Clinical Approach to Infection in the Compromised Host

Fourth Edition

Clinical Approach to Infection in the Compromised Host
Fourth Edition

Edited by
ROBERT H. RUBIN, M.D.

Associate Director, Division of Infectious Disease
Brigham & Women's Hospital
Boston, Massachusetts
and Director of the Center for Experimental Pharmacology and Therapeutics
Harvard University–Massachusetts Institute of Technology
Division of Health Sciences and Technology
Cambridge, Massachusetts

and
LOWELL S. YOUNG, M.D.

Director of Kuzell Institute for Arthritis and Infectious Diseases
Chief of the Division on Infectious Diseases
Pacific California Medical Center
San Francisco, California

With a Foreword by
RALPH VAN FURTH, M.D.

KLUWER ACADEMIC/PLENUM PUBLISHERS
NEW YORK, BOSTON, LONDON, DORDRECHT, MOSCOW

Library of Congress Cataloging-in-Publication Data

Clinical approach to infection in the compromised host/edited by Robert H. Rubin and
Lowell S. Young.—4th ed.
        p.   ; cm.
    Includes bibliographical references and index.
    ISBN 0-306-46693-7
        1. Opportunistic infections.   2. Communicable diseases.   3. Immunological deficiency
syndromes—Complications.   I. Rubin, Robert H., 1941–   II. Young, Lowell S.
    [DNLM: 1. Immunologic Deficiency Syndromes—complications.   2. Acquired
Immunodeficiency Syndrome—immunology.   3. Immunocompromised
Host—immunology.   WD 308 C6408 2002]
    RC112 .C59 2002
    616.9′0479—dc21
                                                                                          2001053922

ISBN 0-306-46693-7

©2002 Kluwer Academic / Plenum Publishers
233 Spring Street, N.Y., New York 10013

http://www.wkap.nl/

10   9   8   7   6   5   4   3   2   1

A C.I.P. record for this book is available from the Library of Congress

Printed in the United States of America

Contributors

Sarah W. Alexander • Division of Hematology and Oncology, Rainbow Babies and Children's Hospital, University Hospital of Cleveland, Cleveland, Ohio 44106

Constance A. Benson • Division of Infectious Diseases, Department of Medicine, University of Colorado Health Sciences Center, Denver, Colorado 80262

Michael Boeckh • Program in Infectious Diseases, Fred Hutchinson Cancer Research Center, and Department of Medicine, University of Washington School of Medicine, Seattle, Washington 98109

John C. Christenson • Department of Pediatrics, Division of Infectious Diseases and Geographic Medicine, University of Utah School of Medicine, Salt Lake City, Utah 84132

A. Benedict Cosimi • Department of Surgery, Massachusetts General Hospital, Boston, Massachusetts 02114

Jules L. Dienstag • Gastrointestinal Unit, Medical Services, Massachusetts General Hospital, and Department of Medicine, Harvard Medical School, Boston, Massachusetts 02114

Jay Alan Fishman • Infectious Disease Division and Transplantation Unit, Massachusetts General Hospital, Harvard Medical School, Boston, Massachusetts 02114

Mitchell Goldman • Indiana University School of Medicine Division of Infectious Diseases, Indianapolis, Indiana 46202

Reginald Greene • Chest Division, Radiology Service, Massachusetts General Hospital, and Department of Radiology, Harvard Medical School, Boston, Massachusetts 02114

Paul D. Griffiths • Department of Virology, Royal Free and University College School of Medicine, London, NW3 2PF, United Kingdom

Harry R. Hill • Departments of Pediatrics, Internal Medicine, and Pathology, Division of Clinical Immunology and Allergy, University of Utah School of Medicine, Salt Lake City, Utah 84132

Richard Allen Johnson • Harvard Medical School, Infectious Disease Unit, Massachusetts General Hospital, and Department of Dermatology, Beth Israel-Deaconess Medical Center, Harvard Medical School, Boston, Massachusetts 02114

Bart Jan Kullberg • Department of Medicine, Nijmegen University Medical Center St. Radboud, 6500 HB Nijmegen, Netherlands

Kieren A. Marr • Program in Infectious Diseases, Fred Hutchinson Cancer Research Center, and Department of Medicine, University of Washington School of Medicine, Seattle, Washington 98109

Brigitta U. Mueller • Deparment of Medicine, Children's Hospital, Boston, Massachusetts, and Department of Pediatrics, Harvard Medical School, Boston, Massachusetts 02115

Philip A. Pizzo • Deparment of Medicine, Children's Hospital, Boston, Massachusetts, and Department of Pediatrics, Harvard Medical School, Boston, Massachusetts 02115

Peter M. Rosenberg • St. John's Health Center, Santa Monica, California 90404

Robert H. Rubin • Division of Infectious Disease, Brigham and Women's Hospital; Center for Experimental Pharmacology and Therapeutics, Harvard–Massachusetts Institute of Technology Division of Health Sciences and Technology; and Department of Medicine, Harvard Medical School, Boston, Massachusetts 02114

George A. Sarosi • Roudebush VA Medical Center, Indianapolis, Indiana 46202

W. Michael Scheld • Departments of Internal Medicine (Infectious Diseases) and Neurosurgery, University of Virginia School of Medicine, Charlottesville, Virginia, 22908

Robert T. Schooley • Division of Infectious Diseases, Department of Medicine, University of Colorado Health Sciences Center, Denver, Colorado 80262

Arthur Sober • Harvard Medical School, Department of Dermatology, Massachusetts General Hospital, Boston, Massachusetts 02115

Allan R. Tunkel • Department of Internal Medicine, MCP Hahnemann University, Philadelphia, Pennsylvania 19129

Jos W. M. van der Meer • Department of Medicine, Nijmegen University Medical Center, St. Radboud, 6500 HB Nijmegen, Netherlands

Ralph van Furth • University of Leiden Medical School, 2341 NL Oegstgees, Netherlands

L. Joseph Wheat • Department of Medicine, Indiana University School of Medicine, and Department of Veterans Affairs Hospital, Indianapolis, Indiana 46202

Lowell S. Young • Kuzell Institute for Arthritis and Infectious Diseases, Division of Infectious Diseases, California Pacific Medical Center, University of California at San Francisco, San Francisco, California 94115

Foreword

At the beginning of the new millennium, it is opportune to review what has been accomplished in the field of infectious diseases during the last decades of the previous century. The paradigm of the immunocompromised host has taught much about the pathophysiology of infectious diseases, particularly with regard to immunological aspects of host defense. In the beginning, Robert Good called immunodeficiency syndromes "experiments of nature." In the 1960s and subsequent decades, the clinical and immunological aspects of immune deficiencies were studied and adequate treatment attempted. A reflection of these developments were the three successful meetings on these topics in Veldhoven, The Netherlands (1980), Stirling, Scotland (1982) and Toronto, Canada (1984) and the foundation of the International Immunocompromised Host Society in 1985. Since then, the IIHS has organized meetings every two years.

The main concern of clinicians is to understand the ways in which the courses of infectious processes differ in individuals with impaired host defense from that of healthy individuals. Scientists attracted to this field want to unravel the pathophysiology of immune deficiencies. These aspects of this field—particularly the clinical pictures of inborn and acquired immune deficiency, the therapeutic interventions, and the modes of prevention of infections in immunocompromised patients—are all covered in the fourth edition of *Clinical Approach to Infection in the Compromised Host*. This volume includes the knowledge accumulated during the last 50 years and it is timely that this new edition appeared.

An infection can be defined as a defense reaction of the host upon the invasion and multiplication of microorganisms. A compromised host can have damaged integument, i.e., skin or mucosa, allowing easy penetration of bacteria, or a defective humoral and/or cellular immune function, as a result of which microorganisms cannot be eliminated optimally. During the last decade, research on immune deficiency was directed mainly toward the un-raveling of the molecular pathways of impaired host defense mechanisms and the characterization of the genetic mutations involved, with the prospect of novel strategies for therapeutic interventions and possible corrective gene therapy. In this foreword, I will take a helicopter view of the various aspects of host defense mechanisms with special emphasis on genetic factors, because of their relevance for the course and outcome of infections.

During life, there exist phases of age-related compromised immune functions. After birth there is a physiological immune deficiency because the production of antibodies commences slowly upon contact of the neonate with microorganisms and upon vaccination. The production of IgM commences at the fifth month of gestation and reaches mature levels between the second and fourth years of life. The production of IgG antibodies starts immediately after birth and is at a mature level between the ages of 6 and 10 years. The production of serum IgA and secretory IgA commences during the first trimester of life, and reaches mature levels at 12–16 years and 2–4 years of age. Thus, during the first years of life the humoral immune system is deficient, which might help explain the frequent occurrence of respiratory and gastrointestinal infections in children under 12 years of age.

At the other end of life, in the elderly, the immune functions wane gradually. In individuals older than 65 years antibody production—for example, after influenza or pneumococcal vaccination—is less than in younger adults. Attenuation of neutrophil functions and T lymphocyte-mediated immune responses also occurs with age. This decrease of host defense mechanisms explains in part why the elderly suffer a higher morbidity and mortality from infections. Another explanation for the frequent occurrence of (viral) respiratory and other infections in the elderly is their contact with children who suffer from such infections, which they have acquired in day care centers, kindergarten, or primary school.

Most microorganisms enter the body via mucosal

membranes, which implies a sequence of events from adherence of microorganisms to epithelial cells, their penetration of epithelial cells or passage between these cells, multiplication of microorganisms within host cells or connective tissues, and their subsequent dissemination via lymphatics or blood vessels to other sites of the body.

The knowledge of the structures at the surface of microorganisms by which they adhere to binding sites of epithelial host cells, such as protein or carbohydrate structures, is rather limited. It is of interest that several classes of host-cell receptors, which are involved in immunological processes, also serve as receptors for some viruses; for example, ICAM-1 for rhinoviruses, epidermal growth factor receptor for reovirus and vaccinia virus, CD46 (membrane cofactor receptor) for measles virus, CD2 receptor for Epstein–Barr virus, VLA2 receptor for echovirus, $\alpha_v\beta_3$ and $\alpha_v\beta_5$ integrin receptors for adenovirus, epidermal growth factor receptor for vaccinia, IgG superfamily receptor for poliovirus, CD4, and the chemokine receptors CXCR4 and CCR5 for the human immunodeficiency virus (HIV). Since the expression of these receptors can be modulated by, for example, cytokines and chemokines such changes can also affect the host cell–virus interaction. Virus–receptor interactions will lead to signal transduction within the host cells and subsequent biochemical and functional changes.

It is often not well appreciated that viruses can alter the immune responses by infecting immunocompetent cells, such as lymphocytes, NK cells, monocytes/macrophages, and dendritic cells. The most ardent immunosuppressive virus is HIV, but other viruses like HTLV I, measles, rubella, cytomegalovirus, Epstein–Barr virus, rubella, and hepatitis B can cause immunosuppression as well. Viruses can also evade the immune system by suppressing the expression of MHC class I and II molecules or cell-adhesion molecules, or form defense molecules that interfere with cytokine or chemokine functions. Whether there is a relation between the genetic makeup of the infected patient and the occurrence of immunosuppression by the virus is not yet known.

Individuals suffering from primary immune deficiency disorders, which are due to mutations in genes involved in the formation of components involved in the immune response, have an impaired host defense against infections. At the present time a large number of mutations involved in primary immune deficiency have been identified.

In the case of agammaglobulinemia, the absence of antibodies will preclude bactericidal serum activity and the elimination of microorganisms by phagocytes. Subtler

are the subclass deficiencies with the consequent increased susceptibility for infections with encapsulated bacteria and defective response to polysaccharide antigens upon vaccination, which occur among Caucasians, particularly in several ethnic groups. IgA deficiency is the most common form of subclass deficiency in Caucasians and is less frequent in other ethnic groups. Although individuals with IgA deficiency are at risk for mucosal infections, they are usually healthy, probably because of a compensatory increased synthesis of IgM, including secretory IgM.

How can we respond adequately to prevent infections in individuals with inborn errors of antibody formation? Substitution therapy with gammaglobulin is probably the best option for patients with severe hypo- and agammaglobulinemia, as well as for those with serious infection in the setting of IgG subclass deficiency. In some cases bone-marrow transplantation might be indicated, but this is infrequently applied. Immune globulins specific for varicella–zoster or cytomegalovirus can be given as prophylaxis or treatment in immunocompromised patients. Parenteral vaccination of individuals with defective antibody formation will often not lead to an adequate antibody response. However, in the case of IgG deficiency, stimulation of the synthesis of secretory antibodies is of utmost importance since they can prevent invasion of microorganisms. Therefore, the development of new vaccines—for example, DNA vaccines—and of new formulations which can be used for mucosal immunization is warranted. Vaccine delivery by transgenic plants or fruits is an attractive option.

Individuals genetically deficient in components of the classic or alternative complement pathways are particularly susceptible to infections with *Neisseria* species. When such infection cannot be controlled with antimicrobial therapy, plasma transfusion can be considered.

Phagocytes are the most important cells for the elimination of microorganisms from the body. These cells, formed in the bone marrow, are transported via the circulation to areas of infection, where they adhere to endothelial cells. Transendothelial migration occurs in the postcapillary venules by squeezing between endothelial cells. Local accumulation of phagocytes is determined by adhesion molecules on the luminal surface of endothelial cells (ICAM-1, ICAM-2, or VCAM-1) which serve as receptors for complementary adhesion molecules (the β_2 integrins LFA-1, CR3, p150,95, and the β_1 integrin VLA-4) on circulating cells. During infection upregulation of expression and activation of adhesion molecules on endothelial cells and leukocytes occurs by stimulation with cytokines and other inflammatory mediators and by intact bacteria or their products, such as endotoxin, thus enhancing the adhesion of leukocytes to endothelial cells and their subsequent migration.

In children with leukocyte adhesion deficiency (LAD I)—a congenital defect of β_2 integrins—neutrophils do not adhere to endothelial cells, and as a consequence are unable to migrate to tissues with a resulting lack of pus formation at sites of infection. A similar gene mutation encoding for β_2 integrins occurs in Holstein cattle and dogs.

Phagocytosis of microorganisms is facilitated by opsonins, i.e., immunoglobulins and components of the complement system, that are recognized by respective immunoglobulin receptors (FcγR I, FcγR II, and FcγR III and also FcαR) and complement receptors (CR1, CR3, and CR4). Recently, allotypic polymorphism of Fcγ receptors, which affect the affinity for IgG2, has been discovered. Consequently, IgG2-opsonized bacteria and IgG2-immune complexes, which can determine the course and outcome of infections, are less efficiently phagocytosed by individuals who are homozygous for the low-affinity receptors. On the other hand, during infection cytokines can up- or downregulate the expression and function of Fc and complement receptors of neutrophils and monocytes/macrophages, which will also affect the elimination of opsonized microorganisms.

Cytokines and chemokines are humoral mediators that act in a network and determine the outcome of infections as favorable, chronic, or fatal. These factors stimulate or inhibit the recruitment, movement, and function of cells involved in inflammation, thus being essential in the containment or removal of microorganisms. In contrast to extracellular bacteria that are eliminated mainly by neutrophils, intracellular bacteria (e.g., mycobacteria, *Listeria*, salmonellae, and brucellae) are eliminated optimally only by cytokine-activated macrophages, a condition called cell-mediated immunity. Recently, a number of individual patients and families suffering from recurrent or chronic infections with mycobacteria and/or salmonellae were described, who did not recover despite adequate antimicrobial therapy. In these patients, mutations in genes coding for interleukin-12 (IL-12) receptors or interferon gamma (IFN-γ) receptors have been identified. These receptors are functionally defective and the production of IFN-γ and IL-12 is usually reduced in these patients. Patients with inherited IL-12 deficiency are described as well, but inherited deficiencies of other cytokines are not reported as yet. Possibly, such deficiencies are not compatible with normal embryonic development.

The chronic course or fatal outcome of infections in all these kinds of patients can be explained by their inability to develop adequate cell-mediated immunity.

The total absence of neutrophils is in essence not compatible with life, but cyclic granulocytopenia is a condition accompanied with recurrent infections. There is a multitude of congenital disorders of neutrophils and other leukocytes that are caused by mutation in genes coding for enzymes involved in the functions of these cells. Patients with such disorders, which are relatively rare, suffer frequently from severe infections. Possibilities for preventive measures are limited, with the exception of IFN-γ treatment of chronic granulomatous disease patients; other options are bone marrow transplantation and, in the future, gene therapy.

To my knowledge, individuals without monocytes/macrophages do not exist. It is probable that these cells play a pivotal role in embryonic development, and in the absence of these cells no viable fetus will develop.

During the last decade, we have learned more about genetic factors that determine the predisposition for certain infections. It has been known for some time that individuals heterozygous for sickle cell anemia are less susceptible to malaria. Recently, it has been reported that homozygous mutation of the cystic fibrosis gene increases the susceptibility for typhoid fever. Polymorphism of HLA class II immune response genes determines differences in immune responsiveness between individuals. For example, the susceptibility and/or course of mycobacterial infections, such as leprosy and tuberculosis, typhoid fever and other salmonellosis, paralytic poliomyelitis, and dengue fever, are linked with HLA class II alleles. More recently a correlation between the susceptibility or outcome of meningococcal disease and genes for mannose-binding, lectin, and factors influencing the production of TNF or IL-10 has been reported.

Pharmacogenomics is a new discipline that is also relevant for the treatment of the immunocompromised host. It concerns the interindividual variability in drug response based on the genetic polymorphism of multiple genes involved in metabolic and physiologic pathways. For example, the genetic polymorphism of the cytochrome P-450 system determines the pharmacodynamics of some antimicrobial drugs. The poor or extensive enzyme activity of cytochrome P-450 affects the acetylation of isoniazid and, thus, the therapeutic levels of this drug; antimicrobials like rifampicin, erythromycin, antimycotic drugs, and antiretroviral protease inhibitors increase or decrease cytochrome P-450 isoenzymes which results in a decrease or increase of their pharmacodynamic effects. In the future, the detection of genetic polymorphism of individuals may have great potential for the development of new drugs and avoidance of drug toxicity.

In this foreword, I had the intention of triggering the attention of clinicians for some immunological and genetic aspects which have an impact on the clinical approach to infections in the immunocompromised host. I trust this will be an adjunct during the reading of this excellent volume.

Ralph van Furth
Oegstgeest, The Netherlands

Preface

As we open the freshly minted pages of the fourth edition of *Clinical Approach to Infection in the Compromised Host*, we look back in amazement at the more than two decades that have passed since this project was initiated. Who would have thought? It all began in the spring of 1977, at the annual Epidemic Intelligence Service (EIS) Conference of the Centers for Disease Control when two friends from medical school days (both ex-EISers), one now in California and one still in Boston, met and began to discuss their current clinical and research interests. By happy coincidence, we had both become enthralled in a very specialized aspect of infectious disease—the infectious disease problems of the immunocompromised host. Even the term "compromised host" was a relatively novel one. We belonged to a small community of clinical investigators who were part infectious disease clinicians, part microbiologists, part hematologists, oncologists, and organ transplanters, as well as clinical immunologists and white cell biologists. As a community we were convinced that these compromised hosts provided an ideal opportunity for deciphering the nuances of host–pathogen interaction. In addition, even in the 1970s, it was already clear that for a number of diseases previously untreatable, disease modifying therapy was on the immediate horizon *provided we could learn to prevent and treat the invasive infections* made possible by this disease-modifying therapy.

In the more than two decades that have passed, the number of immunocompromised patients has increased exponentially due to two factors, one predictable and positive while the other was unexpected and has had catastrophic consequences. The first of these has been the remarkable transformation that has occurred in the prognosis of patients who formerly succumbed rapidly to their malignancy, their end-stage renal, cardiac, hepatic, and pulmonary disease, and to such autoimmune processes as vasculitis, inflammatory bowel disease, and other conditions. Today, these individuals have reasonable hopes of

years, if not decades, of productive life because of advances in the use of cancer chemotherapeutic agents and immunosuppressive therapy, in immunology and transplantation, and in the linkage of effective antimicrobial therapy to prevent and treat the infectious complications of the disease-modifying therapy. For example, the prophylactic use of trimethoprim–sulfamethoxazole in a variety of immunocompromised patients has essentially eradicated *Pneumocystis carinii* pneumonia, nocardiosis, listeriosis, and, perhaps, toxoplasmosis from many compromised patient populations; and the preemptive use of intravenous, followed by oral, ganciclovir has decreased the incidence of symptomatic cytomegalovirus disease in the transplant patient receiving antilymphocyte antibody therapy from ~65% to near zero. Thus, the appropriate deployment of antimicrobial programs can make safe immunosuppressing therapies that can control the underlying disease. Our task is to delineate other prophylactic and preemptive strategies to prevent the infections that are still occurring in these patients.

The second factor, the unexpected and catastrophic, is obviously the advent of the epidemic of human immunodeficiency virus (HIV) infection that results in the acquired immunodeficiency syndrome (AIDS). From a cluster of cases of opportunistic infection in gay males, this has been transformed into an epidemic of worldwide proportions, threatening the existence of whole societies in Africa and Asia, and causing great morbidity and mortality throughout the world. Today, although the greatest burden of this disease is borne by the poor, all segments of society are at risk. Despite the catastrophe, remarkable progress has been made in dealing with this plague in the past two decades. This program has included: the identification of the causative virus, the dynamics of its replication, and the modulation of this by cytokines and chemokines elaborated by the host in response to other processes; a whole array of diagnostic tests, the most important being reliable anti-HIV antibody testing and viral load measurements by molecular diagnostic techniques (these not only allow early diagnosis but guide therapy, as well as provide protection of the blood and allograft supply); a clear understanding of the epidemiology of the AIDS epidemic and what public health measures can help alleviate the spread of HIV; an increased understanding of virus–host interactions; the definition of the clinical manifestations of AIDS and the correlation of these with viral load and CD4 lymphocyte measurements; and last, but far from least, the development of an armamentarium of anti-HIV drugs and the knowledge of how to use them. The advent of highly active antiretroviral therapy (HAART) has transformed AIDS from a progressive illness with an early death sentence into a manageable chronic disease, with issues now being quality of life and therapeutic side effects rather than the immediacy of death itself. The biggest problem now is the cost of these drugs and how to get them to the impoverished, particularly in the developing areas of the world.

The net result of these two factors—successful immunosuppressing therapy and HAART therapy for AIDS—has meant that many of these patients are being returned to relatively normal lives in the general community. No longer are their infectious disease problems the sole province of ivory tower academicians who communicated chiefly with each other. It is fair to say that the infectious disease problems of the immunocompromised host are now the concern of all practitioners of medicine, of as much concern to the primary care physician as to the tertiary care specialist. Gratifyingly, there has been an accompanying explosion of information on the science and practice of caring for the infectious disease problems of these patients. Despite the medical conferences, new journals, newsletters, cassettes, videotapes, and other multimedia attempts to convey the necessary information to the expanding group of physicians who need such knowledge, we continue to believe that there is a compelling need for the clinician to have access in the dead of night or the heat of day to well-written, sage advice from veterans of battles similar to those they are now undertaking. We have previously stated that the best way to learn is to sit at the feet of a master—such as Mark Hopkins and the Log—for several years. It is our hope that we have been able to bring together a group of Mark Hopkinses, all veterans of these battles and distinguished contributors to the field, with this book serving as a Log for all those with a need to know. All credit for achieving these objectives is owed to our contributors. We accept responsibility for any inadequacies.

As we peruse this edition, we are also reminded that, as in any dynamic field of medicine and science, we stand on the shoulders of those who have gone before. These include our teachers, our patients, and our colleagues. Unfortunately, much of what we have learned has come from the study of patients who, despite the best efforts that could be made at the time, have succumbed to their infections. Memories of them and their courage in the face of extreme adversity continue to inspire us, and this book is in part dedicated to them and honors them. It is also dedicated to the memory of two of our teachers: Dr. Louis Weinstein and Dr. Alex Langmuir, who taught us the joys, challenges, science, and art that illuminate the practice of infectious disease care, epidemiology, and research. Giants surely did once walk the Earth …

Robert H. Rubin
Lowell S. Young

Boston and San Francisco

Preface to the First Edition

The science and practice of infectious disease cut across all medical disciplines, from medicine to surgery, and from cardiology to neurology. Because of the diverse nature of infection and the clinical settings in which it occurs, the acquisition of the skills needed to become expert in clinical infectious diseases has usually required a lengthy apprenticeship. As one of us has noted, "The practice of infectious disease is akin to many primitive arts, being handed down by oral traditions from generation to generation. The best way to learn is to sit at the feet of a master for several years, asking, observing, and studying—the medical equivalent of Mark Hopkins and the Log."

However, increasingly it has become apparent that a more efficient means of communicating the art and science of clinical infectious disease to the general medical community is necessary. The infections themselves, the potential therapeutic modalities, the clinical settings in which they occur, and the occurrence of such infections far away from the academic medical center—all these have put a new emphasis on disseminating the most up-to-date information available to diagnose and treat clinical infection. This is particularly true when one considers the gamut of infections that afflict the patient with a defect in host defense. Those of us with a particular interest in this area of medicine and infectious diseases are painfully aware of the special nature of these patients and their problems and the rather extended apprenticeship we have served in learning to deal with these problems. We have been impressed that although great strides have been taken in general infectious disease in moving beyond the Log and the oral tradition, in this area of infectious disease such efforts are just beginning. Thus, the idea for this book was conceived to attempt to meld the scientific advances in this area with the experience we had in dealing with such patients to construct a useful, practical guide to the problem of infection in the compromised host. We wanted to share the fruits of our apprenticeship with the rest of the medical community who increasingly are being called on to deal with these clinical problems.

The next step was to find out whether a publisher would be interested. Ms. Hilary Evans of Plenum was quickly recruited to the effort. She has been a bulwark of strength and encouragement during the lengthy gestation period. Finally, there comes the recruitment of the other contributors. Perhaps the most pleasant surprise in this whole experience was the enthusiasm with which our contributors brought their expertise to the endeavor—all of us agreeing that a need existed for a practical guide to patient management in the immunosuppressed host that was based on firm scientific data whenever this was available and on the art and judgment of medicine when such data was unavailable. With admiration and gratitude we thank our contributors, who have taught us so much in the preparation of this book.

Finally, it is fitting that we express our gratitude to three different groups of individuals who have made this book possible—our teachers, Mort Swartz, Louis Weinstein, Alex Langmuir, Don Armstrong, and Don Louria, who have served as our models in their ability to blend the sciences of microbiology, immunology, and epidemiology with the art of clinical medicine; our families, who have supported us in this effort and whose time has been stolen to prepare this work; and perhaps most of all, our patients, the immunocompromised patients with life-threatening infections who continue to teach us and inspire us with their courage and faith as we painfully learn how best to deal with infection in the compromised host.

Robert H. Rubin
Lowell S. Young

Boston and Los Angeles

Contents

5 Central Nervous System Infection in the Immunocompromised Host

ALLAN R. TUNKEL and W. MICHAEL SCHELD

6 Fungal Infections in the Immunocompromised Host

L. JOSEPH WHEAT, MITCHELL GOLDMAN, and GEORGE A. SAROSI

7   Mycobacterial and Nocardial Diseases in the Compromised Host

LOWELL S. YOUNG and ROBERT H. RUBIN

8 *Pneumocystis carinii* and Parasitic Infections in the Immunocompromised Host 265

JAY ALAN FISHMAN

9 Viral Hepatitis in the Compromised Host

PETER M. ROSENBERG and JULES L. DIENSTAG

13 Infectious Complications in Children with Cancer and Children with Human Immunodeficiency Virus Infection 441
SARAH W. ALEXANDER, BRIGITTA U. MUELLER, and PHILIP A. PIZZO

14 Infections Complicating Congenital Immunodeficiency Syndromes 465
JOHN C. CHRISTENSON and HARRY R. HILL

15 Management of Infections in Leukemia and Lymphoma 497
LOWELL S. YOUNG

16 Infection in Hematopoietic Stem Cell Transplantation 527
MICHAEL BOECKH and KIEREN A. MARR

17   Infection in the Organ Transplant Recipient ..
ROBERT H. RUBIN

18 Surgical Aspects of Infection in the Compromised Host 681

A. BENEDICT COSIMI

Clinical Approach to Infection in the Compromised Host

Fourth Edition

1

Introduction

ROBERT H. RUBIN and LOWELL S. YOUNG

Two terms that have become common in the contemporary literature of internal medicine, surgery, and pediatrics are the *immunocompromised host* and *opportunistic infection*. The first of these terms, *immunocompromised host* (or such variants as the *compromised host*), describes a group of individuals with impairment of either or both natural and specific immunity to infection (*impaired host defenses*) such that they are at increased risk for infection by a variety of microorganisms. The microorganisms that invade these individuals may be grouped into three major categories:

1. *True pathogens* are the classic plagues of mankind (e.g., influenza, typhoid fever, bubonic plague, diphtheria), invading normal and abnormal host alike. The organisms causing these infections possess virulence factors that are capable of overcoming the natural resistance mechanisms of the nonimmune host, with survival being dependent on either the rapid development of a specific immune response or the institution of effective antimicrobial therapy or both. The virulence factors involved include the production of toxins and/or the ability to traverse mucocuta-

neous barriers, evade phagocytosis, and resist extra- and intracellular microbicidal systems.

2. *Sometime pathogens* are commonly present as colonizers of the mucocutaneous surfaces of the body, causing clinical disease only when they are introduced into normally sterile tissues following a break in the integrity of a mucocutaneous surface. When such a break occurs, these organisms then possess sufficient virulence characteristics to cause lethal infection, again unless either specific immunity develops or effective antimicrobial therapy is introduced. Examples of this form of infection include staphylococcal and Group A streptococcal sepsis following breaks in skin integrity and gram-negative and *Bacteroides fragilis* sepsis following bowel perforation.

3. *Nonpathogens* are generally susceptible to nonspecific (natural) resistance supplemented by specific immunity. They usually have no impact on the normal host, being capable of invading and causing disease only in individuals with impairment of either nonspecific or specific host defenses or both. Examples of this form of infection include *Pneumocystis carinii* pneumonia and invasive aspergillosis.

The term *opportunistic infection* is used to denote invasive infection due to nonpathogens or to infections with sometime or even true pathogens of a type and/or severity rarely encountered in the normal host [e.g., disseminated zoster in the lymphoma patient, hepatosplenic candidiasis in the leukemic patient, and recurrent *Salmonella* bacteremia in the acquired immunodeficiency syndrome (AIDS) patient]. Although opportunistic infection appropriately has been the focus of attention in the compromised host, it

Robert H. Rubin • Division of Infectious Disease, Brigham and Women's Hospital; Center for Experimental Pharmacology and Therapeutics, Harvard–Massachusetts Institute of Technology Division of Health Sciences and Technology; and Department of Medicine, Harvard Medical School, Boston, Massachusetts 02114. **Lowell S. Young** • Kuzell Institute for Arthritis and Infectious Diseases, Division of Infectious Diseases, California Pacific Medical Center, University of California at San Francisco, San Francisco, California 94115.

Clinical Approach to Infection in the Compromised Host (Fourth Edition), edited by Robert H. Rubin and Lowell S. Young. Kluwer Academic/Plenum Publishers, New York, 2002.

is important to emphasize the impact of both true and sometime pathogens in this patient population.

As the number of immunocompromised patients has exploded in recent years, responsibility for their care has spread from the academic medical center to practitioners at every level—the primary care physician, the general internist and surgeon, as well as the subspecialist. It is incumbent on all of us to become familiar both with the unusual infectious disease problems that occur in these patients—the prevention, diagnosis, and treatment of these problems—and with the ways in which underlying disease and/or its therapy can modify the clinical presentation and management of common conditions. For example, too often we have seen the diagnosis of a perforated abdominal viscus in a patient on immunosuppressive therapy missed and the patient succumb because of the absence of the classic signs of an acute abdomen, a not uncommon event in such patients.

This volume was conceived as an attempt to deal with this issue; to summarize directly and succinctly the major issues and controversies involving the medical and surgical management of immunocompromised patients. We hope that the dominant characteristics of this assembly of views are candor and the presentation of a particular approach to the clinical management of infection in the compromised host. The contributors have been asked to meet the important issues "head-on" and to identify those areas in which useful knowledge does or does not exist. After giving a fair summary of the published literature, they elaborate on their views about the most expeditious, economically sound, logical approach to the challenging manifestations of infection in what are likely to be very ill patients. Where data are insufficient to enable one to choose among several clinical alternatives, the contributors have said so, thereby distinguishing established principle from opinion. This effort is not intended as an encyclopedic compendium of the recent medical knowledge on this complicated subject, but rather as a practical guide to clinical decision making.

The guiding theme of this volume is to identify the epidemiologic, pathophysiologic, and clinical clues that will lead to early diagnosis and effective therapy or, better yet, prevention, rather than to discuss the way a disease looks at the autopsy table. This volume has been purposely organized with this perspective in mind. On the one hand, there are detailed chapters on host defenses, as well as particularly important infections (e.g., herpesviruses, fungi, parasites). On the other hand, since the patient does not present with a label stating the name of the infection, other chapters are devoted to particular organ system infections (e.g., of the skin, lungs, and central nervous system) and infection in particular patient groups (e.g., pediatric and adult cancer, bone marrow and organ transplant, and AIDS patients). The clinician needs both kinds of information; e.g., an approach to the patient with pneumonia as well as a detailed analysis of what to do and what to anticipate once the cause of the pneumonia in this particular patient population has been determined. Although such a multifaceted approach results in some repetition of material, we believe that such repetition is both warranted and useful ("creative redundancy"). Throughout the volume, great reliance has been placed on actual case examples to illustrate important clinical points.

There clearly has been a great deal of ferment and progress in this field since the last edition of this volume. The challenges, however, remain:

1. A myriad of infectious agents can cause potentially lethal infection in these patient groups. Although treatment of established disease has improved, prevention should be the first goal of the clinician: we need to develop better epidemiologic protection, as well as additional preemptive and prophylactic strategies against such important pathogens as Epstein–Barr virus, fungi, antimicrobial-resistant bacteria, and the community-acquired respiratory viruses.

2. The impaired inflammatory response of immunosuppressed patients, particularly those with profound neutropenia and those receiving corticosteroids, can greatly attenuate the signs and symptoms of microbial invasion until the disease is relatively far advanced. Since the chance of recovery from most infections in this population is directly related to how early the diagnosis is made and effective therapy instituted, subtle signs and symptoms require constant attention, expert radiological evaluation [particularly computed tomographic (CT) scanning for indications not acceptable in the normal host], and invasive biopsies.

3. Diagnostic techniques adequate for the care of the nonimmunosuppressed population are not optimal for these patients, at least in part since the nature of the infections can be so different. For example, conventional blood culturing techniques are not adequate for the diagnosis of invasive fungal and nocardial infection. Even when an isolate is made, usually after an invasive procedure, standardized antimicrobial techniques are not widely available.

4. The nature of many of the infections observed is

such that prolonged therapy is often necessary, increasing the risk of toxicity due to such drugs as amphotericin, high-dose trimethoprim–sulfamethoxazole, ganciclovir, and others.

5. In addition to the risk of infection, the differential diagnosis in many of these patients includes such noninfectious etiologies as recurrent neoplastic disease, allograft rejection, graft-versus-host disease, and relapsing collagen vascular disease, thus complicating the clinician's task.

As we face these and other challenges, controversy is inevitable. Table 1 delineates some of the most important questions facing the clinician today in dealing with the compromised host. Gratifyingly, these are distinctly different than those outlined in the previous edition, indicating progress. Answering these questions will clearly advance the field significantly in the next decade.

TABLE 1. Areas of Controversy in the Management of Infection in the Immunocompromised Host

1. What is the best strategy for preventing bacterial and fungal infection in the neutropenic cancer or bone marrow transplant patient?
2. What is the optimal empiric antimicrobial therapy for suspected bacterial sepsis in the neutropenic patient? Monotherapy versus potentially synergistic combination therapy? Vancomycin as primary therapy with the gram-negative regimen or only after the diagnosis of gram-positive infection is made?
3. What is the relative efficacy of the different antifungal drugs for the different fungi at different sites? Conventional amphotericin versus the different lipid-associated amphotericin preparations, and these versus the new drugs such as voriconazole and the echinocandins? Do combinations offer any advantage?
4. What are the indications for initiating and stopping empiric antifungal therapy? Which of the drugs is best for what patient populations?
5. What are the indications for continued empiric antimicrobial therapy in the culture-negative but persistently febrile neutropenic patient?
6. What is the most cost-effective program for preventing the infectious disease consequences of cytomegalovirus and Epstein–Barr virus infection in transplant patients?
7. Will antiviral strategies decrease the incidence of allograft injury and/or graft-versus-host disease in transplant patients?
8. What is the best strategy for managing hepatitis B and C virus infection in patients already or about to be immunosuppressed (e.g., organ transplant candidates and recipients)?
9. What is optimal anti-HIV therapy for the different stages of this infection such that the measurable antiviral load is undetectable and resistance is avoided?
10. Other than *Pneumocystis carinii* infection, are there opportunistic infections that can be effectively prevented with prophylactic or preemptive strategies in the AIDS patient and other immunocompromised patient populations?

There are two competing major developments that will affect our ability as well as Europe's ability to meet these challenges and settle the controversies that characterize this field. One development is the governmental and societal pressures for the implementation of cost-effective measures in clinical practice. How we define *cost-effectiveness* per se is controversial. Anyone who treats patients is aware, however, of the increased consciousness relating to reducing unnecessary tests, the use of less expensive antimicrobial agents, and the desire to discharge patients early following chemotherapy or pharmacological therapy with the potential for persisting side effects. One would hope that many of these cost-cutting measures have been of benefit. In certain situations, however, they may limit our ability to understand fully the complexities of the biological phenomena that are occurring in our more complicated immunosuppressed patients.

Clinically, early discharge may spell early readmission. By their very nature, immunocompromised patients tend to require expensive care; one example of this being the frequent need for CT scanning when conventional radiography is adequate for the normal host. It is likely that the trend emphasizing economics will persist, and many of the changes in this text are aimed at providing the clinician with useful guidelines that will enable the practice of first-rate medicine that is responsive to the demands of cost control. If economics is an issue in the United States and Europe, it is a reality in the developing world where the care of AIDS patients is nonexistent. How the economic issues are solved will determine greatly what progress will be made in the care of the immunocompromised host.

A happier development is that we are beginning to enjoy the fruits of the revolution in science that has occurred in the pharmaceutical and biotechnology industries. Never before have we had such a richness of approved drugs and drugs in development as we have now. The molecular biology revolution, monoclonal antibody technology, combinatorial chemistry, and rapid throughput screening have brought us effective AIDS drugs, new immunosuppressive and chemotherapeutic agents, and some new antibacterial, antifungal, and antiviral agents. It is paradoxical that a traditional area of pharmaceutical strength—the development of new antibiotics—has been lagging behind such areas as recombinant proteins, hematopoietic growth factors, and therapeutic monoclonal antibodies at a time when antimicrobial resistance is burgeoning. Clearly, particularly for the immunocompromised host new antimicrobial agents in all categories are needed if we are not to be overwhelmed with drug-resistant tuberculosis, azole-resistant fungi, ganciclovir-resistant herpesviruses, lamivudine-resistant hepatitis B, vancomycin-resistant enterococci, methicillin-resistant *Staphylococcus aureus*, penicillin-resistant pneumococci, and highly resistant gram-negative bacilli, virtually all of which have a higher attack rate in the immunocompromised patient than in the general population. Finally, there is a compelling need both for new paradigms for developing and deploying these new agents (in a cost-effective way) and for trained clinical investigators for facilitating this process.

It is hoped that the material in this volume will both lead to the optimal management of immunocompromised patients today and help to show the way to new approaches for tomorrow. This volume is primarily an attempt to summarize the many developments that have occurred in a rapidly evolving field and to direct the thoughts of investigators to where major opportunities lie. The volume should be regarded as serving three purposes: (1) a status report as to where we are in the understanding and management of an increasingly important group of patients, (2) a practical guide to the clinician for the everyday management of these patients, and (3) a charge to the future to investigators who constantly seek to improve what we can for a courageous and deserving group of human beings: the increasing population of individuals who are immunocompromised by disease and/or its therapy.

2

Defects in Host Defense Mechanisms

JOS W. M. VAN DER MEER and BART JAN KULLBERG

1. Colonization

Under normal conditions, large areas of the human body surface are colonized with microorganisms. The skin and the mucous membranes of the oropharynx, nasopharynx, intestinal tract, and parts of the genital tract each have their own microflora.[1] These patterns of colonization are determined by microbial factors, exogenous factors, and host factors. An important microbial factor is adhesion to epithelial cells, often with distinct specificity for a certain type of epithelial cell.[2] Microorganisms also can influence the patterns of colonization by producing bacteriocins and other products that inhibit the growth of other microorganisms[3] or by competing for essential nutrients.[4] By means of such mechanisms, colonizing microorganisms form a barrier against microorganisms from the outside world. This type of barrier has been designated *colonization resistance*.[5] In the nose,[6] diphtheroids hamper the local growth of *Staphylococcus aureus*, whereas in the gastrointestinal tract, anaerobic microorganisms inhibit the colonization and outgrowth of aerobic gram-negative rods.[4,5] By contrast, certain pathogenic microorganisms (e.g., respiratory viruses) facilitate bacterial colonization.[7,8] Of the exogenous factors that influence the normal flora, diet[1,9] and, more significantly, disinfectants and antimicrobial drugs[10] should be mentioned. The host factors that play a role in colonization are complex. The adherence of bacteria to epithelial cells is dependent on specific receptors on somatic cells.[2,11] Host factors like fibrinogen and fibronectin (see Section 3.1.3) may play a role. Interindividual differences in bacterial adherence to epithelial cells may be based on certain blood group glycoproteins[12] or human leukocyte antigen (HLA) types[13] that are expressed on epithelial cells and serve as receptors. The precise mechanisms by which race,[14] hormonal status,[15] pregnancy,[16] alcoholism,[17] intravenous drug abuse,[18] hemodialysis,[19] and underlying diseases such as cancer[20–22] and diabetes[17] alter colonization are unknown. Several other factors, considered part of the host defense in a stricter sense (Table 1), also influence colonization of the body surface.

2. First Line of Defense

2.1. Skin

Normal skin is not very hospitable to most microorganisms.[23] By desquamation of the horny layer, bacteria are constantly eliminated from the skin. Dependent on their chemical structure, skin surface lipids produced by the sebaceous glands as well as keratinocytes either inhibit the growth of bacteria, such as streptococci, or promote the growth of diphtheroids.[23] The lactic acid in sweat can be used as a source of energy by staphylococci, although inhibition of growth occurs at higher concentrations.[24] When the skin is humid and the pH increases, higher bacterial counts are found.[25]

2.2. Mucosa

At the mucosal level, a number of effective mechanisms are encountered. Ciliary motion, ventilation, and coughing maintain sterile airways below the vocal cords

Jos W. M. van der Meer and Bart Jan Kullberg • Department of Medicine, Nijmegen University Medical Center St Radboud, 6500 HB Nijmegen, The Netherlands.

Clinical Approach to Infection in the Compromised Host (Fourth Edition), edited by Robert H. Rubin and Lowell S. Young. Kluwer Academic / Plenum Publishers, New York, 2002.

TABLE 1. Host Defense Mechanisms

Line of defense	Nonspecific	Specific
Surface defense (skin, mucous membranes)	Mechanical barrier Secretory barrier Ciliary motion Movement	Immunoglobulins
Humoral defense	Lysozyme and lactoferrin Complement system Fibronectin Interferons Interleukins	Immunoglobulins

Cellular defense	Phagocytic cells
	    Neutrophilic granulocytes
	    Eosinophilic granulocytes
	    Mononuclear phagocytes
	    Natural killer cells
		Cell-mediated immunity (T lymphocytes and macrophages)

in the healthy human being.[26] Gastric acid is an effective barrier against bacteria from outside; thus, hypochlorhydria (e.g., as induced by H_2 receptor antagonists) is associated with colonization of the stomach and even intestinal infections.[27] Unconjugated bile has antibacterial properties and probably helps restrict the number of bacteria in the small intestine.[28] Intestinal motility also inhibits bacterial outgrowth. Colonization of the urinary tract is largely prevented by regular voiding.[29] In the female genital tract, production of antimicrobial factors (such as H_2O_2) by the commensal *Lactobacillus* species creates an environment that is hostile to many pathogens.[30] Lysozyme and lactoferrin in tears and saliva display antibacterial activity: lysozyme through inhibition of the cell wall synthesis of gram-positive bacteria and lactoferrin through interference with bacterial iron metabolism[31,32] (see Section 3.1.1). More important are the immunoglobulins, especially secretory IgA, on these surfaces; by coating microorganisms, secretory IgA prevents adherence to mucosal cells (see Section 3.1.6).

Symbiosis of the host with colonizing microflora depends in the first instance on the integrity of the mechanical barrier of skin and mucous membranes. Damage to this first line of defense, even by trivial injuries such as a puncture, can turn colonizing microorganisms into pathogens that impose great demands on the other defense mechanisms. Moreover, pathogens from the outside world then may gain entrance and cause infection. At this point, it should be stressed that a number of microorganisms are capable of penetrating intact epithelia. If the microorganisms succeed in passing the first line of defense, then the second line of defense, consisting of hu-

moral and cellular defense mechanisms (Table 1), is needed for elimination of the invaders.

3. Second Line of Defense

The second line of defense is made up of humoral and cellular defense mechanisms (Table 1) that act in close cooperation. Thus, discussing humoral and cellular defense mechanisms separately is a somewhat artificial approach. However, for reasons of clarity and because quite a number of pathological states involve either the humoral or the cellular defense mechanisms, these mechanisms are dealt with in separate sections.

3.1. Humoral Defense Mechanisms

The humoral defense system consists of a number of nonspecific and specific factors that interact with microorganisms (Table 1).

3.1.1. Lysozyme and Lactoferrin

Lysozyme, one of the nonspecific factors (see also Section 2.2), is an enzyme found in many body fluids. It is present in high concentrations in the azurophilic and specific granules of polymorphonuclear phagocytes and is constitutively secreted by mononuclear phagocytes.[33] By cleaving the linkage between *N*-acetyl-muramic acid, lysozyme interferes with cell wall synthesis, especially that of gram-positive bacteria[31]; it also markedly amplifies the effector mechanism of the complement cascade[34] (see Section 3.1.2). Lactoferrin is an iron-binding protein that, when not fully saturated with iron, inhibits the growth of microorganisms, such as gram-negative and gram-positive bacteria and various *Candida* species.[32] It then acts as an iron-chelating agent, thereby depriving the microorganisms of the iron necessary for their growth. It is an important enzyme of neutrophil-specific granules (see Section 3.2.1h).

3.1.2. Complement

More important in the humoral host defense is the complement system, which consists of at least 19 plasma proteins that are able to react in a cascade.[33,34] This cascade has four essential elements: (1) classic activation, (2) alternative activation, (3) amplification, and (4) the effector mechanism (Fig. 1). Antigen–antibody complexes can bind the first component of the complement system (C1) at the exposed Fc portion of the antibody; in this way,

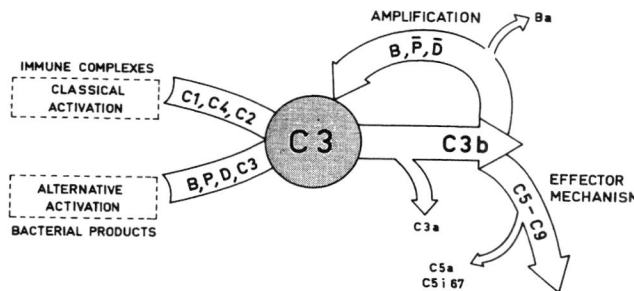

FIGURE 1. Schematic representation of the complement system. From Daha.[35]

classic activation is initiated. The classic route also can be started by the acute-phase reactant mannan-binding protein (MBP) to microorganisms or by activated factor XII. Once activated, C1 activates C4 and C2; the next step is activation of C3, which is converted into C3a and C3b.

Bacterial products, especially molecules with repeating chemical structures (lipopolysaccharides, teichoic acid, and endogenous molecules like CRP and fibrillar β-amyloid), can activate the alternative route via factors B, P, and D. When C3b, formed by either the classic or the alternative route, reacts with factors B, P, and D, complement activation may be amplified. The effector mechanism is then initiated by the conversion of C5 into C5b and C5a and subsequent formation of the multimolecular complex C5b6789.

A variety of inhibitors, also part of the complement system, regulate the system. For a more detailed account of these factors, the reader is referred to the reviews by Joiner *et al.*[34] and Figueroa and Densen.[36] The biological effects of complement are essentially as follows: Factors C3a and C5a are able to liberate histamine and serotonin from mast cells and basophils, thereby increasing vascular permeability. Factors C5a, Ba, and C5b67, and to a much lesser extent, C3a, have a chemotactic effect on leukocytes; in addition, C5a induces adhesiveness, oxidative burst, and degranulation of neutrophils (see Section 3.2.1). Since C3b-bound complexes can bind to C3b receptors on granulocytes and monocytes–macrophages, C3b acts as an opsonin (see Section 3.2.1e) and promotes phagocytosis. The effector complex C5b6789 is able to penetrate cell membranes, leading to lysis of the microorganisms as well as of erythrocytes and tumor cells.

Thus, the complement system provides protection with a potent host defense system against microorganisms that operates even before a specific immune response has developed via the alternative pathway and via the classic

route once there is an acute phase response or a specific antibody response. Moreover, it appears that certain antibodies also amplify the alternative route. From the deficiencies of individual complement components in humans, the relative role and importance of these factors in host defense can be assessed. Infections are relatively rare in patients with a deficiency of C1, C4, and C2, probably because the alternative pathway is intact. Recurrent infections, especially those caused by encapsulated bacteria, have been reported in patients lacking C3.[37,38] Deficiency of C5 is associated with an impaired capacity to generate chemotactic activity in serum and with recurrent pyogenic infections.[39] Leiner's syndrome (eczema, diarrhea, and recurrent gram-negative bacteremia) has been qualified as a C5-dysfunction syndrome; generation of chemotactic factors is defective and there is a concomitant, as yet unexplained, opsonic defect.[40] Among those with a deficiency of one of the terminal components (C5–C8), a high incidence of chronic or recurrent *Neisseria* infections has been found.[36,41] The mortality rate of the latter infections in these patients is remarkably low, suggesting that an intact terminal pathway contributes to a poor outcome.

Deficiencies in alternative pathway proteins are rare. A combined deficiency of factors B and C2 predisposes to serious infections with encapsulated bacteria.[42] Similar infectious conditions occur in the event of factor D deficiency[43] and also in C3b-inactivator deficiency.[44] In sickle cell anemia (see Section 5.3), reduced activation of alternative pathway factors leads to impaired opsonization of pneumococci and salmonellae.[45,46] A more detailed discussion of the primary complement deficiencies is given in Chapter 14.

During various disease states, such as gram-negative or gram-positive bacteremia, massive complement activation may occur in association with hypotension, respiratory distress syndrome, and disseminated intravascular coagulation (DIC).[47] Under these circumstances, acquired complement deficiency can develop, leading to an inability to cope with infectious agents.[48,49] Acquired complement deficiency, as in systemic lupus erythematosus (SLE), has been reported to be associated with meningococcal infection.[41,50,51] In Felty's syndrome, acquired hypocomplementemia appears to predispose to serious infectious conditions[52] (see Section 5.7). Nonenzymatic glycosylation of C3 may occur in diabetes mellitus and may lead to impaired opsonization[53] (see Section 5.4).

That virus infections are not an overt problem in complement-deficiency states does not imply that the complement system does not play a role in the defense against viruses. In fact, there is considerable evidence indicating that the complement system has a central role in host defense mechanisms, as reviewed by Lachmann.[54]

3.1.3. Fibronectin

Fibronectin originally has been described as cold-insoluble globulin in fibrinogen-rich plasma. It has a molecular weight of 440,000 and is detectable in normal plasma at a concentration of approximately 0.35 g/liter.[55] Most of the soluble, circulating fibronectin is produced by hepatocytes.[56] This soluble form is secreted on mucosal surfaces. When the molecule binds to cellular receptors and is incorporated into extracellular and basement membrane matrix, it forms an insoluble protein.[55] Fibronectin has an important function in adherence of cells to other cells, to basement membranes, and to microorganisms. Although mononuclear phagocytes have receptors for fibronectin,[57] it is questionable whether fibronectin should be regarded as an opsonin,[58] i.e., a ligand between a particle and a phagocyte that mediates ingestion (see Section 3.2.1). It is more likely that fibronectin acts by inducing more receptors for endocytosis on phagocytic cells.[57,58] It also enhances bactericidal activity of mononuclear phagocytes.[58] In experimental animal studies, plasma fibronectin has been shown to promote clearance of injected particles.[58] Deficiency of plasma fibronectin has been reported in DIC, septicemia, trauma, and shock.[59] Fibronectin-rich cryoprecipitate infusions are not beneficial in patients with sepsis.[60] Congenital fibronectin deficiency does not seem to be associated with impaired host defense.[61]

3.1.4. Interferons

The interferons (IFNs) are a group of species-specific glycoproteins that exert a wide array of biological effects. Three main classes of IFN are distinguished on the basis of their antigenic specificities: interferon-α, interferon-β, and interferon-γ. IFN-α, formerly called "leukocyte interferon," is highly heterogeneous: There are more than a dozen genes that code for human IFN-α.[62] IFN-γ-β, formerly called "fibroblast interferon," exists as two subtypes, whereas only one type of IFN-γ (formerly called "immune interferon" or "acid-labile interferon") is known.

Generally speaking, IFN-γα and β act mainly as antiviral agents, whereas IFN-γ is more active as an immunomodulator. The overall effects of the IFNs are shown schematically in Fig. 2. The antiviral effects are not virus specific; they are mediated partly by intracellular changes leading to inhibition of viral replication in a

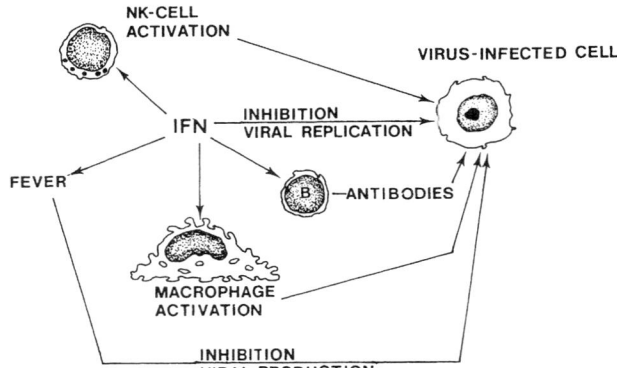

FIGURE 2. Schematic representation of the actions of the interferons. No efforts have been made to separate the functions of the various types of interferon. Modified after a model proposed by Dr. H. Schellekens.

number of ways.[63] *In vivo*, the antiviral effects also seem to be mediated by stimulation of host defense mechanisms, especially natural killer (NK) cells (see Section 3.2.2) and macrophages (see Section 3.2.3c).[62,64] IFN-γ is a major cytokine, and as such it is an important activator of macrophages[65] (see Section 3.2.3c). After exposure to IFN-γ, macrophages are more sensitive to bacterial endotoxin, and these cells will produce more of the proinflammatory cytokines[66] (see Section 3.1.5).

It is very likely that the IFNs are designed to work at short range, mainly at the site of infection. Nevertheless, serum IFN activity can be measured during viral infections such as influenza.[67] An important question with therapeutic implications is whether deficiency of IFNs occurs, e.g., at the cellular level. Such deficiencies indeed have been reported for IFN-γ in newborns[68] and in patients with primary immunodeficiencies[69–71] or the acquired immunodeficiency syndrome (AIDS).[72–74] For allograft recipients, deficient production of both IFN-γ and IFN-α has been reported.[75,76] Congenital defects in the IFN-γ receptor[77,78] will be discussed under Section 3.2.3d. IFNs recently have become available on a large scale, and their clinical use to treat the compromised host is discussed in Section 9.

3.1.5. Cytokines

3.1.5a. Interleukins and Tumor Necrosis Factor. "Interleukins" is the collective name for a series of humoral factors that are produced mainly (but not exclusively) by leukocytes, especially mononuclear phagocytes and lymphocytes, and transfer signals between cells of the immune system and usually also between other cells. As such, they act as local and systemic mediators of the host response to infection. Formerly, they were designated as "lymphokines" if they were primarily produced by lymphocytes and as "monokines" if mononuclear phagocytes were the major source of production. At present, the term "interleukin" (IL) or the more general term "cytokine" is preferred. According to an international convention, these cytokines are now numbered (IL-1, IL-2, ...).[79] The nomenclature is not consequent, in a sense; e.g., tumor necrosis factor α (TNF-α, synonym: "cachectin") and lymphotoxin (TNF-β) should have been assigned an interleukin number. The same holds for the interferons, especially IFN-γ (see Section 3.1.4). Although not all these factors have been shown to have a sizable impact on host defense, this rapidly expanding field is briefly reviewed.

"Interleukin-1" is the collective name for two 17-kDa proteins (IL-1-β and -α) that are produced and released mainly by mononuclear phagocytes (see Section 3.2.1c) when these cells encounter bacterial lipopolysaccharide, microbial cell wall substances (peptidoglycan, muramyl peptides), and bacterial exotoxins (e.g., toxic shock syndrome toxin 1)[80]; other cells (e.g., endothelial cells, keratinocytes, astrocytes) also are able to produce IL-1. IL-1-β largely is released extracellularly and acts as an autocrine, a paracrine, and an endocrine substance, whereas IL-1-α stays mainly cell associated and is believed to exert its function in cell–cell interaction. IL-1 has a large number of important biological effects: It acts as an endogenous pyrogen and stimulates endothelial cells, neutrophils, T and B lymphocytes, hematopoietic cells, and mesenchymal cells (Fig. 3).[80] These biological effects are produced either directly by interaction of IL-1 with specific receptors on the membrane of the target cell or indirectly by the induction of intermediary cytokines like IL-6 and IL-8. Inside the effector cells, the effects of IL-1 are largely mediated via the liberation of arachidonic acid, via the cyclooxygenase pathway or the lipoxygenase pathway.[80]

TNF-α also is a 17-kDa protein that shares many of its effects with IL-1,[81] although it binds to different receptors. Like IL-1, this cytokine is produced and secreted by stimulated mononuclear phagocytes and by other cells. Many of the effects of these cytokines are definitely beneficial in host defense against infection. This beneficence is underscored by experiments showing that a low dosage of exogenous IL-1 protects against lethal bacterial infection even in granulocytopenic animals[82] and by the observations in experimental animals that the absence of TNF has a negative impact on granuloma formation in mycobacterial infection, formation of abscesses in intra-abdominal infection, and fungal infection.[83,84]

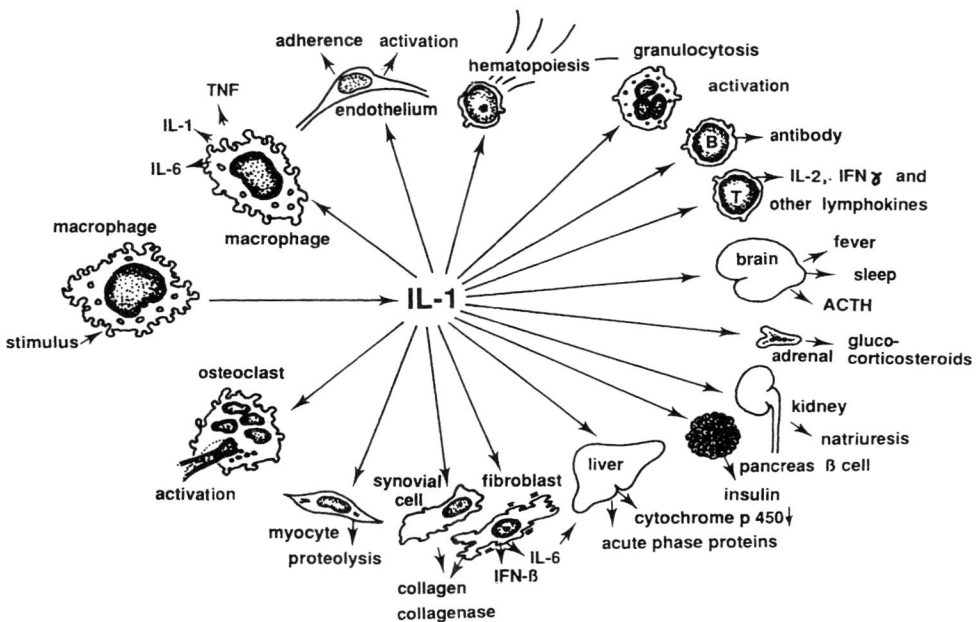

FIGURE 3. Multiple biological activities of interleukin-1 (IL-1). The effects of tumor necrosis factor (TNF) are similar, with the restriction that the effects on lymphocytes are less pronounced.

Under certain circumstances, however, these cytokines may be deleterious for the host. During serious bacterial infection, TNF, IL-1-β, and IL-6 have been detected in the circulation; high concentrations of these cytokines, particularly of TNF, in the circulation appear to indicate a poor prognosis.[85,86] In a rabbit model, IL-1 and TNF produce a state with the clinical characteristics of septic shock and respiratory distress syndrome.[87] In other serious infections, such as cerebral malaria, and in inflammatory states (e.g., the capillary leak syndrome induced by IL-2 treatment), the aforementioned cytokines play an important pathophysiologic role.[88,89] Interference with deleterious cytokinemia, e.g., by administration of antibodies against TNF, so far has failed as a therapeutic option in septic shock and in severe malaria in humans (see also Chapter 4).

IL-2, a 15-kDa glycoprotein originally described as a T-cell growth factor, is produced and secreted by a subset of T lymphocytes; it is the major mediator of the proliferation of T lymphocytes and stimulates production of other cytokines by T cells.[90] In addition, IL-2 activates B lymphocytes and NK cells (see Section 3.2.2). The activation of blood lymphocytes by IL-2 and the induction of other cytokines are the bases for the anticancer therapy with this cytokine (IL-2 with or without *ex vivo* activated killer cells).[91] Probably due to liberation of cytokines such as TNF and IL-1,[89] serious side effects (capillary leak syndrome) occur with high-dose IL-2 therapy.

IL-3 is a multilineage hematopoietic growth factor produced mainly by T lymphocytes.[92] It synergizes with other hematopoietic growth factors (see Section 3.1.5b).

IL-4, a product of T-helper lymphocytes, acts on many types of cells and has a role in the activation, proliferation, and differentiation of B lymphocytes. The molecule plays a crucial stimulatory role in immunoglobulin E (IgE) production,[93] an effect that can be blocked by IFNs, and also plays a role in the activation of T lymphocytes and NK cells. On mononuclear phagocytes, IL-4 induces the expression of class II antigens of the major histocompatibility complex, and it downregulates the expression of various Fc receptors; in addition, IL-4 blocks the production of IL-1, IL-6, IL-8, and TNF by these cells.[93] Because of the latter effects, IL-4 can be considered an anti-inflammatory cytokine. In experimental leishmaniasis, IL-4 plays a beneficial role.[94]

IL-5, produced by T lymphocytes, is the major growth and differentiation factor for eosinophils; it also activates eosinophils and is chemotactic for these cells.[95] IL-5 plays a key role in states of eosinophilia (see Section 3.2.1f).

IL-6 is a 22-kDa cytokine produced by many types of cells (macrophages, fibroblasts, endothelial cells, and smooth muscle cells), especially if these cells are stimulated by IL-1 and TNF.[96] IL-6 is a growth factor for B lymphocytes (see Section 3.1.6), stimulates T cells together with IL-1, and is a major stimulant for the synthesis of acute phase proteins by the hepatocytes.[97] The molecule also stimulates hematopoiesis and is a relatively weak but probably crucial endogenous pyrogen; its action should probably be considered more anti-inflammatory that proinflammatory.[98]

IL-7, a cytokine produced by bone marrow stroma, seems to be an important cytokine in the control of lymphocytopoiesis (for both T and B lymphocytes).[99]

IL-8 is produced by many types of cells after stimulation with a variety of exogenous inducers (e.g., endotoxin) and physiological inducers (IL-1 and TNF).[100] In fact, this molecule is responsible for neutrophil-activating effects originally ascribed to IL-1. These neutrophil-activating effects are induction of directed migration, degranulation of specific granules, and superoxide production.[100]

IL-9, a 30- to 40-kDa glycoprotein produced by stimulated T-helper lymphocytes, stimulates proliferation of a restricted number of T-helper lymphocytes and of hematopoietic progenitor cells, especially those of the erythroid lineage.[101–103]

IL-10 is a 17- to 20-kDa cytokine produced by a number of cell types.[104] It is a major anti-inflammatory cytokine, since it inhibits the macrophage-induced production of other cytokines, especially IFN-γ and TNF, and enhances the proliferative response of thymocytes and mast cells to IL-2 and IL-4.[104] IL-10 also is involved in B-lymphocyte proliferation and function.

IL-11 is a 20-kDa cytokine that resembles IL-6: it stimulates hematopoiesis and induces hepatic acute phase proteins. It has less of an effect on T and B lymphocytes. An interesting function is the protection of intestinal epithelium.[105]

IL-12 is a heterodimer composed of a 35-kDa and a 40-kDa subunit. It is one of the major IFN-γ-inducing cytokines. Thereby it plays a crucial role in host defense against intracellular pathogens such as mycobacteria and *Salmonella* spp.[106] Its role in these infections has become apparent from recent reports of patients with a defective IL-12 receptor, who suffer from recalcitrant infections caused by these pathogens.[107,108]

IL-13 is a cytokine that resembles IL-4, in that it plays a role in IgE production by B lymphocytes. It stimulates differentiation of macrophages from precursors, but at the same time inhibits proinflammatory cytokine production.[109]

IL-14 is a 55-kDa molecule that stimulates the proliferation of B lymphocytes; it also has a role in the maintenance of B-cell memory.[110] IL-15 induces proliferation and differentiation of B cells, T cells, NK cells, and mast cells.[111,112] IL-16 is a product of CD8+ lymphocytes and attracts CD4+ lymphocytes. It seems to play a role in viral infection.[113,114] IL-17 is a product of CD4+ cells and induces production of IL-6, IL-8, G-CSF, as well as prostaglandin E2 by a variety of cells. IL-17 also plays a role in neutrophil differentiation.[115]

IL-18 is related to the IL-1 family in terms of both structure and function. Similar to IL-1, IL-18 participates in both innate and acquired immunity. Formerly called IFN-γ-inducing factor, IL-18 plays an important role in the TH1 response, primarily by its ability to induce IFN-γ production in T cells and natural killer cells.[116] Similar to IL-1β, IL-18 is synthesized as a biologically inactive precursor molecule, and the IL-18 precursor requires cleavage into an active, mature molecule by the protease called IL-1 beta-converting enzyme (ICE), which is also known as caspase-1.[117] Mature IL-18 induces a variety of other proinflammatory cytokines and chemokines, e.g., TNF-α, IL-1β, IL-8, macrophage inflammatory protein-1α, and monocyte chemotactic protein-1 in mononuclear cells.[118] There are accumulating data that IL-18 plays an important role in the innate host response to infection.

3.1.5b. Colony-Stimulating Factors. Colony-stimulating factors (CSFs) are hematopoietic growth factors that are capable of stimulating the clonal growth of hematopoietic precursor cells *in vitro*.[119,120] One of the CSFs, IL-3, has been discussed in the preceding section. Also available in recombinant form and for clinical use are three other growth factors: granulocyte–macrophage CSF (GM-CSF), granulocyte CSF (G-CSF) (see Section 9), and macrophage CSF (M-CSF). These CSFs stimulate proliferation and differentiation of cells from bone marrow and blood.

GM-CSF is a 22-kDa glycoprotein produced by many types of cells, especially macrophages, endothelial cells, and T lymphocytes. With increasing concentrations of GM-CSF *in vitro*, growth of the precursor cells of mononuclear phagocytes, neutrophils, eosinophils, and megakaryocytes is stimulated.[119,120] In addition, GM-CSF induces differentiation of granulocytes and mononuclear phagocytes and at high concentrations potentiates the microbicidal function of these cells.[121] *In vivo*, GM-CSF is used to combat the neutropenia in AIDS and in hematologic disorders and to ameliorate the neutropenia induced during cancer chemotherapy and after bone marrow transplantation.[119,120] Although enhanced neutrophil function has been found in humans after administration of GM-CSF,[121,122] the clinical relevance of this finding has not been established.

G-CSF, a 19-kDa protein produced by mononuclear phagocytes, fibroblasts, and endothelial cells, acts primarily as a terminal differentiation and activation factor for neutrophils.[119] G-CSF can induce oxygen metabolism in neutrophils[123] (see Section 3.2.1f) and enhances survival of these cells. In humans, it has been used for the same indications as GM-CSF. By virtue of its selective effect on neutrophils, rG-CSF is an important therapeutic agent for patients with congenital or acquired neutropenia.[120,124] In addition, rG-CSF is used to elicit white blood cells from donors, raising renewed interest in white blood cell transfusions in patients with neutropenia-related infections (see Section 9).[125] Because of its capacity to activate neutrophils, rG-CSF is under investigation as a therapeutic agent in severe pneumonia and in disseminated candidiasis in patients without neutropenia.[126,127] In addition to the proinflammatory effects on granulocytes, G-CSF has anti-inflammatory effects, since it inhibits proinflammatory cytokine production.[128]

M-CSF is a 70- to 90-kDa glycoprotein, which is a product of mononuclear phagocytes, fibroblasts, and endothelial cells. It promotes the growth of mononuclear phagocytes and stimulates macrophage function.[119] The clinical experience with this molecule has been limited.

3.1.5c. Cytokine Network. The cytokines and growth factors discussed above interact with each other and with the interferons (see Section 3.1.4) in a complicated network. Many of these cytokines are able to induce each other *in vivo*. For example, IL-1 readily induces IL-6 and IL-8 (see Section 3.1.5a) and also induces itself in an autocrine and paracrine fashion (Fig. 3); in addition, it induces TNF.[129] Apart from these positive-feedback loops, inhibitory loops also are activated. The most important inhibitors recognized so far for the cytokine network are the IL-1 receptor antagonist (IL-1ra), the soluble receptors for TNF, and the induction of cytokines with anti-inflammatory properties, such as IL-4, IL-10, and transforming growth factor-β (TGF-β).[130–132] The potential of these inhibitors of the cytokine network to interfere with deleterious cytokinemia and local cytokine responses (see Section 3.1.5) has been demonstrated by a large number of preclinical studies.[133–136]

In experimental infections in mice (especially leishmaniasis) two major cytokine patterns have been recognized and linked to production by certain types of T lymphocytes (see Section 3.2.3b): the so-called T-helper-1 (TH1 or type 1) response is characterized by production of IL-2, IL-12, and IFN-γ, whereas the T-helper-2 (TH2 or type 2) response produces IL-4, IL-5, and IL-10.[137,138] The TH1 response enhances host defense and immunity and the TH2 response dampens the immune response. In humans, immune responses to several chronic infections appear to polarize along TH1/TH2 lines as well.[139] Although the TH1/TH2 concept is useful to give a simple description of the prevalent cytokine pattern, it is not very useful in infectious diseases in which a neutrophil response is crucial. Also a strict dichotomy, as originally proposed, is an oversimplification.[140,141] In many clinical situations both patterns are found and thereby the concept loses its significance. It also does not take into account that many types of cells other than T-helper lymphocytes produce cytokines: macrophages, granulocytes, NK cells, endothelial cells, mast cells, and epithelial cells. Finally, the concept does not deal much with sequential production of pro- and anti-inflammatory cytokines.

3.1.6. Immunoglobulins

The immunoglobulins, which make up the specific humoral response, are products of the B-lymphocyte system. During the development of the pre-B cells, clonal diversity is generated as a consequence of a series of gene rearrangements.[142] In this way, the potential to generate millions of B-cell clones is created. On encounter with an antigen, certain B-cell clones expand to produce specific antibodies. Early in this response, IgG is secreted; during differentiation to the plasma cell stage, recombination or deletion of DNA may occur so that the other Ig classes (IgG, IgA, IgE, IgD) are produced and finally secreted. The functioning of B lymphocytes is controlled by regulatory T cells (see Section 3.2.3b) and various humoral B-cell growth and differentiation factors such as IL-1, IL-2, IL-6, IL-11, and IL-14 (see Section 3.2.3a). Some B cells do not differentiate into plasma cells but become long-lived memory B cells, which enable the body to produce an immediate antibody response to secondary exposure to the antigen.

Immunoglobulins carry out their functions by means of antigen binding at the Fab sites on the Ig molecule. These functions include neutralization and agglutination of the antigen, complement activation and binding (see Section 3.1.2), prevention of epithelial attachment of the antigen, and mediation of endocytosis, i.e., opsonization (see Section 3.2.1e). Not all the different classes and subclasses of immunoglobulins carry out all these functions to the same extent.

IgM is not in itself an opsonin; only when combined with complement factor C3b is this complex opsonic. Because of its pentameric structure, IgM is very efficient as an agglutinin. There are four IgG subclasses in humans; of these, IgG1 and IgG3 bind C3b and are especially important as opsonins (see Section 3.2.1e). IgG2 and possibly IgG4 are thought to play a role as antibodies against microbial polysaccharides, e.g., those of the capsules of type B *Haemophilus influenzae* and *Streptococcus pneumoniae*.[143,144]

The IgA found on mucosal surfaces consists of equal amounts of the two subclasses IgA1 and IgA2, produced locally by plasma cells in the mucosa; this secretory IgA is a dimer of two IgA molecules coupled by a small polypeptide, the J chain, with a secretory component that is a polypeptide produced by epithelial cells.[145] Secretory IgA is not very opsonic; it prevents adherence of bacteria to the mucosal surface, inhibits the motility of bacteria, may agglutinate bacteria, and neutralizes enterotoxins and viruses.[145] Of the two subclasses of IgA on the mucosa, only IgA1 is cleaved by the IgA proteases produced by such bacteria as *Neisseria gonorrhoeae*, *N. meningitidis*, *H. influenzae*, *S. pneumoniae*, and *S. sanguis*.[145] About 90% of the IgA in serum, which is produced by plasma cells in bone marrow, is of the IgA1 subclass.[146] The few IgA-bearing B lymphocytes in peripheral blood are probably on their way to the mucosal surfaces. The role of circulating IgA in host defense is unclear.

IgE is normally present in the circulation in very low concentrations.[147] This immunoglobulin plays a role in acute allergic reactions and helminthic infestation. Its production is under the control of IL-4[93] (see Section 3.1.5a). When antigen combines with IgE on mast cells and basophils, these degranulate and subsequently release a variety of amines (e.g., histamine). These products are responsible for increased vascular permeability and the influx of eosinophils. IgE also acts as a ligand for the killing of schistosomes by macrophages[148]; the ligand for killing of these organisms by eosinophils is not IgE, but IgG and complement.[149] The role of circulating IgD is unclear. As an immunoglobulin on the surface of B cells (like membrane-bound IgM), it serves as a receptor for antigens.[150]

Clearly, the various immunoglobulin classes and subclasses play different roles in the handling of antigens. Therefore, a deficiency of all immunoglobulins, i.e., hypogammaglobulinemia, is associated with undue susceptibility not only to encapsulated bacteria, such as *S. pneumoniae* and *H. influenzae* (both of which exhibit tropism in the respiratory tract), but also to enteric pathogens such as species of *Salmonella* and *Campylobacter*.[151–153] In addition, there is susceptibility to the protozoa *Giardia lamblia*.[154] These patients also find it difficult to cope with viruses such as poliovirus, echovirus, and rotavirus.[155–157] Severe *Mycoplasma* and *Ureaplasma* infections have also been reported.[158,159]

Patients with selective IgA deficiency may suffer from recurrent respiratory infections and protracted giardiasis.[160] Those with IgG subclass deficiencies, which may be associated with an IgA deficiency, exhibit a relatively high incidence of *H. influenzae* infections.[144,161] Patients with IgM deficiency are especially at risk for meningococcal infections.[162] Remarkably, not all cases of selective immunoglobulin isotype deficiencies are associated with repeated infections; many IgA-deficient persons are asymptomatic.[160] For instance, an asymptomatic familial complete deficiency of IgA1, IgG1, IgG2, and IgG4 has also been reported.[163]

Primary immunoglobulin deficiency states are due to (1) B-cell defects (as in X-linked hypogammaglobulinemia, in which functional Bruton tyrosin kinase is lacking[164]), (2) abnormalities of T-cell subpopulations or macrophages (as in some cases of late-onset hypogammaglobulinemia) and their soluble products,[165,166] (3) deficiency of molecules that are responsible for intercellular signaling between T and B lymphocytes (e.g., the crucial signaling molecule CD40 ligand),[167,168] or (4) mechanisms that have not yet been elucidated.

A more detailed account of the primary immunoglobulin-deficiency states and their treatment is given in Chapter 20. Secondary immunoglobulin deficiencies are defined as (1) disorders in which immunoglobulin synthesis is decreased, as occurs in chronic lymphocytic leukemia, multiple myeloma, and other lymphoproliferative diseases and to some extent after splenectomy (see Section 5.2), or (2) disorders with increased immunoglobulin catabolism associated with severe burns, protein-losing enteropathies, and nephrotic syndrome. Administration of immunoglobulin to treat disorders with immunoglobulin losses is not beneficial in most cases.

Tuftsin is a tetrapeptide (Thr-Lys-Pro-Arg) that binds covalently to the Fc portion of IgG.[169] Its principal function is thought to be activation of phagocytes. To do this, tuftsin first must be freed from the immunoglobulin by the action of a splenic enzyme (tuftsin endocarboxypeptidase) and by the enzyme leukokinase, which is bound to the membrane of a neutrophil or mononuclear phagocyte. Its biological relevance is suggested by the occurrence of a congenital deficiency state associated with recurrent infections of the respiratory tract, skin, and lymph nodes.[169] The infectious conditions associated with splenectomy can be explained in part by a failure to free tuftsin from its carrier, the immunoglobulin (see Section 5.2). Deficiency of tuftsin has been reported in AIDS; this finding could partially explain the human immunodeficiency virus (HIV)-infected patient's susceptibility to encapsulated bacteria.[170]

3.2. Cellular Defense Mechanisms

3.2.1. Phagocytic Cells

Phagocytic cells are the cells of the granulocytic series (neutrophilic granulocytes and eosinophilic granu-

locytes) and the mononuclear phagocytes. The development of these cells is shown in Fig. 4.

3.2.1a. Kinetics of Neutrophilic Granulocytes. The proliferation of neutrophil precursors is under the control of hematopoietic growth factors such as IL-3, GM-CSF, and G-CSF (see Section 3.1.5).[171] In the neutrophilic granulocyte series, it takes approximately 6 days for metamyelocytes to form by sequential division and another 6 days for the metamyelocytes to mature into polymorphonuclear granulocytes.[172] A large number of neutrophils (approximately 10 times the circulating population) remain in the bone marrow as a reserve that can be released into the circulation when there is an inflammatory stimulus. The neutrophils that enter the circulation are distributed over two compartments: one consisting of free circulating neutrophils (circulating pool) and the other of neutrophils that adhere loosely to the vascular endothelium (marginating pool).[171] Under normal circumstances, the two pools are of approximately equal size and in dynamic equilibrium. Several factors can influence the pool sizes by disturbing adherence to the endothelium. Epinephrine and the glucocorticosteroids are potent inhibitors of margination.[173] C5a (the chemotactically active cleavage product of complement factor C5) plays a key role in the margination of neutrophils.[174]

The process of neutrophil extravasation requires a cascade of sequential events between neutrophils and endothelial cells, that usually are divided into four steps (Fig. 5).[175,176] After an inflammatory stimulus, the first step of the neutrophil adhesion cascade is principally mediated by the selectins. L-selectin, expressed on neutrophils, E-selectin and P-selectin, both expressed by activated endothelial cells, interact with their mucinlike ligands and circulating neutrophils begin to roll over the endothelial cell surface.[177] In step 2, neutrophils become activated by locally produced tissue-derived factors (e.g., chemokines). This process leads to both shedding of L-selectin and activation of specific leukocyte integrins, such as very late activation antigen 4 (VLA-4), Mac-1, and leukocyte function-associated molecule 1 (LFA-1). In the third step, the interaction of these activated integrins on neutrophils with their ligands on endothelial cells—vascular cell adhesion molecule 1 (VCAM-1) with VLA-4 and intercellular adhesion molecule 1 (ICAM-1) with Mac-1 and LFA-1—causes the firm adhesion of neutrophils to the endothelium. Finally, leukocytes migrate between the endothelial cells and invade the tissues.[175,176] The magnitude of the marginating pool is also principally regulated by the selectins.[171] The transient neutropenia that occurs during hemodialysis is the result of complement activation by the dialyzer membrane and subsequent neutrophil margination and sequestration in the pulmo-

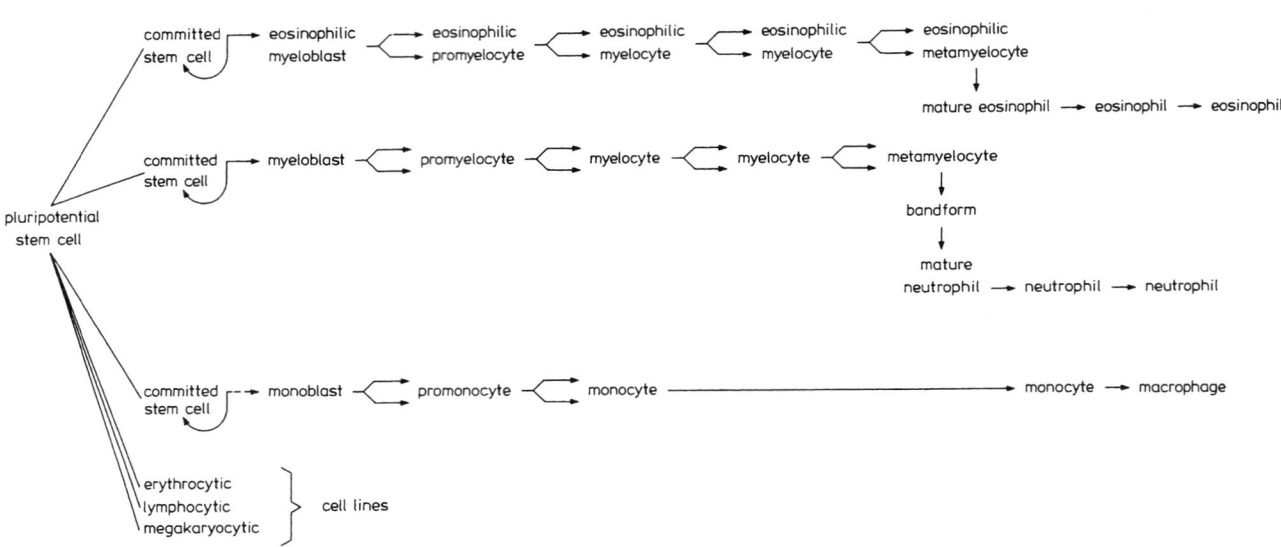

FIGURE 4. Development of the phagocytic cells.

nary circulation.[178] Neutrophils disappear randomly from the circulation into the tissues with a half-life of approximately 6 hr; in the tissues, they are estimated to survive approximately 1–3 days.[172]

3.2.1b. Kinetics of Eosinophils. The eosinophilic granulocytes most probably develop from their own com-

mitted stem cells (Fig. 4).[179] The proliferation and maturation of eosinophilic precursors is under the control of IL-3, GM-CSF, and IL-5.[180] The time needed for proliferation, maturation, and circulation seems to be similar to that described for neutrophils.[180] Eosinophils become activated by exposure to IL-5 and activated eosinophils

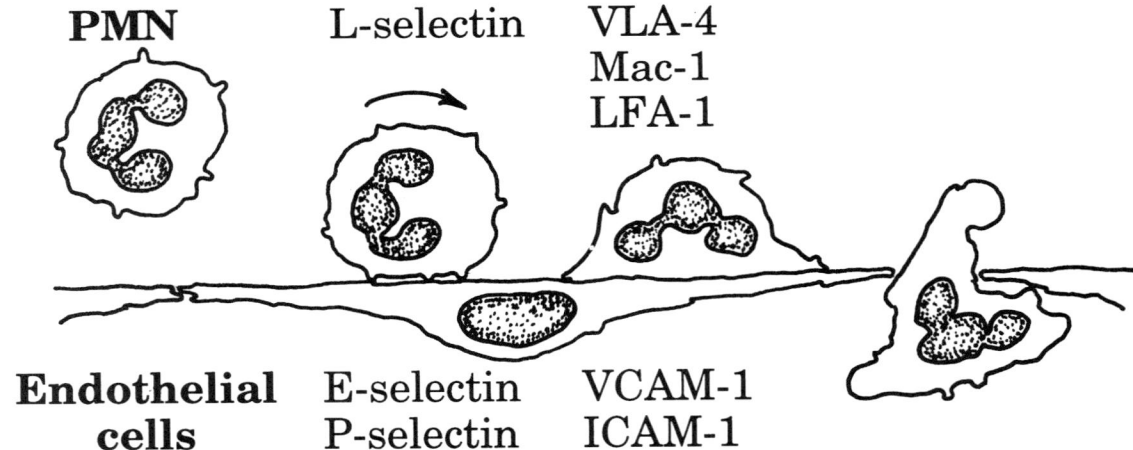

FIGURE 5. The process of neutrophil extravasation. The first step of neutrophil adhesion is mediated by selectins, leading to rolling of the neutrophil over the endothelial cell surface. In the second step, neutrophils are activated, leading to both shedding of L-selectin and activation of specific integrins, such as very late activation antigen 4 (VLA-4), Mac-1, and leukocyte function-associated molecule 1 (LFA-1). In step 3, interaction of vascular cell adhesion molecule 1 (VCAM-1) with VLA-4 and intercellular adhesion molecule 1 (ICAM-1) with Mac-1 and LFA-1 causes adhesion of neutrophils. In step 4, leukocytes migrate between the endothelial cells and invade the tissues.

themselves are capable of producing a variety of cytokines, such as IL-3, IL-5, GM-CSF, IL-6, IL-8, as well as mediators such as leukotriene C4 and platelet-activating factor.[181] Autocrine cytokines inhibit apoptosis of eosinophils, leading to a prolonged survival in the tissues of 12–14 days, as opposed to a normal cell survival of 48 hr.[182]

The cause of persistent eosinophilia and the hypereosinophilic syndrome is unknown. Recent studies suggest that in patients with the hypereosinophilic syndrome, a clone of abnormal T cells produces large amounts of IL-5, which may cause the eosinophilia.[183]

3.2.1c. Kinetics of Mononuclear Phagocytes. Mononuclear phagocytes also derive from the bone marrow.[184] Monoblasts[185,186] divide to form promonocytes, which divide and form monocytes. Without further maturation, the monocytes enter the circulation, which they leave again with a half-life of approximately 70 hr.[187] In the tissues where these cells differentiate into one of the various types of macrophages, they survive for several weeks.[184] The macrophages, together with their precursor cells, form the mononuclear phagocyte system (MPS)[188] depicted in Fig. 6. The reticuloendothelial system, a grouping of macrophages, reticulum cells, fibroblasts, endothelial cells, and other cells of divergent origin proposed by Aschoff, is considered an outdated concept. It is uncertain whether the Langerhans cell in the skin, the veiled cell in the afferent lymph, and the interdigitating cell in the lymph nodes and spleen are mononuclear phagocytes.[188]

3.2.1d. Kinetics of Phagocytes during Inflammation. During an acute inflammation, an increase in phagocytes is observed at the site of inflammation. Neutrophils and sometimes eosinophils first appear in the inflammatory field, followed by an increasing number of macrophages. The formation of this inflammatory exudate is the result of various mechanisms. Activation of several inflammatory humoral factors (e.g., kinins, cytokines, prostaglandins, complement factors) leads to increased local blood flow and increased vascular permeability. Humoral factors (especially C5a, leukotriene B, IL-8 and related cytokines, and bacterial products) attract neutrophils and mononuclear phagocytes (chemotaxis).[190] In the blood, neutrophilia develops as a result of the release of the marrow reserve as well as increased

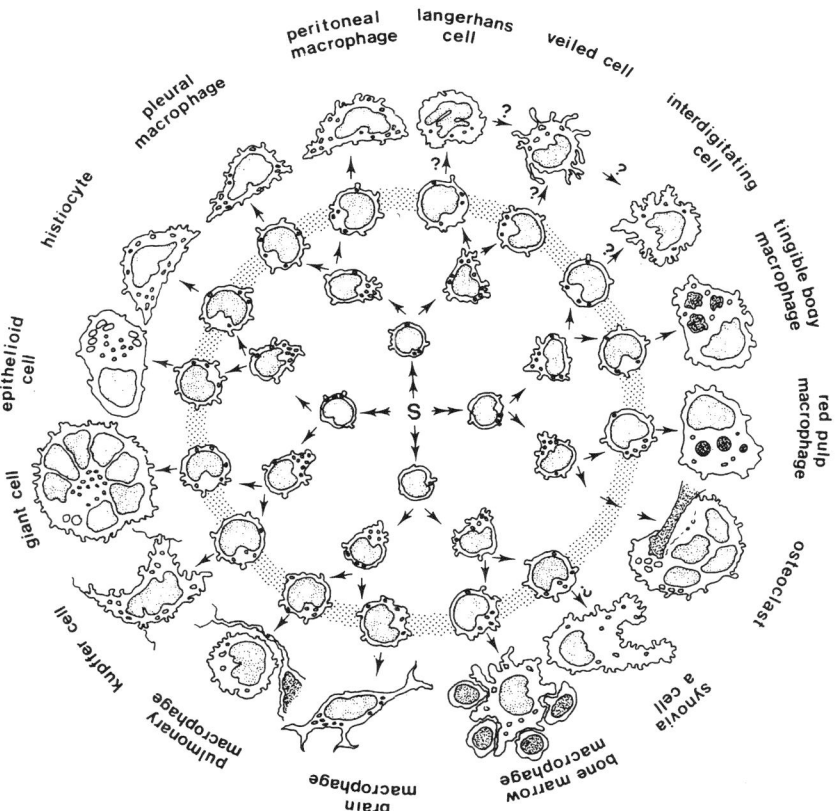

FIGURE 6. The mononuclear phagocyte system (MPS).[188] The hematopoietic stem cell (S) is the precursor for the monoblasts (shown in the inner ring). By division of monoblasts, promonocytes (second ring) are formed, which then form monocytes that enter the bloodstream (shaded area). Monocytes differentiate into macrophages, which have different names in different tissues (outer ring). Those cells for which the evidence to include them in the MPS is inconclusive are indicated with a question mark. Based on experiments in the mouse, the monocyte is not considered a precursor of the osteoclast.[189]

granulocytopoiesis. Several humoral factors (e.g., IL-1) seem to play a role in the induction of this granulocytosis.[80] Monocytopoiesis also is stimulated; humoral factors produced by tissue macrophages seem to be responsible.[191] In humans IL-10 injection is followed by monocytosis.[192] Generally, eosinopenia is observed initially during the inflammatory response.[179] Under certain conditions (e.g., metazoal infection), eosinophilia and eosinophil accumulation in the tissues occur; IL-5 appears to be a major factor in this process.[95]

3.2.1e. Endocytosis.   Endocytosis is the cellular uptake of materials, such as microorganisms, debris, immune complexes, effete red blood cells, and tumor cells by engulfment via the cell membrane. It can be subdivided into (1) pinocytosis (drinking by cells) and (2) phagocytosis (eating by cells). Pinocytosis is the uptake of small particles in small vacuoles. For the uptake of many kinds of molecules, pinocytosis is a receptor-mediated process. Monocytes and macrophages, fibroblasts, endothelial cells, hepatocytes, and various other cells are able to perform pinocytosis; granulocytes are believed incapable of pinocytosis.

Phagocytosis is the uptake of particles larger than 1 μm. It is a receptor-mediated process, which means that the particles that are taken up bind to receptors on the cell membrane either directly or by means of ligands. The most efficient uptake occurs via the Fc-γ receptor and the C3b receptor with IgG1, IgG3, and C3b as ligands or opsonins (Fig. 7). Thus, the first step is the binding of the opsonins to the particle. IgM is not in itself an opsonin,

since there are no receptors for the Fc portion of IgM on phagocytic cells (see Section 3.1.6). However, C3b can bind to IgM and then to the C3b receptors on the phagocyte, thereby mediating endocytosis. Fibronectin may be an opsonin as well (see Section 3.1.3). Binding of the opsonins to a series of receptors in a zipperlike fashion[193] leads to engulfment of the particle (Fig. 7). The pseudopods of the cell close around the particle until it is enclosed in a vacuole, the phagosome. The rate of ingestion by neutrophils is greater than that by monocytes and macrophages.[194] Eosinophils also are slower phagocytes than are neutrophils.[179]

3.2.1f. Intracellular Killing.   As soon as the particle (or the opsonins) makes contact with the cell membrane, oxidases in the membrane are triggered to activate oxygen-dependent microbicidal mechanisms.[195] Substances such as superoxide (O_2^-), hydrogen peroxide (H_2O_2), and hydroxyl radical (^-OH) are formed. During and after ingestion, the lysosomes (granules) fuse with the phagosome and pour their enzymes into the vacuole (degranulation) (Fig. 7). The lysosomal enzymes[196] react with the ingested particle. One lysosomal enzyme, myeloperoxidase, triggers the reaction of H_2O_2 with chloride, which yields the potent microbicidal product hypochlorite. A number of antimicrobial polypeptides in the lysosomes (e.g., cathepsin G, elastase, lysozyme, and defensins) contribute to microbial killing in an oxygen-independent way.[197] The formation of reactive nitrogen products, especially nitric oxide (NO), from L-arginine represents another potent microbicidal mechanism. The production of these products is under the control of inducible nitric oxide synthase (iNOS); iNOS is stimulated by cytokines such as TNF. The evidence regarding the role of NO in intracellular killing, especially of mycobacteria, comes from studies in rodents; its status in humans is less clear.[198] Peroxynitrite ($ONOO^-$) is a combined product of $NO-O_2^-$, and has microbicidal activity, e.g., against *Candida* species.[199,200] Inhibition of the pathways for either O_2^- or NO production inhibits microbial killing by $ONOO^-$. Production of $ONOO^-$ by macrophages is under the influence of IFN-γ.[200]

During the process of phagocytosis and intracellular killing of microorganisms, neutrophils usually die and are taken up by macrophages, which have a greater ability to survive and which digest much of the endocytosed material enzymatically. Certain microorganisms are not readily killed by normal macrophages, but can be digested when the latter are activated by products of stimulated T lymphocytes. These mechanisms are discussed in Section 3.2.3c.

Eosinophils are able to kill several species of meta-

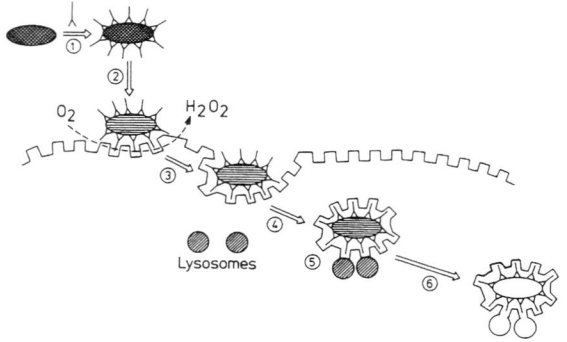

FIGURE 7. Phagocytosis and intracellular killing. A particle is opsonized (1) by opsonins (λ). The next step is recognition and binding (2) of the opsonins to a series of receptors in the membrane of the phagocytic cell. This binding triggers an oxidase in the membrane to activate oxygen-dependent microbicidal mechanisms. Meanwhile, the particle is engulfed by the cell membrane in a zipperlike fashion (3), and a phagosome is formed (4). Lysosomes fuse with the phagosome and pour their enzymes into the phagosomes (5,6).

zoa, this killing being largely an extracellular event. After attachment of the eosinophil to the parasite, a process mediated by IgE and probably also by complement, the parasite is damaged by exocytosed products of the eosinophil; in particular, major basic protein, eosinophil cationic protein, eosinophil peroxidase, and oxidative products.[180] IL-5 (see Section 3.1.5a) activates the effector function of eosinophils.[95]

3.2.1g. Clearance of Particles from the Bloodstream. The preceding sections described phagocyte function in the inflammatory field. However, there remains an important question: What happens to foreign particles (e.g., bacteria) that have gained access to the bloodstream? Circulating neutrophils and monocytes do not seem to play an important role in the elimination of such substances; macrophages in the liver (Kupffer cells) and spleen are especially important, however, for the uptake of this material.[201] The rate of clearance by these macrophages is dependent on the nature of the material, i.e., their charge and chemical composition, which determine binding to macrophage receptors or ligands (opsonins). In addition, the availability of opsonins and the functional state of the macrophages are important for the rate of clearance.[202,203] Opsonins not only increase the rate of clearance but also determine the site of uptake; particles opsonized by IgG are taken up mainly in the spleen, whereas C3b causes binding by Kupffer cells.[201,204]

3.2.1h. Other Functions of Phagocytes. Neutrophils contain two types of lysosomes, i.e., primary (azurophilic) granules and secondary (specific) granules, both of which contain different enzymes.[196] The primary granules tend to fuse with phagosomes (see Section 3.2.1f). Degranulation of specific granules, however, occurs earlier, at the time of nonspecific surface contact (e.g., during diapedesis), and the enzymes (e.g., lysozyme, lactoferrin, and vitamin B_{12}-binding protein) are released extracellularly by exocytosis.[196] In addition, cytokines like IL-1 and IL-1ra are produced and secreted by neutrophils. The secretory capacity of macrophages, however, is much greater. IL-1, TNF, IL-6, IL-8, IL-10, and other cytokines (see Section 3.1.5a), procoagulant factor, various complement factors, prostaglandins and other arachidonate metabolites, growth factors, and many other substances, in addition to the lysosomal enzymes, are secretory products of macrophages.[205] Apart from this secretory function, macrophages play a role in the processing and presentation of antigen to T cells[206,207] (see Section 3.2.3a).

3.2.1i. Deficient Phagocyte Function. The phagocytic cells constitute such a major defense system against infections that when there is a numeric or functional deficiency of these cells, bacterial or fungal infec-

tions are almost inevitable. A shortage of circulating neutrophils, as in idiopathic neutropenia, aplastic anemia, drug-induced agranulocytosis, and leukemia, predisposes to bacterial and fungal infections. There is a quantitative relationship between the number of circulating neutrophils and the incidence of infection[208,209]: at concentrations below 500 granulocytes/mm^3, and even more so below 100 granulocytes/mm^3, there is a high risk for these infections among patients with leukemia and aplastic anemia and to a lesser extent idiopathic granulocytopenia.[210] This difference in susceptibility to infection can probably be explained by the presence of damaged mucosa resulting from cytotoxic therapy for leukemia, such damaged mucosa providing an easy route of entry for microorganisms. In idiopathic granulocytopenic patients, the normal or even elevated monocyte counts also may compensate for the defect. In neutropenia, infectious complications usually arise insidiously, with little or no inflammatory signs and without formation of pus. Such infections usually run a fulminant course (see Chapter 15).

Qualitative defects can affect the various steps of the phagocytic process (see Fig. 6), each of which can be assessed in the laboratory. Congenital defects of adhesion molecules exist, e.g., LAD type 1 syndrome, which is caused by a defect in the β-integrins,[211] and LAD type 2, which is caused by a defect in the fucosylation of the ligands for the selectins.[212] As in the case of deficiency of chemotactic factors (e.g., C3 or C5) and defective chemotaxis (either congenital or acquired), these syndromes also are associated with recurrent bacterial and fungal infections. The congenital defects are more extensively reviewed in Chapter 14; the acquired defects are discussed in Sections 4–9.

During the late 1950s, it was shown that a delay in the migration of neutrophils of no more than 2 hr has a devastating effect on experimental infections.[213] In patients with defective chemotaxis, purulent infections of the skin, subcutaneous tissues, lymph nodes, and lungs are commonly encountered, whereas septicemia is a rare event. *Staphylococcus*, *Streptococcus*, *Candida* species, and *Escherichia coli* are the major pathogens. Opsonization of *S. pneumoniae* and *H. influenzae* in particular is impaired in hypogammaglobulinemia (see Section 3.1.6). C3 deficiency also leads to impaired opsonization and clinically to purulent infections caused by gram-positive and gram-negative bacteria as well as fungi. Similar opsonic defects may develop during massive complement consumption (see Section 3.1.2). Low plasma concentrations of fibronectin, as observed in shock, septicemia, and trauma, probably should not be considered indicative of a hypo-opsonic state[60] (see Section 3.1.3).

Primary defects of endocytosis by granulocytes and mononuclear phagocytes are rare, but a number of disease states and drugs have been reported to affect endocytosis (see Sections 4–7). Moreover, it has been well established in *in vivo* experiments that the uptake of particles from the circulation by macrophages can easily become saturated.[203] Patients with systemic lupus erythematosus (SLE) and other diseases in which circulating immune complexes are present[214–218] exhibit impaired clearance, probably due to saturation of receptors by the circulating complexes. Decreased Fc-mediated clearance has been found in end-stage renal disease[219] and also in AIDS.[220] Certain histocompatibility antigens (especially HLAB8/DR3 and DR2 haplotypes) have been associated with an Fc-receptor defect.[221,222] In primary biliary cirrhosis, defective C3b-mediated clearance has been demonstrated.[223] Impaired clearance of injected material (e.g., labeled aggregated albumin) has been described in alcoholic liver cirrhosis.[224–226] Fat emulsions administered intravenously (Intralipid) have been shown to produce mononuclear phagocyte blockade[227]; this blockade could contribute to greater susceptibility to generalized infection and possibly to a poor outcome of infection as well. High doses of intravenous gammaglobulin preparations may block Fc receptors in idiopathic thrombocytopenic purpura.[228] Such a mechanism could have a negative impact in compromised hosts receiving such preparations (e.g., hypogammaglobulinemic patients and bone marrow transplant recipients). Clinically, these gammaglobulin preparations seem to cause no such problems. Moreover, no evidence of Fc-receptor blockade was found in hypogammaglobulinemic patients treated with intravenous immunoglobulin.[229]

Defects of intracellular killing may be congenital or acquired; the congenital abnormalities such as chronic granulomatous disease are discussed in Chapter 14. Some of the acquired abnormalities of intracellular killing are discussed in Sections 4–8.

3.2.2. Natural Killer Cells

NK cells, which originate in bone marrow and probably belong to the lymphocyte series, are able to kill certain virus-infected cells as well as tumor cells.[230] These cells are not phagocytic and can carry out their cytotoxic function without previous sensitization, without involvement of the major histocompatibility system, and in the absence of antibody and complement. Phenotypically, they are described as large granular lymphocytes.[230] Killing of target cells is mediated by so-called perforins, originating from the granules. These substances are similar to the membrane attack complex of the terminal pathway of the complement cascade (see Section 3.1.2), in that they penetrate cell membranes and produce cell lysis.[231] The relative role of NK cells in host defense *in vivo* is not precisely known, but susceptibility to certain viruses [e.g., herpes simplex virus, cytomegalovirus (CMV)] can be correlated with NK-cell activity.[232,233] Deficient NK cell function is found in AIDS[234] (see Section 5.8).

An important interaction is that between the IFNs, IL-2, and NK cells, in which the activity of the latter is enhanced.[62,230] NK cells produce cytokines like IL-1 and IFN-γ. Since these killer cells have receptors for IgG, they also may function as killer cells in the presence of antibody, i.e., antibody-dependent cellular cytotoxicity.[230]

3.2.3. Cellular Immunity

Cellular immunity consists of a number of effector mechanisms in which T lymphocytes and macrophages interact.

3.2.3a. Initiation of Cellular Immune Response. To initiate an immune response in T lymphocytes (and B lymphocytes) (see Section 3.1.5), most antigens have to be processed and presented by accessory cells.[206] Mononuclear phagocytes act as accessory cells; in addition, dendritic cells[235] and Langerhans cells in the skin[236] (see Fig. 6) also present antigen efficiently. Whether the latter two cell types belong to the mononuclear phagocyte system is uncertain.[188] T cells of the helper–inducer type recognize antigen only when it is presented on the membrane of the antigen-presenting cell together with an HLA-class II molecule: HLA-D(R), HLA-DP, HLA-DQ (Fig. 8). For the activation and proliferation of T cells that respond specifically to a certain antigen, a number of cytokines, adhesion molecules (LFA-1, ICAM), and other accessory molecules (CD40, Fas) are required. For instance, antigen-stimulated macrophages release IL-1, IL-6, IL-12, and IL-18, which trigger the activation and proliferation of T lymphocytes.[80,96,116,237] Once triggered by antigen, IL-1, and IL-6, T lymphocytes produce IL-2 and IFN-γ,[90,237] which induces clonal expansion of activated T lymphocytes that respond to that specific antigen (Fig. 8).

3.2.3b. T Lymphocytes. T lymphocytes can be divided into effector T cells and regulatory cells[238] (see Fig. 8). Of the effector T cells, the cytotoxic T cells have been well characterized. These cells are able to kill virus-infected cells and tumor cells in the absence of antibody. For recognition of the target by cytotoxic T cells, class I (HLA-A, HLA-B, HLA-C) molecules must appear together with the antigen.[239] Since HLA class I molecules are expressed on every nucleated cell, this effector mech-

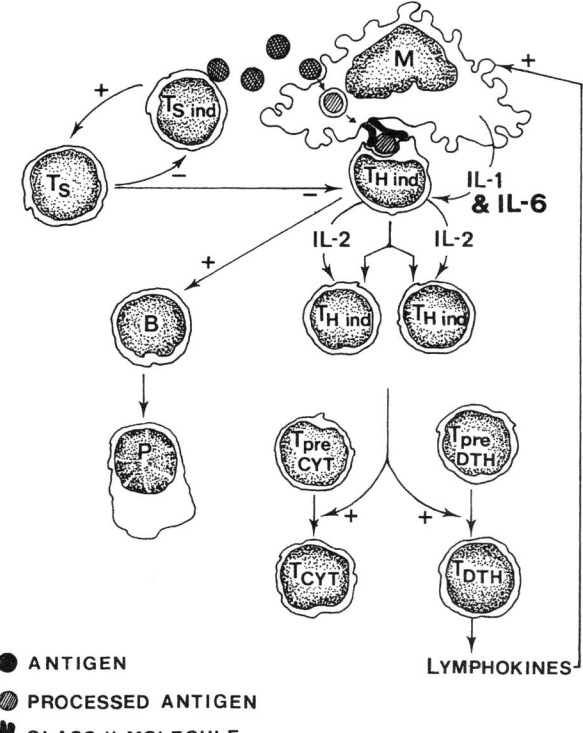

● ANTIGEN

◉ PROCESSED ANTIGEN

⊀ CLASS II MOLECULE

FIGURE 8. Macrophage–T-lymphocyte interaction. The macrophage (M) ingests the antigen; the antigen is then processed and presented on the cell membrane together with a class II molecule of the major histocompatibility complex. The presented antigen and the class II molecule are recognized by a T helper-inducer lymphocyte (T_{Hind}). The latter cell is activated by IL-1 and IL-6 secreted by the antigen-stimulated macrophage. The T lymphocyte produces IL-2, proliferates, stimulates B cells, and induces differentiation of effector T cells. The latter consist of cytotoxic T cells (T_{CYT}) and T cells that play a role in delayed-type hypersensitivity (T_{DTH}) and their precursors. The T_{DTH} cells produce lymphokines (mainly IFN-α), which activates the macrophage. T suppressor-inducer cells (T_{Sind}) respond to soluble antigen and induce suppressor T lymphocytes (T_S), which in turn inhibit T_{Hind} cells. Scheme designed by B. J. M. Zegers and J. W. M. van der Meer.

anism could be operational against all kinds of virus-infected cells. Another effector T cell is the T cell that plays a role in delayed-type hypersensitivity (DTH) reactions and produces lymphokines that activate macrophages.

Regulatory T cells can be subdivided into T-helper–inducer cells (synonym: T-helper cells) and suppressor T cells.[238] The helper–inducer cells regulate the proliferation and differentiation of not only effector T cells but also B cells (Fig. 8). During differentiation, T lymphocytes express phenotypically stable antigens that are easily recognizable with commercially available monoclonal

antibodies. However, it should be kept in mind that these phenotypes cannot be directly associated with function. For example, the CD4-positive population contains not only helper cells but also effector cells as well as suppressor–inducer cells.[240] T lymphocytes are important producers of cytokines and this property has been the basis for the TH1/TH2 concept, discussed in Section 3.1.5c.

3.2.3c. Activated Macrophages. With the microbicidal systems (see Section 3.2.1f), normal macrophages will not be able to kill a number of microorganisms. These microorganisms, which include protozoa (*Toxoplasma gondii*, *Leishmania* spp., *Trypanosoma* spp.), fungi (*Histoplasma capsulatum*, *Cryptococcus neoformans*, *Pneumocystis carinii*), bacteria (*Mycobacteria*, *Listeria monocytogenes*, *Salmonella* spp., *Brucella* spp., *Legionella* spp.), *Chlamydia*, and viruses, survive and even replicate inside the macrophages of a nonimmune individual.[241,242] When T-cell immunity arises, activated T-helper cells will produce cytokines, such as IL-12,[237] IFN-γ,[65] and TNF[243] (see Sections 3.1.4 and 3.1.5a), that activate the macrophages[65,244] (Fig. 8). On activation, several events take place in the macrophages; the oxygen-dependent and oxygen-independent (e.g., nitrogen products) microbicidal mechanisms (see Section 3.2.1f) become fully activated, leading to killing of the intracellular pathogens.

3.2.3d. Defective Cellular Immunity. All components of the cellular immune system are crucial to optimal functioning. Thus, abnormalities of, or even an imbalance among, regulatory T cells, effector T-cell deficiencies, deficient signaling between cells, and macrophage dysfunction may lead to decreased cellular immunity. Quite a number of molecular defects leading to cellular immunodeficiency have been discovered in recent years.[245] Examples are defects of the intracellular signaling pathways in T cells,[246] or intercellular signaling molecules, such as CD40/CD40 ligand[167] and Fas–Fas ligand,[247,248] or receptor defects, such as defects of the IL-12 and IFN-γ receptor.[77,78,107,108,249] (see also Section 3.1.5a). It should be stressed that not all disturbances of this delicate network necessarily lead to a state of cellular immunodeficiency.

The most impressive acquired deficiency of helper T cells is AIDS, in which the retrovirus HIV infects and destroys these T lymphocytes (see Section 5.8 and Chapter 16). In severe, chronic, inflammatory disease states (e.g., sarcoidosis[250]), an imbalance among regulatory T cells has been demonstrated. In some immunodeficiency states (congenital as well as acquired disorders), both the cellular and humoral immunities are defective. The quality of cellular immunity in patients can be measured to some extent with both *in vitro* and *in vivo* tests. An important *in vitro* test is still the lymphocyte transformation study using T-cell mitogens such as anti-CD3, phytohemagglutinin, and concanavalin A as well as anamnestic antigens. To sort out the nature of a defect, more sophisticated tests can be performed. Assessment of the intra- and intercellular signaling pathways is becoming increasingly important in the workup of presumed cellular immune defects.[251] Measurement of macrophage activation in patient material is difficult. *In vivo* tests include measurement of DTH skin reactions to recall antigens such as tuberculin, varidase, trichophyton antigen, *Candida* antigen, and mumps antigen, as well as a primary antigen such as *Helix pomata* hemocyanin,[252] after sensitization. A person who does not respond to the antigens is called "anergic."

In the face of true cellular immunodeficiency, infections caused by viruses and the intracellular pathogens (see Section 3.2.3c) may ensue. The relative importance of these pathogens is reflected by the fact that many are discussed in separate chapters of this volume. The congenital immunodeficiency diseases with major deficiencies of cell-mediated immunity (CMI) are discussed in detail in Chapter 14. Acquired disturbances of cellular immunity occur as a result of (1) malignant diseases involving the lymphoid system or the mononuclear phagocyte system, e.g., Hodgkin's disease and non-Hodgkin's lymphoma (see Section 5.1); (2) treatment with immunosuppressive drugs, e.g., glucocorticosteroids, azathioprine, cyclophosphamide, and cyclosporin A (see Section 6.2.2); (3) viral infections, such as cytomegalovirus (CMV) infection, Epstein–Barr virus (EBV) infection, probably hepatitis C, and especially HIV infection (see Section 5.8); (4) pregnancy[253]; (5) high estrogen status[254]; (6) protein energy malnutrition (see Section 4.2); and (7) aging (see Section 4.3). The mechanisms behind these states of cellular immunodeficiency are rather complex and differ in the various conditions mentioned, since they may involve either regulatory T cells, effector T cells, macrophages, soluble factors such as cytokines (production as well as response), or combinations thereof.

4. Factors that Influence the Quality of Host Defense

4.1. Genetic Control of Host Defense against Infection

The quality of the defense against a variety of microorganisms appears to be under genetic control. Animal experiments have demonstrated that genetic susceptibility and resistance involves various aspects of the humoral and cellular immune responses.[255] Also in humans, it is likely that susceptibility to most microorganisms is determined by a large number of polymorphic genes.[256] The degree of the antibody response to a series of antigens is associated with HLA and also with certain immunoglobulin allotypes (i.e., genetically variable markers within immunoglobulin molecules[257]). There is circumstantial evidence that survival during epidemics is under control of the HLA system.[258] In this context the association between certain common West African HLA antigens and protection from severe malaria may be relevant.[259] The HLA-DR background may directly control the quality of antigen presentation; however, since the genes for cytokines like TNF-α and lymphotoxin are located closely to those of HLA-DR, some of the effects of HLA on disease susceptibility may be due to genetic polymorphism of these cytokines.[260] For susceptibility to severe malaria and fulminant meningococcal infection, genetic polymorphism of cytokines such as TNF-α and IL-10 has been investigated directly.[261] From the published data no clear picture has emerged so far and more studies are awaited.

The genetic polymorphisms of Fc receptors and the control of the Fc-receptor function of phagocytic cells by HLA haplotypes are discussed in Section 3.2.1i. Another example of genetic surveillance of immune reactivity is encountered in mycobacterial diseases, notably tuberculosis and leprosy. Already many years ago it was found that the HLA type seems to determine the type of disease that develops.[262]

More recently a number of exciting discoveries have addressed the role of the NRAMP1 gene in host defense to infections. In mice, natural resistance to infection with some mycobacteria is influenced by the gene for natural-resistance-associated macrophage protein 1 (Nramp1), but the exact role of the human homologue of this gene, NRAMP1, in tuberculosis is unknown. Four NRAMP1 polymorphisms were each significantly associated with susceptibility to tuberculosis in West Africa.[263] Genetic polymorphisms of molecules such as the vitamin D receptor,[264] as well as the IFN-γ receptor and the IL-12 receptor determine severity and outcome of mycobacterial infection.[77,78,107,108,249] Despite these developments, our knowledge of the genetic control of host defense is still too limited to be taken into account clinically,[265] with the exception of administering recombinant cytokines such as IFN-γ in case of clear deficiency of IFN-γ or its inducers.[107]

4.2. Nutritional Status

Quantitatively inadequate protein-energy nutrition is an important problem in the world, not only for inhabitants of developing countries but also in many patients with a severe underlying illness requiring intensive treatment. In these patients, host defense mechanisms may deteriorate further, because malnutrition has an important impact on host defense and may lead to infectious complications. Chandra[266] has provided an impressive list of infections that are influenced by nutritional status. Of these, tuberculosis, bacterial diarrhea, bacterial and viral respiratory infections, *Pneumocystis carinii* infection, candidiasis, and aspergillosis should be mentioned. Host defense defects that are produced by protein-energy malnutrition are dependent on the degree of malnutrition and can be summarized from the vast amount of literature as follows[266,267]:

1. The skin and mucous membrane barriers may become somewhat impaired by thinning of the mucosa with a low lysozyme concentration and decreased secretory IgA levels.
2. Both pathways of the complement system are invariably impaired, which affects chemotaxis and especially opsonization.
3. Apart from the previously mentioned decrease in secretory IgA, the immunoglobulin levels and the antibody response to most protein antigens are normal.
4. Although the microbicidal function of neutrophils has been reported to be reduced, it is doubtful whether this reduction contributes to the increased susceptibility to infection.
5. Depressed NK cell activity has been found in malnourished children.[268]
6. Distorted thymic morphology, deficiency of some thymic factors, and an insufficiency of regulatory T cells (especially helper T cells) are probably responsible for the observed disorder of CMI.

A shortage of some vitamins may affect CMI.[266] There is now good evidence that vitamin A supplementation to children with mild vitamin A deficiency has an important impact on mortality from infection.[269–272]

Deficiencies of trace elements also may further impair host defense in already compromised patients. Zinc deficiency produces disturbed T-cell function, at least in experimental animals, and increases their susceptibility to *Listeria* and *Salmonella* species.[273] It is important to remember that zinc deficiency—believed to be responsible, at least in part, for the immunosuppression in protein-energy malnutrition—has been reported during total parenteral nutrition[274] and in sickle cell anemia (see Section 5.3). Supplements of zinc have been shown to affect the functions of phagocytic cells as well as those of T cells.[274]

Whether iron deficiency leads to a significantly increased susceptibility to infection in humans is uncertain.[275] However, killing of microorganisms by neutrophils and abnormal function of T lymphocytes have been found in patients with iron deficiency.[275,276] In contrast, iron repletion and iron overload predispose to infection, probably by making iron available to microorganisms as a nutrient.[277] This phenomenon had been observed already by Trousseau in the 19th century (cited by Murray *et al.*[277]). Iron is an essential growth factor for most bacteria. The organisms obtain iron by producing chelators called "siderophores," which bind iron for uptake into the bacteria. Several microorganisms, such as *Yersinia enterocolitica*, are thought to be of low virulence because they lack siderophores. During iron overload, listeriosis and severe yersiniosis have been reported, as well as the rare but severe infection with the fungi of the class of Zygomycetes, zygomycosis or mucormycosis.[278,279] These infections are seen particularly in patients treated with deferoxamine, since this agent can be utilized by the microorganisms as an exogenous siderophore, leading to increased iron uptake and enhanced outgrowth.

A phosphate deficit, which may occur during intravenous hyperalimentation, is associated with a decrease in the chemotactic, phagocytic, and microbicidal functions of granulocytes and clinically with fungal and bacterial infections.[280] Intravenous fat emulsions, which are associated with coagulase-negative staphylococcal bacteremia and Malassezia furfur fungemia,[281,282] are able to suppress the oxidative metabolism in neutrophils[283]; their effects on the mononuclear phagocyte system are discussed in Section 3.2.1i. In addition, lipoproteins enhance the growth of *Candida albicans* both directly and by interfering with plasma candidastatic factors.[284]

4.3. Age and Host Defense

Human beings are most susceptible to infection at the two extremes of life. Neonates can be considered compromised hosts, since they have an important route of entry for microorganisms (i.e., the umbilicus), a blood–cerebrospinal fluid barrier that allows bacteria easy access to the meninges, and immature host defense mechanisms. Of the latter, the initial absence of IgM and IgA, the weak antibody response to polysaccharide antigens,[285] the lower levels of complement factors (especially factors of the alternative pathways[286,287]), and the inadequate functioning of the phagocytic cells[288–291] are crucial. The chemotactic activity of both polymorphs and monocytes is low.[292,293] Cord blood serum does not provide optimal opsonization of microorganisms that require activation of the alternate complement pathway (e.g., *E. coli*).[287] Phagocytosis is generally normal, with the exception of the phagocytosis of group B streptococci by neonatal monocytes.[291] The microbicidal function of granulocytes and monocytes in newborns is a controversial subject. Marodi et al.[291] showed normal killing of *E. coli*, *S. aureus*, and group B streptococcus by neonatal granulocytes; for monocytes, they found defective killing of *S. aureus* and group B streptococcus and normal killing of *E. coli*. The relative importance of abnormal antigen presentation by monocytes, defective IFN-γ production, and NK cell function in neonates is difficult to assess.[68,294–296]

In the elderly, the decline in the quality of the first line of defense (i.e., atrophy and dryness of skin and mucous membranes), reduced vitality, and increased risk for trauma, together with retardation of the repair process, should probably be regarded as the major causes of increased susceptibility to infections. In addition, the primary and secondary humoral responses[297,298] have been reported to be reduced. Also, there is evidence for reduced oxidative metabolism of the neutrophils, which interestingly can be improved by the addition of recombinant growth hormone.[299] In aged individuals, CMI is suboptimal, probably due to impaired T-cell function.[300,301] The extent to which these changes in the second line of defense play a role in susceptibility to infection and outcome is unclear.

4.4. Body Temperature and Host Defense

Elevated body temperature directly affects the susceptibility to infection. First, the replication of certain viruses and bacteria is inhibited at higher temperatures.[302] Microorganisms may require more iron for growth at elevated temperatures.[275] Antibody synthesis as well as T-cell proliferation and function increase at higher temperatures.[303,304] These effects are explained by the greater efficacy of pyrogenic cytokines like IL-1 (see Section 3.1.5a) at higher temperatures. By contrast, NK cell activity seems to be impaired at higher temperatures.[305] Both production and activity of IFNs appear to be enhanced at higher temperatures.[306] The effects of elevated temperatures on the function of phagocytic cells are less straightforward, but in general, phagocytosis and intracellular killing are more efficient at higher temperatures.[306]

Thus, whether fever is beneficial in humans is clearly a difficult judgment; the answer most likely depends to a great extent on the type of infection. Appropriate management of body temperature could be important for compromised hosts, but our knowledge in this respect is still too limited to make this statement with certainty.

4.5. Stress and Host Defense

It is a common belief that psychological stress suppresses host defense against infection. Indeed, there is evidence that psychological stress has an influence on the function of T cells and NK cells.[307,308] The effects are most likely mediated by endogenous opioids, the hypothalamic–pituitary–adrenal axis, catecholamines, heat-shock proteins, and cytokines like IL-1.[308–311] The question whether psychological stress leads to an increased susceptibility to infection has been the subject of several studies (reviewed by Swartz[309]). It was demonstrated that psychological stress was associated with an increased risk of acute viral respiratory illness, a risk that was related to the amount of stress.[312] The effects of stress were consistent but small. There is no information on nonviral infections and on susceptibility in the already compromised host.

5. Diseases that Affect Host Defense

Throughout this chapter, diseases that affect the particular host defense mechanisms have been mentioned. In this section, a number of disease states associated with suppressed host defense mechanisms are discussed.

5.1. Leukemia and Lymphoma

The number of granulocytes is not the only factor that determines the susceptibility to infection in acute leukemia (see Section 3.2.1i); other factors include mucosal damage due to cytotoxic therapy, concomitant monocytopenia and lymphopenia, tissue infiltration leukemic cells, leukostasis in the vasculature, and use of intravascular catheters.

Bone marrow transplantation (BMT), an important therapeutic approach to acute leukemia, aplastic anemia, severe combined immunodeficiencies, and a number of inborn errors of metabolism, is accompanied by a series of infectious complications, depending on the phase of the BMT procedure.[313] Soon after BMT, there are infections associated with granulocytopenia and indwelling intravascular catheters; herpes simplex infections may also occur. If graft-versus-host disease of the intestinal tract develops, bacteremia and fungemia may arise in the gut. After BMT, there is a state of severe T-cell dysfunction,[314] during which CMV and varicella–zoster virus (VZV) infections are frequently encountered. Since the B-cell function is suboptimal[314] and there is little humoral immunity left after BMT, severe pneumococcal infections may develop after the patient is released from the hospital. A more detailed account of these infectious conditions and the management of such patients is given in Chapter 15.

In chronic lymphocytic leukemia, hypogammaglobulinemia may become so severe that recurrent infections caused by *Pneumococcus* and *Haemophilus influenzae* develop (see Section 3.1.6). Despite these infectious complications, immunoglobulin substitution in these cases does not seem to be cost-effective.[315] Similar problems may be encountered in other B-cell malignancies, such as myeloma. In hairy cell leukemia, infections caused by intracellular microorganisms, especially mycobacteria, are prominent and point to T-cell dysfunction.[316] In Hodgkin's disease, T-cell function is disturbed because of an excess of T-suppressor cells, particularly in the advanced stages and in certain histologic types of the disease.[317] T-cell dysfunction decreases CMI, which can be measured both *in vitro* and *in vivo*. Infections with intracellular microorganisms, especially VZV, thus can be explained. In Hodgkin's disease, the significance of a circulating inhibitor of chemotaxis[318] for the susceptibility to infection is unclear. Chemotherapy and radiotherapy cause not only impairment of T-cell function (see Sections 6 and 8) but also a decrease in the ability to mount specific antibodies against *H. influenzae* B.[319] In addition, splenectomy may enhance the risk of fulminant pneumococcal and *H. influenzae* infections (see Section 5.2). Although not yet studied as extensively, the host defense defects in other lymphoproliferative disorders are very likely to be quite similar to those found in Hodgkin's disease.

5.2. Splenectomy

For patients undergoing splenectomy during a staging procedure for hematologic malignancy or for treatment of acquired hemolytic anemia or thalassemia, the risk that they will develop overwhelming septicemia some time during their life is approximately 5%.[320–322] This risk is much lower after splenectomy due to traumatic rupture of the spleen. The septicemia is caused mainly by *S. pneumoniae* and *H. influenzae*, but the presence of such microorganisms as *Meningococcus*, *Staphylococcus*, and *Capnocytophaga canimorsus* (formerly DF2 bacillus) also has been reported.[320–323] The reasons for this increased susceptibility seem to be manifold. First, the architecture of the spleen is such that it can be considered as a sieve, in which macrophages are present at strategic positions. It is the most important organ for the removal of particles that are not opsonized by complement (see Section 3.2.1g). In fact, an increased amount of antibody has been found to be necessary for efficient clearance of opsonized particles after splenectomy.[324] The spleen also plays an important role in the humoral response: The primary immunoglobulin response takes place in the spleen,[325] and low levels of IgM have been observed after splenectomy in children.[326] Impaired antibody production against pneumococcal and other thymus-independent antigens has been reported in splenectomized adults.[327,328] A reduced level of the complement factor properdin, which could lead to suboptimal opsonization, has been demonstrated after splenectomy.[329] Furthermore, postsplenectomy there is a decrease in functional tuftsin (see Section 3.1.6), which also may be important.[330]

Because of the risk of pneumococcal infection after splenectomy, immunization with polyvalent pneumococcal vaccine is recommended. For patients undergoing elective splenectomy, it is good practice to administer the vaccine in the weeks before the splenectomy to obtain a better immune response. The polysaccharide vaccine is T-cell-independent and therefore does not induce immunologic memory. The protection provided, especially in the compromised host, is probably not longer than a couple of years. Revaccination is presumably necessary after 3–5 years. Although vaccination has been shown to be effective, infection may still occur, especially in patients with underlying disease and in small children.[330] Thus, antibiotic prophylaxis is still recommended for small children. For adults splenectomized for Hodgkin's disease, we advocate patient-initiated treatment with oral amoxicillin at the onset of a febrile illness. Now that a vaccine against *Haemophilus influenzae* type B (HIB) has become available, HIB vaccination in patients with splenectomy may be considered. However, protective antibodies against HIB are usually acquired naturally during childhood. Indeed, in a recent study, 100% of 561 patients splenectomized before 1993 (i.e., before HIB vaccine was introduced) had natural protective antibodies, which was in sharp contrast to their immunity to pneumococci.[331] Nevertheless, while studies assessing the value of HIB vaccination in this patient group are pending, it seems prudent to vaccinate splenectomized patients against HIB.[332]

5.3. Sickle Cell Anemia

Infections, especially pneumococcal septicemia, meningitis, and *Salmonella osteomyelitis*, are a major cause of morbidity and mortality in patients with sickle cell anemia.[333] The risk of infectious complications is highest for children with a palpable spleen before 6 months of age. Functional splenectomy, the result of repeated splenic infarctions, appears to be an important host defense defect. Johnston et al.[45] demonstrated deficient opsonization due to a defect in the alternate pathway of complement. These defective host defense mechanisms enhance the risk of pneumococcal complications. Susceptibility to *Salmonella* infections can be explained at least in part by a similar mechanism.[46] Suppressed CMI with zinc deficiency and decreased nucleoside phosphorylase activity was described in sickle cell disease.[334] The efficacy of the pneumococcal vaccine in patients with sickle cell disease has been demonstrated, but pneumococcal infections still occur[330,335] (see also Section 5.2).

5.4. Diabetes Mellitus

Because of the assumption that diabetes is associated with an increased susceptibility to infection, many investigators have studied host defense in this disease. A major question is whether diabetics should be regarded as truly compromised hosts.[336] Breaches in the first line of defense due to injections, diabetic vascular disease, and neuropathy are important determinants of infections. High concentrations of glucose in urine and secretions may promote colonization by *Candida* species and other microorganisms. Still, it is difficult to explain in this manner the association of diabetes mellitus with, for instance, malignant external otitis (caused by *Pseudomonas aeruginosa*).[337]

Rare but rather specific for the diabetic patient is rhinocerebral zygomycosis, which is especially manifested during diabetic ketoacidosis.[338] Zygomycosis (also called mucormycosis) is caused by fungi that are strongly dependent on iron supply and compete with serum apotransferrin for the uptake of iron.[339] It has been demonstrated that the change of pH during diabetic ketoacidosis reduces the affinity of serum apotransferrin to bind iron. Thus, during acidosis, increased amounts of iron may be available to the microorganism, leading to enhanced outgrowth.[340]

Abnormalities in the second line of defense in diabetics can be summarized as follows:

1. Immunoglobulin response appears to be normal in diabetic patients. Nonenzymatic glycosylation of complement factor 3 has been reported in diabetics; this may lead to impaired opsonization of, for example, *Candida albicans*.[53]
2. In a series of studies, the chemotactic activity of granulocytes was shown to be impaired[341–343]; these abnormalities are not associated with keto-

acidosis. The chemotactic responsiveness of monocytes has been found to be depressed, possibly as a consequence of autooxidative cell damage.[344] The delayed inflammatory response already noted in experimental diabetic animals many years ago[345] thus can be explained.

3. Phagocytic adherence and phagocytosis have been shown to be reduced at high glucose concentrations, high osmolarity, and low pH.[346,347] Abnormalities of the bactericidal function of granulocytes have also been described[348,349]; the impaired glucose metabolism of the phagocytes could well be the basis of the observed abnormalities.
4. Clearance studies with aggregated albumin demonstrate no abnormalities in patients with diabetes.[350]
5. The T-cell function can be considered normal in diabetic patients, with the possible exception of patients with poorly controlled diabetes.[351,352] Animal experiments have provided some evidence for altered CMI,[353] but the consequences for host defense in diabetic patients are unclear.

An interesting observation is the high frequency of diabetes among patients with myeloperoxidase deficiency and fungal infections.[354,355] Although the clinical importance of myeloperoxidase for the host defense against invasive candidiasis is not clear, the myeloperoxidase activity in leukocytes of diabetic patients with severe infections should be determined.

5.5. Chronic Renal Failure

Patients with chronic renal failure are said to be at increased risk for bacterial infection,[356] but it is questionable whether the chronic renal failure itself is an important factor.[357] Chronic renal failure per se does not seem to impair humoral defense mechanisms. However, discrete abnormalities of the cellular defense mechanisms have been reported. A decreased bone marrow pool reserve of granulocytes has been found in uremic patients.[358] Reduced neutrophil accumulation *in vivo*[359] and impaired neutrophil chemotaxis *in vitro*[360] have been observed. These defects can be attributed largely to impaired generation of chemotactic factors in uremic serum and are partially corrected by peritoneal dialysis, but not by hemodialysis.[360] Both granulocytic function, as measured by surface adherence, phagocytosis, and intracellular killing, and opsonization were shown to be normal in patients with chronic renal failure.[361] Phagocytosis by mononuclear phagocytes appears to be abnormal both *in vitro*[362] and *in vivo*.[219,363] CMI, as measured by the DTH skin reaction, is diminished in chronic renal failure and does not improve with hemodialysis.[364] Lymphocytopenia, which occurs in uremia, and suppression of the proliferative response by uremic serum are in agreement with this observation.[365] *In vitro*, the mitogenic and antigenic responses of lymphocytes from uremic patients were found to be normal by some investigators and abnormal by others.[366,367]

The situation becomes different when a treatment such as hemodialysis is taken into account. The first line of defense is damaged by multiple punctures, which may lead to intravascular infection, especially when prosthetic material is present. The effects of hemodialysis on complement activation and neutrophil kinetics are discussed in Section 3.2.1a. The fact that antibody response is abnormal in hemodialysis patients was suggested by hepatitis B vaccine trials.[368–371] This response can be enhanced by coadministration of IL-2.[372] In addition, abnormal CMI has also been observed in patients on hemodialysis.[373–375] Iron overload in hemodialysis patients increases the risk of bacteremia[376] (see Section 4.2). On the other hand, dialysis patients who are treated with deferoxamine to chelate aluminum or iron have been shown to be at risk for zygomycosis. The fungus requires iron as a growth factor and is able to utilize the exogenous deferoxamine as an exogenous siderophore in place of its endogenous siderophore rhizoferrin to facilitate iron uptake, leading to enhanced outgrowth.[338,377,378] In patients on chronic ambulatory peritoneal dialysis (CAPD), peritonitis is the most frequently encountered infectious complication,[379] and gram-positive microorganisms (especially *Staphylococcus epidermidis*) are the most common microorganisms. Exogenous contamination, the catheter, and intestinal disease (such as diverticulitis) play an important role in the pathogenesis. The phagocytic and microbicidal functions of peritoneal macrophages from CAPD patients appear to be adequate, whereas a serious deficiency of opsonins (IgG as well as C3) has been found in the peritoneal dialysis effluent.[380,381] In a prospective study, low heat-stable opsonic activity correlated with a high incidence of *S. epidermidis* peritonitis.[381] Interestingly, diabetic CAPD patients are not at increased risk for peritonitis,[382] and their peritoneal effluent exhibits high opsonic activity.[381]

5.6. Alcoholism and Hepatic Cirrhosis

Alcoholism is associated with a number of infectious complications, such as pneumonia and tuberculosis. In alcoholism, the first line of defense easily may become severely compromised. Repeated aspiration of stomach contents, a depressed cough reflex, reduced glottal closure, trauma, and lifestyle play a role in the pathogenesis of infections. Many disorders also are present in the second line of defense. The antibody response to new antigens is impaired, whereas the secondary response is normal.[383] Depressed complement activity sometimes found in alcoholics may be responsible for decreased chemotactic responsiveness.[341] Inhibition of granulocyte adherence can be attributed to alcohol,[384,385] and this may explain the leukopenia and a blunted granulocyte response to bacterial infection that have often been noted.[385] Phagocytosis and the microbicidal function of granulocytes are generally considered normal, whereas macrophage function (including clearance of particles from the bloodstream) and response to cytokines are abnormal[385,386]; also, CMI has been found to be depressed in severe alcoholism.[385,387] Fewer T cells and a reduced mitogenic responsiveness also have been described.[385] Alcoholism has an inhibitory effect on NK cell function.[388] In alcoholism, other factors usually contribute to the increased susceptibility to infection; often, there will be a concomitant protein-energy malnutrition and vitamin deficiency (see Section 4.2) as well as liver cirrhosis. In liver cirrhosis, septicemia and spontaneous bacterial peritonitis are the major infectious complications.[389] In patients with cirrhosis, serum has been shown to be less chemotactic for neutrophils than normal serum, probably due to the presence of a specific inhibitor.[390,391] Although usually normal, reduced complement activity and abnormal opsonization have been observed in some patients with alcoholic cirrhosis.[391,392] The concentration of complement factor C3 in ascitic fluid appears to be related to the risk of developing spontaneous bacterial peritonitis.[393] Phagocytosis and intracellular killing by granulocytes and monocytes are normal.[394] Clearance of particles from the bloodstream is impaired, partly as the result of a diminished hepatic blood flow.[223,224] Furthermore, the T-cell responses in cirrhosis may be depressed.[394] The relative roles of the reported abnormalities as well as their relevance for clinical practice are unclear.

5.7. Systemic Lupus Erythematosus, Rheumatoid Arthritis, and Felty's Syndrome

It is uncertain whether SLE per se predisposes to infection.[395,396] Such factors as immunosuppressive therapy may well account for the clinical impression of increased susceptibility to infection.[397] Herpes zoster, fungal, and CMV infections are not rare in SLE patients on immunosuppressive therapy.[398] A number of host defense defects have been found in SLE. Acquired complement deficiencies with defective generation of chemotactic factors and poor opsonization as well as a C5a inhibitor have been reported in SLE.[399–401] A serum inhibitor of both phagocytosis and degranulation could explain the decreased particle uptake *in vitro*,[402] while immune complexes are probably responsible for impaired particle clearance *in vivo*[214,215] (see Section 3.2.1g). Furthermore, granulocytopenia and possibly a decreased granulocyte reserve[403] could further contribute to defective host defense. The same holds for the frequently occurring lymphopenia (both T and B cells) and depressed CMI, which lead to anergy.[404]

Whether there is increased susceptibility to infection in rheumatoid arthritis is even more questionable, although a case can be made for infectious arthritis,[396] which may develop as a hematogenous infection or as a complication of intra-articular injections. Hypocomplementemia, the presence of immune complexes and rheumatoid factor, an impaired chemotactic response,[405] and depression of the bactericidal function of monocytes[406] have been suggested as explanations for increased susceptibility to infection. Local defense in joints may be impaired; a factor that impedes phagocytosis has been found in synovial fluid.[407]

The frequent infections encountered in Felty's syndrome (rheumatoid arthritis with splenomegaly and granulocytopenia) cannot be related solely to the degree of granulocytopenia.[408] The cause of the granulocytopenia is still controversial; impaired granulocytopoiesis, decreased granulocyte reserve, excessive margination, and increased neutrophil destruction have all been suggested.[396] The defects in the host defense mechanisms are similar to those seen in rheumatoid arthritis.[396] Severe hypocomplementemia may be a major factor in the susceptibility to infection in Felty patients.[52,408] There is no consensus regarding the therapeutic value of splenectomy in Felty's syndrome, but the risk of postsplenectomy septicemia should be taken into account (see Section 5.2).

5.8. Infections

Many microorganisms are able to diminish the resistance of the host, thus opening a route for infection with secondary pathogens. Via numerous mechanisms, microorganisms are able to compromise the first line of defense as well as humoral and cellular defense mechanisms (extensively reviewed by Mackowiak[409] and O'Grady and Smith[410]). Hypocomplementemia and blockading of the mononuclear phagocyte system are dealt with in Sections 3.1.2 and 3.2.1i. Whether the function of neutrophils and monocytes is impaired by viral or bacterial infection is a controversial subject.[411,412] An interesting pathogen in this

respect is *Capnocytophaga* species. *Capnocytophaga gingivalis*, *C. ochracea*, and *C. sputigena* are dental microorganisms that may cause invasive infection in the compromised host.[413] Reportedly, these *Capnocytophaga* species are able to induce a disorder of neutrophil migration *in vivo*.[414] Impairment of CMI during the infectious disease (by affecting either the macrophages or T-cell function) also is important. Viruses that are able to weaken cellular immunity include CMV, EBV, hepatitis C virus, and last but not least HIV. It is well known that CMV infection (which is more extensively discussed in Chapter 13) is accompanied by secondary infections with bacteria, fungi, and protozoa.[415–417] A low level of helper T cells and an increased number of suppressor T cells, as seen in CMV mononucleosis, may be an explanation.[418] However, others have found that such T-cell abnormalities are not produced by CMV in recipients of renal transplants treated with low-dose immunosuppressives.[419] The association of non-A, non-B hepatitis with prolonged kidney graft survival and increased susceptibility to infection, suggesting impairment of CMI, has been reported.[420] This notion is sustained by the observation that liver transplant recipients with HCV hepatitis have significantly more episodes of major infections than all other types of patients. The incidence of major bacterial infections is not higher in HCV patients, but the risk of fungal infections and CMV disease is significantly increased.[421]

During infection with HIV, the T lymphocytes that express the CD4 molecule (the receptor for the virus) disappear by formation of syncytia of these cells and by other cytotoxic mechanisms that have not been fully elucidated.[422] Macrophages, which bear the CD4 molecule, are less susceptible to killing by HIV. The macrophages may become functionally impaired to some extent (decreased chemotaxis and impaired killing)[423]; it is unclear whether this effect is directly due to the infection of these cells.

In symptomatic HIV-infected patients, the number of CD4-positive T lymphocytes decreases over time. When CD4-positive cell counts fall below 200/mm, opportunistic infections indicative of AIDS tend to occur (see Chapter 12). These are, first of all, the types of infections related to severely impaired CMI: *Pneumocystis carinii* pneumonia (and extrapulmonary pneumocystosis), CMV infection, VZV infection, herpes simplex infection, mycobacterial infection, toxoplasmosis, cryptococcosis, histoplasmosis, papovavirus infection, and salmonellosis (see Chapter 12). When patients respond to highly active antiretroviral therapy (HAART) with a clinically significant rise in CD4 cells, host defense against these opportunistic pathogens is restored.[424] The HAART-induced restoration of cell-mediated immunity may lead to paradoxical worsening of inflammation through increasing numbers and reactivity of CD4 cells, e.g., in mycobacterial infection.[425]

Interestingly, some opportunistic infections regularly seen in patients with impaired CMI, e.g., listeriosis, are relatively rare in AIDS patients.[426] A variety of infections occur in symptomatic HIV infection that cannot be easily explained by the impaired CMI. First, the exact defense defect underlying the severe and persistent mucosal candidiasis is not very well understood.[427] Even though neutropenia (due to the HIV infection itself or to drugs like zidovudine) and impaired neutrophil function[428,429] occur in AIDS, candidemia and disseminated candidiasis are remarkably rare in AIDS.

The host defense defects that permit infections with a number of intestinal pathogens, such as *Isospora belli*, *Cryptosporidium*, and *Microsporidium*, also are not elucidated, but probably the T-cell defect is responsible here.

The increased incidence of infections caused by encapsulated bacteria, such as pneumococci and *H. influenzae*,[430,431] points to defective humoral immunity. Indeed, antigen-specific B-lymphocyte responses are impaired, despite polyclonal B-cell activation.[234] This result also becomes clear from the impaired antibody response to protein and polysaccharide vaccines.[432–434] The reported deficiency of tuftsin (see Section 3.1.6) may contribute to the development of these infections.[170] An additional explanation for the occurrence of these bacterial infections and possibly also those caused by *S. aureus*[435] is the impaired Fc-receptor-mediated clearance in AIDS[220] (see Section 3.2.1i).

The defense against *Aspergillus* species also may be impaired, although it is unclear to what extent the deficient T-cell function contributes to this impairment[436,437]; factors that may contribute to the relatively rare *Aspergillus* infections in AIDS patients are neutropenia, corticosteroid therapy, marijuana use, previous *Pneumocystis* pneumonia and CMV disease, the development of cystic lung lesions, and antibiotics.[438] The impact of the impaired NK cell function that has been noted in AIDS is not clear.[234]

6. Immunosuppressive Drugs

6.1. Glucocorticosteroids

Although there is consensus that both excessive endogenous production[439] and therapeutically administered[440] glucocorticosteroids lead to increased susceptibility to infection, the magnitude of this problem for the individual patient is difficult to predict. A thorough meta-analysis revealed that the rate of infectious complications was not increased in patients given prednisone at a daily dose of less than 10 mg or a cumulative dose of less than 700 mg.[441] The glucocorticosteroids are able to affect many aspects of the host defense. The skin and mucous membrane barriers suffer relatively little damage, although atrophy and delayed healing after injury may play a role in the pathogenesis of infections. Of the humoral defense mechanisms, the complement system does not seem to be affected, whereas some changes in immunoglobulin levels have been observed.[442,443]

The major problems involve the cellular defense mechanisms. Although granulocytopoiesis seems to be enhanced by glucocorticosteroids[444] and the marrow pool reserve is mobilized by these drugs,[445] the negative effects on neutrophilic granulocytes seem to outweigh these advantages. At the site of inflammation, there is reduced accumulation of neutrophils as a result of impaired margination, probably due to reduced stickiness of the granulocytes and diminished chemotactic activity.[384,446,447] Glucocorticosteroids affect phagocytosis and intracellular killing by neutrophils only at very high concentrations; most investigators therefore agree that these effects are not of clinical importance. A fall in the number of eosinophils in the blood is seen after glucocorticosteroid administration.[179]

The mononuclear phagocyte system (MPS) also is affected by glucocorticosteroids. First, profound monocytopenia occurs after administration of these drugs.[446,448] Monocyte chemotaxis may become impaired,[449] but the effects on intracellular killing by mononuclear phagocytes are controversial.[449–451] The clearance of particles from the bloodstream has been shown to be reduced.[452] The production of a number of macrophage products (e.g., IL-1, TNF) is inhibited by glucocorticosteroids.[453,454] Preexposure of macrophages to IFN-γ or granulocyte colony-stimulating factor prevents such inhibition.[455,456] The reduced production of cytokines such as IL-1 may have important consequences for T-cell function, febrile response, and other aspects of the acute-phase response. Also, macrophage activation is impaired, probably mainly by inhibition of the response to cytokines.[457] Glucocorticosteroids also have effects on T lymphocytes (reviewed by Cupps and Fauci[458]). Within hours after administration of glucocorticosteroids, a profound lymphocytopenia (also involving B cells to some extent) occurs due to redistribution of these cells.[440] In contrast to the situation in mice, rats, and rabbits, lymphocytolysis is not an important mechanism in humans.[440]

The redistribution of lymphocytes as well as the effects on the MPS are thought to be the major reasons for altered functioning of lymphocytes during corticosteroid therapy.[458] From the foregoing, the anti-inflammatory and immunosuppressive effects of glucocorticosteroids are clear and the consequences for susceptibility to infection can be understood. These effects may even become enhanced by their influence on vascular permeability, wound healing, and a number of metabolic processes. The infections that occur are those associated with impaired phagocyte function, such as infections caused by staphylococci, gram-negative bacteria, *Candida*, and *Aspergillus* species,[459] as well as with suppressed CMI (see Sections 3.2.3c and 3.2.3d).

To avoid infectious complications, patients who need chronic glucocorticosteroid treatment should be converted to a single, alternate-day dose whenever possible.[440,446]

6.2. Other Immunosuppressive Agents

6.2.1. Cytotoxic Drugs

Cytotoxic drugs such as cyclophosphamide, azathioprine, and methotrexate (MTX) interfere with host defense mainly through their effects on cell proliferation. Thus, by inducing neutropenia, monocytopenia, and lymphocytopenia (the degree depending on the drug given), such immunosuppressive drugs may give rise to infectious complications. Cyclophosphamide is a potent lymphocytotoxic drug; by interference with the B-cell system, it is able to suppress especially the primary but also the secondary antibody responses,[460] and by inhibition of the T-cell response, the drug impairs CMI.[461,462] The thiopurines azathioprine and 6-mercaptopurine produce moderate suppression of both the humoral and the cellular immune responses. The views on the effects of thiopurines on immunoglobulin synthesis are conflicting.[463] Granulocyte function does not seem to be affected, whereas monocyte exudation is impaired,[464,465] NK-cell function is somewhat suppressed, and the effect of T-cell function is not pronounced.[466,467] MTX decreases immunoglobulin synthesis[468,469] and inhibits phagocytosis and intracellular killing by granulocytes.[470] In addition, MTX affects monocyte production and macrophage activation.[471,472] The production of IL-1 is decreased both *in vitro* and in patients with rheumatoid arthritis receiving their first dose of MTX, but not after prolonged treatment.[472,473] There is little effect on T-lymphocyte function.[468] The combination of immunosuppressive cytotoxic agents and glucocorticosteroids greatly affects the susceptibility to infections as well as the outcome of these infections.

6.2.2. Cyclosporin and Tacrolimus

Cyclosporin is a potent immunosuppressive drug that is widely used in transplantation medicine and for treatment of immunologic disorders. Its main site of action is the T-helper-inducer lymphocyte subpopulation; cyclosporin has no direct effect on the functions of B cells, macrophages, neutrophils, or NK cells.[474] Tacrolimus (FK506) has a similar mechanism of action.[475] Both drugs bind to cytoplasmatic T-cell receptors, leading to inactivation of calcineurin.[476] Calcineurin is a crucial enzyme in the signaling pathway inducing IL-2 gene transcription, and treatment with either of these calcineurin inhibitors blocks production of IL-2 (see Section 3.1.5) and lymphokines such as IFN-γ (see Section 3.1.4).[476,477] Controlled clinical studies have shown that in transplant recipients, cyclosporin therapy leads to increased graft survival and is associated with equal or lower incidence of bacterial or viral infections than conventional immunosuppressive agents (reviewed by Kim and Perfect[474]). Important side effects of cyclosporin include acute and chronic nephrotoxicity, which contributes not only to renal vasoconstriction and ischemia, but also to induction of transforming growth factor-β (TGF-β).[478] A discussion of the numerous interactions of cyclosporin with other drugs is beyond the scope of this chapter. Tacrolimus appears to be more potent than cyclosporin, leading to a lower incidence of acute and long-term rejection of both kidney and liver transplants than with cyclosporin-based immunosuppression.[479,480]

6.2.3. Mycophenolate Mofetil, Sirolimus, and IL-2 Receptor Antibodies

Mycophenolate mofetil (MMF) is an antiproliferative agent, which interrupts the DNA replication of T cells that proliferate after stimulation of the IL-2 receptor, analogous to the mechanism of action of azathioprine.[481] MMF now has largely replaced azathioprine in renal transplant patients, due to its potency to reduce the incidence of acute rejection.[482,483] However, MMF reportedly increases the risk of invasive cytomegalovirus infection.[482–484] Since the IL-2 receptor is expressed only on activated T cells, blocking the IL-2 receptor may provide highly specific immunosuppression. Two IL-2 receptor-blocking monoclonal antibodies, daclizumab and basiliximab, have been developed. Initial clinical trials have suggested that these agents are effective in preventing transplant rejection, without occurrence of opportunistic infections.[485–487] Sirolimus is a pharmacologic inhibitor of the intracellular signaling pathway of the IL-2 receptor.

Preliminary data suggest that sirolimus may lead to further reduction of allograft rejection when added to a cyclosporin-based regimen.[488] Other strategies currently under investigation include blocking of ligand–receptor interactions between antigen-presenting cells and T cells, e.g., aimed at blockade of CD28/B7, CD40/154, ICAM-1/LFA1, and LFA3/CD2.[489] The impact of these modalities on host defense to infection is still unclear.

6.2.4. Antimicrobials

In vitro studies have shown that numerous pharmacologic agents, e.g., antimicrobial drugs (reviewed by Hauser and Remington[490]), negatively affect the function of cells involved in the defense against microorganisms. The clinical significance of these findings is doubtful. The same holds for claims that certain antimicrobials are immunostimulating. Such claims have been published for macrolides,[491] certain cephalosporins (e.g., cefodizime),[492] ciprofloxacin,[493] fusidin,[494] and amphotericin B.[495,496] So far, these immunostimulating effects do not have practical implications.

6.2.5. Antilymphocyte Antibodies

Polyclonal antithymocyte globulin (ATG) and monoclonal antilymphocyte antibodies (such as OKT3, WT32) are potent immunosuppressive agents used mainly for prevention and treatment of transplant rejection.[497–500] Through the destruction of T lymphocytes, these agents lead to lymphocytopenia and suppression of CMI, and thus to increased susceptibility to infections (especially CMV infections).[497,501] The polyclonal ATG products also act on non-T cells (depending on brand and batch), which may further depress host defenses.[497] Using OKT3, Thistlethwaite *et al.*[498] reported a 22% incidence of infections after single exposure to the monoclonal antibodies and more than a 50% incidence after multiple exposures, and others have also reported that the incidence of infection is higher with the use of these polyclonal or monoclonal antibodies than with immunosuppressive drugs, such as glucocorticosteroids, azathioprine, and cyclosporin.[500–506]

6.3. Blood Products

The immunosuppressive effects of blood products have received much attention for two reasons: (1) survival of organ transplants can be improved by pretransplantation blood transfusion when the transfusion is matched for HLA-DR antigen,[507,508] and (2) immunologic abnormalities occur in hemophiliacs receiving repeated doses of plasma protein concentrates.[218,509] A detailed discussion of these observations goes beyond the scope of this chapter. Suffice it to say that depressed cellular immunity has been observed in HIV-negative hemophiliacs (abnormal T-cell subsets, defective NK cell activity) and in patients who have received multiple blood transfusions (reduced NK cell activity).[510] The significance for susceptibility to infection is unknown. Transmission by transfusion of infectious agents that modulate the immune response (e.g., CMV, hepatitis C) also should be taken into account (see Section 5.8).

7. Plasmapheresis

Plasma exchange is a treatment modality used in a variety of disorders; plasmapheresis is especially applied in diseases that seem to be immunologically mediated, and this treatment therefore often is instituted in patients who already are receiving immunosuppressive therapy. On the basis of a small retrospective study in rapidly progressive glomerulonephritis, it has been suggested that this procedure increases the incidence of life-threatening opportunistic infection.[511] However, this suggestion was not confirmed in a relatively large, randomized, controlled trial in severe lupus nephritis.[512] It is a common assumption that one of the modes of action of plasmapheresis is through the removal of circulating cytokines. However, in view of the short half-life of these molecules, this concept is not tenable.[513,514] The removal of mediators that have a lower endogenous clearance, such as soluble receptors for TNF-α and IL-6 as well as C-reactive protein, is well established.[515] Also, it has been suggested that anti-inflammatory mediators may be supplemented by plasmapheresis, such as C1-inhibitor, antithrombin III, and the anticoagulant factors protein S and protein C.[515]

8. Radiation

Radiation damages both proliferating and nonproliferating cells in a dose-dependent fashion. Thus, host defense mechanisms may be affected in a number of ways. The effects are dependent not only on the dose, but also on the time course and the radiosensitivity of the cells in the area of the body that is receiving radiation. At the level of the first line of defense, the epithelial barrier may be damaged; the rapidly proliferating epithelium of the gastrointestinal tract in particular is highly sensitive to radiation injury, and invasion by intestinal bacteria may ensue.

When B cells are irradiated, the primary antibody response is depressed for several weeks.[516] Secondary antibody response is much less sensitive to radiation damage, while plasma cells that produce antibody are relatively radioresistant. Cellular defense may be affected because of the radiosensitivity of the hematopoietic tissues. Granulocytopenia, which occurs after destruction of the mitotic pool and after depletion of the marrow pool reserve (i.e., not until 48–72 hr after total-body irradiation), is the major factor leading to infectious complications. After meningeal irradiation for acute lymphocytic leukemia, neutrophils acquire a transient microbicidal defect that probably reflects damage to the granulocyte precursors.[517] Mononuclear phagocyte precursors in bone marrow also are radiosensitive, whereas monocytes and macrophages are not. Since the latter cells are relatively long-lived, a deficiency of mononuclear phagocytes (e.g., in CMI) may not be immediately apparent.

Of the counterpart in cellular immunity—the T cells—it is known that thymocytes are radiosensitive, whereas mature T cells are radioresistant.[516] CMI may become suppressed when radiation precedes exposure to the antigen, whereas existing cellular immunity reactions are not hampered until large dosages are given.[518] Relapses of infections caused by dormant intracellular pathogens (e.g., *Mycobacterium tuberculosis*, VZV) may occur, especially when the focus of infection is within the area that received radiation. Total lymphoid irradiation,[519] which has been a part of the treatment of Hodgkin's disease and non-Hodgkin's lymphoma for more than two decades, is not accompanied by significant infectious complications. Nevertheless, there is profound and sustained immunosuppression,[519] which has also been exploited for the treatment of nonhematologic diseases.

9. Attempts to Strengthen Host Defense

When a patient exhibits a defective host defense, the prevention of infection is the main concern, followed by optimal treatment once infection has developed. Prevention of both damage to the first line of defense and colonization that could lead to infection with (multiresistant) microorganisms is important, as are the prevention and control of the adverse effects of treatment. Thanks to modern biotechnology (recombinant DNA technology and monoclonal antibody techniques), a number of modalities are now emerging for the substitution of host defense defects, for the strengthening of certain defense

mechanisms, and for the treatment of a number of specific infections. Before we briefly review these strategies, it should be stressed that such measures as physiotherapy, improvement of nutritional status, and immunization with relevant vaccines remain crucial. The most important vaccines that are used in the compromised host are hepatitis B vaccine (see Section 5.5), pneumococcal polysaccharide vaccine (see Section 5.2), and *Haemophilus influenzae* type B conjugate vaccine.[520]

Still, very little can be done to improve the first line of defense, although we can find some pointers in the treatment of burns, which involves not only creams such as silver sulfadiazine to strengthen the barriers, but also skin grafting and even covering the injury with *in vitro* cultured epithelium.[509] Acceleration of wound healing using epidermal growth factor and maybe other cytokines are potentially important treatment modalities.[521,522] An elegant way of strengthening the first line of defense against urinary tract infections in postmenopausal women is through the local administration of estrogens. In this way, the glycogen in the vaginal epithelium is restored, allowing recolonization with lactobacilli.[523,524]

Various options are available to improve the second line of defense. Administration of fresh plasma or supplementation of individual complement components is worthwhile in primary complement-deficiency states.[36] Whether such supplementation is beneficial in secondary complement deficiencies is not known. Substitution with fibronectin-rich cryoprecipitates during sepsis and trauma does not seem to be effective[60] (see Section 3.1.3). IFN-α has an important role as antiviral agent in the therapy of chronic viral diseases, such as for genital warts[525] and for chronic hepatitis B and C. In patients with active hepatitis B, IFN-α treatment leads to viral eradication and seroconversion in 25–40%.[526,527] Long-term follow-up now has established a significant improval of survival in IFN-α-treated patients with chronic hepatitis B and cirrhosis.[528] In patients with hepatitis C, of whom 70 to 85% will go on to develop chronic hepatitis, treatment with IFN-α alone leads to a sustained response in 15 to 20%.[529] With the addition of ribavirin to IFN-α monotherapy, the response rate increases to approximately 50%.[530]

IFN-γ increases the resistance to a number of infections, probably by enhancing cellular defense (see Section 3.2.3c). Treatment with IFN-γ had a beneficial effect when administered to patients with lepromatous leprosy in some,[531] but not all studies.[532,533] In combination with pentavalent antimony, IFN-γ is effective in visceral leishmaniasis.[534,535] Impressive results have been obtained with this cytokine used prophylactically and therapeutically in chronic granulomatous disease.[536,537] The incidence of invasive aspergillosis in the group of CGD patients who received IFN-γ prophylaxis was significantly reduced from 24 to 4% in 2 years[536] (see Chapter 14). Although the mode of action of IFN-γ at the molecular level in this disorder is not understood, neutrophils from patients on IFN-γ were found to produce significantly more damage to *Aspergillus* hyphae *in vitro*.[538] IFN-γ also has been used successfully in refractory tuberculosis and atypical mycobacterial infection.[539,540] It is likely that more indications for IFN-γ as a therapeutically useful immunostimulator for infections in the compromised host will emerge during the next decade.

Animal models have demonstrated that early treatment with IL-1 protects against death from bacterial and candidal infection, even in neutropenic animals.[82,541] Recombinant TNF seems to be somewhat less active in such infections,[542] but may be therapeutically useful in mycobacterial infections.[83,244] These cytokines have not yet been used for the treatment of infections in humans because of toxicity.

The proinflammatory cytokines may become useful as vaccine adjuvants. IL-2 already has been used as such for low responders to hepatitis B vaccine.[372] IL-2 also has been applied therapeutically in a few patients with severe immunodeficiency whose T lymphocytes did not produce IL-2.[543,544] Furthermore, intradermal injections of this cytokine in lepromatous leprosy have led to a reduction in the number of bacilli in the skin.[545] It should be kept in mind that a high incidence of bacteremia due to *Staphylococcus aureus* and *S. epidermidis* has been encountered in patients treated with high dosages of IL-2; this result may be explained by an acquired chemotactic defect in the neutrophils of these patients.[546]

Promising results have been obtained with the colony-stimulating factors GM-CSF and G-CSF (see Section 3.1.5b). In cancer patients, colony-stimulating factors have been used both to shorten the duration of chemotherapy-induced granulocytopenia[547] and as adjunctive therapy in patients with febrile neutropenia.[548] The potential beneficial role of rGM-CSF in the prevention of fungal infections has been suggested in a placebo-controlled study of patients with acute myelogenous leukemia.[549] rGM-CSF was associated with a higher rate of complete response, longer overall survival, and lower fungal infection-related mortality.[549] In spite of the concept that neutropenia is a strong risk factor for infections, all other studies have failed to demonstrate a benefit of CSF-induced accelerated bone marrow recovery on either the incidence of invasive infections or mortality in cancer patients. This is in sharp contrast with the beneficial results of prolonged rG-CSF treatment in patients with

congenital neutropenia, leading to a significant reduction in infections.[124,550] Several studies have addressed the beneficial role of rG-CSF on neutrophil function in nonneutropenic patients.

Treatment with rG-CSF accelerates radiological improvement of severe community-acquired pneumonia and reduces the rate of serious complications (e.g., empyema, adult respiratory distress syndrome, and disseminated intravascular coagulation).[126] Preclinical studies have demonstrated the benefit of G-CSF in the treatment of disseminated candidiasis in nonneutropenic animals.[551] Subsequently, a randomized clinical study has suggested that the combination of rG-CSF with fluconazole is beneficial in nonneutropenic patients with invasive or disseminated candidiasis.[127]

Substitution with IgG, given either intravenously[552] or subcutaneously,[553] is indicated in severe hypogammaglobulinemia (see Chapter 14). The latter method is particularly useful in patients who experience serious side effects with intravenous or intramuscular immunoglobulins or who have circulating anti-IgA.[553,554] High-titered specific antibody preparations played an important role in the treatment of varicella infections in immunocompromised children before acyclovir became available.[555] At present, anti-CMV IgG is given prophylactically to prevent CMV pneumonitis after bone marrow transplantation and therapeutically in conjunction with the antiviral drug ganciclovir[556–558] (see Chapter 17). In bone marrow transplant recipients, intravenous immunoglobulins appear to be beneficial for other infections as well.[559] In neutropenic patients, it is important to also consider plasma supplementation, since a deficiency of opsonins has been noted in such patients.[555,560] Administration of immunoglobulin can block T-cell activation by staphylococcal and streptococcal superantigens *in vitro*.[561] High doses of intravenous immunoglobulin have been suggested to be beneficial in patients with streptococcal toxic shock syndrome caused by group A streptococcus.[562] Randomized studies are under way to further explore this modality. In infants with prematurity or bronchiopulmonary dysplasia, who are at high risk for acquiring respiratory syncytial virus (RSV) infection, prophylaxis with RSV-IgG leads to a reduction in incidence and severity of RSV infection.[563,564] This modality is now replaced by monthly injections of palivizumab, a monoclonal IgG antibody that binds to RSV, resulting in a significant reduction of RSV infections in high-risk patient groups.[565]

Treatment of life-threatening gram-negative infection with passive immunization with monoclonal antibodies against endotoxin core so far has failed[566,567] (see

Chapter 4). Even in patients with fulminant meningococcal septicemia, which is considered the prototype of severe endotoxinemia, monoclonal antibodies against endotoxin did not alter the course of the disease.[568]

On the basis of experimental animals,[130,569] clinical trials have attempted to modulate the proinflammatory cytokine response in patients with severe sepsis, using monoclonal antibodies against TNF-α, recombinant soluble TNF-α receptor, or recombinant IL-1 receptor antagonist. In contrast to the favorable results in animal models of endotoxin challenge or serious infection, the clinical studies so far have been disappointing.[570–572] The reasons for these failures are multifactorial: the clinical syndrome of sepsis is very heterogeneous with a vast number of confounders; the intervention is started relatively late in the course of the illness; the optimal dose and duration of treatment are unknown; a single intervention may not be sufficient; and the 28-day mortality is an unrealistic endpoint.[573,574] In sharp contrast with these negative results are those obtained in rheumatoid arthritis[575] and in Crohn's disease, in which the anticytokine strategies (especially anti-TNF treatments) have met with spectacular results.[576] The possible downside of anti-TNF treatment is that such strategies have a negative effect on infections that are to be contained in granulomas, such as those caused by mycobacteria and fungi, and in abscesses.[577]

There are uncontrolled data in the literature to support plasma exchange in fulminant meningococcal septicemia (see Section 7).[513,578,579]

In severely neutropenic patients with bacterial infection, granulocyte transfusions have been given as an adjunct to antimicrobial therapy for a number of years. The enthusiasm for this approach has vanished for cost-benefit as well as risk-benefit reasons[580] and also because expectations were that treatment of neutropenic patients with rCSF would have a great impact on the incidence and severity of infections. For some patients with phagocyte dysfunction (as in chronic granulomatous disease), granulocyte transfusions have been beneficial,[581] but in these patients, treatment with IFN-γ is the primary therapy of choice.[536] In refractory infection in patients with phagocyte disorders, as well as in neutropenic patients with refractory infection, many centers now have begun to administer granulocyte transfusions derived from donors pretreated with rG-CSF.[125] This pretreatment of donors with daily injections of rG-CSF leads to a four- to tenfold increase of their absolute neutrophil count, greatly enhances the yield of neutrophils, and by preserving the granulocytes by inhibition of apoptosis, the 1- and 24-hr posttransfusion neutrophil counts in the recipients are substantially higher.[582]

There also is some experience with the local instillation of donor granulocytes, e.g., in abscesses. Transfusions of monocytes, which have a longer life span, have not yet been given. For treatment of phagocyte disorders, quite a number of drugs (e.g., ascorbic acid, levamisole, cimetidine) have been tried, mostly without success.[581] Bone marrow transplantation cannot be considered first-line treatment for these disorders.[581]

To restore T-cell function, thymus transplantation or the administration of either thymic hormones or (antigen-specific) transfer factor has been undertaken in selected patients.[583] Lymphocyte transfusions, which are not a common form of treatment, have not been successful in AIDS patients. Adenosine deaminase (ADA) deficiency results in severe combined immune deficiency disease, which is fatal without treatment. The cause for this is believed to be the accumulation of one of the substrates for ADA, 2′-deoxyadenosine to which especially T cells are hypersensitive. For ADA deficiency and other disorders like Wiskott–Aldrich syndrome, bone marrow transplantation has been shown to be a solution (see Chapter 14). Weekly injections of ADA coupled to polyethylene glycol (to prolong the half-life of the enzyme) have corrected the immune defect in ADA deficiency.[584] Efficient transfer of a recombinant ADA gene into hematopoietic stem cells is a therapeutic option if it results in the outgrowth of a "genetically repaired" lymphoid system. Several patients have now been treated with ADA gene therapy, although persistence of the gene has been limited until now.[585,586]

In animal experiments, resistance to infection has been enhanced by immunomodulators such as endotoxin, detoxified endotoxin, *Corynebacterium parvum*, BCG and muramyl peptides, and glucan.[542] It is likely that these therapies work by induction of cytokines. It is our expectation that treatment with cytokines will have a greater future than is currently realized.

References

1. Mackowiak PA: The normal microbial flora. *N Engl J Med* **307:** 83–93, 1982.
2. Abraham SN: Host defenses against adhesion of bacteria to mucosal surfaces. In Gallin JI, Fauci AS (eds): *Advances in Host Defense Mechanisms*, Vol. 4. Raven Press, New York, 1985, pp. 63–88.
3. Savage DC: Colonization by and survival of pathogenic bacteria on intestinal mucosal surfaces. In Britton G, Marschall KC (eds): *Adsorption of Microorganisms to Surfaces*. John Wiley, New York, 1980, pp. 175–206.
4. Guiot HF: Role of competition for substrate in bacterial antagonism in the gut. *Infect Immun* **38:**887–892, 1982.

5. Van der Waay D, Berghuis-de Vries JM, Lekkerkerk-van der Wees JEC: Colonization resistance of the digestive tract and the spread of bacteria to the lymphatic organs in mice. *J Hyg* **70:**335–342, 1972.

6. Heczko PB, Pryjma J, Kasprowicz J: Influence of host and parasite factors on the nasal carriage of staphylococci. In Jeljaszawicz J (ed): *Staphylococci and Staphylococcal Infection*. Karger, Basel, 1973, pp. 581–594.

7. Young LS, LaForce FM, Head JJ, Feeley JC, Bennett JV: A simultaneous outbreak of meningococcal and influenza infections. *N Engl J Med* **287:**5–9, 1972.

8. Sanford BA, Shelokov A, Ramsay MA: Bacterial adherence to virus-infected cells: A cell culture model of bacterial superinfection. *J Infect Dis* **137:**176–181, 1978.

9. Kominos SD, Copeland CE, Grosiak B, Postic B: Introduction of *Pseudomonas aeruginosa* into a hospital via vegetables. *Appl Microbiol* **24:**567–570, 1972.

10. Pollack M, Charache P, Nieman RE, Jett MP, Reimhardt JA, Hardy PH Jr: Factors influencing colonisation and antibiotic-resistance patterns of gram-negative bacteria in hospital patients. *Lancet* **2:**668–671, 1972.

11. Beachey EH: Bacterial adherence: Adhesin-receptor interactions mediating the attachment of bacteria to mucosal surface. *J Infect Dis* **143:**325–345, 1981.

12. Johnson JR: Virulence factors in *Escherichia coli* urinary tract infection. *Clin Microbiol Rev* **4:**80–128, 1991.

13. Kinsman OS, McKenna R, Noble WC: Association between histocompatability antigens (HLA) and nasal carriage of *Staphylococcus aureus*. *J Med Microbiol* **16:**215–220, 1983.

14. Noble WC: Carriage of *Staphylococcus aureus* and beta haemolytic streptococci in relation to race. *Acta Derm Venereol* **54:**403–405, 1974.

15. Winkler J, Block C, Leibovici L, Faktor J, Pitlik SD: Nasal carriage of *Staphylococcus aureus*: Correlation with hormonal status in women. *J Infect Dis* **162:**1400–1402, 1990.

16. Thadepalli H, Chan WH, Maidman JE, Davidson EC Jr: Microflora of the cervix during normal labor and the puerperium. *J Infect Dis* **137:**568–572, 1978.

17. Mackowiak PA, Martin RM, Smith JW: The role of bacterial interference in the increased prevalence of oropharyngeal gram-negative bacilli among alcoholics and diabetics. *Am Rev Respir Dis* **120:**589–593, 1979.

18. Tuazon CU, Sheagren JN: Increased rate of carriage of *Staphylococcus aureus* among narcotic addicts. *J Infect Dis* **129:**725–727, 1974.

19. Chow JW, Yu VL: *Staphylococcus aureus* nasal carriage in hemodialysis patients. Its role in infection and approaches to prophylaxis. *Arch Intern Med* **149:**1258–1262, 1989.

20. Klein RS, Recco RA, Catalano MT, Edberg SC, Casey JI, Steigbigel NH: Association of *Streptococcus bovis* with carcinoma of the colon. *N Engl J Med* **297:**800–802, 1977.

21. Alpern RJ, Dowell VR Jr.: *Clostridium septicum* infections and malignancy. *JAMA* **209:**385–388, 1969.

22. Black PH, Kunz LJ, Swartz MN: Salmonellosis—A review of some unusual aspects. *N Engl J Med* **262:**811–817, 864–870, 921–927, 1960.

23. Reichert U, Saint Leger D, Schaefffer H: Skin surface chemistry and microbial infection. *Semin Dermatol* **1:**91–100, 1982.

24. Smith RF: Lactic acid utilization by the cutaneous Micrococcaceae. *Appl Microbiol* **21:**777–779, 1971.

25. Blank I, Oawes RK: The water content of *Stratum corneum*: The importance of water in promoting bacterial multiplication on cornified epithelium. *J Invest Dermatol* **31:**141–145, 1958.

26. Newhouse M, Sanchis J, Bienenstock J: Lung defense mechanisms (first of two parts). *N Engl J Med* **295:**990–998, 1976.

27. Giannella RA, Broitman SA, Zamcheck N: Influence of gastric acidity on bacterial and parasitic enteric infections. A perspective. *Ann Intern Med* **78:**271–276, 1973.

28. Binder HJ, Filburn B, Floch M: Bile acid inhibition of intestinal anaerobic organisms. *Am J Clin Nutr* **28:**119–125, 1975.

29. Hinman F Jr, Cox CE: The voiding vesical defense mechanism: The mathematical effect of residual urine, voiding interval and volume on bacteriuria. *J Urol* **96:**491–498, 1966.

30. Klebanoff SJ, Hillier SL, Eschenbach DA, Waltersdorph AM: Control of the microbial flora of the vagina by H_2O_2-generating lactobacilli. *J Infect Dis* **164:**94–100, 1991.

31. Strominger JL, Tipper DJ: Structure of bacterial cell walls: The lysozyme substrate. In Osserman E, Canfield W (eds): *Lysozyme*. Academic Press, New York, 1974, pp. 169–184.

32. Masson PL, Heremans JF, Schonne E: Lactoferrin, an iron-binding protein in neutrophilic leukocytes. *J Exp Med* **130:**643–658, 1969.

33. McClelland DB, van Furth R: *In vitro* synthesis of lysozyme by human and mouse tissues and leucocytes. *Immunology* **28:**1099–1114, 1975.

34. Joiner KA, Brown EJ, Frank MM: Complement and bacteria: Chemistry and biology in host defense. *Annu Rev Immunol* **2:**461–491, 1984.

35. Daha MR: Biological properties of immune complexes. *Neth J Med* **27:**375–379, 1984.

36. Figueroa JE, Densen P: Infectious diseases associated with complement deficiencies. *Clin Microbiol Rev* **4:**359–395, 1991.

37. Alper CA, Colten HR, Gear JS, Rabson AR, Rosen FS: Homozygous human C3 deficiency. The role of C3 in antibody production, C-1s-induced vasopermeability, and cobra venom-induced passive hemolysis. *J Clin Invest* **57:**222–229, 1976.

38. Roord JJ, Daha M, Kuis W, *et al*: Inherited deficiency of the third component of complement associated with recurrent pyogenic infections, circulating immune complexes, and vasculitis in a Dutch family. *Pediatrics* **71:**81–87, 1983.

39. Snyderman R, Durack DT, McCarty GA, Ward FE, Meadows L: Deficiency of the fifth component of complement in human subjects. Clinical, genetic and immunologic studies in a large kindred. *Am J Med* **67:**638–645, 1979.

40. Miller ME, Koblenzer PJ: Leiner's disease and deficiency of C5. *J Pediatr* **80:**879–880, 1972.

41. Ellison RTD, Kohler PF, Curd JG, Judson FN, Reller LB: Prevalence of congenital or acquired complement deficiency in patients with sporadic meningococcal disease. *N Engl J Med* **308:**913–916, 1983.

42. Newman SL, Vogler LB, Feigin RD, Johnston RB Jr: Recurrent septicemia associated with congenital deficiency of C2 and partial deficiency of factor B and the alternative complement pathway. *N Engl J Med* **299:**290–292, 1978.

43. Kluin-Nelemans HC, van Velzen-Blad H, van Helden HP, Daha MR: Functional deficiency of complement factor D in a monozygous twin. *Clin Exp Immunol* **58:**724–730, 1984.

44. Ziegler JB, Alper CA, Rosen RS, Lachmann PJ, Sherington L: Restoration by purified C3b inactivator of complement-mediated function *in vivo* in a patient with C3b inactivator deficiency. *J Clin Invest* **55:**668–672, 1975.

45. Johnston RB Jr, Newman SL, Struth AG: An abnormality of the alternate pathway of complement activation in sickle-cell disease. *N Engl J Med* **288:**803–808, 1973.

46. Hand WL, King NL: Serum opsonization of salmonella in sickle cell anemia. *Am J Med* **64:**388–395, 1978.

47. Fearon DT, Ruddy S, Schur PH, McCabe WR: Activation of the properdin pathway of complement in patients with gram-negative bacteremia. *N Engl J Med* **292:**937–940, 1975.

48. Rytel MW, Dee TH, Ferstenfeld JE, Hensley GT: Possible pathogenetic role of capsular antigens in fulminant pneumococcal disease with disseminated intravascular coagulation (DIC). *Am J Med* **57:**889–896, 1974.

49. Greenwood BM, Onyewotu II, Whittle HC: Complement and meningococcal infection. *Br Med J* **1:**797–799, 1976.

50. Dance DAB, Smith CL: Complement deficiency and sporadic meningococcal disease. *N Engl J Med* **309:**615–616, 1983.

51. Lehman TJ, Bernstein B, Hanson V, Kornreich H, King K: Meningococcal infection complicating systemic lupus erythematosus. *J Pediatr* **99:**94–96, 1981.

52. Breedveld FC, Lafeber GJ, van den Barselaar MT, van Dissel JT, Leijh PC: Phagocytosis and intracellular killing of *Staphylococcus aureus* by polymorphonuclear cells from synovial fluid of patients with rheumatoid arthritis. *Arthritis Rheum* **29:**166–173, 1986.

53. Hostetter MK: Handicaps to host defense. Effects of hyperglycemia on C3 and *Candida albicans*. *Diabetes* **39:**271–275, 1990.

54. Lachmann PJ: Antibody and complement in viral infections. *Br Med Bull* **41:**3–6, 1985.

55. Proctor RA: Fibronectin: A brief overview of its structure, function, and physiology. *Rev Infect Dis* **9**(Suppl 4)**:**S317–S321, 1987.

56. Tamkun JW, Hynes RO: Plasma fibronectin is synthesized and secreted by hepatocytes. *J Biol Chem* **258:**4641–4647, 1983.

57. Bevilacqua MP, Amrani D, Mosesson MW, Bianco C: Receptors for cold-insoluble globulin (plasma fibronectin) on human monocytes. *J Exp Med* **153:**42–60, 1981.

58. Proctor RA: Fibronectin: An enhancer of phagocyte function. *Rev Infect Dis* **9**(Suppl 4)**:**S412–S419, 1987.

59. Mosher DF, Williams EM: Fibronectin concentration is decreased in plasma of severely ill patients with disseminated intravascular coagulation. *J Lab Clin Med* **91:**729–735, 1978.

60. Grossman JE: Plasma fibronectin and fibronectin therapy in sepsis and critical illness. *Rev Infect Dis* **9**(Suppl 4)**:**S420–S430, 1987.

61. Shirakami A, Shigekiyo T, Hirai Y, *et al*: Plasma fibronectin deficiency in eight members of one family. *Lancet* **1:**473–474, 1986.

62. Balkwill FR: Interferons. *Lancet* **1:**1060–1063, 1989.

63. Joklik WK: The molecular basis of the antiviral activity of interferons: Introductory remarks. *Ann NY Acad Sci* **350:**432–440, 1980.

64. Schellekens H, Weimar W, Cantell K, Stitz L: Antiviral effect of interferon *in vivo* may be mediated by the host. *Nature* **278:**742, 1979.

65. Murray HW: Interferon-γ, the activated macrophage, and host defense against microbial challenge. *Ann Intern Med* **108:**595–608, 1988.

66. Arenzana-Seisdedos F, Virelizier JL, Fiers W: Interferons as macrophage-activating factors. III. Preferential effects of interferon-gamma on the interleukin 1 secretory potential of fresh or aged human monocytes. *J Immunol* **134:**2444–2448, 1985.

67. Ennis FA, Meager A, Beare AS, *et al*: Interferon induction and increased natural killer-cell activity in influenza infections in man. *Lancet* **2:**891–893, 1981.

68. Bryson YJ, Winter HS, Gard SE, Fischer TJ, Stiehm ER: Deficiency of immune interferon production by leukocytes of normal newborns. *Cell Immunol* **55:**191–200, 1980.

69. Epstein LB, Ammann AJ: Evaluation of T lymphocyte effector function in immunodeficiency diseases: Abnormality in mitogen-stimulated interferon in patients with selective IgA deficiency. *J Immunol* **112:**617–626, 1974.

70. Lipinski M, Virelizier JL, Tursz T, Griscelli C: Natural killer and killer cell activities in patients with primary immunodeficiencies or defects in immune interferon production. *Eur J Immunol* **10:** 246–249, 1980.

71. Virelizier JL, Lenoir G, Griscelli C: Persistent Epstein–Barr virus infection in a child with hypergammaglobulinaemia and immunoblastic proliferation associated with a selective defect in immune interferon secretion. *Lancet* **2:**231–234, 1978.

72. Murray HW, Rubin BY, Masur H, Roberts RB: Impaired production of lymphokines and immune (gamma) interferon in the acquired immunodeficiency syndrome. *N Engl J Med* **310:**883–889, 1984.

73. Buimovici-Klein E, Lange M, Ramey WG, Grieco MH, Cooper LZ: Cell-mediated immune responses in AIDS. *N Engl J Med* **311:** 328–329, 1984.

74. Rossol S, Voth R, Laubenstein HP, *et al*: Interferon production in patients infected with HIV-1. *J Infect Dis* **159:**815–821, 1989.

75. Rytel MW, Balay J: Impaired production of interferon in lymphocytes from immunosuppressed patients. *J Infect Dis* **127:**445–449, 1973.

76. Weimar W, Van Ruyven CM, Geerlings W: Gamma interferon production capacity after renal transplantation. *Transplant Proc* **15:**421–423, 1983.

77. Newport MJ, Huxley CM, Huston S, *et al*: A mutation in the interferon-gamma-receptor gene and susceptibility to mycobacterial infection. *N Engl J Med* **335:**1941–1949, 1996.

78. Dorman SE, Holland SM: Mutation in the signal-transducing chain of the interferon-gamma receptor and susceptibility to mycobacterial infection. *J Clin Invest* **101:**2364–2369, 1998.

79. Aarden L, Brunner TK, Cerotini JC: Revised nomenclature for antigen-non-specific T cell proliferation and helper factors. *J Immunol* **123:**2928–2929, 1979.

80. Dinarello CA: Interleukin-1. *FASEB J* **2:**108–115, 1988.

81. Beutler B, Cerami A: Cachectin: More than a tumor necrosis factor. *N Engl J Med* **316:**379–385, 1987.

82. Van der Meer JWM, Barza M, Wolff SM, Dinarello CA: A low dose of recombinant interleukin 1 protects granulocytopenic mice from lethal gram-negative infection. *Proc Natl Acad Sci USA* **85:**1620–1623, 1988.

83. Kindler V, Sappino AP, Grau GE, Piguet PF, Vassali P: The inducing role of tumor necrosis factor in the development of bactericidal granulomas during BCG infection. *Cell* **56:**731–740, 1989.

84. Netea MG, Van Tits LHJ, Curfs JAHJ, *et al*: Increased susceptibility of TNFaLTa double knockout mice to systemic candidiasis through impaired recruitment of neutrophils and phagocytosis of *Candida albicans*. *J Immunol* **163:**1498–1505, 1999.

85. Waage A, Brandtzaeg P, Halstensen A, Kierulf P, Espevik T: The complex pattern of cytokines in serum from patients with meningococcal septic shock. Association between interleukin 6, interleukin 1, and fatal outcome. *J Exp Med* **169:**333–338, 1989.

86. Cannon JG, Thompkins RG, Gelfand JA, *et al*: Circulating interleukin-1 and tumor necrosis factor in septic shock and experimental endotoxin fever. *J Infect Dis* **161:**79–84, 1990.

87. Okusawa S, Gelfand JA, Ikejima T, Connolly RJ, Dinarello CA: Interleukin-1 induces a shock-like state in rabbits. *J Clin Invest* **81:**1162–1172, 1988.

88. Grau GE, Taylor TE, Molyneux ME, *et al*: Tumor necrosis factor

and disease severity in children with falciparum malaria. *N Engl J Med* **320:**1586–1591, 1989.

89. Mier JW, Vachino G, van der Meer JW, *et al*: Induction of circulating tumor necrosis factor (TNF alpha) as the mechanism for the febrile response to interleukin-2 (IL-2) in cancer patients. *J Clin Immunol* **8:**426–436, 1988.

90. Kuziel WA, Greene WC: Interleukin-2. In Thomson A (ed): *The Cytokine Handbook.* Academic Press, San Diego, 1991, pp. 83–102.

91. Rosenberg SA, Lotze MT, Muul LM, *et al*: A progress report on the treatment of 157 patients with advanced cancer using lymphokine-activated killer cells and interleukin-2 or high-dose interleukin-2 alone. *N Engl J Med* **316:**889–897, 1987.

92. Schrader JW: Interleukin-3. In Thomson A (ed): *The Cytokine Handbook.* Academic Press, San Diego, 1991, pp. 103–118.

93. Banchereau J: Interleukin-4. In Thomson A (ed): *The Cytokine Handbook.* Academic Press, San Diego, 1991, pp. 119–148.

94. Carter KC, Gallagher G, Baillie AJ, Alexander J: The induction of protective immunity to *Leishmania major* in the BALB/c mouse by interleukin 4 treatment. *Eur J Immunol* **19:**779–782, 1989.

95. Sanderson CJ: Interleukin-5. In Thomson A (ed): *The Cytokine Handbook.* Academic Press, San Diego, 1991, pp. 149–167.

96. Van Snick J: Interleukin-6: An overview. *Annu Rev Immunol* **8:** 253–278, 1990.

97. Gauldie J, Richards C, Harnish D, Lansdorp P, Baumann H: Interferon beta 2/B-cell stimulatory factor type 2 shares identity with monocyte-derived hepatocyte-stimulating factor and regulates the major acute phase protein response in liver cells. *Proc Natl Acad Sci USA* **84:**7251–7255, 1987.

98. Schindler R, Mancilla J, Endres S, Ghorbani R, Clark SC, Dinarello CA: Correlations and interactions in the production of interleukin-6 (IL-6), IL-1, and tumor necrosis factor (TNF) in human blood mononuclear cells: IL-6 suppresses IL-1 and TNF. *Blood* **75:**40–47, 1990.

99. Goodwin RG, Namen AE: Interleukin-7. In Thomson A (ed): *The Cytokine Handbook.* Academic Press, San Diego, 1991, pp. 192–200.

100. Van Damme J: Interleukin-8. In Thomson A (ed): *The Cytokine Handbook.* Academic Press, San Diego, 1991, pp. 201–214.

101. Renauld JC, Goethals A, Houssiau F, Merz H, Van Roost E, Van Snick J: Human P40/IL-9. Expression in activated CD4+ T cells, genomic organization, and comparison with the mouse gene. *J Immunol* **144:**4235–4241, 1990.

102. Donahue RE, Yang YC, Clark SC: Human P40 T-cell growth factor (interleukin-9) supports erythroid colony formation. *Blood* **75:**2271–2275, 1990.

103. Williams DE, Morrissey PJ, Mochizuki DY, *et al*: T-cell growth factor P40 promotes the proliferation of myeloid cell lines and enhances erythroid burst formation by normal murine bone marrow cells *in vitro. Blood* **76:**906–911, 1990.

104. Zlotnik A, Moore KW: Interleukin 10. *Cytokine* **3:**366–371, 1991.

105. Opal SM, Jhung JW, Keith JC Jr, *et al*: Recombinant human interleukin-11 in experimental *Pseudomonas aeruginosa* sepsis in immunocompromised animals. *J Infect Dis* **178:**1205–1208, 1998.

106. Lamont AG, Adorini L: IL-12: A key cytokine in immune regulation. *Immunol Today* **17:**214–217, 1996.

107. De Jong R, Altare F, Haagen IA, *et al*: Severe mycobacterial and *Salmonella* infections in interleukin-12 receptor-deficient patients. *Science* **280:**1435–1438, 1998.

108. Altare F, Durandy A, Lammas D, *et al*: Impairment of mycobacterial immunity in human interleukin-12 receptor deficiency. *Science* **280:**1432–1435, 1998.

109. Chomarat P, Bancherau J: Interleukin-4 and interleukin-13: Their similarities and discrepancies. *Int Rev Immunol* **17:**1–52, 1998.

110. Ambrus JL Jr, Pippin J, Joseph A, *et al*: Identification of a cDNA for a human high-molecular-weight B-cell growth factor [published erratum appears in *Proc Natl Acad Sci USA* **93**(15):8154, 1996]. *Proc Natl Acad Sci USA* **90:**6330–6334, 1993.

111. Trentin L, Zambello R, Facco M, Sancetta R, Agostini C, Semenzato G: Interleukin-15: A novel cytokine with regulatory properties on normal and neoplastic B lymphocytes. *Leuk Lymphoma* **27:**35–42, 1997.

112. Waldmann TA, Tagaya Y: The multifaceted regulation of interleukin-15 expression and the role of this cytokine in NK cell differentiation and host response to intracellular pathogens. *Annu Rev Immunol* **17:**19–49, 1999.

113. Cruikshank WW, Kornfeld H, Center DM: Signaling and functional properties of interleukin-16. *Int Rev Immunol* **16:**523–540, 1998.

114. Center DM, Kornfeld H, Cruikshank WW: Interleukin 16 and its function as a CD4 ligand. *Immunol Today* **17:**476–481, 1996.

115. Fossiez F, Bancherau J, Murray R, Van Kooten C, Garrone P, Lebecque S: Interleukin-17. *Int Rev Immunol* **16:**541–551, 1998.

116. Dinarello CA: IL-18: A TH1-inducing, proinflammatory cytokine and new member of the IL-1 family. *J Allergy Clin Immunol* **103:** 11–24, 1999.

117. Fantuzzi G, Reed DA, Dinarello CA: IL-12-induced IFN-gamma is dependent on caspase-1 processing of the IL-18 precursor. *J Clin Invest* **104:**761–767, 1999.

118. Puren AJ, Fantuzzi G, Gu Y, Su MS, Dinarello CA: Interleukin-18 (IFNγ-inducing factor) induces IL-8 and IL-1β via TNFα production from non-CD14+ human blood mononuclear cells. *J Clin Invest* **101:**711–721, 1998.

119. Groopman JE, Molina JM, Scadden DT: Hematopoietic growth factors. Biology and clinical applications. *N Engl J Med* **321:** 1449–1459, 1989.

120. Moore MA: The clinical use of colony stimulating factors. *Annu Rev Immunol* **9:**159–191, 1991.

121. Baldwin GC, Gasson JC, Quan SG, *et al*: Granulocyte-macrophage colony-stimulating factor enhances neutrophil function in acquired immunodeficiency syndrome patients. *Proc Natl Acad Sci USA* **85:**2763–2766, 1988.

122. Roilides E, Walsh TJ, Pizzo PA, Rubin M: Granulocyte colony-stimulating factor enhances the phagocytic and bactericidal activity of normal and defective human neutrophils. *J Infect Dis* **163:** 579–583, 1991.

123. Nathan CF: Respiratory burst in adherent human neutrophils: Triggering by colony-stimulating factors CSF-GM and CSF-G. *Blood* **73:**301–306, 1989.

124. Dale DC, Bonilla MA, Davis MW, *et al*: A randomized controlled phase III trial of recombinant human granulocyte colony-stimulating factor (filgrastim) for treatment of severe chronic neutropenia. *Blood* **81:**2496–2502, 1993.

125. Dignani MC, Anaissie EJ, Hester JP, *et al*: Treatment of neutropenia-related fungal infections with granulocyte colony-stimulating factor-elicited white blood cell transfusions: A pilot study. *Leukemia* **11:**1621–1630, 1997.

126. Nelson S, Belknap SM, Carlson RW, *et al*: A randomized controlled trial of filgrastim as an adjunct to antibiotics for treatment of hospitalized patients with community-acquired pneumonia. *J Infect Dis* **178:**1075–1080, 1998.

127. Kullberg BJ, Vandewoude K, Herbrecht R, Jacobs F, Aoun M, Kujath P: A double-blind, randomized, placebo-controlled phase

II study of filgrastim (recombinant granulocyte colony-stimulating factor) in combination with fluconazole for treatment of invasive candidiasis and candidemia in nonneutropenic patients. 38th Interscience Conference on Antimicrobial Agents and Chemotherapy. Washington, DC, 1998, p. 479.

128. Görgen I, Hartung T, Leist M, *et al*: Granulocyte colony-stimulating factor treatment protects rodents against lipopolysaccharide-induced toxicity via suppression of systemic tumor necrosis factor-α. *J Immunol* **149:**918–924, 1992.

129. Ikejima T, Okusawa S, Ghezzi P, Van der Meer JWM, Dinarello CA: Interleukin-1 induces tumor necrosis factor (TNF) in human peripheral blood mononuclear cells *in vitro* and a circulating TNF-like activity in rabbits. *J Infect Dis* **162:**215–223, 1990.

130. Dinarello CA: Interleukin-1 and interleukin-1 antagonism. *Blood* **77:**1627–1652, 1991.

131. Engelmann H, Novick D, Wallach D: Two tumor necrosis factor-binding proteins purified from human urine. Evidence for immunological cross-reactivity with cell surface tumor necrosis factor receptors. *J Biol Chem* **265:**1531–1536, 1990.

132. Chantry D, Turner M, Abney E, Feldmann M: Modulation of cytokine production by transforming growth factor-beta. *J Immunol* **142:**4295–4300, 1989.

133. Wakabayashi G, Gelfand JG, Burke JF, Thompson RC, Dinarello CA: A specific receptor antagonist for interleukin-1 prevents *E. coli*-induced shock in rabbits. *FASEB J* **5:**338–343, 1991.

134. Dinarello CA, Thompson RC: Blocking IL-1—Interleukin-1 receptor antagonist *in vivo* and *in vitro*. *Immunol Today* **12:**404–410, 1991.

135. Lesslauer W, Tabuchi H, Gentz R, *et al*: Recombinant soluble tumor necrosis factor proteins protect mice from lipopolysaccharide-induced lethality. *Eur J Immunol* **21:**2883–2886, 1991.

136. Mohler KM, Torrance DS, Smith CA, *et al*: Soluble tumor necrosis factor (TNF) receptors are effective therapeutic agents in lethal endotoxemia and function simultaneously as both TNF carriers and TNF antagonists. *J Immunol* **151:**1548–1561, 1993.

137. Mosmann TR, Cherwinski H, Bond MW, Giedlin MA, Coffman RL: Two types of murine helper T cell clone. I. Definition according to profiles of lymphokine activities and secreted proteins. *J Immunol* **136:**2348–2357, 1986.

138. Abbas AK, Murphy KM, Sher A: Functional diversity of helper T lymphocytes. *Nature* **383:**787–793, 1996.

139. Romagnani S: Lymphokine production by human T cells in disease states. *Annu Rev Immunol* **12:**227–257, 1994.

140. Kelso A: Th1 and Th2 subsets: Paradigms lost? *Immunol Today* **16:**374–379, 1995.

141. Allen JE, Maizels RM: Th1-Th2: Reliable paradigm or dangerous dogma? *Immunol Today* **18:**387–392, 1997.

142. Tonegawa S: Somatic generation of antibody diversity. *Nature* **302:**575–581, 1983.

143. Oxelius VA: Chronic infections in a family with hereditary deficiency of IgG2 and IgG4. *Clin Exp Immunol* **17:**19–27, 1974.

144. Siber GR, Schur PH, Aisenberg AC, Weitzman SA, Schiffman G: Correlation between serum IgG-2 concentrations and the antibody response to bacterial polysaccharide antigens. *N Engl J Med* **303:**178–182, 1980.

145. Tomasi TB Jr, Plaut AG: Humoral aspects of mucosal immunity. In Gallin JI, Fauci AS (eds): *Advances in Host Defense Mechanisms*. Raven Press, New York, 1982, pp. 31–61.

146. Andre C, Andre F, Fargier C: Distribution of IgA 1 and IgA 2 plasma cells in various normal human tissues and in the jejunum of plasma IgA-deficient patients. *Clin Exp Immunol* **33:**327–331, 1978.

147. Buckley RH, Becker WG: Abnormalities in the regulation of human IgE synthesis. *Immunol Rev* **41:**288–314, 1978.

148. Dessaint JP, Capron A, Joseph M: Interaction of schistosomiasis and macrophages. In Van Furth R (ed): *Mononuclear Phagocytes: Characteristics, Physiology, and Function*. Martinus Nijhoff, The Hague, 1985, pp. 593–598.

149. Vadas MA, Butterworth AE, Sherry B, *et al*: Interactions between human eosinophils and schistosomula of *Schistosoma mansoni*. I. Stable and irreversible antibody-dependent adherence. *J Immunol* **124:**1441–1448, 1980.

150. Thorbecke GJ, Leski AA: Immunoglobulin D: Structure and function. *Ann NY Acad Sci* **399:**1–410, 1982.

151. Rosen FS, Cooper MD, Wedgwood RJ: The primary immunodeficiencies (1). *N Engl J Med* **311:**235–242, 1984.

152. Melamed I, Bujanover Y, Igra YS, Schwartz D, Zakuth V, Spirer Z: *Campylobacter enteritis* in normal and immunodeficient children. *Am J Dis Child* **137:**752–753, 1983.

153. van der Meer JW, Mouton RP, Daha MR, Schuurman RK: *Campylobacter jejuni* bacteraemia as a cause of recurrent fever in a patient with hypogammaglobulinaemia. *J Infect* **12:**235–239, 1986.

154. Ochs HD, Ament ME, Davis SD: Giardiasis with malabsorption in X-linked agammaglobulinemia. *N Engl J Med* **287:**341–342, 1972.

155. Wright PF, Hatch MH, Kasselberg AG, Lowry SP, Wadlington WB, Karzon DT: Vaccine-associated poliomyelitis in a child with sex-linked agammaglobulinemia. *J Pediatr* **91:**408–412, 1977.

156. McKinney RE Jr, Katz SL, Wilfert CM: Chronic enteroviral meningoencephalitis in agammaglobulinemic patients. *Rev Infect Dis* **9:**334–356, 1987.

157. Saulsbury FT, Winkelstein JA, Yolken RH: Chronic rotavirus infection in immunodeficiency. *J Pediatr* **97:**61–65, 1980.

158. So AK, Furr PM, Taylor-Robinson D, Webster AD: Arthritis caused by *Mycoplasma salivarium* in hypogammaglobulinaemia. *Br Med J* **286:**762–763, 1983.

159. Taylor-Robinson D, Furr PM, Webster AD: *Ureaplasma urealyticum* in the immunocompromised host. *Pediatr Infect Dis* **5:**S236–S238, 1986.

160. Ammann AJ, Hong R: Selective IgA deficiency: Presentation of 30 cases and a review of the literature. *Medicine* **50:**223–236, 1971.

161. Oxelius VA, Laurell AB, Lindquist B, *et al*: IgG subclasses in selective IgA deficiency: Importance of IgG2-IgA deficiency. *N Engl J Med* **304:**1476–1477, 1981.

162. Hobbs JR, Milner RD, Watt PJ: Gamma-M deficiency predisposing to meningococcal septicaemia. *Br Med J* **4:**583–586, 1967.

163. Lefranc MP, Lefranc G, de Lange G, *et al*: Instability of the human immunoglobulin heavy chain constant region locus indicated by different inherited chromosomal deletions. *Mol Biol Med* **1:**207–217, 1983.

164. Saffran DC, Parolini O, Fitch-Hilgenberg ME, *et al*: Brief report: A point mutation in the SH2 domain of Bruton's tyrosine kinase in atypical X-linked agammaglobulinemia. *N Engl J Med* **330:**1488–1491, 1994.

165. Reinherz EL, Geha R, Wohl ME, Morimoto C, Rosen FS, Schlossman SF: Immunodeficiency associated with loss of T4+ inducer T-cell function. *N Engl J Med* **304:**811–816, 1981.

166. Eibl MM, Mannhalter JW, Zlabinger G, *et al*: Defective macrophage function in a patient with common variable immunodeficiency. *N Engl J Med* **307:**803–806, 1982.

167. Van Kooten C, Banchereau J: CD40-CD40 ligand: A multifunctional receptor-ligand pair. *Adv Immunol* **61:**1–77, 1996.

168. Kroczek RA, Graf D, Brugnoni D, et al: Defective expression of CD40 ligand on T cells causes "X-linked immunodeficiency with hyper-IgM (HIGM1)." Immunol Rev 138:39–59, 1994.

169. Najjar VA, Fridkin M: Antineoplastic, immunogenic and other effects of the tetrapeptide tuftsin: A natural macrophage activator. Ann NY Acad Sci 419:1–273, 1983.

170. Corazza GR, Zoli G, Ginaldi L, et al: Tuftsin deficiency in AIDS. Lancet 337:12–13, 1991.

171. Opdenakker G, Fibbe WE, Van Damme J: The molecular basis of leukocytosis. Immunol Today 19:182–189, 1998.

172. Cronkite EP, Fliedner TM: Granulocytopoiesis. N Engl J Med 270:1347–1352, 1964.

173. Joyce RA, Boggs DR, Hasiba U, Srodes CH: Marginal neutrophil pool size in normal subjects and neutropenic patients as measured by epinephrine infusion. J Lab Clin Med 88:614–620, 1976.

174. Gallin JI, Wright DG, Malech HL, Davis JM, Klempner MS, Kirkpatrick CH: Disorders of phagocyte chemotaxis. Ann Intern Med 92:520–538, 1980.

175. Butcher EC: Leukocyte-endothelial cell recognition: Three (or more) steps to specificity and diversity. Cell 67:1033–1036, 1991.

176. Springer TA: Traffic signals for lymphocyte recirculation and leukocyte emigration: The multistep paradigm. Cell 76:301–314, 1994.

177. Hogg N: The leukocyte integrins. Immunol Today 10:111–114, 1989.

178. Arnaout MA, Hakim RM, Todd RFd, Dana N, Colten HR: Increased expression of an adhesion-promoting surface glycoprotein in the granulocytopenia of hemodialysis. N Engl J Med 312:457–462, 1985.

179. Bass DA: Eosinophil behavior during host defense reactions. In Gallin JI, Fauci AS (eds): Advances in Host Defense Mechanisms. Raven Press, New York, 1982, pp. 211–241.

180. Weller PF: The immunobiology of eosinophils. N Engl J Med 324:1110–1118, 1991.

181. Capron M, Desreumaux P: Immunobiology of eosinophils in allergy and inflammation. Res Immunol 148:29–33, 1997.

182. Simon HU, Yousefi S, Schranz C, Schapowal A, Bachert C, Blaser K: Direct demonstration of delayed eosinophil apoptosis as a mechanism causing tissue eosinophilia. J Immunol 158:3902–3908, 1997.

183. Simon HU, Plotz SG, Dummer R, Blaser K: Abnormal clones of T cells producing interleukin-5 in idiopathic eosinophilia. N Engl J Med 341:1112–1120, 1999.

184. Van Furth R, Diesselhoff-Den Dulk MMC, Sluiter W: New perspectives on the kinetics of mononuclear phagocytes. In Van Furth R (ed): Mononuclear Phagocytes: Characteristics, Physiology, and Function. Martinus Nijhoff, The Hague, 1985, pp. 201–208.

185. Goud TJ, Schotte C, van Furth R: Identification and characterization of the monoblast in mononuclear phagocyte colonies grown in vitro. J Exp Med 142:1180–1199, 1975.

186. van der Meer JW, van de Gevel JS, Beelen RH, Fluitsma D, Hoefsmit EC, van Furth R: Culture of human bone marrow in the Teflon culture bag: Identification of the human monoblast. J Reticuloendothel Soc 32:355–369, 1982.

187. van Furth R, Raeburn JA, van Zwet TL: Characteristics of human mononuclear phagocytes. Blood 54:485–500, 1979.

188. Van Furth R: Cells of the mononuclear phagocyte system: Nomenclature in terms of sites and conditions. In Van Furth R (ed): Mononuclear Phagocytes: Functional Aspects. Martinus Nijhoff, The Hague, 1980, pp. 1–30.

189. Burger EH, Van der Meer JW, van de Gevel JS, Gribnau JC, Thesingh GW, van Furth R: In vitro formation of osteoclasts from long-term cultures of bone marrow mononuclear phagocytes. J Exp Med 156:1604–1614, 1982.

190. Snyderman R, Goetzl EJ: Molecular and cellular mechanisms of leukocyte chemotaxis. Science 213:830–837, 1981.

191. Sluiter W, Van Waarde D, Hulsing-Hesselink E: Humoral control of monocyte production during inflammation. In Van Furth R (ed): Mononuclear Phagocytes: Functional Aspects. Martinus Nijhoff, The Hague, 1980, pp. 325–339.

192. Chernoff AE, Granowitz EV, Shapiro L, et al: A randomized, controlled trial of IL-10 in humans. Inhibition of inflammatory cytokine production and immune responses. J Immunol 154:5492–5499, 1995.

193. Griffin FM Jr, Griffin JA, Leider JE, Silverstein SC: Studies on the mechanism of phagocytosis. I. Requirements for circumferential attachment of particle-bound ligands to specific receptors on the macrophage plasma membrane. J Exp Med 142:1263–1282, 1975.

194. Leijh PC, van den Barselaar MT, van Furth R: Kinetics of phagocytosis and intracellular killing of Staphylococcus aureus and Escherichia coli by human monocytes. Scand J Immunol 13:159–174, 1981.

195. Clark RA: The human neutrophil respiratory burst oxidase. J Infect Dis 161:1140–1147, 1990.

196. Wright DG: The neutrophil as a secretous organ of host defense. In Gallin JI, Fauci AS (eds): Advances in Host Defense Mechanisms. Raven Press, New York, 1982, pp. 75–110.

197. Lehrer RI, Ganz T: Antimicrobial polypeptides of human neutrophils. Blood 76:2169–2181, 1990.

198. Moncada S, Higgs EA: Endogenous nitric oxide: Physiology, pathology and clinical relevance. Eur J Clin Invest 21:361–374, 1991.

199. Vazquez-Torres A, Balish E: Macrophages in resistance to candidiasis. Microbiol Mol Biol Rev 61:170–192, 1997.

200. Vazquez-Torres A, Jones-Carson J, Balish E: Peroxynitrite contributes to the candidacidal activity of nitric oxide-producing macrophages. Infect Immun 64:3127–3133, 1996.

201. Arend WP, Mannik M: Studies on antigen-antibody complexes. II. Quantification of tissue uptake of soluble complexes in normal and complement-depleted rabbits. J Immunol 107:63–75, 1971.

202. Van Es LA, Daha MR, Kijlstra A: Clearance of soluble immune complexes and aggregates. In Peeters H (ed): Protides of the Biological Fluids. Pergamon Press, Oxford, 1979, pp. 159–162.

203. Haakenstad AO, Mannik M: Saturation of the reticuloendothelial system with soluble immune complexes. J Immunol 112:1939–1948, 1974.

204. Atkinson JP, Frank MM: Studies on the in vivo effects of antibody. Interaction of IgM antibody and complement in the immune clearance and destruction of erythrocytes in man. J Clin Invest 54:339–348, 1974.

205. Nathan CF: Secretory products of macrophages. J Clin Invest 79:319–326, 1987.

206. Unanue ER: Antigen-presenting function of the macrophage. Annu Rev Immunol 2:395–428, 1984.

207. Unanue ER, Allen PM: The basis for the immunoregulatory role of macrophages and other accessory cells. Science 236:551–557, 1987.

208. Bodey GP, Buckley M, Sathe YS, Freireich EJ: Quantitative relationships between circulating leukocytes and infection in patients with acute leukemia. Ann Intern Med 64:328–340, 1966.

209. van der Meer JW, Alleman M, Boekhout M: Infectious episodes in severely granulocytopenic patients. Infection 7:171–175, 1979.

210. Dale DC, Guerry Dt, Wewerka JR, Bull JM, Chusid MJ: Chronic neutropenia. Medicine 58:128–144, 1979.

211. Anderson DC, Schmalsteig FC, Finegold MJ, *et al*: The severe and moderate phenotypes of heritable Mac-1, LFA-1 deficiency: Their quantitative definition and relation to leukocyte dysfunction and clinical features. *J Infect Dis* **152:**668–689, 1985.

212. Etzioni A, Frydman M, Pollack S, *et al*: Brief report: Recurrent severe infections caused by a novel leukocyte adhesion deficiency. *N Engl J Med* **327:**1789–1792, 1992.

213. Miles AA, Miles EM, Burke J: The value and duration of defense reactions of the skin to the primary lodgment of bacteria. *Br J Exp Pathol* **38:**79–96, 1957.

214. Frank MM, Hamburger MI, Lawley TJ, Kimberly RP, Plotz PH: Defective reticuloendothelial system Fc-receptor function in systemic lupus erythematosus. *N Engl J Med* **300:**518–523, 1979.

215. Lobatto S, Daha MR, Breedveld FC, *et al*: Abnormal clearance of soluble aggregates of human immunoglobulin G in patients with systemic lupus erythematosus. *Clin Exp Immunol* **72:**55–59, 1988.

216. Hamburger MI, Moutsopoulos HM, Lawley TJ, Frank MM: Sjögren's syndrome: A defect in reticuloendothelial system Fc-receptor-specific clearance. *Ann Intern Med* **91:**534–538, 1979.

217. Hamburger MI, Gerardi EN, Fields TR, Bennett RS: Lympho-plasmapheresis and reticuloendothelial system Fc receptor function in rheumatoid arthritis. *Arthritis Rheum* **24:**S399–S404, 1981.

218. Kimberly RP, Inman RD, Bussel JB, Polk JR, Hilgartner MW: Modulation of mononuclear phagocyte system function and circulating immune complexes by lyophilized concentrates in patients with classic hemophilia. *Clin Immunol Immunopathol* **31:**321–330, 1984.

219. Ruiz P, Gomez F, Schreiber AD: Impaired function of macrophage Fc gamma receptors in end-stage renal disease. *N Engl J Med* **322:**717–722, 1990.

220. Bender BS, Frank MM, Lawley TJ, Smith WJ, Brickman CM, Quinn TC: Defective reticuloendothelial system Fc-receptor function in patients with acquired immunodeficiency syndrome. *J Infect Dis* **152:**409–412, 1985.

221. Lawley TJ, Hall RP, Fauci AS, Katz SI, Hamburger MI, Frank MM: Defective Fc-receptor functions associated with the HLA-B8/DRw3 haplotype: Studies in patients with dermatitis herpetiformis and normal subjects. *N Engl J Med* **304:**185–192, 1981.

222. Kimberly RP, Gibofsky A, Salmon JE, Fotino M: Impaired fc-mediated mononuclear phagocyte system clearance in HLA-DR2 and MT1-positive healthy young adults. *J Exp Med* **157:**1698–1703, 1983.

223. Jaffe CJ, Vierling JM, Jones EA, Lawley TJ, Frank MM: Receptor specific clearance by the reticuloendothelial system in chronic liver diseases. Demonstration of defective C3b-specific clearance in primary biliary cirrhosis. *J Clin Invest* **62:**1069–1077, 1978.

224. Biozzi G, Benacerraf B, Halpern BN: Exploration of the phagocyte function of the reticuloendothelial system with heat denatured human serum albumin labeled with I-131 and applications to the measurement of liver blood flow in normal man and in some pathological conditions. *J Lab Clin Med* **51:**230–238, 1958.

225. Lahnborg G, Friman L, Berghem L: Reticuloendothelial function in patients with alcoholic liver cirrhosis. *Scand J Gastroenterol* **16:**481–489, 1981.

226. Rimola A, Soto R, Bory F, Arroyo V, Piera C, Rodes J: Reticuloendothelial system phagocytic activity in cirrhosis and its relation to bacterial infections and prognosis. *Hepatology* **4:**53–58, 1984.

227. Seidner DL, Mascioli EA, Istfan NW, *et al*: Effects of long-chain triglyceride emulsions on reticuloendothelial system function in humans. *J Parenter Enteral Nutr* **13:**614–619, 1989.

228. Dwyer JM: Manipulating the immune system with immune globulin. *N Engl J Med* **326:**107–116, 1992.

229. Halma C, Daha MR, van der Meer JW, *et al*: Effect of monomeric immunoglobulin G (IgG) on the clearance of soluble aggregates of IgG in man. *J Clin Lab Immunol* **35:**9–15, 1991.

230. Herberman RB: Natural killer cells. *Annu Rev Med* **37:**347–352, 1986.

231. Krahenbuhl O, Tschopp J: The mechanism of lymphocyte-mediated killing. Perforin-induced pore formation. *Immunol Today* **12:**399–402, 1991.

232. Ritz J: The role of natural killer cells in immune surveillance. *N Engl J Med* **320:**1748–1749, 1989.

233. Biron CA, Byron KS, Sullivan JL: Severe herpesvirus infections in an adolescent without natural killer cells. *N Engl J Med* **320:**1731–1735, 1989.

234. Lane HC, Fauci AS: Immunologic abnormalities in the acquired immunodeficiency syndrome. *Annu Rev Immunol* **3:**477–500, 1985.

235. Van Voorhis WC, Valinsky J, Hoffman E, Luban J, Hair LS, Steinman RM: Relative efficacy of human monocytes and dendritic cells as accessory cells for T cell replication. *J Exp Med* **158:**174–191, 1983.

236. Stingl G, Katz SI, Clement L, Green I, Shevach EM: Immunologic functions of Ia-bearing epidermal Langerhans cells. *J Immunol* **121:**2005–2013, 1978.

237. Locksley RM: Interleukin-12 in host defense against microbial pathogens. *Proc Natl Acad Sci USA* **90:**5879-5880, 1993.

238. Ballieux RE, Heijnen CJ: Immunoregulatory T cell subpopulations in man: Dissection by monoclonal antibodies and Fc-receptors. *Immunol Rev* **74:**5–28, 1983.

239. Meuer SC, Schlossman SF, Reinherz EL: Clonal analysis of human cytotoxic T lymphocytes: T4+ and T8+ effector T cells recognize products of different major histocompatibility complex regions. *Proc Natl Acad Sci USA* **79:**4395–4399, 1982.

240. Thomas Y, Rogozinski L, Irigoyen OH, *et al*: Functional analysis of human T cell subsets defined by monoclonal antibodies. IV. Induction of suppressor cells within the OKT4+ population. *J Exp Med* **154:**459–467, 1981.

241. Hahn H, Kaufmann SH: The role of cell-mediated immunity in bacterial infections. *Rev Infect Dis* **3:**1221–1250, 1981.

242. Murray HW: How protozoa evade intracellular killing. *Ann Intern Med* **98:**1016–1018, 1983.

243. Bermudez LE, Young LS: Tumor necrosis factor, alone or in combination with IL-2, but not IFN-γ, is associated with macrophage killing of *Mycobacterium avium* complex. *J Immunol* **140:**3006–3013, 1988.

244. Adams DO, Hamilton TA: The cell biology of macrophage activation. *Annu Rev Immunol* **2:**283–318, 1984.

245. Siegel RL, Issekutz T, Schwaber J, Rosen FS, Geha RS: Deficiency of T helper cells in transient hypogammaglobulinemia of infancy. *N Engl J Med* **305:**1307–1313, 1981.

246. Notarangelo LD: Immunodeficiencies caused by genetic defects in protein kinases. *Curr Opin Immunol* **8:**448–453, 1996.

247. Itoh N, Yonehara S, Ishii A, *et al*: The polypeptide encoded by the cDNA for human cell surface antigen Fas can mediate apoptosis. *Cell* **66:**233–243, 1991.

248. Smith CA, Farrah T, Goodwin RG: The TNF receptor superfamily of cellular and viral proteins: Activation, costimulation, and death. *Cell* **76:**959–962, 1994.

249. Holland SM, Dorman SE, Kwon A, *et al*: Abnormal regulation of interferon-gamma, interleukin-12, and tumor necrosis factor-alpha

in human interferon-gamma receptor 1 deficiency. *J Infect Dis* **178:** 1095–1104, 1998.

250. Hunninghake GW, Crystal RG: Pulmonary sarcoidosis: A disorder mediated by excess helper T-lymphocyte activity at sites of disease activity. *N Engl J Med* **305:**429–434, 1981.

251. Rosen FS, Wedgwood RJP, Eibl M, *et al*: Primary immunodeficient diseases. Report of a WHO scientific group. *Clin Exp Immunol* **112**(Suppl 1)**:**1–28, 1998.

252. Kallenberg CGM, Torensma R, The TH: The immune response to primary immunogens in man. In Reeves WG (ed): *Recent Developments in Clinical Immunology.* Elsevier, Amsterdam, 1984, pp. 1–26.

253. Weinberg ED: Pregnancy-associated depression of cell-mediated immunity. *Rev Infect Dis* **6:**814–831, 1984.

254. Styrt B, Sugarman B: Estrogens and infection. *Rev Infect Dis* **13:**1139–1150, 1991.

255. Skamene E (ed): *Genetic Control of Host Response to Infection and Malignancy.* Alan Liss, New York, 1985.

256. Hill AV: The immunogenetics of human infectious diseases. *Annu Rev Immunol* **16:**593–617, 1998.

257. van Eden W, de Vries RR, van Rood JJ: The genetic approach to infectious disease, with special emphasis on the MHC. *Disease Markers* **1:**221–242, 1983.

258. de Vries RR, Meera Khan P, Bernini LF, van Loghem E, van Rood JJ: Genetic control of survival in epidemics. *J Immunogenet* **6:**271–287, 1979.

259. Hill AV, Allsopp CE, Kwiatkowski D, *et al*: Common west African HLA antigens are associated with protection from severe malaria. *Nature* **352:**595–600, 1991.

260. Mira JP, Cariou A, Grall F, *et al*: Association of TNF2, a TNF-alpha promoter polymorphism, with septic shock susceptibility and mortality: A multicenter study. *JAMA* **282:**561–568, 1999.

261. Westendorp RG, Langermans JA, Huizinga TW, *et al*: Genetic influence on cytokine production and fatal meningococcal disease. *Lancet* **349:**170–173, 1997.

262. de Vries RR, Ottenhoff TH, van Schooten WC: Human leukocyte antigens (HLA) and mycobacterial disease. *Springer Semin Immunopathol* **10:**305–318, 1988.

263. Bellamy R, Ruwende C, Corrah T, McAdam KP, Whittle HC, Hill AV: Variations in the NRAMP1 gene and susceptibility to tuberculosis in West Africans. *N Engl J Med* **338:**640–644, 1998.

264. Bellamy R, Ruwende C, Corrah T, *et al*: Tuberculosis and chronic hepatitis B virus infection in Africans and variation in the vitamin D receptor gene. *J Infect Dis* **179:**721–724, 1999.

265. Pollack MS, Rich RR: The HLA complex and the pathogenesis of infectious diseases. *J Infect Dis* **151:**1–8, 1985.

266. Chandra RK: Nutrition, immunity, and infection: Present knowledge and future directions. *Lancet* **1:**688–691, 1983.

267. Keusch GT, Wilson CS, Waksal S: Nutrition, host defenses and the lymphoid system. In Gallin JI, Fauci AS (eds): *Advances in Host Defense Mechanisms,* Vol. 2. Raven Press, New York, 1983, pp. 275–359.

268. Salimonu LS, Ojo-Amaize E, Williams AI, *et al*: Depressed natural killer cell activity in children with protein-calorie malnutrition. *Clin Immunol Immunopathol* **24:**1–7, 1982.

269. Sommer A, Tarwotjo I, Djunaedi E, *et al*: Impact of vitamin A supplementation on childhood mortality. A randomised controlled community trial. *Lancet* **1:**1169–1173, 1986.

270. Hussey GD, Klein M: A randomized, controlled trial of vitamin A in children with severe measles. *N Engl J Med* **323:**160–164, 1990.

271. Rahmathullah L, Underwood BA, Thulasiraj RD, *et al*: Reduced mortality among children in southern India receiving a small weekly dose of vitamin A. *N Engl J Med* **323:**929–935, 1990.

272. West CE, Rombout JH, van der Zijpp AJ, Sijtsma SR: Vitamin A and immune function. *Proc Nutr Soc* **50:**251–262, 1991.

273. Sugarman B: Zinc and infection. *Rev Infect Dis* **5:**137–147, 1983.

274. Tucker SB, Schroeter AL, Brown PW Jr, McCall JT: Acquired zinc deficiency. Cutaneous manifestations typical of acrodermatitis enteropathica. *JAMA* **235:**2399–2402, 1976.

275. Bullen JJ, Griffiths E (eds): *Iron and Infection.* John Wiley, New York, 1987.

276. Chandra RK, Au B, Woodford G, Hyam P: Iron status, immune response and susceptibility to infection. *Ciba Found Symp* 249–268, 1976.

277. Murray MJ, Murray AB, Murray MB, Murray CJ: The adverse effect of iron repletion on the course of certain infections. *Br Med J* **2:**1113–1115, 1978.

278. van Asbeck BS, Verbrugh HA, van Oost BA, Marx JJ, Imhof HW, Verhoef J: *Listeria monocytogenes* meningitis and decreased phagocytosis associated with iron overload. *Br Med J (Clin Res Ed)* **284:**542–544, 1982.

279. Boelaert JR, van Landuyt HW, Valcke YJ, *et al*: The role of iron overload in *Yersinia enterocolitica* and *Yersinia pseudotuberculosis* bacteremia in hemodialysis patients. *J Infect Dis* **156:**384–387, 1987.

280. Craddock PR, Yawata Y, VanSanten L, Gilberstadt S, Silvis S, Jacob HS: Acquired phagocyte dysfunction. A complication of the hypophosphatemia of parenteral hyperalimentation. *N Engl J Med* **290:**1403–1407, 1974.

281. Freeman J, Goldmann DA, Smith NE, Sidebottom DG, Epstein MF, Platt R: Association of intravenous lipid emulsion and coagulase-negative staphylococcal bacteremia in neonatal intensive care units. *N Engl J Med* **323:**301–308, 1990.

282. Long JG, Keyserling HL: Catheter-related infection in infants due to an unusual lipophilic yeast—Malassezia furfur. *Pediatrics* **76:**896–900, 1985.

283. Robin AP, Arain I, Phuangsab A, Holian O, Roccaforte P, Barrett JA: Intravenous fat emulsion acutely suppresses neutrophil chemiluminescence. *J Parenter Enteral Nutr* **13:**608–613, 1989.

284. Netea MG, Curfs JHAJ, Demacker PMN, Meis JFGM, Van der Meer JWM, Kullberg BJ: Infusion of lipoproteins into human volunteers enhances the growth of *Candida albicans. Clin Infect Dis* **28:**1148–1151, 1999.

285. Smith DH, Peter G, Ingram DL, Harding AL, Anderson P: Responses of children immunized with the capsular polysaccharide of *Hemophilus influenzae,* type b. *Pediatrics* **52:**637–644, 1973.

286. Adamkin D, Stitzel A, Urmson J, Farnett ML, Post E, Spitzer R: Activity of the alternative pathway of complement in the newborn infant. *J Pediatr* **93:**604–608, 1978.

287. Marodi L, Leijh PC, Braat A, Daha MR, van Furth R: Opsonic activity of cord blood sera against various species of microorganism. *Pediatr Res* **19:**433–436, 1985.

288. Park BH, Holmes B, Good RA: Metabolic activities in leukocytes of newborn infants. *J Pediatr* **76:**237–241, 1970.

289. McCracken GH Jr, Eichenwald HF: Leukocyte function and the development of opsonic and complement activity in the neonate. *Am J Dis Child* **121:**120–126, 1971.

290. Schuit KE, Powell DA: Phagocytic dysfunction in monocytes of normal newborn infants. *Pediatrics* **65:**501–504, 1980.

291. Marodi L, Leijh PC, van Furth R: Characteristics and functional capacities of human cord blood granulocytes and monocytes. *Pediatr Res* **18:**1127–1131, 1984.

292. Marodi L, Jezerniczky J, Csorba S, Karmazsin L, Leijh PC, van Furth R: Chemotactic and random movement of cord-blood granulocytes. *Experientia* **40:**1407–1410, 1984.

293. Marodi L, Csorba S, Nagy B: Chemotactic and random movement of human newborn monocytes. *Eur J Pediatr* **135:**73–75, 1980.

294. van Tol MJ, Zijlstra J, Thomas CM, Zegers BJ, Ballieux RE: Distinct role of neonatal and adult monocytes in the regulation of the *in vitro* antigen-induced plaque-forming cell response in man. *J Immunol* **133:**1902–1908, 1984.

295. Kohl S, Frazier JJ, Greenberg SB, Pickering LK, Loo LS: Interferon induction of natural killer cytotoxicity in human neonates. *J Pediatr* **98:**379–384, 1981.

296. Wakasugi N, Virelizier JL: Defective IFN-gamma production in the human neonate. I. Dysregulation rather than intrinsic abnormality. *J Immunol* **134:**167–171, 1985.

297. Phair J, Kauffman CA, Bjornson A, Adams L, Linnemann C Jr: Failure to respond to influenza vaccine in the aged: Correlation with B-cell number and function. *J Lab Clin Med* **92:**822–828, 1978.

298. Roberts-Thomson IC, Whittingham S, Youngchaiyud U, Mackay IR: Ageing, immune response, and mortality. *Lancet* **2:**368–370, 1974.

299. Wiedermann CJ, Niedermuhlbichler M, Beimpold H, Braunsteiner H: *In vitro* activation of neutrophils of the aged by recombinant human growth hormone. *J Infect Dis* **164:**1017–1020, 1991.

300. Gardner ID: The effect of aging on susceptibility to infection. *Rev Infect Dis* **2:**801–810, 1980.

301. Saltzman RL, Peterson PK: Immunodeficiency of the elderly. *Rev Infect Dis* **9:**1127–1139, 1987.

302. Mackowiak PA: Direct effects of hyperthermia on pathogenic microorganisms: Teleologic implications with regard to fever. *Rev Infect Dis* **3:**508–520, 1981.

303. Jampel HD, Duff GW, Gershon RK, Atkins E, Durum SK: Fever and immunoregulation. III. Hyperthermia augments the primary *in vitro* humoral immune response. *J Exp Med* **157:**1229–1238, 1983.

304. Hanson DF, Murphy PA, Silicano R, Shin HS: The effect of temperature on the activation of thymocytes by interleukins I and II. *J Immunol* **130:**216–221, 1983.

305. Dinarello CA, Dempsey RA, Allegretta M, *et al*: Inhibitory effects of elevated temperature on human cytokine production and natural killer activity. *Cancer Res* **46:**6236–6241, 1986.

306. Roberts NJ Jr.: Impact of temperature elevation on immunologic defenses. *Rev Infect Dis* **13:**462–472, 1991.

307. Shavit Y, Lewis JW, Terman GW, Gale RP, Liebeskind JC: Opioid peptides mediate the suppressive effect of stress on natural killer cell cytotoxicity. *Science* **223:**188–190, 1984.

308. Young RA: Stress proteins and immunology. *Annu Rev Immunol* **8:**401–420, 1990.

309. Swartz MN: Stress and the common cold. *N Engl J Med* **325:**654–656, 1991.

310. Ironson G, Wynings C, Schneiderman N, *et al*: Posttraumatic stress symptoms, intrusive thoughts, loss, and immune function after Hurricane Andrew [see comments]. *Psychosom Med* **59:**128–141, 1997.

311. Kiecolt-Glaser JK, Marucha PT, Malarkey WB, Mercado AM, Glaser R: Slowing of wound healing by psychological stress [see comments]. *Lancet* **346:**1194–1196, 1995.

312. Cohen S, Tyrrell DA, Smith AP: Psychological stress and susceptibility to the common cold. *N Engl J Med* **325:**606–612, 1991.

313. van der Meer JW, Guiot HF, van den Broek PJ, van Furth R: Infections in bone marrow transplant recipients. *Semin Hematol* **21:**123–140, 1984.

314. Witherspoon RP, Lum LG, Storb R: Immunologic reconstitution after human marrow grafting. *Semin Hematol* **21:**2–10, 1984.

315. Weeks JC, Tierney MR, Weinstein MC: Cost effectiveness of prophylactic intravenous immune globulin in chronic lymphocytic leukemia. *N Engl J Med* **325:**81–86, 1991.

316. Bennett C, Vardiman J, Golomb H: Disseminated atypical mycobacterial infection in patients with hairy cell leukemia. *Am J Med* **80:**891–896, 1986.

317. Engleman EG, Benike CJ, Hoppe RT, Kaplan HS, Berberich FR: Autologous mixed lymphocyte reaction in patients with Hodgkin's disease. Evidence for a T cell defect. *J Clin Invest* **66:**149–158, 1980.

318. Ward PA, Berenberg JL: Defective regulation of inflammatory mediators in Hodgkin's disease. Supernormal levels of chemotactic-factor inactivator. *N Engl J Med* **290:**76–80, 1974.

319. Weitzman SA, Aisenberg AC, Siber GR, Smith DH: Impaired humoral immunity in treated Hodgkin's disease. *N Engl J Med* **297:**245–248, 1977.

320. Chilcote RR, Baehner RL, Hammond D: Septicemia and meningitis in children splenectomized for Hodgkin's disease. *N Engl J Med* **295:**798–800, 1976.

321. Weitzman S, Aisenberg AC: Fulminant sepsis after the successful treatment of Hodgkin's disease. *Am J Med* **62:**47–50, 1977.

322. Schwartz PE, Sterioff S, Mucha P, Melton LJd, Offord KP: Postsplenectomy sepsis and mortality in adults. *JAMA* **248:**2279–2283, 1982.

323. Kullberg BJ, Westendorp RG, van 't Wout JW, Meinders AE: Purpura fulminans and symmetrical peripheral gangrene caused by *Capnocytophaga canimorsus* (formerly DF-2) septicemia—A complication of dog bite. *Medicine* **70:**287–292, 1991.

324. Hosea SW, Brown EJ, Hamburger MI, Frank MM: Opsonic requirements for intravascular clearance after splenectomy. *N Engl J Med* **304:**245–250, 1981.

325. Benner R, Hijmans W, Haaijman JJ: The bone marrow: The major source of serum immunoglobulins, but still a neglected site of antibody formation. *Clin Exp Immunol* **46:**1–8, 1981.

326. Schumacher MJ: Serum immunoglobulin and transferrin levels after childhood splenectomy. *Arch Dis Child* **45:**114–117, 1970.

327. Di Padova F, Durig M, Harder F, Di Padova C, Zanussi C: Impaired antipneumococcal antibody production in patients without spleens. *Br Med J (Clin Res Ed)* **290:**14–16, 1985.

328. Amlot PL, Hayes AE: Impaired human antibody response to the thymus-independent antigen, DNP- Ficoll, after splenectomy. Implications for post-splenectomy infections. *Lancet* **1:**1008–1011, 1985.

329. Carlisle HN, Saslaw S: Properdin levels in splenectomized persons. *Proc Soc Exp Biol Med* **102:**150–155, 1959.

330. Broome CV, Facklam RR, Fraser DW: Pneumococcal disease after pneumococcal vaccination: An alternative method to estimate the efficacy of pneumococcal vaccine. *N Engl J Med* **303:**549–552, 1980.

331. Konradsen HB, Rasmussen C, Ejstrud P, Hansen JB: Antibody levels against *Streptococcus pneumoniae* and *Haemophilus influenzae* type b in a population of splenectomized individuals with varying vaccination status. *Epidemiol Infect* **119:**167–174, 1997.

332. Teare L, O'Riordan S: Is splenectomy another indication for *Haemophilus influenzae* type b vaccination? *Lancet* **340:**1362, 1992.

333. Barret-Connor E: Infection and sickle cell anemia. In Allen JC (ed): *Infection in the Compromised Host: Clinical Correlations and Therapeutic Approaches*. Williams & Wilkins, Baltimore, 1981, pp. 107–120.

334. Ballester OF, Prasad AS: Anergy, zinc deficiency, and decreased nucleoside phosphorylase activity in patients with sickle cell anemia. *Ann Intern Med* **98:**180–182, 1983.

335. Ammann AJ, Addiego J, Wara DW, Lubin B, Smith WB, Mentzer WC: Polyvalent pneumococcal–polysaccharide immunization of patients with sickle-cell anemia and patients with splenectomy. *N Engl J Med* **297:**897–900, 1977.

336. Allen JC: The diabetic as a compromised host. In Allen JC (ed): *Infection in the Compromised Host: Clinical Correlations and Therapeutic Approaches.* Williams & Wilkins, Baltimore, 1981, pp. 229–270.

337. Rubin J, Yu VL: Malignant external otitis: Insights into pathogenesis, clinical manifestations, diagnosis, and therapy. *Am J Med* **85:**391–398, 1988.

338. Kullberg BJ, Van der Meer JWM: Zygomycosis. In Armstrong D, Cohen J (eds): *Infectious Diseases.* Mosby, London, 1999, pp. 4.9.14–4.9.15.

339. Meyers BR, Wormser G, Hirschman SZ, Blitzer A: Rhinocerebral mucormycosis: Premortem diagnosis and therapy. *Arch Intern Med* **139:**557–560, 1979.

340. Artis WM, Fountain JA, Delcher HK, Jones HE: A mechanism of susceptibility to mucormycosiss in diabetic ketoacidosis: Transferrin and iron availability. *Diabetes* **31:**1109–1114, 1982.

341. Brayton RG, Stokes PE, Schwartz MS, Louria DB: Effect of alcohol and various diseases on leukocyte mobilization, phagocytosis and intracellular bacterial killing. *N Engl J Med* **282:**123–128, 1970.

342. Mowat A, Baum J: Chemotaxis of polymorphonuclear leukocytes from patients with diabetes mellitus. *N Engl J Med* **284:**621–627, 1971.

343. Miller ME, Baker L: Leukocyte functions in juvenile diabetes mellitus: Humoral and cellular aspects. *J Pediatr* **81:**979–982, 1972.

344. Hill HR, Augustine NH, Rallison ML, Santos JI: Defective monocyte chemotactic responses in diabetes mellitus. *J Clin Immunol* **3:**70–77, 1983.

345. Sheldon WH, Bauer H: The development of the acute inflammatory response to experimental cutaneous mucormycosis in normal and diabetic rabbits. *J Exp Med* **110:**845–859, 1959.

346. Bagdade JD, Root RK, Bulger RJ: Impaired leukocyte function in patients with poorly controlled diabetes. *Diabetes* **23:**9–15, 1974.

347. Chernew I, Braude AI: Depression of phagocytosis by solutes in concentrations found in the kidney and urine. *J Clin Invest* **41:**1945–1951, 1962.

348. Nolan CM, Beaty HN, Bagdade JD: Further characterization of the impaired bactericidal function of granulocytes in patients with poorly controlled diabetes. *Diabetes* **27:**889–894, 1978.

349. Tan JS, Anderson JL, Watanakunakorn C, Phair JP: Neutrophil dysfunction in diabetes mellitus. *J Lab Clin Med* **85:**26–33, 1975.

350. Berken A, Sherman AA: Reticuloendothelial system phagocytosis in diabetes mellitus. *Diabetes* **23:**218–220, 1974.

351. MacCuish AC, Urbaniak SJ, Campbell CJ, Duncan LJ, Irvine WJ: Phytohemagglutinin transformation and circulating lymphocyte subpopulations in insulin-dependent diabetic patients. *Diabetes* **23:**708–712, 1974.

352. Eliashiv A, Olumide F, Norton L, Eiseman B: Depression of cell-mediated immunity in diabetes. *Arch Surg* **113:**1180–1183, 1978.

353. Mahmoud AA, Rodman HM, Mandel MA, Warren KS: Induced and spontaneous diabetes mellitus and suppression of cell-mediated immunologic responses. Granuloma formation, delayed dermal reactivity and allograft rejection. *J Clin Invest* **57:**362–367, 1976.

354. Parry MF, Root RK, Metcalf JA, Delaney KK, Kaplow LS, Richar WJ: Myeloperoxidase deficiency: Prevalence and clinical significance. *Ann Intern Med* **95:**293–301, 1981.

355. Cech P, Stalder H, Widmann JJ, Rohner A, Miescher PA: Leukocyte myeloperoxidase deficiency and diabetes mellitus associated with *Candida albicans* liver abscess. *Am J Med* **66:**149–153, 1979.

356. Montgomerie JZ, Kalmanson GM, Guze LB: Renal failure and infection. *Medicine* **47:**1–32, 1968.

357. Clarke IA, Ormrod DJ, Miller TE: Host immune status in uremia. V. Effect of uremia on resistance to bacterial infection. *Kidney Int* **24:**66–73, 1983.

358. Perescenschi G, Zakouth V, Spirer Z, Aviram A: Leukocyte mobilization by epinephrine and hydrocortisone in patients with chronic renal failure. *Experientia* **33:**1529–1530, 1977.

359. Perillie PE, Nolan JP, Finch SC: Studies of the resistance to infection in diabetes mellitus: Local exudative cellular response. *Lab Clin Med* **59:**1008–1015, 1962.

360. Salant DJ, Glover AM, Anderson R, *et al*: Depressed neutrophil chemotaxis in patients with chronic renal failure and after renal transplantation. *J Lab Clin Med* **88:**536–545, 1976.

361. Abrutyn E, Solomons NW, St. Clair L, MacGregor RR, Root RK: Granulocyte function in patients with chronic renal failure: Surface adherence, phagocytosis, and bactericidal activity *in vitro*. *J Infect Dis* **135:**1–8, 1977.

362. Urbanitz D, Sieberth HG: Impaired phagocytic activity of human monocytes in respect to reduced antibacterial resistance in uremia. *Clin Nephrol* **4:**13–17, 1975.

363. Nelson J, Ormrod DJ, Miller TE: Host immune status in uremia. IV. Phagocytosis and inflammatory response *in vivo*. *Kidney Int* **23:**312–319, 1983.

364. Kirkpatrick CH, Wilson WEC, Talmage DW: Immunologic studies in human organ transplantation. I. Observation and characterization of suppressed cutaneous reactivity in uremia. *J Exp Med* **119:**727–742, 1964.

365. Newberry WM, Sanford JP: Defective cellular immunity in renal failure: Depression of reactivity of lymphocytes to phytohemagglutinin by renal failure serum. *J Clin Invest* **50:**1262–1271, 1971.

366. Daniels JC, Sakai H, Remmers AR Jr, *et al*: *In vitro* reactivity of human lymphocytes in chronic uraemia: Analysis and interpretation. *Clin Exp Immunol* **8:**213–227, 1971.

367. Miller TE, Stewart E: Host immune status in uraemia. I. Cell-mediated immune mechanisms. *Clin Exp Immunol* **41:**115–122, 1980.

368. Stevens CE, Alter HJ, Taylor PE, Zang EA, Harley EJ, Szmuness W: Hepatitis B vaccine in patients receiving hemodialysis. Immunogenicity and efficacy. *N Engl J Med* **311:**496–501, 1984.

369. Crosnier J, Jungers P, Courouce AM, *et al*: Randomised placebo-controlled trial of hepatitis B surface antigen vaccine in French haemodialysis units: II, Haemodialysis patients. *Lancet* **1:**797–800, 1981.

370. Desmyter J, Colaert J, De Groote G, *et al*: Efficacy of heat-inactivated hepatitis B vaccine in haemodialysis patients and staff. Double-blind placebo-controlled trial. *Lancet* **2:**1323–1328, 1983.

371. Lelie PN, Reesink HW, de Jong-van Manen ST, Dees PJ, Reerink-Brongers EE: Immune response to a heat-inactivated hepatitis B vaccine in patients undergoing hemodialysis. Enhancement of the response by increasing the dose of hepatitis B surface antigen from 3 to 27 micrograms. *Arch Intern Med* **145:**305–309, 1985.

372. Meuer SC, Dumann H, Meyer zum Buschenfelde KH, Kohler H: Low-dose interleukin-2 induces systemic immune responses against HBsAg in immunodeficient non-responders to hepatitis B vaccination. *Lancet* **1:**15–18, 1989.

373. Ruddy MC, Rubin AL, Novogrodsky A, Stenzel KH: Decreased macrophage-mediated suppression of lymphocyte activation in chronic renal failure. *Am J Med* **75:**571–579, 1983.

374. Sengar DP, Rashid A, Harris JE: *In vitro* reactivity of lymphocytes obtained from uraemic patients maintained by heamodialysis. *Clin Exp Immunol* **21:**298–305, 1975.

375. Langhoff E, Ladefoged J: Cellular immunity in renal failure: Depression of lymphocyte transformation by uraemia and methyl-prednisolone. Intra-individual consistency of lymphocyte responses to the *in vitro* suppressive effect of steroid. *Int Arch Allergy Appl Immunol* **74:**241–245, 1984.

376. Boelaert JR, Daneels RF, Schurgers ML, Matthys EG, Gordts BZ, Van Landuyt HW: Iron overload in haemodialysis patients increases the risk of bacteraemia: A prospective study. *Nephrol Dial Transplant* **5:**130–134, 1990.

377. Boelaert JR, Vergauwe PL, Vandepitte JM: Mucormycosis infection in dialysis patients. *Ann Intern Med* **107:**782–783, 1987.

378. De Locht M, Boelaert JR, Schneider YJ: Iron uptake from ferrioxamine and from ferrirhizoferrin by germinating spores of *Rhizopus microsporus*. *Biochem Pharmacol* **47:**1843–1850, 1994.

379. Gloor HJ, Nichols WK, Sorkin MI, *et al*: Peritoneal access and related complications in continuous ambulatory peritoneal dialysis. *Am J Med* **74:**593–598, 1983.

380. Verbrugh HA, Keane WF, Hoidal JR, Freiberg MR, Elliott GR, Peterson PK: Peritoneal macrophages and opsonins: Antibacterial defense in patients undergoing chronic peritoneal dialysis. *J Infect Dis* **147:**1018–1029, 1983.

381. Keane WF, Peterson PK: Host defense mechanisms of the peritoneal cavity and continuous ambulatory peritoneal dialysis. *Peritoneal Dialysis Bull* **4:**122–127, 1984.

382. Amair P, Khanna R, Leibel B, *et al*: Continuous ambulatory peritoneal dialysis in diabetics with end-stage renal disease. *N Engl J Med* **306:**625–630, 1982.

383. Gluckman SJ, Dvorak VC, MacGregor RR: Host defenses during prolonged alcohol consumption in a controlled environment. *Arch Intern Med* **137:**1539–1543, 1977.

384. MacGregor RR, Spagnuolo PJ, Lentek AL: Inhibition of granulocyte adherence by ethanol, prednisone, and aspirin, measured with an assay system. *N Engl J Med* **291:**642–646, 1974.

385. MacGregor RR: Alcohol and immune defense. *JAMA* **256:**1474–1479, 1986.

386. Bermudez LE, Young LS: Ethanol augments intracellular survival of *Mycobacterium avium* complex and impairs macrophage responses to cytokines. *J Infect Dis* **163:**1286–1292, 1991.

387. Berenyi MR, Straus B, Cruz D: *In vitro* and *in vivo* studies of cellular immunity in alcoholic cirrhosis. *Am J Dig Dis* **19:**199–205, 1974.

388. Saxena QB, Mezey E, Adler WH: Regulation of natural killer activity *in vivo*. II. The effect of alcohol consumption on human peripheral blood natural killer activity. *Int J Cancer* **26:**413–417, 1980.

389. Wilcox CM, Dismukes WE: Spontaneous bacterial peritonitis. A review of pathogenesis, diagnosis, and treatment. *Medicine* **66:**447–456, 1987.

390. DeMeo AN, Andersen BR: Defective chemotaxis associated with a serum inhibitor in cirrhotic patients. *N Engl J Med* **286:**735–740, 1972.

391. Blussé van Oud Alblas A, Janssens AR, Leijh PC, van Furth R: Functions of granulocytes and monocytes in primary biliary and alcoholic cirrhosis. *Clin Exp Immunol* **62:**724–731, 1985.

392. Wyke RJ, Rajkovic IA, Williams R: Impaired opsonization by serum from patients with chronic liver disease. *Clin Exp Immunol* **51:**91–98, 1983.

393. Runyon BA: Patients with deficient ascitic fluid opsonic activity are predisposed to spontaneous bacterial peritonitis [published erratum appears in *Hepatology* **8(5):**1184, 1988]. *Hepatology* **8:**632–635, 1988.

394. Sorrell MF, Leevy CM: Lymphocyte transformation and alcoholic liver injury. *Gastroenterology* **63:**1020–1025, 1972.

395. Staples PJ, Gerding DN, Decker JL, Gordon RS Jr: Incidence of infection in systemic lupus erythematosus. *Arthritis Rheum* **17:**1–10, 1974.

396. Abeles M: The rheumatic patient as a compromised host. In Allen JC (ed): *Infection in the Compromised Host: Clinical Correlations and Therapeutic Approaches*. Williams & Wilkins, Baltimore, 1981, pp. 197–227.

397. Ginzler E, Diamond H, Kaplan D, Weiner M, Schlesinger M, Seleznick M: Computer analysis of factors influencing frequency of infection in systemic lupus erythematosus. *Arthritis Rheum* **21:**37–44, 1978.

398. Moutsopoulos HM, Gallagher JD, Decker JL, Steinberg AD: Herpes zoster in patients with systemic lupus erythematosus. *Arthritis Rheum* **21:**789–802, 1978.

399. Clark RA, Kimball HR, Decker JL: Neutrophil chemotaxis in systemic lupus erythematosus. *Ann Rheum Dis* **33:**167–172, 1974.

400. Perez HD, Lipton M, Goldstein IM: A specific inhibitor of complement (C5)-derived chemotactic activity in serum from patients with systemic lupus erythematosus. *J Clin Invest* **62:**29–38, 1978.

401. Jasin HE, Orozco JH, Ziff M: Serum heat labile opsonins in systemic lupus erythematosus. *J Clin Invest* **53:**343–353, 1974.

402. Zurier RB: Reduction of phagocytosis and lysosomal enzyme release from human leukocytes by serum from patients with systemic lupus erythematosus. *Arthritis Rheum* **19:**73–78, 1976.

403. Kimball HR, Wolff SM, Talal N, Plotz PH, Decker JL: Marrow granulocyte reserves in the rheumatic diseases. *Arthritis Rheum* **16:**345–352, 1973.

404. Rosenthal CJ, Franklin EC: Depression of cellular-mediated immunity in systemic lupus erythematosus. Relation to disease activity. *Arthritis Rheum* **18:**207–217, 1975.

405. Mowat AG, Baum J: Chemotaxis of polymorphonuclear leukocytes from patients with rheumatoid arthritis. *J Clin Invest* **50:**2541–2549, 1971.

406. Bar-Eli M, Ehrenfeld M, Litvin Y, Gallily R: Monocyte function in rheumatoid arthritis. *Scand J Rheumatol* **9:**17–23, 1980.

407. Turner RA, Schumacher R, Myers AR: Phagocytic function of polymorphonuclear leukocytes in rheumatic diseases. *J Clin Invest* **52:**1632–1635, 1973.

408. Breedveld FC, Fibbe WE, Hermans J, van der Meer JW, Cats A: Factors influencing the incidence of infections in Felty's syndrome. *Arch Intern Med* **147:**915–920, 1987.

409. Mackowiak PA: Microbial synergism in human infections. *N Engl J Med* **298:**21–26, 1978.

410. O'Grady F, Smith H (eds): *Microbial Perturbation of Host Defences*. Academic Press, New York, 1981.

411. Solberg CO, Hellum KB: Neutrophil granulocyte function in bacterial infections. *Lancet* **2:**727–730, 1972.

412. Barbour AG, Allred CD, Solberg CO, Hill HR: Chemiluminescence by polymorphonuclear leukocytes from patients with active bacterial infection. *J Infect Dis* **141:**14–26, 1980.

413. Parenti DM, Snydman DR: Capnocytophaga species: Infections in nonimmunocompromised and immunocompromised hosts. *J Infect Dis* **151:**140–147, 1985.

414. Shurin SB, Socransky SS, Sweeney E, Stossel TP: A neutrophil disorder induced by capnocytophaga, a dental micro-organism. *N Engl J Med* **301**:849–854, 1979.

415. Rubin RH, Cosimi AB, Tolkoff-Rubin NE, Russell PS, Hirsch MS: Infectious disease syndromes attributable to cytomegalovirus and their significance among renal transplant recipients. *Transplantation* **24**:458–464, 1977.

416. Chatterjee SN, Fiala M, Weiner J, Stewart JA, Stacey B, Warmer N: Primary cytomegalovirus and opportunistic infections. Incidence in renal transplant recipients. *JAMA* **240**:2446–2449, 1978.

417. Rand KH, Pollard RB, Merigan TC: Increased pulmonary superinfections in cardiac-transplant patients undergoing primary cytomegalovirus infection. *N Engl J Med* **298**:951–953, 1978.

418. Carney WP, Rubin RH, Hoffman RA, Hansen WP, Healey K, Hirsch MS: Analysis of T lymphocyte subsets in cytomegalovirus mononucleosis. *J Immunol* **126**:2114–2116, 1981.

419. Van Es A, van Gemert GW, Baldwin WK: Viral infection and T-lymphocyte subpopulations in renal transplant recipients. *N Engl J Med* **309**:110–111, 1983.

420. LaQuaglia MP, Tolkoff-Rubin NE, Dienstag JL, *et al*: Impact of hepatitis on renal transplantation. *Transplantation* **32**:504–507, 1981.

421. Singh N, Gayowski T, Wagener MM, Marino IR: Increased infections in liver transplant recipients with recurrent hepatitis C virus hepatitis. *Transplantation* **61**:402–406, 1996.

422. Greene WC: The molecular biology of human immunodeficiency virus type 1 infection. *N Engl J Med* **324**:308–317, 1991.

423. Smith PD, Ohura K, Masur H, Lane HC, Fauci AS, Wahl SM: Monocyte function in the acquired immune deficiency syndrome. Defective chemotaxis. *J Clin Invest* **74**:2121–2128, 1984.

424. Schneider MM, Borleffs JC, Stolk RP, Jaspers CA, Hoepelman AI: Discontinuation of prophylaxis for *Pneumocystis carinii* pneumonia in HIV-1-infected patients treated with highly active antiretroviral therapy. *Lancet* **353**:201–203, 1999.

425. Foudraine NA, Hovenkamp E, Notermans DW, *et al*: Immunopathology as a result of highly active antiretroviral therapy in HIV-1-infected patients. *AIDS* **13**:177–184, 1999.

426. Berenguer J, Solera J, Diaz MD, Moreno S, Lopez-Herce JA, Bouza E: Listeriosis in patients infected with human immunodeficiency virus. *Rev Infect Dis* **13**:115–119, 1991.

427. Sobel JD: Controversial aspects of candidiasis in the acquired immunodeficiency syndrome. In Vandenbossche H, Mackenzie D, Cauwenbergh G, Drouet E, Dupont B (eds): *Mycoses in AIDS Patients*. Plenum Press, New York, 1990, pp. 93–100.

428. Lazzarin A, Uberti Foppa C, Galli M, *et al*: Impairment of polymorphonuclear leucocyte function in patients with acquired immunodeficiency syndrome and with lymphadenopathy syndrome. *Clin Exp Immunol* **65**:105–111, 1986.

429. Ellis M, Gupta S, Galant S, *et al*: Impaired neutrophil function in patients with AIDS or AIDS-related complex: A comprehensive evaluation. *J Infect Dis* **158**:1268–1276, 1988.

430. Redd SC, Rutherford GWd, Sande MA, *et al*: The role of human immunodeficiency virus infection in pneumococcal bacteremia in San Francisco residents. *J Infect Dis* **162**:1012–1017, 1990.

431. Schlamm HT, Yancovitz SR: *Haemophilus influenzae* pneumonia in young adults with AIDS, ARC, or risk of AIDS. *Am J Med* **86**:11–14, 1989.

432. Steinhoff MC, Auerbach BS, Nelson KE, *et al*: Antibody responses to *Haemophilus influenzae* type B vaccines in men with human immunodeficiency virus infection. *N Engl J Med* **325**:1837–1842, 1991.

433. Kroon FP, van Dissel JT, Rijkers GT, Labadie J, van Furth R: Antibody response to *Haemophilus influenzae* type b vaccine in relation to the number of CD4+ T lymphocytes in adults infected with human immunodeficiency virus. *Clin Infect Dis* **25**:600–606, 1997.

434. Kroon FP, van Dissel JT, Ravensbergen E, Nibbering PH, van Furth R: Antibodies against pneumococcal polysaccharides after vaccination in HIV-infected individuals: 5-year follow-up of antibody concentrations. *Vaccine* **18**:524–530, 1999.

435. Jacobson MA, Gellermann H, Chambers H: *Staphylococcus aureus* bacteremia and recurrent staphylococcal infection in patients with acquired immunodeficiency syndrome and AIDS-related complex. *Am J Med* **85**:172–176, 1988.

436. Stevens DA, Denning DW: Pulmonary aspergillosis in AIDS. *N Engl J Med* **325**:354–357, 1991.

437. Schaffner A: Pulmonary aspergillosis in AIDS. *N Engl J Med* **325**:354–357, 1991.

438. Denning DW, Follansbee SE, Scolaro M, Norris S, Edelstein H, Stevens DA: Pulmonary aspergillosis in the acquired immunodeficiency syndrome. *N Engl J Med* **324**:654–662, 1991.

439. Graham BS, Tucker WS Jr: Opportunistic infections in endogenous Cushing's syndrome. *Ann Intern Med* **101**:334–338, 1984.

440. Fauci AS, Dale DC, Balow JE: Glucocorticosteroid therapy: Mechanisms of action and clinical considerations. *Ann Intern Med* **84**:304–315, 1976.

441. Stuck AE, Minder CE, Frey FJ: Risk of infectious complications in patients taking glucocorticosteroids. *Rev Infect Dis* **11**:954–963, 1989.

442. Sneiderman CA, Wilson JW: Effects of corticosteroids on complement and the neutrophilic polymorphonuclear leukocyte. *Transplant Proc* **7**:41–48, 1975.

443. Butler WT, Rossen RD: Effects of corticosteroids on immunity in man. I. Decreased serum IgG concentration caused by 3 or 5 days of high doses of methylprednisolone. *J Clin Invest* **52**:2629–2640, 1973.

444. Suda T, Miura Y, Ijima H, Ozawa K, Motoyoshi K, Takaku F: The effect of hydrocortisone on human granulopoiesis *in vitro* with cytochemical analysis of colonies. *Exp Hematol* **11**:114–121, 1983.

445. Dale DC, Fauci AS, Guerry DI, Wolff SM: Comparison of agents producing a neutrophilic leukocytosis in man. Hydrocortisone, prednisone, endotoxin, and etiocholanolone. *J Clin Invest* **56**:808–813, 1975.

446. Dale DC, Fauci AS, Wolff SM: Alternate-day prednisone. Leukocyte kinetics and susceptibility to infections. *N Engl J Med* **291**:1154–1158, 1974.

447. Wiener SL, Wiener R, Urivetzky M, *et al*: The mechanism of action of a single dose of methylprednisolone on acute inflammation *in vivo*. *J Clin Invest* **56**:679–689, 1975.

448. Thompson J, Furth Rv: The effect of glucocorticosteroids on the kinetics of mononuclear phagocytes. *J Exp Med* **131**:429–442, 1970.

449. Rinehart JJ, Balcerzak SP, Sagone AL, LoBuglio AF: Effects of corticosteroids on human monocyte function. *J Clin Invest* **54**:1337–1343, 1974.

450. Rinehart JJ, Sagone AL, Balcerzak SP, Ackerman GA, LoBuglio AF: Effects of corticosteroid therapy on human monocyte function. *N Engl J Med* **292**:236–241, 1975.

451. Van Zwet TL, Thompson J, Van Furth R: Effect of glucocorticosteroids on the phagocytosis and intracellular killing by peritoneal macrophages. *Infect Immun* **12**:699–705, 1975.

452. Atkinson JP, Frank MM: Complement-independent clearance of

IgG-sensitized erythrocytes: Inhibition by cortisone. *Blood* **44:** 629–637, 1974.

453. Werb ZA: Hormone receptors and hormonal regulation of macrophage physiological functions. In Van Furth R (ed): *Mononuclear Phagocytes: Functional Aspects.* Martinus Nijhoff, The Hague, 1980, pp. 809–829.

454. Knudsen PJ, Dinarello CA, Strom TB: Glucocorticoids inhibit transcriptional and posttranscriptional expression of interleukin 1 in U937 cells. *J Immunol* **139:**4129–4134, 1987.

455. Luedke CE, Cerami A: Interferon-gamma overcomes glucocorticoid suppression of cachectin/tumor necrosis factor biosynthesis by murine macrophages. *J Clin Invest* **86:**1234–1240, 1990.

456. Roilides E, Uhlig K, Venzon D, Pizzo PA, Walsh TJ: Prevention of corticosteroid-induced suppression of human polymorphonuclear leukocyte-induced damage of *Aspergillus fumigatus* hyphae by granulocyte colony-stimulating factor and gamma interferon. *Infect Immun* **61:**4870–4877, 1993.

457. Dannenberg AM Jr: The antinflammatory effects of glucocorticosteroids. A brief review of the literature. *Inflammation* **3:**329–343, 1979.

458. Cupps TR, Fauci AS: Corticosteroid-mediated immunoregulation in man. *Immunol Rev* **64:**134–155, 1982.

459. Gustafson TL, Schaffner W, Lavely GB, Stratton CW, Johnson HK, Hutcheson RH Jr: Invasive aspergillosis in renal transplant recipients: Correlation with corticosteroid therapy. *J Infect Dis* **148:**230–238, 1983.

460. Gershwin ME, Goetzl EJ, Steinberg AD: Cyclophosphamide: Use in practice. *Ann Intern Med* **80:**531–540, 1974.

461. Balow JE: Cyclophosphamide suppression of established cell-mediated immunity. *J Clin Invest* **56:**65–70, 1975.

462. Fauci AS, Wolff SM, Johnson JS: Effect of cyclophosphamide upon the immune response in Wegener's granulomatosis. *N Engl J Med* **285:**1493–1496, 1971.

463. Ten Berge RJ, Schellekens PT: A critical analysis of the use of azathioprine in clinical medicine. *Neth J Med* **26:**164–171, 1983.

464. Losito A, Williams DG, Cooke G, Harris L: The effects on polymorphonuclear leucocyte function of prednisolone and azathioprine *in vivo* and prednisolone, azathioprine and 6-mercaptopurine *in vitro.* *Clin Exp Immunol* **32:**423–428, 1978.

465. Hersh EM, Wong VG, Freireich EJ: Inhibition of the local inflammatory response in man by antimetabolites. *Blood* **27:**38–48, 1966.

466. Cseuz R, Panayi GS: The inhibition of NK cell function by azathioprine during the treatment of patients with rheumatoid arthritis. *Br J Rheumatol* **29:**358–362, 1990.

467. Förre O, Bjerkhoel F, Salvesen CF, *et al:* An open, controlled, randomized comparison of cyclosporine and azathioprine in the treatment of rheumatoid arthritis: A preliminary report. *Arthritis Rheum* **30:**88–92, 1987.

468. Andersen PA, West SG, O'Dell JR, Via CS, Claypool RG, Kotzin BL: Weekly pulse methotrexate in rheumatoid arthritis. Clinical and immunologic effects in a randomized, double-blind study. *Ann Intern Med* **103:**489–496, 1985.

469. Boerbooms AM, Jeurissen ME, Westgeest AA, Theunisse H, Van de Putte LB: Methotrexate in refractory rheumatoid arthritis. *Clin Rheumatol* **7:**249–256, 1988.

470. Hyams JS, Donaldson MH, Metcalf JA, Root RK: Inhibition of human granulocyte function by methotrexate. *Cancer Res* **38:** 650–655, 1978.

471. Johnston CA, Russell AS, Kovithavongs T, Dasgupta M: Measures of immunologic and inflammatory responses *in vitro* in rheumatoid patients treated with methotrexate. *J Rheumatol* **13:** 294–296, 1986.

472. Johnson WJ, DiMartino MJ, Meunier PC, Muirhead KA, Hanna N: Methotrexate inhibits macrophage activation as well as vascular and cellular inflammatory events in rat adjuvant induced arthritis. *J Rheumatol* **15:**745–749, 1988.

473. Barrera P, Boerbooms AM, Demacker PN, van de Putte LB, Gallati H, van der Meer JW: Circulating concentrations and production of cytokines and soluble receptors in rheumatoid arthritis patients: Effects of a single dose methotrexate. *Br J Rheumatol* **33:**1017–1024, 1994.

474. Kim JH, Perfect JR: Infection and cyclosporine. *Rev Infect Dis* **11:**677–690, 1989.

475. Spencer CM, Goa KL, Gillis JC: Tacrolimus. An update of its pharmacology and clinical efficacy in the management of organ transplantation. *Drugs* **54:**925–975, 1997.

476. Kahan BD: Cyclosporin. *N Engl J Med* **321:**1725–1738, 1989.

477. Granelli-Piperno A, Inaba K, Steinman RM: Stimulation of lymphokine release from T lymphoblasts. Requirement for mRNA synthesis and inhibition by cyclosporin A. *J Exp Med* **160:**1792–1802, 1984.

478. Shin GT, Khanna A, Ding R, *et al: In vivo* expression of transforming growth factor-beta1 in humans: Stimulation by cyclosporine. *Transplantation* **65:**313–318, 1998.

479. European FK506 Multicentre Liver Study Group: Randomised trial comparing tacrolimus (FK506) and cyclosporin in prevention of liver allograft rejection. *Lancet* **344:**423–428, 1994.

480. Jensik SC, and the FK 506 Kidney Transplant Study Group: Tacrolimus (FK 506) in kidney transplantation: Three-year survival results of the US multicenter, randomized, comparative trial. *Transplant Proc* **30:**1216–1218, 1998.

481. Sievers TM, Rossi SJ, Ghobrial RM, *et al:* Mycophenolate mofetil. *Pharmacotherapy* **17:**1178–1197, 1997.

482. The Tricontinental Mycophenolate Mofetil Renal Transplantation Study Group: A blinded, randomized clinical trial of mycophenolate mofetil for the prevention of acute rejection in cadaveric renal transplantation. *Transplantation* **61:**1029–1037, 1996.

483. US Renal Transplant Mycophenolate Mofetil Study Group: Mycophenolate mofetil in cadaveric renal transplantation. *Am J Kidney Dis* **34:**296–303, 1999.

484. Moreso F, Seron D, Morales JM, *et al:* Incidence of leukopenia and cytomegalovirus disease in kidney transplants treated with mycophenolate mofetil combined with low cyclosporine and steroid doses. *Clin Transplant* **12:**198–205, 1998.

485. Vincenti F, Kirkman R, Light S, *et al:* Interleukin-2-receptor blockade with daclizumab to prevent acute rejection in renal transplantation. *N Engl J Med* **338:**161–165, 1998.

486. Vincenti F, Grinyo J, Ramos E, *et al:* Can antibody prophylaxis allow sparing of other immunosuppressives? *Transplant Proc* **31:** 1246–1248, 1999.

487. Nashan B, Moore R, Amlot P, *et al:* Randomised trial of basiliximab versus placebo for control of acute cellular rejection in renal allograft recipients. *Lancet* **350:**1193–1198, 1997.

488. Kahan BD, Julian BA, Pescovitz MD, Vanrenterghem Y, Neylan J, Group atRS: Sirolimus reduces the incidence of acute rejection episodes despite lower cyclosporine doses in Caucasian recipients of mismatched primary renal allografts: A phase II trial. *Transplantation* **68:**1526–1532, 1999.

489. Sayegh MH, Turka LA: The role of T-cell costimulatory activation pathways in transplant rejection. *N Engl J Med* **338:**1813–1821, 1998.

490. Hauser WE, Remington JS: The effect of antibiotics on the humoral and cell-mediated immune responses. In Sabath LD (ed): *Action of Antibiotics in Patients.* Huber, Bern, 1982, pp. 127–147.

491. Morikawa K, Watabe H, Araake M, Morikawa S: Modulatory effect of antibiotics on cytokine production by human monocytes *in vitro. Antimicrob Agents Chemother* **40:**1366–1370, 1996.

492. Bergeron Y, Deslauriers AM, Ouellet N, Gauthier MC, Bergeron MG: Influence of cefodizime on pulmonary inflammatory response to heat-killed *Klebsiella pneumoniae* in mice. *Antimicrob Agents Chemother* **43:**2291–2294, 1999.

493. Riesbeck K, Sigvardsson M, Leanderson T, Forsgren A: Super-induction of cytokine gene transcription by ciprofloxacin. *J Immunol* **153:**343–352, 1994.

494. Genovese F, Mancuso G, Cuzzola M, *et al*: Improved survival and antagonistic effect of sodium fusidate on tumor necrosis factor alpha in a neonatal mouse model of endotoxin shock. *Antimicrob Agents Chemother* **40:**1733–1735, 1996.

495. Thomas MZ, Medoff G, Kobayashi GS: Changes in murine resistance to *Listeria monocytogenes* infection induced by amphotericin B. *J Infect Dis* **127:**373–377, 1973.

496. Vonk AG, Netea MG, Denecker NEJ, Verschueren ICMM, Van der Meer JWM, Kullberg BJ: Modulation of the pro- and anti-inflammatory cytokine balance by amphotericin B. *J Antimicrobial Chemother* **42:**469–474, 1998.

497. Cosimi AB: The clinical value of antilymphocyte antibodies. *Transplant Proc* **13:**462–468, 1981.

498. Thistlethwaite JR Jr, Stuart JK, Mayes JT, *et al*: Complications and monitoring of OKT3 therapy. *Am J Kidney Dis* **11:**112–119, 1988.

499. Frenken LA, Hoitsma AJ, Tax WJ, Koene RA: Prophylactic use of anti-CD3 monoclonal antibody WT32 in kidney transplantation. *Transplant Proc* **23:**1072–1073, 1991.

500. Burk ML, Matuszewski KA: Muromonab-CD3 and antithymocyte globulin in renal transplantation. *Ann Pharmacother* **31:**1370–1377, 1997.

501. Rubin RH, Cosimi AB, Hirsch MS, Herrin JT, Russell PS, Tolkoff-Rubin NE: Effects of antithymocyte globulin on cytomegalovirus infection in renal transplant recipients. *Transplantation* **31:**143–145, 1981.

502. Macris MP, Van Buren CT, Sweeney MS, Frazier OH, Duncan JM: Selective use of OKT3 in heart transplantation with the use of risk factor analysis. *J Heart Transplant* **8:**296–302, 1989.

503. Gordon RD, Tzakis AG, Iwatsuki S, *et al*: Experience with Orthoclone OKT3 monoclonal antibody in liver transplantation. *Am J Kidney Dis* **11:**141–144, 1988.

504. Renlund DG, O'Connell JB, Gilbert EM, *et al*: A prospective comparison of murine monoclonal CD-3 (OKT3) antibody-based and equine antithymocyte globulin-based rejection prophylaxis in cardiac transplantation. Decreased rejection and less corticosteroid use with OKT3. *Transplantation* **47:**599–605, 1989.

505. Millis JM, McDiarmid SV, Hiatt JR, *et al*: Randomized prospective trial of OKT3 for early prophylaxis of rejection after liver transplantation. *Transplantation* **47:**82–88, 1989.

506. Grino JM, Alsina J, Sabater R, *et al*: Antilymphoblast globulin, cyclosporine, and steroids in cadaveric renal transplantation. *Transplantation* **49:**1114–1117, 1990.

507. van Rood JJ, Claas FH: The influence of allogeneic cells on the human T and B cell repertoire. *Science* **248:**1388–1393, 1990.

508. Lagaaij EL, Hennemann IP, Ruigrok M, *et al*: Effect of one-HLA-DR-antigen-matched and completely HLA-DR-mismatched blood transfusions on survival of heart and kidney allografts. *N Engl J Med* **321:**701–705, 1989.

509. Lederman MM, Ratnoff OD, Scillian JJ, Jones PK, Schacter B: Impaired cell-mediated immunity in patients with classic hemophilia. *N Engl J Med* **308:**79–83, 1983.

510. Gascon P, Zoumbos NC, Young NS: Immunologic abnormalities in patients receiving multiple blood transfusions. *Ann Intern Med* **100:**173–177, 1984.

511. Wing EJ, Bruns FJ, Fraley DS, Segel DP, Adler S: Infectious complications with plasmapheresis in rapidly progressive glomerulonephritis. *JAMA* **244:**2423–2426, 1980.

512. Pohl MA, Lan SP, Berl T: Plasmapheresis does not increase the risk for infection in immunosuppressed patients with severe lupus nephritis. The Lupus Nephritis Collaborative Study Group. *Ann Intern Med* **114:**924–929, 1991.

513. van Deuren M, Santman FW, van Dalen R, Sauerwein RW, Span LF, van der Meer JW: Plasma and whole blood exchange in meningococcal sepsis. *Clin Infect Dis* **15:**424–430, 1992.

514. Van der Meer JWM, Van Deuren M, Kullberg BJ: The interplay of pro-inflammatory cytokines and anti-inflammatory mediators during severe infection. In Vincent JL (ed): *Yearbook of Intensive Care and Emergency Medicine.* Springer, Brussels, 1995, pp. 377–383.

515. van Deuren M, Frieling JT, van der Ven-Jongekrijg J, *et al*: Plasma patterns of tumor necrosis factor-alpha (TNF) and TNF soluble receptors during acute meningococcal infections and the effect of plasma exchange. *Clin Infect Dis* **26:**918–923, 1998.

516. Doria G, Agarossi G, Adorini L: Selective effects of ionizing radiations on immunoregulatory cells. *Immunol Rev* **65:**23–54, 1982.

517. Baehner RL, Neiburger RG, Johnson DE, Murrmann SM: Transient bactericidal defect of peripheral blood phagocytes from children with acute lymphoblastic leukemia receiving craniospinal irradiation. *N Engl J Med* **289:**1209–1213, 1973.

518. Slater JM, Ngo E, Lau BH: Effect of therapeutic irradiation on the immune responses. *Am J Roentgenol* **126:**313–320, 1976.

519. Strober S, Dhillon M, Schubert M, *et al*: Acquired immune tolerance to cadaveric renal allografts. A study of three patients treated with total lymphoid irradiation. *N Engl J Med* **321:**28–33, 1989.

520. Eskola J, Kayhty H, Takala AK, *et al*: A randomized, prospective field trial of a conjugate vaccine in the protection of infants and young children against invasive Haemophilus influenzae type b disease. *N Engl J Med* **323:**1381–1387, 1990.

521. Brown GL, Nanney LB, Griffen J, *et al*: Enhancement of wound healing by topical treatment with epidermal growth factor. *N Engl J Med* **321:**76–79, 1989.

522. Greenhalgh DG: The role of growth factors in wound healing. *J Trauma* **41:**159–167, 1996.

523. Raz R, Stamm WE: A controlled trial of intravaginal estriol in postmenopausal women with recurrent urinary tract infections. *N Engl J Med* **329:**753–756, 1993.

524. Eriksen B: A randomized, open, parallel-group study on the preventive effect of an estradiol-releasing vaginal ring (Estring) on recurrent urinary tract infections in postmenopausal women. *Am J Obstet Gynecol* **180:**1072–1079, 1999.

525. Reichman RC, Oakes D, Bonnez W, *et al*: Treatment of condyloma acuminatum with three different interferon-alpha preparations administered parenterally: A double-blind, placebo-controlled trial. *J Infect Dis* **162:**1270–1276, 1990.

526. Perrillo RP, Schiff ER, Davis GL, *et al*: A randomized, controlled trial of interferon alfa-2b alone and after prednisone withdrawal for the treatment of chronic hepatitis B. The Hepatitis Interventional Therapy Group. *N Engl J Med* **323:**295–301, 1990.

527. Wong DK, Cheung AM, O'Rourke K, Naylor CD, Detsky AS,

Heathcote J: Effect of alpha-interferon treatment in patients with hepatitis B e antigen-positive chronic hepatitis B. A meta-analysis. *Ann Intern Med* **119:**312–323, 1993.

528. de Jongh FE, Janssen HL, de Man RA, Hop WC, Schalm SW, van Blankenstein M: Survival and prognostic indicators in hepatitis B surface antigen-positive cirrhosis of the liver. *Gastroenterology* **103:**1630–1635, 1992.

529. Davis GL, Balart LA, Schiff ER, *et al*: Treatment of chronic hepatitis C with recombinant interferon alfa. A multicenter randomized, controlled trial. Hepatitis Interventional Therapy Group. *N Engl J Med* **321:**1501–1506, 1989.

530. Davis GL, Esteban-Mur R, Rustgi V, *et al*: Interferon alfa-2b alone or in combination with ribavirin for the treatment of relapse of chronic hepatitis C. International Hepatitis Interventional Therapy Group. *N Engl J Med* **339:**1493–1499, 1998.

531. Nathan CF, Kaplan G, Levis WR, *et al*: Local and systemic effects of intradermal recombinant interferon-gamma in patients with lepromatous leprosy. *N Engl J Med* **315:**6–15, 1986.

532. Barral-Netto M, Santos S, Santos I, *et al*: Immunochemotherapy with interferon-gamma and multidrug therapy for multibacillary leprosy. *Acta Trop* **72:**185–201, 1999.

533. Sampaio EP, Moreira AL, Sarno EN, Malta AM, Kaplan G: Prolonged treatment with recombinant interferon gamma induces erythema nodosum leprosum in lepromatous leprosy patients. *J Exp Med* **175:**1729–1737, 1992.

534. Badaro R, Falcoff E, Badaro FS, *et al*: Treatment of visceral leishmaniasis with pentavalent antimony and interferon gamma. *N Engl J Med* **322:**16–21, 1990.

535. Squires KE, Rosenkaimer F, Sherwood JA, Forni AL, Were JB, Murray HW: Immunochemotherapy for visceral leishmaniasis: A controlled pilot trial of antimony versus antimony plus interferon-gamma. *Am J Trop Med Hyg* **48:**666–669, 1993.

536. Gallin JI, Malech HL, Melnick DA, and the International Chronic Granulomatous Disease Cooperative Study Group: A controlled trial of interferon gamma to prevent infection in chronic granulomatous disease. *N Engl J Med* **324:**509–516, 1991.

537. Bernhisel-Broadbent J, Camargo EE, Jaffe HS, Lederman HM: Recombinant human interferon-γ as adjunct therapy for *Aspergillus* infection in a patient with chronic granulomatous disease. *J Infect Dis* **163:**908–911, 1991.

538. Rex JH, Bennett JE, Gallin JI, Malech HL, Decarlo ES, Melnick DA: *In vivo* interferon-γ therapy augments the *in vitro* ability of chronic granulomatous disease neutrophils to damage *Aspergillus* hyphae. *J Infect Dis* **163:**849–852, 1991.

539. Condos R, Rom WN, Schluger NW: Treatment of multidrug-resistant pulmonary tuberculosis with interferon-gamma via aerosol. *Lancet* **349:**1513–1515, 1997.

540. Holland SM, Eisenstein EM, Kuhns DB, *et al*: Treatment of refractory disseminated nontuberculous mycobacterial infection with interferon gamma. A preliminary report. *N Engl J Med* **330:**1348–1355, 1994.

541. Kullberg BJ, Van 't Wout JW, Poell RJM, Van Furth R: Combined effect of fluconazole and recombinant human interleukin-1 on systemic candidiasis in mice. *Antimicrob Agents Chemother* **36:**1225–1229, 1992.

542. Vogels MTE, Van der Meer JWM: Use of immune modulators in nonspecific therapy of bacterial infections. *Antimicrob Agents Chemother* **36:**1–5, 1992.

543. Doi S, Saiki O, Hara T, *et al*: Administration of recombinant IL-2 augments the level of serum IgM in an IL-2 deficient patient. *Eur J Pediatr* **148:**630–633, 1989.

544. Pahwa R, Chatila T, Pahwa S, *et al*: Recombinant interleukin 2 therapy in severe combined immunodeficiency disease. *Proc Natl Acad Sci USA* **86:**5069–5073, 1989.

545. Cohn ZA, Kaplan G: Hansen's disease, cell-mediated immunity, and recombinant lymphokines. *J Infect Dis* **163:**1195–1200, 1991.

546. Klempner MS, Noring R, Mier JW, Atkins MB: An acquired chemotactic defect in neutrophils from patients receiving interleukin-2 immunotherapy. *N Engl J Med* **322:**959–965, 1990.

547. Crawford J, Ozer H, Stoller R, *et al*: Reduction by granulocyte colony-stimulating factor of fever and neutropenia induced by chemotherapy in patients with small-cell lung cancer. *N Engl J Med* **325:**164–170, 1991.

548. Maher DW, Lieschke GJ, Green M, *et al*: Filgrastim in patients with chemotherapy-induced febrile neutropenia. A double-blind, placebo-controlled trial. *Ann Intern Med* **121:**492–501, 1994.

549. Rowe JM, Andersen JW, Mazza JJ, *et al*: A randomized placebo-controlled phase III study of granulocyte-macrophage colony-stimulating factor in adult patients (> 55 to 70 years of age) with acute myelogenous leukemia: A study of the Eastern Cooperative Oncology Group (E1490). *Blood* **86:**457–462, 1995.

550. Boxer LA, Hutchinson R, Emerson S: Recombinant human granulocyte-colony-stimulating factor in the treatment of patients with neutropenia. *Clin Immunol Immunopathol* **62:**S39–S46, 1992.

551. Kullberg BJ, Netea MG, Curfs JHAJ, Keuter M, Meis JFGM, Van der Meer JWM: Recombinant murine granulocyte colony-stimulating factor protects against acute disseminated *Candida albicans* infection in non-neutropenic mice. *J Infect Dis* **177:**175–181, 1998.

552. Buckley RH, Schiff RI: The use of intravenous immune globulin in immunodeficiency diseases. *N Engl J Med* **325:**110–117, 1991.

553. Van der Meer JWM, De Windt GE, Van den Broek PJ: Subcutaneous immunoglobulin substitution in hypogammaglobulinemia. In Krijnen HW, Strengers PFW, Van Aken WG (eds): *Immunoglobulins.* CLB, Amsterdam, 1988, pp. 71–76.

554. Gardulf A, Hammarstrom L, Smith CI: Home treatment of hypo-gammaglobulinaemia with subcutaneous gammaglobulin by rapid infusion. *Lancet* **338:**162–166, 1991.

555. Winston DJ, Young LS: Immunization of the compromised host against infectious complications. In Allen JC (ed): *Infection in the Compromised Host: Clinical Correlations and Therapeutic Approaches.* Williams & Wilkins, Baltimore, 1981, pp. 37–89.

556. Winston DJ, Ho WG, Lin CH, *et al*: Intravenous immune globulin for prevention of cytomegalovirus infection and interstitial pneumonia after bone marrow transplantation. *Ann Intern Med* **106:**12–18, 1987.

557. Reed EC, Bowden RA, Dandliker PS, Lilleby KE, Meyers JD: Treatment of cytomegalovirus pneumonia with ganciclovir and intravenous cytomegalovirus immunoglobulin in patients with bone marrow transplants. *Ann Intern Med* **109:**783–788, 1988.

558. Emanuel D, Cunningham I, Jules-Elysee K, *et al*: Cytomegalovirus pneumonia after bone marrow transplantation successfully treated with the combination of ganciclovir and high-dose intravenous immune globulin. *Ann Intern Med* **109:**777–782, 1988.

559. Sullivan KM, Kopecky KJ, Jocom J, *et al*: Immunomodulatory and antimicrobial efficacy of intravenous immunoglobulin in bone marrow transplantation. *N Engl J Med* **323:**705–712, 1990.

560. Keusch GT, Ambinder EP, Kovacs I, Goldberg JD, Phillips DM, Holland JF: Role of opsonins in clinical response to granulocyte transfusion in granulocytopenic patients. *Am J Med* **73:**552–563, 1982.

561. Norrby-Teglund A, Kaul R, Low DE, *et al*: Plasma from patients with severe invasive group A streptococcal infections treated with normal polyspecific IgG inhibits streptococcal superantigen-induced T cell proliferation and cytokine production. *J Immunol* **156**:3057–3064, 1996.

562. Kaul R, McGeer A, Norrby-Teglund A, *et al*: Intravenous immunoglobulin therapy for streptococcal toxic shock syndrome—A comparative observational study. *Clin Infect Dis* **28**:800–807, 1999.

563. Groothuis JR, Simoes EA, Levin MJ, *et al*: Prophylactic administration of respiratory syncytial virus immune globulin to high-risk infants and young children. The Respiratory Syncytial Virus Immune Globulin Study Group. *N Engl J Med* **329**:1524–1530, 1993.

564. The PREVENT Study Group: Reduction of respiratory syncytial virus hospitalization among premature infants and infants with bronchopulmonary dysplasia using respiratory syncytial virus immune globulin prophylaxis. *Pediatrics* **99**:93–99, 1997.

565. The IMpact-RSV Study Group: Palivizumab, a humanized respiratory syncytial virus monoclonal antibody, reduces hospitalization from respiratory syncytial virus infection in high-risk infants. *Pediatrics* **102**:531–537, 1998.

566. Ziegler EJ, Fisher CJ Jr, Sprung CL, *et al*: Treatment of gram-negative bacteremia and septic shock with HA-1A human monoclonal antibody against endotoxin. A randomized, double-blind, placebo-controlled trial. *N Engl J Med* **324**:429–436, 1991.

567. Greenberg RN, Wilson KM, Kunz AY, Wedel NI, Gorelick KJ: Randomized, double-blind phase II study of anti-endotoxin antibody (E5) as adjuvant therapy in humans with serious gram-negative infections. *Prog Clin Biol Res* **367**:179–186, 1991.

568. Derkx B, Wittes J, McCloskey R, and the Euroean Pediatric Meningococcal Septic Shock Trial Study Group: Randomized, placebo-controlled trial of HA-!A, a human monoclonal antibody to endotoxin, in children with meningococcal septic shock. *Clin Infect Dis* **28**:770–777, 1999.

569. Tracey KJ, Fong Y, Hesse DG, *et al*: Anti-cachectin/TNF monoclonal antibodies prevent septic shock during lethal bacteraemia. *Nature* **330**:662–664, 1987.

570. Fisher CJ Jr, Agosti JM, Opal SM, *et al*: Treatment of septic shock with the tumor necrosis factor receptor:Fc fusion protein. *N Engl J Med* **334**:1697–1702, 1996.

571. Abraham E, Wunderink R, Silverman H, *et al*: Efficacy and safety of monoclonal antibody to human tumor necrosis factor alpha in patients with sepsis syndrome. A randomized, controlled, double-blind, multicenter clinical trial. *JAMA* **273**:934–941, 1995.

572. Opal SM, Fisher CJ Jr, Dhainaut JF, *et al*: Confirmatory interleukin-1 receptor antagonist trial in severe sepsis: A phase III, randomized, double-blind, placebo-controlled, multicenter trial. The Interleukin-1 Receptor Antagonist Sepsis Investigator Group. *Crit Care Med* **25**:1115–1124, 1997.

573. Bone RC: Why sepsis trials fail. *JAMA* **276**:565–566, 1996.

574. Zeni F, Freeman B, Natanson C: Anti-inflammatory therapies to treat sepsis and septic shock: A reassessment. *Crit Care Med* **25**:1095–1100, 1997.

575. Elliott MJ, Maini RN, Feldmann M, *et al*: Randomised double-blind comparison of chimeric monoclonal antibody to tumour necrosis factor alpha (cA2) versus placebo in rheumatoid arthritis. *Lancet* **344**:1105–1110, 1994.

576. Targan SR, Hanauer SB, Van Deventer SHJ, *et al*: A short-term study of chimeric monoclonal antibody Ca2 to tumor necrosis factor-α for Crohn's disease. *N Engl J Med* **337**:1029–1035, 1997.

577. Echtenacher B, Falk W, Mannel DN, Krammer PH: Requirement of endogenous tumor necrosis factor/cachectin for recovery from experimental peritonitis. *J Immunol* **145**:3762–3766, 1990.

578. Scharfman WB, Tillotson JR, Taft EG, Wright E: Plasmapheresis for meningococcemia with disseminated intravascular coagulation. *N Engl J Med* **300**:1277–1278, 1979.

579. Bjorvatn B, Bjertnaes L, Fadnes HO, *et al*: Meningococcal septicaemia treated with combined plasmapheresis and leucapheresis or with blood exchange. *Br Med J (Clin Res Ed)* **288**:439–441, 1984.

580. Winston DJ, Ho WG, Gale RP: Therapeutic granulocyte transfusions for documented infections. A controlled trial in ninety-five infectious granulocytopenic episodes. *Ann Intern Med* **97**:509–515, 1982.

581. van der Meer JW, van den Broek PJ: Present status of the management of patients with defective phagocyte function. *Rev Infect Dis* **6**:107–121, 1984.

582. Feldman E, Hester JP, Vartivarian SE. The use of granulocyte colony-stimulating factor-enhanced granulocyte transfusions from normal donors in patients with neutropenia-related fungal infections. 33rd Interscience Conference on Antimicrobial Agents and Chemotherapy. Washington, DC, 1993, Abstract 711.

583. Hassner A, Adelman DC: Biologic response modifiers in primary immunodeficiency disorders. *Ann Intern Med* **115**:294–307, 1991.

584. Hershfield MS, Buckley RH, Greenberg ML, *et al*: Treatment of adenosine deaminase deficiency with polyethylene glycol-modified adenosine deaminase. *N Engl J Med* **316**:589–596, 1987.

585. Hoogerbrugge PM, van Beusechem VW, Fischer A, *et al*: Bone marrow gene transfer in three patients with adenosine deaminase deficiency. *Gene Ther* **3**:179–183, 1996.

586. Onodera M, Ariga T, Kawamura N, *et al*: Successful peripheral T-lymphocyte-directed gene transfer for a patient with severe combined immune deficiency caused by adenosine deaminase deficiency. *Blood* **91**:30–36, 1998.

3

Mucocutaneous Infections in the Immunocompromised Host

RICHARD ALLEN JOHNSON and ARTHUR SOBER

1. Introduction

Among the most formidable challenges to the clinician is the care of the patient with an impaired immune system—the compromised host. Two characteristics in particular contribute to the complexity of management of infection in these patients: (1) the exceptionally broad variety of potential microbial pathogens and (2) the wide spectrum of clinical manifestations of disease resulting from the abnormal immune response.

In the compromised patient, cutaneous and subcutaneous tissues may be expected to be an important aspect of infection, for three reasons.[1,2] First, the skin together with the mucosal surfaces represents the first line of defense of the body against the external environment. These barriers assume an even greater importance when secondary defenses, such as phagocytosis, cell-mediated immunity, and antibody production, are impaired. Second, the rich blood supply of the skin provides a route of spread of infection both from the skin to other bodily locations and to the skin from infected sites. In the latter case, a skin lesion may serve as an early warning system to alert the patient and the clinician to the existence of a systemic infection. These cutaneous lesions may be benign in ap-

pearance, presumably because of the impaired host immune response, and therefore may be easily missed or dismissed as insignificant. Third, skin infections are common, occurring in up to one-third of significantly compromised hosts.

This chapter will give an overview of infection of the cutaneous and subcutaneous tissues in compromised hosts. Topics of discussion are the skin as a barrier to infection, a four-part classification of skin infection in compromised patients, dermatologic lesions associated with human immune deficiency virus (HIV) disease, and diagnostic considerations. In that opportunistic neoplasms and various paraneoplastic inflammatory disorders can enter into the differential diagnosis of opportunistic infection, these also will be discussed.

2. Skin as a Barrier to Infection

The skin is usually quite resistant to infection. The mechanisms by which the resistance occurs are not well understood. Three important components that contribute to microbial resistance are nonspecific: (1) intact keratinized layers of the skin, which prevent penetration of microorganisms; (2) dryness of the skin, which retards the growth of certain organisms such as aerobic gram-negative bacilli and *Candida* species; and (3) the suppressant effect of the normal skin flora, which appears to reduce colonization of pathogens, a phenomenon known as "bacterial interference." Within this framework, then, one might expect potentially serious skin infections to develop under the following circumstances: (1) destruction by trauma or bypass by introduction of intravascular

Richard Allen Johnson • Harvard Medical School, Infectious Disease Unit, Massachusetts General Hospital, and Department of Dermatology, Beth Israel-Deaconess Medical Center, Harvard Medical School, Boston, Massachusetts 02115.   **Arthur Sober** • Harvard Medical School Department of Dermatology, Massachusetts General Hospital, Boston, Massachusetts 02115.

Clinical Approach to Infection in the Compromised Host (Fourth Edition), edited by Robert H. Rubin and Lowell S. Young. Kluwer Academic/Plenum Publishers, New York, 2002.

TABLE 1. Types of Skin Infection by Pathophysiologic Events

Type of infection	Pathogen	Site of infection	Healthy host	Compromised host
Primary skin infections with common pathogens	*S. aureus* Group A streptococcus	Epidermis, hair follicles, dermis, subcutaneous tissues	Impetigo, ecthyma, folliculitis, abscess, intertrigo	Soft-tissue infection, necrotizing soft-tissue infection, septicemia
Unusually widespread cutaneous infection	Dermatophytes *Candida* spp. Herpes simplex virus, varicella–zoster virus, cytomegalovirus, Epstein–Barr virus Molluscum contagiosum virus Human papillomavirus	Epidermis, intertriginous sites, hair follicles Oropharynx, esophagus, genitalia	Dermatophytosis: epidermal (limited), folliculitis Candidiasis: intertrigo, genital Localized herpes; resolves spontaneously Herpes zoster (mild) Molluscum contagiosum (localized, nonfacial) Common and mucosal warts	Dermatophytosis: epidermal (extensive), folliculitis Candidiasis: intertrigo, folliculitis, mucosal Chronic herpetic ulcers Extensive herpes zoster ± hematogenous dissemination to skin Hairy leukoplakia Widespread MC, resistant to therapy Widespread warts: squamous cell carcinoma (*in situ* and invasive)
Opportunistic primary cutaneous infection	Atypical mycobacteria *Nocardia* Molds *Prototheca*	Dermis, subcutaneous tissues	Swimming pool granuloma	Soft-tissue infection ± necrosis; septicemia
Systemic infection metastatic to cutaneous and subcutaneous sites	Bacteria Fungal pneumonitis with fungemia	Dermis, subcutaneous tissues	Soft-tissue infection ± necrosis Nodules	Soft-tissue infection ± necrosis Nodules

catheters of the previously intact keratinized layer of skin; (2) moistening of the skin, such as under occlusive dressings; and (3) alteration of the normal colonizing flora, such as after administration of antimicrobial agents. These types of events would represent some risk to the normal patient, but are considerably more threatening to the compromised patient with impaired immunologic defenses that are likely to be more readily overwhelmed when the primary cutaneous barrier breaks down.

An example of these phenomena is the development of invasive fungal infection in compromised patients whose skin has been traumatized by tape holding intravascular lines in place. Infection with *Rhizopus* species has been associated with use of Elastoplast tape to secure intravascular catheters.[3] Skin infection with *Aspergillus* species has occurred at the site of boards to stabilize arms to protect intravenous lines.[4] Because of the occurrence of these types of infections, the following approach would seem warranted: Occlusive dressings in immunocompromised patients should be avoided when possible, and skin covered by such dressings should be routinely inspected. Paper tape should be used in preference to cloth tape, and surgical dressings might be secured with girdles of elasticized netting rather than tape whenever possible.

The effect of chronic administration of cortico-

steroids on the skin is another factor that may contribute to increased susceptibility of compromised patients to infection. Steroid therapy appears to inhibit proliferation of fibroblasts, synthesis of mucopolysaccharides, and deposition of collagen. The net effect is thin and atrophic skin that heals poorly. Minor trauma generates lesions that tend to persist, providing potential portals of entry for pathogens. An example of the phenomenon that has been observed is recurrent staphylococcal cellulitis about the elbow in patients receiving chronic immunosuppressive therapy after renal transplantation. These patients exhibited two adverse effects of chronic corticosteroid administration: (1) thinning of the skin leading to enhanced susceptibility of the tissue to trauma and (2) steroid-induced proximal myopathy. Because of the myopathy, the patients tended to rise from the sitting position by pushing off with their elbows, and thus traumatizing them. Cellulitis about the elbows recurs in these patients until protection is provided to their elbows and the steroid dose is decreased.

3. Types of Skin Infection

Infection of the cutaneous and subcutaneous tissues in compromised patients can be classified in a variety of ways: by pathogen, by underlying immunologic defect, or by pace of illness. An additional categorization considers pathophysiologic events and consists of four groups (Fig. 1): (1) infection originating in skin and being typical of that which occurs in immunocompetent persons, albeit with the potential for more serious illness; (2) extensive cutaneous involvement with pathogens that normally produce trivial or well-localized disease in immunocompetent patients; (3) infection originating from a cutaneous source and caused by opportunistic pathogens that rarely cause disease in immunocompetent patients but may cause either localized or widespread disease in compromised persons; and (4) cutaneous or subcutaneous infection that represents metastatic spread from a noncutaneous site. Cutaneous and subcutaneous infections in compromised patients are discussed in this section within the framework of these four groups.

3.1. Primary Skin Infections with Common Pathogens

The incidence and severity of conventional forms of infections originating in the skin often are increased in the compromised host. Gram-positive organisms, such as *Staphylococcus aureus* and group A streptococci, most commonly cause these infections. Patients with granulocytopenia are more susceptible to cellulitis caused by less virulent (for the skin) bacterial pathogens, such as Enterobacteriaceae and *Pseudomonas* species, and by anaerobic bacteria. Patients with leukemia or diminished cellmediated immunity may have erysipelas-like infection, caused by such organisms as *Cryptococcus neoformans* or *Candida* species, mimicking cellulitis caused by common gram-positive bacteria.

When evaluating cellulitis in a compromised patient, common as well as uncommon/rare pathogens must be considered as potential pathogens. If a patient does not respond to conventional antimicrobial therapy, an aggressive approach to diagnosis is warranted, with biopsy of lesions for Gram and other stains, cultures, and dermatopathology, to correctly identify the pathogen.

3.2. Unusually Widespread Cutaneous Infection

Nonvirulent skin fungi and viruses constitute the two major causes of infection in this category. These patho-

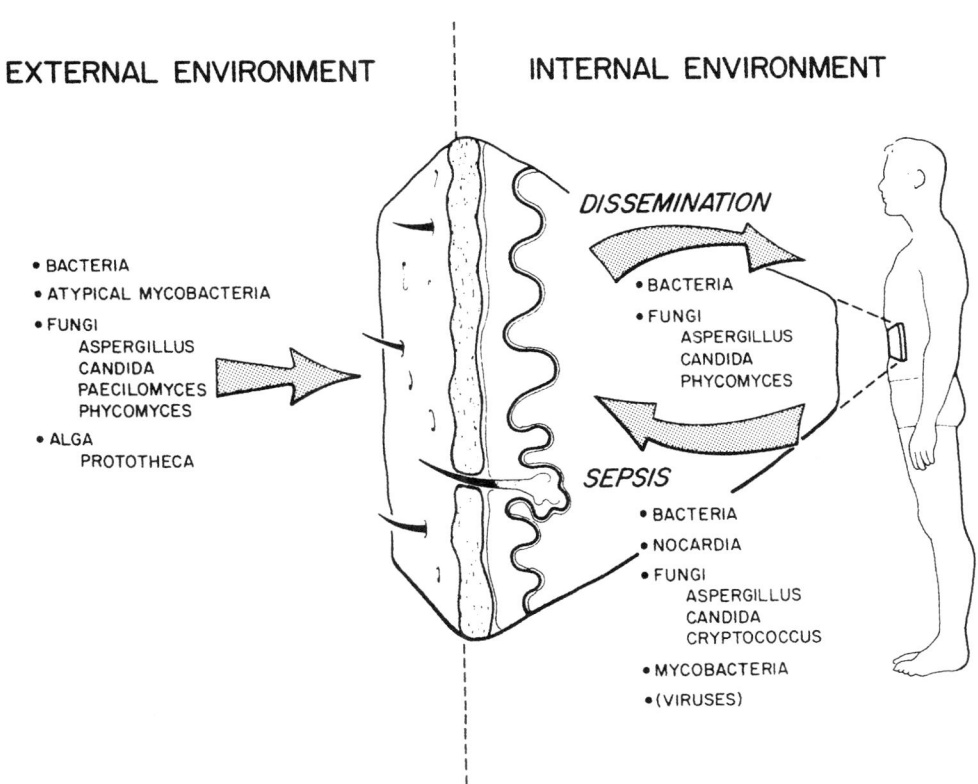

FIGURE 1. Schematic representation of the role of skin in the occurrence of localized and disseminated infection in the compromised host.

TABLE 2. Pathogens and Types of Skin Infections[a]

Pathogen	Clinical infection
Type 2A infections (common pathogens causing aggressive/extensive local infection)	
Bacteria	
Staphylococcus aureus	Primary infection (folliculitis, furuncles, carbuncles, abscess, impetigo, bullous impetigo, and ecthyma)
	Secondary infection of dermatoses/breaks in integrity of skin
	Soft-tissue infections
	Toxin syndromes (SSSS, SSF, STSS)
Group A streptococcus (*Streptococcus pyogenes*)	Primary infection (impetigo, ecthyma)
	Secondary infection of dermatoses/breaks in integrity of skin
	Soft-tissue infections (erysipelas, cellulitis, necrotizing STI)
	Toxin syndromes (SSSS, SSF, STSS)
Group B streptococcus	Primary infection (intertrigo of anogenital region)
	STI (anogenital region, lower extremities)
Treponema pallidum	Chancre: multiple, large
Fungi	
Dermatophytes	Extensive epidermal dermatophytosis
	Proximal subungual onychomycosis
	Majocchi's granuloma
Candida species	Oropharyngeal candidiasis
	Esophageal/tracheobronchial candidiasis
	Recurrent vulvovaginal candidiasis
	Invasive candidiasis
Viruses	
Herpes simplex virus	Eczema herpeticum
	Chronic (>1 month), large ulcers (orofacial, anogenital, digital, esophageal)
Varicella-zoster virus	Severe varicella
	Herpes zoster ± severe necrotizing infection
Molluscum contagiosum virus	Umbilicated papules/nodules: facial, multiple, confluent, large/giant; unresponsive to therapy
Human papillomavirus	Infection of keratinized skin; verruca vulgaris, verruca plana, verruca plantaris
	Mucosal infection: condyloma acuminatum; SIL, SCCIS
Infestations	
Scabies	Crusted on hyperkeratotic scabies
Type 2B infections (opportunistic pathogens causing aggressive/extensive local infection)	
Bacteria	
Bacillus cereus	Necrotizing lesion with bulla formation
Corynebacterium jeikeium	Cellulitis at skin puncture sites (bone marrow biopsy, IV catheters)
M. fortuitum complex (*M. fortuitum*, *M. chelonae*, *M. abscessus*)	Wound infections at sites of puncture wound (injection or traumatic), surgical wounds: nodules, ulcers, granulomas, verrucous plaques
Nocardia species	Abscess, ulcers, granulomas, STI, mycetoma, lymphocutaneous infection
Fungi	
Aspergillus species	Erythematous/purpuric papules, ulcer/eschar
	Red plaque—± vesicles, pustules, ulcerations under elastasized adhesive tape
Mucormycosis	STI beginning at breaks in skin (IV lines, injection sites, macerated skin, burns, insect bites)
	Rhinocerebral mucormycosis: infection begins in nasal, paranasal, oropharyngeal mucosa; the infections then extend to skin with orbital cellulitis, edema
Phaeohyphomycosis (dematiaceous fungi) (*Exophiala, Phialophora, Fonsecaea, Cladosporium, Alternaria, Bipolaris, Curvularia,* and *Exserohilum*)	Low-grade, chronic subcutaneous soft-tissue infection
Chromomycosis	Plaques (verrucous, nodular, tumorous with pseudoepitheliomatous hyperplasia)
Mycetoma	Tumor with draining sinuses
Hyalohyphomycosis (*Fusarium, Penicillium, Paecilomyces, Acremonium,* and *Scopulariopsis* species)	Local infection of thermal burns, punctures, IV lines

TABLE 2. (Continued)

Pathogen	Clinical infection
Type 2C infections (opportunistic pathogens causing aggressive invasive local infections ± metastatic dissemination)	
Gram-negative rods (*E. coli*, *P. aeruginosa*, *S. maltophila*, *K. pneumonia*, *P. mirabilis*, *A. hydrophila*)	STI ± hemorrhagic bullae, ± abscess, ± necrosis, ± ulceration
Type 2D infections (opportunistic/common pathogens with blood vessel invasion and metastatic cutaneous lesions)	
Bacteria,	
Corynebacterium jeikeium	Papular eruption (erythematous hemorrhagic)
	Soft-tissue abscess
	Necrotizing infection
Streptococcus pneumoniae (pneumococcus)	STI
V. vulnificus	STI ± necrosis
H. cinaedi	STI
M. tuberculosis	Disseminated small red-brown papules/papulovesicles/papulopustules
M. avium-intracellulare complex	Disseminated lesions (papules, nodules, pustules, abscesses, ulcers)
M. haemophilum	STI, nodules, abscess
B. henselae, *B. quintana*	Localized or disseminated red-violaceous papules, nodules, plaques; subcutaneous nodules/tumors
Nocardia species	Disseminated pustules, abscesses, nodules
T. pallidum	Secondary syphilis: atypical lues maligna
	Rapid progression to neurosyphilis
	Seronegative
	Treatment failure
Aspergillus species	Erythematous macules/papules; subcutaneous nodules/abscess
Sporotrix schenkii	Papules, nodules, erosions, ulcers; arthritis; eye involvement
Endemic fungi	
Blastomyces, Coccidioides, Histoplasma	Disseminated papules
Penicillium marneffei	Disseminated molluscum contagiosumlike lesions; genital ulcers
Phaeohyphomycosis	
Dematiaceous fungi	Erythematous nodules, papules, pustules, plaques, ± ulceration, ± hemorrhage
Hyalohyphomycosis	Disseminated red papules to necrotic lesions
Viruses	
Herpes simplex virus	Cutaneous dissemination with vesicles, erosions, ulcers
	Visceral dissemination (hepatitis, pneumonitis, encephalitis)
Varicella–zoster virus	Varicella, ± visceral involvement
	Herpes zoster with cutaneous dissemination
	Herpes zoster with visceral dissemination
HHV-8	Kaposi's sarcoma: papules/nodules; edema ± visceral involvement
Infestations	
Strongyloidiasis	Disseminated infection presenting as periumbilical purpura
Leishmaniasis	Erythrodermic or dermatomyositislike eruption
American trypanosomiasis	Visceral infection, i.e., infection resembling cellulitis in cardiac transplant recipient with history of trypanosomal cardiomyopathy
Acanthamoebiasis	Disseminated nodules, abscesses, ulcers

[a]Abbreviations: STI, soft-tissue infection; SSSS, staphylococcal scaled skin syndrome; SSF, staphylococcal scarlet fever; STSS, staphylococcal toxic shock syndrome.

gens typically cause minor infections in immunocompetent persons, but in compromised patients they tend to cause more extensive disease that may lead to more serious systemic illness. Viruses that cause exanthems (e.g., those caused by rubella, measles, or enterovirus) do occur in immunocompromised patients, but the more problematic pathogens include the family of herpesviruses and human papillomaviruses (HPV).

Nonvirulent fungi include the dermatophytes (*Trichophyton* species, *Microsporum* species, and *Epidermophyton*), *Candida* species, *Pityrosporum* species, *Fusarium solani*, and *Alternaria alternata*. These fungi frequently colonize human skin and cause localized, superficial skin infection in immunocompetent persons, particularly when the skin has been traumatized. The incidence and severity of infection may be increased in compromised patients. Topical corticosteroid preparations prescribed mistakenly for epidermal dermatophytoses compromise local immunity, facilitating growth of the fungus causing extensive local epidermal infection (so-called tinea incognito in that the diagnosis of dermatophytosis is missed); dermatophytic folliculitis (Majocchi's granuloma) commonly is seen as an associated finding. Systemic corticosteroid therapy also can cause widespread epidermal dermatophytosis. These dermatomycoses are best treated with oral agents such as terbinafine, itraconazole, or fluconazole; secondary prophylaxis is often necessary.

In chronically immunosuppressed patients, HPV-induced lesions, i.e., verrucae and condylomata, either may be extremely numerous or may form large confluent lesions. Up to 40% of renal transplant recipients develop warts following transplantation, half of these have more than ten warts, and up to 1% have extensive disease. The incidence and severity of warts seem to be related to immunosuppression, with previously acquired latent virus reactivating with institution of immunosuppressive therapy. In compromised patients, HPV-induced lesions have the potential for malignant transformation, particularly in sun-exposed areas of the body. Squamous cell carcinoma (SCC) arising in sites of chronic sun exposure occur 36 times more frequently in renal transplant recipients than in the general population, some clearly arising within warts; HPV DNA is demonstrable within the tumors. Management of patients with extensive warts should include avoidance of sun exposure, use of strong sunscreens, reduction in immunosuppressive therapy when possible, and careful observation for the development of malignant lesions. HPV-induced anogenital *in situ* and invasive SCC also is more common in transplant recipients and HIV-infected individuals; these persons also should be screened for *in situ* and invasive SCC with Papanicolaou test of the anus and cervix and lesional biopsy when indicated.

Skin infections with members of the herpesvirus family, particularly herpes simplex virus (HSV) and varicella–zoster virus (VZV), are very common in compromised patients. Nasolabial or anogenital infections due to HSV occur in as many as half of renal transplant recipients, patients with malignancy, those receiving chemotherapy, and HIV-infected individuals. Immunocompromised patients may have more serious forms of HSV infection including chronic herpetic ulcers, esophageal or respiratory tract infection, or disseminated infection [patients with lymphoma, transplant recipients (bone marrow, renal, cardiac, or liver), and neonates].

In compromised patients, reactivation of VZV is common, occurring in 14% of persons with Hodgkin's disease, 8% of non-Hodgkin's lymphoma and renal transplant recipients, and 2% with solid tumors. Visceral dissemination occurs in 15 to 30% of patients with Hodgkin's disease with zoster; systemic dissemination, however, is uncommon in renal transplant recipients. Reactivated VZV infection is particularly problematic for bone marrow transplant recipients, of whom one-half will develop herpes zoster. In one-third of these, VZV will disseminate, and in one-fourth a generalized atypical recurrent varicellalike illness will develop.

Reactivated cytomegalovirus (CMV) causes hepatitis, pneumonitis, chorioretinitis, encephalitis, and gastroenteritis in transplant recipients and HIV-infected individuals; cutaneous CMV infections are rare. Cutaneous CMV infections are reported to present as nodules, ulcers, indurated plaques, vesicles, petechiae, or a maculopapular exanthem. Reactivation of Epstein–Barr virus (EBV) results in oral hairy leukoplakia on the lateral aspects of the tongue, a lesion nearly pathognomonic for HIV disease.

3.3. Opportunistic Primary Cutaneous Infection

Following inoculation into the skin, organisms of low virulence can cause local or disseminated infections in some persons with impaired immune defenses. Localized disease can be caused by *Paecilomyces*, atypical mycobacteria, and *Prototheca*. Localized disease with life-threatening systemic spread may be caused by *Pseudomonas aeruginosa*, *Aspergillus* species, *Candida* species, and *Rhizopus* species.

P. aeruginosa causes a necrotizing soft-tissue infection, i.e., ecthyma gangrenosum, which occurs at the portal of entry (most commonly in naturally occluded sites such as the anogenital region or axillae), especially in granulocytopenia patients. Histologically, the organism causes a necrotizing vasculitis that results in tissue infarction and hematogenous dissemination.

Paecilomyces is a saprophytic fungus that may cause an ulcerating soft-tissue infection following laceration of the pretibial region in renal transplant recipients. Environmental mycobacteria such as *M. marinum*, *M. chelonae*, *M. kansasii*, and *M. haemophilum* may cause cutaneous infection following inoculation. In the normal host, the infection is localized and may resolve without therapeutic intervention; however, in the compromised host, more extensive local infection may occur as well as lymphatic or hematogenous dissemination. *Prototheca wickerhamii* is an algalike organism ubiquitous in nature that may cause localized infection following trauma or surgery.

Primary infection caused by *Aspergillus*, *Rhizopus*,

or *Candida* species arises at localized cutaneous sites, but has the potential for disseminated disease in the compromised host. Primary cutaneous infection with these fungi has been associated with use of adhesive or Elastoplast tape, cardiac electrode leads, or extravasation of intravenous fluids. *Aspergillus* and *Rhizopus* species can invade blood vessels, resulting in infarction, hemorrhage, and hematogenous dissemination.

3.4. Systemic Infection Metastatic to Cutaneous and Subcutaneous Sites

In a report of dermatologic manifestations of infection in compromised patients, 8 of 31 patients (26%) had apparent spread of systemic infection to cutaneous and subcutaneous tissues.[2] In 6 of these 8 patients, cutaneous or subcutaneous lesions were the first clinical sign of disseminated infection. In compromised hosts, cutaneous lesions resulting from hematogenous spread of infection are caused in general by three classes of pathogens: (1) *Pseudomonas aeruginosa* and other bacteria; (2) the endemic systemic mycoses (*Histoplasma capsulatum*, *Coccidioides immitis*, *Blastomyces dermatitidis*, and *Penicillium marneffei* (Southeast Asia); and (3) the ubiquitous opportunistic organisms *Aspergillus* species, *Cryptococcus neoformans*, *Candida* species, *Rhizopus* species, and *Nocardia* species.

Hematogenous dissemination of *P. aeruginosa* to the skin of a compromised patient can result in subcutaneous nodules, cellulitis, or necrotizing soft-tissue infection (pyoderma gangrenosum). The usual setting is profound granulocytopenia, often with acute leukemia.

Fungi endemic to geographical regions often cause asymptomatic primary pulmonary infection, followed by prolonged latency. In the setting of deficiency of cell-mediated immunity, latent fungi can cause local pulmonary infection with subsequent hematogenous dissemination, commonly to mucocutaneous sites.

Ubiquitous opportunistic fungi and *Nocardia* species can cause asymptomatic pulmonary infections that disseminate hematogenously in the compromised host. *Candida* usually disseminates from the gastrointestinal tract or an infected intravascular line. Disseminated cryptococcosis often presents with cutaneous lesions (molluscum contagiosumlike facial lesions), subcutaneous nodules, or cellulitis prior to the clinical presentation of meningitis. Disseminated histoplasmosis also presents on the skin with molluscum contagiosumlike facial lesions, guttate psoriasislike lesions, as well as other morphologies. Nocardia disseminates from pulmonary infection, resulting in subcutaneous nodules.

4. Diagnostic Aspects of Skin Infections in the Compromised Patient

In the immunocompetent patient, the gross appearance of a skin lesion is an important aspect of diagnosis. By contrast, the clinical value of the gross appearance of a cutaneous lesion in a compromised host is likely to be limited, for two reasons: First, in compromised patients, the variety of organisms that may cause infection is substantially greater than in immunocompetent persons. Second, in compromised patients, the inflammatory response to infection may be altered. A cutaneous lesion results not only from the invading pathogen itself but also from the inflammatory response of the body to the microbe. Thus, with an impaired inflammatory response, the prediction is that the diagnostic usefulness of gross appearance would be suboptimal. For these two reasons, it is key to realize that the differential diagnosis of a particular skin lesion in a compromised patient is extensive.

The approach to biopsy of a cutaneous lesion suspected to be infectious should include two considerations: (1) The most rapid and most sensitive methods for detecting microbes both histologically and immunologically should be used and (2) appropriate cultures and stains should be obtained to optimize the chance of identifying the pathogen. A 6- or 8-mm punch biopsy is usually adequate. Half the tissue is sent for histopathologic evaluation by routine methods and also by special stains for fungi, mycobacteria, and bacteria. The other half is sent to the microbiology laboratory for culture for aerobic and anaerobic bacteria, mycobacteria, and fungi (at 25°C and 37°C) and also for Gram's stain, acid-fast, modified acid-fast, and direct fungal stains of touch preparations or ground tissue.

5. Opportunistic Infections by Pathogen

5.1. Bacterial Infections

5.1.1. *Staphylococcus aureus*

Staphylococcus aureus causes the greater majority of all pyodermas and soft-tissue infections. Although not one of the cutaneous resident flora, it colonizes the anterior nares in up to 25% of healthy individuals at any one time and more than 50% of chronically ill individuals. The incidence of *S. aureus* nasal carriage is higher in chronically ill individuals, especially those with HIV disease, diabetes mellitus, cancer (especially hematologic malignancies), neutropenia, abnormal leukocyte func-

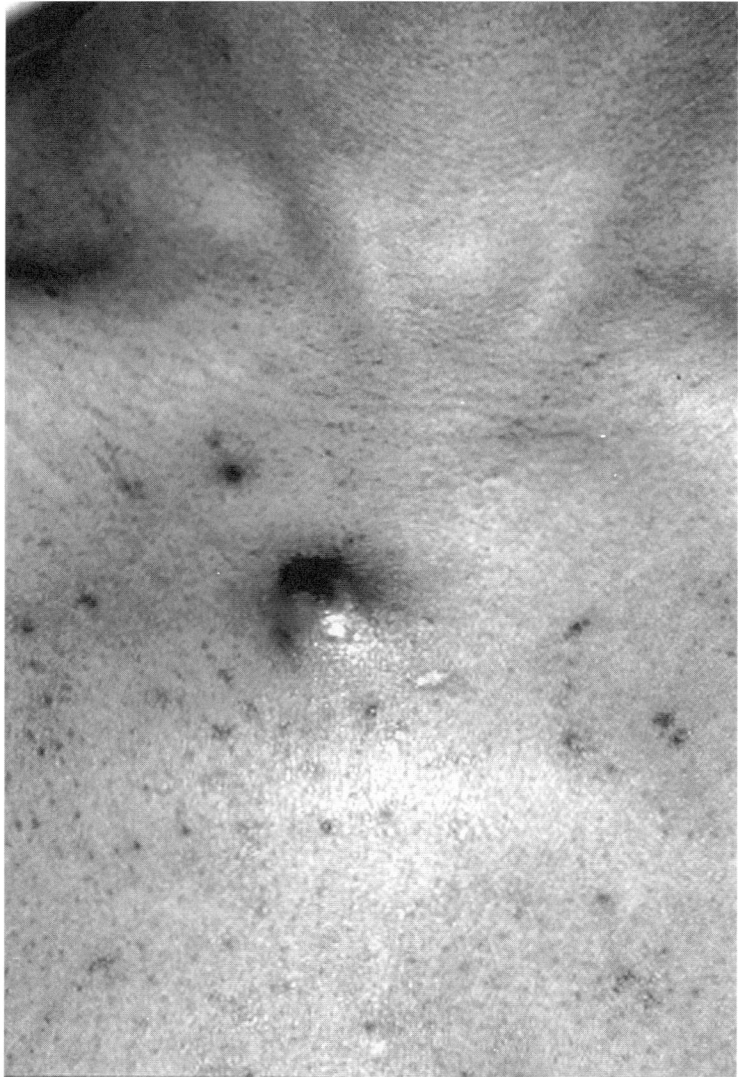

FIGURE 2. *Staphylococcus aureus*: Abscess arising in excoriation. This individual with HIV disease has pruritic eosinophilic folliculitis, excoriated chest lesions, resulting in secondary *S. aureus* infection with subsequent abscess formation. Color representation of Figure 2 follows page 708.

tion, chronic granulomatous disease, hyperimmunoglobulinemia E syndrome, and interleukin-2 therapy.

Ensconced in the nares, *S. aureus* is able to colonize and infect superficial skin lesions by entering hair follicles or small breaks in the epidermis, by secondary infection (Fig. 2) of dermatologic disorders (scabies, eczematous dermatitis, herpetic ulcer, Kaposi's sarcoma, molluscum contagiosum), drug injection sites (Fig. 3), or via vascular access line and drainage tubes, resulting in pyodermas (folliculitis, furuncles, carbuncles, abscess, impetigo, bullous impetigo, and ecthyma). Once established in the skin, *S. aureus* is able to invade more deeply into the soft tissue with resultant erysipelas (horizontal spread in lymphatics) and cellulitis (vertical spread into subcutaneous fat) (Fig. 4). *S. aureus* is the most common cause of wound infections. Risk factors for surgical wound infection are dependent on the following: host factors (immune status, diabetes mellitus); surgical factors (disruption of tissue perfusion that accompanies the surgical procedure, foreign body use); staphylococcal factors (substances that mediate tissue adherence and invasion or that enable staphylococci to survive host defenses and antibiotics in tissues, and antimicrobial prophylaxis). Bacteremia can result in deposition of *S. aureus* in the skin, resulting in petechiae, hemorrhages, subcutaneous nodules, soft-tissue infections, and pyomyositis.

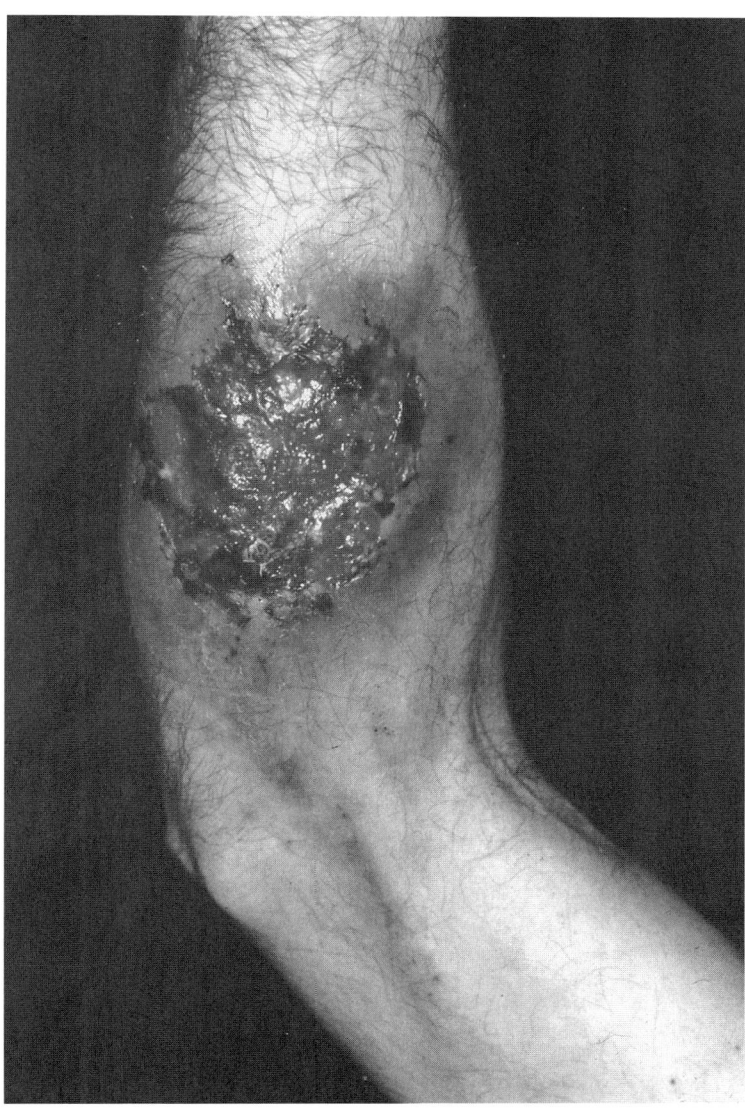

FIGURE 3. *S. aureus* cellulitis in site of drug injection. Advanced soft-tissue infection on the forearm at the site of heroin injection. Lesion biopsy specimen also showed foreign body reaction. Color representation of Figure 3 follows page 708.

Various strains of *S. aureus* are capable of producing a variety of toxins, which cause the clinical syndromes of staphylococcal scalded skin syndrome (rare in infants greater than 2 years of age), staphylococcal scarlet fever, and toxic shock syndrome (TSS). TSS is a febrile, multiorgan disease caused by the elaboration of staphylococcal toxins, characterized by a generalized scarlatiniform eruption, hypotension, functional abnormalities of three or more organ systems, and desquamation in the evolution of the exanthem. Cellulitis caused by *S. aureus* that produce TSS toxins can be accompanied by the cutaneous and systemic findings of staphylococcal scarlet fever or TSS.[5]

5.1.2. β-Hemolytic *Streptococcus*

Group A β-hemolytic streptococci (*Streptococcus pyogenes*) (GAS) commonly colonize the upper respiratory tract and secondarily infect (impetiginize) minor skin lesions from which invasive infection can arise.[6–8] Certain strains of group A streptococci have a higher affinity for the skin than the respiratory tract and can colonize the skin, subsequently causing superficial pyodermas or soft-tissue infections. Lymphatic obstruction/lymphedema predisposes to erysipelas or cellulitis. Individuals at higher risk are those who have had radical mastectomy with axillary node dissection or saphenous vein harvest. Absence

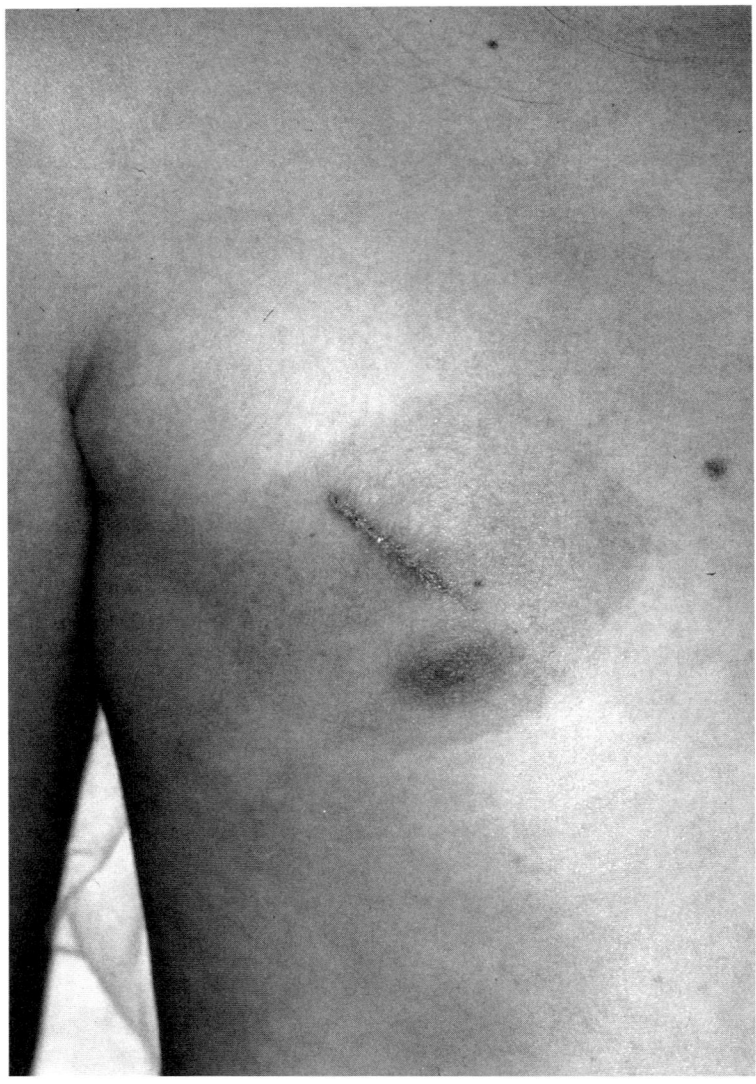

FIGURE 4. *S. aureus* cellulitis in scratch. Early cellulitis arising on the chest at the site of a cat scratch in an 8-year-old HIV-infected girl. Color representation of Figure 4 follows page 708.

of antibodies against pyrogenic exotoxins A and/or B has been reported as a risk factor for developing invasive streptococcal disease; toxic shock and mortality were associated with a lack of antipyrogenic exotoxin A antibodies.[9] Antibodies against pyrogenic exotoxin A were vital for mediating the outcome of invasive GAS disease.

Group B streptococci commonly colonize the perineum and may cause soft-tissue infections at this site. Advanced age, cirrhosis, diabetes, stroke, breast cancer, surgical wounds, decubitus ulcer, neurogenic bladder, and foreign bodies (breast or penile implants) have all been associated with a significantly increased risk of acquiring group B streptococcus infection. An unusual form of recurrent cellulitis of the lower extremities in women results from impaired lymphatic drainage due to neoplasia, radical vulvectomy or pelvic surgery, or radiation therapy. Morbidity and mortality are relatively high for group B streptococcus infections, with a high incidence of bacteremia. Other streptococci such as the enterococcus can cause invasive infections with septicemia, endocarditis (Fig. 5), and soft-tissue infections.

5.1.3. *Streptococcus pneumoniae* (Pneumococcus)

Streptococcus pneumoniae is a rare cause of cellulitis, occurring in individuals predisposed by connective

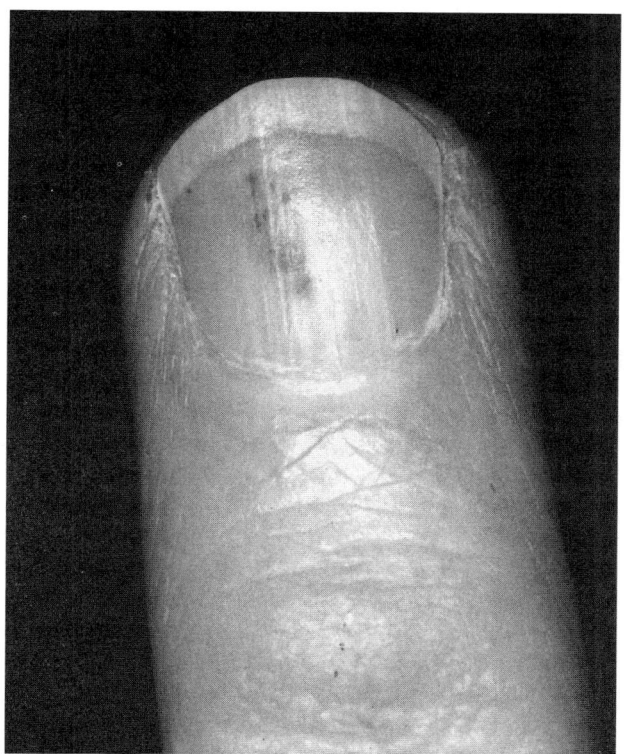

FIGURE 5. Enterococcus: Endocarditis with nail bed splinter hemorrhages. Splinter hemorrhages in the midportion of the nail bed in a diabetic with urinary tract infection, complicated by sepsis, endocarditis, and peripheral embolization. Color representation of Figure 5 follows page 708.

tissue disease, alcoholism, drug abuse, HIV disease, or corticosteroid therapy.[10,11] Clinically, infected areas are characterized by bullae, brawny erythema, and a violaceous hue. Approximately 50% of cases are the result of pneumococcal bacteremia, often from a pulmonary source. Because of underlying medical conditions and often pneumonia, the morbidity is high.

5.1.4. *Bacillus cereus*

Primary cutaneous *Bacillus cereus* infection presents as a single bulla with necrosis on the extremity of an immunocompromised patient.[12–14] The large gram-positive rods of *B. cereus* may be mistaken for *Clostridium* species in lesional biopsy specimens and smears.

5.1.5. *Corynebacterium jeikeium*

Skin and soft-tissue infections due to *Corynebacterium jeikeium* occur in granulocytopenic patients and

take either of two forms: (1) primary infections [cellulitis at bone marrow biopsy sites, infection at insertion sites of intravascular catheters, skin fissures (perianal)][12–14] and (2) secondary infections (erythematous or hemorrhagic papular rash, soft-tissue abscess, or necrotic lesions) following bacteremia from primary infection sites.[15]

5.1.6. *Escherichia coli* and Other Gram-Negative Bacilli

Escherichia coli and other gram-negative bacilli rarely cause soft-tissue infection (Fig. 6) or hemorrhagic plaques associated with hematogenous dissemination, in individuals with cirrhosis, neutropenia, or leukocyte dysfunction.[16–22] In a report of seven patients with gram-negative bacillary cellulitis and cirrhosis, soft-tissue infections were characterized by bullous lesions, ulcers, abscesses, or extensive cutaneous necrosis. Bacteremia occurred in six patients and patients eventually died. Isolates from the skin included *Klebsiella pneumoniae*, *E. coli*, *P. aeruginosa*, *Proteus mirabilis*, and *A. hydrophila*.

5.1.7. *Pseudomonas aeruginosa*

Pseudomonas aeruginosa causes the necrotizing soft-tissue infection ecthyma gangrenosum (EG), which occurs as a primary skin infection (Fig. 7)[23–25] or as a complication of pseudomonal bacteremia.[26] EG occurs commonly as a nosocomial infection, especially in immunocompromised patients with diabetes, neutropenia, or poor neutrophil function. *P. aeruginosa* is the most common pathogen causing gangrenous cellulitis in childhood. *P. aeruginosa* gains entry into the dermis and subcutaneous tissues via adnexal epidermal structures or areas of loss of epidermal integrity (pressure ulcers, thermal burns, and trauma). EG occurs most frequently in the axillae or anogenital regions but can arise at any cutaneous site. Clinically, EG presents initially as an erythematous, painful plaque that quickly undergoes necrosis. Established lesions show bulla formation, hemorrhage, necrosis, and surrounding erythema. If effective antibiotic therapy is not initiated promptly, the necrosis often may extend rapidly. Bacteremia occurs soon after the onset of EG and may result in metastatic spread of *P. aeruginosa* with subcutaneous nodules and abscesses. Histologically, EG is characterized by a distinctive septic vasculitis.

Hematogenous dissemination of *P. aeruginosa* to the skin can result in multiple subcutaneous nodules, hemorrhagic bullae, multiple small hemorrhagic papules, and/or EG.[27]

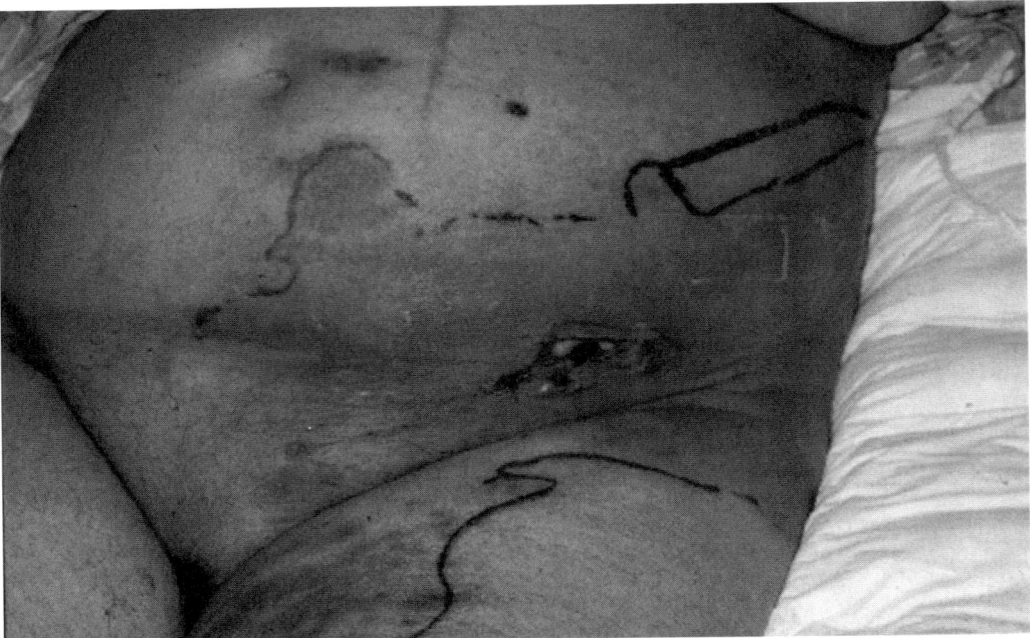

FIGURE 6. *Escherichia coli:* Cellulitis. Enlarging red, hot, tender plaque arising at a lymph node biopsy site in elderly patient with lymphoma. Color representation of Figure 6 follows page 708.

5.1.8. *Stenotrophomonas (Xanthomonas) maltophilia*

Stenotrophomonas [formerly classified as *Xanthomonas* (formerly classified as *Pseudomonas*)] *maltophilia* is a significant cause of morbidity and mortality in hospitalized patients with neutropenia, cancer, and undergoing chemotherapy. Primary cellulitis (often necrotizing), disseminated cutaneous nodules, and mucocutaneous ulcers caused by *Xanthomonas* are often associated with underlying malignancies.[28]

5.1.9. *Aeromonas* Species

Aeromonas hydrophila, a gram-negative facultative rod that is found naturally in aqueous environments, causes soft-tissue infections in healthy individuals and more serious infections in the compromised host.[29,30] *A. hydrophila* soft-tissue infections occur following injuries sustained in a contaminated aquatic environment or the "outdoors." *A. hydrophila* (a normal inhabitant of the foregut of leeches) cellulitis has also followed the therapeutic use of leeches (in 7–20% of patients) following reimplantation or flap surgery. In the compromised host, *Aeromonas* causes severe cellulitis and necrotizing soft-tissue infections.

5.1.10. *Vibrio* Species

Vibrio vulnificus is a free-living gram-negative rod, occurring naturally in the marine environment, occasionally contaminating oysters and other shellfish. Marine *Vibrio* species can cause sepsis and soft-tissue infections (Fig. 8), particularly in patients with cirrhosis and/or diabetes mellitus.[31] *V. damselae* may cause fulminant necrotizing soft-tissue infections in immunocompetent patients. Either ingestion of raw seafood or exposure of open wounds to seawater can result in *Vibrio* bacteremia and soft-tissue infections. Individuals with cirrhosis, hemochromatosis, and diabetes mellitus and other patients with chronic disease are advised to avoid eating raw seafood.

Marine *Vibrio* soft-tissue infections occur by direct inoculation into a superficial wound or by bacteremic spread to the skin (metastatic infection). Following ingestion of *V. vulnificus* in contaminated seafood, the organism is capable of crossing the gut mucosa rapidly, invading the bloodstream without causing gastrointestinal symptoms. The clinical picture is one of abrupt onset of chills and fever, often followed by hypotension, usually complicated by development of metastatic cutaneous lesions within 36 hr after onset. The cutaneous lesions

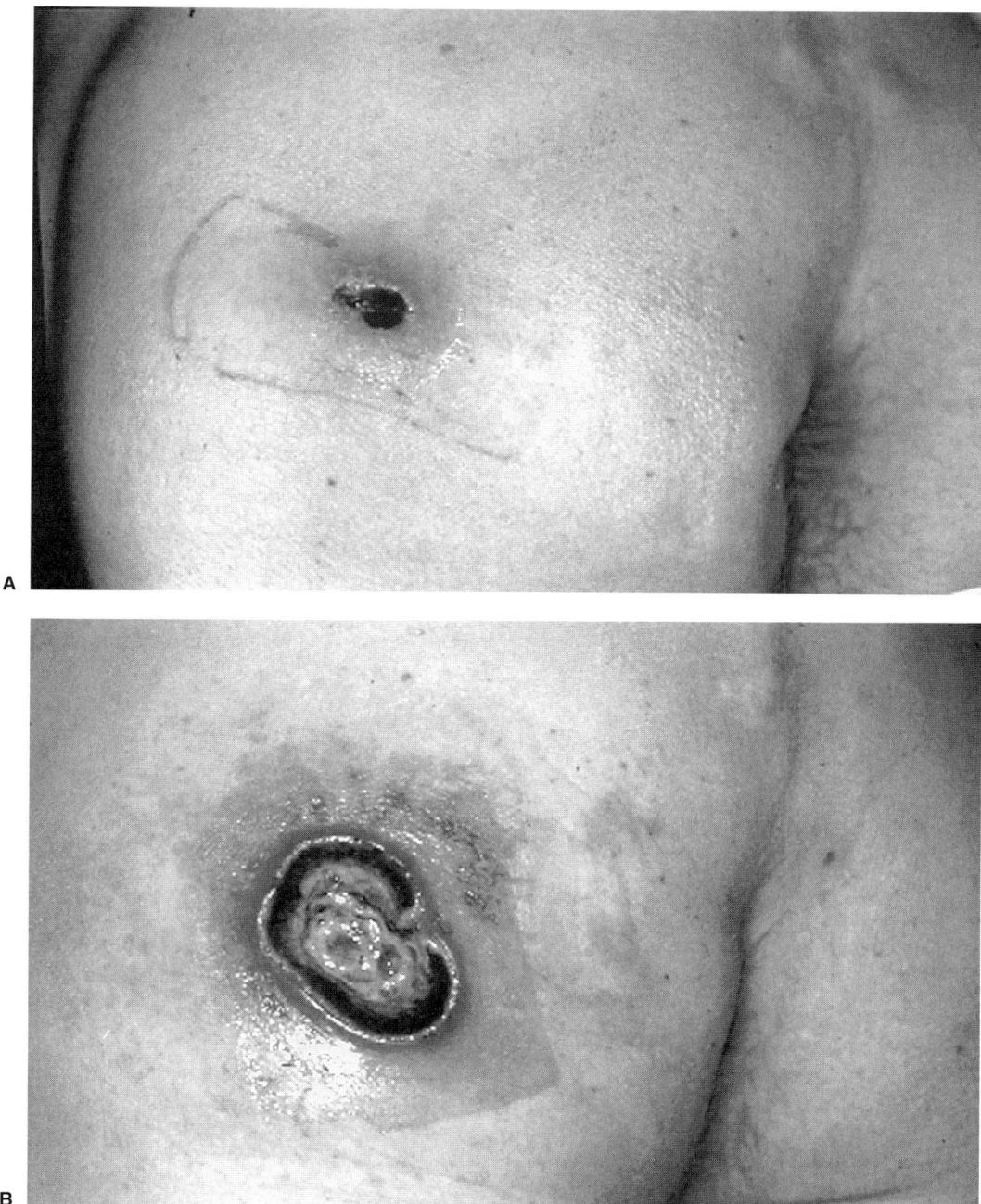

FIGURE 7. *Pseudomonas aeruginosa*: Ecthyma gangrenosum. (A) A very painful, ulcerated plaque on the buttock of a man with advanced HIV disease and neutropenia. Oral ciprofloxacin was given; an adverse drug eruption occurred and the drug was discontinued. (B) The ulcer enlarged and was associated with bacteremia. Neutropenia persisted and the patient subsequently died of *Pseudomonas* pneumonitis. Color representation of Figure 7 follows page 708.

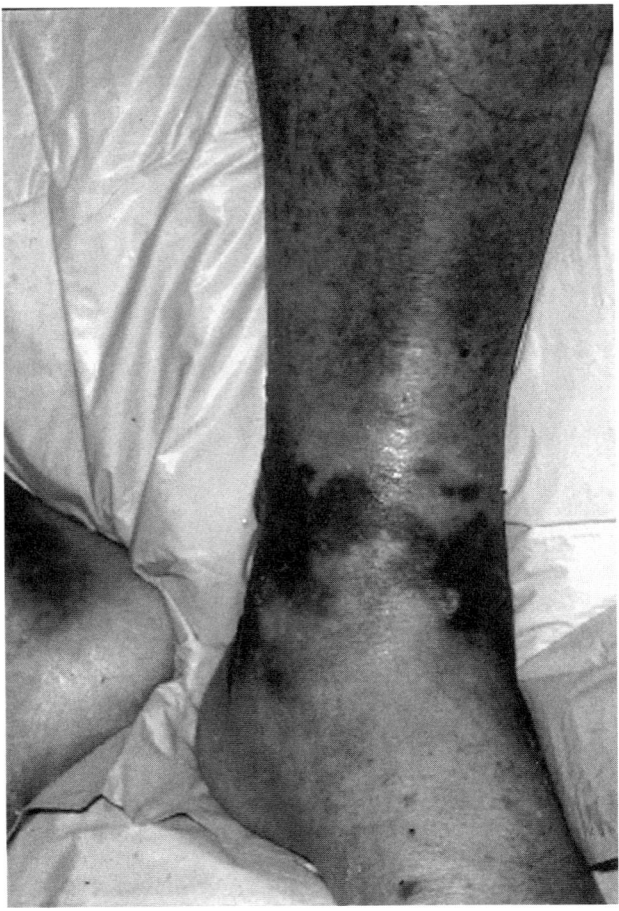

FIGURE 8. *Vibrio vulnificus*: Cellulitis. Hemorrhage plaques and bullae on the lower legs and feet of a diabetic with cirrhosis. The patient had ingested raw clams with subsequent *V. vulnificus* gastroenteritis (asymptomatic), bacteremia, and bilateral lower leg soft-tissue infection. Color representation of Figure 8 follows page 708.

begin as erythematous plaques, rapidly evolving to hemorrhagic bullae and then to necrotic ulcers. The lesions arise commonly on the extremities, occasionally bilaterally. Soft-tissue infections also can arise following inoculation of *V. vulnificus* or *V. alginolyticus* directly into a site of soft-tissue injury. Infection by either of these *Vibrio* species can be life threatening in compromised hosts.

5.1.11. *Helicobacter cinaedi*

Helicobacter cinaedi causes a syndrome characterized by fever, bacteremia, and recurrent and/or chronic cellulitis (resembling erythema nodosum) in compromised patients. In a series of 23 febrile patients with *H. cinaedi* bacteremia (11 were HIV-infected; the others had underlying alcoholism, diabetes, or malignancy), 9 had cellulitis (some with a distinctive red brown or copper discoloration with minimal warmth). In a series of 7 patients (6 HIV-infected, 1 with a history of alcoholism) with *H. cinaedi* soft-tissue infections, cellulitis with adjacent arthritis occurred in 2 patients. The organism is carried as bowel flora in 10% of homosexual men (no carriage in other groups). Diagnosis is made by considering it in immunocompromised individuals, demonstrating cellulitis on lesional biopsy (excluding panniculitis), and failure to isolate other pathogens. Bacteremia is intermittent; the organism is difficult to isolate, requiring hydrogen in the culture vial. Ciprofloxacin 500 mg bid or clarithromycin 500 mg bid is effective, given for 6 to 8 weeks to prevent relapse.

5.1.12. Mycobacteria Infections

5.1.12a. *Mycobacterium tuberculosis*. In developing countries, tuberculosis is the most common opportunistic infection in HIV disease; however, cutaneous tuberculosis is relatively uncommon. As has been true in the past, most cases of symptomatic tuberculosis represent reactivation of latent infection. In non-HIV-infected persons who have tuberculosis in some form, the incidence of extrapulmonary tuberculosis is 15%; in HIV disease, 20 to 40%. In advanced HIV disease, the incidence of extrapulmonary disease increases to 70%.

The etiologic agents of human tuberculosis include *M. tuberculosis*, *M. bovis*, and occasionally bacillus Calmette–Guérin (BCG). Cutaneous tuberculosis is highly variable in its clinical presentation. Cutaneous tuberculosis occurs following *M. tuberculosis* exposure to an exogenous source or by autoinoculation or endogenous spread from another site. Modes of endogenous spread to skin include: direct extension from underlying tuberculous infection, i.e., lymphadenitis or tuberculosis of bones and joints results in scrofuloderma; lymphatic spread to skin results in lupus vulgaris; hematogenous dissemination results in either acute miliary tuberculosis,[32–34] lupus vulgaris, or metastatic tuberculosis abscess. Tuberculosis cutis miliaris disseminata is the lesion of miliary tuberculosis, presenting as 1- to 4-mm red–brown papules–papulovesicles–papulopustules.[35,36] On acid-fast stain of lesional skin biopsy specimens, numerous acid-fast bacilli are seen; however, giant cells and granulomas are absent. In HIV-infected individuals, prior BCG immunization can be followed by reactivation of and infection by BCG at the site of immunization, dissemination of BCG, or lymphadenitis.

5.1.12b. Environmental Mycobacteria. Mycobacteria other than tuberculosis (MOTT) (also known as nontuberculous mycobacteria, environmental mycobacteria, potentially pathogenic environmental mycobacteria, atypical mycobacteria, nonleprous mycobacteria) are widely distributed in soil, dust, and water. They are classified by Runyon groups: group I (photochromogens), *M. marinum, M. kansasii*; group II (scotochromogens), *M. szulgai, M. gordonae, M. malmoense*; group III (nonphotochromogens), *M. avium-intracellulare* complex (MAC); group IV (rapid growers), *M. fortuitum* complex [*M. fortuitum, M. chelonae, M. abscessus* (MFC)]. In the immunocompetent host, injury (trauma or surgery) is followed by the development of localized cellulitis or abscess formation in four to six weeks. In the compromised host, the usual presentation is of multiple erythematous or violaceous subcutaneous nodules without history of trauma; lesions may progress to abscesses that drain and ulcerate.[37]

M. kansasii,[38,39] *M. marinum*,[40–42] *M. gordonae*,[43] *M. malmoense*,[44] *M. fortuitum*,[45,46] and *M. chelonae*[47,48] (Fig. 9) can cause infection in the healthy as well as the compromised host. Inoculation has occurred via puncture wounds (injection or traumatic) or surgical procedures (augmentation mammoplasty, median sternotomy,[47,48] percutaneous catheterization), or rarely primary cellulitis occurs without recognizable skin trauma. Contaminated gentian violet used for skin marking has been the source in some outbreaks. MFC soft-tissue infections characteristically occur several weeks after the injury and appear as indolent wound infections (nodules, ulcers, granulomatous papules, or verrucous plaques). In immunocompromised individuals, MFC can disseminate hematogenously to skin (multiple recurring abscesses on the extremities) and joints.[49]

In advanced HIV disease, MAC commonly causes systemic infection; however, it rarely causes cutaneous infection.[50–52] Cutaneous MAC infections usually are complications of disseminated disease; lesions vary from papules, nodules, pustules, and soft-tissue abscesses to ulcerations; localized infection without apparent disseminated infection has been reported.[51] Cutaneous ulcerations have occurred at the sites of underlying MAC-associated lymphadenitis. Subcutaneous abscesses and ulcers due to localized MAC infection also have been described.

Interpretation of isolation of MAC and/or demonstration of acid-fast bacilli in skin biopsy specimens from patients with advanced HIV disease is difficult in that approximately 40% of these individuals have MAC bacteremia (if not on prophylaxis for MAC). In most cases in whom MAC is demonstrated on lesional biopsy speci-

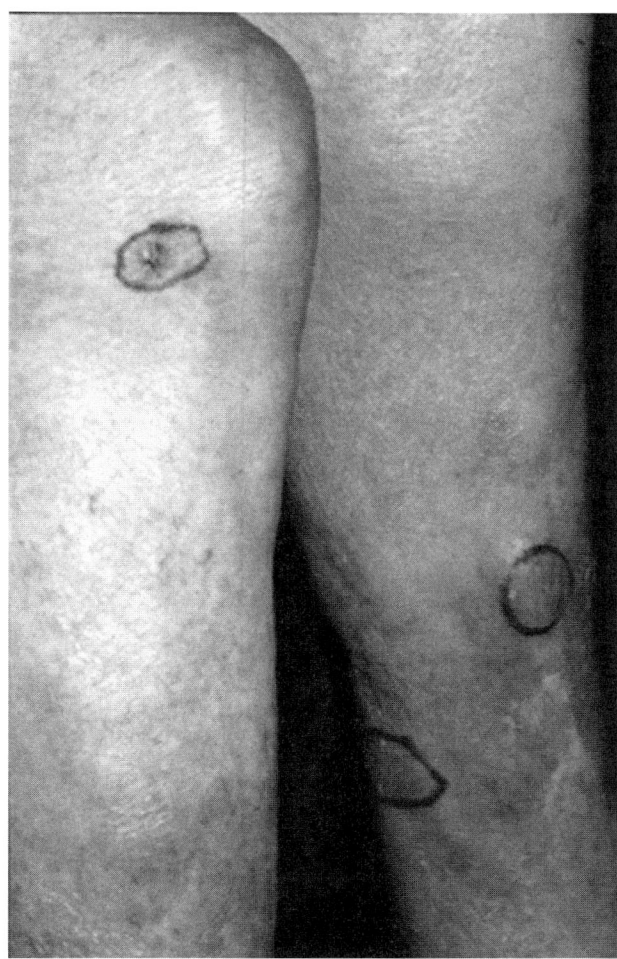

FIGURE 9. *Mycobacterium chelonae*: Cellulitis. Inflammatory, tender nodules on the legs of an 84-year-old woman treated with prednisone for asthma. Color representation of Figure 9 follows page 708.

men, the presence of MAC is incidental, having no part in the pathogenesis of the cutaneous lesion. MAC infection has been reported in three patients who presented with either submandibular, axillary, or inguinal lymphadenitis.[53] Following incision and drainage or spontaneous rupture, scrofuloderma occurred with the formation of deep ulcerative lesions; resolution occurred after a short course of routine antituberculous therapy. Subcutaneous masses also can represent underlying osteomyelitis.

Cutaneous *M. haemophilum* infection has been reported to induce erythema, swelling, painful nodules, and abscess formation and disseminated cutaneous lesions with systemic involvement of bones, joints, lymphatics, and lungs.[54–58] Recovery of *M. haemophilum* requires a high level of clinical suspicion and special handling of mycobacterial cultures by the microbiology laboratory, including cultivation on enriched chocolate agar or heme-supplemented media and incubation at 30°C for up to 8 weeks. Response to antimycobacterial therapy has been poor; disease tends to recur and progress.

5.1.12c. *Mycobacterium leprae*. The interrelationship of *M. leprae* and HIV in dually infected persons has not been adequately studied to date.[59–62] Tropical areas such as Africa and India that have a high prevalence of leprosy are expected to bear the brunt of the HIV epidemic during the next decade. It is probable that leprosy will accelerate the course of HIV disease and that HIV infection will result in a higher ratio of cases of lepromatous versus tuberculoid leprosy and resistance to antilepromatous therapy.[63,64] The natural history of leprosy in HIV disease has been reported in 275 patients from Haiti; 6.5% of the entire cohort was HIV-seropositive. No difference in HIV seropositivity was detected in patients with either lepromatous or tuberculoid types of leprosy. Of the HIV-seropositive patients, 22% developed new skin lesions and lepromin anergy during the course of dapsone–rifampin leprosy therapy, as compared to 0.8% of HIV-seronegative patients.

5.1.13. Bacillary Angiomatosis

Bacillary angiomatosis (BA) and bacillary peliosis (BAP) occur most commonly in the setting of HIV-induced immunodeficiency, characterized by angioproliferative lesions resembling pyogenic granulomas or Kaposi's sarcoma.[65] BAP is caused by infection with fastidious gram-negative bacilli of the genus *Bartonella*, *B. henselae*, and *B. quintana*. The vascular lesions are referred to as BA and those occurring in the liver or spleen as peliosis (BAP). In immunocompetent individuals, *B. henselae* also causes cat scratch disease. HIV-infected individuals with BAP usually have moderate to advanced disease; rarely, BA occurs in immunocompetent, non-HIV-infected individuals. The varied tissue response to *Bartonella* infection in the immunocompetent individual is analogous to the clinical patterns occurring in leprosy. Individuals with intact cellular immunity develop cat scratch disease or tuberculous leprosy; those with impaired cellular immunity develop BAP or lepromatous leprosy. Currently, the prevalence of BA is very low due to prophylaxis given for infections such as *Mycobacterium avium* complex (MAC) and improved immune function with highly active antiretroviral therapy (HAART).

In a study of 49 individuals with BAP, 53% were infected with *B. henselae* and 47% were infected with *B.*

quintana.[66] *B. henselae* and *B. quintana* were equally likely to cause cutaneous BA; only *B. henselae* was associated with hepatosplenic peliosis. Patients with *B. henselae* infection were epidemiologically linked to cat and flea exposure. Those with *B. quintana* infections were linked to low income, homelessness, and exposure to head or body lice. Prior treatment with macrolide (erythromycin, clarithromycin, azithromycin) antibiotics appeared to be protective against infection with either species.

The domestic cat serves as a major persistent reservoir for *B. henselae*. Cats experience prolonged, asymptomatic bacteremia, and can transmit the infection to humans.[67] The cat flea is the vector of *B. henselae* among cats. The domestic cat, however, appears to be a major vector (by scratch or bite) from cat to humans. Antibiotic treatment of infected cats and control of flea infestation are potential strategies for decreasing exposure to *Bartonella*.

Whether asymptomatic or latent infection occurs in humans is not known. The incubation period is unknown, but is probably days to weeks. Patients with localized infection may be free of systemic symptoms. Cutaneous BA may be painful; in contrast, similar-appearing lesions of Kaposi's sarcoma usually are not painful unless ulcerated or secondarily infected. Individuals with more widespread disseminated *Bartonella* infection often experience fever, malaise, and weight loss.

Clinically, the cutaneous lesions of BA are red-to-violaceous, dome-shaped papules, nodules, or plaques, ranging in size from a few millimeters up to 2 to 3 cm in diameter (dermal vascular lesions with thinned or eroded epidermis). Less commonly, domed subcutaneous masses occur without the characteristic red color of more superficial vascular lesions.[68] Lesions are soft to firm and may be tender to palpation. The number of lesions ranges from solitary lesions to hundreds. Nearly any cutaneous site may be involved, but the palms, soles, and oral cavity are usually spared. Following hematogenous or lymphatic dissemination, the spectrum of internal disease caused by *B. henselae* and *B. quintana* includes soft-tissue masses, bone marrow, lymphadenopathy, splenomegaly, and hepatomegaly; internal involvement can occur with or without cutaneous lesions.

The differential diagnosis of the cutaneous papulonodular lesions includes Kaposi's sarcoma, pyogenic granuloma, epithelioid (histiocytoid) angioma, cherry angioma, sclerosing hemangioma, angiokeratomas, and disseminated deep fungal infections. Subcutaneous BA nodules and tumor must be differentiated from enlarged lymph nodes and subcutaneous masses.

The histopathology of lesional skin biopsy specimens of BA is characterized by two patterns of lobular proliferations of capillaries and venules. Pyogenic granuloma-like lesions are characterized by proliferation of small round blood vessels with plump endothelial cells. The stroma is edematous and loose. The inflammatory infiltrate is composed of lymphocytes, histiocytes, and neutrophils. The overlying epidermis may show collarette formation, thinning, or ulceration. Few if any bacteria are visualized by silver stain. The second type of lesion arising deeper in the dermis or subcutis appears more cellular, made up of myriad small, round blood vessels lined by plump endothelial cells. The interstitium shows a granular amphophilic material. Abundant clusters of bacilli, corresponding to sites of granular material, are visualized by silver stain. Percutaneous liver biopsy in patients with peliosis hepatis may be contraindicated because of the vascular nature of the lesions and the risk for uncontrolled bleeding. Histology of liver lesions shows blood-filled cysts with clusters of bacilli in the connective tissue rims of the cysts.[69]

The infecting *Bartonella* species can be identified by molecular techniques on tissue samples. Isolation of *Bartonella* is possible from lesional tissue biopsy specimens and/or blood. The diagnosis also can be confirmed by detection of anti-*Bartonella* antibodies. The diagnosis of BAP usually is made by the demonstration of pleomorphic bacilli on a Warthin–Starry or similar silver stain, or by electron microscopy.

The course of BA is variable. In some patients, lesions regress spontaneously. BA infection may spread hematogenously or via lymphatics to involve bone marrow, bone, spleen, and liver. Death may occur secondary to laryngeal obstruction, liver failure, or pulmonary infection. As with other opportunistic infections in HIV disease, BAP can recur. During the past decade, the incidence of BA has decreased due to the use of antibiotics for MAC prophylaxis and improved immune function following HAART.

BAP is preventable. *B. henselae* is contracted from cats; avoidance should prevent infection. *B. quintana* occurs among homeless people; infection can be prevented by improved hygiene.[70] The antibiotics of choice are erythromycin 250–500 mg PO qid or doxycycline 100 mg bid, continued until the lesions resolve, usually in 3 to 4 weeks. Secondary prophylaxis is indicated in patients with recurrent BAP, especially if immune restoration is not possible.

5.1.14. Nocardiosis

Cutaneous nocardiosis can occur as a primary cutaneous infection [abscesses, ulcers, granulomas, soft-tissue infection, mycetoma, sporotrichoid (lymphocuta-neous) infection][71,72] or secondary cutaneous infection (pustules, abscesses, nodules) complicating hematogenous dissemination from the lungs.[73–76] Nocardiosis in HIV disease is rare; prophylaxis for *Pneumocystis carinii* pneumonia (PCP) with sulfonamides also may provide primary prophylaxis for nocardiosis.[77] Primary cutaneous nocardiosis in HIV disease has been reported to occur at the site of heroin injection; abscesses appeared initially that evolved into large ulcerations. Primary cutaneous nocardiosis may result in lymphangitic proximal extension (sporotrichoid pattern).[78,79]

5.2. Fungal Infections

5.2.1. Superficial Fungal Infections (Dermatomycoses)

5.2.1a. Dermatophytoses. Dermatophytes, i.e., *Trichophyton*, *Microsporum*, and *Epidermophyton*, may occur on any keratinized epidermal structure, i.e., epidermis (stratum corneum), nails, and hair. This group of fungi infect nonviable tissue in otherwise healthy individuals; however, in the compromised host, direct invasion of the dermis may occur. Dermatophytoses are of importance for three reasons: the morbidity and disfigurement caused by the dermatophyte infection itself, which can be quite extensive; the breakdown in the integrity of the skin that can occur, providing a portal of entry for other pathogens, particularly *S. aureus*; and such infections can cause clinical manifestations that mimic other dermatologic conditions.[80] Dermatophyte infections in the compromised host are more frequent, often widespread, atypical in appearance, or invasive.[81]

Epidermal dermatophytosis often is widespread in HIV-infected individuals (Fig. 10) and in transplant recipients. Local immunity to dermatophytic infection is commonly suppressed in patients who have been misdiagnosed as having an inflammatory dermatosis such as eczema or psoriasis and treated with topical corticosteroid preparations, so-called tinea incognito. Clinically this presents as one or many plaques, in some cases with sharply marginated borders, in some cases without scaling, and variable degrees of erythema. Papules or nodules within tinea incognito represent a dermatophytic folliculitis (Majocchi's granuloma). Majocchi's granuloma does occur in the absence of topical corticosteroid use.[82] Inflammatory plaques, ± abscess formation, ± hemorrhage associated with dermal invasion also have been reported to occur in hairy and glabrous sites.[83–85] Epidermal dermatophytosis also can occur in sites of irradiation in which local immunosuppression has occurred.[86]

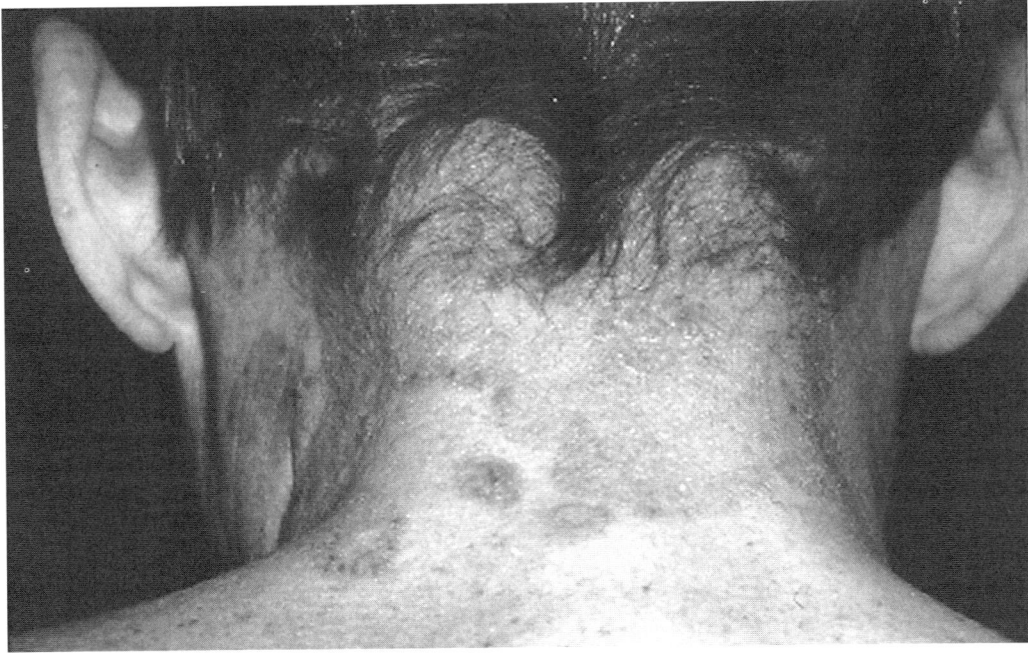

FIGURE 10. Epidermal dermatophytosis: Red scaling plaque on the neck is seen on the neck; lesions were widespread, involving 25% of the body surface area in a patient with advanced HIV disease. Tinea pedis and toenail onychomycosis was also present. Color representation of Figure 10 follows page 708.

In immunocompromised individuals, *Trichophyton rubrum* causes proximal subungual onychomycosis (PSO), and infection of the undersurface of the proximal nail plate (Fig. 11). PSO occurs most often in HIV-infected individuals, and its diagnosis is an indication for HIV testing. PSO also can be seen in transplant recipients and Waldenström's macroglobulinemia. Clinically, PSO initially appears as a chalky white discoloration of the proximal nail plate. KOH preparation of the dorsal nail plate is negative for fungal elements; the undersurface of the nail plate obtained from a core of nail with a skin punch reveals fungal elements on the undersurface of the involved nail plate. Unless immunocompromise is restored, dermatophyte infections are chronic and recurrent.[87] In that many patients are taking oral imidazoles such as fluconazole or itraconazole for candidiasis or cryptococcosis, dermatophytoses are inadvertently treated and kept under control. Terbinafine, which is highly efficacious for dermatophytic infection, is not predictably effective for nondermatophytic fungal infections.

5.2.1b. Pityrosporiasis. *Malassezia furfur* (*P. orbiculare*, *Pityrosporum ovale*) can cause extensive pityriasis (tinea) versicolor, especially in diabetics and individuals treated with topical or systemic corticosteroids. *Pityrosporum* folliculitis, which occurs more commonly in HIV disease, transplant recipients, pregnancy, malignancy, and chronic renal failure, presents as multiple small folliculocentric papules and pustules (acneform) on the upper trunk.[88,89] Diagnosis is made by demonstration of yeast in the follicular infundibulum. *Pityrosporum* folliculitis must be differentiated from both cutaneous candidiasis and hematogenously disseminated candidiasis to skin in the compromised host.[90] *Pityrosporum* may have some role in the pathogenesis of seborrheic dermatitis, which is common in HIV disease.

5.2.1c. Candidiasis. Oropharyngeal candidiasis (OPC) associated with *Pneumocystis carinii* pneumonia in young homosexual men marked the advent of the HIV epidemic.[91] OPC occurs in the majority of HIV-infected individuals during the natural course of HIV disease as a result of impaired cell-mediated immunity. The oropharynx is the most common site of mucosal candidiasis, which may extend into the esophagus and/or tracheobronchial tree in advanced HIV disease. Recurrent candidal vulvovaginitis is common in HIV-infected women, and may be the first clinical expression of immunodeficiency.[92] In contrast, *Candida* intertrigo, which is more common than mucosal candidiasis in the normal host, is uncommon in adults with HIV disease. Candidiasis of moist, keratinized cutaneous sites such as the anogenital

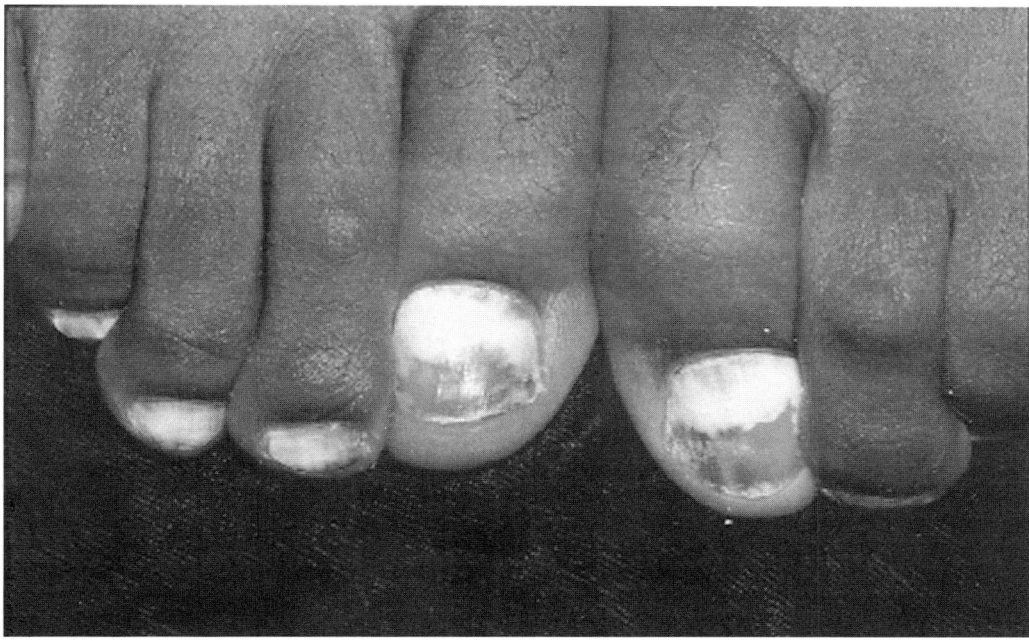

FIGURE 11. Onychomycosis: Proximal subungual. White nails are seen proximally on toenails of this HIV-infected black male. The dermatophyte *Trichophyton rubrum* track proximally over the dorsal nail plate to the undersurface of the nail. This variant of onychomycosis indicates significant immunocompromise, and usually occurs in moderately advanced HIV disease. Color representation of Figure 11 follows page 708.

region occurs with some frequency. Systemic infection originating in the bowel occurs in individuals with prolonged neutropenia. OPC in the absence of predisposing local or systemic causes always should raise the issue of HIV serotesting.

Candida colonization of the oropharynx is common in HIV-infected individuals.[93] In a study of HIV-infected outpatients (median CD4 cell count 113/μl), *Candida* species were isolated from the oral swabs in 60% of individuals, in the absence of any clinical findings of thrush.[94] *C. albicans* was the most prevalent colonizing species isolated from each individual. Five other species were also isolated; 22% of patients were colonized with two different *Candida* species. Isolation of non-*albicans* species alone correlated with advanced HIV diseases with very low CD4 cell counts.

OPC is a marker of HIV disease progression.[95] In a study of the onset of oropharyngeal candidiasis following documented dates of HIV seroconversion, candidiasis was noted in 4% at 1 year after seroconversion, 8% at 2 years, 15% at 3 years, 18% at 4 years, 26% at 5 years; the median CD4 cell count was 392/μl when OPC was first detected. OPC and esophageal candidiasis have been reported to occur as manifestations of primary HIV infection.[96] Esoph-

ageal candidiasis, an AIDS-defining condition, occurs only with advanced CD4 count reduction (<100/μl).

Although OPC often is asymptomatic, the presence of white curdlike colonies of *Candida* within the mouth is a constant reminder of HIV disease to the patient. When symptomatic, common complaints associated with OPC include a soreness or burning sensation in the mouth, sensitivity eating spicy foods, and/or reduced or altered sense of taste. Symptomatic esophageal candidiasis is less common than oropharyngeal infection, and is usually, but not invariably, associated with oropharyngeal disease. The most common symptoms associated with esophageal candidiasis include retrosternal burning and odynophagia. Female patients with HIV infection are increasingly subject to vulvovaginal candidiasis associated with vulvar pruritus, dysuria, dyspareunia, and vaginal discharge.

On physical examination and esophagoscopy, oropharyngeal and esophageal candidiasis present most commonly with a *pseudomembranous* (thrush) pattern (Fig. 12) and less often with a chronic *hyperplastic* and/or *atrophic* pattern. Pseudomembranous candidiasis is characterized by white to creamy curdlike plaques on any surface of the oral mucosa, with these white areas being colonies of *Candida*. The "curds" are easily removed

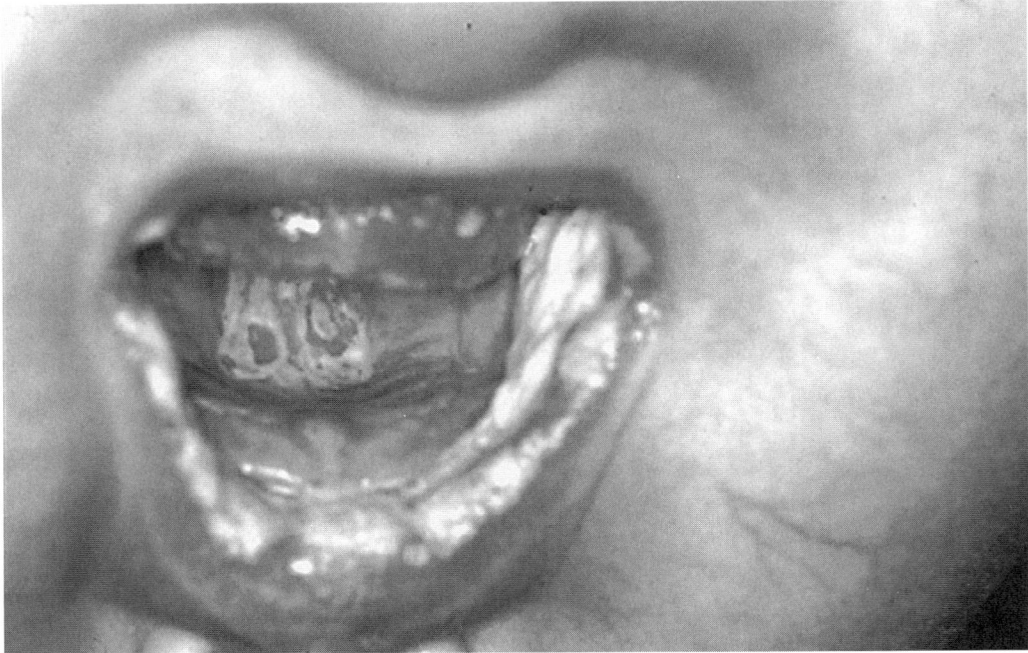

FIGURE 12. Mucosal candidiasis: Pseudomembranous or thrush. White colonies of *Candida* are seen on the oral mucosa. The infant was been treated with ACTH. Color representation of Figure 12 follows page 708.

with a dry gauze (in contrast to the lesions of oral hairy leukoplakia, which are relatively fixed to the underlying mucosa), with some bleeding of the mucosa sometimes occurring. Atrophic candidiasis often is overlooked on examination of the mouth and often is the initial presentation of oropharyngeal candidiasis; it appears as patches of erythema, most commonly occurring in the vault of the mouth on the hard and/or soft palate. On the dorsal surface of the tongue, atrophic candidiasis causes areas of depapillation resulting in a smooth red mucosa. There may be areas of pseudomembranous involvement at some sites, while others manifest the atrophic pattern. Chronic hyperplastic candidiasis presents as both red and white patches at any site in the oropharynx. In edentulous patients with dentures, pseudomembranous and/or atrophic candidiasis typically is seen under the mucosa occluded by dentures.

Candidal angular cheilitis occurs at the corners of the mouth as an intertrigo, unilaterally or bilaterally, and is more common in edentulous patients; it may occur in conjunction with oropharyngeal or esophageal disease or as the only manifestation of candidal infection.

Chronic and/or recurrent candidal vulvovaginitis is a common opportunistic infection in HIV-infected women with moderate to advanced immunodeficiency. Children

with HIV infection commonly experience candidiasis in the diaper area and intertrigo in the axillae and neck fold.

The incidence of cutaneous candidiasis, i.e., *Candida* intertrigo, may be somewhat increased in immunocompromised adults. Fingernail chronic *Candida* paronychia with secondary nail dystrophy (onychia) is common in HIV-infected children.[97]

Candidemia occurs in HIV-infected individuals undergoing total parenteral nutrition, intravenous antibiotic therapy through a central venous catheter, or cancer chemotherapy, having a central venous catheter for >90 days. In a study of HIV-infected children with fungemia, non-*Candida albicans* species and *Candida* (*Torulopsis*) *glabrata* were isolated relatively commonly.[98] Clinically, cutaneous dissemination of candidiasis presents as multiple erythematous papules on ears, extremities, and the trunk; lesions may be ecchymotic in the setting of thrombocytopenia.

The diagnosis of mucocutaneous candidal infection requires the presence of clinical manifestations of infection plus the demonstration of the organism on microscopic examination of a potassium hydroxide preparation of material taken from the lesion. Since *Candida* is a commensal organism in the oral cavity, isolation of the organism on culture in the absence of a clinically overt abnormality is not very meaningful. At times, lesional biopsy is required for the diagnosis of hyperplastic candidiasis. The differential diagnosis of oropharyngeal candidiasis includes oral hairy leukoplakia, migratory glossitis (geographic tongue), lichen planus, bite line irritation, and smoker's leukoplakia.

Management of mucosal candidiasis should be directed at control of symptomatic candidiasis, which may be followed by secondary prophylaxis (Table 3). Prolonged prophylaxis with topical or systemic agents increases the risk of azole-resistant infection. Topical treatments rely on high patient compliance in that they require administration 4 to 5 times daily, but they usually are preferred over systemic drugs for initial treatment. Agents for topical therapy of OPC include nystatin (suspension, tablets, pastilles),[99] clotrimazole (troche), itraconazole solution, fluconazole solution, and amphotericin B solution. The imidazoles, fluconazole (oral solution, tablets, IV solution), itraconazole (capsules, oral solution[100]), and ketoconazole (tablets) are available for systemic therapy. Terbinafine is an excellent agent for dermatophytoses, but not for candidal infections.[101] OPC relapses in approximately 40% of cases within 4 weeks of discontinuing therapy.

Secondary prophylaxis of OPC and esophageal candidiasis often is indicated unless the immunocompromise is restored. Fluconazole-resistant oropharyngeal and/or esophageal candidiasis occurs relatively frequently in chronically treated patients.[102,103] HIV-infected individuals taking fluconazole 200 mg/day prophylactically experienced reduction in the frequency of cryptococcal infection, esophageal candidiasis, and superficial fungal infection, especially those with CD4 cell counts of $\leq$50/μl; but they did not experience reduction in overall mortality rate.[104] Despite fluconazole efficacy in preventing fungal infections, daily routine prophylaxis is not recommended for all individuals with advanced HIV disease because of cost, possible emergence of drug-resistant candidiasis, and potential drug interactions.[105]

5.2.2. Invasive Fungal Infections Involving the Skin

The major importance of the cutaneous manifestations of systemic mycotic infection is that these manifestations may be the first clue to the presence of such disseminated infection. The most important examples of this phenomenon are disseminated cryptococcal infection, which occurs in approximately 10% of patients with untreated advanced HIV disease and infection due to *Histoplasma capsulatum* or *Coccidioides immitis*. These

TABLE 3. Treatment of Oropharyngeal and/or Esophageal Candidiasis

Episodic	May be effective for patients with mild to moderate immunocompromise
Topical therapy	These preparations are effective in the immunocompetent individual but relatively ineffective with decreasing cell-mediated immunity
Clotrimazole	For OPC, oral tablets (troche), 10 mg, one tablet 5 times daily may be effective; one tablet in a single application
Nystatin	For OPC, vaginal tablets, 100,000 units qid dissolved slowly in the mouth, are the most effective preparation. The oral suspension, 1 to 2 teaspoons, held in mouth for 5 min and then swallowed may be effective
Oral therapy	Usually required for patients with moderate to severe immunocompromise
Fluconazole	150 mg PO as a single dose
Itraconazole (capsules or oral solution)	100 mg PO bid for 2 weeks
Recurrent candidiasis	In immunocompromised individuals, response to topical and/or system therapy; however, recurrence is the rule. May become refractory to intermittent therapy, requiring daily chemoprophylaxis, with either topical or systemic treatment. Increase dose with resistant disease
Fluconazole	200 mg PO once followed by 100 mg daily for 2 to 3 weeks, then discontinue. Increase dose to 400 to 800 mg in resistant infection. Also available in IV form
Itraconazole	100 mg PO bid for 2 weeks
Ketoconazole	200 mg PO qid for 1 to 2 weeks
Fluconazole-resistant candidiasis	Fluconazole-resistant candidiasis defined as clinical persistence of infection following treatment with fluconazole 100 mg/day PO for 7 days. Occurs most commonly in HIV-infected individuals with CD4 cell counts <50/μl who have had prolonged fluconazole exposure
Fluconazole	Increase dose to 200 mg PO bid; chronic low-dose fluconazole treatment (50 mg/day) facilitates emergence of resistant strains
Itraconazole	200 mg PO bid; 50% of fluconazole-resistant *Candida* strains sensitive to itraconazole
Amphotericin B	For severe resistant disease. New liposomal preparations are effective and less toxic. Recurrence is the rule; maintenance therapy is often required

last are seen much less frequently, since exposures to these organisms occur only in geographically restricted areas (*H. capsulatum* being found in the east central portion of the United States, Ohio and Mississippi River valleys, Virginia and Maryland, as well as parts of Central America and, in the case of *H. capsulatum* var. *duboisii*, Africa; *C. immitis* being found in the desert soil of the southwestern United States, Mexico, and parts of Central and South America). In Southeast Asia, systemic infection with *Penicillium marneffei* in HIV disease has been reported.[106]

The pathogenesis of these infections resembles that of tuberculosis: primary infection occurs in the lungs following the inhalation of air contaminated with these organisms. The initial response to this event is a polymorphonuclear leukocyte one, which serves to limit the extent of primary infection. However, the definitive host response is a cell-mediated immune response, which both limits the impact of postprimary systemic dissemination of the organisms and prevents the subsequent breakdown of sites of dormant infection. Thus, patients with advanced HIV disease are at risk for three patterns of infection: (1) progressive, primary infection with systemic spread due to a failure of the normal cell-mediated immune response; (2) reactivation of dormant sites of infection, with secondary systemic dissemination of the organisms; and (3) reinfection in a patient who has lost the protective immunity engendered years previously on exposure to this same organism, with such reinfection resulting in a pattern of disease akin to that seen in patients with progressive primary infection. Whenever systemic dissemination of these organisms occurs, there is an approximately 10% incidence of mucocutaneous disease, often as the first recognizable manifestation of systemic infection.[107]

Skin lesions occurring in disseminated mycotic infections for the most part are asymptomatic apart from their cosmetic appearance. Thus, the symptom complex that the patient presents with is determined primarily by the other sites of involvement: symptoms referable to the respiratory tract in patients with active lung infection; symptoms referable to the central nervous system in patients with seeding of the central nervous system; systemic complaints of fever, chills, sweats, weight loss, and so forth dependent on the organism load and the intensity of the host inflammatory response.[108] Oral and/or esophageal ulcerations due to *Histoplasma capsulatum* may, however, be painful.

The most common appearence of skin lesions due to systemic fungal infection in the HIV-infected individual is that of multiple molluscum contagiosumlike lesions,

papules, or nodules occurring on the face.[109,110] On occasion these lesions may become ulcerated, taking on a herpetiform appearance. Other reported cutaneous findings include erythematous macules; necrotic or keratin-plugged papules and nodules; pustules, folliculitis, or acneform lesions; vegetative plaques; or a panniculits.[111] Facial lesions are most common, but lesions also are seen on the trunk and extremities. Oral mucosal lesions occurring in HIV-infected patients with disseminated mycotic infections include nodules and vegetations; ulcerations may occur on the soft palate, oropharynx, epiglottis, and nasal vestibule. These occur most commonly with histoplasmosis, occasionally in cryptococcosis, but not in coccidioidomycosis. Hepatosplenomegaly and/or lymphadenopathy occur commonly in patients with disseminated histoplasmosis.

The cornerstone of diagnosis in this clinical situation is skin biopsy for culture and pathologic examination. It is important to recognize that such biopsies have two purposes: diagnosis of a particular skin lesion and early recognition of disseminated infection in a compromised host. Because of these dual objectives, any unexplained skin lesion in these patients should be considered for biopsy. Histologically, diagnosis is made by demonstration of fungal forms with hematoxylin–eosin, periodic acid–Schiff (PAS), or methenamine silver stain of the lesional biopsy specimens or of a touch preparation. Tzanck smears obtained by scraping the top of a lesion, placing the material on a glass slide, fixing with methyl alcohol, and staining with rapid Giemsa technique show multiple encapsulated and budding yeast. India ink preparation of lesional skin scraping also can be used to demonstrate encapsulated and budding cryptococcal yeast forms. Fungi also can be isolated on culture of the skin biopsy specimen.

The differential diagnosis of patients with skin lesions possibly due to systemic fungal infection includes molluscum contagiosum, verruca vulgaris, verruca plana, disseminated herpetic or varicella infection, bacillary angiomatosis, and furunculosis. The diagnosis of cutaneous infection with *Cryptococcus neoformans*, *Coccidioides immitis*, or *Histoplasma capsulatum* is prima facie evidence of disseminated infection and must be treated as such.

5.2.2a. Ubiquitous Fungi.

1. Candidiasis, invasive. Invasive candidiasis is the most common invasive fungal infection. *Candida albicans* causes 75% of cases of disseminated candidal infections; *C. tropicalis*, 20% of cases. *C. tropicalis*, however, causes 60% of cases of dissemiated candidal infections with cutaneous involvement and *C. albicans* causes only 20%.[112] The gastrointestinal tract is the site of primary invasion; *Candida* enters blood vessels and disseminates widely. Diagnosis often is delayed because of the nonspecific clinical manifestations and difficulty in identifying *Candida* isolated on culture as a pathogen. The usual clinical scenario is of a patient with fever, neutropenia, clinical deterioration, and failure of response to multiple antimicrobial agents.

The cutaneous findings usually are subtle, with erythematous papules (single, multiple, localized, or diffuse) (Fig.

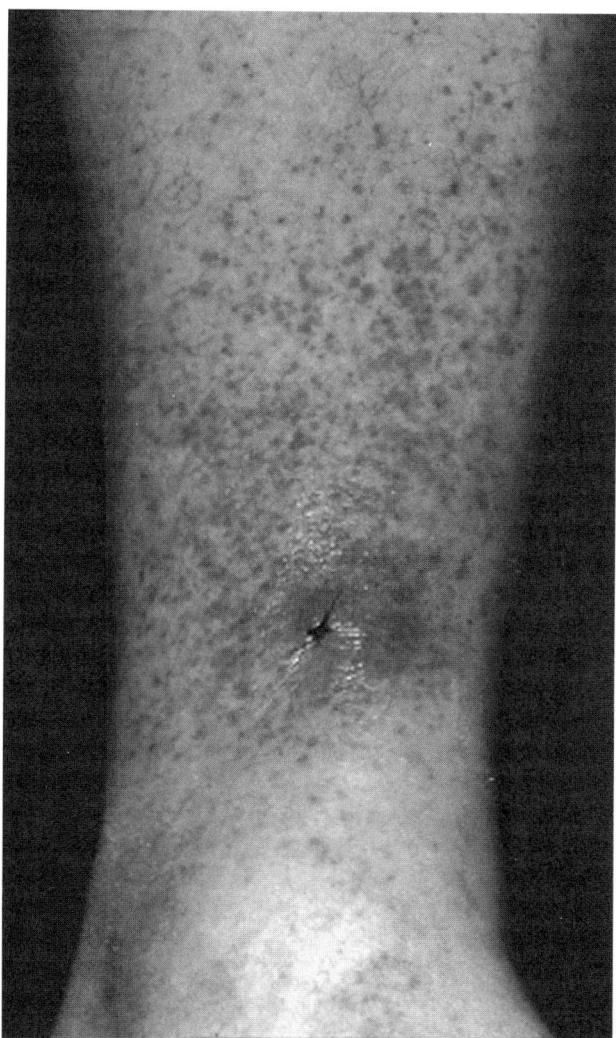

FIGURE 13. Candidemia: Disseminated *C. tropicalis* infection. A hemorrhagic nodule on the leg with multiple petechiae in an elderly patient with acute myelogenous leukemia and thrombocytopenia. *C. tropicalis* was isolated from the lesional skin biopsy specimen and blood. Color representation of Figure 13 follows page 708.

13) arising on the trunk and proximal extremities,[113,114] and in some cases, papulonodules, purpura, central necrosis, ecthyma gangrenosum-like lesions, and nodular folliculitis in hair-bearing areas (heroin users).[115–118] Other clinical findings include endophthalmitis seen on fundoscopic examination, arthritis, and muscle abscess.[119,120] A touch preparation of a punch biopsy specimen examined with KOH or Gram's stain may show *Candida*, allowing rapid diagnosis.[121] Lesional skin biopsy specimens show a range of inflammatory changes ranging from a sparse perivascular mononuclear infiltrate to a leukocytoclastic vasculitis; *Candida* may be sparse or present in large numbers, in and around dermal blood vessels.

2. Cryptococcosis. *Cryptococcus neoformans* is the second most common fungal opportunist (*C. albicans* being the most common, usually causing mucocutaneous infection as well as invasive disease) in the compromised host.[122,123] Disseminated cryptococcosis is by far the most common life-threatening fungal infection in HIV disease. Cutaneous cryptococcosis occurs in 5–10% of individuals with disseminated infection[124] and essentially is always associated with systemic infection in advanced HIV disease (CD4 cell count <50/μl). Cutaneous manifestations can present 2 to 6 weeks before signs of systemic infection.

Hematogenous dissemination of *C. neoformans* to the skin results in lesions of various morphologies that generally are asymptomatic. The most common morphology of cutaneous cryptococcosis is of molluscum contagiosumlike lesions (Fig. 14), i.e., umbilicated skin-colored or pink papules or nodules (54%); other types of cutaneous lesions include pustules, cellulitis,[125] ulceration, panniculitis, palpable purpura, subcutaneous abscesses, and vegetating plaques.[126,127] Lesions commonly occur on the face, but may be widespread. Oral nodules and ulcers also occur alone or with cutaneous lesions. The papules and nodules of cryptococcosis, ranging from solitary to greater than 100 in number, are usually skin-colored, with little if any inflammatory erythema, lacking the central umbilication or keratotic plug characteristic of mollusca. Occasionally, crusting or ulceration occurs resembling chronic herpetic ulcers. In more darkly pigmented individuals, lesions may be hypo- or hyperpigmented. Cutaneous cryptococcosis in some cases occurs in the absence of demonstrable fungal infection in the lung or meninges. Hematogenous dissemination of *Histoplasma capsulatum* or *Coccidioides immitis* can produce identical skin lesions on the face.

Cryptococcal cellulitis occurs following fungemia and most often occurs in the compromised host (corticosteroid therapy, systemic lupus, chronic lymphocytic leukemia, myeloma, chronic active hepatitis, cervical

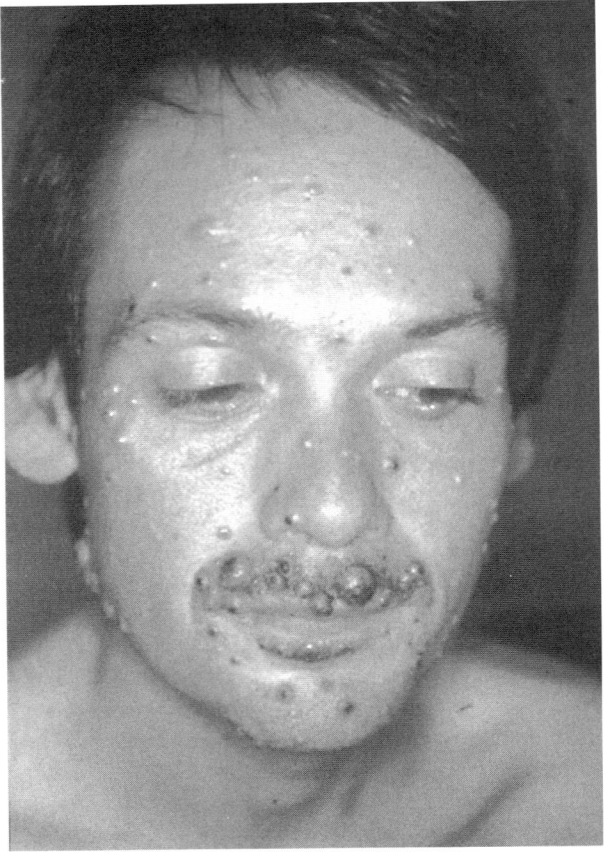

FIGURE 14. Cryptococcosis, disseminated. Multiple large molluscum contagiosumlike lesions on the face of a male with advanced HIV disease. Disseminated histoplasmosis and penicillinosis can also present with similar clinical findings. Color representation of Figure 14 follows page 708.

medullary tumor, congenital lymphedema, congenital lymphedema with lymphopenia, liver transplantation, inflammatory bowel disease, and kidney transplantation). This fungal cellulitis presents with a single or multiple red, hot, tender plaques.[128,129] Primary cutaneous cryptococcosis has been reported, presenting as a nodule, plaque, or ulcer, or in some cases with sporotrichoid spread.[130–132]

3. Aspergillosis. *Aspergillus* species are ubiquitous in the environment in food, water, soil, plants, and decaying vegetation. Aspergillosis can involve the skin either as a primary cutaneous infection or secondarily via invasion into the skin from an underlying infected site (nose, sinuses, and orbit) or hematogenous dissemination to the skin (usually from a primary lung infection). Aspergillosis occurs in the setting of severe or prolonged granulocytopenia due to cytotoxic therapy for leukemia or lym-

phoma, high-dose or prolonged systemic corticosteroid therapy in transplant recipients or those with collagen vascular disease. Primary cutaneous aspergillosis occurs at sites of intravascular catheters and drainage tubes, use of arm boards, and extensive trauma or burns.[4,133–136]

Primary cutaneous aspergillosis has been reported more commonly in children with acute lymphocytic leukemia. Lesions arise most commonly on the palm, initially with erythematous or purpuric papules that progress to violaceous plaques with hemorrhagic bullae, which in turn ulcerate and form necrotic eschars. Secondary cutaneous or disseminated aspergillosis presents as erythematous macules or papules that evolve to purpuric or necrotic lesions, hemorrhagic bullae, and subcutaneous nodules or abscesses.

Potassium hydroxide preparation of scraping from the inner aspect of the bulla roof shows large hyphae. Lesional skin biopsy shows regularly septate, dichotomously branching hyphal elements that may be angioinvasive, which must be differentiated from *Scopulariopsis, Pseudoallescheria, Fusarium,* and *Penicillium.* The majority of patients with invasive disseminated aspergillosis die, despite treatment with amphotericin B.[137]

4. Sporotrichosis. *Sporotrix schenkii* is ubiquitous in the environment in rotting organic matter. Percutaneous inoculation results in limited forms of cutaneous sporotrichosis in immunocompetent individuals. In the compromised host, dissemination of local infection to other organs occurs from lung or skin foci. Dissemination of sporotrichosis is associated with severe malnutrition, sarcoidosis, malignancy, diabetes mellitus, alcoholism, organ transplantation, and HIV disease. A range of cutaneous lesions includes papules to nodules, which may become eroded, ulcerated, crusted, or hyperkeratotic (Fig. 15).[138,139] Individual lesions may remain discrete or become confluent. Lesions are often disseminated, but sparing the palms, soles, and oral mucosa. Ocular involvement results in hypopeon, scleral perforation, and prolapse of the uvea.[140] Joint infection with frank arthritis also is common in the disseminated form of sporotrichosis occurring in HIV disease. Other organs involved in disseminated sporotrichosis in HIV disease include joints, lung, liver, spleen, intestine, and meninges.

5. Mucormycosis. Mucormycosis is a group of infectious syndromes caused by *Mucor, Rhizopus, Absidia,* and *Cunninghamella* species, which have identical presentations and appear morphologically identical in tissue. Conditions predisposing to cutaneous mucormycosis include diabetes, severe thermal burns, trauma, leukemia, organ transplantation, and use of Elastoplast dressing. Superficial cutaneous mucormycosis presents as erythe-

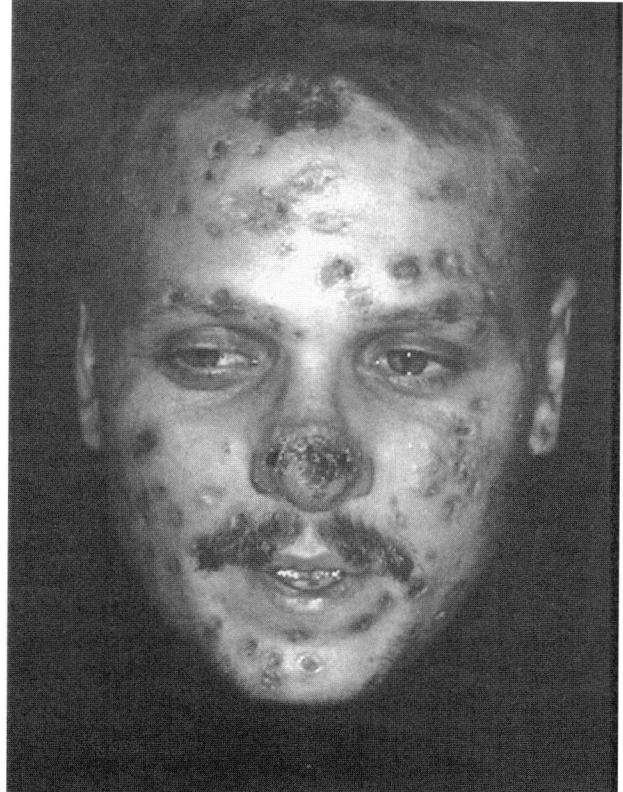

FIGURE 15. Sporotrichosis, disseminated. Multiple crusted ulcers on the face of an HIV-infected male with advanced HIV disease. Cutaneous lesions were disseminated, and associated sporotrichoid infectious arthritis. Color representation of Figure 15 follows page 708.

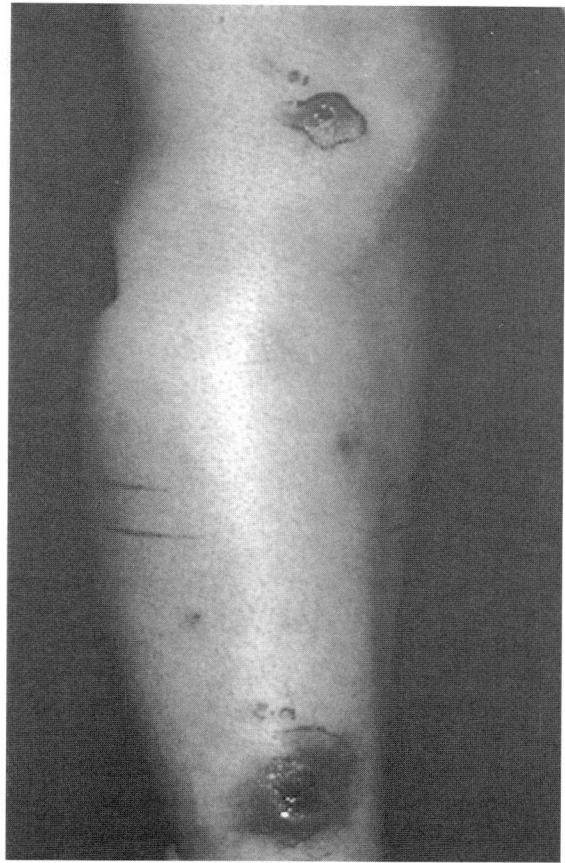

FIGURE 16. Mucormycosis, primary cutaneous. Two large crusted ulcers are seen on the leg. The patient was being treated for acute myelogenous leukemia. Color representation of Figure 16 follows page 708.

matous plaque with subsequent vesicles, pustules, and ulceration at sites dressed with contaminated elasticized adhesive tape. The fungus also can enter the skin via breaks caused by intravascular lines, injections, macerated skin, burns, or insect bites (Fig. 16). The fungus causes a necrotizing infection with subsequent vascular invasion, soft-tissue infection, necrosis, and ulceration.

Rhinocerebral mucormycosis occurs in the setting of poorly controlled diabetes (especially with ketoacidosis) and systemic corticosteroid therapy, especially in patients with leukemia, lymphoma, or organ transplant recipients. Infection begins on the palate or in the nose and paranasal sinuses, spreading rapidly to the central nervous system via the orbit and cribriform plate. Infected mucosal sites are black and necrotic. With extension of infection through the nasal turbinates, orbital cellulitis occurs associated with extraocular muscle paresis, proptosis, chemosis, and eyelid edema. Histologically, broad, irregularly shaped

nonseptate hyphae with right-angle branching are seen in the infected tissue.

6. Phaeohyphomycosis. The phaeohyphomycoses present as a clinically varied group of fungal infections caused by dematiaceous (pigmented or black) fungi that in tissue appear as yeastlike cells, branched or unbranched septate hyphae, or a combination of these forms. Chromomycosis appears clinically as verrucous, nodular, or tumorous plaques with pseudoepitheliomatous hyperplasia, showing large, pigmented, round, thick-walled cells with septation (sclerotic or Medlar bodies). Mycetoma appears clinically as a tumor with draining sinuses and granules in the abscesses. The clinical presentation varies with the host's immune response. The causative fungi include *Exophiala*, *Phialophora*, *Fonsecaea*, *Cladosporium*, *Alternaria*, *Bipolaris*, *Curvularia*, and *Exserohilum* species.

In the compromised host, infection with dematiaceous

fungi can present as primary subcutaneous phaeohyphomycosis, localized inoculation cutaneous phaeohyphomycosis, and systemic or invasive phaeohyphomycosis with secondary or metastatic cutaneous lesions. Primary inoculation phaeohyphomycosis may present with a spectrum of clinical lesions: ulcer with necrosis, subcutaneous cyst, boggy verrucous plaque, crusted nodule with pustules, subcutaneous abscesses with sinus tracts, dermatomal vesicles and crusts, fluctuant abscess with surrounding cellulitis, and a necrotic ulcerated plaque resembling ecthyma gangrenosum. Disseminated phaeohyphomycosis involving the skin presents as tender erythematous nodules, ulcerated papules, hemorrhagic pustules, and scaling hyperpigmented plaques.

7. Hyalohyphomycosis. The hyalohyphomycoses are a variety of opportunistic fungal infections caused by nondematiaceous molds or yeasts, having a nonseptated tissue form. The fungi in this group include *Fusarium, Penicillium, Paecilomyces, Acremonium,* and *Scopulariopsis* species.

Fusarium species are ubiquitous soil saprophytes. Risk factors for infection are granulocytopenia and systemic corticosteroid therapy. Localized cutaneous infection is associated with thermal burns, punctures, and intravascular lines and can be the source of hematogenous dissemination. *Fusarium* onychomycosis can be the source of disseminated *Fusarium* infection in the compromised host.[141,142] Disseminated fusariosis is characterized by evolution of widespread cutaneous lesions from erythematous to necrotic lesions, propensity for vascular invasion and thrombosis with tissue infarction, and acute branching broad septate hyphae.[143] Tissue infarction also is seen in the lesions of *Aspergillus* and *Mucor* infections.

The dimorphic fungus *Penicillium marneffei* is the third most common opportunistic infection in HIV-infected residents of countries of Southeast Asia and the southern part of China.[144,145] In a report of 92 patients,[146] the clinical presentation included fever, weight loss, cough, anemia, and disseminated papular skin lesions (71%). The most common skin lesions were umbilicated papules, occurring most frequently on the face, ears, upper trunk, and arms. Genital ulcers also were reported, ranging in size from <1 cm to 3 cm in diameter. Oral lesions included papules and ulcers. *P. marneffei* preferentially disseminates to lung and liver as well.

5.2.2b. Endemic Fungi. Fungi endemic to geographic regions cause primary pulmonary infection following inhalation of spores. Primary infection is usually asymptomatic, resolving undiagnosed. The organisms are normally contained by the immune system but not eradicated; immunosuppression late in life may lead to reactivation of the disease. In the immunocompromised host,

the latent fungi reactivate in the pulmonary focus, causing infection in the lungs; invasion of blood vessels results in hematogenous dissemination to multiple organ systems. Alternatively, progressive primary infection can occur in the compromised host, again with systemic dissemination.

1. Blastomycosis. Blastomyces dermatitidis is endemic to the south-central and Midwestern United States, the central provinces of Canada, and Africa. Blastomycosis occurs most often in apparently immunocompetent individuals; more severe or disseminated infections occur in compromised persons. Disseminated blastomycosis presents with hundreds of erythematous papules that progress to pustules.[147–149]

2. Coccidioidomycosis. Coccidioides immitis is limited to the semiarid areas of the southwestern United States, Mexico, and Central and South America. Latent pulmonary infection may become active and disseminate in pregnancy, those with African and Filipino ancestry, and immunocompromised individuals with advanced HIV disease, hematologic malignancies, or organ transplant. Cutaneous lesions of disseminated coccidioidomycosis are usually asymptomatic, beginning as papules, evolving to pustules, plaques, or nodules with minimal surrounding erythema; lesions often resemble molluscum contagiosum.[150] In time, lesions may enlarge and become confluent, with formation of abscess, multiple draining sinus tracts, ulcers, subcutaneous cellulitis, verrucous plaques, granulomatous nodules, and with healing, scars.[151] Lesional biopsy specimens show sporangia, hyphal forms, and arthroconidia. Disseminated coccidioidomycosis is diagnosed culturally by isolating the fungus from infected tissues. Serum complement fixation titers are often helpful in diagnosis, but may be negative in the setting of immunocompromise.

3. Histoplasmosis. Histoplasma capsulatum is restricted to the Ohio and Mississippi River valleys, Virginia, and Maryland as well as parts of Central America; *H. capsulatum* var. *duboisii* occurs in Africa. In endemic geographic areas, e.g., Indiana, disseminated histoplasmosis is the leading opportunistic infection in HIV disease. In a report from Kansas City of HIV-infected individuals, the annual incidence of histoplasmosis was 4.7%[152]; the following were associated with an increased risk for histoplasmosis: a history of exposure to chicken coops, a positive baseline serology for complement-fixing antibodies to *Histoplasma* mycelium antigen, and a baseline CD4 of <150/μl. Disseminated histoplasmosis also occurs in young children, those with Hodgkin's disease, and patients treated with systemic corticosteroids or chemotherapy.

Disseminated histoplasmosis presents with a variety of cutaneous findings in approximately 10% of cases:

erythematous macules; necrotic or keratin-plugged molluscum contagiosumlike papules and nodules; pustules, folliculitis, acneform lesions, a rosacealike eruption, guttate psoriasislike eruption; ulcers; vegetative plaques; or panniculitis.[111,153,154] Several different morphologic lesions may occur on a patient. Lesions occur most commonly on the face, followed by the extremities and trunk. Oral mucosal lesions include nodules and vegetations; ulceration occurs on the soft palate, oropharynx, epiglottis, and nasal vestibule. A subtle, widespread, exanthematous or psoriasiform eruption may occasionally develop in HIV-infected patients already on systemic antifungal therapy, in whom systemic symptoms are completely lacking. Oral nodular and ulcerative lesions also occur in disseminated histoplasmosis.[155] Hepatosplenomegaly and/or lymphadenopathy occur commonly in patients with disseminated histoplasmosis.

5.3. Viral Infections

Viruses are major pathogens causing opportunistic infections (OIs) in the compromised host, many of which are manifested at mucocutaneous sites, ranging from cosmetically disfiguring facial molluscum contagiosum (MCV) to extensive common or genital warts to life-threatening or invasive human papillomavirus (HPV)-induced squamous cell carcinoma. In the great majority of cases, viral OIs represent reactivation of latent viral infection, i.e., herpes family of viruses, or of subclinical infection with HPV or MCV.

5.3.1. Measles

Measles had been rare in the industrialized nations because of childhood immunization; in third world countries such as in Africa, measles is common and is associated with significant morbidity and mortality. In the United States, focal epidemics have occurred due to failure of immunization. Measles occurring in the setting of HIV disease in unvaccinated persons has high morbidity and mortality rates.[156] The immunogenicity of measles vaccine in children with HIV infection is low, with only 25% of immunized HIV-infected children developing antibody.[157] Clinically, measles occurring in HIV disease may be atypical with a prolonged period of rash or absence of exanthem or enanthem.[158,159] Diagnosis of measles is usually made on clinical findings; however, in cases in which the exanthem is atypical, documentation of seroconversion is helpful. In some cases, seroconversion does not occur due to abnormal B-cell function; lesional skin biopsy is helpful, showing multinucleated keratino-

cytes. Children who develop measles may have severe or occasionally fatal infection.

5.3.2. Herpetoviridae (Human Herpesviruses)

Human herpesviruses (HHV), i.e., herpes simplex virus (HSV) types 1 and 2, cytomegalovirus (CMV),[160] varicella–zoster virus (VZV), Epstein–Barr virus (EBV), and human herpesvirus-6, -7, -8 (HHV-6, HHV-7, HHV-8), share three characteristics in common that make them particularly effective pathogens in the compromised host: *latency* (once infected with the virus, the individual remains infected for life, with immunosuppression being the major factor responsible for reactivation of the virus from a latent state); *cell association* (these viruses are highly cell associated, rendering humoral immunity inefficient as a host defense and cell-mediated immunity paramount in the control of these infections); and *oncogenicity* (all herpes group viruses should be regarded as potentially oncogenic, with the clearest demonstration of this being EBV-related lymphoproliferative disease).[161] Of the herpes group viruses, those with the greatest impact on the mucocutaneous tissues of the compromised host are HSV, VZV, HHV-8, and, to a lesser extent, EBV.

Reactivated HHV infection can be particularly severe in compromised hosts, resulting in chronic persistent disease and in some cases life-threatening disease. Reactivated HHV infections may contribute to increased HIV expression and potentially modify and accelerate the course of HIV disease. Acute or chronic infections caused by HSV or VZV are usually treated until lesions have resolved, secondary prophylaxis not usually being required. CMV infections, especially retinitis, may be devastating and require lifelong prophylaxis.

Corticosteroid therapy in HIV disease is of concern because of the possible risk of reactivation of HHV. In a report of patients treated with prednisone, the incidence of clinically active infections due to CMV, HSV, and VZV that occurred within a 30-day period of therapy was compared for each group; the median total dose of prednisone was 1600 mg.[162] No statistically significant differences between the cases and controls were detected in terms of the incidence of clinically active herpesvirus infections, i.e., CMV infection (2.5% vs. 5.0%), HSV (1.6% vs. 1.5%), and VZV (0 vs. 0.3%). Only CD4 cell count of $<50/\mu l$ was a significant risk factor for the development of any herpesvirus infection or for the development of a clinically active CMV infection. The risk of HHV infections was related to the stage of HIV infection and was not influenced by corticosteroid therapy.

5.3.2a. Herpes Simplex Virus-1 and -2 Infections. The great majority of HSV-1 and HSV-2 infections occurring in the compromised host are reactivations of latent infections. With mild immunocompromise, lesions are self-limited, resolving within a week or two. Extensive, local cutaneous infections, i.e., eczema herpeticum (Kaposi's varicelliform eruption), can occur with locally impaired immune function (e.g., atopic dermatitis) or in the systemically compromised host.[163]

With more advanced compromise, lesions tend to be subacute or chronic, indolent, or atypical, responding less promptly to oral antiviral therapy. Clinically, reactivated infections (ulcers) are larger and deeper. Ulcerated, crusted lesions at perioral, anogenital, or digital locations are usually HSV in etiology, in spite of atypical clinical appearances. HSV is a treatable cause of intraoral ulcers and always should be identified and treated in the compromised host. Disseminated HSV infection can involve the skin only or of more concern the viscera (lungs, liver, and brain), which has significant associated morbidity and mortality.

In 1981, chronic perianal herpetic ulcers associated with a severe, previously undetected, acquired immune deficiency were an early harbinger of the impending HIV epidemic.[164] Genital herpes and other genital ulcerative diseases are risk factors for acquisition of HIV infection during sexual intercourse.[165] The seroprevalence of HSV-2 infection in the United States is indicative of unprotected sexual intercourse and increased risk of HIV transmission.[166] Reactivated latent HSV infection is one of the most common OIs in HIV disease. Reactivation of latent HSV infection has been documented to increase HIV plasma viral load level[167]; however, acyclovir use was not associated with prolonged survival.[168] Herpetic as well as other genital ulcers are risk factors for transmission of HIV.[165]

During the asymptomatic phase of HIV infection, the clinical manifestations of HSV infection are no different from those occurring in the normal host: orolabial lesions, usually due to HSV-1, or anogenital lesions, usually due to HSV-2, triggered by such factors as fever, stress, other viral infections, or exposure to intense ultraviolet light. The lesions heal within 1 to 2 weeks with or without antiviral treatment. However, with progression of the immunodeficiency, recurrence, even after effective therapy, delayed healing, and chronic ulcers, occasionally due to treatment-resistant strains of HSV, become common. Intermittent asymptomatic shedding of HSV is common. In a report of patients with advanced HIV disease, HSV was isolated on periodic culture of the perianal region in 24% of patients in the absence of erosive or ulcerative lesions, even among those without a history of perianal HSV lesions.[169] Shedding was short-lived, inter-

mittent, and not associated with early subsequent development of perianal ulcers.

With increasing immunocompromise, recurrent HSV infection may become persistent and progressive. Erosions occurring at the typical sites (perioral, anogenital, and digital) enlarge and deepen into painful ulcers.[170,171] Oropharyngeal herpetic ulcers can occur alone or in association with lesions of the lip(s). Untreated, these ulcers may become confluent, forming large lesions.[172,173] Herpetic infection of one or more fingers can form severely painful, large whitlows.[174] HSV can be inoculated into nearly any site including the ears and toes.[174,175] In addition to ulceration, chronic HSV infections also can present as proliferative lesions of the epidermis with or without scale. Herpetic ulcers on the buttocks, perineum, and anus can be associated with painful intra-anal or rectal HSV ulcerations. Genital herpetic ulcers are common. Untreated in individuals with advanced HIV disease, the ulcers persist and enlarge. Hematogenous dissemination with visceral infection is less common in HIV disease than in other compromised states. Large atrophic scars may follow healing of deep herpetic ulcers.

Symptomatically, recurrent herpetic infection is characterized by an itching or tingling sensation at the site, often prior to any visible alteration. Vesicles and pustules frequently rupture leaving superficial erosions and ulcers, which are associated with varying degrees of discomfort and pain. A herpetic whitlow, for example, occurring on the distal finger often is associated with excruciating pain due to the closed tissue spaces involved and the extensive enervation of the site.[176] Herpetic ulcers on the anorectal mucosa cause pain on defecation, tenesmus, and constipation, and at times are associated with HSV colitis and diarrhea. HSV from labial or oropharyngeal ulcers is swallowed in saliva and can infect the esophageal epithelium, usually in patients with very low CD4 counts. Esophageal herpetic ulcers present with odynophagia and/or chest pain.[177] Predisposing factors include nasogastric procedures, corticosteroid therapy, and cancer chemotherapy. Extraesophageal herpetic lesions (labial, oropharyngeal) are present in one-third of patients with esophageal ulcers.

In patients with advanced immunodeficiency, recurrent lesions may fail to heal and continue to enlarge, forming large, chronic ulcers.[178] In the absence of other causes of immunocompromise, chronic herpetic ulcers present for more than a month's duration is an AIDS-defining opportunistic infection. Such ulcers may develop on any mucocutaneous epithelium, but are seen most commonly in perineal, anal, buttock, genital, perioral, and digital sites (Fig. 17). Untreated or acyclovir-resistant

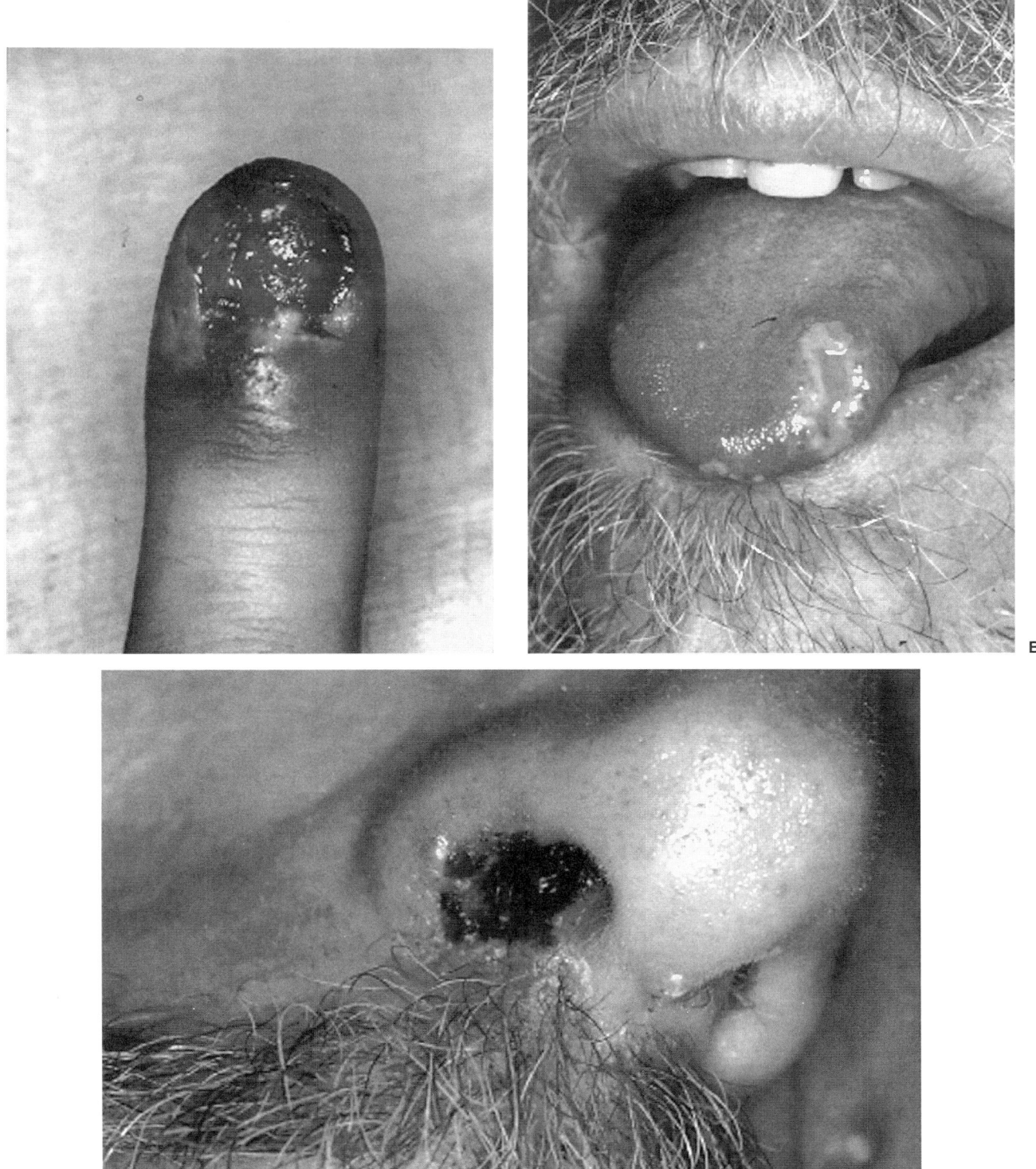

FIGURE 17. HSV infection: Distal digit, tongue, face. Inflammation of the (A) distal digit and nail bed was associated with chronic ulcers of the (B) tongue and (C) face in a patient with advanced HIV disease. All lesions resolved with oral valaciclovir. Color representation of Figure 17 follows page 708.

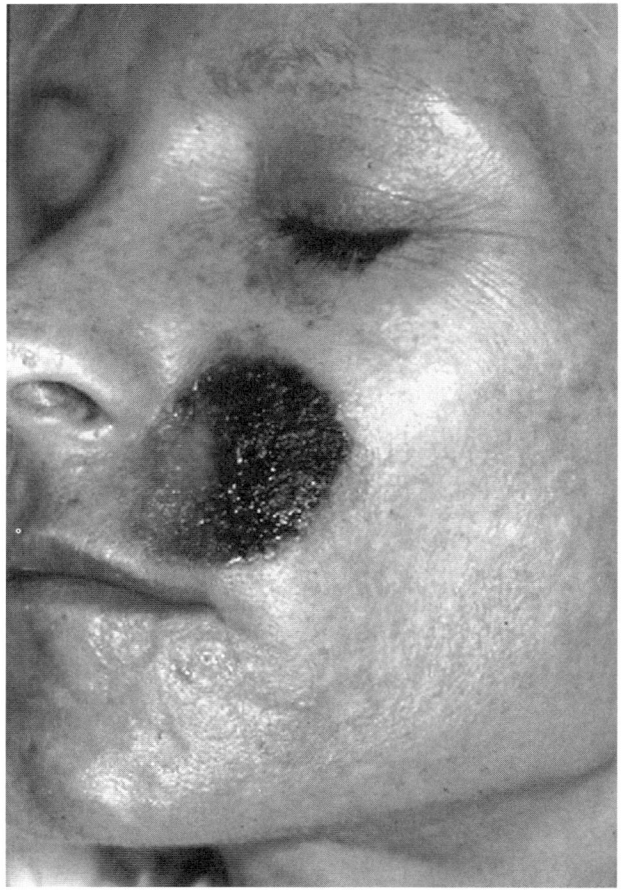

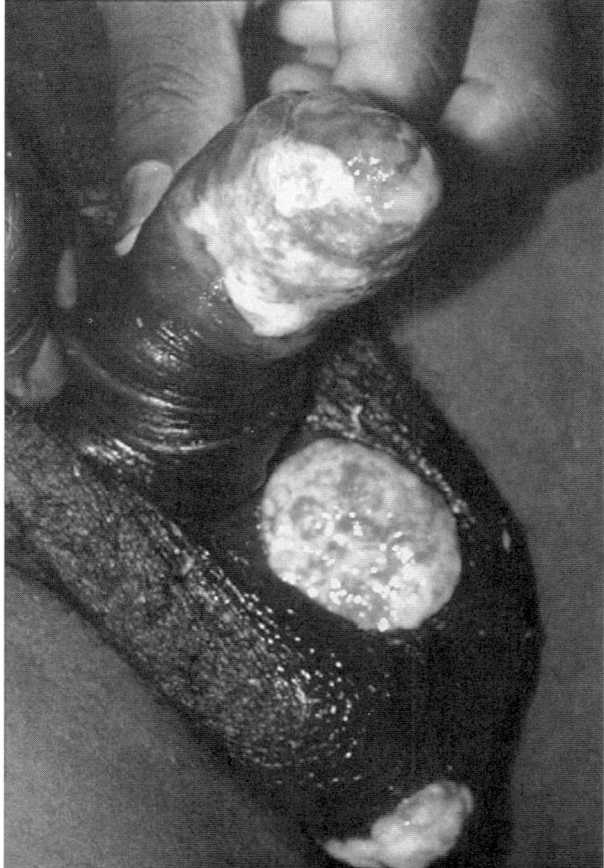

FIGURE 18. HSV infection, acyclovir-resistant: Chronic ulcer on face. A large ulcer painted with gentian violet and atrophic scars on the face of a woman with advanced HIV disease. Ulcers had been present for 1 year; the HSV isolated on culture was acyclovir-resistant. All lesions resolved with intravenous foscarnet. Color representation of Figure 18 follows page 708.

FIGURE 19. HSV infection, acyclovir-resistant: Chronic ulcer on penis. Large ulcers on the penis and scrotum of a male with advanced HIV disease failed to respond to intravenous acyclovir, foscarnet, or cidofovir. Color representation of Figure 19 follows page 708.

HSV ulcers may become confluent, forming lesions up to 20 cm in diameter, involving half the face (Fig. 18), genitalia (Fig. 19), or perineum (Fig. 20).

HSV infections can be diagnosed by isolation of the virus or identification of HSV antigen in lesional smears or biopsy specimens. If indicated, the isolate can be tested for sensitivity to various antiviral agents. Histology shows multinucleated giant epidermal cells indicative of HSV or VZV infection. The Tzanck test, which looks for giant epithelial or adnexal cells, preferably multinucleated, in smears of lesional exudate, is useful but is not always positive even in frank herpetic lesions; its reliability is completely dependent on the skill of the microscopist. Lesional biopsy is helpful when giant epidermal cells are detected, but cannot distinguish HSV from VZV infection. Viral culture of a lesion has a high yield in making the diagnosis. The polymerase chain reaction can detect VZV and HSV DNA sequences from a variety of sources including formalin-fixed tissue specimens.[179]

Currently, three drugs are available for oral therapy of HSV infections; famciclovir[180] and valaciclovir are absorbed much better than acyclovir (Table 4). These agents can be given to treat primary or reactivated infection or to suppress reactivation. In the management of chronic herpetic ulcers, immunosuppressive therapy should be reduced when possible. Intravenous acyclovir (5 mg/kg every 8 hr) may be given for severe infections; the improved blood levels of famciclovir and valaciclovir make oral therapy more effective than with oral acyclovir. Foscarnet and cidofovir are administered intravenously

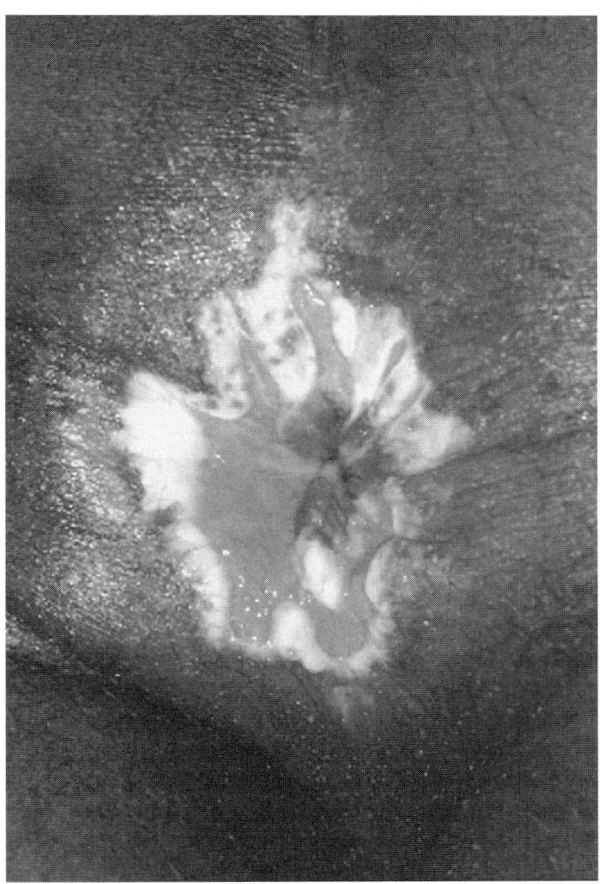

FIGURE 20. HSV infection, acyclovir-resistant: Chronic ulcer on perineum. A large chronic ulcer on the perineum of a male with advanced HIV disease failed to respond to oral and intravenous acyclovir. Color representation of Figure 20 follows page 708.

for infections caused by acyclovir-resistant HSV. Cidofovir gel has been effective as a topical therapy of acyclovir-resistant HSV infections.[181] The use of chronic HSV suppression is controversial. The oral antiviral agents are indicated for frequently recurring HSV infection.

5.3.2b. Varicella–Zoster Virus Infections. Primary VZV infection manifests as varicella (chickenpox); reactivation of VZV from a dorsal root ganglion or cranial nerve ganglion manifests as herpes zoster (HZ).[182] In the compromised host, VZV infection can present as severe varicella, dermatomal HZ, disseminated HZ (sometimes without dermatomal HZ), and chronic or recurrent HZ. Disseminated HZ is defined as cutaneous involvement by greater than three contiguous dermatomes or >20 lesions scattered outside the initial dermatome, or systemic infection (hepatitis, pneumonitis, encephalitis). Disseminated

VZV infection in an individual harboring latent VZV can present with clinical pattern of scattered vesicles in the absence of dermatomal HZ. The compromised host previously infected with VZV is subject to exogenous reinfection with VZV.[183] Underlying conditions of immunocompromise that are associated with an increased incidence of HZ include Hodgkin's disease, non-Hodgkin's lymphoma, solid tumors (e.g., small cell carcinoma of the lung), solid organ and bone marrow transplantation, and HIV disease. Local immunosuppression related to irradiation, nodal involvement by tumor, or surgical sites also is associated with an increased incidence of HZ.[184] Rarely, the reactivation of VZV can produce pain without any cutaneous lesions (zoster sine zoster). In immunocompetent individuals, the main complication of zoster is postherpetic neuralgia (defined as pain persisting more than 6 weeks after the development of cutaneous lesions). Rare complications are myelitis or large-vessel granulomatous arteritis. In the compromised host, central nervous system complications are more serious in the form of progressive small-vessel encephalitis or myelitis.

Primary VZV infection is nearly always symptomatic (i.e., varicella or chickenpox)[185]; reactivated infection presents as HZ. Children with HIV disease represent the largest reservoir of VZV-susceptible immunodeficient children in the world, numbering several million in Africa. Varicella occurring in HIV-infected children and adults can be severe, prolonged,[186] and complicated by VZV dissemination (pneumonia, hepatitis, encephalitis, and pancreatitis), disseminated intravascular coagulation, bacterial superinfection, and death. Primary, recurrent, and persistent VZV infections are a frequent cause of morbidity and hospitalization for HIV-infected children. Rather than resolving, persistent crusted lesions can occur at sites of initial vesicle formation, lasting for weeks or months. In a report of HIV-infected children with varicella, the most common complication was recurrence of VZV infection in 53% of cases.[187–189] Sixty-one percent of children experienced herpes zoster during the first episode of reactivated VZV infection; 32% had dissemination of herpes zoster, associated with a low CD4 cell count. A second episode of varicella can occur, presumably following exposure to a different VZV strain than that which caused varicella initially. In a study of 30 cases of varicella in HIV-infected children, 27% developed HZ an average of 1.9 years after varicella (range, 0.8–3.7).[187–189] Children with <15% CD4 cell levels at onset of varicella were at very high risk of reactivation. Of the children with HZ, 50% developed recurrent HZ episodes.

HIV-infected individuals who are seronegative for VZV, and hence are at risk for primary infection, should

TABLE 4. Treatment of Herpes Simplex Viral Infections

Prevention	Skin-to-skin contact should be avoided during outbreak of cutaneous HSV infection
Topical antiviral therapy	
Penciclovir 1% cream	Apply q2h while awake for recurrent orolabial infection in immunocompetent individual
Oral antiviral therapy	Currently, available drugs for HSV infections include acyclovir, valaciclovir, and famciclovir. Valaciclovir, the prodrug of acyclovir, has a better availability and is nearly 85% absorbed after oral administration. Famciclovir is equally effective for cutaneous HSV infections
Genital herpes	
First episode	Antiviral agents are more effective in treating primary infections than recurrences
Acyclovir	400 mg tid or 200 mg 5×/day × 7–10 days
Valaciclovir	1 g bid × 7–10 days
Famciclovir	250 mg tid × 5–10 days
Recurrences	Most episodes of recurrent herpes do not benefit from pulse therapy with oral acyclovir. In severe recurrent disease, patients who start therapy at the beginning of the prodrome or within 2 days after onset of lesions may benefit from therapy by shortening and reducing severity of eruption; however, recurrences cannot be prevented
Acyclovir	400 mg PO tid for 5 days *or* 800 mg PO bid for 5 days
Valaciclovir	500 mg bid × 5 days
Famciclovir	125 mg bid for 5 days
Chronic suppression	Decreases frequency of symptomatic recurrences and asymptomatic HSV shedding. After 1 year of continuous daily suppressive therapy, acyclovir should be discontinued to determine the recurrence rate
Acyclovir	400 mg bid
Valaciclovir	500–1000 mg qd
Famciclovir	250 mg bid
Nongenital mucocutaneous herpes	Although not as well studied, the dosing is identical to that of genital herpes in most cases
Mucocutaneous disease in immunocompromised	Neither the need for nor the proper increased dosage of acyclovir has been established conclusively. Patients with herpes who do not respond to the recommended dose of acyclovir may require a higher oral dose of acyclovir, IV acyclovir, or be infected with an acyclovir-resistant strain, requiring IV foscarnet. The roles of valaciclovir and famciclovir are not yet established
Acyclovir	5 g/kg IV q8h for 7–14 days *or* 400 mg 5×/day × 7–14 days
Oral valaciclovir or famciclovir	Have reduced the necessity for IV acyclovir therapy
Neonatal	
Acyclovir	20 mg/kg IV q8h × 14–21 days
Acyclovir-resistant	In HIV-infected patients, chronic HSV infections are mucocutaneous, rarely invasive
Foscarnet	40 mg/kg IV q8h × 14–21 days
Aldara cream	tid with occlusion may be effective

promptly receive zoster immune globulin on exposure to the virus and high-dose intravenous acyclovir (10 mg/kg every 8 hr) instituted at the earliest signs of primary infection. This disastrous illness is of primary concern in HIV-infected children, as >90% of the adult population in the United States is seropositive for VZV. Acute varicella does not appear to worsen the course of HIV infection with regard to CD4 or CD8 cell levels.

The more common problem in the compromised host is reactivated infection, where latent virus present in the dorsal nerve root ganglia becomes reactivated.[190] Typically zoster occurs for the first time relatively early in the course of HIV disease, even before oral hairy leukoplakia and oropharyngeal candidiasis.[191,192] Thus, among a group of 112 HIV-infected homosexual men with zoster, 23% had developed AIDS 2 years later, 46% 4 years later, and 73% 6 years later.[193] The more extensive the dermatomal

involvement, the greater the pain, and the more extensive the involvement of cranial and cervical dermatomes, the more advanced the HIV infection is found to be. HZ is also common in other compromised hosts as well.[194,195]

The first manifestation of zoster is often pain in the dermatome that subsequently manifests the classical grouped vesicles on an erythematous base. At times, multidermatomal involvement, either contiguous (Fig. 21) or noncontiguous, may occur. Zoster also may recur within the same dermatome or persist chronically for many months.[196] Occasionally, in these patients, the dermatomal eruption may be bullous, hemorrhagic, or necrotic and be accompanied by severe pain. Not infrequently, patients with zoster experience hematogenously borne cutaneous dissemination, without visceral involvement or inordinate morbidity. Persistent disseminated VZV lesions are often very painful and appear as crusted to

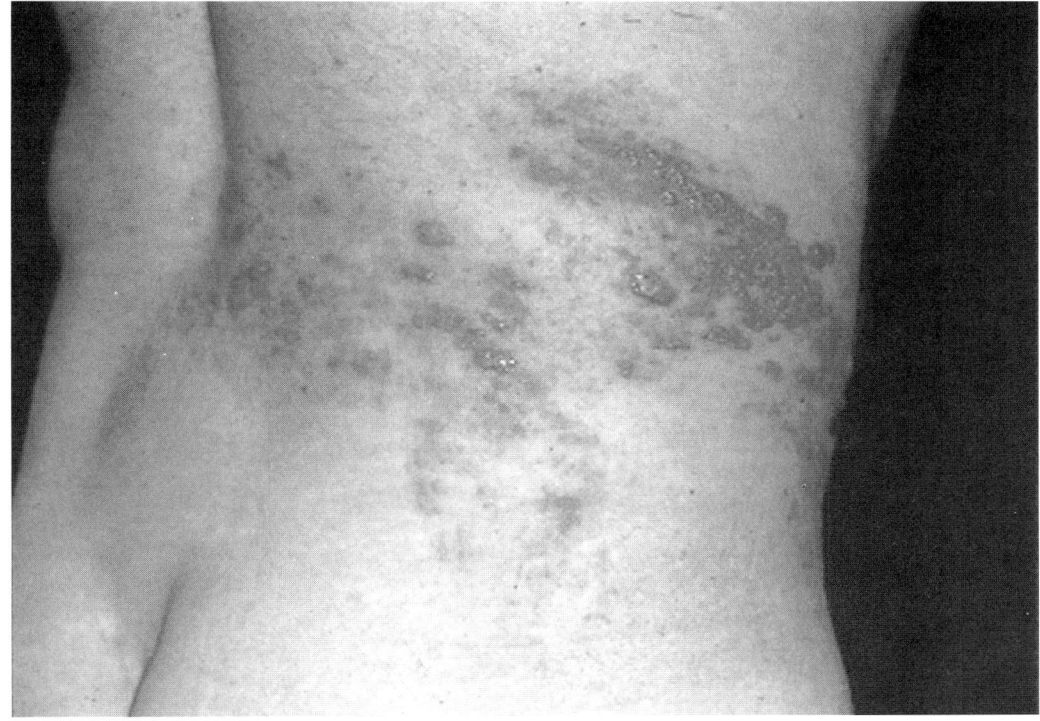

FIGURE 21. VZV infection: Herpes zoster. This typical zosteriform vesiculobullous eruption involving three contiguous dermatomes on the flank was the indication for HIV testing in this HIV-infected male. Color representation of Figure 21 follows page 708.

hyperkeratotic plaques, 1–2 cm in diameter, occasionally with marginal vesicles.[197] These ecthymatous or chronic VZV infections are sparse, typically 5 to 15 in number, on the trunk or proximal extremities.[198,199] The majority of HIV-infected patients with zoster have an uneventful recovery.

Herpes zoster can be the initial clinical presentation of HIV disease. Those who acquire HIV infection sexually are reported to experience HZ more commonly than do those who acquire it by injecting drug use.[200] In a report from Kenya, 85% of patients (16–50 years old) presenting with HZ were HIV-infected.[201] The duration of illness was longer in HIV-infected patients compared with non-HIV-infected cases of HZ (32 vs. 22 days). Seventy-four percent of HIV-infected individuals with HZ had generalized lymphadenopathy compared with only 3% in the noninfected group. Severe pain (69% vs. 39%), bacterial superinfection (15% vs. 6%), more than one affected dermatome (38% vs. 18%), and cranial nerve involvement were all more common in HIV-infected individuals with HZ. The mean CD4 cell count at presentation was 333/μl in the HIV-infected group and 777/μl in the HIV-negative group. Recovery was generally complete and uncomplicated. In Ethiopia, 95% of patients (mean age 35 years) with HZ ophthalmicus were reported to be HIV-infected.[202] Severe eyelid involvement occurred in 25%, ocular involvement in 78%, visual loss in 56%, and postherpetic neuralgia in 55%. Severity of HZ ophthalmicus was associated with delay in presentation, lack of antiviral therapy, and advanced HIV disease.

In a report of homosexual men followed after HIV-1 seroconversion, 20% had an episode of HZ after a mean follow-up of 54 months; 10% of those experienced one recurrence.[203] In a report of patients with advanced HIV disease (CD4 cell count <25/μl) treated with zidovudine, 16% had a history of HZ on enrollment and 13% of these had a recurrence during the 2-year follow-up.[204] HZ was not associated with a more rapid progression to AIDS.

Major complications of HZ occur in one-quarter of cases and include blindness (HZ ophthalmicus), neurological complications, chronic cutaneous infection, postherpetic neuralgia, and bacterial superinfection, all of which occur more commonly if the CD4 cell count is <200/μl.

HZ is a clinical indicator of faltering immunity, and its occurrence always should raise the issue of HIV serotesting. The incidence of HZ in HIV disease is approximately 25%. In a cohort study of 287 homosexual men with well-defined dates of HIV seroconversion and 419 HIV-seronegative homosexual men, the incidence of HZ was 15 times greater in HIV-seropositive men [29.4 cases/1000 person-years (PY)] than in HIV-seronegative men (2.0 cases/1000 PY).[205] The overall age-adjusted relative risk was 16.9.

Herpes zoster often occurs early in the course of HIV disease and can occur soon after varicella in children. In several studies, HZ was not predictive of faster progression to advanced HIV disease. Extent of dermatomal involvement, severity of pain, and involvement of cranial or cervical dermatomes have been correlated with a poor outcome of HIV disease.

In a study of men with HZ, the incidence of first episode was 52 per 1000 PY in HIV-infected men and 3.3 per 1000 PY in non-HIV-infected men.[206] HZ recurred in 26% of HIV-infected men. The incidence of HZ increased by 31.2 per 1000 PY at CD4 cells $\geqslant$500/μl, 47.2 per 1000 PY at CD4 cells 200–499/μl, and 97.5 per 1000 PY at CD4 cells <200/μl. The incidence of HZ increases with the decrease in CD4 cell counts and T-cell reactivity, but HZ is not an independent predictor of disease progression.

Clinically, HZ is typical in most cases. In advanced HIV disease, the spectrum of lesions is much wider. Solitary or few ulcerations can occur. Epidermal proliferative lesions, either solitary or few, resemble basal cell or squamous cell carcinomas.[207] Scattered or zosteriform verrucous lesions also occur.[208] In advanced HIV disease, VZV can also infect the neural tissue of the central nervous system (encephalitis), retina (acute retinal necrosis), or the spinal cord, with or without cutaneous lesions.[209–212] HZ precedes the onset of acute retinal necrosis by several days in 60–90% of cases.

In that zoster often occurs early in HIV disease, the course is fairly uneventful for the majority of patients. It is most often unidermatomal, but may be multidermatomal (Fig. 21), recurrent within the same dermatome, or disseminated. The eruption may be bullous, hemorrhagic, and/or necrotic and may be accompanied by severe pain. The majority of HIV-infected patients with HZ experience an uneventful recovery, but atypical clinical courses are not uncommon. Limited cutaneous dissemination of zoster secondary to viremia is common in some patients with zoster, but uneventful recovery is the rule. Ophthalmic zoster has the highest incidence of serious complications, which include corneal ulceration, variable decrease of visual acuity,[213,214] and retinal necrosis.[215] Viral encephalitis can occur via entry into the brain by VZV infection of the optic nerve,[216] or follow hematogenous dissemina-

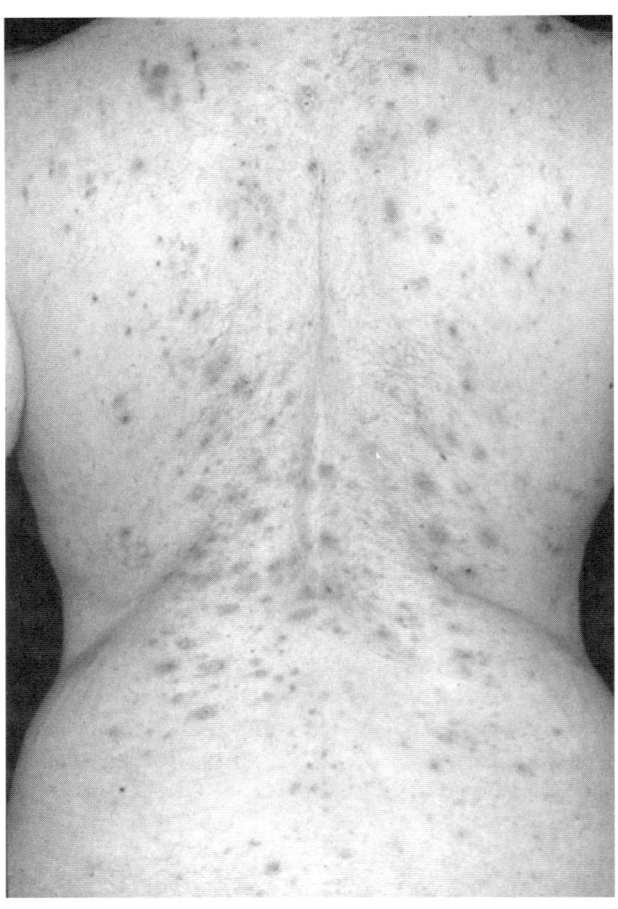

FIGURE 22. VZV infection: Disseminated infection. Disseminated VZV infection with >100 vesicles is seen on the back of a male with herpes zoster on the flank (not seen) and with advanced HIV disease. Color representation of Figure 22 follows page 708.

tion.[217] Cutaneous dissemination of VZV in patients with HZ is relatively common (Figs. 22, 23)[218]; however, significant visceral involvement is rare. Viscerally disseminated HZ in a compromised host can be life-threatening.

In HIV-diseased individuals, cutaneous VZV lesions can persist[199] for months following either primary[198] or reactivated VZV infection with a pattern of zoster[196,219] or disseminated infection,[197,220] referred to as chronic verrucous or ecthymatous VZV infection. Lesions may persist for months, either in the localized or disseminated form, appearing as hyperkeratotic, ulcerated, painful lesions (Fig. 24) often with central crusting and/or ulceration with a border of vesicles.[190] A rare complication of zoster is the occurrence of a granulomatous vasculitis in the involved dermatome, without persistence of the VZV genome, possibly as a reaction to minute amounts of viral

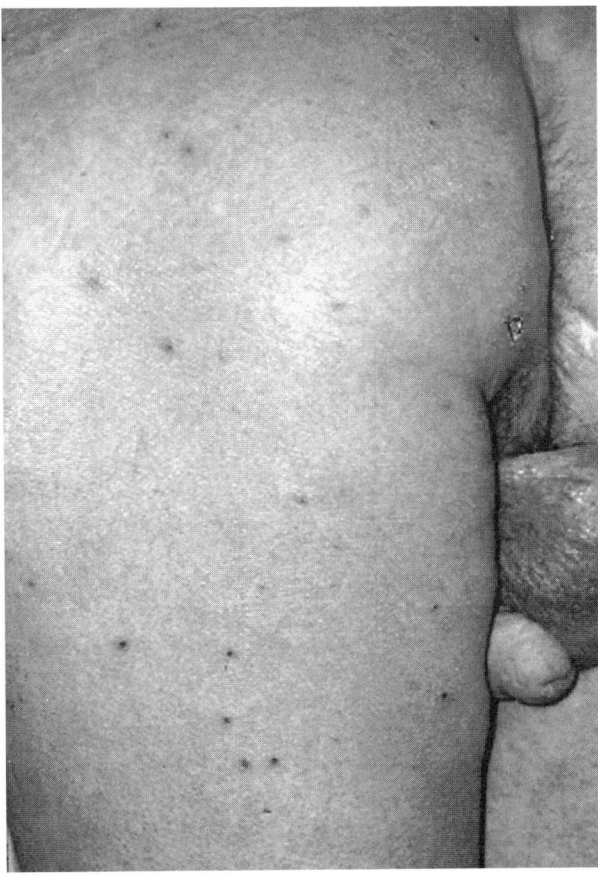

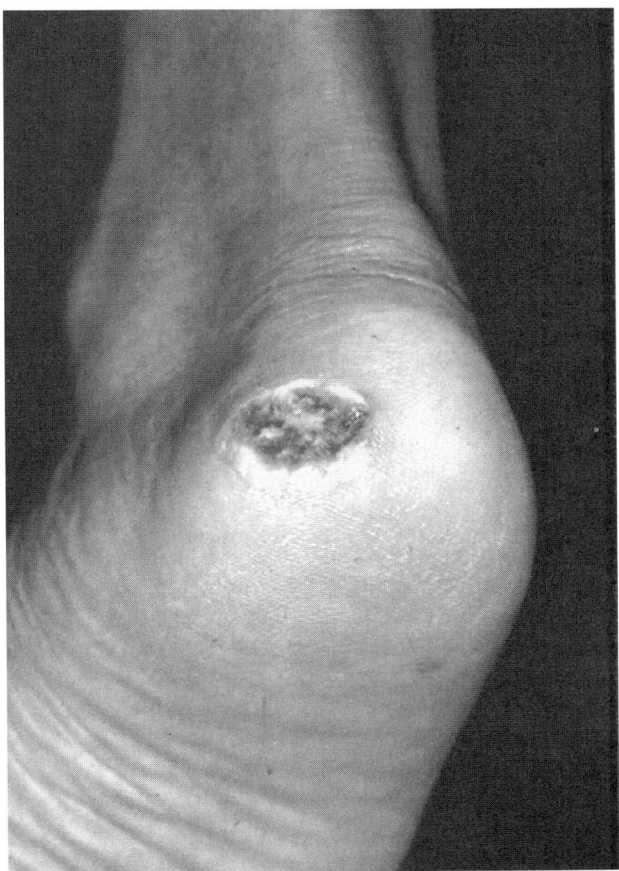

FIGURE 23. VZV infection: Disseminated infection. This patient with advanced HIV disease had approximately 40 vesicles and crusted erosion on the buttocks and thighs. Color representation of Figure 23 follows page 708.

FIGURE 24. VZV infection: Local infection. A solitary very painful ulcer of the heel had been present for 3 weeks in a male with advanced HIV disease. The lesion resolved with oral acyclovir. Color representation of Figure 24 follows page 708.

proteins.[221] Small-vessel encephalitis caused by VZV is more common in the compromised host.[182]

The diagnosis of varicella and HZ can be confirmed by detection of viral antigen on a smear of the base of a vesicle or erosion or in a section of a lesional biopsy specimen. A positive Tzanck test confirms the diagnosis of either VZV or HSV. Isolation of VZV on culture is more difficult than isolation of HSV. Lesional biopsy also is helpful to establish a diagnosis, especially in unusual manifestations of VZV infection such as ecthymatous or chronic verrucous lesions; the diagnosis is confirmed by detection of VZV antigen.

Administration of varicella vaccine in early HIV disease in children appears safe and beneficial. HIV-infected children exposed to VZV, whether varicella or zoster, may benefit by treatment with varicella–zoster immune globulin prophylactically, as well as acyclovir.

Most persons with zoster occurring in early HIV disease do well without antiviral therapy. The same drugs approved for treatment of HSV are approved for treatment of VZV infection: famciclovir, valaciclovir, and acyclovir. Intravenous acyclovir (10 mg/kg every 8 hr) is given for severe infections. Because of the risk of visual impairment following ophthalmic zoster, intravenous acyclovir is usually given. As with HSV infections, acyclovir-resistant strains emerge following prolonged acyclovir treatment; most of these resistant strains respond to foscarnet therapy. Secondary prophylaxis usually is not indicated after VZV infection resolves.

The diagnosis of VZV infection usually is made on a clinical basis, supported by the finding of giant and/or multinucleated acanthocytes on cytologic study of vesicle fluid. Alternatively, the lesions can be biopsied or cultured. The differential diagnosis of varicella includes disseminated HSV infection, cutaneous dissemination of zoster, eczema herpeticum, disseminated vaccinia, bullous impetigo, and various vesicular viral exanthems such as enterovirus infection. The prodromal pain of herpes zoster can mimic cardiac or pleural disease, an acute abdomen, or vertebral disk disease. The rash must be distinguished from zosteriform HSV infection. Ecthymatous VZV lesions must be differentiated from impetigo, ecthyma, or deep mycotic infections.

The management of VZV infections in patients with mild to moderate immunocompromise is identical to that in the immunocompetent host (Table 5). The cornerstone of treatment for severe VZV infection and/or VZV infection in the severely immunocompromised host is intravenous acyclovir. As with HSV infection, acyclovir-resistant VZV has been reported following chronic acyclovir therapy for persistent or recurrent VZV infection.[222,223]

5.3.2c. Epstein–Barr Virus Infections. Epstein–Barr virus (EBV) selectively infects cells of the B-lymphocyte lineage and certain types of squamous epithelium. The majority of adults have been infected with EBV and harbor the latent virus. EBV infection plays an important role in the pathogenesis of three important clinical conditions in the HIV-infected patient population: oral hairy leukoplakia (OHL), classic Burkitt's lymphoma, and EBV-positive large-cell lymphoma. Of these, OHL is the one with primary effects on the mucocutaneous tissues and correlates with moderate to advanced HIV-induced immunodeficiency.[224,225] EBV DNA can be detected in the oral epithelium in HIV-infected patients without clinical signs of OHL, and its detection may be a marker for symptomatic HIV disease.[226] Whether OHL develops after reactivation of latent EBV or superinfection is uncertain.[227]

OHL is a lesion specific to HIV disease. OHL has

TABLE 5. Management of Varicella–Zoster Virus Infection

Prevention	
Immunization	VZV immunization is now available (Varivax) and is 80% effective in preventing symptomatic primary VZV infection; 5% of newly immunized children develop rash. Those at high risk for varicella, who should be immunized, include normal VZV-negative adults, children with leukemia, and immunocompromised individuals (treatment, HIV infection, cancer). VZV vaccine results in both cell-mediated immunity and antibody production against the virus
Symptomatic	Directed at reducing pruritis
Lotions	Little if any efficacy. May symptomatically help pruritis by cooling the involved skin
Oral antipruritic agents	Agents causing sedation are most effective, such as doxepin in the evening
Caution	Antipyretic administration is of concern because of a possible link between aspirin and Reye's syndrome in children with varicella
Antiviral agents	Indicated in herpes zoster if patient older than 60 years
Otherwise healthy patient	If begun wihin 24 hr after onset of rash, decreases the severity of varicella and reduces secondary cases
Acyclovir	20 mg/kg (800 mg max) 5/day × 7–10 days
Valaciclovir	1000 mg tid × 7–10 days
Famciclovir	500 mg tid × 7–10 days
VZV infection (varicella or zoster) in immunocompromised host with severe infections	
Acyclovir	10 mg/kg IV q8h × 7–10 days
Acyclovir-resistant	
Foscarnet	40 mg/kg IV q8h × 7 days
Pain management	Required for many patients with acute zoster as well as postherpetic neuralgia. The pain may be severe, disturb sleep, and often give rise to depression
Acute herpes zoster	Extrastrength acetaminophen
	Codeine 30–60 mg q4–6h when necessary
	Prednisone 40–60 mg qd may reduce the incidence of postherpetic neuralgia
Postherpetic neuralgia (pain that persists more than 6 weeks after development of rash)	Amitriptyline or nortriptyline 25–75 mg qd
	Phenytoin 300–400 mg qd
	Gabapentin 300 mg tid
Treatment of bacterial superinfection	Directed at *S. aureus* and/or group A streptococcus
Mupirocin ointment	Applied twice daily to lesions
Oral antibiotic	

been reported in up to 28% of HIV-infected patients, and is more common in males.[228,229] In a study of OHL as a clinical marker of HIV disease progression, OHL was detected in 9% of individuals 1 year after seroconversion, 16% at 2 years, 15% at 3 years, 35% at 4 years, and 42% at 5 years.[95] The median CD4 cell count when OHL was first detected was 468/μl. In individuals without an AIDS-defining illness when OHL is first detected, the probability of developing AIDS (without HAART) has been reported to be 48% by 16 months after detection and 83% by 31 months.[224] Persons with OHL and a history of hepatitis B virus infection have a fourfold risk for early progression to AIDS; those with syphilis have a nearly threefold risk for early AIDS diagnosis.[230] In a study of HIV-associated mucocutaneous disorders in 456 patients (1982–1992),[231] OHL was diagnosed in 16% of cases, when the median CD4 cell count was 235/μl. Median survival time after OHL diagnosis was 20 months. In patients with a CD4 cell count of ≥300/μl, the detection of OHL was associated with shorter median survival time of 25 months compared with 52 months in those without OHL.

Oral hairy leukoplakia typically presents as hyperplastic, verrucous, whitish, epithelial plaques on the lateral aspects of the tongue (Fig. 25), frequently extending onto the contiguous dorsal or ventral surfaces.[232] Usually, a single lesion or three to six discrete plaques separated by normal-appearing mucosa are observed. Much less commonly, OHL occurs on the buccal mucosa opposing the tongue and on the soft palate. Although described as hairy, the most frequently noted appearance of the lesion occurring on the tongue is a corrugated appearance, with parallel white rows arranged nearly vertically. Useful diagnostic criteria for OHL are a white lesion, involvement of the lateral aspect of the tongue, lack of change in appearance with rubbing, and the lesion does not respond to antifungal therapy. Differential diagnostic considerations include hyperplastic oral candidiasis, condyloma acuminatum, geographic (migratory) glossitis, lichen planus, tobacco-associated leukoplakia, mucous patch of secondary syphilis, squamous cell carcinoma, and occlusal trauma.

The differential diagnosis includes hyperplastic oral candidiasis, HPV-induced neoplasia [condyloma acuminatum, squamous intraepithelial lesion (SIL) (also known as dysplasia or intraepithelial neoplasia), squamous cell carcinoma (SCC) *in situ*, or invasive SCC], geographic tongue, lichen planus, tobacco-associated leukoplakia, mucous patch of secondary syphilis, and bite trauma. The histologic findings of OHL are acanthosis,

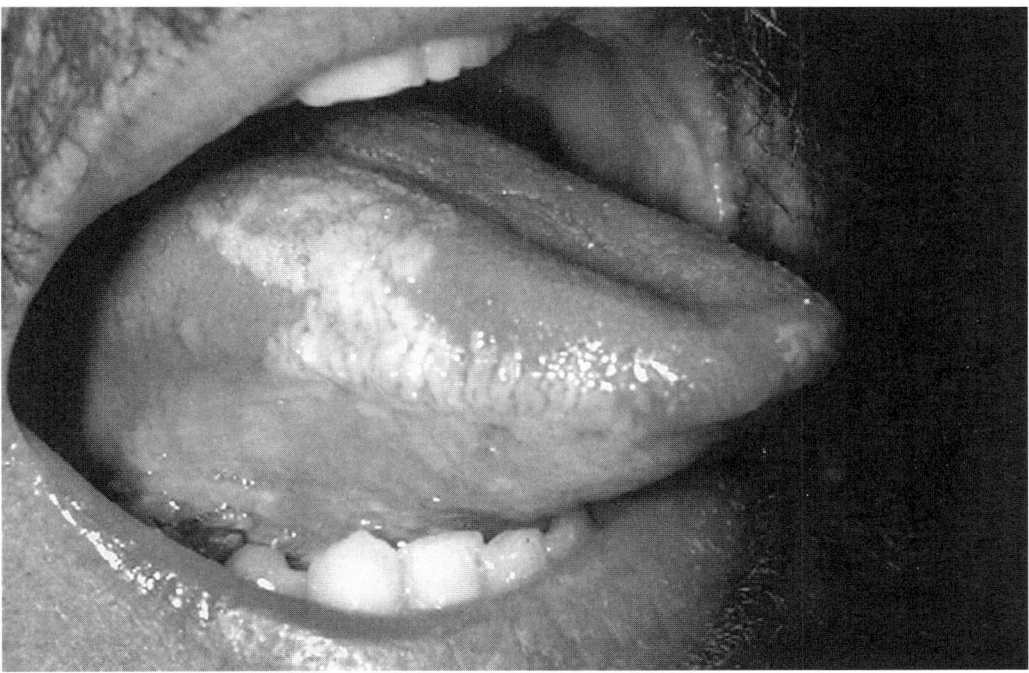

FIGURE 25. EBV infection: Oral hairy leukoplakia. A large asymptomatic plaque on the lateral, dorsal, and ventral tongue of a patient with moderately advanced HIV disease. Color representation of Figure 25 follows page 708.

marked parakeratosis with the formation of ridges and keratin projections, areas of ballooning cells resembling HPV-induced koilocytosis, and little or no dermal inflammatory reaction. Electron microscopy demonstrates 100-nm intranuclear virions and 240-nm encapsulated virus particles. Using *in situ* hybridization, EBV DNA can be demonstrated within nuclei in the upper portions of the epithelium. Keratinocytes are widely infected with EBV; however, expression of viral antigens and replication appears dependent on some process in the epithelial cell maturation–differentiation phase. The diagnosis of OHL is usually made on clinical findings; however, if the diagnosis is uncertain, confirmatory lesional biopsy is advised.

For the most part, OHL is asymptomatic, but its presence may be associated with some degree of anxiety. Patients should be reassured and advised that OHL is not thrush. With HAART, OHL usually resolves without additional interventions. In concerned patients with persistent lesions, topically applied podophyllin in benzoin is effective; recurrence within weeks to months is common. Acyclovir, valaciclovir, famciclovir, ganciclovir, or foscarnet, given for other indications, are often effective therapies for OHL.

Posttransplant cutaneous B-cell lymphoma, associated with EBV infection, is an uncommon complication of solid organ transplantation.[233] Findings are usually confined to the limited regions of the skin; systemic involvement is not common. Treatment is usually directed at the lesions, with surgery or radiotherapy.

5.3.2d. Cytomegalovirus. Seroprevalence studies of cytomegalovirus (CMV) infection indicate that 50% of the general population is infected by age 50 years. The seroprevalence is much higher in lower socioeconomic groups and among injecting drug users. Most cases of primary CMV infection are asymptomatic; following primary infection, CMV enters a latent phase of infection, during which asymptomatic viral shedding in saliva, semen, and/or urine is extremely common. In the compromised host, CMV disease occurs via primary CMV infection, reactivation of latent CMV infection, or reinfection with a new CMV subtype. In most cases, CMV disease represents reactivation of latent virus. CMV infection often occurs soon after immunosuppression associated with cancer chemotherapy, organ transplantation, and systemic corticosteroid therapy. CMV is the most common viral pathogen in patients with advanced HIV-induced immunodeficiency. In a study of 82 HIV-1-seropositive persons, 51.7% of those with either AIDS-related complex (ARC) or AIDS had evidence of CMV infection of circulating polymorphonuclear cells, whereas no infection was detected among the 50 asymptomatic HIV-infected persons.

Manifestations of CMV infection include retinitis, esophagitis, colitis, gastritis, hepatitis, and encephalitis. In a multicenter study of 1002 persons with AIDS or ARC, median survival after diagnosis of CMV disease was 173 days, and CMV was an independent predictor of death.[234] Disseminated CMV has been demonstrated in 93% of patients with AIDS; at autopsy, however, association with skin lesions was not reported. CMV reactivation and dissemination are common events as immunodeficiency worsens. As an opportunistic organism CMV commonly infects the retina, causing a sight-threatening retinitis,[235] and the large intestine, causing colitis manifested by intractable diarrhea. Widespread infection is associated with a generalized wasting syndrome, pneumonitis, and encephalitis. CMV infection, although present within various organs as documented by viral culture, is not necessarily the cause of the tissue dysfunction.[160]

Specific CMV-induced skin lesions have not been identified in compromised individuals. Cutaneous ulcers and a morbilliform eruption are the most common presentation of cutaneous CMV involvement.[31,236–238] Evidence for CMV infection in a variety of mucous membrane lesions, implied by specific cytopathic changes in biopsy specimens by light and electron microscopy, immunofluorescence, immunoperoxidase, and *in situ* hybridization techniques, has been reported[239,240]; however, the role of CMV in pathogenesis of the lesions is not certain. Perianal ulceration caused by CMV occurs as the infection spreads from contiguous gastrointestinal sites. CMV was considered to be the cause of perianal and oral ulceration in five patients with advanced disease studied, based on typical histologic changes and positive fluorescent monoclonal anti-CMV antibody studies. Empirical treatment with acyclovir failed, but all ulcers healed with either foscarnet or ganciclovir treatment. Other reported presentations of CMV infection in skin of HIV-infected individuals include macular purpura of the extremities associated with leukocytoclastic vasculitis and small, keratotic, verrucous lesions, 1 to 3 cm in diameter, scattered on the trunk, limbs, and face.

5.3.2e. Human Herpesvirus-8 (Kaposi's Sarcoma-Associated Herpesvirus). Kaposi's sarcoma (KS) is a hemangiomalike proliferation of endothelial-derived cells, first reported by Moritz Kaposi in 1872 as "idiopathic pigmented sarcoma of the skin." Classic KS occurs in men of Mediterranean or Eastern European ancestry and clinically presents as slowly growing tan to violaceous papules, nodules, and tumors on the lower extremities. In the 1950s, African KS was described in young persons from equatorial Africa. In the 1960s, a third variant of KS was described in patients on long-term immunosuppressive therapy (i.e., recipients of organ transplants). Most recently, in 1981, epidemic or HIV-associated KS was one of the first disorders to be identified in patients with AIDS. Most cases of HIV-associated KS occurred in homosexual or bisexual males. Early in the HIV epidemic, approximately 40% of homosexual or bisexual men with AIDS had KS; more recently the prevalence of KS appears to have declined. In all other groups at risk for HIV infection, the prevalence of KS has been and remains at the 1 to 3% level.

HHV-8 is a lymphotropic, oncogenic γ-herpesvirus that was first detected in KS in 1994.[241,242] HHV-8 also has been detected in other neoplasms such as body cavity-based lymphoma and Castleman's tumor. HHV-8 has been identified in every variant of KS, with its DNA being confined to the nuclei of proliferating spindle cells. The HHV-8 genome also was detected in peripheral blood mononuclear cells of approximately 50% of HIV-KS patients, occasionally months before the occurrence of lesions.[243,244] The role of HHV-8 as a true infectious agent was confirmed in several serological studies that provided evidence for a humoral immune response to HHV-8; seroconversion has been detected months before the clinical appearance of KS.[245,246] After infection with HHV-8, en-

dothelial cells change their biological behavior, resulting in the production of and susceptibility to various inflammatory and angiogenic cytokines. Once morphologic transformation into spindle cells has occurred, these cells start to produce autocrine and paracrine growth factors such as basic fibroblast growth factor (bFGF), interleukin-8 (IL-8), platelet-derived growth factor (PDGF), and interleukin-6 (IL-6), which stimulate KS cell proliferation and regular angiogenesis.

Although the clinical course of HIV-associated KS may be quite variable, it is far more aggressive than other variants of KS, with the frequent occurrence of widespread cutaneous and visceral lesions.[247] Early lesions of HIV-associated KS present as slight discolorations of the skin, usually barely palpable papules, and if very early, macules. These lesions arise within the dermis, lacking any epidermal change, and are pinkish in color, with faint hues of tan, yellow, and green (biliverdin), giving the appearance of a bruise. Over a period of weeks to months to years, these early lesions enlarge into nodules (Fig. 26) or frank tumors, and the color darkens to a violaceous, Concord grape color, often with a yellow–green halo. As lesions enlarge, epidermal changes may occur, showing a shiny, atrophic appearance if stretched, or at times hyperkeratosis with scale formation. In late lesions, tumor necrosis may occur with erosion or ulceration of the surface.

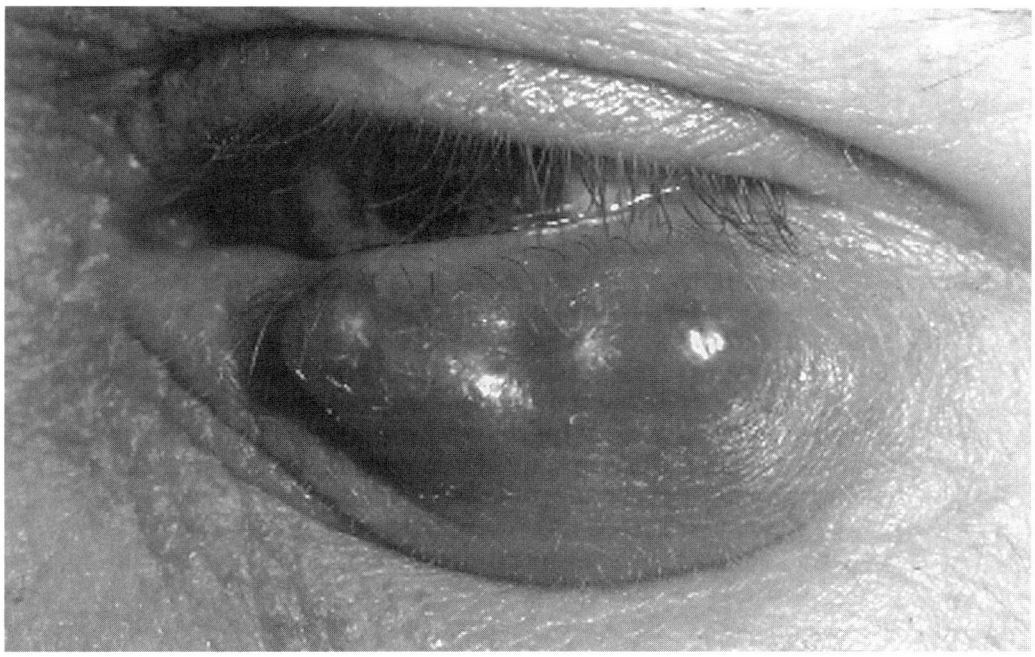

FIGURE 26. Kaposi's sarcoma. A large asymptomatic plaque on the lower eyelid of an HIV-infected male, cosmetically disfiguring. Color representation of Figure 26 follows page 708.

Oral lesions are common and may be the first site of involvement, occurring typically on the hard palate as a violaceous stain of the mucosa.

Although HIV-associated KS can occur at any site on the skin, certain sites are much more common than others. On the head and neck, lesions commonly develop on the tip of the nose, the cheeks (Fig. 27), eyelids, and ears. In time, discrete lesions may coalesce, forming large plaques (Fig. 28). Occasionally, KS lesions may form on the bulbar conjunctiva, appearing as a subconjunctival hemorrhage. Facial edema is common, due to lymphatic obstruction by the KS lesions, and is sometimes extreme, causing gross distortion of the patient's appearance. In some patients, visible lesions are scanty or absent and the presentation of the KS is predominantly that of edema isolated to the face and/or one or more of the extremities.

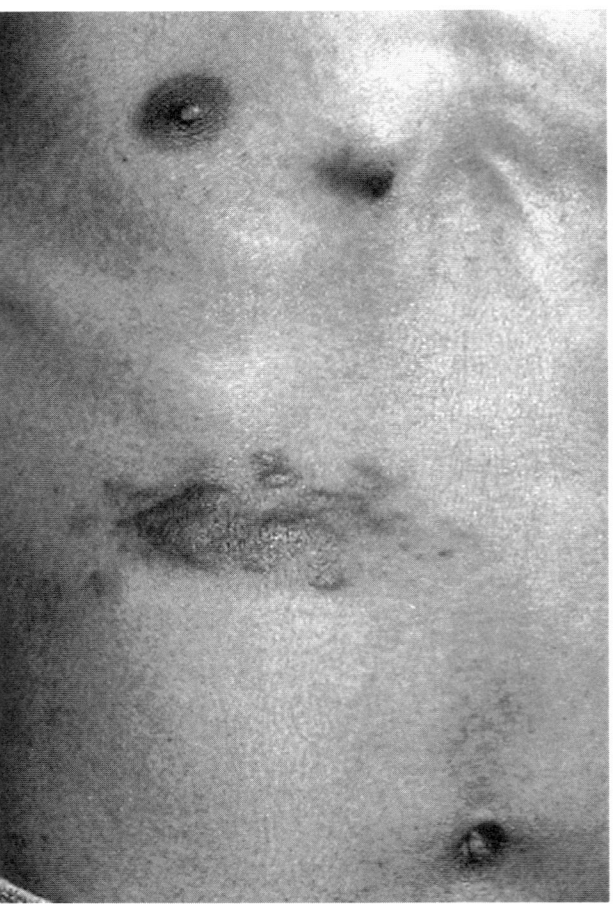

FIGURE 28. Kaposi's sarcoma. Large, confluent violaceous plaques on the trunk of an HIV-infected individual. Color representation of Figure 28 follows page 708.

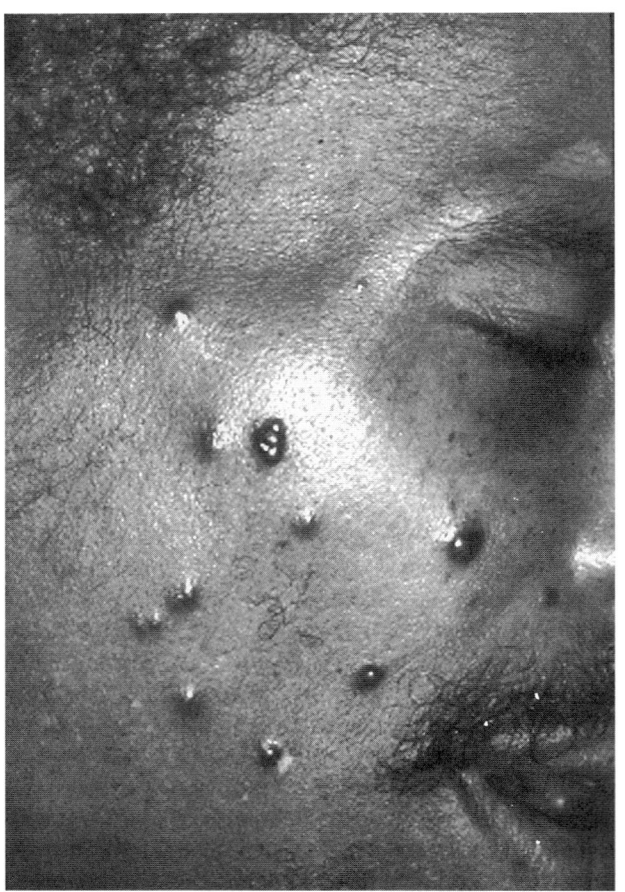

FIGURE 27. Kaposi's sarcoma. Multiple purple nodules on the face of a black HIV-infected male; asymptomatic but causing significant cosmetic disfigurement and stigmatization. Color representation of Figure 27 follows page 708.

In a study of 173 patients with epidemic KS the distribution of mucocutaneous lesions was trunk, 52%; legs, 45%; arms, 38%; face, 33%: and oral cavity (Figs. 29, 30), 40%. Koebnerization of KS lesions has been reported at sites of venipuncture, BCG injection, abscess formation, and contusions.

Although the diagnosis of KS usually can be suspected clinically, in most instances histologic diagnosis on a lesional punch biopsy specimen should be accomplished. The differential diagnosis of possible KS lesions depends on the stage of disease encountered. An early, nearly macular (patch stage) lesion can be mistaken for a bruise, hemangioma, dermatofibroma, insect bite, or benign nevus. More advanced, nodular, or plaque KS lesions must be differentiated from psoriasis, lichen planus, secondary syphilis, insect bites, benign nevi, nonmelanoma cancer, melanoma, and metastatic visceral malig

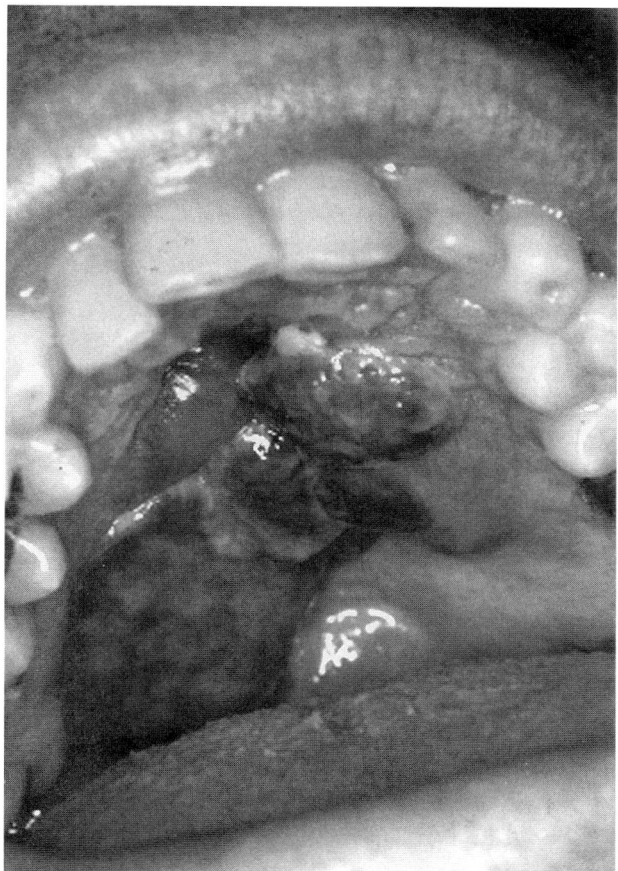

FIGURE 29. Kaposi's sarcoma. Confluent, eroded, and painful tumors on the palate of an HIV-infected individual. Color representation of Figure 29 follows page 708.

nancies. Once the individual KS plaque or tumorous lesions have coalesced to form tuberous lesions, the major alternative diagnosis is lymphoma.

The course of KS depends on restitution of immune function. Few patients die of complications directly related to KS. Currently, individuals with untreated HIV disease still present with KS. With effective response to HAART, KS is relatively uncommon, and if present usually regresses or resolves without specific therapy for KS. Without immune restitution by HAART, however, established KS lesions tend to enlarge in size and deepen in color, at times coalescing, while more and more lesions appear. An occasional patient will develop KS lesions involving internal organs in the absence of any visible mucocutaneous involvement. At postmortem examination, most patients with KS will be found to have lesions in the gastrointestinal tract, lymph nodes, liver, lung, spleen, and/or kidneys. In addition, there appears to be

an increased incidence of second malignancies in patients with AIDS-associated KS.

In the management of KS, the initial therapeutic focus is reduction in immune compromise by changing immunosuppressive drug therapies or by effective treatment of HIV disease with HAART. In many cases, KS will regress or resolve. For persistent KS, a spectrum of increasingly aggressive therapies are available.

Localized cutaneous disease is best treated with application of a panretin gel, intralesional injection of vinblastine,[248] cryotherapy,[249] surgical excision, or radiation. Indolent, disseminated cutaneous KS is best treated with systemic immunotherapy or chemotherapy. Systemic alpha-interferon is effective in indolent or slowly growing KS in patients with CD4− lymphocyte counts >400/μl, who have few systemic symptoms and no opportunistic infections (40 to 50% response rate, although this rate may be higher when zidovudine also is administered). Because of its slow onset of action, usually requiring 6 to 8 weeks for an initial response, systemic alpha-interferon is inappropriate therapy for rapidly growing KS. In such patients and those who fail to respond to interferon therapy, chemotherapy akin to that utilized for the more aggressive forms of the disease may then be employed. Systemic chemotherapy is indicated for aggressive disseminated and/or visceral KS. The chemotherapeutic agents of choice are vincristine and bleomycin, both of which are marrow-sparing agents, usually in combination with adriamycin. Such regimens yield a significant response in 79% of cases. Liposome-encapsulated doxorubicin has enhanced tumor uptake and lesser systemic side effects. Paclitaxel may be effective as a single agent. Antiviral agents, effective in treatment of HHV-8 infection, may be effective in treatment of KS.[250,251]

5.3.3. Molluscum Contagiosum

Molluscum contagiosum virus (MCV) commonly infects keratinized skin subclinically and can cause lesions at sites of minor trauma and in the infundibular portion of the hair follicle.[252] Transmission is usually via skin-to-skin contact, occurring commonly in children and sexual partners. The clinical course of MCV infection in HIV disease differs significantly from that in the normal host and is an excellent clinical marker of the degree of immunodeficiency.[253] In adults with multiple mollusca occurring outside of the genital area, especially head and neck lesions, HIV infection should be considered. Large and confluent lesions cause significant morbidity and disfigurement. Extensive MCV infection is uncommon in other compromised adults such as those with atopic der-

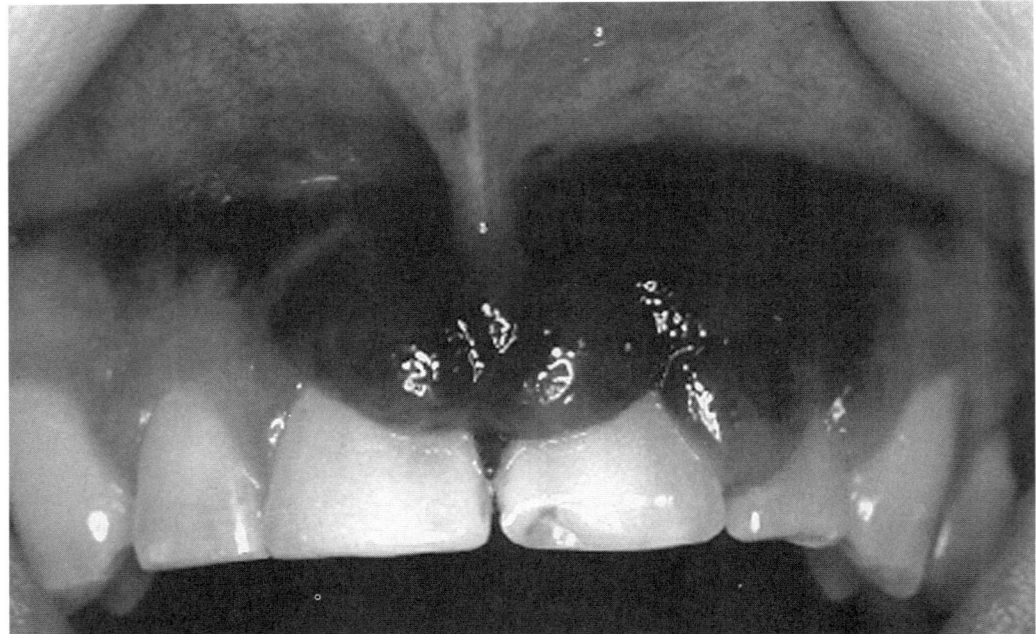

FIGURE 30. Kaposi's sarcoma. A large confluent plaque on the upper anterior gingiva of an HIV-infected individual. Color representation of Figure 30 follows page 708.

matitis,[254] sarcoidosis,[255] cutaneous T-cell lymphoma,[256] lymphatic leukemia, lymphoma,[257] and thymoma.[258]

Prior to HAART, MCV infections were detected in 10% of individuals with HIV disease, and in 30% of those with CD4 cell counts of <100/μl, the number of lesions was inversely related to the CD4 cell count.[259] In a study of 27 HIV-infected patients with MCV infection,[253] mean CD4 cell count was 85.7 cells/μl within 60 days of mollusca diagnosis; 52% of patients had facial and neck lesions alone, and 26% had lesions in areas associated with sexual transmission. *Pneumocystis carinii* pneumonia previously had occurred in 30% of individuals and KS had been diagnosed in 56%.

Clinically, MCV infection presents as skin-colored papules or nodules, often with a characteristic central umbilicated keratotic plug. Lesions are usually 2 to 6 mm in diameter, but may be >1 cm in diameter (giant molluscum). Large lesions may mimic epidermal inclusion cysts, arising on or about the ear or on the trunk. Shortly after their appearance, lesions may be solitary; in time, multiple lesions are more typical (50+ lesions). With persistent and progressive immunodeficiency, mollusca may continue to enlarge and proliferate, resulting in confluent masses of lesions, e.g., involving the entire beard area. Large and/or multiple confluent facial lesions cause significant cosmetic disfigurement. The most characteristic sites of occurrence in HIV-infected adults are on the face (Fig. 31), beard area, neck, and scalp; anogenital and intertriginous [axillae, groin (Fig. 32)] involvement also is common. In males, lesions are often confined to the beard area, the skin having been inoculated during the process of shaving. Occasionally, lesions become secondarily infected with *S. aureus*, resulting in abscess formation, or with *P. aeruginosa* with resultant necrotizing cellulitis. Significant postinflammatory hyper- or hypopigmentation (Fig. 31), more pronounced in more heavily melanized skin, may occur following cryosurgery of lesions, adding to the cosmetic disfigurement of the mollusca.

The diagnosis of MCV infection in the HIV-infected patient is usually made on clinical grounds, but histologic confirmation is required in some patients. The differential diagnosis of solitary molluscum contagiosum includes verruca vulgaris, condyloma acuminatum, basal cell carcinoma, keratoacanthoma, and squamous cell carcinoma. The differential diagnosis of multiple facial mollusca contagiosa includes hematogenous dissemination to the skin of invasive fungal infections (cryptococcosis, histoplasmosis, coccidioidomycosis, and penicilliosis). Lesional skin biopsy is indicated in patients with sudden appearance of molluscalike facial papules associated with fever, headache, confusion, or pulmonary infiltrate to rule

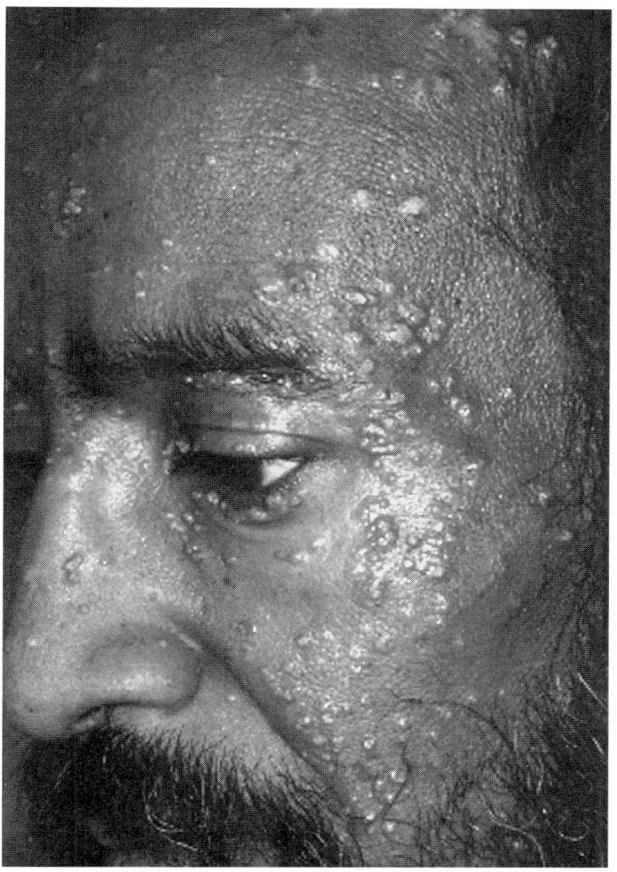

FIGURE 31. Molluscum contagiosum. Multiple, skin-colored papules on the face of an HIV-infected male. Color representation of Figure 31 follows page 708.

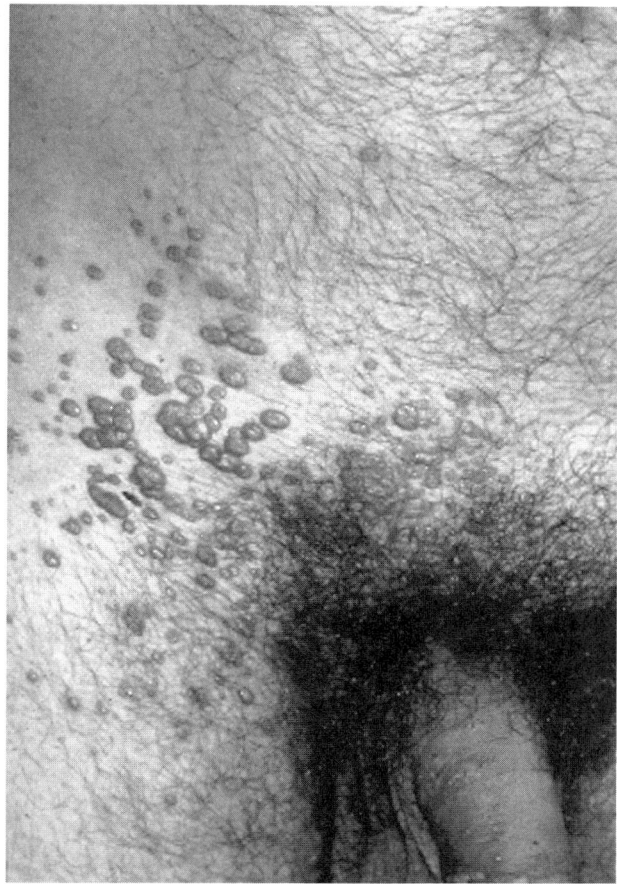

FIGURE 32. Molluscum contagiosum. Multiple, skin-colored papules and nodules, becoming confluent on the lower abdomen of a male with advanced HIV disease. Color representation of Figure 32 follows page 708.

out deep mycosis with hematogenous dissemination to the skin.

In HIV-infected individuals, MCV infection tends to be progressive and recurrent after the usual therapies. In HIV-infected individuals, MCV has been demonstrated within clinically normal epidermis surrounding lesions, suggesting the mechanism by which new lesions recur at treatment sites.[260,261] With response to HAART, MCV infections regress or resolve completely, associated with increased CD4 cell counts and reduced viral load level.[262]

Therapeutically, the most efficacious approach toward MCV infection is correction of the underlying immunodeficiency; if this can be accomplished, lesions regress. If correction of immunodeficiency is not possible, treatment is directed at controlling the numbers and bulk of cosmetically disturbing lesions rather than at eradica-

tion of all lesions. Liquid nitrogen cryospray is the most convenient therapy and usually must be repeated every 2 to 4 weeks. Electrosurgery is more effective than cryosurgery; local anesthesia is required by most subjects with either injected lidocaine or EMLA cream. CO_2 or pulse-dye laser ablation also is effective but relatively costly. Five percent imiquimod cream applied three times a week is an effective patient-administered therapy in children and adults. Cidofovir, a nucleotide analogue with activity against several DNA viruses, given either intravenously or topically as a cream, may be an effective therapy.[263]

5.3.4. Human Papillomavirus Infections

Subclinical infection with human papillomavirus (HPV) is nearly universal in humans. With immunocompromise, cutaneous and/or mucosal HPV infection (re)emerges from latency, presenting clinically as verruca, condyloma acuminatum, squamous intraepithelial lesion (SIL), squamous cell carcinoma *in situ* (SCCIS), or invasive squamous cell carcinoma (SCC).[264]

Human papillomavirus colonizes keratinized skin of all humans, producing common warts (verruca vulgaris, verruca plantaris, verruca plana) in many healthy individuals during the course of a lifetime. The greater majority of sexually active individuals are subclinically infected with one or multiple HPV types. HPV-6 and -11 infect mucosal sites (genitalia, anus, perineum, oropharynx) and cause genital warts (condyloma acuminatum); HPV-16 and -18 cause precancerous lesions SIL, SCCIS, and invasive SCC.

In organ transplant recipients, HIV-infected individuals, and other compromised hosts, verrucae are not unusual in morphology, number, or response to treatment; however, with advancing disease, verrucae can enlarge (Fig. 33), become confluent, and become unresponsive to therapy.[265] HPV-5 can cause an unusual pattern of extensive verruca plana and pityriasis (tinea) versicolorlike warts, similar to the pattern seen in epidermodysplasia verruciformis. With moderate or advanced immunodeficiency, warts and condyloma (Fig. 34) may become much more numerous, confluent, and refractory to usual treatment modalities. Precancerous lesions identical to mucosal lesions, namely, SIL and SCCIS, can occur periungually on the fingers (Fig. 35). In some cases, invasive SCC can arise at one or multiple sites on the fingers and/or nail bed. These tumors are aggressive and invade to the underlying periosteum and bone relatively early due to the proximity of the underlying bony structures.

HPV-induced SIL, SCCIS, and invasive SCC of mucosal sites, especially the transitional epithelium of the cervix and anus, are increasingly more common in HIV disease. SIL and SCCIS occur in spite of maintained or improved immune function as a result of HAART. Invasive cervical SCC has been added to the list of AIDS-defining conditions. HPV-induced neoplasms may well be diagnosed more frequently as the life span of persons with HIV diseases is increased with treatment.

HPV DNA is 2 to 3 times as frequent in cervicovaginal lavage specimens and almost 15 times as common in anal swab specimens from HIV-infected women than in those from HIV-seronegative women.[266,267] HIV-seropositive women are 5 times as likely as HIV-seronegative women to have vulvovaginal condyloma and oral or anal SIL.[268,269] The increased prevalence of HPV-induced lesions in HIV disease probably is related to deficient cell-mediated immunity rather than impaired

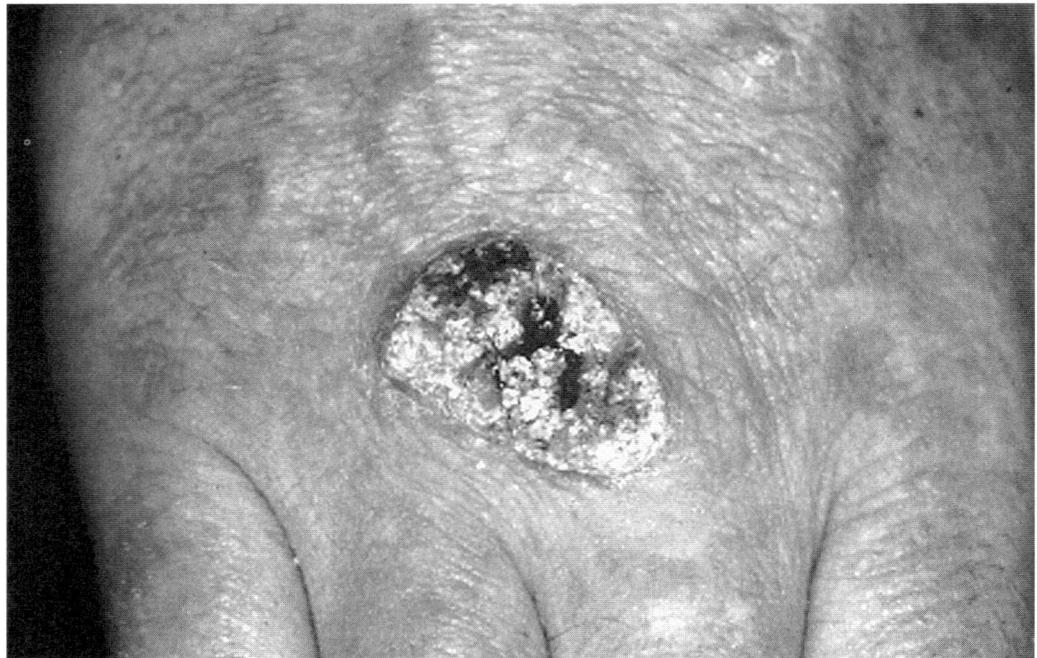

FIGURE 33. HPV infection: Verruca. A huge verruca vulgaris (common wart) on the dorsal hand of a renal transplant recipient. Color representation of Figure 33 follows page 708.

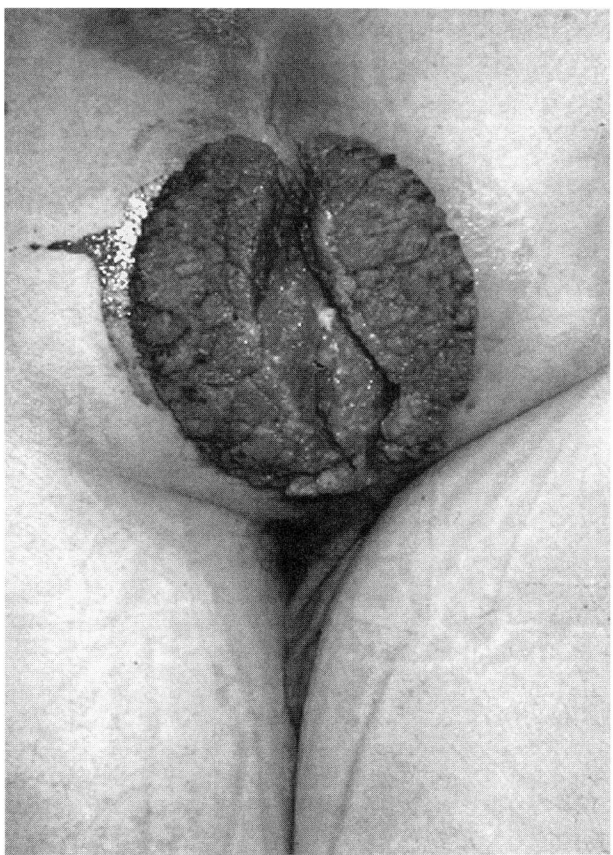

FIGURE 34. HPV infection: Gigantic perineal/perianal condyloma. A huge, enlarging, recurring tumor in the perineum of a 21-year-old liver transplant recipient. The lesion was removed surgically, regrowing less rapidly, after immunosuppressive therapy was reduced. Color representation of Figure 34 follows page 708.

specific antibody formation.[270] Increased HPV replication of the more oncogenic HPV types occurs with more advanced immunosuppression.[271]

In a study of the association between anal SIL, HPV infection, and immunosuppression among HIV-seropositive and HIV-seronegative homosexual men, anal HPV DNA was detected in 55% of HIV-seropositive and 23% of HIV-seronegative men by Southern transfer hybridization and in 92 and 78% by polymerase chain reaction (PCR).[272] Anal SIL was noted in 26% of HIV-seropositive and in 8% of HIV-seronegative men; high-grade SIL was noted in 4% of HIV-seropositive and in 0.5% of HIV-seronegative men. Among HIV-infected men, anal SIL, detection of specific anal HPV types, and detection of high levels of anal HPV DNA were all associated with advanced HIV disease. The risk of anal SIL among HIV-

seropositive men with CD4 cell counts $<500/\mu l$ was increased 2.9-fold over that of HIV-seropositive men with CD4 counts $>500/\mu l$.

In a study of HPV infection in HIV-seronegative and seropositive women, HPV DNA was detected in 83% of the seropositive and 62% of the seronegative women.[266] Twenty percent of seropositive women and 3% of seronegative women had persistent infections with HPV-16-associated viral types (16, 31, 33, 35, 58) or HPV-18-associated types (18 or 45), which are most strongly associated with cervical cancer. HIV-infected women were noted to have a high rate of persistent HPV infections with the types of HPV that are strongly associated with the development of high-grade SIL and invasive SCC.

The degree of immunosuppression correlates with the presence of HPV DNA, extent of HPV infection, and potential for malignant transformation, with individuals with CD4 cell counts $<200/\mu l$ being at greatest risk.[273] The potential for malignant transformation varies considerably according to the type and site of HPV-infected epithelium, being greatest for the transitional epithelium of the cervix and anus, lesser for vulvar epithelium, and least for the epithelium of the male genitalia, perineum, inguinal folds, and perianal regions. The immune mechanisms underlying the increased rates of anogenital neoplasia in HIV-infected individuals[274–277] are not well understood but are thought to be related to high prevalence of HPV infection, impairment of cellular immunity, activation of latent HPV replication, and local suppression of cytokine production.

With advanced immunodeficiency, low- or high-grade SIL caused most frequently by HPV-16 and -18 can arise on epithelium of the cervix, external genitalia, perineum, anus, oropharynx, or keratinized skin, especially the nail beds. Although differentiation from condyloma cannot be made on clinical grounds alone, SIL and SCCIS often present as multiple smooth, pink to skin color to tan/brown macules or papules, which may form confluent cobblestoned, well-demarcated plaques. Lesions are usually multifocal but may be unifocal. In some cases, multiple foci of epithelial erosion occur and concomitant herpetic infection must be ruled out. Massive HPV-induced lesions (older terminology "giant condyloma of Buschke–Löwenstein") have a much greater chance of showing foci of SIL, SCCIS, or invasive SCC histologically. Extragenital HPV-induced SIL, SCCIS, and invasive SCC also occur in keratinized skin such as the nail beds.[278,279]

Oropharyngeal HPV-induced lesions resemble anogenital condyloma, and are pink or white in color but

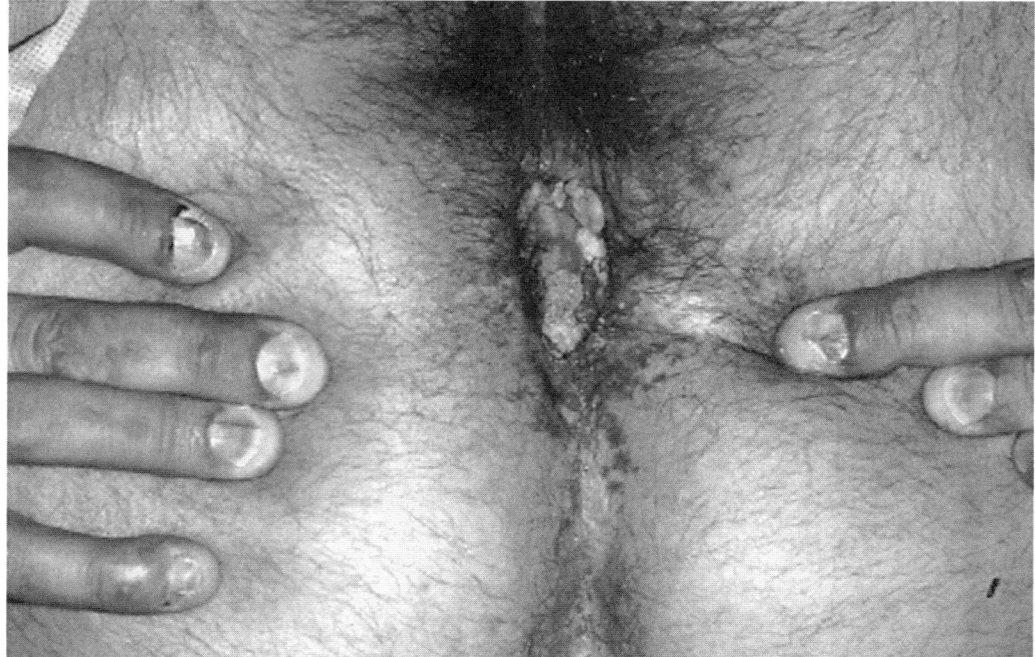

FIGURE 35. HPV infection: Squamous cell carcinoma. Squamous cell carcinoma *in situ* is seen on the perineum and periungual area on multiple fingers of male with advanced HIV disease. The squamous cell carcinoma on the left thumb and index finger had become invasive, necessitating amputation. Color representation of Figure 35 follows page 708.

never the tan to brown color of some genital lesions. Extensive intraoral condyloma acuminatum (oral florid papillomatosis) presents as multiple large plaques, analogous to anogenital giant condylomata acuminata of Buschke–Löwenstein, and can also transform to verrucous carcinoma.

The diagnosis of verrucae and condylomata usually is made on clinical findings. In individuals with advanced HIV disease, biopsy of suspected HPV infections of the anogenital region is recommended because of the high prevalence of SIL and SCCIS. Exfoliative cytology is very effective at detecting SIL or SCC of the cervix and also may be helpful with anal involvement.[280] External anogenital SIL and SCCIS, unlike cervical and anal lesions, cannot be screened for by the Papanicolaou test using exfoliative cytology. Lesional biopsy specimens should be obtained from several sites, especially in individuals at higher risk for malignant transformation. In most patients, histologic findings are relatively uniform at multiple biopsy sites, ranging from low- or high-grade SIL to SCCIS. At one point, however, invasive SCC usually is unifocal within a field of SCCIS. A larger nodule within a field of SIL or SCCIS should be excised to rule out invasive SCC. Over the course of HPV-

induced neoplasia, multiple invasive SCCs may arise at various anogenital epithelial sites.

The natural history of external anogenital HPV-induced neoplasia is probably similar to that of condyloma acuminatum. Prolonged, severe immunodeficiency provides the necessary milieu for the emergence of HPV-induced anogenital neoplasia. The incidence of transformation of SCCIS to invasive SCC appears to be low. The relative risk for HPV-related anal SCC is much higher in HIV-infected than in non-HIV-infected homosexual men and is more likely in advanced HIV disease.

Invasive cervical SCC is an AIDS-defining condition; however, a documented increase in incidence has not yet been reported. Cervical and anal neoplasias are likely to become more common manifestations of HIV disease as patients with profound immunodeficiency, who previously would have succumbed to opportunistic infections, are now surviving for extended periods because of increasingly effective antiretroviral prophylaxis of opportunistic infections and newer antimicrobial therapies. External anogenital SIL and SCCIS also may become more common in long-term survivors of HIV disease.

Low- or high-grade SIL of the external anogenital

epithelium can be treated by several methods: topical chemotherapy (5% 5-fluorouracil or 5% imiquimod cream, especially for extensive multifocal lesions); surgical excision of single or several lesions; or focal destruction of lesions by cryosurgery, electrosurgery, or laser surgery. Unlike topical 5-fluorouracil or imiquimod, surgical modalities treat only clinically detectable lesions and not subclinical infection. For minimally invasive SCC arising in an area of external anogenital SCCIS, surgical excision is recommended with adequate borders around the lesion. The role for adjunctive radiotherapy has not yet been defined, nor has the use of combined modality therapy with external beam radiotherapy plus chemotherapy.

Individuals with documented external anogenital high-grade SIL or SCCIS should be followed by periodic follow-up examinations (every 3 to 4 months), noting the appearance of new lesions at these sites or an enlarging nodule or ulcerated site; biopsy of these sites is recommended. In that HPV-induced neoplasia may extend to the cervix and/or anus, direct examination by speculum and anoscope also should be performed; samples for cytology should be obtained using a cervical brush and cytofix solution.

The number of individuals with SIL/SCCIS of the external anogenital region is expected to grow with the increasing numbers of long-term survivors of HIV disease. The prevalence of HPV infection of nonkeratinized and mucous epithelium in HIV-infected homosexual men is high, up to 20% having clinically detectable anogenital condylomata and up to 50% having infection detectable by cytologic smears of anal canal mucosa.[281] HPV types 6 and 11, which in the normal host infect the anogenital epithelium and are of low oncogenic risk, can be demonstrated in extensive verrucae on the hands and feet and in invasive anogenital SCC of HIV-infected men. Similarly, HIV-infected women have rates of cervical dysplasia five to ten times higher than non-HIV-infected women. Thus, in one group of HIV-infected women, three-fourths had vulvar HPV infection, two-thirds of these having condylomata and one-third vulvar intraepithelial neoplasia.[282,283]

Verrucae occurring in HIV-infected individuals usually are asymptomatic, the most common complaint being a cosmetic one. Warts on the plantar aspect of the foot can become large and painful. Verruca vulgaris and verruca plantaris appear as well-demarcated keratotic papules or nodules, usually with multiple tiny red-brown dots representing thrombosed capillaries; palmar and plantar warts characteristically interrupt the normal dermatoglyphics. They may be very numerous and confluent, giving the appearance of a mosaic. Verruca plana appears as a well-demarcated, flat-topped papule, which lacks the dots seen in other types of verrucae.[284,285] When present in the beard area, hundreds of flat warts may be present. All types of verrucae may have a linear arrangement due to koebnerization or autoinoculation.

Condylomata acuminata are usually asymptomatic, although voluminous lesions may be painful and bleed.[286] Pain associated with HPV infection most commonly is due to therapy, but also is sensed in voluminous verrucous masses. Condylomata appear as well-demarcated papular or nodular lesions arising anywhere on the anogenital, vaginal, cervical, rectal, or oropharyngeal epithelium. Lesions may be very numerous and become confluent. The prevalence of cervical intraepithelial neoplasia is increased in HIV disease; however, prevalence of invasive SCC of the cervix is not certain at this time.[283] Intraepithelial neoplasia involving the anogenital skin also is increased; there may be an increase in the incidence of invasive SCC as well.[287,288]

The diagnosis of verrucae and condylomata usually is made on a clinical basis. Acetowhitening, the appearance of white micropapules or macules after the application of 5% acetic acid (white vinegar) to the anogenital epithelium, can be helpful in defining the extent of HPV infection. The diagnosis of intraepithelial neoplasia and invasive SCC can be made only histologically and atypical lesions should be biopsied. The differential diagnoses of verrucae include molluscum contagiosum, various benign and malignant epidermal neoplasms, KS deep mycotic lesions hematogenously disseminated to the skin, and bacillary angiomatosis. The differential diagnoses of condylomata acuminata include various benign and malignant mucocutaneous neoplasms, condylomata lata of secondary syphilis, and molluscum contagiosum.

Efficacy of treatment of verruca vulgaris and condyloma acuminatum in HIV disease varies with the degree of immunocompromise. In patients with early disease, these lesions should be managed as in the normal host. In patients with advanced HIV-induced immunodeficiency, complete eradication of benign HPV-induced lesions is unlikely, and aggressive treatment such as laser surgery is contraindicated.[289] Cytologic smears and/or lesional biospies should be obtained to monitor the evolution from cytologic atypia to intraepithelial neoplasia or invasive SCC.

5.4. Sexually Transmitted Diseases

Genital ulcer disease caused by syphilis, genital herpes, and chancroid has been associated with increased transmission of HIV.

Syphilis

Coinfection with *Treponema pallidum* and HIV interacts at several levels.[290,291] Both are sexually transmitted diseases and may be acquired from a dually infected sex partner. Primary syphilis (characterized by chancres or ulcers on the vulva, cervix, penis, anus, rectum, or oropharynx) facilitates acquisition of HIV because of the break in the integrity of the epithelium. Syphilis occurring in more advanced HIV disease can present with highly atypical findings (Table 6): the immunologic defects associated with HIV infection may block the appearance of the usual antibody response to *T. pallidum* infection, so that false-negative serologic tests for syphilis can be observed in the face of active and even progressive infection; the clinical manifestations of the disease may be altered, the response to therapy decreased; the duration of the stages of syphilis may be greatly telescoped, all due to the immunocompromised state of the host.

Shortly after *T. pallidum* penetrates the intact mucous membrane or abraded skin, it spreads via lymphatics and the systemic circulation. Healing of the primary chancre, control of the infection in draining lymph nodes, the development of lesions at metastatic sites, the rapidity with which normally late manifestations of syphilis develop, the clinical manifestations of the infection, and the response to antimicrobial therapy are all influenced by the

TABLE 6. Variations in Syphilis Occurring in HIV Disease

Clinical findings	Increased severity of clinical findings and more rapid clinical course
	Atypical primary syphilis: giant chancres
	Atypical findings in secondary syphilis: cervical lymphadenopathy, lues maligna (secondary syphilis with vasculitis manifested by fever, malaise, headache, and nodules, indurated plaques ± hyperkeratosis and/or ulceration, sclerosis)
	Early onset of tertiary syphilis
	Greater likelihood of ocular (retrobulbar optic neuritis) and neurologic disease
Course of syphilis	Rapid progression to tertiary disease within the first year of infection, i.e., meningovascular syphilis
Serologic response to syphilis	Limited, delayed, or absent antibody responses to syphilis with repeatedly negative reagin and treponemal antibody testing in serum and CSF
Response to therapy	Greater likelihood of treatment failure
	Relapse without reexposure despite "adequate" treatment

degree of immunocompromise present. Thus, the greater the immunocompromise, the slower the healing, the greater the organism burden, the more common systemic spread, the more rapid the development of "late-stage" disease, and the more likely the failure of conventional antimicrobial therapy, particularly if bacteriostatic drugs such as doxycycline or erythromycin are utilized.

The majority of HIV-infected persons who acquire syphilis have the expected clinical course of disease and the expected serologic findings in serum and cerebrospinal fluid, and they respond to the recommended therapeutic regimens. In a small percentage, however, the clinical manifestations, clinical course, serologic response, and response to antibiotic treatment are unusual, especially with moderate to advanced HIV-induced immunodeficiency. In that unusual clinical presentations or failures of treatment are reported as single or a few cases, the percentage of cases of syphilis in HIV-infected patients with an unusual clinical course of disease is unknown. An inadequate immune response to *T. pallidum* is considered to cause the various abnormalities in the course of syphilis in the HIV-infected patient.[292,293]

Not uncommonly, however, the extent of ulceration is far greater than normally seen, and multiple chancres may be observed.[294] Instead of healing within 3–6 weeks, persistent ulceration may occur.

Any of the clinical manifestations of secondary syphilis that occur in the normal host may occur in the HIV patients, with the most important mucocutaneous abnormalities outlined in Table 6.[295] In addition, lues maligna, a rare form of secondary syphilis characterized by pleomorphic skin lesions including pustules, nodules, ulcers, and a necrotizing vasculitis, may be seen.[296] However, the most notable aspect of syphilis in the HIV-infected individual is the rapid progression of disease, such that normally late sequelae of *T. pallidum* infection may be observed less than 6 months after primary infection, even in the face of normally adequate therapy.[297,298]

In general, HIV-infected individuals with syphilis have a higher incidence of systemic symptoms, simultaneous multiorgan involvement, and atypical rashes and are particularly prone to the development of neurosyphilis and uveitis. Because of the malignant course of the disease and the unreliability of traditional diagnostic and therapeutic approaches, both an aggressive biopsy and treatment program are essential.

The cornerstone of the diagnostic approach to syphilis in non-HIV-infected individuals is serologic testing.[299,300] However, since both false-positive and false-negative results occur in HIV-infected persons,[301,302] such testing must be supplemented by biopsies of suspicious lesions.[303] Fluid from such lesions should be examined immediately by dark-field microscopy. In addition, specific immunofluorescent or immunoperoxidase staining of the pathologic specimens can lead to the definitive diagnosis. The differential diagnosis of syphilitic cutaneous lesions is quite broad, depending on the particular lesion being considered. Thus, the differential diagnosis of primary lesions of syphilis includes mainly herpes simplex infection, chancroid, and bacterially infected genital lesions of any cause, although under appropriate epidemiologic conditions such entities as donovanosis, lymphogranuloma venereum, mycobacterial infection, and tularemia must be considered. The differential diagnosis of secondary syphilis includes drug eruption (e.g., captopril), pityriasis rosea, infectious exanthems, infectious mononucleosis, tinea corporis, tinea versicolor, scabies, "id" reaction, condylomata acuminata, acute guttate psoriasis, and lichen planus. The differential diagnosis of tertiary syphilis includes lymphoma, tuberculosis, sarcoidosis, and deep fungal infections.

In HIV-infected patients with syphilis, two rules of clinical management apply: (1) disseminated infection, particularly with central nervous system involvement, should be assumed, and therapy appropriate for neurosyphilis prescribed in any patient coinfected with syphilis and HIV who has evidence of compromised immune function, regardless of the apparent clinical stage of syphilis observed; and (2) close follow-up, with repeated clinical, serologic, and cerebrospinal fluid examinations, is necessary, as even the best of regimens will sometimes fail in HIV-infected patients with significant immunocompromise.[304,305]

In a report of syphilis and HIV infection, 23% of individuals who presented with syphilis were concurrently HIV-infected.[306] The clinical presentation of syphilis in patients with HIV infection differs from that of patients without HIV infection in that patients with HIV infection present more often in the secondary stage (53% vs. 33%) and those with secondary syphilis are more likely to have chancres (43% vs. 15%).

HIV testing is advised for all sexually active patients with syphilis. Although uncommon, seronegative primary and secondary syphilis have been reported in HIV-infected individuals.[301] Nearly all HIV-infected individuals with symptomatic neurosyphilis have positive syphilis serologies.[302] Normally, treponemal tests remain positive throughout life. However, 7% of asymptomatic HIV-infected patients with a history of syphilis and 38% of those with symptomatic HIV infection with a history of syphilis have been reported to lose reactivity of treponemal tests.[307]

Neurosyphilis should be considered in the differential diagnosis of neurologic disease in HIV-infected persons. When clinical findings suggest syphilis but serologic tests are negative or confusing, alternative tests such as biopsy of lesions, dark-field examination, and direct fluorescent antibody staining of lesion material should be used.

In comparison with HIV-seronegative individuals, HIV-infected patients who have early syphilis may be at increased risk for neurologic complications and may have higher rates of treatment failure with currently recommended regimens. The magnitude of these risks, although not defined precisely, is probably minimal. No treatment regimens for syphilis are demonstrably more effective in preventing neurosyphilis in HIV-infected patients than the syphilis regimens recommended for HIV-seronegative individuals. Careful follow-up after therapy is essential. The current Centers for Disease Control recommendations for treating early syphilis appear adequate for most patients, whether or not HIV infection is present.[308,309]

Penicillin regimens should be used whenever possible for all stages of syphilis in HIV: benzathine penicillin G, 2.4 million units IM, as for HIV-seronegative individuals. Some authorities advise cerebrospinal fluid examination and/or treatment with a regimen appropriate for neurosyphilis for all patients coinfected with syphilis and HIV, regardless of the clinical stage of syphilis. Patients should be followed clinically and with quantitative nontreponemal serologic tests (VDRL, RPR) at 1, 2, 3, 6, 9, and 12 months after treatment. Patients with early syphilis whose titers increase or fail to decrease fourfold within 6 months should undergo cerebrospinal fluid examination and be re-treated. In such patients, cerebrospinal fluid abnormalities could be due to HIV-related infection, neurosyphilis, or both.[290] In that *T. pallidum* may persist in the central nervous system of the HIV-infected patient in spite of adequate antibiotic treatment, the possibility of chronic maintenance treatment, analogous to secondary prophylaxis of cryptococcal meningitis, has been raised.

5.5. Parasitic Infestations

Protozoan infections are among the most common opportunistic infections in HIV disease; mucocutaneous involvement, however, is uncommon.[310]

5.5.1. Extrapulmonary Pneumocystosis

Pneumocystis carinii is a common opportunistic pathogen in untreated HIV disease with CD4 counts <200 cells/μl, most commonly causing pneumonia. Extrapulmonary pneumocystosis can be rare in the initial presentation of HIV infection, manifested by unilateral or bilateral polyploid masses, and may be accompanied by loss of hearing. Similar lesions may occur at the tympanic membrane, middle ear, and mastoid air cells, associated with retrograde spread via the eustachian tube. Gangrene of the foot has been reported in a patient with widespread pneumocystosis; microemboli containing *P. carinii* were present in smaller arterioles and capillaries within necrotic skin of the toes. Widespread violaceous papules and nodules arising on the torso, arms, and legs, resembling KS have been reported.[311]

5.5.2. Strongyloidiasis

Once infected with *Strongyloides stercoralis*, the organism persists in the host via autoinoculation. In the compromised host with underlying conditions such as chronic infection (HIV disease, tuberculosis, lepromatous leprosy), neoplasms (lymphoma, leukemia), or organ transplantation, the number of filariform larvae can increase tremendously resulting in disseminated strongyloidiasis or hyperinfection syndrome.

Disseminated strongyloidiasis should be considered in the following circumstances[312]: (1) eosinophilia may be modest or absent; systemic corticosteroids and host debilitation may suppress this characteristic finding; the presence of eosinophilia should initiate a vigorous search for parasites; (2) unexplained and/or persistent bacteremia with enteric organisms despite administration of appropriate antibiotics; (3) serious infection (pneumonia, meningitis, bacteremia) from a suspected intra-abdominal source; (4) nonspecific gastrointestinal symptoms (abdominal pain and distention, diarrhea, nausea, and vomiting); (5) nonspecific pulmonary symptoms and signs (cough, wheezing, hemoptysis, transient interstitial infiltrates); (6) concurrent infection or prior therapy for other intestinal parasites; and (7) history of residence or travel to an endemic area even many years previously.

Clinically, periumbilical purpura is suggestive of disseminated strongyloidiasis. The ecchymoses are said to resemble multiple thumbprints on the abdominal wall, radiating from the umbilicus to the flanks and lower extremities.[313] Fine petechiae are also seen having a reticulated pattern of linear and serpiginous purpuric streaks.[314]

5.5.3. Leishmaniasis

Following asymptomatic or symptomatic primary infection, *Leishmania* often remains latent in the reticulo-

endothelial system. Subclinical *Leishmania* infection is common in Mediterranean countries; 5–15% of adults in parts of Italy have a positive leishmanin skin test.[315] In previously infected individuals, antigen-specific T cells and NK cells interact with parasitized phagocytes in an equilibrium such that only a very low level of replication of *Leishmania* occurs. In HIV-infected persons, the equilibrium is lost.[316] As immunodeficiency progresses, the protozoans may escape confinement by immune surveillance and cause visceral leishmaniasis (VL) (kala-azar). Reactivated leishmaniasis also occurs in organ transplant recipients.

Coinfection with HIV and *Leishmania* has been reported in >700 patients living in the Mediterranean basin, with the greatest number in Spain. In southern Europe, 50% of adult VL cases have occurred in HIV-infected individuals; 1.5 to 9% of HIV-infected individuals have either newly acquired or reactivated VL. More than 400 cases of coinfection with HIV and *Leishmania* have been reported from Spain, 85% in injecting drug users (IDU). Person-to-person transmission of *Leishmania* as well as HIV has been suggested in IDUs. VL may be the presenting manifestation of HIV disease. The course of persons who harbor *Leishmania* as well as HIV remains poorly defined. Coinfection of HIV and *Leishmania* in other sites of endemic leishmaniasis such as Kenya, Sudan, India, and Brazil is not well defined.

No characteristic skin lesions have been described in HIV disease. Normal skin also may be parasitized. Cutaneous leishmaniasis (CL) usually represents primary infection presenting in multiple crusted papulonodules in sites exposed to insect vectors, disseminated nodules,[317] as well as an erythrodermic and dermatomyositislike eruption.[318] A generalized psoriasiform eruption has been reported in a patient with VL.[319] Leishmaniasis also can present at sites of HSV or VZV infection or of KS in HIV-infected individuals with CL or VL.[320] Digital necrosis has been reported associated with leishmanial vasculitis.[321]

Diagnosis may be confirmed by demonstration of *Leishmania* on lesional skin biopsy and/or bone marrow aspiration. Leishmanial serology is often negative. The incidence of relapse of visceral leishmaniasis is high in HIV-infected individuals.

5.5.4. American Trypanosomiasis (Chagas' Disease)

American trypanosomiasis can reactivate in patients with cardiomyopathy treated with cardiac transplantation, and present with lesions resembling soft-tissue infection.[322–326] In patients who have received transplants, trypanosomiasis reactivates with clinical presentation of fever, heart failure, and soft-tissue infection on the trunk and/or lower extremities. *Trypanosoma cruzi* can be detected on lesional skin biopsy.

5.5.5. Acanthamoebiasis

Acanthamoebae are free-living amoebae, which can enter the upper respiratory tract, disseminate hematogenously, and cause encephalitis and disseminated cutaneous lesions in advanced HIV disease. Cutaneous lesions appear as initially erythematous dermal to subcutaneous papules and/or nodules that suppurate, forming abscesses and ulcerations.[327–330] Acanthamoebic cysts and trophozoites, which resemble macrophages, can be visualized in lesional biopsy specimens with PAS or Gomori's methenamine silver stain and immunofluorescence techniques.[331] A leukocytoclastic vasculitis can also occur. The organism can be isolated on culture of a biopsy specimen.

5.5.6. Prototothecosis

Prototheca species are algae, found in water, sewage, soil, and trees; *Prototheca wickerhamii* is the only one capable of infecting humans. Human infection occurs at sites of traumatic inoculation, producing localized infection in the olecranon bursa in the normal host. In the compromised host, lesions occur at site of inoculation and have a widely varied clinical appearance, ranging from papules, vesicles, ulcers, or verrucous plaques.[332,333] Disseminated infection can occur following localized cutaneous lesions such as insect bite in the compromised host.[334] *Prototheca* can be identified by PAS and silver stains of lesional biopsy specimens and isolated on Sabouraud's dextrose agar.

5.6. Arthropod Infestations

Crusted (Norwegian) Scabies

Crusted or hyperkeratotic or Norwegian scabies occurs in compromised hosts. Currently in the United States, HIV disease is the most common associated immunocompromised state; crusted scabies also occurs in leprosy (the original report of crusted or Norwegian scabies was in lepers from Norway), Down's syndrome, transplant recipients,[335] chronic lymphocytic leukemia,[336] adult T-cell leukemia,[337] solid tumors, and vasculitis.

In obtunded or compromised individuals, pruritus may be diminished or absent in crusted scabies. Scabetic infestation can be severe, with millions of mites infesting the skin, presenting as a hyperkeratotic dermatitis but resembling atopic erythroderma, psoriasis vulgaris, keratoderma blennorrhagicum, keratosis follicularis (Darier's disease), or seborrheic dermatitis (in infants).[338] Thickly crusted plaques occur on the ears, buttocks, and extensor surfaces of the extremities, palms, and soles. Heavy infestation occurs around the nails with nail dystrophy and subungual and periungual scale-crust.[339,340] Scabetic infestation, which usually spares the head and neck in adults, can be generalized. *Staphylococcus aureus* and gram-negative superinfection can occur, which has been complicated by septicemia and death.[295,341,342] Because of the number of organisms in crusted scabies, recurrences are common and hospital epidemics may occur.

Use of potent topical corticosteroids for such previously diagnosed pruritic conditions may mask the presence of scabetic infestation. Eradication of the infestation is difficult because of the number of organisms. Topical treatment with gamma benzene hexachloride, permethrin lotion, or 10% sulfur ointment is effective; total-body application is required. Keratolytic agents are needed to debride hyperkeratotic areas in conjunction with debridement of involved nails. Orally administered ivermectin has been reported to be effective in scabies.[343,344]

6. Diagnostic Aspects of Skin Infections in the Compromised Patient

In the immunocompetent patient, the gross appearance of a skin lesion is an important aspect of diagnosis. By contrast, the clinical value of the gross appearance of a cutaneous lesion in a compromised person is likely to be limited, for two reasons: First, in compromised patients, the variety of organisms that may cause infection is substantially greater than in immunocompetent persons. Second, in compromised patients, the inflammatory response to infection may be altered. It is noteworthy that a cutaneous lesion results not only from the invading pathogen itself but also from the inflammatory response of the body to the microbe. Thus, with an impaired inflammatory response, the prediction is that the diagnostic usefulness of gross appearance would be suboptimal. For these two reasons, it is key to realize that the differential diagnosis of a particular skin lesion in a compromised patient is extensive. This point is emphasized in Table 1, which indicates that no gross morphology of a skin lesion in a compromised person is pathognomonic for a single etiologic agent. All these considerations stress the value of skin biopsy in establishing a diagnosis.

The approach to biopsy of a cutaneous lesion suspected to be infectious should include two considerations: (1) the most rapid and sensitive methods for detecting microbes both histologically and immunologically should be used and (2) appropriate cultures and stains should be obtained to optimize the chance of identifying the pathogen. Biopsy, if possible, should be a generous wedge excision. Half the tissue is then sent for histopathologic evaluation by routine methods and also by special stains for fungi, mycobacteria, and bacteria. The other half is sent to the microbiology laboratory for culture for aerobic and anaerobic bacteria, mycobacteria (at 25°C and 37°C), and fungi, as well as for Gram's, acid-fast, modified acid-fast, and direct fungal stains of touch preparations or ground material.

In conclusion, to take advantage of the skin as an early warning system of serious infection in compromised patients, both physician and patient should search for cutaneous and subcutaneous lesions. Unexplained skin lesions then should be evaluated by biopsy for culture and histologic examination.

7. Inflammatory Disorders that Simulate Infection in the Compromised Host

Several inflammatory cutaneous disorders occur in the compromised host that can be mistaken for infections. These disorders are treated with anti-inflammatory agents such as corticosteroids. Antibiotics are ineffective and surgical debridement contraindicated.

7.1. HIV-Associated Eosinophilic Folliculitis

Eosinophilic folliculitis is a relatively common pruritic eruption of sterile, pruritic papules and pustules on the face, trunk, and extremities.[345] The eruption occurs nearly exclusively in HIV disease, presenting in either advanced HIV disease or during immune reconstitution following initiation of HAART.[346] Histologically, neutrophilic and eosinophilic infiltration of the hair follicles is observed. A clinical entity (Ofuji's disease) with identical histology occurs in non-HIV-infected individuals. Clinically, however, nonfollicular pustules, coalescing plaques, and urticarial lesions are seen. A 1- to 2-week course of oral corticosteroids is highly effective in providing symptomatic relief of eosinophilic folliculitis. Agents used for long-term suppression of eosinophilic folliculitis

included topical corticosteroid preparations, ultraviolet B phototherapy, and oral agents such as isotretinoin and itraconazole.

7.2. Neutrophilic Dermatoses

7.3.1. Acute Febrile Neutrophilic Dermatosis (Sweet's Syndrome)

Acute febrile neutrophilic dermatosis presents as painful inflammatory plaques often accompanied by fever, arthralgia, and peripheral leukocytosis (neutrophilia). Multiple lesions arise acutely, are tender and/or painful, and occur most commonly on the face, neck, arms, and legs. Systemic symptoms of fever, headache, arthralgia, and malaise often accompany the cutaneous manifestations. Sweet's syndrome occurs as a paraneoplastic reaction pattern as well as associated with various infections and inflammatory disorders (parainflammatory). Approximately 10% of cases of Sweet's syndrome are associated with malignancies, i.e., preleukemias (myelodysplastic syndrome), leukemias, and solid tumors. When associated with an underlying malignancy, Sweet's syndrome persists for months or years. The treatment of choice for Sweet's syndrome is prednisone 70 mg tapered over 1 to 2 weeks. Lesions recur unless the underlying malignancy is effectively treated.

7.3.2. Pyoderma Gangrenosum

Pyoderma gangrenosum is a rapidly evolving, idiopathic, chronic, and severely disabling skin disease, characterized by sudden occurrence of single or multiple inflammatory plaques. Approximately half of cases are idiopathic, while others are associated with paraproteinemia, myeloma, and leukemia. Surgical debridement is contraindicated, for any trauma to the skin of these individuals can cause the lesions to flare at that site. The most effective agents for management are oral corticosteroids and cyclosporin A.

8. Conclusion

To take advantage of the skin as an early warning system of serious infection in compromised patients, both the physician and the patient should routinely search for cutaneous and subcutaneous lesions. Unexplained skin lesions should be evaluated by biopsy for culture and histologic examination.

Bibliography

Bowden RA, Ljungman P, Paya CV (eds): *Transplantation Infections.* Lippincott-Raven, Philadelphia, 1998.

Grossman ME: *Cutaneous Infections in the Immunocompromised Host.* William & Wilkins, Baltimore, 1995.

References

1. Wolfson JS, Sober AJ, Rubin RH: Dermatologic manifestations of infection in the compromised host. *Annu Rev Med* **34:**205–217, 1983.

2. Wolfson JS, Sober AJ, Rubin RH: Dermatologic manifestations of infections in immunocompromised patients. *Medicine* **64:**115–133, 1985.

3. Dennis JE, Rhodes KH, Cooney DR, Roberts GD: Nosocomial *Rhizopus* infection (zygomycosis) in children. *J Pediatr* **96:**824–828, 1980.

4. Grossman ME, Fithian EC, Behrens C, Bissinger J, Fracaro M, Neu HC: Primary cutaneous aspergillosis in six leukemic children. *J Am Acad Dermatol* **12:**313–318, 1985.

5. Cone LA, Woodard DR, Byrd RG, Schulz K, Kopp SM, Schlievert PM: A recalcitrant, erythematous, desquamating disorder associated with toxin-producing staphylococci in patients with AIDS. *J Infect Dis* **165:**638–643, 1992.

6. Cupps TR, Cotton DJ, Schooley RT, Fauci AS: Facial erysipelas in the immunocompromised host. Report of two cases. *Arch Dermatol* **117:**47–49, 1981.

7. Hewitt WD, Farrar WE: Bacteremia and ecthyma caused by *Streptococcus pyogenes* in a patient with acquired immunodeficiency syndrome. *Am J Med Sci* **295:**52–54, 1988.

8. Janssen F, Zelinsky-Gurung A, Caumes E, Decazes JM: Group A streptococcal cellulitis-adenitis in a patient with acquired immunodeficiency syndrome. *J Am Acad Dermatol* **24:**363–365, 1999.

9. Mascini EM, Jansze M, Schellekens JFP, *et al*: Invasive group A streptococcal disease in The Netherlands: Evidence for a protective role of anti-exotoxin A antibodies. *J Infect Dis* **181:**631–638, 2000.

10. Patel M, Ahrens JC, Moyer DV, DiNubile MJ: Pneumococcal soft-tissue infections: A problem deserving more recognition [see comments]. *Clin Infect Dis* **19:**149–151, 1994.

11. Rodriguez Barradas MC, Musher DM, Hamill RJ, Dowell M, Bagwell JT, Sanders CV: Unusual manifestations of pneumococcal infection in human immunodeficiency virus-infected individuals: The past revisited. *Clin Infect Dis* **14:**192–199, 1992.

12. Jaruratanasirikul S, Kalnauwakul S, Lekhakula A: Traumatic wound infection due to *Bacillus cereus* in an immunocompromised patient: A case report. *Southeast Asian J Trop Med Public Health* **18:**112–114, 1987.

13. Khavari PA, Bolognia JL, Eisen R, Edberg SC, Grimshaw SC, Shapiro PE: Periodic acid-Schiff-positive organisms in primary cutaneous *Bacillus cereus* infection. Case report and an investigation of the periodic acid-Schiff staining properties of bacteria. *Arch Dermatol* **127:**543–546, 1991.

14. Henrickson KJ, Shenep JL, Flynn PM, Pui CH: Primary cutaneous *Bacillus cereus* infection in neutropenic children. *Lancet* **1:**601–603, 1989.

15. Jerdan MS, Shapiro RS, Smith NB, Virshup DM, Hood AF: Cuta-

neous manifestations of Corynebacterium group JK sepsis. *J Am Acad Dermatol* **16:**444–447, 1987.

16. Chapman RA, Van Slyck EJ, Madhavan T: Skin lesions associated with *E. coli* sepsis in a patient with acute leukemia. *Henry Ford Hosp Med J* **28:**47–48, 1980.

17. Dreizen S, McCredie KB, Bodey GP, Keating MJ: Unusual mucocutaneous infections in immunosuppressed patients with leukemia—Expansion of an earlier study. *Postgrad Med* **79:**287–294, 1986.

18. Livingston W, Grossman ME, Garvey G: Hemorrhagic bullae in association with *Enterobacter cloacae* septicemia. *J Am Acad Dermatol* **27:**637–638, 1992.

19. DiGioia RA, Kane JG, Parker RH: Crepitant cellulitis and myonecrosis caused by *Klebsiella.* *JAMA* **237:**2097–2098, 1977.

20. Bagel J, Grossman ME: Hemorrhagic bullae associated with *Morganella morganii* septicemia. *J Am Acad Dermatol* **12:**575–576, 1985.

21. Bonner MJ, Meharg JG Jr: Primary cellulitis due to *Serratia marcescens.* *JAMA* **250:**2348–2349, 1985.

22. Hartmann F, Gheorghiu T, Leupold H, Baer F, Diehl V: *Serratia* infections in patients with neutropenia. *Klin Wochenschr* **69:**491–494, 1991.

23. Blumenthal NC, Sood UR, Aronson PJ, Hashimoto K: Facial ulcerations in an immunocompromised patient. Ecthyma gangrenosum. *Arch Dermatol* **126:**529–532, 1990.

24. Huminer D, Siegman-Igra Y, Morduchowicz G, Pitlik SD: Ecthyma gangrenosum without bacteremia. Report of six cases and review of the literature. *Arch Intern Med* **147:**299–301, 1987.

25. Sangeorzan JA, Bradley SF, Kauffman CA: Cutaneous manifestations of *Pseudomonas* infection in the acquired immunodeficiency syndrome [letter]. *Arch Dermatol* **126:**832–833, 1990.

26. Collini FJ, Spees EK, Munster A, Dufresne C, Millan J: Ecthyma gangrenosum in a kidney transplant recipient with Pseudomonas septicemia. *Am J Med* **80:**729–734, 1986.

27. Bagel J, Grossman ME: Subcutaneous nodules in *Pseudomonas* sepsis. *Am J Med* **80:**528–529, 1986.

28. Pham BN, Aractingi S, Dombret H, Arlet G, Hunault M, Degos L: *Xanthomonas* (formerly *Pseudomonas*) *maltophilia*—Induced cellulitis in a neutropenic patient [letter] [published erratum appears in *Arch Dermatol* **128**(10):1378, 1992]. *Arch Dermatol* **128:**702–704, 1992.

29. Francis YF, Richman S, Hussain S, Schwartz J: Aeromonas hydrophila infection: Ecthyma gangrenosum with aplastic anemia. *NY State J Med* **82:**1461–1464, 1982.

30. Gold WL, Salit IE: *Aeromonas hydrophila* infections of skin and soft tissue: Report of 11 cases and review. *Clin Infect Dis* **16:**69–74, 1993.

31. Bhawan J, Gellis S, Ucci A, Chang TW: Vesiculobullous lesions caused by cytomegalovirus infection in an immunocompromised adult. *J Am Acad Dermatol* **11:**743–747, 1984.

32. Antinori S, Galimberti L, Tadini GL, *et al*: Tuberculosis cutis miliaris disseminata due to multidrug-resistant *Mycobacterium tuberculosis* in AIDS patients. *Eur J Clin Microbiol Infect Dis* **14:**911–914, 1995.

33. Libraty DH, Byrd TF: Cutaneous miliary tuberculosis in the AIDS era: Case report and review [see comments]. *Clin Infect Dis* **23:**706–710, 1996.

34. Hudson CP, Wood R, O'Keefe EA: Cutaneous miliary tuberculosis in the AIDS era [letter; comment]. *Clin Infect Dis* **25:**1484, 1997.

35. Stack RJ, Bickley LK, Coppel IG: Miliary tuberculosis presenting

as skin lesions in a patient with acquired immunodeficiency syndrome. *J Am Acad Dermatol* **23:**1031–1035, 1990.

36. Rohatgi PK, Palazzolo JV, Saini NB: Acute miliary tuberculosis of the skin in acquired immunodeficiency syndrome [see comments]. *J Am Acad Dermatol* **26:**356–359, 1992.

37. Patel R, Roberts GD, Keating MR, Paya CV: Infections due to nontuberculous mycobacteria in kidney, heart, and liver transplant recipients. *Clin Infect Dis* **19:**263–273, 1994.

38. Stellbrink HJ, Koperski K, Albrecht H, Greten H: *Mycobacterium kansasii* infection limited to skin and lymph node in a patient with AIDS. *Clin Exp Dermatol* **15:**457–458, 1990.

39. Nandwani R, Shanson DC, Fisher M, Nelson MR, Gazzard BG: *Mycobacterium kansasii* scalp abscesses in an AIDS patient [letter]. *J Infect* **31:**79–80, 1995.

40. Gombert ME, Goldstein EJ, Corrado ML, Stein AJ, Butt KM: Disseminated *Mycobacterium marinum* infection after renal transplantation. *Ann Intern Med* **94:**486–487, 1981.

41. Enzenauer RJ, McKoy J, Vincent D, Gates R: Disseminated cutaneous and synovial *Mycobacterium marinum* infection in a patient with systemic lupus erythematosus. *South Med J* **83:**471–474, 1990.

42. Lambertus MW, Mathisen GE: *Mycobacterium marinum* infection in a patient with cryptosporidiosis and the acquired immunodeficiency syndrome. *Cutis* **42:**38–40, 1988.

43. Rusconi S, Gori A, Vago L, Marchetti G, Franzetti F: Cutaneous infection caused by *Mycobacterium gordonae* in a human immunodeficiency virus-infected patient receiving antimycobacterial treatment [letter] [see comments]. *Clin Infect Dis* **25:**1490–1491, 1997.

44. Claydon EJ, Coker RJ, Harris JR: *Mycobacterium malmoense* infection in HIV positive patients [see comments]. *J Infect* **23:**191–194, 1991.

45. Sack JB: Disseminated infection due to *Mycobacterium fortuitum* in a patient with AIDS. *Rev Infect Dis* **12:**961–963, 1990.

46. Shafer RW, Sierra MF: *Mycobacterium xenopi, Mycobacterium fortuitum, Mycobacterium kansasii,* and other nontuberculous mycobacteria in an area of endemicity for AIDS. *Clin Infect Dis* **15:**161–162, 1992.

47. Wallace RJ Jr, Brown BA, Onyi GO: Skin, soft tissue, and bone infections due to *Mycobacterium chelonae chelonae*: Importance of prior corticosteroid therapy, frequency of disseminated infections, and resistance to oral antimicrobials other than clarithromycin. *J Infect Dis* **166:**405–412, 1992.

48. Jopp-McKay AG, Randell P: Sporotrichoid cutaneous infection due to *Mycobacterium chelonei* in a renal transplant patient. *Australas J Dermatol* **31:**105–109, 1990.

49. Swetter SM, Kindel SE, Smoller BR: Cutaneous nodules of *Mycobacterium chelonae* in an immunosuppressed patient with pre-existing pulmonary colonization. *J Am Acad Dermatol* **28:**352–355, 1993.

50. Friedman BF, Edwards D, Kirkpatrick CH: *Mycobacterium avium-intracellulare*: Cutaneous presentations of disseminated disease. *Am J Med* **85:**257–263, 1988.

51. Esteban J, Gorgolas M, Fernandez-Guerrero ML, Soriano F: Localized cutaneous infection caused by *Mycobacterium avium* complex in an AIDS patient. *Clin Exp Dermatol* **21:**230–231, 1996.

52. Meadows JR, Carter R, Katner HP: Cutaneous *Mycobacterium avium* complex infection at an intramuscular injection site in a patient with AIDS. *Clin Infect Dis* **24:**1273–1274, 1997.

53. Barbaro DJ, Orcutt VL, Coldiron BM: *Mycobacterium avium-Mycobacterium intracellulare* infection limited to the skin and

lymph nodes in patients with AIDS. *Rev Infect Dis* **11:**625–628, 1989.

54. Rogers PL, Walker RE, Lane HC, *et al*: Disseminated *Mycobacterium haemophilum* infection in two patients with the acquired immunodeficiency syndrome. *Am J Med* **84:**640–642, 1988.

55. Kristjansson M, Bieluch VM, Byeff PD: *Mycobacterium haemophilum* infection in immunocompromised patients: Case report and review of the literature. *Rev Infect Dis* **13:**906–910, 1991.

56. Dever LL, Martin JW, Seaworth B, Jorgensen JH: Varied presentations and responses to treatment of infections caused by *Mycobacterium haemophilum* in patients with AIDS. *Clin Infect Dis* **14:**1195–1200, 1992.

57. Mycobacterium haemophilum infections—New York City metropolitan area, 1990–1991. *Morb Mortal Wkly Rep* **40:**636–637, 643, 1991.

58. Straus WL, Ostroff SM, Jernigan DB, *et al*: Clinical and epidemiologic characteristics of *Mycobacterium haemophilum*, an emerging pathogen in immunocompromised patients. *Ann Intern Med* **120:**118–125, 1994.

59. Lamfers EJ, Bastiaans AH, Mravunac M, Rampen FH: Leprosy in the acquired immunodeficiency syndrome [letter]. *Ann Intern Med* **107:**111–112, 1987.

60. Meeran K: Prevalence of HIV infection among patients with leprosy and tuberculosis in rural Zambia. *Br Med J* **298:**364–365, 1989.

61. Turk JL, Rees RJ: AIDS and leprosy [editorial]. *Lepr Rev* **59:**193–194, 1988.

62. Leonard G, Sangare A, Verdier M, *et al*: Prevalence of HIV infection among patients with leprosy in African countries and Yemen. *J Acquir Immune Defic Syndr* **3:**1109–1113, 1990.

63. Gormus BJ, Murphey-Corb M, Martin LN, *et al*: Interactions between simian immunodeficiency virus and *Mycobacterium leprae* in experimentally inoculated rhesus monkeys. *J Infect Dis* **160:**405–413, 1989.

64. Kennedy C, Lien RA, Stolz E, van Joost T, Naafs B: Leprosy and human immunodeficiency virus infection. A closer look at the lesions. *Int J Dermatol* **29:**139–140, 1990.

65. Wong R, Tappero J, Cockerell CJ: Bacillary angiomatosis and other *Bartonella* species infections. *Semin Cutan Med Surg* **16:**188–199, 1997.

66. Koehler JE, Sanchez MA, Garrido CS, *et al*: Molecular epidemiology of bartonella infections in patients with bacillary angiomatosis-peliosis [see comments]. *N Engl J Med* **337:**1876–1883, 1997.

67. Koehler JE, Glaser CA, Tappero JW: *Rochalimaea henselae* infection. A new zoonosis with the domestic cat as reservoir [see comments]. *JAMA* **271:**531–535, 1994.

68. Schwartz RA, Nychay SG, Janniger CK, Lambert WC: Bacillary angiomatosis: Presentation of six patients, some with unusual features. *Br J Dermatol* **136:**60–65, 1997.

69. Perkocha LA, Geaghan SM, Yen TS, *et al*: Clinical and pathological features of bacillary peliosis hepatis in association with human immunodeficiency virus infection [see comments]. *N Engl J Med* **323:**1581–1586, 1990.

70. Tompkins LS: Of cats, humans, and *Bartonella* [editorial; comment]. *N Engl J Med* **337:**1916–1917, 1997.

71. Nishimoto K, Ohno M: Subcutaneous abscesses caused by *Nocardia brasiliensis* complicated by malignant lymphoma. A survey of cutaneous nocardiosis reported in Japan. *Int J Dermatol* **24:**437–440, 1985.

72. Moeller CA, Burton CSD: Primary lymphocutaneous *Nocardia brasiliensis* infection. *Arch Dermatol* **122:**1180–1182, 1986.

73. Kalb RE, Kaplan MH, Grossman ME: Cutaneous nocardiosis. Case reports and review. *J Am Acad Dermatol* **13:**125–133, 1985.

74. Shapiro PE, Grossman ME: Disseminated *Nocardia asteroides* with pustules. *J Am Acad Dermatol* **20:**889–892, 1989.

75. Christoph I: Pulmonary *Cryptococcus neoformans* and disseminated *Nocardia brasiliensis* in an immunocompromised host. Case report. *NC Med J* **51:**219–220, 1990.

76. Hodohara K, Fujiyama Y, Hiramitu Y, *et al*: Disseminated subcutaneous *Nocardia asteroides* abscesses in a patient after bone marrow transplantation. *Bone Marrow Transplant* **11:**341–343, 1993.

77. Javaly K, Horowitz HW, Wormser GP: Nocardiosis in patients with human immunodeficiency virus infection. Report of 2 cases and review of the literature. *Medicine* **71:**128–138, 1992.

78. Shelkovitz-Shilo I, Feinstein A, Trau H, Kaplan B, Sofer E, Schewach-Millet M: Lymphocutaneous nocardiosis due to *Nocardia asteroides* in a patient with intestinal lymphoma. *Int J Dermatol* **31:**178–179, 1992.

79. Yang LJ, Chan HL, Chen WJ, Kuo TT: Lymphocutaneous nocardiosis caused by *Nocardia caviae*: The first case report from Asia. *J Am Acad Dermatol* **29:**639–641, 1993.

80. Almeida L, Grossman M: Widespread dermatophyte infections that mimic collagen vascular disease. *J Am Acad Dermatol* **23:**855–857, 1990.

81. Elewski BE, Sullivan J: Dermatophytes as opportunistic pathogens. *J Am Acad Dermatol* **30:**1021–1022, 1994.

82. Sequeira M, Burdick AE, Elgart GW, Berman B: New-onset Majocchi's granuloma in two kidney transplant recipients under tacrolimus treatment. *J Am Acad Dermatol* **38:**486–488, 1998.

83. Squeo RF, Beer R, Silvers D, Weitzman I, Grossman M: Invasive *Trichophyton rubrum* resembling blastomycosis infection in the immunocompromised host. *J Am Acad Dermatol* **39:**379–380, 1998.

84. Demidovich CW, Kornfeld BW, Gentry RH, Fitzpatrick JE: Deep dermatophyte infection with chronic draining nodules in an immunocompromised patient. *Cutis* **55:**237–240, 1995.

85. Novick NL, Tapia L, Bottone EJ: Invasive *Trichophyton rubrum* infection in an immunocompromised host. Case report and review of the literature. *Am J Med* **82:**321–325, 1987.

86. Maor MH: Dermatophytosis confined to irradiated skin. A case report [letter]. *Int J Radiat Oncol Biol Phys* **14:**825–826, 1988.

87. Wright DC, Lennox JL, James WD, Oster CN, Tramont EC: Generalized chronic dermatophytosis in patients with human immunodeficiency virus type I infection and CD4 depletion [letter]. *Arch Dermatol* **127:**265–266, 1991.

88. Bufill JA, Lum LG, Caya JG, *et al*: Pityrosporum folliculitis after bone marrow transplantation. Clinical observations in five patients. *Ann Intern Med* **108:**560–563, 1988.

89. Yohn JJ, Lucas J, Camisa C: Malassezia folliculitis in immunocompromised patients. *Cutis* **35:**536–538, 1985.

90. Klotz SA, Drutz DJ, Huppert M, Johnson JE: Pityrosporum folliculitis. Its potential for confusion with skin lesions of systemic candidiasis. *Arch Intern Med* **142:**2126–2129, 1982.

91. Gottlieb MS, Schroff R, Schanker HM, *et al*: *Pneumocystis carinii* pneumonia and mucosal candidiasis in previously healthy homosexual men: Evidence of a new acquired cellular immunodeficiency. *N Engl J Med* **305:**1425–1431, 1981.

92. Imam N, Carpenter CC, Mayer KH, Fisher A, Stein M, Danforth SB: Hierarchical pattern of mucosal candida infections in HIV-seropositive women [see comments]. *Am J Med* **89:**142–146, 1990.

93. Carlin EM, Hannan M, Walsh J, *et al*: Nasopharyngeal flora in HIV seropositive men who have sex with men. *Genitourin Med* **73:**477–480, 1997.

94. McNeil J, Kan V: Oral yeast colonization of HIV-infected outpatients [letter]. *AIDS* **9:**301–302, 1995.

95. Lifson AR, Hilton JF, Westenhouse JL, *et al*: Time from HIV seroconversion to oral candidiasis or hairy leukoplakia among homosexual and bisexual men enrolled in three prospective cohorts. *AIDS* **8:**73–79, 1994.

96. Decker CF, Tiernan R, Paparello SF: Esophageal candidiasis associated with acute infection due to human immunodeficiency virus [letter; comment]. *Clin Infect Dis* **14:**791, 1992.

97. Prose NS: HIV infection in children. *J Am Acad Dermatol* **22:**1223–1231, 1990.

98. Walsh TJ, Gonzalez C, Roilides E, *et al*: Fungemia in children infected with the human immunodeficiency virus: New epidemiologic patterns, emerging pathogens, and improved outcome with antifungal therapy. *Clin Infect Dis* **20:**900–906, 1995.

99. MacPhail LA, Hilton JF, Dodd CL, Greenspan D: Prophylaxis with nystatin pastilles for HIV-associated oral candidiasis. *J Acquir Immune Defic Syndr Hum Retrovirol* **12:**470–476, 1996.

100. Graybill JR, Vazquez J, Darouiche RO, *et al*: Randomized trial of itraconazole oral solution for oropharyngeal candidiasis in HIV/AIDS patients. *Am J Med* **104:**33–39, 1998.

101. Nandwani R, Parnell A, Youle M, *et al*: Use of terbinafine in HIV-positive subjects: Pilot studies in onychomycosis and oral candidiasis. *Br J Dermatol* **134**(Suppl 46)**:**22–24 (discussion 39), 1996.

102. Laguna F, Rodriguez-Tudela JL, Martinez-Suarez JV, *et al*: Patterns of fluconazole susceptibility in isolates from human immunodeficiency virus-infected patients with oropharyngeal candidiasis due to *Candida albicans*. *Clin Infect Dis* **24:**124–130, 1997.

103. Maenza JR, Merz WG, Romagnoli MJ, Keruly JC, Moore RD, Gallant JE: Infection due to fluconazole-resistant *Candida* in patients with AIDS: Prevalence and microbiology. *Clin Infect Dis* **24:**28–34, 1997.

104. Powderly WG, Finkelstein D, Feinberg J, *et al*: A randomized trial comparing fluconazole with clotrimazole troches for the prevention of fungal infections in patients with advanced human immunodeficiency virus infection. NIAID AIDS Clinical Trials Group [see comments]. *N Engl J Med* **332:**700–705, 1995.

105. Schuman P, Capps L, Peng G, *et al*: Weekly fluconazole for the prevention of mucosal candidiasis in women with HIV infection. A randomized, double-blind, placebo-controlled trial. Terry Beirn Community Programs for Clinical Research on AIDS [see comments]. *Ann Intern Med* **126:**689–696, 1997.

106. Supparatpinyo K, Chiewchanvit S, Hirunsri P, Uthammachai C, Nelson KE, Sirisanthana T: *Penicillium marneffei* infection in patients infected with human immunodeficiency virus [see comments]. *Clin Infect Dis* **14:**871–874, 1992.

107. Diamond RD: The growing problem of mycoses in patients infected with the human immunodeficiency virus. *Rev Infect Dis* **13:**480–486, 1991.

108. Wheat LJ, Connolly-Stringfield PA, Baker RL, *et al*: Disseminated histoplasmosis in the acquired immune deficiency syndrome: Clinical findings, diagnosis and treatment, and review of the literature. *Medicine* **69:**361–374, 1990.

109. Rico MJ, Penneys NS: Cutaneous cryptococcosis resembling molluscum contagiosum in a patient with AIDS. *Arch Dermatol* **121:**901–902, 1985.

110. Picon L, Vaillant L, Duong T, *et al*: Cutaneous cryptococcosis

resembling molluscum contagiosum: A first manifestation of AIDS. *Acta Derm Venereol* **69:**365–367, 1989.

111. Cohen PR, Bank DE, Silvers DN, Grossman ME: Cutaneous lesions of disseminated histoplasmosis in human immunodeficiency virus-infected patients. *J Am Acad Dermatol* **23:**422–428, 1990.

112. Wingard JR, Merz WG, Saral R: *Candida tropicalis*: A major pathogen in immunocompromised patients. *Ann Intern Med* **91:** 539–543, 1979.

113. Bodey GP, Luna M: Skin lesions associated with disseminated candidiasis. *JAMA* **229:**1466–1468, 1974.

114. Grossman ME, Silvers DN, Walther RR: Cutaneous manifestations of disseminated candidiasis. *J Am Acad Dermatol* **2:**111–116, 1980.

115. File TM Jr, Marina OA, Flowers FP: Necrotic skin lesions associated with disseminated candidiasis. *Arch Dermatol* **115:**214–215, 1979.

116. Fine JD, Miller JA, Harrist TJ, Haynes HA: Cutaneous lesions in disseminated candidiasis mimicking ecthyma gangrenosum. *Am J Med* **70:**1133–1135, 1981.

117. Benson PM, Roth RR, Hicks CB: Nodular subcutaneous abscesses caused by *Candida tropicalis* [letter]. *J Am Acad Dermatol* **16:** 623–624, 1987.

118. Collignon PJ, Sorrell TC: Disseminated candidiasis: Evidence of a distinctive syndrome in heroin abusers. *Br Med J* **287:**861–862, 1983.

119. Jarowski CI, Fialk MA, Murray HW, *et al*: Fever, rash, and muscle tenderness. A distinctive clinical presentation of disseminated candidiasis. *Arch Intern Med* **138:**544–546, 1978.

120. Marcus J, Grossman ME, Yunakov MJ, Rappaport F: Disseminated candidiasis, *Candida* arthritis, and unilateral skin lesions. *J Am Acad Dermatol* **26:**295–297, 1992.

121. Held JL, Berkowitz RK, Grossman ME: Use of touch preparation for rapid diagnosis of disseminated candidiasis. *J Am Acad Dermatol* **19:**1063–1066, 1988.

122. Currie BP, Casadevall A: Estimation of the prevalence of cryptococcal infection among patients infected with the human immunodeficiency virus in New York City. *Clin Infect Dis* **19:**1029–1033, 1994.

123. Dromer F, Mathoulin S, Dupont B, Laporte A: Epidemiology of cryptococcosis in France: A 9-year survey (1985–1993). French Cryptococcosis Study Group. *Clin Infect Dis* **23:**82–90, 1996.

124. Murakawa GJ, Kerschmann R, Berger T: Cutaneous *Cryptococcus* infection and AIDS. Report of 12 cases and review of the literature. *Arch Dermatol* **132:**545–548, 1996.

125. Anderson DJ, Schmidt C, Goodman J, Pomeroy C: Cryptococcal disease presenting as cellulitis [see comments]. *Clin Infect Dis* **14:**666–672, 1992.

126. Manfredi R, Mazzoni A, Nanetti A, Mastroianni A, Coronado O, Chiodo F: Morphologic features and clinical significance of skin involvement in patients with AIDS-related cryptococcosis. *Acta Derm Venereol* **76:**72–74, 1996.

127. Dimino-Emme L, Gurevitch AW: Cutaneous manifestations of disseminated cryptococcosis. *J Am Acad Dermatol* **32:**844–850, 1995.

128. Hall JC, Brewer JH, Crouch TT, Watson KR: Cryptococcal cellulitis with multiple sites of involvement. *J Am Acad Dermatol* **17:** 329–332, 1987.

129. Goldman M, Pottage JC Jr: Cryptococcal infection of the breast. *Clin Infect Dis* **21:**1166–1169, 1995.

130. Shuttleworth D, Philpot CM, Knight AG: Cutaneous cryptococcosis: Treatment with oral fluconazole. *Br J Dermatol* **120:**683–687, 1989.

131. Cusini M, Cagliani P, Grimalt R, Tadini G, Alessi E, Fasan M: Primary cutaneous cryptococcosis in a patient with the acquired immunodeficiency syndrome [letter]. *Arch Dermatol* **127:**1848–1849, 1991.

132. Antony SA, Antony SJ: Primary cutaneous *Cryptococcus* in non-immunocompromised patients. *Cutis* **56:**96–98, 1995.

133. Hunt SJ, Nagi C, Gross KG, Wong DS, Mathews WC: Primary cutaneous aspergillosis near central venous catheters in patients with the acquired immunodeficiency syndrome. *Arch Dermatol* **128:**1229–1232, 1992.

134. Allo MD, Miller J, Townsend T, Tan C: Primary cutaneous aspergillosis associated with Hickman intravenous catheters. *N Engl J Med* **317:**1105–1108, 1987.

135. Arikan S, Uzun O, Cetinkaya Y, Kocagoz S, Akova M, Unal S: Primary cutaneous aspergillosis in human immunodeficiency virus-infected patients: Two cases and review [see comments]. *Clin Infect Dis* **27:**641–643, 1998.

136. Thakur BK, Bernardi DM, Murali MR, McClain SA, Clark RA: Invasive cutaneous aspergillosis complicating immunosuppressive therapy for recalcitrant pemphigus vulgaris. *J Am Acad Dermatol* **38:**488–490, 1998.

137. Pursell KJ, Telzak EE, Armstrong D: *Aspergillus* species colonization and invasive disease in patients with AIDS. *Clin Infect Dis* **14:**141–148, 1992.

138. Oscherwitz SL, Rinaldi MG: Disseminated sporotrichosis in a patient infected with human immunodeficiency virus [letter]. *Clin Infect Dis* **15:**568–569, 1992.

139. Dong JA, Chren MM, Elewski BE: Bonsai tree: Risk factor for disseminated sporotrichosis. *J Am Acad Dermatol* **33:**839–840, 1995.

140. Heller HM, Fuhrer J: Disseminated sporotrichosis in patients with AIDS: Case report and review of the literature. *AIDS* **5:**1243–1246, 1991.

141. Girmenia C, Iori AP, Boecklin F, *et al*: *Fusarium* infections in patients with severe aplastic anemia: Review and implications for management. *Haematologica* **84:**114–118, 1999.

142. Girmenia C, Arcese W, Micozzi A, Martino P, Bianco P, Morace G: Onychomycosis as a possible origin of disseminated *Fusarium solani* infection in a patient with severe aplastic anemia [letter]. *Clin Infect Dis* **14:**1167, 1992.

143. June CH, Beatty PG, Shulman HM, Rinaldi MG: Disseminated *Fusarium moniliforme* infection after allogeneic marrow transplantation. *South Med J* **79:**513–515, 1986.

144. Chariyalertsak S, Sirisanthana T, Supparatpinyo K, Praparattanapan J, Nelson KE: Case-control study of risk factors for *Penicillium marneffei* infection in human immunodeficiency virus-infected patients in northern Thailand. *Clin Infect Dis* **24:**1080–1086, 1997.

145. Duong TA: Infection due to *Penicillium marneffei*, an emerging pathogen: Review of 155 reported cases [see comments]. *Clin Infect Dis* **23:**125–130, 1996.

146. Supparatpinyo K, Khamwan C, Baosoung V, Nelson KE, Sirisanthana T: Disseminated *Penicillium marneffei* infection in southeast Asia. *Lancet* **344:**110–113, 1994.

147. Butka BJ, Bennett SR, Johnson AC: Disseminated inoculation blastomycosis in a renal transplant recipient. *Am Rev Respir Dis* **130:**1180–1183, 1984.

148. Pappas PG: Blastomycosis in the immunocompromised patient. *Semin Respir Infect* **12:**243–251, 1997.

149. Pappas PG, Threlkeld MG, Bedsole GD, Cleveland KO, Gelfand MS, Dismukes WE: Blastomycosis in immunocompromised patients. *Medicine* **72:**311–325, 1993.

150. Prichard JG, Sorotzkin RA, James RED: Cutaneous manifestations of disseminated coccidioidomycosis in the acquired immunodeficiency syndrome. *Cutis* **39:**203–205, 1987.

151. Vartivarian SE, Coudron PE, Markowitz SM: Disseminated coccidioidomycosis. Unusual manifestations in a cardiac transplantation patient. *Am J Med* **83:**949–952, 1987.

152. McKinsey DS, Spiegel RA, Hutwagner L, et al: Prospective study of histoplasmosis in patients infected with human immunodeficiency virus: Incidence, risk factors, and pathophysiology. *Clin Infect Dis* **24:**1195–1203, 1997.

153. Carme B, Ngaporo AI, Ngolet A, Ibara JR, Ebikili B: Disseminated African histoplasmosis in a Congolese patient with AIDS. *J Med Vet Mycol* **30:**245–248, 1992.

154. Bellman B, Berman B, Sasken H, Kirsner RS: Cutaneous disseminated histoplasmosis in AIDS patients in south Florida. *Int J Dermatol* **36:**599–603, 1997.

155. Eisig S, Boguslaw B, Cooperband B, Phelan J: Oral manifestations of disseminated histoplasmosis in acquired immunodeficiency syndrome: Report of two cases and review of the literature. *J Oral Maxillofac Surg* **49:**310–313, 1991.

156. Kaplan LJ, Daum RS, Smaron M, McCarthy CA: Severe measles in immunocompromised patients. *JAMA* **267:**1237–1241, 1992.

157. Krasinski K, Borkowsky W: Measles and measles immunity in children infected with human immunodeficiency virus. *JAMA* **261:**2512–2516, 1989.

158. Markowitz LE, Chandler FW, Roldan EO, et al: Fatal measles pneumonia without rash in a child with AIDS. *J Infect Dis* **158:**480–483, 1988.

159. McNutt NS, Kindel S, Lugo J: Cutaneous manifestations of measles in AIDS. *J Cutan Pathol* **19:**315–324, 1992.

160. Horn TD, Hood AF: Cytomegalovirus is predictably present in perineal ulcers from immunosuppressed patients. *Arch Dermatol* **126:**642–644, 1990.

161. Stewart JA, Reef SE, Pellett PE, Corey L, Whitley RJ: Herpesvirus infections in persons infected with human immunodeficiency virus. *Clin Infect Dis* **21**(Suppl 1):S114–S120, 1995.

162. Keiser P, Jockus J, Horton H, Smith JW: Prednisone therapy is not associated with increased risk of herpetic infections in patients infected with human immunodeficiency virus. *Clin Infect Dis* **23:**201–202, 1996.

163. Masessa JM, Grossman ME, Knobler EH, Bank DE: Kaposi's varicelliform eruption in cutaneous T cell lymphoma. *J Am Acad Dermatol* **21:**133–135, 1989.

164. Siegal FP, Lopez C, Hammer GS, et al: Severe acquired immunodeficiency in male homosexuals, manifested by chronic perianal ulcerative herpes simplex lesions. *N Engl J Med* **305:**1439–1444, 1981.

165. Hook EWd, Cannon RO, Nahmias AJ, et al: Herpes simplex virus infection as a risk factor for human immunodeficiency virus infection in heterosexuals [see comments]. *J Infect Dis* **165:**251–255, 1992.

166. Fleming DT, McQuillan GM, Johnson RE, et al: Herpes simplex virus type 2 in the United States, 1976 to 1994 [see comments]. *N Engl J Med* **337:**1105–1111, 1997.

167. Mole L, Ripich S, Margolis D, Holodniy M: The impact of active herpes simplex virus infection on human immunodeficiency virus load. *J Infect Dis* **176:**766–770, 1997.

168. Torres RA, Neaton JD, Wentworth DN, et al: Acyclovir use and survival among human immunodeficiency virus-infected patients with CD4 cell counts of 500/mm³. The Terry Beirn Community Programs for Clinical Research on AIDS (CPCRA). *Clin Infect Dis* **26:**85–90, 1998.

169. Pannuti CS, Cristina M, Finck DS, et al: Asymptomatic perianal shedding of herpes simplex virus in patients with acquired immunodeficiency syndrome. *Arch Dermatol* **133:**180–183, 1997.

170. Burgoyne M, Burke W: Atypical herpes simplex infection in patients with acute myelogenous leukemia recovering from chemotherapy. *J Am Acad Dermatol* **20:**1125–1126, 1989.

171. Kalb RE, Grossman ME: Chronic perianal herpes simplex in immunocompromised hosts. *Am J Med* **80:**486–490, 1986.

172. Toback AC, Grossman ME: Chronic intraoral herpes simplex infection in renal transplant recipients. *Transplant Proc* **18:**966–969, 1986.

173. Grossman ME, Stevens AW, Cohen PR: Brief report: Herpetic geometric glossitis [see comments]. *N Engl J Med* **329:**1859–1860, 1993.

174. Zuretti AR, Schwartz IS: Gangrenous herpetic whitlow in a human immunodeficiency virus-positive patient. *Am J Clin Pathol* **93:**828–830, 1990.

175. Weaver G, Kostman JR: Inoculation herpes simplex virus infections in patients with AIDS: Unusual appearance and location of lesions. *Clin Infect Dis* **22:**141–142, 1996.

176. Norris SA, Kessler HA, Fife KH: Severe, progressive herpetic whitlow caused by an acyclovir-resistant virus in a patient with AIDS [letter]. *J Infect Dis* **157:**209–210, 1988.

177. Genereau T, Lortholary O, Bouchaud O, et al: Herpes simplex esophagitis in patients with AIDS: Report of 34 cases. The Cooperative Study Group on Herpetic Esophagitis in HIV Infection. *Clin Infect Dis* **22:**926–931, 1996.

178. Stone WJ, Scowden EB, Spannuth CL, Lowry SP, Alford RH: Atypical herpesvirus hominis type 2 infection in uremic patients receiving immunosuppressive therapy. *Am J Med* **63:**511–516, 1977.

179. Nahass GT, Mandel MJ, Cook S, Fan W, Leonardi CL: Detection of herpes simplex and varicella-zoster infection from cutaneous lesions in different clinical stages with the polymerase chain reaction. *J Am Acad Dermatol* **32:**730–733, 1995.

180. Schacker T, Hu HL, Koelle DM, et al: Famciclovir for the suppression of symptomatic and asymptomatic herpes simplex virus reactivation in HIV-infected persons. A double-blind, placebo-controlled trial. *Ann Intern Med* **128:**21–28, 1998.

181. Lalezari J, Schacker T, Feinberg J, et al: A randomized, double-blind, placebo-controlled trial of cidofovir gel for the treatment of acyclovir-unresponsive mucocutaneous herpes simplex virus infection in patients with AIDS. *J Infect Dis* **176:**892–898, 1997.

182. Gilden DH, Kleinschmidt-DeMasters BK, LaGuardia JJ, Mahalingam R, Cohrs RJ: Neurologic complications of the reactivation of varicella-zoster virus. *N Engl J Med* **342:**635–645, 2000.

183. McNamara MP, LaCrosse S, Piering WF, Rytel MW: Exogenous reinfection with varicella-zoster virus [letter]. *N Engl J Med* **317:**511, 1987.

184. Rusthoven JJ, Ahlgren P, Elhakim T, et al: Varicella-zoster infection in adult cancer patients. A population study. *Arch Intern Med* **148:**1561–1566, 1988.

185. Kelley R, Mancao M, Lee F, Sawyer M, Nahmias A, Nesheim S: Varicella in children with perinatally acquired human immunodeficiency virus infection. *J Pediatr* **124:**271–273, 1994.

186. Baran J Jr, Khatib R: Recrudescence of initial cutaneous lesions after crusting of chickenpox in an adult with advanced AIDS

suggests prolonged local viral persistence. *Clin Infect Dis* **24:** 741–742, 1997.

187. Leibovitz E, Cooper D, Giurgiutiu D, *et al*: Varicella-zoster virus infection in Romanian children infected with the human immuno-deficiency virus. *Pediatrics* **92:**838–842, 1993.

188. Gershon AA, Mervish N, LaRussa P, *et al*: Varicella-zoster virus infection in children with underlying human immunodeficiency virus infection. *J Infect Dis* **176:**1496–1500, 1997.

189. von Seidlein L, Gillette SG, Bryson Y, *et al*: Frequent recurrence and persistence of varicella-zoster virus infections in children infected with human immunodeficiency virus type 1. *J Pediatr* **128:**52–57, 1996.

190. Gulick RM, Heath-Chiozzi M, Crumpacker CS: Varicella-zoster virus disease in patients with human immunodeficiency virus in-fection [comment]. *Arch Dermatol* **126:**1086–1088, 1990.

191. Colebunders R, Mann JM, Francis H, *et al*: Herpes zoster in African patients: A clinical predictor of human immunodeficiency virus infection. *J Infect Dis* **157:**314–318, 1988.

192. Friedman-Kien AE, Lafleur FL, Gendler E, *et al*: Herpes zoster: A possible early clinical sign for development of acquired immuno-deficiency syndrome in high-risk individuals. *J Am Acad Derma-tol* **14:**1023–1028, 1986.

193. Melbye M, Grossman RJ, Goedert JJ, Eyster ME, Biggar RJ: Risk of AIDS after herpes zoster. *Lancet* **1:**728–731, 1987.

194. Dolin R, Reichman RC, Mazur MH, Whitley RJ: NIH conference. Herpes zoster-varicella infections in immunosuppressed patients. *Ann Intern Med* **89:**375–388, 1978.

195. Feld R, Evans WK, DeBoer G: Herpes zoster in patients with small-cell carcinoma of the lung receiving combined modality treatment. *Ann Intern Med* **93:**282–283, 1980.

196. Hoppenjans WB, Bibler MR, Orme RL, Solinger AM: Prolonged cutaneous herpes zoster in acquired immunodeficiency syndrome [see comments]. *Arch Dermatol* **126:**1048–1050, 1990.

197. Gilson IH, Barnett JH, Conant MA, Laskin OL, Williams J, Jones PG: Disseminated ecthymatous herpes varicella-zoster virus in-fection in patients with acquired immunodeficiency syndrome. *J Am Acad Dermatol* **20:**637–642, 1989.

198. Leibovitz E, Kaul A, Rigaud M, Bebenroth D, Krasinski K, Bor-kowsky W: Chronic varicella zoster in a child infected with human immunodeficiency virus: Case report and review of the literature. *Cutis* **49:**27–31, 1992.

199. LeBoit PE, Limova M, Yen TS, Palefsky JM, White CR Jr, Berger TG: Chronic verrucous varicella-zoster virus infection in patients with the acquired immunodeficiency syndrome (AIDS). Histo-logic and molecular biologic findings. *Am J Dermatopathol* **14:** 1–7, 1992.

200. Alliegro MB, Dorrucci M, Pezzotti P, *et al*: Herpes zoster and progression to AIDS in a cohort of individuals who seroconverted to human immunodeficiency virus. Italian HIV Seroconversion Study. *Clin Infect Dis* **23:**990–995, 1996.

201. Tyndall MW, Nasio J, Agoki E, *et al*: Herpes zoster as the initial presentation of human immunodeficiency virus type 1 infection in Kenya. *Clin Infect Dis* **21:**1035–1037, 1995.

202. Bayu S, Alemayehu W: Clinical profile of herpes zoster ophthal-micus in Ethiopians. *Clin Infect Dis* **24:**1256–1260, 1997.

203. McNulty A, Li Y, Radtke U, *et al*: Herpes zoster and the stage and prognosis of HIV-1 infection. *Genitourin Med* **73:**467–470, 1997.

204. Glesby MJ, Moore RD, Chaisson RE: Herpes zoster in patients with advanced human immunodeficiency virus infection treated with zidovudine. Zidovudine Epidemiology Study Group. *J Infect Dis* **168:**1264–1268, 1993.

205. Buchbinder SP, Katz MH, Hessol NA, *et al*: Herpes zoster and human immunodeficiency virus infection [see comments]. *J Infect Dis* **166:**1153–1156, 1992.

206. Veenstra J, Krol A, van Praag RM, *et al*: Herpes zoster, immuno-logical deterioration and disease progression in HIV-1 infection. *AIDS* **9:**1153–1158, 1995.

207. Tsao H, Tahan SR, Johnson RA: Chronic varicella zoster infection mimicking a basal cell carcinoma in an AIDS patient. *J Am Acad Dermatol* **36:**831–833, 1997.

208. Nikkels AF, Rentier B, Pierard GE: Chronic varicella-zoster virus skin lesions in patients with human immunodeficiency virus are related to decreased expression of gE and gB. *J Infect Dis* **176:** 261–264, 1997.

209. Batisse D, Eliaszewicz M, Zazoun L, Baudrimont M, Pialoux G, Dupont B: Acute retinal necrosis in the course of AIDS: Study of 26 cases [see comments]. *AIDS* **10:**55–60, 1996.

210. Galindez OA, Sabates NR, Whitacre MM, Sabates FN: Rapidly progressive outer retinal necrosis caused by varicella zoster virus in a patient infected with human immunodeficiency virus. *Clin Infect Dis* **22:**149–151, 1996.

211. Manian FA, Kindred M, Fulling KH: Chronic varicella-zoster virus myelitis without cutaneous eruption in a patient with AIDS: Report of a fatal case. *Clin Infect Dis* **21:**986–988, 1995.

212. Lionnet F, Pulik M, Genet P, *et al*: Myelitis due to varicella-zoster virus in two patients with AIDS: Successful treatment with acyclovir. *Clin Infect Dis* **22:**138–140, 1996.

213. Kestelyn P, Stevens AM, Bakkers E, Rouvroy D, Van de Perre P: Severe herpes zoster ophthalmicus in young African adults: A marker for HTLV-III seropositivity. *Br J Ophthalmol* **71:**806–809, 1987.

214. Shuler JD, Engstrom RE Jr., Holland GN: External ocular dis-ease and anterior segment disorders associated with AIDS. *Int Ophthalmol Clin* **29:**98–104, 1989.

215. Seiff SR, Margolis T, Graham SH, O'Donnell JJ: Use of intra-venous acyclovir for treatment of herpes zoster ophthalmicus in patients at risk for AIDS. *Ann Ophthalmol* **20:**480–482, 1988.

216. Rostad SW, Olson K, McDougall J, Shaw CM, Alvord EC Jr: Transsynaptic spread of varicella zoster virus through the visual system: A mechanism of viral dissemination in the central nervous system [published erratum appears in *Hum Pathol* **20:**820, 1989]. *Hum Pathol* **20:**174–179, 1989.

217. Ryder JW, Croen K, Kleinschmidt-DeMasters BK, Ostrove JM, Straus SE, Cohn DL: Progressive encephalitis three months after resolution of cutaneous zoster in a patient with AIDS. *Ann Neurol* **19:**182–188, 1986.

218. Cohen PR, Grossman ME: Clinical features of human immuno-deficiency virus-associated disseminated herpes zoster virus infection—A review of the literature. *Clin Exp Dermatol* **14:**273–276, 1989.

219. Disler RS, Dover JS: Chronic localized herpes zoster in the ac-quired immunodeficiency syndrome [letter] [see comments]. *Arch Dermatol* **126:**1105–1106, 1990.

220. Janier M, Hillion B, Baccard M, *et al*: Chronic varicella zoster infection in acquired immunodeficiency syndrome [letter]. *J Am Acad Dermatol* **18:**584–585, 1988.

221. Langenberg A, Yen TS, LeBoit PE: Granulomatous vasculitis occurring after cutaneous herpes zoster despite absence of viral genome. *J Am Acad Dermatol* **24:**429–433, 1991.

222. Safrin S, Berger TG, Gilson I, *et al*: Foscarnet therapy in five patients with AIDS and acyclovir-resistant varicella-zoster virus infection. *Ann Intern Med* **115:**19–21, 1991.

223. Safrin S, Assaykeen T, Follansbee S, Mills J: Foscarnet therapy for acyclovir-resistant mucocutaneous herpes simplex virus infection in 26 AIDS patients: Preliminary data. *J Infect Dis* **161:**1078–1084, 1990.

224. Greenspan D, Greenspan JS, Hearst NG, *et al:* Relation of oral hairy leukoplakia to infection with the human immunodeficiency virus and the risk of developing AIDS. *J Infect Dis* **155:**475–481, 1987.

225. Greenspan D, Greenspan JS: Oral manifestations of HIV infection. *Dermatol Clin* **9:**517–522, 1991.

226. Webster-Cyriaque J, Edwards RH, Quinlivan EB, Patton L, Wohl D, Raab-Traub N: Epstein–Barr virus and human herpesvirus 8 prevalence in human immunodeficiency virus-associated oral mucosal lesions. *J Infect Dis* **175:**1324–1332, 1997.

227. Triantos D, Porter SR, Scully C, Teo CG: Oral hairy leukoplakia: Clinicopathologic features, pathogenesis, diagnosis, and clinical significance. *Clin Infect Dis* **25:**1392–1396, 1997.

228. Shiboski CH, Hilton JF, Neuhaus JM, Canchola A, Greenspan D: Human immunodeficiency virus-related oral manifestations and gender. A longitudinal analysis. The University of California, San Francisco Oral AIDS Center Epidemiology Collaborative Group. *Arch Intern Med* **156:**2249–2254, 1996.

229. Shiboski CH, Hilton JF, Greenspan D, *et al:* HIV-related oral manifestations in two cohorts of women in San Francisco. *J Acquir Immune Defic Syndr* **7:**964–971, 1994.

230. Greenspan D, Greenspan JS, Overby G, *et al:* Risk factors for rapid progression from hairy leukoplakia to AIDS: A nested case-control study. *J Acquir Immune Defic Syndr* **4:**652–658, 1991.

231. Husak R, Garbe C, Orfanos CE: Oral hairy leukoplakia in 71 HIV-seropositive patients: Clinical symptoms, relation to immunologic status, and prognostic significance [see comments]. *J Am Acad Dermatol* **35:**928–934, 1996.

232. Alessi E, Berti E, Cusini M, *et al:* Oral hairy leukoplakia [see comments]. *J Am Acad Dermatol* **22:**79–86, 1990.

233. McGregor JM, Yu CC, Lu QL, Cotter FE, Levison DA, MacDonald DM: Posttransplant cutaneous lymphoma. *J Am Acad Dermatol* **29:**549–554, 1993.

234. Gallant JE, Moore RD, Richman DD, Keruly J, Chaisson RE: Incidence and natural history of cytomegalovirus disease in patients with advanced human immunodeficiency virus disease treated with zidovudine. The Zidovudine Epidemiology Study Group [see comments]. *J Infect Dis* **166:**1223–1227, 1992.

235. Heinemann MH: Characteristics of cytomegalovirus retinitis in patients with acquired immunodeficiency syndrome. *Am J Med* **92:**12S–16S, 1992.

236. Pariser RJ: Histologically specific skin lesions in disseminated cytomegalovirus infection. *J Am Acad Dermatol* **9:**937–946, 1983.

237. Patterson JW, Broecker AH, Kornstein MJ, Mills AS: Cutaneous cytomegalovirus infection in a liver transplant patient. Diagnosis by *in situ* DNA hybridization. *Am J Dermatopathol* **10:**524–530, 1988.

238. Curtis JL, Egbert BM: Cutaneous cytomegalovirus vasculitis: An unusual clinical presentation of a common opportunistic pathogen. *Hum Pathol* **13:**1138–1141, 1982.

239. Walker JD, Chesney TM: Cytomegalovirus infection of the skin. *Am J Dermatopathol* **4:**263–265, 1982.

240. Swanson S, Feldman PS: Cytomegalovirus infection initially diagnosed by skin biopsy. *Am J Clin Pathol* **87:**113–116, 1987.

241. Chang Y, Cesarman E, Pessin MS, *et al:* Identification of herpesvirus-like DNA sequences in AIDS-associated Kaposi's sarcoma. *Science* **266:**1865–1869, 1994.

242. Moore PS, Chang Y: Detection of herpesvirus-like DNA sequences in Kaposi's sarcoma in patients with and those without HIV infection. *N Engl J Med* **332:**1181–1185, 1995.

243. Whitby D, Howard MR, Tenant-Flowers M, *et al:* Detection of KS-associated herpesvirus in peripheral blood of HIV-infected individuals and progression to Kaposi's sarcoma. *Lancet* **346:**799–802, 1995.

244. Moore PS, Kingsley LA, Holmberg SD, *et al:* Kaposi's sarcoma associated herpesvirus infection prior to onset of Kaposi's sarcoma. *AIDS* **10:**175–180, 1996.

245. Kedes DH, Operskalski E, Busch M, *et al:* The seroepidemiology of human herpesvirus 8 (Kaposi's sarcoma associated herpesvirus): Distribution of infection in KS risk groups and evidence for sexual transmission. *Nat Med* **2:**918–924, 1996.

246. Gao SJ, Kingsley L, Li M, *et al:* KSHV antibodies among Americans, Italians, and Ugandans with and without Kaposi's sarcoma. *Nat Med* **2:**925–928, 1996.

247. Friedman-Kien AE, Saltzman BR: Clinical manifestations of classical, endemic African, and epidemic AIDS-associated Kaposi's sarcoma. *J Am Acad Dermatol* **22:**1237–1250, 1990.

248. Boudreaux AA, Smith LL, Cosby CD, Bason MM, Tappero JW, Berger TG: Intralesional vinblastine for cutaneous Kaposi's sarcoma associated with acquired immunodeficiency syndrome. A clinical trial to evaluate efficacy and discomfort associated with infection. *J Am Acad Dermatol* **28:**61–65, 1993.

249. Tappero JW, Berger TG, Kaplan LD, Volberding PA, Kahn JO: Cryotherapy for cutaneous Kaposi's sarcoma (KS) associated with acquired immune deficiency syndrome (AIDS): A phase II trial. *J Acquir Immune Defic Syndr* **4:**839–846, 1991.

250. Morfeldt L, Torssander J: Long-term remission of Kaposi's sarcoma following foscarnet treatment in HIV-infected patients. *Scand J Infect Dis* **26:**749–752, 1994.

251. Hammoud Z, Parenti DM, Simon GL: Abatement of cutaneous Kaposi's sarcoma associated with cidofovir treatment [see comments]. *Clin Infect Dis* **26:**1233, 1998.

252. Weinberg JM, Mysliwiec A, Turiansky GW, Redfield R, James WD: Viral folliculitis. Atypical presentations of herpes simplex, herpes zoster, and molluscum contagiosum [see comments]. *Arch Dermatol* **133:**983–986, 1997.

253. Myskowski PL: Molluscum contagiosum. New insights, new directions [editorial; comment]. *Arch Dermatol* **133:**1039–1041, 1997.

254. Pauly CR, Artis WM, Jones HE: Atopic dermatitis, impaired cellular immunity, and molluscum contagiosum. *Arch Dermatol* **114:**391–393, 1978.

255. Ganpule M, Garretts M: Molluscum contagiosum and sarcoidosis: Report of a case. *Br J Dermatol* **85:**587–589, 1971.

256. Rosenberg EW, Yusk JW: Molluscum contagiosum. Eruption following treatment with prednisone and methotrexate. *Arch Dermatol* **101:**439–441, 1970.

257. Redfield RR, James WD, Wright DC, *et al:* Severe molluscum contagiosum infection in a patient with human T cell lymphotrophic (HTLV-III) disease. *J Am Acad Dermatol* **13:**821–824, 1985.

258. Cotton DW, Cooper C, Barrett DF, Leppard BJ: Severe atypical molluscum contagiosum infection in an immunocompromised host. *Br J Dermatol* **116:**871–876, 1987.

259. Koopman RJ, van Merrienboer FC, Vreden SG, Dolmans WM: Molluscum contagiosum; a marker for advanced HIV infection [letter]. *Br J Dermatol* **126:**528–529, 1992.

260. Smith KJ, Skelton HGD, Yeager J, James WD, Wagner KF: Molluscum contagiosum. Ultrastructural evidence for its presence in

skin adjacent to clinical lesions in patients infected with human immunodeficiency virus type 1. Military Medical Consortium for Applied Retroviral Research [see comments]. *Arch Dermatol* **128:** 223–227, 1992.

261. Smith KJ, Yeager J, Skelton H: Molluscum contagiosum: Its clinical, histopathologic, and immunohistochemical spectrum. *Int J Dermatol* **38:**664–672, 1999.

262. Hicks CB, Myers SA, Giner J: Resolution of intractable molluscum contagiosum in a human immunodeficiency virus-infected patient after institution of antiretroviral therapy with ritonavir. *Clin Infect Dis* **24:**1023–1025, 1997.

263. Meadows KP, Tyring SK, Pavia AT, Rallis TM: Resolution of recalcitrant molluscum contagiosum virus lesions in human immunodeficiency virus-infected patients treated with cidofovir [see comments]. *Arch Dermatol* **133:**987–990, 1997.

264. Cohen LM, Tyring SK, Rady P, Callen JP: Human papillomavirus type 11 in multiple squamous cell carcinomas in a patient with subacute cutaneous lupus erythematosus. *J Am Acad Dermatol* **26:** 840–845, 1992.

265. Viac J, Chardonnet Y, Euvrard S, Chignol MC, Thivolet J: Langerhans cells, inflammation markers and human papillomavirus infections in benign and malignant epithelial tumors from transplant recipients. *J Dermatol* **19:**67–77, 1992.

266. Sun XW, Kuhn L, Ellerbrock TV, Chiasson MA, Bush TJ, Wright TC Jr: Human papillomavirus infection in women infected with the human immunodeficiency virus [see comments]. *N Engl J Med* **337:**1343–1349, 1997.

267. Sun XW, Ellerbrock TV, Lungu O, Chiasson MA, Bush TJ, Wright TC Jr: Human papillomavirus infection in human immunodeficiency virus-seropositive women. *Obstet Gynecol* **85:**680–686, 1995.

268. Hillemanns P, Ellerbrock TV, McPhillips S, *et al:* Prevalence of anal human papillomavirus infection and anal cytologic abnormalities in HIV-seropositive women. *AIDS* **10:**1641–1647, 1996.

269. Chiasson MA, Ellerbrock TV, Bush TJ, Sun XW, Wright TC Jr: Increased prevalence of vulvovaginal condyloma and vulvar intraepithelial neoplasia in women infected with the human immunodeficiency virus. *Obstet Gynecol* **89:**690–694, 1997.

270. Hagensee ME, Kiviat N, Critchlow CW, *et al:* Seroprevalence of human papillomavirus types 6 and 16 capsid antibodies in homosexual men. *J Infect Dis* **176:**625–631, 1997.

271. Palefsky JM, Holly EA, Ralston ML, Jay N: Prevalence and risk factors for human papillomavirus infection of the anal canal in human immunodeficiency virus (HIV)-positive and HIV- negative homosexual men. *J Infect Dis* **177:**361–367, 1998.

272. Kiviat NB, Critchlow CW, Holmes KK, *et al:* Association of anal dysplasia and human papillomavirus with immunosuppression and HIV infection among homosexual men. *AIDS* **7:**43–49, 1993.

273. Chopra KF, Tyring SK: The impact of the human immunodeficiency virus on the human papillomavirus epidemic. *Arch Dermatol* **133:**629–633, 1997.

274. Palefsky JM, Shiboski S, Moss A: Risk factors for anal human papillomavirus infection and anal cytologic abnormalities in HIV-positive and HIV-negative homosexual men. *J Acquir Immune Defic Syndr* **7:**599–606, 1994.

275. Maiman M, Fruchter RG, Guy L, Cuthill S, Levine P, Serur E: Human immunodeficiency virus infection and invasive cervical carcinoma. *Cancer* **71:**402–406, 1993.

276. Palefsky JM: Anal human papillomavirus infection and anal cancer in HIV-positive individuals: An emerging problem [editorial]. *AIDS* **8:**283–295, 1994.

277. Carter PS, de Ruiter A, Whatrup C, *et al:* Human immunodeficiency virus infection and genital warts as risk factors for anal intraepithelial neoplasia in homosexual men. *Br J Surg* **82:**473–474, 1995.

278. Fader DJ, Stoler MH, Anderson TF: Isolated extragenital HPV-thirties-group-positive bowenoid papulosis in an AIDS patient. *Br J Dermatol* **131:**577–580, 1994.

279. Tosti A, La Placa M, Fanti PA, *et al:* Human papillomavirus type 16-associated periungual squamous cell carcinoma in a patient with acquired immunodeficiency syndrome [letter]. *Acta Derm Venereol* **74:**478–479, 1994.

280. Palefsky JM, Holly EA, Hogeboom CJ, Berry JM, Jay N, Darragh TM: Anal cytology as a screening tool for anal squamous intraepithelial lesions. *J Acquir Immune Defic Syndr Hum Retrovirol* **14:**415–422, 1997.

281. Critchlow CW, Holmes KK, Wood R, *et al:* Association of human immunodeficiency virus and anal human papillomavirus infection among homosexual men. *Arch Intern Med* **152:**1673–1676, 1992.

282. Schrager LK, Friedland GH, Maude D, *et al:* Cervical and vaginal squamous cell abnormalities in women infected with human immunodeficiency virus. *J Acquir Immune Defic Syndr* **2:**570–575, 1989.

283. Feingold AR, Vermund SH, Burk RD, *et al:* Cervical cytologic abnormalities and papillomavirus in women infected with human immunodeficiency virus. *J Acquir Immune Defic Syndr* **3:**896–903, 1990.

284. Berger TG, Sawchuk WS, Leonardi C, Langenberg A, Tappero J, Leboit PE: Epidermodysplasia verruciformis-associated papillomavirus infection complicating human immunodeficiency virus disease. *Br J Dermatol* **124:**79–83, 1991.

285. Prose NS, von Knebel-Doeberitz C, Miller S, Milburn PB, Heilman E: Widespread flat warts associated with human papillomavirus type 5: A cutaneous manifestation of human immunodeficiency virus infection. *J Am Acad Dermatol* **23:**978–981, 1990.

286. Laraque D: Severe anogenital warts in a child with HIV infection [letter]. *N Engl J Med* **320:**1220–1221, 1989.

287. Kiviat N, Rompalo A, Bowden R, *et al:* Anal human papillomavirus infection among human immunodeficiency virus-seropositive and -seronegative men. *J Infect Dis* **162:**358–361, 1990.

288. Ampel NM, Stout ML, Garewal HS, Davis JR: Persistent rectal ulcer associated with human papillomavirus type 33 in a patient with AIDS: Successful treatment with isotretinoin. *Rev Infect Dis* **12:**1004–1007, 1990.

289. Beck DE, Jaso RG, Zajac RA: Surgical management of anal condylomata in the HIV-positive patient. *Dis Colon Rectum* **33:** 180–183, 1990.

290. Musher DM: Syphilis, neurosyphilis, penicillin, and AIDS. *J Infect Dis* **163:**1201–1206, 1991.

291. Hicks CB: Syphilis and HIV infection. *Dermatol Clin* **9:**493–501, 1991.

292. Tramont EC: Syphilis in the AIDS era [editorial]. *N Engl J Med* **316:**1600–1601, 1987.

293. Tramont EC: Syphilis in adults: From Christopher Columbus to Sir Alexander Fleming to AIDS. *Clin Infect Dis* **21:**1361–1369, 1995.

294. Gregory N, Sanchez M, Buchness MR: The spectrum of syphilis in patients with human immunodeficiency virus infection. *J Am Acad Dermatol* **22:**1061–1067, 1990.

295. Glover RA, Piaquadio DJ, Kern S, Cockerell CJ: An unusual presentation of secondary syphilis in a patient with human immunodeficiency virus infection. A case report and review of the literature. *Arch Dermatol* **128:**530–534, 1992.

296. Tosca A, Stavropoulos PG, Hatziolou E, *et al*: Malignant syphilis in HIV-infected patients [see comments]. *Int J Dermatol* **29:**575–578, 1990.

297. Johns DR, Tierney M, Felsenstein D: Alteration in the natural history of neurosyphilis by concurrent infection with the human immunodeficiency virus. *N Engl J Med* **316:**1569–1572, 1987.

298. Holtom PD, Larsen RA, Leal ME, Leedom JM: Prevalence of neurosyphilis in human immunodeficiency virus-infected patients with latent syphilis [see comments]. *Am J Med* **93:**9–12, 1992.

299. Recommendations for diagnosing and treating syphilis in HIV-infected patients. *Morb Mortal Wkly Rep* **37:**600–602, 607–608, 1988.

300. Rompalo AM, Cannon RO, Quinn TC, Hook EWD: Association of biologic false-positive reactions for syphilis with human immunodeficiency virus infection. *J Infect Dis* **165:**1124–1126, 1992.

301. Hicks CB, Benson PM, Lupton GP, Tramont EC: Seronegative secondary syphilis in a patient infected with the human immunodeficiency virus (HIV) with Kaposi sarcoma. A diagnostic dilemma [published erratum appears in *Ann Intern Med* **107:**946, 1987] [see comments]. *Ann Intern Med* **107:**492–495, 1987.

302. Matlow AG, Rachlis AR: Syphilis serology in human immunodeficiency virus-infected patients with symptomatic neurosyphilis: Case report and review. *Rev Infect Dis* **12:**703–707, 1990.

303. Tikjob G, Russel M, Petersen CS, Gerstoft J, Kobayasi T: Seronegative secondary syphilis in a patient with AIDS: Identification of *Treponema pallidum* in biopsy specimen. *J Am Acad Dermatol* **24:**506–508, 1991.

304. Fiumara N: Human immunodeficiency virus infection and syphilis. *J Am Acad Dermatol* **21:**141–142, 1989.

305. Musher DM, Hamill RJ, Baughn RE: Effect of human immunodeficiency virus (HIV) infection on the course of syphilis and on the response to treatment. *Ann Intern Med* **113:**872–881, 1990.

306. Hutchinson CM, Hook EW 3rd, Shepherd M, Verley J, Rompalo AM: Altered clinical presentation of early syphilis in patients with human immunodeficiency virus infection. *Ann Intern Med* **121:**94–100, 1994.

307. Haas JS, Bolan G, Larsen SA, Clement MJ, Bacchetti P, Moss AR: Sensitivity of treponemal tests for detecting prior treated syphilis during human immunodeficiency virus infection [see comments]. *J Infect Dis* **162:**862–866, 1990.

308. 1998 guidelines for treatment of sexually transmitted diseases. Centers for Disease Control and Prevention. *Morb Mortal Wkly Rep* **47:**1–111, 1998.

309. Rolfs RT, Joesoef MR, Hendershot EF, *et al*: A randomized trial of enhanced therapy for early syphilis in patients with and without human immunodeficiency virus infection. The Syphilis and HIV Study Group [see comments]. *N Engl J Med* **337:**307–314, 1997.

310. Curry A, Turner AJ, Lucas S: Opportunistic protozoan infections in human immunodeficiency virus disease: Review highlighting diagnostic and therapeutic aspects [see comments]. *J Clin Pathol* **44:**182–193, 1991.

311. Litwin MA, Williams CM: Cutaneous *Pneumocystis carinii* infection mimicking Kaposi sarcoma. *Ann Intern Med* **117:**48–49, 1992.

312. Scowden EB, Schaffner W, Stone WJ: Overwhelming strongyloidiasis: An unappreciated opportunistic infection. *Medicine* **57:**527–544, 1978.

313. Bank DE, Grossman ME, Kohn SR, Rabinowitz AD: The thumbprint sign: Rapid diagnosis of disseminated strongyloidiasis. *J Am Acad Dermatol* **23:**324–326, 1990.

314. Kalb RE, Grossman ME: Periumbilical purpura in disseminated strongyloidiasis. *JAMA* **256:**1170–1171, 1986.

315. Davidson RN: AIDS and leishmaniasis [editorial]. *Genitourin Med* **73:**237–239, 1997.

316. Desjeux P: Global control and *Leishmania* HIV co-infection. *Clin Dermatol* **17:**317–325, 1999.

317. Da-Cruz AM, Machado ES, Menezes JA, Rutowitsch MS, Coutinho SG: Cellular and humoral immune responses of a patient with American cutaneous leishmaniasis and AIDS. *Trans R Soc Trop Med Hyg* **86:**511–512, 1992.

318. Dauden E, Penas PF, Rios L, *et al*: Leishmaniasis presenting as a dermatomyositis-like eruption in AIDS. *J Am Acad Dermatol* **35:**316–319, 1996.

319. Rubio FA, Robayna G, Herranz P, *et al*: Leishmaniasis presenting as a psoriasiform eruption in AIDS [letter]. *Br J Dermatol* **136:**792–794, 1997.

320. Barrio J, Lecona M, Cosin J, Olalquiaga FJ, Hernanz JM, Soto J: *Leishmania* infection occurring in herpes zoster lesions in an HIV-positive patient [see comments]. *Br J Dermatol* **134:**164–166, 1996.

321. Caballero-Granado FJ, Lopez-Cortes LF, Borderas F, Regordan C: Digital necrosis due to *Leishmania* species infection in a patient with AIDS. *Clin Infect Dis* **26:**198–199, 1998.

322. Stolf NA, Higushi L, Bocchi E, *et al*: Heart transplantation in patients with Chagas' disease cardiomyopathy. *J Heart Transplant* **6:**307–312, 1987.

323. Libow LF, Beltrani VP, Silvers DN, Grossman ME: Post-cardiac transplant reactivation of Chagas' disease diagnosed by skin biopsy. *Cutis* **48:**37–40, 1991.

324. Kirchhoff LV, Gam AA, Gilliam FC: American trypanosomiasis (Chagas' disease) in Central American immigrants. *Am J Med* **82:**915–920, 1987.

325. Kirchhoff LV: Chagas disease. American trypanosomiasis. *Infect Dis Clin North Am* **7:**487–502, 1993.

326. Kirchhoff LV: American trypanosomiasis (Chagas' disease). *Gastroenterol Clin North Am* **25:**517–533, 1996.

327. Murakawa GJ, McCalmont T, Altman J, *et al*: Disseminated acanthamebiasis in patients with AIDS. A report of five cases and a review of the literature. *Arch Dermatol* **131:**1291–1296, 1995.

328. Tan B, Weldon-Linne CM, Rhone DP, Penning CL, Visvesvara GS: Acanthamoeba infection presenting as skin lesions in patients with the acquired immunodeficiency syndrome. *Arch Pathol Lab Med* **117:**1043–1046, 1993.

329. Helton J, Loveless M, White CR Jr: Cutaneous acanthamoeba infection associated with leukocytoclastic vasculitis in an AIDS patient. *Am J Dermatopathol* **15:**146–149, 1993.

330. Chandrasekar PH, Nandi PS, Fairfax MR, Crane LR: Cutaneous infections due to Acanthamoeba in patients with acquired immunodeficiency syndrome. *Arch Intern Med* **157:**569–572, 1997.

331. May LP, Sidhu GS, Buchness MR: Diagnosis of Acanthamoeba infection by cutaneous manifestations in a man seropositive to HIV. *J Am Acad Dermatol* **26:**352–355, 1992.

332. Woolrich A, Koestenblatt E, Don P, Szaniawski W: Cutaneous prototheosis and AIDS. *J Am Acad Dermatol* **31:**920–924, 1994.

333. Goldstein GD, Bhatia P, Kalivas J: Herpetiform prototheosis. *Int J Dermatol* **25:**54–55, 1986.

334. Wirth FA, Passalacqua JA, Kao G: Disseminated cutaneous prototheosis in an immunocompromised host: A case report and literature review. *Cutis* **63:**185–188, 1999.

335. Anolik MA, Rudolph RI: Scabies simulating Darier disease in an immunosuppressed host. *Arch Dermatol* **112:**73–74, 1976.

336. Tibbs CJ, Wilcox DJ: Norwegian scabies and herpes simplex in a patient with chronic lymphatic leukemia and hypogammaglobulinemia [letter]. *Br J Dermatol* **126:**523–524, 1992.

337. Suzumiya J, Sumiyoshi A, Kuroki Y, Inoue S: Crusted (Norwegian) scabies with adult T-cell leukemia. *Arch Dermatol* **121:**903–904, 1985.

338. Donabedian H, Khazan U: Norwegian scabies in a patient with AIDS. *Clin Infect Dis* **14:**162–164, 1992.

339. Portu JJ, Santamaria JM, Zubero Z, Almeida-Llamas MV, Aldamiz-Etxebarria San Sebastian M, Gutierrez AR: Atypical scabies in HIV-positive patients. *J Am Acad Dermatol* **34:**915–917, 1996.

340. Arico M, Noto G, La Rocca E, Pravata G, Bivona A: Localized crusted scabies in the acquired immunodeficiency syndrome. *Clin Exp Dermatol* **17:**339–341, 1992.

341. Hulbert TV, Larsen RA: Hyperkeratotic (Norwegian) scabies with gram-negative bacteremia as the initial presentation of AIDS [letter]. *Clin Infect Dis* **14:**1164–1165, 1992.

342. Glover A, Young L, Goltz AW: Norwegian scabies in acquired immunodeficiency syndrome: Report of a case resulting in death from associated sepsis [letter]. *J Am Acad Dermatol* **16:**396–399, 1987.

343. Meinking TL, Taplin D, Hermida JL, Pardo R, Kerdel FA: The treatment of scabies with ivermectin. *N Engl J Med* **333:**26–30, 1995.

344. Taplin D, Meinking TL: Treatment of HIV-related scabies with emphasis on the efficacy of ivermectin. *Semin Cutan Med Surg* **16:**235–240, 1997.

345. Rosenthal D, LeBoit PE, Klumpp L, Berger TG: Human immunodeficiency virus-associated eosinophilic folliculitis. A unique dermatosis associated with advanced human immunodeficiency virus infection. *Arch Dermatol* **127:**206–209, 1991.

346. Bachmeyer C, Cordier F, Cazier A, Blum L, Mougeot-Martin M: [Eosinophilic folliculitis associated with AIDS after antiretroviral tri-therapy (letter)]. *Presse Med* **28:**2226, 1999.

4

Clinical Approach to the Compromised Host with Fever and Pulmonary Infiltrates

ROBERT H. RUBIN and REGINALD GREENE

1. The Febrile Pneumonitis Syndrome and Its Importance

The immunocompromised patient in whom fever and pneumonitis develop presents a formidable challenge to the clinician for several reasons:

1. First and foremost, a legion of microbial invaders, ranging from common viral and bacterial pathogens to exotic fungal and protozoan agents, can cause pulmonary infection in these patients.[1–9]
2. Dual or sequential infection is common; hence, the clinician must be alert to the possibility that a single diagnosis may not adequately characterize a patient or lead to optimal therapy.[2,3]
3. The differential diagnosis of the febrile pneumonitis syndrome includes not only infection but also a multitude of noninfectious causes of pulmonary inflammation; e.g., radiation lung injury, drug reactions, the underlying neoplasm, pulmonary embolic disease, leukoagglutinin transfusion reactions, pulmonary hemorrhage, atypical pulmonary edema, and alveolar proteinosis.[2,3,5,9–11]

Robert H. Rubin • Division of Infectious Disease, Brigham and Women's Hospital; Center for Experimental Pharmacology and Therapeutics, Harvard–Massachusetts Institute of Technology Division of Health Sciences and Technology; and Department of Medicine, Harvard Medical School, Boston, Massachusetts 02114. Reginald Greene • Chest Division, Radiology Service, Massachusetts General Hospital, and Department of Radiology, Harvard Medical School, Boston, Massachusetts 02114.

Clinical Approach to Infection in the Compromised Host (Fourth Edition), edited by Robert H. Rubin and Lowell S. Young. Kluwer Academic/Plenum Publishers, New York, 2002.

In addition to the broad differential diagnosis that must be considered, the clinician's task is further complicated in many patients by the subtlety of the clinical presentation. The impaired inflammatory response that is characteristic of so many immunocompromised states may greatly alter the clinical presentation of the pulmonary process. Since physical findings, presenting symptoms, radiologic patterns, and even tissue pathology are largely determined by the inflammatory response to the inciting agent, it is axiomatic that all these phenomena can be greatly modified in the compromised host. In particular, the manifestations of microbial invasion can be greatly attenuated until late in the disease process in these patients. As a consequence, the microbial load present at the time of diagnosis is usually significantly greater than in the normal host, thus providing a greater challenge to the clinician prescribing antimicrobial therapy.[3,6,12,13] It is not surprising, then, that the survival of the immunocompromised host with pulmonary infection is determined in large part by the speed with which diagnosis is made and effective therapy instituted. Accordingly, even subtle clinical and radiologic findings must be carefully evaluated. Thus, an unexplained cough in an individual with the acquired immunodeficiency syndrome (AIDS), even in the absence of physical findings or abnormalities on the chest roentgenogram, can represent significant *Pneumocystis carinii* pneumonia, demonstrable by nuclear medicine or computerized tomographic (CT) scan and diagnosable by examination of respiratory secretions. Similarly, an unexplained fever in a leukemic patient with prolonged chemotherapy-induced neutropenia can be due to invasive pulmonary aspergillosis that may be delineated on a CT scan of the chest, in the face of negative conventional chest radiography, and diagnosed by percutaneous needle aspiration biopsy.

Of all the host defense defects, severe granulocytopenia will have the most profound effect in this regard (although high-dose corticosteroids and advanced AIDS can have a similar impact). The incidence of cough and purulent sputum production, rate of development and progression of radiologic findings, occurrence of cavitation, and pleural space involvement are all markedly diminished in patients with profound granulocytopenia. At the same time, the unopposed microbial invasion is proceeding at a far more rapid rate than is normally observed. As a consequence, the effects of the pulmonary process in granulocytopenic patients may be modified.[3,6,12,13] For example, in children with acute leukemia and pneumonia who have absolute granulocyte counts of less than 1000/mm^3, the incidence of positive blood cultures has been reported to be 64%, with the incidence of positive blood cultures falling to 0% in children with leukemia, pneumonia, and absolute granulocyte counts greater than 1000/mm^3. Similarly, although septic shock is a rare occurrence in an immunologically normal patient with pneumonia, it is not unusual in a severely granulocytopenic patient.[3]

The clinical importance of the febrile pneumonitis syndrome is illustrated by the following observations: The lungs are involved in more than 50% of immunocompromised patients who develop febrile complications, with autopsy evidence of pulmonary infection in more than 90% of patients who succumb.[5] As many as 58% of patients with cancer and profound granulocytopenia who die have been shown at autopsy to have clinically unrecognized, and hence inadequately treated, pneumonia.[9,14] Patients with significant lung injury from noninfectious processes, particularly pulmonary infarction, have a high rate of secondary infection, which is often the immediate cause of death. This result is particularly likely if the primary lung injury required prolonged intubation and HEPA filtered air was not being provided.[2] Mortality rates of 35–90% have been reported in immunocompromised patients with pneumonia, with the exact incidence depending on the underlying disease, the severity of the pneumonia at the time of diagnosis, and the degree of host defense impairment.[10,15] For example, Poe et al.[16] have reported mortality rates that approached 100% in immunocompromised patients with pneumonia who demonstrated the following characteristics: a Pao$_2$ <50 mm Hg within 72 hr of admission, corticosteroid administration at the time of presentation, and the need for mechanical ventilation.

The foundation of the approach to the febrile pneu-
monitis syndrome in this patient population is the recognition of the clinical importance of this syndrome, an awareness of the subtlety of its clinical presentation, and the need for more intensive and invasive diagnostic procedures than in the normal host. Despite the obvious difficulties, a variety of clues are available to the clinician in approaching this clinical problem: (1) the clinical and epidemiologic setting in which the pulmonary process is occurring; (2) an understanding of the host defense defects present; (3) the rate of progression of the illness; (4) the pattern of radiologic abnormality produced on chest radiography; and (5) the proper deployment and application of information gained from a series of increasingly invasive diagnostic techniques. A logical approach based on these elements will enable the clinician to arrive rapidly at the appropriate diagnosis. Such rapid diagnosis and appropriate institution of therapy can result, even in this population, in a gratifying rate of clinical response and meaningful patient survival.[2,3]

2. Overview of the Infectious Causes of the Febrile Pneumonitis Syndrome in the Immunocompromised Host

The first concern of the clinician confronted with an immunocompromised patient with fever and pulmonary infiltrates is infection. Overall, 75–90% of episodes are due to infection, with particularly high rates of noninfectious causes of the febrile pneumonitis syndrome occurring in cancer patients and transplant recipients, especially the former (Table 1).[2,3,5,8–11] Most published series devoted to pulmonary infections in immunocompromised patients have emphasized the importance of opportunistic gram-negative, fungal, nocardial, and herpes group viral infections. Such series have been primarily concerned with infections in patients with acute leukemia or other illnesses being treated with intensive immunosuppressive therapy within the hospital environment.[17,18] This emphasis on opportunistic infection is also appropriate in patients with AIDS.[19] However, in all immunocompromised patients, even the AIDS patient, common infectious and noninfectious causes of pneumonitis must be considered as well (Table 1).[1,7–9,20,21]

It also should be emphasized that in some patients who are less intensively immunosuppressed, opportunistic infections are uncommon. The number one cause of respiratory infections in these patients, when they are not undergoing acute, immunosuppressive therapy, are the community-acquired respiratory viruses, most commonly influenza but also including parainfluenza, respiratory

TABLE 1. Etiology of Febrile Pneumonitis Syndrome in 100 Cancer Patients and 51 Renal Transplant Patients at Massachusetts General Hospital[a]

Etiology (Percent)	Number of patients		
	Cancer	Renal transplant	Total
Infectious causes			
Conventional bacterial infection (23.8)	26	10	36
Viral infection (13.2)	11	9	20
Fungal infection (10.6)	10	6	16
Nocardia asteroides (8.6)	5	8	13
Pneumocystis carinii (5.3)	6	2	8
Mycobacterium tuberculosis (0.7)	1	0	1
Mixed infections (9.9)	14	1[b]	15
Total (72.2)	73	36	109
Noninfectious causes			
Pulmonary emboli (7.9)	3	9	12
Recurrent tumor (5.3)	8	0	8
Radiation pneumonitis (4.6)	7	0	7
Pulmonary edema (4.6)	1	6	7
Drug-induced pneumonitis (3.3)	5	0	5
Leukoagglutinin reaction (1.3)	2	0	2
Pulmonary hemorrhage (0.7)	1	0	1
Total (27.8)	27	15	42

[a]Data taken and modified from Rubin[3] and Ramsey *et al.*[2]

[b]This one case was that of an aspiration pneumonia from whom mixed oropharyngeal floras were grown from a transtracheal aspirate. In addition, 23 renal transplant patients with a primary pulmonary process developed superinfection.

syncytial virus, adenoviruses, and even rhinoviruses. In the setting of communitywide respiratory virus activity, immunosuppressed patients are at particular risk, with higher rates of viral pneumonia and bacterial (and even fungal) superinfection than what is observed in the general population.[22–28] Similarly, *Streptococcus pneumoniae* is the single most common bacterial infection in the cancer, organ transplant, and AIDS populations, especially those that acquire their infections in the community.[2–4,20,21]

Illustrative Case 1

A 42-year-old male cardiac transplant patient was admitted with a 4-hr history of fever, rigors, pleurisy, purulent sputum production, and shortness of breath. He had received a heart transplant for a congestive cardiomyopathy 3 years previously. Since then, he had had no episodes of rejection and was currently maintained on cyclosporine 250 mg, azathioprine 100 mg, and prednisone 10 mg/day. One week prior to admission, he developed an upper respiratory infection characterized by low-grade fever, malaise, anorexia, myalgias, and nonproductive cough. A number of individuals at work, his wife, and one of his children had similar illnesses. He had appeared to be getting somewhat better, when he was awakened early on the morning of admission with a shaking

chill, pleuritic chest pain, an increased cough now productive of purulent blood-tinged sputum, and shortness of breath. Physical examination revealed a toxic, tachypneic gentleman with a temperature of 103.4°F (39.7°C) and a respiratory rate of 35. Herpes labialis was evident and signs of consolidation were present at the left lung base. Laboratory data revealed Hct of 43%, WBC of 14,000/mm³, with 80% polys, 14% bands, 3% lymphs, and 3% monos. Sputum examination revealed sheets of polymorphonuclear leukocytes and gram-positive diplococci. Sputum and two blood cultures grew *Streptococcus pneumoniae*. Chest radiography (Fig. 1) demonstrated a focal air space consolidation of the lung.

Comment: This is the classic presentation of community-acquired pneumococcal pneumonia following a viral upper respiratory infection. Indeed, a communitywide influenza outbreak was taking place, with immunosuppressed individuals such as this bearing a particularly heavy burden from this infection. In this case, bacterial superinfection with *Streptococcus pneumoniae* further complicated the clinical course. In this stable heart transplant patient receiving maintenance immunosuppressive therapy, the pathogenesis, clinical presentation, physical findings, chest radiography, and response to therapy are virtually identical to those observed in the general population.

2.1. Factors that Determine the Risk of Pulmonary Infection

The risk in the compromised host of invasive infection in general and of pulmonary infection in particular is determined largely by the interaction of two factors: the patient's net state of immunosuppression and the epidemiologic exposures he or she encounters. Thus, if the infecting inoculum is great enough, even a normal individual can develop life-threatening infection; conversely, if the degree of immunosuppression is great enough, even the most innocuous of commensal organisms can pose a severe threat.[29]

Certain defects in host defense render the individual particularly susceptible to infection with particular classes of microorganisms (Table 2). These correlations are especially useful in pediatric patients with congenital defects that are relatively "pure," such as isolated defects in antibody formation, complement function, or granulocyte function (see Chapters 2 and 14). In most individuals, however, the situation is far more complicated, with a variety of defects present in the same individual due to the effects of acquired disease and its therapy. For example, following a combination of splenectomy, radiation, and chemotherapy, patients with Hodgkin's disease, which itself is associated with profound defects in T-lymphocyte function and cell-mediated immunity, will develop marked B-lymphocyte dysfunction as manifested by low levels of serum immunoglobulin M and specific antibody against *Haemophilus influenzae* type B, poor response to pneumococcal vaccine, and an increased risk of life-threatening systemic and pulmonary infection with these two organ-

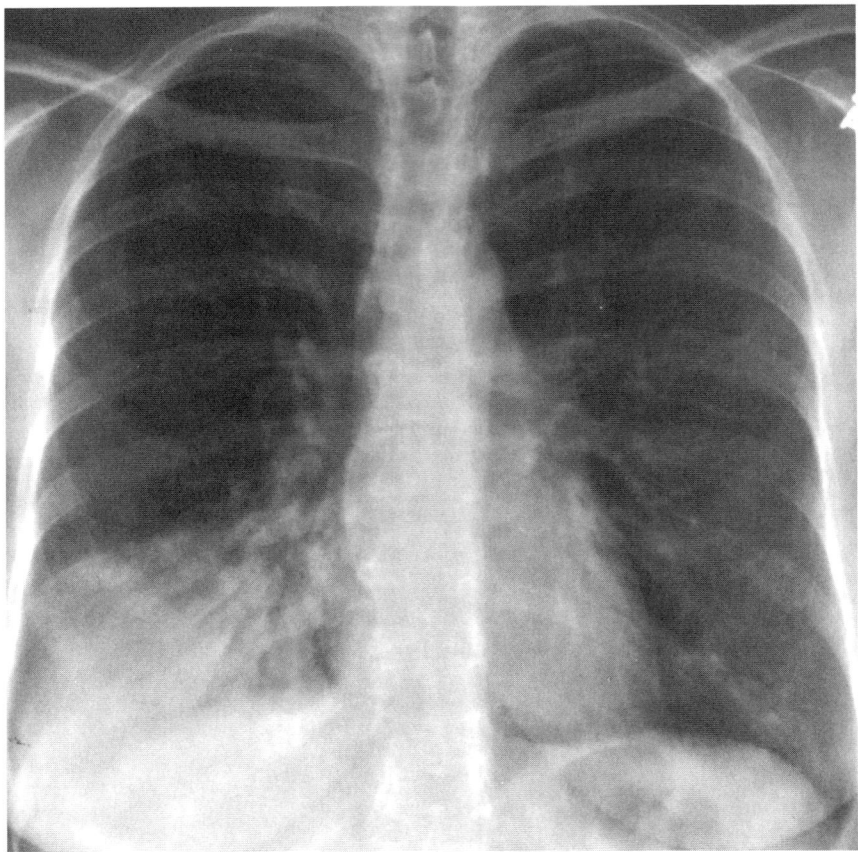

FIGURE 1. Community-acquired *Streptococcus pneumoniae* pneumonia. (A, B) Frontal and lateral chest radiographs show a focal opacity in the right lower lobe consisting of dense consolidation with prominent air bronchograms and a tiny sympathetic pleural effusion along the lateral costophrenic sulcus.

A

isms.[30–32] Thus, in the majority of immunocompromised patients, the concern is the "net state of immunosuppression," rather than a single defect.

2.1.1. Net State of Immunosuppression

The net state of immunosuppression is a complex function determined by the interaction of a number of factors[29]:

1. Host defense defects caused by the disease process itself.
2. Dose, duration, and temporal sequence of immunosuppressive therapy employed.
3. Presence or absence of neutropenia.
4. Anatomic integrity of the tracheobronchial tree (including the presence or absence of such "foreign bodies" as endotracheal tubes and obstructing tumor masses).
5. "Functional" integrity of the oropharyngeal and gastric mucosae in terms of their ability to resist the adherence of potential pathogens to these mucosal surfaces.
6. Such metabolic factors as protein–calorie malnutrition, uremia, and, perhaps, hyperglycemia.
7. Presence of infection with one or more of the immunomodulating viruses: human immunodeficiency virus (HIV), the hepatitis viruses B and C (HBV and HCV), cytomegalovirus (CMV), Epstein–Barr virus (EBV), and probably human herpesvirus 6 (HHV-6).

The most important determinant of the net state of immunosuppression is the nature of the immunosuppressive therapy that has been and continues to be administered. A few general points regarding immunosuppressive therapy bear special emphasis. The first and perhaps most important is that the particular dose of immunosuppression being administered on a given day is less important than the dosages employed over a sustained period of time (the area under the curve). For example, although the highest daily doses of immunosuppressive therapy that

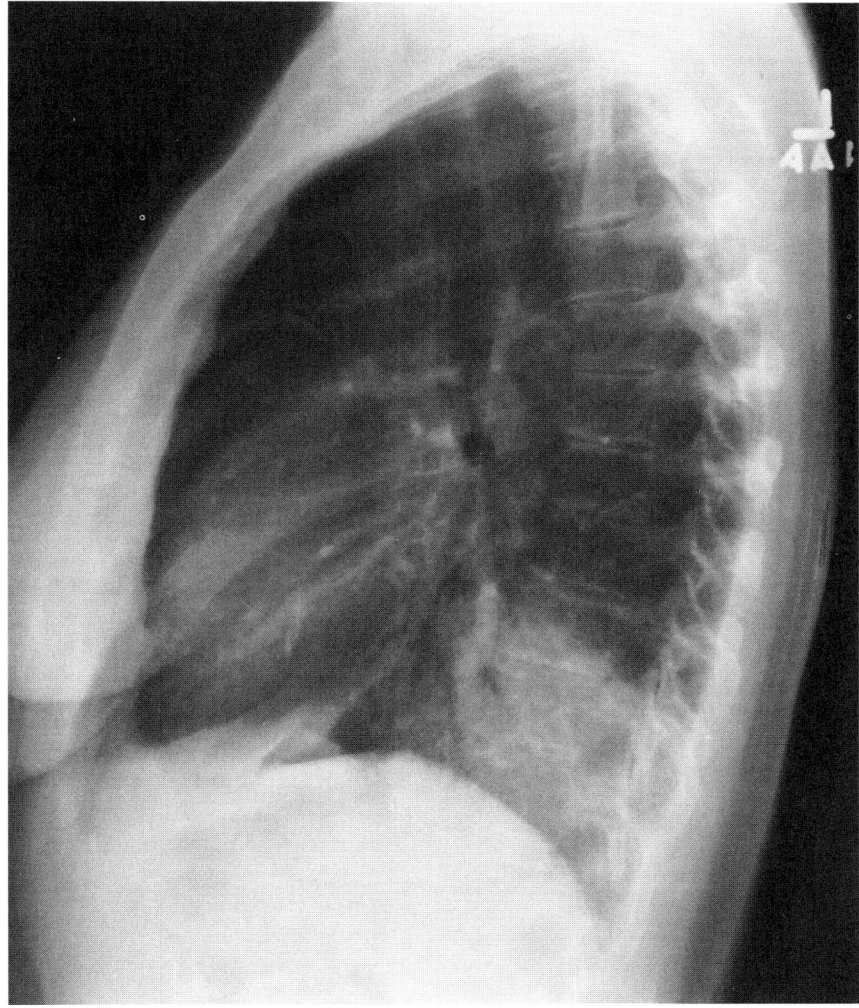

B

FIGURE 1. (*Continued*)

are employed in organ transplantation are administered in the first 2–3 weeks posttransplant, the risk of opportunistic infection is extremely low until more than 4 weeks posttransplant, at which time the daily doses of immunosuppressive drugs have fallen significantly. We have suggested that immunosuppressive therapy is like buying something by credit card: if you increase immunosuppression, you have the immediate gratification of improved allograft function; the bill (in terms of infection) comes due 3–4 weeks later.[29,33]

The second general point is that the effects of the entire immunosuppressive regimen must be considered; the net state of immunosuppression is determined not just by the summation of the effects of the individual agents,

but also by the interactions of these different effects. For example, whereas such antilymphocyte antibody therapies as antithymocyte globulin (polyclonal, of either equine or rabbit origin) or OKT3 (monoclonal, of murine origin) are capable of reactivating such viruses as CMV and EBV from latency, cyclosporine, tacrolimus, and rapamycin are not. However, once virus is reactivated, cyclosporine, tacrolimus, and rapamycin will greatly amplify the effects of these viruses by specifically blocking the virus-specific cytotoxic T-cell response, the key host defense. Thus, from the point of view of the host, the worst possible scenario is to reactivate virus with antilymphocyte antibody therapy and then accentuate the effects of the virus that has been reactivated with the other

TABLE 2. Pulmonary Infections to Which Patients with Specific Host Defense Defects Are Predisposed[a]

Host defense defect	Pulmonary infections to which patient is predisposed
Oral and tracheobronchial ulceration or obstruction or both	Oral bacterial flora Enterobacteriaceae
Decrease in the number of fully functional granulocytes	Oral bacterial flora Enterobacteriaceae *Aspergillus* species *Pseudomonas aeruginosa*
Hypogammaglobulinemia	*Streptococcus pneumoniae* *Haemophilus influenzae* type B (*Pneumocystis carinii*)
Depressed cell-mediated immunity	Typical and atypical mycobacteria Fungi Viruses (CMV, varicella zoster virus, herpex simplex, measles virus) *Pneumocystis carinii* *Toxoplasma gondii* *Strongyloides stercoralis*
Complement defects	*Streptococcus pneumoniae* *Haemophilus influenzae* type B

[a]Modified from Rubin.[3] Infections uncommonly associated with a particular defect are listed in parentheses.

immunosuppressive drugs. Indeed, in this instance specific preemptive antiviral therapy needs to be administered concomitantly (see Chapter 17).[34–36]

Finally, it is now apparent that cytokines, chemokines, and other factors elaborated by the host in response to a variety of insults (e.g., infection, allograft rejection, graft vs. host disease, and antilymphocyte antibody therapy) will modulate the course of different infections and interact with the effects of exogenous immunosuppressive therapy on the host. For example, it is clear that there are at least three signaling pathways that can result in the reactivation of CMV from latency: tumor necrosis factor (TNF) interacting with the TNF receptor on latently infected cells; catecholamines through cyclic AMP; and certain proinflammatory cytokines. The TNF pathway appears to be the most important of these. Thus, replicating CMV can appear in any circumstance where TNF is released (e.g., sepsis, rejection, antilymphocyte antibody therapy). The CMV will then be modulated by the immunosuppression being administered and in turn the level of CMV will help determine the susceptibility to a variety of opportunistic pulmonary pathogens, ranging from *Pneumocystis carinii* to *Aspergillus* species.[29,37]

More than 95% of the pneumonias that occur in the

immunocompromised patient follow the delivery of sufficient numbers of virulent microorganisms to the lower respiratory tract via the tracheobronchial tree. In most cases, this infection occurs following the aspiration of gastric or oropharyngeal flora, thus emphasizing the importance of microbial colonization patterns in the pathogenesis of pneumonia,[38,39] particularly bacterial pneumonia. Even in the case of invasive pulmonary aspergillosis, however, there is evidence that preceding nasopharyngeal colonization may play a role in the pathogenesis of this form of infection as well.[40–43] In addition, inhalation of aerosolized pathogens that are present in unusually high amounts in the patient's environment can be responsible for the microbial challenge to the lungs (see Section 2.1.2).

The normal bacterial flora of the oropharynx is predominantly gram positive and sensitive to a broad range of antibiotics. In particular, the prevalence of gram-negative oropharyngeal colonization in a normal population is on the order of 2–6%, and even challenging normal individuals with gram-negative organisms will rarely induce sustained colonization with these antibiotic-resistant, virulent bacterial species. This remains true for "physically healthy" hospitalized patients (e.g., psychiatric patients), but the rate of *Staphylococcus aureus* and/or gram-negative carriage rises to 35% in the moderately ill hospitalized patients and to 73% in critically ill patients.[38,39,44–46] Sustained colonization with one or any combination of Enterobacteriaceae, *Pseudomonas aeruginosa*, *Staphylococcus aureus*, or *Candida* species is the rule in many immunocompromised patient populations, particularly cancer patients being treated with cytoreductive chemotherapy. Although environmental exposures, the use of antimicrobial therapies, the deployment of respiratory therapy equipment, and other factors will contribute to the rate of nasopharyngeal colonization, the crucial factor is the alteration in the surface characteristics of the mucosal epithelium. In the face of chemotherapy-induced mucositis, steroid therapy, an advanced state of illness, and other factors, the adherence of the more virulent flora, the first step in the pathogenesis of pneumonia in many immunocompromised individuals, is promoted.[4,44,47,48]

Adherence of bacteria to epithelial cells is mediated through specific interactions between adhesins on the surfaces of bacteria (particularly specialized appendages termed "pili" or "fimbriae") and receptors on the surfaces of the epithelial cells. Oropharyngeal epithelial cells from individuals colonized with gram-negative organisms have been shown to support an increased rate of gram-negative adherence when compared to epithelial cells from uncolonized individuals.[44–48] The mechanisms involved in this change in adherence patterns are still being delineated, although the ubiquitous glycoprotein fibronectin appears to be a critical determinant of the microbial flora of the oropharynx. Normally, fibronectin coats this epithelial surface, promoting the attachment of gram-positive organisms and enhancing the bactericidal function of phagocytic cells directed against other microbial species. This causes a selective disadvantage for gram-negative bacteria, impeding colonization with these organisms. Destruction of this fibronectin coat or inadequate production, as occurs in a variety of disease states, malnutrition, dehydration, and in response to chemotherapy, will have at least two adverse effects: (1) It will impede the attachment of the desirable normal flora and (2) it will expose epithelial cell surface receptors for a variety of gram-negative bacterial surface adhesins. The net result is gram-negative overgrowth of the oropharynx.[4,44–54] Once gram-negative oropharyngeal colonization is established, it tends to persist. The next step is aspiration of these organisms into the lower respiratory tract. Pharyngeal secretions are aspirated in the majority of normal individuals during sleep and in 70% of individuals with a decreased level of consciousness. The impact of the aspiration episode(s) is increased by depressed gag and cough reflexes, with further amplification in the presence of endotracheal and nasogastric tubes. Two statistics emphasize the importance of these observations: (1) Approximately 90% of patients with gram-negative pneumonia have had prior oropharyngeal colonization with the same organism and (2) pneumonia develops five to eight times more frequently in individuals colonized with these virulent gram-negative organisms. Hence, the ability of the host's epithelial surfaces to resist colonization with virulent microbial species is an important determinant of the net state of immunosuppression.[4,44–54]

An additional source of gram-negative bacillary pneumonia is the stomach in the compromised host. The normal production of acid by the stomach maintains a low bacterial count. However, as gastric pH rises, the level of microbial contamination (particularly with gram-negative organisms) rises, serving as a reservoir for both the oropharynx and the respiratory tract.[4,55–57] The use of drugs such as proton pump inhibitors, histamine-2 blockers, and antacids, which can raise gastric pH, have been linked in some, but not all, studies to an increased risk of gram-negative pneumonia (with sucralfate for gastric mucosal protection in some, but not all, studies being free of this added risk). The use of enteral feedings, many of which have a pH between 6.4 and 7.0, also will favor gram-negative colonization. It also has been suggested that impaired gastric and small intestinal motility result in

bacterial overgrowth in the duodenum. This overgrowth is followed by duodenal–gastric reflux, with further amplification of bacterial growth caused by elevated gastric pH. Once these events have occurred, oropharyngeal and tracheal colonization can occur quite readily, thus facilitating the development of pneumonia.[55–61]

2.1.2. Epidemiologic Aspects

The epidemiologic aspects of pulmonary infection in the immunocompromised host can be divided into two general categories: pulmonary infections related to exposures occurring within the community and pulmonary infections related to exposures within the hospital environment.

2.1.2a. Community-Acquired Infection. There are four major considerations when considering the possibility of community-acquired pulmonary infection in the compromised host[1–4,22–29]:

1. The geographically restricted, systemic mycoses (e.g., blastomycosis, coccidioidomycosis, and histoplasmosis).
2. Tuberculosis.
3. *Strongyloides stercoralis* infection.
4. Community-acquired respiratory viruses (e.g., influenza, respiratory syncytial virus, parainfluenza, adenoviruses, and rhinoviruses).

The systemic mycoses and tuberculosis share a similar pathogenesis. The portal of entry is the lung, with primary infection following inhalation of a sufficient inoculum that escapes an initial nonspecific inflammatory response and postprimary dissemination to other bodily sites being a not infrequent event. Limitation of the extent of both the pulmonary and sites of metastatic infection is dependent on the development of a specific cell-mediated immune response. At a later date, reactivation of latent infection can occur, with the possibility of secondary dissemination, due to the waning of immunity and/or local anatomic factors (e.g., erosion of an infected node into the lung). Because of the central role of specific cell-mediated immunity in the control of these infections, it is not surprising that immunocompromised individuals with deficits in this aspect of host defense are susceptible to these infections, with three patterns of disease being observed: progressive primary infection of the lungs; reactivation infection, usually of the lungs, with the potential for secondary dissemination; and reinfection following a new exposure, with a high probability of dissemination, in individuals whose immunity has been ablated due to disease or its therapy. The end result in many immunocompromised patients (especially AIDS patients, but also transplant patients, those with lymphoma, and those receiving intensive immunosuppressive therapy) is disseminated infection, with either or both extensive pulmonary disease and evidence of metastatic infection to such sites as the skin, central nervous system, and skeletal system. In particular, the coexistence of tuberculosis, especially drug-resistant tuberculosis, in populations with a high incidence of HIV infection is now recognized as a particular threat not only to these patients and other immunocompromised patients but also to the community at large.[29,33,62–67]

Illustrative Case 2

A 37-year-old Hispanic male, who worked at a municipal hospital as an emergency ward attendant, presented with fevers, a nonproductive cough, and increasing shortness of breath of several weeks' duration. Although several friends and lovers had previously been diagnosed with AIDS, he had refused HIV testing for himself. Over the past year, he had sought treatment outside his place of employment for recurrent anogenital herpes and oral thrush. On physical examination, his temperature was 102.2°F (39°C), his respiratory rate was 24, and his blood pressure was 110/70. He appeared thin and chronically ill, with oral thrush, a painful 2 × 3 cm ulceration adjacent to his anus, and a slightly enlarged liver and spleen. Laboratory studies revealed the following: Hct 29%, WBC 3400/mm^3 with 65% polys, 14% bands, 12% monocytes, and 9% lymphs; SGOT 110; alkaline phosphatase three times the upper limits of normal; bilirubin 2.4/3.0. Chest X-ray, which was initially interpreted as negative, revealed a faint, diffuse miliary infiltrate (Fig. 2). Subsequently, his HIV antibody test was shown to be positive, and he was found to have a CD4 count of 61. *Mycobacterium tuberculosis*, subsequently found to be resistant to isoniazid and rifampin, was isolated from an induced sputum and from Dupont isolator blood cultures. On admission, he was placed in isolation and following evaluation begun on therapy with isoniazid, pyrazinamide, rifampin, ethambutol, and streptomycin, as well as anti-HIV therapy. Unfortunately, he continued to decline and expired 2 weeks after admission, even before the suspected diagnosis of miliary tuberculosis could be confirmed.

Comment. This case was part of a cluster of cases of drug-resistant tuberculosis involving immunocompromised individuals at this municipal hospital and dramatically delineates a current public health dilemma. This Hispanic man with undiagnosed AIDS could have contracted his drug-resistant tuberculosis within his community or occupationally. The extent and pace of his illness were greatly amplified by the level of immunocompromise engendered by his HIV infection, and both within

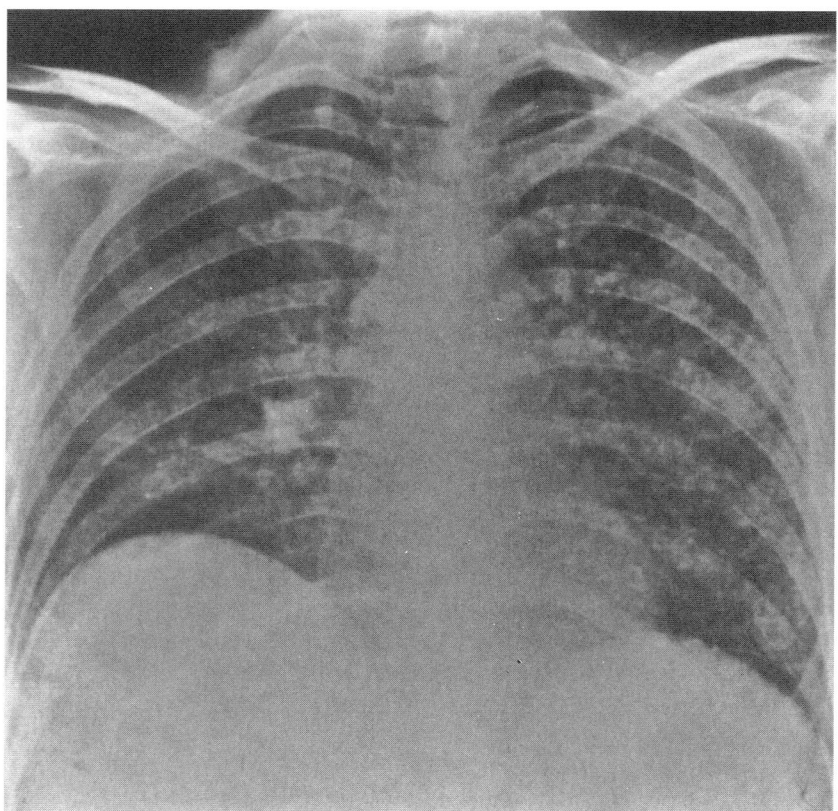

A

FIGURE 2. Miliary tuberculosis. Diffuse miliary opacities of 2–4 mm diameter due to miliary tuberculosis. The diffuse, tiny, nodular opacities are characteristic of blood-borne mycobacterial or fungal infection. (A) Conventional chest radiograph. (B) CT section in another patient through the level of the carcina demonstrates multiple pinhead-sized nodules distributed diffusely throughout the lung due to miliary tuberculosis in an AIDS patient. (Courtesy of Paul Stark, MD)

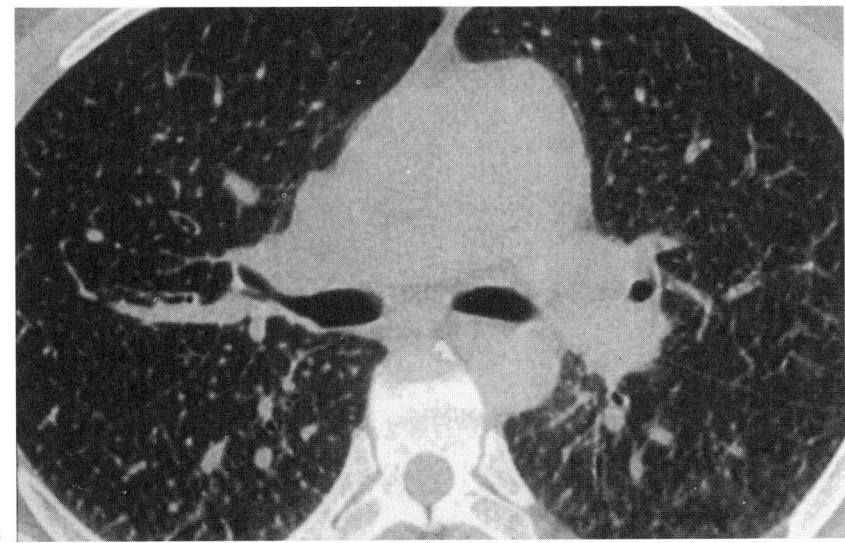

B

FIGURE 2. (*Continued*)

the community and within the hospital he posed a significant risk to other individuals, particularly other immunocompromised individuals, with whom he came into contact. The extent of his illness was underestimated by the initial evaluation of his chest radiograph (presumably due to the limited inflammatory response he was able to mount), thereby increasing the hazard to both himself and others.

Illustrative Case 3

A 46-year-old man with non-Hodgkin's lymphoma previously treated with radiotherapy and chemotherapy, but for the past 6 months with prednisone and weekly vincristine, entered with a 2-week history of a "cold." He had been relatively well until the gradual onset of fever, night sweats, anorexia, a 10-pound weight loss, increased fatiguability, and headaches. He did not complain of chest pain but had noted some dyspnea on exertion. There was no history of past tuberculosis or tuberculous exposure. Although he had been a resident of New England for the last 10 years, he had lived and worked for more than 30 years of his life on a farm in the San Joaquin valley of California. Physical examination revealed a chronically ill man with a respiratory rate of 18, temperature of 101°F (38.3°C), coarse rhonchi over the right upper chest posteriorly, and an enlarged spleen. Laboratory data revealed Hct of 32% and WBC count of 11,000/mm³ with 85% polys, 3% bands, and 12% lymphs. Skin testing revealed anergy (including testing with coccidioidin). Sputum examination revealed abundant polymorphonuclear leukocytes and normal throat flora. Cerebrospinal fluid examination revealed 52 leukocytes/mm³, 89% lymphs and 11% polys, a sugar of 31 mg/dl (simultaneous blood sugar of 110 mg/dl), and a protein of 96 mg/dl. Chest radiograph revealed a cavitary lesion in the right upper lobe. Complement-fixing antibody to *Coccidioides immitis* was positive in both the serum and the cerebrospinal fluid.

Comment. This is an example of disseminated coccidioidomycosis years after the patient had been primarily infected, with reactivation of an old pulmonary focus and secondary dissemination. The impaired cell-mediated immunity of this patient, induced primarily by the treat-

ment of his lymphoma, made these events possible. The important clue to the diagnosis lay in the patient's remote epidemiologic history, with the diagnosis established by serologic testing. A negative coccidioidin skin test and a positive serologic test constitute the characteristic pattern observed in this circumstance.

Strongyloides stercoralis is of particular concern, because of its unique life cycle. Alone among the intestinal nematodes that affect humans, *S. stercoralis* possesses an autoinfection cycle that can take place entirely within a person's gastrointestinal (GI) tract. As a result, chronic asymptomatic GI infestation can be maintained for decades after the person has been exposed in an endemic area. With the onset of depressed cell-mediated immune function, due either to disease or its therapy, overwhelming systemic invasion by this organism can occur. Hemorrhagic pulmonary consolidation or diffuse, bilateral alveolar opacities, often with accompanying GI complaints (including hemorrhagic enterocolitis), may develop. Alternatively, systemic strongyloidiasis may be accompanied by the adult respiratory distress syndrome (see Chapter 8).[68,69]

Perhaps the most important community exposure for immunocompromised patients is that related to respiratory viruses in the household, workplace, and general community (alternatively, these same viruses can be brought into the hospital setting from the community, with rapid spread among patients and staff). The community-acquired respiratory viruses causing disease in the immunocompromised patient population are diverse, including the

Orthomyxoviridae (influenza viruses), the Paramyxoviridae (respiratory syncytial virus and the parainfluenza viruses), the Picornaviridae (rhinoviruses and enteroviruses), the Coronaviridae (coronaviruses), and the Adenoviridae (adenoviruses). Transmission of the virus is through contact with virus-laden respiratory secretions: person to person in aerosols or on the hands, or through contact with fomites. Infection in immunocompromised individuals generally reflects the level of virus transmission in the general community. The importance of these infections is underlined by the following statistics: At the MD Anderson Cancer Center, during two 6-month surveillance periods in the early 1990s, 67 (31%) of 217 hematopoietic stem cell patients admitted with an acute respiratory illness had infection with one or more of these respiratory viruses (49% respiratory syncytial virus, 18% influenza, 18% picornavirus, 9% parainfluenza viruses, and 6% adenoviruses); 58% of these infections were complicated by pneumonia and 51% of the pneumonias were fatal.[70] Similar data are available from the Huddinge University Hospital in Sweden, where a prospective surveillance study among a comparable population of patients was carried out between 1989 and 1996 and involved 545 patients. This study revealed an incidence of community-acquired respiratory virus disease of 9.3% in allogeneic and 1.8% in autologous hematopoietic stem cell transplant patients (38% influenza viruses, 21% respiratory syncytial virus, 21% parainfluenza viruses, and 21% adenoviruses).[71] Less complete information is available in other populations, but what information is available is consistent with that cited for the hematopoietic stem cell transplant population.[72–75] For example, during a recent communitywide epidemic of influenza A infection, 28 kidney and liver transplant recipients were admitted to the Massachusetts General Hospital over a 6-week period, with more than half of these having pneumonia. As a general rule then, immunosuppressed patients with these infections have a higher attack rate for viral pneumonia, superinfecting bacterial pneumonia, rate of transmitting the infection to others, and prolonged illness than the general population.

At present, prevention and treatment of community-acquired respiratory virus infection in immunocompromised patients is in its infancy. Although yearly influenza immunization is recommended for these patients, efficacy is significantly attenuated when compared with the general population.[72,76–79] There is inadequate information available to comment regarding amantadine, rimantadine, or the new neuraminidase inhibitors in the prevention or treatment of influenza in immunosuppressed patients. There are anecdotal reports of success with ribavirin in the management of both respiratory syncytial virus and adenovirus infection, but its true efficacy (with or without immunoglobulin) remains to be defined.[72] In sum, avoidance of exposure and traditional infection control procedures are the cornerstones of our present efforts to control this important class of infection.

2.1.2b. Hospital-Acquired Infection. As important as community exposures are in the evaluation of immunocompromised hosts with the febrile pneumonitis syndrome, exposures within the hospital are even more important. Nosocomial exposures to air contaminated with *Pseudomonas aeruginosa*, *Klebsiella pneumoniae* and other Enterobacteriaceae, and *Aspergillus* species, as well as to potable water laden with *Legionella* species, have resulted in epidemic pulmonary infection in immunocompromised patients.[1–3,40,41,80–88] As far as *Legionella* infection is concerned, *L. pneumophila*, as in the general population, accounts for the majority of cases in the immunocompromised patients, but in addition the burden of disease from unusual *Legionella* species (e.g., *L. micdadei*, *L. longbeachae*, *L. bozemanii*) is borne most heavily in these patients.[89–91] The source of nosocomial legionellosis is a contaminated potable water system, with patients aspirating the organisms thus presented.[84] The extent of the problem is underlined by a recent survey undertaken in the United Kingdom, in which *Legionella* species were isolated from 55% of the water supplies of 69 transplant units, with *L. pneumophila* being isolated from 45%.[85] Prevention of nosocomial legionellosis requires surveillance and then disinfection of infected water supplies. Although hyperchlorination was recommended in the past for this purpose, superheat and flush or copper–silver ionization are currently the preferred methods.[84]

Nosocomial outbreaks of pulmonary infection have been observed among immunocompromised patients, with two epidemiologic patterns having been identified: domiciliary and nondomiciliary. The term "domiciliary" is used to describe outbreaks in which the patient is infected on the ward or in his or her hospital room. Such epidemics of *Pseudomonas*, *Aspergillus*, and *Klebsiella* (among others) infection have been not uncommon and are relatively easily identified because of temporal and spatial clustering of cases of opportunistic infection. The exposures have been due to construction, aerosolization of standing water laden with gram-negative bacilli or *Legionella*, or contamination of the air-handling system for the ward or the patient's room. Such outbreaks are effectively prevented by the provision of high-efficiency particulate air (HEPA) filtering in hospital locations where immunocompromised patients are housed. As one measure of the effectiveness of this approach, our experience with intubated transplant patients is very striking. In the absence of HEPA-filtered air, the incidence of gram-negative or *Aspergillus* superinfection in transplant patients requiring more than 3 days of tracheal intubation and ventilatory support approached 100%; at present, when care is administered in HEPA-filtered rooms, the incidence of secondary infection of the lungs does not become significant for at least a week, and we have successfully cared for immunosuppressed patients who required intubation for more than a month. Thus, an adequately filtered air supply is directly translated into a lower incidence of pulmonary infection and a decreased mortality due to opportunistic pulmonary infection.[2,23,29,92]

Although reports of domiciliary outbreaks of nosocomial pulmonary infection among immunocompromised patients are not uncommon, nondomiciliary exposures and outbreaks are actually a greater problem. These infections occur when patients are taken from their rooms through the hospital to the radiology, endoscopy, or operating suites for essential procedures, with exposures occurring along the way or at the site of the procedure. Construction within the hospital environment has been the major cause of air contamination leading to nondomiciliary outbreaks. Thus, we have observed an outbreak of invasive pulmonary aspergillosis among transplant patients, leukemic patients, and patients receiving immunosuppressive therapy for collagen vascular disease due to construction in the central radiology suite.[92] Similarly, cardiac transplant patients have acquired this infection while waiting outside the cardiac catheterization laboratory for a routine endomyocardial biopsy procedure. The use of routine surgical masks to protect these patients has not been effective. Although the ideal equipment has yet to be designed for protecting patients being transported in the hospital environment, efforts to develop prototype portable HEPA-filtered transport equipment to protect patients when they must travel out of the protected environment of the transplant or oncology units are to be commended.[93] In the meantime attention to transport routes in the hospital, with the avoidance of areas of construction, must receive a high priority for these patients.

Illustrative Case 4

An 11-year-old girl in remission after induction chemotherapy for acute myelogenous leukemia entered the hospital for a scheduled round of chemotherapy. Five days postadmission, she spiked a fever and an infiltrate was noted on chest CT scan. Needle aspiration biopsy yielded a pure growth of *Aspergillus fumigatus*. Amphotericin therapy was instituted, with a good response after a prolonged hospitalization. That month, five other immunocompromised patients (three renal transplant patients, an adult oncology patient, and one patient being treated with high-dose corticosteroids for systemic lupus erythematosus) from different locations within the hospital developed invasive pulmonary aspergillosis. The epidemic was traced to construction in the central

radiology suite. With cessation of the construction, the epidemic came to an end.

Comment. This is a well-documented nondomiciliary outbreak of invasive pulmonary aspergillosis. Such nondomiciliary exposures are probably quantitatively more important than domiciliary exposures, but are more difficult to identify because of the lack of temporal and spatial clustering. Constant surveillance of immunocompromised patients for such excessive environmental hazards is essential. Suspicion regarding such a nondomiciliary, nosocomial exposure should be aroused whenever opportunistic pulmonary infection is identified in a patient whose net state of immunosuppression should not be great enough for such an infection to occur unless the patient had experienced an unusually intense exposure. For example, the occurrence of invasive aspergillosis in the first 3 weeks postorgan transplantation or in the first week of cancer chemotherapy should be cause for an epidemiologic investigation.

Immunocompromised patients, then, can be compared to "sentinel chickens" placed in the swamps surrounding major population centers to monitor the level of mosquito-borne arbovirus infection. In this context, it is our contention that immunocompromised patients are sentinel chickens placed in the swamps of our hospital (and community) environment. Any excess traffic in microbes will be seen first and most severely in these individuals and constant vigilance is necessary to protect these patients. The lesson for the clinician is twofold: Constant surveillance to effect early identification and correction of hazards is important to prevent infection; in addition, while correcting circumstances that have epidemic potential, the clinician must be aware of the prevalent endemic microbial flora at any point in time to facilitate the care of the individual patient.[94]

2.2. Pathology of Pulmonary Infections in the Immunocompromised Host

The evaluation of pulmonary biopsy material for possibly treatable infection is an important part of the assessment of many immunocompromised patients with the febrile pneumonitis syndrome. The histologic patterns of pulmonary injury observed following microbial invasion in this patient population have been classified by Nash into six general patterns, each with its own differential diagnostic considerations: acute nonnecrotizing pneumonia, acute necrotizing pneumonia, diffuse alveolar damage, diffuse alveolar damage with foamy alveolar exudate, granulomatous pneumonitis, and bronchiolitis obliterans-organizing pneumonia (BOOP).[95]

2.2.1. Acute Nonnecrotizing Pneumonia

Acute nonnecrotizing pneumonia is caused by a variety of bacterial agents, most commonly *Streptococcus pneumoniae*. It is characterized by a fibrinopurulent exudate filling airways and air spaces, without destruction of the alveoli. In most instances, the causative agent can be isolated from the respiratory secretions and lung tissue and effective antimicrobial therapy can result in complete healing of the process. Indeed, lung biopsy is usually not necessary to make this diagnosis.[95]

2.2.2. Acute Necrotizing Pneumonia

Acute necrotizing pneumonia adds the element of pulmonary tissue destruction to the previous pattern of acute inflammation. The distribution of the lesions depends on the route by which the organisms invade the lung: Organisms that reach the lung via the tracheobronchial tree and invade at the level of the bronchi and adjacent airspaces produce a pattern of bronchopneumonia (e.g., *Staphylococcus aureus*), whereas those that invade the lung at the level of the distal airspaces produce an airspace consolidation that can progress to a full-blown lobar pneumonia (e.g., *Klebsiella pneumoniae*). Hematogenous spread of organisms to the lung typically produces a nodular focus of necrotizing pneumonia that is located in the periphery of the lung and has no relationship to the segmental anatomy of the tracheobronchial tree. Such lesions are caused by a variety of virulent organisms once they enter the bloodstream, including *S. aureus*, *Pseudomonas aeruginosa*, *Candida albicans* and *tropicalis*, and *Aspergillus fumigatus*.[95] The most common bacterial infections that produce acute necrotizing pneumonia in the immunocompromised host are *S. aureus*, *K. pneumoniae*, *P. aeruginosa*, and *Legionella* species. In the case of *P. aeruginosa* infection, in addition to the nonspecific necrotizing features, so-called *Pseudomonas* vasculitis characterized by masses of bacteria surrounding and invading blood vessels is a common finding. In the case of *Legionella* infection, the inflammatory exudate usually contains numerous macrophages in addition to the neutrophils seen with the other bacterial processes, and extensive leukocytoclasis of alveolar inflammatory cells may be observed.[95,96] *Candida*, *Aspergillus*, and *Zygomycetes* are the fungal species most commonly associated with pulmonary necrosis, both because of direct effects of these organisms and because of their propensity for invading pulmonary arteries, causing infarction and hemorrhage (as well as a propensity for hematogenous dissemination). The opportunistic bacterial pathogen, *Nocardia asteroides*, the clinical effects of which resemble those of *Aspergillus*, can likewise produce necrotizing changes in the lung.[95,97] Finally, the herpes group viruses and adenoviruses can cause necrotizing pneumonias associated with a neutrophilic exudate. Cytomegalovirus (CMV) and adenovirus infection may be associated with two pathologic patterns: a necrotizing pneumonia (in the case of adenoviruses, bronchiolitis and bronchitis as well) and diffuse alveolar damage. In the case of CMV, enlarged cells exhibiting the characteristic cytopathic CMV changes may be present, although the absence of these cells does not rule out this diagnosis. Viral culture, immunofluorescent staining, and *in situ* polymerase chain reaction (PCR) are more sensitive in diagnosing CMV pulmonary infection than the demonstration of the cytomegaly cells with their intranuclear inclusions.[95,97]

2.2.3. Diffuse Alveolar Damage

Diffuse alveolar damage is a nonspecific response to a variety of insults and is the pathologic finding seen in the adult respiratory distress syndrome (ARDS) and its less severe variant, acute lung injury (ALI), such as that occurring in sepsis and septic shock. The term "diffuse alveolar damage" is given to a pathologic pattern of damaged alveolar capillary endothelium and alveolar epithelium, interstitial edema, and hyaline membranes. Over a few weeks, this acute exudative picture evolves into an organizing (proliferative) phase characterized by interstitial fibrosis, chronic inflammation, and regenerating alveolar epithelium. This lesion is the characteristic one caused by a variety of viruses, most notably influenza, but also including CMV and adenoviruses.[95,97]

2.2.4. Diffuse Alveolar Damage with Foamy Alveolar Exudate

Diffuse alveolar damage with eosinophilic alveolar foam is the classic pathologic finding with *Pneumocystis carinii* pneumonia.[98] Typically, at the time of biopsy evidence of the organizing phase of alveolar damage is already present, with the organisms being demonstrable within the alveolar foam with such special stains as the methenamine–silver stain or by immunohistochemical methods. It also should be emphasized, however, that the absence of the alveolar foam does not rule out pneumocystosis, with organisms still being demonstrable in association with nonspecific diffuse alveolar damage with the special stains. As with other infections that occur in the AIDS patient, the burden of *Pneumocystis* organisms is far higher in this group of immunosuppressed individuals and the consequences of the *P. carinii* infection are often greater in AIDS patients than in other immunocompromised patients with this infection: a higher likelihood of developing changes of ARDS, a higher incidence of severe interstitial fibrosis, the occurrence of metastatic infection to other bodily sites, and the not uncommon occurrence of pneumatoceles or bullae as a consequence of the *Pneumocystis* infection—unusual events in the non-AIDS patient.[95,97]

2.2.5. Granulomatous Pneumonitis

Granulomatous pneumonitis is characterized by a nodular infiltrate composed of epithelioid histiocytes, multinucleated giant cells, and surrounding areas of fibrosis and chronic inflammation. Varying degrees of central necrosis may be observed. This is the classic pathologic lesion observed with *Mycobacterium tuberculosis*, *Cryptococcus neoformans*, *Histoplasma capsulatum*, *Coccidioides immitis*, and *Blastomyces dermatitidis*. In saying this, however, two points must be particularly emphasized: First, prior to the development of the established granuloma, an acute neutrophilic response is the first line of defense against invasion with these microbes. Therefore, the exact mix of acute and granulomatous inflammation seen will depend on the age of the lesion biopsied. Second, the most important factor in determining the exact pathologic picture observed with the organisms is the ability of the particular host to generate an inflammatory response. For example, depending on the level of cell-mediated immune response possible, the pathologic pattern may range from a fully developed granulomatous response to the presence of organisms in the lung without any inflammatory response. Hence, when an etiologic diagnosis is not immediately apparent after careful examination of the biopsy material, a broad range of special stains and probes aimed at delineating specific organisms should be employed, recognizing that the more immunocompromised the individual, the greater the likelihood of an atypical pathologic response.[95,97]

2.2.6. Bronchiolitis Obliterans-Organizing Pneumonia

The term "bronchiolitis obliterans-organizing pneumonia" (BOOP) encompasses a pathologic entity consisting of fibrous organization of an inflammatory exudate. As such, it is the end result of a variety of infectious and noninfectious inflammatory processes that affect the lung and is a not uncommon finding on biopsy, particularly if the biopsy is carried out relatively late in the disease process. Cultures are usually negative from tissue with this pathologic pattern. Since this pattern is a "final common pathway" for a variety of processes, this finding is of little use in terms of etiologic diagnosis and therapy, although the extent of the fibrotic reaction may have

important prognostic implications as to the reversibility of the pulmonary injury.[95,97,99]

2.3. Clinical Clues to the Diagnosis of Pulmonary Infection

An important component in constructing a differential diagnosis for an immunosuppressed patient with possible pneumonia is an understanding of the "temporal" aspects of the underlying disease. That is, whether the patient's immunocompromised state is due to antileukemia chemotherapy or HIV infection or cyclosporine-based immunosuppression postorgan transplantation, the duration of the immunosuppressed state is an important determinant of what microorganisms are likely to be present; there is a timetable for each type of patient that defines when in the postimmunosuppressed state different microbes are likely to invade the lung (exceptions to the timetable usually are due to unusual environmental exposures within the hospital).[29] Thus, patients with acute leukemia who have pneumonia on first presentation almost assuredly have either bacterial infection or leukemic infiltrates rather than opportunistic infection and almost never require a lung biopsy before initiating effective antimicrobial therapy. In contrast, this same patient with fever and pneumonitis after 3 weeks of chemotherapy-induced neutropenia and broad-spectrum antibacterial therapy is at high risk for invasive fungal infection. Thus, the understanding of the expected timetable of infection for each of the immunocompromising illnesses (see Chapters 13–17) can play an extremely useful role in constructing the differential diagnosis.

Perhaps the most useful clue to the correct diagnosis in patients with the febrile pneumonitis syndrome comes from an assessment of the mode of onset and rate of progression of the pulmonary process. Thus, an acute onset over less than 24 hr of symptoms severe enough to bring the patient to medical attention would suggest conventional bacterial infection (and, of the noninfectious causes, pulmonary embolic disease, pulmonary edema, a leukoagglutinin reaction, or pulmonary hemorrhage). A subacute onset over a few days to a week would suggest viral or *Mycoplasma* infection, *Pneumocystis*, or in some instances *Aspergillus*, *Nocardia*, or *Rhodococcus*. A more chronic course over one or more weeks would suggest fungal, nocardial, rhodococcal, or tuberculous infection (as well as tumor or radiation or drug-induced pneumonitis). When the mode of clinical presentation is combined with the radiologic finding, the range of etiologic possibilities becomes considerably smaller and much more manageable for the clinician (Table 3).[2,3]

**TABLE 3. Differential Diagnosis of Fever
and Pulmonary Infiltrates in the Compromised Host
According to Roentgenographic Abnormality and
the Rate of Progression of the Symptoms**[a]

Chest radiographic abnormality	Etiology according to the rate of progression of the illness	
	Acute	Subacute-chronic
Consolidation	Bacterial (including Legionnaires' disease) Thromboembolic Hemorrhage (pulmonary edema)	Fungal Nocardial Tuberculous Tumor (Viral, drug-induced, radiation, *Pneumocystis*)
Peribronchovascular	Pulmonary edema Leukoagglutinin reaction (bacterial)	Viral *Pneumocystis* Radiation Drug-induced (fungal, nocardial, tuberculous, tumor)
Nodular infiltrate[b]	(Bacterial, pulmonary edema)	Tumor Fungal Nocardial Tuberculous (*Pneumocystis*)

[a]Modified from Rubin.[3] An acute illness is one that develops and requires medical attention in a matter of relatively few hours (<24). A subacute-chronic process develops over several days to weeks. Note that unusual causes of a process are in parentheses.
[b]A nodular infiltrate is defined as one or more large (>1 cm² on chest radiography) focal defects with well-defined, more or less rounded edges, surrounded by aerated lung. Multiple tiny nodules of smaller size, as sometimes caused by such an agent as CMV or varicella-zoster virus, are not included here.

Additional useful information may also be obtained by measuring the arterial partial pressure of oxygen (Pao_2) while the patient is breathing room air. Most of the disease processes that cause the febrile pneumonitis syndrome in the compromised host are associated with significant impairment in oxygenation early in the clinical course (room air Pao_2 <65 mm Hg). By contrast, most patients with pulmonary disease caused by fungi, tuberculosis, *Nocardia*, *Rhodococcus*, and tumor will have relatively well-maintained oxygenation (room air Pao_2 >70 mm Hg) until very late in the course, despite extensive consolidation on chest radiography. Although a rare patient with these forms of infection will have an acute overwhelming pneumonia resembling conventional acute bacterial infection in both clinical presentation and arterial blood gas findings, the great majority will have subacute or chronic presentations associated with well-preserved oxygenation. For example, among more than

25 organ transplant recipients with primary fungal or nocardial pulmonary infections, all but 2 had a Pao_2 value of >70 mm Hg. Both exceptions had concomitant congestive heart failure and chronic obstructive pulmonary disease to explain their low Pao_2 values.[2,3]

The hypoxemia in patients with acute bacterial and viral infection and the noninfectious causes of the febrile pneumonitis syndrome results from a large shunt combined with regions of low ventilation–perfusion (V/Q) ratios in the involved lung tissues. Maintenance of the Pao_2 in the fungal, nocardial, tuberculous, rhodococcal, and presumably tumor patients appears to be due to diversion of obstructed blood flow to the involved lung, thus minimizing V/Q mismatch. The pathogenetic mechanism for this obstruction may be thrombosis, vascular obstruction, direct invasion of the blood vessels, or a strong, unopposed reflex arteriolar vasoconstriction in response to regional alveolar hypoxia.[2,3]

3. Overview of Noninfectious Causes of the Febrile Pneumonitis Syndrome

The occurrence of noninfectious forms of the febrile pneumonitis syndrome in the compromised host is determined by the underlying disease and how it is treated. Thus, in the patient with malignant disease, the major causes of this syndrome are radiation pneumonitis, drug-induced pulmonary injury, parenchymal tumor invasion, and rarely an unusual form of alveolar proteinosis. In the organ transplant patient, in the patient receiving corticosteroids, and in other groups of patients immunocompromised by nonmalignant disease, the major considerations are pulmonary emboli and pulmonary edema. In the patient with HIV infection, noninfectious causes of pulmonary infiltrates of particular importance include Kaposi's sarcoma, non-Hodgkin's lymphoma, and two unusual forms of interstitial lung disease of unclear etiology: lymphoid interstitial pneumonitis (particularly in children with AIDS) and nonspecific interstitial pneumonitis. AIDS patients also can develop drug-induced pulmonary disease as a consequence of the use of such drugs as bleomycin in the treatment of the secondary malignancies that can complicate the course of the AIDS patient (Table 4). In addition, critically ill patients with all forms of immunocompromise can develop ARDS as a result of systemic sepsis. Less commonly, any patient with a major clotting or platelet disorder can develop pulmonary hemorrhage and any transfused patient is at risk for the development of a leukoagglutinin reaction (Table 1).[3,5,8,10,11]

TABLE 4. Cytotoxic and Noncytotoxic Chemotherapeutic Agents Known to Induce Pulmonary Disease[a]

Cytotoxic	Noncytotoxic
Azathioprine	Bleomycin sulfate
Bleomycin sulfate	Cytosine arabinoside
Busulfan	Methotrexate sodium
Chlorambucil	Procarbazine
Cyclophosphamide	hydrochloride
Hydroxyurea	
Melphalan	
Mitomycin	
Nitrosourea (BCNU, CCNU, methyl-CCNU)	
Procarbazine hydrochloride	

[a]Modified from Rosenow et al.[5] Note that although both bleomycin and procarbazine are associated mainly with cytotoxic reactions, noncytotoxic reactions have also been observed, albeit uncommonly.

3.1. Radiation Pneumonitis

Radiation lung injury is of two types: an acute form, radiation pneumonitis, which begins at the end of a course of radiation therapy or up to 6 months later, and a chronic form, radiation fibrosis, which may follow acute disease or begin without premonitory symptoms 6 or more months following the completion of therapy. Pathologically, radiation pneumonitis is characterized by the desquamation of bronchiolar and alveolar cells and by the formation of protein-rich hyaline membranes as a result of the exudation of plasma into the alveolar spaces through injured pulmonary capillaries. Engorgement and thrombosis of capillaries and arterioles are evident and the alveolar septa are thickened by lymphocytic infiltrates and immature collagen deposition. Changes in surfactant production and metabolism may be particularly striking during this phase of radiation injury. This exudative phase is manifested on CT scan by findings of a homogeneously increased attenuation that progresses over time to patchy and then more dense consolidation. Particularly in the early exudative phase, high-dose corticosteroid administration is associated with rapid resolution of these CT scan findings (although moderate doses of prednisone, <40 mg/day ± azathioprine, do not prevent radiation lung injury).[100–107] Radiation fibrosis is characterized by the replacement of normal pulmonary parenchyma and architecture by dense connective tissue. Clinically, radiation pneumonitis begins insidiously with fever without rigors, nonproductive cough, progressive dyspnea, and a characteristic pattern of pulmonary infiltrates on chest radiography (see Illustrative Case 5). Radiation fibrosis, as its name suggests, is not associated with symptoms of ongoing inflammation; it is usually asymptomatic, and when signs and symptoms are present, they are related to the progressive pulmonary fibrosis (decreased exercise tolerance, dyspnea, orthopnea, cyanosis, clubbing, and cor pulmonale).[100–104]

The incidence and severity of radiation lung injury are largely determined by the characteristics of the radiation administered: the volume of lung exposed, the higher the dose, and the shorter the period of time over which the therapy is administered (fractionation of therapy permits repair of sublethal damage between doses), the higher is the incidence of radiation lung disease. It has been suggested that with the therapy protocols in current use, radiation pneumonitis is rarely seen at doses below 30 Gy, may develop at doses between 30 and 40 Gy, and is almost always evident with doses above 40 Gy. Evidence of radiation pneumonitis can be found an average of 8 weeks after a 40-Gy dose, appearing 1 week earlier for each 10-Gy increase in dose above 40. Because of the nature of the radiation therapy administered, symptomatic radiation injury of the lung is most common in patients receiving radiotherapy for breast and lung cancer and lymphoma, with clinical manifestations developing in 3–15% of these individuals.[100,108–117]

Certain modifying factors greatly enhance the risk of radiation lung injury, the most important of such factors being previous radiotherapy to the lung, abrupt withdrawal of corticosteroid treatment, and the concomitant administration of cytotoxic cancer chemotherapy. Surprisingly, age does not appear to play an important role. Finally, there appears to be individual variability in susceptibility to radiation injury, as there are several reports of severe pneumonitis developing after relatively small doses of radiation.[109,110,118–123] One might question whether such increased susceptibility is related pathogenetically to the uncommon occurrence of extensive radiation pneumonitis beyond the radiation field. A trivial explanation for these events is that they result from technical errors involving radiation port placement. However, particularly in the rare cases of bilateral pneumonitis following unilateral irradiation, it has been suggested that a delayed hypersensitivity reaction to an antigen generated or released by radiation injury is responsible. Consistent with this hypothesis are several observations: When bronchoalveolar lavage is carried out in patients undergoing unilateral radiation therapy, an increase in the number of lymphocytes can be demonstrated from both lungs; gallium scans carried out in the same time period demonstrate increased uptake in both the irradiated and the nonirradiated lung. The long latent period, the involvement of nonirradiated tissue, the idiosyncratic occurrence of the process, the striking clinical response to corticosteroid therapy that is observed, and the experimental observation that a monoclonal antibody that blocks the interaction between the CD40 ligand and the CD40 receptor is protective against radiation lung injury are all consistent with the hypothesis that a hypersensitivity reaction can play a role in the pathogenesis of radiation pneumonitis. In addition, nonspecific inflammation appears to play a role, as shown by the observation that nonsteroidal anti-inflammatory drugs can provide significant protection in experimental models of radiation pneumonitis.[124–131]

It is clear, moreover, that other processes are involved in the pathogenesis of radiation lung injury. Nitric oxide appears to be an inflammatory mediator in the process, with an inhibitor of nitric oxide synthase being shown in an experimental model of radiation pneumonitis to be protective.[132] There is an increasing body of evidence that suggests that oxygen free radicals are the key mediators of radiation lung injury, with administration of human superoxide dismutase to experimental animals by gene therapy offering significant protection against injury. It is likely that the fibrosis that develops, the final common pathway of the different processes involved, is secondary to the release of transforming growth factor beta (TGF-β). The overall complexity of the pathogenesis of radiation lung injury is further illustrated by observations that thrombomodulin is involved, and that in the experimental animal vitamin A administration is protective.[123,133–138]

In sum, then, the clinician should be highly suspicious of the possibility of radiation pneumonitis in a patient with one of the tumor types noted who has a subacute–chronic onset of respiratory symptoms during or following the completion of a course of radiotherapy and from whom a history of one or more of the adjunctive factors can be obtained. Early diagnosis by biopsy can lead to effective therapy with corticosteroids, although such therapy will be ineffective if significant delay occurs.

Illustrative Case 5

A 41-year-old woman with known Hodgkin's disease presented with a 3-week history of increasing dyspnea, nonproductive cough, fever, night sweats, and malaise. Stage IIIb Hodgkin's disease had been diagnosed 6 months previously and was treated with 5000 rads to the mediastinum and para-aortic lymph nodes followed by cyclic MOPP (nitrogen mustard, vincristine, prednisone, and procarbazine) chemotherapy, the most recent cycle having been completed 2 weeks previously. The patient noted the insidious onset of fever, nonproductive cough, malaise, and dyspnea on exertion. Although she was quite dyspneic by the time she sought medical attention, she could not designate any one time when a major change in her clinical status occurred. There were no significant travel exposures and there were no illnesses in her family. Physical examination revealed a dyspneic woman with a temperature of 101°F (38.3°C) and a respiratory rate of 35. General physical examination was unrevealing. Laboratory data included Hct 34%, WBC 5400, with 82% polys, 11% lymphs, and 7% monos. Room air arterial blood gases revealed a Pa_{O_2} of 42 mm Hg, Pa_{CO_2} 28 mm Hg, and pH 7.52. Chest radiography revealed diffuse consolidative and peribronchovascular opacities predominantly central in distribution (Fig. 3). The diagnosis of radiation pneumonitis was made by transbronchial biopsy through the fiberoptic bronchoscope. The patient was treated with corticosteroids with marked improvement. Two weeks after the initiation of steroids, room air arterial blood gases were Pa_{O_2} 90 mm Hg, Pa_{CO_2} 36 mm Hg, and pH 7.45.

Comment. This is a classic case of radiation pneumonitis: an insidious onset of the febrile pneumonitis syndrome in the appropriate clinical setting, with a typical chest radiograph. Because of the need for corticosteroid therapy, concomitant infection needed to be ruled out, particularly *P. carinii* and a variety of viruses, including CMV. This was done and the patient had a gratifying response to therapy.

3.2. Drug-Induced Pneumonitis

Several chemotherapeutic agents, most notably bleomycin, busulfan, mitomycin (with and without vinca alkaloids), cyclophosphamide, and chlorambucil, produce pulmonary injury akin to that caused by radiation (see Table 4). It should be assumed that all alkylating agents have this ability because of their radiomimetic and mutagenic capabilities. In the lung, these effects cause injury, particularly to the lining epithelium of the alveoli and to the alveolar capillary endothelium, resulting, as with radiation lung injury, in two clinical syndromes: a progressive interstitial pneumonitis characterized by fever without chills, dyspnea, nonproductive cough, and progressive hypoxia usually beginning weeks to months after significant amounts of the drug have been administered; and a chronic interstitial fibrosis that may follow symptomatic inflammatory lung disease or occur insidiously without previous warning. Some patients may have symptoms due to drug-induced lung injury but negative chest radiographs. In these instances, a decrease in diffusing capacity, a positive Ga-67 scan, or finding on chest CT scan can be very helpful.[3,5,10,11]

The inflammatory manifestations, like early radiation pneumonitis, may be responsive to the cessation of the provoking drug and the initiation of steroid therapy, whereas the chronic fibrotic process is not. Again, as with radiation pneumonitis, both the dose of the agent administered (and the time course over which it is administered) and individual susceptibility to lung injury appear to be important in the pathogenesis of this process. Given the similarities between drug-induced and radiation lung injury in presumed pathogenesis, radiographic and histologic

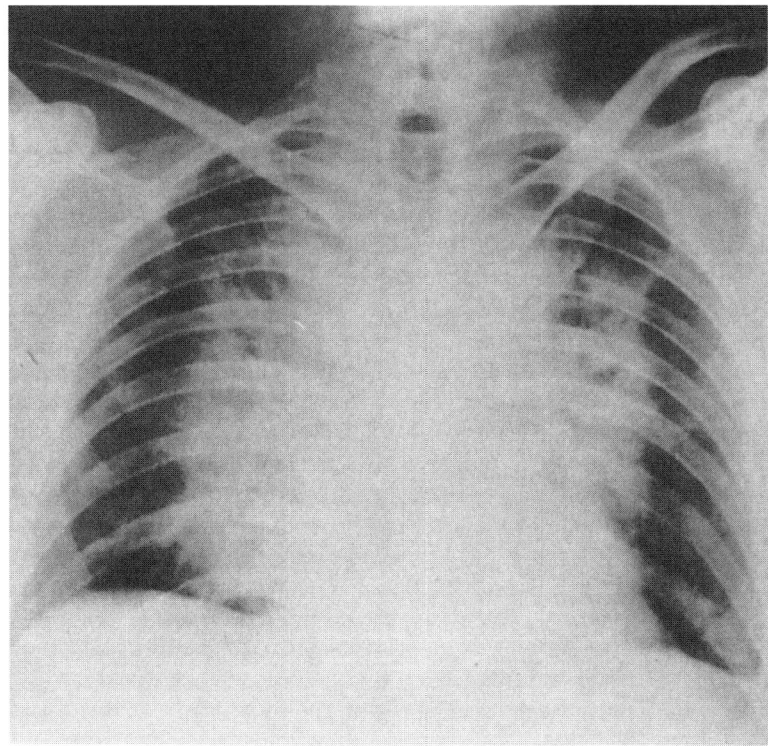

FIGURE 3. Radiation pneumonitis. Nonanatomic perimediastinal lung opacity 2 months after cessation of radiotherapy corresponds to the shape of irradiation portals and is typical of radiation pneumonitis.

appearances, and clinical presentation, it is not surprising that the combination of these treatment modalities is associated with a greater risk of pulmonary disease than when either type of agent is used alone.[109–111,118–120,122–124,139–142] It has been suggested that both these drugs and radiation induce the local production of oxygen radicals such as hydrogen peroxide and superoxide anions, which then produce lung injury. Consistent with this hypothesis is the observation that administration of oxygen to a patient receiving bleomycin will accelerate drug-induced pulmonary toxicity.[143–147] Conversely, the administration of the antioxidant vitamin E and superoxide dismutase to animal models of bleomycin lung disease have been shown to be protective.[148–150] Both types of lung injury are associated with restrictive defects and lowered diffusing capacities on pulmonary function testing.[151–153] Studies of bleomycin toxicity in animal models suggest that disturbances in surfactant production play a role in the genesis of the pulmonary functional defect and that the administration of artificial surfactant, perhaps through the interference with neutrophil adherence and superoxide production in the injured lung, can provide protection.[154,155] Given the subacute onset of symptoms, the

interstitial pulmonary infiltrate, and the clinical setting, the major differential diagnostic considerations are *P. carinii* infection and viral pneumonitis. Usually, these conditions can be distinguished only by lung biopsy.[156–163]

3.2.1. Busulfan Pulmonary Injury

Of all the cancer chemotherapeutic agents, busulfan and bleomycin are the drugs most commonly implicated as causes of pulmonary injury. In the case of busulfan, pulmonary injury is usually observed in patients placed on maintenance long-term therapy with this agent in the treatment of chronic myelogenous leukemia or, less commonly, polycythemia vera. In these patients, pulmonary disease may develop as early as 1 year after the initiation of such therapy, but more commonly requires up to 4 years.[164,165] The incidence of busulfan lung injury is not well established but appears to be less than 5%. For example, in one series of 23 well-studied patients followed for an average period of approximately 2 years, only one patient developed clinical pulmonary disease and this patient received a smaller dose of busulfan than did several of the other patients. Pathologically, intra-alveolar fibrosis and large atypical alveolar mononuclear cells may be seen in a much higher percentage of patients receiving chronic busulfan therapy.[166–169] Figure 4 demonstrates the typical radiologic findings in a patient with this form of lung injury.

3.2.2. Bleomycin Pulmonary Injury

In part because of its efficacy as a chemotherapeutic agent for a wide variety of tumors, bleomycin is the single most common cause of drug-induced pulmonary injury. In an estimated 2.5–13% of patients receiving this agent, symptomatic pulmonary disease develops, and the reported mortality has been as high as 50%.[158–163,170–175] Clinically manifest lung injury will occur in most patients receiving a cumulative bleomycin dose greater than 500 mg.[171] In addition, there have now been several reports of life-threatening pulmonary disease when as little as 50–180 mg has been administered.[174,176,177] Early pulmonary injury may be detected by the demonstration of a decrease in diffusing capacity and vital capacity at a time when the chest radiograph is still normal.[172,176] An important clue for the clinician is the relatively high rate of bleomycin toxicity in patients who have received prior radiotherapy. For example, in one study of 101 patients receiving bleomycin therapy, 5 of 12 who had received previous radiotherapy developed pulmonary disease, whereas only 4 of 89 not receiving radiotherapy developed comparable levels of pulmonary injury.[173]

The nitrosourea compounds (BCNU, CCNU, and methyl-CCNU) also have emerged as important causes of cytotoxic drug-associated pulmonary injury akin to that caused by busulfan and bleomycin.[139,178,179] Less commonly, cyclophosphamide and chlorambucil appear to be associated with the same process.[140,180–186] The recent report of a high incidence of *Pneumocystis carinii* pneumonia in patients with necrotizing vasculitis or leukemia being treated with cyclophosphamide and prednisone underlines the clinician's dilemma. The clinical and radiologic presentations of *P. carinii* pneumonia and cytotoxic drug-induced pneumonitis are identical. Invasive studies for a precise diagnosis rather than empiric therapy therefore are essential in this clinical situation.[187]

Illustrative Case 6

A 51-year-old man with Hodgkin's disease was admitted with a 2-week history of fever, night sweats, nonproductive cough, and increasing shortness of breath. Two years earlier, Hodgkin's disease was diagnosed and was treated with splenectomy, total nodal irradiation, and several cycles of MOPP therapy for stage IIIA disease. Eight months previously, recurrent disease involving the lung, liver, and bone was diagnosed and bleomycin was begun. A total dose of 340 mg had been

administered by the time of this admission. On physical examination, his temperature was 100.6°F (38.1°C), respiratory rate was 30, and pulse rate was 90. He was a chronically ill-appearing man in mild respiratory distress. Fine rales were heard over both lung bases. Laboratory evaluation revealed Hct 34%, WBC 3200 with 83% polys, 4% lymphs, and 13% monos. Room air arterial blood gases revealed a Pa_{O_2} of 54 mm Hg, Pa_{CO_2} of 32 mm Hg, and pH 7.50. Chest radiography demonstrated a diffuse peribronchovascular infiltrate in both lung fields, most prominent at the bases. Transbronchial biopsy via the fiberoptic bronchoscope yielded characteristic changes of bleomycin-induced lung disease, with no evidence of infection. Therapy with corticosteroids was associated with symptomatic improvement and an increase in room air Pa_{O_2} to approximately 60–65 mm Hg. Over the next 3 months, he remained stable with respect to pulmonary function and radiographic findings. However, he succumbed to progressive Hodgkin's disease. At postmortem examination, a mixed pulmonary picture consisting of extensive intra-alveolar fibrosis and interstitial pneumonitis with frequent large atypical alveolar mononuclear cells was observed; the usual picture of bleomycin (and busulfan) lung injury.

Comment. This was a typical presentation of bleomycin-induced lung disease in a patient predisposed to its development by the previous radiotherapy he had received. The dose of bleomycin administered, although not excessive, was clearly in the range associated with at least a 5–10% risk of pulmonary toxicity. Again, the treatable differential diagnostic possibilities lay chiefly between bleomycin lung disease and *Pneumocystis.* Transbronchial biopsy provided an easy, well-tolerated means of diagnosis. Only a moderate response to corticosteroid therapy was observed in this case.

3.2.3. Methotrexate Pulmonary Injury

Methotrexate (MTX), a folic acid antagonist (antimetabolite), is widely used in the treatment of leukemia, lymphoma, and other neoplastic conditions, in bone marrow transplantation, and increasingly in the management of such nonmalignant conditions as psoriasis, rheumatoid arthritis, and ectopic pregnancy. MTX also can produce subacute and chronic pulmonary injury syndromes, but in most instances these syndromes differ from those caused by radiation and the alkylating agents: MTX appears to cause an acute, allergic, granulomatous reaction that may subside despite the continuation of therapy (or not reappear after cessation of the drug and subsequent rechallenge), although progressive, chronic interstitial fibrosis also may develop. The duration of MTX therapy before the onset of symptoms can range from a few days (in at least one reported case, with a total dose of 12.5 mg) to more than 5 years with an average weekly dose during this period of 25–50 mg. In addition, it can occur weeks after the drug has been discontinued.[188–204] Clinical disease and radiographic findings resemble those seen with the other forms of drug-induced pulmonary injury. However, eosinophilia is commonly present, and bronchoalveolar lavage in patients with MTX-induced pneumonitis (as opposed to those receiving MTX without

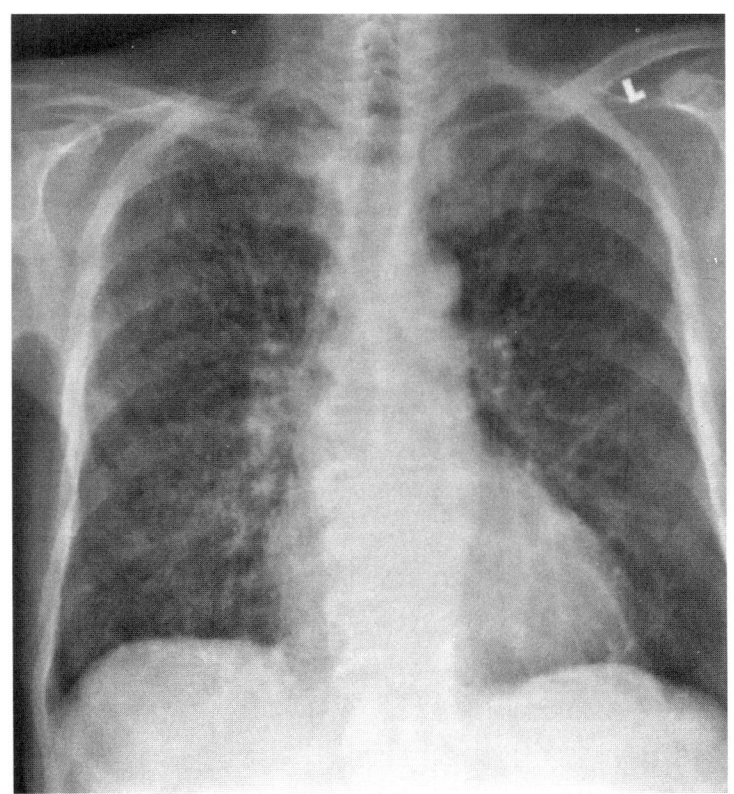

A

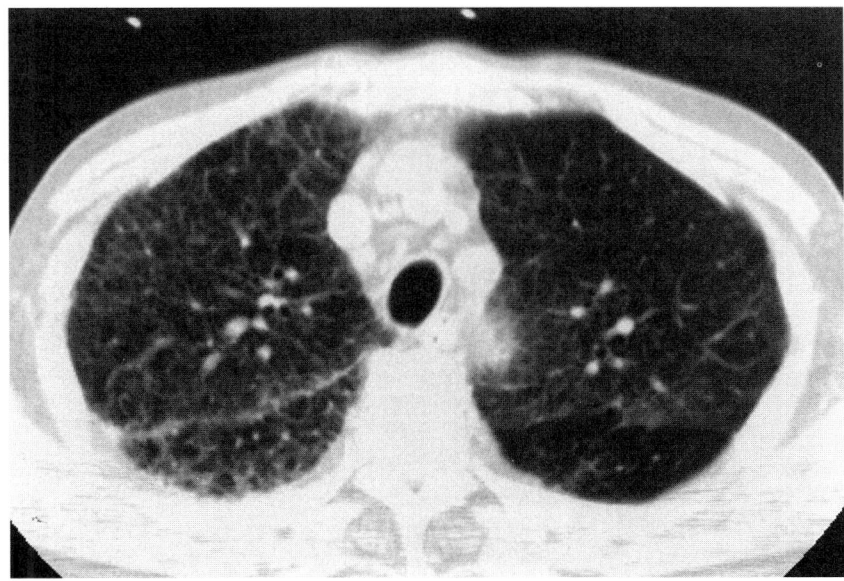

B

FIGURE 4. Bronchiolitis obliterans-organizing pneumonia (BOOP) secondary to busulfan. (A) Chest radiograph shows diffuse reticular lung opacities in a patient with acute respiratory decompensation during busulfan therapy for myelofibrosis. (B) A CT section through the upper chest demonstrates nonspecific diffuse ground-glass opacities, septal lines, and bronchiolectasis.

pulmonary effects) reveals evidence of a lymphocytic alveolitis with a predominance of CD4 cells, both findings suggesting an immunologic rather than a toxic reaction. Not surprisingly, preceding radiation therapy appears to play no role in predisposing patients to this form of pulmonary injury.[205] In addition to corticosteroids, cyclophosphamide has been reported useful in the management of this condition (as it is for other forms of granulomatous tissue injury, e.g., Wegener's granulomatosis).[206] Rarely, bleomycin, cytosine arabinoside, and procarbazine can produce lung injury with an MTX-like pathologic picture, presumably due to a similar allergic rather than cytotoxic mechanism. Conversely, there is an occasional case of pneumonitis associated with MTX use in which the pathology and clinical course more closely resemble a bleomycinlike toxic reaction than the more common, presumably allergic disease.[5,195,205]

3.2.4. Uncommon Causes of Drug-Induced Pulmonary Injury

Other agents to which compromised patients may be exposed that may be associated with the production of either (or both) the febrile pneumonitis syndrome and chronic interstitial fibrosis include melphalan,[207] azathioprine,[208] diphenylhydantoin,[209] amitriptyline,[210] parenteral gold therapy,[211] d-penicillamine,[212] nitrofurantoin,[213] mitozantrone,[214] and amiodarone.[215] This last, a potent antiarrhythmic agent, has been a particular problem in heart transplantation, where we have observed patients who were receiving amiodarone at the time of transplant develop an acute ARDS syndrome, probably due to the combination of amiodarone and cyclosporine, and the other circumstances that obtain during cardiac transplantation. Our policy at present is to attempt to wean patients from amiodarone prior to transplantation.

3.3. Neoplastic Pulmonary Invasion

Fever and a clinical presentation suggesting pneumonia caused by neoplastic invasion of the lung may sometimes occur, particularly in patients with lymphoma. Pulmonary involvement occurs in 20–30% of patients with Hodgkin's disease, with most having the nodular sclerosis type of tumor. Almost invariably, Hodgkin's disease of the lung is associated with mediastinal lymph node involvement (or at least a history of previously treated mediastinal disease). In the rare Hodgkin's disease patient with pulmonary invasion in the absence of mediastinal adenopathy, there almost always is evidence of extrathoracic disease.[216–219]

Primary intrathoracic disease is uncommon in patients with non-Hodgkin's lymphoma (4% in one series of 1269 patients), but in approximately half of these patients, thoracic disease will eventually develop.[220] Primary pulmonary disease accounts for only 10% of these thoracic cases. Unlike the situation in Hodgkin's disease, a significant proportion of these patients may have pulmonary involvement in the absence of mediastinal nodal disease.[219–222]

Leukemia patients occasionally may have leukemic infiltrates in their lungs with associated fever, especially patients with acute monocytic and chronic lymphatic leukemia. Necropsy studies would suggest that as many as 25% of such patients will have leukemic pulmonary invasion. Occasionally, particularly in leukemic patients with white blood cells greater than 200,000/mm^3 and a high percentage of blast cells, leukostasis and occlusion of pulmonary vessels may produce pulmonary symptoms and even infiltrates on chest radiographs. Superinfection of such areas is quite common. Because of decreased compliance within small blood vessels, leukemic blast cells may be particularly important in the pathogenesis of the intravascular leukostasis.[5,223–225] A variation of this process is what has been termed "leukemic cell lysis pneumopathy," in which persons with large numbers of circulating leukemic blasts develop fever, respiratory distress, and patchy lung infiltrates within a few days of rapid chemotherapy-induced destruction of blasts. It has been suggested that this occurrence represents either diffuse alveolar or pulmonary capillary damage due to enzymes released locally by the destroyed blast cells.[5,226] However, infiltrates in such patients sufficient to cause radiographic abnormalities are much more frequently the result of infection, hemorrhage, or heart failure, and the clinician should proceed on this basis rather than pass off such infiltrates as being caused by the leukemia.[227,228]

Perhaps the most common association between neoplastic pulmonary invasion and the febrile pneumonitis syndrome is related to endobronchial lesions from primary or metastatic cancer that may cause bronchial obstruction, distal atelectasis, and bacterial infection. Bronchoscopic demonstration of such lesions can be quite useful and lead to effective surgical or radiation therapy.[219]

3.4. Other Noninfectious Causes of the Febrile Pneumonitis Syndrome

Even in immunosuppressed patients without neoplastic disease, noninfectious causes of the febrile pneumonitis syndrome account for as many as 25% of such cases. Here, pulmonary emboli and atypical pulmonary edema are the major causes. Such difficulties have been particularly prominent in renal transplant patients.[2,229,230] In these patients, surgical manipulation of pelvic or lower extremity venous structures is especially associated with pulmonary emboli, acute allograft failure, oliguria, and fluid overload (rather than primary cardiac disease) in patients with pulmonary edema. The administration of such antilymphocyte antibody therapies as OKT3 can lead to a febrile pulmonary edema picture due to cytokine release.[231,232]

A major difference between immunosuppressed patients and normal patients with these forms of primary pulmonary disease is the high rate of superinfection in the immunosuppressed patient. For example, in one series of renal transplant patients,[2] eight of nine patients with pulmonary embolic disease developed life-threatening superinfection, with rates of superinfection nearly as high being noted in patients on high-dose corticosteroids, those with lymphoma, and those with other causes of significant immunosuppression. We have found that the restriction of immunocompromised patients to HEPA-filtered rooms after primary lung injury has had a major impact in preventing secondary superinfection of the injured lungs, particularly if the patient is intubated.

An unusual cause of fever and pulmonary infiltrates in the compromised host is a leukoagglutinin reaction. Leukoagglutinin reactions result in a syndrome of febrile pulmonary edema of noncardiac origin characterized by the abrupt onset of fever, chills, tachypnea, nonproductive cough, and respiratory distress in the first 24 hr following blood transfusion (and most commonly during the transfusion or in the first few hours following it). Such reactions are initiated by the interaction of preformed agglutinating antibodies with antigens on leukocyte surfaces, probably of both HLA and non-HLA type. The antibodies are usually present in the patient's serum because of sensitization by past transfusions or pregnancies and are directed against leukocytes transfused with the unit of blood; rarely, the antibodies may be present in the plasma of the blood being transfused, and they then act against the patient's leukocytes.[233–236]

Illustrative Case 7

A 22-year-old woman with a relapse of acute myelogenous leukemia developed fever and acute respiratory distress 4 hr after a blood transfusion. The patient had been well until 7 months previously, when she presented with fever and ecchymoses, and a diagnosis of acute myelogenous leukemia was made. Full remission was induced with cytosine arabinoside, daunorubicin, and thioguanine therapy after a stormy course requiring multiple red cell and platelet transfusions. Once in remission, she remained well until a few days prior to admission, when a bone marrow aspiration revealed recurrent disease and she was found to be anemic and thrombocytopenic with a guaiac-positive stool. She was admitted to the hospital and given 2 units of packed red blood

cells and 15 units of platelets. Approximately 4 hr after the initiation of the transfusion therapy, she complained of a shaking chill, spiked a fever, and developed progressive respiratory distress. Physical examination revealed an acutely dyspneic young woman with a temperature of 103.4°F (39.7°C), respiratory rate of 40, and diffuse rales and rhonchi over both lung fields. Cardiac examination revealed a regular tachycardia, no gallops, and a grade 2/6 pulmonic flow murmur. Room air arterial blood gases demonstrated a Pao_2 of 39 mm Hg, $Paco_2$ of 28 mm Hg, and pH of 7.56. Chest radiography revealed a normal-sized heart with a diffuse patchy infiltrate consistent with pulmonary edema (Fig. 5A). Sputum examination revealed no polymorphonuclear leukocytes or organisms. The patient was treated with intubation, positive end expiratory pressure, and oxygen supplementation as needed. At 12 hr after intubation, she was extubated, and by 24 hr, her chest radiograph had returned to normal (Fig. 5B).

Comment. Noncardiogenic pulmonary edema secondary to a leukoagglutinin reaction is evident in this case. The therapy here does not include digitalis or diuretics, but rather is centered on adequate respiratory support with oxygen, positive end expiratory pressure, and intubation as needed. As in any immunosuppressed host, extubation should be carried out as soon as possible.

Pulmonary alveolar proteinosis is characterized by the accumulation within the alveoli of eosinophilic, proteinaceous, periodic acid–Schiff-positive material that is surfactantlike in character.[237–239] It has been suggested that there are two forms of alveolar proteinosis: a primary form, in which the proteinaceous material is derived from surfactant, and a secondary one, in which cellular debris and fibrin are significant components of the intra-alveolar material. The secondary form occurs in association with leukemia, lymphoma, metastatic melanoma, chemotherapy, primary immunodeficiencies, AIDS, and as a consequence of infection.[240,241] A defect in the macrophage clearance of surfactant components from the alveoli has been postulated as the cause of this condition (with secondary forms of alveolar proteinosis developing as a consequence of macrophage injury). Conversely, defects in alveolar macrophage migration and macrophage metabolism are inducible by bronchoalveolar lavage material obtained from patients with alveolar proteinosis. These macrophage defects probably play a role in the relatively high rate of secondary infection, particularly with *Nocardia asteroides*, that occurs in patients with this condition.[237–244]

Symptomatically, patients with alveolar proteinosis commonly present with the subacute complaints of nonproductive cough, fever, and slight dyspnea on exertion. More acute symptoms raise the possibility of secondary microbial invasion. The usual radiographic manifestations are those of airspace disease, with a minimum of air bronchograms. Typically, the chest radiograph reveals a bilateral, diffuse, perihilar, or central infiltrate that is more prominent at the lung bases, although atypical radiographs can occur.[242]

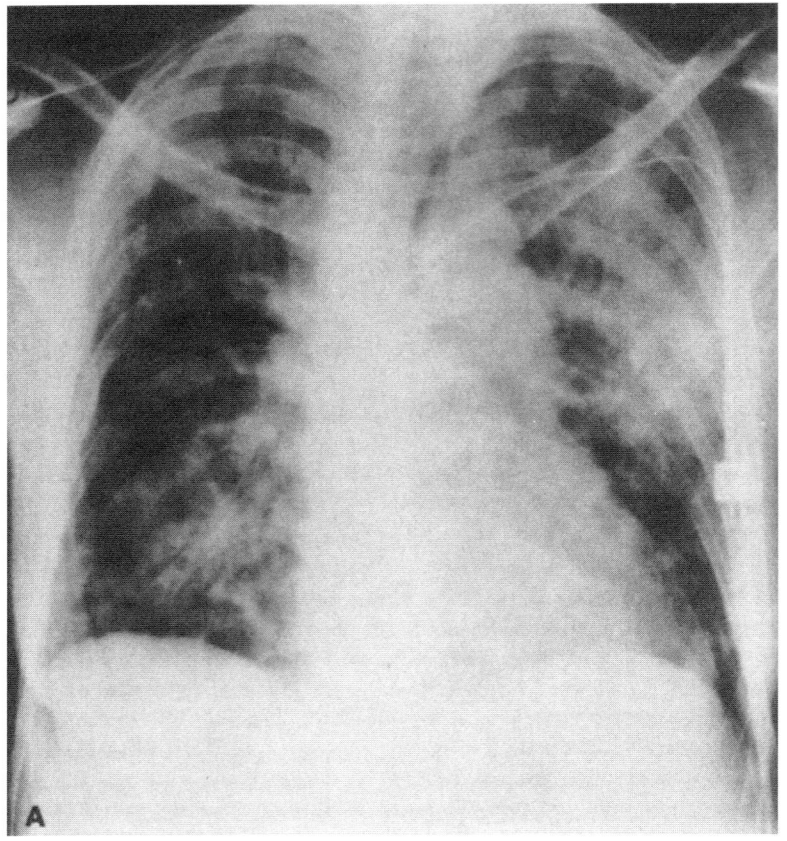

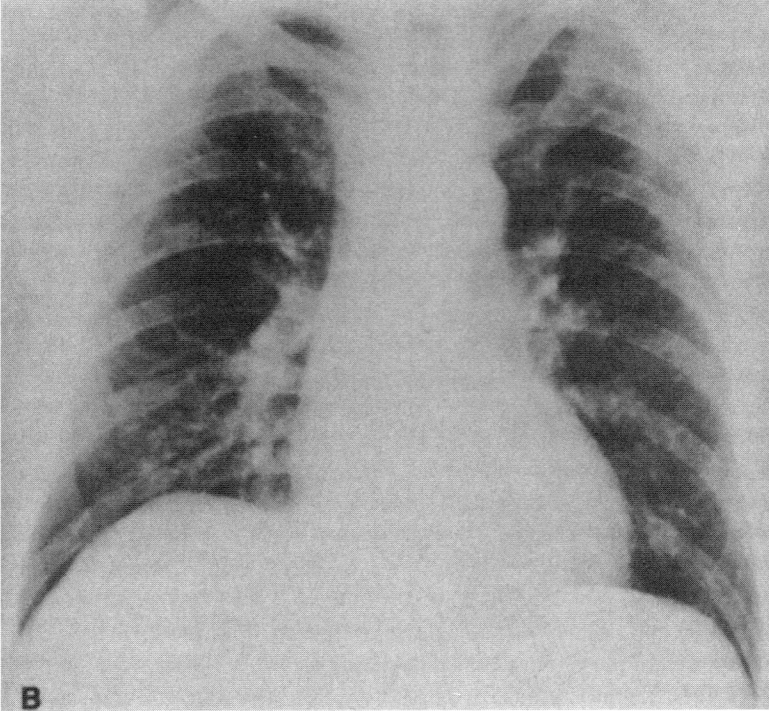

FIGURE 5. Leukoagglutinin reaction. (A) The appearance of patchy multifocal consolidation several hours after an intravenous transfusion is characteristic of the leukoagglutinin reaction. (B) At 24 hr later, the lung spontaneously cleared.

4. Radiologic Clues to the Diagnosis of the Febrile Pneumonitis Syndrome

Although no particular chest radiologic pattern is specific for a given pathologic process or microbial invader, particularly in the immunosuppressed patient, certain patterns are more characteristic of some processes than of others. Recognition of particular patterns can aid greatly in narrowing the range of differential diagnostic possibilities. The following radiographic characteristics are useful descriptors of radiologic patterns for clinical–radiologic–pathologic correlations[2]:

1. Time of appearance, progression, or resolution of new chest-imaging abnormalities as correlated with clinical findings.

2. Distribution and location of radiologic abnormalities. An opacity or small group of opacities confined to one anatomic area (e.g., segment or lobe) is considered focal (Fig. 1), whereas widespread or innumerable lesions are considered diffuse (Fig. 2 and Fig. 5A). Abnormalities that are distributed in more than one area but are not so abundant as to be too numerous to count are termed multifocal (Fig. 4). As visualized on CT, abnormalities may be located centrally (Fig. 4) or peripherally (Fig. 6) or both.

3. Lung opacities, which are divided into three major groups:

a. In the first group are consolidations, in which there is substantial replacement of alveolar air by tissue density material. Air bronchograms and peripheral location are characteristic of consolidative lesions. On CT, a dense consolidation often exhibits air bronchograms. Incomplete consolidation can result in ground-glass opacification. On CT, ground-glass opacification is recognized as regions with a slight increase in lung attenuation where the underlying vasculature remains visible and air bronchograms are absent. On radiography, ground-glass opacification results in subtle or barely visible lung opacity.

b. In the second group are linear and peribronchovascular (or interstitial) opacities. Interstitial opacities are predominantly oriented along the peribronchial or perivascular bundles. On CT and radiography, interstitial opacities are recognized by bronchovascular irregularity, or by thickened or irregular septa, or by both.

c. In the third group are nodular opacities that are subdivided into the following types: miliary, sublobular, lobular, and macronodules. Tiny nodules with diameters of less than 3 mm are often termed miliary and can be indicative of interstitial granulomas; e.g., miliary tuberculosis (Fig. 2B). Sublobular opacities are usually ill-defined nodules 4–8 mm in diameter, are often centrilobular in location, and have a "tree-in-bud" appearance (Fig. 7). These findings are indicative of airspace disease in small lung units prior to frank consolidation. Sublobular nodules may be caused by acute or chronic pneumonias, metastases, or large granulomas. Lobular nodular opacities (10–15 mm) have septal margins and often are seen in infectious consolidations of secondary pulmonary lobules. These lesions may also be seen in pulmonary infarcts or aspiration (Fig. 6). Macronodules (i.e., >10–15 mm) include lobular nodular opacities and are often space-occupying, nonanatomic lesions with well-defined, more or less rounded edges surrounded by aerated lung. These are often seen in fungal infection. Occasionally, small, well-defined, peripheral consolidations may take on the appearance of macronodules. The nodules of Kaposi's sarcoma are variable, ranging from sublobular to lobular and to macronodular size (Fig. 8). The peribronchial location of nodules in Kaposi's sarcoma is a clue to their identity and interstitial location.

4. Other characteristics that should be looked for include pleural fluid, atelectasis, cavitation, lymphadenopathy, cardiac enlargement, and pericardial abnormality. Pleural fluid is a clue to congestive heart failure and fluid overload when bilateral and to necrotizing or granulomatous infection (or subdiaphragmatic inflammation), especially when associated with lymphadenopathy or cavitation, when unilateral.

5. In the case of pulmonary nodules seen on CT in the severely neutropenic transplant patient, a surrounding halo of ground-glass opacification is an important characteristic early CT lesion of invasive aspergillosis attributed to hemorrhage associated with local infarction[2–5] (Figs. 9, 10), and an air crescent cavitation is an important characteristic CT lesion of late invasive aspergillosis attributed to ischemic necrosis[6–8] (Fig. 11). The ground-glass halo is identified as a substantial surround of ground-glass opacity and should be differentiated from the faint indistinct margination that might be caused by technical volume averaging (Figs. 9, 10). An air crescent is identified by a distinct crescent of gas capping a ball of necrotic debris in a healing focus of invasive aspergillosis and needs to be differentiated from a thick-walled cavity lacking a distinct crescent (Fig. 11).

4.1. Correlation of Radiologic Findings, Rate of Progression, and Clinical Signs

By combining this classification of radiologic findings with information concerning the rate of progression of the illness (acute vs. subacute–chronic), as outlined in Table 3 (Section 2.3), a useful differential diagnosis is

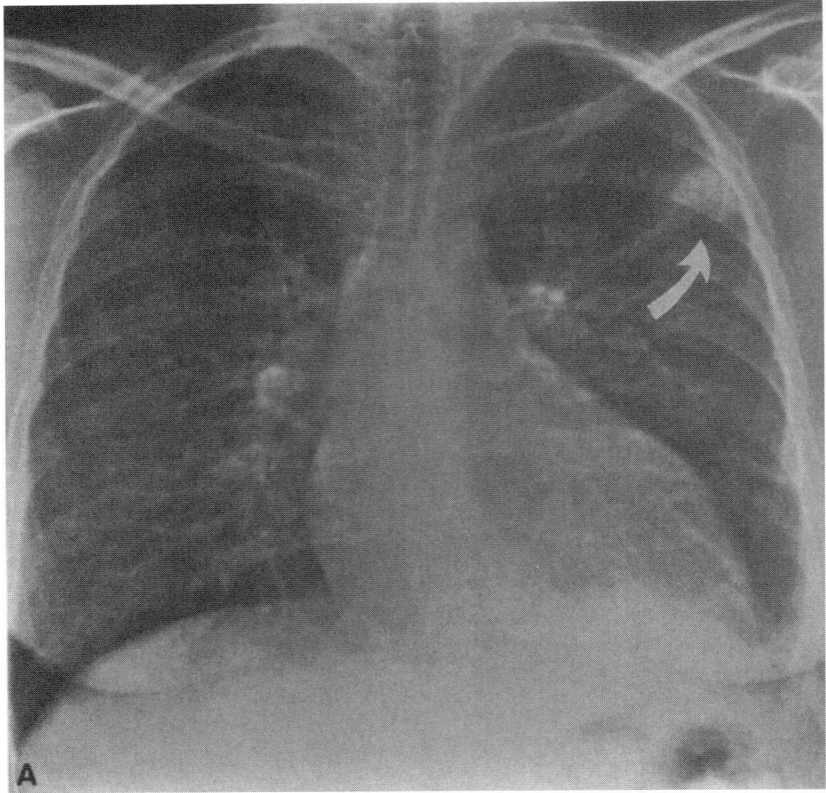

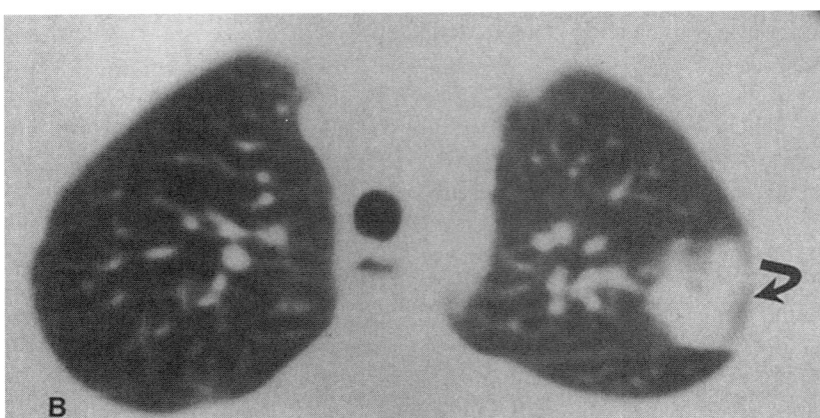

FIGURE 6. Pulmonary infarct. (A) A small opacity in the periphery of the left upper lobe of a diabetic patient with fever and a foot ulcer is due to a pulmonary infarct. The opacity is caused by a hemorrhagic infarct in a secondary pulmonary lobule. A straight interface along the bottom of the opacity is caused by a marginating interlobular septum. (B) A CT scan confirms the presence of the peripheral opacity. Other scan sections demonstrated additional bilateral infarcts and a left pleural effusion.

then generated. Thus, focal or multifocal consolidation of acute onset will quite likely be caused by bacterial infection; similar lesions with subacute-chronic histories are most likely secondary to fungal, tuberculous, or nocardial infections. Macronodules are usually a sign of fungal or nocardial disease, particularly if they are subacute in onset. Subacute disease with diffuse abnormalities, either of the peribronchovascular type or miliary micronodules, are often caused by viruses or *Pneumocystis* (although in the AIDS patient, disseminated tuberculosis and systemic fungal infection are also considerations). As noted in Table 3, noninfectious causes are added to the differential diagnosis when the history is appropriate, the radiologic findings are consistent, and ancillary radiographic signs (such as hilar adenopathy in patients with Hodgkin's disease) are present.[2,3]

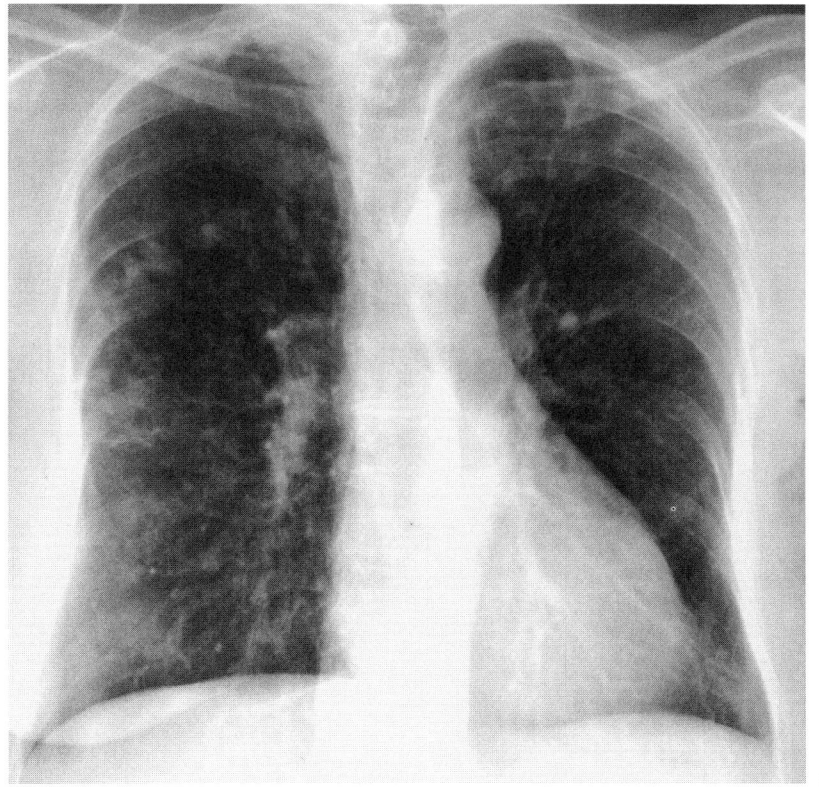

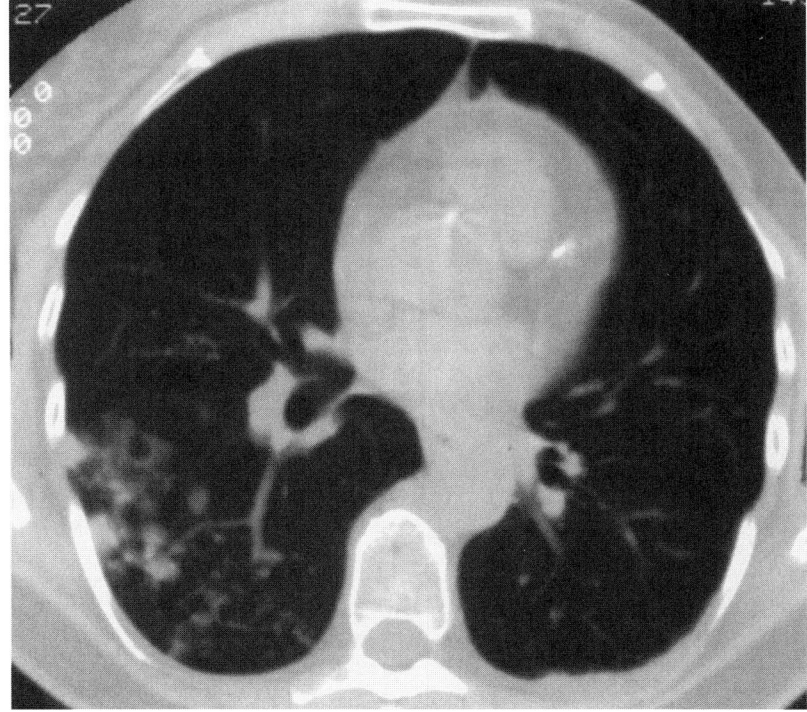

FIGURE 7. Tree-in-bud opacities due to *Mycobacterium avium-intracellulare* infection in an AIDS patient. (A) Frontal chest radiograph shows multiple small patchy opacities at the mid-level of the right lung. (B) A CT section shows multiple tree-in-bud opacities in the right lower lobe due to *M. avium-intracellulare* infection.

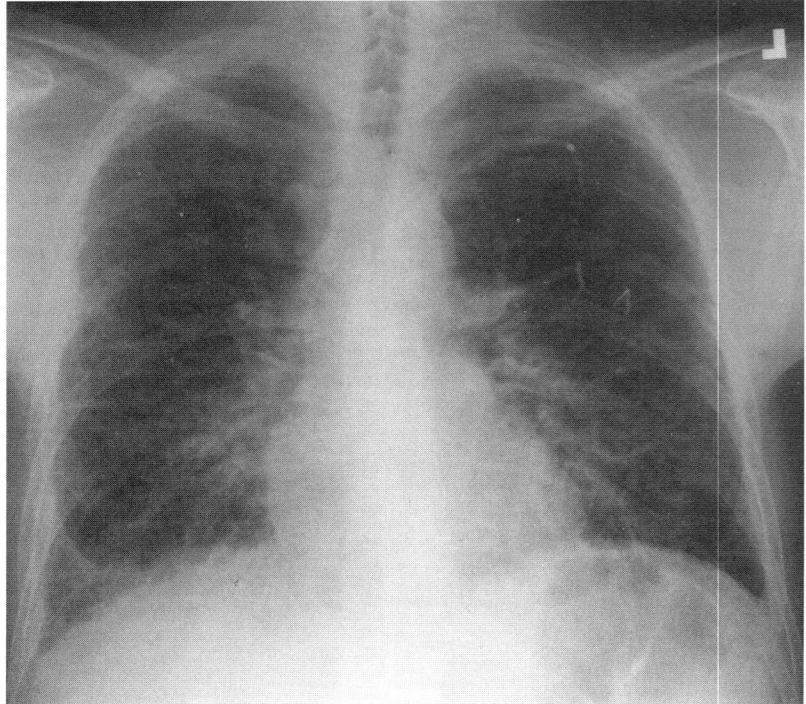

A

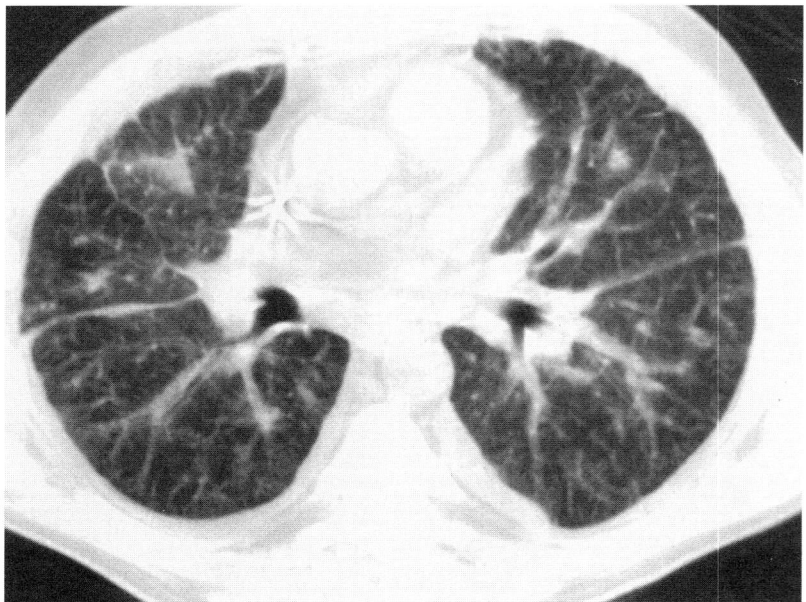

B

FIGURE 8. AIDS and Kaposi's sarcoma. Diffuse bronchovascular thickening, septal lines, and peribronchial nodules due to AIDS-related Kaposi's sarcoma. (A) Frontal chest radiograph shows diffuse reticular peribronchovascular thickening and septal lines extending to the pleural surface associated with bilateral pleural thickening. (B) A CT section through the lower chest shows diffuse bronchial wall thickening, peribronchial nodules, septal lines at the pleural surface, and bilateral pleural effusions.

Additional clues can be found by examining the pulmonary lesion for the development of cysts or cavitation or by carefully delineating the location of the opacity or opacities. Cavitation suggests necrotizing infection such as that caused by fungi, *Nocardia, Rhodococcus,* certain gram-negative bacilli (most commonly *Klebsiella* and *Pseudomonas*), and *Staphylococcus aureus,* or necrotic tumor.[2,3,95,245,246] Atypical upper lobe cysts may be found in AIDS patients treated with inhaled pentamidine who develop *Pneumocystis carinii* pneumonia (Fig. 12).

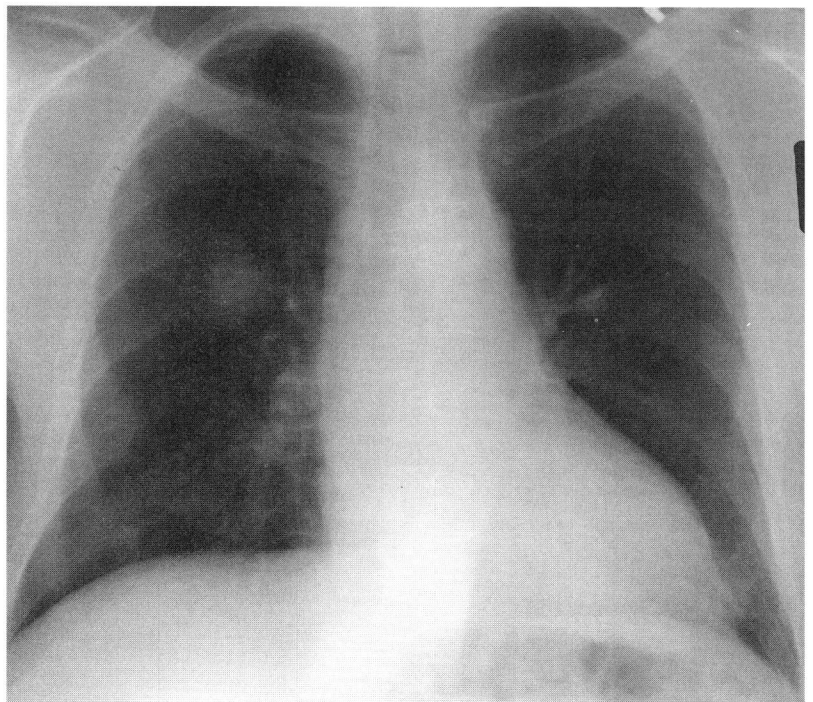

A

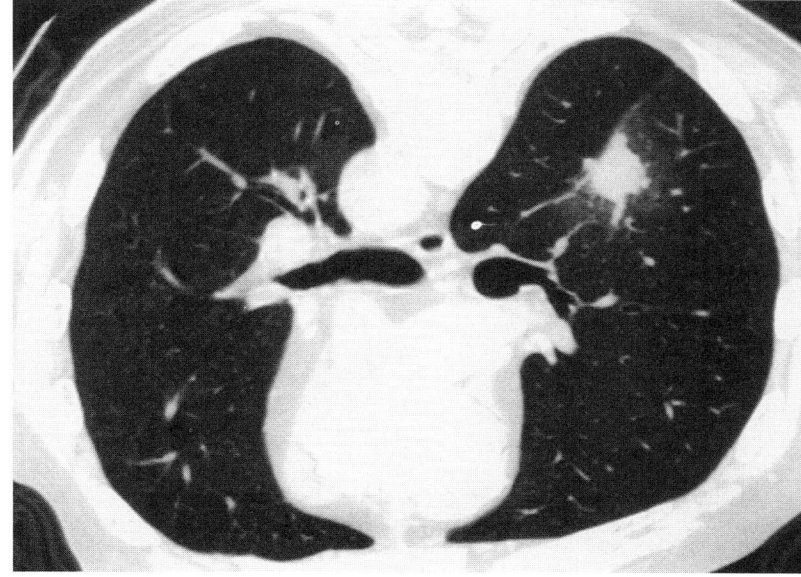

B

FIGURE 9. Nocardial infection in nonneutropenic cardiac transplant patient. (A) Frontal chest radiograph shows a new 2-cm nodule in the right upper lobe. (B) A CT section through the tracheal carina level shows a peripheral right upper lobe nodule adjacent to the major fissure surrounded by a ring of ground-glass opacity (a "halo"). (C) Prone CT during percutaneous needle aspiration biopsy 5 days later shows the nodule has cavitated. *Nocardia asteroides* was isolated from the lung aspirate. (*Continued on next page*)

In an appropriate setting, air crescent cavitation may suggest late invasive aspergillosis.

The best clues to the radiologic diagnosis of radiation pneumonitis are the timing of onset with respect to radiation treatment and the location and configuration of the pulmonary infiltrate, which is almost always confined to the outlines of the radiation portals. Thus, the diagnosis of radiation pneumonitis should be suspected when imaging demonstrates an infiltrate (particularly a peribronchovascular one) with relatively sharp margins that do not

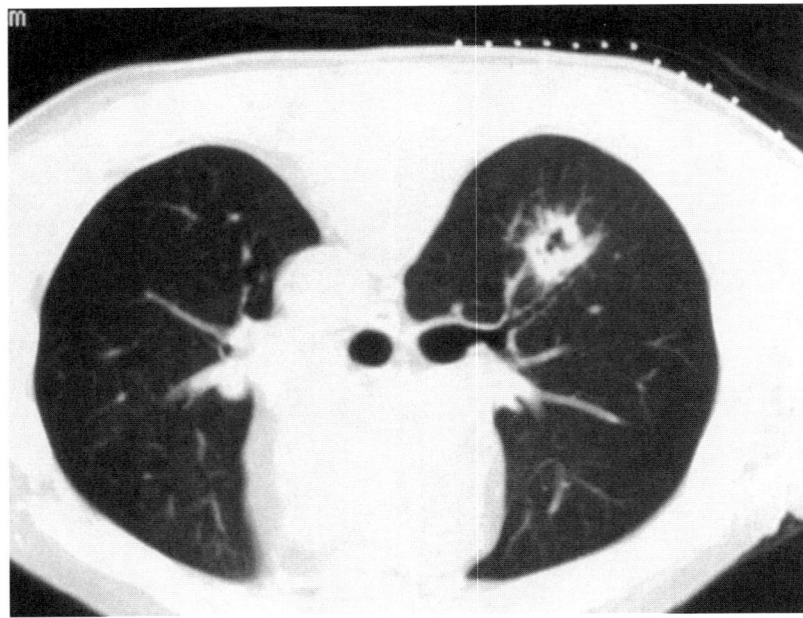

c

FIGURE 9. (*Continued*)

correspond to the bronchopulmonary anatomy but adjoin the edges of the radiation field (Fig. 3). Changes that occur outside this area should be minor. Since many cases of radiation pneumonitis follow mediastinal irradiation, the infiltrates are often central in location, in contrast to the usual peripheral location of most other processes that affect this population.[100–104,108,124]

The depressed inflammatory response of the immunocompromised host may greatly modify or delay the appearance of a pulmonary lesion on images. This depression is most frequently seen in patients with severe neutropenia (particularly those with an absolute granulocyte count <100/mm³),[247–249] but is also seen with steroid treatment.[250] When such severe neutropenia is present, atelectasis may be the only radiologic clue to the presence of clinically important pulmonary infection. In particular, radiologic evidence of fungal invasion, which normally excites a less exuberant inflammatory response than does

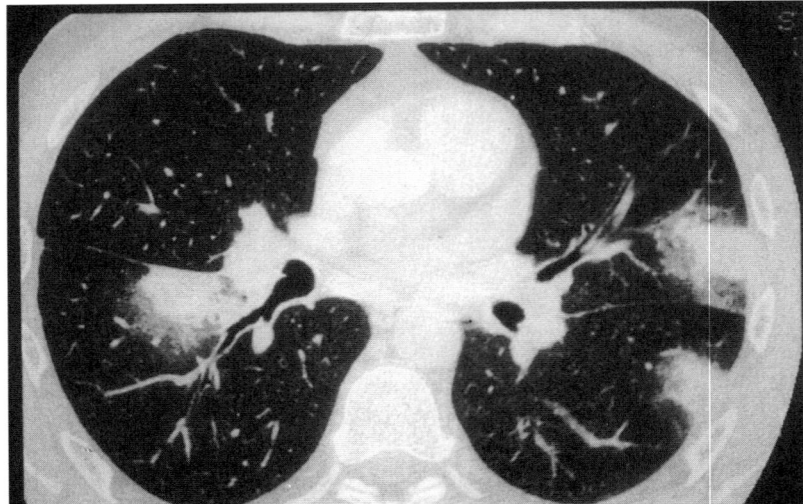

FIGURE 10. Early invasive aspergillosis. CT scan of severely neutropenic allogeneic bone marrow transplant patient shows three subpleural pulmonary nodules with surrounding halos of ground-glass opacity due to invasive pulmonary aspergillosis.

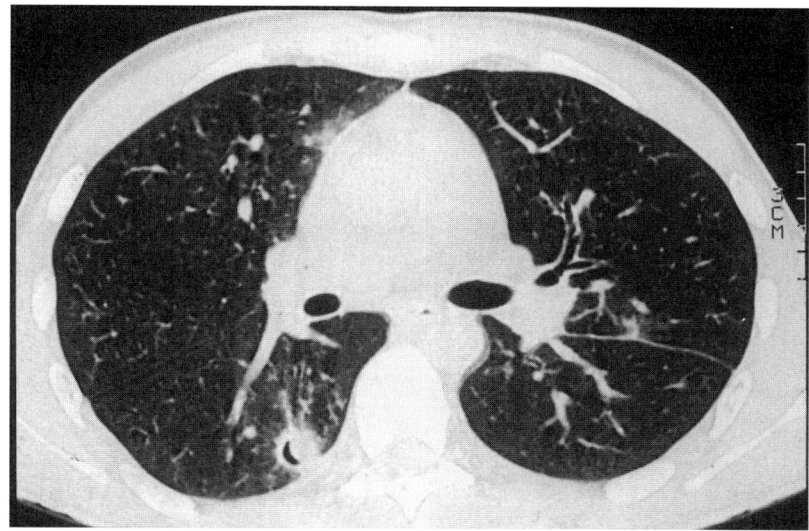

FIGURE 11. Late invasive aspergillosis. Nodule with air crescent cavity late in the course of invasive pulmonary aspergillosis. A CT section through the lower chest shows a nodule with a crescentic gas collection partially outlining an infarcted lung mass that has pulled away from viable lung.

bacterial invasion, will often be very slow to appear. By contrast, in patients recovering from neutropenia there may be a paradoxical increase in the radiologic findings (and sputum production) as the granulocyte count recovers, despite a good clinical response to antimicrobial therapy.[247–249]

CT of the chest has revolutionized the evaluation of the immunocompromised host with the febrile pneumonitis syndrome. This has become even more true as technical advances have resulted in more rapid acquisition of higher-quality CT images, thus reducing breathing artifacts associated with tachypnea. Although conventional chest radiography remains the first procedure for evaluating immunocompromised patients for possible pulmonary disease, the high sensitivity and precise anatomic localization possible with CT scanning can result in earlier localization, diagnosis, and treatment. Thus, CT provides a greater chance of survival from opportunistic infection.[251,252] CT is more sensitive and effective than chest radiography in diagnosing disease in the immunocompromised patient.[253] It is especially useful when the chest radiograph is negative and when the radiographic findings are subtle or nonspecific. Since localizing pulmonary symptoms are often absent, CT is a great aid in localizing disease for biopsy. CT can be more specific than radiography in the diagnosis of opportunistic infection. For example, under appropriate clinical circumstances, the CT halo sign often can differentiate nodules caused by *Aspergillus* species from infections caused by conventional bacterial or viral species[9,254] (Fig. 10). Similarly, high-resolution CT can be extremely useful in the

evaluation of patients with possible *Pneumocystis* infection.[255] Thus, CT is now of primary importance in the evaluation of the immunosuppressed patient in the following situations:

1. In the evaluation of febrile, severely neutropenic patients with negative or subtle chest radiographic findings, CT is far more sensitive in the detection of potentially treatable opportunistic infection, particularly that due to fungal pathogens such as *Aspergillus*. Intrathoracic complications of bone marrow transplantation, for instance, are found with CT in 57% of patients with clinical symptoms and negative radiographs.[256] Thus, CT has become the established imaging method of choice for the diagnosis of occult or subtle disease because of the limited sensitivity of conventional radiography, especially when the inflammatory response is impaired by severe neutropenia.

2. Similarly, equivocal chest radiographs of patients receiving exogenous immunosuppressive therapy (e.g., organ transplant patients) are best reevaluated with chest CT. For example, we have seen a number of patients with pulmonary nodules due to such organisms as *Cryptococcus neoformans* that were clearly seen on chest CT evaluation after no abnormality or very minimal findings were detected on chest radiography (Fig. 13). Cure of such individuals is far more easily accomplished by the earlier diagnosis afforded by CT than after systemic dissemination and CNS seeding has occurred by which time radiography may have become positive. Nodules detected with CT are suggestive of fungal, nocardial, rhodococcal,

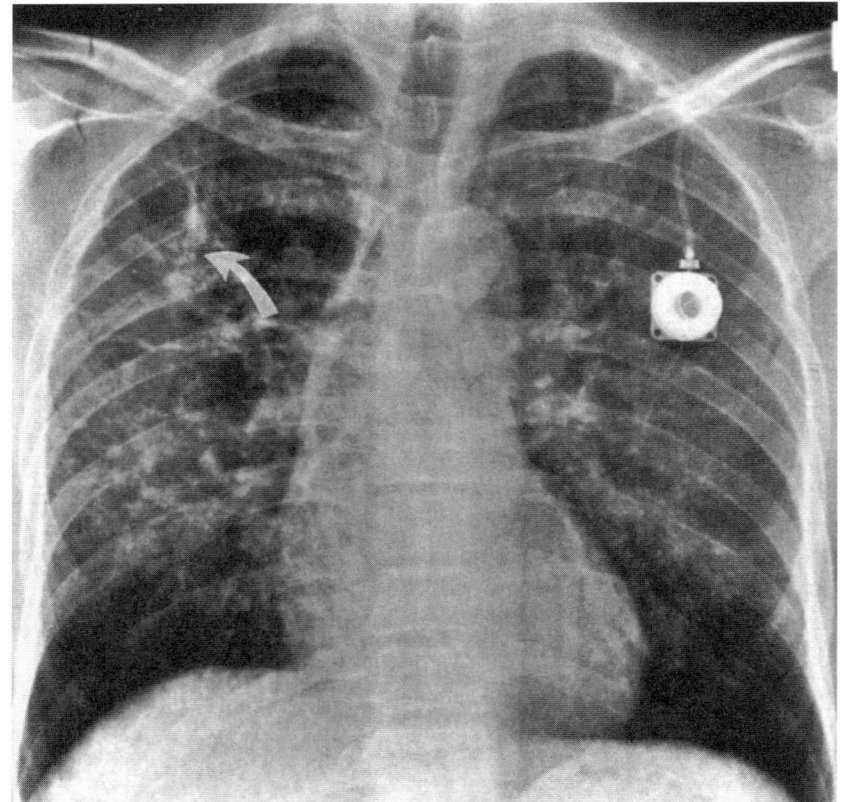

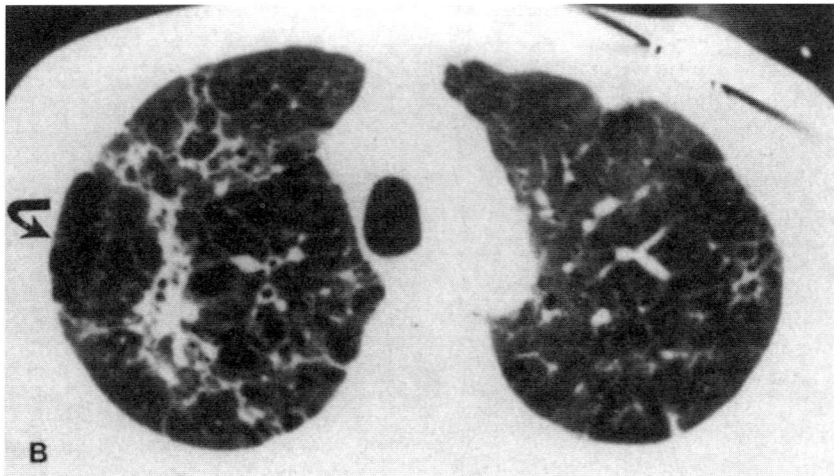

FIGURE 12. Cystic apical *Pneumocystis carinii* pneumonia. (A) Cystic lung regions of *Pneumocystis* pneumonia are most evident in the right apex of an HIV-positive patient treated with aerosolized pentamidine. (B) A CT scan confirms cystic lesions with well-formed walls, characteristic of postaerosolized pentamidine *Pneumocystis* pneumonia and unlike bullous emphysema.

or tuberculous infection in the transplant patient. By the time conventional radiography becomes definitely abnormal, the disease process is often far advanced.

3. In the AIDS patient, with a negative chest radiograph and a clinically compatible syndrome (e.g., nonproductive cough, fever, and hypoxemia), CT is of value in detecting *Pneumocystis carinii* pneumonia. Since 10–20% of *Pneumocystis* patients have normal chest radiographs, CT may be useful both in detecting subtle disease and in ruling out the possibility of this infection (Fig. 14).[255,257]

4. Although an abnormal chest radiograph may lead

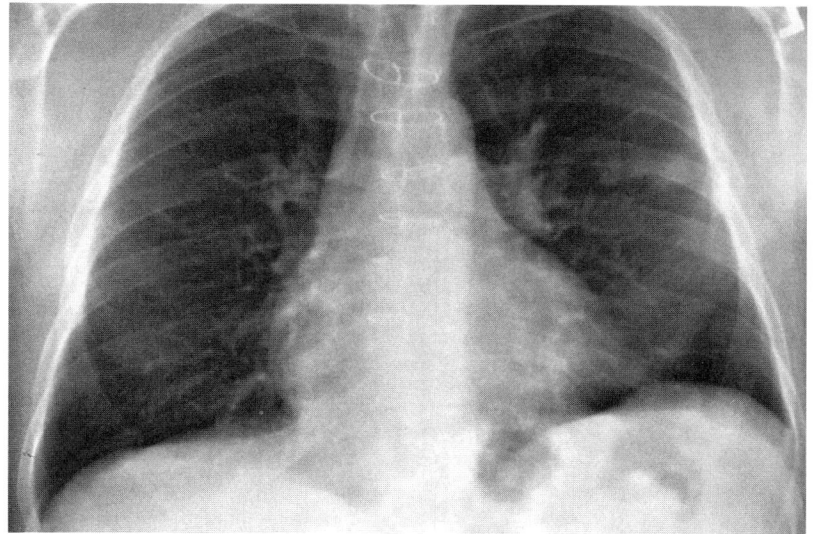

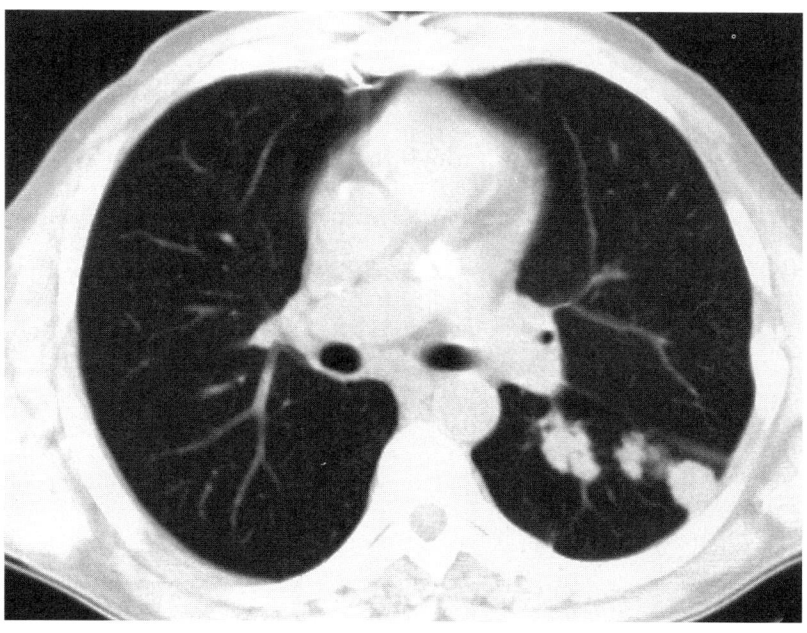

FIGURE 13. Pulmonary cryptococcosis. (A) Frontal chest radiograph shows large nodular opacities in the left midlung in a minimally symptomatic, nonneutropenic cardiac transplant patient. (B) A CT section through the midchest confirms the lung nodules. Other CT sections (not shown) showed additional nodules of varying sizes.

to the diagnosis of infection, it greatly underestimates the extent of the disease process compared with CT. Particularly with opportunistic fungal and nocardial infection, precise knowledge of the extent of the infection at diagnosis and the response of all sites of disease to therapy will lead to the best therapeutic outcome. A general rule of thumb in the treatment of opportunistic infection in the immunocompromised host is that the best clinical results will be obtained if the clinician continues therapy at least until all evidence of clinical disease has resolved. This goal can best be achieved with CT guidance.

5. Since dual or sequential pulmonary infections are not uncommon in the immunocompromised host, more than one etiologic agent may be responsible for a given clinical episode.[2,3,258,259] In patients who have responded slowly or poorly to what should be appropriate therapy, CT may provide clues that additional diagnostic possibilities should be considered. For example, in AIDS pa-

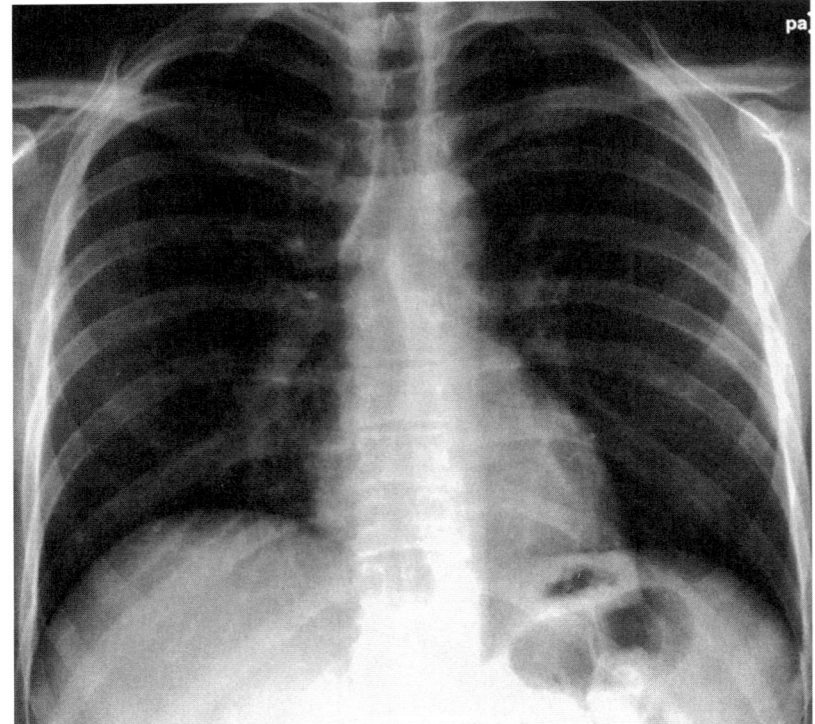

A

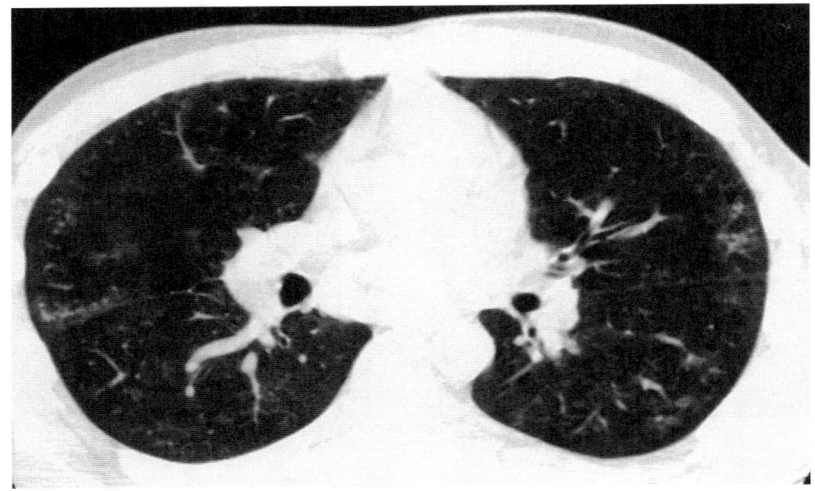

B

FIGURE 14. *Pneumocystis carinii* pneumonia. (A) Frontal chest radiograph in AIDS patient with very low CD4 count and subacute cough shows clear lungs with no convincing evidence of focal or diffuse abnormality. (B) A CT section through the lower chest of this patient taken on the same day as the chest radiograph shows a faint diffuse ring of subpleural ground-glass opacities. *Pneumocystis carinii* in large number were recovered from the bronchoalveolar lavage. (C) A CT scan through another AIDS patient with *P. carinii* pneumonia shows more dense diffuse ground-glass opacities. (Courtesy of Paul Stark, MD)

tients with *Pneumocystis* pneumonia, characteristic CT and radiographic findings usually reflect the diffuse interstitial and alveolar abnormalities (Fig. 14). Since acinar and macronodular opacities are relatively unusual manifestations of *P. carinii* pneumonia,[255,260] their identification in a patient not responding to appropriate therapy should raise the possibility of concomitant Kaposi's sarcoma (Fig. 8) or infection with other agents (Fig. 15). In particular, the identification of a thick-walled cavitary nodule is highly suggestive of a new or coexisting fungal or bacterial infection in the patient with microbiologically or pathologically confirmed *P. carinii* infection.[7]

6. CT also can help in defining which invasive diagnostic procedure is most likely to yield a diagnosis and where the disease is most likely to be found for successful biopsy.[7] CT can provide precise guidance for percutane-

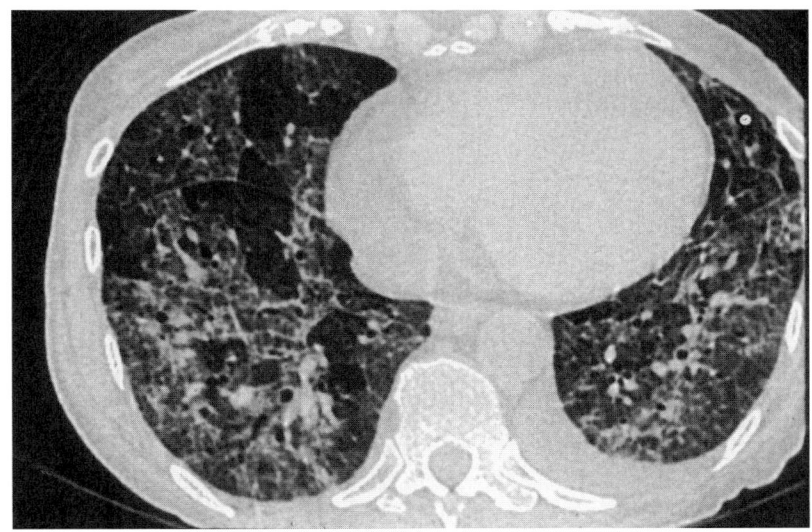

C

FIGURE 14. (*Continued*)

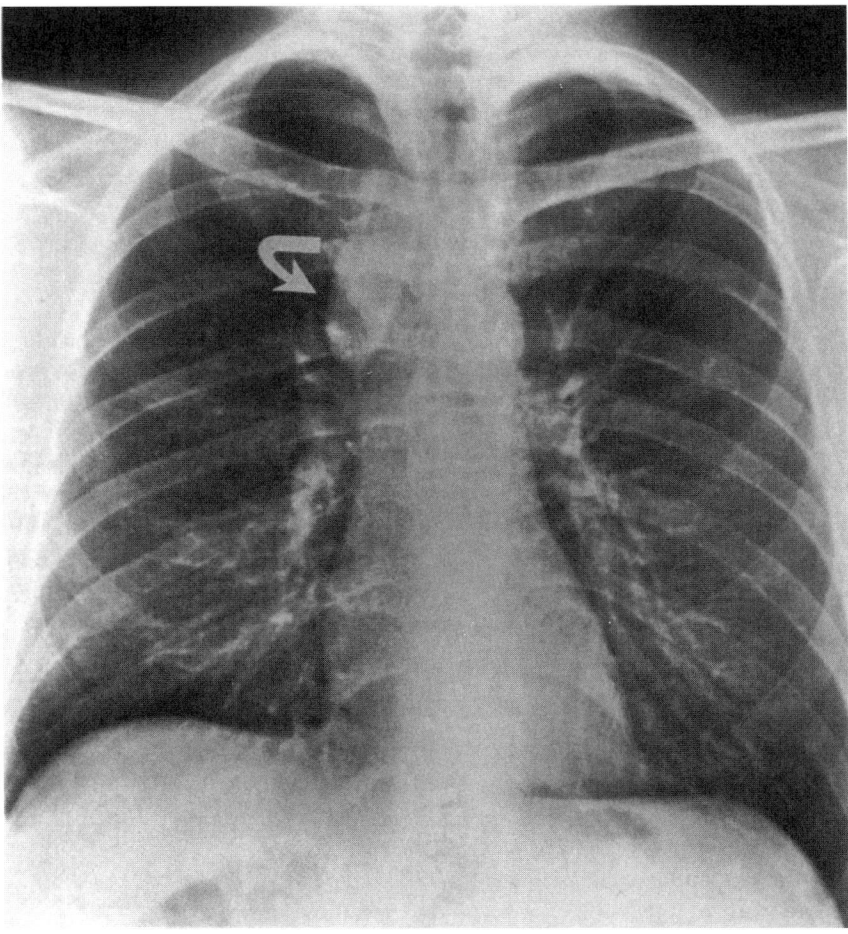

FIGURE 15. Atypical mycobacterial infection. Right-sided mediastinal adenopathy is evident in an AIDS patient with dyspnea, fever, and cough due to *Mycobacterium avium-intracellulare* infection. The adenopathy is also consistent with *M. tuberculosis* and lymphoma. On CT scan, low attenuation centers with enhancing rims of enlarged mediastinal lymph nodes favor mycobacterial infection over lymphoma.

ous needle biopsy, thoracoscopic, or open lung excision in the case of peripheral lung nodules.[261] CT also is the best means of predicting whether bronchoscopy is likely to be the most appropriate diagnostic modality for a particular patient. Thus, in patients with pulmonary nodules, CT demonstration of the feeding bronchus correlates with a 60% diagnostic yield with bronchoscopy, as opposed to a 30% yield if this finding is not present.[262] If CT demonstrates centrally located diffuse opacifications, a bronchoscopic approach is the modality of choice for diagnosis.[263]

7. CT also can narrow the differential diagnosis in the patient with suspected opportunistic infection. Cavitary CT lesions are suggestive of infections with *Nocardia*, *Rhodococcus*, *Cryptococcus*, and *Aspergillus*. Opacified secondary pulmonary lobules in the lung periphery are suggestive of bland pulmonary infarcts and of cavitated septic or hemorrhagic *Aspergillus* infarcts. Diffuse peribronchial opacities are suggestive of fluid overload and graft-versus-host disease (Fig. 16). Dense regional or lobar consolidation on CT is most suggestive of bacterial pneumonia or fungal infection (Fig. 1). Apical cystic CT lesions are suggestive of *P. carinii* pneumonia, especially in AIDS patients treated with prophylactic aerosolized pentamidine.[264] High-resolution CT can aid in differentiating between apical lung opacities that are due to *M. tuberculosis* scars and benign apical "cap."[265]

4.2. Radiologic Aspects of Thoracic Disease in AIDS Patients

Since the early days of the HIV epidemic, pulmonary disease has been recognized as a major cause of morbidity and mortality. Prior to the advent of *Pneumocystis* prophylaxis, 70% of patients had *Pneumocystis* pneumonia as their AIDS defining illness, with *Pneumocystis* also being the most common cause of death. Two major advances in the care of AIDS patients have markedly changed the nature of the intrathoracic complications observed: The first of these was the widespread deployment of anti-*Pneumocystis* prophylaxis, resulting in a significant decrease in the incidence of this infection. The second advance was the development of highly active antiretroviral therapy (HAART), which has greatly improved the prognosis for HIV-infected individuals, with a corresponding fall in all forms of pulmonary infection in these patients. The three most common causes of respiratory infection in patients with HIV infection are acute bronchitis (~13 episodes/100 patient years), bacterial pneumonia (~5 episodes/100 patient years), and *P. carinii* (~5 episodes/100 patient years). When one follows HIV patients over a 5-year period, there is a gradual increase in the incidence of both bacterial pneumonia and *Pneumocystis* (from 3–4 in the first year to 7–10 episodes/100 patient years in the fifth year), as well as other opportunis-

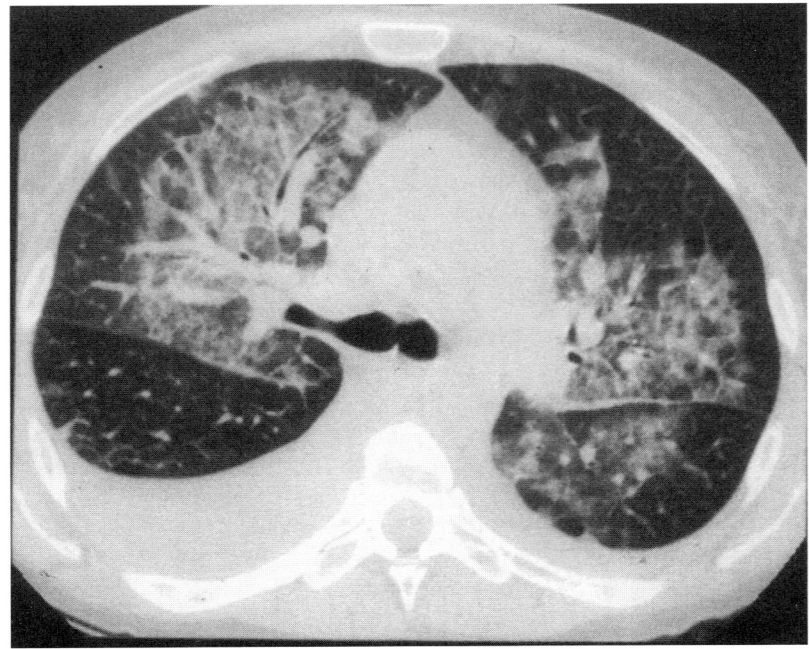

FIGURE 16. Hydrostatic "batwing" pulmonary edema. A CT section through the level of the carina shows perihilar ground-glass opacification, subpleural thickening, and bilateral pleural effusion due to congestive heart failure.

tic infections and tuberculosis.[266,267] Not surprisingly, the nature of the pulmonary infection occurring can be correlated with the CD4 count of the patient: acute bronchitis being the predominant lower respiratory infection of a cohort of patients who enter with a CD4 count $\geq$ 200 cells/mm^3; in members of the cohort with CD4 counts <200 cells/mm^3 acute bronchitis, bacterial pneumonia, and *Pneumocystis* all occurred at high rates.[267] Once either bacterial pneumonia or *Pneumocystis* pneumonia occur, the mean survival rate is approximately 2 years.[268] Other causes of pulmonary disease in AIDS patients include mycobacterial infection, Kaposi's sarcoma, fungus infection, lymphoma, and lymphocytic interstitial pneumonitis. Other less common AIDS-related conditions include septic and nonseptic pulmonary emboli and viral infection, particularly that due to CMV and herpes simplex virus.[7,269] Typical radiologic findings in AIDS patients with these conditions are presented in Table 5.

4.2.1. *Pneumocystis carinii* Pneumonia

The two factors that determine risk of *Pneumocystis* infection in HIV-infected individuals are CD4 count and whether or not effective anti-*Pneumocystis* prophylaxis is being administered. This infection is quite uncommon with a CD4 count >200/mm^3. With patients not receiving prophylaxis, the median CD4 count of those developing *Pneumocystis* is ~100/mm^3, while for those receiving prophylaxis the median count is ~20/mm^3.[266,270,271]

The disease most often presents as diffuse, perihilar, bronchovascular lung opacities, but focal and multifocal opacities also occur. The radiographic findings are often quite subtle in the early stages of the disease and may be overlooked if the AIDS history is unknown or if no prior chest radiographs are available for comparison. Approximately 10–20% of *Pneumocystis* infections are totally occult by conventional radiography.[272] CT is not generally required or employed to make the diagnosis in typical cases. However, when radiographic findings are absent, CT may demonstrate more typical or extensive diffuse ground-glass lung opacities characteristic of *P. carinii* pneumonia. In more chronic *Pneumocystis* infection, or at the conclusion of therapy, CT may show evidence of pulmonary fibrosis and septal lines.[273] Atypical findings of *Pneumocystis* in AIDS patients include focal lung opacities, cavities, miliary nodules, and very rarely intrathoracic lymphadenopathy and pleural effusions.[260,274] Radiographic detection of pleural effusion and intrathoracic lymphadenopathy, however, are not characteristic of *Pneumocystis*,[275] but massive pleural effusion and massive lymphadenopathy have been rarely reported.[276,277] Small pleural effusions and small intrathoracic lymph nodes are more likely to be detected with CT.[278] *Pneumocystis* pneumonia often presents as apical lung opacities with cysts or cavities in those previously treated with aerosolized pentamidine (Fig. 8).[279] The cavities and subpleural cysts tend to have thicker walls than true subpleural areas of emphysema.[279–281] Some *Pneumocystis* patients may first present with pneumothorax.[282–284] Rarely, *Pneumocystis* pulmonary lesions have been described to increase off steroids and to decrease if steroids are reapplied.[250] Gallium-67 lung uptake in AIDS-related *Pneumocystis* pneumonia is usually diffuse and intense, while thallium lung uptake is negative. Calcifications in affected abdominal viscera may be seen with CT after treatment for *Pneumocystis* pneumonia.[277] Although

TABLE 5. Radiologic Findings in Acquired Immunodeficiency Syndrome Patients with Pulmonary Disease of Different Etiologies[a]

Cause of pulmonary disease	Lung[b]				Hilar and mediastinal lymph nodes		Pleura
	D	MF	F	G/T	Enlarged/CT attenuation	Gallium[c]	Fluid
Pneumocystis carinii	+++	++	+	+++/0	0/	0	0
Mycobacteria	+	++	++	++/0	+++/low	+++	+++
Bacteria	+	+	+++	+/0	0	0	+++
Kaposi's sarcoma	+++	+++	+	0/+	++/high	0	+++
Fungus	+	+++	+++	++/0	++/low	++	+++
Lymphoma		++	++	+/+	++/high	+++	+++
Lymphocytic interstitial pneumonitis	+++			+/	0	0	0
Cytomegalovirus	+++	++	+	+/	0	0	0
Septic emboli		+++	+	+/	0	0	+++

[a]Results are expressed as degrees of positivity.
[b]D, diffuse; MF, multifocal; F, focal; G/T, gallium/thallium.
[c]Nuclear scans.

magnetic resonance imaging (MRI) has not yet established itself as an important diagnostic modality in *Pneumocystis*, higher total T1 and T2 signal has been found in patients with *Pneumocystis* when compared with normals.[285] In order of frequency, CT images of *Pneumocystis* pneumonia can show diffuse opacification, patchy opacities with spared regions, and focal peripheral opacities. *Pneumocystis* pneumonia often heals with pulmonary fibrosis, cystic spaces, and zones of emphysema, especially in the upper lung zones.[252] Radiographic and CT images of CMV pneumonia are indistinguishable from those of *Pneumocystis*.

4.2.2. Mycobacterial Infection

Tuberculosis is an infection of great importance in the AIDS patient. In these individuals, tuberculosis is most often the result of reactivation of a dormant focus, but the radiologic appearance more closely resembles that of primary tuberculosis. Thus, tuberculosis in AIDS patients more often presents with intrathoracic lymphadenopathy, miliary lung nodules, pleural fluid, and extrapulmonary disease, rather than the apical consolidation and cavitary disease characteristic of reactivation disease.[286–288] Once again, CD4 count can be used to stratify the appearance of tuberculosis, with counts of <200/mm^3 being associated with the miliary form of the disease and higher counts with more typical reactivation disease.[266] Approximately 85% of AIDS patients with tuberculosis have abnormal chest radiographs.[289] Tuberculosis and infection with atypical mycobacteria are by far the most common infectious causes of intrathoracic lymph node enlargement in the AIDS patient (Fig. 10). Hilar and mediastinal lymphadenopathy may be evident on conventional radiographs, but CT is significantly more sensitive. Enlarged intrathoracic lymph nodes with low CT attenuation centers and enhancing rims are characteristic of tuberculosis,[290] but are occasionally observed in patients with disseminated fungal infection. The low CT attenuation of tuberculous lymph nodes is different from the higher attenuation of enlarged nodes in Kaposi's sarcoma and lymphoma. Gallium-67 uptake in the lung tends to be more patchy and less intense than in *Pneumocystis*.

Atypical mycobacterial infection of the lungs, especially with *Mycobacterium avium-intracellulare* (MAI), occurs in the AIDS patient but is much less common than pulmonary infection with *M. tuberculosis* and tends to occur late in the course of AIDS, tending to occur in the AIDS patient with a CD4 count <50/mm^3. MAI infection, unlike *M. tuberculosis*, usually has a gastrointestinal portal of entry and is disseminated by the time it is first detected in the chest. Radiologic findings are not distinguishable from disseminated *M. tuberculosis* infection and include intrathoracic adenopathy and miliary nodules or tree-in-bud opacities (Fig. 2).[10,291] Other atypical mycobacterial infections of the lungs, including the newly described *M. celatum*,[292] occasionally can be seen in patients with advanced AIDS.

4.2.3. Pyogenic Bacterial Pneumonia

As previously stated, community-acquired bacterial pneumonia is common in AIDS and is most commonly due to *Streptococcus pneumoniae*, *Haemophilus influenzae*, and in the late stages of disease a variety of gram-negative organisms, including *Pseudomonas aeruginosa*.[266] Of importance, prophylaxis for *Pneumocystis* with trimethoprim–sulfamethoxazole and for MAI with azithromycin or clarithromycin is associated with a lower risk of bacterial pneumonia.[293] The radiologic manifestations of bacterial pneumonia are similar in AIDS and non-AIDS patients, with pulmonary lesions that are likely to be focal, consolidative, cavitary, and associated with pleural fluid. However, bacterial pneumonia can present with regional or even diffuse interstitial-appearing opacities that can mimic *Pneumocystis*.[294] Under appropriate therapy, bacterial infections are much more likely to clear within 3 weeks than is *Pneumocystis*. Gallium-67 uptake in the lung is less likely to be positive in bacterial pneumonia.

4.2.4. Kaposi's Sarcoma of the Lung

Kaposi's sarcoma involves the lung in about one fifth of AIDS patients with skin lesions. The lesions tend to be diffuse interstitial or airspace opacities and associated with small peribronchial nodular opacities.[295] The peribronchovascular lung lesions are often associated with subpleural nodules that are particularly well seen on CT (Fig. 8).[296] The lesions can simulate lymphatic spread of tumor. Asymptomatic lung lesions of Kaposi's sarcoma are often found at postmortem examination in AIDS patients with normal chest radiographs.[297] Hilar lymph node enlargement and pleural effusions are common accompaniments of lung lesions. When the mucosa of the tracheobronchial tree is involved, atelectasis may develop.[298] The lymph nodes harboring Kaposi's sarcoma tend to take up thallium.[299] When AIDS patients with Kaposi's sarcoma of the skin develop diffuse lung opacities, gallium scanning can be helpful in differentiating between *Pneumocystis* and Kaposi's sarcoma of the lung. The lung lesions of Kaposi's sarcoma do not take up gallium, while the lungs of patients with *Pneumocystis* pneumonia avidly take up the gallium.[257]

4.2.5. Fungus Infection of the Lung

The most common fungal agent invading the lungs of the AIDS patient (other than *Pneumocystis*, which is now classified as a fungus) is *Cryptococcus neoformans* (Fig. 9). Other fungi that commonly affect the lungs of AIDS patients include *Aspergillus* species, *Histoplasma capsulatum*, and *Coccidioides immitis* (these last two being relatively common in the geographic regions where this fungus grows in the soil, e.g., the midwestern section of the United States for *Histoplasma* and the southwestern section of the United States for *Coccidioides*). In a substantial minority of patients (~25%) disseminated fungal infection is first discovered in the lungs, especially when CT scanning is employed.[300] In cryptococcal disease, chest radiographs usually are negative or reveal focal disease, with or without cavitation.[301] Diffuse lung disease is often found in patients with *H. capsulatum* infection. *Aspergillus* species tend to cause lung infection late in the course of AIDS and are not as common in AIDS as in other immunocompromised patients (e.g., transplant patients, patients with neutropenia, patients receiving prolonged courses of corticosteroids). Radiographs of pulmonary fungal infection can reveal consolidations, nodules, nodules with halos, cavities, cavities with air crescents, and pleural effusions. In AIDS patients, *Aspergillus* infections produce varied imaging findings, including upper lobe cavities similar to tuberculosis, pleural-based lung nodules and infarcts, and diffuse infiltrates.[302]

4.2.6. Lymphoma of the Lung

A small but important group of AIDS patients develop B-cell or other forms of non-Hodgkin's lymphoma in the central nervous system and/or abdominal viscera. The lesions tend to occur as macronodules in the lung parenchyma, often in association with pleural fluid.[303] Lymphoma in AIDS patients causes enlargement of mediastinal and hilar lymph nodes only rarely.[272,304,305] Lung involvement and pleural effusions are more common in AIDS-related lymphomas than in non-AIDS lymphomas. CT is extremely useful in detecting the focal or multifocal solid lung masses of AIDS-related lymphomas of the lung.[306] Nuclear scanning is said to differentiate among pulmonary infection, lymphoma, and Kaposi's sarcoma.[299] In infection the 3-hr delayed thallium scan is negative, while the gallium scan is positive in the lungs. In lym-

phoma of the lung, both thallium and gallium scans are positive. In Kaposi's sarcoma, the thallium scan is positive, while the gallium scan is negative. Lymph node uptake of gallium is avid in mycobacterial infection and lymphoma.[306] Potential causes of lymphadenopathy in the AIDS patient include follicular hyperplasia, non-Hodgkin's lymphoma, mycobacterial infection, metastatic cancer, Kaposi's sarcoma, multiple infectious agents, and drug-induced disease (e.g., that due to trimethoprim–sulfamethoxazole).[307] By far the most common cause of radiologically identifiable lymph node enlargement in the AIDS patient is tuberculosis. High CT attenuation of enlarged lymph nodes in AIDS is suggestive of Kaposi's sarcoma. Low CT attenuation of enlarged lymph nodes is more often associated with mycobacterial infection.[308]

4.2.7. Lymphocytic Interstitial Pneumonitis

Lymphocytic interstitial pneumonitis is not uncommonly found in pediatric patients with AIDS. The lung lesions tend to be nonspecific linear interstitial opacities admixed with patchy areas of airspace consolidation. In contradistinction to *Pneumocystis*, gallium uptake in the lung is mild or negative in AIDS-related lymphocytic interstitial pneumonia.[257,306]

5. Specific Diagnosis

The effective therapy of the febrile pneumonitis syndrome in immunocompromised patients requires rapid and precise diagnosis. Although the diagnostic clues discussed thus far may greatly limit the differential diagnostic possibilities, the specific diagnosis should be sought whenever possible. An aggressive approach to diagnosis will limit drug toxicity and the risk of potentially lethal superinfection without exposing the patient to potentially inadequate therapy. Not surprisingly, several studies have shown that the rapidity with which the diagnosis is made has a major impact in determining the outcome of therapy, whether one is dealing with a noninfectious disease, a conventional bacterial infection, or invasive fungal, nocardial, or rhodococcal disease.[1–3,140,177,309] Therefore, great emphasis must now be placed on the techniques available for making a precise diagnosis: immunologic studies, conventional examination of expectorated or induced sputum specimens, and a variety of invasive procedures designed to sample either lower respiratory tract secretions or lung tissue or both. Indeed, it is largely the skill of the clinician in utilizing the specific diagnostic techniques available that will determine the rate of survival in immunocompromised patients with the febrile pneumonitis syndrome.

As we discuss this aggressive approach to diagnosis, we also must emphasize that the clinician must constantly keep in mind both the risks and the benefits that are involved. The critical question that must be asked before undertaking any invasive procedure in this patient population is what is the likelihood that this procedure will result in a major change in therapy and prolongation of patient survival. Thus, invasive procedures are rarely indicated in patients with advanced leukemia, AIDS, or metastatic cancer. On the other hand, an aggressive approach for precise diagnosis is clearly indicated in organ transplant recipients, patients with treatable Hodgkin's disease or other forms of cancer with a reasonable expectation of a meaningful response to therapy, and patients with such conditions as collagen vascular disease or inflammatory bowel disease being treated with immunosuppressive therapy. In addition to the ethical considerations involved in this distinction between these two groups of individuals, the practical matter is that the diagnostic yield of invasive procedures is much higher in the second group of patients. In the poor-prognosis patient, even open lung biopsy may not lead to any diagnosis in as many as 20% of patients and in others the findings may include diagnoses such as hemorrhage, tumor, and bronchiolitis obliterans that do not lead to effective therapy. Therefore, risk–benefit analysis must be a careful part of the clinician's diagnostic approach to the febrile pneumonitis syndrome in the immunocompromised patient.[1–3,5,10]

5.1. Immunologic Techniques for Specific Diagnosis

Measurement of antibody and delayed hypersensitivity skin test responsiveness to microbial antigens are time-honored techniques for diagnosing invasive infection. However, such methods have limited applicability when caring for immunosuppressed patients with the febrile pneumonitis syndrome, for several reasons:

1. Even under the best of circumstances in the normal host, there is a delay between the onset of infection and the development of a measurable immune response. In the immunocompromised patient, such responses may be further delayed or totally abrogated. To wait for the development of such a response can interfere greatly with the need and desire to arrive at a rapid diagnosis. For example, we have cared for a liver transplant patient who contracted HIV infection at the time of transplantation (prior to the availability of HIV testing) but who did not develop a positive HIV antibody test until more than 2½ years later, at which time he already had overt AIDS.[306]

In contrast to the lack of sensitivity of the antibody response in diagnosing clinical disease, a variety of antibody measurements are helpful in predicting the risk of disease if exposed: Thus, a negative test for circulating antibody to *Toxoplasma gondii* is very useful in ruling out the possibility of encephalitis due to this organism in AIDS patients, as well as in predicting susceptibility to systemic toxoplasmosis and the need for prophylaxis in a recipient of a cardiac allograft from a seropositive donor. Similarly, a negative antibody test for varicella zoster virus in an immunocompromised individual delineates an individual very susceptible to disseminated visceral infection on exposure to this virus. Finally, the attack rate for clinical disease due to CMV in transplant patients can be predicted by knowing the antibody status of the donor and recipient (see Chapters 17 and 18).

2. Since many of the opportunistic infections that cause life-threatening disease in the compromised host cause asymptomatic subclinical infection in the normal population, the presence of a positive result may have little meaning. The classic examples of this phenomenon are the many attempts to make the diagnosis of invasive candidiasis or aspergillosis on the basis of the presence of precipitating or agglutinating antibodies directed against these organisms. It is now clear that because of the failure of development of such antibodies in many compromised patients and because of their presence in many normals, these tests are of limited diagnostic value.[307–309]

3. Appropriate serologic or skin tests are not available for many of the disease processes under consideration.

Therefore, the effectiveness of such traditional immunologic techniques for diagnosis has been disappointing. Antibody testing has been useful in patients with histories of possible exposure to *Coccidioides immitis* or *Histoplasma capsulatum*, in whom the demonstration of elevated or rising titers of complement-fixing antibody is an excellent clue to the presence of active infection with these agents (see Illustrative Case 3 and Chapter 6).

5.2. Sputum Examination

The usual clinical approach to the diagnosis of pneumonia is based on the Gram's stain and cultural examination of expectorated sputum specimens. It should be emphasized that strict criteria should be employed when viewing the Gram's stain of an expectorated sputum specimen before trusting the validity of the specimen: few squamous epithelial cells (<10 per low-power field) and many polymorphonuclear leukocytes (>25 per low-power field). If such criteria are not met, the validity of the specimen is in question. For a variety of reasons, such an approach is often of little diagnostic value in the compromised patient. First, many of these patients, particularly those with significant leukopenia, fail to produce sputum.[12,13,310–313] Second, the upper respiratory tract of many of these patients is frequently colonized with a large number of potential pathogens, particularly gram-negative bacilli and fungi. Expectorated sputum specimens therefore will be contaminated by these potential pathogens and differentiation between organisms truly invading the lung and those that colonize the pharynx may be quite difficult. Third, certain organisms that commonly cause pneumonia in this population, particularly the fungi, rarely shed sufficient organisms into the sputum to permit diagnosis by cultural or microscopic examination. Finally, the noninfectious causes of pulmonary infiltrates will not be diagnosed by examination of expectorated sputum specimens. Therefore, although the clinician should always initiate the diagnostic evaluation of the patient with possible pneumonia by an examination of an expectorated sputum specimen, more invasive diagnostic procedures are usually necessary.

Because of the large number of AIDS patients presenting with possible *P. carinii* pneumonia and the need for noninvasive tests for diagnosis, attention has been focused on the examination of induced sputum. Provided skilled respiratory therapists are carrying out the sputum-induction procedure and equally skilled microbiologists are evaluating the material obtained, the diagnostic yield can be high. Thus, in AIDS patients, approximately two thirds of *Pneumocystis* pneumonias can be diagnosed following staining with Giemsa and other conventional stains. When immunofluorescent staining utilizing a monoclonal antibody specific for *P. carinii* is used, sensitivity and specificity of greater than 90% can be obtained. One caution must be noted, however. Because the organism burden is far greater in AIDS patients with *Pneumocystis* pneumonia than in other immunocompromised patients such as transplant or lymphoma patients with this infection, the sensitivity of this procedure is considerably less (at least 25% less in our experience) in these other patient groups.[314–317]

With the increased ability to carry out induced-sputum examinations, many centers are looking to utilize this procedure in the diagnosis of other forms of pneumonia, particularly those due to bacteria and fungi. Although precise data regarding sensitivity and specificity for other infections in different patient populations are not currently available, there is general agreement that examination of an induced-sputum specimen is an improvement over conventional expectorated-sputum examination. Indeed, it has largely replaced transtracheal aspiration in the initial evaluation of patients with pneumonia of unclear cause. Problems with the transtracheal approach have included hemorrhage, cervical cellulitis, and oropharyngeal contamination (particularly in nonexpert hands).[5,318] This having been said, however, it is fair to say that in the hands of an experienced operator, transtracheal aspiration can be particularly useful in the evaluation of the occasional patient with possible bacterial infection, particularly anaerobic bacterial infection, provided the following guidelines are followed[2,3,319,320]:

1. There are three absolute contraindications to transtracheal aspiration: (a) an uncooperative patient, (b) a patient with unsuitable anatomic characteristics (such as an obese person, a child, or one who has undergone surgery or radiation therapy of the neck) that make the procedure technically difficult, and (c) those with uncorrectable bleeding diatheses.
2. In the process of anesthetizing the area over the cricothyroid membrane where the lavage needle and catheter will be inserted, a small 25-gauge needle is employed both to deliver the local anesthesia and to delineate the track that the larger needle will follow.
3. In neutropenic patients, broad-spectrum antibacterial therapy is initiated immediately after the procedure and continued for a minimum of 48 hr postprocedure.

5.3. Invasive Diagnostic Techniques

If the diagnosis has not been made by sputum examination, a more invasive procedure in which direct sampling of lower respiratory secretions, pulmonary tissue, or both may be accomplished is then required. The choice of procedure is dependent on several factors: the patient's degree of illness, the rate of progression of the disease, the type of imaging finding, and the relative expertise and experience of personnel at the institution. If the presumed pneumonitis and the degree of hypoxia are progressing rapidly, the definitive diagnostic procedure—the open lung biopsy—should be carried out immediately. This urgency is particularly true when the imaging pattern is nonspecific and diffuse or multifocal. Despite the need for general anesthesia, thoracotomy, and a postoperative chest tube, it is remarkable how well this procedure is tolerated, especially if the treatable process is identified. Increasingly, video-assisted transthoracoscopic biopsy in skilled hands is proving to be a satisfactory alternative to the traditional open lung biopsy, with considerably less morbidity for the patient. If the pulmonary process is progressing at a more desultory pace, progressive hypoxia is not an immediate problem, and the clinical problem is more of a diagnostic dilemma than a therapeutic emergency, then less invasive techniques can be attempted, with the open or transthoracoscopic lung biopsy held in reserve if these techniques fail.[3,5,318]

5.3.1. Bronchoscopic Diagnostic Techniques

Fiberoptic bronchoscopy has become a cornerstone of invasive diagnostic studies in the immunocompromised host because it provides opportunities for bronchoalveolar lavage, transbronchial biopsy, bronchial brushing, and inspection of the anatomy of the tracheobronchial tree. Two cautions should be emphasized, however, regarding bronchoscopy in this susceptible patient population. First, even uncomplicated bronchoalveolar lavage will cause a fall in oxygen saturation of 5–10% during and immediately after the procedure, a decline that can be clinically significant in terms of the subsequent need for assisted ventilation.[321] Second, particularly in the elderly debilitated patient, too much local anesthesia from the procedure will leave the patient with an impaired gag reflex that is inadequate to protect the airway from aspiration pneumonia. In addition, contamination of the diagnostic material obtained by the bactericidal anesthetic agents can lower the diagnostic yield.[322–324]

Fiberoptic bronchoscopy and the ancillary procedures it makes possible is the diagnostic procedure of choice in the immunocompromised patient with diffuse lung disease. Bronchoalveolar lavage by itself, which has the lowest rate of complications of any of the bronchoscopic procedures, is particularly useful in diagnosing *P. carinii* infection and pulmonary hemorrhage (>95% sensitivity in experienced hands) and is moderately effective in diagnosing cryptococcal pneumonia in the AIDS patient. In the transplant patient, the diagnostic yield of lavage for isolated cryptococcal nodules is far less (<20%). Because of success in diagnosing *Pneumocystis*, bronchoalveolar lavage has become the diagnostic procedure of choice in AIDS patients with diffuse pulmonary opacities. Additional information regarding the prognosis of an AIDS patient with *Pneumocystis* can be obtained by analyzing the bronchoalveolar lavage fluid for the presence of neutrophils. If present, they predict a much poorer outcome of therapy. In diagnosing *Pneumocystis*, trans-

bronchial biopsy and bronchial brushing add significantly to the complication rate without improving the diagnostic yield. Therefore, in AIDS patients, biopsy procedures are usually restricted to patients with focal disease suggesting malignancy or with invasive fungal infection such as aspergillosis.[325–328]

In contrast, bronchoalveolar lavage is of little value in the diagnosis of pulmonary infiltrates in neutropenic leukemic patients. It has a very low yield in diagnosing invasive pulmonary aspergillosis and a high rate of false-positive bacterial isolations due to contamination from the upper airway. Bronchoalveolar lavage also will fail to diagnose more than 50% of malignancies and drug-induced pulmonary processes. An additional problem in the neutropenic patient is the potential for life-threatening bacteremia and postbronchoscopy pneumonia with any bronchoscopic procedure. In order to avoid these problems, we routinely begin broad-spectrum intravenous antimicrobial therapy with ceftazidime or imipenem, with or without vancomycin (depending on cultures of the upper airway), as soon as adequate specimens are obtained in the neutropenic patient. Such antibiotics are continued for 48 hr after the procedure unless some untoward event has occurred that requires further antibiotic therapy.[3,329–332]

Transbronchial biopsy is particularly useful in the diagnosis of allograft rejection in lung transplant recipients, leukemic infiltrates, radiation- and drug-induced pneumonitis, various forms of interstitial pneumonitis, and some viral infections.[1,5,8,10] Tumor invasion is often missed, however, as the biopsy forceps tends to slide off the tumor. Transbronchial needle aspiration through the bronchoscope, utilizing an 18-gauge needle, has been reported to increase the diagnostic yield, providing valuable tissue for histologic examination from paratracheal, peribronchial, and carinal areas without a significant increase in complications.[333]

The yield from transbronchial biopsy is clearly operator-dependent. At many centers, including our own, there has been increasing concern regarding the quality of the biopsy material obtained, particularly in non-AIDS patients. Because of this concern, there has been a revival of interest in an older procedure—transpleural lung biopsy via the thoracoscope (increasingly, this procedure has been greatly improved by video assistance, permitting the operator greater control in carrying out this procedure). Transthoracoscopic biopsy appears to be particularly useful in the diagnosis of patients with diffuse interstitial pulmonary disease or with focal, pleural-based disease in the periphery of the lung. Although an artificial pneumothorax is induced to carry out the procedure and a chest tube is needed for 1–3 days postprocedure, it may be carried out under local anesthesia and appears to be well tolerated, even in bone marrow transplant patients. This approach yields a higher-quality diagnostic specimen than does transbronchial biopsy and is usually better tolerated.[334–337]

Bronchial brushing has become less popular in recent years, primarily because it adds little to bronchoalveolar lavage in the evaluation of the AIDS patient. There is still considerable interest, however, in utilizing protected catheters (telescoped plugged catheters) to sample the lower respiratory flora without contamination from the oropharynx. Utilizing this approach and quantitative microbiology, there is some evidence that a finding of more than 10^3 colony-forming units of a particular bacterial species is associated with invasive disease. At many centers, however, there is as yet a significant false-positive and false-negative rate, and the use of protected catheters therefore should be regarded as a research procedure at present.[325,326,338–340]

5.3.2. Percutaneous Needle Biopsy of the Lung

The diagnostic procedure of choice for small focal lesion in the lung periphery is some form of percutaneous, transthoracic needle biopsy technique (just as the diagnostic procedure of choice for central or diffuse lung infiltrates involves either bronchoscopic or thoracoscopic approaches). This is not to say that focal lung disease cannot be diagnosed by bronchoscopic biopsies, but that the diagnostic yield of needle biopsy is greater for focal diseases and the procedure is usually less uncomfortable for the patient.[5]

Extensive experience has been obtained in recent years with CT-guided transthoracic needle biopsy. CT localization has increased the diagnostic yield and the safety of the procedure.[341–343] The procedures employed can be divided into two general categories: fine-gauge needle aspiration procedures (<18 gauge) that provide small amounts of material for microbiologic and cytologic examination and larger-bore (>18 gauge) biopsy techniques aimed at providing larger pieces of tissue for histologic as well as cultural analysis. Some investigators have been successful in obtaining tissue core samples with a 19/22-gauge coaxial system that uses a fine, circumferentially beveled aspiration needle.[344,345] The advantage of the coaxial system is that it permits multiple samples to be obtained through a single small-gauge puncture of the pleural surface of the lung.

Aspiration-needle biopsy is the most widely used procedure for the diagnosis of focal pulmonary processes in the immunocompromised host, particularly when infection is the primary consideration.[345] Percutaneous needle aspiration is particularly well suited for diagnosing focal peripheral lung lesions, e.g., those due to *Nocardia*, fungi, or tuberculosis.[346] A sensitivity greater than 80% has been noted for infection and greater than 90% for malignancy. At our institution, we have successfully diagnosed more than 90% of fungal or nocardial infections occurring in immunosuppressed patients approached in this manner and regard this procedure as the one of choice for such peripherally placed, focal pulmonary lesions not diagnosed by induced-sputum examination. The diagnostic yield is particularly high if cavitation is present in the lesion. In contrast, the diagnostic yield of this procedure in patients with diffuse lung disease is quite low and transthoracic needle aspiration should not be carried out in such individuals.[341,342] In the immunocompromised patient with a focal lung lesion, transthoracic needle aspiration with an ultrathin needle can give decisive information and low false-positive results with a low incidence of complications (Fig. 17).[11] CT fluoroscopy is a new rapid CT-imaging technique similar to conventional fluoroscopy that appears to be promising as a control method for transthoracic needle biopsy of the focal pulmonary lesions.[12]

Needle-biopsy-induced hemoptysis and focal pulmonary hemorrhage in patients with adequate coagulation factors (e.g., platelet count >75,000/mm^3 prior to and for at least 24 hr postprocedure, and normal prothrombin and partial thromboplastin times) are partly related to the skill of the operator and the size of the needle employed. Avoidance of visible bronchi, accompanying arteries, and associated veins can minimize bleeding complications. With 22- and 23-gauge thin-wall needles, the incidence of postprocedure hemoptysis is less than 1%. Such complications as air embolism, implantation of malignant cells into the needle tract, spread of tumor or infection into the pleural space, and bleeding into the chest wall, although reported, are quite rare with fine-gauge needle biopsy.[341,342]

The most common major complication of percutaneous fine-needle aspiration of the lung is pneumothorax, with a wide range of reported rates of 15–40% with the higher rates detected when CT rather than radiography is used as the standard for detection.[13] About one third of pneumothoraces require chest tube insertion.[14] Both the incidence of pneumothorax and the requirement for a chest tube appear to be particularly great in individuals with chronic obstructive pulmonary disease (three times the risk of those without emphysema), the FEV$_1$ being the best predictor of risk.[341,342,347–349] Other potential factors

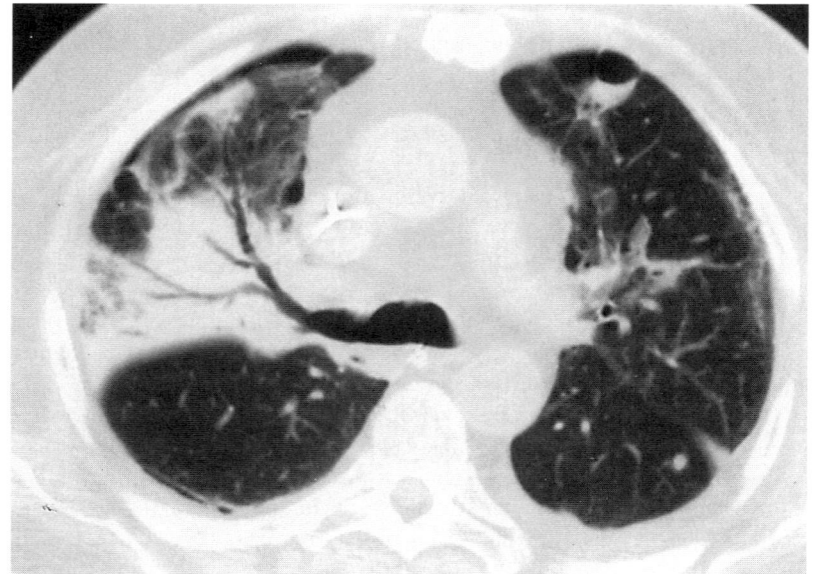

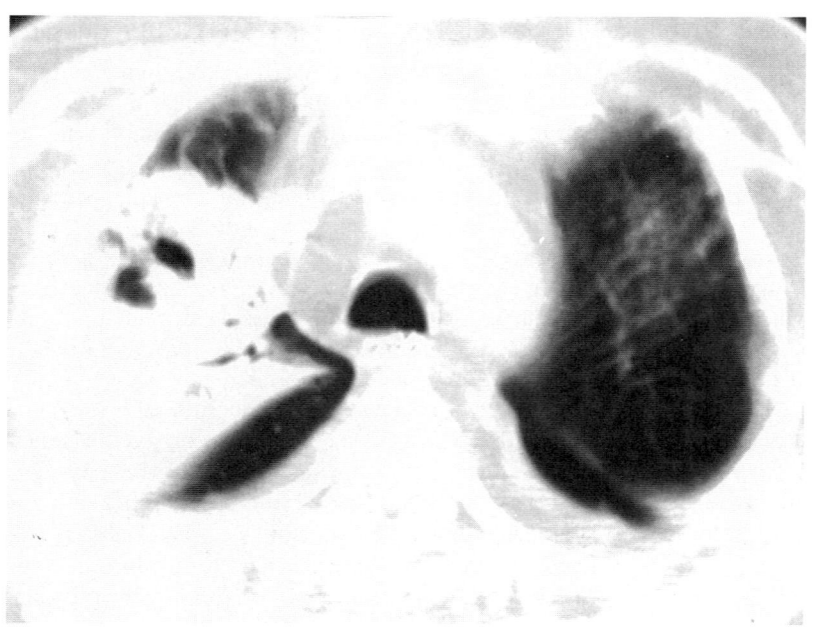

FIGURE 17. Legionnaire's disease. (A) A CT section through the right upper lobe in a cardiac transplant patient with respiratory failure due to *Legionella pneumophila* demonstrates a dense consolidation with a permanent air bronchogram. (B) A CT scan ten days later demonstrates a cavitated right upper lobe mass. Diagram was made by percutaneous needle aspiration.

increasing the risk of pneumothorax include small nodule size and penetration of aerated lung. Moore *et al.*,[352] have suggested that careful attention to certain precautions can decrease the requirement for chest tube placement to approximately 1%. Among others, the precautions include making a single pleural puncture at the biopsy site.[349–352]

5.3.3. Open Lung Biopsy

The definitive diagnostic procedure in the immunocompromised patient with the febrile pneumonitis syndrome remains open lung biopsy, which should be seriously considered if arterial hypoxemia is intensifying, the pulmonary infiltrates are spreading rapidly, and the patient has a hopeful prognosis from his underlying disease. Despite the need for general anesthesia, thoracotomy, and a postoperative chest tube, the procedure is remarkably well tolerated, especially in those patients in whom a treatable disorder is identified. A specific diagnosis is made in approximately 80% of immunosuppressed patients who come to open lung biopsy. The undiagnosed cases probably represent instances of unrecognized pulmonary drug toxicity, the effects of antecedent antimicrobial therapy in modifying the disease process, or even some new form of pulmonary infection (the examples of *Legionella pneumophila* and *L. micdadei*, both of which had their greatest impact on immunosuppressed patients, come readily to mind in this regard). There is a false-negative rate with open lung biopsy of less than 5%, these instances presumably being related to sampling error or inappropriate handling of specimens.[353–363]

A more important question is how frequently the knowledge obtained results in meaningful therapy. The answer to this question is clearly different for different patient populations. Thus, there is now general consensus that open lung biopsy is rarely indicated in patients with AIDS or in febrile neutropenic patients with acute leukemia. In contrast, in patients with Hodgkin's disease, organ transplant patients, or other immunosuppressed patients whose life expectancies from their underlying conditions can be measured in years, it is very clear that when the febrile pneumonitis syndrome develops, open lung biopsy can provide lifesaving information. In these latter individuals, open lung biopsy should be carried out if rapid clinical deterioration is occurring as outlined above or if less invasive diagnostic procedures have failed to yield a diagnosis.[353–363]

The final issue in need of discussion is the question of the circumstances under which empiric therapy can be carried out without an invasive diagnostic procedure. In our experience, the following are the most common situations in which empiric therapy is necessary: (1) far-advanced AIDS, relapsing acute myelogenous leukemia, and other advanced malignances that limit life expectancy because of the severity of the underlying illness; (2) leukemia prior to therapy, since there is an exceedingly low probability of opportunistic infection and antibacterial therapy has a high probability of success; (3) the presence of either an uncorrectable bleeding diathesis or such impaired pulmonary function that invasive diagnostic techniques would not be tolerated; and (4) patient refusal of invasive diagnostic studies. In such patients, the choice of empiric therapy is made on the basis of the indirect clues outlined previously: the epidemiologic and clinical setting, the nature of the immune defect(s) present, the pace of the pulmonary process, and the radiographic pattern.

6. Superinfection

Implicit in this review and in most published reports concerning the febrile pneumonitis syndrome in the immunocompromised patient is that a single etiology is responsible for the disease syndrome. That this is not always the case is shown in Table 1, in which 10% of the subjects, particularly the cancer patients, were shown to have mixed infections. Particular combinations of agents that are likely to be present together are CMV with *Pneumocystis*, gram-negative, or fungal infection; *Nocardia* and *Aspergillus*; *Cryptococcus* and *Nocardia*; *Cryptococcus* and *Pneumocystis*; mycobacterial and fungal infection; and radiation pneumonitis with gram-negative bacillary infection.[3,5]

Even more important is the occurrence of superinfection. For example, in one series of renal transplant patients, it was noted that pulmonary superinfection accounted for 81% of the fatalities.[2] Superinfection appears to be most common in the following situations: following an intense environmental exposure within the hospital to *Aspergillus* species or such gram-negative bacilli as *Pseudomonas aeruginosa*, the previously injured lung being particularly susceptible to environmental hazards; patients who are intubated; patients on high-dose corticosteroids; severely neutropenic patients; and those with pulmonary infarcts or severe chemical injury from aspiration.[1–6,8] The pathogens responsible for superinfection are somewhat different from those responsible for primary infection. Virtually all the instances of bacterial superinfection are due to gram-negative bacilli. Two organisms that rarely produce primary pulmonary infection—herpes simplex virus and *Candida* species—are not uncommon causes of superinfection, as are *Aspergillus fumigatus* and *Torulopsis glabrata*, again, particularly in the intubated patient.[2,6]

The clinician should be particularly alert to the possibility of superinfection in the following circumstances: in the patient who has shown an initial clinical response to therapy in terms of temperature curve, arterial blood gases, well-being, leukocyte count, and sputum production, but who now shows deterioration in one or more of these parameters; in any patient with significant leukopenia who continues to receive high-dose corticosteroid therapy; and in the patient with progressive deterioration despite apparently effective treatment. In any of these instances, aggressive diagnostic techniques should be again undertaken. At postmortem examination in too many patients, the primary cause of fever and pneumonitis is no longer present, but instead, multiple superinfecting microorganisms can be demonstrated. Earlier recognition and better prevention are both necessary in dealing with this unsolved problem complicating the febrile pneumonitis syndrome in the compromised host.[1-5,8] Concern for the possibility of superinfection should be at a particularly high level if there are any questions regarding the quality of the air supply to which the patient is being exposed (e.g., non-HEPA-filtered air for an intubated patient). Immunocompromised patients with lung injury of diverse etiologies are the most susceptible of individuals to pulmonary superinfection.

References

1. Williams DM, Krick JA, Remington JS: Pulmonary infection in the compromised host. *Am Rev Respir Dis* **114:**359–394, 593–627, 1976.
2. Ramsey PG, Rubin RH, Tolkoff-Rubin NE, *et al*: The renal transplant patient with fever and pulmonary infiltrates: Etiology, clinical manifestations, and management. *Medicine* **59:**206–222, 1980.
3. Rubin RH: The cancer patient with fever and pulmonary infiltrates: Etiology and diagnostic approach. In Remington JS, Swartz MN (eds): *Current Clinical Topics in Infectious Disease*, Vol. I. McGraw-Hill, New York, 1980, pp. 288–303.
4. Bergen GA, Shelhamer JH: Pulmonary infiltrates in the cancer patient: New approaches to an old problem. *Infect Dis Clin North Am* **10:**297–325, 1996.
5. Rosenow EC III, Wilson WR, Cockerill FR III: Pulmonary disease in the immunocompromised host. *Mayo Clin Proc* **60:**473–487, 610–631, 1985.
6. Crawford SW: Noninfectious lung disease in the immunocompromised host. *Respiration* **66:**385–395, 1999.
7. Murray JF, Mills J: Pulmonary infectious complications of human immunodeficiency virus infection: Parts I and II. *Am Rev Respir Dis* **141:**1356–1372, 1990.
8. Ettinger NA, Trulock EP: Pulmonary considerations of organ transplantation: Parts I–III. *Am Rev Respir Dis* **143:**1386–1405, 1991; **144:**213–224, 433–451, 1991.
9. Hildebrand FL, Rosenow EC, Haberman TM, Tazelaar HD: Pulmonary complications of leukemia. *Chest* **9:**1233–1239, 1990.
10. Rosenow EC III: Diffuse pulmonary infiltrates in the immunocompromised host. *Clin Chest Med* **11:**55–64, 1990.
11. White DA, Matthay RA: Noninfectious pulmonary complications of infection with the human immunodeficiency virus. *Am Rev Respir Dis* **140:**1763–1787, 1989.
12. Sickles EA, Greene WH, Wiernik PH: Unusual presentation of infection in granulocytopenic patients. *Arch Intern Med* **135:**715–719, 1975.
13. Bodey GP, Buckley M, Sathe YS, *et al*: Quantitative relationships between circulating leukocytes and infection in patients with acute leukemia. *Am J Roentgenol* **64:**328–340, 1966.
14. Heussel CP, Kauczor HU, Heussel GE, *et al*: Pneumonia in febrile neutropenic patients and in bone marrow and blood stem-cell transplant recipients: Use of high resolution computed tomography. *J Clin Oncol* **17:**796–805, 1999.
15. Carratala J, Roson B, Fernandez-Sevilla A, *et al*: Bacteremic pneumonia in neutropenic patients with cancer: Causes, empirical antibiotic therapy, and outcome. *Arch Intern Med* **158:**868–872, 1998.
16. Poe RH, Wahl GW, Qazi R, *et al*: Predictors of mortality in the immunocompromised patient with pulmonary infiltrates. *Arch Intern Med* **146:**1304–1308, 1986.
17. Ewig S, Glasmacher A, Ulrich B, *et al*: Pulmonary infiltrates in neutropenic patients with acute leukemia during chemotherapy: Outcome and prognostic factors. *Chest* **114:**444–451, 1998.
18. Collin BA, Ramphal R: Pneumonia in the compromised host including cancer patients and transplant patients. *Infect Dis Clin North Am* **12:**781–805, 1998.
19. Conces DJ Jr: Pulmonary infections in immunocompromised patients who do not have acquired immunodeficiency syndrome: A systematic approach. *J Thoracic Imaging* **13:**234–246, 1998.
20. Bartlett JG: Pneumonia in the patient with HIV infection. *Infect Dis Clin North Am* **12:**807–820, 1998.
21. Baril L, Astagneau P, Nguyen J, *et al*: Pyogenic bacterial pneumonia in human immunodeficiency virus infected inpatients: A clinical, radiological, microbiological, and epidemiological study. *Clin Infect Dis* **26:**964–971, 1998.
22. Raballa N, Rodrigues P, Lebeaga R, *et al*: Conventional respiratory viruses recovered from immunocompromised patients: Clinical considerations. *Clin Infect Dis* **28:**1043–1048, 1999.
23. Yousuf HM, Englund J, Couch R, *et al*: Respiratory syncytial virus-associated infections in adult recipients of solid organ transplants. *J Heart Lung Transplant* **17:**202–210, 1998.
24. Krinzman S, Basgoz N, Kradin R, *et al*: Influenza among hospitalized adults with leukemia. *Clin Infect Dis* **24:**1095–1099, 1997.
25. McGrath D, Falagas ME, Freeman R, *et al*: Adenovirus infection in adult orthotopic liver transplant recipients: Incidence and clinical significance. *J Infect Dis* **177:**459–462, 1998.
26. Simsir A, Greenebaum E, Nuovo G, Schulman LL: Late fatal adenovirus pneumonitis in a lung transplant recipient. *Transplantation* **65:**592–594, 1998.
27. Hale GA, Heslop HE, Krance RA, *et al*: Adenovirus infection after pediatric bone marrow transplantation. *Bone Marrow Transplant* **23:**277–282, 1999.
28. Ghosh S, Champlin R, Cough R, *et al*: Rhinovirus infections in myelosuppressed adult blood and marrow transplant recipients. *Clin Infect Dis* **29:**528–532, 1999.
29. Fishman JA, Rubin RH: Infection in organ transplant recipients. *N Engl J Med* **338:**1741–1751, 1998.
30. Weitzman SA, Aisenberg AC: Fulminant sepsis after the successful treatment of Hodgkin's disease. *Am J Med* **62:**47–50, 1977.
31. Weitzman SA, Aisenberg AC, Siber GR, *et al*: Impaired humoral immunity in treated Hodgkin's disease. *N Engl J Med* **297:**245–248, 1977.
32. Siber GR, Weitzman SA, Aisenberg AC, *et al*: Impaired antibody response to pneumococcal vaccines after treatment for Hodgkin's disease. *N Engl J Med* **299:**442–446, 1978.
33. Rubin RH: Infectious disease complications of renal transplantation. *Kidney Int* **44:**221–236, 1993.
34. Hibberd PL, Tolkoff-Rubin NE, Cosimi AB, *et al*: Symptomatic cytomegalovirus disease in the cytomegalovirus antibody seropositive renal transplant recipient treated with OKT3. *Transplantation* **53:**68–72, 1992.
35. Hibberd PL, Tolkoff-Rubin NE, Conti D, *et al*: Preemptive gan-

ciclovir therapy to prevent cytomegalovirus disease in cytomegalovirus antibody positive renal transplant recipients: A randomized controlled trial. *Ann Intern Med* **123:**18–25, 1995.

36. Turgeon N, Fishman JA, Basgoz N, *et al*: Effect of acyclovir or ganciclovir therapy after preemptive intravenous ganciclovir therapy to prevent cytomegalovirus disease in cytomegalovirus seropositive renal and liver transplant recipients receiving antilymphocyte antibody therapy. *Transplantation* **66:**1780–1786, 1998.

37. Reinke P, Prosch S, Kern F, Volk HD: Mechanisms of human cytomegalovirus (HCMV) (re)activation and its impact on organ transplant patients. *Transplant Infect Dis* **1:**157–164, 1999.

38. Johanson WG Jr, Pierce AK, Sanford JP: Changing pharyngeal bacterial flora of hospitalized patients. *N Engl J Med* **281:**1137–1140, 1969.

39. Johanson WG Jr, Higuchi JJ, Chadhuri TR, *et al*: Bacterial adherence to epithelial cells in bacterial colonization of the respiratory tract. *Am Rev Respir Dis* **121:**55–63, 1980.

40. Aisner J, Schimpff SC, Bennett JE, *et al*: Aspergillus infections in cancer patients: Association with fireproofing materials in a new hospital. *JAMA* **235:**411–413, 1976.

41. Sarubbi FA Jr, Kopf HB, Wilson MB, *et al*: Increased recovery of *Aspergillus flavius* from respiratory secretions during hospital construction. *Am Rev Respir Dis* **125:**33–38, 1982.

42. Hadley S, Karchmer AW: Fungal infections in solid organ transplant recipients. *Infect Dis Clin North Am* **9:**1045–1074, 1995.

43. Wald A, Lesenring W, Burik J, *et al*: Epidemiology of *Aspergillus* infections in a large cohort of patients undergoing bone marrow transplantation. *J Infect Dis* **175:**1459–1466, 1999.

44. LaForce FM: Lower respiratory tract infections. In Bennett JV, Brachman PS (eds): *Hospital Infections*, 3rd ed. Little, Brown, Boston, 1992, pp. 611–639.

45. LaForce FM, Hopkins J, Trow R, *et al*: Human oral defenses against gram-negative rods. *Am Rev Respir Dis* **114:**929–935, 1976.

46. Reynolds HY: Bacterial adherence to respiratory tract mucosa: A dynamic interaction leading to colonization. *Semin Respir Infect* **2:**8–19, 1987.

47. Faling LJ: Advances in preventing nosocomial pneumonia. *Am Rev Respir Dis* **137:**256–258, 1988.

48. Proctor RA: Fibronectin: A brief overview of its structure, function, and physiology. *Rev Infect Dis* **9:**S317–S321, 1987.

49. Proctor RA: Fibronectin: An enhancer of phagocyte function. *Rev Infect Dis* **9:**S412–S419, 1987.

50. Dal Nogare AR, Toews GB, Pierce AK: Increased salivary elastase precedes gram-negative bacillary colonization in postoperative patients. *Am Rev Respir Dis* **135:**671–675, 1987.

51. Niederman MS, Merrill WW, Ferranti RD, *et al*: Nutritional status and bacterial binding in the lower respiratory tract in patients with chronic tracheostomy. *Ann Intern Med* **100:**795–800, 1984.

52. Martin TR: The relationship between malnutrition and lung infections. *Clin Chest Med* **8:**359–372, 1987.

53. Woods DE, Straus DC, Johanson WG, *et al*: Role of fibronectin in prevention of adherence of *Pseudomonas aeruginosa* to buccal cells. *J Infect Dis* **143:**784–790, 1981.

54. Johanson WG Jr, Pierce AK, Sanford JP, *et al*: Nosocomial respiratory infections with gram-negative bacilli. *Ann Intern Med* **77:**701–706, 1972.

55. Atherton ST, White DJ: Stomach as source of bacteria colonizing respiratory tract during artificial ventilation. *Lancet* **2:**968–969, 1978.

56. du Moulin GC, Hedley-Whyte J, Paterson DG, *et al*: Aspiration of gastric bacteria in antacid-treated patients: A frequent cause of postoperative colonization of the airways. *Lancet* **1:**242–245, 1982.

57. Donowitz LG, Page MC, Mileur GL, *et al*: Alterations of normal gastric flora in critical care patients receiving antacid and cimetidine therapy. *Infect Control* **7:**23–26, 1986.

58. Ruddell WSJ, Axon ATR, Finlay JM, *et al*: Effect of cimetidine on gastric bacterial flora. *Lancet* **1:**672–674, 1990.

59. Driks MR, Craven DE, Celli BR, *et al*: Nosocomial pneumonia in intubated patients given sucralfate as compared with antacids or histamine type 2 blockers. *N Engl J Med* **317:**1376–1382, 1987.

60. Pingleton SK, Hinthorn DR, Liu C: Enteral nutrition in patients receiving mechanical ventilation: Multiple sources of tracheal colonization include the stomach. *Am J Med* **80:**827–832, 1986.

61. Inglis TJJ, Sherratt MJ, Sproat LJ, *et al*: Gastroduodenal dysfunction and bacterial colonization of the ventilated lung. *Lancet* **1:**911–913, 1993.

62. Centers for Disease Control: Nosocomial transmission of multidrug-resistant tuberculosis among HIV-infected persons—Florida and New York, 1988–1991. *Morb Mortal Wkly Rep* **40:**585–591, 1991.

63. Dooley SW, Villarino ME, Lawrence M, *et al*: Nosocomial transmission of tuberculosis in a hospital unit for HIV-infected patients. *JAMA* **257:**2632–2634, 1992.

64. Edlin BR, Tokars JL, Grieco MH, *et al*: An outbreak of multidrug-resistant tuberculosis among hospitalized patients with the acquired immunodeficiency syndrome. *N Engl J Med* **326:**1514–1521, 1992.

65. Fischl MA, Uttamchandani RB, Daikos GL, *et al*: An outbreak of tuberculosis caused by multiple drug resistant tubercle bacilli among patients with HIV infection. *Ann Intern Med* **117:**177–183, 1992.

66. Small PM, Shafer RW, Hopewell PC, *et al*: Exogenous reinfection with multidrug-resistant *Mycobacterium tuberculosis* in patients with advanced HIV infection. *N Engl J Med* **328:**1137–1144, 1993.

67. Tollemar JG; Fungal infections in solid organ transplant recipients. In Bowden RA, Ljungman P, Paya CV (eds): *Transplant Infections*. Lippincott-Raven Publishers, Philadelphia, 1998, pp. 339–350.

68. Scowden EB, Schaffner W, Stone WJ: Overwhelming strongyloidiasis: An unappreciated opportunistic infection. *Medicine* **57:**527–544, 1978.

69. Morgan JS, Schaffner W, Stone WJ: Opportunistic strongyloidiasis in renal transplant recipients. *Transplantation* **42:**518–524, 1986.

70. Whimbey E, Champlin R, Couch R, *et al*: Community respiratory virus infections among hospitalized adult bone marrow transplant recipients. *Clin Infect Dis* **22:**778–782, 1996.

71. Ljungman P: Respiratory virus infections in bone marrow transplant recipients: The European perspective. *Am J Med* **102**(3A)**:**44–47, 1997.

72. Whimbey EE, Englund JA: Community respiratory virus infections in transplant recipients. In Bowden RA, Ljungman P, Paya CV (eds): *Transplant Infections*. Lippincott-Raven Publishers, Philadelphia, 1998, pp. 295–308.

73. Yousuf HM, Englund J, Cough R, *et al*: Influenza among hospitalized adults with leukemia. *Clin Infect Dis* **24:**1095–1099, 1997.

74. Bridges ND, Spray TL, Collins MH, *et al*: Adenovirus infection in the lung results in graft failure after lung transplantation. *J Thorac Cardiovasc Surg* **116:**617–623, 1998.

75. Krinzman S, Basgoz N, Kradin R, *et al*: Respiratory syncytial

virus-associated infections in allograft recipients of solid organ transplants. *J Heart Lung Transplant* **17:**202–210, 1998.

76. Adman D, Engelhard D, Strauss N, *et al*: Antibody response to influenza immunization in patients after heart transplantation. *Vaccine* **15:**1518–1522, 1997.

77. Dengler TJ, Strnad N, Buhring I, *et al*: Differential immune response to influenza and pneumococcal vaccination in immunosuppressed patients after heart transplantation. *Transplantation* **66:** 1340–1347, 1998.

78. Blumberg EA, Fitzpatrick J, Stutman PC, *et al*: Safety of influenza vaccine in heart transplant recipients. *J Heart Lung Transplant* **17:** 1075–1080, 1998.

79. Fraund S, Wagner D, Pethig K, *et al*: Influenza vaccine in heart transplant recipients. *J Heart Lung Transplant* **18:**220–225, 1999.

80. Gump DW, Frank RO, Winn WC Jr, *et al*: Legionnaires' disease in patients with associated serious disease. *Ann Intern Med* **90:**538–542, 1979.

81. Haley CE, Cohen ML, Halter J, *et al*: Nosocomial Legionnaires' disease: A continuing common-source epidemic at Wadsworth Medical Center. *Ann Intern Med* **90:**583–586, 1979.

82. Tkatch LS, Kusne S, Irish WD, *et al*: Epidemiology of legionella pneumonia and factors associated with legionella-related mortality at a tertiary care center. *Clin Infect Dis* **27:**1479–1486, 1998.

83. Kool JL, Fiore AE, Kioski CM, *et al*: More than 10 years of unrecognized nosocomial transmission of Legionnaires' disease among transplant patients. *Infect Control Hosp Epidemiol* **19:**898–904, 1998.

84. Chow JW, Yu VL: Legionella: A major opportunistic pathogen in transplant recipients. *Semin Respir Infect* **13:**132–139, 1998.

85. Patterson WJ, Hay J, Seal DV, McLuckie JC: Colonization of transplant unit water supplies with Legionella and protozoa: Precautions required to reduce the risk of legionellosis. *J Hosp Infect* **37:**7–17, 1997.

86. Rose HD: Mechanical control of hospital ventilation and aspergillus infections. *Am Rev Respir Dis* **105:**306–307, 1972.

87. Rhame FS, Streifel AJ, Kersey JH Jr, *et al*: Extrinsic risk factors for pneumonia in the patient at high risk of infection. *Am J Med* **75**(5A):42–52, 1984.

88. Opal SM, Asp AA, Cannady PB Jr, *et al*: Efficacy of infection control measures during a nosocomial outbreak of disseminated aspergillosis associated with hospital construction. *J Infect Dis* **153:**634–637, 1986.

89. Korman TM, Fuller A, Ibrahim J, *et al*: Fatal *Legionella longbeachae* infection following heart transplantation. *Eur J Clin Microbiol Infect Dis* **17:**53–55, 1998.

90. Gordon EA, Gordon FD, Hayek J, *et al*: Lung abscess complicating *Legionella micdadei* pneumonia in an adult liver transplant recipient: Case report and review. *Transplantation* **65:**130–134, 1998.

91. Harris A, Lally M, Albracht M: *Legionella bozemanii* pneumonia in three patients with AIDS. *Clin Infect Dis* **27:**97–99, 1998.

92. Hopkins C, Weber DJ, Rubin RH: Invasive aspergillus infection: Possible nonward common source within the hospital environment. *J Hosp Infect* **12:**19–25, 1989.

93. Kacmarek RM, Kratohuil J, Dashevsky Y, *et al*: Performance of prototype portable HEPA-filtered positive pressure enclosures. *Respir Care* **37:**1368, 1992.

94. Rubin RH: The compromised host as sentinel chicken. *N Engl J Med* **317:**1151–1153, 1987.

95. Nash G: Pathology of pulmonary infections: Immune compromised vs. normal host. *Chest* **95**(Suppl):176S–180S, 1989.

96. Myerowitz RL (ed): *The Pathology of Opportunistic Infections with Pathogenetic, Diagnostic, and Clinical Correlations*. Raven Press, New York, 1983, pp. 83–94.

97. Nash G: Pathologic features of the lung in the immunocompromised host. *Hum Pathol* **13:**841–858, 1982.

98. Weber WR, Askin FB, Dehner LP: Lung biopsy in *Pneumocystis carinii* pneumonia: A histopathologic study of typical and atypical features. *Am J Clin Pathol* **67:**11–19, 1977.

99. Epler GR, Colby TV, McLoud TC, *et al*: Bronchiolitis obliterans organizing pneumonia. *N Engl J Med* **312:**152–158, 1985.

100. Gross NJ: Pulmonary effects of radiation therapy. *Ann Intern Med* **86:**81–92, 1977.

101. Jennings FL, Arden A: Development of radiation pneumonitis: Time and dose factors. *Arch Pathol Lab Med* **74:**351–360, 1962.

102. Teates CD: The effects of unilateral thoracic irradiation on pulmonary blood flow. *Am J Roentgenol* **102:**875–882, 1968.

103. Margolis LW, Phillips TL: Whole-lung irradiation for metastatic tumor. *Radiology* **93:**1173–1179, 1969.

104. Deeley TJ: The effects of radiation on the lungs in the treatment of carcinoma of the bronchus. *Clin Radiol* **11:**33–39, 1960.

105. Gross NJ: Surfactant subtypes in experimental lung damage: Radiation pneumonitis. *Am J Physiol* **260**(4 Pt 1):L302–L310, 1991.

106. Hallman M, Maasilta P, Kivisaari L, *et al*: Changes in surfactant in bronchoalveolar lavage fluid after hemithorax irradiation in patients with mesothelioma. *Am Rev Respir Dis* **141:**998–1005, 1990.

107. Kwok E, Chan CK: Corticosteroids and azathioprine do not prevent radiation-induced lung injury. *Can Respir J* **5:**211–214, 1998.

108. Ikezoe J, Takashima S, Morimoto S, *et al*: CT appearance of acute radiation-induced injury in the lung. *Am J Roentgenol* **150:**765f–770f, 1988.

109. Lingos TI, Recht A, Vicini F, *et al*: Radiation pneumonitis in breast cancer patients treated with conservative surgery and radiation therapy. *Int J Radiat Oncol Biol Phys* **21:**355–360, 1991.

110. Tarbell NJ, Thompson L, Mauch P: Thoracic irradiation in Hodgkin's disease: Disease control and long-term complications. *Int J Radiat Oncol Biol Phys* **18:**275–281, 1990.

111. Segawa Y, Takigawa N, Kataoka M, *et al*: Risk factors for development of radiation pneumonitis following radiation therapy with or without chemotherapy for lung cancer. *Int J Radiat Oncol Biol Phys* **39:**91–98, 1997.

112. Monson JM, Stark P, Reilly JJ, *et al*: Clinical radiation pneumonitis and radiographic changes after thoracic radiation therapy for lung carcinoma. *Cancer* **82:**842–850, 1998.

113. Marks LB, Munley MT, Bentel GC, *et al*: Physical and biological predictors of changes in whole lung function following thoracic radiation. *Int J Radiat Oncol Biol Phys* **39:**563–570, 1997.

114. Cheng SH, Jian JJ, Chan KY, *et al*: The benefit and risk of postmastectomy radiation therapy in patients with high risk breast cancer. *Am J Radiat Oncol Biol Phys* **21:**12–17, 1998.

115. Kwa SL, Lebesque JV, Theuws JC, *et al*: Radiation pneumonitis as a function of mean lung dose: An analysis of pooled data of 540 patients. *Int J Radiat Oncol Biol Phys* **42:**1–9, 1998.

116. Halme M, Hallman M, Ruotsalainen T, *et al*: Tumor response and radiation-induced lung injury in patients with recurrent small cell lung cancer treated with radiotherapy and concomitant interferon-alpha. *Lung Cancer* **23:**39–52, 1999.

117. Graham MV, Purdy JA, Emami B, *et al*: Clinical dose-volume histogram analysis for pneumonitis after 3D treatment for non-small cell lung cancer (NSCLC). *Int J Radiat Oncol Biol Phys* **45:**323–329, 1999.

118. Castellino RA, Glatstein E, Turbow MM, *et al*: Latent radiation injury of lung or heart activated by steroid withdrawal. *Ann Intern Med* **80:**593–599, 1974.

119. Cohen IJ, Loven D, Schoenfeld T, *et al*: Dactinomycin potentiation of radiation pneumonitis: A forgotten interaction. *Pediatr Hematol Oncol* **8:**187–192, 1991.

120. Blomgrist C, Tiusaneu K, Elomaa I, *et al*: The combination of radiotherapy, adjuvant chemotherapy (cyclophosphamide-doxorubicin-ftorafin) and tamoxifen in stage II breast cancer: Long-term follow-up results of a randomized trial. *Br J Cancer* **66:**1171–1176, 1992.

121. Quon H, Shepherd FA, Payne DG, *et al*: The influence of age on the delivery, tolerance, and efficacy of thoracic radiation on the combined modality treatment of limited stage small cell lung cancer. *Int J Radiat Oncol Biol Phys* **43:**39–45, 1999.

122. Yamada M, Kudoh S, Hirata K, *et al*: Risk factors on pneumonitis following chemoradiotherapy for lung cancer. *Eur J Cancer* **34:**71–75, 1998.

123. Beinert T, Binder D, Stuschke M, *et al*: Oxidant-induced lung injury in anticancer therapy. *Eur J Med Res* **4:**43–53, 1999.

124. Roswit B, White DC: Severe radiation injuries of the lung. *Am J Roentgenol* **129:**127–136, 1977.

125. Gibson PG, Bryant DH, Morgan GW, *et al*: Radiation-induced lung injury: A hypersensitivity pneumonitis? *Ann Intern Med* **109:**288–291, 1988.

126. Gross NJ, Holloway NO, Narine KR: Effects of some nonsteroidal anti-inflammatory agents on experimental radiation pneumonitis. *Radiat Res* **127:**317–324, 1991.

127. Kataoka M, Kawamura M, Itoh H, *et al*: Ga-67 citrate scintigraphy for the early detection of radiation pneumonitis. *Clin Nucl Med* **17:**27–31, 1992.

128. Kataoka M, Kawamura M, Ueda N, *et al*: Diffuse gallium-67 uptake in radiation pneumonitis. *Clin Nucl Med* **15:**707–711, 1990.

129. Arbetter KR, Prakash UB, Tazelaar HD, Douglas WW: Radiation-induced pneumonitis in the "nonirradiated" lung. *Mayo Clinic Proc* **74:**27–36, 1999.

130. Martin C, Ronero S, Sanchez-Paya J, *et al*: Bilateral lymphocytis alveolitis: A common reaction after unilateral thoracic radiation. *Eur Respir J* **13:**727–732, 1999.

131. Adawi A, Zhang Y, Baggs R, *et al*: Blockade of CD40-CD40 ligand interreactions protects against radiation-induced pulmonary inflammation and fibrosis. *Clin Immunol Immunopathol* **89:**222–230, 1998.

132. Nozaki Y, Hasegawa Y, Takeuchi A, *et al*: Nitric oxide as an inflammatory mediator of radiation pneumonitis in rats. *Am J Physiol* **272:**L651–L658, 1997.

133. Epperly MW, Bray JA, Krager S, *et al*: Intratracheal infection of adenovirus containing the human MnSOD transgene protects athymic nude mice from radiation-induced organizing alveolitis. *Int J Radiat Oncol Biol Phys* **43:**169–181, 1999.

134. Epperly MW, Travis EL, Sikora C, Greenberger JS: Magnesium superoxide dismutase (MnSOD) plasmid/liposome pulmonary radioprotective gene therapy: Modulation of irradiation-induced mRNA for IL-1, TNF-alpha, and TGF-beta correlates with delay of organizing alveolitis/fibrosis. *Biol Blood Marrow Transplant* **5:**204–214, 1999.

135. Anscher MS, Kong FM, Marks LB, *et al*: Changes in plasma transforming growth factor beta during radiotherapy and the risk of symptomatic radiation-induced pneumonitis. *Int J Radiat Oncol Biol Phys* **37:**253–258, 1997.

136. Burger A, Loffler H, Bamberg M, Rodemann HP: Molecular and cellular basis of radiation fibrosis. *Int J Radiat Biol* **73:**401–408, 1998.

137. Hayer-Jensen M, Kong FM, Fink LM, Anscher MS: Circulating thrombomodulin during radiation therapy of lung cancer. *Radiat Oncol Invest* **7:**238–242, 1999.

138. Redlich CA, Rockwell S, Chung JS, *et al*: Vitamin A inhibits radiation-induced pneumonitis in rats. *J Nutr* **128:**1661–1664, 1998.

139. Rubio C, Hill ME, Milan S, *et al*: Idiopathic pneumonia syndrome after high dose chemotherapy for relapsed Hodgkin's disease. *Br J Cancer* **75:**1044–1048, 1997.

140. Rosenow EC III: The spectrum of drug-induced pulmonary disease. *Ann Intern Med* **77:**977–991, 1972.

141. Brettner A, Heitzman ER, Woodin WG: Pulmonary complications of drug therapy. *Radiology* **96:**31–38, 1970.

142. Whitcomb ME: Drug-induced lung disease. *Chest* **63:**418–422, 1973.

143. Goldiner PL, Schweizer O: The hazards of anesthesia and surgery in bleomycin-treated patients. *Semin Oncol* **6:**121–124, 1979.

144. Tryka AF, Skornik WA, Godleski JJ, *et al*: Potentiation of bleomycin-induced lung injury by exposure to 70% oxygen: Morphologic assessment. *Am Rev Respir Dis* **126:**1074–1079, 1982.

145. Einhorn L, Krause M, Hornback N, *et al*: Enhanced pulmonary toxicity with bleomycin and radiotherapy in oat cell lung cancer. *Cancer* **37:**2414–2416, 1976.

146. Bloor AJ, Scarle JR, Marcus RE: Two cases of fatal bleomycin pneumonitis complicating the treatment of non-Hodgkin's lymphoma. *Clin Lab Hematol* **20:**119–121, 1998.

147. Luis M, Ayuso A, Martinez G, *et al*: Intraoperative respiratory failure in a patient after treatment with bleomycin: Previous and current intraoperative exposure to 50% oxygen. *Eur J Anaesthesiol* **16:**66–68, 1999.

148. Nakamura H, Sato S, Takahashi K: Effects of vitamin E deficiency on bleomycin-induced pulmonary fibrosis in the hamster. *Lung* **166:**161–176, 1988.

149. Yamazaki C, Hoshino J, Hori T, *et al*: Effect of lecithinized superoxide dismutase on the interstitial pneumonia model induced by bleomycin in mice. *Jpn J Pharmacol* **75:**97–100, 1997.

150. Yamazaki C, Hoshino J, Sekiguchi T, *et al*: Production of superoxide and nitric oxide by alveolar macrophages in the bleomycin-induced interstitial pneumonia mice model. *Jpn J Pharmacol* **78:**69–73, 1998.

151. Littler WA, Ogilvie C: Lung function in patients receiving busulphan. *Br Med J* **4:**530–532, 1970.

152. Rodman T, Karr S, Close HP: Radiation reaction in the lung: Report of a fatal case in a patient with carcinoma of the lung, with studies of pulmonary function before and during prednisone therapy. *N Engl J Med* **262:**431–434, 1960.

153. Brady LW, Germon PA, Cander L: The effects of radiation therapy on pulmonary function in carcinoma of the lung. *Radiation* **85:**130–134, 1965.

154. Horiuchi T, Mason RJ, Kuroki Y, *et al*: Surface and tissue forces, surfactant protein A, and the phospholipid components of pulmonary surfactant in bleomycin-induced pulmonary fibrosis in the rat. *Am Rev Respir Dis* **141:**1006–1013, 1990.

155. Suwabe A, Otake K, Yakuwa N, *et al*: Artificial surfactant (Surfactant TA) modulates adherence and superoxide production of neutrophils. *Am J Respir Crit Care Med* **158:**1890–1899, 1998.

156. Sostman HD, Matthay RA, Putman CE: Cytotoxic drug-induced lung disease. *Am J Med* **62:**608–615, 1977.

157. Willson JVK: Pulmonary toxicity of antineoplastic drugs. *Cancer Treat Rep* **62:**2003–2008, 1978.

158. Holoye PY, Luna MA, MacKay B, *et al*: Bleomycin hypersensitivity pneumonitis. *Ann Intern Med* **88:**47–49, 1978.

159. Rosenow EC III: Chemotherapeutic drug-induced pulmonary disease. *Semin Respir Med* **2:**89–96, 1980.

160. Collis CH: Lung damage from cytotoxic drugs. *Cancer Chemother Pharmacol* **4:**17–27, 1980.

161. Weiss RB, Muggia FM: Cytotoxic drug-induced pulmonary disease. *Am J Med* **68:**259–266, 1980.

162. Batist G, Andrews JL Jr: Pulmonary toxicity of antineoplastic drugs. *JAMA* **246:**1449–1453, 1981.

163. Ginsberg SJ, Comis RL: The pulmonary toxicity of antineoplastic agents. *Semin Oncol* **9:**34–51, 1982.

164. Oliner H, Schwartz R, Rubio F Jr, *et al*: Interstitial pulmonary fibrosis following busulfan therapy. *Am J Med* **31:**134–139, 1961.

165. Leake E, Smith WG, Woodiff HK: Diffuse interstitial pulmonary fibrosis after busulphan therapy. *Lancet* **2:**432–434, 1963.

166. Heard BE, Cooke RA: Busulphan lung. *Thorax* **23:**187–193, 1968.

167. Kirschner RH, Esterly JR: Pulmonary lesions associated with busulfan therapy of chronic myelogenous leukemia. *Cancer* **27:**1074–1080, 1971.

168. Manning DM, Strimlan CV, Turbiner EH: Early detection of busulfan lung: Report of a case. *Clin Nucl Med* **5:**412–414, 1980.

169. Hankins DG, Sanders S, MacDonald FM, *et al*: Pulmonary toxicity recurring after a six week course of busulfan therapy and after subsequent therapy with uracil mustard. *Chest* **73:**413–416, 1978.

170. Horowitz AL, Friedman M, Smither J, *et al*: The pulmonary changes of bleomycin toxicity. *Radiology* **106:**65–68, 1973.

171. Blum RH, Carter SK, Agre K: A clinical review of bleomycin—A new antineoplastic agent. *Cancer* **31:**903–914, 1973.

172. Pascual RS, Mosher MB, Sikand RS, *et al*: Effects of bleomycin on pulmonary function in man. *Am Rev Respir Dis* **108:**211–217, 1973.

173. Samuels ML, Johnson DE, Itoloye PY, *et al*: Large-dose bleomycin therapy, and pulmonary toxicity: A possible role of prior radiotherapy. *JAMA* **235:**1117–1120, 1976.

174. Iacovino JR, Leitner J, Abbas AK, *et al*: Fatal pulmonary reaction from low doses of bleomycin: An idiosyncratic tissue response. *JAMA* **235:**1253–1255, 1976.

175. Dearnaley DP, Horwich A, Ahern R, *et al*: Combination chemotherapy with bleomycin, etoposide, and cisplatin (BEP) for metastatic testicular teratoma: Long-term follow-up. *Eur J Cancer* **27:**684–691, 1991.

176. Perez-Guerra F, Harkleroad LE, Walsh RE, *et al*: Acute bleomycin lung. *Am Rev Respir Dis* **106:**909–913, 1972.

177. Brown WG, Hasan FM, Barbee RA: Reversibility of severe bleomycin-induced pneumonitis. *JAMA* **239:**2012–2014, 1978.

178. Aronin PA, Mahaley MS Jr, Rudnick SA, *et al*: Prediction of BCNU pulmonary toxicity in patients with malignant gliomas: An assessment of risk factors. *N Engl J Med* **303:**183–188, 1980.

179. Durant JR, Norgard MJ, Murad TM, *et al*: Pulmonary toxicity associated with bischloromethyl nitrosourea (BCNU). *Ann Intern Med* **90:**191–194, 1979.

180. Rodin AE, Haggard ME, Travis LB: Lung changes and chemotherapeutic agents in childhood: Report of a case associated with cyclophosphamide therapy. *Am J Dis Child* **120:**337–340, 1970.

181. Dohner VA, Ward HP, Standard RE: Alveolitis during procarbazine, vincristine and cyclophosphamide therapy. *Chest* **62:**636–639, 1972.

182. Patel AR, Shah PC, Rhee HL, *et al*: Cyclophosphamide therapy and interstitial pulmonary fibrosis. *Cancer* **38:**1542–1549, 1976.

183. Rubio FA: Possible pulmonary effects of alkylating agents. *N Engl J Med* **287:**1150–1151, 1972.

184. Rose MA: Busulphan toxicity syndrome caused by chlorambucil. *Br Med J* **2:**123–127, 1975.

185. Godard P, Marty JP, Michel FB: Interstitial pneumonia and chlorambucil. *Chest* **76:**471–473, 1979.

186. Twohig KJ, Matthay RA: Pulmonary effects of cytotoxic agents other than bleomycin. *Clin Chest Med* **11:**31–54, 1990.

187. Sen RP, Walsh TE, Fisher W, *et al*: Pulmonary complications of combination therapy with cyclophosphamide and prednisone. *Chest* **99:**143–146, 1991.

188. Whitcomb ME, Schwartz MI, Tormey DC: Methotrexate pneumonitis: Case report and review of the literature. *Thorax* **27:**636–639, 1972.

189. Goldman GC, Moschella SL: Severe pneumonitis occurring during methotrexate therapy: Report of two cases. *Arch Dermatol* **103:**194–197, 1971.

190. Everts CS, Westcott JL, Bragg DG: Methotrexate therapy and pulmonary disease. *Radiology* **107:**539–543, 1973.

191. Lisbona A, Schwartz J, Lachance C, *et al*: Methotrexate-induced pulmonary disease. *J Can Assoc Radiol* **24:**215–220, 1973.

192. Sostman HD, Matthay RA, Putman CE, *et al*: Methotrexate-induced pneumonitis. *Medicine* **55:**371–388, 1976.

193. Gutin PH, Green MR, Bleyer WA, *et al*: Methotrexate pneumonitis induced by intrathecal methotrexate therapy: A case report with pharmacokinetic data. *Cancer* **38:**1529–1534, 1976.

194. Lascari AD, Strano AJ, Johnson WW, *et al*: Methotrexate-induced sudden fatal pulmonary reaction. *Cancer* **40:**1393–1397, 1977.

195. Rosenow EC III, Unni KK: Drug-induced pulmonary granulomas. *Lung Biol Health Dis* **20:**469–484, 1983.

196. Green L, Schattner A, Berkenstadt H: Severe reversible interstitial pneumonitis induced by low dose methotrexate: Report of a case and review of the literature. *J Rheumatol* **16:**1007–1008, 1989.

197. Ridley MG, Wolfe CS, Mathews JA: Life-threatening acute pneumonitis during low dose methotrexate treatment for rheumatoid arthritis: A case report and review of the literature. *Ann Rheum Dis* **47:**784–788, 1988.

198. Shapiro CL, Yeap BY, Godleski J, *et al*: Drug-related pulmonary toxicity in non-Hodgkin's lymphoma: Comparative results with three different treatment regimens. *Cancer* **68:**699–705, 1991.

199. Cook NJ, Carroll GJ: Successful reintroduction of methotrexate after pneumonitis in two patients with rheumatoid arthritis. *Ann Rheum Dis* **51:**272–274, 1992.

200. Hargreaves MR, Mowat AG, Benson MK: Acute pneumonitis associated with low dose methotrexate treatment for rheumatoid arthritis: Report of five cases and review of published reports. *Thorax* **47:**628–633, 1992.

201. Kremer JM, Phelps CT: Long-term prospective study of the use of methotrexate in the treatment of rheumatoid arthritis: Update after a mean of 90 months. *Arthritis Rheum* **35:**138–145, 1992.

202. Elsasser S, Dalquen P, Soler M, *et al*: Methotrexate-induced pneumonitis: Appearance four weeks after discontinuation of treatment. *Am Rev Respir Dis* **140:**1089–1092, 1989.

203. Ohosone Y, Okano Y, Kameda H, *et al*: Clinical characteristics of patients with rheumatoid arthritis and methotrexate induced pneumonitis. *J Rheumatol* **24:**2299–2303, 1997.

204. Horrigan TJ, Fanning J, Marcotte MP: Methotrexate pneumonitis after systemic treatment for ectopic pregnancy. *Am J Obstet Gynecol* **176:**714–715, 1997.

205. White DA, Rankin JA, Stover DE, *et al*: Methotrexate pneu-

monitis: Bronchoalveolar lavage findings suggest an immunologic disorder. *Am Rev Respir Dis* **139:**18–21, 1989.

206. Suwa A, Kirakata M, Satoh S, *et al*: Rheumatoid arthritis associated with methotrexate-induced pneumonitis: Improvement with i.v. cyclophosphamide therapy. *Clin Exp Rheumatol* **17:**355–358, 1999.

207. Codling BW, Chakera TM: Pulmonary fibrosis following therapy with melphalan for multiple myeloma. *J Clin Pathol* **25:**668–673, 1972.

208. Rubin G, Baume P, Vandenberg R: Azathioprine and acute restrictive lung disease. *Aust NZ J Med* **2:**272–274, 1972.

209. Hazlett DR, Ward GW, Madison DS: Pulmonary function loss in diphenylhydantoin therapy. *Chest* **66:**660–664, 1974.

210. Marshall A, Moore K: Pulmonary disease after amitriptyline overdosage. *Br Med J* **1:**716–717, 1973.

211. Winterbauer RH, Wikske KR, Wheelis RF: Diffuse pulmonary injury associated with gold treatment. *N Engl J Med* **294:**919–921, 1976.

212. Zitnik RJ, Cooper JA Jr: Pulmonary disease due to antirheumatic agents. *Clin Chest Med* **11:**139–150, 1990.

213. Jick SS, Jick H, Walker AM, *et al*: Hospitalizations for pulmonary reactions following nitrofurantoin use. *Chest* **96:**512–515, 1989.

214. Tomlinson J, Tighe M, Johnson S, *et al*: Interstitial pneumonitis following mitozantrone, chlorambucil and prednisolone (MCP) chemotherapy. *Clin Oncol (R Coll Radiol)* **11:**184–186, 1999.

215. Wilson BD, Clarkson CE, Lippmann ML: Amiodarone-induced pulmonary inflammation: Correlation with drug dose and lung levels of drug, metabolite, and phospholipid. *Am Rev Respir Dis* **143:**1110–1114, 1991.

216. Whitcomb ME, Schwartz MI, Keller AR, *et al*: Hodgkin's disease of the lung. *Am Rev Respir Dis* **106:**79–85, 1972.

217. Martin JJ: The Nisbet Symposium: Hodgkin's disease—Radiological aspects of the disease. *Australas Radiol* **11:**206–218, 1967.

218. Strickland B: Intra-thoracic Hodgkin's disease. Part II. Peripheral manifestations of Hodgkin's disease in the chest. *Br J Radiol* **40:**930–938, 1967.

219. Fraser RF, Pare JAP: Neoplastic diseases of the lungs. In Fraser RG, Pare JAP (eds): *Diagnosis of Diseases of the Chest*, 2nd ed, Vol II. Saunders, Philadelphia, 1978, pp. 981–1134.

220. Rosenberg SA, Diamond HD, Jaslowitz B, *et al*: Lymphosarcoma: A review of 1269 cases. *Medicine* **40:**31–84, 1961.

221. Rose HA: Primary lymphosarcoma of the lung. *J Thorac Cardiovasc Surg* **33:**254–263, 1957.

222. Baron MG, Whitehouse WM: Primary lymphosarcoma of the lung. *Am J Roentgenol* **85:**294–308, 1961.

223. Vernant JP, Brun B, Mannoni P, *et al*: Respiratory distress of hyperleukocytic granulocytic leukemias. *Cancer* **44:**264–268, 1979.

224. McKee LC Jr, Collins RD: Intravascular leukocyte thrombi and aggregates as a cause of morbidity and mortality in leukemia. *Medicine* **53:**463–478, 1974.

225. Myers TJ, Cole SR, Klatsky AU, *et al*: Respiratory failure due to pulmonary leukostasis following chemotherapy of acute non-lymphocytic leukemia. *Cancer* **51:**1808–1813, 1983.

226. Tryka AF, Godleski JJ, Fanta CH: Leukemic cell lysis pneumonopathy: A complication of treated myeloblastic leukemia. *Cancer* **50:**2763–2770, 1982.

227. Green RA, Nichlos NJ: Pulmonary involvement in leukemia. *Am Rev Respir Dis* **80:**833–844, 1959.

228. Blank N, Castellino RA, Shah V: Radiographic aspects of pulmonary infection in patients with altered immunity. *Radiol Clin North Am* **11:**175–190, 1973.

229. Simmons RL, Uranga VM, LaPlante ES, *et al*: Pulmonary complications in transplant recipients. *Arch Surg* **105:**260–268, 1972.

230. Friedman M, Libert R, Michaelson ED: Unilateral pulmonary edema after renal transplantation. *N Engl J Med* **293:**343–344, 1975.

231. Cosimi AB, Cho SI, Delmonico FL, *et al*: A randomized clinical trial comparing OKT3 and steroids for treatment of hepatic allograft rejection. *Transplantation* **43:**91–95, 1987.

232. Ortho Multicenter Transplant Study Group: A randomized clinical trial of OKT3 monoclonal antibody for acute rejection of cadaveric renal transplants. *N Engl J Med* **313:**337–342, 1985.

233. Ward HN: Pulmonary infiltrates associated with leukoagglutinin transfusion reactions. *Ann Intern Med* **73:**689–694, 1970.

234. Thompson JSA, Severson CD, Parmely MJ, *et al*: Pulmonary "hypersensitivity" reactions induced by transfusion of non-HL-A leukoagglutinins. *N Engl J Med* **284:**1120–1125, 1971.

235. Tenholder MF, Hooper RG: Pulmonary infiltrates in leukemia. *Chest* **78:**468–473, 1980.

236. Popovsky MA, Abel MD, Moore SB: Transfusion-related acute lung injury associated with passive transfer of antileukocyte antibodies. *Am Rev Respir Dis* **128:**185–189, 1983.

237. Schiller V, Aberle DR, Aberle AM: Pulmonary alveolar proteinosis: Occurrence with metastatic melanoma to lung. *Chest* **96:**466–467, 1989.

238. Honda Y, Takahashi H, Shijubo N, *et al*: Surfactant protein A concentration in bronchoalveolar lavage fluids of patients with pulmonary alveolar proteinosis. *Chest* **103:**496–499, 1993.

239. Crouch E, Persson A, Chang D: Accumulation of surfactant protein D in human pulmonary alveolar proteinosis. *Am J Pathol* **142:**241–248, 1993.

240. Singh G, Katyal SL, Bedrossian CW, *et al*: Pulmonary alveolar proteinosis: Staining for surfactant apoprotein in alveolar proteinosis and in conditions simulating it. *Chest* **83:**82–86, 1983.

241. Ruben FL, Talamo TS: Secondary pulmonary alveolar proteinosis occurring in two patients with acquired immune deficiency syndrome. *Am J Med* **80:**1187–1190, 1986.

242. Godwin JD, Muller NL, Takasugi JE: Pulmonary alveolar proteinosis: CT findings. *Radiology* **169:**609–613, 1988.

243. Carre PC, Didier AP, Pipy BR, *et al*: The lavage fluid from a patient with alveolar proteinosis inhibits the *in vitro* chemiluminescence response and arachidonic acid metabolism of normal guinea pig alveolar macrophages. *Am Rev Respir Dis* **142:**1068–1072, 1990.

244. Hoffman RM, Dauber JH, Rogers RM: Improvement in alveolar macrophage migration after therapeutic whole lung lavage in pulmonary alveolar proteinosis. *Am Rev Respir Dis* **134:**1030–1032, 1989.

245. Muntaner L, Leyes M, Payeras A, *et al*: Radiologic features of *Rhodococcus equi* pneumonia in AIDS. *Eur J Radiol* **24:**66–70, 1997.

246. Capdevila JA, Bujan S, Gavalda J, *et al*: *Rhodococcus equi* pneumonia in patients infected with the human immunodeficiency virus. Report of 2 cases and review of the literature. *Scand J Infect Dis* **29:**535–541, 1997.

247. Sickles EA, Greene WH, Wiernik PH: Unusual presentation of infection in granulocytopenic patients. *Arch Intern Med* **135:**715–719, 1975.

248. Sickles EA, Young VM, Greene WH, *et al*: Pneumonia in acute leukemia. *Arch Intern Med* **79:**528–534, 1973.

249. Levine AS, Schimpff SC, Graw RG, *et al*: Hematologic malignancies and other marrow failure states: Progress in the management of complicating infections. *Semin Hematol* **11:**141–202, 1974.

250. Groskin SA, Stadnick ME, DuPont PG: *Pneumocystis carinii* pneumonia: Effect of corticosteroid treatment on radiographic appearance in a patient with AIDS. *Radiology* **180:**423–425, 1991.

251. Kuhlman JE, Fishman EK, Hruban RH, *et al*: Disease of the chest in AIDS: CT diagnosis. *RadioGraphics* **9(5):**827–857, 1989.

252. Kuhlman JE, Knowles M, Fishman EK, *et al*: Premature bullous damage in AIDS: CT diagnosis. *Radiology* **173:**23–26, 1989.

253. Barloon TJ, Galvin JR, Mori M, *et al*: High-resolution ultrafast chest CT in the clinical management of febrile bone marrow transplant patients with normal or nonspecific chest roentgenograms. *Chest* **99:**928–933, 1991.

254. Kuhlman JE, Fishman EK, Siegelman SS: Invasive pulmonary aspergillosis in acute leukemia: Characteristic findings on CT, the CT halo sign and the role of CT in early diagnosis. *Radiology* **157:**611–614, 1985.

255. Gruden JF, Huang L, Turner J, *et al*: High resolution CT in the evaluation of clinically suspected *Pneumocystis carinii* pneumonia in AIDS patients with normal, equivocal, or nonspecific findings. *Am J Roentgenol* **169:**967–975, 1997.

256. Graham NJ, Muller NL, Miller RR, *et al*: Intrathoracic complications following allogeneic bone marrow transplantation: CT findings. *Radiology* **181:**153–156, 1991.

257. Golden JA, Sollitto RA: The radiology of pulmonary disease: Chest radiography, computed tomography, and gallium scanning. In White DA, Stover DE (eds): *Pulmonary Effects of AIDS*. W. B. Saunders, St. Louis, MO, 1988, pp. 481–495.

258. Glatman-Freedman A, Ewig JM, Dobroszycki J, *et al*: Simultaneous *Pneumocystis carinii* and pneumococcal pneumonia in human immunodeficiency virus-infected children. *J Pediatr* **132:**169–171, 1998.

259. Clark TM, Burman WJ, Cohn DL, Mehler PS: Septic shock from *Mycobacterium tuberculosis* after therapy for *Pneumocystis carinii*. *Arch Intern Med* **158:**1033–1035, 1998.

260. Barrio JL, Suarez M, Rodriguesz JL, *et al*: *Pneumocystis carinii* pneumonia presenting as cavitating and noncavitating solitary pulmonary nodules in patients with the acquired immunodeficiency syndrome. *Am Rev Respir Dis* **134:**1094–1096, 1986.

261. Plunkett MB, Peterson MS, Landereneau RJ, *et al*: Peripheral pulmonary nodules: Preoperative percutaneous needle localization with CT guidance. *Radiology* **185:**274–276, 1992.

262. Naidich DP, Sussman R, Kutcher WL, *et al*: Solitary pulmonary nodules: CT-bronchoscopic correlation. *Chest* **3:**595–598, 1988.

263. Janzen DL, Adler BD, Padley SPG, *et al*: Diagnostic success of bronchoscopic biopsy in immunocompromised patients with acute pulmonary disease: Predictive value of disease distribution as shown on CT. *Am J Roentgenol* **160:**21–24, 1993.

264. Srivatsa SS, Burger CD, Douglas WW: Upper lobe pulmonary parenchymal calcification in a patient with AIDS and *Pneumocystis carinii* pneumonia receiving aerosolized pentamidine. *Chest* **101:**266–267, 1992.

265. Im JG, Webb WR, Han MC, *et al*: Apical opacity associated with pulmonary tuberculosis: High-resolution CT findings. *Radiology* **178:**727–731, 1991.

266. Bartlett JG: Pneumonia in the patient with HIV infection. *Infect Dis Clin North Am* **12:**807–820, 1998.

267. Wallace JM, Hansen NI, Lavange L, *et al*: Respiratory disease trends in the pulmonary complications of HIV infection study cohort. *Am J Respir Crit Care Med* **155:**72–80, 1997.

268. Osmond DH, Chin DP, Glassroth J, *et al*: Impact of bacterial pneumonia and *Pneumocystis carinii* pneumonia on human immunodeficiency virus disease progression. Pulmonary Complications of HIV Study Group. *Clin Infect Dis* **29:**536–543, 1999.

269. Carson PJ, Goldsmith JC: Atypical pulmonary diseases associated with AIDS. *Chest* **100:**675–677, 1991.

270. Stansell JD, Osmond DH, Charlebois E, *et al*: Predictors of *Pneumocystis carinii* pneumonia in HIV-infected persons. *Am J Respir Crit Care Med* **155:**60–66, 1997.

271. Bochini A, Smacchia C, DiFine M, *et al*: Community-acquired pneumonia in a cohort of former injection drug users with and without human immunodeficiency virus infection: Incidence, etiologies, and clinical aspects. *Clin Infect Dis* **23:**107–113, 1996.

272. Naidich DP: Pulmonary manifestations of HIV infection. In Greene R, Muhm JR (eds): *Syllabus: A Categorical Course in Chest Radiology*. Radiological Society of North America, Chicago, 1992, pp. 135–155.

273. Bergin CJ, Wirth RL, Berry GJ, *et al*: *Pneumocystis carinii* pneumonia: CT and HRCT observations. *J Comput Assist Tomogr* **14:**756–759, 1990.

274. Wasser LS, Brown E, Talavera W: Miliary PCP in AIDS. *Chest* **96:**693–695, 1989.

275. Cohen BA, Pmeranz S, Rabinowitz JG, *et al*: Pulmonary complications of AIDS: Radiologic features. *Am J Roentgenol* **143:**115–122, 1984.

276. Jayes RL, Kamerow HN, Hasselquist SM, *et al*: Disseminated pneumocystosis presenting as a pleural effusion. *Chest* **103:**306–308, 1993.

277. Radin DR, Baker EL, Klatt EC, *et al*: Visceral and nodal calcification in patients with AIDS-related *Pneumocystis carinii* infection. *Am J Roentgenol* **154:**27–31, 1990.

278. Kuhlman JE, Kavuru M, Fishman EK, *et al*: *Pneumocystis carinii* pneumonia: Spectrum of parenchymal CT findings. *Radiology* **175:**711–714, 1990.

279. Goodman PC, Daley C, Minagi H: Spontaneous pneumothorax in AIDS patients with *Pneumocystis carinii* pneumonia. *Am J Roentgenol* **147:**29–31, 1986.

280. Travis WD, Pittaluga S, Lipschik GY, *et al*: Atypical pathologic manifestations of *Pneumocystis carinii* pneumonia in acquired immune deficiency syndrome. *Am J Surg Pathol* **14:**615–625, 1990.

281. Gurney JW, Bates FT: Pulmonary cystic disease: Comparison of *Pneumocystis carinii* pneumatoceles and bullous emphysema due to intravenous drug abuse. *Radiology* **173:**27–31, 1989.

282. Smith RL, Berkowitz KA, Aranda CP: Bronchoalveolar lavage neutrophila seen in *Pneumocystis* pneumonia presenting pneumothorax. *Chest* **100:**865–867, 1991.

283. McClellan MD, Miller SB, Parsons PE, *et al*: Pneumothorax with *Pneumocystis carinii* pneumonia in AIDS. *Chest* **100:**1224–1228, 1991.

284. Pinsk R, Rogers LF: Cystic parenchymal changes associated with spontaneous pneumothorax in an HIV-positive patient. *Chest* **97:**1471–1472, 1990.

285. McFadden RG, Carr TJ, Mackie IDF: Thoracic magnetic resonance imaging in the evaluation of HIV-1/AIDS pneumonitis. *Chest* **101:**371–374, 1992.

286. Hill AR, Premkumar S, Brustein S, *et al*: Disseminated tuberculosis in the acquired immunodeficiency syndrome era. *Am Rev Respir Dis* **144:**1164–1170, 1991.

287. Long R, Maycher B, Scalcini M, *et al*: The chest roentgenogram in pulmonary tuberculosis patients seropositive for human immunodeficiency virus type 1. *Chest* **99:**123–127, 1991.

288. Flora GS, Modilevsky T, Antoniskis D, *et al*: Undiagnosed tuber-

culosis in patients with human immunodeficiency virus infection. *Chest* **98:**1056–1059, 1990.

289. Barnes PF, Bloch AB, Davidson PT, *et al*: Tuberculosis in patients with human immunodeficiency virus infection. *N Engl J Med* **324:**1644–1650, 1991.

290. Pastores SM, Naidich DP, Aranda C, *et al*: Intrathoracic adenopathy associated with pulmonary tuberculosis in patients with human immunodeficiency virus infection. *Chest* **103:**1433–1437, 1993.

291. Horsburgh CR: *Mycobacterium avium* complex infection in the acquired immunodeficiency syndrome. *N Engl J Med* **324:**1332–1338, 1991.

292. Zurawski CA, Cage GD, Rimland D, Blumberg HM: Pneumonia and bacteremia due to *Mycobacterium celatum* masquerading as *Mycobacterium xenopi* in patients with AIDS: An underdiagnosed problem? *Clin Infect Dis* **24:**140–143, 1997.

293. Redd SC, Rutherford GW III, Sande MA, *et al*: The role of human immunodeficiency virus infection in pneumococcal bacteremia in San Francisco residents. *J Infect Dis* **162:**1012–1017, 1990.

294. Magnenat JL, Nicod LP, Auckenthaler R, *et al*: Mode of presentation and diagnosis of bacterial pneumonia in human immunodeficiency virus-infected patients. *Am Rev Respir Dis* **144:**917–922, 1991.

295. Sadaghdar H, Eden E: Pulmonary Kaposi's sarcoma presenting as fulminant respiratory failure. *Chest* **100:**858–860, 1991.

296. Naidich DP, Tarras M, Garay SM, *et al*: Kaposi's sarcoma: CT-radiographic correlation. *Chest* **96:**723–728, 1989.

297. Garay SM, Belenko M, Fazzine I, *et al*: Pulmonary manifestations of Kaposi's sarcoma. *Chest* **91:**39–43, 1987.

298. Nathan S, Vaghaiwalla R, Mohsenifar Z: Use of Nd:YAG laser in endobronchial Kaposi's sarcoma. *Chest* **98:**1299–1300, 1990.

299. Lee VW, Fuller JD, O'Brien MJ, *et al*: Pulmonary Kaposi sarcoma in patients with AIDS: Scintigraphic diagnosis with sequential thallium and gallium scanning. *Radiology* **180:**409–412, 1991.

300. Chechani V, Kamholz SL: Pulmonary manifestations of disseminated cryptococcosis in patients with AIDS. *Chest* **98:**1060–1066, 1990.

301. Zuger A, Louie E, Holzman RS, *et al*: Cryptococcal disease in patients with the acquired immunodeficiency syndrome. *Ann Intern Med* **104:**234–240, 1986.

302. Denning DW, Follansbee SE, Scolaro M, *et al*: Pulmonary aspergillosis in the acquired immunodeficiency syndrome. *N Engl J Med* **324:**654–662, 1991.

303. Sider L, Weiss AJ, Smith MD, *et al*: Varied appearance of AIDS-related lymphoma in the chest. *Radiology* **171:**629–632, 1989.

304. Heitzman ER: Pulmonary neoplastic and lymphoproliferative disease in AIDS: A review. *Radiology* **177:**347–351, 1990.

305. Bazot M, Cadranel J, Benayoun S, *et al*: Primary pulmonary AIDS-related lymphoma: Radiographic and CT findings. *Chest* **116:**1282–1286, 1999.

306. Townsend RR: CT of AIDS-related lymphoma. *Am J Roentgenol* **156:**969–974, 1991.

307. Bottles K, McPhaul LW, Volberding P: Fine-needle aspiration biopsy of patients with acquired immunodeficiency syndrome (AIDS): Experience in an outpatient clinic. *Ann Intern Med* **108:**42–45, 1988.

308. Herts BR, Megibow AJ, Birnbaum BA, *et al*: High-attenuation lymphadenopathy in AIDS patients: Significance of findings at CT. *Radiology* **185:**777–781, 1992.

309. Aisner J, Schimpff SC, Wiernik PH: Treatment of invasive aspergillosis: Relation of early diagnosis and treatment to response. *Ann Intern Med* **86:**539–543, 1977.

310. Rubin RH, Tolkoff-Rubin NE: The problem of human immuno-

deficiency virus (HIV) infection and transplantation. *Transplant Int* **1:**36–42, 1988.

311. Schaefer JC, Yu B, Armstrong D: An *Aspergillus* immunodiffusion test in the early diagnosis of aspergillosis in adult leukemia patients. *Am Rev Respir Dis* **113:**325–329, 1976.

312. Filice G, Yu B, Armstrong D: Immunodiffusion and agglutination tests for *Candida* in patients with neoplastic disease: Inconsistent correlation of results with invasive disease. *J Infect Dis* **135:**349–357, 1977.

313. Edwards JE Jr, Lehrer RI, Stiehm ER, *et al*: Severe candidal infections: Clinical perspective, immune defense mechanisms, and current concepts of therapy. *Ann Intern Med* **89:**91–106, 1978.

314. Murray PR, Washington JA II: Microscopic and bacteriologic analysis of expectorated sputum. *Mayo Clin Proc* **50:**339–344, 1975.

315. Bodey GP, Powell RD, Hersh EM, *et al*: Pulmonary complications of acute leukemia. *Cancer* **19:**781–793, 1966.

316. Sickles EA, Young VM, Greene WH, *et al*: Pneumonia in acute leukemia. *Ann Intern Med* **79:**528–534, 1973.

317. Aisner J, Kuols LK, Sickles EA, *et al*: Transtracheal selective bronchial brushing for pulmonary infiltrates in patients with cancer. *Chest* **69:**367–371, 1976.

318. Bigby TD, Margolskee D, Curtis JL, *et al*: The usefulness of induced sputum in the diagnosis of *Pneumocystis carinii* pneumonia in patients with the acquired immunodeficiency syndrome. *Am Rev Respir Dis* **133:**515–518, 1986.

319. Pitchenik AE, Ganjei P, Torres A, *et al*: Sputum examination for the diagnosis of *Pneumocystis carinii* pneumonia in AIDS. *Am Rev Respir Dis* **133:**226–229, 1986.

320. Kovacs JA, Ng JL, Masur H, *et al*: Diagnosis of *Pneumocystis carinii* pneumonia: Improved detection in sputum with use of monoclonal antibodies. *N Engl J Med* **318:**589–593, 1988.

321. O'Brien RF, Quinn JL, Miyahara BT, *et al*: Diagnosis of *Pneumocystis carinii* pneumonia by induced sputum in a city with moderate incidence of AIDS. *Chest* **95:**136–138, 1989.

322. Masur H, Shelhamer J, Parrillo JE: The management of pneumonias in immunocompromised patients. *JAMA* **253:**1769–1773, 1985.

323. Barlett JG: Diagnostic accuracy of transtracheal aspiration: Bacteriologic studies. *Am Rev Respir Dis* **115:**777–782, 1977.

324. Matthay RA, Moritz ED: Invasive procedures for diagnosing pulmonary infection: A critical review. *Clin Chest Med* **2:**3–19, 1981.

325. Verra F, Mouda H, Rauss A, *et al*: Bronchoalveolar lavage in immunocompromised patients: Clinical and functional consequences. *Chest* **101:**1215–1220, 1992.

326. Thiede WH, Banaszak GF: Selective bronchial catheterization. *N Engl J Med* **286:**525–528, 1972.

327. Repsher LH, Schroter G, Hammon WS: Diagnosis of *Pneumocystis carinii*. *N Engl J Med* **287:**340–341, 1972.

328. Finley R, Kieff E, Thompson S, *et al*: Bronchial brushing in the diagnosis of pulmonary disease in patients at risk for opportunistic infection. *Am Rev Respir Dis* **109:**379–386, 1974.

329. Xaubet A, Torres A, Marco F, *et al*: Pulmonary infiltrates in immunocompromised patients: Diagnostic value of telescopic plugged catheter and bronchoalveolar lavage. *Chest* **95:**130–135, 1989.

330. Weldon-Linne CM, Rhone DP, Bourassa R: Bronchoscopy specimens in adults with AIDS: Comparative yields of cytology, histology, and cultures for diagnosis of infectious agents. *Chest* **98:**24–28, 1990.

331. Malabonga VM, Basti J, Kamholz SL: Utility of bronchoscopic sampling techniques for cryptococcal disease in AIDS. *Chest* **99:**370–372, 1991.

332. Mason GR, Hashimoto CH, Dickman PS, *et al*: Prognostic implications of bronchoalveolar lavage neutrophilia in patients with *Pneumocystis carinii* pneumonia and AIDS. *Am Rev Respir Dis* **139:**1336–1342, 1989.

333. Saito H, Anaissie GE, Morice RC, *et al*: Bronchoalveolar lavage in the diagnosis of pulmonary infiltrates in patients with acute leukemia. *Chest* **94:**745–749, 1988.

334. Kovalski R, Hansen-Flaschen J, Lodato RF, *et al*: Localized pulmonary infiltrates: Diagnosis by bronchoscopy and resolution with therapy. *Chest* **97:**674–678, 1990.

335. Beyt BE Jr, King DK, Glew RH: Fatal pneumonitis and septicemia after fiberoptic bronchoscopy. *Chest* **72:**105–107, 1977.

336. Robbins H, Goldman AL: Failure of a "prophylactic" antimicrobial drug to prevent sepsis after fiberoptic bronchoscopy. *Am Rev Respir Dis* **116:**325–326, 1977.

337. Mehta AC, Kavuru MS, Meeker DP, *et al*: Transbronchial needle aspiration for histology specimens. *Chest* **96:**1228–1232, 1989.

338. Dijkman JH, van der Meer JWM, Bakker W, *et al*: Transpleural lung biopsy by the thoracoscopic route in patients with diffuse interstitial pulmonary disease. *Chest* **82:**76–83, 1982.

339. Lloyd MS: Thoracoscopy and biopsy in the diagnosis of pleurisy with effusion. *Q Bull Sea View Hosp* **14:**128–133, 1953.

340. DeCamp PT, Mosley PW, Scott ML, *et al*: Diagnostic thoracoscopy. *Ann Thorac Surg* **16:**79–84, 1973.

341. Oldenburg FA, Newhouse MT: Thoracoscopy: A safe, accurate diagnostic procedure using the rigid thoracoscope and local anesthesia. *Chest* **75:**45–50, 1979.

342. Faling LJ: New advances in diagnosing nosocomial pneumonia in intubated patients. Part I. *Am Rev Respir Dis* **137:**253–255, 1988.

343. Marquette CH, Ramon P, Courcol R, *et al*: Bronchoscopic protected catheter brush for the diagnosis of pulmonary infections. *Chest* **93:**746–750, 1988.

344. Wimberley N, Faling LJ, Bartlett JG: A fiberoptic bronchoscopy technique to obtain uncontaminated lower airway secretions for bacterial culture. *Am Rev Respir Dis* **119:**337–343, 1979.

345. Westcott JL: Percutaneous transthoracic needle biopsy: State of the art. *Radiology* **169:**593–601, 1988.

346. Perlmutt LM, Johnston WW, Dunnick NR: Percutaneous transthoracic needle aspiration: A review. *Am J Roentgenol* **152:**451–455, 1989.

347. Conces DJ Jr, Clark SA, Tarver RD, *et al*: Transthoracic aspiration needle biopsy: Value in the diagnosis of pulmonary infections. *Am J Roentgenol* **152:**31–34, 1989.

348. Greene R, Szyfelbein W, Isler RJ, *et al*: Supplementary tissue core histology from fine needle transthoracic aspiration biopsy. *Am J Roentgenol* **144:**787–792, 1985.

349. Greene R: Transthoracic needle aspiration biopsy. In Athanasoulis C, Pfister R, Greene R, *et al* (eds): *Interventional Radiology*. Saunders, Philadelphia, 1981, pp. 587–634.

350. Scott WW, Kuhlman JE: Focal pulmonary lesions in patients with AIDS: Percutaneous transthoracic needle biopsy. *Radiology* **180:**419–421, 1991.

351. Miller KS, Fish GB, Stanley JH, *et al*: Prediction of pneumothorax rate in percutaneous needle aspiration of the lung. *Chest* **93:**742–745, 1988.

352. Moore EH, Shepard JO, McCloud TC, *et al*: Positional precautions in needle aspiration lung biopsy. *Radiology* **175:**733–735, 1990.

353. Zavala DC, Bedell GN, Rossi NP: Trephine lung biopsy with a high-speed air drill. *J Thorac Cardiovasc Surg* **64:**220–228, 1972.

354. McCartney RL: Hemorrhage following percutaneous lung biopsy. *Radiology* **112:**305–307, 1974.

355. Clore F, Virapongse C, Saterfiel J: Low-risk large-needle biopsy of chest lesions. *Chest* **96:**538–541, 1989.

356. Goralnick CH, O'Connell DM, El Yousef SJ, *et al*: CT-guided cutting-needle biopsies of selected chest lesions. *Am J Roentgenol* **151:**903–907, 1988.

357. Toledo-Pereyra LH, DeMeester TR, Kinealey A, *et al*: The benefit of open lung biopsy in patients with previous non-diagnostic transbronchial lung biopsy: A guide to appropriate therapy. *Chest* **77:**647–650, 1980.

358. Jaffe JP, Maki DG: Lung biopsy in immunocompromised patients: One institution's experience and an approach to management of pulmonary disease in the compromised host. *Cancer* **48:**1144–1153, 1981.

359. Haverkos HW, Dowling JN, Pasculle AW, *et al*: Diagnosis of pneumonitis in immunocompromised patients by open lung biopsy. *Cancer* **52:**1093–1097, 1983.

360. McKenna RJ Jr, Mountain CF, McMurtey MJ: Open lung biopsy in immunocompromised patients. *Chest* **86:**671–674, 1984.

361. Cockerill FR III, Wilson WR, Carpenter HA, *et al*: Open lung biopsy in immunocompromised patients. *Arch Intern Med* **145:**1398–1404, 1985.

362. Cheson BD, Samlowski WE, Tang TT, *et al*: Value of open-lung biopsy in 87 immunocompromised patients with pulmonary infiltrates. *Cancer* **55:**453–459, 1985.

363. Catterall JR, McCabe RE, Brooks RG, *et al*: Open lung biopsy in patients with Hodgkin's disease and pulmonary infiltrates. *Am Rev Respir Dis* **139:**1274–1279, 1989.

5

Central Nervous System Infection in the Immunocompromised Host

ALLAN R. TUNKEL and W. MICHAEL SCHELD

1. Introduction

The risk of an infection of the central nervous system (CNS) in an immunocompromised patient depends on the underlying disease and its treatment, the duration of immunosuppression, and the type of immune abnormality.[1,2] The four major types of host defense abnormalities encountered in the immunosuppressed patient are defects in cell-mediated immunity, humoral immunity, the number and function of neutrophils, and loss of splenic function (i.e., from surgery, disease, or radiotherapy). Knowledge of the underlying host defense abnormality is often helpful in predicting the infecting microorganism (Table 1).

Patients with defects in cell-mediated immunity include but are not limited to those with lymphoma, organ transplant recipients, patients treated with daily corticosteroid therapy, and patients with human immunodeficiency virus type 1 (HIV-1) infection. These patients are more susceptible to infection by intracellular microorganisms (e.g., *Listeria monocytogenes*, *Toxoplasma gondii*), the eradication of which is dependent on an intact T-lymphocyte–macrophage system. Patients with defective humoral immunity (e.g., chronic lymphocytic leukemia, multiple myeloma, and Hodgkin's disease posttreatment) are unable to mount a normal antibody response to

bacterial infection and are subject to infection with encapsulated bacteria such as *Streptococcus pneumoniae* and *Haemophilus influenzae* type B. These same microorganisms may be responsible for bacteremia or CNS infection or both in patients who have undergone splenectomy or in patients with diminished splenic function. Neutropenic patients are particularly prone to CNS infections caused by gram-negative bacilli (e.g., *Pseudomonas aeruginosa*) and fungal species such as *Candida* and *Aspergillus*.

Certain aspects of CNS infections in the immunosuppressed patient make their identification and management challenging: (1) The number of potential invading microorganisms is large; (2) infection with more than one agent, or sequential infection, is not uncommon; (3) multiple organ systems may be simultaneously involved; (4) the manifestations of infection may be masked because of the host's diminished inflammatory response; and (5) survival is dependent on rapid diagnosis and initiation of therapy.[3] The following sections briefly review the spectrum of CNS infections in the immunocompromised patient, emphasizing the epidemiology, clinical presentation, diagnosis, and treatment of these disorders in this patient population.

2. Viral Infections

2.1. Human Immunodeficiency Virus Type 1

2.1.1. Epidemiology and Etiology

Infection with the human immunodeficiency virus type 1 (HIV-1) is a global problem. By 1991, 2.5 million persons worldwide had developed the acquired immuno-

Allan R. Tunkel • Department of Internal Medicine, MCP Hahnemann University, Philadelphia, Pennsylvania 19129. W. Michael Scheld • Departments of Internal Medicine (Infectious Diseases) and Neurosurgery, University of Virginia School of Medicine, Charlottesville, Virginia 22908.

Clinical Approach to Infection in the Compromised Host (Fourth Edition), edited by Robert H. Rubin and Lowell S. Young. Kluwer Academic/Plenum Publishers, New York, 2002.

TABLE 1. Common Etiologic Agents of Central Nervous System Infection in Immunocompromised Patients

Class of organism	Cellular immunodeficiency	Humoral immunodeficiency	Neutropenia
		Type of immune defect	
Viruses	Cytomegalovirus Herpes simplex virus Varicella-zoster virus JC virus	Cytomegalovirus	—
Bacteria	*Listeria monocytogenes* *Mycobacterium avium-intracellulare* *Mycobacterium tuberculosis* *Nocardia* species *Salmonella* species	*Streptococcus pneumoniae* *Haemophilus influenzae* *Neisseria meningitidis* *Staphylococcus aureus* Other streptococci	*Escherichia coli* *Klebsiella* species *Pseudomonas aeruginosa* *Staphylococcus aureus*
Fungi	*Aspergillus* species *Candida* species *Cryptococcus neoformans* *Histoplasma capsulatum* *Coccidioides immitis* Mucoraceae family *Pseudallescheria boydii*		*Aspergillus* species *Candida* species Mucoraceae family *Pseudallescheria boydii* *Fusarium* species
Protozoa	*Toxoplasma gondii* *Strongyloides stercoralis*		

deficiency syndrome (AIDS); an estimated 1 of every 250 persons 15–49 years of age is now infected with HIV-1.[4,5] As of January 1995, the Centers for Disease Control and Prevention (CDC) had estimated that the number of HIV-1-infected persons in the United States with severe immunosuppression had increased to more than 200,000. HIV-1 is spread by sexual intercourse, inoculation of infected blood or blood products (e.g., factor VIII concentrate), and perinatal exposure either *in utero*, during delivery, or through breast milk.[6] The patient populations most commonly infected with HIV-1 are persons with multiple sexual partners (particularly homosexual males), injection drug users, recipients of blood products, persons from endemic areas (e.g., the Caribbean and Central and West Africa), and children of HIV-1-infected mothers. There remains no epidemiologic evidence for transmission of HIV-1 by insect bite, saliva, tears, urine, or casual contact.[7] Health care workers also can become infected, although this is an unusual mode of transmission (risk of seroconversion of approximately 0.3% after a single percutaneous exposure to HIV-1-infected blood).

2.1.2. Clinical Presentation

CNS pathology is a common manifestation of HIV-1 infection.[4,5,8] The virus enters the CNS early in the course of infection and has been shown to be present in all stages of disease, irrespective of neurologic symptoms, although most CNS diseases complicating HIV-1 infection occur in its late or AIDS phase. From 40 to 70% of patients with AIDS or symptomatic HIV-1 infection have prominent neurologic symptoms; as many as 90% of patients with AIDS have abnormalities of the nervous system identified at autopsy.[9–14] The CNS syndromes associated with HIV-1 infection may result directly from HIV-1 infection or from opportunistic infections or neoplasms. The following sections review the neurologic manifestations directly related to HIV-1, with discussion of specific CNS opportunistic infections encountered in AIDS patients in other parts of this chapter.

Approximately 5–10% of persons newly infected with HIV-1 develop an acute aseptic meningoencephalitis syndrome,[15,16] occurring just before seroconversion and during or after the "mononucleosislike syndrome." Patients may manifest headache, fever, meningismus, cranial neuropathies (most often involving cranial nerves V, VII, and VIII), altered mental status, and focal or generalized seizures.

AIDS dementia complex (also called HIV-1 encephalitis, HIV-1 encephalopathy, or HIV-1-associated cognitive/motor complex) occurs almost exclusively during the AIDS phase of infection and is characterized by the triad of cognitive, motor, and behavioral dysfunction[4,8,17–19]; it may be the first sign of AIDS or HIV-1 infection. Recent prospective data from the Multicenter AIDS Cohort Study[20] demonstrated an incidence rate over a 5-year period of AIDS dementia complex of 7.3 cases per 100 person-years for individuals with CD4 lymphocyte counts of $\leqslant 100 \times 10^6$/liter, 3.0 cases in those with counts of 101–200 $\times 10^6$/liter, 1.3–1.7 cases for counts of 201–500 $\times 10^6$/liter, and 0.5 case for counts >500 $\times 10^6$/liter. AIDS dementia complex is a dementia of the subcortical type, with a predilection for frontal white matter. The term "subcortical dementia" has been applied because clinically symptoms associated with cortical involvement, such as aphasia and apraxia, are uncommon and because pathologically white matter and subcortical gray matter are most prominently involved. Initially, patients present with behavioral symptoms of apathy, inattention, forgetfulness, impaired concentration, mental slowing, and social withdrawal; these symptoms are found in about 40% of cases. Difficulty in reading is common and is usually due to problems in concentration rather than difficulty in understanding the printed word. There is often striking abulia, in which the patient takes long pauses before answering questions and exhibits a general torpidity. Other presentations include acute confusion, hallucinations, and psychosis. These early signs often may be attributed to depression and may be ignored until they progress to more dramatic deficits, including severe dementia and psychomotor retardation. Seizures occur as an early symptom in about 10% of cases and eventually in 20–50% of patients with more advanced illness. Patients with AIDS dementia complex may seem normal by bedside mental status testing, but neuropsychological testing reveals impaired fine and rapid motor control, diminished verbal fluency and short-term memory, impaired visual–motor and visual–spatial abilities, and deficiencies in complex problem solving. Rapid thinking and quick reactions give patients the greatest difficulty. Motor disturbances are seen in nearly 50% of patients and usually lag behind intellectual impairment. Initially, there is loss of coordination, tremors, and unsteady gait that may progress to severe ataxia and paraplegia. Incontinence usually occurs late in the disease course and is progressive over several months, at a time when patients are nearly vegetative and unable to ambulate. This syndrome may progress gradually over a period of several months to more than 1 year, or it may fluctuate with sudden deterioration, sometimes in association with systemic illness.

At least 75% of HIV-1-infected children have abnormal neurologic development.[21,22] Approximately one fourth have a static encephalopathy, most likely due to fetal or perinatal complications. Previous studies documented that almost half of patients developed a progressive encephalopathy that was clinically similar to that seen in adults with AIDS dementia complex.[23] However,

two recent longitudinal studies of 172 and 766 children both estimated that 23% of perinatally infected children with AIDS developed progressive encephalopathy.[5] Symptoms and signs of progressive encephalopathy occur from as young as 2 months of age to 5.5 years (mean age of 18 months); for some children, this picture is the presenting manifestation of HIV-1 infection.[24] Children with progressive encephalopathy frequently have seizures and lethargy, followed by spastic paraparesis or quadriparesis, neurologic deterioration, and dementia with loss of previously attained developmental milestones. Examination reveals intellectual decline and symmetrical motor impairment. The untreated child with progressive encephalopathy deteriorates without improvement; the estimated mean survival after diagnosis is 11–22 months.[5]

Vacuolar myelopathy is a syndrome associated with vacuolar degeneration primarily affecting the lateral and posterior columns of the spinal cord, causing a syndrome of progressive spastic paraparesis evolving over a period of weeks to months.[25,26] There is gait ataxia, leg weakness, upper motor neuron signs, incontinence, and posterior column deficits; a discrete sensory level is distinctly unusual. In one study, 90% of patients had evidence of AIDS dementia complex.[27] Vacuolar myelopathy is found in 22–55% of AIDS patients at autopsy.[4,5]

2.1.3. Diagnosis

HIV-1 can affect a host response at the meninges early in infection, and CNS persistence after the initial infection is common. Most often, there are no symptoms or signs following initial infection, although some patients may present with neurologic findings (see Section 2.1.2). In persons with asymptomatic HIV-1 infection, electrophysiologic tests (electroencephalography, multimodal evoked-potential tests, and otoneurologic tests) may be the most sensitive indicators of subclinical neurologic impairment[28]; abnormalities of electrophysiologic tests are far more common in asymptomatic carriers of HIV-1 than in controls and tend to progress over time. Cerebrospinal fluid (CSF) examination can be normal; therefore, CSF analysis in HIV-1-infected patients cannot prove or disprove that HIV-1 is causing a particular neurologic syndrome. However, at least 40% of HIV-1-infected patients have abnormalities of CSF, although the findings are nonspecific.[16,29–31] There is a mild lymphocytic pleocytosis in 20% of patients, with cell counts ranging from 5 to 50 $\times 10^6$ cells/liter, increased protein from 0.5 to 1.0 g/liter (50–100 mg/dl) in 60% of patients, and a normal glucose. HIV-1 is grown from approximately 20% to over 60% of CSF cultures.[32,33] HIV-1 can be recovered from

CSF in all stages of virus infection independent of the degree of clinically apparent immune suppression. However, the value of CSF virus isolation in predicting neurologic involvement is poor.[4] If the CD4–CD8 ratio is determined in CSF, it is low, especially late in infection, paralleling the pattern seen in the peripheral blood.[34] High titers of anti-HIV-1 antibodies also may be detected in CSF[32,35,36]; calculation of relative CSF and serum titers may indicate the presence of anti-HIV-1 antibody synthesis within the CNS. Oligoclonal bands directed against HIV-1 epitopes also appear in CSF.[37] However, neither culture of HIV-1 from CSF nor the finding of intrathecal synthesis of antibodies to HIV-1 is specific for AIDS dementia complex; both findings may be present in the absence of neurologic symptoms, although they are more common in patients with full-blown AIDS. CSF p24 antigen also may be detected,[38] with CSF concentrations often higher than those found in serum. Increased CSF concentrations of β_2-microglobulin[39] and quinolinic acid[40] also have been reported in HIV-1-infected patients; high CSF concentrations of β_2-microglobulin correlate with the severity of symptoms attributable to the AIDS dementia complex. Recently, a number of investigators have measured CSF concentrations of HIV-1 RNA (the "viral load") and have found a correlation between CSF viral load and severity of AIDS dementia.[41,42] However, there was considerable overlap in CSF values of HIV-1 RNA in relation to clinical neurologic severity, so that CSF viral load cannot be used in isolation as a diagnostic marker for AIDS dementia complex or even as an independent predictor of its severity.[43] Another study found no significant differences in CSF concentrations of HIV-1 RNA in patients with or without HIV encephalitis, but found that CSF concentrations of HIV-1 RNA correlated with concentrations of HIV-1 RNA found in plasma.[44] Further studies are needed to determine the importance of CSF viral load as a marker for neurologic disease in HIV-1-infected patients.

Imaging studies may be useful in the diagnosis of AIDS dementia complex,[4,8,19] although in patients with this clinical presentation, brain imaging is most often performed to exclude other diagnoses. In patients with AIDS dementia complex, computed tomography (CT) may be normal or reveal cerebral atrophy with increase in ventricular and sulcal size and hypodense lesions of cerebral white matter. Magnetic resonance imaging (MRI) also shows cerebral atrophy, although later in the course of illness, bilateral, diffuse, homogeneous, confluent white matter abnormalities are found (best seen on T2-weighted images). Children with progressive encephalopathy often have basal ganglia calcifications and atrophy.[45,46] Electroencephalography usually shows differing degrees of generalized slowing in a nonspecific pattern. HIV-1-related vacuolar myelopathy is seldom associated with abnormalities on MRI; on rare occasions, spinal cord atrophy and hyperintense signals on T2-weighted images may be observed.[4]

2.1.4. Treatment

Several studies have suggested that zidovudine has therapeutic and prophylactic efficacy in patients with AIDS dementia complex[47–49]; improvement has been detected clinically, neuropsychologically, and radiographically.[50] Zidovudine also is a promising treatment for progressive encephalopathy of children,[51] in which a continuous intravenous zidovudine infusion leads to improvement in cognition, language, socialization, affect, coordination, and gait. In the only prospective controlled trial of symptomatic AIDS dementia complex in adults,[52] two high doses of zidovudine (1000 mg/day and 2000 mg/day) were better than placebo with respect to improved performance on neuropsychological tests, although conventional doses of zidovudine were not tested. The efficacy of highly active antiretroviral therapy (HAART) in patients with AIDS dementia complex has not been assessed, although use of combination therapies with agents that penetrate into the CNS is an appropriate strategy pending further study; certainly the current standard of care is combination therapy. The use of antiretroviral therapy in patients with HIV-1 infection is discussed in detail in Chapter 12.

Adjuvant therapies (to attenuate the toxic pathways leading to brain dysfunction) also are under investigation for the treatment of AIDS dementia complex.[8] These therapies can be divided according to their target of action: immunosuppressive acting on immunopathology, neuroprotective acting on the primary target cell, and compensatory acting on the neural networks. Ongoing clinical trials will determine whether these strategies are efficacious in patients with AIDS dementia complex.

2.2. Herpesviruses

2.2.1. Epidemiology and Etiology

All herpesviruses contain double-stranded DNA surrounded by a capsid that consists of 262 capsomeres and provides icosapentahedral symmetry to the virus. The main herpesviruses that infect the CNS are herpes simplex virus (HSV), varicella–zoster virus (VZV), and cytomegalovirus (CMV); these viruses are discussed in the following sections.

The pathogenesis of human disease caused by HSV depends on intimate, personal contact of a susceptible individual with someone excreting the virus; HSV must come in contact with mucosal surfaces or abraded skin for infection to be initiated.[53] Viral replication occurs at the site of infection, and the capsid is transported within neurons to the dorsal root ganglia, where another round of viral replication occurs and latency is established. Infection with HSV-1 generally is limited to the oropharynx and can be transmitted via respiratory droplets or through direct contact with infectious secretions. Acquisition of HSV-2 usually is the consequence of transmission via genital routes. It is estimated that over one third of the world's population has recurrent HSV infections and is capable of transmitting HSV during episodes of productive infection. The epidemiology of infection has been clarified by methodological advances in antibody detection. Serologic studies performed in New Orleans demonstrated acquisition of antibodies to HSV-1 in over 90% of children by age 15 years. However, there is a significantly lower prevalence of antibodies during childhood, adolescence, and even later in life in the relatively middle and upper socioeconomic classes. HSV infections of the CNS are among the most severe of all human viral infections of the brain. Herpes simplex encephalitis is associated with significant morbidity and mortality (>70% mortality with no or ineffective therapy).[54] In the United States, herpes simplex encephalitis is thought to account for about 10–20% of encephalitic viral infections of the CNS[55]; the majority of cases (94–96%) are caused by HSV-1. Herpes simplex encephalitis occurs throughout the year and in patients of all age groups[53]; Caucasians account for 95% of patients with biopsy-proven disease. HSV encephalitis does not appear to be more common in immunosuppressed patients, although its incidence may be increased among persons infected with HIV-1 (6% in one series).[56]

VZV causes two clinically distinct diseases: (1) varicella, characterized by a generalized vesicular rash that occurs in epidemics, and (2) zoster, a common reactivation infection usually seen in the elderly.[57] Humans are the only known reservoir for VZV. Varicella is presumed to be transmitted via the respiratory route, with initial viral replication in the nasopharynx or upper respiratory tract.[58] Varicella is endemic in the population, but becomes epidemic among susceptible individuals during the late winter and early spring; 90% of cases occur in children less than 10 years of age. The actual incidence of CNS complications during active varicella infection is unknown, although the observed incidence ranges from 0.1 to 0.75% in some series.[59] Herpes zoster is a consequence of reactivation of latent VZV. Persons at greatest risk for developing herpes zoster, as well as those at increased risk for complications, are individuals with deficiencies in cell-mediated immunity.[57] A direct correlation exists between cutaneous dissemination and the appearance of visceral complications, including meningoencephalitis.[60,61] A variety of VZV-induced neurologic disorders have been described in AIDS patients, including multifocal leukoencephalitis, ventriculitis, myelitis, myeloradiculitis, and focal brain stem lesions.[57] A multifocal VZV leukoencephalitis grossly resembling progressive multifocal leukoencephalopathy also has been described in patients with underlying malignancies.[57,62] In general, the CNS complications of herpes zoster are associated with a higher morbidity and mortality than are those of acute varicella, possibly due in part to the patient's advanced age and underlying disease status.[59]

CMV infections also are ubiquitous, although these viruses are highly species-specific, linked only to infection in humans. Most individuals experience CMV infection at some point during their lifetime; the prevalence of antibodies indicating infection increases with advancing age.[63,64] Routes of transmission include sexual, salivary, close contact with an infected person excreting virus, parenteral via blood transfusion or organ transplantation, and transplacental leading to intrauterine infection.[65,66] The most common form of CNS disease occurs early in life as a consequence of intrauterine infection; only 1% of newborns excrete CMV at birth, 10% of whom develop clinical evidence of disease. Immunocompromised patients (e.g., patients with AIDS or following bone marrow or organ transplantation) represent a common group that can present with life- and sight-threatening disease due to CMV[67–69]; CNS complications, however, are uncommon. In patients with AIDS, CNS infection with CMV results in two distinct neuropathologic patterns: microglial nodular encephalitis and ventriculoencephalitis.[70]

2.2.2. Clinical Presentation

The majority of patients with biopsy-proven herpes simplex encephalitis present with a focal encephalopathic process characterized by altered mentation and decreasing levels of consciousness with focal neurologic findings (e.g., dysphasia, weakness, paresthesias) (Table 2).[53,71] These patients nearly always present with fever and personality changes. Seizures, either focal or generalized, occur in approximately two thirds of patients with proven disease. The clinical course may evolve slowly or with alarming rapidity; progressive loss of consciousness leading to coma, unfortunately, is common. Although clinical evidence of a localized temporal lobe lesion often is

**TABLE 2. Clinical Findings
in Patients with Brain-Biopsy-Proven
Herpes Simplex Encephalitis[a]**

Finding	Patients with finding
Historic findings	
Alteration of consciousness	97%
Fever	90%
Headache	81%
Personality change	71%
Seizures	67%
Vomiting	46%
Hemiparesis	33%
Memory loss	24%
Clinical findings at presentation	
Fever	92%
Personality change	85%
Dysphasia	76%
Autonomic dysfunction	80%
Ataxia	40%
Hemiparesis	38%
Seizures	38%
Cranial nerve deficits	32%
Visual field loss	14%
Papilledema	14%

[a]Adapted from Whitley.[53]

thought to be herpes simplex encephalitis, a variety of other diseases can be shown to mimic this condition.[72] Immunocompromised patients with herpes simplex encephalitis may develop a more diffuse nonnecrotizing encephalitis involving the cerebral hemispheres and brain stem.[73]

Several categories of CNS infection are caused by varicella virus. Cerebellar ataxia is the most common neurologic abnormality associated with varicella; the frequency of cerebellar dysfunction is approximately 1 in 4000 cases.[57] Symptoms include nausea, vomiting, headache, nuchal rigidity, and ataxia; seizures are rare. The cerebellar manifestations of varicella are usually self-limited, resolving within several weeks. Meningoencephalitis or cerebritis is a less common but frequently more severe CNS complication of varicella.[57] Neurologic symptoms may occur from 11 days before to several weeks after the onset of the varicella rash. Headache, fever, and vomiting often are accompanied by an altered sensorium; seizures occur in 29–52% of cases. Focal neurologic abnormalities include cranial nerve dysfunction, aphasia, and hemiplegia.[60] Encephalitis is the most common CNS abnormality associated with herpes zoster. It is seen most commonly in patients of advanced age, in patients following immunosuppression, and in patients

with disseminated cutaneous zoster.[57,74] Altered mentation without other explanation in patients with either localized or disseminated herpes zoster may be the sole manifestation; fever may not be present. Other symptoms and signs include hallucinations, meningismus, ataxia, seizures, and motor paralysis. A distinctive CNS process that sometimes is seen in cases of ophthalmic zoster is contralateral hemiplegia[57]; it accounts for up to one third of cases of CNS abnormalities in herpes zoster. In typical cases, zoster ophthalmicus precedes the appearance of hemiplegia by several weeks or more,[75] although the onset of hemiplegia may be as late as 6 months after the rash has resolved.[76]

CNS complications due solely to CMV infection are uncommon and usually manifest as meningitis, encephalitis, or meningoencephalitis. A form of CNS disease also occurs in the immunocompromised host following organ transplantation (usually documented on postmortem examination of brain tissue)[77,78] and in patients with AIDS.[79,80] CMV encephalitis in AIDS usually presents in a subacute or chronic course, with cortical dysfunction leading to confusion, disorientation, and perhaps seizures[81]; if present, brain stem lesions may produce focal signs. In patients with necrotizing ventriculoencephalitis, a rapidly fatal form of CMV encephalitis, cranial nerve defects, nystagmus, and cognitive disturbances (mental slowness and memory deficit) often are observed.[70,82]

2.2.3. Diagnosis

The diagnosis of herpes simplex encephalitis is established by brain biopsy, which currently remains the most specific means of diagnosis. However, proper localization of the major disease focus by clinical, electrical, and radiographic techniques is mandatory before a biopsy is attempted. In the early stages of herpes simplex encephalitis, congestion of capillaries and other small vessels in the cortex and subcortical white matter is evident, as are petechiae.[83,84] Perivascular cuffing becomes prominent in the second and third weeks of infection. The microscopic appearance then becomes dominated by evidence of necrosis and inflammation, with widespread areas of hemorrhagic necrosis. The presence of intranuclear inclusions (Cowdry type A) supports the diagnosis of viral infection, but is found in only about 50% of patients. Immunofluorescence studies of brain tissue provide a rapid, sensitive, and reliable method for detecting herpes antigen, provided that sufficient antigen is present in the specimen and nonspecific immunofluorescence can be minimized.[85] Virus also can be isolated from brain biopsy specimens.[86] The decision to perform a brain biopsy in patients with presumed herpes simplex encephalitis is discussed in Section 2.2.4.

CSF examination with routine studies in patients with herpes simplex encephalitis is nondiagnostic.[53,87] The white cell count is invariably elevated (mean of 100 × 10⁶/liter) in 97% of patients with brain-biopsy-proven disease, with a predominance of lymphocytes. The CSF protein is similarly elevated, averaging approximately 1.0 g/liter (100 mg/dl). The presence of red blood cells in CSF is not diagnostic for herpes simplex encephalitis, but suggests this diagnosis in the appropriate clinical setting. About 5–10% of patients with herpes simplex encephalitis have completely normal CSF on first evaluation. Routine attempts to isolate the virus from CSF are rarely successful (about 4% positive). An assay technique for the detection of HSV antigen in the CSF of patients with herpes simplex encephalitis has been developed and is 80% sensitive and 90% specific if performed within 3 days of the onset of illness.[88] Recent studies suggest that detection of HSV DNA within CSF cells by the polymerase chain reaction (PCR) is highly sensitive and specific for the diagnosis of herpes simplex encephalitis.[89] In one study, the sensitivity and specificity of PCR were 91% and 92%, respectively, in patients with biopsy-proven disease[90]; the specificity would have been higher except that some tissue specimens were fixed in formalin, which killed infectious virus. PCR detection of HSV DNA in CSF has become the diagnostic procedure of choice in patients with herpes simplex encephalitis.[53]

Noninvasive neurodiagnostic studies may support a diagnosis of herpes simplex encephalitis. Electroencephalography (EEG) appears to be the most sensitive (about 84%) for diagnosis, exhibiting characteristic spike-and-slow-wave activity and periodic lateralized epileptiform discharges (PLEDs), which arise predominantly over the temporal and frontotemporal regions[91,92]; the specificity of this test is only 32.5%, however. CT scans initially show low-density areas with mass effect localized to the temporal lobe, which can progress to either or both radiolucent and hemorrhagic areas[93,94]; these areas are seen in 50–75% of patients at some time during the illness. Magnetic resonance imaging (MRI) (with enhancement) demonstrates lesions earlier and is superior to CT in localizing these lesions to the orbital–frontal and temporal lobes.[95,96]

In varicella-associated cerebritis, the CSF is often abnormal, with a mild to moderate lymphocytic pleocytosis and elevated protein.[57] The EEG usually is diffusely abnormal, although focal abnormalities may occur even without clinical seizure activity. Postmortem studies of the brain in fatal varicella cerebritis reveal a lack of distinctive histopathologic findings. There usually is diffuse cerebral edema; intranuclear inclusions rarely have been observed. In herpes-zoster-associated encephalitis, lumbar puncture frequently yields an abnormal CSF formula, with a lymphocytic pleocytosis, elevated protein, and normal glucose. It should be noted, however, that as many as 40–50% of patients with uncomplicated herpes zoster without CNS symptoms have a mild CSF pleocytosis or elevated CSF protein concentration.[57] A diffuse slowing may be evident on EEG without detection of a specific abnormal focus.[57,60] VZV has been cultured from brain and CSF in a number of cases of herpes-zoster-associated encephalitis,[97–99] and viral inclusions in glial cells, neurons, and arteries of the brain are well described in fatal cases[100]; VZV antibodies and lymphocyte-associated VZV antigens also have been demonstrated in the CSF of these patients.[98,101,102] In zoster ophthalmicus with contralateral hemiplegia, cerebral angiography often demonstrates unilateral arteritis or thrombosis of individual vessels[75,76]; CT may show evidence of cerebral infarction in some cases.[103]

The gold standard for the diagnosis of CMV disease in any site is isolation of the organism; however, it may take as long as 2–4 weeks before evidence of cytopathic effect is observed in cell culture systems. Even if brain biopsy specimens are obtained, they may not reveal typical CMV histopathology or a positive culture. Diagnosis of CNS infection caused by CMV is best made by PCR of CSF, which has a high sensitivity and specificity for CNS involvement.[104–107] Focal abnormalities on CT also have been described in CNS infections caused by CMV,[108] including one patient with a ring-enhancing lesion that proved to be a CMV abscess.[109]

2.2.4. Treatment

The therapy of herpes simplex encephalitis has undergone a major evolution in recent years. The first antiviral drug available (in the late 1960s and early 1970s) was idoxuridine[110–112]; unfortunately, severe toxicity (bone marrow suppression and secondary bacterial infections) limited its usefulness.[54] Later studies documented the efficacy of vidarabine in biopsy-proven herpes simplex encephalitis. Initial double-blind, placebo-controlled studies revealed that vidarabine decreased mortality in herpes simplex encephalitis from 70 to 28% 1 month after disease onset and from 70 to 44% 6 months later.[113] A subsequent open, uncontrolled trial of nearly 100 patients with proven disease defined a long-term mortality rate of 40%.[114] Younger patients (<30 years old) and those with a more normal level of consciousness were more likely to return to a normal level of functioning following vidarabine therapy than were older patients who were semi-comatose or comatose. For best results, antiviral therapy must be instituted prior to the onset of hemorrhage or before widespread bilateral disease ensues.

More recent trials conducted by the NIAID Collaborative Antiviral Study Group have documented that acyclovir is superior to vidarabine for the treatment of biopsy-proven herpes simplex encephalitis.[115,116] One study[116] demonstrated that acyclovir decreased the mortality rate to 19% 6 months after initiation of therapy (vs. 55% for vidarabine); 38% of patients, regardless of age, returned to a normal level of functioning or were found to have only minor impairment after treatment with acyclovir. This outcome was better than that for vidarabine recipients, of whom only 15–20% were judged to be normal on long-term follow-up. In addition, the ease of administration and side effect profile were more favorable with acyclovir than with vidarabine. In a recent study of 42 patients with herpes simplex encephalitis treated with acyclovir, 30% either died or had a severe neurologic deficit[117]; of the remaining 70%, most had persistent neurologic symptoms and/or signs. On the basis of the clinical studies cited above and the fact that herpes simplex remains the only viral CNS infection for which therapy has proved useful in rigorous controlled trials,[87] an empiric course of acyclovir should be considered for presumed herpes simplex encephalitis. Brain biopsy should be performed in unclear cases or when the patient has not responded to adequate acyclovir therapy. Acyclovir should be administered at a dosage of 10 mg/kg intravenously every 8 hr in patients with normal renal function; therapy should be continued for 14 days. Relapse of herpes simplex encephalitis has been documented following administration of acyclovir,[118–120] and some patients may require a longer duration of therapy. On the basis of the demonstration of antiviral synergy *in vitro*, combinations of antiviral agents may prove useful for the therapy of herpes simplex encephalitis.[53]

Intravenous acyclovir has become the antiviral agent of choice for immunocompromised patients at high risk for progressive disease caused by VZV.[121,122] No clinical trial has established the value of an antiviral agent for herpes-zoster-associated encephalitis; however, in a patient with herpes zoster who has clinical evidence of encephalitis, acyclovir should be used empirically. Indeed, the most striking feature in patients with herpes-zoster-associated encephalitis who were treated with acyclovir was the rapid and dramatic clinical resolution of the encephalopathic state, usually within 72 hr.[123] Acyclovir also was efficacious in two HIV-1-infected patients with herpes zoster leukoencephalitis.[124]

The management of patients with CNS infections

caused by CMV is complicated by the inability to accurately diagnose the infection. Three antiviral agents (ganciclovir, foscarnet, and cidofovir) have been licensed for the treatment of life- and sight-threatening CMV infections in immunocompromised patients, although there are no data regarding the utility of these compounds for the management of CNS infections caused by CMV. Fulminant CMV encephalitis has developed in patients being treated with maintenance ganciclovir for CMV retinitis.[125]

2.3. Progressive Multifocal Leukoencephalopathy

2.3.1. Epidemiology and Etiology

Progressive multifocal leukoencephalopathy (PML) is an opportunistic, demyelinating infection that occurs exclusively in immunocompromised patients.[126,127] Infection is known to occur particularly in patients with defects in cell-mediated immunity, most often with lymphoproliferative disorders, but also in patients receiving antineoplastic chemotherapy for myeloproliferative disorders and malignancies, bone marrow transplantation, renal transplantation, autoimmune diseases, sarcoidosis, tuberculosis, Whipple's disease, nontropical sprue, hypogammaglobulinemia, and idiopathic $CD4^+$ T lymphocytopenia.[127–130] Within the last decade, however, it has become apparent that infection with HIV-1 greatly increases the risk of PML,[127,131–133] which occurs in up to 4% of patients with AIDS. As with other neurologic complications of HIV-1, PML may be the presenting manifestation of the immunodeficient state.

In virtually all cases, PML has been associated with CNS invasion by JC virus,[126,127] a member of the polyoma subgroup of the genus Papovaviridae; however, in a small number of cases, a second polyoma virus, similar to SV40, has been recovered.[134] The designation JC virus is taken from the initials of the patient from whom it was recovered. JC virus is a small, icosahedral agent with a genome composed of supercoiled, double-stranded DNA. The virus is ubiquitous; acquisition of antibody to JC virus begins in infancy. By late adult life, the prevalence of antibody in the general population is over 70%. Urinary excretion of JC virus, believed to represent reactivated JC virus infection, is common under conditions of immunosuppression; the virus has been detected in the urine of 12.9% of patients with leukemia, 7% of bone marrow transplant patients, and 18% of renal transplant recipients at some time during the period of immunosuppression.[132,135] Urinary excretion also occurs in 0.4% of pregnant women. A fourfold rise in antibody titer to JC virus has been observed in 9–15% of pregnant women, most commonly at the end of the second trimester.[136–138]

Virtually all patients with PML have impaired T-lymphocyte function. The pathogenesis of CNS infection is unknown, although it has been suggested that JC virus may infect B lymphocytes, which may then penetrate into the brain and initiate virus infection in the perivascular spaces.[139,140]

2.3.2. Clinical Presentation

PML usually begins insidiously with initial symptoms reflecting cerebral involvement.[126] Early in the disease course, alterations in personality are common, followed by blunting of intellect and frank dementia as the disease progresses. Involvement of the dominant hemisphere may produce expressive or receptive aphasia or both. Ataxic gait, limb dysmetria, and dysarthria often indicate cerebellar involvement.[141] Visual abnormalities (e.g., quadrantanopsias or hemianopsias) occur in approximately 50% of patients. In AIDS patients, limb weakness, gait abnormalities (typically ataxia), visual loss, and altered mental status are the most common initial complaints. Brain stem and cerebellar involvement are uncommon as initial presentations. In the majority of cases of PML, death occurs within 1 year, although the disease may be rapidly progressive, with death occurring within 2 months, or more prolonged, with reported survivals of 8–10 years. In AIDS patients, the clinical course of PML is similar to that seen in PML in other immunocompromised persons and characteristically is steadily and rapidly progressive, with an average survival of 4 months (range 0.3–18 months).[131] Spontaneous remission of PML has been observed in the setting of AIDS; two HIV-1-infected patients with PML exhibited dramatic, though incomplete, recovery of neurologic function and have survived more than 30 months since the onset of symptoms.[142]

2.3.3. Diagnosis

PML should be considered in any immunocompromised patient who develops neurologic findings. CSF is usually normal or contains increased protein; in rare cases, a lymphocytic pleocytosis may be present.[143] CT scanning shows demyelination in most cases, manifested as hypodense, nonenhancing white matter lesions without mass effect or evidence of edema[144,145]; CT findings, however, may be proportionally less abnormal than clinical findings. The CT lesions usually remain nonenhancing even with double-dose contrast-delayed scanning. MRI is the diagnostic method of choice because it is more sensitive than CT and can detect injury to myelin[145–147]; abnormalities noted on MRI are chiefly high-intensity signal

lesions located in the centrum semiovale, periventricular areas, and cerebellum on T2-weighted images (Fig. 1). Serologic tests of blood or CSF for antibody to JC virus are unreliable because antibody to JC virus is common in the general population and because many patients with PML fail to develop a significant rise in antibody titers.[148] Urinary excretion of JC virus particles occurs frequently in immunosuppressed patients,[132,135] so detection of virus in the urine is not indicative of PML.

Confirmation of the diagnosis of PML during life requires brain biopsy.[149,150] Pathologically, there is virus-induced lysis of oligodendrocytes with resultant loss of myelin. There also is abortive infection of astrocytes, which undergo morphologic alteration to develop features suggestive of neoplastic transformation. The cerebrum, cerebellum, or brain stem may be affected; significant involvement of the spinal cord does not occur. Nuclei of infected oligodendrocytes contain JC viral nucleic acids, express early and late viral proteins, and contain typical polyomavirus virions. The diagnostic sensitivity of brain biopsy may be increased by use of immunocytochemical or *in situ* nucleic acid hybridization methods.[151,152] Recently, PCR has been utilized as a diagnostic test for detection of JC virus DNA in CSF samples from patients with PML,[153–155] with a sensitivity and specificity of 82% and 100%, respectively, in one study.[153] Pending further studies, PCR is likely to prove to be a sensitive and highly specific diagnostic test for confirming the diagnosis of PML and may reduce the need for brain biopsy to establish the diagnosis.

2.3.4. Treatment

Proven antiviral therapy for PML does not currently exist. Cytarabine has been used in AIDS and non-AIDS patients with PML,[56,126,156–161] with variable results. A recently published trial of cytarabine (administered either intravenously or intrathecally) in 57 HIV-1-infected patients with biopsy-confirmed PML found no significant difference in survival in patients receiving cytarabine,[162] suggesting that this drug has no role in the treatment of AIDS-associated PML. Other agents utilized, with anecdotal success, have included adenine arabinoside,[163,164] acyclovir,[165] cidofovir,[166,167] and camptothecin,[168] but no large-scale trials have been performed with any of these agents. Several cases have been documented in which PML has remitted, either spontaneously or in response to withdrawal of immunosuppressive drugs.[150] High-dose zidovudine therapy (1200 mg/day) led to a dramatic response in an AIDS patient with biopsy-proven PML[169]; this patient had clinical deterioration with reduction of the zidovudine dosage and seemed to stabilize after the

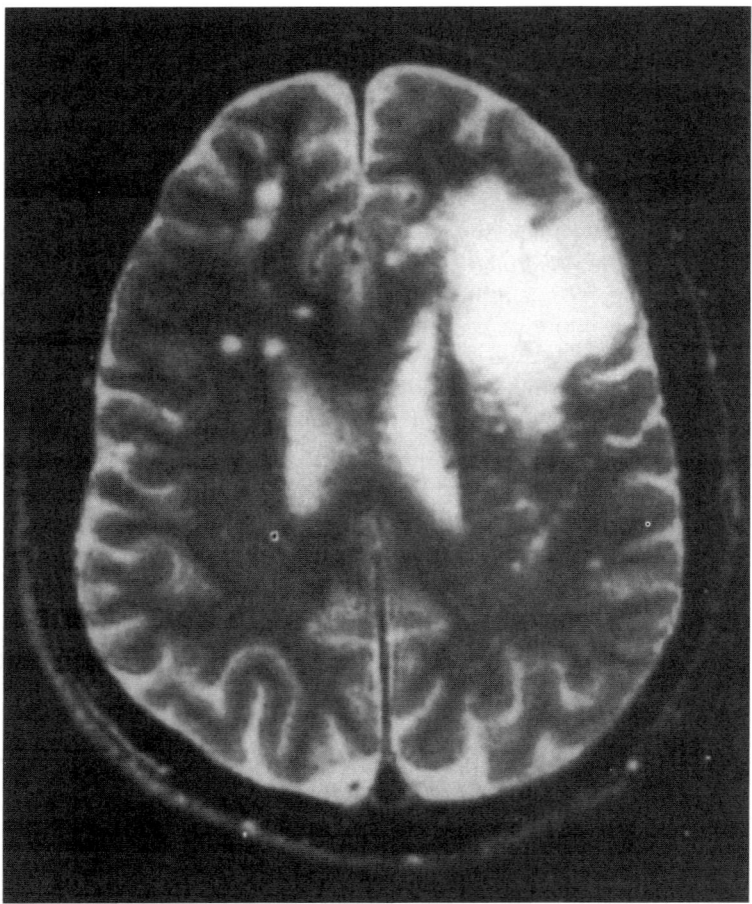

FIGURE 1. T2-weighted MRI of the brain in an AIDS patient with progressive multifocal leukoencephalopathy, revealing increased signal intensity in the left frontoparietal region with several smaller lesions in the right frontal lobe. These lesions are subcortical in location, have discrete borders, and are without mass effect.

higher dose was resumed. Use of highly active antiretroviral therapy has also shown efficacy in the therapy of AIDS patients with PML,[170–173] suggesting that optimal control of HIV-1 infection may lead to regression of PML in patients with AIDS. However, a recent study found that PML may not improve despite an adequate virologic response to highly active antiretroviral therapy,[174] indicating that definitive therapy is still needed for PML.

3. Bacterial Infections

3.1. Meningitis

3.1.1. Epidemiology and Etiology

Bacterial meningitis remains an important and devastating disease. In a surveillance study of all cases of bacterial meningitis in 27 states of the United States from 1978 through 1981, the overall yearly attack rate was 3.0 cases/100,000 population, although there was variability depending on geographic locale, sex, and race[175]; over 80% of cases were caused by *Haemophilus influenzae*, *Neisseria meningitidis*, or *Streptococcus pneumoniae*. In a subsequent surveillance study conducted during 1995 in laboratories serving all the acute care hospitals of 22 counties of four states (>10 million population),[176] the incidence of bacterial meningitis decreased dramatically as a result of a vaccine-related decline in meningitis caused by *H. influenzae* type B such that, in the United States, bacterial meningitis is now a disease predominantly of adults rather than infants and children. In the subgroup of patients 16 years of age and older, most cases of bacterial meningitis are caused by *S. pneumoniae*, *N. meningitidis*, and *L. monocytogenes*.[177] The isolation of particular bacterial species varies depending on the patient's age, underlying disease status, and other predisposing conditions.

Haemophilus influenzae is a gram-negative bacillus

that was previously isolated in 45–48% of all cases of bacterial meningitis in the United States[175,178]; this organism is now isolated in only 7% of cases.[176] The overall mortality rate is 3–6%.[175,176,178] The vast majority of cases previously occurred in children less than 6 years of age, with 90% of cases caused by capsular type B strains. The use of *H. influenzae* type B conjugate vaccines has reduced the incidence of *H. influenzae* type B meningitis more than 90% among infants and children.[179,180] *Haemophilus influenzae* represents only about 5% of total CSF isolates after 6 years of age; isolation of the organism in this age group should suggest the presence of certain predisposing conditions such as sinusitis, otitis media, epiglottitis, pneumonia, head trauma with CSF leak, diabetes mellitus, alcoholism, splenectomy or asplenic states, and immune deficiency (e.g., hypogammaglobulinemia).[181,182]

Neisseria meningitidis is a non-spore-forming, oxidase-positive, gram-negative coccus that may appear as biscuit-shaped diplococci in clinical specimens. It most often causes meningitis in children and young adults in the United States, with an overall mortality rate of 3–13%.[175,176,178] Infection with the meningococcus may occur in epidemics (usually due to serogroups A and C), although in sporadic cases, serogroup B strains account for many isolates in this country. Serogroup Y strains may be associated with pneumonia. Several outbreaks of disease caused by serogroup C meningococci have been reported in the United States and Canada, with most caused by one strain of electrophoretic type 37 (ET-37) complex termed ET-15.[183,184] *Neisseria meningitidis* infections are more likely in patients who have deficiencies of the terminal complement components (C5, C6, C7, C8, and perhaps C9),[185] the so-called membrane attack complex, in which there is a greater than 8000-fold increased incidence of neisserial infection, although mortality rates from neisserial infections are lower than in patients with an intact complement system.

Streptococcus pneumoniae is a non-spore-forming, nonmotile, gram-positive coccus that typically appears lancet-shaped in clinical specimens. Pneumococcal meningitis is now the most common etiologic agent of bacterial meningitis in the United States, accounting for 47% of total cases, with a mortality rate of 19–26%.[175,176,178] Infection often is associated with distant or contiguous foci of infection (e.g., pneumonia, otitis media, mastoiditis, sinusitis, endocarditis). Serious infections with *S. pneumoniae* may be observed in patients with predisposing conditions such as splenectomy or asplenic states, multiple myeloma, hypogammaglobulinemia, and alcoholism.[186,187] The pneumococcus is the most common

CSF isolate in head trauma patients who have suffered basilar skull fracture with subsequent CSF leak.[188]

Other bacterial species are less common causes of bacterial meningitis, but may be more frequently observed in immunocompromised patients. Meningitis due to *Listeria monocytogenes*, a gram-positive, catalase-positive bacillus, represents only about 8% of cases of bacterial meningitis in the United States, but carries a high mortality rate (15–29%).[175,176,178] Infection with *Listeria* is more common in neonates, the elderly, alcoholics, cancer patients, and immunosuppressed adults (e.g., renal transplant patients).[189–193] Despite the increased incidence of *Listeria* infection in patients with deficiencies in cell-mediated immunity, however, *Listeria* meningitis is infrequently found in patients with HIV-1 infection.[194,195] In addition, up to 30% of adults and 54% of children and young adults who have listeriosis have no apparent underlying disease. Outbreaks of listerial infection, including meningitis, have been associated with the consumption of contaminated cole slaw, milk, and cheese, with sporadic cases traced to contaminated cheese, turkey franks, and alfalfa sprouts, pointing to the intestinal tract as the usual portal of entry.[189,192,193]

Aerobic gram-negative bacilli have become increasingly important as etiologic agents of meningitis.[196,197] *Klebsiella* species, *Escherichia coli*, and *Pseudomonas aeruginosa* may be isolated from CSF of patients after head trauma, neurosurgical procedures, the elderly, immunosuppressed patients, and patients with gram-negative septicemia. *Staphylococcus epidermidis* is the most common cause of meningitis in patients with CSF shunts.[198] Meningitis due to *Staphylococcus aureus* usually is found in the early postneurosurgical period as well as in patients with CSF shunts.[199] Underlying diseases in patients with *S. aureus* meningitis also include diabetes mellitus, alcoholism, chronic renal failure requiring dialysis, and malignancies. The group B streptococcus (*S. agalactiae*) is a common cause of meningitis in neonates,[200] but also occasionally is observed to cause meningitis in adults, especially in those more than 60 years of age; additional risk factors for group B streptococcal meningitis include diabetes mellitus, malignancy, alcoholism, hepatic failure, renal failure, and corticosteroid therapy.[201,202]

Rarely, meningitis due to *Nocardia* species has been reported. *Nocardia* are non-spore-forming, filamentous, gram-positive branching rods that are partially acid-fast. In a recent review of 28 patients who met the criteria for nocardial meningitis,[203] predisposing conditions (immunosuppressive drug therapy, malignancy, head trauma, prior CNS procedures, chronic granulomatous disease, sarcoidosis) were noted in approximately 75% of cases.

CNS disease is thought to occur via hematogenous dissemination from a primary site, usually the lungs.

3.1.2. Clinical Presentation

The classic clinical presentation in patients with bacterial meningitis includes fever, headache, meningismus, and signs of cerebral dysfunction[204–206]; these are found in more than 85% of patients. The meningismus may be subtle or marked, or accompanied by either or both Kernig's and Brudzinski's signs[207]; however, these signs are elicited in only about 50% of adults with bacterial meningitis, and their absence does not rule out this diagnosis. Cerebral dysfunction is manifested by confusion, delirium, or a declining level of consciousness ranging from lethargy to coma. Cranial nerve palsies and focal cerebral signs are uncommon (10–20% of cases), while seizures occur in about 30% of cases. Cranial nerve palsies (especially involving cranial nerves III, IV, VI, and VII), when they occur, likely develop as the nerve becomes enveloped by exudate in the arachnoid sheath surrounding the nerve. Alternatively, cranial nerve palsies may be a sign of increased intracranial pressure. Focal neurologic deficits and seizure activity arise from cortical and subcortical ischemia and infarction, which is the result of inflammation and thrombosis of blood vessels, often within the subarachnoid space. Hemiparesis also may be a sign of a large subdural effusion, which arises when infection in the adjacent subarachnoid space leads to an increase in permeability of the thin-walled capillaries and veins of the inner layer of the dura. This effusion is usually a self-limited process in which the fluid in the subdural space is resorbed, although an enlarging effusion can lead to mass effect with resultant hemiparesis. Papilledema is rare (<1% of bacterial meningitis cases) and should suggest an alternative diagnosis (e.g., an intracranial mass lesion). Later in the disease course, patients may develop signs of increased intracranial pressure including coma, hypertension, bradycardia, and third nerve palsy; these are ominous prognostic signs.

Certain symptoms and signs may suggest an etiologic diagnosis in patients with bacterial meningitis. Meningococcemia, with or without meningitis, may present with a prominent rash, located principally on the extremities, in about 50% of cases.[204,206] The rash is typically erythematous and macular early in the disease course, but quickly evolves into a petechial phase with further coalescence into a purpuric form. The rash often matures rapidly, with new petechial lesions appearing during the physical examination. A similar rash also may be seen in rapidly overwhelming sepsis due to *S. pneu-moniae* or *H. influenzae*, occurring primarily in splenectomized patients. An additional suppurative focus of infection (e.g., otitis media, sinusitis, or pneumonia) may be seen in about 30% of patients with pneumococcal or *H. influenzae* meningitis. In patients who have suffered a basilar skull fracture in which a dural fistula is produced between the subarachnoid space and the nasal cavity, paranasal sinuses, or middle ear, meningitis usually is caused by the pneumococcus[188]; these patients commonly present with rhinorrhea or otorrhea due to a CSF leak, and a persistent defect is a common explanation for recurrent bacterial meningitis. *Listeria monocytogenes* meningitis has an increased tendency to cause seizures and focal deficits early in the course of infection[189–193]; some patients may present with ataxia, cranial nerve palsies, or nystagmus (due to rhomboencephalitis). However, patients with *Listeria* meningitis may present with no evidence of parenchymal brain involvement.

Certain subgroups of patients may not manifest many of the classic signs and symptoms of bacterial meningitis.[208] Elderly patients, especially those with underlying conditions such as diabetes mellitus or cardiopulmonary disease, may present insidiously with lethargy or obtundation and variable signs of meningeal irritation and without fever.[209] In this subgroup of patients, an altered mental status should not be ascribed to other causes until bacterial meningitis has been excluded by CSF examination. Neonates with bacterial meningitis usually do not demonstrate meningismus or fever and the only clinical clues to meningitis are listlessness, high-pitched crying, fretfulness, refusal to feed, irritability, or respiratory distress.[200,210] The diagnosis of bacterial meningitis in neutropenic patients requires a high index of suspicion, as symptoms and signs may initially be subtle due to the patient's impaired ability to mount a local inflammatory response.[1,2]

3.1.3. Diagnosis

The diagnosis of bacterial meningitis rests on CSF examination by lumbar puncture. The typical CSF findings in patients with acute bacterial meningitis are shown in Table 3.[204,206] The opening pressure is elevated in virtually all cases; values over 600 mm H_2O suggest cerebral edema, the presence of intracranial suppurative foci, or communicating hydrocephalus. The CSF white cell concentration usually is elevated in untreated bacterial meningitis, ranging from 100 to $10,000 \times 10^6$/liter. There most often is a neutrophilic predominance, although about 10% of patients with bacterial meningitis present with a predominance of lymphocytes (>50%) in

TABLE 3. Typical Cerebrospinal Fluid Findings in Acute Bacterial Meningitis

Test	Finding
White cell count	$1000-5000 \times 10^6$/liter (range: <100 to >10,000 $\times 10^6$/liter)
Neutrophils	≥80%
Protein	1–5 g/liter (100–500 mg/dl)
Glucose	≤2.2 mmoles/liter (≤40 mg/dl)
Gram stain	Positive in 60–90%
Culture	Positive in 70–85%
Bacterial antigens[a]	Positive
Lactate	Positive (≥35 mg/dl)
C-reactive protein	Positive
Limulus lysate[b]	Positive

[a]See the text for a complete description of available bacterial antigen tests.
[b]Positive only in meningitis due to gram-negative bacteria.

CSF; this finding is more common in neonatal gram-negative meningitis and in meningitis due to *L. monocytogenes*. Occasionally, patients may have very low CSF white cell counts ($0-20 \times 10^6$/liter) despite high bacterial concentrations in CSF, a finding associated with a poor prognosis. Therefore, a Gram stain and culture always should be performed on CSF specimens even if the white cell count is normal. A CSF glucose concentration of less than 2.2 mmoles/liter (40 mg/dl) is found in about 60% of patients with bacterial meningitis and a CSF/serum glucose ratio of less than 0.31 in about 70% of patients.[211] The CSF protein is elevated in virtually all cases of bacterial meningitis, presumably due to disruption of the blood–brain barrier. A recent analysis found that a CSF glucose of less than 1.9 mmoles/liter (34 mg/dl), a CSF/blood glucose ratio of less than 0.23, a CSF protein of more than 2.2 g/liter (220 mg/dl), a CSF leukocyte count of more than 2000×10^6/liter, or a CSF neutrophil count of more than 1180×10^6/liter was an individual predictor of bacterial meningitis, as compared to viral meningitis, with a certainty of 99% or better.[212] A normal CSF white cell count and protein may be seen in specimens obtained at the onset of meningitis, in some cases of neonatal meningitis, and in severely immunocompromised patients.

CSF examination by Gram stain may permit a rapid, accurate identification of the causative microorganism in 60–90% of cases of bacterial meningitis.[211] The probability of detecting the organism correlates with bacterial concentrations in CSF; concentrations less than or equal to 10^3 colony-forming units (CFU)/ml are associated with poor microscopic results (organisms seen 25% of the time), whereas microscopy is positive in 97% of cases in

which CSF bacterial concentrations are equal to or greater than 10^5 CFU/ml.[213] The probability of identifying an organism may decrease in patients who have received prior antimicrobial therapy.[214,215] The CSF Gram stain is positive in less than half of cases of meningitis due to *L. monocytogenes*.[190] Several rapid diagnostic tests also have been developed to aid in the diagnosis of bacterial meningitis.[216] Counterimmunoelectrophoresis (CIE) may detect specific antigens in CSF due to meningococci (serogroups A and C), *H. influenzae* type B, pneumococci (83 serotypes), type III group B streptococci, and *E. coli* K1. The sensitivity of CIE ranges from 62 to 95%, although the test is highly specific. Newer tests employing staphylococcal coagglutination or latex agglutination are more rapid and sensitive than CIE, with the ability to detect much lower concentrations (~1 ng/ml) of bacterial antigen. However, none of the tests currently available detects group B meningococcal antigens. One of these rapid diagnostic tests (preferably latex agglutination) should be performed on all CSF specimens from patients in whom bacterial meningitis is suspected if the CSF indices are consistent with the diagnosis and the CSF Gram stain is negative. However, it must be emphasized that a negative test does not rule out infection due to a particular meningeal pathogen. Lysate prepared from amoebocytes of the horseshoe crab, *Limulus polyphemus*, is useful in suspected cases of gram-negative meningitis in which a positive test is due to the presence of endotoxin,[217] although the test does not distinguish among gram-negative organisms that may be present in CSF. Recently, procalcitonin has been evaluated as a marker to differentiate bacterial from viral meningitis. In one study, a serum procalcitonin concentration >0.2 ng/ml had a sensitivity and specificity of up to 100% in the diagnosis of bacterial meningitis[218]; further studies are needed, however, to determine the utility of this test in distinguishing bacterial from viral meningitis.

CSF studies in nocardial meningitis usually reveal a neutrophilic pleocytosis that tends to persist in the presence of empiric antimicrobial therapy.[203] *Nocardia* species typically grow on routine laboratory media, but growth may require several weeks of incubation. Although culture of large CSF volumes and use of enhanced concentration techniques before staining may improve the yield, the utility of these maneuvers for nocardial meningitis has not been well defined.

3.1.4. Treatment

3.1.4a. Specific Antimicrobial Agents. The initial management of patients with presumed bacterial meningitis includes performance of an emergent lumbar puncture.[206,219] If the CSF formula is consistent with that diagnosis, empiric antimicrobial therapy should be initiated rapidly, often before results of Gram stain or CSF bacterial antigen tests are available. If no etiologic agent can be identified by results of Gram stain or rapid diagnostic tests, empiric therapy should be initiated on the basis of the patient's age and underlying disease status. In patients who present with a focal neurologic examination, a CT scan should be performed immediately to exclude the presence of an intracranial mass lesion. However, if there is any delay in obtaining the CT scan, empiric antimicrobial therapy should be started immediately and before the lumbar puncture because of the high mortality rate in patients with bacterial meningitis in whom antimicrobial therapy is delayed. Choices for empiric antibiotic therapy in patients with presumed bacterial meningitis based on age and underlying disease are shown in Table 4. Once the infecting microorganism is isolated, antimicrobial therapy can be modified for optimal treatment (Table 5); choices of specific agents are reviewed in greater detail below. Recommended doses of antimicrobial agents for CNS infections in adults with normal renal function are shown in Table 6.

For meningitis caused by *S. pneumoniae*, penicillin

TABLE 4. Empiric Therapy of Purulent Meningitis[a]

Predisposing condition	Therapy
Age	
0–4 weeks	Ampicillin plus cefotaxime; ampicillin plus an aminoglycoside
4–12 weeks	Ampicillin plus a third-generation cephalosporin[b]
3 months to 18 years	Third-generation cephalosporin[b]; ampicillin plus chloramphenicol
18–50 years	Third-generation cephalosporin[b,c]
>50 years	Ampicillin plus a third-generation cephalosporin[b]
Immunocompromise	
Cellular immunodeficiency	Ampicillin plus a third-generation cephalosporin[b]
Humoral immunodeficiency	Third-generation cephalosporin[b]
Asplenia	Third-generation cephalosporin[b]
Neutropenia	Ampicillin plus ceftazidime or cefepime
Postneurosurgical or CSF shunt	Vancomycin plus ceftazidime or cefepime

[a]Vancomycin should be added to empiric therapeutic regimens when highly penicillin or cephalosporin-resistant pneumococci are suspected.
[b]Cefotaxime or ceftriaxone.
[c]Add ampicillin if meningitis caused by *Listeria monocytogenes* is suspected.

G and ampicillin are equally efficacious for susceptible strains. However, on the basis of recent studies of pneumococcal susceptibility patterns, these agents are not recommended for empiric therapy when the pneumococcus is considered a likely infecting pathogen. In the past, pneumococci were uniformly susceptible to penicillin with minimal inhibitory concentrations (MICs) of 0.06 µg/ml or less. Several reports have now documented pneumococcal strains that are relatively resistant (MIC range: 0.1–1.0 µg/ml) or highly resistant (MIC ≥2.0 µg/ml) to penicillin.[206,219–222] Pneumococcal strains resistant to the third-generation cephalosporins (MIC ≥2 µg/ml) also have been reported; these agents have been associated with treatment failure in patients with pneumococcal meningitis.[206,219] Because sufficient CSF concentrations of penicillin are difficult to achieve with standard parenteral dosages (initial CSF concentrations of ~1 µg/ml), penicillin never should be used as empiric therapy for suspected or proven pneumococcal meningitis. Empiric therapy should consist of the combination of vancomycin plus a third-generation cephalosporin (either cefotaxime or ceftriaxone), pending susceptibility testing of the pneumococcal isolate. Once susceptibility testing of the isolate is performed, antimicrobial therapy can be modified for optimal treatment (Table 5). Of concern is the report of 11 consecutive patients with CSF-culture-proven pneumococcal meningitis who were treated with intravenous vancomycin.[223] All patients initially improved and 10 were ultimately cured of their infection, although 4 patients experienced a therapeutic failure, leading to a change in vancomycin therapy. These data indicate the need for careful monitoring of adult patients receiving vancomycin therapy for pneumococcal meningitis. In patients not responding, use of intrathecal or intraventricular vancomycin is a reasonable option.[224] The addition of rifampin to vancomycin, with or without the third-generation cephalosporin, also has been recommended by some authorities, although clinical data are not available; rifampin should only be added if the organism is susceptible and the expected clinical or bacteriologic response is delayed.[225]

Penicillin G and ampicillin also are efficacious for meningitis due to *N. meningitidis*. However, these recommendations may change in the future as meningococcal strains that are relatively resistant to penicillin (MIC range: 0.1–0.7 µg/ml) have been described in several areas, particularly Spain[226]; 20% of meningococcal isolates were relatively resistant to penicillin in 1989.[227] This resistance appears to be mediated by a reduced affinity of the antibiotic for penicillin-binding proteins 2 and 3. In the United States, meningococcal strains relatively resis-

TABLE 5. Antimicrobial Therapy for Central Nervous System Infections

Organism	Standard therapy	Alternative therapies
Aspergillus species	Amphotericin B[a]	Itraconazole[b]; amphotericin B lipid complex[b]; liposomal amphotericin B[b]; voriconazole[a]
Bacteroides fragilis	Metronidazole	Chloramphenicol; clindamycin; meropenem
Candida species	Amphotericin B[a]	Fluconazole[b]; amphotericin B lipid complex[b]; liposomal amphotericin B[b]
Coccidioides immitis	Fluconazole	Amphotericin B[c]
Cytomegalovirus	Ganciclovir	Foscarnet
Enterobacteriaceae	Third-generation cephalosporin[d]	Aztreonam; fluoroquinolone[b]; trimethoprim–sulfamethoxazole; meropenem
Fusobacterium species	Penicillin G	Metronidazole
Haemophilus influenzae		
β-Lactamase-negative	Ampicillin	Third-generation cephalosporin[d]; chloramphenicol; cefepime
β-Lactamase-positive	Third-generation cephalosporin[d]	Aztreonam; cefepime; chloramphenicol, fluoroquinolone
Herpes simplex virus	Acyclovir	Vidarabine
Histoplasma capsulatum	Amphotericin B	Fluconazole[b]; itraconazole[b]
Listeria monocytogenes	Ampicillin or penicillin G[e]	Trimethoprim–sulfamethoxazole
Mycobacterium tuberculosis	Isoniazid, rifampin, pyrazinamide[f]	—
Neisseria meningitidis	Penicillin G or ampicillin	Third-generation cephalosporin[d]
Nocardia asteroides	Trimethoprim–sulfamethoxazole or sulfadiazine	Minocycline; imipenem; third-generation cephalosporin[d]; amikacin
Pseudallescheria boydii	Voriconazole[h]	Itraconazole[b]; fluconazole[b]; miconazole[c]
Pseudomonas aeruginosa	Ceftazidime[e] or cefepime[e]	Aztreonam[e]; fluoroquinolone[e]; meropenem[e]
Staphylococcus aureus		
Methicillin-sensitive	Nafcillin or oxacillin	Vancomycin
Methicillin-resistant	Vancomycin[g]	—
Staphylococcus epidermidis	Vancomycin[g]	—
Streptococcus agalactiae	Ampicillin or penicillin G[e]	Third-generation cephalosporin[d]; vancomycin
Streptococcus milleri; other streptococci	Penicillin G	Third-generation cephalosporin[d]; vancomycin
Streptococcus pneumoniae		
Penicillin MIC <0.1 μg/ml	Penicillin G or ampicillin	Third-generation cephalosporin[d]; vancomycin
Penicillin MIC 0.1–1.0 μg/ml	Third-generation cephalosporin[d]	Vancomycin[g]; meropenem
Penicillin MIC ≥2.0 μg/ml	Vancomycin + a third-generation cephalosporin[d,g]	Meropenem
Strongyloides stercoralis	Thiabendazole	—
Toxoplasma gondii	Pyrimethamine + sulfadiazine	Pyrimethamine + clindamycin; pyrimethamine + either clarithromycin, azithromycin, dapsone, or atovaquone[b]
Varicella-zoster virus	Acyclovir	Vidarabine

[a]Addition of flucytosine should be considered.
[b]Effectiveness has not been clearly documented.
[c]Intravenous and intrathecal administration.
[d]Cefotaxime or ceftriaxone.
[e]Addition of an aminoglycoside should be considered.
[f]Ethambutol or streptomycin should be added for presumably resistant strains.
[g]Addition of rifampin should be considered.
[h]Not yet approved by the FDA for clinical use.

tant to penicillin also have been described, accounting for 3% of isolates in one population-based surveillance study.[228] In addition, chloramphenicol resistance among meningococci has recently been reported.[229] The clinical significance of these resistant isolates is unclear because patients with meningococcal meningitis have recovered with standard penicillin therapy, although some authorities would treat patients with meningococcal meningitis

with a third-generation cephalosporin pending results of susceptibility testing.

Therapy of meningitis due to *H. influenzae* type B has been markedly altered due to the emergence of β-lactamase-producing strains, accounting for approximately 24% and 32% of CSF isolates overall in the United States in 1981[175] and 1986,[178] respectively. Resistance to chloramphenicol also has been described, al-

TABLE 6. Recommended Doses of Antimicrobial Agents for Central Nervous System Infections in Adults with Normal Renal and Hepatic Function[a]

Antimicrobial agent	Total daily dose	Dosing interval (hr)
Acyclovir	30 mg/kg	8
Amikacin	15 mg/kg	8
Amphotericin B	0.6–1.0 mg/kg[b]	24
Amphotericin B lipid complex	5 mg/kg	24
Ampicillin	12 g	4
Atovaquone	3000 mg	6
Azithromycin	1200–1500 mg	24
Aztreonam	6–8 g	6–8
Cefepime	4–6 g	8–12
Cefotaxime	8–12 g	4–6
Ceftazidime	6 g	8
Ceftriaxone	4 g	12
Chloramphenicol	4–6 g[c]	6
Ciprofloxacin	800–1200 mg	8–12
Clindamycin	2400–4800 mg	6
Ethambutol[d]	15 mg/kg	24
Fluconazole	400–800 mg	12
Flucytosine[d]	100 mg/kg	6
Foscarnet	180 mg/kg	8
Ganciclovir	10 mg/kg	12
Gentamicin	3–5 mg/kg	8
Isoniazid[d]	300 mg[e]	24
Itraconazole[d]	800 mg	12
Liposomal amphotericin B	3–5 mg/kg	24
Meropenem	6 g	8
Metronidazole	30 mg/kg	6
Miconazole	1.5–3.0 g	8
Nafcillin	9–12 g	4
Oxacillin	9–12 g	4
Penicillin G	24 million units	4
Pyrazinamide[d]	15–30 mg/kg	24
Pyrimethamine[d]	25–100 mg[f]	24
Rifampin[d]	600 mg	24
Sulfadiazine[d]	4–6 g	6
Thiabendazole[d]	50 mg/kg	12
Tobramycin	3–5 mg/kg	8
Trimethoprim–sulfamethoxazole	10–20 mg/kg[g]	6–12
Vancomycin	2–3 g	8–12
Vidarabine	15 mg/kg	24

[a]Unless otherwise indicated, therapy is administered intravenously.
[b]Higher doses are recommended for aspergillosis or mucormycosis; the dose can be increased to 1.5 mg/kg per day in severely ill patients.
[c]Higher dose is recommended for pneumococcal meningitis.
[d]Oral administration.
[e]Initiate therapy at a dose of 10 mg/kg.
[f]Higher dose is used in AIDS patients with toxoplasmic encephalitis.
[g]The dose is based on trimethoprim component.

though it occurs more commonly in Spain (>50% of isolates)[230,231] than in the United States (<1% of isolates).[232] In addition, a prospective study found chloramphenicol to be bacteriologically and clinically inferior to ampicillin, ceftriaxone, or cefotaxime in childhood bacterial meningitis due predominantly to *H. influenzae* type B.[233] On the basis of these findings and of studies documenting efficacy of the third-generation cephalosporins (cefotaxime or ceftriaxone) similar to that of the combination of ampicillin plus chloramphenicol for bacterial meningitis, the American Academy of Pediatrics has recommended the use of the third-generation cephalosporins as empiric antibiotic therapy in children with bacterial meningitis.[234] Cefuroxime, despite initial studies suggesting that it was as efficacious as ampicillin plus chloramphenicol for childhood bacterial meningitis, should not be used for treatment of bacterial meningitis. A prospective randomized study comparing ceftriaxone to cefuroxime for the treatment of childhood bacterial meningitis documented the superiority of ceftriaxone, with which patients had milder hearing impairment and more rapid CSF sterilization than the patients who received cefuroxime.[235] Cefepime also has been studied in the therapy of bacterial meningitis in infants and children and was found to be safe and therapeutically equivalent to cefotaxime.[236]

The third-generation cephalosporins have revolutionized the treatment of meningitis due to enteric gram-negative bacilli.[237] Cure rates with the use of these agents have ranged from 78% to 94%,[238,239] compared to previous mortality rates of 40–90% with standard regimens (usually an aminoglycoside with or without chloramphenicol). One particular agent, ceftazidime, has been shown to be efficacious for the treatment of *P. aeruginosa* meningitis, resulting in cure of 19 of 24 patients when administered either alone or in combination with an aminoglycoside.[240] Another study in 10 pediatric patients with meningitis due to *Pseudomonas* species revealed that 7 patients were cured clinically and 9 were cured bacteriologically when treated with ceftazidime-containing regimens.[241] Intrathecal or intraventricular aminoglycoside therapy should be considered only if there is no response to systemic therapy; this mode of treatment is rarely needed at present. In a study of 21 children with bacterial meningitis, imipenem was efficacious in eradication of bacteria from CSF[242] but was associated with a high rate (33%) of seizure activity, limiting its usefulness in bacterial meningitis; meropenem, a new carbapenem with less seizure proclivity than imipenem, is a better alternative for the therapy of CNS infections. The fluoroquinolones (e.g., ciprofloxacin or pefloxacin) have been used successfully in some patients with gram-negative bacillary meningitis.[243] At present, the fluoroquinolones should be used only in adult patients with bacterial meningitis who are failing conventional therapy or when the causative organism is resistant to standard drugs.

Despite the broad range of activity of the third-generation cephalosporins, they are ineffective in meningitis caused by *L. monocytogenes*. In this situation, therapy should consist of ampicillin or penicillin G,[189,191–193] with the addition of an aminoglycoside considered in proven infection due to documented *in vitro* synergy. Alternatively, trimethoprim–sulfamethoxazole (TMP-SMZ), which is bactericidal against *Listeria in vitro*, can be used.[244] Chloramphenicol shows varying activity against *L. monocytogenes in vitro*, although its use has been associated with an unacceptably high failure rate in *Listeria* meningitis. Vancomycin is unsatisfactory for *Listeria* meningitis despite favorable *in vitro* sensitivity results.

S. aureus meningitis should be treated with nafcillin or oxacillin,[199] with vancomycin reserved for patients who are allergic to penicillin or when methicillin-resistant organisms are suspected or isolated. Meningitis due to coagulase-negative staphylococci, most commonly encountered in CSF shunt infections, should be treated with vancomycin. Rifampin should be added if the patient fails to improve[198,245]; shunt removal is often necessary to optimize therapy. A combination of ampicillin plus an aminoglycoside has been standard therapy for neonatal meningitis due to *S. agalactiae*[200] and also is recommended for meningitis in adults due to this organism.[201,202] This combination is recommended due to demonstrated *in vitro* synergy and recent reports documenting the presence of penicillin-tolerant strains. The third-generation cephalosporins are alternative agents, with vancomycin reserved for patients with significant penicillin allergy. Nocardial meningitis should be treated with a sulfonamide, with or without trimethoprim[203,246,247]; alternative agents include minocycline, amikacin, imipenem, the third-generation cephalosporins, and the fluoroquinolones, which are all active *in vitro* against *Nocardia*.[248–250]

The duration of therapy for bacterial meningitis has traditionally been 10–14 days for most cases of nonmeningococcal meningitis and 3 weeks for meningitis due to gram-negative enteric bacilli.[251] Certain subsets of patients, however, may respond to shorter courses of therapy. Meningococcal meningitis can be treated for 7 days with intravenous penicillin, and some authors have even suggested that 4 days is adequate therapy[252]; however, this study requires confirmation, as only 50 patients were studied and no control group was included. Several studies comparing 7 days of treatment to 10 days of treatment have documented that 7 days of therapy are effective and safe in infants and children with *H. influenzae* meningitis.[253,254] However, therapy must be individualized, and some patients may require longer courses of treatment. In adults with meningitis due to enteric gram-negative bacilli, treatment with appropriate regimens should be continued for 3 weeks due to the high rate of relapses with shorter courses of therapy. A period of at least 21 days is the recommended duration of therapy for *Listeria* meningitis.[192,193]

3.1.4b. Adjunctive Therapy. Despite the availability of effective bactericidal antibiotics for bacterial meningitis, the morbidity and mortality from this disorder remain unacceptably high. Over the last several decades, an explosion of new information detailing the pathogenic and pathophysiologic mechanisms operable in bacterial meningitis has led to the development of innovative treatment strategies for this disorder.[255–258] Studies in experimental animal models have demonstrated that once bacteria enter the subarachnoid space, an intense inflammatory response is generated. This response can be augmented by antimicrobial therapy, which leads to rapid lysis of organisms and release of virulence factors such as cell wall or lipopolysaccharide, which in turn increases the CSF release of inflammatory cytokines. Corticosteroids have been evaluated as agents to reduce this inflammatory response and the subsequent pathophysiologic consequences of bacterial meningitis (e.g., cerebral edema, increased intracranial pressure) in experimental animal models, leading to numerous studies of adjunctive dexamethasone in patients with bacterial meningitis.[206,219] A recently published meta-analysis of these studies confirmed the benefit of adjunctive dexamethasone (0.15 mg/kg every 6 hr for 2–4 days) for *H. influenzae* type B meningitis, and if commenced with or before parenteral antimicrobial therapy, suggested benefit for pneumococcal meningitis.[259] In contrast, a retrospective, nonrandomized study of children with pneumococcal meningitis published after the meta-analysis demonstrated that use of adjunctive dexamethasone was not associated with a beneficial effect,[260] although the dexamethasone was administered before or within 1 hr of the first antimicrobial dose and the children in the dexamethasone group had a higher severity of illness.

When using adjunctive dexamethasone, timing of administration is crucial. Administration concomitant with or just before the first dose of the antimicrobial agent is optimal for attenuating the subarachnoid space inflammatory response. In adults or in patients with meningitis caused by bacterial pathogens other than *H. influenzae* type B, the routine use of adjunctive dexamethasone is not recommended, pending results of ongoing studies.[206,219] The use of adjunctive dexamethasone is of particular concern in patients with pneumococcal meningitis caused by highly penicillin- and cephalosporin-resistant strains, in which patients may require antimicrobial therapy with vancomycin; a diminished CSF inflammatory response after dexamethasone administration might significantly reduce vancomycin penetration into CSF and delay CSF sterilization. For any patient with pneumococcal meningitis receiving adjunctive dexamethasone who is not improving as expected or who has a pneumococcal isolate for which the cefotaxime or ceftriaxone MIC is ⩾2.0 μg/ml, a repeat lumbar puncture 36–48 hr after initiation of antimicrobial therapy is recommended to document the sterility of CSF.[261]

3.2. Brain Abscess

3.2.1. Epidemiology and Etiology

Brain abscess is one of the most serious complications of head and neck infections. Case–fatality rates, even in the antibiotic era, ranged from 30 to 60% (similar to the preantibiotic era) until recently, when the overall mortality rate has decreased to 0–24%[262,263]; this improvement is likely due to recent developments in diagnosis and treatment that are discussed in Sections 3.2.3 and 3.2.4. When evaluating the likely bacterial species responsible for brain abscess formation, isolation frequency is highest for certain microorganisms and depends on the pathogenic mechanisms involved.[263,264] Streptococci (aerobic, anaerobic, and microaerophilic) are the bacteria most commonly isolated (in 60–70% of abscesses). These bacteria, especially the *S. milleri* group, normally reside in the oral cavity, appendix, and female genital tract and have a proclivity for abscess formation.[265,266] *S. aureus* accounts for 10–15% of isolates, usually in patients with cranial trauma or endocarditis. The attention to proper techniques has increased the isolation of anaerobes from brain abscesses, with *Bacteroides* species isolated in 20–40% of cases, often in mixed culture.[267,268] Enteric gram-negative bacilli (e.g., *Proteus* species, *E. coli*, *Klebsiella* species, and *Pseudomonas* species) are isolated in 23–33% of patients, often in patients who have otitic foci of infection or who are immunosuppressed.

Other bacterial pathogens may be isolated from patients who are immunocompromised,[3] although these organisms make up a small proportion of all bacterial brain abscesses (<1%). These organisms include *H. influenzae*, *S. pneumoniae*, *L. monocytogenes*, and *Nocardia asteroides*.[3,269–273] The latter more often is isolated in patients with T-lymphocyte or mononuclear phagocyte defects (secondary to corticosteroid therapy, in organ transplant patients, and in patients with neoplastic disease),[274–277] although up to 48% of patients with nocardiosis have no underlying conditions. One case of nocardial brain abscess has been seen in a pregnant female[278] and other cases have been described in patients with AIDS.[279,280] *Nocardia* usually spreads to the brain via hematogenous dissemination from distant foci of infection (e.g., the lung). In patients with HIV-1 infection, bacterial causes of brain abscess also include *L. monocytogenes*[194] and *Salmonella* group B.[281]

3.2.2. Clinical Presentation

The clinical course of brain abscess may be indolent or fulminant.[262,264,269,282,283] Most of the clinical manifestations are due to the presence of a space-occupying lesion within the brain, not to the systemic signs of infection. Headache is observed in over 70% of patients and may be moderate to severe and hemicranial or generalized in location. Other findings include nausea and vomiting (half the cases), presumably due to increased intracranial pressure; nuchal rigidity and papilledema occur in about one fourth of cases. Mental status changes, ranging from lethargy to coma, are seen in the majority of patients. Approximately 50% of patients develop focal neurologic deficits (usually hemiparesis), which vary depending on the location of brain involved. Seizures, usually generalized, occur in 25–35% of cases, but fever is found in only 45–50% of cases. Less than half of patients with brain abscess present with the classic triad of fever, headache, and focal neurologic deficit.

The location of the brain abscess may define the clinical presentation.[262–264,269] Patients with a frontal lobe abscess often present with headache, drowsiness, inattention, and deterioration of mental status (commonly seen is hemiparesis with unilateral motor signs and a motor speech disorder). The clinical presentation of cerebellar abscesses is ataxia, nystagmus, vomiting, and dysmetria. Temporal lobe abscesses may cause ipsilateral headache and aphasia, if the lesion is in the dominant hemisphere. In addition, a visual field defect (e.g., upper homonymous quadrantanopsia) may be the only presenting sign of a temporal lobe abscess. Abscesses of the brain stem usually present with facial weakness, fever, headache, hemiparesis, dysphagia, and vomiting.[284] In addition, the clinical presentation of brain abscess in the immunocompromised patient may be masked by the diminished inflammatory response.[3]

In nocardial brain abscess, the presentation is generally nonspecific (fever, headache, and focal deficits determined by the site and size of the lesion).[274–280] The clinical suspicion of nocardial brain abscess may be increased by the presence of pulmonary, skin, or muscle lesions, which are present concurrently in many, but not all, cases.

3.2.3. Diagnosis

CT has revolutionized the diagnosis of brain abscess. It is not only an excellent means of examination of the brain parenchyma but also is superior to standard radiologic procedures for examination of the paranasal sinuses, mastoids, and middle ear.[285] The availability of CT has significantly improved the mortality rate in pa-

tients with bacterial brain abscess. Data from the University of California in San Francisco demonstrated a decrease in the overall mortality rate from 44% during the 3 years before CT scanning to zero for the 3 years following the introduction of CT scanning in 1977[286]; this lower mortality rate was principally related to early diagnosis and an accurate method of postoperative follow-up. The characteristic CT appearance of brain abscess is that of a hypodense center with a peripheral uniform ring enhancement following the injection of contrast material; this ring is surrounded by a variable hypodense area of brain edema (Fig. 2). Other CT findings include nodular enhancement and areas of low attenuation without enhancement, the latter being observed during the early cerebritis stage prior to abscess formation; as the abscess progresses, contrast enhancement is observed. Once the abscess becomes encapsulated in the later stages, contrast

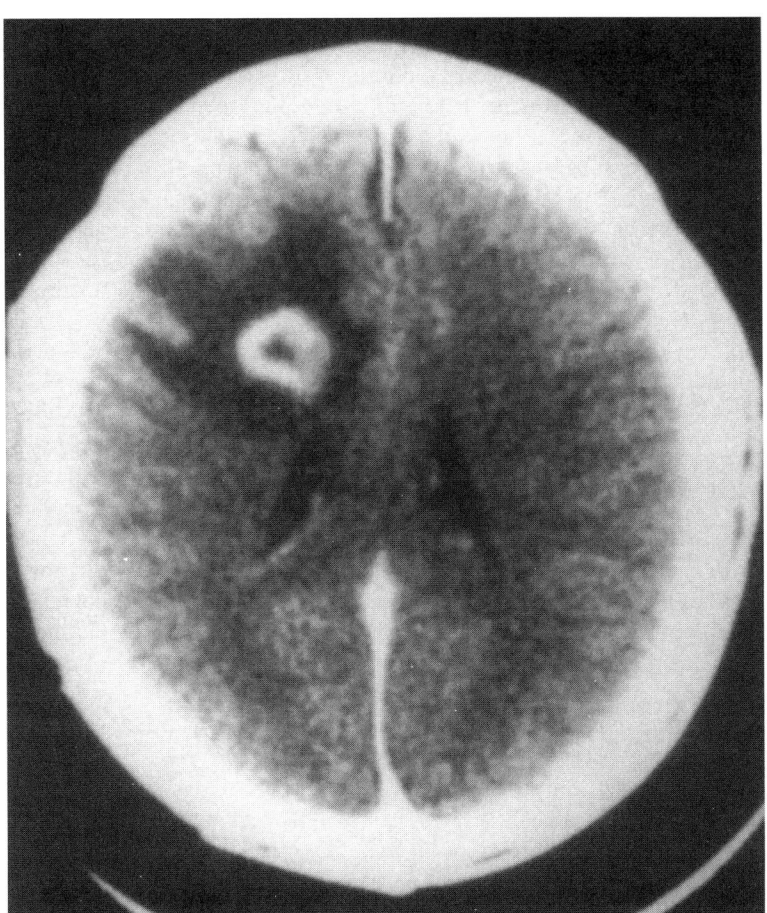

FIGURE 2. CT of the head revealing a ring-enhancing frontal lobe abscess and surrounding edema. Cultures of the abscess grew *Streptococcus milleri.*

no longer differentiates the lucent center, and the CT appearance is similar to the early cerebritis stage. The absence of contrast likely indicates a well-encapsulated lesion. CT scanning also is useful for following the course of brain abscess, although after aspiration improvement in CT appearance may not be seen for up to 5 weeks or longer.

MRI has been extensively studied in the diagnosis of brain abscess and is now the first imaging procedure of choice in patients suspected of having this disorder.[287,288] MRI offers significant advantages over CT in the early detection of cerebritis, detection of cerebral edema with greater contrast between edema and adjacent brain, more conspicuous spread of inflammation into the ventricles and subarachnoid space, and earlier detection of satellite lesions. Contrast enhancement with the paramagnetic agent gadolinium diethylenetriamine penta-acetic acid (Gd-DTPA) provides the added advantage of clearly differentiating the central abscess, surrounding enhancing rim, and cerebral edema surrounding the abscess.

A major advance in the use of CT scanning is the ability to perform stereotactic CT-guided aspiration of the brain abscess to facilitate bacteriologic diagnosis and guide antimicrobial therapy. Aspiration during the cerebritis stage, however, may be complicated by hemorrhage. All brain abscess lesions >2.5 cm in diameter should be stereotactically aspirated or excised for optimal management.[263,289] At the time of aspiration, specimens should be sent for Gram stain, routine culture, and anaerobic culture. In addition, other special stains such as Ziehl–Neelsen for mycobacteria, modified acid-fast for *Nocardia*, and silver stains for fungi should be performed; specimens should also be sent to the microbiology laboratory for culture of these organisms.

3.2.4. Treatment

3.2.4a. Antimicrobial Therapy.   When a diagnosis of bacterial brain abscess is made either presumptively by radiologic studies or by aspiration of the abscess, antimicrobial therapy should be initiated. Aspiration may provide an etiologic diagnosis based on Gram's or other special stains, but if the aspirate is unrevealing or cannot be performed, empiric therapy should be initiated on the basis of the patient's predisposing condition and presumed pathogenic mechanism of abscess formation.[263,264] Due to the high rate of isolation of streptococci (particularly the *S. milleri* group) from brain abscesses of various etiologies, high-dose intravenous penicillin G or another drug (e.g., a third-generation cephalosporin, either cefotaxime or ceftriaxone) active against this organism

should be included in the initial therapeutic regimen. Penicillin G also is active against most anaerobic species, with the notable exception of *Bacteroides fragilis*, which may be isolated in a high percentage of brain abscess cases; metronidazole should be included in the initial regimen when this organism is suspected. Metronidazole has bactericidal activity against *B. fragilis* and attains high concentrations in brain abscess pus, and its entry into cerebral abscesses is not affected by concomitant corticosteroid therapy.[290,291] When *S. aureus* is considered a likely pathogen (e.g., following cranial trauma or postneurosurgery), nafcillin should be used, with vancomyin reserved for patients allergic to penicillin or when methicillin-resistant organisms are suspected or isolated.[292] For empiric therapy when members of the Enterobacteriaceae family are suspected, either a third-generation cephalosporin or trimethoprim-sulfamethoxazole should be used.[293,294] If *P. aeruginosa* is a likely infecting pathogen, ceftazidime is the agent of choice. When a brain abscess due to *Nocardia* is suspected or proven, the sulfonamides, with or without trimethoprim,[203,246,247] are a reasonable first choice for treatment. Alternative agents include minocycline, amikacin, imipenem, the third-generation cephalosporins and the fluoroquinolones, which are among the most active agents against *Nocardia in vitro*,[248–250] although clinical information is lacking. Combination therapies have been utilized in patients with *Nocardia* brain abscess[295–299] and combination regimens containing the third-generation cephalosporins or imipenem along with a sulfonamide should be considered for immunocompromised patients or those in whom therapy fails.[283] For *L. monocytogenes* brain abscess, combination therapy with ampicillin and gentamicin should be used.

Once the infecting pathogen is isolated, antimicrobial therapy can be modified for optimal therapy. Our recommendations for standard therapy, with alternative agents, are shown in Table 5. Doses of these agents used for CNS infections are shown in Table 6. Antimicrobial therapy with high-dose intravenous antibiotics should be continued for 6–8 weeks and often is followed by oral antibiotic therapy for 2–6 months if one or more appropriate agents are available.[263,264] Shorter courses (3–4 weeks) may be adequate for patients who have undergone surgical excision of the abscess. Surgical therapy often is required for brain abscess, although certain subsets of patients may be treated with medical therapy alone,[284,300–302] these being patients with (1) medical conditions that increase the risk of surgery, (2) multiple abscesses, (3) abscesses in a deep or dominant location, (4) concomitant meningitis or ependymitis, (5) early abscess reduction with clinical improvement after antimicrobial therapy, and (6) abscess size under 3 cm. However, in one series, no abscess larger than 2.5 cm resolved without surgical therapy.[300] Furthermore, patients treated with antimicrobial therapy alone may require prolonged (up to 12 weeks) courses of parenteral therapy, and must be carefully followed clinically and radiographically. Antimicrobial therapy for nocardial brain abscess has ranged from 3 to 12 months.[248,276] However, therapy in immunosuppressed patients should probably be continued for up to 1 year, with careful follow-up to monitor for relapse.

3.2.4b. Surgical Therapy. Most patients with brain abscess require surgical management for optimal therapy. The two procedures judged equivalent by outcome are aspiration of the abscess after burr hole placement and complete excision after craniotomy.[269,303,304] The choice of procedure must be individualized for each patient. Aspiration may be performed by stereotactic CT guidance, which affords the surgeon rapid, accurate, and safe access to virtually any intracranial point[305–307]; aspiration also can be used for swift relief of increased intracranial pressure. The major disadvantage of aspiration is incomplete drainage of multiloculated lesions; these patients frequently require excision. Complete excision after craniotomy is most often employed in patients in a stable neurologic condition. Surgery should also be performed when abscesses exhibit gas on radiologic evaluation. If the patient develops worsening neurologic deficits (e.g., deteriorating consciousness or signs of increased intracranial pressure), surgery should be performed emergently. Excision is contraindicated in the early stages of brain abscess formation before a capsule is formed and occasionally due to abscess loculation.

3.3. Tuberculosis

3.3.1. Epidemiology and Etiology

Virtually all tuberculous infections of the CNS are caused by the human tubercle bacillus, *Mycobacterium tuberculosis*. Much of the data on the incidence of CNS tuberculosis were obtained in the first half of the 20th century, when approximately 5–15% of individuals exposed to tuberculosis developed symptomatic disease; of this number, 5–10% of patients ultimately had CNS involvement.[308] CNS tuberculosis is particularly common in less developed areas of the world (e.g., India and Africa).[309] Tuberculous meningitis accounts for approximately 15% of extrapulmonary cases or about 0.7% of all clinical tuberculosis in the United States.[310] Factors such as advanced age, immunosuppressive drug therapy, transplantation, lymphoma, gastrectomy, pregnancy, diabetes

mellitus, and alcoholism are known to compromise the immune response in patients with smoldering chronic organ tuberculosis, leading to reactivation of latent foci and progression to the clinical syndrome of late generalized tuberculosis.[311,312] The advent of HIV-1 infection also has influenced the epidemiology of tuberculosis in the United States, with an estimated 6000–9000 new cases annually.[313] Although the majority of tuberculosis cases in HIV-1-infected patients are pulmonary, extrapulmonary tuberculosis (including CNS disease) occurs in more than 70% of patients with AIDS or AIDS discovered soon after the diagnosis of tuberculosis, but in only 24–45% of patients with tuberculosis and less advanced HIV-1 infection,[314] suggesting that extrapulmonary tuberculosis appears to be more common in patients with more severe HIV-1-induced immunosuppression.

Mycobacterium tuberculosis is a nonmotile bacillus, an obligate aerobic parasite the only natural reservoir of which is humans.[308] The thick cell wall of the mycobacteria can be visualized on Ziehl–Neelsen, Kinyoun, or fluorochrome staining. Infection begins with inhalation of infectious particles, with hematogenous infection occurring 2–4 weeks following infection.[308,311] At distant sites of infection (e.g., the CNS), the primary tubercles grow, the caseous centers liquefy, organisms proliferate, and the lesion ultimately ruptures, discharging organisms and their antigenically potent products into the surrounding tissues. The specific CNS syndromes associated with infection are a function of the original location of the tubercle. Foci located on the surface of the brain or ependyma rupture into the subarachnoid space or the ventricular system, causing meningitis. Foci deep within the brain or spinal cord parenchyma enlarge to form tuberculomas or, more rarely, tuberculous abscesses. Tuberculomas can occur throughout the cerebral hemispheres, the basal ganglia, the cerebellum, and the brain stem.

The nontuberculous or atypical mycobacteria also may be causes of CNS infection, although there are very few reported cases even in the presence of disseminated disease.[315] Isolated cases have occurred in both immunosuppressed and immunocompetent patients,[316–318] in whom CNS infection was presumed to be a result of hematogenous dissemination of the organism.

3.3.2. Clinical Presentation

The clinical picture of tuberculous meningitis is quite variable.[308] Children commonly develop nausea, vomiting, and behavioral changes; headache is seen in fewer than 25% of cases. Seizures are infrequent (seen in 10–20% of children prior to hospitalization), although more than 50% of patients may develop seizures during hospitalization. An encephalitic course has been described in children characterized by stupor, coma, and convulsions without signs of meningitis. In adults, the clinical presentation of tuberculous meningitis tends to be more indolent.[308,319,320] In the usual patient, an insidious prodrome characterized by malaise, lassitude, low-grade fever, intermittent headache, and changing personality ensues. Within 2–3 weeks, there is development of a meningitis phase manifested as protracted headache, meningismus, vomiting, and confusion. In some adults, the initial prodromal stage may take the form of a slowly progressive dementia over several months or years characterized by personality changes, social withdrawal, and memory deficits. In contrast, patients also may present with a rapidly progressive meningitis syndrome indistinguishable from pyogenic bacterial meningitis.[311] A history of prior clinical tuberculosis is infrequent (<20% of cases).[319,321]

On physical examination, children and adults present with more uniform findings, although considerable variation does exist.[308,311,316,319–325] Fever is an inconstant finding on physical examination (50–98% of cases). Meningismus and signs of meningeal irritation are not uniform findings (absent in 25–80% of children and adults). Focal neurologic signs most frequently consist of unilateral or, less commonly, bilateral cranial nerve palsies, which are seen in up to 30% of patients on presentation; the most frequently affected is cranial nerve VI, followed by cranial nerves III, IV, and VIII. Hemiparesis may result from ischemic infarction in the anterior cerebral circulation, most commonly in the territory of the middle cerebral artery.[326] Abnormal movements (chorea, hemiballismus, athetosis, myoclonus, and cerebellar ataxia) are less frequently seen on neurologic examination. Funduscopic examination may reveal choroidal tubercles, but these are seen in only about 10% of cases of tuberculous meningitis.

The clinical manifestations of tuberculous meningitis do not seem to be modified by HIV-1 infection.[327,328] Most patients present with fever, headache, and altered mentation. Meningeal signs are absent in up to 50% of patients, although in a recent review of patients with tuberculous meningitis admitted to an intensive care unit, 88% had meningeal stiffness.[325]

Symptoms of tuberculomas often are limited to seizures and correlates of increased intracranial pressure.[308] Papilledema is seen in most cases, accompanied by neurologic deficits reflecting the location of the lesions. Fever and signs of systemic infection are rarely present. The mean duration of symptoms is weeks to months, and some observers note that the patient's symptoms are less dramatic than would be expected from the radiologic or surgical size of the lesion. Only about 30% of patients have evidence of tuberculous infection outside the CNS.[311]

3.3.3. Diagnosis

Routine laboratory tests are not helpful in the diagnosis of CNS tuberculosis.[308,311] A syndrome of inappropriate antidiuretic hormone secretion has been seen in tuberculous meningitis, manifested as hyponatremia and hypochloremia.[329] Chest radiograph abnormalities are common in children with CNS tuberculosis, usually reflecting primary tuberculosis, occasionally with superimposed calcifications or with miliary spread.[308,322] In adults, the adenopathy and dense infiltrates of primary tuberculosis are less frequently found; changes include apical scarring, calcified Ghon complexes, and nodular upper lobe disease. Miliary disease has been documented radiologically by chest radiograph in 25–50% of cases of tuberculous meningitis in adults.[330,331] Tuberculin skin tests are usually positive in patients with CNS tuberculosis; rates of 85–90% positivity occur in children.[308,322] In adults, approximately 35–60% of individuals thought to have tuberculous meningitis do not react to first- or second-strength tuberculin testing.[319,323,324] In HIV-1-infected patients with tuberculosis, rates of tuberculin reactivity in active disease are low, but not negligible, usually from 33% to 71%.[314]

CSF abnormalities are traditionally seen in tuberculous meningitis.[308,311] The fluid is typically clear or opalescent, but when the CSF is allowed to stand at room temperature or in the refrigerator for a short time, a cobweblike clot may form that is the classic "pellicle" of tuberculosis. It occurs secondary to the high fibrinogen concentration in the fluid along with the presence of inflammatory cells; this presence is characteristic, although by no means invariable. A moderate pleocytosis is characteristic of tuberculous meningitis, with 90–100% of patients having more than 5×10^6 white cells/liter. The number of cells seldom exceeds 300×10^6/liter, although exceptions do occur (between 500×10^6 and 1500×10^6/liter in about 20% of patients).[324] Initially, both lymphocytes and neutrophils predominate, with rapid conversion into a lymphocytic predominance over several weeks. The converse can be seen following the introduction of antituberculous chemotherapy, in which an initial lymphocytic predominance shifts to a neutrophilic predominance on subsequent CSF examinations, the so-called "therapeutic paradox." There is usually a modest depression of CSF glucose, with a median of 2.2 mmoles/liter

(40 mg/dl) reported in most series. Hypoglycorrachia has correlated with more advanced stages of clinical disease.[308] CSF protein is elevated in the majority of cases, with median values of 1.5–2.0 g/liter (150–200 mg/dl). Occasionally, CSF protein values in excess of 10–20 g/liter are reported, usually in conjunction with spinal block.[321] The identification of tuberculous organisms in CSF by specific stains is difficult because of the small population of organisms. In many series, fewer than 25% of specimens were smear-positive,[308,316,319,325] although one review demonstrated positive smears in 52% of CSF specimens.[320] The yield may be increased by staining the pellicle (if present) as well as layering the centrifuged sediment of large CSF volumes onto a single slide with repeated applications until the entire pellet can be stained at once. Obtaining repeated specimens also may increase the yield. An 86% rate of acid-fast smear positivity was demonstrated when up to four separate specimens were examined for each patient[324]; this rate has not been consistently duplicated in the literature, however. Proof of infection requires isolation of the organism from CSF, although false-negative CSF cultures are common, with mycobacteria isolated from less than 50% of patients with a clinical diagnosis of tuberculous meningitis.[308] Higher culture yields may be obtained by processing multiple specimens for each patient, although even with as many as four CSF specimens, almost 20% of patients with a clinical diagnosis of tuberculous meningitis have negative CSF cultures.[324]

Based on the findings presented above, several newer diagnostic modalities are under development for the diagnosis of tuberculous meningitis (Table 7).[308,332] Some tests utilize biochemical assays to measure some feature of the organism or the host response to it (e.g., bromide partition test, adenosine deaminase assay). Other modalities are immunologic tests that detect mycobacterial antigen or antibody in the CSF [e.g., tuberculostearic acid antigen, enzyme-linked immunosorbent assay (ELISA), latex agglutination]. These immunodiagnostic tests have some promise for rapid and sensitive diagnosis of tuberculous meningitis, although there are problems with the presence of cross-reacting antibodies against nonpathogenic mycobacteria, as well as with the presence of bacterial or fungal antigenic moieties. The technique of PCR for detecting fragments of mycobacterial DNA in CSF specimens appears to be an equally promising tool.[333–336] Before these tests can be considered useful in the diagnosis of tuberculous meningitis, large-scale confirmatory studies first must be performed.

There are no radiologic changes that are pathognomonic for tuberculous infection of the CNS. Angiographic evaluation may show a characteristic triad consisting of (1) a hydrocephalic pattern to the vessels, (2) narrowing of the vessels at the base of the brain, and (3) narrowed or occluded small and medium-sized vessels with scanty collaterals; these changes are not invariably present, however,[308] and angiography is rarely employed at present. On CT scanning, hydrocephalus is frequently present at diagnosis or develops during the course of infection. With the addition of intravenous contrast material, enhancement of the basal cisterns results, with widening and blurring of the basilar arterial structures. Periventricular lucencies may be evident, reflecting the presence of periventricular tuberculous exudate and tubercle formation adjacent to the ependyma and choroid (Fig. 3). MRI with gadolium enhancement has been shown to be more sensitive than CT in detecting the anatomic abnormalities of tuberculous meningitis.[337] MR angiography also has been utilized to detect the characteristic vascular narrowing and the rare complication of aneurysm formation in patients with tuberculous meningitis.[338]

On CT scanning, tuberculomas may appear as isodense or hypodense areas with uniform contrast enhancement.[339] With high-resolution scanning, a thick ring of contrast enhancement may surround a characteristic punctate area of central clearing, signaling the presence of caseation within the lesion. Surrounding edema may or may not be present. On MRI, noncaseating granulomas are often bright on T2-weighted images whereas caseating granulomas are isointense to markedly hypointense. Following aspiration of these lesions, approximately 60% of tissue specimens have been smear-positive for acid-fast bacilli, and 50–60% of specimens grow the organism in culture.[308]

With regard to the nontuberculous or atypical mycobacteria, stricter criteria are needed to define CNS disease.[316–318] Repeated isolation of multiple colonies of nontuberculous mycobacteria must be demonstrated in the absence of other pathogens to implicate these agents as the cause of CNS disease.

TABLE 7. Newer Diagnostic Tests for Tuberculous Meningitis[a]

Test	Sensitivity	Specificity
Radiolabeled bromide partition	90–94%	88–96%
Adenosine deaminase	73–100%	71–99%
Tuberculostearic acid	95%	99%
Mycobacterial antigen	79–94%	95–100%
Mycobacterial antibody	27–100%	94–100%
Polymerase chain reaction	83–100%	80–100%

[a]Adapted from Zugar and Lowy.[308]

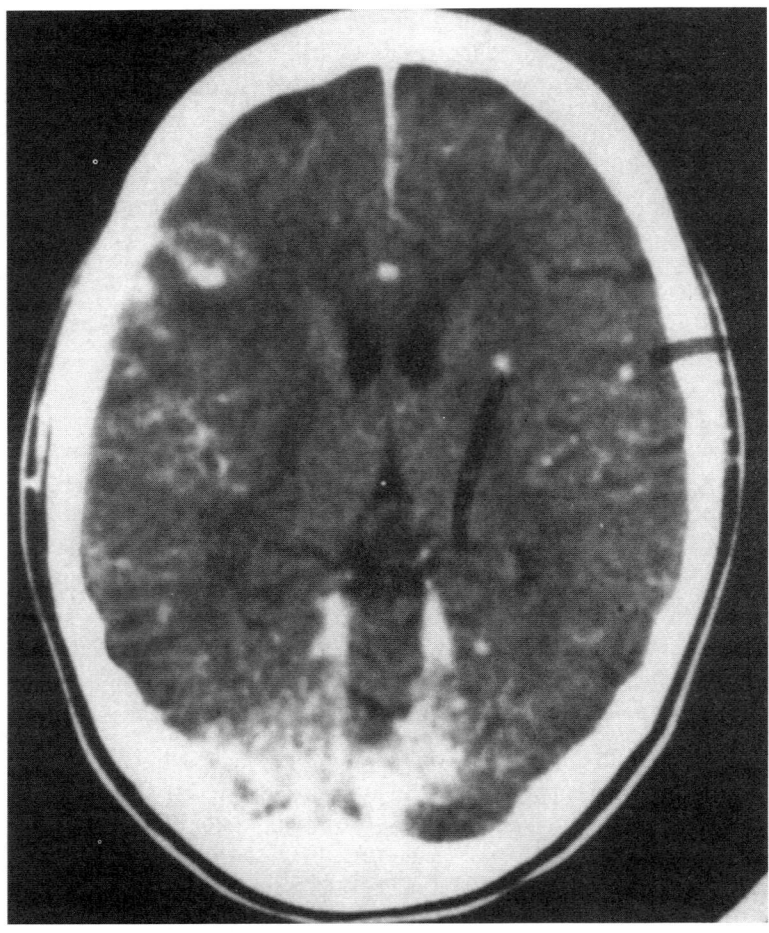

FIGURE 3. CT of the head with contrast enhancement in a patient with tuberculous meningitis, revealing enhancement of the basal meninges.

3.3.4. Treatment

3.3.4a. Antimicrobial Therapy. CNS tuberculosis was a uniformly fatal disease until the advent of effective chemotherapy. However, the optimal drug regimen, dose, route of administration, and duration of treatment still remain undefined. The most important principle of therapy is that it should be initiated on the basis of strong clinical suspicion and not delayed until proof of infection has been obtained, since outcome of tuberculous meningitis is determined by the clinical stage at admission and the delay in initiating therapy.[325] In general, the principles of treatment of CNS tuberculosis do not differ significantly from those for other forms of tuberculosis.[308,311] In all modern regimens, isoniazid and rifampin are the mainstays of treatment. More recently, regimens have taken advantage of the newly appreciated merits of pyrazin-amide and its intracellular microbicidal activity. All these drugs have good CNS penetration; in the presence of meningeal inflammation, the CSF concentrations of isoniazid, rifampin, and pyrazinamide are 90%, 20%, and 100%, respectively, of peak serum concentrations. For treatment of tuberculous meningitis, the World Health Organization recommends the combination of isoniazid, rifampin, pyrazinamide, and ethambutol for 2 months, followed by 6–7 months of isoniazid and rifampin.[340] The American Thoracic Society recommends 2 months of isoniazid, rifampin, and pyrazinamide followed by 4 months of isoniazid and rifampin for adults and 10 months for children,[341] although tuberculous meningitis in children has been successfully treated with a 6-month course of therapy.[342] Other authors recommend a total treatment duration of 9 months for CNS tuberculosis.[311] However, therapy for tuberculous meningitis may need to be individualized, with longer durations of therapy utilized in patients with a higher severity of illness.[343] In addition, HIV-1-infected patients may require longer courses of treatment.[344] Doses of these agents used in CNS infections are shown in Table 6. Ethambutol or streptomycin should be added in cases of suspected drug resistance. Clinical disease with resistant organisms occurs with higher frequency among immigrants from countries in Asia, Africa, and the Americas; known contacts of drug-resistant cases; homeless and impoverished individuals; and residents of certain geographic areas in the United States (particularly adjacent to the Mexican border).[308,344]

Secondary resistance may develop during chemotherapy if compliance with a multidrug regimen is poor. In patients with multidrug-resistant strains of *M. tuberculosis*, the optimal therapeutic regimen is unknown, but should be guided by susceptibility testing and drug penetration into the CSF. Ethambutol penetrates poorly into CSF except when the meninges are inflamed. Ethionamide crosses both healthy and inflamed meninges, with peak CSF concentrations comparable to those achieved in serum. The fluoroquinolones (ciprofloxacin and ofloxacin) also penetrate well into CSF and have good *in vitro* activity against *M. tuberculosis*. Amoxicillin–clavulanate or imipenem has been used in individual cases of meningitis caused by multidrug-resistant *M. tuberculosis*. Preliminary data have suggested that HIV-1-infected patients with meningitis caused by rifampin- or multidrug-resistant *M. tuberculosis* were more likely to die in the 60 days after collection of CSF than HIV-1-infected patients with fully susceptible or other resistant isolates.[345]

3.3.4b. Adjunctive Therapy. Despite the availability of effective antituberculous chemotherapy, tuberculous meningitis is associated with persistent morbidity and mortality; this circumstance has intensified the search for adjunctive treatment agents. Although chemotherapy may halt bacterial growth, inflammation often continues at the base of the brain, with organization of necrotic tissue and exudate, fibroblastic proliferation, and formation of dense fibrocollagenous tissue compressing adjacent structures and impeding spinal fluid circulation. The most persistently advocated adjuvants for tuberculous meningitis are corticosteroids, although the place of steroids in the treatment of tuberculous meningitis remains unclear.[308,311] Corticosteroids have been demonstrated to abrogate the signs and symptoms of disease in which patients frequently defervesce, with clearing of sensorium and improvement in well-being, even after only a few doses. A number of studies have suggested that the primary value of corticosteroids may be in their ability to

treat or avert the development of spinal block, possibly by lowering the protein content of CSF; an improvement in overall mortality has been ascribed to their specific impact on this poor prognostic sign.[346] Although corticosteroids reduce mortality, a concomitant almost compensatory increase in significant neurologic sequelae has been seen in survivors with tuberculous meningitis. However, two recent studies did not observe this paradoxical finding. One study from Egypt found reduced neurologic complications and case–fatality rates in patients with tuberculous meningitis who received dexamethasone in conjunction with antituberculous chemotherapy.[347] Another prospective, controlled, randomized study from the Republic of South Africa found that corticosteroids significantly improved survival rate and intellectual outcome in children with tuberculous meningitis.[348] Most authorities now advocate the use of corticosteroids in tuberculous meningitis in selected cases with extreme neurologic compromise, elevated intracranial pressure, impending herniation, or impending or established spinal block.[308,311,349] Some authors also recommend corticosteroids in patients with CT evidence of either hydrocephalus or basilar meningitis. Prednisone, 1 mg/kg per day, tapered over 1 month, often is recommended, although varying doses of dexamethasone or hydrocortisone also have been used.

3.3.4c. Surgery. Surgical intervention in tuberculous meningitis is primarily aimed at relief of hydrocephalus; clinical response to ventriculoatrial or ventriculoperitoneal shunt placement in children has ranged from 66 to 100% in small series.[350] However, criteria for selection of patients for shunt surgery for hydrocephalus complicating tuberculous meningitis remain unclear. One study demonstrated that neurologic status on admission was the most important factor in determining outcome in patients who had undergone shunt surgery for hydrocephalus[351]; patients with normal sensorium with or without neurologic deficits were recommended for early shunt surgery, whereas for those with altered sensorium or who were deeply comatose with or without decerebrate–decorticate posturing, external drainage was recommended. A subsequent study by the same authors recommended that patients with altered sensorium undergo shunt surgery, whereas those with severe neurologic compromise (deep coma with or without decerebrate–decorticate posturing) have a trial of external drainage with their response determining the need for shunt surgery.[352] With the advent of effective antituberculous chemotherapy, surgery is no longer recommended for tuberculomas of the CNS,[353] unless the lesions are critically located, or producing obstructive hydrocephalus or compression of the brain stem.

4. Fungal Infections

4.1. Meningitis

4.1.1. Epidemiology and Etiology

The incidence of fungal meningitis has increased dramatically in recent years because of the increased numbers of immunosuppressed patients.[354] While pathogenic fungi may cause only mild symptoms in the immunologically intact person, they may cause severe neurologic disease in the patient with altered immunity. Prognosis in patients with fungal meningitis depends on severity of immunosuppression, fungal virulence factors, inoculum size of the organism, and timeliness of initiation of therapy.

Cryptococcus neoformans is the most common fungal cause of clinically recognized meningitis. Cryptococcal meningitis occurs most commonly in persons who are immunosuppressed[355–357]; these include patients with reticuloendothelial malignancies (e.g., lymphoma), sarcoidosis, organ transplantation, collagen vascular diseases, diabetes mellitus, chronic hepatic failure, chronic renal failure, and patients receiving corticosteroids. Currently, patients with AIDS are in the highest-risk group[358–363]; clinical studies suggest that 5–10% of AIDS patients develop cryptococcal meningitis.[364] Cases also have been documented, however, in apparently healthy individuals. *Cryptococcus neoformans* is a yeastlike fungus that is ubiquitous in nature and is the only encapsulated yeast known to be pathogenic in humans. It is associated with bird droppings, although it can also be found in fruits, vegetables, milk, and soil.[365] Disease develops following inhalation of the unencapsulated yeast, which is followed by a primary pulmonary infection that is usually asymptomatic or self-limited. The CNS is the most frequent site of extrapulmonary involvement with *Cryptococcus* via hematogenous dissemination. CNS disease is characterized as a meningitis or meningoencephalitis, although rarely an isolated granulomatous lesion (cryptococcoma) may develop.

The etiologic agent of coccidioidomycosis is *Coccidioides immitis*, a thermal dimorphic fungus, which grows as a mold in the external environment and as a yeastlike structure in human tissues.[366–368] The organism is endemic in the semiarid regions of the Americas and desert areas of the southwestern United States (California, Arizona, New Mexico, Texas), where about one third of the population is infected. Disease begins with aerosolization of arthroconidia from the soil, which are inhaled into the lungs. In the tissues, arthroconidia swell and form large, hard-walled structures called "spherules." The ini-

tial pulmonary infection is usually asymptomatic or self-limited, with fewer than 1% of patients developing disseminated disease within the first 3–6 months after initial infection[369,370]; one third to one half of patients with disseminated disease have meningeal involvement. Predisposition to the development of disseminated disease has been associated with infancy and old age, male gender, non-Caucasian race, pregnancy, and immunosuppression; corticosteroid therapy, antineoplastic chemotherapy, immune suppression for organ transplantation, and diseases that impair cellular immunity (e.g., AIDS) profoundly increase the susceptibility to life-threatening coccidioidal infections.[368]

Histoplasma capsulatum is a thermal dimorphic fungus that grows as a mycelium in the soil where a humid environment, moderate temperatures, and shady conditions facilitate its growth.[371] The organism is endemic to fertile river valleys, principally the Mississippi and Ohio river basins. The spores from the fungus may be aerosolized and inhaled. In endemic areas, humans are readily and presumably repeatedly infected; rarely, systemic disease develops. Hematogenous dissemination tends to occur in immunocompromised individuals with impaired cell-mediated immunity.[372–375] With dissemination, multiple organs are involved, with CNS involvement occurring in 10–20% of cases.[376] CNS disease may occur more frequently in patients with AIDS.

Candida species are ubiquitous organisms that are normal commensals of humans.[377] The organism usually colonizes mucosal surfaces in the yeast form, but with true infection, pseudohyphae or hyphae frequently are found. When present, tissue invasion invariably occurs in patients with altered host defenses, including patients receiving corticosteroid or broad-spectrum antibiotic therapy or hyperalimentation, in premature infants, in patients with malignancy, neutropenia, chronic granulomatous disease, diabetes mellitus, or thermal injuries, and in patients with a central venous catheter in place.[378–381] Widespread dissemination of *Candida* to multiple sites may ensue, including the CNS[382–384]; CNS candidiasis, however, is rarely diagnosed antemortem. Candidal meningitis is uncommon, occurring in fewer than 15% of patients with CNS candidiasis. *Candida albicans* is the species most commonly found in CNS disease.

4.1.2. Clinical Presentation

The time course of fungal meningitis may be variable, depending on the clinical setting; cases may present acutely, subacutely, or chronically. Some of the fungal meningitides may cause symptoms that persist for years in the absence of antifungal treatment. In contrast, the same organisms can produce severe symptoms and signs within a few days and without clinical signs of meningeal irritation in the immunocompromised patient.

The clinical presentation of cryptococcal meningitis in non-AIDS and AIDS patients is shown in Table 8.[355-365,385] In non-AIDS patients, cryptococcal meningitis typically manifests as a subacute process after days to weeks of symptoms. Headache is the most frequent complaint. Fever, meningismus, and personality changes also may occur; confusion, irritability, and other personality changes reflecting meningoencephalitis are found in about one half of patients. Ocular abnormalities occur in about 40% of patients and include papilledema (with or without loss of visual acuity) and cranial nerve palsies; *Cryptococcus* also has the propensity to directly invade the optic nerve. Rare findings include seizures and focal neurologic deficits. In AIDS patients, the presentation of cryptococcal meningitis can be very subtle, with minimal if any symptoms. AIDS patients may present with only headache and lethargy. Although fever is common, meningeal signs occur in only a minority of patients. Photophobia and cranial nerve palsies are often absent. On the African continent, AIDS patients with cryptococcal meningitis have been observed to have higher rates of neurologic compromise,[386,387] possibly because of the advanced stage of illness at the time of presentation.

Meningeal coccidioidomycosis may present acutely, although it most often follows a subacute or chronic course, and is invariably fatal if not treated.[369,370] Patients generally complain of headache, low-grade fever, weight loss, and mental status changes; about one half of patients develop disorientation, lethargy, confusion, or memory loss. Nausea, vomiting, focal neurologic deficits, and sei-

zures also may develop; signs of meningeal irritation are usually absent, although this symptom has been reported in as many as one third of cases. Atypical presentations include subarachnoid hemorrhage, spinal cord compression, and severe spinal arachnoiditis.

The presenting symptoms of *Histoplasma* meningitis are nonspecific.[371,375] Symptoms usually include headache and fever. Only about one half of patients have neurologic symptoms; seizures or focal neurologic deficits occur in 10–30% of cases.[376] Mental status abnormalities include reduced level of consciousness, confusion, personality changes, and memory impairment.

The clinical presentation of candidal meningitis is nonspecific.[382-384] The onset of symptoms may be abrupt or insidious. The most common symptoms are fever, headache, and meningismus; some patients have depressed mental status, confusion, cranial neuropathies, and other focal neurologic signs.

4.1.3. Diagnosis

Although the laboratory findings in fungal meningitis have been well studied, there are no pathognomonic features associated with any specific fungal organism. The laboratory differences in non-AIDS and AIDS patients with cryptococcal meningitis are shown in Table 9.[355-364,385] On CSF examination, most patients with fungal meningitis have a pleocytosis with white cell counts ranging from 20×10^6 to 500×10^6 cells/liter; the proportion of neutrophils usually is less than 50%. Severely immunosuppressed patients (e.g., AIDS) may have very low or even normal CSF leukocyte counts during active infection, particularly in cryptococcal meningitis; normal CSF indices were found in 17% of HIV-1-infected patients

---

TABLE 8. Clinical Presentation of Cryptococcal Meningitis in Non-AIDS and AIDS Patients[a]

Symptom or sign	Patients affected	
	Non-AIDS	AIDS
Headache	87%	81%
Fever	60%	88%
Nausea, vomiting, malaise	53%	38%
Mental status changes	52%	19%
Meningeal signs	50%	31%
Visual changes, photophobia	33%	19%
Seizures	15%	8%
No symptoms or signs	10%	12%

[a]Adapted from Patterson and Andriole.[385]

TABLE 9. Laboratory Findings in Non-AIDS and AIDS Patients with Cryptococcal Meningitis[a]

Parameter	Patients with finding	
	Non-AIDS	AIDS
Positive blood culture	—	30–63%
Positive serum cryptococcal antigen	66%	99%
CSF opening pressure >200 mm H_2O	72%	62%
CSF glucose <2.2 mmoles/liter (40 mg/dl)	73%	33%
CSF protein >0.45 g/liter (45 mg/dl)	89%	58%
CSF leukocytes >20 × 10^6/liter	70%	23%
Positive CSF India ink	60%	74%
Positive CSF culture	96%	95%
Positive CSF cryptococcal antigen	86%	91–100%

[a]Adapted from Chuck and Sande[361] and Patterson and Andriole.[385]

with cryptococcal meningitis in a study from South Africa.[386] As many as 65% of AIDS patients with cryptococcal meningitis exhibit fewer than 5×10^6 white blood cells/liter in CSF. Occasionally in coccidioidal meningitis, CSF examination may reveal prominent eosinophilia.[370,388] CSF protein concentrations are generally elevated, with concentrations above 10 g/liter suggesting subarachnoid block. CSF glucose concentrations are often reduced, but may be normal, as is seen in two thirds of AIDS patients with cryptococcal meningitis.

Conclusive proof that a fungal organism is causing the meningitis rests on identification of the fungus in CSF or brain tissue. However, CSF cultures are not always positive in fungal meningitis.[354,389] For example, only 25–50% of patients with coccidioidal meningitis have positive CSF cultures. For these reasons, a minimum of 5 ml CSF should be cultured when a mycosis is suspected; repeated culture of large volumes of CSF also is recommended. The yield of CSF culture in cryptococcal meningitis is excellent for both non-AIDS and AIDS patients (Table 9). In patients with cryptococcal meningitis, CSF India ink examination remains a rapid, effective test that is positive in 50–75% of cases; the yield increases up to 88% in patients with AIDS. Recovery of *H. capsulatum* from the CSF is accomplished in 25–65% of documented cases.[366,371,376] In *Candida* meningitis, CSF pleocytosis is commonly seen, with a mean of 600×10^6 cells/liter; lymphocytes or neutrophils may predominate. Yeast cells are detected in about half of cases on direct microscopy of CSF. *Candida* organisms can be readily grown from CSF in the majority of cases; a single positive culture from a patient with risk factors or symptoms is considered significant when CSF indices are compatible with meningitis and the fungus is isolated in pure culture.[390]

As cultures may be negative or require long periods before positive results are noted in patients with fungal meningitis, adjunctive studies (particularly serologic tests) may be helpful in the diagnosis. The latex agglutination test for cryptococcal polysaccharide antigen is both sensitive and specific for the diagnosis of cryptococcal meningitis when samples are first heated to eliminate rheumatoid factor.[389,391,392] False-positive tests may occur if surface condensation from agar plates contaminates the assay slide or with other infections (e.g., disseminated *Trichosporon beigelii* and paravertebral bacterial infections). False-negative tests are unusual; in most cases, they are due to early infection with a low burden of organisms in the CSF or may represent a prozone phenomenon resulting from antigen excess. The polysaccharide antigen test may be positive early in infection, even when the CSF culture is negative; titers of 1:8 or more by latex agglutination indicate a presumptive diag-

nosis of cryptococcal meningitis. Cryptococcal polysaccharide antigen also can be found in the serum as well as the CSF, usually in severely immunosuppressed patients such as those with AIDS. Serum cryptococcal antigen detection has been used as a screen for possible CNS infection in AIDS patients,[389] although the value of serum antigen in screening patients suspected of having meningeal disease has not been definitively established. Cryptococcal polysaccharide antigen titers are generally higher in serum than in CSF in patients with cryptococcal meningitis. In addition, extremely high CSF antigen titers have been reported in patients with AIDS; early investigators suggested that CSF titers of 1:10,000 or more predicted a poor outcome,[359] although some patients have clearly responded to therapy even with high initial titers.

For *C. immitis* infection, elevated concentrations of complement-fixing antibodies are the hallmark of disseminated disease; serum titers above 1:32–1:64 suggest dissemination.[389] However, the titers may be low when other body sites are not involved. CSF complement-fixing antibodies are present in at least 70% of cases of early coccidioidal meningitis and from virtually all patients as the infection progresses. In coccidioidal meningitis, the complement-fixing antibody titers appear to parallel the course of meningeal disease,[369] although patients who relapse after an initial response to antifungal therapy generally develop CSF pleocytosis or abnormal protein or glucose concentrations before detectable CSF antibody recurs. Patients with immunodeficiencies also may fail to develop complement-fixing antibodies in either serum or CSF.

Antibody detection in CSF also has been used for the diagnosis of *Histoplasma* meningitis.[393] These tests [complement fixation and radioimmunoassay (RIA)] have excellent sensitivity for diagnosis, but are less specific. Cross-reactivity with other fungal pathogens occurs in approximately 50% of cases. *Histoplasma* antigen detection in urine, CSF, and serum may be useful[394] and support the use of empiric antifungal therapy. However, a recent study utilizing an RIA for detection of *H. capsulatum* antigen in CSF from patients with meningitis revealed positive tests in only 4 of 12 patients[395]; cross-reactions may occur in patients with coccidioidal meningitis. In addition to CSF, fungal cultures of blood, bone marrow, sputum, and urine should be obtained where appropriate.[371,376] The value of antigen or antibody tests in the diagnosis of CNS candidiasis has not been established.

4.1.4. Treatment

4.1.4a. Antifungal Therapy   The antimicrobial agents of choice, with alternatives, for the treatment of fungal meningitis are shown in Table 5 (Section 3.1.4a); doses for CNS infections are shown in Table 6 (Section 3.1.4a). Before the availability of amphotericin B, cryptococcal meningitis was nearly always fatal. Although the prognosis improved dramatically with the introduction of amphotericin B, morbidity, mortality, and relapse rates remained high in immunocompromised patients (cure rates ≤52% after a first course of therapy).[356] With the discovery of the *in vitro* synergy between amphotericin B and flucytosine, a large prospective collaborative trial was designed to compare the combination of amphotericin B (0.3 mg/kg per day) plus flucytosine (150 mg/kg per day) for 6 weeks to amphotericin B (0.4 mg/kg per day) alone for 10 weeks in the treatment of acute cryptococcal meningitis.[396] Combination therapy produced fewer failures, fewer relapses, and more rapid sterilization of CSF and was associated with decreased nephrotoxicity in comparison to therapy with amphotericin B alone. Cure or improvement occurred in 67% of patients on combination therapy versus 41% of patients receiving low-dose amphotericin B. There were no significant differences in mortality rates between the two groups. This study has been criticized, however, for the dose of amphotericin B in the single-agent arm of the trial. A subsequent study demonstrated that a 4-week regimen of combination amphotericin B plus flucytosine therapy could be used in a subset of patients who had at presentation no neurologic complications, no underlying diseases, no immunosuppressive therapy, a pretreatment CSF white cell count greater than 20×10^6/liter, and serum cryptococcal antigen titer less than 1:32, and at 4 weeks a negative CSF India ink and a CSF cryptococcal antigen titer less than 1:8.[397] The patients treated with combination therapy had a high rate of toxicity due to flucytosine, with one or more toxic reactions noted in 38% of patients, mainly hematologic, indicating the need to monitor flucytosine serum concentrations during therapy (maintain serum concentrations at 50–100 μg/ml).[398] In AIDS patients, no differences in survival were noted in a retrospective analysis if patients were treated with amphotericin B alone or amphotericin B combined with flucytosine,[361] although in patients receiving combination therapy, flucytosine had to be discontinued in over half the patients due to toxicity, primarily cytopenias.

In AIDS patients with cryptococcal meningitis, there had been a poor response to standard antifungal therapy[358,359] consisting of amphotericin B either with or without flucytosine. Fluconazole, a triazole antifungal agent, also has been studied for the management of this condition. Fluconazole has excellent oral absorption, a long half-life (about 30 hr), and a very good penetration into the CSF (CSF concentrations about 70–80% of peak serum concentrations). Fluconazole was initially shown to be effective in small, uncontrolled studies of cryptococcal meningitis in AIDS patients.[399,400] Two subsequent trials were done to compare fluconazole to standard antifungal therapy for cryptococcal meningitis in patients with AIDS. In one study,[401] fluconazole (400 mg/day) was compared to the combination of amphotericin B (0.7 mg/kg per day) plus flucytosine (150 mg/kg per day). The failure rate was 57% (8 of 14 patients) in the group receiving fluconazole and 0% (0 of 6 patients) in the patients receiving standard amphotericin B plus flucytosine combination therapy. Combination therapy had superior mycologic and clinical efficacy in this study. A subsequent study done by the Mycoses Study Group examined fluconazole (initial dose of 400 mg followed by 200 mg/day) versus amphotericin B (at least 0.3 mg/kg per day) for acute cryptococcal meningitis in AIDS patients.[402] There were no significant differences in the number of patients who were cured or improved in either treatment group. In addition, the overall case–fatality rates were not significantly different, although there was a trend to early mortality (within the first 2 weeks) in the patients treated with fluconazole. Following a *post hoc* analysis of the data, however, it appeared that fluconazole was inferior to amphotericin B only in patients with certain negative prognostic signs such as a positive blood culture for *C. neoformans*, a CSF cryptococcal antigen titer greater than 1:128, a positive CSF India ink smear, or altered mentation.[403] This study has been criticized, however, because the drug dosages used in both arms of the study may have been too low for the optimal treatment of cryptococcal meningitis. Despite these concerns, the results supported the initial use of amphotericin B (with or without flucytosine) in AIDS patients with cryptococcal meningitis for a period of approximately 2 weeks, followed by fluconazole (400 mg/day) to complete a 10-week course.

Based on the need for more data to better define the optimal antifungal therapy for AIDS patients with cryptococcal meningitis, several additional trials were undertaken. In a double-blind multicenter trial, patients with a first episode of AIDS-associated cryptococcal meningitis were randomly assigned to treatment with amphotericin B (0.7 mg/kg per day) without or with flucytosine (100 mg/kg per day) for 2 weeks, followed by 8 weeks of treatment with either itraconazole (400 mg/day) or fluconazole (400 mg/day).[404] At 2 weeks, there were no statistically significant differences in clinical outcome or CSF sterilization in the patients who received amphotericin B alone versus those who received combination therapy. At 10 weeks, 72% of the fluconazole recipients versus 60% of the itraconazole recipients had negative CSF cultures, although the proportion of patients with clinical responses was similar in both groups. In a multivariate analysis, the addition of flucytosine during the initial 2 weeks and treatment with fluconazole for the next 8 weeks were independently associated with CSF sterilization. The authors recommended induction therapy with higher-dose amphotericin B (0.7 mg/kg per day) plus flucytosine followed by consolidation therapy with fluconazole as the treatment of choice for AIDS patients with acute cryptococcal meningitis.

In recent years, commercially available lipid formulations of amphotericin B have been developed for use in invasive fungal infections[405] and have been studied in patients with cryptococcal meningitis.[406] One prospective study compared liposomal amphotericin B (4 mg/kg per day) with amphotericin B (0.7 mg/kg per day) in the treatment of cryptococcal meningitis.[407] The median time to CSF sterilization was shorter in patients treated with liposomal amphotericin B, and at 14 days a significantly greater number of patients had CSF culture conversion. However, the time to clinical response as well as the clinical failure rate during initial therapy did not differ between groups. Further comparative trials are needed before liposomal amphotericin B can be recommended over standard therapy for cryptococcal meningitis.

In non-AIDS patients with cryptococcal meningitis, the optimal use of fluconazole is unclear. Several anecdotal reports and small case series have reported the benefit of fluconazole monotherapy for non-AIDS patients with cryptococcal meningitis,[408,409] although no comparative trials with amphotericin B have been performed. In a recently published abstract of a retrospective review of 158 HIV-1-negative patients with CNS cryptococcosis,[410] patients with CNS cryptococcosis were more likely to receive an induction regimen containing amphotericin B (89%) and subsequent therapy with fluconazole (65%), suggesting that fluconazole has an important role for consolidation therapy in this patient population. However, pending further data, non-AIDS patients with cryptococcal meningitis should continue to receive standard therapy with amphotericin B plus flucytosine for 4–6 weeks.

One important issue in AIDS patients with cryptococcal meningitis is the high rate of relapse if antifungal therapy is discontinued after a course of acute therapy.[358,359] In AIDS patients, long-term suppressive therapy with ketoconazole or amphotericin B was associated with improved median survival (238 vs. 141 days).[361] Several recent studies have demonstrated that fluconazole is the antifungal agent of choice for prevention of relapse in AIDS patients with cryptococcal meningitis.[411,412] A placebo-controlled trial[412] found that the rate of relapse was markedly diminished in patients receiving fluconazole suppressive therapy (3% vs. 37% in patients receiving placebo). Similarly, a trial completed by the Mycoses Study Group revealed that fluconazole (200 mg/day) was superior to amphotericin B (1 mg/kg per week) in prevention of relapse in AIDS patients with crytococcal meningitis (2% vs. 18%).[413] Fluconazole (200 mg/day) also has been compared to itraconazole (200 mg/day) as maintenance therapy in AIDS-associated cryptococcal meningitis.[414] However, this study had to be stopped prematurely because the rate of culture-positive relapse was greater among patients receiving itraconazole (23%) than fluconazole (4%), indicating that fluconazole remains the treatment of choice for maintenance therapy of AIDS-associated cryptococcal disease. Antifungal therapy needs to be continued for life in patients with AIDS. It has been suggested that the prostate gland represents a sequestered reservoir from which systemic relapse can occur.[415,416]

The treatment of choice for coccidioidal meningitis in immunosuppressed patients was amphotericin B administered both intravenously and intrathecally.[369,417] The intrathecal administration of amphotericin B may be via the lumbar, cisternal, or ventricular routes, the latter usually through an Ommaya reservoir. The usual dose of intrathecal amphotericin B is 0.5 mg, although doses of 1.0–1.5 mg can be used if combined with hydrocortisone; the duration is three times weekly for 3 months. Mortality rates of about 50% have been reported,[369] but one study found an overall reported survival of 91%, over a followup of 75 months, when larger doses of intrathecal amphotericin B (1.0–1.5 mg) were used.[418] Therapy is discontinued after the CSF has been normal for at least 1 year on a regimen of intrathecal administration once every 6 weeks. However, intrathecal amphotericin B is poorly tolerated and often leads to arachnoiditis. A recent collaborative study examined the efficacy of fluconazole (400 mg once daily) in the treatment of coccidioidal meningitis.[419] Of 47 evaluable patients, 79% responded to treatment, with most of the improvement occurring within 4–8 months of initiation of therapy; the authors recommended lifelong treatment with fluconazole because of the high relapse rate when azole therapy is discontinued. Based on these results, fluconazole is recommended as first-line therapy for coccidioidal meningitis.

For CNS disease due to *H. capsulatum*, therapy with amphotericin B (at least 30–35 mg/kg total dose) should be used.[371,376] In AIDS patients, the initial response rate is more than 80%. Relapse is common, however, necessitating maintenance amphotericin B or ketoconazole therapy; amphotericin B has been more effective than ketoconazole for maintenance therapy.[375] The triazole antifungal agents itraconazole and fluconazole have been studied as maintenance therapy to prevent relapse in AIDS patients with disseminated histoplasmosis[376]; itraconazole appears superior to fluconazole in this regard. CNS involvement with progressive disseminated histoplasmosis is a grim prognostic sign and does not appear to respond to conventional antifungal therapy with amphotericin B; the mortality rate was 61.5% in AIDS patients with CNS histoplasmosis in a recent review.[375]

Amphotericin B, alone or in combination with flucytosine, also is the treatment of choice for *Candida* meningitis; there are no studies comparing the efficacy of single or combination therapy. Cure rates with amphotericin B ranged from 67 to 89% in adults and 71 to 100% in neonates with *Candida* meningitis. Combination amphotericin B plus flucytosine has been recommended by some investigators,[354,420] on the basis of more rapid sterilization of CSF cultures and possible reduction of long-term neurologic sequelae in newborns. There are no studies, however, comparing the efficacy of amphotericin B alone to the combination of amphotericin B plus flucytosine for the treatment of *Candida* meningitis. Isolated case reports have suggested that fluconazole may be an acceptable alternative agent for the therapy of *Candida* meningitis.[421,422]

4.1.4b. Adjunctive Therapy.   Patients with cryptococcal meningitis may have several complications despite appropriate antifungal therapy. In a recent study of 236 AIDS patients with cryptococcal meningitis, only 55% were alive with negative CSF cultures at 10 weeks.[423] Increased intracranial pressure and/or hydrocephalus has been noted in some AIDS patients with cryptococcal meningitis, with an association between elevated intracranial pressure and early mortality.[404] Therapeutic modalities for these complications include shunting of CSF, frequent high-volume lumbar punctures, and acetazolamide. However, the precise role of these adjunctive measures in the treatment of cryptococcal meningitis remains to be established.

Several recent reports have examined the role of reduction of intracranial pressure in AIDS patients with cryptococcal meningitis. In one report of five AIDS patients with cryptococcal meningitis who developed severe intracranial hypertension with progressive neurologic deterioration, CSF shunting led to a dramatic improvement.[424] However, this report was anecdotal and did not indicate whether less invasive means would have led to an appropriate clinical response. In another report of 10 HIV-1-infected patients with cryptococcal meningitis who presented with symptoms of elevated intracranial pressure and had an opening pressure greater than 200

mm H_2O during lumbar puncture, all returned to their premorbid level of consciousness following placement of a lumbar drain and normalization of intracranial pressure[425]; eight patients eventually required placement of lumbar peritoneal shunts for persistently elevated intracranial pressure. A third report also demonstrated the safety and efficacy of shunting procedures for hydrocephalus in ten non-HIV-1-infected patients with cryptococcal meningitis[426]; nine of the patients had noticeable improvement in dementia and gait following placement of a ventriculoatrial or ventriculoperitoneal shunt. These data indicate that shunting procedures can ameliorate the sequelae of elevated intracranial pressure in patients with cryptococcal meningitis, although further studies are needed to determine the optimal management strategy for reduction of elevated intracranial pressure in this patient population.

4.2. Brain Abscess

4.2.1. Epidemiology and Etiology

The incidence of fungal brain abscesses has increased in recent years due to the prevalent administration of immunosuppressive therapy, broad-spectrum antibiotic therapy, and corticosteroids.[354,427–429] Unfortunately, the diagnosis of fungal brain abscess often is unexpected and many cases are not discovered until autopsy. Many of the etiologic agents of fungal meningitis also may cause brain abscess (e.g., *Cryptococcus neoformans*, *Coccidioides immitis*, *Histoplasma capsulatum*); the epidemiologic and etiologic characteristics of these organisms are described in Section 4.1.1. Although *Candida* species may produce meningitis, focal CNS lesions are more common with this organism.[377,383,430,431] In autopsy studies, *Candida* has emerged as the most prevalent etiologic agent of fungal brain abscess; neuropathologic lesions include microabscesses, macroabscesses, noncaseating granulomas, and diffuse glial nodules. However, several other fungal organisms should be considered in the differential diagnosis of fungal brain abscess in the immunosuppressed patient; these organisms are described in detail below.

Cases of intracranial infection due to *Aspergillus* species have been reported worldwide, with most cases occurring in adults. The lungs are the usual site of primary infection, and intracranial seeding occurs during dissemination of the organism or by direct extension from an area anatomically adjacent to the brain (e.g., the paranasal sinuses).[427,432] Most cases of invasive aspergillosis occur in neutropenic patients who have an underlying hematologic malignancy.[433,434] Other risk groups include patients with hepatic disease, Cushing's syndrome, diabetes mellitus, or chronic granulomatous disease, intravenous drug abusers, postcraniotomy patients, organ transplant recipients, HIV infection, and patients receiving chronic corticosteroid therapy[3,354,435–440]; some patients have no discernible risk factors.[441] Concurrent intracranial involvement is seen in 13–16% of patients with pulmonary aspergillosis; of patients with disseminated disease, the brain is involved in 40–70% of cases. *Aspergillus* species are ubiquitous molds found in soil, water, and decaying vegetation. The organism has a characteristic microscopic appearance in which the hyphae develop terminal buds that crown the organism with multiple small conidiae.

Mucormycosis (zygomycosis, phycomycosis) is one of the most acute, fulminant fungal infections known. Many conditions that predispose to mucormycosis have been described,[427,442–447] including diabetes mellitus (70% of cases), usually in association with acidosis, acidemia from profound systemic illnesses (e.g., sepsis, severe dehydration, severe diarrhea, chronic renal failure), hematologic neoplasms, renal transplant recipients, injection drug users, and use of deferoxamine; fewer than 5% of cases are found in normal hosts. CNS disease may occur from direct extension of the rhinocerebral form of mucormycosis or by hematogenous dissemination from other sites of primary infection. The Mucoraceae are ubiquitous fruit and bread molds that thrive in soil, manure, and decaying material.[447,448] They include the genera *Rhizopus, Absidia,* and *Mucor;* the genus *Rhizopus* is responsible for most cases of cerebral mucormycosis.

Pseudallescheria boydii is a common mold readily isolated from soil that may cause CNS disease in both normal and immunocompromised hosts (e.g., neutropenia, cellular immunodeficiency).[427,449–452] This organism is being increasingly referred to as *Scedosporium apiospermum,* the asexual form of *P. boydii. Pseudallescheria boydii* may enter the CNS by direct trauma, by hematogenous dissemination from a pulmonary route, via an intravenous catheter, or by direct extension from infected sinuses. Brain abscess is the usual CNS manifestation, although meningitis and ventriculitis also have been reported. There is an association between near-drowning and subsequent illness, due to the pathogen's presence in contaminated water and manure.

4.2.2. Clinical Presentation

The usual presenting signs and symptoms of fungal brain abscess are similar to those of patients with a bacterial brain abscess or tumor, and they relate to the fulminant nature of the particular infectious agent and the intracranial location of the abscess. However, certain fungal pathogens may present with specific characteristics following CNS infection. Patients with *Aspergillus* brain abscess most commonly manifest signs of a stroke referable to the involved area of brain.[427] Headache, encephalopathy, and seizures also may occur. Fever is not a consistent feature, and signs of meningeal irritation are rare. Patients with *Aspergillus* brain abscess commonly have evidence of aspergillosis involving other organs.[434,438,441]

Patients with rhinocerebral mucormycosis present initially with complaints referable to the eyes or sinuses including headache (often unilateral), facial pain, diplopia, lacrimation, and nasal stuffiness or discharge[427,442,443,453]; fever and lethargy also may occur. Initial signs include development of a nasal ulcer, facial swelling, nasal discharge, proptosis, and external ophthalmoplegia as the infection begins to spread posteriorly to involve the orbit; orbital involvement occurs in two thirds of patients. Cranial nerve abnormalities are common (including cranial nerves II–VII, IX, and X) and blindness may occur as a result of vascular compromise. Thrombosis is a striking feature of this disease because the organism has a proclivity for blood vessel invasion. Focal neurologic deficits such as hemiparesis, seizures, or monocular blindness suggest far-advanced disease. With further progression, invasion and occlusion of the cavernous sinus and internal carotid artery can occur.[454] This clinical syndrome of enlarging areas of black mucosal or even facial necrosis is often quite dramatic.

The development of CNS infection due to *P. boydii* tends to become manifest 15–30 days after an episode of near-drowning.[450,451] Brain abscesses can be located in the cerebrum, cerebellum, or the brain stem; clinical presentations include seizures, altered consciousness, headache, meningeal irritation, focal neurologic deficits, abnormal behavior, and aphasia. Metastatic skin lesions may herald the fungemia as the organism spreads to the CNS. The clinical manifestations of CNS disease due to *Cryptococcus, Histoplasma, Coccidioides,* and *Candida* depend on the intracranial location of the abscess.

4.2.3. Diagnosis

Diagnosis of fungal brain abscesses often is difficult. CSF results are usually abnormal, but the findings are nonspecific. CT and MRI are quite sensitive in defining the lesions but seldom show changes specific for fungal brain abscess. Some exceptions do exist. The finding of a cerebral infarct in a patient with risk factors for invasive aspergillosis should suggest that diagnosis.[427] The areas of infarction typically develop into either single or mul-

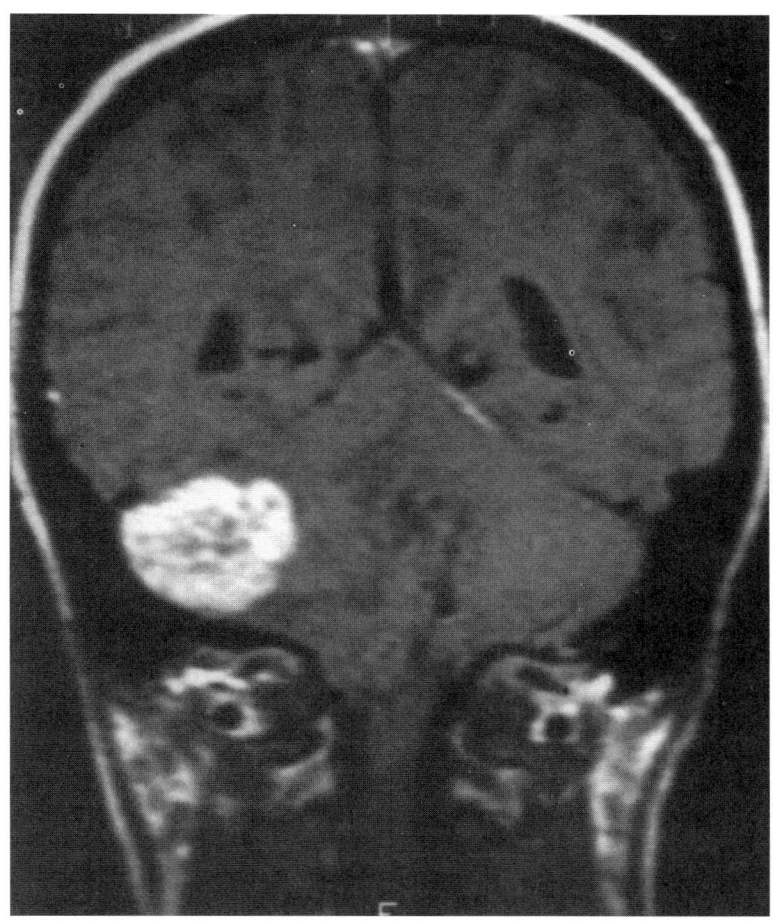

FIGURE 4. Tl-weighted MRI of the head in a patient with chronic granulomatous disease revealing an enhancing mass in the right cerebellum. Biopsy of the lesion revealed invasive aspergillosis.

tiple abscesses involving the cerebrum (usually frontal or temporal lobes) or cerebellum (Fig. 4). In patients with rhinocerebral mucormycosis, CT and MRI may show characteristic changes, including sinus opacification, erosion of bone, and obliteration of deep fascial planes.[455,456] Frontal lobe involvement may show little or no ring enhancement; the lack of contrast enhancement is a poor prognostic sign, as it indicates failure of host defense mechanisms to isolate or encapsulate the offending organisms. In intravenous drug abusers with cerebral mucormycosis, the basal ganglia are the most frequent site of CNS disease.[445] Cavernous sinus involvement may be seen on MRI.

Definitive diagnosis of fungal brain abscess requires biopsy of the lesion and examination by Gram's stain, potassium hydroxide wet mount, and appropriate fungal stains (mucicarmine, methenamine–silver) and cultures.[427] The stains highlight the characteristic morphology of fungi and yeasts. For example, in tissue sections, cells of

C. neoformans can be specifically identified with the mucicarmine stain. *Aspergillus* species manifest as septate hyphae with acute-angle, dichotomous branching in tissue sections. Examination of scrapings or biopsies of necrotic nasal turbinates may be helpful in the diagnosis of rhinocerebral mucormycosis, revealing the typical nonseptate hyphae. Biopsy of cerebral lesions in mucormycosis usually demonstrate irregular hyphae, right-angle branching, and lack of septae. *Pseudallescheria boydii* appears as septate hyphae in clinical specimens, although the hyphae are narrower and as a rule do not show the characteristic dichotomous branching encountered in invasive aspergillosis. *Pseudallescheria boydii* can also be identified by staining of biopsy tissue with fluorescent antibodies, a sensitive method that does not cross-react with hyphae of other common pathogenic fungi.

Diagnosis of fungal brain abscess may be difficult even with biopsy; occasionally, the organism may be seen on direct examination but may not grow in culture. In these situations, serologic testing may be helpful. Serologic tests for cryptococcal polysaccharide antigen are very useful in diagnosis of CNS disease (see Section 4.1.3).[389,391,392] Serologic tests for other fungal infections, such as coccidioidomycosis[369,389] and histoplasmosis,[394,395] also may be helpful, although their routine use in diagnosis of CNS infections remains to be established. Serologic assays for aspergillosis, mucormycosis, and candidiasis are under development and at present cannot be used to reliably establish or exclude the presence of CNS disease due to these organisms.[427]

4.2.4. Treatment

The optimal therapy of fungal brain abscesses usually requires a combined medical and surgical approach. Surgery includes either excision or drainage of the abscess. The mainstay of medical therapy for *Aspergillus* brain abscess is amphotericin B (0.8–1.25 mg/kg per day), with doses up to 1.5 mg/kg per day depending on the clinical response.[457] Few instances of survival have been recorded in CNS aspergillosis despite administration of amphotericin B.[435,436] Most patients have required a total of more than 3 g for eradication of CNS disease. Concomitant therapy with flucytosine or rifampin has been tried, although no controlled trials have been done to examine the efficacy of this approach. A recent literature review has suggested that success rates in CNS aspergillosis may be greater with the addition of flucytosine to amphotericin B. Itraconazole has *in vitro* activity against *Aspergillus*,[458] and high-dose therapy (800 mg daily for 5 months followed by 400 mg daily for 4.5 months) resulted in complete resolution of cerebral abscesses caused by *Aspergillus fumigatus* in an elderly asthmatic patient who was treated with corticosteroids.[459] Excisional surgery or drainage was a key factor in the successful management of several cases of CNS aspergillosis.[460–462]

Mucormycosis also should be treated with amphotericin B, along with correction of underlying metabolic derangements and aggressive surgical debridement.[427,443,444,463] The role of surgery in the treatment of cerebral mucormycosis cannot be overemphasized. Because of their propensity to invade blood vessels, the Mucoraceae cause extensive tissue infarction, thereby impairing delivery of antifungal agents to the site of infection. This damage often leaves surgery as the only modality that may effectively eliminate the invading microorganisms. Hyperbaric oxygen therapy has been reported to be a useful adjunct in cerebral mucormycosis,[464,465] although no prospective, controlled trials have been performed to adequately assess its efficacy.

For *P. boydii* brain abscess, surgical drainage is the cornerstone of effective therapy.[450] The organism demonstrates *in vitro* resistance to amphotericin B; the antifungal treatment of choice is miconazole [see Tables 5 and 6 (Section 3.1.4a)].[450,451] The drug often must be given by the intravenous or intrathecal route, or by both, and relapses are common.[466] In addition, dosage increments often must be made during therapy because miconazole serum concentrations fall while patients are receiving stable doses.[427] *Pseudallescheria boydii* strains with *in vitro* resistance to miconazole also have been reported. A partial response to fluconazole (600–800 mg/day) has been shown in a single case report.[467] Itraconazole has been used successfully to treat pulmonary infection,[468] but its efficacy in CNS disease is unknown. Itraconazole oral suspension has better absorption than the capsules and is probably the de facto drug of choice. Voriconazole, in conjunction with surgical drainage, has been successfully used in one patient with *P. boydii* brain abscess.[469]

Medical therapy for fungal brain abscesses due to *C. neoformans*, *C. immitis*, *H. capsulatum*, and *Candida* species is detailed in Section 4.1.4. Drugs of choice, with alternative agents, are shown in Table 5; recommended doses in CNS infections are shown in Table 6.

5. Protozoal Infections

5.1. Toxoplasmosis

5.1.1. Epidemiology and Etiology

Toxoplasma gondii is an obligate intracellular protozoan of humans and animals. The incidence of human infection depends on dietary habits (especially the amount of meat consumed and whether eaten raw, rare, or well done), the number of stray cats living in close proximity to humans, climatic conditions (moderate temperatures and high humidity favor oocyst survival in soil), and the overall level of sanitation and hygiene.[470] In most cases, the tissue cyst form of the organism persists, but the person has no clinical manifestations. *Toxoplasma gondii* can infect the CNS in a variety of syndromes, but is usually associated with development of intracerebral mass lesions or encephalitis in immunocompromised hosts. In the past, most cases of CNS toxoplasmosis occurred from reactivation of disease in patients with reticuloendothelial malignancies (e.g., lymphoma, leukemia),[470–473] either due to the malignancy itself or to associated immunosuppressive or cytotoxic drug therapy. CNS disease also has occurred in patients receiving immunosuppressive therapy after organ transplantation and for treatment of collagen vascular disorders[472,474–476]; one case has been documented in a patient with untreated Hodgkin's disease who was receiving no cytotoxic drug therapy.[477] Disease in organ transplant recipients not only occurs secondary to reactivation but also may occur after transfer of infected cysts in the allograft, most commonly in heart transplant recipients.[474,478,479] Disseminated acute acquired disease also has been reported in renal and liver transplant recipients. The number of cases of CNS toxoplasmosis has increased dramatically since 1981, specifically in association with the AIDS epidemic.[480–484] The use of trimethoprim–sulfamethoxazole prophylaxis[485] and highly active antiretroviral therapy[486] has led to a decrease in the incidence of toxoplasmic encephalitis in this risk group.

Toxoplasma gondii is a sporozoan of the order Coccidia that exists in several distinct forms: (1) the tachyzoite or endozoite, which is the cell-invasive, rapidly proliferating form; (2) the tissue cyst, which contains intracystic bradyzoites; and (3) the oocyst, which produces infectious sporozoites.[470,487] The only definitive hosts for *T. gondii* are cats. Cats become infected by eating animals (usually rodents) that contain cysts in their tissues or by ingesting oocysts passed in the feces of other cats. When the cat ingests either oocysts or tissue cysts, the sexual phase begins in the feline gut, in which the protozoa infect epithelial cells in the small intestine and develop into merozoites, which infect other epithelial cells. Some merozoites develop into gametocytes that fuse and form diploid oocysts; millions of oocysts are excreted in the feces over a period of 1–3 weeks. The time to first appearance of oocysts in cat feces depends on the form of *Toxoplasma* that infected the cat: 3–5 days after ingestion of tissue cysts, 7–10 days after ingestion of tachyzoites, and 20–24 days after ingestion of oocysts. These oocysts may remain infectious in the environment for more than a year, but can be rendered noninfectious by boiling water or by dry heat at higher than 66°C. The asexual phase of the parasite's life cycle occurs in incidental hosts as well as in felines and is largely extraintestinal. After oocyst ingestion, sporozoites are released into the small intestine, penetrate the gut wall, replicate, and are spread hematogenously throughout the body. Once inside macrophages or other cells, the sporozoite transforms into a tachyzoite that multiplies until many organisms accumulate, at which time the host cell ruptures. The released tachyzoites invade adjacent cells, and the cycle is repeated until host immunity develops, limiting further multiplication. Control of toxoplasmic infection in humans depends primarily on intact cell-mediated immunity, which successfully limits replication of tachyzoites, although it fails to kill all the organisms. Some tachyzoites transform into bradyzoites and become dormant tissue cysts, which retain the ability to resume multiplication if cell-mediated immunity wanes.

5.1.2. Clinical Presentation

The clinical manifestations of CNS toxoplasmosis in the immunocompromised patient may be variable, ranging from an insidious process evolving over several weeks to acute onset with a confusional state; the initial symptoms and signs may be focal, nonfocal, or both.[470,487] Focal abnormalities depend on the intracranial location of the lesions and include homonymous hemianopsia, diplopia, cranial nerve palsies, hemiparesis, hemiplegia, hemisensory loss, aphasia, focal seizures, personality changes, movement disorders, and cerebellar dysfunction. *Toxoplasma gondii* has a predilection to localize in the basal ganglia and brain stem, producing extrapyramidal symptoms resembling Parkinson's disease. Generally, patients who present with nonfocal abnormalities develop signs of focal neurologic disease as the infection progresses, although a few patients develop a diffuse, rapidly fatal encephalopathic process. Nonfocal evidence of neurologic dysfunction may predominate; findings include generalized weakness, headache, confusion, lethargy, alteration of mental status, personality changes, and coma. Due to the predominant CNS parenchymal involvement of toxoplasmosis, evidence of meningeal inflammation is rarely observed.

The clinical presentation of CNS toxoplasmosis may vary depending on the pathogenesis of infection. Transplant recipients often have nonfocal disease in which the disease is diffuse and disseminated.[472,474,475,478,479] Early signs and symptoms include weakness, lethargy, confusion, decreased responsiveness, generalized seizures, and headache; localizing neurologic signs tend to occur late in the course of infection in transplant recipients or not at all. In patients with underlying malignancies (e.g., Hodgkin's disease), the presentation of toxoplasmic encephalitis is evenly distributed between focal and nonfocal manifestations of encephalitis.[470–473] Patients with AIDS often present with nonspecific symptoms such as neuropsychiatric complaints, headache, disorientation, confusion, and lethargy; associated fever and weight loss are also common.[480–484] Typically, the course is subacute, progressing over 2–8 weeks. Patients then develop evidence of focal CNS mass lesions with ataxia, aphasia, hemiparesis, visual field loss, and vomiting, or a more generalized encephalitis with increasing confusion, dementia, and stupor; seizures are common and may be the presenting clinical manifestation of CNS toxoplasmosis.

5.1.3. Diagnosis

The role of serologic testing for the diagnosis of toxoplasmic encephalitis depends on the type of patient.[470,487] In the immunocompetent patient, the diagnosis of acute acquired toxoplasmosis is usually established by a fourfold rise in IgG antibody titer. Although the presence of a single high antibody titer also suggests acute infection, such titers can persist for years after acute infection.[488] Therefore, the presence of IgM antibody directed against *T. gondii* may be useful for the diagnosis of acute acquired toxoplasmosis; generally, a positive result indicates infection within the preceding 3–4 months, although exceptions do occur.

In the immunosuppressed host, the value of serologic testing for the diagnosis of toxoplasmosis depends on the pathogenesis of infection. For example, in heart transplant recipients, toxoplasmic encephalitis most often follows acute acquisition of the organism from the transplanted allograft[474,478,479,489]; seronegative recipients prior to transplant who receive an organ from a seropositive donor seroconvert and generally develop severe symptomatic disease. In contrast, seropositive recipients prior to transplantation often develop significant IgM and IgG antibody titers to *T. gondii* after transplantation but remain asymptomatic. Significant rises in antibody titers also have been demonstrated in a variety of immunocompromised patients without specific evidence of active infection.[490] However, in many immunocompromised patients (e.g., AIDS, bone marrow transplant recipients), toxoplasmic encephalitis occurs as a result of a recrudescence of a latent infection.[480–484] In this situation, the presence of antitoxoplasmic antibody can almost be demonstrated uniformly prior to the development of the encephalitis; rises in antibody titer are observed in only a minority of cases and titers may even decline as the encephalitis progresses. In AIDS patients, more than 97% of patients with toxoplasmic encephalitis have serum antibody titers against *T. gondii* ranging from 1:8 to higher than 1:1024[470]; the predictive value of a positive serology in patients with characteristic abnormalities on radiographic studies may be as high as 80% in the United States.[482,491] In contrast, in a retrospective review of 115 patients with AIDS and CNS toxoplasmosis at San Francisco General Hospital between 1981 and 1990,[484] 4 of 18 patients with pathologically confirmed disease had undetectable anti-*Toxoplasma* IgG antibody by an indirect immunofluorescence assay. Despite these conflicting data, many physicians in the United States initiate a therapeutic trial of antitoxoplasmic chemotherapy in an AIDS patient who is seropositive for *T. gondii* and has characteristic neuroradiographic abnormalities.[492] This approach generally is valid in AIDS patients with presumed CNS toxoplasmosis. However, in populations in whom other CNS processes are more prevalent, the predictive value of a positive serology may be much lower.[9,493] In addition, in populations wherein the overall seroprevalence for *T. gondii* is very high, there is a lower predictive value of a positive serology in distinguishing toxoplasmic encephalitis from other infectious and noninfectious etiologies that cause similar neuroradiologic abnormalities. Determination of antitoxoplasmic antibodies in the CSF, reflecting intrathecal antibody synthesis, may be a useful adjunctive test.[494] The formula utilized is as follows:

$$\frac{\text{CSF dye test titer (reciprocal)} \times \text{total serum IgG}}{\text{Serum dye test titer (reciprocal)} \times \text{total CSF IgG}}$$

Using this formula, an index greater than 1 is indicative of intrathecal production of antitoxoplasmic antibody in the immunocompromised host. The utility of this test, however, decreases in patients with high concentrations of serum antibody.

CT and MRI are both extremely useful in the diagnosis of CNS toxoplasmosis.[470,482,495,496] The characteristic CT appearance (seen in 90% of patients) is that of rounded isodense or hypodense lesions with ring enhancement after the administration of contrast material; however, homogeneous enhancement or no enhancement also can be seen. There are multiple lesions in approximately 75% of cases, often involving the corticomedullary junction and the basal ganglia, although any part of the CNS may be involved. Marked edema and mass effect also are frequently observed. A double-dose delayed-contrast study may be a more sensitive method for delineating the true extent of disease. Unfortunately, CT usually underestimates the number of lesions documented pathologically at autopsy.[481] MRI has a greater sensitivity than CT (Fig. 5) and has detected lesions in patients with active toxoplasmic encephalitis whose CT scans were normal. Therefore, MRI should be performed in AIDS patients with neurologic symptoms and antibody to *T. gondii* in which CT shows no abnormality. CT and MRI also may be useful in following response to therapy, in which most patients have radiographic evidence of improvement within 10–14 days of initiation of antitoxoplasmic therapy; resolution of most abnormalities may take as long as 6 months.

Definitive diagnosis of toxoplasmic encephalitis re-

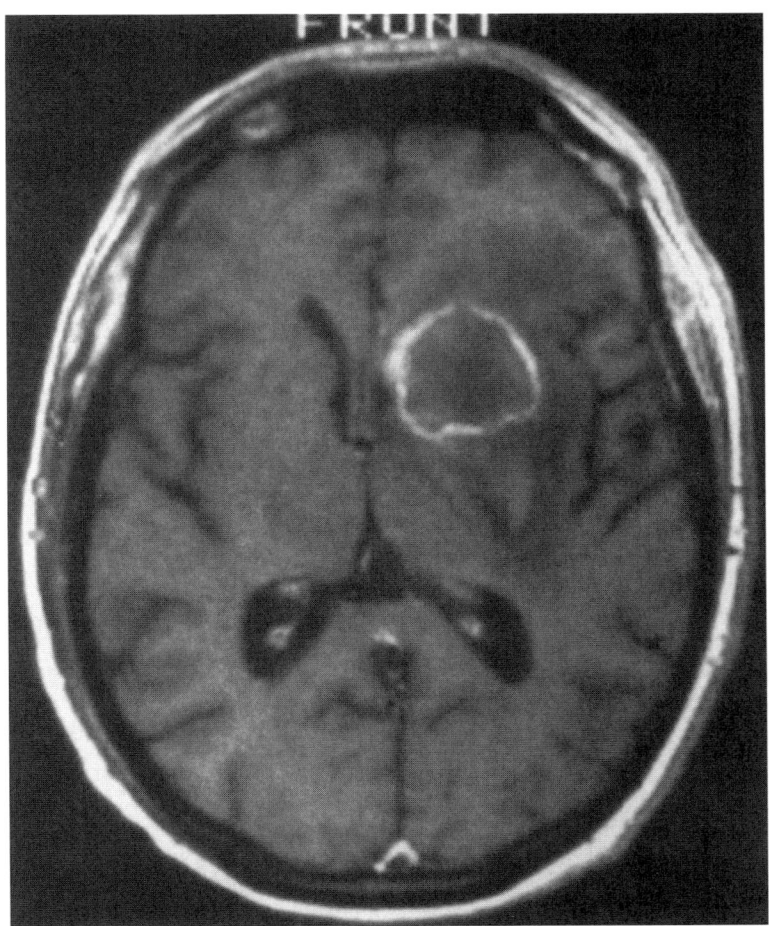

FIGURE 5. T1-weighted MRI of the head in an AIDS patient with toxoplasmosis, revealing a large mass in the left frontal lobe with enhancement after the administration of gadolinium. Note the surrounding edema and effacement of the frontal horn of the lateral ventricle.

quires the demonstration of the organism in clinical specimens. Some patients may have a concomitant toxoplasmic pneumonitis, in which the pathogen is detected in bronchoalveolar lavage fluid, as was recently shown in Giemsa-stained specimens from three organ transplant recipients with disseminated toxoplasmosis.[497] Pseudocysts and tachyzoites, which are easily identifiable by histopathologic stains, may not be found in the center of the necrotic lesion and are best identified at the periphery of the lesion or within normal brain tissue. A sensitive test for rapid diagnosis is the immunofluorescence technique, using monoclonal antitoxoplasmic antibodies on brain tissue touch preparations.[498]

In HIV-1-infected patients with the presumptive diagnosis of toxoplasmic encephalitis, the approach is somewhat different. When contrast CT or MRI reveals the presence of multiple ring-enhancing lesions and the patient has positive anti-*Toxoplasma* IgG serologic tests,

empiric therapy for toxoplasmic encephalitis should be initiated; clinical and radiographic improvement should be observed within 10–14 days in patients with toxoplasmic encephalitis.[499] For patients with positive anti-*Toxoplasma* IgG serologic tests and a single lesion identified by MRI, consideration should be given to thallium-201 single-photon emission computed tomography (^{201}Tl-SPECT) scanning[500,501] or to positron-emission tomography scanning employing ^{18}F-fluorodeoxyglucose ([18]FDG-PET),[502] which are highly specific for the diagnosis of primary CNS lymphoma and would warrant stereotactic brain biopsy. In patients with mass lesions and negative anti-*Toxoplasma* IgG serologic tests, the diagnosis of toxoplasmic encephalitis is possible but unlikely; brain biopsy is optimal in this situation, although some experts have recommended an empiric trial for toxoplasmic encephalitis, with brain biopsy performed in patients who fail to respond.[503] Patients with single lesions on MRI and negative serologic tests should undergo a stereotactic brain biopsy.

5.1.4. Treatment

The mainstay of therapy for toxoplasmic encephalitis is the combination of pyrimethamine and sulfadiazine [see Table 5 (Section 3.1.4a)],[470,487] which are inhibitors of dihydrofolate reductase and dihydrofolate synthetase, respectively. These drugs act synergistically against *T. gondii* by blocking folic acid metabolism. Current recommendations for dosing [see Table 6 (Section 3.1.4a)] are to administer pyrimethamine as a 200-mg loading dose, followed by 50–75 mg/day for the first 3–6 weeks of therapy in AIDS patients and for a minimum of 4–6 weeks at a dose of 25–50 mg/day for other immunodeficient patients. Sulfadiazine is given at a dose of 4–6 g/day in four divided doses by mouth. These drugs are highly effective against the tachyzoites, but have no effect on cyst forms. Patients who are immunocompromised often relapse when therapy is discontinued; in AIDS patients, the relapse rate is at least 50%.[504] Antitoxoplasmic therapy should be continued until adequate cell-mediated immunity has been restored, or if it cannot be, maintenance therapy should be continued for life,[505] utilizing lower doses of pyrimethamine (25–50 mg) plus sulfadiazine (2–4 g), given daily or two or three times weekly. However, both pyrimethamine and sulfadiazine are associated with significant toxicity (up to 60% in AIDS patients), often necessitating discontinuation of therapy.[504] Skin rash and drug-induced hematologic toxicity are most commonly observed. Folinic acid (e.g., leucovorin), at dosages of 10–20 mg daily (up to 50 mg/day), is often employed in hopes of decreasing the hematologic side effects seen when pyrimethamine and sulfadiazine are used in combination.

The toxicity associated with standard antitoxoplasmic therapy has led to a search for new therapeutic agents. Clindamycin (at dosages of 1200–4800 mg/day), in combination with pyrimethamine, has been efficacious in several reports, with a trend to lower toxicity than observed with pyrimethamine–sulfadiazine combinations.[506–509] One published trial[510] comparing pyrimethamine plus clindamycin to pyrimethamine plus sulfadiazine as treatment for toxoplasmic encephalitis in AIDS suggested that the relative efficacy of pyrimethamine plus clindamycin was approximately equal to that of pyrimethamine plus sulfadiazine, although there was a trend to greater survival in patients randomized to the pyrimethamine–sulfadiazine combination. In another study, the response rate in AIDS patients with toxoplasmic encephalitis to the combination of pyrimethamine plus clindamycin was 71%.[499] In a subsequent trial comparing pyrimethamine plus clindamycin to pyrimethamine plus sulfadiazine,[511] no statistically significant differences in efficacy were noted during acute therapy, although the rate of crossover, motivated by a lack of response, was higher among the recipients of pyrimethamine plus clindamycin. Trimethoprim–sulfamethoxazole also has been suggested as a valuable alternative to pyrimethamine plus sulfadiazine in AIDS patients with toxoplasmic encephalitis,[512] although further studies with the use of this agent are needed. Other agents under investigation include macrolides and azalides such as roxithromycin, clarithromycin, and azithromycin; atovaquone; dapsone; the purine analogue arprinocid; and dihydrofolate reductase inhibitors such as trimetrexate or piritrexin.[470,487] However, it remains to be determined whether these agents will have a role in the treatment of toxoplasmic encephalitis.

5.2. Strongyloidiasis

5.2.1. Epidemiology and Etiology

Many species of helminths can invade or involve the CNS, although *Strongyloides stercoralis* is the one most frequently associated with the immunocompromised patient.[513–515] Under normal conditions, *Strongyloides* does not involve the CNS, although in the hyperinfection syndrome, CNS involvement may be prominent.[516] *Strongyloides stercoralis* is a small nematode that can live free in moist soil or can parasitize the small bowel of humans. The hermaphroditic adult females live buried in the intestinal crypts of the duodenum and jejunum, where they produce up to 40 eggs daily. The eggs release rhabditiform larvae that undergo molts to yield filariform or infective larvae; these infective larvae can penetrate the skin of mammals to establish intestinal parasitism. These invading larvae are highly motile, first moving rapidly through the skin and then traveling into the lymphatic system to reach the venous system. Next, there is migration from pulmonary blood vessels into alveoli, up the airways to the glottis, and then down the esophagus to reach the small bowel. The larvae also can accomplish an endogenous cycle within the body of a single host, reinvading the same host's tissues by passing through the bowel wall or the skin around the anus without reaching the environment; this process is called *autoinfection*. Only small numbers of larvae are usually involved, but if cellular immunity is compromised, the number of recycling larvae increases enormously, producing the hyperinfection syndrome in which the larvae can disseminate to many sites throughout the body.[515,516] The diseases that predispose to hyperinfection include chronic infections (tuberculosis and syphilis), chronic renal disease, malnutrition, alcoholism, lymphomas, burns, renal transplantation, systemic lupus erythematosus, and other debilitating conditions. In most cases, immunosuppressive or cytotoxic chemotherapy, especially corticosteroids, has been used. HIV-1-infected patients also can develop disseminated strongyloidiasis in the absence of corticosteroids or other immunosuppressive therapies.[517]

5.2.2. Clinical Presentation

In the hyperinfection syndrome with disseminated strongyloidiasis, almost all patients have abdominal pain, distention, or ileus, although these signs can be subtle in patients receiving corticosteroids. The leading CNS manifestation is bacterial meningitis caused by enteric bacteria,[516,518] which occurs secondary to seeding of the meninges during persistent or recurrent bacteremias associated with the migration of infective larvae; the larvae may carry organisms such as *E. coli*, *Klebsiella*, *Serratia*, and enterococci on their surfaces as they exit the intestine, or they may carry the enteric organisms within their own gastrointestinal tracts. Larvae are rarely found in the CNS, suggesting that meningitis is secondary to seeding during bacteremias rather than secondary to direct invasion. However, larvae occasionally have been found in the meninges[518] and within the parenchyma of the brain.[519]

5.2.3. Diagnosis

Diagnosis of the hyperinfection syndrome is proved by demonstration of larvae in the gastrointestinal tract or elsewhere. An increase in blood eosinophils or serum IgE concentration may be a useful indicator of infection in patients who do not have concurrent infection with other tissue helminths, although the blood eosinophil count is frequently normal in patients with depressed cell-mediated immunity and disseminated disease. Serologic assays for IgG antibodies to *Strongyloides* may be positive but are of limited value.[515] CNS infection with *S. stercoralis* is difficult to document because the organism is only rarely found in the meninges or the parenchyma of the brain.[516,518,519] Larvae may reach the CNS via hematogenous dissemination with impaction in small blood vessels, causing microinfarction. In one case inspected at autopsy,[519] the brain was edematous and a focal hemorrhage with necrosis was found in the cerebellum. Microinfarcts with slight inflammatory infiltrates were scattered throughout the brain. Serial sections occasionally showed degenerating larvae

within microinfarcts, sometimes within capillaries. Apparently viable larvae were found throughout the brain without evidence of necrosis or inflammation; larvae also were found in the perivascular spaces, the dura, and the epidural, subdural, and subarachnoid spaces.

5.2.4. Treatment

Most immunocompromised patients with disseminated strongyloidiasis have died. However, if the diagnosis is established, treatment should be initiated with thiabendazole (25 mg/kg twice daily)[513]; therapy may need to be continued for at least 2–3 weeks in the hyperinfection syndrome. Thiabendazole has been shown to penetrate into the CSF of a patient with disseminated strongyloidiasis and *Serratia* meningitis.[520]

References

1. Rubin RH, Hooper DC: Central nervous system infections in the compromised host. *Med Clin North Am* **69**:281–293, 1985.
2. Armstrong D, Wong B: Central nervous system infections in immunocompromised hosts. *Annu Rev Med* **33**:291–308, 1982.
3. Hooper DC, Pruitt AA, Rubin RH: Central nervous system infection in the chronically immunosuppressed. *Medicine* **61**:166–188, 1982.
4. Berger JR, Simpson DM: Neurologic complications of AIDS. In Scheld WM, Whitley RJ, Durack DT (eds): *Infections of the Central Nervous System*, 2nd ed. Lippincott-Raven Publishers, Philadelphia, 1997, pp. 255–271.
5. Janssen RS: Epidemiology and neuroepidemiology of human immunodeficiency virus infection. In Berger JR, Levy RM (eds): *AIDS and the Nervous System*, 2nd ed. Lippincott-Raven Publishers, Philadelphia, 1997, pp. 13–37.
6. Friedland GH, Klein RS: Transmission of the human immunodeficiency virus. *N Engl J Med* **317**:1125–1134, 1987.
7. Lifson AR: Do alternate modes for transmission of human immunodeficiency virus exist? *JAMA* **259**:1353–1356, 1988.
8. Price RW: Neurologic complications of HIV infection. *Lancet* **348**:445–452, 1996.
9. Levy RM, Bredesen DE, Rosenblum ML: Neurological manifestations of the acquired immunodeficiency syndrome (AIDS): Experience at UCSF and review of the literature. *J Neurosurg* **62**:475–495, 1985.
10. McArthur JC: Neurologic manifestations of AIDS. *Medicine* **66**:407–437, 1987.
11. Gabuzda DH, Hirsch MS: Neurologic manifestations of infection with human immunodeficiency virus: Clinical features and pathogenesis. *Ann Intern Med* **107**:383–391, 1987.
12. Levy RM, Janssen RS, Bush TJ, *et al*: Neuroepidemiology of acquired immunodeficiency syndrome. *J Acquir Immune Defic Syndr* **1**:31–40, 1988.
13. Lantos PL, McLaughlin JE, Scholtz CL, *et al*: Neuropathology of the brain in HIV infection. *Lancet* **1**:309–311, 1989.
14. Malouf R, Jacquette G, Dobkin J, *et al*: Neurologic disease in

human immunodeficiency virus-infected drug abusers. *Arch Neurol* **47:**1002–1007, 1990.

15. Carne CA, Tedder RS, Smith A, *et al*: Acute encephalopathy coincident with seroconversion for anti-HTLV-III. *Lancet* **2:**1206–1208, 1985.

16. Hollander H, Stringari S: Human immunodeficiency virus-associated meningitis: Clinical course and correlations. *Am J Med* **83:**813–816, 1987.

17. Navia BA, Jordan BD, Price RW: The AIDS dementia complex. I. Clinical features. *Ann Neurol* **19:**517–524, 1986.

18. Navia BA, Price RW: The acquired immunodeficiency syndrome dementia complex as the presenting or sole manifestation of human immunodeficiency virus infection. *Arch Neurol* **44:**65–69, 1987.

19. Bredesen DE: Clinical features (pp. 400–402). In Ho DD (moderator): The acquired immunodeficiency syndrome (AIDS) dementia complex. *Ann Intern Med* **111:**400–410, 1989.

20. Bacellar H, Munoz A, Miller E, *et al*: Temporal trends in the incidence of HIV-1–related neurologic diseases: Multicenter AIDS Cohort Study, 1985–1992. *Neurology* **44:**1892–1900, 1994.

21. Belman AL, Ultmann MH, Horoupian D, *et al*: Neurologic complications in infants and children with acquired immune deficiency syndrome. *Ann Neurol* **18:**560–566, 1985.

22. Epstein LG, Sharer LR, Joshi VV, *et al*: Progressive encephalopathy in children with acquired immune deficiency syndrome. *Ann Neurol* **17:**488–496, 1985.

23. Epstein LG, Sharer LR, Goudsmit J: Neurological and neuropathological features of human immunodeficiency virus infection in children. *Ann Neurol* **23:**S19–S23, 1988.

24. Davis SL, Halsted CC, Levy N, *et al*: Acquired immune deficiency syndrome presenting as progressive infantile encephalopathy. *J Pediatr* **110:**884–888, 1987.

25. Petito CK, Navia BA, Cho ES, *et al*: Vacuolar myelopathy pathologically resembling subacute combined degeneration in patients with the acquired immunodeficiency syndrome. *N Engl J Med* **312:**874–879, 1985.

26. Goldstick L, Mandybur TI, Bode R: Spinal cord degeneration in AIDS. *Neurology* **35:**103–106, 1985.

27. Navia BA, Cho ES, Petito CK, *et al*: The AIDS dementia complex. II. Neuropathology. *Ann Neurol* **19:**525–535, 1986.

28. Koralnik IJ, Beaumanoir A, Hausler R, *et al*: A controlled study of early neurologic abnormalities in men with asymptomatic human immunodeficiency virus infection. *N Engl J Med* **323:**864–870, 1990.

29. Hollander H: Cerebrospinal fluid normalities and abnormalities in individuals infected with human immunodeficiency virus. *J Infect Dis* **158:**855–858, 1988.

30. Appelman ME, Marshall DW, Brey RL, *et al*: Cerebrospinal fluid abnormalities in patients without AIDS who are seropositive for human immunodeficiency virus. *J Infect Dis* **158:**193–199, 1988.

31. Chalmers AC, April BS, Shephard H: Cerebrospinal fluid and human immunodeficiency virus: Findings in healthy, asymptomatic, seropositive men. *Arch Intern Med* **150:**1538–1540, 1990.

32. Resnick L, Berger JR, Shapshak P, *et al*: Early penetration of the blood–brain-barrier by HIV. *Neurology* **38:**9–14, 1988.

33. Hollander H, Levy JA: Neurologic abnormalities and recovery of human immunodeficiency virus from cerebrospinal fluid. *Ann Intern Med* **106:**692–695, 1987.

34. McArthur JC, Sipos E, Cornblath DR, *et al*: Identification of mononuclear cells in CSF of patients with HIV infection. *Neurology* **39:**66–70, 1989.

35. Resnick L, diMarzo-Veronese F, Schüpbach J, *et al*: Intra-blood-brain-barrier synthesis of HTLV-III-specific IgG in patients with neurologic symptoms associated with AIDS or AIDS-related complex. *N Engl J Med* **313:**1498–1504, 1985.

36. Chiodi F, Norkrans G, Hagberg L, *et al*: Human immunodeficiency virus infection of the brain. II. Detection of intrathecally synthesized antibodies by enzyme linked immunosorbent assay and imprint immunofixation. *J Neurol Sci* **87:**37–48, 1988.

37. Grimaldi LME, Castagna A, Lazzarein A, *et al*: Oligoclonal IgG bands in cerebrospinal fluid and serum during asymptomatic human immunodeficiency virus infection. *Ann Neurol* **24:**277–279, 1988.

38. Portegies P, Epstein LG, Hung ST, *et al*: Human immunodeficiency virus type 1 antigen in cerebrospinal fluid: Correlation with clinical neurologic status. *Arch Neurol* **46:**261–264, 1989.

39. Brew BJ, Bhalla RB, Fleisher M, *et al*: Cerebrospinal fluid β_2–microglobulin in patients infected with human immunodeficiency virus. *Neurology* **39:**830–834, 1989.

40. Heyes MP, Rubinow D, Lane C, *et al*: Cerebrospinal fluid quinolinic acid concentrations are increased in acquired immune deficiency syndrome. *Ann Neurol* **26:**275–277, 1989.

41. Ellis RJ, Hsia K, Spector SA, *et al*: Cerebrospinal fluid human immunodeficiency virus type 1 RNA levels are elevated in neurocognitively impaired individuals with acquired immunodeficiency syndrome. *Ann Neurol* **42:**679–688, 1997.

42. McArthur JC, McClernon DR, Cronin MF, *et al*: Relationship between human immunodeficiency virus-associated dementia and viral load in cerebrospinal fluid and brain. *Ann Neurol* **42:**689–698, 1997.

43. Price RW, Staprans S: Measuring the ''viral load'' in cerebrospinal fluid in human immunodeficiency virus infection: Window into brain infection? *Ann Neurol* **42:**675–678, 1997.

44. Bossi P, Dupin N, Coutellier A, *et al*: The level of human immunodeficiency virus (HIV) type 1 RNA in cerebrospinal fluid as a marker of HIV encephalitis. *Clin Infect Dis* **26:**1072–1073, 1998.

45. Mintz M, Epstein LG, Koenigsberger MR: Neurological manifestations of pediatric acquired immunodeficiency syndrome: Clinical features and therapeutic approaches. *Semin Neurol* **12:**51–56, 1992.

46. Belman AL, Lantos G, Horoupian D, *et al*: AIDS: Calcifications of the basal ganglia in infants and children. *Neurology* **36:**1192–1199, 1986.

47. Fiala M, Cone LA, Cohen N, *et al*: Responses of neurologic complications of AIDS to 3'-azido-3'-deoxythymidine and 9-(1,3-dihydroxy-2-propoxymethyl)guanine. I. Clinical features. *Rev Infect Dis* **10:**250–256, 1988.

48. Portegies P, de Gans J, Lange JMA, *et al*: Declining incidence of AIDS dementia complex after introduction of zidovudine treatment. *Br Med J* **299:**819–821, 1989.

49. Vago L, Castagna A, Lazzarin A, *et al*: Reduced frequency of HIV-induced brain lesions in AIDS patients treated with zidovudine. *J Acquir Immune Defic Syndr* **6:**42–45, 1993.

50. Schmitt FA, Bigley JW, McKinnes R, *et al*: Neuropsychological outcome of zidovudine (AZT) treatment of patients with AIDS and AIDS-related complex. *N Engl J Med* **319:**1573–1578, 1988.

51. Pizzo PA, Eddy J, Falloon J, *et al*: Effect of continuous intravenous infusion of zidovudine (AZT) in children with symptomatic HIV infection. *N Engl J Med* **319:**889–896, 1988.

52. Sidtis JJ, Gatsonis C, Price RW, *et al*: Zidovudine treatment of the AIDS dementia complex: Results of a placebo-controlled trial. AIDS Clinical Trials Group. *Ann Neurol* **33:**343–349, 1993.

53. Whitley RJ: Herpes simples virus. In Scheld WM, Whitley RJ, Durack DT (eds): *Infections of the Central Nervous System*, 2nd ed. Lippincott-Raven Publishers, Philadelphia, 1997, pp. 73–89.

54. Boston Interhospital Virus Study Group and the NIAID-Sponsored Cooperative Antiviral Clinical Study: Failure of high dose 5-iodo-2′-deoxyuridine in the therapy of herpes simplex virus encephalitis: Evidence of unacceptable toxicity. *N Engl J Med* **292:**599–603, 1975.

55. Corey L, Spear PG: Infection with herpes simplex viruses (second of two parts). *N Engl J Med* **314:**749–757, 1986.

56. Snider WD, Simpson DM, Nielsen S, *et al*: Neurological complications of acquired immune deficiency syndrome: Analysis of 50 patients. *Ann Neurol* **14:**403–418, 1983.

57. Gnann JW Jr, Whitley RJ: Neurologic manifestations of varicella and herpes zoster. In Scheld WM, Whitley RJ, Durack DT (eds): *Infections of the Central Nervous System*, 2nd ed. Lippincott-Raven Publishers, Philadelphia, 1997, pp. 91–105.

58. Weller TH: Varicella and herpes zoster. Changing concepts of the natural history, control, and importance of a not-so-benign virus (first of two parts). *N Engl J Med* **309:**1362–1367, 1983.

59. Barnes DW, Whitley RJ: CNS diseases associated with varicella zoster virus and herpes simplex virus infection: Pathogenesis and current therapy. *Neurol Clin* **4:**265–283, 1986.

60. Mazur MH, Dolin R: Herpes zoster at the NIH: A 20-year experience. *Am J Med* **65:**738–744, 1978.

61. Jemsek J, Greenberg SB, Taber L, *et al*: Herpes zoster-associated encephalitis: Clinicopathologic report of 12 cases and review of the literature. *Medicine* **62:**81–97, 1983.

62. Horton B, Price RW, Jimenez D: Multifocal varicella-zoster virus leukoencephalitis temporally remote from herpes zoster. *Ann Neurol* **9:**251–266, 1981.

63. Krech U: Complement-fixing antibodies against cytomegalovirus in different parts of the world. *Bull WHO* **49:**103–106, 1973.

64. Krech U, Tobin J: A collaborative study of cytomegalovirus antibodies in mothers and young children in 19 countries. *Bull WHO* **59:**605–610, 1981.

65. Onorato IM, Morens DM, Martone WJ, *et al*: Epidemiology of cytomegalovirus infections: Recommendations for prevention and control. *Rev Infect Dis* **7:**479–496, 1988.

66. Pass RF: Epidemiology and transmission of cytomegalovirus. *J Infect Dis* **153:**243–248, 1988.

67. Glenn J: Cytomegalovirus infections following renal transplantation. *Rev Infect Dis* **3:**1151–1178, 1981.

68. Winston DJ, Gale RP, Meyer DV, *et al*: Infectious complications of human bone marrow transplantation. *Medicine* **58:**1–31, 1979.

69. Jacobson MA, Mills J: Serious cytomegalovirus disease in the acquired immunodeficiency syndrome (AIDS): Clinical findings, diagnosis, and treatment. *Ann Intern Med* **108:**585–594, 1988.

70. Grassi MP, Clerici F, Perin C, *et al*: Microglial nodular encephalitis and ventriculoencephalitis due to cytomegalovirus infection in patients with AIDS: Two distinct clinical patterns. *Clin Infect Dis* **27:**504–508, 1998.

71. Whitley RJ, Tilles J, Linneman C, *et al*: Herpes simplex encephalitis: Clinical assessment. *JAMA* **247:**317–320, 1982.

72. Whitley RJ, Cobbs CG, Alford CA Jr, *et al*: Diseases that mimic herpes simplex encephalitis: Diagnosis, presentation, and outcome. *JAMA* **262:**234–239, 1989.

73. Johnson M, Valyi-Nagy T: Expanding the clinicopathologic spectrum of herpes simplex encephalitis. *Hum Pathol* **29:**207–210, 1998.

74. Tenser RB: Herpes simplex and herpes zoster: Nervous system involvement. *Neurol Clin* **2:**215–240, 1984.

75. Patresi R, Freemon FR, Lowry JL: Herpes zoster ophthalmicus with contralateral hemiplegia. *Arch Neurol* **34:**640–641, 1977.

76. Walker RJ, Gammal TE, Allen MB: Cranial arteritis associated with herpes zoster. *Radiology* **107:**109–110, 1973.

77. Schneck SA: Neuropathological features of human organ transplantation. I. Possible cytomegalovirus infection. *J Neuropathol Exp Neurol* **24:**415–429, 1965.

78. Schober R, Herman MM: Neuropathology of cardiac transplantation. *Lancet* **1:**962–967, 1973.

79. Moskowitz LB, Gregorios JB, Hensley GT, *et al*: Cytomegalovirus induced demyelination associated with acquired immune deficiency syndrome. *Arch Pathol Lab Med* **108:**873–877, 1984.

80. Post MJD, Hensley GT, Moskowitz LB, *et al*: Cytomegalic inclusion virus encephalitis in patients with AIDS: CT, clinical and pathologic correlation. *Am J Neuroradiol* **7:**275–280, 1986.

81. Holland NR, Power C, Mathews VP, *et al*: Cytomegalovirus encephalitis in acquired immunodeficiency syndrome (AIDS). *Neurology* **44:**507–514, 1994.

82. Kalayjian RC, Cohen ML, Bonomo RA, Flanigan TP: Cytomegalovirus ventriculoencephalitis in AIDS. A syndrome with distinct clinical and pathologic features. *Medicine* **72:**67–77, 1993.

83. Boos J, Esiri MM: Biopsy histopathology in herpes simplex encephalitis and in encephalitis of undefined etiology. *Yale J Biol Med* **57:**751–755, 1984.

84. Garcia JH, Colon LE, Whitley RJ, *et al*: Diagnosis of viral encephalitis by brain biopsy. *Semin Diagn Pathol* **1:**71–80, 1984.

85. Nahmias AJ, Whitley RJ, Visintine AN, *et al*: Herpes simplex virus encephalitis: Laboratory evaluations and their diagnostic significance. *J Infect Dis* **145:**829–836, 1982.

86. Griffith JF, Ch'ien LT: Herpes simplex virus encephalitis: Diagnostic and treatment considerations. *Med Clin North Am* **647:**991–1008, 1983.

87. Whitley RJ: Viral encephalitis. *N Engl J Med* **323:**242–250, 1990.

88. Lakeman FD, Koga J, Whitley RJ: Detection of antigen to herpes simplex virus in cerebrospinal fluid from patients with herpes simplex encephalitis. *J Infect Dis* **155:**1172–1178, 1987.

89. Rowley A, Lakeman F, Whitley RJ, *et al*: Diagnosis of herpes simplex encephalitis by DNA amplification of cerebrospinal fluid cells. *Lancet* **335:**440–441, 1990.

90. Lakeman FD, Whitley RJ, CASG NIAID: Diagnosis of herpes simplex encephalitis: Application of polymerase chain reaction to cerebrospinal fluid from brain biopsied patients and correlation with disease. *J Infect Dis* **171:**857–863, 1995.

91. Ch'ien LT, Boehm RM, Robinson H, *et al*: Characteristic early electroencephalographic changes in herpes simplex encephalitis. *Arch Neurol* **34:**361–364, 1977.

92. Smith JB, Westmoreland BF, Reagan TJ, *et al*: A distinctive clinical EEG profile in herpes simplex encephalitis. *Mayo Clin Proc* **50:**469–474, 1975.

93. Enzmann DR, Ransom B, Norman D, *et al*: Computed tomography of herpes simplex encephalitis. *Radiology* **129:**419–425, 1978.

94. Zimmerman RD, Russell EJ, Leeds N, *et al*: CT in the early diagnosis of herpes simplex encephalitis. *Am J Roentgenol* **134:**61–66, 1980.

95. Schlesinger Y, Buller RS, Brunstrom JE, *et al*: Expanded spectrum of herpes simplex encephalitis in childhood. *J Pediatr* **126:**234–241, 1995.

96. Johnson RT: Acute encephalitis. *Clin Infect Dis* **23:**219–226, 1996.

97. Ophir O, Seigman-Igra Y, Vardinon N, *et al*: Herpes zoster encephalitis: Isolation of virus from cerebrospinal fluid. *Isr J Med Sci* **20:**1189–1192, 1984.

98. Andiman WA, White-Greenwald M, Tinghitella T: Zoster encephalitis: Isolation of virus and measurement of varicella-zoster specific antibodies in cerebrospinal fluid. *Am J Med* **73:**769–772, 1982.

99. Steele RW, Keeney RE, Bradsher RW, *et al*: Treatment of varicella-zoster meningoencephalitis with acyclovir: Demonstration of virus in cerebrospinal fluid by electron microscopy. *Am J Clin Pathol* **80:**57–60, 1983.

100. Jemsek J, Greenberg SB, Taber L, *et al*: Herpes zoster-associated encephalitis: Clinicopathologic report of 12 cases and review of the literature. *Medicine* **62:**81–97, 1983.

101. Bieger RC, Van Scoy RE, Smith TF: Antibodies to varicella zoster in cerebrospinal fluid. *Arch Neurol* **34:**489–491, 1977.

102. Mathiesen T, Linde A, Olding-Stenvisk E, Wahren B: Antiviral IgM and IgG subclasses in varicella-zoster-associated neurological syndromes. *J Neurol Neurosurg Psychiatry* **52:**578–582, 1989.

103. Kuroiwa Y, Furukawa T: Hemispheric infarction after herpes zoster ophthalmicus: Computed tomography and angiography. *Neurology* **31:**1030–1032, 1981.

104. Clifford DB, Buller RS, Mohammed S, *et al*: Use of polymerase chain reaction to demonstrate cytomegalovirus DNA in CSF of patients with human immunodeficiency virus infection. *Neurology* **43:**75–79, 1993.

105. Cinque P, Vago L, Brytting M, *et al*: Cytomegalovirus infection of the central nervous system in patients with AIDS: Diagnosis by DNA amplification from cerebrospinal fluid. *J Infect Dis* **166:**1408–1411, 1992.

106. Wolf DG, Spector SA: Diagnosis of human cytomegalovirus central nervous system disease in AIDS patients by DNA amplification from cerebrospinal fluid. *J Infect Dis* **166:**1412–1415, 1992.

107. Wildemann B, Haas J, Lynen N, *et al*: Diagnosis of cytomegalovirus encephalitis in patients with AIDS by quantitation of cytomegalovirus genomes in cells of cerebrospinal fluid. *Neurology* **50:**693–697, 1998.

108. Masdeu JC, Small CB, Weiss L, *et al*: Multifocal cytomegalovirus encephalitis in AIDS. *Ann Neurol* **23:**97–99, 1988.

109. Levy RM, Bredesen DE: Central nervous system dysfunction in acquired immunodeficiency syndrome. *J Acquir Immune Defic Syndr* **1:**41–64, 1988.

110. Breeden CJ, Hall TC, Tyler HR: Herpes simplex encephalitis treated with systemic 5-iodo-2' deoxyuridine. *Ann Intern Med* **65:**1050–1056, 1966.

111. Nolan DC, Carruthers MM, Lerner AM: Herpesvirus hominis encephalitis in Michigan: Report of thirteen cases, including six treated with idoxuridine. *N Engl J Med* **282:**10–13, 1970.

112. Nolan DC, Lauter CB, Lerner AM: Idoxuridine in herpes simplex virus (type 1) encephalitis: Experience with 29 cases in Michigan, 1966 to 1971. *Ann Intern Med* **78:**243–246, 1973.

113. Whitley RJ, Soong S-J, Dolin R, *et al*: Adenine arabinoside therapy of biopsy-proved herpes simplex encephalitis: National Institute of Allergy and Infectious Diseases Collaborative Antiviral Study. *N Engl J Med* **297:**289–294, 1977.

114. Whitley RJ, Soong S-J, Hirsch MS, *et al*: Herpes simplex encephalitis: Vidarabine therapy and diagnostic problems. *N Engl J Med* **304:**313–318, 1981.

115. Skolderberg B, Firsgren M, Alestig K, *et al*: Acyclovir versus vidarabine in herpes simplex encephalitis: A randomized multicentre study of consecutive Swedish patients. *Lancet* **2:**707–711, 1984.

116. Whitley RJ, Alford CA, Hirsch MS, *et al*: Vidarabine versus acyclovir therapy of herpes simplex encephalitis. *N Engl J Med* **314:**144–149, 1986.

117. McGrath N, Anderson NE, Croxson MC, Powell KF: Herpes simplex encephalitis treated with acyclovir: Diagnosis and long term outcome. *J Neurol Neurosurg Psychiatry* **63:**321–326, 1998.

118. Van Landingham KE, Marsteller HB, Ross GW: Relapse of herpes simplex encephalitis after conventional acyclovir therapy. *JAMA* **259:**1051–1053, 1988.

119. Rothman A, Cheeseman SH, Lehrman SN: Herpes simplex encephalitis in a patient with lymphoma: Relapse following acyclovir therapy. *JAMA* **259:**1056–1057, 1988.

120. Kimura H, Aso K, Kuzushima K, *et al*: Relapse of herpes simplex encephalitis in children. *Pediatrics* **89:**891–894, 1992.

121. Balfour HH Jr, Bean B, Laskin O, *et al*: Acyclovir halts progression of herpes zoster in immunocompromised patients. *N Engl J Med* **308:**1448–1453, 1983.

122. Shepp D, Dandliker PS, Meyers JD: Treatment of varicella-zoster virus in severely immunocompromised patients: A randomized comparison of acyclovir and vidarabine. *N Engl J Med* **314:**208–212, 1986.

123. Johns DR, Gress DR: Rapid response to acyclovir in herpes zoster-associated encephalitis. *Am J Med* **82:**560–562, 1987.

124. Otero J, Ribera E, Gavalda J, *et al*: Response to acyclovir in two cases of herpes zoster leukoencephalitis and review of the literature. *Eur J Clin Microbiol Infect Dis* **17:**286–289, 1998.

125. Berman SM, Kim RC: The development of cytomegalovirus encephalitis in AIDS patients receiving ganciclovir. *Am J Med* **96:**415–419, 1994.

126. Greenlee JE: Progressive multifocal leukoencephalopathy. In Remington JS, Swartz MN (eds): *Current Clinical Topics in Infectious Diseases*. Blackwell, Boston, 1989, pp. 140–156.

127. Greenlee JE: Progressive multifocal leukoencephalopathy—Progress made and lessons relearned. *N Engl J Med* **338:**1378–1380, 1998.

128. Seong D, Bruner JM, Lee KH, *et al*: Progressive multifocal leukoencephalopathy after autologous bone marrow transplantation in a patient with chronic myelogenous leukemia. *Clin Infect Dis* **23:**402–403, 1996.

129. Chikezie PU, Greenberg AL: Idiopathic CD4⁺ T lymphocytopenia presenting as progressive multifocal leukoencephalopathy: Case report. *Clin Infect Dis* **24:**526–527, 1997.

130. Bezrodnik L, Samara R, Krasovec S, *et al*: Progressive multifocal leukoencephalopathy in a patient with hypogammaglobulinemia. *Clin Infect Dis* **27:**181–184, 1998.

131. Berger JR, Kaszovitz B, Post MJD, *et al*: Progressive multifocal leukoencephalopathy associated with human immunodeficiency virus infection: A review of the literature with a report of sixteen cases. *Ann Intern Med* **107:**78–87, 1987.

132. Chaisson RE, Griffin DE: Progressive multifocal leukoencephalopathy in AIDS. *JAMA* **264:**79–82, 1990.

133. Berger JR, Concha M: Progressive multifocal leukoencephalopathy: The evolution of a disease once considered rare. *J Neurovirol* **1:**5–18, 1995.

134. Weiner LP, Herndon RM, Narayan O, *et al*: Isolation of virus related to SV40 from patients with progressive multifocal leukoencephalopathy. *N Engl J Med* **286:**385–390, 1972.

135. Arthur RR, Shah KV, Yolken RH, *et al*: Detection of human papovaviruses BKV and JCV in urines by ELISA. In Sever JL, Madden DL (eds): *Polyomaviruses and Human Neurological Diseases*. Alan R. Liss, New York, 1983, pp. 169–176.

136. Andrews CA, Daniel RW, Shah KV: Serological studies of papovavirus infections in pregnant women and renal transplant recipients. In Sever JL, Madden DL (eds): *Polyomaviruses and*

Human Neurological Diseases. Alan R. Liss, New York, 1983, pp. 133–141.

137. Daniel R, Shah K, Madden D, *et al*: Serological investigation of the possibility of congenital transmission of papovavirus JC. *Infect Immun* **33:**319–321, 1981.

138. Coleman DV, Gardner SD, Mulholland C, *et al*: Human polyoma-viruses in pregnancy: A model for the study of defense mecha-nisms to virus reactivation. *Clin Exp Immunol* **53:**289–296, 1983.

139. Houff SA, Major EO, Katz DA, *et al*: Involvement of JC virus-infected mononuclear cells from the bone marrow and spleen in the pathogenesis of progressive multifocal leukoencephalopathy. *N Engl J Med* **318:**301–305, 1988.

140. Houff SA, Major EO, Katz DA, *et al*: JC virus infection of lym-phocytes in the pathogenesis of progressive multifocal leuko-encephalopathy. *Ann Neurol* **24:**142, 1988.

141. Parr J, Horoupian DS, Winkelman AC: Cerebellar form of pro-gressive multifocal leukoencephalopathy. *Can J Neurol Sci* **6:**123–128, 1979.

142. Berger JR, Mucke L: Prolonged survival and partial recovery in AIDS-associated progressive multifocal leukoencephalopathy. *Neurology* **38:**1060–1065, 1988.

143. Walker DL: Progressive multifocal leukoencephalopathy: An op-portunistic viral infection of the central nervous system. In Vinken PJ, Bruyn GW (eds): *Handbook of Clinical Neurology*, Vol. 34. Elsevier, Amsterdam, 1978, pp. 307–329.

144. Krupp LB, Lipton RB, Swerdlow ML, *et al*: Progressive multi-focal leukoencephalopathy: Clinical and radiographic findings. *Ann Neurol* **17:**344–349, 1985.

145. Koeppen S: Progressive multifocal leukoencephalopathy: Neuro-logical findings and evaluation of magnetic resonance imaging and computed tomography. *Neurosurg Rev* **10:**127–132, 1987.

146. Guilleux MH, Seiner RE, Young I: MR imaging in progressive multi-focal leukoencephalopathy. *Am J Neuroradiol* **7:**1033–1035, 1986.

147. Mark AS, Atlas SW: Progressive multifocal leukoencephalopathy in patients with AIDS: Appearance on MR images. *Radiology* **173:**517–520, 1989.

148. Padgett BL, Walker DL: Virologic and serologic studies of pro-gressive multifocal leukoencephalopathy. In Sever JL, Madden DL (eds): *Polyomaviruses and Human Neurological Diseases*. Alan R Liss, New York, 1983, pp. 107–117.

149. Richardson EP, Webster HD: Progressive multifocal leukoenceph-alopathy: Its pathological features. In Sever JL, Madden DL (eds): *Polyomaviruses and Human Neurological Diseases*. Alan R. Liss, New York, 1983, pp. 191–203.

150. Schlitt M, Morawetz RB, Bonnin J, *et al*: Progressive multifocal leukoencephalopathy: Three patients diagnosed by brain biopsy, with prolonged survival in two. *Neurosurgery* **18:**407–414, 1986.

151. Aksamit AJ, Sever JL, Major EO: Progressive multifocal leuko-encephalopathy: JC virus detection by in situ hybridization com-pared with immunohistochemistry. *Neurology* **36:**499–504, 1986.

152. Aksamit A, Major EO, Ghatak NR, *et al*: Diagnosis of progressive multifocal leukoencephalopathy by brain biopsy with biotin la-beled DNA:DNA in situ hybridization. *J Neuropathol Exp Neurol* **46:**556–566, 1987.

153. Weber T, Turner RW, Frye S, *et al*: Specific diagnosis of progres-sive multifocal leukoencephalopathy by polymerase chain reac-tion. *J Infect Dis* **169:**1138–1141, 1994.

154. Hammarin AL, Bogdanovic G, Svedhem V, *et al*: Analysis of PCR as a tool for detection of JC virus DNA in cerebrospinal fluid for diagnosis of progressive multifocal leukoencephalopathy. *J Clin Microbiol* **34:**2929–2932, 1996.

155. Bogdanovic G, Priftakis P, Hammarin AL, *et al*: Detection of JC virus in cerebrospinal fluid (CSF) samples from patients with progressive multifocal leukoencephalopathy but not in CSF sam-ples from patients with herpes simplex encephalitis, enteroviral meningitis, or multiple sclerosis. *J Clin Microbiol* **36:**1137–1138, 1998.

156. Bauer WR, Turel AP, Johnson KP: Progressive multifocal leuko-encephalopathy and cytarabine. *JAMA* **226:**174–176, 1973.

157. Conomy JP, Beard NS, Matsumoto H, *et al*: Cytarabine treatment of PML. *JAMA* **229:**1313–1316, 1974.

158. Marriott PJ, O'Brien MD, MacKenzie ICK, *et al*: Progressive multifocal leukoencephalopathy: Remission with cytarabine. *J Neurol Neurosurg Psychiatry* **38:**205–209, 1975.

159. Smith CR, Sima AAF, Salit IE, *et al*: Progressive multifocal leukoencephalopathy: Failure of cytarabine therapy. *Neurology* **32:**200–203, 1982.

160. Portegies P, Algra PR, Hollak CEM, *et al*: Response to cytarabine in progressive multifocal leucoencephalopathy in AIDS. *Lancet* **337:**680–681, 1991.

161. Moreno S, Miralles P, Diaz MD, *et al*: Cytarabine therapy for progressive multifocal leukoencephalopathy in patients with AIDS. *Clin Infect Dis* **23:**1066–1068, 1996.

162. Hall CD, Dafni U, Simpson D, *et al*: Failure of cytarabine in progressive multifocal leukoencephalopathy associated with hu-man immunodeficiency virus infection. *N Engl J Med* **338:**1345–1351, 1998.

163. Miller JR, Barrett RE, Britton CB, *et al*: Progressive multifocal leukoencephalopathy in a male homosexual with T-cell immune deficiency. *N Engl J Med* **307:**1436–1437, 1982.

164. Rand KH, Johnson KP, Rubinstein LJ, *et al*: Adenine arabinoside in the treatment of progressive multifocal leukoencephalopathy: Use of virus-containing cells in the urine to assess response to therapy. *Ann Neurol* **1:**458–462, 1977.

165. Bedri J, Weinstein W, DeGregorio P, *et al*: Progressive multifocal leukoencephalopathy in acquired immunodeficiency syndrome. *N Engl J Med* **309:**492–493, 1983.

166. Sadler M, Chinn R, Healy J, *et al*: New treatments for progressive multifocal leukoencephalopathy in HIV-1–infected patients. *AIDS* **12:**533–535, 1998.

167. Blick G, Whiteside M, Griegor P, *et al*: Successful resolution of progressive multifocal leukoencephalopathy after combination therapy with cidofovir and cytosine arabinoside. *Clin Infect Dis* **26:**191–192, 1998.

168. Vollmer-Haase J, Young P, Ringelstein EB: Efficacy of campto-thecin in progressive multifocal leucoencephalopathy. *Lancet* **349:**1366, 1997.

169. Conway B, Halliday WC, Brunham RC: Human immunodefi-ciency virus-associated progressive multifocal leukoencephalopa-thy: Apparent response to 3′-azido-3′-deoxythymidine. *Rev Infect Dis* **12:**479–482, 1990.

170. Baqi M, Kucharczyk W, Walmsley SL: Regression of progressive multifocal leukoencephalopathy with highly active antiretroviral therapy. *AIDS* **11:**1526–1527, 1997.

171. Elliot B, Aromin I, Gold R, *et al*: Two to five year remission of AIDS-associated progressive multifocal leukoencephalopathy with combined antiretroviral therapy. *Lancet* **349:**850, 1997.

172. Albrecht H, Hoffmann C, Degen O, *et al*: Highly active anti-retroviral therapy significantly improves the prognosis of patients with HIV-associated progressive multifocal leukoencephalopathy. *AIDS* **12:**1149–1154, 1998.

173. Cinque P, Casari S, Bertelli D: Progressive multifocal leuko-

encephalopathy, HIV, and highly active antiretroviral therapy. *N Engl J Med* **339:**848–849, 1998.

174. Tantisiriwat W, Tebas P, Clifford DB, *et al*: Progressive multifocal leukoencephalopathy in patients with AIDS receiving highly active antiretroviral therapy. *Clin Infect Dis* **28:**1152–1154, 1999.

175. Schlech WF, Ward JI, Band JD, *et al*: Bacterial meningitis in the United States, 1978 through 1981: The national bacterial meningitis surveillance study. *JAMA* **253:**1749–1754, 1985.

176. Schuchat A, Robinson K, Wenger JD, *et al*: Bacterial meningitis in the United States in 1995. *N Engl J Med* **337:**970–976, 1997.

177. Durand ML, Calderwood SB, Weber DJ, *et al*: Acute bacterial meningitis in adults. A review of 493 episodes. *N Engl J Med* **328:**21–28, 1993.

178. Wenger JD, Hightower AW, Facklam RR, *et al*: Bacterial meningitis in the United States, 1986: Report of a multistate surveillance study. *J Infect Dis* **162:**1316–1323, 1990.

179. Robbins JB, Schneerson R, Anderson P, Smith DH: Prevention of systemic infections, especially meningitis, caused by *Haemophilus influenzae* type b. *JAMA* **276:**1181–1185, 1996.

180. Steinhoff MC: *Haemophilus influenzae* type b infections are preventable everywhere. *Lancet* **349:**1186–1187, 1997.

181. Spagnuolo PT, Ellner JJ, Lerner PI, *et al*: *Haemophilus influenzae* meningitis: The spectrum of disease in adults. *Medicine* **61:**74–84, 1982.

182. Farley MM, Stephens DS, Brachman PS, *et al*: Invasive *Haemophilus influenzae* disease in adults. A prospective, population-based surveillance. *Ann Intern Med* **116:**806–812, 1992.

183. Jackson LA, Schuchat A, Reeves MW, *et al*: Serogroup C meningococcal outbreaks in the United States. An emerging threat. *JAMA* **273:**383–389, 1995.

184. Whalen CM, Hockin JC, Ryan A, Ashton R: The changing epidemiology of invasive meningococcal disease in Canada, 1985 through 1992. Emergence of a virulent clone of *Neisseria meningitidis*. *JAMA* **273:**390–394, 1995.

185. Ross SC, Densen P: Complement deficiency states and infection: Epidemiology, pathogenesis and consequences of neisserial and other infections in an immune deficiency. *Medicine* **63:**243–273, 1984.

186. Geiseler PJ, Nelson KE, Levin S, *et al*: Community-acquired purulent meningitis: A review of 1,316 cases during the antibiotic era, 1954–1976. *Rev Infect Dis* **2:**725–745, 1980.

187. Olopoenia L, Frederick W, Greaves W, *et al*: Pneumococcal sepsis and meningitis in adults with sickle cell disease. *South Med J* **83:**1002–1004, 1990.

188. Tunkel AR, Scheld WM: Acute infectious complications of head trauma. In Braakman R (ed): *Handbook of Clinical Neurology: Head Injury*. Elsevier, Amsterdam, 1990, pp. 317–326.

189. Gellin BG, Broome CV: Listeriosis. *JAMA* **261:**1313–1320, 1989.

190. Kessler SL, Dajani AS: *Listeria* meningitis in infants and children. *Pediatr Infect Dis J* **9:**61–63, 1990.

191. Cherubin CE, Appleman MD, Heseltine PNR, *et al*: Epidemiological spectrum and current treatment of listeriosis. *Rev Infect Dis* **13:**1108–1114, 1991.

192. Lorber B: Listeriosis. *Clin Infect Dis* **24:**1–11, 1997.

193. Mylonakis E, Hohmann EL, Calderwood SB: Central nervous system infection with *Listeria monocytogenes*: 33 years' experience at a general hospital and review of 776 cases from the literature. *Medicine* **77:**313–336, 1998.

194. Berenguer J, Solera J, Diaz MD, *et al*: Listeriosis in patients infected with human immunodeficiency virus. *Rev Infect Dis* **13:**115–119, 1991.

195. Decker CF, Simon GL, DiGioia, *et al*: *Listeria monocytogenes* infections in patients with AIDS: Report of five cases and review. *Rev Infect Dis* **13:**413–417, 1991.

196. Mangi RJ, Quintiliani R, Andriole VT: Gram-negative bacillary meningitis. *Am J Med* **59:**829–836, 1975.

197. Cherubin CE, Marr JS, Sierra MF, *et al*: *Listeria* and gram-negative bacillary meningitis in New York City, 1972–1979: Frequent causes of meningitis in adults. *Am J Med* **71:**199–209, 1981.

198. Kaufman BA, Tunkel AR, Pryor JC, *et al*: Meningitis in the neurosurgical patient. *Infect Dis Clin North Am* **4:**677–701, 1990.

199. Schlessinger LS, Ross SC, Schaberg DR: *Staphylococcus aureus* meningitis: A broad-based epidemiologic study. *Medicine* **66:**148–156, 1987.

200. Saez-Llorens X, McCracken GH Jr: Bacterial meningitis in neonates and children. *Infect Dis Clin North Am* **4:**623–644, 1990.

201. Dunne DW, Quagliarello V: Group B streptococcal meningitis in adults. *Medicine* **72:**1–10, 1993.

202. Domingo P, Barquet N, Alvarez M, *et al*: Group B streptococcal meningitis in adults: Report of twelve cases and review. *Clin Infect Dis* **25:**1180–1187, 1997.

203. Bross JE, Gordon G: Nocardial meningitis: Case reports and review. *Rev Infect Dis* **13:**160–165, 1991.

204. Roos KL, Tunkel AR, Scheld WM: Acute bacterial meningitis in children and adults. In Scheld WM, Whitley RJ, Durack DT (eds): *Infections of the Central Nervous System*, 2nd ed. Lippincott-Raven Publishers, Philadelphia, 1997, pp. 355–401.

205. Carpenter RR, Petersdorf RG: The clinical spectrum of bacterial meningitis. *Am J Med* **33:**262–275, 1962.

206. Tunkel AR, Scheld WM: Acute meningitis. In Mandell GL, Bennett JE, Dolin R (eds): *Principles and Practice of Infectious Diseases*, 5th ed. Churchill-Livingstone, Philadelphia, 2000, pp. 959–997.

207. Verghese A, Gallemore G: Kernig's and Brudzinski's signs revisited. *Rev Infect Dis* **9:**1187–1192, 1987.

208. Geiseler PJ, Nelson KE: Bacterial meningitis without clinical signs of meningeal irritation. *South Med J* **75:**448–450, 1982.

209. Gorse GJ, Thrupp LD, Nudleman KL, *et al*: Bacterial meningitis in the elderly. *Arch Intern Med* **144:**1603–1607, 1984.

210. Ashwal S, Perkin RM, Thompson JR, *et al*: Bacterial meningitis in children: Current concepts of neurologic management. *Curr Probl Pediatr* **24:**267–284, 1994.

211. Marton KI, Gean AD: The spinal tap: A new look at an old test. *Ann Intern Med* **104:**840–848, 1986.

212. Spanos A, Harrell FE Jr, Durack DT: Differential diagnosis of acute meningitis: An analysis of the predictive value of initial observation. *JAMA* **262:**2700–2707, 1989.

213. La Scolea LJ Jr, Dryja D: Quantitation of bacteria in cerebrospinal fluid and blood of children with meningitis and its diagnostic significance. *J Clin Microbiol* **19:**187–190, 1984.

214. Kaplan SL, O'Brian Smith E, Wills C, *et al*: Association between preadmission oral antibiotic therapy and cerebrospinal fluid findings and sequelae caused by *Haemophilus influenzae* type b meningitis. *J Pediatr Infect Dis* **5:**626–632, 1986.

215. Blazer S, Berant M, Alon U: Bacterial meningitis: Effect of antibiotic treatment on cerebrospinal fluid. *Am J Clin Pathol* **80:**386–387, 1983.

216. Gray LD, Fedorko DP: Laboratory diagnosis of bacterial meningitis. *Clin Microbiol Rev* **5:**130–145, 1992.

217. Saubolle MA, Jorgensen JH: Use of the *Limulus* amebocyte lysate test as a cost-effective screen for gram-negative agents of meningitis. *Diagn Microbiol Infect Dis* **7:**177–183, 1987.

218. Viallon A, Zeni F, Lambert C, *et al*: High sensitivity and specificity of serum procalcitonin levels in adults with bacterial meningitis. *Clin Infect Dis* **28**:1313–1316, 1999.

219. Tunkel AR, Scheld WM: Acute bacterial meningitis. *Lancet* **346**:1675–1680, 1995.

220. Appelbaum PC: Antimicrobial resistance in *Streptococcus pneumoniae*: An overview. *Clin Infect Dis* **15**:77–83, 1992.

221. Austrian R: Confronting drug-resistant pneumococci. *Ann Intern Med* **121**:807–809, 1994.

222. Paris MM, Ramilo O, McCracken GH Jr: Management of meningitis caused by penicillin-resistant *Streptococcus pneumoniae*. *Antimicrob Agents Chemother* **39**:2171–2175, 1995.

223. Viladrich PF, Gudiol F, Linares J, *et al*: Evaluation of vancomycin for therapy of adult pneumococcal meningitis. *Antimicrob Agents Chemother* **35**:2467–2472, 1991.

224. Ahmed A: A critical evaluation of vancomycin for treatment of bacterial meningitis. *Pediatr Infect Dis J* **16**:895–903, 1997.

225. American Academy of Pediatrics, Committee on Infectious Diseases: Therapy for children with invasive pneumococcal infections. *Pediatrics* **99**:289–299, 1997.

226. Campos J, Mendelman PM, Sako MU, *et al*: Detection of relatively penicillin G-resistant *Neisseria meningitidis* by disk susceptibility testing. *Antimicrob Agents Chemother* **31**:1478–1482, 1987.

227. Saez-Nieto JA, Lujan R, Berron S, *et al*: Epidemiology and molecular basis of penicillin-resistant *Neisseria meningitidis* in Spain: A 5-year history (1985–1989). *Clin Infect Dis* **14**:394–402, 1992.

228. Jackson LA, Tenover FC, Baker C, *et al*: Prevalence of *Neisseria meningitidis* relatively resistant to penicillin in the United States, 1991. *J Infect Dis* **169**:438–441, 1994.

229. Galimand M, Gerbaud G, Guibourdenche M, *et al*: High-level chloramphenicol resistance in *Neisseria meningitidis*. *N Engl J Med* **339**:868–874, 1998.

230. Campos J, Garcia-Tornel S, Sanfeliu I: Susceptibility studies of multiply resistant *Haemophilus influenzae* isolated from pediatric patients and contacts. *Antimicrob Agents Chemother* **25**:706–709, 1984.

231. Campos J, Garcia-Tornel S, Gairi JM, *et al*: Multiply resistant *Haemophilus influenzae* type b causing meningitis: Comparative clinical and laboratory study. *J Pediatr* **108**:897–902, 1986.

232. Givner LB, Abramson JS, Wasilauskas B: Meningitis due to *Haemophilus influenzae* type b resistant to ampicillin and chloramphenicol. *Rev Infect Dis* **11**:329–334, 1989.

233. Peltola J, Anttila M, Renkonen OV, *et al*: Randomized comparison of chloramphenicol, ampicillin, cefotaxime, and ceftriaxone for childhood bacterial meningitis. *Lancet* **1**:1281–1287, 1989.

234. American Academy of Pediatrics Committee on Infectious Diseases: Treatment of bacterial meningitis. *Pediatrics* **81**:904–907, 1988.

235. Schaad UB, Suter S, Gianella-Borradori A, *et al*: A comparison of ceftriaxone and cefuroxime for the treatment of bacterial meningitis in children. *N Engl J Med* **322**:141–147, 1990.

236. Saez-Llorens X, Castano E, Garcia R, *et al*: Prospective randomized comparison of cefepime and cefotaxime for treatment of bacterial meningitis in infants and children. *Antimicrob Agents Chemother* **39**:937–940, 1995.

237. Cherubin CE, Eng RHK, Norrby R, *et al*: Penetration of newer cephalosporins into cerebrospinal fluid. *Rev Infect Dis* **11**:526–548, 1989.

238. Cherubin CE, Corrado ML, Nair SR, *et al*: Treatment of gram-negative bacillary meningitis: Role of new cephalosporin antibiotics. *Rev Infect Dis* **4**:S453–S464, 1982.

239. Landesman SH, Corrado ML, Shah PM, *et al*: Past and current roles of cephalosporin antibiotics in treatment of meningitis: Emphasis on use in gram-negative bacillary meningitis. *Am J Med* **71**:693–703, 1981.

240. Fong IW, Tomkins KB: Review of *Pseudomonas aeruginosa* meningitis with special emphasis on treatment with ceftazidime. *Rev Infect Dis* **7**:604–612, 1985.

241. Rodriguez WJ, Khan WN, Cocchetto DM, *et al*: Treatment of *Pseudomonas* meningitis with ceftazidime with or without concurrent therapy. *Pediatr Infect Dis J* **9**:83–87, 1990.

242. Wong VK, Wright HT Jr, Ross LA, *et al*: Imipenem/cilastatin treatment of bacterial meningitis in children. *Pediatr Infect Dis J* **10**:122–125, 1991.

243. Tunkel AR, Scheld WM: Treatment of bacterial meningitis. In Wolfson JS, Hooper DC (eds): *Quinolone Antimicrobial Agents*. American Society for Microbiology, Washington, DC, 1993, pp. 381–395.

244. Levitz RE, Quintiliani R: Trimethoprim-sulfamethoxazole for bacterial meningitis. *Ann Intern Med* **100**:881–890, 1984.

245. Gombert ME, Landesman SH, Corrado ML, *et al*: Vancomycin and rifampin therapy for *Staphylococcus epidermidis* meningitis associated with CSF shunts. *J Neurosurg* **55**:633–636, 1981.

246. Wallace RJ Jr, Septimus EJ, Williams TW Jr, *et al*: Use of trimethoprim-sulfamethoxazole for treatment of infections due to *Nocardia*. *Rev Infect Dis* **4**:315–325, 1982.

247. Smego RA Jr, Moeller MB, Gallis HA: Trimethoprim-sulfamethoxazole therapy for *Nocardia* infections. *Arch Intern Med* **143**:711–718, 1983.

248. Filice GA, Simpson GL: Management of *Nocardia* infections. In Remington JS, Swartz MN (eds): *Current Clinical Topics in Infectious Diseases*. McGraw-Hill, New York, 1984, pp. 49–64.

249. Wallace RJ Jr, Steele LC, Sumter G, *et al*: Antimicrobial susceptibility patterns of *Nocardia asteroides*. *Antimicrob Agents Chemother* **32**:1776–1779, 1988.

250. Berkey P, Moore D, Rolston K: *In vitro* susceptibilities of *Nocardia* species to newer antimicrobial agents. *Antimicrob Agents Chemother* **32**:1078–1079, 1988.

251. Radetsky M: Duration of treatment in bacterial meningitis: A historical inquiry. *Pediatr Infect Dis J* **9**:2–9, 1990.

252. Viladrich PF, Pallares R, Ariza J, *et al*: Four days of penicillin therapy for meningococcal meningitis. *Arch Intern Med* **146**:2380–2382, 1986.

253. Jadavji T, Biggar WD, Gold R, *et al*: Sequelae of acute bacterial meningitis in children treated for seven days. *Pediatrics* **78**:21–25, 1985.

254. Lin TY, Chrane DF, Nelson JD, *et al*: Seven days of ceftriaxone therapy is as effective as ten days' treatment for bacterial meningitis. *JAMA* **253**:3559–3563, 1985.

255. Tunkel AR, Wispelwey B, Scheld WM: Bacterial meningitis: Recent advances in pathophysiology and treatment. *Ann Intern Med* **112**:610–623, 1990.

256. Quagliarello V, Scheld WM: Bacterial meningitis: Pathogenesis, pathophysiology, and progress. *N Engl J Med* **327**:864–872, 1992.

257. Pfister HW, Fontana A, Tauber MG, *et al*: Mechanisms of brain injury in bacterial meningitis: Workshop summary. *Clin Infect Dis* **19**:463–479, 1994.

258. Pfister HW, Koedel U, Paul R: Acute meningitis. *Curr Infect Dis Rep* **1**:153–159, 1999.

259. McIntyre PB, Berkey CS, King SM, *et al*: Dexamethasone as

adjunctive therapy in bacterial meningitis. A meta-analysis of randomized clinical trials since 1988. *JAMA* **278:**925–931, 1997.

260. Arditi M, Mason EO Jr, Bradley JS, *et al:* Three-year multicenter surveillance of pneumococcal meningitis in children: Clinical characteristics, and outcome related to penicillin susceptibility and dexamethasone use. *Pediatrics* **102:**1087–1097, 1998.

261. Kaplan SL, Mason EO Jr: Management of infections due to antibiotic-resistant *Streptococcus pneumoniae. Clin Microbiol Rev* **11:**628–644, 1998.

262. Kaplan K: Brain abscess. *Med Clin North Am* **69:**345–360, 1985.

263. Tunkel AR, Wispelwey B, Scheld WM: Brain abscess. In Mandell GL, Bennett JE, Dolin R (eds): *Principles and Practice of Infectious Diseases,* 5th ed. Churchill-Livingstone, Philadelphia, 2000, pp. 1016–1028.

264. Heilpern KL, Lorber B: Focal intracranial infections. *Infect Dis Clin North Am* **10:**879–898, 1996.

265. Murray HW, Gross KC, Masur H, *et al:* Serious infections caused by *Streptococcus milleri. Am J Med* **64:**759–764, 1978.

266. Shlaes DM, Lerner PI, Wolinsky E, *et al:* Infections due to Lancefield group F and related streptococci (*S. milleri, S. anginosus*). *Medicine* **60:**197–207, 1981.

267. de Louvois J, Gortvai P, Hurley R: Bacteriology of abscesses of the central nervous system: A multicentre prospective study. *Br Med J* **2:**981–984, 1977.

268. de Louvois J: The bacteriology and chemotherapy of brain abscess. *J Antimicrob Chemother* **4:**395–413, 1978.

269. Chun CH, Johnson JD, Hofstetter M, *et al:* Brain abscess: A study of 45 consecutive cases. *Medicine* **65:**415–431, 1986.

270. Lechtenberg R, Sierra MF, Pringle GF, *et al: Listeria monocytogenes:* Brain abscess or meningoencephalitis? *Neurology* **29:** 86–90, 1979.

271. Nieman RE, Lorber B: Listeriosis in adults: A changing pattern— Report of eight cases and review of the literature. *Rev Infect Dis* **2:**207–227, 1980.

272. Curry WA: Human nocardiosis: A clinical review with selected case reports. *Arch Intern Med* **140:**818–826, 1980.

273. Grigoriadis E, Gold WL: Pyogenic brain abscess caused by *Streptococcus pneumoniae:* Case report and review. *Clin Infect Dis* **25:** 1108–1112, 1997.

274. Palmer DL, Harvey RL, Wheeler JK: Diagnostic and therapeutic considerations in *Nocardia asteroides* infection. *Medicine* **53:** 391–401, 1974.

275. Wilson JP, Turner HR, Kirchner KA, *et al:* Nocardial infections in renal transplant recipients. *Medicine* **68:**38–57, 1989.

276. Berkey P, Bodey GP: Nocardial infection in patients with neoplastic disease. *Rev Infect Dis* **11:**407–412, 1989.

277. Lerner PI: Nocardiosis. *Clin Infect Dis* **22:**891–905, 1996.

278. Braun TI, Kerson LA, Eisenberg FP: Nocardial brain abscesses in a pregnant woman. *Rev Infect Dis* **13:**630–632, 1991.

279. Holtz HA, Lavery DP, Kapila R: Actinomycetales infection in the acquired immunodeficiency syndrome. *Ann Intern Med* **102:**203–205, 1985.

280. Kim J, Minamoto GY, Grieco MH: Nocardial infection as a complication of AIDS: Report of six cases and review. *Rev Infect Dis* **13:**624–629, 1991.

281. Sharer LR, Kapila R: Neuropathologic observations in acquired immunodeficiency syndrome (AIDS). *Acta Neuropathol* **66:**188–198, 1985.

282. Brewer NS, MacCarty CS, Wellman WE: Brain abscess: A review of recent experience. *Ann Intern Med* **82:**571–576, 1985.

283. Mathisen GE, Johnson JP: Brain abscess. *Clin Infect Dis* **25:**763–781, 1997.

284. Carpenter JL: Brain stem abscesses: Cure with medical therapy, case report, and review. *Clin Infect Dis* **18:**219–226, 1994.

285. Zimmerman RD, Weingarten K: Neuroimaging of cerebral abscesses. *Neuroimaging Clin North Am* **1:**1–16, 1991.

286. Rosenblum ML, Joff JT, Norman D, *et al:* Decreased mortality from brain abscesses since advent of computerized tomography. *J Neurosurg* **49:**658–668, 1978.

287. Zimmerman RD, Haimes AB: The role of MR imaging in the diagnosis of infections of the central nervous system. In Remington JS, Swartz MN (eds): *Current Clinical Topics in Infectious Diseases.* Blackwell, Boston, 1989, pp. 82–108.

288. Zimmerman RA, Girard NJ: Imaging of intracranial infections. In Scheld WM, Whitley RJ, Durack DT (eds): *Infections of the Central Nervous System,* 2nd ed. Lippincott-Raven Publishers, Philadelphia, 1997, pp. 923–944.

289. Mamerlak AN, Mampalam TJ, Obana WG, Rosenblum ML: Improved management of multiple brain abscesses: A combined surgical and medical approach. *Neurosurgery* **36:**76–86, 1995.

290. Holm S, Kourtopoulos H: Penetration of antibiotics into brain tissue and brain abscesses: An experimental study in steroid treated rats. *Scand J Infect Dis* **44**(Suppl):68–70, 1985.

291. Alderson D, Strong AJ, Ingham MR, *et al:* Fifteen year review of the mortality of brain abscess. *Neurosurgery* **8:**1–6, 1981.

292. Levy RM, Gutin PH, Baskin DS, *et al:* Vancomycin penetration of a brain abscess: Case report and review of the literature. *Neurosurgery* **18:**633–636, 1986.

293. Greene BM, Thomas FE Jr, Alford RH: Trimethoprim-sulfamethoxazole and brain abscess. *Ann Intern Med* **82:**812–813, 1975.

294. Sjolin J, Eriksson N, Arneborn P, *et al:* Penetration of cefotaxime and desacetylcefotaxime into brain abscesses in humans. *Antimicrob Agent Chemother* **35:**2606–2610, 1991.

295. Kim J, Minamoto GY, Hoy CD, *et al:* Presumptive cerebral *Nocardia asteroides* infection in AIDS: Treatment with ceftriaxone and minocycline. *Am J Med* **90:**656–658, 1991.

296. Krone A, Schaal KP, Brawanski A, Schuknecht B: Nocardial cerebral abscess cured with imipenem/amikacin and enucleation. *Neurosurg Rev* **12:**333–340, 1989.

297. Garlando F, Bodmer T, Lee C, *et al:* Successful treatment of disseminated nocardiosis complicated by cerebral abscess with ceftriaxone and amikacin: Case report. *Clin Infect Dis* **15:**1039–1040, 1992.

298. Jansen C, Frenay HM, Vandertop WP, Visser MR: Intracerebral *Nocardia asteroides* abscess treated by neurosurgical aspiration and combined therapy with sulfadiazine and cefotaxime. *Clin Neurol Neurosurg* **93:**253–255, 1991.

299. Gombert ME, du Bouchet L, Aulicino TM, Berkowitz LB: Antimicrobial synergism in the therapy of experimental cerebral nocardiosis. *J Antimicrob Chemother* **24:**39–43, 1989.

300. Rosenblum ML, Hoff JT, Norman D, *et al:* Nonoperative treatment of brain abscesses in selected high-risk patients. *J Neurosurg* **52:**217–225, 1980.

301. Boom WH, Tuazon CU: Successful treatment of multiple brain abscesses with antibiotics alone. *Rev Infect Dis* **7:**189–199, 1985.

302. Fulgham JR, Wijkicks EFM, Wright AJ: Cure of a solitary brainstem abscess with antibiotic therapy: Case report. *Neurology* **46:**1451–1454, 1996.

303. Stephanov S: Surgical treatment of brain abscess. *Neurosurgery* **22:**724–730, 1988.

304. Mampalam TJ, Rosenblum ML: Trends in the management of bacterial brain abscesses: A review of 102 cases over 17 years. *Neurosurgery* **23:**451–458, 1988.

305. Skrap M, Melatini A, Vassallo A, Sidoti C: Stereotactic aspiration

and drainage of brain abscesses. Experience with 9 cases. *Minim Invasive Neurosurg* **39:**108–112, 1996.

306. Shahzadi S, Lozano AM, Bernstein M, *et al*: Stereotactic management of bacterial brain abscess. *Can J Neurol Sci* **23:**34–39, 1996.

307. Laborde G, Klimek L, Harders A, Gilsbach J: Frameless stereotactic drainage of intracranial abscesses. *Surg Neurol* **40:**16–21, 1993.

308. Zugar A, Lowy FD: Tuberculosis of the central nervous system. In Scheld WM, Whitley RJ, Durack DT (eds): *Infections of the Central Nervous System*, 2nd ed. Lippincott-Raven Publishers, Philadelphia, 1997, pp. 417–443.

309. Gracey DR: Tuberculosis in the world today. *Mayo Clin Proc* **63:**1251–1255, 1988.

310. Farer LS, Lowell AM, Meador MP: Extrapulmonary tuberculosis in the United States. *Am J Epidemiol* **19:**205–217, 1979.

311. Leonard JM, Des Prez RM: Tuberculous meningitis. *Infect Dis Clin North Am* **4:**769–787, 1990.

312. Singh N, Paterson DL: *Mycobacterium tuberculosis* infection in solid-organ transplant recipients: Impact and implications for management. *Clin Infect Dis* **27:**1266–1277, 1998.

313. Markowitz N, Hansen NI, Hopewell PC, *et al*: Incidence of tuberculosis in the United States among HIV-infected persons. *Ann Intern Med* **126:**123–132, 1997.

314. Barnes PF, Bloch AB, Davidson PT, *et al*: Tuberculosis in patients with human immunodeficiency virus infection. *N Engl J Med* **324:**1644–1650, 1991.

315. Horsburgh CR, Mason UG, Farhi DC, *et al*: Disseminated infection with *Mycobacterium avium-intracellulare*: A report of 13 cases and a review of the literature. *Medicine* **64:**36–48, 1985.

316. Klein NC, Damsker B, Hirschman SZ: Mycobacterial meningitis: Retrospective analysis from 1970–1983. *Am J Med* **79:**29–34, 1985.

317. Wolinsky E: Nontuberculous mycobacteria and associated diseases. *Am Rev Respir Dis* **119:**107–159, 1979.

318. Cegielski JP, Wallace RJ Jr: Infections due to nontuberculous mycobacteria. In Scheld WM, Whitley RJ, Durack DT (eds): *Infections of the Central Nervous System*, 2nd ed. Lippincott-Raven Publishers, Philadelphia, 1997, pp. 445–461.

319. Ogawa SK, Smith MA, Brennessel DJ, *et al*: Tuberculous meningitis in an urban medical center. *Medicine* **63:**317–326, 1987.

320. Kent SJ, Crowe SM, Yung A, *et al*: Tuberculous meningitis: A 30–year review. *Clin Infect Dis* **17:**987–994, 1993.

321. Alvarez S, McCabe WR: Extrapulmonary tuberculosis revisited: A review of experience at Boston City and other hospitals. *Medicine* **63:**25–54, 1984.

322. Idriss ZH, Sinno AA, Kronfol NM: Tuberculous meningitis in childhood. *Am J Dis Child* **130:**364–367, 1976.

323. Haas EJ, Madhavan T, Quinn EL, *et al*: Tuberculous meningitis in an urban general hospital. *Arch Intern Med* **137:**1518–1521, 1977.

324. Kennedy DH, Fallon RJ: Tuberculous meningitis. *JAMA* **241:**264–268, 1979.

325. Verdon R, Chevret S, Laissy JP, Wolff M: Tuberculous meningitis in adults: Review of 48 cases. *Clin Infect Dis* **22:**982–988, 1996.

326. Leiguarda R, Berthier M, Starkstein S, *et al*: Ischemic infarction in 25 children with tuberculous meningitis. *Stroke* **19:**200–204, 1988.

327. Berenguer J, Moreno S, Laguna F, *et al*: Tuberculous meningitis in patients infected with human immunodeficiency virus. *N Engl J Med* **326:**668–672, 1992.

328. Dube MP, Holtom PD, Larsen RA: Tuberculous meningitis in patients with and without human immunodeficiency virus infection. *Am J Med* **93:**520–524, 1992.

329. Smith J, Godwin-Austen R: Hypersecretion of anti-diuretic hormone due to tuberculous meningitis. *Postgrad Med J* **56:**41–44, 1980.

330. Clark WC, Metcalf JC, Muhlbauer MS, *et al*: *Mycobacterium tuberculosis* meningitis: A report of twelve cases and a literature review. *Neurosurgery* **18:**604–610, 1986.

331. Stockstill MT, Kauffman CA: Comparison of cryptococcal and tuberculous meningitis. *Arch Neurol* **40:**81–85, 1983.

332. Daniel TM: New approaches to the rapid diagnosis of tuberculous meningitis. *J Infect Dis* **155:**599–602, 1987.

333. Kaneko K, Onodera O, Miyatake T, *et al*: Rapid diagnosis of tuberculous meningitis by polymerase chain reaction (PCR). *Neurology* **40:**1617–1618, 1990.

334. Liu PY, Shi Z, Lau Y, Hu B: Rapid diagnosis of tuberculous meningitis by a nested amplification protocol. *Neurology* **44:**1161–1164, 1994.

335. Folgueira L, Delgado R, Palenque E, Noriega AR: Polymerase chain reaction for rapid diagnosis of tuberculous meningitis in AIDS patients. *Neurology* **44:**1336–1338, 1994.

336. Bonington A, Strang JIG, Klapper PE, *et al*: Use of Roche AMPLICOR *Mycobacterium tuberculosis* PCR in early diagnosis of tuberculous meningitis. *J Clin Microbiol* **36:**1251–1254, 1998.

337. Offenbacher H, Fazekas F, Schmidt R, *et al*: MRI in tuberculous meningoencephalitis: Report of four cases and review of the neuroimaging literature. *J Neurol* **238:**340–344, 1991.

338. Gupta RK, Gupta S, Singh D, *et al*: MR imaging and angiography in tuberculous meningitis. *Neuroradiology* **36:**87–92, 1994.

339. Whiteman MLH: Neuroimaging of central nervous system tuberculosis in HIV-infected patients. *Neuroimag Clin North Am* **7:**199–214, 1997.

340. Raviglione MC, Narain JP, Kochi A: HIV-associated tuberculosis in developing countries: Clinical features, diagnosis and treatment. *Bull WHO* **70:**515–526, 1992.

341. American Thoracic Society: Treatment of tuberculosis and tuberculosis infection in adults and children. *Am J Respir Crit Care Med* **149:**1359–1374, 1994.

342. Alarcon F, Escalante L, Perez Y, *et al*: Tuberculous meningitis: Short course of chemotherapy. *Arch Neurol* **47:**1313–1317, 1990.

343. Humphries M: The management of tuberculous meningitis. *Thorax* **47:**577–581, 1992.

344. Havlir DV, Barnes PF: Tuberculosis in patients with human immunodeficiency virus infection. *N Engl J Med* **340:**367–373, 1999.

345. Tunkel AR: Chronic meningitis. *Curr Infect Dis Rep* **1:**160–165, 1999.

346. Escobar JA, Belsey MA, Duenas A, *et al*: Mortality from tuberculous meningitis reduced by steroid therapy. *Pediatrics* **56:**1050–1055, 1975.

347. Girgis NI, Farid Z, Kilpatrick ME, *et al*: Dexamethasone adjunctive treatment for tuberculous meningitis. *Pediatr Infect Dis J* **10:**179–183, 1991.

348. Schoeman JF, Van Zyl LE, Laubscher JA, Donald PR: Effect of corticosteroids on intracranial pressure, computed tomographic findings, and clinical outcome in young children with tuberculous meningitis. *Pediatrics* **99:**226–231, 1997.

349. McGowan JE, Chesney PJ, Crossley KB, *et al*: Guidelines for the use of systemic glucocorticosteroids in the management of selected infections. *J Infect Dis* **165:**1–13, 1992.

350. Chitale VR, Kasaliwal GT: Our experience of ventriculoatrial shunt using Upadhyaya valve in cases of hydrocephalus associated with tuberculous meningitis. *Prog Pediatr Surg* **15:**223–236, 1982.

351. Palur R, Rajshekhar V, Chandy MJ, *et al*: Shunt surgery for hydrocephalus in tuberculous meningitis: A long-term follow up study. *J Neurosurg* **74:**64–69, 1991.

352. Mathew JM, Rajshekhar V, Chandy MJ: Shunt surgery in poor grade patients with tuberculous meningitis and hydrocephalus:

Effects of response to external ventricular drainage and other variables on long-term outcome. *J Neurol Neurosurg Psychiatry* **65:**115–118, 1998.

353. Bagga A, Kaira V, Ghai OP: Intracranial tuberculoma: Evaluation and treatment. *Clin Pediatr* **27:**487–490, 1988.

354. Salaki JS, Louria DB, Chmel H: Fungal and yeast infections of the central nervous system: A clinical review. *Medicine* **63:**108–132, 1984.

355. Diamond RD, Bennett JE: Prognostic factors in cryptococcal meningitis: A study of 111 patients. *Ann Intern Med* **80:**176–181, 1974.

356. Sabetta JR, Andriole VT: Cryptococcal infection of the central nervous system. *Med Clin North Am* **69:**333–345, 1985.

357. Perfect JR, Durack DT, Gallis HA: Cryptococcemia. *Medicine* **62:**98–109, 1983.

358. Kovacs JA, Kovacs AA, Polis M, *et al*: Cryptococcosis in the acquired immunodeficiency syndrome. *Ann Intern Med* **103:**533–538, 1985.

359. Zugar A, Louie E, Holzman RS, *et al*: Cryptococcal disease in patients with the acquired immunodeficiency syndrome: Diagnostic features and outcome of treatment. *Ann Intern Med* **104:**234–240, 1986.

360. Eng RHK, Bishburg E, Smith SM, *et al*: Cryptococcal infections in patients with acquired immune deficiency syndrome. *Am J Med* **81:**19–23, 1986.

361. Chuck SL, Sande MA: Infections with *Cryptococcus neoformans* in the acquired immunodeficiency syndrome. *N Engl J Med* **321:**794–799, 1989.

362. Clark RA, Greer D, Atkinson W, *et al*: Spectrum of *Cryptococcus neoformans* infection in 68 patients with human immunodeficiency virus. *Rev Infect Dis* **12:**768–777, 1990.

363. Mitchell TG, Perfect JR: Cryptococcosis in the era of AIDS 100 years after the discovery of *Cryptococcus neoformans*. *Clin Microbiol Rev* **8:**515–548, 1995.

364. Powderly WG: Cryptococcal meningitis and AIDS. *Clin Infect Dis* **17:**837–842, 1993.

365. Levitz SM: The ecology of *Cryptococcus neoformans* and the epidemiology of cryptococcosis. *Rev Infect Dis* **13:**1163–1169, 1991.

366. Treseler CB, Sugar AM: Fungal meningitis. *Infect Dis Clin North Am* **4:**789–808, 1990.

367. Kirkland TN, Fierer J: Coccidioidomycosis: A reemerging infectious disease. *Emerg Infect Dis* **2:**192–199, 1996.

368. Galgiani JN: Coccidioidomycosis: A regional disease of national importance. Rethinking approaches to control. *Ann Intern Med* **130:**293–300, 1999.

369. Bouza E, Dreyer JS, Hewitt WL, *et al*: Coccidioidal meningitis: An analysis of thirty-one cases and review of the literature. *Medicine* **60:**139–172, 1981.

370. Ampel NM, Wieden MA, Galgiani JN: Coccidioidomycosis: Clinical update. *Rev Infect Dis* **11:**897–911, 1989.

371. Wheat LJ: Histoplasmosis. *Infect Dis Clin North Am* **2:**841–859, 1988.

372. Kauffman CA, Israel KS, Smith JW, *et al*: Histoplasmosis in immunosuppressed patients. *Am J Med* **64:**923–932, 1978.

373. Wheat LJ, Slama TG, Norton JA, *et al*: Risk factors for disseminated or fatal histoplasmosis. *Ann Intern Med* **96:**159–163, 1982.

374. Johnson PC, Khardori N, Najjar AF, *et al*: Progressive disseminated histoplasmosis in patients with acquired immunodeficiency syndrome. *Am J Med* **85:**152–158, 1988.

375. Wheat LJ, Connolly-Stringfield PA, Baker RL, *et al*: Disseminated histoplasmosis in the acquired immune deficiency syndrome: Clinical findings, diagnosis and treatment, and review of the literature. *Medicine* **69:**361–374, 1990.

376. Wheat J: Histoplasmosis. *Medicine* **76:**339–354, 1997.

377. Crislip MA, Edwards JE Jr: Candidiasis. *Infect Dis Clin North Am* **3:**103–133, 1989.

378. Meunier-Carpentier I, Kiehn TE, Armstrong D: Fungemia in the immunocompromised host: Changing patterns, antigenemia, high mortality. *Am J Med* **71:**363–370, 1981.

379. Gold JWM: Opportunistic fungal infections in patients with neoplastic disease. *Am J Med* **76:**458–463, 1984.

380. Wey SB, Mori M, Pfaller MA, *et al*: Risk factors for hospital-acquired candidemia: A matched case-control study. *Arch Intern Med* **149:**2349–2353, 1989.

381. Lee BE, Cheung PY, Robinson JL, *et al*: Comparative study of mortality and morbidity in premature infants (birth weight, <1,250 g) with candidemia or candidal meningitis. *Clin Infect Dis* **27:**559–565, 1998.

382. Bayer AS, Edwards JE Jr, Seidel JS, *et al*: *Candida* meningitis: Report of seven cases and review of the English literature. *Medicine* **55:**477–486, 1976.

383. Lipton SA, Hickey WF, Morris JH, *et al*: Candidal infection in the central nervous system. *Am J Med* **76:**101–108, 1984.

384. Walsh TJ, Hier DB, Caplan LP: Fungal infections of the central nervous system: Comparative analysis of risk factors and clinical signs in 57 patients. *Neurology* **35:**1654–1657, 1985.

385. Patterson TF, Andriole VT: Current concepts in cryptococcosis. *Eur J Clin Microbiol Infect Dis* **8:**457–465, 1989.

386. Moosa MYS, Coovadia YM: Cryptococcal meningitis in Durban, South Africa: A comparison of clinical features, laboratory findings, and outcome for human immunodeficiency virus (HIV)-positive and HIV-negative patients. *Clin Infect Dis* **24:**131–134, 1997.

387. Heyderman RS, Gangaidzo IT, Hakim JG, *et al*: Cryptococcal meningitis in human immunodeficiency virus-infected patients in Harare, Zimbabwe. *Clin Infect Dis* **26:**284–289, 1998.

388. Schermoly MJ, Hinthorn DR: Eosinophilia in coccidioidomycosis. *Arch Intern Med* **148:**895–896, 1988.

389. Perfect JR: Diagnosis and treatment of fungal meningitis. In Scheld WM, Whitley RJ, Durack DT (eds): *Infections of the Central Nervous System*, 2nd ed. Lippincott-Raven Publishers, Philadelphia, 1997, pp. 721–739.

390. Geers TA, Gordon SM: Clinical significance of *Candida* species isolated from cerebrospinal fluid following neurosurgery. *Clin Infect Dis* **28:**1139–1147, 1999.

391. Goodman JS, Kaufman L, Loening MG: Diagnosis of cryptococcal meningitis: Detection of cryptococcal antigen. *N Engl J Med* **285:**434–436, 1971.

392. Snow RM, Dismukes WE: Cryptococcal meningitis: Diagnostic value of cryptococcal antigen in cerebrospinal fluid. *Arch Intern Med* **135:**1155–1157, 1975.

393. Wheat LJ, French M, Batteiger B, *et al*: Cerebrospinal fluid *Histoplasma* antibodies in central nervous system histoplasmosis. *Arch Intern Med* **145:**1237–1240, 1985.

394. Wheat LJ, Kohler RB, Tewari RP: Diagnosis of disseminated histoplasmosis by detection of *Histoplasma capsulatum* antigen in serum and urine specimens. *N Engl J Med* **314:**83–88, 1986.

395. Wheat LJ, Kohler RB, Tewari RP, *et al*: Significance of *Histoplasma* antigen in the cerebrospinal fluid of patients with meningitis. *Arch Intern Med* **149:**302–304, 1989.

396. Bennett JE, Dismukes WE, Duma RJ, *et al*: A comparison of amphotericin B alone and combined with flucytosine in the treatment of cryptococcal meningitis. *N Engl J Med* **301:**126–131, 1979.

397. Dismukes WE, Cloud G, Gallis HA, *et al*: Treatment of cryptococcal meningitis with combination amphotericin B and flucytosine for four as compared with six weeks. *N Engl J Med* **317:**334–341, 1987.

398. Stamm AM, Diasio RB, Dismukes WE, *et al*: Toxicity of amphotericin B plus flucytosine in 194 patients with cryptococcal meningitis. *Am J Med* **83:**236–242, 1987.

399. Byrne WR, Wajszczuk CP: Cryptococcal meningitis in the acquired immunodeficiency syndrome (AIDS): Successful treatment with fluconazole after failure of amphotericin B. *Ann Intern Med* **108:**384–385, 1988.

400. Stern JJ, Hartman BJ, Sharkey P, *et al*: Oral fluconazole therapy for patients with acquired immunodeficiency syndrome and cryptococcosis: Experience with 22 patients. *Am J Med* **85:**477–480, 1988.

401. Larsen RA, Leal MAE, Chan LS: Fluconazole compared with amphotericin B plus flucytosine for cryptococcal meningitis in AIDS: A randomized trial. *Ann Intern Med* **113:**183–187, 1990.

402. Saag MS, Powderly WG, Cloud GA, *et al*: Comparison of amphotericin B with fluconazole in the treatment of acute AIDS-associated cryptococcal meningitis. *N Engl J Med* **326:**83–89, 1992.

403. Bennett JE: Current therapy of deep mycoses. In Mandell GL, Douglas RG Jr, Bennett JE (eds): *Principles and Practice of Infectious Diseases, Update 11.* Churchill Livingstone, New York, 1991.

404. Van der Horst CM, Saag MS, Cloud GA, *et al*: Treatment of cryptococcal meningitis associated with the acquired immunodeficiency syndrome. *N Engl J Med* **337:**15–21, 1997.

405. Wong-Beringer A, Jacobs RA, Guglielmo BJ: Lipid formulations of amphotericin B: Clinical efficacy and toxicities. *Clin Infect Dis* **27:**603–618, 1998.

406. Sharkey PK, Graybill JR, Johnson ES, *et al*: Amphotericin B lipid complex compared with amphotericin B in the treatment of cryptococcal meningitis in patients with AIDS. *Clin Infect Dis* **22:**315–321, 1996.

407. Leendera ACAP, Reiss P, Portegies P, *et al*: Liposomal amphotericin B (AmBisome) compared to amphotericin B both followed by oral fluconazole in the treatment of AIDS-associated cryptococcal meningitis. *AIDS* **11:**1463–1471, 1997.

408. Yamaguchi H, Ikemoto H, Watanabe K, *et al*: Fluconazole monotherapy for cryptococcosis in non-AIDS patients. *Eur J Clin Microbiol Infect Dis* **15:**787–792, 1996.

409. Anthony SJ, Patel A, Leonard J: Use of fluconazole in the treatment of non-AIDS cryptococcal meningitis. *J Natl Med Assoc* **89:**694–695, 1997.

410. Pappas PG, Perfect J, Larsen RA, *et al*: Cryptococcosis in HIV-negative patients: Analysis of 306 cases (abstract 101). *Clin Infect Dis* **27:**939, 1998.

411. Sugar AM, Saunders C: Oral fluconazole as suppressive therapy of disseminated cryptococcosis in patients with acquired immunodeficiency syndrome. *Am J Med* **85:**481–489, 1988.

412. Bozzette SA, Larsen RA, Chiu J, *et al*: A placebo-controlled trial of maintenance therapy with fluconazole after treatment of cryptococcal meningitis in the acquired immunodeficiency syndrome. *N Engl J Med* **32:**580–584, 1991.

413. Powderly WG, Saag MS, Cloud GA, *et al*: A controlled trial of fluconazole or amphotericin B to prevent relapse of cryptococcal meningitis in patients with the acquired immunodeficiency syndrome. *N Engl J Med* **326:**793–798, 1992.

414. Saag MS, Cloud GA, Graybill JR, *et al*: A comparison of itraconazole versus fluconazole as maintenance therapy for AIDS-associated cryptococcal meningitis. *Clin Infect Dis* **28:**291–296, 1999.

415. Larsen RA, Bozzette S, McCutchan JA, *et al*: Persistent *Cryptococcus neoformans* infection of the prostate after successful treatment of meningitis. *Ann Intern Med* **111:**125–128, 1989.

416. Bozzette SA, Larsen RA, Chiu J, *et al*: Fluconazole treatment of persistent *Cryptococcus neoformans* prostatic infection in AIDS. *Ann Intern Med* **115:**285–286, 1991.

417. Sobel RA, Ellis WG, Nielsen SL, *et al*: Central nervous system coccidioidomycosis: A clinicopathologic study of treatment with and without amphotericin B. *Hum Pathol* **15:**980–995, 1984.

418. Labadie EL, Hamilton RH: Survival improvement in coccidioidal meningitis by high-dose intrathecal amphotericin B. *Arch Intern Med* **146:**2013–2018, 1986.

419. Galgiani JN, Catanzaro A, Cloud G, *et al*: Fluconazole therapy for coccidioidal meningitis. *Ann Intern Med* **119:**28–35, 1993.

420. Smego RA, Perfect JR, Durack DT: Combined therapy with amphotericin B and 5-flucytosine for *Candida* meningitis. *Rev Infect Dis* **6:**791–801, 1984.

421. Casado JL, Quereda C, Oliva J, *et al*: Candidal meningitis in HIV-infected patients: Analysis of 14 cases. *Clin Infect Dis* **25:**673–676, 1997.

422. Rodriguez-Arrondo F, Aguirrebengoa K, De Arce A, *et al*: Candidal meningitis in HIV-infected patients: Treatment with fluconazole. *Scand J Infect Dis* **30:**417–418, 1998.

423. Robinson PA, Bauer M, Leal MAE, *et al*: Early mycological treatment failure in AIDS-associated cryptococcal meningitis. *Clin Infect Dis* **28:**82–92, 1999.

424. Bach MC, Tally PW, Godofsky EW: Use of cerebrospinal fluid shunts in patients having acquired immunodeficiency syndrome with cryptococcal meningitis and uncontrollable intracranial hypertension. *Neurosurgery* **41:**1280–1283, 1997.

425. Fessler RD, Sobel J, Guyot L, *et al*: Management of elevated intracranial pressure in patients with cryptococcal meningitis. *J Acquir Immune Defic Syndr* **17:**137–142, 1998.

426. Park MK, Hospenthal DR, Bennett JE: Treatment of hydrocephalus secondary to cryptococcal meningitis by use of shunting. *Clin Infect Dis* **28:**629–633, 1999.

427. Sepkowitz K, Armstrong D: Space-occupying fungal lesions of the central nervous system. In Scheld WM, Whitley RJ, Durack DT (eds): *Infections of the Central Nervous System*, 2nd ed. Lippincott-Raven Publishers, Philadelphia, 1997, pp. 741–762.

428. Hagensee ME, Bauwens JE, Kjos B, Bowden RA: Brain abscess following marrow transplantation: Experience at the Fred Hutchinson Cancer Center, 1984–1992. *Clin Infect Dis* **19:**402–408, 1994.

429. Selby R, Ramirez CB, Singh R, *et al*: Brain abscess in solid organ transplant recipients receiving cyclosporine-based immunosuppression. *Arch Surg* **132:**304–310, 1997.

430. Parker JC Jr, McCloskey JJ, Lee RS: The emergence of candidosis: The dominant postmortem cerebral mycosis. *Am J Clin Pathol* **70:**31–36, 1978.

431. Burgert SJ, Classen DC, Burke JP, Blatter DD: Candidal brain abscess associated with vascular invasion: A devastating complication of vascular catheter-related candidemia. *Clin Infect Dis* **21:**202–205, 1995.

432. Denning DW: Invasive aspergillosis. *Clin Infect Dis* **26:**781–805, 1998.

433. Young RC, Bennett JE, Vogel CL, *et al*: Aspergillosis: The spectrum of the disease in 98 patients. *Medicine* **49:**147–173, 1970.

434. Meyer RD, Young LS, Armstrong D, *et al*: Aspergillosis complicating neoplastic disease. *Am J Med* **54:**6–15, 1973.

435. Weiland D, Ferguson RM, Peterson PK, *et al*: Aspergillosis in 25 renal transplant patients: Epidemiology, clinical presentation, diagnosis, and management. *Ann Surg* **198:**622–629, 1983.

436. Britt RH, Enzmann DR, Remington JS: Intracranial infection in cardiac transplant recipients. *Ann Neurol* **9:**107–119, 1981.

437. Green M, Wald ER, Tzakis A, *et al*: Aspergillosis of the CNS in a pediatric liver transplant recipient: Case report and review. *Rev Infect Dis* **13:**653–657, 1991.

438. Beal MF, O'Carroll P, Kleinman GM, *et al*: Aspergillosis of the nervous system. *Neurology* **32:**473–479, 1982.

439. Denning DW, Follansbee SE, Scolaro M, *et al*: Pulmonary aspergillosis in the acquired immunodeficiency syndrome. *N Engl J Med* **324:**654–662, 1991.

440. Minamoto GY, Barlam TF, Vander Els NJ: Invasive aspergillosis in patients with AIDS. *Clin Infect Dis* **14:**66–74, 1992.

441. Walsh TJ, Hier DB, Caplan LR: Aspergillosis of the central nervous system: Clinicopathological analysis of 17 patients. *Ann Neurol* **18:**574–582, 1985.

442. Meyer RD, Rosen P, Armstrong D: Phycomycosis complicating leukemia and lymphoma. *Ann Intern Med* **77:**871–879, 1972.

443. Parfrey NZ: Improved diagnosis and prognosis of mucormycosis. *Medicine* **65:**113–123, 1986.

444. Morduchowicz G, Shmueli D, Shapira Z, *et al*: Rhinocerebral mucormycosis in renal transplant recipients. *Rev Infect Dis* **8:**441–446, 1986.

445. Stave GM, Heimberger T, Kerkering TM: Zygomycosis of the basal ganglia in intravenous drug users. *Am J Med* **86:**115–117, 1989.

446. Daly AL, Velazquez LA, Bradley SF, *et al*: Mucormycosis: Association with deferoxamine therapy. *Am J Med* **87:**468–471, 1989.

447. Sugar AM: Mucormycosis. *Clin Infect Dis* **14:**S126–S129, 1992.

448. Rinaldi MG: Zygomycosis. *Infect Dis Clin North Am* **3:**19–41, 1989.

449. Travis LB, Roberts GD, Wilson WR: Clinical significance of *Pseudallescheria boydii*: A review of ten years' experience. *Mayo Clin Proc* **60:**531–537, 1985.

450. Berenguer J, Diaz-Mediavilla J, Urra D, *et al*: Central nervous system infection caused by *Pseudallescheria boydii*. *Rev Infect Dis* **11:**890–896, 1989.

451. Dworzack DL, Clark RB, Borkowski WJ, *et al*: *Pseudallescheria boydii* brain abscess: Association with near-drowning and efficacy of high-dose, prolonged miconazole therapy in patients with multiple abscesses. *Medicine* **68:**218–224, 1989.

452. Kershaw P, Freeman R, Templeton D, *et al*: *Pseudallescheria boydii* infection of the central nervous system. *Arch Neurol* **47:**468–472, 1990.

453. Rangel-Guerra R, Martinez HR, Saenz C: Mucormycosis: Report of 11 cases. *Arch Neurol* **42:**578–581, 1985.

454. Anaissie EJ, Shikhani AH: Rhinocerebral mucormycosis with internal carotid occlusion: Report of two cases and review of the literature. *Laryngoscope* **95:**1107–1113, 1985.

455. Anderson D, Matick H, Naheedy MH, *et al*: Rhinocerebral mucormycosis with CT scan findings. *Comput Radiol* **8:**113–117, 1984.

456. Press GA, Weindling SM, Hesselink JR, *et al*: Rhinocerebral mucormycosis: MR manifestations. *J Comput Assist Tomogr* **12:**744–749, 1988.

457. Denning DW, Stevens DA: Antifungal and surgical treatment of invasive aspergillosis: Review of 2,121 published cases. *Rev Infect Dis* **12:**1147–1201, 1990.

458. Zuckerman J, Tunkel AR: Itraconazole: A new triazole antifungal agent. *Infect Control Hosp Epidemiol* **15:**397–410, 1994.

459. Sanchez C, Mauri E, Dalmau D, *et al*: Treatment of cerebral aspergillosis with itraconazole: Do high doses improve the prognosis? *Clin Infect Dis* **21:**1485–1487, 1995.

460. Goodman ML, Coffey RJ: Stereotactic drainage of *Aspergillus* brain abscess with long-term survival: Case report and review. *Neurosurgery* **24:**96–99, 1989.

461. Venugopal PV, Venugopal TV, Thiruneelakantan K, *et al*: Cerebral aspergillosis: Report of two cases. *Sabouraudia* **15:**225–230, 1977.

462. Klein HJ, Richter HP, Schachenmayr W: Intracerebral *Aspergillus* abscess: Case report. *Neurosurgery* **13:**306–309, 1983.

463. Ochi JW, Harris JP, Feldman JI, *et al*: Rhinocerebral mucormycosis: Results of aggressive surgical debridement and amphotericin B. *Laryngoscope* **98:**1339–1342, 1988.

464. Couch L, Theilen F, Mader JT: Rhinocerebral mucormycosis with cerebral extension successfully treated with adjunctive hyperbaric oxygen therapy. *Arch Otolaryngol Head Neck Surg* **114:**791–794, 1988.

465. Ferguson BJ, Mitchell TG, Moon R, *et al*: Adjunctive hyperbaric oxygen for treatment of rhinocerebral mucormycosis. *Rev Infect Dis* **10:**551–559, 1988.

466. Fisher JF, Shadomy S, Teabeaut JR, *et al*: Near-drowning complicated by brain abscess due to *Petriellidium boydii*. *Arch Neurol* **39:**511–513, 1982.

467. Bailey T, Graham MB, Powderly W: Disseminated *Pseudallescheria boydii* infection treated with fluconazole. Abstract 71. In Sixth International Symposium on Infections in the Immunocompromised Host, Peebles, Scotland, 1990.

468. Nomdedeu J, Brunet S, Martino R, *et al*: Successful treatment of pneumonia due to *Scedosporium apiospermum* with itraconazole: Case report. *Clin Infect Dis* **16:**731–733, 1993.

469. Nasky MA, McDougal, Peacock JE Jr: *Pseudallescheria boydii* brain abscess successfully treated with voriconazole and surgical drainage: Case report and literature review of central nervous system pseudallescheriasis. *Clin Infect Dis* **31:**673–677, 2000.

470. Dukes CS, Luft BJ, Durack DT: Toxoplasmosis of the central nervous system. In Scheld WM, Whitley RJ, Durack DT (eds): *Infections of the Central Nervous System*, 2nd ed. Lippincott-Raven Publishers, Philadelphia, 1997, pp. 785–806.

471. Carey RM, Kimball AC, Armstrong D, *et al*: Toxoplasmosis: Clinical experiences in a cancer hospital. *Am J Med* **54:**30–38, 1973.

472. Ruskin J, Remington JS: Toxoplasmosis in the compromised host. *Ann Intern Med* **84:**193–199, 1976.

473. Hakes TB, Armstrong D: Toxoplasmosis: Problems in diagnosis and treatment. *Cancer* **52:**1535–1540, 1983.

474. Hakim M, Esmore D, Wallwork J, *et al*: Toxoplasmosis in cardiac transplantation. *Br Med J* **292:**1108, 1986.

475. Reynolds ES, Walls KW, Pfeiffer RI: Generalized toxoplasmosis following renal transplantation. *Arch Intern Med* **118:**401–405, 1966.

476. Deleze M, Mintz G, Carmen Majia MD: *Toxoplasma gondii* encephalitis in systemic lupus erythematosus: A neglected cause of treatable nervous system infection. *J Rheumatol* **12:**994–996, 1985.

477. Green JA, Spruance SL, Cheson BD: Favorable outcome of central nervous system toxoplasmosis occurring in a patient with untreated Hodgkin's disease. *Cancer* **45:**808–810, 1980.

478. Luft BJ, Naot Y, Araujo FG, *et al*: Primary and reactivated *Toxoplasma* infection in patients with cardiac transplants. *Am Coll Physicians* **99:**27–31, 1983.

479. Nagington J, Martin AL: Toxoplasmosis and heart transplantation. *Lancet* **2:**679, 1983.

480. Luft BJ, Brooks RG, Conley FK, *et al*: Toxoplasmic encephalitis in patients with acquired immune deficiency syndrome. *JAMA* **252:**913–917, 1984.

481. Navia BA, Petito CK, Gold JWM, *et al*: Cerebral toxoplasmosis complicating the acquired immune deficiency syndrome: Clinical and neuropathological findings in 27 patients. *Ann Neurol* **19:**224–238, 1986.

482. Luft BJ, Remington JS: Toxoplasmic encephalitis in AIDS. *Clin Infect Dis* **15:**211–222, 1992.

483. Renold C, Sugar A, Chave JP, *et al*: *Toxoplasma* encephalitis in patients with acquired immunodeficiency syndrome. *Medicine* **71:**224–239, 1992.

484. Porter SB, Sande MA: Toxoplasmosis of the central nervous system in the acquired immunodeficiency syndrome. *N Engl J Med* **327:**1643–1648, 1992.

485. Carr A, Tindall B, Brew BJ, *et al*: Low-dose trimethoprim sulfamethoxazole prophylaxis for toxoplasmic encephalitis in patients with AIDS. *Ann Intern Med* **117:**106–111, 1992.

486. Furrer H, Egger M, Opravil M, *et al*: Discontinuation of primary prophylaxis against *Pneumocystis carinii* pneumonia in HIV-1–infected adults treated with combination antiretroviral therapy. *N Engl J Med* **340:**1301–1306, 1999.

487. Montoya JG, Remington JS: *Toxoplasma gondii*. In Mandell GL, Bennett JE, Dolin R (eds): *Principles and Practice of Infectious Diseases*, 5th ed. Churchill Livingstone, Philadelphia, 2000, pp. 2858–2888.

488. Brooks RG, McCabe RE, Remington JS: Role of serology in the diagnosis of toxoplasmic lymphadenopathy. *Rev Infect Dis* **9:**775–782, 1987.

489. Rose AG, Uys CJ, Novitsky D, *et al*: Toxoplasmosis of donor and recipient hearts after heterotopic cardiac transplantation. *Arch Pathol Lab Med* **107:**368–373, 1990.

490. Peacock JE Jr, Folds J, Orringer E, *et al*: *Toxoplasma gondii* and the compromised host. *Arch Intern Med* **143:**1235–1237, 1983.

491. Cohn JA, McMeeking A, Cohen W, *et al*: Evaluation of the policy of empiric treatment of suspected *Toxoplasma* encephalitis in patients with the acquired immunodeficiency syndrome. *Am J Med* **86:**521–527, 1989.

492. Cimino C, Lipton RB, Williams A, *et al*: The evaluation of patients with human immunodeficiency virus-related disorders and brain mass lesions. *Arch Intern Med* **151:**1381–1384, 1991.

493. Bishburg E, Eng RHK, Slim J, *et al*: Brain lesions in patients with acquired immunodeficiency syndrome. *Arch Intern Med* **149:**941–943, 1989.

494. Potasman I, Resnick L, Luft BJ, *et al*: Intrathecal production of antibodies against *Toxoplasma gondii* in patients with toxoplasmic encephalitis and the acquired immunodeficiency syndrome (AIDS). *Ann Intern Med* **108:**49–51, 1988.

495. Elkin CM, Leon E, Grenell SL, *et al*: Intracranial lesions in the acquired immunodeficiency syndrome: Radiological (computed tomographic) features. *JAMA* **253:**393–396, 1985.

496. Post MJD, Sheldon JJ, Hensley GT, *et al*: Central nervous system disease in acquired immunodeficiency syndrome: Prospective correlation using CT, MR imaging and pathologic studies. *Radiology* **158:**141–148, 1986.

497. Jacobs F, Depierreux M, Goldman M, *et al*: Role of bronchoalveolar lavage in diagnosis of disseminated toxoplasmosis. *Rev Infect Dis* **13:**637–641, 1991.

498. Sun T, Greenspan J, Tenenbaum M, *et al*: Diagnosis of cerebral toxoplasmosis using fluorescein-labeled antitoxoplasma monoclonal antibodies. *Am J Surg Pathol* **10:**312–316, 1986.

499. Luft BJ, Hafner R, Korzun AH, *et al*: Toxoplasmic encephalitis in patients with the acquired immunodeficiency syndrome. *N Engl J Med* **329:**995–1000, 1993.

500. Ruiz A, Ganz WI, Post MJD, *et al*: Use of thallium-201 brain SPECT to differentiate cerebral lymphoma from *Toxoplasma* encephalitis in AIDS patients. *Am J Neuroradiol* **15:**1885–1894, 1994.

501. Berry I, Gaillard JF, Guo Z, *et al*: Cerebral lesions in AIDS. What can be expected from scintigraphy? Cerebral tomographic scintigraphy using thallium-201: A contribution to the differential diagnosis of lymphomas and infectious lesions. *J Neuroradiol* **22:**218–228, 1995.

502. Pierce MA, Johnson MD, Maciunas RJ, *et al*: Evaluating contrast-enhancing brain lesions in patients with AIDS by using positron emission tomography. *Ann Intern Med* **123:**594–598, 1995.

503. Quality Standards Subcommittee of the American Academy of Neurology: Evaluation and management of intracranial mass lesions in AIDS. *Neurology* **50:**21–26, 1998.

504. Haverkos HW: Assessment of therapy for *Toxoplasma* encephalitis: The TE study group. *Am J Med* **82:**907–914, 1987.

505. Leport C, Raffi F, Matheron S, *et al*: Treatment of central nervous system toxoplasmosis with pyrimethamine/sulfadiazine combination in 35 patients with the acquired immunodeficiency syndrome: Efficacy of long-term continuous therapy. *Am J Med* **84:**94–100, 1988.

506. Rolston KVI, Hoy J: Role of clindamycin in the treatment of central nervous system toxoplasmosis. *Am J Med* **83:**551–554, 1987.

507. Dannemann BR, Israelski DM, Remington JS: Treatment of toxoplasmic encephalitis with intravenous clindamycin. *Arch Intern Med* **148:**2477–2482, 1988.

508. Westblom TO, Belshe RB: Clindamycin therapy of cerebral toxoplasmosis in an AIDS patient. *Scand J Infect Dis* **20:**561–563, 1988.

509. Podzamczer D, Gudiol F: Clindamycin in cerebral toxoplasmosis. *Am J Med* **84:**800, 1988.

510. Dannemann B, McCutchan JA, Israelski D, *et al*: Treatment of toxoplasmic encephalitis in patients with AIDS: A randomized trial comparing pyrimethamine plus clindamycin to pyrimethamine plus sulfadiazine. *Ann Intern Med* **116:**33–43, 1992.

511. Katlama C, De Wit S, O'Doherty E, *et al*: Pyrimethamine-clindamycin vs. pyrimethamine-sulfadiazine as acute and long-term therapy for toxoplasmic encephalitis in patients with AIDS. *Clin Infect Dis* **22:**268–275, 1996.

512. Torre D, Casari S, Speranza F, *et al*: Randomized trial of trimethoprim-sulfamethoxazole versus pyrimethamine-sulfadiazine for therapy of toxoplasmic encephalitis in patients with AIDS. *Antimicrob Agents Chemother* **42:**1346–1349, 1998.

513. Scowden EB, Schaffner W, Stone WJ: Overwhelming strongyloidiasis: An unappreciated opportunistic infection. *Medicine* **57:**527–544, 1978.

514. Igra-Siegman Y, Kapila R, Sen P, *et al*: Syndrome of hyperinfection with *Strongyloides stercoralis*. *Rev Infect Dis* **3:**397–407, 1981.

515. Capello M, Hotez P: Disseminated strongyloidiasis. *Semin Neurol* **13:**169–174, 1993.

516. Cameron ML, Durack DT: Helminthic infections of the central nervous system. In Scheld WM, Whitley RJ, Durack DT (eds): *Infections of the Central Nervous System*, 2nd ed. Lippincott-Raven Publishers, Philadelphia, 1997, pp. 845–878.

517. Schainberg L, Scheinberg MA: Recovery of *Strongyloides ster-coralis* by bronchoalveolar lavage in a patient with acquired immunodeficiency syndrome. *Am J Med* **87:**486, 1989.

518. Owor R, Wamukota WM: A fatal case of strongyloidiasis with *Strongyloides* larvae in the meninges. *Trans R Soc Trop Med Hyg* **70:**497–499, 1976.

519. Neefe LI, Pinilla O, Garagusi VF, *et al*: Disseminated strongyloidiasis with cerebral involvement. *Am J Med* **55:**832–838, 1973.

520. Arroyo JC, Brown A: Concentrations of thiabendazole and parasite-specific IgG antibodies in the cerebrospinal fluid of a patient with disseminated strongyloidiasis. *J Infect Dis* **156:**520–523, 1987.

6

Fungal Infections in the Immunocompromised Host

L. JOSEPH WHEAT, MITCHELL GOLDMAN, and GEORGE A. SAROSI

1. Introduction

Candida, *Aspergillus*, and Mucoraceae are the most common causes of serious fungal infection in granulocytopenic patients, while *Cryptococcus neoformans*, *Histoplasma capsulatum*, and *Coccidioides immitis* are important pathogens in patients with impaired cellular immunity. Infections with less virulent fungi, such as *Trichosporon*, *Fusarium*, *Alternaria*, *Pseudallescheria*, and dematiaceous fungi, are being recognized more frequently.

Neutrophils, monocytes, and macrophages provide the major host defense against *Candida*, *Aspergillus*, and Mucoraceae.[1,2] Diseases or medications that damage phagocytic cells predispose to infection with these fungi. Rapidly progressive illnesses characterize infections in patients with severe granulocytopenia, while a more indolent course may be seen in those with less profound granulocytopenia or with diseases that impair granulocyte function. Cellular immunity provides the most important defense against *C. neoformans*, *H. capsulatum*, and *C. immitis*. They behave as opportunistic pathogens in patients with defective cellular immunity, causing progres-

sive disseminated disease rather than self-limited pulmonary infections. Helper T cells specific for the invading pathogen elaborate cytokines that arm macrophages to inhibit or kill fungal pathogens. Mucosal, but not systemic, *Candida* infections also are characteristic of cellular immunodeficiency.

1.1. Fungal Infections in Granulocytopenia

Fungal infections are major causes of morbidity and mortality in granulocytopenic patients.[3] Fungal infections occur less frequently in granulocytopenic patients with solid tumors or lymphoma than with leukemia. Duration of granulocytopenia and use of broad-spectrum antibiotics or corticosteroids are risk factors for invasive fungal disease in these patients.[4]

Candida and *Aspergillus* account for most of the fungal infections following bone marrow transplantation,[5–7] while other fungi are increasing in importance.[8,9] Infections following bone marrow transplantation are classified as those developing before engraftment (up to 30 days) and those in the postengraftment period (day 30 to day 100), with the highest incidence occurring after engraftment.[7] Also, a period of vulnerability exists following transplantation during the interval between loss of native immunity and development of passively transferred immunity from the donor. Pretransplant characteristics including underlying myelodysplastic syndrome and unrelated donor status are risk factors for invasive fungal infection.[7] In the postengraftment period, use of corticosteroids for chronic graft-versus-host disease (GVHD) is a significant factor predisposing to fungal infection.[7]

L. Joseph Wheat • Departments of Medicine and Pathology, Indiana University School of Medicine, and Department of Veterans Affairs Hospital, Indianapolis, Indiana 46202. **Mitchell Goldman** • Department of Medicine, Indiana University School of Medicine, Indianapolis, Indiana 46202. **George A. Sarosi** • Department of Medicine, Indiana University School of Medicine, and Department of Veterans Affairs Hospital, Indianapolis, Indiana 46202.

Clinical Approach to Infection in the Compromised Host (Fourth Edition), edited by Robert H. Rubin and Lowell S. Young. Kluwer Academic/Plenum Publishers, New York, 2002.

1.2. Fungal Infections in Organ Transplantation

Serious fungal infection occurs in from 5% of patients following renal transplantation to over 20% following liver transplantation, most presenting within 2 months of transplantation. Risk factors include severe underlying disease, prolonged operative times, extended intensive care unit stays, use of broad-spectrum antibiotics, intense immunosuppression with frequent administration of high-dose pulse corticosteroid therapy, and use of devices that breach skin and mucous membrane barriers. More recent studies have identified additional risk factors for fungal infection following liver transplantation, including pre-transplantation anemia, return to surgery, prolonged use of ciprofloxacin[10]; intra-abdominal bleeding, fulminant hepatitis, cytomegalovirus infection[11]; and creatinine >3 mg/dl, operative time >11 hr, retransplantation and early fungal colonization.[12]

The spectrum of fungal diseases varies during the early versus late time period after organ transplantation. For example, Candida and Aspergillus account for most of the early fungal infections following liver transplantation (Candida, 70%, and Aspergillus, 25%). Impaired phagocytic cell function caused by use of high-dose corticosteroid therapy predisposes to these early infections. Later, chronic low-dose corticosteroids, cyclosporin A, or tacrolimus and mycophenolate mofetil impair cellular immunity, predisposing to mucocutaneous candidiasis and systemic mycoses.

1.3. Fungal Infections in Acquired Immunodeficiency Syndrome

Fungal infections are common in patients with acquired immunodeficiency syndrome (AIDS). Mucocutaneous candidiasis occurs in up to 90% of patients but systemic candidiasis is rare. Cryptococcal meningitis occurs in 5 to 12% of patients, histoplasmosis in 2 to 5%, and coccidioidomycosis in 5% in the southwestern United States. Aspergillosis and infections with Mucoraceae, *Pseudallescheria*, *Alternaria*, *Blastomyces dermatitidis*, *Paracoccidioides brasiliensis*, and *Sporothrix schenckii* have been reported.

The incidence and severity of systemic fungal infections increase with progression of HIV infection and reduction in CD4 counts. While mucocutaneous candidiasis may develop in patients with CD4 counts between 200 and 500/μl, most patients have lower counts. Esophagitis occurs after CD4 counts fall below 100 cells/μl.[13] Systemic mycoses usually occur in patients with CD4 counts below 100 (median below 50 cells/μl), and aspergillosis in those with counts below 50 cells/μl,

often in conjunction with granulocytopenia or corticosteroid therapy.[14] Although the incidence of mucosal and systemic mycoses has fallen since introduction of highly active antiretroviral therapies,[15] they persist as problems in persons who have not accessed health care, who are nonadherent to therapy, or who have failed antiretroviral therapy.

2. Antifungal Prophylaxis in the Immunocompromised Host

2.1. Prophylaxis in Hematology

A recent meta-analysis evaluating nearly 6000 subjects enrolled in randomized clinical trials has concluded that antifungal prophylaxis with fluconazole, itraconazole, or low-dose amphotericin B has reduced the morbidity and mortality related to fungal infections in neutropenic cancer patients, particularly among those receiving hematopoietic stem cell or bone marrow transplants.[16] While a number of different regimens of antifungal prophylaxis have resulted in reductions in fungal colonization and superficial fungal infections, a more important goal is the prevention of serious life-threatening systemic fungal infections. To date, the greatest benefits of antifungal prophylaxis in reducing systemic fungal infections have been realized in patients following allogeneic or high-risk autologous bone marrow transplantation where significant decreases in *Candida albicans* infections have been realized.[17,18] As not all patients receiving chemotherapy for acute leukemia are at high risk for invasive fungal infections, it has been more difficult to demonstrate benefit in this population using antifungal prophylaxis.[19,20] Most presently available antifungals when used in prophylaxis regimens have significant limitations, including low efficacy, toxicity, drug interactions, and cost.[21–23] Acknowledging these limitations, antifungal prophylaxis appears to be appropriate at institutions where there is a high incidence of invasive fungal infections in patients expected to be profoundly granulocytopenic for >7 days and for those requiring significant doses of immunosuppressant medications following bone marrow or stem cell transplantation.

2.1.1. Fluconazole

Fluconazole has several advantages over other presently available antifungal agents including its overall safety profile, excellent bioavailability in both oral and intravenous preparations, and minor drug interaction potential. Large placebo-controlled studies have demonstrated a useful role for fluconazole prophylaxis in bone marrow transplant recipients. In addition to reductions in superficial fungal infections, fluconazole prophylaxis at a dose of 400 mg daily was associated with a reduction in systemic candidiasis from 18% to 7% in one study and from 15.8% to 2.8% in another study of patients undergoing allogeneic or autologous bone marrow transplantation.[17,18] A more recent study has shown reductions in proven and probable invasive fungal infections in patients receiving autologous bone marrow transplants not supported with hematopoietic growth factors, while reductions in fungal infections were not seen for those treated with hematopoietic growth factors.[20] Fluconazole prophylaxis was associated with a reduction in overall mortality in one large study in bone marrow transplant recipients,[18] presumably because of the continuation of fluconazole during the 2 months of high risk following transplantation. As expected, fluconazole prophylaxis has had no effect on the incidence of aspergillosis. Fluconazole 100–200 mg daily also may be effective.[24]

Fluconazole prophylaxis during therapy for acute leukemia reduced fungal colonization and superficial infections but did not prevent invasive fungal infections.[19,23] The lack of consistent efficacy of fluconazole prophylaxis in leukemic populations may be attributed to inclusion of heterogeneous populations with differing risks for serious mycoses. Fluconazole prophylaxis is most efficacious in patients with acute myelogenous leukemia undergoing induction therapy with cytarabine plus anthracycline-based regimens.[20]

Fluconazole is less active against non-*albicans Candida* species[25] and is inactive against filamentous fungi. In particular, *C. krusei* is resistant to fluconazole and *C. glabrata* isolates are frequently less susceptible or resistant. Retrospective review of fungal infections after bone marrow transplantation showed a sevenfold increase in *C. krusei* infection in patients receiving fluconazole and an increase in colonization with *C. glabrata*,[26] as recently confirmed by others.[27] Also, fluconazole-resistant *C. albicans* has been noted in patients receiving prophylaxis. While the emergence of colonization and infection with fluconazole-resistant organisms is concerning, the reduction in serious *Candida* infection and its attributable mortality justifies its use for prophylaxis.[27]

Illustrative Case 1

A 34-year-old female underwent a T-cell-depleted allogeneic bone marrow transplantation for chronic myelocytic leukemia. Antifungal prophylaxis with fluconazole, 200 mg daily, was started concurrently with chemotherapy. On the 8th posttransplant day *C. krusei* was isolated from blood, showing a minimum inhibitory concentration (MIC) to fluconazole of >64 μg/ml. Amphotericin B was initiated, the central

venous catheters were removed, and the patient defervesced on the following day.

2.1.2. Itraconazole

Itraconazole is attractive as prophylaxis because of its activity against *Aspergillus*. Most studies have used the capsule formulation, which exhibits variable absorption and drug interactions with agents that reduce gastric acidity. An uncontrolled study of patients with hematologic malignancies reported reduction in proven fatal fungal infections, mostly aspergillosis, in patients who had received itraconazole 400 mg daily compared to those who received ketoconazole in a prior study.[28] Other uncontrolled studies also showed reductions in systemic yeast infections[29] or fungal mortality,[30] compared to historical controls. In a small placebo-controlled study in patients with hematologic malignancies, however, itraconazole prophylaxis when added to oral amphotericin B prophylaxis did not prevent systemic fungal infections or mortality.[31] A study comparing itraconazole 100 mg and fluconazole 50 mg, each given twice daily, in patients undergoing chemotherapy or autologous stem cell transplantation observed no differences in the rate of fungal infections or mortality,[32] but dosages of both agents were too low to assess their efficacies. Blood concentrations of itraconazole of at least 1000 ng/ml are needed to prevent aspergillosis,[33] requiring doses of up to 600 mg daily.

The oral solution formulation of itraconazole in cyclodextrin is better absorbed and yields higher blood concentrations than does the capsule. In a controlled study in patients with hematologic malignancies, itraconazole solution at a dose of 2.5 mg/kg twice daily was associated with a reduction in fungemias due to *Candida* species (0.5% vs. 4% in placebo recipients) and in proven and suspected deep fungal infections from 33% to 24%.[34] No differences in the rate of aspergillosis or mortality were observed. Experience with itraconazole solution 2.5 mg/kg twice daily compared to fluconazole 100 mg orally in a nonblinded randomized trial of 581 patients with hematologic malignancies showed no significant difference in proven invasive fungal infections.[35] There were six (2%) proven fungal infections in the fluconazole group and one (0.3%) in the itraconazole group. *Post hoc* analysis of subjects who developed fungal infections beyond the specified period of study observation did reveal a lower rate of proven fungal infections and aspergillosis in the itraconazole group. A trial is in progress in bone marrow transplant patients comparing fluconazole to itraconazole given intravenously and then orally in the solution formulation.

2.1.3. Intravenous Amphotericin B

The use of amphotericin B as primary antifungal prophylaxis has been less well studied than fluconazole prophylaxis. Uncontrolled trials have reported significant reductions in systemic fungal infections including aspergillosis in allogeneic marrow transplant recipients after instituting low-dose amphotericin B.[36] Low-dose amphotericin B 0.1 mg/kg per day resulted in reduction in fungal colonization and empiric use of high-dose amphotericin B, compared to placebo in autologous transplant recipients, but was not associated with reductions in proven fungal infections.[37] Amphotericin B prophylaxis 0.5 mg/kg per day three times weekly was more toxic but no more effective than fluconazole in patients undergoing chemotherapy for leukemia.[38] In a separate study comparing low-dose amphotericin B at a dose of 2 mg/kg daily to fluconazole 400 mg daily in bone marrow transplant patients, low-dose amphotericin B was as effective as fluconazole but was associated with greater toxicity.[39] Overall, the efficacy of low-dose amphotericin B as prophylaxis is unproven and the potential for toxicity is considerable.

2.1.4. Lipid Preparations of Amphotericin B

The lipid preparations of amphotericin B are less nephrotoxic than the deoxycholate formulation.[40] Liposomal amphotericin B (AmBisome) also has been shown to have less infusion-related side effects than standard amphotericin B.[41] In a placebo-controlled study in allogeneic bone marrow transplant recipients, liposomal amphotericin B 1 mg/kg per day reduced fungal colonization but did not prevent invasive fungal infections.[42] A more recent study comparing liposomal amphotericin B at 2 mg/kg three times weekly to placebo also failed to prevent fungal infections.[43] Finally, a trial comparing amphotericin B colloidal dispersion (ABCD, Amphocil) with fluconazole for antifungal prophylaxis was stopped early due to the severe side effects caused by this lipid preparation.[44]

2.2. Antifungal Prophylaxis in Liver Transplant Recipients

Recent studies have indicated that liver transplant recipients at highest risk for serious invasive fungal infections include those with a number of risk factors as indicated in Section 1.2. To date, published studies of antifungal prophylaxis in this patient group have included those patients at variable risk for fungal infections.[45–48]

In a multicenter comparison of fluconazole 100 mg daily compared to oral–nasogastric administration of nystatin suspension 1,000,000 units every 6 hr for 28 days after transplantation, fluconazole was associated with a 50% reduction in superficial fungal infections and colonization attributed to *Candida*.[45] Invasive *Candida* infections occurred in 9% of the nystatin group and 2.6% of the fluconazole group, but the difference did not reach statistical significance. Finally, there was no difference in mortality comparing nystatin to fluconazole. More recently, fluconazole 400 mg daily for 10 weeks after transplantation was compared to placebo in patients undergoing liver transplantation.[46] Fluconazole significantly reduced superficial (28% with placebo vs. 4% with fluconazole) and invasive fungal infections (23% with placebo vs. 6% with fluconazole). Fluconazole was well tolerated without apparent hepatotoxicity, but cyclosporine levels were higher and adverse neurologic events were more common in the fluconazole group. While there were fewer deaths attributed to fungal infections in the fluconazole group, overall mortality was not affected. Fluconazole prevented infections by most *Candida* species except *C. glabrata*.

Itraconazole oral solution 2.5 mg/kg twice daily also has been used in one small placebo-controlled trial in liver transplant recipients and was associated with significant reductions in the combined end point of superficial, proven deep, and suspected deep fungal infections.[49] All infections in this study were attributed to *Candida* species.

In a retrospective study in high-risk liver transplant recipients, low-dose amphotericin B (10–20 mg daily) prevented *Candida* infections and reduced empiric amphotericin B use, when compared to historical controls.[47] Liposomal amphotericin B (AmBisome) has been evaluated in a randomized placebo-controlled trial.[48] At a dose of 1 mg/kg per day for 5 days, started during transplant operation, invasive fungal infections occurred in none of 40 (0%) patients receiving prophylaxis and 6 of 37 (16%) placebo recipients ($P < .01$).[48] No difference in mortality was observed. While the cost of prophylaxis was about $5000 per patient, it was less than the cost to treat the five invasive *Candida* infections and one case of aspergillosis that occurred in the placebo patients. A large trial comparing fluconazole and liposomal amphotericin B as prophylaxis in patients at high risk for invasive fungal infection is in progress.

2.3. Antifungal Prophylaxis in Lung Transplant Recipients

One retrospective study reported that itraconazole reduced the incidence of aspergillosis from 16% in historical controls to 6%.[50] In an attempt to reduce both *Aspergillus* and *Candida* infections, one center used aerosolized amphotericin B 0.2 mg/kg every 8 hr combined with oral or intravenous fluconazole 400 mg daily for a minimum of 30 days and reported a reduction in invasive fungal infections in the early postoperative period from 23% in historical controls to 0% in those receiving prophylaxis.[51] Based on initial promising results using aerosolized amphotericin B lipid complex (Abelcet), a randomized study is under way.[52]

2.4. Ongoing Investigations in Antifungal Prophylaxis

Ongoing evaluations of investigational agents as antifungal prophylaxis in bone marrow transplant patients at high risk for fungal infection include study of SCH 56592 (posaconazole),[53,54] a new triazole agent with activity against *Candida* and *Aspergillus* species, as well as activity against fluconazole-resistant *Candida*, and study of FK 463, a lipopeptide echinocandinlike agent with activity against *Candida* and *Aspergillus* species.[55]

2.5. Recommendations for Prophylaxis

Prophylaxis with fluconazole is recommended in patients who will undergo allogeneic bone marrow transplantation. Patients receiving induction therapy for acute myelogenous leukemia with cytarabine plus anthracycline-based derivatives and patients undergoing autologous bone marrow or stem cell transplantation without the support of hematopoietic growth factors also are appropriate candidates for fluconazole prophylaxis (Table 1). Itraconazole might be considered as an alternative in

TABLE 1. Indications for Antifungal Prophylaxis

Allogeneic bone marrow or stem cell transplant
Induction therapy for acute myelogenous leukemia with cytarabine and anthracycline
Autologous bone marrow or stem cell transplant *not* supported with hematopoietic growth factors
Liver transplant at high risk for fungal infection with ≥2 of the following[103]:
  Retransplantation
  Creatinine >2 mg/dl
  Choledochojejunostomy
  Intraoperative use of ≥40 units of blood products
  Fungal colonization detected within the first 3 days after transplantation

institutions experiencing a high incidence of aspergillosis. Routine antifungal prophylaxis is not recommended for patients undergoing chemotherapy for solid tumors, patients with leukemia receiving less intensive chemotherapy, or those undergoing autologous bone marrow or stem cell transplant expected to have a shortened duration of neutropenia following treatment with hematopoietic growth factors.

The role of prophylaxis in organ transplantation is less well understood because adequate studies have not been conducted. The problems and cost associated with antifungal prophylaxis caution against its liberal use, in the absence of factors placing the patient at an inordinate risk for serious mycotic infection. Thus, prophylaxis is not routinely recommended. Appropriate candidates for antifungal prophylaxis include patients undergoing liver transplantation at high risk for fungal infection (Table 1). Awaiting comparative trials, fluconazole 400 mg daily or liposomal amphotericin B at a dose of 1 mg/kg daily should be considered.

Secondary prophylaxis for patients who have recently recovered from a systemic or invasive mycosis, such as aspergillosis or chronic disseminated candidiasis, is recommended during subsequent immunosuppression. Chronic maintenance therapy also is important in those with ongoing, severe immunosuppression.

3. Empiric Antifungal Therapy in Neutropenic Patients with Persistent Fever

In the 1970s to early 1980s, invasive fungal infections were increasingly recognized as a cause of morbidity and mortality in neutropenic cancer patients, particularly those undergoing intensive cytotoxic chemotherapy for acute myelogenous leukemia.[56] Unfortunately, due to the lack of sensitivity of fungal cultures accompanied by the difficulties of performing invasive procedures in these patients, fungal infections often were undiagnosed in the antemortem period and only recognized at autopsy.[56] Undiagnosed fungal infections were documented in 24 to 64% of cancer patients at autopsy.[56–58] Further observations indicated that the mortality could be improved by early therapy,[59,60] supporting empiric antifungal therapy without documentation of fungal infection.

Early studies showed that empiric amphotericin B treatment reduced the incidence of invasive fungal infections and improved mortality.[56,61] More recent studies have examined the use of the newer lipid formulations and triazoles, and clinical trials are in progress evaluating extended-spectrum azoles and echinocandins.

3.1. Amphotericin B Deoxycholate

The earliest reported randomized trial of empiric amphotericin B in neutropenic pediatric cancer patients was performed by Pizzo and co-workers[56] with intravenous amphotericin B at a dose of 0.5 mg/kg per day. Patients without a known source of infection who had persistent fever despite antibacterial treatment were randomized either to continue antibacterial therapy alone or to have amphotericin B added. Only 6% of subjects randomized to receive amphotericin B developed a systemic fungal infection, while invasive fungal infections were documented in 31% of controls. While these results did not achieve statistical significance, an analysis including patients from a nonrandomized portion of the study did show significant benefit in the amphotericin B arm compared to the controls (2.5% vs. 31%, respectively). No difference in mortality was seen between groups in the randomized study, however.

A similar study of neutropenic cancer patients with persistent fever despite antibiotics reported significantly fewer fungal infections in those randomized to amphotericin B than in controls who continued antibacterials alone (1.4% vs. 9.4%, respectively).[62] Furthermore, death attributed to fungal infection was lower in the amphotericin group (1.4%) than controls (6%) ($P = .05$). Empiric treatment was most beneficial in patients who had not received antifungal prophylaxis, those with neutrophil counts <100/μl, patients with clinically documented fungal infection, and those >16 years of age.

3.2. Lipid Preparation of Amphotericin B as Empiric Antifungal Therapy

A report combining two prospective randomized clinical trials showed that liposomal amphotericin B at 1 or 3 mg/kg daily was significantly better tolerated than conventional amphotericin B at a dose of 1 mg/kg daily.[63] Success of therapy, defined by resolution of fever, lack of need to add antifungal therapy, and lack of development of a systemic fungal infection, was more common in subjects receiving liposomal amphotericin B 3 mg/kg per day (64%) than in those receiving conventional amphotericin B (49%) ($P = .03$). The efficacy of liposomal amphotericin B 1 mg/kg daily was no different than conventional amphotericin B, however.

Another study showed liposomal amphotericin B 3 mg/kg daily and conventional amphotericin B 0.6 mg/kg daily to be equally effective with respect to resolution of fever and mortality, but there were significantly fewer break-through fungal infections with the lipid preparation (3.2%) than conventional amphotericin B (7.8%).[41] Nephrotoxicity and infusion-related side effects also were reduced with liposomal amphotericin B.

Liposomal amphotericin B (AmBisome) at 3 or 5 mg/kg daily has been compared to amphotericin B lipid complex (ABLC, Abelcet) 5 mg/kg.[64] No differences in breakthrough fungal infections or mortality were reported. Nephrotoxicity and treatment discontinuations due to intolerance were less common with liposomal amphotericin B, however.

Amphotericin B colloidal dispersion 4 mg/kg daily has been compared to conventional amphotericin B 0.8 mg/kg daily.[65] The efficacy of this lipid formulation was comparable to that of conventional amphotericin B, while renal dysfunction was significantly less common for those receiving the colloidial dispersion. Infusion-related events were significantly more common in those treated with the colloidial dispersion than with the deoxycholate preparation, however.

3.3. Triazole Agents as Empiric Antifungal Therapy

3.3.1. Fluconazole as Empiric Antifungal Therapy

Fluconazole has been studied as an alternative to amphotericin B as empiric antifungal for febrile neutropenia.[66–68] As fluconazole lacks activity against *Aspergillus*, its empiric use in patients at risk for aspergillosis has remained a significant concern, however. Due to the potential for infection with fluconazole-resistant *Candida*, studies of fluconazole in this setting have excluded patients who had received prior systemic azole prophylaxis.

In one small study comparing empiric fluconazole therapy to amphotericin B in patients with hematologic malignancies, enrollment was limited to those with normal chest radiographs, no previous aspergillosis, and negative surveillance cultures for *Aspergillus*.[66] In this patient population, fluconazole at a dose of 400 mg daily was as effective as amphotericin B 0.8 mg/kg when given after at least 96 hr of prior antibacterial therapy. The largest randomized study comparing fluconazole to amphotericin B to date enrolled 317 patients, including 120 who had undergone bone marrow or stem cell transplantation.[67] Nearly 60% had abnormal chest radiographs at randomization and surveillance studies for *Aspergillus* were not performed. Empiric antifungal therapy was initiated after at least 96 hr of fever unresponsive to antibacterial therapy. Fluconazole was given at a loading dose of

800 mg followed by 400 mg daily and compared to amphotericin B 0.5 mg/kg daily. Fluconazole was as effective as amphotericin B with responses to therapy seen in 68% of fluconazole recipients and 67% of amphotericin B recipients. No differences in overall mortality or mortality due to fungal infections were found. Persistent or new fungal infections were documented in 8% of fluconazole-treated patients and 6% of those receiving amphotericin B. As expected, fluconazole was better tolerated than amphotericin B. While this study demonstrated equal efficacy of fluconazole and amphotericin B in the setting of neutropenic fever, the study was characterized by a low incidence of invasive aspergillosis and relatively short duration of neutropenia, mostly attributed to support with hematopoietic growth factors. The authors of this study indicated that careful assessment including chest computerized tomography (CT) for patients at risk for aspergillosis be performed before considering fluconazole as empiric therapy.

3.3.2. Itraconazole as Empiric Antifungal Therapy

As itraconazole has antifungal activity against both *Candida* and *Aspergillus*, following the development of both intravenous and a more reliably absorbed oral cyclodextrin formulation, there has been increased interest in the use of itraconazole as empiric antifungal therapy for persistently febrile neutropenic patients. While itraconazole has not been extensively studied in this setting, preliminary results from a recently completed randomized trial comparing itraconazole to amphotericin B in patients with underlying hematological malignancies have been reported.[69] In this investigation, amphotericin B at a dose of 0.7–1.0 mg/kg daily was compared with intravenous itraconazole 200 mg twice daily for 2 days followed by 200 mg daily for a maximum of 12 days before converting to itraconazole oral solution at a dose of 200 mg twice daily. This study excluded those with evidence of a fungal infection at baseline. Five patients in each study arm developed a deep fungal infection; overall itraconazole was at least as effective as amphotericin B and less toxic.

3.4. Ongoing Investigations in Empiric Therapy

Ongoing evaluations of investigational agents as empiric antifungal therapy for persistent fever and neutropenia include study of voriconazole,[70] a new triazole agent with activity against *Aspergillus* and *Candida*, including fluconazole-resistant species, and study of cas-

pofungin acetate (MK-0991), an echinocandin agent with activity against *Candida* and *Aspergillus*.[71]

3.5. Recommendations for Empiric Therapy

Amphotericin B 0.7 mg/kg per day is recommended as empiric antifungal therapy for neutropenic patients with granulocyte counts below 500/µl for more than 7 days who have persistent fever of undefined etiology after at least 96 hr of antibiotic therapy or recurrent fever despite antibacterial therapy (Table 2). A higher dose of 1.0 mg/kg per day may be preferred for patients who have received prior antifungal prophylaxis, because of the increased risk for infection with non-*albicans Candida* or *Aspergillus*. Liposomal amphotericin B is at least as effective and better tolerated, but its higher cost limits its use except in special situations, such as renal insufficiency or treatment with medications that increase the risk for renal insufficiency. Fluconazole 400 to 800 mg daily could be considered if the likelihood of aspergillosis was low and the patient had not received a systemic azole for prophylaxis or treatment of a presumed or proven fungal infection. While less well studied, itraconazole is an alternative in subjects who have not received prior systemic azole therapy for prophylaxis or treatment.

4. Candidiasis

4.1. Epidemiology

Candidiasis occurs in 9 to 25% of patients following bone marrow transplantation, 1 to 13% with granulocytopenia caused by chemotherapy for hematologic malignancy, 1 to 2% following chemotherapy for lymphoma, and 0.5% following chemotherapy for solid tumors. Sys-

---

**TABLE 2. Guidelines for Empiric
Antifungal Therapy during Neutropenia**

Neutropenia and persistent or recurrent unexplained fever despite 96 hr of antimicrobials

Especially those not receiving prior systemic antifungal prophylaxis

Amphotericin B 0.7 mg/kg per day *or*

Liposomal amphotericin B (AmBisome) 3 mg/kg per day (though may be cost prohibitive)

Fluconazole 400–800 mg daily for those at low risk for aspergillosis also effective

Itraconazole intravenous followed by oral solution appears to be an effective alternative

temic candidiasis occurs following liver transplantation in up to 18% of patients and less commonly following other types of organ transplantation.

C. albicans is the most common species isolated from immunocompromised host, causing 40 to 50% of cases of systemic candidiasis. *C. glabrata* is responsible for 5 to 35%, and *C. tropicalis* for 8 to 43% of cases. Less common *Candida* species, including *guilliermondi*, *lusitaniae*, *krusei*, and *parapsilosis*, have accounted for 4 to 12% of cases. Non-*albicans* species may be more common in patients who have received antifungal prophylaxis.[26]

Risk factors for systemic candidiasis following bone marrow transplantation included older age, impaired donor match, use of total-body irradiation, presence of underlying disease, occurrence of acute GVHD, and presence of *Candida* colonization. Prolonged antibiotic treatment and long duration of granulocytopenia also are important predisposing factors.

4.2. Clinical Findings

Fever unresponsive to antibiotic therapy is the usual mode of presentation of systemic candidiasis in the neutropenic patient, usually occurring during the second week of antibiotic therapy or at the time of bone marrow recovery. Dissemination is common, more so with *C. tropicalis* than with *C. albicans*.[72] Common sites of dissemination, in descending order, include the liver, kidney, spleen, heart, gastrointestinal tract, skin, lungs, brain, eye, pancreas, and thyroid. Pulmonary manifestations include local, diffuse, or miliary infiltrates. Skin manifestations include nodules, papules, ecthyma gangrenosum, verrucous plaques, and a picture of purpura fulminans associated with coagulopathy and shock.[73] Candidiasis may cause intra-abdominal abscesses and peritonitis following organ transplantation.[74]

Hepatic involvement is common in disseminated candidiasis in neutropenic subjects, but lesions may not be seen until resolution of granulocytopenia.[75] Alkaline phosphatase elevation may be a clue to the presence of hepatic candidiasis. Abdominal CT scan shows focal lesions in the liver and the spleen. The kidneys, muscle, and other organs also may be involved. White nodules, 1–2 mm in size, stud the liver at laparoscopy, and larger lesions measuring several centimeters in diameter may be seen on CT scans. Biopsy shows microabscesses or granulomas, but organisms may be missed unless multiple sections are examined. Despite demonstration of organisms in fungal stains, cultures are positive in only 30 to

60% of patients. Of note, hepatic lesions may disappear during granulocytopenia and recur after granulocyte recovery, complicating assessment of response to therapy.[76]

4.3. Diagnosis

4.3.1. Mycologic Methods

Despite widespread dissemination, fungemia can be documented in only a third to half of patients. If skin lesions are present, organisms can be seen by fungal stain and isolated in cultures. Isolation of *Candida* species from surveillance cultures may provide a clue to the diagnosis of candidiasis. Surveillance cultures were positive in 91% of cases of systemic candidiasis in one study.[77] While *Candida* species often can be isolated from routine culture media, use of special fungal media for blood cultures and mycotic media for isolation of fungi from other specimens may improve the diagnostic yield.

4.3.2. Serologic Methods

Diagnosis by detection of antigenic materials or metabolic products of *Candida* has been reviewed.[78] Measurement by gas chromatography of ratios of D-arabinitol/L-arabinitol in the urine of neutropenic patients yielded positive results in all ten patients with proven candidiasis and negative results in 94% of controls.[79] Detection of antigen in sera by latex agglutination was not accurate, however.[80] Enzyme immunoassay (EIA) methods show promise,[81] but have not been proven to be useful in patient management. Diagnosis by detection of $(1\rightarrow3)$-β-D-glucan[82] and polymerase chain reaction (PCR) amplification of gene products[83] requires confirmation.

Antibody tests are not useful for several reasons. First, several weeks are required to mount an antibody response.[84] Second, colonization leads to development of antibodies in individuals without significant *Candida* infection, detracting from the significance of a positive test. Third, immunocompromised individuals may not mount an antibody response. Tests for *Candida* antibodies have been falsely negative in half of all patients with serious candidiasis.[85]

4.4. Treatment

Challenges in treatment of candidiasis result from the difficulty with recognition and diagnosis and increasing problems with resistance to antifungal therapy. The widespread usage of fluconazole for prophylaxis or empiric therapy has resulted in more infections with non-*albicans* species and fluconazole-resistant *C. albicans*.[8] Polyene resistance has been reported.[86]

Catheter management is an important aspect of treatment of candidemia.[87,88] Most experts recommend changing all nonsurgically placed catheters, but would attempt to sterilize the bloodstream without removal of tunneled catheters unless the tunnel appeared infected[88] or the fungemia persisted despite treatment.

4.4.1. Amphotericin B

Amphotericin B dosage should be at least 0.7 mg/kg per day,[88] but higher doses of 1 mg/kg per day may be needed for non-*albicans* species, including *C. glabrata* and *C. krusei*. Mortality is higher in patients infected by strains with MICs above 0.8 μg/ml.[89] Therapy should be continued until at least 2 weeks after the last positive blood culture and resolution of the clinical manifestation of candidiasis,[88] but amphotericin B may be replaced by fluconazole.

4.4.2. Lipid Formulations of Amphotericin B

Response rates of 50 to 80% in neutropenic patients were observed with AmBisome,[90] Amphotec,[91] or Abelcet.[92] Similar findings were reported in bone marrow and liver transplant recipients.[93,94] A single, unpublished randomized study comparing Abelcet with amphotericin B noted comparable efficacy but improved tolerability with the lipid formulation.[95] In the absence of evidence for improvement in outcome, the main reason to choose a lipid preparation is concern for nephrotoxicity.

4.4.3. 5-Flucytosine

One study reporting a higher mortality in patients receiving amphotericin B alone (84%) compared to amphotericin B and 5-flucytosine (34%) acknowledged that the patients receiving amphotericin B alone were sicker and that the outcomes were similar if the groups were matched for baseline severity of illness.[96] Some experts recommend combined therapy in patients with candidemia who are clinically unstable or have evidence for disseminated infection, but at reduced 5-flucytosine doses of 100 mg/kg per day.[88] Flucytosine blood concentrations should be measured and maintained between 75 and 100 μg/ml and dosage should be adjusted for renal function. However, the poor tolerability and lack of an intravenous preparation reduce the feasibility of this regimen and its utility has not been established.

4.4.4. Fluconazole

Fluconazole has been used for treatment of disseminated candidiasis in those who failed amphotericin B treatment.[97,98] Expert opinion supports fluconazole use in patients who are stable clinically, without evidence for disseminated involvement, infected with *C. albicans*, and who have not received prophylaxis with a triazole antifungal agent.[88] Fluconazole also may play a role for continued therapy in patients who were initially treated with amphotericin B and subsequently experienced clinical improvement and resolution of granulocytopenia.[88] A loading dosage of 12 mg/kg followed by 6 mg/kg daily in patients with normal renal function is recommended. A dosage of 12 mg/kg per day is advised if fluconazole is used for treatment of *C. glabrata* infection and measurement of blood concentrations is suggested in patients receiving this high dosage.

Fluconazole should not be combined with amphotericin B because of absence of data showing that combined therapy improves outcome, but a study evaluating combination therapy of candidemia in the nongranulocytopenic host is under analysis (Dr. John Rex for the Mycoses Study Group). There is potential for antagonism when these agents are combined.

4.4.5. Adjunctive Cytokine Therapy

Cases of hepatosplenic candidiasis have been described that benefited from interferon-γ and granulocyte–monocyte colony-stimulating factor (GM-CSF).[99] GM-CSF also has shown benefit in treating esophageal candidiasis.[100,101] Some experts recommend administration of GM-CSF or granulocyte colony-stimulating factor (G-CSF) in persistently granulocytopenic patients with candidiasis.[88]

4.4.6. Prevention of Recurrence during Subsequent Granulocytopenia

Patients who have recovered from candidiasis are at risk for recurrence during future granulocytopenia.[97,102] Prophylactic use of amphotericin B or fluconazole during subsequent chemotherapy is advised.[102] In one report, none of 15 consecutive patients with prior hepatosplenic candidiasis experienced relapse following bone marrow transplantation under the protection of intravenous amphotericin B begun before transplantation and continued until engraftment.[102] Lesions may not clear despite good clinical responses to therapy, however.[102]

4.4.7. Treatment Recommendations

Recommendations for treatment of candidiasis have been reviewed.[88,103] Amphotericin B 0.7 to 1 mg/kg per day, or one of the lipid formulations (3–5 mg/kg per day) in patients at increased risk for renal impairment is indicated for treatment of patients with more severe manifestations and for those who have recently received a triazole for antifungal prophylaxis (Table 3). Fluconazole, 6 mg/kg daily, adjusted for renal function, is appropriate for those with milder manifestations and for continuation of treatment after response to amphotericin B. Fluconazole should not be used as initial therapy if the patient has received it frequently in the past, because of the risk for infection with azole-resistant strains. If fluconazole is used for *C. glabrata* infection, the dose should be increased to 12 mg/kg daily and blood concentrations should be determined to avoid levels above 200 μg/ml. A minimum duration of 2 weeks is recommended in patients with fungemia who lack evidence for deep tissue involvement or dissemination. For others, treatment should continue until the lesions have disappeared or calcified as determined by CT scan.

5. Aspergillosis

5.1. Epidemiology

Most infections are acquired in the hospital[104] but patients may become colonized before admission. Genotyping may be used to show the relationship of the patient's strain to that of environmental isolates, assisting in the evaluation of possible nosocomial acquisition.[105] Air sampling is not useful for predicting exposure, however.[106] Exposure also has been attributed to smoking marijuana.[107] Reactivation following a prior episode of aspergillosis also occurs.[108] Use of protective environments reduces the risk for aspergillosis.[109]

Invasive aspergillosis has been reported in 5 to 24% of patients during chemotherapy-induced granulocytopenia. Duration of granulocytopenia during chemotherapy for leukemia, receipt of an allogeneic marrow, positive cytomegalovirus serostatus, delayed marrow engraftment, and age greater than 18 years are important risk factors for aspergillosis following bone marrow transplantation.[110,111] Most cases occur following engraftment during treatment for GVHD.[7,104]

The incidence following solid organ transplantation is lower than following bone marrow transplantation.[112] In a recent study, invasive aspergillosis occurred in only 1% of liver transplant recipients, suggesting that advances in transplant immunology and the use of immunosuppressive drugs have lowered the risk for aspergillosis.[113] The median time to onset was 17 days following transplantation and 75% of cases occurred within 90 days. *A. fumigatus* and *A. flavus* cause the majority of cases, 73% and 15%, respectively. Risk factors include use of high-dose corticosteroids, OKT3 monoclonal antibodies, renal dysfunction, and cytomegalovirus infection.[113]

TABLE 3. Treatment Guidelines for the More Common Fungal Infections

Fungus	Initial treatment	Subsequent treatment[a]
Candida	Amphotericin B[b]	Fluconazole
Aspergillosis	Amphotericin B	Itraconazole
Zygomycosis	Amphotericin B lipid complex	Posaconazole may be active for some strains
Fusariosis	Amphotericin B	Posaconazole, voriconazole, BMS 207147
Cryptococcosis	Amphotericin B + 5-flurocytosine	Fluconazole
Histoplasmosis		
Severe/moderately severe	Liposomal amphotericin B (AmBisome)	Itraconazole[a]
Mild	Itraconazole	
Coccidioidomycosis	Amphotericin B	Itraconazole or fluconazole
Meningitis	Amphotericin B + fluconazole	
Blastomycosis	Amphotericin B	Itraconazole[a]
Paracoccidioidomycosis	Amphotericin B	Itraconazole
Sporotrichosis	Amphotericin B	Itraconazole
Penicilliosis marneffei	Amphotericin B	Itraconazole

[a]Chronic maintenance therapy in patient with AIDS or other ongoing immunosuppression.
[b]One of the lipid preparations may be indicated in those with renal impairment or receiving medications that may increase the toxicity of amphotericin B, such as cyclosporine, pentamidine, aminoglycosides, foscarnet, or cytosine arabinoside.

5.2. Pathogenesis

Inhalation is the source for invasive rhinosinusitis and tracheopulmonary aspergillosis, while cutaneous inoculation may cause skin disease. *Aspergillus* invades blood vessels causing infarction and hematogenous dissemination, but fungemia is rare. However, necrosis may not develop until granulocytopenia resolves.

5.3. Clinical Manifestations

Aspergillosis is progressive and usually fatal in the severely neutropenic patient, but may be more indolent in those with less severe immunosuppression.[114] Invasive pulmonary aspergillosis is the most common clinical manifestation, occurring in about 75% of cases. Necrotizing tracheobronchitis may precede pulmonary involvement.[115] Rhinosinusitis[116] or cutaneous disease from direct inoculation occurs in 5 to 10% of cases. Dissemination to extrapulmonary sites is common.

5.3.1. Invasive Pulmonary Aspergillosis

Pulmonary complaints or hypoxia should raise suspicion of invasive pulmonary aspergillosis. Pleuritic chest pain, pleural friction rub, and nodular or wedge-shaped infiltrates occur in 30% of cases. Chest radiograms show focal or diffuse infiltrates but may be normal.[117] Chest CT scans are more sensitive that radiograms and may reveal nodular lesions surrounded by a zone of attenuation producing a halo effect.[118] Frequently a vague infiltrate first noted during granulocytopenia progresses to a classic wedge-shaped infarct or nodular lesion with cavitation following bone marrow recovery.[118]

5.3.2. Rhinosinusitis and Otitis

Patients may experience nasal congestion, epistaxis, nasal discharge, and sinus and eye pain[116] and examination shows facial tenderness, crusting at the inferior turbinate or cartilaginous septum, nasal or palatal ulcers, and necrotic lesions. Infection may extend to the soft tissues of the face, orbit, or mastoid bone. Invasive external otitis has been observed.[119] This diagnosis should be suspected in immunocompromised patients with ear pain, hearing loss, and otorrhea.

5.3.3. Cutaneous Aspergillosis

Redness and induration that progresses to necrosis with eschar formation characterize inoculation aspergillosis.[116] Patients also may experience embolic skin lesions,[118] which evolve from a macule or papule to a pustule with ulceration and eschar formation.

5.3.4. Disseminated Aspergillosis

Dissemination occurs in up to half of patients. Sites of involvement include the brain in 50 to 60%, gastrointestinal tract in 40 to 50%, kidney in 30%, liver in 30%, thyroid in 25%, heart in 15%, and spleen in 15%. Endocardium and myocardium may be involved, causing emboli to large blood vessels. Rapidly progressive multiorgan failure may occur. *Aspergillus* is a common cause of central nervous system infection, manifested by meningitis, encephalitis, abscesses, and granulomas.[120] Other less common manifestations have been reviewed.[114]

Illustrative Case 2

A 41-year-old male who underwent liver transplantation received 2 g of methylprednisolone for rejection on posttransplant days 3 and 8. Obtundation and multiorgan failure developed on posttransplant day 7. Chest roentgenogram showed diffuse infiltrates and head CT scan showed multiple masses. Hyphal elements were seen in a fungal smear from the nasopharynx and *A. flavus* was isolated. Amphotericin B was initiated but he died later that day. Autopsy revealed aspergillosis of the tracheobronchial tree, lungs, myocardium, endocardium, and brain.

5.4. Diagnosis

Diagnosis of aspergillosis before death may be difficult. Chest CT may assist in early diagnosis, improving the outcome of treatment.[121] Bronchoalveolar lavage is useful in patients with CT abnormalities.[122] Similarly, CT of the sinuses and endoscopic examination of the nares may facilitate early diagnosis of rhinosinusitis.

5.4.1. Mycologic Methods

Aspergillus can be isolated from sputum in fewer than half of cases of invasive pulmonary aspergillosis,[118,123] but a higher yield (60–75%) may be achieved by bronchoscopy.[104,117] Several other fungi resemble *Aspergillus* histopathologically, including *Fusarium*, *Pseudallescheria*, *Penicillium*, and other less common molds. Isolation of *Aspergillus* should be regarded as presumptive evidence for aspergillosis.[124] In one study, 55% of patients with invasive aspergillosis but <1% of controls had positive nose cultures.[125] Following lung transplantation, however, isolation of *Aspergillus* from bronchoalveolar lavage fluid was associated with invasive aspergillosis in only 27% of cases.[126]

5.4.2. Serologic Methods

Diagnosis based on detection of antigens or metabolic products has been proposed. The sensitivity for antigen detection is greater using EIA than latex agglutination, but false-positive results are common with both.[127] A diagnostic approach using CT scan and antigen detection has been proposed.[128] Biochemical detection of $(1\rightarrow3)$-β-D-glucan in serum also has been described,[129] but cross-reactions may occur with other fungi and *Pneumocystis carinii*, limiting the usefulness of this test. PCR methods also are under investigation.[83] None of these methods have been adequately validated for diagnosis of aspergillosis and are not recommended.

5.5. Treatment

5.5.1. Amphotericin B

The mortality is high, up to 90% in some studies.[130] Early diagnosis and aggressive treatment improves outcome. Administration of high doses of amphotericin B (1 to 1.5 mg/kg per day) is recommended, as rapid progression occurs commonly in patients treated with lower doses. A total of at least 35 mg/kg should be given if it is used exclusively, but treatment may be changed to itraconazole or one of the newer triazoles with activity against *Aspergillus* (voriconazole or posaconazole) if patients respond clinically and immunosuppression can be reduced. Resolution of neutropenia is essential for recovery.[118,131]

5.5.2. Lipid Formulations of Amphotericin B

The lipid formulations of amphotericin B have been incompletely investigated for treatment of aspergillosis. In one retrospective review in liver transplant patients, mortality rates were similar with amphotericin B (100%) and the lipid preparations (89%).[113] About one third of patients with invasive aspergillosis responded to amphotericin B colloidal dispersion (ABCD, Amphotec) in a second study.[132] Another report showed a better outcome in patients treated with Amphotec than in historical controls treated with amphotericin B, but baseline characteristics favored the Amphotec group,[133] and no efficacy advantage was observed in a subsequent comparative trial (Jo-Anne van Burik, unpublished data, 2000). Wingard[93] reported a 38% response to amphotericin B lipid complex (ABLC, Abelcet) in bone marrow transplant patients and Walsh[92] reported a 42% response. Higher responses of 60 to 66% were reported with liposomal amphotericin B (AmBisome).[90,134] A study that compared AmBisome 1

mg/kg per day versus 5 mg/kg per day reported comparable outcomes at both doses, but the study was underpowered and failed to include an amphotericin B deoxycholate arm.[135]

5.5.3. 5-Flucytosine

Although some have recommended concurrent use of 5-flucytosine, one study failed to demonstrate benefit.[118] 5-Flucytosine causes neutropenia and gastrointestinal upset, delays bone marrow recovery, and cannot be given parenterally, complicating its use.

5.5.4. Itraconazole

Response rates over 70% have been reported,[25,136–138] but trials comparing itraconazole to amphotericin B have not been conducted. Doses of up to 800 mg daily may be needed to achieve high success rates,[139] although failure caused by resistance to itraconazole has been reported.[140,141] Itraconazole's primary role is consolidation therapy once immunosuppression has been reduced and clinical remission has been induced with amphotericin B.[117] Blood concentration should be measured because of itraconazole's erratic absorption.

5.5.5. Newer Antifungal Agents

Voriconazole[70] and posaconazole (SCH 56592),[142] new triazoles produced by Pfizer and Schering Plough, respectively, are active against *Aspergillus in vitro* and in a murine model.[143] Caspofungin, an echinocandin antifungal agent produced by Merck, also is active in aspergillosis.[144] Some clinical experience has been obtained using voriconazole and caspofungin in patients with aspergillosis. These new agents are discussed in a subsequent section.

5.5.6. Hematopoietic Growth Factors

Granulocyte and granulocyte–monocyte colony-stimulating factor may improve the outcome of invasive fungal disease.[145] However, in the European survey, use of growth factors did not appear to improve survival in aspergillosis.[117] Of note, resolution of neutropenia is clearly associated with an improved outcome,[117] supporting a role for such therapy in the neutropenic patient. GM-CSF has been shown to enhance the penetration of antimicrobial agents into activated phagocytes and improve the function of antigen-presenting cells,[146] supporting other mechanisms of action as adjuvants to antifungal therapy in patients with serious fungal infections.

5.5.7. Prevention of Recurrence of Aspergillosis during Subsequent Granulocytopenia

Patients who recover from invasive aspergillosis are at a high risk for relapse at the time of future immunosuppression.[147] Prophylactic amphotericin B or itraconazole at the time of relapse of leukemia or initiation of immunosuppressive therapy may prevent such recurrences.[108]

5.5.8. Surgery

Resection of residual pulmonary lesions also has been recommended to prevent recurrence during subsequent granulocytopenia. In one report, the operative mortality of prophylactic surgery was 11%, relapse rate 14%, and 3-month survival 77%.[148] This experience supports resection of unilateral, isolated pulmonary lesions before marrow transplantation. In rhinosinusitis and cutaneous aspergillosis, necrotic tissue should be debrided.

5.5.9. Treatment Recommendations

Guidelines recently have been published for treatment of aspergillosis.[149] Amphotericin B 1–1.5 mg/kg per day, or one of the lipid formulations (5–10 mg/kg per day) in patients with underlying renal impairment or who are receiving nephrotoxic medications is recommended (Table 3). Itraconazole 6 mg to 12 mg/kg daily by mouth is an alternative for those with milder manifestations and for continuation of treatment after response to amphotericin B, but problems with absorption and drug interactions must be recognized. If the intravenous formulation is used, the dosage is 3 to 6 mg/kg daily. If the patient has received itraconazole recently, potential infection with itraconazole-resistant *Aspergillus* also must be considered. A minimum duration of 3 months is recommended, but treatment should not be stopped until evidence of invasive infection has resolved.

5.6. Prevention

5.6.1. Use of Protective Environments

Protective air environments reduce the risk for[150] and reduce the prevalence of aspergillosis.[125,151–153] Travel outside of the protective environment should be discouraged and masks should be worn for such travel. Flowers should not be brought into patient rooms, food should be cooked, and smoking prohibited.

5.6.2. Barrier Containment and Decontamination

Construction and remodeling increases the concentration of *Aspergillus* spores in hospital air[154] and outbreaks of aspergillosis may be associated with such activities. Use of barriers around the construction site and decontamination of air ducts may prevent exposure.[150]

6. Zygomycosis

6.1. Epidemiology

Zygomycosis (mucormycosis) is the third leading invasive mycosis in patients with hematologic malignancies and is caused by fungi of the order Mucorales (class Zygomycetes). Members of the genus *Rhizopus* are the most common causes of human disease, but other human pathogens in this group include *Absidia*, *Mucor*, and *Rhizomucor*. The incidence of zygomycosis following organ transplantation has ranged from 1 to 9%,[155] compared to 1% following bone marrow transplantation.[156,157] Mucoraceae can be found in hospital air and zygomycosis can be acquired by inhalation of spores or by direct inoculation.

6.2. Clinical Manifestations

Zygomycosis causes vascular invasion with infarction and hemorrhage. Nasal discharge, intranasal ulcers, facial swelling, proptosis, and ophthalmoplegia[158] characterize rhinocerebral zygomycosis. Central nervous system involvement is especially common, resulting from either hematogenous dissemination or from local extension and may be manifested by parenchymal lesions or meningitis.[157] Pulmonary findings mimic those described with aspergillosis and chest roentgenograms show patchy infiltrates, consolidation, cavity formation, fungus balls, and pleural effusion. Disseminated disease involving the spleen, kidney, liver, heart, stomach, thyroid, and other organs occurs in up to half of fatal cases. Mortality exceeds 50% and is highest in patients with disseminated disease.[155–157]

6.3. Diagnosis

Diagnosis usually is based on identification of broad-based, nonseptate hyphae branching at 90° angles in tissue biopsies. Cultures may be falsely negative. Zygomycetes and *Aspergillus* may be mistaken for one another histopathologically. Serologic tests for antibodies or antigens have not been investigated for diagnosis of zygomycosis.

6.4. Treatment

Successful outcome has occurred in <25% of cases in immunocompromised hosts. Debridement of necrotic tissue and therapy with amphotericin B are essential to recovery. Resection of lung lesions may be helpful in patients with pulmonary zygomycosis.[159] Hyperbaric oxygen treatment was suggested to improve outcome of zygomycosis[160] but has not been adequately evaluated and is not recommended. Use of colony-stimulating factors may improve the outcome in patients with persistent neutropenia.[161] Aggressive treatment with high doses of amphotericin B is recommended in conjunction with reduction of immunosuppression (Table 3). A new antifungal agent, posaconazole, is active *in vitro* against some zygomycetes, but has not been studied in animal models or patients.[162]

7. Fusariosis

7.1. Epidemiology and Pathogenesis

Fusarium species are widely distributed in nature, found in soil and on plant debris, and infection is acquired by inhalation of spores or direct deposition at cutaneous sites. *Fusarium* also causes onychomycosis, which may serve as a portal of entry in the immunocompromised host. Hematogenous dissemination is common following inhalation or cutaneous infection, including onychomycosis. Aggressive management of nail infection is recommended to prevent this complication.

7.2. Clinical Manifestations

Two thirds of patients demonstrate skin lesions.[163,164] The cutaneous manifestations include granulomas, ulcers, pustules, vesicles, nodules, necrosis, mycetomas, and panniculitis. Lesions may resemble ecthyma gangrenosum. Localized pulmonary disease may occur but less frequently than in aspergillosis. Multiorgan dissemination is common and any organ may be involved. The mortality is about 70%, recovery occurring only if neutropenia resolves.[163,164]

7.3. Diagnosis

Definitive diagnosis is based on culture. *Fusarium* species are the only opportunistic molds that can be readily recovered by blood culture. Identification of hyphae in the tissues does not distinguish *Fusarium* from other molds, such as *Aspergillus* or *Scedosporium* species. Serologic methods are not available.

7.4. Treatment

Most strains of *Fusarium* are susceptible to amphotericin B, which remains the treatment of choice. Fluconazole and itraconazole are inactive against *Fusarium* spp., illustrating a limitation of use of these agents for empiric antifungal therapy. Posaconazole demonstrates moderate *in vitro* activity and *in vivo* efficacy in a murine model of fusariosis.[165] Voriconazole also is active against *Fusarium* spp. *in vitro*.[70] Surgical debridement of necrotic tissue and use of colony-stimulating factors may be helpful.[163,164] Treatment with high doses of amphotericin B or one of the lipid formulations is recommended, with optimization of immune function through reduction in immunosuppressants (Table 3).

8. Cryptococcosis

8.1. Epidemiology and Pathogenesis

Cryptococcosis occurs in 5 to 12% of patients with AIDS and 1 to 26% of those who have undergone organ transplantation, but is rare following bone marrow transplantation. Cryptococcosis results from inhalation of *C. neoformans* yeasts and rarely by cutaneous inoculation. Although infection also has been suggested to result from reactivation, its late occurrence following organ transplantation favors exogenous exposure. In two studies, cryptococcal meningitis occurred more than 6 months after transplantation in over 90% of cases, often following treatment of acute rejection.[166,167]

8.2. Clinical Manifestations

Although cryptococcal infection is acquired by inhalation, pulmonary involvement is uncommonly recognized. Radiographic abnormalities include interstitial or nodular infiltrates most commonly and occasionally cavities or pleural effusions.[168] Meningoencephalitis is the characteristic manifestation of cryptococcosis. Symptoms evolve over a few months in most cases, but rapid progression may occur in those with severe immunodeficiencies. Symptoms include fever, headache, nausea, and vomiting, but less than one third exhibit meningismus, altered mentation, or focal neurologic abnormalities. Elevated intracranial pressure is common and may cause death from brain stem herniation. Focal brain lesions occur in 10% of cases, as isolated manifestations or in combination with meningoencephalitis.[169] CT or MRI shows meningeal enhancement, hydrocephalus, cerebral edema or atrophy, and visual loss, which may necessitate aggressive management of intracranial hypertension or decompression of the optic nerve sheath.[170]

Extraneural dissemination occurs in a quarter to half of patients, manifested by hepatosplenomegaly and bone marrow suppression most commonly and by lesions in the eyes, bones, or joints, each occurring in about 5% of patients. Skin involvment occurs in up to a quarter of patients. The prostate may be infected, serving as a source for reactivation following discontinuation of therapy.[171] Other sites of dissemination include the heart, pericardium, muscle, gastrointestinal tract, peritoneum, thyroid, larynx, breast, placenta, urinary tract, and organ of Corti.

8.3. Diagnosis

8.3.1. Mycologic Methods

Cryptococcal infection in the immunocompromised patient usually can be diagnosed without great difficulty. India ink stain or culture of cerebrospinal fluid are positive in 50 to 90% of cases, respectively. Rarely, examination of cerebrospinal fluid obtained by cisternal or ventricular aspiration yields a diagnosis in patients with negative examinations of lumbar fluid.[172]

8.3.2. Antigen Detection

The diagnosis of meningitis can be made initially by detection of cryptococcal polysaccharide antigen in cerebrospinal fluid. Antigen also can be detected in serum, often providing a clue to the diagnosis before lumbar puncture is performed. False-positive results occur in <1% of controls. Enzyme immunoassay for cryptococcal antigen is more sensitive and at least as specific as the latex agglutination test but is more expensive, and thus has not gained wide acceptance.[173] Antigen testing for monitoring therapy or identification of relapse has not been adequately assessed.

8.4. Treatment

8.4.1. Antifungal Medications

Amphotericin B plus 5-flucytosine appears to be the most effective regimen. The response to this combination (68%) was better than to amphotericin B alone (47%), but the dosage of amphotericin B (0.3 mg/kg per day) was not optimal.[174] Successful combination treatment of cryptococcal meningitis with amphotericin B and 5-flucytosine has been reported in 70 to 90% of organ transplant patients.[166,167] Comparing amphotericin B 0.7 mg/kg per day alone or combined with 5-flucytosine for the first 2 weeks, followed by itraconazole versus fluconazole for the next 10 weeks in patients with AIDS, the combined regimen was well tolerated and appeared to reduce the risk for relapse.[175] Furthermore, the mycological response appeared to be superior (~90%) to that with fluconazole or amphotericin B alone (~50%).[176] Although intraventricular treatment has been used, such treatment often causes debilitating arachnoiditis.

Liposomal preparations of amphotericin B are less toxic than the standard formulation and may be administered more aggressively because of reduced nephrotoxicity. Among the lipid formulations, none achieve detectable concentrations in the CSF, but AmBisome reaches the highest levels in brain tissue.[177] Although cerebrospinal fluid sterilization occurred more rapidly with AmBisome than with amphotericin B,[178] a randomized, double-blind comparison failed to confirm clinical superiority (Dr. Richard Hamill for the Mycoses Study Group, unpublished observation).

Fluconazole or itraconazole are inadequate initial treatment of meningitis or widely disseminated infection. When used alone for treatment of meningitis, complete or partial response occurred in 62% patients with fluconazole[176] and 41% with itraconazole.[179] Used as "consolidation" therapy following amphotericin B with or without 5-flucytosine for 2 weeks, itraconazole 200 mg twice daily or fluconazole 400 mg once daily showed a similar clinical outcome, but with slower cerebrospinal fluid sterilization with itraconazole.[175] Resistance to fluconazole may be a cause for treatment failure in some cases.[180–183]

Higher doses of fluconazole (800 to 2000 mg per day) in combination with 5-flucytosine are more effective than fluconazole alone, inducing a response in over 80% of cases,[184] but this regimen has not been compared to amphotericin B plus 5-flucytosine. Voriconaozole is more active than fluconazole for *C. neoformans* and crosses the blood–brain barrier, but has not been studied in meningitis. Posaconazole does not cross the blood–brain barrier.

Maintenance treatment is indicated in patients with AIDS and cryptococcal meningitis. The risk for relapse can be reduced from 25% to below 3% with fluconazole 200 mg daily.[185] Fluconazole is superior to amphotericin B once weekly[186] or itraconazole for chronic maintenance therapy.[187] Whether maintenance therapy can be stopped following improvement in CD4 count in response to antiretroviral therapy requires investigation. The need for maintenance therapy in other immunocompromised hosts is unknown.

8.4.2. Treatment Recommendations

Amphotericin B 0.7 to 1 mg/kg per day and 5-flucytosine 100 mg/kg per day are recommended for the initial threatment of meningitis (Table 3). Guidelines for therapy have been reviewed.[188] After 2 weeks, in patients who respond to therapy, treatment may be changed to fluconazole 6 to 12 mg/kg daily for at least 6 to 12 months. Chronic maintenance therapy with 3 mg/kg daily is recommended in those with AIDS or who have relapsed despite recommended therapy in the presence of ongoing severe immunosuppression. Some immunosuppressed patients, however, may be unable to tolerate 5-flucytosine because of gastrointestinal side effects, bone marrow suppression, or inability to take medications orally. Itraconazole is not recommended for the treatment of meningitis because of suboptimal efficacy,[187] presumably caused by its failure to achieve detectable concentrations in the cerebrospinal fluid. For nonmeningeal disease, fluconazole or itraconazole, 3 to 6 mg/kg per day may be used in those with mild to moderate symptoms, reserving amphotericin B for severe cases.

8.4.3. Cytokines

In vitro and animal model experiments demonstrate synergy between GM-CSF and interferon-γ (IFN-γ) and certain antifungal agents in treatment of cryptococcosis,[189,190] but clinical trials have not been reported. A study is in progess, however, evaluating IFN-γ as adjunctive therapy in patients with AIDS who have cryptococcal meningitis (InterMune Pharmaceuticals, Palo Alto, CA).

8.4.4. Management of Immunosuppressive Medications

Most patients have responded to antifungal treatment while receiving azathioprine 75 to 125 mg/day and

prednisone 15 to 20 mg/day.[167,191–193] Immunosuppressants were stopped and dialysis was resumed in renal allograft patients who have developed acute rejection during treatment of cryptococcal meningitis.[167] If possible, immunosuppression should be reduced; if not possible, chronic antifungal maintenance therapy may be required to prevent relapse.

8.4.5. Management of Intracranial Hypertension

Early recognition and aggressive treatment for elevated intracranial pressure may improve outcome.[175,194] Intracranial hypertension should be suspected in patients with progressive headache, nausea, or vomiting, worsening mental status, or visual impairment. Initial treatment should include lumbar puncture to remove large volumes (25 ml) of cerebrospinal fluid and administration of acetohexamide. Placement of lumbar drains into the subarachnoid space may be needed in patients who require frequent lumbar puncture, but it carries the risk of bacterial superinfection. Cerebrospinal fluid shunting may be needed in patients failing these measures, but should be delayed if possible because shunts placed early in the course of infection may become obstructed. Decompression of the optic nerve sheath may be needed in some patients with visual loss caused by edema of the optic nerve.

8.4.6. Prophylaxis to Prevent Acquisition in AIDS

Fluconazole or itraconazole 200 mg daily in persons with CD4 counts below 100/μl reduce the incidence of cryptococcal meningitis but do not improve survival benefit.[195,196] Because of high cost, lack of survival benefit, and concern about induction of resistance, prophylaxis is not recommended.

8.4.7. Transplantation Following Treatment for Cryptococcosis

Transplantation is not contraindicated in patients who have recovered from cryptococcal infection, and a few patients have been successfully retransplanted without reactivation of cryptococcal infection.[197] The two who were successfully retransplanted had completed antifungal treatment at least 6 months before retransplantation, while a third who was retransplanted only one month after completion of treatment relapsed. Another patient who

11.3.2. Serology

The standard complement fixation test is not specific or sensitive, and thus cannot be relied on. The immuno-diffusion test is more specific but less sensitive. Enzyme immunoassay has increased sensitivity but decreased specificity. The negative serodiagnostic test does not exclude the diagnosis and a positive result does not establish the diagnosis; but rather prompts the need for further diagnostic testing.[242] An antigen cross-reacting with *H. capsulatum* may be detected in the body fluids, providing a rapid diagnosis in patients with disseminated or extensive pulmonary blastomycosis.[243]

11.4. Treatment

11.4.1. Amphotericin B

Treatment guidelines recently have been reviewed.[244] There are no prospective studies evaluating the various treatment options in immunosuppressed patients, but available data, based on small series and case reports, support a recommendation for amphotericin B given until clinical stabilization, followed by an azole.[238] Some authorities recommend a minimum dose of amphotericin B of 1000 mg, before switching to an azole, usually itraconazole. For meningitis, a total course of 2000 to 2500 mg of amphotericin B is recommended.[245]

11.4.2. Itraconazole and Fluconazole

Itraconazole[246] can be used after the patient improves in response to amphotericin B. While ketoconazole is highly effective,[247] its unfavorable side effect profile makes itraconazole the more acceptable option. Fluconazole also is effective therapy (85 to 89% response) for treatment of blastomycosis, but requires use of higher daily doses of 6 to 12 mg/kg.[248]

11.4.3. Chronic Maintenance Therapy

Chronic maintenance therapy has not been studied in blastomycosis. Nevertheless, lifelong suppressive therapy with itraconazole 3 mg/kg once or twice daily and adjusted to achieve blood levels of at least 1 μg/ml is recommended if the immunosuppressive regimen cannot be reduced or the immunosuppressive illness will not improve.[239]

11.4.4. Treatment Recommendations

Treatment guidelines have been reviewed.[244] Amphotericin B 0.7 to 1 mg/kg per day is recommended for initial therapy, because of the poor outcome of blastomycosis in the immunocompromised host (Table 3). Itraconazole 3 mg/kg twice daily can be substituted for amphotericin B after the patient has received about 1 g, and continued for at least one year. A higher total dose of amphotericin B of at least 2 g may be appropriate in those with CNS involvement, followed by fluconazole 12 mg/kg daily. Chronic suppressive therapy is recommended in those who have AIDS or ongoing severe immunosuppression.

12. Unusual Fungal Pathogens

The more common unusual fungal pathogens[9] include *Trichosporon beigelii* and the dematiaceous fungi, which cause phaeohyphomycosis.[249] These infections have been reviewed recently by Walsh and Groll.[250] The dematiaceous fungi include *Alternaria, Bipolaris, Cladosporium,*

TABLE 4. Treatment Guidelines for the Less Common Opportunistic Fungal Infections

Fungus	Initial treatment	Subsequent treatment
Acremonium	Amphotericin B	Posaconazole, voriconazole, BMS 207147
Paecilomyces lilacinus	Amphotericin B or terbinafine	
Paecilomyces variotii	Itraconazole or newer triazole	
Trichophyton	Itraconazole or fluconazole	
Microsporon	Terbinafine	
Trichoderma longibrachiatum	Amphotericin B	Posaconazole, voriconazole, BMS 207147
Pseudallescheria boydii	Amphotericin B + itraconazole	Itraconazole
Scedosporium prolificans	Abelcet	Posaconazole, voriconazole, BMS 207147
Bipolaris	Itraconazole	Posaconazole, voriconazole, BMS 207147
Cladophialophora bantiana	Amphotericin B + itraconazole	Itraconazole
Alternaria	Amphotericin B + itraconazole	Itraconazole
Wangiella (Exophilia) dermatidis	Amphotericin B + itraconazole	Itraconazole

Curvularia, and a few other genera. Less commonly isolated fungi include *Geotrichum, Pseudallescheria, Saccharomyces, Rhodotorula, Blastoschizomyces capitum, Malassezia furfur*, and *Penicillium*.[251] Penicilliosis marneffei is common in patients with AIDS from Southeast Asia,[252] and paracoccidioidomycosis in those from Latin America.[253]

Most infections caused by these fungi occur in granulocytopenic patients and are widely disseminated,[251] or occur in those with AIDS. Clinical findings and morphologic appearance in histopathologic sections of tissues from patients with infections caused by dematiaceous fungi and *Pseudallescheria* resemble those of aspergillosis.[251] Antigen detection also can be used for diagnosis of penicilliosis marneffei and paracoccidioidomycosis.[243]

Treatment for the unusual fungal infections recently has been reviewed and treatment guidelines are summarized in Table 4. Notably, however, the optimal treatment is unclear since prospective treatment trials have not been conducted and relatively few data are available from observational studies. Many of these fungi may be resistant to amphotericin B and respond poorly to treatment.[250,251] Some of the newer triazoles and terbinafine have been useful for treatment of such infections.[250,254–261] Treatment guidelines for penicilliosis marneffei and paracoccidioidomycosis parallel those for histoplasmosis. Optimization of immune status and resection of necrotic tissue or residual lesions is important to a successful outcome.

13. Antifungal Agents

13.1. Amphotericin B

Properties of amphotericin B deoxycholate (Fungizone) have been reviewed.[262] It has a broad spectrum of antifungal activity. A few strains of *C. albicans* and other *Candida* species,[89] *Pseudallescheria boydii, Trichosporon beigelii*, and certain dematiaceous fungi are resistant to amphotericin B, however. Toxicity of amphotericin B may pose major problems in patients receiving other nephrotoxic agents, including cyclosporin A, tacrolimus, adenine arabinoside, foscarnet, cidofovir, and aminoglycosides.[263] Anemia is common and leukopenia or thrombocytopenia has been reported. Rare toxicities include anaphylaxis, cardiomyopathy, diabetes insipidus, and encephalopathy.

13.2. Lipid Formulations of Amphotericin B

Incorporation of amphotericin B into lipid complexes [amphotericin B lipid complex (ABLC, Abelcet), amphotericin B colloidal dispersion (ABCD, Abelcet), or liposomes (L-AmB, AmBisome)] reduces toxicity and may improve efficacy.[40,264,306] Results of studies evaluating the effectiveness of these preparations for treatment of infections in the immunocompromised host were reviewed by Walsh.[265] The biochemical and pharmacokinetic properties of these lipid formulations differ and have been reviewed elsewhere[40,264,266,267] (Table 5).

TABLE 5. Toxicity and Pharmacokinetics of Lipid Formulations of Amphotericin B[a]

Generic name	Amphotericin B deoxycholate	Amphotericin B colloidal dispersion	Amphotericin B lipid complex	Liposomal amphotericin B
Trade name (abbreviation)	Fungizone (AmBd)	Amphotec (ABCD)	Abelcet (ABLC)	AmBisome (L-AmB)
Infusion toxicity relative to AmBd	NA[b]	Higher	Comparable	Less
Nephrotoxicity relative to AmBd	NA	Lower	Lower	Lower
Blood C_{max}				
Dose	0.6 mg/kg	5 mg/kg	5 mg/kg	5 mg/kg
C_{max}	1.1 µg/ml	1.7 µg/ml	3.1 µg/ml	83 µg/ml
Volume of distribution relative to AmBd	NA	Increased	Increased	Decreased
Concentration in CNS of rabbits				
Dose	1 mg/kg per day	5 mg/kg per day	5 mg/kg per day	5 mg/kg per day
Brain	0.53 µg/ml	0.30 µg/ml	0.36 µg/ml	2.33 µg/ml
Cerebrospinal fluid	0 µg/ml	0 µg/ml	0 µg/ml	0 µg/ml
Clearance relative to amphotericin B	NA	Increased	Increased	Decreased
Dosage	0.7–1.5 mg/kg per day	3–6 mg/kg per day	5 mg/kg per day	3–5 mg/kg per day
Infusion rate	2 to 4 hr	1 mg/kg per hr	2.5 mg/kg per hr	30 to 60 min
Need to test dose	Yes	Yes	No	No

[a]Data for this table can be found in Wong-Beringer,[264] Heimenz,[40] Groll,[177] Dix,[266] and Boswell.[306]
[b]NA, not applicable.

ABCD and ABLC achieve lower serum concentrations than does amphotericin B, while L-AmB yields much higher concentrations. L-AmB reaches higher concentrations in brain tissue, but none reach therapeutic concentrations in the CSF.[177] Unfortunately the high costs of the lipid formulations discourage their liberal use.

The best information comes from a large study comparing amphotericin B to L-AmB for empiric therapy of neutropenic fever.[265] In that study, hydrocortisone was prescribed to prevent chills and fever threefold more often with amphotericin B than with L-AmB. Cardiorespiratory events (hypotension, hypertension, tachycardia, hypoxia) also were much less frequent (2- to 20-fold) with L-AmB than with amphotericin B. Skeletal pain, commonly in the spine, has been observed with L-AmB and may be related to rapid infusion (over 30 to 60 min). The reduction in infusion toxicity and cardiorespiratory compromise with the liposomal preparation was attributed to reduced release of inflammatory cytokines (TNF-α, IL-1, and IL-6). Infusion toxicities do not appear to be reduced with ABCD, however.[264] Hepatotoxicity was not increased with L-AmB in comparison to amphotericin B.[265]

Creatinine elevation and hypokalemia were twofold more common with amphotericin B than with L-AmB.[265] Similar data are available from smaller studies comparing the other lipid preparations to amphotericin B, but adequately powered studies comparing one lipid agent to another have not been conducted. The improved renal tolerance was thought to result from reduced transfer of amphotericin B into the renal tissues with the liposomal formulation.[265] Reduced nephrotoxicity is the major reason to use the lipid formulations rather than deoxycholate amphotericin B (Table 6).

A major area of concern in the immunocompromised host is the combined toxicity with coadministered nephrotoxins. Comparing ABCD to amphotericin B in patients receiving cyclosporin A, nephrotoxicity occurred in 28% receiving the lipid preparation versus 68% receiving amphotericin B.[65] Similar findings were observed when

TABLE 6. Indications for Lipid Formulations of Amphotericin B

Underlying renal dysfunction or high risk for renal dysfunction
Intolerance of amphotericin B because of nephrotoxicity
Treatment with other nephrotoxic medications:
• Cyclosporin A, tacrolimus
• Foscarnet, aminoglycosides, pentamidine
• Cytosine arabinoside
Progressive infection despite amphotericin B

comparing L-AmB to amphotericin B in patients receiving cyclosporin A, aminoglycosides, or foscarnet.[265] However, even using L-AmB the risk for nephrotoxicity increased with coadministration of other nephrotoxins, illustrating that this problem has not been eliminated. Nevertheless, concurrent treatment with other nephrotoxic medications is an indication for use of one of the lipid formulations (Table 6).

13.3. 5-Flucytosine

5-Flucytosine is active *in vitro* against *Candida* and *Cryptococcus* and has been proven useful in combination with amphotericin B or fluconazole for treatment of cryptococcal meningitis.[268] 5-Flucytosine should not be used alone because of rapid emergence of resistance, however. Gastrointestinal upset, bone marrow suppression, and hepatitis are its major toxicities and correlate with blood levels exceeding 100 μg/ml.[269] Use of 100 mg/kg per day instead of 150 mg/kg per day may reduce toxicity. Serum levels, blood count, and creatinine should be monitored and dosage adjusted appropriately for renal function to reduce toxicity. The target peak serum concentration should be between 50 and 75 μg/ml.

13.4. Fluconazole

Fluconazole has a broad spectrum of activity and has been useful for treatment of local and systemic candidiasis, cryptococcosis, histoplasmosis, blastomycosis, and coccidioidomycosis (Table 7). Treatment failures related to inherent resistance or development of resistance have been reported with *C. albicans*, *C. glabrata*, and *C. krusei* species,[26,270,271] *C. neoformans*,[180–183] and *H. capsulatum*; however, *C. parapsilosis*, *C. tropicalis*, and *C. guillermondii* also may be resistant.[25] Fluconazole is inactive and an ineffective treatment for aspergillosis or other molds.

Fluconazole is highly bioavailable, yielding predictable blood levels.[272] Because of its long half-life, a higher dose, or so-called loading dose, is given on the first day to achieve therapeutic concentrations more rapidly. Fluconazole is fully water-soluble and largely (80%) excreted unchanged in the urine. Absorption does not require gastric acid and is not improved by administration with food. Absorption is only slightly affected by H_2 blockers.[272,273] Fluconazole concentrations in the cerebrospinal fluid are 60 to 80% of corresponding blood concentrations.

Fluconazole causes fewer drug interactions than do other azoles.[273] Coadministration of rifampin reduced the

TABLE 7. Comparison of Azole Antifungal Agents[a]

Parameter	Ketoconazole	Itraconazole	Fluconazole
Absorption	Unpredictable	Variable	Excellent
Requires gastric acidity	Yes	Yes	No
Intravenous form	No	No	Yes
Hepatic metabolism	Yes	Yes	No (20%)
Renal excretion	No	No	Yes
CSF penetration	No	No	Yes (80%)
Effect on human P450	Moderate	Reduced	Reduced
Drug interactions	Most	Intermediate	Least
GI upset	29%	10%	10%
Hepatotoxicity	Yes	Yes	Yes
Skin rash	10%	2%	4%
Androgen suppression	8%	1%	No

[a]References for the data in this table are reviewed in the sections describing each agent.

area under the curve (AUC) and half-life by about 20% but did not reduce serum levels.[273] Phenytoin did not alter fluconazole pharmacokinetics.[274] Although fluconazole is more specific than ketoconazole for fungal P450 enzymes, fluconazole impairs the metabolism of several medications, causing potential drug interactions. Fluconazole increases cyclosporin A levels twofold, necessitating that concentrations be monitored to minimize nephrotoxicity.[275] Gastrointestinal upset is uncommon except in patients receiving higher doses (400 mg or more daily). Reversible thrombocytopenia, Stevens–Johnson syndrome, fatal hepatitis, and anaphylaxis have been reported.

13.5. Itraconazole

Itraconazole is effective for treatment of oral and esophageal candidiasis, histoplasmosis,[210] cryptococcosis,[175] coccidioidomycosis,[276–280] sporotrichosis,[137,281] aspergillosis,[131,136–138] and phaeohyphomycosis.[259] Some species of *Candida* demonstrate high MICs to itraconazole,[282] and its role for treatment of serious *Candida* infections has not been adequately studied.[283] It remains active against many strains of fluconazole-resistant *Candida* and has proven useful for treatment of fluconazole-refractory thrush and esophagitis.[284–286] Itraconazole's role for treatment of serious fungal infections in the immunocompromised host, especially aspergillosis, remains controversial and has not been adequately studied.

Itraconazole is poorly water-soluble and requires gastric acid for absorption (Table 7). Absorption is re-

duced 15 to 20% by H$_2$ blockers.[282] Blood levels vary widely and may be inadequate in some patients. To achieve therapeutic concentrations more rapidly, a higher dose should be given for the first 3 days of treatment, usually 3 mg/kg three times daily. Blood levels do not increase proportionally with dosages above 3 mg/kg, supporting recommendations to increase the frequency of administrations rather than the individual dose in patients taking more than 3 mg/kg daily. The solution formulation of itraconazole does not require gastric acid for absorption, and thus may be taken by patients who must receive medications that reduce gastric acidity or fail to absorb the capsule formulation. An intravenous formulation overcomes problems associated with poor absorption or inability to take oral medications, permitting use of itraconazole in patients who could not be treated orally.

Itraconazole reaches high tissue levels because of its lipophilic nature, but does not achieve therapeutic concentrations in the cerebrospinal fluid or urine.[272] It is extensively metabolized in the liver by CYP 3A4 enzymes and undergoes rapid degradation in patients treated with enzyme inducers.[282,287] It also inhibits CYP 3A4 enzymes, causing major drug interactions with a number of medications that are eliminated by hepatic metabolism. Itraconazole reduces cyclosporine clearance, requiring careful monitoring and dosage adjustments.[288]

Hypokalemia, edema, and hypertension have been reported in a few patients receiving high doses (800 mg/day).[289] Reversible adrenal corticosteroid suppression also may complicate high-dose therapy.[289] Hepatitis and skin rash occur infrequently.[259,289,290]

13.6. Voriconazole

Voriconazole, a new triazole with a greater spectrum of activity, has been studied for treatment of aspergillosis and candidiasis.[70] It is more active than fluconazole against *Candida*, including strains that are resistant to other azoles,[291,292] *Cryptococcus*,[293] and the dimorphic fungi and expands the spectrum to include *Aspergillus*,[294–297] *Trichosporon beigelii*,[70] *Scedosporium apiospermum*,[298] and other filamentous fungi.[299,300] It penetrates the cerebrospinal fluid, similar to fluconazole, but is eliminated by hepatic metabolism, like itraconazole. Visual disturbances have been recognized in 10 to 40% of patients.

13.7. Posaconazole (SCH 56592)

Posaconazole also is a new triazole with an expanded spectrum of activity against *Aspergillus*, zygomycetes,

Fusarium, some dematiaceous and most dimorphic fungi.[143,165,214,301,302] It does not penetrate the cerebrospinal fluid and is eliminated by hepatic metabolism. It has been studied for treatment of thrush, including that caused by fluconazole-resistant strains.

13.8. Echinocandins

The echinocandins are a new class of antifungal agents that work by inhibition of β (1-3)-glucan synthase, preventing incorporation of glucans into the fungal cell wall.[162,303] These agents are active against *Candida*, *Aspergillus*, and *Pneumocystis carinii* but not against other fungi.[71,144,304,305] One preparation is in clinical investigation, caspofungin (LY-743872, MK-991, Merck). It is well tolerated but must be given intravenously. It may have a role for treatment and prevention of fungal infections in the immunocompromised host, but its narrow spectrum may limit its role for empiric therapy.

13.9. Nikkomycin Z

Nikkomycin Z is a chitin synthase inhibitor and is active against the endemic mycoses. Animal studies have demonstrated efficacy in histoplasmosis, coccidioidomycosis, and blastomycosis. However, it no longer is undergoing clinical investigation, and may never become available for treatment of patients.

References

1. Romani L, Howard DH: Mechanisms of resistance to fungal infections. *Curr Opin Immunol* **7**:517–523, 1995.
2. Romani L: The T cell response against fungal infections. *Curr Opin Immunol* **9**:484–490, 1997.
3. Williamson EC, Millar MR, Steward CJ, *et al*: Infections in adults undergoing unrelated donor bone marrow transplantation. *Br J Haematol* **104**:560–568, 1999.
4. Walsh TJ, Hiemenz J, Pizzo PA: Editorial response: Evolving risk factors for invasive fungal infections—All neutropenic patients are not the same. *Clin Infect Dis* **18**:793–798, 1994.
5. Sable C, Donowitz G: Infections in bone marrow transplant recipients. *Clin Infect Dis* **18**:273–284, 1994.
6. LaRocco MT, Burgert SJ: Infection in the bone marrow transplant recipient and role of the microbiology laboratory in clinical transplantation. *Clin Microbiol Rev* **10**:277–297, 1997.
7. Jantunen E, Ruutu P, Niskanen L, *et al*: Incidence and risk factors for invasive fungal infections in allogeneic BMT recipients. *Bone Marrow Transplant* **19**:801–808, 1997.
8. Perfect JR, Schell WA: The new fungal opportunists are coming. *Clin Infect Dis* **22**(Suppl 2):S112–S118, 1996.
9. Hazen KC: New and emerging yeast pathogens. *Clin Microbiol Rev* **8**:462–478, 1995.
10. Wade JJ, Rolando N, Hayllar K, *et al*: Bacterial and fungal infections after liver transplantation: An analysis of 284 patients. *Hepatology* **21**:1328–1336, 1995.
11. Patel R, Portela D, Badley AD, *et al*: Risk factors of invasive *Candida* and non-*Candida* fungal infections after liver transplantation. *Transplantation* **62**:926–934, 1996.
12. Collins LA, Samore MH, Roberts MS, *et al*: Risk factors for invasive fungal infections complicating orthotopic liver transplantation. *J Infect Dis* **170**:644–652, 1994.
13. Imam N, Carpenter CCJ, Mayer KH, *et al*: Hierarchical pattern of mucosal *Candida* infections in HIV-seropositive women. *Am J Med* **89**:142–146, 1990.
14. Denning DW, Follansbee SE, Scolaro M, *et al*: Pulmonary aspergillosis in the acquired immunodeficiency syndrome. *N Engl J Med* **324**:654–662, 1991.
15. Martins MD, Lozano-Chiu M, Rex JH: Declining rates of oropharyngeal candidiasis and carriage of *Candida albicans* associated with trends toward reduced rates of carriage of fluconazole-resistant *C. albicans* in human immunodeficiency virus-infected patients. *Clin Infect Dis* **27**:1291–1294, 1998.
16. Bow EJ, Laverdiere M, Lussier N, Rotstein C: Antifungal prophylaxis in neutropenia cancer patients: A meta-analysis of randomized controlled trials. Abst #1516. In *Abstracts of the 41st Annual Meeting of the American Society of Hematology*, New Orleans, LA, 1999.
17. Goodman JL, Winston DJ, Greenfield RA, *et al*: A controlled trial of fluconazole to prevent fungal infections in patients undergoing bone marrow transplantation. *N Engl J Med* **326**:845–851, 1992.
18. Slavin MA, Osborne B, Adams R, *et al*: Efficacy and safety of fluconazole prophylaxis for fungal infections after bone marrow transplantation—A prospective, randomized, double-blind study. *J Infect Dis* **171**:1545–1552, 1995.
19. Winston DJ, Chandrasekar PH, Lazarus HM, *et al*: Fluconazole prophylaxis of fungal infections in patients with acute leukemia. *Ann Intern Med* **118**:495–503, 1993.
20. Rotstein C, Bow EJ, Laverdiere M, *et al*: Randomized placebo-controlled trial of fluconazole prophylaxis for neutropenic cancer patients: Benefit based in purpose and intensity of cytotoxic therapy. *Clin Infect Dis* **28**:331–340, 1999.
21. Gubbins PO, Bowman JL, Penzak SR: Antifungal prophylaxis to prevent invasive mycoses among bone marrow transplant recipients. *Pharmacotherapy* **18**:549–564, 1998.
22. Böhme A, Karthaus M, Hoelzer D: Antifungal prophylaxis in neutropenic patients with hematologic malignancies: Is there a real benefit? *Chemotherapy* **45**:224–232, 1999.
23. Schaffner A, Schaffner M: Effect of prophylactic fluconazole on the frequency of fungal infections, amphotericin B use, and health care costs in patients undergoing intensive chemotherapy for hematologic neoplasias. *J Infect Dis* **172**:1035–1041, 1995.
24. Alangaden G, Chandrasekar PH, Bailey E, *et al*: Antifungal prophylaxis with low-dose fluconazole during bone marrow transplantation. *Bone Marrow Transplant* **14**:919–924, 1994.
25. Guinet R, Marlier R: Sensibilite comparee des levures aux ketoconazole, itraconazole et fluconazole en micromethode standardisee en milieu liquide. *Pathol Biol* **38**:575–578, 1990.
26. Wingard JR, Merz WG, Rinaldi MG, *et al*: Increase in *Candida krusei* infection among patients with bone marrow transplantation and neutropenia treated prophylactically with fluconazole. *N Engl J Med* **325**:1274–1277, 1991.
27. Marr KA, Seidel K, White TC, Bowden R: Candidemia in allogeneic blood and bone marrow transplant recipients: Evolution of

risk factors after adoption of prophylactic fluconazole. *J Infect Dis* **181**:309–316, 2000.

28. Tricot G, Joosten E, Boogaerts MA, *et al*: Ketoconazole vs. itraconazole for antifungal prophylaxis in patients with severe granulocytopenia: Preliminary results of two nonrandomized studies. *Rev Infect Dis* **9**(Suppl 1):S94–S99, 1987.

29. Böhme A, Just-Nübling G, Bergmann L, *et al*: Itraconazole for prophylaxis of systemic mycoses in neutropenic patients with haematological malignancies. *J Antimicrob Chemother* **38**:953–961, 1996.

30. Glasmacher A, Molitor E, Hahn C, *et al*: Antifungal prophylaxis with itraconazole in neutropenic patients with acute leukemia. *Leukemia* **12**:1338–1343, 1998.

31. Vreugdenhil G, Van Dijke BJ, Donnelly JP, *et al*: Efficacy of itraconazole in the prevention of fungal infections among neutropenic patients with hematologic malignancies and intensive chemotherapy. A double-blind, placebo controlled study. *Leuk Lymphoma* **11**:353–358, 1993.

32. Huijgens PC, Simoons-Smit AM, Van Loenen AC, Prooy E: Fluconazole versus itraconazole for the prevention of fungal infections in haemato-oncology. *Clin Pathol* **52**:376–380, 1999.

33. Lamy T, Bernard M, Courtois A, *et al*: Prophylactic use of itraconazole for the prevention of invasive pulmonary aspergillosis in high risk neutropenic patients. *Leuk Lymphoma* **30**:163–174, 1998.

34. Menichetti F, Del Favero A, Martino P, *et al*: Itraconazole oral solution as prophylaxis for fungal infections in neutropenic patients with hematologic malignancies: A randomized, placebo-controlled, double-blind, multicenter trial. *Clin Infect Dis* **28**:250–255, 1999.

35. Morgenstern GR, Prentice AG, Prentice HG, *et al*: A randomized controlled trial of itraconazole versus fluconazole for the prevention of fungal infections in patients with haematological malignancies. *Br J Haematol* **105**:901–911, 1999.

36. O'Donnell MR, Schmidt GM, Tegtmeier BR, *et al*: Prediction of systemic fungal infection in allogeneic marrow recipients: Impact of amphotericin prophylaxis in high-risk patients. *J Clin Oncol* **12**:827–834, 1994.

37. Perfect JR, Klotman ME, Gilbert CC, *et al*: Prophylactic intravenous amphotericin B in neutropenic autologous bone marrow transplant recipients. *J Infect Dis* **165**:891–897, 1992.

38. Bodey GP, Anaissie EJ, Elting LS, *et al*: Antifungal prophylaxis during remission induction therapy for acute leukemia fluconazole versus intravenous amphotericin B. *Cancer* **73**:2099–2106, 1994.

39. Wolff SN, Fay J, Stevens D, *et al*: Fluconazole vs low-dose amphotericin B for the prevention of fungal infections in patients undergoing bone marrow transplantation: A study of the North American Marrow Transplant Group. *Bone Marrow Transplant* **25**:853–859, 2000.

40. Hiemenz JW, Walsh TJ: Lipid formulations of amphotericin B: Recent progress and future directions. *Clin Infect Dis* **22**:S133–S144, 1996.

41. Walsh TJ, Finberg RW, Arndt C, *et al*: Liposomal amphotericin B for empirical therapy in patients with persistent fever and neutropenia. *N Engl J Med* **340**:764–771, 1999.

42. Tollemar J, Ringdén O, Andersson S, *et al*: Randomized double-blind study of liposomal amphotericin B (AmBisome) prophylaxis of invasive fungal infections in bone marrow transplant recipients. *Bone Marrow Transplant* **12**:577–582, 1993.

43. Kelsey SM, Goldman JM, McCann S, *et al*: Liposomal amphotericin (AmBisome) on the prophylaxis of fungal infections in neu-

tropenic patients: A randomized, double-blind, placebo-controlled study. *Bone Marrow Transplant* **23**:163–168, 1999.

44. Timmers GJ, Zweegman S, Simoons-Smit AM, *et al*: Amphotericin B colloidal dispersion (Amphocil) vs fluconazole for the prevention of fungal infections in neutropenic patients: Data of a prematurely stopped clinical trial. *Bone Marrow Transplant* **25**:879–884, 2000.

45. Lumbreras C, Cuervas-Mons V, Jara P, *et al*: Randomized trial of fluconazole versus nystatin for the prophylaxis of *Candida* infection following liver transplantation. *J Infect Dis* **174**:583–588, 1996.

46. Winston DJ, Pakrasi A, Busuttil RW: Prophylactic fluconazole in liver transplant recipients—A randomized, double-blind, placebo-controlled trial. *Ann Intern Med* **131**:729–737, 1999.

47. Linden P, Kramer LP, Mazariegos G, *et al*: Low-dose amphotericin B for the prophylaxis of serious *Candida* infections in high-risk liver recipients. Abst J47. In *Program and Abstracts of the 36th Interscience Conference on Antimicrobial Agents and Chemotherapy*, New Orleans, LA, 1996, p. 227.

48. Tollemar J, Höckerstedt K, Ericzon B-G, *et al*: Liposomal amphotericin B prevents invasive fungal infections in liver transplant recipients. A randomized, placebo-controlled study. *Transplantation* **59**:45–50, 1995.

49. Colby WD, Sharpe MD, Ghent CN, *et al*: Efficacy of itraconazole prophylaxis against systemic fungal infection in liver transplant recipients. Abst 1650. In *Abstracts of the 39th Interscience Conference on Antimicrobial Agents and Chemotherapy*, 1999.

50. Nadeem I, Yeldandi V, Sheridan P, *et al*: Efficacy of itraconazole prophylaxis for aspergillosis in lung transplant recipients. In *Program and Abstracts of the 32nd Annual Meeting of the Infectious Disease Society of America*, p. 58A, 1994.

51. Calvo V, Borro JM, Morales P, *et al*: Antifungal prophylaxis during the early postoperative period of lung transplantation. *Chest* **115**:1301–1304, 1999.

52. Lau CL, Palmer SM, D'Amico TA, *et al*: Lung transplantation at Duke University Medical Center. *Clin Transplant* 327–340, 1998.

53. Barchiesi F, Arzeni D, Fothergill AW, *et al*: *In vitro* activities of the new antifungal triazole SCH 56592 against common and emerging yeast pathogens. *Antimicrob Agents Chemother* **44**:226–229, 2000.

54. Kirkpatrick WR, McAtee RK, Fothergill AW, *et al*: Efficacy of SCH56592 in a rabbit model of invasive aspergillosis. *Antimicrob Agents Chemother* **44**:780–782, 2000.

55. Ikeda F, Wakai Y, Matsumoto S, *et al*: Efficacy of FK463, a new lipopeptide antifungal agent, in mouse models of disseminated candidiasis and aspergillosis. *Antimicrob Agents Chemother* **44**:614–618, 2000.

56. Pizzo PA, Ribichaud KJ, Gill FA, Witebsky FG: Empiric antibiotic and antifungal therapy for cancer patients with prolonged fever and granulocytopenia. *Am J Med* **72**:101–111, 1982.

57. Krick JA, Remington JS: Opportunistic invasive fungal infections in patients with leukaemia or lymphoma. *Clin Haematol* **5**:249–310, 1976.

58. Fraser DW, Ward JI, Ajello L, Plikaytis BD: Aspergillosis and other systemic mycoses: The growing problem. *JAMA* **242**:1631–1635, 1979.

59. Pennington JE: Successful treatment of *Aspergillus* pneumonia in hematologic neoplasia. *N Engl J Med* **295**:426–427, 1976.

60. Aisner J, Wiernik PH, Schimpf SC: Treatment of invasive aspergillosis: Relation of early diagnosis and treatment to response. *Ann Intern Med* **86**:539–543, 1977.

61. Stein RS, Kayser J, Flexner JM: Clinical value of empirical am-

photericin B in patients with acute myelogenous leukemia. *Cancer* **50**:2247–2251, 1982.

62. European Organization for Research on Treatment of Cancer International Antimicrobial Therapy Cooperative Group: Empiric antifungal therapy in febrile granulocytopenic. *Am J Med* **86**:668–672, 1989.

63. Prentice HG, Hann IM, Herbrecht R, *et al*: A randomized comparison of liposomal versus conventional amphotericin B for the treatment of pyrexia of unknown origin in neutropenic patients. *Br J Haematol* **98**:711–718, 1997.

64. Wingard JR, White MH, Anaissie EJ, Rafailli JT: A randomized double-blind safety study of AmBisome and Abelcet in febrile neutropenic patients. *Proc 9th Focus Fungal Infect Meet* **98**:711–718, 1999.

65. White MH, Bowden RA, Sandler ES, *et al*: Randomized, double-blind clinical trial of amphotericin B colloidal dispersion vs. amphotericin B in the empirical treatment of fever and neutropenia. *Clin Infect Dis* **27**:296–302, 1998.

66. Viscoli C, Castagnola E, Van Lint MT, Moroni C: Fluconazole versus amphotericin B as empirical antifungal therapy of unexplained fever in granulotopenic cancer patients: A pragmatic, multicentre, prospective and randomized clinical trial. *Eur J Cancer* **32A**:814–820, 1996.

67. Winston DJ, Hathorn JW, Schuster MG, *et al*: A multicenter, randomized trial of fluconazole versus amphotericin B for empiric antifungal therapy of febrile neutropenic patients with cancer. *Am J Med* **108**:282–289, 2000.

68. Malik IA, Moid I, Aziz A, Kham S: A randomized comparison of fluconazole with amphtericin B as empiric anti-fungal agents in cancer patients with prolonged fever and neutropenia. *Am J Med* **105**:478–483, 1998.

69. Boogaerts M, Garber G, Winston D, Reboli A: Itraconazole compared with amphotericin B as empirical therapy for persistent fever of unknown origin in neutropenic patients. *Bone Marrow Transplant* **23**:S111, 1999.

70. Sheehan DJ, Hitchcock CA, Sibley CM: Current and emerging azole antifungal agents. *Clin Microbiol Rev* **12**:40–79, 1999.

71. Pfaller MA, Marco F, Messer SA, Jones RN: *In vitro* activity of two echinocandin derivatives, LY303366 and MK-0991 (L-743,792), against clinical isolates of *Aspergillus*, *Fusarium*, *Rhizopus*, and other filamentous fungi. *Diagn Microbiol Infect Dis* **30**:251–255, 1998.

72. Marina NM, Flynn PM, Rivera GK, Hughes WT: *Candida tropicalis* and *Candida albicans* fungemia in children with leukemia. *Cancer* **68**:594–599, 1991.

73. Silverman RA, Rhodes AR, Dennehy PH: Disseminated intravascular coagulation and purpura fulminans in a patient with candida sepsis. Biopsy of purpura fulminans as an aid to diagnosis of systemic candida infection. *Am J Med* **80**:679–684, 1986.

74. Castaldo P, Stratta RJ, Wood RP, *et al*: Clinical spectrum of fungal infections after orthotopic liver transplant. *Arch Surg* **126**:149–156, 1991.

75. Thaler M, Pastakia B, Shawker TH, *et al*: Hepatic candidiasis in cancer patients: The evolving picture of the syndrome. *Ann Intern Med* **108**:88–100, 1988.

76. Pestalozzi BC, Krestin GP, Schanz U, *et al*: Hepatic lesions of chronic desseminated candidiasis may become invisible during neutropenia. *Blood* **90**:3858–3864, 1997.

77. Karp JE, Merz WG, Charache P: Response to empiric amphotericin B during antileukemic therapy-induced granulocytopenia. *Rev Infect Dis* **13**:592–599, 1991.

78. Walsh TJ, Chanock SJ: Laboratory diagnosis of invasive candidiasis: A rationale for complementary use of culture- and nonculture-based detection systems. *Int J Infect Dis* **1**(Suppl 1): S11–S19, 1997.

79. Christensson B, Wiebe T, Pehrson C, Larsson L: Diagnosis of invasive candidiasis in neutropenic children with cancer by determination of D-arabinitol/L-arabinitol ratios in urine. *J Clin Microbiol* **35**:636–640, 1997.

80. Phillips P, Dowd A, Jewesson P, *et al*: Nonvalue of antigen detection immunoassays for diagnosis of candidemia. *J Clin Microbiol* **28**:2320–2326, 1990.

81. Morhart M, Rennie R, Ziola B, *et al*: Evaluation of enzyme immunoassay for *Candida* cytoplasmic antigens in neutropenic cancer patients. *J Clin Microbiol* **32**:766–776, 1994.

82. Hossain MA, Miyazaki T, Mitsutake K, *et al*: Comparison between Wako-WB003 and Fungitec G tests for detection of (1→3)-β-D-glucan in systemic mycosis. *J Clin Lab Anal* **11**:73–77, 1997.

83. Einsele H, Hebart H, Roller G, *et al*: Detection and identification of fungal pathogens in blood by using molecular probes. *J Clin Microbiol* **35**:1353–1360, 1997.

84. Martínez JP, Gil ML, López-Ribot JL, Chaffin WL: Serologic response to cell wall mannoproteins and proteins of *Candida albicans*. *Clin Microbiol Rev* **11**:121–141, 1998.

85. Preisler HD, Hasenclever HF, Henderson ES: Anti-*Candida* antibodies in patients with acute leukemia: A prospective study. *Am J Med* **51**:352–361, 1971.

86. Wingard JR: Infections due to resistant *Candida* species in patients with cancer who are receiving chemotherapy. *Clin Infect Dis* **19** (Suppl 1):S49–S53, 1994.

87. Rex J: Editorial response: Catheters and candidemia. *Clin Infect Dis* **22**:467–470, 1996.

88. Edwards JE Jr, Bodey GP, Bowden RA, *et al*: International conference for the development of a consensus on the management and prevention of severe candidal infections. *Clin Infect Dis* **25**:43–59, 1997.

89. Powderly W, Kobayashi G, Herzig GP: Amphotericin B-resistant yeast infection in severely immunocompromised patients. *Am J Med* **4**:826–832, 1988.

90. Ringdén O, Meunier F, Tollemar J, *et al*: Efficacy of amphotericin B encapsulated in liposomes (AmBisome) in the treatment of invasive fungal infections in immunocompromised patients. *J Antimicrob Chemother* **28**(Suppl B):73–82, 1991.

91. Noskin GA, Pietrelli L, Coffey G, *et al*: Amphotericin B colloidal dispersion for treatment of candidemia in immunocompromised patients. *Clin Infect Dis* **26**:461–467, 1998.

92. Walsh TJ, Hiemenz JW, Seibel NL, *et al*: Amphotericin B lipid complex for invasive fungal infections: Analysis of safety and efficacy in 556 cases. *Clin Infect Dis* **26**:1383–1396, 1998.

93. Wingard JR: Efficacy of amphotericin B lipid complex injection (ABLC) in bone marrow transplant recipients with life-threatening systemic mycoses. *Bone Marrow Transplant* **19**:343–347, 1997.

94. Merhav H, Mieles L: Amphotericin B lipid complex in the treatment of invasive fungal infections in liver transplant patients. *Transplant Proc* **29**:2670–2674, 1997.

95. Anaissie EJ, White MH, Uzun O, *et al*: Amphotericin B lipid complex vs. amphotericin B for treatment of invasive candidiasis: A prospective, randomized multicenter trial. Abst #LM21. In *Program and Abstracts of the 35th Interscience Conference on Antimicrobial Agents and Chemotherapy*, 1995, p. 330.

96. Horn R, Wong B, Kiehn TE, Armstrong D: Fungemia in a cancer

hospital: Changing frequency, earlier onset, and results of therapy. *Rev Infect Dis* **7:**646–655, 1985.

97. Kauffman CA, Bradley SF, Ross SC, Weber DR: Hepatosplenic candidiasis: Successful treatment with fluconazole. *Am J Med* **91:**137–141, 1991.

98. Anaissie E, Bodey GP, Kantarjian H, *et al*: Fluconazole therapy for chronic disseminated candidiasis in patients with leukemia and prior amphotericin B therapy. *Am J Med* **91:**142–150, 1991.

99. Poynton CH, Barnes RA, Rees J: Interferon γ and granulocyte-macrophage colony-stimulating factor for the treatment of hepatosplenic candidosis in patients with acute leukemia. *Clin Infect Dis* **26:**239–240, 1998.

100. Capetti A, Bonfanti P, Magni C, Milazzo F: Employment of recombinant human granulocyte-macrophage colony stimulating factor in oesophageal candidiasis in AIDS patients. *AIDS* **9:**1378, 1995.

101. Vazquez JA, Gupta S, Villanueva A: Potential utility of recombinant human GM-CSF as adjunctive treatment of refractory oropharyngeal candidiasis in AIDS patients. *Eur J Clin Microbiol Infect Dis* **17:**781–783, 1998.

102. Bjerke JW, Meyers JD, Bowden RA: Hepatosplenic candidiasis—A contraindication to marrow transplantation. *Blood* **84:**2811–2814, 1994.

103. Rex JH, Walsh TJ, Sobel JD, *et al*: Practice guidelines for the treatment of candidiasis. *Clin Infect Dis* **30:**662–678, 2000.

104. Ribaud P, Chastang C, Latgé JP, *et al*: Survival and prognostic factors of invasive aspergillosis after allogeneic bone marrow transplantation. *Clin Infect Dis* **28:**322–330, 1999.

105. Leenders A, Van Belkum A, Janssen S, *et al*: Molecular epidemiology of apparent outbreak of invasive aspergillosis in a hematology ward. *J Clin Microbiol* **34:**345–351, 1996.

106. Hospenthal D, Kwon-Chung KJ, Bennett JE: Concentrations of airborne *Aspergillus* compared to the incidence of invasive aspergillosis: Lack of correlation. *Med Mycol* **36:**165–168, 1998.

107. Hamadeh R, Ardehali A, Locksley RM, York MK: Fatal aspergillosis associated with smoking contaminated marijuana, in a marrow transplant recipient. *Chest* **94:**432–433, 1984.

108. Karp E, Burch PA, Merz WG: An approach to intensive antileukemia therapy in patients with previous invasive aspergillosis. *Am J Med* **85:**203–206, 1988.

109. Armstrong D: Problems in management of opportunistic fungal diseases. *Rev Infect Dis* **II:**S1591–S1599, 1989.

110. Morrison VA, Haake RJ, Weisdorf DJ: Non-*Candida* fungal infections after bone marrow transplantation: Risk factors and outcome. *Am J Med* **96:**497–503, 1994.

111. Meyers JD: Fungal infection in bone marrow transplant patients. *Semin Oncol* **3:**10–13, 1990.

112. Paterson DL, Singh N: Invasive aspergillosis in transplant recipients. *Medicine* **78:**123–138, 1999.

113. Singh N, Arnow PM, Bonham A, *et al*: Invasive aspergillosis in liver transplant recipients in the 1990s. *Transplantation* **64:**716–720, 1997.

114. Denning DW: Invasive aspergillosis. *Clin Infect Dis* **26:**781–803, 1998.

115. Clark A, Skeleton J, Fraser RS: Fungal tracheobronchitis report of 9 cases and review of the literature. *Medicine* **70:**1–14, 1991.

116. Talbot GH, Huang A, Provencher M: Invasive aspergillus rhinosinusitis in patients with acute leukemia. *Rev Infect Dis* **13:**219–232, 1991.

117. Denning DW, Marinus A, Cohen J, *et al*: An EORTC multicentre prospective survey of invasive aspergillosis in haematological

patients: Diagnosis and therapeutic outcome. *J Infect* **37:**173–180, 1998.

118. Burch PA, Karp JE, Merz WG, *et al*: Favorable outcome of invasive aspergillosis in paitents with acute leukemia. *J Clin Oncol* **5:**1985–1993, 1987.

119. Reiss P, Hadderingh R, Schot LJ, Dannder SA: Invasive external otitis caused by *Aspergillus fumigatus* in two patients with AIDS. *AIDS* **5:**605–606, 1991.

120. Bodey GP, Vartivarian S: Aspergillosis. *Eur J Clin Microbiol Infect Dis* **8:**413–437, 1989.

121. Caillot D, Casasnovas O, Bernard A, *et al*: Improved management of invasive pulmonary aspergillosis in neutropenic patients using early thoracic computed tomographic scan and surgery. *J Clin Oncol* **15:**139–147, 1997.

122. Brown MJ, Worthy SA, Flint JD, Muller NL: Invasive aspergillosis in the immunocompromised host: Utility of computed tomography and bronchoalveolar lavage. *Clin Radiol* **53:**255–257, 1998.

123. Albelda SM, Talbot GH, Gerson SL, *et al*: Role of fiberoptic bronchoscopy in the diagnosis of invasive pulmonary aspergillosis in patients with acute leukemia. *Am J Med* **76:**1027–1034, 1984.

124. Horvath J, Dummer S: The use of respiratory tract cultures in the diagnosis of invasive pulmonary aspergillosis. *Am J Med* **100:**171–178, 1996.

125. Aisner J, Murillo J, Schimpff SC, Steere AC: Invasive aspergillosis in acute leukemia: Correlation with nose cultures and antibiotic use. *Ann Intern Med* **90:**4–9, 1979.

126. Nunley DR, Ohori NP, Grgurich WF, *et al*: Pulmonary aspergillosis in cystic fibrosis lung transplant recipients. *Chest* **114:**1321–1329, 1998.

127. Verweij PE, Stynen D, Rijs AJMM, *et al*: Sandwich enzyme-linked immunosorbent assay compared with Pastorex latex agglutination test for diagnosing invasive aspergillosis in immunocompromised patients. *J Clin Microbiol* **33:**1912–1914, 1995.

128. Severens JL, Donnelly JP, Meis JFGM, *et al*: Two strategies for managing invasive aspergillosis: A decision analysis. *Clin Infect Dis* **25:**1148–1154, 1997.

129. Yuasa K, Goto H, Iguchi M, *et al*: Evaluation of the diagnostic value of the measurement of $(1\rightarrow3)$-β-D-glucan in patients with pulmonary aspergillosis. *Respiration* **63:**78–83, 1996.

130. Kaiser L, Huguenin T, Lew PD, *et al*: Invasive aspergillosis—Clinical features of 35 proven cases at a single institution. *Medicine* **77:**188–194, 1998.

131. Allo MD, Miller J, Townsend T, Cissy T: Primary cutaneous aspergillosis associated with Hickman intravenous catheters. *N Engl J Med* **317:**1105–1108, 1987.

132. Oppenheim BA, Herbrecht R, Kusne S: The safety and efficacy of amphotericin B colloidal dispersion in the treatment of invasive mycoses. *Clin Infect Dis* **21:**1145–1153, 1995.

133. White A, Anaissie E, Kusne S, *et al*: Amphotericin B colloidal dispersion vs. amphotericin B as therapy for invasive aspergillosis. *Clin Infect Dis* **24:**635–642, 1997.

134. Mills W, Chopra R, Linch DC, Goldstone AH: Liposomal amphotericin B in the treatment of fungal infections in neutropenic patients: A single-centre experience of 133 episodes in 116 patients. *Br J Haematol* **86:**754–760, 1994.

135. Ellis M, Spence D, De Pauw B, *et al*: An EORTC international multicenter randomized trial (EORTC number 19923) comparing two dosages of liposomal amphotericin B for treatment of invasive aspergillosis. *Clin Infect Dis* **27:**1406–1412, 1998.

136. Denning DW, Tucker RM, Hanson LH, Stevens FA: Treatment of

invasive aspergillosis with itraconazole. *Am J Med* **86:**791–800, 1989.

137. Viviani MA, Tortorano AM, Pagano A, *et al:* European experience with itraconazole in systemic mycoses. *J Am Acad Dermatol* **9**(Suppl 1)**:**S77–S86, 1987.

138. Dupont B: Itraconazole therapy in aspergillosis: Study in 49 patients. *J Am Acad Dermatol* **23:**607–614, 1990.

139. Sánchez C, Mauri E, Dalmau D, *et al:* Treatment of cerebral aspergillosis with itraconazole: Do high doses improve the prognosis? *Clin Infect Dis* **21:**1485–1487, 1995.

140. Denning DW, Venkateswarlu K, Oakley KL, *et al:* Itraconazole resistance in *Aspergillus fumigatus. Antimicrob Agents Chemother* **41:**1364–1368, 1997.

141. Chryssanthou E: *In vitro* susceptibility of respiratory isolates of *Aspergillus* species to itraconazole and amphotericin B. Acquired resistance to itraconazole. *Scand J Infect Dis* **29:**509–512, 1997.

142. Oakley KL, Moore CB, Denning DW: *In vitro* activity of SCH-56592 and comparison with activities of amphotericin B and itraconazole against *Aspergillus* spp. *Antimicrob Agents Chemother* **41:**1124–1126, 1997.

143. Graybill JR, Bocanegra R, Najvar LK, *et al:* SCH56592 treatment of murine invasive aspergillosis. *J Antimicrob Chemother* **42:** 539–542, 1998.

144. Petraitis V, Petraitiene R, Groll AH, *et al:* Antifungal efficacy, safety, and single-dose pharmacokinetics of LY303366, a novel echinocandin B, in experimental pulmonary aspergillosis in persistently neutropenic rabbits. *Antimicrob Agents Chemother* **42:** 2898–2905, 1998.

145. Jones TC: Use of granulocyte-macrophage colony stimulating factor (GM-CSF) in prevention and treatment of fungal infections. *Eur J Cancer [A]* **35:**S8–S10, 1999.

146. Armitage J: Emerging applications of recombinant human granulocyte-macrophage colony-stimulating factor. *J Am Soc Hematol* **92:**4491–4508, 1998.

147. Viollier AF, Peterson DE, DeJongh CA, *et al: Aspergillus* sinusitis in cancer patients. *Cancer* **58:**366–371, 1986.

148. Reichenberger F, Habicht J, Kaim A, *et al:* Lung resection for invasive pulmonary aspergillosis in neutropenic patients with hematologic diseases. *Am J Respir Crit Care Med* **158:**885–890, 1998.

149. Stevens DA, Kan VL, Judson MA, *et al:* Practice guidelines for diseases caused by *Aspergillus. Clin Infect Dis* **30:**696–709, 2000.

150. Opal SM, Asp AA, Canady PB Jr, *et al:* Efficacy of infection control measure during a nosocomial outbreak of disseminated aspergillosis associated with hospital construction. *J Infect Dis* **153:**634–637, 1986.

151. Meunier F: Prevention of mycoses in immunocompromised patients. *Rev Infect Dis* **9:**408–416, 1987.

152. Sherertz RJ, Belani A, Kramer BS, *et al:* Impact of air filtration on nosocomial *Aspergillus* infections unique risk of bone marrow transplant recipients. *Am J Med* **83:**709–718, 1987.

153. Conneally E, Cafferkey MT, Daly PA, *et al:* Nebulized amphotericin B as prophylaxis against invasive aspergillosis in granulocytopenic patients. *Bone Marrow Transplant* **5:**403–406, 1990.

154. Bodey GP, Johnston D: Microbiological evaluation of protected environments during patient occupancy. *Appl Microbiol* **22:**828–836, 1999.

155. Singh N, Gayowski T, Singh J, Yu VL: Invasive gastrointestinal zygomycosis in a liver transplant recipient: Case report and review of zygomycosis in solid-organ transplant recipients. *Clin Infect Dis* **20:**617–620, 1995.

156. Gaziev D, Baronciani D, Galimberti M, *et al:* Mucormycosis after bone marrow transplantation: Report of four cases in thalassemia and review of the literature. *Bone Marrow Transplant* **17:**409–414, 1996.

157. Morrison VA, McGlave PB: Mucormycosis in the BMT population. *Bone Marrow Transplant* **11:**383–388, 1993.

158. Blitzer A, Lawson W, Meyers BR, Biller HF: Patient survival factors in paranasal sinus zygomycosis. *Laryngoscope* **90:**635–648, 1980.

159. Lehrer RI, Howard DH, Sypherd PS, *et al:* Zygomycosis. *Ann Intern Med* **93:**93–108, 1980.

160. Ferguson BJ, Mitchell TG, Moon R, *et al:* Adjunctive hyperbaric oxygen for treatment of rhinocerebral zygomycosis. *Rev Infect Dis* **10:**551–559, 1988.

161. Gonzalez CE, Couriel DR, Walsh TJ: Disseminated zygomycosis in a neutropenic patient: Successful treatment with amphotericin B lipid complex and granulocyte colony-stimulating factor. *Clin Infect Dis* **24:**192–196, 1997.

162. Groll AH, Walsh TJ: Potential new antifungal agents. *Curr Opin Infect Dis* **10:**449–458, 1997.

163. Nelson PE, Dignani MC, Anaissie EJ: Taxonomy, biology, and clinical aspects of *Fusarium* species. *Clin Microbiol Rev* **7:**479–504, 1994.

164. Hennequin C, Lavarde V, Poirot JL, *et al:* Invasive *Fusarium* infections: A retrospective survey of 31 cases. *J Med Vet Mycol* **35:**107–114, 1997.

165. Lozano-Chiu M, Arikan S, Paetznick VL, *et al:* Treatment of murine fusariosis with SCH 56592. *Antimicrob Agents Chemother* **43:**589–591, 1999.

166. Schroter GPJ, Temple DR, Husberg BS, *et al:* Cryptococcosis after renal transplantation: Report of ten cases. *Surgery* **79:**268–277, 1976.

167. Watson AJ, Russell RP, Cabreja RF, Braverman WA: Cure of cryptococcal infection during continued immunosuppressive therapy. *Q J Med* **55:**169–172, 1985.

168. Aberg JA, Mundy LM, Powderly WG: Pulmonary cryptococcosis in patients without HIV infection. *Chest* **115:**734–740, 1999.

169. Andreula CF, Burdi N, Carella A: CNS cryptococcosis in AIDS: Spectrum of MR findings. *J Comput Assist Tomogr* **17:**438–441, 1993.

170. Powderly WG, Cloud GA, Dismukes WE, Saag MS: Measurement of cryptococcal antigen in serum and cerebrospinal fluid: Value in the management of AIDS-associated cryptococcal meningitis. *Clin Infect Dis* **18:**789–792, 1994.

171. Larsen RA, Bozzette S, McCutchan JA, *et al:* Persistent *Cryptococcus neoformans* infection of the prostate after successful treatment of meningitis. *Ann Intern Med* **111:**125–128, 1989.

172. Chan KH: Neurosurgical aspects of cerebral cryptococcosis. *Neurosurgery* **25:**44–48, 1989.

173. Gade W, Hinnefeld SW, Babcock LS, *et al:* Comparison of the premier cryptococcal antigen enzyme immunoassay and the latex agglutination assay for detection of cryptococcal antigens. *J Clin Microbiol* **29:**1616–1619, 1991.

174. Bennett JE, Dismukes WE, Duma RJ, *et al:* A comparison of amphotericin B alone and combined with flucytosine in the treatment of cryptococcal meningitis. *N Engl J Med* **301:**126–131, 1979.

175. Van der Horst CM, Saag MS, Cloud GA, *et al:* Treatment of cryptococcal meningitis associated with the acquired immunodeficiency syndrome. *N Engl J Med* **337:**15–21, 1997.

176. Saag MS, Powderly WG, Cloud GA, *et al:* Comparison of amphotericin B with fluconazole in the treatment of acute AIDS-associated cryptococcal meningitis. *N Engl J Med* **326:**83–89, 1992.

177. Groll A, Giri N, Petraitis V, *et al*: Comparative efficacy and distribution of lipid formulations of amphotericin B in experimental *Candida albicans* infections of the central nervous system. *J Infect Dis* **182**:274–282, 2000.

178. Leenders A, Reiss P, Portegies P, *et al*: Lipsomal amphotericin B (AmBisome) compared with amphotericin B both followed by oral fluconazole in the treatment of AIDS-associated cryptococcal meningitis. *AIDS* **11**:1463–1471, 1997.

179. DeGans J, Portegies P, Tiessens G, *et al*: Itraconazole compared with amphotericin B plus flucytosine in AIDS patients with cryptococcal meningitis. *AIDS* **6**:185–190, 1990.

180. Armengou A, Porcar C, Mascaró J, García-Bragado F: Possible development of resistance to fluconazole during suppressive therapy for AIDS-associated cryptococcal meningitis. *Clin Infect Dis* **23**:1337–1338, 1996.

181. Witt MD, Lewis RJ, Edwards JE Jr, Ghannoum MA: Possible development of resistance to fluconazole during suppressive therapy for AIDS-associated cryptococcal meningitis—Reply. *Clin Infect Dis* **23**:1338, 1996.

182. Paugam A, Dupouy-Camet J, Blanche P, *et al*: Increased fluconazole resistance of *Cryptococcus neoformans* isolated from a patient with AIDS and recurrent meningitis. *Clin Infect Dis* **19**:975–976, 1994.

183. Coker RJ, Harris JR: Failure of fluconazole treatment in cryptococcal meningitis despite adequate CSF levels. *J Infect* **23**:101–103, 1991.

184. Larsen RA, Bozzette SA, Jones BE, *et al*: Fluconazole combined with flucytosine for treatment of cryptococcal meningitis in patients with AIDS. *Clin Infect Dis* **19**:741–745, 1994.

185. Bozzette SA, Larsen RA, Chiu J, *et al*: A placebo-controlled trial of maintenance therapy with fluconazole after treatment of cryptococcal meningitis in the acquired immunodeficiency syndrome. *N Engl J Med* **324**:580–584, 1991.

186. Powderly WG, Saag MS, Cloud GA, *et al*: A controlled trial of fluconazole or amphotericin B to prevent relapse of cryptococcal meningitis in patients with the acquired immunodeficiency syndrome. *N Engl J Med* **326**:793–798, 1992.

187. Saag MS, Cloud GA, Graybill JR, *et al*: A comparison of itraconazole versus fluconazole as maintenance therapy for AIDS-associated cryptococcal meningitis. *Clin Infect Dis* **28**:291–296, 1999.

188. Saag MS, Graybill RJ, Larsen RA, *et al*: Practice guidelines for the management of cryptococcal disease. *Clin Infect Dis* **30**:710–718, 2000.

189. Jahangir M, Hussain I, Ul H, Haroon TS: A double-blind, randomized, comparative trial of itraconazole versus terbinafine for 2 weeks in tinea capitis. *Br J Dermatol* **139**:672–674, 1998.

190. Brummer E, Stevens DA: Macrophage colony-stimulating factor induction of enhanced macrophage anticryptococcal activity: Synergy with fluconazole for killing. *J Infect Dis* **170**:173–179, 1994.

191. Bach MC, Adler JL, Breman J, *et al*: Influence of rejection therapy on fungal and nocardia infections in renal-transplant recipients. *Lancet* **1**:180–184, 1973.

192. Watson AJ, Whelton A, Russell RP: Cure of cryptococcemia and preservation of graft function in a renal transplant recipient. *Arch Intern Med* **144**:1877–1878, 1984.

193. Kong NCT, Shaariah W, Morad Z, *et al*: Cryptococcosis in a renal unit. *Aust NZ J Med* **20**:645–649, 1990.

194. Graybill JR, Sobel J, Saag M, *et al*: Diagnosis and management of increased intracranial pressure in patients with AIDS and cryptococcal meningitis. *Clin Infect Dis* **30**:47–54, 2000.

195. Powderly WG, Finkelstein DM, Feinberg J, *et al*: A randomized trial comparing fluconazole with clotrimazole troches for the prevention of fungal infections in patients with advanced human immunodeficiency virus infection. *N Engl J Med* **332**:700–705, 1995.

196. McKinsey DS, Wheat LJ, Cloud GA, *et al*: Itraconazole prophylaxis for fungal infections in patients with advanced human immunodeficiency virus infection: Randomized, placebo-controlled, double-blind study. *Clin Infect Dis* **28**:1049–1056, 1999.

197. Iitaka K, McEnery PT, West CD: Successful renal transplantation after generalized cryptococcosis. *J Pediatr* **92**:422–423, 1978.

198. Mills SA, Seigler HF, Wolfe WG: The incidence and management of pulmonary mycosis in renal allograft patients. *Ann Surg* **182**:617–626, 1975.

199. Wheat L: Histoplasmosis in the acquired immunodeficiency syndrome. *Curr Top Med Mycol* **7**:7–18, 1996.

200. Wheat LJ, Smith EJ, Sathapatayavongs B, *et al*: Histoplasmosis in renal allograft recipients: Two large urban outbreaks. *Arch Intern Med* **143**:703–707, 1983.

201. Hughes WT: Hematogenous histoplasmosis in the immunocompromised child. *J. Pediatr* **105**:569–575, 1984.

202. Vail G, Young R, Filo RS, *et al*: Incidence of histoplasmosis following allogeneic bone marrow transplant (aBMT) or solid organ transplant (SOT) in a hyperendemic area. Abst # 3465a. In *Program and Abstracts of the 36th Infectious Diseases Society of America Meeting*, Denver, CO, 1998, p. 141.

203. Hood AB, Inglis FG, Lowenstein L, *et al*: Histoplasmosis and thrombocytopenic purpura: Transmisson by renal homotransplantation. *Can Med Assoc J* **93**:587–592, 1965.

204. Wheat J: Histoplasmosis: Experience during outbreaks in Indianapolis and review of the literature. *Medicine* **76**:339–354, 1997.

205. Sridhar NR, Tchervenkov JI, Weiss MA, *et al*: Disseminated histoplasmosis in a renal transplant patient: A cause of renal failure several years following transplantation. *Am J Kidney Dis* **17**:719–721, 1991.

206. Wheat LJ, Kohler RB, Tewari RP: Diagnosis of disseminated histoplasmosis by detection of *Histoplasma capsulatum* antigen in serum and urine specimens. *N Engl J Med* **314**:83–88, 1986.

207. Sathapatayavongs B, Batteiger BE, Wheat LJ, *et al*: Clinical and laboratory features of disseminated histoplasmosis during two large urban outbreaks. *Medicine* **62**:263–270, 1983.

208. Wheat LJ, Connolly-Stringfield PA, Baker RL, *et al*: Disseminated histoplasmosis in the acquired immune deficiency syndrome: Clinical findings, diagnosis and treatment, and review of the literature. *Medicine* **69**:361–374, 1990.

209. Johnson P, Wheat LJ, Cloud G, *et al*: A multicenter randomized trial comparing amphotericin B (AmB) and liposomal amphotericin B (AmBisome, LAmB) as induction therapy of disseminated histoplasmosis (DH) in AIDS patients. *Ann Intern Med*, in press, 2002.

210. Wheat J, Hafner R, Korzun AH, *et al*: Itraconazole treatment of disseminated histoplasmosis in patients with the acquired immunodeficiency syndrome. *Am J Med* **98**:336–342, 1995.

211. Vullo V, Mastroianni CM, Ferone U, *et al*: Central nervous system involvement as a relapse of disseminated histoplasmosis in an Italian AIDS patient. *J Infect* **35**:83–84, 1997.

212. Wheat J, MaWhinney S, Hafner R, *et al*: Treatment of histoplasmosis with fluconazole in patients with acquired immunodeficiency syndrome. *Am J Med* **103**:223–232, 1997.

213. Wheat J, Marichal P, Vanden Bossche H, *et al*: Hypothesis on the

mechanism of resistance to fluconazole in *Histoplasma cap-sulatum. Antimicrob Agents Chemother* **41:**410–414, 1997.

214. Connolly P, Wheat J, Schnizlein-Bick C, *et al*: Comparison of a new triazole antifungal agent, Schering 56592, with itraconazole and amphotericin B for treatment of histoplasmosis in immuno-competent mice. *Antimicrob Agents Chemotherapy* **43:**322–328, 1999.

215. McKinsey DS, Gupta MR, Riddler SA, *et al*: Long-term ampho-tericin B therapy for disseminated histoplasmosis in patients with the acquired immune deficiency syndrome. *Ann Intern Med* **111:** 655–659, 1989.

216. Wheat J, Hafner R, Wulfson M, *et al*: Prevention of relapse of histoplasmosis with itraconazole in patients with the acquired immunodeficiency syndrome. *Ann Intern Med* **118:**610–616, 1993.

217. Hecht FM, Wheat J, Korzun AH, *et al*: Itraconazole maintenance treatment for histoplasmosis in AIDS: A prospective, multicenter trial. *J Acquir Immun Defic Syndr Hum Retrovirol* **16:**100–107, 1997.

218. Wheat J, Sarosi G, McKinsey D, *et al*: Practice guidelines for the management of patients with histoplasmosis. *Clin Infect Dis* **30:** 688–695, 2000.

219. Goldman M, Cloud GA, Smedema M, *et al*: Does long-term itraconazole prophylaxis result in *in vitro* azole resistance in mu-cosal *Candida albicans* isolates from persons with advanced hu-man immunodeficiency virus infection? *Antimicrob Agents Che-mother* **44:**1585–1587, 2000.

220. Ampel NM, Ryan KJ, Carry PJ, *et al*: Fungemia due to *Cocci-dioides immitis. Medicine* **65:**312–321, 1999.

221. Galgiani JN, Ampel NM: Coccidioidomycosis in human immuno-deficiency virus-infected patients. *J Infect Dis* **162:**1165–1169, 1990.

222. Drutz D, Catanzaro A: Coccidioidomycosis Parts 1 and 2. *Am Rev Respir Dis* **117:**559–585, 727–771, 1978.

223. Singh VR, Smith DK, Lawrence J, *et al*: Coccidioidomycosis in patients infected with human immunodeficiency virus: Review of 91 cases at a single institution. *Clin Infect Dis* **23:**563–568, 1996.

224. Galgiani JN: Coccidioidomycosis in the immunosuppressed host. In Einstein HE, Catanzaro A (eds): *Proceedings of the 5th International Conference on Coccidioidomycosis, National Foun-dation for Infectious Diseases*, Washington, DC, 1996, pp. 312–318.

225. Pappagianis D, Zimmer BL: Serology of coccidioidomycosis. *Clin Microbiol Rev* **3:**247–268, 1990.

226. Smith CE, Saito MT, Beard RR, *et al*: Serological tests in the diagnosis and prognosis of coccidioidomycosis. *Am J Hyg* **52:** 1–21, 1950.

227. Fish DG, Ampel NM, Galgiani JN, *et al*: Coccidioidomycosis during human immunodeficiency virus infection: A review of 77 patients. *Medicine* **69:**384–391, 1980.

228. Stevens D: Coccidioidomycosis. *N Engl J Med* **332:**1077–1082, 1995.

229. Labadie ELMD, Hamilton RHMD: Survival improvement in coc-cidioidal meningitis by high-dose intrathecal amphotericin B. *Arch Intern Med* **146:**2013–2018, 1985.

230. Galgiani JN, Catanzaro A, Cloud GA, *et al*: Fluconazole therapy for coccidioidal meningitis. *Ann Intern Med* **119:**28–35, 1993.

231. Galgiani J, Stevens D, Graybill J, Cloud G: Ketoconazole therapy of progressive coccidioidomycosis. *Am J Med* **84:**603–610, 1988.

232. Catanzaro A, Galgiani JN, Levine BE, *et al*: Fluconazole in the treatment of chronic pulmonary and nonmeningeal disseminated coccidioidomycosis. *Am J Med* **98:**249–256, 1995.

233. Classen DC, Burke JP, Smith CB: Treatment of coccidioidal men-ingitis with fluconazole. *J Infect Dis* **158:**903–904, 1988.

234. Galgiani JN, Catanzaro A, Cloud GA, *et al*: Comparison of oral fluconazole and itraconazole for progressive, nonmeningeal coc-cidioidomycosis: A randomized, double-blind trial. *Ann Intern Med* **133:**676–686, 2000.

235. Dewsnup DH, Galgiani JN, Graybill JR, *et al*: Is it ever safe to stop azole therapy for *Coccidioides immitis* meningitis? *Ann Intern Med* **124:**305–310, 1996.

236. Galgiani JN, Ampel NM, Catanzaro A, *et al*: Practice guidelines for the treatment of coccidioidomycosis. *Clin Infect Dis* **30:**658–661, 2000.

237. Schroter GPJ, Bakshandeh K, Husberg BS, Weil R III: Coccidi-oidomycosis and renal transplantation. *Transplantation* **23:**485–489, 1977.

238. Pappas PG, Threlkeld MG, Bedsole GD, *et al*: Blastomycosis in immunocompromised patients. *Medicine* **72:**311–325, 1993.

239. Pappas PG, Pottage JC, Powderly WG, *et al*: Blastomycosis in patients with the acquired immunodeficiency syndrome. *Ann In-tern Med* **116:**847–853, 1992.

240. Recht LD, Davies SF, Eckman MR, Sarosi GA: Blastomycosis in immunosuppressed patients. *Am Rev Respir Dis* **125:**359–362, 1982.

241. Laine L, Dretler RH, Conteas CN, *et al*: Fluconazole compared with ketoconazole for treatment of *Candida* esophagitis in AIDS. *Ann Intern Med* **117:**655–660, 1992.

242. Sarosi GA, Davies SF: Blastomycosis. *Am Rev Respir Dis* **120:** 911–937, 1979.

243. Wheat J, Wheat H, Connolly P, *et al*: Cross-reactivity in *Histo-plasma capsulatum* variety *capsulatum* antigen assays of urine samples from patients with endemic mycoses. *Clin Infect Dis* **24:** 1169–1171, 1997.

244. Chapman SW, Bradsher RW Jr, Campbell GD Jr, *et al*: Practice guidelines for the management of patients with blastomycosis. *Clin Infect Dis* **30:**679–683, 2000.

245. Gonyea EF: The spectrum of primary blastomycotic meningitis: A review of central nervous system blastomycosis. *Ann Neurol* **3:** 26–39, 1978.

246. Dismukes WE, Bradsher RW Jr, Cloud GC, *et al*: Itraconazole therapy for blastomycosis and histoplasmosis. *Am J Med* **93:**489–497, 1992.

247. Dismukes WE, Cloud G, Bowles C, *et al*: Treatment of blasto-mycosis and histoplasmosis with ketoconazole: Results of a pro-spective randomized clinical trial. *Ann Intern Med* **103:**861–872, 1985.

248. Pappas PG, Bradsher RW, Kauffman CA, *et al*: Treatment of blastomycosis with higher doses of fluconazole. *Clin Infect Dis* **25:**200–205, 1997.

249. Vartivarian SE, Anaissie EJ, Bodey GP: Emerging fungal patho-gens in immunocompromised patients: Classification, diagnosis, and management. *Clin Infect Dis* **17**(Suppl 2)**:**S487–S491, 1993.

250. Walsh TJ, Groll AH: Emerging fungal pathogens: Evolving chal-lenges to immunocompromised patients for the twenty-first cen-tury. *Transplant Infect Dis* **1:**247–261, 1999.

251. Anaissie E, Bodey GP, Rinaldi MG: Emerging fungal pathogens. *Eur J Clin Microbiol Infect Dis* **8:**323–330, 1989.

252. Supparatpinyo K, Khamwan C, Baosoung V, *et al*: Disseminated *Penicillium marneffei* infection in Southeast Asia. *Lancet* **334:** 110–113, 1994.

253. Goldani LZ, Sugar AM: Paracoccidioidomycosis and AIDS: An overview. *Clin Infect Dis* **21:**1275–1281, 1995.

254. Walsh TJ, Peter J, McGough DA, et al: Activities of amphotericin B and antifungal azoles alone and in combination against *Pseudallescheria boydii. Antimicrob Agents Chemother* **39:**1361–1364, 1995.

255. Patterson TF, Andriole VT, Zervos MJ, et al: The epidemiology of pseudallescheriasis complicating transplantation: Nosocomial and community-acquired infection. *Mycoses* **33:**297–302, 1990.

256. Goldberg SL, Geha DJ, Marshall WF, et al: Successful treatment of simultaneous pulmonary *Pseudallescheria boydii* and *Aspergillus terreus* infection with oral itraconazole. *Clin Infect Dis* **16:**803–805, 1993.

257. Ruxin TA, Steck WD, Helm TN, et al: *Pseudallescheria boydii* in an immunocompromised host—Successful treatment with debridement and itraconazole. *Arch Dermatol* **132:**382–384, 1996.

258. Verweij PE, Cox NJM, Meis JFG: Oral terbinafine for treatment of pulmonary *Pseudallescheria boydii* infection refractory to itraconazole therapy. *Eur J Clin Microbiol Infect Dis* **16:**26–28, 1997.

259. Sharkey PK, Graybill JR, Rinaldi MG, et al: Itraconazole treatment of phaeohyphomycosis. *J Am Acad Dermatol* **23:**577–586, 1990.

260. Anaissie E, Bodey GP, Kantarjian H, et al: New spectrum of fungal infection in patients with cancer. *Rev Infect Dis* **11:**369–378, 1989.

261. Lowenthal RM, Atkinson K, Challis DR, et al: Invasive *Trichosporon cutaneum* infection: An increasing problem in immunosuppressed patients. *Bone Marrow Transplant* **2:**321–327, 1987.

262. Gallis HA, Drew RH, Pickard WW: Amphotericin B: 30 years of clinical experience. *Rev Infect Dis* **12:**308–329, 1987.

263. Walsh TJ, Pizzo A: Treatment of systemic fungal infections: Recent progress and current problems. *Eur J Clin Microbiol Infect Dis* **7:**460–475, 1988.

264. Wong-Beringer A, Jacobs RA, Guglielmo BJ: Lipid formulations of amphotericin B: Clinical efficacy and toxicities. *Clin Infect Dis* **27:**603–618, 1998.

265. Walsh TJ, Finberg RW, Arndt C, et al: Liposomal amphotericin B for empirical therapy in patients with persistent fever and neutropenia. *N Engl J Med* **340:**764–771, 1999.

266. Dix SP: Pharmacology of lipid formulations of amphotericin B. *Infect Dis Clin Pract* **7:**S8–S15, 1998.

267. Storm G, Van Etten E: Biopharmaceutical aspects of lipid formulations of amphotericin B. *Eur J Clin Microbiol Infect Dis* **16:**64–73, 1997.

268. Petersen D, Demertzis S, Freund M, et al: Individualization of 5–fluorocytosine therapy. *Chemotherapy* **40:**149–156, 1994.

269. Stamm AM, Diasio RB, Dismukes WE, et al: Toxicitiy of amphotericin B plus flucytosine in 194 patients with cryptococcal meningtitis. *Am J Med* **83:**236–242, 1987.

270. Kitchen VS, Savage M, Harris JRW: *Candida albicans* resistance in AIDS [letter to the editor]. *J Infect* **22:**204–205, 1991.

271. Fox R, Neal KR, Leen CLS, et al: Fluconazole resistant candida in AIDS. *J Infect* **22:**201–212, 1991.

272. Brammer KW, Farrow PR, Faulkner JK: Clinical pharmacology pharmacokinetics and tissue penetration of fluconazole in humans. *Rev Infect Dis* **12:**S318–S326, 1990.

273. Apseloff G, Hilligoss DM, Gardner MJ, et al: Induction of fluconazole metabolism by rifampin: *In vivo* study in humans. *J Clin Pharmacol* **31:**358–361, 1991.

274. Blum RA, Wilton JH, Hilligoss DM, et al: Effect of fluconazole on the disposition of phenytoin. *Clin Pharmacol Ther* **49:**420–425, 1991.

275. Canafax DM, Graves NM, Hilligoss DM, et al: Increased cyclo-sporine levels as a result of simultaneous fluconazole and cyclosporine therapy in renal transplant recipients: A double-blind, randomized pharmacokinetic and safety study. *Transplant Proc* **23:**1041–1042, 1991.

276. Graybill JR, Stevens DA, Galgiani N, et al: Itraconazole treatment of coccidioidomycosis. *Am J Med* **89:**282–290, 1990.

277. Stevens DA: Itraconazole and fluconazole for treatment of coccidioidomycosis. *Clin Infect Dis* **18:**470–470, 1994.

278. Tucker RM, Denning DW, Arathoon EG, et al: Itraconazole therapy for nonmeningeal coccidioidomyosis: Clinical and laboratory observation. *J Am Acad Dermatol* **23:**593–601, 1990.

279. Tucker RM, Haq Y, Denning DW, Stevens DA: Adverse events associated with itraconazole in 189 patients on chronic therapy. *J Antimicrob Chemother* **26:**561–566, 1990.

280. Diaz M, Puente R, DeHoyos LA, Cruz S: Itraconazole in the treatment of coccidioidomycosis. *Chest* **100:**682–684, 1990.

281. Sharkey-Mathis PK, Kauffman CA, Graybill JR, et al: Treatment of sporotrichosis with itraconazole. *Am J Med* **95:**279–285, 1993.

282. Grant SM, Clissold SP: Itraconazole A: A review of its pharmacodynamic and pharmacokinetic properties, and therapeutic use in superficial and systemic mycoses. *Drugs* **37:**310–344, 1989.

283. Van't Wout JW, Novakova I, et al: The efficacy of itraconazole against systemic fungal infections in neutropenic patients: A randomized comparative study with amphotericin B. *J Infect* **22:**45–52, 1991.

284. Barchiesi F, Colombo AL, McGough DA, et al: *In vitro* activity of itraconazole against fluconazole-susceptible and -resistant *Candida albicans* isolates from oral cavities of patients infected with human immunodeficiency virus. *Antimicrob Agents Chemother* **38:**1530–1533, 1994.

285. Phillips P, Zemcov J, Mahmood W, et al: Itraconazole cyclodextrin solution for fluconazole-refractory oropharyngeal candidiasis in AIDS: Correlation of clinical response with *in vitro* susceptibility. *AIDS* **10:**1369–1376, 1996.

286. Cartledge JD, Midgley J, Gazzard BG: Itraconazole cyclodextrin solution: The role of *in vitro* susceptibility testing in predicting successful treatment of HIV-related fluconazole-resistant and fluconazole-susceptible oral candidosis. *AIDS* **11:**163–168, 1997.

287. Blomley M, DeBelder A, Thyway Y, Weston M: Itraconazole and anti-tuberculosis drugs. *Lancet* **336:**1255, 1991.

288. Kramer MR, Marshall SE, Denning DW, et al: Cyclosporine and itraconazole interaction in heart and lung transplant recipients. *Ann Intern Med* **113:**327–329, 1990.

289. Sharkey PK, Rinaldi MG, Dunn JF, et al: High-dose itraconazole in the treatment of severe mycoses. *Antimicrob Agents Chemother* **35:**707–713, 1991.

290. Tucker RM, Denning DW, Dupont B, Stevens DA: Itraconazole therapy for chronic coccidioidal meningitis. *Ann Intern Med* **112:**108–112, 1990.

291. Belanger P, Nast CC, Fratti R, et al: Voriconazole (UK-109,496) inhibits the growth and alters the morphology of fluconazole-susceptible and -resistant *Candida* species. *Antimicrob Agents Chemother* **41:**1840–1842, 1997.

292. Hegener P, Troke PF, Fätkenheuer G, et al: Treatment of fluconazole-resistant candidiasis with voriconazole in patients with AIDS. *AIDS* **12:**2227–2228, 1998.

293. Pfaller MA, Zhang J, Messer SA, et al: *In vitro* activities of voriconazole, fluconazole, and itraconazole against 566 clinical isolates of *Cryptococcus neoformans* from the United States and Africa. *Antimicrob Agents Chemother* **43:**169–171, 1999.

294. Cuenca-Estrella M, Rodríguez-Tudela JL, Mellado E, et al: Com-

parison of the *in vitro* activity of voriconazole (UK-109,496), itraconazole and amphotericin B against clinical isolates of *Aspergillus fumigatus*. *J Antimicrob Chemother* **42:**531–533, 1998.

295. Murphy M, Bernard EM, Ishimaru T, Armstrong D: Activity of voriconazole (UK-109,496) against clinical isolates of *Aspergillus* species and its effectiveness in an experimental model of invasive pulmonary aspergillosis. *Antimicrob Agents Chemother* **41:**696–698, 1997.

296. Van't Hek LG, Verweij PE, Weemaes CM, *et al*: Successful treatment with voriconazole of invasive aspergillosis in chronic granulomatous disease. *Am J Respir Crit Care Med* **157:**1694–1696, 1998.

297. Schwartz S, Milatovic D, Thiel E: Successful treatment of cerebral aspergillosis with a novel triazole (voriconazole) in a patient with acute leukaemia. *Br J Haematol* **97:**663–665, 1997.

298. Girmenia C, Luzi G, Monaco M, Martino P: Use of voriconazole in treatment of *Scedosporium apiospermum* infection: Case report. *J Clin Microbiol* **36:**1436–1438, 1998.

299. Johnson EM, Szekely A, Warnock DW: *In-vitro* activity of voriconazole, itraconazole and amphotericin B against filamentous fungi. *J Antimicrob Chemother* **42:**741–745, 1998.

300. Radford SA, Johnson EM, Warnock DW: *In vitro* studies of activity of voriconazole (WK-109,496), a new triazole antifungal agent, against emerging and less-common mold pathogens. *Antimicrob Agents Chemother* **41:**841–843, 1997.

301. Lutz JE, Clemons KV, Aristizabal BH, Stevens DA: Activity of the triazole SCH 56592 against disseminated murine coccidioidomycohsis. *Antimicrob Agents Chemother* **41:**1558–1561, 1997.

302. Law D, Moore CB, Denning DW: Activity of SCH 56592 compared with those of fluconazole and itraconazole against *Candida* spp. *Antimicrob Agents Chemother* **41:**2310–2311, 1997.

303. Bartizal K, Gill CJ, Abruzzo GK, *et al*: *In vitro* preclinical evaluation studies with the echinocandin antifungal MK-0991 (L-743,872). *Antimicrob Agents Chemother* **41:**2326–2332, 1997.

304. Graybill JR, Bocanegra R, Luther M, *et al*: Treatment of murine *Candida krusei* or *Candida glabrata* infection with L-743,872. *Antimicrob Agents Chemother* **41:**1937–1939, 1997.

305. Bartlett MS, Current WL, Goheen MP, *et al*: Semisynthetic echinocandins affect cell wall deposition of *Pneumocystis carinii in vitro* and *in vivo*. *Antimicrob Agents Chemother* **40:**1811–1816, 1996.

306. Boswell GW, Buell D, Bekersky I: AmBisome (liposomal amphotericin B): A comparative review. *J Clin Pharmacol* **38:**583–592, 1998.

7

Mycobacterial and Nocardial Diseases in the Compromised Host

LOWELL S. YOUNG and ROBERT H. RUBIN

1. Introduction

Grouping together mycobacteria and *Nocardia* may appear to be a hasty gesture. Nonetheless, both groups of organisms are acid-fast and the *Nocardia* may cause clinical manifestations that could be mistaken for reactivation tuberculosis or the spread of mycobacterial disease to bone, brain, and soft tissues from an initial pulmonary site. It is true that *Nocardia* grow relatively more rapidly in comparison to the causative agent of tuberculosis or the most important of the nontuberculous mycobacteria, *Mycobacterium avium*. Further, the *Nocardia* are generally susceptible to a variety of antimicrobial agents, particularly the sulfonamides, that are generally inactive against mycobacteria. In terms of the compromised host, both newly acquired and reactivated disease by both groups of organisms can be associated with immunosuppressive therapy such as corticosteroids. Therefore, from the perspective of syndromic classifications, the overlap in disease caused by mycobacteria and *Nocardia* justifies grouping these organisms together. With regard to histopathological straining characteristics, *Nocardia* may re-

semble *Actinomycetes*, an occasional cause of oral cavity and head and neck infections. However, *Actinomycetes* are strict penicillin-susceptible anaerobes and are an uncommon cause of disease in immunocompromised hosts.

Mycobacterium tuberculosis ranks first among all potentially fatal infectious disorders of the human race.[1] The annual incidence of malaria may likely be higher, but mortality from tuberculosis, estimated at between 2 and 3 million individuals per year, is greater than that associated with malaria.[2] A rising association of tuberculosis and human immunodeficiency virus (HIV) disease in many developing countries is likely to propel the death rates from tuberculosis to even higher levels in the next decade.[3,4] While in North America, Japan, and Western Europe tuberculosis incidence rates are on the decline, this is not the case in areas severely afflicted by the acquired immunodeficiency syndrome (AIDS) pandemic. Another noteworthy area where multiply drug-resistant tuberculosis strains have appeared in great numbers are in the countries of the former Soviet Union. As HIV-related diseases begin to involve the latter, the potential for even more serious drug-resistant disease seems to be increasing.

Nontuberculous mycobacteria were considered a curiosity until the advent of the AIDS pandemic. Indeed, descriptions of chronic lung disease complicated by "atypical tuberculosis" (atypical in the sense that they were not the same as tuberculosis or bovine tuberculosis) were initially associated with syndromes clinically indistinguishable from tuberculosis.[5,6] During the height of the AIDS epidemic in the United States and Western Europe and prior to the introduction of highly active antiretroviral therapy, one particular species, the organisms of the *Mycobacterium avium–intracellulare* complex (hereafter re-

Lowell S. Young • Kuzell Institute for Arthritis and Infectious Diseases, Division of Infectious Diseases, California Pacific Medical Center; University of California at San Francisco; San Francisco, California 94115.   **Robert H. Rubin** • Division of Infectious Disease, Brigham and Women's Hospital; Center for Experimental Pharmacology and Therapeutics, Harvard–Massachusetts Institute of Technology Division of Health Sciences and Technology and Department of Medicine, Harvard Medical School, Boston, Massachusetts 02114.

Clinical Approach to Infection in the Compromised Host (Fourth Edition), edited by Robert H. Rubin and Lowell S. Young. Kluwer Academic/Plenum Publishers, New York, 2002.

ferred to as MAC) became the most common bacteremic infection in patients with advanced AIDS.[7] These organisms are generally resistant to conventional antituberculosis medications, thus triggering an urgent search for alternative therapies.[8] As a result of the AIDS pandemic, several additional species of nontuberculous mycobacteria were first identified after organisms were isolated from bacteremic cases of human disease and these isolates did not correspond phenotypically and genotypically with any other known species.[9,10] These "new" species of nontuberculous mycobacteria include *M. genovense*,[11] *M. malmoense*,[12] and *M. haemophilum*.[13] These organisms may cause clinical syndromes indistinguishable from MAC and often are susceptible to the same types of therapeutic agents. Nonetheless, they can be quite fastidious and may require special media growth supplements (e.g., hemin for *M. haemophilum*) and incubation conditions at low temperatures (e.g., 30–32°C) for optimal recovery. This emphasizes the need for close interaction between clinicians and the microbiology laboratory to optimize care of the immunocompromised patient.

Historically, the second most important mycobacterial disease after tuberculosis was leprosy.[14] Tens of millions of cases of leprosy persist, primarily in Africa, Asia, and South and Central America. Overall, however, the incidence appears to be on the decline. It is unclear whether the spread of the AIDS epidemic to the aforementioned areas will result in a net increase in leprosy, but such an epidemiologic development might be anticipated. Important information about the principal mycobacterial pathogens—*M. tuberculosis*, MAC, and *M. leprae*—has come recently from the complete sequencing of the genomes of representative pathogenic isolates of these mycobacterial species. Already an extraordinary amount of information has been obtained that casts a major light about pathogenic mechanisms and offers clues to intracellular persistence and latency, as well as tissue trophism. This information may well be translated into improved chemotherapeutic agents, vaccines, and other major steps toward infection control.

2. Microbiologic Classification and Clinical Syndromes

Table 1 summarizes the classic Runyon classification of mycobacteria, which was based on phenotypic and growth characteristics.[5] This scheme is still useful for laboratory identification purposes but has less relevance clinically. Table 2 represents a more clinically oriented classification by targeted tissue trophism (i.e., type of organ involvement).

TABLE 1. Classification of Mycobacteria[a]

Group	Representative species of mycobacteria
Typical, slow-growing, strict pathogens	*M. tuberculosis, M. bovis, M. leprae*
Photochromogens[b]	*M. kansasii*
	M. marinum
Scotochromogens[c]	*M. scrofulaceum*
	M. szulgai
	M. xenopi
Nonphotochromogens	*M. avium*
	Mycobacterium avium complex
	M. intracellulare
	M. malmoense
	M. haemophilum
Rapidly growing organisms	*M. abscessus*
	M. fortuitum
	M. chelonei

[a]Based on Runyon.[5]
[b]Photochromogen: isolate is buff-colored in the dark but turns yellow with brief exposure to light.
[c]Scotochromogen: isolate is yellow-orange or orange even when grown in the dark.

M. tuberculosis (and closely related *M. bovis*) are probably the most virulent organisms, which reflect their ability to cause disease in normal hosts. *M. bovis* is still a problem in children who are apparently immunologically intact but are exposed to the natural reservoir of bovine tuberculosis.[15] While we associate *M. tuberculosis*, *M. kansasii*, and *M. bovis* with pulmonary disease, they have the capacity to cause disease in virtually any organ of the body including brain, skin, and bone. So it is with the MAC, although systemic infection caused by the latter almost exclusively occurs in immunocompromised individuals.[7] In young children, MAC is a cause of lymphadenopathy, as is *M. scrofulaceum*. Cutaneous, soft-tissue and wound infections caused by the rapid-growing species *M. marinum*, *M. fortuitum*, and *M. chelonei* are particularly vexing problems that may have to be dealt with by a combination of chemotherapy and surgical excision of infected tissue site. One major review has called attention to the fact that all the new species of mycobacteria are environmental in origin; indeed, there are reports of rare previously undetected or undescribed mycobacterial species whose initial "debut" in a human host has been as a complication of AIDS.[9]

3. Pathogenesis and Epidemiology

M. tuberculosis is transmitted primarily as an airborne disease, although occasionally contact transmission has been reported. The only reservoir for humans is an-

TABLE 2. Clinical Disease Caused by Mycobacteria

Clinical disease	Common etiologic species	Growth rate	Morphologic growth features
Pulmonary	M. avium complex	Slow (>7 days)	Usually not pigmented
	M. kansasii	Slow	Photochromogen
	M. chelonae subspecies	Rapid (<7 days)	Not pigmented
	M. xenopi	Slow	Pigmented, grows at 45°C
	M. terrae	Slow	Not pigmented
	M. tuberculosis	Slow	Not pigmented
Lymphadenitis	M. avium complex	Slow	Usually not pigmented
	M. scrofulaceum	Slow	Scotochromogen
Cutaneous	M. leprae	Very slow	Not culturable ex vivo
	M. marinum	Rapid	Photochromogen; requires low temperatures (28° to 30°C) for isolation
	M. fortuitum	Rapid	Not pigmented
	M. chelonae	Rapid	Not pigmented
	M. ulcerans	Slow	Usually a scotochromogen; requires low temperatures for isolation
Disseminated	M. avium complex	Slow	Isolates from patients with AIDS, often pigmented (80%)
	M. bovis	Slow	
	M. genovense	Very slow (>6 wk)	Growth better in broth than agar
	M. kansasii	Slow	Photochromogen
	M. chelonae	Rapid	Not pigmented
	M. haemophilum	Slow	Not pigmented; requires hemin, often needs low temperature and CO_2
	M. simiae	Slow	Photochromogen
	M. malmoense	Very slow (>6 wk)	Prefers low pH
	M. tuberculosis	Slow	Not pigmented

other human subject, yet airborne spread can involve infection by droplet nuclei that are spread over a large volume of air at a considerable distance, e.g., the classical tuberculosis sanitoria and closed quarters such as prisons or submarines.[16] After inhalation, the organisms establish localized disease in the lung and may spread by the bloodstream route. In the great majority of individuals initially infected by *M. tuberculosis* through the airborne route, containment of the infection to the lung parenchyma occurs and this may or may not be represented by a visible evidence on the chest X ray (scars, calcification, Ghon complex). It is believed that the majority of clinical cases of tuberculosis (i.e., tuberculosis disease) occur via the reactivation mechanism, but a central unresolved puzzle in tuberculosis pathogenesis is how reactivation disease occurs. This may be related to genetic characteristics of both the host and the pathogen.[17–19] Steroid therapy, cytolytic cancer chemotherapy, radiation treatment of a neoplasm, or use of antisera that impairs T-lymphocyte function have been classically associated with reactivation tuberculosis, but what causes reactivation in "normals" is unclear.

Skin test conversion (Koch old tuberculin, Mantoux testing) is a standardized diagnostic approach for determining tuberculosis exposure. Viewed in the light that most immunologically intact (i.e., nonimmune suppressed) individuals convert their tuberculin test from negative to positive has been a valuable epidemiologic tool. However, interpretation of a positive test has difficulties and has proved challenging. In the light of immune suppression associated with AIDS and other conditions, criteria for a positive test have been recently revised.[20] Many patients with active tuberculosis and particularly those with concurrent AIDS are often totally anergic and no amount of antigen can prompt a delayed hypersensitivity response. So-called anergy testing and "two-stage" tuberculin testing also are difficult to interpret and should be discouraged in the evaluation of immunocompromised patients.[21,22] The most important factor in the clinical approach is a high index of suspicion: occult febrile illnesses, unexplained weight loss or wasting, "culture negative" microbiologic specimens or tissue biopsies, or inflammatory processes that have no easy explanation should prompt studies aimed at identifying a possible mycobacterial process.

Confounding the interpretation of skin tests is the fact that many individuals have received antecedent BCG (live attenuated *M. bovis* vaccine) vaccination. Efficacy of the various vaccine preparations (there are dozens available worldwide) has been controversial and may be related to wide variations in the antigens present in various preparations of BCG vaccines that have been used.[23,24] Skin test conversion often follows BCG immunization but there is no correlation between the magnitude of the tuberculin reactivity that follows such and the degree of protection. Therefore, it has been recommended that a prior history of BCG immunization be ignored in interpreting a tuberculin skin test. Reactivity appears to wane with time after immunization.

Presently, the annual number of new cases of tuberculosis in the United States is approximately 20,000 per year and is decreasing.[25] It has been noted that while patients who are homeless, immunosuppressed, and HIV positive continue to be a substantial proportion of cases of active disease, 40% of all cases of tuberculosis occurring in the United States are now in individuals born outside of this country. Converted to rates, the incidence is five times higher in foreign- as compared with native-born individuals, and the decrease since 1992 in the incidence of U.S. tuberculosis has been limited to those born in North America (but *not* the foreign born).

As underscored by Table 2, almost any organ can be involved by mycobacterial disease. The most virulent organism even in normal hosts is *M. tuberculosis*. Extrapulmonary manifestations have been well described in the literature and the organism that comes closest to *M. tuberculosis* in terms of clinical disease involvement is MAC. The systemic disease, however, is primarily seen in immune suppressed hosts as opposed to normal hosts (MAC disease). *M. leprae*'s tissue trophism is remarkable in affecting primarily the skin and Schwann cells of the peripheral nervous system. However, while the skin changes are disfiguring and yield the characteristic clinical appearance of patients with leprosy or Hansen's disease, the real danger to the patient with leprosy is nerve involvement presenting initially as numbness and loss of proprioception, which leads to traumatic tissue destruction as the patient becomes unaware of diminished sensation in peripheral limbs. In this sense, the damage due to leprosy can mimic that seen in patients with severe diabetic neuropathy. The classic association of a cutaneous abnormality with sensoneural loss of course should be the clue to leprosy, but relatively few cases are now seen in the United States. Most patients live in the southwestern region and have been born outside of the United States.

The most important of the nontuberculous mycobacteria in developed countries is MAC. Very early in the AIDS pandemic evidence became available that systemic disease caused by these organisms was extraordinarily common.[4,26] Prospective studies prior to the advent of anti-MAC chemoprophylaxis indicated that after an AIDS patient's CD4 count dropped below 50 cells/mm[3] the likelihood of bacteremic MAC disease was 50% in the ensuing year.[27] The conclusion of these studies is that MAC disease appeared to be inevitable unless some type of chemotherapeutic intervention was invoked.

Occasionally, patients with high CD4 counts have had fever and fleeting lung infiltrates or lymphadenopathy due to MAC organisms. Since 1995 and the introduction of triple drug, so-called highly active antiretroviral chemotherapy (HAART), some patients who have ongoing MAC disease but high (>200) CD4 counts may have inflammatory masses wherein MAC organisms are isolated. This picture, however, is a reflection of immune reconstitution and a responsive host dealing with the organism as contrasted with the findings of progressive MAC disease in the face of increasing T-cell lymphopenia. The typical features of the syndrome in patients who receive neither active antiretroviral treatment nor prophylaxis was that of fever, progressive weight loss associated with drenching night sweats, recurrent chills, diarrhea, abdominal organomegaly, and abdominal pain. Laboratory findings have almost always included anemia, abnormalities of liver function (in particular, alkaline phosphatase) and neutropenia. Occasionally granulomas are detected in aspirate bone marrow or liver biopsies. Untreated patients with such findings and positive cultures for MAC in the blood rarely survived more than a year.[28]

Inasmuch as MAC is an aquatic organism, the likely route of acquisition in most AIDS patients appears to be the gastrointestinal tract. Diarrhea and malabsorption have been characteristic of the early stages of the disease, but dissemination in patients with severe immune deficiency inevitably occurs. Hepatomegaly, splenomegaly, and retroperitoneal lymphadenopathy are the likely causes of the severe abdominal pains that many of these patients have experienced.

Until the advent of the AIDS epidemic routine blood cultures to document the presence of mycobacteria were not commonly obtained. Nonetheless, the crucial procedure in documenting nontuberculous mycobacterial blood infections is an initial laboratory step in which the leukocytes (mononuclear cells or monocytes) are lysed, thereby releasing viable intracellular organisms. Subculture onto enriched media will usually reveal the organism (even if it is not one commonly used to culture mycobacterium), but cultures should be held on the order of 2–6 weeks inasmuch as organisms are slow growing.[8]

3.1. Association of Tuberculosis with HIV Disease

Beginning in the 1980s it was recognized that a large proportion of individuals being treated in tuberculosis clinics were HIV seropositive.[4] This association continues to this day. In fact, in sub-Saharan Africa between two thirds and three quarters of all individuals being treated in tuberculosis clinics have underlying HIV seropositivity. Viewed in another context, from the standpoint of host immune response, interesting associations have been noted between the appearance of clinical disease and CD4 counts. Pulmonary tuberculosis has been documented in individuals with CD4 counts above 300 cells/mm^3. Inasmuch as a CD4 count of 200 cells/mm^3 or less is AIDS defining and is associated with risk of opportunistic infections such as *Pneumocystis carinii* pneumonia, what this association for tuberculosis suggests is that tuberculosis is a more virulent disease afflicting individuals who have relatively higher CD4 counts. Miliary tuberculosis not surprisingly has been associated with lower CD4 counts in the range of 100 or lower and the same also has been observed for the involvement of the central nervous system.[4,29] This emphasizes the importance of knowing an individual's tuberculin status because the progression of an underlying HIV disease is likely to result in decreased tuberculin reactivity.

Illustrative Case 1

The patient was a 57-year-old black male who was admitted to the hospital with fevers as high as 105°F. Thirty years before, the patient was inducted into the U.S. Army at which time a screening PPD tuberculin test was found to be positive at 15 mm of induration. He received no preventative chemotherapy and subsequently after discharge from the army was employed as a chef. By the 1990s, the patient was known to be HIV positive, yet he had repeated CD4 counts in excess of 1000 cells/mm^3 and normal helper:suppressor T-cell ratios and CD4 percentages. Approximately 3 months before admission to hospital the individual consulted his primary care physician and because of a CD4 count of 925 cells/mm^3 elected not to begin antiretroviral chemotherapy. A few weeks thereafter he began to experience progressive weight loss, which exceeded 30 pounds over a 2.5-month period. Weight loss was accompanied by abrupt onset of shaking chills, temperatures as high as 105°F, night sweats, and diarrhea. Although a screening chest X ray was felt to be normal, he developed progressive shortness of breath. When O$_2$ saturation declined below 92% he was admitted to hospital. Within a day of that hospitalization the patient was found to have a CD4 count of 9 cells/mm^3 and a percent T-helper lymphocytes of 1%. Chest X ray on admission revealed fine, diffuse, bilateral lower lobe infiltrates, but multiple induced cough specimens were negative for *P. carinii* or any bacterial pathogen. The patient was initiated on high-dose IV trimethoprim–sulfamethoxazole for presumed *P. carinii* pneumonia. Peripheral blood antibody measurements revealed high levels (in excess of 320) of complement-fixing antibodies against cytomegalovirus but no titers were elevated against *Legionella pneumophila* or *Mycoplasma pneumoniae*. Despite IV trimethoprim–sulfamethoxazole, the patient continued to decline and both lung fields became involved with bilateral, dense, fluffy infiltrates. Progressive arterial O$_2$ desaturation was accompanied by marked deterioration in the patient's liver function tests. After 10 days of hospitalization, the patient became progressively jaundiced with a striking rise in alkaline phosphatase in excess of 2000 units with accompanying rises of transaminase values (ALTs) into the 1000s. He was intubated and despite broad-spectrum antibacterial agents (imi-

penem, amikacin) and the addition of empiric antifungal and antituberculous chemotherapy (which also covered MAC), he expired after approximately 3.5 weeks of hospitalization. Postmortem examination revealed macroscopic and microscopic necrotic foci in lungs, liver, and spleen. All organs cultured grew pan-susceptible *Mycobacterium tuberculosis* as did three of three blood cultures inoculated into mycobacterial broth media.

Comment. This case illustrates the rapid progression of disseminated tuberculosis in the AIDS patient. At the time that he was initially found to be seropositive against HIV, he elected not to begin antiretroviral chemotherapy, an option that remains available to this day. Nonetheless, the precipitous decline in the CD4 helper lymphocyte count is noteworthy and is suggestive of systemic tuberculosis with rapid clinical progression. *M. tuberculosis* is probably the most virulent of all opportunistic pathogens and can cause miliary disease and death even in normal hosts. In the case of HIV disease the virus has a specific tropism for CD4 lymphocytes. In the process of activating host defense against mycobacteria, cytokines are released by macrophages that cause further T-cell activation and enhanced expression of CD4 and HIV coreceptors on the surface of lymphocytes. This accelerates the course of the underlying HIV because there are far more targets (surface receptors for virus) expressed on CD4 lymphocytes than on normal lymphocytes. If the patient is not receiving antiretroviral chemotherapy, rapid CD4 depletion ensues. The decline in T-cell-mediated immunity will further enhance host susceptibility to the opportunistic mycobacterial disease.

In this case, blood O$_2$ desaturation and liver function abnormalities predated the onset of gross pulmonary infiltrates. While consideration of *P. carinii* pneumonia was logical, the presentation of wasting and fever is something that should alert clinicians to the possibility of an underlying mycobacterial process. Admittedly, the low CD4 helper lymphocyte count would have more likely presaged a nontuberculous mycobacterial disease such as MAC rather than *M. tuberculosis* in North American and European patients. CD4 counts are often high (e.g., >200 cells/mm^3) in North American and European AIDS patients, because *M. tuberculosis* can cause pulmonary and systemic disease, while host defenses are more relatively intact. A higher index of suspicion should have led to earlier institution of antituberculous chemotherapy. In this case, the patient received only preterminal quadruple therapy against *M. tuberculosis* and the organism was fully susceptible to each agent. After the patient's demise, the organism grew from blood cultures as well as all tissues cultured. Of note from an epidemiologic perspective, this patient worked as a chef for a large family, yet none of the immediate family contacts or social contacts were found to be tuberculin positive. This emphasizes that specific host susceptibility is far and away the most important determinant leading to the expression of clinical disease.

3.2. Revised Guidelines for Tuberculin Reactivity and Preventative Chemotherapy

With the availability of agents such as isoniazid, considerable optimism was expressed about the potential for eradication of tuberculosis in many developed countries. This was based on the postulate that humans are the only reservoir of disease and the identification of tuberculin-positive individuals allows identification of individuals who harbor viable organisms, and thus are susceptible to reactivation disease (to say nothing of spread to other individuals). The problematic nature of isoniazid (not a totally safe drug for "preventative chemotherapy") led to generally poor acceptance of so-called "preventative chemotherapy," which was hoped to be the basis of tuberculosis "eradication." Recently, more focused guidelines have been published for tuberculosis screening and control.[20,30] The highest clinical priority for using preventative chemotherapy is in individuals who have some identified basis for decreased immune response: immunosuppressive therapy, organ transplantation, and HIV seropositivity. Further, recent contacts of active cases and individuals whose chest X ray shows scars and calcific lesions consistent with old, untreated tuberculosis would fall into this group 1. In these individuals a test reactivity of 5 mm or greater would be considered positive. It was recognized (and was published some time ago) that more than 10 mm of induration was associated with increased likelihood of disease progression, but on a statistical basis the relative advancement to active disease in people with such types of tuberculin reactions was generally low (group 2). Individuals who were considered to be at low risk of developing active tuberculosis, i.e., not immune suppressed, not in recent contact with an active case, and not having any lesions on chest X ray suggestive of prior exposure to tuberculosis, should be viewed as having a positive tuberculin test only if greater or equal to 15 mm of reactivity is unequivocally documented (group 3).

Traditionally, individuals with "positive" tuberculin tests have been treated with 12 months of isoniazid (300 mg qd). The major benefit for those who are compliant with these regimens has been treatment for at least 6 months.[30] As a compromise the new guidelines recommend 9 months of isoniazid prophylaxis. More recently, there has been evidence that an alternative regimen may be shorter and equally active as the traditional approach to preventative chemotherapy with isoniazid. This is a combination with pyrazinamide and rifampin in the conventional daily dose taken for 2 months or rifampin alone for 4 months.[30]

4. Treatment of Tuberculosis, Drug-Resistant Tuberculosis, and Alternative Antituberculosis Chemotherapy

Conventional treatment of pulmonary tuberculosis has been based on the combination of oral isoniazid, rifampin, ethambutol for 6 months, and pyrazinamide for the initial 2 months.[31] Duration of therapy in the AIDS patient and other immunocompromised hosts should be more than 6 months and should be linked to clinical response and any subsequent improvement in host immune status.

The problem of drug-resistant tuberculosis became strikingly apparent during the late 1980s when certain major American cities witnessed epidemics of drug-resistant disease in AIDS patients. While the definition of resistance may be an arbitrary laboratory criterion, reports prior to 1990 indicated that the incidence of either isoniazid or rifampin-resistance alone was usually no more than 10%. The definition of multiple drug-resistant (MDR) tuberculosis was set at resistance against at least two major antituberculous medications, usually isoniazid or rifampin, but could include other "first-line agents" such as streptomycin, pyrazinamide, and ethambutol. The incidence of MDR tuberculosis in some of the AIDS patients reported in the last decade has approached 20% (reviewed in Ref. 4). Additionally, a high incidence of resistant strains has been reported from areas such as Russia where combination therapy in the initial treatment of tuberculosis may not be as widely practiced as in the United States.

In terms of antituberculous effect, streptomycin is as potent as isoniazid and rifampin, and thus is considered to be a "major" therapeutic agent. Because of the need for parenteral administration, however, and its attendant toxicities, this drug is usually reserved for retreatment. Alternative therapies have included the fluoroquinolones of which the most experience has been with oxfloxacin and its enantiomer, levofloxacin. However, by in vitro testing, gatifloxacin and moxifloxacin appear to be more potent (but human clinical data are lacking). Other alternative regimens include cycloserine, clofazimine, and other aminoglycosides. Selection of secondary regimens, i.e., for clinical failure or for drug resistance, should be based on in vitro susceptibility testing. Assuming that patients have been started on the initial recommended four-drug regimen for pulmonary disease (isoniazid, rifampin, pyrazinamide, and ethambutol) and assuming initial susceptibility to at least half the agents in the original regimen, there still will be time for laboratory isolation and susceptibility testing in order to derive an appropriate therapeutic regimen even in the majority of situations where drug resistance is detected. Stated in another way, there does not appear to be compelling evidence for assuming multidrug resistance in most cases unless there is a major outbreak of MDR disease in the community or institution and clinical evidence suggests that more than the conventional four-drug starting regimen should be employed.

4.1. Chemotherapy of MAC Disease

Much has been learned about the treatment and prevention of bacteremic MAC disease in AIDS patients. There are, however, no controlled or comparative studies in any other types of immune-suppressed patients. There are no large-scale studies of a comparative nature in which some of the drugs identified as active in AIDS patients have been employed to treat conditions such as pulmonary MAC in the adult or for that matter MAC lymphadenopathy as occurs not uncommonly in children. Nonetheless, the weight of the clinical evidence suggests that agents identified as active in the AIDS population are likely to be active in non-HIV-positive individuals.[10,32]

The principle in the chemotherapeutic approach to MAC disease is the recognition that standard antituberculous chemotherapy is usually ineffective, particularly the agents isoniazid and pyrazinamide. Rifampin in doses commonly used probably inhibits no more than 15 to 20% of MAC strains.[8] Ethambutol, on the other hand, has a mild potentiating effect for drugs that have intrinsic anti-MAC activity and its use also may tend to deter the emergence of resistance to other drugs. The aminoglycoside group, including streptomycin, amikacin, and kanamycin, is likely to be effective if used in high dose. The most experience has been with IV amikacin.[33] The California Collaborative Treatment Group studies have suggested that amikacin exerts a bactericidal effect *in vivo* simply because the other drugs used in combination with amikacin in that trial would not have been expected to yield a 2 log reduction in bacteremia.[33] Further, amikacin in experimentally infected beige mice (the most commonly used test system to identify active agents) was found to be active *in vivo*, lowering the magnitude of bacteremia as well as the tissue burden.[34] The long-term untoward consequences of such treatment are obvious, namely, damage to renal function and to VIII nerve function.

Monotherapeutic studies were actually carried out in patients with MAC bacteremia and the only conventional antimycobacterial agent found to have a modest impact on the bacteremia was ethambutol.[35] Rifampin and clofazimine as monotherapy were inactive. No studies using aminoglycosides have compared them with an alternative anti-MAC regimen.

The single most important advance in the management of MAC disease is the use of modern macrolides such as azithromycin, clarithromycin, and possibly roxithromycin to treat human disease. The first evidence came from animal studies[36,37] and then pilot studies in humans in which quantitative blood cultures were used to document significant logarithmic reductions in the magnitude of MAC bacteremia.[38,39] Quantitation of the decline in bacterial load also was accompanied by evidence of clinical response: defervescence, reduction or amelioration of night sweats, stabilization of weight, and improved quality of life. Nonetheless, the monotherapy with

azithromycin or clarithromycin did lead to breakthrough with resistant organisms.[40] In experimental studies in the murine test system, the concomitant use of at least one agent, usually ethambutol, tended to reduce the development of resistance.[41] Therapeutic doses of azithromycin have ranged from 500 to 1200 mg per day and for clarithromycin 1–4 g/day in divided doses (bid). The recommendation is to use clarithromycin at the lower dose. Comparative trials have indicated a comparable microbiologic effect of 600 mg of azithromycin and 1 g of clarithromycin per day.[42] Roxithromycin has never been compared with the modern macrolides and no major therapeutic study with this agent exists. Nonetheless, it does possess significant *in vivo* activity in the experimental test system.[43]

After the encouraging results in the treatment of MAC bacteremia, a number of trials were undertaken in which azithromycin and clarithromycin were used prophylactically to prevent MAC bacteremia in AIDS patients with CD4 counts less than 100 mm^3.[44,45] These studies were preceded by well-designed studies that showed that a modern rifamycin derivative, rifabutin, also had prophylactic efficacy in humans.[46] However, rifabutin is subject to drug interactions with a number of compounds including protease inhibitors, and thus is not recommended for use in patients receiving HAART. Additionally, several studies have looked at the increased benefit of either azithromcyin or clarithromycin, each with rifabutin.[44,45] While there was additional reduction in the incidence of MAC bacteremia, there also was a proportionate increase in untoward effects and the issue of drug interactions persists. Thus, the recommendation is for patients who are still receiving prophylaxis not to combine azithromycin or clarithromycin (and perhaps roxithromycin) with a rifamycin whether it be rifampin or rifabutin.

The availability of HAART since the end of 1995 has greatly diminished the indications for either macrolide treatment or prophylaxis of MAC disease in AIDS patients. Inasmuch as MAC disease appears to occasionally complicate cancer chemotherapy and organ transplantation, the lessons learned from the extensive clinical trials in AIDS can probably be extrapolated to other immune-suppressed hosts.[10] The minimum effective regimen is a modern macrolide plus ethambutol. Since there is a basis for treating mycobacterial diseases with several potent drugs to minimize the emergence of resistance, it may be desirable to add additional agents. Further, in patients who are quite critically ill at the onset of therapy, a so-called "induction regimen" aimed at maximum bactericidal effect at the beginning of therapy may accelerate clinical improvement and defervescence (although it must be recognized that defervescence on solo macrolide treatment can occur within 7–14 days of the initiation of even oral therapy). The most potent parenteral agent for induction in addition to a macrolide is probably amikacin IV, given in a total daily dose of 15–20 mg/kg.

Besides amikacin, choice of companion anti-MAC drugs in the initial regimen is limited. For reasons indicated earlier, a rifamycin probably should be avoided in HAART recipients because of the drug interaction issues (although azithromycin is less subject to interaction with rifabutin than clarithromycin). New agents that offer promise derive basically from *in vitro* studies, and studies in experimentally immune-suppressed animals suggest the following might be useful companion drugs in addition to aminogylocides: a fluoroquinolone such as moxifloxacin,[47] an oxazolidinone such as linazolid,[48] and an antimalarial agent such as mefloquine.[49] It must be recognized that there are no human studies of treating MAC in humans that validate the inclusion of these agents at present. Nonetheless, a significant effect has been observed using monotherapy with moxifloxacin, linezolid, and mefloquine as monotherapy in experimental infection.

Illustrative Case 2

The patient was a 38-year-old cosmetician who lived in an agricultural area of the Western United States. Two years prior to admission she had an abnormal chest X ray taken at local chest clinic where she had been referred by her primary care physician. While she had no history of smoking or exposure to tuberculosis, she was initially diagnosed as having pulmonary tuberculosis based on the recovery of acid-fast organisms from random sputum samples. After several months, a state reference laboratory identified the organism as belonging to the MAC, not *M. tuberculosis*.

Chest X ray revealed extensive bilateral cavitary and fibrotic disease. Her HIV serology was negative, but the CD4 helper T-cell count was 194/mm³ with 1064 CD8 suppressor cells/mm³. The patient denied a history of smoking, ethanolism, or illicit drug intake, but she did note a variety of hypersensitivity reactions to some of the beauty products that she worked with. Her hematologic examination was unremarkable except for mild anemia, which had not been responsive to iron supplementation. Serum protein electrophoresis was unremarkable. The patient was begun on oral chemotherapy with clarithromycin (500 mg bid), ethambutol (800 mg qd), and oxfloxacin (400 mg qd). After an initial response for 6 months, her symptoms of cough and weight loss recurred. Treatment subsequently was altered to include parenteral amikacin, but without an effect despite her isolate's *in vitro* susceptibility. After a progressive downhill course extending over 2.5 years, the patient expired in respiratory failure.

Comment. This patient was not overtly immune compromised, yet laboratory testing revealed a significant decrease in CD4 T-helper lymphocytes. A syndrome has been reported by Holland and colleagues,[50] so-called idiopathic T-cell lymphopenia, in which systemic MAC disease has been one of the most frequent opportunistic pathogens. There was initial concern that these patients might reflect another variant of "acquired immune deficiency" and in this respect the patient technically could be categorized as such. Cases reported by Holland and colleagues[50] from the National Institutes of Health appear to have more severe CD4 lymphopenia. The syndrome represented by this patient is reminiscent of some of the original cases reported by Runyon[5] of nontuberculous mycobacterial disease focusing on patients hospitalized at the Battey State Hospital in Georgia after World War II. However, the agents from whom "the Battey bacillus" or *M. avium–intracellulare* complex were cultured usually were male and had a history of smoking. As reported by several authors, the pulmonary disease as represented in cases such as this appears to be localized to the lung and often is indistinguishable from pulmonary tuberculosis: multiple cavities, fibrotic progressive processes with consolidation are typical.[51,52] The majority of patients appear to be women and most do not have a strong history of smoking. In our experience of more than 50 patients referred to this center with pulmonary MAC disease, CD4 counts are abnormally low but not as strikingly low as those with systemic MAC and idiopathic CD4 lymphopenia. The treatment is the same, but symptoms may recur and the duration of therapy is unclear. If the lung disease is restricted to one lobe, then surgical resection is a possible consideration. More usual, however, is for the pattern of disease to be diffuse. Some patients probably can have treatment discontinued after 9 months if they respond; others have persistent disease with consistently positive sputum cultures. Extensive studies using modern macrolide therapy such as clarithromycin or azithromycin reveal that patients will respond to these macrolide-containing regimens, but the long-term prognosis is unclear.[10] Relapse is common. Further, the number of agents that can be used in addition to either azithromycin or clarithromycin should be two or more, but choices are limited. Rifamycins (rifampin/rifabutin) are among the choice because non-AIDS patients usually will not be receiving medications subject to major drug interactions. The nature of the underlying defect that predisposes to pulmonary MAC disease requires further investigation. The disease is limited to the lungs (as opposed to severe idiopathic CD4 lymphopenia where dissemination can occur) yet abnormalities in numbers of CD4 lymphocytes have been noted as in this patient and many others. Functional abnormality such as the patient's response to proinflammatory cytokines or a reduced number of cytokine receptors or leukocytes in these patients are possibilities that could explain host susceptibility.

4.2. Role of *in Vitro* Susceptibility Testing

Susceptibility testing usually is routinely carried out on initial isolates of *M. tuberculosis*. Patients who have persisting disease or apparently relapsed disease also should have isolates tested. In many communities the testing of drugs against *M. tuberculosis* is carried out in municipal or state laboratories and significant delays may occur before results are known. It is clear that close collaboration between clinicians and public health officials is needed if evidence of drug-resistant strains of *M. tuberculosis* appears. The alternative regimens for treating disease-resistant strains clearly should be identified by appropriate panels of drug susceptibility tests.

For the nontuberculous mycobacteria, it can be assumed reasonably that initial isolates of MAC are macrolide, ethambutol, and aminoglycoside susceptible. However, for the challenging cases and in individuals who have recurrent disease with biopsy-confirmed culture data, susceptibility tests are advisable. These should be done in accordance to recently published guidelines.[53]

The large and apparently increasing number of nontuberculous mycobacteria that can cause disease in individuals with defects in cellular immunity presents a considerable clinical challenge. By and large, agents such as *M. genovense*, *M. malmoense*, and *M. haemophilum* are susceptible to modern macrolides, but no major chemotherapeutic trials have been carried out for disease caused by these specific nontuberculous mycobacteria. Additionally, the so-called rapid-growing nontuberculous mycobacteria consist of three distinct species: *M. fortuitum*, *M. chelonae*, and *M. abscessus* (these had been grouped together, yet it is recognized that versus *M. chelonae* tobramycin is significantly more active than amikacin). Agents that should be tested against the rapid growers include amikacin, cefoxitin, a quinolone such as moxifloxacin, a macrolide such as clarithromycin, doxycycline, and probably imipenem.

M. kansasii is more predictably susceptible to rifampin than to isoniazid, and indeed some strains from AIDS patients have been found to be isoniazid resistant. In other respects the clinical disease tends to mimic tuberculosis. *M. marinum* also is usually susceptible to agents such as azithromycin and clarithromycin.

5. Nocardiosis

The term "nocardiosis" encompasses a group of clinical syndromes caused by the genus *Nocardia*, with a number of species of varying virulence for humans causing disease, particularly in the immunocompromised host. Nocardial infection was first described in 1888 by Edward Nocard,[54] who described bovine farcy, a granulomatous process characterized by pulmonary disease, abscesses, sinus tracts, and spread to other organs. The causative organism, now called *Nocardia farcinica*, was found to be very similar to actinomycetes and mycobacteria both clinically and microbiologically. In 1943, Waksman[55] and Henrici described a separate species, *Nocardia asteroides*, the cause of the majority of human cases of invasive disease. Nocardial species can be described as belonging to a group of bacterial organisms termed the "aerobic actinomycetes." Their subgroup, the aerobic nocardiform actinomycetes, includes *Mycobacterium*, *Corynebacterium*, *Nocardia*, *Rhodococcus*, *Gordona*, and *Tsukamurella*. These organisms all have cell walls containing, among other constituents, mycolic acids of

various chain lengths, which are responsible for the varying degrees of acid fastness observed with these organisms under the microscope after appropriate staining. Since the description of *N. farcinica* and *N. asteroides*, more than a dozen different species have been identified. Of these, *N. asteroides*, *N. farcinica*, *N. nova*, *N. brasiliensis*, *N. otitidiscaviarum*, and *N. transvalensis* are the most important causes of human infection. Speciation is useful in predicting antimicrobial susceptibility; virtually all *N. asteroides* and *N. brasiliensis* isolates being sulfonamide sensitive, with more variable susceptibility for the other species.[56]

Although true bacteria, the clinical syndromes and pathogenesis of nocardial infection are quite similar to those observed with both mycobacteria and such fungal infections as those caused by *Aspergillus* species: infection occurring via inhalation or through direct inoculation of skin and soft tissues; once primary infection occurs, bloodstream dissemination occurs with the potential for metastatic infection throughout the body, but most commonly to the central nervous system (CNS), the skin and soft tissues, bone, and lungs; and treatment needs to be prolonged, with eradication of both the primary site of infection and also all areas of metastatic disease.[56]

5.1. Microbiology and Epidemiology

Nocardia are ubiquitous saprophytes in the environment, being found in decaying organic material, standing pools of water, and soil. Infection occurs due to inhalation of an aerosol or inoculation of material from one of these sites. Indeed, such activities as gardening by immunocompromised patients can lead to nocardiosis. Animal-to-human and human (infected patient)-to-human transmission have not been documented to occur, although fomites on the hands of medical personnel have been suspected on occasion. Nosocomial outbreaks of invasive nocardiosis have been reported among immunosuppressed patients, presumably due to the inhalation of aerosolized organisms from the environment-contaminated dust, hospital construction, or even groundswork around the hospital that liberated large amounts of dust. Disinfection with formaldehyde of hospital units for the care of immunosuppressed transplant patients has been reported to interrupt a nosocomial outbreak of nocardiosis.[56–60]

Nocardiae grow in an aerobic environment as branching, filamentous gram-positive rods that are positive by modified acid-fast stain (e.g., Kinyoun stain). This latter is accomplished by substituting sulfuric acid for the more potent hydrochloric acid used in the decolorization step in staining for mycobacteria (e.g., Ziehl–Neelsen stain).

Search for such organisms under the microscope in sputum and other clinical materials can allow the probable diagnosis to be made several days before the organism can be identified in cultures. *Nocardia* species can be grown on most nonselective media, if the diagnostic laboratory is alerted that nocardiosis is part of the differential diagnosis. However, in such specimens as sputum, in which heavy overgrowth with conventional bacteria can obscure the presence of *Nocardia*, the diagnostic yield is greatly increased by the use of such selective media as Thayer–Martin agar with antibiotics, paraffin agar, or buffered charcoal-yeast extract medium (commonly used for the isolation of *Legionella*). Once a colony of nocardial organisms is identified, speciation is then accomplished by a battery of biochemical tests, although modern molecular techniques are starting to come into vogue for this purpose.[56,60–64]

5.2. Pathogenesis and Pathology

The predominant initiating event for nocardiosis is inhalation of an aerosol of organisms, with delivery of viable organisms to the lower respiratory tract. Alternatively, direct inoculation of skin and soft tissues can occur with injury of the primary cutaneous barrier being an essential first step in introducing the infection, either preexisting skin injury as with a surgical wound or injury at the time of inoculation. In all cases, the initial response to nocardial invasion is a polymorphonuclear leukocyte (PMN) one. PMNs have been shown to effect growth inhibition by an oxidant-independent mechanism and to inhibit proliferation. However, PMNs are not able to kill these organisms. Nocardial superoxide dismutase and catalase appear to be of critical importance in protecting these bacteria from PMN-induced killing. It is not surprising then that patients with chronic granulomatous disease are at risk for this infection. Similarly, nonactivated macrophages are unable to kill pathogenic strains, although avirulent strains are easily killed by such resident macrophage populations as the alveolar macrophages.[65–70]

The key host defense against nocardiosis is a developing cell-mediated immune response; humoral immunity appears to offer little protection. In the successful response, the T-lymphocyte response is the orchestrator of the host defense, liberating cytokines that activate macrophages, which then are capable of both phagocytosis and intracellular killing. This bactericidal effect is accomplished by lysosomal enzymes, with virulent organisms being more resistant to phagocytosis and phagosome–lysosome fusion. It is not surprising then that in murine models the transfer of only T lymphocytes from immune animals protected naïve animals from lethal challenge. Thus, nocardiosis predominantly is seen in patients with depressed cell-mediated immunity (and to a significantly lesser extent, those with profound defects in leukocyte number and function); e.g., those on steroids, organ transplant recipients, lymphoma patients, and those being treated with systemic therapy for collagen vascular disease. Notable by its absence from this list are patients with AIDS, in whom nocardiosis is an uncommon infection, occurring predominantly late in the course of such patients when it does occur. This lack of nocardial infection in AIDS patients is only partially explained by the wide use of trimethoprim–sulfamethoxazole prophylaxis in these individuals. In addition, nocardial infection is a particular problem in patients with chronic pulmonary disease, particularly alveolar proteinosis.[56,65,66,68,71–73]

Bloodstream invasion by nocardial species from the initial portal of entry is a not uncommon event, as demonstrated by the frequent recognition of metastatic sites of infection. However, isolation of *Nocardia* from blood cultures, even when Dupont isolators are employed, is uncommon unless an endovascular foreign body is present.[74] Sites of metastatic spread via the bloodstream commonly include the skin, CNS (including the epidural and subarachnoid spaces and brain parenchyma), and skeletal system (including joint prostheses), but essentially any site can be involved, from the eye to the thyroid, from the adrenal to an abnormal cardiovascular surface. In addition, contiguous spread to adjoining tissues is not uncommon. In the case of the most common form of nocardiosis seen in compromised hosts, pulmonary infection, spread to the pleural space and/or pericardium is not uncommon (with the result being the possibility of nocardial empyema and/or purulent pericarditis).[56,65,75–78]

Pathologically, the hallmark signs of nocardial infection are inflammation, necrosis, and abscess formation. The histology of this lesion runs the gamut from a suppurative to a granulomatous response, often in the same patient. The branching, beaded, gram-positive rods that are observed microscopically in smears of colonies from an agar plate can be demonstrated in these abscesses. These forms usually will stain acid-fast positive when tissue sections stained with such stains as the Fite–Faraco are examined. So-called ''sulfur granules'' (bacterial macrocolonies) can be found in drainage from nocardial mycetomas (essentially identical to those observed in similar lesions produced by *Actinomycetes*).[56,65]

5.3. Clinical Manifestations

As previously stated, non-AIDS patients with defects in cell-mediated immunity (e.g., those on steroids, lymphoma patients, and transplant patients), patients with chronic granulomatous disease, and those with alveolar proteinosis and other chronic lung diseases are the individuals at greatest risk of this infection. In addition, such systemic illnesses as alcoholism (and presumably other forms of advanced liver disease), diabetes, and metastatic cancer can provide enough host compromise to permit this infection to flourish. Overall, a summary of more than 1000 cases in the world's literature found that less than 40% of patients with nocardiosis were thought to be relatively normal hosts, with a significantly smaller percentage having the pulmonary form of infection, and virtually none of these having disseminated infection (as opposed to compromised hosts, in whom disseminated infection is exceedingly common).[56,57,65,66,76,77,79–85]

With the ubiquity of *Nocardia* species in our environment, inoculation injury following trauma to the skin is not uncommon and is almost assuredly underdiagnosed. *N. asteroides* infection is often self-limited; in contrast, *N. brasiliensis* is the most common cause of a progressive sporotrichoidlike process involving both the skin and the lymphatics. Systemic disease can occur due to bloodstream invasion from these sites. Because of this and because metastatic infection to the skin is a common event with pulmonary infection, in the immunocompromised host the diagnosis of nocardial infection of the skin should be assumed to be a case of disseminated infection and treated as such, rather than as a trivial, superficial problem treatable with excision and/or a short course of antimicrobial therapy.[56,65,86–90] An uncommon form of nocardial infection of the skin and soft tissues in the compromised host is a mycetoma, a chronically progressive, destructive process usually located in the distal portions of the extremities. It is likely that in the compromised host dissemination would occur before a fully formed mycetoma could develop.

The most important clinical syndrome produced by *Nocardia* is that which follows invasion of the lung. Approximately 90% of pulmonary nocardial cases are caused by members of the *N. asteroides* complex. The clinical symptoms include fever, chills, night sweats, cough, purulent sputum production, pleuritic chest pain, malaise, easy fatigability, and dyspnea on exertion, usually with a subacute course over days to weeks. Alternatively, the infection in the lung and systemically can be relatively silent, with the presenting symptoms bringing the patient to medical attention being a site of metastatic infection, most commonly of the skin or CNS. This is a necrotizing infection (akin to that produced by tuberculosis and aspergillosis), so that there is far less ventilation–perfusion mismatch in most patients than is observed with conventional pyogenic infection (e.g., pneumococcal pneumonia). As a result, oxygenation usually is relatively

preserved at rest until disease is relatively far advanced. The chest radiologic picture is predominantly one of focal or multifocal disease, nodules, and/or areas of consolidation, which frequently progress to cavitation. The so-called "halo sign," described as being characteristic of neutropenic patients with invasive aspergillosis, can be duplicated by nocardial infection. Nocardial infection has a strong element of contiguous spread, so pleural effusions and/or empyema are common (and purulent pericarditis is not uncommon). Endobronchial inflammatory masses can be noted (as with aspergillosis) and fistulous tracts to adjoining structures may occur (as with actinomycosis). Healing with therapy will usually result in fibrosis, with retraction of the involved lung tissue in a fashion akin to that often seen with pulmonary tuberculosis. Microbial activity of such an area of the lung cannot be reliably determined by radiologic examination.[56,65,66,81,91–95]

Occurring commonly in conjunction with pulmonary infection or as the first manifestation of nocardiosis is CNS infection. In one large series of systemic nocardiosis, almost 50% of patients had evidence of CNS disease. Although *Nocardia* can produce meningitis alone, its most frequent effects are due to space-occupying lesions in the brain and less frequently the spinal cord and epidural space. Again, as with the pulmonary infection, clinical presentation is usually subacute–chronic. The exact presenting signs and symptoms will depend on the precise localization of the nocardial lesion, with a broad variety of clinical syndromes having been described, including the following: seizures, focal motor and/or sensory deficits, unilateral or bilateral paresis, cerebellar disturbance, and even behavioral and psychiatric disturbances. Alternatively, the neurologic examination can be essentially normal in a patient with pulmonary nocardiosis, but with significant disease found with brain imaging studies, either CT or MRI.[56,65,66,73,75–78,82,96–100]

Illustrative Case 3

A 56-year-old man had successfully undergone cardiac transplantation 3 years previously. His posttransplant course had gone extremely well, and he now was fully rehabilitated: working full time, golfing, and raising roses. His immunosuppressive program at this time included the following: cyclosporine 200 mg/day, mycophenolate 1 g twice daily, and prednisone 15 mg/day. He presented now for a routine clinic visit with a 3-week history of low-grade cough, occasional night sweats, and a subcutaneous nodule on his right arm. In retrospect, these symptoms began a few weeks after he undertook a major revision of his rose garden. Physical examination revealed a temperature of 99.6°F, blood pressure 150/90, pulse 86 and regular, and respirations 20. He was a ruddy complected, slightly cushingoid man in no acute distress. Examination of his skin revealed not one but three subcutaneous nodules: a 2 × 2 cm, slightly tender, indurated nodule on the volar surface of his right forearm; a 1 × 1.5 cm nontender nodule on his left thigh; and a 2 × 3 cm nontender nodule in the middle of his left buttock. The first of these had some erythema of the skin overlying the nodule, although the skin over the others was normal in appearance. The induration of all three nodules could be appreciated by the examiner by palpation. There was no adenopathy. Chest and cardiac examination were within normal limits, as was the abdominal exam. Neurologic examination was likewise normal. Laboratory evaluation revealed: Hct 34%, WBC 8700 (84% polys, 5% monos, 11% lymphs), negative urinalysis; BUN 32, creatinine 1.9 mg/dl, and normal liver function tests. Chest X ray and then CT demonstrated a 3 × 4 cm cavitary mass in the right lower lobe and two smaller nodules in the left lower lobe. CT scan of the brain revealed a 3 × 3 cm mass lesion in the right frontal cortex and three smaller nodules on the left side. An induced sputum and Dupont isolator blood cultures were negative for any pathogens. A nuclear scan of the skeletal system was negative.

An excisional biopsy of the right forearm lesion yielded a suppurative mass that on microscopic exam had demonstrable acid-fast (and gram-positive) branching, filamentous rods that grew *Nocardia asteroides* on culture. A needle aspiration of the cavitary lung lesion yielded identical results. The patient was treated initially with intravenous imipenem/cilastin for one month, then with trimethoprim–sulfamethoxazole at recommended doses for an additional year, and then maintained on low-dose trimethoprim–sulfamethoxazole indefinitely. Clinically, he has done well, with clearing of his subcutaneous nodules and the smaller lung nodules, with marked shrinkage of the cavitary lesion. The brain lesions have cleared completely without surgical drainage.

Comment. This is a classical case of nocardial infection in a susceptible host, caused by the most likely species to cause this syndrome, *Nocardia asteroides*. The exposure that led to the infection presumably occurred during his extensive digging of the ground for his rose garden (such exposures should be avoided by the chronically immunosuppressed). As with 10–20% of these cases the manifestation of the infection that brought the patient to medical attention was an area of metastatic infection. The clinically silent brain involvement also is not uncommon and needs to be sought with appropriate imaging studies. Diagnosis was made by skin biopsy. The aspiration of the lung lesion was carried out to rule out another concomitant infection (in transplant patients a number of patients with dual *Nocardia* and fungal infection have been seen and invasive pulmonary aspergillosis is another infection that can be acquired by working with the soil).[101,102] Because of the cerebral disease, treatment initially was with imipenem/cilastin, and then was switched to standard therapy with trimethoprim–sulfamethoxazole, but greatly prolonged because of the large microbial burden and evidence of metastatic disease. Treatment was continued until all evidence of active pulmonary and metastatic infection was gone.

Other sites of metastatic infection have been well described and include the bones and joints, the eye (particularly the retina), the thyroid, the adrenal, the kidney, and cardiovascular sites.[56,65,75–78,96–100,103–107]

5.4. Diagnosis

As demonstrated by the illustrative case, most cases of nocardiosis require some form of invasive diagnostic procedure to facilitate diagnosis. These include skin, lung, brain, and bone biopsy. All such biopsies should be subjected to direct microscopic examination (with both Gram's and modified acid-fast stains), culture, and pathology. An area of some confusion is the approach to the isolation of *Nocardia* from an induced or expectorated

sputum. It is clear that patients with chronic lung disease (particularly bronchitis and bronchiectasis) not on steroids can have transient nocardial carriage with no recognized consequences. In patients who are not immunocompromised and have no evidence of disease on radiologic evaluation, it is reasonable to follow these patients without therapy. However, in patients who are immunocompromised, even in the absence of radiologic findings or symptoms, we would advocate a course of preemptive antimicrobial therapy until the colonization is eliminated.[56,66,108,109]

There is presently no skin test or serologic test to aid in the diagnosis of nocardiosis. As far as the latter is concerned, after multiple failures by a number of investigators, Angeles and Sugar,[110] using greatly purified nocardial-specific antigen, have reported success using an enzyme immunoassay, including a dot–blot adaptation of this technique; similarly, using a similar antigen preparation, Boiron and Provost[111] have reported success with a Western blot technique. These tests report excellent sensitivity and specificity in patients with *N. asteroides* and *N. brasiliensis*. Unfortunately, these tests are not commercially available.

5.5 Clinical Management

When *Nocardia* is first demonstrated in human secretions or tissue, the next step is to define the extent of the disease in the body; that is, the assumption should be made that systemic spread has occurred unless the clinician can rule it out. A standard workup would include a careful palpation of the skin, chest CT scan, brain CT or MRI scan, and bone scan. In patients with positive blood cultures, evidence of cardiovascular disease should be sought with a transesophageal echocardiogram and an MR angiogram (regarding the possible presence of either endocarditis or a mycotic aneurysm) that might require surgical attack.

Medical therapy with sulfonamides is the mainstay of therapy, with surgical excision being ancillary and primarily utilized for diagnosis. Today, most clinicians utilize trimethoprim–sulfamethoxazole as the treatment of choice, with there being evidence *in vitro* of synergistic activity between these two drugs against most strains of *Nocardia*. However, it bears emphasis that there is no published clinical experience demonstrating a clear advantage for the combination. In patients with normal renal function, doses of 5–15 mg/kg of trimethoprim (25–75 mg of sulfamethoxazole) are utilized, with the higher doses being utilized for patients with a high microbial burden, cerebral disease, or other evidence of systemic spread.[56,112–114] Most experts advocate monitoring serum sulfonamide levels (aiming to achieve a level of 100–150 mg/l).[57] This last is particularly important in patients with renal dysfunction, those failing to respond to therapy, and patients receiving cyclosporine or tacrolimus, which can interact with the trimethoprim–sulfamethoxazole to cause synergistic nephrotoxicity. In addition, myelotoxicity has to be looked for. With trimethoprim–sulfamethoxazole therapy, or one of the alternative regimens, a discernible clinical response should be apparent within 10–14 days of initiating therapy, with some patients reporting improvement in 3–5 days.[56]

Alternatives to sulfonamide therapy are available. Whereas virtually all isolates of the *N. asteroides* complex and *N. brasiliensis* are sensitive to trimethoprim–sulfamethoxazole (unless the isolate represents a break through in a patient receiving trimethoprim–sulfamethoxazole prophylaxis), other species are more variable. If one of these species is isolated, particularly *N. otitidiscaviarum*, antimicrobial sensitivity testing is obligatory. Alternative regimens include intravenous imipenem/cilastin, cefotaxime, and ceftriaxone, with or without amikacin. The greatest experience probably has been with imipenem/cilastin, particularly in patients not tolerating trimethoprim–sulfamethoxazole and those with CNS disease (as in Illustrative Case 3). Other alternatives include minocycline, ciprofloxacin, ampicillin or amoxicillin–clavunate, and erythromycin. However, whereas there is a large published experience with sulfonamides and lesser experience with imipenem/cilastin and minocycline, for most of the others there are only scattered case reports.[115–129] Accordingly, our practice is to devise a two-drug regimen based on *in vitro* antimicrobial susceptibility testing, if we cannot use trimethoprim–sulfamethoxazole, imipenem/cilastin, or minocycline. We also would be inclined to use surgical excision of the infection, if possible, in conjunction with such treatment. The analogy to the management of atypical mycobacterial infection is clear.

The optimal duration of therapy has never been defined. Although relatively short courses (1–3 months) can be curative in nonimmunosuppressed patients or those with inoculation cutaneous infection, in immunosuppressed patients durations of ≥4–6 months are recommended to prevent relapse. In patients with CNS infection a minimum of 12–15 months is appropriate.[112,113,130,131] Our approach is to utilize these suggestions as only that—suggestions. Rather, we will treat "long enough," a period of time determined by the extent and location of the disease, the clinical response of the patient to the therapy, and what can be done to decrease the patient's net state of immunosuppression. In general, we would not begin to consider cessation of treatment until all evidence of clinical disease has disappeared.

Prognostic factors that determine the response to therapy include microbial load, the tissue involvement (e.g., CNS involvement worse than pulmonary worse than skin or skeletal, with multiple abscesses worse than a single lesion), decreasing immunosuppression, duration of therapy, and the promptness with which the diagnosis is made and therapy initiated.[56,65,112,132,133] This is a life-threatening disease in compromised hosts and is better prevented with low-dose trimethoprim–sulfamethoxazole prophylaxis and appropriate epidemiologic precautions.[134]

References

1. Dye C, Scheele S, Dolin P, *et al*: Global burden of tuberculosis: Estimated incidence, prevalence, and mortality by country. WHO Global Surveillance and Monitoring Project. *JAMA* **282:**677, 1999.
2. Bloom BR, Small PM: The evolving relation between humans and *Mycobacterium tuberculosis*. *N Engl J Med* **338:**677, 1998.
3. Snider DE, Castro KG: The global threat of drug-resistant tuberculosis. *N Engl J Med* **338:**1689, 1998.
4. Young LS: The Garrod Lecture. Mycobacterial diseases in the 1990s. *J Antimicrob Chemother* **32:**179, 1993.
5. Runyon E: Anonymous mycobacteria in pulmonary disease. *Med Clin North Am* **43:**273, 1959.
6. Wolinsky E: Nontuberculous mycobacteria and associated diseases. *Am Rev Respir Dis* **119:**107, 1979.
7. Korvick JA, Benson CA: *Mycobacterium avium Complex Infection: Progress in Research and Treatment*. American Society for Microbiology, New York, 1996.
8. Inderlied CB, Nash KA: Microbiology and *in vitro* susceptibility testing. In Korvick JA, Benson CA (eds): *Mycobacterium avium Complex Infection: Progress in Research and Treatment*. Marcel Dekker, New York, 1996, pp. 109–121.
9. Falkinham JO: Epidemiology of infection by nontuberculous mycobacteria. *Clin Microbiol Rev* **9:**177, 1996.
10. Wallace RJ, Glassroth J, Griffith DE: Diagnosis and treatment of disease caused by non-tuberculous mycobacteria. *Am J Respir Crit Care Med* **156:**S1, 1997.
11. Matsiota-Bernard P, Thierry D, De Truchis P, *et al*: *Mycobacterium genovense* infection in a patient with AIDS who was successfully treated with clarithromycin. *Clin Infect Dis* **20:**1565, 1995.
12. Buchholz UT, McNeil MM, Keyes LE, Good RC: *Mycobacterium malmoense* infections in the United States, January 1993 through June 1995. *Clin Infect Dis* **27:**551, 1998.
13. Kiehn TE, White M: *Mycobacterium haemophilum*: An emerging pathogen. *Eur J Clin Microbiol Infect Dis* **13:**925, 1994.
14. Gelber RH: Progress in the chemotherapy of leprosy: Status, issues and prospects. *Prog Drug Res* **34:**421, 1990.
15. Dankner WM, Waecker NJ, Essey MA, *et al*: *Mycobacterium bovis* infections in San Diego: A clinicoepidemiologic study of 73 patients and a historical review of a forgotten pathogen. *Medicine* **72:**11, 1993.

16. Snider DE, Raviglione M, Kochi A: Global burden of tuberculosis. In Bloom B (ed): *Tuberculosis Pathogenesis: Protection and Control*. American Society for Microbiology, Washington, DC, 1994, pp. 3–12.

17. Bellamy R, Ruwende C, Corrah T, *et al*: Variations in the NRAMP1 gene and susceptibility to tuberculosis in West Africans. *N Engl J Med* **338:**640, 1998.

18. Ordway DJ, Sonnenberg MG, Donahue SA, *et al*: Drug-resistant strains of *Mycobacterium tuberculosis* exhibit a range of virulence for mice. *Infect Immun* **63:**741, 1995.

19. Valway SE, Sanchez MP, Shinnick TF, *et al*: An outbreak involving extensive transmission of a virulent strain of *Mycobacterium tuberculosis*. *N Engl J Med* **338:**633, 1998.

20. American Thoracic Society and Centers for Disease Control: Targeted tuberculin testing and treatment of latent tuberculosis infection. *Am J Respir Crit Care Med* **161:**S221, 2000.

21. Slovis BS, Plitman JD, Haas DW: The case against anergy testing as a routine adjunct to tuberculin skin testing. *JAMA* **283:**2003, 2000.

22. Webster CT, Gordin FM, Matts JP, *et al*: Two-stage tuberculin skin testing in individuals with human immunodeficiency virus infection. Community Programs for Clinical Research on AIDS. *Am J Respir Crit Care Med* **151:**805, 1995.

23. Young DB, Robertson BD: TB vaccines: Global solutions for global problems. *Science* **284:**1479, 1999.

24. Behr MA, Wilson MA, Gill WP, *et al*: Comparative genomics of BCG vaccines by whole-genome DNA microarray. *Science* **284:**1520, 1999.

25. Talbot EA, Moore M, McCray E, Binkin NJ: Tuberculosis among foreign-born persons in the United States, 1993–1998. *JAMA* **284:**2894, 2000.

26. Young LS, Inderlied CB, Berlin OG, Gottlieb MS: Mycobacterial infections in AIDS patients, with an emphasis on the *Mycobacterium avium* complex. *Rev Infect Dis* **8:**1024, 1986.

27. Nightingale SD, Byrd LT, Southern PM, *et al*: Incidence of *Mycobacterium avium-intracellular* complex bacterium in human immunodeficiency virus-positive patients. *J Infect Dis* **165:**1082, 1992.

28. Young LS: Mycobacterial infections in immunocompromised patients. *Curr Opin Infect Dis* **9:**240, 1996.

29. Kim JH, Langston AA, Gallis HA: Miliary tuberculosis: Epidemiology, clinical manifestations, diagnosis, and outcome. *Rev Infect Dis* **12:**583, 1990.

30. Cohn DL, El-Sadr WM: Treatment of latent tuberculosis infection. In Reichman LB, Hershfield E (eds): *Tuberculosis: A Comprehensive International Approach*. Marcel Dekker, New York, 2000, pp. 471–478.

31. Bass JB, Farer LS, Hopewell PC, *et al*: Treatment of tuberculosis and tuberculosis infection in adults and children. American Thoracic Society and The Centers for Disease Control and Prevention. *Am J Respir Crit Care Med* **149:**1359, 1994.

32. Wallace RJ, Brown BA, Griffith DE, *et al*: Initial clarithromycin monotherapy for *Mycobacterium avium-intracellulare* complex lung disease. *Am J Respir Crit Care Med* **149:**1335, 1994.

33. Chiu J, Nussbaum J, Bozzette S, *et al*: Treatment of disseminated *Mycobacterium avium* complex infection in AIDS with amikacin, ethambutol, rifampin, and ciprofloxacin. *Ann Intern Med* **113:**358, 1990.

34. Inderlied CB, Kolonoski PT, Wu M, Young LS: Amikacin, ciprofloxacin and imipenem treatment for disseminated *Mycobacterium avium* complex infection of beige mice. *Antimicrob Chemother* **33:**176, 1989.

35. Kemper CA, Havlir D, Haghighat D, *et al*: The individual microbiologic effect of three antimycobacterial agents, clofazimine, ethambutol, and rifampin, on *Mycobacterium avium* complex bacteremia in patients with AIDS. *J Infect Dis* **170:**157, 1994.

36. Fernandes PB, Hardy DJ, McDaniel D, *et al*: *In vitro* and *in vivo* activities of clarithromycin against *Mycobacterium avium*. *Antimicrob Agents Chemother* **33:**1531, 1989.

37. Inderlied CB, Kolonoski PT, Wu M, Young LS: *In vitro* and *in vivo* activity of azithromycin (CP 62, 993) against the *Mycobacterium avium* complex. *J Infect Dis* **159:**994, 1989.

38. Dautzenberg B, Truffot C, Legris S, *et al*: Activity of clarithromycin against *Mycobacterium avium* infection in patients with the acquired immune deficiency syndrome. *Am Rev Respir Dis* **144:**564, 1991.

39. Young LS, Wiviott L, Wu M, *et al*: Azithromycin reduces *Mycobacterium avium* complex bacteremia and relieves the symptoms of disseminated disease in patients with AIDS. *Lancet* **338:**1107, 1991.

40. Chaisson RE, Benson CA, Dube MP, *et al*: Clarithromycin therapy for bacteremic *Mycobacterium avium* complex. *Ann Intern Med* **121:**905, 1994.

41. Bermudez LE, Nash KA, Petrofsky M, *et al*: Effect of ethambutol on emergence of clarithromycin-resistant *Mycobacterium avium* complex in the beige mouse model. *J Infect Dis* **174:**1218, 1996.

42. Dunne M, Fessel J, Kumar P, *et al*: A randomized, double-blind trial comparing azithromycin and clarithromycin in the treatment of disseminated *Mycobacterium avium* infection in patients with human immunodeficiency virus. *Clin Infect Dis* **31:**1245, 2000.

43. Bermudez LE, Kolonoski P, Young LS: Roxithromycin alone or in combination with either ethambutol or levofloxacin for disseminated *Mycobacterium avium* infection in beige mice. *Antimicrob Agents Chemother* **40:**1033, 1996.

44. Pierce M, Crampton S, Henry D, *et al*: A randomized trial of clarithromycin as prophylaxis against disseminated *Mycobacterium avium* complex infection in patients with advanced acquired immunodeficiency syndrome. *N Engl J Med* **335:**384, 1996.

45. Havlir DV, Dube MP, Sattler FR, *et al*: Prophylaxis against disseminated *Mycobacterium avium* complex with weekly azithromycin, daily rifabutin, or both. California Collaborative Treatment Group. *N Engl J Med* **335:**392, 1996.

46. Nightingale SD, Cameron DW, Gordin FM, *et al*: Two placebo controlled trials of rifabutin prophylaxis against *Mycobacterium avium* complex infection in AIDS. *N Engl J Med* **329:**828, 1993.

47. Bermudez LE, Inderlied CB, Kolonoski P, *et al*: Activity of moxifloxacin by itself and in combination with ethambutol, rifabutin, and azithromycin *in vitro* and *in vivo* against *Mycobacterium avium*. *Antimicrob Agents Chemother* **45:**217, 2001.

48. Wu M, Aralor P, Nash K, *et al*: Linezolid, a new oxazoladinone, has activity *in vitro* and in macrophage culture system against *Mycobacterium avium* complex (MAC). In *38th Interscience Conference on Antimicrobial Agents and Chemotherapy*. American Society for Microbiology, San Diego, California, 1998, p. 210.

49. Bermudez LE, Kolonoski P, Wu M, *et al*: Mefloquine is active *in vitro* and *in vivo* against *Mycobacterium avium* complex. *Antimicrob Agents Chemother* **43:**1870, 1999.

50. Holland SM, Eisenstein EM, Kuhns DB, *et al*: Treatment of refractory disseminated nontuberculous mycobacterial infection with interferon gamma. A preliminary report. *N Engl J Med* **330:**1348, 1994.

51. Prince DS, Peterson DD, Steiner RM, *et al*: Infection with *Mycobacterium avium* complex in patients without predisposing conditions. *N Engl J Med* **321:**863, 1989.

52. Reich JM, Johnson RE: *Mycobacterium avium* complex pulmonary disease presenting as an isolated lingular or middle lobe pattern. The Lady Windermere syndrome. *Chest* **101:**1605, 1992.

53. Woods GL: Disease due to the *Mycobacterium avium* complex in patients infected with human immunodeficiency virus: Diagnosis and susceptibility testing. *Clin Infect Dis* **18:**S227, 1994.

54. Nocard E: Note sur la maladie de bonefs de la guadeloupe, connue sous le nom de farcin. *Ann Inst Pasteur Paris* **2:**293, 1888.

55. Waksman S: The nomenclature and classification of the actinomytes. *J Bacteriol* **46:**337, 1943.

56. Sorrell T: *Nocardia Species*, Vol. 2. Iredell J (ed). Churchill Livingstone, New York, 2000, p. 2637.

57. McNeil MM, Brown JM: The medically important aerobic actinomycetes: Epidemiology and microbiology. *Clin Microbiol Rev* **7:**357, 1994.

58. Houang ET, Lovett IS, Thompson FD, *et al*: *Nocardia asteroides* infection—A transmissible disease. *J Hosp Infect* **1:**31, 1980.

59. Sahathevan M, Harvey FA, Forbes G, *et al*: Epidemiology, bacteriology and control of an outbreak of *Nocardia asteroides* infection on a liver unit. *J Hosp Infect* **18**(Suppl. A):473, 1991.

60. Provost F, Laurent F, Blanc MV, Boiron P: Transmission of nocardiosis and molecular typing of *Nocardia* species: A short review. *Eur J Epidemiol* **13:**235, 1997.

61. Ashdown LR: An improved screening technique for isolation of *Nocardia* species from sputum specimens. *Pathology* **22:**157, 1990.

62. Shawar RM, Moore DG, LaRocco MT: Cultivation of *Nocardia* spp. on chemically defined media for selective recovery of isolates from clinical specimens. *J Clin Microbiol* **28:**508, 1990.

63. Vickers RM, Rihs JD, Yu VL: Clinical demonstration of isolation of *Nocardia asteroides* on buffered charcoal-yeast extract media. *J Clin Microbiol* **30:**227, 1992.

64. Wilson RW, Steingrube VA, Brown BA, Wallace RJ Jr: Clinical application of PCR-restriction enzyme pattern analysis for rapid identification of aerobic actinomycete isolates. *J Clin Microbiol* **36:**148, 1998.

65. Lerner PI: Nocardiosis. *Clin Infect Dis* **22:**891, 1996.

66. Beaman BL, Beaman L: *Nocardia* species: Host–parasite relationships. *Clin Microbiol Rev* **7:**213, 1994.

67. Filice GA, Beaman BL, Krick JA, Remington JS: Effects of human neutrophils and monocytes on *Nocardia asteroides*: Failure of killing despite occurrence of the oxidative metabolic burst. *J Infect Dis* **142:**432, 1980.

68. Davis-Scibienski C, Beaman BL: Interaction of *Nocardia asteroides* with rabbit alveolar macrophages: Association of virulence, viability, ultrastructural damage, and phagosome–lysosome fusion. *Infect Immun* **28:**610, 1980.

69. Beaman L, Beaman BL: Monoclonal antibodies demonstrate that superoxide dismutase contributes to protection of *Nocardia asteroides* within the intact host. *Infect Immun* **58:**3122, 1990.

70. Shetty AK, Arvin AM, Gutierrez KM: *Nocardia farcinica* pneumonia in chronic granulomatous disease. *Pediatrics* **104:**961, 1999.

71. Deem RL, Doughty FA, Beaman BL: Immunologically specific direct T lymphocyte-mediated killing of *Nocardia asteroides*. *J Immunol* **130:**2401, 1983.

72. Deem RL, Beaman BL, Gershwin ME: Adoptive transfer of immunity to *Nocardia asteroides* in nude mice. *Infect Immun* **38:**914, 1982.

73. Oerlemans WG, Jansen EN, Prevo RL, Eijsvogel MM: Primary cerebellar nocardiosis and alveolar proteinosis. *Acta Neurol Scand* **97:**138, 1998.

74. Kontoyiannis DP, Ruoff K, Hooper DC: *Nocardia* bacteremia. Report of 4 cases and review of the literature. *Medicine* **77:**255, 1998.

75. Palmer SM Jr, Kanj SS, Davis RD, Tapson VF: A case of disseminated infection with *Nocardia brasiliensis* in a lung transplant recipient. *Transplantation* **63:**1189, 1997.

76. Sakai C, Takagi T, Satoh Y: *Nocardia asteroides* pneumonia, subcutaneous abscess and meningitis in a patient with advanced malignant lymphoma: Successful treatment based on *in vitro* antimicrobial susceptibility. *Intern Med* **38:**683, 1999.

77. Wong KM, Chak WL, Chan YH, *et al*: Subcutaneous nodules attributed to nocardiosis in a renal transplant recipient on tacrolimus therapy. *Am J Nephrol* **20:**138, 2000.

78. Ogg G, Lynn WA, Peters M, *et al*: Cerebral *Nocardia* abscesses in a patient with AIDS: Correlation of magnetic resonance and white cell scanning images with neuropathological findings. *J Infect* **35:**311, 1997.

79. Kim J, Minamoto GY, Grieco MH: *Nocardial* infection as a complication of AIDS: Report of six cases and review. *Rev Infect Dis* **13:**624, 1991.

80. Forbes GM, Harvey FA, Philpott-Howard JN, *et al*: Nocardiosis in liver transplantation: Variation in presentation, diagnosis and therapy. *J Infect* **20:**11, 1990.

81. Farina C, Boiron P, Goglio A, Provost F: Human nocardiosis in northern Italy from 1982 to 1992. Northern Italy Collaborative Group on Nocardiosis. *Scand J Infect Dis* **27:**23, 1995.

82. Boiron P, Provost F, Chevrier G, Dupont B: Review of nocardial infections in France 1987 to 1990. *Eur J Clin Microbiol Infect Dis* **11:**709, 1992.

83. Rinaldi S, D'Argenio P, Fiscarelli E, *et al*: Fatal disseminated *Nocardia farcinica* infection in a renal transplant recipient. *Pediatr Nephrol* **14:**111, 2000.

84. Magee CC, Halligan RD, Milford EL, Sayegh MH: Nocardial infection in a renal transplant recipient on tacrolimus and mycophenolate mofetil. *Clin Nephrol* **52:**44, 1999.

85. Reddy SS, Holley JL: Nocardiosis in a recently transplanted renal patient. *Clin Nephrol* **50:**123, 1998.

86. Merigou D, Beylot-Barry M, Ly S, *et al*: Primary cutaneous *Nocardia asteroides* infection after heart transplantation. *Dermatology* **196:**246, 1998.

87. Bhalodia AM, Lertzman BH, Kantor GR, Granick MS: Localized cutaneous *Nocardia brasiliensis* mimicking foreign body granuloma. *Cutis* **61:**161, 1998.

88. Folgaresi M, Ferdani G, Coppini M, Pincelli C: Primary cutaneous nocardiosis. *Eur J Dermatol* **8:**430, 1998.

89. Lee MS, Sippe JR: Primary cutaneous nocardiosis. *Australas J Dermatol* **40:**103, 1999.

90. Wenger PN, Brown JM, McNeil MM, Jarvis WR: *Nocardia farcinica* sternotomy site infections in patients following open heart surgery. *J Infect Dis* **178:**1539, 1998.

91. Casty FE, Wencel M: Endobronchial nocardiosis. *Eur Respir J* **7:**1903, 1994.

92. Pickles RW, Malcolm JA, Sutherland DC: Endobronchial nocardiosis in a patient with AIDS. *Med J Aust* **161:**498, 1994.

93. Martinez-Marcos FJ, Viciana P, Canas E, *et al*: Etiology of solitary pulmonary nodules in patients with human immunodeficiency virus infection. *Clin Infect Dis* **24:**908, 1997.

94. Kim Y, Lee KS, Jung KJ, *et al*: Halo sign on high resolution CT: Findings in spectrum of pulmonary diseases with pathologic correlation. *J Comput Assist Tomogr* **23:**622, 1999.

95. Gaeta M, Blandino A, Scribano E, *et al*: Computed tomography

halo sign in pulmonary nodules: Frequency and diagnostic value. *J Thorac Imaging* **14:**109, 1999.

96. Sabeel A, Alrabiah F, Alfurayh O, Hassounah M: Nocardial brain abscess in a renal transplant recipient successfully treated with triple antimicrobials. *Clin Nephrol* **50:**128, 1998.

97. Harvey AL, Myslinski J, Ortiz L: A case of *Nocardia* epidural abscess. *J Emerg Med* **16:**579, 1998.

98. Torres OH, Domingo P, Pericas R, *et al*: Infection caused by *Nocardia farcinica*: Case report and review. *Eur J Clin Microbiol Infect Dis* **19:**205, 2000.

99. Mogilner A, Jallo GI, Zagzag D, Kelly PJ: *Nocardia* abscess of the choroid plexus: Clinical and pathological case report. *Neurosurgery* **43:**949, 1998.

100. Hiller R, Singh H, Crone M: Left leg paralysis in a renal transplant. *Am J Kidney Dis* **33:**E4, 1999.

101. McCown HF, Sahn EE: Subcutaneous phaeohyphomycosis and nocardiosis in a kidney transplant patient. *J Am Acad Dermatol* **36:**863, 1997.

102. Sartoris KE, Baillie GM, Tiernan R, Rajagopalan PR: Phaeohyphomycosis from *Exphiala jeanselmei* with concomitant *Nocardia asteroides* infection in a renal transplant recipient: Case report and review of the literature. *Pharmacotherapy* **19:**995, 1999.

103. Yap EY, Fam HB, Leong KP, Buettner H: *Nocardia* choroidal abscess in a patient with systemic lupus erythematosus. *Aust NZ J Ophthalmol* **26:**337, 1998.

104. Dinulos JG, Darmstadt GL, Wilson CB, *et al*: *Nocardia asteroides* septic arthritis in a healthy child. *Pediatr Infect Dis J* **18:**308, 1999.

105. Arnal C, Man H, Delisle F, *et al*: *Nocardia* infection of a joint prosthesis complicating systemic lupus erythematosus. *Lupus* **9:**304, 2000.

106. Carriere C, Marchandin H, Andrieu JM, *et al*: *Nocardia* thyroiditis: Unusual location of infection. *J Clin Microbiol* **37:**2323, 1999.

107. Midiri M, Finazzo M, Bartolotta TV, Maria MD: Nocardial adrenal abscess: CT and MR findings. *Eur Radiol* **8:**466, 1998.

108. Georghiou PR, Blacklock ZM: Infection with *Nocardia* species in Queensland. A review of 102 clinical isolates. *Med J Aust* **156:**692, 1992.

109. Cremades MJ, Menendez R, Santos M, Gobernado M: Repeated pulmonary infection by *Nocardia asteroides* complex in a patient with bronchiectasis. *Respiration* **65:**211, 1998.

110. Angeles AM, Sugar AM: Rapid diagnosis of nocardiosis with an enzyme immunoassay. *J Infect Dis* **155:**292, 1987.

111. Boiron P, Provost F: Use of partially purified 54-kilodalton antigen for diagnosis of nocardiosis by Western blot (immunoblot) assay. *J Clin Microbiol* **28:**328, 1990.

112. Smego RA Jr, Moeller MB, Gallis HA: Trimethoprim-sulfamethoxazole therapy for *Nocardia* infections. *Arch Intern Med* **143:**711, 1983.

113. Wallace RJ Jr, Septimus EJ, Williams TW Jr, *et al*: Use of trimethoprim-sulfamethoxazole for treatment of infections due to *Nocardia*. *Rev Infect Dis* **4:**315, 1982.

114. Smego RA Jr, Gallis HA: The clinical spectrum of *Nocardia brasiliensis* infection in the United States. *Rev Infect Dis* **6:**164, 1984.

115. Gombert ME, Aulicino TM, duBouchet L, *et al*: Therapy of experimental cerebral nocardiosis with imipenem, amikacin, trimetho-prim-sulfamethoxazole, and minocycline. *Antimicrob Agents Chemother* **30:**270, 1986.

116. Threlkeld SC, Hooper DC: Update on management of patients with *Nocardia* infection. *Curr Clin Top Infect Dis* **17:**1, 1997.

117. Goldstein FW, Hautefort B, Acar JF: Amikacin-containing regimens for treatment of nocardiosis in immunocompromised patients. *Eur J Clin Microbiol* **6:**198, 1987.

118. Gombert ME, Aulicino TM: Synergism of imipenem and amikacin in combination with other antibiotics against *Nocardia asteroides*. *Antimicrob Agents Chemother* **24:**810, 1983.

119. Gombert ME, duBouchet L, Aulicino TM, Berkowitz LB: Antimicrobial synergism in the therapy of experimental cerebral nocardiosis. *J Antimicrob Chemother* **24:**39, 1989.

120. Menendez R, Cordero PJ, Santos M, *et al*: Pulmonary infection with *Nocardia* species: A report of 10 cases and review. *Eur Respir J* **10:**1542, 1997.

121. Wren MV, Savage AM, Alford RH: Apparent cure of intracranial *Nocardia asteroides* infection by minocycline. *Arch Intern Med* **139:**249, 1979.

122. Bach MC, Monaco AP, Finland M: Pulmonary nocardiosis. Therapy with minocycline and with erythromycin plus ampicillin. *JAMA* **224:**1378, 1973.

123. Hall WA, Martinez AJ, Dummer JS, Lunsford LD: Nocardial brain abscess: Diagnostic and therapeutic use of stereotactic aspiration. *Surg Neurol* **28:**114, 1987.

124. Wortman PD: Treatment of a *Nocardia brasiliensis* mycetoma with sulfamethoxazole and trimethoprim, amikacin, and amoxicillin and clavulanate. *Arch Dermatol* **129:**564, 1993.

125. Wallace RJ Jr, Nash DR, Johnson WK, *et al*: Beta-lactam resistance in *Nocardia brasiliensis* is mediated by beta-lactamase and reversed in the presence of clavulanic acid. *J Infect Dis* **156:**959, 1987.

126. Bath PM, Pettingale KW, Wade J: Treatment of multiple subcutaneous *Nocardia asteroides* abscesses with ciprofloxacin and doxycycline. *Postgrad Med J* **65:**190, 1989.

127. Yew WW, Wong PC, Kwan SY, *et al*: Two cases of *Nocardia asteroides* sternotomy infection treated with ofloxacin and a review of other active antimicrobial agents. *J Infect* **23:**297, 1991.

128. Meier B, Metzger U, Muller F, *et al*: Successful treatment of a pancreatic *Nocardia asteroides* abscess with amikacin and surgical drainage. *Antimicrob Agents Chemother* **29:**150, 1986.

129. Leitersdorf I, Silver J, Naparstek E, Raveh D: Tetracycline derivatives, alternative treatment for nocardiosis in transplanted patients. *Clin Nephrol* **48:**48, 1997.

130. Byrne E, Brophy BP, Perrett LV: *Nocardia* cerebral abscess: New concepts in diagnosis, management, and prognosis. *J Neurol Neurosurg Psychiatry* **42:**1038, 1979.

131. Geiseler PJ, Andersen BR: Results of therapy in systemic nocardiosis. *Am J Med Sci* **278:**188, 1979.

132. Simpson GL, Stinson EB, Egger MJ, Remington JS: Nocardial infections in the immunocompromised host: A detailed study in a defined population. *Rev Infect Dis* **3:**492, 1981.

133. Arduino RC, Johnson PC, Miranda AG: Nocardiosis in renal transplant recipients undergoing immunosuppression with cyclosporine. *Clin Infect Dis* **16:**505, 1993.

134. Munoz P, Munoz RM, Palomo J, *et al*: *Pneumocystis carinii* infection in heart transplant recipients. Efficacy of a weekend prophylaxis schedule. *Medicine* **76:**415, 1997.

8

Pneumocystis carinii and Parasitic Infections in the Immunocompromised Host

JAY ALAN FISHMAN

1. Introduction

International travel and shifting patterns of immigration have increased the importance of awareness of the major clinical syndromes associated with infections due to parasites. In the immunocompromised individual, life-threatening infection may emerge decades after a forgotten exposure in an endemic area. Most clinicians have some familiarity with the major clinical syndromes associated with malaria, Chagas' disease, giardiasis, amebiasis, or the helminthic diseases. The presence, progression, and manifestations of some of the common parasitic diseases are altered by immune compromise. Prior to the recognition of the acquired immunodeficiency syndrome (AIDS), important parasites in the immunocompromised host were largely limited to infections with *Toxoplasma gondii*, *Pneumocystis carinii*, *Strongyloides stercoralis*, and occasionally babesiosis or malaria related to transfusions in splenectomized patients. Recently, however, the pattern of parasitic infection has been altered by a number of important trends in clinical medicine (see Table 1):

- The growing population of immunocompromised individuals.
- Prolonged survival with immune deficits.
- Increased use of immunosuppressive therapies in underdeveloped regions.

- Highly active antiretroviral therapies (HAART) for human immunodeficiency virus (HIV) infection have resulted in immune reconstitution in many individuals with reduced susceptibility to common parasites and to *Pneumocystis carinii*.
- New immunosuppressive agents are employed in a broader range of patients including generic cyclosporine, tacrolimus, mycophenylate mofetil, sirolimus, costimulatory blockade, antilymphocyte antibodies, and broader application of intensive chemotherapy and hematopoietic transplantation for malignancy.
- Broader use of routine prophylactic strategies for common infections in compromised individuals.

Successful parasitism is defined by the adaptation of an organism to the host environment. In the absence of an immunologic niche for the organism, the parasite will either fail to establish infection or overwhelm the host. The effects of immune compromise on the manifestations of parasitic infections are defined by the organisms' "natural" mode of evasion/interaction with the host's immune system and by the nature of the immune lesion(s) (see Table 1). Thus, an organism of low native virulence (e.g., *P. carinii*) causes great morbidity in patients with AIDS or following organ transplantation.

1.1. Parasite-Specific Factors: Development and Distribution

Each year, parasites cause over 2 billion infections worldwide. It is predictable that some of these infections occur in individuals immunocompromised by malnutri-

JAY ALAN FISHMAN • Infectious Disease Division and Transplantation Unit, Massachusetts General Hospital, Harvard Medical School, Boston, Massachusetts 02114.

Clinical Approach to Infection in the Compromised Host (Fourth Edition), edited by Robert H. Rubin and Lowell S. Young. Kluwer Academic/Plenum Publishers, New York, 2002.

TABLE 1. Mechanisms for the Evasion of Host Immune Response in Parasites

Host response	Mediator	Mechanism	Examples
Nonspecific inflammation	Anti-inflammatory molecules	—	Amebae, *T. taeniaeformis*
Humoral	Complement	Surface resistance	*T. cruzi*, schistosomes, *Leishmania*
Humoral	Antibody	Shedding antigen	*Trichinella*, schistosomes
		Antigenic variation	Trypanosomes, *Giardia*
		Antigenic mimicry	Schistosomes
		Antibody destruction	Filaria, *T. cruzi*
		Host antigen coat	*P. carinii*, *T. vivax*
		Polyclonal stimulation	Trypanosomes
Cellular	Macrophage	Block fusion/acidification	*T. gondii*
Phagocytosis	—	Escape phagolysosome	*T. cruzi*
		Evade oxidative burst	*T. gondii*, *L. donovani*
		Alter macrophage function[a]	*Leishmania*, *T. brucei*
	Eosinophil	Inhibit attachment	Schistosomes
None	Privileged site	Escape into gut	*Ascarius*, hookworms
		Eye	*Ochocercus*
		Lymphoblast	*Theileria*
		Liver	Malaria
		Muscle	*Sarcocystis*
		Intestinal epithelia	Coccidia

[a]IL-1, interleukin 1; MHC, gene products of the major histocompatibility locus.

tion, by the epidemic of AIDS in developing regions, or by immunosuppressive therapy. A history of travel to endemic areas (recent or distant) or of exposures to food, water, animals, blood products, or other vectors of parasitic disease should suggest acute or reactivated infection in symptomatic individuals. Some infections are prevalent in subgroups of immunologically normal hosts in developed regions. Thus, homosexual males have an increased incidence of infections with intestinal parasites (including *Trichuris* and pathogenic *Entamoeba histolytica*, *Giardia lamblia*, *Strongyloides stercoralis*) and nonpathogenic protozoa. Infection with other pathogens (cytomegalovirus, *Salmonella*, *Shigella*) in this population may contribute to the pathogenesis or severity of concomitant infections. Day care and chronic care centers are common sources for infections with *Giardia*, amebae, and *Cryptosporidium* species.

A few features of parasitic infections merit emphasis.

- The life cycle of the parasite determines the nature and duration of the exposure of the organism to the host's immune system (see Section 2.1) and the clinical manifestations of infection.
- Immune suppression has the greatest effect on the life cycles of organisms that are normally suppressed or regulated by the host's immune response.
- In general, only those infections due to organisms that can complete their life cycle within the human

host are amplified in the immunocompromised host. *Strongyloides stercoralis* is unusual as a nematode because of the ability to complete its life cycle within the human host. Thus, the inflammatory response to all of the helminths are reduced in patients treated with corticosteroids, but only *Strongyloides* infection is significantly exacerbated by such treatment.

- The protozoa as a group have the capability to complete their life cycles within the human host and are common pathogens in the setting of immune deficiency. Accelerated growth of protozoans derepressed by immune dysfunction produces systemic diseases that reflect both the initial sites (organs) of infection and the metastatic spread of infection. Thus, depression of immune barriers may predispose to invasive disease by gastrointestinal parasites, as in amebiasis, or to dissemination of intracellular organisms that overwhelm the reticuloendothelial system as may be seen in leishmaniasis in AIDS or solid organ transplantation.
- Latency is a critical feature of most parasitic infections of compromised hosts. Primary infections (e.g., due to *Strongyloides*, *Leishmania donovani*, *T. cruzi*, and *T. gondii*) may be observed in endemic regions, but are generally less common than reactivation syndromes due to more distant exposures. Transplanted organs are commonly impli-

cated as the source of latent infection in immunologically naive allograft recipients.

- The impact of infections due to organisms that require maturation outside the host and subsequent penetration (often ingestion) into the host is limited by the size of the initial inoculum. The burden of this group of organisms cannot increase during the course of disease. As a result, helminths (worms) tend to cause mechanical obstruction due to size, location, and nonspecific inflammatory responses (fibrosis) and are generally limited to the gastrointestinal (GI) tract. These pathogens (e.g., *Schistosoma* or the liver flukes) may cause organ failure (renal or hepatic), which necessitate transplantation, but are not significantly exacerbated by immune suppression.
- Coinfection is a critical feature of these infections. Thus, invasive disease of the gastrointestinal tract or lungs is more common in the setting of simultaneous viral infection (particularly cytomegalovirus) of these organs. In the compromised individual, all active infections must be treated for successful clinical resolution.

1.2. Host–Parasite Interactions and Mechanisms of Immune Evasion

Significant infections due to parasites occur when the balance between host protective mechanisms and parasite growth is disrupted. Specific immune lesions (e.g., hypogammaglobulinemia) may not predispose to parasites normally controlled by other mechanisms (e.g., T lymphocytes). Some infections (e.g., malaria, amebae) are not appreciably exacerbated by immune suppression. A second group causes little or no disease (subclinical or mild, commensal, or latent infection) until activated in the setting of immune compromise. Some of the most "successful" parasites have the ability to avoid detection or killing or both by the immune system. Some of the common mechanisms of immune evasion are listed in Table 1. A number of parasites are resistant to antibody- and complement-mediated lysis (e.g., *Schistosoma, Trypanosoma cruzi*).[1–3] Others vary or shed surface antigens to avoid detection. Malarial parasites may alter the surface of host cells and release toxins, which induce the production of cytokines, enhancing display of receptors needed for cellular penetration. Still others become coated with host proteins to diminish immune detection. Filaria, *Leishmania*, and trypanosomes are capable of inducing defects in cell-mediated immunity.[4] Patients with defects in cell-mediated immunity are particularly susceptible to

TABLE 2. Parasitic Infections of Importance in the Immunocompromised Host[a]

Mechanism	Organisms
Neutrophil inflammation	*P. carinii*[b]
Humoral immunity	*G. lamblia, Cryptosporidium*
Cellular immunity	*P. carinii,*[b] *T. gondii, Cryptosporidium, Strongyloides, Leishmania, Microsporidia, Isospora belli, G. lamblia, E. histolytica, Cyclospora*

[a]This table does not list common parasitic infections that are not increased in severity or frequency in immunocompromised hosts.
[b]*Pneumocystis carinii* is included for purposes of this discussion, but is considered to be a fungus.

infections due to *Pneumocystis, T. gondii, Cryptosporidium* species, *Leishmania* species, and *S. stercoralis* (see Table 2). Subgroups of organisms of particular importance to the compromised host are the intracellular parasites (*T. gondii, Leishmania*, and *T. cruzi*) that evade killing by host macrophages. All share the need to evade the humoral (complement and antibody) immune response and intracellular oxidative killing mechanisms prior to establishing intracellular residence.

Perhaps the best example of the interaction of a parasite with the immune system is that of *Leishmania* species. The manifestations of cutaneous leishmaniasis (*L. mexicana* complex, *L. braziliensis* complex, *L. tropica, L. major*) range from localized cutaneous disease to diffuse cutaneous or mucocutaneous ("espundia") involvement. Disseminated disease (visceral leishmaniasis or kala-azar) involving the liver, spleen, bone marrow, and reticuloendothelial system also occurs (*L. donovani, L. chagasi, L. infantum*). In the presence of a normal cell-mediated immune response and in the absence of specific antibody, cutaneous lesions often heal spontaneously. Patients with diffuse cutaneous leishmaniasis generally have high levels of specific antibody without antigen-specific delayed-type hypersensitivity (DTH). Relapsing (recidivans) and mucocutaneous disease occurs in the presence of DTH, but macrophage dysfunction is suggested by the paucity of granulomata in affected tissues. Visceral disease occurs in the absence of cell-mediated immunity and in the presence of specific antibody. Fatal disease has been reported in AIDS patients in the setting of marked T-lymphocyte deficiency. In normal hosts, secondary *Leishmania* infections produce smaller lesions and lower parasite burdens than primary infections. The intensity of inflammation in Chagas' disease is similarly dictated by the intensity of the host response. The role of autoimmunity in the clinical manifestations of Chagas' disease remains unresolved.

1.3. Missing Infections in Compromised Hosts

Immunity to parasites is complex. Generally, multiple components of the immune system contribute to the prevention or resolution of infection. Thus, common knowledge about the nature of immune deficiency in individuals is often incorrect. Patients with AIDS who undergo organ transplantation have been shown to require immune suppression to prevent graft rejection. This observation suggests that although the prime immune deficiencies of AIDS and in transplantation are "T-cell mediated," some aspects of immune function are well preserved in HIV infection. Thus, it is predictable that specific infections would occur with differing frequencies in transplantation, neutropenia, and AIDS. A subgroup of common parasitic infections has not increased substantially in frequency or severity in individuals infected with HIV.[5] These infections include *S. stercoralis*, malaria, *E. histolytica*, and trypanosomiasis. Reports of strongyloidiasis in AIDS generally have been in individuals also receiving immunosuppressive therapies or with underlying malignancy. Given the relative absence of disseminated strongyloidiasis in this population, the immune deficits of AIDS must not include some of the relevant host defenses (e.g., of the intestinal mucosa) that are altered in organ transplantation or in neutropenia. Perhaps of greater importance is the relative absence of *Strongyloides* in transplant recipients on corticosteroid-sparing regimens. It has been suggested that steroids may mimic the effect of naturally occurring ecdysteroids that accelerate the maturation (molting) of rhabditiform larvae.

Multiple potential pathogens often are found in diarrheal stools from individuals infected with HIV.[6,7] The organism(s) causing disease (e.g., *Microsporidia* or *Cryptosporidia* species) may be demonstrated microscopically in the small intestine but often are undetected in stool samples. These diarrheal pathogens are now being recognized in hematopoietic and solid organ transplant recipients, particularly in agricultural areas subject to episodic flooding of potable water supplies. Gastrointestinal cytomegalovirus is an important cofactor to many of these agents.

2. *Pneumocystis carinii*

Pneumocystis carinii was described in 1909 by Chagas and again in 1910 by Carini. *P. carinii* was not recognized as a pathogen of humans until 1942. The first clear association of *P. carinii* with human disease was in 1951, when Vanek and Jirovec[8] found the organism in the lungs of malnourished infants and neonates with an "interstitial plasma cell pneumonitis."[9–16] This unusual disease had been associated with epidemics of pneumonia in malnourished children in the aftermath of each of the major wars.[13,17,18] *Pneumocystis carinii* was first recognized in patients receiving corticosteroids and chemotherapeutic drugs in the 1950s with clusters of cases in clinical oncology centers in the 1970s.[9,10,18–25] The emergence of *P. carinii* as a major pathogen of individuals with AIDS has revolutionized the approach to diagnosis and management of patients with *Pneumocystis* pneumonia.[26–30] *Pneumocystis* also has emerged as a major pathogen in solid organ transplant recipients.[29,30] The approach to the prevention of *Pneumocystis* pneumonia has evolved based on experience in AIDS and due to the large number of individuals intolerant of first-line therapies (due to sulfa drug intolerance).[29,30] Recommendations for the prevention of *Pneumocystis* infection have changed with the development of highly active antiretroviral therapies (HAART) for HIV infection.

2.1. The Organism: Taxonomy and Life Cycle

The taxonomic position of *P. carinii* remains uncertain. The organism bears resemblance to both the fungi and the protozoan parasites.[26,31–34] The appearance of the organism *in vivo* is most similar to that of the protozoa, including the thick-walled cyst form with multiple internal sporozoites and the small, thin-walled trophozoites (Fig. 1). Antimicrobial agents used to treat protozoan infections including *T. gondii* and malaria have been successful in the treatment of *Pneumocystis* pneumonia. By contrast, the cyst wall contains β-1,3-glucans and stains with both methenamine silver and the periodic acid–Schiff (PAS) stains typically used for fungi. Two important enzymes of folate metabolism (dihydrofolate reductase and thymidylate synthase) are encoded on separate genes encoding distinct proteins. This contrasts with one gene encoding a bifunctional protein (both enzymatic activities) in the protozoa.[35,36] The airborne spread of infection supports identification with the dimorphic fungi. Molecular studies of the organism suggest that *Pneumocystis* is more closely related to the fungi than to the protozoan parasites.[31–34] Phylogenetic mapping based on ribosomal messenger RNA sequences also places the organism more closely with the yeasts than with the protozoa.[31,34] On the basis of the common derivation of both the fungi and the protozoa from the classic *Protista*, it may well be that *Pneumocystis* represents a unique phylogenetic niche and will bear relationships with multiple

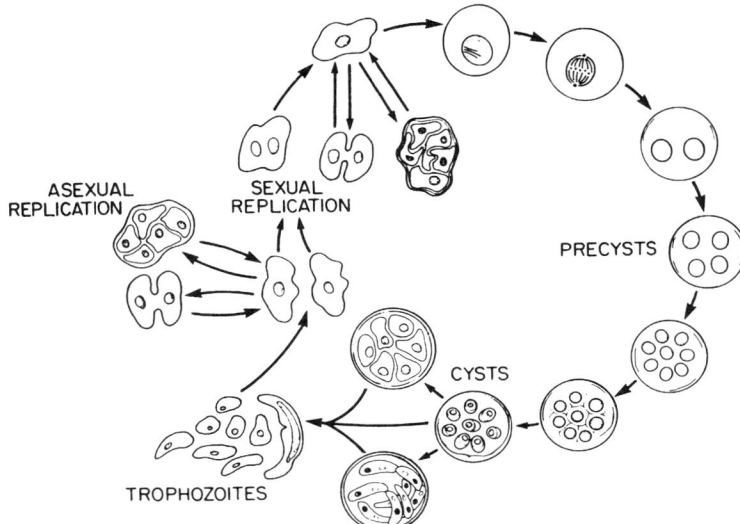

FIGURE 1. Life cycle of *Pneumocystis carinii*. Both sexual and asexual reproduction have been postulated. Glucan synthesis is necessary for cyst wall formation. Only the cyst form stains with methenamine–silver stains.

groups of organisms including the fungi, protozoa, algae, and slime molds (*Dictyostelium*). Further genetic data may clarify these questions.

Pneumocystis carinii is an extracellular organism that appears in three forms in the pulmonary alveolus (Fig. 1). Up to 95% of the organisms are trophozoites: motile, pleomorphic, thin-walled (20 nm) nucleated organisms 2–6 μm in diameter with pseudopodia and a dense covering of small filopodia ("tubular expansions") of uncertain functions.[37–39] The trophozoite contains a nucleolus, nuclear pores, primitive mitochondria, endoplasmic reticulum (ER), and ribosomes, but apparently lacks Golgi, flagellae, and cilia. The "cyst" form is a thick-walled (100–200 nm) sphere 4–7 μm in diameter that contains up to eight internal daughter cells called "intracystic bodies" or "sporozoites." The cyst wall has an electron-lucent middle layer that is stained by the methenamine–silver technique and is absent in trophozoites (Fig. 2). Thus, the silver stains commonly used to detect *P. carinii* in tissue samples or in sputum samples will detect only 3–10% of the organism burden. The sporozoites each have a nucleus, mitochondrion, and large numbers of ribosomes and ER. Intermediate forms between the trophozoite and the cysts have been termed "precysts." These have an oval shape, intermediate cell wall thickness, loss of tubular expansions, and occasionally a nuclear "synaptonemal complex" consistent with meiotic division. It is postulated that the eight daughter nuclei are the product of two meiotic divisions and one mitosis (see Fig. 1). While some reports exist of intracytoplasmic location for *P. carinii*, this observation has not been common. In general, the organisms are

embedded in a layer of alveolar material along the epithelial surface. Some organisms are seen surrounded by cytoplasmic protrusions from type I epithelial cells and within vesicles of alveolar macrophages. The trophozoites are often closely adherent to the epithelial surface with interdigitation of the cell membranes.[38,40] In areas of epithelial cell loss, organisms adhere to the basement membrane. The mechanisms of cell injury and the nature of the interaction between the lung cells and *P. carinii* are unknown. Productive culture *in vitro* has not been achieved without cell contact.

The absence of a continuous *in vitro* cultivation system for *P. carinii* has made studies of the life cycle difficult.[38,41,42] Observation of the organism in tissue sections or with a feeder layer of mammalian tissue culture cells *in vitro* suggests the scheme diagrammed in Fig. 1. Both sexual and asexual replication have been postulated. Trophozoites mature into the aforementioned early cyst forms ("precysts") with up to eight visible nuclei and a thick outer cell wall. It is the cell wall maturation step that appears to be blocked by inhibitors of glucan synthesis. This group of glucan synthase inhibitors (echinocandins and others) blocks the production of cysts *in vivo* and inhibits increases in organism burden during exposure to these agents. Cell membranes form around each of the internal nuclei of cysts, forming the internal "sporozoites." The cysts rupture to release immature trophozoites of a variety of shapes and sizes to restart the cycle. *In vitro* cultivation for periods up to 10 days has been achieved using rat-lung-derived organisms cultured on a variety of mammalian cell lines. Despite extensive efforts, the system has been improved very little since the

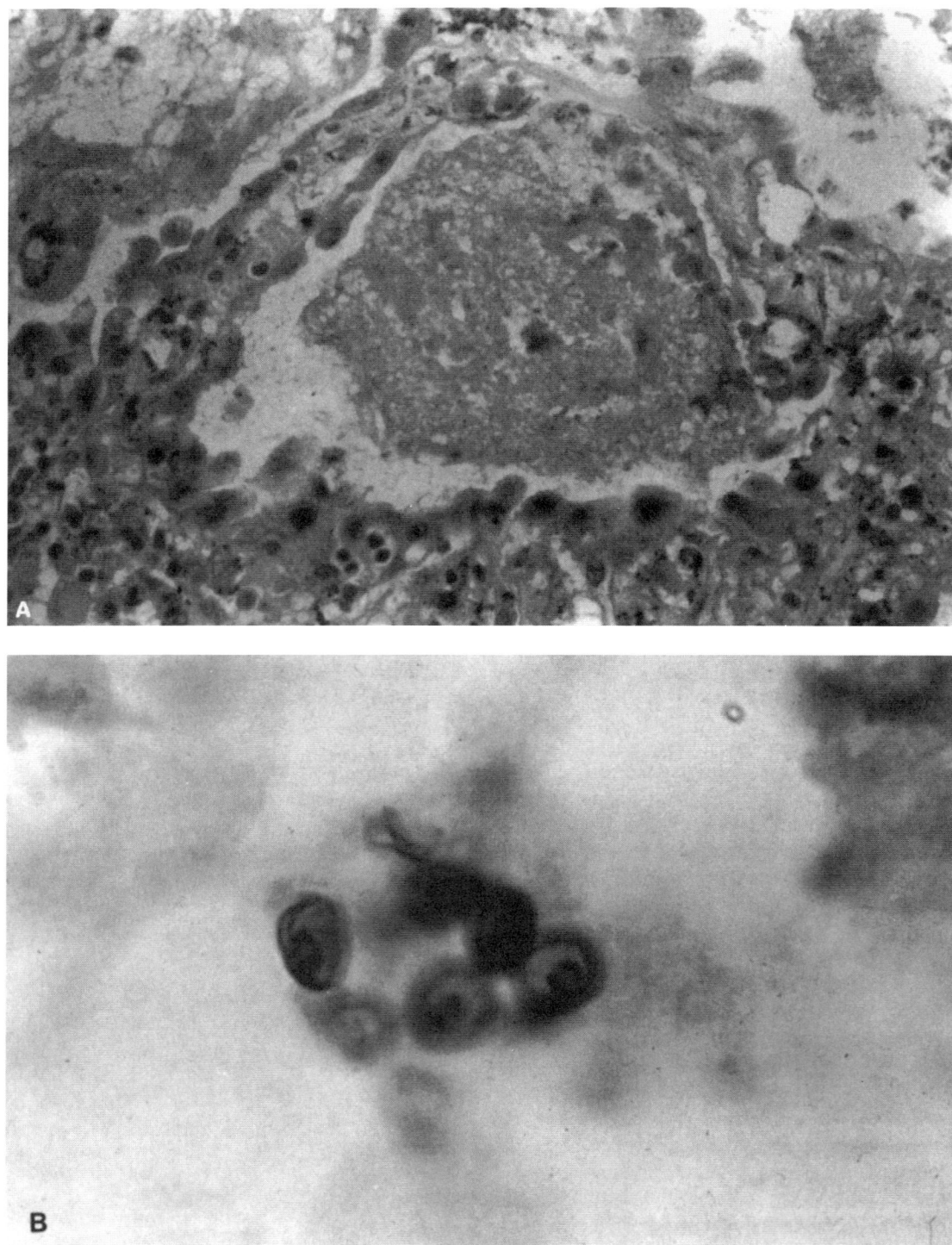

FIGURE 2. (A) Hematoxylin–eosin stain of *Pneumocystis*-infected lung demonstrating pathognomonic intra-alveolar "frothy" material, interstitial widening, and minimal inflammatory cell infiltration. The biopsy was obtained from a patient with AIDS. (B) Induced sputum examination reveals multiple cyst forms stained with a rapid methenamine–silver stain. Immunofluorescent staining is preferred for rapidity and because both cyst and trophozoite forms of *P. carinii* are detected.

first *in vitro* cultivation by Pifer *et al.*[41,43] and Latorre *et al.*[44] in 1977.[45,46] Continuous axenic culture of organisms has not been achieved.[42] The culture systems have assisted in "cleaning up" preparation of organisms that are contaminated by host lung cells and proteins and have been useful in drug screening assay systems. Viability tests for *P. carinii* are in their infancy. Reviews of the biology of *P. carinii* have been published.[25,26,40,45]

2.2. Epidemiology

2.2.1. Animal Studies and Serology

The natural reservoir for *P. carinii* is unknown. The association of protein-calorie malnutrition and of immune suppression with the development of *P. carinii* pneumonitis has been documented in the rat and mouse models of this disease.[40,46–48] The rat model has changed little since the description of the induction of *P. carinii* pneumonia in animals treated with cortisone acetate by Weller in 1955 and by Frenkel *et al.*[49] in 1966.[50,51] The animal model has been modified to utilize "virus-free" rats using transtracheal inoculation of *P. carinii*; this modification results in fewer infections in these immunosuppressed animals due to other pathogens and a more consistent level of infection.[52] These studies suggested that pneumonitis was the result of the emergence of latent infection during immune suppression. Hughes and others also have demonstrated aerosol transmission of the organism in the animal model. Clusters of infection in clinical oncology centers and serologic studies support the aerosol transmission of infection from environmental or human sources or both. Recent studies using the corticosteroid-treated animal model of infection suggest that few organisms (<100 cysts) are needed to cause infection in the immunocompromised host.

The role of T lymphocytes in protection against infection is best illustrated by the use of cyclosporin A in rats and the use of antibodies to T-helper lymphocytes (CD4+) in mice to deplete the host immune response.[53,54] The protective effect of the passive transfer of T lymphocytes in the mouse model also supports the primacy of the T-cell response in prevention of *P. carinii* infection.[55,56] Augmentation of the macrophage response using interferon-γ (IFN-γ) appears to reduce the amount of antibiotic needed to clear infection.[57,58] The roles of colony-stimulating factors [granulocyte- and granulocyte-macrophage-stimulating factors (G-, GM-, and M-CSF)] in the clearance of infection are not yet known. However, mice deficient in GM-CSF have enhanced susceptibility to infection. Passive immunization with monoclonal anti-bodies is partially protective against *P. carinii* infection, suggesting a role for both cellular and humoral immune mechanisms. A major role for animal models has been the evaluation of therapies for the treatment of *Pneumocystis* pneumonia. The efficacy of antibiotics in the rat model has been shown to correlate with successful outcome in clinical applications. Most of the alternative agents available for treatment of *P. carinii* (aerosolized pentamidine, clindamycin–primaquine, dapsone–trimethoprim, atovaquone, echinocandin and other glucan synthase inhibitors, azithromycin, trimetrexate, erythromycin–sulfa) have been developed and tested using the rat and mouse models.

A limited number of antigens have been detected on *P. carinii*.[59–63] Monoclonal antisera raised to these moieties have been useful in the development of immunofluorescent staining of clinical specimens. The major antigens detected in human organisms by Western immunoblotting are of molecular weights 110–116, 50–55, 60–65, 35–45, and 22–25 kDa. These molecules are poorly soluble and very "sticky," accounting for the low level of antigenemia seen in *Pneumocystis* pneumonia and the clumping of organisms. A number of other antigens are variably detected. There is variation in the pattern of glycosylation of *P. carinii* antigens isolated from different species. The role of these antigens in immunity to *P. carinii* is not known.[60,63,64] As was noted, passive transfer of T cells, but not serum, from immune animals is protective against *Pneumocystis* infection.[55] Up to 87% of adults have lymphocyte proliferation in response to stimulation with *P. carinii* antigens.[65,66] Solubilized (and particulate) glycoproteins of the 55–60 kDa and 100–116 kDa ranges stimulate T-lymphocyte proliferation from sensitized hosts.[60,65] In the lungs, antigenic processing by accessory cells (dendritic cells and macrophages) is needed for the generation of *Pneumocystis*-specific T-cell proliferation. IFN-γ and dapsone appear to enhance intracellular killing of *P. carinii* by macrophages. Opsonization by immune serum is not essential but improves phagocytosis by nonimmune macrophages.[67] The organisms are subsequently degraded without evidence of intracellular replication. Infection by HIV decreases internalization but not adhesion of *P. carinii* to alveolar macrophages. The production of cytokines (TNF-α and IL-1β) by macrophages in response to *P. carinii* also is blocked by infection with HIV.

2.2.2. The Susceptible Host

Pneumocystis has been documented as a cause of pneumonia in a broad range of immunocompromised pa-

**TABLE 3. Conditions Associated
with *Pneumocystis carinii* Pneumonia**

Acquired immunodeficiency syndrome (AIDS)
Chemotherapy (especially corticosteroids)
Radiation therapy
Organ transplantation
Prematurity
Malnutrition (protein and calorie)
Malignancies (especially hematopoietic)
Congenital immune deficiency diseases (cellular, humoral, combined)
Collagen vascular disease
Hematologic disorders
Cushing's syndrome
Nephrotic syndrome

tients (Table 3). In the non-AIDS patient, the propensity for the development of *Pneumocystis* pneumonia is related to three factors: (1) the duration of immune suppression or neutropenia, (2) the specific drugs to which a patient has been exposed, and (3) the nature of the underlying disease.[68] The presentation of disease will vary based on the underlying predisposing condition. As a general rule, more severe disease is seen in T-lymphocyte deficiencies or hematopoietic malignancies.[69] Individuals who are malnourished or treated with corticosteroids tend to have greater susceptibility than do patients with other induced immune deficiencies.[70–72] The manifestations of disease are frequently muted in patients receiving a range of immunosuppressive agents or in AIDS patients with advanced disease. Symptoms may emerge with the cessation of immune suppression or with the return of immune function in neutropenic, HIV-infected, or bone marrow transplant or organ transplant recipients.

Serologic studies suggest that seroconversion to *P. carinii* usually occurs some time after the third year of life. The earliest studies of *Pneumocystis* occurring in the epidemic form as "interstitial plasma cell pneumonitis" in malnourished children in orphanages demonstrate that low serum immunoglobulin and low serum albumin levels are associated with both the occurrence and the poor outcome of this form of the disease.[10,11,40] The absence of immunity to *P. carinii* in babies is illustrated in children with congenital HIV infection. In the pre-HAART era, these patients would be expected to survive for less than 1 year. Patients developing AIDS or HIV infection after 1 year of age do somewhat better with pneumocystosis. Both the malnourished infants and congenitally infected AIDS patients develop *Pneumocystis* pneumonia on average by 6 months of age.

Pneumocystis carinii is an organism of low native

virulence. Apparent enhanced virulence of *P. carinii* in some individuals may be a function of coinfection alone, immune suppression due to coexistent viral [cytomegalovirus (CMV), HIV] infection, or the possibility that coinfection with certain agents may enhance the virulence of infection due to *P. carinii*. The association of CMV with *Pneumocystis* is commonly observed.[73,74] This association is largely due to the frequency of CMV infection in the population of immunocompromised individuals. CMV is well known as a systemic immunosuppressive agent, but its effects on the pathogenesis of *Pneumocystis* remain unclear. *In vitro*, CMV infection enhances adhesion of organisms to the feeder cell monolayer. Asymptomatic CMV is often found in respiratory secretions from compromised individuals, but does not appear to increase the morbidity or mortality due to *P. carinii* pneumonia in AIDS. Invasive CMV pneumonitis may increase the severity of *Pneumocystis* pneumonia and requires treatment of both entities.[75]

Reports of *Pneumocystis* pneumonia in immunologically normal hosts have raised suspicions about an increased environmental exposure, possibly due to *P. carinii* in AIDS.[21,76] However, these case reports lack clear documentation of *P. carinii* infection and of normal immune function. Autopsy studies do not support the existence of the organism as a commensal. However, serologic studies suggest that subclinical exposure occurs in most individuals before the age of 5.[62,77–81] Various tests of immunoglobulin G (IgG) serum antibodies [immunofluorescence, enzyme-linked immunosorbent assay (ELISA)] have detected infection in 1–100% of normal adults and in 30–100% of infected adults.[77,80] Significant titers of antibody to *P. carinii* are detected in most patients at the time of diagnosis of *Pneumocystis* pneumonia.[78,79] Detection of circulating antigen would be preferred for establishing the presence of *P. carinii*. Antigen detection systems have low specificity due to impure antigen preparations used to generate the detector antibodies. Improvements in antigen isolation (gel electrophoresis, molecular cloning) and in antibody development (i.e., monoclonal antibodies) may make clinical antigen detection feasible. Polymerase chain reaction (PCR) assays have not been applied routinely for clinical diagnosis.

2.3. Changing Patterns of *Pneumocystis*

Prior to the use of prophylactic antimicrobial agents, *P. carinii* pneumonia was the major complication and diagnostic manifestation of AIDS and a common complication of immune suppression of transplantation and cancer therapies. Without prophylaxis, over 80% of individuals infected with HIV and up to 15% of other immunocompromised hosts would be expected to develop significant *Pneumocystis* pneumonia. Four factors have contributed to the reduction in morbidity associated with *Pneumocystis* infection in developed regions: (1) the incidence of *Pneumocystis* pneumonia and of other opportunistic infections have been reduced in AIDS patients receiving successful antiretroviral therapies (HAART); (2) recognition of *P. carinii* infection as a common presentation of AIDS; (3) improvements in the treatment of viral coinfections (i.e., CMV); and (4) routine use of anti-*Pneumocystis* prophylactic therapies. The incidence of pneumocystosis in AIDS patients is greatest in individuals with fewer than 200 CD4+ lymphocytes/mm^3 or in whom fewer than 20% of circulating lymphocytes are CD4+.[54,69] It is likely that alveolar macrophage activity against the organism also is decreased by HIV and CMV infections. Recent data suggest that primary prophylaxis may not be needed in individuals with persistent improvement in immune functions during HAART therapy as indicated by the absence of detectable HIV and CD4+ lymphocyte counts above 200/ml for 6 months or more. This observation has led to modification of the guidelines for prophylaxis. By contrast, prophylaxis is generally underutilized in patients receiving immune suppressive therapies for autoimmune and connective tissue diseases resulting in preventable infection in these individuals.

2.4. Clinical Manifestations of *Pneumocystis* Infection

The clinical manifestations of *Pneumocystis* pneumonia depend on the patient's condition: preexisting lung injury, immune function, concomitant infections, or drug therapies (Table 4).[61] In the adult without AIDS, *P. carinii* pneumonia is usually subacute to acute in onset, developing over a few days to weeks (Fig. 3). The patient develops progressive dyspnea, tachypnea, cyanosis, and a nonproductive cough. Patients may report low-grade fevers, sweats, or systemic flulike symptoms. Auscultatory findings at the onset are minimal, generally no more than scattered rales and somewhat diminished breath sounds. In the adult with AIDS, the manifestations of the initial episode of *P. carinii* pneumonia, usually dyspnea and fever, evolve more slowly, often over 2–5 weeks.[29,82] Subsequent relapses may evolve more rapidly, especially in the setting of other infections (e.g., CMV) or fibrosis or emphysematous changes from previous infections (Fig. 3).

By the time of hospitalization, arterial hypoxemia is generally moderate to severe and the alveolar–arterial O$_2$ gradient is considerably widened: The degree of arterial

TABLE 4. Clinical Presentation of *Pneumocystis carinii* Pneumonia[a,b]

Clinical signs	Non-AIDS	AIDS (low CD4+ count)
Dyspnea	Common	Common
Cough	Common	Common
Fever	Common	Common
Progression[a]	Rapid (7–21 days)	Gradual (2–5 weeks)
Hypoxemia	Severe	Moderate to severe
Leukocytosis	Often absent (neutropenic)	Often absent (lymphopenic)
Chest radiograph	Diffuse bilateral interstitial infiltrate (variable)	Asymmetric or bilateral interstitial infiltrate (often normal)
Response: initial Rx	Rapid (3–5 days)	Slow (5–9 days)
Recurrence	Unusual	Common[b]
Side effects of therapy	Usually mild	Common; some severe

[a]Altered by type and duration of immunodeficiency. More common/severe with CMV coinfection.
[b]Patients responding to HAART therapies may have fewer or less severe infections.

hypoxemia is out of proportion to the physical and radiologic findings. Dyspnea and arterial hypoxemia often occur in the face of a normal chest radiograph. Pleurisy and pneumothorax may occur acutely. In the patient undergoing chemotherapy, clinical manifestations of pulmonary disease often intensify after the immunosuppressive agents are discontinued, and pulmonary infiltrates appear on the chest radiograph as the host's inflammatory response reemerges (Fig. 4). Conversely, the use of corticosteroids or cyclosporine or tacrolimus therapy may mask the signs and symptoms of *Pneumocystis* pneumonia until late in the course of disease.

Manifestations of extrapulmonary disease due to *P. carinii* depend on the location of infection.[83] Mass lesions of the liver or spleen may be silent. Colonic and omental lesions have caused obstruction, and embolic phenomena have been seen in virtually every organ system including retinal lesions.

In the organ transplant recipient, *P. carinii* pneumonia will occur approximately 6–8 weeks after the initiation of immunosuppressive therapy or during periods of increased immunosuppression for treatment of episodes of graft rejection. Unprophylaxed liver transplant recipients treated with corticosteroids for autoimmune hepatitis prior to surgery, may develop *Pneumocystis* infection within days of transplantation. The incidence of *Pneumocystis* pneumonia depends on the center where

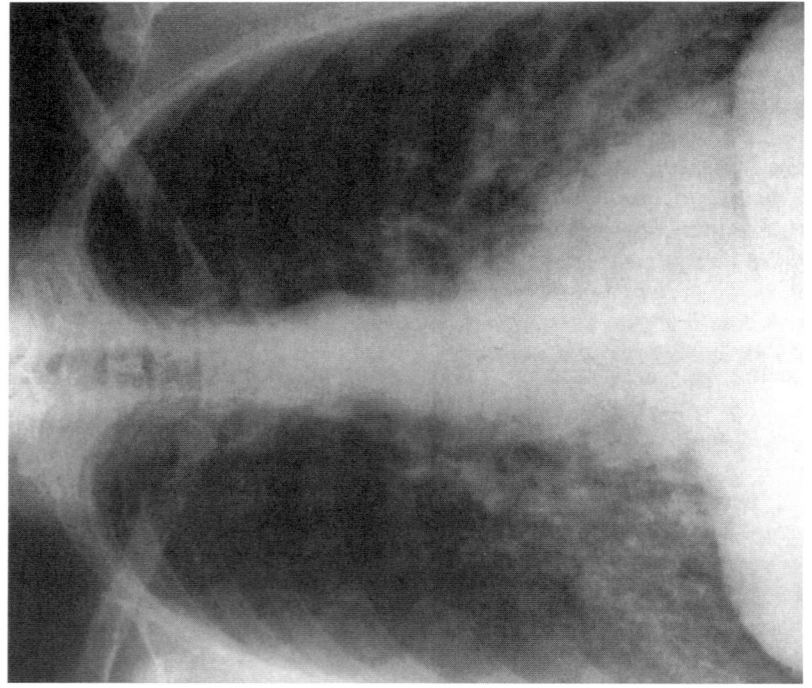

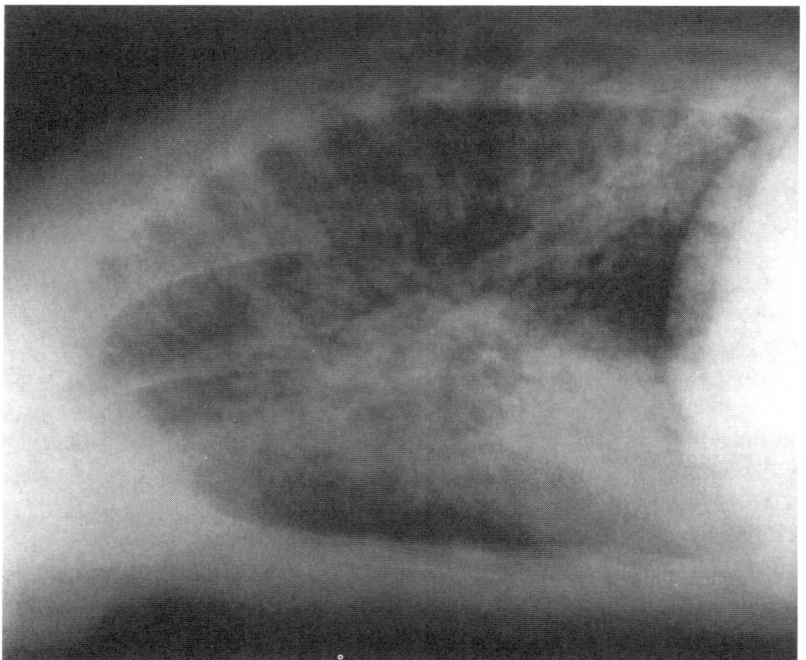

FIGURE 3. Chest radiograph of a 36-year-old man with *P. carinii* pneumonia following chemotherapy for a non-Hodgkin's lymphoma. A complete clinical history appears in Illustrative Case 1 (Section 2.9.2). Typical diffuse and bilateral, fine, interstitial infiltrates are observed.

transplantation is performed and the immunosuppression regimens and prophylactic regimens employed. In patients receiving heart–lung and single-lung transplants, the incidence of asymptomatic *Pneumocystis* isolation from these organs approaches two thirds of the total number of patients in some centers. Of these, approximately half will be expected to develop symptomatic disease in the absence of treatment or prophylaxis. By contrast, among other organ transplant recipients, including heart transplants, only 5–10% will be expected to carry or develop *Pneumocystis* infections. Lung transplant recipients are instructive in terms of the pulmonary inflammatory response to *Pneumocystis* infection. They tend to have a lymphocyte-predominant response to the acute infection, with the recruitment of macrophages during and after therapy. Despite therapy with cyclosporine, lymphocytes are found in the infected transplanted lung in large numbers. These are primarily T lymphocytes with normal helper–suppressor ratios. Over half of this group of patients with *Pneumocystis* pneumonia also will have a secondary bacterial or viral infection. Heart and heart–lung transplant recipients are particularly susceptible to coinfection with CMV. The cytotoxic T-lymphocyte-mediated response to pulmonary CMV may be difficult to separate from organ rejection. A number of centers have noted that patients with *Pneumocystis* pneumonia while on cyclosporine have an increased mortality over other immunocompromised patients with *Pneumocystis*. Bacterial infection of the lung remains more common than *Pneumocystis* pneumonia in the pediatric immunocompromised population.[84] Early signs of pneumocystosis include diarrhea, poor feeding, and coryza. The respiratory manifestations progress to nasal flaring, intercostal retraction, and cyanosis. Fever may be absent. As in the adult, arterial hypoxemia is generally present along with respiratory alkalosis (pH 7.45–7.6; P_{CO_2} 20–40 mm Hg). *Pneumocystis* infection in HIV-infected children less than 1 year of age is a predictor of very poor short-term survival.[84]

2.5. Radiology of *Pneumocystis carinii* Pneumonia

Variability of the radiographic picture matches that of the clinical presentation of *P. carinii*. Like many of the "atypical" pneumonias (pulmonary infection without sputum production), no diagnostic pattern exists for *Pneumocystis* pneumonia on routine chest radiograph. The chest radiograph may be entirely normal despite significant hypoxemia and diffuse parenchymal involvement.[85,86] Diffuse, fine, "ground-glass" interstitial infiltrates with a perihilar predominance are common (Fig.

3).[87] These infiltrates may progress to involve the entire lung with progressive consolidation. "Atypical features" are often seen: small effusions, asymmetry or focal consolidation, small nodules or cavities, linear opacities, pneumothoraces, lymphadenopathy.[88–90] Distortions of the radiographic pattern may occur in the presence of preexisting pulmonary disease (e.g., radiation or cytotoxic injury). Accentuation may be noted in the presence of superimposed viral (CMV) infection or after weaning of immunosuppressive agents (Fig. 3). Abscess formation may be due to *Pneumocystis* alone, when *P. carinii* develops in a preexisting cavity, or with bacterial or fungal superinfection (Fig. 4).

In the AIDS patient, radiographic disease will commonly progress despite appropriate therapy. While this progression may reflect superinfection, it is more often an indication of the greater organism load seen in these patients. The intravenous drug abuser often will have small cysts and bulli in the peripheral lung fields; these changes are more often perihilar with pneumocystosis.[90] The use of aerosolized pentamidine for prophylaxis against *P. carinii* in AIDS patients and in non-AIDS patients has resulted in a series of otherwise unusual radiologic presentations of *Pneumocystis* pneumonia (Fig. 5). Maldistribution of drug may account for the development of *Pneumocystis* only in the upper lobes. This distribution of disease, coupled with the apparent tendency of these patients to develop cystic changes in the parenchyma, also explains a predilection for spontaneous pneumothoraces. Pneumothorax also may complicate the therapy of intubated patients with infection or residual fibrosis from previous *Pneumocystis* infection.[90] The development of extrapulmonary pneumocystosis, while rarely seen in non-AIDS patients, is probably due to the reduced systemic absorption of pentamidine during aerosol administration. The clinical presentation is generally a mass lesion of the liver or spleen.

In transplanted lungs, infection must be differentiated from rejection of the transplanted organ. Rejection may cause nodular and interstitial infiltrates indistinguishable from *P. carinii* pneumonia. These changes are more common in the period 6–8 weeks after transplantation. Infection in these hosts is more often due to CMV or CMV with *P. carinii* or other agents than to any other single pathogen.

Children with epidemic "interstitial plasma cell pneumonitis" on the basis of malnutrition, crowding, and institutional living quarters have a more gradual progression of the chest radiograph. Vascular markings and atelectasis are commonly seen, with hyperinflation and intercostal widening preceding consolidation. In AIDS,

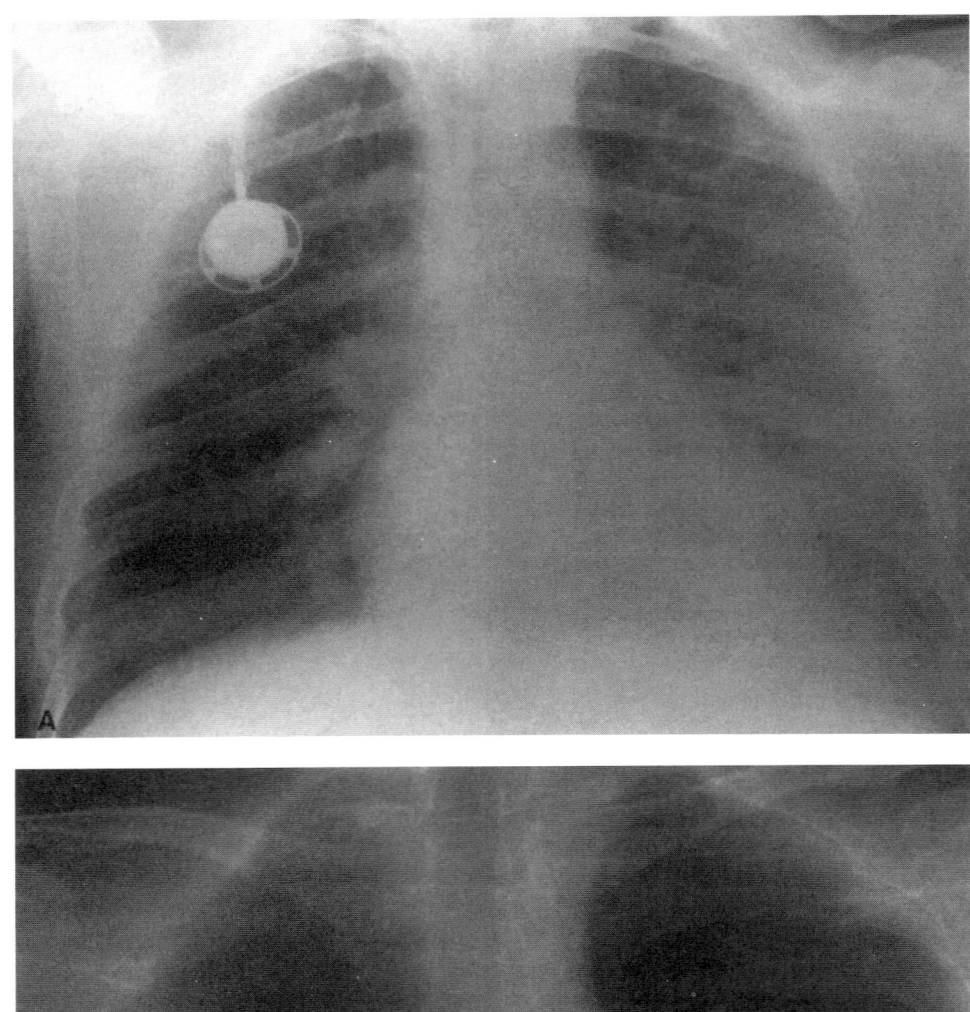

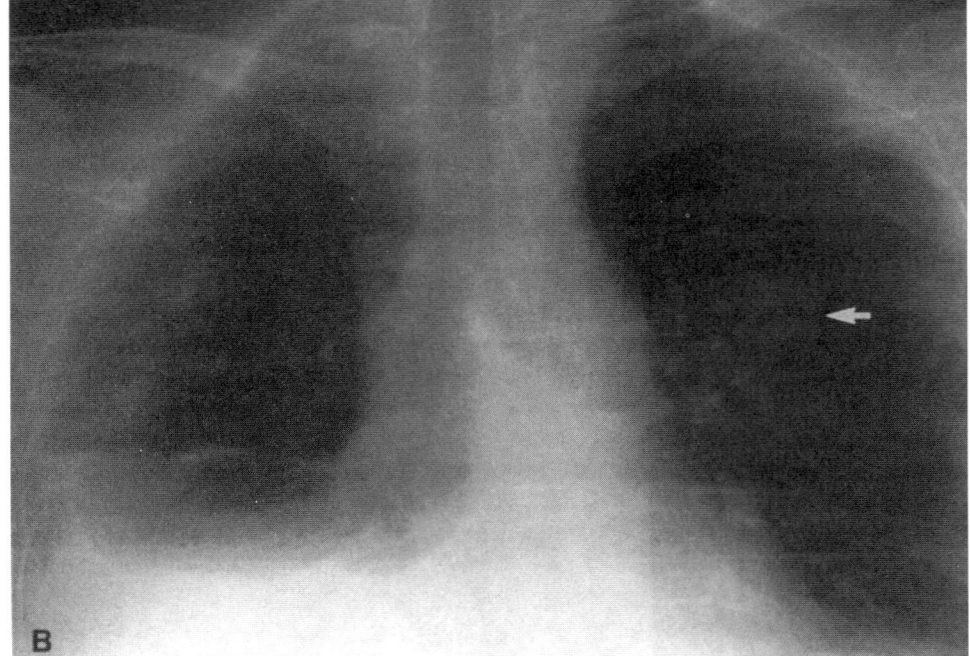

FIGURE 4. (A) Chest radiograph of a 38-year-old man with AIDS who presented with fever, cough, and malaise of 3 weeks' duration. His CD4+-lymphocyte count was 87 at the time of admission. An abscess cavity was seen in the left upper lobe. Bronchoscopic biopsy revealed only *P. carinii*. (B) Chest radiograph of a 43-year-old woman who became febrile and dyspneic 6 weeks following liver transplantation. A small abscess cavity was seen (←) in addition to a benign right-sided pleural effusion. Percutaneous needle aspiration of the abscess cavity revealed *P. carinii*. (C) A CT scan of the chest demonstrates progression of the abscess despite therapy with intravenous pentamidine. The patient recovered completely after 4 weeks of therapy.

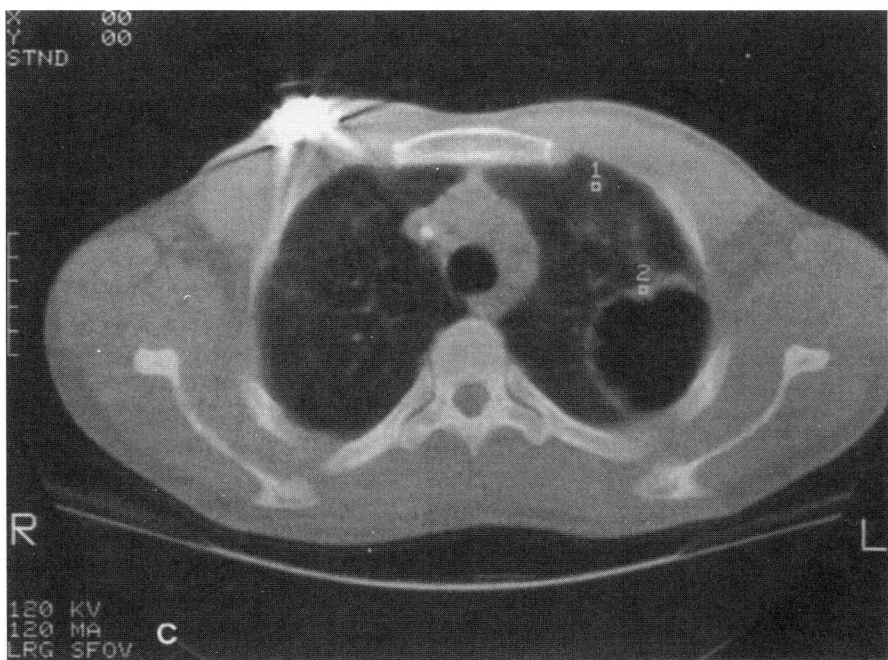

FIGURE 4. (*Continued*)

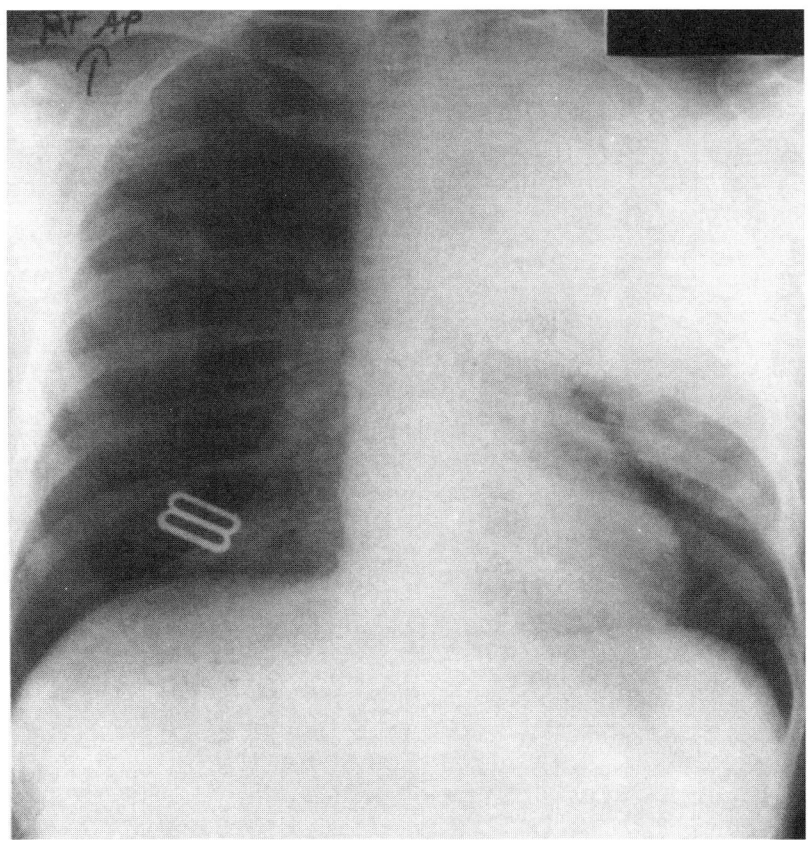

FIGURE 5. Simultaneous upper lobe infection with *P. carinii* and *Legionella pneumophila* in a 46-year-old patient with AIDS while on prophylaxis with aerosolized pentamidine.

lymphocytic interstitial pneumonitis (LIP) may mimic the radiologic appearance of *Pneumocystis* infection. This complication involves a diffuse lymphoid hyperplasia and infiltration of the interstitial space with lymphocytes. Alternatives to plain radiographic imaging include the computerized tomography (CT) and nuclear magnetic resonance imaging (MRI) scans, ultrasound, and nuclear medicine imaging including gallium, radiolabeled immunoglobulin, and white blood cell scans. The tissue–air interface is poorly imaged by MRI and makes this modality less useful. In patients on corticosteroids and in AIDS, the CT scan often will reveal diffuse interstitial and nodular parenchymal involvement of the *Pneumocystis*-infected lungs in the setting of normal or nearly normal routine chest radiographs (Fig. 6). The CT scan is sensitive to emerging or atypical patterns of lung injury, including cysts and microabscesses (Figs. 4 and 6). The correlation of CT scans with histopathology is quite good; imaging demonstrates the patchy distribution of lung involvement and the apposition of normal parenchyma with consolidated tissue. Thus, CT also is useful to direct biopsy procedures. Ultrasound and CT scanning are both useful in the evaluation of extrapulmonary masses due to *P. carinii*. This presentation needs to be separated from other infections (e.g., fungi, mycobacteria) and from lymphoma or metastatic tumor. Multiple small lesions may be seen in the liver or spleen with punctate or rim calcifications. These foci are often better identified by ultrasound than by CT scan. They are clumped hypoechoic masses that develop an echogenic rim during therapy. Biopsy can be performed using ultrasound or fluoroscopic guidance.

Each of the nuclear medicine imaging techniques is limited by the need for tissue inflammation to accumulate the imaging agent and to produce a localized image. In marked neutropenia or uremia or in infections that do not induce much local inflammation, images may not develop. Conversely, the diffuse inflammation that is often observed in the lungs of patients with AIDS (possibly due to cytotoxic lymphocytes for HIV or CMV) may produce false-positive images. Nuclear medicine imaging may detect inflammation earlier than other techniques. Furthermore, the ability to scan the entire body reveals unexpected findings in up to 15% of scans. Gallium citrate (^{67}Ga scintography) scanning, radiolabeled human serum immunoglobulin ([^{111}In]-IgG) imaging, ^{99m}Tc, and diethylenetriamine pentacetic acid (DTPA) scans are abnormal in *Pneumocystis* pneumonia.[86,91–93] The diffuse uptake of ^{67}Ga in the lungs coupled with hypoxemia and a decreased diffusion capacity ($D_L CO$) to carbon monoxide have been used in many centers to make a presumptive

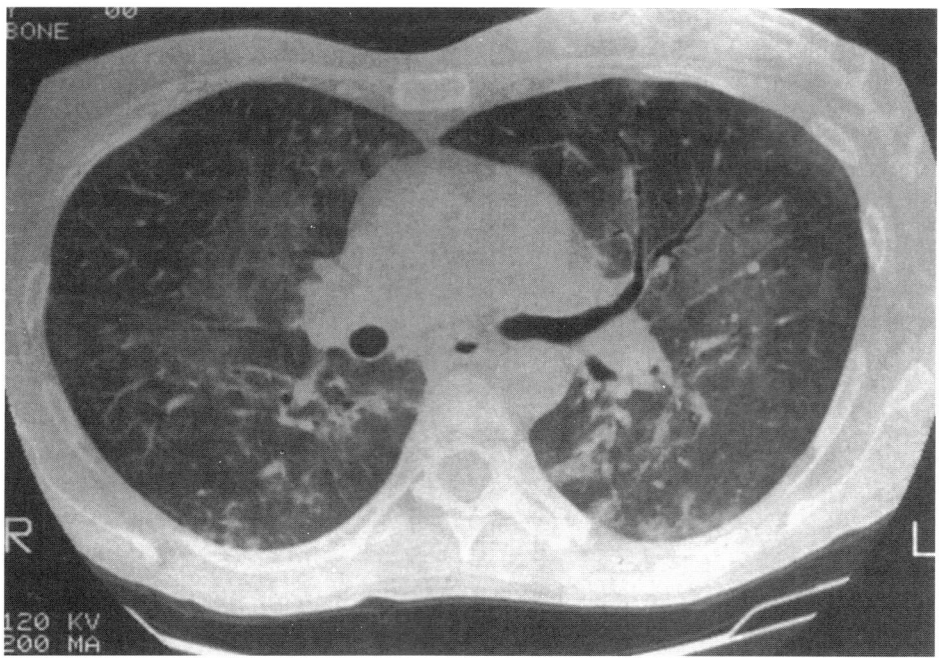

FIGURE 6. CT scan of the chest of an AIDS patient with *P. carinii* pneumonia and a normal chest radiograph. Multiple small interstitial densities and areas of parenchymal consolidation are seen.

diagnosis of *Pneumocystis* pneumonia in AIDS patients. These tests are also abnormal in non-AIDS patients with *P. carinii* infection. The main deficiency of this method, as with all noninvasive imaging techniques, is a lack of specificity. Half the positive images seen in pulmonary gallium scans of AIDS patients will be due to *P. carinii*. Lymph node uptake (as seen in AIDS-associated adenopathy) is common. Drug reactions, adult respiratory distress syndrome (ARDS), CMV, mycobacteria, radiation injury, and other insults may provoke a positive image. However, the image will precede demonstrable infection in many patients by as much as 4–6 weeks. A negative gallium scan is rarely seen (<7%) in pneumocystosis. Normal images should be seen by 3–5 weeks after the start of therapy in the absence of other processes. DTPA scans are a function of fluid movement and label clearance out of the alveolar space. While abnormal in *Pneumocystis* pneumonia, DTPA scans are nondiagnostic. The positron emission tomographic (PET) scan may provide useful information about the course of infection as metabolic labels for the growth of *P. carinii* are developed.

2.6. Laboratory Evaluation

Laboratory evaluation of the immunocompromised patient with pulmonary symptoms provides information about the susceptibility and the prognosis of the patient with *Pneumocystis* pneumonia. The level of serum lactic dehydrogenase (LDH) is elevated in most patients with *Pneumocystis* pneumonia [>300 international units (IU)/ ml]). Very high LDH levels indicate that large amounts of lung tissue are involved, and levels over 600 or 700 IU/ml

carry a poor prognosis. Other diffuse pulmonary processes, including pulmonary emboli with infarction, lymphoma, other pneumonias, and LIP, also raise serum LDH levels. The characteristic hypoxemia of *Pneumocystis* pneumonia produces a broad alveolar–arterial Po_2 gradient; gradients in excess of 30 mm Hg tend to have a higher mortality. Another indicator of diffuse lung injury is an elevation in the level of angiotensin-converting enzyme. This level is also increased by smoking and by sarcoidosis. Pulmonary function testing is not useful diagnostically but may indicate abnormalities in oxygen exchange. However, arterial blood gas measurements are very useful in the management of patients in making decisions in regard to intubation and the use of corticosteroids as adjunctive therapy to antimicrobial agents. Corticosteroids have been shown to be of benefit in hastening improvement in oxygenation in nonintubated patients with a Pao_2 between 35 and 75 mm Hg while breathing room air or a hypoxemia ratio (Pao_2/Fio_2) between 75 and 350. In the markedly neutropenic or lymphopenic patient, consideration also may be given to the use of colony-stimulating factors to augment the host response. There are few prospective clinical data to support use of these agents.

2.7. Histopathologic Diagnosis

Identification of *P. carinii* as a specific etiologic agent of pneumonia in an immunocompromised patient should lead to successful treatment (Table 5). Given the frequent coexistence of multiple processes or infections, the potential toxicity of the agents currently available for the treatment of *Pneumocystis* pneumonia, and the impor-

TABLE 5. Diagnostic Techniques for *Pneumocystis carinii*[a]

Technique	Yield	Complications	Comments
Routine sputum	Poor	Rare	Cultures needed
Induced sputum	30–55%	Rare	First choice; excellent in AIDS
Transtracheal aspiration	Fair	Common; bleeding, subcutaneous air	Rarely worthwhile
Gallium scan, D_Lco	Nonspecific	Injection site	Positive in >95% of infected patients
BAL[b]	>50% (>95% in AIDS)	Bleeding, aspiration fever, bronchospasm	Wedged terminal BAL with immunofluorescence
BAL/brushing	As for BAL alone	As for BAL	Not useful for *P. carinii*
BAL/transbronchial biopsy	Over 90% (all patients)	See BAL; pneumothorax	Impression smears; cultures; pathology
Open lung biopsy	Over 95% (all patients)	Anesthesia, air leakage, altered respiration, wound infection	
Needle aspirate	Up to 60%	Pneumothorax, bleeding	Best in localized disease

[a]All samples should be cultured and stained for bacteria (including mycobacteria), fungi, viruses, and protozoa and handled with caution. Optimal procedures will depend on the locally available expertise.
[b]BAL, bronchoalveolar lavage.

tance of a specific etiologic diagnosis in the compromised host, it is advantageous to have histopathologic confirmation of the diagnosis prior to initiating therapy.[37,94] In the absence of data suggesting that antibiotic-resistant organisms contribute significantly to failures of clinical therapy, the absence of a clinical response to first-line therapy in the setting of known *Pneumocystis* infection should suggest the presence of another simultaneous process. Further, in the non-AIDS patients, no more than 15–25% of pulmonary infiltrates are caused by *Pneumocystis*. The broad antibacterial spectrum of trimethoprim-sulfamethoxazole (TMP-SMX) may delay or obscure the ability to make an alternative diagnosis. It may be undesirable in some patients to use invasive techniques to obtain samples for the diagnosis of *Pneumocystis* pneumonia. However, in compromised hosts, empiric therapy must be balanced against the possibility of misdiagnosis, suboptimal or delayed therapy, and/or the avoidable toxicities of antimicrobial agents. In AIDS patients with depressed CD4+ lymphocyte counts, the frequency of *Pneumocystis* pneumonia in patients not receiving prophylaxis may make a therapeutic trial more appropriate than invasive diagnostic tests. The optimal approach therefore must be based on the patient's clinical condition. For patients treated empirically, the physician must have a low threshold to adopt a more invasive posture should the clinical situation deteriorate. A distinction should be made between the diagnosis of *Pneumocystis* infection in AIDS and in non-AIDS patients (see Table 4).[82,95,96] The burden of organisms in infected AIDS patients appears greater than that of other immunocompromised hosts. Thus, the identification of organisms by noninvasive techniques is more often achieved in the AIDS patient (see Fig. 2). In general, noninvasive testing should be attempted to make the initial diagnosis of *Pneumocystis* pneumonia, but invasive techniques should be used when necessary and clinically feasible to identify problems such as carcinoma impinging on the airway, viral or fungal coinfection, pulmonary embolism, or congestive heart failure. The most commonly used techniques in order of increasing invasiveness are outlined in Table 5. It is important to adapt these recommendations for the techniques available at a given institution and for the relative skill of the practitioners involved in providing these diagnostic techniques. The yield of diagnostically useful material is generally greater from tissue biopsies than from induced sputa or bronchoscopy specimens. Suspicion of *Pneumocystis* pneumonia should lead to early, invasive diagnosis in the non-AIDS-compromised host. The response to therapy decreases over time and empiric therapy (e.g., TMP-SMX-induced nephrotoxicity and hemato-

poietic suppression in patients receiving cyclosporine) may introduce avoidable toxicities.

2.7.1. Histology of Infection

The diagnosis of *P. carinii* infection has been improved by the use of induced sputum samples and of immunofluorescent monoclonal antibodies to detect the organism in clinical specimens.[61] The recognition of small numbers of organisms is of uncertain diagnostic value in an asymptomatic individual or without a history of prior PCP. This is to say that given the presence of both cellular and serologic exposure to *Pneumocystis* in the general population, it might be expected that *Pneumocystis* could be isolated in some nonimmunocompromised host as either a nonpathogen or during a minor infection in an immunologically normal host. However, therapy should be initiated with the isolation of this organism in an individual with altered immune function, especially T-lymphocyte function.[69] Conversely, the identification of this organism in a normal host should initiate a search for immune deficiency.

In the lungs, *P. carinii* produces a characteristic interstitial and alveolar infiltrate (see Fig. 2).[38] This infiltrate is diagnostic even in the absence of clearly identifiable organisms. In the malnourished infant or neonate, the reaction to *Pneumocystis* pneumonia ("epidemic pneumocystosis") is primarily a disease of the interstitium. The pathology of interstitial plasma cell pneumonia of the neonate includes interstitial edema with infiltration of plasma cells and lymphocytes with a characteristic frothy exudate in the alveolar space. In the immunosuppressed adult and child, the disease tends to be more alveolar. The alveolar space is filled with a frothy eosinophilic material that contains organisms and debris of macrophages and alveolar epithelial cells as well as edema fluid and protein.[67] The distribution of disease is often patchy, with normal lung adjacent to areas of dense consolidation. Identification of organisms requires special stains. The most commonly used tissue stain is the methenamine–silver, which stains only the cyst forms of the organism (Fig. 2). Because cysts represent only 5–10% of the total infectious burden of *Pneumocystis* in the lungs, the silver stain greatly underestimates the organism load. To identify the presence of trophozoites, a polychrome stain needs to be done, usually on impression smears made from the cut surface of a lung biopsy specimen or from sputum smears. These are discussed in some greater detail below.

The primacy of the interstitial injury probably accounts for the marked hypoxemia seen in *Pneumocystis* pneumonia. While early disease is characterized by clumps of organisms at the alveolar epithelial surface, progressive infection causes epithelial injury and sloughing with interstitial cellular infiltration. In normal animals, *P. carinii* elicits primarily a polymorphonuclear leukocyte response in addition to alveolar macrophages early in disease. In the T-cell-deficient host, the inflammatory response is muted. The nature of the infiltrate depends on the nature of either the underlying immune defect or the immune suppressive regimen that is being used. The pathognomonic frothy alveolar infiltrate should be distinguished from hyaline membranes that may line alveoli in oxygen toxicity, alveolar proteinosis, or the adult respiratory distress syndrome (ARDS). All these conditions can coexist with *Pneumocystis*. In pediatric AIDS patients, LIP (see Section 2.5) may occur in the absence of clear infectious etiology. This is a systemic proliferation of lymphocytes and of lymphoid tissue, but may produce the same radiologic picture as *Pneumocystis* pneumonia in children with AIDS.

In tissue sections or on smears, *P. carinii* may be demonstrated by a variety of staining methods. Direct immunofluorescent staining of organisms using monoclonal antibodies is very useful for screening-induced sputum specimens. These antibodies generally bind both cysts and trophozoites. The cyst wall can be displayed by a variety of staining techniques; of these, the Gomori methenamine–silver nitrate method (which stains organisms brown or black) is most reliable, even though it is susceptible to artifacts. Sporozoites and trophozoites are stained by polychrome stains, particularly the Giemsa stain. The Giemsa, Wright's, toluidine blue O, or Grocott's rapid silver stain technique is most useful in dealing with the lung imprints, bronchial lavage fluid, or pulmonary aspirates. Rapid polychrome staining (Diff-Quick, American Scientific Products, Inc.) and a rapid silver-staining technique are useful in screening smears. When a silver stain is used, a counterstain such as Gram's, Wright's, Giemsa, hematoxylin, or trichrome may be required to identify intracystic bodies and to distinguish cysts from red blood cells and yeasts.

Following the resolution of acute infection, interstitial fibrosis and small areas of emphysema are often seen. The relative roles of *P. carinii*, drug therapy, and concomitant infection (e.g., HIV, CMV) in this pulmonary picture are unclear. In AIDS patients, residual organisms are commonly detected months after the completion of successful therapy. These organisms do not correlate with the incidence of recurrent disease and are not thought to represent "resistant" organisms.

Extrapulmonary disease has been reported in both AIDS and non-AIDS patients. Extrapulmonary organisms occasionally have been identified in lymphoid tissue, blood, bone marrow, liver, spleen, heart, kidney, pancreas, adrenal, thyroid, thymus, mesentery, ear, and eye tissue.[97–100] In extrapulmonary sites, care must be taken to avoid confusing yeast forms with *Pneumocystis*. In the AIDS patient population, dissemination is most often associated with prophylactic therapy with aerosolized pentamidine or with the absence of prophylaxis against *Pneumocystis* pneumonia. The patients present with mass lesions in the liver or spleen and may develop ischemic injury when clumps of organisms embolize to small blood vessels.[99] These lesions must be biopsied to distinguish them from metastatic tumor, lymphoma, or focal fungal infections.

2.7.2. Sputum Examination

Sputum collected for routine bacterial and fungal stains and cultures is rarely usable for the diagnosis of *Pneumocystis* pneumonia.[37,101] The technique of sputum induction has been very useful in the diagnosis of *Pneumocystis* infection in all immunocompromised individuals when coupled with the use of immunofluorescent antibodies for the detection of *Pneumocystis* in these specimens.[95,102–106] Sputum induction has become the diagnostic technique of choice for *P. carinii*. It should be noted that many bacteria do not grow *in vitro* after exposure to hypertonic saline, so it is important that routine sputum collection be utilized for bacterial and fungal diagnosis. Patients are exposed to aerosolized hypertonic saline or water for up to 30 min and smears are prepared from the mucoid portion of the collected specimens. Smears can be prepared in a number of ways, including after treatment of the specimen with a mucolytic agent (acetylcysteine, Mucomyst) or dithiothreitol just prior to making the smear. The cytocentrifuge also has been useful for this purpose. Smears should be stained with Giemsa or Diff-Quik stains for the intercystic bodies or with toluidine blue O or rapid silver stain, which stain the cyst wall (Fig. 2). Because cyst stains detect only 5–10% of the total organisms, the Giemsa stain is preferred over the more complex silver stain. However, the Giemsa stains are difficult to read. This problem has been overcome by the use of monoclonal antibodies directed against surface epitopes from *P. carinii*.[101,104] With some experience, these commercially available kits are easy to use with a relatively low level of background. The use of immunofluorescent microscopy should increase the detection of *Pneumocystis* by up to 10–20% over conventional staining. The same techniques are used to process bronchoalveolar lavage specimens. It is advantageous to concentrate these specimens using a cytocentrifuge or a microcentrifuge prior to preparing smears due to the effect of large fluid volumes associated with bronchoalveolar lavage.

2.7.3. Fiberoptic Bronchoscopy

In experienced hands, pulmonary bronchoscopy with multiple biopsies will provide the diagnosis of *Pneumocystis* pneumonia in over 90% of all patients.[94,96,107–112] Wedged terminal lavage in aliquots of at least 50 cm^3 in at least three aliquots should be sufficient to detect *Pneumocystis* infection without biopsy in over 80% of all patients and in up to 95% of patients with AIDS. The presence of other pathogens in lavage specimens is often difficult to interpret. For example, the frequent colonization of the upper airway with *Candida* and the frequent isolation of CMV from such samples is of uncertain importance without histopathologic confirmation. Further, the ability to use bronchoscopic lavage for diagnosis is completely dependent on the skill of the laboratories handling the specimens. Biopsies are not generally needed to make the diagnosis of *P. carinii* pneumonia in AIDS, but will often provide useful information about the patient's status in regard to interstitial injury after chemo- or radiotherapy, viral infection, ARDS, or response to therapy. The complication rate is institution-dependent, but generally low. Biopsies (open or bronchoscopic) may be preferred if the clinical laboratories lack experience with *P. carinii*.

2.7.4. Transtracheal Aspiration and Percutaneous Needle Aspiration

Transtracheal aspiration for the diagnosis of *Pneumocystis* infection is probably unnecessary given the advantages of immunofluorescent staining coupled to induced sputum or bronchoscopy. The incidence of complications outweighs the potential benefit of the rapid production of a diagnostic specimen. Even in experienced hands, the diagnostic yield of tracheal aspiration is lower than that achieved by induced-sputum examination when both are coupled to immunofluorescence microscopy.

Radiologically guided percutaneous needle aspiration of the lung produces diagnostic specimens in up to 60% of patients with *P. carinii* pneumonia.[113] This technique is useful both in diffuse lung disease and in the evaluation of focal and peripheral processes seen on chest radiograph or CT scan. Pneumothorax is common (up to a third of cases in some series); 20% of pneumothoraces require chest-tube insertion. Inadequate specimens and bleeding are the other main complications of this technique.

2.7.5. Open Lung Biopsy and Video-Assisted Thoracoscopy

Surgical open lung biopsy remains the "gold standard" for the evaluation of pulmonary processes in the immunocompromised host.[114] When skilled surgeons perform the procedure, the complication rate of thoracotomy and biopsy should be low. This approach provides the best specimen for cultures and histopathology. The use of video-assisted thoracoscopic biopsy (VATS) directed by radiologic CT scanning generally provides adequate samples for both histologic and microbiologic assessment with minimal morbidity in experienced hands. Impression smears taken from the cut surface of the lung biopsy are often adequate for the diagnosis and treatment of *P. carinii* pneumonia.

2.8. Therapy of *Pneumocystis carinii* Infections

Due to the frequency of *P. carinii* pneumonia in patients with AIDS, the number of therapeutic options available for the treatment of *Pneumocystis* pneumonia has increased.[25,29,30,115] For most of the available antimicrobial agents, the potential for side effects such as rash, hepatitis or pancreatitis, or GI intolerance must be balanced against the potential for bone marrow suppression, which is common to almost all the agents discussed in this section and in Table 6.[29,30,116] A few general points may be made about therapy:

1. The most effective systemic therapy for the treatment of *P. carinii* pneumonia in all patients remains trimethoprim–sulfamethoxazole (cotrimoxazole, TMP-SMX).[117–120] This consideration includes such factors as the rapidity of clinical response and the ease of administration (oral bioavailability).

2. The use of adjunctive therapies (colony-stimulating factors, immune modulators, aerosolized pentamidine, corticosteroids, antibodies) must be tailored to the individual patient. Reduction in immune suppression is key.

3. Experience in AIDS patients suggests that treatment can be continued through the occurrence of mild side effects including rash, mild elevation of serum liver function tests, and slight bone marrow depression. Such treatment may require adjustments in dosage, the interval of administration, or the form of the antimicrobial given. Antimicrobial side effects are reduced in individuals treated with corticosteroids during antimicrobial therapy. These guidelines are not universally applicable to non-AIDS patients, in part because adverse reactions are less common but persistent despite dose adjustments.

4. *Pneumocystis carinii* that is resistant to anti-

TABLE 6. The Treatment of *Pneumocystis carinii*[a]

Agent(s) (route)[b]	Dose	Options/comments[b]
First line[b]		
Trimethoprim and sulfamethoxazole (TMP-SMX) (IV/PO)	15–20 mg/kg per day TMP 75–100 mg/kg per day SMX	Treat through rash with reduced dose or desensitize in AIDS; alternate agents in non-AIDS
Second line		
Dapsone (PO) with	100 mg/day	Methemoglobinemia; G6PD; *may* be tolerated in sulfa allergy
TMP (PO/IV)	15–20 mg/kg per day	
Atovaquone, suspension	750 mg liquid PO tid	Variable absorbance, with fatty food; few side effects
Pentamidine isethionate (IV)	4 mg/kg per day 300 mg/day maximum	Lower dose (2–3 mg/kg); IM not advised; breakthrough in transplant and with CMV
Third line[b]		
Trimetrexate (IV)	30–45 mg/m² per day	Efficacy = pentamidine; anemia
with folinic acid	80–100 mg/m² per day	Marrow toxicity; early relapse
Clindamycin (IV/PO)	450–600 mg q6h	Methemoglobinemia; diarrhea (pyrimethamine for primaquine)
and primaquine	15–30 mg base qd	
Others		
Pyrimethamine	Load 50 mg bid ×2d, then 25–50 mg qd	Not studied fully
with sulfadiazine	Load 75 mg/kg, then 100 mg/kg per day	Maximum 4 g in two doses; up to 8 g (maximum)
Fansidar	Not standardized	Long half-life; not in sulfa allergy
Piritrexim/folinic acid	Under study	Like trimetrexate
8-aminoqunoline	Under study	
Macrolide/sulfonamide	Under study	Synergy; macrolides alone inactive

[a]Adjunctive therapies (see text); corticosteroids (high dose with rapid taper); possibly interferon gamma; granulocyte-macrophage colony-stimulating factor.
[b]Based on clinical judgement of physician; some agents not FDA approved for this indication (ranking of therapies based on author's experience).

microbial agents has been described by a number of authors but in the absence of standardized microbiologic assays has not yet been demonstrated as a cause of clinical therapeutic failure. The apparent failure of an individual patient to respond to therapy may reflect either inadequate serum or tissue levels of antimicrobial, greater degrees of lung injury, or concomitant processes. However, resistance to sulfa drugs has been indicated by mutations in the dihydropteroate synthase gene, which appear to be more common in individuals failing sulfa or sulfone prophylaxis. Switching agents for reasons other than toxicity are not generally recommended unless adequate time (minimum 7 days) with appropriate serum drug levels has been achieved.[116,120,121] While there are patients who appear to "do better" on one agent instead of another, it is much more common to recognize a second process (infection, tumor, allergy, ARDS) complicating *Pneumocystis* pneumonia than "resistant" infection.

5. Coinfection with pathogens in addition to *P. carinii* is common.

6. The duration of therapy in the immunocompromised patient with *Pneumocystis* has not been studied carefully. The use of 14 days of therapy in non-AIDS patients and 21 days in AIDS patients is arbitrary. Shorter courses may well be effective, especially in the setting of antimicrobial agents with long serum half-lives, but only

if secondary prophylaxis is initiated for patients with persistent immune deficiency. Residual organisms present in bronchoalveolar lavage specimens at the completion of therapy are of uncertain importance, but are largely nonviable. These organisms do not correlate with the incidence of recurrent disease. Non-AIDS patients respond to therapy and prophylaxis as well as or better than patients with AIDS, and with fewer adverse reactions.

2.8.1. Trimethoprim–Sulfamethoxazole (TMP-SMX, Cotrimoxazole)

Trimethoprim–sulfamethoxazole is the drug of first choice for the treatment of *P. carinii* pneumonia in patients who tolerate this agent.[24,26,29,30,118–120,122–125] This preference is based on (1) the availability of both intravenous and oral formulations of the drug, which enhances the ease of administration; (2) the ability to follow serum levels of the sulfa component; and (3) the efficacy of this drug for both therapy and prophylaxis. The onset of action for cotrimoxazole is rapid; clinical responses are seen as early as 3–4 days into therapy. Serum levels with orally administered drug are equivalent to intravenous levels, given normal GI function. Therapy is generally initiated with a total dose of 20 mg/kg per day of TMP coupled with 100 mg/kg per day of SMX divided into four daily doses. Peak serum levels are reached within 2 hr after oral administration and probably in the range of 5–15 μg/ml of TMP and 100–150 μg/ml of SMX. Serum SMX levels of over 200 μg/ml are associated with a somewhat higher incidence of drug toxicity, in particular bone marrow suppression. The rapid onset of action of this agent may provide the margin necessary to avoid intubation of the critically ill patient. In the non-AIDS patient, TMP-SMX evokes many fewer adverse reactions than does pentamidine. The rates of adverse reactions in the AIDS patient population are roughly equivalent and amount to 30–50% of all patients who take either agent. A course of 14 days of therapy is adequate if immune suppression can be reduced or reversed; 21 days of therapy is preferred in AIDS patients or patients on chronic immune suppression (organ transplant recipients) in whom immune suppression cannot be varied. Chronic immune suppression requires prophylactic antibiotic therapy in patients who have had an episode of *Pneumocystis* pneumonia.

The proper dosing of TMP-SMX in adults has not been studied.[121,123] The dosing regimens in common use were developed in children with leukemia and have not been reevaluated in adults with any form of underlying immune deficiency. While successful, these levels may be excessive, and it is worth monitoring serum levels at some point during the course of hospitalization. In the immunocompromised host, TMP-SMX covers a broad spectrum of organisms, including *Listeria*, many *Nocardia*, and many common bacterial pathogens including both encapsulated and unencapsulated gram-negative and gram-positive organisms. Thus, there may be unexpected beneficial effects when treating a patient for *Pneumocystis* with this agent. The toxic side effects of TMP-SMX are generally those of sulfa allergy. In the AIDS patient population, the adverse reactions to this drug include reactions to TMP and to the carriers and dyes present in various formulations. Some of the allergies in AIDS patients can be quite severe, including Stevens–Johnson's syndrome, hepatotoxicity with eosinophilia and cell necrosis, erythema multiforme exudativum, and nephrotoxicity. Both components in the combination can produce bone marrow suppression, including thrombocytopenia and neutropenia; these side effects are frequently reversible by reducing the total dose of the drug or supplementing with folinic acid. The bone marrow-suppressive effects are greater in patients with underlying hematologic disorders or those receiving cytotoxic chemotherapy. Folinic acid probably should not be used in patients with acute leukemias. Solid organ transplant recipients, particularly with renal and hepatic transplants, frequently suffer

nephrotoxicity with intravenous TMP-SMX treatment (and often with pentamidine) and may not recover normal renal function subsequent to drug-induced renal injury. In the HIV-infected, transplant, or chemotherapy patient receiving a variety of hematopoietic-suppressive therapies, drug-related toxicities are common and may necessitate the switching of antimicrobial agents.

2.8.2. Pentamidine

Pentamidine isethionate was the first agent employed for the successful therapy of *P. carinii* pneumonia.[10,40,117,126,127] Pentamidine was first administered intramuscularly in an epidemic of infantile *Pneumocystis* pneumonia. In this population, it reduced mortality from 50% to 3.5%. The success of pentamidine therapy in subsequent trials varied widely, with survival rates of 25–85% of affected individuals. While pentamidine therapy is generally successful in up to 75% of individuals, over half will have adverse effects when receiving this drug via the intramuscular route, the major complications being sterile abscesses at the site of injection. Currently, pentamiine isethionate is given intravenously by infusion over 1–2 hr in a 5% glucose solution at a dose between 2 and 4 mg/kg per day. Because of the prolonged half-life of this drug and the high levels of tissue binding, it may be advantageous to begin therapy at the 4 mg/kg per day level and to use lower doses subsequently. Therapeutic efficacy is achieved more slowly than with other agents, often requiring 5–7 days before clinical improvement is observed. Therapeutic levels persist in the lungs long after treatment is completed. Pentamidine may play a role in the reduction of symptoms due to reduced secretion of tumor necrosis factor by macrophages involved in the phagocytosis of *P. carinii*. In Europe, pentamidine methanesulfonate is also available and requires different dosing than the isethionate form available in the United States. Pentamidine is a very useful drug in the treatment of AIDS patients allergic to cotrimoxazole, but is probably a drug of second choice in the non-AIDS patients who tolerate TMP-SMX.

The incidence of side effects to intravenously administered pentamidine is roughly equivalent to that seen with cotrimoxazole in AIDS patients.[120] Adverse reactions are both idiosyncratic and dose-related. Administration of pentamidine in the presence of renal dysfunction is associated with an increased incidence of most of the side effects of pentamidine therapy. These adverse reactions include hypoglycemia, hyperglycemia, neutropenia, thrombocytopenia, azotemia, pancreatitis, nausea, and altered taste sensation. Pancreatic dysfunction is more common after a total dose exceeding 3 g pentamidine. This injury may occur after the cessation of therapy because of the prolonged half-life of the drug, which is frequently over 2 months. Hypoglycemia or hyperglycemia may precede permanent insulin dependence. Pentamidine should be avoided in pancreas transplant recipients due to the potential for islet cell necrosis. Despite initial enthusiasm, aerosolized pentamidine has not proven useful for the initial therapy of *Pneumocystis* pneumonia, but may be useful as an adjunct to therapy. A hepatic metabolite of pentamidine may be responsible for most of the toxicities of this agent. New diamidines with fewer side effects and greater efficacies are under development.

2.8.3. Alternative Regimens

Multiple new drugs and drug combinations are available for the treatment of *Pneumocystis*. Because of the frequency of adverse reactions in AIDS patients, most of these regimens have been used almost exclusively in the subpopulation of AIDS patients with allergies both to TMP-SMX and to pentamidine. Dapsone (100 mg PO per day) has been used in combination therapy with trimethoprim (15 mg/kg per day PO divided into three doses) as an effective alternative oral therapy.[129–132] Many AIDS patients intolerant of sulfamethoxazole will tolerate dapsone (4,4′-diaminodiphenyl sulfone), which is metabolized by the liver ($t_{1/2} \geqslant 30$ hr). However, the long half-life and side-effect profile (neutropenia in 19%, anemia, fever, hemolysis in G6PD-deficiency, rash, hepatitis) may be particularly disadvantageous in the marrow or organ transplant recipient. The absorption of dapsone from the GI tract may be reduced by antiviral therapy with DDI (2′,3′-dideoxyinosine).

Atovaquone (750 mg suspension PO tid) is FDA-approved for the treatment of mild to moderately severe *Pneumocystis* pneumonia. Side effects of atovaquone are relatively uncommon and are generally mild. Comparative trials between atovaquone (tablets) and TMP-SMX suggest that TMP-SMX should be preferred in patients who tolerate this therapy.[141] Bioavailability of atovaquone has been improved by reformulation as a suspension. Up to 7% of HIV-infected patients develop limiting toxicity on atovaquone during therapy (vs. 20% for TMP-SMX); however, significantly more patients failed therapy due to lack of response in the atovaquone group than in the TMP-SMX group. When pentamidine was compared to atovaquone for therapy of mild to moderate infection, lack of response was observed in 29% of atovaquone patients and 19% of pentamidine patients. However, atovaquone was better tolerated with treatment-limiting

side effects in 9% versus 24% for pentamidine. The incidence of rash, the most common side effect of atovaquone, correlates with increasing serum drug levels.[141] Other toxicities include diarrhea, nausea, vomiting, fever, and increased liver function tests.[61,141] Atovaquone will irreversibly stain clothes (yellow), which limits use for prophylaxis in children. Preliminary data in stable organ transplantation patients suggest that there is no interaction between atovaquone and cyclosporine or tacrolimus. Atovaquone is useful for both therapy and prophylaxis in the BMT and solid organ transplant populations. Animal data suggest the possible presence of an interaction between atovaquone and erythromycin that merits further study.

Trimetrexate (45 mg/m^2 per day) *with* folinic acid (80 mg/m^2 per day) has been approved for use in moderately severe pneumonia.[137–140] Trimetrexate is a dihydrofolate reductase inhibitor and is lipid soluble with a serum half-life up to 34 hr. It will produce severe neutropenia in the absence of folinic acid supplementation (which should be continued for 3 to 5 days after cessation of trimetrexate) and in some patients with simultaneous infections due to HIV, CMV, or during therapy with antiviral antimicrobial agents. Side effects include fever, rash, leukopenia, and transaminase elevation. Relapsed infection in AIDS patients has been somewhat more frequent than with other therapies. The survival rate following therapy in AIDS patients is higher with TMP-SMX than with trimetrexate for moderately severe *Pneumocystis* pneumonia. Piritrexim is pharmacologically similar to trimetrexate but has been most useful in combination with a sulfonamide. A variety of chemical modifications of the methotrexate molecule may lead to more selective activity against *P. carinii*.[61]

The combination of clindamycin (600–900 mg IV or PO q6–8h) and primaquine (15–30 mg of base/day PO) is effective in mild to moderate infection.[133–135] No significant differences were observed among treatment groups receiving TMP-SMX, dapsone–trimethoprim, or clindamycin–primaquine for mild-to-moderate *Pneumocystis* pneumonia in AIDS in terms of survival, dose-limiting toxicity, therapeutic failure, or the ability to complete 21 days of therapy.[131] For all three regimens, dose-limiting toxicity was experienced by 30.9% of patients and 6.1% were considered to be therapeutic failures by day 7.[131] Thus, for an individual patient, the side-effect profile is the main determinant of the choice of therapy. The main toxicities of clindamycin include rash (16%), methemoglobinemia, anemia, neutropenia, and the development of *Clostridium difficile* colitis. Pyrimethamine (50–100 mg/day PO after 100–200 mg load) with sulfadiazine or trisulfapyrimidines (4–8 g/day) also are effective but require folinic acid (10 mg/day) supplementation. Pyrimethamine will decrease the renal clearance of creatinine without altering the glomerular filtration rate. The macrolides (azithromycin, clarithromycin) have little efficacy as monotherapy, but appear to enhance the efficacy of sulfamethoxazole. However, this combination provides little benefit over TMP-SMX.

The clinical utility of DFMO (α-difluoromethylornithine) has not been well established. The presence of the target enzyme in *P. carinii* (ornithine decarboxylase, ODC) and activity against polyamine biosynthesis *in vitro* have been demonstrated.[142] Because humans and *P. carinii* share the target enzyme ODC, the differential sensitivity of the organism to DFMO with rapid depletion of polyamines in *P. carinii in vitro* suggests a mechanism of action beyond ODC inhibition. Clinical experience with DFMO as primary therapy for *Pneumocystis* pneumonia has not been encouraging. Newer agents under study include the echinocandins (glucan synthase inhibitors) which block formation of cysts, the 8-aminoquinolines, which have entered clinical trials, the dicationic substituted *bis*-benzimidazoles (antimicrotubule pentamidine derivatives), terbinafine, isoprinosine, bilobalide (a sesquiterpene from Gingko biloba leaves), biguanide inhibitors (PS-15) of dihydrofolate reductase, quinghaosu, albendazole, proguanil, terbinafine, guanylhydrazones, and some nonquinolone topoisomerase inhibitors.

2.8.4. Adjunctive Therapies

Many patients with *P. carinii* pneumonia will suffer disease progression despite appropriate antimicrobial therapy. In this setting, the initial delay of 4–7 days before responding to therapy may necessitate intubation with mechanical ventilation. It may be that the successful killing of intra-alveolar organisms may contribute to the local inflammatory process and further diminish oxygenation. These patients often will require supplemental oxygen and are at risk for bacterial and fungal superinfection after intubation. One approach to this problem has been the judicious use of corticosteroids in selected patients with *Pneumocystis* pneumonia with hypoxemia and prior to intubation early in the patient's course. Clinical trials have demonstrated that corticosteroids administered in the first 72 hr of therapy for *Pneumocystis* pneumonia are of significant benefit in AIDS patients in terms of morbidity, mortality, and the avoidance of intubation in patients with an arterial Po_2 on room air between 35 and 72 mm Hg or with a hypoxemia ratio (Po_2/Fio_2) between 75 and 350.[143–146] Experience with both neutropenic and organ transplant patients with *Pneumocystis* pneumonia has

been equally gratifying. In AIDS patients, the benefits of early steroid therapy were as follows: up to a 50% reduction in patients requiring intubation, a marked reduction in the number of patients experiencing deterioration in oxygenation during the first 7 days of therapy, a reduction in the number of side effects due to antimicrobial agents observed, a significant decrease in patient mortality in the first 84 days after hospitalization (to 50%), and a persistent improvement in exercise tolerance after the completion of therapy. In addition, many of these patients were found to eat better. Patients with undiagnosed CMV adrenalitis also may benefit. The incidence of side effects was surprisingly low when the patient was given a maximum of 14 days of tapering steroid therapy. Patients in whom steroid therapy is not tapered are prone to recrudescence of hypoxemia and of acute pulmonary symptoms. The optimal dose of steroids has not been established. One useful regimen is a dose of 40–60 mg prednisone or prednisolone given orally or intravenously twice a day. After 5–7 days, the steroids are tapered over a period of 7 days to 2 weeks. Predictable side effects of steroid excess are rarely seen with a short course of modest steroid doses. An excess incidence of opportunistic infection, gastric irritation, or acceleration of the underlying disease due to HIV was not observed. Patients did observe an increased incidence of oral herpes simplex with oral thrush, both of which are improved with careful attention to oral care.

An alternative approach to the suppression of the acute inflammatory response is the augmentation of the immune response to *Pneumocystis* using immune modulators.[58] Animal studies and some clinical anecdotes have suggested that interferon-γ (IFN-γ) has the effect of reducing the amount of *Pneumocystis* found in infected lungs, probably by enhancement of the macrophage response. IFN-γ administered either intravenously or by aerosol may accelerate the clearance of organisms without greatly enhancing the local inflammatory response.[57,58] In leukopenic patients or in AIDS, an alternative would be the utilization of macrophage or granulocyte–macrophage colony-stimulating factors (M-CSF or GM-CSF). Unlike the effect of IFN-γ, the effects of colony-stimulating factors on local immunity may require T-lymphocyte function. There is controversy regarding the use of colony-stimulating factors in patients infected with HIV due to increased viral replication in the presence of M-CSF and GM-CSF. It is not yet clear whether or not these effects are of physiologic significance. G-CSF has been used without complication to support leukocyte counts in solid organ transplant recipients in the presence of CMV infection or drug-induced neutropenia.

The role of aerosolized pentamidine in the acute management of patients with *Pneumocystis* pneumonia remains unclear. The distinct advantage of having an agent that does not disseminate systemically and obtains high local concentrations appears obvious. It may be that aerosolization of pentamidine is a useful supplement to intravenous pentamidine therapy in the first few days prior to obtaining good lung tissue levels of this agent. Other immune modulators are being evaluated for the treatment of *Pneumocystis*. Preliminary data suggest that the fluoroquinolones and possibly the echinocandins and pneumocandins will also be useful.

2.8.5. Response to Therapy

The initial manifestations of *Pneumocystis* pneumonia have been altered by prophylactic antimicrobial therapy, by successful highly active antiretroviral therapy (HAART) for AIDS, and by new immunosuppressive regimens, including those with calcineurin inhibitors, mycophenylate mofetil, and antithymocyte globulins. The response to therapy also has changed with these new regimens. While the incidence of *Pneumocystis* pneumonia has declined in response to HAART, when infected, AIDS patients are presenting with more advanced disease and with atypical manifestations (upper lobe disease, multiple infections, extrapulmonary disease) or with a history of adverse reactions to one or another of the primary therapeutic agents. In general, the response of the non-AIDS immunosuppressed patient is determined by the ability to reduce exogenous immunosuppressive regimens. Most non-AIDS immunosuppressed patients do better with initial therapy for *P. carinii* pneumonia than do patients with low CD4+ lymphocyte counts and HIV infection: The response is more rapid and recurrence is relatively uncommon. Failure of a patient to respond to cotrimoxazole therapy in 5–7 days is unusual. However, little benefit is likely to be seen if a switch to pentamidine is made before 7 days into the course of therapy. The chest radiograph may progress while oxygenation and nonspecific indicators of lung injury gradually improve. Failure to improve more often is due to other factors than it is to failure of a given antimicrobial agent. Adding pentamidine to TMP-SMX offers no advantage over simply switching agents, and there may be antagonism between these agents when used in combination. Patients switched from cotrimoxazole to pentamidine for reasons of therapeutic failure generally do less well than patients who can be treated for 2–3 weeks on either agent. Of the newer agents, there appears to be antagonism in the animal model between erythromycin and atovaquone. The echi-

nocandins (β-1,3 glucan synthase inhibitors) inhibit cell wall synthesis, preventing cyst formation. This class of agents may prove to be useful adjuncts to therapy. The clinical assessment of the patient is usually the best guide to subsequent therapy. Patients who fail to respond to antimicrobial agents within 7 days or so are good candidates for bronchoscopy with biopsy or lavage or both to clarify the nature of their progressive pulmonary disease.

The survival of patients on anti-*Pneumocystis* therapy has improved to between 80% and 90% at most medical centers. In the pre-HAART era, AIDS patients receiving zidovudine (AZT) have fewer and less severe episodes of *Pneumocystis* pneumonia. However, of patients developing opportunistic infections, the fraction with *Pneumocystis* remains about the same, at 50–60%. The benefits of AZT therapy appeared to diminish somewhat over time, reflecting viral resistance. Complications are more common in the AIDS population, including a tenfold increase in the incidence of significant skin rash and fever. Hepatic toxicity occurs in up to 20% of these patients. Minor adverse reactions (skin rash or transient liver function test abnormalities) may be due to either component, and these may be reversed by continuing the drug at reduced levels. It is worth checking serum SMX levels if side effects occur. Serious toxicity requires switching to an alternative regimen.

2.9. Prevention of *Pneumocystis carinii* Pneumonia in the Susceptible Host

2.9.1. Patients Requiring Prophylaxis

The spectrum of patients who will require anti-*Pneumocystis* prophylaxis has changed dramatically with newer immunosuppressive regimens for organ transplantation and the treatment of graft-versus-host disease, the use of more intensive chemotherapeutic regimens for malignancy, intensive immune suppression for connective tissue diseases, and conversely, the decreasing incidence of opportunistic infection in AIDS.[392] There are few clear rules regarding patients who "should" be on prophylaxis. However, experience dictates that these might include:

• AIDS patients with CD4+ lymphocyte counts below 200 CD4+ lymphocytes/mm³ blood or less than 20% CD4+ lymphocytes total, rising HIV viral loads, persistent cytomegalovirus infection, or recurrent opportunistic infections suggestive of persistent T-cell defects despite HAART therapy. In individuals who are noncompliant with HAART therapy or for whom HAART is not available, anti-*Pneumocystis* prophylaxis should be attempted.
• Individuals receiving anti-T-cell therapies or cor-

ticosteroids over 20 mg/day of prednisone for a period of over 2–3 weeks (an arbitrary duration consistent with the life cycle of the organism).

• Solid organ transplant recipients, depending on the incidence of infection in the institution (>5–10% without prophylaxis) but up to lifelong for heart, liver, and liver recipients and 6 months to a year posttransplant for kidney recipients. These recommendations are consistent with the periods of greatest risk due to intensity of immune suppression. Any transplant recipient with a history of *Pneumocystis* pneumonia or frequent opportunistic infections, who is receiving prophylaxis or therapy for CMV infection or treatment of acute rejection, merits consideration of *Pneumocystis* prophylaxis. Individuals with chronic graft dysfunction and who are receiving higher than usual levels of immune suppression also merit prophylaxis.

• Use in neutropenic cancer patients is controversial, given the marrow suppression that may result from TMP-SMX use. However, intensive chemotherapy with neutropenia of more than 7–10 days may justify prophylaxis with alternative agents.

In AIDS patients, the incidence of *Pneumocystis* is greatest in patients with less than 200 CD4+ lymphocytes/mm^3 blood or less than 20% CD4+ lymphocytes total. The incidence of primary *Pneumocystis* pneumonia and other opportunistic infections (e.g., CMV and *M. avium-intracellulare* complex) is reduced in proportion to the control achieved of viral infection. Prophylaxis with TMP-SMX reduces the incidence of pneumonia by over tenfold. While these guidelines are useful, the incidence of *Pneumocystis* pneumonia is also high in patients with rapidly progressive immune deterioration or a prior history of *Pneumocystis* infection, or both. Prophylaxis against *Pneumocystis* and other opportunistic infections has changed due to HAART therapy. Recent data suggest that individuals with CD4+ T-cell counts above 200/mm^3 and with viral replication suppressed to undetectable levels by antiretroviral therapy for a period of 3 or more months do not appear to require primary prophylaxis for *P. carinii* pneumonitis (and toxoplasmosis). The caveats to this recommendation are that the follow-up period is relatively brief. Anecdotal reports and nonrandomized trials suggest that sustained counts over 200 CD4+ lymphocytes/mm^3 in individuals with a history of *Pneumocystis* pneumonia may also be protected, but sufficient data do not yet exist. It does appear that significant immune reconstitution occurs in most individuals with AIDS after more than 12 weeks of successful antiviral therapy. However, it is expected that the degree of immune function achieved is inconsistent and in some individuals recurrence of viral infection and opportunistic infections have been observed. Thus, firm recommendations for secondary prophylaxis with HAART must be individualized. A number of AIDS patients have experienced complications of immune reconstitution (*Pneumocystis* pneumonitis, CMV retinitis, MAC lymphadenitis) upon initiation of HAART. These cases probably reflect subclinical infection with symptoms reflecting recrudescence of the immune response. A similar picture may occur if HAART is introduced before completion of therapy for active infection due to *Pneumocystis*, CMV, or tuberculosis.

2.9.2. Antimicrobial Prophylaxis—Agents

Pioneering studies in children with hematopoietic malignancies, especially those on corticosteroid therapy, led to the development of TMP-SMX for the prevention of *Pneumocystis* pneumonia.[71,72,119,124,147,148] Successful experience with prophylaxis at the Saint Jude's Children's Research Hospital (Dr. Walter Hughes) was broadened to include immunocompromised adults and patients with severe combined immunodeficiency syndrome (SCID).[147] The advantages of daily cotrimoxazole (single or double strength) were the prevention of infection due to *Pneumocystis* as well as most *Toxoplasma gondii*, *Listeria monocytogenes*, *Nocardia asteroides*, and common urinary, gastrointestinal, and pulmonary bacterial pathogens. The incidence of side effects was fairly modest with relatively short periods of prophylaxis. Oral administration of 5 mg/kg per day of TMP as cotrimoxazole divided into two daily doses or 150 mg/m^2 per day successfully prevents the development of *Pneumocystis* infection. Even on this modest dosage, bone marrow suppression is relatively common but usually mild. It is equally effective for the prevention of *Pneumocystis* to administer cotrimoxazole 3 days each week (either consecutive or alternative) with one double-strength tablet at bedtime. Lower doses of antimicrobial have also been effective.[147] These doses of TMP-SMX are usually well tolerated, with a minimal incidence of side effects and relatively modest bone marrow suppression. It should be noted that we have observed pulmonary and central nervous system infections due to *T. gondii* and *Nocardia* species in individuals receiving 3-day/week regimens in adult solid organ and hematopoietic stem cell transplantation recipients. Alternative regimens are occasionally necessary in non-AIDS patients requiring prophylaxis. These patients will tolerate atovaquone (1500 mg PO qd), pentamidine 300 mg by aerosol or intravenously every 3–4 weeks, dapsone (100 mg orally a day), or weekly Fansidar (pyrimethamine–

sulfadoxine).[149] The use of these latter two agents should be guided by knowledge of their side effects.

TMP-SMX is the agent of choice for the prevention of *Pneumocystis* infection in patients who tolerate this agent.[150,151] Aerosolized pentamidine has become less popular for prophylaxis due to inconvenience of aerosol administration, atypical presentations of pulmonary and occasionally extrapulmonary pneumocystosis, the lack of activity against *T. gondii* and other potential pathogens, and a lower success rate in preventing *Pneumocystis* infection than TMP-SMX. Lower-dose regimens of TMP-SMX are generally well tolerated (in up to 75–80% of AIDS patients); potential side effects are those seen in therapy. In patients completing a course of cotrimoxazole therapy for pneumonia, prophylaxis should be initiated immediately to prevent sensitization. In patients who do not tolerate cotrimoxazole, pentamidine aerosol has been very useful. This mode of drug administration has been associated with atypical presentations of *Pneumocystis* pneumonia, including extrapulmonary *Pneumocystis* infection and pneumothoraces (i.e., apical *Pneumocystis* infection) in some patients. Successful administration of aerosolized pentamidine depends not only on the dosage and schedule of administration, but also on the nebulizer used to create the aerosol. It has been found that health care personnel may be exposed to both pentamidine and infectious agents from the lungs of patients during the nebulizer treatments. The nebulizer must produce a mist of 1–3 μm droplets to be successful. Careful positioning of the patient assists in the proper distribution of this drug. A tightly closed nebulizer system and dedicated room must be considered for the safe administration of this agent. Patients who are intolerant of aerosolized pentamidine may benefit from the use of bronchodilators prior to administration or may be treated intravenously. The side effects of pentamidine will develop more slowly during aerosolized treatment but may occur after a total dose of 3 g. Aerosolized pentamidine also has the effect of reducing the number of organisms in sputum such that the sensitivity of microscopic diagnosis on induced sputum samples may be reduced. However, this agent has proved very useful in large population studies.

Some of the longer-acting sulfa-derived agents including Fansidar and dapsone have been useful in preventing infection at relatively low cost.[149–152] Breakthrough infections have been seen with dapsone at 50 mg/day doses and side effects are more common at the recommended 100 mg/day dose. This drug is occasionally associated with nausea, asymptomatic methemoglobinemia, and hemolytic anemia, especially in glucose-6-phosphate dehydrogenase (G6PD)-deficient patients. There is some

suggestion that DDI use in the treatment of HIV infection may interfere with the absorption of dapsone from the GI tract. Fansidar administered weekly also has been effective in clinical trials. Concerns about this agent are derived from its long half-life and the occasional episode of severe hepatitis in some individuals taking Fansidar for prophylaxis against malaria. Atovaquone may also be useful for prevention of *Pneumocystis* infection, especially in patients intolerant of sulfa drugs. Atovaquone is also active against *Toxoplasma gondii* and is used at a dose of 1500 mg PO daily.[392] Concerns about the prevention of *Pneumocystis* also have led to considerations about the possible person-to-person spread of this organism. While serologic studies have suggested that such transmission is possible, the strongest suggestion of person-to-person transmission is the increased incidence of *Pneumocystis* pneumonia in non-AIDS patients in institutions caring for AIDS patients with *Pneumocystis* infection. Patients infected with *Pneumocystis* should not share rooms with other immunocompromised patients. This problem merits further study.

Illustrative Case 1

A 36-year-old man with non-Hodgkin's lymphoma was treated with high-dose corticosteroids and cytotoxic chemotherapy. He presented with fevers, a diffuse pulmonary process, and hypoxemia. His chest radiograph [Fig. 3 (Section 2.4)] had both alveolar and interstitial infiltrates at the time of transfer from an outside hospital. The patient has been followed for a stage 1 high-grade B-cell lymphoma diagnosed 2 years prior to admission. The pathology on this tumor included cells varying from small noncleaved, non-Burkitt's cells to immunoblasts. He was initially irradiated in the left groin area but had a recurrence in the right groin 1 year later with the same histology. At the time, his abdominal CT and bone marrow biopsy did not reveal tumor and he started treatment with ProMACE-cyta-BOM: multiple courses of cytoxan, adriamycin, and VP 16, followed by bleomycin, vincristine, methotrexate, and cytosine arabinoside, followed by high-dose prednisone. He had completed four cycles of therapy and was on high-dose prednisone at the time of his admission to the Massachusetts General Hospital. His presentation followed the development of diffuse bilateral pulmonary infiltrates over a period of 4 days. He had been treated with oral amoxicillin–clavulanate. He developed a rapid deterioration of pulmonary function. He was markedly hypoxemic, with diffuse rales and rhonchi, fever to 102°F, with a total white blood count of 2000 (42% polymorphonuclear leukocytes).

After transfer, initial antibiotic coverage was broad. Antimicrobial agents included cotrimoxazole (TMP-SMX) for the presumed diagnosis of *Pneumocystis carinii* pneumonia as well as erythromycin, ticarcillin, gentamicin, and vancomycin. Induced sputum was noted to contain *P. carinii* by direct immunofluorescence. Initial blood, urine, and sputum cultures were negative for other pathogens, including CMV.

When first seen in consultation, he had completed approximately 14 days of pentamidine isethionate after 7 days of TMP-SMX and was on a gradual steroid taper. His course had been complicated by a skin rash and fever, probably due to TMP-SMX, and thrombocytopenia due to pentamidine. Notable in his course was that despite therapy for *Pneumo-*

cystis for 21 days, his chest radiograph had failed to improve, with diffuse infiltrates consistent with adult respiratory distress syndrome. The effects of past chemotherapy also were considered. It was recommended that the patient be taken for open lung biopsy due to the failure of his chest X ray and pulmonary functions to improve and the inability to isolate further pathogens. At the same time, his anti-*Pneumocystis* therapy was stopped (other than prophylactic pentamidine, 300 mg intravenously every 3 weeks). The lung specimen grew CMV overnight via Shell vial. Pathology demonstrated an interstitial process consistent with lung toxicity due to Cytoxan in addition to intracytoplasmic and intranuclear inclusions consistent with CMV infection and rare *P. carinii* cysts. On the basis of these results, he was treated with intravenous ganciclovir, at an initial dose of 450 mg intravenously every 12 hr. His steroid taper was continued. Despite some early radiologic improvements with lysis of his fevers, the chest X ray failed to improve greatly; however, his oxygenation improved markedly during therapy. He completed 21 days of antiviral therapy and was discharged home. He has received ganciclovir and dapsone prophylaxis for subsequent periods of chemotherapy. Bacterial and fungal cultures remain negative.

Comments. This case illustrates the complexity of the management of immunocompromised patients with the "febrile pneumonitis syndrome." The likelihood of significant infection rises with the amount and the duration of immune suppression. The rapid progression of *P. carinii* is typical for this infection in a non-AIDS patient. The frequency of side effects of therapy for *P. carinii* is illustrated by the leukopenia (due to cotrimoxazole) and thrombocytopenia (due to pentamidine) seen in this host. Initial management with oral amoxicillin–clavulanate in a sick, compromised patient and in the absence of microbiologic (Gram's stain) evidence for bacterial infection is worrisome. Delays in the recognition and treatment of *Pneumocystis* pneumonia contributed to this patient's progressive deterioration. The use of corticosteroids in chemotherapy proved to be a microbiologic double-edged sword. Steroid therapy was probably the main factor in the development of infection in this host but it may decrease pulmonary inflammation and transiently improve oxygenation in the acute setting. While well studied in AIDS patients, steroid therapy for *P. carinii* pneumonia has not been examined in the heterogeneous non-AIDS population.

Failure to improve pulmonary function by day 7–10 of therapy for *P. carinii* infection should suggest additional evaluation. Congestive heart failure, pulmonary embolus, and bacterial superinfection [Fig. 5 (Section 2.5)] are common cofactors. Because clinically significant antimicrobial resistance has not been clearly demonstrated in *P. carinii*, switching therapy for reasons other than toxicity is *not* recommended. An aggressive approach to histopathologic diagnosis is often required in the absence of a reasonable microbiologic diagnosis. The ability of the patient to tolerate invasive procedures is best earlier in the course of illness. In this patient, the successful treatment of *P. carinii* pneumonia was masked by CMV pneumonitis and by pulmonary toxicity due to chemotherapy. Treatment of CMV pneumonia with ganciclovir (not an FDA-approved indication) produced a rapid clinical improvement. The likelihood of recrudescence of infection due to *P. carinii* or CMV or both necessitates prophylaxis for subsequent periods of neutropenia and corticosteroid therapy.

3. *Toxoplasma gondii*

Serologic evidence suggests that up to 70% of all individuals are exposed to *Toxoplasma gondii* at some point during their lives. Thus, it is surprising that *Toxoplasma* causes significant infection only infrequently outside the immunocompromised host. In immunocompromised adults and in fetuses, this relatively benign organism causes significant morbidity. Toxoplasmosis represents a spectrum of diseases caused by infection with *T. gondii*. The presence or absence of the organism in tissues ("infection"), particularly in the cyst form, is not indicative of clinical disease ("toxoplasmosis") in the absence of an appropriate clinical presentation. Acute infection in the immunocompromised individual requires prompt diagnosis and therapy to avoid significant injury. The important role of infection of the central nervous system (CNS) by this organism complicates both the diagnosis and the treatment of toxoplasmosis. This problem is further compounded by the lack of reliable and reproducible serologic tests for use in immunocompromised individuals. Since its description in the gondii, a rodent from North Africa, and in rabbits, it has become apparent that there are strain differences among *T. gondii* isolates from various regions of the world.[153,154] This has complicated the development of molecular diagnostic tests.

3.1. The Organism

3.1.1. Life Cycle

Toxoplasma gondii is an obligate intracellular protozoan of the order Coccidia. The multiple developmental stages of *T. gondii* are relevant to the clinical manifestations of toxoplasmosis (Fig. 7). Oocysts initiate the life cycle of *T. gondii* and are produced only in the intestines of members of the cat family. This form of organism is oval in shape, measuring 10–15 μm in diameter, and is produced in the small intestine following both asexual (schizogeny) and sexual (gametogeny) reproduction.

Self-propagating infection (the enteroepithelial cycle) within the intestinal epithelium generates millions of oocysts per day in cat feces. Fecal oocysts will survive for over a year in moist soil. These oocysts undergo sporulation over 2–3 days after excretion and become infective. This form can be ingested directly by humans (fecal–oral route) to initiate infection in the human (intermediate) host. Infected oocysts are killed by boiling or adequately cooking (over 65°C) meats or vegetables.

Proteolytic disruption of oocysts or of tissue cysts occurs after ingestion. The released sporozoites are motile and penetrate nucleated cells throughout the body via direct invasion of local intestinal epithelial cells or to distant sites via the bloodstream and lymphatics. Sporozoites mature into trophozoites within vacuoles of retic-

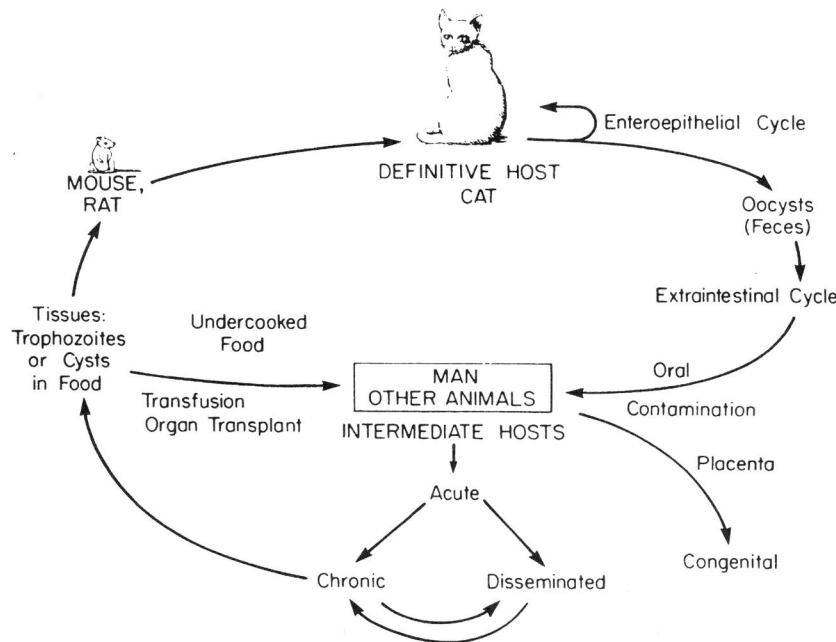

FIGURE 7. Life cycle of *Toxoplasma gondii.*

uloendothelial cells and of other tissues.[155] Trophozoites are 4- to 8-μm crescentic, nucleated organisms that move by body flexion without a flagellum. These trophozoites multiply within a cell by internal division (endodyogeny) until the cell ruptures, initiating a new cycle of invasion.[155] In the face of an immune response and for unclear reasons, infectious spread may be interrupted by the development of tissue cysts. These cysts are large (up to 200 μm) and contain thousands of slowly dividing, relatively inactive trophozoites (sometimes called "bradyzoites"). These cysts act as a reservoir of infection throughout the life of the host. Rupture of cysts due to ingestion of undercooked meats or due to immunosuppression of the host initiates another cycle of infection. Tissue cysts can be disrupted by freezing and thawing or by cooking. Bradyzoites are killed by heat and by normal gastric secretions.

3.1.2. Epidemiology

The infected cat produces millions of oocysts each day for a period of up to 3 weeks. The organism has been found in most animal species that consume either plants or meat. Up to 25% of lamb and pork have been shown to contain tissue cysts. Uncommonly, *Toxoplasma* has been isolated from beef, goat's milk, and eggs, as well as from vegetables consumed by seropositive vegetarians. *Toxoplasma gondii* is found worldwide. Seropositivity increases with age.[153] Infectious oocysts may be carried by coprophagous organisms including cockroaches, flies, worms, and snails. Significant infection has occurred in laboratory workers who have become inoculated with organisms. *Toxoplasma* has prolonged survival in refrigerated blood samples and may be transmitted by transfusion of either whole blood or white blood cells.[156] Transmission of toxoplasmosis has occurred in the setting of organ transplantation into a seronegative recipient as well as by reactivation of latent disease by immune suppression.[157] There are multiple subpopulations of *Toxoplasma* that vary in virulence and may have differing developmental characteristics, including a predilection for different organ systems. The determinants of specific organ involvement are unclear. Parasitization is most common in the brain, heart, lungs, pericardium, and lymphoid tissues.[158] Trophozoites and cysts coexist during active infection and tissue cysts persist after the clearance of an acute infection.

The incidence and virulence of *Toxoplasma* infections vary by region. In Europe, seroprevalence approaches 75%, while estimates in the United States vary from 5% to 40%. Clinically apparent infection occurs in up to 30% of *T. gondii*-seropositive individuals with untreated AIDS. While patients on maintenance therapy for toxoplasmosis do not develop *Pneumocystis* pneumonia, cotrimoxazole prophylaxis is largely, but incompletely, protective against activation of *T. gondii*.

T lymphocytes mediate much immunity to *T. gondii*, largely via the mechanism of macrophage activation.[159,160] Without specific activation, *Toxoplasma* blocks the fusion and acidification of phagolysosomes. Agents that interfere with either T-cell or monocyte function (steroids, HIV, antithymocyte globulin, lymphoma) predispose to toxoplasmosis.[159,161–174] Stimulation of monocytes (IFN, interleukin-2, colony-stimulating factors) by cytokines may enhance killing of *T. gondii*.[159,173] In the non-AIDS patient, the majority of cases of toxoplasmosis occur in patients with hematopoietic malignancy (lymphoma, leukemia) during chemotherapy (especially regimens including corticosteroids) and in organ, particularly heart, and hematopoietic transplantation recipients.[167,168,175–179] The predilection for CNS involvement suggests that immune clearance of *T. gondii* is less effective in the CNS. The roles of antibody and complement in the killing clearance of *T. gondii* remain uncertain, although the combination can kill extracellular trophozoites. The development of T- and B-cell immunity to *T. gondii* coincides with the clearance of extraneural organisms, the regression of tissue invasion, and the development of tissue cysts.

3.2. Clinical Presentations of Toxoplasmosis

Toxoplasmosis occurs with four distinct clinical presentations: (1) congenital, (2) acquired in immunocompetent individuals, (3) disseminated in immunocompromised individuals, and (4) as reactivation of latent infection within the eye (ocular toxoplasmosis).

3.2.1. Congenital Toxoplasmosis

Congenital infections result from acute infections during pregnancy. They are usually asymptomatic infections occurring in immunocompetent individuals. However, infected women who become immunocompromised may reactivate infection and transmit the organism to the fetus. Women who are infected and seroconvert prior to conception have a low risk of transmission to the fetus. The incidence of fetal infection resulting in abortion (stillbirth) or in significant congenital disease rises during the course of pregnancy: from 25% in the first trimester to two thirds of third-trimester fetuses. While the incidence of infection is quite high, the majority of infants infected during the latter trimesters do not show signs of infection, and treatment of the mother with specific antibiotics sig-

nificantly reduces the incidence of congenital infection (by over 50%). The primary manifestations of congenital disease are of CNS disease, particularly chorioretinitis or hydrocephalus. Sequelae of infection may not be immediately evident at birth. Children may have blindness, psychomotor or mental retardation, jaundice, thrombocytopenia, anemia, encephalitis, microcephaly, hypothermia, or pneumonitis. Most of these congenital malformations are not specific to *T. gondii*, but are related to inflammation and scarring in the sites of greatest infection. Mild disease is seen restricted to hepatosplenomegaly or lymphadenopathy. These individuals may ultimately develop CNS disease. It is currently thought that most children infected *in utero* will develop *some* disease related to congenital toxoplasmosis.

3.2.2. Acquired Toxoplasmosis in Immunocompetent Individuals

Over 80% of individuals with *Toxoplasma* infection will be asymptomatic. Alternatively, *Toxoplasma* will present with nontender lymphadenopathy affecting cervical lymph nodes or systemic lymph glands.[180–184] In this setting, toxoplasmosis may be confused with a "flulike" syndrome or infectious mononucleosis and is generally benign and self-limited. Atypical features such as those seen in lymphoma may include fever, hepatosplenomegaly, atypical lymphocytosis, sweats, muscle aches, sore throat, and maculopapular rashes. The lymph nodes may be tender but rarely become fluctuant without superinfection. Rarely, chorioretinitis may occur in acute infection. While the clinical course is relatively benign, symptoms may persist for up to a year. Fluctuating adenopathy and persistence of lymphadenopathy is occasionally seen. More severe disease involving the heart, lungs, and CNS is rarely observed.[185] Myocarditis, myositis, and polymyositis may occur in normal individuals and in those on modest amounts of immune suppression, occasionally in association with dermatomyositis. Because of the broad differential diagnosis of common lymphadenopathy syndromes, including viral illnesses and cat-scratch disease, the diagnosis of toxoplasmosis is dependent on serologic testing or the identification of organisms. Occasionally, lymphadenopathy is not significant in the setting of multisystem involvement including hepatitis, myositis, or fever of unknown etiology.

3.2.3. Ocular Toxoplasmosis

Toxoplasma gondii is a common cause of chorioretinitis throughout the world, usually as a result of reactivation of latent congenital infection.[169,170,186] This form of infection occurs in immunologically normal young adults, affecting the retina and underlying choroid and presenting clinically as unilateral or bilateral chorioretinitis with visual loss, glaucoma, photophobia, and pain. Congenital ocular toxoplasmosis may cause developmental abnormalities of the optic neuraxis, including such common manifestations as strabismus, cataracts, nystagmus, or optic neuritis. The characteristic lesion on funduscopic exam is a cluster of areas of focal necrotizing retinitis, giving a white-to-yellow raised cotton patch with blurred margins. Healing lesions are initially pale and gradually take on black pigment. Occasionally, associated acute panuveitis, papillitis, or optic nerve atrophy may occur. Recurrent disease is common (up to 30% after specific chemotherapy), often in areas of scars from previous infection. Ocular toxoplasmosis usually occurs in the absence of systemic symptoms. Bilateral infection is characteristic of congenital disease, while acute disease is usually unilateral. Eye infection in the immunologically normal host produces severe inflammation and necrosis. Granulomatous inflammation of the choroid accompanies retinal injury. Distinctive characteristics of ocular toxoplasmosis include the clarity of vitreous and aqueous humors, the presence of bilateral macular involvement, and the appearance of a normal retinal exam in the presence of focal areas of retinal degeneration. AIDS-associated *Toxoplasma* chorioretinitis is uncommon; it will often occur (~30–60%) in patients with concurrent CNS lesions. These patients present with acute loss of visual acuity, eye pain, and with panophthalmitis and areas of coagulative necrosis containing cysts and tachyzoites. Occasionally, retinal vascular thrombosis may occur. The characteristic retinal lesions often are raised with healing and may be unilateral or bilateral. The posterior uveitis of toxoplasmosis may be confused with that of syphilis, tuberculosis, histoplasmosis, or leprosy.

3.2.4. Toxoplasmosis in Immunocompromised Hosts

The true incidence of toxoplasmosis in immunocompromised hosts is unclear. Toxoplasmosis of the immunocompromised individual is usually due to the reactivation of latent infection in the absence of a limiting immune response. Acute infection in this population generally results from organ transplantation or the direct infusion of contaminated blood or blood products.[171,175,177,187–192] The occurrence of toxoplasmosis in a seronegative individual in the absence of organ or blood transfusion is uncommon. Blood products from patients with chronic myelogenous leukemia (CML) tend to maintain high levels of parasitemia despite high antibody titers. Because of the survival of organisms in stored citrated blood for up to 2

months at 4°C, patients with CML should not be used for blood or organ donation. The serologic status of organ donors and recipients, particularly cardiac, should be assessed prior to transplantation.

Because of the protean manifestations of toxoplasmosis, this infection must be considered in the differential diagnosis of many systemic illnesses in the immunocompromised host. The non-AIDS immunosuppressed patient is more likely to display systemic manifestations of toxoplasmosis similar to those seen in the immunocompetent host than is the individual with AIDS. A "mono"-like prodrome with fever and lymphadenopathy may precede other manifestations. In this setting, disseminated infection will involve the brain, liver, bone marrow, heart, omentum, spleen, and other organs. Multiple brain lesions are common. The presentation of toxoplasmosis in the immunodeficient host may be that of focal mass lesions (e.g., seizures), "diffuse" infection with widespread microglial nodules of the gray matter, or encephalitis with diffuse CNS impairment or multiple focal neurologic deficits.[186] The diffuse form suggests a degree of immune impairment inadequate for abscess formation and often progresses rapidly to death. Some 10% (United States) to 25% (Europe) of acutely infected immunocompromised individuals will have a neurologic presentation.

Toxoplasmosis in AIDS may also present with pneumonitis, myositis, myocarditis, diffuse gastrointestinal involvement with pain, diarrhea, hepatic dysfunction, ascites, orchitis, panhypopituitarism, diabetes insipidus, and SIADH. Pulmonary infection may also be significant in both AIDS and non-AIDS compromised hosts.[193,194] The clinical onset of pneumonia is associated with the gradual onset of fever, dyspnea, cough, and occasionally hemoptysis. Chest X rays reveal a diffuse bilateral pulmonary infiltrate with atypical features including interstitial infiltrates and small nodules. The serum LDH is often elevated. In this setting, it should be possible to isolate organisms from lung washings.[195] In both AIDS and transplant patients, asymptomatic myocarditis and myositis are common, associated histologically with myocyte necrosis.

In the organ transplant recipient, the clinical presentation is similar, with the exception that the disease is generally due to dissemination from the transplanted organ.[157,168] In contrast to AIDS, allograft recipients who develop infections are generally seronegative. The greater severity of disease in the seronegative recipient of a seropositive organ than in reactivation disease in the seropositive individual suggests that antibody is partially protective. Given the predilection of the organism for myocardium, toxoplasmosis occurs most often in heart transplantation. Transmission in large series exceeds 80% in cardiac transplantation into seronegative individuals. In part, this may reflect the intensity of infection (i.e., number of cysts) from regions with high endemicity (e.g., France, Haiti) as well as immune compromise. Such individuals may present with heart failure or arrhythmia, which may be indistinguishable from graft rejection or primary cardiac CMV infection. Identification of organisms, including tissue cysts, in a seronegative individual should raise the specter of acute disseminated disease.

The incidence of toxoplasmosis in AIDS appears to be decreasing with the introduction of more effective antiviral therapies. Without HAART, *Toxoplasma* encephalitis has been a common presentation for patients with AIDS. In untreated AIDS patients, as in most immunocompromised individuals, toxoplasmosis is usually disseminated at the time of presentation.[196,197] In the AIDS population, lymphadenopathy may be seen in primary HIV infection and with opportunistic infections including mycobacteria and CMV. Lymph node biopsies are generally nondiagnostic. In AIDS, the occurrence of disseminated *Toxoplasma* infection is usually correlated with low circulating CD4+ lymphocyte levels (<50). Between 5% and 50% of *Toxoplasma*-seropositive HIV-infected individuals may develop CNS toxoplasmosis, based on autopsy studies. AIDS patients from endemic

areas due to consumption of undercooked pork or lamb or contaminated vegetables are at particular risk. In patients from such areas (e.g., Haiti, much of Africa), *T. gondii* infection is commonly diagnosed. Signs of meningitis are uncommon, while seizures, confusion, depression, visual changes, and hydrocephalus are more common. For HIV-infected individuals whose CD4+ counts remain above 200 lymphocytes/mm^3 for at least 3 months, studies suggest that primary prophylaxis may be discontinued without short-term (less than a year) detriment. Definitive data are lacking regarding the long-term persistence and efficacy of immune reconstitution during HAART.

The presentation of CNS toxoplasmosis can be as mild as slightly altered mental status, minor neurologic signs, or fever with headache. Seizures and gross motor deficits may occur in up to 10% of individuals. The CNS lesions of toxoplasmosis must be distinguished from other infections or tumor (Fig. 8).[175,197–205] The infection can present as focal abscesses or diffuse meningoencephalitis, based on the distribution of the preexisting latent lesions. Acute, primary toxoplasmosis has also been reported. The progression of *Toxoplasma*-induced mental status changes is subacute when compared to that due to HIV alone. HIV-associated leukoencephalopathy or progressive multifocal leukoencephalopathy forms part of the "AIDS dementia complex," which also can present

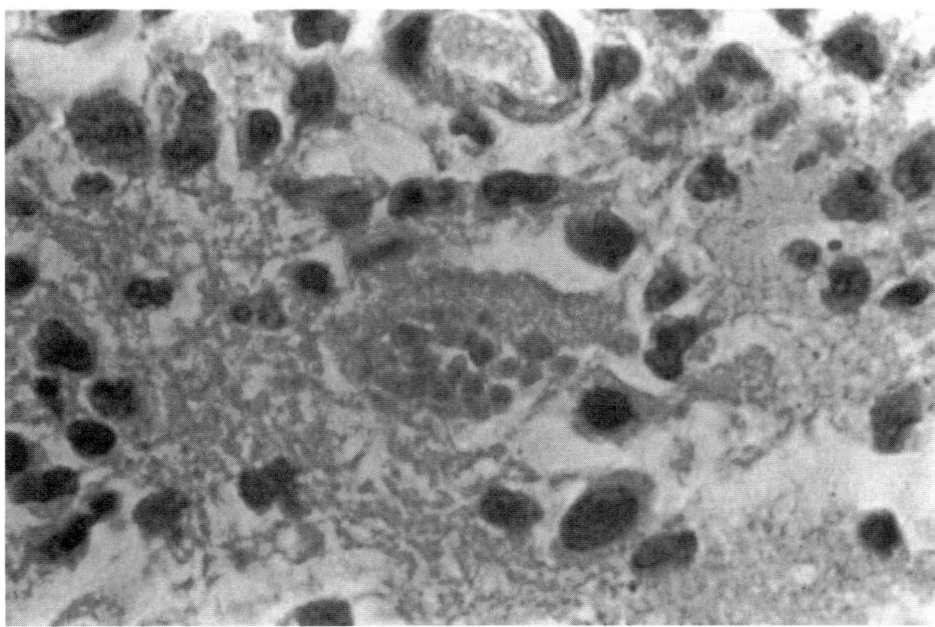

FIGURE 8. Hematoxylin-eosin stain of a brain biopsy specimen from the edge of a large anterior brain abscess from a patient with AIDS. A tissue cyst contains multiple intracystic bodies or bradyzoites consistent with a diagnosis of toxoplasmosis.

with focal or global CNS deterioration and may coexist with acute infections of the brain. Some of our patients have presented with multiple organisms in a single small brain abscess, including *T. gondii* with anaerobic bacteria, *Aspergillus* species, and mycobacteria, including tuberculosis and *M. avium* complex. Because serologic testing in immunodeficient patients is not often helpful in establishing a diagnosis of toxoplasmosis, biopsies of infected areas may be required to confirm the diagnosis.

3.3. Diagnosis

3.3.1. Laboratory Evaluation for *Toxoplasma gondii*

The diagnosis of infection due to *Toxoplasma* is difficult, most notably in the immunocompromised individual in whom early diagnosis is most important and serologic tests are least helpful.[206] Demonstration of the tissue cyst form of the organism or of a positive serum IgG level suggests the possibility of infection due to *T. gondii*, but does not prove ongoing disease, i.e., toxoplasmosis. Tissue cysts persist in the brain, lung, liver, lymph node, heart, and spleen for years after acute infection. The presence of trophozoites in lymph node, blood, brain, cerebrospinal fluid (CSF), or other tissues during acute infection, or the demonstration of an acute (IgM) immune response to the organism, or the amplification of *T. gondii*-specific DNA from body fluids is needed for confirmation.

Brain disease features prominently in the presentation of immunocompromised patients with toxoplasmosis (Figs. 8 and 9).[202] Decisions about the management of suspected toxoplasmosis involving the brain are complicated by the risk of performing definitive diagnostic procedures, including lesion aspirates or open brain biopsies, and the potential for toxicity of antimicrobial agents (notably pyrimethamine or sulfa), which approaches 50% with the antimicrobial agents commonly used in AIDS patients. The lumbar puncture is nondiagnostic in most cases. The CSF will have slightly elevated protein levels and few white blood cells. These patients will have a normal to slightly decreased glucose concentration in the CSF; hypoglycorrhachia is usually associated with the rupture of organisms directly into the CSF. Elevations in antibodies to *T. gondii* in the CSF are highly suggestive of acute infection.

In AIDS, empiric therapeutic trials frequently are the best approach in stable patients. In organ or hematopoietic transplant recipients, given the risk of the rapid progression of fungal infection of the CNS, the threshold for invasive diagnosis to distinguish potential etiologies of

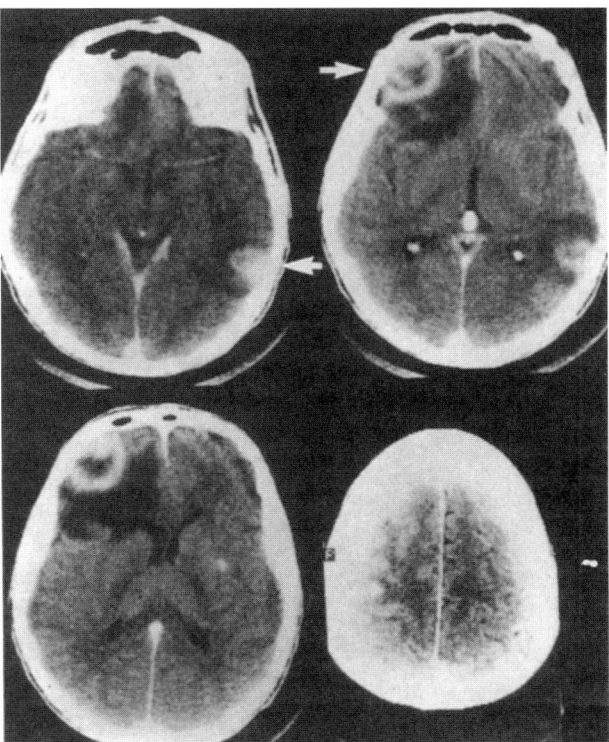

FIGURE 9. CT scan from the patient in Fig. 8 reveals multiple and bilateral contrast-enhancing brain abscesses surrounded by edema. This presentation is seen in *T. gondii* infection more often than the large solitary lesions often associated with CNS lymphoma.

infection must be lower. In either group, the presence of rapidly progressive brain lesions, clinical deterioration, lesions unresponsive to empiric therapy, or serologic testing not consistent with the clinical picture, a brain biopsy may allow the definitive separation of toxoplasmosis from other processes. Biopsy of brain lesions has the added advantage of identifying other potentially treatable processes. In addition to meningoencephalitis, encephalomyelitis, or mass effect due to brain abscess, the patient may present with pneumonitis, myocarditis, or signs of hepatitis. *Toxoplasma gondii* trophozoites may be found in lung lavage or lung biopsy sample from patients with pneumonitis.

3.3.2. Radiology

Radiologic evaluation will generally begin with a head CT scan and chest radiograph. The chest radiograph is nonspecific, the picture being that of prominent hilar lymphadenopathy with diffuse interstitial infiltrates. Like other atypical pneumonias, atypical patterns including nodules and asymmetric patches are common. Cavitation or pleural effusions are uncommon. Chest radiographs do not tend to improve over a 14-day course of therapy unless corticosteroid therapy is employed.

In AIDS patients, the presence of multiple and bilateral intraparenchymal lesions that are contrast-enhancing on CT scans is often considered diagnostic of CNS toxoplasmosis (Fig. 9).[207,208] CT studies of AIDS patients with solitary CNS lesions show that about half will be due to *T. gondii*, with the balance being lymphoma, progressive multifocal leukoencephalopathy, and less commonly other infectious or malignant processes.[208] Non-AIDS-immunocompromised patients also will have multiple CNS lesions, but the differential diagnosis must be broadened and invasive diagnosis considered earlier. Up to 20% of patients will have single lesions without contrast enhancement. Occasionally, uptake of contrast dye may be delayed, particularly in neutropenic patients and in those on corticosteroids. Repeat scans after the administration of increased doses of contrast dye may demonstrate lesions not previously observed on routine scanning. CT scans without contrast are usually negative. The majority of lesions are nodular or have ring enhancement with a predilection for the basal ganglia and the gray–white junction. Most lesions, especially those of the posterior fossa, are better seen on MRI scan, which often reveal small, bilateral lesions not seen by CT. Similar lesions of the posterior fossa also may be seen with cryptococcal and mycobacterial infections. Large focal abscesses may be seen that cannot be distinguished from necrotic tumor or abscesses due to other pathogens. Serologic testing and CSF antibody levels may be useful in this group. Radiologic evidence of encephalitis is seen less frequently. Thallium single photon emission computed tomography (SPECT) scanning or positron emission tomography (PET) generally demonstrate greater uptake of methionine or glucose than lymphomatous lesions, but overlap (false-positive and -negative scans) does occur.

3.3.3. Histopathology and Culture

The identification of *T. gondii* in clinical specimens is difficult. In general, organisms are not found in body fluids; they may be found in cytology specimens from lung washings or CSF or as "contaminants" of viral tissue cultures.[209,210] Impression smears of the tissues can be stained with Giemsa stain to demonstrate both the cyst and trophozoite forms. Tissue cyst walls are stained by PAS stain (see Fig. 8). Trophozoites are stained with either Wright or Giemsa stains and by fluorescent or peroxidase-tagged antibody to *T. gondii* on histologic sections. Touch preparations of tissues and cytocentrifuged fluid samples may be stained with Wright–Giemsa. The presence of multiple tissue-cyst forms in areas of acute tissue inflammation in the absence of other pathogens may be used as presumptive evidence for the presence of acute infection. Tissues or fluids inoculated into mice also have been used for detection of *T. gondii* infection. This method is not practical for routine clinical diagnosis. Further, mice generally do not die from human *Toxoplasma* infection; serologic testing and pathology are generally necessary to confirm the diagnosis of toxoplasmosis using this system. Tissue culture methods have been improved with the identification of plaques in cell culture monolayers and are more generally available. Growth of *T. gondii* from blood in tissue culture can be considered evidence of disseminated infection. Isolation of viable organisms from tissue does not assist in determining the acuteness of infection. *Toxoplasma gondii* organisms are uncommonly found in lymph nodes outside acute infection, but may be found in other tissues for months after therapy. The identification of organisms in fetal tissue or from the placenta is diagnostic of congenital infection. The detection of *Toxoplasma* in infected tissues may be improved by immunostaining amplification using peroxidase–antiperoxidase, fluorescent antibody (often with significant nonspecific staining of tissues), or ELISAs using unfixed tissues. All have proved useful clinically, although the immunoperoxidase is the most consistently sensitive and specific.

In the normal individual, the histopathologic changes seen in the lymph nodes are often diagnostic.[181,183,211] Epithelioid histiocytes appear in clusters, monocytes infiltrate the sinuses, and reactive hyperplasia is observed. Organisms and giant cells are rarely seen. In the eye, retinitis with necrosis, vascular proliferation, and granulomata are seen. Myositis may be seen in any muscle but is most prominent in the heart. Diffuse patches of mononuclear infiltration may occur in the presence or absence of organisms. Diffuse meningoencephalitis will produce multiple areas of necrosis, particularly of the gray matter and microglial nodules. Organisms are found within blood vessel walls, in the periphery of abscesses and surrounding areas of necrosis, and in normal tissues. Periventricular lesions are more often seen in infants, accounting for a high incidence of hydrocephalus. In the immunocompromised host, many small areas of coagulation necrosis may be seen; large abscesses occur in some patients.

In contrast to the lesions produced in other organs, *T.*

gondii usually evokes diffuse disease in the lungs. This result may reflect the role played by the lungs as a filter for circulating organisms and the extensive resident phagocyte population.[193,194] The pneumonitis is predominantly interstitial and characterized by an infiltrate of macrophages, lymphocytes, plasma cells, and occasionally polymorphonuclear leukocytes. The local inflammatory response is usually mild. Areas of bronchopneumonia, endarteritis, and necrosis may develop. Discrete areas of interstitial infiltration may progress to areas of consolidation with necrosis or infarction. Proliferating stages of *T. gondii* are seen inside alveolar macrophages and in the alveolar space. With active myocarditis, organisms also can be found in the endocardium. Death may result from hepatic necrosis, cardiomyopathy, or secondary pulmonary edema. As in the brain, toxoplasmosis may occur in the setting of other opportunistic infections including *P. carinii* pneumonia, bacterial pneumonia, or fungal abscess.

3.3.4. Skin Testing and Cellular Immunity

Skin testing for delayed hypersensitivity is not useful in establishing the diagnosis of acute toxoplasmosis. Positive skin tests may not develop for months after acute infection and may never develop in immunocompromised patients, including those with AIDS. Skin testing may provide an alternative to serologic testing for population screening. Similarly, lymphocyte transformation using patients' cells stimulated with *Toxoplasma* antigens also is an indicator of previous exposure to *T. gondii*. Inversion of the T-lymphocyte helper/suppressor ratio (an excess of suppressor T cells) occasionally is observed in the presence of acute toxoplasmosis. Such an inversion also occurs in the presence of viral infection, including HIV

and CMV, and in a variety of other conditions. The demonstration of circulating antigen from *T. gondii* in sera is not routinely available, but may be helpful in establishing the acuteness of exposure.[212,213]

3.3.4a. Antibody Detection Tests and Polymerase Chain Reaction. In the immunologically intact individual, serologic testing is diagnostic for *T. gondii* infection (Table 7).[184,214] The immunocompromised host will often fail to generate a specific antibody response to acute infection, or this response will be much delayed. In the patient receiving immunosuppressive therapy, false-positive serologic tests may occur in the setting of organ transplantation into seropositive organ recipients. These include both IgM and IgG antibodies. The absence of diagnostic serology may increase the need for a tissue diagnosis in the immunocompromised host. In the immunodeficient individual, the presence of a positive test is still of clinical importance. Because much of acute toxoplasmosis occurs in seropositive individuals, the value of serologic diagnosis in the absence of an elevation of the serum IgM titer is questionable. However, conversion from seronegative to seropositive or the presence of a fourfold rise in titer can be taken as indicative of acute toxoplasmosis for most of the tests currently available. The currently available tests are summarized in Table 7. The presence of a differential concentration of specific antibodies in CSF when compared with serum may be used to indicate the presence of primary infection of the CNS.[186,215,216]

Specific diagnostic tests merit comment. Positive values must be established for each laboratory. The standard serologic assay is the Sabin–Feldman dye test, which can be standardized to reference sera available from the World Health Organization.[214,217] Like most of

TABLE 7. Serologic Assays for *Toxoplasma gondii* Infection[a]

Test	Titer		Comments
	Acute	Chronic	
Sabin–Feldman dye test (IgG)	1:1000	1:4–1:2000	Remains elevated; onset 1–2 weeks
Indirect fluorescent antibody (IFA)-IgG[b]	1:1000	1:4–1:2000	Remains elevated; onset 2–3 weeks; some false-positives[b]
IFA-IgM[b]	1:64	0–1:20	Negative in months; first positive acutely (1 week)
Direct agglutination test (IgG)	1:1000–1:20	1:64,000	Remains elevated; BME[c] to block IgM agglutination
Indirect hemagglutination (IHA) (IgG)	1:1000	1:16–1:256	Remains elevated; delayed onset
Complement fixation (IgG)	1:32	0–1:8	Remains elevated; onset 2–3 weeks
ELISA[b]-double sandwich-IgM	1:256 (or 1.7 units)	1.7–3.0	Remains elevated; early onset; sensitive
Immunosorbent (IgM) (latex bead)	Positive	Positive	Simple, sensitive (vs. IFA); fewer false-positives

[a]A two-tube or fourfold rise in titer to "acute" level is diagnostic for any test. Positive/diagnostic values for the various tests will vary among clinical laboratories and some patients will fall outside these ranges. These represent adult values.
[b]May give a false-positive value in the presence of rheumatoid, antinuclear, or other autoantibodies; a single high titer is diagnostic of acute infection.
[c]BME = β-mercaptoethanol.

the tests that measure primarily IgG antibodies, positive tests do not occur until after 2 weeks of infection, with peak titers occurring at times up to 2 months. While titers decline over 1–2 years, low titers persist for life. As with all serologic tests, the titer of antibody does not correlate with the severity of illness, but may provide information about the ability of the host to respond to new antigenic challenge. Because immunity to toxoplasmosis is largely T-cell-mediated, the presence or absence of antibody will not determine the ability of the host to respond to acute infection. The IgM titer correlates most closely with acute infection; a number of tests have been developed for this purpose.[218,219] The direct agglutination test using either fixed whole trophozoites or antigen-coated latex particles, the IgM immunofluorescent antibody test, and the conventional IgM ELISA all give false-positive results in the case of high endogenous levels of nonspecific IgM, rheumatoid factor, and antinuclear antibodies.[220–223] The double-sandwich IgM ELISA and IgM immunosorbent assay using solid phase do not suffer from such false-positive results and have much greater sensitivity than the alternatives.[218,222–224] In the immunocompetent individual, acute infection should be accompanied by seroconversion from negative to positive or by the demonstration of a fourfold (two-tube) rise from a low chronic titer to a high acute titer in sera drawn at least 3 weeks apart and run simultaneously. If initial sera are drawn too late in the course of infection, such diagnostic increases may be missed. Panels of serologic tests are often more useful than single assays.

Polymerase chain reaction (PCR) assays are available on both a research and commercial basis. These nonstandardized tests have been useful in the diagnosis of intrauterine infections, ocular, congenital, acute disseminated, and CNS infections particularly when brain biopsy is not feasible. The tests are highly specific, but have lacked sensitivity (varying from 11 to 85%) in part because some patients with *Toxoplasma* encephalitis or ocular infection lack parasitemia or dissemination outside the blood–brain barrier. Tests based on amplification of genomic repetitive elements and nested PCR systems have been somewhat more sensitive than those based on the B1 element.

3.3.4b. Diagnosis in the Immunodeficient Patient.   In the immunodeficient host, rapid progression of disease or atypical presentations may necessitate tissue diagnosis.[225] Many patients will fail to demonstrate a rise in IgM serum titer. In AIDS patients, there may be some advantage to the commercial solid-phase IgM immunosorbent assay. The failure of serologic testing points out the importance of obtaining baseline titers of *Toxoplasma*

antibodies in AIDS or organ transplant patients. Demonstration of relatively enhanced antibody production in the CSF has been associated with *Toxoplasma* encephalitis. Local antibody production should be associated with a specific antibody level (as a fraction of total local IgG) greater than that fraction of specific antibody present in serum (as a fraction of total serum IgG). A similar concept has been applied to the diagnosis of ocular toxoplasmosis, in which serum titers are typically low. In some hosts, notably in newborns and in recent organ transplant (especially cardiac) recipients, elevated titers of anti-*Toxoplasma* antibody may be normal. IgG antibody titers in newborns may reflect maternal antibody titers. In congenital infection, the child may not produce specific antibodies before 2–9 months of age. Serial tests may be helpful in this situation. Cardiac and heart–lung transplant recipients may demonstrate significant titers even without acute infection. Elevated CSF IgM levels should always be taken as indication of possible brain involvement. PCR testing of blood or buffy coat may detect cases of disseminated infection in advance of seroconversion. PCR of CSF also may be useful if positive. Negative PCR tests do not prove the absence of toxoplasmosis.

3.4. Therapy of *Toxoplasma gondii* Infection

3.4.1. Acute Infection

Initial therapy for *T. gondii* infection should include a reduction in the immunosuppressive therapy whenever possible. Cellular immunity is needed to eradicate intracellular organisms. Extracellular organisms are killed by antibody in conjunction with complement, while cyst forms are largely resistant to antibiotic therapy with the possible exception of atovaquone. Empiric therapy should be initiated in patients with the appropriate clinical syndrome and who are at risk for disseminated toxoplasmosis. Such conditions include CNS, pericardial, pulmonary, hepatic, splenic, or bloodborne disease in patients with AIDS, patients receiving high-dose corticosteroid and/or calcineurin-inhibitor therapies, cardiac transplant recipients at risk for primary infection, bone marrow transplantation recipients, or acute hematopoietic malignancy.[206,209,225–229]

Antimicrobial therapy is outlined in Table 8. A synergistic combination of antibiotics including pyrimethamine (200 mg load, then 25–75 mg/day with folinic acid 5–15 mg/day) and sulfonamide or clindamycin is favored for *Toxoplasma* infection in all patients. Data in AIDS patients suggest that oral clindamycin (600 mg PO q6h) may be as effective as high-dose intravenous drug (i.e.,

TABLE 8. Antimicrobial Therapy for *Toxoplasma gondii* Infections[a]

Condition/duration	Drug[b]	Dose	Alternatives	Comments
Disseminated/CNS acute encephalitis; 6 weeks	Pyrimethamine with folinic acid and sulfadiazine *or* trisulfapyrimidine	100–200 mg PO load; then 25–50 mg PO qd or qod 10 mg/day 4 g PO; then 1–1.5 g PO qid 75–100 mg/kg per day	Pyrimethamine/folinic acid plus clindamycin (900–1200 mg mg/d IV q6h *or* 300–600 PO qid), *or* azithromycin (1200–1500 mg/d); or clarithromycin (1 g bid), *or* atovaquone (750 mg tid-qid)	Bone marrow suppression; decrease dose for neutropenia; may avoid folinic acid in leukemia; lifelong suppression; role of HAART in prevention of CNS relapse not clear; sulfa allergy common; *C. difficile* common
Suppression (after acute infection to 6–12 months after Toxo + heart transplant); lifelong	Pyrimethamine	25–75 mg/day; folinic acid; *and* sulfadiazine 500–1000 mg/d or clindamycin 300–450 mg PO qid	Replace clindamycin with atovaquone 750 mg PO tid-qid *or* dapsone 100 mg/d *or* azithromycin 600 mg/d PO	Alternatives without documented efficacy; no data to support discontinuation of HAART
Prophylaxis (AIDS with + serology); lifelong	TMP-SMX	DS PO qd	SS qd or DS tiw *or* dapsone 50 mg/d + pyrimethamine 50 mg/wk + folinic acid *or* atovaquone 750 mg bid	Breakthroughs in nondaily regimens in transplant; secondary prophylaxis essential; need for primary prophylaxis after recovery of CD4+ counts (HAART)?
Congenital (1st 18 weeks of gestation to term)	Spiramycin (FDA)	1 g PO tid or qid		In pregnancy or sulfa allergy with pyrimethamine; Rx needed for neonate
Ocular or transfusion	As for CNS infection			Steroids for inflammation in meningitis/eye infections
Acute normal host	None			

900–1200 mg IV q6h).[227-231] The most common sulfonamide used is sulfadiazine (6–8 g/day after a 4-g load), but it is interchangeable with trisulfapyrimidine for this purpose.[228] Other sulfonamides are not equally effective. It must be noted that solid organ and hematopoietic transplant recipients (renal, or those on calcineurin inhibitors) will rarely tolerate high-dose intravenous sulfa or TMP-SMX. Hydration and adequate urine flow are critical. As an alternative to pyrimethamine–sulfa or clindamycin–pyrimethamine, pyrimethamine may be used with atovaquone (750 mg PO tid or qid), an effective oral agent in individuals with normal gastrointestinal function, and which has unique activity against the intracystic form of *T. gondii*. Alternate strategies include pyrimethamine with azithromycin (1000–1500 mg/day), trimetrexate, or minocycline (100 mg PO bid).[232,233] The newer macrolides roxithromycin, azithromycin, and clarithromycin in combination with sulfonamide or pyrimethamine may have the advantage of excellent tissue penetration in excess of serum levels.

Therapy should be adjusted to the underlying immune disorder.[225] For most immunodeficient patients, pyrimethamine is given for up to 6 weeks at 25 or 50 mg/day. In AIDS patients, 50–100 mg/day is preferred. In most patients, this drug will induce bone marrow suppression, which may be relieved by calcium leucovorin. Some patients will experience an altered taste sensation, headaches, or GI upset while taking pyrimethamine. The patient must be well hydrated to prevent crystalluria. Alternative therapies are less toxic and less active.[234] Therapy is generally continued at high dose for 6 weeks with reduced dosing thereafter for chronic suppression. Seronegative recipients of heart transplants from seropositive donors should receive 6–8 weeks of pyrimethamine (50 mg/day) with sulfadiazine (2–4 g/day) or clindamycin (1200–1800 mg/day).[178,235] For CNS infection, no significant changes are seen in radiologic studies before 2–3 weeks; the neurologic examination is more sensitive to progression or response to therapy. Non-AIDS immunodeficient patients do better in general than AIDS patients suffering from disseminated toxoplasmosis. The main limitation of therapy in AIDS has been drug toxicity, which occurs in at least half of individuals. Discontinuation of therapy in the absence of effective HAART for at least 3 months is generally associated with a relapse of brain disease. Acute neurologic deterioration may occur with the initiation of HAART or reduction in immune suppression.

3.4.2. Prophylaxis

In seropositive AIDS patients with CD4+ lymphocyte counts <200/mm³ and in seropositive neutropenic patients or those on high-dose corticosteroids, primary prophylaxis is recommended using TMP-SMX (DS qd). The incidence of toxoplasmosis is reduced in AIDS patients receiving cotrimoxazole prophylaxis. The use of pyrimethamine (50 mg/day) and sulfadiazine (2 g/day) also prevents both diseases.[228] Given the lower sensitivity of *T. gondii* to sulfa compared with *P. carinii* and/or the role of latent CNS infection, daily drug regimens are preferred. For seronegative recipients of cardiac transplants, more intensive prophylaxis with pyrimethamine and sulfa (2–4 g/day) is suggested for the first 3 months posttransplantation with lifelong TMP-SMX (1 DS tab/day) thereafter. Alternatives include dapsone (50–100 mg/day) with pyrimethamine (50 mg biw-tiw), and folinic acid (10 mg/week), atovaquone (1500 mg/day ± pyrimethamine), or one of the macrolide regimens (none well studied). Given the long serum half-life of pyrimethamine, a regimen of 3 days a week is generally adequate for the prevention of disease. Alternatively, Fansidar can be given at a dose of 1 or 2 tablets per week in many patients. Breakthrough infection has occurred during therapy with this antimicrobial combination. In solid organ transplant recipients, atovaquone (1500 mg/day) prevents both infections.

After acute infection, lifelong suppressive therapy is recommended in the absence of return of normal immune function.[206] Data on secondary prophylaxis for *T. gondii* in HAART or after transplantation are lacking. Success is directly proportional to the penetration of the agent into the CNS and the patient's ability to tolerate drug side effects. Relapse in AIDS has been observed during prophylactic therapy with pyrimethamine, trimethoprim-sulfamethoxazole, and spiramycin.[206,231,236] Higher-dose pyrimethamine (50–75 mg/day) has been used in some AIDS and transplant patients for therapy and prophylaxis to increase serum levels. In general, folinic acid supplementation is needed. In AIDS patients with toxoplasmosis, some early relapses have been seen with the macrolide prophylactic regimens.[236] In combinations with pyrimethamine, skin rash (38%), GI (38%), liver function test (77%), or hematologic (54%) toxicities are seen. In the setting of other immunosuppressive therapies or antiviral therapies, some patients will tolerate slightly reduced sulfonamide doses or use of another sulfonamide preparation in therapy or prophylaxis, rather than needing to discontinue an agent that is causing minor side effects. While the use of corticosteroids may reduce mass effect and brain edema in the acute phase, it is not clear that there is a role for these agents in the long-term management of toxoplasmosis.[170] Immune modulators, including IFN-γ, may be useful as adjuncts to antimicrobial agents in clearing intracellular organisms.

Illustrative Case 2

A 37-year-old man was brought to the Massachusetts General Hospital emergency department because of progressive inability to care for himself. The patient was a homosexual whose single sexual partner had died of AIDS 15 months previously. At that time, the patient was serum HIV-negative by ELISA. By history, the patient had been well until 3 months prior to admission, holding full-time employment as a computer programmer. He had complained of some decreased "ability to concentrate" and of mild fatigue. These symptoms were attributed by the patient to depression following the anniversary of his partner's death. Two weeks before presentation, he had seen his personal physician for a general examination. This physician had detected mild diffuse and nontender lymphadenopathy and noted that the patient seemed "tired" but otherwise normal. A serum HIV screening test had been positive by ELISA and confirmed by Western blot; the patient had not yet been notified. A serologic test for toxoplasmosis revealed a positive IgG titer (1:1024) and a negative IgM titer. On the day of admission, the patient had not appeared at work and had been found comfortable but somewhat confused at home. He had no pets and no other known infectious exposures.

On examination, the patient was thin but in no acute distress. He complained of a mild headache but denied fevers, photophobia, or meningismus. On neurologic examination, the patient's short-term memory was impaired and he forgot the ends of some sentences. His speech was fluent. He was not able to recognize some objects or written words. His general examination revealed mild lymphadenopathy and a palpable spleen tip but was otherwise unremarkable. Laboratory evaluation revealed a total white blood cell count of 2300 with a normal differential and 137 CD4+ lymphocytes. His hematology, chemistries, and chest radiograph were otherwise within normal limits. A head CT scan was performed that demonstrated multiple (at least three), bilateral, contrast-enhancing, ring-shaped lesions with surrounding edema (Fig. 9). A lymph node biopsy revealed only reactive hyperplasia.

On the basis of the presumed diagnosis of toxoplasmosis, the patient was treated empirically with pyrimethamine and sulfadiazine and switched to clindamycin and pyrimethamine when a drug rash developed on day 5 of therapy. Because of progressive neurologic deterioration in the absence of improvement by CT scan, a stereotactic brain biopsy of one of the brain lesions was performed. *Toxoplasma gondii* was demonstrated (Fig. 8). On antimicrobial agents, the patient improved only slightly over 6 weeks of therapy. A repeat CT scan revealed some improvement in some of the lesions, but a large frontal lesion remained unchanged. Biopsy of the anterior lesion revealed B-cell lymphoma. This tumor progressed rapidly despite therapy. The patient died 3 months later.

Comment. The presentation of toxoplasmosis of the CNS can be subtle. The HIV-positive patient is often well-appearing, without fever or headache. Up to 60% will have focal neurologic deficits, fever, altered mental status, or seizures. The geographic distribution of infection varies with the incidence of *T. gondii* infection in the general population. While toxoplasmosis is an uncommon presenting manifestation of AIDS in the United States (3–5%), up to a third of seropositive AIDS patients will eventually develop disease without prophylaxis. The clinical presentation of toxoplasmosis in the AIDS patient is often indistinguishable from HIV-encephalitis, CNS lymphoma, or progressive multifocal leukoencephalopathy (PML).

The CT scan is often used diagnostically for *T. gondii* infection; however, the MRI scan is more sensitive. The presence of multiple and bilateral contrast-enhancing (nodular or ring) lesions by CT or MRI scan is highly correlated with toxoplasmosis. However, lymphoma can have the same appearance or, as in this case, can coexist with *T. gondii* infection. These tumors are often aggressive and poorly responsive to treatment. By MRI scan, PML lesions usually cause multiple or diffuse subcortical changes of high intensity without gadolinium enhancement. A response to therapy for CNS toxoplasmosis (encephalitis or brain abscess) is often seen clinically within a week and radiographically in 2–3 weeks. Failure to improve on empiric therapy may necessitate further investigation.

Toxoplasmosis is generally seen in AIDS patients with CD4 t lymphocyte counts of less than 200/ml blood. The presentation can be similar in non-AIDS immunocompromised patients, although fever and systemic signs are common. PML and lymphoma also are seen in organ transplantation recipients and following intensive chemotherapy for carcinoma. Infection occurs almost exclusively in IgG-seropositive individuals. An IgM response is often absent. Lifelong suppressive therapy is needed in AIDS patients after the initial treatment. Breakthrough infection has been seen in AIDS patients on prophylaxis for *P. carinii*. Clinically, patients with CNS toxoplasmosis may worsen with the initiation of HAART therapy. This is thought to reflect the recrudescence of immune function with declining HIV viral loads. Such patients, as for reconstitution syndrome in *Pneumocystis* infection, may benefit from the transient addition of corticosteroids for suppression of the inflammatory response. However, other potential pathogens also should be excluded.

4. *Cryptosporidium* Species

Cryptosporidium is a protozoan parasite that can cause severe and persistent diarrhea in immunocompromised patients, particularly those with AIDS, but also in individuals with immunoglobulin deficiency, solid organ and hematopoietic transplant recipients, and neutropenic hosts. This is a common form of self-limited gastroenteritis in less severely immunodeficient individuals and in the immunocompetent host.[6,237–244] The organism also has been associated with biliary disease and uncommonly with respiratory infection in some individuals. The relative absence of effective therapy for cryptosporidiosis increases the impact of this pathogen. Despite the description of this organism in mice by Tyzzer in 1907, the organism was not linked to significant disease in animals until the 1950s. It is recognized as a major pathogen of turkeys, calves, and lambs. Severe cryptosporidial diarrhea was detected early in the AIDS epidemic.[7,245] In

Africa and Haiti, AIDS-associated diarrhea presents as the constellation of findings referred to as "slim disease": diarrhea, weight loss, fatigue, fever, and, eventually, death.[6,246–248] Up to 40% of these patients' clinical manifestations may be due to *Cryptosporidium*.[7,240,249] In the United States, carriage of *Cryptosporidium* in AIDS patients is common (up to 5%).[238] Some 10–15% of patients with AIDS and diarrhea will have cryptosporidiosis.[250] More recently, it has been recognized as a common cause of enteritis worldwide in immunocompetent hosts.[244,251–255] *Cryptosporidium* has been detected in up to 5% of immunologically normal individuals experiencing gastroenteritis or chronic malabsorption.[244] In the individual with diarrhea containing *Cryptosporidium*, other organisms are often detected, including amebae, CMV, *Giardia lamblia*, *Isospora belli*, and adenovirus.[250] Outbreaks of cryptosporidial infection have occurred in day-care centers and subsequently in the families of affected children. Immunocompromised individuals, including bone marrow transplantation recipients, organ transplant recipients, individuals with primary immunoglobulin deficiencies, as well as patients with AIDS, are candidates for severe and generally unremitting gastrointestinal and gallbladder infection.[254–256] Despite its growing importance, the pathogenesis of cryptosporidiosis has not been clarified.[257]

4.1. The Organism

Cryptosporidium is a coccidian protozoan parasite (phylum Apicomplexa, class Sporozasida, subclass Coccidiasina) of which two other members (*Isospora belli* and *Eimeria* sp.) cause significant infection of gastrointestinal and respiratory epithelia in compromised hosts. Many species of cryptosporidia have been identified in most vertebrate species. The ability to transmit infection between species, most notably between animals and man, indicates that this zoonotic organism lacks host specificity.[258] Two species appear to cause significant disease in mammals: *C. parvum* and *C. muris*. *C. parvum* is somewhat smaller (2- to 3-μm diameter oocsts) than *C. muris* (5- to 8-μm oocysts). *C. parvum* is primarily responsible for the diarrheal disease of humans and of cattle, while *C. muris* infects primarily the stomach of nonhuman mammals. The variability of the ability of clinical isolates of *C. parvum* to cause disease in humans and to respond to therapy suggests that further subspecies of *C. parvum* exist. Some additional subspecies have been identified in stools of patients with AIDS. Additional antigenically distinct subspecies are beginning to be identified in animals. The coccidia complete their life cycle within the human host. The exacerbation of infection by immune suppression reflects the role of the immune system in controlling replication. Infection is initiated by ingestion (fecal–oral route) or occasionally by inhalation of oocysts with an average of 7–10 days elapsing before clinical symptoms emerge. Disease may be delayed for up to a month following exposure. It is likely that the oocyst can also initiate an autoinfectious cycle. Excystation, usually in the presence of bile or digestive enzymes, releases four motile sporozoites that attach to and penetrate the host epithelial cell. These motile sporozoites mature within a unique four-layered parasitophorous vacuole where they mature into trophozoites (Fig. 10). Asexual (schizogeny) division produces up to eight meronts, which mature into merozoites that can either reinfect host epithelial cells or initiate a sexual cycle (gametogeny). Sexual reproduction completes the cycle by producing mature oocysts. A fraction of the oocysts have thin walls and rupture within the intestinal lumen to reinitiate the cycle within the host. The majority are excreted in feces. Shedding of cysts occurs in individuals with and without immune compromise or symptomatic diarrhea.

Little is known about the pathogenesis of this infection. Symptomatic disease is generally associated with infection of the proximal small bowel. The voluminous water diarrhea suggests a hypersecretory mechanism the basis of which remains unknown. No toxin production has been demonstrated and the degree of villous injury is modest compared with the intensity of the diarrheal illness. The hepatobiliary tree may serve as a reservoir for reinfection in the immunocompromised host. Carriage in the gallbladder has been associated with failure to clear infections both in AIDS patients and in children with hypogammaglobulinemia (Fig. 10). The unique intracellular vacuole places the organism in an extracytoplasmic location, which may contribute to the inability of antimicrobial agents to clear infection.

4.2. Epidemiology

Seroprevalence studies for *Cryptosporidium* have demonstrated the presence of organisms in over 65% of individuals in rural and urban slum areas of underdeveloped nations and 20–30% of people in more developed countries. In regions of the United States with high levels of dairy and cattle farming, seroprevalence approaches 50%. The worldwide distribution of this disease is confirmed by the high incidence of this infection in patients with AIDS: up to 20% will have identifiable organisms and up to 5% will develop cryptosporidial enteritis.[250,251] Cryptosporidia also have been detected in symptomatic and asymptomatic patients with hemato-

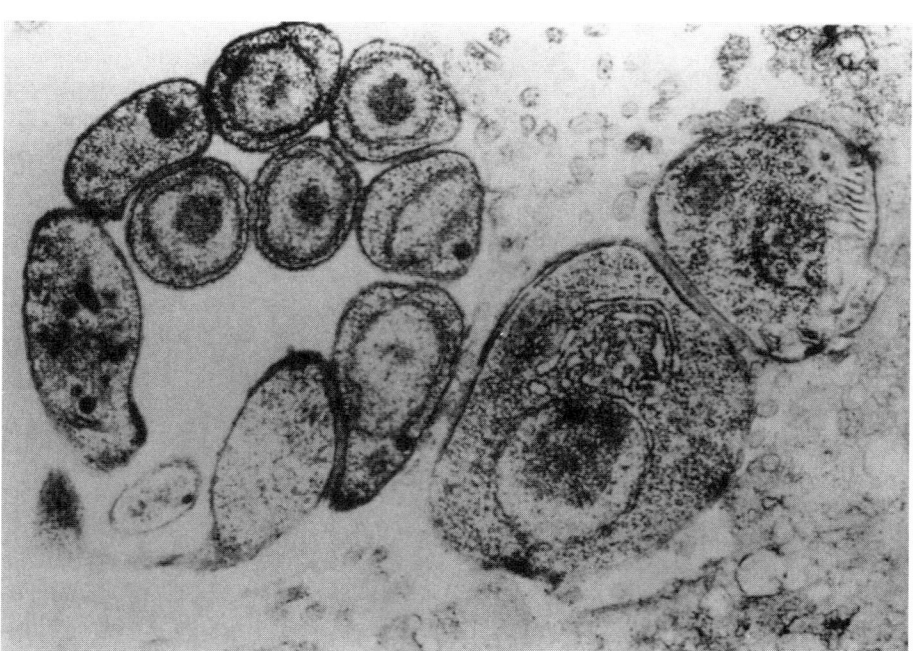

FIGURE 10. Cryptosporidia from the gallbladder of a patient with AIDS. Electron micrograph. ×7500.

logic malignancies and other immune deficits.[259,260] Shedding is sporadic, making epidemiologic studies less reliable. *Cryptosporidium* is probably a common cause of travelers' diarrhea.[261] Organisms are generally transmitted by fecal–oral contamination. Waterborne transmission has also been demonstrated.[262,263] Spread occurs between animals and between man and animals. Human-to-human spread causes the epidemics seen occasionally in day-care centers and in families.[253,264] Studies in immunocompetent patients demonstrated a range of 1.5–10% of diarrheal diseases is caused by cryptosporidiosis.[252] The Massachusetts General Hospital experience has noted that about 3% of individuals with significant diarrheal illness will have *Cryptosporidium* identified in their stools in the absence of other known pathogens. There is no gender preference, but about half are children younger than 5 years of age. Clustering of cases appears to occur during later summer and the fall. Animal-to-person spread is common in animal handlers. *Cryptosporidium* oocysts have been found in rivers throughout the western United States and in sewage. It is likely that the contamination is due in part to the relative resistance of this organism to traditional means of sterilization of common source water supply, including chlorination, iodophors, hypochlorite, and formaldehyde. Water filtration appears to be more effective than is treatment with disinfectants. The relative role of person-to-person spread compared with contamination of public water supplies is unclear. However, water-borne outbreaks have been documented with compromised hosts serving as sentinels for inadequacy of the treatment of public water supplies. Clusters of cases have been documented in rural areas after flooding of well water and with cattle contaminating public water supplies. An outbreak, which may have affected 403,000 residents of Milwaukee, Wisconsin in 1993, was attributed to heavy rain and snow runoff carrying contaminants into the rivers supplying Lake Michigan prior to water treatment that met federal standards. Outbreaks also have been associated with fruits and vegetables washed in contaminated water. Organisms survive freezing and require heating above 72°C for 1 min for inactivation.

Susceptible patients are those with the entire range of immune dysfunction. Young children and newborns may be more susceptible due to immaturity of immune function or to fecal–oral contamination. Infections have been reported after viral infections (measles, chicken pox, infectious mononucleosis). Elderly individuals may be similarly at risk. Significant cryptosporidial diarrhea has been seen in diabetics after gastrointestinal surgery, in organ transplant recipients, in children with primary im-

mune deficiencies, in AIDS patients, and in normal individuals.[239,256,264,265] Both humoral and cellular immune mechanisms appear to be involved in and necessary for protection against *Cryptosporidium*.[266] A syndrome similar to cryptosporidiosis is seen in many patients with AIDS-associated enteropathy who have malabsorption and villous atrophy in the presence or absence of identifiable pathogens.[147,238,249] The significance of small numbers of organisms or of HIV itself in the pathogenesis of this syndrome remains unclear.

4.3. Diagnosis

Cryptosporidium infection should be suspected in any patient with profuse watery diarrhea, but primarily in those in whom an underlying immunodeficiency has been identified. It is probably worthwhile to exclude other possible etiologies of infection. More common infections are those due to toxigenic *E. coli*, *Salmonella*, *Campylobacter*, *Shigella*, and antimicrobial-associated *Clostridium difficile*.[6,237] *Cholera* and *Yersinia* may have a similar presentation.[267] Viral agents are more difficult to identify, but it is likely that adenovirus, rotavirus, and Norwalk virus are more common pathogens.[268] In the immunocompromised host, cytomegalovirus is an important differential consideration as a predisposing factor for *Cryptosporidium* as well as causing primary invasive gastrointestinal infection.[210] Coinfection with Microsporidia is occasionally seen. Mycobacterial infection and histoplasmosis also may cause diarrheal illness in these patients. Many parasites will cause diarrheal syndromes, including *Giardia lamblia*, *Entamoeba histolytica*, *Cyclospora cayetensis*, Microsporidia, and *Isospora belli*.[269,270]

Diagnosis may be made noninvasively by stool testing. Because oocysts are similar in size to yeasts, identification of *Cryptosporidium* requires special staining. Oocysts may be concentrated by the Sheather sugar coverslip flotation method that allows quantitation and identification of infectious organisms. Modified acid-fast staining, Kinyoun's carbol–fuchin negative staining, auramine staining, and safranin staining have largely replaced this laborious method. Also available are indirect immunofluorescent staining with monoclonal antibodies and rapid methods including direct immunofluorescence and ELISA.[271] Serologic tests are under development but have not been refined to allow the diagnosis of acute infection. The immunocompromised host with cryptosporidiosis will not develop a significant serologic response to the organism. It may be necessary and worthwhile to proceed to small or large bowel biopsy. Organisms and typical histopathology are better seen in the small bowel.[270,272] On biopsy, infection may be patchy. Typically, there is a loss or blunting of the villi; crypt abscesses develop in immunologically normal hosts. There is an acute and chronic inflammatory response in the lamina propria. Standard hematoxylin–eosin-stained specimens will reveal small organisms at the tips and between microvilli. Ultrastructural studies demonstrate the presence of an extracytoplasmic but intracellular parasitophorous vacuole in immunocompromised individuals. The range of histopathology is quite broad. Mild inflammation may progress to focal necrosis. Our experience suggests a synergistic injury between cytomegalovirus and *Cryptosporidium*. This affects colonic mucosa and also the esophagus, stomach, appendix, pancreatic and bile ducts, the respiratory tract, and the gallbladder. A few AIDS patients have had cryptosporidial cholangitis in the setting of simultaneous cytomegaloviral infection of the gallbladder. When *Cryptosporidium* is detected in the lungs, it is usually found in association with gastrointestinal cryptosporidiosis.[273,274] In our experience, this has been seen only in AIDS patients and probably reflects aspiration of organisms, rather than acute, primary pulmonary infection.[194]

As noted above, *Cryptosporidium* is detected by either wet mount or fixed preparation of stool or other excretions. The oocysts do not stain with iodine and are orange with Truant's auramine–rhodamine stains. Yeasts are brown with iodine and do not stain with Truant's. On fixed smear, oocysts stain red with dense internal granules on Kinyoun stain, while yeasts stain green. The typical morphology of *Cryptosporidium* also will be seen on Giemsa stain, using a light green counterstain. Improvements in direct immunofluorescent antibody staining may enhance detection. Antigen detection and PCR assays on serum samples have been reported. Specimens should be handled carefully due to the possibility of aerosolizing infectious organisms.

4.4. The Patient

The patient presents with watery diarrhea, abdominal pain, anorexia, nausea and vomiting, fever, and myalgias.[275] They often have been treated for diarrhea in the recent past with incomplete resolution. Less commonly, cholangitis, pancreatitis, reactive arthritis (Reiter's syndromelike illness), and pneumonia may be recognized.[276] Stool examination reveals watery stool without blood or white cells, with intermittent shedding of a large number of cryptosporidial oocysts. Severe diarrhea may be associated with malabsorption, as measured by D-xylose absorption studies, vitamin B_{12} malabsorption, and steator-

rhea. Mucosal thickening and small bowel dilatation may be noted in radiographic studies. Infection of the gallbladder is common. In the immunocompetent patient, the syndrome should resolve in 1 to 3 weeks. Organisms may continue to be shed after the resolution of symptoms. The ability to resolve infection depends on reversal of immune compromise in addition to therapy. Immunity seems to include both B- and T-cell responses, although humoral immunity is incompletely protective. CD4+ lymphocytes appear to control susceptibility and duration of infection while interferon-γ reduces the intensity of infection in animal models.

In the immunocompromised patient, recurrent disease may occur in the absence of therapy. Diarrhea may be significant enough to require hospitalization for dehydration or wasting. The right upper quadrant localization of abdominal symptoms may suggest acute cholecystitis. The gallbladder may be dilated, with thickened walls and dilated bile ducts. In the absence of cholangitis, the syndrome may be mimicked by a number of enteropathies in AIDS patients, including that of primary HIV infection.[6,7] The severe diarrhea associated with cytomegalovirus infection in AIDS patients or after transplantation may be bloody and merits separate therapy.

4.5. Therapy of Cryptosporidiosis

There is no consistently useful therapy for cryptosporidiosis (Table 9). Support with fluids and electrolytes and added nutrition may be necessary. In immunosuppressed patients, the disease will resolve if immunosuppressive regimens can be reduced or eliminated. Antimotility agents have not been demonstrated to be effective. Preliminary encouraging results with the macrolide spiramycin (2–3 gm/day) have not been consistently reproducible.[277,278] Spiramycin does appear to have some efficacy in cryptosporidiosis in the non-AIDS immunocompromised individuals. The drug is poorly absorbed with food. Anecdotal reports of adverse effects during therapy with high-dose spiramycin (1.5 g every 8 hr IV) suggest that some patients have had increased stool output and volume loss with the development of fecal leukocytes, protein loss, and progressive loss of mucosal folds in the presence of very few organisms. Occasionally, apoptosis with loss of columnar epithelium and vacuolization has been observed. In AIDS patients treated with spiramycin, most continue to excrete organisms, but some have had remission of symptoms. Controlled studies of this agent do not exist but are being conducted. Other macrolides have been studied including clarithromycin, which appears to be useful in prophylaxis, and azithromycin, which was

TABLE 9. Therapy for Common Intestinal Parasites of Compromised Hosts

Infecting organism	Primary	Alternative	Comments
Cryptosporidia No therapy proven efficacious. Self-limited in immunocompetent patients	Paromomycin 500–750 mg tid or qid to 100 mg bid with food × 14–28 days, ± azithromycin 600 mg/day × 4 weeks, followed by paramomycin for 8 weeks (*JID* **178**:900, 1998), *or* nitrazoxanide 500 mg bid-qid	Octreotide 50–500 μg tid SC or IV at μg/hr *or* azithromycin 1200 mg PO qd × 28, then 500 mg/day (azithromycin alone not very effective); atovaquone 750 mg bid-tid with food	HAART is most effective (*Lancet* **351**:256, 1998); nitrazoxanide up to 2.0 g/day promising but not FDA approved; Unimed: 1-800-864-6330; diclazuril (not FDA approved) Janssen 800-521-2437; symptomatic Rx; nutritional support; hyperimmune colostrum; benefit of rifabutin? water: boiled 1μ filter
Isospora belli	TMP-SMX 2 DS tab PO bid or 1 DS tid × 10 days, then bid × 2–4 weeks	Pyrimethamine 50–75 mg/d PO + folinic acid 10 mg/d PO × 14–28 days ± sulfadiazine if resistant	Chronic suppression in AIDS patients without HAART response; 1 TMP-SMX 3×/wk *or* (pyrimethamine 25 mg/d PO + folinic acid 5 mg/d PO)
Microsporidiosis (*CID* **27**:1, 1998) Ocular: *Encephalitozoon hellum* or *cuniculi*, *Vittaforma corneae* or *Nosema* sp. Intestinal: *E. bieneusi*, *E. (Septata) intestinalis*. Disseminated: *E. hellum* or *cuniculi*, *E. intestinalis*, *Pleistophora* sp.	Albendazole 400–800 mg PO bid × 21 days to 3 months	Metronidazole 500 mg tid; atovaquone 750 mg PO tid; thalidomide ?(HIV, males) 100 mg qd (teratogen); in HIV+ patients, ?*E. hellum* with fumagillin eyedrops; for *V. corneae*, may need keratoplasty	To obtain fumagillin: 1-800-547-1392. Thrombocytopenia with fumagillin in *E. bieneusi*; HAART most successful (*Lancet* **351**:256, 1998); Dx: Most labs use modified trichrome stain; need electron micrographs for species identification; FA and PCR methods in development

not effective in clinical trials. Diclazuril sodium, a benzeneacetonitrile derivative, and letrazuril, which has greater bioavailability, did not prove effective in clinical trials. Paromomycin has reduced parasite carriage in some patients but does not appear to produce cure in trials. Nitrazoxanide also is in trials with some data suggesting clinical improvement and reduction in oocyte shedding in small numbers of patients. Alpha-difluoromethylornithine (DFMO) also has demonstrated some palliative effect on infection, but toxicities (largely bone marrow depression) have limited its use. Preliminary animal studies using hyperimmune sera against *Cryptosporidium* or bovine colostrum are encouraging, but efficacy in humans has not yet been demonstrated.[279,280] Bovine colostrum is inhibitory for cryptosporidial growth *in vitro*. A few patients have been treated with transfer factor, diclazuril, and leclazuril, but these agents have not proved useful in clinical trials.

Somatostatin has been useful in reducing the severity of the secretory-type diarrhea in a few patients.[281] Octreotide is a somatostatin analogue that also has been useful in treating secretory diarrhea.[282,283] In AIDS patients with refractory symptoms, stool frequency and volume often decrease significantly during therapy with octreotide. Unfortunately, patients with no identifiable organisms appear to do better with this agent than do individuals with documented cryptosporidiosis.

It is worth remembering that organisms shed by patients are infectious. Precautions are necessary to prevent spread from infected individuals within the hospital setting.

5. *Isospora belli*

Isospora belli is a coccidian protozoan that infects the GI tract of immunocompromised individuals. *Isospora* was described in 1915, but is still poorly understood. Isosporiasis is a common cause of diarrheal disease in tropical regions but is found worldwide.[284] The common forms of disease are due to *I. belli* and occasionally *I. hominis*. Both are normally of low native virulence. These organisms also can cause or contribute to disease in immunocompromised individuals.

5.1. The Organism: Life Cycle and Epidemiology

Isospora is epidemic in tropical and subtropical areas, including parts of South America, Africa, and Southeast Asia. Person-to-person spread probably accounts for outbreaks seen in the institutional setting. *Isospora* has been demonstrated in 0.2–1% of patients with AIDS in the United States and in up to 15% in Haiti, but its true prevalence is not known. The mechanism of acquisition of *Isospora* infection is not known. It is likely that many carriers of the disease are asymptomatic.

The organism completes its life cycle within the human host, and both sexual and asexual cycles can continue indefinitely within the GI tract. The infection is transmitted by an elliptical oocyst approximately 25–30 μm × 10–15 μm in size. There are two internal sporocysts containing four sporozoites each. Ingestion of sporulated oocysts releases infectious sporozoites that invade the intestinal epithelium and undergo asexual and sexual reproduction. Unsporulated oocysts also are formed and are shed intermittently and mature outside the infected individual. Oocysts can remain viable in the environment for months.

5.2. The Patient

Isospora belli causes diarrhea in immunocompetent patients, especially young children and residents of chronic psychiatric care facilities. *I. belli* also causes chronic diarrhea in malnourished individuals and in AIDS patients who are not receiving cotrimoxazole prophylaxis for *Pneumocystis* and in those without access or response to immune reconstitution with antiviral therapies (HAART).[241,285] In the immunocompetent patient, *I. belli* infection has an incubation period of 7–10 days and causes self-limited diarrheal illness. The patient presents with watery, nonbloody diarrhea with nausea, abdominal pain, and weight loss. Systematic signs (headache, malaise, myalgias) are often present, although high-grade fever is unusual. The patient continues to secrete infectious oocysts for weeks after acute infection. Some individuals have presented with a more chronic abdominal pain or diarrhea syndrome. Prolonged infection may cause malabsorption. The infection is self-limited in the normal host, clearing in 4–6 weeks.

5.3. Histopathology and Diagnosis

Diagnosis of isosporiasis is established by identification of *Isospora* oocysts in fecal specimens in the appropriate clinical setting. Organisms can be identified by the modified acid-fast stain, sugar flotation, auramine–rhodamine stain, or intestinal biopsy. Histopathology of the biopsy specimen reveals mucosal atrophy, villous blunting, crypt hypertrophy, and inflammation of the lamina propria, primarily with eosinophils, lymphocytes, and plasma cells. The organism is found in vacuoles within the cytoplasm of epithelial cells. Extraintestinal dissemination of the organism is rare. Multiple stool specimens may be necessary to demonstrate organisms because of small numbers of shed oocysts. Fecal leukocytes may be absent. Charcot–Leyden crystals are often seen in stool samples. String tests, duodenal aspirates, and small bowel biopsies have been useful in detecting organisms. Electron microscopy may be necessary to find organisms on colonic biopsy. There are no useful serologic tests at present. A mild leukocytosis and eosinophilia may be seen in the peripheral blood smear.

5.4. Therapy

Isosporiasis responds to therapy with oral trimethoprim-sulfamethoxazole (TMP-SMX) at a dose of 160 mg TMP component four times a day for 10 days (Table 9).[285,286] In AIDS patients, prolonged additional therapy with TMP-SMX (twice a day for an additional 3 weeks) has been necessary for clearance of oocysts.[287] Prophylactic therapy with TMP-SMX (once a day) or pyrimethamine–sulfadoxine (Fansidar) has been useful in preventing relapses. Other "successful" therapies may be useful in part because of the treatment of other concomitant infections.[286] Patients may continue to excrete organisms long after the successful completion of therapy. This observation probably supports the use of prophylactic therapy in AIDS patients or other symptomatic immunocompromised hosts. In transplant patients with isosporiasis and in AIDS patients, reduction in immune suppression or antiviral therapy with immune reconstitution is needed for complete resolution.

6. Microsporidia

The Microsporidia make up a phylum consisting of approximately 80 genera and over 700 species of organisms. These are obligate, intracellular, spore-forming, protozoal parasites that were first identified in 1857 as causing disease in insects, fish, snails, rodents, and some primates. They are occasionally found in irrigation and drainage ditches and in surface water. These organisms were rarely implicated in clinical disease prior to the advent of AIDS.[288–290] Case reports describe a series of children with seizure disorders and children and adults with corneal ulcerations, keratitis, or iritis. Keratitis appears to be a rare manifestation of microsporidiosis of the immunocompetent host. Disseminated disease has been described in a young boy with thymic aplasia. Autopsy of this child revealed disseminated Microsporidia involving the lungs, stomach, colon, kidneys, adrenal glands, heart, liver, and other muscles. The organism was identified as *Nosema connori*. Cases of microsporidiosis have been elicited by treatment with corticosteroids as well as in other immunodeficient states. Corneal infections with Microsporidia have been identified in both AIDS and non-AIDS immunocompromised patients.

Microsporidial infection has been identified in patients with AIDS since the mid-1980s as a cause of chronic diarrhea with weight loss.[267,289] In general, the cause of this form of infection has been identified as *Enterocytozoon bieneusi*, for which man is the definitive host. Other common forms in humans include *Encephalitozoon hellum*, *Enc.* (formerly *Septata*) *intestinalis*, *Pleistophora* spp., *Trachipleistophora hominis* and *T. anthropophthera*, *Vittaforma corneae* (formerly *Nosema corneum*), and *Brachiola vesicularum*. Further genera have been identified in small numbers of patients. These relatively undefined species have been attributed to groupings called either "*Nosema*-like" or "collective groups" of Microsporidia based on the region of origin. Up to 30% of AIDS patients with weight loss and chronic diarrhea are infected with Microsporidia in some series. Disseminated disease in AIDS patients has involved skeletal muscle, kidney, liver, eye, and intestinal wall as well as intestinal epithelial cells. In solid organ transplant recipients (heart, lung, liver, kidney–pancreas) and BMT recipients, microsporidial diarrhea, pneumonitis, keratitis, and encephalitis have been observed.

6.1. The Organism

Microsporidia have spores ranging in size from 1 to 20 μm.[291] Spores infecting mammals are generally 1–2 μm in diameter, requiring electron microscopy for identification, using biopsy specimens or aspirates of affected large or small bowel. These spores have thick walls, allowing them to persist outside the host and also making them difficult to stain. They are generally gram-positive and contain PAS-positive granules. Birefringent spores will be found throughout the fibrous stroma but will not stain on hematoxylin–eosin stain. Organisms are found in the supranuclear cytoplasm of cells as binucleated spores containing a coiled polar tube. Small-bowel biopsy reveals the greatest number of organisms. All the develop-

mental forms of the organism's life cycle occur within the epithelial cells of the small bowel.[292] Infection of the liver produces a granulomatous hepatitis. Organisms can spread via blood, lymph, or infected macrophages throughout the body. Significant accumulations have been found most often in brain or kidney. Brain infection results in focal seizure disorders, while kidney involvement may produce interstitial nephritis of some severity. Organisms have been identified in the prostate and urethra of infected individuals although formal demonstration of sexual transmission has not been made. In non-AIDS patients, a granulomatous vasculitis of cerebral vessels may produce a meningoencephalitis. It appears that cell-mediated immunity is critical for protection against significant disease due to the Microsporidia.[288]

6.2. The Patient

Like individuals with cryptosporidiosis and isosporiasis, patients with microsporidial infection present with chronic watery, nonbloody, nonmucoid diarrhea without fever. Patients have weight loss despite continuing to eat well. Intestinal mucosae are normal to minimally inflamed endoscopically. Enterocyte degeneration accompanies partial villous atrophy and modest inflammation in the lamina propria. Person-to-person spread seems likely. By contrast, patients with "AIDS-associated enteropathy" without identifiable infection are more likely to have malabsorption, including lactase deficiency, with minimal changes in villous morphology. The possible mechanisms of this syndrome include bacterial overgrowth secondary to local immune deficiency. Some patients have presented with myositis (*Pleistophora* sp.), bronchiolitis and nephritis (*Enc. hellum*), sinusitis (*Enc. intestinalis*), or hepatitis (*Enc. cuniculi*). The mechanism for dissemination remains uncertain.

6.3. Diagnosis

Suspicion of the presence of Microsporidia may be raised by the presence of organisms seen on Brown–Brenn-stained tissue sections (Fig. 11). The organism is occasionally seen on Giemsa-stained impression smears and on thin sections stained with methylene blue–azure II with a basic fuschin or toluidine blue counterstain. Species identification requires electron microscopy.[291] Unconcentrated stool specimens or duodenal aspirates can be screened after fixation in 3 volumes of 10% formalin or on thin smears made with methanol fixation. Small bowel biopsy (distal duodenum) provides the optimal specimen for histologic demonstration of enteric infection. Staining

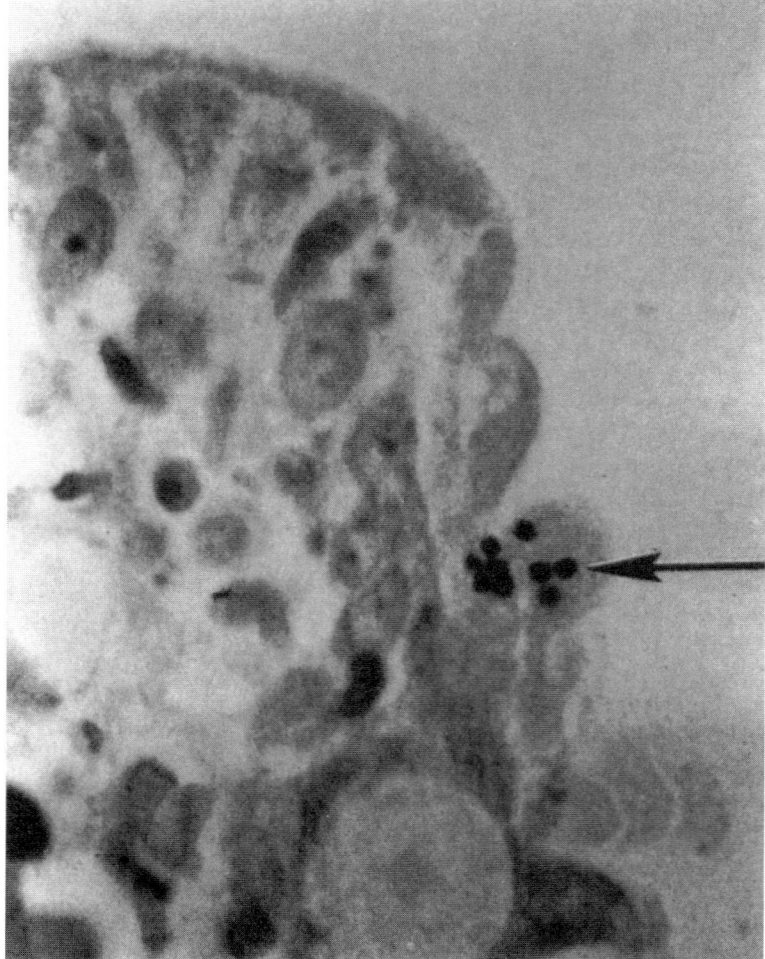

FIGURE 11. Microsporidia within the jejunal epithelium of a patient with AIDS (←) are gram-positive spores by Brown–Brenn stain. ×1750. Courtesy of Drs. R. Weber and R. T. Bryan, Parasitic Diseases Branch, Centers for Disease Control.

with a mixture including chromotrope 2R (Harleco, Inc.) has been used by some workers in place of Giemsa, toluidine blue O, Gram's stains, stool concentration, or electron microscopic examination.[293] Intestinal biopsies reveal villus atrophy, crypt hyperplasia, and variably increased numbers of intraepithelial lymphocytes usually with minimal neutrophilic infiltration. *Enc. intestinalis* is often more widespread and may be found in macrophages of the lamina propria. Non-AIDS patients have a more severe inflammatory response and greater destructive ulceration of corneal tissues during microsporidial infection. The diagnosis of microsporidium infection can be made by electron microscopy of corneal or conjunctival scrapings. A variety of serologic tests have been developed to detect antibodies binding to spores and extracts of organisms. Immunofluorescence microscopy is not generally useful for stool diagnosis but may be useful with

intestinal biopsies and with tissues in disseminated disease. Molecular studies of these organisms have also led to the development of a variety of nucleic acid tests that are available on a research basis. These tests are being refined for clinical use.

6.4. Therapy

Therapy for microsporidiosis depends on reversal of immune deficits (Table 9). In AIDS, immune reconstitution in response to HAART therapy has been associated with remission of chronic disease. Albendazole (400 mg PO bid for 3–6 weeks) has had some efficacy in reducing disease activity and reducing or eradicating parasites from stool, sputum, eye scrapings, sinuses, and other sites. These infections generally have been due to *Enc. intestinalis*, *Enc. hellum*, or *Enc. cuniculi* but not *Ent.*

bieneusi. Thalidomide, 100 mg per day (a teratogen), has been reported to be useful in some HIV-infected patients infected with *Ent. bieneusi*. Fumagillin (for *Enc. cuniculi*) and an analogue TNP-470 have limited activity against *V. corneae* and *Enc. intestinalis* in cell cultures. Furizolidone and fumagillin have been used in small numbers of AIDS and organ transplant patients with *Ent. bieneusi*. Many agents including thiabendazole, 5-fluorouracil, sparfloxacin, nifedipine, and itraconazole have limited activity. Other agents with reported efficacy may have activity against co-pathogens: azithromycin, atovaquone, metronidazole, and thiabendazole. Symptomatic therapy and nutritional support are important adjuncts to therapy. The frequency of isolation of Microsporidia (0–30%) varies with the region from which the patients are derived and with travel to tropical regions. It is likely that improved diagnostic techniques will further increase recognition of this phylum in many immunocompromised hosts.

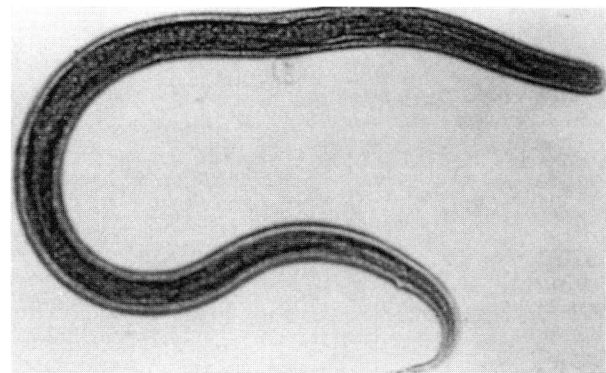

FIGURE 12. *Strongyloides stercoralis* larva isolated from the sputum of a patient with disseminated strongyloidiasis due to an ACTH-secreting tumor.

7. *Strongyloides stercoralis*

Since the original association of *Strongyloides stercoralis* with "Cochin China diarrhea" in 1876, strongyloidiasis has been recognized as an important human intestinal pathogen. This nematode currently infects almost 100 million people worldwide.[294,295] This parasitic helminth can complete its entire life cycle within the human host, allowing for persistent and occasionally lifelong infection. In the immunocompromised host, with increased numbers of larvae completing the autoinfection cycle, large numbers of worms enter the systemic circulation producing the "hyperinfection syndrome."[296,297] When organisms are found in organs not generally associated with the life cycle of *Strongyloides*, "disseminated" or overwhelming infection is said to have occurred. The presence of a persistent carrier state greatly enlarges the at-risk population for severe disease during periods of immune suppression.

7.1. The Organism

Strongyloides stercoralis is a nematode found worldwide in tropical and subtropical regions in warm, moist soil often contaminated by human feces. The organism completes a complex life cycle within the human host.[295] The infective form is the filariform larvae that penetrate exposed skin areas that come into contact with the soil. These larvae follow the venous circulation to the right heart and to the pulmonary alveolar capillary bed (Fig. 12). The worms then penetrate into the alveolar space, are carried up the bronchial tree, and are swallowed. Some larvae migrate through other tissues, especially muscles, producing local symptoms. Male worms cannot penetrate the mucosa and are excreted. Beneath the small intestine mucosa, female larvae mature through two molts, producing adult female worms that produce fertilized eggs through parthenogenesis. Eggs mature to first-stage rhabditiform larvae within the intestinal mucosa. Maturation of the rhabditiform larvae occurs over 24–48 hr. These larvae are passed in the stool or enter an autoinfective cycle within the gastrointestinal tract. The autoinfective reproductive cycle allows for enhanced growth in the absence of immunologic controls. Some of the rhabditiform larvae mature within the intestinal lumen and penetrate into the vascular tree via the wall of the bowel or perianal skin (external autoinfection) to reinitiate the cycle. Larvae passed with the stool may become infective either via direct maturation or via intermediate sexual development into male and female forms. The filariform larva is approximately 600 μm long, while the rhabditiform larva is approximately 200–300 μm long. The free-living female worm is 1 mm long; the adult male worm is slightly shorter.

7.2. Epidemiology

Strongyloides has been found in temperate and tropical regions. It is an unusual parasite in that the human is the major host, although some other animals, including cats, dogs, and subhuman primates, may harbor active infections. The frequency of infection of these animals in endemic areas is not known. The parasite is found worldwide and is most common in the tropics and subtropics.

Approximately 1% of dogs in the eastern United States may be infected. Patients may have no history of travel to an endemic region. Chronic strongyloidiasis is a condition of relatively low worm burden restricted to an auto-infective cycle between the skin and the intestinal tract, but without a sufficient immune response to clear the infection. Many give a remote history of rash associated with febrile or diarrheal illness. Chronic infection has persisted for over 30 years in some patients.

The exact components of the immune system responsible for prevention of disease or the reduction of the severity of infection are not known.[295] Disseminated infection has been reported in people with a broad array of immune defects.[298] This population includes individuals with hematopoietic malignancies or connective tissue disease being treated with immunosuppressive therapies (notably corticosteroids); hosts with congenital or acquired hypogammaglobulinemia, chronic malignancies, malnutrition, severe burns, or alcoholism with hepatic cirrhosis; and persons with occupational exposure to contaminated feces.[299–306] Increased corticosteroid dosages used for the treatment of *Strongyloides*-induced bronchospasm or for organ graft rejection have been associated with the development of disseminated infection.[307–312] Renal and hepatic transplantation have been associated with strongyloidiasis. Disseminated disease has been reported only in renal transplant recipients and anecdotally in hepatic transplant recipients receiving tacrolimus. All reported cases were in patients receiving corticosteroid therapy. Cyclosporin A, but not tacrolimus, has been reported to exert an inhibitory effect on *Strongyloides*.[313] Whether corticosteroids play a direct role in the maturation of rhabditiform larvae (via mimicking of parasite ecdysteroid hormones) is uncertain. The graft may also transmit the organism. Abdominal surgery or steroid therapy for apparent ulcer disease, ulcerative colitis, or Crohn's disease can exacerbate underlying infections.[309,314] The hyperinfection syndrome has also been reported in normal individuals without apparent predisposing immune defects.

Strongyloidiasis has not been associated with AIDS in the absence of other factors contributing to susceptibility (e.g., chemotherapy, malignancy, malnutrition). Strongyloidiasis has been reported to be more common in sexually active homosexual men with normal immune function. Direct transmission of the parasite occurs via rectal intercourse, by oral–anal exposure, or by contact with skin in the perianal area. Few patients with AIDS are reported to have developed disseminated strongyloidiasis.[7,302,306] This finding contradicts the perception that cellular immunity is solely responsible for protection against *Strongyloides*. The relative absence of this finding in AIDS may reflect underdiagnosis or unreported cases of disease, but it is apparent that this infection is still more common in other classes of immunocompromised individuals.

7.3. Pathogenesis

The role of the immune system in the modulation of infection due to *Strongyloides* is demonstrated by changes in the course of disease in the presence of immune suppression (Figs. 12 and 13). In the normal host, repenetration of the gut wall by maturing filariform larvae is limited. Long-standing infection may be associated with minimal fibrosis, with adult worms in the crypts of Lieberkühn and with eggs and larvae in the bowel lumen. Chronic infection produces minimal inflammation of the bowel wall with some villous blunting. With immune suppression, especially in the setting of corticosteroids, penetration by the larvae through the gut wall increases. In this setting, worms accumulate in the lungs in significant numbers. This penetration produces bowel wall edema with mucosal ulceration and mucous secretion. Depending on the level of immune suppression, the inflammatory response on the wall of both the small and large intestine may be severe or minimal. The inflammatory response is both acute and chronic and includes plasma cells, eosinophils, histiocytes, giant cells, lymphocytes, and neutrophils. Granuloma formation may occur around degenerating larvae, but is often absent during immune suppression.

Bacterial or fungal infection carried from the lumen of the GI tract may cause acute infection of the bowel wall, peritonitis, or sepsis. Involvement of the CNS is common, with meningitis due to *Strongyloides* and to accompanying organisms. Gram-negative bacteremia, pneumonia, and meningitis are common features of disseminated disease. Bacteria are thought to escape via mucosal breaks or attached to migrating worms. We also have observed recurrent episodes of polymicrobial sepsis and peritonitis in renal transplant patients with hyperinfection syndrome.

7.4. The Patient

Acute GI infection generally will produce epigastric fullness or pain and in some individuals diarrhea and malabsorption.[315] Passage of larvae through the lungs may produce eosinophilic pneumonia or "Loeffler's syndrome" or milder manifestations of dyspnea, cough, bronchospasm, and fever.[316] Most of these individuals

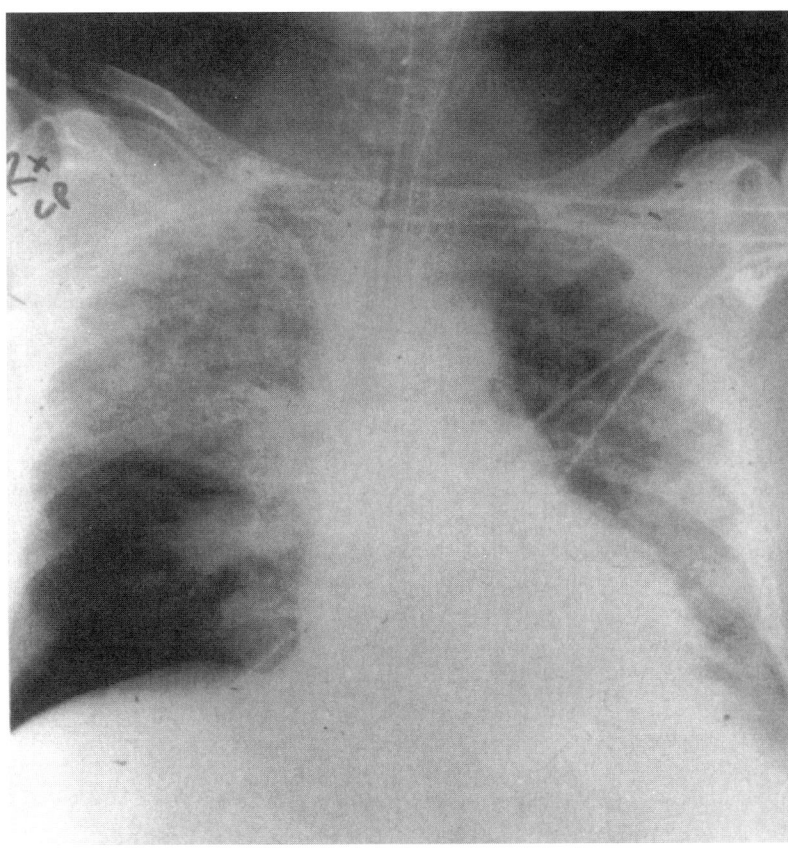

FIGURE 13. Chest radiograph of a patient with disseminated strongyloidiasis. The patient had fever, hemoptysis, hypoxemia, and gram-negative bacteremia with fluctuating, dense pulmonary infiltrates.

will have a peripheral blood eosinophilia. Gram-negative septicemia and necrotizing pneumonia may occur numerous times without specific treatment of the underlying worm infection. Gram-negative meningitis is a common complication of strongyloidiasis; larvae are infrequently detected in the meninges or CSF of affected patients. Bacterial superinfection as a complication of strongyloidiasis is equally common in normal and in immunocompromised individuals, supporting the presumption that bacterial infection is a function of worm penetration, rather than of the underlying immunodeficiency state. The mechanism of such superinfection is unknown, but superinfection occurs in one third to one half of individuals with disseminated infection. Most individuals with chronic infection are asymptomatic. Others have abdominal pain, diarrhea, and urticaria.[301] In chronic infection, respiratory symptoms are less prominent; however, immune complex disease may cause arthritis. The two common skin manifestations of strongyloidiasis include (1) a migratory, pruritic, raised, linear rash called "creeping eruption" or "larva currens" and (2) crops of urticarial eruptions that appear to be manifestations of immediate

hypersensitivity reactions to migrating worms. Two thirds of individuals with chronic infection will develop transient and recurrent urticarial eruptions on the skin of the waist and buttocks. The migratory rash may move across the skin at a rate of up to 10 cm/hr. It is unclear whether these eruptions are a reaction to the worms themselves, secreted antigens, or components of the skin or gut flora that are carried along with the organism. Visceral or cutaneous migration appears to be much slower but may produce a similar rash.

7.5. "Hyperinfection Syndrome" and Disseminated Strongyloidiasis

The predilection of this organism for the lungs and the CNS is manifested most impressively with disseminated infection in immunocompromised individuals (Figs. 12 and 13).[296,303] The complications associated with dissemination reflect both a large worm burden and the effects of organisms accompanying the migrating nematodes. Local and systemic infections and allergic responses may be seen to both the worms and the "passenger" bacteria from the gut. Unlike the transient eosinophilic pneumonia that may be seen with acute infection, the hyperinfection syndrome is accompanied by significant pneumonitis.[312] Pulmonary bacterial superinfection occurs in the setting of small-airway obstruction secondary to entrapped worms. Pneumonitis is generally accompanied by abdominal crisis: severe abdominal pain with ileus, small-bowel obstruction, and occasionally septic shock. Hepatic failure has been reported; CNS involvement may include eosinophilic meningitis, altered mental status, coma, or focal neurologic deficits. Polymicrobial bloodstream infection may be seen and includes the entire range of gut flora, including *Candida*. Cavitary pulmonary lesions may develop, and transient rashes or skin swelling of the buttocks or lower abdomen may be noted. Peripheral blood eosinophilia is variably observed. Mortality with disseminated infection generally exceeds 75% and is usually due to gram-negative sepsis. Because the consequences of disseminated *Strongyloides* are so grave, *preemptive* treatment should be instituted prior to elective immune suppression in patients with exposures to endemic regions, even without identification of organisms.[294] Disseminated infection has been observed up to 3 years after solid organ transplantation, but generally during periods of prolonged neutropenia following chemotherapy or hematopoietic transplantation.

7.6. Diagnosis

Early diagnosis and therapy are the main determinants of outcome. The diagnosis of strongyloidiasis should be suspected in the presence of GI symptoms accompanied by urticaria or eosinophilia or in individuals who have lived in endemic areas. Because systemic manifestations may be altered by the presence of immunodeficiency, clinical suspicion in the presence of rhabditiform larvae found in stool or in duodenal aspirates should be considered sufficient for the diagnosis in the appropriate host. Strongyloidiasis needs to be considered in the appropriate patient whose pneumonia does not respond to therapy based on sputum examination and culture data. GI symptoms are frequently missed because of the preeminent pulmonary or CNS symptoms.[315] Negative stool examinations may be misleading. Eosinophilia is frequently absent in disseminated infections and in patients receiving corticosteroids.

The diagnosis of strongyloidiasis is based on the demonstration of filariform larvae in stools, sputum, or CSF. In chronic infection, small numbers of larvae may be hard to find. Larvae may be found by use of the Enterotest capsule ("string test"), which contains a long nylon thread. The swallowed capsule releases the thread, which is withdrawn after several hours and may be coated with mucus containing larvae. Duodenal aspiration or purged stool specimens also may reveal organisms not detected in routine specimens. Sputum examination may reveal bacterial or fungal pneumonia in the absence of identifiable larvae. Diagnosis is occasionally made on unstained wet mounts of bronchoalveolar lavage specimens or transtracheal aspirates from infected lungs.[306] Formal-ether concentrates of sputum may be of use if the larvae are few in number. The worm also will be seen by Papanicolaou stain or by experienced observers on Gram's and acid-fast stains of concentrated specimens. A modified agar plate is useful in stool examination. Serologic testing is rarely available in a time frame useful for clinical diagnosis, but should be used for screening of patients from endemic regions. Molecular diagnostics are under development.

Neither chest nor abdominal radiologic studies are diagnostic of infections. Chest X rays show patchy or diffuse bilateral pulmonary infiltrates (Fig. 13). Pulmonary processes may be transient or progress to consolidation, especially in the presence of bacterial superinfection. These processes clear with appropriate therapy. Barium swallows demonstrate duodenal and jejunal dilatations and bowel wall edema with narrowing in areas of fibrosis. Disseminated disease is frequently accompanied by dilatation of the small bowel with air–fluid levels.

In the immunologically normal individual with active infection, higher titers to *Strongyloides* antigens have been demonstrated. The utility of these tests in individuals with disseminated infection has not been established. Both immunofluorescence assays and ELISAs have been described. These assays may become useful in detecting at-risk populations before the initiation of immunosuppressive therapy. Antigen detection tests are under development.

7.7. Therapy

All patients infected with *S. stercoralis* should be treated. Uncomplicated GI infections may be treated successfully with thiabendazole, 50 mg/kg per day (PO), divided into two doses, to a maximum dose of 3 g/day over 2 or 3 days. Thiabendazole has many toxicities, including nausea, vomiting, dizziness, and occasionally a sense of disembodiment and of urine odor. In uncomplicated disease, lower doses are likely to be effective, and confirmation of the clearance of worms should be made by stool examination at 6 and 12 months after therapy. Disseminated disease is treated with the same drug for a period of 5–7 days. In our experience, thiabendazole is effective but poorly tolerated in cirrhosis or transplant recipients. A number of drugs are being studied but are of less certain efficacy, including mebendazole, cambendazole, and albendazole in high doses. Ivermectin (200 μg/kg per day for 1–2 days) is useful and better tolerated than thiabendazole in clinical trials and our experience but is not FDA approved for this indication.[317] In most centers, ivermectin is considered the drug of choice for disseminated disease. Albendazole (400 mg PO qd × 3 days) also has been useful in some individuals. Side effects of therapy may include hypotension, neurotoxicity, and leukopenia with mild elevations of liver function tests. Patients with disseminated infection are treated for 5–7 days for strongyloidiasis, but may require more prolonged therapy for secondary bacterial infections of the lungs, DNS, or abdomen. Immunosuppressive agents should be reduced as much as possible and areas of focal infection drained. Repeated courses of therapy may be necessary in chronically immunosuppressed individuals (i.e., those with AIDS or organ transplants). Reports of failures have been made with each of these regimens. Therapeutic failures with thiabendazole may be successfully treated with ivermectin. Patients at risk for dissemination due to immune suppression should be evaluated for the presence of the carriage state.

Illustrative Case 3

A 66-year-old man immigrated to the United States from the Dominican Republic approximately 30 years prior to this admission. He had a history of heavy smoking and alcohol abuse and had a sister with a history of active pulmonary tuberculosis. During an evaluation for hypertension and angina 3 years prior to admission, the patient was noted to have a 5-mm nodule in his left chest on chest X ray. No further evaluation of this nodule was performed at the time. A few days prior to admission to Massachusetts General Hospital, he presented to an outside hospital with chest pain, nausea, and vomiting. He was ruled out for myocardial infarction but was found to have exercise-induced electrocardiographic changes. His laboratory evaluation included the following values: potassium, 1.9; chloride, 89; bicarbonate, 49; glucose, 150–180; a fasting cortisol of 72.3 (8 AM) and 56.4 (12 PM). A review of systems revealed an anxious man with bilateral lower extremity edema and increased abdominal girth with progressive dyspnea on exertion. His appetite was increased but he had lost 7 kilograms in weight over the past few weeks. The patient had new-onset diarrhea without abdominal pain and a cough with occasional hemoptysis but with little sputum production. He did note intermittent fevers as high as 101°F and penile paresthesias when urinating. He was allergic (hives) to sulfa drugs.

On physical examination, the patient was a Hispanic man who appeared cushingoid in body habitus. His skin examination revealed hyperpigmented knuckles and nail beds. He had three-plus pitting edema to the knees without clubbing. His chest X ray (Fig. 13) revealed a right middle lobe infiltrate with mediastinal widening and a generalized interstitial pattern. His oral examination revealed thrush.

His white blood count was normal without eosinophils. His ACTH was 665 (normal range, 10–56). His amylase was 269 (45–113); his ionized calcium was 0.94 (1.14–1.30). Stool examination was negative

for pathogenic bacteria or parasites. Sputum examination and cultures were unremarkable.

The patient developed progressive respiratory distress requiring intubation and transfer to the medical intensive care unit. Deep-suctioned sputum examination was positive for both *P. carinii* and *S. stercoralis* (Fig. 12). He was treated with thiabendazole, intravenous pentamidine, and broad-spectrum antimicrobial agents. His admission blood cultures grew enteric gram-negative rods overnight and sputum cultures grew CMV. The patient failed to clear his *Strongyloides* infection with thiabendazole and was treated with ivermectin. Nonetheless, he progressed to frank perforation of his colon. Liver biopsy during laparotomy revealed a malignant small-cell neoplasm consistent with small-cell carcinoma of the lung. The patient expired on the 14th hospital day.

Comments. This case demonstrates the importance of a careful epidemiologic history in the management of the immunocompromised patient. Patients from areas endemic for *S. stercoralis* remain at risk for disseminated infection ("hyperinfection") for many years after the initial exposure. This common presentation included a negative stool examination for *S. stercoralis* and a normal sputum Gram's stain. The patient was immunocompromised due to an ACTH-secreting small-cell carcinoma presumed to be lung-derived. Pulmonary infection with *Strongyloides* is probably less important than the frequency of bronchial obstruction with subsequent bacterial superinfection. The cataclysmic event in this patient's course was bowel perforation and gram-negative sepsis *despite* appropriate antibiotic therapy. Sustained bacteremia even without perforation is a common complication of strongyloidiasis in the compromised host. Patients may have paralytic ileus with or without perforation, hemorrhagic and fluctuating pulmonary infiltrates progressing to adult respiratory distress syndrome, bacteremia or bacterial pneumonia, and parasitic or bacterial meningitis, *or both*. While eosinophilia (>25–30%) is common, this finding is usually absent in patients receiving corticosteroids. Larvae are detected in sputum or pulmonary aspirates, especially when concentrated prior to examination.

In patients known to be at risk for *Strongyloides* infection, aggressive evaluation or preemptive therapy is mandatory *prior to* initiating immunosuppression, e.g., organ transplantation or chemotherapy. Larvae have been detected in purged stools when routine examinations are unremarkable. Ivermectin has been used successfully in many such patients and is generally preferred over thiabendazole. The presence of *P. carinii* speaks to the chronicity of immune suppression. CMV is an important cofactor to many systemic infections, including *Pneumocystis*, and may contribute to bowel ulceration and/or perforation in the compromised host. In general, such patients will have a negative CMV antigenemia assay in the peripheral blood and will require endoscopy with biopsy for diagnosis.

8. *Entamoeba histolytica* and Amebiasis

Entamoeba histolytica is a parasite of worldwide distribution that lives in the intestinal lumen of over 500 million people.[318] The organism is usually a benign commensal with disease occurring in about 10% of exposed individuals. Under appropriate conditions, *E. histolytica* invades the intestinal mucosa, causing dysentery, mass lesions (ameboma), or extraintestinal lesions including liver abscesses. Because the basis of the conversion from commensal to invasive parasite is unclear, the mechanism by which amebiasis is exacerbated by immune suppression also is not understood. Invasive amebae (*E. histolytica*) can be distinguished from nonpathogenic forms (*E. dispar*) using antibody staining, zymodeme analysis, genetic probes (e.g., for adhesion molecules and enzymes), and restriction mapping of amebic strains. The significant clinical features of amebiasis include the separation of nonpathogenic from pathogenic amebae, the distinction of lumenal infection from invasive disease, and the consideration of this diagnosis prior to the initiation of immunosuppressive therapies.

8.1. The Organism

E. histolytica is a pseudopod-forming protozoan of the family Endamoebidae, order Amoebida, class Lobosea, which also includes other human parasites: *E. hartmanni* ("small amebae"), *E. polecki*, *E. coli*, and *E. gingivalis*. Rarely, amebae common in animals have been found in humans; these organisms have not been implicated in disease. The motile form of the organism is the trophozoite, which lives in the colonic lumen. The trophozoites divide and encyst, producing thick-walled cysts with four nuclei. Unlike other parasites, these cysts do not persist in tissues and are found only in the lumen of the human bowel. Trophozoites are fragile and have not been implicated in the transmission of disease. Trophozoites are found in tissues during invasion. Strain differences appear to be common in *E. histolytica*. These differences account for geographic variation in the pathogenicity of the organism. There are antigenic, enzymatic, and genetic (cDNA) differences between pathogenic and nonpathogenic strains of amebae. Characteristic electrophoretic patterns of isoenzymes or "zymodemes" may distinguish pathogenic strains.[319] *In vitro*, it was thought to be possible to change a nonpathogenic organism into a pathogenic organism by altering the bacterial flora surrounding the organisms in culture. However, these studies have not been reproducible and it is not clear that this conversion occurs *in vivo*. The determinants of pathogenicity are not clear. *E. histolytica* and *E. dispar* are indistinguishable morphologically.

8.2. Life Cycle

The life cycle of the organism begins with the ingestion of food contaminated by infective cysts, which can survive for weeks in the environment. Excystment in the small bowel releases motile trophozoites covered with filopodia used for epithelial attachment in the colon. The organism grows in a low-oxygen environment. Encyst-

ment of trophozoites occurs only in the large bowel. The characteristic of hematophagia is considered diagnostic of invasive amebae and active disease. The nonpathogenic cysts of *E. hartmanni* are much smaller in diameter (10 μm) than those from *E. histolytica* (8–20 μm). Humans are definitive hosts for *E. histolytica*. The development of good animal models has been difficult.[320] However, characteristic intestinal and hepatic disease has been produced by coinfection using pathogenic *E. histolytica* with bacteria and by direct inoculation of organisms into the hepatic parenchyma or circulation.[320,321]

8.3. Epidemiology

Infection due to *E. histolytica* is common in tropical regions, in parts of Central and South America, India, and western and South Africa. Up to 50% of the residents of some areas are infected. People in these hyperendemic regions are subject to constant reinfection. Occasional epidemics of amebiasis are related to contamination of a water supply or to unsanitary conditions surrounding institutionalized individuals. Individuals exposed while traveling usually have resided in endemic areas for a month or longer. Amebiasis is the third leading cause of mortality due to parasitic infection, following malaria and schistosomiasis.

Infection is most common in persons of lower socioeconomic status, with crowded living conditions and poor sanitation, in institutionalized individuals, and in promiscuous male homosexuals. Male homosexuals in the United States have a fixed high incidence of carriage of *E. histolytica*, with some urban areas reporting infection rates approaching 30% of sexually active individuals.[322,323] AIDS patients do not appear to be at increased risk for the development of amebiasis. The incidence of invasive disease has not corresponded to the presence of luminal infection in homosexual AIDS patients in either the United States or Latin America. Reports of the increased incidence of amebiasis in homosexual patients with AIDS in the United States do not demonstrate an increased prevalence of pathogenic strains or of invasive GI disease.[7,324] It appears that most of the parasites recovered from homosexual men are nonpathogenic and probably are not relevant to GI symptoms. Invasive disease is most common in children, pregnant women and immediately postpartum, malnutrition, and immunocompromised individuals. Among compromised hosts, patients receiving chemotherapy, corticosteroid therapy, or immunosuppression for organ transplantation are at a greater risk for the development of fulminant colitis.[325,326]

Serologic assays are useful for the detection of invasive disease. These tests generally revert to negative within 6–12 months after the acute invasive disease. However, the indirect hemagglutination assay remains positive for years.

8.4. Immunology

Antibodies against *E. histolytica* that develop during acute infection do not appear to be protective.[327,328] Such antibodies may persist for up to 10 years after acute infection. The presence of serum antibodies correlates with the presence of both invasive amebiasis and hepatic disease. Antibody can lyse trophozoites *in vitro*. Complement also can produce partial lysis. Complement-resistant amebae may "cap off" bound surface antibody or may shed surface antigens to limit the efficacy of the humoral immune response. Amebae isolated from liver abscesses appear to be resistant to antibody-mediated lysis. The cellular immune response appears to be depressed acutely during infection and returns after treatment of disease.[326,327] Sera from infected individuals may contain a factor that is suppressive for cell-mediated immune responses. Recurrent infection is uncommon and resistance to subsequent infection appears to be mediated by the cellular immune response activating killing by macrophages.

8.5. Pathogenesis

The pathology of amebic disease is important. The characteristics of inflammatory changes in the bowel may be easily confused with other more benign processes. Invasion of the intestinal mucosa produces a local loss of mucin from the surface of epithelial cells, with underlying edema and hyperemia. Superficial ulceration develops with minimal local inflammation. Ulceration progresses superficially, with penetration into the mucosa producing the characteristic "flask ulcer" extending into the submucosa of the intestinal wall. The ulcers themselves are usually small with raised borders and a necrotic base and with normal mucosa between adjacent ulcers. Extensive disease may involve large segments of the intestinal mucosa. Organisms are found superficially at the edge of the epithelial lesion. Trophozoites are found in tissues in an amorphous, eosinophilic matrix. Given the ability of the organism to lyse neutrophils, leukocytes are found only at the edge of established amebic ulcers. Chronic amebic infection will result in thickening of the colonic mucosa.

Progressive disease within the wall of the intestine can produce a pseudotumor consisting of necrotic tissue with acute and chronic inflammation and granulation and fibrosis. This mass lesion is called an "ameboma." Further complications occur locally with perforation of the GI tract or by penetration into the portal circulation, allowing seeding of the liver.[319] Liver involvement may occur in the absence of clinically important intestinal disease. The liver is involved by local inflammation, followed by focal necrosis and granuloma formation. Periportal inflammation and fibrosis may be seen with few if any organisms in the ports. Progressive necrosis and parenchymal reaction provide a thin capsule to the enlarging amebic abscess. Organisms are found primarily at the edge of the abscess. Amebic abscesses are single lesions in over 80% of individuals and are generally found in the right lobe in the posterior portion adjacent to the diaphragm. Large abscesses may decompress into the right or left chest or into the bronchial tree, producing catastrophic disease. The necrotic debris has been characterized as "anchovy paste" exudate, which may be found in the sputum after rupture of an abscess into the bronchi. Rupture of abscesses of the left lobe of the liver may be associated with acute pericardial tamponade. Bacterial superinfection is surprisingly uncommon. With appropriate therapy, pathologic changes in the intestines or the liver resolve without fibrosis.

8.6. The Patient

The presentation of amebic infection depends on the extent of disease and the condition of the host. Clinical presentations of this common disease can be divided into subcategories: asymptomatic cyst carriers; a possible chronic, nondysenteric colitis syndrome; acute rectocolitis; toxic megacolon with fulminant colitis; ameboma; and painless rectal bleeding. The interaction of immune depression with the presentation of disease appears to determine the severity and rapidity of disease progression.[326] Immune compromise does not appear to affect the development of either invasive or noninvasive disease. Since local inflammation usually is modest, fibrosis is not characteristic of disease. The alterations produced by immune suppression may be subtle. It should be anticipated that bacterial superinfection and peritonitis are more common in the setting of broad-spectrum immune suppression such as with corticosteroid therapy. The important factors in infection are the strain of *E. histolytica* and the nutritional status of the host.

Only 10–20% of individuals infected with *E. histolytica* will develop clinically significant disease. Most individuals will spontaneously eradicate the parasites. Symptomatic individuals are at risk of further complications of disease if not adequately treated. It is possible that the majority of these infections are with nonpathogenic strains. A syndrome of irritable bowel disease in the presence of *E. histolytica* has been termed "chronic non-dysenteric infection." There is some controversy as to whether or not the organisms play a significant role in the development of the syndrome of chronic intermittent abdominal pain with diarrhea.[329]

The basis of this pattern of infection remains unclear. In general, the presentation of acute disease depends on the extent of colonic involvement and the rapidity with which the disease develops. "Acute rectocolitis" usually involves the gradual onset of diarrhea and acute abdominal pain. Watery stools may become blood-stained. Tenesmus is common. Many small, superficial, mucosal ulcerations with segmental distribution are seen by colonoscopy. The ulcers contain necrotic debris and trophozoites, with underlying hyperemia and some submucosal hemorrhage. Fecal leukocytes are uncommon. While adults are generally well compensated, children may be toxic, with high fevers, and become rapidly dehydrated.[330] Right lower quadrant pain may be due to acute appendicitis (which is uncommon) or to "typhlitis," which usually occurs in the presence of mucosal thickening in the ileocecal region. Complications are common in individuals developing amebic appendicitis. Liver abscesses, bleeding, and perforation with fistula formation are common complications in patients with undiagnosed amebic appendicitis.

Host factors seem to determine the progression of acute dysenteric disease to the fulminant colitis.[326,327] Fulminant colitis is a rapidly progressive syndrome. The patient may present with bloody diarrhea with foul odor and diffuse abdominal pain. High fever is present, with signs of ileus, dehydration, and shock developing early. Tenesmus and rectal bleeding are present, and perforation with peritonitis occurs in up to two thirds of these individuals. Toxic megacolon may occur but is uncommon. Extraintestinal disease may occur. The syndrome is more common in the presence of malnutrition, in older individuals, and in the immunocompromised patient, particularly those receiving corticosteroids. Up to two thirds of these individuals will die due to complications of intestinal injury. Surgical debridement is of uncertain value in the absence of acute peritonitis or toxic megacolon. Both antiamebic and antibacterial therapies are necessary.

Up to 2% of patients with invasive disease will develop ameboma, usually of the cecum or ascending colon. The diagnosis is frequently incidental to the evaluation of acute dysenteric amebiasis or with an asymptomatic abdominal mass lesion. Diffuse thickening of the gastric wall with mucosal ulceration may give the appearance of Crohn's disease. Treatment of this complication with corticosteroids may have disastrous side effects.[325] The presence of rectal bleeding without pain may be chronic and is a reflection of congestive colitis, usually without ulceration.

Extraintestinal disease is generally restricted to amebic liver abscess. This complication is more common in lower socioeconomic groups, men than in women, and alcoholics than in nondrinkers. In addition to poor hygiene, host immunity also may be a contributing factor. Amebic liver abscess usually occurs in the absence of acute rectocolitis. Intestinal involvement can be demonstrated in only 38% of individuals with hepatic abscess. The clinical presentation is usually acute, with abdominal pain, fever, and symptoms localizing in the subdiaphragmatic region. Right shoulder pain is common and is increased by coughing. Fever, rigors, and sweats are common, associated with cachexia. Some patients will present with a more gradual evolution with hepatomegaly and anemia. In most individuals, liver function tests are mildly abnormal and the patients may be slightly jaundiced. The patient often will have an elevated white blood count. Other extraintestinal foci of infection may include the skin of the perianal region or the skin or tissues overlying the involved intestines or thoracic regions. Amebic brain abscess is generally catastrophic. Genitourinary disease is described in the form of rectovaginal fistulae or as ulceration with granulomatous disease of the vagina or penis.

8.7. Diagnosis

The diagnosis of amebiasis is made by the demonstration of trophozoites of *E. histolytica* in smears made from colonic samples. Motile organisms containing red blood cells in the presence of small superficial ulcers of the colonic mucosa are the most common findings. Differentiation of pathogenic amebae from nonpathogenic organisms can be done on wet-mounted smears using fresh or preserved specimens. The addition of iodine will allow the differentiation of *E. histolytica* cysts from yeasts and other amebic species. The addition of methylene blue will allow the differentiation of cysts from leukocytes, which stain blue. Areas with both mucus and blood are optimal for examination. Cyst morphology is best seen on fixed specimens stained with iron–hematoxylin or a trichrome stain. This morphology is more easily observed after concentration of a specimen using sedimentation or flotation on formalin–ethyl acetate.

Serologic diagnosis is useful for the diagnosis of invasive GI disease or hepatic disease but is rarely useful acutely. Individuals with positive serologies and negative stool examinations or with a syndrome consistent with amebiasis (liver abscess, mass lesions, acute colitis) should undergo endoscopic evaluation (unless the patient has toxic megacolon and fulminant colitis). Radiologic evaluation will reveal the presence of thickened intestinal walls in severe amebic colitis, with narrowing and loss of folds.[331] Ameboma may be confused with chronic infection due to tuberculosis, *Yersinia* infection, carcinoma, lymphoma, or regional enteritis. Infection by *E. histolytica* must be excluded before initiating corticosteroid therapy for colitis.

In the absence of localizing signs, the diagnosis of a liver abscess may be overlooked. Ultrasonography, CT, or liver–spleen scintigraphy will demonstrate liver abscesses in the majority of cases. The presence of a liver abscess with a positive amebic serology should be considered diagnostic of amebic liver abscess. In immunosuppressed individuals or in malnourished patients with chronic abscesses, negative serologies may be slow to convert to positive. Organisms are infrequently found on stool examination to confirm the presence of *E. histolytica* infection. Serologic tests are particularly useful because they peak 2–3 months after acute infection and generally return to low levels by 1 year after infection. In endemic areas, positive serologic tests are less useful. Percutaneous drainage of a large abscess carries the risk of rupture and spread of infection. However, percutaneous drainage in the setting of disease unresponsive to antimicrobial agents or of anticipated rupture or in the presence of mixed bacterial and amebic infection may be indicated. Occasionally, percutaneous drainage has provided a diagnosis in the setting of negative serologic tests. Catheters placed into liver abscesses should be removed immediately after draining the lesion.

8.8. Therapy

A broad range of antiamebic drugs is available and their use is based both on their site of action (tissue vs. colonic lumen) and the potential toxicities of each agent (Table 9). Agents with actions within the lumen of the bowel include diiodohydroxyquin, diloxanide furoate, and paromomycin. The most effective drugs available at present have activity both in the tissue and in the intestinal lumen; they include metronidazole and the nitroimidazole derivatives (tinidazole, ornidazole, secnidazole, and nimorazole). High-dose metronidazole (750 mg tid for 5–10 days) is needed and often poorly tolerated. Emetine and dehydroemetine must be given intramuscularly and have depressant effects on the myocardium. Erythromycin and

tetracycline have activity only in the bowel wall. Chloroquine works only in hepatic disease.

Treatment for asymptomatic cyst passers is controversial due to the frequency of infection with nonpathogenic strains of amebae.[329] The luminal agents such as diloxanide are more effective than metronidazole in the absence of invasive disease. Invasive disease should be treated with a nitroimidazole with a luminal agent. In the presence of colonic perforation, antibacterial therapy should be added, using metronidazole to cover anaerobic flora as well as the amebic infection. The need for surgery with microperforation is controversial. However, with acute appendicitis or toxic megacolon, surgery will be necessary. Ameboma may be cured with antibiotic therapy alone and surgery should be avoided if possible.[332] Similarly, liver abscess is rarely an indication for surgery. Antimicrobial therapy is generally successful and abscesses resolve without fibrosis. Because treatment failures with metronidazole have been observed, liver abscesses are generally treated with combination therapy using a nitroimidazole agent and dehydroemetine.

9. Primary Amebic Meningoencephalitis

Primary amebic meningoencephalitis (PAM) is an uncommon infection of the CNS produced by the amebae *Naegleria* and *Acanthamoeba* and rarely *Balamuthia mandrillaris*.[333] Depending on the species of organism, the progression of disease may be chronic or acute. PAM is usually fatal.

9.1. The Organism

While secondary involvement of the nervous system with amebic infection can occur with *Entamoebae histolytica*, primary infections are due to *Naegleria fowleri* and a number of species of *Acanthamoeba*. *Naegleria* trophozoites are amorphous or sluglike in shape. The most easily identifiable form of the organism is the flagellate induced from trophozoites placed in distilled water. Flagellates do not occur during human infection. *Acanthamoeba* trophozoites have pseudopodia surrounding one end. A number of *Acanthamoeba* species are pathogenic for humans including: *A. castellani*, *A. hatchetti*, *A. griffini*, *A. divonensis*, *A. palestinensis*, *A. culbertsoni*, *A. astronyxis*, and *A. rhysodes*. Both *Naegleria* and *Acanthamoeba* may be seen containing ingested red blood cells; both occur in a thick-walled cyst form. The trophozoites of these organisms vary from 7 to 20 μm in size. *B. mandrillaris* trophozoites are 30 μm in diameter and are indistinguishable from those of *Acanthamoeba*. *B. mandrillaris* cysts have a wavy, three-layered outer wall which averages 15 μm in diameter. These organisms are capable of producing PAM in many animal species by intranasal or intravenous inoculation. There does not appear to be an intermediate host. The virulence of a given strain will depend on the organism and on the host's immune status.

9.2. Epidemiology

Amebic meningoencephalitis has been reported worldwide. The majority of cases are due to *Naegleria* species. The natural habitat of these amebae is soil and fresh warm water. Common-source epidemics have been detected as a result of contaminated water supplies. However, these organisms are commonly found in pools of warm water such as aquariums or hospital hydrotherapy tanks. *Acanthamoeba* has been isolated in pharyngeal swabs of normal individuals with respiratory viral illnesses and positive serologic assays found in healthy individuals. However, other than corneal disease, granulomatous amebic encephalitis due to *Acanthamoeba* is generally restricted to debilitated individuals or those with immune defects. These defects include individuals with AIDS, diabetes mellitus, cirrhosis, corticosteroid therapy, and cancer chemotherapy and in renal, hepatic, and bone marrow transplant recipients. Corneal infection by *Acanthamoeba* is often associated with contamination of contact lens solutions. *B. mandrillaris* causes infection in both normal and immunocompromised individuals. Reported cases of PAM have come from throughout the United States and Europe, central Africa, India and other parts of Asia, and Australia. Pathogenic organisms are easily missed in the evaluation of acutely ill patients.

9.3. Pathogenesis

Naegleria generally enters the CNS via the olfactory neuroepithelium when water or dust particles enter the nose. By contrast, *Acanthamoeba* causes infection of other organ systems, including skin, lung, or eye, and then spreads to the CNS via the bloodstream. The clinical presentation of *Naegleria* infection is acute and rapidly progressive meningitis that is fatal in less than 4 days without therapy. The inflammatory cell response is primarily with polynuclear leukocytes. As *Naegleria* invades the superficial cortex, the olfactory and frontotemporal areas of the brain quickly develop hemorrhagic necrosis. Trophozoites are prominent.

The pace of meningoencephalitis due to *Acanthamoeba* is generally slower, producing ultimately fatal disease in 2–3 weeks. The inflammatory response is usually lymphocytic, with macrophages, granuloma formation, giant cells, and vasculitis. The pathology and clinical picture are those of multiple space-occupying brain abscesses. These abscesses are located in the white matter in deep midline and midbrain structures. Organisms are found in tissues, but in general not in CSF. *Acanthamoeba* keratitis is usually associated with corneal trauma and the use of contact lenses. Both cysts and trophozoites are found in affected corneas in the center of an inflammatory ring. Giant cells may be present in any inflammatory exudate. Involvement of the posterior segment of the eye may occur. Because of the effects on deep structures, *Acanthamoeba* may cause a picture of subacute or chronic encephalitis before focal deficits develop related to brain abscesses. *B. mandrillaris* causes a subacute-to-chronic granulomatous meningoencephalitis similar to that of *Acanthamoeba* in both compromised and normal hosts. Involved areas may contain cysts and trophozoites in a perivascular location, with vasculitis and hemorrhagic necrosis of meninges and brain tissues. Organisms also may be found in other tissues (kidney, skin, adrenals). The inflammatory exudate includes lymphocytes, monocytes, plasma cells, and giant cells.

9.4. The Patient and Diagnosis

While acute PAM will present in a fashion similar to cases of severe bacterial meningitis, the patient will rapidly progress from headache, fever, and vomiting to seizures, coma, or paralysis. *Naegleria* infection will generally present within 1 week after exposure to contaminated water or possibly to inhaled cysts. *Naeglaria* may produce false sensations of taste or smell similar to those seen in some patients with heart failure and early brain stem herniation. *Naegleria* trophozoites will be found in CSF, with markedly low glucose levels and a neutrophil count as high as 15,000. The CSF protein level may be normal or slightly elevated. Progression is even more rapid in immunocompromised hosts.

Patients with *Acanthamoeba* infection also may present with headache, fever, and meningismus. The infectious exposure is generally uncertain. Nodular or ulcerative skin lesions may be present for months before clinical disease emerges. Infections involving the sinuses, bone, and lungs have also been described. Focal neurologic deficits or seizures occur early in the course of this subacute disease. The patient will undergo a gradual neurologic deterioration over the course of 2–4 weeks. The clinical picture exceeds what would be expected from a focal brain abscess. The CSF generally does not contain organisms and the glucose is normal or slightly depressed. The CSF contains 100–400 cells, which include both lymphocytes and neutrophils. Organisms are found in brain biopsy specimens. Serologic tests are useful diagnostically but not clinically. *Acanthamoeba* infection may produce disease of the eye or respiratory tract in normal hosts. Eye injury has been associated with infections carried by contact lenses or local trauma.

While infection due to *B. mandrillaris* occurs in normal individuals, individuals with AIDS, diabetes, and renal failure are notably susceptible. Disease has been reported as acute necrotizing meningoencephalitis or as subacute, granulomatous meningoencephalitis. *B. mandrillaris* may be seen in wet mounts of CSF that generally has a mononuclear pleocytosis. Imaging studies of the brain demonstrate multiple hypodense mass lesions. The organism can be grown on tissue culture cells or in axenic media.

9.5. Therapy

Amebic keratitis can be cured with early diagnosis, generally before the characteristic corneal ring develops. Corneal scrapings are often contaminated with common bacterial pathogens; *Acanthamoeba* must be considered in the differential diagnosis and appropriate parasitologic evaluation made. Motile trophozoites may be observed in wet mounts of corneal scrapings. Cysts and trophozoites may be seen in Giemsa, hematoxylin and eosin, periodic acid–Schiff, and Calcofluor white-stained specimens. Molecular assays are under development.

Few individuals have survived primary amebic meningoencephalitis.[333,334] Early diagnosis and therapy of *Naeglaria* infection are critical to survival. Those few patients who survived received intrathecal and systemic treatment with amphotericin B at high dose (≥ 1 mg/kg per day). Given the rapid progression of PAM, there is no opportunity for gradual escalation of dosing. Passive immunotherapy may be useful when antisera become available clinically. There is no useful therapy for *Acanthamoeba*, although *in vitro* sensitivity of *Acanthamoeba* to polymyxin, pentamidine, propamidine, ketoconazole, miconazole, 5-fluorocytosine, paromomycin, neomycin, and ivermectin has been demonstrated. These agents, coupled with debridement, have had some efficacy in the treatment of keratitis. Effective therapies for *B. mandrillaris* have not yet been described.

10. Leishmaniasis

Leishmaniasis encompasses a variety of syndromes, including cutaneous, mucocutaneous, and visceral disease. The manifestations of infection vary depending on the species of *Leishmania* and the immune status of the host. Because all species of *Leishmania* are intracellular parasites of macrophages, the functional status of T lymphocytes and of cytokines affecting macrophage function will determine whether or not the organism disseminates locally or systemically. In endemic regions and travelers to such areas, leishmaniasis has been recognized in a broad spectrum of both normal and immunocompromised individuals.

10.1. The Organism

Species of *Leishmania* can be separated by geographic distribution and the usual clinical manifestations of infection. The organisms that cause cutaneous and mucocutaneous leishmaniasis include *L. braziliensis*, *L. major*, *L. mexicana*, *L. tropica*, *L. peruviana*, and *L. aethiopica*. However, most species may be associated with more invasive disease also. *Leishmania tropica* can cause chronic, relapsing cutaneous disease (recidivans form) and was associated with visceral disease in U.S. soldiers in the Middle East after Operation Desert Storm. Similarly, *L. mexicana* has been found in patients with visceral disease. *Leishmania donovani* (India and Africa), *L. infantum* (Mediterranean), and *L. chagasi* (South America) can cause visceral leishmaniasis. Consensus about species differentiation is being addressed using sequence and hybridization analysis using kinetoplast DNA probes.

Leishmania exists in two forms. Within the sand fly (phlebotomus) vector or in culture, they develop from amastigotes into single-celled, flagellated extracellular promastigotes. The promastigotes migrate to the sand fly pharynx in a nondividing or stationary phase, a form that has high infectivity for macrophages. Human infection is initiated by the bite of an infected female phlebotomine sand fly. Within the cells of vertebrate hosts, the organisms become small, rounded amastigotes (without flagellae) of 2–3 μm in diameter. Organisms enter macrophages at the site of the bite and replicate by binary fission within the macrophage. Each strain of *Leishmania* has a unique and complex interaction with the phagolysosome; in general, the parasite is resistant to the lysosomal acidic pH. Complement activation may assist in the penetration of promastigotes into macrophages.

10.2. Epidemiology

The spread of leishmaniasis is dependent on the presence of the appropriate species of sand fly.[335] Other than *L. donovani* and *L. tropica*, the organism is generally maintained in wild animals, including rodents, dogs, marsupials, and other wild animals. *Leishmania donovani* and

L. tropica appear to be able to use the human as a definitive host. Various forms of *Leishmania* are found in the southern United States, Central and South America, and throughout Africa, southern Asia, Europe, and the Middle East. Virtually all the cases seen in the United States are acquired outside the country. The exact frequency of these infections is unclear, because the pathogen has caused disease as late as 30 years after the initial infection. Distant exposure must be excluded before the diagnosis is excluded. In endemic areas, the annual incidence is 0.1–1% and may go as high as 5% during epidemics. Malnourished or immunocompromised individuals are most susceptible to symptomatic and severe infections.[336]

Malnutrition is probably the most important immunosuppressive mechanism predisposing to severe visceral leishmaniasis.[335,337,338] While many individuals in endemic areas are infected but develop relatively mild disease, individuals with malnutrition are much more likely to develop classic, advanced, visceral disease. This difference is most easily seen in children, in whom untreated symptomatic infection will be fatal in 75–90% of those affected. Only 10–20% of affected individuals will develop clinically apparent disease.[339]

In immunocompromised individuals, especially those receiving corticosteroids, with hematologic malignancies, or after organ transplants, disease may progress more rapidly and be more difficult to diagnose.[336,340–345] Leishmaniasis may not present for years after exposure and/or initiation of immune suppression. In these individuals, the disease is more often chronic and the response to therapy less rewarding. Chronic, relapsing, visceral leishmaniasis has been described as a complication of AIDS in patients from Spain, France, and Italy.[346,347] Some AIDS patients do not develop antibodies to *Leishmania*, in contrast to the nearly uniform detection of antibodies in immunocompetent individuals and in non-AIDS immunocompromised patients with visceral disease. In solid organ transplant recipients, pulse-dose steroids, antilymphocyte antibodies, and intensified immune suppression used to treat graft rejection may accelerate disease. At least one renal allograft recipient with exposure in the pretransplant period developed clinical disease more than 7 years after transplantation, although the average time is closer to 8 months.[345,348] It is possible that *Leishmania* may reduce immune responsiveness to infections due to other organisms. Antibodies against a broad range of antigens have been observed in early visceral disease.

10.3. Pathogenesis and Immunology

Cutaneous and mucocutaneous forms of leishmaniasis begin when promastigotes are injected subcutaneously by the sand fly and enter local host cells. Local inflammation is primarily lymphocytic and granulomatous, with necrosis of the skin occurring early. Organisms spread via the bloodstream or lymphatics to the mucosal surfaces of the nose, mouth, pharynx, and larynx. Inflammation is generally modest. Dermal necrosis is probably due to the local immune reaction. Subsequently, hyperkeratosis and acanthosis may occur. The number of organisms in infected cells is variable.

In visceral leishmaniasis, infected macrophages from the skin serve as a reservoir for organisms that infect spleen, lymph nodes, liver, bone marrow, and intestinal mucosa. This infection causes hyperplasia of focal lymphoid tissue with granulomata. Ulceration of mucosal surfaces may occur, and endothelial proliferation may occur in pulmonary alveolar capillaries and in blood vessels of the renal glomeruli. The spleen and liver are enlarged due to parasitization of macrophages and Kupffer cells. Some areas of skin around the initial bite will have nodules containing parasites; some areas that appear normal will also contain parasites.

Immunity to *Leishmania* is thought to be mediated by CD4+ T lymphocytes, with lymphokines from these cells enhancing the killing of intracellular organisms.[347,349] The level of immune response determines the manifestations of the disease. In cutaneous leishmaniasis, patients lacking immune responsiveness to the parasite may develop tissue cutaneous leishmaniasis (DCL) with little lymphocyte infiltration of areas of involvement. By contrast, leishmaniasis recidivans is the result of an exuberant lymphocytic response to low numbers of parasites that persist within macrophages. In visceral leishmaniasis, the host is completely lacking a cellular immune response to leishmanial antigens.[350] Parasitization of the reticuloendothelial system is uncontrolled. Despite high levels of circulating antibodies in immune complexes, disease may spread rapidly without therapy. Suppressor lymphocytes appear to play a role in the loss of antigen reactivity seen in visceral leishmaniasis.

10.4. The Patient

Infection in normal and immunocompromised individuals share most clinical characteristics. Cutaneous and mucocutaneous leishmaniasis may cause clinical symptoms weeks to years after initial infection. In cutaneous disease, a small papule at the site of the bite will develop into a nodule. Necrosis leaves a painless ulcer with raised firm edges. Multiple lesions may occur in the same area or along lymphatics. These lesions will generally heal spontaneously. Specific strains of *Leishmania* may involve the ears, upper face, and nose or cause painful lesions. Immune suppression may cause a relapse in previously healed areas. Experience with leishmaniasis in soldiers serving in the Middle East suggests that systemic manifestations of cutaneous disease are common and may reflect organisms found in the bone marrow and in other visceral locations in the immunologically normal host.

Mucocutaneous leishmaniasis ("espundia") is usually a complication of cutaneous disease occurring years after the initial skin lesions have healed. Organisms that spread to mucosal tissues may produce symptoms similar to sinusitis or nosebleeds. Granulomatous inflammation with necrosis may destroy the nasal septum and surrounding tissues. This inflammation is frequently complicated by bacterial superinfection.

Uncommon forms of leishmaniasis include a chronic relapsing or recidivans disease and DCL. Chronic relapsing disease occurs in or surrounding the area of original cutaneous involvement. Diffuse cutaneous disease is associated with lack of immune reactivity to leishmanial antigens. Patients develop nonulcerative nodules across the skin, with little inflammatory reaction. Visceral leishmaniasis will become symptomatic weeks to months after infection, with the gradual onset of fever, sweats, and weight loss. Nonspecific abdominal complaints accompany hepatosplenomegaly and wasting of large muscle groups. Pancytopenias and hypergammaglobulinemia are common. Complications are the results of anemia, thrombocytopenia, liver failure, or secondary bacterial infection. Atypical manifestations include involvement of the gastrointestinal tract and peripheral neuropathies (axonal degeneration and demyelination) described in patients with AIDS.

10.5. Diagnosis

Patients with symptoms of leishmaniasis are modest problems in differential diagnosis. The cutaneous lesions, especially in the immunocompromised host, might be seen in diseases due to mycobacteria, including *M. leprae* or *M. marinum*, cutaneous diphtheria, sarcoidosis, histoplasmosis, yaws, and fungi including sporotrichosis. Mucocutaneous disease may be confused with the destructive lesions of lethal midline granulomatous disease or blastomycosis. Visceral leishmaniasis with hepatosplenomegaly can mimic or complicate lymphoma, malaria, schistosomiasis, endocarditis, brucellosis, leukemia, immunosuppression for organ transplantation or with corticosteroids, or advanced histoplasmosis.[345,346]

Diagnosis is based on the identification of organisms from tissues. In cutaneous disease, needle aspiration or punch biopsy should be performed at the edge of the lesion. In visceral kala-azar, tissue aspirate from spleen, bone marrow, lymph node, or liver is often diagnostic. Parasites are occasionally seen in peripheral blood smears or sputum samples. Parasites can be cultured in special media or in animals. Serologic tests are useful in non-AIDS patients but may be negative in the first week to 10 days after infection.[351]

10.6. Therapy

Leishmaniasis is treated with pentavalent antimony: stibogluconate sodium (Pentostam) or meglumine antimoniate (Glucantime). Stibogluconate is used at a dose of 20 mg/kg body weight for 2–3 weeks. Controversy exists as to the maximum daily dose, but it should generally not exceed 850 mg to 1 g daily in adults. Relapse with mucocutaneous and visceral disease is not uncommon and may necessitate multiple courses of treatment for cure. Mortality approaches 28% in immunocompromised individuals. Occasional abnormalities in liver function tests and electrocardiograms may be seen, with elevations of the total white blood count as side effects of therapy. Pancreatitis is common after pentavalent antimony therapy.[348] Patients also may complain of arthralgias. Meglumine is given by injection for American cutaneous disease (20 mg/kg per day × 15 days). Allopurinol (20 mg/kg per day in four divided doses) may have additive efficacy with pentavalent antimony or in patients with unresponsive disease. Allopurinol monotherapy may be less toxic, less costly, and more effective than the antimony compounds. Antimony resistance has been documented in India. Interferon has been used with antimony as primary therapy in some patients. Amphotericin B and lipid-associated formulations of amphotericin B have been used as alternatives for therapy of cutaneous, mucocutaneous, and visceral leishmaniasis. Amphotericin should not be given with antimony compounds. Pentamidine is another alternative for therapy of visceral disease. Relapses are apparently related to failure to develop an immune response during the course of therapy. Newer therapies under study include topical therapies for skin disease and ketoconazole for systemic disease due to *L. mexicana*.[352]

11. Other Parasitic Diseases of the Immunocompromised Host

The presentations and frequencies of some of the most common parasitic diseases of man are relatively unchanged in the patient with altered immunity. Others are important in the presence of specific immune lesions. The failure to detect enhanced infection may relate to variability in the duration or severity of these diseases in normal individuals. More often, the failure of immune suppression to exacerbate infection suggests either a lack of involvement of the immune system in the control of infection in the normal host or the inability of the organism to complete its life cycle in man. A number of important infections are occasionally problematic in the patient with immune dysfunction. These infections are discussed below.

11.1. *Giardia lamblia*

Giardiasis is the most common protozoan disease in the United States and an important cause of enteric disease worldwide. Infection with *G. lamblia* is caused by ingestion of food or water contaminated with cysts. Trophozoites develop in the duodenum after exposure of cysts to acid in the stomach. This free-swimming flagellate causes disease by attachment to intestinal epithelial cells and may cause prolonged infection despite therapy. The patient may develop diarrhea, malabsorption, and weight loss of varying degrees.[353] Children are more frequently infected than adults and prior infection appears to confer protection against subsequent attacks.[354] Immunocompromised individuals appear to be at greater risk, perhaps most often related to malnutrition.[355] Infection appears to be more common in patients with hypogammaglobulinemia or dysgammaglobulinemia, although circulating antibody does not appear to be protective.[277,356–361] Mucosal IgA may provide a barrier against giardial infection.[362] The role of T lymphocytes may be in enhancing the production of mucosal IgA. Secretory IgA of breast milk may also be protective.[363]

11.1.1. Epidemiology

Despite the frequency of *G. lamblia* in the stools of homosexual males, giardiasis is not a common pathogen of unusual severity in patients with AIDS.[364–367] *Giardia* is found in contaminated water, in animals including beavers and dogs, and after person-to-person spread in daycare centers.[263,354,368–372] Infection rates are high in institutions for the mentally disabled and in patients with achlorhydria. The cyst is hardy and survives for months in fresh cool water.[373] Small numbers of cysts can cause infection.

11.1.2. Diagnosis and Therapy

Giardia is a common cause of diarrhea and nonspecific abdominal complaints. Unexplained malabsorption or lactose intolerance may be observed.[353,374–376] Colonoscopy will be entirely normal.[377] The detection of organisms in stool or on duodenal aspirate or using the Enterotest vial will provide the diagnosis. Antigen detection tests allow the detection of *Giardia* by immunofluorescence or ELISA. Serologic testing is useful in epidemiology. Patients are generally treated with metronidazole, 250 mg three or four times a day for 7–10 days. This drug should be used with caution in children or pregnant women. Repeat therapy may be necessary in up to 20% of individuals. Malabsorptive symptoms may be slower to resolve than the infection itself. Alternatively, quinacrine hydrochloride can be used for therapy at 100 mg orally three times a day for 5 days (2 mg/kg in young children). Furazolidone, paromomycin, and tinidazole have also been used successfully. Albendazole has some activity but is not recommended for routine use.

11.2. Malaria

Given the importance of malaria as a pathogen worldwide, it is striking that this intracellular protozoan has not emerged as an opportunistic pathogen in immunocompromised hosts. Malaria has been transmitted with renal, heart, and liver allografts derived from endemic regions. It is unclear whether infection acquired other than at the time of transplantation is due to new primary infection (mosquito bite) or recrudescence of old infection (e.g., hepatic *Plasmodium vivax*) in an immunocompromised host. While immunity to *P. falciparum* develops in individuals from endemic regions, the association of immune suppression with enhanced severity or incidence of infection has not been made.

The life cycle of malaria begins when the female anopheline mosquito inoculates sporozoites into the host during a blood meal. These organisms enter liver cells and proliferate, releasing merozoites that then invade erythrocytes. Immunity occurs in endemic areas after repeated infection and is both species- and strain-specific. Passive transfer of antibodies is protective, blocking infection of red blood cells and killing intracellular organisms via antibody-dependent cellular cytotoxicity (ADCC). ADCC requires normal splenic function. The development of immunity requires normal T-cell function. Immunity to sporozoites appears to be mediated at least in part by T cells and not by antibody.

Malaria is primarily a tropical disease and is caused by four major species, with each having different clinical manifestations. *Plasmodium falciparum* causes the most rapidly progressive disease, including anemia, renal failure, cerebral disease, pulmonary edema, liver failure, and death. *Plasmodium vivax* causes anemia and splenic rupture in severe cases. *Plasmodium malariae* may persist as an asymptomatic infection for many years and may cause nephrotic syndrome in children. *Plasmodium ovale* causes the acute infectious syndrome seen in all forms of malaria. The acute presentations will be of rigors with high fever and sweats. Headache and nausea and occasionally seizures may be seen in all forms, but are particularly important symptoms with *P. falciparum*. As the life cycle of the organism become synchronized, cycles of chills, fever, and sweats become characteristic for the species of malaria. Most patients will have nonspecific complaints, including myalgias, cough, and diarrhea, and may have anemia, jaundice, or abdominal tenderness. Relapse in malaria (*P. ovale* and *P. vivax*) may present years after initial infection. It is likely that infection is more severe in patients with functional or anatomic asplenia, during pregnancy, and possibly during immune suppression.[378] Corticosteroids administered for renal transplantation tend to normalize serum creatinine values, perhaps by reversal of graft rejection, but do not affect the characteristic fevers associated with malaria.

Therapy is based on the type of malaria seen on the peripheral blood smear and on the pattern of antibiotic resistance in the area in which it was acquired. If malaria is likely, the blood smears should be repeated after a negative examination, as the level of parasitemia may fluctuate. Significant infection in immunocompromised individuals has been associated with transfusions of infected blood with *P. malariae* into immunocompromised individuals. Transfusional disease will cause significant infection in immunocompromised individuals.[378]

11.3. Babesiosis

The babesiae are protozoan parasites of animals transmitted by the ixodid tick to humans as an incidental host. There are over 70 species of *Babesia*, several of which cause human disease, including *B. microti* and *B. divergens*. Most of the cases of human babesiosis have been described from the northeastern United States. A few cases of *B. divergens* have been reported from Europe. Occasionally, babesiosis has been seen as a complication of blood transfusion or in organ transplantation. On the basis of serologic studies, it is likely that babesial infections occur worldwide but are asymptomatic or mildly symptomatic.

The organism reproduces within erythrocytes, causing hemolysis and hemoglobinuria. Hypotension may result from the release of a kallikrein activator by the organism. Splenectomy or abnormal splenic function and

T-lymphocyte dysfunction have been associated with especially severe disease. Recurrent and apparently relapsing infection has been reported in a single AIDS patient. The level of parasitemia is markedly enhanced by administration of corticosteroids in animal models.

Most patients will present with fever, chills, diarrhea, vomiting, anemia, myalgias, and fatigue.[379–383] The clinical manifestations will depend on the species of *Babesia* causing infection. *Babesia microti* causes mild disease that appears to remit spontaneously.[383] Patients will have mild elevations in liver function tests and parasitemia of less than 10%. Symptoms and parasitemia may exist for up to 4 months and are increased after splenectomy. *Babesia divergens* has been seen in splenectomized hosts, with severe disease culminating in renal failure, hypotension, and severe anemia. These infections have all been fatal. Diagnosis is by blood smear. Antibody-based tests are also available.[384]

Treatment has been controversial, but therapy with clindamycin (1.2 g bid IV or 600 mg PO tid × 7 days) and quinine (650 mg PO tid) has been effective in some patients.[385,386] Alternative therapies include TMP-SMX with pentamidine or quinine with azithromycin. The high-grade parasitemia seen in splenectomized patients or in those receiving corticosteroids has responded to exchange blood transfusion in a few patients. It appears that pentamidine isethionate also reduces parasitemia.

11.4. American Trypanosomiasis (Chagas' Disease)

Trypanosoma cruzi is a protozoan that infects man during a blood meal by the reduvid bug. Endemic regions for Chagas' disease (*T. cruzi*) and for African sleeping sickness (caused by *T. brucei brucei* and *T. brucei gambiense*) do not overlap. *T. cruzi* develop in the bloodsucking Triatominae or kissing bug, releasing metacyclic trypamastigotes, which are discharged with feces during blood meals. The parasite infects a variety of cell types and transforms intracellularly as amastigotes, differentiates into trypamastigotes, and enters new cells after rupture of the host cell. The vectors are commonly found in dilapidated housing and in the wild in endemic regions where the organism is maintained in many mammalian host intracellularly as trypomastigotes. *Trypanosoma cruzi* have a particular predilection for muscle (including cardiac) and neuroglial cells and produce local inflammation with lymphocytes, macrophages, and plasma cells. Within heart muscle, pseudocysts are formed that are clumps of intracellular amastigotes, often in association with myocarditis. In addition to reduvid infection, infection has occurred as a result of organ transplantation and via blood transfusion. Heart failure due to Chagas' disease is associated with biventricular enlargement, thinning, aneurysms, and mural thrombi and is a common basis for transplantation in endemic regions. Conduction abnormalities also are common. Dilatation of the esophagus and colon also is associated with loss of innervation of the mesenteric plexi in association with local inflammation.

Up to 20 million people are infected worldwide, including up to 40–50% of the population of endemic areas of Central and South America. Cases related to blood transfusion have been reported in the United States and Canada, as well as in endemic areas.

Approximately 10–30% of infected individuals develop clinical disease. The patient will present with fever, lymphadenopathy, hepatosplenomegaly, and headache. A small, painful indurated area (chagoma) or unilateral orbital edema with conjunctivitis (Romana's sign) will be present in many patients. The patient may develop symptoms of myocarditis or meningoencephalitis. Sequelae of the acute infection may be seen many years after symptomatic or asymptomatic initial infection. The major complications of Chagas' disease are cardiac arrhythmias or conduction defects with congestive heart failure. GI involvement may appear as megacolon or megaesophagus.

Latent infection may be reactivated by immune suppression, including individuals who are the recipients of infected organ grafts.[1] Immunosuppression of individuals with chronic infection due to *T. cruzi* often is of greater severity than the original disease. In renal allograft recipients, cases of central nervous system involvement have been described. Activation of Chagas' disease has occurred in the setting of AIDS. In both groups, brain abscesses have developed, which is not a part of the syndrome in normal hosts. In Brazil and elsewhere in South America, heart transplants have been performed in patients with known Chagas' disease, some of whom developed cardiac allograft infection. In some, disease progressed despite treatment with benznidazole. Reduced levels of immune suppression with tacrolimus or cyclosporine and prophylaxis with nifurtimox have allowed successful transplantation in some of these individuals. Disease may relapse with cessation of therapy or due to CMV infection. Transfusion-associated Chagas' disease is reported in renal and hepatic allograft recipients.

Diagnosis is based on detection of circulating parasites in blood or buffy coat and later the detection of IgM antibody to the parasite. Chronic carriers may have elevated IgG titers; these antibodies may cross-react with those for syphilis, malaria, leishmania, and autoimmune antigens. We have seen parasites in pericardial fluid from a liver transplant recipient, who also had parasites detected in bone marrow and lymph node biopsies. Molecular, axenic, and murine (xenodiagnosis) cultivation also have been achieved. Treatment is with nifurtimox or benznidazole. Side effects are common and treatment often fails to eradicate the parasite.

11.5. African Trypanosomiasis

African trypanosomiasis is caused by *Trypanosoma brucei*, which causes African sleeping sickness. Infection is initiated by the bite of the tsetse fly of Africa. After being injected into the human host, the parasites multiply locally, producing a chancre at the site of replication. Once in the bloodstream or within tissues, the parasite evades immune detection through a process called "antigenic variation." A hemolymphatic phase of the disease occurs, with bloodstream invasion weeks or months after the initial chancre. This phase is characterized by fever, lymphadenopathy, fleeting rashes, edema of face or legs, ascites, or pleural and pericardial effusion. Jaundice and myocarditis may progress to rapidly fatal complications. Trypanosomal invasion of the basal ganglia produces meningeal inflammation extending into the brain cortex with perivascular cuffing. Persistent headache and altered mental function will develop, with a decreased level of consciousness commonly termed "sleeping sickness."

This disease affects over 20,000 people a year. In contrast to most of the pathogens of importance to the immunocompromised individual, African trypanosomiasis causes immune suppression sufficient to allow the development of opportunistic infection, especially pneumonia. Because the treatment for infection is often toxic (suramin or melarsoprol), malnourished patients in endemic areas need to have their nutrition and general clinical status optimized before they will tolerate treatment. Patients will often relapse after therapy. Many of the manifestations of the disease appear to be immune-mediated. Generalized B-cell activation results in an increase in serum immunoglobulins (including autoantibodies) and immune complexes. Patients with African trypanosomiasis also may have diminished reactivity to vaccination or to skin testing.

Diagnosis is based on the detection of parasites in blood, on aspiration of chancres, from lymph nodes, or from organisms found in CSF. Therapy is effective if meningoencephalitis has not developed. The drugs used in therapy include suramin and pentamidine. Both drugs have toxic side effects. Late-stage disease is treated with melarsoprol. Lethal encephalopathy occurs in up to 10% of patients treated with melarsoprol. Some success with α-difluoromethylornithine (DFMO) has been reported.

11.6. *Cyclospora*

Cyclospora was first described in Papua New Guinea in 1997.[387] In the interim this group of organisms has been attributed to the blue-green algae, cyanobacteria, and fungi and has been termed a "large cryptosporidium"; they have been implicated as the causative agent of a severe diarrheal illness in travelers and in patients with AIDS.[388–391] Electron microscopic studies have suggested that these "cyanobacteriumlike" bodies represent a new protozoan pathogen of the coccidian genus *Cyclospora*, and named *Cyclospora cayetanensis*. Cyclospora and *Eimeria* species appear to belong to the same genus and are found in the gastrointestinal tracts of snakes, moles, rodents, chickens, and humans. Like *Cryptosporidium*, this organism causes self-limited but relapsing diarrheal illness in travelers, children, and other normal hosts (up to 3 months) and possibly persistent diarrhea in the cases described in patients with AIDS. Infection is most often described in travelers after contact with contaminated water supplies in tropical regions. *Cyclospora* does not appear to be a major cause of travelers' diarrhea; it is found in many (up to 0.5%) urban water supplies worldwide. Person-to-person spread was suggested by an outbreak occurring among medical house staff in Chicago. Asymptomatic infection has been demonstrated.

The organism is 8–10 μm in diameter with a cluster (morula) of refractile, membrane-bound globules within a limiting membrane. The organism is a thick-walled sphere with a fibrillar coat and granular cytoplasm. The organism contains two sporocysts that contain sporozoites with nuclei and micronemes. The organism is variably acid-fast, although the organism can be identified often in wet mounts of diarrheal stool or by recognizing the characteristic blue autofluorescence under the ultraviolet epifluorescence microscope. Safranin staining and sporulation tests also may be used for identification. In jejunal biopsies, *Cyclospora* are found in a characteristic supranuclear location of the cytoplasm of jejunal epithelial cells. Oocysts are long-lived in the environment and are resistant to chlorination, formalin, and freezing. Sporulation occurs in the environment, as for *Isospora*, suggesting that the replication of this organism may not be intensively regulated by the hosts' immune system.

Infected individuals develop symptoms within 7–10 days of exposure. Patients present with a flulike illness with the abrupt onset of watery diarrhea accompanied by prominent systemic signs including myalgia, headache, nausea, vomiting, and low-grade fever. The intensity of diarrhea fluctuates over 4–8 weeks without therapy. Treatment is limited to cotrimoxazole (DS bid), which should be followed with prophylaxis in immunodeficient individuals.

References

1. Joiner K, Sher A, Gaither T, *et al*: Evasion of alternative complement pathway of *Trypanosoma cruzi* results from inefficient binding of factor B. *Proc Natl Acad Sci USA* **83**:6593–6597, 1986.
2. Ruppel A, McLaren DJ, Diesfield HJ, *et al*: *Schistosoma mansoni*: Escape from complement-mediated parasiticidal mechanisms following percutaneous primary infection. *Eur J Immunol* **14**:702–708, 1984.
3. Sher A, Hieny S, Joiner K: Evasion of the alternative complement pathway by metacyclic trypomastigotes of *Trypanosoma cruzi*: Dependence on the developmentally regulated synthesis of surface proteins and *N*-linked carbohydrate. *J Immunol* **137**:2961–2967, 1986.
4. Piessens WF, Partono F, Hoffmann SL, *et al*: Antigen-specific suppressor T-lymphocytes in human lymphatic filariasis. *N Engl J Med* **307**:144–148, 1982.
5. Lucas SB: Missing infections in AIDS. *Trans R Soc Trop Med Hyg* **84**:34–38, 1990.
6. Greenson JK, Belitsos PC, Yardley JH, *et al*: AIDS enteropathy: Occult enteric infections and duodenal mucosal alterations in chronic diarrhea. *Ann Intern Med* **114**:366–372, 1991.
7. Smith PD, Quinn TC, Strober W, *et al*: Gastrointestinal infections in AIDS. *Ann Intern Med* **116**:63–77, 1992.
8. Vanek J, Jirovec O: Parasitare Pneumonie, "Interstitielle" Plasmazell Pneumonie der Frühgeburten verursachte durch *Pneumocystis carinii*. *Zentralbl Bakteriol* **158**:120–127, 1952.
9. Dutz W: *Pneumocystis carinii* pneumonia. *Pathol Annu* **5**:309–341, 1970.
10. Gajdusek DC: *Pneumocystis carinii*—Etiologic agent of interstitial plasma cell pneumonia of young and premature infants. *Pediatrics* **19**:543–545, 1957.
11. Perera DR, Western KA, Johnson HD, *et al*: *Pneumocystis carinii* pneumonia in a hospital for children. *JAMA* **214**:1074–1078, 1970.
12. Robbins JB, DeVita VT, Dutz W: *Symposium on* Pneumocystis carinii *infection*. NCI Monograph No. 43. National Cancer Institute, Washington, DC, 1976.
13. Stagno S, Pifer LL, Hughes WT, *et al*: *Pneumocystis carinii* pneumonitis in young immunocompetent infants. *Pediatrics* **66**:56–62, 1980.
14. Van Der Meer G, Brug SL: Infection par *Pneumocystis* chez l'homme et chez les animaux. *Ann Soc Belg Med Trop* **22**:301–307, 1942.
15. Walzer PD, Shultz MG, Western KA: *Pneumocystis carinii* pneumonia and primary immune deficiency disease of infancy and childhood. *J Pediatr* **82**:416–422, 1973.
16. Watanabe JM, Chinchinian H, Weitz C, *et al*: *Pneumocystis carinii* pneumonia in a family. *JAMA* **193**:119–120, 1965.
17. Redman JC: *Pneumocystis carinii* pneumonia in an adopted Vietnamese infant. *JAMA* **230**:1561–1563, 1973.
18. Walzer PD, Perl DP, Krogstad DJ, *et al*: *Pneumocystis carinii* pneumonia in the United States: Epidemiologic, diagnostic, and clinical features. *Ann Intern Med* **80**:83–93, 1974.
19. Dutz W, Post C, Jennings-Khodadad E, *et al*: Therapy and prophylaxis of *Pneumocystis carinii*. In Robbins JB, DeVita VT Jr, Dutz W (eds): *Symposium on Pneumocystis carinii* Infection. NCI

Monograph 43. National Cancer Institute, Washington, DC, 1976, pp. 179–185.

20. Brazinsky JH, Phillips JE: *Pneumocystis* pneumonia transmission between patients with lymphoma. *JAMA* **209:**1527, 1969.

21. Ruebush TK II, Weinstein RA, Baehner RL, *et al*: An outbreak of *Pneumocystis* pneumonia in children with acute lymphocytic leukemia. *Am J Dis Child* **132:**143–148, 1978.

22. Singer C, Armstrong D, Rosen PP, *et al*: *Pneumocystis carinii* pneumonia: A cluster of eleven cases. *Ann Intern Med* **82:**772–777, 1975.

23. Walzer PD, Perl DP, Krogstad DJ, *et al*: *Pneumocystis carinii* pneumonia in the United States: Epidemologic, diagnostic and clinical features. In Robbins JB, DeVita VT Jr, Dutz W (eds): *Symposium on Pneumocystis carinii Infection*. NCI Monograph 43. National Cancer Institute, Washington, DC, 1976, pp. 55–63.

24. Winston DJ, Gale RP, Meyer DV, *et al*: Infectious complications of human bone marrow transplantation. *Medicine* **58:**1–31, 1979.

25. Walzer PD: *Pneumocystis carinii Pneumonia*. Marcel Dekker, New York, 1994.

26. Fishman JA: *Pneumocystis carinii* pneumonia. In Fishman AP (ed): *Pulmonary Diseases and Disorders*. McGraw-Hill, New York, 1992, pp. 263–286.

27. Gottlieb MS, Schroff R, Shander HM: *Pneumocystis carinii* pneumonia and mucosal candidiasis in previously healthy homosexual men: Evidence of a new acquired cellular immunodeficiency. *N Engl J Med* **305:**1425–1431, 1981.

28. Moskowitz LB, Kory P, Chan JC, *et al*: Unusual causes of death in Haitians residing in Miami: High prevalence of opportunistic infections. *JAMA* **250:**1187–1191, 1983.

29. Fishman JA: Prevention of infection due to *Pneumocystis* carinii. *Antimicrob Agents Chemother* **42:**995–1004, 1998.

30. Fishman JA: Treatment of infection due to *Pneumocystis* carinii. *Antimicrob Agents Chemother* **42:**1309–1314, 1998.

31. Lundgren B, Cotton R, Lundgren JD, *et al*: Identification of *Pneumocystis carinii* chromosomes and mapping of five genes. *Infect Immun* **58:**1705–1710, 1990.

32. Edman JC, Kovacs JA, Masur H, *et al*: Ribosomal RNA sequence shows *Pneumocystis carinii* to be a member of the fungi. *Nature* **334:**519–522, 1988.

33. Sogin ML, Edman JC: A self-splicing intron in the small subunit rRNA gene of *Pneumocystis carinii*. *Nucleic Acids Res* **17:**5349–5359, 1989.

34. Stringer SL, Stringer JR, Blase MA, *et al*: *Pneumocystis carinii*: Sequence from ribosomal RNA implies a close relationship with fungi. *Exp Parasitol* **68:**450–461, 1989.

35. Edman JC, Edman U, Cao M, *et al*: Isolation and expression of the *Pneumocystis carinii* dihydrofolate reductase gene. *Proc Natl Acad Sci USA* **86:**8625–8629, 1989.

36. Edman U, Edman JC, Lundren B, *et al*: Isolation and expression of the *Pneumocystis carinii* thymidylate synthase gene. *Proc Natl Acad Sci USA* **86:**6503–6507, 1989.

37. Smith JW, Bartlett MS: Diagnosis of *Pneumocystis* pneumonia. *Lab Med* **10:**430–435, 1979.

38. Campbell WG: Ultrastructure of *Pneumocystis* in human lung. *Arch Pathol Lab Med* **93:**312–324, 1972.

39. Vossen MEMH, Beckers PJA, Meuwissen JHETL, *et al*: Developmental biology of *Pneumocystis carinii*: An alternative view on the life cycle of the parasite. *Z Parasitenkd* **55:**101–118, 1978.

40. Hughes WT: *Pneumocystis carinii Pneumonitis*. CRC Press, New York, 1987.

41. Pifer LL, Hughes WT, Murphy MJ: Propagation of *Pneumocystis carinii in vitro*. *Pediatr Res* **11:**305–316, 1977.

42. Cushion MT: *In vitro* studies of *Pneumocystis carinii*. *J Protozool* **36:**45S–52S, 1989.

43. Murphy MJ, Pifer LL, Hughes WT: *Pneumocystis carinii in vitro*. *Am J Pathol* **86:**387, 1977.

44. Latorre CR, Sulzer AJ, Norman LG: Serial propagation of *Pneumocystis carinii* in cell line cultures. *Appl Env Microbiol* **33:**1204–1206, 1977.

45. Smith JW, Bartlett MS, Queener SF: *In vitro* cultivation of *Pneumocystis*. Development of models and their use to discover new drugs for therapy and prophylaxis of *Pneumocystis carinii* pneumonia. In Walzer PD (ed): *Pneumocystis carinii Pneumonia*. Marcel Dekker, New York, 1994, pp. 487–510.

46. Walzer PD, Schnelle V, Armstrong D, *et al*: A new experimental model for *Pneumocystis carinii* infection. *Science* **197:**177–179, 1977.

47. Hughes WT, Bartley DL, Smith BM: A natural source of infection due to *Pneumocystis carinii*. *J Infect Dis* **147:**595, 1983.

48. Veda K, Goto Y, Yamazaki S, *et al*: Chronic fatal pneumocystosis in nude mice. *Jpn J Exp Med* **47:**475–482, 1977.

49. Frenkel JK, Good JT, Shultz JA: Latent *Pneumocystis* infection of rats: Relapse and chemotherapy. *Lab Invest* **15:**1559–1577, 1966.

50. Frenkel JK: *Pneumocystis jiroveci* n. sp. In Robbins JB, DeVita VT Jr, Dutz W (eds): *Symposium on Pneumocystis carinii Infection*. NCI Monograph 43. National Cancer Institute, Washington, DC, 1976, pp. 13–30.

51. Hendley JO, Weller TH: Activation and transmission in rats of infection with *Pneumocystis*. *Proc Soc Exp Biol Med* **137:**1401–1404, 1971.

52. Bartlett MS, Fishman JA, Durkin MM, *et al*: *Pneumocystis carinii*: Improved models to study efficacy of drugs for treatment or prophylaxis of *Pneumocystis* pneumonia in the rat (*Rattus* spp.). *Exp Parasitol* **70:**100–106, 1990.

53. Shellito J, Suzara VV, Blumenfeld W, *et al*: A new model of *Pneumocystis carinii* infection in mice selectively depleted of help of T lymphocytes. *J Clin Invest* **85:**1686–1693, 1990.

54. Harmsen AG, Stankiewicz M: Requirement for CD4+ cells in resistance to *Pneumocystis carinii* pneumonia in mice. *J Exp Med* **172:**937–946, 1990.

55. Furuta T, Veda K, Fujiwara K, *et al*: Cellular and humoral immune response of mice subclinically infected with *Pneumocystis carinii*. *Infect Immun* **47:**544–548, 1985.

56. Furuta TK, Ueda K, Fujiwara K: Effect of T-cell transfer on *Pneumocystis carinii* infection in nude mice. *Jpn J Exp Med* **54:**57–64, 1984.

57. Beck JM, Liggitt HD, Brunette EN, *et al*: Reduction in intensity of *Pneumocystis carinii* pneumonia in mice by aerosol administration of gamma interferon. *Infect Immun* **59:**3859–3862, 1991.

58. Murray HW, Gellene RA, Libby DM: Activation of tissue macrophages from AIDS patients: *In vitro* response of AIDS alveolar macrophages to lymphokines and interferon-gamma. *J Immunol* **135:**2374–2377, 1985.

59. Brzosko WJ, Nowoslawski A: Identification of *Pneumocystis carinii* antigens in tissues. *Bull Acad Pol Sci* **13:**49–54, 1965.

60. Graves DC: Immunologic studies of *Pneumocystis carinii*. *J Protozool* **36:**60–69, 1989.

61. Fishman JA: Prevention of infection due to *Pneumocystis carinii* in transplant recipients. *Clin Infect Dis* **33:**1397–405, 2001.

62. Hughes WT: Recent advances in serodiagnosis of *Pneumocystis carinii*. *Chest* **89:**764–765, 1986.

63. Fisher DJ, Gigliotti F, Zauderer M, *et al*: Specific T-cell response to a *Pneumocystis carinii* surface glycoprotein (gp120) after immunization and natural infection. *Infect Immun* **59**:3372–3376, 1991.

64. Herrod HG, Valenski WR, Woods DR, *et al*: The *in vitro* response of human lymphocytes to *Pneumocystis carinii* antigen. *J Immunol* **126**:59–61, 1981.

65. Giggliotti F, Hughes WT: Passive immunoprophylaxis with specific monoclonal antibody confers partial protection against *Pneumocystis carinii* pneumonitis in animal models. *J Clin Invest* **81**:1666–1668, 1988.

66. Hagler DN, Deepe GS, Pogue CL, *et al*: Blastogenic responses to *Pneumocystis carinii* among patients with human immunodeficiency (HIV) infection. *Clin Exp Immunol* **74**:7–13, 1988.

67. Masur H, Jones TC: The interaction *in vitro* of *Pneumocystis carinii* with macrophages and L-cells. *J Exp Med* **147**:157–170, 1978.

68. Hughes WT, Feldman S, Aur RJA, *et al*: Intensity of immunosuppressive therapy and the incidence of *Pneumocystis carinii* pneumonitis. *Cancer* **36**:2004–2009, 1975.

69. Masur H, Ognibene FP, Yarchoan R, *et al*: CD4 counts as predictors of opportunistic pneumonias in human immunodeficiency virus (HIV) infection. *Ann Intern Med* **111**:223–231, 1989.

70. Hughes WT, Johnson WW: Recurrent *Pneumocystis carinii* pneumonia following apparent recovery. *J Pediatr* **79**:755–759, 1971.

71. Hughes WT, Price RA, Sisko F, *et al*: Protein calorie malnutrition: A host determinant for *Pneumocystis carinii* infection. *Am J Dis Child* **128**:44–550, 1974.

72. Hughes WT, Sanyal SK, Price RA: Signs, symptoms and pathophysiology of *Pneumocystis carinii* pneumonitis. In Robbins JB, DeVita VT Jr, Dutz W (eds): *Symposium on Pneumocystis carinii Infection*. NCI Monograph 43. National Cancer Institute, Washington, DC, 1976, pp. 77–88.

73. Jacobson MA, Mills J: Serious cytomegalovirus disease in acquired immunodeficiency syndrome (AIDS): Clinical findings, diagnosis, and treatment. *Ann Intern Med* **108**:585–594, 1988.

74. Wang NS, Huang SN, Thurlbeck WM: Combined *Pneumocystis carinii* and cytomegalovirus infection. *Arch Pathol Lab Med* **90**:529–535, 1970.

75. Bozzette SA, Arcia J, Bartok AE, *et al*: Impact of *Pneumocystis carinii* and cytomegalovirus on the course and outcome of atypical pneumonia in advanced human immunodeficiency virus disease. *J Infect Dis* **165**:93–98, 1992.

76. Lyons HA, Vinijchaikul K, Hennigar GR: *Pneumocystis carinii* pneumonia unassociated with other disease. *Arch Intern Med* **108**:173–180, 1961.

77. Meuwissen JHET, Leeuwenberg ADEM: A microcomplement fixation test applied to infection with *Pneumocystis carinii*. *Trop Geogr Med* **24**:282–291, 1972.

78. Meyer JD, Pifer LL, Sale GE, *et al*: The value of *Pneumocystis carinii* antibody and antigen detection for diagnosis of *Pneumocystis carinii* pneumonia after marrow transplantation. *Am Rev Respir Dis* **120**:181–182, 1979.

79. Meuwissen JHET, Tauber I, Leeuwenberg ADEM, *et al*: Parasitologic and serologic observations of infection with *Pneumocystis* in humans. *J Infect Dis* **136**:43–49, 1977.

80. Norman L, Kagan IG: Some observations on the serology of *Pneumocystis carinii* infections in the United States. *Infect Immunol* **8**:317–321, 1973.

81. Pifer LL: Serodiagnosis of *Pneumocystis carinii*. *Chest* **87**:698–700, 1985.

82. Kovacs JA, Hiemenz JW, Macher AM, *et al*: *Pneumocystis carinii* pneumonia: A comparison between patients with the acquired immunodeficiency syndrome and patients with other immunodeficiencies. *Ann Intern Med* **100**:663–671, 1984.

83. Kwok S, O'Donnell JJ, Wood IS: Retinal cotton-wool spots in a patient with *Pneumocystis carinii* infection. *N Engl J Med* **307**:184–185, 1982.

84. Scott GB, Hutto C, Makuch RW, *et al*: Survival in children with perinatally acquired human immunodeficiency virus type 1 infection. *N Engl J Med* **321**:1971–1976, 1989.

85. Sirotzky L, Memoli V, Roberts JL, *et al*: Recurrent *Pneumocystis* pneumonia with normal chest roentgenograms. *JAMA* **240**:1513–1515, 1978.

86. Turbiner EH, Yeh SDJ, Rosen PP, *et al*: Abnormal gallium scintigraphy in *Pneumocystis carinii* pneumonia with a normal chest radiography. *Radiology* **127**:437–438, 1978.

87. Goodell B, Jacobs JB, Powell RD, *et al*: *Pneumocystis carinii*: The spectrum of diffuse interstitial pneumonia in patients with neoplastic disease. *Ann Intern Med* **72**:337–340, 1970.

88. Doppman JL, Geelhoed GW: Atypical radiographic features in *Pneumocystis carinii* pneumonia. In Dutz W, Robbins JB, DeVita VT Jr (eds): *Symposium on Pneumocystis carinii Infection*. NCI Monograph 43. National Cancer Institute, Washington, DC, 1976, pp. 89–97.

89. Cross AS, Steigbigel RT: *Pneumocystis carinii* pneumonia presenting as localized nodular densities. *N Engl J Med* **291**:831–832, 1974.

90. Luddy RE, Champion LAA, Schwartz AD: *Pneumocystis carinii* pneumonia with pneumatocoele formation. *Am J Dis Child* **131**:470–471, 1973.

91. Levenson SM, Warren RD, Richman SD, *et al*: Abnormal pulmonary gallium accumulation in *P. carinii* pneumonia. *Radiology* **119**:395–398, 1976.

92. Fishman JA, Strauss HW, Fischman AJ, *et al*: Imaging of *Pneumocystis carinii* pneumonia with ^{111}In-labelled non-specific polyclonal IgG: An experimental study in rats. *Nucl Med Comm* **12**:175–187, 1991.

93. Fishman JA: Radiologic approach to the diagnosis and management of *Pneumocystis carinii*. In Walzer P (ed): *Pneumocystis carinii Pneumonia*. Marcel Dekker, New York, 1994, pp. 415–438.

94. Hopewell PC: Bronchoalveolar lavage and transbronchial biopsy for the diagnosis of pulmonary infections in the acquired immunodeficiency syndrome. *Ann Intern Med* **102**:747–752, 1985.

95. Kovacs JA, Ng VL, Masur H, *et al*: Diagnosis of *P. carinii* pneumonia: Improved detection in sputum with use of monoclonal antibiotics. *N Engl J Med* **318**:589–593, 1988.

96. Stover DE, Zaman MB, Hajdu SI, *et al*: Bronchoalveolar lavage in the diagnosis of diffuse pulmonary infiltrates in the immunocompromised host. *Ann Intern Med* **101**:1–7, 1984.

97. Awen C, Baltzan M: Systemic dissemination of *Pneumocystis carinii* pneumonia. *Can Med Assoc J* **104**:809–812, 1971.

98. Coulman CU, Greene I, Archibald RWR: Cutaneous *Pneumocystis*. *Ann Intern Med* **106**:396–398, 1987.

99. Fishman JA: Case records of the Massachusetts General Hospital. *N Engl J Med* **332**:249–258, 1995.

100. Raviglionne MC: Extrapulmonary pneumocystosis: The first fifty cases. *Rev Infect Dis* **12**:1127–1138, 1990.

101. Lau WK, Young LS, Remington JS: *Pneumocystis carinii* pneumonia: Diagnosis by examination of pulmonary secretions. *JAMA* **236**:2399–2402, 1976.

102. Bigby PD, Margolskee D, Curtis J, *et al*: Usefulness of induced

sputum in diagnosis of pneumonia in patients with acquired immunodeficiency syndrome. *Am Rev Respir Dis* **133:**515–518, 1986.

103. Bigby TD, Margolskee D, Curtis JL, *et al*: The usefulness of induced sputum in the diagnosis of *Pneumocystis carinii* pneumonia in patients with the acquired immunodeficiency syndrome. *Am Rev Respir Dis* **133:**515–518, 1986.

104. Kovacs JA, Gill V, Swann JC, *et al*: Prospective evaluations of monoclonal antibody in diagnosis of *Pneumocystis carinii* pneumonia. *Lancet* **2:**1–3, 1986.

105. Lim SK, Eveland WC, Porter RJ: Direct fluorescent-antibody method for the diagnosis of *Pneumocystis carinii* pneumonitis from sputa or tracheal aspirates from humans. *Appl Microbiol* **27:**144–149, 1974.

106. Baughman RP: Current methods of diagnosis. In PD Walzer (ed): *Pneumocystis carinii Pneumonia*. Marcel Dekker, New York, 1994, pp. 381–402.

107. Pennington JE, Feldman NT: Pulmonary infiltrates and fever in patients with hematologic malignancy: Assessment of transbronchial biopsy. *Am J Med* **62:**581–587, 1977.

108. Blumenfeld W, Wager E, Hadley WK: Use of transbronchial biopsy for diagnosis of opportunistic pulmonary infections in acquired immunodeficiency syndrome (AIDS). *Am J Clin Pathol* **81:**1–5, 1984.

109. Broaddus C, Dake MD, Stulbarg MS, *et al*: Bronchoalveolar lavage and transbronchial biopsy for the diagnosis of pulmonary infections in the acquired immunodeficiency syndrome. *Ann Intern Med* **102:**747–752, 1985.

110. Burt ME, Flye WW, Webber BL, *et al*: Prospective evaluation of aspiration needle, cutting needle transbronchial, and open lung biopsy in patients with pulmonary infiltrates. *Ann Thorac Surg* **32:**146–153, 1981.

111. Coleman DL, Dodek PM, Luce JM, *et al*: Diagnostic utility of fiberoptic bronchoscopy in patients with *Pneumocystis carinii* pneumonia and the acquired immunodeficiency syndrome. *Am Rev Respir Dis* **128:**795–799, 1983.

112. Springmeyer SC, Silvestri, RC, Sale GE, *et al*: Role of transbronchial biopsy for the diagnosis of diffuse pneumonia in immunocompromised marrow transplant recipients. *Am Rev Respir Dis* **126:**763–765, 1982.

113. Chaudhary S, Hughes WT, Feldman S, *et al*: Percutaneous transthoracic needle aspiration of the lung. *Am J Dis Child* **131:**902–906, 1977.

114. Rossiter SJ, Miller DC, Churg AM, *et al*: Open lung biopsy in the immunosuppressed patient. *J Thorac Cardiovasc Surg* **77:**338–345, 1979.

115. Young RC, DeVita VT Jr: Treatment of *Pneumocystis carinii* pneumonia. In Robbins JB, DeVita VT Jr, Dutz W (eds): *Symposium on Pneumocystis carinii Infection*. NCI Monograph 43. National Cancer Institute, Washington, DC, 1976, pp. 193–198.

116. Haverkos HW: PCP therapy project group: Assessment of therapy for *Pneumocystis carinii* pneumonia. *Am J Med* **76:**501–508, 1984.

117. Hughes WT, Feldman S, Chaudhary SC, *et al*: Comparison of pentamidine isethionate and trimethoprim-sulfamethoxazole in the treatment of *Pneumocystis carinii* pneumonia. *J Pediatr* **92:**285–291, 1978.

118. Hughes WT, Feldman S, Sanyal SK: Treatment of *Pneumocystis carinii* pneumonitis with trimethoprim/sulfamethoxazole. *Can Med Assoc J* **112:**47S-50S, 1975.

119. Hughes WT, McNabb PC, Makres TD, *et al*: Efficacy of trimethoprim and sulfamethoxazole in the prevention and treatment of *Pneumocystis carinii* pneumonitis. *Antimicrob Agents Chemother* **5:**289–293, 1974.

120. Wharton JM, Coleman DL, Wofsy CB, *et al*: Trimethoprim-sulfamethoxazole or pentamidine for *Pneumocystis carinii* pneumonia in the acquired immunodeficiency syndrome. *Ann Intern Med* **105:**37–44, 1986.

121. Kluge RM, Spaulding DM, Spain JA: Combination of pentamidine and trimethoprim-sulfamethoxazole in the therapy of *Pneumocystis carinii* pneumonia in rats. *Antimicrob Agents Chemother* **13:**975–978, 1978.

122. Hughes WT, Kuhn S, Chaudhary S, *et al*: Successful chemoprophylaxis of *Pneumocystis carinii* pneumonia pneumonitis. *N Engl J Med* **297:**1419–1426, 1977.

123. Lau WK, Young LS: Trimethoprim-sulfamethoxazole treatment of *Pneumocystis carinii* pneumonia in adults. *N Engl J Med* **295:**716–718, 1976.

124. Winston DJ, Lau WK, Gale RP, *et al*: Trimethoprim-sulfamethoxazole for the treatment of *Pneumocystis carinii* pneumonia. *Ann Intern Med* **97:**762–769, 1980.

125. Young LS: Treatment of *Pneumocystis carinii* pneumonia in adults with trimethoprim/sulfamethoxazole. *Rev Infect Dis* **4:**608–613, 1982.

126. Ivady G, Paldy L: Ein neues Behandlungserfarhen der interstitiellen plasmazelligen Pneumonie Frühgeborener mit fünfwertigen Stibium und aromatischen Diamidinen. *Monatsschr Kinderheilkd* **106:**10–14, 1958.

127. Western KA, Perera DR, Schultz MG: Pentamidine isethionate in the treatment of *Pneumocystis carinii* pneumonia. *Ann Intern Med* **73:**695–702, 1970.

128. Hughes WT, Smith BL: Efficacy of diaminodiphenyl sulfone and other drugs in murine *Pneumocystis carinii* pneumonitis. *Antimicrob Agents Chemother* **26:**436–440, 1984.

129. Hughes WT, Smith-McCain BL: Effects of sulfonyl urea compounds on *Pneumocystis carinii*. *J Infect Dis* **153:**944–947, 1984.

130. Leoung GS, Mills J, Hopewell PC, *et al*: Dapsone-trimethoprim for *Pneumocystis carinii* pneumonia in the acquired immunodeficiency syndrome. *Ann Intern Med* **105:**45–48, 1986.

131. Medina I, Mills J, Leoung G, *et al*: Oral therapy for *Pneumocystis carinii* pneumonia in the acquired immunodeficiency syndrome: A controlled trial of trimethoprim-sulfamethoxazole versus trimethoprim-dapsone. *N Engl J Med* **323:**776–782, 1990.

132. Metroka CE, McMechan MF, Andrada R, *et al*: Failure of prophylaxis with dapsone in patients taking dideoxyinosine. *N Engl J Med* **325:**737, 1991.

133. Noskin GA, Murphy RL, Black JR, *et al*: Salvage therapy with clindamycin/primaquine for *Pneumocystis carinii* pneumonia. *Clin Infect Dis* **14:**183–188, 1992.

134. Post C, Fakoughi T, Dutz W, *et al*: Prophylaxis of epidemic infantile pneumocystosis with a 20:1 sulfadoxine and pyrimethamine combination. *Curr Ther Res* **13:**273–279, 1971.

135. Toma E, Fournier S, Poisson M, *et al*: Clindamycin with primaquine for *Pneumocystis carinii* pneumonia. *Lancet* **1:**1046–1048, 1989.

136. Whisnant JK, Buckley RH: Successful pyrimethamine-sulfadiazine therapy of *Pneumocystis* pneumonia in infants with X-linked immunodeficiency with hyper IgM. In Robbins JB, DeVita VT Jr, Dutz W (eds): *Symposium on Pneumocystis carinii Infection*. NCI Monograph 43. National Cancer Institute, Washington, DC, 1975, pp. 211–216.

137. Allegra CJ, Chabner BA, Tuazon CU, *et al*: Trimetrexate for the treatment of *Pneumocystis carinii* pneumonia in patients with the

acquired immunodeficiency syndrome. *N Engl J Med* **317:**978–985, 1987.

138. Polsen DC, Kovacs JA, Lipschik GY: Folate antagonists in the treatment of *Pneumocystis carinii* pneumonia. In PD Walzer (ed): *Pneumocystis carinii Pneumonia.* Marcel Dekker, New York, 1994, pp. 545–560.

139. Sattler FR, Allegra CJ, Verdegam TD, *et al:* Trimetrexate-leucovorin dosage evaluation study for treatment of *Pneumocystis carinii* pneumonia. *J Infect Dis* **161:**91–96, 1990.

140. Allegra CJ, Kovacs JA, Chabner BA, *et al:* Potent *in vivo* and *in vitro* activity of a lipid soluble antifolate, trimetrexate, against *Pneumocystis carinii. Clin Res* **34:**674A, 1986.

141. Falloon J, Kovacs J, Hughes W, *et al:* A preliminary evaluation of 566C80 for the treatment of *Pneumocystis* pneumonia in patients with the acquired immunodeficiency syndrome. *N Engl J Med* **325:**1534–1538, 1991.

142. Golden JA, Sjoerdsma A, Santi DV: *Pneumocystis carinii* pneumonia treated with alpha-difluoromethylornithine. *West J Med* **141:**613–623, 1984.

143. Bozzette SA, Sattler FR, Chiu J, *et al:* A controlled trial of early adjunctive treatment with corticosteroids for *Pneumocystis carinii* pneumonia in the acquired immunodeficiency syndrome. *N Engl J Med* **323:**1451–1457, 1990.

144. Gagnon S, Boota AM, Fisch MA, *et al:* Corticosteroids as adjunctive therapy for severe *Pneumocystis carinii* pneumonia in the acquired immunodeficiency syndrome. *N Engl J Med* **323:**1444–1450, 1990.

145. McGowan JE Jr, Chesney PJ, Crossley KB, *et al:* Guidelines for the use of systemic glucocorticosteroids in the management of selected infections. *J Infect Dis* **165:**1–13, 1992.

146. Montaner JSG, Lawson LM, Levitt N, *et al:* Corticosteroids prevent early deterioration in patients with moderately severe *Pneumocystis carinii* pneumonia and the acquired immunodeficiency syndrome (AIDS). *Ann Intern Med* **113:**14–20, 1990.

147. Hughes WT, Rivera GK, Schell MJ, *et al:* Successful intermittent prophylaxis for *Pneumocystis carinii* pneumonitis. *N Engl J Med* **316:**1627–1632, 1987.

148. Hughes WT, Price RA, Kim HK, *et al: Pneumocystis carinii* pneumonitis in children with malignancies. *J Pediatr* **82:**404–415, 1973.

149. Fischl MA: Fansidar prophylaxis of *Pneumocystis* pneumonia in the acquired immunodeficiency syndrome. *Ann Intern Med* **105:**629, 1986.

150. Hughes WT: Limited effect of trimethoprim sulfamethoxazole prophylaxis on *Pneumocystis carinii. Antimicrob Agents Chemother* **16:**333–335, 1979.

151. Hughes WT: Five-year absence of *Pneumocystis carinii* pneumonitis in a pediatric oncology center. *J Infect Dis* **150:**305–306, 1984.

152. Gottlieb M, Knight S, Mitsuyasu R, *et al:* Prophylaxis of *Pneumocystis carinii* infection in acquired immunodeficiency syndrome (AIDS) with pyrimethamine/sulfadoxine (Fansidar). *Lancet* **2:**398–399, 1984.

153. Feldman HA: Toxoplasmosis: An overview. *Bull NY Acad Med* **50:**110–127.

154. Janku J: Pathogenesis and pathologic anatomy of coloboma of macula lutea in eye of normal dimensions, and in microphthalmic eye with parasites in return. *Cas Lek Clsk* **62:**1021–1027, 1923.

155. Jones TC, Hirsch JG: The interaction between *Toxoplasma gondii* and mammalian cells. II. The absence of lysosome fusion with phagocytic vacuoles containing living parasites. *J Exp Med* **136:**1173–1194, 1972.

156. Raisanen S: Toxoplasmosis transmitted by blood transfusions. *Transfusion* **18:**329–332, 1978.

157. Michaels MG, Wald ER, Ficker FJ, *et al:* Toxoplasmosis in pediatric recipients of heart transplants. *Clin Infect Dis* **14:**847–851, 1992.

158. Leak D, Meghji M: Toxoplasmic infection in cardiac disease. *Am J Cardiol* **43:**841–849, 1979.

159. Anderson SE Jr, Remington JS: Effect of normal and activated human macrophages on *Toxoplasma gondii. J Exp Med* **139:**1154–1174, 1974.

160. Araujo FG: Depletion of L3T4+ (CD4+) T lymphocytes prevents development of resistance to *Toxoplasma gondii* in mice. *Infect Immun* **59:**1614–1619, 1991.

161. McLeod R, Remington JS: Influence of infection with toxoplasma on macrophage function, and role of macrophages in resistance to toxoplasma. *Am J Trop Med Hyg* **26:**170–186, 1977.

162. Lindberg RE, Frenkel JK: Toxoplasmosis in nude mice. *J Parasitol* **63:**210–221, 1977.

163. Frenkel JK: Effects of cortisone, total body radiation and nitrogen mustard on chronic latent toxoplasmosis. *Am J Pathol* **33:**618–619, 1957.

164. Strannegard O, Lycke E: Effect of antithymocyte serum on experimental toxoplasmosis in mice. *Infect Immun* **5:**769–774, 1972.

165. Remington JS, Krahenbuhl JL, Mendenhall JW: A role for activated macrophages in resistance to infection with toxoplasma. *Infect Immun* **6:**829–834, 1972.

166. Jones TC, Len L, Hirsch J: Assessment *in vitro* of immunity against *Toxoplasma gondii. J Exp Med* **171:**466–482, 1975.

167. Hirsch R, Burke BA, Kersey JH: Toxoplasmosis in bone marrow transplant recipients. *J Pediatr* **105:**426–428, 1984.

168. Jehn V, Fink M, Gundlach P: Lethal cardiac and cerebral toxoplasmosis in a patient with acute myeloid leukemia after successful allogenic bone marrow transplantation. *Transplantation* **38:**430–433, 1984.

169. Nicholdon DH, Wolchok EB: Ocular toxoplasmosis in an adult receiving long-term corticosteroid therapy. *Arch Ophthalmol* **94:**248–257, 1976.

170. O'Connor GR, Frenkel JK: Dangers of steroid treatment in toxoplasmosis. *Arch Ophthalmol* **94:**213, 1976.

171. Rose AG, Uys CJ, Novitsky D, *et al:* Toxoplasmosis of donor and recipient hearts after heterotopic cardiac transplantations. *Arch Pathol Lab Med* **107:**368–373, 1983.

172. Ryning FW, McLeod R, Madox JC, *et al:* Probable transmission of *Toxoplasma gondii* by organ transplantation. *Ann Intern Med* **84:**47–49, 1979.

173. Swartzberg JE, Krahenbuhl JL, Remington JS: Dichotomy between macrophage activation and degree of protection against *Listeria monocytogenes* and *Toxoplasma gondii* in mice stimulated with *Corynebacterium parvum. Infect Immun* **12:**1037–1043, 1975.

174. Wong B, Gold JWM, Brown AE, *et al:* Central nervous system toxoplasmosis in homosexual men and parenteral drug abusers. *Ann Intern Med* **100:**36–42, 1984.

175. Ruskin J, Remington JS: Toxoplasmosis in the compromised host. *Ann Intern Med* **84:**193–199, 1976.

176. Barlotta FM, Odhoa M Jr, Neu HC, *et al:* Toxoplasmosis, lymphoma or both? *Ann Intern Med* **70:**517–528, 1979.

177. Herb HM, Jontofsoh R, Loffler HD, *et al:* Toxoplasmosis after renal transplantation. *Clin Nephrol* **8:**529–532, 1978.

178. Luft BJ, Naot Y, Araujo FG, *et al:* Primary and reactivated toxoplasma infection in patients with cardiac transplant: Clinical spec-

trum and problems in diagnosis in a defined population. *Ann Intern Med* **99:**27–31, 1983.

179. Whiteside JD, Begent RHJ: Toxoplasma encephalitis complicating Hodgkin's disease. *J Clin Pathol* **28:**443–445, 1975.

180. Brooks RG, McCabe RE, Remington JS: Role of serology in the diagnosis of toxoplasmic lymphadenopathy. *Rev Infect Dis* **9:**1055–1062, 1987.

181. McCabe RE, Brooks RG, Dorfman RF, *et al*: Clinical spectrum in 107 cases of toxoplasmic lymphadenopathy. *Rev Infect Dis* **9:**754–774, 1987.

182. Siim JC: Acquired toxoplasmosis: Report of seven cases with strongly positive serologic reactions. *JAMA* **147:**1651–1654, 1951.

183. Gard S, Magnusson HJ: Glandular form of toxoplasmosis in connection with pregnancy. *Acta Med Scand* **141:**59–64, 1951.

184. Welch PC, Masur H, Jones TC, *et al*: Serologic diagnosis of acute lymphadenopathic toxoplasmosis. *J Infect Dis* **142:**256–264, 1980.

185. Wolf A, Cowen D, Paige B: Human toxoplasmosis: Occurrence in infants with encephalomyelitis—Verification by transmission to animals. *Science* **89:**226–227, 1939.

186. Rollins DF, Tabbara KF, O'Connor GR, *et al*: Detection of toxoplasma antigen and antibody in ocular fluids in experimental ocular toxoplasmosis. *Arch Ophthalmol* **101:**455–457, 1983.

187. Siegel S, Lunde M, Gelderman A, *et al*: Transmission of toxoplasmosis by leukocyte transfusion. *Blood* **37:**388–394, 1971.

188. Stinson EB, Biber CP, Griepp RB, *et al*: Infectious complications after transplantation in man. *Ann Intern Med* **74:**22–36, 1971.

189. Dummer JS, Bahnson HT, Griffith BP, *et al*: Infections in patients on cyclosporine and prednisone following cardiac transplantation. *Transplant Proc* **14:**2779–2781, 1983.

190. Luft BJ, Naot Y, Araujo FG, *et al*: Primary and reactivated toxoplasma infection in patients with cardiac transplants. *Ann Intern Med* **99:**27–31, 1983.

191. Hakin M, Wreghitt TG, English TAH, *et al*: Significance of donor transmitted disease in cardiac transplantation. *Heart Transplant* **4:**302–306, 1985.

192. Hakin M, Esmore D, Wallwork J, *et al*: Toxoplasmosis in cardiac transplantation. *Br Med J* **292:**1108–1109, 1986.

193. Pomeroy C, Filice GA: Pulmonary toxoplasmosis: A review. *Clin Infect Dis* **14:**863–870, 1992.

194. Fishman J: *Pneumocystis carinii* pneumonia. In Fishman AP, Elias JA, Fishman JA, Grippi MA, Kaiser LR, Senior RM (eds): *Fishman's Pulmonary Diseases and Disorders*, 3rd ed. McGraw-Hill, New York, 1998, pp. 2313–2332.

195. Jacobs F, Depierreux M, Goldman M, *et al*: Role of bronchoalveolar lavage in diagnosis of disseminated toxoplasmosis. *Rev Infect Dis* **13:**637–641, 1990.

196. Oksenhendler E, Cadranel J, Sarfati C, *et al*: *Toxoplasma gondii* pneumonia in patients with the acquired immunodeficiency syndrome. *Am J Med* **88:**18N–21N, 1990.

197. Snider WD, Simpson DM, Neilsen S, *et al*: Neurological complications of acquired immunodeficiency syndrome: Analysis of 50 patients. *Ann Neurol* **14:**403–418, 1983.

198. Ghatak NR, Sawyer DR: A morphologic study of opportunistic cerebral toxoplasmosis. *Acta Neuropathol* **42:**217–221, 1978.

199. Kerstin F, Newmann J: "Malignant lymphoma" of the brain following renal transplantation. *Acta Neuropathol* **6:**131–133, 1975.

200. Luft BJ, Brooks RG, Conley FK, *et al*: Toxoplasmic encephalitis in patients with acquired immune deficiency syndrome. *JAMA* **252:**913–917, 1984.

201. Luft BJ, Remington JS: Toxoplasmosis of the central nervous system. In Remington JS, Swartz M (eds): *Current Clinical Topics in Infectious Disease*, Vol. 5. McGraw-Hill, New York, 1985, pp. 315–358.

202. Luft BJ, Remington JS: Toxoplasmic encephalitis. *J Infect Dis* **157:**1–6, 1988.

203. Powell HC, Gibbs JC Jr, Lorenzo AM, *et al*: Toxoplasmosis of the central nervous system in the adult: Electron microscopic observations. *Acta Neuropathol* **41:**211–216, 1978.

204. Schulkof LA, Russell JR: Intracerebral toxoplasmosis presenting as a mass lesion. *Surg Neurol* **4:**9–11, 1975.

205. Slavick HE, Lipman IJ: Brain stem toxoplasmosis complicating Hodgkin's disease. *Arch Neurol* **34:**636–637, 1977.

206. Wanke C, Tuazon CU, Kovacs A, *et al*: Toxoplasma encephalitis in patients with acquired immune deficiency syndrome: Diagnosis and response to therapy. *Am J Trop Med Hyg* **36:**509–516, 1987.

207. Menges HW, Fischer E, Valavanis A, *et al*: Cerebral toxoplasmosis in the adult. *J Comput Assist Tomogr* **3:**413–416, 1979.

208. Ciricillo SF, Rosenblum ML: Use of CT and MR imaging to distinguish intracranial lesions and to define the need for biopsy in AIDS patients. *J Neurosurg* **73:**720–724, 1990.

209. Shepp DH, Hackman RC, Conley FK, *et al*: *Toxoplasma gondii* reactivation identified by detection of parasitemia in tissue culture. *Ann Intern Med* **103:**218–221, 1985.

210. Hofflin JM, Remington JS: Tissue culture isolation of toxoplasma from blood of a patient with AIDS. *Arch Intern Med* **145:**925–926, 1985.

211. Dorfman RF, Remington JS: Value of lymph-node biopsy in the diagnosis of acute acquired toxoplasmosis. *N Engl J Med* **289:**878–881, 1973.

212. Araujo FG, Remington JS: Antigenemia in recently acquired acute toxoplasmosis. *J Infect Dis* **141:**144–150, 1980.

213. Brooks RG, Sharma SD, Remington JS: Detection of *Toxoplasma gondii* antigens by a dot-immunobinding technique. *J Clin Microbiol* **21:**113–116, 1985.

214. Sabin AV, Ruchman I: Characteristics of toxoplasma neutralizing antibody. *Proc Soc Exp Biol Med* **51:**1–6, 1942.

215. Desmonts G: Definitive serologic diagnosis of ocular toxoplasmosis. *Arch Ophthalmol* **76:**839–851, 1966.

216. Potasman I, Resnick L, Luft BJ, *et al*: Intrathecal production of antibodies against *Toxoplasma gondii* in patients with toxoplasmic encephalitis and the acquired immunodeficiency syndrome (AIDS). *Ann Intern Med* **108:**49–51, 1988.

217. Sabin A, Feldman HA: Dyes as microchemical indicators of a new immunity phenomenon affecting a protozoan parasite (Toxoplasma). *Science* **108:**660–663, 1978.

218. Remington JS, Eimstad WM, Araujo FG: Detection of immunoglobulin M antibodies with antigen-tagged latex particles in an immunosorbent assay. *J Clin Microbiol* **17:**939–941, 1983.

219. Remington JS, Miller MJ, Brownlee I: IgM antibodies in acute toxoplasmosis. II. Prevalence and significance in acquired cases. *J Lab Clin Med* **71:**855–866, 1968.

220. Desmonts G, Remington JS: Direct agglutination test for diagnosis of toxoplasma infection: Method for increasing sensitivity and specificity. *J Clin Microbiol* **11:**562–568, 1980.

221. McCabe RE, Gibbons D, Brooks RG, *et al*: Agglutination test for diagnosis of toxoplasmosis in AIDS. *Lancet* **2:**680, 1983.

222. Siegel JP, Remington JS: Comparison of methods for quantitating antigen specific immunoglobulin M antibody with a reverse enzyme linked immunoabsorbent assay. *J Clin Microbiol* **18:**63–70, 1983.

223. Walls KW, Bullock SL, English DK: Use of the enzyme-linked immunosorbent assay (ELISA) and its microadaptation for the

serodiagnosis of toxoplasmosis. *J Clin Microbiol* **5:**273–277, 1977.

224. Wielaard F, van Gruighuigsen H, Duermeyer W, *et al*: Diagnosis of acute toxoplasmosis by an enzyme immunoassay for specific immunoglobulin M antibodies. *J Clin Microbiol* **17:**981–987, 1983.
225. Hakes TB, Armstrong D: Toxoplasmosis: Problems in diagnosis and treatment. *Cancer* **52:**1535–1540, 1983.
226. Cohn JA, McMeeking A, Cohen W, *et al*: Evaluation of the policy of empiric treatment of suspected toxoplasma encephalitis in patients with the acquired immunodeficiency syndrome. *Am J Med* **86:**521–527, 1989.
227. Dannemann B, McCutchan JA, Israelski D, *et al*: Treatment of toxoplasmic encephalitis in patients with AIDS. *Ann Intern Med* **116:**33–43, 1992.
228. Haverkos HW: Assessment of therapy for toxoplasma encephalitis. *Am J Med* **82:**907, 1987.
229. Danneman BR, Israelski DM, Remington JS: Treatment of toxoplasmic encephalitis with intravenous clindamycin. *Arch Intern Med* **148:**2477–2482, 1988.
230. Katlama C: Evaluation of the efficacy and safety of clindamycin plus pyrimethamine for induction and maintenance therapy of toxoplasmic encephalitis in AIDS. *Eur J Clin Microbiol Infect Dis* **10:**189–191, 1991.
231. Heald A, Flepp M, Chave J-P, *et al*: Treatment for cerebral toxoplasmosis protects against *Pneumocystis carinii* pneumonia in patients with AIDS. *Ann Intern Med* **115:**760–763, 1991.
232. Kovacs JA, Allegra CJ, Chabner BA, *et al*: Potent effect of trimetrexate, a lipid-soluble antifolate, on *Toxoplasma gondii*. *J Infect Dis* **155:**1027–1032, 1987.
233. Araujo FG, Prokocimer P, Remington JS: Clarithromycin-minocycline is synergistic in a murine model of toxoplasmosis. *J Infect Dis* **165:**788, 1992.
234. Grossman P, Remington J: The effect of trimethoprim and sulfamethoxazole on *Toxoplasma gondii in vitro* and *in vivo*. *Am J Trop Med Hyg* **28:**445–455, 1979.
235. Hakim M: Toxoplasmosis in cardiac transplantation. *Br Med J* **92:**1108, 1986.
236. Leport C, Vilde JL, Katlama C, *et al*: Failure of spiramycin to prevent neurotoxoplasmosis in immunosuppressed patients. *JAMA* **255:**2290, 1986.
237. Guerrant RL, Bobak DA: Bacterial and protozoal gastroenteritis. *N Engl J Med* **325:**327–340, 1991.
238. Laughon BE, Druckman DA, Vernon A, *et al*: Prevalence of enteric pathogens in homosexual men with and without acquired immunodeficiency syndrome. *Gastroenterology* **94:**984–993, 1988.
239. Miller RA, Holmberg R Jr, Clausen CR: Life-threatening diarrhea caused by *Cryptosporidium* in a child undergoing therapy for acute lymphocytic leukemia. *J Pediatr* **103:**256–259, 1983.
240. Moura H, Fernandes O, Viola JPB, *et al*: Enteric parasites and HIV infection: Occurrence in AIDS patients in Rio de Janeiro, Brazil. *Mem Inst Oswaldo Cruz Rio de Janeiro* **84:**527–533, 1989.
241. Soave R, Armstrong D: *Cryptosporidium* and cryptosporidiosis. *Rev Infect Dis* **8:**1012–1023, 1986.
242. Soave R, Johnson WD Jr: *Cryptosporidium* and *Isospora belli* infections. *J Infect Dis* **157:**225–229, 1988.
243. Vuorio A, Jokipii AMM, Jokipii L: *Cryptosporidium* in asymptomatic children. *Rev Infect Dis* **13:**261–264, 1991.
244. Wolfson JS, Richter JM, Waldron MA, *et al*: Cryptosporidiosis in immunocompetent patients. *N Engl J Med* **213:**1278–1282, 1985.

245. Connolly GM, Dryden MS, Shanson DC, *et al*: Cryptosporidial diarrhea in AIDS and its treatment. *Gut* **29:**593–597, 1988.
246. Moore JD, Buster SH: Intestinal parasites in Haitian entrants. *J Infect Dis* **150:**965, 1984.
247. Fleming AF: Opportunistic infections in AIDS in developed and developing countries. *Trans R Soc Trop Med Hyg* **84:**1–6, 1990.
248. Colebunders R, Lusakumuni K, Nelson AM, *et al*: Persistent diarrhoea in Zairian AIDS patients: An endoscopic and histological study. *Gut* **29:**1687–1691, 1988.
249. Whiteside ME, Barkin JS, May RG, *et al*: Enteric coccidiosis among patients with the acquired immunodeficiency syndrome. *Am J Trop Med Hyg* **33:**1065–1072, 1984.
250. Smith PD, Lane HC, Gill VJ, *et al*: Intestinal infections in patients with the acquired immunodeficiency syndrome (AIDS). *Ann Intern Med* **108:**328–333, 1988.
251. Bogaerts J, Lepage P, Rouvroy D, *et al*: *Cryptosporidium* spp., a frequent cause of diarrhea in central Africa. *J Clin Microbiol* **20:**874–876, 1984.
252. Shepherd RC, Reed CL, Sinha GP: Shedding of oocysts of *Cryptosporidium* in immunocompetent patients. *J Clin Pathol* **41:**1104–1106, 1988.
253. Caprioli A, Gentile G, Baldassarri L, *et al*: Cryptosporidium as a common cause of childhood diarrhoea in Italy. *Epidemiol Infect* **102:**537–540, 1989.
254. Current WL, Reese NC, Ernst JV, *et al*: Human cryptosporidiosis in immunocompetent and immunodeficient persons: Studies of an outbreak and experimental transmission. *N Engl J Med* **308:**1252–1257, 1983.
255. DuPont HL: Cryptosporidiosis and the healthy host. *N Engl J Med* **312:**1319–1320, 1985.
256. Weisburger WR, Hutcheon DF, Yardley JH, *et al*: Cryptosporidiosis in an immunosuppressed renal-transplant recipient with IgA deficiency. *Am J Clin Pathol* **72:**473–478, 1979.
257. Darban H, Enriquez J, Sterling CR, *et al*: Cryptosporidiosis facilitated by murine retroviral infection with LP-BM5. *J Infect Dis* **164:**741–745, 1991.
258. Tzipori S: Cryptosporidiosis in animals and humans. *Microbiol Rev* **47:**84–96, 1983.
259. Gentile G, Venditti M, Micozzi A, *et al*: Cryptosporidiosis in patients with hematologic malignancies. *Rev Infect Dis* **13:**842–846, 1991.
260. Gentile G, Caprioli A, Donelli G, *et al*: Asymptomatic carriage of *Cryptosporidium* in two patients with leukemia [letter]. *Am J Infect Control* **18:**127–128, 1990.
261. Soave R, Ma P: Cryptosporidiosis: Traveler's diarrhea in two families. *Arch Intern Med* **145:**7–72, 1983.
262. Horowitz MA, Hughes JM, Craun GF: Outbreaks of waterborne disease in the United States. *J Infect Dis* **133:**588–593, 1974.
263. Hayes EB, Matte TD, O'Brien TR, *et al*: Large community outbreak of cryptosporidiosis due to contamination of a filtered public water supply. *N Engl J Med* **320:**1372–1376, 1989.
264. Martino P, Gentile G, Caprioli A, *et al*: Hospital-acquired cryptosporidiosis in a bone marrow transplantation unit. *J Infect Dis* **158:**647–648, 1988.
265. Roncoroni AJ, Gomez MA, Mera J, *et al*: *Cryptosporidium* infection in renal transplant. *J Infect Dis* **160:**559, 1989.
266. Rehg JE, Hancock ML, Woodmansee DB: Characterization of a dexamethasone-treated model of cryptosporidial infection. *J Infect Dis* **158:**1406–1407, 1988.
267. Dryden MS, Shanson DC: The microbial causes of diarrhea in

patients infected with the human immunodeficiency virus. *J Infect Dis* **17:**107–114, 1988.

268. Kaljot KT, Ling JP, Gold JW, *et al*: Prevalence of acute enteric viral pathogens in acquired immunodeficiency syndrome patients with diarrhea. *Gastroenterology* **97:**1031–1032, 1989.

269. Knight R, Wright SG: Progress report: Intestinal protozoa. *Gut* **19:**940, 1978.

270. Kotler DP, Francisco A, Clayton F, *et al*: Small intestinal injury and parasitic diseases in AIDS. *Ann Intern Med* **113:**444–449, 1990.

271. Ungar B: Enzyme-linked immunoassay for detections of *Cryptosporidium* antigens in fecal specimens. *J Clin Microbiol* **28:**2491–2495, 1990.

272. Erlandsen SL, Chase DG: Morphological alterations in the microvillous border of villous epithelial cells produced by intestinal microorganisms. *Am J Clin Nutr* **27:**1277–1286, 1974.

273. Kibbler CC, Smith A, Hamilton-Dutoit SJ, *et al*: Pulmonary cryptosporidiosis occurring in a bone marrow transplant patient. *Scand J Infect Dis* **19:**581–584, 1987.

274. Ma P, Villanueva TG, Kaufman D, *et al*: Respiratory cryptosporidiosis in the acquired immune deficiency syndrome: Use of modified cold Kinyoun and Hemacolor stains for rapid diagnosis. *JAMA* **252:**1298–1301, 1984.

275. Meisel JL, Perera DR, MeLigro C, *et al*: Overwhelming watery diarrhea associated with *Cryptosporidium* in an immunosuppressed patient. *Gastroenterology* **70:**1156–1160, 1976.

276. Cron R, Sherry D: Reiter's syndrome associated with cryptosporidial gastroenteritis. *J Rheumatol* **10:**1062–1063, 1995.

277. Moskovitz BL, Stanton TL, Kusmierek JJ: Spiramycin therapy for cryptosporidial diarrhoea in immunocompromised patients. *J Antimicrob Chemother* **22:**189–191, 1988.

278. Portnoy D, Whiteside ME, Buckley E III, *et al*: Treatment of intestinal cryptosporidiosis with spiramycin. *Ann Intern Med* **101:**202–204, 1984.

279. Louie E, Borkowsky W, Klesius PK, *et al*: Treatment of cryptosporidiosis with oral bovine transfer factor. *Clin Immunol Immunopathol* **44:**329–334, 1987.

280. Saxon A, Weinstein W: Oral administration of bovine colostrum anticryptosporidia antibody fails to alter the course of human cryptosporidiosis. *J Parasitol* **73:**413–415, 1987.

281. Cook DJ, Kelton JG, Stanisz AM, *et al*: Somatostatin treatment for cryptosporidial diarrhea in a patient with the acquired immunodeficiency syndrome (AIDS). *Ann Intern Med* **108:**708–709, 1988.

282. Cello JP, Grendell JH, Basuk P, *et al*: Effect of octreotide on refractory AIDS-associated diarrhea. *Ann Intern Med* **115:**705–710, 1991.

283. Nousbaum JB, Robaszkiewicz M, Cauvin JM, *et al*: Treatment of intestinal cryptosporidiosis with zidovudine and SMS 201–995, a somatostatin analog. *Gastroenterology* **101:**874, 1991.

284. Godiwala T, Yaeger R: *Isospora* and traveler's diarrhea, *Ann Intern Med* **106:**908–909, 1987.

285. DeHovitz A, Pape JW, Boney M, *et al*: Clinical manifestations and therapy of *Isospora belli* infections in patients with the acquired immunodeficiency syndrome. *N Engl J Med* **315:**87–90, 1986.

286. DeHovitz J: Management of *Isospora belli* infections in AIDS patients. *Infect Med* **5:**437–440, 1988.

287. Pape JW, Verdier RI, Johnson WD Jr: Treatment and prophylaxis of *Isospora belli* infection in patients with the acquired immunodeficiency syndrome. *N Engl J Med* **320:**1044–1047, 1989.

288. Bryan RT, Cali A, Owen RL, *et al*: Microsporidia: Opportunistic pathogens in patients with AIDS. In Sun T (ed): *Progress in Clinical Parasitology*, Vol. II. Field & Wood, New York, 1991, pp. 1–26.

289. Desportes I, Charpentier Y, Gallian A, *et al*: Occurrence of a new microsporidian: *Enterocytozoon bieneusi* n.g., n.sp., in the enterocysts of a human patient with AIDS. *J Protozool* **32:**250–254, 1985.

290. Orenstein JM, Chiang J, Steinberg W, *et al*: Intestinal microsporidiosis as a cause of diarrhea in HIV-infected patients: A report of 20 cases. *Hum Pathol* **21:**475–481, 1990.

291. Cali A, Owen RL: Microsporidiosis. In Balows A, Hausler WJ Jr, Ohashi M *et al* (eds): *Laboratory Diagnosis of Infectious Diseases: Principles and Practice*. Springer-Verlag, New York, 1988, pp. 929–950.

292. Cali A, Owen RL: Intracellular development of *Enterocytozoon*, a unique microsporidian found in the intestine of AIDS patients. *J Protozool* **37:**145–155, 1990.

293. Weber R, Bryan RT, Owen RL, *et al* (Enteric Opportunistic Infections Working Group): Improved light-microscopical detection of *Microsporidia* spores in stool and duodenal aspirates. *N Engl J Med* **326:**161–166, 1992.

294. Genta RM: Global prevalence of strongyloidiasis: Critical review with epidemiologic insight into the prevention of disseminated disease. *Rev Infect Dis* **11:**755–767, 1989.

295. Neva FA: Biology and immunology of human strongyloidiasis. *J Infect Dis* **153:**397–406, 1986.

296. Igra-Siegman Y, Kapila R, Sen P, *et al*: Syndrome of hyperinfection with *Strongyloides stercoralis*. *Rev Infect Dis* **3:**397–407, 1981.

297. Scowden EB, Schaffner W, Stone WJ: Overwhelming strongyloidiasis: An unappreciated opportunistic infection. *Medicine* **57:**527–544, 1978.

298. Rivera E, Maldonado N, Velez-Garcia E, *et al*: Hyperinfection with *Strongyloides stercoralis*. *Ann Intern Med* **72:**199–204, 1970.

299. Nucci M, Portugal R, Pulcheri W, *et al*: Fatal strongyloidiasis in patients with hematologic malignancies. *Clin Infect Dis* **21:**675–677, 1995.

300. Rogers W, Nelson B: Strongyloidiasis and malignant lymphoma. *JAMA* **195:**173–175, 1966.

301. Purtilo DT, Meyers WM, Connor DH: Fatal strongyloidiasis in immunosuppressed patients. *Am J Med* **56:**488–493, 1974.

302. Maayan S, Wormser GP, Widerhorn J, *et al*: *Strongyloides stercoralis* hyperinfection in a patient with the acquired immune deficiency syndrome. *Am J Med* **83:**945–948, 1987.

303. Kuberski TT, Gabor EP, Bourdreaux D: Disseminated strongyloidiasis—A complication of the immunosuppressed host. *West J Med* **122:**504–508, 1975.

304. Cohen J, Spry CJF: *Strongyloides stercoralis* infection and small intestinal lymphoma. *Parasitol Immunol* **1:**167–169, 1979.

305. De Oliviera RB, Voltarelli JC, Meneghelli VG: Severe strongyloidiasis associated with hypogammaglobulinemia. *Parasitol Immunol* **3:**165–169, 1981.

306. Kramer MR, Gregg PA, Goldstein M, *et al*: Disseminated strongyloidiasis in AIDS and non-AIDS immunocompromised hosts: Diagnosis by sputum and bronchoalveolar lavage. *South Med J* **83:**1226–1229, 1990.

307. Berger R, Kraman S, Paciotti M: Pulmonary strongyloidiasis complicating therapy with corticosteroids. *Am J Trop Med Hyg* **29:**31–34, 1980.

308. Higenbotham TW, Heard BE: Opportunistic pulmonary strongyloidiasis complicating asthma treated with steroids. *Thorax* **31:**226–233, 1976.

309. Meltzer RS, Singer C, Armstrong D, *et al*: Pulmonary strongyloidiasis complicating therapy with corticosteroids. *Am J Med Sci* **277:**91–98, 1979.

310. Meyers AM, Shapiro DJ, Milne FJ, *et al*: *Strongyloides stercoralis* hyperinfection in a renal allograft recipient. *South Afr Med J* **50:**1301–1302, 1976.

311. Nwokolo C, Imohiosen EAE: Strongyloidiasis of respiratory tract presenting as "asthma." *Br Med J* **2:**153–154, 1973.

312. Scoggin CH, Call NB: Acute respiratory failure due to disseminated strongyloidiasis in a renal transplant recipient. *Ann Intern Med* **87:**456–458, 1977.

313. Nolan TJ, Schad GA: Tacrolimus allows autoinfective development of the parasitic nematode *Strongyloides stercoralis*. *Transplantation* **62:**1038, 1996.

314. Cruz T, Reboucas G, Rocha H: Fatal strongyloidiasis in patients receiving corticosteroids. *N Engl J Med* **275:**1093–1096, 1966.

315. Cookson JB, Montgomery RD, Morgan HV, *et al*: Fatal paralytic ileus due to strongyloidiasis. *Br Med J* **4:**771–772, 1972.

316. Rassiga AL, Lawry JL, Forman WB: Diffuse pulmonary infection due to *Strongyloides stercoralis*. *JAMA* **230:**426–430, 1974.

317. Naquira C, Jiminez G, Guerra JG, *et al*: Ivermectin for human strongyloidiasis and other intestinal helminths. *Am J Trop Med Hyg* **40:**304–309, 1989.

318. Ravdin JI: Amebiasis. In *Human Infection by Entamoeba histolytica*. John Wiley, New York, 1988.

319. Sargeaunt PG, Jackson TFHG, Simjee AE: Biochemical homogeneity of *Entamoeba histolytica* isolates, especially those from liver abscess. *Lancet* **1:**1386–1388, 1982.

320. Villarejos SVM: Corticosteroid and experimental amoebiasis in rats. *J Parasitol* **48:**194, 1962.

321. Anaya-Velazquez F, Tsutsumi V, Gonzolez-Robles A, *et al*: Intestinal invasive amebiasis: An experimental model in rodents using axenic or monoxenic strains of *Entamoeba histolytica*. *Am J Trop Med Hyg* **34:**723–730, 1985.

322. Allason-Jones E, Mindel A, Sargeaunt P, *et al*: *Entamoeba histolytica* as a commensal intestinal parasite in homosexual men. *N Engl J Med* **315:**353–356, 1986.

323. Sullam PM: *Entamoeba histolytica* in homosexual men. *N Engl J Med* **316:**690, 1987.

324. Antony MA, Brandt LJ, Klein RS, *et al*: Infectious diarrhea in patients with AIDS. *Dig Dis Sci* **33:**1141–1146, 1988.

325. El-Hannawy M, Abd-Rabbo H: Hazards of cortisone therapy in hepatic amebiasis. *J Trop Med Hyg* **81:**71–73, 1978.

326. Denis M, Chadee K: Immunopathology of *Entamoeba histolytica* infections. *Parasitol Today* **4:**247–252, 1988.

327. Kretschmer RR: Immunology of amebiasis. In Martinez-Palomo A (ed): *Amebiasis*. Elsevier, Amsterdam, 1986, pp. 95–167.

328. Trissl D: Immunology of *Entamoeba histolytica* in human and animal hosts. *Rev Infect Dis* **4:**1154–1184, 1982.

329. Nanda R, Baveja U, Annand BS: *Entamoeba histolytica* cyst passers: Clinical features and outcome in untreated subjects. *Lancet* **2:**301–303, 1984.

330. Balikian JP, Bitar JG, Rishani KK, *et al*: Fulminant necrotizing amebic colitis in children. *Am J Protocol* **28:**69–78, 1977.

331. Cardoso JM, Kimura K, Stoopen M, *et al*: Radiology of invasive amebiasis of the colon. *J Roentgenol* **128:**935–946, 1977.

332. Ralls PW, Barnes PF, Johnson MB, *et al*: Medical treatment of hepatic amebic abscess: Rare need for percutaneous drainage. *Radiology* **165:**805–807, 1987.

333. Duma RJ, Ferrell HW, Nelson CE, *et al*: Primary amebic meningoencephalitis. *N Engl J Med* **281:**1315–1323, 1969.

334. Duma RJ, Finley R: *In vitro* susceptibility of pathogenic *Naegleria* and *Acanthamoeba* species to a variety of therapeutic agents. *Antimicrob Agents Chemother* **10:**370–376, 1976.

335. Grimaldi G Jr, Tesh RB, McMahon-Pratt D: A review of the geographical distribution and epidemiology of leishmaniasis in the New World. *Am J Trop Med Hyg* **41:**687–725, 1989.

336. Ma DDF, Concannon AJ, Hayes J: Fatal leishmaniasis in renal-transplant patients. *Lancet* **2:**311–312, 1979.

337. Cerf BJ, Jones TC, Badaro R, *et al*: Malnutrition as a risk factor for severe visceral leishmaniasis. *J Infect Dis* **156:**1030–1033, 1987.

338. Harrison LH, Naidu TG, Drew JS, *et al*: Reciprocal relationship between undernutrition and the parasitic disease visceral leishmaniasis. *Rev Infect Dis* **8:**447–453, 1986.

339. Pampiglione S, Manson-Bahr PEC, Giungu F, *et al*: Studies on Mediterranean leishmaniasis. II. Asymptomatic cases of visceral leishmaniasis. *Trans R Soc Trop Med Hyg* **68:**447–453, 1974.

340. Aguardo JM, Gomez J, Figuera A, *et al*: Visceral leishmaniasis complicating acute leukemia. *J Infect Dis* **7:**272–274, 1983.

341. Aguardo JM, Plaza J, Escudero A: Visceral leishmaniasis in renal-transplant recipient. *J Infect* **13:**301–302, 1986.

342. Badaro R, Carvalho EM, Rocha H, *et al*: *Leishmania donovani*: An opportunistic microbe associated with progressive disease in three immunocompromised patients. *Lancet* **1:**647–649, 1986.

343. Broeckaert A, Michielsen P, Vandepitte J: Fatal leishmaniasis in renal-transplant patient. *Lancet* **2:**740–741, 1979.

344. Fernandez-Guerrero ML, Aguado JM, Buzon L, *et al*: Visceral leishmaniasis in immunocompromised hosts. *Am J Med* **83:**1098–1102, 1987.

345. Ma DDF, Concannon AJ, Hayes J: Fatal leishmaniasis in a renal transplant patient. *Lancet* **2:**311–312, 1979.

346. Montalban C, Martinez-Fernandez R, Calleja JL, *et al*: Visceral leishmaniasis (kala-azar) as an opportunistic infection in patients infected with the human immunodeficiency virus in Spain. *Rev Infect Dis* **11:**655–660, 1989.

347. Yebra M, Segovia J, Manzano L, *et al*: Disseminated-to-skin kala-azar and the acquired immunodeficiency syndrome. *Ann Intern Med* **108:**490–491, 1988.

348. Berenguer J, Gomez-Campdera F, Padilla B, *et al*: Visceral leishmaniasis (kala-azar) in transplant recipients: Case report and review. *Transplantation* **65:**1401–1404, 1998.

349. Weiser WY, Van Neil A, Clark SC, *et al*: Recombinant human granulocyte/macrophage colony stimulating factor activates intracellular killing of *Leishmania donovani* by human monocyte-derived macrophages. *J Exp Med* **166:**1436–1446, 1987.

350. Carvalho EM, Teixeira R, Johnson WD: Cell mediated immunity in American visceral leishmaniasis: Reversible immunosuppression during acute infection. *Infect Immun* **33:**498–502, 1981.

351. Badaro R, Reed S, Barral A, *et al*: Evaluation of the micro enzyme-linked immunosorbent assay (ELISA) for antibodies to American visceral leishmaniasis: Standardization of parasite antigen to detect infection-specific responses. *Am J Trop Med Hyg* **35:**72–78, 1986.

352. Navin TR, Arana BA, Arana FE, *et al*: Placebo-controlled clinical trial of sodium stibogluconate (Pentostam) versus ketoconazole for treating cutaneous leishmaniasis in Guatemala. *J Infect Dis* **165:**528–534, 1992.

353. Hoskins LC, Winawer SJ, Broitman SA, *et al*: Clinical giardiasis and intestinal malabsorption. *Gastroenterology* **53:**265–279, 1967.

354. Keysteon JS, Krajden S, Warren MR: Person-to-person transmission of *Giardia lamblia* in day care nurseries. *Can Med Assoc J* **119:**241–248, 1978.

355. Smith PD, Keister DB, Elson CO: Human host response to *Giardia lamblia* trophozoites. II. Antibody-dependent killing *in vitro*. *Cell Immunol* **82:**308–315, 1983.

356. Hermans PE, Huizenga KA, Hoffman HN, *et al*: Dysgamma-globulinemia associated with nodular lymphoid hyperplasia of the small intestine. *Am J Med* **40:**78–89, 1966.

357. Jones EG, Brown WR: Serum and intestinal fluid immunoglobulin in patients with giardiasis. *Am J Dig Dis* **19:**791–796, 1974.

358. Roberts-Thomson IC, Stevens DP, Mahmoud AA, *et al*: Acquired resistance to infection in an animal model of giardiasis. *J Immunol* **117:**2036–2037, 1976.

359. Roberts-Thomson IC, Mitchell GF: Giardiasis in mice. I. Prolonged infections in certain mouse strains and hypothymic (nude) mice. *Gastroenterology* **75:**42–50, 1978.

360. Stevens DP, Frank DM, Mahmoud AAF: Thymus dependency of host resistance to *Giardia muris* infection: Studies in nude mice. *J Immunol* **120:**680–682, 1978.

361. Zinneman HH, Kaplan AP: The association of giardiasis with reduced intestinal secretory immunoglobulin A. *Am J Dig Dis* **17:**793–797, 1972.

362. Ament ME, Rubin CE: Relation of giardiasis to abnormal intestinal structure and function in gastrointestinal immunodeficiency syndromes. *Gastroenterology* **62:**216–226, 1972.

363. Stevens DP, Frank DM: Local immunity in murine giardiasis: Is milk protective at the expense of maternal gut? *Trans Assoc Am Physicians* **91:**268–272, 1978.

364. Owen RL: Sexually related intestinal disease. In Sleisenger MH, Fordtran JS (eds): *Gastrointestinal Disease*, 3rd ed. Saunders, Philadelphia, 1983, pp. 966–985.

365. Philips SC, Mildvan D, William DC, *et al*: Sexual transmission of enteric protozoa and helminths in a venereal disease clinical population. *N Engl J Med* **305:**603–606, 1981.

366. Schmerin MJ, Jones TC, Klein H, *et al*: Giardiasis: Association with homosexuality. *Ann Intern Med* **88:**801–803, 1978.

367. Janoff EN, Smith PD, Blaser MJ: Acute antibody responses to *Giardia lamblia* are depressed in patients with the acquired immunodeficiency syndrome. *J Infect Dis* **157:**798–804, 1988.

368. Balck RE, Dykes AC, Sinclair SP, *et al*: Giardiasis in day care centers: Evidence of person-to-person transmission. *Pediatrics* **60:**486–491, 1977.

369. Barbour AG, Nichols CR, Fukushima T: An outbreak of giardiasis in a group of campers. *Am J Trop Med* **25:**384–389, 1976.

370. Brodsky RE, Spencer HC, Schultz MG: Giardiasis in American travelers in the Soviet Union. *J Infect Dis* **130:**319–323, 1974.

371. Butler T, Middleton FG, Earnest DL, *et al*: Chronic and recurrent diarrhea in American servicemen in Vietnam. *Arch Intern Med* **132:**373–377, 1973.

372. Dykes AC, Juanek DD, Lorenz RA: Municipal water-borne giardiasis: An epidemiologic investigation: Beavers implied as a possible reservoir. *Ann Intern Med* **92:**165–170, 1980.

373. Shaw PK, Brodsky RE, Lyman DO, *et al*: A community-wide outbreak of giardiasis with documented transmission by municipal water. *Ann Intern Med* **87:**426–432, 1977.

374. Morecki R, Parker JG: Ultrastructure studies of the human *Giardia lamblia* and subjacent jejunal mucosa in a subject with steatorrhea. *Gastroenterology* **52:**51–164, 1967.

375. Sheehy TW, Holley HP Jr: *Giardia*-induced malabsorption in pancreatitis. *JAMA* **233:**1373–1375, 1975.

376. Tandon BN, Tandon RK, Satpathy BK, *et al*: Mechanism of malabsorption in giardiasis: A study of bacterial flora and bile salt deconjugation in upper jejunum. *Gut* **18:**176–181, 1977.

377. Saha TK, Ghosh TK: Invasion of small intestinal mucosa by *Giardia lamblia* in man. *Gastroenterology* **72:**402–405, 1977.

378. Tapper ML, Armstrong D: Malaria complicating neoplastic disease. *Arch Intern Med* **136:**807–810, 1976.

379. Healy GR: *Babesia* infections in man. *Hosp Pract* **13:**107–116, 1979.

380. Healy GR, Wlazer PD, Sulzer AJ: A case of asymptomatic babesiosis in Georgia. *Am J Trop Med Hyg* **25:**376–378, 1976.

381. Ruebush TK II, Spielman A: Human babesiosis in the United States. *Ann Intern Med* **88:**263, 1978.

382. Ruebush TK II, Cassaday PB, March JH, *et al*: Human babesiosis on Nantucket Island: Clinical features. *Ann Intern Med* **86:**6–9, 1977.

383. Ruebush TK II, Juranek DD, Chisholm ES, *et al*: Human babesiosis on Nantucket Island: Evidence for self-limited and subclinical infections. *N Engl J Med* **297:**825–827, 1977.

384. Chisholm ES, Ruebush TK II, Sulzer AJ, *et al*: *Babesia microti* infection in man: Evaluation of an indirect immunofluorescent antibody test. *Am J Trop Med Hyg* **7:**14–19, 1978.

385. Miller LH, Neva FH, Gill F: Failure of chloroquine in human babesiosis (*Babesia microti*). *Ann Intern Med* **88:**200–202, 1978.

386. Rowin KS, Tanowitz HB, Wittner M: Therapy of experimental babesiosis. *Ann Intern Med* **97:**556–558, 1982.

387. Ashford RW: Occurrence of an undescribed coccidian in man in Papua New Guinea. *Ann Trop Med Parasitol* **73:**497–500, 1979.

388. Ortega YR, Sterling CR, Gilmann RH, *et al*: *Cyclospora* species—A new protozoan pathogen of humans. *N Engl J Med* **328:**1308–1312, 1993.

389. Hoge CW, Shlim DR, Rahaj R, *et al*: Epidemiology of diarrhoeal illness associated with coccidian-like organism among travellers and foreign residents in Nepal. *Lancet* **341:**1175–1179, 1993.

390. Hart AS, Ridinger MT, Soundarajan R, *et al*: Novel organism associated with chronic diarrhoea in AIDS. *Lancet* **335:**169–170, 1990.

391. Long EG, White EH, Carmichael WW, *et al*: Morphologic and staining characteristics of cyanobacterium-like organism associated with diarrhea. *J Infect Dis* **164:**199–202, 1991.

392. Fishman JA: Prevention of infection due to *P. carinii* in transplant recipients. *Clin Infect Dis* **33:**1397–1405, 2001.

9

Viral Hepatitis in the Compromised Host

PETER M. ROSENBERG and JULES L. DIENSTAG

1. Introduction

Five categories of viral hepatitis have been recognized: hepatitis A, hepatitis B, hepatitis C, hepatitis B-associated delta hepatitis (hepatitis D), and hepatitis E. All have the potential to produce similar illnesses; however, clinical severity and outcome as well as the contribution of immunologic mechanisms to the clinical expression of infection and illness differ among the various types. Observations of viral hepatitis in immunosuppressed persons have taught us important lessons about the biology of these viral agents, on the one hand, and about the approach to immunologically compromised patients with hepatitis, on the other. Unlike susceptibility to opportunistic infections, susceptibility to viral hepatitis per se is not increased in the immunosuppressed; however, because patients with immunologic derangements are very likely to require transfusion of blood products and because hepatitis B and C viruses are transmitted by similar parenteral routes, the frequency of viral hepatitis in this group of patients is increased. In addition, severity of acute illness, likelihood of chronic infection, infectivity, serologic responsiveness, and the early and late consequences of chronic infection differ distinctly between immunocompetent and immunosuppressed hosts.

Among the five categories of viral hepatitis, only hepatitis B and hepatitis C agents appear to be more

Peter M. Rosenberg • Saint John's Health Center, Santa Monica, California 90404.   **Jules L. Dienstag** • Gastrointestinal Unit, Medical Services, Massachusetts General Hospital, and Department of Medicine, Harvard Medical School, Boston, Massachusetts 02114.

Clinical Approach to Infection in the Compromised Host (Fourth Edition), edited by Robert H. Rubin and Lowell S. Young. Kluwer Academic/Plenum Publishers, New York, 2002.

frequent and to alter clinical expression in immunocompromised persons. Hepatitis A is primarily an enterically spread virus, unassociated with appreciable viremia, chronic infection, or chronic hepatitis. Although the frequency of infection with hepatitis A virus (HAV) is high in persons with Down's syndrome, in whom immunologic deficiencies have been documented, the frequency of serologic markers of HAV infection is just as high in institutionalized persons with other forms of mental retardation, unassociated with compromised immune function.[1] Moreover, the clinical features and benign outcome of acute HAV infection in institutionalized mentally retarded persons are indistinguishable from those of comparably aged normal, immunocompetent persons. In other categories of immunocompromised persons, such as those requiring chronic hemodialysis, neither the prevalence nor the incidence of HAV infection is higher than those found in immunocompetent persons.[2] Hepatitis E virus, like hepatitis A, is transmitted enterically and has been associated with epidemics in developing countries including Pakistan, India, and Mexico. While hepatitis E infection does not lead to chronic hepatitis, it is associated with a 10% mortality rate in pregnant women. Thus far, in the United States, cases of hepatitis E have been confined to persons returning or immigrating from areas of known endemicity.[3] "Hepatitis G virus," also known as GB virus C, suggested initially as a putative hepatitis virus, is highly prevalent in several groups of immunocompromised patients, including hemodialysis patients,[4] organ transplant recipients,[5] and patients infected with human immunodeficiency virus (HIV),[6] as well as in immunocompetent patients with hepatitis C virus infection, but has not been associated with liver disease and is no longer considered a true liver pathogen. Although clinical and histological observations hint that both HAV and the delta agent are directly toxic to hepatocytes and do not require participation by the host immune system for cytopathology,[7,8] in all likelihood liver injury in all the viral hepatitides involves host cell-mediated immune responses to virus-infected hepatocytes. In considering viral hepatitis in the immunocompromised host, most attention has been focused on hepatitis B and hepatitis C, but new information on delta hepatitis has appeared.

2. Role of Immunologic Mechanisms in the Pathogenesis of Viral Hepatitis

2.1. Hepatitis B

Hepatitis B virus (HBV) is a DNA virus that belongs to the hepadnavirus group, characterized by three morphologic forms, a 42-nm genome-containing virion, and 22-nm spherical and tubular forms composed of the virion surface protein, hepatitis B surface antigen (HBsAg); an association with acute and chronic liver disease; and a predominantly double-stranded DNA genome with a single-stranded region. Within the virion is the nucleocapsid, on the surface of which is expressed hepatitis B core antigen (HBcAg) and on the inside of which is the DNA, DNA polymerase, and hepatitis B e antigen (HBeAg), a nonparticulate, internal nucleocapsid protein. In persons infected with HBV, serologic markers detectable routinely include HBsAg, HBeAg, their respective antibodies, anti-HBs and anti-HBe, and antibody to HBcAg (anti-HBc). In general, those with current HBV infection have circulating HBsAg and anti-HBc, whereas those who have recovered have circulating anti-HBc and anti-HBs. When HBV infection is accompanied by circulating HBeAg, HBV DNA is readily detectable in serum by relatively insensitive hybridization assays with a sensitivity of $\geq 10^5$ to 10^6 virions/ml and the levels of replication and infectivity are high; in those with HBsAg in serum who are anti-HBe positive, serum HBV DNA is undetectable with blotting assays (levels of $<10^5$/ml, as measured by sensitive amplification assays) and virus replication and infectivity are quite limited. Among the several HBV antibodies, anti-HBs is considered the protective antibody; the goal of immunoprophylaxis is to provide the susceptible host with circulating anti-HBs. The three primary modes of hepatitis B transmission are via percutaneous–transmucosal inoculation, intimate contact, and perinatal exposure.

In immunocompetent persons infected with HBV, clinical expression of infection is quite variable. In most cases, infection is accompanied by liver cell necrosis and followed by virus elimination and recovery. In rare instances, most or all hepatocytes are destroyed, leading to fulminant hepatitis, with a mortality rate of approximately 80%. In <5% of immunocompetent adults with clinically apparent acute hepatitis B, infection remains chronic. Some in this category remain free of liver morphologic abnormalities and are classified as chronic carriers, while others continue to have liver cell necrosis and inflammation; those with chronic inflammatory activity may have mild to severe chronic hepatitis. In addition, both the level of viral replication and the rate of progression may vary considerably among persons with chronic hepatitis B. Because asymptomatic chronic hepatitis B carriers have normal liver morphology despite ongoing virus replication in liver cells, liver cell necrosis in hepatitis B appears not to be the result of a direct cytopathic injury by HBV. The differences in expression and outcome of HBV infection therefore have been attributed to variability in host immune responsiveness.[9,10]

Observations that support the contribution of immunologic mechanisms include the following:

1. Chronic hepatitis B carriage is more likely to follow acute HBV infection in those with immunologic immaturity (e.g., neonates) or with immunologic deficiencies (e.g., persons with HIV infection, Down's syndrome, lepromatous leprosy, chronic renal failure, and those receiving cytotoxic or immunosuppressive chemotherapy).[11–15]
2. Chimpanzees treated with cyclophosphamide during acute experimental infection with HBV remain chronically infected with the virus.[16]
3. A close spatial relationship exists between necrotic hepatocytes and mononuclear inflammatory cells, presumed to be the immunologic effectors of hepatocytolysis.
4. When cytotoxic chemotherapy is withdrawn in immunosuppressed persons with circulating HBsAg, massive hepatocellular necrosis and severe hepatitis may occur,[17–19] presumably as a result of the restoration of immune competence.
5. Similarly, in patients with chronic hepatitis B, immunosuppressive therapy allows an increase in the level of HBV replication, and a transient elevation in serum aminotransferase levels may follow withdrawal of immunosuppressive therapy.[20–24]

The cellular rather than humoral immune response is the major contributor to the immunopathogenesis of viral hepatitis. Experiments in transgenic mouse models of

HBV infection have demonstrated that transfer of HBV-specific cytolytic T lymphocytes (CTLs) can result in liver cell necrosis.[25,26] In addition, in patients with chronic hepatitis B, manipulation of humoral immunity by infusion of large doses of anti-HBs does not lead to hepatocellular necrosis,[27,28] while cellular immune stimulation or reconstitution is followed by elevation of aminotransferase levels, a reflection of liver cell injury.[28–30] Also cited to acquit humoral immune responses in the pathogenesis of liver injury in viral hepatitis is the observation that both acute and chronic hepatitis, often quite severe, can occur in the absence of intact humoral immune responsiveness, e.g., in patients with agammaglobulinemia.[31]

Failure to mount a vigorous CD4+ lymphocyte response against HBcAg in the setting of acute HBV infection appears to be at least partially responsible for development of chronic infection, even in immunocompetent individuals.[32,33] The CD8+ CTL response against HBsAg epitopes, also relatively weak in patients with chronic HBV infection, may be sufficient to cause liver injury but is ineffective at viral clearance.[34] In addition to differences in cell-mediated cytolysis, differences in cytokine profiles distinguish between those with acute, self-limited hepatitis B and those with chronic infection; Th1 cytokines [interleukin-2, tumor necrosis factor-alpha (TNF-α), and interferon-γ] predominate in the former, and Th2 cytokines (interleukins-4, -5, -6, and -10) predominate in the latter.[35] Adding even more to the complexity of the relationship between immunologic function and hepatitis B is the detection of HBV DNA and HBsAg in bone marrow cells and peripheral blood lymphocytes from patients with hepatitis B,[36] suggesting that HBV may have a direct effect on immunologic competence, mediated by infection of immunocytes. In managing immunologically compromised patients, therefore, we should remain cognizant of the relationship between immunologic integrity and response to hepatitis B.

2.2. Hepatitis C

Hepatitis C virus (HCV) is an RNA virus that belongs to the flavivirus family. Its genome consists of a single open reading frame encoding three structural and at least seven nonstructural proteins. The envelope proteins, especially a short domain of the E2 protein, are highly variable, which may contribute significantly to the virus' ability to evade the host immune system. HCV is transmitted primarily by percutaneous inoculation and, unlike HBV, very inefficiently by sexual or perinatal routes. Also unlike HBV, HCV is rarely eradicated by the host immune response even in immunocompetent individuals and infection becomes chronic in approximately 85% of those infected. Antibodies to HCV, which are found in more than 90% of patients with chronic HCV infection,[37] are ineffective at viral clearance but have not been implicated in the pathogenesis of HCV-related liver injury (although they contribute to many of the extrahepatic manifestations of HCV infection). On the other hand, HCV-specific CD4+ T lymphocytes[38] and CTLs[39] have been identified in livers of patients with chronic HCV infection, suggesting that the cellular immune response contributes to liver injury in chronic hepatitis C. The identification of healthy carriers with normal liver histology and normal serum aminotransferase levels despite detectable HCV RNA in the blood suggests that HCV is not directly cytopathic to hepatocytes in immunocompetent hosts.[40] As is the case for hepatitis B, immunologic properties of the host probably affect the clinical expression and outcome of hepatitis C infection.

3. Viral Hepatitis in the Immunocompromised Host

Following the identification of HBsAg, then known as Australia antigen, by Blumberg and associates,[41] the first patients found to have a high frequency (11%) of antigenemia were leukemics and for a brief time the antigen was even thought to be a leukemia antigen. When the frequency of chronic HBs antigenemia in other patient groups was tested, Blumberg and colleagues found that not only leukemics but also institutionalized persons with Down's syndrome (30%), patients with Hodgkin's disease (85%), patients with lepromatous leprosy (20%), and chronically hemodialyzed patients with chronic renal failure (10%) had high HBsAg prevalences.[12–15] In each of these patient groups, cellular immune defects have been postulated or documented.[41,42] Common to all these groups of immunocompromised patients are an increased rate of exposure to blood products and/or other risk factors for viral hepatitis, a tendency toward subclinical acute infection, a greatly increased risk of chronic infection, and a tendency to support high levels of viral replication, and therefore to be highly infectious. Neonates share with these groups the tendencies toward subclinical acute infection and greater chronicity, likely due to their immunologic immaturity at the time of infection. Organ transplant recipients, patients with HIV infection, and patients who have undergone cytotoxic therapy are somewhat different from the aforementioned populations in that they were in many cases immunocompetent at the time of infection with HBV.

A set of classic studies on hepatitis B in patients with Down's syndrome provides a paradigm for the response to HBV infection in the immunocompromised patient. In this population, defects in cellular immune function have been invoked to explain their high frequency of infections in general and of chronic infection after acute hepatitis B in particular.[43] As in neonates and other immunosuppressed persons, in those with Down's syndrome, acute HBV infection is almost invariably subclinical. Chronicity has followed acute infection in 20–38% of persons with Down's syndrome.[44–49] Not only are they more likely to be exposed because of circumstances of their care (institutionalization at an early age), to have subclinical acute HBV infections, and to harbor infection chronically, they also are more likely to support higher levels of virus replication, and therefore to be more infectious for others. In a comparison by Szmuness et al.,[50] persons with Down's syndrome and chronic HBV infection were found to have a higher prevalence of HBeAg than that of HBsAg-positive blood donors or HBsAg-positive persons with non-Down's types of mental retardation. The frequency of HBeAg in Down's syndrome was even higher, 61%, in a study from Australia,[45] figures comparable to the HBeAg prevalence found by Szmuness et al.[50] in hemodialysis patients (Table 1). Patients with Down's syndrome thus serve as a reservoir for dissemination of infection to others institutionalized with them and to those who care for them; they become a focus for the amplification of HBV infection.

As the prevalence of hepatitis B in the United States has declined, the importance of hepatitis C in various populations of immunosuppressed individuals has been increasingly recognized. These populations include organ transplant recipients and patients with hypogammaglobulinemia and HIV infection. It is clear that these conditions are associated with higher levels of HCV viremia, but the overall impact of immunodeficiency on the clinical course of HCV infection has been somewhat more difficult to determine than that on the course of HBV infection. Unlike the immune response to HBV, the immune response to HCV is rarely effective in clearing the virus even in immunocompetent individuals, making the contrast between the clinical courses of HCV in immunocompetent versus immunodeficient persons less marked. Of note, immunosuppression blunts the serologic response to HCV, as measured by enzyme immunoassay for anti-HCV. Therefore, in immunocompromised patients, a negative serologic test for anti-HCV does not exclude HCV infection. The most sensitive test for HCV infection is the detection of HCV RNA in serum by polymerase chain reaction (PCR). Therefore, negative tests for anti-HCV must be interpreted cautiously in immunosuppressed patients with chronic hepatitis.

From a practical perspective, clinically significant immunosuppression and hepatitis are likely to overlap and to be encountered primarily in five categories of patients: hemodialyzed patients with chronic renal failure, recipients of organ transplants, patients with lymphoproliferative and myeloproliferative malignancies, patients with hypogammaglobulinemia, and patients with AIDS.

3.1. Hemodialyzed Patients with Chronic Renal Failure

3.1.1. Hepatitis B in Hemodialysis Units

During the late 1960s and early 1970s, hepatitis B was recognized as a major hazard for patients and staff of hemodialysis units.[51,52] A substantial body of literature documented the high risk observed during that period, the tendency for HBsAg to be of subtype *ay* rather than the more prevalent *ad*, and the mechanisms of transmission in such units. In a study by Szmuness et al.,[53] the point prevalence of current or past HBV infection was 50% in patients and 34% in staff. During the first year in hemodialysis units, the incidence of new infections ranged between 40 and 47% in patients and staff. Recognition soon emerged of the relationship between duration of treatment and intensity of blood exposure in dialysis units and hepatitis B.

Since the institution of specific infection control practices in 1977 and of hepatitis B vaccination in 1982, the rate of HBV transmission by hemodialysis has fallen markedly (see Section 4). Nevertheless, five hepatitis B outbreaks were reported as recently as 1994.[54] The following risk factors for HBV transmission have been consistently demonstrated in dialysis units: presence of a

TABLE 1. Frequency of Replicative Hepatitis B Virus Infection in HBsAg Carrier Populations, as Reflected by HBeAg Status[a]

Carrier population	Percent with HBeAg
Volunteer blood donors	9
Mentally retarded persons	
Down's syndrome	39
Other mentally retarded	10
Hemodialysis patients	71

[a]Adapted from Szmuness et al.[50]

chronically infected patient; failure to separate an infected patient by room, machine, and staff; and failure to vaccinate patients against HBV.[55–57] In these units, extracorporeal blood is present almost continuously and little imagination is required to comprehend that HBV can be transmitted via blood leaks in dialysis machines as well as via contamination by blood of gloves, clamps, other instruments, dialysis machine surfaces and control knobs, and other environmental surfaces.[58] Although the modes of inoculation differ, the risk to patients (exposed via blood transfusion and contaminated instruments and surfaces) and staff (exposed primarily via needlesticks and other percutaneous–transmucosal penetrations) are quite comparable.

The clinical course of HBV in hemodialysis patients, however, differs significantly from that in hemodialysis staff (and other immunocompetent persons). On the one hand, hemodialysis patients are much more likely than hemodialysis staff to have an asymptomatic, anicteric acute illness associated with HBV infection, while, on the other hand, hemodialysis patients are substantially more susceptible to chronic infection (Table 2).[51,59–63] Once chronically infected, hemodialysis patients are also more likely to remain HBeAg-positive, i.e., to have high levels of ongoing virus replication and to be infectious.[50,64,65] Thus, although specific immunologic defects have not been demonstrated consistently in hemodialysis patients with chronic HBV infection,[60,61] hemodialysis patients do have evidence of depressed cellular immune function[42] and do respond to HBV infection as immunosuppressed hosts do: with failure to destroy virus-infected hepatocytes, and therefore with a paucity of hepatitis symptoms; with failure to clear infection; and with failure to contain virus replication and infectivity. Therefore, once HBV infection is introduced into a dialysis unit, unless steps are taken to control or eliminate it (see Section 4), dialysis patients who are infected tend to be very infectious and to propagate the cycle of infection within the hemodialysis unit to other patients and staff and to amplify the spread of infection outside the unit to family and other household contacts.[53,66,67]

TABLE 2. Hepatitis B Virus Infection in Hemodialysis Units: Comparison between Patients and Staff

	Patients (%)	Staff (%)
Prevalence of current or past infection[2,54,60]	50–67	34
Incidence of new infections during first year[2,54]	40	47
Clinically apparent acute illness[52]	32	85
Remain chronically infected[61–63]	60–90	10
HBeAg-positive chronic infection[51,65,66]	53–61	10

Limited information is available about the long-term consequences of chronic HBV infection in hemodialysis patients. Many come eventually to renal transplantation, which may have a dramatic impact on their clinical courses (see Section 3.2), and few studies have been done to assess hepatic morphology and the course of liver disease in hemodialysis patients who do not receive renal homografts. What data exist indicate that histologic progression is quite variable in this group of HBsAg-positive patients.[68,69] Clinical expression of HBV infection, however, remains silent. In one study, for example,[69] liver histology ranged from normal to mild chronic hepatitis to moderate chronic hepatitis to severe chronic hepatitis. Follow-up biopsies done in a limited number of these patients showed progression from the milder forms of chronic hepatitis to more severe chronic hepatitis with fibrosis, despite the absence of necrosis and inflammation. Despite such progression, symptoms were minimal or absent and aminotransferase levels were no higher than 1.5–3 times the upper limits of normal. Thus, as observed in other immunosuppressed patients (see Section 3.2), histologic liver lesions that are nonprogressive in immunocompetent hosts may deteriorate silently to more severe lesions in hemodialysis patients with chronic HBV infection.[70]

3.1.2. Hepatitis C in Hemodialysis Units

Long before the discovery of HCV, what was recognized as non-A, non-B hepatitis was found to occur with increased frequency in hemodialysis units. As control measures have reduced substantially the risk of HBV infection in hemodialysis units (see Section 4), the occurrence of HCV infection in patients and staff of dialysis units has been recognized with increasing frequency. Attention to hemodialysis-associated non-A, non-B hepatitis was generated initially by Galbraith et al.,[71] who described two outbreaks in a London hospital of HBsAg-negative hepatitis in 1966–1967 and 1968–1970 involving 29 hemodialysis patients. Originally labeled incorrectly as cases of hepatitis A,[72] these illnesses were shown by retrospective serologic analysis to be cases of non-A, non-B hepatitis. Seven patients had two discrete episodes of non-A, non-B hepatitis, and in eight, chronic liver disease developed. Other outbreaks and sporadic cases of non-A, non-B hepatitis have been reported from other hemodialysis units.[73–75]

The institution of infection control practices in dialysis units and screening of blood donors have substantially reduced the risk of HCV infection in dialysis patients. Currently, the prevalence of HCV seropositivity among chronic hemodialysis patients averages 10% and is as high as 60% in some centers, though many of these patients were infected prior to screening of the blood pool for HCV.[56] Risk factors for HCV seropositivity in the dialysis population include history of blood transfusion, amount of blood transfused, number of years on dialysis, and intravenous drug use.[76–79] Because their humoral immune responses to HCV may be blunted, however, up to 20 to 40% of patients with end-stage renal disease and chronic hepatitis C may have false-negative results on serologic testing for HCV.[80] The number of years on dialysis has been the most consistently identified risk factor, with prevalence of HCV seropositivity increasing more than threefold with increasing duration of dialysis.[76] This association is independent of history of blood transfusion or intravenous drug use, suggesting true nosocomial transmission of HCV via hemodialysis. Recently, HCV RNA has been detected in dialysis ultrafiltrate samples from a patient with chronic hepatitis C, further supporting the possibility of direct patient-to-patient HCV transmission via dialysis.[81]

Few reliable data exist to define the natural history of chronic hepatitis C in hemodialysis patients, because many long-term survivors with end-stage renal disease go on to renal transplantation and many others die of causes unrelated to liver disease. Further complicating natural history studies is the fact that the duration of HCV infection can be rarely established with certainty in this population. Most authors have concluded that HCV infection is, for the most part, clinically and biochemically silent in the dialysis population; however, a recent report identified a strong correlation between hepatitis C viremia and elevated aminotransferase levels.[82] Furthermore, biochemical markers do not always correlate with histologic severity of hepatitis; at least one study found that the majority of HCV-infected dialysis patients with normal aminotransferase levels had histopathologic changes consistent with hepatitis.[83] A single study found HCV infection to be an independent predictor of mortality in the dialysis population after correction for age, transplantation, time on hemodialysis, and race.[84] Given that many dialysis patients go on to receive renal allografts, the forthcoming discussion of HCV in the transplant setting also will be relevant to the prognosis of HCV infection in the dialysis population.

Finally, coinfection with the so-called hepatitis G virus (HGV, or GB virus C) is common in HCV-infected dialysis patients, as it is in transfusion recipients, occurring in 29% of HCV-infected dialysis patients in a recent report.[85] Hepatitis G, however, has no pathogenic effect on the liver, either alone or in combination with HCV infection. Thus, HGV coinfection does not worsen the course of HCV infection in dialysis patients.

3.2. Hepatitis in Recipients of Organ Transplants

Hepatitis occurs in recipients of transplanted kidneys, livers, hearts, and bone marrow. In liver transplant recipients, other factors, such as graft rejection, are a cause for abnormal biochemical tests, and for histologic changes that obscure the presence of suspected transplantation-associated viral hepatitis. Virologic markers alone can be relied on with confidence to define the role of hepatitis viruses in liver injury after liver transplantation. Therapy with immunosuppressive drugs such as azathioprine, cyclosporin A, and tacrolimus, not to mention other potentially hepatotoxic drugs, may also contribute to abnormal liver tests in transplant recipients. On the other hand, evidence points to a role for viral hepatitis agents in a substantial proportion of the hepatic dysfunction seen in transplanted patients. Pharmacologic immunosuppression in transplant patients with viral hepatitis leads to predictably elevated levels of viral replication, and the clinical implications of this enhanced viral replication will be discussed in Sections 3.2.3 and 3.2.4. Data generated in the study of renal and hepatic transplant patients should be applicable to recipients of other solid organ transplants as well.

3.2.1. Hepatitis B in Renal Transplant Recipients

Exposure to hepatitis B is common among renal transplant recipients. Hemodialysis preceding transplantation, intentional pretransplantation blood transfusions, and additional percutaneous exposures in the peritransplant period contribute to the potential for HBV infection. Rarely, the transplanted organ may derive from an HBsAg-positive donor, or even more rarely from an HBsAg-negative, anti-HBc-positive donor.[86,87] Among these risk factors, pretransplantation hemodialysis is considered the largest contributor by some investigators,[88] but others find that the frequencies of pre- and posttransplantation HBV infection are similar.[89] In 1974, Luby et al.[90] followed 45 consecutive renal transplant recipients prospectively and identified 9 cases (20%) of hepatitis B. Fortunately, the prevalence of HBV infection in renal transplant patients has fallen markedly since the implementation of specific infection control measures for the prevention of HBV transmission in dialysis units.

In addition to the natural immunosuppression of chronic renal failure and hemodialysis that antedate transplantation, transplant recipients are subjected to pharmacologic immunosuppression with high-dose cortico-steroids and azathioprine, mycophenolate, cyclosporin A, or tacrolimus, not to mention occasional requirements for other forms of immunosuppressive therapy, all designed to abrogate cellular immune function sufficiently to prevent allograft rejection. As expected, like other immuno-suppressed patients, once infected with HBV, transplant recipients are less likely to experience a recognizable acute icteric illness, but they are more likely to remain chronically infected, to retain high levels of virus replication, and to be highly infectious for their contacts.[68] The presence of HBsAg in their urine is quite common[91] and has also been linked circumstantially to HBV exposure in their family contacts.[92] Reports exist as well of hepatitis B reactivation after renal transplantation in a number of patients who became HBsAg-positive after transplantation but who had anti-HBc in an earlier serum sample[24,93–95] or reactivation of virus replication in those with relatively nonreplicative HBV infection at the time of transplantation.[94] These patterns have been observed more commonly in oncology patients receiving cytotoxic chemotherapy (see Section 3.3). Although their frequency in transplant recipients is unknown, their occurrence raises the possibility that some apparent *de novo* cases of hepatitis B after transplantation actually represent reactivation.

The effect of HBV infection on patient survival after renal transplantation remains controversial, but a growing body of evidence favors a significant decrease in survival in HBsAg-positive patients (Fig. 1). Pirson *et al.*[96] reported that, whether HBsAg was present at the time of transplantation or acquired thereafter, mortality, resulting from progressive liver disease, was fivefold higher between 6 months and 4 years after transplantation in HBsAg-positive renal allograft recipients. Similarly, Hillis *et al.*[97] found that renal transplant recipients who were HBsAg positive at the time of transplantation had a dramatically reduced survival, although most deaths in this series were not related to liver disease. In a subsequent prospective study of 22 HBsAg-positive renal transplant recipients, all of whom had benign histologic diagnoses on initial biopsy, Parfrey *et al.*[98] found progression to more severe chronic hepatitis in 6 patients (27%) and progression to cirrhosis in 7 patients (32%). Eight of the 13 with advanced chronic hepatitis or cirrhosis had clinical evidence of jaundice, portal hypertension, and/or liver failure; 3 died of liver failure, and 2 died of hepato-cellular carcinoma. In contrast, a cohort of 10 HBsAg-positive patients with chronic renal failure managed with maintenance hemodialysis instead of transplantation experienced no liver-related deaths, and 4 ultimately cleared

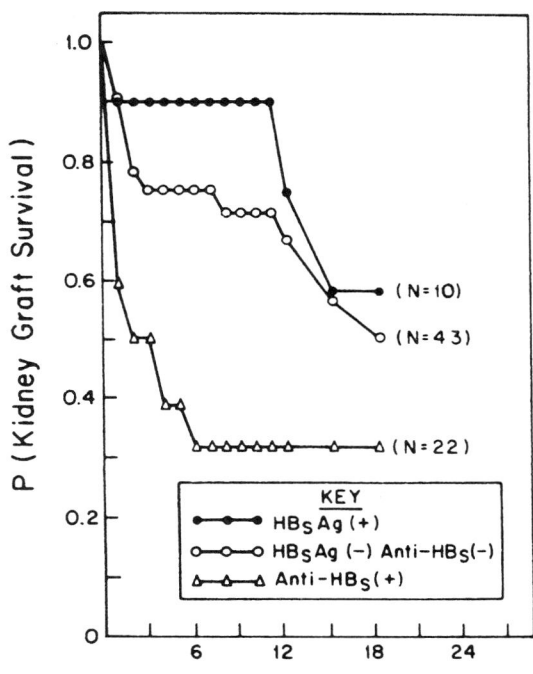

FIGURE 1. Renal allograft survival curves for patients who were known HBsAg carriers, who had no hepatitis B markers, or who had anti-HBs prior to transplantation. The numbers in parentheses represent the number of patients in each category. (From London *et al.*[61] with permission.)

HBsAg. These authors suggested that transplantation of HBsAg-positive patients with end-stage chronic renal failure may be inadvisable. Recently, Mathurin *et al.*[99] reported the results of a case–control study designed to determine the effect of viral hepatitis on patient and graft survival after renal transplantation. These authors found a significant decrease in 10-year survival (55%) in 128 HBsAg-positive patients, compared with 80% in 128 matched controls; most of the excess mortality in the HBV-infected group was attributable to liver disease. Presence of biopsy-proven cirrhosis also was found to be an independent predictor of mortality in this series, and the authors suggested that the presence of cirrhosis on a pretransplant liver biopsy should be a contraindication to renal transplantation. Unfortunately, pretransplant liver biopsies were not performed routinely on these patients; therefore, whether progression to cirrhosis occurred before or after transplantation cannot be determined.

As noted above for hemodialysis patients and as emphasized by the findings of Parfrey *et al.*,[98] transplant recipients with chronic hepatitis B may have insidiously progressive chronic hepatitis and cirrhosis in the absence of symptoms.[68,100] Degos *et al.*[88] found that in 25% of patients with mild chronic hepatitis at the time of transplantation histologic lesions evolved to more severe and progressive chronic hepatitis; Parfrey *et al.*[98] found that 42% of patients with normal liver histology or mild chronic hepatitis on initial biopsy progressed to more severe chronic hepatitis or cirrhosis. The risk of progressive hepatitis B correlates with older age at transplantation and age at acquisition of HBV infection.[101] Patients who acquire *de novo* HBV infection shortly after renal transplantation are at especially high risk of developing decompensated liver disease, resulting from either acute or chronic hepatitis.[102] In rare cases, a syndrome of fibrosing cholestatic hepatitis caused by hepatitis B has been reported following renal transplantation.[103,104] This syndrome, characterized by cholestasis, rapidly progressive liver failure, and exuberant viral replication in the setting of immunosuppression, was initially described in HBV-infected liver-transplant recipients and will be discussed further in Section 3.2.3.

Other factors besides HBV may contribute to hepatic dysfunction in renal transplant patients and should be included in differential diagnostic considerations. Azathioprine may be hepatotoxic; other hepatotropic (hepatitis C) and systemic virus infections, especially herpes simplex, zoster, and cytomegalovirus (CMV), may cause liver dysfunction; sepsis too can result in secondary hepatocellular dysfunction. As hepatitis B has been reduced in hemodialysis populations by control measures, so too has the role of HBV in posttransplant hepatic disease declined. Currently, hepatitis C is the cause of most cases of transplantation-associated hepatitis.

3.2.2. Hepatitis C in Renal Transplant Recipients

Prior to the identification of hepatitis C, non-A, non-B hepatitis was recognized as a source of appreciable morbidity and even mortality in renal transplant recipients. Ware *et al.*[105] described 72 episodes of acute hepatitis or elevations in serum aminotransferase activity in 62 (38%) of 162 renal transplant recipients. Although drug hepatotoxicity and infections with CMV, HBV, Epstein–Barr virus (EBV), or varicella zoster could be incriminated in as many as 74% of the 34 cases of acute hepatitis, drug hepatotoxicity or infection with these agents was not linked to the chronic liver disease encountered in this group of patients. Of the 38 cases of chronic liver disease, 27 (71%) were attributed by serologic exclusion to non-A, non-B hepatitis agent(s). Sixteen of the patients with

chronic liver disease, most of whom had non-A, non-B hepatitis, had progressive deterioration; cirrhosis developed ultimately in 11, 4 experienced a transition from mild to more severe chronic hepatitis, and 1 died of liver failure 1 year after the onset of disease.

Degos *et al.*[88] found that most cases of liver disease in renal transplant recipients began prior to transplantation and were acquired during hemodialysis. In contrast, analyses by others have shown that most of the liver disease that occurs in renal transplant recipients occurs after transplantation. LaQuaglia *et al.*[106] reviewed the 10-year experience of 405 consecutive renal transplant recipients at the Massachusetts General Hospital. Despite reliance on frozen washed red blood cells (RBCs) as the exclusive source of transfused blood, 10.4% of this cohort of patients had biochemical or clinical evidence of acute hepatitis after transplantation and 62% of the patients with hepatitis (or 6.5% of the total) were categorized by serologic and clinical exclusion as having non-A, non-B hepatitis. Five cases were caused by HBV, but all five occurred during the first 3 years of the observation period, 1970–1973, a trend that has been recognized almost universally in American transplantation programs. Therefore, the relative proportion of non-A, non-B hepatitis cases during the last 7 years of observation was even higher. After acute infection, chronicity was very common; chronic hepatitis developed in 93% of those whose acute hepatitis occurred during the first year after transplantation and in 64% of those affected after the first year. Not only was chronicity likely, but morbidity and mortality were unexpectedly common; those with hepatitis had a significantly higher mortality rate (45%) than patients without hepatitis (16%) (Fig. 2, Table 3). An important observation emerging from this study was the fact that 80% of the deaths were not related to liver disease but to extrahepatic sepsis. Even among survivors, life-threatening extrahepatic infections were significantly more common in patients with hepatitis (52%) than in patients without hepatitis (20%). These findings, in addition to a significantly increased 1-year allograft survival in the patients with hepatitis (73%) compared with that in patients without hepatitis (50%), suggested that non-A, non-B hepatitis infection, like infection with CMV, had an immunosuppressive effect on transplant recipients.

Illustrative Case 1

A 62-year-old man required hemodialysis for intravenous-contrast-dye-associated renal failure; after a year of hemodialysis, he underwent cadaver-kidney transplantation with a then-standard prednisone–azathioprine immunosuppressive regimen (in the precyclosporine era). His immediate posttransplantation course was uneventful, but his

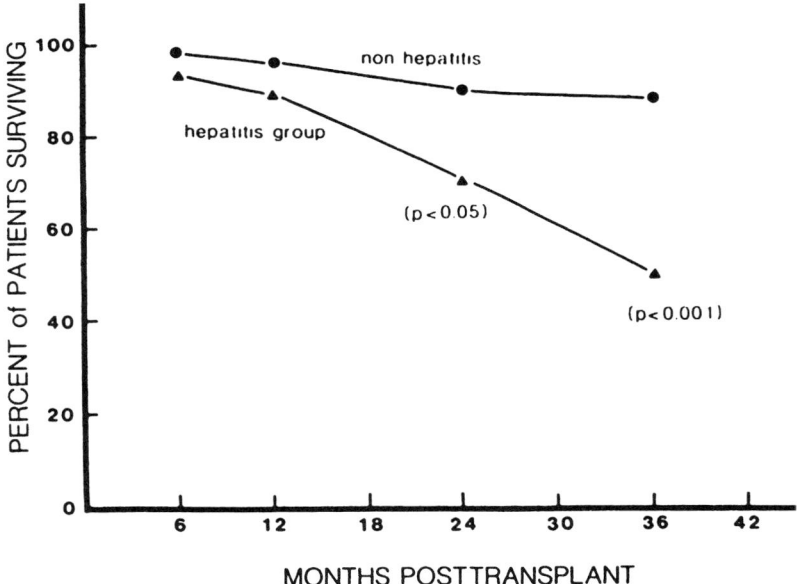

FIGURE 2. Survival of patients with and without hepatitis following renal transplantation. (From LaQuaglia *et al.*[106])

EL30aminotransferase levels, normal during hemodialysis, became abnormal several months after transplantation and a diagnosis of acute non-A, non-B hepatitis (hepatitis C) was made. Although a liver biopsy done 2 years after transplantation for nafcillin-associated acute toxic hepatitis demonstrated the presence of underlying cirrhosis, he never experienced any complications of chronic liver disease or clinical manifestations of hepatic decompensation. On the other hand, he had staphylococcal olecranon bursitis at 2 years and fatal septic vasculitis 6 months thereafter, attributed at postmortem examination to disseminated (skin, lungs, small bowel) *Legionella* infection.

Comment. This patient's course provides an example of acute hepatitis C after renal transplantation, progressing histologically within 2 years to cirrhosis but without clinical manifestations of liver disease. His clinical course was dominated instead by life-threatening infections, first staphylococcal bursitis, then disseminated legionellosis, the cause of death. Although the organism identified at autopsy was a surprise, the case is a vivid example of the type of extrahepatic life-threatening infections to which patients with posttransplantation hepatitis C are vulnerable.

TABLE 3. Impact of Non-A, Non-B Hepatitis on Renal Transplant Recipients[a]

Clinical feature	Hepatitis (%)	No hepatitis (%)
Graft survival	73	50
Mortality	45	16
Life-threatening infections among survivors	52	20

[a]From LaQuaglia *et al.*[106]

Since the development of assays for antibodies to HCV, HCV infection has been documented as being quite prevalent in the renal transplant population, with prevalences ranging from 20% to 30%.[107,108] That immunosuppression increases the level of HCV replication in renal transplant recipients has been well documented[109]; however, elevated serum aminotransferase levels are uncommon following renal transplantation in HCV-infected persons who had normal levels prior to transplantation. In one study, sustained aminotransferase elevations developed in only 8 of 38 (22%) HCV-infected patients over a mean follow-up of 65 months after renal transplantation.[110] In another study, elevated serum aminotransferase levels developed in 60% of HCV-infected patients within 4 years following renal transplantation, but no mortality was attributable to liver disease.[107] In fact, the long-term consequences of HCV infection in the renal transplant population remain controversial. Pereira *et al.*[111] found a fivefold increase in liver disease in HCV-infected renal transplant patients compared with HCV-negative recipients as well as significantly increased mortality, largely related to sepsis. Unlike those in most other studies, however, patients in the latter study had a high rate of pretransplant aminotransferase elevations. Most studies have failed to show any decrease in patient or allograft survival in HCV-infected renal transplant recipients, with follow-up observations as long as 18 years.[112–115] In one recent study, a comparison of liver histology in dialysis patients to that in renal transplant recipients revealed a similar histologic activity score and stage in the two groups, despite longer duration of end-stage renal disease and of HCV infection in the transplant group.[116] This observation suggests that immunosuppression after renal transplantation may not have a significant effect on the course of hepatitis C, despite the existence of higher levels of viral replication. Only the recent study by Mathurin et al.[99] found decreased patient survival, graft survival, and liver-specific mortality in HCV-infected renal transplant recipients. In this study, 10-year survival after renal transplantation was 65% in HCV-infected patients, compared with 85% in matched controls. Multivariate analysis confirmed that HCV-seropositivity was independently associated with patient survival, as were biopsy-proven cirrhosis, age at transplantation, and year of transplantation. These authors suggested that inadequate duration of follow-up in previous studies may have been responsible for the failure by others to detect an impact of HCV infection on mortality after renal transplantation. Also likely to explain the difference between this study and others is the fact that patients in this study had more advanced liver disease prior to renal transplantation than those in the aforementioned studies, athough pretransplant liver biopsies were not performed. Given that the degree of pretransplant liver disease is probably the most important predictor of liver-related morbidity and mortality following renal transplantation, pretransplantation liver biopsy is an important tool in the selection of patients for renal transplantation.

Several unusual HCV-related syndromes affecting renal transplant patients also bear mention. Analogous to the syndrome mentioned in the discussion of posttransplant HBV infection, rapidly progressive fibrosing cholestatic hepatitis has been described in HCV-infected patients following renal transplantation.[117,118] Most of these patients have succumbed to liver failure, but a few have been salvaged with either reduction of immunosuppression[117] or interferon therapy.[118] Another syndrome, HCV-related glomerulonephritis, has been well-described in immunocompetent patients and has recently been described in HCV-infected renal transplant recipients.[119,120] In the context of renal transplantation, this syndrome assumes particular importance, challenging the clinician to distinguish between virus-associated and transplant-associated glomerulopathy.

3.2.3. Hepatitis B in Liver Transplant Recipients

Both donor organs and transfused blood are screened for evidence of HBV infection; because such screening is so effective, the likelihood of acquiring hepatitis B from an allograft or transfused blood is extraordinarily low. Of note, transmission of HBV via hepatic allografts obtained from HBsAg-negative but anti-HBc-positive donors has been described, and such organs should probably be transplanted only into recipients already infected with HBV.[121] On the other hand, patients who undergo liver transplantation for liver failure caused by hepatitis B—those with chronic hepatitis B or rarely those with acute fulminant hepatitis B—are at risk of reinfection of the new liver with residual HBV.[122–124] Whether the source of reinfection is circulating virions or extrahepatic reservoirs of HBV (e.g., bone marrow, lymphocytes, pancreas),[125] reinfection of the allograft (in the absence of specific therapy to prevent it) occurs almost universally.[126,127] The rare exception is the patient who undergoes liver transplantation for acute fulminant hepatitis B.[123] In some cases of fulminant hepatitis B, the destruction of the liver and consequently the substrate for virus replication is presumed to be so complete that no virus remains at the time of transplantation. These fortunate few may not experience reinfection of the allograft with HBV.

Active HBV replication (as indicated by the presence of HBeAg and HBV DNA in serum) in the pretransplant period appears to increase the likelihood of clinically significant HBV recurrence in the posttransplant period, although pretransplantation high-level virus replication does not necessarily predict the recurrence or the severity of recurrence with great accuracy.[123,127,128] Almost inevitably, the immunosuppressive agents required for control of rejection, including glucocorticosteroids (mediated via a glucocorticoid-responsive element in the HBV genome), provide a potent stimulus for virus replication.[129,130] Therefore, almost all patients who undergo liver transplantation for hepatitis B will have high levels of HBV replication, regardless of their pretransplant level of replication. Whereas in normal immunocompetent persons, increased levels of HBV replication are associated with liver injury, in immunosuppressed transplant recipients the exogenous immunosuppression is usually sufficient to blunt liver injury substantially. Therefore, a majority of such patients, approximately 50%, become asymptomatic but highly replicative "carriers" of HBV with little or no virus-induced liver injury, despite the presence of intrahepatic HBV antigens.

Unfortunately and unpredictably a proportion of patients will have ongoing virus-induced liver injury after transplantation. The spectrum of clinical disease is broad and variable, ranging from mild, minimally progressive liver injury, reflected by limited hepatocellular inflammation and elevation of liver chemistry tests, to a recapitulation of chronic hepatitis, to a more rapidly progressive syndrome known as fibrosing cholestatic hepatitis. As a result, the overall success rate (e.g., 1-year survival) of liver transplantation in patients with hepatitis B is approximately 20% lower than that in patients without hepatitis B.[123] This decay in relative success is associated not only with recurrent liver disease but also with increased graft loss, sepsis, and such other unexplained systemic complications as severe pancreatitis.

Fibrosing cholestatic hepatitis (FCH) was originally described by the Cambridge-King's College Hospital transplant group in patients with recurrent HBV infection following liver transplantation.[131] In the Cambridge–King's College series, FCH occurred in 24% of HBsAg-positive transplant patients who survived more than 2 months; in other series, the frequency of FCH was below 10%.[131,132] The syndrome manifests clinically as rapidly progressive liver failure, often within months of transplantation, and usually culminates in death within 4 to 6 weeks of onset. Unlike fulminant hepatitis in immunocompetent patients, FCH is characterized by a unique histologic pattern consisting of hepatocyte ballooning, cholestasis, serpiginous periportal fibrosis, and prominent HBV antigen expression in hepatocytes, with a relative paucity of inflammatory cells. The presence of unusually large quantities of HBV antigens within infected hepatocytes and the relative absence of inflammatory cells suggest that liver injury in FCH, unlike that in nearly all other forms of viral hepatitis, results from a direct cytopathic effect of HBV on hepatocytes, rather than from a cytolytic T-cell response to virus-infected hepatocytes.[133] This concept is supported by evidence from both transgenic mouse[134] and cell culture[135] models, in which overexpression of HBV antigens has been found to induce cytopathic changes in hepatocytes. In a single patient with HBV-induced FCH following liver transplantation, a pre-S mutation resulting in intracellular virus accumulation has been identified.[136]

In the early weeks and even months after transplantation commonly the level of circulating HBsAg is beneath the detection threshold. Therefore, HBsAg may be undetectable for several months after transplantation. To distinguish between the effects of rejection and HBV, pathologists rely on the distribution of necroinflammatory activity. As noted, detection of intrahepatic HBV antigens is almost invariable; however, liver injury associated with rejection tends to be portal–periportal, while liver injury associated with HBV (or other hepatotropic viruses) tends to be more lobular.[137]

Neither hepatitis B vaccination nor interferon ther-

apy has been effective in preventing reinfection of the allograft with HBV after transplantation.[138] Similarly, interferon has not been effective in treating hepatitis B after transplantation.[123,139,140] In immunocompetent persons, the efficacy of interferon in hepatitis B is felt to be mediated by a combination of its antiviral properties as well as its immunomodulatory properties. Immunosuppressed organ recipients are similar to other immunocompromised hosts in failing to respond to interferon, probably because of the blunting of cell-mediated responsiveness. In fact, because interferon can enhance cellular expression of HLA class I, theoretically interferon treatment could precipitate or perpetuate allograft rejection. In practice, however, interferon therapy for viral hepatitis has not been associated with rejection after organ transplantation.

High-anti-HBs-titer polyclonal and monoclonal hepatitis B immune globulins (HBIG) have been used in an attempt to prevent recurrent hepatitis B after transplantation.[126,127,140–143] The long-term use of high-dose HBIG appears to decrease rates of both HBV recurrence and mortality in HBV-infected liver transplant recipients; survival rates have improved to approximately 80% at 1 year and 65% at 3 years after transplantation.[127] Consequently, all liver transplantation centers rely on posttransplantation HBIG, administered daily for a week beginning in the anhepatic phase and continuing at 4- to 6-week intervals indefinitely thereafter, in anyone undergoing liver transplantation for end-stage chronic hepatitis B. Still, HBV recurs despite immunoprophylaxis in 15 to 50% of patients, whether secondary to saturation of HBIG's binding capacity in the setting of high-level viral replication or to escape mutations in the surface antigen.[127,128,143–147] In addition, HBIG therapy is expensive, costing approximately $20,000 per year. Therefore, better strategies for the prevention of recurrent HBV infection after transplantation are needed.

Development of the nucleoside analogue lamivudine as therapy for hepatitis B has changed dramatically the approach to prevention and treatment of recurrent HBV following liver transplantation. Lamivudine, a potent inhibitor of HBV DNA polymerase, is highly effective in suppressing HBV replication in both immunocompetent and immunosuppressed patients. In a pilot study, lamivudine given prior to transplantation prevented HBV recurrence in 11 of 12 patients transplanted for HBV-related cirrhosis.[148] Larger studies have confirmed that lamivudine, whether given before or after transplantation, effectively suppresses HBV replication.[149–151] Development of lamivudine-resistant HBV, however, has been described in as many as 25% of patients on lamivudine for 1 year after liver transplantation.[151] Resistance results from mutations in a highly conserved portion of the HBV DNA polymerase gene encoding an amino acid motif with the sequence YMDD.[152–154] These mutations were first described in liver transplant patients and occur earlier and more frequently in this setting, presumably because of the immunosuppression-driven increase in viral replication. Although these resistant HBV variants appear to be somewhat attenuated, hepatic decompensation, including death from liver failure, has been described in approximately a third of patients with such viral breakthrough after liver transplantation.

The optimal timing of lamivudine therapy remains somewhat controversial. Given that the presence of HBV DNA at the time of transplantation portends a worse outcome, a rational approach would be to reduce viral replication by treating with lamivudine prior to transplantation. On the other hand, given that the availability of donor organs is unpredictable, this strategy might be associated with prolonged therapy prior to transplantation in a proportion of patients, and prolonged lamivudine therapy may contribute to the development of YMDD mutations after transplantation. The issue of the timing of therapy will require clarification over the next few years. Still, preliminary observations suggest that patients with lamivudine-associated YMDD-variant hepatitis B prior to transplantation respond to a combination of HBIG plus lamividine after transplantation with suppression of recurrent HBV infection. Long-term observations of the impact of pre- and posttransplantation lamivudine therapy are anticipated, and early results suggest that lamivudine therapy will have a beneficial effect on graft and patient survival rates after liver transplantation for hepatitis B.

Given the limitations of both HBIG and lamivudine in preventing recurrence of HBV after liver transplantation, attention has focused on the combination of the two. Recently, the combination of lamivudine and HBIG was reported to be successful in preventing HBV recurrence in every one of a group of 14 patients.[155] Although the study was uncontrolled and no patient was followed for more than 18 months, many transplantation centers have already adopted this combination approach for routine prophylaxis after liver transplantation in patients with hepatitis B. Large-scale, multicenter studies are in progress to test the relative efficacy of monotherapy and combination therapy regimens and to determine how long to use each of the two components of combination therapy. In addition, new antiviral agents, including adefovir dipovoxil, effective against both wild-type and YMDD-variant HBV are being studied in combination with lami-

vudine. Inevitably, future strategies will include multiple nucleoside analogues used in combination.

3.2.4. Hepatitis D in Liver Transplant Recipients

The outcome of liver transplantation in patients with end-stage liver disease secondary to infection with the HBV-associated delta hepatitis agent (hepatitis D virus) appears to be better than that observed in patients with hepatitis B alone; a 1-year survival of almost 80% can be expected, indistinguishable from that seen in patients without hepatitis B. Although HDV reinfection is quite likely, if not invariable, reexpression of HBV markers after transplantation may not occur in such patients. Thus, HDV infection after liver transplantation is unique in two ways: (1) this is the only situation observed in which HDV infection occurs in its human host without the help of HBV and (2) this is the only scenario besides fulminant hepatitis B in which liver transplantation is not invariably accompanied by HBV reinfection of the allograft. In fact, up to 20% of patients who receive liver allografts for hepatitis D lose both HBV and HDV after transplantation. In a study of 27 Italian and Belgian patients who underwent liver transplantation for hepatitis D, 21 (78%) survived, 5 (18%) lost both HDV and HBV, 11 (41%) had recurrent HDV without HBV, and 11 (41%) had recurrent HDV plus HBV. In the group with recurrent HDV alone, none of the 11 had recurrent liver injury.[156] In contrast, among the 11 with recurrent HDV plus HBV, recurrent liver disease occurred 3 to 8 (mean 6) months after transplantation in all 11, and 2 died. Thus, liver injury was confined exclusively to the group with both HBV and HDV infection. In short, although a proportion of patients who undergo liver transplantation for hepatitis D will have recurrent, often severe, liver disease, overall the prognosis is better for such patients than for patients with hepatitis B alone.

3.2.5. Hepatitis C in Liver Transplant Patients

End-stage liver disease resulting from chronic hepatitis C is currently the leading indication for orthotopic liver transplantation. Reinfection with HCV is universal in the allograft; however, in rare cases, HCV infection after liver transplantation may derive from other sources: from the large volume of blood transfused during the transplant procedure and in the peritransplant period or from a donor liver obtained from an HCV-infected person.

The availability of serologic tests for anti-HCV has been of limited value in assessing the frequency of new

HCV infection and reinfection after liver transplantation. Studies based on amplification of HCV RNA by PCR suggest that HCV reinfection after liver transplantation is the rule rather than the exception; however, studies based on enzyme immunoassay testing for anti-HCV fail to identify antibody in a substantial proportion of HCV-infected allograft recipients.[157–163] The humoral immune response to HCV is relatively feeble compared with that mounted against other hepatitis viruses and immunosuppression may blunt the immune response. Therefore, pretransplantation anti-HCV may become undetectable after transplantation and immunosuppression, despite the persistence of HCV infection, as demonstrated by the presence of HCV RNA.

Amplification assays for HCV RNA, such as PCR and branched DNA methods, have provided valuable insight into posttransplant HCV infection. HCV RNA remains detectable at a low level even in the immediate posttransplant period and the level rises gradually thereafter, reaching the pretransplant level on posttransplant day 9.[164] Thereafter, the level of viremia is greatly influenced by therapy given for acute rejection, such as high-dose methylprednisolone, which leads to 4- to 100-fold elevations in the HCV RNA level.[165] Based on these data, many centers strive to minimize immunosuppressive drug dosages in patients undergoing liver transplantation for hepatitis C. By approximately 1 month posttransplant, however, HCV RNA levels begin to rise rapidly in most patients, even in the absence of antirejection therapy. The level peaks between 1 and 6 months after transplantation, correlating with onset of an acute lobular hepatitis and intrahepatic expression of HCV antigens.[165] Acute hepatitis resulting from recurrent HCV infection occurs in 57% of patients undergoing liver transplantation for HCV-related disease within the first year after transplantation.[166] This acute hepatitis is usually self-limited and its resolution correlates with a decrease in hepatitis C viremia.[165] By the end of the first year, the HCV RNA level is 3 to 112 times the pretransplant level and higher levels of viremia appear to correlate with more severe chronic graft damage.[165]

The long-term impact of recurrent HCV infection on the outcome of liver transplantation is somewhat controversial. Feray et al.[166] found that in 42% of patients with recurrent HCV infection, chronic hepatitis developed; 25% of those with chronic hepatitis had some degree of fibrosis and one patient had documented cirrhosis within 1 year of transplantation. These authors concluded that patients undergoing liver transplantation for hepatitis C have a good prognosis, despite the likelihood of recurrent infection. The King's College group reported that by 5 years posttransplantation, 20% of those undergoing liver transplantation for chronic hepatitis C had progressed histologically to cirrhosis and only 5% had no evidence of chronic hepatitis.[167] Five patients in this cohort lost their hepatic allografts to HCV-related cirrhosis and three to progressive cholestatic hepatitis. Nevertheless, 5-year graft and patient survival rates did not differ between those undergoing transplantation for hepatitis C and those undergoing transplantation for other causes of liver disease. Yet another group of investigators found that even after up to 10 years of posttransplantation follow-up, no difference in survival was observed between patients undergoing transplantation for hepatitis C or for other reasons.[168] In this study, significant fibrosis developed in only 10% of HCV-infected patients within 10 years of transplantation.

Despite the lack of evidence for an overall decrease in survival in patients with recurrent HCV infection after transplantation, recurrent HCV can affect mortality and morbidity dramatically in individual cases. The most severe form of recurrent disease is a progressive cholestatic hepatitis similar to the fibrosing cholestatic hepatitis observed in HBV-infected patients after transplantation. In one report, the frequency of this syndrome in HCV-reinfected patients was 6%.[169] Mean serum bilirubin level in affected patients was 24.7 mg/dl, and retransplantation was required in all patients within 5 months of the onset of cholestasis. Similar to HBV-associated fibrosing cholestatic hepatitis (FCH), this syndrome is characterized histologically by centrilobular ballooning degeneration, bridging fibrosis, and cholestasis.[170] Interestingly, in a longitudinal study of HCV replication by Gane et al.,[165] the level of HCV RNA rose dramatically after an initial episode of acute lobular hepatitis in a patient in whom FCH developed subsequently, whereas viremia levels decreased within a month of the lobular hepatitis in the remaining patients. This observation supports but does not prove the theory that HCV, like HBV, may exert a cytopathic effect under conditions of high-level replication and suppressed immune responsiveness.

Efforts to treat hepatitis C after liver transplantation have focused on recombinant interferon alpha, either alone or in combination with ribavirin. In immunocompetent persons with chronic hepatitis C, 3 million units of interferon administered subcutaneously three times a week for 6 months is likely to achieve biochemical remission in approximately 50% of cases; however, the relapse rate after cessation of therapy is 50%.[171] For this minimal regimen, the likelihood of a sustained virologic response is below 10%. Results of treatment for recurrent hepatitis C after liver transplantation have been even more disappointing; however, most trials have been small, nonrandomized if controlled at all, and lacking in power. In a group of 18 patients who had biochemical, virologic, and histologic evidence of recurrent hepatitis C, Wright *et al.*[172] described treatment with a regimen of interferon, 3 million units, three times a week for at least 4 months; an initial virologic response occurred in only 28% of patients and virologic relapse occurred in all responders. Feray *et al.*[173] reported a sustained response in only 1 of 14 patients with recurrent hepatitis C treated with a similar regimen of interferon. More concerning, the same group found a 36% incidence of chronic rejection in the interferon-treated patients. Vargas *et al.*[174] applied the same regimen to patients with acute lobular hepatitis, the first clinical indicator of recurrent hepatitis C, but this strategy also failed to clear viremia, to improve aminotransferase levels, or to prevent progression to chronic hepatitis.

The nucleoside analogue ribavirin is the only other currently available antiviral agent with potential activity against HCV. Similar to experience in the nontransplant setting, short-term treatment with ribavirin alone has been shown to improve biochemical activity of posttransplantation hepatitis C but to have no effect on viremia.[175] Continuation of ribavirin therapy for at least 6 months has resulted in complete normalization of aminotransferases in 26% of patients as well as an improvement in the degree of necroinflammatory activity on liver biopsy.[176] The most recent report, however, identified no improvement in liver histology, despite improvement in aminotransferase activity.[177] Therefore, the main role of ribavirin is likely to be as an adjunct to interferon therapy. Bizollon *et al.*[178] reported dramatic results when treating clinically recurrent hepatitis C with the combination of interferon 3 million units three times a week and ribavirin 1200 mg/day for 6 months, followed by maintenance therapy with ribavirin alone. All 21 patients treated achieved a return to normal of aminotransferase activity during combination therapy and only one relapsed during maintenance therapy. Moreover, reduction in viremia and improvement in histology were seen in all patients and this therapy did not result in a single episode of rejection. Therapy had to be stopped in 14% of patients because of ribavirin-induced hemolytic anemia.

Illustrative Case 2

A 42-year-old abstinent alcoholic man underwent liver transplantation in July 1989, for complications of end-stage liver disease and portal hypertension. He received 18 units of packed red blood cells at the time of transplantation; prior to surgery, his serum was reactive for isolated anti-HBs. Six months after discharge from the hospital he was living independently and had returned to work. Ongoing immunosuppression included cyclosporine, azathioprine, and prednisone and his aminotransferase levels and bilirubin were normal.

Nine months following liver transplantation, routine follow-up monitoring unearthed an AST of 122 IU, ALT of 170 IU, and bilirubin of 1.8 mg/dl. A liver biopsy showed mild rejection as well as histologic findings consistent with viral hepatitis. After increased immunosuppression with steroid pulses, aminotransferase levels fell somewhat but not to normal, and within another 3 months his AST was up to 362, SGPT 223, and bilirubin 6.3. Tests for HBsAg and anti-HCV were negative and the patient's stored pretransplant serum was also anti-HCV-negative. When pre- and posttransplant serum samples were tested for HCV RNA by PCR, however, the posttransplantation sample, not the pretransplantation sample, contained HCV RNA.

A course of recombinant interferon-alfa-2b was begun at a dose of 3 million units subcutaneously three times weekly. Aminotransferase levels did not fall and a modest decrease in bilirubin was followed by a return to pretreatment, elevated levels. The patient's liver function began to deteriorate with progressive jaundice and coagulopathy. An episode of encephalopathy led to hospitalization and he underwent a second transplant 24 months after receiving his first liver. He made a steady recovery thereafter and maintained normal aminotransferase levels.

Comment. This patient had no evidence of hepatitis C before his first liver transplant. Abnormal liver tests after transplantation raised the question of rejection as well as hepatitis, and the liver biopsy was consistent with both diagnoses. Serologic testing for anti-HCV was negative following liver transplantation, a frequent occurrence in immunosuppressed patients; however, hepatitis C viremia was demonstrated by PCR. This infection was acquired at the time of transplantation when the patient received multiple blood products. His clinical course was characterized by progressive deterioration of liver function and an unremitting illness reminiscent of FCH, for which interferon treatment is usually ineffective.

3.3. Hepatitis in Oncology Patients

3.3.1. Hepatitis B in Oncology Patients Receiving Chemotherapy

Hepatitis B has been shown to be a hazard for oncology patients who require immunosuppressive therapy.[13,14,18,179–188] Reports of the frequency of circulating HBsAg in oncology patients with leukemia and lymphoma ranged from a low of 1% to a high of 33%; the frequency of antibodies to HBV in these patients has been reported in the range of 19 to 55%.[13,14,181,182,189–192] Therefore, total current or past exposure to HBV has been observed in 29–69% of patients with myeloproliferative and lymphoproliferative malignancies. In patients with solid tumors, the frequencies are lower; HBsAg has been found in approximately 1% and HBV antibodies in an additional 1–16%.[182,192,193] These high frequencies of HBV infection result from the exposure of such patients to blood products, especially when bone marrow is replaced by malignant cells and during therapy-induced marrow aplasia, and high-risk hospital environments. Moreover, their immunosuppressed status subjects these patients to a high likelihood of remaining chronically infected with HBV after acute infection.

During cytotoxic chemotherapy, immunosuppression allows an often dramatic increase in the level of circulating viremia, as reflected by substantial rises in the titer of HBsAg during therapy and return to pretreatment titers when the marrow is repopulated[182]; conversion from nonreplicative (HBeAg/HBV DNA negative) to replicative (HBeAg/HBV DNA positive) HBV infection[186]; and even occasionally reexpression of hepatitis B surface antigenemia in patients whose blood contained anti-HBs prior to chemotherapy,[182,183,189,194] i.e., a reactivation of HBV infection.[182,186,195] Because levels of viremia are high and because most of the infections are subclinical, these patients provide a poorly appreciated reservoir of HBV infection and serve as a source of infection for the medical personnel who care for them as well as family members.[183,189,196] Falls in the level of anti-HBs also have

been observed during chemotherapy-induced marrow aplasia in these patients,[182] and many oncology patients who are immunosuppressed fail to mount any antibody response (anti-HBs or anti-HBc) to HBV.[189]

Concern for these patients has been raised by the observation that withdrawal of chemotherapy in asymptomatic HBsAg carrier oncology patients has been associated with acute hepatitislike exacerbations (Fig. 3). Although in some patients this may be followed by recovery and elimination of the virus,[186] in others massive hepatic necrosis with fulminant, often fatal, hepatitis follows within 1–6 weeks,[18,197–199] up to 3 months in one case.[186] In a prospective study, Lok et al.[200] found that 22% of HBV carriers receiving cytotoxic therapy for lymphoma experienced acute, icteric hepatitis; hepatic failure developed in 7.4%, with a mortality of 50%. These events have been interpreted as a reflection of the absence of a direct cytopathic effect of HBV during periods of intense virus replication in the permissive environment of chemotherapy but recurrence of presumably immunologically

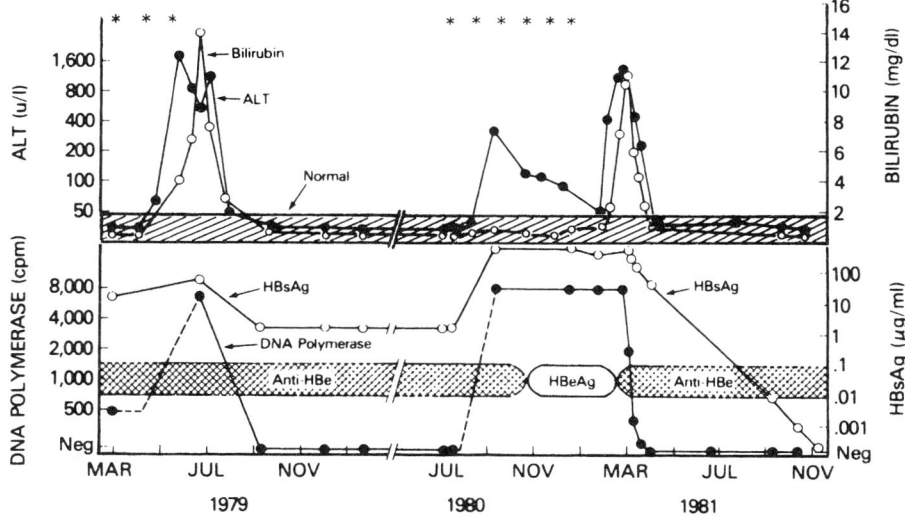

FIGURE 3. Clinical, biochemical, and serologic findings in a 35-year-old HBsAg-positive patient with lymphoma who underwent chemotherapy. Asterisks indicate courses of chemotherapy. Clinical and histologic evaluation prior to chemotherapy suggested that the patient was an asymptomatic HBsAg carrier. Prior to chemotherapy, she lacked serum HBV DNA and had circulating anti-HBe, but she had low-level DNA polymerase activity. During the third course of combination chemotherapy with cyclophosphamide, vincristine, prednisone, and procarbazine, she became icteric, and serum alanine aminotransferase (ALT) increased dramatically. During her severe episode of acute hepatitis, her HBsAg titer and DNA polymerase level increased, HBV DNA became detectable in serum, and her prothrombin time increased to 4 sec above that of control. With recovery from this severe episode of hepatitis, her ALT, bilirubin, and HBsAg titer fell to prechemotherapy levels and both HBV DNA and DNA polymerase became undetectable. Approximately 1 year later, when her lymphoma recurred and she was treated again with chemotherapy, she had an increase in expression of HBV replication, reflected by an asymptomatic increase in HBsAg titer, reappearance in serum of HBV DNA and DNA polymerase, and conversion from anti-HBe positive to HBeAg positive. Chemotherapy was discontinued after six cycles, and 3 months later severe hepatitis recurred, with high ALT and bilirubin, coagulopathy (prothrombin time 12 sec above control), and encephalopathy. At the same time, markers of high HBV replication disappeared. Prednisone therapy was instituted and the severe hepatitis resolved. Nine months later, her HBsAg became undetectable. HBsAg, hepatitis B surface antigen; HBeAg, hepatitis B e antigen. (From Hoofnagle et al.[186])

mediated lysis of an abundance of HBV-infected and virus-antigen-expressing hepatocytes once immune competence is restored by withdrawal of immunosuppressive therapy.

The response of HBsAg-positive oncology patients to chemotherapy is not uniform, however; the acute, severe hepatitis described in oncology patients receiving chemotherapy does not always occur as chemotherapy is being withdrawn. Included in reports of this observation are instances of acute severe hepatitis, acute hepatitislike events, and biochemical evidence of hepatocellular damage that begin during chemotherapy, at a time of active virus replication and peak immunosuppression.[182,184–187] Thus, two points can be made. First, acute HBV infection is not always asymptomatic in immunosuppressed oncology patients. Second, acute hepatitis can occur either during chemotherapy or after its withdrawal. Because the course of events in HBsAg-positive oncology patients receiving chemotherapy is unpredictable, both administration and withdrawal of chemotherapy should be accompanied by careful monitoring of aminotransferase levels and viral replication. In the preantiviral era, some authorities suggested that in these patients withdrawal of chemotherapy be done gradually rather than abruptly and that immunosuppressive chemotherapy be reinstituted at the first sign of acute hepatitis during withdrawal; the effectiveness of this strategy was never established. With the current availability of lamivudine, the simplest and most effective strategy is to pretreat with lamivudine in anticipation of beginning chemotherapy, suppressing HBV replication during and after withdrawal of cytotoxic drugs. Alternatively and especially if hepatitis B reactivation is recognized after chemotherapy is begun or withdrawn, lamivudine therapy can be initiated at the first sign of a return of replicative hepatitis B. Although there are no published data on this approach, anecdotal reports and experiences are beginning to accumulate, and data from the use of lamivudine in other settings suggest that it would be quite effective in oncology patients.

Chronicity of hepatitis B after acute infection, the expected pattern,[201] is not invariable either. Of six children with acute leukemia in whom HBsAg became detectable, all six cleared antigenemia[189] and 86% of a group of oncology patients with acute hepatitis B following injections with an HBV-contaminated tumor vaccine recovered completely and cleared HBsAg.[184]

In summary, then, the frequency of HBV infection in oncology patients is increased primarily as a result of exposure through transfused blood products. With more sensitive, contemporary blood screening for HBsAg, this route of exposure is less likely than was the case a decade or two ago. During chemotherapy in patients with HBV infection, levels of HBV replication tend to increase and withdrawal of chemotherapy may be associated with severe acute hepatitis, as immunologic competence is restored and virus-expressing hepatocytes are destroyed. As a rule, acute HBV infection acquired during chemotherapy or disease-induced immunosuppression tends to be mild and asymptomatic and chronicity of HBV infection can be anticipated in such patients. Finally, exceptions to all of these generalizations occur with some regularity.

3.3.2. Hepatitis C in Oncology Patients

HCV infection occurs in oncology patients, primarily as a result of exposure through transfusions[187,188,201–203] but is less well characterized in this group of patients than hepatitis B. Malone and Novak[201] found that in more than one half of a group of 31 children with acute leukemia and acute transfusion-associated hepatitis (18 type non-A, non-B, 13 type B), chronic hepatitis, primarily moderate-to-severe chronic hepatitis, developed. Hepatitis is undesirable in these patients not only because of the high frequency of chronic liver disease, but also because of the resulting limitation on the amount of additional chemotherapy that can be given. Unfortunately, relatively few data are available on the effect of cytotoxic chemotherapy on the course of hepatitis C. Poor survival due to the underlying malignancy makes the effect of an insidious, chronic disease like hepatitis C both difficult to measure and relatively less important. Unlike hepatitis B, acute flares of hepatitis or fibrosing cholestatic hepatitis have rarely been reported in patients with hepatitis C undergoing cytotoxic chemotherapy. Only one report has been published of fulminant hepatitis following withdrawal of chemotherapy in a patient with chronic HCV infection.[204]

3.4. Hepatitis in Patients with AIDS

3.4.1. Hepatitis B in Patients with AIDS

Because patients in high-risk groups for AIDS also are at high risk for other blood-borne viral infections, their exposure to hepatitis B is increased.[205] The acquired immunosuppression associated with human immunodeficiency virus (HIV) infection has the same impact on hepatitis B as other categories of immunocompromise. The level of replication and viremia tend to be very high, while the degree of hepatitis virus-induced liver injury tends to be blunted by impairment of host cytolytic T-cell activity.[206,207] Therefore, patients with HIV infection and a depressed CD4 level tend to harbor high levels of hepatitis B viremia but to remain relatively free of HBV-induced liver injury. Like other immunocompromised patients, HIV-infected patients are much more likely to acquire chronic infection following acute HBV infection, with a rate as high as 50%.[208] In addition, patients with AIDS have multiple potential factors that can lead to liver injury—atypical mycobacterial, fungal, and/or protozoan liver disease, CMV-induced liver injury and cholangiopathy, drug hepatotoxicity (antifungal drugs, antituberculous drugs, zidovudine, sulfonamides, protease inhibitors, etc.), fatty liver, peliosis, malignant infiltration (lymphoma, Kaposi's sarcoma), extrahepatic biliary obstruction, nonspecific cholestasis—providing an often confusing differential diagnosis.[209] Another potentially interesting interaction between HBV and HIV relates to the fact that the product of the X gene of HBV has been shown to transactivate transcription of HIV.[210]

Despite higher levels of hepatitis B viremia in patients with HIV infection, HIV coinfection does not seem to alter the clinical course of chronic hepatitis B; Scharschmidt et al.[211] found no difference in survival between patients with HIV/HBV coinfection and patients with HIV infection alone. An isolated case of fibrosing cholestatic hepatitis has been reported in the setting of HIV/HBV coinfection, however.[212] Finally, because of their level of immunosuppression, patients with AIDS do not respond to interferon therapy for chronic hepatitis B; both the antiviral and immunomodulatory effects of interferon appear to be necessary for interferon efficacy, and immunosuppressed AIDS patients are unlikely to be sufficiently immunologically competent to mount the cell-mediated immune attack on HBV-infected hepatocytes felt to be required for successful interferon therapy.[213,214] Persons with HIV infection who have a normal CD4 count may be treated with interferon, for this subgroup may retain the immunologic competency to respond to interferon. Recently, however, lamivudine has eclipsed interferon as a treatment for HIV/HBV coinfection. The activity of lamivudine against HBV, in fact, was initially discovered in a group of coinfected patients receiving lamivudine for HIV infection.[215,216] The potency of lamivudine against both viruses makes it an excellent option for treating coinfected individuals. Long-term data as to lamivudine's effect on HBV in this setting are lacking, however, and the propensity of HBV to develop resistance to this agent must be considered.

Delta hepatitis occurs in multiply transfused hemo-philiacs and in intravenous drug users, groups at high risk as well for HIV infection.[217,218] As is the case for HBV, delta hepatitis does not appear to have a dramatic clinical impact on this group, but detailed studies are lacking.

3.4.2. Hepatitis C in Patients with AIDS

In view of their shared routes of transmission, HCV and HIV infections commonly occur together. Until recently, complications of HIV infection had largely eclipsed those of HCV infection in coinfected patients. With the development of highly active antiretroviral therapy (HAART), however, patients with HIV infection now survive longer and HCV has become clinically important in HIV/HCV-coinfected patients. Recent evidence suggests that HIV coinfection worsens the course of chronic hepatitis C.[219–222] Coinfection with HIV and HCV has been reported to result in increased liver-related mortality and in more rapid progression to cirrhosis than infection with HCV alone.[219,220] Whereas the mean time from initial HCV infection to cirrhosis appears to be approximately 20 years in immunocompetent persons, the mean time to cirrhosis in HIV/HCV-coinfected persons was found in one study to be as short as 6.9 years.[223] Martin et al.[224] described three HIV-infected patients in whom transfusion-associated non-A, non-B hepatitis progressed from the onset of hepatitis to cirrhosis within 3 years. Similarly, several case reports confirm the strikingly rapid progression of liver disease in patients who acquire HIV and HCV simultaneously, culminating in death from liver failure within 3 years of infection.[225,226] Low CD4 counts ($<200/mm^3$) and advanced age at onset of HCV infection, in addition to simultaneous acquisition of HCV and HIV infections, have been suggested as variables associated with more rapid progression of HCV-associated liver disease in coinfected patients.[219] Rapidly progressive, fatal fibrosing cholestatic hepatitis resulting from HCV infection has been reported recently in two HIV-infected patients.[227]

4. Prevention

An increased exposure to blood-borne hepatitis viruses and a high likelihood of chronic infection and of highly replicative infection and infectivity are themes that have emerged often during the consideration of hepatitis in immunosuppressed persons. These observations translate into a high risk of infection among immunosuppressed patients and their contacts, primarily health workers and family members. A simple program of education, serologic surveillance, and commonsense precautionary measures can reduce markedly the frequency of new hepatitis virus infections in both staff and patients. For example, in hemodialysis units, recommended procedures include patient and staff education about the risks of hepatitis and its potential avenues of spread within high-risk units; periodic (every 1–3 months) screening of staff and patients for HBsAg and anti-HBs as well as for elevations of serum aminotransferase levels; segregation of HBsAg-positive from HBV-susceptible patients and designation of a separate dialysis machine, or even dialysis unit, for HBsAg-positive patients; assignment of anti-HBs-positive, i.e., immune personnel to care for HBsAg-positive patients; adherence to strict standards of personal and environmental hygiene; and chemical disinfection (with formalin, glutaraldehyde, hypochlorite, or iodophors) of surfaces contaminated by spilled blood or secretions. This approach has been credited with markedly reducing the frequency of hepatitis B as a hemodialysis-acquired infection.[55,56,58] Every attempt should be made to avoid sharing of secretions with or ungloved-hand exposure to blood and body fluids from patients with hepatitis. This admonition applies in the hospital and in the home of such patients. Certainly, organ donors should be screened for HBsAg and anti-HCV and organs from HBsAg-positive and anti-HCV-positive donors should not be used for transplantation, except in very unusual circumstances (e.g., the recipient is already infected with the virus).

Still, although precautions and physical barriers have made an important contribution to limiting the spread of HBV in high-risk areas (hemodialysis, transplantation, and oncology units), occasional cases of hepatitis B continue to surface.[54] In addition to general hygienic measures, immunoprophylaxis of susceptible patients, staff, and family members to prevent hepatitis B infection is recommended. Hepatitis B vaccine has been shown to be safe, immunogenic, and protective against HBV infection. Its use is recommended preexposure for patients and staff of hemodialysis, transplantation, and oncology units as well as for family members who share a household with a chronic HBsAg carrier. Three intramuscular (deltoid, not gluteal) injections of 1 ml, containing 10–20 mg of HBsAg (differs by manufacturer), are administered at time 0, 1, and 6 months to immunocompetent adults; half-dose injections are recommended for infants and children under the age of 10. In immunosuppressed persons, a double-dose regimen, in which each injection consists of 40 mg of HBsAg, is recommended.[228]

Unfortunately, despite the higher vaccine dose, immunocompromised persons do not respond optimally to hepatitis B vaccine. Whereas 95–97% of immunocompetent children and adults acquire protective anti-HBs after vaccination with standard vaccine doses and are protected, the immunogenicity of the vaccine is substantially reduced in hemodialysis patients given double the dose, down to approximately 60–80%.[229-232] Conflicting results have emerged from studies of the protective efficacy of hepatitis B vaccine in hemodialysis patients. The French (Pasteur)[233] and Dutch (Netherlands Red Cross)[234] vaccines were shown to be protective in controlled trials, while the vaccine prepared in the United States (Merck Sharp & Dohme) was not effective in this group of patients.[230] The discrepancy and failure to demonstrate vaccine efficacy in the American study may have resulted in part from the much lower than anticipated hepatitis B attack rate among placebo recipients in the participating American hemodialysis units or perhaps from the increased immunogenicity of the European vaccines, which are not subjected to as many inactivation steps as the American vaccine.[235] An adequate explanation for the differences in outcomes of the two sets of studies, however, is not established. Certainly, immunogenicity is lower in hemodialysis patients, and therefore protection of nonresponders to the vaccine would not be expected. In the same vein, the argument could be advanced that the European vaccines are slightly more immunogenic in hemodialysis patients, and therefore more protective. In the United States study, however, HBs antigenemia and even clinical hepatitis B occurred in vaccine *responders* with circulating anti-HBs documented prior to HBV infection.[230] In all likelihood, suboptimal immunogenicity and protective efficacy of currently available hepatitis B vaccines in hemodialysis patients should be anticipated. Higher rates of response may be achievable by vaccinating patients with end-stage renal disease before initiation of hemodialysis.[236]

Similarly, in renal transplant recipients, only 18–32% have been shown to acquire anti-HBs after double-dose vaccination, and the levels of anti-HBs achieved are substantially lower than those in immunocompetent persons. Moreover, the appearance of anti-HBs is often delayed and of limited durability.[231,235] An important point emanating from studies of hepatitis B vaccine in renal transplant patients, however, is the safety of vaccination in this group. Data from some transplant centers have suggested that anti-HBs-positive hemodialysis patients had an increased rate of graft rejection. Studies of hepatitis B vaccine in renal transplant recipients, however, have shown no association between vaccine-induced anti-HBs and graft rejection.[237]

In oncology patients, immunosuppressed both by their underlying diseases and by immunosuppressive, cytotoxic chemotherapy, responsiveness to hepatitis B vaccine appears to be linked with survival. Among surviving adults with solid tumors, the anti-HBs response rates to three 40-μg doses of vaccine have been shown to approach 70%; however, responsiveness was negligible, 9%, among those who did not survive, and even among survivors, antibody levels tended to be low and poorly sustained.[193] Similarly, only 25% of children with solid tumors have responded to three 40-μg doses of hepatitis B vaccine, four times the recommended dose in this age group.[238] Expected immunogenicity rates to hepatitis B vaccine in immunosuppressed groups relative to immunocompetent controls are shown in Fig. 4.[193,230,237,239] Whether vaccinated oncology patients in whom anti-HBs appears will be protected remains to be seen. Occasionally, even naturally acquired anti-HBs can become undetectable in oncology patients during cytotoxic chemotherapy.[182,189]

Despite these limitations, however, and because of the consequences of HBV infection in immunosuppressed persons, most authorities recommend prophylactic administration of three 40-μg doses of hepatitis B vaccine in these patients. Adequate prevention of hepatitis B in immunosuppressed hemodialysis and renal transplant patients probably will require vaccination of patients with chronic renal failure prior to transplantation, preferably even before initiating hemodialysis, as renal failure evolves. In oncology patients, the earlier that the vaccine is administered the better; for those who are likely to acquire exposure to blood products during the course of their therapy, the first dose of vaccine should be given before chemotherapy is begun.

For unvaccinated health workers without natural immunity to HBV in dialysis, transplant, and oncology units who sustain a percutaneous or transmucosal exposure to HBsAg-positive material, a combination of passive and active immunization is recommended (postexposure prophylaxis). Passive immunization with high-anti-HBs-titer hepatitis B immune globulin (HBIG), 0.06 ml/kg IM, should be accomplished as soon after the accidental exposure as possible. This should be accompanied or followed shortly thereafter by initiation of a complete course of hepatitis B vaccine.[240] Similarly, when immunosuppressed patients suffer an accidental, identifiable percutaneous HBV exposure, HBIG should be administered.

Efforts to prevent the spread of HCV infection in immunosuppressed patients, as in immunocompetent patients, have been hampered by inability to develop a vaccine against HCV. In settings such as dialysis centers, however, standard precautions should be adequate to prevent HCV transmission; neither periodic HCV testing in this population nor isolation of HCV-infected patients is recommended currently.[241] Screening of blood products and donor organs for HCV has reduced the incidence of HCV infection in patients with malignancies and those undergoing organ transplantation.

5. Summary

Because they are exposed to blood products, immunosuppressed patients have an increased risk of infection with the blood-borne hepatitis viruses, HBV and HCV. Included in the category of immunocompromised patients most likely to be encountered by clinicians are chronically hemodialyzed persons with end-stage renal failure, organ transplant recipients, oncology patients receiving cytotoxic chemotherapy, and persons with AIDS. Although the precise pathogenesis of viral hepatitis has not been defined, most evidence points toward a central role for cytolysis of virus-infected hepatocytes by host cellular immune mechanisms. Deficient in cellular immune competence, these patients are more likely not only to be exposed and infected, but also to remain chronically infected after acute infection, to maintain high levels of virus replication, and to be highly infectious for their

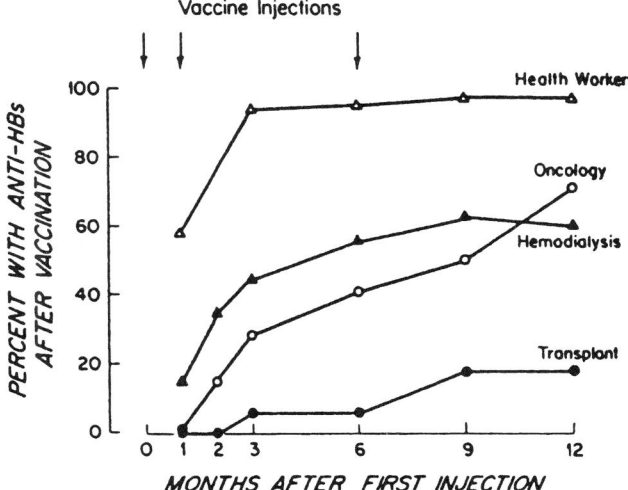

FIGURE 4. Relative immunogenicity of triply inactivated, plasma-derived hepatitis B vaccine (Hepatvax-B, Merck Sharp & Dohme) in renal transplant recipients,[187] oncology patients receiving chemotherapy,[147] and hemodialysis patients[180] who received three 40-μg injections compared with immunocompetent health workers[188] who received three 20-μg injections.

contacts. Despite the fact that acute infection may not be accompanied by severe illness, late life-threatening consequences of chronic hepatitis are observed in immunosuppressed patients. Even mild chronic hepatitis, which is more benign and less progressive in immunocompetent persons, can deteriorate to the more severe and progressive forms of chronic hepatitis in the immunosuppressed patient. Moreover, reactivation of hepatitis virus replication and clinical hepatitis can occur in immunocompromised persons. Such hepatitis reactivation can be especially severe during withdrawal of cytotoxic chemotherapy in oncology patients who are hepatitis B carriers, presumably as a result of sudden restoration of cellular immune cytotoxic potential. In renal transplant recipients, controversy exists over the impact of HBV and HCV infections on graft and patient survival. In liver transplant recipients, an ever-increasing burden of recurrent hepatitis C is being observed, and this may turn out over time to have grave consequences for graft and patient survival. The ramifications of viral hepatitis in HIV-infected patients are also being recognized with increasing frequency as survival in these patients is extended by contemporary antiretroviral regimens. Prevention of viral hepatitis in patients, their household contacts, and health workers is based on programs of mechanical intervention, education, surveillance, and other commonsense hygienic precautions. In addition, preexposure immunoprophylaxis with hepatitis B vaccine and postexposure immunoprophylaxis with a combination of hepatitis B immune globulin and vaccine for accidental inoculation are recommended.

References

1. Szmuness W, Purcell RH, Dienstag JL, et al: Antibody to hepatitis A antigen in institutionalized mentally retarded patients. JAMA 237:1702–1705, 1977.

2. Szmuness W, Dienstag JL, Purcell RH, et al: Hepatitis type A and hemodialysis: A seroepidemiologic study in 15 U.S. centers. Ann Intern Med 87:8–12, 1977.

3. Balayan MS: HEV infection: Historical perspectives, global epidemiology, and clinical features. In Hollinger FB, Lemon SM, Margolis H (eds): Viral Hepatitis and Liver Disease. Williams & Wilkins, Baltimore, 1991, pp. 498–501.

4. Umlauft F, Wong DT, Underhill PA, et al: Hepatitis G virus infection in hemodialysis patients and the effects of interferon treatment. Am J Gastroenterol 92:1986–1991, 1997.

5. Bizollon T, Guichard S, Ahmed SN, et al: Impact of hepatitis G virus co-infection on the course of hepatitis C virus infection before and after liver transplantation. J Hepatol 29:893–900, 1998.

6. Rey D, Fraize S, Vidinic J, et al: High prevalence of GB virus C/hepatitis G virus RNA in patients infected with human immunodeficiency virus. J Med Virol 57:75–79, 1999.

7. Popper H, Thung SN, Gerber MA, et al: Histologic studies of severe delta agent infection in Venezuelan Indians. Hepatology 3:906–912, 1983.

8. Friedman LS, Dienstag JL: The disease and its pathogenesis. In Gerety RJ (ed): Hepatitis A. Academic Press, New York, 1984, pp 55–79.

9. Lee WM: Hepatitis B virus infection. N Engl J Med 337:1733–1745, 1997.

10. Koziel MJ: The immunopathogenesis of hepatitis B infection. In Schinazi RF, Sommadossi JP, Thomas HC (eds): Therapies for Viral Hepatitis. International Medical Press, London, 1998, pp 53–64.

11. Horvath J, Raffanti SP: Clinical aspects of the interactions between human immunodeficiency virus and the hepatotropic viruses. Clin Infect Dis 18:339–347, 1994.

12. Blumberg BS, Sutnick AI, London WT: Australia antigen as a hepatitis virus: Variation in host response. Am J Med 48:1–8, 1970.

13. Blumberg BS, Gerstley BJS, Hungerford DA, et al: A serum antigen (Australia antigen) in Down's syndrome, leukemia and hepatitis. Ann Intern Med 66:924–931, 1967.

14. Blumberg BS, Sutnick AI, London WT: Hepatitis and leukemia: Their relation to Australia antigen. Bull NY Acad Med 44:1566–1586, 1968.

15. Szmuness W, Harley EJ, Ikram H, et al: Sociodemographic aspects of the epidemiology of hepatitis B. In Vyas GN, Cohen SN, Schmid R (eds): Viral Hepatitis. Franklin Institute, Philadelphia, 1978, pp 297–320.

16. Markenson JA, Gerety RJ, Hoofnagle JH, et al: Effects of cyclophosphamide on hepatitis B virus infection and challenge in chimpanzees. J Infect Dis 131:79–87, 1975.

17. Bird GL, Smith H, Portmann B, Alexander GH, Williams R: Acute liver decompensation on withdrawal of cytotoxic chemotherapy and immunosuppressive therapy in hepatitis B carriers. Q J Med 73:895–902, 1989.

18. Galbraith RM, Eddleston ALWF, Williams R, et al: Fulminant hepatic failure in leukemia and choriocarcinoma related to withdrawal to cytotoxic drug therapy. Lancet 2:528–530, 1975.

19. Thung SN, Gerber MA, Klion F, Gilbert H: Massive hepatic necrosis after chemotherapy withdrawal in a hepatitis B virus carrier. Arch Intern Med 145:1313–1314, 1985.

20. Lok ASF, Liang RHS, Chiu EKW, et al: Reactivation of hepatitis B virus replication in patients receiving cytotoxic therapy. Gastroenterology 100:182–188, 1991.

21. Perrillo R, Schiff E, Davis G, et al: A randomized, controlled trial of interferon alfa-2b alone and after prednisone withdrawal for the treatment of chronic hepatitis B. N Engl J Med 323:295–301, 1990.

22. Sagnelli E, Manzillo G, Maio G, et al: Serum levels of hepatitis B surface and core antigens during immunosuppressive treatment of HBsAg-positive chronic active hepatitis. Lancet 2:395–397, 1980.

23. Scullard GH, Smith CI, Merigan TC, et al: Effects of immunosuppressive therapy on viral markers in chronic active hepatitis B. Gastroenterology 81:987–991, 1981.

24. Villa E, Theodossi A, Portmann B, et al: Reactivation of hepatitis B virus infection in two patients: Immunofluorescence studies of liver tissue. Gastroenterology 80:1048–1053, 1981.

25. Chisari FV: Hepatitis B virus transgenic mice: Insights into the virus and the disease. Hepatology 22:1316–1325, 1995.

26. Moriyama T, Guilhot S, Klopchin K, et al: Immunobiology and pathogenesis of hepatocellular injury in hepatitis B virus transgenic mice. Science 248:361–364, 1990.

27. Reed WD, Eddleston ALWF, Cullens H, *et al*: Infusion of hepatitis B antibody in antigen-positive active chronic hepatitis. *Lancet* **2:**1347–1351, 1975.

28. Kohler PF, Trembath J, Merrill DA, *et al*: Immunotherapy with antibody, lymphocytes and transfer factor in chronic hepatitis B. *Clin Immunol Immunopathol* **2:**465–471, 1974.

29. Thomas HC, Chadwick RG, Jain S, *et al*: Levamisole therapy for HBs antigen positive chronic liver disease (abstract). *Gastroenterology* **73:**A52, 1977.

30. Jain S, Thomas HC, Sherlock S: Transfer factor in the attempted treatment of patients with HBsAg-positive chronic liver disease. *Clin Exp Immunol* **30:**10–15, 1977.

31. Good RA, Page AR: Fatal complications of virus hepatitis in two patients with agammaglobulinemia. *Am J Med* **29:**804–810, 1960.

32. Ferrari C, Penna A, Bertoletti A, *et al*: Cellular immune response to hepatitis B virus-encoded antigens in acute and chronic hepatitis B virus infection. *J Immunol* **145:**3442–3449, 1990.

33. Ferrrari C, Penna A, Giuberti T, *et al*: Intrahepatic, nucleocapsid antigen-specific T cells in chronic active hepatitis B. *J Immunol* **139:**2050–2058, 1987.

34. Barnaba V, Franco A, Alberti A, *et al*: Recognition of hepatitis B virus envelope proteins by liver-infiltrating lymphocytes in chronic HBV infection. *J Immunol* **143:**2650–2655, 1989.

35. Bertoletti A, D'Elios MM, Boni C, *et al*: Different cytokine profiles of intrahepatic T cells in chronic hepatitis B and hepatitis C virus infections. *Gastroenterology* **112:**193–199, 1997.

36. Romet-Lemonne JL, McLane MF, Elfassi E, *et al*: Hepatitis B virus infection in cultured human lymphoblastoid cells. *Science* **221:**667–669, 1983.

37. Ray R, Khanna A, Lagging LM, *et al*: Peptide immunogen mimicry of putative E1 glycoprotein-specific epitopes in hepatitis C virus. *J Virol* **68:**4420–4426, 1994.

38. Lohr HF, Schlaak JF, Kollmannsperger S, *et al*: Liver-infiltrating and circulating CD4+ T cells in chronic hepatitis C: Immunodominant epitopes, HLA restriction, and functional significance. *Liver* **16:**174–182, 1996.

39. Koziel MJ, Dudley D, Wong JT, *et al*: Intrahepatic cytotoxic T lymphocytes specific for hepatitis C virus in persons with chronic hepatitis. *J Immunol* **149:**3339–3344, 1992.

40. Shindo M, Arai K, Sokawa Y, Okuno T: The virological and histological states of anti-hepatitis C virus-positive subjects with normal liver biochemical values. *Hepatology* **22:**418–425, 1995.

41. Blumberg BS, Alter HJ, Visnich S: A "new" antigen in leukemia sera. *JAMA* **191:**541–546, 1965.

42. Goldblum SE, Reed WP: Host defenses and immunologic alterations associated with chronic hemodialysis. *Ann Intern Med* **93:**597–613, 1980.

43. Agarwal SS, Blumberg BS, Gerstley BS, *et al*: DNA polymerase activity as an index of lymphocyte stimulation: Studies in Down's syndrome. *J Clin Invest* **49:**161–169, 1970.

44. Boughton CR, Hawkes RA, Schroeter DR, *et al*: The epidemiology of hepatitis B in a residential institution for the mentally retarded. *Aust NZ J Med* **6:**521–529, 1976.

45. Hawkes RA, Boughton CR, Schroeter DR, *et al*: Hepatitis B infection in institutionalized Down's syndrome inmates: A longitudinal study with five hepatitis B virus markers. *Clin Exp Immunol* **40:**478–486, 1980.

46. Madden DL, Dietzman DE, Matthew EB, *et al*: Epidemiology of hepatitis B virus in an institution for mentally retarded persons. *Am J Ment Defic* **80:**369–375, 1976.

47. Chaudhary RK, Perry E, Cleary TE: Prevalence of hepatitis B infection among residents of an institution for the mentally retarded. *Am J Epidemiol* **105:**123–126, 1977.

48. Szmuness W, Pick R, Prince AM: The serum hepatitis virus specific antigen (SH): A preliminary report of epidemiologic studies in an institution for the mentally retarded. *Am J Epidemiol* **92:**51–61, 1970.

49. Hollingsworth DR, Hollingsworth JW, Roeckel I, *et al*: Immunologic reactions and Australia antigenemia in Down's syndrome. *J Chronic Dis* **27:**483–490, 1974.

50. Szmuness W, Neurath AR, Stevens CE, *et al*: Prevalence of hepatitis B "e" antigen and its antibody in various HBsAg carrier populations. *Am J Epidemiol* **113:**113–121, 1981.

51. Garibaldi RA, Forrest JN, Bryan JA, *et al*: Hemodialysis-associated hepatitis. *JAMA* **225:**384–389, 1973.

52. Snydman DR, Bryan JA, Hanson B: Hemodialysis-associated hepatitis in the United States—1972. *J Infect Dis* **132:**109–113, 1975.

53. Szmuness W, Prince AM, Grady GF, *et al*: Hepatitis B infection: A point-prevalence study in 15 US hemodialysis centers. *JAMA* **227:**901–906, 1974.

54. Centers for Disease Control and Prevention: Outbreaks of hepatitis B virus infection among hemodialysis patients—California, Nebraska, and Texas, 1994. *MMWR* **45:**285–289, 1996.

55. Alter MJ, Favero MS, Maynard JE: Impact of infection control strategies on the incidence of dialysis-associated hepatitis in the United States. *J Infect Dis* **153:**1149–1151, 1986.

56. Tokars JI, Miller E, Alter MJ, Arduino MJ: National surveillance of dialysis-associated diseases in the United States, 1995. *ASAIO J* **44:**98–107, 1998.

57. Alter MJ, Ahtone J, Maynard JE: Hepatitis B virus transmission associated with a multiple-dose vial in a hemodialysis unit. *Ann Intern Med* **99:**330–333, 1983.

58. Centers for Disease Control: Hepatitis—Control measures for hepatitis B in dialysis centers. Viral hepatitis investigations and control series. US Department of Health, Education and Welfare, Phoenix, November 1977, pp. 1–9.

59. Szumness W: Large-scale efficacy trials of hepatitis B vaccines in the USA: Baseline data and protocols. *J Med Virol* **4:**327–340, 1979.

60. Sengar DPS, Rashid A, McLeish WA, *et al*: Hepatitis B surface antigen (HBsAg) infection in a hemodialysis unit. II. Factors affecting host immune response to HBsAg. *Can Med Assoc J* **113:**945–948, 1975.

61. London WT, Drew JS, Lustbader ED, *et al*: Host responses to hepatitis B infection in patients in a chronic hemodialysis unit. *Kidney Int* **12:**51–58, 1977.

62. Snydman DS, Bregman D, Bryan JA: Hemodialysis-associated hepatitis in the United States, 1974. *J Infect Dis* **135:**687–691, 1977.

63. Ribot S, Rothstein M, Goldblat M, *et al*: Duration of hepatitis B surface antigenemia (HBsAg) in hemodialysis patients. *Arch Intern Med* **139:**178–180, 1979.

64. Gahl GM, Hess G, Arnold W, *et al*: Hepatitis B virus markers in 97 long-term hemodialysis patients. *Nephron* **24:**58–63, 1979.

65. Beorchia S, Trepo C, Betuel H, *et al*: Interest of HBeAg in the study of immunogenetic factors influencing HBV infection in hemodialysed and kidney transplanted patients. In Touraine JL, Traeger J, Betuel H, *et al* (eds): *Transplantation and Clinical Immunology*, Vol. 10. Excerpta Medica, Amsterdam, 1979, pp. 44–49.

66. Snydman DR, Bryan JA, Macon EJ, *et al*: Hemodialysis-associated hepatitis: Report of an epidemic with further evidence on mechanisms of transmission. *Am J Epidemiol* **104:**563–570, 1976.

67. Garibaldi RA, Hatch FE, Bisno AL, *et al*: Nonparenteral serum hepatitis: Report of an outbreak. *JAMA* **220:**963–966, 1972.

68. Coughlin GP, Van Deth AG, Disney APS, *et al*: Liver disease and the e antigen in HBsAg carriers with chronic renal failure. *Gut* **21:**118–122, 1980.

69. Degott C, Degos F, Jungers P, *et al*: Relationship between liver histopathological changes and HBsAg in 111 patients treated by long-term hemodialysis. *Liver* **3:**377–384, 1983.

70. Jungers P, Naret C, Degott C, *et al*: Histological and immunological survey of chronic active hepatitis in 650 hemodialyzed patients. In Touraine JL, Traeger J, Betuel H, *et al* (eds): *Transplantation and Clinical Immunology*, Vol. 10. Excerpta Medica, Amsterdam, 1979, pp. 38–43.

71. Galbraith RM, Dienstag JL, Purcell RH, *et al*: Non-A, non-B hepatitis associated with chronic liver disease in a hemodialysis unit. *Lancet* **1:**951–953, 1979.

72. Galbraith RM, Portmann B, Eddleston ALWF, *et al*: Chronic liver disease developing after outbreak of HBsAg-negative hepatitis in haemodialysis unit. *Lancet* **2:**886–890, 1975.

73. Marmion BP, Burrell CJ, Tonkin RW, *et al*: Dialysis-associated hepatitis in Edinburgh, 1969–1978. *Rev Infect Dis* **4:**619–637, 1982.

74. Coursaget P, Maupas P, Dubois F, *et al*: Hepatitis non-A, non-B chez six malades hemodialyses. *Nouv Presse Med* **7:**3515–3519, 1978.

75. Mery JP, Simon N, Courouce AM: Hepatite non-A, non-B chez les hemodialyses chroniques: 5 observations. *Nouv Presse Med* **8:**3973, 1979.

76. Moyer LA, Alter MJ: Hepatitis C virus in the hemodialysis setting: A review with recommendations for control. *Semin Dialysis* **7:**124–127, 1994.

77. Hardy NM, Sandroni S, Danielson S, Wildon WJ: Antibody to hepatitis C virus increases with time on hemodialysis. *Clin Nephrol* **38:**44–48, 1992.

78. Niu MT, Coleman PH, Alter MJ: Multicenter study of hepatitis C virus infection in chronic hemodialysis patients and hemodialysis center staff members. *Am J Kidney Dis* **22:**568–573, 1993.

79. Niu MT, Alter MJ, Kristensen C, Margolis HS: Outbreak of hemodialysis associated non-A, non-B hepatitis and correlation with antibody to hepatitis C virus. *Am J Kidney Dis* **19:**345–352, 1992.

80. Pereira BJG, Levey AS: Hepatitis C infection in dialysis and renal transplantation. *Kidney Int* **51:**981–999, 1997.

81. Valtuille R, Fernandez JL,Berridi J, *et al*: Evidence of hepatitis C virus passage across dialysis membrane. *Nephron* **80:**194–196, 1998.

82. Fabrizi F, Lunghi G, Andrulli S, *et al*: Influence of HCV viremia upon serum aminotransferase activity in chronic dialysis patients. *Nephrol Dial Transplant* **12:**1394–1398, 1997.

83. Al-Wakeel J, Malik GH, Al-Mohaya S, *et al*: Liver disease in dialysis patients with antibodies to hepatitis C virus. *Nephrol Dial Transplant* **11:**2265–2268, 1996.

84. Stehman-Breen CO, Emerson S, Gretch D, Johnson RJ: Risk of death among chronic dialysis patients infected with hepatitis C virus. *Am J Kidney Dis* **32:**629–634, 1998.

85. De Medina M, Ashby M, Schluter V, *et al*: Prevalence of hepatitis C and G virus infection in chronic hemodialysis patients. *Am J Kidney Dis* **31:**224–226, 1998.

86. Wolf JL, Perkins HA, Schreeder MT, *et al*: The transplanted kidney as a source of hepatitis B infection. *Ann Intern Med* **91:**412–413, 1979.

87. Lutwick LI, SyWassink JM, Corry RJ, *et al*: The renal transplant as a source of hepatitis B virus (HBV) (abstract). *Clin Res* **29:**258A, 1981.

88. Degos F, Degott C, Bedrossian J, *et al*: Is renal transplantation involved in posttransplantation liver disease? A prospective study. *Transplantation* **29:**100–102, 1980.

89. Toussaint C, Thiry L, Kinnaert P, *et al*: Prognostic significance of hepatitis B antigenemia in kidney transplantation. *Nephron* **17:**335–342, 1976.

90. Luby JP, Vurnett W, Hull AR, *et al*: Relationship between cytomegalovirus and hepatic function abnormalities in the period after renal transplant. *J Infect Dis* **129:**511–518, 1974.

91. Kaiser L, Kelly TJ, Patterson MJ, *et al*: Hepatitis B surface antigen in urine of renal transplant recipients. *Ann Intern Med* **94:**783–784, 1981.

92. Mayor GH, Kelly TJ, Hourani MR, *et al*: Intermittent hepatitis B surface antigenuria in a renal transplant recipient. *Am J Med* **68:**305–307, 1980.

93. Nagington J, Cossart YE, Cohen BJ: Reactivation of hepatitis B after transplantation operations. *Lancet* **1:**558–560, 1977.

94. Dusheiko G, Song E, Bowyer S, *et al*: Natural history of hepatitis B virus infection in renal transplant recipients—A fifteen-year follow-up. *Hepatology* **3:**330–336, 1983.

95. Marcellin R, Giostra E, Martinot-Peignoux M, *et al*: Redevelopment of hepatitis B surface antigen after renal transplantation. *Gastroenterology* **100:**1432–1434, 1991.

96. Pirson Y, Alexandre GPJ, van Ypersele de Strihou C: Long-term effect of HBs antigenemia on patient survival after renal transplantation. *N Engl J Med* **296:**194–196, 1977.

97. Hillis WD, Hillis A, Walker WG: Hepatitis B surface antigenemia in renal transplant recipients: Increased mortality risk. *JAMA* **242:**329–332, 1979.

98. Parfrey PS, Forbes RDC, Hutchinson TA, *et al*: The clinical and pathological course of hepatitis B liver disease in renal transplant recipients. *Transplantation* **37:**461–466, 1984.

99. Mathurin P, Mouquet C, Poynard T, *et al*: Impact of hepatitis B and C virus on kidney transplantation outcome. *Hepatology* **29:**257–263, 1999.

100. Anuras S, Piros J, Bonney WW, *et al*: Liver disease in renal transplant recipients. *Arch Intern Med* **137:**42–48, 1977.

101. Rao KV, Kasiske BL, Anderson WR: Variability in the morphological spectrum and clinical outcome of chronic liver disase in hepatitis B-positive and B-negative renal transplant recipients. *Transplantation* **51:**391–396, 1991.

102. Scott D, Mijch A, Lucas CR, *et al*: Hepatitis B and renal transplantation. *Transplant Proc* **19:**2159–2160, 1987.

103. Lam PWY, Wachs ME, Somberg KA, *et al*: Fibrosing cholestatic hepatitis in renal transplant recipients. *Transplantation* **61:**378–381, 1996.

104. Chen CH, Chen PJ, Chu JS, *et al*: Fibrosing cholestatic hepatitis in a hepatitis B surface antigen carrier after renal transplantation. *Gastroenterology* **107:**1514–1518, 1994.

105. Ware AJ, Luby JP, Hollinger B, *et al*: Etiology of liver disease in renal-transplant patients. *Ann Intern Med* **91:**364–371, 1979.

106. LaQuaglia MP, Tolkoff-Rubin NE, Dienstag JL, *et al*: Impact of hepatitis on renal transplantation. *Transplantation* **32:**504–507, 1981.

107. Chan TM, Lok ASF, Cheng IKP, Chan RT: A prospective study of hepatitis C virus infection among renal transplant recipients. *Gastroenterology* **104:**862–868, 1993.

108. Rhor MS, Lesniewski RR, Rubin CA, *et al*: Risk of liver disease in HCV-seropositive kidney transplant recipients. *Ann Surg* **217:** 512–517, 1993.

109. Genesca J, Vila J, Cordoba J, *et al*: Hepatitis C virus infection in renal transplant recipients: Epidemiology, clinical impact, serological confirmation, and viral replication. *J Hepatol* **22:**272–277, 1995.

110. Kazi S, Prasad S, Pollak R, *et al*: Hepatitis C infection in potential recipients with normal liver biochemistry does not preclude renal transplantation. *Dig Dis Sci* **39:**961–964, 1994.

111. Pereira BJG, Wright TL, Schmid CH, *et al*: The impact of pre-transplantation hepatitis C infection on the outcome of renal transplantation. *Transplantation* **60:**799–805, 1995.

112. Lau JYN, Davis GL, Brunson ME, *et al*: Hepatitis C virus infection in kidney transplant recipients. *Hepatology* **18:**1027–1031, 1993.

113. Roth D, Zucker K, Cirocco R, *et al*: A prospective study of hepatitis C virus infection in renal allograft recipients. *Transplantation* **61:**886–889, 1996.

114. Stempel CA, Lake J, Kuo G, *et al*: Hepatitis C—Its prevalence in end-stage renal failure patients and clinical course after kidney transplantation. *Transplantation* **55:**273–276, 1993.

115. Pol S, Legendre C, Saltiel C, *et al*: Hepatitis C virus in kidney recipients: Epidemiology and impact on renal transplantation. *J Hepatol* **15:**202–206, 1992.

116. Glicklich D, Thung SN, Kapoian T, *et al*: Comparison of clinical features and liver histology in hepatitis C-positive dialysis patients and renal transplant recipients. *Am J Gastroenterol* **94:**159–163, 1999.

117. Dellatdetsima JK, Boletis JN, Makris F, *et al*: Fibrosing cholestatic hepatitis in renal transplant recipients with HCV infection (abstract). *Hepatology* **26:**154A, 1997.

118. Rosenberg PM, Toth CM, Pascual M, *et al*: HCV-associated fibrosing cholestatic hepatitis after renal transplantation: Response to interferon alfa therapy. *Hepatology* **28:**846A, 1998.

119. Cruzado JM, Gil-Vernet S, Ercilla G, *et al*: Hepatitis C virus-associated membranoproliferative glomerulonephritis in renal allografts. *J Am Soc Nephrol* **7:**2469–2475, 1996.

120. Morales JM, Pascual-Capdevila J, Campistol JM, *et al*: Membranous glomerulonephritis associated with hepatitis C virus infection in renal transplant patients. *Transplantation* **63:**1634–1639, 1997.

121. Dickson RC, Everhart JE, Lake JR, *et al*: Transmission of hepatitis B by transplantation of liver from donors positive for antibody to hepatitis B core antigen. *Gastroenterology* **113:**1668–1673, 1997.

122. Wright TL, Lake JR: Liver transplantation for patients with hepatitis B: What have we learned from our results? *Hepatology* **13:**796–799, 1991.

123. Todo S, Demetris AJ, Van Thiel D, *et al*: Orthotopic liver transplantation for patients with hepatitis B virus-related liver disease. *Hepatology* **13:**619–626, 1991.

124. Eason JD, Freeman RB, Rohrer RJ, *et al*: Should liver transplantation be performed for patients with hepatitis B? *Transplantation* **57:**1588–1593, 1994.

125. Yoffe B, Burns DK, Bhatt HS, *et al*: Extrahepatic hepatitis B virus DNA sequences in patients with acute hepatitis B infection. *Hepatology* **12:**188–192, 1990.

126. Feray C, Zignego AL, Samuel D, *et al*: Persistent hepatitis B virus infection of mononuclear blood cells without concomitant liver infection: The liver transplantation model. *Transplantation* **49:** 1155–1158, 1991.

127. Samuel D, Bismuth A, Mathieu D, *et al*: Passive immunoprophylaxis after liver transplantation in HBsAg-positive patients. *Lancet* **337:**813–815, 1991.

128. Samuel D, Muller R, Alexander G, *et al*: Liver transplantation in European patients with the hepatitis B surface antigen. *N Engl J Med* **329:**1842–1847, 1993.

129. Miller RH, Kaneko S, Chung CT, *et al*: Compact organization of the hepatitis B virus genome. *Hepatology* **9:**322–327, 1989.

130. Farza H, Salmon AM, Hadchouel M, *et al*: Hepatitis B surface antigen gene expression is regulated by sex steroids and glucocorticoids in transgenic mice. *Proc Natl Acad Sci USA* **84:**1187–1191, 1987.

131. Davies SE, Portman BC, O'Grady JG, *et al*: Hepatic histological findings after transplantation for chronic hepatitis B virus infection, including a unique pattern of fibrosing cholestatic hepatitis. *Hepatology* **13:**150–157, 1991.

132. O'Grady JG, Smith HM, Davies SE, *et al*: Hepatitis B virus reinfection after orthotopic liver transplantation: Serological and clinical implications. *J Hepatol* **14:**104–111, 1992.

133. Lau JYN, Bain VG, Davies SE, *et al*: High-level expression of hepatitis B viral antigens in fibrosing cholestatic hepatitis. *Gastroenterology* **102:**956–962, 1992.

134. Chisari FV, Filippi P, Buras J, *et al*: Structural and pathological effects of synthesis of hepatitis B virus large envelope polypeptide in transgenic mice. *Proc Natl Acad Sci USA* **84:**6909–6913, 1987.

135. Roingeard P, Romet-Lemonne JL, Leturcq D, *et al*: Hepatitis B virus core antigen accumulation in an HBV nonproducer clone of HepG2–transfected cells is associated with cytopathic effect. *Virology* **179:**113–120, 1990.

136. Bock CT, Tillmann HL, Maschek HJ, *et al*: A pre-S mutation isolated from a patient with chronic hepatitis B infection leads to virus retention and misassembly. *Gastroenterology* **113:**1976–1982, 1997.

137. Demetris AJ, Jaffe R, Sheahan DG, *et al*: Recurrent hepatitis B in liver allograft recipients: Differentiation between viral hepatitis B and rejection. *Am J Pathol* **125:**161–172, 1986.

138. Carey W, Pimentel R, Westveer MK, *et al*: Failure of hepatitis B immunization in liver transplant recipients: Results of a prospective trial. *Am J Gastroenterol* **85:**1590–1592, 1990.

139. Rakela J, Wooten RS, Batts KP, *et al*: Failure of interferon to prevent recurrent hepatitis infection in hepatic allograft. *Mayo Clin Proc* **64:**429–432, 1989.

140. Neuhaus P, Steffen R, Blumhardt G, *et al*: Experience with immunoprophylaxis and interferon therapy after liver transplantation in HBsAg positive patients. *Transplant Proc* **23:**1522–1524, 1991.

141. Lauchart W, Muller R, Pichlmayr R: Immunoprophylaxis of hepatitis B reinfection in recipients of human liver allografts. *Transplant Proc* **19:**2387–2389, 1987.

142. Lauchart W, Muller R, Pichlmayr R: Long-term immunoprophylaxis of hepatitis B virus reinfection in recipients of human liver allografts. *Transplant Proc* **19:**4051–4053, 1987.

143. Mora NP, Klintmalm GB, Poplawski SS, *et al*: Recurrence of hepatitis B after transplantation: Does hepatitis B immunoglobulin modify the recurrent disease? *Transplant Proc* **22:**1547–1548, 1990.

144. Lemmens HP, Langrehr JM, Blumhardt G, *et al*: Outcome following orthotopic liver transplantation in HBsAg-positive patients using short- or long-term immunoprophylaxis. *Transplant Proc* **26:**3622–3623, 1994.

145. Langrehr JM, Lemmens HP, Keck H, *et al*: Liver transplantation in

hepatitis B surface antigen positive patients with postoperative long-term immunoprophylaxis. *Transplant Proc* **27**:1215–1216, 1995.

146. Grazi GL, Mazziotti A, Sama C, *et al*: Liver transplantation in HBsAg-positive HBV DNA-negative cirrhotics: Immunoprophylaxis and long-term outcome. *Liver Transplant Surg* **2**:418–425, 1996.

147. Terrault NA, Zhou S, Combs C, *et al*: Prophylaxis in liver transplant recipients using a fixed dosing schedule of hepatitis B immunoglobulin. *Hepatology* **24**:1327–1333, 1996.

148. Grellier L, Mutimer D, Ahmed M, *et al*: Lamivudine prophylaxis against reinfection in liver transplantation for hepatitis B cirrhosis. *Lancet* **348**:1212–1215, 1996.

149. Nery JR, Weppler D, Rodriguez M, *et al*: Efficacy of lamivudine in controlling hepatitis B virus recurrence after liver transplantation. *Transplantation* **65**:1615–1621, 1998.

150. Bain VG, Kneteman NM, Ma MM, *et al*: Efficacy of lamivudine in chronic hepatitis B patients with active viral replication and decompensated cirrhosis requiring liver transplantation. *Transplantation* **62**:1456–1462, 1996.

151. Perillo R, Rakela J, Martin P, *et al*: Lamivudine for suppression and/or prevention of hepatitis B when given pre/post liver transplantation (abstract). *Hepatology* **26**:260A, 1997.

152. Tipples GA, Ma MM, Rischer KP, *et al*: Mutation in HBV RNA-dependent DNA polymerase confers resistance to lamivudine in vivo. *Hepatology* **24**:714–717, 1996.

153. Ling R, Mutimer D, Ahmed M, *et al*: Selection of mutations in the hepatitis B virus polymerase during therapy of transplant recipients with lamivudine. *Hepatology* **24**:711–713, 1996.

154. Bartholomew MM, Jansen RW, Jeffers LJ, *et al*: Hepatitis B virus resistance to lamivudine given for recurrent infection after orthotopic liver transplantation. *Lancet* **349**:20–22, 1997.

155. Markowitz JS, Martin P, Conrad AJ, *et al*: Prophylaxis against hepatitis B recurrence following liver transplantation using combination lamivudine and hepatitis B immune globulin. *Hepatology* **28**:585–589, 1998.

156. Ottobrelli A, Marzano A, Smedile A, *et al*: Patterns of hepatitis delta virus reinfection and disease in liver transplantation. *Gastroenterology* **101**:1649–1655, 1991.

157. Read AE, Donegan E, Lake J, *et al*: Hepatitis C in patients undergoing liver transplantation. *Ann Intern Med* **114**:282–284, 1991.

158. Martin P, Munoz SJ, Di Bisceglie AM, *et al*: Recurrence of hepatitis C infection following orthotopic liver transplantation. *Hepatology* **13**:719–721, 1991.

159. Wright TL, Hsu H, Greenberg H, *et al*: Recurrence of hepatitis C viral (HCV) infection and disease following liver transplantation (OLTx) (abstract). *Gastroenterology* **100**:812, 1991.

160. Poterucha J, Rakela J, Tarwell H, *et al*: Diagnosis of chronic hepatitis C after liver transplantation using polymerase chain reaction (abstract). *Gastroenterology* **100**:786, 1991.

161. Chung RT, Katkov WN, Dienstag JL, *et al*: Dynamics of hepatitis C viremia in patients before and after orthotopic liver transplantation (abstract). *Gastroenterology* **100**:730, 1991.

162. Wright TL, Donegan E, Hsu HH, *et al*: Recurrent and acquired hepatitis C viral infection in liver transplant recipients. *Gastroenterology* **103**:317–322, 1992.

163. Shah G, Demetris AJ, Gavaler JS, *et al*: Incidence, prevalence, and clinical course of hepatitis C following liver transplantation. *Gastroenterology* **103**:323–329, 1992.

164. Fukumoto T, Berg T, Ku Y, *et al*: Viral dynamics of hepatitis C early after orthotopic liver transplantation: Evidence for rapid turnover of serum virions. *Hepatology* **24**:1351–1354, 1996.

165. Gane EJ, Naoumov NV, Qian KP, *et al*: A longitudinal analysis of hepatitis C virus replication following liver transplantation. *Gastroenterology* **110**:167–177, 1996.

166. Feray C, Gigou M, Samuel D, *et al*: The course of hepatitis C virus infection after liver transplantation. *Hepatology* **20**:1137–1143, 1994.

167. Gane EJ, Portmann BC, Naoumov NV, *et al*: Long-term outcome of hepatitis C infection after liver transplantation. *N Engl J Med* **334**:815–820, 1996.

168. Boker KHW, Dalley G, Bahr MJ, *et al*: Long-term outcome of hepatitis C virus infection after liver transplantation. *Hepatology* **25**:203–210, 1997.

169. Schluger LK, Sheiner PA, Thung SN, *et al*: Severe recurrent cholestatic hepatitis C following orthotopic liver transplantation. *Hepatology* **23**:971–976, 1996.

170. Dickson RC, Caldwell SH, Ishitani MB, *et al*: Clinical and histologic patterns of early graft failure due to recurrent hepatitis C in four patients after liver transplantation. *Transplantation* **61**:701–705, 1996.

171. Davis GL, Balart LA, Schiff ER, *et al*: Treatment of chronic hepatitis C with recombinant interferon alfa: A multicenter randomized, controlled trial. *N Engl J Med* **321**:1501–1506, 1989.

172. Wright TL, Combs C, Kim M, *et al*: Interferon alfa therapy for hepatitis C virus infection after liver transplantation. *Hepatology* **20**:773–779, 1994.

173. Feray C, Samuel D, Gigou M, *et al*: An open trial of interferon alfa recombinant for hepatitis C after liver transplantation: Antiviral effects and risk of rejection. *Hepatology* **22**:1084–1089, 1995.

174. Vargas V, Charco R, Castells L, Esteban R, Margarit C: Alpha-interferon for acute hepatitis C in liver transplant patients. *Transplant Proc* **27**:1222–1223, 1995.

175. Cattral MS, Krajden M, Wanless IR, *et al*: A pilot study of ribavirin therapy for recurrent hepatitis C virus infection after liver transplantation. *Transplantation* **61**:1483–1488, 1996.

176. Aljumah AA, Cattral MS, Greig PD, *et al*: Long-term ribavirin therapy for recurrent hepatitis C after liver transplantation. *Transplant Proc* **29**:514, 1997.

177. Feray C, Roche B, Dussaiz E, *et al*: An open trial of ribavirin in chronic HCV infection after liver transplantation (abstract). *Hepatology* **26**:157A, 1997.

178. Bizollon T, Palazzo U, Ducerf C, *et al*: Pilot study of the combination of interferon alfa and ribavirin as therapy of recurrent hepatitis C after liver transplantation. *Hepatology* **26**:500–504, 1997.

179. Sutnick AI, London WT, Blumberg BS, *et al*: Austalia antigen (a hepatitis associated antigen) in leukemia. *J Natl Cancer Inst* **44**:1241–1249, 1970.

180. Sutnick AI, Levine PH, London WT, *et al*: Frequency of Australia antigen in patients with leukemia in different countries. *Lancet* **1**:1200–1202, 1971.

181. Grange MJ, Erlinger S, Teilletd F, *et al*: A possible relationship to treatment between hepatitis-associated antigen and chronic persistent hepatitis in Hodgkin's disease. *Gut* **14**:433–437, 1973.

182. Wands JR, Chura CM, Roll FJ, *et al*: Serial studies of hepatitis-associated antigen and antibody in patients receiving antitumor chemotherapy for myeloproliferative and lymphoproliferative disorders. *Gastroenterology* **68**:105–112, 1975.

183. Wands JR, Walker JA, Davis TT, *et al*: Hepatitis B in an oncology unit. *N Engl J Med* **291**:1371–1375, 1974.

184. Schulman AN, Fagen ND, Brezina M, et al: HBe-antigen in the course and prognosis of hepatitis B infection: A prospective study. Gastroenterology 78:253–258, 1980.

185. Trinchet JC, Beugrand M, Hecht Y, et al: Hépatite fulminante à virus B survenue au cours d'un traitement immunolodepresseur. Gastroenterol Clin Biol 4:59–62, 1980.

186. Hoofnagle JH, Dusheiko GM, Schafer DF, et al: Reactivation of chronic hepatitis B virus infection by cancer chemotherapy. Ann Intern Med 96:447–449, 1982.

187. Vergani D, Locasciulli A, Masera G, et al: Histological evidence of hepatitis-B-virus infection with negative serology in children with acute leukemia who develop chronic liver disease. Lancet 1:361–364, 1982.

188. Wade JC, Gaffey M, Wiernik PH, et al: Hepatitis in patients with acute nonlymphocytic leukemia. Am J Med 75:413–422, 1983.

189. Locasciulli A, Santamaria M, Masera G, et al: Hepatitis B virus markers in children with acute leukemia: The effect of chemotherapy. J Med Virol 15:29–33, 1985.

190. Cowan DH, Kouroupis GM, Leers W-D: Occurrence of hepatitis and hepatitis B surface antigen in adult patients with acute leukemia. Can Med Assoc J 112:693–697, 1975.

191. Sauerbruch T, Frosner G, Theml H, et al: Hepatitis B-virus-marker and rotelnantikorper bei patienten mit hodgkin-non-hodgkin-lymphomen und allgemein internistischen erkrankungen. Blut 40:259–266, 1980.

192. Tabor E, Gerety RJ, Mott M, et al: Prevalence of hepatitis B in a high risk setting: A serologic study of patients and staff in a pediatric oncology unit. Pediatrics 61:711–715, 1978.

193. Weitberg AG, Weitzman SA, Watkins E, et al: Immunogenicity of hepatitis B vaccine in oncology patients receiving chemotherapy. J Clin Oncol 3:718–722, 1985.

194. Schulman AN, Fagen ND, Ling CM, et al: Repeated type B hepatitis infections: Ten cases studied prospectively (abstract). Gastroenterology 72:1182, 1977.

195. Lightdale CJ, Ikram H, Pinsky C: Primary hepatocellular carcinoma with hepatitis B antigenemia: Effects of chemotherapy. Cancer 46:1117–1122, 1980.

196. Steinberg SC, Alter HJ, Leventhal BG: The risk of hepatitis transmission to family contacts of leukemia patients. J Pediatr 87:753–757, 1975.

197. Wands JR: Subacute and chronic active hepatitis after withdrawal of chemotherapy (letter). Lancet 2:979, 1975.

198. Thung SN, Gerber MA, Klion F, et al: Massive hepatic necrosis after chemotherapy withdrawal in a hepatitis B virus carrier. Arch Intern Med 145:1313–1314, 1985.

199. Lau JYN, Lai CL, Lin HJ, et al: Fatal reactivation of chronic hepatitis B virus infection following withdrawal of chemotherapy in lymphoma patients. Q J Med 73:911–917, 1989.

200. Lok ASF, Liang RHS, Chiu EKW, et al: Reactivation of hepatitis B virus replication in patients receiving cytotoxic therapy. Gastroenterology 100:182–188, 1991.

201. Malone W, Novak R: Outcome of hepatitis in children with acute leukemia. Am J Dis Child 134:584–587, 1980.

202. Locasciulli A, Alberti A, Barbieri R, et al: Evidence of non-A, non-B hepatitis in children with acute leukemia and chronic liver disease. Am J Dis Child 173:354–356, 1983.

203. Barton JB, Conrad ME: Beneficial effects of hepatitis in patients with acute myelogenous leukemia. Ann Intern Med 90:188–190, 1979.

204. Vento S, Cainelli F, Mirandola F, et al: Fulminant hepatitis on withdrawal of chemotherapy in carriers of hepatitis C virus. Lancet 347:92–93, 1996.

205. De Cock KM, Niland JC, Lu HP, et al: Experience with human immunodeficiency virus infection in patients with hepatitis B virus and hepatitis delta virus infections in Los Angeles. Am J Epidemiol 127:1250–1260, 1988.

206. Trent Mills C, Lee E, Perrillo R: Relationship between histology, aminotransferase levels, and viral replication in chronic hepatitis B. Gastroenterology 99:519–524, 1990.

207. Rector WG, Govindarajan S, Horsburgh CR, et al: Hepatic inflammation, hepatitis B replication, and cellular immune function in homosexual males with chronic hepatitis B and antibody to human immunodeficiency virus. Am J Gastroenterol 83:262–266, 1988.

208. Krogsgaard K, Lindhardt B, Nielson J: The influence of HTLV-III infection on the natural history of hepatitis B virus infection in male homosexual HBsAg carriers. Hepatology 7:37–41, 1987.

209. Bonacini M: Hepatobiliary complications in patients with human immunodeficiency virus infection. Am J Med 92:404–411, 1992.

210. Seto E, Benedict Yen TS, Matija Peterlin B, et al: Transactivation of the human immunodeficiency virus long terminal repeat by the hepatitis B virus X protein. Proc Natl Acad Sci USA 85:8286–8290, 1988.

211. Scharschmidt BF, Held MJ, Hollander HH, et al: Hepatitis B in patients with HIV infection: Relationship to AIDS and patient survival. Ann Intern Med 117:837–838, 1992.

212. Fang JWS, Wright TL, Lau JYN: Fibrosing cholestatic hepatitis in a patient with HIV and hepatitis B. Lancet 342:1175, 1993.

213. McDonald JA, Caruso L, Karayiannis P, et al: Diminished responsiveness of male homosexual chronic hepatitis B carriers with HTLV-III antibodies to recombinant alfa-interferon. Hepatology 4:719–723, 1987.

214. Wong D, Colina Y, Naylor C: Interferon alfa treatment of chronic hepatitis B: Randomized trial in a predominantly homosexual male population. Gastroenterology 108:165–171, 1996.

215. Benhamou Y, Katlama C, Lunel F, et al: Effects of lamivudine on replication of hepatitis B virus in HIV-infected men. Ann Intern Med 125:705–712, 1996.

216. Schnittman SM, Pierce PF: Potential role of lamivudine in the clearance of chronic hepatitis B virus infection in patients co-infected with human immunodeficiency virus type 1. Clin Infect Dis 23:638–639, 1996.

217. Novick DM, Farci P, Croxson TS, et al: Hepatitis D virus and human immunodeficiency virus antibodies in parenteral drug abusers who are hepatitis B surface antigen positive. J Infect Dis 158:795–803, 1988.

218. Salomon RE, Kaslow RA, Phair JP, et al: Human immunodeficiency virus and hepatitis delta virus in homosexual men. A study of four cohorts. Ann Intern Med 108:51–54, 1988.

219. Collier J, Heathcote J: Hepatitis C viral infection in the immunosuppressed patient. Hepatology 27:2–6, 1998.

220. Eyster ME, Diamondstone LS, Lien JM, et al: Natural history of hepatitis C virus infection in multitransfused hemophiliacs: Effect of coinfection with human immunodeficiency virus. J AIDS 6:602–610, 1993.

221. Bierhoff E, Fischer HP, Pffifer U, et al: Hepatitis und posthepatitische Zirrhose bei AIDS. Verh Dtsch Ges Path 79:249–253, 1995.

222. Rockstroh JK, Spengler U, Sudhop T, et al: Immunosuppression may lead to progression of hepatitis C virus-associated liver dis-

ease in hemophiliacs coinfected with HIV. *Am J Gastoenterol* **91:**2563–2568, 1996.

223. Soto B, Sanchez-Quijano A, Rodrigo L, *et al*: Human immunodeficiency virus infection modifies the natural history of chronic parenterally-acquired hepatitis C with an unusually rapid progression to cirrhosis. *J Hepatol* **26:**1–5, 1997.

224. Martin P, DiBisceglie AM, Kassianides C, *et al*: Rapidly progressive non-A, non-B hepatitis in patients with human immunodeficiency virus infection. *Gastroenterology* **97:**1559–1561, 1989.

225. Ridzon R, Gallagher K, Ciesielski C, *et al*: Simultaneous transmission of human immunodeficiency virus and hepatitis C virus from a needle-stick injury. *N Engl J Med* **336:**919–922, 1997.

226. Bjoro K, Froland SS, Yun Z, *et al*: Hepatitis C infection in patients with primary hypogammaglobulinemia after treatment with contaminated immune globulin. *N Engl J Med* **331:**1607–1611, 1994.

227. Rosenberg PM, Graeme-Cook FM, Farrell JJ, *et al*: HCV-induced fibrosing cholestatic hepatitis in HIV infection: A newly recognized syndrome. *Am J Gastroenterol*, **97**(2)**:** in press, 2002.

228. Immunization Practices Advisory Committee: Recommendations for protection against viral hepatitis. *MMWR* **34:**313–324, 329–335, 1985.

229. Stevens CE, Szmuness W, Goodman AI, *et al*: Hepatitis B vaccine: Immune response in hemodialysis patients. *Lancet* **2:**1211–1213, 1980.

230. Stevens CE, Alter HJ, Taylor PE, *et al*: Hepatitis B vaccine in patients receiving hemodialysis: Immunogenicity and efficacy. *N Engl J Med* **311:**496–501, 1984.

231. Grob PJ, Binswanger U, Zaruba K, *et al*: Immunogenicity of a hepatitis B subunit vaccine in hemodialysis and in renal transplant recipients. *Antiviral Res* **3:**43–52, 1983.

232. Bergamini F, Zanetti AR, Ferroni P, *et al*: Immune response to hepatitis B vaccine in staff and patients in renal dialysis units. *J Infect* **7**(Suppl 1)**:**35–40, 1983.

233. Crosnier J, Jungers P, Courouce A-M, *et al*: Randomized placebo-controlled trial of hepatitis B surface antigen vaccine in French haemodialysis units. II. Haemodialysis patients. *Lancet* **2:**797–800, 1981.

234. Desmyter J, Colaert J, De Groot G, *et al*: Efficacy of heat-inactivated hepatitis B vaccine in hemodialysis patients and staff: Double-blind placebo-controlled trial. *Lancet* **2:**1323–1327, 1983.

235. Desmyter J, Colaert J: Comparative immunogenicity of MSD, Pasteur and CLB hepatitis B vaccines in 245 hemodialysis patients (abstract). In Vyas GN, Dienstag JL, Hoofnagle JH (eds): *Viral Hepatitis and Liver Disease*. Grune & Stratton, New York, 1984, pp. 709–710.

236. Seaworth B, Drucker J, Starling J, *et al*: Hepatitis B vaccines in patients with chronic renal failure before dialysis. *J Infect Dis* **157:**332–337, 1988.

237. Jacobson IM, Jaffers G, Dienstag JL, *et al*: Immunogenicity of hepatitis B vaccine in renal transplant recipients. *Transplantation* **39:**393–395, 1985.

238. Arnold W, Baumann W: Vaccination of children with malignant diseases with alum-absorbed hepatitis B vaccine—Immunogenicity studies. *Scand J Infect Dis* **38**(Suppl)**:**33–36, 1983.

239. Dienstag JL, Werner BF, Polk BF, *et al*: Hepatitis B vaccine in health care personnel: Safety, immunogenicity, and indicators of efficacy. *Ann Intern Med* **101:**34–40, 1984.

240. Centers for Disease Control: Postexposure prophylaxis of hepatitis B: Recommendations of the Immunization Practices Advisory Committee. *Ann Intern Med* **101:**351–354, 1984.

241. Kellerman S, Alter MJ: Preventing hepatitis B and hepatitis C virus infections in end-stage renal disease patients: Back to basics (editorial). *Hepatology* **29:**291, 1999.

10

The Herpesviruses

PAUL D. GRIFFITHS

1. General Characteristics of Herpesviruses

"Herpesviruses have always been with us, will always remain with us, and we must learn to live with them."

1.1. Herpesviruses Have Coevolved with Humans

The aphorism cited above can be justified by the results of detailed molecular biologic studies tracking the rate of genetic change in human and animal herpesviruses.[1] Back-projection indicates that a putative ancestral herpesvirus was present around 200 million years ago (see Fig. 1). The persistence of herpesviruses since that time, despite our acquisition of sophisticated immune mechanisms, indicates that they must be among the most successful group of viruses ever to infect humans. To achieve this, herpesviruses have evolved ways of interacting with human cells and avoiding immune responses, some of which are discussed later, and whose study has educated us about previously unknown aspects of cell biology. Note that success is defined as transmission of virus between individuals so that they persist; the ability to induce disease is not a prerequisite for success and may indeed be detrimental if people with obvious symptoms are shunned by potential contacts. Indeed, it is likely that herpesviruses have evolved strategies to persist while causing minimal disturbance to their hosts by means of a standoff between the virus's replicative ability and the

host's immune responses. Thus, when viewed from the enormous time scale in Fig. 1, the current problems caused by herpesviruses may be attributed to very recent changes in medical practice, particularly transplantation, together with emergence of the human immunodeficiency virus (HIV) pandemic, which have each produced large numbers of individuals with impaired immune responses.

There are eight herpesviruses that infect humans, classified into the three subfamilies alpha-, beta-, and gamma-Herpesvirinae. Table 1 shows both their common and systematic names. These subfamilies were originally defined on biological criteria. Thus, alphaherpesviruses are tropic for neural cells and replicate rapidly in cell culture; betaherpesviruses replicate slowly in cells of fibroblast origin, remaining largely cell associated; and gammaherpesviruses are tropic for lymphocytes and are potentially oncogenic. More recently, these biological differences were shown to segregate with genetic diversity (see Fig. 1) and so the classification system has been retained, with new members allocated to subfamilies solely on molecular biological criteria (reviewed in Ref. 2).

1.1.1. Discovery of Human Herpesviruses

Figure 2 shows the years when each human herpesvirus was first propagated, together with the techniques used.

The herpes simplex viruses (HSV) are the most difficult to pinpoint in history. They appear to have been propagated in the 1920s using the then-current approach of inoculation into experimental animals. However, it was not until the 1960s that HSV-1 was clearly distinguished from HSV-2 using serologic techniques.[2] The availability of the "new" technology in the form of cell cultures rapidly led to the identification of cytomegalovirus (CMV)[3]

Paul D. Griffiths • Department of Virology, Royal Free and University College School of Medicine, London NW3 2PF, England.

Clinical Approach to Infection in the Compromised Host (Fourth Edition), edited by Robert H. Rubin and Lowell S. Young. Kluwer Academic/Plenum Publishers, New York, 2002.

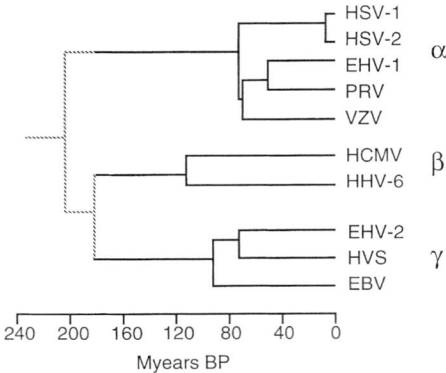

FIGURE 1. Evolutionary time scale of herpesviruses. HVS, Herpesvirus saimiri; EHV, equine herpesvirus; PRV, pseudorabies virus; M years BP, millions of years before present. (Reprinted with permission from McGeoch *et al.*[1])

and varicella–zoster virus (VZV)[4] in 1956 and 1958, respectively.

The discovery of Epstein–Barr virus (EBV) is a unique story. A surgeon, Dennis Burkitt, was in Africa studying the epidemiology of the tumor that was to bear his name. He observed that Burkitt's lymphoma did not occur above a certain altitude and hypothesized that an insect-transmitted virus may be the cause. He sent biopsy material to Epstein, who propagated the cells *in vitro* and visualized a herpesvirus by electron microscopy in 1964.[5] Virus-infected cells were used to detect immunoglobulin G (IgG) antibodies and surprisingly EBV was shown to be a common infection worldwide. A technician in Henle's laboratory in Philadelphia acted as a negative control for EBV serology until she returned to work after an illness and was found to have seroconverted to EBV. Her illness was infectious mononucleosis and this seren-

dipitous observation identified the most common disease association of EBV (reviewed in Ref. 6). A quarter of a century later, the "new" technology that allowed lymphocytes to be propagated in the laboratory led to the discovery[7] of human herpesvirus 6 (HHV-6) (during a search for viruses associated with lymphadenopathy) and HHV-7 (from a healthy member of laboratory staff).[8] In 1994, the "new" technique of representational difference analysis allowed a segment of the HHV-8 genome to be identified,[9] leading rapidly to sequencing of the entire molecule.[10]

It will be evident from this truncated history that herpesviruses are common and that our ability to detect their presence is limited by the availability of specialized techniques. Given the time scales in Figs. 1 and 2, only a brave investigator would state categorically that we have discovered all of the herpesviruses that can infect humans. Clearly, the study of herpesviruses is a rapidly moving subject and I will attempt in this chapter to emphasize basic principles and refer readers to seminal articles and recent reviews as a way of summarizing the vast amount of information available.

1.2. Herpesviruses Have Large Complex Genomes

Herpesviruses possess some of the largest genomes found in viruses. The genomes of different viruses are arranged in distinct ways according to the topology of repeat regions and the presence of inverted segments that produce genetic isomers. These isomers have no pathogenic significance but can be useful in typing individual strains from patients.

Strains of each of the eight human herpesviruses have been sequenced in their entirety and the number of open reading frames (or genes) predicted. The function of these genes will be discussed later. The nomenclature used to describe these genes can be arcane and confusing. Wherever possible, this chapter will employ the system of numbering genes sequentially in particular segments (e.g., UL30 is the 30th gene in the unique long region).

1.2.1. Genome Structures/Nomenclature

As shown in Fig. 3, all herpesviruses have repeat sequences at their termini. These allow the genome to circularize and be copied via rolling circle replication, but also can facilitate virus evolution through recombination events (reviewed in Ref. 11). Repeat sequences of varying lengths also are found internal to the termini. These again can facilitate virus evolution and can be helpful in typing the origin of particular virus strains.

TABLE 1. The Human Herpesviruses

Subfamilies and common names	Systematic nomenclature
Alphaherpesviruses	
Herpes simplex virus type 1	HHV-1
Herpes simplex virus type 2	HHV-2
Varicella–zoster virus	HHV-3
Betaherpesviruses	
Human cytomegalovirus	HHV-5
Human herpesvirus type 6	HHV-6
Human herpesvirus type 7	HHV-7
Gammaherpesviruses	
Epstein–Barr virus	HHV-4
Human herpesvirus type 8	HHV-8

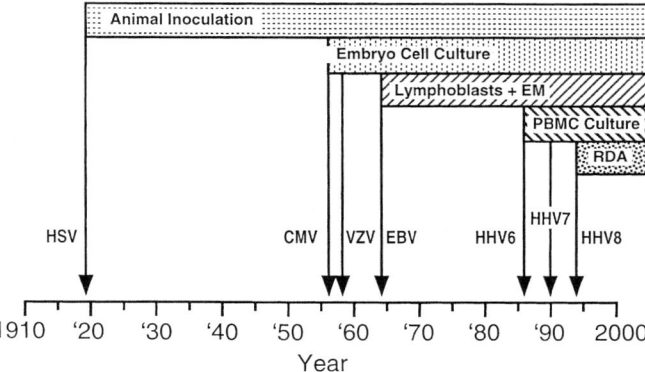

FIGURE 2. Year of first propagation of human herpesviruses together with methods used. RDA, representational difference analysis; EM, electron microscopy.

The regions of the genomes that do not contain repeat sequences are termed unique. For four viruses, two unique (U) regions [termed long (L) and short (S)] are present, each flanked by terminal repeats. When the VZV genome is copied, the US can be in either orientation leading to two possible isomeric forms. For the HSV-1, HSV-2, and CMV genomes, both the US and UL regions can invert, leading to four possible isomeric forms. Each isomer is biologically identical but can be differentiated by typing methods.

1.2.2. Cascade Sequence of Gene Expression

The genomes of herpesviruses bring the potential of introducing a large number of virus-encoded proteins into the environment of the infected cell. This potential is actively regulated by expressing preferentially those gene products that are required at particular stages of the virus's intracellular life cycle. These stages are divided into alpha, beta, or gamma (known alternatively as immediate-early, early, and late, respectively) and summarized in Fig. 4.

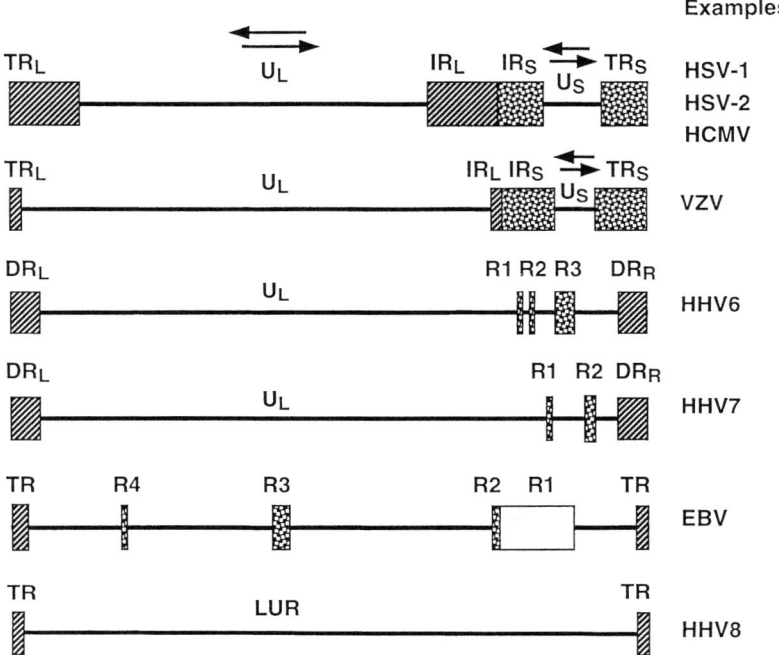

FIGURE 3. Genomic organization of human herpesviruses. Each genome is drawn to a standard length with repeat sequences shown as boxes (R1, R2, ...). Horizontal arrows indicate regions which can invert, producing genomic isomers. U, unique; L, long; S, short; IR, inverted repeat; TR, terminal repeat; LUR, long unique region. (Figure drawn from data provided in Roizman,[2] Mocarski,[197] Gompels,[204] Nicholas,[205] and Russo.[10])

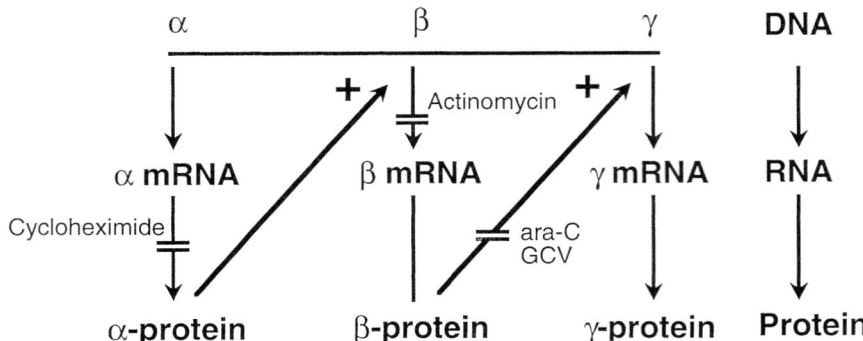

FIGURE 4. Cascade of herpesvirus gene expression. α, alpha or immediate-early; β, beta or early; γ, gamma or late; mRNA, messenger RNA.

Following infection, the first set of genes to be transcribed are the alpha genes. Their protein products enable the virus to take control of cellular macromolecular synthesis and counteract cellular and immune defenses designed to protect the host against invading viruses. For example, some proteins interact with p53 to prevent the cell triggering apoptosis. Some alpha proteins also are transactivators responsible for activating the promoter of the next class of genes once the cell has been successfully taken over.

The beta genes encode enzymes and other proteins required for replication. These genes ensure that the components required for macromolecular synthesis are available to allow DNA replication to occur, which is especially important when the host cell is terminally differentiated and so not destined to divide. DNA replication is an important landmark in the replicative cycle of herpesviruses and must be completed before gamma proteins can be produced.

The gamma genes encode the structural proteins which make up the capsid, tegument, and envelope of the virus (see Figs. 5 and 6). They are regulated at both transcriptional and translational stages. Thus, although some gamma genes are transcribed at beta times, they are not translated until gamma times and are sometimes termed "delayed-early" to reflect this. The term "true-late" is also sometimes used to describe genes that are strictly transcribed only after DNA replication (reviewed in Ref. 2).

From this conventional description of the cascade sequence of herpesvirus replication in a single cell, it might be thought that the first herpesvirus-encoded proteins to appear in the cell would be alpha, followed by beta, then gamma. This is true when a herpesvirus *reactivates* from DNA latent within an individual cell. However, herpesviruses have evolved an additional mecha-

nism to optimize the rapid appearance of virus-encoded proteins in *newly infected cells*. As shown in Fig. 7, as soon as a virion penetrates the plasma membrane of an uninfected cell it uncoats and releases virus-encoded proteins into the cell. It is thus structural proteins, particularly of the tegument, which are the first to appear in the newly infected cell. Since these were synthesized in the previous cell as gamma proteins, the appearance of herpesvirus-encoded proteins *in a newly infected cell* follows the order gamma, alpha, beta, gamma. For clarity, I will use the term "input gamma" to indicate the tegument proteins introduced into the cell at the time of uncoating. These incoming input-gamma tegument proteins contain transactivators[12] and play important roles in regulating the response of the cell to the incoming virus.

1.2.3. There Is a Core Set of Replicative Genes

Despite the amazing biological diversity of herpesviruses, they share a core set of genes required for DNA replication. These are listed in Table 2 for representative members of the three subfamilies and are beta genes that produce a series of enzymic activities. They are important because they represent molecular targets for the design of antiviral drugs. Genetic studies have also revealed other blocks of genes that are conserved among herpesviruses but a detailed discussion is beyond the scope of this chapter (reviewed in Ref. 2).

1.2.4. Some Genes Are Structural

Given the large coding capacity of herpesviruses, it is perhaps surprising that relatively few genes are required to produce the structural proteins of the capsid, tegument, or envelope (summarized in Table 3 for HSV). Sets of core genes have been expressed as recombinant

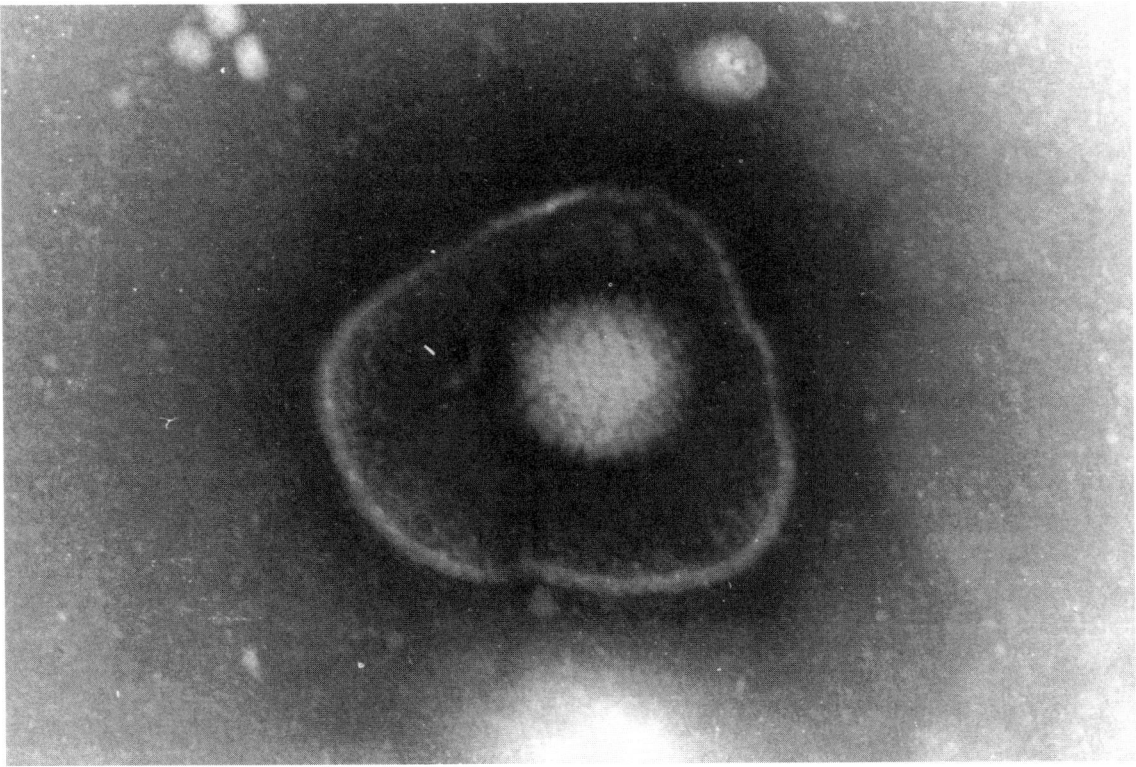

FIGURE 5. Electron micrograph of a herpes virion.

proteins and have been shown to assemble into capsids (reviewed in Ref. 13). Note that some proteolytic processing is required to produce the mature forms of these proteins so that herpesvirus-encoded proteases represent additional molecular targets for the design of antiviral compounds.

1.2.5. Some Genes Are Targets for Antiviral Chemotherapy

This section will be limited to the discussion of the mechanism of action of compounds which are licensed or in clinical trials. The indications for using these compounds in clinical practice will be discussed later.

1.2.5a. Nucleosides.   Acyclovir (ACV) is the prototype for this class of compound.[14] ACV diffuses into and out of cells. In cells infected with a herpesvirus, viral enzymes can recognize ACV and phosphorylate it. For alpha- and gammaherpesviruses, the enzyme that phosphorylates ACV is thymidine kinase (Tk). For beta-herpesviruses, the phosphorylation is achieved by the UL97 family of protein kinases.

Once ACV is phosphorylated within the virus-infected cell, it is charged and so unable to diffuse out of the cell. A concentration gradient is thereby formed across the plasma membrane, aiding diffusion of more ACV into the infected cell. Cellular enzymes convert ACV monophosphate to the triphosphate, which is a potent inhibitor of herpesvirus DNA polymerase and an obligate chain terminator and a suicide inhibitor of the enzyme.[15] In combination, these characteristics potently inhibit herpesvirus replication. Selectivity for virus-infected cells is achieved because ACV is a better substrate for herpesvirus Tk than cellular Tk and because ACV triphosphate is a better inhibitor of virus-encoded DNA polymerase than cellular DNA polymerase.

Ganciclovir (GCV) and penciclovir (PCV) are structurally related to ACV (see Fig. 8). They are anabolized by the same pathway as ACV so that either drug can inhibit HSV or VZV. Note that ACV can also be activated by UL97 so that this compound can also inhibit CMV. Both GCV and PCV possess a free hydroxyl at a position equivalent to the 3′ of the open sugar ring and so can allow DNA elongation. This means that they are not

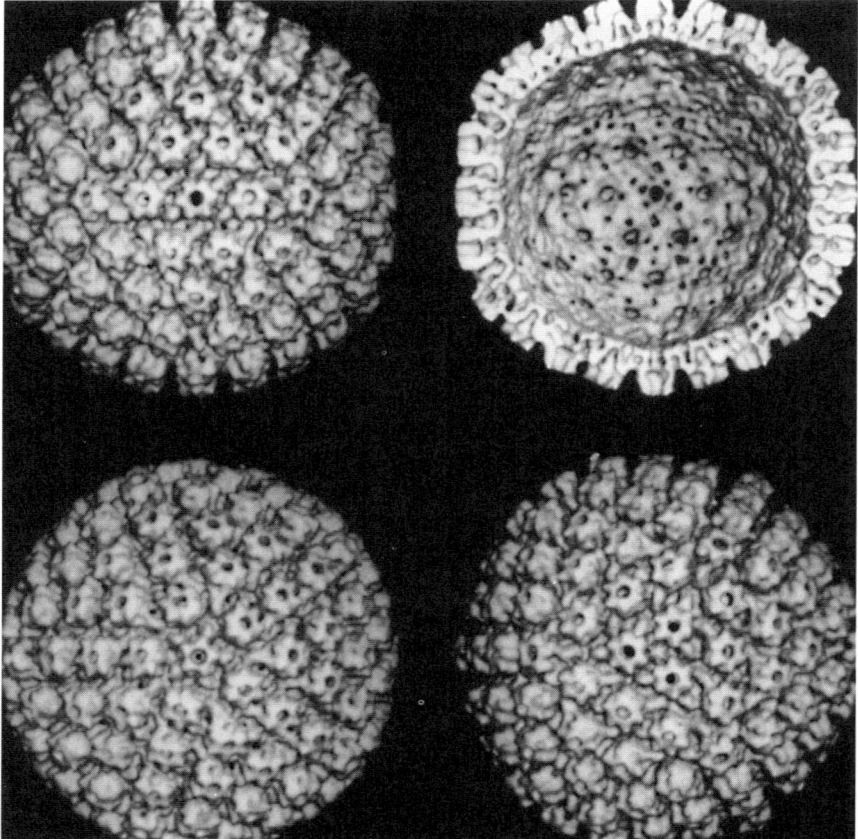

FIGURE 6. Structure of the HSV-1 capsid. Prepared by cryoelectron microscopy and three-dimensional image reconstruction to a resolution of approximately 2.5 nm. The two-fold (upper left), fivefold (lower left), and threefold (lower right) axes of symmetry are shown. The upper right panel shows an equatorial section illustrating the inner surface of the capsid and the thickness of the shell (approximately 15 nm). (Reprinted with permission from Homa and Brown.[13])

obligate chain terminators and are not suicide inhibitors of DNA polymerase, although chain termination usually occurs after incorporation of one or more molecules. The ability to allow chain elongation is theoretically undesirable because it might occur in uninfected cells leading to a mutagenic event in cellular DNA; GCV in particular is oncogenic at low dosage in rodents due to its incorporation into cellular DNA, whereas this effect is only seen with PCV after lifetime exposure of rodents to high doses. Lobucavir is structurally related to PCV, but with a closed carbocyclic ring at the position equivalent to a sugar in a natural nucleoside. It is probably anabolized by cellular enzymes to the triphosphate that inhibits herpesviral DNA polymerase. It is not an obligate chain terminator and is oncogenic after lifetime exposure of rodents to high doses.

Sorivudine (BVaraU; see Fig. 8) also is anabolized by the same pathway but with an important difference. Sorivudine monophosphate is not a good substrate for cellular enzymes, although sorivudine diphosphate is readily converted to the triphosphate. The enzymatic activity (thymidylate kinase) required to convert sorivudine monophosphate to the disphosphate is provided by the Tk of HSV-1 and VZV but not HSV-2. Sorivudine therefore is a poor inhibitor of HSV-2 relative to its potency against the other two alphaherpesviruses.

Sorivudine is well absorbed orally, but ACV, GCV, and PCV, in decreasing order, are less well absorbed. Oral bioavailability of these compounds has been improved via esters which are absorbed and then cleaved in the intestinal wall and/or liver to release free compound. Valaciclovir is the valine ester of ACV. Valganciclovir is the valine ester of GCV and famciclovir (FCV) is an ester of PCV, which requires oxidation of the purine ring in the liver as well as ester cleavage in order to release PCV.

1.2.5b. Nucleotides. Nucleotide compounds are phosphonates, structurally equivalent to the nucleoside monophosphate but without the charge that would prevent cellular uptake. They bypass the Tk (or UL97) step and are converted to their diphosphates (equivalent to the nucleoside triphosphate) (see Fig. 9) by cellular enzymes. Their selectivity therefore resides in the preferential inhi-

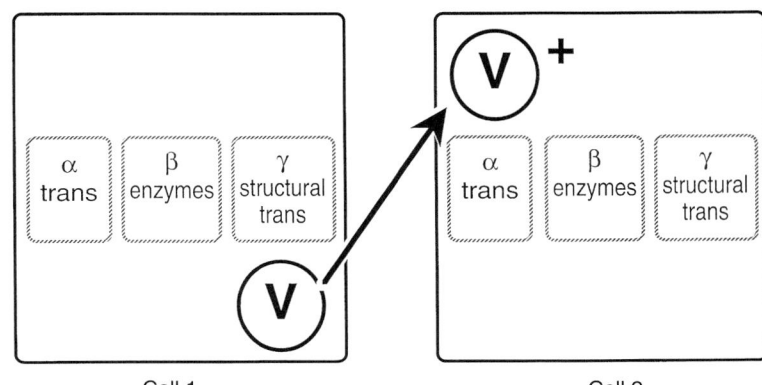

FIGURE 7. Sequential appearance of proteins of different classes within cells. In the left-hand cell, herpesvirus reactivation from latency followed by cascade gene expression produces proteins in the sequence alpha, beta, and gamma. When the newly formed herpesvirus infects a new cell, the sequence of proteins is input gamma, alpha, beta, gamma.

Cell 1   Cell 2

bition of herpesvirus DNA polymerase rather than cellular DNA polymerase by the phosphonate diphosphate. Examples include cidofovir (Fig. 8) and adefovir.

1.2.5c. Foscarnet. When natural nucleoside triphosphates are incorporated into a growing DNA chain, pyrophosphate is eliminated as the phosphodiester bond is formed. Foscarnet is a structural analog of pyrophosphate and inhibits herpesviral DNA polymerase by occupying the enzymatic site for pyrophosphate.

1.2.5d. Antisense Compounds. Molecules complementary (antisense) to mRNA can bind to mRNA and prevent expression of the encoded gene. This approach

TABLE 3. Structural Proteins of the HSV Virion

Capsid	Tegument	Envelope
UL18	UL11	UL1
UL19	UL13	UL10
UL26	UL21	UL20
UL26.5	UL36	UL22
UL35	UL37	UL27
UL38	UL41	UL34
UL6	UL46	UL44
UL12.5	UL47	UL45
UL15	UL48	US4
UL25	UL49	US6
	US9	US7
	US11	US8
	RL1	
	RS1	

[a]Adapted from Homa and Brown.[13]

has been applied to CMV with the design of a molecule (fomivirsen) antisense to the major alpha gene and a different molecule (GEM 132) antisense to another alpha gene, both of which are important transactivators. The half-life of these compounds can be very long because they have been modified, to varying degrees, to contain components not readily degraded by host cells such as phosphorothioates (substitution of sulfur into the phosphodiester background) and/or modified sugar residues.

1.2.5e. Benzimidivir. This is a novel compound with a chemical structure and mode of action unlike any other antiherpes drug. It does not require phosphorylation or any other modification to become active. It has activity against CMV only and appears to act by blocking the switch into rolling circle replication that normally pro-

TABLE 2. Herpesvirus Core Replication Genes

	HSV	HCMV	HHV-8
Proteins required to replicate virion DNA			
DNA polymerase	UL30	UL54	ORF9
Polymerase processivity factor	UL42	UL44	
Single-stranded DNA binding protein	UL29	UL57	ORF6
Origin binding protein	UL9		
Helicase-primase complex	UL5	UL70	ORF40
	UL8	UL101/2	ORF41
	UL52	UL105	ORF44
Enzymes required for nucleic acid metabolism			
Alkaline exonuclease	UL12	UL98	ORF37
Thymidine kinase	UL23		ORF21
Ribonucleotide reductase	UL39		ORF60
	UL40	UL45	ORF61
Uracil glycosylase	UL2	UL114	ORF46
DUTPase	UL50	UL72	ORF54

[a]Summarized from data given in Roizman,[2] Russo,[10] and Mocarski.[197]

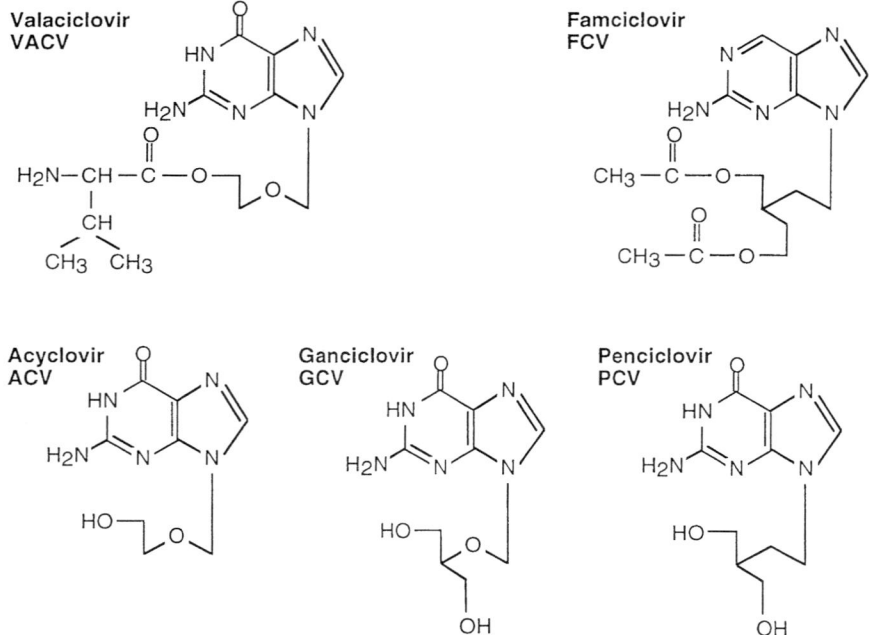

FIGURE 8. Chemical structures of antiviral drugs with activity against herpesviruses *in vivo.*

duces an exponential increase in viral DNA. Recent results demonstrate that this is achieved by inhibition of UL97.

1.2.6. These Genes Are Mutated in Resistant Strains

This statement is best exemplified by studies of strains of HSV resistant to ACV (reviewed in Ref. 16).

This drug exerts profound control over HSV replication, but thereby produces selective pressure for the emergence of resistance at the Tk and DNA polymerase loci. Three phenotypes are recognized. The easiest for the virus to produce is the Tk-minus (Tk−) phenotype where a point mutation in the Tk gene produces a premature stop codon and a truncated, nonfunctional protein. Such a virus can

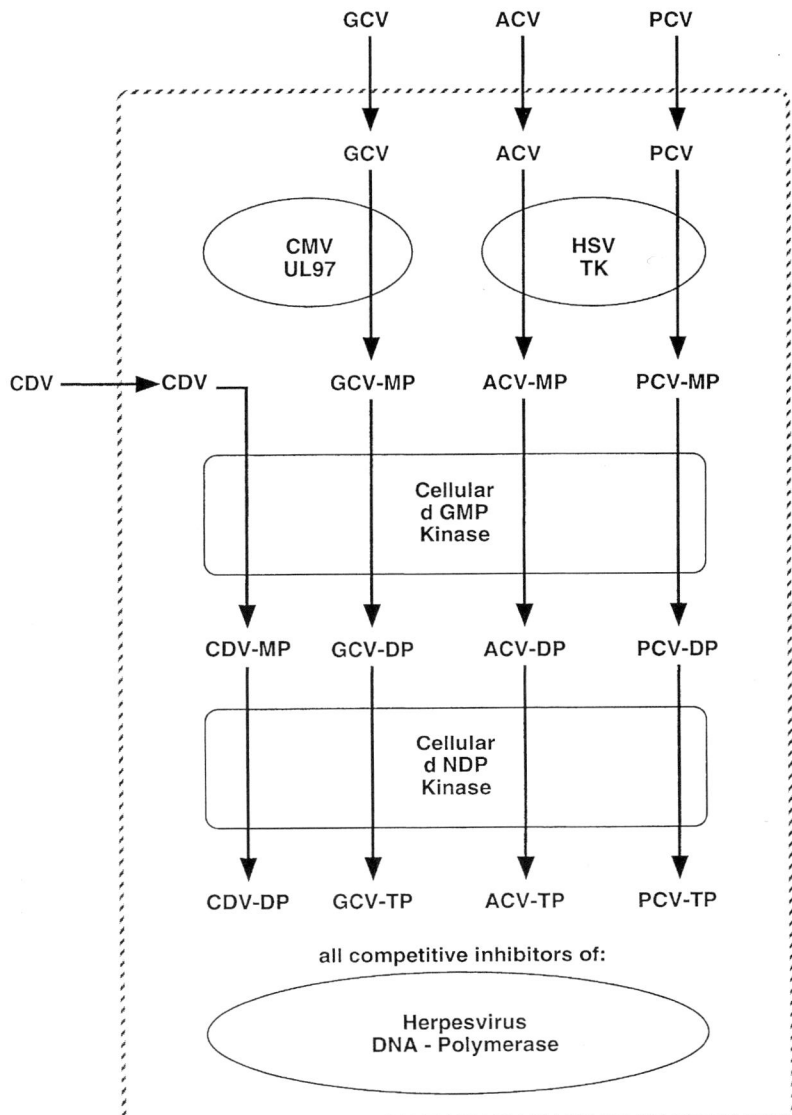

FIGURE 9. Anabolism of nucleoside analogues to their phosphorylated active forms. Enzymes encoded by herpesviruses are shown as ovals. Enzymes encoded by mammalian cells are shown as rectangles. The curved arrows indicate that GCV is phosphorylated by HSV-Tk and that ACV is phosphorylated by CMV-UL97. ACV, acyclovir; CDV, cidofovir; GCV, ganciclovir; TK, thymidine kinase; PCV, penciclovir; UL97, 97th gene in unique long region; MP, monophosphate; DP, diphosphate; TP, triphosphate. (Drawn from data provided by DeClercq.[206])

escape from ACV control but fortunately has debilitated replicative ability. In particular, it cannot reactivate from latency because Tk is required in the environment of the terminally differentiated dorsal root ganglion cell to produce natural nucleosides required for replication of a DNA virus. Note that a Tk− virus can be rescued by superinfection with a wild-type HSV,[17] but fortunately this seems to be exceedingly uncommon in nature. The debilitated nature of the Tk− phenotype means that it is readily controlled by individuals with normal immunity when it replicates in the skin.[18] However, in the immunocompromised host, the balance between inherent patho-

genicity of virus and immune defense is altered so that Tk− strains can nevertheless cause chronic infection and disease. Note also that genetic changes elsewhere in the genome may potentially facilitate replication of a Tk− virus and this appears to be the explanation for a unique strain of HSV recently identified in an immunocompromised patient.[19]

The Tk-altered (Tka) phenotype occurs when the Tk gene is mutated to reduce its phosphorylation of ACV, but the Tk enzyme can still activate natural nucleosides. As would be expected, this phenotype confers normal patterns of replication on the virus, which can reactivate from latency. Fortunately, very few genetic changes can produce this combination of effects so the Tka phenotype remains rare.

The third phenotype occurs when mutations in DNA polymerase reduce its affinity for ACV triphosphate but still allow the enzyme to recognize natural nucleoside triphosphates. The mutated enzyme therefore functions at normal levels and is associated with a fully virulent phenotype able to reactivate from latency.

1.2.6a. Patterns of Cross-resistance. Given the structural similarity of ACV, GCV, and PCV (Fig. 8), it is not surprising that virus resistant to one compound exhibits cross-resistance to the others. In practice, ACV has been used so widely that the small number of resistant strains available for study have been selected *in vitro* through the use of this drug. Virtually all of these viruses exhibit cross-resistance to PCV and/or GCV. The same almost certainly applies to VZV, although the number of strains available for analysis at this time is much smaller.

If strains of HSV (or VZV) resistant to ACV are encountered in clinical practice and alternative treatment is required, then foscarnet is the first choice.[20] Cidofovir is an alternative which, by bypassing Tk, has the advantage of avoiding the genetic change commonly selected for under ACV pressure.[21]

1.2.6b. Relevance to Future Compounds. The principles learned from ACV/HSV will almost certainly be applicable to other antiviral compounds as they reach clinical trials. We should thus prepare for reports of herpesvirus strains resistant to antisense compounds and protease inhibitors, even before such drugs are licensed.

The basic principles to help avoid the development of resistance are to reduce herpesvirus replication in the presence of drug (i.e., treat with potent compounds and adequate doses for a short time and then stop therapy) and to treat promptly at an early stage (when the virus load is low).

1.2.7. Most Genes Are Dispensable for Growth in Cell Culture and Probably Interact with the Host Cell

When tested for their ability to replicate in cell cultures, most herpesviruses with gene deletions have no apparent replicative defect. Such genes are sometimes termed "nonessential." This description cannot be extrapolated to the human host for two reasons: (1) cell cultures provide an excess of nutrients not necessarily found in normal habitats, e.g., dorsal root ganglia; and (2) cell cultures are not subject to immune effectors that can destroy virus-infected cells *in vivo*. Since many genes

in herpesviruses can be described as "nonessential" for replication in cell cultures yet are conserved in nature, it is likely that they have been selected and retained over time because they provide a useful interaction with the host. A definite function for most genes cannot yet be assigned, but some have recently yielded to investigation.

1.3. Herpesviruses Avoid Immune Responses

The immune system recognizes virus-encoded proteins either through the humoral or cell-mediated arms of the immune system. Herpesviruses avoid stimulating the immune system through a variety of tactics.

1.3.1. They Establish Latency

Following initial infection, the cell may be activated into full productive infection or may harbor a latent herpesvirus genome. From the perspective of an individual cell, latency can be envisaged as a "default setting" for the virus in the absence of stimulation. Latency has been best studied for HSV but many aspects still remain to be defined mechanistically (reviewed in Ref. 2).

Cells with latent HSV genomes express an RNA latency-associated transcript (LAT), which maps to a region of alpha gene expression. The LAT facilitates the establishment and maintenance of the latent state. The LAT does not encode a protein of normal size but may produce small peptides with possible regulatory function from small open reading frames.[22]

It is evident that this minimal expression of genetic information is insufficient to trigger responses by the immune system, so that the virus genome can persist for the life of the individual in this state. However, in order to transmit to others, the virus must reactivate at some stage. As shown in Fig. 7, alpha proteins are produced which render the cell a target for cell-mediated immune responses. If the virus can complete its replicative cycle and produce daughter virions before the cell is destroyed, then transmission may occur. Since attempts to reactivate in individuals with normal immunity are frequently met by prompt immune responses, reactivations are typically more commonly recognized in individuals who are immunocompromised.

1.3.2. They Interfere with Immune Responses

1.3.2a. Humoral Immunity. Free virions in the extracellular fluid are susceptible to neutralization by antibodies directed against surface glycoproteins. To reduce this effect, HSV has evolved a tactic to prevent activation of effectors utilizing the Fc portion of the bound antibody. Two HSV genes (gE and gI) together form an Fc receptor on the virion surface. Antibodies bound to neutralizing epitopes on glycoproteins are then captured and bound back onto the virion via the receptor, thereby rendering the Fc portion inaccessible to the immune system.[23]

1.3.2b. Cell-Mediated Immunity. The cell-mediated immune (CMI) system of cytotoxic T cells normally recognizes virus-encoded epitopes in the context of HLA class I proteins. As shown in Fig. 10, peptides are degraded from ubiquitinated virus-encoded proteins in the proteosome and transferred via transporter associated with antigen presentation (TAP) into the lumen of the endoplasmic reticulum (ER). There they associate with the heavy chains of class I HLA molecules plus β_2-m and are transported through the ER and trans-Golgi to reach the plasma membrane. In the ER, a mechanism exists to identify misfolded proteins and reexport them back to the cytoplasm for ubiquitination and subsequent proteosomal degradation. This system is remarkably effective and herpesviruses have evolved multiple mechanisms to interfere at several distinct points (Fig. 10):

1. EBV requires EBNA1 to maintain its episomal DNA form but peptides from EBNA1 are not presented by infected cells because this protein has an unusual structure with multiple glycine–alanine repeats that cannot readily be digested in the proteosome. Note that this glycine–alanine repeat structure can be inserted into other proteins to decrease their immune recognition.[24] The CMV protein ppUL83 (pp65) is one of the input-gamma tegument proteins that is revealed as the incoming virion is uncoated. It phosphorylates the major alpha protein of CMV, and thereby prevents its degradation in the proteosome.[25]
2. HSV encodes a protein, ICP47, which blocks the activity of TAP.[26] CMV has a similar protein, Us6.
3. The Us3 gene of CMV and the ml52 gene of murine CMV bind mature HLA class I complexes and sequester them in the ER.[27] In this respect, they have a similar function to the E3-19K protein of adenovirus where this phenomenon was first described.
4. The CMV genes, Us2 and Us11, mimic the effect

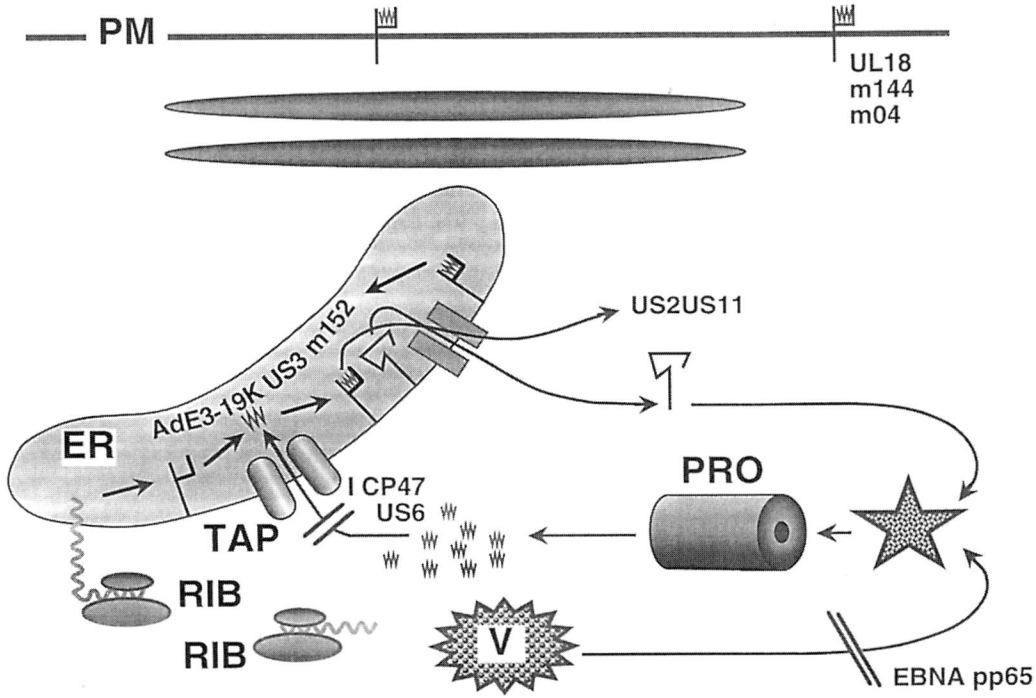

FIGURE 10. Intracellular pathways leading to presentation of viral peptide epitopes at the plasma membrane by HLA class I molecules. Letters and numbers indicate virus proteins which can interfere with antigen presentation (as discussed in the text). V, virus-encoded protein; *, protein modified by ubiquitin; RIB, ribosome; PRO, proteasome; TAP, transporter associated with antigen presentation; ER, endoplasmic reticulum; PM, plasma membrane.

of cellular genes and reexport mature HLA class I proteins back into the cytosol for degradation.[28]

5. If all these mechanisms were effective, then the virus-infected cell would lack display of mature HLA class I complexes. Such a situation is detected by natural killer (NK) cells, which sense the absence of these molecules through several types of receptors. To prevent NK-mediated lysis, some herpesviruses thus encode HLA-decoy molecules (U$_L$18 and U$_L$40 in human CMV; m144, m04 in murine CMV) that pass an inhibitory signal to NK cells, presumably without presenting viral peptides which could be recognized by cytotoxic T lymphocytes (CTL).[29]

One can only marvel at these multiple adaptive mechanisms, presumably acquired over the last 100 million years as jawed vertebrates developed CMIR.[30] Furthermore, the U$_L$18 gene of CMV lacks introns but is otherwise remarkably similar to normal cellular class I molecules. This suggests that U$_L$18 was acquired from the cell, perhaps through retransposition of spliced mRNA sometime during the last 90 million years, since the related HHV-6 and -7 lack this gene (Fig. 1). The temporal control of these multiple genes is also remarkable, ranging from input-gamma (U$_L$83), -alpha (U$_S$3), -beta (U$_S$6), and -gamma (U$_L$18) times.

Note that these inhibiting effects of class I display are not absolute and can be overcome by up-regulation of host defense molecules. Thus, CD4+ T cells and macrophages infiltrating an HSV lesion can produce interferon-gamma (IFN-γ) and up-regulate class I display sufficiently to help CTL recognize virus-infected cells and terminate the infection.[31]

1.4. Herpesviruses Are Important in the Immunocompromised

1.4.1. They Act as Opportunistic Agents

Since attempts to reactivate in individuals with normal immunity are frequently met by prompt immune responses, reactivations are typically more commonly recognized in individuals who are immunocompromised. Thus, in addition to the immune modulatory effects at the level of the infected cell described above, immunocompromised patients have a relative deficiency of functional CTL.

1.4.2. They Have a Characteristic Temporal Appearance

1.4.2a. Allograft Recipients. Following organ transplantation, herpesviruses appear in a predictable pat-

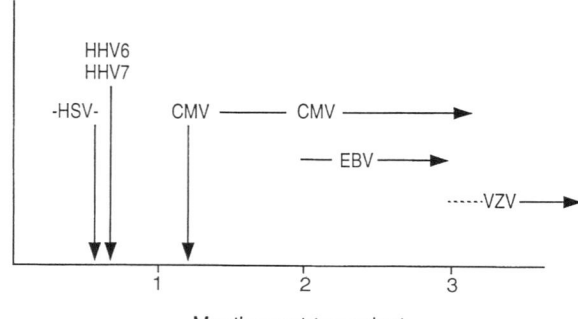

FIGURE 11. Time of typical appearance of herpesviruses and herpesvirus diseases post-transplant. Vertical arrows indicate median times at which virus excretion was detected; horizontal arrows indicate when disease due to each virus usually appears. (Drawn from data in Saral *et al.*,[32] Griffiths *et al.*,[33] and Kidd *et al.*[149])

tern, as shown in Fig. 11. HSV appears soon after transplant during the stage of maximal immunosuppression.[32] However, the other herpesviruses are significantly delayed in appearance, which does not accord with the simple interpretation that removal of immune surveillance triggers their immediate reactivation. Even with sensitive techniques such as polymerase chain reaction (PCR), the median time to CMV detection is 36 days posttransplant.[33] Even more remarkably, VZV is not a major problem in the immediate posttransplant weeks but becomes significant months later when most other herpesviruses are no longer detectable.[34]

These patterns have not been explained satisfactorily, but are consistent and reproducible. One possibility is that different alpha herpesviruses have a natural frequency of attempted reactivation from dorsal root ganglia, most of which are clinically unrecognized because of effective immune control. Removal of immune surveillance would then require a wait of varying length before the next reactivation was due, i.e., HSV reactivation would occur before VZV reactivation. Another possibility for betaherpesviruses is that transmission with the donated organ[35] may require activation of latent genomes and productive infection in that organ before virus can spread viremically.

1.4.2b. Patients with HIV. As shown in Fig. 12, patients with HIV also have a characteristic temporal appearance of herpesviruses. There are parallels with transplant patients, shown in Fig. 11, e.g., HSV reactivates when the patient is profoundly immunocompromised and zoster tends to occur at relatively high CD4 counts and may be the initial AIDS-indicator disease.[36] However, there also are major differences. HHV-6 is detected when CD4 counts are high and less frequently as the CD4 count

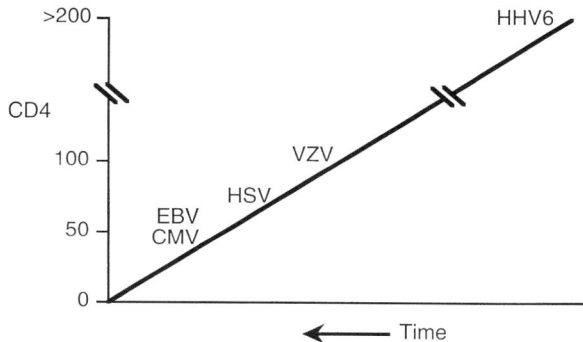

FIGURE 12. Time of typical appearance of herpesviruses in HIV-positive patients according to CD4 count. (Drawn from data in Glesby *et al.*,[36] Fairfax *et al.*,[37] and Fabio *et al.*[38])

declines.[37,38] This remarkable paradoxical relationship may reflect the fact that HHV-6 replicates in CD4 cells so that loss of these cells decreases the chance of detecting virus *in the blood*. However, it should be noted that HHV-6 is readily detected in multiple organs from AIDS patients at autopsy,[39] so that results from blood may underestimate the pathologic significance of this virus.

1.4.3. The Incidence and Severity of Disease Are Increased

Through the use of frequent cell culture and sensitive PCR methods, we now recognize that HSV-2 is a common infection in people with normal immunity, although most episodes of active infection are asymptomatic.[40,41] Studies of immuncompromised patients conducted before the routine use of ACV prophylaxis indicate a high frequency of HSV infection with a high risk of symptomatic disease.[32]

With the exception of some cases of HSV, most individuals with normal immunity have infrequent episodes of active infection with herpesviruses and the severity of disease is usually mild, if present at all. In contrast, immunocompromised patients have a high incidence of documented infection with multiple herpesviruses. The clinical consequences of such infections will be discussed later, but in general the severity of herpesvirus disease is increased, with involvement of structures deeper than the mucosa (HSV), involvement of multiple dermatomes (VZV), multiple end-organ diseases (CMV), or the appearance of tumors (EBV, HHV-8).

1.4.4. Some Disease May Be Immunopathologic

In this context, immunopathology implies a contribution from the host toward the production of disease. For example, infectious mononucleosis is caused by the T-cell response to EBV-infected B cells; HSV may induce stromal keratitis because a viral epitope recognized by CD4+ T cells cross-reacts with an epitope in the normal cornea.[42]

Despite the fact that patients are immunocompromised, they may still develop immunopathologic disease. Thus, CMV is tropic for the lung but CMV interstitial pneumonitis rarely occurs without coincidental graft-versus-host disease (GVHD).[43] CMV retinitis in AIDS patients became complicated by vitritis once highly active antiretroviral therapy (HAART) became available and restored some immune responses against CMV antigens. The immunocompromised appear to be relatively spared from the greatest ravages of HSV encephalitis, perhaps because they cannot mount an effective inflammatory response that can raise intracranial pressure and cause hypoxic necrosis in the brain. These examples illustrate that the immune system is not always "the good guy" and that the terms "immunocompromised" or "immunodysregulated" should be used in preference to "immunosuppressed."

1.4.5. They May Trigger Other Pathologic Phenomena

End-organ diseases caused by herpesviruses are termed their "direct" effects. In addition, herpesviruses may cause *indirect* effects by triggering a variety of pathologic phenomena after organ transplantation.[44] For example, by damaging the normal mucosal barrier, HSV may provide a portal of entry for bacterial pathogens. CMV may facilitate bacteremias or fungal infections by adding to the total level of immunosuppression. Activation of CMV and other betaherpesviruses may trigger graft rejections. CMV is associated with accelerated atherosclerosis post-heart transplant, possibly because of effects on smooth muscle cells in the intima.[45]

1.4.6. They Can Interact with HIV

Given that active herpesvirus infections are widespread in patients with AIDS who also have extensive HIV replication, the possibility exists that HIV may interact with one or more herpesviruses (reviewed in Ref. 46). Such interactions could take place in a single coinfected cell or the two viruses could infect neighboring cells (Table 4). Such interactions could have negative effects on HIV if both viruses compete for the same metabolic pathway or could lead to up-regulation of HIV. The results of molecular or cellular biological experiments to assess these possibilities are shown in Table 5.

TABLE 4. Interactions between HIV and Potential Cofactor Viruses that Could Lead to Increased HIV Replication[a]

Effect of interaction	HIV and cofactor virus infect	
	Same cell	Neighboring cells
Activate proviral HIV DNA	Transactivation	Cytokine release Antigen presentation
Alter tropism of HIV RNA	Pseudotype formation	Up-regulation CD4/ coreceptor Induction alternative receptor

[a]Reprinted from Griffiths,[46] with permission.

1.4.7. Treatment Benefits May Be Profound

Prevention of herpesvirus infection may have clinical benefits unrecognized by those unaware of the possible indirect effects of herpesviruses. Since herpesvirus infections occur in the context of complex medical cases, there is no role for believing anecdotal cases, irrespective of their plausibility. However, if the results of randomized placebo-controlled trials of antiherpes drugs (or vaccines) show significant reductions in conditions identified as possible indirect effects, then we should accept these associations as causal and update our understanding of the pathogenic mechanisms involved. For example, it is interesting to note that a placebo-controlled trial of low-dose (200 mg qid) ACV in neutropenic patients significantly reduced the incidence of bacteremias.[47] Unrecognized HSV-induced mucosal ulceration therefore may be an important source of bacterial colonization leading to bacteremias. Likewise, a placebo-controlled trial of valaciclovir has recently reported reduced incidence of biopsy-proven graft rejection after renal transplantation.[48] A herpesvirus therefore may be triggering some cases of allograft rejection, with CMV and the other betaherpesviruses being prime candidates. Prolonged

follow-up of heart allograft recipients who participated in a placebo-controlled trial of GCV has revealed decreased fungal infections[49] and decreased accelerated posttransplant atherosclerosis,[50] supporting the causal association of CMV with these syndromes.

In HIV-positive patients, the clinical consequences of interactions described *in vitro* are likely to be complex given the number of possibilities involved (see Table 5). Those that have been studied *in vivo* are listed in Table 6. Colleagues should consider the possible effects on HIV of giving antiherpes drugs to AIDS patients and the possible effects on herpesviruses of giving HAART. Given the complexities, placebo-controlled trials are essential and it is interesting that, despite initial controversy, a meta-analysis of ACV trials has confirmed a small but significant reduction in mortality from the chronic suppressive use of ACV.[51] Since in another controlled trial higher doses of ACV conferred no additional benefits,[52] this suggests the effect was due to inhibition of a very sensitive herpesvirus such as HSV. This possibility is supported by the finding that both HSV and HIV are shed from genital HSV ulcers.[53] Clearly, there also is a public health implication for HIV-positive individuals with genital ulcer disease due to HSV who may have increased HIV infectivity for others.[53] Indeed, it may be argued that unrecognized genital HSV infection may underlie the heterosexual epidemic of AIDS in developed countries, which is a provocative hypothesis that could be tested by controlled intervention trials with prophylactic ACV.

2. Characteristics of Particular Herpesviruses

2.1. Herpes Simplex 1 and 2

2.1.1. Epidemiology

It should be noted that the term "genital herpes" refers to infection of the genital area and may be caused

TABLE 5. *In Vitro* Evidence for Cofactor Herpesviruses Stimulating HIV Replication[a]

Virus	Transactivation[b]	Cytokine release	Alternative HIV receptor	CD4 or coreceptor up-regulation	Antigen presentation	Pseudotype formation
HSV	✓	×				✓
VZV		×				
EBV	✓	✓				×
CMV	✓	✓	✓	✓	✓	✓
HHV-6	✓	×		✓		
HHV-7			×	×		
HHV-8		×		×		

[a]Reprinted from Griffiths,[46] with permission.
[b]✓, data support stimulation; ×, data reject stimulation (or show inhibition); blank, no data.

TABLE 6. *In Vivo* Evidence for Cofactor Herpesviruses Stimulating HIV Replication[a]

Virus	Virus increases HIV RNA or DNA[b]	Antibody increases AIDS	Antibody increases death	Cofactor virus increases AIDS	Cofactor virus increases death	Autopsy	
						Organ	Cell
HSV	✓	×				✓	
VZV						×	
EBV						✓	
CMV		✓/×	✓	✓	✓	✓	✓
HHV-6	✓					✓	
HHV-7						✓	
HHV-8							

[a]Reprinted from Griffiths,[46] with permission.
[b]✓, data support stimulation; ×, data reject stimulation; blank, no data.

by either HSV-2 or HSV-1. Likewise, oral herpes is an anatomic description of the condition usually but not always caused by HSV-1.

Most seroepidemiologic studies have not distinguished between HSV-1 and HSV-2. They report that antibodies against either (or both) viruses are acquired during early childhood and increase progressively until approximately 80% of adults in developed countries and over 95% in developing countries are seropositive.[54] Routes of acquisition of virus include salivary transmission from mother to child, salivary transmission from relative/friends to child, salivary transmission among adolescents, and transmission from saliva/genital secretions among the sexually active.

As expected, type-specific serology shows that most infections acquired by children are HSV type 1 (although neonates may rarely acquire HSV-2 perinatally). Among sexually active adolescents, HSV-1 still predominates, with HSV-2 acquisition typically beginning from 20 years onward. Overall, approximately 22% of the U.S. population was HSV-2 seropositive in a 1988–1994 survey,[55] this figure having risen by a third since the first national survey performed between 1976 and 1980.

2.1.2. Pathogenesis

Following primary infection and replication of virus in epithelial cells, HSV (of either type) is taken up by sensory nerve endings supplying the affected dermatome and establishes latency in corresponding dorsal root ganglia. Latency-associated transcripts (LAT) are produced from the latent genome but the virus remains invisible to the immune system because proteins are not expressed from these noncoding transcripts. The genome can be activated to produce infectious virions by diverse stimuli, including ultraviolet light, stress, and menstruation. These stimuli activate the dorsal root ganglia so that the

full complement of virus-encoded proteins is produced. These are packaged to form capsids that travel down the sensory nerve by axonal transport. They are enveloped at the cell membrane of the nerve and released to initiate infection in contiguous epithelial cells (reviewed in Ref. 54).

The presence of genes able to interfere with the presentation of viral epitopes in the context of class I HLA molecules (see Fig. 10) permits viral replication for a few hours without an effective immune response. However, the production of IFN-γ by macrophages and NK cells leads to up-regulation of class I display on the infected cells, allowing their ultimate destruction by CTL.[31] The episode of virus reactivation thus is eventually terminated by a coordinated immune response requiring components of the innate and adaptive immune systems.

Following solid organ transplantation, lymphocyte blastogenic responses return to normal within 6 months, although recipients of bone marrow require HSV reactivation to stimulate these responses in the newly engrafted marrow.[56]

2.1.2a. Direct Effects. Most of the HSV infections in the immunocompromised result from reactivation of latent infection. Much of the morbidity in the immunocompromised host can be attributed to local spread of HSV (see Table 7). For example, oral HSV may cause esophagitis or focal pneumonitis and spread may be facilitated by the presence of nasogastric or endotracheal tubes, respectively. HSV may involve multiple sites of the skin, especially if there is a preexisting skin condition such as eczema. Such autoinoculation may include keratitis or whitlow, and the patient should be advised of the potential infectivity of HSV lesions. Extensive replication of HSV-2 can involve tissues deeper than the mucosa and clinically resemble "bed sores." Local involvement of the sacral nerves also can precipitate meningitis, which may be recurrent. Viremic dissemination of HSV plays

TABLE 7. Pathogenesis of HSV Diseases

Pathogenic feature	Examples of disease
Local infection	Mucosal lesions
Local spread	Esophagitis, focal pneumonitis, "bed sores," eczema herpeticum, autoinoculation (keratitis/whitlow), meningitis (HSV-2)
Damaged epithelial layer	Bacteremia
Viremic spread	Diffuse pneumonitis, hepatitis
Inflammatory response	Activate HIV

only a minor part in pathogenesis but occasional cases of HSV hepatitis, gastrointestinal disease, or diffuse pneumonitis are seen after primary HSV infection. Rarely, primary HSV infections may be transmitted via the donated organ.[57]

2.1.2b. Indirect Effects. The damaged and denuded epithelium of an HSV lesion represents a potential portal of entry for secondary bacterial pathogens. In HIV-positive patients, the inflammatory response to HSV can stimulate cells with latent HIV proviral DNA to produce new HIV virions (see Table 7). This may manifest as a small increase in the HIV plasma RNA level or the appearance of a novel quasi-species previously quiescent.[53]

2.1.3. Clinical Manifestations

In individuals with normal immune responses, most HSV infections are medically trivial, although they may have marked cosmetic and/or psychosexual effects on the patient. Symptoms associated with primary infection of the oral mucosa include stomatitis, although most infections are asymptomatic. Genital infection due to HSV-2 or HSV-1 may be asymptomatic or may present with vesicles/ulcers. Primary HSV-2 genital infection may be accompanied by meningitis, especially in females. Primary HSV-1 infection also can involve the cornea, causing keratitis, or the brain, causing severe encephalitis.

Symptoms associated with reactivation of infection include recurrent episodes of crops of vesicles at the mucosal/epithelial border of the mouth, termed "cold sores" or "fever blisters," which are usually caused by HSV-1. Repeated reactivation of HSV-1 also can cause keratitis with dendritic or geographical ulcers. Reactivation from sacral dorsal root ganglia produces vesicles and/or ulcers on the genital mucosa and is usually caused by HSV-2. It is now known that most reactivations of HSV-2 are either totally asymptomatic or produce such

minor signs that they are not noticed by the patient.[41] This is true also of males who are HIV-positive.[58] This lack of symptomatology coupled with frequent reactivation presumably facilitates transmission of HSV-2 and explains its current high prevalence.

The diseases caused by HSV in immunocompromised patients are summarized in Table 8. Fortunately, most of these are now rare due to the use of acyclovir prophylaxis and prompt treatment of early disease. Without treatment, virus excretion may be prolonged for weeks instead of days in the immunocompetent. Ulceration of oral or genital mucosa may be extensive and persist for weeks and may rarely lead to HSV dissemination (Table 8).

2.1.4. Diagnosis

Typical vesicular lesions of recurrent HSV infection may be readily diagnosed clinically. Atypical lesions or ulcers are more common and should be cultured for HSV.

HSV also should be included in the differential diagnosis of esophagitis or pneumonitis. Cells obtained by endoscopy should be stained using monoclonal antibodies for evidence of HSV infection. In cases of meningitis or encephalitis, cerebrospinal fluid (CSF) should be obtained and tested for HSV using PCR.[59]

If episodes of HSV fail to respond to acyclovir, the drug should be given intravenously. If there is still no response, the possibility of resistance should be considered and cultures sent for sensitivity testing.

Detection of IgG antibodies against HSV (type 1 or 2) is used at baseline to identify patients with latent infection. A variety of methods are widely available. Serology has no role to play once the patient is immunocompromised.

**TABLE 8. HSV Diseases
in the Immunocompromised**

Symptoms	Solid transplantation	Bone marrow transplantation	AIDS
Oral	+	++	+
Genital	+	+	++
Bed sores			++
Keratitis	+	+	+
Esophagitis	+	+	+
Pneumonitis	+	+	+
Fever/viremia	+	+	+
Meningitis	+	+	+
Encephalitis	+	+	+

2.1.5. Management

After transplantation, approximately 70% of seropositive patients will excrete HSV.[32] Although many will not suffer serious disease, oral infections are painful, add to the total burden of disease, reduce fluid and nutritional intake, and may reduce compliance with multiple oral drug regimens. For all these reasons, prophylaxis with acyclovir is given routinely. Natural history studies show that those most likely to excrete HSV posttransplant are those with high IgG antibodies pretransplant.[60] High antibody titers probably identify individuals with frequently reactivating HSV who have boosted their antibody levels in the past and who are likely to reactivate again once immunocompromised. A case could be made for reserving acyclovir prophylaxis for those with high baseline levels of IgG antibodies, but in practice prophylaxis is advised for all seropositives because acyclovir is so safe and effective.

Occasional patients will develop breakthrough episodes of HSV infection or disease despite receiving acyclovir prophylaxis. Such isolates should be cultured but are invariably sensitive to acyclovir. Breakthrough episodes therefore should prompt reassessment of patient compliance and the possibility of gastrointestinal disease impairing absorption of the drug.

For treatment of established disease, acyclovir should be given intravenously to ensure adequate tissue penetration of drug. If the patient has recovered sufficiently to be considered for discharge home, then a full course can be completed by giving valaciclovir.

The treatment of HSV strains proven to have acyclovir resistance is with foscarnet.[20] Once the lesions have resolved, the wild-type HSV latent in the dorsal root ganglia may reactivate again, so prophylaxis with acyclovir should be restarted once the foscarnet is stopped. This cycle may repeat itself several times in a profoundly immunocompromised AIDS patient, with selection of resistance at the epithelial level but failure of this debilitated virus to establish latency and reactivate (see Section 1.2.6). An alternative to consider is cidofovir, either systemically or in a topical form, which is currently undergoing clinical trials.[21]

Controlled clinical trials of other antiviral compounds such as famciclovir support their use for the treatment of HSV infections in patients with normal immunity. They could be considered for the use in the immunocompromised host once ongoing controlled trials demonstrating efficacy and safety in each population have been published. However, it should be noted that they offer little real advantage over the well-established acyclovir.

In particular, resistant strains of HSV will be cross-resistant to famciclovir, so this drug has no role to play in the management of HSV resistance.

2.2. Varicella–Zoster Virus

2.2.1. Epidemiology

Primary infection with VZV causes the well-known systemic vesicular rash of chickenpox. This is highly communicable so most children acquire chickenpox. Accordingly, the prevalence of IgG antibodies against VZV approximates 90% of adolescents in developed countries. For reasons that remain unexplained, chickenpox in certain tropical communities is a disease of adults, not children.[61]

Reactivation of latent VZV causes an outbreak of vesicles in a dermatomal distribution, termed herpes zoster or shingles. Zoster increases in frequency as age increases, with small numbers of cases in children and young adults and a high attack rate in the elderly (attack rate, 10/1000 patient years in those 75 years of age or older).[62]

This epidemiologic picture may change with the recent licensure of a live-attenuated Oka strain VZV vaccine. Originally designed to protect immunocompromised children from chickenpox, this vaccine has been recommended for all children in the United States (except those who are immunocompromised). At present, only a small proportion of children are receiving the vaccine, but if the proportion increases substantially, it could alter the natural history of VZV disease. Using a mathematical model, Garnett and Ferguson[63] have illustrated how VZV vaccine could increase or decrease the severity of disease due to chickenpox or to zoster (see Table 9).

2.2.2. Pathogenesis

Primary VZV infection is acquired through the respiratory tract, usually after contact with a case of chickenpox. Following virus replication in the upper respiratory tract, VZV establishes a primary viremia and reaches the lymphoreticular tissue of the spleen. While the patient remains asymptomatic, VZV replicates to a high level and then is released to cause a secondary viremia. This marks the end of the incubation period and disseminates virus to multiple sites (reviewed in Ref. 64). In the skin, VZV replicates in dermal perivascular tissues producing edema fluid, which splits the epithelial layer (acantholysis) causing vesicle formation. The vesicle fluid contains a high concentration of virus particles which are progressively

**TABLE 9. Immunization of Children
to Prevent Chickenpox: Possible Effects
at the Population Level on the Severity of VZV Disease**

Possible effects observed	Explanation for observation
Decreased chickenpox severity	Direct effect of the reduced number of cases of chickenpox
Increased chickenpox severity	Vaccine induces a shift in age distribution of natural chickenpox so that adolescents/adults are predominantly affected, with consequential increased severity
Decreased zoster severity	Direct effect of the reduced number of people entering mature decades with latent VZV
Increased zoster severity	If repeated natural exposure to chickenpox normally maintains cell-mediated immunity keeping VZV suppressed into latency, removal of chickenpox cases could lead to an increased incidence of zoster

aFrom Garnett and Ferguson.[63]

degraded by polymorphonuclear leukocytes and macrophages to form pustules (bacteriologically sterile). The tissue fluid is absorbed to form a crust that provides protection for lateral ingrowth of epithelial cells. When the epithelial defect is repaired, the crust is sloughed off to reveal normal-looking skin. This orderly process can be impaired if the child scratches the lesions, introducing secondary bacterial infections. Healing may then be delayed and may produce scarring.

During VZV replication in the skin, some virus reaches the dorsal root ganglia via axonal transport. There it establishes latency similar to that described earlier for HSV, although details of the molecular and cell biology are less well defined. The dorsal root ganglia involved are those that serve the area of skin affected by the chickenpox rash. Since this involves the face and trunk predominantly, latent VZV, and so future zoster, is found mainly in the thoracic and cranial nerves, although all parts of the body can be affected. Indeed, it is instructive to review that the dermatomes of the body were described by studying the areas of skin involved by zoster and matching this with the inflammation seen in the corresponding dorsal root ganglia at autopsy (reviewed in Ref. 65).

While reactivation of VZV usually produces zoster, some cases can occur without a visible rash (zoster sine herpete). This has been inferred by finding cases of VZV PCR viremia and/or appearance of new immune reactivity to VZV in the form of IgM antibodies or lymphocyte proliferation responses.[66,67]

In the immunocompromised host, zoster also can be followed by viremic dissemination of VZV to multiple sites, including the skin. This is termed disseminated zoster but in clinical terms resembles chickenpox in its severity. The pathogenic features of the diseases associated with VZV infection are listed in Table 10.

Direct Effects. In addition to cutaneous dissemination, viremia can produce encephalitis or acute retinal necrosis. Local spread to the neighboring dermatome is often seen in zoster and is not a manifestation of viremic dissemination. Occasionally, severe cases of cranial zoster produce nectrotizing arteritis which causes contralateral hemiparesis.[68]

2.2.3. Clinical Manifestations

In children with normal immunity, chickenpox is a self-limiting disease. The number of vesicles provides an estimate of the severity of the rash, with the median varying between 200 and 500.[69] The number of vesicles is increased in younger children and in those who acquire chickenpox from a sibling, presumably because prolonged contact allows a greater inoculum to be transmitted than when contact is more fleeting. The vesicles usually appear in a series of crops representing repeated episodes of viremic release from the spleen over 2 to 4 days. Examination of an acute case therefore reveals some lesions at the vesicular stage, with others pustular or forming crusts.

In immunocompromised children, the number of vesicles is typically increased.[70] Their character also may alter, with a hemorrhagic base heralding a poor prognosis. Lesions also may coalesce, producing bullae or purpura fulminans. Crops of vesicles may continue to appear with a prolonged time of up to a week until new vesicle formation is halted. The increased quantity and duration of viremia explains why visceral involvement is increased in the immunocompromised host.[71] Pneumonitis and en-

TABLE 10. Pathogenesis of VZV Diseases

Pathogenic feature	Examples of disease
Local infection	Skin lesions
Local spread	Neighboring dermatome, eczema herpeticum
Damaged epithelial layer	Bacterial superinfection, scarring
Viremic spread	Diffuse pneumonitis, encephalitis, cutaneous dissemination, acute retinal necrosis
Inflammatory response	

TABLE 11. VZV Diseases in the Immunocompromised

Symptoms	Solid transplantation	Bone marrow transplantation	AIDS
Chickenpox	+	+	
Zoster	+	+ +	+ +
Disseminated zoster	+	+	
Atypical chickenpox			+
Pneumonitis	+	+	+
Encephalitis	+	+	+
Motor neuropathy	+	+	+
Acute retinal necrosis			+

cephalitis are the most common clinical manifestations (see Table 11).

In the individual with normal immunity, zoster is usually heralded by unilateral pain that can be sufficient to make the patient seek medical attention. The appearance of characteristic vesicles and the progression along the route of a single dermatome both provides a diagnosis and gives this virus family its name (Greek: *Herpes* = creeping). Each vesicle follows the typical stages described earlier for chickenpox, so the skin is typically recovered within 10–20 days, although there may be residual scarring or altered skin pigmentation. However, the pain may still persist despite healing of the skin [postherpetic neuralgia (PHN)]. Several stages of pain have been described (see Fig. 13) but as discussed elsewhere[72] it is probably best to consider pain as a continuum termed "zoster-associated pain" (ZAP), without making artificial time boundaries. ZAP is increased in severity and persistence in the elderly for reasons that are not clear but relate more to the repair ability of neural tissue rather than to virologic factors.

In the immunocompromised host, zoster may be self-limiting and involve the same dermatomes as in the immunocompetent, i.e., thoracic > lumbar = cranial > sacral,[73] may be more extensive and certainly has a greater disposition to disseminate.[74,75] Indeed, patients may present with a florid varicella rash without clear evidence of a distinct dermatomal phase. Patients may also present with nonspecific features of fever, nausea, abdominal pain, and vomiting 1–3 days before the onset of rash.[73] Rarely, patients may have visceral involvement without cutaneous manifestations.[73] Before antiviral therapy was available, cases of extensive, deep, necrotic zoster were described. The incidence of PHN depends on the age of the patient being studied and bears no apparent relationship to the extent of the rash.[76] For example, in AIDS patients who are relatively young, persisting neural pain is rare, despite the extensive zoster rash they may have recently recovered from. Dissemination of VZV to other dermatomes, to lung, or to brain, should always be considered.

2.2.4. Diagnosis

The diagnosis of chickenpox or zoster is frequently made clinically. Patients with atypical lesions should have skin scrapes taken from vesicles and stained with monoclonal antibodies to detect VZV antigens.[77] Cases of "chickenpox" in a patient with a previous history of chickenpox should be investigated by direct staining for VZV antigens and by checking for the presence of IgG antibodies from the previous episode. Patients with organ involvement should have tests for VZV as part of the differential diagnosis, e.g., monoclonal antibody staining of cells obtained from the lung by bronchoalveolar lavage in the case of pneumonitis or CSF tested by PCR for VZV DNA in the case of encephalitis.

2.2.5. Management

2.2.5a. Prevention. Patients without a history of chickenpox should be offered VZV vaccine if sufficiently immunocompetent. They should be advised to avoid people with rashes and their seronegative family members

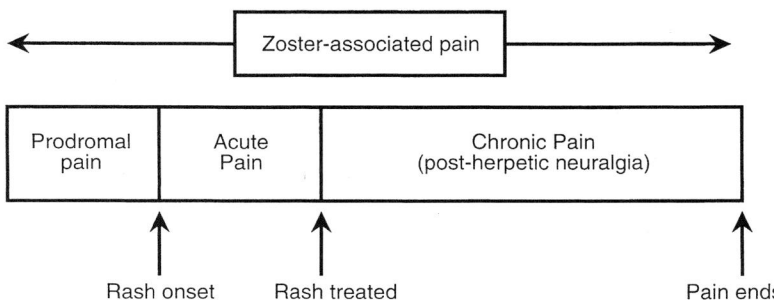

FIGURE 13. Terms used to describe the pain caused by reactivation of varicella–zoster virus. (Reprinted with permission from Wood.[72])

should be given VZV vaccine. If they have contact with a suspected case of chickenpox, then VZV IgG should be checked with a rapid test, e.g., latex agglutination. If seronegative, they should be offered[70] zoster immune globulin (ZIG). All contacts of chickenpox should be followed closely, advised to check their temperature twice daily for 21 days, and told to report for examination if they develop fever. Intravenous acyclovir should be given at the first sign of a vesicle.

Acyclovir has been used in small studies for the prophylaxis of chickenpox contacts.[78] Formal studies are required in the immunocompromised before a recommendation can be made, but if it is to be used, e.g., where ZIG is unavailable or contraindicated, then it should be given for 1 week starting on day 7 after contact, i.e., to coincide with the anticipated appearance of secondary viremia.[78]

2.2.5b. Treatment.   All cases of chickenpox in the immunocompromised should be given acyclovir intravenously as soon as the diagnosis is made unless all vesicles have already scabbed. Unlike the case with normal children, there is no maximum time limit from disease onset because immunocompromised children may continue to have prolonged new vesicle formation and ongoing viremia. The IV route is often changed to oral valaciclovir once the patient's condition improves sufficiently for hospital discharge to be considered, but oral medication should not be used in the initial phase unless controlled clinical trials subsequently demonstrate its efficacy and safety.

Likewise, all cases of zoster in the immunocompromised host should be treated promptly as above. Given the pathogenesis described earlier, it is difficult to see why practitioners would give topical therapy to the skin. The nervous system is the major target for antiviral intervention because it contains the cells forming new virus particles, so systemic medication with good tissue penetration is required. The therapeutic goals in the immunocompromised host for zoster are prevention of viremic dissemination and accelerated skin healing rather than reduction of ZAP, which is the major goal in the elderly patient with normal immunity. Due to the risk of VZV dissemination, steroids should not be given for zoster in the immunocompromised host, although they may have a minor role in accelerating local healing in carefully selected elderly patients.[79]

Resistance to acyclovir occurs rarely in AIDS patients who are profoundly immunocompromised and unable to control VZV replication, despite high-dose IV acyclovir administration. The lesions progressively take on a thickened, verrucous form. Most of these viruses are Tk− and so debilitated, although DNA polymerase mutants have been described. Therapy is with foscarnet.[80]

Throughout, acyclovir has been emphasized for the therapy of VZV infections, with its prodrug, valaciclovir, substituted if necessary on clinical criteria. Famciclovir also could be used for treatment once controlled clinical trials reporting its efficacy and safety in the immunocompromised have been published. Sorivudine is the most potent drug against VZV *in vitro* and clinical trials confirm its potency in the immunocompromised host.[81] However, it is not licensed because a minor breakdown product (bromovinyluracil) inhibits an enzyme responsible for the catabolism of 5-fluorouracil and 5-fluorocytosine and deaths occurred when sorivudine was co-administered to patients receiving cancer chemotherapy regimens containing either of these two drugs. This contraindication was recognized and publicized before the drug was released in Japan, but apparently ignored.

2.3. Cytomegalovirus

2.3.1. Epidemiology

Seroepidemiologic studies show that, by age 30 years, 50–70% of adults in developed countries and nearly 100% in developing countries have IgG antibodies to CMV.[82] Virtually all such infections are acquired asymptomatically, although rare cases of CMV hepatitis or CMV mononucleosis do occur in people with normal immunity.

CMV acquisition is more common in lower socioeconomic groups.[83] Natural routes of transmission include *intrauterine* (presumably through maternal viremia in 0.3–1% of cases); *perinatal* (through contact with infected maternal genital secretions and/or breast milk in 10–20% of cases); *horizontal* in childhood (through saliva); and *horizontal* in the sexually active (through saliva and/or genital secretions). In addition, iatrogenic sources of CMV include donated solid organ, blood transfusion, and semen used for *in vitro* fertilization.

In all cases, seropositive individuals should be regarded as possessing latent CMV capable of reactivation. Thus, if they become immunocompromised, CMV may reactivate from latency and may cause disease (reactivation infection). In addition, if their organs are harvested for transplantation, they may transmit virus,[35] irrespective of whether the recipient is seronegative (primary infection) or seropositive (reinfection). It has been noted that all the organs harvested from seropositive donors are concordant (i.e., either all transmit or do not transmit to recipients). This indicates that either CMV is latent in most organs or that passenger leukocytes contain CMV and are transmitted passively with the donor organ. Given the ability of CMV to replicate in most cells *in vivo* and

TABLE 12. Pathogenesis of CMV Diseases

Pathogenic feature	Examples of disease
Local infection	
Local spread	Progression of retinitis
Damaged endothelial layer	Accelerated atherosclerosis
Viremic spread	Hepatitis, gastrointestinal disease, adrenalitis, retinitis, encephalitis, polyradiculopathy, pneumonitis, bone marrow suppression
Inflammatory response	Allograft rejection, pneumonitis (BMT), vitritis (AIDS)

mouse transgenic studies showing activation of the major immediate-early promoter in multiple tissues,[84] the former possibility is more likely.

2.3.2. Pathogenesis

2.3.2a. Viremia and Viral Load. Most CMV disease in the immunocompromised is attributable to viremic spread to multiple organs (Table 12). In all populations, the risk of disease correlates strongly with high CMV viral loads. This phenomenon was first described in 1975 by Stagno and colleagues[85] who compared serial viruria titrations in neonates with symptomatic congenital, asymptomatic congenital, or perinatal infection. The group with CMV disease had on average one log higher viruria than those with asymptomatic congenital infection, who in turn had an average one log higher viruria than those with perinatal infection. This observation suggested that there might be a threshold viral load above which CMV disease became common, and this possibility has been investigated using quantitative-competitive PCR (QCPCR). After renal transplant, a significant correlation is seen[86] between the median values of maximum viruria posttransplant and the presence of CMV disease (see Fig. 14) and the same is true for viral loads in the blood.[87] Similar results are found in liver transplant and bone marrow transplant patients, with CMV viral loads in the blood significantly greater in patients with CMV disease in each case.[88,89]

For each transplant population, it is well known[90] that donor–recipient serostatus at the time of transplant identifies patients at risk of CMV disease (e.g., see Fig. 14). For recipients of solid organs, the group with highest risk are D+R− (i.e., donor seropositive, recipient seronegative), followed by D+R+, then D−R+. These groups correspond to primary infection, reinfection plus reactivation, and reactivation infections, respectively. For

bone marrow transplant recipients, the highest risk is D−R+, followed by D+R+, then D+R−. These groups correspond to reactivation, reactivation in the presence of marrow from immune donors, and possible transmission of virus from donor marrow, respectively. In addition, multiple studies in all transplant patient groups have shown that the detection of viremia is a risk factor for CMV disease.[91] To determine whether donor–recipient serostatus, viremia, and high viral load were independent markers of high-risk patients or were different ways of measuring the same pathogenesis factor, multivariate statistical analyses were performed.

For all three patient populations, high viral load remained a risk factor for CMV disease after viremia and donor–recipient serostatus had been controlled for statistically. In contrast, donor–recipient serostatus and viremia were no longer statistically significant once viral load had been controlled for.[86,89] Thus, high viral load is the major *determinant* of CMV disease and the classically defined risk factors of donor–recipient serostatus and viremia are *markers* of CMV disease by virtue of their association with high viral load. Indeed, it was shown that the relationship between increasing viral load and disease is nonlinear, such that a threshold value exists above which CMV disease becomes more common,[88] implying that marked prevention of disease could be obtained if drugs were deployed to prevent viral load reaching these critical values. Furthermore, serial measurements of viremia in several groups of immunocompromised patients demonstrated that CMV replicates with rapid dynamics, approximating to a half-life of 1 day.[92] This means that its reputation as a "slowly growing" virus is undeserved and that drugs of high potency are required to interfere with its replication. In addition, this high rate of replication explains how CMV variants resistant to GCV can evolve and provides a basis for calculating their relative fitness compared to wild type.[92] These mathematical modeling techniques also can be used to explain and predict the circumstances under which resistant strains become prominent.[93] In summary, short courses of GCV are unlikely to select resistant strains, but repeated courses, especially with oral GCV, provide ideal opportunities for resistant strains to flourish. They also demonstrate why resistant strains are cultured infrequently in practice; the process of incubating mixed populations of strains for 3–4 weeks in cell cultures lacking GCV allows the wild-type virus to outcompete the mutant strain, leading to the conclusion that resistance is not present. Thus, viral load measurements explain much of the pathogenesis of CMV disease and are important for understanding disease processes, for targeting the deployment of antiviral drugs, and for measuring the success of antiviral therapy and

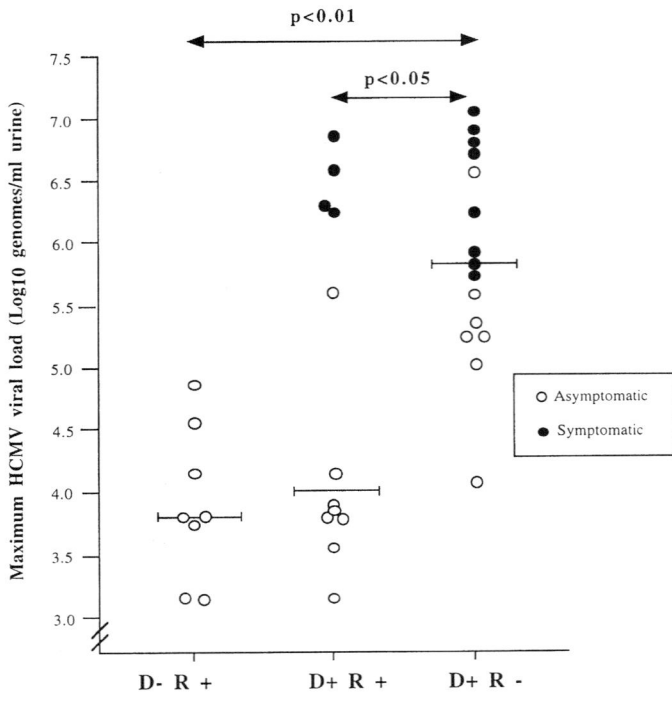

FIGURE 14. Interrelationships between peak CMV viral load, donor-recipient serostatus and CMV disease in 35 patients with active CMV infection after renal transplantation. (Reprinted with permission from Cope *et al.*[86])

predicting the development of resistance. However, this does not mean that knowledge of the other factors is without value. In particular, information about donor–recipient serostatus is available at the time of transplant and so can be used to decide which therapeutic option is most appropriate for individual patients. Furthermore, because high viral loads coincide with CMV disease, their detection cannot give prognostic value, whereas the detection of viremia can. For all these reasons, measurements of CMV viral load have been given prominence in this section on understanding pathogenic processes in cohorts of patients, whereas viremia is emphasized later as an important diagnostic tool for deciding which individual should be treated.

2.3.2b. Disease Processes within Infected Organs.

1. Direct Effects. Much of the end-organ disease caused by CMV can be attributed to *lysis*, i.e., destruction of cells as a direct result of virus replication. This can be seen clearly in the special case of the retinal cells destroyed by CMV retinitis, but similar processes probably account for hepatitis, adrenalitis, gastrointestinal tract ulceration, encephalitis, and polyradiculopathy. In all these cases, CMV can be seen histopathologically in biopsies

(through the owl's eye inclusions formed when cells are producing CMV) (see Fig. 15), can be cultured from biopsies (showing productive replication), and disease responds to antiviral therapy. In contrast, some other diseases associated with CMV may be triggered by the virus but be caused by immune responses.

In CMV pneumonitis after bone marrow transplant, disease does not present when the patient is most profoundly immunocompromised but appears typically in the second month posttransplant. Patients may have CMV viremia before this and may have asymptomatic CMV lung infection as shown by bronchial lavage at day 35,[94] but they will not develop disease until the marrow has engrafted. In addition, the presence of GVHD represents a risk factor for CMV pneumonitis. Once established, CMV pneumonitis responds very poorly to ganciclovir alone,[95] but the addition of immunoglobulin may give an improved response rate.[96] Taken together with parallel findings in mouse models, these results suggest that CMV pneumonitis is a disease triggered by the virus but caused by an aberrant host response (reviewed in Ref. 43). Although no subsequent studies have identified the host factors responsible, a cytokine-driven disease caused by

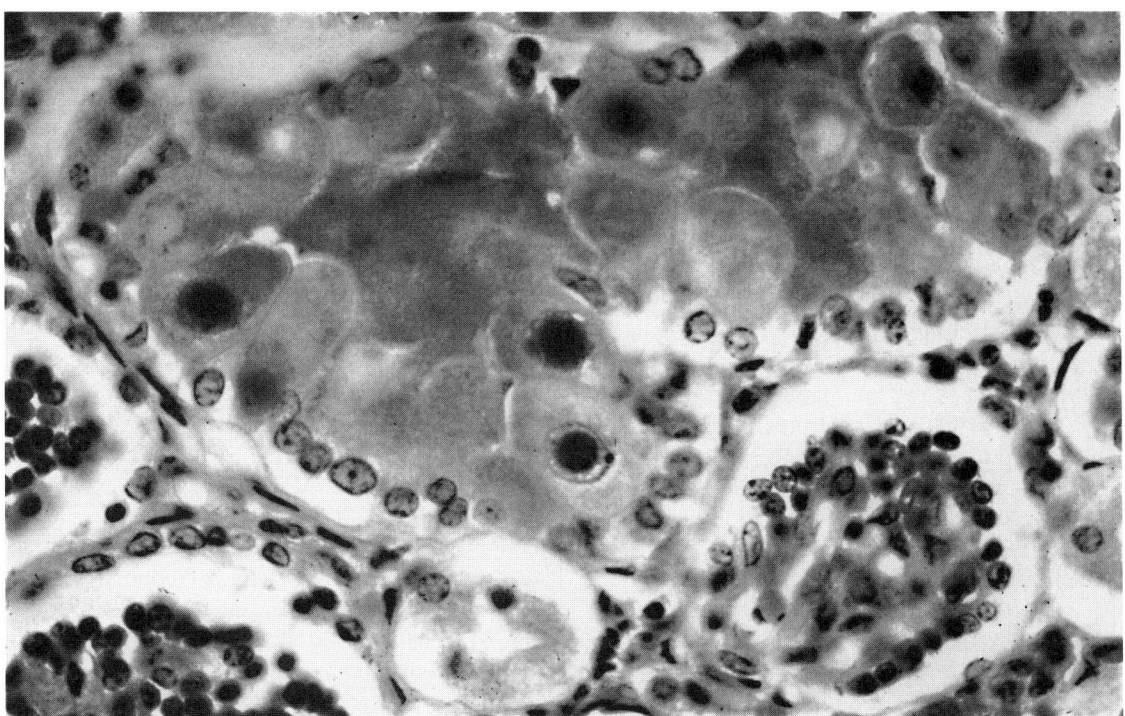

FIGURE 15. Histologic preparation of kidney showing owl's eye inclusions typical of CMV.

abnormal cell-mediated effectors is suspected. If correct, this hypothesis would explain why CMV pneumonitis is uncommon in AIDS patients, because by the time HIV has damaged the cell-mediated immune system sufficiently to allow CMV viremic spread to the lungs patients may have insufficient cell-mediated immune effectors to cause the aberrant inflammatory lung disease. A corollary to the hypothesis was that if the immune deficit in AIDS patients could be restored with antiretroviral medication, then CMV pneumonitis might appear as a "new" disease in AIDS patients.[43] At the time of writing, the author is unaware of any cases of CMV pneumonitis in AIDS patients given HAART, an observation that would suggest three possibilities. First, HAART may not have repaired the type of cell-mediated immune response necessary (or may have repaired also a set of immune regulators that normally control the aberrant response but which are absent soon after bone marrow transplant). Second, asymptomatic CMV lung infection may be uncommon in AIDS patients so that HAART provides the immune effectors but that there is no stimulus for them to home to the lungs. Third, the hypothesis may be incorrect. Although no cases of *lung* disease have been seen under

HAART administration, an inflammatory response to CMV has been seen in the *eye*, with corresponding increased levels of patient morbidity.

Patients with quiescent CMV retinitis receiving maintenance ganciclovir have developed vitritis after administration of HAART, which has dramatically altered the patient's perception of CMV eye disease. Before the availability of HAART, early cases of CMV retinitis were often identified through ophthalmologic screening of patients with low CD4 counts and/or CMV viremia. In many such cases, patients were asymptomatic and received lifelong ganciclovir to prevent CMV affecting their vision. Once given HAART, the presumed inflammatory response to persisting CMV antigens produces a cloudy vitreous, adversely affecting vision. Thus, for some patients, HAART has turned an asymptomatic disease into a symptomatic one.

2. Indirect Effects. CMV is associated with an increased incidence of acute graft rejection. The presumed pathogenesis involves CMV infection of the transplanted organ acting like a transplantation antigen, marking the organ for immune attack. Evidence for CMV playing this pathogenic role includes statistical association,[97] detection of CMV in organs undergoing rejection, apparent response of late rejection to ganciclovir therapy in an uncontrolled study,[98] and a significant reduction in acute graft rejection among patients randomized to high-dose valaciclovir in a placebo-controlled trial of prophylaxis after renal transplant.[48]

CMV is also associated with accelerated atherosclerosis after heart transplantation.[97] Several potential pathogenic mechanisms which fall under the heading of damaged endothelial surface (Table 12) could explain this association. CMV is found in monocytes/macrophages that could be attracted to sites of graft atheroma, either bringing CMV to that site or facilitating the formation of foam cells laden with oxidized lipids.[99] CMV major immediate-early protein binds p53 and has been found bound to p53 in arterial smooth muscle cells.[45] This suggests that CMV could reduce apoptosis leading to proliferation of such cells. The Us28 gene of CMV encodes a chemokine receptor that, once transferred experimentally to smooth muscle cells, confers on them the ability to migrate toward a source of chemokines. Thus, CMV infection might stimulate chemotactic mobility of these cells toward a site of inflammation.[100] Finally, CMV stimulates the formation of reactive oxidized intermediates and could contribute further to the progression of atherosclerosis.[101] It should be noted that follow-up of heart allograft patients who took part in a placebo-controlled trial of GCV has recently found reduced accelerated atherosclerosis among those allocated the drug.[50]

CMV is also associated with bacterial or fungal superinfection and follow-up of the heart allograft patients mentioned above has demonstrated reduced fungal infection in those randomized to GCV.[49] This implies that CMV is functionally immunosuppressive, although no immunologic mechanism has ever been confirmed. If CMV does contribute to the net level of immuosuppression, this could explain why it is associated with EBV-induced lymphoproliferative disease. It also could explain why AIDS patients with first-episode CMV retinitis have a significantly increased rate of death if their CMV viral load in blood is above the median of the whole group of patients[102] and why the death rate (in the pre-HAART era) was associated more strongly with CMV viral load than HIV viral load.[103]

2.3.3. Clinical Manifestations

The major clinical manifestations of CMV disease in different groups of immunocompromised patients are summarized in Table 13. These should be defined with reference to the criteria laid down at the International

TABLE 13. CMV Diseases in the Immunocompromised

Symptoms	Solid transplantation	Bone marrow transplantation	AIDS
Fever/hepatitis	+ +	+	+
Gastrointestinal	+	+	+
Retinitis	+	+	+ +
Pneumonitis	+	+ +	
Myelosuppression		+ +	
Encephalopathy			+
Polyradiculopathy			+
Addisonian			+
Immunosuppression	+		
Rejection/GVHD	+	?	
Atherosclerosis	+		
Death		+	+

CMV Workshop,[104] which include compatible clinical features plus signs of end-organ dysfunction plus demonstration of CMV in the affected organ (exception retina). In particular, diseases should be described in terms of the body system affected and the term "CMV syndrome" avoided.

2.3.3a. Fever/Leukopenia.   CMV viremia often is associated with prolonged spiking fever (e.g., >38°C on 3 consecutive days), with or without leukopenia. These constitutional symptoms may resolve spontaneously or may lead to end-organ disease. Other causes of fever (e.g., bacteremia) and leukopenia (e.g., doses of immunosuppressive drugs) must be excluded.

2.3.3b. Hepatitis.   Transaminases are raised (e.g., >2.5 × upper limit of normal), with or without alkaline phosphatase representing an obstructive component. Hyperbilirubinemia may be present but frank jaundice is uncommon. Hepatitis usually resolves spontaneously but may herald other end-organ disease.

2.3.3c. Gastrointestinal Disease.   CMV may involve the gastrointestinal tract anywhere from the mouth to the anus. The presentation is usually with pain, often accompanied by fever. Esophagitis, odynophagia, and abdominal pain mimicking perforation indicate involvement of the esophagus/colon. Endoscopy reveals mucosal ulcerations, with or without *Candida* superinfection. The ulcers respond slowly to treatment and may perforate and/or hemorrhage.

2.3.3d. Retinitis.   This can occur in any immunocompromised patient but is most common in AIDS. Symptoms, if present, include "floaters," flashing lights, and loss of central vision. Small peripheral lesions may be unnoticed by the patient; lesions involving the macula may be imminently sight-threatening and demand immediate treatment. Involvement of a large proportion of the retina interrupts retinal/scleral attachment and represents a risk factor for retinal detachment. Before the availability of HAART, most patients had progression of retinitis and the goal of treatment was to preserve vision (reviewed in Ref. 105). Treatment with HAART may be followed by vitritis (see Section 2.3.2), with paradoxical impaired vision despite better control of retinitis.

2.3.3e. Encephalitis. In AIDS patients, CMV reaches the brain by one of two routes: extension of a neighboring endotheliitis or by the choroid plexus. In the former case, the encephalitis follows a subacute course, difficult to distinguish from HIV dementia. In the latter, necrotizing ventricular encephalitis produces cranial nerve defects, nystagmus, and ventriculomegaly (reviewed in Ref. 106). In both cases, response to treatment is poor.

2.3.3f. Polyradiculopathy. An AIDS patient with a very low CD4 count presents with subacute weakness of the lower limbs, with or without bladder paralysis. Lumbar puncture reveals abundant polymorphonuclear leukocytes in the CSF. Immediate treatment is indicated but the clinical response is poor.

2.3.3g. Pneumonitis. Most cases occur after bone marrow transplant with concurrent GVHD (see Section 2.3.2). There is rapid onset of dyspnea plus hypoxia. Chest X ray may be relatively clear initially but progresses to show interstitial infiltrates. There is a high mortality, with poor response to treatment.[96]

2.3.4. Diagnosis

2.3.4a. Detection of Viremia. This can be performed by any published method that has been shown to provide a good positive predictive value for CMV disease, e.g., 50–60% for the patient population to be followed. Thus, the rapid diagnostic techniques using cell culture amplification testing of virus (termed DEAFF testing in Europe[107] and shell vial in the United States[108]) are no longer sufficiently sensitive and should be replaced with newer methods. Examples include PCR in whole blood,[109] PCR in plasma,[110] and antigenemia.[111] Note that a randomized trial has shown PCR to be superior to conventional cell culture for deciding when to initiate preemptive therapy.[112] Laboratory protocols differ and it is important that all aspects of each method are followed in detail including sample processing and virus detection. These have been optimized to avoid the detection of latent virus while providing good sensitivity for predicting disease. It should be noted that these methodologies have not been optimized for the detection of asymptomatic infection. Thus, it is not possible to "mix and match" different

aspects of PCR protocols. Whichever method is chosen, the results must be audited at regular intervals to ensure that the anticipated positive predictive values are being attained. This also should be done when any changes to immunosuppressive or antiviral protocols are being contemplated.

2.3.4b. CNS Involvement. PCR of CSF is the method of choice for diagnosing CMV CNS infection.[113]

2.3.4c. DEAFF/Shell Vial. This method is still sufficiently sensitive and robust to diagnose CMV lung infection using bronchoalveolar lavage fluid. Cells from this fluid also can be cytocentrifuged and stained with monoclonal antibodies; but this approach, while more rapid, lacks sensitivity compared to DEAFF/shell vial amplification.

2.3.4d. Histopathology. This is performed on tissue biopsies to detect classic Cowdry type A intranuclear "owl's eye" inclusion bodies (see Fig. 15). It is insensitive and so has a high specificity for disease.[114]

2.3.4e. Cell Culture. This is performed on tissue biopsies after tissue is minced and inoculated directly onto cells. It is slow but sensitive.

2.3.4f. Serology. Many enzyme immunoassays are commercially available for the detection of CMV IgG antibodies pretransplant in both donor and recipient. Serologic testing has no role to play posttransplant.

2.3.5. Management

The principles of managing CMV infection and disease in the immunocompromised host are to anticipate their development, to define policies for monitoring patients routinely for the presence of viremia (and/or detection of CMV in urine or saliva) according to their baseline risk of CMV disease, and to enhance surveillance if patients develop a condition likely to increase their risk of CMV disease. Using the principles of evidence-based medicine, the patient then will be offered prophylaxis or preemptive therapy based on an assessment of their individualized risk of disease, together with data from controlled clinical trials in the same patient group supporting the efficacy and safety of possible antiviral interventions.

Strategies for Deploying Antiviral Agents. Different strategies have been devised for controlling CMV disease based on the efficacies and toxicities of the drugs available at present (summarized in Table 14).

1. True Prophylaxis. This strategy may be used where an assessment at baseline shows that the risk of disease is high, the chance of severe disease is also high, and at least one double-blind, randomized, placebo-controlled trial supports the efficacy and safety of pro-

TABLE 14. Strategies for Chemotherapy of CMV

Term used	When drug given	Risk of disease	Acceptable toxicity	Treatment decision prompted by
Prophylaxis	Before active infection	Low	None	Clinician
Delayed prophylaxis	Before active infection but after rejection	Medium	Low	Clinician
Suppression	After peripheral detection of virus	Medium	Low	Laboratory
Preemptive therapy	After systemic detection of virus	High	Medium	Laboratory
Treatment	Once disease apparent	Established	High	Both

phylaxis in the target population. The patient then will be given the drug from the time of transplant onward and continued for the duration studied in the controlled clinical trial that provided evidence for its use. This is termed "true prophylaxis," because from a virologic perspective it administers the drug before there is active virus replication.

2. Delayed Prophylaxis. At baseline, a decision was made that true prophylaxis was not indicated. However, the patient's situation has changed, e.g., because augmented immunosuppression is required to control an episode of graft rejection and so it is decided to start prophylaxis now.[115] This is still termed "prophylaxis" because the drug is given before there is active virus replication.

3. Suppression. The patient has been monitored by collecting weekly samples of urine and/or saliva and processing them by a laboratory method shown to provide a moderate positive predictive value for CMV disease, e.g., 30%.[91,109] It is decided to give an antiviral drug with the intention of suppressing virus replication below the level needed to cause viremia. This strategy therefore is termed "suppression" and requires a drug more potent than may be required for prophylaxis because it has been given when the virus already has a "headstart" in the race for control of replication.

4. Preemptive Therapy. This term describes intervention when the results of laboratory tests indicate that a patient is at imminent risk of CMV disease.[116] It is used in two circumstances: detection of viremia in any immunocompromised patient and detection of asymptomatic lung infection after bone marrow transplant.[94] In either case, a highly potent drug is required.

In the first example, the patient has been monitored by collecting weekly samples of blood processed by laboratory methods known to provide a high positive predictive value for CMV disease, e.g., 50–60%.[91,109] It is decided to give an antiviral drug with the intention of halting CMV viremia before it reaches the high viral loads required to cause disease. In the second example, bone marrow transplant patients are tested for evidence of CMV lung infection by performing bronchial lavage at day 35. If infection is found, patients are treated in the hope of preventing the immunopathologic pneumonitis. Although successful in one non-placebo-controlled study,[94] this approach has now been superseded by less invasive testing for viremia.

Decision points for starting preemptive therapy must be based on the results of clinicopathologic studies with the assay under evaluation. Examples include detection of viremia by PCR or antigenemia above a cutoff value associated with a high risk of disease or two consecutive samples PCR-positive. More recently, the results of viral dynamic assessments have been applied to this problem; patients at risk of disease can be identified by the absolute value of viral load found in the first PCR-positive sample, coupled with an assessment of individual viral dynamics by calculating the rate of increase from the last PCR-negative sample.[117]

5. Treatment of Established Disease. The patient meets the case definition of CMV because he or she has compatible symptoms and signs, together with detection of CMV in the affected organ. A highly potent drug is required that will penetrate the affected organ and resolve the disease, including any immunopathologic components.

Results of Double-Blind, Randomized, Placebo-Controlled Trials. Results of published trials defined according to these criteria are given in Table 15, parts A–D. It will be seen that the most potent drug *in vitro*, ganciclovir, has been subjected to several such clinical trials, but the other licensed compounds, foscarnet and cidofovir, have not. Other agents such as IFN-α, acyclovir, valaciclovir, and immunoglobulin also have been evaluated, although they may not conventionally be thought of as having anti-CMV activity.

Table 15A shows that, in addition to ganciclovir,[118–123] IFN-α[124–126] and acyclovir[48,127,128] have activity against CMV *in vivo*. The only two studies not to show an effect were the two studies of immunoglobulin,[129,130] which suggests that if immunoglobulin has a role in the

TABLE 15. Double-Blind, Placebo-Controlled, Randomized Trials of CMV

Strategy	Drug	Bone marrow	Renal	Heart	Liver
A. Significantly reduced CMV infection (marked with asterisk)					
Treatment	GCV	Reed[118]*			
Suppressive	GCV	Goodrich[119]*			
Prophylaxis	Interferon		Cheeseman[124]*		
			Hirsch[125]*		
			Lui[126]*		
	ACV	Prentice[128]*	Balfour[127]*		
	VACV		Lowance[48]*		
	Ig		Metselaar[130]*		Snydman[129]
	GCV	Winston[121]*		Merigan[122]	Gane[123]*
		Goodrich[120]*		Macdonald[131]*	
B. Significantly reduced CMV disease (marked with asterisk)					
Treatment	GCV	Reed[118]			
Suppressive	GCV	Goodrich[119]*			
Prophylaxis	IFN		Cheeseman[124]		
			Hirsch[125]*		
			Lui[126]		
	ACV	Prentice[128]	Balfour[127]*		
	VACV		Lowance[48]*		
	Ig		Metselaar[130]		Snydman[129]*
	GCV	Winston[121]		Merigan[122]*	Gane[123]*
		Goodrich[120]*		Macdonald[131]*	
C. Significantly reduced death (marked with asterisk)					
Treatment	GCV	Reed[118]			
Suppressive	GCV	Goodrich[119]*			
Prophylaxis	IFN		Cheeseman[124]		
			Hirsch[125]		
			Lui[126]		
	ACV	Prentice[128]*	Balfour[127]		
	VACV		Lowance[48]		
	Ig		Metselaar[130]		Snydman[129]
	GCV	Winston[121]		Merigan[122]	Gane[123]
		Goodrich[120]		Macdonald[131]	
D. Significantly improved indirect effects (marked with asterisk)					
Treatment	GCV	Reed[118]			
Suppressive	GCV	Goodrich[119]			
Prophylaxis	IFN		Cheeseman[124]		
			Hirsch[125]		
			Lui[126]		
	ACV	Prentice[128]	Balfour[127]		
	VACV		Lowance[48]*		
	Ig		Metselaar[130]		Snydman[129]
	GCV	Winston[121]		Merigan[122]*	Gane[123]
		Goodrich[120]		Macdonald[131]	

prophylaxis of CMV disease, it may not be working through inhibition of CMV replication.

Table 15B shows that ganciclovir failed to demonstrate a significantly better resolution of established CMV disease than placebo.[118] Part of this disappointing outcome may be attributed to the low dose (2.5 mg/kg tid) and/or short duration used (14 days) to treat gastrointestinal disease in bone marrow transplant patients.[118] Nevertheless, it illustrates the difficulty of treating established CMV disease and so argues that the other strategies, which aim to prevent CMV disease, always should be pursued in preference to waiting for disease to establish itself. Ganciclovir did reduce CMV disease when used in the suppressive mode of bone marrow transplant patients.[119] It also had a significant benefit when used in one[120] of two trials of prophylaxis after bone marrow transplant; the second study[121] showed a strong trend in favor of ganciclovir, which just failed to reach conventional statistical significance. Ganciclovir also significantly reduced CMV disease following prophylaxis given orally to liver transplant patients[123] and intravenously to heart transplant patients.[122,131] However, benefit after heart transplant was seen in the low-risk group only, with no effect in the D+R− group of one study,[122] whereas the opposite outcome was seen in a second.[131] This difference might result from the longer treatment course in the later study, together with a design difference such that patients with rejection were given additional doses of GCV. Prophylactic acyclovir significantly reduced CMV disease after renal transplant,[127] as did prophylactic valaciclovir.[48] In a prophylaxis trial after bone marrow transplant, acyclovir significantly decreased CMV viremia and showed a nonsignificant trend toward reduced CMV disease.[128] A trial of immunoglobulin prophylaxis showed reduced "CMV syndrome," despite having no significant effect on CMV infection.[129] Subgroup analysis showed an effect on CMV-associated fungal superinfection (part of the predefined "CMV syndrome") so it remains possible that the immunoglobulin predominantly reduces fungal rather than CMV infection.

Table 15C examines whether these drugs demonstrated a survival benefit subsequent to their inhibition of CMV. The number of deaths in the solid organ transplant populations is too low to provide the statistical power to address this issue. After bone marrow transplant, ganciclovir significantly improved survival when used suppressively.[119] However, when used prophylactically, no effect was seen.[120,121] This was not a problem of small sample size and neither study demonstrated even a trend in favor of ganciclovir. The most likely explanations are that some patients with viremia still received preemptive therapy,[120] so reducing CMV-induced mortality in both arms, but that ganciclovir-induced neutropenia induced by prophylactic GCV predisposed patients to succumb to bacterial or fungal superinfections, so mitigating the potential benefits of this drug. Overall, these studies indicate that ganciclovir is too toxic a compound to be used for prophylaxis after bone marrow transplant, although it is literally lifesaving when used in suppressive mode.[119] This illustrates that in prophylaxis all patients are exposed to side effects and that suppression, by limiting the number of patients exposed to the drug, can produce better therapeutic ratios. In contrast, acyclovir prophylaxis after bone marrow transplant produced a survival benefit,[128] presumably because its more modest efficacy was not offset by serious toxicity.

Table 15D summarizes the studies that so far have reported significantly reduced indirect effects of CMV. After renal transplant, valaciclovir produced a marked reduction in biopsy-proven acute graft rejection corresponding to a 50% decrease in incidence among seronegative recipients at risk of primary infection.[48] The effects in seropositive recipients were smaller, implying that CMV (rather than another herpesvirus susceptible to the drug) is responsible for this indirect effect and that most CMV-induced graft rejection occurs in the subset of patients with primary infection. Following heart transplantation, GCV significantly reduced fungal infections[49] and accelerated atherosclerosis.[50]

Conclusions

1. Decisions about which drugs to recommend for particular treatment strategies must draw on evidence-based medicine provided by the results of controlled clinical trials. Decisions must be based on considerations of toxicity as well as on efficacy.

2. For bone marrow transplant patients, if you wish to use prophylaxis, give acyclovir[128]; if you wish to use suppression, give ganciclovir.[119] Since the latter study showed that of the samples of urine, saliva, and blood processed, only blood provided prognostic value, this conclusion probably applies to preemptive therapy.

3. The toxicity problems of ganciclovir apply specifically to the bone marrow transplant population. A head-to-head comparison of low-dose ganciclovir versus acyclovir for prophylaxis after liver transplant shows that ganciclovir is superior.[132]

4. Similar data from AIDS patients cannot be presented, because remarkably no trials have been designed based on these virologic criteria. Two trials[52,133] have been conducted of clinical prophylaxis but these are not the same as true prophylaxis, because at the time of randomization, although no patients had CMV disease, some of them would be expected to have asymptomatic infection. Indeed, the subsequent virologic studies show that oral ganciclovir had its greatest effect when given to the subset receiving true prophylaxis and had little effect in the subset receiving preemptive therapy.[134] This illustrates clearly that the therapeutic implications of viral dynamics are not simply abstract concepts but have real-world benefits to offer patients. In contrast, the virologic studies of valaciclovir show that this drug had its greatest effect when given for preemptive therapy.[135] This might

be thought to imply that valaciclovir is more potent *in vivo* than oral ganciclovir, but a randomized head-to-head comparison would be needed to test this possibility. Such a trial is unlikely to be conducted because the high dose of valaciclovir chosen (2 g qid) was poorly tolerated by the AIDS patients,[52] although the same dose was safe when studied in renal transplant patients.[48]

5. The acyclovir and valaciclovir studies do show that potent inhibition of DNA polymerase by acyclovir triphosphate can have clinical utility under some circumstances. In the renal transplant study, the authors provide evidence that plasma levels of acyclovir were higher than expected because of poor renal clearance but were still lower than would be required to inhibit CMV based on *in vitro* data,[136] which demonstrates clearly that the IC_{50} levels produced by fibroblast cell cultures are misleadingly high. Pharmaceutical companies screening for other anti-CMV compounds in this cell line should consider whether they risk rejecting a clinically useful compound. Likewise, those seeking therapeutic compounds for other betaherpesviruses should consider whether their results for acyclovir (or other compounds in development) are being fairly represented.

2.4. Human Herpesvirus 6 and 7

2.4.1. Epidemiology

Seroepidemiologic studies show that over 90% of children acquire each of these viruses during the first 2 years of life.[137] The presumed source of infection is the mother, with saliva a likely means of transmission. Studies suggest infection is acquired from 6 months onward, presumably once passive protective maternal immunity declines.[137] Among the minority of children who experience exanthem subitum, their median ages are 7 months for HHV-6 and 10 months for HHV-7.[138]

2.4.2. Pathogenesis

Since these viruses were only identified in 1986 and 1990, respectively, less hard data are available about their pathogenesis (Table 16). Both viruses are tropic for CD4+ lymphocytes. HHV-7 uses CD4 as its receptor,[139] whereas HHV-6 uses the CD46 molecule.[140] HHV-6 also replicates in macrophages and NK cells, leading to the suggestion that it has the potential to be immunomodulatory (reviewed in Ref. 141). HHV-6 encodes a chemokine gene, termed U83.[142] When expressed as an Fc-fusion protein, it showed transient mobilization of ionized calcium and chemotactic activation, although to a lower

TABLE 16. Pathogenesis of HHV-6/HHV-7 Diseases

Pathogenic feature	Examples of disease
Local infection	
Local spread	
Damaged epithelial layer	
Viremic spread	Encephalitis, hepatitis, bone marrow suppression, pneumonitis?
Inflammatory response	Allograft rejection, GVHD?, up-regulates HIV (HHV-6)

level than with RANTES. The corresponding gene in HHV-6 type A lacks a signal sequence, and so is probably not secreted, which is possibly an important clue to help explain the apparent biological differences between types A and B of HHV-6. Both HHV-6A and -B induce T-cell apoposis *in vitro*, which is augmented by tumor necrosis factor-alpha (TNF-α) and which affects uninfected bystander cells.[143] Studies of HHV-6A and -B in the thymus/liver-implanted SCID-hu mouse model show progressive thymocyte depletion, especially of T-cell progenitors.[144]

Both HHV-6 and HHV-7 have the potential to interact with HIV *in vitro*, HHV-6 can transactivate the HIV LTR but also competes with HIV when both viruses replicate fully in cell culture.[145] HHV-7 inhibits HIV by virtue of competition for the CD4 receptor.[139]

In patients with or without exanthem subitum, HHV-6 has demonstrated clear neurotropism and is strongly implicated as the cause of febrile fits,[137] being detected by PCR in the CSF of such cases.[146] It also has been detected by CSF PCR in cases of focal encephalitis, clinically indistinguishable from herpes simplex encephalitis.[147] HHV-6 also has been implicated as an etiologic agent of multiple sclerosis, persisting within the white matter. There are certainly attractions in a model of multiple sclerosis pathogenesis where a herpesvirus able to reactivate periodically could render oligodendrocytes susceptible to immune attack, but the data available to date fall far short of proof.

2.4.2a. Direct Effects. In the immunocompromised host, case reports have appeared of encephalitis, pneumonitis, and bone marrow suppression where HHV-6 was detected in the affected organ. Although some of these cases seem plausible, in view of the ubiquity of HHV-6 infection, parallel investigations are required in control patients before it can be concluded that this virus plays an etiologic role. For example, a recent case series shows that HHV-6 detection by PCR in CSF correlates with CNS symptoms compared to a control group of asymptomatic immunocompromised patients.[148]

TABLE 17. Detection of HHV-6 and HHV-7 after Solid Organ Transplantation

Author	Virus studied	Detection method	Number patients	Number bloods	Antiviral prophylaxis	Observed diseases[a]	Case reports
Schmidt[198]	HHV-6	PCR	46 liver	287	Ig	None	None
Herbein[199]	HHV-6	Virus isolation	32 liver, renal	NG[b]	NG	None	None
Osman[150]	HHV-6 HHV-7	PCR	56 renal	NG	NG	None	↑CMV disease
Griffiths[33]	HHV-6 HHV-7	PCR plus QCPCR	60 liver	536	None	Rejection	
Kidd[149]	HHV-6 HHV-7	PCR plus QCPCR	52 renal	596	None	Rejection	↑CMV disease

[a]Effects on whole population.
[b]NG, not given.

Prospective virologic studies have been conducted in transplant patients and the results are summarized in Table 17 for recipients of solid organs and in Table 18 for recipients of bone marrow. Overall, these studies showed a high incidence of infection but low incidence of the diseases indicated in the case reports. It also was not clear from many studies whether patients were receiving prophylaxis with antiherpes drugs.

2.4.2b. Indirect Effects.   Two studies aimed to identify sufficient patients who had not received antiviral prophylaxis and to determine whether active infection with HHV-6 or HHV-7 was associated with graft rejection in the whole population and whether any association was confounded by CMV. After liver transplant, biopsy-proven graft rejection was significantly associated with CMV and HHV-6, but not with HHV-7.[33] In multivariate analyses, the effects of CMV and HHV-6 were independent of each other. After renal transplant, biopsy-proven

graft rejection was significantly associated with HHV-7.[149] Interestingly, these pathologic associations followed the pattern of virus replication as documented by QCPCR for HHV-6 and HHV-7. Specifically, the median HHV-7 virus load was significantly greater than that of HHV-6 after renal transplant, whereas after liver transplant the trend was in the opposite direction.[33,149] In each case, the median quantity of CMV viral load was significantly greater than that of either HHV-6 or HHV-7. These results imply that HHV-6 and HHV-7 have differing abilities to replicate in hosts with different transplanted organs and that their replication may trigger graft rejections. The results also suggest that some of the association between CMV and graft rejection may be confounded by the presence of HHV-7.

Interestingly, CMV disease was significantly increased in renal patients coinfected with HHV-7,[149] but neither HHV-7 nor HHV-6 increased the incidence of

TABLE 18. Detection of HHV-6 and HHV-7 after Bone Marrow Transplantation

Author	Virus studied	Detection method	Number patients	Number bloods	Antiviral prophylaxis	Observed diseases[a]	Case reports
Wilborn[200]	HHV-6	PCR	57	415	NG[b]	None	GVHD
Kadakia[201]	HHV-6	Virus isolation	26	NG	NG	No association with engraftment	
Wang[162]	HHV-6 HHV-7	PCR	37	270	High-dose ACV (9)[c]	None	Engraftment
Appleton[202]	HHV-6	PCR, IHC	57	NG	Moderate-dose ACV	None	GVHD (skin not blood)
Chan[203]	HHV-6 HHV-7	PCR	61	563	NG	None	Engraftment

[a]Effects on whole population.
[b]NG, not given
[c](9) indicates the number of patients given prophylaxis.

CMV disease after liver transplant. Another study of renal transplant patients first reported that CMV disease was associated with HHV-7 coinfection.[150] Another publication in liver transplant patients, using serology to detect HHV-6 infections, found increased CMV disease among those with immune responses to HHV-6,[151] and this association has been confirmed using PCR assays (Dr. C. Paya; personal communication). Another recent publication using PCR in liver transplant patients also has reported a significant association between detection of HHV-6 and CMV disease.[152] Further detailed studies are required to determine whether these viruses truly interact or whether they share a common risk factor, such as graft rejection. One possible source of confounding can be excluded; HHV-6 and HHV-7 do not appear to cause owl's eye inclusions, which could cause misdiagnosis of CMV disease.[114]

Several studies have shown that HHV-6 can be detected in multiple tissues from AIDS patients at autopsy.[39,153,154] Given that betaherpesviruses can have either positive or negative cofactor effects on HIV replication *in vitro*, multiple tissues collected at autopsy were studied to determine whether the presence of a given virus is associated with an increased HIV viral load measured by QCPCR of proviral DNA. The median quantity of proviral DNA was increased significantly in the presence of HHV-6 but not CMV or HHV-7.[155] It also was interesting to note that the median viral load of HHV-6 was increased in the presence of HIV, suggesting a bidirectional mutual interaction between these viruses *in vivo*.[155] There was no evidence of competition between HHV-7 and HIV, as has been described *in vitro*.[156]

2.4.3. Clinical Manifestations

Exanthem subitum is characterized by a high fever (38–39°C) for 2–3 days, which ceases suddenly (hence, subitum), and is followed by a pale maculopapular rash. The clinical course is usually benign, but complications of convulsions, hepatitis, encephalitis, or infectious mononucleosis may ensue.

2.4.4. Diagnosis

Serology for IgG antibodies is available but rarely used due to the high prevalence of antibodies.

Suspected cases of HHV-6 infection should be investigated by a test for virus or its components, which has been shown to give negative results in patients with normal immunity, i.e., will not detect latent HHV-6. Exam-

ples include virus isolation, PCR, and methods corresponding to DEAFF/shell vial and antigenemia for CMV. It is interesting to note that HHV-6 is preferentially detected in HIV-positive patients with high CD4 counts.[37] This surprising relationship for a herpesvirus probably reflects the number of CD4+ cells that are required to produce detectable levels of HHV-6/7 in the blood.[37]

When PCR methods are used, it is important to ensure that the DNA extracted from leukocytes is not simply detecting HHV-6 genome integrated into the chromosomes of these cells. Such DNA integration was first reported for chromosome 17 and subsequently has been found in chromosomes 1 and 22.[157–160] The biological significance of chromosomal integration is uncertain, although there is clear evidence of chromosomal inheritance of what appears to be full-length HHV-6 genome. From the diagnostic standpoint, QCPCRs are useful because the quantity of HHV-6 DNA is high.[33] However, the easiest way to distinguish chromosomal integration from true infection is to examine serial samples; a strong HHV-6 PCR reaction pretransplant that persists posttransplant is not consistent with the expected and described pattern of HHV-6 infection posttransplant.[33]

2.4.5. Management

HHV-6 and HHV-7 do not form plaques in cell culture, so standard *in vitro* assays for drug sensitivity cannot be employed. Nevertheless, available *in vitro* data show that HHV-6 is sensitive to GCV and foscarnet but not ACV.[161] Controlled trials in humans are required to determine whether these results accurately mirror those found *in vivo*, especially since the *in vitro* data for CMV/ACV gave misleading results. Indeed, one study reports that HHV-6 was detected less frequently when bone marrow transplant patients were receiving ACV prophylaxis.[162]

In the immunocompromised host, HHV-6 should be considered in the differential diagnosis of encephalitis, pneumonitis, and bone marrow failure after successful engraftment. In general, empiric treatment for suspected CMV disease should be expected to also provide some inhibition of HHV-6.

When reviewing protocols for antiviral prophylaxis for CMV, consider that some drugs may also have activity against HHV-6 and HHV-7, so that some of the clinical benefit attributed to inhibition of CMV may be due to inhibition of these newer betaherpesviruses. Consider that acute treatment for suspected CMV disease may also suppress HHV-6 or HHV-7 so that clinical responses may be heterogeneous.

2.5. Epstein–Barr Virus

2.5.1. Epidemiology

Serologic assays show that 90% of adults in developed and virtually 100% in developing countries have been infected with EBV. The virus is spread by saliva and can be detected in the mouth washings of most seropositive adults.[163] Neonates are protected from infection by maternal antibody but acquire infection from 6 months onward. Most infections at this age are asymptomatic, although occasional cases of glandular fever do occur. In contrast, primary infection in adolescents is frequently accompanied by infectious mononucleosis.

There are two different types of EBV, termed EBV-1 and EBV-2.[164] This distinction is based upon genetic polymorphisms in the EBNA2, -3A, -3B, and -3C nuclear antigens. Individual strains of EBV can be distinguished by polymorphisms produced by the number of terminal repeats present in the genome.

2.5.2. Pathogenesis

EBV in saliva infects a new individual via B cells situated in the crypts of the oral mucosa. The EBV gp350 envelope protein binds onto the CD21 (C3d) receptor protein. EBV then establishes a persistent infection of B cells, which circulate throughout the body at a frequency ranging between 1 and 60 per million B cells in the blood.[165,166] These cells then return to the oral mucosa where they represent a source of productive infection. Originally, it was thought that EBV also persistently infected epithelial cells, but the need for chronic recirculating B cells to maintain persistent infection of the oral epithelium has been demonstrated under informative circumstances in a bone marrow transplant patient. When B cells of recipient origin were ablated by total-body irradiation, salivary excretion ceased but salivary excretion was reestablished years later with a different EBV strain, presumably acquired exogenously.[167] Furthermore, patients with a congenital absence of B lymphocytes have no detectable EBV infection in saliva or blood and have no T-cell recognition of EBV epitopes.[168] These results indicate that salivary excretion of EBV probably originates from B cells and that epithelial cells are not infected with EBV (reviewed in Ref. 168).

2.5.2a. Latency.   During latency, a single protein (EBNA1) is expressed. The function of EBNA1 is to maintain the episomal form of EBV, ensuring that the genome is replicated in tandem with cell division so that each daughter cell contains EBV. The question then arises as to why cells expressing EBNA1 are not destroyed in immunocompetent hosts. EBNA1 contains an unusual glycine-alanine repeat structure that interferes with the ability of proteosomal enzymes to process the protein into immunogenic peptides (see Fig. 10). As a result, cells expressing EBNA1 can evade the cell-mediated immune response of immunocompetent people.

EBV-encoded proteins (see Table 19) are highly immunogenic, so virus-producing cells are rapidly eliminated in immunocompetent persons. During acute infectious mononucleosis, CTL directed against lytic and latent antigens can be detected using limiting dilution assays.[169,170] In the acute phase, more cells recognize lytic antigens, but cells recognizing latent antigens become dominant during convalescence. These assays identified between 1:100 and 1:500 T cells as being specific for an individual EBV epitope during the acute phase and between 1:500 and 1:2500 during convalescence. More recently, the use of tetramer technology has forced a reappraisal of these estimates. The use of tetramers recognizing EBV epitopes in the context of common HLA proteins produces an estimate of CTL frequency approximately 10-fold higher than that noted above.[171] This high frequency indicates that most activated, proliferating T cells seen during the acute phase of infectious mononucleosis have been stimulated directly by EBV and recognize EBV-encoded epitopes.

In individuals with distinct forms of immunodeficiency, cells expressing other EBV proteins can survive without destruction. They are categorized according to the type of latency found in cells maintained *in vitro*.

1. Latency I.   Cells with this type of latency are typically found in Burkitt's lymphoma. Patients have chronic antigenic stimulation of B cells (due to malaria or to HIV infection), which presumably drives B-cell proliferation. Cells are selected by their ability to escape cell-mediated immune control manifest by constitutive decreased display of HLA molecules and/or decreased cell adhesion molecules.

**TABLE 19. EBV-Encoded Proteins
Expressed during Latency**

Protein	Importance	CMI recognition
EBNA-1	Maintains episome	Absent
EBNA-2	Transformation/proliferation	Weak
EBNA-3A	Transformation/proliferation	Strong
EBNA-3B		
EBNA-3C	Transformation/proliferation	Strong
EBNA-LP	Leader protein	Weak
LMP-1	Transformation/proliferation	Weak
LMP-2	Transformation/proliferation	Weak

2. Latency II. These cells are found in nasopharyngeal carcinoma, in EBV-associated Hodgkin's disease (75% of Hodgkin's disease in developing countries; 25% in developed countries are EBV-associated) or T-cell lymphomas, or in cases of hemophagocytosis syndrome.

3. Latency III. Cells expressing this larger number of EBV proteins can only persist when cell-mediated immune responses are profoundly compromised. Cells with this pattern therefore are found in lymphoproliferation after transplant or AIDS.

2.5.2b. Lymphoproliferation. Proliferation of these cells leads to selection of variants with genetic changes that facilitate their ability to persist. The LMP1 protein of EBV is a strong candidate for transforming activity encoded by the virus and acts like a constitutively active TNF-α receptor, driving the availability of NFKB.[172] In addition, the cellular proto-oncogene, *myc*, may be activated by transfer from chromosome 8 to 14 (the heavy chain immunoglobulin locus), chromosome 2 (kappa chain locus), or chromosome 22 (lambda chain locus). Furthermore, tumor suppressor genes such as p53 may be inactivated or cellular adhesion molecules downregulated. This sequential acquisition of phenotypic changes leads ultimately to a cell with highly malignant behavior. Initially, the lymphoproliferation therefore is polyclonal but may ultimately become monoclonal. Likewise, the EBV genome may be monoclonal, showing its derivation from the progenitor cell of the ultimately dominant tumor type, or polyclonal, representing EBV infection of several coevolving lineages. The mono- and polyclonality of EBV can be demonstrated by using restriction enzymes to cut the EBV episome in regions flanking terminal repeats (see Fig. 3).

2.5.2c. X-Linked Lymphoproliferation. Massive lymphoproliferation, frequently fatal, was initially described in the Duncan kindred and subsequently identified in many families.[173] It represents an X-linked inheritance of selective inability to control lymphoproliferation caused by EBV. Studies of affected families mapped the gene to the long arm of the X chromosome. Two groups then independently identified the defective gene as SAP (SLAM-associated protein). One group achieved this through positional cloning of the X chromosome,[174] whereas the second[175] identified SAP while studying proteins that can bind to SLAM (signaling lymphocyte-activation molecule).

SLAM is present on both B and T lymphocytes. It acts as a self-ligand so that SLAM–SLAM interactions on neighboring cells can mediate cell–cell signaling events. In T cells, SLAM activation mediates proliferation, induces IFN-γ, and switches the phenotype toward Th1 (i.e., toward CTL effector function). In B cells, SLAM increases proliferation and differentiation.

This recent work will lead to a mechanistic explanation for X-linked lymphoproliferation (XLP); at present, it is clear only that the normal SAP–SLAM interactions are dysfunctional (reviewed in Ref. 176). Patients with XLP exhibit proliferation of both B and T cells. It can be assumed that the T cells are ineffective, because the disease may be controlled by passive adoptive transfer of functional T cells. Nevertheless, these T cells could be pathogenic to the host and death is frequently attributed to hepatic necrosis and bone marrow failure accompanied by infiltration of these organs by CTL.

2.5.3. Clinical Manifestations

2.5.3a. Infectious Mononucleosis. Primary EBV infection in adolescence is accompanied by a 50% rate of infectious mononucleosis; a high attack rate attributed to a large virus inoculum acquired through deep kissing. A typical case would be a college freshman presenting with fever, malaise, and sore throat. Most cases resolve spontaneously, but symptoms may persist for weeks or months and can be debilitating to academic studies due to persistent tiredness and malaise. Occasional cases are life threatening in the acute phase, with respiratory embarrassment secondary to enlarged tonsillar tissue or with splenic rupture due to massive engorgement with lymphocytes.

2.5.3b. Lymphoproliferative Syndrome. This term encompasses a wide variety of clinical presentations with varying prognoses. In recipients of solid organs, it frequently results from primary EBV infection derived from the donor (reviewed in Ref. 177). Although the virus is of donor origin, the proliferating cells are derived from the recipient. The tumor may be localized to the donated organ, especially in recipients of kidneys. Alternatively, the tumor may involve a lymph node (including intra-abdominal sites) or be extranodal. Lymphoproliferative syndrome should be considered in the differential diagnosis of fever posttransplant (including rejection where the donated organ is involved).

2.5.3c. X-Linked Lymphoproliferative Syndrome. This condition presents at an average age of 2.5 years in boys who have had normal responses to other infectious agents and vaccines. It represents a selective inability to control EBV-driven proliferation of B and T lymphocytes. The clinical course of XLP is variable, with three major phenotypes that may overlap (reviewed in Ref. 178). Approximately 50% of boys have severe infectious mononucleosis, 25–30% have malignant lymphomas (often extranodal in the ileocecal region), and 25–30% have acquired hypogammaglobulinemia. There is a very high ultimate mortality, with death due to hepatic necrosis and bone marrow failure.

2.5.4. Diagnosis

2.5.4a. Serologic Response to EBV. Cells expressing EBNA or the late structural protein virus capsid antigen (VCA) can be propagated in the laboratory and used to measure antibodies of IgG or IgM class. The typical response to primary EBV infection is illustrated schematically in Fig. 16. IgM anti-VCA appear first and are transient, being replaced by IgG class antibodies. The response to EBNA is delayed so that it first appears during convalescence.

The profile of VCA-IgM positive plus EBNA negative therefore is consistent with primary EBV infection, while a profile of VCA-IgM negative, VCA-IgG positive, and EBNA positive accords with infection in the past. These tests can be helpful in identifying donor–recipient pairs at high risk of lymphoproliferative disease (donor seropositive–recipient seronegative).

2.5.4b. Infectious Mononucleosis. The clinical presentation may be obvious, with confirmation made by testing for heterophile agglutinins (e.g., Paul–Bunnel test, Monospot). If ampicillin has been given inadvertently, the resultant florid maculopapular rash is itself

diagnostic. In atypical cases, specific EBV serology can be ordered, with the presence of VCA IgM and IgG antibodies, together with the absence of EBNA antibodies characteristic of primary infection (see Fig. 16). Specific serology should also be requested in suspected cases of primary EBV infection in childhood, since tests for heterophile agglutinins usually give negative results.

2.5.4c. Lymphoproliferative Syndrome. A biopsy of the affected tissue is required. EBV infection should be sought by staining sections for the noncoding transcripts termed EBERs or for EBNA. Studies of the mono- and polyclonality of the tumor may be requested, but these do not always correlate directly with clinical outcome. Recent results of testing blood for EBV DNA by PCR show a correlation with disease.[179] Specific serology may also be requested and is most useful when the patient remains seronegative and the donor is shown to be seronegative, in which case the diagnosis can be excluded. In cases of CNS involvement, PCR for EBV DNA can be performed on CSF.

2.5.4d. X-Linked Lymphoproliferative Syndrome. The diagnosis is made when a male presents with severe infectious mononucleosis, lymphoma, or acquired hypogammaglobulinemia (reviewed in Ref. 178). A full family history may reveal the premature deaths of multiple males in several generations on the maternal side. The recent identification of the XLP gene as SAP should prompt genetic studies of this locus in the presenting case and

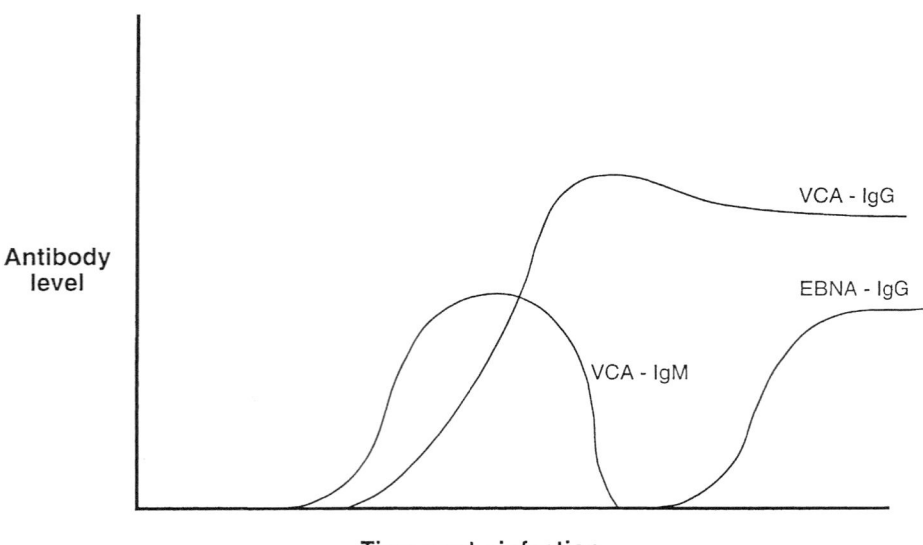

FIGURE 16. Schematic representation of the temporal appearance of EBV antibodies following primary infection. VCA, virus capsid antigen; EBNA, Epstein-Barr nuclear antigen.

family members. However, not all cases of XLP have mutations in the SAP exons; it may be that mutations elsewhere in the gene may be responsible or the diagnosis in some earlier cases of "XLP" may have been incorrect.[174]

2.5.5. Management

2.5.5a. Infectious Mononucleosis. Controlled trials show that acyclovir decreases EBV excretion from the saliva but does not affect the progression of disease.[180] This is presumably because the immunopathologic effect of T-cell recognition of infected B cells, once established, does not require ongoing viral replication for its maintenance. Steroids have also been tried but do not affect the clinical course.[181] Management of an acute case of infectious mononucleosis therefore consists of ensuring that tonsillar enlargement does not threaten the airway, avoidance of ampicillin and the rash it produces, together with supportive care and advice about prognosis.

2.5.5b. Lymphoproliferative Syndrome. Once the diagnosis has been made in a transplant patient, it is important to reduce the level of immunosuppression and follow carefully to determine whether the tumor regresses. Complete resolution is more likely if the tumor is polyclonal and if it appeared soon after transplantation.

In transplant patients who do not respond or in AIDS patients the next approach is adoptive immunotherapy with T cells reactive against EBV.[182] If patients do not respond, or if appropriately matched cells are not available, cancer chemotherapy should be considered, although the results in patients who are already profoundly immunocompromised can be disappointing. For CNS lymphomas, radiotherapy also may be considered.

Some clinicians also advocate giving acyclovir to patients with lymphoproliferative syndrome. Although there are no data from controlled trials, this may be worth trying in patients with polyclonal tumors in an attempt to limit EBV proliferation. However, once a monoclonal tumor has become established, it will propagate using the cellular DNA polymerase alpha and delta; acyclovir is a selective antiviral drug partly because it is a poor inhibitor of such enzymes. Therefore, there seems to be no rationale for giving acyclovir in these circumstances. In contrast, acyclovir prophylaxis might have a role in preventing EBV lymphoproliferation, but its low incidence precludes a controlled trial designed to assess this.

2.5.5c. X-Linked Lymphoproliferative Syndrome. Lymphoproliferation in a male patient without known causes of immunosuppression requires a family history to examine the possibility of X-linked lymphoproliferative syndrome, together with genetic studies of the SAP gene (reviewed in Ref. 176). Identification of the affected gene offers new approaches to managing XLP. Affected families may be offered genetic counseling and screening of fetuses, coupled with therapeutic termination of males bearing the affected gene. Alternatively, male newborns may be tested to determine whether they have the defective gene. This would provide prognostic information to the parents and allow prophylactic use of immunoglobulin infusions which are given with the aim of preventing primary EBV infection (although there are no controlled trial data to demonstrate efficacy). Finally, males carrying the affected gene could be offered bone marrow transplantation. This has been performed in males affected by XLP with some success, but a better outcome might be expected if transplantation was performed before EBV was acquired.

2.5.5d. Vaccine Prospects. Primary EBV infection potentially could be prevented by a vaccine that induced neutralizing antibodies. Most neutralizing antibodies are directed against gp350. This has been prepared in various ways and considered as a potential immunogen. In experimental symptoms, it protects cotton-top tamarins against EBV-induced lymphomas. At present, no vaccine preparations are undergoing extensive clinical trial, although prevention of infectious mononucleosis among an initially seronegative college freshman remains a practical target population for evaluating efficacy.

2.6. Human Herpesvirus 8

2.6.1. Epidemiology

Since serologic assays are insensitive in patients with proven Kaposi's sarcoma (KS), it is reasonable to conclude that they probably underestimate the prevalence of infection in the general community. Nevertheless, among HIV-negative individuals, 1% of blood donors, and 8% of sexually transmitted disease (STD) clinic attendees have detectable antibodies.[183] Infection appears to be more common in central Africa. In Cameroon, infection was first detected at age 4 years.[184] In Uganda,[185] the adult level of seropositivity was achieved before puberty and was significantly associated with hepatitis B transmission. Such figures suggest that HHV-8 may be acquired during childhood in African countries, whereas in the United States, the prevalence of HHV-8 more closely resembles that of HSV-2 than the remaining six human herpesviruses. Among male homosexuals in the United States, the incidence of antibodies increases in parallel with the number of lifetime sexual partners, suggesting that HHV-8 is transmitted sexually.[186] Earlier reports of detection of HHV-8 by PCR in semen have not been confirmed. The virus is readily detected in saliva,[187] but the epidemiology of KS and of HHV-8 seropositivity does not support ready transmission by this route.

Areas with a high incidence of KS include sub-Saharan African and Italy. Although the absolute seroprevalence of HHV-8 may be underestimated, the apparent proportion seropositive in a community increases in parallel with its incidence of KS.[183]

2.6.2. Pathogenesis

HHV-8 possesses a series of genes likely to facilitate progression of the infected cell through the cell cycle (Table 20). Many of these genes also are present in herpesvirus saimiri, an oncogenic virus of the squirrel monkey. The EBV genome does not contain all these genes, but it is interesting to note that EBV activates the equivalent cellular genes when it infects cells (Table 20). This suggests that the gammaherpesviruses share a common pathway for activating cells, with EBV stimulating the host genes, while HHV-8 brings them into target cells. Presumably, the genes in HHV-8 and herpesvirus saimiri were captured from the host by the virus genome in the last 80 million years, since EBV diverged at this time (see Fig. 1).

The K12 gene of HHV-8 transforms cells so that the virus may be directly oncogenic.[188] In addition, the release of cytokines from HHV-8-infected cells may facilitate progression of KS so that the tumor may be a cytokine-driven proliferation of target cells, which are thought to be endothelial lymphatics.

Serologic data in HIV-positive patients suggest that HHV-8 seroconversion occurs around the time that KS

TABLE 21. HHV-8 after Renal Transplantation

Possible reactivation[a]	Possible primary infection[b]
11 patients with posttransplant KS compared to 11 controls without KS	220 patients tested at baseline and 12 months later
HHV-8 IgG-positive prior to transplant in 10/11 KS versus 2/17 controls	25 seroconverters identified
	6 donors tested, 5 HHV-8 IgG-positive
	2/25 developed KS

[a]From Parravicini et al.[190]
[b]From Regamey et al.[191]

appears.[189] It seems improbable that AIDS patients will have acquired a new sexually transmitted infection for the first time once they are immunocompromised. Alternative explanations include reactivation of latent HHV-8 with boosting of antibody levels recorded as seroconversion by insensitive assays, and the possibility that exposure to HHV-8 may frequently not lead to infection, unless the patient is immunocompromised.

In transplant patients, one study reports the development of KS following nonprimary infection,[190] while a second implicates transmission from the donor[191] based on serial serologic assays (Table 21). These differences may reflect selection biases inherent in the different study designs (case–control study of disease vs. cohort study of infection, respectively). Given the experience from studying other herpesviruses, it is reasonable to conclude that HHV-8 may be acquired either from the donor or may reactivate from latency soon after transplantation. Additional detailed studies will be required to show if reinfection occurs.

2.6.3. Clinical Features

2.6.3a. Kaposi's Sarcoma. KS was originally described as a tumor of elderly males, frequently affecting the lower limbs. It follows an indolent course and rarely disseminates (termed "classic KS"). African KS occurs in younger people, is more aggressive, and often involves internal organs.

In the early 1980s, KS was recognized as an important AIDS indicator disease. In clinical course, it resembles African KS, with multicentric involvement of skin, often associated with KS in the gastrointestinal tract, lungs, and eye.

2.6.3b. Body Cavity Lymphomas. HHV-8 is also associated with a particular type of lymphoma that does not form aggregating masses of cells.[192] The B cells populate the body cavities (such as peritoneum or pleura),

TABLE 20. Gammaherpesvirus Genes (or Their Cellular Homologs) Likely to Facilitate Cell Proliferation[a]

HHV-8 ORF	Herpesvirus saimiri	EBV[b]	Function
16	✓	+	Bcl-2
72	✓	+	Cyclin-D
4	✓	+	Complement-binding protein
74	✓	+	IL-8 receptor
K2		+	IL-6
K14		+	Adhesion
2	✓		Dihydrofolate reductase

[a]From Russo[10] and Muralidhar et al.[188]
[b]+, cellular gene activated.

producing effusions. They are rare but occur in patients who are profoundly immunocompromised, particularly secondary to HIV infection.

2.6.3c. Other Associations. Reported detection of HHV-8 DNA in biopsies taken from a wide variety of skin tumors as well as from normal skin from transplant patients has not been confirmed. A reported association with multiple myeloma or with sarcoidosis also has not been confirmed.

2.6.4. Diagnosis

HHV-8 DNA can be detected by PCR, but particular care to avoid cross-contamination is essential. The presence of HHV-8 DNA in the blood is associated with a high risk of future KS.[193] Often, HHV-8 DNA only becomes detectable just before the appearance of KS,[189] suggesting that clinically inapparent lesions may be the source of virus.

Diagnosis of KS is by biopsy and histopathologic examination. HHV-8 DNA has been detected in all types of KS.

IgG antibodies can be detected by a variety of methods, which focus on antigens expressed during the lytic or latent phases of the HHV-8 replication cycle. There is a poor interassay correlation between the available methods.[194]

2.6.5. Management

If it is possible to reduce immunosuppression, e.g., posttransplant or by giving HAART to a previously untreated HIV-positive patient, then the lesions of KS may regress spontaneously. Indeed, the incidence of KS has declined markedly with the availability of HAART.

Small, indolent lesions of KS may not require treatment. There are no controlled trials of different treatments, but agents used include retinoic acid, local vinblastine injections, radiotherapy, or systemic chemotherapy. The latter is often poorly tolerated in profoundly immunocompromised AIDS patients.

There is no evidence from controlled trials that the administration of antiherpes drugs has any effect on the course of KS. Anecdotal reports of benefit are unconvincing, because, as discussed earlier for EBV, these drugs have selective action against the viral DNA polymerase, whereas malignant cells use the host DNA polymerases alpha and delta. However, if the pathogenesis of KS involves the paracrine secretion of cytokines and if such production from cells infected with HHV-8 can be inhibited by antiherpes compounds, then they may be able to affect KS pathogenesis. Placebo-controlled trials are required to address this possibility.

In contrast, it is plausible that prophylaxis with antiherpes compounds might reduce the occurrence of future KS by interfering with paracrine secretion of cytokines. No controlled trials have been conducted but cohort studies report that ganciclovir, foscarnet, and acyclovir (in decreasing order of potency) are associated with decreased incidence of KS in AIDS patients.[195,196]

ACKNOWLEDGMENTS. I would like to thank the following colleagues for many stimulating discussions and collaborations concerning herpesviruses in the immunocompromised host: Professor Vince Emery and Dr. Duncan Clark (Virology), Professor Grant Prentice (Bone Marrow Transplant), Dr. Paul Sweny (Renal Transplant), Dr. Andy Burroughs (Liver Transplant), Dr. Margaret Johnson (AIDS), and Dr. Pauline Wilson (Ophthalmology/AIDS); Dr. Judith Feinberg (Infectious Diseases Center, University of Cincinnati), as well as the graduate students whose work I have quoted (Drs. Frances Bowen, Lea Cope, Dee Gor, and Jane Deayton).

References

1. McGeoch DJ, Cook S, Dolan A, Jamieson FE, Telford EA: Molecular phylogeny and evolutionary timescale for the family of mammalian herpesviruses. *J Mol Biol* **247**(3):443–458, 1995.
2. Roizman B: The family herpesviridae: A brief introduction. In Roizman B, Whitley RJ, Lopez C (eds): *The Human Herpesviruses.* Raven Press, New York, 1993, pp. 1–9.
3. Rowe WP, Hartley JW, Waterman S, Turner HC, Huebner RJ: Cytopathic agent resembling human salivary gland virus recovered from tissue cultures of human adenoids. *Proc Soc Exp Biol Med* **92**(2):418, 1956.
4. Weller TH, Witton HM, Bell EJ: The etiologic agents of varicella and herpes zoster. Isolation, propagation and cultural characteristics *in vitro. J Exp Med* **108**(6):843–868, 1958.
5. Epstein MA, Barr YM: Cultivation *in vitro* of human lymphoblasts from Burkitt's malignant lymphoma. *Lancet* **1**(7327):252–253, 1964.
6. Griffin BE: Relation of Burkitt's tumor-associated herpes-type virus to infectious mononucleosis. *Rev Med Virol* **8**(2):61–66, 1998.
7. Salahuddin SZ, Ablashi DV, Markham PD, *et al*: Isolation of a new virus, HBLV, in patients with lymphoproliferative disorders. *Science* **234**(4776):596–601, 1986.
8. Frenkel N, Schirmer EC, Wyatt LS, *et al*: Isolation of a new herpesvirus from human CD4+ T cells. *Proc Natl Acad Sci USA* **87**(2):748–752, 1990.
9. Chang Y, Cesarman E, Pessin MS, *et al*: Identification of herpesvirus-like DNA sequences in AIDS-associated Kaposi's sarcoma. *Science* **266**(5192):1865–1869, 1994.
10. Russo JJ, Bohenzky RA, Chien MC, *et al*: Nucleotide sequence of the Kaposi sarcoma-associated herpesvirus (HHV-8). *Proc Natl Acad Sci USA* **93**(25):14862–14867, 1996.

11. Umene K: Mechanism and application of genetic recombination in herpesviruses. *Rev Med Virol* **9**(3):171–182, 1999.

12. Post LE, Mackem S, Roizman B: Regulation of alpha genes of herpes simplex virus: Expression of chimeric genes produced by fusion of thymidine kinase with alpha gene promoters. *Cell* **24**(2):555–565, 1981.

13. Homa FL, Brown JC: Capsid assembly and DNA packaging in herpes simplex virus. *Rev Med Virol* **7**(2):107–122, 1997.

14. Elion GB, Furman PA, Fyfe JA, *et al*: Selectivity of action of an antiherpetic agent, 9-(2-hydroxyethoxymethyl) guanine. *Proc Natl Acad Sci USA* **74**(12):5716–5720, 1977.

15. Furman PA, St. Clair MH, Spector T: Acyclovir triphosphate is a suicide inactivator of the herpes simplex virus DNA polymerase. *J Biol Chem* **259**(15):9575–9579, 1984.

16. Collins P, Darby G: Laboratory studies of herpes simplex virus strains resistant to acyclovir. *Rev Med Virol* **1**(1):19–28, 1991.

17. Efstathiou S, Kemp S, Darby G, Minson AC: The role of herpes simplex virus type 1 thymidine kinase in pathogenesis. *J Gen Virol* **70**(Pt 4):869–879, 1989.

18. Kost RG, Hill EL, Tigges M, Straus SE: Brief report: Recurrent acyclovir-resistant genital herpes in an immunocompetent patient. *N Engl J Med* **329**(24):1777–1782, 1993.

19. Horsburgh BC, Chen SH, Hu A, *et al*: Recurrent acyclovir-resistant herpes simplex in an immunocompromised patient: Can strain differences compensate for loss of thymidine kinase in pathogenesis? *J Infect Dis* **178**(3):618–625, 1998.

20. Safrin S, Crumpacker C, Chatis P, *et al*: A controlled trial comparing foscarnet with vidarabine for acyclovir-resistant mucocutaneous herpes simplex in the acquired immunodeficiency syndrome. The AIDS Clinical Trials Group. *N Engl J Med* **325**(8):551–555, 1991.

21. Safrin S, Cherrington J, Jaffe HS: Clinical uses of cidofovir. *Rev Med Virol* **7**(3):145–156, 1997.

22. Randall G, Lagunoff M, Roizman B: The product of ORF O located within the domain of herpes simplex virus 1 genome transcribed during latent infection binds to and inhibits *in vitro* binding of infected cell protein 4 to its cognate DNA site. *Proc Natl Acad Sci USA* **94**(19):10379–10384, 1997.

23. Frank I, Friedman HM: A novel function of the herpes simplex virus type 1 Fc receptor: Participation in bipolar bridging of antiviral immunoglobulin G. *J Virol* **63**(11):4479–4488, 1989.

24. Powis SH: Lessons from an age-old war. *Nat Med* **4**(8):887–888, 1998.

25. Gilbert MJ, Riddell SR, Plachter B, Greenberg PD: Cytomegalovirus selectively blocks antigen processing and presentation of its immediate-early gene product. *Nature* **383**(6602):720–722, 1996.

26. Hill A, Jugovic P, York I, *et al*: Herpes simplex virus turns off the TAP to evade host immunity. *Nature* **375**(6530):411–415, 1995.

27. Jones TR, Wiertz EJ, Sun L, *et al*: Human cytomegalovirus US3 impairs transport and maturation of major histocompatibility complex class I heavy chains. *Proc Natl Acad Sci USA* **93**(21):11327–11333, 1996.

28. Wiertz EJ, Jones TR, Sun L, *et al*: The human cytomegalovirus US11 gene product dislocates MHC class I heavy chains from the endoplasmic reticulum to the cytosol. *Cell* **84**(5):769–779, 1996.

29. Reyburn HT, Mandelboim O, Vales-Gomez M, *et al*: The class I MHC homologue of human cytomegalovirus inhibits attack by natural killer cells. *Nature* **386**(6624):514–517, 1997.

30. Agrawal A, Eastman QM, Schatz DG: Transposition mediated by RAG1 and RAG2 and its implications for the evolution of the immune system. *Nature* **394**(6695):744–751, 1998.

31. Posavad CM, Koelle DM, Corey L: Tipping the scales of herpes simplex virus reactivation: The important responses are local. *Nat Med* **4**(4):381–382, 1998.

32. Saral R, Ambinder RF, Burns WH, *et al*: Acyclovir prophylaxis against herpes simplex virus infection in patients with leukemia. A randomized, double-blind, placebo-controlled study. *Ann Intern Med* **99**(6):773–776, 1983.

33. Griffiths PD, Ait-Khaled M, Bearcroft CP, *et al*: Human herpesviruses 6 and 7 as potential pathogens after liver transplantation: Prospective comparison with the effect of cytomegalovirus. *J Med Virol* **59**(4):496–501, 1999.

34. Perren TJ, Powles RL, Easton D, Stolle K, Selby PJ: Prevention of herpes zoster in patients by long-term oral acyclovir after allogeneic bone marrow transplantation. *Am J Med* **85**(2A):99–101, 1988.

35. Grundy JE, Lui SF, Super M, *et al*: Symptomatic cytomegalovirus infection in seropositive kidney recipients: Reinfection with donor virus rather than reactivation of recipient virus. *Lancet* **2**(8603):132–135, 1988.

36. Glesby MJ, Moore RD, Chaisson RE: Herpes zoster in patients with advanced human immunodeficiency virus infection treated with zidovudine. Zidovudine Epidemiology Study Group. *J Infect Dis* **168**(5):1264–1268, 1993.

37. Fairfax MR, Schacker T, Cone RW, Collier AC, Corey L: Human herpesvirus 6 DNA in blood cells of human immunodeficiency virus-infected men: Correlation of high levels with high CD4 cell counts. *J Infect Dis* **169**(6):1342–1345, 1994.

38. Fabio G, Knight SN, Kidd IM, *et al*: Prospective study of human herpesvirus 6, human herpesvirus 7, and cytomegalovirus infections in human immunodeficiency virus-positive patients. *J Clin Microbiol* **35**(10):2657–2659, 1997.

39. Clark DA, Ait-Khaled M, Wheeler AC, *et al*: Quantification of human herpesvirus 6 in immunocompetent persons and postmortem tissues from AIDS patients by PCR. *J Gen Virol* **77**(Pt 9):2271–2275, 1996.

40. Wald A, Corey L, Cone R, *et al*: Frequent genital herpes simplex virus 2 shedding in immunocompetent women. Effect of acyclovir treatment. *J Clin Invest* **99**(5):1092–1097, 1997.

41. Wald A, Zeh J, Selke S, Ashley RL, Corey L: Virologic characteristics of subclinical and symptomatic genital herpes infections. *N Engl J Med* **333**(12):770–775, 1995.

42. Zhao ZS, Granucci F, Yeh L, Schaffer PA, Cantor H: Molecular mimicry by herpes simplex virus-type 1: Autoimmune disease after viral infection. *Science* **279**(5355):1344–1347, 1998.

43. Grundy JE, Shanley JD, Griffiths PD: Is cytomegalovirus interstitial pneumonitis in transplant recipients an immunopathological condition? *Lancet* **2**(8566):996–999, 1987.

44. Rubin RH: The indirect effects of cytomegalovirus infection on the outcome of organ transplantation. *JAMA* **261**(24):3607–3609, 1989.

45. Speir E, Modali R, Huang ES, *et al*: Potential role of human cytomegalovirus and p53 interaction in coronary restenosis. *Science* **265**(5170):391–394, 1994.

46. Griffiths PD: Studies to further define viral co-factors for human immunodeficiency virus. *J Gen Virol* **79**(Pt 2):213–220, 1998.

47. Lonnqvist B, Palmblad J, Ljungman P, *et al*: Oral acyclovir as prophylaxis for bacterial infections during induction therapy for acute leukaemia in adults. The Leukemia Group of Middle Sweden. *Supp Care Cancer* **1**(3):139–144, 1993.

48. Lowance D, Neumayer H-H, Legendre C, *et al*: Valaciclovir reduces the incidence of cytomegalovirus disease and acute rejec-

tion in renal allograft recipients. *N Engl J Med* **340**(19):1462–1470, 1999.

49. Wagner JA, Ross H, Hunt S, *et al*: Prophylactic ganciclovir treatment reduces fungal as well as cytomegalovirus infections after heart transplantation. *Transplantation* **60**(12):1473–1477, 1995.

50. Valantine HA, Gao S-Z, Menon SG, *et al*: Impact of prophylactic immediate posttransplant ganciclovir on development of transplant atherosclerosis: A post-hoc analysis of a randomised, placebo-controlled study. *Circulation* **100**(1):61–66, 1999.

51. Ioannidis JP, Collier AC, Cooper DA, *et al*: Clinical efficacy of high-dose acyclovir in patients with human immunodeficiency virus infection: A meta-analysis of randomized individual patient data. *J Infect Dis* **178**(2):349–359, 1998.

52. Feinberg JE, Hurwitz S, Cooper D, *et al*: A randomized, double-blind trial of valaciclovir prophylaxis for cytomegalovirus disease in patients with advanced human immunodeficiency virus infection. *J Infect Dis* **177**(1):48–56, 1998.

53. Schacker T, Ryncarz AJ, Goddard J, *et al*: Frequent recovery of HIV-1 from genital herpes simplex virus lesions in HIV-1–infected men. *JAMA* **280**(1):61–66, 1998.

54. Corey L, Spear PG: Infections with herpes simplex viruses (1). *N Engl J Med* **314**(11):686–691, 1986.

55. Fleming DT, McQuillan GM, Johnson RE, *et al*: Herpes simplex virus type 2 in the United States, 1976 to 1994. *N Engl J Med* **337**(16):1105–1111, 1997.

56. Meyers JD, Flournoy N, Thomas ED: Infection with herpes simplex virus and cell-mediated immunity after marrow transplant. *J Infect Dis* **142**(3):338–346, 1980.

57. Koneru B, Tzakis AG, DePuydt LE, *et al*: Transmission of fatal herpes simplex infection through renal transplantation. *Transplantation* **45**(3):653–656, 1988.

58. Schacker T, Zeh J, Hu HL, Hill E, Corey L: Frequency of symptomatic and asymptomatic herpes simplex virus type 2 reactivations among human immunodeficiency virus-infected men. *J Infect Dis* **178**(6):1616–1622, 1998.

59. Lakeman FD, Whitley RJ: Diagnosis of herpes simplex encephalitis: Application of polymerase chain reaction to cerebrospinal fluid from brain-biopsied patients and correlation with disease. National Institute of Allergy and Infectious Diseases Collaborative Antiviral Study Group. *J Infect Dis* **171**(4):857–863, 1995.

60. Berry NJ, Grundy JE, Griffiths PD: Radioimmunoassay for the detection of IgG antibodies to herpes simplex virus and its use as a prognostic indicator of HSV excretion in transplant recipients. *J Med Virol* **21**(2):147–154, 1987.

61. Sinha DP: Chickenpox—A disease predominantly affecting adults in rural West Bengal, India. *Int J Epidemiol* **5**(4):367–374, 1976.

62. Donahue JG, Choo PW, Manson JE, Platt R: The incidence of herpes zoster. *Arch Intern Med* **155**(15):1605–1609, 1995.

63. Garnett GP, Ferguson NM: Predicting the effect of varicella vaccine on subsequent cases of zoster and varicella. *Rev Med Virol* **6**(3):151–161, 1996.

64. Mims C, Fenner F: The pathogenesis of the acute exanthems. *Rev Med Virol* **6**(1):1–8, 1996.

65. Head M, Campbell AW, Kennedy PGE: The pathology of herpes zoster and its bearing on sensory localisation. *Rev Med Virol* **7**(3):131–143, 1997.

66. Ljungman P, Lonnqvist B, Gahrton G, *et al*: Clinical and subclinical reactivations of varicella-zoster virus in immunocompromised patients. *J Infect Dis* **153**(5):840–847, 1986.

67. Wilson A, Sharp M, Koropchak CM, Ting SF, Arvin AM: Sub-clinical varicella-zoster virus viremia, herpes zoster, and T lymphocyte immunity to varicella-zoster viral antigens after bone marrow transplantation. *J Infect Dis* **165**(1):119–126, 1992.

68. Doyle PW, Gibson G, Dolman CL: Herpes zoster ophthalmicus with contralateral hemiplegia: Identification of cause. *Ann Neurol* **14**(1):84–85, 1983.

69. Ross AH: Modification of chicken pox in family contacts by administration of gamma globulin. *N Engl J Med* **267**(8):369–376, 1962.

70. Zaia JA, Levin MJ, Preblud SR, *et al*: Evaluation of varicella-zoster immune globulin: Protection of immunosuppressed children after household exposure to varicella. *J Infect Dis* **147**(4):737–743, 1983.

71. Atkinson K, Storb R, Prentice RL, *et al*: Analysis of late infections in 89 long-term survivors of bone marrow transplantation. *Blood* **53**(4):720–731, 1979.

72. Wood MJ: Management strategies in herpes: How can the burden of zoster-associated pain be reduced? PPS, Worthing, UK, 1993.

73. Locksley RM, Flournoy N, Sullivan KM, Meyers JD: Infection with varicella-zoster virus after marrow transplantation. *J Infect Dis* **152**(6):1172–1181, 1985.

74. Ruckdeschel JC, Schimpff SC, Smyth AC, Mardiney MR Jr: Herpes zoster and impaired cell-associated immunity to the varicella-zoster virus in patients with Hodgkin's disease. *Am J Med* **62**(1):77–85, 1977.

75. Atkinson K, Meyers JD, Storb R, Prentice RL, Thomas ED: Varicella-zoster virus infection after marrow transplantation for aplastic anemia or leukemia. *Transplantation* **29**(1):47–50, 1980.

76. Whitley RJ, Shukla S, Crooks RJ: The identification of risk factors associated with persistent pain following herpes zoster. *J Infect Dis* **178**(Suppl 1):S71–S75, 1998.

77. Drew WL, Mintz L: Rapid diagnosis of varicella-zoster virus infection by direct immunofluorescence. *Am J Clin Pathol* **73**(5):699–701, 1980.

78. Suga S, Yoshikawa T, Ozaki T, Asano Y: Effect of oral acyclovir against primary and secondary viraemia in incubation period of varicella. *Arch Dis Child* **69**(6):639–642 (discussion 642–643), 1993.

79. Whitley RJ, Weiss H, Gnann JWJ, *et al*: Acyclovir with and without prednisone for the treatment of herpes zoster. A randomized, placebo-controlled trial. The National Institute of Allergy and Infectious Diseases Collaborative Antiviral Study Group. *Ann Intern Med* **125**(5):376–383, 1996.

80. Breton G, Fillet AM, Katlama C, Bricaire F, Caumes E: Acyclovir-resistant herpes zoster in human immunodeficiency virus-infected patients: Results of foscarnet therapy. *Clin Infect Dis* **27**(6):1525–1527, 1998.

81. Gnann JWJ, Crumpacker CS, Lalezari JP, *et al*: Sorivudine versus acyclovir for treatment of dermatomal herpes zoster in human immunodeficiency virus-infected patients: Results from a randomized, controlled clinical trial. Collaborative Antiviral Study Group/AIDS Clinical Trials Group, Herpes Zoster Study Group. *Antimicrob Agents Chemother* **42**(5):1139–1145, 1998.

82. Griffiths PD, Baboonian C: A prospective study of primary cytomegalovirus infection during pregnancy: Final report. *Br J Obstet Gynaecol* **91**(4):307–315, 1984.

83. Stagno S, Cloud GA: Working parents: The impact of day care and breast-feeding on cytomegalovirus infections in offspring. *Proc Natl Acad Sci USA* **91**(7):2384–2389, 1994.

84. Baskar JF, Smith PP, Ciment GS, *et al*: Developmental analysis of the cytomegalovirus enhancer in transgenic animals. *J Virol* **70**(5):3215–3226, 1996.

85. Stagno S, Reynolds DW, Tsiantos A, *et al*: Comparative serial virologic and serologic studies of symptomatic and subclinical congenitally and natally acquired cytomegalovirus infections. *J Infect Dis* **132**(5):568–577, 1975.

86. Cope AV, Sweny P, Sabin C, *et al*: Quantity of cytomegalovirus viruria is a major risk factor for cytomegalovirus disease after renal transplantation. *J Med Virol* **52**(2):200–205, 1997.

87. Hassan-Walker AF, Kidd IM, Sabin C, *et al*: Quantity of human cytomegalovirus (CMV) DNAemia as a risk factor for CMV disease in renal allograft recipients: Relationship with donor/recipient CMV serostatus, receipt of augmented methylprednisolone and anti-thymocyte globulin (ATG). *J Med Virol* **58**(2):182–187, 1999.

88. Cope AV, Sabin C, Burroughs A, *et al*: Interrelationships among quantity of human cytomegalovirus (HCMV) DNA in blood, donor–recipient serostatus, and administration of methylprednisolone as risk factors for HCMV disease following liver transplantation. *J Infect Dis* **176**(6):1484–1490, 1997.

89. Gor D, Sabin C, Prentice HG, *et al*: Longitudinal fluctuations between peak virus load, donor/recipient serostatus, acute GvHD and CMV disease. *Bone Marrow Transplant* **21**(6):597–605, 1998.

90. Betts RF, Freeman RB, Douglas RG Jr, Talley TE: Clinical manifestations of renal allograft derived primary cytomegalovirus infection. *Am J Dis Child* **131**(7):759–763, 1977.

91. Meyers JD, Ljungman P, Fisher LD: Cytomegalovirus excretion as a predictor of cytomegalovirus disease after marrow transplantation: Importance of cytomegalovirus viremia. *J Infect Dis* **162**(2):373–380, 1990.

92. Emery VC, Cope AV, Bowen EF, Gor D, Griffiths PD: The dynamics of human cytomegalovirus replication *in vivo*. *J Exp Med* **190**(2):177–182, 1999.

93. Emery VC, Griffiths PD: Prediction of cytomegalovirus load and resistance patterns after antiviral chemotherapy. *Proc Natl Acad Sci USA* **97**(14):8039–8044, 2000.

94. Schmidt GM, Horak DA, Niland JC, *et al*: A randomized, controlled trial of prophylactic ganciclovir for cytomegalovirus pulmonary infection in recipients of allogeneic bone marrow transplants; The City of Hope-Stanford-Syntex CMV Study Group. *N Engl J Med* **324**(15):1005–1011, 1991.

95. Safrit JT, Koup RA: The CD4 loss in AIDS patients is not immunopathologically mediated. *Rev Med Virol* **6**(1):13–16, 1996.

96. Ljungman P, Engelhard D, Link H, *et al*: Treatment of interstitial pneumonitis due to cytomegalovirus with ganciclovir and intravenous immune globulin: Experience of European Bone Marrow Transplant Group. *Clin Infect Dis* **14**(4):831–835, 1992.

97. Grattan MT, Moreno-Cabral CE, Starnes VA, Oyer PE, Stinson EB: Cytomegalovirus infection is associated with cardiac allograft rejection and atherosclerosis. *JAMA* **261**(24):3561–3566, 1989.

98. Reinke P, Fietze E, Ode-Hakim S, *et al*: Late acute renal allograft rejection and symptomless cytomegalovirus infection. *Lancet* **344**(8939–8940):1737–1738, 1996.

99. Guetta E, Guetta V, Shibutani T, Epstein SE: Monocytes harboring cytomegalovirus: Interactions with endothelial cells, smooth muscle cells, and oxidized low-density lipoprotein. Possible mechanisms for activating virus delivered by monocytes to sites of vascular injury. *Circ Res* **81**(1):8–16, 1997.

100. Streblow DN, Soderberg-Naucler C, Vieira J, *et al*: The human cytomegalovirus chemokine receptor US28 mediates vascular smooth muscle cell migration. *Cell* **99**(5):511–520, 1999.

101. Speir E, Shibutani T, Yu ZX, Ferrans V, Epstein SE: Role of reactive oxygen intermediates in cytomegalovirus gene expression and in the response of human smooth muscle cells to viral infection. *Circ Res* **79**(6):1143–1152, 1996.

102. Bowen F, Wilson P, Cope A, *et al*: Cytomegalovirus retinitis in AIDS patients: Influence of cytomegaloviral load on response to ganciclovir, time to recurrence and survival. *AIDS* **10**(13):1515–1520, 1996.

103. Spector SA, Hsia K, Crager M, *et al*: Cytomegalovirus (CMV) DNA load is an independent predictor of CMV disease and survival in advanced AIDS. *J Virol* **73**(8):7027–7030, 1999.

104. Ljungman P, Plotkin SA: Workshop of CMV disease: Definitions, clinical severity scores, and new syndromes. *Scand J Infect Dis* **99**(Suppl):87–89, 1995.

105. Jacobson MA: Treatment of cytomegalovirus retinitis in patients with the acquired immunodeficiency syndrome. *N Engl J Med* **337**(2):105–114, 1997.

106. Griffiths PD, McLaughlin JE: Cytomegalovirus. In Scheld WM, Whitley RJ, Durack DT (eds): *Infections of the Central Nervous System*. Lippincott-Raven, Philadelphia, 1997, pp. 107–115.

107. Griffiths PD, Panjwani DD, Stirk PR, *et al*: Rapid diagnosis of cytomegalovirus infection in immunocompromised patients by detection of early antigen fluorescent foci. *Lancet* **2**(8414):1242–1245, 1984.

108. Gleaves CA, Smith TF, Shuster EA, Pearson GR: Rapid detection of cytomegalovirus in MRC-5 cells inoculated with urine specimens by using low-speed centrifugation and monoclonal antibody to an early antigen. *J Clin Microbiol* **19**(6):917–919, 1984.

109. Kidd IM, Fox JC, Pillay D, *et al*: Provision of prognostic information in immunocompromised patients by routine application of the polymerase chain reaction for cytomegalovirus. *Transplantation* **56**(4):867–871, 1993.

110. Spector SA, Merrill R, Wolf D, Dankner WM: Detection of human cytomegalovirus in plasma of AIDS patients during acute visceral disease by DNA amplification. *J Clin Microbiol* **30**(9):2359–2365, 1992.

111. The TH, van der Bij W, van den Berg AP, *et al*: Cytomegalovirus antigenemia. *Rev Infect Dis* **12**(Suppl 7):S734–S744, 1990.

112. Einsele H, Ehninger G, Hebart H, *et al*: Polymerase chain reaction monitoring reduces the incidence of cytomegalovirus disease and the duration and side effects of antiviral therapy after bone marrow transplantation. *Blood* **86**(7):2815–2820, 1995.

113. Shinkai M, Spector SA: Quantitation of human cytomegalovirus (HCMV) DNA in cerebrospinal fluid by competitive PCR in AIDS patients with different HCMV central nervous system diseases. *Scand J Infect Dis* **27**(6):559–561, 1995.

114. Mattes FM, McLaughlin JE, Emery VC, Clark DA, Griffiths PD: Histopathological detection of owl's eye inclusions is still specific for cytomegalovirus in the era of human herpesviruses 6 and 7. *J Clin Pathol* **53**(8):612–614, 2000.

115. Hibberd PL, Tolkoff-Rubin NE, Conti D, *et al*: Preemptive ganciclovir therapy to prevent cytomegalovirus disease in cytomegalovirus antibody-positive renal transplant recipients. A randomized controlled trial. *Ann Intern Med* **123**(1):18–26, 1995.

116. Rubin RH: Preemptive therapy in immunocompromised hosts. *N Engl J Med* **324**(15):1057–1059, 1991.

117. Emery VC, Sabin CA, Cope AV, *et al*: Application of viral-load kinetics to identify patients who develop cytomegalovirus disease after transplantation. *Lancet* **355**(9220):2032–2036, 2000.

118. Reed EC, Wolford JL, Kopecky KJ, *et al*: Ganciclovir for the treatment of cytomegalovirus gastroenteritis in bone marrow transplant patients. A randomized, placebo-controlled trial. *Ann Intern Med* **112**(7):505–510, 1990.

119. Goodrich JM, Mori M, Gleaves CA, *et al*: Early treatment with ganciclovir to prevent cytomegalovirus disease after allogeneic bone marrow transplantation. *N Engl J Med* **325**(23):1601–1607, 1991.

120. Goodrich JM, Bowden RA, Fisher L, *et al*: Ganciclovir prophylaxis to prevent cytomegalovirus disease after allogeneic marrow transplant. *Ann Intern Med* **118**(3):173–178, 1993.

121. Winston DJ, Ho WG, Bartoni K, *et al*: Ganciclovir prophylaxis of cytomegalovirus infection and disease in allogeneic bone marrow transplant recipients. Results of a placebo-controlled, double-blind trial. *Ann Intern Med* **118**(3):179–184, 1993.

122. Merigan TC, Renlund DG, Keay S, *et al*: A controlled trial of ganciclovir to prevent cytomegalovirus disease after heart transplantation. *N Engl J Med* **326**(18):1182–1186, 1992.

123. Gane E, Saliba F, Valdecasas GJ, *et al*: Randomised trial of efficacy and safety of oral ganciclovir in the prevention of cytomegalovirus disease in liver-transplant recipients. The Oral Ganciclovir International Transplantation Study Group. *Lancet* **50** (9093):1729–1733, 1997.

124. Cheeseman SH, Rubin RH, Stewart JA, *et al*: Controlled clinical trial of prophylactic human-leukocyte interferon in renal transplantation. Effects on cytomegalovirus and herpes simplex virus infections. *N Engl J Med* **300**(24):1345–1349, 1979.

125. Hirsch MS, Schooley RT, Cosimi AB, *et al*: Effects of interferon-alpha on cytomegalovirus reactivation syndromes in renal-transplant recipients. *N Engl J Med* **308**(25):1489–1493, 1983.

126. Lui SF, Ali AA, Grundy JE, *et al*: Double-blind, placebo-controlled trial of human lymphoblastoid interferon prophylaxis of cytomegalovirus infection in renal transplant recipients. *Nephrol Dial Transplant* **7**(12):1230–1237, 1992.

127. Balfour HHJ, Chace BA, Stapleton JT, Simmons RL, Fryd DS: A randomized, placebo-controlled trial of oral acyclovir for the prevention of cytomegalovirus disease in recipients of renal allografts. *N Engl J Med* **320**(21):1381–1387, 1989.

128. Prentice HG, Gluckman E, Powles RL, *et al*: Impact of long-term acyclovir on cytomegalovirus infection and survival after allogeneic bone marrow transplantation. European Acyclovir for CMV Prophylaxis Study Group. *Lancet* **343**(8900):749–753, 1994.

129. Snydman DR, Werner BG, Dougherty NN, *et al*: Cytomegalovirus immune globulin prophylaxis in liver transplantation. A randomized, double-blind, placebo-controlled trial. The Boston Center for Liver Transplantation CMVIG Study Group. *Ann Intern Med* **119**(10):984–991, 1993.

130. Metselaar HJ, Rothbarth PH, Brouwer RM, *et al*: Prevention of cytomegalovirus-related death by passive immunization. A double-blind placebo-controlled study in kidney transplant recipients treated for rejection. *Transplantation* **48**(2):264–266, 1989.

131. Macdonald PS, Keogh AM, Marshman D, *et al*: A double-blind placebo-controlled trial of low-dose ganciclovir to prevent cytomegalovirus disease after heart transplantation. *J Heart Lung Transplant* **14**(1):32–38, 1995.

132. Winston DJ, Wirin D, Shaked A, Busuttil RW: Randomised comparison of ganciclovir and high-dose acyclovir for long-term cytomegalovirus prophylaxis in liver-transplant recipients. *Lancet* **346**(8967):69–74, 1995.

133. Spector SA, McKinley GF, Lalezari JP, *et al*: Oral ganciclovir for the prevention of cytomegalovirus disease in persons with AIDS. Roche Cooperative Oral Ganciclovir Study Group. *N Engl J Med* **334**(23):1491–1497, 1996.

134. Spector SA, Wong R, Hsia K, Pilcher M, Stampien MJ: Plasma cytomegalovirus (CMV) DNA load predicts CMV disease and survival in AIDS patients. *J Clin Invest* **101**(2):497–502, 1998.

135. Griffiths PD, Feinberg J, Fry J, *et al*: The effect of valaciclovir on cytomegalovirus viremia and viruria detected by polymerase chain reaction in patients with advanced human immunodeficiency virus disease. *J Infect Dis* **177**(1):57–64, 1998.

136. Fletcher CV, Englund JA, Edelman CK, *et al*: Pharmacologic basis for high-dose oral acyclovir prophylaxis of cytomegalovirus disease in renal allograft recipients. *Antimicrob Agents Chemother* **35**(5):938–943, 1991.

137. Hall CB, Long CE, Schnabel KC, *et al*: Human herpesvirus-6 infection in children. A prospective study of complications and reactivation. *N Engl J Med* **331**(7):432–438, 1994.

138. Tanaka K, Kondo T, Torigoe S, *et al*: Human herpesvirus 7: Another causal agent for roseola (exanthem subitum). *J Pediatr* **125**(1):1–5, 1994.

139. Lusso P, Secchiero P, Crowley RW, *et al*: CD4 is a critical component of the receptor for human herpesvirus 7: Interference with human immunodeficiency virus. *Proc Natl Acad Sci USA* **91**(9): 3872–3876, 1994.

140. Santoro F, Kennedy PE, Locatelli G, *et al*: CD46 is a cellular receptor for human herpesvirus 6. *Cell* **99**(7):817–827, 1999.

141. Lusso P: Human herpesvirus 6 (HHV-6). *Antiviral Res* **31**(1–2):1–21, 1996.

142. Zou P, Isegawa Y, Nakano K, *et al*: Human herpesvirus 6 open reading frame U83 encodes a functional chemokine. *J Virol* **73**(7):5926–5933, 1999.

143. Inoue Y, Yasukawa M, Fujita S: Induction of T-cell apoptosis by human herpesvirus 6. *J Virol* **71**(5):3751–3759, 1997.

144. Gobbi A, Stoddart CA, Malnati MS, *et al*: Human herpesvirus 6 (HHV-6) causes severe thymocyte depletion in SCID- hu Thy/Liv mice. *J Exp Med* **189**(12):1953–1960, 1999.

145. Levy JA, Landay A, Lennette ET: Human herpesvirus 6 inhibits human immunodeficiency virus type 1 replication in cell culture. *J Clin Microbiol* **28**(10):2362–2364, 1990.

146. Caserta MT, Hall CB, Schnabel K, *et al*: Neuroinvasion and persistence of human herpesvirus 6 in children. *J Infect Dis* **170**(6): 1586–1589, 1994.

147. McCullers JA, Lakeman FD, Whitley RJ: Human herpesvirus 6 is associated with focal encephalitis. *Clin Infect Dis* **21**(3):571–576, 1995.

148. Wang FZ, Linde A, Hagglund H, *et al*: Human herpesvirus 6 DNA in cerebrospinal fluid specimens from allogeneic bone marrow transplant patients: Does it have clinical significance? *Clin Infect Dis* **28**(3):562–568, 1999.

149. Kidd IM, Clark DA, Andrew DA, *et al*: Prospective study of betaherpesvirus infections following renal transplantation: Association of human herpesvirus 7 with CMV disease. *Transplantation* **56**(4):867–871, 1993.

150. Osman HK, Peiris JS, Taylor CE, *et al*: "Cytomegalovirus disease" in renal allograft recipients: Is human herpesvirus 7 a cofactor for disease progression? *J Med Virol* **48**(4):295–301, 1996.

151. Dockrell DH, Prada J, Jones MF, *et al*: Seroconversion to human herpesvirus 6 following liver transplantation is a marker of cytomegalovirus disease. *J Infect Dis* **176**(5):1135–1140, 1997.

152. Humar A, Malkan G, Moussa G, *et al*: Human herpesvirus-6 is associated with cytomegalovirus reactivation in liver transplant recipients. *J Infect Dis* **181**(4):1450–1453, 2000.

153. Knox KK, Carrigan DR: Disseminated active HHV-6 infections in patients with AIDS. *Lancet* **343**(8897):577–578, 1994.

154. Corbellino M, Lusso P, Gallo RC, *et al*: Disseminated human herpesvirus 6 infection in AIDS. *Lancet* **342**(8881):1242, 1993.

155. Emery VC, Atkins MC, Bowen EF, *et al*: Interactions between

β-herpesviruses and human immunodeficiency virus *in vivo*: Evidence for increased human immunodeficiency viral load in the presence of human herpesvirus 6. *J Med Virol* **57**(3):278–282, 1999.

156. Crowley RW, Secchiero P, Zella D, *et al*: Interference between human herpesvirus 7 and HIV-1 in mononuclear phagocytes. *J Immunol* **156**(5):2004–2008, 1996.

157. Daibata M, Taguchi T, Sawada T, Taguchi H, Miyoshi I: Chromosomal transmission of human herpesvirus 6 DNA in acute lymphoblastic leukaemia. *Lancet* **352**(9127):543–544, 1998.

158. Daibata M, Taguchi T, Nemoto Y, Taguchi H, Miyoshi I: Inheritance of chromosomally integrated human herpesvirus 6 DNA. *Blood* **94**(5):1545–1549, 1999.

159. Luppi M, Barozzi P, Morris C, *et al*: Human herpesvirus 6 latently infects early bone marrow progenitors *in vivo*. *J Virol* **73**(1):754–759, 1999.

160. Torelli G, Barozzi P, Marasca R, *et al*: Targeted integration of human herpesvirus 6 in the p arm of chromosome 17 of human peripheral blood mononuclear cells *in vivo*. *J Med Virol* **46**(3):178–188, 1995.

161. Burns WH, Sandford GR: Susceptibility of human herpesvirus 6 to antivirals *in vitro*. *J Infect Dis* **162**(3):634–637, 1990.

162. Wang FZ, Dahl H, Linde A, *et al*: Lymphotropic herpesviruses in allogeneic bone marrow transplantation. *Blood* **88**(9):3615–3620, 1996.

163. Gerber P, Lucas S, Nonoyama M, Perlin E, Goldstein LI: Oral excretion of Epstein–Barr virus by healthy subjects and patients with infectious mononucleosis. *Lancet* **2**(7785):988–989, 1972.

164. Sample J, Young L, Martin B, *et al*: Epstein-Barr virus types 1 and 2 differ in their EBNA-3A, EBNA-3B, and EBNA-3C genes. *J Virol* **64**(9):4084–4092, 1990.

165. Lam KM, Syed N, Whittle H, Crawford DH: Circulating Epstein–Barr virus-carrying B cells in acute malaria. *Lancet* **337**(8746):876–878, 1991.

166. Miyashita EM, Yang B, Lam KM, Crawford DH, Thorley-Lawson DA: A novel form of Epstein–Barr virus latency in normal B cells *in vivo*. *Cell* **80**(4):593–601, 1995.

167. Gratama JW, Oosterveer MA, Zwaan FE, *et al*: Eradication of Epstein–Barr virus by allogeneic bone marrow transplantation: Implications for sites of viral latency. *Proc Natl Acad Sci USA* **85**(22):8693–8696, 1988.

168. Faulkner GC, Burrows SR, Khanna R, *et al*: X-linked agammaglobulinemia patients are not infected with Epstein–Barr virus: Implications for the biology of the virus. *J Virol* **73**(2):1555–1564, 1999.

169. Steven NM, Annels NE, Kumar A, *et al*: Immediate early and early lytic cycle proteins are frequent targets of the Epstein–Barr virus-induced cytotoxic T cell response. *J Exp Med* **185**(9):1605–1617, 1997.

170. Steven NM, Leese AM, Annels NE, Lee SP, Ricksinson AB: Epitope focusing in the primary cytotoxic T cell response to Epstein–Barr virus and its relationship to T cell memory. *J Exp Med* **184**(5):1801–1813, 1996.

171. Callan MF, Tan L, Annels N, *et al*: Direct visualization of antigen-specific CD8+ T cells during the primary immune response to Epstein–Barr virus *in vivo*. *J Exp Med* **187**(9):1395–1402, 1998.

172. Liebowitz D: Epstein–Barr virus and a cellular signalling pathway in lymphomas from immunosuppressed patients. *N Engl J Med* **338**(20):1413–1421, 1998.

173. Purtilo DT, Cassel CK, Yang JP, Harper R: X-linked recessive progressive combined variable immunodeficiency (Duncan's disease). *Lancet* **1**(7913):935–940, 1975.

174. Coffey AJ, Brooksbank RA, Brandau O, *et al*: Host response to EBV infection in X-linked lymphoproliferative disease results from mutations in an SH2–domain encoding gene. *Nat Genet* **20**(2):129–135, 1998.

175. Sayos J, Wu C, Morra M, *et al*: The X-linked lymphoproliferative-disease gene product SAP regulates signals induced through the co-receptor SLAM. *Nature* **395**(6701):462–469, 1998.

176. Klein G, Klein E: Sinking surveillance's flagship. *Nature* **395**(6701):441–444, 1998.

177. Haque TH, Crawford DH: Transmission of Epstein–Barr virus during transplantation. *Rev Med Virol* **6**(2):77–84, 1996.

178. Purtilo DT: X-linked lymphoproliferative disease manifests immune deficiency to Epstein–Barr virus which results in diverse diseases. *Rev Med Virol* **2**(3):153–160, 1992.

179. Lucas KG, Burton RL, Zimmerman SE, *et al*: Semiquantitative Epstein–Barr virus (EBV) polymerase chain reaction for the determination of patients at risk for EBV-induced lymphoproliferative disease after stem cell transplantation. *Blood* **91**(10):3654–3661, 1998.

180. Andersson J, Britton S, Ernberg I, *et al*: Effect of acyclovir on infectious mononucleosis: A double-blind, placebo-controlled study. *J Infect Dis* **153**(2):283–290, 1986.

181. Tynell E, Aurelius E, Brandell A, *et al*: Acyclovir and prednisolone treatment of acute infectious mononucleosis: A multicenter, double-blind, placebo-controlled study. *J Infect Dis* **174**(2):324–331, 1996.

182. Emanuel DJ, Lucas KG, Mallory GBJ, *et al*: Treatment of post-transplant lymphoproliferative disease in the central nervous system of a lung transplant recipient using allogeneic leukocytes. *Transplantation* **63**(11):1691–1694, 1997.

183. Kedes DH, Operskalski E, Busch M, *et al*: The seroepidemiology of human herpesvirus 8 (Kaposi's sarcoma-associated herpesvirus): Distribution of infection in KS risk groups and evidence for sexual transmission. *Nat Med* **2**(8):918–924, 1996.

184. Gessain A, Mauclere P, van Beveren M, *et al*: Human herpesvirus 8 primary infection occurs during childhood in Cameroon, Central Africa. *Int J Cancer* **81**(2):189–192, 1999.

185. Simpson GR, Schulz TF, Whitby D, *et al*: Prevalence of Kaposi's sarcoma associated herpesvirus infection measured by antibodies to recombinant capsid protein and latent immunofluorescence antigen. *Lancet* **348**(9035):1133–1138, 1996.

186. Martin JN, Ganem DE, Osmond DH, *et al*: Sexual transmission and the natural history of human herpesvirus 8 infection. *N Engl J Med* **338**(14):948–954, 1998.

187. Koelle DM, Huang ML, Chandran B, *et al*: Frequent detection of Kaposi's sarcoma-associated herpesvirus (human herpesvirus 8) DNA in saliva of human immunodeficiency virus-infected men: Clinical and immunologic correlates. *J Infect Dis* **176**(1):94–102, 1997.

188. Muralidhar S, Pumfery AM, Hassani M, *et al*: Identification of kaposin (open reading frame K12) as a human herpesvirus 8 (Kaposi's sarcoma-associated herpesvirus) transforming gene. *J Virol* **72**(6):4980–4988, 1998.

189. Gao SJ, Kingsley L, Hoover DR, *et al*: Seroconversion to antibodies against Kaposi's sarcoma-associated herpesvirus-related latent nuclear antigens before the development of Kaposi's sarcoma. *N Engl J Med* **335**(4):233–241, 1996.

190. Parravicini C, Olsen SJ, Capra M, *et al*: Risk of Kaposi's sarcoma-associated herpes virus transmission from donor allografts among Italian posttransplant Kaposi's sarcoma patients. *Blood* **90**(7):2826–2829, 1997.

191. Regamey N, Tamm M, Wernli M, *et al*: Transmission of human herpesvirus 8 infection from renal-transplant donors to recipients. *N Engl J Med* **339**(19):1358–1363, 1998.

192. Cesarman E, Chang Y, Moore PS, Said JW, Knowles DM: Kaposi's sarcoma-associated herpesvirus-like DNA sequences in AIDS-related body-cavity-based lymphomas. *N Engl J Med* **332**(18):1186–1191, 1995.

193. Whitby D, Howard MR, Tenant-Flowers M, *et al*: Detection of Kaposi sarcoma associated herpesvirus in peripheral blood of HIV-infected individuals and progression to Kaposi's sarcoma. *Lancet* **346**(8978):799–802, 1995.

194. Rabkin CS, Schulz TF, Whitby D, *et al*: Interassay correlation of human herpesvirus 8 serologic tests. HHV-8 Interlaboratory Collaborative Group. *J Infect Dis* **178**(2):304–309, 1998.

195. Mocroft A, Youle M, Gazzard B, *et al*: Anti-herpesvirus treatment and risk of Kaposi's sarcoma in HIV infection. Royal Free/Chelsea and Westminster Hospitals Collaborative Group. *AIDS* **10**(10):1101–1105, 1996.

196. Glesby MJ, Hoover DR, Weng S, *et al*: Use of antiherpes drugs and the risk of Kaposi's sarcoma: Data from the Multicenter AIDS Cohort Study. *J Infect Dis* **173**(6):1477–1480, 1996.

197. Mocarski ES: Cytomegalovirus biology and replication. In Roizman B, Whitley RJ, Lopez C (eds): *The Human Herpesviruses.* Raven Press, New York, 1993, pp. 173–226.

198. Schmidt CA, Wilbron F, Weiss K, *et al*: A prospective study of human herpesvirus type 6 detected by polymerase chain reaction after liver transplantation. *Transplantation* **61**(4):662–664, 1996.

199. Herbein G, Strasswimmer J, Altieri M, *et al*: Longitudinal study of human herpesvirus 6 infection in organ transplant recipients. *Clin Infect Dis* **22**(1):171–173, 1996.

200. Wilborn F, Brinkmann V, Schmidt CA, *et al*: Herpesvirus type 6 in patients undergoing bone marrow transplantation: Serologic features and detection by polymerase chain reaction. *Blood* **83**(10):3052–3058, 1994.

201. Kadakia MP, Rybka WB, Stewart JA, *et al*: Human herpesvirus 6: Infection and disease following autologous and allogeneic bone marrow transplantation. *Blood* **87**(12):5341–5354, 1996.

202. Appleton AL, Sviland L, Peiris JS, *et al*: Human herpes virus-6 infection in marrow graft recipients: Role in pathogenesis of graft-versus-host disease. Newcastle upon Tyne Bone Marrow Transport Group. *Bone Marrow Transplant* **16**(6):777–782, 1995.

203. Chan PK, Peiris JS, Yuen KY, *et al*: Human herpesvirus-6 and human herpesvirus-7 infections in bone marrow transplant recipients. *J Med Virol* **53**(3):295–305, 1997.

204. Gompels UA, Nicholas J, Lawrence G, *et al*: The DNA sequence of human herpesvirus-6: Structure, coding content, and genome evolution. *Virology* **209**(1):29–51, 1995.

205. Nicholas J: Determination and analysis of the complete nucleotide sequence of human herpesvirus. *J Virol* **70**(9):5975–5989, 1996.

206. DeClercq E: Trends in the development of new antiviral agents for the chemotherapy of infections caused by herpesviruses and retroviruses. *Rev Med Virol* **5**(3):149–164, 1995.

11

Morbidity in Compromised Patients Related to Viruses Other than Herpes Group and Hepatitis Viruses

ROBERT T. SCHOOLEY

1. Introduction

The past decade has witnessed greatly increased awareness of the multiple roles played by viral agents in immunocompromised patients. This interest, which has been stimulated by the development of effective antiviral chemotherapy for herpes group viruses, in particular, has been greatly facilitated by the development of more direct approaches to diagnosis using molecular biological and monoclonal antibody technology. Lessened reliance on standard serologic techniques for viral diagnosis has been especially useful in immunocompromised patients in whom humoral responses may be delayed or absent.

It has become increasingly apparent that viral agents may contribute to morbidity in immunocompromised patients by several mechanisms in addition to traditional cytopathology and end-organ damage. Immunomodulation, which has been extensively studied in the herpes group virus system, is a feature of other viral agents, such as measles and mumps. In addition, immunocompromised patients may prove to be the population in which the biological relevance of our increasing understanding of the transforming properties of several viral agents,

Robert T. Schooley • Division of Infectious Diseases, Department of Medicine, University of Colorado Health Sciences Center, Denver, Colorado 80262.

Clinical Approach to Infection in the Compromised Host (Fourth Edition), edited by Robert H. Rubin and Lowell S. Young. Kluwer Academic/Plenum Publishers, New York, 2002.

such as BK virus, adenoviruses, and papillomaviruses, is tested. This chapter attempts to outline the morbidity caused by viruses other than herpes group and hepatitis viruses in immunocompromised patients. Although immunocompromised patients are susceptible to the same spectrum of infections that cause morbidity for the general populations, certain agents appear to be particularly devastating to immunocompromised individuals. Only agents for which increased morbidity and mortality in immunocompromised patients have been clearly demonstrated are discussed (Table 1).

2. DNA Viruses

2.1. Adenoviruses

Adenoviruses are a common cause of acute febrile illness in immunocompetent adults. In children, however, up to 10% of hospitalizable pneumonias may be caused by adenovirus infection.[1] Over the past decade, it has become apparent that adenoviruses may cause major morbidity in immunocompromised patients.[2–6] Adenovirus-associated morbidity has been reported in patients with both cellular and humoral immune defects with predisposing conditions that have included severe combined immunodeficiency, malignancy, and solid organ or bone marrow allograft transplantation. In such patients, adenovirus infections may be subclinical or they may be associated with severe morbidity or death.[2,6–9]

TABLE 1. Organ System Involvement with Viruses Other Than Herpes Group and Hepatitis Viruses in Immunocompromised Patients

Virus	Organ system						
	CNS	Pulmonary	Hepatic	GI	Renal	Musculoskeletal	Skin
Adenovirus	+	+ +	+ + +	+ +	+		
Papovaviruses							
JC	+						
BK					+		
Papillomaviruses							+ +
Vaccinia							+ + +
Polio	+ +						
Echo	+ +						+ +
Coxsackie	+			+			
Measles	+ +	+ +					+
RSV		?					
Rotavirus				+ +			

Illustrative Case 1

A 19-year-old man with renal failure caused by focal sclerosing glomerulonephritis received a renal allograft from a cadaveric donor.[6] Eight weeks after transplantation, fever, leukopenia, and thrombocytopenia developed. Cytomegalovirus (CMV) was isolated from his saliva. An eightfold rise in complement-fixation antibody titers to CMV was documented. The syndrome resolved spontaneously over 2–3 weeks. Six months later, he returned to hemodialysis because of recurrent focal sclerosis.

One year after the initial transplantation procedure, a second allograft from a cadaveric donor was implanted. He received prednisone, azathioprine, and antithymocyte globulin, but progressive azotemia, hypertension, and oliguria developed. Two weeks following transplantation, fever and dyspnea developed. Bilateral basilar rales were present. A chest radiograph revealed diffuse interstitial pulmonary infiltrates. Leukopenia with a relative lymphocytosis was noted. The immunosuppressive regimen was rapidly tapered. Hemodialysis was reinstituted 3 weeks following transplantation. With withdrawal of the immunosuppressive regimen, the fever and pulmonary interstitial infiltrates resolved over a 2-week period. Adenovirus type 34 and CMV were isolated from three urine specimens. Adenovirus antibodies measured by hemagglutinin inhibition rose from <8 to 32 in conjunction with the clinical syndrome. The CMV antibody titer did not rise.

Comment. This patient with a previous renal allograft procedure that had been complicated by CMV reactivation developed fever, leukopenia, and interstitial pneumonitis in conjunction with isolation of adenovirus type 34 and CMV from urine. The brisk humoral immune response to adenovirus type 34 coupled with the lack of an antibody titer rise to CMV suggests that the febrile illness during the second posttransplant period was caused by adenovirus rather than CMV. Management consisted of supportive care and of withdrawal of immunosuppressive therapy.

Adenovirus hepatitis is the most frequently recognized clinical entity in this group of patients. Patients with adenovirus hepatitis present with fever, anorexia, nausea, and vomiting. The hepatitis may be fulminant with rapidly rising hepatocellular enzymes, and swiftly deteriorating hepatic synthetic function.[10–15] Hepatic tissue obtained by biopsy or at autopsy generally reveals necrotic hepatocytes with amphophilic or basophilic nuclear inclusion bodies. Portal tracts and hepatic blood vessels are usually spared, as compared with the marked hepatocellular involvement. Electron microscopy of hepatic tissue reveals crystalline arrays of hexagonal adenovirus virions.[2–4] Although some patients with adenovirus-associated hepatitis recover, many succumb to hepatic or extrahepatic involvement with the virus.

Pneumonia is also frequently recognized as a manifestation of adenovirus infection in the immunocompromised host.[4,5,14–22] Although patients may present with isolated pneumonia, many patients have hepatic involvement as well. In most cases, the pneumonia is bilateral and interstitial, although cases of unilateral involvement have been reported.[2] Pleural effusions may appear in up to 20% of cases.[2,18] Lung biopsy reveals interstitial pneumonia with intranuclear inclusions. The course of adenovirus pneumonia is variable and may be dependent on the degree of immunosuppression of the host, the prior experience of the host with adenoviruses, and perhaps the serotype of the isolate. Patients have been reported with rapidly progressive pneumonia with hypoxemia and death; others have manifested a much more indolent course.[2,18]

A number of patients have been reported in whom the predominant clinical manifestation of adenovirus infection was gastroenteritis.[23–26] Adenovirus was isolated from the stools of 12 of 78 bone marrow allograft recipients participating in a prospective study of gastrointestinal (GI) pathogens during the early posttransplant period. In 9 of these patients, adenovirus was the sole viral pathogen. These patients presented with vomiting (8 of 9), diarrhea (5 of 9), and abdominal cramps (4 of 9). Four also had respiratory symptoms. Among this group of patients there were four deaths. Most of the adenovirus isolates were recovered during a 3-month period, during which adenovirus was prevalent among the pediatric population of the hospital, suggesting that patients acquired the pathogen from hospital visitors and staff. Colitis also has been associated with adenovirus infection in patients with AIDS,[25] and in a single patient infected with both adenovirus type 2 and Epstein–Barr virus.[26] Adenoviruses may be isolated relatively frequently from the urinary tracts of patients with human immunodeficiency virus-1 (HIV-1) infection.[27–29] Although adenovirus, on occasion, may contribute to pulmonary pathology in patients with AIDS,[30] urinary isolation is usually without obvious symptoms. In the setting of organ transplantation, however, adenovirus-associated hemorrhagic cystitis has been frequently reported.[31–36] Adenovirus isolates associated with hemorrhagic cystitis are often of serotype 11.

Renal parenchymal involvement by adenovirus also has been documented in one large carefully performed study of bone marrow allograft recipients and in one case report of a patient with Hodgkin's disease.[37] In this study, comprising 1051 patients over a 5-year period, 6 adenoviruses were isolated from 51 patients. In all but 10 patients, either no clinical manifestations were apparent in conjunction with adenoviral isolation, or other clinical problems, including herpes group virus activity, precluded attribution of clinical findings to the isolated adenovirus. Adenoviruses were isolated from renal parenchyma of 5 of the 10 patients for whom organ damage in association with adenovirus infection could be documented. Four of the five patients developed renal impairment; three required dialysis. Although each of the five patients received potentially nephrotoxic drugs, the demonstration of adenovirus inclusions in association with tubular epithelial necrosis in two of the patients supports the hypothesis that the virus also contributed to renal functional impairment. In addition to hepatic, pulmonary, and GI involvement, adenovirus may occasionally involve the central nervous system (CNS) in immunocompromised patients.[38–40] This complication appears to be relatively infrequent even in patients with disseminated adenovirus infection. Other rare complications of adenovirus infection in immunocompromised patients include a report of granuloma annulare in a patient with acquired immunodeficiency syndrome (AIDS),[41] and hemophagocytic syndrome in a bone marrow allograft recipient.[42]

Diagnosis is greatly facilitated by a familiarity with the clinical syndromes associated with adenovirus infection in the immunocompromised and by attention to ongoing viral activity in the community. Definitive diagnosis rests with the demonstration of virus in an involved organ by viral isolation, electron microscopy, cytologic examination,[43,44] demonstration of adenoviral antigens in infected cells, or the demonstration of a fourfold rise in adenovirus-specific antibodies during the clinical episode.[45] The development of enzyme immunoassays (EIA) for adenoviral antigens in stool[45] and of polymerase chain reaction (PCR) technology for identification of adenovirus DNA in perpiheral blood mononuclear cells[46,47] will provide for easier diagnosis of adenovirus infections and undoubtedly will lead to an enhanced appreciation of the clinical spectrum of adenovirus infections in the immunocompromised host.

Although immunoglobulin (Ig) therapy has been advocated for patients with humoral immune deficiency,[17] there is no convincing evidence from controlled trials that this approach is beneficial. At present, no chemotherapeutic agents have demonstrated efficacy against adenovirus *in vivo*. However, several novel nucleoside analogues have recently demonstrated activity against certain serotypes of adenovirus *in vitro*.[48]

As in other situations in which the degree of immunosuppression can be manipulated, it would be prudent to decrease the exogenous immunosuppression to as great a degree as possible during acute infection. Because adenoviruses are relatively highly contagious and infection can have significant consequences for immunocompromised persons, respiratory and enteric isolation should be employed for patients with recognized adenovirus infection in units with immunosuppressed patients.

2.2. Papovaviruses

The relatively recently described human papovaviruses, BK and JC, have received increasing attention as pathogens in immunosuppressed patients. Serologic surveys have demonstrated that these agents are ubiquitous in the adult population, with up to 90% of adults having evidence of prior BK virus infection and 70% of adults having evidence of prior JC virus infection.[49–51] There have been reports of morbidity in association with these viruses in immunocompetent patients.

2.2.1. JC Virus

Illustrative Case 2

A 68-year-old man developed renal failure in December of 1981 as a result of rapidly progressive glomerulonephritis despite a course of therapy with prednisone and cyclophosphamide. He received a cadaveric renal allograft in November of 1982 following 6 months of hemodialysis. His early posttransplant course was characterized by repeated bouts of acute rejection, for which he received three courses of antithymocyte globulin in addition to azathioprine and prednisone.

Seven months after transplantation, his family noted the onset of forgetfulness and emotional lability. Progressive dementia was noted over the ensuing 6 weeks. He returned for evaluation following a grand mal seizure. On examination, he was an afebrile elderly man in no acute distress. His general physical examination was remarkable for a cushingoid habitus and a nontender renal allograft in the right iliac fossa. He had no focal neurologic findings. Short-term memory was markedly impaired. Long-term memory was moderately impaired. He could name the three most recent presidents but could not perform serial 7 subtractions.

A computed tomographic (CT) study of the cranium revealed an area of decreased density with indistinct margins in the left frontal lobe. No mass effect was noted (Fig. 1). The initial radiographic interpretation was that the CT scan was most compatible with an infarct of several weeks' duration, although a low-grade infiltrating tumor could not be ruled out. A lumbar puncture revealed a normal opening pressure. Two lymphocytes were present. The glucose and protein were normal. An electroencephalogram (EEG) was diffusely abnormal, with slowing throughout both hemispheres.

A biopsy of the left frontal lobe lesion revealed replacement of normal brain tissue with large, bizarre astrocytes surrounded by degenerating oligodendrocytes. Electron microscopy, which revealed papovavirus particles within nuclei of the degenerating oligodendrocytes, established the diagnosis of progressive multifocal leukoencephalopathy.

A second CT study revealed the development of additional areas of decreased attenuation within the white matter of the brain. Following the brain biopsy, the dementia progressed rapidly. No antiviral therapy was administered. Increasing seizure activity led to the development of aspiration pneumonia. The patient died 10 months following transplantation, after a 4-month course of progressive multifocal leukoencephalopathy.

JC virus, a member of the human polyomavirus group, is the etiologic agent for the vast majority of cases of progressive multifocal leukoencephalopathy (PML). This rare clinical entity, initially described by Richardson and co-workers[52] in 1958, had been reported in fewer than 250 patients prior to 1989. With the rapid increase in the number of individuals with HIV-associated cellular immunodeficiency, the number of reported cases of PML has increased significantly over the past decade.[53] Almost all patients have demonstrable immunologic defects, with cellular immune defects being much more common than defects in humoral immunity. The combination of the immunocompromised patient population in which the disease was initially recognized and the morphology of involved oligodendrocytes led Richardson to suggest a

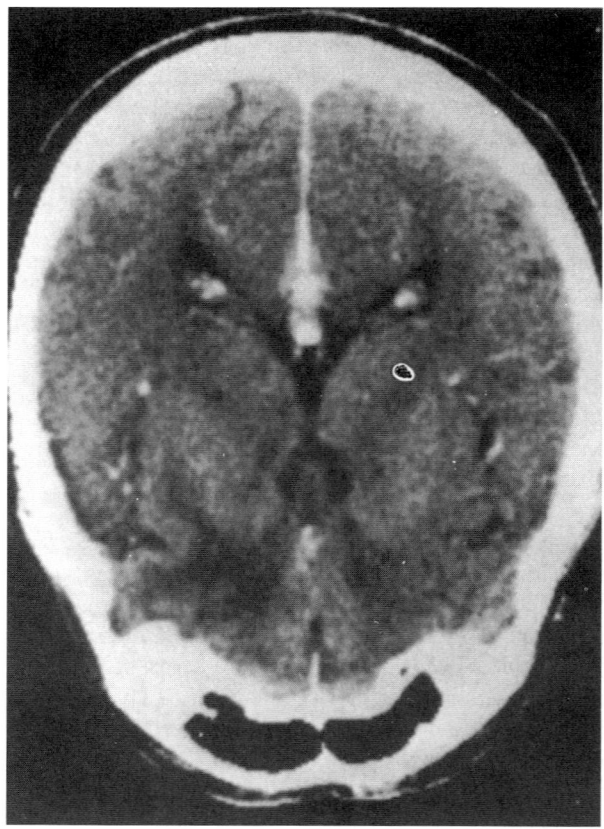

FIGURE 1. Computed tomographic study of the patient in Illustrative Case 2, demonstrating areas of decreased attenuation, particularly involving the left frontal lobe.

viral etiology for this clinical entity.[52,54] Following a decade during which viral particles compatible with polyomavirus were demonstrated in oligodendrocytes by electron microscopy,[55] two groups reported the isolation of viral agents from PML brains in tissue culture.[56,57] One group isolated a polyomavirus, designated JC virus, from a PML brain, using primary human fetal glial cell cultures; a second group reported the isolation of a virus identical to, or very closely related to, simian virus 40 (SV40) using an independent technique. All subsequent isolates have been of the JC type.

Serologic studies have demonstrated that JC virus is a ubiquitous agent. Antibodies to JC virus can be demonstrated in sera from 70% of the adult population.[50] Primary infection occurs most frequently during the second decade of life and is generally subclinical. Following primary infection, the virus is thought to remain within the host in latent form. With immunosuppression such as that used in renal transplantation or in pregnancy, reactivation of JC virus is evident in some persons both by serologic changes and by shedding of the virus in the urine.[58,59] With the application of more sensitive techniques such as PCR technology, shedding of JC virus is detected on occasion from nonimmunocompromised individuals.[60–62] In general, these bouts of reactivation are also clinically inapparent. For reasons that are unclear, JC virus invades the CNS of a very small minority of patients and results in PML.[63,64] Southern blot analysis and PCR technology also have demonstrated JC virus DNA in brain tissue obtained from immunocompromised individuals without PML.[65–67]

Patients with PML have generally experienced defects in cellular immunity for months to years prior to onset of neurologic symptoms. The chronicity of the cellular immune defect associated with HIV infection places patients with AIDS at particular risk for PML.[53,68–70] HIV is itself neurotropic.[71,72] Recently, it has been demonstrated additionally that the HIV-1 regulatory gene *tat* also up-regulates the JC virus late promoter.[73] In addition, the JC virus early promoter stimulates expression of the HIV-1 long terminal repeat sequence.

A number of neurologic syndromes, including dementia, aseptic meningitis, peripheral neuropathy, and vacuolar myelopathy, have been reported in association with HIV infection.[71,72,74] Thus, not all neurologic syndromes in AIDS patients clinically compatible with PML are the result of JC virus. Presentation is usually insidious, with neurologic signs reflecting the multifocal nature of the process.[53,54,63,64] In most patients, progression is relatively rapid with early deterioration in cognitive function, speech, and vision. Motor deficits, cortical blindness, and sensory abnormalities follow shortly thereafter. Patients rarely complain of headaches or exhibit fever or seizures. Occasional cases have been described in which involvement of the spinal cord has been clinically prominent, but these cases are very uncommon. Most patients succumb to the illness within 6 months of the onset of neurologic signs. In one relatively large series of patients with HIV-1-associated PML, the median survival after diagnosis was only 3 months; 90% of patients were dead within a year of diagnosis.[53]

Definitive diagnosis requires brain biopsy. Nonetheless, in the appropriate setting, an extremely strong clinical diagnosis can be established without resorting to an invasive procedure. Lumbar puncture usually reveals a normal opening pressure with a normal glucose and protein level. A few lymphocytes may be present, but in general the hallmark of this illness in both the cerebrospinal fluid (CSF) and the brain is the lack of inflammatory reaction. The demonstration of JCV DNA in CSF by PCR

is strong evidence that the illness is PML.[75] EEGs are diffusely abnormal but nondiagnostic. The individual lesions are usually too small to be detectable by radionuclide scanning. Pneumoencephalography and cerebral angiography are usually normal. CT, however, is often capable of demonstrating multiple demyelinative lesions. The ubiquity of antibodies to JC virus in the adult population and the relative frequency of serologic evidence of JC virus activity in immunocompromised hosts in the absence of PML[76] make serologic diagnosis impossible.

In atypical cases, brain biopsy may be required to rule out other treatable entities. The central pathologic feature of PML is demyelination, which occurs as a result of lytic infection of oligodendrocytes by JC virus. In contrast to multiple sclerosis, very few inflammatory cells are demonstrable. With disease progression, confluent demyelination results in widely distributed plaques of increasing size. Electron microscopic evaluation of involved oligodendrocytes reveals crystalline arrays of papovavirus particles.[77] Viral antigens also can be demonstrated by other techniques, such as immunofluorescence or immunoperoxidase staining. Late in disease, as oligodendrocytes are progressively depleted, viral particles and antigen become more difficult to demonstrate. JC virus can be isolated in tissue culture, using primary human fetal glial cells. Typical cytopathic effects of loss of contact inhibition, destruction of spongioblasts, and the appearance of multinucleate astrocytes are usually apparent 10–12 days after initiation of the culture. Recently, several new techniques have been applied to the diagnosis of PML. These include *in situ* hybridization,[78–81] and immunohistochemical staining of brain tissue.[82,83] PCR is perhaps the most powerful technique in that JC virus DNA has been detected in brain tissue of 80–100% of patients with PML.[84–86] JC virus DNA is also detected in the CSF of patients with PML.[75,85,86] However, the results of PCR-based diagnostic approaches must be interpreted with caution, since on occasion JCV DNA can be detected in CSF from patients with other CNS diagnoses. In view of the demonstration of JC virus sequences in brain tissue, urine, and peripheral blood mononuclear cells from patients without PML, amplifying JCV DNA from individuals without PML should not be unexpected.[65–67,87,88]

There is no effective treatment of PML. There have been a number of reports of therapy with interferon-β (IFN-β) and with nucleoside analogues, including IUDR, cytosine arabinoside, and adenine arabinoside.[89–102] Although occasional case reports have maintained that stabilization or improvement was coincident with the use of these agents, these observations are uncontrolled and anecdotal. One case has been reported with apparent improvement with zidovudine therapy.[103] The recent handful of reports that intravenous and/or intrathecal cytarabine therapy has been associated with improvement in some[104–107] but not all[108] PML patients led to the initiation of a randomized, blinded trial of this agent within the AIDS Clinical Trials Group of the NIAID. In this study cytarabine was shown to be ineffective.[109] The advent of highly active antiretroviral therapy (HAART) has provided an opportunity to reverse HIV-1-associated immunosuppression in many patients. Case reports of the association of a reversal of PML in patients with HIV-1 infection with the initiation of antiretroviral chemotherapy are increasingly common.[110]

2.2.2. BK Virus

BK virus initially was isolated from the urine of a renal allograft recipient with ureteral stenosis.[111] Subsequent case reports also have made note of the association between BK virus infection and ureteral stenosis.[112] One carefully performed pathologic study documented the presence of papovavirus inclusions in ureteral epithelial cells in an area of ureteral stenosis in two renal allograft recipients. BK virus excretion is much more frequently observed, however, than is ureteral stenosis.[60,61,113,114] As in the case of CMV, BK virus may be transmitted to BK-seronegative allograft recipients by kidneys from BK-seropositive donors.[114] An association between BK virus excretion and the development of pancreatitis postrenal transplantation has also been demonstrated.[115] Whether this association implies an etiologic relationship and whether it might be generalizable to other clinical settings is unclear.

Outside the setting of renal transplantation, reports of BK virus-associated morbidity have been sparse. One 6-year-old boy was reported with tubulointerstitial nephritis associated with BK virus infection.[116] This child with congenital dysgammaglobulinemia developed renal failure and subsequently died after a series of infections including *Cryptococcus neoformans* and *Enterobacter cloacae sepsis*. At postmortem examination, renal glomeruli were relatively normal, but renal tubular cells were focally necrotic with large amounts of polyomavirus antigen in both tubular lining cells and cast material. DNA hybridization studies revealed a disseminated BK virus infection with BK sequences in kidney lymph nodes, spleen, and lungs. It is likely from the data generated since the initial description of BK virus 25 years ago that this agent will be increasingly implicated as a pathogen in immunocompromised persons *in vitro*. Neither seroconversion nor viral excretion was affected by prophylactic

IFN-β therapy in a randomized study of renal allograft recipients.[115]

2.2.3. Papillomavirus

For several hundred years, warts have been postulated to be of an infectious etiology. Human inoculation experiments in the first decade of the 20th century established the viral etiology of human warts, but further progress was greatly hampered by the lack of suitable tissue culture techniques for propagation of the etiologic agents. The recent application of molecular biologic techniques to the study of human warts has confirmed the viral etiology of these common skin tumors and has provided insight into the plurality of animal papillomaviruses. These agents have been shown to be small DNA viruses and are included in the same family (Papovaviridae) as polyomaviruses. These recent studies also have delineated the associations of specific strains of papillomavirus with specific clinical types of warts.[117,118]

Clinicians have long been aware of the increased frequency and severity of warts in patients with immunodeficiency states.[119–122] This increased frequency has been noted in patients with a wide variety of primary and secondary immunodeficiency states but is most notable in patients with defects in cellular immunity. It has been the impression of many clinicians that warts are particularly likely to be induced by sun exposure. Scattered reports of possible degeneration of warts to squamous cell tumors have appeared.[123,124] In the setting of HIV infection, in particular, a strong association has been made between the presence of genital condylomata and human papillomavirus.[125] In addition, an association between the presence of human papillomavirus and cervical intraepithelial neoplasia is emerging in HIV-1-infected women.[126,127]

Therapy is generally based on local destruction of the involved area by cryosurgery or electrodesiccation. These modes of therapy are much less likely to meet with success in immunocompromised persons than in immunologically normal hosts. Current clinical trials are under way in which the potential value of interferons is being addressed in the management of warts. If these studies suggest that interferons are useful in immunologically normal persons, extension of the studies to immunocompromised patients appears likely.

2.3. Vaccinia

Prior to the cessation of routine smallpox vaccination, vaccinia necrosum was a major cause of morbidity and mortality with immunocompromised vaccinees.[128]

This entity, also known as progressive vaccinia, occurred when patients with predominantly cellular immune defects were inadvertently vaccinated for smallpox with vaccinia virus.[128–131] The vaccination site initially appears normal but, rather than resolution, progressive destruction is observed at the site of vaccination. Eventually metastatic cutaneous lesions are evident. In general, very little lymphadenopathy or local reaction is noted in the absence of bacterial superinfection. Treatment has included vaccinia immune globulin and thiosemicarbazone.[132,133]

3. RNA Viruses

3.1. Picornaviruses

3.1.1. Poliomyelitis

Paralytic poliomyelitis has become an extremely rare disease in industrialized countries. In the United States, cases of paralytic poliomyelitis declined 20-fold with the introduction of the Salk vaccine in 1955 and an additional 100-fold with the widespread use of the live attenuated (Sabin) vaccine.[134] The gains in control of paralytic poliomyelitis for the general population with the Sabin vaccine have been complicated by the fact that the live attenuated strains capable of inducing lasting immunity in the immunocompetent host can induce paralytic poliomyelitis in the immunocompromised host.[135,136] This is particularly true for the type 2 vaccine strain. Since 1969, 7% of cases of paralytic poliomyelitis in the United States were in immunocompromised persons in association with live attenuated vaccine strains.[135] The risk of paralytic poliomyelitis following administration of oral polio vaccine strain is increased approximately 10,000-fold in hypogammaglobulinemics.[136] By contrast, paralysis is seen in association with much more comparable proportions (about 2.5%) of immunocompetent and immunocompromised persons in association with infection with wild-type strains.

An analysis of cases of paralytic poliomyelitis in immunocompromised patients in association with live attenuated vaccine strains has revealed that patients with humoral, cellular defects and combined effects are at increased risk from oral polio vaccine.[137–144] Nonetheless, not all immunodeficient patients without prior immunity who receive oral polio vaccine develop paralytic poliomyelitis; it has been estimated that the risk for hypogammaglobulinemics is that 1 in 40 will develop paralytic disease following exposure to the oral vaccine.[136] The incubation period between vaccination and onset of para-

lytic manifestations is typically prolonged in immunocompromised persons but ranged from 2 weeks to 7 months in one survey.[135] The disease in immunodeficient persons is frequently characterized by a stuttering prolonged onset of neurologic symptoms, which include both upper and lower motor neuron signs. GI shedding of virus frequently becomes chronic in immunodeficient persons.

Vaccination with live attenuated polio vaccine of patients with known or suspected immunodeficiency of any variety should be avoided. Vaccination of family members living with such persons with live attenuated vaccine also should be avoided, given the demonstrable spread of vaccine strains among family members and contacts as the result of focal shedding of virus. Once neurologic manifestations have become apparent, no effective therapy is available. An excellent case can be made, however, for long-term immunoglobulin therapy of hypogammaglobulinemic patients following inadvertent exposure to live attenuated vaccine if the exposure is recognized prior to the onset of neurologic manifestations. Because of the prolonged fecal shedding of virus by immunocompromised persons, care should be exercised to prevent nosocomial spread if such patients are hospitalized among other immunodeficient persons.

3.1.2. Coxsackievirus and Echoviruses

In addition to being at increased risk for paralytic poliomyelitis in association with live attenuated polio vaccine, hypogammaglobulinemic patients are at increased risk for chronic meningoencephalitis from other enteroviruses.[145–158] Chronic meningoencephalitis is most frequently associated with echovirus infections (types 2, 3, 5, 9, 11, 19, 24, 25, 30, 33) but has occasionally been reported in association with coxsackie B infection.[159,160] The onset may be relatively acute but is frequently insidious. Severity ranges from subclinical involvement to more obvious neurologic findings including headaches, nuchal rigidity, cognitive dysfunction, lethargy, seizures, tremor, motor weakness, or ataxia. Many but not all patients also manifest a prominent dermatomyositislike syndrome. The pathogenesis of the myositis is not certain, but isolation of virus from muscle of one of these cases suggests that direct muscle involvement by virus may contribute to this aspect of the illness in some patients.[151] The recent application of PCR technology to the diagnosis of enterovirus infections[158] has provided an important additional diagnostic approach in this clinical setting.

CSF usually includes a predominantly lymphocytic pleocytosis and a mildly elevated protein with a normal glucose. Virus can be isolated from CSF repeatedly or intermittently but is usually not isolatable from extraneural sites.

No specific antiviral therapy for echoviruses or coxsackieviruses is currently available. An agent has recently been developed that binds to the virion capsid protein of a number of picornaviruses.[161] The drug (WIN 54954) has been used successfully in a murine coxsackievirus A9 myocarditis model.[162] Although most hypogammaglobulinemic patients with chronic meningoencephalitis resulting from enteroviruses are receiving supplemental immunoglobulin therapy at presentation, clinical improvement has been reported in several patients in association with therapy with immunoglobulin preparations with neutralizing activity to the specific implicated enterovirus.[147–151,158,163]

In addition to the occasional involvement of the CNS of hypogammaglobulinemic patients by coxsackievirus, two recent reports have implicated coxsackieviruses in the causation of acute gastroenteritis among hospitalized immunocompromised patients.[164,165] Both reports involved the same bone marrow transplant unit.

Illustrative Case 3

A 9-year-old boy with renal failure caused by obstructive uropathy received four renal allografts between 1976 and 1979. The first three allografts were lost to rejection. He underwent splenectomy in January 1981, in preparation for a fourth allograft procedure. He received a cadaveric allograft in July 1982. Immunosuppression consisted of cyclophosphamide, prednisone, and total lymphoid irradiation. During the first 3 months following transplantation, he had several episodes of rejection that were managed with boluses of solumedrol. Antithymocyte globulin was added to the immunosuppressive regimen in the fourth postoperative month.

In November 1982, he awoke one morning with a headache, fever, and vomiting. He was initially managed with Tylenol. That evening, he awoke with a temperature of 103°F (39.4°C) and had a grand mal seizure similar to seizures he had experienced with bouts of fever in the past. In the emergency room, he was found to be febrile to 102.6°F (39.2°C) and to have a blood pressure of 200/100. The fundi were benign. The neck was supple. The abdomen was distended, with active bowel sounds. No localized tenderness or rebound was appreciated. Stool was guaiac negative. The neurologic examination was normal.

Laboratory studies included a serum sodium level of 138 mEq/liter, potassium 4.8 mEq/liter, chloride 111 mEq/liter, and bicarbonate 21 mEq/liter. The BUN and creatinine were unchanged from the previous week at 30 and 1.3 mg/dl, respectively. The hematocrit was 24%, the white blood cell (WBC) count 4500/mm³, with 67% polymorphonuclear cells, 4% bands, 8% lymphocytes, 16% monocytes, and 5% basophils. A traumatic lumbar puncture revealed 8800 red blood cells (RBCs)/mm³ and 8 WBCs/mm³, of which 80% were lymphocytes and 20% were polymorphonuclear cells. The CSF glucose was 64 mg/dl; the protein was 79 mg/dl. CSF was sent for bacterial, fungal, and viral cultures and for cryptococcal antigen.

Over the next several days, his headache resolved, but he remained febrile to 101–102°F. Vomiting ceased but he had several loose stools per day that were trace guaiac positive. A percutaneous renal biopsy revealed acute and chronic rejection. Allograft irradiation was added;

the antithymocyte globulin was continued. A lumbar puncture performed 4 days after admission revealed 10 RBCs/mm³ and 22 WBCs/mm³. Twenty percent of the WBCs were polymorphonuclear cells; the remainder were mononuclear cells. The CSF glucose was 55 mg/dl; the protein had fallen to 34 mg/dl.

By the tenth hospital day he was afebrile. His headaches, nausea, and diarrhea had resolved. Following discharge, cytopathic effect consistent with that induced by an enterovirus was identified in viral cultures of the initial CSF sample. During hospitalization, three other patients in the transplant unit experienced unexplained bouts of diarrhea.

In summary, this heavily immunosuppressed renal allograft recipient experienced a bout of fever, headaches, nausea, vomiting, and diarrhea at a time during which enterovirus was isolated from CSF. The possibility of nosocomial spread was raised by the simultaneous occurrence of unexplained diarrhea in three other concurrently hospitalized allograft recipients.

In one of the two reports, 7 of 14 patients transplanted during a 3-week period developed gastroenteritis related to coxsackievirus, type A1.[164] Infected patients had significantly larger stool volumes than noninfected patients. Six of the seven infected patients died. At postmortem examination, pronounced GI lymphoid atrophy was evident. This was associated with overlying foamy vacuolated GI epithelium. The extremely high mortality rate led to closing of the bone marrow transplantation unit to new admissions and to the institution of enteric precautions for all patients. These maneuvers were associated with cessation of the outbreak.

3.2. Paramyxoviruses

3.2.1. Measles

Measles (rubeola) is generally a self-limited disease of childhood that has decreased in incidence by 95% since widespread vaccination began in 1961.[166] However, it is not infrequently associated with major morbidity in immunocompromised hosts. In addition to the direct effects on the host, measles virus has the potential to induce more profound immunosuppression in a fashion analogous to that documented for the herpesviruses.[167–169]

The most frequently encountered forms of excess morbidity associated with measles infection in immunocompromised hosts are giant cell pneumonia[170–174] and encephalitis.[173–177] The mortality rate of measles in patients with malignancy or advanced HIV infection may be as high as 30–40%.[172–175] The association between measles virus and giant cell pneumonia in immunocompromised children was first proved by Enders and colleagues,[170] who isolated the virus from the lungs of three immunocompromised children with fatal pneumonia. The pathologic hallmark of this illness is the multinucleated giant cells, which are widespread in lungs and other organs. It should be noted that these patients often exhibit no rash.[172] Diagnosis is facilitated by the recognition of the clinical syndrome of fever and diffuse interstitial pneumonia in an immunocompromised person who has no history of measles vaccination, particularly if a history of exposure to measles is elicited.

Subclinical involvement of the CNS in up to 50% of immunocompetent patients with measles is suggested by the frequent observation of abnormal EEGs in acute measles.[178–181] The occasional isolation of measles virus from the brain of patients with fatal measles encephalitis and the demonstration of the association between measles and subacute sclerosing panencephalitis (SSPE)[179–182] further strengthen the evidence that the CNS may be targeted by measles virus in the normal host. Measles encephalitis in immunocompromised patients has been increasingly recognized over the past three decades.[176,177,181–185] Measles encephalitis in immunocompromised patients may initially present as a focal process with motor deficits or focal seizures, but progression over a relatively brief period of time is quite common. In contrast to patients with measles virus-associated giant cell pneumonia, immunocompromised patients with measles encephalitis usually manifest a viral exanthem that may precede the onset of neurologic findings by several days to up to 6 months. Initial CSF examination may reveal a normal glucose and protein level and few or no cells. Most patients succumb to this complication of measles virus infection. Although IFN-α therapy was associated with a brief stabilization of one patient,[186] failure of IFN-α therapy has also been reported.[187] Recently, there have been several anecdotal reports of the use of intravenous and/or associated ribavirin in the management of immunocompromised patients with measles.[172,188,189] The reports of convulsions and pneumonia in immunocompromised patients receiving the live attenuated measles vaccine underscore the rationale for avoiding this live vaccine in immunocompromised patients.[190]

3.2.2. Respiratory Syncytial Virus

Respiratory syncytial virus (RSV) is the most frequent cause of pneumonia, bronchiolitis, or tracheobronchitis among young children.[191] Among persons over the age of 3, RSV infection is usually manifest as only tracheobronchitis or more commonly as an upper respiratory illness.[192] RSV infection has been well recognized as a nosocomial pathogen[193–195] and has been demonstrated to be associated with significantly increased mortality in infants with congenital heart disease.[196] Although animal models have demonstrated prolonged RSV shedding in immunosuppressed animals,[197] only scattered reports of increased morbidity resulting from RSV among immunosuppressed children have appeared.[198,199] The extent of the increased morbidity caused by RSV among immuno-

suppressed adults appears to be extremely limited, except in the setting of bone marrow transplantation. Ribavirin has been used successfully in the treatment of RSV infection.

3.3. Rotaviruses

Rotaviruses are a major cause of infantile diarrhea of worldwide distribution. Clinical manifestations range from mild diarrhea to a severe dehydrating illness that may be fatal.[200,201] Although most cases of outbreaks of viral gastroenteritis among adults and school-age children appear to be the result of Norwalk-like agents, experimental transmission of rotavirus to adults has been associated with diarrhea.[202] With the increasingly frequent application of the enzyme-linked immunosorbent assay (ELISA) for detection of rotavirus antigen in stool, the agent is becoming increasingly recognized as a pathogen for immunocompromised children and adults.

Rotavirus-associated diarrhea in immunocompromised patients may present as a relatively chronic process or the onset may be more acute.[165,203,204] Rotavirus infection was documented in 9 of 78 hospitalized bone marrow allograft recipients.[165] In 8 of these patients, rotavirus was the only viral pathogen isolated in association with gastroenteritis. Although not as striking as with coxsackie A1-associated gastroenteritis patients investigated in the same unit, rotavirus-infected patients exhibited increased mortality, as compared with patients who did not experience viral gastroenteritis.

Rotavirus does not grow efficiently using conventional tissue culture techniques. Although rotavirus can be demonstrated in stool by conventional or immune electron microscopy, the development of ELISA and radioimmunoassay has greatly facilitated diagnosis of rotavirus gastroenteritis.[205–209] The value of serologic responses to rotaviruses has not yet been demonstrated in immunocompromised individuals.

Therapy is primarily supportive. In immunocompetent infants, most of the severe morbidity is directly attributable to fluid and electrolyte imbalance.[189] The mechanism by which rotavirus infection appears to contribute to increased mortality in bone marrow allograft recipients is unclear. Although one could hypothesize that additional insults to the GI mucosa could potentiate the development of bacteremia, this was not documented in the bone marrow transplant prospective study.[165] In infants with severe combined immunodeficiency, oral therapy with human milk containing antibodies to rotavirus may be useful.[203] As in the situation with other outbreaks of nosocomial diarrhea, rotavirus-associated diarrhea should result in the institution of enteric precautions, particularly in settings in which other immunocompromised patients are present. Although most immunocompetent adults do not develop diarrhea in association with rotavirus exposure, serologic responses of adult contacts of children with rotavirus diarrhea and the occasional outbreak of rotavirus diarrhea in settings such as nursing homes suggest that hospital staff may facilitate nosocomial spread.[210–212] Therapy of rotavirus infection is primarily supportive, although hyperimmune bovine colostrum has recently been reported to exhibit modest benefits.[213–215]

References

1. Chanock RM: Impact of adenovirus in human disease. *Prev Med* **3:**466–472, 1974.
2. Zahradnik JM, Spencer MJ, Parker DD: Adenovirus infection in the immunocompromised patient. *Am J Med* **68:**725–732, 1980.
3. Carmichael GP, Zahradnik JM, Moyer GM, *et al*: Adenovirus hepatitis in an immunosuppressed adult patient. *Am J Clin Pathol* **71:**352–355, 1979.
4. Rodriguez FH, Liuzza GE, Gohd RH: Disseminated adenovirus serotype 31 infection in an immunocompromised host. *Am J Clin Pathol* **82:**615–618, 1979.
5. Myerowitz RL, Stadler H, Oxman MN, *et al*: Fatal disseminated adenovirus infection in a renal transplant recipient. *Am J Med* **59:**591–598, 1975.
6. Michaels MG, Green M, Wald ER, *et al*: Adenovirus infection in pediatric liver transplant recipients. *J Infect Dis* **165:**170–174, 1992.
7. Salt A, Sutehall G, Sargaison M, *et al*: Viral and *Toxoplasma gondii* infections in children after liver transplantation. *J Clin Pathol* **43:**63–67, 1990.
8. Wasserman R, August CS, Plotkin SA: Viral infections in pediatric bone marrow transplant patients. *Pediatr Infect Dis J* **7:**109–115, 1988.
9. Strickler JG, Singleton TP, Copenhaver CM, *et al*: Adenovirus in the gastrointestinal tracts of immunosuppressed patients. *Am J Clin Pathol* **97:**555–558, 1992.
10. Wreghitt TG, Gray JJ, Ward KN, *et al*: Disseminated adenovirus infection after liver transplantation and its possible treatment with ganciclovir (letter). *J Infect* **19:**88–89, 1989.
11. Norris SH, Butler TC, Glass N, *et al*: Fatal hepatic necrosis caused by disseminated type 5 adenovirus infection in a renal transplant recipient. *Am J Nephrol* **9:**101–105, 1989.
12. Janner D, Petru AM, Belchis D, *et al*: Fatal adenovirus infection in a child with acquired immunodeficiency syndrome (see comments). *Pediatr Infect Dis J* **9:**434–436, 1990.
13. Johnson PR, Yin JA, Morris DJ, *et al*: Fulminant hepatic necrosis caused by adenovirus type 5 following bone marrow transplantation. *Bone Marrow Transplant* **5:**345–347, 1990.
14. Ljungman P, Ehrnst A, Bjorkstrand B, *et al*: Lethal disseminated adenovirus type 1 infection in a patient with chronic lymphocytic leukemia. *Scand J Infect Dis* **22:**601–605, 1990.
15. Keller EW, Rubin RH, Black PH, *et al*: Isolation of adenovirus type 34 from a renal transplant recipient with interstitial pneumonia. *Transplantation* **23:**188–191, 1977.
16. Neiman PE, Reeves W, Ray G, *et al*: A prospective analysis of interstitial pneumonia and opportunistic viral infections among recipients of allogeneic bone marrow grafts. *J Infect Dis* **136:**754–767, 1977.
17. Lecatsas G, Van Wyk JAC: DNA viruses in urine after renal transplantation. *S Afr Med J* **53:**787–788, 1978.
18. Wigger HJ, Blanc WA: Fatal hepatic and bronchial necrosis in adenovirus infection with thymic alymphoplasia. *N Engl J Med* **275:**870–874, 1968.
19. Dagan R, Schwartz RH, Insel RA, *et al*: Severe diffuse adenovirus 7a pneumonia in a child with combined immunodeficiency: Possible therapeutic effect of human serum immune globulin containing specific neutralizing antibody. *Pediatr Infect Dis* **3:**246–249, 1984.
20. Siegal FP, Dikman SH, Arayatu RB, *et al*: Fatal disseminated adenovirus pneumonia in an agammaglobulinemic patient. *Am J Med* **71:**1062–1067, 1981.
21. Shields AF, Hackman RC, Fife HK, *et al*: Adenovirus infection in patients undergoing bone marrow transplantation. *N Engl J Med* **312:**529–533, 1985.
22. Valteau D, Hartmann O, Benhamou E, *et al*: Nonbacterial nonfungal interstitial pneumonitis following autologous bone marrow transplantation in children treated with high-dose chemotherapy without total-body irradiation. *Transplantation* **45:**737–740, 1988.
23. Yolken RH, Bishop CA, Townsend TR, *et al*: Infectious gastroenteritis in bone marrow transplant recipients. *N Engl J Med* **306:**1009–1012, 1982.
24. Johansson ME, Wirgart BZ, Grillner L, *et al*: Severe gastroenteritis in an immunocompromised child caused by adenovirus type 5. *Pediatr Infect Dis J* **9:**449–450, 1990.
25. Janoff EN, Orenstein JM, Manischewitz JF, *et al*: Adenovirus colitis in the acquired immunodeficiency syndrome. *Gastroenterology* **100:**976–979, 1991.
26. Okano M, Thiele GM, Davis JR, *et al*: Adenovirus type-2 in a patient with lethal hemorrhagic colonic ulcers and chronic active Epstein–Barr virus infection. *Ann Intern Med* **108:**693–699, 1988.
27. deJong PJ, Valderrama G, Spigland I, *et al*: Adenovirus isolates from urine of patients with acquired immunodeficiency syndrome. *Lancet* **1:**1293–1296, 1983.
28. Horwitz MS, Valderrama G, Hatcher V, *et al*: Characterization of adenovirus isolates from AIDS patients. *Ann NY Acad Sci* **437:**161–174, 1984.
29. Hierholzer JC, Wigand R, Anderson LJ, *et al*: Adenoviruses from patients with AIDS. A plethora of serotypes and a description of five new serotypes of subgenus D types (types 43-47). *J Infect Dis* **154:**804–813, 1988.
30. Valainis GT, Carlisle JT, Daroca PJ, *et al*: Respiratory failure complicated by adenovirus type 29 in a patient with AIDS. *J Infect Dis* **160:**349–351, 1989.
31. Lecatsas G, Prozesky OW, Van Wyk J: Adenovirus type 11 associated with hemorrhagic cystitis after renal transplantation. *S Afr Med J* **48:**1932–1935, 1974.
32. Cassano WF: Intravenous ribavirin therapy for adenovirus cystitis after allogeneic bone marrow transplantation. *Bone Marrow Transplant* **7:**247–248, 1991.
33. Buchanan W, Bowman JS, Jaffers G: Adenoviral acute hemorrhagic cystitis following renal transplantation (letter). *Am J Nephrol* **10:**350–351, 1990.
34. Yagisawa T, Takahashi K, Yamaguchi Y, *et al*: Adenovirus induced nephropathy in kidney transplant recipients. *Transplant Proc* **21:**2097–2099, 1989.

35. Ambinder R, Burns W, Forman M, *et al*: Hemorrhagic cystitis associated with adenovirus infection in bone marrow transplantation. *Arch Intern Med* **146:**1400–1401, 1986.

36. Miyamura K, Takeyama K, Kojima S, *et al*: Hemorrhagic cystitis associated with urinary excretion of adenovirus type 11 following allogeneic bone marrow transplantation. *Bone Marrow Transplant* **4:**533–535, 1989.

37. Teague MW, Glick AD, Fogo AB: Adenovirus infection of the kidney: Mass formation in a patient with Hodgkin's disease. *Am J Kidney Dis* **18:**499–502, 1991.

38. Chou SM, Roos R, Burrell R, *et al*: Subacute focal adenovirus encephalitis. *J Neuropathol Exp Neurol* **32:**34–50, 1973.

39. Lau YL, Levinsky RJ, Morgan G, *et al*: Dual meningoencephalitis with echovirus type 11 and adenovirus in combined (common variable) immunodeficiency. *Pediatr Infect Dis J* **7:**873–876, 1988.

40. Davis D, Henslee PJ, Markesbery WR: Fatal adenovirus meningoencephalitis in a bone marrow transplant patient. *Ann Neurol* **23:**385–389, 1988.

41. Coldiron BM, Freeman RG, Beaudoing DL: Isolation of adenovirus from a granuloma annulare-like lesion in the acquired immunodeficiency syndrome-related complex. *Arch Dermatol* **124:**654–655, 1988.

42. Levy J, Wodell RA, August CS, *et al*: Adenovirus-related hemophagocytic syndrome after bone marrow transplantation. *Bone Marrow Transplant* **6:**349–352, 1990.

43. Bayon MN, Drut R: Cytologic diagnosis of adenovirus bronchopneumonia. *Acta Cytol* **35:**181–182, 1991.

44. Mahafzal Am, Landry ML: Evaluation of immunofluorescent reagents, centrifugation, and conventional cultures for the diagnosis of adenovirus infection. *Diagn Microbiol Infect Dis* **12:**407–411, 1989.

45. Wood DJ, Bijlsma K, deJong JC, *et al*: Evaluation of a commercial monoclonal antibody-based enzyme immunoassay for detection of adenovirus types 40 and 41 in stool specimens. *J Clin Microbiol* **27:**1155–1158, 1989.

46. Allard A, Girones R, Juto P, *et al*: Polymerase chain reaction for detection of adenoviruses in stool samples (published erratum appears in *J Clin Microbiol* **29:**2683). *J Clin Microbiol* **28:**2659–2667, 1990.

47. Flomenberg P, Gonzales E, Piaskowski V, Casper JT: Detection of adenovirus DNA in peripheral blood mononuclear cells by polymerase chain reaction assay. *J Med Virol* **51:**182–188, 1997.

48. Gordon YJ, Romanowski E, Araullo-Cruz T, *et al*: Inhibitory effect of (S)-HPMPC, (S)-HPMPA, and 2′-non-cyclic GMP on clinical ocular adenovirus isolates is serotype-dependent in vitro. *Antiviral Res* **16:**11–16, 1991.

49. Shah KV, Daniel RW, Warszawski RM: High prevalence of antibodies of BK virus, on SV 40 related papovavirus, in residents of Maryland. *J Infect Dis* **127:**784–787, 1973.

50. Padgett B, Walker D: Natural history of human polyomavirus infections. In Stevens JG, Todaro GJ, Fox CF (eds): *Persistent Viruses*. Academic Press, New York, 1978, pp. 751–758.

51. Shah KV, Daniel RW, Zeigel KK, Murphy GP: Search for BK and SV 40 virus reactivation in renal transplant recipients. *Transplantation* **17:**131–134, 1974.

52. Astrom KE, Mancall EL, Richardson EP Jr: Progressive multifocal leukoencephalopathy: Hitherto unrecognized complication of chronic lymphatic leukemia and Hodgkin's disease. *Brain* **81:**93–111, 1958.

53. Gillespie SM, Chang Y, Lemp G, *et al*: Progressive multifocal leukoencephalopathy in persons infected with human immunodeficiency virus, San Francisco, 1981–89. *Ann Neurol* **30:**597–604, 1991.

54. Richardson EP Jr: Progressive multifocal leukoencephalopathy. *N Engl J Med* **265:**815–823, 1961.

55. ZuRhein GM, Chow SM: Particles resembling papovaviruses in human cerebral demyelinating disease. *Science* **148:**1477–1479, 1965.

56. Padgett BL, Walker DL, ZuRhein GM, *et al*: Cultivation of papova-like virus from human brain with progressive multifocal leukoencephalopathy. *Lancet* **1:**1257–1260, 1971.

57. Weiner LP, Herndon RM, Narayan O, *et al*: Isolation of virus related to SV40 from patients with progressive multifocal leukoencephalopathy. *N Engl J Med* **286:**385–390, 1972.

58. Hogan T, Borden E, McBain J, *et al*: Human polyomavirus infections with JC virus and BK virus in renal transplant patients. *Ann Intern Med* **92:**373–378, 1980.

59. Coleman D, Wolfendale M, Daniel R, *et al*: A prospective study of human polyomavirus infection in pregnancy. *J Infect Dis* **142:**1–8, 1980.

60. Arthur RR, Dagostin S, Shah KV: Detection of BK virus and JC virus in urine and brain tissue by the polymerase chain reaction. *J Clin Microbiol* **27:**1174–1179, 1989.

61. Mori M, Aoki N, Shimada H, *et al*: Defection of JC virus in the brains of aged patients without progressive multifocal leukoencephalopathy by the polymerase chain reaction and Southern hybridization analysis. *Neurosci Lett* **41:**151–155, 1992.

62. Myers C, Frisque RJ, Arthur RR: Direct isolation and characterization of JC virus from urine samples of renal and bone marrow transplant patients. *J Virol* **63:**4445–4449, 1989.

63. Lyon LW, McCormick WF, Schochet SS Jr: Progressive multifocal leukoencephalopathy. *Neurology* **21:**72–77, 1971.

64. Horte-Barbosa L, Hamilton R, Fucillo DA, *et al*: Progressive multifocal leukoencephalopathy. *N Engl J Med* **286:**1060, 1972.

65. Elsner C, Dorries K: Evidence of human polyomavirus BK and JC infection in normal brain tissue. *Virology* **191:**72–80, 1992.

66. Quinlivan EB, Norris M, Bouldin TW, *et al*: Subclinical central nervous system infection with JC virus in patients with AIDS. *J Infect Dis* **166:**80–85, 1992.

67. White FA 3d, Ishaq M, Stoner GL, *et al*: JC virus DNA is present in many human brain samples from patients without progressive multifocal leukoencephalopathy. *J Virol* **66:**5726–5734, 1992.

68. Miller JR, Barrett RE, Britton CB: Progressive multifocal leukoencephalopathy in a male homosexual with T-cell immune deficiency. *N Engl J Med* **307:**1436–1438, 1982.

69. Snider WD, Simpson DM, Nielsen S, *et al*: Neurological complications of acquired immune deficiency syndrome: Analysis of 50 patients. *Ann Neurol* **14:**403–417, 1983.

70. Levy RM, Bredesen ED, Rosenblum ML: Neurological manifestation of the acquired immune deficiency syndrome (AIDS): Experience at UCSF and the review of the literature. *J Neurosurg* **62:**475–495, 1985.

71. Shaw GM, Harper ME, Hahn BH, *et al*: HTLV-III infection in brains of children and adults with AIDS encephalopathy. *Science* **227:**177–181, 1985.

72. Ho DD, Rota TR, Schooley RT, *et al*: Isolation of HTLV-III from CSF and neural tissues of patients with AIDS related neurological syndromes. *N Engl J Med* **313:**1493–1497, 1985.

73. Tada H, Rappaport J, Lashgari M, *et al*: Trans-activation of the JC virus late promoter by the *tat* protein of type 1 human immunodeficiency virus in glial cells. *Proc Natl Acad Sci USA* **87:**3479–3483, 1991.

74. Petito CK, Navia BA, Cho E-S, *et al*: Vacuolar myelopathy patho-logically resembling subacute combined degeneration in patients with the acquired immune deficiency syndrome. *N Engl J Med* **312**:874–879, 1985.

75. Matsiota-Bernard P, de Truchis P, Gray F, Flament-Saillour M, Voyatzakis E, Nauciel C: JC virus detection in the cerebrovascular fluid of AIDS patients with progressive multifocal leukoencepha-lopathy and monitoring of the antiviral treatment by a PCR method. *J Med Microbiol* **46**:256–259, 1997.

76. Hogan TF, Borden EC, McBain JA, *et al*: Human polyomavirus infections with JC virus and BK virus in renal transplant patients. *Ann Intern Med* **92**:373–378, 1980.

77. Woodhouse MA, Dayan AD, Burston J, *et al*: Progressive multi-focal leukoencephalopathy: Electron microscope study of four cases. *Brain* **90**:863–870, 1967.

78. Teo CG, Wong SY, Best PV: JC virus genomes in progressive multifocal leukoencephalopathy: Detection using a sensitive non-radioisotopic in situ hybridization method. *J Pathol* **157**:135–140, 1989.

79. Boerman RH, Arnoldus EP, Raap AK, *et al*: Diagnosis of progres-sive multifocal leukoencephalopathy by hybridization techniques. *J Clin Pathol* **42**:153–161, 1989.

80. Houff SA, Katz D, Kufta CV, *et al*: A rapid method for *in situ* hybridization for viral DNA in brain biopsies from patients with AIDS. *AIDS* **3**:843–845, 1989.

81. Hulette CM, Downey BT, Burger PC: Progressive multifocal leu-koencephalopathy. Diagnosis by in situ hybridization with a bio-tinylated JC virus DNA probe using an automated Histomatic Code-On slide stainer. *Am J Surg Pathol* **15**:791–797, 1991.

82. Schmidbauer M, Budka H, Shah KV: Progressive multifocal leu-koencephalopathy (PML) in AIDS and in the pre-AIDS era. A neuropathological comparison using immunocytochemistry and in situ DNA hybridization for virus detection. *Acta Neuropathol* **80**:375–380, 1990.

83. Knowles WA, Sharp IR, Efstratiou L, *et al*: Preparation of mono-clonal antibodies to JC virus and their use in the diagnosis of progressive multifocal leukoencephalopathy. *J Med Virol* **34**:127–131, 1991.

84. Telenti A, Aksamit AJ Jr, Proper J, *et al*: Detection of JC virus DNA by polymerase chain reaction in patients with progressive multifocal leukoencephalopathy. *J Infect Dis* **162**:858–861, 1990.

85. Henson J, Rosenblum M, Armstrong D, *et al*: Amplification of JC virus DNA from brain and cerebrospinal fluid of patients with progressive multifocal leukoencephalopathy. *Neurology* **41**:1967–1971, 1991.

86. Brouqui P, Bollet C, Delmont J, *et al*: Diagnosis of progressive multifocal leukoencephalopathy by PCR detection of JC virus from CSF (letter). *Lancet* **339**:1182, 1992.

87. Kitamura T, Aso Y, Kuniyoshi N, *et al*: High incidence of urinary JC virus excretion in nonimmunosuppressed older patients. *J In-fect Dis* **161**:1128–1133, 1990.

88. Tornatore C, Berger JR, Houff SA, *et al*: Detection of JC virus DNA in peripheral multifocal leukoencephalopathy. *Ann Neurol* **31**:454–462, 1992.

89. ZuRhein GM, Varakis J: Progressive multifocal luekoencephalop-athy in a renal allograft recipient. *N Engl J Med* **291**:798, 1974.

90. Bauer WR, Turel AP, Johnson KP: Progressive multifocal leuko-encephalopathy and cytarabine: Remission with treatment. *JAMA* **226**:174–176, 1973.

91. Smith CR, Sima AAF, Salit IE, *et al*: Progressive multifocal leukoencephalopathy: Failure of cytarabine therapy. *Neurology* **32**:200–203, 1982.

92. Conomy JP, Beard NS, Matsumoto H, *et al*: Cytarabine treatment of PML. *JAMA* **229**:1313–1316, 1974.

93. VanHorn G, Bastian FO, Moate JL: Progressive multifocal leuko-encephalopathy: Failure of response to transfer factor and cytara-bine. *Neurology* **28**:744–747, 1978.

94. Peters ACB, Vertsteeg J, Bots GTA, *et al*: Progressive multifocal leukoencephalopathy: Immunofluorescent demonstration of SV 40 antigen on CSF cells and response to cytarabine therapy. *Arch Neurol* **37**:497–501, 1980.

95. Rand RH, Johnson KP, Rubenstein LJ, *et al*: Adenine arabinoside in the treatment of progressive multifocal leukoencephalopathy: Use of virus containing cells in the urine to assess response to therapy. *Ann Neurol* **1**:458–462, 1977.

96. Rockwell D, Ruben FL, Windlestein A, *et al*: Absence of immune deficiency in a case of progressive multifocal leukoencephalopa-thy. *Am J Med* **61**:433–436, 1976.

97. Marriott PS, O'Brien MD, MacKenzie IC, *et al*: Progressive multi-focal leukoencephalopathy: Remission with cytarabine. *J Neurol Neurosurg Psychiatry* **38**:205–209, 1975.

98. Castleman G, Scully RE, McNeeley BU: Clinicopathological con-ference. *N Engl J Med* **286**:1047–1054, 1972.

99. Tarsy D, Holden EM, Segarra JM, *et al*: 5-Iodo-2'-deoxyuridine given intraventricularly in the treatment of progressive multifocal leukoencephalopathy. *Cancer Chemother Rep* **57**:73–78, 1973.

100. Hedley Whyte ET, Smith BP, Tyler HR, *et al*: Multifocal leuko-encephalopathy with remission and five year survival. *J Neuro-pathol Exp Neurol* **25**:107–116, 1966.

101. Holden EM, Tarsy D, Calabresi P, *et al*: Use of 5-iodo-2'-deoxyuridine in progressive multifocal leukoencephalopathy. *Neurology* **21**:448, 1971.

102. Tashiro K, Doi S, Moriwaka F, *et al*: Progressive multifocal leuko-encephalopathy with magnetic resonance imaging verification and therapeutic trials with interferon. *J Neurol* **234**:427–429, 1987.

103. Conway B, Halliday WC, Brunham RD: Human immunodefi-ciency virus-associated progressive multifocal leukoencephalopa-thy: Apparent response to 3'-azido-3'-deoxythymidine. *Rev Infect Dis* **12**:479–482, 1990.

104. O'Riordan T, Daly PA, Hutchinson M, *et al*: Progressive multi-focal leukoencephalopathy—Remission with cytarabine. *J Infect* **20**:51–54, 1990.

105. Lidman C, Lindquist L, Mathiesen T, *et al*: Progressive multifocal leukoencephalopathy in AIDS (letter). *AIDS* **5**:1039–1041, 1991.

106. Portegies P, Algra PR, Hollak CE, *et al*: Response to cytarabine in progressive multifocal leukoencephalopathy in AIDS (letter). *Lancet* **337**:680–681, 1991.

107. Nicoli F, Chave B, Peragut JC, *et al*: Efficacy of cytarabine in progressive multifocal leukoencephalopathy in AIDS (letter; com-ment). *Lancet* **339**:306, 1992.

108. Smith CR, Sima AA, Salit IE, *et al*: Progressive multifocal leuko-encephalopathy: Failure of cytarabine therapy. *Neurology* **32**:200–203, 1982.

109. Vandersteenhoven JJ, Dbaibo G, Boyko OB, *et al*: Progressive multifocal leukoencephalopathy in pediatric acquired immuno-deficiency syndrome. *Pediatr Infect Dis J* **11**:232–237, 1992.

110. Inui K, Miyagawa H, Sashihara J, Miyoshi H, Tanaka-Taya K, Nishigaki T, Teraoka S, Mano T, Ono J, Okada S: Remission of progressive multifocal encephalopathy following highly active antiretroviral therapy in a patient with HIV infection. *Brain Dev* **21**:416–419, 1999.

111. Gardner SD, Field AM, Coleman DV, *et al*: New human papova-virus (BK) isolated from urine after renal transplantation. *Lancet* **1**:1253–1257, 1971.

112. Coleman DV, MacKenzie EFD, Gardner SD, *et al*: Human poly-omavirus (BK) infection and ureteric stenosis in renal allograft recipients. *J Clin Pathol* **31:**338–347, 1978.

113. Lecatsas A, Prozesky OW, Van Wyk J, *et al*: Papovavirus in urine after renal transplantation. *Nature* **241:**343–344, 1973.

114. Andrews C, Shah KV, Rubin RH, *et al*: BK papovavirus in renal transplant recipients: Contribution of donor kidneys. *J Infect Dis* **145:**276, 1982.

115. Cheeseman SH, Black PH, Rubin RH, *et al*: Interferon and BK papovavirus. Clinical and laboratory studies. *Infect Dis* **41:**157–161, 1980.

116. Rosen S, Harmon W, Krensky AM, *et al*: Tubulointerstitial nephritis associated with polyomavirus (BK type) infection. *N Engl J Med* **308:**1192–1196, 1983.

117. Gross G, Pfister H, Hagedorn M, *et al*: Correlation between human papillomavirus (HPV) type and histology of warts. *J Invest Dermatol* **78:**160–164, 1982.

118. Favre M, Obalek S, Jablonska S, *et al*: Human papillomavirus type 49, a type isolated from flat warts of renal transplant patients. *J Virol* **63:**4909, 1989.

119. Morrison WL: Viral warts, herpes simplex, and herpes zoster in patients with secondary immune deficiencies and neoplasms. *Br J Dermatol* **92:**625–630, 1975.

120. Barnett N, Mak H, Winkelstein J: Extensive verrucosis in primary immunodeficiency diseases. *Arch Dermatol* **119:**5–7, 1983.

121. Spencer ES, Anderson HK: Clinically evident, non-terminal infections with herpesviruses and the wart virus in immunosuppressed renal allograft recipients. *Br Med J* **3:**251–254, 1970.

122. Perry TL, Harman L Jr: Warts in diseases with immune defects. *Cutis* **13:**359–362, 1974.

123. Mullen DL, Silverberg SG, Penn I, *et al*: Squamous cell carcinoma of the skin and lip in renal homograft recipients. *Cancer* **37:**729–734, 1976.

124. Barr BB, Benton EC, McLaren K, *et al*: Human papilloma virus infection and skin cancer in renal allograft recipients. *Lancet* **1:** 124–129, 1989.

125. Matorras R, Ariceta JM, Rementeria A, *et al*: Human immuno-deficiency virus-induced immunosuppression: A risk factor for human papillomavirus infection. *Am J Obstet Gynecol* **164:**42–44, 1991.

126. Henry MJ, Stanley MW, Cruikshank S, *et al*: Association of human immunodeficiency virus-induced immunosuppression with human papillomavirus infection and cervical intraepithelial neoplasia. *Am J Obstet Gynecol* **160:**352–353, 1989.

127. Alloub MI, Barr BB, McLaren KM, *et al*: Human papillomavirus infection and cervical intraepithelial neoplasia in women with renal allografts. *Br Med J* **298:**153–156, 1989.

128. Lane JM, Ruben FL, Abrutyn E, *et al*: Deaths attributable to smallpox vaccination 1959 to 1966, and 1968. *JAMA* **212:**441–444, 1970.

129. Lane JM, Ruben FL, Neff JM, *et al*: Complications of smallpox vaccination, 1968. II Results of the ten statewide surveys. *J Infect Dis* **122:**303–308, 1970.

130. Lane JM, Ruben RL, Neff JM, *et al*: Complications of smallpox vaccination, 1968: 1. National surveillance in the United States. *N Engl J Med* **281:**1201–1208, 1969.

131. Neff JM, Lane JM, Pert JH, *et al*: Complications of smallpox vaccination. I. National surveillance in the United States, 1963. *N Engl J Med* **276:**125–132, 1967.

132. Turner W, Bauer DJ, Nimmo-Smith RH: Eczema vaccinatum treated with N-methylisatin β-thiosemicarbazone. *Br Med J* **1:** 1317–1319, 1962.

133. Brainerd HD, Hanna L, Jawetz E: Methisazone in progressive vaccinia. *N Engl J Med* **276:**620–622, 1967.

134. Centers for Disease Control: *Poliomyelitis Surveillance Summary, 1980–1981.* CDC, Atlanta, 1982.

135. Moore M, Katona P, Kaplan JE, *et al*: Poliomyelitis in the United States, 1969–1981. *J Infect Dis* **146:**558–563, 1982.

136. Wyatt HV: Poliomyelitis in hypogammaglobulinemics. *J Infect Dis* **128:**802–806, 1973.

137. Chang TW, Weinstein L, MacMahon HE: Paralytic poliomyelitis in a child with hypogammaglobulinemia: Probable implications of type I vaccine strain. *Pediatrics* **37:**630–636, 1966.

138. Riker JB, Brandt CD, Chandra R, *et al*: Vaccine-associated polio-myelitis in a child with thymic abnormality. *Pediatrics* **48:**923–929, 1971.

139. Feigin RD, Guggenheim MA, Johnson SD: Vaccine related para-lytic poliomyelitis in an immunodeficient child. *J Pediatr* **79:**642–647, 1971.

140. Saulsbury FT, Winkelstein JA, Davis LE, *et al*: Combined immuno-deficiency and vaccine related poliomyelitis in a child with cartilage hair hypoplasia. *J Pediatr* **86:**868–872, 1975.

141. Centers for Disease Control: *Neurotropic Disease Surveillance.* Annual Summary 1969. US Department of Health, Education and Welfare, Public Health Service, Washington, DC, 1969.

142. Lopez C, Biggar WD, Park BH, *et al*: Nonparalytic poliovirus infections in patients with severe combined immunodeficiency disease. *J Pediatr* **84:**447–502, 1974.

143. Davis LE, Bodian D, Price D, *et al*: Chronic progressive polio-myelitis secondary to vaccination of an immunodeficient child. *N Engl J Med* **297:**241–245, 1977.

144. Wright PF, Hatch MH, Kasselberg AG, *et al*: Vaccine associated poliomyelitis in a child with sex linked agammaglobulinemia. *J Pediatr* **91:**408–412, 1977.

145. Ziegler JB, Penny R: Fatal echo 30 virus infection and amyloidosis in X-linked hypogammaglobulinemia. *Clin Immunol Immuno-pathol* **3:**347–352, 1975.

146. Bardelas JA, Winkelstein JA, Seto DSY, *et al*: Fatal ECHO 24 infection in a patient with hypogammaglobulinemia: Relationship to dermatomyositis-like syndrome. *J Pediatr* **90:**396–399, 1977.

147. Wilfert CM, Buckley RH, Mohanakumar T, *et al*: Persistent and fatal central nervous system echovirus infections in agamma-globulinemia. *N Engl J Med* **296:**1485–1489, 1977.

148. Webster ADB, Tripp JH, Hayward AR, *et al*: Echovirus encepha-litis and myositis in primary immunoglobulin deficiency. *Arch Dis Child* **53:**33–37, 1978.

149. Bodensteiner JB, Morris HH, Howell JT, *et al*: Chronic echo type 5 virus meningoencephalitis in X-linked hypogammaglobuline-mia: Treatment with immune plasma. *Neurology* **29:**815–819, 1979.

150. Weiner LS, Howell JT, Langford MP, *et al*: Effect of specific anti-bodies on chronic echovirus type 5 encephalitis in a patient with hypogammaglobulinemia. *J Infect Dis* **140:**858–863, 1979.

151. Mease PJ, Ochs HD, Wedgewood RJ: Successful treatment of echovirus meningoencephalitis and myositis-fascitis with intra-venous immune globulin therapy in a patient with X-linked agammaglobulinemia. *N Engl J Med* **304:**1278–1281, 1981.

152. Hodes DS, Espinoza DV: Temperature sensitivity of isolate of echovirus type II causing chronic meningoencephalitis in an agammaglobulinemic patient. *J Infect Dis* **144:**377, 1981.

153. Wagner DK, Marti GE, Jaffe ES, *et al*: Lymphocyte analysis in a patient with X-linked agammaglobulinemia and isolated growth hormone deficiency after development of echovirus dermatomyo-sitis and meningoencephalitis. *Int Arch Allergy Appl Immunol* **89:**143–148, 1989.

154. Van Maldergem L, Mascart F, Ureel D, *et al*: Echovirus meningo-encephalitis in X-linked hypogammaglobulinemia. *Acta Paediatr Scand* **78:**325–326, 1989.

155. Smith JK, Chi DS, Guarderas J, *et al*: Disseminated echovirus infection in a patient with multiple myeloma and a functional defect in complement. Treatment with intravenous immunoglobulin. *Arch Intern Med* **149:**1455–1457, 1989.

156. Biggs DD, Toorkey BC, Carrigan DR, *et al*: Disseminated echovirus infection complicating bone marrow transplantation. *Am J Med* **88:**421–425, 1990.

157. Rotbart HA, Kinsella JP, Wasserman RL: Persistent enterovirus infection in culture-negative meningoencephalitis: Demonstration by enzymatic RNA amplification. *J Infect Dis* **161:**787–791, 1990.

158. Misbah SA, Spickett GP, Ryba PC, *et al*: Chronic enteroviral meningoencephalitis in agammaglobulinemia: Case report and literature review. *J Clin Immunol* **12:**266–270, 1992.

159. Cooper JB, Pralt WR, English BK, *et al*: Coxsackievirus B3 producing fatal meningoencephalitis in a patient with X-linked agammaglobulinemia. *Am J Dis Child* **137:**82–83, 1983.

160. Hertel NT, Pedersen FK, Heilmann C: Coxsackie B3 virus encephalitis in a patient with agammaglobulinemia. *Eur J Pediatr* **148:**642–643, 1989.

161. Woods MG, Diana GD, Rogge MC, *et al*: *In vitro* and *in vivo* activities of WIN 54954, a new broad-spectrum antipicornavirus drug. *Antimicrob Agents Chemother* **33:**2069–2074, 1989.

162. See DM, Tilles JG: Treatment of coxsackievirus A9 myocarditis in mice with WIN 54954. *Antimicrob Agents Chemother* **36:**425–428, 1992.

163. Jantausch BA, Luban NLC, Duffy L, Rodriguez WJ: Maternal plasma transfusion in the treatment of disseminated neonatal echovirus 11 infection. *Pediatr Infect Dis J* **14:**154–155, 1995

164. Townsend TR, Yolken RH, Bishop CA, *et al*: Outbreak of coxsackie A1 gastroenteritis: A complication of bone marrow transplantation. *Lancet* **1:**820–823, 1982.

165. Yolken RH, Bishop CA, Townsend TA, *et al*: Infectious gastroenteritis in bone marrow transplant recipients. *N Engl J Med* **306:**1009–1012, 1982.

166. Krugman S: Present status of measles and rubella immunization in the United States: A medical progress report. *J Pediatr* **90:**1–12, 1977.

167. Smithwick EM, Berkovich S: *In vitro* suppression of lymphocyte response to tuberculosis by live measles virus. *Proc Soc Exp Biol Med* **123:**276–278, 1966.

168. Schooley RT, Hirsch MS, Colvin RB, *et al*: Association of herpesvirus infections with T-lymphocyte subset alterations, glomerulopathy, and opportunistic infections after renal transplantation. *N Engl J Med* **308:**313–318, 1983.

169. Blumberg RS, Schooley RT: Lymphocyte markers and infectious diseases. *Semin Hematol* **22:**81–114, 1985.

170. Enders JF, McCarthy K, Mitus A, *et al*: Isolation of measles virus at autopsy in cases of giant cell pneumonia without rash. *N Engl J Med* **261:**875–881, 1959.

171. Breitfeld V, Hashida Y, Sherman FE, *et al*: Fatal measles infection in children with leukemia. *Lab Invest* **29:**279–291, 1973.

172. Kaplan LJ, Daum RS, Smaron M, *et al*: Severe measles in immunocompromised patients. *JAMA* **267:**1237–1241, 1992.

173. Gray MM, Hann IM, Glass S, *et al*: Mortality and morbidity caused by measles in children with malignant disease attending four major treatment centers: A retrospective view. *Br Med J* **295:**19–22, 1987.

174. Kernahan J, McQuillin J, Craft AW: Measles in children who have malignant disease. *Br Med J* **295:**15–18, 1987.

175. Sension MG, Quinn TC, Markowitz LE, *et al*: Measles in hospitalized African children with human immunodeficiency virus. *Am J Dis Child* **142:**1271–1272, 1988.

176. Aicardi J, Goutieres F, Arseni-Nunes ML, *et al*: Acute measles encephalitis in children with immunosuppression. *Pediatrics* **59:**232–239, 1977.

177. Simpson R, Eden OB: Possible interferon response in a child with measles encephalitis during immunosuppression. *Scand J Infect Dis* **16:**315–319, 1984.

178. Gibbs FA, Gibbs EL, Carpenter PR, *et al*: Electrocephalographic changes in uncomplicated childhood diseases. *JAMA* **171:**1050–1059, 1959.

179. Meulen VT, Muller D, Kackell Y, *et al*: Isolation of infectious measles virus in measles encephalitis. *Lancet* **2:**1172–1175, 1972.

180. Shaffer MF, Rake G, Hodes HL: Isolation of virus from a patient with fatal encephalitis complicating measles. *Am J Dis Child* **64:**815, 1982.

181. Connolly JH, Allen IV, Hurwitz LJ, *et al*: Measles virus antibody and antigen in subacute sclerosing panencephalitis. *Lancet* **1:**542–544, 1967.

181. Barbosa LH, Fucciloo DA, Sever JL, *et al*: Subacute sclerosing panencephalitis: Isolation of measles virus from a brain biopsy. *Nature* **221:**974, 1969.

182. Editorial: Measles encephalitis during immunosuppressive treatment. *Br Med J* **1:**1552, 1976.

183. Krasinski K, Borkowsky W: Measles and measles immunity in children infected with human immunodeficiency virus. *JAMA* **261:**2512–2517, 1989.

184. Murphy JV, Yunis EJ: Encephalopathy following measles infection in children with chronic illness. *J Pediatr* **88:**937–942, 1976.

185. Agamanolis DP, Tun JS, Parker DL: Immunosuppressive measles encephalitis in a patient with a renal transplant. *Arch Neurol* **36:**686–690, 1979.

186. Pullen CR, Noble TC, Scott DJ, *et al*: Atypical measles infections in leukaemic children on immunosuppressive treatment. *Br Med J* **1:**1562–1565, 1976.

187. Olding-Stenkvist E, Forsgren M, Henley D, *et al*: Measles encephalopathy during immunosuppression: Failure of interferon treatment. *Scand J Infect Dis* **14:**1–4, 1982.

188. Fouillard L, Mouthon L, Laporte JP, *et al*: Severe respiratory syncytial virus pneumonia after autologous bone marrow transplantation: A report of three cases and review. *Bone Marrow Transplant* **9:**97–100, 1992.

189. Win N, Mitchell D, Pugh S, *et al*: Successful therapy with ribavirin of late onset respiratory syncytial virus pneumonitis complicating allogeneic bone transplantation. *Clin Lab Haematol* **14:**29–32, 1992.

190. Mitus A, Holloway A, Evans AE, *et al*: Attenuated measles vaccine in children with leukemia. *Am J Dis Child* **103:**243–248, 1962.

191. Glezen WP, Denhy FW: Epidemiology of acute lower respiratory disease in children. *N Engl J Med* **288:**498–505, 1973.

192. Hall CB, Geiman JM, Biggar R, *et al*: Respiratory syncytial virus infections within families. *N Engl J Med* **294:**414–419, 1976.

193. Hall GB, Douglas RG Jr: Modes of transmission of respiratory syncytial virus. *J Pediatr* **99:**100–103, 1981.

194. Hall CB: Nosocomial viral respiratory infections. Perennial weeds on pediatric wards. *Am J Med* **70:**670–677, 1981.

195. Sims DG, Downham MAPS, Webbs JKG, *et al*: Hospital cross-infection on children's wards with respiratory syncytial virus and the role of adult carriage. *Acta Paediatr Scand* **64:**541–545, 1975.

196. MacDonald NE, Hall CB, Suffin SC, *et al*: Respiratory syncytial

viral infection in infants with congenital heart disease. *N Engl J Med* **307:**397–400, 1982.

197. Johnson RA, Prince GA, Suffin SC, *et al*: Respiratory syncytial virus infection in cyclophosphamide-treated cotton rats. *Infect Immun* **37:**369–373, 1982.

198. Milder JE, McDearmon SC, Walzer PD: Presumed respiratory syncytial virus pneumonia in an adolescent compromised host. *South Med J* **72:**1195–1198, 1979.

199. Hall CB, MacDonald NE, Klemperer MR, *et al*: Respiratory syncytial virus in immunosuppressed children. *Pediatr Res* **15:**613, 1981.

200. Rodriguez WJ, Kim HW, Arrobio JO, *et al*: Clinical features of acute gastroenteritis associated with human reovirus like agent in infants and young children. *J Pediatr* **91:**188–193, 1977.

201. Carlson JAK, Middleton PJ, Ssymanski MT, *et al*: Fatal rotavirus gastroenteritis. An analysis of 21 cases. *Am J Dis Child* **132:**477–479, 1978.

202. Kapikian AZ, Wyatt RG, Levin MM, *et al*: Oral administration of human rotavirus to volunteers: Induction of illness and correlates of resistance. *J Infect Dis* **147:**95–106, 1983.

203. Saulsbury FT, Winklestein JA, Yolken RH: Chronic rotavirus infection in immunodeficiency. *J Pediatr* **97:**61–65, 1980.

204. Jarvis WR, Middleton PJ, Gilford EW: Significance of viral infections in severe combined immunodeficiency disease. *Pediatr Infect Dis* **2:**187–192, 1983.

205. Flewett TH, Bryden AS, Davies H: Virus particles in gastroenteritis. *Lancet* **2:**1497, 1973.

206. Kapikian AZ, Kim HW, Wyatt RG, *et al*: Reovirus like agent in stools: Association with infantile diarrhea and development of serologic tests. *Science* **185:**1049–1053, 1974.

207. Brendt CD, Kim HW, Rodriguez WH, *et al*: Comparison of direct electron microscopy, immune electron microscopy, and rotavirus enzyme linked immunosorbent assay for detection of gastroenteritis viruses in children. *J Clin Microbiol* **13:**976–988, 1981.

208. Middleton PJ, Holdaway MD, Petore M, *et al*: Solid phase radioimmunoassay for the detection of rotavirus. *Infect Immun* **16:**439–444, 1977.

209. Yolken RH, Kim HW, Clem T, *et al*: Enzyme linked immunosorbent assay (ELISA) for detection of human reovirus-like agent of infantile gastroenteritis. *Lancet* **2:**263–267, 1977.

210. Kim HW, Brandt CD, Kaoikain AZ, *et al*: Human reovirus like agent (HRLVA) infection. Occurrence in adult contacts of pediatric patients with gastroenteritis. *JAMA* **238:**404–407, 1977.

211. Halvrsrud J, Ostavik I: An epidemic of rotavirus associated gastroenteritis in a nursing home for the elderly. *Scand J Infect Dis* **12:**161–164, 1980.

212. Holzel H, Cubett DW, McSwiggan DA, *et al*: An outbreak of rotavirus infection among adults in a cardiology ward. *J Infect* **2:**33, 1980.

213. Sarker SA, Casswall TH, Mahalanabis D, Alam NH, Albert MJ, Brussow H, Fuchs GJ, Hammerstrom L: Successful treatment of rotavirus diarrhea in children with immunoglobulin from immunized bovine colostrum. *Pediatr Infect Dis J* **17:**1149–1154, 1998.

214. Mitra AK, Mahalanabis D, Ashraf H, *et al*: Hyperimmune cow colostrum reduces diarrhoea due to rotavirus: A double-blind, controlled clinical trial. *Acta Paediatr* **84:**996–1001, 1995.

215. Ylitalo S, Uhari M, Rasi S, Pudas J, Leppaluoto J: Rotaviral antibodies in the treatment of acute rotaviral gastroenteritis. *Acta Paediatr* **87:**264–267, 1998.

12

Pathogenesis and Clinical Manifestations of HIV-1 Infection

ROBERT T. SCHOOLEY and CONSTANCE A. BENSON

Human immunodeficiency virus-1 (HIV-1) is a member of the lentivirus family of human retroviruses.[1] As such, it shares several biologic and molecular properties with a number of related agents affecting nonhuman primates, goats, sheep, and horses.[2–4] In addition to HIV-1, a related human pathogen, HIV-2 also has been identified.[5,6] Although this agent also has been associated with acquired immunodeficiency syndrome (AIDS) in humans, it is less pathogenic than HIV-1. These agents are enveloped viruses that carry their genetic information in the form of two identical strands of RNA that must be reverse transcribed to DNA following infection of mammalian cells.[7] This process is the basis for the nomenclature of the retrovirus family. Following reverse transcription, the double-stranded DNA product may remain within the cytoplasm or it may be transported to the nucleus where it integrates within the genome of the host cell and establishes lifelong infection of that cell and its progeny.[8] Although the complement of regulatory genes varies somewhat from one retrovirus to another, these agents share the property of having a complex set of genes that regulate viral expression within the host cell and that are clearly integral components of the pathogenesis of retroviral infection.[9] Clinical manifestations of infection likewise are host and virus dependent and include both direct manifestations of the viral infection and indirect effects mediated through effects of the viruses on specific organ systems. In the case of HIV-1, the major target organs are the immune and central nervous systems. In addition, in late stages of the disease process an increasing array of metabolic disturbances are encountered that contribute to the wasting syndrome frequently observed in advanced HIV-1 infection. This chapter will focus primarily on clinical manifestations of HIV-1 infection that are directly attributable to HIV-1 infection.

1. Epidemiology of HIV-1 Infection

The initial recognition of a syndrome that was ultimately defined as the acquired immunodeficiency syndrome occurred in the summer of 1981 when a cluster of cases of *Pneumocystis carinii* pneumonia was identified among homosexual men in Los Angeles.[10] The syndrome was further delineated 6 weeks later in a report of cases of Kaposi sarcoma among gay men in New York and California. With the description of the etiologic agent in 1983, it became possible to understand the clinical spectrum of illness, the modes of transmission, and the distribution of the etiologic agent.[11] As noted above, AIDS is caused by a pathogenic human retrovirus that is transmitted by sexual intercourse, blood-borne contamination, and perinatally. Although the clinical syndrome of AIDS was initially described in the United States, HIV-1 was likely only introduced into this country relatively recently. Current evidence suggests that this virus arose from closely related primate lentiviruses and was initially transmitted to man within this century in West Africa.[12] Over the past 30 years, HIV-1 has spread to all populated areas of the world

Robert T. Schooley and Constance A. Benson • Division of Infectious Diseases, Department of Medicine, University of Colorado Health Sciences Center, Denver, Colorado 80262.

Clinical Approach to Infection in the Compromised Host (Fourth Edition), edited by Robert H. Rubin and Lowell S. Young. Kluwer Academic/Plenum Publishers, New York, 2002.

and now affects over 40 million people. The predominant modes of transmission, and hence, the affected patient populations, vary from region to region. Although the predominant mode of spread of the virus was initially among homosexual men in North America and Europe, the major route of spread on these continents is among heterosexual adults, as it is in the remainder of the world.[13] In the developed world a significant amount of the transmission is among injecting drug users. In the developing world perinatal transmission accounts for a large number of new cases.

2. Primary HIV-1 Infection

Although the initial recognition of the AIDS epidemic arose from identification of individuals with opportunistic infections associated with profound cellular immunodeficiency, it was initially thought that infection with HIV-1 was characterized by many years of silent infection and that the infection became clinically manifest only when sufficient immunodeficiency had developed to allow expression of any one of a number of opportunistic infectious pathogens or of one of several AIDS-associated malignancies. After it became possible to identify HIV-1 infection serologically prior to the onset of immunodeficiency, appreciation developed of a clinical syndrome associated with initial infection.[14,15] This syndrome is a febrile illness that occurs 3–12 weeks after infection and varies in severity from being easily mistaken as a mild "viral" syndrome to being associated with a meningoencephalitis that leads to hospital admission. The frequency with which the syndrome is recognized depends in part on the index of suspicion of the clinician and by recognition of a recent exposure to HIV-1 infection. Depending on these factors, primary infection may be clinically apparent up to 70% of the time.[15] In most series, however, significantly lower fractions of HIV-1 seroconverters are recognized clinically. Clinical manifestations of primary HIV-1 infection include fever, malaise, headache, lymphadenopathy, and a rash.[15–17] Diagnosis is made by the recognition of a compatible clinical syndrome and the demonstration of seroconversion to HIV-1 coincident with the onset of the illness. This is one of the few clinical settings in which serum HIV-1 p24 antigen determinations are useful in the management of individual patients.[18–20]

It is likely that many or all of the clinical manifestations of the primary infection syndrome are related to a high level of viremia[19,20] and to an array of cytokines elaborated as part of the brisk humoral and cellular immune response to the infection.[21,22] During primary infection, levels of virus similar to those encountered during advanced disease are encountered in plasma and in peripheral blood mononuclear cells.[19,20] This high level of viremia is associated with wide dissemination of the virus to lymphoid tissues and to the central nervous system (CNS).[14] HIV-1 replication declines in association with the evolution of HIV-1-specific cell-mediated immune responses.[23–26] Primary infection may be associated with significant declines in CD4 cell counts and with the other manifestations of cell-mediated immunodeficiency.[27] In occasional individuals, the cell-mediated immunodeficiency associated with primary infection has been so severe that oral candidiasis or *Pneumocystis carinii* pneumonia has been observed. Under most circumstances the clinical manifestations resolve and the CD4 cell counts return toward preinfection levels over a 2- to 12-week period.[16] Those with more severe symptoms and/or immunologic activation during primary HIV-1 infection are more likely to progress more rapidly to symptomatic HIV-1 disease than are those with subclinical primary infection.[28,29]

Management of primary infection includes both supportive care and antiretroviral chemotherapy. Although concerns have been raised that administration of antiretroviral agents during primary infection might compromise the evolution of effective HIV-1-specific humoral and cell-mediated immune responses, most recommend the early initiation of antiretroviral chemotherapy in patients identified during primary infection.[30,31] These recommendations are derived both from a controlled, randomized clinical trial that demonstrated a reduction in HIV-1-associated clinical manifestations in patients receiving zidovudine (ZDV) compared to those receiving placebo, as well as from observations that early treatment might preserve cell-mediated immune responses to the virus.[25,26,32]

3. Pathogenesis of HIV-1 Infection

Following resolution of primary infection, HIV-1 enters a phase of clinical latency. This period is of variable duration but is on the average in the range of 8–12 years. During this period, lower amounts of virus are demonstrable in the peripheral blood of most individuals[33] and vigorous cellular and humoral immune responses to the virus can be detected.[34–36] Initially it was thought that lymphocytes of the CD4 surface phenotype were the exclusive reservoir of the virus; however, it has become apparent that the virus is capable of infecting monocytes,[37–39] follicular dendritic cells,[40] epidermal Langerhans cells,[41] and alveolar macrophages.[42] In addition, the virus is readily demonstrable within a variety of cells in the CNS.

As molecular techniques for detecting HIV-1 have improved, it has become apparent that even during the so-called latent period of infection there is an extremely vigorous viral infection at a number of sites, particularly among lymphoid organs.[43–48] Studies of viral dynamics indicate that 10 billion viral particles are produced per day (Fig. 1). Virus circulating in plasma has a half-life of less than 6 hr. Ninety-nine percent of this virus is produced by activated CD4 cells. This high rate of viral replication in CD4 cells is associated with a major increase in the daily production of CD4 cells and is associated with a gradual decrease in both cellular and humoral immune functions as CD4 cells decline in number. Although enumeration of CD4 cells is the most convenient approach to monitoring HIV-1-associated immunodeficiency, it is clear that a wide array of functional abnormalities precedes and accompanies the numerical depletion of CD4 cells. These defects include but are not restricted to a decrease in responsiveness to mitogens and antigens,[49–51] decreased natural killer and cytotoxic T-cell activity,[52,53] decreased proliferative responses to specific pathogens,[54] as well as decreases in humoral immune responsiveness.[55–57] These abnormalities are associated with a general activation and disordered regulation of both the humoral and cellular arms of the immune response.[57–60] The rate at which these abnormalities progress is variable from patient to patient and may reflect contributions from both viral[61,62] and immunologic factors. Although much has been written about mechanisms whereby the progressive immunodeficiency is an indirect consequence of HIV-1-specific immune responses,[63–66] at this point most evidence implicates the cytopathic properties of the virus as the major contributor to immune dysfunction.[43–48] Although the high rate of viral replication in activated CD4 cells accounts for most of the acute immunological manifestations of HIV-1 infection, the 1% of viral replication that occurs in resting CD4 cells and in monocytes plays a critical role in maintaining a reservoir of virus that is capable of reinfection of the CD4 cell compartment whenever viral suppression is suboptimal.[67,68]

From the standpoint of clinical management, CD4 cell enumeration remains the most practical measure of HIV-1-associated immunodeficiency. Additional prognostic information may be gained by monitoring parameters of viral load such as plasma viremia[69] or of immune activation by measuring coexpression of T-cell surface activation markers such as HLA-DR or CD38.[70] However, these parameters are generally more useful in stratifying individuals participating in clinical trials than in managing individual patients. It should be recognized, however, that CD4 cell enumeration provides only a general approximation of the immunologic responsiveness of an HIV-1-infected individual and that major functional differences in immunity may be unrecognized by CD4 cell enumeration alone. A number of factors contribute to interpatient variability in terms of the CD4 cell count as it relates to disease prognosis. First, there is significant variability in CD4 cells, even in the same individual,

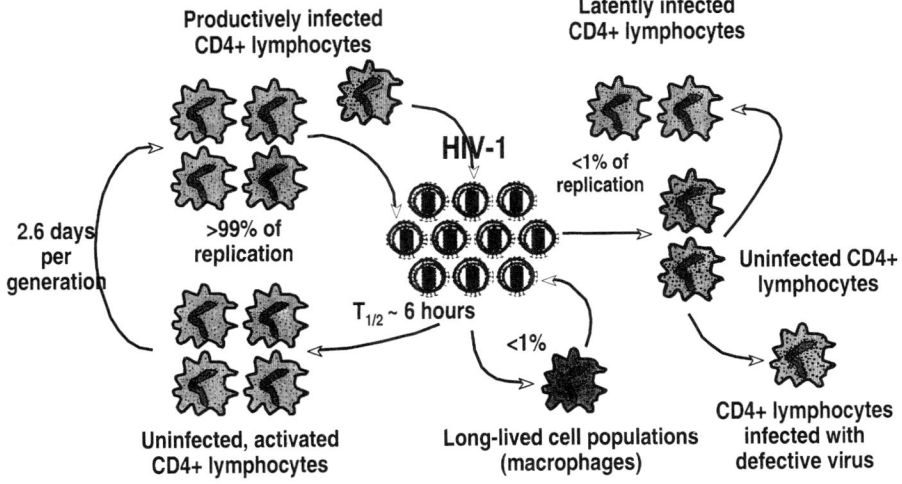

FIGURE 1. Dynamics of HIV-1 *in vivo.*

when measured repetitively. Second, CD4 cell enumeration does not capture information about lymphocyte function or about the competence of other arms of the immune response. Finally, the peripheral blood compartment comprises only 2–5% of the total body lymphoid tissue. Thus, relatively small shifts in CD4 cells among lymphoid compartments may not be reflected by peripheral blood CD4 cell enumeration. Splenectomized individuals provide perhaps the most extreme example of this situation. CD4 cells rise significantly in the peripheral blood following splenectomy, but an HIV-infected, splenectomized individual is much more immunocompromised than is reflected by peripheral CD4 cell enumeration. Indeed, in this setting, it is appropriate to initiate *P. carinii* prophylaxis at levels of 500–800 CD4 cells. Thus, although CD4 cell enumeration is widely used for prognostication in HIV-1 infection, there remains significant room for improvement in this regard.

As noted above, HIV-1-infected individuals manifest significant variability in terms of clinical course. In general, once primary infection has passed, most individuals are not symptomatic with HIV-1 infection at peripheral blood levels of CD4 cells above 500/mm^3. An increasing fraction of individuals develops signs and symptoms of HIV-1 infection as CD4 cells decline from 550/mm^3 to the range of 300/mm^3. These symptoms and signs include fatigue, weight loss, adenopathy, night sweats, as well as infections such as herpes zoster and oral and vaginal thrush. As CD4 cells in the peripheral blood approach 200/mm^3, individuals are at increasing risk for *P. carinii* pneumonia.[71,72] It should be noted, however, that unexplained fever and oral thrush are independent predictors of *P. carinii* pneumonia.[72] In general, other than for *P. carinii* pneumonia and *Mycobacterium tuberculosis* infection, most HIV-1-infected individuals do not develop additional HIV-1-associated opportunistic infections until CD4 cells decline to 50/mm^3 or less. At or below this level, HIV-1-infected individuals are at increasing risk for a wide array of opportunistic infections such as *Toxoplasma gondii* encephalitis, cryptococcal meningitis, and cytomegalovirus (CMV) retinitis. Below 50 CD4 cells/mm^3, there is so much variability in clinical course that continued monitoring of CD4 cells has little to offer under most circumstances. Management of specific HIV-1-associated opportunistic infections is outlined in the other chapters of this book.

4. Neurologic Manifestations of HIV-1 Infection

Early in the HIV-1 epidemic it was recognized that neurologic complications are a frequent manifestation of HIV-1 disease.[73] Early analysis of CNS manifestations of HIV-1 infection focused on the propensity of several organisms such as cytomegalovirus, *T. gondii*, and *Cryptococcus neoformans* to involve the CNS.[73,74] However, with the demonstration that HIV-1 itself could be readily detected within the CNS, investigation of the pathogenesis of HIV-1-associated neurologic dysfunction began to focus increasingly on the neurotropism of HIV-1.[14,75–78] HIV-1-associated neurologic abnormalities may occur at any time during HIV-1 infection.[14] AIDS dementia complex may be the heralding manifestation of AIDS[79,80] or it may occur as a late manifestation of advanced disease. Up to two-thirds of patients with HIV-1 infection may experience neurologic abnormalities at some point during their illness.[81]

4.1. Pathogenesis of the Neurologic Complications of HIV-1

HIV-1 is readily detected within the CNS.[14,75–78] Entrance of HIV-1 into the CNS is demonstrable during primary infection,[14,76] and has been demonstrated *in vitro* within fetal neural tissues obtained from abortuses of HIV-1-infected women.[82] Immunohistochemical staining and *in situ* hybridization studies suggest that the monocyte/macrophage and glial cells are the predominant cell type within which HIV-1 is localized in the brain.[78,83–87] HIV-1 has on occasion been demonstrated within oligodendrocytes and astroglial cells *in vivo*[88] and to infect glial cells *in vitro*.[89–91] Thus, it is possible that some of the CNS pathology associated with HIV infection is the direct result of lytic infection of neurons. The relative infrequency with which HIV-1 is found within neurons *in vivo*, however, has led to speculation that other mechanisms might contribute to the pathogenesis of AIDS dementia complex. These indirect mechanisms of pathogenesis include the possibility that viral[92] or cellular products[93–103] elaborated by HIV-1-infected monocytes/macrophages and glial cells might lead to neuronal dysfunction. Much attention has focused on the possible role of tumor necrosis factor-alpha (TNF-α) in the pathogenesis of AIDS dementia complex.[93–95,97–100] Interest in TNF-α is the result of the detection of this cytokine in cerebrospinal fluids (CSFs) obtained from some[95,97,99,100] but not all[104,105] HIV-1-infected individuals and from evidence that TNF-α may damage neuronal cells *in vitro*.[93,94] Elevated levels of quinolinic acid also have been detected in CSF of patients with advanced HIV-1 infection.[96,101] This metabolite of the kynurenine pathway is an agonist of *N*-methyl-D-aspartate receptors and is toxic *in vitro* to neuronal cells at concentrations below those observed in some patients with advanced HIV-1 infection.[106]

In addition to TNF-α and quinolinic acid, several other cytokines and other markers of immunologic activation including interleukin-1 (IL-1), IL-6, β$_2$-microglobulin, and neopterin are found in increased concentrations within the CSF of individuals with HIV-1 infection.[99,105,107–110] In and of themselves, these molecules may not be neurotoxic, but the role of these molecules in recruiting an ongoing inflammatory response within the CNS may play a key role in the pathogenesis of AIDS dementia complex. In addition, the finding of a vigorous and broadly directed HIV-1-specific cytotoxic T-cell response within the CSF[111] and of brain-reactive antibodies in the patients with AIDS dementia complex[112] has raised the possibility that autoimmune mechanisms also may contribute to the pathogenesis of AIDS dementia complex. Given the multiplicity of clinical manifestations of AIDS dementia complex, it is likely that the pathogenesis of the neurologic complications of HIV-1 infection involves more than a single mechanism. The rapid response to antiretroviral therapy of many of the clinical manifestations of AIDS dementia complex argues strongly, however, that neuronal destruction is not the predominant mechanism whereby HIV-1 infection leads to CNS dysfunction.

5. Clinical Manifestations of AIDS Dementia Complex

Multiple neurologic syndromes have been associated with HIV-1 infection. These include acute and chronic aseptic meningitis, dementia, myelopathy, inflammatory neuropathy, mononeuritis, sensory neuropathy, and granulomatous angiitis.[14,76,113–122] The diversity of the neurologic manifestations of HIV-1 reflects both the multifactorial nature of the pathogenesis of HIV-1-associated neurologic complications, and the complex nature of the underlying disease.

5.1. Aseptic Meningitis

HIV-1 infection of the CNS occurs early during primary infections.[14,123,124] Involvement may be subclinical or it may be associated with signs and symptoms of aseptic meningitis.[14,124] Such patients may present with fever, headache, confusion, and photophobia with or without other clinical manifestations of primary HIV-1 infection.[14,76] CSF findings may include a mild CSF lymphocytosis and an elevated CSF protein. In most patients the CSF glucose is normal or slightly depressed. It is unusual to observe CSF proteins in excess of 200 mg/dl or markedly decreased CSF glucose in the absence of other CNS processes. Diagnosis may be facilitated by the rec-

ognition of an exposure history and/or a clinical syndrome suggestive of primary HIV-1 infection. In that HIV-1 antibodies may be absent at the time of clinical presentation of primary HIV-1 infection, diagnostic yield may be increased by the inclusion of studies designed to detect the presence of HIV-1 itself. These studies include serum HIV-1 p24 antigen, viral cultures of the peripheral blood or CSF, or detection of viral nucleic acid sequences by PCR technology.

In most cases, the aseptic meningitis syndrome associated with primary HIV-1 infection is self-limited and subsides after a period of several weeks to a few months. On occasion, the meningitis may smolder along for a longer period of time and/or may become chronic. Therapy of HIV-associated aseptic meningitis is dictated primarily by other manifestations of the illness. Studies of the efficacy of antiretroviral agents in primary infection are currently under way. Until these studies have been completed, therapy will remain largely supportive.

5.2. Subacute Encephalitis

HIV infection of the brain may be associated with progressive cognitive impairment characterized by motor and/or behavioral findings. This syndrome has been termed subacute encephalitis,[73] AIDS encephalopathy,[125] or the AIDS dementia complex.[116] The prevalence of this syndrome in HIV-1-infected individuals depends on the sensitivity of the tools used to make this diagnosis and the severity of the underlying HIV-1 disease in the population under study. Several early studies employing neuropsychological testing instruments and noninvasive neurodiagnostic studies emphasized the high prevalence of subtle neurocognitive deficits in patients with early asymptomatic HIV infection.[125–127] These reports led to unwarranted concerns that subacute encephalitis in individuals otherwise asymptomatic from HIV infection could be a major problem in the workplace. The United States military cited such concerns as the basis for the imposition of a number of restrictions on HIV-1-infected service personnel.[128] Subsequent larger studies in which appropriately matched control subjects were included, however, have demonstrated that significant neurocognitive impairment in early phases of HIV-1 infection during which individuals are otherwise asymptomatic is the exception rather than the rule.[129–136] These studies strongly suggest that some of the prior restraints on occupational activities by HIV-1-infected individuals were premature. In children, particularly those infected *in utero*, CNS manifestations may be more prominent.[137–139]

As the underlying disease worsens, individuals with HIV-1 infection are more likely to manifest neurocognitive impairment.[131–133] Clinically, such individuals exhibit forgetfulness, concentration deficits, and slowness of thought.[118] In addition to these abnormalities, many patients exhibit mood disturbances such as blunted affect, apathy, and depression. Motor abnormalities are less frequently observed and may be difficult to distinguish from the effects of coexisting wasting observed in late disease.

Much effort has been expended in an attempt to develop reliable neuropsychometric tests of predictive value in the assessment of subacute encephalitis.[140–145] These efforts have been more significant in terms of implications for clinical trials[143] than from the standpoint of clinical practice. In general, extensive neuropsychometric testing is beyond the capability of most practicing clinicians and adds little to management of individual patients.

A wide array of noninvasive neurodiagnostic studies have been employed to detect CNS abnormalities in HIV-1-infected individuals. These approaches include electroencephalography,[146–149] transcutaneous electrical nerve stimulation,[150] computed axial tomography,[151–153] magnetic resonance imaging,[153–155] proton magnetic resonance spectroscopy,[156,157] single-photon emission computed tomography (SPECT scanning),[157–161] and positron emission tomography.[162] Each technique has advantages and disadvantages in terms of cost, sensitivity, and specificity. Electroencephalography may be abnormal during asymptomatic HIV infection, but the specificity and clinical significance of electroencephalographic abnormalities in this setting are marginal. Similarly, individuals with HIV infection, especially those with advanced disease, may exhibit evidence of cerebral atrophy and/or ventricular enlargement by computed tomography or by magnetic resonance imaging. However, in that individuals with HIV-1 infection without AIDS dementia complex also may exhibit similar findings and in that some patients with advanced dementia may show little evidence of cerebral atrophy, such studies are of limited use in evaluating individual patients with dementia. At present, these approaches are most appropriately applied in the context of investigations of the pathogenesis or drug therapy of AIDS dementia complex.[163] In addition to neuropsychometric and noninvasive neurodiagnostic studies, a number of cytokines and other molecules associated with immunologic activation may be found in the CSF of patients with AIDS dementia complex.[96–100,105,107–110,164–168] Although quantitation of these substances in CSF may be useful in clinical research settings, such studies are currently of limited utility in clinical practice.

The clinical course of subacute encephalitis is variable. Factors that account for the presence and/or the natural history of subacute encephalopathy have not yet been fully delineated. It is possible that viral variation plays an important role in the neuropathogenesis of HIV infection.[169–172] It is also possible that host factors play a critical role in determining the natural history of the AIDS dementia complex. In some patients, clinical manifestations, once established, may progress relentlessly. In others, the course may be stuttering.[158] In many patients, however, cognitive dysfunction closely parallels the course of the underlying disease with clear-cut deterioration occurring in conjunction with the onset of infectious complications of HIV-1 infection.

Therapy of HIV-1-associated subacute encephalitis involves both supportive measures and antiretroviral chemotherapy. Since the advent of antiretroviral chemotherapy, the prevalence of cases of AIDS dementia complex has lagged behind that predicted in the era before widespread antiretroviral chemotherapy.[173,174] Although penetration of nucleoside analogue reverse transcriptase inhibitors into the brain is limited, several studies have documented the efficacy of zidovudine and dideoxyinosine in the management of AIDS dementia complex.[175–181] In addition to the effects of antiretroviral therapy in clinical manifestations of AIDS dementia complex, zidovudine administration has been associated with decreases in viral antigens and cytokine abnormalities in CSF.[107,110,177,182] It has been suggested that these cytokines might serve as useful surrogate markers of antiretroviral activity in the CNS. Although controlled trials with other antiretroviral agents have not been performed, it has been noted that the incidence of AIDS dementia complex has been declining as antiretroviral chemotherapy has improved.

5.3. Peripheral Neuropathy

Peripheral neuropathy also is common in patients with HIV-1 infection.[14,183,184] This manifestation of HIV-1 infection may precede the onset of AIDS, but rarely occurs in individuals who are otherwise asymptomatic from the standpoint of HIV-1 infection. Most patients experience a distal symmetric predominantly sensory neuropathy manifested clinically as painful paresthesias or numbness.[183,185] HIV-1-associated peripheral neuropathy may be difficult to distinguish on a clinical basis from the peripheral neuropathy associated with dideoxycytidine (ddC) or dideoxyinosine therapy.[186] Although HIV-1 has been isolated from peripheral nerves,[14,185] the pathogenesis of HIV-1-associated peripheral neuropathy has not been completely elucidated. Occasional HIV-1-infected patients may present with a chronic demyelinating polyneuropathy, or a Guillain–Barré-like syndrome.[183,184] Even less frequently, mononeuritis multiplex has been described in association with HIV-1 infection. The course of HIV-1-associated peripheral neuropathy is variable.[185]

In some patients, initially painful paresthesias evolve into a less troublesome hypesthesia. Management consists primarily of the use of antiretroviral therapy as dictated by the underlying disease. Some patients benefit clinically from the use of phenytoin, nonnarcotic analgesics, and tricyclic antidepressants, although controlled clinical trials have not demonstrated clear benefit of these approaches.[186]

5.4. Vacuolar Myelopathy

Vacuolar myelopathy has also been reported in association with HIV-1 infections.[187,188] This process involves primarily the lateral and posterior columns of the spinal cord and is often accompanied by subacute encephalitis.[74,187] Patients present with a progressive spastic paraparesis manifest by ataxia of gait, leg weakness, incontinence, and upper motor neuron and posterior column signs.[187,188] As in the case of peripheral neuropathy, the course is variable. Although no controlled clinical trials focusing on therapy of this complication of HIV-1 infection have been reported, antiretroviral therapy is a quite reasonable approach.

6. Cachexia and Wasting in HIV-1 Infection

Cachexia and wasting were major complications of advanced HIV-1 infection prior to the advent of more effective antiretroviral chemotherapy.[189,190] Weight loss in HIV-1 infection may arise from several causes including anorexia, malabsorption, and hypermetabolism.[191–194] Weight loss may occur gradually in association with the underlying features noted above or it may occur in a stepwise fashion in association with opportunistic complications of the illness.[195] Decreased levels of testosterone also have been noted in advanced disease. In any given patient the degree to which each of these factors contributes may vary. Appropriate management of individuals with HIV-1 infection requires attention to wasting and cachexia with careful consideration of factors playing a role in individual patients. In addition to appropriate management of opportunistic infections and nutritional counseling, some patients may benefit modestly from human growth hormone, megesterol, or nandrolone decanoate.[196–199]

7. Metabolic Disturbances

A number of metabolic disturbances have been documented in HIV-1-infected individuals.[200] Hypertriglyceridemia is perhaps the most prominent metabolic disturbance in advanced HIV-1 infection.[201,202] Elevated levels of triglycerides are more pronounced in patients with AIDS than in those with subclinical HIV-1 infection.[201] Hypertriglyceridemia in HIV infection is associated with decreased clearance of triglycerides from the circulation caused by diminished lipoprotein lipase activity.[203] In addition, hepatic fatty acid synthesis is increased.[200,204]

Several hypotheses have been advanced to explain these abnormalities in lipid metabolism. Most of these hypotheses focus on a central role for one or more cytokines including TNF,[202,205–207] IFN-γ,[206,208,209] and IL-6.[210–213] Increased serum levels of TNF-α have been demonstrated in some[205,206] but not all[202] studies of patients with advanced HIV-1 infection. In studies such as these, given the autocrine and paracrine nature of most cytokines, it is difficult to extrapolate from measures of serum levels of a given cytokine to its role in the pathogenesis of AIDS. Indeed, measures of messenger RNA in appropriate cells may provide more insight than serum levels of cytokines in situations in which cytokines may be cleared rapidly by binding with serum or cell-associated ligands. Acid-labile IFN-α also is elevated in individuals with HIV-1 infections.[202,208,209] Serum levels of IFN-α correlate both with the severity of HIV-1 infection[208] and with serum triglyceride levels.[202,203] In addition, therapy with antiretroviral drugs results in a reduction of both serum interferon levels and serum triglycerides.[202,209] IL-6 also enhances hepatic lipogenesis.[203] Serum levels of IL-6 are elevated in HIV-1 infection,[211–214] and correlate with advancing disease.

HIV infection is also associated with increases in resting energy expenditure.[215–218] This abnormality is detectable in early HIV-1 infection[218] and may be exacerbated in advanced disease with the onset of infections with organisms such as *Mycobacterium avium-intracellulare*.[219] Finally, patients with HIV-1 infection exhibit decreases in protein synthesis and breakdown and in glucose cycling.[220] The role of metabolic abnormalities in the pathogenesis of AIDS is the subject of intense investigation. Nonetheless, at this point no one of these abnormalities has been shown to be the predominant mediator of cachexia in AIDS.

8. Principles of Antiretroviral Chemotherapy

Insights into the pathogenesis of HIV-1 infection coupled with the availability of an array of more potent antiretroviral chemotherapeutic agents have drastically modified the prognosis of HIV-1 infection in patient populations with access to medical care.[221,222] A detailed description of each of the currently available antiretroviral chemotherapeutic agents is beyond the scope of this chapter. Nonetheless, a description of the general features of the classes of drugs currently available and an outline of the current approach to their use follow.

8.1. Classes of Agents

8.1.1. Reverse Transcriptase Inhibitors

At present nine reverse transcriptase inhibitors have been approved for use in the United States. Six of these (zidovudine, didanosine, zalcitabine, stavudine, lamivudine, and abacavir) are nucleoside analogues; three are classified as nonnucleoside analogue reverse transcriptase inhibitors. Each agent in this class inhibits the reverse transcription of HIV-1 RNA into DNA by the virally encoded enzyme, reverse transcriptase. The nucleoside analogues mediate their effects as chain terminators after being inappropriately incorporated into viral DNA. The nonnucleoside analogues possess diverse chemical structures but bind the same pocket in the reverse transcriptase enzyme and sterically inhibit the enzyme. The nucleoside analogues include four agents of moderate potency (zidovudine, didanosine, zalcitabine, and stavudine) that each inhibit HIV-1 replication in $vivo$ by approximately 0.5 $\log_{10}$. The remaining two nucleoside analogues (abacavir and lamivudine) are approximately tenfold more potent and inhibit susceptible viruses by approximately 1.5 $\log_{10}$ in $vivo$. The nonnucleoside analogues (delavirdine, nevirapine, and efavirenz) are generally about as potent as abacavir and lamivudine in that they also inhibit HIV-1 in $vivo$ approximately 1.5 $\log_{10}$.

Toxicities of these agents are somewhat agent specific. Zidovudine is primarily associated with marrow toxicity; didanosine and stavudine are more frequently associated with neurotoxicity and pancreatitis. Zalcitabine is used infrequently but it also is associated with neurotoxicity. Lamivudine and abacavir are generally well tolerated, although 3–5% of patients taking abacavir experience an unusual hypersensitivity reaction during the first several weeks of initiating the drug. This multisystem disorder is generally progressive and includes various combinations of fever, rash, gastrointestinal, and pulmonary symptoms. The mechanism of this complication of abacavir therapy has not been delineated. The syndrome is fully reversible if the drug is stopped and not restarted, but can be fatal if it is continued or interrupted and restarted. Patients initiating abacavir therapy should be fully educated about this complication and counseled to contact their physicians immediately if a compatible clinical syndrome develops. They should be particularly warned about the danger of stopping and restarting therapy with the drug, since this is frequently associated with an accelerated clinical syndrome that can be fatal. The nonnucleoside reverse transcriptase inhibitors may induce a pruritic eruption during the first several weeks of therapy. These rashes are more frequently observed with nevirapine and delavirdine than with efavirenz and usually are self-limiting and do not require interruption of therapy. Patients may benefit from a short course of an antihistamine when such a rash is observed. Efavirenz is not infrequently associated with CNS disturbances, especially manifest as sleep disturbances with vivid dreams and/or insomnia. Since efavirenz induces its own metabolism, these complications are generally maximal in the first 2 weeks of therapy and may subside if patients are managed supportively.

8.1.2. HIV-1 Protease Inhibitors

At this writing five HIV-1 protease inhibitors have been approved in the United States (saquinavir, ritonavir, indinavir, nelfinavir, and amprenavir). These agents competitively inhibit the ability of the HIV-1 protease enzyme to cleave the long polyprotein precursor initially transcribed by the virus into smaller viral core components. When taken in full doses, the HIV-1 protease inhibitors generally inhibit HIV-1 in $vivo$ in the range of 1.5 $\log_{10}$. Although each agent may be associated with gastrointestinal disturbances, the dose-limiting toxicities of each agent are not identical. Full doses of ritonavir are often associated with circumoral paresthesia, diarrhea, and nausea. Nelfinavir is associated with diarrhea in approximately 50% of patients. Although this may be improved by supportive measures such as lomotil, this side effect may cause as many as 10–20% of patients to seek an alternative agent. Indinavir is more water soluble than the other protease inhibitors, and thus, is renally excreted to a much greater extent. This requires patients to pay assiduous attention to hydration in order to avoid nephrolithiasis. Amprenavir and saquinavir may be associated with nausea and/or diarrhea, although these complications are generally less severe than observed with nelfinavir or ritonavir.

Over the past 24 months, it also has become apparent that the use of these agents also may be associated with significant metabolic abnormalities including diabetes mellitus, hypercholesterolemia, hypertriglyceridemia, and fat redistribution syndromes.[223–229] HIV-1 protease inhibitor-associated fat redistribution syndromes include various combinations of peripheral wasting, central adiposity, gynecomastia, and/or dorsocervical fat deposition.

These complications also are observed in patients with HIV-1 infection who are not on protease inhibitors, but there is no question but that the prevalence of these complications is more frequent in patients who receive these agents chronically. The mechanism(s) responsible for these complications have not yet been delineated.

8.2. Clinical Use of Antiretroviral Agents

Two consensus panels have provided excellent guidance as to the appropriate use of antiretroviral chemotherapeutic agents.[223–229] Although the recommendations vary in some detail, in general there is substantial agreement among the two sets of guidelines. The contemporary approach to the use of antiretroviral chemotherapeutic agents is based both on the current understanding of retroviral pathogenesis and on a large number of controlled clinical trials (Table 1). Although there are many nuances in antiretroviral chemotherapy that govern the approach to individual clinical situations (as there are in all areas of medicine), a working knowledge of these general principles is extremely helpful in patient management. In particular, these principles are extremely helpful in approaching decisions regarding when to initiate therapy, what agents to use, and when to change therapy.[230–232]

8.2.1. When Should Therapy Be Started?

This question is one that has been posed to patients and physicians since the initial use of zidovudine in 1986. The answer to this question is not the same for every patient and it has changed over time as additional agents have become available and as additional insights into HIV pathogenesis have emerged. As outlined in Principle 1, HIV-1 replication is the principal driving feature of HIV-1-associated morbidity and mortality. HIV-1-driven CD4 cell destruction begins during primary infection and continues throughout the lifetime of the infected host. It is clear that effective control of viral replication by an efficacious antiretroviral regimen is associated with substantial immune reconstitution and with greatly reduced morbidity and mortality.[233–235] If there were no acute or chronic toxicities of antiretroviral chemotherapy, the most prudent recommendation would be that antiretroviral chemotherapy should be initiated whenever infection is discovered. However, the toxicity, cost, and inconvenience of antiretroviral chemotherapy must be balanced against the benefits of therapy. Furthermore, since antiretroviral regimens are complex and since incomplete adherence is associated with a high risk of regimen failure, unless patients are fully committed to therapy, premature initia-

TABLE 1. Principles of Antiretroviral Chemotherapy[a]

Principle 1	Ongoing HIV replication leads to immune system damage and disease progression. HIV infection is always harmful, and true long-term survival free of clinically significant immune dysfunction is unusual.
Principle 2	Plasma HIV RNA levels indicate the magnitude of HIV replication and its associated rate of CD4 cell destruction, while CD4 cell counts indicate the extent of HIV-1-induced immune damage already suffered.
Principle 3	Treatment decisions should be individualized by level of risk indicated by plasma HIV RNA levels and CD4 cell counts.
Principle 4	The use of potent combination antiretroviral therapy to suppress HIV replication to below the levels of detection of sensitive plasma HIV RNA assays limits the potential for selection of antiretroviral-resistant HIV-1 variants.
Principle 5	The most effective means to accomplish durable suppression of HIV replication is the simultaneous initiation of combinations of effective anti-HIV drugs that are not cross-resistant with drugs with which the patient has been treated previously.
Principle 6	Antiretroviral drugs should always be used according to optimum schedules and dosages.
Principle 7	Antiretroviral drugs are limited in number and cross-resistance between drugs has been documented; changes in antiretroviral therapy constrain future therapeutic options.
Principle 8	Women should receive optimal antiretroviral therapy regardless of pregnancy status.
Principle 9	The same principles of antiretroviral therapy apply to HIV-infected children and adults, although the treatment of HIV-infected children involves unique pharmacologic, virologic, and immunologic considerations.
Principle 10	Persons with acute primary HIV infections should be treated with antiretroviral agents that suppress virus replication to levels below the limit of detection of sensitive plasma HIV-1 RNA assays.
Principle 11	HIV-1-infected persons, even those with plasma HIV-1 RNA levels below detectable limits, should be considered infectious and should be counseled to avoid sexual and drug-use behaviors that are associated with transmission or acquisition of HIV-1 and other infectious pathogens.

[a]Adapted from the Report of the United States Public Health Service Panel on Antiretroviral Chemotherapy.[265]

tion of therapy runs a great risk of the emergence of resistant virus. With these counterbalancing features in mind, it is essential to accurately assess the likelihood of disease progression in individual patients if appropriate advice is to be given. As stated in Principles 2 and 3, the level of viral replication that can be accurately assessed by measuring HIV-1 RNA in the plasma is an extremely strong predictor of the rate of disease progression and CD4 cell loss in the individual patient.[236,237] The CD4 cell count is the best predictor of the instantaneous risk of HIV-1-associated morbidity and mortality. One should individualize the decision to initiate therapy with these considerations in mind. Although the time of initiation of therapy will vary from patient to patient, most practitioners recommend initiation of therapy in patients with fewer than 350–500 CD4 cells/mm^3 and/or more than 5000 to 10,000 copies of HIV-1 RNA/ml. If there are aspects of a patient's course that suggest more rapid progression of disease such as the presence of clinical manifestations or a rapidly falling CD4 cell count or rising plasma HIV-1 RNA level, earlier therapy is advisable. If a patient is not convinced that he or she will be able to strictly adhere to a complex regimen, it is generally advisable to delay therapy.

8.2.2. What Antiretroviral Agents Should Be Used?

When therapy is initiated, the goal of therapy should be to maximally suppress HIV-1 for as long as possible with the least toxicity in that patient. This is best achieved by working with the patient to craft a well-tolerated regimen that is both potent and poses a high genetic barrier to resistance (Principles 4–7). With currently available agents, such a regimen is usually composed of at least three agents. These agents should be chosen on the basis of their potency, their degree of cross-resistance, and their tolerability in the patient under care. Most physicians initiate a regimen that includes two nucleoside analogue reverse transcriptase inhibitors (such as zidovudine or stavudine and lamivudine, or didanosine and stavudine) and a potent third agent such as a protease inhibitor, a nonnucleoside reverse transcriptase inhibitor, or abacavir. The specific agents to be included in a regimen should be chosen after a thorough discussion with the patient about the potential toxicities of the agents in the regimen and about issues such as convenience. No single regimen is appropriate for all patients. Perhaps the most important thing to emphasize to patients is that full adherence to the regimen chosen is critical and that the current array of agents allows for a great degree of flexibility if the regimen initially selected is not optimally tolerated.

8.2.3. When Should Therapy Be Changed?

The extent to which a durable response can be expected is a function of the degree of viral suppression. Physicians should endeavor to drive HIV-1 RNA levels below 20–50 copies/ml over the first 12–16 weeks of therapy. When sensitive techniques are employed, smoldering viral replication can be detected in peripheral blood and lymphoid tissues of many patients even when such success is achieved.[68,238,239] The rate of regimen failure is a function of many variables that include the degree of residual viral replication and the genetic barrier posed to the virus. Although many patients with lesser degrees of viral suppression do extremely well for prolonged periods of time with respect to sustained viral suppression, rising CD4 cell counts, and clinical stability, these patients are at increased risk for regimen failure as residual viral replication in the presence of selective pressure favors the outgrowth of drug-resistant virus.[240–242] Since drug resistance generally occurs sequentially rather than simultaneously in multidrug regimens, the short- and intermediate-term benefits of continued incomplete suppression of viral replication must be balanced against the longer-term risk of the accumulation of increasing drug resistance.[243–245] Attempts to modify or change regimens in individual patients must take these considerations into account as well as the extent to which additional therapeutic options are available to the patient.

8.2.4. What Agents Should Be Used When Regimens Fail?

As patients experience regimen failure, the accumulation of drug resistance in the viral quasispecies significantly limits distal therapeutic options (Principle 7). The general goals of therapy in salvage situations remain the same; the extent to which these goals can be achieved, however, is more limited. Until recently the conventional wisdom was that all drugs in a failing regimen must be changed and should never be reused. With the realization that drugs fail sequentially, it often is possible to preserve the use of several drugs in a failing regimen in the salvage regimen or subsequently.[243–245] This requires that patients be followed closely for failure and that contemporary molecular tools for viral characterization be employed to monitor the virus for drug susceptibility. Genotypic testing has the advantages of being somewhat less expensive and slightly more rapid but the disadvantage of being nonquantitative. Phenotypic susceptibility testing techniques are developing rapidly and have the major advantage of being quantitative. Several prospective and retrospective analyses have demonstrated the utility of susceptibility testing in constructing new regimens after failure of one or more prior regimens.[246–251] By making use of these tools, clinicians have an opportunity to construct combination regimens in salvage situations that are reasonably likely to result in high levels of viral suppression.

9. Primary and Secondary Prophylaxis of Opportunistic Infections in HIV-1-Infected Individuals

Prior to the availability of potent antiretroviral chemotherapy, most decisions about prophylaxis of opportunistic infections were irreversible. The general approach was to identify patient populations at increased risk for specific infections and to place patients on prophylaxis for life. In most cases opportunistic infections required lifelong therapy, since attempts to withdraw therapy after resolution of the acute infection were invariably complicated by relapse. The potential for improvement in pathogen-specific immune recovery was first reported by Autran and colleagues,[233] who reported that six patients treated with ZDV, ddC, and followed for 6 months experienced a substantial increase in CD4 count and decline in viral load. All six had no lymphoproliferative responses to recall or neoantigens at baseline; however, after 3–6 months of treatment all had developed responses to CMV and antigens.[233] Subsequently, Torriani and colleagues[252] reported eight patients with quiescent CMV retinitis who immunologically and virologically improved after initiation of potent antiretroviral therapy and had their CMV maintenance therapy discontinued. Four of eight patients who had a rise in CD4 count to levels above 200 cells/μl developed robust LPA responses to CMV. Those who had lower CD4 count increases and higher viral loads on therapy did not develop LPA responses.[252] Komanduri and colleagues[253] reported that after treatment of active CMV retinitis with and initiation of potent antiretroviral therapy, substantial increases in CMV-specific CD4 T-cell responses could be demonstrated, with proportions similar to those seen in HIV-seropositive, CMV-seronegative controls.[253] Havlir *et al.*[254] similarly showed that patients with HIV infection and disseminated *Mycobacterium avium* complex (MAC) disease, who began potent antiretroviral therapy and had an increase in CD4 T-lymphocyte count and decline in viral load, developed LPA responses to MAC Ags that were similar in magnitude to those who were HIV-seronegative, and were considerably greater than those who were HIV-seropositive without MAC disease who were not receiving antiretroviral therapy. While these descriptive data are useful in demonstrating that pathogen-specific *in vitro* responses are generated following initiation of potent antiretroviral therapies, there are

scant data available to address the predictive value of these responses in providing protection against opportunistic infections. To address this point, Torriani and colleagues[252] followed their original observations with a report demonstrating that the failure to develop LPA responses to CMV antigens was associated not only with HIV virologic failure and immunologic failure but also with reactivation of CMV disease. Additional studies evaluating the predictive value of recovery of LPA and other immunologic responses to specific antigens are in progress. Taking these and other data into account, the United States Public Health Service and the Infectious Diseases Society of America have recently revised their guidelines for the use of prophylactic agents directed at opportunistic pathogens in HIV-1-infected individuals.[255] The changes in these recommendations are summarized below.

9.1. *Pneumocystis carinii* Pneumonia and *Toxoplasma gondii*

The guidelines for initiation of primary or secondary prophylaxis for *Pneumocystis carinii* pneumonia (PCP) and toxoplasmosis have not substantively changed with the advent of potent antiretroviral chemotherapy. Based on data from three prospective observational studies, one retrospective review, and one randomized clinical trial, it is recommended that discontinuation of primary prophylaxis for PCP be considered for patients responding to highly active antiretroviral therapy (HAART) who have a sustained increase in CD4 T-cell count from <200 cells/μl to >200 cells/μl for 3–6 months.[256–258] While many of these studies also evaluated the risk of toxoplasmosis, there were insufficient numbers of patients at risk to support a recommendation for routine discontinuation of primary prophylaxis for toxoplasmosis. Prophylaxis should be reinstituted if the CD4 T-cell count again declines to <200 cells/μl or symptoms related to HIV disease intervene. Although patients with a prior episode of PCP or toxoplasmosis receiving secondary prophylaxis also may be at substantially lower risk for relapse when they respond immunologically to HAART as above, there are insufficient data from available trials at present to recommend discontinuation of secondary prophylaxis for PCP or toxoplasmosis.

9.2. Tuberculosis

There are no changes in recommendations related to those who are considered to be candidates for the use of chemoprophylaxis for prevention of tuberculosis. It is now recommended that all HIV-infected persons who have a positive tuberculin skin test, have no evidence of active tuberculosis, and no history of prior treatment or prophylaxis should receive 9 months of chemoprophylaxis with either INH 300 mg qd, INH 900 mg twice weekly, or rifampin or rifabutin plus pyrazinamide daily for 2 months. Rifampin should not be used with protease inhibitors or nonnucleoside reverse transcriptase inhibitors. Rifabutin is an acceptable alternative but should not be used with ritonavir or saquinavir. It may be used at one-half the daily dose (150 mg qd) with indinavir, nelfinavir, or amprenavir. Rifabutin should not be used with delavirdine, but may be used with efavirenz if the dose is adjusted to 450 mg qd. There are insufficient data regarding the use of rifabutin with nevirapine to make a formal recommendation. One should be cautioned that all these recommendations are made on the basis of few clinical data and best clinical judgment should guide management in individual cases. For those who have an increase in CD4 T-cell count from $<$200 cells/μl to $>$200 cells/μl, a repeat PPD may be considered if it had previously been negative.

9.3. *Mycobacterium avium* Complex

There also is no change in the recommendation for when to initiate primary MAC prophylaxis. Based on data from one prospective observational study and two randomized clinical trials, one may consider discontinuation of primary prophylaxis for MAC disease in patients who have a sustained increase in CD4 T-cell count from $<$100 cells/μl to $>$100 cells/μl for 3–6 months in response to potent antiretroviral therapy.[259] Primary prophylaxis should be reinstituted if the CD4 T-cell count again declines to $<$50 cells/μl. Although a prospective clinical trial is in progress evaluating discontinuation of secondary maintenance therapy, there are insufficient data to routinely recommend this approach at present, thus full therapeutic doses of antimycobacterial agents for treatment of MAC disease should be continued for life, pending further information. At least one anecdotal experience has appeared that suggests that it is possible to stop secondary prophylaxis in some patients.[260] The same drug interaction issues should be considered when using rifabutin in the context of treatment for MAC disease.

9.4. Cytomegalovirus Disease

No substantive changes have been made to the recommendations for primary CMV prophylaxis. With regard to secondary maintenance therapies, the regimens noted to be effective for chronic suppression now include parenteral cidofovir and the combination of ganciclovir via intraocular implant combined with oral ganciclovir. A number of studies have demonstrated that for patients with CMV retinitis receiving HAART who have had an increase in their CD4 T-cell count from $<$100 cells/μl to $>$100–150 cells/μl and whose plasma HIV RNA level has been suppressed, the risk of progression of CMV retinitis is substantially reduced.[261–263] Thus, discontinuation of CMV maintenance therapy may be considered for those who meet these criteria and sustain a rise in CD4 T-cell count for 3–6 months. Factors that should be taken into account when deciding to discontinue maintenance therapy include magnitude and duration of suppression of plasma HIV RNA levels, anatomic location of the retinitis, vision in the contralateral eye, and the feasibility of regular ophthalmologic monitoring. Maintenance therapy should be reinstituted if the CD4 T-cell count declines to $<$50–100 cells/μl.

9.5. Invasive Fungal Infections

The recommendations related to primary prophylaxis for invasive fungal infections have not changed, with the following exceptions: (1) Due to the recognition that itraconazole is embryotoxic and teratogenic and that craniofacial and skeletal abnormalities have been reported following prolonged *in utero* exposure to fluconazole,[264] primary prophylaxis for invasive fungal infections using systemic azoles should not be administered during pregnancy and azoles should be discontinued for HIV-infected women who become pregnant; effective birth control is recommended for all women on azole therapy; (2) discontinuation of secondary prophylaxis/maintenance therapy for invasive fungal infections is not currently recommended, although the risk may be low for patients who have an increase in CD4 T-cell counts in response to potent antiretroviral therapy.

9.6. Bacterial Respiratory Infections

The principal changes in the guidelines related to bacterial respiratory infections are the following: (1) periodic revaccination with pneumococcal vaccine may be considered; intervals of 5 years are recommended; (2) revaccination one time should be considered if the initial pneumococcal vaccine immunization was administered when the CD4 T-cell count was $<$200 cells/μl and the CD4 T-cell count has subsequently increased to $>$200 cells/μl on HAART.

9.7. Human Herpesvirus 8

The following new recommendation related to human herpesvirus 8 has been added: because potent antiretroviral therapy that suppresses HIV replication reduces the frequency of Kaposi's sarcoma and may prevent the progression of or development of new Kaposi's sarcoma lesions, HAART should be considered in all persons who are candidates for such and all persons with Kaposi's sarcoma.

9.8. Hepatitis C Virus

Guidelines related to hepatitis C virus (HCV) added to the recently revised guidelines[265] include: (1) HIV-infected patients should be screened for HCV infection using an enzyme immunoassay for detection of HCV antibody. Positive results should be confirmed with additional testing (RIBA or reverse transcriptase–polymerase chain reaction). (2) HIV-HCV coinfected persons should be advised to avoid excessive amounts of alcohol. (3) Patients with chronic hepatitis C should be vaccinated against hepatitis A. (4) HIV-HCV coinfected persons should be evaluated for chronic liver disease and for possible treatment; treatment should be done in conjunction with experts experienced in the treatment of HIV and HCV. (5) Elevated liver enzymes are common among those coinfected with HCV, and while they should be carefully monitored, HAART should not be withheld; those who do receive HAART may experience an inflammatory reaction mimicking an exacerbation of hepatitis and careful monitoring of liver function and possible temporary cessation of antiretroviral therapy may be necessary.

References

1. Chiu I-M, Yaniv A, Dahlberg JE, *et al*: Nucleotide sequence evidence for the relationship of AIDS retrovirus to lentiviruses. *Nature* **317**:366–368, 1985.
2. Narayan O, Cork LC: Lentiviral diseases of sheep and goats: Chronic pneumonia, leukoencephalomyelitis and arthritis. *Rev Infect Dis* **7**:89–98, 1985.
3. Stephens RM, Casey JW, Rice NR: Equine infectious anemia virus *gag* and *pol* genes: Relatedness to visna and AIDS virus. *Science* **231**:589–594, 1986.
4. Daniel MD, Letvin NL, King NW, *et al*: Isolation of T-cell tropic HTLV-III-like retrovirus from macaques. *Science* **228**:1201–1204, 1985.
5. Clavel F, Mansinho K, Chamaret S, *et al*: Human immunodeficiency virus type 2 infection associated with AIDS in West Africa. *N Engl J Med* **316**:1180–1185, 1987.
6. Brun-Vezinet F, Rey MA, Katlama C, *et al*: Lymphadenopathy-associated virus type 2 in AIDS and AIDS-related complex: Clinical and virological features in four patients. *Lancet* **1**:128–132, 1987.
7. Varmus H: Retroviruses. *Science* **240**:1427–1435, 1988.
8. Bushman FD, Fujiwara T, Craigie R: Retroviral DNA integration directed by HIV integration protein *in vitro*. Science **249**:1555–1558, 1990.
9. Haseltine WA: The molecular biology of HIV-1. In Devita Jr, Hellman S, Rosenberg SA (eds): AID*S: Etiology, Diagnosis, Prevention, and Treatment*. Lippincott, Philadelphia, 1992, pp. 39-59.
10. Gottlieb MS, Schroff R, Schanker HM, *et al*: *Pneumocystis carinii* pneumonia and mucosal candidiasis in previously healthy homosexual men: Evidence of a new acquired cellular immunodeficiency. *N Engl J Med* **305**:1425–1431, 1981.
11. Barre-Sinoussi F, Chermann JC, Rey F, *et al*: Isolation of a T-lymphotropic retrovirus from a patient at risk for acquired immune deficiency syndrome. *Science* **220**:868–871, 1983.
12. Gao F, Bailes E, Robertson DL, *et al*: Origin of HIV-1 in the chimpanzee Pan troglodytes troglodytes. *Nature* **397**:436–441, 1999.
13. Centers for Disease Control and Prevention: Update: Trends in AIDS incidence—United States, 1996. *MMWR* **46**:861–868, 1997.
14. Ho DD, Sarngadharan MG, Resnick L, *et al*: Primary human T-lymphotropic virus type III infection. *Ann Intern Med* **103**:880–883, 1985.
15. Tindall B, Barker S, Donovan B, *et al*: Characterization of the acute clinical illness associated with human immunodeficiency virus infection. *Arch Intern Med* **148**:945–949, 1988.
16. Fox R, Eldred LJ, Fuchs EJ, *et al*: Clinical manifestations of acute infection with human immunodeficiency virus in a cohort of gay men. *AIDS* **1**:35–38, 1987.
17. Gaines H, von Sydow M, Pehrson PO, *et al*: Clinical picture of primary HIV infection presenting as a glandular-fever-like illness. *Br Med J* **297**:1363–1368, 1988.
18. Gaines H, Albert J, von Sydow M, *et al*: HIV antigenaemia and virus isolation from plasma during primary HIV infection. *Lancet* **1**:1317–1318, 1987.
19. Daar ES, Mougdil T, Meyer RD, *et al*: Transient high levels of viremia in patients with primary human immunodeficiency virus type 1 infection. *N Engl J Med* **324**:961–964, 1991.
20. Clark SJ, Saag MS, Decker WD, *et al*: High titers of cytopathic virus in plasma of patients with symptomatic primary HIV-1 infection. *N Engl J Med* **324**:954–960, 1991.
21. Gaines H, von Sydow MAE, von Stedingk LV, *et al*: Immunological changes in primary HIV-1 infection. *AIDS* **4**:995–999, 1990.
22. Tindal B, Cooper DA: Primary HIV infection: Host responses and intervention strategies. *AIDS* **5**:1–14, 1991.
23. Koup RA, Safrit JT, Cao Y, Andrews CA, McLeod G, Borkowsky W, Farthing C, Ho DD: Temporal association of cellular immune responses with the initial control of viremia in primary human immunodeficiency virus type 1 syndrome. *J Virol* **68**:4650–4655, 1994.
24. Borrow P, Lewicki H, Hahn BH, Shaw GM, Oldstone MB: Virus-specific CD8+ cytotoxic T-lymphocyte activity associated with control of viremia in primary human immunodeficiency virus type 1 infection. *J Virol* **68**:6103–6110, 1994.
25. Rosenberg ES, Billingsley JM, Caliendo AM, Boswell SL, Sax PE, Kalams SA, Walker BD: Vigorous HIV-1 specific CD4+ T cell responses associated with control of viremia. *Science* **278**:1447–1450, 1997.
26. Kahn JO, Walker BD: Acute human immunodeficiency virus type 1 infection. *N Engl J Med* **339**:33–39, 1998.

27. Pederson C, Dickmeiss E, Gaub J, *et al*: T-cell subset alterations and lymphocyte responsiveness to mitogens and antigen during severe primary infection with HIV: A case series of seven consecutive HIV seroconverters. *AIDS* **4:**523–526, 1990.

28. Sheppard HW, Ascher MS, McRae B, *et al*: The initial immune response to HIV and immune system activation determine the outcome of HIV disease. *J AIDS* **4:**704–712, 1991.

29. Sinicco A, Sciandra M, Lucchini A, *et al*: *Acute Primary HIV-1 Infection: Risk of Developing AIDS*. VIII International Conference on AIDS, Amsterdam, 1992.

30. Tindall B, Carr A, Goldstein D, *et al*: Administration of zidovudine during primary HIV-1 infection may be associated with a less vigorous immune response. *AIDS* **7:**127–128, 1993.

31. Tindall B, Gaines H, Imrie A, *et al*: Zidovudine in the management of primary HIV-1 infection. *AIDS* **5:**477–484, 1991.

32. Kinloch-De Loes S, Hirschel BJ, Hoen B, *et al*: A controlled trial of zidovudine in primary human immunodeficiency virus infection. *N Engl J Med* **333:**408–413, 1995.

33. Ho DD, Moudgil T, Alam M: Quantitation of human immunodeficiency virus type 1 in the blood of infected persons. *N Engl J Med* **321:**1621–1625, 1989.

34. Walker BD, Chakrabarti S, Moss B, *et al*: HIV-specific cytotoxic T lymphocytes in seropositive individuals. *Nature* **328:**345–348, 1987.

35. Weiss RA, Clapham RP, Cheingsong-Popov R, *et al*: Neutralization of human T-lymphotropic virus type III by sera of AIDS and AIDS-risk patients. *Nature* **316:**69–72, 1985.

36. Robert-Guroff M, Brown M, Gallo RC: HTLV-III-neutralizing antibodies in patients with AIDS and AIDS-related complex. *Nature* **316:**72–74, 1985.

37. McElrath MJ, Pruett JE, Cohn ZA: Mononuclear phagocytes of blood and bone marrow: Comparative roles as viral reservoirs in human immunodeficiency virus type 1 infections. *Proc Natl Acad Sci USA* **86:**675–679, 1989.

38. Ho DD, Rota TR, Hirsh MS: Infection of monocyte/ macrophages by human T lymphotropic virus type III. *J Clin Invest* **77:**1712–1715, 1986.

39. Popovic M, Gartner S: Isolation of HIV-1 from monocytes but not T lymphocytes. *Lancet* **2:**916, 1987.

40. Tenner-Racz K, Racz P, Dietrich M, *et al*: Altered follicular dendritic cells and virus-like particles in AIDS and AIDS-related lymphadenopathy. *Lancet* **1:**105–106, 1985.

41. Tschaler E, Groh V, Popovic M, *et al*: Epidermal Langerhans' cells: A target for HTLV-III/LAV infection. *J Invest Dermatol* **88:**233–237, 1987.

42. Salahuddin SZ, Rose RM, Groopman JE, *et al*: Human T lymphotropic virus type III infection of human alveolar macrophages. *Blood* **68:**281–284, 1986.

43. Pantaleo G, Graziosi C, Demarest JF, *et al*: HIV infection is active and progressive in lymphoid tissue during the clinically latent stage of disease. *Nature* **362:**355–358, 1993.

44. Embretson J, Zupancic M, Ribas JL, *et al*: Massive covert infection of helper T lymphocytes and macrophages by HIV during the incubation period of AIDS. *Nature* **362:**359–362, 1993.

45. Piatak M Jr, Saag MS, Yang LC, *et al*: High levels of HIV-1 in plasma during all stages of infection determined by competitive PCR. *Science* **259:**1749–1754, 1993.

46. Wei X, Ghosh SK, Taylor ME, *et al*: Viral dynamics in human immunodeficiency virus type 1 infection. *Nature* **373:**117–122, 1995.

47. Ho DD, Neumann AU, Perelson AS: Rapid turnover of plasma virions and CD4 lymphocytes in HIV-1 infection. *Nature* **373:**123–126, 1995.

48. Perelson AS, Neumann AU, Markowitz M, Leonard JM, Ho DD: HIV-1 dynamics *in vivo*: Virion clearance rate, infected cell lifespan, and viral generation time. *Science* **271:**1582–1586, 1996.

49. Ammann AJ, Abrams D, Conant M, *et al*: Acquired immune dysfunction in homosexual men: Immunologic profiles. *Clin Immunol Immunopathol* **27:**315–325, 1983.

50. Lane HC, Depper JM, Green WC, *et al*: Qualitative analysis of immune function in patients with the acquired immunodeficiency syndrome: Evidence for a selective defect in soluble antigen recognition. *N Engl J Med* **313:**79–84, 1985.

51. Giorgi JV, Fahey JL, Smith DC, *et al*: Early effects of HIV on CD4 lymphocytes *in vivo*. *J Immunol* **138:**3725–3730, 1987.

52. Rook AH, Masur H, Lane HC, *et al*: Interleukin-2 enhances the depressed natural killer and cytomegalovirus-specific cytotoxic activities of lymphocytes from patients with the acquired immune deficiency syndrome. *J Clin Invest* **72:**398–403, 1983.

53. Shearer GM, Salahuddin SZ, Markham PD, *et al*: Prospective study of cytotoxic T lymphocyte responses to influenza and antibodies to human T lymphotropic virus-III in homosexual men: Selective loss of an influenza-specific, human leukocyte antigen-restricted cytotoxic T lymphocyte response in human T lymphotropic virus-III positive individuals with symptoms of acquired immunodeficiency syndrome and in a patient with acquired immunodeficiency syndrome. *J Clin Invest* **76:**1699–1704, 1985.

54. Hoy JF, Lewis DE, Miller GG: Functional versus phenotypic analysis of T cells in subjects seropositive for the human immunodeficiency virus: A prospective study of *in vitro* responses to Cryptococcus neoformans. *J Infect Dis* **158:**1071–1078, 1988.

55. Reinherz E, Schlossman SF: The differentiation and function of human T lymphocytes. *Cell* **19:**821–827, 1980.

56. Engleman EG, Benike C, Grumet C, *et al*: Activation of human T lymphocyte subsets: Helper and suppressor/cytotoxic T cells recognize and respond to distinct histocompatibility antigens. *J Immunol* **127:**2124–2129, 1981.

57. Meuer SC, Scholssman SF, Reinherz E: Clonal analysis of human cytotoxic T lymphocytes: T4+ and T8+ effector T cells recognize products of different major histocompatibility complex regions. *Proc Natl Acad Sci USA* **79:**4395–4399, 1982.

58. Klatzmann D, Champagne E, Chamaret S, *et al*: T-lymphocyte T4 molecule behaves as the receptor for human retrovirus LAV. *Nature* **312:**767–768, 1984.

59. McDougal JS, Mawle A, Cort SP, *et al*: Cellular tropism of the human retrovirus HTLV-III/LAV: 1. Role of T-cell activation and expression of the T4 antigen. *J Immunol* **135:**3151, 1985.

60. McDougal JS, Kennedy MS, Sligh JM, *et al*: Binding of HTLV/LAV to T4+ T cells by a complex of the 110 K viral protein and the T4 molecule. *Science* **231:**382–385, 1986.

61. Groenik M, Fouchier RAM, De Goede REY, *et al*: Phenotypic heterogeneity in a panel of infectious molecular human immunodeficiency virus type 1 clones derived from a single individual. *J Virol* **65:**1968–1975, 1991.

62. Koot M, Keet IPM, Vos AHV, *et al*: Prognostic value of HIV-1 syncytium-inducing phenotype for rate of CD4+ cell depletion and progression to AIDS. *Ann Intern Med* **118:**681–688, 1993.

63. Weinhold KJ, Lyerly HK, Stanley SD, *et al*: HIV-1 GP120-mediated immune suppression and lymphocyte destruction in the absence of viral infection. *J Immunol* **142:**3091–3097, 1989.

64. Lanzavecchia A, Roosnek E, Gregory T, *et al*: T cells can present antigens such as HIV gp120 targeted to their own surface molecules. *Nature* **334:**530–532, 1988.

65. Siliciano RF, Lawton T, Knall C, *et al*: Analysis of host-virus

interactions in AIDS with anti-gp120 T cell clones: Effect of HIV sequence variation and a mechanism for CD4+ cell depletion. *Cell* **54:**561–575, 1988.

66. Ziegler JL, Stites DP: Hypothesis: AIDS is an autoimmune disease directed at the immune system and triggered by a lymphotropic retrovirus. *Clin Immunol Immunopathol* **41:**305–313, 1986.

67. Perelson AS, Essunger P, Cao Y: Decay characteristics of HIV-1 infected compartments during combination therapy. *Nature* **387:** 188–191, 1997.

68. Finzi D, Blankson J, Siliciano JD: Latent infection of CD4+ T cells provides a mechanism for lifelong persistence of HIV-1, even in patients on effective combination therapy. *Nat Med* **5:**512–517, 1995.

69. Coombs RW, Collier AC, Allain JP, *et al*: Plasma viremia in human immunodeficiency virus infection. *N Engl J Med* **321:** 1626–1631, 1989.

70. Giorgi JV, Hultin LE, McKeating JA, *et al*: Shorter survival in advanced human immunodeficiency virus type 1 infection is more closely associated with T lymphocyte activation than with plasma virus burden or virus chemokine coreceptor usage. *J Infect Dis* **179:**859–870, 1999.

71. Masur H, Ognibene FP, Yarchoan R, *et al*: CD4 counts as predictors of opportunistic pneumonias in human immunodeficiency virus (HIV) infection. *Ann Intern Med* **111:**223–231, 1989.

72. Phair J, Munoz A, Detels R, *et al*: The risk of *Pneumocystis carinii* pneumonia among men infected with human immunodeficiency virus type 1. *N Engl J Med* **322:**161–165, 1990.

73. Snider WD, Simpson DM, Nielson S, *et al*: Neurological complications of acquired immune deficiency syndrome; analysis of 50 patients. *Ann Neurol* **14:**403–418, 1983.

74. Nielson SL, Petito CK, Urmacher CD, *et al*: Subacute encephalitis in acquired immune deficiency syndrome: A postmortem study. *Am J Clin Pathol* **82:**678–682, 1984.

75. Shaw GM, Harper ME, Hahn BH, *et al*: HTLV-III infection in brains of children and adults with AIDS encephalopathy. *Science* **227:**177–182, 1985.

76. Ho DD, Rota TR, Schooley RT, *et al*: Isolation of HTLV-III from cerebrospinal fluid and neural tissues of patients with neurologic syndromes related to the acquired immune deficiency syndrome. *N Engl J Med* **313:**1493–1497, 1985.

77. Levy JA, Shimabukuro J, Hollander H, *et al*: Isolation of AIDS-associated retroviruses from cerebrospinal fluid and brain of patients with neurological symptoms. *Lancet* 2:586–588, 1985.

78. Koenig S, Gendelman HE, Orenstein JM, *et al*: Detection of AIDS virus in macrophages in brain tissue from AIDS patients with encephalopathy. *Science* **233:**1089–1093, 1986.

79. Janssen RS, Nwanyanwu OC, Selik RM, *et al*: Epidemiology of human immunodeficiency virus encephalopathy in the United States. *Neurology* **42:**1472–1476, 1992.

80. Navia BA, Price RW: The acquired immunodeficiency syndrome dementia complex as the presenting and sole manifestation of human immunodeficiency syndrome virus infection. *Arch Neurol* **44:**65–69, 1987.

81. Simpson DM, Tagliati R: Neurological manifestations of HIV infection. *Ann Intern Med* **121:**769–785, 1994.

82. Lyman WD, Kress Y, Kure K, *et al*: Detection of HIV in fetal central nervous system tissue. *AIDS* **4:**917–920, 1990.

83. Wiley CA, Schrier RD, Nelson JA, *et al*: Cellular localization of human immunodeficiency virus infection within the brains of acquired immune deficiency syndrome patients. *Proc Natl Acad Sci USA* **83:**7089–7093, 1986.

84. Gabuzda DH, Ho DD, de la Monte SM, *et al*: Immunohistochemical identification of HTLV-III antigen in brains of patients with AIDS. *Ann Neurol* **20:**289–295, 1986.

85. Vazeux R, Brousse N, Jarry A, *et al*: AIDS subacute encephalitis. Identification of HIV-infected cells. *Am J Pathol* **20:**289–295, 1986.

86. Stoler MH, Eskin TA, Benn S, *et al*: Human T-cell lymphotropic virus type III infection of the central nervous system. A preliminary *in situ* analysis. *JAMA* **256:**2360–2364, 1986.

87. Pumarola-Sune T, Navia BA, Cordon-Cardo C, *et al*: HIV antigen in the brains of patients with the AIDS dementia complex. *Ann Neurol* **21:**490–496, 1987.

88. Gyorkey F, Melnick JL, Gyorkey P: Human immunodeficiency virus in brain biopsies of patients with AIDS and progressive encephalopathy. *J Infect Dis* **155:**870–876, 1987.

89. Chiodi F, Fuerstenberg S, Gidlund M, *et al*: Infection of brain-derived cells with the human immunodeficiency virus. *J Virol* **61:** 1244–1247, 1987.

90. Koyanagi Y, Miles S, Mitsuyasu RT, *et al*: Dual infection of the central nervous system by AIDS viruses with distinct cellular tropisms. *Science* **236:**819–822, 1987.

91. Cheng-Mayer C, Rutka JT, Rosenblum ML, *et al*: Human immunodeficiency virus can productively infect cultured human glial cells. *Proc Natl Acad Sci USA* **84:**3526–3530, 1987.

92. Brenneman DE, Westbrook CL, Fitzgerald SP, *et al*: Neuronal cell killing by the envelope protein of HIV and its prevention by vasoactive intestinal peptide. *Nature* **335:**639–642, 1988.

93. Rutka JT, Giblin JR, Berens ME, *et al*: The effects of human recombinant tumor necrosis factor on glioma-derived cell lines: Cellular proliferation, cytotoxicity, morphological and radioreceptor studies. *Int J Cancer* **41:**573–582, 1988.

94. Selmaj KW, Raine CS: Tumor necrosis factor mediates myelin and oligodendrocyte damage *in vitro*. *Ann Neurol* **23:**339–346, 1988.

95. Mintz M, Rapaport R, Oleske JM, *et al*: Elevated serum levels of tumor necrosis factor are associated with progressive encephalopathy in children with acquired immunodeficiency syndrome. *Am J Dis Child* **143:**771–774, 1989.

96. Heyes MP, Brew BJ, Martin A, *et al*: Quinolinic acid in cerebrospinal fluid and serum in HIV-1 infection: Relationship to clinical and neurological status. *Ann Neurol* **29:**202–209, 1991.

97. Grimaldi LM, Martino GV, Franciotta DM, *et al*: Elevated alpha-tumor necrosis factor levels in spinal fluid from HIV-1 infected patients with central nervous system involvement. *Ann Neurol* **29:**21–25, 1991.

98. Merrill JE, Chen IS: HIV-1, macrophages, glial cells, and cytokines in AIDS nervous system disease. *FASEB J* **5:**23391–23397, 1991.

99. Perrella O, Carrieri PB, Guarnaccia D, *et al*: Cerebrospinal fluid cytokines in AIDS dementia complex. *J Neurol* **239:**387–388, 1992.

100. Grimaldi LME, Martino GV, Franciotta DM, *et al*: Elevated alpha-tumor necrosis factor levels in spinal fluid from HIV-1 infected patients with central nervous system involvement. *Neurology* **29:** 21–25, 1991.

101. Heyes MP, Brew B, Martin A, *et al*: Cerebrospinal fluid quinolinic acid concentrations are increased in acquired immune deficiency syndrome. *Adv Exp Med Biol* **294:**687–690, 1991.

102. Cvetkovich TA, Lazar E, Blumberg BM, *et al*: Human immunodeficiency virus type 1 infection of neural xenografts. *Proc Natl Acad Sci USA* **89:**5162–5166, 1992.

103. Pulliam L, Herndier BG, Tang NM, *et al*: Human immunodefi-

ciency virus-infected macrophages produce soluble factors that cause histological and neurochemical alterations in cultured human brains. *J Clin Invest* **87**:503–512, 1991.

104. Shaskan EG, Thompson RM, Price RW: Undetectable tumor necrosis factor-alpha in spinal fluid from HIV-1 infected patients (letter). *Ann Neurol* **31**:687–689, 1992.

105. Gallo P, Laverda AM, De Rossi A, *et al*: Immunological markers in the cerebrospinal fluid of HIV-1 infected children. *Acta Paediatr Scand* **80**:659–666, 1991.

106. Whetsell WO, Schwarcz R: Prolonged exposure to submicromolar concentrations of quinolinic acid causes excitotoxic damage in organotypic cultures of rat corticostriatal system. *Neurosci Lett* **97**:271–275, 1989.

107. Brew BJ, Bhalla RB, Paul M, *et al*: Cerebrospinal fluid neopterin in human immunodeficiency virus type 1 infection. *Ann Neurol* **28**:556–560, 1990.

108. Achim CL, Morey MK, Wiley CA: Expression of major histocompatibility complex and HIV antigens within the brains of AIDS patients. *AIDS* **5**:535–541, 1991.

109. McArthur JC, Nance-Spronson TE, Griffin DE, *et al*: The diagnostic utility of elevation in cerebrospinal fluid beta 2-microglobulin in HIV-1 dementia. Multicenter AIDS cohort study. *Neurology* **42**:1707–1712, 1992.

110. Brew BJ, Bhalla RB, Paul M, *et al*: Cerebrospinal fluid beta 2-microglobulin in patients with AIDS dementia complex: An expanded series including response to zidovudine treatment. *AIDS* **6**:461–465, 1992.

111. Jassoy C, Johnson RP, Navia BA, *et al*: Detection of a vigorous HIV-1-specific cytotoxic T lymphocyte response in cerebrospinal fluid from infected persons with AIDS dementia complex. *J Immunol* **149**:3113–3119, 1992.

112. Kumar M, Resnick L, Loewenstein DA, *et al*: Brain-reactive antibodies and the AIDS dementia complex. *J AIDS* **2**:469–471, 1989.

113. Lipkin WI, Parry G, Kiprox D, *et al*: Inflammatory neuropathy in homosexual men with lymphadenopathy. *Neurology* **35**:1479–1483, 1985.

114. Petito CK, Navia BA, Cho ES, *et al*: Vacuolar myelopathy pathologically resembling subacute combined degeneration in patients with acquired immunodeficiency syndrome. *N Engl J Med* **312**:874–879, 1985.

115. Levy RM, Bredesen DE, Rosenblum ML: Neurological manifestations of the acquired immunodeficiency syndrome (AIDS): Experience at UCSF and review of the literature. *J Neurosurg* **62**:475–495, 1985.

116. Navia BA, Jordan BD, Price RW: The AIDS dementia complex. Clinical features. *Ann Neurol* **19**:517–524, 1986.

117. Yankner BA, Skolnick P, Shoukimas GM, *et al*: Cerebral granulomatous angiitis associated with acute HTLV-III infection of the central nervous system. *Ann Neurol* **20**:362–364, 1986.

118. Cornblath DR, McArthur JC, Kennedy PG, *et al*: Inflammatory demyelinating peripheral neuropathies associated with human T-cell lymphotropic virus type III infection. *Ann Neurol* **21**:32–40, 1987.

119. Gabuzda D, Hirsch MS: Neurologic manifestations of infection with human immunodeficiency virus. *Ann Intern Med* **107**:383–391, 1987.

120. de la Monte SM, Ho DD, Schooley RT, *et al*: Subacute encephalomyelitis of AIDS and its relation to HTLV-III infection. *Neurology* **37**:562–569, 1987.

121. de la Monte SM, Gabuzda DH, Ho DD, *et al*: Peripheral neuropa-

thy in the acquired immunodeficiency syndrome. *Ann Neurol* **28**:485–492, 1988.

122. Ho DD, Bredesen DE, Vinters HV, *et al*: The acquired immunodeficiency syndrome (AIDS) dementia complex (clinical conference). *Ann Intern Med* **111**:400–410, 1989.

123. Goudsmit J, Wolters ED, Bakker M, *et al*: Intrathecal synthesis of antibodies to HTLV-III in patients without AIDS or AIDS-related complex. *Br Med J* **292**:1231–1234, 1986.

124. Carne CA, Tedder RS, Smith A, *et al*: Acute encephalopathy coincident with seroconversion for anti-HTLV-III. *Lancet* **2**:1206–1208, 1985.

125. Grant I, Atkinson JH, Hesselink JR, *et al*: Evidence for early central nervous system involvement in the acquired immunodeficiency syndrome (AIDS) and other human immunodeficiency virus (HIV) infections. Studies with neuropsychologic testing and magnetic resonance imaging. *Ann Intern Med* **107**:828–836, 1987.

126. Saykin AJ, Janssen RS, Sprehn GC, *et al*: Neuropsychological dysfunction in HIV-infection: Characterization in a lymphadenopathy cohort. *Int J Clin Neuropsychol* **10**:81–95, 1988.

127. Marshall DW, Goethe KE, Mitchell JE, *et al*: Neurological and neuropsychological status of human immunodeficiency virus (HIV) serum antibody positive asymptomatic patients. *Neurology* **38**:247, 1988.

128. Clark JB: Policy considerations of human immunodeficiency virus (HIV) infection in US naval aviation personnel. *Aviat Space Environ Med* **61**:165–181, 1990.

129. Goethe KE, Mitchell JE, Marshall DW, *et al*: Neuropsychological and neurological function of human immunodeficiency virus seropositive asymptomatic individuals. *Arch Neurol* **46**:129–133, 1989.

130. Tross S, Price RW, Navia B, *et al*: Neuropsychological characterization of the AIDS dementia complex: A preliminary report. *AIDS* **2**:81–88, 1989.

131. McArthur JC, Cohen BA, Selnes OA, *et al*: Low prevalence of neurological and neuropsychological abnormalities in otherwise healthy HIV-1-infected individuals: Results from the multicenter AIDS Cohort Study. *Ann Neurol* **26**:601–611, 1989.

132. Janssen RS, Saykin AJ, Cannon L, *et al*: Neurological and neuropsychological manifestations of HIV-1 infection: Association with AIDS-related complex but not asymptomatic HIV-1 infection. *Ann Neurol* **26**:592–600, 1989.

133. Gibbs A, Andrewes DG, Szmukler G, *et al*: Early HIV-related neuropsychological impairment: Relationship to stage of viral infection. *J Clin Exp Neuropsychol* **12**:766–780, 1990.

134. Egan VG, Crawford JR, Brettle RP, *et al*: The Edinburgh cohort of HIV-positive drug users: Current intellectual function is impaired, but not due to early AIDS dementia complex. *AIDS* **4**:651–656, 1990.

135. Clifford DB, Jacoby RG, Miller JP, *et al*: Neuropsychometric performance of asymptomatic HIV-infected subjects. *AIDS* **4**:767–774, 1990.

136. McAllister RH, Herns MV, Harrison MJ, *et al*: Neurological and neuropsychological performance in HIV seropositive men without symptoms. *J Neurol Neurosurg Psychiatry* **55**:143–148, 1992.

137. The European Collaborative Study: Neurologic signs in young children with human immunodeficiency virus infection. *Pediatr Infect Dis J* **9**:402–406, 1990.

138. Civitello LA: Neurologic complications of HIV infection in children. *Pediatr Neurosurg* **17**:104–112, 1991.

139. Blanche S, Tardieu M, Duliege A, *et al*: Longitudinal study of 94 symptomatic infants with perinatally acquired human immunodeficiency virus infection. Evidence for a bimodal expression of

clinical and biological symptoms. *Am J Dis Child* **144**:1210–1215, 1990.

140. Perdices M, Cooper DA: Simple and choice reaction time in patients with human immunodeficiency virus infection. *Ann Neurol* **25**:460–467, 1989.

141. Van Gorp WG, Miller EN, Satz P, *et al*: Neuropsychological performance in HIV-1 immunocompromised patients: A preliminary report. *J Clin Exp Neuropsychol* **11**:763–773, 1989.

142. Butters N, Grant I, Haxby J, *et al*: Assessment of AIDS-related cognitive changes: Recommendations of the NIMH Workshop on Neuropsychological Assessment Approaches. *J Clin Exp Neuropsychol* **12**:963–978, 1990.

143. Price RW, Sidtis JJ: Evaluation of the AIDS dementia complex in clinical trials. *J AIDS* **3**(Suppl 2):S51–S60, 1990.

144. Karlsen NR, Reinvang I, Froland SS: Slowed reaction time in asymptomatic HIV-positive patients. *Acta Neurol Scand* **86**:242–246, 1992.

145. Kovner R, Lazar JW, Lesser M, *et al*: Use of the Dementia Rating Scale as a test for neuropsychological dysfunction in HIV-positive IV drug abusers. *J Subst Abuse Treat* **9**:133–137, 1992.

146. Tinuper P, de Carolis P, Galeotti M, *et al*: Electroencephalogram and HIV infection: A prospective study in 100 patients. *Clin Electroencephalogr* **21**:145–150, 1990.

147. Itil TM, Ferracuti S, Freedman AM, *et al*: Computer-analyzed EEG (CEEG) and dynamic brain mapping in AIDS and HIV related syndrome: A pilot study. *Clin Electroencephalogr* **21**:140–144, 1990.

148. Elovaara I, Saar P, Valle SL, *et al*: EEG in early HIV-1 infection is characterized by anterior dysrhythmicity of low maximal amplitude. *Clin Electroencephalogr* **22**:131–140, 1991.

149. Schmitt B, Seeger J, Jacobi G: EEG and evoked potentials in HIV-infected children. *Clin Electroencephalogr* **23**:111–117, 1992.

150. Taylor DN, Wallace JG, Masdeu JC: Perception of different frequencies of cranial transcutaneous electrical nerve stimulation in normal and HIV-positive individuals. *Percept Mot Skills* **74**:259–264, 1992.

151. Balakrishnan J, Becker PS, Kumar AJ, *et al*: Acquired immunodeficiency syndrome: Correlation of radiologic and pathologic findings in the brain. *Radiographics* **10**:201–215, 1990.

152. Elovaara I, Poutiainen E, Raininko R, *et al*: Mild brain atrophy in early HIV infection: The lack of association with cognitive deficits and HIV-specific intrathecal immune response. *J Neurol Sci* **99**:121–136, 1990.

153. Post MJ, Berger JR, Quencer RM: Asymptomatic and neurologically symptomatic HIV-seropositive individuals: Prospective evaluation with cranial MR imaging. *Radiology* **178**:131–139, 1991.

154. Flowers CH, Mafee MF, Crowell R, *et al*: Encephalopathy in AIDS patients: Evaluation with MR imaging. *Am J Neuroradiol* **11**:1235–1245, 1990.

155. Sonnerborg A, Saaf J, Alexius B, *et al*: Quantitative detection of brain aberrations in human immunodeficiency virus type 1-infected individuals by magnetic resonance imaging. *J Infect Dis* **162**:1245–1251, 1990.

156. Menon DK, Baudouin CJ, Tomlinson D, *et al*: Proton MR spectroscopy and imaging of the brain in AIDS: Evidence of neuronal loss in regions that appear normal with imaging. *J Comput Assist Tomogr* **14**:882–885, 1990.

157. Menon DK, Ainsworth JG, Cox IJ, *et al*: Proton MR spectroscopy of the brain in AIDS dementia complex. *J Comput Assist Tomogr* **16**:538–542, 1992.

158. Maini CL, Pigorini F, Pau FM, *et al*: Cortical cerebral blood flow in HIV-1-related dementia complex. *Nucl Med Commun* **11**:639–648, 1990.

159. Bottomley PA, Hardy CJ, Cousins, JP, *et al*: AIDS dementia complex: Brain high-energy phosphate metabolite deficits. *Radiology* **176**:407–411, 1990.

160. Masdeu JC, Yudd A, Van Heertum RL, *et al*: Single-photon emission computed tomography in human immunodeficiency virus encephalopathy: A preliminary report. *J Nucl Med* **32**:1471–1475, 1991.

161. Ajmani A, Habte-Gabr E, Zarr M, *et al*: Cerebral blood flow SPECT with Tc-99m exametzine correlates in AIDS dementia complex stages. A preliminary report. *Clin Nucl Med* **16**:656–659, 1991.

162. Kuni CC, Rhame FS, Meier MJ, *et al*: Quantitative I-123-IMP brain SPECT and neuropsychological testing in AIDS dementia. *Clin Nucl Med* **16**:174–177, 1991.

163. Brunetti A, Berg G, DiChiro G, *et al*: Reversal of brain metabolic abnormalities following treatment of AIDS dementia complex with 3′-azido-2′,3′-dideoxythymidine (AZT, zidovudine): A PET-FDG study. *J Nucl Med* **30**:581–590, 1989.

164. Larsson M, Hagberg L, Forsman A, *et al*: Cerebrospinal fluid catecholamine metabolites in HIV-infected patients. *J Neurosci Res* **28**:406–409, 1991.

165. Liuzzi GM, Mastroianni CM, Vullo V, *et al*: Cerebrospinal fluid myelin basic protein as predictive marker of demyelination in AIDS dementia complex. *J Neuroimmunol* **36**:251–254, 1992.

166. Heyes MP, Brew BJ, Saito K, *et al*: Inter-relationships between quinolinic acid, neuroactive kynurenines, neopterin and beta 2-microglobulin in cerebrospinal fluid and serum of HIV-1-infected patients. *J Neuroimmunol* **40**:71–80, 1992.

167. Bogner JR, Junge-Hulsing B, Kronawitter U, *et al*: Expansion of neopterin and beta 2-microglobulin in cerebrospinal fluid reaches maximum levels early and late in the course of human immunodeficiency virus infection. *Clin Invest* **70**:665–669, 1992.

168. Berger JR, Tornatore C, Major EO, *et al*: Relapsing and remitting human immunodeficiency virus-associated leukoencephalomyelopathy. *Ann Neurol* **31**:34–38, 1992.

169. Chiodi F, Valentin A, Keys B, *et al*: Biological characterization from blood and cerebrospinal fluid. *Virology* **173**:178–187, 1989.

170. O'Brien WA, Koyanagi Y, Namazie A, *et al*: HIV-1 tropism for mononuclear phagocytes can be determined by regions of gp120 outside the CD4-binding domain. *Nature* **348**:69–73, 1990.

171. Li Y, Kappes JC, Conway JA, *et al*: Molecular characterization of human immunodeficiency virus type 1 cloned directly from uncultured human brain tissue: Identification of replication-competent and -defective viral genomes. *J Virol* **65**:3973–3985, 1991.

172. Pang S, Vinters HV, Akashi T, *et al*: HIV-1 *env* sequence variation in brain tissue of patients with AIDS-related neurologic disease. *J AIDS* **4**:1082–1092, 1991.

173. Portegies P, deGans J, Lange JM, *et al*: Declining incidence of AIDS dementia complex after introduction of zidovudine treatment. *Br Med J* **299**:819–821, 1989.

174. Vago L, Castagna A, Lazzanin A, *et al*: Reduced frequency of HIV-induced brain lesions in AIDS patients treated with zidovudine. *J AIDS* **6**:42–45, 1993.

175. Yarchoan R, Berg G, Brouwer P, *et al*: Response of human immunodeficiency virus associated neurological diseases to 3′-azido-3′-deoxythymidine. *Lancet* **1**:132–135, 1987.

176. Schmitt FA, Bigley JW, McKinnis R, *et al*: Neuropsychological outcome of zidovudine (AZT) treatment of patients with AIDS and AIDS-related complex. *N Engl J Med* **319**:1573–1578, 1988.

177. deGans J, Lange JMA, Derix MMA, et al: Decline of HIV antigen levels in cerebrospinal fluid during treatment with low-dose zidovudine. AIDS 2:37–40, 1988.

178. Brew BJ: Acute encephalopathy caused by human immunodeficiency virus apparently responsive to zidovudine. Med J Aust 151:725–726, 1989.

179. Yarchoan R, Pulda JM, Thomas RV, et al: Long-term toxicity/activity profile of 2′,3′-dideoxyinosine in AIDS or AIDS-related complex. Lancet 336:526–529, 1990.

180. Brouwers P, Moss H, Wolters P, et al: Effect of continuous-infusion zidovudine therapy on neuropsychologic functioning in children with symptomatic human immunodeficiency virus infection. J Pediatr 117:980–985, 1990.

181. Hollweg M, Riedel RR, Goebel FD, et al: Remarkable improvement of neuropsychiatric symptoms in HIV-infected patients after AZT therapy. Klin Wochenschr 69:409–412, 1991.

182. Sidtis JJ, Gatsonis C, Price RW, et al: Zidovudine treatment of the AIDS dementia complex: Results of a placebo controlled trial. Ann Neurol 33:343–349, 1992.

183. Lipkin WI, Parry G, Kiprov D, et al: Inflammatory neuropathy in homosexual men with lymphadenopathy. Neurology 35:1479–1483, 1985.

184. Cornblath DR, McArthur JC, Kennedy PG, et al: Inflammatory demyelinating peripheral neuropathies associated with human T-cell lymphotropic virus type III infection. Ann Neurol 21:32–40, 1987.

185. Simpson DM, Tagliati M: Nucleoside analogue-associated peripheral neuropathy in HIV infection. J AIDS 9:153–161, 1995.

186. Kieburtz K, Simpson DM, Yiannoutsos C, ACTG 242 Study Team: A randomized trial of amitriptiline and mexiletine for painful neuropathy in HIV. Neurology 51:1682–1688, 1998.

187. Petito CK, Navia BA, Cho ES, et al: Vacuolar myelopathy pathologically resembling subacute combined degeneration in patients with the acquired immunodeficiency syndrome. N Engl J Med 312:874–879, 1985.

188. DeLaMonte SM, Ho DD, Schooley RT, et al: Subacute encephalomyelitis of AIDS and its relation to HTLV-III infection. Neurology 37:562–569, 1987.

189. Kotler DP, Wang I, Pierson RN: Body composition studies in patients with the acquired immunodeficiency syndrome. Am J Clin Nutr 42:1255–1265, 1985.

190. Kotler DP, Tierney AR, Wang J, et al: Magnitude of body-cell-mass depletion and the timing of death from wasting in AIDS. Am J Clin Nutr 50:444–447, 1989.

191. Dobs A, Few W, Blackman M, Harman S, Hoover D, Graham N: Serum hormones in men with human immunodeficiency virus-associated wasting. J Clin Endocrinol Metab 81:4108–4112, 1996.

192. Coodley GO, Loveless MO, Nelson HD, Coodley MK: Endocrine function in the HIV wasting syndrome. J AIDS 7:46–51, 1994.

193. Grinspoon S, Corcoran C, Lee K, et al: Loss of lean body and muscle mass correlates with androgen levels in hypogonadal men with acquired immunodeficiency syndrome and wasting. J Clin Endocrinol Metab 81:4051–4058, 1996.

194. Grunfeld C, Pang M, Shimizu L, Shigenaga JK, Jensen P, Feingold KR: Resting energy expenditure, caloric intake and short-term weight change in human immunodeficiency virus infection and the acquired immunodeficiency syndrome. Am J Clin Nutr 55:455–460, 1992.

195. Macallan DC, Noble C, Baldwin C, Foskett M, McManus T, Griffin GE: Prospective analysis of patterns of weight change in stage IV HIV infection. Am J Clin Nutr 58:417–424, 1993.

196. Oster MH, Enders SR, Samuels SJ, Cone LA, Hooton TM, Browder HP, Flynn NM: Megestrol acetate in patients with AIDS and cachexia. Ann Intern Med 121:400–408, 1994.

197. Von Roenn JH, Armstrong D, Kotler DP, et al: Megestrol acetate in patients with AIDS-related cachexia. Ann Intern Med 121:393–399, 1994.

198. Gold J, High HA, Li Y, et al: Safety and efficacy of nandrolone decanoate for treatment of wasting in patients with HIV infection. AIDS 10:745–752, 1996.

199. Griffin GE, Paton NI, Cofrancesco J Jr, et al: Nutrition and quality of life in HIV infection. The role of growth hormone in HIV-associated wasting. J Clin Res 1:199–218, 1998.

200. Grunfeld C, Feingold KR: Metabolic disturbances and wasting in the acquired immune deficiency syndrome. N Engl J Med 327:329–337, 1992.

201. Grunfeld C, Kotler DP, Hamadeh R, et al: Hypertriglyceridemia in the acquired immunodeficiency syndrome. Am J Med 86:27–31, 1991.

202. Grunfeld C, Kotler DP, Shigenaga JK, et al: Circulating interferon-γ levels and hypertriglyceridemia in the acquired immunodeficiency syndrome. Am J Med 90:154–162, 1991.

203. Grunfeld C, Pang M, Doerrler W, et al: Lipids, lipoproteins, triglyceride clearance and cytokines in human immunodeficiency virus infection and the acquired immunodeficiency syndrome. J Clin Endocrinol Metab 74:1045–1052, 1992.

204. Hellerstein MK, Grunfeld C, Wu K, et al: Increased de novo hepatic lipogenesis in human immunodeficiency virus-infected humans. J Clin Endocrinol Metab 76:559–565, 1993.

205. Lahdevirta J, Maury CPJ, Teppo A-M, et al: Elevated levels of circulating cachectin/tumor necrosis factor in patients with acquired immunodeficiency syndrome. Am J Med 85:289–291, 1988.

206. Reddy MM, Sorrell SJ, Lange M, et al: Tumor necrosis factor and HIV p24 antigen levels in serum of HIV-infected populations. J AIDS 1:436–440, 1988.

207. Dezube BJ, Pardee AB, Beckett LA, et al: Cytokine dysregulation in AIDS: The in vivo overexpression of mRNA of tumor necrosis factor-α and its correlation with that of the inflammatory cytokine GRO. J AIDS 5:1099–1104, 1992.

208. Buimovici-Klein E, Lange M, Klein RJ, et al: Long-term follow-up of serum-interferon and its acid-stability in a group of homosexual men. AIDS Res 2:99–108, 1986.

209. Mildvan D, Machado SG, Wilets I, et al: Endogenous interferon and triglyceride concentrations to assess response to zidovudine in AIDS and advanced AIDS-related complex. Lancet 339:453–456, 1992.

210. Grunfeld C, Adi S, Soued M, et al: Search for mediators of the lipogenic effects of tumor necrosis factor: Potential role for interleukin 6. Cancer Res 50:4233–4238, 1990.

211. Breen EC, Rezai AR, Nakajima K, et al: Infection with HIV is associated with elevated IL-6 levels and production. J Immunol 144:480–484, 1990.

212. Birx DL, Redfield RR, Tencer K, et al: Induction of interleukin 6 during human immunodeficiency virus infection. Blood 76:2303–2310, 1990.

213. Honda M, Yamamoto S, Cheng M, et al: Human soluble Il-6 receptor: Its detection and enhanced release by HIV infection. J Immunol 148:2175–2180, 1992.

214. deWit R, Raasveld MH, tenBerge RJ, et al: Interleukin-6 concentrations in the serum of patients with AIDS-associated Kaposi's sarcoma during treatment with interferon-alpha. J Intern Med 229:539–542, 1991.

215. Hommes MJ, Romijn JA, Godfried MH, *et al*: Increased resting energy expenditure in human immunodeficiency virus-infected men. *Metabolism* **39:**1186–1190, 1990.

216. Grunfeld C, Pang M, Shimizu L, *et al*: Resting energy expenditure, caloric intake, and short-term weight change in human immunodeficiency virus infection and the acquired immunodeficiency syndrome. *Am J Clin Nutr* **55:**455–460, 1992.

217. Melchior JC, Salmon D, Rigaud D, *et al*: Resting energy expenditure is increased in stable, malnourished HIV-infected patients. *Am J Clin Nutr* **53:**437–441, 1992.

218. Hommes MJT, Romijn JA, Endert E, *et al*: Resting energy expenditure and substrate oxidation in human immunodeficiency virus (HIV)-infected asymptomatic men: HIV affects host metabolism in the early asymptomatic stage. *Am J Clin Nutr* **54:**311–315, 1991.

219. Melchiro JC, Raguin G, Rigaud D, *et al*: Acute systemic infection predicted by unusual increase in energy expenditure in HIV infected patients. In *Abstracts of the Seventh International Conference on AIDS*, Istituto Superiore di Sanita, Rome, 1991, Vol. 1, p. 293.

220. Stein TP, Nutinsky C, Condoluci D, *et al*: Protein and energy substrate metabolism in AIDS patients. *Metabolism* **39:**876–881, 1990.

221. Palella FJ, Delaney KM, Moorman AC, Loveless MO, Fuhrer J, Satten GA, Aschman DJ, Holmberg SD: Declining morbidity and mortality among patients with advanced human immunodeficiency virus infection. HIV outpatient study investigators. *N Engl J Med* **338:**853–860, 1998.

222. Hogg RS, Heath KV, Yip B, *et al*: Improved survival among HIV-infected individuals following initiation of antiretroviral therapy. *JAMA* **279:**450–454, 1998.

223. Carr A, Samaras K, Burton S, *et al*: A syndrome of peripheral lipodystrophy, hyperlipidemia, and insulin resistance in patients receiving HIV protease inhibitors. *AIDS* **12:**F51–F58, 1998.

224. Eastone JA, Decker CF: New-onset diabetes mellitus associated with use of protease inhibitor. *Ann Intern Med* **127:**948,1997.

225. Hengel RL, Watts NB, Lennox JL, *et al*: Benign symmetric lipomatosis associated with protease inhibitors. *Lancet* **350:**1596, 1997.

226. Dube MP, Johnson DI, Currier JS, Leedom JM: Protease inhibitor-associated hyperglycaemia. *Lancet* **350:**713–714, 1997.

227. Lo JC, Mulligan K, Tai VW, Algren H, Schambelan M: Buffalo hump in men with HIV-1 infection. *Lancet* **351:**867–870, 1998.

228. Lumpkin MM: Reports of diabetes and hyperglycemia in patients receiving protease inhibitors for the treatment of human immunodeficiency virus. FDA Public Health Advisory, Washington, DC, June 1997.

229. Sullivan AK, Nelson MR: Marked hyperlipidaemia on ritonavir therapy. *AIDS* **11:**938–939, 1997.

230. Centers for Disease Control and Prevention: Report of the NIH panel to define principles of therapy of HIV infection and guidelines for the use of antiretroviral agents in HIV-infected adults and adolescents. *MMWR* **47**(No. RR-5):1–41, 1998.

231. Carpenter CCJ, Fischl MA, Hammer SM, *et al*: Antiretroviral therapy for HIV infection in 1998: Updated recommendations of the International AIDS Society—USA panel. *JAMA* **280:**78–86, 1998.

232. Carpenter CCJ, Cooper DA, Fischl MA, *et al*: Antiretroviral therapy for HIV Infection in 1999: Updated recommendations of the International AIDS Society—USA panel. *JAMA* **283:**2936–2937, 2000.

233. Autran B, Carcelain G, Li TS, *et al*: Positive effects of combined antiretroviral therapy on CD4+ T cell homeostasis and function in advanced HIV disease. *Science* **277:**112–116, 1997.

234. Lederman MM, Connick E, Landay A, *et al*: Immunologic responses associated with 12 weeks of combination antiretroviral therapy consisting of zidovudine, lamivudine, and ritonavir: Results of AIDS Clinical Trials Group Protocol 315. *J Infect Dis* **178:**70–79, 1998.

235. Hammer SM, Squires KE, Hughes MD, *et al*: A controlled trial of two nucleoside analogues plus indinavir in persons with human immunodeficiency virus infection and CD4 cell counts of 200 per cubic millimeter or less. *N Engl J Med* **337:**725–733, 1997.

236. Mellors JW, Rinaldo CR, Gupta P, White RM, Todd JA, Kingsley LA: Prognosis of HIV-1 infection predicted by quantity of virus in plasma. *Science* **272:**1167–1170, 1996.

237. Mellors JW, Munoz A, Giorgi JV, *et al*: Plasma viral load and CD4+ lymphocytes as prognostic markers of HIV-1 infection. *Ann Intern Med* **126:**946–954, 1997.

238. Gunthard HF, Frost SDW, Leigh Brown AJ, *et al*: Evolution of envelope sequences of HIV-1 in cellular reservoirs in the setting of potent antiviral therapy. *J Virol* **73:**9401–9412, 1999.

239. Zhang L, Ramratnam B, Tenner-Racz K, *et al*: Quantifying residual HIV-1 replication in patients receiving combination antiretroviral therapy. *N Engl J Med* **340:**1605–1613, 1999.

240. Ledergerber B, Egger M, Opravil M, *et al*: Clinical progression and virological failure on highly active antiretroviral therapy in HIV-1 patients: A prospective cohort study. *Lancet* **353:**863–868, 1999.

241. Deeks S, Barbour J, Swanson M, *et al*: Sustained CD4 T cell response after virologic failure of protease inhibitor based regimens: Correlation between CD4 and viral load response after two years of therapy (abstract 494). In *6th Conference on Retroviruses and Opportunistic Infections*, Chicago, IL, 1999.

242. Kaufmann D, Pantaleo G, Sudre P, *et al*: CD4+ cell count in HIV-1 infected individuals remaining viraemic with highly active antiretroviral therapy (HAART), Swiss HIV Cohort Study. *Lancet* **351:**723–724, 1998.

243. Havlir D, Hellmann N, Petropoulos C, *et al*: Viral rebound in the presence of indinavir without protease inhibitor resistance (abstract LB12). In *6th Conference on Retroviruses and Opportunistic Infections*, Chicago, IL, 1999.

244. Holder DJ, Condra J, Schlief WA, *et al*: Virologic failure during combination therapy with Crixivan and RT inhibitors is often associated with expression of resistance-associated mutations in RT only (abstract 492). In *6th Conference on Retroviruses and Opportunistic Infections*, Chicago, IL, 1999.

245. Descamps D, Peytavin G, Calvez V, *et al*: Virologic failure, resistance and plasma drug measurements in induction maintenance therapy trial (abstract 493). In *6th Conference on Retroviruses and Opportunistic Infections*, Chicago, IL, 1999.

246. Patick A, Zhang M, Hertogs K, *et al*: Correlation of virological response with genotype and phenotype of plasma HIV-1 variants in patients treated with nelfinavir in the US expanded access program (abstract 178). In *2nd International Workshop on HIV Drug Resistance and Treatment Strategies*, Chicago, IL, 1999.

247. Miller MD, Wulfsohn MS, Margot NA, *et al*: Retrospective analysis of baseline susceptibility to adefovir dipivoxil is predictive of virological response in GS-96-408 (abstract 106). *Antiviral Ther* **4**(Suppl 1):70, 1999.

248. Lanier R, Danehower S, Daluge S: Genotypic and phenotypic correlates of response to abacavir (ABC, 1592) (abstract 52). *Antiviral Ther* **3**(Suppl 1):36, 1998.

249. Deeks SG, Hellman NS, Grant RM, *et al*: Novel four-drug salvage treatment regimens after failure of a human immunodeficiency virus type 1 protease inhibitor-containing regimen: Antiviral activity and correlation of baseline phenotypic drug susceptibility with virologic outcome. *J Infect Dis* **179:**1375-1381, 1999.

250. Durant J, Clevenbergh P, Halfon P, *et al*: Drug-resistance genotyping in HIV-1 therapy: The VIRADAPT randomised controlled trial. *Lancet* **353:**2195–2199, 1999.

251. Baxter JD, Mayers DL, Wentworth DN, *et al*: Final results of CPCRA 046: Pilot study of the short-term effects of antiretroviral management based on plasma genotypic antiretroviral resistance testing (GART) in patients failing antiretroviral therapy. *Antiviral Ther* **4**(Suppl 1)**:**43, 1999.

252. Torriani FJ, Freeman WR, Karavellas O, *et al*: Lymphoproliferative responses to CMV predict CMV retinitis reactivation in patients who discontinued CMV maintenance therapy (abstract 250). *6th Conference on Retroviruses and Opportunistic Infections*, Chicago, IL, 1999.

253. Komanduri KV, Viswanathan MN, Wieder ED, *et al*: Restoration of cytomegalovirus specific CD4+ T-lymphocyte responses after ganciclovir and highly active antiretroviral therapy in individuals infected with HIV-1. *Nature Med* **4:**953–956, 1998.

254. Havlir DV, Schrier R, Torriani F, *et al*: Reconstitution of *M. avium* complex (MAC) immune responses after highly active antiretroviral therapy (HAART) (abstract 248). In *6th Conference on Retroviruses and Opportunistic Infections*, Chicago, IL, 1999.

255. USPHS/IDSA Prevention of Opportunistic Infections Working Group: 1999 USPHS/IDSA guidelines for the prevention of opportunistic infections in persons infected with human immunodeficiency virus. *Ann Intern Med* **131:**873–901, 1999.

256. Schneider MME, Borleffs JC, Stoke RP, *et al*: Discontinuation of prophylaxis for *Pneumocystis carinii* pneumonia in HIV-1-infected patients treated with highly active antiretroviral therapy. *Lancet* **353:**201–203, 1999.

257. Lopez JC, Pena JM, Miro JM, *et al*: Discontinuation of PCP prophylaxis (PRO) is safe in HIV-infected patients (PTS) with immunological recovery with HAART. Preliminary results of an open, randomized and multicentric clinical trial (GESIDA04/98) (abstract 7). In *6th Conference on Retroviruses and Opportunistic Infections*, Chicago, IL, 1999.

258. Furrer H, Egger M, Opravil M, *et al*: Discontinuation of primary prophylaxis against *Pneumocystis carinii* pneumonia in HIV-1-infected adults treated with combination antiretroviral therapy. *N Engl J Med* **340:**1301–1306, 1999.

259. Currier JS, Williams PS, Koletar S, *et al*: A randomized, placebo-controlled trial of azithromycin prophylaxis for the prevention of *Mycobacterium avium* complex (MAC) in subjects with increases in CD4 cells on antiretroviral therapy (abstract LB-23). In *Program and Abstracts of the 39th Interscience Conference on Antimicrobial Agents and Chemotherapy*, San Diego, CA, 1999.

260. Aberg JA, Yajko DM, Jacobson MA, *et al*: Eradication of AIDS-related disseminated *Mycobacterium avium* complex infection after 12 months of antimycobacterial therapy combined with highly active antiretroviral therapy. *J Infect Dis* **178:**1446–1449, 1998.

261. Macdonald JC, Torriani FM, Morse LS, *et al*: Lack of reactivation of cytomegalovirus retinitis after stopping CMV maintenance therapy in AIDS patients with sustained elevations in CD4 T cells in response to highly active antiretroviral therapy. *J Infect Dis* **177:**1182–1187, 1998.

262. Tural C, **2 more authors**, *et al*: Long-lasting remission of cytomegalovirus retinitis without maintenance therapy in human immunodeficiency virus-infected patients. *J Infect Dis* **177:**1080–1083, 1998.

263. Vrabec TR, Baldassano VF, Whitcup SM: Discontinuation of maintenance therapy in patients with quiescent cytomegalovirus retinitis and elevated CD4+ counts. *Ophthalmology* **105:**1259–1264, 1998.

264. Pursley TJ, Blomquist IK, Abraham J, Andersen HF, Bartley JA: Fluconazole-induced congenital anomalies in three infants. *Clin Infect Dis* **22:**336–340, 1996.

265. Centers for Disease Control and Prevention: Recommendations for prevention and control of hepatitis C virus (HCV) infection and HCV-related chronic disease. *MMWR* **47**(No. RR-19)**:**1–392, 1998.

13

Infectious Complications in Children with Cancer and Children with Human Immunodeficiency Virus Infection

SARAH W. ALEXANDER, BRIGITTA U. MUELLER, and PHILIP A. PIZZO

1. Introduction

Many of the infectious complications encountered in adult immune-compromised patients are also observed in children. However, there are some unique differences that make the management of an infection in a child with cancer, human immunodeficiency virus (HIV) infection, or compromised immune system a challenge. This chapter will focus on the problems that are specific for patients in the pediatric age range.

Infection is a very common event during normal childhood, the average child experiencing three to ten febrile illnesses per year during the first few years of life. The spectrum of infections is age-dependent and the evaluation of a child with a suspected infection must take the age of the child into account. In fact, the infectious process may have started *in utero* and have resulted in organ damage by the time of birth, as evidenced by some children with HIV infection or other congenital infections.

The premature or very low birth weight baby has a compromised immune system due to immaturity and several studies have addressed the use of bone marrow stimulation with growth factors to decrease the incidence of infectious complications.[1,2] The newborn and infant are protected to a degree through the passive transfer of maternal antibodies against certain infections. The young child often experiences primary infections with the opportunistic pathogens, such as varicella zoster virus (chickenpox) or *Pneumocystis carinii*, in contrast to the reactivation infections that occur in adults. Otherwise well-tolerated infections, such as bronchitis or otitis, to which children are frequently exposed, can result in serious morbidity in the immunocompromised child.

It may be necessary to modify diagnostic techniques used in adults for infants and children. For example, the volume of blood that can be obtained for a blood culture is necessarily less in a neonate or child. Even the identification of the source of infection can be a greater challenge in a child than in an adult. The younger the child, the fewer "typical" clinical symptoms are recognizable. For example, an infant with meningitis may present with mere "fussiness" or have a decreased appetite and the typical neck stiffness seen in older children and adults usually is not present.

It is not only necessary to adjust the dosage of therapeutic agents on the basis of the growing child's weight or body surface area, but also to take into account the differences in pharmacokinetics and pharmacodynamics between children and adults. The hepatic glucuronidation

Sarah W. Alexander • Division of Hematology and Oncology, Rainbow Babies & Children's Hospital, Cleveland, Ohio 44106. **Brigitta U. Mueller and Philip A. Pizzo** • Department of Medicine, Children's Hospital, Boston, Massachusetts, and Department of Pediatrics, Harvard Medical School, Boston, Massachusetts 02115.

Clinical Approach to Infection in the Compromised Host (Fourth Edition), edited by Robert H. Rubin and Lowell S. Young. Kluwer Academic/Plenum Publishers, New York, 2002.

process, for example, is relatively immature during the first 2–3 months of life, thus decreasing the metabolic clearance of many drugs. Renal excretion reaches adult levels between 6 and 12 months of life, due to slow maturation of glomerular filtration and tubular function, as well as increase in renal blood flow.[3]

Another complicating factor is the inability of the young child to swallow tablets or capsules and the all-too-frequent lack of specific pediatric formulations (or pharmacokinetic/toxicity information in children) of many commonly used drugs. One of the important contributions of the pediatric acquired immunodeficiency syndrome (AIDS) community has been the support of efforts to enhance public awareness for the need for pharmaceutical trials in children.[4,5] The U.S. Food and Drug Administration (FDA) now permits the approval of drugs for use in children based on efficacy data gathered in adults if the disease in children and adults is similar, provided that the pharmaceutical companies submit dosing (pharmacokinetic) and safety (toxicity) data from trials performed in an adequate number of children. Indeed, if the drug is likely to be used in children, it is now mandatory for pharmaceutical companies to provide such data in a timely fashion (even retrospectively for already approved drugs).

A problem commonly encountered in adolescents is the lack of compliance. Adolescents typically believe themselves to be invulnerable and not infrequently have difficulties adhering to a long-term treatment plan, such as a prophylactic regimen against opportunistic infections.

On the other hand, children have in general fewer comorbid conditions than their adult counterparts. Sixty percent of all cancers in adults occur in patients greater than 65 years old, many of whom are likely to have preexisting conditions at the time of their cancer diagnosis which pose additional clinical challenges.[6,7] For some clinical events that are analogous in the pediatric and adult populations, such as febrile neutropenia in oncology patients, children consistently have lower reported incidence of serious morbidity and mortality.

As in adults, the degree of immunosuppression (i.e., loss or diminution of one or more arms of the immune system) directly influences the incidence and severity of infection. The persistent vulnerability resulting from progressive immunosuppression in HIV infection or the ablation of host immunity in bone marrow transplantation leads to a higher incidence of infection and increased morbidity and mortality. However, even a minor deficiency of the host's immune system (e.g., chronic corticosteroid therapy) places the child at an increased risk

for complications of an infection such as varicella zoster (e.g., disseminated disease or pneumonitis).

2. Background

2.1. Malignancies in Childhood

Cancer in childhood is a relatively rare event.[8] It is estimated that 8800 new cases of cancer are diagnosed each year in the United States in children under the age of 15 years.[9] This is in comparison to approximately 1,000,000 new cases per year in the adult population. Despite its rarity, cancer is the leading cause of disease-specific deaths for 3- to 14-year-old children and accounts for approximately 10% of deaths during childhood.[10] The types of cancer are markedly different for children as compared to adults. Leukemia and brain tumors account for almost 50% of pediatric tumors,[8,11] whereas lung, prostate, breast, and colon cancer are the leading causes of adult cancer.[12] There are distinct patterns of disease based on age at diagnosis (Table 1). The majority of the tumors occurring in the first 2 years of life are embryonal, including neuroblastoma, Wilms' tumor, retinoblastoma, primitive neuroectodermal tumors, and hepatoblastoma. Acute lymphoblastic leukemia (ALL) has a sharp incidence peak at age 2 to 4 years. Osteosarcoma, Ewing's sarcoma, Hodgkin's disease, and non-Hodgkin's lymphoma all have increasing incidence later in childhood.[11] The incidence of cancer also varies by race (for example, ALL and Ewing's sarcoma are significantly less common in African-American children than in Caucasians).

There has been a dramatic improvement in the success rates of pediatric cancer therapy over the past three decades, particularly for ALL, lymphomas, and soft-tissue sarcomas. More than 65% of all children diagnosed with cancer survive for more than 5 years. In general, pediatric malignancies are considered to be more responsive than those in adults. Despite this success there still are some pediatric tumor types with extremely poor prognoses, for example, brain stem gliomas and rhabdoid tumor.

The treatment modalities used in adults and children include chemotherapy, surgery, and radiation. Chemotherapy is used extensively in pediatric oncology, and for some malignancies, such as leukemias and lymphomas, it is the primary treatment modality. Chemotherapy also is used as a neoadjuvant modality, for example, in non-metastatic osteosarcoma, Ewing's sarcoma, and hepatoblastoma, to facilitate local control measures either with surgery or radiation. Chemotherapy is most often admin-

TABLE 1. Predominant Pediatric Cancers by Age and Site[a]

Tumors	Newborns (<1 year)	Infants (1–2 years)	Children (3–11 years)	Adolescents and young adults (12–21 years)
Leukemias	Congenital leukemia AML AMMoL CML (juvenile)	ALL AML CML (juvenile)	ALL AML	AML ALL
Lymphomas	Very rare	Lymphoblastic	Lymphoblastic	Lymphoblastic Large cell Burkitt's Hodgkin's disease
CNS	Medulloblastoma	Medulloblastoma	Astrocytoma Medulloblastoma Ependymoma	Astrocytoma Craniopharyngioma Medulloblastoma
Head and neck	Retinoblastoma Rhabdomyosarcoma Neuroblastoma PNET	Retinoblastoma Rhadomyosarcoma Neuroblastoma	Rhabdomyosarcoma Lymphoma	Lymphoma Rhabdomyosarcoma PNET
Thoracic	Neuroblastoma Teratoma	Neuroblastoma Teratoma	Lymphoma Neuroblastoma Rhabdomyosarcoma	Lymphoma Ewing's/PNET Rhabdomyosarcoma
Abdominal	Neuroblastoma Mesoblastic nephroma Hepatoblastoma Wilms' (>6 mo)	Neuroblastoma Wilms' Hepatoblastoma Leukemia Rhabdomyosarcoma	Neuroblastoma Wilms' Lymphoma Hepatoma Rhabdomyosarcoma	Lymphoma Hepatocellular carcinoma Rhabdomyosarcoma PNET
Gonadal	Yolk sac tumor Teratoma Neuroblastoma	Yolk sac tumor	Rhabdomyosarcoma	Rhabdomyosarcoma Dysgerminoma Teratocarcinoma
Extremity	Fibrosarcoma	Fibrosarcoma Rhabdomyosarcoma	Rhabdomyosarcoma Ewing's/PNET	Osteosarcoma Ewing's/PNET Rhabdomyosarcoma

[a]Adapted from Pizzo et al.[10]

[b]AML, acute myelogenous leukemia; AMMoL, acute myelomonocytic leukemia; CML, chronic myelogenous leukemia; PNET, primitive neuro-ectodermal tumor.

istered in combination regimens. Dose intensity—the amount of drug administered in a given period of time—is thought to be important for some pediatric tumors. The doses are calculated per kilogram body weight or per meter squared of body surface area and often are higher than those used in adults.[13] (An exception are neonates who have less well-developed renal and hepatic excretory mechanisms.) This may be one of the reasons why the cure rate of pediatric cancers is generally higher than what is achieved in adult oncology. The higher dose intensity is feasible due to the generally higher tolerance of therapy in children and the liberal use of hematopoietic stem cell factors.

Depending on the nature of the malignancy and the degree, duration, and intensity of therapy, children undergoing therapy for cancer will be immunosuppressed to a variable degree and duration, but generally will experi-ence intervals of immunologic recovery. Fever and infections are expected complications. Skills in the diagnosis and management of infectious complications are imperative for the oncologist or infectious disease specialist caring for these children, especially if the curative potential is to be reached.

2.2. HIV Infection in Childhood

The Joint United Nations Program on AIDS (UN-AIDS) recently published staggering numbers in regard to HIV infection: worldwide an estimated 1.2 million children are currently living with HIV/AIDS, and 590,000 were infected in 1998 alone (reference: UNAIDS website, http://www.unaids.org). Through June 1998, a total of 8280 cases of AIDS in children <13 years of age in the

United States have been reported to the Centers for Disease Control and Prevention (CDC).[14]

HIV infection in children was first recognized in children who had received a transfusion of contaminated blood or coagulation products, but this route of infection has been virtually eliminated in the United States.[14–16] Over 95% of HIV-infected children now acquire the infection vertically, i.e., from their mothers during gestation or birth.[14,17] Children who were infected *in utero* may already have a compromised immune system as neonates or very young infants, whereas children infected during delivery (probably about two thirds of cases) will develop a slower but progressive immune disturbance if they are not recognized and treated.

Without intervention, perinatal transmission rates of 25 to 30% and 13 to 20% have been observed in the United States and in Europe, respectively, compared to almost 40% in African or Asian countries.[18,19] In 1994, the results of the AIDS Clinical Trials Group protocol 076 (PACTG 076) employing zidovudine during pregnancy reported a 67% reduction in perinatal transmission in the zidovudine compared to the placebo group.[20,21] This resulted in new guidelines issued by the CDC.[22,23]

Measurement of viral RNA or DNA copy numbers as well as culture techniques have become the standard tools to make the diagnosis of HIV infection in an infant or child.[24,25] It is assumed that children who have a positive HIV polymerase chain reaction (PCR) result within the first 48 hr after birth were infected *in utero*, while those who are infected during the intrapartum period might not become positive until 2–6 weeks after birth. The diagnosis of HIV infection in an infant born to a seropositive mother by antibody test [enzyme-linked immunosorbent assay (ELISA) or Western blot] is not reliable because of the passive transfer of maternal antibodies. However, in children older than 18 months of age, serologic tests for specific antibodies are still important to establish the diagnosis of HIV infection, especially if PCR assays or culture methods are not available.

The American Academy of Pediatrics (AAP) recommends routine HIV education and testing with consent of *all* women seeking prenatal care so that every woman will

TABLE 2. CDC 1994 Revised Classification System for HIV Infection in Children Less than 13 years of Age[a]

A. Pediatric HIV Classification[b]

Immune categories	Clinical Categories			
	(N) No symptoms	(A) Mild symptoms	(B) Moderate symptoms	(C) Severe symptoms
(1) No suppression	N1	A1	B1	C1
(2) Moderate suppression	N2	A2	B2	C2
(3) Severe suppression	N3	A3	B3	C3

[a]Using this system children are classified according to three parameters: infection status, clinical status, and immunological status. The categories are mutually exclusive. Once classified in a more severe category, a child is **not** reclassified in a less severe category even if the clinical or immunologic status improves.
[b]Children whose HIV infection status is not confirmed are classified by using the above grid with a letter E (for vertically exposed) placed before the appropriate classification code, e.g., EN2.
[c]Both category C and lymphoid interstitial pneumonitis in category B are reportable to state and local health departments as AIDS.

B. Immunologic Categories[a,b]

Immunologic category	Age Groups		
	0–11 months	1–5 years	>6 years
(1) No suppression	>1500 cells/μl (>25%)	>1000 cells/μl (>25%)	>500 cells/μl (>25%)
(2) Moderate suppression	750–1499 cells/μl (15–24%)	500–999 cells/μl (15–24%)	200–499 cells/μl (15–24%)
(3) Severe suppression	<750 cells/μl (<15%)	<500 cells/μl (<15%)	<200 cells/μl (<15%)

[a]The immunologic category classification is based on age-specific CD4+ T-lymphocyte count or percent of total lymphocytes and is designed to determine severity of immunosuppression attributable to HIV for age. If either CD4 count or percent results in classification into a different category, the child should be classified into the more severe category. A value should be confirmed before reclassification of the child into a more severe category. Regardless of subsequent CD4 determinations, children should not be reclassified into a less severe category.
[b]From Centers for Disease Control and Prevention.[22]

(continued)

TABLE 2. (*Continued*)

C. Abbreviated Clinical Categories[a]

Category N: Not symptomatic
  Children who have no signs or symptoms considered to be the result of HIV infection
Category A: Mildly symptomatic ($\geq$2 symptoms required)
  Dermatitis or lymphadenopathy ($>$0.5 cm at more than two sites; bilateral = one site)
  Hepatomegaly or splenomegaly
  Parotitis or recurrent or persistent respiratory infection, sinusitis, or otitis media
Category B: Moderately symptomatic
  Anemia, neutropenia, or thrombocytopenia persisting 30 days
  Bacterial meningitis, pneumonia, or sepsis or diarrhea, recurrent or chronic
  Candidiasis, oropharyngeal thrush, persisting for $>$2 months in children 6 months of age
  Cardiomyopathy, nephropathy, or hepatitis
  Cytomegalovirus infection or recurrent herpes simplex virus stomatitis; bronchitis, pneumonitis, or esophagitis with onset before 1 month of age
  Herpes zoster (shingles) involving $\geq$2 distinct episodes or more than one dermatome
  Leiomyosarcoma
  Lymphoid interstitial pneumonia or pulmonary lymphoid hyperplasia complex
  Nocardiosis or toxoplasmosis, onset before 1 month of age
  Persistent fever (lasting $>$1 month)
  Varicella, disseminated (complicated chickenpox)
Category C: Severely symptomatic
  Any condition listed in the 1987 surveillance case definition for AIDS, with the exception of lymphoid interstitial pneumonia
  Serious bacterial infections, multiple or recurrent
  Candidiasis, esophageal or pulmonary (bronchi, trachea, lungs)
  Coccidiomycosis, disseminated
  Cryptosporidiosis or isosporidiosis with diarrhea persisting $>$1 month
  Cytomegalovirus disease with onset of symptoms at age $>$1 month
  Encephalopathy
  Herpes simplex virus infection causing a mucocutaneous ulcer persisting for 1 month; or bronchitis, pneumonitis, or esophagitis for any duration affecting a child 1 month of age
  Histoplasmosis, disseminated
  Kaposi's sarcoma; primary lymphoma of brain; or small, noncleaved cell, immunoblastic, or large cell lymphoma of B-cell or unknown immunologic phenotype
  Mycobacterium tuberculosis, disseminated or extrapulmonary **or** *Mycobacterium*, other species or unidentified species, disseminated
  Mycobacterium avium complex or *Mycobacterium kansasii*, disseminated
  Pneumocystis carinii pneumonia
  Progressive multifocal leukoencephalopathy
  Salmonella (nontyphoid) septicemia, recurrent
  Toxoplasmosis of the brain with onset 1 month of age
  Wasting syndrome **plus** (1) chronic diarrhea (i.e., at least two loose stools per day for $>$30 days) **or** (2) documented fever (for 30 days, intermittent or constant)

[a]For complete listing, see Centers for Disease Control and Prevention.[22]

know her HIV status and the methods available both to prevent the acquisition and transmission of HIV and to determine whether it is appropriate to breastfeed.[26] Guidelines regarding the monitoring of children born to HIV-infected women also have been published both by the AAP and the CDC.[23,26]

HIV infection in infants and children has a different presentation from adults, and the CDC classifies HIV infection in children under the age of 13 years based on clinical and immunologic parameters (Table 2). Common clinical features seen during the course of HIV infection are lymphadenopathy, fevers, malaise and loss of energy, hepatosplenomegaly, respiratory tract infections, as well as recurrent and chronic otitis and sinusitis. Other typical findings are failure to thrive, sometimes associated with chronic diarrhea, failure to grow, the presence and persistence of mucocutaneous candidiasis and many nonspecific cutaneous manifestations. The initial symptoms may be subtle and sometimes difficult to distinguish from manifestations caused by drug use during pregnancy, or from problems associated with prematurity, or congenital infections other than with HIV.

A major decrease in morbidity and mortality of HIV infection in children and adults has occurred during the last few years. However, infants who are not known to be HIV-infected or do not have access to early intervention are still at high risk for early and severe morbidity and continue to have a high mortality rate.[27,28] Perinatally acquired HIV infection follows a bimodal course, with about one third of the children becoming symptomatic within the first 2 years of life. In a European study of 392 HIV-infected children, 20% of children died or developed an AIDS-defining symptom (CDC category C, Table 2) within the first year of life and 4.7% per year thereafter, reaching a cumulative incidence of 36% by 6 years of age.[27] Two thirds of the children alive at 6 years of age had only minor symptoms and one third had well-preserved CD4 counts ($>25\%$), despite prior clinical manifestations. The early manifestations of clinical symptoms in an infant, especially opportunistic infections, encephalopathy, or hepatosplenomegaly, have been associated with a poor prognosis.[27] Infants with very high HIV RNA copy numbers shortly after birth are presumed to have been infected *in utero* and tend to have early onset of symptoms.[29]

With expanding knowledge of the dynamics of viral replication and its relationship to disease progression and prognosis, it has become clear that early and aggressive therapy offers the potential benefit of a prolonged asymptomatic period. The indications for the initiation and monitoring of antiretroviral therapy developed by panels of experts include clinical, immunologic, and virologic parameters.[23,25,30]

3. Interface between Cancer and Infections

There are important associations between neoplasms and infections. Cancers can occur in patients who are immunocompromised because of an infection (e.g., HIV infection) and infections are common in patients whose immune system is impaired because of cancer or its therapy. Furthermore, the two problems often have very similar presentations. For example, the signs and symptoms heralding the hematopoietic malignancies of childhood frequently include fever and malaise. Several studies have shown that as many as 60% of children with ALL present with fever of greater than $38°C$ at the time of diagnosis not infrequently attributed to a bacterial or viral infection. Similarly, non-Hodgkin's lymphoma is not infrequently misdiagnosed as infectious mononucleosis, while the lytic bone lesions associated with Ewing's sarcoma can be confused with osteomyelitis. Conversely, it should be noted that an infection is occasionally misdiagnosed as a malignancy. For example, histoplasmosis involving the bone marrow has been confused with lymphoma. Acute cryptococcal pneumonia can present in the child with cancer as one or more radiographic nodules and therefore can be misinterpreted as being tumor metastases, and a cerebral lesion in an HIV-infected patient can be due to either toxoplasmosis or lymphoma.

It is also possible that a malignancy and an infection have the same etiologic agent. For example, Epstein–Barr virus (EBV) is the etiologic agent responsible for infectious mononucleosis in normal hosts, but in some families it may result in a fatal X-linked lymphoproliferative syndrome. Primary infection or reactivation of EBV infection is the most common cause for the occurrence of lymphoproliferative disorders in immunocompromised patients.[31–34] Moreover, EBV is etiologically linked with the pathogenesis of African Burkitt's lymphoma, primary central nervous system lymphomas, as well as smooth muscle tumors (leiomyosarcomas) in HIV-infected or posttransplantation patients.[35–39] Interestingly, leiomyosarcomas that occur in nonimmunocompromised patients are not associated with EBV infection and adult patients with HIV infection only rarely develop leiomyosarcomas. A significant increase in the incidence of Kaposi's sarcoma (KS) in children from African countries recently has been published. Kaposi's sarcoma is a tumor closely associated with human herpes virus type 8 (HHV-8) infection.[40,41] In Zambia, KS now comprises almost 20% of all childhood cancers, compared to 6% prior to 1986.

4. Perturbations of Host Defense that Contribute to Infection

Cancer and its therapy cause a spectrum of perturbations to the normal immune system. The underlying disease can have varied and significant effects. Patients with leukemia or metastatic solid tumors can have bone marrow replacement causing prolonged neutropenia. Patients with solid tumors often are at risk for recurrent local infections from disrupted architecture. For example, children with genitourinary rhabdomyosarcoma often have infectious complications related to obstruction of the urinary tract caused by tumor. Malnutrition due to the underlying disease and its therapy can contribute to the degree of immunosuppression.

Therapy for the underlying malignancy broadens the spectrum of host defense perturbations, the most significant of which is chemotherapy or radiation-induced myelosuppression. Chemotherapy also can cause significant disruption of mucosal integrity providing a portal for entry for infectious agents. The majority of pediatric patients receiving chemotherapy will have a permanent central venous access device implanted, representing another site of potential infection. Therapy can render patients hypogammaglobulinemic, particularly after allogeneic bone marrow transplantation. Children with solid tumors requiring surgical therapy are at heightened risk for wound infections if they have received neoadjuvant chemotherapy. Patients, most commonly those with bone tumors undergoing limb-sparing procedures, often have foreign materials left in place that can provide an additional nidus for infection. Splenectomy, no longer part of the routine management of patients with Hodgkin's disease, is still sometimes required in the management of the pediatric abdominal tumors and leaves the patient at higher risk for infections with encapsulated organisms.

Flow cytometric analysis of lymphocyte subpopulations in healthy children has revealed age-related changes in the number of the different subgroups.[42–44] As in adults, infection with HIV results in profound deficiencies in cell-mediated and humoral immunity, secondary to both quantitative and qualitative defects, leading to a progressive dysfunction of the immune system with depletion of CD4-positive T cells. In the absence of early antiretroviral therapy, an abnormal CD4 percentage (less than the 10th percentile for uninfected children) was found in 83% and an abnormally low absolute CD4 count in 67% of HIV-infected children who were less than 2 years of age.[45] A similar depletion of CD4-positive cells has been described in children treated with chemotherapy for cancer, and these changes can persist even after therapy.[46] Not unexpectedly, an inverse correlation was observed between the patients' ages and the number of CD4 lymphocytes 6 months after completion of therapy.[47]

Other immune abnormalities commonly seen in HIV-infected children include decreased lymphocyte proliferation in response to an antigen, polyclonal B-cell activation resulting in hypergammaglobulinemia, and altered function of monocytes and neutrophils.[48–50] In contrast, it is common to see persistent hypogammaglobulinemia in patients undergoing chemotherapy or transplantation.[51,52] Since many children have not yet developed immunity to common infections (due to lack of exposure to natural infection or immunization) and since persistence of preexisting antibodies is not guaranteed, they are at risk for common childhood viral infections, such as varicella and measles. In order to decrease the risk of exposure, it is important to create herd immunity by vaccinating as many children (and sometimes adults) as possible in the community.

5. Fever

Fever is a common problem in children with a compromised immune system. In a prospective review of 1001 episodes of fever in pediatric and young adult patients with cancer being treated at the National Cancer Institute in the late 1970s and early 1980s, approximately one half of all patients became febrile, often requiring hospitalization and treatment for presumed or documented infections.[53] Eighty percent of these episodes occurred when the patients were granulocytopenic and 20% in patients who were not granulocytopenic. Fevers in the nonneutropenic population were most often ascribed to chemotherapy (especially methotrexate, cytarabine arabinoside, cyclophosphamide, and actinomycin D), the underlying malignancy, and viral infections.

Fever also is a common symptom of HIV infection and can be caused by either an infectious organism or a malignancy. In a recent study of 42 HIV-infected children under the age of 36 months, a mean of 1.8 visits for fever per child and year was noted.[54] Blood cultures were obtained in 89% of the visits and were positive in a third of cases. Rapid progression of HIV disease was associated with a fourfold higher incidence of febrile episodes compared to children with nonrapid progression ($P < 0.01$).[54]

5.1. Fever and Neutropenia in Children

It has been more than 30 years since the initial observation that neutropenia was a significant risk factor for infection in patients with cancer receiving myelotoxic therapy.[55] In a seminal study by Schimpff et al.,[56] it was shown that the empiric use of carbenicillin and gentamicin in febrile patients with cancer and neutropenia decreased mortality. Based on this and multiple subsequent studies, empiric antibiotic therapy has become the standard of care for both pediatric and adult patients with fever and neutropenia.

Illustrative Case 1

A 6-month-old infant with metastatic Wilm's tumor was admitted for fever. He had received adriamycin and vincristine 7 days earlier. At home he was noted to have mild rhinorrhea and was slightly fussy. An older sibling had recently been ill with a "cold." His parents took his temperature at home and found it to be 38.6°C. As they had been instructed to do for fevers greater than 38.0°C, they called the pediatric oncologist. They were asked to bring him in immediately for evaluation.

On arrival he was febrile and moderately tachypenic but with an otherwise unremarkable physical exam. He was found to have a total white blood cell count of 540/mm³ and an absolute neutrophil count of 50/mm³. Two sets of blood cultures were drawn and he was hospitalized and placed on broad-spectrum intravenous antibiotics. A chest radiograph was negative. Two days later he was afebrile and clinically well. His preantibiotic cultures were positive for *Klebsiella pneumoniae*.

Fever is most commonly defined as a single temperature greater than 38.5°C or persistent temperature greater than 38°C. Neutropenia is usually defined as an absolute neutrophil count less than 500 cells/mm³. Some studies also include patients with absolute neutrophil counts of less than 1000 cells/mm³ whose counts are expected to fall below 500 cells/mm³ within the first 24 hr after presentation with fever.

The initial assessment for pediatric patients with fever and neutropenia is the same as for adults and is guided by rather simple principles. A careful history and meticulous physical examination are imperative. Special attention should be paid to areas at increased risk for infection in patients receiving cytotoxic therapy, including the oropharynx, perianal region, central line site if present, and any site of recent invasive procedures. Blood cultures should be obtained both peripherally and from central lines. Other sites for culture are based on specific clinical suspicions. Radiologic studies should be ordered based on specific symptoms or physical findings. Baseline chemistries and an assessment of renal function are usually obtained to ensure appropriate organ function.

Initial antimicrobial coverage strategies have been studied extensively in both adult and pediatric populations. Empiric therapy is based on early antibiotic coverage for the most likely infecting organisms as well as those organisms that have the potential of being rapidly lethal if not appropriately treated. There is no single best regimen for all patients with fever and neutropenia, with the choice influenced by local patterns of infecting organisms and their resistance spectra as well as consideration of toxicities, ease of administration, and cost.

The organisms causing bacteremia in patients with fever and neutropenia have continued to change over the past 30 years from an initial predominance of gram-positive organisms in the 1950s to gram-negative organisms, including *Pseudomonas aeruginosa*, *Escherichia coli*, and *Klebsiella pneumoniae* in the 1960s to 1980s and now again back to gram-positive organisms, particularly coagulase-negative staphylococci, streptococci, *Staphylococcus aureus*, *Corynebacterium jeikeium*, and enterococci.[57–65]

The antibiotic regimens used in children in general do not differ from those used in adults. Either combination therapy, usually with a beta-lactam and an aminoglycoside, or single-agent therapy is employed. Ceftazidime was the first agent to undergo significant evaluation as monotherapy for empiric coverage of patients with fever and neutropenia. Use of this agent has been shown to be as safe as combination regimens.[66–68] Awareness of local patterns of infection is important if ceftazidime monotherapy is to be employed. In centers with high rates of *Citrobacter*, *Serratia*, and *Enterobacter*, ceftazidime monotherapy should be avoided because of the risk of emergence of resistant organisms by beta-lactamase induction.[69] Cefepime is a newer broad-spectrum cephalosporin with improved gram-positive coverage in comparison to ceftazidime. It has been shown to be safe and effective as monotherapy in small studies, but whether it is truly better than prior generation cephalosporins has not yet been shown.[70]

The carbapenem family of antibiotics also has been studied as monotherapy for fever and neutropenia. Both imipenem and meropenem are efficacious and safe in most patients.[71–76] However, imipenem does lower the seizure threshold, a consideration that is important in treating children with brain tumors.

Ciprofloxacin (a fluoroquinolone) monotherapy has been evaluated for use in the setting of fever and neutropenia with mixed results, primarily due to inadequate coverage for gram-positive organisms and therefore is not generally recommended as monotherapy.[77] Newer generation quinolones (e.g., trovafloxacin and levofloxacin) with broader gram-positive coverage may prove to be reasonable monotherapy for fever and neutropenia. Quinolones for the most part have been avoided in pediatric patients because of the additional concern of potential cartilage damage, although studies evaluating these agents are under way.

Given the evolution of the spectrum of infecting organisms in patients with fever and neutropenia the question of whether empiric coverage with vancomycin is warranted has been investigated in both adult and pediatric patients. There have been some studies that have shown that the addition of vancomycin increases the "success" of the empiric regimen.[78,79] Others have shown that vancomycin is not essential for initial therapy and that in contrast to infections with gram-negative organisms, initiating treatment of gram-positive bacteria at the time of their identification is a safe and reasonable alternative for most patients.[80–82] The reluctance to use vancomycin more broadly is due to the emergence of vancomycin-resistant organisms, most notably vancomycin-resistant enterococci. There are situations, however, in which empiric vancomycin is generally recommended.[83]

These include institutions where there are high rates of fulminant infections with gram-positive organisms (i.e., *Streptococcus viridans*) as well as for individuals with obvious serious central line infections and with hypotension at presentation.

5.2. Fever and Neutropenia: Differences between Adults and Children

Many of the large clinical trials evaluating different therapies for fever and neutropenia have been conducted in combined pediatric and adult populations. In general, patient age has not been used as one of the criteria when evaluating subgroup outcomes. The largest body of data addressing the question of whether there is a difference between pediatric and adult patients with fever and neutropenia has come from a retrospective analysis of several large trials.[84] This study of 3080 patients with fever and neutropenia compared the pediatric group, which comprised 25% of the study population, to the adult group in terms of infection types and outcomes. The pediatric patient population was different with proportionally more children with ALL than acute myelogenous leukemia (AML) compared to the adult population. There also were significantly more children with solid tumors undergoing intensive myelosuppressive chemotherapy. The rate of bacteremia in the two groups was similar. The organisms causing bacteremia were similar, although children were noted to have a higher rate of streptococcal infections. The rate of clinically documented infections in the pediatric group was lower, and consequently children had a higher incidence of fever of unknown origin. Of the sites of infection documented clinically, children were more likely to have upper respiratory infections compared to lower respiratory tract infections in their adult counterparts. Mortality was lower in children as was the incidence of death related to infection (1% vs. 4%, $P = 0.001$).[84]

5.3. Risk Stratification

It has become increasingly clear that not all patients with fever and neutropenia are at equal risk for serious infection.[85–87] Factors available at the time of presentation that consistently appear to confer low risk are a short duration of neutropenia (less than 7–10 days), being clinically well without significant comorbid medical conditions or significant focal infections, and having cancer that is not progressive.[88–91] These factors have been evaluated in pediatric populations with similar findings.[92]

Risk stratification of patients with fever and neu-

tropenia provides the opportunity to evaluate "less intense" therapy for the low-risk subset. This includes the use of broad-spectrum oral antibiotics and outpatient management. The potential advantages of these modifications to standard therapy include avoiding nosocomial infections and the psychosocial burdens of hospitalization, as well as decreasing costs. There is mounting evidence that the treatment of patients with "low-risk" fever and neutropenia with broad-spectrum oral antibiotics in the outpatient setting may be a reasonable alternative to parenteral therapy.

A series of studies performed by Malik *et al.* in an indigent population of oncology patients in Pakistan has assessed the efficacy of oral ofloxacin for patients with low-risk fever and neutropenia. The initial trial compared oral ofloxacin to combination intravenous therapy (amikacin with carbenicillin, cloxacillin, or pipericillin). All therapy was administered as inpatients. There was no significant difference found in "success" or mortality from infection.[91] Subsequent studies of outpatient therapy with oral ofloxacin, involving both adult and pediatric patients, have shown it to be a reasonable alternative to inpatient intravenous therapy.[93–95]

Comparison of intravenous and oral therapy administered to patients with low-risk fever and neutropenia in the outpatient setting has been evaluated. Rubenstein *et al.*[96] compared therapy with oral ciprofloxacin and clindamycin to intravenous aztreonam and clindamycin. This study was stopped early because of an unexpected incidence of renal toxicity in the oral therapy group. A follow-up study using ciprofloxacin and amoxicillin/clavulanate as the oral regimen showed no difference in efficacy or toxicity between the oral and intravenous therapy groups.[97]

A randomized, double-blind, placebo-controlled study of oral versus intravenous therapy for patients with low-risk fever and neutropenia recently has been completed at the NIH (Freifeld[97a]). This study compared intravenous ceftazidime to oral ciprofloxacin and augmentin in adult and pediatric inpatients. The results provide strong evidence that oral antibiotic therapy for patients with low-risk fever and neutropenia is a safe and effective alternative to standard intravenous therapy.

5.4. Prolonged Fever and Neutropenia: Antifungal Therapy

Children and adults with prolonged neutropenia are at high risk for serious infections and require vigilant care.[98] Persistent fever in the absence of other findings in the neutropenic host is not in itself an indication for changing antibiotic therapy, except for the addition of

TABLE 3. Modifications of Therapy in Patients with Fever and Neutropenia

Clinical event	Possible modification of therapy
Breakthrough bacteremia	If gram-positive isolate (e.g., *S. epidermidis*), add vancomycin
	If gram-negative isolate (i.e., presumably resistant), switch to new regimen
Catheter-associated infection	Add vancomycin (as well as gram negative coverage if not already being given)
Severe oral mucositis or necrotizing gingivitis	Add specific antianaerobic agent (e.g., clindamycin or metronidazole)
Esophagitis	Trial of oral clotrimazole, ketoconazole, fluconazole, or IV amphotericin
Pneumonitis	
Diffuse or interstitial	Trial of trimethoprim–sulfamethoxazole and azithromycin (plus broad-spectrum antibiotics if the patient is neutropenic)
New infiltrate in granulocytopenic patient receiving antibiotics	If granulocyte count is rising, watch and wait
	If granulocyte count is not recovering, biopsy to establish diagnosis; if biopsy cannot be done, add amphotericin B empirically
Perianal tenderness	If patient is already receiving broad-spectrum antibiotics, add specific antianaerobic agents
	If patient is not on antibiotics, begin broad-spectrum therapy, including anaerobic coverage
Persistent fever and neutropenia	Continue antibiotics and after 7 days of persistent fever and neutropenia add systemic antifungal therapy empirically

empiric antifungal therapy.[99,100] Changes in clinical status or new microbiological data should be used to guide antibiotic modifications. Suggested therapies are outlined in Table 3.

Prolonged neutropenia puts the patient at significant risk for invasive fungal infection. The diagnosis of fungal infections in the neutropenia patient remains challenging, with fever often being the only presenting clinical sign. Even in proven disseminated candidal disease, blood cultures often remain negative. Imaging with computer tomography (CT) and magnetic resonance imaging (MRI) has become a routine part of the care of the persistently febrile neutropenic patient to assess for evidence of hepatosplenic candidiasis. Subtle pulmonary findings can be the first signs of invasive aspergillus infection, definitively diagnosed often only by open lung biopsy. The rationale for empiric antifungal therapy is the same as that for antibacterials, mainly to decrease infection-related mortality by early initiation of therapy. There have been several prospective randomized trials that have shown a decreased incidence of death related to fungal infection in patients with prolonged fever and neutropenia who received empiric amphotericin B.[98]

Illustrative Case 2

A 15-year-old adolescent with newly diagnosed ALL was undergoing induction chemotherapy. He had been febrile at the time of diagnosis and therefore had been started on broad-spectrum intravenous antibiotics. He rapidly defervesced and all blood cultures had remained negative.

On day 20 of induction he once again developed fevers. His physical examination was unremarkable. He was neutropenic, though beginning to show evidence of early bone marrow recovery, rising platelet count, and peripheral monocytes. He remained febrile over the next several days and empiric amphotericin B was started. By day 26 of induction he was no longer neutropenic, but remained febrile. Multiple sets of blood cultures all remained negative. A CT scan of his abdomen performed with contrast revealed multiple "bull's-eye" lesions in his liver, spleen, and kidneys. A needle biopsy of his liver confirmed the diagnosis of hepatosplenic candidiasis.

Empiric therapy with amphotericin B has been limited by significant renal toxicity. Recently there have been two trials comparing two of the liposomal preparations of amphotericin to standard amphotericin B for empiric therapy in febrile neutropenic patients, both of which showed equal efficacy but less renal toxicity in those patients receiving the liposomal formulations.[101,102] Empiric therapy with fluconazole also has been shown to be efficacious and less toxic. It should not be used, however, in those patients with clinical signs suggestive of aspergillosis or in those who have received fluconazole prophylaxis and therefore are at higher risk for infection with fluconazole-resistant candidal species.[103,104]

5.5. Viral Infections

Viral infections are extremely common during childhood, both in the healthy and the immunocompromised child. Their diagnosis has remained challenging, and often "nonspecific viral syndromes" are made as diagnoses of exclusion. Advances in PCR technology as well as standardized serological and antigen detection assays have begun to have had a significant impact on the ability of clinicians to assign causality of various infectious syndromes and more importantly, begin appropriate therapy.

Cytomegalovirus (CMV) is a ubiquitous pathogen transmitted both vertically and by close person-to-person contact. The majority of primary infections in immunocompetent individuals are asymptomatic. The prevalence of antibodies to CMV in adult populations varies geographically from 40 to 100%.[105] The virus persists in a latent form after primary infection. Both primary infections and reactivation can cause serious disease in the immunocompromised individual, primarily involving the lungs, gastrointestinal tract, bone marrow, retina, and central nervous system in most cases.

Prophylactic strategies for high-risk bone marrow transplantation patients and improved screening for early infection with PCR and CMV antigenemia assays have had a significant impact on the morbidity and mortality related to CMV infection.[106,107] Prevention of transfusion-acquired CMV for immunosuppressed seronegative patients is very important.[108] Blood banking strategies to provide CMV-negative blood have evolved to the use of leukofiltered products, abrogating the need of dual inventories of (CMV-positive and -negative) blood.[109]

Herpes simplex virus (HSV) type 1 is a common infectious agent in childhood, while HSV-2 is often acquired in adolescence. Infection in immunocompetent children is often asymptomatic, while it can cause severe and recurrent gingivostomatitis, often accompanied by fever, in immunosuppressed individuals. Reactivation of HSV disease in bone marrow transplantation patients is common and prophylaxis for individuals who are seropositive is routinely employed.

Illustrative Case 3

A 3-year-old boy on maintenance therapy for ALL came to clinic for a 1-day history of low-grade fevers and a new rash on his torso. An older brother had been convalescing from chicken pox that had erupted 2 weeks earlier. In the clinic the child played in the waiting room with multiple other children.

When the lesions were examined and determined to be classic for varicella zoster, the child was immediately moved into respiratory isolation and admitted to the hospital for intravenous acyclovir (1500 mg/m^2 per day). His complete blood count was remarkable for a white blood cell count of 2900/mm^3 with 39% polys, 47% lymphocytes, 9% atypical lymphocytes, and 5% monocytes. Serum transaminases were mildly elevated. Oxygen saturation on room air was 100% and a chest radiograph was negative. He was treated for 7 days with intravenous acyclovir and had an uneventful recovery, with his numerous lesions finally crusting prior to his discharge.

Eighteen other children were with him in the waiting room. Fourteen of these children were susceptible to primary varicella and were given prophylactic varicella zoster immune globulin intramuscularly. Within 2 weeks four patients were diagnosed with varicella. All cases were mild but still required hospitalization and intravenous acyclovir.

Varicella zoster virus (VZV) infection (chickenpox) is a common childhood illness with a peak incidence in late winter and early spring. Approximately two thirds of the cases occur in children between the ages of 5 and 9 years. The majority of the cases in normal children are clinically apparent with less than 5% remaining asymptomatic. In the 1970s, primary varicella infection carried a 10% mortality rate in children with cancer.[110] The use of postexposure prophylactic varicella zoster immune globulin (VZIG) as well as the use of the antiviral agent acyclovir has dramatically decreased the incidence of severe disseminated disease as well as overall mortality in immunocompromised individuals. However, approximately 50% of the cases of varicella occur in patients without a known exposure, making VZIG without benefit for this group.[111] Acyclovir therapy is highly effective, but treatment failures have been reported, albeit rarely. An additional problem of the current strategies is the need to withhold chemotherapy pending the evaluation and treatment of the exposed or affected child.[111] Prevention remains an important goal and vaccination strategies have been studied, primarily in patients with ALL.

Herpes zoster is caused by reactivation of the varicella virus often presenting with dermatomal cutaneous disease. In immunosuppressed children there is a risk of progression of focal disease to disseminated visceral disease, most commonly affecting the brain, liver, and lungs. Disseminated disease carries a high mortality and its prevention with early therapy with intravenous acyclovir is imperative. Interestingly, although HIV-infected children can present with either dermatomal zoster or a chronic form of cutaneous chicken pox, it is rarely complicated by systemic dissemination.[112]

Common pediatric respiratory viruses (respiratory syncytial virus, parainfluenza, adenovirus, influenza, and rhinovirus) can cause significant morbidity in immunocompromised children. Because these infections are so ubiquitous they are an extremely common cause of fever in this population and contribute significantly to hospital admissions and delayed chemotherapy. Preventing exposure to these agents is difficult, particularly in families with multiple young children.

5.6. Protozoa, *Pneumocystis carinii*

Pneumocystis carinii pneumonia (PCP) was first described in the 1940s in malnourished neonates in Eastern Europe. In the 1960s, with the early use of immunosuppressive therapy, PCP was recognized more frequently, most commonly in patients with leukemia.[113] In the mid-1970s, prophylaxis for PCP in patients receiving chemotherapy was shown to be efficacious. In the 1980s PCP emerged as the most common opportunistic infection in patients with AIDS.[114] Although infection with PCP has become a relatively common problem, several fundamental issues regarding this organism remain unresolved. The taxonomy remains controversial as does the mode of transmission and whether it should be considered contagious. There is some evidence that primary infection in the immunocompromised host, usually occurring in early childhood, carries a worse prognosis compared to reactivation of disease, though this remains controversial.

Illustrative Case 4

A 9-month-old girl was admitted in acute respiratory distress. She was born 6 weeks prematurely to an HIV-infected mother who was also addicted to cocaine. She was small for age and her growth was at the fifth percentile for height and weight. She was hospitalized at 6 weeks of age for respiratory syncytial virus infection and received ribavirin and supportive care. She had two episodes of otitis media and had had persistent oral candidiasis.

She presented with a 1-day history of poor oral intake, fevers to 39.2°C, tachypnea, and perioral cyanosis. Physical examination revealed a respiratory rate of 60 breaths/min with wheezes and rhonchi on expiration. She had cervical adenopathy and hepatosplenomegaly. A chest radiograph revealed hyperinflation and bilateral interstitial infiltrates. The white blood cell count was $3600/m^3$ with 59% polys, 12% bands, 22% lymphocytes, 4% monocytes, and 3% eosinophils. The absolute CD4 count was $995/mm^3$. The serum alkaline phosphatase and lactate dehydrogenase were minimally elevated. Two blood cultures, a throat culture, and a urine culture were sent. Flexible bronchoscopy was performed and *P. carinii* was identified. High-dose trimethoprim–sulfamethoxazole and prednisone at 2 mg/kg per day was begun. The child's condition gradually worsened, requiring mechanical ventilation. Despite these interventions the patient died. At autopsy there was extensive acute inflammatory changes of the alveoli with *P. carinii* cysts and trophozoites.

The presentation of PCP in the oncology patient is often acute, with hypoxemia and significant respiratory distress. The presentation in children with HIV can be more varied, from an insidious pattern with prolonged cough and low-grade fevers to a syndrome very similar to that of the oncology patient. The diagnosis is made by the demonstration of organisms in lung tissue or lower respiratory tract secretions and is most often made by bronchoscopy. The drug of choice for therapy is trimethoprim–sulfamethoxazole (TMP-SMX), with parenteral pentamidine used as an alternative in those individuals who are intolerant of TMP-SMX. Corticosteroids have been shown to be beneficial in HIV-infected adults with severe PCP and this practice has been adopted in pediatric patients, usually at a dose of 2 mg/kg per day of prednisone. Prophylaxis for PCP is critical in the management of these patients.

5.7. Fever in Pediatric HIV Infection

Since HIV-infected children can be severely immunosuppressed at a very young age, it is not surprising that

infection with *S. pneumoniae* and *H. influenzae*, the organisms that are responsible for most common childhood infections, such as otitis media and bronchitis, commonly cause bacteremia.[62] However, the greatest risk factor for bacteremia in HIV-infected children is the presence of a central venous catheter, which often leads to infections with gram-positive organisms.[54,63,64]

Infections with opportunistic organisms, such as *Mycobacterium avium* complex (MAC) or *P. carinii*, are common in pediatric and adult patients with very low CD4 counts.[115,116] However, the definition of "low" CD4 counts is age-dependent and an infant may present with PCP as the first indication of HIV infection.[117] This has resulted in revised guidelines for PCP prophylaxis in children.[116] The risk for fungemia appears to be highest in HIV-infected children with a central venous catheter.[118]

Illustrative Case 5

A 17-month-old male with perinatally acquired HIV infection was evaluated for 2 weeks of daily fevers to 38.5°C and persistent diarrhea. The child reportedly had had night sweats for the prior 6 weeks but fever only in the last 2 weeks. Three months previously the child was hospitalized for *S. pneumoniae* bacteremia. Since that admission the child had had a poor appetite, intermittent loose stools, and no weight gain. On physical examination the child was below the fifth percentile for height, weight, and head circumference. Peripheral adenopathy and hepatosplenomegaly were palpable. The most recent absolute CD4 cell count was $185/mm^3$. The white blood cell count was $3500/mm^3$, the hematocrit was 28%, and the platelet count was $110,000/mm^3$. Serum transaminases were normal. Blood and urine cultures were sent and demonstrated no growth. Stool cultures for bacterial (including mycobacterial species), viral, and protozoal pathogens were negative on multiple samples.

Fever in children with HIV infection may be a manifestation of HIV infection alone, though intercurrent or infectious events represent the more likely etiology. In children with advanced disease, persistent fever without clear source is more common.

6. Immunizations

Immunizations are an important part of the care of healthy children. The routine vaccination schedule will often be disrupted for younger children undergoing therapy for cancer. There are no universally accepted recommendations for immunizing children undergoing therapy for cancer but some general guidelines can be applied. The American Academy of Pediatrics and the Centers for Disease Control regularly publish updated guidelines regarding immunization practices in healthy and immunocompromised patients.[119,120] Guidelines for vaccination strategies in HIV-infected children have also been published.[121] In considering reasonable vaccine strategies, information about the host's risks for infection needs to be balanced with the safety and efficacy of each vaccine in this population. Two main concerns have to be considered: (1) Will the patient be able to mount (or maintain) an antibody response? and (2) Could the vaccine itself (in the case of attenuated live organisms) cause disease?[122]

6.1. Live Vaccines

Measles–mumps–rubella (MMR) is a live virus vaccine that is contraindicated in patients undergoing active chemotherapy. In the early development of the vaccine one of eight children vaccinated with the Edmonton b strain measles vaccine died of the disease. Immunization has been shown to be safe and is recommended for patients with ALL after completion of therapy, with the usual waiting period of 3–6 months to allow for T-cell reconstitution. It has been recommended that patients having completed therapy for Hodgkin's disease not be vaccinated, given their prolonged T-cell deficits.[123] Household contacts can be safely vaccinated because transmission of the vaccine virus does not occur.[119] Children with severe immunosuppression (for definition, see Table 2) due to HIV infection should not receive the MMR vaccine either; however, HIV-infected children with mild or moderate immune suppression can follow the normal immunization schedules.[119] The risk associated with wild-type measles is considered to be higher than the risk for vaccine-related complications.[124,125]

Oral polio vaccine (OPV) is a live virus vaccine that is contraindicated in immunocompromised patients and their household contacts.[119] Inactivated polio vaccine (IPV) can be used safely during treatment for nonimmunized patients. Reimmunization after the completion of therapy with IPV or OPV is generally recommended.[123] HIV-infected children or their household contacts should not receive OPV. In fact, to avoid inadvertent infection of children with unrecognized HIV infection, the AAP now recommends IPV for the first two doses (at 2 and 4 months of age) for all children.

The varicella vaccine is potentially the most extensively studied vaccine in immunocompromised children. Varicella vaccine is now universally recommended for healthy individuals in early childhood and for susceptible older children and adolescents. It is not contraindicated in household contacts of immunosuppressed individuals.[119] The use of varicella vaccine has been studied extensively in children with ALL. Currently it is not licensed for routine use in children with malignancies receiving active chemotherapy, but its use should be considered in chil-

dren with ALL who have been in remission for more than 1 year. It has been shown to be safe in children with ALL, with the most common toxicity being a mild rash occurring in 50% of those treated.[126] The efficacy of the vaccine has been shown in the degree of protection from acquiring the disease from household contacts, with 14% developing mild disease and complete protection from severe varicella. There has been concern about increasing the incidence of herpes zoster (shingles) in immunocompromised vaccinees, but it has been shown that leukemic vaccinees are less likely to develop zoster than comparable children with leukemia who had wild-type infection.[88,126]

The varicella vaccine is considered contraindicated in all HIV-infected children and adults.[119] Unfortunately, antibody testing to evaluate protective immunity is not reliable in the immunocompromised patient. A negative result does not exclude prior contact with varicella and a positive result could be due to passive transfer through a recent blood transfusion. Passive immunoprophylaxis either with varicella zoster immune globulin within 96 hr of exposure or regular infusions of gammaglobulin may be indicated in the child at high risk.

6.2. Live Bacteria

The only live bacterium that potentially could be used in a childhood immunization schedule is the BCG vaccine. Although not used in the United States, it is still recommended in more than 100 countries worldwide.[119] Since disseminated BCG-itis can occur in immunocompromised children, it is generally contraindicated in that population.[127–130] However, in populations with a high risk for tuberculosis, BCG is recommended for asymptomatic HIV-infected children.[119]

6.3. Inactivated Bacteria, Inactivated Viruses, Polysaccharide–Protein Conjugates, and Toxoids

Diptheria–pertussis–tetanus (DPT) vaccine has been evaluated in young infants with neuroblastoma and in children receiving maintenance therapy for various malignancies showing these children to mount adequate responses.[131,132] It has been suggested that children should be given this vaccine at scheduled times even while undergoing active therapy. The alternative approach is to immunize at the end of therapy. To decrease the risk of seizures (a special concern in children with brain tumors or preexisting seizure disorder) it is currently recommended that all children should receive the acellular form of the pertussis component (DtaP). There are no special exceptions for HIV-infected children; however, in the case of a child with progressive encephalopathy prior discussion with a specialist is recommended.

Patients with ALL as well as those with Hodgkin's disease treated at a time when splenectomy was routine have been shown to be at a higher risk for invasive pneumococcal and *Haemophilus influenzae* type B (HIB) disease.[133–135] Pneumococcal infection contributes significantly to the morbidity and mortality of pediatric HIV infection.[136,137] The polysaccharide pneumococcal vaccine is only moderately immunogenic when studied in oncology or transplantation populations. Evaluation of the newer pneumococcal vaccines needs to be carried out in children with cancer; however, it currently is recommended that patients who are expected to be functionally or anatomically asplenic should receive pneumococcal vaccine. This includes children with HIV infection.[119,121]

The HIB conjugate vaccines have been tested in children who were on treatment or who had completed treatment for ALL, as well as in children with HIV infection. The responses were not normal, but those who did respond had protective antibodies that were measurable for 12 months.[133,138,139] Though the data are not definitive, some authors recommend the vaccination of *all* immunosuppressed individuals with pneumococcal, HIB, and meningococcal vaccines.[120,123]

The efficacy of the influenza vaccine in patients undergoing cancer therapy is controversial. In the pediatric age group it was shown in a small retrospective study of patients with ALL that immunization decreased the incidence of influenza infection compared to unimmunized controls.[140] Despite the lack of definitive data it is generally recommended that all immunocompromised children and their household contacts receive yearly influenza vaccines prior to flu season.[120]

6.4. Passive Immunization

Intravenous immunoglobulin (IVIG) has been shown to prevent serious bacterial infections in patients with congenital immunodeficiencies. The monthly administration of IVIG has been studied in asymptomatic and symptomatic children with HIV infection.[141,142] As long as children (all being treated with antiretroviral therapy) also received TMP-SMX as PCP prophylaxis, there was no difference between the group receiving IVIG and the group receiving placebo. The current recommendation is to use prophylactic IVIG (400 mg/kg per dose every 28 days) in HIV-infected children with hypogammaglobulinemia, poor functional antibody development (i.e., lack of antibody response after immunizations), or significant

recurrent infections despite therapy with appropriate antibiotics.[121] The prophylactic use of IVIG is not generally recommended for oncology patients. There may be a subset of patients, however, who develop hypogammaglobulinemia for whom IVIG may be beneficial.[143]

7. Infection Prophylaxis

The best infection prophylaxis in the care of immunocompromised patients (and others) is diligent hand washing before and after contact with patients and for patients themselves after handling pets. Although children and adults with neutropenia should avoid contact with sick persons, which rarely means staying out of school during the cold season, there is no need to isolate immunocompromised patients except in rare circumstances (e.g., the first few weeks after bone marrow transplantation). Teachers or daycare workers should be made aware of the child's immunocompromised state and asked to notify the parents in case of an outbreak of a contagious disease, such as varicella. HIV-infected patients and other immunocompromised persons should avoid eating raw or undercooked eggs, poultry, meat, or seafood or unpasteurized dairy products. However, the routine prescription of a "neutropenic diet," which excludes salads and other raw food, is no longer recommended.

Although pharmacologic prophylaxis is used widely in both pediatric and adult oncology, there is ongoing concern regarding the emergence of resistant pathogens. Primarily because of this threat and because prophylaxis has never been shown to impact mortality, the 1997 recommendations from the Infectious Diseases Society of America are that the routine use of antimicrobial prophylaxis should be avoided in patients receiving treatment for cancer, except for the institution of PCP prophylaxis in children undergoing prolonged and intensive chemotherapy.[83]

However, in HIV-infected children and adults there is a strong correlation between the degree of the immunosuppression and the risk for certain opportunistic infections. The most commonly used indicator for the need of antimicrobial prophylaxis is the CD4 count.[116,121] In addition, many children and adults receive suppressive therapy (i.e., secondary prophylaxis) to prevent a recurrence or reactivation of certain infections (e.g., CMV disease, MAC infection). There is not yet consensus on whether children and adults whose immune system improves (i.e., CD4 count increases) due to aggressive antiretroviral therapy need to continue prophylactic antimicrobial therapy.

7.1. Prophylaxis during Episodes of Fever and Neutropenia

Prophylactic antimicrobial regimens are used extensively in patients receiving cytotoxic chemotherapy. Early on it was noted that the patient's own enteric flora was often the culprit in documented infections during periods of fever and neutropenia. Initially it was shown that nonabsorbable antibiotics decreased the rate of infectious complications. These agents are in general unpalatable and were limited by difficulty with patient compliance, particularly in children.

The two agents that have received the most attention as prophylactic therapy for bacterial infections in neutropenic cancer patients have been TMP-SMX and the oral quinolones, primarily ciprofloxacin.[79,144–146] Both agents have been shown to decrease documented infections, primarily bacteremia, in neutropenic patients. Documenting an effect on overall mortality has been more difficult. The use of TMP-SMX is complicated by a significant rate of allergic reactions and a risk of bone marrow suppression with the potential for prolongation of neutropenia. Fluoroquinolones are in general well tolerated but have been associated with a higher than expected rate of streptococcal bacteremia, leading some to suggest the addition of penicillin or clindamycin to the prophylactic regimen. The utility of either agent is limited by the potential for the emergence of resistant pathogens.[147,148]

The investigation and routine use of fluoroquinolones in pediatrics has been restricted by the potential for arthropathy. In juvenile animals, exposure to fluoroquinolones has been associated with a risk of arthropathy expressed clinically as lameness and associated with characteristic histologic findings of blisters and erosions of articular cartilage. This finding has been consistent for all fluoroquinolones tested.[149] There is, however, a growing body of evidence supporting the safety of quinolone antibiotics in the pediatric population. Nalidixic acid, a nonfluorinated quinolone, has been used in children for decades. In animal studies it causes the classic cartilage changes but it has not been found to cause any arthropathy in children, including those treated for prolonged periods of time.[150,151] In 1997, Hampel et al.[152] reviewed the worldwide experience with ciprofloxacin in pediatric patients based on its compassionate use and concluded that short courses of ciprofloxacin appear to be safe. No cases of the experimentally induced cartilage damage have been confirmed in humans. However, since the prophylactic use of fluoroquinolones in patients with cancer is of questionable utility, these drugs should best be reserved for justified indications.

7.2. Mycobacterial Prophylaxis

Tuberculosis in childhood is almost always a primary infection acquired from an adult with contagious disease. The HIV epidemic has led to a resurgence of adult and pediatric cases of tuberculosis.[153,154] Children with immune deficiency (due to cancer therapy or HIV infection) who were exposed to a potentially contagious case of tuberculosis within the last 3 months and without evidence of active disease should receive prophylaxis with isoniazid for at least 6 months, even if the skin test is negative. The Mantoux skin test (5 tuberculin units) should always be combined with a test for another common antigen, such as *Candida* or mumps, to assess anergy. The interpretation of a Mantoux test in an immunocompromised individual is more stringent than in otherwise healthy people, and an induration of ≥5 mm is considered positive.[155]

Among the atypical mycobacteria, MAC is most commonly encountered in the pediatric population, followed by *M. scrofulaceum*, *M. kansasii*, and *M. marinum*. Only MAC infection causes frequent and serious enough infection to necessitate prophylaxis in certain populations, especially HIV-infected children.[115,121] Chemoprophylaxis with clarithromycin or azithromycin is recommended for adults and children >6 years of age with a CD4 count <50 cells/mm^3.[121] For children aged 2–6 years, prophylaxis should be initiated with a CD4 count <75 cells/mm^3, for children aged 1–2 years at <500 cells/mm^3, and in children <12 months of age at a CD4 count of <750 cells/mm^3.[121] There are no specific guidelines for prophylaxis in children with cancer and MAC infection is rarely a problem in that population.

7.3. Antifungal and Antiparasitic Prophylaxis

Prophylaxis against fungal organisms also is a part of routine care for many patients receiving chemotherapy but is not recommended in HIV-infected patients, except to prevent recurrence of previously documented severe disease.[121] Fluconazole is the most widely used agent both for adults and children. It has been shown to decrease fungal colonization and in some studies it decreased the rate of invasive infections, but it has not been shown to improve overall survival.[156–158] Like its antibacterial counterparts, the widespread use of fluconazole for prophylaxis raises concerns for the development of resistant *Candida* species.[159,160]

Aspergillus infection still has a high mortality in immunocompromised patients.[161,162] There are no recommendations for pharmacologic prophylaxis. However, patients who received a bone marrow transplant should be cared for in HEPA-filtered rooms for the first few weeks, and all patients should avoid construction and renovation sites, especially within the hospital.

Cryptococcal infection can lead to systemic disease and meningitis in HIV-infected children and adults and sometimes presents as pulmonary nodules that must be differentiated from metastases in cancer patients.[163–165] Although there is no indication for prophylaxis of primary infection, suppressive therapy with fluconazole is indicated in children and adults with systemic disease. Whether this secondary prophylaxis can be stopped if a patient's immune status improves (i.e., an increase in CD4 counts occurs) is unclear.[121]

Prior to the AIDS epidemic, the majority of the cases of PCP reported to the CDC were in patients with leukemia, primarily those with ALL.[113] The disease was fulminant in presentation and carried a near 100% mortality without treatment. In 1977, Hughes et al.[166] demonstrated that prophylaxis with TMP-SMX was highly effective in preventing PCP in high-risk oncology patients.[166,167] The use of PCP prophylaxis has since become a routine part of the management of most childhood cancers.

Defining which patients are most likely to benefit from PCP prophylaxis has become complex and involves knowledge of the underlying disease as well as the chemotherapeutic regimen. The patients at highest risk included those with ALL and non-Hodgkin's lymphoma, those undergoing bone marrow transplantation, and infants with severe combined immunodeficiency.[166,168] More recently, patients with brain tumors receiving steroids for greater than 2 months have been shown to be at significant risk and accordingly should receive prophylaxis.[169–171] The risk in patients undergoing intensive chemotherapeutic regimens for treatment of solid tumors is less clear, however. In some series the rate of infection without prophylaxis is high (for example, 23% for patients undergoing therapy for rhabdomyosarcoma in the original series by Hughes et al.[166]) and PCP has been reported as a major toxicity in many therapeutic trials of solid tumors, including neuroblastoma, Ewing's sarcoma, rhabdomyosarcoma, and osteosarcoma.[172–174]

Most patients with cancer are able to tolerate prophylaxis with TMP-SMX. Alternative regimens used in oncology patients are based on data from prophylaxis trials in AIDS patients and include the use of aerosolized pentamidine (which is difficult to administer to the very young child), as well as dapsone or atovaquone.

Pneumocystis carinii pneumonia was the AIDS indicator disease in almost 40% of the pediatric cases reported to the CDC until recently.[175] The peak incidence of PCP in infancy occurs during the first 3 to 6 months of life, often as the first symptom of HIV infection, and has been associated with a mortality of 39 to 65% in infants, in spite of improved diagnosis and treatment. In 1991, the CDC issued guidelines for PCP prophylaxis in children, taking into account the age-dependent levels of normal CD4 cell numbers.[176] However, a survey published in 1995 revealed basically no change between 1988 and 1992 in the incidence of PCP among infants born to HIV-infected mothers.[177] This was partly due to the fact that many children were not known to be HIV-infected or their immune status (i.e., CD4 count) was not known when they presented with PCP. Furthermore, among the infants known to be HIV infected who had a CD4 count performed within 1 month of PCP diagnosis, 18% had a CD4 count over 1500 cells/mm^3, the recommended threshold for initiation of PCP prophylaxis.[177] Simultaneously it was noticed that primary prophylaxis during the first year of life was highly effective in the prevention of PCP.[178] The following recommendations were developed based on these data[116]:

- birth to 4–6 weeks, HIV exposed or infected: no prophylaxis, since PCP is rare and due to concerns regarding kernicterus with TMP-SMX
- 4–6 weeks to 4 months, HIV exposed or infected: prophylaxis for all children
- 4–12 months, HIV infected or indeterminate: prophylaxis for all children
- 4–12 months, HIV infection reasonably excluded: no prophylaxis
- 1–5 years, HIV infected: prophylaxis if CD4$^+$ count is <500 cells/mm^3 or CD4$^+$ percentage is <15%
- over 6 years, HIV infected: prophylaxis if CD4$^+$ count is <200 cells/mm^3 or CD4$^+$ percentage is <15%
- all children with a history of PCP: prophylaxis

The recommended prophylactic regimen is TMP-SMX with TMP 150 mg/m^2 per day and SMX 750 mg/m^2 per day given orally in divided doses twice a day during 3 consecutive days per week. Alternative regimens, if TMP-SMX is not tolerated, are dapsone orally (2 mg/kg per day) or aerosolized pentamidine. However, breakthrough infections can occur with every regimen and appear to be most frequent with intravenous pentamidine and least common with TMP-SMX.[179,180]

Toxoplasmosis is a rare opportunistic infection in children but sometimes can be seen as a congenitally acquired disease.[181] The use of TMP-SMX for PCP prophylaxis appears to decrease the incidence of toxoplasmosis as well.[182]

Infection with Cryptosporidium can cause severe

and persistent diarrhea in immunocompromised patients and can be acquired due to the contamination of municipal water supplies.[183–185] Preventive measures for children are no different than for adults; however, it may be more difficult to prevent children from swallowing water during swimming and to avoid touching stray dogs or cats.[121]

7.4. Antiviral Prophylaxis

Immunocompromised children, including those with HIV infection, are at risk for severe varicella infections and therefore should receive VZIG as soon as possible after exposure but at minimum within 96 hr of exposure. Significant exposures include household contacts, shared hospital rooms, or "face-to-face" indoor play. The time of interaction considered significant for exposure differs, with some experts suggesting 5 min as significant and others defining close contact as more than 1 hr. Oral acyclovir is generally not recommended for prevention of acquisition of infection.

Respiratory syncytial virus (RSV) infection is common in pediatric oncology and bone marrow transplantation patients. In otherwise healthy children greater than 2 years of age RSV primarily causes upper airway disease. In patients receiving chemotherapy it can cause pneumonia,[186] and in those undergoing bone marrow transplantation RSV pneumonia carries a 50–80% mortality rate.[187,188] Prophylaxis for RSV has been studied extensively in premature neonates and the monthly administration of either RSV-IVIG (Respigam) or the humanized monoclonal antibody, palivizumab (Synergis), during the RSV season has been shown to decrease the frequency and severity of infections.[189–191] Neither agent has been evaluated in controlled trial in immunodeficient children. However, the current recommendations from the AAP suggest that children with severe immunodeficiencies may benefit from RSV-IVIG. If these infants are receiving IVIG monthly, providers may consider substituting RSV-IVIG during the RSV season.

Prophylaxis for HSV and CMV infection has become an important component of the care of bone marrow transplant patients. Routine prophylaxis for patients receiving chemotherapy is not recommended. However, patients with recurrent HSV lesions may benefit from a suppressive regimen of oral acyclovir.

8. Use of Growth Factors

Growth factors, primarily filgrastim [granulocyte colony-stimulating factor (G-CSF)] and sargramostin [granulocyte–macrophage colony-stimulating factor (GM-CSF)], are used extensively in the care of pediatric cancer patients and have also been studied in children with HIV disease.[192–195] These agents increase the number and phagocytic function of polymorphonuclear cells in the peripheral blood.[196–200] Given the association of the degree and duration of neutropenia with the risk of serious infection, there was initially great excitement about the potential impact growth factors might have in patients receiving myelosuppressive chemotherapy. Guidelines for the use of cytokines in patients have been developed by the American Society of Clinical Oncology (ASCO).[201,202]

Two strategies for growth factor use have received the greatest attention. The first is initiating therapy at the time of diagnosis with fever and neutropenia. Studies in both pediatric and adult patients have had variable results, with some investigators reporting a moderate decrease in number of days with fever, neutropenia, and antibiotic administration, as well as a decrease in length of hospitalization.[203–207] No study has shown a reduction in the rate of serious infection or infection-related mortality. Overall, there is no strong evidence supporting the initiation of growth factors at the time of fever and neutropenia.

The second strategy for growth factor use has been one of primary prophylaxis, initiating therapy after completion of each cycle of chemotherapy. Despite encouraging early clinical trials,[208] multiple subsequent studies have shown decreased duration of neutropenia and variable effect on the incidence of fever and neutropenia and of documented infection but no discernible effect on infection-related mortality.[194,209–212]

The ASCO guidelines recommend primary prophylaxis with growth factors when the expected rate of fever and neutropenia is greater than 40%. This guideline is difficult to apply, particularly in pediatrics, where the data regarding incidence of febrile neutropenia for standard treatment regimens have not been systematically evaluated. The patterns of growth factor use have been explored in both adults and children.[201,202] Pediatric oncologists are much more likely than their adult counterparts to use growth factors as primary prophylaxis, often due to protocol requirements but also due to the fact that most chemotherapeutic regimens used for pediatric malignancies will predictably result in periods with marked neutropenia.

Although an absolute neutrophil count (ANC) less than 500 cells/mm^3 is commonly associated with an increased risk for infectious complications in children and adults with cancer, this association is not clearly defined for HIV-infected patients.[213–215] Cytokines have been used to treat neutropenia in patients with HIV infection.[195,216,217] Although effective in raising the neutrophil count, there is concern that GM-CSF, but not G-CSF, will enhance replication of HIV.[218] Since there are now a growing number of antiretroviral agents available, it is usually possible to avoid the use of cytokines by choosing a less marrow-toxic regimen.

9. Conclusions

Many of the principles of chemoprophylaxis or treatment of infections apply to both children and adults. However, there are some notable differences. Most opportunistic infections present as primary infections in children compared to reactivation in adults, which may have an impact on immunization strategies and other prophylactic measures. Commonly used antibiotics and antiviral agents often are not produced in a form suitable for small children and child-specific dosages and toxicity data may not be available. Cancer therapy in children commonly results in severe neutropenia, and strict isolation is neither feasible nor desirable in the young child.

The chances for a cure (in the case of cancer) or good quality of life (in the case of cancer or HIV disease) have improved greatly during the last decades, thanks in large part to better antimicrobial prevention and treatment strategies. However, a continued effort to address child-specific issues such as the ones mentioned above is necessary in order to further decrease the morbidity and mortality associated with infections of immunocompromised patients.

References

1. Schibler KR, Osborne KA, Leung LY, Le TV, Baker SI, Thompson DD: A randomized, placebo-controlled trial of granulocyte colony-stimulating factor administration to newborn infants with neutropenia and clinical signs of early-onset sepsis. *Pediatrics* **102**:6–13, 1998.

2. Cairo MS, Agosti J, Ellis R, *et al*: A randomized, double-blind, placebo-controlled trial of prophylactic recombinant human granulocyte-macrophage colony-stimulating factor to reduce nosocomial infections in very low birth weight neonates. *J Pediatr* **134**:64–70, 1999.

3. Reed MD, Besunder JB: Developmental pharmacology: Ontogenic basis of drug disposition. *Pediatr Clin North Am* **36**:1053–1074, 1989.

4. Committee on Drugs: Unapproved uses of approved drugs: The physician, the package insert, and the Food and Drug Administration: Subject review. *Pediatrics* **98**:143–145, 1996.

5. Kauffman RE: Status of drug approval processes and regulation of medications for children. *Curr Opin Pediatr* **7**:195–198, 1995.

6. Yancik R: Epidemiology of cancer in the elderly. Current status and projections for the future. *Rays* **22**:3–9, 1997.

7. Yancik R, Wesley MN, Ries LA, *et al*: Comorbidity and age as predictors of risk for early mortality of male and female colon carcinoma patients: A population-based study. *Cancer* **82**:2123–2134, 1998.

8. Robison LL: General principles of the epidemiology of childhood cancer. In Pizzo PA, Poplack DG (eds): *Principles and Practice of Pediatric Oncology.* Lippincott-Raven, Philadelphia, 1997, pp. 1–10.

9. Grovas A, Fremgen A, Rauck A, *et al*: The National Cancer Data Base report on patterns of childhood cancers in the United States. *Cancer* **80**:2321–2332, 1997.

10. Pizzo P, Poplack DG (eds): *Principles and Practice of Pediatric Oncology.* Lippincott-Raven, Philadelphia, 1997.

11. Gurney JG, Severson RK, Davis S, Robison LL: Incidence of cancer in children in the United States. Sex-, race-, and 1-year age-specific rates by histologic type. *Cancer* **75**:2186–2195, 1995.

12. Wingo PA, Landis S, Ries LA: An adjustment to the 1997 estimate for new prostate cancer cases. *Cancer* **80**:1810–1813, 1997.

13. Balis FM, Holcenberg JS, Poplack DG: General principles of chemotherapy. In Pizzo PA, Poplack DG (eds): *Principles and Practice of Pediatric Oncology.* Lippincott-Raven, Philadelphia, 1997, pp. 215–272.

14. Centers for Disease Control and Prevention: US HIV and AIDS cases reported through December 1997. HIV/AIDS surveillance report. Year-end edition. *MMWR* **9**(2):1–44, 1997.

15. Ammann AJ, Cowan MJ, Wara DW, *et al*: Acquired immunodeficiency in an infant: Possible transmission by means of blood products. *Lancet* **1**:956–958, 1983.

16. Centers for Disease Control: Unexplained immunodeficiency and opportunistic infections in infants—New York, New Jersey, California. *MMWR* **31**:665–667, 1982.

17. Centers for Disease Control and Prevention: Update: Trends in AIDS incidence, deaths and prevalence—United States, 1996. *MMWR* **46**:165–173, 1997.

18. The European Collaborative Study: Vertical transmission of HIV-1: Maternal immune status and obstetric factors. *AIDS* **10**:1675–1681, 1996.

19. The Working Group on Mother-to-Child Transmission of HIV: Rates of mother-to-child transmission of HIV-1 in Africa, America, and Europe: Results from 13 perinatal studies. *J Acquir Immune Defic Syndr Hum Retrovirol* **8**:506–510, 1995.

20. Connor EM, Sperling RS, Gelber R, *et al*: Reduction of maternal–infant transmission of immunodeficiency virus type 1 with zidovudine treatment. *N Engl J Med* **331**:1173–1180, 1994.

21. Connor EM, Mofenson LK: Zidovudine for the reduction of perinatal human immunodeficiency virus transmission: Pediatric AIDS Clinical Trials Group protocol 076—Results and treatment recommendations. *Pediatr Infect Dis J* **14**:536–541, 1995.

22. Centers for Disease Control and Prevention: Recommendations of the US Public Health Service Task Force on the use of zidovudine to reduce perinatal transmission of human immunodeficiency virus. *MMWR* **43**:1–20, 1994.

23. Centers for Disease Control and Prevention: Guidelines for the use of antiretroviral agents in pediatric HIV infection. *MMWR* **47**:1–44, 1998.

24. Kline MW, Lewis DE, Hollinger FB, *et al*: A comparative study of human immunodeficiency virus culture, polymerase chain reaction and anti-human immunodeficiency virus immunoglobulin A antibody detection in the diagnosis during early infancy of vertically acquired human immunodeficiency virus infection. *Pediatr Infect Dis J* **13**:90–94, 1994.

25. Centers for Disease Control and Prevention: Report of the NIH panel to define principles of therapy of HIV infection and guide-

lines for the use of antiretroviral agents in HIV-infected adults and adolescents. *MMWR* **47:**1–83, 1998.

26. American Academy of Pediatrics Committee on Pediatric AIDS: Human milk, breastfeeding, and transmission of human immunodeficiency virus in the United States. *Pediatrics* **96:**977–979, 1995.

27. Blanche S, Newell ML, Mayaux MJ, *et al*: Morbidity and mortality in European children vertically infected by HIV- 1. The French Pediatric HIV Infection Study Group and European Collaborative Study. *J Acquir Immune Defic Syndr Hum Retrovirol* **14:**442–450, 1997.

28. Palella FJ, Delaney KM, Moorman AC, *et al*: Declining morbidity and mortality among patients with advanced human immunodeficiency virus infection. *N Engl J Med* **338:**853–860, 1998.

29. Dickover RE, Dillon M, Leung KM, *et al*: Early prognostic indicators in primary perinatal human immunodeficiency virus type 1 infection: Importance of viral RNA and the timing of transmission on long-term outcome. *J Infect Dis* **178:**375–387, 1998.

30. Centers for Disease Control and Prevention: Public Health Service task force recommendations for the use of antiretroviral drugs in pregnant women infected with HIV-1 for maternal health and for reducing perinatal HIV-1 transmission in the United States. *MMWR* **47:**1–31, 1998.

31. Boyle GJ, Michaels MG, Webber SA, *et al*: Posttransplantation lymphoproliferative disorders in pediatric thoracic organ recipients. *J Pediatr* **131:**309–313, 1997.

32. Kornstein MJ, Weber J, Luck JB, Massey GV, Strom S, McWilliams NB: Epstein–Barr virus-associated lymphoproliferative disorder. *Arch Pathol Med* **113:**481–484, 1989.

33. Newell KA, Alonso EM, Whitington PF, *et al*: Posttransplant lymphoproliferative disease in pediatric liver transplantation. Interplay between primary Epstein–Barr virus infection and immunosuppression. *Transplantation* **62:**370–375, 1996.

34. Purtilo DT, Tatsumi E, Manolov G, *et al*: Epstein–Barr virus as an etiological agent in the pathogenesis of lymphoproliferative and aproliferative diseases in immune deficient patients. *Int Rev Exp Pathol* **27:**113–183, 1985.

35. Liebowitz D: Epstein–Barr virus and a cellular signaling pathway in lymphomas from immunosuppressed patients. *N Engl J Med* **338:**1413–1421, 1998.

36. Dockrell DH, Strickler JG, Paya CV: Epstein–Barr virus-induced T cell lymphoma in solid organ transplant recipients. *Clin Infect Dis* **26:**180–182, 1998.

37. Manez R, Breinig MC, Linden P, *et al*: Posttransplant lymphoproliferative disease in primary Epstein–Barr virus infection after liver transplantation: The role of cytomegalovirus disease. *J Infect Dis* **176:**1462–1467, 1997.

38. Timmons CF, Dawson DB, Richards CS, Andrews WS, Katz JA: Epstein–Barr virus-associated leiomyosarcomas in liver transplantation recipients. *Cancer* **76:**1481–1489, 1995.

39. McClain KL, Leach CT, Jenson HB, *et al*: Association of Epstein–Barr virus with leiomyosarcomas in young people with AIDS. *N Engl J Med* **332:**12–18, 1995.

40. Athale UH, Patil PS, Chintu C, Elem B: Influence of HIV epidemic on the incidence of Kaposi's sarcoma in Zambian children. *J Acquir Immune Defic Syndr Hum Retrovirol* **8:**96–100, 1995.

41. Chintu C, Athale UH, Patil PS: Childhood cancers in Zambia before and after the HIV epidemic. *Arch Dis Child* **73:**100–105, 1995.

42. Erkeller-Yuksel FM, Deneys V, Hannet I, *et al*: Age-related changes in human blood lymphocyte subpopulations. *J Pediatr* **120:**216–222, 1992.

43. The European Collaborative Study: Age-related standards for T lymphocyte subsets based on uninfected children born to human immunodeficiency virus 1-infected mothers. *Pediatr Infect Dis J* **11:**1018–1026, 1992.

44. Comans-Bitter WM, de Groot R, van den Beemd R, *et al*: Immunophenotyping of blood lymphocytes in childhood. Reference values for lymphocyte subpopulations. *J Pediatr* **130:**388–393, 1997.

45. McKinney RE, Wilfert CM: Lymphocyte subsets in children younger than 2 years old: Normal values in a population at risk for human immunodeficiency virus infection and diagnostic and prognostic application to infected children. *Pediatr Infect Dis J* **11:**639–644, 1992.

46. Mackall CL, Fleisher TA, Brown MR, *et al*: Lymphocyte depletion during treatment with intensive chemotherapy for cancer. *Blood* **84:**2221–2228, 1994.

47. Mackall CL, Fleisher TA, Brown MR, *et al*: Age, thymopoiesis, and CD4+ T-lymphocyte regeneration after intensive chemotherapy. *N Engl J Med* **332:**143–149, 1995.

48. Roilides E, Black C, Reimer C, Rubin M, Venzon D, Pizzo PA: Serum immunoglobulin G subclasses in children infected with human immunodeficiency virus type 1. *Pediatr Infect Dis J* **10:**134–139, 1991.

49. Borkowsky W, Rigaud M, Krasinski K, Moore T, Lawrence R, Pollack H: Cell-mediated and humoral immune responses in children infected with human immunodeficiency virus during the first four years of life. *J Pediatr* **120:**371–375, 1992.

50. Luzuriaga K, Koup RA, Pikora CA, Brettler DB, Sullivan JL: Deficient human immunodeficiency virus type 1-specific cytotoxic T cell responses in vertically infected children. *J Pediatr* **119:**230–236, 1991.

51. Kelsey SM, Lowdell MW, Newland AC: IgG subclass levels and immune reconstitution after T cell-depleted allogeneic bone marrow transplantation. *Clin Exp Immunol* **80:**409–412, 1990.

52. Pollock CA, Mahony JF, Ibels LS, *et al*: Immunoglobulin abnormalities in renal transplant recipients. *Transplantation* **47:**952–956, 1989.

53. Pizzo PA, Robichaud KJ, Wesley R, Commers JR: Fever in the pediatric and young adult patient with cancer. A prospective study of 1001 episodes. *Medicine* **61:**153–165, 1982.

54. Lichenstein R, King JC Jr, Farley JJ, Su P, Nair P, Vink PE: Bacteremia in febrile human immunodeficiency virus-infected children presenting to ambulatory care settings. *Pediatr Infect Dis J* **17:**381–385, 1998.

55. Bodey GP, Buckley M, Sathe YS, Freireich EJ: Quantitative relationships between circulating leukocytes and infections in patients with acute leukemia. *Ann Intern Med* **64:**328–340, 1966.

56. Schimpff S, Satterlee W, Young VM, Serpick A: Empiric therapy with carbenicillin and gentamicin for febrile patients with cancer and granulocytopenia. *N Engl J Med* **284:**1061–1065, 1971.

57. Pizzo PA, Ladisch S, Simon RM, Gill F, Levine AS: Increasing incidence of gram-positive sepsis in cancer patients. *Med Pediatr Oncol* **5:**241–244, 1978.

58. Gunther G, Bjorkholm M, Bjorklind A, Engervall P, Stiernstedt G: Septicemia in patients with hematological disorders and neutropenia. A retrospective study of causative agents and their resistance profile. *Scand J Infect Dis* **23:**589–598, 1991.

59. Koll BS, Brown AE: Changing patterns of infections in the immunocompromised patient with cancer. In Pizzo PA (ed): *Hematology/ Oncology Clinics of North America*, Vol. 7:4. Saunders, Philadelphia, 1993, pp. 753–770.

60. Aquino VM, Pappo A, Buchanan GR, Tkaczewski I, Mustafa MM: The changing epidemiology of bacteremia in neutropenic children with cancer. *Pediatr Infect Dis J* **14:**140–143, 1995.

61. Dayan PS, Chamberlain JM, Arpadi SM, Farley JJ, Stavola JJ, Rakusan TA: *Streptococcus pneumoniae* bacteremia in children infected with HIV: Presentation, course, and outcome. *Pediatr Emerg Care* **14:**194–197, 1998.

62. Andiman WA, Mezger J, Shapiro E: Invasive bacterial infections in children born to women infected with human immunodeficiency virus type 1. *J Pediatr* **124:**846–852, 1994.

63. Fichtenbaum CJ, Dunagan WC, Powderly WG: Bacteremia in hospitalized patients infected with the human immunodeficiency virus: A case–control study of risk factors and outcome. *J Acquir Immune Defic Syndr Hum Retrovirol* **8:**51–57, 1995.

64. Roilides E, Marshall D, Venzon D, Butler K, Husson R, Pizzo PA: Bacterial infections in human immunodeficiency virus type 1-infected children: The impact of central venous catheters and antiretroviral agents. *Pediatr Infect Dis J* **10:**813–819, 1991.

65. Roilides E, Butler KM, Husson RN, Mueller BU, Lewis LL, Pizzo PA: *Pseudomonas* infections in children with human immunodeficiency virus infection. *Pediatr Infect Dis J* **11:**547–553, 1992.

66. Pizzo PA, Hathorn JW, Hiemenz J, *et al*: A randomized trial comparing ceftazidime alone with combination antibiotic therapy in cancer patients with fever and neutropenia. *N Engl J Med* **315:**552–558, 1986.

67. Sanders JW, Powe NR, Moore RD: Ceftazidime monotherapy for empiric treatment of febrile neutropenic patients: A meta-analysis. *J Infect Dis* **164:**907–916, 1991.

68. De Pauw BE, Deresinski SC, Feld R, Lane-Allman EF, Donnelly JP, The Intercontinental Antimicrobial Study Group: Ceftazidime compared with piperacillin and tobramycin for the empiric treatment of fever in neutropenic patients with cancer. A multicenter randomized trial. *Ann Intern Med* **120:**834–844, 1994.

69. Pizzo PA: Management of fever in patients with cancer and treatment-induced neutropenia. *N Engl J Med* **328:**1323–1332, 1993.

70. Eggimann P, Glauser MP, Aoun M, Meunier F, Calandra T: Cefepime monotherapy for the empirical treatment of fever in granulocytopenic cancer patients. *J Antimicrob Chemother* **32** (Suppl B):151–163, 1993.

71. Liang R, Yung R, Chiu E, *et al*: Ceftazidime versus imipenem-cilastin as initial monotherapy for febrile neutropenic patients. *Antimicrob Agents Chemother* **34:**1336–1341, 1990.

72. Freifeld AG, Walsh T, Marshall D, *et al*: Monotherapy for fever and neutropenia in cancer patients: A randomized comparison of ceftazidime versus imipenem. *J Clin Oncol* **13:**165–176, 1995.

73. Riikonen P: Imipenem compared with ceftazidime plus vancomycin as initial therapy for fever in neutropenic children with cancer. *Pediatr Infect Dis J* **10:**918–923, 1991.

74. Rolston KV, Berkey P, Bodey GP, *et al*: A comparison of imipenem to ceftazidime with or without amikacin as empiric therapy in febrile neutropenic patients. *Arch Intern Med* **152:**283–291, 1992.

75. The Meropenem Study Group of Leuven LaN: Equivalent efficacies of meropenem and ceftazidime as empirical monotherapy of febrile neutropenic patients. *J Antimicrob Chemother* **36:**185–200, 1995.

76. Cometta A, Glauser MP: Empiric antibiotic monotherapy with carbapenems in febrile neutropenia: A review. *J Chemother* **8:**375–381, 1996.

77. Meunier F, Zinner SH, Gaya H, *et al*: Prospective randomized evaluation of ciprofloxacin versus piperacillin plus amikacin for empiric antibiotic therapy of febrile granulocytopenic cancer patients with lymphomas and solid tumors. The European Organization for Research on Treatment of Cancer International Antimicrobial Therapy Cooperative Group. *Antimicrob Agents Chemother* **35:**873–878, 1991.

78. Shenep JL, Hughes WT, Roberson PK, *et al*: Vancomycin, ticarcillin, and amikacin compared with ticarcillin-clavulanate and amikacin in the empirical treatment of febrile, neutropenic children with cancer. *N Engl J Med* **319:**1053–1058, 1988.

79. Karp JE, Dick JD, Angelopulos C, *et al*: Empiric use of vancomycin during prolonged treatment-induced granulocytopenia. Randomized, double-blind, placebo-controlled clinical trial in patients with acute leukemia. *Am J Med* **81:**237–242, 1986.

80. European Organization for Research and Treatment of Cancer (EORTC) International Antimicrobial Therapy Cooperative Group: National Cancer Institute of Canada-Clinical Trials Group. Vancomycin added to empirical combination antibiotic therapy for fever in granulocytopenic cancer patients. *J Infect Dis* **163:**951–958, 1991.

81. Rubin M, Hathorn JW, Marshall D, Gress J, Steinberg SM, Pizzo PA: Gram-positive infections and the use of vancomycin in 550 episodes of fever and neutropenia. *Ann Intern Med* **108:**30–35, 1988.

82. Ramphal R, Bolger M, Oblon DJ, *et al*: Vancomycin is not an essential component of the initial empiric treatment regimen for febrile neutropenic patients receiving ceftazidime: A randomized prospective study. *Antimicrob Agents Chemother* **36:**1062–1067, 1992.

83. Hughes WT, Armstrong D, Bodey GP, *et al*: 1997 guidelines for the use of antimicrobial agents in neutropenic patients with unexplained fever. Infectious Diseases Society of America. *Clin Infect Dis* **25:**551–573, 1997.

84. Hann I, Viscoli C, Paesmans M, Gaya H, Glauser M: A comparison of outcome from febrile neutropenic episodes in children compared with adults: Results from four EORTC studies. International Antimicrobial Therapy Cooperative Group (IATCG) of the European Organization for Research and Treatment of Cancer (EORTC). *Br J Haematol* **99:**580–588, 1997.

85. Freifeld AG, Pizzo PA: The outpatient management of febrile neutropenia in cancer patients. *Oncology* **10:**599–606, 611–612; discussion, 615–616.

86. Buchanan GR: Approach to treatment of the febrile cancer patient with low-risk neutropenia. *Hematol Oncol Clin North Am* **7:**919–935, 1993.

87. Elting LS, Rubenstein EB, Rolston KV, Bodey GP: Outcomes of bacteremia in patients with cancer and neutropenia: Observations from two decades of epidemiological and clinical trials. *Clin Infect Dis* **25:**247–259, 1997.

88. Hardy I, Gershon AA, Steinberg SP, LaRussa P: The incidence of zoster after immunization with live attenuated varicella vaccine. A study in children with leukemia. Varicella Vaccine Collaborative Study Group. *N Engl J Med* **325:**1545–1550, 1991.

89. Talcott JA, Finberg R, Mayer RJ, Goldman L: The medical course of cancer patients with fever and neutropenia. Clinical identification of a low-risk subgroup at presentation. *Arch Intern Med* **148:**2561–2568, 1988.

90. Talcott JA, Siegel RD, Finberg R, Goldman L: Risk assessment in cancer patients with fever and neutropenia: A prospective, two-center validation of a prediction rule. *J Clin Oncol* **10:**316–322, 1992.

91. Malik IA, Abbas Z, Karim M: Randomised comparison of oral ofloxacin alone with combination of parenteral antibiotics in neutropenic febrile patients. *Lancet* **339:**1092–1096, 1992.

92. Lucas KG, Brown AE, Armstrong D, Chapman D, Heller G: The identification of febrile, neutropenic children with neoplastic disease at low risk for bacteremia and complications of sepsis. *Cancer* **77:**791–798, 1996.

93. Malik IA, Khan WA, Aziz Z, Karim M: Self-administered antibiotic therapy for chemotherapy-induced, low-risk febrile neutropenia in patients with nonhematologic neoplasms. *Clin Infect Dis* **19:**522–527, 1994.

94. Malik IA, Khan WA, Karim M, Aziz Z, Khan MA: Feasibility of outpatient management of fever in cancer patients with low-risk neutropenia: Results of a prospective randomized trial. *Am J Med* **98:**224–231, 1995.

95. Malik IA: Out-patient management of febrile neutropenia in indigent paediatric patients. *Ann Acad Med Singapore* **26:**742–746, 1997.

96. Rubenstein EB, Rolston K, Benjamin RS, *et al*: Outpatient treatment of febrile episodes in low-risk neutropenic patients with cancer. *Cancer* **71:**3640–3646, 1993.

97. Rolston K, Rubenstein EB, Etling L, *et al*: Ambulatory management of febrile episodes in low risk neutropenic patients. In *Programs and Abstracts of the 35th Interscience Conference of Antimicrobial Agents and Chemotherapy (San Francisco)* (abstract 2235). American Society for Microbiology, 1995, p. 333.

97a. Freifeld A, Walsh T, Channock S, *et al*: A double blind comparison of empirical oral and intravenous antibiotic therapy for low risk patients with febrile neutropenia during cancer chemotherapy. *N Engl J Med* **341:**305–311, 1999.

98. Pizzo PA, Robichaud KJ, Gill FA, Witebsky FG: Empiric antibiotic and antifungal therapy for cancer patients with prolonged fever and granulocytopenia. *Am J Med* **72:**101–111, 1982.

99. Lee JW, Pizzo PA: Management of the cancer patient with fever and prolonged neutropenia. *Hematol Oncol Clin North Am* **7:**937–960, 1993.

100. De Pauw BE, Raemaekers JM, Schattenberg T, Donnelly JP: Empirical and subsequent use of antibacterial agents in the febrile neutropenic patient. *J Intern Med Suppl* **740:**69–77, 1997.

101. Walsh TJ, Finberg RW, Arndt C, *et al*: Liposomal amphotericin for empirical therapy in patients with persistent fever and neutropenia. *N Engl J Med* **340:**764–771, 1999.

102. White MH, Bowden RA, Sandler ES, *et al*: Randomized, double-blind clinical trial of amphotericin B colloidal dispersion vs. amphotericin B in the empirical treatment of fever and neutropenia. *Clin Infect Dis* **27:**296–302, 1998.

103. Viscoli C, Castagnola E, Van Lint MT, *et al*: Fluconazole versus amphotericin B as empirical antifungal therapy of unexplained fever in granulocytopenic cancer patients: A pragmatic, multicentre, prospective and randomised clinical trial. *Eur J Cancer* **32A:**814–820, 1996.

104. Viscoli C, Castagnola E, Machetti M: Antifungal treatment in patients with cancer. *J Intern Med Suppl* **740:**89–94, 1997.

105. van der Meer JT, Drew WL, Bowden RA, *et al*: Summary of the International Consensus Symposium on Advances in the Diagnosis, Treatment and Prophylaxis and Cytomegalovirus Infection. *Antiviral Res* **32:**119–140, 1996.

106. Gondo H, Minematsu T, Harada M, *et al*: Cytomegalovirus (CMV) antigenaemia for rapid diagnosis and monitoring of CMV-associated disease after bone marrow transplantation. *Br J Haematol* **86:**130–137, 1994.

107. Einsele H, Ehninger G, Hebart H, *et al*: Polymerase chain reaction monitoring reduces the incidence of cytomegalovirus disease and the duration and side effects of antiviral therapy after bone marrow transplantation. *Blood* **86:**2815–2820, 1995.

108. Bowden RA: Transfusion-transmitted cytomegalovirus infection. *Hematol Oncol Clin North Am* **9:**155–66, 1995.

109. Bowden RA, Slichter SJ, Sayers M, *et al*: A comparison of filtered leukocyte-reduced and cytomegalovirus (CMV) seronegative blood products for the prevention of transfusion-associated CMV infection after marrow transplant. *Blood* **86:**3598–3603, 1995.

110. Feldman S, Hughes WT, Daniel CB: Varicella in children with cancer: Seventy-seven cases. *Pediatrics* **56:**388–397, 1975.

111. Buda K, Tubergen DG, Levin MJ: The frequency and consequences of varicella exposure and varicella infection in children receiving maintenance therapy for acute lymphoblastic leukemia. *J Pediatr Hematol Oncol* **18:**106–112, 1996.

112. von Seidlein L, Gillette SG, Bryson Y, *et al*: Frequent recurrence and persistence of varicella-zoster virus infections in children infected with human immunodeficiency virus type 1. *J Pediatr* **128:**52–57, 1996.

113. Walzer PD, Perl DP, Krogstad DJ, Rawson PG, Schultz MG: *Pneumocystis carinii* pneumonia in the United States. Epidemiologic, diagnostic, and clinical features. *Ann Intern Med* **80:**83–93, 1974.

114. Selik RM, Starcher ET, Curran JW: Opportunistic diseases reported in AIDS patients: Frequencies, associations, and trends. *AIDS* **1:**175–182, 1987.

115. Lewis LL, Butler KM, Husson RN, *et al*: Defining the population of human immunodeficiency virus-infected children at risk for *Mycobacterium avium-intracellulare* infection. *J Pediatr* **121:**677–683, 1992.

116. Centers for Disease Control and Prevention: 1995 Revised guidelines for prophylaxis against *Pneumocystis carinii* pneumonia for children infected with or perinatally exposed to human immunodeficiency virus. *MMWR* **44:**1–12, 1995.

117. Maldonado YA, Araneta RG, Hersh AL: The Northern California Pediatric HIV Consortium: *Pneumocystis carinii* pneumonia prophylaxis and early clinical manifestations of severe perinatal human immunodeficiency virus type 1 infection. *Pediatr Infect Dis J* **17:**398–402, 1998.

118. Gonzales CE, Venson D, Lee S, Mueller BU, Pizzo PA, Walsh TJ: Risk factors for fungemia in children infected with human immunodeficiency virus: A case–control study. *Clin Infect Dis* **23:**515–521, 1996.

119. American Academy of Pediatrics: Immunization in Special Clinical Circumstances. In Pickering LK (ed): *2000 Red Book: Report of the Committee on Infectious Diseases*, 25th ed., American Academy of Pediatrics, Elk Grove Village, IL, 2000, pp. 56–66.

120. Centers for Disease Control and Prevention: Recommendations of the Advisory Committee on Immunization Practices (ACIP): Use of vaccines and immune globulins in persons with altered immunocompetence. *MMWR* **42:**1–18, 1993.

121. Centers for Disease Control and Prevention: 1997 USPHS/IDSA guidelines for the prevention of opportunistic infections in persons infected with human immunodeficiency virus. *Ann Intern Med* **127:**922–946, 1997.

122. Ridgway D, Wolff LJ: Active immunization of children with leukemia and other malignancies. *Leuk Lymphoma* **9:**177–192, 1993.

123. Ambrosino DM, Molrine DC: Critical appraisal of immunization strategies for prevention of infection in the compromised host. *Hematol Oncol Clin North Am* **7:**1027–1050, 1993.

124. Palumbo P, Hoyt L, Demasio K, Oleske J, Connor E: Population-based study of measles and measles immunization in human immunodeficiency virus-infected children. *Pediatr Infect Dis J* **11:**1008–1014, 1992.

125. Kaplan LJ, Daum RS, Smaron M, McCarthy CA: Severe measles in immunocompromised patients. *JAMA* **267:**1237–1241, 1992.

126. LaRussa P, Steinberg S, Gershon AA: Varicella vaccine for immunocompromised children: Results of collaborative studies in the United States and Canada. *J Infect Dis* **174**(Suppl 3)**:**S320–S323, 1996.

127. Abramowsky C, Gonzalez B, Sorensen RU: Disseminated bacillus Calmette–Guérin infections in patients with primary immunodeficiencies. *Am J Clin Pathol* **100:**52–56, 1993.

128. Skinner R, Appleton AL, Sprott MS, *et al*: Disseminated BCG infection in severe combined immunodeficiency presenting with severe anaemia and associated with gross hypersplenism after bone marrow transplantation. *Bone Marrow Transplant* **17:**877–880, 1996.

129. Talbot EA, Perkins MD, Fagundes S, Silva M, Frothingham R: Disseminated bacille Calmette–Guérin disease after vaccination: Case report and review. *Clin Infect Dis* **24:**1139–1146, 1997.

130. Besnard M, Sauvionb S, Offredo C, *et al*: Bacillus Calmette–Guérin infection after vaccination of human immunodeficiency virus-infected children. *Pediatr Infect Dis J* **12:**993–997, 1993.

131. Orgel HA, Hamburger RN, Mendelson LM, Miller JR, Kung FH: Antibody responses in normal infants and in infants receiving chemotherapy for congenital neuroblastoma. *Cancer* **40:**994–997, 1977.

132. Kung FH, Orgel HA, Wallace WW, Hamburger RN: Antibody production following immunization with diphtheria and tetanus toxoids in children receiving chemotherapy during remission of malignant disease. *Pediatrics* **74:**86–89, 1984.

133. Feldman S, Gigliotti F, Shenep JL, Roberson PK, Lott L: Risk of *Haemophilus influenzae* type b disease in children with cancer and response of immunocompromised leukemic children to a conjugate vaccine. *J Infect Dis* **161:**926–931, 1990.

134. Chilcote RR, Baehner RL, Hammond D: Septicemia and meningitis in children splenectomized for Hodgkin's disease. *N Engl J Med* **295:**798–800, 1976.

135. Siber GR: Bacteremias due to *Haemophilus influenzae* and *Streptococcus pneumoniae:* their occurrence and course in children with cancer. *Am J Dis Child* **134:**668–672, 1980.

136. Farley JJ, King JC, Nair P, Hines SE, Tressler RL, Vink PE: Invasive pneumococcal disease among infected and uninfected children of mothers with human immunodeficiency virus infection. *J Pediatr* **124:**853–858, 1994.

137. Janoff EN, Breiman RF, Daley CL, Hopewell PC: Pneumococcal disease during HIV infection. Epidemiology, clinical, and immunologic perspectives. *Ann Intern Med* **117:**314–324, 1992.

138. Kristensen K: Antibody response to a *Haemophilus influenzae* type b polysaccharide tetanus toxoid conjugate vaccine in splenectomized children and adolescents. *Scand Infect Dis J* **24:**629–632, 1992.

139. Shenep JL, Feldman S, Gigliotti F, *et al*: Response of immunocompromised children with solid tumors to a conjugate vaccine for *Haemophilus influenzae* type b. *J Pediatr* **125:**581–584, 1994.

140. Brydak LB, Rokicka-Milewska R, Machala M, Jackowska T, Sikorska-Fic B: Immunogenicity of subunit trivalent influenza vaccine in children with acute lymphoblastic leukemia. *Pediatr Infect Dis J* **17:**125–129, 1998.

141. The National Institute of Child Health and Human Development Intravenous Immunoglobulin Study Group: Intravenous immune globulin for the prevention of bacterial infections in children with symptomatic human immunodeficiency virus infection. *N Engl J Med* **325:**73–80, 1991.

142. Mofenson LM, Moye J Jr: Intravenous immune globulin for the prevention of infections in children with symptomatic human immunodeficiency virus infection. *Pediatr Res* **33**(Suppl):S80–S89, 1993.

143. Yap PL: Prevention of infection in patients with B cell defects: Focus on intravenous immunoglobulin. *Clin Infect Dis* **17**(Suppl 2)**:**S372–375, 1993.

144. Gualtieri RJ, Donowitz GR, Kaiser DL, Hess CE, Sande MA: Double-blind randomized study of prophylactic trimethoprim/sulfamethoxazole in granulocytopenic patients with hematologic malignancies. *Am J Med* **74:**934–940, 1983.

145. Kauffman CA, Liepman MK, Bergman AG, Mioduszewski J: Trimethoprim/sulfamethoxazole prophylaxis in neutropenic patients. Reduction of infections and effect on bacterial and fungal flora. *Am J Med* **74:**599–607, 1983.

146. Cruciani M, Rampazzo R, Malena M, *et al*: Prophylaxis with fluoroquinolones for bacterial infections in neutropenic patients: A meta-analysis. *Clin Infect Dis* **23:**795–805, 1996.

147. Carratala J, Fernandez-Sevilla A, Tubau F, Callis M, Gudiol F: Emergence of quinolone-resistant *Escherichia coli* bacteremia in neutropenic patients with cancer who have received prophylactic norfloxacin. *Clin Infect Dis* **20:**557–560; discussion 561–563, 1995.

148. Donnelly JP: Is there a rationale for the use of antimicrobial prophylaxis in neutropenic patients? *J Intern Med Suppl* **740:**79–88, 1997.

149. Lietman PS: Fluoroquinolone toxicities. An update. *Drugs* **49:**159–163, 1995.

150. Schaad UB, Wedgwood-Krucko J: Nalidixic acid in children: Retrospective matched controlled study for cartilage toxicity. *Infection* **15:**165–168, 1987.

151. Nuutinen M, Turtinen J, Uhari M: Growth and joint symptoms in children treated with nalidixic acid. *Pediatr Infect Dis J* **13:**798–800, 1994.

152. Hampel B, Hullmann R, Schmidt H: Ciprofloxacin in pediatrics: Worldwide clinical experience based on compassionate use—Safety report. *Pediatr Infect Dis J* **16:**127–129; discussion 160–162, 1997.

153. Dumois JA: Tuberculosis in children with HIV infection. *Pediatr AIDS HIV Infect Fetus Adolesc* **3:**177–182, 1992.

154. Chan SP, Birnbaum J, Rao M, Steiner P: Clinical manifestations and outcome of tuberculosis in children with acquired immunodeficiency syndrome. *Pediatr Infect Dis J* **15:**443–447, 1996.

155. Committee on Infectious Diseases: Update on tuberculosis skin testing of children. *Pediatrics* **97:**282–284, 1996.

156. Winston DJ, Chandrasekar PH, Lazarus HM, *et al*: Fluconazole prophylaxis of fungal infections in patients with acute leukemia. Results of a randomized placebo-controlled, double-blind, multicenter trial. *Ann Intern Med* **118:**495-503, 1993.

157. Schaffner A, Schaffner M: Effect of prophylactic fluconazole on the frequency of fungal infections, amphotericin B use, and health care costs in patients undergoing intensive chemotherapy for hematologic neoplasias. *J Infect Dis* **172:**1035–1041, 1995.

158. Gotzsche PC, Johansen HK: Meta-analysis of prophylactic or empirical antifungal treatment versus placebo or no treatment in patients with cancer complicated by neutropenia. *Br Med J* **314:**1238–1244, 1997.

159. Maenza JR, Keruly JC, Moore RD, Chaisson RE, Merz WG, Gallant JE: Risk factors for fluconazole-resistant candidiasis in human immunodeficiency virus-infected patients. *J Infect Dis* **173:** 219–225, 1996.
160. Wingard JR, Merz WG, Rinaldi MG, Johnson TR, Karp JE, Saral R: Increase in *Candida krusei* infection among patients with bone marrow transplantation and neutropenia treated prophylactically with fluconazole. *N Engl J Med* **325:**1274–1277, 1991.
161. Shetty D, Giri N, Gonzalez CE, Pizzo PA, Walsh TJ: Invasive aspergillosis in human immunodeficiency virus-infected children. *Pediatr Infect Dis J* **16:**216–221, 1997.
162. Weinberger M, Elattar I, Marshall D, et al: Patterns of infections in patients with aplastic anemia and the emergence of *Aspergillus* as a major cause of death. *Medicine* **71:**24–42, 1992.
163. Leggiadro RJ, Barrett FF, Hughes WT: Extrapulmonary cryptococcosis in immunocompromised infants and children. *Pediatr Infect Dis J* **11:**43–47, 1992.
164. Allende M, Pizzo PA, Horowitz M, Pass HI, Walsh TJ: Pulmonary cryptococcosis presenting as metastases in children with sarcomas. *Pediatr Infect Dis J* **12:**240–243, 1993.
165. Gonzales GE, Shetty D, Lewis LL, Mueller BU, Pizzo PA, Walsh TJ: Cryptococcosis in human immunodeficiency virus-infected children. *Pediatr Infect Dis J* **15:**796–800, 1996.
166. Hughes WT, Kuhn S, Chaudhary S, et al: Successful chemoprophylaxis for *Pneumocystis carinii* pneumonitis. *N Engl J Med* **297:**1419–1426, 1977.
167. Hughes WT: Five-year absence of *Pneumocystis carinii* pneumonitis in a pediatric oncology center. *J Infect Dis* **150:**305–306, 1984.
168. Leggiadro RJ, Winkelstein JA, Hughes WT: Prevalence of *Pneumocystis carinii* pneumonitis in severe combined immunodeficiency. *J Pediatr* **99:**96–98, 1981.
169. Sepkowitz KA, Brown AE, Telzak EE, Gottlieb S, Armstrong D: *Pneumocystis carinii* pneumonia among patients without AIDS at a cancer hospital. *JAMA* **267:**832–837, 1992.
170. Henson JW, Jalaj JK, Walker RW, Stover DE, Fels AO: *Pneumocystis carinii* pneumonia in patients with primary brain tumors. *Arch Neurol* **48:**406–409, 1991.
171. Schiff D: Pneumocystis pneumonia in brain tumor patients: Risk factors and clinical features. *J Neurooncol* **27:**235–240, 1996.
172. Cangir A, Morgan SK, Land VJ, Pullen J, Starling SA, Nitschke R: Combination chemotherapy with adramycin (NSC-123127) and dimethyl triazeno imidazole carboxamide (DTIC) (NSC-45388) in children with metastatic sold tumors. *Med Pediatr Oncol* **2:**183–190, 1976.
173. Chusid MJ, Heyrman KA: An outbreak of *Pneumocystis carinii* pneumonia at a pediatric hospital. *Pediatrics* **62:**1031–1035, 1978.
174. Cangir A, Vietti TJ, Gehan EA, et al: Ewing's sarcoma metastatic at diagnosis. Results and comparisons of two intergroup Ewing's sarcoma studies. *Cancer* **66:**887–893, 1990.
175. Simonds RJ, Oxtoby MJ, Caldwell B, Gwinn ML, Rogers MF: *Pneumocystis carinii* pneumonia among US children with perinatally acquired HIV infection. *JAMA* **270:**470–473, 1993.
176. Centers for Disease Control: Guidelines for prophylaxis against *Pneumocystis carinii* pneumonia for children infected with human immunodeficiency virus. *MMWR* **40:**1–13, 1991.
177. Simonds RJ, Lindegren ML, Thomas P, et al: Prophylaxis against *Pneumocystis carinii* pneumonia among children with perinatally acquired human immunodeficiency virus infection in the United States. *N Engl J Med* **332:**786–790, 1995.
178. Thea DM, Lambert G, Weedon J, et al: Benefit of primary pro-
phylaxis before 18 months of age in reducing the incidence of *Pneumocystis carinii* pneumonia and early death in a cohort of 112 human immunodeficiency virus-infected infants. *Pediatrics* **97:** 59–64, 1996.
179. Mueller BU, Butler KM, Husson RN, Pizzo PA: *Pneumocystis carinii* pneumonia despite prophylaxis in children with human immunodeficiency virus infection. *J Pediatr* **119:**992–994, 1991.
180. Nachman SA, Mueller BU, Mirochnik M, Pizzo PA: High failure rate of dapsone and pentamidine as *Pneumocystis carinii* pneumonia prophylaxis in human immunodeficiency virus-infected children. *Pediatr Infect Dis J* **13:**1004–1006, 1994.
181. Mitchell CD, Erlich SS, Mastrucci MT, Hutto JC, Scott GB, Park WB: Congenital toxoplasmosis occurring in infants infected with human immunodeficiency virus 1. *Pediatr Infect Dis J* **9:**512–518, 1990.
182. Carr A, Tindall B, Brew BJ, et al: Low-dose trimethoprim-sulfamethoxazole prophylaxis for toxoplasmic encephalitis in patients with AIDS. *Ann Intern Med* **117:**106–111, 1992.
183. Gentile G, Venditti M, Micozzi A, et al: Cryptosporidiosis in patients with hematologic malignancies. *Rev Infect Dis* **13:**842–846, 1991.
184. Flanigan T, Whalen C, Turner J, et al: *Cryptosporidium* infection and CD4 counts. *Ann Intern Med* **116:**840–842, 1992.
185. Mac Kenzie WR, Hoxie NJ, Proctor ME, et al: A massive outbreak in Milwaukee of *Cryptosporidium* infection transmitted through the public water supply. *N Engl J Med* **331:**161–167, 1994.
186. Hall CB, Powell KR, MacDonald NE, et al: Respiratory syncytial viral infection in children with compromised immune function. *N Engl J Med* **315:**77–81, 1986.
187. Hertz MI, Englund JA, Snover D, Bitterman PB, McGlave PB: Respiratory syncytial virus-induced acute lung injury in adult patients with bone marrow transplants: A clinical approach and review of the literature. *Medicine* **68:**269–281, 1989.
188. Harrington RD, Hooton TM, Hackman RC, et al: An outbreak of respiratory syncytial virus in a bone marrow transplant center. *J Infect Dis* **165:**987–993, 1992.
189. Groothuis JR, Simoes EA, Levin MJ, et al: Prophylactic administration of respiratory syncytial virus immune globulin to high-risk infants and young children. The Respiratory Syncytial Virus Immune Globulin Study Group. *N Engl J Med* **329:**1524–1530, 1993.
190. The PREVENT Study Group: Reduction of respiratory syncytial virus hospitalization among premature infants and infants with bronchopulmonary dysplasia using respiratory syncytial virus immune globulin prophylaxis. *Pediatrics* **99:**93–99, 1997.
191. The IMpact-RSV Study Group: Palivizumab, a humanized respiratory syncytial virus monoclonal antibody, reduces hospitalization from respiratory syncytial virus infection in high-risk infants. *Pediatrics* **102:**531–537, 1998.
192. Nemunaitis J, Rosenfeld CS, Ash R, et al: Phase III randomized, double-blind placebo-controlled trial of rhGM-CSF following allogeneic bone marrow transplantation. *Bone Marrow Transplant* **15:**949–954, 1995.
193. Piguet D, Chapuis B: Recombinant human granulocyte-macrophage colony-stimulating factor in acquired or chemotherapy-induced neutropenia. An open clinical trial. *Acta Oncol* **33:**639–643, 1994.
194. Pui C-H, Boyett JM, Hughes WT, et al: Human granulocyte colony-stimulating factor after induction chemotherapy in children with acute lymphoblastic leukemia. *N Engl J Med* **336:**1781–1787, 1997.
195. Zuccotti GV, Plebani A, Biasucci G, et al: Granulocyte-colony

stimulating factor and erythropoietin therapy in children with human immunodeficiency virus infection. *J Intern Med Res* **24:** 115–121, 1996.

196. Wang JM, Chen ZG, Colella S, *et al*: Chemotactic activity of recombinant human granulocyte colony-stimulating factor. *Blood* **72:**1456–1460, 1988.

197. Welte K, Zeidler C, Reiter A, *et al*: Differential effects of granulocyte-macrophage colony-stimulating factor and granulocyte colony-stimulating factor in children with severe congenital neutropenia. *Blood* **75:**1056–1063, 1990.

198. Weisbart RH, Golde DW, Clark SC, Wong GG, Gasson JC: Human granulocyte-macrophage colony-stimulating factor is a neutrophil activator. *Nature* **314:**361–363, 1985.

199. Sieff CA, Emerson SG, Donahue RE, *et al*: Human recombinant granulocyte-macrophage colony-stimulating factor: A multilineage hematopoietin. *Science* **230:**1171–1173, 1985.

200. Roilides E, Walsh TJ, Pizzo PA, Rubin M: Granulocyte colony-stimulating factor enhances the phagocytic and bactericidal activity of normal and defective human neutrophils. *J Infect Dis* **163:** 579–583, 1991.

201. American Society of Clinical Oncology: American Society of Clinical Oncology recommendations for the use of hematopoietic colony-stimulating factors: Evidence-based, clinical practice guidelines. *J Clin Oncol* **12:**2471–2508, 1994.

202. American Society of Clinical Oncology: 1997 update of recommendations for the use of hematopoietic colony-stimulating factors: Evidence-based, clinical practice guidelines. *J Clin Oncol* **15:** 3288, 1997.

203. Maher DW, Lieschke GJ, Green M, *et al*: Filgrastim in patients with chemotherapy-induced febrile neutropenia. A double-blind, placebo-controlled trial. *Ann Intern Med* **121:**492–501, 1994.

204. Riikonen P, Saarinen UM, Mäkipernaa A, *et al*: Recombinant human granulocyte-macrophage colony-stimulating factor in the treatment of febrile neutropenia: a double blind placebo-controlled study in children. *Pediatr Infect Dis J* **13:**197–202, 1994.

205. Mitchell PL, Morland B, Stevens MC, *et al*: Granulocyte colony-stimulating factor in established febrile neutropenia: A randomized study of pediatric patients. *J Clin Oncol* **15:**1163–1170, 1997.

206. Vellenga E, Uyl-de Groot CA, de Wit R, *et al*: Randomized placebo-controlled trial of granulocyte-macrophage colony-stimulating factor in patients with chemotherapy-related febrile neutropenia. *J Clin Oncol* **14:**619–627, 1996.

207. Anaissie EJ, Vartivarian S, Bodey GP, *et al*: Randomized compari-son between antibiotics alone and antibiotics plus granulocyte-macrophage colony-stimulating factor (*Escherichia coli*-derived in cancer patients with fever and neutropenia. *Am J Med* **100:**17–23, 1996.

208. Crawford J, Ozer H, Stoller R, *et al*: Reduction by granulocyte colony-stimulating factor of fever and neutropenia induced by chemotherapy in patients with small-cell lung cancer. *N Engl J Med* **325:**164–170, 1991.

209. Pettengell R, Gurney H, Radford JA, *et al*: Granulocyte colony-stimulating factor to prevent dose-limiting neutropenia in non-Hodgkin's lymphoma: A randomized controlled trial. *Blood* **80:** 1430–1436, 1992.

210. Gerhartz HH, Engelhard M, Meusers P, *et al*: Randomized, double-blind, placebo-controlled, phase III study of recombinant human granulocyte-macrophage colony-stimulating factor as adjunct to induction treatment of high-grade malignant non-Hodgkin's lymphomas. *Blood* **82:**2329–2339, 1993.

211. Ohno R, Tomonaga M, Ohshima T, *et al*: A randomized controlled study of granulocyte colony stimulating factor after intensive induction and consolidation therapy in patients with acute lymphoblastic leukemia. *Int J Hematol* **58:**73–81, 1993.

212. Heil G, Chadid L, Hoelzer D, *et al*: GM-CSF in a double-blind randomized, placebo controlled trial in therapy of adult patients with de novo acute myeloid leukemia (AML). *Leukemia* **9:**3–9, 1995.

213. Farber BF, Lesser M, Kaplan MH, Woltmann J, Napolitano B, Armellino D: Clinical significance of neutropenia in patients with human immunodeficiency virus infection. *Infect Control Hosp Epidemiol* **12:**429–434, 1991.

214. Moore RD, Keruly JC, Chaisson RE: Neutropenia and bacterial infection in acquired immunodeficiency syndrome. *Arch Intern Med* **155:**1965–1970, 1995.

215. Meynard JL, Guiguet M, Arsac S, Frottier J, Meyohas MC: Frequency and risk factors of infectious complications in neutropenic patients infected with HIV. *AIDS* **11:**995–998, 1997.

216. Mueller BU, Jacobsen F, Butler KM, Husson RN, Lewis LL, Pizzo PA: Combination treatment with azidothymidine and granulocyte colony-stimulating factor in children with human immunodeficiency virus infection. *J Pediatr* **121:**797–802, 1992.

217. Hermans P: Haematopoietic growth factors as supportive therapy in HIV-infected patients. *AIDS* **9**(suppl 2):S9–S14, 1995.

218. Kalter DC, Nakamura M, Turpin JA, *et al*: Enhanced HIV replication in macrophage colony-stimulating factor-treated monocytes. *J Immunol* **146:**298–306, 1991.

14

Infections Complicating Congenital Immunodeficiency Syndromes

JOHN C. CHRISTENSON and HARRY R. HILL

1. Introduction

Nothing is more challenging to the physician than the management of a serious infection in an immunocompromised host. Under such circumstances, the clinician must bring to bear all of his or her knowledge of microbiology and antimicrobial therapy, but in addition he or she must have a firm understanding of the host defense abnormality underlying that individual patient's disease. As indicated by Stollerman,[1] we can no longer have a primarily parasite-oriented approach to infectious diseases but must include host factors in the data on which we make potentially life-saving decisions.

Nowhere is this truer than in the management of the patient with a congenital defect in their host defense system. Moreover, no clinician (pediatrician, internist, otolaryngologist, surgeon) can afford to ignore this area of medicine, since many such patients are being discovered and are being kept alive and functioning almost normally well into adulthood. It is absolutely essential that the diagnosis of a congenital immunodeficiency disease be made as soon as possible. The reader might ask, "Why," since there are few definitive procedures for correcting serious defects in the host defense mechanism.

John C. Christenson • Department of Pediatrics, Division of Infectious Disease and Geographic Medicine, University of Utah School of Medicine, Salt Lake City, Utah 84132.   **Harry R. Hill** • Departments of Pediatrics, Internal Medicine, and Pathology, Division of Clinical Immunology and Allergy, University of Utah School of Medicine, Salt Lake City, Utah 84132.

Clinical Approach to Infection in the Compromised Host (Fourth Edition), edited by Robert H. Rubin and Lowell S. Young. Kluwer Academic/Plenum Publishers, New York, 2002.

The answer to that question lies in the fact that survival and often the quality of life have improved in almost all instances in which the basic underlying pathophysiology of an immune defect has been elucidated. A prime example of this is chronic granulomatous disease, once known as fatal granulomatous disease. Since the basic defect in phagocyte intracellular bactericidal activity was discovered in this disease, more appropriate antimicrobial therapy combined with judicious use of surgical drainage in combination with the use of immunomodulator agents such as interferon-gamma have resulted in less morbidity and mortality. Similarly, appropriate use of antibiotics along with immunoglobulin therapy has had an extremely beneficial effect in patients with hypogammaglobulinemia.

A critical factor in the detection of immunodeficient patients is to maintain a high degree of suspicion in order to discover such individuals as early as possible. In one study of patients with acquired common variable hypogammaglobulinemia, a period of approximately 10 years lapsed between onset of recurrent infections and diagnosis of the antibody deficiency.[2] This is unacceptable by present-day standards, since simple tests are available in most hospital and commercial laboratories to adequately screen for hypogammaglobulinemia. Although there are several excellent reviews on detecting the immunodeficient patient,[1,3–9] we shall briefly review the factors that should alert the clinician to the possibility of a defect and then discuss the readily available tests that are useful in screening suspected patients.

Deciding which patients are suffering from a host defense abnormality and which require further investigation is difficult in many instances. A careful history detailing the number, type, and severity of infections is critical. Particular attention should be addressed to determining whether the infection and its etiology were documented by culture results, molecular diagnostic tests, roentgenograms, scans, antibody titer rises, or other means. This is essential since often a diagnosis of pneumonia or blood "poisoning" is mentioned by the patient without any documentation made available. A complete family history, concentrating on recurrent infections, early deaths, malignancies, and consanguinity, also may yield valuable data. A thorough physical examination and appropriate laboratory tests should be performed to rule out physical or anatomic defects that might lead to recurrent infections. Recurrent meningitis because of dermal sinus or basilar skull fracture and recurrent urinary tract infections secondary to ureteral problems are prime examples of such anatomical abnormalities.

Moreover, one must exclude those patients who are normal but who are exposed in their environment to a number of respiratory illnesses, and thus are often ill. It has been shown that preschool and school-aged children may have 6 and up to 12 respiratory infections per year.[10] If one assumes a 2-week period for each individual infection, including prodrome, disease, and convalescence, then it would not be unusual for a child to be sick 12 (3 months) to 24 weeks (6 months) per year. An adult probably averages 2 to 4 respiratory infections per year. These figures should be kept in mind, as they are very useful in explaining recurrent infections to the individual who turns out on testing to have a normal immune system. We would point out that we have yet to see pharyngitis, tonsillitis, or colds alone be a serious problem in any of our true immunodeficient patients.

After determining that a patient's recurrent infections are not the result of anatomical defects or epidemiologic exposure, then it is important to divide the host defense mechanism into its major components and consider the type of infection generally seen with defects in each system. Generally, patients with abnormalities in the thymus-dependent portions of their immune systems (cell-mediated immunity) tend to have severe or recurrent viral infections caused by varicella zoster virus (VZV), herpes simplex virus (HSV), cytomegalovirus (CMV), or fungal and yeast infections such as those caused by *Candida albicans*.[11–14] A variety of intracellular bacterial pathogens and *Pneumocystis carinii* infections also occur. Abnormalities in T-lymphocyte numbers and function can be screened for fairly simply employing skin tests for delayed hypersensitivity, lymphocyte mitogenic responses, and enumeration of T cells and T-cell subsets.[15] We currently employ *C. albicans* antigen and diphtheria–tetanus

toxoid. Tritiated thymidine incorporation is used to determine the mitogenic response of lymphocytes exposed to the nonspecific mitogens phytohemagglutinin or concanavalin A or to tetanus toxoid or *Candida* antigen. More recently, analysis of the production of T-cell cytokines such as interleukin-2 (IL-2), IL-12, or interferon-gamma (IFN-γ) or the receptors for these and other critical cytokines can be employed in the workup of these patients.

One of the major functions of antibody in the host defense mechanism is to opsonize or coat bacteria so that phagocytic cells may ingest and kill them. Antibody deficiency syndromes are often characterized by severe and recurrent pyogenic bacterial infections. Almost all such patients have recurrent respiratory infections that include draining otitis media, mastoiditis, sinusitis, bronchitis, and multiple episodes of pneumonia.[2,16] This often leads to the development of bronchiectasis and chronic respiratory problems.[17] These patients also suffer from chronic diarrhea and may have systemic infections such as sepsis, meningitis, or osteomyelitis.[2,16,17]

The diagnosis of hypogammaglobulinemia is dependent on (1) the history of severe or recurrent bacterial infections, (2) a low level of immunoglobulins, and (3) the inability to make good specific antibodies. Hypogammaglobulinemia is best documented by determining IgG, IgM, and IgA levels by nephelometry. These tests are far more sensitive and specific in quantitating immunoglobulins than is protein electrophoresis or immunoelectrophoresis. Recurrent sinopulmonary infections suggest an immunoglobulin deficiency. In those patients with normal IgG concentrations, a selective deficiency in IgG subclasses is still a possibility.[18–21]

The ability of a patient to make good specific antibody also can be assessed fairly easily using common serologic techniques that are available in most hospital or commercial laboratories. Anti-blood group A and B titers, or isohemagglutinins, can be run wherever blood typing is performed. These are predominantly IgM antibodies directed against cross-reacting polysaccharide antigens on bacteria normally present in the gastrointestinal (GI) flora. By 6 months of age, a child should have a titer of 1:8 or greater against A or B substance unless the blood type is AB.[7] Adults generally have titers of 1:32 to 1:128. Other serologic tests that can be used to assess specific antibody production include the antistreptolysin O, anti-DNAase B, or Streptozyme test if a patient has had a past documented streptococcal infection or a rubella titer if the patient has already received this vaccine. (Note that immunization with live virus vaccines is not recommended in patients with immune deficiencies, including antibody deficiencies, since severe infections may occur.) Alternatively, one can measure influenza, diphtheria, or tetanus titers following immunization with influenza vaccine or diphtheria–tetanus toxoid. In addition, specific antibody responses can be measured after immunization with *Haemophilus influenzae* type B conjugate and pneumococcal vaccines. The assessment of an immune response to polysaccharides is important for the detection of those patients with normal immunoglobulin levels who have an impaired response to these antigens.[22–24] Finally, one can enumerate the number of peripheral blood lymphocytes that have immunoglobulin on their surface or surface markers for B cells.

Congenital deficiencies in the complement system are often associated with infections similar to those observed in hypogammaglobulinemia and include respiratory infections, sepsis, and meningitis.[25–32] Commonly available tests for measuring complement activity and levels include nephelometry, radial immunodiffusion (commercial kits are available for measuring C3, C4, C5, and factor B) and hemolytic assays or enzyme-linked immunoassays for determining total hemolytic complement.[33] Through the use of these tests, it is often possible to detect total absence of a single complement component and to define activation of one or more pathways during specific infections or other disease processes.

Infections in patients with phagocyte abnormalities often manifest as abscesses or episodes of cellulitis.[34–42] Staphylococci predominate as the etiologic agent; this is probably because this organism is the most numerous among the skin flora. Streptococci and some gram-negative infections as well as those caused *by C. albicans* also occur in these patients.[35,39,43] The metabolic activity of phagocytes, which relates to their ability to activate the hexose monophosphate shunt and generate toxic oxygen radicals essential to microbicidal activity, can be assessed with the nitroblue tetrazolium (NBT) dye-reduction test,[44,45] detection of chemiluminescence,[46,47] or through flow cytometry using dihydrorhodamine 123 fluorescence.[48,49] In the NBT test, a yellow dye solution is mixed with a drop of blood and often with a substance such as endotoxin that stimulates phagocyte metabolism. After an appropriate interval, a smear is made and the cells that have activated their respiratory burst and developed a black deposit of NBTH within their cytoplasm are enumerated.[44] Patients with defects in this important microbicidal pathway fail to reduce the dye and therefore do not form black deposits in their cells.

In the chemiluminescence (CL) assay, polymorphonuclear leukocytes (PMNs) are incubated with a phagocytizable particle in a dark-adapted scintillation vial.[46,47] The mixture is placed in a liquid scintillation counter, out of phase, with one photomultiplier tube disconnected, and counted at intervals for 60–90 min. Following particle ingestion, the microbicidal mechanism of both PMNs and monocytes are activated and excited molecular oxygen species are generated.[50,51] On decaying to the ground state, these emit photons that can be measured as light in the scintillation system. Patients with defects in the generation of these oxygen species either fail to generate or produce lower levels of chemiluminescence than do controls.[52–54] This technique represents a simple and sensitive means for screening for defects in phagocyte oxidative metabolism. Myeloperoxidase deficiency also may be detected using this technique as well as by a simple histochemical stain.[52] All suspected defects in microbicidal activity should be confirmed by more classic phagocytosis and killing assays.[55]

At this time, flow cytometry utilizing dihydrorhodamine 123 appears to be the most sensitive assay for the diagnosis of patients with chronic granulomatous disease. Patients with recessive disease also may be detected using this methodology.[48,49]

Defects in leukocyte motility are best screened for *in vitro* using the Boyden chamber or a migration under agarose technique.[56–59] These assays have been found to correlate more closely with clinical abnormalities than do the results of skin window techniques.[60] These are more difficult tests, of course, and are not commonly available except at larger medical centers. One additional laboratory test, the serum level of IgE, may be of value in the diagnosis of a number of patients with the syndrome of hyperimmunoglobulinemia E, recurrent infections, and defective PMN chemotaxis.[35,36] Extremely high levels of IgE are usually present in these patients[35] as well as in the closely related or identical syndrome of Job.[61] Table 1 summarizes the appropriate tests to be employed in the patient suspected of an immune deficiency.

2. Aim of Therapy in Congenital Immunodeficiency Diseases

2.1. Treatment of Life-Threatening Infections

After the diagnosis of immunodeficiency disease is established, what are the therapeutic goals in the management of infections in these patients? These can be grouped as follows: (1) prevention or therapy of potentially life-threatening infections; (2) therapy of less severe acute infections; and (3) management of chronic infection so as to minimize the development of long-term sequelae.

Patients with congenital immune deficiency are es-

TABLE 1. Immunologic Screening Tests

Portion of the immune system	Tests
Antibody	Serum IgG, IgM, IgA levels
	Serum IgG subclass levels
	IgG response to proteins (diphtheria, tetanus) and polysaccharides (pneumococcus, *Haemophilus influenzae* type B, meningococcus)
	Isohemagglutinin titers for IgM antibody response
Cell-mediated	Total lymphocyte count (complete blood count, differential)
	Skin tests (*C. albicans*, diphtheria, tetanus)
	T cells and T-cell subsets
	Mitogen responses
	HIV antibody, if disease is suspected
Complement	Total hemolytic complement (CH50)
	Quantitation of serum complement components (C2, C3, C4, C6, and factor B)
Phagocyte	WBC and differential count
	NBT or dihydrorhodamine fluorescence test for respiratory burst activity
	Serum IgE levels
	Flow cytometric analysis for CD11/18
	Chemotaxis assay

pecially prone to develop sudden overwhelming and often fatal infection. In such cases, it is absolutely essential that the clinician have an understanding of the basic underlying defect in the host defense mechanism, the most likely etiologic agents in infection, and the available therapeutic modalities. The following illustrate a few examples:

1. Patients with the Wiskott–Aldrich syndrome are particularly susceptible to the development of serious overwhelming infections with *Streptococcus pneumoniae*, *Haemophilus influenzae*, and other encapsulated bacteria because they fail to make good antibodies to polysaccharide antigens.[62,63] A number of individuals with this syndrome have died of such infection. Thus, in an acutely ill patient with this particular immunodeficiency, therapy would logically include the intravenous administration of an antibiotic that would be effective against these organisms and intravenous immunoglobulin in an attempt to supply missing antibodies. Severe viral infections such as those due to respiratory syncytial virus and rotavirus also occur in these patients and can be life threatening.

2. Patients with chronic granulomatous disease (CGD) often develop severe pneumonias and other infections caused by staphylococci and a variety of other pathogens such as *Nocardia* and *Aspergillus* because their leukocytes fail to kill these organisms following inges-

tion.[34,37,39,40] A CGD patient with severe pneumonia should be worked up completely to establish a diagnosis but treated initially with antimicrobial agents directed against staphylococci and especially drugs that are known to penetrate into cells. Thus, nafcillin or oxacillin along with agents such as rifampin or an aminoglycoside may be employed for best results.[64,65]

3. Patients with congenital absence of their terminal complement components (C6, C7, C8) often suffer repeated disseminated infections with *Neisseria meningitidis* or *N. gonorrhoeae*.[28,29,31,32] An acutely toxic patient known to have such a defect in his complement system should receive appropriate intravenous antimicrobial therapy for these organisms. Fresh frozen plasma therapy also might conceivably be of value in supplying the missing complement components. No data on the clinical efficacy of such therapy are available, however.

Such dramatic examples point out the need for a thorough knowledge of host defense abnormalities, likely microbial pathogens, and newer therapeutic modalities in the successful therapy of these life-threatening infections in the congenital immunodeficient syndromes.

2.1.1. Early Diagnosis of Infection

Early diagnosis of infection and correct definition of the etiologic agent are extremely important in managing severe life-threatening infections in immune-deficient hosts. Again, a high degree of suspicion must be maintained for infection in these patients. A variety of diagnostic tests may lend additional support in suggesting the presence of infection. These include routine and more specialized cultures such as transtracheal aspirates,[66] lung aspirates,[67,68] bronchoalveolar lavage,[69,70] or open biopsies[71] in diagnosing pneumonias and cerebrospinal fluid (CSF) taps or even brain biopsies in central nervous system (CNS) infections. Polymerase chain reaction (PCR) assays of CSF, serum, or tissue also may be useful in the diagnosis of certain viral infections.[72,73] A variety of scans are now available that have proven useful in defining infections in these patients and include technetium (^{99m}Tc)-labeled bone scans, computed tomography (CT), and magnetic resonance imaging (MRI). Although a positive result with these latter techniques may be quite helpful, a negative one does not rule out the presence of infection in an immunocompromised host. On several occasions, we have seen osteomyelitis progress significantly in the face of negative bone scans in such patients or the presence of abscesses without being detected on routine scanning procedures. Again, the most likely problem in these patients is infection, and one should not rule it out on the basis of laboratory tests or radiographs alone.

Additional tests for the rapid diagnosis of bacterial meningitis that may be of value in the immunodeficient patient should be mentioned. These include the limulus lysate test for gram-negative endotoxin in CSF and CSF lactate levels.[74–76] The limulus lysate test has been reported to give positive results in up to 99% of patients with gram-negative meningitis.[75] The CSF lactate, which increases during bacterial but not viral CNS infection, is generally elevated in almost all patients with culture-proven bacterial meningitis.[75,77] Some laboratories can measure levels of tumor necrosis factor and interleukins in CSF. When compared to patients with viral meningitis, patients with bacterial meningitis will have higher levels of these cytokines, which may help in the differentiation between bacterial or viral etiologies.

In recent years, molecular assays such as PCR have been a great addition to the diagnostic armamentarium. PCRs are commercially available for the diagnosis of infections caused by CMV, human immunodeficiency virus (HIV), HSV, human herpesvirus 6 (HHV-6), and enterovirus. Newer PCRs are being developed for the diagnosis of infections caused by bacteria such as *Mycobacterium tuberculosis*, *Bordetella pertussis*, *Streptococcus pneumoniae*, and *Bartonella hensalae*.[78]

In summary, the key to early diagnosis of infection in the congenitally immunodeficient patient is a high index of suspicion for infection. On the first sign of such a problem, appropriate routine and specialized diagnostic cultures and antibody studies should be obtained. If the illness appears less than immediately life threatening, radiographs and scans may be helpful. If the illness does appear life threatening, appropriate antimicrobial therapy should be administered immediately and should be based on a clear understanding of the patient's host defense abnormality and the most likely etiologic agents.

2.1.2. Appropriate Antimicrobial Therapy

Selecting an appropriate antimicrobial agent in the severely ill immunocompromised host is a difficult task. We must often forsake much of what is taught in infectious disease training programs and initiate antimicrobial therapy with the broadest, most potent antimicrobial agents. Thus, initial therapeutic regimens should be designed to quickly stem the progression of infection and prevent a fatal outcome. To illustrate, we were recently challenged by the presentation of one of our Wiskott–Aldrich syndrome patients in the emergency room with a history of sudden onset of fever, chills, and extreme leth-

argy. On physical examination, he had a fever of 104°F and the usual eczema and petechiae seen in these patients. Knowing that such patients have great difficulty in making antibodies to polysaccharide antigens, we initially suspected infection with *S. pneumoniae*, *H. influenzae*, or *N. meningitidis*, but we could not rule out other organisms such as *E. coli* or *Klebsiella pneumoniae*. Since this was obviously a life-threatening infection, we drew appropriate cultures, did a lumbar puncture, and initiated therapy with vancomycin and ceftriaxone. In this critically ill patient, we believe we were justified in treating all or most of the potential pathogens immediately. The next day, the patient was alert, active, and clamoring to go home when a blood culture from the preceding day grew out *S. pneumoniae*. Again, this course of therapy was initiated for a brief period of time to give us time to make a specific diagnosis. As soon as the etiology of his infection was known and antimicrobial susceptibility information was available, a narrower spectrum agent was continued.[79] Thus, only a very limited course of broad-spectrum antimicrobial therapy was given, which would be highly unlikely to result in abnormal colonization or superinfection. The antimicrobial agents selected for treating life-threatening infections in the immunocompromised host should be chosen based on the patient's host defense defect and the most likely pathogen involved, as will be discussed later in this chapter. In addition, an attempt should be made to select bactericidal rather than bacteriostatic antibiotics when possible, since the host's immune system may not be capable of microbial killing. Intravenous therapy is indicated for all such serious infections.

2.1.3. Immunologic Adjuncts to Therapy

After appropriate cultures and laboratory tests have been obtained and antimicrobial therapy has been instituted in the critically ill congenitally immunodeficient patient, the clinician should consider what he might do to enhance host defenses.

In the antibody-deficient patient, an effort should be directed toward supplying missing antibodies. In the acutely ill individual, this can be done with intravenous gammaglobulin (IVIG), 400–600 mg/kg infused over 2–4 hr. Its use has been shown to ameliorate the severity of the illness and improve survival.[80–82] IVIG is preferred over fresh frozen plasma because the risk of HIV infection or hepatitis is now almost nonexistent. In addition, IVIG is now readily available in all hospital pharmacies for use when necessary.

There are now a number of immunoglobulin preparations modified for intravenous use. These preparations are altered in some way so that they do not contain immunoglobulin aggregates that are capable of activating the complement system. Intramuscular immunoglobulin preparations containing aggregates can trigger the complement system if inadvertently injected intravenously and thereby cause anaphylacticlike reactions.[83,84] Such intravenous immunoglobulin preparations have been found to be superior to intramuscular globulin preparations in reducing specific acute illnesses and reducing the number of days that hypogammaglobulinemic individuals required antibiotics.[82,85] Furthermore, they also have been found to be effective in preventing symptomatic CMV infection following transplantation[86] and treating chronic echovirus encephalitis.[87] Hyperimmune IVIG preparations and monoclonal antibodies also may have a role in preventing and even treating severe disease in the immunocompromised host much as has been shown with hyperimmune preparations to respiratory syncytial virus. Plasma therapy also has been used with some success in Wiskott–Aldrich syndrome patients,[88] in patients with ataxia telangiectasia,[89] and in patients with complement component deficiencies.[84]

Additional immediate immunotherapeutic maneuvers that one might employ in the severely infected immunocompromised host include the use of granulocyte transfusions in patients with marked neutropenia or a serious granulocytopathy. Good results have been reported in such patients treated with daily infusions of three to four units of granulocytes by Fudenberg and co-workers.[90] Attempts to correct T-cell abnormalities in the acutely ill patient have not met with much success and are best carried out after infection is brought under control. Efforts at transplantation during such episodes have not often resulted in survival. More recently immunomodulatory therapy with IL-2 or IFN-γ has been employed with some success.

2.2. Minimizing the Effects of Less Severe Acute Infections

Immunocompromised patients are seldom entirely free from pyogenic infections. Thus, the antibody-deficient patient often suffers recurrent or chronic episodes of sinusitis, otitis media, mastoiditis, or bronchitis. Patients with absence of C3 or C5 have similar problems, whereas those with phagocyte movement or killing defects suffer recurrent abscesses, episodes of cellulitis, and pneumonias. These recurrent acute infections detract significantly from the patient's quality of life and may evolve into either the severe life-threatening infections discussed

above or chronic, indolent infections that have serious sequelae as discussed in Section 2.3. Every attempt should be made to control these infections by using both antimicrobial therapy and immunologic regimens. In many cases, the etiologic agents may be different from those seen in normal hosts and may have antibiotic resistance, in part because of the chronic use of antimicrobial agents by these patients. These acute infectious processes may be divided as follows.

2.2.1. Respiratory Infections

Acute sinusitis is not uncommon in the patient with deficiency of antibodies, C3, or C5 or with phagocyte dysfunction. This type of infection is characterized by low-grade fever, congestion, postnasal drip, and tenderness over the involved sinus. Such infection is associated in many instances with recurrent otitis media. The organisms involved are often similar to those in the uncompromised host and include *H. influenzae*, *S. pneumoniae*, and occasionally *S. pyogenes*, or *Neisseria* species. Anaerobes also may be involved in sinusitis.[91] In contrast to normal patients, immune-deficient hosts may have organisms such as *S. aureus* or gram-negative bacilli such as *P. aeruginosa* isolated from the middle ear in association with symptoms of otitis media. While generally not recommended, we have found cultures taken with a nasopharyngeal swab to be very useful in selecting antibiotics for these patients. It is surprising how many times immunodeficient patients will yield essentially pure cultures of a pathogen such as *H. influenzae* or *S. pneumoniae* from nasopharyngeal specimens. In the face of clinically apparent disease, antimicrobial agents should be administered, but in the absence of infection, we have elected to withhold treatment of the carrier state, since this would probably only result in replacement of one pathogen by another or in the emergence of antibiotic resistance.

Therapy of sinusitis and otitis media should be initiated with an antibiotic such as amoxicillin, although this agent may not be effective if *Staphylococcus* is involved or if other resistant organisms are present. Nasopharyngeal, sinus, or middle ear fluid cultures should be taken in the immunodeficient patient and antimicrobial susceptibility patterns determined on significant isolates. This is in contrast to the usual practice in the normal host with acute otitis or sinusitis where such cultures are really not indicated and are not cost effective. Failure to evaluate the immune-deficient host with appropriate culture and sensitivity determinations often results in significant delay in the initiation of appropriate therapy. Moreover, should infection disseminate and become life threatening, the initial culture and sensitivity results would be of great benefit in selecting appropriate therapy.

Acute bronchitis and pneumonia also are common occurrences in the patient with a congenital defect in the immune system. Low-grade fever, cough, and sputum production with bronchitis are common in antibody and complement deficiencies and also may be seen in T-lymphocyte and phagocyte abnormalities. Typeable and nontypeable *H. influenzae* often are isolated from sputum specimens in these patients. A host of other organisms also may be involved. A Gram's stain of expectorated sputum combined with culture results should be the guide to therapy. Initially, a drug such as amoxicillin or in the adult a tetracycline such as doxycycline or a macrolide is often helpful in bronchitis. Therapy should be altered when the results of cultures and sensitivities are returned. Amoxicillin–clavulanic acid, dicloxacillin, or an oral cephalosporin may be used if staphylococci are involved. While respiratory viruses are more common, in older children and adults, *Mycoplasma pneumoniae*, *Chlamydia pneumoniae*, and *Bordetella* spp. are common causes of bronchitis-like illnesses. Antimicrobial therapy for bronchitis is generally limited to 7–10 days.

Recurrent pneumonia is one of the most common problems faced by physicians caring for immunodeficient patients. Pyogenic infections caused by *H. influenzae* and *S. pneumoniae* as well as a host of other pathogens occur in antibody or complement deficiencies. In patients with phagocyte dysfunction, *S. aureus* is more prevalent as an etiologic agent. Sputum or transtracheal cultures should be obtained as well as blood, urine, and pleural fluid when possible for culture and examination for microbial antigens. If the patient appears toxic and if empyema or pneumatoceles are present, the patient should be hospitalized and treated with parenteral therapy. In the less ill patient, we have been successful on numerous occasions with oral therapy combined with chest physical therapy, postural drainage, and close follow-up evaluation. High-dose amoxicillin or cefuroxime axetil usually have been selected based on the results of cultures and Gram's stains. In the nonhospitalized patient, close follow-up is required, including radiographs to rule out progression, effusion, empyema, or pneumatocele. If such complications should develop in the face of antimicrobial therapy, the patient should be hospitalized and additional diagnostic procedures such as transtracheal aspirate or lung biopsy should be carried out in order to document the etiology. Intravenous gammaglobulin (400 mg/kg) should be administered to antibody-deficient patients in addition to their regular maintenance doses when pneumonia develops.

2.2.2. Gastrointestinal Infections

The respiratory and GI tracts are the prime sites for microbial challenge in the normal and immunodeficient patient. Thus, it is not unusual that these two areas should be the major focus of infection in the compromised individual. Most patients with combined T- and B-cell defects have some degree of diarrhea and malabsorption.[92]

Patients with isolated IgA deficiency or IgA deficiency associated with other immunoglobulin abnormalities often suffer chronic diarrhea because of *Giardia lamblia* infection. This agent is probably the leading cause of infectious diarrhea in immunodeficient patients. Symptoms have their onset within 6–8 days of infection and usually consist of cramping abdominal pain, nausea, diarrhea, and occasionally low-grade fever. The disease may last for weeks or months in the immunodeficient individual. The diagnosis may be established by examining multiple stools for cysts and trophozoites in approximately 50% of patients. More reliable means for establishing the diagnosis consist of examining duodenal or jejunal aspirates or biopsies. Since this disease is so common in our IgA- and combined immunoglobulin-deficient patients, we often will initiate therapy with metronidazole after a stool specimen has been collected for examination (*Giardia* antigen assay) but before a specific diagnosis has been made. Therapy is continued for 1 week and results in an approximate cure rate of 70%. The symptomatic patient may be treated with the alternative antimicrobial agent such as quinacrine if no response is observed following initial therapy (see Chapter 7, this volume). If the young child is unable to swallow tablets, a suspension of furazolidone (Furoxone) can be used with similar results.

Bacterial overgrowth in the small intestine also has been suspected as a cause of diarrhea in immunodeficient patients. Good data are not available, correlating bacterial counts in the small bowel and symptomatology. On occasion, an immunodeficient patient with marked diarrhea and no specific isolatable pathogen will respond to antimicrobial therapy. A nonabsorbable drug such as neomycin may be employed, or occasionally an absorbable drug such as tetracycline will be of benefit to the older patient. Although the data on such therapy are practically nonexistent, a trial in the patient with marked diarrhea may be indicated after specific pathogens have been ruled out as the etiology. Other important GI pathogens commonly reported as cause of disease in the immunodeficient host are rotavirus and *Cryptosporidium* sp. Both have been documented as causative agents of chronic diarrhea in this population. Unfortunately, there is no consistent effective therapy for either one. Some clinicians have used with some success oral immunoglobulin therapy on patients with chronic rotavirus diarrhea.[93] Paromomycin, clarithromycin, azithromycin, and bovine colostrum have been advocated by some clinicians to treat infections by *Cryptosporidium*. Results have been generally disappointing. Because of the frequent exposure to antimicrobial agents, the immunodeficient patient is also susceptible to infection by *Clostridium difficile*.

Plasma infusions or IVIG have often been found to decrease the severity and frequency of diarrhea in the immunodeficient patient.[84,88,94] The reason for this is unclear, since the IgA in plasma does not appear to cross into the GI tract. Perhaps the plasma contains other factors such as lymphocyte products that are helpful in enhancing local secretory immunity. Recurrent episodes of perirectal abscesses in a young child should suggest a possible immunodeficiency, especially neutrophil disorders. Commonly patients with leukocyte adherence defects may present with perirectal abscesses requiring broad-spectrum antibiotics and surgical drainage.[95]

2.2.3. Cutaneous Infections

Recurrent cutaneous infections are the hallmark of patients with phagocyte defects.[96] Abscess formation is common in patients with neutrophil and macrophage chemotactic, phagocytic, and killing defects. These may consist of small "pimples," larger boils, or huge abscesses. Interestingly, the patients with the largest abscesses often have phagocyte motility defects, leading to their recurrent infections. The reason for this may be related to observations made in animals by Miles and co-workers.[97,98] This group found that there was a critical 2–4 hr period during which phagocytes must arrive at the site of bacterial invasion if infection is to be contained. If phagocyte accumulation was delayed past this time interval, larger local lesions or systemic infection occurred. Thus, the large abscesses in the patients with chemotactic defects probably result from delayed accumulation of the first wave of phagocytes. This allows further bacterial multiplication with subsequent production of an increased amount of inflammatory mediators via the complement system and other pathways. This added stimulus continues to call in additional cells and results in the large abscesses observed.

Microbicidal defects such as those seen in chronic granulomatous disease allow intracellular growth of bacteria, lysis of PMNs and other cells, and release of important inflammatory mediators that also result in accumulations of bacteria, phagocytes, and debris. The abscesses in these patients are often caused by staphylococci but may

involve a whole host of organisms.[38] Needle aspiration or open drainage usually yields an etiologic agent on which sensitivity testing can be carried out. Many of the patients with CGD have serious problems in wound healing, so that recently we have not recommended incision and drainage except where absolutely necessary. When possible, aspiration with a large-bore needle has been successful in relieving pressure and obtaining an etiologic diagnosis. Therapy is then instituted with an appropriate bactericidal agent. Such medical management of what used to be considered surgical cases has resulted in far less overall morbidity in CGD patients.

Cutaneous and mucocutaneous candidiasis also have been observed in patients with phagocyte abnormalities.[43,99] Some also may have T-lymphocyte problems so that the exact role of each abnormality in the overall clinical picture is unknown. The candidiasis in these patients is usually quite difficult to treat and may require prolonged topical and even systemic therapy for a brief period of time. Nystatin has generally been used in topical therapy, and agents such as amphotericin B, 5-flucytosine, and miconazole have occasional use in a severely affected patient to decrease the overall numbers of these organisms in lesions. An antifungal agent, ketoconazole, has been used to treat a variety of fungal infections. Whereas it is quite effective against superficial infections,[100] it has been less successful when used against deep infections such as those in bones and joints.[101]

Newer antifungal agents such as fluconazole[102,103] and itraconazole[104] have been shown to be promising agents in the treatment of fungal infections in the immunocompromised host. The latter agent has *in vitro* activity against *Aspergillus* spp.

A recent comparative study in the treatment of cryptococcal meningitis in AIDS patients between amphotericin B plus 5-flucytosine and fluconazole demonstrated that the combination regimen was superior to fluconazole.[105] However, fluconazole has been found to be a highly effective agent in preventing recurrent cryptococcal infection.[106] Further comparative studies will be necessary to determine the precise role of these agents in patients with primary immunodeficiencies. They seem to be quite appropriate for mild to moderate cases of mucocutaneous candidiasis. Unfortunately, the routine prophylactic use of fluconazole in high-risk patients has selected resistant pathogens, such as *Candida krusei*, which may be more difficult to treat.[107]

The formulations of amphotericin B combined with liposomes have demonstrated improved pharmacokinetics of the agent with better penetration and fewer side effects and are more effective in the treatment of infections caused by *Aspergillus* and in hepatosplenic candidiasis.[108,109]

Cellulitis caused by streptococci, staphylococci, and other agents also has occasionally been a problem in patients with phagocyte immune-deficiency diseases.[35,36] These generally require rapid diagnosis based on blood cultures or local needle aspirates of the infected area and are treated initially with parenteral bactericidal antimicrobial agents, following which we can usually switch to oral agents.

2.3. Prevention of Chronic Infections and Their Sequelae

2.3.1. Antimicrobial and Other Therapy

A third and perhaps most important objective in the management of the congenital immune deficient patient is to prevent the long-term sequelae of infections. Acute, recurrent, and chronic infections often lead to the development of serious sequelae. Death in the antibody-deficiency syndromes, for instance, is usually the result of respiratory failure secondary to bronchiectasis and recurrent pneumonias. Similar respiratory problems may occur in ataxia telangiectasia and even in diseases associated with phagocyte disorders such as chronic granulomatous disease. On occasion, recurrent staphylococcal pneumonias with pneumatocele formation and chronic pulmonary fibrosis have resulted in lobectomy and the attending complications. When infection becomes manifested in other lobes, the patient with a history of such surgery is even more compromised. For this reason, such procedures are now discouraged in most centers. Postural drainage and physiotherapy cannot be overstressed in these patients, since adequate removal of plugs, inflammatory cells, and bacterial debris contributes greatly to the prevention of long-term pulmonary complications. All of our patients with pulmonary disease are followed with at least yearly chest roentgenograms and pulmonary function studies. Some chest physicians employ the intermittent use of oral antibiotics (amoxicillin or tetracyclines) in these patients on a 1-week-on, 1-week-off basis. We do not do this and only employ antibiotics when clinical or radiographic findings point to an increasingly purulent bronchitis or pneumonia.

Additional Therapy.   The patient with chronic sinusitis may benefit from drainage procedures such as the Caldwell–Luc procedure, whereas recurrent otitis media patients may benefit from adenoidectomy and insertion of tympanic membrane drainage tubes. Patients with recurrent abscesses, especially those caused by staphylococci, may respond to vigorous washing with an agent containing povidone-iodine or chlorhexidine. These preparations tend to dry out the skin, however, and may actually increase abscess formation. In other individuals with severe, essentially incapacitating abscess formation, we have had to resort to chronic long-term antimicrobial prophylaxis. One patient with Job's syndrome with severe staphylococcal abscesses was treated for more than 8 years with cloxacillin with an excellent response. As soon as the medication was stopped, abscesses recurred. We do not advocate the indiscriminate use of antibiotic prophylaxis, but in a few instances it can be quite useful. In general, the choice of such therapy should be limited to situations in which (1) one etiologic agent predominates, (2) therapy is not associated with significant toxicities, and (3) resistance is not likely to readily develop. In general, we have tended to use close to full therapeutic doses rather than the low-dose therapy commonly associated with "prophylaxis." Although there are few hard data on this point, we believe that low-dose therapy might favor the selection of resistant organisms.

Trimethoprim–sulfamethoxazole combinations may have several uses in the therapy or prevention of infection in patients with recurrent infections. Such preparations have been used in children with recurrent episodes of otitis media[110] and recurrent urinary tract infections with some success.[111] Hughes and co-workers[112] showed that 150 mg trimethoprim combined with 750 mg sulfamethoxazole/m^2 per day was 100% effective in preventing the development of *Pneumocystis carinii* pneumonia in immunosuppressed patients. This is an exciting observation, since infection with this agent is a very common complication in immunodeficiency disease. Johnston and co-workers[113] also indicated that sulfonamides may have a beneficial effect in patients with chronic granulomatous disease (CGD). This stems from the observation that a number of these patients have had fewer infectious complications while on long-term sulfonamide therapy, and their PMNs have shown slightly better bactericidal activity in the presence of sulfonamides. A review by Gonzalez and Hill[114] summarizes the advantages and disadvantages of antimicrobial prophylaxis in patients with CGD. Agents that concentrate within the polymorphonuclear cell such as rifampin and clindamycin are attractive agents for patients with phagocytic disorders. The potential for emergence of resistance and side effects need to be considered when making the decision to initiate a prophylactic agent.

2.3.2. Immunotherapy

In the antibody-deficient patient, prevention of chronic infections and sequelae is aided significantly by the use of immunoglobulin. In spite of the availability of IVIG, which has been shown to have increased efficacy in preventing infections, some individuals because of their insurance or convenience or even availability of product choose to use intramuscular immunoglobulin. Intramuscular immunoglobulin (IMIG) is administered prophylactically in a dose of approximately 0.6 ml/kg every 3 weeks after an initial loading dose of twice that amount. In the larger person, this can result in the requirement for a substantial injection volume (70 kg × 0.6 ml = 42 ml). For this reason, many of the larger patients prefer weekly doses amounting to approximately one third of the dose; 10 ml/week usually offers adequate protection against pyogenic infections in most patients. We do not routinely check immunoglobulin levels following injection, since the level seldom correlates specifically with protection against infection. An attempt is made, however, to adjust the dose according to the patient's symptoms. Approximately 10% of patients will develop reactions following IM immunoglobulin administration. Most will be caused by the inadvertent injection of the material into small veins in the muscle. The aggregates contained within the IM preparation will result in complement activation and an anaphylaxislike picture. Patients also may develop IgE or IgG antibodies directed against various proteins in the immunoglobulin preparations. Patients with a total absence of IgA are likely to develop such antibodies to this immunoglobulin that may result in anaphylactoid reactions after IM administration. Such patients may be skin tested for such reactivity using low concentrations of the preparation. Both types of reactions and especially those caused by aggregates may be managed subsequently through the use of IV gammaglobulin or plasma therapy. This must be administered carefully with emergency equipment readily available in the event that a reaction should occur.

The IVIG preparations represent a significant advance in the therapy of patients with antibody deficiency disease. The use of IV gammaglobulin as a prophylactic agent has clearly resulted in a reduction of morbidity in patients with X-linked agammaglobulinemia and hypogammaglobulinemia. Clear advantages over plasma therapy and IM gammaglobulin are minimal if any risk of hepatitis or transmission of CMV or HIV; easy access and no need for blood typing; relatively predictable pharmacokinetics resulting in predictable immunoglobulin levels; and little problem with aggregates and their associated adverse reactions. IVIG also can be easily infused at home with few complications. This potentially reduces the cost of medical care.[115]

As mentioned previously, multiple studies have demonstrated the efficacy of IVIG as a therapeutic and prophylactic agent. More recently, hyperimmune CMV IVIG has been used during bone marrow transplantation to reduce the incidence of CMV pneumonia.[80–82,86,116] Hyperimmune respiratory syncytial virus (RSV) immunoglobulin (RSVIG) has recently been shown to be effective in reducing the incidence and morbidity of RSV in infants with chronic pulmonary disease. Although studies are lacking, anecdotal experience may point to the beneficial effect in selected immunocompromised hosts and use of this preparation should be considered for immune-deficient infants during the RSV season.[117] In fact, all immune-deficient patients under 2–3 years of age should likely receive RSV hyperimmune IVIG or RSV monoclonal antibody, especially during the respiratory viral season.

A number of substances derived from human or animal sources or synthesized chemically have been used in attempts to enhance the host defense mechanism of immunodeficient patients. Transfer factor, a small-molecular-weight (<10,000), nonimmunogenic protein derived from human leukocyte lysates, has been used with some success to prevent infections in patients with Wiskott–Aldrich syndrome or chronic mucocutaneous candidiasis.[118–120] Unfortunately, results have been quite variable and the use of this agent has been associated with renal toxicity and the development of hemolytic anemia in a small number of patients.[118]

Thymosin, a partially purified extract of beef thymic tissue, has been reported to increase T-cell numbers in a variety of congenital and acquired immunodeficiency syndromes.[121] Conversion to positive mitogenic responses and mixed leukocyte reactions has followed *in vitro* incubation with the agent, whereas *in vivo* therapy has been reported to cause clinical improvement in patients with (1) Nezelof's syndrome of cellular immunodeficiency with immunoglobulins, (2) ataxia telangiectasia, (3) DiGeorge's syndrome, and (4) Wiskott–Aldrich syndrome.[122] In general, these studies have not always been impressive. Moreover, allergic reactions and hepatitis have occurred during the use of this agent. Neither transfer factor nor thymosin is available in a standardized form. It seems clear that their use at the present time should be restricted to investigational settings and considerably more experience will be required before their clinical value can be established.

Patients with enzyme defects in their purine metabolic pathways including adenosine deaminase and nucleoside phosphorylase deficiency have a picture similar to that of patients with severe combined immunodeficiency disease, although the symptomatology may be less severe initially. Polmar and co-workers[123] have treated one such individual with repeated administrations of glycerol-frozen packed human erythrocytes that contain high levels of adenosine deaminase. This has been followed by return of lymphocyte function to normal *in vitro* and positive test reactivity and the patient has remained well. Adenosine deaminase in polyethylene glycol (PEG-ADA) with prolonged half-life is now also available to treat these patients. Similar therapy has been attempted in a patient with nucleoside phosphorylase deficiency with some success. Recently, gene therapy of these disorders has been attempted with at least some success.

Bone marrow, fetal thymus, and fetal liver transplant have been used in a variety of immunodeficiency diseases.[124–126] Marrow transplantation between HLA-matched siblings (especially HLA-D matched, mixed lymphocyte culture nonreactive pairs) has been successful in a number of cases of severe combined immunodeficiency disease and in several cases with Wiskott–Aldrich syndrome.[127] Excellent results have been obtained when HLA-matched donors have been available. Almost all patients who receive marrow from an individual not matched at the D locus have died of severe graft-versus-host disease and overwhelming infection within the first 1–2 months after transplant. Recent attempts at transplanting mismatched marrow have employed lectin or monoclonal antibody removal of mature T cells from donor marrow. These have met with considerable success. Gram-negative enteric pathogens, viruses, or *P. carinii* have usually been the etiologic agents in the severe infections developing in these patients. This had led to attempts aimed at suppressing the GI flora of patients after transplantation with a variety of systemic and oral, nonabsorbable antibiotics.[128,129] In addition, laminar flow rooms have been used to manage these susceptible individuals. More recently HLA-identical or haploidentical stem cell transplants have been employed to correct a number of the congenital immunodeficiencies.

Fetal thymic and hepatic tissue obtained before immunocompetence has been established have been used in cases of severe combined immunodeficiency disease for which no HLA-D-matched marrow donor has been available. Reconstitution has been successful in several instances but usually has not been of a long-lasting nature. Transplantation of fetal thymus or thymic epithelial cells maintained *in vitro* has been attempted in a variety of immunodeficiency disorders including DiGeorge's syndrome,[130] Nezelof's syndrome,[131] chronic mucocutaneous candidiasis,[132] and ataxia telangiectasia.[131] Transient improvement in skin test reactions and *in vitro* lymphocyte responses and a decrease in infectious complications have been reported following such therapy, but permanent reconstitution has been rare.

Recently, a number of substances with possible immune-potentiating effects have been investigated in patients with congenital immune deficiencies. Some years ago, we and others[133,134] showed that neutrophil chemotaxis and lysosomal enzyme release were modulated in part by cyclic $3',5'$-guanosine monophosphate (cGMP) and cyclic $3',5'$-adenosine monophosphate (cAMP). Moreover, microtubular polymerization and function appeared to be dependent on the intracellular levels of these cyclic nucleotides. Oliver and Zurier[135] showed that patients with Chediak–Higashi syndrome, who have PMNs filled with large lysosomal granules and that have chemotactic and bactericidal defects, have disordered microtubular function that can be corrected *in vitro* with cGMP or its acetylcholine (Ach). Boyer and co-workers[136] subsequently treated a patient with this disease with ascorbic acid, an agent that has been shown to alter cGMP and cAMP and to alter leukocyte function.[137] Following such treatment, the patient improved clinically and had partial reversal of his *in vitro* leukocyte function abnormalities. More recently, we have employed moderately high doses of ascorbic acid in several patients with hyperimmunoglobulinemia E, recurrent infections, and defective PMN chemotaxis. In several such patients, we have observed rather dramatic results with a marked decrease in infections after *in vivo* therapy with 1000–2000 mg ascorbic acid per day. In others, no effect has been observed. Thus, as in the other congenital immunodeficiency syndromes that have a variable course, it is extremely difficult to evaluate therapy.

A dramatic improvement in the health status of patients with chronic granulomatous disease has been achieved with the use of IFN-γ. A reduction in serious infections in recipients of IFN-γ was observed in a recent study published by a large collaborative study group.[138] In addition, Jeppson and associates[139] have recently shown that recombinant human IFN-γ significantly improves the chemotactic responsiveness of neutrophils of patients with hyperimmunoglobulinemia E. Petrak and colleagues[140] report that IFN-γ administered subcutaneously three times weekly in a dose similar to that used in the CGD study improved neutrophil chemotaxis and probably benefited three of four patients in a noncontrolled pilot study. Granulocyte colony-stimulating factor also has been of some benefit to these patients in isolated cases. Patients with common variable hypogammaglobulinemia who failed to make IL-2 have been treated with this cytokine and have shown clinical and immunologic improvement.[141]

3. Specific Infections in Immunodeficiency Syndromes

This section reviews the major congenital immunodeficiency syndromes, discusses their major host defense abnormalities, and underscores the most common infec-

TABLE 2. Congenital Immunodeficiencies and Associated Infections and Pathogens

X-linked agammaglobulinemia-hypogammaglobulinemia [B-lymphocyte disorders]	Sepsis/meningitis *Streptococcus pneumoniae* *Haemophilus influenzae* *Pseudomonas aeruginosa* Persistent viral infections Gastroenteritis—*Giardia lamblia*, rotavirus, *Cryptosporidium parvum* Chronic enteroviral infections Recurrent otitis media Recurrent sinopulmonary infections
DiGeorge's syndrome, other T-lymphocyte defects	Candidiasis *Pneumocystis carinii* pneumonia Pneumonia Aspergillosis Salmonellosis Cytomegalovirus infections Severe respiratory viral infections
Complement disorders	Infections by *Neisseria meningitidis*, *Streptococcus pneumoniae*, *Haemophilus influenzae*
Phagocytic cells/chemotactic disorders (chronic granulomatous disease)	*Staphylococcus aureus* Gram-negative bacilli *Aspergillus*

tions that complicate each disorder. The discussion is confined to the most serious and most common etiologic agents in these patients. A summary of these appears in Table 2.

3.1. Combined B- and T-Cell Defects

3.1.1. Severe Combined Immunodeficiency Disease

This disease occurs as an X-linked lymphopenic form and as autosomal-recessive forms. The disease is characterized by lymphopenia, a marked decrease in T- and B-cell numbers, low serum immunoglobulins, no antibody responses following immunization, negative skin test reactions, and severe and recurrent infections. Desquamating skin rashes are common and should lead one to suspect immune deficiency if associated with diarrhea, failure to thrive, or pulmonary infiltrates. Most patients become symptomatic within the first few months of life, but a few will do relatively well until 6–9 months of age. Persistent cutaneous *Candida* infection, chronic diarrhea, and failure to thrive should suggest the need for T-cell quantitation or a study of the lymphocyte mitogenic response. Signs and symptoms suggestive of severe combined immunodeficiency disease (SCID) are similar to those in acquired immunodeficiency syndrome caused by HIV. HIV antibody testing of the patient is recommended as part of the evaluation. Respiratory infection is also very common and may be caused by *P. carinii*. A chronic,

pertussis-like cough is often present and may be associated with tachypnea, retractions, and a markedly decreased arterial PO$_2$. When *Pneumocystis* pneumonia is suspected, a bronchoalveolar lavage is recommended. If negative, an open lung biopsy is indicated to confirm the diagnosis. Although sputum, transtracheal aspirates or needle aspirates of the lung are useful if positive, false-negative results are frequent in non-AIDS patients,[69] and should not delay the confirmation of diagnosis of *Pneumocystis carinii* pneumonia (PCP) by means of open lung biopsy. It is essential to make an appropriate etiologic diagnosis of pneumonia in SCID, since other bacterial and viral pathogens may produce very similar findings. To begin treatment without documentation of the etiology usually results in a therapeutic dilemma several days later if the response is less than dramatic. One is then forced to add or subtract antibiotics at random, since routine cultures can no longer be relied on. Therefore, if at all possible, the patient with SCID with severe pneumonia should have an open lung biopsy to define the etiology.

Trimethoprim–sulfamethoxazole is the treatment of choice for both children[142] and adults[143] with *Pneumocystis* pneumonia. This combination antibiotic has replaced pentamidine as first-line therapy, which had a high adverse events rate despite being fairly effective.[143] Hughes and co-workers[142] first reported on the successful therapy of 16 of 20 patients with pneumonia caused by *Pneumocystis*. Subsequently, other investigators have confirmed and extended these results in both children[144]

TABLE 3. Congenital Immunodeficiencies Suggested Initial Empiric Therapy for Specific Infections

Disorder	Type of infection	Suggested therapy
Severe combined immunodeficiency (SCID)	Bacterial pneumonia or sepsis	Cefotaxime or ceftriaxone
	If *Staphylococcus aureus* suspected	Add nafcillin or oxacillin
	If *Pseudomonas aeruginosa* suspected	Ceftazidime plus aminoglycoside or Extended-spectrum penicillin plus aminoglycoside
	If methicillin-resistant *S. aureus* or drug-resistant *S. pneumoniae* suspected	Cefotaxime or ceftriaxone plus vancomycin
	Herpes simplex virus or varicella zoster virus infections	Acyclovir
	Cytomegalovirus infection	Ganciclovir, foscarnet
	Mucocutaneous candidiasis	Fluconazole
Purine pathway enzyme deficiencies	Similar to SCID	Similar to SCID
Wiskott–Aldrich	Bacterial pneumonia or sepsis	Cefotaxime or ceftriaxone
	If methicillin-resistant *S. aureus* or drug-resistant *S. pneumoniae* suspected	Add vancomycin
Ataxia telangiectasia	Chronic sinopulmonary infections, mild to moderate	Amoxicillin–clavulanic acid
	Severe	Cefotaxime or ceftriaxone
Mucocutaneous candidiasis	Mild to moderate fungal disease	Fluconazole, ketoconazole, nystatin, topical amphotericin B
	Systemic fungal disease	Amphotericin B
	Hepatosplenic candidiasis	Amphotericin B lipid complex
Antibody deficiency disorders, common variable hypogammaglobulinemia	Pneumonia, sepsis[a]	Third-generation cephalosporin or Piperacillin–tazobactam plus aminoglycoside or Meropenem plus aminoglycoside
	Otitis media, sinusitis	Amoxicillin–clavulanic acid, cefuroxime[b]
	Chronic enteroviral meningoencephalitis	Intravenous immunoglobulin, pleconaril [investigational agent]
Selective IgA deficiency	Upper respiratory tract infections	Amoxicillin–clavulanic acid, cefuroxime
	Giardiasis	Metronidazole or quinacrine
Complement disorders	Pneumonia, sepsis, meningitis	Cefotaxime or ceftriaxone[a]
Neutrophil disorders	Perirectal abscesses	Amoxicillin–clavulanic acid or Piperacillin–tazobactam or Meropenem or Clindamycin plus aminoglycoside
Chemotactic disorders	Pneumonia	Clindamycin, dicloxacillin or Nafcillin or oxacillin plus aminoglycoside
Hyperimmunoglobulinemia E	Cutaneous infections	
Chronic granulomatous disease	Musculoskeletal or soft-tissue infections	Clindamycin plus aminoglycoside or Nafcillin or Oxacillin plus aminoglycoside

[a]If drug-resistant *S. pneumoniae* is suspected, add vancomycin.
[b]If drug-resistant *S. pneumoniae* is suspected, add rifampin or clindamycin.

and adults.[143] The recommended dose in children is 20 mg/kg trimethoprim and 100 mg/kg sulfamethoxazole, whereas doses in adults have ranged from 906 to 1200 mg trimethoprim and 4800 to 6000 mg sulfamethoxazole per day. Survival rates of 80% in children and 83% in adults have been reported in small series compared with an overall reported cure rate of 42% using pentamidine.[145] In general, a response is seen within 2 to 4 days with defervescence of fever if present, lowered respiratory rate, and gradual clearing of the chest radiographs. Therapy is continued for 14–21 days. Recently, researchers have documented the beneficial effects of corticosteroids in preventing the early deterioration observed in patients being treated for PCP.[146] Pentamidine has been used in combination with trimethoprim–sulfamethoxazole in several critically ill patients, but no evidence of a synergistic or additive effect has been observed. Pentamidine preferably should be given intramuscularly and is associated with immediate reactions including hypotension, tachycardia, nausea, and vomiting. In addition, microscopic hematuria, azotemia, granulocytopenia, and hypoglycemia have occurred with its use. Inhaled pentamidine using a nebulizer has been shown to be effective in the treatment of mild cases of PCP.[147] In addition, it also has been used for prophylaxis, resulting in fewer side effects than when used systemically.[148] This mode of administration should be reserved for patients older than 8 years of age.

In the patient with severe combined immunodeficiency disease with suspected bacterial pneumonia or sepsis, therapy should initially be broad spectrum and should include a third-generation cephalosporin such as cefotaxime, ceftriaxone, or ceftazidime. Should penicillin-resistant staphylococci subsequently be cultured, then a penicillinase-resistant derivative such as nafcillin or oxacillin should be added. If *Pseudomonas aeruginosa* is suspected or isolated, a combination of an extended-spectrum penicillin such as piperacillin or ticarcillin, plus an aminoglycoside should be utilized. Ceftazidime plus an aminoglycoside would be an acceptable alternative. Later, after the staphylococcal etiology is defined, a combination of nafcillin plus the aminoglycoside would be expected to have a synergistic effect against most penicillin-susceptible as well as penicillin-resistant staphylococci.[149]

Systemic candidiasis also may occur in such patients and especially those subjected to hyperalimentation for failure to thrive.[150,151] The presence of one or more positive blood cultures coupled with the presence of *Candida* in the urine or *Candida* lesions in the eye grounds should suggest systemic involvement. Therapy should logically consist of amphotericin B therapy with or without 5-flucytosine. This latter agent has the advantage that oral administration results in excellent serum and CSF levels of the drug shortly after therapy is initiated (in contrast to amphotericin B).[152] For this reason, we elect to use it in the seriously ill immunodeficient patient, even though a number of *Candida* isolates will be resistant. It should always be combined with amphotericin B or other antifungals to prevent the development of resistance. Other agents, such as miconazole, have been used with some success in candidiasis in normal individuals and may also be of value in the immune-deficient patient. Fluconazole also appears to be beneficial.

Disseminated viral infections also are common in patients with severe combined immunodeficiency diseases. HSV, VZV, and CMV may produce disseminated disease. Acyclovir has been shown to be effective in the treatment of severe HSV infections. In addition, acyclovir at a dose of 1500 mg/m^2 per day divided every 8 hr has been used to treat disseminated varicella in immunodeficient hosts, resulting in a reduction in morbidity and mortality.[153]

Ganciclovir has been used in the treatment of CMV chorioretinitis, pneumonitis, and gastrointestinal disease in patients with AIDS with mixed results.[154] Further studies will be needed to evaluate its clinical utility in other immunodeficiency states.

Live virus vaccines should not be given to these individuals or household family members as paralytic poliomyelitis or encephalitis has developed following poliovirus, measles, or mumps vaccines.

Following bone marrow transplant in patients with SCID, graft-versus-host disease (GVHD) often leads to severe infection. Two weeks after transplant, a severe reaction often ensues that is characterized by blood diarrhea, desquamating skin rash, and abnormal liver function. At this time the patient is most susceptible to overwhelming sepsis and death from enteric organisms, including *E. coli*, *Enterobacter*, *Klebsiella*, and *Pseudomonas aeruginosa*. Broad-spectrum antimicrobial therapy is indicated in such patients when signs of infection develop.

Because of the combined nature of their defect, patients with SCID have a profound defect in antibody production, and thus critically need to be supplied with immunoglobulin. Immunoglobulin (IVIG) should routinely be administered. Because of the marked propensity of these patients to develop GVHD, all blood products should be irradiated (3000–5000 rad) prior to administration in order to eliminate immunocompetent lymphocytes.

Permanent reconstitution in SCID is generally attempted after acute infection has been controlled and is best carried out with an HLA-matched sibling bone marrow transplant as indicated earlier.[124] Alternatively, marrow from a parent that has been treated to remove mature T cells may be employed. Alternative procedures include fetal liver transplantation with or without fetal thymus transplant.[125–127] More recently, stem cell transplants or marrow transplants from an HLA-identical or haploidentical parent have been performed with considerable success.[155]

3.1.2. Purine Pathway Enzyme Deficiencies

Patients with defects in their purine metabolic pathways have infections identical to those with SCID described above except they may have later onset and are often somewhat less severe because a degree of immune function remains in some instances.[156–158] The immune deficiency in these patients probably results from the accumulation of substances toxic to the developing immune system because of an enzyme deficiency. Adenosine deaminase (ADA) and purine nucleoside phosphorylase (PNP) deficiencies have both been associated with combined B- and T-cell abnormalities.[11] Generally, some immune functions may remain, so that initial symptoms of infection may not appear until 6–12 months of age. Examination of the patients' RBCs or WBCs for ADA or nucleoside phosphorylase levels is diagnostic, since affected patients' levels are quite low. Serum or urine uric acid concentrations are also quite low because of the purine metabolic block and may be used as a rapid screening test for the disorder. The gene for ADA deficiency has been mapped to chromosome 20q13, while that for PNP deficiency maps to chromosome 14q13.1.[159]

Therapy for the common infections in these patients would be similar to those reported above for SCID. In addition, these patients may respond with partial immunologic reconstitution to the administration of glycerol-frozen packed erythrocytes containing high levels of the missing enzymes. Polyethylene glycol ADA with an extended half-life in the body is also used to treat these patients. Immunoglobulin therapy should also be included when evidence of defective antibody production is present. More recently, patients with ADA deficiency have been the first to undergo gene therapy employing autologous bone marrow following retroviral insertion of the gene for this enzyme.

3.1.3. Wiskott–Aldrich Syndrome

These patients have as their major host defense abnormality an inability to make antibody directed at poly-saccharide (and to a lesser degree protein) antigens.[62,63] The basic defect is unknown, but many believe it results from abnormal macrophage processing of antigen.[63] Later, these patients seem to lose T-cell function, making this a combined B- and T-cell abnormality. Along with eczema, thrombocytopenia in the 50,000 to 100,000 or lower range often suggests the diagnosis. Infections seldom are a problem until maternal immunoglobulin disappears in the infant 4–6 months after delivery. Recently, a novel gene termed WASP, which encodes a Wiskott–Aldrich syndrome protein, has been found to have point mutations or single base deletions in patients with the disorder.[159] The Wiskott–Aldrich syndrome patient then suffers repeated episodes of otitis media, pneumonia, and sinusitis, mostly caused by *S. pneumoniae* and *H. influenzae*. Polysaccharide-encapsulated enteric organisms such as *E. coli* and *K. pneumoniae* also may be a problem. The most severe infectious problem in the Wiskott–Aldrich patient results from overwhelming sepsis with *S. pneumoniae*. In such individuals, less than 6 hr may elapse between onset of symptoms and death. It is critical that clinicians managing such patients be aware of this and that they inform the patients or the parents of the possible consequences of such infection. Our patients with Wiskott–Aldrich syndrome are cautioned to call us immediately and/or report to an emergency room if they have onset of chills, fever, or other signs of systemic infection. Furthermore, they are instructed to warn the clinician they see of the potential life-threatening nature of their disease. After one such patient died of pneumococcemia on a hurried 4-hr trip in from a rural area, we have started giving large doses of amoxicillin or a cephalosporin to such patients to take immediately if they cannot reach a medical facility or physician within an hour. There is no proof that such oral therapy will be beneficial and it may make subsequent cultures invalid, but the fulminant nature of infection in these patients is the basis for this recommendation. Immediately upon seeing such a patient, we would recommend blood, urine, and CSF cultures followed within minutes by the administration of high intravenous doses of a third-generation cephalosporin in addition to vancomycin or rifampin.[79] *Streptococcus pneumoniae* and *H. influenzae* are by far the most likely pathogens so that this combination should cover both ampicillin-susceptible and -resistant strains of *H. influenzae* and *S. pneumoniae*. Since antibody deficiency is a prime component of this disease as well as shock, we would also give IVIG (400 mg/kg) to the critically ill Wiskott–Aldrich syndrome patient.

Unfortunately, splenectomy, which greatly aids the thrombocytopenia observed in this disease, may contribute to the incidence of overwhelming infections. Some investigators have been resorting to such therapy in severely symptomatic thrombocytopenia patients, however.[160] These patients have been placed on oral antibiotic prophylaxis with excellent results. Many later develop lymphomas, however.

Less severe acute and chronic infections in Wiskott–Aldrich syndrome patients need to be treated promptly with appropriate antimicrobial therapy to prevent the subsequent development of serious infection and to inhibit the development of local complications such as mastoiditis or osteomyelitis underlying otitis media or sinusitis. We also have seen *S. pneumoniae* meningitis develop in a Wiskott–Aldrich patient in the hospital while undergoing oral therapy for otitis media. These cases serve to point out how serious a problem such infections can be in these individuals and how rapidly they can develop.

Long-term immunotherapy in the Wiskott–Aldrich syndrome would consist of regular IVIG infusions every 3–4 weeks. Recently bone marrow or stem cell transplantation has been used with good success in these patients.

3.1.4. Ataxia Telangiectasia

These patients develop cerebellar ataxia, telangiectasia (most prominent on the bulbar conjunctiva, nose, ears, and antecubital fossae), and recurrent sinopulmonary infections. Many have depressed skin test reactivity and lymphocyte responses and also may have absent IgA and IgE or well-depressed IgG2 and/or IgG4. A single ataxia telangiectasia gene has been located on chromosome 11q22.23.[159] Chronic sinopulmonary infections with *H. influenzae*, *S. aureus*, and *S. pneumoniae* are common and bronchiectasis develops in most cases.

Antimicrobial agents should be selected on the basis of nasopharyngeal, tympanocentesis, or sputum cultures and be used on an intermittent rather than continuous basis. Plasma therapy or more recently IVIG has resulted in beneficial results in several instances.[89,161] Since chronic respiratory infection and failure are often the cause of death in this disease, every emphasis should be made to ensure that adequate pulmonary physiotherapy and postural drainage are accomplished.

3.2. Congenital Pure T-Cell Immunodeficiencies

3.2.1. Cellular Immunodeficiency with Immunoglobulins

Cellular immunodeficiency with immunoglobulins is primarily a T-cell defect, although abnormalities in

immunoglobulin synthesis also may occur. These are felt to be secondary to the primary defect in T cells. The disease may be inherited in an autosomal recessive pattern or may be sporadic and is characterized by failure to develop a thymus. The time of onset of infections varies but is usually between 6 months and 1 year. More recent evidence suggests that a number of these patients may have PNP deficiency. Recurrent otitis media, monilial diaper rash, and diarrhea often are presenting complaints. Viral infections such as varicella or reactions to live virus vaccines may result in severe disease in these individuals. Gram-negative sepsis with agents such as *E. coli* and *Pseudomonas aeruginosa* or PCP[162,163] may result in fatalities. Therapy is similar for these agents to that described for SCID.

As mentioned, these patients may have a deficiency of one or more immunoglobulin classes, but in addition may fail to make adequate antibodies against T-dependent antigens. For this reason, such patients should receive immunoglobulin replacement therapy and should receive IVIG on admission with serious systemic infection.

Thymosin has been used in several patients with this disorder with variable results. One patient was reported to have an increase in T-cell rosettes, positive delayed hypersensitivity skin tests, and a decrease in infectious complications following such therapy.[164] Others have seen little response to this preparation.

3.2.2. DiGeorge's Syndrome

This syndrome, which results from abnormal embryologic development of the third and fourth pharyngeal pouches, often results in absence or partial absence of the thymus and a marked deficiency in the T-lymphocyte system. Abnormalities of the aortic arch, the parathyroids, and occasionally the thyroid gland also may occur along with dysmorphic facial features. The diagnosis should be suspected in the presence of seizures, hypocalcemia, and cardiac anomalies in a newborn. Roentgenograms may reveal absence of thymic tissues and T-cell quantitation will be quite low. Thus, cardiac and parathyroid complications may alert the clinician to the diagnosis before infections occur. Later, chronic mucocutaneous candidiasis, chronic rhinitis, and recurrent pneumonias develop. *Pneumocystis carinii* pulmonary infection is not uncommon. Diarrhea and failure to thrive also may occur. Therapy for these infections is as outlined for SCID.

Although immunoglobulins are usually normal in these patients, there may be some problems in making specific antibodies to T-cell-dependent antigen. Thus, immunoglobulin therapy may be beneficial. Monosomy of chromosome 22 can be demonstrated by fluorescent *in situ* hybridization in most of the patients.[159]

3.2.3. Chronic Mucocutaneous Candidiasis

Chronic mucocutaneous candidiasis probably represents a spectrum of diseases, and in fact defective function has been observed at different sites in the immune system in different patients. An autosomal recessive pattern has been detected as well as sporadic cases. The disease also may occur in association with multiple endocrine abnormalities including hypoparathyroidism, diabetes mellitus, and Addison's disease. Abnormal immune parameters may include (1) decreased skin test and *in vitro* lymphocyte mitogenic responses to *Candida* and other antigens and (2) a specific defect in skin test and *in vitro* mitogenic responses to *Candida* but absent production of migration inhibition factor (MIF) by lymphocytes challenged with *Candida* antigen. We[43] and others[99] have also observed chronic candidiasis in association with both neutrophil and monocyte chemotactic defects.

Infections in these patients are almost entirely limited to candidiasis[11] unless several defects in the immune system are present.[43] Infection is usually confined to the skin and mucous membranes, with systemic involvement being extremely rare. We have seen one case with the granulomatous form of the disease (in which large granulomas develop) in which the patient developed *Candida* endocarditis and one patient who had associated anorexia nervosa who had a blood culture taken shortly before death that grew *Candida*. Otherwise, *Candida* infection has not been invasive.

Topical therapy with gentian violet, nystatin, or miconazole may be helpful in the patient with limited disease. Applications should be made several times (3–4) daily in a viscous suspension that will hold the antimicrobial agent in contact with the organism for as long as possible. Nystatin suppositories or pastilles have been most useful in treating both vaginal and oral infections, since they are in a viscous, slow-release form. With more severe disease, it is often necessary to use systemic therapy to decrease the overall load of organisms so that immunotherapy or topical antimicrobial may be effective. Amphotericin B, 5-flucytosine, and miconazole have been employed with limited success, but it is not clear whether clinical failures have resulted from development of drug resistance that can be documented *in vitro*. Oral antifungal agents such as ketoconazole and fluconazole appear to be particularly useful in this condition.

Transfer factor as well as thymosin have been re-ported to cause at least temporary improvement in some patients with chronic mucocutaneous candidiasis[119,122] but cannot be accepted as standard therapy at this time.

3.3. B-Cell Immunodeficiency

Congenital abnormalities of the B-cell system are associated with severe and recurrent infections and may be classified as follows.

3.3.1. Transient Hypogammaglobulinemia of Infancy

A small number of infants have a lag in antibody production that may or may not be associated with some increase in infections.[165] Unfortunately, this condition is grossly overdiagnosed and is often used as the pretext for the inappropriate use of immunoglobulin. A study by Tiller and Buckley[165] suggested that most such patients do not have serious bacterial infections and most can make specific antibody when immunized. Such individuals, therefore, would seldom appear to require routine administration of immunoglobulin. If the clinician would confine immunoglobulin use to those individuals who (1) have IgG levels in the ≤200 mg/dl range, (2) suffer severe and recurrent bacterial infections, and (3) do not produce antibody in response to immunization, then the inappropriate use of this agent would be greatly reduced.

3.3.2. Sex-Linked Hypogammaglobulinemia

This disease, which is X-linked, occurs exclusively in males and is associated with extremely low levels (<100 mg/dl) of IgG, IgM, and IgA, low numbers of B lymphocytes, absent germinal centers in lymph nodes, and a marked decrease in plasma cells. Infections usually begin at 4–6 months of age, when maternal antibody disappears from the infant. The disorder is caused by the absence of what is termed Bruton's tyrosine kinase (Btk), which is encoded on the X chromosome and which is apparently critical for B-cell maturation past the pre-B-cell stage.[166] Recurrent pyogenic infections including otitis media, sinusitis, conjunctivitis, pneumonia, and sepsis occur and are commonly caused by *S. pneumoniae*, *H. influenzae*, *S. aureus*, *N. meningitidis*, and *P. aeruginosa*.[167,168] Such patients are especially prone to overwhelming infection and usually develop chronic pulmonary disease secondary to recurrent respiratory infections. Initial antimicrobial therapy of the seriously ill patient should be broad spectrum and include a third-generation cephalosporin and a penicillinase-resistant penicillin. Intravenous immunoglobulin (400 mg/kg) also should be administered. More localized acute infections also should be treated promptly. These usually respond to oral therapy with penicillin, amoxicillin, dicloxacillin, or an oral cephalosporin derivative, depending on the results of cultures and sensitivity patterns. Therapy of sinusitis may require 3–4 weeks, whereas otitis will usually respond within 14 days. As described previously, an attempt should be made in immune-deficient patients to determine the etiologic agent and its sensitivity through the use of nasopharyngeal, middle ear fluid, sinus, or sputum cultures. Amoxicillin/clavulanic acid seems to be an excellent choice for those infections of the respiratory tract that could be caused by beta-lactamase-producing organisms such as *H. influenzae*, *S. aureus*, and *Moraxella catarrhalis*.

Prevention of chronic pulmonary disease is dependent on prompt treatment of acute infections as well as attention to pulmonary physiotherapy and drainage. Judicious use of oral antibiotics in treating exacerbations of bronchitis and decreasing purulence in chronic bronchiectasis is also indicated. Oral amoxicillin, cephalosporins, or tetracyclines (in the adult) may be used.

Although viral infections are not usually a problem in antibody-deficient patients, such patients may develop serious disease from live virus vaccines and also may develop chronic CNS infection with enteroviruses.[169] These may be treated effectively with high-dose IVIG. An investigational agent, pleconaril, may be useful for the treatment of these chronic infections.[170]

3.3.3. Hypogammaglobulinemia Associated with Hyperimmunoglobulinemia M

This is a sex-linked disorder characterized by normal to increased levels of IgM and IgM-producing B and plasma cells but with low levels of IgG and IgA. The disorder has been shown to be due to a primary dysfunction of isotype switching.[171] In the majority of cases, which occur in a sex-linked pattern, the disorder results from the absence on T cells of the ligand for CD40 on B cells. A minority of cases have an autosomal recessive pattern of inheritance in which signal transduction through CD40 on B cells is impaired. The patients also have neutropenia and thrombocytopenia and may have hemolytic anemias and lymphomas. Antibodies directed against polysaccharide antigens such as isohemagglutinins or opsonins for *S. pneumoniae* and *H. influenzae* are present in serum and are of the IgM variety. Low to absent levels of IgG and IgA antibodies are present in serum, and respiratory and GI secretions, however. These patients therefore have respiratory, soft-tissue, and GI infections

but do not often suffer overwhelming episodes of sepsis. Therapy should be like that described under sex-linked hypogammaglobulinemia (Section 3.3.2) and should include immunoglobulin replacement therapy to supply missing IgG antibodies. Cryptosporidia infection may result in severe liver disease, so avoidance of contaminated water is important. Many of these patients develop lymphomas later in life. Attempts have recently been made to use bone marrow or stem cell transplants in these patients.

3.3.4. Selective IgM Deficiency

This disease is probably of genetic origin, but the exact inheritance pattern has not been elucidated. It is characterized by extremely low levels (<20 mg/dl) of IgM without other detectable abnormalities. Because these patients lack production of IgM antibodies directed against the polysaccharide capsules of many pyogenic bacteria, they tend to have severe systemic infections caused by *S. pneumoniae*, *H. influenzae*, and *E. coli* as well as other encapsulated pathogens. These often present as sepsis or meningitis and should be treated with broad-spectrum antimicrobial regimens such as ceftriaxone or cefotaxime along with plasma infusions to supply missing IgM. Long-term immunoglobulin is not indicated, since commercial preparations contain low levels of this immunoglobulin and the half-life is quite short.

3.3.5. Selective IgA Deficiency

This immune deficiency is one of the most prevalent abnormalities in the host defense mechanism, occurring in approximately 1 in 700 individuals.[172] These patients usually have absent serum and secretory IgA but normal to only slightly decreased IgA-bearing B cells. There appears to be a terminal block in B-cell differentiation into plasma cells and subsequently a block in actual IgA synthesis. At least one individual has also been described who lacked secretory piece and had undetectable IgA in external secretions but normal serum levels of IgA.[173] Absence of IgA has been found in a number of relatively normal individuals who do not appear to have an increased incidence of infection.[174] Such patients, however, may have an increased incidence of cutaneous, respiratory, and GI allergies, which has led to the speculation that secretory IgA has a major role in preventing absorption of potential allergens. Autoimmune diseases such as rheumatoid arthritis and systemic lupus erythematosus (SLE) are also more common in these patients. Other individuals with absent IgA do appear to have recurrent viral and bacterial respiratory infections. These may present with upper respiratory infections, chronic bronchitis leading to bronchiectasis, or pneumonias. Such infections should be treated with antimicrobial agents based on the specific pathogens isolated and treatment should include strict attention to pulmonary therapy. Allergic therapy for rhinitis or asthma also may greatly benefit certain individuals with IgA deficiency. Plasma has on occasion had a beneficial effect in some of these patients despite the fact that little infused IgA can be demonstrated in external secretions.[175]

Gastrointestinal symptoms also are common in IgA deficiency and often include the development of lymphonodular hyperplasia. Chronic diarrhea secondary to *Giardia* is quite common and should be treated. Recently, a common susceptibility gene located in the MHC class III region on chromosome 6 has been implicated in both IgA deficiency and common variable immunodeficiency.[176]

3.3.6. Common Variable Hypogammaglobulinemia

This immune deficiency is usually classified as acquired; however, studies have indicated a possible genetic basis.[177] This is the most common form of serious immune deficiency that we see. It has its onset several years to several decades after birth and is characterized by a variable incidence of immunoglobulin deficiency. IgG is usually low with or without low concentrations of IgM and IgA. Approximately one third of patients have defective T-cell function and 25% eventually develop malignancies including thymomas and lymphoreticular tumors.[2] The exact pathogenesis of this disorder is not known, although deficiency of production of T-cell cytokines such as IL-2 has been implicated in some patients.[159] Respiratory infections including otitis media, sinusitis, bronchitis, and recurrent pneumonias are present in up to 98% of patients. These are usually caused *by H. influenzae, S. pneumoniae, S. pyogenes,* and *S. aureus.* Bacteremia and meningitis are less common but do occur. Chronic diarrhea in association with giardiasis occurs in up to one third of individuals with this syndrome. A number of associated problems such as thyroid abnormalities, achlorhydria, nodular lymphoid hyperplasia, and arthritis are common.

Respiratory infections are managed in a manner similar to that described for sex-linked hypogammaglobulinemia. Immunoglobulin replacement therapy is usually beneficial and may markedly limit chronic diarrhea along with therapy aimed at giardiasis. Pulmonary therapy as well as sinus drainage procedures may be beneficial in preventing long-term sequelae or respiratory infections.

Selective use of antimicrobial prophylaxis with trimethoprim–sulfamethoxazole may be useful as well.

3.4. Complement Component Deficiencies

The complement system represents an important aspect of the humoral portion of the host defense mechanism. Individual components play a major role in bacterial and yeast opsonization (C3b and C5b), viral neutralization (C4b), phagocyte chemotaxis (C5a), and lysis of some microorganisms (C6, C7, C8, and C9). Total or partial deficiency of individual components is genetically determined by autosomal inheritance. The homozygous state generally results in serious disease and complete absence of the component, whereas the heterozygous state leads to little in the way of symptomatology, but the component is present in approximately half the normal concentration.[25–32] Infections in patients with complement deficiencies vary significantly according to the nature of the missing component. The classic complement pathway involving C1, 4, 2, 3, 5, 6, 7, 8, and 9 is triggered predominantly by antigen–antibody complexes. In the absence of antibody, another system, the alternative pathway, which involves properdin, factor A or C3b, and factor B or C3 proactivator, is triggered and acts directly on C3 to activate the terminal part of the system. In general, therefore, the classic pathway, which is somewhat more efficient, is important in the immune host, whereas the alternative pathway functions in the individual lacking antibodies to certain bacteria. Two of the most important functions of the complement system appear to be to generate chemotactic factors and inflammatory mediators (C3a and C5a) and to opsonize microorganisms (C3b, C3bi, and C5b).

3.4.1. Early Classic Pathway Component Deficiencies

Deficiencies of Clq, r, s, C2, and C4 do not usually result in an increased incidence of infections, probably because the alternative complement pathway remains intact. Rather, these patients suffer from disease that resembles collagen vascular disorders.[178–181] Nephritis is particularly common and may be associated with rashes, Raynaud's phenomena, and arthritis. Severe infections occasionally occur, and these may result in part from the fact that the alternative pathway is somewhat less efficient than the classic one. Such patients have suffered from meningitis, sepsis, pneumonia, otitis media, paronychia, and sinusitis caused by agents that include *S. aureus, S. pneumoniae,* and *Salmonella.* In several cases, there have been underlying conditions that would predispose an individual to infection (including a skull fracture) or the individual suffered only one serious infection. Thus, it is quite difficult to relate unusual or recurrent infections to a deficiency of the early complement components. Of interest is a report by Newman et al.[182] in which two patients with C2 deficiency who developed repeated episodes of *S. pneumoniae* sepsis are described. In addition to C2 deficiency, however, both individuals were found to have low levels of factor B of the alternative complement pathway and deficient functional activity of this pathway. Each of these individuals as well as patients with other early component deficiencies have responded to appropriate antimicrobial therapy without added therapeutic maneuvers.

The gene for C2 is located with the major histocompatibility complex on the short arm of chromosome 6. Deficiency of C2 as well as of the components of the C1 complex (C1q, r, s) and C4 are all inherited in an autosomal recessive pattern.

3.4.2. C3 and C5 Deficiency

As mentioned, C3 and C5 breakdown products play a major role in the host defense mechanism, assisting in immune adherence, opsonization, and chemotaxis and being important as inflammatory mediators. Absence of C3 leads to infections that are similar to and even more severe than those in patients with hypogammaglobulinemia. Recurrent otitis media, sinusitis, pneumonias, paronychia, and impetigo are common.[25,183] Moreover, sepsis is not uncommon.[25,183] Etiologic agents include *S. pneumoniae*, *S. pyogenes*, *Klebsiella*, *N. meningitidis*, *H. influenzae*, and *E. coli*. Of interest is the fact that at least one patient had little response in the way of a leukocytosis despite repeated pyogenic infections. This may be related to the role of C3b breakdown products in releasing bone marrow PMN reserves. In general, these patients have responded well to appropriate antimicrobial agents alone. Initial therapy of the critically ill patient should include agents effective against the organisms listed above. In the critically ill patient, a unit or two of fresh-frozen plasma might also be of some benefit, especially if shock is present.

Functional deficiency of C5 was initially reported by Miller and associates.[29,184] These infants each had severe seborrheic dermatitis, intractable diarrhea, and recurrent infections with yeast and gram-negative bacilli. All had marked failure to thrive. Normal levels of C5 were detected by immunochemical means in each instance, but the patient's serum did not support phagocytosis of baker's yeast particles *in vitro*. Moreover, addition of

purified C5 to these sera corrected the defect in opsonization, suggesting that the patient's own C5 was not functionally active. Administration of fresh plasma to these individuals resulted in significant clinical improvement and restoration of opsonic activity *in vitro*. Thus, this entity became the first complement component deficiency that was found to be at least partially responsive to plasma therapy. Such individuals therefore should be treated with appropriate antimicrobial agents or topical antifungal drugs but should also receive a trial of fresh plasma therapy. It should be pointed out that plasma or blood stored at refrigerator temperature for 24 hr or more has little remaining functional C5 and will not correct this abnormality. Miller and Kablenzer[185] have subsequently pointed out the similarity between this disease and that described by Leiner[186] and have suggested that these two entities are the same.

Total deficiency of the fifth component of complement has also been described.[32] This patient suffered severe and repeated infections including chronic oral and vaginal candidiasis, infected cutaneous ulcers, subcutaneous abscesses, otitis media, and sepsis. Etiologic agents have included staphylococci, *Proteus*, *Pseudomonas*, *Enterobacter*, enterococci, and *S. pyogenes*. The patient also had a lupuslike syndrome and Raynaud's phenomena. Persistent shedding of cytomegalovirus occurred in her urine. Response to antimicrobial agents was not dramatic and other therapy was not attempted according to the report. It would seem that appropriate antimicrobial agents combined with fresh plasma therapy might have been of some benefit in this individual and in other patients with this disorder.

3.4.3. C6, C7, and C8 Deficiency

Absence of the sixth, seventh, and eighth components of complement have been reported in man.[27,28,30] Petersen and co-workers[31] indicated that 13 of 24 patients with absence of one of these components had at least one and usually several episodes of disseminated *Neisseria gonorrhoeae* or *N. meningitidis* infection. Most had joint or skin lesions associated with dissemination and few had problems with other infections except for one patient with C6 deficiency who also had four episodes of pneumococcal pneumonia.[28] Associated problems included Raynaud's phenomena, sclerodactylia, telangiectasia, ankylosing spondylitis, and SLE. Subsequently, sera from such individuals have been shown to lack bactericidal activity for *Neisseria* species. All patients with disseminated neisserial disease, particularly those with recurrent episodes, should probably be screened for complement component deficiencies using a total hemolytic complement assay. Therapy of the acutely ill patient with such a defect should include high-dose parenteral penicillin, since this would seem to cover most of the pathogens causing serious disease in these individuals.

3.4.4. Alternative Pathway Defects

Absence or deficiency of alternative pathway components may be congenital, developmental, or acquired. Newborn infants, for instance, have low levels of factor B and impaired nonspecific serum opsonic activity.[187] This is apparently a developmental problem, since levels subsequently become normal. Patients with the nephrotic syndrome apparently lose factor B in their urine, and thus also have defective nonspecific opsonic activity. Congenital deficiency of factor B also has been described in association with C2 deficiency as mentioned above.[182] Absence of properdin associated with recurrent infections also has been reported.[188] Patients with defects in the alternative complement pathway have in common an unusual incidence of severe and often overwhelming infections. The organisms involved are often polysaccharide-coated ones that can activate the alternative pathway in normal serum and include *N. meningitidis*, *S. pneumoniae*, *H. influenzae*, and *C. albicans*. The overwhelming nature of the infections in such patients demands that therapy be instituted promptly and be directed toward the organisms mentioned above. A useful adjunct in preventing serious infection might be through the use of vaccines against *S. pneumoniae*, *H. influenzae*, and *N. meningitidis*.

3.5. Phagocyte Abnormalities

The importance of phagocytes including PMNs, macrophages, and other fixed tissue histiocytes is illustrated best in the individual who has a marked deficiency of these cells. Profound neutropenia with a variety of causes including congenital, autoimmune, toxic, or those associated with malignancy usually result in severe infections and often death. Once the absolute PMN count drops below 500/mm³, one out of four individuals dies of overwhelming infection.[189] Moreover, the infections in these patients are ones not commonly observed in individuals with adequate numbers of phagocytes and include severe gram-negative and gram-positive sepsis, severe pneumonia, perirectal and abdominal abscesses, cutaneous and systemic candidiasis, and *Aspergillus* infection. Many of these patients have normal antibody and complement levels as well as adequate T-cell function but still contract these serious infections indicating the critical role that phagocytes play in the host defense mechanism against a number of bacterial, fungal, and even viral pathogens.[190,191]

3.5.1. Congenital Neutropenias

Several forms of congenital neutropenia occur which may or may not result in serious infection. These are briefly discussed below.

3.5.1a. Infantile Lethal Agranulocytosis. This is an autosomal recessive disease associated with profound neutropenia, eosinophilia, and monocytosis. The bone marrow reveals a striking absence of neutrophilic precursors. Steroids, splenectomy, or other maneuvers have little effect and the patients suffer severe infections with staphylococci, *E. coli*, *Proteus mirabilis*, *P. aeruginosa*, streptococci, and *Candida* organisms.[192] Despite antibiotic prophylaxis or prompt therapy of individual infectious episodes, death almost always results early in life. It would seem that bone marrow transplantation should be attempted early if an HLA-D-matched marrow donor is available. Kostmann[193] first described these patients in several Swedish families, but similar cases have been described elsewhere. In a recent report, five patients with congenital agranulocytosis were given recombinant human granulocyte colony-stimulating factor (G-CSF). All patients experienced an increase in their absolute neutrophil counts, from less than 100 to between 1300 and 9500 cells/μl.[194] This resulted in a decrease in morbidity and a reduction in need for antimicrobial therapy. Additional studies have indicated that G-CSF is lifesaving in these patients.

3.5.1b. Chronic Benign Neutropenia. This is usually a sporadic disease, but autosomal recessive inheritance has been reported. Unlike the fatal neutropenia described above, these patients have a much milder course.[195] They also have an eosinophilia and often a marked monocytosis, which may help to explain their ability to overcome infections. Bone marrow examination reveals arrest at the myelocyte or metamyelocyte stage. During acute infection, some of these patients appear to be able to mount a neutrophilic response. Infections include cutaneous and deep abscesses caused by staphylococci, streptococci, and gram-negative bacteria, recurrent pneumonias, and mouth ulcerations. Prompt antimicrobial therapy based on the results of aspirate cultures and limited incision and drainage is indicated. In general, these patients do not require therapeutic granulocyte transfusions, since they have some neutrophilic response. Moreover, as they grow older, infectious complications become fewer and most live an essentially normal life span. Initial antimicrobial therapy should include a penicillinase-resistant penicillin and in life-threatening infections a third-generation cephalosporin. Oral antimicrobial therapy should be used for cutaneous abscesses to prevent dissemination and limited drainage should be accomplished.

3.5.1c. Neutropenia with Hypogammaglobulinemia with Increased IgM. As mentioned under antibody-deficiency syndromes, patients with this sex-linked disease have a rather profound neutropenia that contributes to their infectious complications including pneumonias, episodes of sepsis, and occasional cervical adenitis and abscesses. The neutropenia in this disease may be cyclic or constant in nature. Therapy should include IgG replacement but rarely requires white cell transfusion except in the severely ill neutropenic patient.

Recent articles by Boxer and Blackwood[196,197] provide an excellent review of all these clinical disorders of neutropenia and enumerate their workup and therapy.

3.5.2. Chemotactic Defects

A number of individuals suffer severe and repeated infections secondary to phagocyte chemotactic defects. As mentioned, animal studies have shown that a critical 2- to 4-hr period exists during which phagocytes must arrive at the site of microbial invasion if infection is to be suppressed or contained.[97,98] Individuals with chemotactic defects have a variety of infections that are usually confined to the skin, lymph nodes, mucous membranes, and respiratory tract. Staphylococci predominate, probably because they make up a large part of the normal skin flora. Other common pathogens observed in patients with chemotactic defects include streptococci, *C. albicans*, *E. coli*, and *Trichophyton rubrum*.[35,36,39,43,99,198] The following disorders are probably congenital and are associated with chemotactic defects and recurrent infections.

3.5.2a. The "Lazy-Leukocyte" Syndrome. Miller and associates[39] described two patients in 1971 who had gingivitis, recurrent otitis media, rhinitis, and stomatitis. Each was found to have marked neutropenia, but in addition PMN random migration and chemotactic function were markedly abnormal. Both patients grew normally and severe life-threatening infections were not reported. Therapy of such patients would best be directed at the etiologic agent involved with oral antibiotics.

3.5.2b. Congenital Ichthyosis. Subsequently, Miller and co-workers[198] described two kindreds with congenital ichthyosis who suffered chronic, recurrent *Trichophyton rubrum* infections as well as otitis media, re-

current upper respiratory infections, deep abscesses, and generalized impetigo. Peripheral neutrophil counts were normal as was the random motility of their PMNs. Chemotaxis was markedly depressed, however. Griseofulvin therapy of the dermatoses, incision and drainage of abscesses, and oral therapy of otitis media and sinusitis should be reasonably effective in these individuals.

3.5.2c. Hyperimmunoglobulinemia E, Defective Chemotaxis, and Recurrent Infection. Hill and Quie,[35] Clark and co-workers,[99] and others[43,199,200] have described a whole host of patients with a syndrome of hyperimmunoglobulinemia E, allergic manifestations, recurrent infection, and defective PMN and monocyte chemotaxis. Several aspects of the syndrome including its association with bone disease have been reviewed.[201] In at least one report,[43] there appears to be a familial pattern in the inheritance of this syndrome. These patients have usually had severe eczema,[35,43,99,198] although urticaria[202] and allergic rhinitis[203] have also been reported. The infections suffered by these patients vary from multiple superficial cutaneous abscesses to deep-seated abscesses in the buttocks, scalp, or other tissues.[35,43,99,200,202,203] Chronic rhinitis, bronchitis, and otitis media also occur. Furthermore, serious systemic infections such as sepsis and pneumonia have been reported.[202] These are almost always caused by staphylococci, but streptococci and *P. aeruginosa* have also been incriminated on occasion. Chronic cutaneous candidiasis may also be a problem.[43,99] These patients may be related to those described by Buckley and associates[204] and are most certainly a variant of Job's syndrome described by Davis and associates.[205] In fact, we studied the original Job's syndrome patients, who are red-haired females with severe eczema, hyperimmunoglobulinemia E, and recurrent staphylococcal abscesses and found them to have a profound defect in chemotactic function. The defect in these patients is not always present and may be related to the release of allergic mediators that can depress PMN chemotaxis.[35,186,202] Of interest are findings suggesting that allergen-induced reactions can depress chemotactic function and that several of these patients have high levels of IgE antibody directed against staphylococci, their most prominent pathogen.[206,207] Thus, in the individual with the appropriate genetic factors, staphylococci may evoke a strong IgE response that on subsequent challenge leads to allergic release of mediators, thereby depressing the phagocyte system. Depressed neutrophil chemotaxis has even been described in patients with cow's milk or soy protein intolerance.[208] These patients are prone to recurrent respiratory infections. Lymphocytes of patients with the hyper-IgE syndrome have an imbalance in IFN-γ production in relation to IL-4 production. This would drive IgE levels up and possibly result in poor activation of PMNs and macrophages, since IFN-γ is a major activator of these cells.[209]

Therapy of these patients should be undertaken after appropriate cultures are obtained from abscesses, areas of cellulitis, or otitis media. Initial therapy should be directed at staphylococci and streptococci, as these are by far the most common pathogens. In abscesses and localized cellulitis, an oral agent such as dicloxacillin usually is sufficient. Limited incision and drainage is often required from which large volumes of purulent material will often be obtained. Healing from such surgical procedures is usually excellent, in contrast to that of patients with chronic granulomatous disease. Cutaneous *Candida* infection is quite difficult to treat and often requires extremely prolonged therapy with nystatin or clotrimazole. Systemic therapy with coverage aimed predominantly at *S. aureus* is indicated for pneumonia and sepsis in these patients.[202] Nafcillin and gentamicin should be used initially in such patients because of the synergism reported for these agents against both penicillin-sensitive and penicillin-resistant strains.[149] Levamisole is capable of enhancing chemotactic function both *in vitro*[210] and *in vivo*[211] in these patients. A controlled study with this agent failed, however, to decrease the number of serious infections, even though chemotactic function was increased.[212] We have observed some response to oral ascorbic acid in several individuals with hyper-IgE and recurrent infections. Following *in vitro* studies of ours[202] suggesting that histamine H2-blocking agents might improve chemotaxis in these patients, Mawhinney and associates[213] treated such a patient with 200 mg of cimetidine four times a day. The patient's chemotaxis remained normal throughout the treatment period. Soderberg-Warner *et al.*[214] suggested that trimethoprim–sulfamethoxazole may be useful in preventing infections in these patients. Jeppson and associates[139] have reported that IFN-γ improves the chemotaxis of the PMNs of these patients *in vitro*. IFN-γ therapy in the hyper-IgE patients is still experimental but may be of value.

3.5.2d. Defective Chemotaxis, Neutropenia, and Hyperimmunoglobulinemia A.

Björksten and Lundmark[215] described four siblings with recurrent bacterial infections, a defect in PMN chemotaxis, neutropenia, eosinophilia, and hyperimmunoglobulinemia A. These patients had recurrent cutaneous abscesses, otitis media, and pneumonia. *S. aureus* and *C. albicans* were isolated from cultures and responded to therapy like that described above for the hyper-IgE syndrome.

3.5.2e. Actin Dysfunction.

Boxer and co-workers[216] described an infant with blepharitis, vesicular skin lesions, abscesses, and sepsis with organisms including *S. aureus*, *C. albicans*, *Enterococcus faecalis*, and *E. coli*. This patient had a profound defect in PMN chemotaxis and phagocytosis and was found to have actin that polymerized poorly after treatment with potassium chloride. The patient eventually received a bone marrow transplant. Subsequent outcome is unknown.

3.5.2f. Monocyte Chemotactic Deficiency.

Defective monocyte chemotactic responsiveness on occasion also has been observed in the patients with hyperimmunoglobulinemia E. In addition, Snyderman *et al.*[217] and Gallin[218] have reported patients with chronic mucocutaneous candidiasis who had defective monocyte chemotaxis. In at least one case, transfer factor therapy significantly enhanced function and improved the patient's clinical condition.[218]

3.5.3. Microbicidal Defects

Microbicidal defects in phagocytes also result in serious infections and may lead to rather marked sequelae. The major microbicidal mechanism of the PMN involves the production of toxic oxygen products or radicals including hydrogen peroxide, superoxide, and perhaps singlet oxygen via the hexose monophosphate shunt.[190] Additional factors important in microbicidal activity include the lysosomal enzyme, myeloperoxidase, and a halide.[219] Several defects in this system exist that result in the intracellular survival or even multiplication of bacteria. The specific syndromes associated with microbicidal defects are discussed below.

3.5.3a. Chronic Granulomatous Disease.

Chronic granulomatous disease was the first granulocyte defect to be described and have its mechanism elucidated. This disease is usually sex-linked but may occur in autosomal recessive form or in association with severe G6PD deficiency.[37,38,220] Following phagocytosis, the cells of these patients fail to undergo the respiratory burst, do not activate their hexose monophosphate shunt, and do not produce those toxic oxygen products necessary for microbicidal activity. In the sex-linked form of the disease, this was thought to be caused by the absence of an NADH or NADPH oxidase required to activate the cell. Evidence, however, suggests that the abnormality is due to the absence of the 91-kDa heavy chain of cytochrome *b* 558 that is responsible for electron transfer in the initial stages of the respiratory burst.[208] Glutathione peroxidase deficiency was once believed to be behind the autosomal recessive form of the disease. It appears that these autosomal recessive forms are due to the absence of a 47- or

67-kDa cytosolic protein necessary for activity of the oxidase or the absence of 22-kDa light chain of the cytochrome. The gene for the 91-kDa cytochrome *b* 558 heavy chain is located at Xp21.1 (gp 91phox), while that for the 22-kDa light chain is at 16q24 (gp 22phox).[159] The gene for the 47-kDa cytosolic factor is at 7q11.23, while that for the 67-kDa factor is at 1q25 (gp 67phox).[159] These patients usually have early onset of recurrent abscesses, especially about the nose and mouth, hepatic abscesses, pneumonias, and osteomyelitis. The organisms involved are either catalase positive or they do not make hydrogen peroxide themselves. *S. pneumoniae*, group A and D streptococci, and α-streptococci are killed normally by these patients' cells.[221] By contrast, *S. aureus, S. epidermidis, E. coli, Serratia marcescens*, and *C. albicans* are not killed by the PMNs of these patients. *Aspergillus* as well as disseminated bacille Calmette–Guérin (BCG) infection has also been reported in these individuals.[222,223] Fungal infections were found in 20.4% of a large series of 245 cases of chronic granulomatous disease.[224] In addition to the infections mentioned, many of these patients have GI symptoms including diarrhea and stomach outlet obstruction because of granulomas. Granulomas also form in the abdomen and urinary tract and can lead to obstruction and additional infectious complications. Infection is usually caused by staphylococci, with *Klebsiella, E. coli, Serratia marcescens, C. albicans, Pseudomonas, Aspergillus, Proteus*, and *Salmonella* also being involved on occasion. A multiply-resistant pathogen, *Burkholderia cepacia*, has been recognized as an emerging pathogen in patients with CGD, causing necrotizing pneumonia and lymphadenitis.[224]

Therapy of these patients can be extremely difficult, since both PMNs and monocytes lack the ability to kill the organisms described. It is essential whenever possible to make an etiologic diagnosis by needle aspirate of abscesses, osteomyelitis, liver abscesses, or pneumonias. The use of extensive drainage procedures should be limited because these individuals heal very poorly. We have seen an inguinal incision and drainage site require 6 months to heal even with constant topical care. Moreover, the use of lobectomy in these patients should be discouraged, since the patient simply goes on to develop disease in other lobes and then is even more compromised.

Initial antimicrobial therapy in deep-seated bone or tissue infections (sepsis is very rare) should always include a penicillinase-resistant penicillin such as nafcillin. It may be combined with gentamicin for synergism or to cover other enteric gram-negative bacilli. Several investigators have used agents such as rifampin, clindamycin, or chloramphenicol which appear to penetrate WBCs more efficiently.[64,65,225] Such agents have been shown to have better activity *in vitro* in the presence of PMNs. Therapy must be continued for long periods, especially when incision and drainage are not carried out. After an initial 2- to 3-week period, oral antimicrobial agents may be substituted for parenteral therapy.

Antimicrobial prophylaxis per se was not previously recommended in this disease because of the large number of organisms that can be involved and because *Candida* and other fungal and bacterial pathogens may emerge and cause even more serious infection. Some success has been observed, however, through the use of sulfonamides or trimethoprim–sulfamethoxazole. This agent as well as rifampin and clindamycin are extremely effective in penetrating the cell membrane of PMNs and concentrating within the cell. Chronic use of these agents has been reported to result in some reduction in infectious complications and a slight increase in *in vitro* PMN bactericidal activity.[113] Granulocyte transfusion therapy has been used on occasion in severe infection, but results overall have not been dramatic. In one large study, patients with untreated fungal pneumonia or systemic disease usually succumbed.[224] More than one half the patients who were treated with appropriate antifungal agents survived, but granulocyte administration had no significant effect.[221] A number of older individuals are now known to have the disease, and it appears that infections become somewhat less severe with age.

A large, multicenter international trial of IFN-γ in patients with chronic granulomatous disease has recently shown a 70% overall reduction in the risk of serious infection. This is the first documented benefit of a recombinant human cytokine in preventing infections in an immunodeficient patient population.

3.5.3b. Myeloperoxidase Deficiency. Hereditary deficiency of the lysosomal enzyme myeloperoxidase has been described in association with *Candida* infections.[226] In addition, patients with Chediak–Higashi syndrome who have recurrent infections, oculocutaneous albinism, and giant PMN lysosomal granules also have a relative deficiency of myeloperoxidase, since their granules do not readily discharge myeloperoxidase into phagocytic vacuoles. The gene for this autosomal recessive disorder is on chromosome 1. Absence or deficiency of myeloperoxidase results in delayed killing of bacteria and may play a major role in killing yeasts such as *C. albicans*. A simple histochemical stain can be used to diagnose this disorder. Infections include those caused by *Candida*, which may be cutaneous or systemic, but also may include abscesses caused by *S. aureus*. Systemic *Candida* infection should be treated with low-dose amphotericin

B[227] combined with 5-flucytosine initially, whereas topical therapy with nystatin should control cutaneous infections. Fluconazole may be an appropriate alternative.

3.5.3c. Down's Syndrome. Patients with Down's syndrome may have defective PMN bactericidal activity against *S. aureus* associated with recurrent infections by this agent.[228,229] Nitroblue tetrazolium dye reduction was somewhat less than in controls in these patients, but other metabolic parameters have not been systemically examined. Patients with Down's syndrome probably have an increased incidence of staphylococcal abscesses and pneumonia, and this agent should be suspected as the etiologic agent in such patients. Therapy with a penicillinase-resistant penicillin should be aggressive. Incision and drainage is often indicated in abscesses.

3.5.4. Neonates with Combined Defects in Chemotaxis and Phagocytosis

Newborn infants have a rather profound defect in PMN and perhaps monocyte chemotactic responsiveness.[230–232] In addition, the PMNs of stressed but not normal neonates also seem to have a defect in oxidative metabolism and intracellular killing.[47,233,234] Thus, the neonate has abnormalities in both major functional activities of the PMN and suffers infections that are compatible with such defects. Cutaneous abscesses caused by *S. aureus* or gram-negative organisms or cellulitis caused by group A or group B streptococci are not uncommon in such infants. Chronic candidiasis is also quite common and may be related to abnormal PMN function since the T-lymphocyte system is usually normal at this age. Systemic or pulmonary infection with group B streptococci, *E. coli, Klebsiella pneumoniae, S. aureus, H. influenzae* (nontypeable), and *S. pneumoniae* may occur and have a very high incidence of morbidity and mortality.[235–240] Such serious infections must be treated early with appropriate antimicrobial therapy that probably should consist of ampicillin and an aminoglycoside such as gentamicin. Such a drug combination has synergism against many organisms including group B streptococci and *E. coli*, the most common pathogens isolated. Despite appropriate therapy, however, the mortality rate in neonatal sepsis and meningitis remains quite high. We have turned to evaluating the results of transfusion with fresh whole blood on mortality in early onset group B streptococcal sepsis.[241–243] In limited studies, transfusion of relatively large amounts of blood containing antibodies against the infecting organism has dramatically improved survival. Further studies have been carried out in neonatal animals to confirm the efficacy of passive antibody administration in protection and to determine whether white cell transfusions will also be of benefit. Experiments indicated that the administration of antibody and functional phagocytes offers significant protection against overwhelming infection in these immune-compromised neonatal animals.[238] Two review articles have been published recently that summarize the current knowledge pertaining to the use of intravenous gammaglobulin and granulocyte transfusions in the neonate.[244,245] These results underscore the importance of the statement attributed to George Bernard Shaw that "There is at bottom only one genuinely scientific treatment for all diseases and that is to stimulate the phagocyte."[190] This may be especially true in the congenital immunodeficient patient with infection in whom antimicrobial therapy alone may not suffice.

3.5.5. Leukocyte Adhesion Deficiency

The deficiency of CD11/CD18 (LAD Type I) complex is transmitted as an autosomal recessive trait. The gene encoding CD18, the common beta subunit, has been mapped to chromosome 21 at 21q22.3.[159] This deficiency is characterized by abnormal neutrophil mobilization frequently associated with marked leukocytosis, frequent infections such as gingivitis, perirectal abscesses, otitis media, sinusitis, sepsis, and pneumonia. A history of delayed umbilical cord separation also may be observed in these patients along with markedly elevated peripheral leukocyte counts. These infections are commonly caused by *S. aureus*, group A streptococci, *Proteus mirabilis, P. aeruginosa*, and *E. coli*. This syndrome is associated with absent or deficient expression of the plasma membrane glycoprotein LFA-1 (CD11a/CD18), Mac-1 (CD11b/CD18), and p150/95 (CD11c/CD18) due to defective production of the common CD18 component. The function of these surface glycoproteins is to promote a series of leukocyte adhesion-dependent interactions including binding to iC3b, aggregation, adhesion to endothelial cell surfaces, chemotaxis, phagocytosis and particle-induced respiratory burst activity of phagocytes, lymphoproliferative response of lymphocytes, and cytotoxicity mediated by T lymphocytes.[246]

There appear to be two variants of this disease: a severe form with a complete absence of CD11/CD18 expression and a moderate form with 5 to 20% of normal expression. The severity of the patient's clinical manifestations are closely related to the variant of the disease. Patients with severe deficiency are likely to die in the first year of life, while those with mild to moderate disease may live into adulthood.

In the second form of the disease sialyl-Lewis X,

LAD type II, the ligand on neutrophils for E-selectin is missing.[247,248] These patients' cells are unable to roll along the endothelium due to the lack of interaction with the endothelial cell selectins. Clinical manifestations are similar to LAD type I.

References

1. Stollerman GH: Immunologic deficiencies in the training of physicians. *J Chronic Dis* **26:**679–688, 1973

2. Hermans PE, Diaz-Buxo JA, Stobo JD: Idiopathic late-onset immunoglobulin deficiency. *Am J Med* **61:**221–237, 1976.

3. Albano EA, Pizzo PA: The evolving population of immunocompromised children. *Pediatr Infect Dis J* **7:**S79–S86, 1988.

4. Hill HR: Evaluating the patient with recurrent infections. *South Med J* **70:**230–235, 1977.

5. Hill HR: Laboratory aspects of immune deficiency in children. *Pediatr Clin North Am* **27:**805–830, 1980.

6. Hill HR: Immunodeficiency diseases. In Stefanini M, Benson E (eds): *Progress in Clinical Pathology*. Grune & Stratton, New York, 1981, pp. 205–238.

7. Johnston RB Jr, Janeway CA: The child with frequent infections: Diagnostic considerations. *Pediatrics* **43:**596–600, 1969.

8. Johnston RB Jr, Lawton AR III, Cooper MD: Disorders of host defence against infection: Pathophysiologic and diagnostic considerations. *Med Clin North Am* **57:**421–440, 1973.

9. Shyur S-D, Hill HR: Immunodeficiency in the 1990s. *Pediatr Infect Dis J* **10:**595–611, 1991.

10. Dingle JH, Badger GF, Jordan WS Jr: *Illness in the Home*. Press of Case Western Reserve University, Cleveland, 1964.

11. Ammann AJ: T cell and T-B cell immunodeficiency disorders. In Miller ME (ed): *The Child with Recurrent Infection*. Saunders, Philadelphia, 1977, pp. 293–311.

12. Groshong T, Horowitz S, Lovchik J, et al: Chronic cytomegalovirus infection, immunodeficiency, and monoclonal gammopathy-antigen-driven malignancy. *J Pediatr* **88:**217–223, 1976.

13. Kretschmer R, Say B, Brown D, et al: Congenital aplasia of the thymus gland (DiGeorge's syndrome). *N Engl J Med* **279:**1295–1301, 1968.

14. Lawlor GJ Jr, Ammann AJ, Wright WC Jr, et al: The syndrome of cellular immunodeficiency with immunoglobulins. *J Pediatr* **84:**183–192, 1974.

15. Jondal M, Holm G, Wigzell H: Surface markers on human T and B lymphocytes. I. A large population of lymphocytes forming nonimmune rosettes with sheep red blood cells. *J Exp Med* **136:**207–215, 1972.

16. Goldman AS, Goldblum RM: Primary deficiencies in humoral immunity. In Miller ME (ed): *The Child with Recurrent Infection*. Saunders, Philadelphia, 1977, pp. 277–291.

17. Hecht F, McCaw BK, Koler RD: Ataxia-telangiectasia-clonal growth of translocation lymphocytes. *N Engl J Med* **289:**286–291, 1973.

18. Shackelford PG, Polmar SH, Mayus JL, et al: Spectrum of IgG subclass deficiency in children with recurrent infections: Prospective study. *J Pediatr* **108:**647–653, 1986.

19. Heiner DC: Recognition and management of IgG subclass deficiencies. *Pediatr Infect Dis J* **6:**235–238, 1987.

20. Umetsu DT, Ambrosino DM, Quinti I, et al: Recurrent sino-

21. Berger M: Immunoglobulin G subclass determination in diagnosis and management of antibody deficiency syndromes. *J Pediatr* **110:**325–328, 1987.

22. Ambrosino DM, Siber GR, Chilmonczyk BA: An immunodeficiency characterized by impaired antibody responses to polysaccharides. *N Engl J Med* **316:**790–793, 1987.

23. Gigliotti F, Herrod HG, Kalwinsky DK, et al: Immunodeficiency associated with recurrent infections and an isolated in vivo inability to respond to bacterial polysaccharides. *Pediatr Infect Dis J* **7:**417–420, 1988.

24. Wasserman RL, Sorensen RU: Evaluating children with respiratory tract infections: The role of immunization with bacterial polysaccharide vaccine. *Pediatr Infect Dis J* **18:**157–163, 1999.

25. Alper CA, Colten HR, Gear JSS, et al: Homozygous human C3 deficiency. *J Clin Invest* **57:**222–229, 1976.

26. Ballow M, Shira JE, Harden L, et al: Complete absence of the third component of complement in man. *J Clin Invest* **56:**703–710, 1975.

27. Boyer JT, Gall EP, Norman ME, et al: Hereditary deficiency of the seventh component of complement. *J Clin Invest* **56:**905–913, 1975.

28. Leddy JP, Frank MM, Gaitner I, et al: Hereditary deficiency of the sixth component of complement in man. *J Clin Invest* **53:**544–553, 1974.

29. Miller ME, Seals J, Kaye R, et al: A familial, plasma-associated defect of phagocytosis: A new cause of recurrent bacterial infections. *Lancet* **2:**60–63, 1968.

30. Petersen BH, Graham JA, Brooks GF: Human deficiency of the eighth component of complement. *J Clin Invest* **57:**283–290, 1976.

31. Petersen BH, Lee TJ, Snyderman RJ, et al: *Neisseria meningitidis* and *Neisseria gonorrheae* bacteremia associated with C6, C7, or C8 deficiency. *Ann Intern Med* **90:**917–920, 1979.

32. Rosenfeld SI, Baum J, Steigbigel RT, et al: Hereditary deficiency of the fifth component of complement in man. *J Clin Invest* **57:**1635–1643, 1976.

33. Jaskowski TD, Martins TB, Litwin CM, Hill HR: Comparison of three different methods for measuring classical pathway complement activity. *Clin Diagn Lab Immunol* **6:**137–139, 1999.

34. Curnutte JT, Whitten DM, Babior BM: Defective superoxide production by granulocytes from patients with chronic granulomatous disease. *N Engl J Med* **290:**593–596, 1974.

35. Hill HR, Quie PG: Raised serum-IgE levels and defective neutrophil chemotaxis in three children with eczema and recurrent bacterial infections. *Lancet* **1:**183–187, 1974.

36. Hill HR, Quie PG: Defective neutrophil chemotaxis associated with hyperimmunoglobulinemia E. In Bellanti JA, Dayton DH (eds): *The Phagocytic Cell in Host Resistance*. Raven, New York, 1975, pp. 249–266.

37. Holmes B, Park BH, Malawista SE, et al: Chronic granulomatous disease in females: A deficiency of leukocyte glutathione peroxidase. *N Engl J Med* **283:**217–221, 1970.

38. Johnston RB Jr, Baehner RL: Chronic granulomatous disease: Correlation between pathogenesis and clinical findings. *Pediatrics* **48:**730–739, 1971.

39. Miller ME, Oski FA, Harris MB: Lazy-leukocyte syndrome. *Lancet* **1:**665–669, 1971.

40. Oh MK, Rodey GE, Good RA, et al: Defective candicidal capacity

of polymorphonuclear leukocytes from patients with chronic granulomatous disease of childhcod. *J Pediatr* **75:**300–303, 1969.

41. Quie PG, Hill HR: Granulocytopathies. *Dis Mon* **1:**1–32, 1973.

42. Quie PG: Bactericidal functicn of human polymorphonuclear leukocytes. *Pediatrics* **50:**264–270, 1972.

43. Van Scoy RE, Hill HR, Ritts RE Jr, *et al*: Familial neutrophil chemotaxis defect, recurrent bacterial infections, mucocutaneous candidiasis and hyperimmunoglobulinemia E. *Ann Intern Med* **82:**766–771, 1975.

44. Baehner RL, Nathan DG: Quantitative nitroblue tetrazolium test in chronic granulomatous disease. *N Engl J Med* **278:**971–980, 1968.

45. Park BH: The use and limitations of the nitroblue tetrazolium test as a diagnostic aid. *J Pediatr* **78:**376–378, 1971.

46. Cheson BD, Christensen RL, Sperling R, *et al*: The origin of the chemiluminescence of phagocytizing granulocytes. *J Clin Invest* **58:**789–796, 1976.

47. Shigeoka AO, Santos JI, Hill HR: Functional analysis of neutrophil granulocytes from healthy, infected and stressed neonates. *J Pediatr* **95:**454–469, 1979.

48. Emmendorffer A, Nakamura M, Rothe G, *et al*: Evaluation of flow cytometric methods for diagnosis of chronic granulomatous disease variants under routine laboratory conditions. *Cytometry* **18:**147–155, 1994.

49. Crockard AD, Thompson JM, Boyd NA, *et al*: Diagnosis and carrier detection of chronic granulomatous disease in five families by flow cytometry. *Int Arch Allergy Immunol* **114:**144–152, 1997.

50. Allen RC, Stjernholm RL, Steele RH: Evidence for the generation of an electronic excitation state(s) in human polymorphonuclear leukocytes and its participation in bactericidal activity. *Biochem Biophys Res Commun* **47:**679–684, 1972.

51. Allen RC, Yevich SJ, Orth RW, *et al*: The superoxide anion and singlet molecular oxygen: Their role in the microbicidal activity of the polymorphonuclear leukocyte. *Biochem Biophys Res Commun* **60:**909–917, 1974.

52. Rosen H, Klebanoff SJ: Chemiluminescence and superoxide production by myeloperoxidase-deficient leukocytes. *J Clin Invest* **58:**50–60, 1976.

53. Shigeoka AO, Hill HR: Recurrent pseudomonas infection associated with neutrophil dysfunction. *Scand J Infect Dis* **10:**307–311, 1978.

54. Stevens P, Winston DJ, Van Dyke K: *In vitro* evaluation of opsonic and cellular granuloctye function by luminol-dependent chemiluminescence: Utility in patients with severe neutropenia and cellular deficiency states. *Infect Immun* **22:**41–51, 1978.

55. Quie PG, White JG, Holmes B, *et al*: *In vitro* bactericidal capacity of human polymorphonuclear leukocytes: Diminished activity in chronic granulomatous disease of childhood. *J Clin Invest* **46:**668–679, 1967.

56. Ward PA, Cochrane CG, Muller-Eberhard HJ: The role of serum complement in chemotaxis of leukocytes *in vitro*. *J Exp Med* **122:**327–346, 1965.

57. Hill HR, Hogan NA, Mitchell TG, *et al*: Evaluation of a cytocentrifuge method for measuring neutrophil granulocyte chemotaxis. *J Lab Clin Med* **86:**703–710, 1975.

58. Cutler JE: A simple *in vitro* method for studies on chemotaxis. *Proc Soc Exp Biol Med* **147:**471–474, 1974.

59. Nelson RD, Quie PG, Simmons RL: Chemotaxis under agarose: A new and simple method for measuring chemotaxis and spontaneous migration of human polymorphonuclear leukocytes and monocytes. *J Immunol* **115:**1650–1656, 1975.

60. Miller ME: Leukocyte movement—*in vitro* and *in vivo* correlates. *J Pediatr* **83:**1104–1106, 1973.

61. Hill HR, Quie PG, Pabst HF, *et al*: Defect in neutrophil granulocyte chemotaxis in Job's syndrome or recurrent "cold" staphylococcal abscesses. *Lancet* **2:**617–619, 1974.

62. Ayoub EM, Dudding BA, Cooper MD: Dichotomy of antibody response to group A streptococcal antigen in Wiskott–Aldrich syndrome. *J Lab Clin Med* **72:**183–187, 1968.

63. Blaese RM, Strober W, Waldmann TA: Immunodeficiency in Wiskott–Aldrich syndrome. *Birth Defects* **11:**250–254, 1975.

64. Ezer G, Soothill JF: Intracellular bactericidal effects of rifampin in both normal and chronic granulomatous disease polymorphs. *Arch Dis Child* **49:**463–466, 1974.

65. Philippart AI, Colodny AH, Baehner RL: Continuous antibiotic therapy in chronic granulomatous disease: Preliminary communication. *Pediatrics* **50:**923–925, 1972.

66. Hahn HH, Beaty HN: Transtracheal aspiration in the evaluation of patients with pneumonia. *Ann Intern Med* **72:**183–187, 1970.

67. Hughes JR, Sinha DP, Cooper MR, *et al*: Lung tap in childhood. *Pediatrics* **41:**477–484, 1969.

68. Finaldn M: Diagnostic lung puncture. *Pediatrics* **44:**471–485, 1969.

69. Kahn FW, Jones JM: Analysis of bronchoalveolar lavage specimens from immunocompromised patients with a protocol applicable in the microbiology laboratory. *J Clin Microbiol* **26:**1150–1155, 1988.

70. Pattishall EN, Noyes BE, Orenstein DM: Use of bronchoalveolar lavage in immunocompromised children with pneumonia. *Pediatr Pulmonol* **5:**1–5, 1988.

71. Wolff LJ, Bartlett MS, Baehner RL, *et al*: The causes of interstitial pneumonitis in immunocompromised children: An aggressive systematic approach to diagnosis. *Pediatrics* **60:**41–49, 1977.

72. Byington CL, Taggart EW, Carroll KC, Hillyard DR: A polymerase chain reaction-based epidemiologic investigation of the incidence of nonpolio enteroviral infections in febrile and afebrile infants 90 days and younger. *Pediatrics* **103:**e1–e7, 1999.

73. Cunningham CK, Bonville CA, Ochs HD, *et al*: Enteroviral meningoencephalitis as a complication of X-linked hyper IgM syndrome. *J Pediatr* **134:**584–588, 1999.

74. Berman NS, Siegel SE, Nachum P, *et al*: Cerebrospinal fluid endotoxin concentrations in gram-negative bacterial meningitis. *J Pediatr* **88:**553–556, 1976.

75. McCracken GH Jr: Rapid identification of specific etiology in meningitis. *J Pediatr* **88:**706–708, 1976.

76. Nachum R, Lipsey A, Siegel SE: Rapid detection of gram-negative bacterial meningitis by the limulus lysate test. *N Engl J Med* **289:**931–934, 1973.

77. Controni G, Rodrigues WJ, Hicks JM, *et al*: Cerebrospinal fluid lactic acid levels in meningitis. *J Pediatr* **91:**379–384, 1977.

78. Haisch CE: The use of polymerase chain reaction in clinical infectious diseases. *Infect Dis Clin Pract* **5:**180–184, 1996.

79. Committee on Infectious Diseases: Therapy for children with invasive pneumococcal infections. *Pediatrics* **99:**289–299, 1997.

80. NIH Consensus Conference: Intravenous immunoglobulin. Prevention and treatment of disease. *JAMA* **264:**3189–3193, 1990.

81. Roifman CM, Gelfand EW: Replacement therapy with high dose intravenous gamma-globulin improves chronic sinopulmonary disease in patients with hypogammaglobulinemia. *Pediatr Infect Dis J* **5:**S92–S96, 1988.

82. Stiehm ER: New pediatric indications for IVIG. *Contemp Pediatr* **8:**29, 1991.

83. Henney CS, Ellis EF: Antibody production to aggregated human gammaglobulin in acquired hypogammaglobulinemia. *N Engl J Med* **278:**1144–1146, 1968.

84. Miller ME: Uses and abuses of plasma therapy in the patient with recurrent infections. *J Allergy Clin Immunol* **51**:45–56, 1973.

85. Cunningham-Rundles C, Siegel FP, Smithwick EM, *et al*: Efficacy of intravenous immunoglobulin in primary humoral immunodeficiency disease. *Ann Intern Med* **101**:435–439, 1984.

86. Condie RM, O'Reilly RJ: Prevention of cytomegalovirus infection by prophylaxis with an intravenous, hyperimmune, native unmodified cytomegalovirus globulin. Randomized trial in bone marrow transplant recipients. *Am J Med* **76**(3A):134–141, 1984.

87. Mease PJ, Ochs HD, Wedgewood RJ: Successful treatment of echovirus meningoencephalitis and myositis-fasciitis with intravenous immune globulin therapy in a patient with X-linked agammaglobulinemia. *N Engl J Med* **304**:1278–1281, 1981.

88. Stiehm ER, Vaerman JP, Fudenberg HH: Plasma infusions in immunologic deficiency states: Metabolic and therapeutic studies. *Blood* **28**:918–937, 1966.

89. Ammann AJ, Good RA, Bier D, *et al*: Long-term plasma infusions in a patient with ataxia-telangiectasia and deficient IgA and IgE. *Pediatrics* **44**:672–676, 1969.

90. Fudenberg HH, Spitter LE, Levin AS: Treatment of immune deficiency. *Am J Pathol* **69**:529–535, 1972.

91. Editorial: Chronic sinusitis. *Lancet* **1**:442–443, 1974.

92. Ament ME: Immunodeficiency syndromes and gastrointestinal disease. *Pediatr Clin North Am* **22**:807–825, 1975.

93. Guarino A, Guandalini S, Albano F, *et al*: Enteral immunoglobulin for treatment of protracted rotaviral diarrhea. *Pediatr Infect Dis J* **10**:612–614, 1991.

94. Binder HJ, Reynolds RD: Control of diarrhea in secondary hypogammaglobulinemia by fresh plasma infusions. *N Engl J Med* **277**:802–803, 1967.

95. Schmalstieg FC: Leukocyte adherence defect. *Pediatr Infect Dis J* **7**:867–872, 1988.

96. Hill HR: Clinical disorders of leukocyte function. *Contemp Top Immunobiol* **1984**:345–393, 1984.

97. Miles AA, Miles EM, Burke J: The value and duration of defense reactions of the skin to primary lodgement of bacteria. *Br J Exp Pathol* **38**:79–96, 1957.

98. Miles AA: The acute reaction of injury as an antimicrobial defense. In Thomas L, Uhr JW, Grant L (eds): *International Symposium on Injury, Inflammation and Immunity*. Williams & Wilkins, Baltimore, 1964, pp. 162–182.

99. Clark RA, Root RK, Kimball HR, *et al*: Defective neutrophil chemotaxis and cellular immunity in a child with recurrent infections. *Ann Intern Med* **78**:515–519, 1973.

100. Horsburgh CR Jr, Kirkpatrick CH: Long-term therapy of chronic mucocutaneous candidiasis with ketoconazole: Experience with twenty-one patients. *Am J Med* **74**:23–29, 1983.

101. Horsburgh CR, Cannody PB, Kirkpatrick CH: Treatment of fungal infections in the bones and joints with ketoconazole. *J Infect Dis* **147**:1064–1069, 1983.

102. Viscoli C, Castagnola E, Fioredda F, *et al*: Fluconazole in the treatment of candidiasis in immunocompromised children. *Antimicrob Agents Chemother* **35**:365–367, 1991.

103. Ikemoto H: A clinical study of fluconazole for the treatment of deep mycoses. *Diagn Microbiol Infect Dis* **12**:239S–247S, 1989.

104. Sharkey PK, Rinaldi MG, Dunn JF, *et al*: High-dose itraconazole in the treatment of severe mycoses. *Antimicrob Agents Chemother* **35**:707–713, 1991.

105. Larsen RA, Leal MAE, Chan LS: Fluconazole compared with amphotericin B plus flucytosine for cryptococcal meningitis in AIDS. *Ann Intern Med* **113**:183–187, 1990.

106. Bozette SA, Larsen RA, Chiu J, *et al*: A placebo-controlled trial of maintenance therapy with fluconazole after treatment of cryptococcal meningitis in the acquired immunodeficiency syndrome. *N Engl J Med* **324**:580–584, 1991.

107. Wingard JR, Merz WG, Rinaldi MG, *et al*: Increase in *Candida krusei* infection among patients with bone marrow transplantation and neutropenia treated prophylactically with fluconazole. *N Engl J Med* **325**:1274–1277, 1991.

108. Hiemenz JW, Walsh TJ: Lipid formulations of amphotericin B: Recent progress and future directions. *Clin Infect Dis* **22**:S133–S144, 1996.

109. Walsh TJ, Hiemenz JW, Seibel NL, *et al*: Amphotericin B lipid complex for invasive fungal infections: Analysis of safety and efficacy in 556 cases. *Clin Infect Dis* **26**:1383–1396, 1998.

110. Perrin JM, Charney E, MacWhinney JB, *et al*: Sulfasoxazole as chemoprophylaxis for recurrent otitis media. *N Engl J Med* **291**:664–667, 1974.

111. Hardin GM, Ronald AR: A controlled study of antimicrobial prophylaxis of recurrent urinary infections in women. *N Engl J Med* **291**:597–601, 1974.

112. Hughes WT, Kuhn S, Chaudhary S, *et al*: Successful chemoprophylaxis for *Pneumocystis carinii* pneumonitis. *N Engl J Med* **297**:1419–1426, 1977.

113. Johnston RB Jr, Wilfert CM, Buckley RH, *et al*: Enhanced bactericidal activity of phagocytes from patients with chronic granulomatous disease in the presence of sulphisoxazole. *Lancet* **1**:824–827, 1975.

114. Gonzalez LA, Hill HR: Advantages and disadvantages of antimicrobial prophylaxis in chronic granulomatous disease of childhood. *Pediatr Infect Dis J* **7**:83–85, 1988.

115. Kobayashi RH, Kobayashi AD, Lee N, *et al*: Home self-administration of intravenous immunoglobulin therapy in children. *Pediatrics* **85**:705–709, 1990.

116. Stiehm ER, Ashida E, Kim KS, *et al*: Intravenous immunoglobulins on therapeutic agents. *Ann Intern Med* **107**:367–382, 1987.

117. Meissner HC, Welliver RC, Chartrand SA, *et al*: Prevention of respiratory syncytial virus infection in high risk infants: Consensus opinion on the role of immunoprophylaxis with RSV hyperimmune globulin. *Pediatr Infect Dis J* **15**:1059–1068, 1996.

118. Ballow M, Dupont B, Good RA: Autoimmune hemolytic anemia in Wiskott–Aldrich syndrome during treatment with transfer factor. *J Pediatr* **83**:772–780, 1973.

119. Pabst HF, Swanson R: Successful treatment of candidiasis with transfer factor. *Br Med J* **2**:442–443, 1972.

120. Wybran J, Levin AS, Spitter LE, *et al*: Rosette-forming cells, immunologic deficiency diseases and transfer factor. *N Engl J Med* **288**:710–713, 1973.

121. Wara DW, Ammann AJ: Activation of T-cell rosettes in immunodeficient patients by thymosin. *Ann NY Acad Sci* **249**:308–314, 1975.

122. Goldstein AL, Cohen GH, Rossio JL, *et al*: Use of thymosin in the treatment of primary immunodeficiency diseases and cancer. *Med Clin North Am* **60**:591–606, 1976.

123. Polmar SH, Stern RC, Schwartz AL, *et al*: Enzyme replacement therapy for adenosine deaminase deficiency and severe combined immunodeficiency. *N Engl J Med* **295**:1337–1343, 1976.

124. Bortin MM, Rimm AA, *et al*: Severe combined immunodeficiency disease: Characterization of the disease and results of transplantation. *JAMA* **238**:591–600, 1977.

125. Hong R: Thymus transplants: A look to the future. *Birth Defects* **11**:357–360, 1975.

126. Buckley RH, Whisnant JK, Schiff RI, *et al*: Correction of severe combined immunodeficiency by fetal liver cells. *N Engl J Med* **294:**1076–1081, 1976.

127. Buckley RH: Immunoreconstitution. In Miller ME (ed): *The Child with Recurrent Infection.* Saunders, Philadelphia, 1977, pp. 313–328.

128. Thomas FD, Stork R, Clift RA, *et al*: Bone-marrow transplantation. *N Engl J Med* **292:**832–843, 1975.

129. Levine AS, Siegel SE, Schreiber AD, *et al*: Protected environments and prophylactic antibodies. *N Engl J Med* **288:**477–483, 1973.

130. Cleveland WW: Immunologic reconstitution in the Di-George syndrome by fetal thymic transplant. *Birth Defects* **11:**352–356, 1975.

131. Ammann AJ, Wara DW, Doyle NE, *et al*: Thymus transplantation in patients with thymic hypoplasia and abnormal immunoglobulin synthesis. *Transplantation* **20:**457–466, 1975.

132. Kirkpatrick CH, Wells SA, Burdick JF, *et al*: Effects of fetal thymus transplantation on defective cellular immunity. *Transplantation* **20:**367–369, 1975.

133. Hill HR, Estensen RD, Quie PG, *et al*: Modulation of human neutrophil chemotactic responses by cyclic 3'5'-guanosine monophosphate and cyclic 3'5'-adenosine monophosphate. *Metabolism* **24:**447–456, 1975.

134. Zurier RB, Weissmann G, Hoffstein S, *et al*: Mechanism of lysosomal enzyme release from human leukocytes. II. Effects of cAMP and cGMP, autonomic agonists, and agents which affect microtubule function. *J Clin Invest* **53:**297–309, 1974.

135. Oliver JM, Zurier RB: Correction of characteristic abnormalities of microtubule function and granule morphology in Chediak–Higashi syndrome with cholinergic agonists. *J Clin Invest* **57:**1239–1247, 1976.

136. Boyer LA, Watanabe AM, Rister M, *et al*: Correction of leukocyte function in Chediak–Higashi syndrome by ascorbate. *N Engl J Med* **295:**1041–1045, 1976.

137. Goetzel EJ, Wasserman SI, Gigli I, *et al*: Enhancement of random migration and chemotactic response of human leukocytes by ascorbic acid. *J Clin Invest* **53:**813–818, 1974.

138. The International Chronic Granulomatous Disease Cooperative Study Group: A controlled trial of interferon gamma to prevent infection in chronic granulomatous disease. *N Engl J Med* **324:**509–516, 1991.

139. Jeppson JD, Jaffe HS, Hill HR: Use of recombinant human interferon gamma to enhance neutrophil chemotactic responses in Job syndrome of hyperimmunoglobulinemia E and recurrent infections. *J Pediatr* **118:**383–387, 1991.

140. Petrak BA, Augustine NH, Hill HR: Recombinant human interferon-gamma treatment of patients with Job syndrome of hyperimmunoglobulinemia E and recurrent infections. *Clin Res* **42:**1A, 1994.

141. Cunningham-Rundles C, Kazbay K, Hassett J, *et al*: Brief report: Enhanced humoral immunity in common variable immunodeficiency after long-term treatment with polyethylene glycol-conjugated interleukin-2. *N Engl J Med* **331:**918–921, 1994.

142. Hughes WT, Feldman S, Chaundhary SC, *et al*: Comparison of pentamidine isethionate and trimethoprim-sulfamethoxazole in the treatment of *Pneumocystis carinii* pneumonia. *J Pediatr* **92:**285–291, 1978.

143. Lau WK, Young LS: Trimethoprim-sulfamethoxazole treatment of *Pneumocystis carinii* pneumonia in adults. *N Engl J Med* **295:**716–718, 1976.

144. Larter WE, John TJ, Sieber OF Jr, *et al*: Trimethoprim-sulfamethoxazole treatment of *Pneumocystis carinii* pneumonitis. *J Pediatr* **92:**826–828, 1978.

145. Walzer PD, Perl DP, Krogstad PJ, *et al*: *Pneumocystis carinii* pneumonia in the United States: Epidemiologic, diagnostic, and clinical features. *Ann Intern Med* **80:**83–93, 1974.

146. Montaner JSF, Lawson LM, Levitt N, *et al*: Corticosteroids prevent early deterioration in patients with moderately severe *Pneumocystis carinii* pneumonia and the acquired immunodeficiency syndrome (AIDS). *Ann Intern Med* **113:**14–20, 1990.

147. Soo Hoo GW, Mohserifar Z, Meyer RD: Inhaled or intravenous pentamidine therapy for *Pneumocystis carinii* pneumonia in AIDS. A randomized trial. *Ann Intern Med* **113:**195–202, 1990.

148. Hirschel B, Lazzarin A, Chopard P, *et al*: A controlled study of inhaled pentamidine for primary prevention of *Pneumocystis carinii* pneumonia. *N Engl J Med* **324:**1079–1083, 1991.

149. Watanakunakorn C, Glotzbecker C: Enhancement of the effects of antistaphylococcal antibiotics by aminoglycosides. *Antimicrob Agents Chemother* **6:**802–806, 1974.

150. Curry CR, Quie PG: Fungal septicemia in patients receiving parenteral hyperalimentation. *N Engl J Med* **285:**1221–1225, 1971.

151. Montgomerie JZ, Edwards JE Jr: Association of infection due to *Candida albicans* with intravenous hyperalimentation. *J Infect Dis* **1978:**197–201, 1978.

152. Hill HR, Mitchell TG, Matsen JM, *et al*: Recovery from disseminated candidiasis in a premature neonate. *Pediatrics* **53:**748–752, 1974.

153. Shepp DH, Dandliker PS, Meyers JD: Current therapy of varicella zoster virus infection in immunocompromised patients. A comparison of acyclovir and vidarabine. *Am J Med* **85**(Suppl 2A):96–98, 1988.

154. Buhles WC Jr., Mastre BJ, Tinker AJ, *et al*: Ganciclovir treatment of life- or sight-threatening cytomegalovirus infection: Experience in 314 immunocompromised patients. *Rev Infect Dis* **10:**S495–S506, 1988.

155. Buckley RH, Schiff SE, Schiff RI, *et al*: Hematopoietic stem-cell transplantation for the treatment of severe combined immunodeficiency. *N Engl J Med* **340:**508–516, 1999.

156. Parkman R, Gelfand EW, Rosen FS, *et al*: Severe combined immunodeficiency and adenosine deaminase deficiency. *N Engl J Med* **292:**714–719, 1975.

157. Stoop JW, Zegers BJM, Hendricks GFM, *et al*: Purine nucleoside phosphorylase deficiency associated with selective cellular immunodeficiency. *N Engl J Med* **96:**651–655, 1977.

158. Ackeret C, Pluss HJ, Hitzig WH: Hereditary severe combined immunodeficiency and adenosine deaminase deficiency. *Pediatr Res* **10:**67–70, 1976.

159. Shyur S-D, Hill HR: Recent advances in the genetics of primary immunodeficiency syndromes. *J Pediatr* **129:**8–24, 1996.

160. Lum LG, Tubergen DG, Corasti L, *et al*: Splenectomy in the management of the thrombocytopenia of the Wiskott–Aldrich syndrome. *N Engl J Med* **302:**892–896, 1980.

161. Stiehm ER: Plasma therapy: An alternative to gamma globulin injections in immunodeficiency. *Birth Defects* **11:**343–346, 1975.

162. Greenberg AH, Ray M, Tsai YT: Thymic alymphoplasia and dysgammaglobulinemia type I. Clinical, immunologic, and pathologic studies of one case. *J Pediatr* **75:**95–103, 1969.

163. Allibone EC, Goldie W, Marmon BP: *Pneumocystis carinii* pneumonia and progressive vaccines in siblings. *Arch Dis Child* **39:**26–34, 1964.

164. Wara DW, Goldstein AL, Doyle NE, *et al*: Thymosin activity in

patients with cellular immunodeficiency. *N Engl J Med* **292:**70–74, 1975.

165. Tiller TL Jr, Buckley RN: Transient hypogammaglobulinemia of infancy: Review of the literature, clinical and immunologic features of 11 new cases, and long-term follow-up. *J Pediatr* **92:**347–353, 1978.

166. Tsukada S, Saffran DC, Rawlings DJ, *et al*: Deficient expression of a B cell cytoplasmic tyrosine kinase in human X-linked agammaglobulinemia. *Cell* **72:**279–290, 1993.

167. Bruton OC: Agammaglobulinemia. *Pediatrics* **9:**722–728, 1952.

168. Davis SD: Antibody deficiency diseases. In Stiehm ER, Fulginiti VA (eds): *Immunologic Disorders in Infants and Children*. Saunders, Philadelphia, 1973, pp. 184–198.

169. Wilfert CM, Buckley RH, Mohammakeimar T, *et al*: Persistent and fatal central-nervous-system echovirus infections in patients with agammaglobulinemia. *N Engl J Med* **26:**1485–1489, 1977.

170. Abdel-Rahman SM, Kearns GL: Single-dose pharmacokinetics of a pleconaril (VP63843) oral solution and effect of food. *Antimicrob Agents Chemother* **42:**2706–2709, 1998.

171. Levitt D, Haber P, Rich K, *et al*: Hyper IgM immunodeficiency. *J Clin Invest* **72:**1650–1657, 1983.

172. Ammann AJ, Hong R: Selective IgA deficiency. In Stiehm ER, Fulginiti VA (eds): *Immunologic Disorders in Infants and Children*. Saunders, Philadelphia, 1973, pp. 199–214.

173. Strober W, Krakauer R, Klaeveman HL, *et al*: Secretory component deficiency: A disorder of the IgA immune system. *N Engl J Med* **294:**351–356, 1976.

174. Ammann AJ, Hong R: Selective IgA deficiency: Presentation of 30 cases and a review of the literature. *Medicine* **50:**223–226, 1971.

175. South MA, Cooper MD, Wollheim FA, *et al*: The IgA system. II. The clinical significance of IgA deficiency: Studies in patients with agammaglobulinemia and ataxia-telangiectasia. *Am J Med* **44:**168–178, 1978.

176. Sneller MC, Strober W, Eisenstein E, *et al*: New insights into common variable immunodeficiency. *Ann Intern Med* **118:**720–730, 1993.

177. Douglas SD, Goldberg LS, Fudenberg HH: Clinical, serologic and leukocyte function studies on patients with idiopathic "acquired" agammaglobulinemia and their families. *Medicine* **48:**48–53, 1970.

178. Day NK, Geiger H, Stroud R, *et al*: C1r deficiency: An inborn error associated with cutaneous and renal disease. *J Clin Invest* **51:**1102–1108, 1972.

179. Klemperer MR, Woodworth HC, Rosen FS, *et al*: Hereditary deficiency of the second component of complement (C'2) in man. *J Clin Invest* **45:**880–890, 1966.

180. Osterland CK, Espinoza L, Parker LP, *et al*: Inherited C2 deficiency and systemic lupus erythematosus studies on a family. *Ann Intern Med* **82:**323–328, 1975.

181. Gilliland BC, Schaller JG, Leddy JP, *et al*: Lupus syndrome in a C4-deficient child. *Arthritis Rheum* **18:**401, 1975.

182. Newman SL, Vogler LB, Feigin RD, *et al*: Recurrent septicemia associated with congenital deficiency of C2 and partial deficiency of factor B and the alternative complement. *N Engl J Med* **299:**290–292, 1978.

183. Alper CA, Colton HR, Rosen FS, *et al*: Homozygous deficency of C3 in a patient with repeat infections. *Lancet* **2:**1179–1181, 1972.

184. Miller ME, Nilsson UR: A familial deficiency of the phagocytosis-enhancing activity of serum related to a dysfunction of the fifth component of complement (C5). *N Engl J Med* **282:**354–358, 1970.

185. Miller ME, Kablenzer PG: Leiner's disease and deficiency of C5. *J Pediatr* **80:**879–880, 1972.

186. Leiner C: Uber Erythrodermia desquamativa, eine Eigenartige universelle Dermatose der Brustkinder. *Arch Dermatol Syph* **89:**163, 1908.

187. Hill HR, Hogan NA, Bale JF, *et al*: Evaluation of non-specific (alternative pathway) opsonic activity by neutrophil chemiluminescence. *Int Arch Allergy Appl Immunol* **53:**490–497, 1977.

188. Neu RL, Stockman JA III, Spitzer RE, *et al*: 46,XY/46,XY,21q-mosaicism in an infant with neutropenia and properdin deficiency. *J Med Genet* **13:**332–334, 1976.

189. Bodey GP, Buckley M, Sathe YS, *et al*: Quantitative relationships between circulating leukocytes and infection in patients with acute leukemia. *Ann Intern Med* **64:**328–340, 1966.

190. Stossel TP: Phagocytosis. *N Engl J Med* **290:**717–723, 1974.

191. Van De Meer JWM, Van Den Brock PJ: Present status of the management of patients with defective phagocyte function. *Rev Infect Dis* **6:**107–121, 1984.

192. Kauder E, Mauer AM: Neutropenias of childhood. *J Pediatr* **69:**147–157, 1966.

193. Kostmann R: Infantile genetic agranulocytosis. *Acta Paediatr* **45:**1–78, 1956.

194. Borilla MA, Gillio AP, Ruggeiro M, *et al*: Effects of recombinant human granulocyte colony-stimulating factor on neutropenia in patients with congenital agranulocytosis. *N Engl J Med* **320:**1574–1580, 1989.

195. Zuelzer WW, Bajoghli M: Chronic granulocytopenia in childhood. *Blood* **23:**359–374, 1964.

196. Boxer LA, Blackwood RA: Leukocyte disorders: Quantitative and qualitative disorders of the neutrophil, Part 1. *Pediatr Rev* **17:**19–28, 1996.

197. Boxer LA, Blackwood RA: Leukocyte disorders: Quantitative and qualitative disorders of the neutrophil, Part 2. *Pediatr Rev* **17:**47–50, 1996.

198. Miller ME, Norman ME, Koblenzer PJ, *et al*: A new familial defect of neutrophil movement. *J Lab Clin Med* **82:**1–8, 1973.

199. Dahl MV, Greene WH Jr, Quie PG: Infection, dermatitis, increased IgE, and impared neutrophil chemotaxis. *Arch Dermatol* **112:**1387–1390, 1976.

200. Jacobs JC, Norman ME: A familial defect of neutrophil chemotaxis with asthma, eczema, and recurrent skin infections. *Pediatr Res* **11:**732–736, 1977.

201. Hill HR: The syndrome of hyperimmunoglobulinemia E and recurrent infections. *Am J Dis Child* **136:**767–771, 1982.

202. Hill HR, Estensen RD, Hogan NA, *et al*: Severe staphylococcal disease associated with allergic manifestations, hyperimmunoglobulinemia E, and defective neutrophil chemotaxis. *J Lab Clin Med* **88:**796–806, 1976.

203. Hill HR, Williams PB, Krueger GG, *et al*: Recurrent staphylococcal abscesses associated with defective neutrophil chemotaxis and allergic rhinitis. *Ann Intern Med* **85:**39–43, 1976.

204. Buckley RH, Wray BB, Belmaker EZ: Extreme hyperimmunoglobulinemia E and undue susceptibility to infection. *Pediatrics* **49:**59–69, 1972.

205. Davis SD, Schaller J, Wedgwood RJ: Job's syndrome: Recurrences, "cold," staphylococcal abscesses. *Lancet* **1:**1013–1017, 1966.

206. Rubin JL, Griffiths RW, Hill HR: Allergen induced depression of neutrophil chemotaxis in allergic individuals. *J Allergy Clin Immunol* **62:**301–308, 1978.

207. Schopfer K, Baerlocher K, Price P, *et al*: Staphylococcal IgE

antibodies, hyperimmunoglobulinemia E and *Staphylococcus aureus* infections. *N Engl J Med* **300:**835–838, 1979.

208. Butler HL, Byrne WJ, Marmer DJ, *et al*: Depressed neutrophil chemotaxis in infants with cow's milk and/or soy protein intolerance. *Pediatrics* **67:**264–268, 1981.

209. Del Prete G, Tiri A, Maggi E, *et al*: Defective *in vitro* production of gamma interferon and tumor necrosis factor-alpha by circulating T cells from patients with hyperimmunoglobulinemia syndrome. *J Clin Invest* **84:**1830–1835, 1989.

210. Hogan NA, Hill HR: Levamisole enhances PMN chemotaxis and elevates cellular cyclic GMP. *J Infect Dis* **138:**437–444, 1978.

211. Wright DG, Kirkpatrick CH, Gallin JI: Effects of levamisole on normal and abnormal leukocyte locomotion. *J Clin Invest* **59:**941–950, 1977.

212. Donabedian H, Alling DW, Vallin JI: Levamisole is inferior to placebo in the hyperimmunoglobulin E recurrent infection (Job's) syndrome. *N Engl J Med* **307:**290–292, 1982.

213. Mawhinney H, Killen M, Fleming WA, *et al*: The hyperimmunoglobulinemia E syndrome—A neutrophil chemotactic defect reversible by histamine H2 receptor blockade? *Clin Immunol Immunopathol* **17:**483–491, 1980.

214. Soderberg-Warner M, Rice-Mendoza CA, Mendoza GR, *et al*: Neutrophil and T lymphocyte characteristics of two patients with the hyper IgE syndrome. *Pediatr Res* **17:**820–824, 1983.

215. Björksten B, Lundmark KM: Recurrent bacterial infections in four siblings with neutropenia, eosinophilia, hyperimmunoglobulinemia A, and defective neutrophil chemotaxis. *J Infect Dis* **133:**63–71, 1976.

216. Boxer LA, Hedley-Whyte ET, Stossel TP: Neutrophil actin dysfunction and abnormal neutrophil behavior. *N Engl J Med* **291:**1093–1099, 1974.

217. Snyderman R, Altman LC, Frankel A, *et al*: Defective mononuclear leukocyte chemotaxis: A previously unrecognized immune dysfunction. *Ann Intern Med* **78:**509–513, 1973.

218. Gallin JL: Abnormal chemotaxis: Cellular and humoral components. In Bellanti JA, Dayton DH (eds): *The Phagocytic Cell in Host Resistance*. Raven, New York, 1975, pp. 227–248.

219. Klebanoff SJ: Myeloperoxidase-halide-hydrogen peroxidase antimicrobial system. *J Bacteriol* **95:**2131–2138, 1968.

220. Gray GR, Klebanoff SJ, Stamatoyannopoulos G, *et al*: Neutrophil dysfunction, chronic granulomatous disease, and non-spherocytic haemolytic anaemia caused by complete deficiency of glucose-6-phosphate dehydrogenase. *Lancet* **2:**530–534, 1973.

221. Kaplan EL, Laxdal T, Quie PG: Studies on polymorphonuclear leukocytes from patients with chronic granulomatous disease of childhood: Bactericidal capacity for streptococci. *Pediatrics* **41:**591–599, 1968.

222. Raubitschak AA, Levin AS, Stites DP, *et al*: Normal granulocyte infusion therapy for aspergillosis in chronic granulomatous disease. *Pediatrics* **51:**230–233, 1973.

223. Verronen P: Presumed disseminated BCG in a boy with chronic granulomatous disease of childhood. *Acta Pediatr Scand* **63:**627–630, 1974.

224. Cohen MS, Isturiz RE, Malech HL, *et al*: Fungal infection in chronic granulomatous disease. *Am J Med* **71:**59–66, 1981.

225. Quie PG: Infections due to neutrophil malfunction. *Medicine* **52:**411–417, 1973.

226. Salmon SE, Cline MJ, Schultz J, *et al*: Myeloperoxidase deficiency. *N Engl J Med* **282:**250–253, 1970.

227. Medoff G, Dismicker WE, Meade RH, *et al*: A new therapeutic approach to Candida infections. *Arch Intern Med* **130:**241–249, 1972.

228. Gregory L, Williams R, Thompson E: Leukocyte function in Down's syndrome and acute leukemia. *Lancet* **1:**1359–1361, 1977.

229. Rosner F, Kozinn PJ, Jervis GA: Leukocyte function and serum immunoglobulins in Down's syndrome. *NY State J Med* **73:**672–675, 1973.

230. Klein RB, Fischer TJ, Gard SE, *et al*: Decreased mononuclear and polymorphonuclear chemotaxis in human newborns, infants, and young children. *Pediatrics* **60:**467–472, 1977.

231. Miller ME: Chemotactic function in the human neonate: Humoral and cellular aspects. *Pediatr Res* **5:**587–592, 1971.

232. Tono-Oda T, Nakayama M, Uehara H, *et al*: Characteristics of impaired chemotactic function in cord blood leukocytes. *Pediatr Res* **13:**148–151, 1979.

233. Mills EL, Thompson T, Björksten B, *et al*: The chemiluminescence response and bactericidal activity of polymorphonuclear neutrophils from newborns and their mothers. *Pediatrics* **63:**429–434, 1979.

234. Wright WC Jr, Ank BJ, Herbert J, *et al*: Decreased bactericidal activity of leukocytes of stressed newborn infants. *Pediatrics* **56:**579–584, 1975.

235. Bortolussi R, Thompson TR, Ferrieri P: Early-onset pneumococcal sepsis in newborn infants. *Pediatrics* **60:**352–355, 1977.

236. Filice GA, Cantrell HF, Smith AB, *et al*: *Listeria monocytogenes* infection in neonates: Investigation of an epidemic. *J Infect Dis* **138:**17–23, 1978.

237. Headings DL, Overall JC Jr: Outbreak of meningitis in a newborn intensive care unit caused by a single Escherichia coli K1 serotype. *J Pediatr* **90:**99–102, 1977.

238. Hemming VG, McClosky DW, Hill HR: Pneumonia in the neonate associated with group B streptococcal septicemia. *Am J Dis Child* **130:**1231–1233, 1976.

239. Hill HR, Hunt CE, Matsen JM: Nosocomial colonization with *Klebsiella* type 26 in a neonatal intensive care unit associated with an outbreak of sepsis, meningitis and necrotizing enterocolitis. *J Pediatr* **85:**415–419, 1974.

240. Pickering LK, Simon FA: Reevaluation of neonatal *Haemophilus influenzae* infections. *South Med J* **70:**205–208, 1977.

241. Hill HR: Host defenses in the neonate: Prospects for enhancement. *Semin Perinatol* **9:**2–11, 1985.

242. Hill HR: Diagnosis and treatment of sepsis in the neonate. In Root RK, Sande MA (eds): *Septic Shock*. Churchill Livingstone, New York, 1985, pp. 219–232.

243. Shigeoka AO, Hall RT, Hill H: Blood-transfusion in group-B streptococcal sepsis. *Lancet* **1:**636–638, 1978.

244. Gonzalez LA, Hill HR: The current status of intravenous gammaglobulin use in neonates. *Pediatr Infect Dis J* **8:**315–322, 1989.

245. Hill HR: Granulocyte transfusions in neonates. *Pediatr Rev* **12:**298–302, 1991.

246. Anderson DC, Schmalsteig FC, Finegold MJ, *et al*: The severe and moderate phenotypes of heritable Mac-1, LFA-1 deficiency: Their quantitative definition and relation to leukocyte dysfunction and clinical features. *J Infect Dis* **152:**668–689, 1985.

247. Etzioni A, Harlan JM, Pollack S, *et al*: Leukocyte adhesion deficiency (LAD) II: A new adhesion defect due to absence of sialyl Lewis X, the ligand for selectins. *Immunodeficiency* **4:**307–308, 1993.

248. Phillips ML, Schwartz BR, Etzioni A, *et al*: Neutrophil adhesion in leukocyte adhesion deficiency syndrome type 2. *J Clin Invest* **96:**2898–2906, 1995.

15

Management of Infections in Leukemia and Lymphoma

LOWELL S. YOUNG

1. Introduction

Much of the useful clinical information, as well as persisting controversies surrounding the management of infectious problems in the immunocompromised host, has come from studies of patients with leukemias and lymphomas. That these diseases should be the object of great interest is not difficult to fathom: they are rapidly or ultimately fatal if untreated, the magnitude of treatment with cytotoxic and immunosuppressive agents has few parallels in the therapy of any human illness (probably exceeded only by bone marrow transplantation, which is often employed for leukemia), and the infectious problems are acute, life threatening, and sometimes highly unusual. Infectious problems developing in other types of immunocompromised patients often result from utilization of the same medications and therapeutic strategies initially attempted in patients with hematologic malignancies. At many medical centers, individuals with acute leukemia have the most severe impairment of host defense mechanisms and represent the prime example of the immunocompromised host.

It has been gratifying that there has been substantial progress in effectively treating several major types of leukemias and lymphomas to the extent that "cure" can

Lowell S. Young • Kuzell Institute for Arthritis and Infectious Diseases, Division of Infectious Diseases, California Pacific Medical Center; and Department of Medicine, University of California at San Francisco, San Francisco, California 94115.

Clinical Approach to Infection in the Compromised Host (Fourth Edition), edited by Robert H. Rubin and Lowell S. Young. Kluwer Academic/Plenum Publishers, New York, 2002.

be an accurate and honest assessment of some treatment results. This success has been achieved with more intensive chemotherapy and radiation, better supportive care, and improved management of infectious complications. The nature of infectious complications has changed somewhat since reviews of two decades or more ago, but most patients still become neutropenic and develop fevers. Some infectious complications have become more amenable to therapy, like *Escherichia coli* sepsis and pneumocystosis. Effective treatment has become available for chickenpox and chemoprophylaxis for *Pneumocystis carinii* infections. On the other hand, a more ominous development has been the emergence of opportunistic fungal infections that are exceedingly difficult to diagnose and treat. Obviously, nature abhors an "ecologic vacuum," and the upsurge in fungal infections is probably a manifestation of better control of bacterial septicemias. Long periods of neutropenia, improved treatment of bacterial pathogens, coupled with failure to achieve substantial improvement in the underlying disease are the major prelude to fungal superinfection. In spite of this trend, clinicians need to be alert to the fact that some rather common infectious problems can still come back to haunt the patient with acute leukemia or lymphoma who has experienced "successful" treatment of the underlying disease. Probably the best example of this phenomenon is the problem of pneumococcal sepsis in splenectomized patients with Hodgkin's disease[1] or well-engrafted recipients of bone marrow transplants.[2] Not to be ignored at present is the potential for post-transfusion viral disease. These patients receive an enormous number of blood products and the possibility of acquired immunodeficiency syndrome (AIDS) must be considered in any febrile, deteriorating patient who was transfused before 1986 in the United States or Western Europe. Transfusion-associated AIDS is still of concern in some countries where routine screening of blood products for human immunodeficiency virus (HIV) is not equivalent to standard practices in the developed countries. Quiescent hepatitis C disease is a matter of increasing concern, but this chronic and often progressive disorder is usually not the cause of acute or chronic febrile syndromes in patients with hematologic malignancies.

2. Host Defenses against Infection in Leukemias and Lymphomas

There are several well-known shibboleths about the nature of the impairment in host defenses in leukemias and lymphomas and how specific defects relate to the incidence and type of complicating infections.[3] Nevertheless, it would be wise to accept all these traditional associations with some caution. The "classic" concept is that acute leukemia is associated with a paucity of normally functioning neutrophils, a granulocyte killing defect associated with cellular immaturity, or both, whereas patients with lymphomas have abnormalities in cell-mediated defense mechanisms ("delayed" hypersensitivity). Thus, leukemics experience pyogenic infections (staphylococcal, streptococcal, gram-negative), whereas patients with lymphomas experience infections by agents against which delayed hypersensitivity mechanisms are thought to be important (tuberculosis, fungal diseases). Finally, diseases in which humoral immunity is impaired either through hypogammaglobulinemia (chronic lymphatic leukemia) or the production of an abnormal immunoglobulin (multiple myeloma or macroglobulinemia) would be characterized by a defect in opsonizing or microbicidal antibodies. If levels of opsonizing antibodies are depressed, the net result is impaired granulocyte function and an increased susceptibility to acute bacterial infections similar to that seen with the acute leukemias is observed. One of the major problems, however, in attempting to relate types of opportunistic infection to underlying disease involves the major effect of treatment (chemotherapy and radiation being the major factors) on the incidence and nature of infectious complications. Except for those physicians who are caring for patients at the very inception of therapy, it will be rare for a clinician to face an "unmodified" patient. In our experience, the classic association of infections caused by encapsulated bacteria (pneumococci and to a lesser extent *Haemophilus influenzae*) has been observed in untreated patients with multiple myeloma or chronic lymphatic leukemia. These

diseases, if unmodified by therapeutic intervention, may show the "classic" associations. If, however, pharmacologic agents with effects on cell-mediated immune function such as the corticosteroids are used to treat these same diseases, the likelihood of infections by fungal organisms, agents such as *Listeria monocytogenes*, and pneumocystosis will be greatly increased. Likewise, when intensive cytotoxic treatment that depresses the total circulating neutrophil count is given to patients with non-Hodgkin's lymphomas and multiple myelomas, patients can develop overwhelming gram-negative and opportunistic fungal infections analogous to what has been regularly encountered in patients being treated for acute leukemia. Some antineoplastic agents may actually have antimicrobial effects, and these may be additional factors selecting for the types of organisms that cause infectious complications.[4,5]

In severe aplastic anemia, the patterns of infection are virtually identical to those observed in acute leukemia. Patients with acute leukemias and the aplastic anemias usually have decreased numbers of normally functioning neutrophils, but it is interesting that within the group of patients with aplasia whom we have followed, there appear to be at least two subpopulations: a group with fulminant disease who experience rapid decline of circulating neutrophils to negligible levels and a second group who have more indolent disease in whom circulating neutrophil counts of 200 to 600/mm³ persist. Clearly, those in the first group are likely to die within weeks to months unless they spontaneously recover hematopoietic function or receive a bone marrow transplant. In contrast, those patients with higher, stable residual white counts may have a life span of many years. The reduced number of neutrophils that are circulating in that latter group of patients seem adequate defense against pyogenic and gram-negative infections.

No clinician needs to be reminded of the crucial role of neutrophils in defense against bacterial and fungal infections. In acute leukemia, the paucity of mature neutrophils is clearly related to this susceptibility, and a sharp increase in the incidence of opportunistic infection has been observed with levels <500/mm³.[6] However, factors such as the ability of the bone marrow to compensate in times of acute infection and the ability of circulating phagocytic cells to migrate into infected tissues may be of greater importance than the absolute circulating neutrophil count per se.[7] Clearly, it may be hazardous to equate risk of infection with only the absolute concentration of circulating neutrophils and not the trend in these levels. For instance, patients now treated with some of the modern aggressive antileukemic protocols will often experience a rapid plummeting of their granulocyte counts such that these levels will halve on each successive day of cytotoxic treatment. Even though patients may have a circulating level of 2000 neutrophils/mm[3] and some circulating white cells may appear morphologically normal, the production of neutrophils has been virtually halted; such patients with rapidly plummeting neutrophil counts are highly susceptible to acute bacterial infections.

A more interesting issue involves defective functional capacity of phagocytic cells in leukemic patients whose peripheral blood counts may not be at a "neutropenic" level. Techniques for evaluating microbicidal function of individual leukocytes in mixed populations[8,9] suggest that morphologically mature neutrophils of some patients with acute myelogenous leukemia have impaired ability to kill certain species of fungi and bacteria. The usual presumption has been that cells of the neutrophil series more immature than the metamyelocyte do not phagocytize and kill normally. However, it appears that some leukemic cells in the peripheral blood are morphologically normal but are functionally abnormal. The abnormality of impaired intracellular killing may not be present in the neutrophils alone but may be present in the mononuclear phagocytes as well. Thus, intracellular bacteria then may find a "sanctuary" from the bactericidal action of systemically administered antimicrobial agents, and this could explain in part the persistence of bacterial infection in patients receiving antibiotics. Again, there are many problems in the interpretation of these data including the effect of the underlying disease (in this case acute leukemia) and the added effect of chemotherapy. The cause of the defective microbicidal activity has not been determined with certainty, but there is concomitant evidence that some of these morphologically normal cells have low levels of lysozyme.[9] The risk of pneumocytosis in patients who have received high-dose steroids during remission underscores a persisting risk despite recovery of the white count. In addition, it has been demonstrated that children with acute lymphatic leukemia undergoing craniospinal irradiation may have defective leukocyte microbicidal function.[10] These observations link an increased clinical incidence of infectious complications with a demonstrable abnormality in phagocyte function.

It is widely held that thymus-derived lymphocyte (T-cell)-mediated immune functions are important defenses against a variety of fungal diseases. On the other hand, one of the more significant findings during the last two decades is the rather striking increase in opportunistic fungal diseases, particularly candidiasis, aspergillosis, and mucormycosis, in the neutropenic acute leukemic population.[11,12] The most plausible explanation for this

trend is the intensive multiple-agent treatment protocols given for acute leukemia do more than depress granulocyte precursors and have significant effects on cell-mediated immune function. However, it also seems likely that neutrophils are important for host defenses against organisms such as *Aspergillus* and *Candida*, which is supported by some studies.[9–11]

In general, humoral immunity as measured by antibody responses and complement levels seems to be relatively well preserved in the acute stages of therapy for acute lymphatic leukemia or acute myelogenous leukemia. Immunization with antigens such as *Pseudomonas aeruginosa* lipopolysaccharides results in brisk type-specific antibody response early in the course of both types of leukemia.[13] A functional assay of circulating antibodies against *P. aeruginosa*, the opsonophagocytosis test, showed no difference between normal serum and serum from children initially treated for acute lymphatic leukemia.[14] However, with resumption of intensive chemotherapy for relapsed leukemia, humoral levels of antibodies (as measured by passive hemagglutination and by opsonophagocytosis against *P. aeruginosa*) declined. This finding suggests that antibody-synthesizing capacity decreases as the underlying disease progresses and/or as more intensive chemotherapy is given in an attempt to induce remission. Opsonic deficiencies therefore parallel levels of circulating neutrophils,[13] a finding that may be critical for interpretation of the efficacy of transfused granulocytes. In chronic lymphocytic leukemia hypogammaglobulinemia is the rule rather than the exception.[15]

The effect of antineoplastic chemotherapy in enhancing susceptibility to infection can neither be overlooked nor easily estimated. In addition to their quantitative effects on circulating neutrophils and lymphocytes, large doses of corticosteroids and vinca alkaloids such as vincristine and vinblastine affect bactericidal or locomotive function of phagocytes.[16] Even antibiotics of the polyene class or sulfonamides have been shown, in doses commonly used, to depress neutrophil function.[17,18] The use of pharmacological agents alone or in combination can compound an underlying intrinsic defect in host defenses, and such defects may persist for years.[19]

Patients with advanced Hodgkin's disease occasionally exhibit the following immunologic abnormalities: (1) reduced response to a variety of antigens that induce delayed hypersensitivity reactions, (2) delayed homograft rejection, and (3) increased susceptibility to infections caused by varicella zoster virus (VZV), granulomatous infections such as tuberculosis, and *L. monocytogenes*.[20] The blastogenic response of peripheral blood lymphocytes from patients with active Hodgkin's disease is reduced on exposure to common antigens such as phytohemagglutinin and those used in skin testing. Contrary to widespread impression, delayed hypersensitivity responses as evaluated by a battery of skin tests employed in untreated patients in all stages of disease are normally present[20]: only 11.7% of more than 100 patients in one series were anergic to a battery of six skin-test antigens. No patient with stage I disease was anergic, and the incidence of complete anergy increased to 27% of patients with stage IV disease. Skin-test reactivity was found to correlate with the absence of systemic symptoms of disease as well as tumor histologic type but was not a useful prognostic sign. The mean absolute lymphocyte count was inversely proportional to skin-test reactivity and declined in most advanced stages of disease. Thus, patients found to have stage I Hodgkin's disease demonstrated skin-test reactivity similar to normal controls, and in all stages of disease skin-test reactivity correlated with absence of symptoms. It was the conclusion of Young *et al.*[20] that anergy, when it exists, is a result of disease progression and not the cause of the disease. These authors also acknowledge that the prognostic value of the skin test as a measure of cellular immune function may be obscured by successful chemotherapeutic and radiotherapeutic intervention. King *et al.*[19] have pointed out that patients successfully treated for lymphomas usually are not anergic when treated with "recall" antigens. By contrast, most are unable to develop delayed cutaneous hypersensitivity to neoantigens.

There are conflicting data on abnormalities in polymorphonuclear leukocyte, monocyte, and macrophage function in patients with lymphomas. Steigbigel *et al.*[21] studied patients with Hodgkin's disease and other lymphomas in all stages of disease prior to staging laparotomy. These patients were untreated and none had evidence of active infection. Neutrophil function, serum opsonic activity, and monocyte-derived macrophage phagocytosis and killing were all within normal limits. During the course of *in vitro* experiments with macrophages incubated with *L. monocytogenes*, these investigators obtained a larger recovery of *L. monocytogenes* from supernatant fluids after macrophage cell lysis. However, these results could not be interpreted as solely reflecting a defect in intracellular killing, since both these workers and others found that macrophages or reticuloendothelial cells from Hodgkin's disease patients are more rapidly phagocytic than normal macrophages or reticuloendothelial cells. Furthermore, differences in macrophage function tests between cells from Hodgkin's disease patients and normals were observed during the early incubation

period (1–2 hr), but later (3–6 hr) there was no significant difference between the two groups. These results suggest that either macrophages from patients with Hodgkin's disease are more avidly phagocytic than those of normal subjects or that macrophages from patients with the disease kill intracellular organisms such as *Listeria* more slowly following ingestion.

Although the studies of Steigbigel *et al.*[21] failed to clarify an immunologic defect in these untreated patients with Hodgkin's disease, others[19] have reported that patients receiving chemotherapy or radiation demonstrate impaired monocyte function. The possibility remains that those individuals who actually develop evidence of infection (as opposed to the uninfected patients with Hodgkin's disease who were studied by Steigbigel) might manifest defects in intracellular killing, particularly against pathogens such as *Listeria*, mycobacteria, *Nocardia*, or *Salmonella*. Alternatively, some authors propose that defects are present in patients other than those having a nodular sclerosing histological pattern (which was the principal type that Steigbigel *et al.*[21] studied) or that the defect in cell-mediated immune function lies not with the phagocytic cell but perhaps in the capacity to activate macrophages and to engage in the process of chemotaxis.[21] The latter function has been found to be abnormal in patients with Hodgkin's disease because of the presence of a chemotactic factor inhibitor.[23] Thus, although the efferent loop, i.e., postphagocytic killing, may be normal, afferent processes in this immunologic "arc" may be impaired in Hodgkin's disease. *In vitro* tests of bactericidal capacity involve direct mixing of microbes and cells (which may be activated in the process of harvesting or cultivation), whereas a chemotactic factor inhibitor may impair *in vivo* activation of the locomotion of cells that possess microbicidal activity (i.e., monocytes and neutrophils).

Earlier studies of the primary and secondary antibody response in patients with Hodgkin's disease showed these to be unimpaired.[24] It has been stated that serum immunoglobulin levels are generally normal at the inception of treatment. However, more intensive combined treatment modalities that include radiotherapy, chemotherapy, and splenectomy are currently employed.[25] The clinical observation of increased numbers of infections related to pneumococci and *H. influenzae* in patients receiving combined radiotherapy and chemotherapy where single-treatment modalities reduced antibody titers insignificantly suggests a problem with humoral immunity. Although splenectomy had no effect on antibody titers, it potentiated the reduction of IgM levels by chemotherapy. However, IgG and IgA levels and alternate-pathway complement activation were either normal or elevated. These authors concluded that aggressive treatment with chemotherapy and radiation therapy impairs humoral defenses against encapsulated microorganisms, and thus magnifies the risk of postsplenectomy septicemia in patients with Hodgkin's disease. Thus, they concluded that treatment rather than the initial severity of disease is the major determinant responsible for depressed antibody levels. They have called attention to the irony that cure of Hodgkin's disease in certain patients by the combination of splenectomy and intensive chemotherapy has transformed a cellular immune defect in the underlying disorder to a life-threatening deficiency in humoral immunity.[25] It must be repeatedly emphasized that splenectomized patients, particularly those with a hematologic or lymphoproliferative malignancy, are prone to develop overwhelming septicemia due to encapsulated bacteria.

3. The Role of Infection in Mortality from Leukemia and Lymphomas

There is no disputing the widely held view that infection plays the major role in the mortality associated with leukemias and some lymphomas. Although this information emanates from the major cancer treatment centers and therefore represents a population in which particularly vigorous chemotherapeutic, radiotherapeutic, and supportive measures have been given, it would be surprising if the role of infection was significantly different in patients with leukemia and lymphomas who received no treatment for their underlying disease. Table 1 is a summary of some relatively old studies reporting primary causes of death in acute leukemia or hematological malig-

nancy. Five of the series concentrate exclusively on acute leukemia and one of these focuses exclusively on childhood leukemia (primarily of the lymphatic type). Of the series summarized by Levine et al.[26] from the National Cancer Institute, a total of 450 patients had hematologic malignancies, of which the majority were cases of acute leukemia and malignant lymphomas. Although the series summarized in Table 1 are by no means comprehensive, they do point toward some interesting trends observed over the past 40 years. Perhaps the most meaningful juxtaposition is the report by Hersh et al.[27] published in 1965, which summarizes the experience between 1954 and 1963 at the National Cancer Institute. Results for the period analyzed by Hersh contrast significantly with the experience reported by Levine et al.[26] from the same institution for the period between 1965 and 1971. Whereas infection accounted for 38% of the primary causes of death in acute leukemia between 1954 and 1963, it had increased to 69% in the latter series from the same institution. One of the major differences between the two series was the decline in percentage of deaths primarily associated with hemorrhage. It is particularly impressive that hemorrhage as the major coprimary cause of death declined at the National Cancer Institute from 31.7% in the 1954–1963 observation period to only 10% in the series reported by Levine and colleagues.[26] The report of Viola[28] from Yale University School of Medicine encompassed the years 1951–1966, for the most part of a period during which several modern techniques for the control of bleeding were not available; hence, the largest proportion of cases with acute leukemia died of hemorrhage alone. Difficulty must be acknowledged, however, in distinguishing between the roles of infection and hemorrhage as causes of mortality because a number of infectious

TABLE 1. Primary Causes of Death in Acute Leukemia or Hematologic Malignancy

Investigators	Site	Years	Population	Number of patients	Infection (%)	Hemorrhage (%)	Both (%)	Others (%)
Hersh et al.[27]	NCI	1954–1963	Acute leukemia	366	37.7	20.5	31.7	—
Viola[28]	Yale	1951–1966	Acute leukemia	78	40	33	7	20
Hughes[29]	St. Jude's	1962–1969	Acute childhood leukemia	199	45	—	33[a]	22
Levine et al.[70]	NCI	1965–1971	Hematologic malignancy	450	69	11	10	10
Chang et al.[30]	M.D. Anderson	1966–1972	Adult acute leukemia	315	66	15	9	10
Feld and Bodey[32]	M.D. Anderson	1966–1973	Malignant lymphoma	206	51	9	—	40[b]
Estey et al.[137]	M.D. Anderson	1973–1979	Acute leukemia	123	49	11	20	20

[a]Infection and hemorrhage pooled together.
[b]Organ failure and electrolyte imbalance, 23%; neoplasia, 11%; infarction, 6%.

processes can trigger disseminated intravascular coagulation, and conversely a site where bleeding has occurred can become a focus of infection.

The problem in distinguishing between hemorrhage alone and infection plus hemorrhage is emphasized by the experience reported by Hughes[29] at St. Jude's Children's Cancer Research Hospital where the results showed that infection was usually accompanied by hemorrhage, and together they accounted for 33% of all deaths in this pediatric population. Unquestionably, more effective control of bleeding complications with platelet transfusions and clotting factors seems responsible for the decline in the role of hemorrhage and the relative increase in the proportion of cases where infection has been the major cause of death. In the last quarter century, infection alone appears to account for almost 70% of deaths from acute leukemia.[30] Hemorrhage and infection together represent another 10% of cases, so that it seems reasonable to attribute some 75–80% of deaths in acute leukemia to infection-related causes.

Relatively less information is available on the role of infection in mortality from malignant lymphoma. Casazza et al.[31] reviewed the experience at the National Cancer Institute between 1953 and 1964 and reported that 99 of 139 patients who died had infections demonstrable at autopsy. However, 124 of these 139 patients were believed to have died primarily of the disseminated neoplasm, albeit with serious associated microbial infection. Only two patients in the survey died of infection while the neoplastic disease was apparently in remission. These authors attributed only 10% of deaths in patients with Hodgkin's disease and mycoses fungoides primarily to infectious causes. They concluded that even fewer patients, less than 4% with lymphocytic lymphoma and no patients with histiocytic lymphoma, died primarily of infection but rather had a refractory neoplasm as the cause of death.

A larger more recent study has summarized the experience from the M.D. Anderson Hospital between 1966 and 1973 and is based on 206 patients with malignant lymphomas who had complete postmortem examinations.[32] The most common cause of death was thought to be infection, which accounted for 51% of the overall mortality. Death was thought to result from "organ failure" (hepatic and renal) in 25% of patients, hemorrhage in 9%, disseminated neoplastic disease in 8%, infarction of tissue in 6%, a second neoplasm in 3%, and electrolyte abnormalities in 1%. Finally, the experience in the treatment of 300 patients with Hodgkin's disease has been reported from Stanford University Medical Center by Notter et al.[33] Although the overall mortality in the series

of 300 patients was relatively low (19%), histologic evidence of Hodgkin's disease was present at death in more than two thirds of patients who came to autopsy, and infection contributed to death in half of all autopsied patients. It was believed that few of these deaths were related to infection in the absence of active neoplasia.

4. Problems with the Interpretation of Fever and Infection Incidence Data in Neutropenic States

There are several major problems that must be acknowledged in any analytic approach to the scope of the infection problem in patients with leukemias and lymphomas: In recognition of these problems, the information summarized in the preceding and ensuing two sections should be accepted in only the broadest of terms, with stress placed on general relationships rather than the precision and accuracy of the information reported:

1. The limitations of extrapolating from autopsy data should be abundantly clear because infections present at autopsy and/or identified as the cause of death may have little relationship to a febrile episode experienced several weeks before. In many referral centers the overall rate at which autopsies have been performed (percentage of total deaths) has been on a marked decline. Thus, infectious complications whose diagnosis is based on tissue examination may be significantly underestimated.

2. Multiple agents may be recovered from blood or histologic material obtained by biopsy or autopsy and the relative importance of each isolate may be difficult to ascertain.

3. Tumor may coexist with infection and the causes of death may be multiple, including persistent neoplasm.

4. Different criteria have been used for defining neutropenic states and most of the lymphomas have been reclassified. Therefore, some series cannot be directly compared.

5. Antibiotic usage has changed over the last 30 years with the introduction of newer systemic agents and the increasing tendency to use prophylactic oral antibiotics. This may influence the recovery of pathogens from disease sites and could be responsible for the decreasing ability to establish a link between fever and documented infection in recent studies.

6. Varying criteria have been employed for defining infection. Some studies have included a category such as "infection caused by unknown organisms,"[34] and others have employed a criterion in which the response to chemotherapy or antimicrobials is used to imply documented

infection. One of the more frustrating aspects of dealing with neutropenic patients is that a significant number may have an infected local site such as soft tissue or lung, but the precise cause of the infectious process is never identified. This lack of verification often results from the inability to undertake diagnostic studies or from concurrent antimicrobial therapy at the time that diagnostic attempts are made. Still, in the analysis of such events, it would seem more appropriate to distinguish microbiologically proven infections from febrile episodes whose microbial etiology is not documented. To compound the nosologic dilemma, further terms such as "sepsis syndrome" have appeared in the literature and cogent arguments for their adoption have appeared.[35] The increasing popularity of these concepts stems from the recognition that frequently febrile patients appear "toxic" and have some compromise in organ function, tissue perfusion, or impaired oxygenation, yet their blood cultures are negative. There clearly are clinical situations where nonbacteremic patients appear to be as ill as those who ultimately are found to have positive blood cultures, and the evolution of concepts like "sepsis syndrome" is appropriate to describe these conditions.

7. The source of bacteremias in some immunocompromised patients often is never found.[3,36] This may be because of failure to culture stool and distinguish normal flora and potential pathogens such as *P. aeruginosa*. As more careful surveillance studies have been carried out, the role of the gastrointestinal (GI) tract as a source of many septicemias in neutropenic patients has been firmly established.[37]

8. Many studies analyzing the total incidence of fever and infection in neoplastic states fail to take into account the stage of the underlying disease and whether patients are in a pretreatment, posttreatment, or consolidation phase. This factor may be critically related to the incidence and type of infecting agent. For instance, pneumocystosis in children is usually a disease of the remission phase. When it occurs in adults with lymphomas, the underlying disease is most often not under chemotherapeutic control.

5. Causes of Fever in Leukemia and Lymphoma

It is well recognized that fever can be a manifestation of underlying neoplastic disease, and this is particularly true of patients with leukemia and lymphoma. On the other hand, the appearance of fever in neutropenic patients necessitates a thorough search for an infectious cause.[38] In both children and adults pneumocystis disease may declare itself clinically as steroid dosing is tapered, suggesting that a decrease in anti-inflammatory therapy triggers in part a host response to the pathogen. Further, it must be recognized that definitions of fever vary but most clinicians accept 38.3°C (101°F) in the absence of environmental causes. Temperatures ranging from 38.0 to 38.3°C represent a febrile state that depending on the clinical circumstances may prompt clinical concern. Table 2 is an attempt to summarize a number of studies primarily carried out in patients with acute leukemia but including some patients with lymphoma. Emphasis is placed on the type of neoplasm, the number of febrile episodes, and the proportion with documented infection versus unexplained fever. Several of these studies are particularly meritorious in that they represent a homogeneous population followed with the same chemotherapeutic protocol; other studies, such as our own, merely

TABLE 2. Documentation of Infection in the Febrile Neutropenic Patients

Investigators	Years	Disease[a]	Total number of episodes/total patients	Documented infection (%)	Other cause (%)	Unexplained fever (%)
Silver[43]	1955–1956	AcLK	92/36	48	16	36
Raab et al.[138]	1960[b]	LK, lymphoma	162/97	40	6	54
Bodey et al.[54]	1966–1972	AcLK	1894/494	64	1	35
Goodall and Vosti[44]	1971–1972	AcLK	45/24	24	—	76
Burke et al.[36]	1965–1973	AcLK	180/104	66	8	26
Atkinson et al.[46]	1974[b]	AcLK, aplasia	100/56	62	6	32
Young[42]	1974–1976	AcLK, aplasia	353/198	33	—	67
Wade et al.[39]	1981[b]	AcLK	121/92	18	—	82
Pizzo et al.[99]	1986[b]	AcLK, solid tumor	612/318	28	—	71

[a]AcLK, acute leukemia.
[b]Year published.

represent a compilation of underlying diseases, number of episodes for which these patients were followed, and the proportion who were found to have microbiologically documented infection. Although generalizations are perilous, it can be seen from the studies summarized in Table 2 that up to 68% of febrile episodes in patients with leukemia and lymphomas have been associated with a documented infection. In approximately 10% of patients, another cause such as a transfusion reaction or a drug allergy has been implicated and the remainder have unexplained fever. On the other hand, some of the more recent studies show a decline in the incidence of documented infection to a range where it is less than 20%.[39–41] Factors that may be responsible for this trend are the increasing tendency to use prophylactic nonabsorbable or systemic antibiotics to prevent infection in neutropenic patients and the aggressive use of empirical antimicrobial therapy given at the very onset of fever in patients with falling neutrophil counts.

The cause of unexplained fever is often the subject of spirited bedside debate. One of the most remarkable findings in our reported experience is that patients who had fever and undocumented infection still responded to a systemic regimen of aggressive antimicrobial therapy in approximately the same proportion as patients who had documented bacterial infection, almost exactly 75%.[42] Defervescence in response to antibiotic treatment even when infection is undocumented in neutropenic subjects is hardly a new observation; for instance, Silver et al.,[43] surveying their experience at the National Cancer Institute between 1955 and 1956, reported that 14 of 33 patients with "fever of undetermined origin" responded to systemic antimicrobial therapy. The systemic therapy in that series was often a tetracycline. The authors concede the possibility that antecedent use of penicillin might have been responsible for drug fever in a few cases. On the other hand, all the patients who responded to tetracycline were subsequently rechallenged with a penicillin and none had recurrent fever, suggesting that the initial febrile elevation while the patient was receiving penicillin was not caused by drug allergy.

For those patients who defervesce after initiation of systemic antimicrobial therapy, the best explanation appears to be effective treatment of a localized infection that has not yet "spilled over" into the bloodstream so that the infecting organism is not recovered from blood cultures or other readily accessible sites. Support for this concept comes from several sources.

1. Those fevers of undetermined origin that persist for several weeks on antibiotics eventually terminate in a serious microbial infection.[43,44] Whether these processes were originally caused by the organism subsequently isolated from blood cannot be proven. Goodall and Vosti[44] noted that roughly one half of patients who were granulocytopenic from leukemia therapy and had no cause discovered prior to the start of antibiotics subsequently developed an infection that they interpreted as a superinfection. Clearly, however, the possibility that an initially localized process that then spread and was responsible for "breakthrough" sepsis could not be excluded in their series.

2. Surveillance cultures reveal GI colonization by organisms such as *P. aeruginosa* and *Klebsiella* species that are not part of the normal bowel flora. They may be suppressed by systemic antibiotic therapy or oral antimicrobials leading to defervescence.

3. We have carried out several studies with paired patient sera and antigens of *P. aeruginosa* and Enterobacteriaceae and demonstrated seroconversion after defervescence and achievement of remission (unpublished data).

Illustrative Case 1

A 36-year-old Hispanic man had newly diagnosed acute myelocytic leukemia. He was admitted to the hospital for intensive induction chemotherapy with a combination of thioguanine, cytosine arabinoside, and daunorubicin. He had been treated by his own physician for symptoms of respiratory infection, cough, and easy bruisability. An antimicrobial, ampicillin, had been one of the agents used. Following hospitalization and initiation of treatment, his white count rapidly plummeted to less than 100 mature granulocytes/mm³ and he developed fever to 104.9°F (40.5°C). A chest radiograph revealed right lower lung field consolidation (Fig. 1), and despite a platelet count of 30,000/mm³, a transtracheal aspirate was performed. Gram's stains of the transtracheal aspirate revealed few neutrophils but moderate numbers of gram-negative bacilli. Aerobic cultures of the aspirate revealed pure growth of *P. aeruginosa*. Six sets of blood cultures taken at the time of temperature elevation eventually showed no growth. Immediately after the transtracheal aspirate, he was started on amikacin and carbenicillin, and the patient gradually defervesced over the ensuing 3 days. Eventually he achieved a hematologic remission and was discharged from the hospital.

Comment. *P. aeruginosa* bacteremia in our experience has been one of the most common causes of gram-negative sepsis in the neutropenic subject. However, variations in institutional incidence do occur, and it is our impression that the problem of *Pseudomonas* is most common when very intense cytotoxic regimens are used. Of great interest in this case was that multiple blood cultures drawn when the patient was febrile failed to grow *P. aeruginosa*, yet transtracheal aspiration revealed abundant growth of *Pseudomonas* organisms. This case underscores our belief that many patients have localized infections caused by *P. aeruginosa* that have not yet "spilled over" into the bloodstream at the time the patient suddenly develops fever and symptoms of sepsis. The nonbacteremic *Pseudomonas* pneumonia in this case was probably caused by aspiration. Prompt treatment prevented dissemination of disease, and better control of opportunistic infection may have provided the basis for early achievement of remission. Had tracheal aspiration not been per-

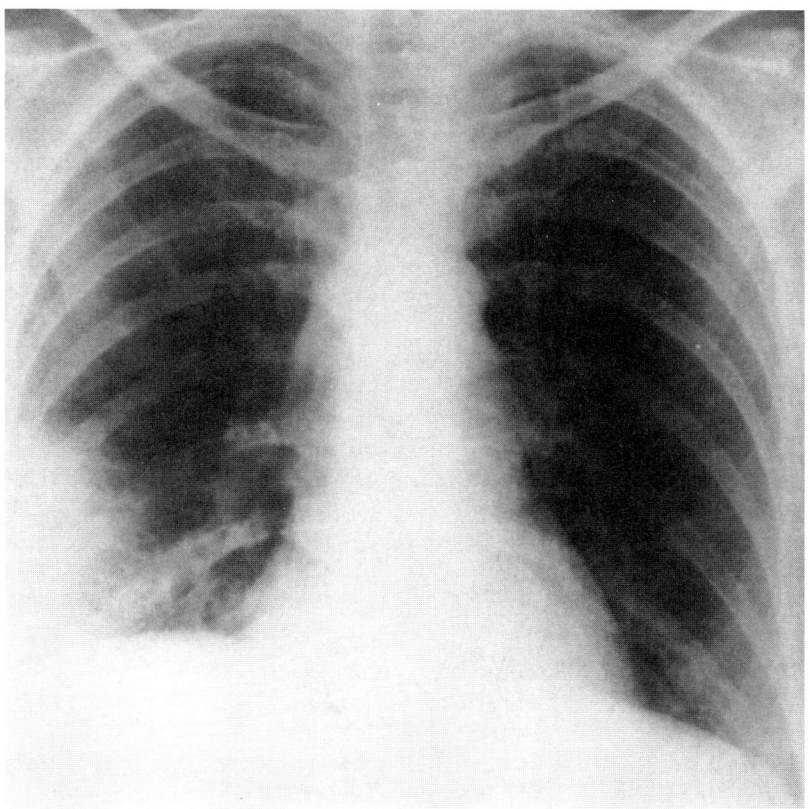

FIGURE 1. PA view of chest in Illustrative Case 1, a febrile neutropenic patient with acute myelocytic leukemia who developed right lower lobe consolidation. All blood cultures were negative, but a transtracheal aspirate grew a pure culture of *Pseudomonas aeruginosa*.

formed, this case would be yet another febrile episode in a neutropenic leukemic, attributed to an infectious process, but of undetermined etiology. In reality, the decision to proceed with transtracheal aspiration was made in ignorance of the platelet count. We do not recommend performing transtracheal aspiration with platelet counts less than 40,000/mm^3 and today the use of this diagnostic approach is rare. In this case, the decision to proceed with transtracheal aspiration was probably ill conceived, yet gave information that was highly valuable in the management of the patient. An alternative to the procedure would be bronchoalveolar lavage with a means for "protecting" the lung specimen from contamination by upper airway flora.

Febrile episodes are the rule rather than the exception in patients being intensely treated for leukemia and lymphoma. Silver et al.[43] observed that only 2 patients of a total of 36 remained afebrile throughout the course of their induction therapy for acute leukemia. Most patients experience more than one episode of fever during the course of induction attempts. In the study of Boggs and Frei,[45] children with acute leukemia experienced 4.6 febrile episodes per 100 days of hospitalization; the figure was slightly lower, 3.9 episodes per 100 days, in adult patients with acute leukemia, and the latter figure was

twice as great for patients with lymphomas. Understandably, most episodes of fever occur while the diseases are in relapse and/or are being intensely treated. In one study, only 4 of 100 febrile episodes occurred while the underlying neoplastic disease was in remission.[46] All these febrile patients were adults with acute lymphoblastic leukemia immediately prior to or during radiation directed at the central nervous system. In our experience, fevers occurring during hematological remission often suggest allergy or a persistent viral disease such as hepatitis or cytomegalovirus infection.

Although the studies summarized in Table 2 are useful in giving a broad view of the likelihood of infection in febrile conditions, it should be recognized that one of the problems with such data is often a lack of a clear-cut relationship to the clinical stage of underlying disease. Stated in another way, the likelihood of serious infection will differ if the patient is presenting for initial chemotherapy, is at the beginning of a consolidation cycle, or has developed the severe complications of cytotoxic treatment. Thus, Burke et al.[36] at the Johns Hopkins Hospital noted that almost three quarters of their patients with acute myelocytic leukemia had fever on presentation, but of these only 15% had severe infection and 43% had infected local sites. As chemotherapy was given, fully 90% of patients developed a life-threatening infection. In the latter study, several different antileukemic regimens were used. By contrast, Goodall and Vosti[44] at Stanford University Hospital reported a series of 24 patients who were all treated in the same manner for their acute leukemia. They reported that fever on admission to the hospital prior to the induction of chemotherapy usually resulted from an identifiable cause, but this observation differs somewhat from the Johns Hopkins experience where 40% of patients did not have a documented infection on presentation. During the initiation of chemotherapy, Goodall and Vosti[44] noted fevers in approximately a third of patients, but most of these were short-lived and some resolved without antimicrobial therapy. As patients became granulocytopenic secondary to treatment, either a cause of fever was found or superinfection subsequently developed. Unfortunately, some infections were not recognized until autopsy. The latter fact illustrates a common dilemma for those investigators attempting to analyze the nature of infection in neutropenic cancer patients, namely, whether autopsy or terminally documented infection can be etiologically linked to earlier febrile episodes(s).

6. Site of Involvement and the Nature of the Microbial Pathogen(s)

Information published on the nature of fatal infection—the major site of involvement and the relative proportion of bacterial, fungal, and viral etiologies—is somewhat similar for hematologic malignancies. Table 3 summarizes four studies: one from the National Cancer Institute (patients with "hematologic malignancy" included those with acute leukemias and lymphomas) and the experience from three studies relating to the treatment

of acute leukemia and malignant lymphoma at the M.D. Anderson Hospital. Only two major sites of involvement are listed: systemic infection and respiratory infections. Obviously, systemic infections must start in some localized area. As pointed out by Burke et al.,[36] the most common primary site of acute leukemia infection is the perirectal/genital area, followed by the urinary tract, skin, and only then the lung. In Burke's studies as well as others, approximately 20% of patients had an unknown source of systemic infection.

From the segregation of pneumonia from systemic infections in Table 3 (a respiratory source excludes those patients whose systemic infections might have originated in the lung), the conclusion is inescapable: bacteria are the most common cause of both systemic infections and pneumonia and nonpulmonary bacterial infections outnumber bacterial respiratory infections. The table, however, does not reflect the current view that the relative proportion of disseminated fungal infections noted at autopsy is increasing. Up to as many as one half of patients with leukemias and lymphomas autopsied in three series had disseminated fungal disease.[11,20,47] Thus, it appears that Table 3 may underestimate the prevalence of fungal disease at autopsy. Clearly, there may be coexistent bacterial and fungal disease and the concurrent presence of viruses and parasites cannot be dismissed as well. Under the category of "other" types of fatal infection are such entities as *Pneumocystis carinii* pneumonia, toxoplasmosis,[48] and several viral diseases. The relatively small proportion of viral isolations also may be a reflection of less adequate techniques for identification of such agents.

With respect to the etiology of fatal bloodstream infection in acute adult and pediatric leukemias, Table 4 summarizes the causative organisms expressed as a percent of the total. Two of these reviews, those of Hersh et al.[27] and Levine et al.,[26] come from the National Cancer Institute and highlight the similarities and differences observed during three periods, 1954–1959, 1960–1963,

TABLE 3. Types of Fatal Infection in Hematologic Disorders

| Investigators | Years | Patients | N | Systemic infection | | | Pneumonia | | | Other |
				Bacterial (%)	Fungal (%)	Both (%)	Bacterial (%)	Fungal (%)	Both (%)	
Levine et al.[26]	1965–1971	Hematologic malignancy	354	37	13	3	19	12	3	13
Chang et al.[30]	1966–1972	Acute leukemia	234	56	10	6	11	3	—	12
Feld and Bodey[32]	1966–1973	Malignant lymphoma	104	44	4	—	36	2	—	13
Estey et al.[137]	1973–1979	Acute leukemia	90	34	15	11	—	b	—	—

[a]Indicates percentage of total in each series with type of fatal infection.
[b]No discrete subdivisions between organisms causing systemic infection and pneumonia.

TABLE 4. Organisms Implicated in Fatal Septicemic Infections[a]

Organism	Hersh et al.[27] 1954–1959	Hersh et al.[27] 1960–1963	Chang et al.[30] 1966–1972	Levine et al.[26] 1965–1971[c]	Hughes[29] 1962–1969[d]	Eting[53] 1980–1993
Pseudomonas aeruginosa	24.6	34.4	18.1	16	36.7	51
Escherichia coli	21.6	12.8	18.8	—	19.3	9
Klebsiella pneumoniae	6.7	12.8	18.7	34	5.1	4
Other bacilli	—	—	18.8	—	9.0	23
Staphylococcus aureus	23.9	4.8	2.0	NA	11.0	24
Fungi	8.2	23.8	NA	13	33.5	—
Multiple organisms	NA[b]	NA	22[e]	3	1.2	—
Total number on which calculation based	366		149	462	155	27

[a]Expressed as percentage of total.
[b]NA, not available.
[c]Patients had negative blood cultures in remainder of infectious deaths.
[d]Total exceeds 100%.
[e]This figure includes E. coli, K. pneumoniae, and other bacilli.

and 1965–1971. Although there are major differences in the way some of these data are tallied, such as the handling of polymicrobial sepsis, all of these studies emphasize the important role that gram-negative rods and particularly *P. aeruginosa* play in fatal infections. The most dramatic decrease in fatal infections caused by a single organism is nicely detailed in the experience of Hersh *et al.*[27] Almost 24% of fatal septicemias between 1954 and 1959 were caused by *S. aureus*, whereas the incidence declined to less than 5% in the ensuing 4-year period. The experience of Chang *et al.*[30] shows an even lower proportion of staphylococcal infections that end with a fatal outcome. Autopsy data are potentially misleading because they may summarize only those infections that have a terminal outcome and might not reflect an actually greater incidence of a specific infection that was more amenable to therapy. Goodall and Vosti[44] noted no cases of staphylococcal infection and a predominance of systemic gram-negative bacillary and fungal infections in their cases. Similarly, in our series of bacteremias complicating acute leukemia and aplastic anemia therapy, we observed only 6 cases of staphylococcal sepsis versus 49 cases of gram-negative septicemia.[42] Perhaps most remarkably there were no deaths associated with *S. aureus* bacteremia, whereas 30% of the gram-negative bacteremias were associated with a fatal outcome. More recent reviews have emphasized resurgence of gram-positive coccal infection in neutropenic subjects[49,50] with the widespread use of indwelling vascular catheters of the Hickman, Broviac, Porta Cath, or similar type. Coagulase-negative staphylococci are often the most common cases of bacteremia in neutropenic cancer, although overall mortality is low. Enterococcal infections also appear to

have increased, perhaps as a result of the selection pressure of cephalosporins. Alpha hemolytic streptococci also have been noted as the cause of serious bacterial infections and seem to be associated with the severe mucositis triggered by some chemotherapeutic regimens.[51] It has been our experience that catheter-associated bacteremia caused by coagulase-negative staphylococci can be treated after the results of the cultures are known and these infections are not fulminant. However, some *S. aureus* infections are now produced by methicillin-resistant strains and aggressive treatment of these complications is still strongly recommended.[52]

Summaries of fatal infection in neutropenic patients could be misleading if no attempt is made to consider the association between underlying diseases and specific pathogens.[53] Although *P. aeruginosa* is a leading cause of mortality in the data summarized in Table 4, an even more striking means for demonstrating the role played by this pathogen in leukemias and lymphomas is demonstrated in Table 5. Although compiled three decades ago, this table summarizes fatal gram-negative bacteremias by neoplastic disease at one institution.[3] More than 60% of acute myelomonocytic leukemias were associated with a fatal gram-negative bacteremia, and of these fatal septicemias 84% were caused by *P. aeruginosa*. The incidence of gram-negative septicemia was relatively low for Hodgkin's disease and non-Hodgkin's lymphomas, 11 and 16%, respectively, but of those fatal bacteremia infections that did occur, 50 and 40%, respectively, were caused by *P. aeruginosa*. In a review published this past decade, *P. aeruginosa* was still associated with the highest case–fatality ratio among bacteremic bloodstream pathogens.[53]

Studies of all bacteremias complicating leukemia,

TABLE 5. Fatal *P. aeruginosa* Bacteremia by Neoplasm[a]

Neoplasm	Number of deaths	All gram-negative bacteremia		*P. aeruginosa*	
		N	Percent	N	Percent fatal bacteremias
Leukemia					
Acute myelo/monocytic	19	12	63	10	84
Acute lymphatic	8	4	50	2	50
Chronic myelocytic	12	2	16	2	100
Chronic lymphatic/lymphocytic lymphoma	37	15	40	11	73
Lymphoma					
Hodgkin's	56	7	11	3	50
Non-Hodgkin's	32	5	16	2	40
Solid tumors	289	7	3	2	28

[a]Modified after Armstrong *et al.*[3]

focusing exclusively on those with fatal outcome, show a relative similarity in the proportion of types of infecting pathogens. Thus, Bodey *et al.*[54] reported that systemic infection and pneumonia together accounted for 60% of total episodes of documented infection in leukemics and the great majority of infecting pathogens were gram-negative bacilli. In their study of all septicemias, Bodey and colleagues noted a relative decline in *E. coli* infections and an increase in the proportion of infections caused by *Klebsiella*, *Enterobacter*, and *Serratia* species. Taken together, these organisms isolated from bacteremic infections in neutropenic leukemics may be more common than *P. aeruginosa*. The increase in enterobacterial infections in patients has been linked to widespread use of third-generation cephalosporins.[55]

In addition to gram-negative bacterial agents commonly isolated from blood culture, several studies of infection in cancer patients show an association between a specific pathogen and a neoplastic disorder. Such tallies of isolates by underlying diseases have been particularly interesting, as shown in Table 6. For instance, listeriosis has traditionally been associated with Hodgkin's disease and lymphomas; indeed, at the one institution from which these data are summarized (Memorial Sloan-Kettering Cancer Center), there were 18 cases of listeriosis in lymphoma patients as contrasted to 3 cases in patients with acute leukemias.[56] Similarly, there were 10 cases of nocardiosis in patients with lymphomas as opposed to 2 in acute leukemia, 24 cases of cryptococcosis in lymphoma patients as opposed to 1 with acute leukemia, and 7 cases of toxoplasmosis in lymphoma patients versus none with leukemia. The large number of cases of *Bacteroides* infection associated with solid tumors is probably the reflection of infectious complications following GI surgery and a large number of "other tumors" associated with tuberculosis were pulmonary in origin.

We believe that the associations shown in Table 6 are valid and consistent with our experience. On the other hand, such information, which might be designated "numerator data" because it refers only to numbers of cases

TABLE 6. Relative Frequency of Some Opportunistic Pathogens by Neoplastic Disease[a]

Organism	Acute leukemia	Hodgkin's disease	Non-Hodgkin's lymphoma	Other tumors
Bacteroides sp.	4	4	2	41
C. neoformans	1	14	10	3
L. monocytogenes	3	12	6	3
M. tuberculosis	1	4	4	72
N. asteroids	2	5	5	7
P. aeruginosa	11	1	14	20
Streptococcus group A sp.	5	5	5	30
K. pneumoniae	8	6	8	2
T. gondii	0	6	1	3

[a]"Numerator" data only (Memorial Hospital). From Armstrong.[56]

rather than to precise relative incidence, must be qualified in the light of the varying incidence of the underlying disorders. During the period when the information summarized in Table 6 was obtained, the estimated annual incidence of solid tumors in the United States was more than 300 times that of acute leukemia; Hodgkin's disease and non-Hodgkin's lymphomas were perhaps 20 times more common than acute leukemia.[57] If we were to take these factors into consideration and the disease incidence data at this particular institution were the same as for the United States as a whole, the classic concept that cryptococcosis, listeriosis, tuberculosis, nocardiosis, and toxoplasmosis are more common in Hodgkin's disease and the lymphomas than in acute leukemia would not hold true.

Clearly, we are not justified in extrapolating from national incidence data to the experience at a single referral institution. Other investigators have also questioned the "classic" associations between underlying disorder and specific opportunistic infection.[58] In future studies it would obviously be desirable to have specific institutional data on the numbers of cases and underlying disease that it treats that can then be used as the appropriate "denominator" figure for calculating relative incidence of infectious complications. Where such information is available, some interesting associations have become apparent: when cases of cryptococcal infection at Memorial Sloan-Kettering Cancer Center prior to the AIDS pandemic were factored for number of patients with an underlying disease admitted during the study period (to determine a case rate per 1000 cases of neoplasm), it was found that the disease most commonly associated with cryptococcosis was chronic lymphatic leukemia.[59] The incidence was 24.3 cases per 1000 in contrast to Hodgkin's disease with an incidence of 13.3 cases per 1000. Chronic myelogenous leukemia interestingly had a case incidence of 10.9 per 1000. These rates for the lymphomas and leukemias were dramatically higher than the incidence of cryptococcosis complicating carcinoma of the breast, where 0.159 case per 1000 was observed.

Applying the same analytic epidemiologic technique to tuberculosis as they did to cryptococcosis, Kaplan *et al.*[60] confirmed the classic association between Hodgkin's disease and tuberculosis. As shown in Table 7, the tuberculosis case rate for patients with Hodgkin's disease was the highest for any neoplasm, 96/10,000 cases, which was followed by carcinoma of the lung (92/10,000) and non-Hodgkin's lymphoma (83/10,000). The association with cancer of the lung has triggered speculation about etiologic links between the two entities as well as suggesting the possible mechanism that a growing carcinoma might lead to breakdown of calcified foci of tuberculous infec-

TABLE 7. Prevalence of Tuberculosis by Neoplastic Disease[a]

Neoplasm	TB cases/10,000
Leukemia	
Acute myelocytic	28
Acute lymphatic	37
Lymphoma	
Hodgkin's disease	96
Non-Hodgkin's	83
Carcinomas	
Lung	92
Stomach	55
Head and neck	51
Colon	6
Bladder	4

[a]Modified from Kaplan *et al.*[60]

tion. Despite the convincingly supported data shown in Table 7, it appears that the incidence of tuberculous disease in immunosuppressed patients is declining in the United States,[61] other than in individuals with AIDS. Where clinical tuberculosis is more common, awareness of this potential complication of antineoplastic chemotherapy remains important.

Consistent with the widely accepted view that Hodgkin's disease is associated with a T-lymphocyte-mediated defect and increased susceptibility to fungal infection, "numerator" data on the incidence of some of the important opportunistic fungal pathogens are summarized in Table 8. Hodgkin's disease has been the most common neoplasm associated with histoplasmosis, cryptococcosis, and coccidioidomycosis. In contrast to the association with Hodgkin's disease, the number of cases of these fungal infections in acute leukemias has been small. There appears to be a strong association between Kaposi's sarcoma seen as part of AIDS and opportunistic mycotic infections caused by cryptococci, *C. immitis*, and *H. capsulatum*. This emphasizes the importance of T-lymphocyte-mediated defenses against these fungal pathogens.

Just as the incidence and the type of infection complicating acute leukemia are dependent on the treatment status and stage of the disease, a similar conclusion can be drawn about the lymphomas. The older report of Casazza *et al.*[31] emphasized that the great majority of serious infections occurred during the advanced stages of the lymphomas. This conclusion has been reaffirmed by more recent reviews. Generally, low fatality rates are associated with infections occurring during the early stages of

TABLE 8. Underlying Diseases in Opportunistic Fungal Infections

Disease	Histoplasmosis[105,139,140]	Cryptococcosis[59]	Coccidicidomycosis[141]
Hodgkin's disease	18	19	6
Chronic lymphatic leukemia	12	5	—
Acute lymphatic leukemia	10	1	—
Other leukemias	6	2	—
Non-Hodgkin's lymphoma	4	9	2
Systemic lupus erythematosus	6	—	—
Renal transplant	7	—	1
Other	3	5	4
Total	66	41	13

disease,[33] but in patients who are unresponsive to further antineoplastic chemotherapy infection is the primary cause of death. Not surprisingly, a variety of bacterial, viral, and fungal processes occur preterminally. In the experience of Feld and Bodey,[32] there was not much difference between Hodgkin's and non-Hodgkin's lymphoma in the incidence of septicemias and pneumonias. There were, however, larger numbers of urinary tract infections in patients with non-Hodgkin's lymphomas, and this may have been related to a higher incidence of genitourinary tract obstruction by tumor.

Another area where there appears to be some difference between Hodgkin's and non-Hodgkin's lymphomas is in the incidence and type of involvement with herpes zoster. Patients vigorously treated with combination chemotherapy and/or radiation therapy for Hodgkin's disease have an incidence of herpes zoster that ranges between 19 and 34.5%.[62–66] This incidence is significantly higher for patients with Hodgkin's disease than for non-Hodgkin's lymphoma by a factor of 3:1. The highest risk (56%) was found in children who received combination chemotherapy plus extensive radiotherapy, but splenectomy did not increase the risk of this viral infection.[66] Most patients developed zoster within a year of initiating chemotherapy and/or radiation therapy, and zoster was most frequently associated with a previously irradiated dermatome.[63] The incidence of dissemination of infection has been quoted as approximately 25%,[65,66] but the definition of dissemination has been quite variable. It is common to see vesicular lesions outside of the primary dermatome without evidence of visceral involvement, and Feldman and co-workers cite an incidence of 50% for this phenomenon, which they term "skin generalization."[63] In general, the mortality from herpes zoster has been low,[63–66] but disseminated disease that includes pneumonic involvement has a mortality that in our experience exceeds 75% and has been reported as high as 100%.[63] With the ready availability of acyclovir and other antiviral agents, however, one would expect the incidence of herpes zoster dissemination to decline and for the case–fatality ratio to fall as well.

The advent of more intensive chemotherapy for non-Hodgkin's lymphoma has made the infectious complications more similar to leukemia. In one series, 83% of documented infections were produced by gram-negative bacilli and staphylococci, with *P. aeruginosa* the major cause of bacteremia and pneumonia.[67] Of 125 patients reviewed, mycobacterial, listerial, nocardial, *P. carinii*, and cryptococcal infections were rare or did not occur at all.

Relatively few modern reports have summarized infection hazards in patients with multiple myeloma. Shaikh and associates[68] reviewed 46 patients with myeloma and found that gram-negative bacilli were the predominant pathogens. This contrasts with earlier studies indicating that *S. pneumoniae* is characteristically associated with this disease.[69]

Finally, the problem of pneumococcal sepsis in patients who are surgically splenectomized or functionally so (after radiation) also should be placed in the following perspective: Even before splenectomy was a routine component of Hodgkin's disease management, cases of overwhelming pneumococcal sepsis were observed in patients with leukemias, lymphomas, and solid tumors.[56] As shown in Table 6, the greatest number of cases in that series occurred in patients with solid tumors rather than lymphomas and acute leukemias. If, however, one "factors" for the known differences in disease incidence, the risk is still considerable in those patients having a primary defect in neutrophil-dependent host defenses. The incidence of postsplenectomy pneumococcal sepsis appears to be particularly significant in the pediatric age population. One of the most interesting observations has been the poor antibody response[70,71] to the pneumococcal type

VI (VIA and VIB in the Danish system, and 6 and 26 in the American system) antigens. Prior to licensure of pneumococcal vaccines, type VI pneumococci were the largest single cause of such bacteremia in hematological malignancy.[72] This suggests that the pneumococcal infection risk since splenectomy became commonly performed represents the continuation of an established trend. Type VI organisms are the most commonly isolated in pneumococcal infections complicating bone marrow transplantation, which may be regarded as creating a "functionally asplenic" state.[2] The major threat that organisms of the type VI serotype(s) pose is probably related to their relatively poor immunogenicity even in normal subjects.[71] Susceptibility to infections by agents with this antigenic composition seems to be increased in immunocompromised subjects.

7. Synthesis

Primarily from information obtained from retrospective or autopsy reviews, clinical trials of antimicrobial therapy, and epidemiologic studies in neutropenic patients, the following "synthesis" is offered as a summary of the changing nature of the problem of fever and infection in the patient with leukemia or lymphoma.

Among neutropenic patients who did not receive prior chemotherapy, approximately one half of patients present with fever (i.e., at time of diagnosis); of these one third are found to have a localized bacterial infection. These localized infections are caused by staphylococci, streptococci, and the more antibiotic-susceptible gram-negative bacilli including *E. coli*. In patients with lymphomas, tuberculosis and cryptococcosis are perhaps the only pathogens unequivocally associated with untreated underlying disease. The remaining two thirds of febrile patients on presentation have fevers caused by their neoplasms or quite localized infections in and around the GI tract that infrequently "spill over" and cause detectable bacteremia. If patients are given oral antibiotics prior to or during chemotherapy, GI colonization by the more antibiotic-resistant gram-negative bacilli such as *P. aeruginosa* and *Klebsiella* species and yeasts may follow. However, severity of underlying disease alone in the absence of a selection pressure from antimicrobials can result in colonization of the host by more drug-resistant organisms.

After antineoplastic treatment has been initiated, fevers secondary to drugs, transfusion reactions, and possibly tumor lysis enter the differential diagnosis. Perhaps one third of these fevers are associated with bacterial infection, but infection should be assumed to be the cause of the temperature elevation and antimicrobials promptly started if the patient is markedly neutropenic (neutrophil count <500/mm³) or the white count is rapidly falling. During first and second induction attempts, documented infections still are caused by more susceptible organisms such as staphylococci and *E. coli*, but as repeated courses of antibiotics are given, the likelihood progressively increases that the patient will be colonized and infected by fungi such as *Candida* and resistant gram-negative bacilli. Many leukemics and lymphoma patients defervesce on empirical antimicrobial treatment during chemotherapy before evidence of remission is achieved. It is likely that the majority of persistent fevers will eventually prove to have a bacterial or fungal cause if the patient is left untreated, although the simultaneous contribution of the tumor to febrile symptoms cannot be excluded. Some but not many of the fevers that persist in the face of antimicrobials may be caused by undiagnosed viral infection. Aggressive use of systemic antimicrobial agents probably terminates localized bacterial infection such as those originating around the GI tract. This could be the explanation of the common observation that many febrile patients without documented bloodstream infection will improve on empirical antimicrobial therapy. Furthermore, attempts to prevent infection by use of oral antimicrobial agents to suppress the fecal flora may contribute to blood culture negativity. Irrespective of the cause of fever, most patients become afebrile with achievement of remission or significant rise in normal neutrophil count to levels exceeding 500 cells/mm³. The exceptions to this probably include chronic viral infections caused by herpesviruses such as cytomegalovirus (CMV).

It is a great oversimplification to regard infection problems in the various leukemias and lymphomas as similar. Considerably more progress has been made in acute lymphocytic leukemia of childhood and some lymphomas than in acute nonlymphocytic leukemia in the adult. Lymphomas complicating AIDS have an especially poor prognosis. At one time steroids were commonly used to treat myelocytic leukemia but probably predisposed to more infection (compared with regimens not incorporating steroids), have little effect against the leukemia, and now are infrequently used. On the other hand, steroids have a well-accepted antitumor effect in lymphocytic leukemias and lymphomas, so by helping to treat the underlying disease, they ultimately reduce overall infection risk. Still, there are infections that occur in patients with lymphoma and lymphatic leukemia that seem related to depressed cell-mediated immunity, and these infections may occur during the remission phase of the disease

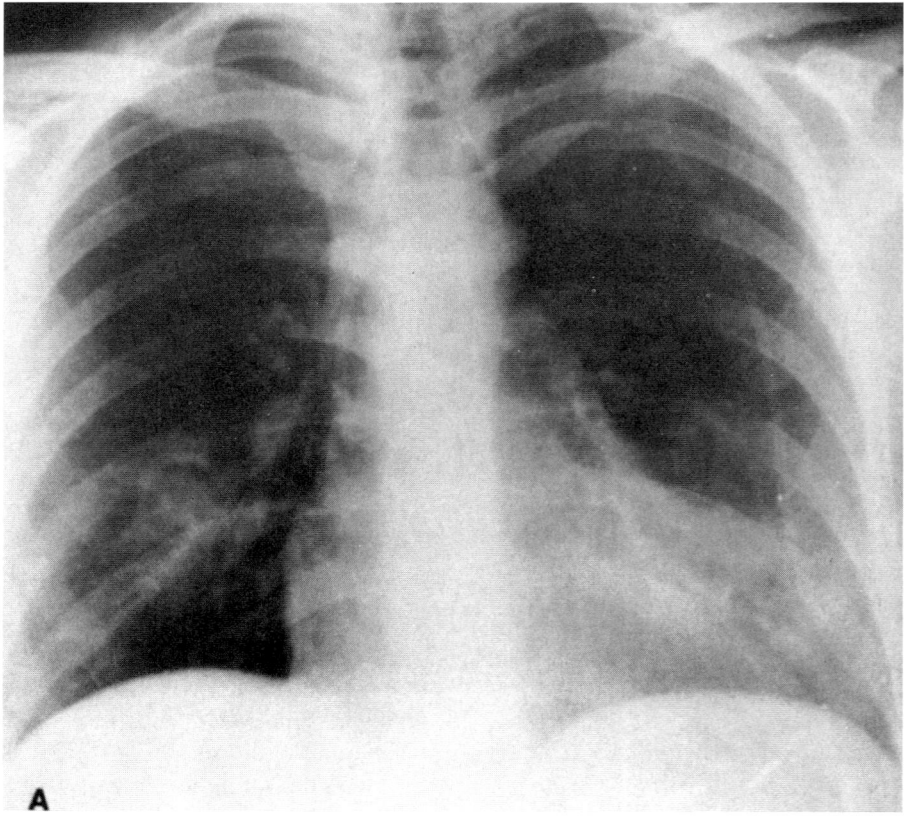

FIGURE 2. (A) Chest radiograph of patient with acute myelocytic leukemia (developing after successful treatment of Hodgkin's disease) who developed cavitating infiltrates in the right middle and left lower lobes after prolonged antibacterial therapy. Although this pattern is most commonly seen with *Aspergillus* lung infections, *Pseudoallescheria boydii* was grown from transbronchial biopsies. (B) Multiple infiltrates, many of them cavitating, developed in this young leukemic male with *S. aureus* bacteremia. The prognosis in this infection is far better than with cavitary fungal disease in patients with hematologic malignancies.

associated with consolidation or "pulses" of maintenance chemotherapy. The best example of such an infection is *P. carinii* pneumonia. If relapse occurs or remission of underlying disease is not achieved, and the patient remains persistently neutropenic, the risk of fatal septicemia caused by *Enterobacter*, *P. aeruginosa*, or the opportunistic fungi increases. Opportunistic fungal infections occur via two primary pathways: the GI route for *Candida* species and the airborne pneumonic route for *Aspergillus* and other molds. (This suggests some measures aimed at prevention and control.) Nonetheless, the pressing need is for sensitive diagnostic methods that will distinguish GI colonization by *Candida* from early systemic invasion. At autopsy, most leukemics and many patients with lymphomas have evidence of infection. Nonetheless, it could be misleading to equate what is found at autopsy with infectious processes or febrile episodes that were evident weeks before death.

Illustrative Case 2

A 36-year-old white woman had a history of stage IVB Hodgkin's disease initially diagnosed 8 years prior to admission. After successful treatment with four cycles of *M*ustard *O*ncovin (Vincristine) *P*rednisone *P*rocarbazine (MOPP) therapy, she was disease-free for almost 7 years but was eventually readmitted to the hospital with increasing fatigue, easy bruisability, and a white count that revealed an acute myelocytic leukemia. She noted daily temperature elevations of 102.2°F (39°C). Physical examination revealed a chronically ill patient with slight lymphadenopathy and hepatomegaly. Peripheral blood showed large numbers of blasts, and she was begun on a 7-day course of thioguanine, cytosine arabinoside, and daunorubicin. Shortly thereafter, she became febrile to 102.9°F (39.4°C) and was started on an aminoglycoside and an antipseudomonal penicillin. After 4 days of therapy, she gradually defervesced but was found to have persistent interstitial markings and a left lower lobe infiltrate on chest radiography. Because of the isolation of *Candida* species from stool and urine, she was started on systemic amphotericin B with therapy increased to a total daily dose of 40 mg. After 2 weeks of amphotericin, marked febrility returned, a right middle lobe infiltrate developed, and amphotericin B was withheld because of

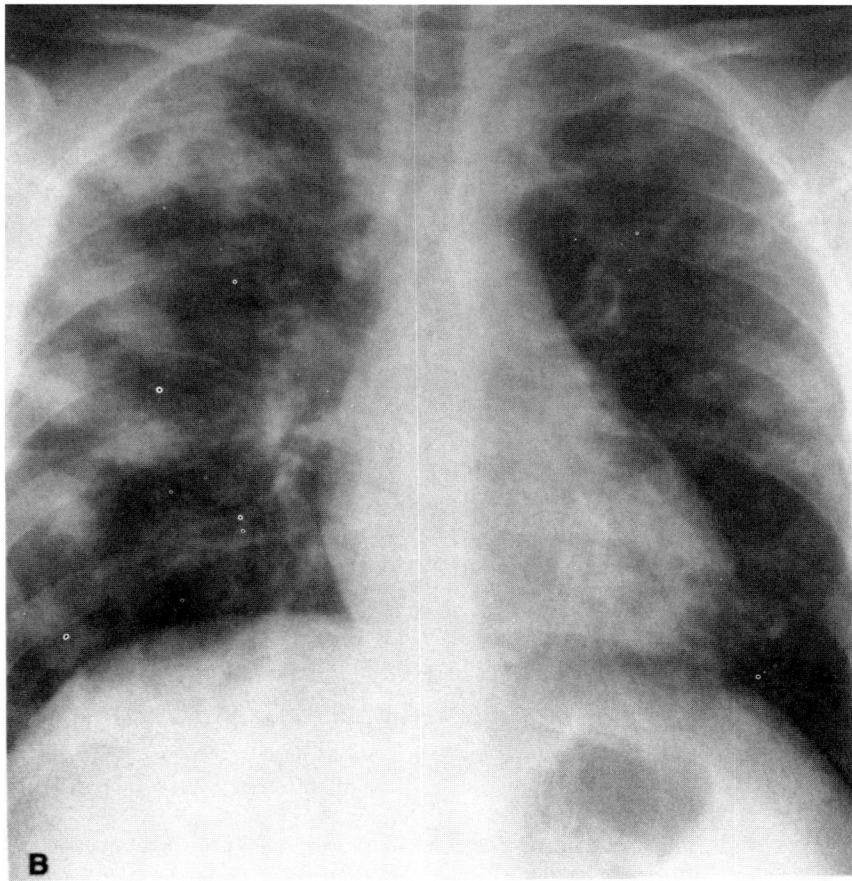

FIGURE 2. (*Continued*)

an increase in serum creatinine. Over the ensuing 4 days, chest radiographs demonstrated both right middle and left lower lobe infiltrates that both contained cavities (Fig. 2A). Fiberoptic bronchoscopy was performed, and brushings revealed multiple branching septate hyphae suggestive of *Aspergillus* species. Systemic antibacterial agents were continued, amphotericin was reinstituted, and a week later *Pseudoallescheria boydii* was grown from bronchial washings. At this point, miconazole, 400 mg q6h, was started. The patient developed a pruritic maculopapular rash over her entire body. Both miconazole and amphotericin were held. At this point, the patient experienced daily intermittent fevers as high as 104°F (40°C), confusion, and disorientation, and amphotericin B was resumed after a 4-day hiatus. Progressive consolidation in both lung fields occurred. The patient developed progressive hypoxia and myoclonic seizures and expired almost 2 months after hospitalization. Autopsy revealed presence of *P. boydii* in both consolidated lungs.

Comment. The patient was aggressively treated with antibacterial and antifungal chemotherapy for fever, although no bacterial cause of infection was documented. Because of concern that she might have underlying systemic candidiasis, a diagnosis suggested by positive stool and urine culture, she was started on amphotericin B. Eventually she developed bilateral cavitary pneumonia secondary to *P. boydii*. This fungus can be mistaken for *Aspergillus* species and causes a similar pathological picture of vascular invasion and infarction of tissue. It is on the latter basis that cavitation occurs, similar to that observed with pulmonary aspergillosis and phycomycosis (mucormycosis).[11] Unfortunately, *P. boydii* is usually resistant to amphotericin B but may be inhibited *in vitro* by miconazole and other azoles. In this case, therapy with amphotericin and miconazole was ineffective, but the major determining factor underlying the progression of fatal outcome in this disease was a refractory leukemia that developed following successful treatment of Hodgkin's disease. Nonetheless, the specific type of superinfection appears to reflect the gradual selecting out of more resistant opportunistic pathogens if the basic disease does not remit. In this case, aggressive diagnostic measures were successful in establishing the diagnosis weeks before death. Autopsy revealed disseminated mold disease.

In contrast to the radiographs from this unfortunate case, Fig. 2B shows the chest films from a young leukemic male who developed overwhelming *S. aureus* septicemia (8/8 positive blood cultures), a shocklike clinical picture, and cutaneous dissemination of staphylococcal lesions. Multiple areas of consolidation in the lung also progressed to cavitation, as shown in the film. Nonetheless, fluids and antibiotic therapy—oxacillin in doses used for endocarditis—resulted in complete clinical cure and almost complete resolution of lung abnormalities prior to achievement of hematologic remission. Indeed, the patient most likely had right-sided staphylococcal endocarditis as the source of septic emboli in both lung fields. This case illustrates the much better prognosis for treatment of systemic staphylococcal infections in acute leukemic patients, particularly when their disease is newly diagnosed and the bacterium is methicillin susceptible. Staphylococci and filamentous fungi can both cause acute cavitary disease in the lung, but the prognosis with the former infection is usually much better than with the latter.

8. Summary of Recommended Therapeutic Strategies

Since 1990, the Infectious Diseases Society of America (IDSA) has published two sets of guidelines on the management of fever in neutropenic patients.[38,73] Basic bedside diagnostic approaches and laboratory studies have been outlined and general approaches to treating and preventing infection based on published evidence have been reviewed. An update of these guidelines, with a review of the recent published literature, was scheduled for publication in early 2002.

The IDSA guidelines show considerable concordance on major issues among clinicians in the field of managing infection in febrile neutropenic subjects. Clearly, however, management approaches will differ depending on different treatment protocols for underlying disease, epidemiologic factors, and local institutional experience. Although there is a gratifying trend to undertake controlled or comparative trials in order to derive more credible and widely applicable guidelines for management, many recommendations still reflect personal experience. It is with that acknowledgment that the following recommendations are made.

8.1. Different Approaches to Leukemia and Lymphoma

There now appears to be markedly divergent infection risk among leukemia (all types) occurring in the adult, acute lymphatic leukemia of childhood, and Hodgkin's disease. Chemotherapy for the latter two groups of disorders is initially effective in the great majority of subjects and is commonly undertaken on an outpatient basis. We see no need for special diets, prophylactic antibiotics during the induction phase, or any type of isolation. If therapy is given to nonhospitalized individuals, each patient should be alerted to immediately report to a physician or emergency room facility with onset of fever (particularly a new pattern), rigors, or other signs suggesting infection. Similar recommendations would apply to most patients with non-Hodgkin's lymphoma

treated outside of the hospital. However, individuals with refractory lymphomatous disease receiving intensive cytotoxic treatment may be viewed as similar to acute leukemia and therefore should be hospitalized. Patients with aplastic anemia (e.g., being treated with high-dose steroid or antithymocyte globulin), multiple myeloma receiving intensive cytotoxic therapy, and chronic myelogenous and lymphatic leukemia in an accelerated disease phase (blast crisis) probably should also be treated like acute leukemic patients.

8.2. Environmental Considerations

Patients whose neutrophil counts are likely to plummet to 500 cells/mm^3 or lower should be placed in a single room and managed with strict handwashing precautions. Numbers of visitors should be held to a minimum. "Reverse isolation" in the traditional sense appears valueless.[74] Masks should be worn by any personnel having upper respiratory infection. Since respiratory infections are often communicable before symptoms appear, it is no major inconvenience to have all personnel wear a nonsterile-type mask on entering the room. This may have an important effect in reminding all persons entering the room to wash their hands.

Sterile sheets are not necessary. Recommended technique for bathing includes sponge baths, but if a tub bath is desired, the tub should be rigorously cleaned and then filled with the hottest possible water from the tap. This is allowed to cool to a level tolerated by the bather. Fresh fruit and vegetables should be excluded from the diet, and none of these materials or flowers are to be brought into a patient's room. All foods should be cooked and rewarmed before serving.

"Protected environments" that employ laminar airflow have been associated with a reduction in infection, but the cost–benefit ratio makes these units hard to justify when the impact on survival and remission rate is assessed.[75] There is good evidence that the incidence of infection is related to the type of leukemia protocol, i.e., some protocols may be associated with a higher rate of infection than others independent of the use of protected environments.[76] We do acknowledge that protected environments employing the laminar airflow principle should reduce true airborne infections. Cases of aspergillosis or mucormycosis developing in a patient managed in a laminar airflow room are exceedingly rare. Interestingly, attempts to reduce airborne fungal spores by air-filtration techniques other than incorporating laminar airflow principles also seem to be successful in reducing *Aspergillus* infection.[77] Overall, based on the totality of evidence protected environments with laminar airflow are not recommended for routine management.[38]

8.3. Prophylactic Antibiotics

Oral nonabsorbable antibiotics have been used to prevent infection during leukemia induction attempts. These regimens included an aminoglycoside (or polymyxin), vancomycin, and nystatin.[78] There are two major problems with this approach. First, unpalatability of the oral medications in addition to the extremely "toxic" clinical condition of some patients are two factors contributing to a situation in which patients most at risk will not or cannot take these medications. Patients most likely to take these medications tend to be less ill and those who are "dropouts" (and sometimes excluded by the analysis) are sicker and at greater risk. To be effective, special effort has to be made to give patients these regimens and they are quite expensive. A second concern is that use of gentamicin as a component of the topical regimen may select for resistant strains of *Pseudomonas* and other gram-negative bacilli.

We have extensive experience with the combination of vancomycin (100 mg q8h), nystatin (400,000 units q4–6h), and colistin or polymyxin B (100 mg q8h) in an attempt to suppress the GI flora. The effect of these efforts has been equivocal at best. Nystatin has had limited impact on fungal colonization and vancomycin is the most unpalatable component of this "cocktail." Our regimen differed from that of Schimpff *et al.*[78] by the substitution of polymyxin B or colistin for gentamicin.

Many investigators have sought an alternative to oral nonabsorbable agents and trimethoprim–sulfamethoxazole has been extensively evaluated. There are now more than a dozen controlled studies in which trimethoprim–sulfamethoxazole has been used in an attempt to prevent bacterial infection in neutropenic patients.[79] Despite the fact that many studies suggest benefit from prophylaxis, the use of trimethoprim–sulfamethoxazole has been reported to be relatively ineffective in some large studies,[80] to be without benefit during consolidation courses of leukemia treatment,[81] to be associated with more prolonged neutropenia in comparison with nalidixic acid,[82] and to be associated with emergence of resistance in gram-negative bacilli.[83,84] The fact that trimethoprim–sulfamethoxazole lacks activity against *P. aeruginosa* is another major drawback. An important placebo-controlled study by Pizzo and colleagues[85] at the National Cancer Institute demonstrated that if one corrected for compliance, only those patients who were highly compliant with trimethoprim–sulfamethoxazole prophylaxis experienced a significant reduction in documented infections and fever. Perhaps equally important, however, was the observation that even patients who took placebo and were highly compliant experienced a significant reduction in infection. This finding strongly suggests that the nature of the compliant patient per se appears to be associated with a significant reduction in documented bacterial infection.

A comprehensive review of the efficacy of prophylaxis with trimethoprim–sulfamethoxazole suggests that overall the benefit from this approach may depend on local epidemiologic circumstances.[79] In settings in which gram-negative enteric bacteria are very susceptible to this agent, the prophylactic use of trimethoprim–sulfamethoxazole may be effective in reducing infections caused by *E. coli*, *Klebsiella*, *Proteus*, *Enterobacter*, and *Serratia* spp. While bacterial infections may be reduced, those that still do occur are more likely to be antibiotic resistant (including resistant to trimethoprim–sulfamethoxazole). Pizzo and associates[85] at the National Cancer Institute have suggested that patients taking prophylactic regimens with trimethoprim–sulfamethoxazole had a significant increase in side effects (skin sensitivity, GI upset) and that these patients experienced significantly more fevers of undetermined origin (FUOs). Use of trimethoprim–sulfamethoxazole prophylaxis still appears justified in protocols specifically aimed at preventing *P. carinii* pneumonia.[38] For patients with acute childhood leukemia, trimethoprim–sulfamethoxazole is virtually 100% effective in preventing pneumocystosis, and such prophylaxis can have an impressive impact on other childhood bacterial infections. Most of these infections occur during remission. Despite the proven efficacy of anti-*Pneumocystis* prophylaxis, we would emphasize its use in populations experiencing a relatively high incidence of pneumocystic infections (more than 5% of patients per year in each diagnostic category). This would include individuals with leukemias and lymphomas treated with large doses of corticosteroids. Patients who have had a proven episode of *P. carinii* pneumonia may be given either continuous or intermittent prophylaxis (160–240 mg q12h of trimethoprim) for the duration of chemotherapy and a period of weeks thereafter.[86] It is only effective as prophylaxis while it is being administered and does not appear to eradicate the carriage of this organism.

We have a long-standing interest in the subject of antimicrobial prophylaxis of infection in neutropenic hosts and have proposed guidelines for improved study design and evaluation of clinical benefit from prophylaxis studies.[87] Excluding pneumocystosis, interest has turned away from trimethoprim–sulfamethoxazole to the new fluoroquinolones for infection prevention in febrile neutropenic patients. Generally similar results have been reported whether the quinolone was compared with placebo,[88] trimethoprim–sulfamethoxazole ± colistin,[89] or vancomycin–polymyxin.[89] Quinolone prophylaxis appears to be associated with significantly fewer gram-negative infections and generally no increase in fungal or gram-positive infection. In fact, the absence of an effect on the latter was observed whether or not the quinolone was poorly absorbed (e.g., norfloxacin) or well absorbed (e.g., ciprofloxacin, ofloxacin) from the gut. A gratifying finding to date is that patients receiving the quinolone had fewer acquired gram-negative bacillary infections and compliance was often better than with the comparison regimen.[83,89] A very large multicenter study has found that ciprofloxacin was more effective in preventing documented gram-negative bacterial infection than norfloxacin, which is less well absorbed systemically.[90] Despite these findings, certain reservations must be expressed about routine use of quinolone prophylaxis during chemotherapy for hematologic malignancy. Though no apparent major drug interactions are limiting factors, the introduction of any new chemotherapeutic agent into a regimen raises this possibility. In spite of an apparently beneficial effect in reducing documented gram-negative infection and bacteremia in particular, quinolone prophylaxis has had no impact on the incidence of gram-positive infections or improved patient survival. While one study using norfloxacin showed a delay to the first febrile episode and initiation of systemic antimicrobial therapy, there has been no consistent reduction in total systemic antibiotic use or reduction in numbers of patients requiring such systemic therapy for "infectious fever."[88] Such conclusions are borne out by several recent reports and meta-analyses of quinolone prophylaxis.[91,92] The finding that prophylaxis fails to reduce systemic antibiotic use again raises the issue of whether quinolone prophylaxis "converts" documentable bacterial infections into cases of FUO, the latter prompting initiation of systemic antibacterials that are given for the same duration as if prophylaxis had not been given. Thus, the cost-effectiveness of routine quinolone prophylaxis must be seriously questioned and our present recommendation, in view of increasing quinolone resistance, is *not* to use them routinely.[38]

Nystatin is not a very effective prophylactic agent against mucosal candidiasis in the markedly neutropenic patient, although many failures may be attributed to poor patient compliance rather than to lack of intrinsic activity. Prophylactic amphotericin B taken orally (not a compound licensed in the United States, but can be prepared by a pharmacy) is preferred by some specialists. There are a paucity of studies demonstrating that ketoconazole is effective in the prophylaxis of systemic candidiasis. Ketoconazole does appear to be useful for the treatment of thrush but lacks activity against *Aspergillus* species.

Fluconazole and itraconazole are widely available antifungal agents with activity directed primarily against yeasts and some dimorphic fungi (e.g., *H. capsulatum* and *C. immitis*). Both agents appear effective in preventing or suppressing oral thrush in cancer patients, and in one large controlled study fluconazole prevented fungal infections in patients undergoing bone marrow transplantation.[93] Itraconazole also was reported to be efficacious in two large studies of patients with hematologic malignancy.[94,95] However, in similar settings an increase in azole-resistant *Candida krusei* has been noted.[96] For this reason routine antifungal prophylaxis is not advocated in the IDSA Guidelines for management of patients with fever and neutropenia.[38]

8.4. Systemic Antimicrobial Agents

In general we subscribe to the patient management guidelines outlined by the IDSA, which expressly addressed the issue of initial antimicrobial therapy in the febrile neutropenic patient.[38] These recommendations reflect in part an extensive summary of studies of empiric antibacterial therapy reviewed by Calandra and Cometta[97] and Pizzo.[98] Essentially, the IDSA Committee recognizes three acceptable initial regimens for the treatment of fever complicating neutropenia: (1) the combination of a beta-lactam agent with antipseudomonal activity such as carbenicillin, ticarcillin, or piperacillin, ceftazidime, or cefepime plus an aminoglycoside (gentamicin, tobramycin, or amikacin); (2) single-agent therapy with a potent beta-lactam compound such as imipenem, meropenem, ceftazidime, or cefepime; and (3) the initial empiric use of vancomycin plus the beta-lactam agent mentioned in option (2). The rationale for using a single potent broad-spectrum beta-lactam agent can be traced to the report of Pizzo and colleagues[99] in 1986, stating that ceftazidime alone was comparable to a regimen incorporating an antipseudomonal penicillin plus an aminoglycoside. Subsequently, a case can be made for monotherapy with carbapenems such as meropenem.[100,101] Cefepime has a broader spectrum of activity than ceftazidime[41] and may be preferred in situations where resistance to ceftazidime has increased. Ceftriaxone has also been used in one large study, but in combination with amikacin.[40]

Severely neutropenic patients may be at greater risk for developing *P. aeruginosa* infections. In such patients we prefer to begin empiric therapy with an aminoglycoside and an antipseudomonal beta-lactam compound, particularly in patients whose absolute neutrophil count is likely to fall below 100 cells/mm³. Modification of the initial drug therapy can clearly be undertaken based on susceptibility of the causative organism isolated from blood or local sites of infection.

Empiric use of vancomycin has stirred considerable controversy.[102] If there is clinical or microbiologic evidence of a venous catheter-associated infection or of methicillin-resistant staphylococcal disease, the initial use of vancomycin may be justified until the results of cultures are obtained. Routine empiric use of vancomycin is discouraged because of concern about increasing resistance among enterococci and staphylococci.[38]

Duration of therapy remains a complex issue. For patients who defervesce and have negative cultures, the discontinuation of systemic antibiotic treatment may be considered within 3–7 days unless there is evidence of persistent localized infection. The major concern in the severely neutropenic patient is the risk of opportunistic fungal infection in patients who do not appear to be responding to the antibacterial regimen. Again, based on bedside clinical assessment, some change in the antibiotic regimen may be justified if there is culture evidence to support such a move. However, if there is persistent fever and negative cultures, then the concern for opportunistic fungal disease increases with the length of neutropenia. A febrile neutropenic patient who develops pulmonary infiltrates while receiving broad-spectrum antibacterial agents most likely has a fungal superinfection and should be started on amphotericin B promptly.

The indications for empiric use of amphotericin in the absence of pulmonary infiltrates remain a subject of continuing debate. The median time for defervescence on antibacterial regimens in febrile neutropenia ranges from 5 to 7 days, so precipitous initiation of antifungal therapy should be discouraged. However, if patients are deteriorating and have no obvious bacterial cause of infection, even in the absence of lung infiltrates or severe esophagitis, then empiric antifungal treatment with an amphotericin preparation should be considered.[98] The reformulation of amphotericin into lipid complexes or liposomes is based on the rationale that these preparations are taken up preferentially by the reticuloendothelial system and released at the site of intracellular infection. The preparations are costly and controversy surrounds their overall efficacy. Nonetheless, a decreased nephrotoxicity of the various preparations makes them a reasonable although expensive alternative to conventional amphotericin B.[103,104]

Azole antifungal agents have been successfully used to treat disseminated or deep-seated *Candida* infections.[105,106] Itraconazole is more active against molds than fluconazole, but experience with its use in severely neutropenic patients is limited. New antifungal agents, e.g., caspofungin, have been used as salvage therapy for proven or suspected aspergillosis. The number of cases reported has been modest, but the agent is fungicidal and has a different mode of action than any other antimycotic agent.

8.5. Risk Assessment and Conversion to Outpatient Treatment

Because of obvious economic constraints much interest has been generated on the feasibility of reducing intravenous treatment by appropriate changes in therapy to an oral regimen. The obvious candidates for such "step down" or "switch" therapy are patients who are not severely neutropenic but may still have fever. This is an area that is still evolving and is based on the recognition that there are both high- and low-risk patients with severely neutropenic patients having persistent neutropenia clearly falling into the former group.[107] Studies reviewed by Rolston[108] and the report of Kern and colleagues[109] give prospective validation to this approach of oral treatment for low-risk patients. There are clearly many potential hazards to early discharge and a change from parenteral to oral treatment in neutropenic patients, but as a general rule of thumb, those individuals who have solid tumors and are likely to recover marrow function within 4–8 days may be considered for such treatment. For a more extensive assessment of scoring systems and the appropriate selection of oral therapy for low-risk febrile neutropenic patients the reviews of Rolston[108] and Klastersky and colleagues[110] are timely.

9. Approach to the Splenectomized Patient

Splenectomized patients, or patients who are functionally asplenic, are candidates for pneumococcal vaccine, but the efficacy of immunization seems limited. An alternative approach is prophylactic oral or parenteral penicillin G (or erythromycin if penicillin allergy is present). However attractive the latter approaches may be, there is no convincing proof of efficacy. Since pneumococcal infections (and *H. influenzae* infections as well) may present with devastating swiftness, we believe that patients should be carefully counseled (1) to seek immediate medical attention with onset of fever accompanied by the first chill and (2) to carry amoxicillin–clavulanate

(Augmentin) and take no less than 1 g PO at onset of upper respiratory infection or chills. The dangers of this latter recommendation are obvious and there are no studies to prove its value. However, it can be argued that the most likely pathogens in the nonneutropenic outpatient are likely to be susceptible to large doses of a penicillin plus a beta-lactamase inhibitor. An alternative treatment would be one of the newer fluoroquinolones such as gatifloxacin or moxifloxacin. Antibiotic suppression of growth in cultures taken subsequent to self-administration of antibiotics may occur, but the most likely pathogens still will be "covered" by this approach.

Illustrative Case 3

A 22-year-old woman was brought to the emergency room with a 6-hr history of fever and shaking chills. The patient had been entirely well until 2 years previously, when she presented with weight loss and night sweats and mediastinal adenopathy was noted on chest radiography. Diagnostic evaluation revealed Hodgkin's disease, nodular sclerosis type, stage IV-B after staging laparotomy, splenectomy, and liver and bone marrow biopsy. She was treated with total lymph node irradiation followed by six courses of MOPP therapy with an excellent clinical response. No further therapy had been given over the last 6 months, during which time she was asymptomatic and without evidence of overt Hodgkin's disease.

She remained well until she awoke at 4:00 AM on the day of admission with shaking chills and fever to 103°F (39.4°C). The patient took two aspirin tablets and returned to sleep until she awoke at 7:00 AM complaining of chills, fever, and a moderately severe bifrontal headache. She was noted to be quite lethargic by her parents and was brought to the hospital after the temperature was found to be 102°F (38.9°C) and she had vomited twice.

On reaching the hospital at 10:00 AM, she was a confused, sleepy but arousable woman with a temperature of 96°F (35.6°C), pulse of 120, and blood pressure of 50 by palpation. She appeared acutely ill and dehydrated. Her skin was cold and clammy. There were no petechiae or ecchymoses. Her neck was supple without clear evidence of pain on movement. Kernig's and Brudzinski's signs were absent. The pharynx was somewhat injected. The chest, cardiac, and abdominal examinations were negative except for the well-healed laparotomy scar. Other than the altered state of consciousness, neurologic examination was normal.

Initial laboratory data revealed HCT 36, WBC 5300/mm³ (with 45% PMNs, 11% bands, 36% lymphocytes, 4% monocytes). Chest radiographs and kidney, ureter, and bladder were within normal limits. Lumbar puncture revealed an opening pressure of 90, no cells, a cerebrospinal fluid (CSF) protein of 30 mg/dl, and CSF sugar of 68 mg/dl. A few pleomorphic gram-negative bacilli were thought to be seen on Gram's stain of the CSF.

Within 30 min of reaching the hospital, the patient received 3 g ampicillin, 3 g nafcillin, and 100 mg gentamicin, as well as 1000 mg methylprednisolone IV. Three liters of normal saline and albumisol were rapidly infused and dopamine was begun. Despite these measures and a full attempt at cardiopulmonary resuscitation, the patient succumbed 2 hr after reaching the hospital. Both sets of blood cultures and the CSF drawn on admission subsequently grew out *H. influenzae* type B. Postmortem examination revealed very early meningitis, no pneumonia, and no evidence of Hodgkin's disease.

Comment. This tragic case illustrates that all-too-common occurrence of effective therapy of the underlying neoplasm but with fatal outcome because of opportunistic infection. Splenectomy and intensive chemo- and radiotherapy probably resulted in marked impairment of host defenses against encapsulated bacteria such as the pneumococcus and, in this case, *H. influenzae* type B. Although the therapy could have been initiated some 15–30 min earlier, when the patient first reached the hospital, without taking the time for the usual extensive evaluation, it is unlikely that at that point such therapy would have been successful. Several important lessons are underlined here:

1. Since there is no effective immunization program to prevent such occurrences, the emphasis at this time must be to initiate earlier therapy. The point to begin therapy is with the first shaking chill. Therefore, our patients who are splenectomized for any reason, but most particularly our patients with Hodgkin's disease who are splenectomized and undergo chemo- and radiotherapy, are told to start treatment at home at appearance of first chill; they should then contact their physician or proceed quickly to the hospital. These patients should have antibiotic at home to be used for this purpose. We recommend initiating therapy with the fixed combination of amoxicillin–clavulanate or one of the newer fluoroquinolones.

2. Speed is of the essence in dealing with this problem, not only speed in initiating therapy at home, but also on reaching the hospital. House officers manning the Emergency Department must be taught that in dealing with these patients a prolonged diagnostic evaluation with delay in initiating therapy is unacceptable. Intravenous antibiotics should be initiated within 10 min of these patients reaching the hospital.

3. The clinicians should not expect to see established pneumonia or sinusitis or severe pharyngitis at the portal of entry in patients such as these. Bacteremia occurs early in the course of such infection, usually without evidence of establishing organ invasion that could be ascertained by physical examination or radiography. Therefore, the possibility of bacteremia must be thoroughly investigated and therapy instituted in this clinical setting even without definite evidence of specific organ invasion.

4. Better methods of preventing this problem are badly needed.

While pneumococcal immunization has a significant effect in preventing disease in immunocompetent persons, no benefit has been observed in asplenic patients or individuals with hematologic malignancies.[111]

10. Neutrophil Transfusions in the Treatment and Prophylaxis of Infection

The primacy of phagocytic leukocytes, either those circulating in the blood or those fixed within the reticuloendothelial system in the defense of the host against bacterial infection, has never been seriously questioned since the work of Metchnikoff at the turn of the century. Almost 60 years ago, neutrophils were injected intramuscularly into neutropenic patients in the hope that their breakdown products would stimulate endogenous neutrophil production.[112] A half a century ago experimental studies in lethally radiated dogs demonstrated that neutrophils harvested by relatively primitive transfusion techniques would circulate and migrate into inflammatory exudates.[113] A realistic dating of neutrophil transfusion therapy in humans, however, goes back only to the mid-1960s when it became possible to harvest granulocytes by a continuous-flow centrifugation technique. In the earlier studies, patients with high circulating white counts secondary to chronic myelogenous leukemia were sought as leukocyte donors, but there always has been a paucity of such patients. Continuous-flow centrifugation techniques required heparinization of the patient. Newer methods have included the harvesting of cells by a discontinuous-flow centrifugation method (which obviates heparinization) and filtration leukopheresis, whereby leukocytes in whole blood are collected by reversible adhesion of cells to nylon wool fibers. Following collection by filtration leukopheresis, cells are eluted and transfused, leading to some functional differences between cells collected by the various techniques.

Both the therapeutic and prophylactic use of exogenous, transfused leukocytes have been largely abandoned for a number of reasons: (1) Even when they were popular and widely used by oncologists, serious questions were raised about their clinical efficacy and safety. A comprehensive review of these controversial issues was presented in a previous edition of this chapter.[114] (2) A strong argument could be raised that the efficacy of therapeutic granulocytes was no better than that of optimal antimicrobial therapy. (3) Serious cost-effectiveness issues were raised about treatment of infection, febrile neutropenic episodes, and prophylactic use in leukemics. In selected situations granulocyte transfusions have still been employed,[115] but their routine use in neutropenia cannot be recommended.[116]

Cytokine Therapy for Stimulation of the Bone Marrow: Colony-Stimulating Factors

The discovery of peptide hormones that stimulate bone marrow stem cell growth and hematopoiesis has stimulated enormous interest and clinical activity in the field of oncologic therapy.[18,117,118] These peptides, also called cytokines, include granulocyte colony-stimulating factor (G-CSF), granulocyte–macrophage colony-stimulating factor (GM-CSF), and interleukin-3 as well as interleukin-6. Each affects bone marrow with differential growth-promoting activity. Some, like G-CSF, are relatively selective for neutrophil production while others have a more "broad-spectrum" effect triggering production of differentiated blood elements such as neutrophils, monocytes, and platelets (GM-CSF). The enthusiasm that has greeted the licensure of such products has led to quite widespread clinical acceptance in the oncology area. There is no doubt that these colony-stimulating factors given in the doses summarized in Table 9 accelerate bone marrow recovery after cytotoxic chemotherapy and may have an impact on severe neutropenic syndromes like aplastic anemia and cyclic neutropenia. The licensed indications in the United States initially have been narrow, but a variety of studies were conducted to determine the broader therapeutic application of these colony-stimulating factors.[119] Despite the abundance of evidence from animal studies that colony-stimulating factors can have a beneficial impact after the development (i.e., treatment) of infection, the actual human clinical data are sparse. For instance, there is no evidence to date that the routine use of these colony-stimulating factors improves patient survival in the treatment of neutropenic states or has any

TABLE 9. Use of Colony-Stimulating Factors in Neutropenia

	GM-CSF, sagramostin	G-CSF, filgastrim
Label indications	To accelerate marrow recovery after autologous marrow transplant	To prevent neutropenia related to treatment of nonmyeloid malignancy
Other uses	To permit myelosuppressive therapies	To correct neutropenia related to AIDS or AZT
Dose, schedule, route	250 μg/m² per day, subcutaneous	5 μ/kg per day, subcutaneous
Expected results	Increase in WBCs (neutrophils, eosinophils, monocytes, platelets)	Increase in WBCs (neutrophils)
Desired results (requires further confirmation)	To prevent neutropenic infection; improved survival	Same as GM
Time to response	2–3 days (longer if marrow suppressed)	Same as GM
Time to relapse after discontinuation	4–7 days	Same as GM
Side effects	Myalgia, fever, flushing, phlebitis; creatinine and liver enzyme abnormalities	Bone pain
Contraindications	Myeloid leukemia or myelodysplasia	Same as GM

beneficial effect when given therapeutically (that is, when added to the best available antimicrobial therapy). A small pilot study has suggested benefit from use of white cells harvested from donors treated with G-CSF.[120] There are some theoretical considerations that suggest that colony-stimulating factors should not be given to patients with acute myelogenous leukemia, hence the contraindications indicated in Table 9.

Most of the studies published to date show that when colony-stimulating factors are given prophylactically to patients with solid tumors, the duration of neutropenia is significantly shortened, the incidence of fever is curtailed, the use of systemic antibiotics is reduced, and the durations of hospitalization may be shortened.[98] All these benefits are highly desirable and could make the use of colony-stimulating factors cost-effective. However, the benefit of these colony-stimulating factors in proven or documented infection remains to be established despite experimental studies suggesting that they enhance neutrophil activity and are beneficial for treating infections such as those caused by *P. aeruginosa*. No study has shown a beneficial effect on overall patient survival when the underlying disease is lymphoma or leukemia, though presumably one might anticipate improved survival if life-threatening infection could be averted or their duration shortened. The most recent guidelines for use of hematopoietic colony-stimulating factors from the American Society of Clinical Oncology do not recommend routine use of these agents in neutropenia.[121] Such a policy concurs with the position of the Infectious Diseases Society of America.[38]

11. Immunoprophylaxis and Immunotherapy of Infection

There are some well-established indications for immunoprophylaxis of infection in the immunocompromised host and these should not be overlooked. In addition, there are areas where licensed vaccines are available, such as pneumococcal vaccine, but doubts persist about their efficacy. Finally, there are several areas of current investigation that involve immunoprophylaxis or immunotherapy of opportunistic infection.

11.1. Childhood and Adult Immunizations

It would be most unfortunate to overlook routine childhood immunizations with antigens such as diphtheria–pertussis–tetanus toxoid. On the other hand, live-virus immunizations should be strictly avoided. Killed polio vaccine (Salk type) should be routinely used and is safe. Guidelines for use of vaccines in adults are routinely updated by the Advisory Committee on Immunization Practices (Centers for Disease Control). As with children, live-virus vaccines should be avoided.

11.2. Passive Antibody

Postexposure immune serum globulin is recommended for hepatitis and for measles following exposure if the patient lacks a history of disease or immunization. Zoster immunoglobulin (but not regular immunoglobulin) can be used for patients without a history of chickenpox within 72 hr of exposure to varicella zoster infection. It may be preferable, however, to wait to see whether varicella zoster infection does develop, in which case the patient can be very effectively treated with acyclovir (10 mg/kg q8h IV). Several studies have suggested that plasma or immunoglobulin with high titers of antibody against CMV might reduce the serious complications of interstitial pneumonitis following marrow transplantation. However, CMV immunoglobulin is not beneficial when given to CMV-seronegative marrow recipients, provided that these patients have been given CMV-seronegative blood products.[122] A long history as well as clinical interest have been associated with use of pooled immunoglobulins for prevention of bacterial infection or its favorable modification in compromised patients.[123,124] Pooled gammaglobulin (serum immune globulin) routinely given to leukemic patients has been of no effect in reducing the incidence of any infection.[125] While some carefully designed studies have shown benefit in patients with low levels of circulating immunoglobulins (e.g., chronic lymphocytic leukemia, some lymphomas), the cost-benefit aspects of routine administration of IgG preparation modified for intravenous use have been challenged.[126] Since hypogammaglobulinemic patients are prone to infections caused by encapsulated respiratory pathogens like streptococci, the advantages derived from routine IV drug administration should be compared with chronic prophylaxis with comparatively safe antimicrobials like penicillin, cephalosporins, or macrolides.

11.3. Influenza Immunization

The Advisory Committee on Immunization Practices of the United States Public Health Service (USPHS) continues to recommend influenza immunization for high-risk groups. Influenza vaccines are generally safe because they contain killed virus, and most vaccine strains engender antibody responses in cancer patients.

However, administration of multiple doses may be necessary for patients receiving immunosuppressive therapy. Aside from antibody responses, there is a paucity of data that influenza immunization reduces active disease, its complications, or influenza-associated mortality in neutropenic patients. Problems associated with influenza infection have been infrequent in our patients over the last few years, but this may be because of a generally low incidence of disease in the community. Although we have not routinely immunized our patients with leukemia and lymphomas against influenza, we have no objections to carrying this out on a routine basis.

Despite its long-standing availability for the chemoprophylaxis of influenza, use of amantadine or rimantidine has never been very popular. Modest but consistent efficacy has been demonstrated in controlled studies, and we would consider its use in unimmunized, immunosuppressed patients during outbreaks of influenza.[127] The same comment may be made about the use of neuramindase inhibitors to treat early influenza, but there are no clinical data on efficacy in patients with hematologic malignancies.

11.4. Pneumococcal Immunization

The humoral antibody response to pneumococcal immunization in patients with lymphomas and myeloma has been disappointing.[111] In addition, a number of cases of infection are caused by serotypes that are not incorporated in the vaccine. Patients with Hodgkin's disease previously treated with radiation or chemotherapy have particularly impaired antibody responses, whereas patients with untreated disease respond in a manner analogous to control. Intensity of treatment rather than splenectomy is directly related to blunted antibody responses, and the impairment may persist for years. Some patients, such as with chronic myelocytic leukemia, do show better antibody responses if immunized prior to splenectomy rather than afterward. Patients with Hodgkin's disease should not be given pneumococcal vaccine during active antineoplastic treatment; it is not certain whether an improved serologic response may result from multiple-dose immunization and delayed therapy of the neoplastic disorder. Those patients in whom a delay in initiating treatment might be considered are likely to have less serious disease (i.e., stage I or II). Since there is no evidence that pneumococcal immunization is effective even in patients with Hodgkin's disease who respond well in terms of a humoral antibody response (splenectomy per se might still be the crucial compromising factor), we would not delay treatment of the tumor in symptomatic patients with progressive disease in the hope of obtaining maximum benefit from pneumococcal immunization.

It may be argued that pneumococcal immunization has little or no "downside risks." Even if the calculated efficacy is extremely modest, the use of a relatively inexpensive vaccine may be moderately beneficial to the host. Thus, in concordance with overall guidelines issued by the USPHS Advisory Committee on Immunization Practices, pneumococcal immunization may be viewed as a prudent move backed by epidemiologic principles but for which confirmatory studies (by randomized prospective trials) in leukemia, lymphoma patients are lacking.[111]

Several experimental approaches to the prevention or treatment of gram-negative bacillary infections have been described.[128] We should recognize, however, that there are major obstacles to immunologic approaches aimed at prevention and treatment of gram-negative infections in patients with hematologic and lymphoreticular malignancies. These problems may be summarized by stating the following questions:

1. What are the antigens that can engender protection?
2. What is the mechanism of this protection?
3. Can comprehensive immunizing preparations be developed?
4. Will the host respond to immunization, i.e., active immunization?
5. If the host is immunologically unresponsive or poorly responsive, will it be better to give passive immunoprophylaxis or therapy?
6. With respect to pulmonary infections, what is the relative importance of "local" immunity as opposed to the systemic immune response?

Active immunization of human subjects has been accomplished mainly by use of a heptavalent *P. aeruginosa* lipopolysaccharide vaccine. The immunogen was a set of seven cell wall endotoxins prepared by standard chemical extraction techniques. In cancer patients, the largest *Pseudomonas* vaccine trial carried out in a single center employed the heptavalent lipopolysaccharide antigen and found limited protection against *P. aeruginosa* infection.[13] However, it is important to note that this study, three decades old, demonstrated no protection against bacteremic infection. Although 67% of subjects with acute leukemia and Hodgkin's disease were found to develop augmented circulating antibody titers against one or more lipopolysaccharide antigens, these elevated titers were short-lived. The immunogenicity of this vaccine in acute leukemics differed in that newly diagnosed patients were usually found to respond with significant antibody increments as opposed to those individuals who are relatively refractory to antileukemic therapy after multiple relapses of their underlying disease. Progression of underlying disease was associated with low antipseudomonal antibody titers and persistent neutropenia. Failures in immunization in that controlled study were related to both low levels of type-specific opsonins and neutropenia.

The experience to date suggests that the most expeditious approach in immunocompromised hosts might be to give passive antibody, but there are unresolved issues about quantity of antibody, mode of delivery, and duration of protection. Only one small study has been reported of the use of *P. aeruginosa* immunoglobulin (IgG) to treat *Pseudomonas* infection and the results of this uncontrolled study are difficult to assess.[129] Even if it were possible to give large quantities of antibody (e.g., by one of the new techniques for IV infusion of IgG), this approach alone might not be successful in the neutropenic patient. Mounting evidence from animal studies suggests that protection will be best achieved by augmenting both granulocyte levels and passive antibody.[130]

A multiplicity of immunizing antigens may be required to obtain type-specific protection against clinically significant gram-negative bacilli. An alternate approach is to engender antibody against a common component of the cell walls of these organisms. Studies from several laboratories have identified important immunochemical similarities between lipopolysaccharide (endotoxin) antigens of the family Enterobacteriaceae. The inner core regions of endotoxins from *E. coli, Proteus, Klebsiella, Serratia, Salmonella,* and so on are structurally similar. Antibodies directed against "core" antigens prepared from "rough" mutants such as *Salmonella minnesota* R595 (a so-called "Re" mutant) appear to have broad cross-protective activity.[131] Anticore antibodies raised in rabbits have given passive protection in mice against a heterologous challenge by *E. coli, Klebsiella,* and *Serratia.*

The advent of monoclonal antibody technology has raised the hope that virtually unlimited amounts of endotoxin core reactive antibody would be available for therapeutic use.[132–134] Despite some initially encouraging results, confirmatory studies were disappointing. Similar disappointing results were obtained with other interventions aimed at blocking the sepsis cascade, including antibodies to cytokines and cytokine receptor-antagonist constructs. The most recent encouragement for interventions in severe sepsis (all cause, not gram-negative alone) is the report of a placebo-controlled study of activated protein C.[135] This preparation has both anti-inflammatory and anticoagulant activities and its use significantly reduced mortality in a study of over 1600 patients. However, neutropenic patients with leukemia and lymphoma were specifically excluded from the trial. It seems unlikely (because of potential bleeding complications) that this agent or similar interventions will be commonly employed to treat septic patients who are neutropenic and thrombocytopenic.[136]

References

1. Chilcote RR, Baehner RL, Hammond D: Septicemia and meningitis in children splenectomized for Hodgkin's disease. *N Engl J Med* **295**(15):798–800, 1976.
2. Winston DJ, Schiffman G, Wang DC, Feig SA, Lin CH, Marso EL, Ho WG, Young LS, Gale RP: Pneumococcal infections after human bone-marrow transplantation. *Ann Intern Med* **91**(6):835–841, 1979.
3. Armstrong D, Young LS, Meyer RD, Blevins AH: Infectious complications of neoplastic disease. *Med Clin North Am* **55**(3):729–745, 1971.
4. Goldschmidt MC, Bodey GP: Effect of chemotherapeutic agents upon microorganisms isolated from cancer patients. *Antimicrob Agents Chemother* **1**(4):348–353, 1972.
5. Schabel FM, Pittillo RF: Screening for and biological characterizations of anti-tumor agents using microorganisms. *Adv Appl Microbiol* **3**:223–256, 1961.
6. Miller SP, Shanbrom E: Infectious syndromes of leukemia and lymphomas. *Am J Med Sci* **246**:420–428, 1963.
7. Malech HL, Gallin JI: Current concepts: Immunology. Neutrophils in human diseases. *N Engl J Med* **317**(11):687–694, 1987.
8. Cline MJ: Defective mononuclear phagocyte function in patients with myelomonocytic leukemia and in some patients with lymphoma. *J Clin Invest* **52**(9):2185–2190, 1973.
9. Lehrer RI, Cline MJ: Leukocyte candidacidal activity and resistance to systemic candidiasis in patients with cancer. *Cancer* **27**:1211–1217, 1972.
10. Baehner RL, Neiburger RG, Johnson DE, Murrmann SM: Transient bactericidal defect of peripheral blood phagocytes from children with acute lymphoblastic leukemia receiving craniospinal irradiation. *N Engl J Med* **289**(23):1209–1213, 1973.
11. Meyer RD, Young LS, Armstrong D, Yu B: Aspergillosis complicating neoplastic disease. *Am J Med* **54**(1):6–15, 1973.
12. Epstein SM, Verney E, Miale TD, Sidransky H: Studies on the pathogenesis of experimental pulmonary aspergillosis. *Am J Pathol* **51**(5):769–788, 1967.
13. Young LS, Meyer RD, Armstrong D: Pseudomonas aeruginosa vaccine in cancer patients. *Ann Intern Med* **79**(4):518–527, 1973.
14. Wollman MW, Young LS, Haghbin M: Anti-*Pseudomonas* heat-stable opsonins in acute lymphoblastic leukemia of childhood. *J Pediatr* **86**:376–381, 1975.
15. Morrison VA: The infectious complications of chronic lymphocytic leukemia. *Semin Oncol* **25**(1):98–106, 1998.
16. Forsgren A, Schmeling D, Banck G: Effect of antibiotics on chemotaxis of human polymorphonuclear leukocytes *in vitro. Infection* **6**:S102–S106, 1978.
17. Björksten B, Ray C, Quie PG: Inhibition of human neutrophil chemotaxis and chemiluminescence by amphotericin B. *Infect Immun* **14**(1):315–317, 1976.

18. Lehrer RI: Inhibition by sulfonamides of the candidacidal activity of human neutrophils. *J Clin Invest* **50**(12):2498–2505, 1971.

19. King GW, Yanes B, Hurtubise PE, Balcerzak SP, LoBuglio AF: Immune function of successfully treated lymphoma patients. *J Clin Invest* **57**(6):1451–1460, 1976.

20. Young RC, Corder MP, Haynes HA, DeVita VT: Delayed hypersensitivity in Hodgkin's disease. A study of 103 untreated patients. *Am J Med* **52**(1):63–72, 1972.

21. Steigbigel RT, Lambert LH Jr, Remington JS: Polymorphonuclear leukocyte, monocyte, and macrophage bactericidal function in patients with Hodgkin's disease. *J Lab Clin Med* **88**(1):54–62, 1976.

22. Sheagren NJ, Block JB, Wolff SM: Reticuloendothelial system phagocytic function in patients with Hodgkin's disease. *J Clin Invest* **46**:855–862, 1967.

23. Ward PA, Berenberg JL: Defective regulation of inflammatory mediators in Hodgkin's disease. Supernormal levels of chemotactic-factor inactivator. *N Engl J Med* **290**(2):76–80, 1974.

24. Brown RS, Haynes HA, Foley HT: Hodgkin's disease: Immunologic, clinical and histologic features in 50 untreated patients. *Ann Intern Med* **67**:291–300, 1967.

25. Weitzman SA, Aisenberg AC, Siber GR, Smith DH: Impaired humoral immunity in treated Hodgkin's disease. *N Engl J Med* **297**(5):245–248, 1977.

26. Levine AS, Schimpff SC, Graw RG Jr, Young RC: Hematologic malignancies and other marrow failure states: Progress in the management of complicating infections. *Semin Hematol* **11**(2):141–202, 1974.

27. Hersh EM, Bodey GP, Nies BA: Cause of death in acute leukemia. *JAMA* **193**:105–109, 1965.

28. Viola MV: Acute leukemia and infection. *JAMA* **201**:923–928, 1967.

29. Hughes WT: Fatal infections in childhood leukemia. *Am J Dis Child* **122**(4):283–287, 1971.

30. Chang HY, Rodriguez V, Narboni G, Bodey GP, Luna MA, Freireich EJ: Causes of death in adults with acute leukemia. *Medicine* **55**(3):259–268, 1976.

31. Casazza AR, Duvall CP, Carbone PP: Summary of infectious complications occurring in patients with Hodgkin's disease. *Cancer Res* **26**(6):1290–1296, 1966.

32. Feld R, Bodey GP: Infections in patients with malignant lymphoma treated with combination chemotherapy. *Cancer* **39**(3):1018–1025, 1977.

33. Notter DT, Grossman PL, Rosenberg SA, Remington JS: Infections in patients with Hodgkin's disease: A clinical study of 300 consecutive adult patients. *Rev Infect Dis* **2**(5):761–800, 1980.

34. Bodey GP, Rodriguez V, Valdivieso M, Feld R: Amikacin for treatment of infections in patients with malignant diseases. *J Infect Dis* **134**(Suppl):S421–S427, 1976.

35. Bone RC, Balk RA, Cerra FB, Dellinger RP, Fein AM, Knaus WA, Schein RM, Sibbald WJ: Definitions for sepsis and organ failure and guidelines for the use of innovative therapies in sepsis. The ACCP/SCCM Consensus Conference Committee. American College of Chest Physicians/Society of Critical Care Medicine. *Chest* **101**(6):1644–1655, 1992.

36. Burke PJ, Braine HG, Rathbun HK, Owens AH Jr: The clinical significance and management of fever in acute myelocytic leukemia. *Johns Hopkins Med J* **139**(1):1–126, 1976.

37. Young LS: Nosocomial infections in the immunocompromised adult. *Am J Med* **70**(2):398–404, 1981.

38. Hughes WT, Armstrong D, Bodey GP, Brown AE, Edwards JE, Feld R, Pizzo P, Rolston KV, Shenep JL, Young LS: 1997 guidelines for the use of antimicrobial agents in neutropenic patients with unexplained fever. Infectious Diseases Society of America. *Clin Infect Dis* **25**(3):551–573, 1997.

39. Wade JC, Schimpff SC, Newman KA, Fortner CL, Standiford HC, Wiernik PH: Piperacillin or ticarcillin plus amikacin. A double-blind prospective comparison of empiric antibiotic therapy for febrile granulocytopenic cancer patients. *Am J Med* **71**(6):983–990, 1981.

40. Efficacy and toxicity of single daily doses of amikacin and ceftriaxone versus multiple daily doses of amikacin and ceftazidime for infection in patients with cancer and granulocytopenia. The International Antimicrobial Therapy Cooperative Group of the European Organization for Research and Treatment of Cancer. *Ann Intern Med* **119**(7 Pt 1):584–593, 1993.

41. Biron P, Fuhrmann C, Cure H, Viens P, Lefebvre D, Thyss A, Viot M, Soler-Michel P, Rollin C, Gres JJ: Cefepime versus imipenem-cilastatin as empirical monotherapy in 400 febrile patients with short duration neutropenia. CEMIC (Study Group of Infectious Diseases in Cancer). *J Antimicrob Chemother* **42**(4):511–518, 1998.

42. Young LS: Amikacin: Experience in a comparative clinical trial with gentamicin in leukopenic subjects. In Luthy R, Siegenthaler W (eds): *Current Chemotherapy*. American Society for Microbiology, Washington, DC, 1978, pp. 246–248.

43. Silver RT, Utz JP, Frei E: Fever, infection and host resistance in acute leukemia. *Am J Med* **24**:25–39, 1958.

44. Goodall PT, Vosti KL: Fever in acute myelogenous leukemia. *Arch Intern Med* **135**(9):1197–1203, 1975.

45. Boggs DR, Frei III E: Clinical studies of fever and infection in cancer. *Cancer* **13**:1240–1253, 1960.

46. Atkinson K, Kay HE, McElwain TJ: Fever in the neutropenic patient. *Br Med J* **3**(924):160–161, 1974.

47. Krick JA, Remington JS: Opportunistic invasive fungal infections in patients with leukaemia lymphoma. *Clin Haematol* **5**(2):249–310, 1976.

48. Israelski DM, Remington JS: Toxoplasmosis in the non-AIDS immunocompromised host. *Curr Clin Top Infect Dis* **13**:322–356, 1993.

49. Wade JC, Schimpff SC, Newman KA, Wiernik PH: *Staphylococcus epidermidis*: An increasing cause of infection in patients with granulocytopenia. *Ann Intern Med* **97**(4):503–508, 1982.

50. Rubin M, Hathorn JW, Marshall D, Gress J, Steinberg SM, Pizzo PA: Gram-positive infections and the use of vancomycin in 550 episodes of fever and neutropenia. *Ann Intern Med* **108**(1):30–35, 1988.

51. Dybedal I, Lamvik J: Respiratory insufficiency in acute leukemia following treatment with cytosine arabinoside and septicemia with Streptococcus viridans. *Eur J Haematol* **42**(4):405–406, 1989.

52. Fowler VG, Jr, Sanders LL, Sexton DJ, Kong L, Marr KA, Gopal AK, Gottlieb G, McClelland RS, Corey GR: Outcome of *Staphylococcus aureus* bacteremia according to compliance with recommendations of infectious diseases specialists: Experience with 244 patients. *Clin Infect Dis* **27**(3):478–486, 1998.

53. Elting LS, Rubenstein EB, Rolston KV, Bodey GP: Outcomes of bacteremia in patients with cancer and neutropenia: Observations from two decades of epidemiological and clinical trials. *Clin Infect Dis* **25**(2):247–259, 1997.

54. Bodey GP, Rodriguez V, Chang HY, Narboni G: Fever and infection in leukemic patients: A study of 494 consecutive patients. *Cancer* **41**(4):1610–1622, 1978.

55. Chow JW, Fine MJ, Shlaes DM, Quinn JP, Hooper DC, Johnson MP, Ramphal R, Wagener MM, Miyashiro DK, Yu VL: Enterobacter bacteremia: Clinical features and emergence of antibiotic resistance during therapy. *Ann Intern Med* **115**(8):585–590, 1991.

56. Armstrong D: Infectious complications of lymphosarcoma. In Molander DW (ed): *Lymphoproliferative Diseases*. Charles C. Thomas, Springfield, IL, 1975, pp. 94–109.

57. Walzer PD, Perl DP, Krogstad DJ: *Pneumocystis carinii* pneumonia in the United States: Epidemiologic, diagnostic, and clinical features. In *Symposium on Pneumocystis carinii Infection*. NCI Monograph 43. National Cancer Institute, Bethesda, 1976, pp. 55–63.

58. Cohen J, Pinching AJ, Rees AJ: Infections and immunosuppression: A study of infective complications of 75 patients with immunologically mediated disease. *Q J Med* **51**:1–15, 1982.

59. Kaplan MH, Rosen PP, Armstrong D: Cryptococcosis in a cancer hospital: Clinical and pathological correlates in forty-six patients. *Cancer* **39**(5):2265–2274, 1977.

60. Kaplan MH, Armstrong D, Rosen P: Tuberculosis complicating neoplastic disease. A review of 201 cases. *Cancer* **33**(3):850–858, 1974.

61. Horsburgh CR Jr, Feldman S, Ridzon R: Practice guidelines for the treatment of tuberculosis. *Clin Infect Dis* **31**(3):633–639, 2000.

62. Wilson JF, Marsa GW, Johnson RE: Herpes zoster in Hodgkin's disease. Clinical, histologic, and immunologic correlations. *Cancer* **29**(2):461–465, 1972.

63. Feldman S, Hughes WT, Kim HY: Herpes zoster in children with cancer. *Am J Dis Child* **126**(2):178–184, 1973.

64. Schimpff SC, O'Connell MJ, Greene WH, Wiernik PH: Infections in 92 splenectomized patients with Hodgkin's disease. A clinical review. *Am J Med* **59**(5):695–701, 1975.

65. Sokal JE, Firat D: Varicella zoster infection in Hodgkin's disease. Clinical and epidemiological aspects. *Am J Med* **39**:452–463, 1965.

66. Reboul F, Donaldson SS, Kaplan HS: Herpes zoster and varicella infections in children with Hodgkin's disease: An analysis of contributing factors. *Cancer* **41**(1):95–99, 1978.

67. Bishop JF, Schimpff SC, Diggs CH, Wiernik PH: Infections during intensive chemotherapy for non-Hodgkin's lymphoma. *Ann Intern Med* **95**(5):549–555, 1981.

68. Shaikh BS, Lombard RM, Appelbaum PC, Bentz MS: Changing patterns of infections in patients with multiple myeloma. *Oncology* **39**(2):78–82, 1982.

69. Twomey JJ: Infections complicating multiple myeloma and chronic lymphocytic leukemia. *Arch Intern Med* **132**(4):562–565, 1973.

70. Levine AM, Overturf GD, Field RF, Holdorf D, Paganini-Hill A, Feinstein DI: Use and efficacy of pneumococcal vaccine in patients with Hodgkin disease. *Blood* **54**(5):1171–1175, 1979.

71. Weibel RE, Vella PP, McLean AA, Woodhour AF, Davidson WL, Hilleman MR: Studies in human subjects of polyvalent pneumococcal vaccines (39894). *Proc Soc Exp Biol Med* **156**(1):144–150, 1977.

72. Folland D, Armstrong D, Seides S, Blevins A: Pneumococcal bacteremia in patients with neoplastic disease. *Cancer* **33**(3):845–849, 1974.

73. Hughes WT, Armstrong D, Bodey GP, *et al*: From the Infectious Diseases Society of America. Guidelines for the use of antimicrobial agents in neutropenic patients with unexplained fever. *J Infect Dis* **161**(3):381–396, 1990.

74. Nauseef WM, Maki DG: A study of the value of simple protective isolation in patients with granulocytopenia. *N Engl J Med* **304**(8):448–453, 1981.

75. Pizzo PA, Levine AS: The utility of protected-environment regimens for the compromised host: A critical assessment. *Prog Hematol* **10**:311–332, 1977.

76. Rodriguez V, Bodey GP, Freireich EJ, McCredie KB, Gutterman JU, Keating MJ, Smith TL, Gehan EA: Randomized trial of protected environment—Prophylactic antibiotics in 145 adults with acute leukemia. *Medicine* **57**(3):253–266, 1978.

77. Rhame FS, Streifel AJ, Kersey JH Jr, McGlave PB: Extrinsic risk factors for pneumonia in the patient at high risk of infection. *Am J Med* **76**(5A):42–52, 1984.

78. Schimpff SC, Greene WH, Young VM, Fortner CL, Cusack N, Block JB, Wiernik PH: Infection prevention in acute nonlymphocytic leukemia. Laminar air flow room reverse isolation with oral, nonabsorbable antibiotic prophylaxis. *Ann Intern Med* **82**(3):351–358, 1975.

79. Young LS: Trimethoprim-sulfamethoxazole and bacterial infections during leukemia therapy. *Ann Intern Med* **95**(4):508–509, 1981.

80. Trimethoprim-sulfamethoxazole in the prevention of infection in neutropenic patients. EORTC International Antimicrobial Therapy Project Group. *J Infect Dis* **150**(3):372–379, 1984.

81. Weiser B, Lange M, Fialk MA, Singer C, Szatrowski TH, Armstrong D: Prophylactic trimethoprim-sulfamethoxazole during consolidation chemotherapy for acute leukemia: A controlled trial. *Ann Intern Med* **95**(4):436–438, 1981.

82. Wade JC, de Jongh CA, Newman KA, Crowley J, Wiernik PH, Schimpff SC: Selective antimicrobial modulation as prophylaxis against infection during granulocytopenia: Trimethoprim-sulfamethoxazole vs. nalidixic acid. *J Infect Dis* **147**(4):624–634, 1983.

83. Dekker AW, Rozenberg-Arska M, Sixma JJ, Verhoef J: Prevention of infection by trimethoprim-sulfamethoxazole plus amphotericin B in patients with acute nonlymphocytic leukaemia. *Ann Intern Med* **95**(5):555–559, 1981.

84. Wells CL, Podzorski RP, Peterson PK, Ramsay NK, Simmons RL, Rhame FS: Incidence of trimethoprim-sulfamethoxazole-resistant enterobacteriaceae among transplant recipients. *J Infect Dis* **150**(5):699–706, 1984.

85. Pizzo PA, Robichaud KJ, Edwards BK, Schumaker C, Kramer BS, Johnson A: Oral antibiotic prophylaxis in patients with cancer: A double-blind randomized placebo-controlled trial. *J Pediatr* **102**(1):125–133, 1983.

86. Hughes WT, Rivera GK, Schell MJ, Thornton D, Lott L: Successful intermittent chemoprophylaxis for *Pneumocystis carinii* pneumonitis. *N Engl J Med* **316**(26):1627–1632, 1987.

87. Young LS: Antimicrobial prophylaxis against infection in neutropenic patients. *J Infect Dis* **147**(4):611–614, 1983.

88. Karp JE, Merz WG, Hendricksen C, Laughon B, Redden T, Bamberger BJ, Bartlett JG, Saral R, Burke PJ: Oral norfloxacin for prevention of gram-negative bacterial infections in patients with acute leukemia and granulocytopenia. A randomized, double-blind, placebo-controlled trial. *Ann Intern Med* **106**(1):1–7, 1987.

89. Young LS: The new fluorinated quinolones for infection prevention in acute leukemia. *Ann Intern Med* **106**(1):144–146, 1987.

90. Prevention of bacterial infection in neutropenic patients with hematologic malignancies. A randomized, multicenter trial comparing norfloxacin with ciprofloxacin. The GIMEMA Infection Program. Gruppo Italiano Malattie Ematologiche Maligne dell'Adulto. *Ann Intern Med* **115**(1):7–12, 1991.

91. Cruciani M, Rampazzo R, Malena M, Lazzarini L, Todeschini G, Messori A, Concia E: Prophylaxis with fluoroquinolones for bacterial infections in neutropenic patients: A meta-analysis. *Clin Infect Dis* 23(4):795–805, 1996.

92. Engels EA, Lau J, Barza M: Efficacy of quinolone prophylaxis in neutropenic cancer patients: A meta-analysis. *J Clin Oncol* 16(3):1179–1187, 1998.

93. Goodman JL, Winston DJ, Greenfield RA, *et al*: A controlled trial of fluconazole to prevent fungal infections in patients undergoing bone marrow transplantation. *N Engl J Med* 326(13):845–851, 1992.

94. Menichetti F, Del Favero A, Martino P, Bucaneve G, Micozzi A, Girmenia C, Barbabietola G, Pagano L, Leoni P, Specchia G, Caiozzo A, Raimondi R, Mandelli F: Itraconazole oral solution as prophylaxis for fungal infections in neutropenic patients with hematologic malignancies: A randomized, placebo-controlled, double-blind, multicenter trial. GIMEMA Infection Program. Gruppo Italiano Malattie Ematologiche dell' Adulto. *Clin Infect Dis* 28(2):250–255, 1999.

95. Rotstein C, Bow EJ, Laverdiere M, Ioannou S, Carr D, Moghaddam N: Randomized placebo-controlled trial of fluconazole prophylaxis for neutropenic cancer patients: Benefit based on purpose and intensity of cytotoxic therapy. The Canadian Fluconazole Prophylaxis Study Group. *Clin Infect Dis* 28(2):331–340, 1999.

96. Wingard JR, Merz WG, Rinaldi MG, Johnson TR, Karp JE, Saral R: Increase in *Candida krusei* infection among patients with bone marrow transplantation and neutropenia treated prophylactically with fluconazole. *N Engl J Med* 325(18):1274–1277, 1991.

97. Calandra T, Cometta A: Antibiotic therapy for gram-negative bacteremia. *Infect Dis Clin North Am* 5(4):817–834, 1991.

98. Pizzo PA: Management of fever in patients with cancer and treatment-induced neutropenia. *N Engl J Med* 328(18):1323–1332, 1993.

99. Pizzo PA, Hathorn JW, Hiemenz J, *et al*: A randomized trial comparing ceftazidime alone with combination antibiotic therapy in cancer patients with fever and neutropenia. *N Engl J Med* 315(9):552–558, 1986.

100. Behre G, Link H, Maschmeyer G, Meyer P, Paaz U, Wilhelm M, Hiddemann W: Meropenem monotherapy versus combination therapy with ceftazidime and amikacin for empirical treatment of febrile neutropenic patients. *Ann Hematol* 76(2):73–80, 1998.

101. Cometta A, Calandra T, Gaya H, Zinner SH, de Bock R, Del Favero A, Bucaneve G, Crokaert F, Kern WV, Klastersky J, Langenaeken I, Micozzi A, Padmos A, Paesmans M, Viscoli C, Glauser MP: Monotherapy with meropenem versus combination therapy with ceftazidime plus amikacin as empiric therapy for fever in granulocytopenic patients with cancer. The International Antimicrobial Therapy Cooperative Group of the European Organization for Research and Treatment of Cancer and the Gruppo Italiano Malattie Ematologiche Maligne dell'Adulto Infection Program. *Antimicrob Agents Chemother* 40(5):1108–1115, 1996.

102. Feld R: Vancomycin as part of initial empirical antibiotic therapy for febrile neutropenia in patients with cancer: Pros and cons. *Clin Infect Dis* 29(3):503–507, 1999.

103. Wong-Beringer A, Jacobs RA, Guglielmo BJ: Lipid formulations of amphotericin B: Clinical efficacy and toxicities. *Clin Infect Dis* 27(3):603–618, 1998.

104. Walsh TJ, Finberg RW, Arndt C, Hiemenz J, Schwartz C, Bodensteiner D, Pappas P, Seibel N, Greenberg RN, Dummer S, Schuster M, Holcenberg JS: Liposomal amphotericin B for empirical therapy in patients with persistent fever and neutropenia. National Institute of Allergy and Infectious Diseases Mycoses Study Group. *N Engl J Med* 340(10):764–771, 1999.

105. Kauffman CA, Bradley SF, Ross SC, Weber DR: Hepatosplenic candidiasis: Successful treatment with fluconazole. *Am J Med* 91(2):137–141, 1991.

106. Anaissie E, Bodey GP, Kantarjian H, David C, Barnett K, Bow E, Defelice R, Downs N, File T, Karam G, *et al*: Fluconazole therapy for chronic disseminated candidiasis in patients with leukemia and prior amphotericin B therapy. *Am J Med* 91(2):142–150, 1991.

107. Talcott JA, Siegel RD, Finberg R, Goldman L: Risk assessment in cancer patients with fever and neutropenia: A prospective, two-center validation of a prediction rule. *J Clin Oncol* 10(2):316–322, 1992.

108. Rolston KV: New trends in patient management: Risk-based therapy for febrile patients with neutropenia. *Clin Infect Dis* 29(3):515–521, 1999.

109. Kern WV, Cometta A, De Bock R, Langenaeken J, Paesmans M, Gaya H: Oral versus intravenous empirical antimicrobial therapy for fever in patients with granulocytopenia who are receiving cancer chemotherapy. International Antimicrobial Therapy Cooperative Group of the European Organization for Research and Treatment of Cancer. *N Engl J Med* 341(5):312–318, 1999.

110. Klastersky J, Paesmans M, Rubenstein EB, Boyer M, Elting L, Feld R, Gallagher J, Herrstedt J, Rapoport B, Rolston K, Talcott J: The Multinational Association for Supportive Care in Cancer risk index: A multinational scoring system for identifying low-risk febrile neutropenic cancer patients. *J Clin Oncol* 18(16):3038–3051, 2000.

111. Shapiro ED, Berg AT, Austrian R, Schroeder D, Parcells V, Margolis A, Adair RK, Clemens JD: The protective efficacy of polyvalent pneumococcal polysaccharide vaccine. *N Engl J Med* 325(21):1453–1460, 1991.

112. Strumia MM: The effect of leukocytic cream injections in the treatment of neutropenias. *Am J Med* 187:527–544, 1934.

113. Brecker G, Wilbur KM, Cronkhite EP: Transfusion of separated leukocytes into irradiated dogs with aplastic marrows. *Proc Soc Exp Biol Med* 84:54–56, 1953.

114. Young LS: Management of infections in leukemia and lymphoma. In *Clinical Approach to Infection in the Compromised Host*. Plenum Press, New York, 1989, pp. 467–502.

115. Bhatia S, McCullough J, Perry EH, Clay M, Ramsay NK, Neglia JP: Granulocyte transfusions: Efficacy in treating fungal infections in neutropenic patients following bone marrow transplantation. *Transfusion* 34(3):226–232, 1994.

116. Hubel K, Dale DC, Engert A, Liles WC: Current status of granulocyte (neutrophil) transfusion therapy for infectious diseases. *J Infect Dis* 183(2):321–328, 2001.

117. Lieschke GJ, Burgess AW: Granulocyte colony-stimulating factor and granulocyte-macrophage colony-stimulating factor (1). *N Engl J Med* 327(1):28–35, 1992.

118. Lieschke GJ, Burgess AW: Granulocyte colony-stimulating factor and granulocyte-macrophage colony-stimulating factor (2). *N Engl J Med* 327(2):99–106, 1992.

119. Rowe JM: Treatment of acute myeloid leukemia with cytokines: Effect on duration of neutropenia and response to infections. *Clin Infect Dis* 26(6):1290–1294, 1998.

120. Dignani MC, Anaissie EJ, Hester JP, O'Brien S, Vartivarian SE, Rex JH, Kantarjian H, Jendiroba DB, Lichtiger B, Andersson BS, Freireich EJ: Treatment of neutropenia-related fungal infections with granulocyte colony-stimulating factor-elicited white blood cell transfusions: A pilot study. *Leukemia* 11(10):1621–1630, 1997.

121. Ozer H, Armitage JO, Bennett CL, Crawford J, Demetri GD, Pizzo PA, Schiffer CA, Smith TJ, Somlo G, Wade JC, Wade JL 3rd, Winn RJ, Wozniak AJ, Somerfield MR: 2000 update of recommendations for the use of hematopoietic colony-stimulating factors: Evidence-based, clinical practice guidelines. American Society of Clinical Oncology Growth Factors Expert Panel. *J Clin Oncol* **18**(20):3558–3585, 2000.

122. Bowden RA, Sayers M, Flournoy N, Newton B, Banaji M, Thomas ED, Meyers JD: Cytomegalovirus immune globulin and seronegative blood products to prevent primary cytomegalovirus infection after marrow transplantation. *N Engl J Med* **314**(16):1006–1010, 1986.

123. Dwyer JM: Manipulating the immune system with immune globulin. *N Engl J Med* **326**(2):107–116, 1992.

124. Buckley RH, Schiff RI: The use of intravenous immune globulin in immunodeficiency diseases. *N Engl J Med* **325**(2):110–117, 1991.

125. Bodey GP, Nies BA, Mohberg NR: Use of gammaglobulins in infection in acute leukemia patients. *JAMA* **190**:1099–1102, 1964.

126. Weeks JC, Tierney MR, Weinstein MC: Cost effectiveness of prophylactic intravenous immune globulin in chronic lymphocytic leukemia. *N Engl J Med* **325**(2):81–86, 1991.

127. Amantadine: does it have a role in the prevention and treatment of influenza? A National Institutes of Health Consensus Development Conference. *Ann Intern Med* **92**(2 Pt 1):256–258, 1980.

128. Young LS, Glauser MP: Gram-negative septicemia and septic shock. *Infect Dis Clin North Am* **5**:739–945, 1992.

129. Jones CE, Alexander JW, Fisher M: Clinical evaluation of *Pseudomonas* hyperimmune globulin. *J Surg Res* **14**(2):87–96, 1973.

130. Harvath L, Andersen BR, Zander AR, Epstein RB: Combined pre-immunization and granulocyte transfusion therapy for treatment of pseudomonas septicemia in neutropenic dogs. *J Lab Clin Med* **87**(5):840–847, 1976.

131. Young LS, Stevens P, Ingram J: Functional role of antibody against "core" glycolipid of Enterobacteriaceae. *J Clin Invest* **56**(4):850–861, 1975.

132. Young LS, Gascon R, Alam S, Bermudez LE: Monoclonal antibodies for treatment of gram-negative infections. *Rev Infect Dis* **11**(Suppl 7):S1564–S1571, 1989.

133. Ziegler EJ, Fisher CJ Jr, Sprung CL, *et al*: Treatment of gram-negative bacteremia and septic shock with HA-1A human monoclonal antibody against endotoxin. A randomized, double-blind, placebo-controlled trial. The HA-1A Sepsis Study Group. *N Engl J Med* **324**(7):429–436, 1991.

134. Greenman RL, Schein RM, Martin MA, *et al*: A controlled clinical trial of E5 murine monoclonal IgM antibody to endotoxin in the treatment of gram-negative sepsis. The XOMA Sepsis Study Group. *JAMA* **266**(8):1097–1102, 1991.

135. Bernard GR, Vincent JL, Laterre PF, LaRosa SP, Dhainaut JF, Lopez-Rodriguez A, Steingrub JS, Garber GE, Helterbrand JD, Ely EW, Fisher CJ Jr: Efficacy and safety of recombinant human activated protein C for severe sepsis. *N Engl J Med* **344**(10):699–709, 2001.

136. Matthay MA: Severe sepsis—A new treatment with both anticoagulant and antiinflammatory properties. *N Engl J Med* **344**(10):759–762, 2001.

137. Estey EH, Keating MJ, McCredie KB, Bodey GP, Freireich EJ: Causes of initial remission induction failure in acute myelogenous leukemia. *Blood* **60**(2):309–315, 1982.

138. Raab SO, Hoeprich PD, Wintrobe MM, *et al*: The clinical significance of fever in acute leukemia. *Blood* **16**:1609–1628, 1960.

139. Davies SF, Khan M, Sarosi GA: Disseminated histoplasmosis in immunologically suppressed patients. Occurrence in a nonendemic area. *Am J Med* **64**(1):94–100, 1978.

140. Cox F, Hughes WT: Disseminated histoplasmosis and childhood leukemia. *Cancer* **33**(4):1127–1133, 1974.

141. Deresinski SC, Stevens DA: Coccidioidomycosis in compromised hosts. *Medicine* **54**:377–395, 1974.

16

Infection in Hematopoietic Stem Cell Transplantation

MICHAEL BOECKH and KIEREN A. MARR

1. Introduction

Hematopoietic stem cell transplantation (HSCT) is being used increasingly as therapy for aplastic anemia, acute leukemia, solid tumors, lymphomas, severe combined immunodeficiency syndrome, and on an experimental basis for some autoimmune disorders.[1] Indeed, recent long-term outcome results in patients with chronic myelogenous leukemia in chronic phase (CML-CP) who received a marrow transplant from an unrelated donor reveal survival rates of more than 75% at 5 years.[2] Effective prevention of infections such as cytomegalovirus (CMV) and invasive candidiasis was an important contributory factor to these survival advantages.[2] Thus, it is likely that HSCT patients will be seen with continued frequency in coming years at medical centers throughout the world. Substantial progress has been made in the understanding of the pathogenesis of infection, developing new diagnostic techniques, and in the development of prevention strategies since the last edition of this chapter. This chapter will review these advances and provide a basic approach to the management of infectious complications following HSCT.

Michael Boeckh and Kieren A. Marr • Program in Infectious Diseases, Fred Hutchinson Cancer Research Center, and Department of Medicine, University of Washington School of Medicine, Seattle, Washington, 98109.

Clinical Approach to Infection in the Compromised Host (Fourth Edition), edited by Robert H. Rubin and Lowell S. Young. Kluwer Academic/Plenum Publishers, New York, 2002.

2. Transplantation Techniques

Advances in therapy, including changes in types of HSCT, conditioning regimens, and methods to prevent and treat graft-versus-host disease (GVHD), affect infectious complications, prevention, and treatment strategies. Traditional methods rely on harvest of marrow for the stem cell source. Recent developments in HSCT include the use of peripheral blood stem cell (PBSC) and umbilical cord blood (UCB) for transplantation. In an effort to eliminate contaminating tumor cells from the PBSC product, a selection of CD34+ cells has recently been advocated for both allogeneic and autologous transplantation. The CD34+ selection process is associated with a significant T cell and monocyte depletion of the PBSC product (i.e., 2–3 log) and may thus affect the posttransplant infection risk. Another novel approach is to perform transplants with nonmyeloablative doses of chemotherapy alone or in combination with nonmyeloablative doses of TBI. This transplant type results in a mixed chimerism that lasts for several months and ultimately in donor chimerism in most cases.[3] After transplant, potent GVHD prophylaxis (e.g., mycophenolate mefotil [MMF] and cyclosporine) is given. Initial results of this transplant modality indicate that the time of granulocytopenia is very short but patients are at risk of GVHD.[4] The impact on infectious complications is currently being studied.

A number of conditioning regimens have been used to prepare patients for allogeneic HSCT. The classic regimen for patients with hematologic malignancy consists of high-dose intravenous cyclophosphamide (CY) with the addition of fractionated total-body irradiation (TBI) but numerous other regimens (e.g., combination chemotherapy without TBI) have been described over the last two decades. Intravenous methotrexate (MTX), corticosteroids, and/or intravenous or oral cyclosporine (CSP), FK506, antithymocyte globulin (ATG), MMF, or rapamycin is given for several months after transplant to prevent acute GVHD. Steroids are also sometimes given for GVHD prophylaxis but the risk of infection is increased with this strategy.[5] In addition, pretreatment of the donor's marrow with monoclonal antibodies or lectins to remove T cells is performed in an effort to prevent acute GVHD. The result is the same for all regimens: patients lose immune reactivity for varying periods after HSCT. Recovery is entirely dependent on the replacement of the hematopoietic and immunologic systems with those of donor origin. Until recovery occurs, patients are subject to a multitude of pathogens.

Patients with identical twin donors or those transplanted for other illnesses, such as severe combined immunodeficiency diseases, have variations on this theme. Patients with syngeneic HSCTs generally do not receive posttransplant MTX, since GVHD cannot occur, whereas patients with severe combined immunodeficiency diseases may need no pretransplant conditioning, since their underlying illness is adequately immunosuppressive. Recipients of syngeneic HSCT have fewer infections than their allogeneic counterparts. In addition, only patients with allogeneic transplants are subject to GVHD, itself immunosuppressive as is its treatment. However, recent changes in transplant techniques such as CD34 selection in autologous transplant recipients as well as the recognition of "pseudo-GVHD"[6] which sometimes leads to use of high-dose steroids may increase the infectious risk in autologous transplant recipients. Knowledge of transplantation techniques is important, as infectious complications evolve with each new modality.

3. Recovery of Host Defenses

Immune reconstitution after HSCT has been reviewed recently by Storek.[7] The most immediate change in host defense that primarily determines early risk of infections in HSCT patients is the precipitous loss of circulating granulocytes, which also occurs during a time of disruption of many anatomic barriers, such as the oral and gastrointestinal (GI) mucosa. The granulocyte count may be low before transplant, depending on the underlying illness and recent therapy, and most patients experience a virtual absence of circulating cells within the first 2 weeks after transplant. Granulocyte engraftment varies with the type of stem cells used for transplantation. Most patients have 500 or more circulating granulocytes/mm^3

by 20–25 days following BMT, compared to 10–20 days after PBSCT and 25–35 days after UCBT. The use of growth factors such as G-CSF or GM-CSF may hasten this recovery. Although the clinical status of patients improves dramatically with recovery of circulating granulocytes, neutrophil function is not yet entirely normal, and defects in neutrophil chemotaxis in patients with GVHD, corticosteroid and/or antithymocyte globulin treatment have been described. The cell input as well as thymic output influence immune recovery.[8,9]

After a myeloablative HSCT, both B and T lymphocyte deficiencies persist for many months or years due to the damage to marrow stroma, thymic stroma, or follicular dendritic cells.[7] Although total lymphocyte count may become normal by the second month after transplant in most patients, there is a quantitative deficiency of CD4 T cells for years after HSCT probably due to insufficient production of T cells in the adult thymus. Also, quantitative deficiencies of B cells occur for months after HSCT, possibly due to insufficient production of B cells in the marrow and/or insufficient delivery of survival signals to newly generated B cells from CD4 T cells or follicular dendritic cells. There is also a quantitative deficiency of dendritic cells during the first several months posttransplant and CD4 T cell and B cell function is deficient for more than 1 year after transplant.[7]

Delayed lymphocyte engraftment is a risk factor for viral infections (e.g., CMV) and is associated with a poor outcome from CMV infection.[10,11] Some *in vitro* lymphocyte functions, including reactivity in mixed leukocyte culture and responses to mitogens or viral antigens, are depressed for 4–5 months after HSCT in all patients, with more prolonged suppression among patients with GVHD.[9,12–16] Related responses, such as *in vitro* interferon production and production of and responsiveness to interleukin-2 (IL-2), are similarly depressed.[17,18] Both NK and cytotoxic T lymphocytes (CTL) mediate activity against virus-infected target cells,[19,20] with recovery of the latter response associated with survival from CMV infection.[15,19] Recovery of virus-specific T cell responses (e.g., against HSV, CMV) may be delayed in patients with acute and chronic GVHD and those requiring corticosteroids, as well as in patients receiving antiviral chemoprophylaxis.[21,22] Lack of recovery of CMV-specific CTL responses are a risk factor for subsequent CMV pneumonia,[15,23] and antiviral prophylaxis with ganciclovir can decrease CMV-specific immune reconstitution.[21,24,25]

Immune reconstitution following unmodified PBSCT is characterized by higher T cell and monocyte counts as well as a transiently improved B cell and NK cell count.[9,26,27] In a randomized trial, this has resulted in lower rates of bacterial, fungal, and to a lesser degree, viral infections in PBSC transplant recipients.[9] CD34 selection of the stem cell product results in a delayed lymphocyte and monocyte engraftment and a higher risk for CMV disease.[28] UCB transplantation is characterized by a prolonged period of granulocytopenia (median approximately 30 days); however, by 1 year immune reconstitution does not appear to be delayed when compared to HLA-identical sibling transplant.[29] Immune reconstitution after nonmyeloablative therapy has not been studied to date.

4. Phases of Infection after HSCT

From the viewpoint of the infectious disease consultant, the posttransplant course may be divided into three periods that correspond broadly to the pattern of immunologic recovery outlined above (Fig. 1). First, the early granulocytopenic period lasts 10–30 days after HSCT in most patients. Without antimicrobial prophylaxis, this period is highlighted by frequent fever and a high risk for bacterial and candidal infection. Second, between recovery of circulating granulocytes and day 100, viral and mold infections become more prominent. A diffuse, apparently noninfectious pneumonia of unknown cause— idiopathic pneumonia syndrome (IPS)—also occurs during this interval. Bacterial infections continue to occur, and fungal (both yeasts and molds) infections overlap in both early phases, especially in the small number of patients with prolonged granulocytopenia because of graft rejection failure. After day 100, the incidence of most infections decreases. Characteristic infections that do occur later include VZV infection and pneumococcal infection, and recently also CMV infection and invasive mold infections.[30,31]

The risk of infection depends in part on the occurrence of GVHD and on concomitant changes in marrow

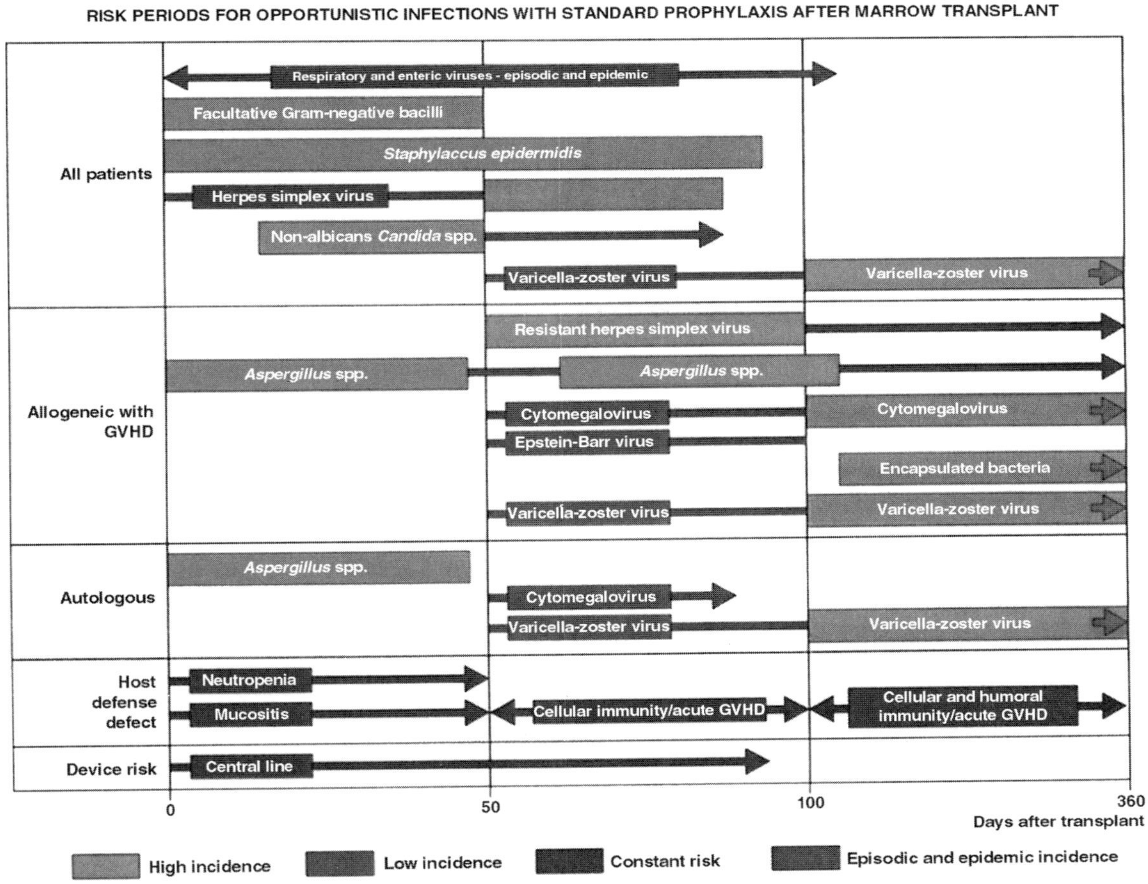

FIGURE 1. Temporal pattern of infections after HSCT with current prevention strategies as outlined in Table 2. (Obtained with permission from Armstrong and Cohen.[302])

function and immune reactivity secondary to its treatment. More recently, the use of effective prevention strategies has modified this risk period, primarily by decreasing the risk of early infections caused by herpes simplex virus (HSV), CMV, and *C. albicans* (Fig. 2). It must be emphasized that individual patients do not always behave quite so predictably, and the consultant's index of suspicion must remain high at all times. New presentations of old infections, and newly recognized infections will certainly continue to be described. Willingness to entertain new possibilities and the use of aggressive diagnostic and therapeutic techniques are important in the care of these patients.

5. Phase I: Infections between Conditioning and Engraftment

Nearly all allogeneic transplant recipients develop fever before or shortly after stem cell infusion, temporally related to declining granulocyte counts. Patients without fever at any time during their course are uncommon, but fever can be seen when the patient is receiving corticosteroids for conditioning-related toxicity or acute GVHD. As in other patients with granulocytopenia, fever may be the best indicator of infection, since common signs of infection may be lacking in the absence of a normal inflammatory response. A recent review of infectious complications prior to engraftment in HSCT (autologous and allogeneic) recipients documented that after fever of unknown origin, which occurred in 50.6% of patients, other common complications included septicemia (12.5%) and pneumonia (11.0%).[32] These complications are discussed in detail below.

5.1. Bacteremia and Candidemia

Bacteremia is temporally associated with the period of profound granulocytopenia.[33] However, late bacteremias also occur among patients with acute GVHD. Bacterial infection as the cause of death has now become unusual, except in patients with either prolonged neutropenia or severe acute GVHD. Recent reviews have noted that gram-negative septicemia and/or pneumonia were contributory causes of death in 5% of allogeneic recipients.[34,35] The organisms recovered from blood cultures have also changed. Whereas aerobic gram-negative organisms were more common prior to 1983, it is now more common to find gram-positive bacteria (especially coagulase-negative *Staphylococcus* and enterococci) as the cause of bacteremia. Although the explanations for this shift are not entirely clear, the universal use of long-term indwelling intravenous catheters, and the aggressive use of broad-spectrum antibiotics effective against more gram-negative aerobes have been implicated. Multiresistant bacterial infections (e.g., vancomycin-resistant enterococci) are also recovered with increased frequency.

Candida species are another major cause of bloodstream infection during this early period after HSCT. Although the incidence and spectrum of *Candida* species have changed since the use of fluconazole for prophylaxis (see Section 5.5.3), a recent study from our institution documented that the median day of candidemia diagnosis remains early after HSCT (day 28, range 12 to 110).[36] Although other yeasts such as *Trichosporon* species can be etiologic agents of fungemia, the vast majority are caused by *Candida* species.[37]

5.2. Early Pneumonia

Bacterial pneumonia is relatively uncommon, except in the setting of ventilator requirement. This may be the result of the early use of empirical antibiotics in patients who cannot develop the usual signs of pulmonary consolidation because of severe granulocytopenia. Nevertheless, bacterial pneumonia related to resistant organisms or those organisms not covered by the empiric regimen may still occur.

Viral and fungal pneumonia are probably the most common causes of infectious pneumonia in the preengraftment period, although the overall incidence remains low. Viruses that can cause pneumonia early after transplant are respiratory viruses (e.g., respiratory syncytial virus [RSV], parainfluenza viruses) and occasionally CMV.[38] Invasive aspergillosis can also occur early after transplant.[32,39] In contrast, pneumonia due to *Pneumocystis carinii* is all but eliminated in the early posttransplant period with pretransplant prophylaxis.[40]

Because of the poor clinical outcome associated with both viral and fungal pneumonia, the diagnosis and treatment of early pulmonary infiltrates are approached aggressively.[41] Computer tomography is often useful for the detection of small pulmonary nodules. Patients in the preengraftment period usually do not produce adequate sputum, and specimens are often difficult to interpret because of contamination with oropharyngeal flora. Bronchoscopy with bronchoalveolar lavage (BAL) is considered the mainstay of diagnosis of pulmonary infiltrates after HSCT and should be performed promptly. While the BAL has excellent sensitivity for viral pathogens, its sensitivity for molds is only about 60%.[41] Thus, more invasive procedures such as thoracoscopically

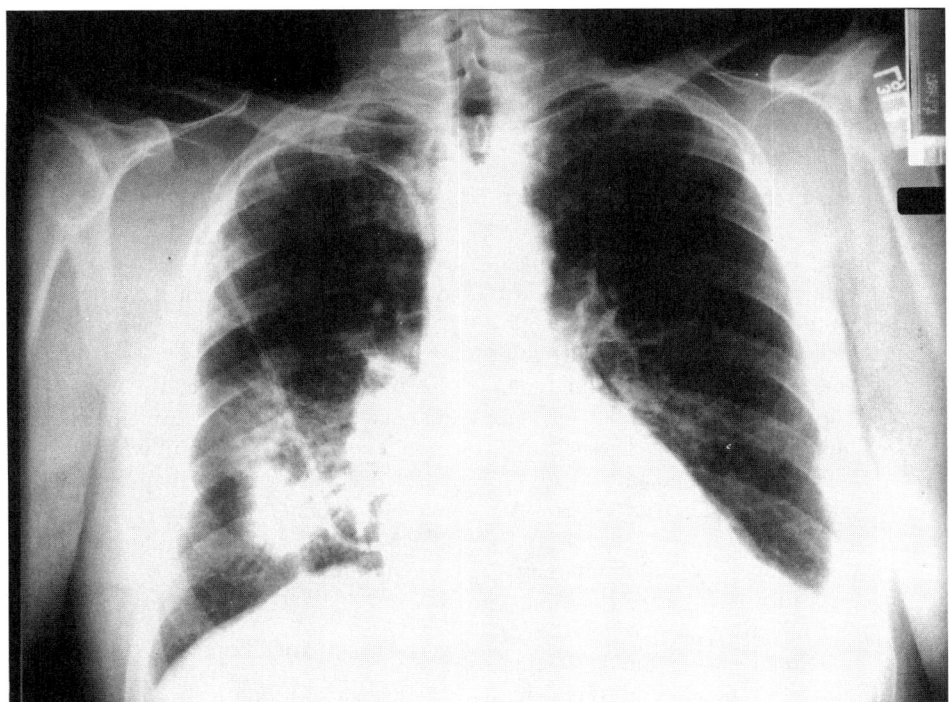

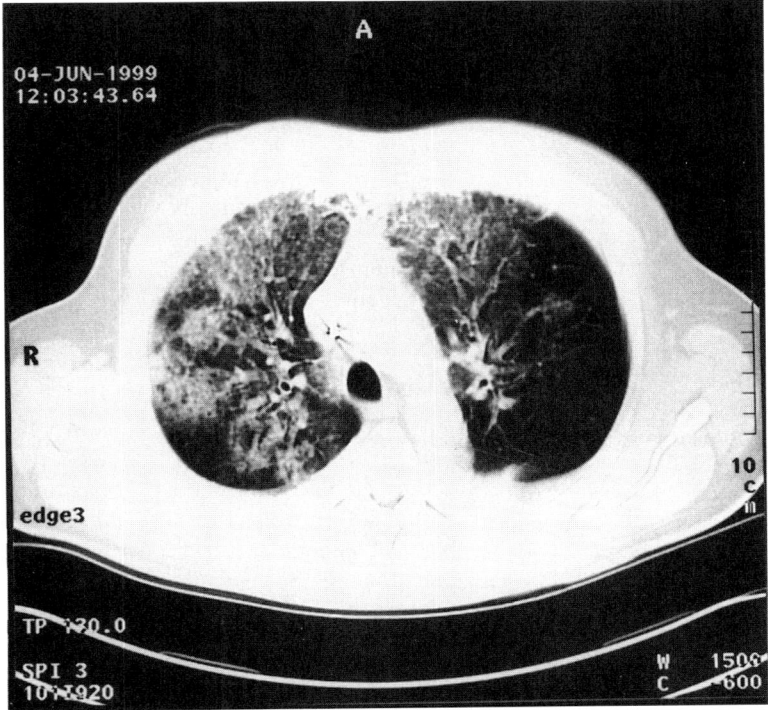

FIGURE 2. Chest radiograph (A) and computed tomography (A) from a 24-year-old patient with pulmonary hemorrhage on day 6 after myeloablative transplant for acute myelogenous leukemia. BAL showed significant bleeding but was negative for bacterial, fungal (including *Pneumocystis carinii*), and viral pathogens. The patient recovered within 2 weeks of steroid therapy. Broad-spectrum antibiotic therapy (imipenem, ciprofloxacin) and amphotericin B (1 mg/kg). All antimicrobials were discontinued within 2 weeks.

guided or open lung biopsies may be required to establish the diagnosis. However, in the preengraftment period, these invasive procedures may be difficult to perform because of low platelet counts. Also, transtracheal aspiration and transbronchial biopsies are generally not performed in the early posttransplant period because of the high bleeding risk.

It should be emphasized that not all infiltrates are of infectious etiology in the early posttransplant period. Noninfectious infiltrates, especially those caused by fluid overload, and conditioning regimen-related pneumonia (i.e., idiopathic pneumonia syndrome; see Section 6.6) can occur. Pulmonary hemorrhage (Fig. 2) occurs mainly in the preengraftment period and is indistinguishable from interstitial pneumonia on regular chest radiograph. Noninfectious diagnoses are made only by exclusion, and empiric treatment with antibiotics, covering gram-negative bacteria and *Legionella*, should be initiated until results of diagnostic tests are available. If fluid overload is suspected, diuresis may be indicated. However, prompt BAL is recommended in CMV seropositive allograft recipients, and during the respiratory virus season one must rule out potentially treatable infections such as RSV lower tract disease.[42] Antifungal treatment should be started only in patients with a high risk (i.e., allograft recipients with characteristic pulmonary findings; see Section 5.5.3). In some patients, a lung biopsy cannot be performed early in their posttransplant course because of refractory thrombocytopenia. Such patients are maintained on empiric treatment without a specific etiological diagnosis. If the infiltrates remain localized, nonbacterial etiologies are less likely, although localized presentations of CMV can occur (Fig. 3). If pulmonary infiltrates become diffuse or hypoxia progresses, then empirical anti-*Pneumocystis* treatment may be considered until such diagnosis has been ruled out in patients who did not receive a pretransplant course of anti-*Pneumocystis* prophylaxis (see Section 5.5.1).

5.3. Intravascular Catheter-Related Infection

Multilumen catheters introduced under sterile conditions through a subcutaneous "tunnel" are used in all patients. This eases the problem of venous access for blood samples as well as hyperalimentation and other infusion requirements. All blood cultures, unless specified otherwise, are obtained through these central lines. This will occasionally cause uncertainty about the interpretation of a positive blood culture, especially for gram-positive organisms. Simultaneous central and "peripheral" blood cultures have not usually aided the interpreta-

tion of positive blood cultures. Catheters also serve as the source of infection.

Although most infections with *Candida* species do not originate from intravascular catheters, two studies have indicated that outcome may be improved if the catheter is removed.[43,44] We typically suggest that lines be removed in the setting of recurrent positive cultures with *Candida* species or *Corynebacterium JK*,[33] and with one positive blood culture of *S. aureus*, atypical mycobacteria, or in a setting of sepsis. Bacteremia with other organisms, in particular a coagulase-negative strain, is usually approached with the lines left in place unless symptoms of sepsis or local signs of infection are present. In particular, infection of the subcutaneous tunnel (e.g., due to atypical mycobacteria[45]) may be difficult to treat without removal of the catheter. Although the catheter should be examined carefully for local signs of infection, purulence and erythema may be notably missing in neutropenic patients. A mild erythema around the exit site of <5 mm is a common finding and does not indicate an exit site infection.

5.4. Fungal Infections

Candidal bloodstream infections that occur prior to engraftment are primarily acquired through the gastrointestinal tract, where these organisms exist as commensal inhabitants. Other nosocomial and catheter-related infections do occur, especially in association with *C. parapsilosis*.[36,46] Prior to the introduction of fluconazole prophylaxis in our institution, during the period between 1980 and 1987, the incidence of invasive candidal infection was 11% and the median time of onset was 2 weeks after HSCT.[47] Most of these infections were caused by *C. albicans* and were associated with tissue invasion and a high mortality (38%). In the last decade, the emergence of fluconazole-resistant non-*albicans* species of *Candida* as pathogens in patients has been well documented.[36] In HSCT patients maintained on fluconazole, these species now constitute the majority of candidal bloodstream infections.[48] A review of patients undergoing HSCT in our institution from 1994 to 1997 revealed a decrease in incidence of candidemia from 11% to 4.6%, and a virtual elimination of *C. albicans* and *C. tropicalis* as pathogens (Fig. 4). In these patients maintained on fluconazole throughout neutropenia and acute GVHD, tissue invasion due to *Candida* species, including hepatosplenic candidiasis (see Section 6.5), has become rare, and attributable mortality has decreased.[36,49]

Invasive aspergillosis has become the most common fungal infection in HSCT recipients maintained on flu-

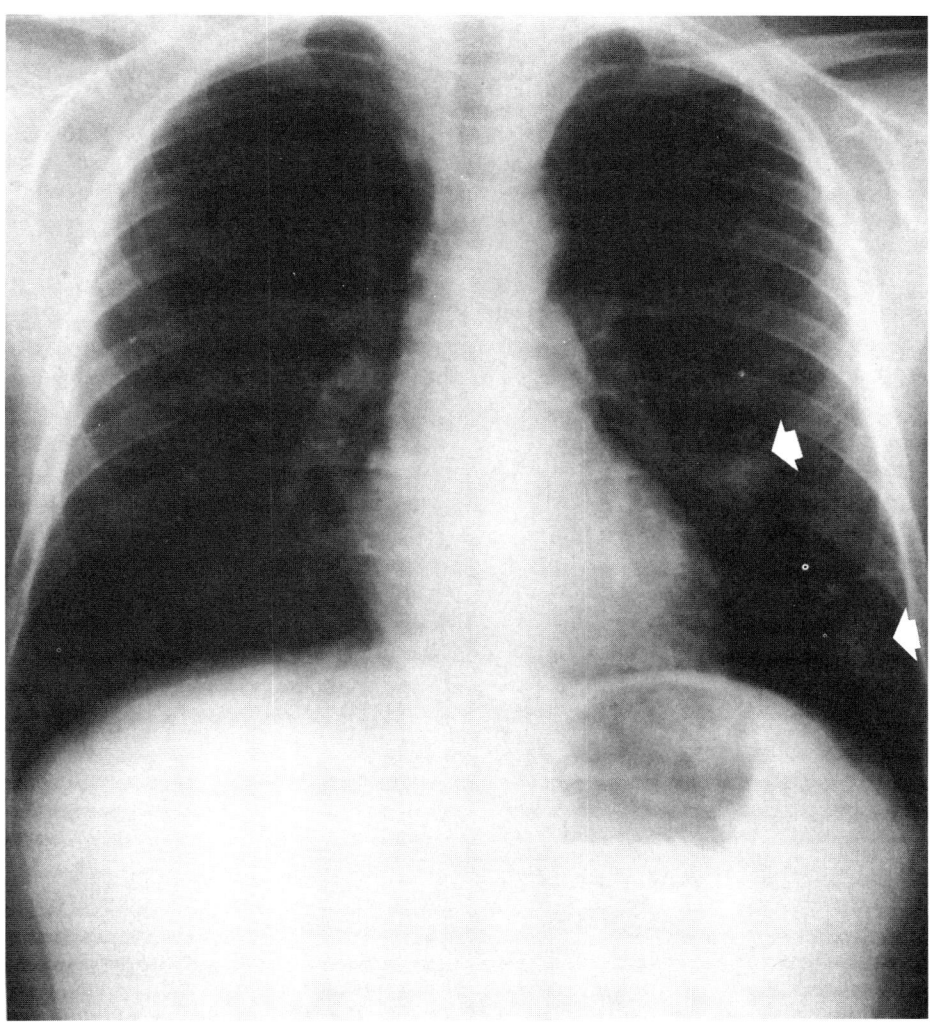

FIGURE 3. "Nodular" cytomegalovirus pneumonia in an 11-year-old boy beginning 70 days after marrow transplant for acute myelocytic leukemia. Two lesions in the left lung are marked by arrows.

conazole prophylaxis (Fig. 5). Several institutions have reported increasing trends of invasive infections over the 1980s and 1990s.[37,39,50,51] This organism is inhaled through the respiratory tract, and as such, may present as localized sinus disease, pulmonary disease, or after dissemination to viscera. An examination of patients undergoing HSCT from 1988 to 1993 revealed that pulmonary aspergillosis has a bimodal timing of diagnosis, with the first peak occurring early, during neutropenia, and a second peak occurring later, corresponding with acute GVHD (Fig. 6).[39] Risk factors associated with each peak are outlined in Table 1. High mortality occurs in HSCT recipients, despite antifungal treatment. Crude mortality approximates 86% for isolated pulmonary disease, 66% in association with sinus disease, and 99% with documented

cerebral invasion.[52] It has been our experience that although overall mortality in allograft recipients approximates 80%, few people survive without prompt reversal in immune dysfunction.[53] One recent study documented that outcome is worst in patients who receive high total doses of corticosteroids.[54] For this reason, attention has turned to establishing effective prevention strategies, methods of early diagnosis, and optimizing therapy.

5.4.1. Diagnosis of Fungal Infections

In the setting of a suspected candidal infection, the physical examination should include a complete search for skin and retinal lesions. Every effort is made to document fungal infections, preferably with culture of the

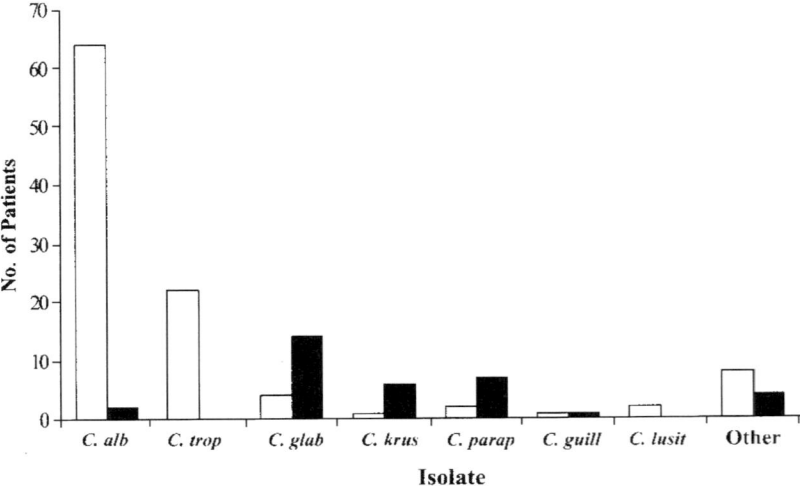

FIGURE 4. Distribution of candidemia in the fluconazole era (black bars) compared to pre-fluconazole era (open bars). (Obtained with permission from Marr *et al.*[36])

organism and pathologic evidence of invasion, if applicable. Blood culture with growth of *Candida* species remains the "gold standard" of diagnosis. Recently developed lysis centrifugation techniques and the BacT/Alert System have increased the sensitivity of cultures.[55,56] Diagnosis of tissue infection can be particularly difficult because blood cultures are frequently negative. Hypodense liver or spleen lesions in a patient at risk for hepatosplenic candidiasis should thus be treated presumptively even in the absence of positive blood cultures. Methods to detect *Candida* antigens, such as cell wall components (i.e., mannan, 1,3-β-glucans, chitin), cytoplasmic enzymes (i.e., enolase), and PCR, have been investigated for use in diagnosing tissue infection and for early diagnosis.[57] Unfortunately, many of these tests have been limited by an inability to differentiate between *Candida* pathogens and commensal isolates. Biopsy of a suspected lesion may thus become necessary if the patient does not respond to appropriate antifungal treatment.

In the setting of a suspected mold infection, physical examination should include a careful examination of the skin, mouth, sinuses, and lungs. Biopsy or BAL, with microbiologic confirmation of the offending organism, is the "gold standard" for diagnosis. Unfortunately, the sensitivity of culture is variable and BAL may reveal *Aspergillus* species in only 50–60% of cases. The diagnosis of invasive aspergillosis prior to an advanced stage in illness is difficult because pulmonary lesions become apparent late in the course of infection and blood cultures are rarely positive.[58] For this reason, detection of *Aspergillus* species in respiratory secretions, even in the absence of other signs of infection, should prompt a complete search for invasive disease and consideration of early antifungal therapy. In the review by Wald and colleagues, "colonization," with any *Aspergillus* species, except for species of *Aspergillus niger* in the stool, was highly predictive of subsequent invasive infection.[39]

For an earlier diagnosis of invasive aspergillosis, investigators have examined the roles of early CT scanning, antigen detection, and PCR. There is no question that CT scans are more sensitive than conventional radiographs for the detection of small lesions. Although investigators have found that CT scan allows for earlier diagnosis and improved prognosis in febrile neutropenic patients,[59] our experience has suggested poor outcome even after detection of such lesions early in HSCT patients. Yet more sensitive methods of diagnosis are needed. Diagnostic methods for detection of *Aspergillus* infection include galactomannan antigen detection and PCR. A sensitive double-sandwich ELISA has been tested and is more widely used in Europe but is not as yet available in the United States. The results of several prospective studies have suggested that this test has utility both for early diagnosis, screening and for follow-up of therapy.[60] Several groups are also investigating the role of PCR for early diagnosis,[61,62] with some studies suggesting utility for early diagnosis using both blood and lower respiratory tract secretions.[61,63]

5.4.2. Treatment of Fungal Infections

All positive blood cultures for *Candida* species should be treated promptly with antifungals. The dose of amphotericin for fungemia and documented candidal in-

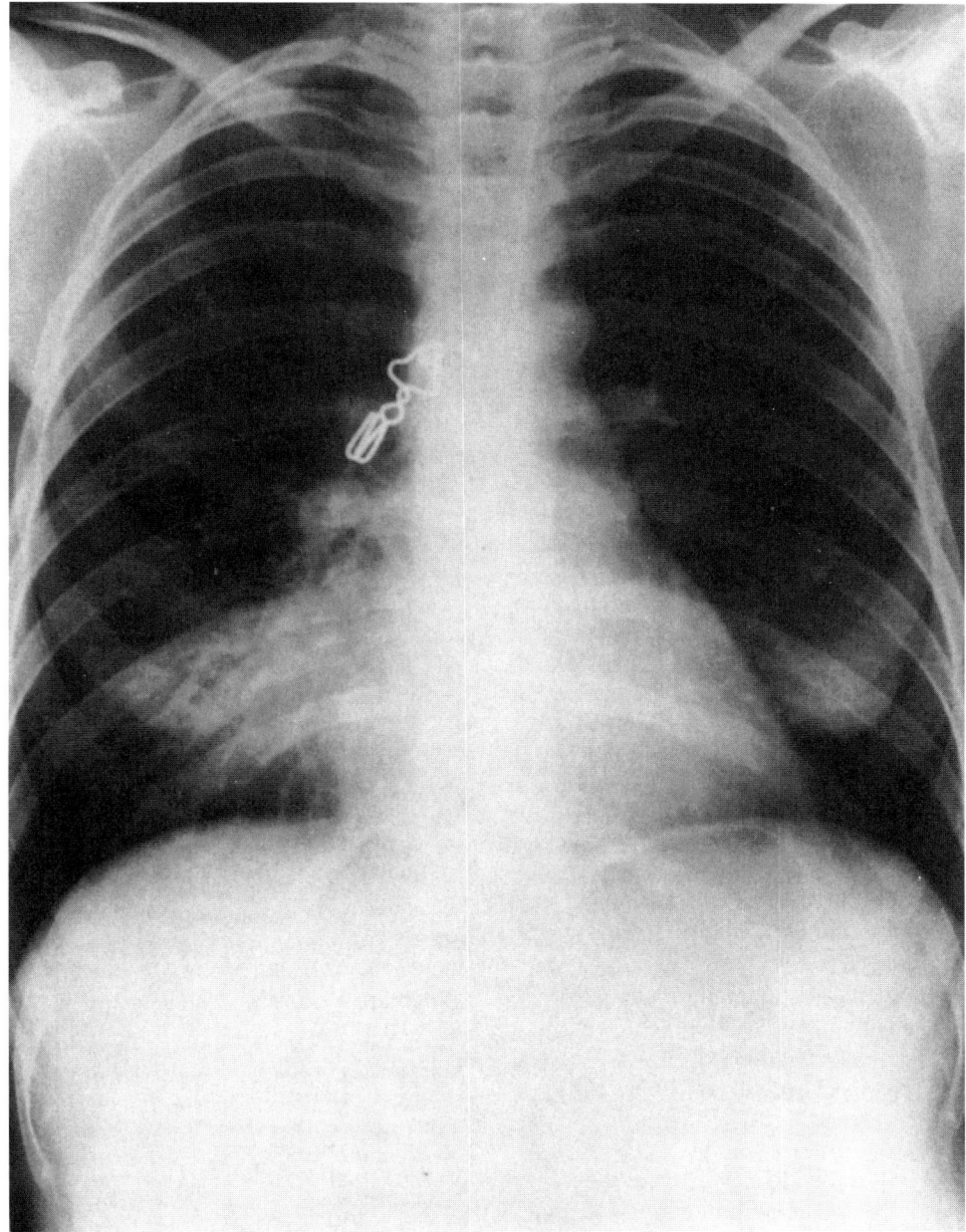

FIGURE 5. *Aspergillus* infection of the right middle lobe proven by open lung biopsy in a 38-year-old woman transplanted for acute nonmyelocytic leukemia in relapse. Involvement of the pericardium was found at the time of biopsy, and at postmortem examination the anterior descending coronary artery was partially occluded.

fections other than mucocutaneous or esophageal infection is at least 0.5 mg/kg per day. Recent evidence indicates that certain non-*albicans Candida* species (e.g., *C. glabrata*, *C. krusei*) warrant higher doses of amphotericin B (1.0 mg/kg).[64] Treatment is continued until resolution of neutropenia, and a minimum of 1.5 g total adult dose

has been delivered. Combination therapy with 5-flucytosine has been advocated in cases of severe invasive candidiasis. When serum levels of 5-flucytosine were maintained below 100 µg/ml, marrow toxicity was generally not observed. Although fluconazole was shown to be efficacious in treating candidemia in nonneutropenic

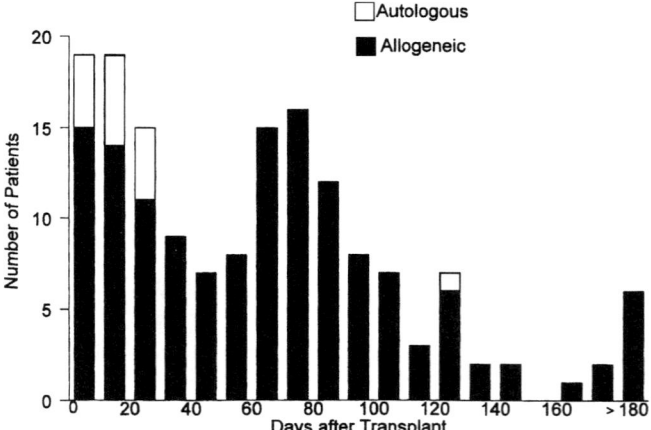

FIGURE 6. Bimodal distribution of aspergillosis after HSCT. (Obtained with permission from Wald *et al.*[39])

hosts,[64] there are currently insufficient data in immuno-compromised patients. It is likely that new, currently investigational agents such as extended-spectrum azole antifungals and echinocandins will have a role in therapy, especially for fluconazole-resistant organisms.

Our experience with the treatment of invasive aspergillosis has been disappointing. Because maximum doses of amphotericin B are still associated with poor outcome and frequent toxicities, attention has turned to the role of lipid amphotericin B formulations, which can deliver more amphotericin while resulting in less nephrotoxicity and infusion-related reactions.[65] Also, other drugs, such as mold-active azole antifungals, echinocandins, and liposomal nystatin, are currently being studied for both combined and monotherapy of aspergillosis in the HSCT setting.

Varying regimens have been used for oral *Candida* infections or for esophageal candidiasis. In patients with esophageal symptoms suggestive of candidiasis, esophagoscopy is performed whenever possible since it has not been clinically possible to distinguish herpetic from *Candida* infection or from other noninfectious causes of esophagitis such as GVHD. Parenteral amphotericin is used instead of oral agents for all fungal infections including esophagitis during the granulocytopenic period.

By contrast, local *Aspergillus* infections of tracheostomy, arteriovenous shunt sites, or infection of the maxilla or sinuses, have been cured with amphotericin and combined resection. The role of surgical resection of isolated pulmonary lesions continues to be debated. While this approach is necessary in the setting of life-threatening hemoptysis, this complication is unusual in HSCT patients. Studies have suggested that outcome after surgical resection may be improved, especially in the set-

ting of pretransplant aspergillosis,[59,66,67] but no randomized trials have been performed. We think that this approach should be considered in patients with a good surgical risk and easily accessible lesions, especially if prolonged neutropenia is anticipated in patients. After successful treatment of invasive aspergillosis, as outlined above, we administer "maintenance" antifungal therapy with an oral mold-active azole drug such as itraconazole oral solution until the patient has discontinued immunosuppressive drugs. Efforts to hasten reconstitution of the immune system should be considered in conjunction with antifungal therapy.

Illustrative Case 1

A 32-year-old male with an underlying disease of acute myelogenous leukemia underwent a matched peripheral blood stem cell transplant from his brother. On day 7 after transplant, he developed neutropenic fever that was unresponsive to ceftazidime and gentamicin. On day 3 of fever, an erythematous, tender 1-cm round lesion developed on the dorsal aspect of his left index finger. Initially firm and nodular, this lesion later became more erythematous and subsequently developed a tiny necrotic center. The patient was started on high doses of amphotericin B (1.5 mg/kg), and cultures of the lesion and blood confirmed infection with *Fusarium* species. Despite receipt of amphotericin B, fever persisted and a left lower lobe nodular infiltrate became apparent on CT scan of his chest. The patient died of respiratory failure on day 28 after transplant.

Comment. This patient had a disseminated infection with one of the "emerging" pathogenic fungi. Although the vast majority of mold infections in this setting are caused by *Aspergillus* species, other molds are becoming more frequent. The rapidly progressive course is typical of infection with this organism, as is the appearance of skin lesions and ultimate fatal outcome. Although amphotericin B has been the predominant agent to treat invasive fungal infections, resistance to this agent is increasingly appreciated in some molds, such as *Fusarium* species and *Pseudallescheria* species, and yeasts such as *Candida lusitaniae*. There

TABLE 1. Risk Factors for Early and Late Aspergillosis[a]

Risk factor	Aspergillosis < day 40		Aspergillosis > day 40	
	Adjusted RR	95% CI	Adjusted RR	95% CI
Age				
≤ 18	1.0	—	1.0	—
19–40	0.98	0.50–1.93	3.03	1.46–6.28
40	1.94	0.98–3.84	5.03	2.41–10.5
Sex				
Female	1.0	—		—
Male	1.73	1.00–2.99	0.91	0.61–1.36
Disease group				
CML chronic phase	1.0	—	1.0	—
Hematologic malignancy, first remission	1.98	0.27–14.2	3.60	1.57–8.28
Hematologic malignancy, non-first remission	8.88	2.13–37.1	3.06	1.62–5.78
Other	5.79	1.26–26.6	3.71	1.77–7.79
Donor type				
HLA matched, related	1.0	—	1.0	—
Autologous/syngeneic	0.86	0.43–1.7	0.09	0.01–0.72
HLA-mismatched, related	2.08	1.08–4.0	0.85	0.44–1.65
Unrelated	1.5	0.67–3.36	1.67	1.04–2.67
Season				
Winter	1.0	—	NS	—
Spring	1.64	0.64–4.18		
Summer	4.45	1.93–10.2		
Fall	2.19	0.89–5.4		
Year of transplant	1.0	0.87–1.16	0.98	0.84–1.17
Construction				
Absent	NS	—	1.0	—
Present			1.84	1.04–3.25
Laminar airflow room				
Yes	1.0	—	NS	—
No	5.58	2.3–13.4		
Neutropenia				
Absent	NS	—	1.0	—
Present			5.97	2.95–12.1
Acute GVHD				
Grade 0–1	NS	—	1.0	—
Grade 2–4			2.60	1.38–4.87
Corticosteroid use				
No	NS	—	1.0	—
Yes			3.14	1.67–5.91

[a]Data obtained from Wald et al.[39] NS, not significant ($p > 0.05$); CML, chronic myelogenous leukemia. Other factors tested but not significant for both early and late aspergillosis were pretransplant CMV serostatus and conditioning regimen.

is preliminary indication that newer mold-active azoles may have better activity against these ampho-resistant molds.[68]

5.5. Antibiotic and Antifungal Treatment

5.5.1. Prophylactic and Empirical Antibiotic Coverage

The care of the HSCT patient with fever and granulocytopenia is similar to that of other patients with granulocytopenia, and many of the considerations raised in Chapter 4 apply. The need for careful and frequent physical examinations and for obtaining a full set of cultures processed for bacterial (including anaerobes), fungal, and viral agents before initiating antibiotic treatment will not be further reiterated.

In the protective environment where antibiotic prophylaxis is routinely given, nonabsorbable antibiotics have been advocated in the past but have largely been replaced by fluoroquinolones and fluconazole due to better tolerability. All approaches of oral antibiotic prophylaxis are hampered by the fact that HSCT patients often have difficulty with oral medications because of mucosal lesions, and many patients cannot take these antibiotics reliably. The use of TMP-SMX for *Pneumocystis* prophylaxis before transplant is advised[40] (Table 2). This regimen also has antibacterial activity. Our regimen consists of the oral administration of TMP-SMX at a dose of 75 mg/m^2 (TMP component) twice daily begun at the start of conditioning and continued up to 48 hr before transplant. TMP-SMX prophylaxis is resumed later in the posttransplant course, and is continued until GVHD prophylaxis is tapered. In patients who do not tolerate TMP-SMX, dapsone is given instead. Finally, prophylactic antibiotics may serve functions other than protection against infection. A recent study performed in patients transplanted in the protective environment suggests that prophylaxis with ciprofloxacin and metronidazole decreases the incidence of acute GVHD grade 2–4 in matched-related transplant recipients, possibly due to reduction in anaerobic bacteria.[69] Other investigators have not confirmed results of this nonblinded study.

Two approaches are being used to initiate systemic antibiotic therapy. The most common empiric strategy is similar to that used in other granulocytopenic patients, i.e., start of systemic intravenous antibiotics at the time of first fever.[70,71] An alternative preemptive strategy is to start systemic antibiotic prophylaxis when the absolute neutrophil count (ANC) drops below 500/mm^3. The rationale for the latter approach is that this practice is associated with a decreased incidence of bacteremia and septicemia compared to when antibiotics are started based on first fever.[72] Also, the ANC usually decreases rapidly below 100/mm^3 and in this situation more than 90% of allograft recipients develop fever. Combined strategies of using oral systemic antibiotic prophylaxis with fluoroquinolones followed by intravenous antibiotics when fever develops have also been used, but there are no controlled studies of the use of prophylactic oral antibiotics for HSCT patients outside of the protective environment and the issue remains controversial.[71]

The choice of specific drugs may depend in part on local usage as well as familiarity with the microbiologic milieu of the particular transplant unit, but in all cases, these drugs should be used in maximum doses for age and weight, with serum levels of aminoglycosides (and other drugs when appropriate) monitored frequently. Based on the spectrum of bacteria recovered from the blood of HSCT patients, initial empirical coverage should be directed toward gram-negative bacteria, including

TABLE 2. Currently Recommended Antimicrobial Prophylaxis Regimens in HSCT Recipients[71,172]

Pathogen	Patients at risk	Prevention strategy	Duration	Alternative	Comments
Pneumocystis carinii	All HSCT recipients	TMP-SMX	Allogeneic: throughout period of immunosuppression Autologous: until day 120	Dapsone, aerosolized pentamidine	Dapsone must be given daily
HSV	HSV seropositive recipients	Acyclovir, valaciclovir	Until day engraftment and resolution		
VZV	VZV seropositive recipients	Acyclovir, valaciclovir	One year after transplant		
CMV	CMV seropositive recipients or donors	Ganciclovir based on antigenemia or DNA detection— or prophylaxis	Preemptive therapy: 2–3 weeks (possibly longer: Ag/PCR must be negative) or until day 100 Prophylaxis: engraftment until day 100	Foscarnet, cidofovir	Acyclovir prophylaxis reduces mortality; continuation of preemptive strategy > day 100 in high-risk patients
	CMV seronegative recipients	Seronegative or leukoreduced blood products	Throughout period of immunosuppression		
Candida albicans and *tropicalis*	All HSCT recipients	Fluconazole	Autologous: until engraftment Allogeneic: until day 75		Allograft recipients benefit from extended prophylaxis
Encapsulated bacteria	Allograft recipients with chronic GVHD	TMP-SMX	Until 1 month after discontinuation of immunosuppression	Penicillin	Daily TMP-SMX also protects against PCP

Pseudomonas aeruginosa. Appropriate options include a third- or fourth-generation cephalosporin with anti-*Pseudomonas* coverage, an antipseudomonal penicillin, imipenem, or meropenem. Little data are available on fluoroquinolones in this situation but preliminary reports suggest that they are equivalent.[73] If patients are treated for fever, a second antibiotic is generally recommended,[70] although initial monotherapy may be sufficient in patients who do not show signs of sepsis or are not colonized with *P. aeruginosa.* The appropriate choice for additional antibiotics is currently debated. In recent years, many centers have seen an increase in renal insufficiency related to the concurrent use of multiple nephrotoxic drugs. Therefore, many patients now receive a combination of fluoroquinolones and a third-generation cephalosporin or a third-generation cephalosporin alone for both empirical coverage and treatment of documented infection. Vancomycin is added either for a positive blood culture for gram-positive cocci or for clinical suspicion of a catheter-related tunnel infection. Controversy remains about the value of adding vancomycin into initial empirical antibiotic regimen,[70,71] especially given the recent emergence of vancomycin-resistant enterococci in many transplant centers.[74] An aminoglycoside is added when a gram-negative organism is isolated from the blood, when clinical shock is present (e.g., hypotension, increasing acidosis), or when fever persists and the patient is colonized with a suspicious gram-negative organism. We have used once-a-day aminoglycoside over the last several years and have seen virtually no renal toxicity with this approach.[34] If renal toxicity is a concern, quinolones can also be added. Amphotericin is added after 3–5 days of persistent fever on a broad-spectrum antibiotic regimen, especially when the patient is colonized with yeast or when *Aspergillus* infections are suspected.[70] Because many patients are receiving fluconazole prophylaxis, the primary concerns with breakthrough fungal infections are those due to molds (*Aspergillus* spp.), so many centers now use higher doses of amphotericin B (i.e., 0.8–1.0 mg/kg) as empirical treatment. Lipid formulations of amphotericin have been shown to be less toxic and associated with fewer breakthrough infections in high-risk patients.[65] Also, extended-spectrum trizoles may be an acceptable alternative.[65a]

5.5.2. Duration of Antibiotic Administration

Various rationales for the use of antibiotics in granulocytopenic patients who do not respond to standard antibiotics after 48–96 hr have been outlined.[70,71] Because the high risk of infection in severely granulocytopenic patients is well known,[75] and studies of the discontinuance of antibiotics in the presence of continued fever and granulocytopenia suggest that bacteremia remains a significant risk,[76] treatment is continued until the patient recovers 500 circulating neutrophils/mm^3 whether or not fever resolves during treatment. One possible exception to this practice is the patient who is clinically stable and afebrile for at least 1 week with neutrophil counts in excess of 200/mm^3. In such patients, discontinuation of empirical antibiotic coverage can be considered. In patients with demonstrated bacterial infection, duration of therapy is decided according to standard practices after recovery of granulocytes. If fever persists despite a broad-spectrum antibacterial and antifungal combination regimen and all cultures and imaging studies remain negative, noninfectious causes of fever such as hyperacute GVHD should be considered, especially in recipients of HLA-mismatched or unrelated donor stem cells. In such case, the use of systemic corticosteroids may be indicated (with continuation of all antimicrobial agents).

Limited information is available on the optimal duration of treatment for documented invasive bacterial infections after HSCT. Bacteremia without a focus is generally treated for at least 2 weeks with single intravenous antibiotics or with combination therapy in cases where synergy can be documented (e.g., *P. aeruginosa*, *K. pneumoniae*). Bacterial pneumonia or other infections with a documented focus may require extended treatment (4–6 weeks), especially if underlying immunosuppression (i.e., high-dose systemic corticosteroids, or severe granulocytopenia) persists.

5.5.3. Prophylactic Antifungal Use

Strategies for the prevention of fungal infections include attempts to augment the host immune response and the use of prophylactic antifungals. The most common fungal pathogens in HSCT recipients include *Candida* species and *Aspergillus* species, but the differing antimicrobial susceptibilities and routes of acquisition of these pathogens have hampered the development of an effective, all-encompassing anti-"fungal" regimen. The use of fluconazole has had a dramatic impact on the incidence and spectrum of candidal infections, but little to no impact on resistant, respiratory-acquired mold infections. Efforts to prevent infections caused by both *Candida* species and *Aspergillus* species will be discussed separately below.

Nonabsorbable antifungals such as oral nystatin, amphotericin B, and clotrimazole troches have been shown in randomized trials to decrease oropharyngeal candidiasis and gastrointestinal colonization, but have no impact on the incidence of hematogenous candidiasis.[77] Well-tolerated oral azole antifungals, especially fluconazole, have virtually replaced all other regimens for the effective prevention of candidiasis. Two randomized, placebo-controlled trials documented that fluconazole administered daily in HSCT recipients (400 mg/day) decreases candidal colonization, disseminated infection, and associated mortality.[78,79] The optimal duration of fluconazole use has been debated. In allogeneic transplant recipients, prolonged duration of antifungal prophylaxis results in overall improved survival (Fig. 7).[80] Lower doses of fluconazole (50 to 200 mg/day) have been shown to decrease candidal colonization with low associated numbers of disseminated infection, but no trials comparing doses have been performed.[77,81] Because of the overall survival benefit, we favor the use of 400 mg/day for an extended period of time (75 days) in allograft recipients and a short duration administered to autograft recipients (30 days) (Table 2).

Prevention of *Aspergillus* infections has been more problematic, in part due to respiratory acquisition of this ubiquitous pathogen and limited antifungal susceptibilities. Attempts at prevention have utilized both amphotericin B and mold-active azoles. Unfortunately, few studies have been randomized and have included patient populations with a high incidence of aspergillosis. Administration of low doses of amphotericin (0.5 mg/kg per day) was associated with a low incidence of aspergillosis versus historic controls, but a randomized study in autograft recipients failed to document a protective benefit.[77,82] In a large, randomized study, aerosolized amphotericin was associated with a low incidence of aspergillosis, but the difference was not statistically significant when compared to placebo.[83] Because of the toxicities associated with the use of amphotericin B, attention has turned to the potential of mold-active azoles for prophylaxis. Studies utilizing the liquid formulation of itraconazole suggest benefit in preventing candidiasis and perhaps aspergillosis.[84,85] To date, the prevention of aspergillosis in HSCT recipients hinges on effective avoidance of exposure, utilizing HEPA filters and laminar airflow containment.[71] As more patients are undergoing HSCT as outpatients, such methods continue to be less efficacious. Prospective, randomized studies utilizing new mold-active compounds are necessary.

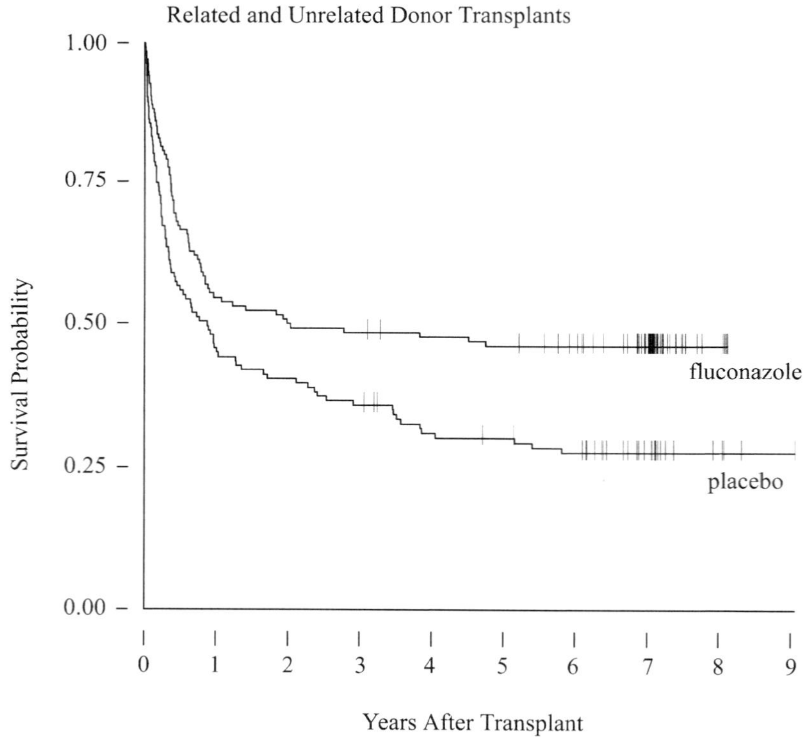

FIGURE 7. Long-term follow-up results of fluconazole prophylaxis. (Obtained with permission from Marr *et al.*[80])

Illustrative Case 2

A 27-year-old male underwent a one-antigen mismatched unrelated donor marrow transplant for T cell non-Hodgkin's lymphoma that had relapsed despite four courses of CHOP chemotherapy. The conditioning regimen included CY and TBI. On day 6 after BMT he developed a fever despite prophylactic fluconazole (400 mg/day). At this time, he was granulocytopenic and ceftazidime and vancomycin were begun empirically. After 3 days of persistent fever, fluconazole was discontinued and amphotericin (0.5 mg/kg per day) was initiated. All blood cultures remained negative and his fever resolved on day 10 with initiation of high doses of corticosteroids for presumed acute GVHD. On day 12, his creatinine increased to 1.7 mg/ml and amphotericin B was discontinued despite an ANC of less than 500/mm³. Blood cultures obtained on day 17 grew *C. albicans*. At this time, his only complaint was persistent diarrhea, thought to be secondary to conditioning regimen-related mucositis. Amphotericin B was restarted at 0.5 mg/kg per day and his double-lumen tunneled catheter was removed. For the following 4 days, blood cultures continued to grow *C. albicans*. On day 27 after BMT the patient sustained a massive gastrointestinal bleed and could not be resuscitated. Autopsy revealed scattered areas of colonic necrosis and colitis with yeast forms and hyphal invasion. MICs to fluconazole for all *C. albicans* obtained from blood cultures were highly resistant (>64 μg/ml).

Comment. This patient developed a disseminated fatal disseminated *C. albicans* infection with a portal of entry through the gastrointestinal mucosa despite prophylaxis with high doses of fluconazole. Although this is a rare event in patients treated as such, fluconazole resistance has been reported both in *C. albicans* and more commonly in non-*C. albicans* species such as *C. krusei* and *C. glabrata*. All candidal infections that break through fluconazole prophylaxis should be treated aggressively with high doses of amphotericin B (1.0 mg/kg/day) or a lipid formulation.

5.5.4. Granulocyte Transfusions and Hematopoietic Growth Factors for Prophylaxis and Therapy

Earlier studies suggest that the use of prophylactic granulocyte transfusions may decrease bacterial and fungal infection after HSCT but survival was not significantly different.[86,87] However, problems with the use of prophylactic granulocytes include lack of availability, CMV matching issues,[88] clotting of the shunt or other symptoms in the granulocyte donor, side effects, and cost of procedure. The combination of these difficulties has led to a virtual discontinuance of the use of prophylactic granulocyte transfusions at most centers, although therapeutic granulocyte transfusions are currently being examined. Recently, there has been renewed interest in therapeutic use of granulocytes. With the availability of hematopoietic growth factors, such as G-CSF, it has become possible to increase the number of granulocytes harvested from a healthy donor tenfold or more over standard pheresis techniques.[89] This approach is currently being studied for therapy of documented infection during the granulocytopenic period.

Hematopoietic growth factors may provide more effective protection during the granulocytopenic period. Several trials utilizing G-CSF or GM-CSF have documented shortened periods of neutropenia and decreased antibiotic utilization, but few have noted a decreased incidence of infectious complications or infection-related mortality (summarized in Ref. 90). There is a potential for the use of growth factors in combination with antimicrobial therapy since increasing neutrophil counts are associated with better outcomes in the setting of both bacterial and fungal infections. Unfortunately, no controlled trials have been performed in patients undergoing HSCT.

6. Phase II: Infections between Engraftment and Day 100

The most important infections occurring during the interval between successful engraftment and day 100 are viral (e.g., CMV, respiratory viruses) and fungal (e.g., invasive mold infections) usually presenting as pneumonia (Fig. 1). The risk of bacterial infections is generally reduced after engraftment with the exceptions of patients who remain granulocytopenic because of graft failure or rejection or those who receive high-dose corticosteroids.

6.1. Interstitial Pneumonia

6.1.1. Etiology and Risk Factors

Interstitial pneumonia remains an important clinical syndrome during the postengraftment period. Etiologies include viral infections (CMV, respiratory viruses), *Pneumocystis carinii* (PCP), and noninfectious causes (idiopathic pneumonia syndrome). With the use of effective prophylaxis, PCP is all but eliminated (Fig. 8). Also, the use of ganciclovir or foscarnet has reduced the incidence of CMV pneumonia during the first 3 months after transplant (see Section 6.2). As prevention strategies are not 100% effective, CMV pneumonia continues to occur, although at much lower incidence (Section 6.2). Respiratory viruses have been appreciated recently as an important cause of interstitial pneumonia.

The risk for CMV pneumonia is increased by any factor that increases the risk of active CMV infection. Thus, patients who are seropositive before transplant have a higher incidence than do seronegative patients. Causality of CMV pneumonia is clearly multifactorial, however. For example, patients who receive TBI for conditioning (i.e., those with leukemia) have a significantly higher incidence of CMV pneumonia than do patients

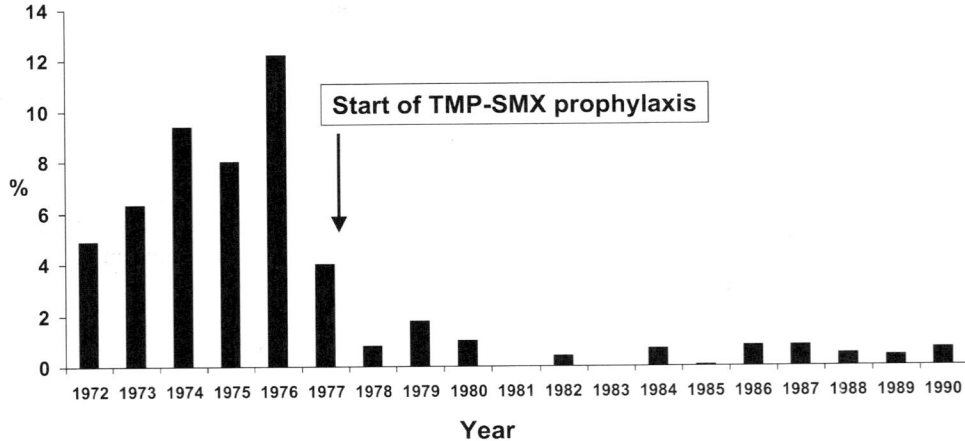

FIGURE 8. Incidence of PCP before and after introduction of TMP-SMX prophylaxis at FHCRC.[34]

who do not receive TBI (e.g., those with aplastic anemia), and if TBI is given to patients with aplastic anemia for conditioning, the incidence of CMV pneumonia increases dramatically.[91] By contrast, syngeneic recipients (twins) transplanted for leukemia rarely develop CMV pneumonia even though they also receive TBI.[92] The other major risk factors for CMV pneumonia are CMV viremia and the occurrence of acute GVHD.[93]

Interstitial pneumonia caused by respiratory viruses (e.g., RSV, parainfluenza viruses) occurs during the respiratory virus season, i.e., when these viruses circulate in the community. Acquisition of respiratory virus infections seems to be mainly related to close contact with an infected individual (i.e., healthcare personnel, family members). Upper respiratory tract infection (URI) is present in the majority of patients with interstitial pneumonia due to respiratory viruses. The most important factor for progression from upper to lower respiratory tract infection is the use of corticosteroids.[94,95]

The occurrence of idiopathic interstitial pneumonia is also increased by the use of TBI and is therefore significantly lower among patients transplanted for aplastic anemia. In this case, however, twin recipients transplanted for leukemia develop idiopathic pneumonia at a rate equivalent to allogeneic transplant recipients, a factor suggesting that idiopathic pneumonia is indeed not an infectious process.[96] The use of fractionated rather than single-exposure TBI has been associated with a significant decrease in the risk for idiopathic interstitial pneumonia.[97] Interstitial pneumonia results in a high case fatality rate (>50%), even for viral pneumonias for which treatment exists.

6.1.2. Clinical Presentation and Diagnosis

Interstitial pneumonia is a syndrome. The designation *interstitial* derives from the histologic picture of interstitial inflammation with a mononuclear cell infiltrate.[98] The clinical syndrome is not distinctive: although some manifestations may precede others by up to 1 week, patients eventually develop fever, nonproductive or poorly productive cough, tachypnea, dyspnea, and occasionally chest pain. Hypoxia is present. Radiological abnormalities may precede or follow symptoms such as cough or fever, and although infiltrates usually become diffuse, both segmental and nodular infiltrates (Fig. 3) have been observed. The risk period for interstitial pneumonia is highest between 30 and 90 days after HSCT. Most, but not all, patients have adequate numbers of circulating granulocytes at onset.

The differential diagnosis has changed over the last decade. While CMV-associated pneumonia was predominant until the early 1990s, it is now less common. Instead, respiratory viruses, idiopathic pneumonia syndrome, and rarely adenovirus pneumonia are prominent causes.

6.2. CMV Infection and Disease

6.2.1. Clinical Manifestations, Epidemiology, and Risk Factors

The most common manifestations of CMV after HSCT are pneumonia and gastrointestinal disease.[99] Hepatitis, retinitis, marrow suppression, and encephalitis are rare manifestations.[100] CMV infection can also cause fever, especially in seronegative patients, but the associa-

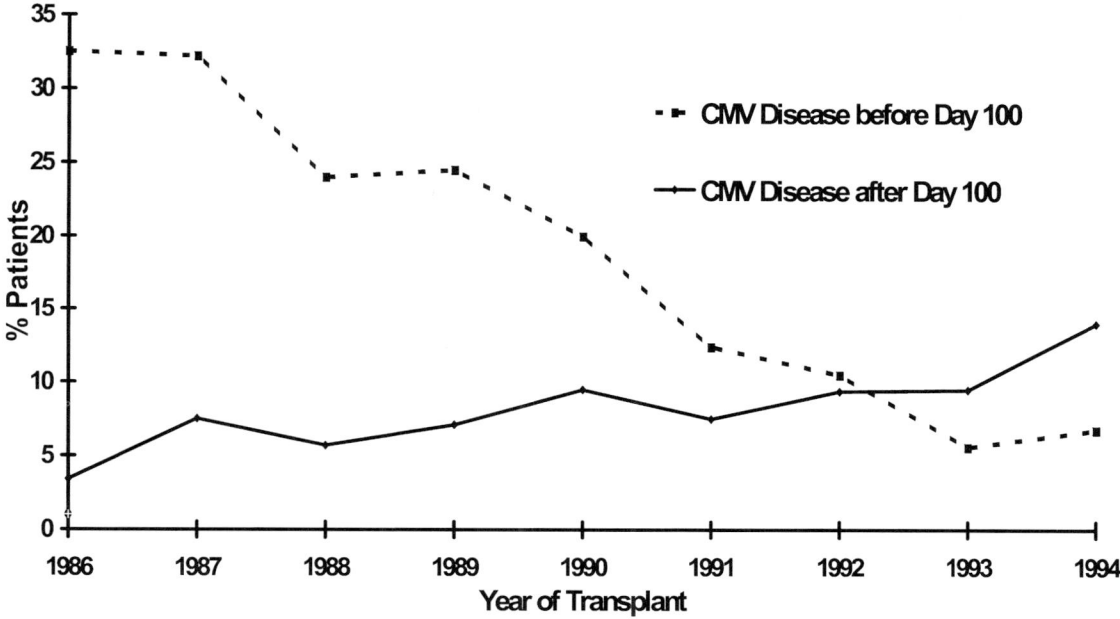

FIGURE 9. Changing pattern of CMV disease in the early 1990s at FHCRC (*N* = 1458 seropositive allograft recipients).[34] (Obtained with permission from Armstrong and Cohen.[302])

tion is often difficult to ascertain after HSCT because of the multiple causes of fever in this setting. CMV seropositive recipients are at highest risk for disease.[101] With ganciclovir prophylaxis or preemptive therapy, the epidemiology of CMV disease has changed (Fig. 9). Recent clinical trials have shown that the majority of CMV disease now occurs after discontinuation of ganciclovir, resulting in a much longer period of high risk.[35,102–107] The median day of onset is now at day 170 after transplant, compared to day 55 without ganciclovir.[108]

There is no difference in the incidence of CMV infection between marrow and peripheral blood stem cell transplantation.[28,109] However, CD34+ selected PBSC transplantation is associated with a higher incidence of CMV infection, higher levels of quantitative pp65 antigenemia, and possibly also disease. This is likely due to the T cell depletion, which is part of the CD34+ selection process.[34] There are no reports on the incidence of CMV infection after UCB transplantation. CMV seropositivity of the recipient is a negative predictor for survival and a risk factor for acute GVHD after UCB transplantation.[110,111] After nonmyeloablative transplantation, CMV infection and disease appears to be delayed but the overall frequency was not different in one study.[112] Differences in the conditioning regimen (e.g., use of T cell depletion) determine the risk and timing of CMV infection. CMV

disease can be fatal after nonmyeloablative transplantation, although the overall fatality rate has not been determined.

The overall impact of ganciclovir prevention strategies (ganciclovir, seronegative or leukocyte-depleted blood products) on survival after transplant has been evaluated in two studies. A study in T-cell-depleted allogeneic transplant recipients indicates that survival remains significantly worse in recipients who are CMV seropositive or have a seropositive donor when compared to seronegative recipients with a seronegative donor.[113] The effect appears to be mediated by an increase of GVHD.[113] In non-T-cell-depleted allogeneic transplant recipients, both recipients and donor positive CMV serostatus remain associated with poor posttransplant outcome.[34] Thus, seronegative recipients should continue to receive stem cells from a seronegative donor whenever possible.

The incidence of CMV pneumonia after unmodified autologous HSCT is lower than that after allogeneic transplant (1–6%),[114–118] but the outcome is similar.[115–118] Available data suggest that the risk of CMV disease after autologous transplant depends on the intensity of treatment prior to transplant in addition to the transplant conditions. Using the antigenemia assay or PCR for CMV DNA, CMV infection rates of 39 and 42%, respectively, have been reported.[119,120] There does not seem to be a difference between marrow and unmodified PBSC with regard to the incidence of CMV infection.[119] However, a significant increase of CMV disease has been observed in CD34+-selected CMV-seropositive autograft recipients: the incidence of CMV disease was 22% and the disease was fatal in 4 of 7 patients in one study.[28]

6.2.2. Diagnosis

Advances in technology have greatly improved the diagnosis of CMV. Currently, CMV pp65 antigenemia or DNA-based detection methods are used for CMV surveillance after HSCT. RNA detection methods seem to be slightly less sensitive than DNA detection methods but similar to pp65 antigenemia assays; however, treatment studies using this assay have not been reported to date. Advantages of antigen-based and molecular assays compared to culture-based assays (i.e., shell vial centrifugation cultures) include a higher sensitivity which leads to earlier detection after transplant. In addition, most of the newer assays allow easy quantification of results.[105,121] Details of quantitative assays for detection of CMV have been reviewed elsewhere.[122] Although earlier studies have documented that CMV DNA can be more readily detected in peripheral blood leukocytes than in plasma,[123,124] our recent experience with a real-time quantitative PCR assay indicates that plasma PCR can also be significantly more sensitive than the antigenemia assay.[34] Because early detection may be critical for early institution of antiviral therapy, especially in CMV-seropositive allograft recipients receiving high-dose steroids, only assays that are at least as sensitive as the pp65 antigenemia assay should be used.[122] During periods of neutropenia, plasma PCR is the method of choice.

Diagnosis of CMV pneumonia requires documentation of CMV in the BAL or biopsy (including autopsy specimens) by culture, immunohistology, or direct fluorescent antibody methods (in BAL specimens) in conjunction with radiographic signs of pneumonia.[125] Symptoms of pneumonia are initially uncharacteristic with dry cough, hypoxemia, and low-degree fever, but it can progress rapidly to respiratory failure. The radiographic picture often shows an interstitial pattern but other patterns including alveolar changes or nodular changes can occasionally be seen (Fig. 3). Severe hypoxemia is a poor prognostic sign and most patients who require ventilator assistance will succumb to the infection.[118] BAL has largely replaced open lung biopsy for the diagnosis of CMV pneumonia,[126] but there is the risk of overdiagnosis, as CMV can be isolated from BAL in patients without any signs or symptoms of respiratory disease.[127,128] CMV detection in BAL from asymptomatic patients at day 35

after BMT is highly predictive for the later development of CMV pneumonia.[128] However, approximately one third of asymptomatic patients who had CMV detected in BAL fluid do not subsequently develop CMV pneumonia. Thus, the finding of CMV in a BAL from asymptomatic patients must be interpreted with caution. The significance of detection of CMV DNA in BAL by PCR is not well studied. Although the negative predictive value is high,[129] positive results are difficult to interpret. Thus, current consensus statements on the definition of CMV disease do not use PCR results.[130] Quantitative PCR assays may provide useful information in the future.[131] Gastrointestinal disease is diagnosed by culture or immunohistology from biopsy specimens. Shell vial centrifugation cultures are particularly useful for rapid diagnosis of CMV in BAL, biopsy, and autopsy specimens, but usually two diagnostic methods are recommended for each specimen for optimal sensitivity.[132] PCR for CMV DNA is presently not used to diagnose CMV gastrointestinal disease.[130] CMV hepatitis occurs only rarely in HSCT recipients. To differentiate from other causes of liver function abnormalities, evaluation of liver tissue is required.[133] Since symptoms of encephalitis are frequently nonspecific, the diagnosis must rely on detection of CMV DNA in the cerebrospinal fluid.[134] The diagnosis of retinitis is based on typical retinal lesions. CMV syndromes with fever and marrow suppression are difficult to differentiate from several other etiologies such as other viral infections, drug toxicity, and marrow rejection. It is therefore currently not accepted as a separate disease entity in the definitions for clinical trials adopted by the international CMV conferences.[133] However, *in vitro* studies support that CMV has an inhibitory effect on hematopoiesis,[135] and there are well-documented cases of late marrow failures after BMT which can also occur with low or undetectable systemic viral load.[136] Fever occurs especially with primary CMV infection.

6.2.3. Prevention

Two strategies are currently used: prophylaxis is based on the pretransplant serostatus and includes administration of antiviral drugs to all patients at risk (Fig. 10). Preemptive therapy restricts antiviral agents to patients who have evidence of CMV infection by antigenemia or DNA detection methods. There are advantages and disadvantages with either strategy, and whether to choose one or the other depends on the availability of a virologic laboratory (Table 3).

Prophylaxis. Two earlier prospective studies have shown a survival benefit of high-dose IV acyclovir in allo-

TABLE 3. CMV Prevention Strategies: Preemptive Therapy versus Prophylaxis

Strategy	Advantages	Disadvantages	References
Prophylaxis			
Intravenous immunoglobulin or CMV-Ig	Low risk for side effects	Efficacy not established High cost Overtreatment	158–165
High-dose acyclovir/ valaciclovir	Low risk for side effects Reduction of CMV infection Survival benefit in patients not receiving ganciclovir prophylaxis or preemptive therapy Activity against other herpesviruses	Low effectiveness for prevention of CMV disease (virologic monitoring still required) Overtreatment	137, 138
Ganciclovir at engraftment	Highly effective Activity against other herpesviruses	High risk of neutropenia High risk of invasive fungal infections High risk of late CMV disease Delay of recovery of CMV-specific T cell function Overtreatment	21, 108, 140–142
Preemptive therapy			
Based on PCR for CMV-DNA	Effective Targeted treatment	Requirement for close virologic monitoring Occasionally missed cases of CMV disease	104, 150
Based on pp65 antigenemia	Effective Targeted treatment Less invasive fungal infections (compared with ganciclovir at engraftment)	Requirement for close virologic monitoring Occasionally missed cases of CMV disease	35

geneic marrow transplant recipients.[137,138] When high-dose IV acyclovir is given from day −5 until engraftment, there does not seem to be a survival benefit in the era of ganciclovir prophylaxis or preemptive therapy.[103] However, a study that compared high-dose IV acyclovir followed by oral acyclovir or valaciclovir resulted in a significant reduction of CMV viremia and thus a reduced need for preemptive therapy with use of valaciclovir;

CMV disease and survival were not different in this study.[139] Three randomized double-blind studies have been published using an early prophylaxis strategy.[35,140,141] All three studies showed a significant reduction of infection and/or disease, however, severe neutropenia was a limiting factor in all studies. There was no benefit in overall survival in any of the studies. An analysis of risk factors for neutropenia in 278 patients who received gan-

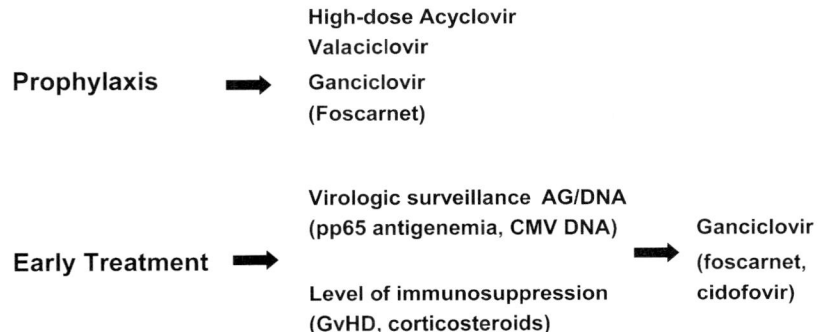

FIGURE 10. CMV prevention strategies: overview of options using antiviral chemotherapy.

ciclovir at engraftment found that early liver failure, renal insufficiency after engraftment, and a low marrow cellularity at day 28 are significantly associated with the development of neutropenia.[142] Ganciclovir prophylaxis using a lower maintenance dose (i.e., 5 mg/kg, 3–5 times a week) starting at engraftment with or without a pretransplant induction course of ganciclovir have also been evaluated in nonrandomized trials.[143–147] Three of these studies, which were performed in unrelated marrow transplant recipients and recipients of T-cell-depleted marrow, respectively, showed unacceptably high rates of CMV disease.[144,145]

Although ganciclovir prophylaxis appears to be highly effective in preventing CMV disease, there are significant disadvantages associated with this strategy. Ganciclovir given at engraftment causes prolonged neutropenia leading to more invasive bacterial and fungal infections.[35,140,142] In addition, a substantial number of patients not at risk for disease (i.e., 60–65%) will unnecessarily receive a potentially marrow-toxic drug. Both factors may contribute substantially to morbidity and financial cost. Furthermore, there is an interference of ganciclovir with the recovery of CMV-specific immune responses.[21] Finally, there appears to be an increased risk of late-onset CMV disease (i.e., after day 100) when ganciclovir is given at engraftment.[35] No controlled trial of foscarnet used as prophylaxis has been reported in HSCT recipients. Two small uncontrolled studies suggest that nephrotoxicity is the major side effect.[148,149] Because foscarnet does not cause marrow toxicity, it is often used as a second-line drug in patients with neutropenia who require antiviral-prophylaxis or treatment.

Preemptive Therapy. Current preemptive therapy strategies are based on virologic monitoring by the pp65 antigenemia assay or DNA detection. Studies using RNA detection methods have not been published. Culture methods are not recommended to start preemptive therapy due to the low sensitivity.[122] CMV disease rates with preemptive therapy strategies range from 3 to 6%.[102,150] Although there is no comparative trial between antigenemia- and PCR-guided strategies, these studies suggest that results are similar. Another risk-adapted strategy that combines both immunologic and virologic factors is to give a short course of ganciclovir to patients who receive high-dose steroids for treatment of acute GVHD or for CMV antigenemia.[151] The strategy was effective in an uncontrolled trial that included HLA matched-related allogeneic transplant recipients.[151] Whether such an approach is useful in unrelated and HLA mismatched patients and whether it is superior to early treatment strategies based on virologic marker only have not been studied in a randomized fashion.

New insight into the dynamics of CMV replication *in vivo* had a major impact on management strategies including dosing and duration of treatment. A recent study shows that the *in vivo* doubling time of CMV in HSCT recipients is only 1 day, contrary to the replication time in fibroblast cell lines where CMV replicates very slowly.[152] Both the initial viral load and the slope of the increase of CMV load are risk factors for CMV disease.[153] Up to 40% of HSCT recipients have increases in CMV load on ganciclovir.[154] Corticosteroid dose is the most important risk factor for this phenomenon during the initial month of antiviral therapy while antiviral resistance was not important.[154] Given the short doubling time of CMV *in vivo* and its apparent correlation with the underlying immunosuppression, twice-a-day induction doses of anti-CMV treatment should be continued beyond 1 week until the CMV load starts declining. Conversely, in patients who show an immediate decline in viral load, 1 week of induction dosing seems to be sufficient.[154] If viral load increases occur during the maintenance-dosing phase within the first 1 or 2 months of therapy, reinduction should be performed. If increases in viral load occur after prolonged exposure to antiviral therapy (>4 weeks), antiviral resistance should also be considered.

Both short- and long-term courses of ganciclovir have been used for preemptive therapy. Short courses are usually given until the virologic marker is negative. Repeated courses are required in up to 30–40% of patients.[35,104] Occasional cases of CMV disease shortly after discontinuation of short-term ganciclovir based on antigenemia have been reported in patients on high-dose steroids,[35,155] although such early rebound has not been seen in PCR-based strategies.[150] Advantages of short-term treatment include lower cost, a lower risk for side effects, and an improved CMV-specific immune reconstitution, which may be associated with a lower risk of late CMV disease.[35] A recent randomized trial demonstrated that foscarnet is as effective as ganciclovir for preemptive treatment of pp65 antigenemia and PCR positivity with less marrow toxicity, with a trend towards increased nephrotoxicity.[156] Cidofovir shows acceptable toxicity,[157] but comparative efficacy data are lacking.

Prophylaxis with Intravenous Immunoglobulin (IVIG), Hyperimmune Globulin (CMV-Ig), or CMV-Specific Monoclonal Antibodies. The use of IVIG or CMV-Ig for the prophylaxis of CMV infection and disease after allogeneic transplants remains controversial. Although the prophylactic use of IVIG is associated with virtually no toxicity, the regimens proposed are costly and most controlled studies do not show a reduction of CMV disease.[158–165] Some studies showed a reduction of bacteremia, interstitial pneumonia, and/or acute GVHD, while other studies did not report such a difference. An improvement of survival has not been reported in any of the studies. Only one small randomized trial of CMV-Ig has been reported in patients who received antigenemia-based preemptive treatment.[166] In this study, there was no reduction of CMV antigenemia compared to pooled IVIG and the incidence of acute GVHD was similar.[166] A randomized study of MSL-109, a human glycoprotein H-specific monoclonal antibody, did not show a reduction of CMV infection or disease in CMV seropositive allogeneic recipients or seronegative recipients with a seropositive donor.[167] There was a transiently reduced mortality in D+/R− patients that was not due to a reduction of CMV disease but the numbers were small and additional studies are needed to confirm this finding. Whether there are indirect effects of CMV-Ig on other opportunistic infections similar to those observed after liver transplantation has not been studied in allogeneic marrow transplant recipients who receive antigenemia- or PCR-guided ganciclovir or ganciclovir prophylaxis.

Prevention of Primary CMV Infection. The risk for CMV transmission in CMV seronegative patients with seronegative stem cell donors is mainly through blood products.[162] Two options exist for reducing this risk of CMV transmission: the use of blood products from CMV seronegative donors or the use of leukocyte-reduced blood products. A recent randomized trial suggests that these two strategies result in similar rates of CMV infection.[168] However, there remains a small risk of CMV disease (up to 2.3% with leukocyte-reduced blood products).[168] Therefore, some centers perform virologic monitoring for antigenemia or CMV DNA with preemptive therapy. There is also a risk of late acquisition of CMV in patients with chronic GVHD. Thus, seronegative or leukocyte-depleted blood should be used beyond day 100 in these patients. The risk of transmission of CMV via the marrow or stem cell product from a seropositive donor to the recipient is approximately 15–20%.[169] Thus, these patients should be considered at risk for CMV disease and antigenemia- or PCR-guided antiviral therapy should be given. IVIG prophylaxis is not effective in preventing CMV disease in this setting,[163,165] although CMV viremia is reduced. Patients who are CMV seronegative before transplant should receive a transplant from a CMV negative donor whenever possible.

Prevention in Autologous Transplantation. High-dose acyclovir given from day −5 until day 30 does not appear to be effective in prevention of CMV disease in seropositive autograft recipients.[117] The low disease incidence would favor a preemptive therapy strategy and CMV quantitation may be useful in identifying patients at risk for disease who might benefit from preemptive treatment. A retrospective analysis suggests that a strategy of ganciclovir based on an antigenemia level of five positive cells per slide reduced the incidence of CMV disease by approximately 60%.[28,119] For CD34-selected seropositive autograft recipients a similar strategy as for allograft recipients has been advocated, e.g., treatment of any level of antigenemia.[28]

6.2.4. Treatment

Treatment of CMV pneumonia with ganciclovir and CMV-Ig or IVIG remains unsatisfactory since fatality rates are still more than 50%.[100,170] Analysis of various antiviral dosing regimens suggests that induction dosing should be continued for 2–3 weeks.[34] Outcome is particularly grim in the presence of copathogens and in mechanically ventilated patients.[170] Whether antiviral combination therapy consisting of ganciclovir and foscarnet or cidofovir results in better outcomes has not been studied. CMV gastrointestinal disease and retinitis usually respond to ganciclovir alone.[171] For CMV gastrointestinal disease prolonged courses of induction doses are required, especially when ulcers are deep on endoscopic examination. Treatment of CMV-related marrow failure consists of foscarnet and G-CSF.[172] Alternatively, ganciclovir, CMV-Ig, or G-CSF can be used.[173]

Illustrative Case 3

A 49-year-old man on day 120 after matched-related donor, nonmyeloablative peripheral blood stem cell transplant for AML was admitted with confusion and somnolence. The patient's pretransplant serologies were positive for HSV and VZV. The CMV status was negative with a positive donor. The patient developed gut GVHD after transplant which was treated with corticosteroids. On admission, he indicated one episode of diarrhea, otherwise no nausea, vomiting, fever, or chills. Medication on admission included cyclosporine, ranitidine, omeprazole, fluconazole, and TMP-SMX. The physical exam was unremarkable. Laboratory values were within normal limits (neutrophil count, creatinine, and electrolytes). A head MRI was negative. CSF evaluation showed a normal protein, no RBC or WBC, bacterial and fungal cultures negative, a negative acid-fast stain, a negative cryptococcal antigen, and no detectable HHV-6 DNA. By day 2 postadmission, mental changes were nearly resolved without specific therapy, but the patient's gastrointestinal symptoms were more pronounced. A flexible sigmoidoscopy was performed. A colon biopsy showed cytomegalic inclusions.

Comment. This case illustrates several points. First, CMV gastrointestinal disease may occur without preceding CMV antigenemia and sometimes even without DNAemia (Table 6). The reasons for this are poorly understood but strain differences have been implicated.[174] Thus, gastrointestinal symptoms in patients at risk for CMV disease should be evaluated by biopsy to initiate appropriate treatment. The major differential diagnosis is GVHD, which requires treatment with corticosteroids. The negative predictive value of the CMV antigenemia assay for gastrointestinal disease is only 85%.[34,175] For gastrointestinal disease, extended ganciclovir treatment is often required. Induction dosing for 2–3 weeks and subsequent maintenance until day 100 is recommended.

TABLE 4. Correlation of CMV Gastrointestinal Disease with CMV pp65 Antigenemia in the Preemptive Therapy Era[34]

Gastrointestinal biopsy	Positive pp65 antigenemia	Number (%)	Maximum levels of pp65 antigenemia (range)
Positive	Never	17 (53)	—
Positive	At or after onset	4 (13)	2 (1–52)
Positive	Before	11 (34)	471 (1–3196)
Total		32 (100)	

The addition of IVIG does not appear to improve the outcome of CMV gastrointestinal disease.[171] Second, this case occurred after nonmyeloablative transplantation. The median day of onset is usually later but by 1 year, the incidence is similar to that observed in myeloablative transplant recipients.[112] Third, the patient was a D+/R− patient. D+/R− patients have only a 15% risk of CMV infection after transplantation when seronegative or leukocyte-reduced blood products are given.[32,169] Importantly, late CMV disease seems to occur predominantly in patients who had signs of CMV infection during the first 100 days after transplant. However, rarely, late CMV disease is seen without CMV infection during the first 100 days. In these cases, patients usually have chronic GVHD and are exposed to CMV via unscreened blood products or through other means of primary acquisition (e.g., toddlers who shed CMV or sexual transmission).

6.3. Infections with Community Respiratory Viruses

Community-acquired respiratory virus infections are an important cause of morbidity and mortality after HSCT. These viruses include RSV, parainfluenza viruses, influenza viruses, and rhinoviruses. The infection epidemiology in HSCT recipients usually parallels that observed in the community, as these viruses circulate in immunocompetent individuals (including healthcare personnel and family members). RSV, influenza viruses, and rhinovirus have a seasonal distribution, while parainfluenza virus infections often occur year-round. The biggest impact on morbidity and mortality after HSCT has been from RSV followed by parainfluenza viruses and influenza viruses.[94,95,176–178] Whether rhinovirus can cause lower tract disease remains controversial.[179,180]

Several methods exist for the diagnosis of respiratory viruses.[181] Since viral load in immunocompromised adults may be very low, appropriate specimen handling is important for recovery of the virus. Nasal wash specimens should be placed on ice or in the refrigerator immediately and transported to the laboratory without delay.[181] Specimen setup in the laboratory should occur within 2–4 hr. Methods available for testing include standard viral cultures (results available in several days), shell vial centrifugation cultures using RSV-specific monoclonal antibodies (results after 1–3 days), direct fluorescent antibody tests (2 hr), enzyme immunoassays (2 hr), and, more recently, RT PCR. A problem with the rapid test methods is a relatively low sensitivity.[181] This can be overcome by combining two rapid tests. On tissue sections from lung biopsy or autopsy specimens, virus-specific monoclonal antibody staining can be used.

RSV. RSV is an RNA virus (paramyxovirus) that causes a wide spectrum of respiratory diseases ranging from life-threatening bronchiolitis in infants and potentially fatal pneumonia in transplant recipients to a mild upper respiratory infection (URI) in immunocompetent adults and older children. The virus is increasingly implicated in a number of respiratory illnesses in immunocompetent or mildly immunosuppressed individuals, such as otitis media, exacerbation of chronic obstructive lung disease, and community-acquired pneumonia. After HSCT and also in severely immunosuppressed nontransplant patients with hematologic malignancies,[182] RSV causes URI, which may progress to fatal pneumonia.[183] During the respiratory virus season (Fig. 11), the incidence is approximately 10% and both allogeneic and autologous transplant recipients may be infected (Table 5). In a large study, no risk factor other than the winter season and male gender could be identified for the acquisition of RSV in HSCT recipients.[95] URI precedes pneumonia in 80% of patients and approximately 40–50% of patients with RSV URI progress to pneumonia after a median of 7 days; however, in 20% of patients with RSV pneumonia, URI is not present or very mild or occurs only concurrently with the onset of pneumonia. Risk factors for progression to pneumonia are older age and transplantation from an HLA mismatched or unrelated donor.[95] Without treatment, RSV pneumonia is almost uniformly fatal.[183] Pulmonary copathogens are detected in one-third of the patients with RSV pneumonia and require aggressive treatment. No controlled trials exist for the treatment of RSV infection and pneumonia in the HSCT setting. Available evidence comes from small uncontrolled cohort studies. The data suggest that treatment of early pneumonia (i.e., prior to mechanical ventilation) is associated with improved outcome. Intermittent short-duration aerosolized ribavirin (2 g over 2 hours three times per day) is considered the treatment of choice for RSV pneumonia. With this regimen, the 30-day all-cause mortality is approximately 45% at our center (Table 5). Systemic ribavirin alone does not seem to be effective, and can cause dermatologic toxicity.[184,185] Whether combined oral and

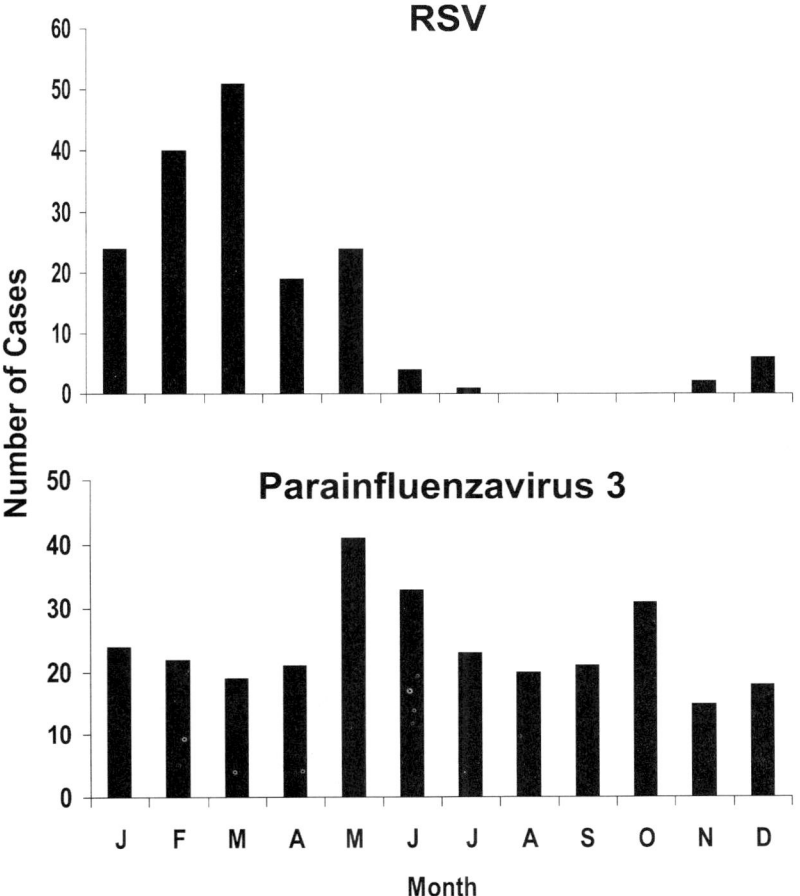

FIGURE 11. Monthly distribution of RSV and parainfluenza virus 3 infections after HSCT over a 9-year period at FHCRC.[34]

aerosolized ribavirin is more effective than aerosolized ribavirin alone has not been studied. The role of concomitant IVIG or RSV-specific immunoglobulin remains unclear. Uncontrolled data suggest that high-titer antibody preparations may be required if such adjunctive therapy is given,[186,187] although this issue has not been studied in a controlled fashion. Whether palivizumab,[188] an RSV-specific monoclonal antibody directed at the F protein of RSV, is effective as adjunctive therapy for RSV pneumonia has not been studied.

Due to the high mortality of RSV pneumonia, much interest has focused on prevention. Possible strategies are similar to those employed for prevention of CMV. While it is well established that RSV URI precedes pneumonia in the majority of cases,[95] limited information is available on the extent of asymptomatic shedding of respiratory

**TABLE 5. Respiratory Virus Infections after HSCT at FHCRC:
Comparison of RSV and Parainfluenza Virus 3**[34,94,95]

Virus	Incidence of infection (during season)	Progression from URI to pneumonia	Time from URI to pneumonia (median)	Proportion of pneumonia without URI	Overall mortality at 1 month after diagnosis of pneumonia
RSV	10%	40%	7 days	20%	45%
Parainfluenza virus 3	7%	18%	7 days	31%	35%

viruses in HSCT recipients.[189,190] Currently, controlled studies are under way to evaluate the efficacy and toxicity of preemptive antiviral therapy based on RSV shedding[190] or based on RSV URI.[95]

Prophylactic measures recommended throughout the respiratory virus season included isolation of infected patients, hand washing prior to, and after, every patient contact, educational efforts targeted at health care personnel and family members, avoiding patient contact of health care personnel and family members with uncontrolled secretions, as well as influenza vaccination of health care personnel and family members.[191,192] Whether pharmacologic prophylaxis (e.g., palivizumab, RSV-Ig) throughout the respiratory virus season is effective in preventing infection and disease in HSCT recipients has not been studied.

Parainfluenza Viruses. Of the four types of parainfluenza viruses, parainfluenza virus 3 is most common (approximately 90%). Parainfluenza virus infection usually does not follow a seasonal pattern (Fig. 11). Overall, the incidence was 7% in two studies.[94,193] Similar to RSV, URI is the predominant presentation, although progression to pneumonia seems to be less common (Table 5).[94] The most important risk factor for the progression from URI to pneumonia is the use of systemic corticosteroids.[94] Although the overall progression rate to pneumonia is only 18%, in allograft recipients receiving >1 mg/kg of prednisone, the risk is 40%, and is 65% with 2 mg/kg.[94] Parainfluenza 3 pneumonia may also occur after autologous transplantation, although mainly in the setting of CD34 selection or use of high-dose steroids.[94] Parainfluenza 3 pneumonia is often associated with serious pulmonary copathogens (53%) such as *Aspergillus fumigatus.* Factors associated with poor outcome after pneumonia include presence of copathogens and mechanical ventilation.[94] Thus, aggressive diagnostic intervention (i.e., BAL) and therapy are indicated in patients with suspected parainfluenza pneumonia. In a large retrospective analysis both URI and pneumonia due to parainfluenza virus 3 were associated with overall mortality in multivariable models.[94] Mortality of pneumonia ranges from 35 to 50%.[94,176] In a retrospective analysis, neither aerosolized ribavirin nor IVIG led to improved outcome of pneumonia or a reduction in viral shedding following pneumonia.[94] Randomized treatment studies have not been performed. Whether earlier treatment is effective in the prevention of pneumonia is unknown.[194]

Influenza Viruses. Influenza virus infections seem to be less common than RSV and parainfluenza virus infections. Both subtypes can cause infection, although type A appears to be more common.[177,178,195] Progression to severe pneumonia can occur similar to RSV and parainfluenza virus.[195] Interestingly, the clinical presentation in HSCT often lacks myalgia, which is commonly seen in immunocompetent individuals. Fortunately, effective prevention is available for influenza, possibly explaining the somewhat lower incidence. As such, healthcare personnel, family members, and visitors are advised to get vaccinated against influenza early in the season. Antiviral therapy is available for influenza virus infection, but these agents have not been studied in HSCT recipients. Amantadine and rimantadine (with a more favorable side effect profile) are effective against influenza A in prevention and treatment of immunocompetent adults. During treatment, rapid emergence of resistance has been described. More recently, neuraminidase inhibitors (i.e., zanamivir, oseltamivir) became available.[196,197] Both agents are active against influenza A and B and are potentially useful alone or in combination with rimantadine.

Illustrative Case 4

A 27-year-old woman presents 24 days after matched-unrelated marrow transplant for AML in remission with bilateral interstitial pneumonia. The patient had a runny nose for 5 days and a cough for 3 days, no shortness of breath, but was otherwise without symptoms. Pretransplant serologies were negative for CMV (donor positive), positive for VZV, HSV, and *Toxoplasma.* The conditioning regimen consisted of cyclophosphamide and TBI, and she received cyclosporin A and methotrexate for GVHD prophylaxis. Engraftment occurred on day 18 after transplant. The patient developed skin GVHD on day 20 and treatment with steroids was initiated (2 mg/kg prednisone). CMV surveillance by the pp65 antigenemia assay has been negative. The patient is currently receiving prednisone (2 mg/kg), cyclosporin A, and daily dapsone for PCP prophylaxis. A BAL was performed and revealed RSV by direct fluorescent antibody staining and shell vial centrifugation cultures. Treatment with aerosolized ribavirin (2 g three times daily) was initiated and continued for 10 days.

Comment. Pneumonia caused by community-acquired respiratory virus should be considered in an HSCT recipient, especially during the winter season. However, parainfluenza virus infection may occur year-round. The patient had upper respiratory symptoms for several days, which offered the first clue to the diagnosis. Since respiratory virus pneumonia is often associated with copathogens, a BAL is always recommended, even if the virus is recovered from the upper respiratory spaces. Institution of treatment before respiratory failure is important.

6.4. Mold Infections

Although mold infections were discussed in the section describing preengraftment infections, it is important to note that the majority of *Aspergillus* pneumonias in allograft recipients actually occur after engraftment, corresponding to the period of acute GVHD (Fig. 6). One survey of *Aspergillus* infections in allograft recipients actually noted a median date of diagnosis of 136 days posttransplantation.[50] Thus, the appearance of nodular pulmonary lesions in an allograft recipient postengraftment should be treated with antifungals and an aggressive approach to securing a diagnosis should be attempted.

Although infections with *Aspergillus* spp. and *Candida* spp. are most common, a variety of fungi cause disease in HSCT recipients. Many are recognized as "emerging" pathogens and will likely continue to attract attention in these immunosuppressed hosts.[198] Most infections are acquired by respiratory inhalation and are dependent on ongoing suppression (i.e., corticosteroids) to cause clinically recognized disease. It is noteworthy that endemic mycoses such as blastomycosis, histoplasmosis, and coccidioidomycosis occur infrequently in this setting.[199] The explanations for this remain unexplained but it is possible that donor-derived cellular immunity is the reason such infections do not occur at a higher frequency.

Other respiratory-acquired molds of importance include those caused by Zygomycetes. A series of HSCT patients revealed an incidence of 0.9% during the period of 1974 to 1989.[37] Of 9 patients with disseminated infection, 6 had invasion into deep tissues and 10 of 13 patients died due to the infection. In this series, most (6 patients) were diagnosed at autopsy. Our recent experience suggests that this infection is currently being diagnosed more frequently antemortem in patients undergoing HSCT.[200] From 1980 to 1999, 29 cases of invasive infection were diagnosed in our institution primarily in patients with severe GVHD. As non-*Aspergillus* mold infections appear to be increasing in frequency (Fig. 12), it has become important to seek microbial diagnoses to direct therapy.

6.5. Hepatosplenic Candidiasis

Hepatosplenic candidiasis is a distinct clinical syndrome of invasive candidiasis, characterized by fever and hypodense liver and/or splenic lesions, usually appearing after engraftment. Histology of biopsied liver shows primarily chronic inflammation, with or without yeast forms.[201] Cultures of blood and biopsy are frequently negative, and thus this diagnosis is primarily made radiographically, sometimes with histologic confirmation. A recent retrospective autopsy study of HSCT patients revealed a low incidence of hepatosplenic candidiasis (3%) in patients treated with prophylactic fluconazole compared to those who did not receive fluconazole (16%).[49]

This emphasizes the importance of *Candida albicans* as an etiologic agent in this syndrome.

Although rare in the azole prophylaxis era, hepatosplenic candidiasis carries a poor prognosis and aggressive treatment is critical. We favor a short treatment course with an amphotericin formulation (e.g., 2–4 weeks), with consideration of alternative oral agents for "maintenance therapy" while the patient has ongoing immunosuppression. Although this infection can be treated successfully with fluconazole in non-HSCT recipients, initial treatment should include amphotericin formulations.[64] Patients who develop infection while on prophylaxis should be treated with antifungals having activity against azole-resistant *Candida* species in molds (i.e., amphotericin B). In refractory cases, combination with 5-flucytosine or fluconazole may be considered. Initial CT or MRI findings clear in the majority of patients but may get worse during therapy and persist as granulomas after successful treatment.[202]

6.6. Idiopathic Pneumonia Syndrome

The most perplexing aspect of the syndrome of interstitial pneumonia has been the inability to identify an infectious agent in one third of patients with histologically demonstrated pneumonia.[98] Idiopathic pneumonia is otherwise indistinguishable clinically and radiologically from CMV or *Pneumocystis* pneumonia. Time of onset is similar to that of other pneumonias, and the mortality rate is about 60%. Etiologies investigated for idiopathic pneumonia have included chlamydial pneumonia, *Mycoplasma*, BK virus, and *Legionella pneumophilia*.[98,203] None of these studies has been revealing. Recently, human herpesvirus type 6 (HHV-6) has been implicated but controversy persists on the pathogenic role of HHV-6 in pneumonia after HSCT.[203,204] Although it remains possible that a heretofore unknown infectious agent is responsible for idiopathic pneumonia, the present data suggest instead that it is the result of radiation and chemotherapy toxicity. These data reveal a lower risk among aplastic anemia patients who do not receive TBI, receipt of fractionated versus single-exposure TBI, and much lower dose rates of radiation. There are no specific recommendations about treatment. Steroids continue to be used without adequate controlled data.[98] Empirical broadspectrum antimicrobial treatment is often given until final results of BAL and biopsies are available. Anti-CMV therapy may be started until results of BAL are available in patients who are at high risk for CMV disease (i.e., seropositive recipients) who did not have adequate surveillance tests performed. Similarly, anti-*Pneumocystis* treatment may be started empirically in patients who did not receive PCP prophylaxis until results of BAL or biopsy are available.

6.7. Herpes Simplex Virus Infection

There was a major shift in the epidemiology of HSV infection in the mid-1980s with the introduction of acyclovir. Without acyclovir prophylaxis, HSV infection usually begins during the granulocytopenic period when it is most difficult to differentiate oral mucosal breakdown (mucositis) from virus infection. Many progress to more typical herpetic lesions involving the lips and nose, although others have HSV recovered from oropharyngeal cultures without such lesions. Without antiviral prophylaxis, 80% of seropositive patients excrete HSV, mostly from the oropharynx, at some time during the first 50 days after transplant (peak excretion during weeks 2–3 after HSCT). By contrast, fewer than 1% of seronegative patients excrete HSV after transplant, and for most purposes (e.g., use of acyclovir prophylaxis), it can be assumed that seronegative patients are not at risk for HSV reactivation. Reactivation of HSV clearly increases the severity of oral mucositis. Untreated patients heal slowly, beginning with recovery of the granulocyte count, and lesions (and excretion) may recur later after HSCT. HSV was a common cause of esophagitis before the availability of acyclovir and has also been recovered from gastric and intestinal ulcers. Patients may develop distant cutaneous lesions, as well as herpes keratitis, through autoinoculation. Cutaneously disseminated HSV infection or encephalitis has been rare, however. One syndrome of note is HSV pneumonia which occurs with either preceding or coincident oral or genital HSV infection, or both.[205] Herpetic whitlow has occurred among patient care personnel, and glove isolation should be used for patients with known oral herpes.

Systemic acyclovir (intravenous or oral) has been highly effective in the treatment of established HSV infection after HSCT.[206] Recommended treatment courses in patients with normal renal function are 250 mg/m^2 every 8 hr IV, 800 mg acyclovir PO three times daily, or 500 mg valaciclovir PO three times daily. In each case, the course should be a minimum of 7 days. Acyclovir treatment has been associated with delay in the specific immune response to HSV after HSCT,[22] as well as with the recovery of HSV strains with reduced sensitivity to acyclovir.[207] Neither of these observations should prevent the use of acyclovir for the treatment of active HSV infection after HSCT when warranted by the clinical situation.

Because of the predictable timing and high frequency of HSV reactivation among seropositive patients, acyclovir is recommended. Intravenous acyclovir given two or three times daily (250 mg/m^2 per dose), oral acyclovir (800 mg twice daily), or valaciclovir (500 mg twice daily) have been shown to be effective, although compliance with oral drugs may be a problem early after transplant due to myositis.[71] Acyclovir prophylaxis is started 1 week before HSCT and continued for 4 weeks after transplant (unless the patient is discharged earlier and has no mucositis) to prevent HSV reactivation among seropositive patients. During the period of severe mucositis, (val)acyclovir is replaced with intravenous acyclovir (250 mg/m^2 twice daily). If prophylaxis is stopped 4 weeks after transplant, subsequent HSV reactivation will occur in approximately 30–40% of seropositive patients, but clinical manifestions are usually mild and treatable with acyclovir.[35] Some patients have repeated reactivations of HSV which often are associated with a development of acyclovir resistance.[207,208] The mechanism of acyclovir resistance is a deficient activity of the viral thymidine kinase (TK) and rarely an altered substrate affinity of the viral TK gene or an altered DNA polymerase. Foscarnet is the drug of choice for acyclovir-resistant HSV infection (40 mg/kg three times per day).[209] It should be noted that subsequent reactivations following acyclovir resistance are usually caused by wild-type HSV strains. Thus, acyclovir may be used in subsequent episodes of reactivation. Long-term suppression with acyclovir or valaciclovir is recommended if a patient has more than two episodes of HSV reactivation after transplant.

7. Phase III: After 100 Days

Infections occurring after 100 days are determined in part by the residual immune deficiency shared by all patients and in part by the additional immunosuppression associated with chronic GVHD and its treatment. The most prominent example of infection determined by the former is VZV infection. A minority of patients with GVHD have an increased incidence of recurrent bacterial infections. The epidemiology of some infections (e.g., CMV) has changed over the last decade due to prevention strategies now routinely administered during the first 100 days after transplant (Fig. 1).

7.1. Varicella–Zoster Virus Infection

Up to 40% of all HSCT patients develop VZV infection.[30,210–213] Median time of onset is 5 months after transplant, and most cases occur within the first year. In a study of 92 patients with VZV infection transplanted during the pre-acyclovir era, 77 had herpes zoster, and 15 had varicellalike infection.[30] In some patients, this was undoubtedly true primary infection, whereas others probably had atypical generalized zoster. One-third of patients with untreated herpes zoster had subsequent cutaneous dissemination. The case–fatality rate for untreated varicella is 35%, and is 30% for treated herpes zoster with dissemination. All deaths have occurred during the first 9 months after HSCT. Other syndromes of importance have included trigeminal zoster with keratitis, postherpetic neuralgia, and local scarring or bacterial superinfection. The incidence of VZV infection is increased among patients with allogeneic (versus autologous and syngeneic) transplants and among those with acute or chronic GVHD. However, if CD34 selection is used in autologous transplantation, the incidence and severity of VZV infection may be similar to those observed after allogeneic transplantation.[214] In the subgroup of patients with chronic GVHD, patients with demonstrable nonspecific suppressor cells have the highest incidence.[215] One study identified VZV seropositive allogeneic transplant recipients of age >10 years who received TBI as a subgroup at particular high risk for VZV infection (cumulative incidence at 3 years after transplant: 44%).[210]

Because of the high mortality rate from VZV infection after HSCT, treatment of all patients with VZV infection with acyclovir (500 mg/m^2 every 8 hr) is recommended. Treatment is continued for 7 days in uncomplicated cases but may be prolonged if patients have persistent new lesions or disseminated disease. In patients with localized zoster, treatment may be continued with valaciclovir (1000 mg three times per day) after 24 hours of IV treatment with acyclovir. VZV infection may recur after the initial episode. Oral acyclovir at a dose of 800 mg twice daily for 1 year prevents VZV infection after HSCT without rebound disease after discontinuation of prophylaxis, and may be particularly useful in patients with ongoing immunosuppression for GVHD after day 100.[216] Strategies that use lower doses and shorter doses of acyclovir were ineffective in preventing VZV infection late after transplant.[217]

Illustrative Case 5

A 32-year-old female patient who underwent an HLA-matched-unrelated marrow transplant for AML presents on day 189 with severe abdominal pain, localizing to the right upper quadrant. The patient has no fever and no diarrhea. Pretransplant serologies were positive for CMV (donor negative), HSV, VZV, and *Toxoplasma*. The conditioning regimen consisted of cyclophosphamide and TBI, and cyclosporin A

and methotrexate were administered for GVHD prophylaxis. The early posttransplant course was complicated by neutropenic fever which resolved after administration of broad-spectrum antibiotics, acute GVHD of the skin (for which systemic corticosteroids were given), and the development of chronic GVHD. On physical exam, the patient had diffuse abdominal pain, maximum in the right upper quadrant, and positive bowel sounds. The rest of the exam was unremarkable. Laboratory findings included an AST level of 540 U/liter (1250 U/liter 4 hr later), and normal values for total bilirubin, serum amylase, absolute neutrophil count, and serum creatinine. Bacterial and fungal blood cultures as well as CMV pp65 antigenemia were all negative. An abdominal radiograph and CT were both normal. Within the first day after admission the patient's status deteriorated rapidly and she died 48 hours after admission. In the hours before death, a few vesicular skin lesions were noted.

Comment. A manifestation of VZV that is particularly difficult to diagnose is the VZV hepatitis that presents without skin lesions. Patients typically are VZV seropositive and present with moderate to severe abdominal pain, often in the right upper quadrant and sometimes only controllable by morphine derivatives, and no apparent skin lesions. Imaging studies are usually negative but transaminases rise rapidly (ALT >500 U/liter). Skin lesions usually occur 48–72 hr after the start of pain, may be low in frequency, and are always disseminated. The recognition of this manifestation of VZV requires a high level of suspicion by the transplant physician or Infectious Diseases consultant because the outcome is invariably fatal without prompt institution of empirical treatment with high-dose intravenous acyclovir (500 mg/m^2 every 8 hr).

7.2. CMV Disease

Late CMV disease is now the major manifestation of CMV disease at many transplant centers.[35,105,218] The majority of late disease occurs during the first year after transplant (median day of onset: 170), but there may be cases up until 3 years after transplant if immunosuppression continues. The clinical manifestation of late CMV disease is similar to that observed early after transplant (Table 6) with the exception of retinitis, which only occurred late after transplant in our experience.[219] Outcome of late CMV disease is poor with pneumonia having the highest mortality. About one third of patients who survive the first episode of late disease will suffer a relapse after a median of 3 months.[220] At 3 months after transplant, late CMV disease can be predicted by the presence of early CMV infection (e.g., pp65 antigenemia), and presence of GVHD.[220] Continued monitoring (pp65 antigenemia, PCR for CMV DNA) of high-risk patients is useful in identifying patients at risk for late CMV disease.[220]

Illustrative Case 6

A 27-year-old woman presented 145 days after matched-unrelated marrow transplant for AML in remission with cough and mild shortness of breath. A chest radiograph shows bilateral interstitial pneumonia. Pretransplant serologies were positive for CMV (donor negative), VZV, HSV, and *Toxoplasma*. The conditioning regimen consisted of cyclo-

TABLE 6. Clinical Manifestations of Late CMV Disease in Seropositive Allograft Recipients at FHCRC (1986–1994)[34]

Site	Number	%
Pneumonia	122	63.8
Alone	110	
Disseminated	12	
Gastrointestinal disease	50	26.2
Retinitis	7	3.7
Alone	5	
Disseminated	2	
Graft failure	5	2.6
Hepatitis	4	2.1
Other	3	1.6
Total	191	100

phosphamide and fractionated TBI, and the patient received cyclosporin A and methotrexate for GVHD prophylaxis. Engraftment occurred on day 20 after transplant. Subsequently, the patient developed skin and gut GVHD (day 28) and treatment with steroids was initiated (2 mg/kg prednisone). She was treated for CMV pp65 antigenemia from day 51 to 99 after transplant. Chronic GVHD was diagnosed on day 86. Current medication consists of prednisone (2 mg/kg), cyclosporin A, and daily TMP-SMX. A BAL showed CMV by shell vial centrifugation cultures, and no other copathogens. The patient died despite immediate start of treatment with ganciclovir and CMV-Ig.

Comment. This is a typical case of late CMV pneumonia. Late CMV disease is an independent cause of late mortality in seropositive recipients. Prevention strategies for late CMV disease are needed. Since virologic markers are highly predictive for late CMV disease, patients who are at high risk for late CMV disease should have continued CMV surveillance by pp65 antigenemia or PCR and preemptive therapy. Since long-term monitoring may not be feasible in many cases, long-term prophylaxis with an oral agent (e.g., valganciclovir or valaciclovir) is currently being studied with regard to efficacy, feasibility, toxicity, and development of resistance.

7.3. Invasive Fungal Infections

Late invasive aspergillosis appears to be an increasingly frequent event in allograft recipients with chronic GVHD and preceding viral infections.[200] In contrast, candidiasis occurs infrequently after day 100. The long-term follow-up of a randomized study comparing fluconazole and placebo revealed that invasive candidiasis (blood or tissue infection) occurred in 8 of 96 patients who received placebo and 1 of 121 patients who received fluconazole for 75 days ($p = 0.0068$).[80] Importantly, clinically extensive chronic GVHD was present in 66% of patients who developed disseminated infection. Thus, although invasive fungal infections are less common "late" after HSCT, invasive aspergillosis and candidiasis do occur, especially in patients with GVHD. The outcome of both mold and candidal infections in this setting remains poor.

7.4. *Pneumocystis carinii* Pneumonia

With the availability of effective prophylaxis (Table 2) early cases are rarely seen and the majority of cases now occur late after transplant (Fig. 8).[40] Most cases occur in patients who do not take the prescribed prophylaxis, are unable to tolerate TMP-SMX due to side effects or allergy, or receive ineffective alternative prophylaxis regimens. The optimal duration of prophylaxis after autologous transplant is currently unknown but recent data from our institution suggest that there is late PCP in autograft recipients as well.[221] Thus, prolonged prophylaxis in some autograft recipients may be required. Approximately 15% of HSCT recipients require alternative prophylaxis regimens at some time after transplantation.[222] Reasons for the need of alternative prophylaxis include allergy to TMP-SMX, gastrointestinal intolerance, increased transaminases, and neutropenia. In our experience, the use of intravenous pentamidine (4 mg/kg monthly or twice monthly up to 3 g total dose) as an alternative form of prophylaxis in TMP-SMX-allergic patients has not prevented PCP.[222] Monthly aerosolized pentamidine has been advocated[71] but there are only small reports on its efficacy in HSCT recipients. Dapsone given twice weekly was not effective in preventing PCP after HSCT.[40] However, dapsone given daily at a dose of 50 mg twice daily provides a more effective alternative.[34] Before starting dapsone, glucose-6-phosphate dehydrogenase deficiency should be ruled out. However, due to the superior results with TMP-SMX in both transplant and HIV-infected patients, it should be given whenever possible. Desensitization is therefore recommended in all patients with allergy to TMP-SMX[40,222] and administration of alternative drugs should only be given until the underlying condition that required the discontinuation of TMP-SMX has improved. There are no data on atovaquone in HSCT recipients.

Although now infrequent, the outcome of PCP remains poor. In a recent review of 10 cases, the fatality rate was 70%.[40] Without prophylaxis, the median time of onset of PCP is 9 weeks after HSCT, similar to that of other nonbacterial pneumonias. The clinical syndrome is indistinguishable both clinically and radiologically from other nonbacterial pneumonias. The diagnosis of *P. carinii* infection is established by either BAL, induced sputum, or thoracoscopic or open lung biopsy. Commonly used staining techniques include methenamine silver or immunofluorescence among others. The treatment of

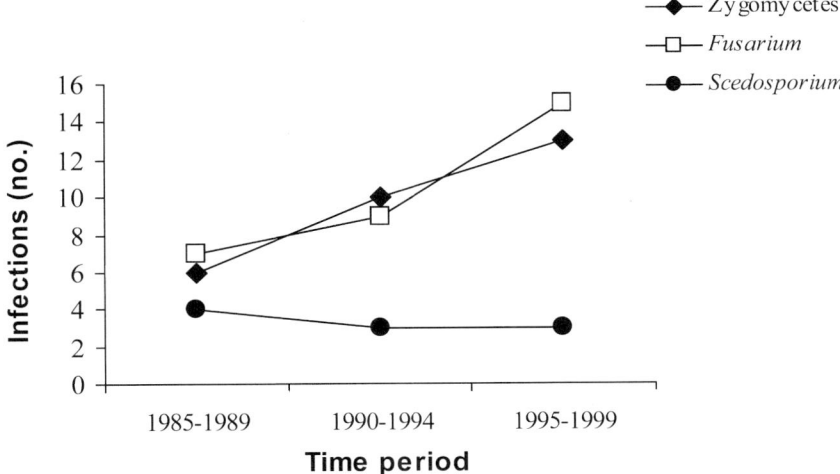

FIGURE 12. Frequency of non-aspergillus mold infections at FHCRC after HSCT.[53]

choice is intravenous TMP-SMX, at a dose of 15–20 mg/kg per day (TMP component) given as divided doses either every 6 or 8 hr among patients with normal renal function. Corticosteriods (40 mg prednisone twice a day for 5 days followed by 40 mg/day for 5 days and 20 mg/day for 11 days) are usually given in the beginning based on results in HIV-infected patients.[223] Intravenous TMP-SMX is preferred for treatment because of uncertainties about absorption. Treatment with TMP-SMX is given for 14 days. Alternative agents used in the HIV setting include intravenous pentamidine, atovaquone, clindamycin/primaquine, dapsone/trimethoprim, and trimetrexate, but no data on treatment outcome are available for HSCT recipients.

7.5. Infections with Encapsulated Bacteria in Patients with Chronic GVHD

Among 89 patients followed 6 months or longer after HSCT, only 30 (34%) remained free of infection.[224] Twenty-five (28%) had three or more infections. Upper respiratory or pulmonary infections were most common. Ten patients had bacterial septicemia. In contrast to the early granulocytopenic period, when gram-negative bacteria were predominant, *Streptococcus pneumoniae* was the most common isolate in proven bacterial infection, followed by *Staphylococcus aureus*. Significant risk factors in the development of late infection were the occurrence of chronic GVHD, inability to respond to dinitrochlorobenzene skin sensitization, and male gender.[31] The occurrence of pneumococcal bacteremia is especially intriguing. In some patients recurrent bacteremias may occur.[34]

The presumed explanation for bacteremic pneumococcal infections is that HSCT patients lose and do not subsequently make opsonizing antibody to encapsulated gram-positive organisms, even after recovery from infection.[31] Patients also respond poorly to immunization with prototype pneumococcal vaccines for the first 1–2 years after transplant, although the response again improves with time.[225] It would seem unlikely that immunization with the available pneumococcal vaccine would provide significant protection for those most in need, patients with chronic GVHD.[226]

Antibacterial prophylaxis is recommended in patients with chronic GVHD to prevent both bacteremic pneumococcal and other infection (Table 2).[71,226] While penicillin appears to work for this indication, the recent emergence of penicillin-resistant pneumococci makes it a less preferable choice.[227] Rather, TMP-SMX given once daily (80 mg TMP component) provides protection against both PCP and encapsulated bacteria. Controlled data are not available to evaluate the efficacy of such prophylaxis, but a retrospective study of nonrandomized treatment groups indicates that patients with chronic GVHD who receive TMP-SMX prophylaxis have a significantly lower incidence of infection. Oral penicillins should be reserved for patients who are unable to tolerate TMP-SMX because of rash, GI side effects, or apparent marrow suppression. There is no reported experience with new quinolones or macrolides for this indication.

Because infection with other organisms including both *Staphylococcus* species and gram-negative aerobic bacteria also occurs, empirical antibiotic treatment of

HSCT patients admitted with clinical sepsis should include broad-spectrum coverage until the identity of the infecting organisms is known.

7.6. Late Poor Graft Function Associated with Infections

Late graft failure may be due to infection. However, the diagnosis is often difficult to establish. Infections that have been associated with impaired graft function are CMV, HHV-6, and human parvovirus B 19. Noninfectious causes such as drug toxicity (e.g., TMP-SMX, mycophenolate, ganciclovir) and hematologic causes such as relapse and rejection must be ruled out. Generally, a bone marrow biopsy is required to make the diagnosis. Techniques used to diagnose organisms in the marrow aspirate include PCR, culture, and immunohistochemistry. Long-term marrow cultures can be used to detect infection of stromal cells.[136] One potential problem in the interpretation of PCR results from marrow samples is the concomitant detection of DNA in the peripheral blood, which might contaminate the marrow sample. However, it should be pointed out that CMV-related graft failure can occur in the absence of a detectable systemic viral load.[136] An operational definition for infection-related graft dysfunction is poor graft function in the absence of other possible causes (i.e., drugs, hematologic causes) and detection of the organism in the marrow aspirate. Since definitive proof of a causative role of the infectious organism is often not possible in a timely fashion, an empiric course of antimicrobial therapy seems reasonable.

8. Other Infections

8.1. Parasitic Infections

Toxoplasmosis appears to be rare after transplantation.[228–230] Disease has occurred exclusively among patients seropositive before HSCT, and thus appears to be caused by reactivation of latent infection. Involvement of heart, lung, and brain has been identified at postmortem examination. As in other immunocompromised hosts, serologic studies are usually unrevealing, and biopsy of implicated organs may be needed for diagnosis. However, cultures or PCR may be used to detect *Toxoplasma* in blood.[231] A review in European centers showed that no clinical cases of posttransplant toxoplasmosis occurred in autologous HSCT recipients, while the incidence was 0.93% in the allogeneic population. Almost one-third of patients had received T-cell-depleted allografts. The brain was the most commonly affected organ (89% of cases)

with lung (49%), liver, heart, and spleen also involved. Dissemination frequently occurs prior to the development of CNS symptoms, which may be associated with fever, pneumonia, or hepatitis.[229] Prophylaxis is warranted for high-risk patient populations such as *Toxoplasma* seropositive HSCT recipients, particularly those receiving manipulated allografts (i.e., T cell depleted or CD34 selected). Primary prophylaxis usually consists of TMP-SMX (at least four double-strength tablets/week), which also provides protection against PCP. Alternatives include pyrimethamine (25 mg/day), though additional measures must be considered for PCP prophylaxis. Prospective screening with PCR may be effective for directing preemptive therapy, but the data are insufficient at this time. Treatment is with pyrimethamine and sulfadiazine. Alternative treatment options include pyrimethamine and clindamycin, or more recently, atovaquone.[230] Treatment for several months may be required.

Cryptosporidiosis also appears to be very rare after HSCT. Clinical manifestations include diarrheal illnesses and sometimes also pneumonia.[232,233] No controlled treatment data exist. Antimicrobials that have been used include spiramycin and paromomycin alone or in combination with azithromycin.[234,235]

8.2. Legionellosis

Legionellosis is a rare but potentially fatal infection after HSCT.[236,237] Non-*pneumophila* species account for about half of the cases. Pneumonia is the principal clinical manifestation but unusual presentations such as pericarditis have been described. Risk factors have not been identified. The hospital water system is a potential source of the infection and should be tested if the incidence changes suddenly.[238] Diagnostic tests include culture, direct fluorescent antibody test, urine antigen tests (*L. pneumophila* only), and the modified Hiemenez stain, which detects acid-fast bacteria on tissue sections.[236] PCR will probably replace some of these techniques in the future.[239] No controlled treatment studies have been published but most patients respond to antimicrobial chemotherapy consisting of quinolones or erythromycin and rifampicin for at least 3 weeks. Longer treatment may be required.[236,237]

8.3. Epstein–Barr Virus

Epstein–Barr virus (EBV) is a DNA virus of the herpesvirus group. It can cause lymphoproliferative disease in transplant recipients (see also Chapter 10). This occurs mainly in T-cell-depleted allograft recipients.[240–243]

However, therapies directed at T cells such as ATG are also associated with a higher risk. Cases have also been described in CD34-selected autograft recipients who received rabbit ATG.[244] Monitoring of EBV by quantitative PCR may prove useful for predicting the disease.[245,246] Initial results with rituximab, an anti-CD20 monoclonal antibody, are encouraging.[247] Donor lymphocyte or EBV-specific T cell therapy appears useful when given prophylactically.[248] However, T cell therapy for established disease may be associated with toxicity.[248,249]

8.4. HHV-6, -7, and -8

HHV-6 and -7 are β herpesviruses (similar to CMV), while HHV-8 is a γ herpesvirus (similar to EBV). HHV-6 and -7 are reviewed in detail in Chapter 10. Several longitudinal studies have examined the role of HHV-6 after HSCT.[250–253] HHV-6 reactivates early after transplant and can be detected in the blood at a median of day 20 after transplant, usually before CMV. Type B is far more common than type A. The search for a clinical syndrome associated with HHV-6 revealed rash, a CNS syndrome consisting of encephalitis and impaired memory,[254] a possible association with interstitial lung disease,[204,255] and an association with delayed platelet engraftment.[253] Secondary graft failure has been documented in case reports.[256] PCR has been used to diagnose HHV-6-associated encephalitis.[254] Whether monitoring for HHV-6 viremia is useful in predicting these clinical syndromes is not known. In solid organ transplant recipients, there seems to be an interaction between HHV-6 and CMV.[257,258] Whether such interaction exists after HSCT has not been studied. Both foscarnet and ganciclovir have *in vitro* activity against HHV-6 and have been used in patients. No controlled treatment data exist. The impact of HHV-7 after HSCT has been examined in two longitudinal studies but no correlation with clinical disease was found. There was, however, a possible association with CMV infection.[250] HHV-8 is associated with Kaposi's sarcoma in HIV-infected individuals. Two recent reports suggest that it can cause a febrile syndrome and marrow failure in HSCT recipients.[259,260]

8.5. Adenovirus

Adenovirus infections are common after HSCT,[261–263] and some recent reports suggest that they may be increasing, possibly related to transplantation practices (e.g., T cell depletion).[264] Clinical manifestations include pneumonia, hepatitis, diarrhea, nephritis, cystitis, and eye infections.[261,263,265] There are no controlled treatment

studies. Intravenous ribavirin has been used, but without success in recent studies.[34,266,267] Recently, cidofovir has been shown to have *in vitro* activity and initial small case series show promising results.[268]

8.6. Hepatitis Viruses

The impact of hepatitis viruses on transplantation outcome has been reviewed extensively.[269] Hepatitis B virus is more likely than hepatitis C virus to cause severe clinical hepatitis and death from posttransplant liver disease.[270] However, these events are still relatively rare. Hepatitis C is a risk factor for venoocclusive disease of the liver and may cause late cirrhosis and liver failure.[271] Screening of donor and recipient is important for appropriate counseling and potential therapy.[269] While antiviral therapy exists for both hepatis B and C virus, these therapies have not been evaluated in the HSCT setting.

8.7. BK and JC Viruses

Polyomaviruses are DNA viruses that can cause disease in HSCT recipients. BK virus is a cause of posttransplant hemorrhagic cystitis.[272] Although BK virus is often detected in urine by PCR, an association with clinical symptoms may be difficult to prove.[273] Detection in blood was not associated with hemorrhagic cystitis in one study.[274] In a study of pediatric HSCT recipients BK virus-associated hemorrhagic cystitis occurred between day 24 and 50 after transplant in 9 of 117 patients. Infection was characterized by a long duration, correlated with use of busulfan, and resulted in bladder tamponade in 2 of 9 patients.[275] One case of BK virus-associated pneumonia has been reported.[276] There is no established treatment. Cidofovir has activity *in vitro* but there are only anecdotal treatment results to date.[277]

JC virus is the cause of progressive multifocal leukoencephalopathy in immunosuppressed patients. Several cases have been reported after HSCT.[278,279] PCR detection of JC virus in the cerebrospinal fluid is used for diagnosis.[280] Cidofovir has activity *in vitro* but little is known about its *in vivo* efficacy.[281]

8.8. Parvovirus B 19

Parvovirus B 19 can infect erythroid progenitor cells. In addition to transient marrow failure, parvovirus B 19 can cause chronic anemia and rarely pancytopenia in HSCT recipients.[282,283] PCR is useful as a diagnostic tool.[283] IVIG is effective in treating parvovirus B 19 symptomatic infection.[284] Prophylactic IVIG seems to have a protective effect against parvovirus B 19, although this has not been established in a randomized study.[285]

8.9. Other Viral Infections

Infections with human papillomavirus (HPV) rarely occur after transplant.[286] In one series, 3 of 238 allogeneic HSCT recipients developed anogenital condylomata associated with HPV.[286]

8.10. Mycobacterial Infections

General aspects of mycobacterial infections are reviewed in Chapter 7. In nonendemic regions for *Mycobacterium tuberculosis* (TB), atypical mycobacteria are significantly more common.[45,287] Atypical mycobacteria occur in both allogeneic and autologous transplant recipients. However, unrelated and HLA mismatched related transplant recipients seem to have the highest incidence.[45] The overall incidence in two large studies was 0.26–0.76% in autologous patients, and 0.37–0.6% in allograft recipients, with a predominance in unrelated transplant recipients (1 versus 0.22% in matched-related allograft recipients). Clinical manifestations of infections with rapid-growing atypical mycobacteria include bacteremia and central line infections with or without fever (Fig. 13). Pulmonary infections are rare but invasive pulmonary disease does occur. For the diagnosis of pulmonary disease, tissue biopsy is required, as colonization is common.[45] Infection responds to combination antibiotic treatment, although removal of the catheter is mandatory. If tunnel infection is present, surgical debridement may be

necessary.[45] The duration of treatment of infection caused by atypical mycobacteria is 4 weeks for line-associated infections with rapid-growing organisms, 2–3 weeks for exit site infection without bacteremia, 6–8 weeks for tunnel infections or bacteremia, and 6 months for biopsy-proven pulmonary disease.[45]

M. tuberculosis only rarely causes disease in nonendemic regions but is a common problem in endemic areas. The incidence of TB after transplantation varies from 0.05% in nonendemic areas to 1.4–5.5% in endemic areas such as Turkey or Hong Kong.[45,287–289] Clinically, pulmonary TB is the most common manifestation, followed by extrapulmonary and disseminated forms. Risk factors include allogeneic transplant, GVHD, and conditioning with TBI. Limited data are available on the management of clinical TB in HSCT recipients. The use of rifampin often causes problems due to drug interactions and liver toxicity. Alternatives to rifampin include quinolones, beta-lactam antibiotics with beta-lactamase inhibitors (not preferred for long-term use), and streptomycin. However, longer duration of treatment is required with these alternative regimens (especially with regimens without streptomycin) due to slow bacterial responses. Prevention strategies recommended in endemic and nonendemic areas consist of tuberculin skin testing of the recipient and donor and administration of prophylaxis with isoniazid for 6–12 months given to positive reactors.[288] Alternative options for prophylaxis include rifampin plus pyrazinamide, rifampin alone, or a combination of quinolones and ethambutol. Healthcare workers in contact with transplant recipients should be regularly tested for tuberculosis.

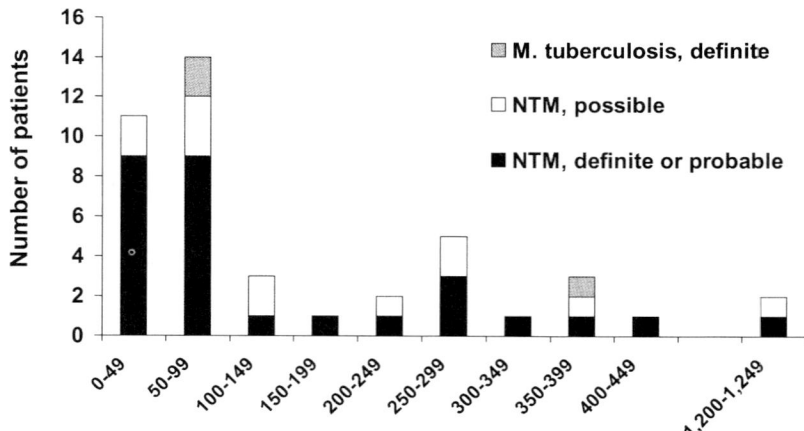

FIGURE 13. Timing of mycobacterial infections after HSCT in a nonendemic area. (Obtained with permission from Gaviria *et al.*[45])

8.11. Nocardiosis

General aspects of nocardial infections are reviewed in Chapter 7. Nocardiosis is a rare late complication after HSCT (median day of onset 226 days).[290,291] *Nocardia asteroides* complex accounted for 96% of the isolates in one study. Infection occurs predominantly in allograft recipients.[291] Nocardiosis has been observed in patients who received intermittent prophylaxis with TMP-SMX for prevention of PCP. The clinical manifestation consists of pulmonary and disseminated disease.[292] Response to treatment with TMP-SMX alone or in combination with synergistic agents is usually good.[291]

9. Pretransplant Infectious Disease Considerations

9.1. Pretransplant Infections

As a general principle, patients with active infections should not be transplanted. However, little specific information has been published regarding the optimal duration of pretransplant treatment and subsequent risk. Often one has to balance the need to treat pretransplant infections with the urgent requirement of transplantation due to the progression of the underlying disease.

9.1.1. Invasive Fungal Infections before Transplantation

Many patients undergoing HSCT have experienced previous episodes of acute candidiasis (i.e., candidemia) and suffer no apparent consequences during the HSCT course. The impact of other, more chronic fungal infections such as hepatosplenic candidiasis and invasive aspergillosis is of more concern. One study has documented that patients with a history of hepatosplenic candidiasis can undergo HSCT while maintained on a low dose (0.5 mg/kg per day) of amphotericin B while neutropenic.[293] In this study, 11 of 15 (73%) patients followed prospectively had persistently abnormal CT scans during HSCT, although only 3 of 15 died with evidence of progressive fungal infection. Currently we administer amphotericin B during neutropenia, followed by fluconazole postengraftment in patients undergoing HSCT.

Although invasive aspergillosis carries a grim prognosis in patients with hematologic malignancies, it is not uncommon to successfully treat this infection in a patient receiving induction or consolidation chemotherapy.[52] One recent retrospective analysis of 48 patients with documented or probable invasive aspergillosis and subsequent HSCT with amphotericin coverage revealed that

the overall incidence of relapsed fungal infection was low (33%); however, the study included mainly good-risk patients.[294] Once aspergillosis relapsed after transplant, the success of treatment was poor (12%).[294] In this series of patients, conditioning with busulfan and cyclophosphamide was associated with a beneficial outcome compared to those patients who received combined TBI. Although the role of surgical resection and the optimal prophylaxis regimen have yet to be defined, previous fungal infection is not an absolute contraindication for subsequent HSCT.[53,294] Secondary treatment and prophylactic regimens should be developed on an individual patient basis, with consideration for previous fungal, viral, and bacterial infections experienced prior to HSCT. Additional use of prophylactic granulocyte infusions from G-CSF-stimulated donors is given on an experimental basis in patients with pretransplant aspergillosis.

9.1.2. Bacterial Infections

Bloodstream infections, sinusitis, pneumonia, and other invasive infections should be treated with appropriate antibiotics for 2 weeks or until recovery. For specific infections (e.g., mycobacterial infections), extended pretransplant treatment courses are required.[45] Whether specific prophylaxis should be given throughout the preengraftment period has not been studied for most bacterial infections. It is our policy to choose antibacterials for empirical treatment of fever that are active against the pathogens documented before transplant. For severe pretransplant infections (e.g., mycobacterial infections), we continue treatment throughout the posttransplant period (i.e., 3–6 months) with no detectable relapse.[45]

9.1.3. Viral Infections

For pretransplant RSV or parainfluenza virus infections, most transplant centers postpone the transplant until resolution and cessation of viral shedding, especially when an allogeneic transplant is planned. However, the issue has not been studied systematically and there is some evidence that low-risk autologous transplant patients may be transplanted without adverse outcome.[295] More studies are needed to define which patients require postponing the transplantation procedure. Use of prophylaxis (e.g., RSV-Ig, palivizumab) during transplant in patients with pretransplant respiratory virus infections has not been studied.

CMV disease has been reported in patients with hematologic malignancies who have received intense chemotherapy and in patients with T cell immunodeficiencies.[34] There is a wide spectrum of severity ranging from mild pneumonia which resolves without treatment, to fatal disease. Although the disease seems to respond to antiviral agents alone in most cases, relapse after transplantation is common and associated with a high fatality rate if the patient is transplanted within 6 months of the disease.[34] Thus, we recommend prophylaxis throughout the posttransplant course (at least 3 months), starting with foscarnet until engraftment followed by ganciclovir or foscarnet.

9.2. Donor Issues

All donors should be in good health at the time of marrow or stem cell donation. However, since stem cell donors are limited, risks and benefits of using a donor who might transmit an infection may have to be weighed in individual cases. Donations following exposure to infectious diseases that have a viremic phase should be avoided within the incubation time of the infection. Serologic testing should be done within 1–2 months before transplant.[71] For a CMV seronegative recipient, a CMV seronegative donor is the preferred choice, if possible, due to an increased mortality with CMV seropositive donors even with CMV antigenemia-guided ganciclovir treatment.[34] Stem cell donations from donors with current or previous hepatitis have been reviewed comprehensively.[269] There is a risk of transmission if the donor has active hepatitis B (HBsAg positive) or C (HCV RNA positive) and alternative donors should be considered. Donors infected with HIV are usually not used. Also, people with active TB should not donate.[71] The stem cell products are screened by culture at the time of harvest but results are often not available at the time of infusion. Antimicrobial therapy of the recipients may be required in rare cases when culture results become positive subsequently. Transplantation during incubation time after proven or potential exposure of the donor should be avoided. Also, stem cell donations directly following vaccination with live-virus vaccine (e.g., polio) should be avoided, although no adverse outcome results have been reported. With regard to travel into malaria-endemic areas, blood bank guidelines for blood donations apply (i.e., donors are deferred for 1 year after travel to malaria-endemic areas and for 3 years donors who have resided in malaria-endemic areas or who have a history of malaria).[71] However, if the only suitable donor does not fit these criteria and immediate transplantation is required, stem cell donation may be considered. In these cases, we use the following guidelines: (1) the team caring for the recipient should be alerted about the stem cell donor's prior travel to malaria-endemic areas, (2) suitable donors can be scheduled for hematopoietic stem cell donation any time after they have completed the 4-week prescribed antimalaria prophylaxis following return from travel, and (3) suitable donors who have traveled to malaria-endemic areas in the preceding 6 months, and have not received antimalaria prophylaxis, should be evaluated for malaria at the time of donor workup. For additional donor issues see the CDC guidelines for preventing opportunistic infections among HSCT recipients.[71]

10. Infection Control Programs and Surveillance

10.1. Isolation Practices

The use of a protective environment (i.e., laminar airflow isolation) is increasingly controversial among transplanters. Our center abandoned the practice several years ago for all patients. General infection control procedures consist of a private room and careful handwashing.[71] Increased isolation methods are used in specific situations, i.e., contact isolation consisting of gowns and gloves for infection or colonization with enteric pathogens (e.g., VRE, *C. difficile*) or highly resistant organisms (e.g., MRSA, multiresistant gram-negative organisms). Respiratory isolation is used for respiratory viruses and VZV (masks, gowns, gloves, and eye protection).[71] For disseminated VZV infection, the patient is placed in a negative pressure room with double doors.

The use of the protective environment remains a controversial issue in 2000, nearly 30 years after inception of the concept. The protective environment is defined as an infection control program consisting of skin cleansing, topical and oral nonabsorbable or absorbable antibiotics, sterile food, and a laminar airflow room.[296] A protective environment has theoretical benefits in addition to infection control for HSCT patients.[296] Both graft rejection and GVHD may be influenced by reduction in endogenous bacterial flora and infection.[297] However, only a few randomized trials have been performed. Among patients transplanted for aplastic anemia, those treated in the protective environment had significantly less acute GVHD compared with patients not transplanted in the protective environment.[298] This effect was also associated with significantly better survival after transplant. Patients with leukemia have not shown such statistically significant benefits, although the incidence of acute GVHD has been lower and survival somewhat better among those transplanted in the protective environment. A recent study by Beelen *et al.* has renewed interest in the question of whether reduction in endogenous bacterial flora and infection can reduce acute GVHD.[69] In a nonblinded, randomized study, patients transplanted under laminar airflow conditions who received ciprofloxacin and metronidazole prophylactically had a lower incidence of acute grade 2–4 GVHD than patients who received ciprofloxacin alone. The effect was only seen in matched-related transplant recipients and was associated with a decrease in the anaerobic flora. Overall survival and the incidence of chronic GVHD were not affected but there was also a trend toward improved survival in matched-related patients.[69]

There are several unresolved questions regarding the use of the protective environment. Although the recent study by Beelen et al.[69] is intriguing and there may be a reduced incidence of early invasive *Aspergillus* infections in patients transplanted in laminar airflow rooms,[39] several recent developments in transplantation practices put the concept into question. These changes include improved methods of prevention of GVHD, availability of HEPA filtration, shorter time to engraftment, and finally cost effectiveness. Indeed, many transplant centers in North America have abandoned the concept entirely (including for patients with aplastic anemia) and use HEPA filtration instead. Perhaps the most important question is whether a laminar airflow room is actually required when HEPA filtration is used and whether effective reduction in endogenous bacterial flora can be accomplished without using laminar airflow room isolation practices. Since oral antibiotics and antifungal agents with good tolerability such as fluoroquinolones, metronidazole, and azoles are now available, this issue should be addressed in a randomized study. Another important issue is the trend to shorten the inpatient stay dramatically. A shorter inpatient stay would thereby further reduce the potential benefit of the laminar airflow room.

10.2. Use of Surveillance Culture

The use of "surveillance" cultures has been debated for some time, and many procedures differ between institutions. It is accepted that granulocytopenic patients usually become infected by resident organisms, either those "endogenous" organisms brought into the hospital with the patient, which include normal flora, or those acquired "exogenously" from the hospital environment and which have become part of the patient's flora after admission (e.g., resistant *Staphylococcus*, *Enterococcus*, *Pseudomonas*, *Candida*, or molds). Thus, with the exception of patients monitored serially for study of specific infection-control programs, we obtain bacterial and fungal cultures

(oropharynx, rectum, vagina) only at admission. Others continue to monitor for resistant organisms.

During febrile granulocytopenic periods, blood cultures are often obtained daily, although a recent study suggests that the continued culturing only rarely provides useful information.[299] We perform blood culture surveillance three times a week in patients who receive systemic corticosteroids regardless of neutrophil counts (due to the fever-masking effect of corticosteroids). Additional cultures from clinically relevant sites are obtained any time infection is suspected. We perform weekly mouth washes for the detection of mold, and administer preemptive therapy with amphotericin drugs if these cultures turn positive.

Viral surveillance testing is performed for CMV in all patients and is the basis for preemptive treatment strategies.[71] The frequency is generally weekly; however, some high-risk patients (e.g., patients after T cell depletion, CD34 selection, during high-dose corticosteroid therapy) may benefit from twice-a-week monitoring during high-risk periods. For respiratory viruses, nasal washes are performed everytime the patient has upper respiratory symptoms (targeted surveillance). Results can be used for preemptive therapy as well as for infection control purposes (duration of isolation). Whether weekly surveillance during the respiratory virus season is useful has not been established.

10.3. Prevention of Exposure by Personnel, Family Members, and Visitors

Personnel, family members, and visitors with communicable diseases should be restricted from contact with HSCT recipients. This includes patients with symptoms of URI and those exposed to or with clinical VZV infection.[71] In addition to the widely recommended handwashing procedures before and after each patient contact, we have recently instituted a "sign-in" procedure during the respiratory virus season in which all members of the personnel (including the medical staff), family members, and visitors are asked daily to confirm with their signature that they are free of URI symptoms (i.e., uncontrolled secretions). All individuals with uncontrolled secretions are restricted from patient contact until symptoms have disappeared. For VZV, vaccination of all family members and visitors with a negative history for chickenpox is advised prior to arrival at the transplant center.[71] This advice should be given during the first contact with the transplant center, thereby allowing necessary time to get the vaccination completed before arrival. Once a contact person has been exposed or infected with VZV, this person will be restricted from contact with the transplant recipient, usually from 8 to 22 days after exposure or until all lesions are crusted, whichever occurs later.

10.4. Postexposure Prophylaxis

Postexposure prophylaxis is recommended in VZV seronegative transplant recipients. VZV-specific immunoglobulin is the standard of care (400 mg/kg). Postexposure use of acyclovir has been reported to ameliorate the disease in immunocompetent patients.[300] Because of the potentially fatal outcome of primary varicella infections, we use postexposure valaciclovir (3 g three times per day) from day 3 to 22 following exposure, in addition to VZV-specific immunoglobulin. After an exposure of seronegative family members or caretakers, we recommend isolation as outlined above. If it is not feasible to isolate the patient during this time period, postexposure valaciclovir or postexposure immunization[301] may be considered. This strategy is aimed at reducing infectivity of the index case and thereby potentially also reducing the risk of transmission to the patient.

10.5. Immunization

Immunization guidelines for HSCT recipients are summarized in the Guidelines for Preventing Opportunistic Infections Among HSCT Recipients, a consensus statement issued by the Centers for Disease Control and Prevention, the Infectious Disease Society of America, and the American Society of Blood and Marrow Transplantation.[71]

11. Conclusions and Future Considerations

Major progress has been made in the diagnosis and prevention of infections in HSCT over the last decade. Perhaps the most impressive examples are the prevention of CMV disease by ganciclovir and of early *Candida albicans* infections by fluconazole. However, many issues in the care of the HSCT recipient remain unresolved. Invasive mold infections are now the leading infectious cause of death at many centers. Thus, there is a great need for improved diagnosis of fungal infections, as well as for prevention and treatment.

Although early CMV disease has been reduced significantly with preemptive treatment or prophylactic strategies, CMV continues to have a significant impact on morbidity and mortality. Late CMV disease remains a clinical problem in patients with continued immunosuppression. Prevention strategies are urgently needed to address this question. Options include long-term antiviral prophylaxis or preemptive treatment strategies based on virologic monitoring as well as immunotherapy strategies such as adoptive transfer of donor-derived T cells. Improved treatment strategies for CMV pneumonia are required due to the persistently poor outcome with ganciclovir and CMV-Ig or IVIG. Thus, antiviral combination strategies, including agents with different mechanisms of action, should be studied. Respiratory viruses have been recognized as cause of fatal pneumonia after HSCT. RSV and parainfluenza viruses can cause severe pneumonias after HSCT. Infection control strategies including vaccination against influenza of susceptible family members and patient care personnel are critical for prevention.

The impact of new transplantation techniques such as CD34 selection and non-myeloablative conditioning regimens on risk for infection requires careful evaluation and possibly adjustment of prevention strategies. Infections occurring in long-term survivors of HSCT are related both to the persistence of immunologic defects and to the additional effects of chronic GVHD and its treatment. Current prevention strategies using chemoprophylaxis are far from perfect, and novel means of augmentation of late immune reconstitution (e.g., vaccination, adoptive transfer of antigen-specific T cells, IL-2) as well as better prevention and treatment for GVHD are needed.

ACKNOWLEDGMENTS.   Earlier editions of this chapter were written by Raleigh Bowden, M.D., and the late Joel D. Meyers, M.D. We thank T. Chauncey, M.D., for providing the figure on the incidence of *Pneumocystis carinii* pneumonia and Julio Gonzales, M.D., for providing Case 3. Michael Boeckh and Kieren Marr were supported in part by grants awarded by the National Institute of Health Cancer Institute (CA 18029 and KO-8 AI1571).

References

1. Thomas E, Storb R, Clift RA, *et al*: Bone-marrow transplantation (first of two parts). *N Engl J Med* **292**:832–843, 1975.
2. Hansen JA, Gooley TA, Martin PJ, *et al*: Bone marrow transplants from unrelated donors for patients with chronic myeloid leukemia [see comments]. *N Engl J Med* **338**:962–968, 1998.
3. Storb R, Yu C, Sandmaier BM, *et al*: Mixed hematopoietic chimerism after marrow allografts. Transplantation in the ambulatory care setting. *Ann NY Acad Sci* **872**:372–375; discussion 375–376, 1999.
4. Sandmaier BM, McSweeney P, Yu C, Storb R: Nonmyeloablative transplants: Preclinical and clinical results. *Semin Oncol* **27**:78–81, 2000.
5. Sayer HG, Longton G, Bowden R, Pepe M, Storb R: Increased risk

of infection in marrow transplant patients receiving methylprednisolone for graft-versus-host disease prevention. *Blood* **84:**1328–1332, 1994.

6. Cooper MH, Hartman GG, Starzl TE, Fung JJ: The induction of pseudo-graft-versus-host disease following syngeneic bone marrow transplantation using FK 506. *Transplant Proc* **23:**3234–3235, 1991.

7. Storek J, Witherspoon RP: Immunologic reconstitution after hematopoietic stem cell transplantation. In Atkinson K (ed): *Clinical Bone Marrow and Blood Stem Cell Transplantation.* Cambridge University Press, Cambridge, 2000, pp. 111–146.

8. Douek DC, Vescio RA, Betts MR, *et al*: Assessment of thymic output in adults after haematopoietic stem-cell transplantation and prediction of T-cell reconstitution [see comments]. *Lancet* **355:** 1875–1881, 2000.

9. Storek J, Dawson MA, Storer B, *et al*: Immune reconstitution after allogeneic marrow versus blood stem cell transplantation. *Blood* **97:**3380–3389, 2001.

10. Einsele H, Ehninger G, Steidle M, *et al*: Lymphocytopenia as an unfavorable prognostic factor in patients with cytomegalovirus infection after bone marrow transplantation. *Blood* **82:**1672–1678, 1993.

11. Fries BC, Khaira D, Pepe MS, Torok-Storb B: Declining lymphocyte counts following cytomegalovirus (CMV) infection are associated with fatal CMV disease in bone marrow transplant patients. *Exp Hematol* **21:**1387–1392, 1993.

12. Meyers JD, Flournoy N, Thomas ED: Cell-mediated immunity to varicella–zoster virus after allogeneic marrow transplant. *J Infect Dis* **141:**479–487, 1980.

13. Ljungman P, Lonnqvist B, Wahren B, Ringden O, Gahrton G: Lymphocyte responses after cytomegalovirus infection in bone marrow transplant recipients—A one-year follow-up. *Transplantation* **40:**515–520, 1985.

14. Meyers JD, Flournoy N, Thomas ED: Cytomegalovirus infection and specific cell-mediated immunity after marrow transplant. *J Infect Dis* **142:**816–824, 1980.

15. Reusser P, Riddell SR, Meyers JD, Greenberg PD: Cytotoxic T-lymphocyte response to cytomegalovirus after human allogeneic bone marrow transplantation: Pattern of recovery and correlation with cytomegalovirus infection and disease. *Blood* **78:** 1373–1380, 1991.

16. Wang FZ, Linde A, Dahl H, Ljungman P: Human herpesvirus 6 infection inhibits specific lymphocyte proliferation responses and is related to lymphocytopenia after allogeneic stem cell transplantation. *Bone Marrow Transplant* **24:**1201–1206, 1999.

17. Levin MJ, Parkman R, Oxman MN, *et al*: Proliferative and interferon responses by peripheral blood mononuclear cells after bone marrow transplantation in humans. *Infect Immun* **20:**678–684, 1978.

18. Azogui O, Gluckman E, Fradelizi D: Inhibition of IL 2 production after human allogeneic bone marrow transplantation. *J Immunol* **131:**1205–1208, 1983.

19. Quinnan GV Jr, Kirmani N, Rook AH, *et al*: Cytotoxic T cells in cytomegalovirus infection: HLA-restricted T-lymphocyte and non-T-lymphocyte cytotoxic responses correlate with recovery from cytomegalovirus infection in bone-marrow-transplant recipients. *N Engl J Med* **307:**7–13, 1982.

20. Bowden RA, Day LM, Amos DE, Meyers JD: Natural cytotoxic activity against cytomegalovirus-infected target cells following marrow transplantation. *Transplantation* **44:**504–508, 1987.

21. Li CR, Greenberg PD, Gilbert MJ, *et al*: Recovery of HLA-

restricted cytomegalovirus (CMV)-specific T-cell responses after allogeneic bone marrow transplant: Correlation with CMV disease and effect of ganciclovir prophylaxis. *Blood* **83:**1971–1979, 1994.

22. Wade JC, Day LM, Crowley JJ, Meyers JD: Recurrent infection with herpes simplex virus after marrow transplantation: Role of the specific immune response and acyclovir treatment. *J Infect Dis* **149:**750–756, 1984.

23. Reusser P, Attenhofer R, Hebart H, *et al*: Cytomegalovirus-specific T-cell immunity in recipients of autologous peripheral blood stem cell or bone marrow transplant. *Blood* **89:**3873–3879, 1997.

24. Bowden RA, Digel J, Reed EC, Meyers JD: Immunosuppressive effects of ganciclovir on *in vitro* lymphocyte responses. *J Infect Dis* **156:**899–903, 1987.

25. Heagy W, Crumpacker C, Lopez PA, Finberg RW: Inhibition of immune functions by antiviral drugs. *J Clin Invest* **87:**1916–1924, 1991.

26. Ottinger HD, Beelen DW, Scheulen B, Schaefer UW, Grosse-Wilde H: Improved immune reconstitution after allotransplantation of peripheral blood stem cells instead of bone marrow [see comments]. *Blood* **88:**2775–2779, 1996.

27. Storek J, Witherspoon RP, Maloney DG, Chauncey TR, Storb R: Improved reconstitution of CD4 T cells and B cells but worsened reconstitution of serum IgG levels after allogeneic transplantation of blood stem cells instead of marrow [letter; comment]. *Blood* **89:**3891–3893, 1997.

28. Holmberg LA, Boeckh M, Hooper H, *et al*: Increased incidence of cytomegalovirus disease after autologous CD34-selected peripheral blood stem cell transplantation [see comments]. *Blood* **94:** 4029–4035, 1999.

29. Talvensaari K, Clave E, Douay CL, *et al*: Immune reconstitution is improved after 1 year in cord blood compared to HLA-identical sibling bone marrow transplanted patients. *Blood* **96:**555a (abstract 2382), 2000.

30. Locksley RM, Flournoy N, Sullivan KM, Meyers JD: Infection with varicella–zoster virus after marrow transplantation. *J Infect Dis* **152:**1172–1181, 1985.

31. Winston DJ, Schiffman G, Wang DC, *et al*: Pneumococcal infections after human bone-marrow transplantation. *Ann Intern Med* **91:**835–841, 1979.

32. Kruger W, Russmann B, Kroger N, *et al*: Early infections in patients undergoing bone marrow or blood stem cell transplantation—A 7 year single centre investigation of 409 cases. *Bone Marrow Transplant* **23:**589–597, 1999.

33. Stamm WE, Tompkins LS, Wagner KF, *et al*: Infection due to Corynebacterium species in marrow transplant patients. *Ann Intern Med* **91:**167–173, 1979.

34. Boeckh M: Unpublished results.

35. Boeckh M, Gooley TA, Myerson D, *et al*: Cytomegalovirus pp65 antigenemia-guided early treatment with ganciclovir versus ganciclovir at engraftment after allogeneic marrow transplantation: A randomized double-blind study. *Blood* **88:**4063–4071, 1996.

36. Marr KA, Seidel K, White TC, Bowden RA: Candidemia in allogeneic blood and marrow transplant recipients: Evolution of risk factors after the adoption of prophylactic fluconazole. *J Infect Dis* **181:**309–316, 2000.

37. Morrison V, Haake R, Weisdorf D: Non-*Candida* fungal infections after bone marrow transplantation: Risk factors and outcome. *Am J Med* **96:**497–503, 1993.

38. Limaye A, Bowden RA, Myerson D, Boeckh M: Cytomegalovirus

disease before engraftment in marrow transplant patients. *Clin Infect Dis* **24:**830–835, 1997.

39. Wald A, Leisenring W, Burik J-A, Bowden RA: Epidemiology of *Aspergillus* infections in a large cohort of patients undergoing bone marrow transplantation. *J Infect Dis* **175:**1459–1466, 1997.

40. Souza JP, Boeckh M, Gooley TA, Flowers ME, Crawford SW: High rates of Pneumocystis carinii pneumonia in allogeneic blood and marrow transplant recipients receiving dapsone prophylaxis. *Clin Infect Dis* **29:**1467–1471, 1999.

41. Marr KA, Bowden RA: New infiltrates in hematopoietic stem cell transplant recipients. In Armstrong D, Cohen J (eds): *Infectious Diseases.* Hartcourt Publishers, London, 1999, pp. 9.4–9.8.

42. Whimbey E, Champlin RE, Englund JA, *et al:* Combination therapy with aerosolized ribavirin and intravenous immunoglobulin for respiratory syncytial virus disease in adult bone marrow transplant recipients. *Bone Marrow Transplant* **16:**393–399, 1995.

43. Anaissie E, Rex J, Uzun O, Vartivarian S: Predictors of adverse outcome in cancer patients with candidemia. *Am J Med* **104:**238–245, 1998.

44. Rex JH, Bennett JE, Sugar AM, *et al:* A randomized trial comparing fluconazole with amphotericin B for the treatment of candidemia in patients without neutropenia. Candidemia Study Group and the National Institute. *N Engl J Med* **331:**1325–1330, 1994.

45. Gaviria JM, Garcia PJ, Garrido SM, Corey L, Boeckh M: Nontuberculous mycobacterial infections in hematopoietic stem cell transplant recipients: Characteristics of respiratory and catheter-related infections [In Process Citation]. *Biol Blood Marrow Transplant* **6:**361–369, 2000.

46. Wingard JR: Importance of *Candida* species other than *C. albicans* as pathogens in oncology patients. *Clin Infect Dis* **20:**115–125, 1995.

47. Goodrich JM, Reed C, Mori M, *et al:* Clinical features and analysis of risk factors for invasive candidal infection after marrow transplantation. *J Infect Dis* **164:**731–740, 1991.

48. Abi Said D, Anaissie E, Uzun O, *et al:* The epidemiology of hematogenous candidiasis caused by different *Candida* species [see comments]. *Clin Infect Dis* **24:**1122–1128, 1997.

49. van Burik JH, Leisenring W, Myerson D, *et al:* The effect of prophylactic fluconazole on the clinical spectrum of fungal diseases in bone marrow transplant recipients with special attention to hepatic candidiasis. *Medicine* **77:**246–254, 1998.

50. Jantunen E, Ruutu P, Niskanen L, *et al:* Incidence and risk factors for invasive fungal infections in allogeneic BMT recipients. *Bone Marrow Transplant* **19:**801–808, 1997.

51. Yamazaki T, Kume H, Murase S, Yamashita E, Arisawa M: Epidemiology of visceral mycoses: Analysis of data in annual of the pathological autopsy cases in Japan. *J Clin Microbiol* **37:**1732–1738, 1999.

52. Denning DW: Therapeutic outcome of invasive aspergillosis. *Clin Infect Dis* **23:**608–615, 1996.

53. Marr KA: Unpublished data.

54. Ribaud P, Chastang C, Latge J, *et al:* Survival and prognostic factors of invasive aspergillosis after allogeneic bone marrow transplantation. *Clin Infect Dis* **28:**322–330, 1999.

55. Henry N, McLimans C, Wright A, *et al:* Microbiological and clinical evaluation of the isolator lysis centrifugation blood culture tube. *J Clin Microbiol* **17:**864–869, 1983.

56. Wilson M, Weinstein M, Reimer L, Mirrett S, Reller L: Controlled comparison of the BacT/Alert and BACTEC 660/730 nonradiometric blood culture systems. *J Clin Microbiol* **1992:**323–329, 1992.

57. Verweij P, Poulain D, Obayashi T, *et al:* Current trends in the detection of antigenaemia, metabolites and cell wall markers for the diagnosis and therapeutic monitoring of fungal infections. *Med Mycol* **36:**146–155, 1998.

58. Duthie R, Denning D: *Aspergillus* fungemia: Report of two cases and review. *Clin Infect Dis* **20:**598–605, 1995.

59. Caillot D, Casasnovas O, Bernard A, *et al:* Improved management of invasive aspergillosis in neutropenic patients using early thoracic computed tomographic scan and surgery. *J Clin Oncol* **15:** 139–147, 1997.

60. Maertens J, Verhaegen J, Demuynck H, *et al:* Autopsy-controlled prospective evaluation of serial screening for circulating galactomannan by a sandwich enzyme-linked immunosorbent assay for hematological patients at risk for invasive aspergillosis. *J Clin Microbiol* **37:**3223–3228, 1999.

61. Hebart H, Loffler J, Meisner C, *et al:* Early detection of aspergillus infection after allogeneic stem cell transplantation by polymerase chain reaction screening [In Process Citation]. *J Infect Dis* **181:** 1713–1719, 2000.

62. Yamakami Y, Hashimoto A, Tokimatsu I, Nasu M: PCR detection of DNA specific for *Aspergillus* species of patients with invasive aspergillosis. *J Clin Microbiol* **34:**2464–2468, 1996.

63. Einsele H, Quabeck K, Muller KD, *et al:* Prediction of invasive pulmonary aspergillosis from colonisation of lower respiratory tract before marrow transplantation [letter]. *Lancet* **352:**1443, 1998.

64. Rex J, Walsh T, Sobel J, *et al:* Practice guidelines for the treatment of candidiasis. *Clin Infect Dis* **30:**662–678, 2000.

65. Walsh T, Finberg R, Arndt C, *et al:* Liposomal amphotericin B for empiric therapy in patients with persistent fever and neutropenia. *N Engl J Med* **340:**764–771, 1999.

65a. Walsh TJ, Pappas P, Winston DJ, *et al:* Voriconazole compared with liposomal amphotericin B for empirical antifungal therapy in patients with neutropenia and persistent fever. *N Engl J Med* **346:** 225–234, 2002.

66. Yeghen T, Kibbler C, Prentice H, *et al:* Management of invasive pulmonary aspergillosis in hematology patients: A review of 87 consecutive cases at a single institution. *Clin Infect Dis* **31:**859–868, 2000.

67. Pidhorecky I, Urschel J, Anderson T: Resection of invasive pulmonary aspergillosis in immunocompromised patients. *Ann Surg Oncol* **7:**312–317, 2000.

68. Sheehan D, Hitchcock C, Sibley C: Current and emerging azole antifungal agents. *Clin Microbiol Rev* **12:**40–79, 1999.

69. Beelen DW, Elmaagacli A, Muller KD, Hirche H, Schaefer UW: Influence of intestinal bacterial decontamination using metronidazole and ciprofloxacin or ciprofloxacin alone on the development of acute graft-versus-host disease after marrow transplantation in patients with hematologic malignancies: Final results and long-term follow-up of an open-label prospective randomized trial. *Blood* **93:**3267–3275, 1999.

70. Hughes WT, Armstrong D, Bodey GP, *et al:* 1997 guidelines for the use of antimicrobial agents in neutropenic patients with unexplained fever. Infectious Diseases Society of America. *Clin Infect Dis* **25:**551–573, 1997.

71. Guidelines for preventing opportunistic infections among hematopoietic stem cell transplant recipients. Recommendations of CDC, the Infectious Disease Society of America, and the American Society of Blood and Marrow Transplantation. *Morbidity Mortality Weekly Rep* **49:**1–128, 2000.

72. Petersen F, Thornquist M, Buckner C, *et al:* The effects of infection prevention regimens on early infectious complications in

marrow transplant patients: A four arm randomized study. *Infection* **16**:199–208, 1988.

73. Winston DJ, Blumer J, Beveridge R, *et al*: Randomized double-blind, multicenter trial of clinafloxacin versus imipenem for empiric therapy of febrile granulocytopenic patients. 10th International Symposium on Infections in the Immunocompromised Host, Davos, Switzerland, 1998.

74. Kapur D, Dorsky D, Feingold JM, *et al*: Incidence and outcome of vancomycin-resistant enterococcal bacteremia following autologous peripheral blood stem cell transplantation. *Bone Marrow Transplant* **25**:147–152, 2000.

75. Bodey GP, Buckley M, Sathe YS, Freireich EJ: Quantitative relationships between circulating leukocytes and infection in patients with acute leukemia. *Ann Intern Med* **64**:328–340, 1966.

76. Gill FA, Robinson R, Maclowry JD, Levine AS: The relationship of fever, granulocytopenia and antimicrobial therapy to bacteremia in cancer patients. *Cancer* **39**:1704–1709, 1977.

77. Uzun O, Anaissie EJ: Antifungal prophylaxis in patients with hematologic malignancies: A reappraisal [see comments]. *Blood* **86**:2063–2072, 1995.

78. Goodman JL, Winston DJ, Greenfield RA, *et al*: A controlled trial of fluconazole to prevent fungal infections in patients undergoing bone marrow transplantation. *N Engl J Med* **326**:845–851, 1992.

79. Slavin MA, Osborne B, Adams R, *et al*: Efficacy and safety of fluconazole prophylaxis for fungal infections after marrow transplantation—A prospective, randomized, double-blind study. *J Infect Dis* **171**:1545–1552, 1995.

80. Marr K, Seidel K, Slavin M, *et al*: Prolonged fluconazole prophylaxis is associated with persistent protection against candidiasis-related death in allogeneic marrow transplant recipients: Long-term follow-up of a randomized, placebo-controlled trial. *Blood* **96**:2055–2061, 2000.

81. Gubbins PO, Bowman JL, Penzak SR: Antifungal prophylaxis to prevent invasive mycoses among bone marrow transplant recipients. *Pharmacotherapy* **18**:549–564, 1998.

82. Perfect J, Klotman M, Gilbert C, *et al*: Prophylactic intravenous amphotericin B in neutropenic autologous bone marrow transplant recipients. *J Infect Dis* **165**:891–897, 1992.

83. Schwartz S, Behre G, Heinemann V, *et al*: Aerosolized amphotericin B inhalations as prophylaxis of invasive aspergillus infections during prolonged neutropenia: Results of a prospective randomized multicenter trial. *Blood* **93**:3654–3661, 1999.

84. Morgenstern G, Prentice A, Prentice H, *et al*: A randomized controlled trial of itraconazole versus fluconazole for the prevention of fungal infections in patients with hematological malignancies. *Br J Haematol* **105**:901–911, 1999.

85. Menichetti F, DelFavero A, Martino P, *et al*: Itraconazole oral solution as prophylaxis for fungal infections in neutropenic patients with hematologic malignancies: A randomized, placebo-controlled, double-blind, multicenter trial. *Clin Infect Dis* **28**:250–255, 1999.

86. Clift RA, Sanders JE, Thomas ED, Williams B, Buckner CD: Granulocyte transfusions for the prevention of infection in patients receiving bone-marrow transplants. *N Engl J Med* **298**:1052–1057, 1978.

87. Strauss RG, Connett JE, Gale RP, *et al*: A controlled trial of prophylactic granulocyte transfusions during initial induction chemotherapy for acute myelogenous leukemia. *N Engl J Med* **305**:597–603, 1981.

88. Winston DJ, Ho WG, Howell CL, *et al*: Cytomegalovirus infections associated with leukocyte transfusions. *Ann Intern Med* **93**:671–675, 1980.

89. Price TH, Bowden RA, Boeckh M, *et al*: Phase I/II trial of neutrophil transfusions from donors stimulated with G-CSF and dexamethasone for treatment of patients with infections in hematopoietic stem cell transplantation. *Blood* **95**:3302–3309, 2000.

90. Wingard J: Growth factors and other immunomodulators. In Bowden R, Ljungman P, Paya C (eds): *Transplant Infections.* Lippincott-Raven, Philadelphia, 1998.

91. Ljungman P, Niederwieser D, Pepe MS, *et al*: Cytomegalovirus infection after marrow transplantation for aplastic anemia. *Bone Marrow Transplant* **6**:295–300, 1990.

92. Applebaum FR, Meyers JD, Fefer A, *et al*: Nonbacterial nonfungal pneumonia following marrow transplantation in 100 identical twins. *Transplantation* **33**:265–268, 1982.

93. Meyers JD, Ljungman P, Fisher LD: Cytomegalovirus excretion as a predictor of cytomegalovirus disease after marrow transplantation: Importance of cytomegalovirus viremia. *J Infect Dis* **162**:373–380, 1990.

94. Nichols WG, Corey L, Gooley T, Davis C, Boeckh M: Parainfluenza virus infections after hematopoietic stem cell transplantation: Risk factors, response to antiviral therapy, and effect on transplant outcome. *Blood* In press, 2001.

95. Boeckh M, Gooley T, Bowden RA, *et al*: Risk factors for progression from respiratory syncytial virus upper respiratory infection to pneumonia after hematopoietic stem cell transplantation. 39th Interscience Conference on Antimicrobial Agents and Chemotherapy, San Francisco, 1999.

96. Meyers JD, Flournoy N, Thomas ED: Nonbacterial pneumonia after allogeneic marrow transplantation: A review of ten years' experience. *Rev Infect Dis* **4**:1119–1132, 1982.

97. Meyers JD, Flournoy N, Wade JC: Biology of interstitial pneumonia after marrow transplantation. In Gale RP (ed): *Recent Advances in Bone Marrow Transplantation.* Liss, New York, 1983, pp. 405–423.

98. Clark JG, Hansen JA, Hertz MI, *et al*: NHLBI workshop summary. Idiopathic pneumonia syndrome after bone marrow transplantation. *Am Rev Respir Dis* **147**:1601–1606, 1993.

99. Boeckh M, Bowden RL: Cytomegalovirus infection in marrow transplantation. *Cancer Treat Res* **76**:97–136, 1995.

100. Boeckh M, Ljungman P: Cytomegalovirus infection after BMT. In Bowden RA, Ljungman P, Paya CV (eds): *Transplant Infections.* Lippincott-Raven, Philadelphia, 1998, pp. 215–227.

101. Meyers JD, Flournoy N, Thomas ED: Risk factors for cytomegalovirus infection after human marrow transplantation. *J Infect Dis* **153**:478–488, 1986.

102. Boeckh M, Bowden RA, Gooley T, Myerson D, Corey L: Successful modification of a pp65 antigenemia-based early treatment strategy for prevention of cytomegalovirus disease in allogeneic marrow transplant recipients [letter]. *Blood* **93**:1781–1782, 1999.

103. Boeckh M, Gooley TA, Bowden RA: Effect of high-dose acyclovir on survival in allogeneic marrow transplant recipients who received ganciclovir at engraftment or for cytomegalovirus pp65 antigenemia [In Process Citation]. *J Infect Dis* **178**:1153–1157, 1998.

104. Einsele H, Ehninger G, Hebart H, *et al*: Polymerase chain reaction monitoring reduces the incidence of cytomegalovirus disease and the duration and side effects of antiviral therapy after bone marrow transplantation. *Blood* **86**:2815–2820, 1995.

105. Zaia JA, Gallez-Hawkins GM, Tegtmeier BR, *et al*: Late cytomegalovirus disease in marrow transplantation is predicted by virus load in plasma. *J Infect Dis* **176**:782–785, 1997.

106. Nguyen Q, Champlin R, Giralt S, *et al*: Late cytomegalovirus

pneumonia in adult allogeneic blood and marrow transplant recipients. *Clin Infect Dis* **28**:618–623, 1999.

107. Zaia JA, Schmidt GM, Chao NJ, *et al*: Preemptive ganciclovir administration based solely on asymptomatic pulmonary cytomegalovirus infection in allogeneic bone marrow transplant recipients: Long-term follow-up. *Biol Blood Marrow Transplant* **1**:88–93, 1995.

108. Boeckh M, Riddell SR, Cunningham T, *et al*: Increased incidence of late CMV disease in allogeneic marrow transplant recipients after ganciclovir prophylaxis is due to a lack of CMV-specific T cell responses. *Blood* **88**(Suppl 1):302a, 1996.

109. Boeckh M, Storer B, Bensinger W, *et al*: CMV infection and disease after marrow vs PBSC transplantation: Results from a randomized trial. Program and abstracts, 43rd Annual Meeting of the American Society of Hematology, Orlando, FL, 2001, abstract 2006.

110. Gluckman E, Rocha V, Boyer-Chammard A, *et al*: Outcome of cord-blood transplantation from related and unrelated donors. *N Engl J Med* **337**:373–381, 1997.

111. Rocha V, Wagner JE, Sobocinski KA, *et al*: Graft-versus-host disease in children who have received a cord-blood or bone marrow transplant from an HLA-identical sibling. Eurocord and International Bone Marrow Transplant Registry Working Committee on Alternative Donor and Stem Cell Sources. *N Engl J Med* **342**:1846–1854, 2000.

112. Junghanss C, Boeckh M, Carter R, *et al*: Incidence of herpesvirus infections following nonmyeloablative allogeneic stem cell transplantation. 42nd Annual Meeting of the American Society of Hematology, San Francisco, 2000.

113. Broers AE, van Der Holt R, van Esser JW, *et al*: Increased transplant-related morbidity and mortality in CMV-seropositive patients despite highly effective prevention of CMV disease after allogeneic T-cell-depleted stem cell transplantation. *Blood* **95**:2240–2245, 2000.

114. Wingard JR, Chen DY, Burns WH, *et al*: Cytomegalovirus infection after autologous bone marrow transplantation with comparison to infection after allogeneic bone marrow transplantation. *Blood* **71**:1432–1437, 1988.

115. Reusser P, Fisher LD, Buckner CD, *et al*: Cytomegalovirus infection after autologous bone marrow transplantation: Occurrence of cytomegalovirus disease and effect on engraftment. *Blood* **75**:1888–1894, 1990.

116. Ljungman P, Biron P, Bosi A, *et al*: Cytomegalovirus interstitial pneumonia in autologous bone marrow transplant recipients. Infectious Disease Working Party of the European Group for Bone Marrow Transplantation. *Bone Marrow Transplant* **13**:209–212, 1994.

117. Boeckh M, Gooley TA, Reusser P, Buckner CD, Bowden RA: Failure of high-dose acyclovir to prevent cytomegalovirus disease after autologous marrow transplantation. *J Infect Dis* **172**:939–943, 1995.

118. Enright H, Haake R, Weisdorf D, *et al*: Cytomegalovirus pneumonia after bone marrow transplantation. Risk factors and response to therapy. *Transplantation* **55**:1339–1346, 1993.

119. Boeckh M, Stevens-Ayers T, Bowden RA: Cytomegalovirus pp65 antigenemia after autologous marrow and peripheral blood stem cell transplantation. *J Infect Dis* **174**:907–912, 1996.

120. Hebart H, Schroder A, Loffler J, *et al*: Cytomegalovirus monitoring by polymerase chain reaction of whole blood samples from patients undergoing autologous bone marrow or peripheral blood progenitor cell transplantation. *J Infect Dis* **175**:1490–1493, 1997.

121. Gerna G, Furione M, Baldanti F, *et al*: Quantitation of human cytomegalovirus DNA in bone marrow transplant recipients. *Br J Haematol* **91**:674–683, 1995.

122. Boeckh M, Boivin G: Quantitation of cytomegalovirus: Methodologic aspects and clinical applications. *Clin Microbiol Rev* **11**:533–554, 1998.

123. Gerna G, Furione M, Baldanti F, Sarasini A: Comparative quantitation of human cytomegalovirus DNA in blood leukocytes and plasma of transplant and AIDS patients. *J Clin Microbiol* **32**:2709–2717, 1994.

124. Boeckh M, Hawkins G, Myerson D, Zaia J, Bowden RA: Plasma PCR for cytomegalovirus DNA after allogeneic marrow transplantation: Comparison with PCR using peripheral blood leukocytes, pp65 antigenemia, and viral culture. *Transplantation* **64**:108–113, 1997.

125. Ljungman P, Plotkin SA: Workshop on CMV disease: Definitions, clinical severity scores, and new syndromes. *Scand J Infect Dis* **99**:S87–S88, 1995.

126. Crawford SW, Bowden RA, Hackman RC, *et al*: Rapid detection of cytomegalovirus pulmonary infection by bronchoalveolar lavage and centrifugation culture. *Ann Intern Med* **108**:180–185, 1988.

127. Ruutu P, Ruutu T, Volin L, *et al*: Cytomegalovirus is frequently isolated in bronchoalveolar lavage fluid of bone marrow transplant recipients without pneumonia. *Ann Intern Med* **112**:913–916, 1990.

128. Schmidt GM, Horak DA, Niland JC, *et al*: A randomized, controlled trial of prophylactic ganciclovir for cytomegalovirus pulmonary infection in recipients of allogeneic bone marrow transplants. The City of Hope-Stanford-Syntex CMV Study Group [see comments]. *N Engl J Med* **324**:1005–1011, 1991.

129. Cathomas G, Morris P, Pekle K, Cunningham I, Emanuel D: Rapid diagnosis of cytomegalovirus pneumonia in marrow transplant recipients by bronchoalveolar lavage using the polymerase chain reaction, virus culture, and the direct immunostaining of alveolar cells. *Blood* **81**:1909–1914, 1993.

130. Ljungman P, Griffith P: *Definitions of Cytomegalovirus Disease: Multidisciplinary Approach to Understanding CMV Disease.* Excerpta Medica, Amsterdam, 1995.

131. Boivin G, Olson CA, Quirk MR, *et al*: Quantitation of cytomegalovirus DNA and characterization of viral gene expression in bronchoalveolar cells of infected patients with and without pneumonitis. *J Infect Dis* **173**:1304–1312, 1996.

132. Hackman RC, Wolford JL, Gleaves CA, *et al*: Recognition and rapid diagnosis of upper gastrointestinal cytomegalovirus infection in marrow transplant recipients. A comparison of seven virologic methods. *Transplantation* **57**:231–237, 1994.

133. Ljungman P, Griffith P: *Definitions of Cytomegalovirus Infection and Disease: Multidiscipinary Approach to Understanding Cytomegalovirus Disease.* Excerpta Medica, Amsterdam, 1993.

134. Cinque P, Vago L, Dahl H, *et al*: Polymerase chain reaction on cerebrospinal fluid for diagnosis of virus-associated opportunistic diseases of the central nervous system in HIV-infected patients. *AIDS* **10**:951–958, 1996.

135. Torok-Storb B, Simmons P, Khaira D, Stachel D, Myerson D: Cytomegalovirus and marrow function. *Ann Hematol* **64**:A128–A131, 1992.

136. Boeckh M, Hoy C, Torok-Storb B: Occult cytomegalovirus infection of marrow stroma. *Clin Infect Dis* **26**:209–210, 1998.

137. Meyers JD, Reed EC, Shepp DH, *et al*: Acyclovir for prevention of cytomegalovirus infection and disease after allogeneic marrow transplantation. *N Engl J Med* **318**:70–75, 1988.

138. Prentice HG, Gluckman E, Powles RL, *et al*: Impact of long-term

acyclovir on cytomegalovirus infection and survival after allogeneic bone marrow transplantation. European Acyclovir for CMV Prophylaxis Study Group. *Lancet* **343:**749–753, 1994.

139. Ljungman P, Camara R, Milpied N, *et al*: A randomized study of valacyclovir as prophylaxis against CMV infection and disease in BMT recipients. 25th Annual Meeting European Group for Blood and Marrow Transplantation, Hamburg, 1999, Vol. Abstract 212.

140. Goodrich JM, Bowden RA, Fisher L, *et al*: Ganciclovir prophylaxis to prevent cytomegalovirus disease after allogeneic marrow transplant. *Ann Intern Med* **118:**173–178, 1993.

141. Winston DJ, Ho WG, Bartoni K, *et al*: Ganciclovir prophylaxis of cytomegalovirus infection and disease in allogeneic bone marrow transplant recipients. Results of a placebo-controlled, double-blind trial. *Ann Intern Med* **118:**179–184, 1993.

142. Salzberger B, Bowden RA, Hackman R, Davis C, Boeckh M: Neutropenia in allogeneic marrow transplant recipients receiving ganciclovir for prevention of CMV disease: Risk factors and outcome. *Blood* **90:**2502–2508, 1997.

143. Atkinson K, Downs K, Golenia M, *et al*: Prophylactic use of ganciclovir in allogeneic bone marrow transplantation: Absence of clinical cytomegalovirus infection. *Br J Haematol* **79:**57–62, 1991.

144. Atkinson K, Arthur C, Bradstock K, *et al*: Prophylactic ganciclovir is more effective in HLA-identical family member marrow transplant recipients than in more heavily immune-suppressed HLA-identical unrelated donor marrow transplant recipients. Australian Bone Marrow Transplant Study Group. *Bone Marrow Transplant* **16:**401–405, 1995.

145. Przepiorka D, Ippoliti C, Panina A, *et al*: Ganciclovir three times per week is not adequate to prevent cytomegalovirus reactivation after T cell-depleted marrow transplantation. *Bone Marrow Transplant* **13:**461–464, 1994.

146. von Bueltzingsloewen A, Bordigoni P, Witz F, *et al*: Prophylactic use of ganciclovir for allogeneic bone marrow transplant recipients [published erratum appears in *Bone Marrow Transplant* **13**(2):232, 1994]. *Bone Marrow Transplant* **12:**197–202, 1993.

147. Yau JC, Dimopoulos MA, Huan SD, *et al*: Prophylaxis of cytomegalovirus infection with ganciclovir in allogeneic marrow transplantation. *Eur J Haematol* **47:**371–376, 1991.

148. Reusser P, Gambertoglio JG, Lilleby K, Meyers JD: Phase I–II trial of foscarnet for prevention of cytomegalovirus infection in autologous and allogeneic marrow transplant recipients [see comments]. *J Infect Dis* **166:**473–479, 1992.

149. Bacigalupo A, Tedone E, Van Lint MT, *et al*: CMV prophylaxis with foscarnet in allogeneic bone marrow transplant recipients at high risk of developing CMV infections. *Bone Marrow Transplant* **13:**783–788, 1994.

150. Einsele H, Hebart H, Kauffmann-Schneider C, *et al*: Risk factors for treatment failures in patients receiving PCR-based preemptive therapy for CMV infection. *Bone Marrow Transplant* **25:**757–763, 2000.

151. Verdonck LF, Dekker AW, Rozenbergarska M, Vandenhoek MR: A risk-adapted approach with a short course of ganciclovir to prevent cytomegalovirus (Cmv) pneumonia in Cmv-seropositive recipients of allogeneic bone marrow transplants. *Clin Infect Dis* **24:**901–907, 1997.

152. Emery VC, Cope AV, Bowen EF, Gor D, Griffiths, PD: The dynamics of human cytomegalovirus replication *in vivo*. *J Exp Med* **190:**177–182, 1999.

153. Emery VC, Sabin CA, Cope AV, *et al*: Application of viral-load kinetics to identify patients who develop cytomegalovirus disease after transplantation. *Lancet* **355:**2032–2036, 2000.

154. Nichols WG, Corey L, Gooley T, *et al*: Rising pp65 antigenemia during preemptive anticytomegalovirus therapy after allogeneic hematopoietic stem cell transplantation: Risk factors, correlation with DNA load, and outcomes. *Blood* **97:**867–874, 2001.

155. Vlieger AM, Boland GJ, Jiwa NM, *et al*: Cytomegalovirus antigenemia assay or PCR can be used to monitor ganciclovir treatment in bone marrow transplant recipients. *Bone Marrow Transplant* **9:**247–253, 1992.

156. Reusser P, Einsele H, Lee J, *et al*: Randomized multicenter trial of foscarnet versus ganciclovir for preemptive therapy of cytomegalovirus infection after allogeneic stem cell transplantation. *Blood* 2002; In press.

157. Ljungman P, Lambertenghi D, Platzbecker U, *et al*: Cidofovir for CMV infection and disease in allogeneic stem cell transplant recipients. *Blood* **97:**388–392, 2001.

158. Meyers JD, Leszczynski J, Zaia JA, *et al*: Prevention of cytomegalovirus infection by cytomegalovirus immune globulin after marrow transplantation. *Ann Intern Med* **98:**442–446, 1983.

159. Winston DJ, Ho WG, Lin CH, *et al*: Intravenous immunoglobulin for modification of cytomegalovirus infections associated with bone marrow transplantation. Preliminary results of a controlled trial. *Am J Med* **76:**128–133, 1984.

160. Winston DJ, Ho WG, Lin CH, *et al*: Intravenous immune globulin for prevention of cytomegalovirus infection and interstitial pneumonia after bone marrow transplantation. *Ann Intern Med* **106:**12–18, 1987.

161. Ringden O, Pihlstedt P, Volin L, *et al*: Failure to prevent cytomegalovirus infection by cytomegalovirus hyperimmune plasma: A randomized trial by the Nordic Bone Marrow Transplant Group. *Bone Marrow Transplant* **2:**299–305, 1987.

162. Bowden RA, Sayers M, Flournoy N, *et al*: Cytomegalovirus immune globulin and seronegative blood products to prevent primary cytomegalovirus infection after marrow transplantation. *N Engl J Med* **314:**1006–1010, 1986.

163. Bowden RA, Fisher LD, Rogers K, Cays M, Meyers JD: Cytomegalovirus (CMV)-specific intravenous immunoglobulin for the prevention of primary CMV infection and disease after marrow transplant [see comments]. *J Infect Dis* **164:**483–487, 1991.

164. Sullivan KM, Kopecky KJ, Jocom J, *et al*: Immunomodulatory and antimicrobial efficacy of intravenous immunoglobulin in bone marrow transplantation [see comments]. *N Engl J Med* **323:**705–712, 1990.

165. Ruutu T, Ljungman P, Brinch L, *et al*: No prevention of cytomegalovirus infection by anti-cytomegalovirus hyperimmune globulin in seronegative bone marrow transplant recipients. The Nordic BMT Group. *Bone Marrow Transplant* **19:**233–236, 1997.

166. Zikos P, Van Lint MT, Lamparelli T, *et al*: A randomized trial of high dose polyvalent intravenous immunoglobulin (HDIgG) vs. cytomegalovirus (CMV) hyperimmune IgG in allogeneic hemopoietic stem cell transplants (HSCT). *Haematologica* **83:**132–137, 1998.

167. Boeckh M, Bowden RA, Storer B, *et al*: Randomized, placebo controlled, double-blind study of a CMV glycoprotein H-specific monoclonal antibody (MSL-109) for prevention of CMV infection after allogeneic hematopoietic stem cell transplant. *Biol Blood Marrow Transplant* **7:**343–351, 2001.

168. Bowden RA, Slichter SJ, Sayers M, *et al*: A comparison of filtered leukocyte-reduced and cytomegalovirus (CMV) seronegative blood products for the prevention of transfusion-associated CMV infection after marrow transplant [see comments]. *Blood* **86:**3598–3603, 1995.

169. Goodrich JM, Boeckh M, Bowden R: Strategies for the prevention of cytomegalovirus disease after marrow transplantation. *Clin Infect Dis* **19**:287–298, 1994.

170. Ljungman P, Engelhard D, Link H, *et al*: Treatment of interstitial pneumonitis due to cytomegalovirus with ganciclovir and intravenous immune globulin: Experience of European Bone Marrow Transplant Group. *Clin Infect Dis* **14**:831–835, 1992.

171. Ljungman P, Cordonnier C, Einsele H, *et al*: Use of intravenous immune globulin in addition to antiviral therapy in the treatment of CMV gastrointestinal disease in allogenic bone marrow transplant patients: A report from the European Group for Blood and Marrow Transplantation (EBMT). Infectious Disease Working Party of the EMBT. *Bone Marrow Transplant* **21**:473–476, 1998.

172. Boeckh M: Current antiviral strategies for controlling cytomegalovirus in hematopoietic stem cell transplant recipients: Prevention and therapy. *Transplant Infect Dis* **1**:165–178, 1999.

173. Bilgrami S, Almeida GD, Quinn JJ, *et al*: Pancytopenia in allogeneic marrow transplant recipients: Role of cytomegalovirus. *Br J Haematol* **87**:357–362, 1994.

174. Boeckh M, Gooley T, Myerson D, Torok-Storb B: Specific CMV genotypes are associated with viremia in patients with CMV disease after allogeneic marrow transplant. 8th European Congress of Clinical Microbiology and Infectious Diseases, Lausanne, Switzerland, 1997.

175. Boeckh M, Bowden RA, Goodrich JM, Pettinger M, Meyers JD: Cytomegalovirus antigen detection in peripheral blood leukocytes after allogeneic marrow transplantation. *Blood* **80**:1358–1364, 1992.

176. Whimbey E, Vartivarian SE, Champlin RE, *et al*: Parainfluenza virus infection in adult bone marrow transplant recipients. *Eur J Clin Microbiol Infect Dis* **12**:699–701, 1993.

177. Whimbey E, Champlin RE, Couch RB, *et al*: Community respiratory virus infections among hospitalized adult bone marrow transplant recipients. *Clin Infect Dis* **22**:778–782, 1996.

178. Ljungman P, Andersson J, Aschan J, *et al*: Influenza A in immunocompromised patients. *Clin Infect Dis* **17**:244–247, 1993.

179. Ghosh S, Champlin R, Couch R, *et al*: Rhinovirus infections in myelosuppressed adult blood and marrow transplant recipients [see comments]. *Clin Infect Dis* **29**:528–532, 1999.

180. Boeckh M, Hayden F, Corey L, Kaiser L: Detection of rhinovirus RNA in bronchoalveolar lavage in hematopoietic stem cell transplant recipients with pneumonia. 40th Interscience Conference on Antimicrobial Agents and Chemotherapy, Toronto, 2000.

181. Englund JA, Piedra PA, Jewell A, *et al*: Rapid diagnosis of respiratory syncytial virus infections in immunocompromised adults. *J Clin Microbiol* **34**:1649–1653, 1996.

182. Whimbey E, Couch RB, Englund JA, *et al*: Respiratory syncytial virus pneumonia in hospitalized adult patients with leukemia. *Clin Infect Dis* **21**:376–379, 1995.

183. Harrington RD, Hooton TM, Hackman RC, *et al*: An outbreak of respiratory syncytial virus in a bone marrow transplant center. *J Infect Dis* **165**:987–993, 1992.

184. Lewinsohn DM, Bowden RA, Mattson D, Crawford SW: Phase I study of intravenous ribavirin treatment of respiratory syncytial virus pneumonia after marrow transplantation. *Antimicrob Agents Chemother* **40**:2555–2557, 1996.

185. Sparrelid E, Ljungman P, Ekelof-Andstrom E, *et al*: Ribavirin therapy in bone marrow transplant recipients with viral respiratory tract infections. *Bone Marrow Transplant* **19**:905–908, 1997.

186. Ghosh S, Champlin RE, Englund J, *et al*: Respiratory syncytial virus upper respiratory tract illnesses in adult blood and marrow

transplant recipients: Combination therapy with aerosolized ribavirin and intravenous immunoglobulin. *Bone Marrow Transplant* **25**:751–755, 2000.

187. DeVincenzo JP, Hirsch RL, Fuentes RJ, Top FH Jr: Respiratory syncytial virus immune globulin treatment of lower respiratory tract infection in pediatric patients undergoing bone marrow transplantation—A compassionate use experience. *Bone Marrow Transplant* **25**:161–165, 2000.

188. Berrey MM, Boeckh M, Bowden RA, *et al*: Phase I evaluation of the RSV-specific humanized monoclonal antibody palivizumab (MEDI-493) in hematopoietic stem cell transplant recipients. *J Infect Dis* 2001.

189. Ljungman P, Gleaves CA, Meyers JD: Respiratory virus infection in immunocompromised patients. *Bone Marrow Transplant* **4**:35–40, 1989.

190. Adams R, Christenson J, Petersen F, Beatty P: Pre-emptive use of aerosolized ribavirin in the treatment of asymptomatic pediatric marrow transplant patients testing positive for RSV. *Bone Marrow Transplant* **24**:661–664, 1999.

191. Raad I, Abbas J, Whimbey E: Infection control of nosocomial respiratory viral disease in the immunocompromised host. *Am J Med* **102**:48–52, 1997; discussion 53–54.

192. Weinstock DM, Eagan J, Malak SA, *et al*: Control of influenza A on a bone marrow transplant unit [In Process Citation]. *Infect Control Hosp Epidemiol* **21**:730–732, 2000.

193. Wendt CH, Hertz MI: Respiratory syncytial virus and parainfluenza virus infections in the immunocompromised host. *Semin Respir Infect* **10**:224–231, 1995.

194. Chakrabarti S, Collingham KE, Holder K, *et al*: Parainfluenza virus type 3 infections in hematopoietic stem cell transplant recipients: Response to ribavirin therapy. *Clin Infect Dis* **31**:1516–1518, 2000.

195. Bowden RA: Respiratory virus infections after marrow transplant: The Fred Hutchinson Cancer Research Center experience. *Am J Med* **102**:27–30, 1997; discussion 42–43.

196. Hayden FG, Osterhaus AD, Treanor JJ, *et al*: Efficacy and safety of the neuraminidase inhibitor zanamivir in the treatment of influenzavirus infections. GG167 Influenza Study Group [see comments]. *N Engl J Med* **337**:874–880, 1997.

197. Nicholson KG, Aoki FY, Osterhaus AD, *et al*: Efficacy and safety of oseltamivir in treatment of acute influenza: A randomised controlled trial. Neuraminidase Inhibitor Flu Treatment Investigator Group. *Lancet* **355**:1845–1850, 2000.

198. Perfect J, Schell W: The new fungal opportunists are coming. *Clin Infect Dis* **22**:S112–S118, 1996.

199. Goldman M: personal communication, 2000.

200. Marr K, Carter R, Crippa F, Boeckh M, Corey L: Epidemiology of invasive fungal infections in hematopoietic stem cell transplant recipients, Infectious Disease Society of America, New Orleans, 2000.

201. Bodey G, Luna M: Disseminated candidiasis in patients with acute leukemia: Two diseases? *Clin Infect Dis* **27**:238, 1998.

202. Anttila V, Lamminen E, Bondestam S, *et al*: Magnetic resonance imaging is superior to computed tomagraphy and ultrasonography in imaging infectious liver foci in acute leukaemia. *Eur J Haematol* **56**:82–87, 1996.

203. Kantrow SP, Hackman RC, Boeckh M, Myerson D, Crawford SW: Idiopathic pneumonia syndrome: Changing spectrum of lung injury after marrow transplantation. *Transplantation* **63**:1079–1086, 1997.

204. Cone RW, Hackman RC, Huang ML, *et al*: Human herpesvirus 6

in lung tissue from patients with pneumonitis after bone marrow transplantation [see comments]. *N Engl J Med* **329:**156–161, 1993.

205. Ramsey PG, Fife KH, Hackman RC, Meyers JD, Corey L: Herpes simplex virus pneumonia: Clinical, virologic, and pathologic features in 20 patients. *Ann Intern Med* **97:**813–820, 1982.

206. Wade JC, Newton B, McLaren C, *et al*: Intravenous acyclovir to treat mucocutaneous herpes simplex virus infection after marrow transplantation: A double-blind trial. *Ann Intern Med* **96:**265–269, 1982.

207. Wade JC, McLaren C, Meyers JD: Frequency and significance of acyclovir-resistant herpes simplex virus isolated from marrow transplant patients receiving multiple courses of treatment with acyclovir. *J Infect Dis* **148:**1077–1082, 1983.

208. Chen Y, Scieux C, Garrait V, *et al*: Resistant herpes simplex virus type 1 infection: An emerging concern after allogeneic stem cell transplantation [In Process Citation]. *Clin Infect Dis* **31:**927–935, 2000.

209. Safrin S, Berger TG, Gilson I, *et al*: Foscarnet therapy in five patients with AIDS and acyclovir-resistant varicella–zoster virus infection. *Ann Intern Med* **115:**19–21, 1991.

210. Han CS, Miller W, Haake R, Weisdorf D: Varicella zoster infection after bone marrow transplantation: Incidence, risk factors and complications. *Bone Marrow Transplant* **13:**277–283, 1994.

211. Koc Y, Miller KB, Schenkein DP, *et al*: Varicella zoster virus infections following allogeneic bone marrow transplantation: Frequency, risk factors, and clinical outcome. *Biol Blood Marrow Transplant* **6:**44–49, 2000.

212. Bilgrami S, Chakraborty NG, Rodriguez-Pinero F, *et al*: Varicella zoster virus infection associated with high-dose chemotherapy and autologous stem-cell rescue. *Bone Marrow Transplant* **23:**469–474, 1999.

213. Schuchter LM, Wingard JR, Piantadosi S, *et al*: Herpes zoster infection after autologous bone marrow transplantation. *Blood* **74:**1424–1427, 1989.

214. Crippa F, Holmberg L, Hooper H, *et al*: Infections after autologous cd34 selected peripheral blood stem cell transplantation. *Blood* **96:**586a, 2000 (abstract 2514).

215. Ljungman P, Bowden RA, Meyers JD: Cytotoxic activity against varicella–zoster virus-infected target cells after marrow transplantation. *J Clin Lab Immunol* **31:**17–21, 1990.

216. Bowden RA, Rogers KS, Meyers JD: Oral acyclovir for the long-term suppression of varicella zoster virus infection after marrow transplantation. 29th Interscience Conference on Antimicrobial Agents and Chemotherapy, Anaheim, CA, 1989.

217. Ljungman P, Wilczek H, Gahrton G, *et al*: Long-term acyclovir prophylaxis in bone marrow transplant recipients and lymphocyte proliferation responses to herpes virus antigens *in vitro*. *Bone Marrow Transplant* **1:**185–192, 1986.

218. Krause H, Hebart H, Jahn G, Muller CA, Einsele H: Screening for CMV-specific T cell proliferation to identify patients at risk of developing late onset CMV disease. *Bone Marrow Transplant* **19:**1111–1116, 1997.

219. Crippa F, Corey L, Chuang EL, Sale G, Borckh M: Virological, clinical, and ophthalmologic features of cytomegalovirus retinitis after hematopoietic stem cell transplantation. *Clin Infect Dis* **32:**214–219, 2001.

220. Boeckh M, Leisenring W, Riddell SR, *et al*: Late cytomegalovirus disease and mortality in allogeneic marrow transplant recipients: Importance of viral load and CMV-specific T cell immunity. *Blood*. In press.

221. Chen CS, Seidel K, Boeckh M, *et al*: The incidence and risk

factors for developing late pneumonia following autologous and allogeneic hematopoietic stem cell transplantation. ASBMT Annual Meeting, Miami, FL, 1998.

222. Walter E, Chauncey T, Boeckh M, *et al*: *Pneumocystis carinii* prophylaxis after marrow transplantation: Efficacy of trimethoprim sulfamethoxazole (TS) and role of TS desensitization and intravenous pentamidine. 33rd Interscience Conference on Antimicrobial Agents and Chemotherapy, New Orleans, LA, 1993.

223. Gagnon S, Boota AM, Fischl MA, *et al*: Corticosteroids as adjunctive therapy for severe Pneumocystis carinii pneumonia in the acquired immunodeficiency syndrome. A double-blind, placebo-controlled trial. *N Engl J Med* **323:**1444–1450, 1990.

224. Atkinson K, Storb R, Prentice RL, *et al*: Analysis of late infections in 89 long-term survivors of bone marrow transplantation. *Blood* **53:**720–731, 1979.

225. Witherspoon RP, Storb R, Ochs HD, *et al*: Recovery of antibody production in human allogeneic marrow graft recipients: Influence of time posttransplantation, the presence or absence of chronic graft-versus-host disease, and antithymocyte globulin treatment. *Blood* **58:**360–368, 1981.

226. Singhal S, Mehta J: Reimmunization after blood or marrow stem cell transplantation [see comments]. *Bone Marrow Transplant* **23:**637–646, 1999.

227. D'Antonio D, Di Bartolomeo P, Iacone A, *et al*: Meningitis due to penicillin-resistant Streptococcus pneumoniae in patients with chronic graft-versus-host disease. *Bone Marrow Transplant* **9:**299–300, 1992.

228. Slavin MA, Meyers JD, Remington JS, Hackman RC: *Toxoplasma gondii* infection in marrow transplant recipients: A 20 year experience. *Bone Marrow Transplant* **13:**549–557, 1994.

229. Martino R, Maertens J, Bretagne S, *et al*: Toxoplasmosis after hematopoietic stem cell transplantation. *Clin Infect Dis* **31:**1188–1195, 2000.

230. Roemer E, Blau W, Basara N, *et al*: Toxoplasmosis, a severe complication in allogeneic hematopoietic stem cell transplantation: Successful treatment strategies during a 5-year single-center experience. *Clin Infect Dis* **32:**1–8, 2001.

231. Held TK, Kruger D, Switala AR, *et al*: Diagnosis of toxoplasmosis in bone marrow transplant recipients: Comparison of PCR-based results and immunohistochemistry. *Bone Marrow Transplant* **25:**1257–1262, 2000.

232. Kibbler CC, Smith A, Hamilton-Dutoit SJ, *et al*: Pulmonary cryptosporidiosis occurring in a bone marrow transplant patient. *Scand J Infect Dis* **19:**581–584, 1987.

233. Manivel C, Filipovich A, Snover DC: Cryptosporidiosis as a cause of diarrhea following bone marrow transplantation. *Dis Colon Rectum* **28:**741–742, 1985.

234. Smith NH, Cron S, Valdez LM, Chappell CL, White AC Jr: Combination drug therapy for cryptosporidiosis in AIDS [see comments]. *J Infect Dis* **178:**900–903, 1998.

235. Nachbaur D, Kropshofer G, Feichtinger H, Allerberger F, Niederwieser D: Cryptosporidiosis after CD34-selected autologous peripheral blood stem cell transplantation (PBSCT). Treatment with paromomycin, azithromycin and recombinant human interleukin-2. *Bone Marrow Transplant* **19:**1261–1263, 1997.

236. Harrington RD, Woolfrey AE, Bowden R, McDowell MG, Hackman RC: Legionellosis in a bone marrow transplant center. *Bone Marrow Transplant* **18:**361–368, 1996.

237. Meletis J, Arlet G, Dournon E, *et al*: Legionnaires' disease after bone marrow transplantation. *Bone Marrow Transplant* **2:**307–313, 1987.

238. Kool JL, Fiore AE, Kioski CM, *et al*: More than 10 years of unrecognized nosocomial transmission of Legionnaires' disease among transplant patients [see comments]. *Infect Control Hosp Epidemiol* **19**:898–904, 1998.

239. Cloud JL, Carroll KC, Pixton P, Erali M, Hillyard DR: Detection of Legionella species in respiratory specimens using PCR with sequencing confirmation. *J Clin Microbiol* **38**:1709–1712, 2000.

240. Zutter MM, Martin PJ, Sale GE, *et al*: Epstein–Barr virus lymphoproliferation after bone marrow transplantation. *Blood* **72**:520–529, 1988.

241. Shapiro RS, McClain K, Frizzera G, *et al*: Epstein–Barr virus associated B cell lymphoproliferative disorders following bone marrow transplantation. *Blood* **71**:1234–1243, 1988.

242. Hale G, Waldmann H: Risks of developing Epstein–Barr virus-related lymphoproliferative disorders after T-cell-depleted marrow transplants. CAMPATH users. *Blood* **91**:3079–3083, 1998.

243. Papadopoulos EB, Carabasi MH, Castro-Malaspina H, *et al*: T-cell-depleted allogeneic bone marrow transplantation as post-remission therapy for acute myelogenous leukemia: Freedom from relapse in the absence of graft-versus-host disease. *Blood* **91**:1083–1090, 1998.

244. Nash RA, Dansey R, Storek J, *et al*: EBV-associated PTLD after high-dose immunosuppressive therapy and autologous CD34–selected stem cell transplantation for severe autoimmune diseases. *Blood* **96**(Suppl):406a (abstract 1747), 2000.

245. Lucas KG, Filo F, Heilman DK, *et al*: Semiquantitative Epstein–Barr virus polymerase chain reaction analysis of peripheral blood from organ transplant patients and risk for the development of lymphoproliferative disease. *Blood* **92**:3977–3978, 1998.

246. Gustafsson A, Levitsky V, Zou JZ, *et al*: Epstein–Barr virus (EBV) load in bone marrow transplant recipients at risk to develop posttransplant lymphoproliferative disease: Prophylactic infusion of EBV-specific cytotoxic T cells. *Blood* **95**:807–814, 2000.

247. Kuehnle I, Huls MH, Liu Z, *et al*: CD20 monoclonal antibody (rituximab) for therapy of Epstein–Barr virus lymphoma after hemopoietic stem-cell transplantation. *Blood* **95**:1502–1505, 2000.

248. Bollard CM, Rooney CM, Huls MH, *et al*: Long term follow-up of patients who received EBV specific CTLs for the prevention or treatment of EBV lymphoma. *Blood* **96**(Suppl):478a, 2000 (abstract 2057).

249. Papadopoulos EB, Ladanyi M, Emanuel D, *et al*: Infusions of donor leukocytes to treat Epstein–Barr virus-associated lymphoproliferative disorders after allogeneic bone marrow transplantation [see comments]. *N Engl J Med* **330**:1185–1191, 1994.

250. Wang FZ, Dahl H, Linde A, *et al*: Lymphotropic herpesviruses in allogeneic bone marrow transplantation. *Blood* **88**:3615–3620, 1996.

251. Cone RW, Huang ML, Corey L, *et al*: Human herpesvirus 6 infections after bone marrow transplantation: Clinical and virologic manifestations. *J Infect Dis* **179**:311–318, 1999.

252. Kadakia MP, Rybka WB, Stewart JA, *et al*: Human herpesvirus 6: Infection and disease following autologous and allogeneic bone marrow transplantation. *Blood* **87**:5341–5354, 1996.

253. Ljungman P, Wang FZ, Clark DA, *et al*: High levels of human herpesvirus 6 DNA in peripheral blood leucocytes are correlated to platelet engraftment and disease in allogeneic stem cell transplant patients. *Br J Haematol* **111**:774–781, 2000.

254. Wang FZ, Linde A, Hagglund H, *et al*: Human herpesvirus 6 DNA in cerebrospinal fluid specimens from allogeneic bone marrow transplant patients: Does it have clinical significance? *Clin Infect Dis* **28**:562–568, 1999.

255. Buchbinder S, Elmaagacli AH, Schaefer UW, Roggendorf M: Human herpesvirus 6 is an important pathogen in infectious lung disease after allogeneic bone marrow transplantation [In Process Citation]. *Bone Marrow Transplant* **26**:639–644, 2000.

256. Carrigan DR, Knox KK: Bone marrow suppression by human herpesvirus-6: Comparison of the A and B variants of the virus [letter; comment]. *Blood* **86**:835–836, 1995.

257. Mendez JC, Dockrell DH, Espy MJ, *et al*: Human beta-herpesvirus interactions in solid organ transplant recipients. *J Infect Dis* **183**:179–184, 2001.

258. Griffiths PD, Ait-Khaled M, Bearcroft CP, *et al*: Human herpesviruses 6 and 7 as potential pathogens after liver transplant: Prospective comparison with the effect of cytomegalovirus. *J Med Virol* **59**:496–501, 1999.

259. Luppi M, Barozzi P, Schulz TF, *et al*: Nonmalignant disease associated with human herpesvirus 8 reactivation in patients who have undergone autologous peripheral blood stem cell transplantation [In Process Citation]. *Blood* **96**:2355–2357, 2000.

260. Luppi M, Barozzi P, Schulz TF, *et al*: Bone marrow failure associated with human herpesvirus 8 infection after transplantation. *N Engl J Med* **343**:1378–1385, 2000.

261. Shields AF, Hackman RC, Fife KH, Corey L, Meyers JD: Adenovirus infections in patients undergoing bone-marrow transplantation. *N Engl J Med* **312**:529–533, 1985.

262. Hale GA, Heslop HE, Krance RA, *et al*: Adenovirus infection after pediatric bone marrow transplantation. *Bone Marrow Transplant* **23**:277–282, 1999.

263. van Kraaij MG, Dekker AW, Verdonck LF, *et al*: Infectious gastroenteritis: An uncommon cause of diarrhoea in adult allogeneic and autologous stem cell transplant recipients [In Process Citation]. *Bone Marrow Transplant* **26**:299–303, 2000.

264. Flomenberg P, Babbitt J, Drobyski WR, *et al*: Increasing incidence of adenovirus disease in bone marrow transplant recipients. *J Infect Dis* **169**:775–781, 1994.

265. Cox GJ, Matsui SM, Lo RS, *et al*: Etiology and outcome of diarrhea after marrow transplantation: A prospective study. *Gastroenterology* **107**:1398–1407, 1994.

266. Liles WC, Cushing H, Holt S, *et al*: Severe adenoviral nephritis following bone marrow transplantation: Successful treatment with intravenous ribavirin [see comments]. *Bone Marrow Transplant* **12**:409–412, 1993.

267. Chakrabarti S, Collingham KE, Fegan CD, Milligan DW: Fulminant adenovirus hepatitis following unrelated bone marrow transplantation: Failure of intravenous ribavirin therapy. *Bone Marrow Transplant* **23**:1209–1211, 1999.

268. Hayashi M, Lee C, de Magalhaes-Silverman M, *et al*: Adenovirus infections in BMT patients successfully treated with cidofovir. *Blood* **96**(Suppl):189a (abstract 810), 2000.

269. Strasser SI, McDonald GB: Hepatitis viruses and hematopoietic cell transplantation: A guide to patient and donor management. *Blood* **93**:1127–1136, 1999.

270. Lau GK, Lie AK, Kwong YL, *et al*: A case-controlled study on the use of HBsAg-positive donors for allogeneic hematopoietic cell transplantation. *Blood* **96**:452–458, 2000.

271. Strasser SI, Sullivan KM, Myerson D, *et al*: Cirrhosis of the liver in long-term marrow transplant survivors. *Blood* **93**:3259–3266, 1999.

272. Bedi A, Miller CB, Hanson JL, *et al*: Association of BK virus with failure of prophylaxis against hemorrhagic cystitis following bone marrow transplantation. *J Clin Oncol* **13**:1103–1109, 1995.

273. Azzi A, Cesaro S, Laszlo D, *et al*: Human polyomavirus BK

(BKV) load and haemorrhagic cystitis in bone marrow transplantation patients. *J Clin Virol* **14**:79–86, 1999.

274. Bogdanovic G, Ljungman P, Wang F, Dalianis T: Presence of human polyomavirus DNA in the peripheral circulation of bone marrow transplant patients with and without hemorrhagic cystitis. *Bone Marrow Transplant* **17**:573–576, 1996.

275. Peinemann F, de Villiers EM, Dorries K, *et al*: Clinical course and treatment of haemorrhagic cystitis associated with BK type of human polyomavirus in nine paediatric recipients of allogeneic bone marrow transplants. *Eur J Pediatr* **159**:182–188, 2000.

276. Sandler ES, Aquino VM, Goss-Shohet E, Hinrichs S, Krisher K: BK papova virus pneumonia following hematopoietic stem cell transplantation. *Bone Marrow Transplant* **20**:163–165, 1997.

277. Held TK, Biel SS, Nitsche A, *et al*: Treatment of BK virus-associated hemorrhagic cystitis and simultaneous CMV reactivation with cidofovir [In Process Citation]. *Bone Marrow Transplant* **26**:347–350, 2000.

278. Coppo P, Laporte JP, Aoudjhane M, *et al*: Progressive multifocal leucoencephalopathy with peripheral demyelinating neuropathy after autologous bone marrow transplantation for acute myeloblastic leukemia (FAB5). *Bone Marrow Transplant* **23**:401–403, 1999.

279. Re D, Bamborschke S, Feiden W, *et al*: Progressive multifocal leucoencephalopathy after autologous bone marrow transplantation and alpha-interferon immunotherapy. *Bone Marrow Transplant* **23**:295–298, 1999.

280. Taoufik Y, Gasnault J, Karaterki A, *et al*: Prognostic value of JC virus load in cerebrospinal fluid of patients with progressive multifocal leukoencephalopathy. *J Infect Dis* **178**:1816–1820, 1998.

281. Houston S, Roberts N, Mashinter L: Failure of cidofovir therapy in progressive multifocal leukoencephalopathy unrelated to human immunodeficiency virus. *Clin Infect Dis* **32**:150–152, 2001.

282. Azzi A, Fanci R, Ciappi S, Zakrzewska K, Bosi A: Human parvovirus B19 infection in bone marrow transplantation patients. *Am J Hematol* **44**:207–209, 1993.

283. Schleuning M, Jager G, Holler E, *et al*: Human parvovirus B19-associated disease in bone marrow transplantation. *Infection* **27**:114–117, 1999.

284. Kurtzman G, Frickhofen N, Kimball J, *et al*: Pure red-cell aplasia of 10 years' duration due to persistent parvovirus B19 infection and its cure with immunoglobulin therapy [see comments]. *N Engl J Med* **321**:519–523, 1989.

285. Frickhofen N, Arnold R, Hertenstein B, Wiesneth M, Young NS: Parvovirus B19 infection and bone marrow transplantation. *Ann Hematol* **64**(Suppl):A121–A124, 1992.

286. Daneshpouy M, Socie G, Clavel C, *et al*: Human papillomavirus infection and anogenital condyloma in bone marrow transplant recipients. *Transplantation* **71**:167–169, 2001.

287. Roy V, Weisdorf D: Mycobacterial infections following bone marrow transplantation: A 20 year retrospective review. *Bone Marrow Transplant* **19**:467–470, 1997.

288. Budak-Alpdogan T, Tangun Y, Kalayoglu-Besisik S, *et al*: The frequency of tuberculosis in adult allogeneic stem cell transplant recipients in Turkey [In Process Citation]. *Biol Blood Marrow Transplant* **6**:370–374, 2000.

289. Aljurf M, Gyger M, Alrajhi A, *et al*: *Mycobacterium tuberculosis* infection in allogeneic bone marrow transplantation patients. *Bone Marrow Transplant* **24**:551–554, 1999.

290. Choucino C, Goodman SA, Greer JP, *et al*: Nocardial infections in bone marrow transplant recipients. *Clin Infect Dis* **23**:1012–1019, 1996.

291. van Burik JA, Hackman RC, Nadeem SQ, *et al*: Nocardiosis after bone marrow transplantation: A retrospective study. *Clin Infect Dis* **24**:1154–1160, 1997.

292. Bhave AA, Thirunavukkarasu K, Gottlieb DJ, Bradstock K: Disseminated nocardiosis in a bone marrow transplant recipient with chronic GVHD. *Bone Marrow Transplant* **23**:519–521, 1999.

293. Bjerke J, Meyers J, Bowden R: Hepatosplenic candidiasis—A contraindication to marrow transplantation? *Blood* **84**:2811–2814, 1994.

294. Offner F, Cordonnier C, Ljungman P, *et al*: Impact of previous aspergillosis on the outcome of bone marrow transplantation. *Clin Infect Dis* **26**:1098–1103, 1998.

295. Aslan T, Fassas AB, Desikan R, *et al*: Patients with multiple myeloma may safely undergo autologous transplantation despite ongoing RSV infection and no ribavirin therapy. *Bone Marrow Transplant* **24**:505–509, 1999.

296. Buckner CD, Clift RA, Sanders JE, *et al*: Protective environment for marrow transplant recipients: A prospective study. *Ann Intern Med* **89**:893–901, 1978.

297. Beelen DW, Haralambie E, Brandt H, *et al*: Evidence that sustained growth suppression of intestinal anaerobic bacteria reduces the risk of acute graft-versus-host disease after sibling marrow transplantation. *Blood* **80**:2668–2676, 1992.

298. Storb R, Prentice RL, Buckner CD, *et al*: Graft-versus-host disease and survival in patients with aplastic anemia treated by marrow grafts from HLA-identical siblings. Beneficial effect of a protective environment. *N Engl J Med* **308**:302–307, 1983.

299. Serody JS, Berrey MM, Albritton K, *et al*: Utility of obtaining blood cultures in febrile neutropenic patients undergoing bone marrow transplantation [In Process Citation]. *Bone Marrow Transplant* **26**:533–538, 2000.

300. Asano Y, Yoshikawa T, Suga S, *et al*: Postexposure prophylaxis of varicella in family contact by oral acyclovir [see comments]. *Pediatrics* **92**:219–222, 1993.

301. Watson B, Seward J, Yang A, *et al*: Postexposure effectiveness of varicella vaccine. *Pediatrics* **105**:84–88, 2000.

302. Armstrong D, Cohen J: *Infectious Diseases*, 1999. Harcourt Publishers Ltd., London.

17

Infection in the Organ Transplant Recipient

ROBERT H. RUBIN

1. Introduction

Over the past 30 years, organ transplantation has been transformed from an interesting experiment in human immunobiology to the most practical means of rehabilitating patients with end-stage disease of the kidney, heart, liver, lungs, and, perhaps, the pancreas and small bowel. As we enter the twenty-first century, the diseases that can be corrected by transplantation continue to increase, the success continues to grow, and the only limitation appears to be the availability of suitable organs for transplantation—a situation hopefully to be corrected in the not too distant future by xenotransplantation. In the United States at the present time, the statistics are as follows: From 1988 to 1996, the 1-year survival rate for renal allografts from living donors increased from 88.8 to 93.9%, and the rate for cadaveric renal allografts from 75.7 to 87.7%. Of equal importance, the long-term survival of functioning allografts has shown similar improvement: the half-life for grafts from living donors increased from 12.7 to 21.6 years, and for grafts from cadaveric donors 7.9 to 13.8 years. If the data are censored to remove patients who died with functioning allografts, the results are even more impressive: from 16.9 to 35.9

years for recipients of allografts from living donors, and from 11.0 to 19.5 years for recipients of allografts from cadaveric donors.[1] For the other organs, the following 1- and 5-year allograft survival statistics are being achieved according to the most recent statistics from the United Network for Organ Sharing (UNOS): liver—80 and 63%, respectively; heart—85 and 68%; and lung—74 and 42%. UNOS patient survival statistics for the same periods at 1 and 5 years are as follows: cadaveric kidney—94 and 82%; living donor kidney—98 and 91%; liver—88 and 74%; heart—86 and 70%; and lung—75 and 44%.[2]

In some centers, even better results are being achieved. For example, the 1-year graft survival rate at Massachusetts General Hospital for cadaveric donor kidneys, livers, hearts, lungs, and pancreases is > 90%, and the 1-year patient survival rate after both cadaveric and living related donor kidney transplantation is > 95%. The remarkable clinical success being achieved at organ transplant centers throughout the world has been accomplished because of progress in the five major areas that contribute to successful transplantation[3]:

1. Optimal tissue typing and matching of donor organ to potential recipient, with a particular emphasis on presensitization testing, thus minimizing the incidence and extent of the rejection process.[4]
2. Careful donor evaluation, meticulous procurement and preservation of the donor organ, and proper preparation of the recipient (in particular, eradicating all treatable infection prior to transplant).
3. Impeccable surgical technique, resulting in a

Robert H. Rubin • Division of Infectious Disease, Brigham and Women's Hospital; Center for Experimental Pharmacology and Therapeutics, Harvard–Massachusetts Institute of Technology Division of Health Sciences and Technology; and Department of Medicine, Harvard Medical School, Boston, Massachusetts 02114.

Clinical Approach to Infection in the Compromised Host (Fourth Edition), edited by Robert H. Rubin and Lowell S. Young. Kluwer Academic/Plenum Publishers, New York, 2002.

minimum of tissue injury, secure vascular, bladder, ureteral, biliary, and bronchial anastomoses, and the prevention and/or aggressive drainage of fluid collections, be they blood, urine, lymphatic, or biliary in origin.

4. Precise, individualized management of the immunosuppressive regimen; on the one hand, effectively preventing or treating allograft rejection, and, on the other, minimizing the severe depression of a broad range of host defenses against infection that can be the consequence of overly aggressive immunosuppressive therapy.

5. Prevention of infection whenever possible with prophylactic or preemptive antimicrobial therapy, and prompt diagnosis and aggressive treatment of microbial invasion when prevention fails.

The net result has been better control of rejection and better prevention and treatment of infection—the two major barriers to successful organ transplantation. These two are closely related, being essentially mirror images of one another, linked by the requirement for immunosuppressive therapy: any intervention that decreases the incidence of infection will permit the safer deployment of more intensive immunosuppressive therapy and thus better management of rejection; and any intervention that decreases the intensity and extent of rejection, thus permitting lesser amounts of immunosuppressive therapy, will be associated with a lower rate of infection. Rejection and infection may be regarded as the two sides of the same problem. Indeed, cytokines, chemokines, and growth factors elaborated in response to either process modulate the occurrence and severity of the other process.

Thus, the *therapeutic prescription* for the organ transplant recipient should be regarded as having two components: an immunosuppressive program to prevent and treat rejection, and an antimicrobial strategy to make this safe. Until specific tolerance becomes a reality in human transplantation, exogenous immunosuppressive therapy will be required. Every immunosuppressive program that has been devised or is presently in development increases the risk of infection. The responsibility of the clinician is to understand this risk and to develop the necessary strategy to reduce the risk, whether the strategy involves active or passive immunization or the targeted deployment of specific antimicrobial agents. In general, the intensity of the immunosuppression, particularly if effective antimicrobial strategies cannot be developed, will determine both the risk of infection and the risk of certain malignancies, particularly lymphoma.[3,5,6]

1.1. The Therapeutic Prescription

1.1.1. Immunosuppressive Agents Employed in Transplantation

The immunosuppressive drugs currently in use for the prevention and treatment of rejection of organ transplants include the following agents: corticosteroids, azathioprine, mycophenolate mofetil, the calcineurin inhibitors (cyclosporine and tacrolimus), rapamycin (sirolimus), and polyclonal and monoclonal antilymphocyte antibodies.

Corticosteroids, particularly prednisone and methylprednisolone, have been key components of the antirejection program since the earliest days of clinical transplantation. One of the major lessons, however, and a cornerstone of modern immunosuppressive therapy is the recognition that there is a limit to the amount of steroids an individual can receive without an undue risk of infection. This has led to the concept of "steroid sparing" therapy, the addition of other drugs that work by other mechanisms in order to achieve the desired net state of immunosuppression while permitting the lowering of the corticosteroid dose. The effects of corticosteroids on the transplant patient can be divided into two general categories: *anti-inflammatory* and *immunosuppressive*. The anti-inflammatory effects are particularly striking[5,7]:

1. *Effects on cytokine production.* The most important anti-inflammatory effects of steroids are secondary to the inhibition of proinflammatory cytokine production (Table 1).

2. *Effects on circulating leukocytes.* Steroids have a complex effect on leukocytes. Although they raise the circulating levels of polymorphonuclear leukocytes (PMNs)—due to both increased production by the bone marrow and demargination of mature PMNs normally adherent to vascular endothelial cells—steroids also decrease PMN accumulation at sites of infection and inflammation. Steroids not only decrease the numbers of circulating lymphocytes, but also change the relationship among the various lymphocyte subpopulations—increasing the ratio of B to T cells, and CD8 to CD4 cells. In addition, steroids cause a decrease in circulating monocytes, eosinophils, and basophils. Many of these effects are due to the impact of steroids on the production of cytokines.

3. *Effects on arachidonic acid pathways.* Steroids inhibit essentially all arachidonic acid metabolites (prostaglandins, thromboxane, and leuko-

TABLE 1. Effects of Glucocorticoids on Cytokine Production and Inflammatory Responses

Mechanism

Following cell entry, steroids bind to the glucocorticoid receptor in the cytosol, activating it.Movement of this complex into the nucleus then occurs, with binding to a particular site of DNA, termed the *glucocorticoid response element*, which then modulates the transcription of specific target genes.

Effects

- Inhibit the *in vivo* expression of interleukin (IL)-1, -2, -3, -4, -10, tumor necrosis factor (TNF), and interferon-γ.
- Accelerate the breakdown of messenger RNA for IL-1, -3, and granulocyte–macrophage colony-stimulating factor
- Inhibit IL-2 receptor expression and signal transduction
- Inhibit T-cell activation by decreasing production of IL-1 by macrophages
- Inhibit macrophage MHC class II expression, IgG receptor expression, and intracellular killing
- Decrease macrophage production of IL-6, TNF, leukotrienes, prostaglandins, platelet activating factor, elastase, collagenase, and histamine-releasing factor
- Inhibit the inducible form of nitric oxide synthose

trienes), as well as platelet-activating factor—potent mediators of inflammation.

4. *Effects on vascular permeability.* Steroids inhibit the production of mediators of vasodilatation and the response to these mediators. Steroids also inhibit the inducible form of nitric oxide synthase, thus decreasing macrophage nitric oxide production, endothelial relaxation, and microvascular leak. The net result is inhibition of microvascular permeability, an important component of the inflammatory response.

The key immunosuppressive effect of corticosteroids is the inhibition of T-cell activation and proliferation (thus blocking clonal expansion in response to antigenic stimulation), which is mediated by its suppression of interleukin-2 (IL-2) and other cytokines required for this to occur. Hence, there is a striking inhibition of cell-mediated immunity. This renders the organ transplant patient particularly susceptible to herpes group viral infections, hepatitis viruses, fungal, nocardial, and mycobacterial processes, *Strongyloides stercoralis*, and infection with such intracellular organisms as *Listeria monocytogenes* and *Salmonella* species. In contrast, although steroids can inhibit the activation of immature B cells (as would be necessary for a primary response to a vaccine), established B-cell responses (or anamnestic responses) are quite resistant to the effects of steroids. Hence, the clinician would be well advised to complete appropriate im-

munizations, whenever possible, prior to the initiation of corticosteroid therapy.[5–8]

From an infectious disease point of view, the most important adverse effects of corticosteroids have to do with their inhibition of the inflammatory response to microbial invasion, particularly in the early stages of the infectious process. The consequences of the impaired inflammatory response are twofold: *the signs and symptoms of infection, as well as the X-ray findings, will be greatly blunted until late in the clinical course; and the microbial burden at the site of infection is likely to be far higher than that observed in normal hosts.*

Illustrative Case 1

A 57-year-old man underwent single lung transplantation for chronic obstructive pulmonary disease. He was immunosuppressed with antilymphocyte antibody induction therapy, cyclosporine, prednisone, and azathioprine. He was maintained on low-dose trimethoprim–sulfamethoxazole prophylaxis. In the first 4 months posttransplant, he required multiple pulse doses of methylprednisolone (500 mg intravenously) and a maintenance prednisone dose of 25 mg/day (usual maintenance dose of ≤ 15 mg/day) because of severe rejection. He presented 4 months posttransplant with complaints of two or three loose bowel movements and mild left lower quadrant discomfort. On physical examination, his temperature was 99.6°F, respiratory rate 18, pulse 85 and regular, and blood pressure 140/90. The patient was obese and somewhat cushingoid in appearance. Positive physical findings were restricted to the abdomen where bowel sounds were slightly hyperactive and there was mild tenderness to deep palpation in the left lower quadrant. There was no guarding, rebound tenderness, or other signs of peritonitis. Pertinent laboratory data included a WBC of 12,500/mm³, Hct 34, normal urinalysis, BUN 36, creatinine 1.4 mg/dl. A CT scan following the administration of rectal contrast revealed a perforation of the sigmoid colon and evidence of inflammation in the surrounding soft tissues. He was taken to the operating room where active diverticulitis with a 2-cm perforation and free peritonitis were noted. A sigmoid colectomy, colostomy, and Hartmann's pouch were performed, and the patient was treated with intravenous ampicillin–sulbactam. He made an uneventful recovery.

Comment: A potentially lethal intra-abdominal catastrophe presented with minimal symptoms because of the anti-inflammatory effects of the immunosuppressive program, particularly the corticosteroids. Despite a free perforation of colonic contents into the peritoneal cavity, the patient did not manifest signs of peritoneal inflammation. If a prompt CT scan with rectal contrast had not been carried out, it is likely the patient would have died of sepsis. The teaching point is that the clinical presentation of life-threatening infection can be very occult, and the clinician must be quick to pursue an aggressive diagnostic workup in the face of subtle signs and symptoms.

Azathioprine has been a cornerstone of steroid sparing therapy for organ transplant recipients for more than 35 years. It is converted into 6-mercaptopurine following administration, with subsequent transformation to a series of intracellularly active metabolites, most notably thiosinic acid and 6-thioguanine nucleotides. These inhibit both an early step in *de novo* purine synthesis and several steps in the purine salvage pathway, thus depleting cellular purine stores and inhibiting DNA and RNA synthesis. The impact of azathioprine is greatest on actively dividing lymphocytes responding to antigenic stimulation, with minimal effects on mature elements of antigenic memory or end-stage lymphocyte function. Azathioprine's major toxicity is on the bone marrow. Its major adverse effects on the course of infection are of two types: it inhibits microbial specific T-cell responses, thus increasing the risk of infection with herpes group viruses, papillomaviruses, fungi, mycobacteria, *S. stercoralis*, and other intracellular organisms; and neutropenia, which can predispose to bacterial sepsis.[5,9] A key step in the metabolism of azathioprine is catalyzed by the enzyme thiopurine methyltransferase. There is considerable genetic heterogeneity in this enzyme in humans, and it is likely that both bone marrow toxicity and the degree of immunosuppression induced by the drug are significantly influenced by whether an individual is a rapid or slow metabolizer of the drug. In general, optimal results are obtained with azathioprine when dosages of $\geq$ 1.5 mg/kg are employed. However, with the recent pharmacogenetic information on the key role of thiopurine methyltransferase, it is reasonable to believe that phenotyping the patient in terms of the activity of this enzyme should greatly improve the therapeutic index of this very useful drug.[9–12]

Mycophenolate mofetil (MMF) is a prodrug, which is administered orally and then cleaved by plasma esterases quite rapidly into the active agent, mycophenolic acid. MMF is a highly selective, noncompetitive, reversible inhibitor of inosine monophosphate dehydrogenase, a crucial enzyme in the *de novo* biosynthesis of guanosine. Proliferating lymphocytes require this pathway, whereas resting lymphocytes and other cell types can make use of the purine salvage pathway if this synthetic option is blocked. MMF is a potent inhibitor of the proliferative response of both B and T lymphocytes to allospecific stimulation. In addition, because guanine nucleotides are necessary for the glycosylation of adhesion molecules, inhibition of lymphocyte migration to sites of rejection or inflammation and decreased adherence to endothelial cells or extracellular protein matrix can be shown. Practically, mycophenolate is often substituted for azathioprine in immunosuppressive regimens, with there seeming to be a more potent antirejection effect without a major increase in either infection or lymphoma (*vide infra*). In terms of the infections that are promoted by this drug, they are essentially identical to those seen with azathioprine, with the notable exception of a lower incidence of neutropenia-related infection. Doses of 1 g two or three times per day are usually employed, with the dose-limiting toxicity usually being gastrointestinal (cramps and diarrhea) rather than bone marrow dysfunction.[5,9,13,14]

Cyclosporine is the immunosuppressive agent that revolutionized organ transplantation. Prior to the advent of cyclosporine-based immunosuppressive regimens in the early 1980s, the 1-year survival rate for cadaveric allografts was barely 50%. Since the deployment of cyclosporine (or its fellow calcineurin inhibitor, tacrolimus—*vide infra*) as the cornerstone of currently used immunosuppression, 1-year cadaveric allograft survival rates of $\geq$ 85% are being achieved throughout the world. Cyclosporine exerts its effects through a complex signaling pathway that results in the inhibition of the transcriptional activation of lymphokine and other genes required for T-cell proliferation, activation, and function. The first step in mediating cyclosporine effects is binding to the cyclophilins, a particular family of immunophilins. The resulting complex has as a target the calcium-dependent serine-threonine phosphatase, calcineurin. Calcineurin, in turn, is involved in the activation of the transcription factor NF-AT, which is required for the transcription of the genes for the cytokines IL-2, IL-3, IL-4, IL-5, interferon-γ (IFN-γ), tumor necrosis factor-α (TNF-α), granulocyte–monocyte colony-stimulating factor (GM-CSF), and the receptors for IL-2 and IL-7. Of these, the IL-2 effects are particularly crucial, resulting in a potent suppression of T-cell function and effect, making cyclosporine a potent inhibitor of allograft rejection. The infectious disease effects of cyclosporine are a direct result of its therapeutic action—a dose-related inhibition of microbial specific T-cell cytotoxic activity being the most important. Thus, cyclosporine will amplify the extent and the effects of any replicating herpes group virus, most notably cytomegalovirus (CMV) and Epstein–Barr virus (EBV) (*vide infra*).[5,15]

Tacrolimus (FK506) has a mechanism of action similar to that of cyclosporine. Thus, the first step occurs with the binding of cyclosporine to a family of immunophilins similar but distinct from those to which cyclosporine binds. These, termed *FK506 binding proteins* (FKBPs), form a complex with tacrolimus, which inhibits the activity of calcineurin. Once again, the effects of T-cell activation are blocked through the inhibition of IL-2 promoter induction. Calcineurin is required for calcium-dependent signal transduction and the activation of the transcription factor NF-AT required for cytokine gene activation. Tacrolimus affects the same array of cytokine genes as cyclosporine, but is 10–100 times more potent. Other immunosuppressive effects of these drugs include the inhibition of

T-cell proliferation and the inhibition of primary or secondary cytotoxic cell proliferation *in vitro*, although direct cytotoxicity and calcium-independent T-cell stimulation are not affected. Both induced immunoglobulin production by B cells and the proliferation of stimulated B cells are inhibited, thus limiting the ability to respond to vaccine administration. Secondary antibody responses, natural killer or antibody-dependent cytotoxic cell function are not inhibited. The effects of tacrolimus on infection are comparable to those seen with cyclosporine, with a particular impact on CMV- and EBV-associated events (*vide infra*).[5,16,17]

Rapamycin (Sirolimus) is structurally quite similar to tacrolimus, but does not exert its immunosuppressive effects by inhibiting calcineurin or the transcription of lymphokine genes. Rather, its targets include RAFT1/FRAP proteins in mammalian cells, which are associated with cell cycle phase G_1. In addition, rapamycin selectively inhibits the synthesis of ribosomal proteins and the induction of mRNA for new ribosomal proteins, which likewise prolongs the cell cycle at the G_1/S interface. Finally, rapamycin inhibits the progression to DNA synthesis and S phase. The net effect is that rapamycin is a less potent inhibitor of cytokine synthesis than cyclosporine or tacrolimus, but is able to inhibit such cyclosporine-resistant immune functions as B-cell immunoglobulin synthesis, antibody-dependent cellular cytotoxicity, lymphocyte-activated killer cells, and natural killer cells. Of perhaps greater importance is rapamycin's ability to inhibit growth factor signaling for both immune and nonimmune cells. This antiproliferative effect may be useful in the prevention and treatment of chronic allograft injury. At present, the primary use of rapamycin is in synergistic combination with cyclosporine, thus permitting lower doses of cyclosporine and, presumably, lesser amounts of nephrotoxicity. In addition, such a combination may permit more steroid sparing therapy. The infectious disease effects of rapamycin, then, are similar to those of cyclosporine, although a particularly high incidence of transient aphthous ulcers and *Pneumocystis carinii* pneumonia have been noted with rapamycin. Because of the latter effect, anti-*Pneumocystis* prophylaxis (e.g., trimethoprim-sulfamethoxazole) should be mandatory whenever significant doses of rapamycin are used.[5,18-21]

Polyclonal antilymphocyte antibodies (antithymocyte globulin, ATG; antilymphocyte serum, ALS). In the United States there are two currently marketed antithymocyte globulins, one of equine and one of rabbit origin. These are potent pan-T-lymphocyte depleting agents that can be employed in two different modes: as induction therapy immediately following transplantation

to prevent rejection; and as a powerful tool to reverse acute rejection that has not responded to pulse doses of methylprednisone (so-called "steroid fast rejection"). The major advantage of induction therapy is that it permits the avoidance of cyclosporine and tacrolimus for 5-14 days (depending on the duration of the ATG therapy), thus avoiding a nephrotoxic insult for kidneys that may be in the throes of ischemia-reperfusion injury and acute tubular necrosis. ATG preparations are not without their problems, however: first is lack of specificity of this polyclonal preparation—antibodies in the preparation can induce neutropenia, thrombocytopenia, and hemolysis; allergic reactions to the foreign protein are common, with these ranging from fevers to anaphylaxis and serum sickness; finally, particularly with the first two or three doses, ATG administration causes the release of IL-1 and TNF in large amounts, resulting in fever, chills, hypotension, and malaise. In addition, these proinflammatory cytokines play an important role in the pathogenesis of CMV and EBV infection (*vide infra*).[5,22]

Monoclonal antilymphocyte antibodies. OKT3, a murine monoclonal antibody to the CD3 T-cell receptor complex, was developed to provide a more specific alternative to polyclonal ATG. Indeed, it has essentially eliminated the nonspecific antibody-induced problems of ATG, while still providing potent induction and antirejection therapy (OKT3 is, at present, the most potent therapy for steroid fast rejection). The adverse effects of OKT3 of infectious disease importance can be grouped into three general categories: (1) elaboration of proinflammatory cytokines, with consequences similar to those observed with the ATG preparations; (2) inhibition of microbial-specific T-cell responses, thus rendering the patient more susceptible to the effects of the herpes group viruses, fungal invasion, and mycobacterial infection; and (3) immune responses to the murine antibody, which will produce fever and allergic manifestations, as well as interfere with its ability to deplete T cells and effect immunosuppression. The use of both OKT3 and ATG, particularly as antirejection therapy, adds significantly to the net state of immunosuppression, requiring an increase in the antimicrobial program needed to prevent infection (*vide infra*).[5,8,23,24]

In an effort to develop more specific immunosuppressive therapies, and in view of the central role of IL-2 and the IL-2 receptor in lymphocyte proliferation and the rejection response, monoclonal antibodies to the IL-2 receptor have been developed. Daclizumab (Zenepax®), which is a humanized IgG, and basilixmab (Simulect®), which is a human and mouse chimeric antibody, have been approved for the prevention of allograft rejection.

Clinical trials with both of these antibodies have shown that these nondepleting, noncytokine-releasing monoclonal antibodies decrease the incidence of acute rejection episodes. Whether or not this will also decrease the incidence of chronic rejection remains to be determined. It is believed that these antibodies have a less marked effect on the course of viral, fungal, and mycobacterial infection than either ATG or OKT3, at least in part due to the absence of the proinflammatory cytokine response with these newer agents but regularly seen with ATG and OKT3.[5,25]

At the present time, standard immunosuppression consists of three-drug therapy: a calcineurin inhibitor (cyclosporine or tacrolimus), azathioprine or mycophenolate, and low-dose prednisone. In patients with acute renal dysfunction, induction ATG or OKT3 is not uncommonly prescribed, in order to avoid nephrotoxic insults from the calcineurin antagonists. Acute rejection is initially treated with two or three daily doses of intravenous methylprednisolone, with OKT3 or ATG utilized if these do not reverse the process. There is at present considerable interest in preventing chronic allograft injury and being able to safely wean patients off prednisone. Newer agents such as rapamycin and the anti-IL-2 receptor antibodies are getting close scrutiny in this regard.

1.1.2. Principles of Antimicrobial Use in Transplantation

Antimicrobial drugs play a major role in the therapeutic prescription for the organ transplant recipient. There are three modes in which antimicrobial agents can be prescribed[5,8,26]:

1. *Therapeutic mode*, in which antimicrobial agents are administered as curative treatment of a patient with clinical infection.
2. *Prophylactic mode*, in which antimicrobial agents are administered to an entire population before an event to prevent infection. For such a strategy to be worthwhile, three criteria must be met: (a) the infection to be prevented must be common enough and important enough to justify the intervention; (b) the antimicrobial regimen is nontoxic enough to make the intervention safe; and (c) the intervention can be shown to be cost effective.
3. *Preemptive mode*, in which antimicrobial agents are administered before clinical infection occurs to a subgroup of individuals shown to be at high risk for life-threatening infection on the basis of a clinical/epidemiologic or a laboratory marker.

The therapeutic use of antimicrobial agents in transplant patients is greatly influenced by the possibility of drug interactions with the immunosuppressive agents, particularly cyclosporine and tacrolimus. There are three general ways in which antimicrobial agents can interact with the calcineurin inhibitors, two of them due to effects on the metabolism of these drugs by hepatic cytochrome P_{450} enzymes and the third due to enhanced nephrotoxicity[5,8,26]:

1. Antimicrobial agents (most notably rifampin, nafcillin, and isoniazid) can *upregulate* the metabolism of the calcineurin inhibitors, resulting in lower blood levels for a given dose of drug, and a high risk for insufficient immunosuppression and resulting allograft rejection.

2. Antimicrobial agents (most notably the macrolides—erythromycin > clarithromycin > azithromycin; and azoles—ketoconazole > itraconazole, voriconazole > fluconazole) can *downregulate* the metabolism of the calcineurin inhibitors, resulting in higher blood levels for a given dose of drug, and a high risk for overimmunosuppression, infection, and nephrotoxicity.

3. Antimicrobial agents given in appropriate doses to patients with therapeutic (not toxic) blood levels of the calcineurin inhibitors can develop *nephrotoxicity* on the initiation of certain antimicrobial drugs, despite the fact that there is no change in the blood level of the calcineurin inhibitor. There are three variants of this form of interaction: (a) dose related—thus, low doses of fluoroquinolones and trimethoprim–sulfamethoxazole are well tolerated, but higher doses are not (e.g., 250 mg of ciprofloxacin or a single-strength trimethoprim–sulfamethoxazole tablet twice daily are well tolerated, higher doses have a dose-related increased risk of renal dysfunction); (b) "accelerated nephrotoxicity"—thus, drugs such as amphotericin or aminoglycosides, which commonly produce nephrotoxicity with sustained exposure (and drugs such as vancomycin, which rarely do), will now manifest nephrotoxicity when added to a calcineurin inhibitor at a much lower cumulative dose (e.g., 100–200 mg of amphotericin in a transplant patient, as opposed to > 500 mg for someone not receiving immunosuppression); (c) "idiopathic nephrotoxicity"—single doses of gentamicin, amphotericin, and intravenous trimethoprim–sulfamethoxazole in the face of therapeutic levels of cyclosporine or tacrolimus have been documented to produce oliguric renal failure in some patients.

Practically, the risk of these interactions leads to the avoidance of aminoglycosides and amphotericin (when possible) in these patients, with the preferential use of advanced-spectrum beta-lactams and azoles in the treatment of bacterial and fungal infections in transplant recipients. When any drug is utilized that affects the hepatic metabolism of the calcineurin antagonists, it is essential that cyclosporine and tacrolimus levels be closely monitored and dosage adjustments made on both the initiation and completion of the course of antimicrobial therapy.[5,8,26]

The most successful antimicrobial prophylactic strategy in transplant patients is the use of low-dose trimethoprim–sulfamethoxazole posttransplant for a minimum of 6–12 months posttransplant (or even longer). This regimen markedly decreases the incidence of urinary tract infection, and essentially eliminates the risk of *Pneumocystis carinii* pneumonia (~15% risk in the first 6 months posttransplant at most transplant centers if anti-*Pneumocystis* prophylaxis is not administered), listeriosis, nocardiosis, and, perhaps, toxoplasmosis. The other prophylactic strategy that has been clearly shown to be cost effective in transplant patients is perioperative (12–72 hr) wound prophylaxis. In addition, many transplant programs utilize antiviral prophylaxis in the prevention of CMV disease (*vide infra*).[5,8,26]

Two forms of preemptive therapy have been defined and are alternative approaches to the prevention of clinical CMV disease: The first is based on clinical/epidemiologic observations: the administration of OKT3 or ATG to patients seropositive for CMV markedly increases the risk of CMV disease; initiating ganciclovir therapy at the time of antilymphocyte antibody therapy preempts this increased risk. Alternatively, a different approach can be taken: 2–7 days prior to the onset of clinical disease, viremia can be demonstrated by polymerase chain reaction (PCR) or antigenemia assay. A preemptive strategy triggered by such a laboratory result can also be successful. Other preemptive strategies to be considered in the transplant patient include the following: candidiuria in a renal transplant patient carries a risk for the development of obstructing fungal balls and ascending candidal pyelonephritis; accordingly, preemptive therapy of asymptomatic candidiuria in renal transplant patients is recommended. Colonization of the respiratory tract of transplant patients with *Aspergillus* species carries an increased risk of subsequent invasive disease. Accordingly, preemptive anti-*Aspergillus* therapy is advocated in such circumstances.[5,8,24,26]

A final point regarding antibacterial and antifungal therapy is the proper deployment of these drugs in patients with infection complicating a surgical misadventure (or other situation in which tissue is devitalized, fluid collections occur, or visceral leaks, e.g., urine or bile, are present). The optimal use of antimicrobial drugs is in conjunction with corrective surgery. The use of antimicrobial therapy without such surgical correction has a high probability of not only failing but of failing with the induction of antimicrobial-resistant infection.[5,8]

2. Risk of Infection in the Organ Transplant Recipient

Table 2 delineates the broad range of infections that occur in organ transplant recipients, classified by the primary factor by which they invade. The risk of these infections, particularly opportunistic infection, in the organ transplant recipient is largely determined by the interaction among three factors: the occurrence of technical mishaps that lead to devitalized tissue, fluid collections, the ongoing need for invasive devices for vascular access, drainage catheters, and other foreign bodies that abridge or otherwise attenuate the primary mucocutaneous barriers to microbial invasion; the epidemiologic exposures the patient encounters; and the patient's net state of immunosuppression. The first of these, the technical/anatomical abnormalities that develop as a consequence of the surgical procedure and the perioperative care, is the most important cause of infection in the first month posttransplant (although the development of such abnormalities at any time in the posttransplant course can increase the risk of infection). Indeed, the incidence of such events is determined by the complexity of the surgery involved—being most common following liver and lung transplantation, somewhat less common following heart transplantation, and least common following kidney transplantation.[3,5,8,26,27]

In those patients free of technical/anatomic abnormalities—the majority of transplant recipients at the most skilled transplant centers—the risk of infection, particularly opportunistic infection, is largely determined by the interaction between the epidemiologic exposures and the net state of immunosuppression. The relationship between these two factors is semiquantitative: if the epidemiologic exposure to a microbial agent is intense enough, even nonimmunosuppressed individuals will become ill; conversely, if the net state of immunosuppression is great enough, then minimal exposure to normally noninvasive, commensal organisms can result in life-threatening infection. This concept is of great practical importance, as the occurrence of invasive infection related to such organisms as *Aspergillus* species, *Legionella* species, and a variety of gram-negative bacilli, at a time when the net state of immunosuppression should not be great enough to permit such an event to occur, can be an important clue

TABLE 2. Classification of Infections Occurring in Transplant Patients[a]

Infections related to technical complications[b]
   Transplantation of a contaminated allograft
   Anastomotic leak or stenosis
   Wound hematoma
   Intravenous line contamination
   Iatrogenic damage to the skin
   Mismanagement of endotracheal tube leading to aspiration
   Infection related to biliary, urinary, and drainage catheters
Infections related to excessive nosocomial hazard
   Aspergillus species
   Legionella species
   Pseudomonas aeruginosa and other gram-negative bacilli
   Nocardia asteroides
Infections related to particular exposures within the community
   Systemic mycotic infections in certain geographic areas
      Histoplasma capsulatum
      Coccidiodes immitis
      Blastomyces dermatitidis
   Stronglyoides stercoralis
   Community-acquired opportunistic infection resulting from ubiquitous saprophytes in the environment[c]
      Cryptococcus neoformans
      Aspergillus species
      Nocardia asteroides
      Pneumocystis carinii
   Respiratory infections circulating in the community
      Mycobacterium tuberculosis
      Influenza
      Adenoviruses
      Parainfluenza
      Respiratory syncytial virus
   Infections acquired by the ingestion of contaminated food/water
      Salmonella species
      Listeria monocytogenes
Viral infections of particular importance in transplant patients
   Herpes group viruses
   Hepatitis viruses
   Papillomavirus
   Human immunodeficieny virus

[a]Modified from Ref. 28.
[b]All lead to infection with gram-negative bacilli, *Staphylococcus* species, and/or *Candida* species.
[c]The incidence and severity of these infections and, to a lesser extent, the other infections listed, are related to the net state of immunosuppression present in a particular patient.

to an excessive environmental hazard that requires immediate attention.[3,5,8,26,27]

2.1. Epidemiologic Exposures of Importance

The epidemiologic exposures of importance for the transplant recipient can be divided into two general categories: those occurring in the community and those occurring within the hospital environment. Community exposures of potential concern include the following: *Mycobacterium tuberculosis*, the geographically restricted systemic mycoses (*Blastomyces dermatitidis*, *Coccidioides immitis*, and *Histoplasma capsulatum*), *Strongyloides stercoralis*, hepatitis B and C, the human immunodeficiency virus (HIV), the enteric bacterial pathogens (particularly *Salmonella* species), and respiratory viruses that can circulate rapidly within households, the workplace, and the community (e.g., influenza, parainfluenza, adenovirus, and respiratory syncytial virus).[3,5,8,26,27]

In the case of *M. tuberculosis* and the mycoses, similar epidemiologic, pathogenetic, and clinical mechanisms are at work. Primary infection occurs via the lungs, after inhalation of an aerosol containing a large inoculum of infectious particles. Primary infection produces a flu-like syndrome, with progressive pulmonary disease and postprimary dissemination being limited by the development of an intense cell-mediated immune response. Late reactivation occurs when such immunity wanes. With posttransplant immunosuppressive therapy particularly aimed at inhibiting cell-mediated immune function, three patterns of clinical disease are observed following infection with these agents in the transplant patient: progressive primary disease with widespread dissemination; reactivation of a long-dormant infection, with progressive local disease (either in the lungs or at sites of previous metastatic spread such as the skeletal system) and/or secondary dissemination; and superinfection, again with dissemination, as immunosuppression causes a waning of previously acquired immunity, with new exposure causing disease. Thus, when evaluating transplant patients with an infectious disease syndrome, careful attention must be paid to the possibility of both recent and remote exposures to these organisms.[3,5,8,26,27]

Strongyloides stercoralis is the one helminth with an autoinoculation cycle in humans; chronic, albeit asymptomatic, infection can be maintained for decades after an individual was initially infected (long after the patient has left endemic areas such as Southeast Asia or South and Central America), and intense and/or disseminated infection can develop after the initiation of immunosuppressive therapy (*vide infra*).[8,27]

Transplant candidates, by the nature of their underlying diseases, are more likely to have received blood products prior to presentation for transplantation, and thus have been at increased risk for acquisition of both hepatitis viruses (hepatitis B, HBV; hepatitis C, HCV) and HIV. Clearly, they are at the same kind of risk for acquiring these infections following such activities as intravenous drug abuse or high-risk sexual practices as the rest of the population. Since these infections can have a major impact on the posttransplant course, with immunosuppression amplifying the extent and effects of these infections (*vide infra*), patients must be screened for their presence prior to transplant, and they should be counseled and protected from acquiring these infections pre- and posttransplant.

The most common form of community-acquired infection occurring in the transplant recipient is that caused by the acute respiratory viruses. Now that successful transplantation and patient rehabilitation is the rule and not the exception, increasing attention is being paid to the occurrence of such common infections as those caused by influenza, parainfluenza, adenoviruses, respiratory syncytial virus, and, probably, even rhinoviruses. Community-wide outbreaks of these infections will have a particular impact on these patients, with more prolonged courses, a higher rate of pneumonia, and an increased rate of bacterial and fungal superinfection compared with the general population. Lung transplant patients appear to be particularly susceptible to these agents, with long-term and progressive lung injury being a too frequent consequence of such infection in lung transplant recipients.[8,28–34] Postinfluenza myocarditis has even been observed in transplant recipients.[35] A variation on this theme is the increasing concern that community-acquired *Chlamydia pneumoniae* infection could be playing a role in the pathogenesis of chronic vasculopathy, particularly in cardiac allograft recipients.[36] Increased attention must be turned to the evaluation and deployment of such preventive and early intervention strategies as vaccination, antiviral strategies (e.g., rimantadine, amantadine, and the neuraminidase inhibitors for influenza), and, most important at this point in time, the isolation of transplant patients from individuals with respiratory infections in the community.[8]

Other exposures of importance to the transplant patient in the community include the enteric pathogens, particularly *Salmonella* species and *Listeria monocytogenes*, in inadequately prepared food, and such recreational pursuits as gardening, in which aerosols of organic material can be created, resulting in exposure to such organisms as *Aspergillus* and *Nocardia* species.[8]

As important as community exposures are to the occurrence of infection in transplant patients, exposures within the hospital are even more important. Nosocomial epidemics of opportunistic infection in transplant patients caused by *Aspergillus* species, *Legionella* species, and gram-negative organisms such as *Pseudomonas aeruginosa* are well recognized. Two epidemiologic patterns of nosocomial exposure have been defined: *domiciliary* and *nondomiciliary*. Domiciliary exposures occur in the room or on the ward where the patient is housed, and are usually caused by contamination of the air supply or potable water by these opportunistic pathogens. Such contamination can occur because of construction; because of the presence of plants contaminated with gram-negative organisms within the patients' rooms; because of showers and toilet facilities that create aerosols of gram negatives or *Legionella* species when used; and because of contaminated water systems and air conditioning or air handling systems. The net result is that these immunosuppressed individuals inhale aerosols contaminated with excessive numbers of these pathogens, producing life-threatening disease. Outbreaks of this type are characterized by clustering of cases in time and space. By prohibiting the entry of plants and flowers into the rooms of these patients, close monitoring of the functioning of showers and toilets, monitoring the hospital water system for *Legionella* contamination, and by the provision of HEPA-filtered air handling systems (particularly in institutions where air supply has been shown to be a problem), this form of excessive epidemiologic exposure can be essentially eliminated.[8,37,38]

Far more problematic are nondomiciliary exposures. These occur when patients are taken from their rooms to such central facilities for essential procedures as the operating room, the radiology suite, the cardiac catheterization laboratory, and others. It is now apparent that excessive exposures of this type are both more common than domiciliary exposures, and more difficult to identify (because of the lack of clustering of cases in time and space). The common denominator in these two forms of nosocomial exposure is usually hospital construction or refurbishing, with the resultant creation of aerosols of potential pathogens. For example, we have observed life-threatening *Aspergillus* infection in renal transplant patients exposed to construction in a radiology suite, in a heart transplant patient exposed while awaiting endomyocardial biopsy outside a cardiac catheterization laboratory, and liver and lung transplant patients exposed in the operating room or intensive care unit. Others have had similar experiences. As a result, increasing emphasis is being placed on protecting patients with special masks and transport devices when they travel within the hospital, and closely monitoring travel routes within the hospital, thus avoiding areas of construction and refurbishment. The operative principle is that the transplant recipient, like other immunosuppressed hosts, is a "sentinel chicken" within the hospital environment—any excess traffic in microbes will be seen first in this patient population, and constant surveillance is essential to prevent catastrophic outbreaks of life-threatening infection.[3,4,8,24,37–40]

2.2. The Net State of Immunosuppression

The net state of immunosuppression is a complex function determined by the interaction of a number of factors. The prime determinant of the net state of immunosuppression is the immunosuppressive regimen prescribed—the dose, the duration, and the temporal sequence in which the drugs that constitute the regimen are deployed. Also to be considered are host defense defects produced by comorbid conditions or the underlying disease that led to the end organ failure that resulted in the need for transplantation.

The second group of factors include neutropenia and the presence or absence of such acquired abnormalities as damage to the mucocutaneous surfaces of the body: foreign bodies that bypass or otherwise compromise the normal functioning of the mucocutaneous barriers, such as bladder, biliary, chest, and drainage catheters, endotracheal tubes, and vascular access devices. As far as neutropenia is concerned, granulocyte colony-stimulating factor (G-CSF) has been shown to be safe and effective in promptly raising the white cell count when it is low, whether this is due to immunosuppressive drugs (e.g., azathioprine or mycophenolate), interferon therapy, or viral infection (e.g., CMV infection). What is less clear is whether or not this translates into less infectious disease morbidity and mortality. In the case of CMV-induced neutropenia, the likely answer is yes, as G-CSF therapy permits the safe use of intravenous ganciclovir therapy of the virus. In the other instances, the evidence is currently lacking. What is apparent is that the administration of G-CSF as an "immune response modifier" to patients with normal white blood cell counts, while safe, is without benefit.[41,42] Recently, the group at the Cleveland Clinic have reported several cases of hypogammaglobulinemia following heart transplantation, with an increased incidence of such infections as nocardiosis and CMV in this setting. The occurrence of hypogammaglobulinemia was related to intensive immunosuppressive therapy with mycophenolate and tacrolimus. The suggestion is that this phenomenon may be more common than is currently recognized, and can be an important contributor to the net state of immunosuppression.[43]

The mucocutaneous surfaces of the body should be regarded as the primary barriers to microbial invasion in transplant patients, particularly important since the secondary defenses are compromised by immunosuppressive therapy. Because the consequences of invasive infection can be so significant in these patients, great emphasis is placed on protection of the skin and mucosal surfaces, the removal of foreign bodies as quickly as possible, and, perhaps, the employment of new technology such as silver, chlorhexidine, and/or antimicrobial-impregnated devices, particularly vascular access catheters.[44,45] Transplant patients are a "leveraged" population in terms of an increased incidence and severity of device-related infections, and are therefore an obvious group for the trial of such new technology.

The third group of factors that help determine the net state of immunosuppression are incompletely understood at present, and these are metabolic abnormalities such as protein-calorie malnutrition, uremia (the effects of uremia on host defenses are delineated in Table 3), and, perhaps, hyperglycemia. Although these by themselves are not sufficiently potent to measurably increase the rate of infection, when such other factors as immunosuppressive therapy are present, they do have an effect. For example, organ transplant patients with a serum albumin level ≤ 2.8 g/dl have a 10-fold greater risk of developing life-threatening infection.[8,26] Consistent with this observation is a report that a low serum albumin level has been correlated with an increased risk of death in dialysis patients as well.[46]

Geriatric (over the age of 60) transplant recipients have an increased risk of infectious disease complications, compared with younger patients, particularly when mycophenolate rather than azathioprine is part of the immunosuppressive regimen.[47] The mechanism for this observation is unclear, with plausible explanations including the following: specific metabolic abnormalities associated with the aging process; decreased metabolism of a particular component of the immunosuppressive regimen associated with age; and an aging immune system that is particularly susceptible to standard doses of immu-

---

TABLE 3. Host Defense Defects Contributing to the Occurrence of Infection in Uremic Patients

Defects inherent to the uremic condition
1. Depressed cell-mediated immunity (delayed rejection of skin and renal allograft and cutaneous anergy)
2. Delayed appearance of leukocytes at sites of inflammation
3. Attenuated antibody response to vaccines
4. Decreased bone marrow pool of granulocytes

Defects related to management of uremic patients
1. Correctable nutritional deficiencies such as protein malnutrition, and zinc and pyridoxine deficiency
2. Mobilization of iron stores by deferoxamine, increasing risk of mucormycosis, listeriosis, and probably other infections
3. Complement activation and leukocyte dysfunction related to hemodialysis (Cuprophane membrane–blood interaction)
4. Chronic peritoneal dialysis patients have a defect in opsonin activity in peritoneal effluent that correlates with risk of *S. epidermidis* peritonitis

nosuppressive drugs. Whatever the mechanism involved, it is clear that age (presumably including the infant as well as the geriatric patient) should be added to the list of factors that contribute to the net state of immunosuppression.

An interesting observation is that race may play a role in the susceptibility to infection posttransplant. It has long been known that African-Americans have an increased incidence of allograft rejection and allograft loss. Recently, Meier-Kriesche et al.[48] reported a decreased incidence of infection posttransplant in African-Americans. This observation, if confirmed, has important implications. First, it suggests that these patients would tolerate more intense immunosuppressive therapy without an undue increase in infectious complications; that is, standard immunosuppressive regimens, primarily defined in Caucasian populations, could be significantly augmented in African-Americans, with clinical benefit. Second, this could be an important clue to the identity of human immune response genes, the definition of which could allow more precise prescription of immunosuppression to each individual.

A major factor in determining the net state of immunosuppression is the presence or absence of infection with one of the immunomodulating viruses—CMV, EBV, HBV, HCV, HIV, and perhaps such viruses as human herpesvirus-6 (HHV-6). It is now clear that, other than the actual immunosuppressive regimen being employed, these are the most important determinants of the net state of immunosuppression. Indeed, an important aspect of the adverse effects of immunosuppressive therapy are their effects in amplifying the impact of these agents (vide infra). One statistic underlines the importance of these viruses in this regard: Over the past two decades, more than 90% of transplant patients at our institution developing opportunistic infection with such organisms as Aspergillus species, Pneumocystis carinii, Nocardia asteroides, and Cryptococcus neoformans did so in the setting of immunomodulating viral infection. Indeed, the few exceptions were patients who acquired their infection as a result of an unusually intense environmental exposure, usually within the hospital environment.[3,5,8,27]

It is also likely that other factors may contribute to the net state of immunosuppression. Among these factors are the immunogenetic makeup of the individual and such previously unconsidered factors as thrombocytopenia. For example, there is evidence that genetic polymorphisms in the genes that control tumor necrosis factor (TNF) production may be important in this regard, with genetically mediated low production of TNF being associated with an increased risk of infection.[49,50] There is also increasing evidence that platelets are specialized inflammatory cells that play a particularly important role in controlling possible fungal infection. Because of the frequency of thrombocytopenia in liver transplant recipients, this is a particular issue in this patient population.[51]

3. Timetable of Infection in the Organ Transplant Recipient

As immunosuppressive regimens have become standardized, it has become apparent that different pathogens affect the transplant recipient at different time points in the posttransplant course. Thus, although infectious disease syndromes can occur at any point in time, the etiology of these syndromes is very different in the first month than at later times. For example, although CMV infection is the most important single cause of clinical infectious disease syndromes in the period 1–4 months posttransplant, it rarely has clinical effects in the first 20 days. Similarly, cryptococcal disease is unusual earlier than 6 months posttransplant, and clinically significant liver disease from newly acquired hepatitis virus infection is usually not observed for several years posttransplant. In addition, although the impact and some of the details are different for the different organs transplanted, the overall pattern is the same for all forms of organ transplantation (Fig. 1). It is useful to divide the posttransplant period into three phases when evaluating the patient for possible infection[3,5,8,27]:

1. *Infection in the first month posttransplant.* The infectious disease problems in this time period are of three types: (a) Infection that was present in the allograft recipient prior to transplant and that continues posttransplant, perhaps exacerbated by posttransplant immunosuppression. The prime concerns here are pneumonia (particularly in the liver, heart, lung, and heart–lung recipient who may have been intubated or whose airway would not be protected because of encephalopathy or severe debility) and bloodstream infection caused by vascular access infection. Because of a shortage of donors, patients coming to extrarenal transplantation often are severely ill, requiring intensive monitoring and support for maintenance care prior to transplant. This results in an increased incidence of aspiration pneumonia and bloodstream infection prior to transplant. In addition, the advanced debility induced by their severe organ failure increases the risk of such processes as tuberculosis, systemic mycoses (both the geographically restricted mycoses and the opportunistic pathogens such as *Cryptococcus neoformans*), and opportunistic pulmonary infection

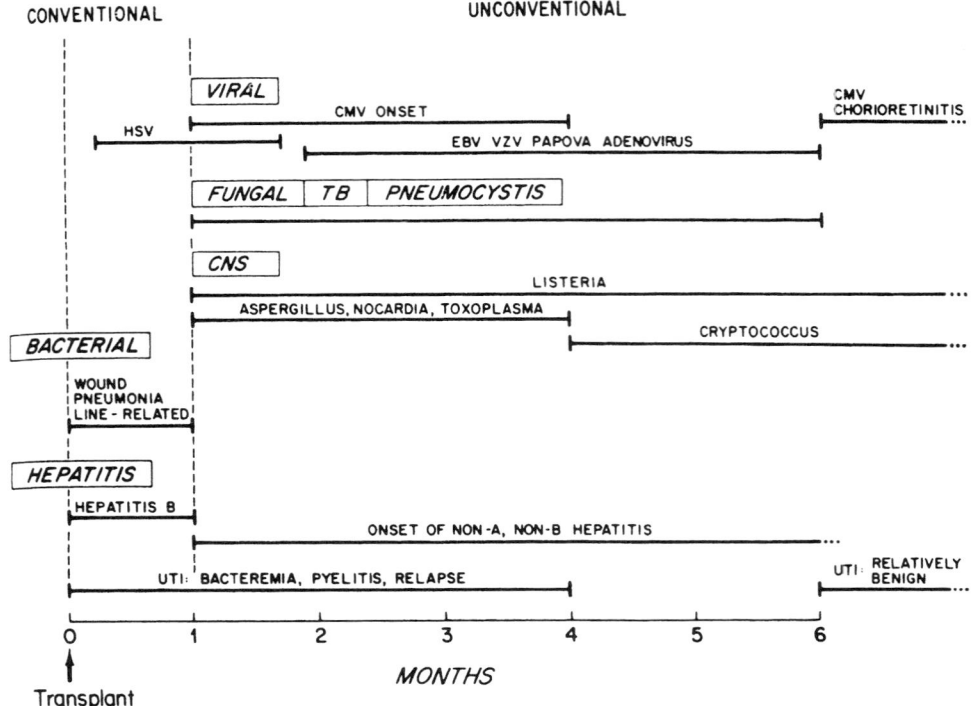

FIGURE 1. Timetable for the occurrence of infection in the renal transplant patient. Exceptions to this timetable should initiate a search for an unusual hazard. CMV, cytomegalovirus; HSV, herpes simplex virus; EBV, Epstein–Barr virus; VZV, varicella–zoster virus; CNS, central nervous system; UTI, urinary tract infection. (Modified from Rubin *et al.*[3])

(including *Pneumocystis carinii* and *Aspergillus* species), particularly if corticosteroids have been used in the management of their primary disease. In general, this is less of an issue for those coming to renal transplantation, since these patients can be maintained usually in a better state of health because of the availability of hemo- and peritoneal dialysis. (b) Infection transmitted via a contaminated allograft, with allograft infection being acquired either from the donor (usually) or in the procurement and preservation process prior to the transplant operation. (c) The same bacterial and candidal infections of the surgical wound, lungs, vascular access, bladder, biliary, chest, and drainage catheters that occur in nonimmunosuppressed patients undergoing comparable amounts of surgery, although the consequences of such infections are much greater in transplant recipients. These last account for greater than 90% of the infections that occur in the first month posttransplant.

It is important to emphasize that infection with such opportunistic pathogens as *Aspergillus* species, *Pneumocystis carinii*, *Legionella* species, and *Nocardia asteroides* is not normally observed in the first month posttransplant. The lack of such infections under normal post-

transplant circumstances at a time when the daily dosage of immunosuppressive therapy is at its highest underlines two important points: (1) The duration of immunosuppression ("the area under the curve") is a more important determinant of the net state of immunosuppression than the particular dose of drug being administered on a given day or over a few days. Indeed, we have likened immunosuppressive therapy to buying a luxury item by credit card—if one raises the dose of immunosuppressive drugs, there is the immediate gratification of improved allograft function, with the bill coming due in 2–4 weeks in terms of infectious disease consequences. (2) The occurrence of opportunistic infection during this first month "golden period" is prima facie evidence of an excessive epidemiologic exposure, since the net state of immunosuppression should not be great enough as yet to permit such an event to occur.

2. Infection 1–6 months posttransplant. Although residual effects of infection acquired earlier may still be noted in this time period, the major infectious disease problems in this time period are predominantly of two different types, with one type contributing to the occurrence of the second: (a) the immunomodulating viruses (particularly CMV, but also including HHV-6, EBV, HBV, HCV, and HIV) which, unless the temporal course is altered by preventive strategies (*vide infra*), exert their primary direct clinical effects in this time period; and (b) such opportunistic pathogens as *Pneumocystis carinii*, *Listeria monocytogenes*, *Aspergillus* species, and *Nocardia asteroides*. The net state of immunosuppression, resulting from the combined effects of sustained immunosuppressive therapy and immunomodulating viral infection, is now great enough for such opportunistic infection to occur, even in the absence of an excessive epidemiologic exposure.[3,5,8,27]

3. *Infection in the late period, more than 6 months posttransplant.* Organ transplant recipients with functioning allografts (and who are thus continuing to receive immunosuppressive therapy) can be divided into three categories in terms of their infectious disease problems: (a) Approximately 75% of transplant recipients will have good graft function, be on maintenance immunosuppression, and be free of chronic viral infection—their infectious disease problems resemble those of the general community (e.g., community-acquired respiratory virus infection, pneumococcal pneumonia, and urinary tract infection); (b) 10–15% of patients have chronic viral infection with CMV, the hepatitis viruses, EBV, papillomavirus, or even HIV, which leads inexorably (unless an effective antiviral program can be prescribed) to destruction of the organ involved (e.g., cirrhosis in the case of HBV and HCV infection), malignancy (hepatocellular carcinoma from the hepatitis viruses, posttransplant lymphoproliferative disease due to the effects of EBV, squamous cell carcinoma from papillomavirus, and, probably, Kaposi's sarcoma from HHV-8), or overt AIDS (from HIV infection); (c) 10–15% of patients with relatively poor allograft function due to both acute and chronic rejection, who have received excessive amounts of immunosuppressive therapy, and who are often infected with one or more of the immunomodulating viruses—these patients, whom we have termed "chronic ne'er-do-wells," are at the highest risk for life-threatening opportunistic infection caused by such agents as *Cryptococcus neoformans*, *Pneumocystis carinii*, *Listeria monocytogenes*, and *Nocardia asteroides*. Ideally, this last group would be better served by decreasing immunosuppression (in the case of renal allograft recipients, returning them to dialysis) and retransplanting. Unfortunately, in part because of the shortage of donors and in part because of the patients' debility at this point, this is often not possible. In that event, prophylactic antimicrobial therapy with trimethoprim–sulfamethoxazole and, possibly, fluconazole should be prescribed for these particularly vulnerable individuals (*vide infra*).[3,5,8,27]

The utility of this timetable as the clinician approaches the issue of infection in the organ transplant recipient is threefold: (1) In the individual patient with a possible infectious disease syndrome the timetable can be quite useful in constructing a differential diagnosis of the possible etiologies. (2) It is a useful infection control device for hospital epidemiologists, as the occurrence of infections not predicted by the timetable is usually a clue to the presence of a previously unidentified environmental exposure of consequence, usually connoting a hazard within the hospital environment. (3) The timetable is the cornerstone on which cost-effective, focused, antimicrobial preventative strategies are based.[3,5,8,27]

Illustrative Case 2

A 56-year-old woman underwent an uncomplicated renal transplant from a cadaveric donor for end-stage renal disease caused by chronic glomerulonephritis. Immunosuppression was accomplished with cyclosporine, mycophenolate, and prednisone, with immediate excellent function. By the third posttransplant day, the serum creatinine level had fallen to 1 mg/dl, where it remained. Ten days posttransplant, fever, chills, and a nonproductive cough developed. Chest radiography revealed a left lower lobe infiltrate. Bronchoscopy with bronchoalveolar lavage was nondiagnostic. Therapy was instituted with piperacillin and tobramycin, but over the next 36 hr increasing dyspnea and fever developed, as well as left-sided pleuritic chest pain. Chest CT scan revealed a pleural-based cavitary lesion in the left lower lobe as well as additional small nodules in both lungs (a total of four, ranging in size from 0.5 to 2.0 cm). Needle aspiration of the cavitary lesion yielded a pure culture of *Aspergillus fumigatus*. Therapy with amphotericin B at a dose of 1.25 mg/kg per day was instituted, and immunosuppression was markedly decreased and then stopped. Unfortunately, progressive disease involving the brain and heart developed, and the patient succumbed. Epidemiologic investigation revealed that this patient and three other immunocompromised patients acquired their invasive aspergillosis through exposure to a radiology suite undergoing reconstruction.

Comment: This is a classic case of a nondomiciliary outbreak of invasive aspergillosis, with a tragic outcome in this woman with a high probability of success from her renal transplant. The lessons here are twofold: rapid, specific diagnosis in transplant patients with progressive pulmonary infection is of critical importance; and opportunistic infection occurring at a point in time when the net state of immunosuppression was relatively low (scarcely 2 weeks posttransplant) is a tip-off to the presence of a significant environmental hazard. This is an example that when the timetable is "wrong," an epidemiologic investigation is in order, even if it is only a single case exception to the timetable.

4. Infection Occurring in the First Month Posttransplant

In the first month posttransplant, the dominant factors determining the risk of infection are technical ones. Because of the technical complexity associated with liver, lung, and heart–lung transplantation, this first month is the critical time period in terms of life-threatening infection for these forms of organ transplantation. Reflecting the technical issues involved, the most important forms of infection are deep wound infection, infection related to vascular insufficiency of the allograft (infarcted tissue being highly prone to microbial invasion), and complications of critical care (pneumonia, if prolonged intubation is required, as well as vascular access and drainage catheter-related sepsis). In the case of renal allograft recipients, the same kinds of technically related infections are of concern, but their incidence is far lower. Cardiac transplant patients have an intermediate risk in terms of the technically related problems of the first month, significantly greater than kidney transplant patients but less than liver and lung recipients. After this first month, the general pattern of infections in terms of incidence, etiology, and clinical significance becomes very similar for all forms of organ transplantation.[3,5,8,27]

When attributing the occurrence of infection to technical issues, this clearly is meant to apply to the surgery itself, as well as the management of the endotracheal tube, vascular access devices, and drainage catheters. One further technical issue that must be stressed is the condition of the organ being transplanted. The result of transplanting damaged, even if uninfected, kidneys, livers, hearts, and lungs is often disastrous. Prolonged allograft nonfunction, usually caused by prolonged ischemia prior to transplantation, has systemic consequences that lead to metabolic derangement, bleeding, encephalopathy, and cardiopulmonary failure—all resulting in increased infection and excessive mortality. Thus, careful attention to the condition of the donor organ is obligatory as part of the technical aspects of transplantation that impact on the incidence of life-threatening infection.[3,5,8]

4.1. Preexisting Infection in the Allograft Recipient

It is axiomatic that a potential organ transplant recipient must be rendered free of active, treatable infection prior to operation and the initiation of immunosuppressive therapy. In particular, ongoing bloodstream infection with *Candida* and bacterial organisms is an absolute contraindication to transplant, as such infection threatens the vascular anastomosis required at the time of transplant. Even in the absence of clear-cut tissue invasion, bloodstream infection should be considered in the following groups of patients: any patient with indwelling vascular access devices, particularly central venous catheters and Swan–Ganz lines; patients with end-stage liver disease and severe portosystemic shunting; and patients with chronic skin disease or compromised bowel mucosal barrier function of any cause. Although chronic infection with such viruses as HBV, HCV, and HIV may be upregu-

lated by immunosuppressive therapy, eradication of the infection prior to transplant is not feasible, and the decision regarding the transplant candidacy of individuals infected with these viruses must be individualized (*vide infra*). In contrast, active bacterial, fungal, and protozoan infections are absolute contraindications to transplantation until they have been eliminated. Among renal transplant recipients, we have observed the following catastrophic forms of infection in the early posttransplant period when active, although asymptomatic, infection was not identified and effectively treated prior to transplant: systemic sepsis and deep wound infection with loss of the allograft caused by smoldering infection of the native kidneys at the time of transplant; miliary tuberculosis in a patient with unsuspected, active tuberculosis at the time of transplant; staphylococcal sepsis from a deep abscess that had been present for months at the site of previous surgery; and progressive lung infection related to a succession of organisms from gram-negative bacilli to fungi, when pneumonia had not been adequately treated prior to transplantation.

The problem is even greater with the extrarenal organs, where the pressure for transplantation in the absence of an artificial life support system akin to dialysis may make the decision-making particularly difficult. A particular concern is the lungs—peritransplant lung injury, whether such injury be due to thromboembolic disease and resulting infarction, chemical injury following aspiration, or microbial invasion, is associated with a high incidence of superinfection posttransplant, with prohibitively high mortality rates resulting. Lung injury must be allowed to heal prior to transplant. An alternative approach can include surgical excision of focal disease, particularly if pulmonary infarction has occurred. This is a particularly important issue in liver transplantation. First, cirrhotic patients are at increased risk for both bacterial and fungal pneumonia.[8] In addition, as failure of the native liver progresses, hepatic encephalopathy ensues, with its associated impairment in the ability to protect the airway against aspiration. We have seen a number of patients with end-stage liver disease lapse into coma, aspirate, and develop diffuse lung injury with infection. Liver transplantation following this event is uniformly unsuccessful, and the ability to clear the lungs of inflammation and infection following this event is very limited; most of these individuals will die before transplantation can safely be performed. Thus, we strongly advocate close monitoring of liver transplant candidates for their ability to protect their airways, and, if this is compromised, intubate prior to aspiration. The patient is then placed on the highest priority emergency transplant list (aiming to transplant within 72 hr of intubation) unless aggressive therapy of the hepatic encephalopathy permits safe extubation and a more leisurely approach to transplantation.[52]

A second issue in liver transplant candidates is intraabdominal infection. The most common untoward event is unrecognized or inadequately treated spontaneous bacterial peritonitis; in addition, a diverticular abscess, appendiceal abscess, or other break in bowel integrity suffered pretransplant can be a significant problem posttransplant. A rare cause of problems is portal vein thrombosis, with retrograde spread into the mesenteric venous system, venous infarction of the bowel, and abscess formation. Again, it is the obligation of the responsible clinician to certify that the patient is free of treatable infection prior to transplantation. As part of this effort, and with the increasing success of liver transplantation, patients with progressive liver disease should be considered, whenever possible, for transplantation weeks to months before the final downhill spiral occurs. Two statements summarize this point of view: If at all possible, the patient should be able to walk to the operating room for the transplant rather than require life support systems. Better to transplant 6 months too early than 1 week too late.[52–54]

In the case of heart, heart–lung, and lung transplant candidates, two particular sites of concern are the respiratory tract and vascular access line-related infection. Active pulmonary infection in these forms of transplantation is likewise associated with a high probability of lethal infection posttransplant; and bacteremia from any source, usually related to venous lines, intra-aortic balloon pumps, or left ventricular assist devices (LVADs) in this situation, threatens the integrity of the vascular anastomoses performed during the transplant procedure. With the increasing use of LVADs as a "bridge" to heart transplantation, there has been particular interest in the infectious complications of these devices and their impact on the subsequent posttransplant course. Currently available information suggests the following[55]: infection complicating the use of the LVAD is common (approximately half of these individuals), including bloodstream, LVAD drivelines, and central venous catheter infection; however, with close surveillance and aggressive bactericidal antimicrobial therapy, these patients can be successfully transplanted, with there being no difference in survival posttransplant in those free of LVAD-associated infection and those with such infection (~60% success rate).

In the case of lung transplant candidates, cystic fibrosis patients are a particular problem, as resistant *Pseudomonas aeruginosa* strains are usually present in the respiratory tract pretransplant, and will usually be present

in the transplanted lungs, despite replacement of both native lungs. Presumably the nasal sinuses are the source of the *Pseudomonas* posttransplant. Two organisms of additional concern are *Aspergillus* species and *Burkholderia cepacia*. Persistent respiratory tract colonization with *Aspergillus* increases the risk of subsequent invasive pulmonary aspergillosis, and attempts to eradicate such carriage with antifungal drugs are recommended by many transplant groups. *B. cepacia* colonization has traditionally been associated with a high rate of lung allograft failure at some, but not all, centers. It is now apparent that *B. cepacia* is not a single species, but rather is a group of closely related bacterial species, all of which are capable of colonizing the respiratory tracts of cystic fibrosis patients. These different species share the characteristic of broad-spectrum antimicrobial resistance, but appear to differ in terms of epidemiology, virulence, and impact on lung transplantation. It is likely that one or more of the *B. cepacia* complex is responsible for a bad outcome following lung transplantation, but that others of the complex have a more benign prognosis. Our view is that until the specifics of these relationships are better delineated, *B. cepacia* colonization should be regarded as a relative, not absolute, contraindication to transplantation.[56–61] In patients with respiratory failure receiving single lung transplants, leaving one native lung in place (e.g., patients with emphysema), the native lung can be a source of posttransplant infection of potentially dire consequence. Hence, careful attention to and eradication of infection in the native lung are essential.[62]

A particular problem of special importance in whole pancreas transplantation is that of candidal infection. Because of the abundance of glucose and glycogen in the body fluids of diabetics, the candidal organism burden is significantly higher in the gut, in the vagina, and on the skin of diabetics. This increases the rate of deep wound infection posttransplant with *Candida* species. Because of this, we advocate fluconazole prophylaxis peritransplant in recipients of pancreatic allografts.[63]

Underlying anatomical or structural weaknesses of important tissues may leave the recipient vulnerable to significant infection posttransplant. Three examples of this phenomenon are especially important: patients with a history of symptomatic diverticulitis pretransplant are at high risk of recurrent diverticulitis with perforation posttransplant; similarly, patients with biliary tract disease pretransplant are at high risk of infectious complications posttransplant; and, finally, patients with significant skin disease (e.g., psoriasis or eczema) are at high risk of staphylococcal bacteremia posttransplant. Recognizing these risks, we advocate sigmoid colectomy for the patients with a history of symptomatic diverticulitis, cholecystectomy for those with cholelithiasis, and aggressive therapy, up to and including pretransplant cyclosporine, to eradicate the skin disease—all to be completed prior to transplant.[8,64]

The 1990s have witnessed an exponential growth in the incidence of methicillin-resistant *Staphylococcus aureus* and vancomycin-resistant *Enterococcus faecium* infection. Transplant recipients, particularly liver transplant recipients, have been particularly affected by these infections, with prolonged in-hospital stays (especially in intensive care units), technical mishaps from the surgery and perioperative care, and exposure to broad-spectrum antibiotics (especially cephalosporins and vancomycin) all playing a role in the epidemiology of these infections. Although the licensure of quinupristin/dalfopristin and linezolid has added significantly to our ability to treat infections due to these organisms, considerable morbidity and mortality, as well as spread to other patients, are still occurring. Surveillance for colonization in the immediate pretransplant period can be helpful in the management of patients posttransplant (e.g., need for infection control precautions, choice of antimicrobial therapy in the initial treatment of infectious disease syndromes, and need for prompt drainage of any fluid collections).[65–69]

The most important parasitic infection that could appear posttransplant, after being asymptomatic pretransplant, is that due to *Strongyloides stercoralis*. *S. stercoralis* is an intestinal nematode that is endemic in many areas of the world. For example, in the United States, it has been found in 36 states.[70,71] The organism has a complex life cycle, the most important aspect of which is an autoinfection component that allows the organism to be maintained in the gastrointestinal (GI) tract of a human host for decades after initial infection had been acquired—long after the host might have moved from an endemic area. Such individuals may be asymptomatic or have only minor GI complaints. Apparently, tissue invasion is prevented by an intact cell-mediated immune response.[72] Following transplantation and the initiation of immunosuppressive therapy, a disastrous hyperinfection syndrome and/or disseminated strongyloidiasis can develop, because of the inhibition of normal cell-mediated immune function. The hyperinfection syndrome represents an exaggeration of the normal life cycle of the parasite, with major impact on the GI tract (a severe, ulcerating, hemorrhagic enterocolitis) and/or lungs (hemorrhagic pneumonia). Disseminated strongyloidiasis consists of extension of the infection outside its normal domain, with the filariform larvae invading all portions of the body. Both of the severe forms of strongy-

loidiasis in the compromised host are associated with recurrent or persistent gram-negative bacteremia and/or meningitis, despite apparently appropriate antibacterial therapy. This appears to be caused by the adherence of gut bacteria to the external surface of the migrating larvae.[70–81] Morgan *et al.*[71] have emphasized that although early diagnosis is the key to effective therapy of these potentially devastating infections in the transplant patient, there are a number of barriers to achieving this: The organism persists long after the individual has left endemic areas; eosinophilia, a cardinal sign of parasitic infection, is commonly absent because of exogenous immunosuppression and concomitant systemic bacterial infection; routine stool examinations for ova and parasites are negative in the overwhelming majority of cases, and more useful diagnostic tests (such as sampling of proximal small bowel contents and sputum cytology) are usually not thought of or performed; and signs and symptoms are nonspecific—complicating bacterial infection will frequently obscure the picture. Therapy of established systemic strongyloidiasis with thiabendazole or ivermectin plus systemic antibacterial therapy aimed at the complicating bacteremia or meningitis is possible (see Chapter 8 for details of management), but mortality remains greater than 50%.[71,73] It is of potential interest that cyclosporine appears to have antiparasitic properties against *S. stercoralis*.[82] However, it is still strongly recommended that the asymptomatic carrier be identified prior to the transplant, and such infection be preemptively eradicated prior to the initiation of immunosuppressive therapy.[83,84] Since routine stool examinations for ova and parasites will diagnose only 27% of asymptomatic carriers,[85] we advocate screening of residents or former residents of *Strongyloides* endemic areas by one of the following approaches pretransplant: examination of Papanicolaou stained smears of sputum and duodenal aspirates and of purged stool specimens; or screen serologically for the presence of antibody to the organism and then making the assumption that a seropositive individual who has not been appropriately treated for this organism harbors it asymptomatically and merits preemptive therapy.

Illustrative Case 3

A 14-year-old boy who had immigrated from Cambodia 2 years previously developed end-stage renal disease caused by reflux nephropathy. While being maintained on chronic hemodialysis, he underwent bilateral nephrectomies in preparation for transplantation. Pretransplant evaluation revealed a negative tuberculin test, no eosinophilia on peripheral blood smear, and three negative stool examinations for ova and parasites. Three weeks after receiving an HLA-identical kidney from his brother, he presented with fever, rigors, cough productive of bloody sputum, abdominal pain, and bloody diarrhea. Chest radiography revealed bilateral patchy densities consistent with bronchopneumonia. Colonoscopy revealed hemorrhagic colitis. Peripheral white blood cell count was 18,000/mm^3 with 85% polymorphonuclear leukocytes and 15% band forms. Two blood cultures grew *E. coli* sensitive to all antibiotics, with persistent bacteremia documented despite therapy with imipenem and gentamicin. He began complaining of a headache 24 hr after admission, and a lumbar puncture revealed an opening pressure of 300 mm H$_2$O, 400 leukocytes (100% polymorphonuclear leukocytes)/mm^3, cerebrospinal fluid (CSF) glucose < 20 mg/dl, and protein 110 mg/dl. CSF cultures grew the same antibiotic-sensitive *E. coli*. Despite continued broad-spectrum antibiotics, the patient died 3 days after admission. Postmortem examination revealed systemic strongyloidiasis involving the GI tract, brain, heart, lung, liver, and renal allograft. There was evidence of an extensive hemorrhagic bronchopneumonia and enterocolitis. Serologic testing on a saved serum specimen drawn pretransplant but tested postmortem was positive for *S. stercoralis* antibody.

Comment. This tragic case of disseminated strongyloidiasis emphasizes several important points. As more transplants are being performed on patients from Southeast Asia and Latin America, the incidence of this entity will continue to increase unless appropriate screening is carried out. Clinical disease usually presents in the first few months posttransplant. Negative routine stool examinations for ova and parasites do not rule out the possibility of this infection, nor does the lack of eosinophilia. The diagnosis of the underlying *S. stercoralis* infection is often obscured by the protean clinical manifestations and the pace of this disseminated syndrome. Treatment of the accompanying bacteremia and bacterial meningitis alone is inadequate if the patient is to be salvaged. This patient would have been best served by preemptive therapy based on a positive serology pretransplant.

4.2. Infection from the Donor

Careful evaluation of the potential organ donor, both living related and cadaveric, is essential in preventing the transmission of life-threatening infection with the allograft. The infections to be considered can be divided into the following categories: (1) active viral infection, particularly with HIV and the hepatitis viruses HBV and HCV; (2) latent infection with such microbial agents as CMV and *Toxoplasma gondii*, which are capable of being reactivated posttransplant, with the possibility of systemic dissemination and clinical disease if appropriate preventive strategies are not utilized; and (3) active infection of the allograft, particularly with bacteria and fungi, but also including certain viruses, that is present as a result of the terminal illness or its care, or as a result of metastatic spread of infection during a preceding illness.[8,27]

Organ transplantation is an extremely efficient means of transmitting HBV, HCV, and HIV. Preventing acute HBV transmission in the peritransplant period by screening of potential donors of both organs and blood is extremely important since acquisition of HBV in this time period is associated with a markedly increased risk of fulminant hepatitis. Presently available methods of screening for hepatitis B surface antigen (HBsAg), the most important marker for the presence of infectious, transmissible virus, are extremely sensitive. It is currently estimated that the risk of transmitting HBV, when adequate testing standards are employed, with blood transfusions is now on the order of 0.002%.[27,86] Transplanting an organ from an HBsAg-positive donor into an anti-HBsAg-negative recipient carries a risk that approaches 100% for infecting the recipient with HBV. Rarely, renal allografts have been transplanted from HBsAg-positive donors into anti-HBsAg-positive recipients without transmitting HBV to the recipient. These successes may have been due to protective effects of previous immunization, despite the absence of a serologic response to the vaccine.[87] Donors who are anti-HBsAg positive may harbor PCR-detectable HBV DNA in their sera and their livers, although the risk of transmitting the virus via the allograft is < 1%.[88,89]

In recent years, there has been considerable interest in the outcome of transplanting organs from donors who are HBsAg negative, but anti-HBc positive. It is now apparent that for renal, and presumably the other extrahepatic organs, the risk of transmitting HBV infection in this fashion is negligible.[90] However, liver allografts procured from HBsAg-negative, anti-HBc-positive donors carry a significant risk of transmitting HBV. Indeed, PCR-detectable virus can be detected in these livers. There is some evidence that the posttransplant HBV infection in these recipients is milder than HBV acquired in other circumstances, and preliminary evidence suggests that posttransplant administration of hepatitis B immune globulin (HBIG) and/or lamivudine is particularly effective in this circumstance. Until more data are available on this point, we would reserve livers from anti-HBc-positive donors for critically ill individuals, and utilize the combination of HBIG and lamivudine posttransplant.[91–97] It is likely that a second group of individuals who would benefit from livers from anti-HBc donors are those who are already anti-HBsAg positive.[97]

The approach to HCV infection remains more controversial. Approximately 5% of potential organ donors are anti-HCV positive, with approximately 50% of these having demonstrable viremia by PCR assay. Organs from donors who are HCV viremic will transmit the virus to a recipient with an efficiency that approaches 100%. Unfortunately, currently available PCR assays do not provide information in a timely enough fashion to impact on the evaluation of a potential cadaveric donor (although, clearly, all possible living donors who are anti-HCV positive should be evaluated for viremia). Hence, the decision-making regarding the suitability of an anti-HCV-positive cadaveric donor cannot be based on knowledge of the true risk of transmission; rather, it has to be made by considering the risks and benefits of providing an organ to a particular individual. Presently available evidence suggests that recipients of organs from an anti-HCV-positive donor have a 50% incidence of becoming anti-HCV positive themselves, ~25% become viremic, and 35% develop liver disease. One suggestion has been to restrict the use of organs from anti-HCV-positive donors to anti-HCV-positive recipients. However, anti-HCV does not prevent reinfection even with the same strain of virus, so this idea has been controversial, although presently available information suggests that liver transplantation from an anti-HCV-positive donor into an anti-HCV-positive recipient does not appear to carry a statistically significant increase in morbidity and mortality. Also, it is clear that adverse effects of HCV infection do not usually become manifest for a number of years after infection acquisition, although chronic liver disease (even requiring liver transplantation) will develop in a significant number of individuals after 5–10 years. Therefore, we have advocated the following approach to organs from anti-HCV-positive donors: We would not use such organs in children and young adults who have the potential for decades of posttransplant life. We would (with informed consent from the recipient) utilize such organs in patients at immediate risk of death without a transplant, the elderly, highly sensitized individuals unlikely to have access to other renal allografts, diabetics, and other individuals with potential life spans of < 10 years. Whether a particular effort in this regard should be devoted to matching anti-HCV-positive donors to anti-HCV-positive recipients remains, in our opinion, unclear at the present time.[86,97–109] The report that certain polyclonal immunoglobulin preparations provide significant protection against HCV infection offers another strategy for safely capturing organs from anti-HCV-positive donors for transplantation.[110]

Although the risks of HBV and HCV infection for transplant patients have been well described, it is clear that as yet undescribed hepatitis viruses are causing chronic hepatitis in transplant recipients; that is, non-A, non-B, non-C, non-D, and non-E hepatitis is present in ~25% of patients with chronic hepatitis. So-called hepatitis G virus, found commonly in both dialysis and transplant recipients, is an "orphan virus," not being linked to any known clinical condition.[111]

HIV infection is also efficiently transmitted by transplantation, with near 100% transmission of the virus from anti-HIV-positive donors. In addition, since there is a window of time in which the virus is present in the blood before the individual seroconverts, there have been a few instances in which HIV has been transmitted from a viremia-positive, antibody-negative donor. There are two safeguards to prevent this uncommon event: First, a careful history should be taken on every potential donor to rule out high-risk behavior that carries a risk for HIV acquisition; i.e., parenteral illegal drug use, male homosexual behavior, promiscuous sexual behavior that could have included contact with HIV-infected individuals, incarceration in a correctional facility, and a history of a clotting disorder (e.g., hemophilia) requiring human-derived clotting factor concentrates. Second, in the near future, it is likely that a rapid test for the HIV p24 antigen will be available, allowing for the routine screening of donors for viremia, especially those in the high-risk group.[97,112,113]

Far more important than concerns regarding false-negative HIV antibody testing is ensuring that the potential donor's own blood and not transfused blood is the specimen being analyzed. In one tragic example, a motor vehicle accident victim who received more than 50 units of blood products prior to being evaluated as a potential donor transmitted HIV infection to the organ recipients. The blood sample that had been judged as HIV antibody negative had been drawn after the administration of the blood products and effective exchange transfusion. In this case, false negativity and a tragic outcome ensued.[113]

Serologic tests for several latent infections that can be carried by the allograft to the recipient, with reactivation and clinical disease posttransplant are well recognized. Important examples of this are CMV, EBV, and, in the case of heart transplant recipients, *Toxoplasma gondii*. Seropositivity for each of these agents connotes latent infection, with this information, plus similar information on the recipient, determining the nature of the posttransplant preventive strategies that will be employed (*vide infra*).

Rarely, dormant fungal or mycobacterial infection that had asymptomatically metastasized to the organ (most commonly the kidney) of the donor during recognized or unrecognized primary infection can be passed to the recipient with potentially catastrophic results. Examples of histoplasmosis[114] and cryptococcosis[115] transmitted in this fashion have been reported. It is thus important in the evaluation of a potential donor that a complete clinical and epidemiologic history be obtained. Although we would not rule out a potential donor on the basis of a positive tuberculin test or a history of residence in a geographic area endemic for histoplasmosis, blastomycosis, or coccidioidomycosis, such information is useful in caring for the recipient posttransplant in terms of the index of suspicion and type of evaluation to which the recipient is subjected for otherwise unexplained febrile illnesses.

The most critical issue in evaluating potential donors, particularly cadaveric donors, is the possibility of transmitting infection related to the terminal illness. There are two considerations here: systemic or isolated organ infection that was the cause of the patient's demise; and bacterial or fungal infection that was acquired in the terminal stage of the individual's illness as part of the care of the patient in the intensive care unit (e.g., vascular access sepsis, pneumonia caused by aspiration, catheter-related urinary tract infection). The latter are usually the most important considerations in evaluating the donor, although the consequences of transplanting an organ from a patient with ongoing bacteremia or fungemia of either type are the same: anastomotic suture lines are threatened, with the vascular suture line being at particular risk for the development of mycotic aneurysms and of catastrophic rupture.[97,116,117]

Clearly, all potential donors with inadequately treated systemic infection are eliminated from consideration. The consequences of failing to do so are illustrated by the report of the cross-Canada spread of the same methicillin-resistant *Staphylococcus aureus* from a cadaveric donor to recipients of the kidneys and liver, with a corneal ring also being positive for the same organism.[118] Similarly, focal infection of the kidneys, lungs, biliary tree, or heart will forestall the use of these organs for transplantation. The key point is not to transplant an already infected organ, or an organ obtained from a patient with ongoing bacteremia or fungemia.[116,117,119–122]

The first group of issues in this area that need to be addressed has to do with the possibility of utilizing organs from donors with systemic infection. Although the individual may have developed irreversible brain injury as a consequence of the infectious process, effective antibiotic therapy may have sterilized the infection such that the organs can be safely used for transplantation. For example, organs from a patient dying of enterococcal endocarditis treated with several days of antibiotics prior to organ procurement have been successfully utilized for transplantation.[123] Because there is considerable interest in increasing the number of organs that might be safely transplanted, we have suggested that the following issues be considered when confronted with this question[117]:

1. *The organism(s).* Not all microbes are of equal virulence in terms of adherence to cardiovascular endothelial surfaces or the ability to metastasize to organs of interest for transplantation. Thus, enterococci, *Staphylococcus aureus*, viridans streptococci, and *Pseudomonas aeruginosa* are noted for their ability to adhere to endothelial surfaces, and *Staphylococcus aureus*, *Salmonella* species, *Aspergillus* species, and *P. aeruginosa* are notable for their ability to establish metastatic infection in organs of interest. In contrast, *E. coli*, *Enterobacter* species, and *Klebsiella* species generally exhibit neither characteristic.

2. *Antimicrobial efficacy.* Not all antimicrobial strategies are of equal efficacy in terms of eliminating bloodstream infection. For the purposes of cleansing the bloodstream and organs of interest of infection, it would seem logical to require bactericidal (as opposed to bacteriostatic) therapy. Because of the far slower response of fungal infection to antimicrobial therapy, patients with candidemia should require far more extensive therapy than those with acute bacteremia or else be eliminated from consideration.

3. *Time course of infection.* The duration and level of bacteremia, and the clinical and microbiologic response to appropriate therapy are important variables to be considered as well. For example, a potential donor with sustained *S. aureus* bacteremia over several days would be an undesirable candidate for this approach. In contrast, acute pneumococcal or meningococcal meningitis, for which appropriate therapy clears the bloodstream in 4–5 days quite reliably, and where metastatic infection to organs such as the kidneys and liver is unusual, might be entities that would lend themselves to an expanded donor pool consideration.

With these considerations in mind, we would propose the following approach, which is aimed at safely expanding the donor pool. It should be emphasized that informed consent on the part of the potential recipient is an integral part of this process.

1. The potential donor has had a bacteremia with a relatively bland organism (e.g., the Enterobacteriaceae with the exception of *Salmonella* species, or viridans streptococci) or with an organism that is rapidly cleared from the bloodstream with effective bactericidal therapy (e.g., penicillin-sensitive pneumococci and meningococci) that has been treated with bactericidal therapy for at least 5 days, and where blood cultures have been shown to become negative. Ideally, some evidence of clinical response to such therapy should be present. Patients with undrained, infected fluid collections ("pus under pressure") require more prolonged treatment regardless of the causative organism.

2. Potential donors with bloodstream infection due to the following organisms would require a minimum of 2 weeks of bactericidal therapy and then "proof of cure" (i.e., negative blood cultures over a week's period off antibiotics): *Staphylococcus aureus*, *Pseudomonas aeruginosa*, and infections due to streptococci that have decreased susceptibility to penicillin.

3. Potential donors with bloodstream or invasive tissue infection due to the following more difficult to treat organisms should be eliminated from consideration for the present: Group A streptococcal infection, vancomycin-resistant enterococcal infection, *Streptococcus milleri* infection, *Salmonella* infection, and fungal, nocardial, or active mycobacterial infection.

4. All recipients receiving organs from this category of expanded donors should receive bactericidal antibiotic therapy directed against the donor's organism for a minimum of 10–14 days posttransplant.

5. An international registry should be established in which all situations in which infected donors were utilized, both successes and failures, are collected, collated, and summarized for the transplant community.

The second group of issues has to do with infection acquired in the terminal care of the donor and/or through contamination during the organ procurement and transport process. In recent years, in an attempt to salvage more organs, several steps have been taken to more closely evaluate these infections in the potential donor and to take steps to limit their consequences:

1. In patients with bladder catheter-associated bacteriuria, as the organs are procured, urine samples are obtained directly from the ureters for urinalysis, Gram stain, and culture. In the absence of evidence of upper urinary tract infection or inflammation, we have successfully utilized such kidneys.

2. In patients with atelectasis and forms of pneumonia unlikely to be associated with bacteremia (e.g., *Haemophilus* species in an adult, *Acinetobacter* and other forms of nosocomial gram-negative pneumonia), if systemic antibiotic therapy effective against the patient's respiratory flora has been administered prior to organ procurement, it may be possible to utilize at least some of the organs for transplantation.

3. In this era of multiorgan procurement from the same donor, great care must be taken to protect each organ from contamination from the other organs. Thus, we have observed contamination of the great vessels of the heart, with subsequent rupture of a mycotic aneurysm at a vascular suture line, caused by spillage of bacteria from the respiratory tract during the procurement procedure.

4. When organs from patients with nosocomial infection distant from the organ utilized are transplanted, after the above precautions are taken, peritransplant prophylaxis is usually extended for 10–14 days to provide an extra level of safety to the recipient.

Other techniques that are employed to protect allograft recipients from contaminated organs include the culturing of organ perfusate and transport media. Indeed, a 4–40% rate of positive cultures has been documented.[119,120,124–127] Most such positive cultures are with nonvirulent skin flora, and these results have correlated poorly with the occurrence of posttransplant allograft infection. Conversely, the occasional instance in which such surveillance cultures have yielded gram-negative bacilli, particularly *P. aeruginosa*, and *Candida* species, have been highly correlated with infection involving the vascular anastomoses of the transplant, with the development of mycotic aneurysms and/or vascular disruption with life-threatening hemorrhage.[119,128–130] The importance of gram-negative infection of the perfusate has been emphasized in a dog transplant model. In these studies, the perfusate was purposely contaminated with *E. coli*, and the kidney was then transplanted. All of the recipients died in approximately 4 days of either vascular anastomotic disruption or generalized sepsis.[130]

Unfortunately, negative perfusate cultures and careful clinical evaluation of the donor prior to organ procurement do not preclude the possibility of serious allograft infection. We have reported a case of unsuspected donor *Pseudomonas* sepsis (donor afebrile prior to transplant, but with premortem donor blood cultures becoming positive several days later for the identical *P. aeruginosa* strain subsequently isolated from the recipient) causing life-threatening infection in both recipients of kidneys from this single donor. In the second posttransplant week, both kidney allograft recipients required emergency graft nephrectomies because of a massive retroperitoneal bleed. At operation, the arterial anastomosis was completely necrotic and disrupted, and grew the same *Pseudomonas*.[119]

We are aware of similar scenarios involving the vascular anastomoses of the other organs as well. Freeman *et al.*[131] have published a reassuring paper in which the outcomes of transplanting organs from donors with positive blood cultures were reviewed retrospectively in the New England region. Of 95 cases of bacteremia in the donor, none of the 212 transplant recipients became infected. However, before we become complacent about this issue, several points bear emphasis: the great majority of the organisms were relatively bland (~15% gram negatives); the duration and level of bacteremia were not well defined; and virtually all of the recipients received appropriate antibiotics posttransplant. We remain particularly concerned about unsuspected donor bacterial or candidal sepsis acquired in the intensive care unit while the donor receives care for the terminal illness, and employ the following approach[119]:

1. Careful culturing of the donor (including preterminal blood cultures) and of the perfusates or organ transplant medium should be continued, with systemic antimicrobial therapy initiated for positive cultures, utilizing shorter durations of therapy (< 7 days) for the nonvirulent organisms and longer durations (> 14 days) if the cultures yield gram-negative bacilli, *Staphylococcus aureus*, or *Candida* species.

2. Certain potential donors at a particularly high risk for occult sepsis should not be used. These include victims of drowning (who may be infected with microorganisms found in water), burn victims, and patients who have been maintained on a respirator with indwelling lines and catheters for a period of more than 7 days.[132,133]

Even these steps will not completely solve the problem of occult sepsis contaminating allografts. The ultimate answer will depend on the development of rapid noncultural diagnostic techniques for detecting microbial DNA or microbial products so that potential donors can be rapidly screened for occult sepsis that could impact on the allograft.

4.3. Wound Infection

The most important form of treatable infection in this time period, in terms of both frequency and clinical impact, is wound-related sepsis. Most of the data in this area have come from studies of renal transplantation, and we will consider this first, with later comments on the special issues that apply to the other forms of organ transplantation.

4.3.1. Wound Infection in Renal Transplant Recipients

The reported incidence of wound infection following renal transplantation has varied from 1.8 to 56%.[134–138] The impact of such infections, particularly when deep in the perinephric space, can be great, with 75% of deep perinephric infections resulting in the need for transplant nephrectomy, and with many lives lost because of systemic sepsis originating from this site or from the development of a mycotic aneurysm in the area of the vascular anastomoses.[139,140] That such infections should be common in these patients is no surprise in view of the effects of chronic uremia, possible protein malnutrition, immunosuppressive therapy, and so forth, on wound healing and resistance to infection. Even more important in the pathogenesis of wound infection is the occurrence of technical complications of the surgery. For example, in one series from the University of Minnesota, the incidence of wound infection was 6.1% among 439 consecutive renal transplant operations. If, however, those occurring in conjunction with the presence of a hematoma or urinary fistula are excluded, the incidence of wound infection in the uncomplicated wounds was only 1.6%, with all being superficial infections. If diabetics and retransplanted patients are excluded, the incidence of wound infection was only 0.7%, again, all superficial.[140] By way of comparison, the rate of wound infection for clean general surgical procedures in nonimmunosuppressed patients is reported as 1.8%.[141] Other transplant groups have noted similarly low rates of wound infection.[142–145]

The unavoidable conclusion is that the most important factor in the prevention of wound sepsis in the transplant patient is the technical quality of the surgery performed. There is probably no other area of general surgery comparable to transplant surgery in which anything less than impeccable surgical technique can have such disastrous consequences. The incidence of wound infection is determined by the ability of the surgeon to prevent urine leaks, wound hematomas, and the development of lymphoceles—all of which markedly increase the risk of infection. One additional factor is the avoidance of unnecessary "dirty" surgery at the time of transplant. For example, a high rate of *Bacteroides fragilis* bacteremia has been observed when elective appendectomy was performed at the time of renal transplantation.[146] This observation is an extension of two general principles: all active infections must be eradicated prior to transplant, and ill-advised surgery has consequences as grave as technically flawed surgery.

Prevention of urinary leaks begins at the time of organ procurement. The primary concern here is to preserve the blood supply of the donor ureter, as ureteral vascular insufficiency resulting in distal ureteral necrosis or fibrosis is a major cause of both urinary extravasation and ureteral obstruction. Damage to this blood supply as a result of stripping of the periureteral adventitial tissue in which the blood vessels run, too extensive dissection of the hilum of the kidney, or the failure to recognize the presence of multiple renal arteries are the major technical problems that must be avoided by the organ procurement surgeon.[147,148] If this is accomplished, attention is then turned to the urinary anastomosis, be it a ureteroneocystostomy or ureteropyelostomy, in which watertight, nonobstructing anastomoses are essential. Although a discussion of the surgical details of the urinary anastomosis and the choice between the types is beyond the scope of this chapter, their importance cannot be minimized. In the best hands, urologic complication rates of less than 2% should be obtained.[149–151] When such complications occur, they must be promptly corrected, with drainage, and provision made for unobstructed urine flow. If unattended, there is a high risk of life-threatening sepsis.

The second major preventable factor in the development of wound sepsis is the formation of a wound hematoma. Any technical problems resulting in bleeding will be exacerbated by the uremic state and by the heparin commonly employed when posttransplant hemodialysis is required. Wound hemostasis at the time of transplantation must be meticulous. When bleeding or other complications require reexploration or transplant nephrectomy, the incidence of wound infection increases 10-fold.[138–140,152]

The final technical consideration here is the prevention of lymphoceles. Such collections of lymph in the retroperitoneal wound of the renal transplant patient occur at reported rates of 2–18% and may result in mechanical obstruction and/or become secondarily infected.[153–155] Lymphoceles result when, at the time of transplantation, lymphatic vessels, especially those crossing the iliac arteries, are cut without ligation or lymph nodes are removed for tissue typing purposes, again without adequate ligation of the lymphatic channels. Lymph collecting in the retroperitoneal space will not be absorbed and must be drained surgically, either externally or into the peritoneal cavity. A clinical clue to the presence of such a lymphocele is the development of unilateral leg edema on the side of the transplant.[139,153–156]

Thus far, the factors leading to the development of wound infection in the renal transplant recipient have been explored. Next to be considered are the questions of how best to prevent (other than with expert surgery), diagnose, and treat wound infections. In the area of prevention, there is general agreement among most transplant surgeons that local irrigation of the transplant wound with antibacterial solutions is beneficial, although such practice has never been subjected to a careful randomized study in the transplant setting. Similarly, the use of open drains, such as Penrose drains, is thought to be associated with the risk of introducing microorganisms from the skin surface.[138,139,157] Many groups, including our own, employ closed suction drainage (as with a Jackson–Pratt drain) in an effort to obliterate dead space and prevent fluid collections, removing the drain when less than 40 ml/day is being delivered into the system. Usually the drain can be removed in less than 5 days, after an ultrasound study is negative for a drainable fluid collection. If copious, nonbloody drainage continues for a longer period, either a lymph leak or a urine leak is present. The latter can be ruled out by administering intravenously the dye indigo carmine, which imparts a blue color to the urine, and checking the color of the drainage.

Perioperative antibiotic prophylaxis has been reported to decrease the incidence of wound infection. A regimen aimed at uropathogens and staphylococci, such as ampicillin–sulbactam or cefazolin, should be used. Such therapy should be initiated on call to the operating room, and continued for no more than 24 hr posttransplant. The exact choice of antibiotics should be guided by knowledge of the prevalent bacterial flora causing wound infection at a particular institution and the antibiotic susceptibility patterns of these organisms. Two points should be emphasized regarding such a prophylactic program: (1) no antibiotic prophylaxis program can take the place of technically expert surgery and (2) such a program is aimed at protecting against wound infection, not later urinary tract infection or other problems. There are better methods for preventing urinary tract infection (*vide infra*), and prolonging broad-spectrum parenteral antibiotic prophylaxis adds little to the care of transplant patients.[5,8,27,143–145,152,157]

The diagnosis of wound infection requires a high index of suspicion. As detailed in Chapter 18, the immunosuppressive therapy being administered will frequently obscure the usual presenting signs of wound infection. Therefore, any transplant patient with an unexplained fever should be subjected to either ultrasonographic or computed tomographic (CT) scanning of the deeper operative sites. In selected patients, sterile needle aspiration of the wound can be helpful in diagnosing more superficial wound collections.

Any collection identified should be promptly drained under broad-spectrum antibiotic coverage. If infection is identified, then appropriate antibiotics are usually continued for 10–14 days or until the patient has been afebrile for 5–7 days. Whenever perinephric infection is identified, the possible need for graft nephrectomy to facilitate drainage and to prevent catastrophic anastomotic leaks should always be kept in mind. If deep sterile collections are found, we once again prefer closed suction drainage to the placement of Penrose drains.

4.3.2. Wound Infection in Liver Transplant Recipients

As with wound infection in renal transplant patients, the major determinants of wound infection (as well as other forms of infection in the first month posttransplant) in the liver transplant recipient are technical aspects of the operation itself. The complexity of the surgery—hepatectomy, and then four vascular anastomoses (the suprahepatic vena caval anastomosis, the infrahepatic vena caval anastomosis, the portal vein anastomosis, and the reconstruction of the hepatic artery), as well as a biliary anastomosis—in a patient with a bleeding diathesis is a daunting challenge, to say the least. Intra-abdominal bleeding and contamination with upper GI microbial flora are not uncommon, leading to a relatively high rate of deep wound infection—probably on the order of 5–15%, which is increased further if reexploration for bleeding or retransplantation are required, or a biliary leak occurs. Intra-abdominal fluid collections, particularly hematomas, must be drained or the risk of secondary infection is prohibitive. This is especially true of hematomas, in which the benefits of removing these nidi for future infection outweigh the risks of reexploration. Typically, such reexploration is best carried out 2–4 days posttransplant, at a time when the coagulopathy that contributed to the intra-abdominal bleeding has been corrected. The nature of the biliary anastomosis also plays a role in determining the frequency of posttransplant wound infection, with choledochojejunostomies carrying an increased risk because of the bowel manipulations necessary to effect the anastomosis.[158–165]

Although peritransplant antimicrobial prophylaxis for the renal transplant patient is more or less optional, most transplant groups regard it as being obligatory for liver transplantation. The controversy concerns the constituents of this prophylactic program. Most groups use perioperative systemic prophylaxis consisting of a cephalosporin (usually cefazolin or cefotaxime) ± ampicillin or gentamicin, beginning on call to the operating room, and continuing for 3–7 days posttransplant. In addition, the transplant groups at Groningen and the Mayo Clinic have championed the concept of selective bowel decontamination, beginning at least 1 week before likely transplantation, and continuing for the first 21 days posttransplant. A common program of selective bowel decontamination that is employed consists of the administration four times per day of the following: a suspension of gentamicin, 80 mg/10 ml; polymyxin E, 100 mg/10 ml; nystatin, 2×106 U/10 ml; plus mucosal paste (Orabase) containing the same antibiotics at 2% concentrations. This program, when combined with perioperative systemic antibiotics, has been associated with a very low rate of infection.[166–170] This stands in direct contrast to reports from other centers of a 16–42% incidence of fungal infections in their liver transplant patients,[170–174] including a 16% incidence of fungemia in one of these studies.[174]

Recent studies by Arnow and colleagues[175,176] have shown that beginning the program of selective bowel decontamination in the perioperative period adds little to the effects of systemic antibiotics, thus emphasizing the need to begin this program a week or more prior to transplantation to reap its optimal benefits. In many patients, this may be logistically difficult to accomplish. Because of the lack of controlled trials in this area, however, it is not clear what the optimal prophylactic regimen should be. For example, at Massachusetts General Hospital, we have a similarly low rate of wound infection and postoperative candidal infection ($< 3\%$), utilizing the following program: oral nystatin or clotrimazole three to four times a day, beginning as soon as it appears likely that the patient is within 3 weeks of transplantation ("high on the waiting list"), continuing this through the transplant period and whenever antibacterial therapy is being administered; cefazolin, 1 g every 8 hr, is administered for 3 days, followed by oral trimethoprim–sulfamethoxazole (or ciprofloxacin, in the trimethoprim–sulfamethoxazole-intolerant patient) for 6–12 months. Despite the difficulties in comparing results of different regimens from different programs, we would suggest that the following principles for preventing deep wound infection in the liver transplant recipient appear reasonable at the present time:

1. The first rule of wound infection prevention is technically perfect surgery.
2. Because of the relatively large numbers of candidal species present in the upper GI tract, the fungal promoting effects of the immunosuppressive therapy (particularly the corticosteroids), and the exposure of the patient to the selective pressures of broad-spectrum antibacterial therapy, it is not surprising that invasive candidal infection is a particular problem in the first month after liver transplantation. Therefore, it is reasonable to believe that an antifungal prophylaxis

program should be undertaken. Our approach at present is to utilize mucosal prophylaxis with oral nystatin or clotrimazole, reserving fluconazole for use in patients likely to have high candidal burdens in their GI tracts and/or to be at significant risk of candidal contamination of the surgical site. Hence, we employ fluconazole in the following instances: in diabetics, in patients who have received broad-spectrum antibacterial therapy in the weeks just prior to transplantation, in patients undergoing choledochojejunostomy biliary anastomoses, and in patients undergoing re-exploration.[177,178] Both oral and parenteral fluconazole are well tolerated in the liver transplant patient when it is clinically indicated, although it is important to remember that fluconazole will downregulate cyclosporine and tacrolimus metabolism. Hence, lower doses of these drugs must be administered in the face of fluconazole therapy or toxic blood levels of the calcineurin inhibitors will be achieved.[179]

3. The best regimen for antibacterial prophylaxis also remains to be defined. One cannot argue against selective bowel decontamination, although there is a need for controlled trials to prove its efficacy. Quinolones, which have shown efficacy in other immunosuppressed patient populations when compared with nonabsorbable antibiotics, merit a trial here as well, probably in combination with an antifungal regimen.[173] Perioperative systemic therapy with such drugs as cefazolin and cefotaxime (with or without other drugs) appears to be well established, although there is a need to establish the best regimen. Although there are conflicting data on this subject, we believe that systemic antimicrobial agents should be redosed during the liver transplant procedure, commensurate with the level of blood replacement that is needed.[175] In other operative situations, the efficacy of prophylactic cephalosporin therapy has been shown to be significantly attenuated when intraoperative blood levels are allowed to fall below therapeutic levels.[180–182]

4.3.3. Wound Infection in Heart, Lung, and Heart–Lung Transplant Recipients

The incidence of wound infection following cardiac transplantation has been reported to vary from 3.6 to 62.5%, but most of these cases are usually related to various catheter sites or vascular cutdown sites and are rarely fatal.[183–185] Lower-dose steroid therapy appears to be associated with a decreased risk of wound infection in these patients.[186] The big concern is infection of the median sternotomy wound, with its potential for mediastinitis and infection spreading directly to the allograft. Sternal wound osteomyelitis and infectious pericarditis are other consequences of wound infection in these patients.[187,188] Fortunately, with careful surgical technique and antimicrobial prophylaxis, this is uncommon. Most cardiac surgical groups utilize a cephalosporin either alone or with vancomycin for perioperative prophylaxis, continuing it until the chest tubes and all catheters have been removed (usually 5–7 days posttransplant).

An uncommon, but usually disastrous, complication is infection at the aortic suture line, resulting in a mycotic aneurysm.[189–193] Whether such complications will increase in patients having left ventricular assist devices placed as a "bridge" to transplantation remains to be determined.[194]

In the case of single, double lung, or heart–lung transplantation, the biggest additional concern is the integrity of the bronchial anastomosis. Otherwise, the incidence of wound infection is related to the technical management of the surgery, and the drainage catheters. We find it useful to monitor chest tube drainage approximately every 3 days prior to removal, aggressively treating organisms that are present in the pleural space or mediastinum. A variety of antimicrobial prophylaxis programs have been utilized in the various forms of lung transplantation. Our own approach is based on two observations: the sputum of many patients coming to lung or heart–lung transplantation is usually colonized with a variety of pathogenic gram-negative bacilli, *Staphylococcus aureus*, and/or fungal species; and the microbial flora of the native respiratory tract is an important source of wound infection (as well as postoperative pneumonia) in lung transplant recipients. Rather than utilizing the same prophylactic regimen for all lung transplant recipients, we prefer to monitor these patients, as they approach transplantation, with weekly sputum cultures, and then design individualized antimicrobial prophylaxis programs based on the results of these cultures.

4.4. Other Causes of Infection in the First Month

The major remaining causes of infection in the renal transplant patient during the first month following transplantation are pneumonia, urinary tract infection, and intravenous line-related sepsis. One general principle underlines the occurrence of each of these—the immunosuppressed patient tolerates poorly the presence of foreign bodies that bypass normal local host defenses; i.e., endotracheal tubes, urinary catheters, and plastic intravenous catheters. These should be used sparingly, removed promptly, and managed with impeccable aseptic technique. In the case of pneumonia, the first concern is prevention with the use of appropriate anesthetic and analgesic management so that the endotracheal tube can be removed as quickly as possible posttransplant, aspiration is prevented, and chest physical therapy and early patient mobilization are employed to prevent atelectasis. An additional factor in preventing pneumonia is the aggressive management of nausea to prevent vomiting, even if a nasogastric tube is required because of gastric dysfunction (particularly in diabetics). Aspiration pneumonia is a frequent consequence of vomiting in the immunosuppressed patient.

Plastic catheters for intravenous use are to be discouraged, particularly central venous pressure catheters or Swan–Ganz pulmonary artery catheters for hemodynamic monitoring. Unless a major complication occurs in the perioperative period, virtually no suitable candidate for a renal transplant should require this type of monitoring. Although the problem of intravenous line-related sepsis is a general one throughout the hospital, its consequences can be particularly disastrous in these patients. For example, although a transient intravenous line-related *Candida* septicemia is associated with metastatic infection in less than 5% of normal individuals, among immunosuppressed patients more than 50% will develop metastatic infections if left untreated.[195]

The same general considerations apply to the other forms of organ transplantation, except even more intensely. In the case of liver transplantation, patency of the vascular anastomoses and the intactness of the biliary anastomosis are more frequently an issue than patency of the renal vascular anastomoses and stability of the ureteral anastomosis. Portal vein thrombosis, hepatic artery thrombosis, and hepatic vein occlusion developing in the first few days posttransplant are well-recognized complications of liver transplantation, particularly in young children whose tiny vessels can pose a major technical problem and in patients who become hypotensive. Manifestations of these complications include ascites, variceal bleeding, and deterioration in liver function tests and clinical status. Not uncommonly, fever and bacteremia may be the major clues. Sepsis is particularly common following interruption in the hepatic arterial circulation, with secondary infection of hepatic infarcts leading to areas of hepatic gangrene, abscess formation, and fulminant sepsis related to bowel flora or candidal organisms. A more insidious consequence of vascular insufficiency can result when the vascular supply to the liver parenchyma remains intact, but the biliary anastomosis is

rendered ischemic. This results in a breakdown of the biliary anastomosis, a bile leak (or late stenosis), and secondary infection. Such secondary infection may take the form of deep wound infection, cholangitis, liver abscess, and/or bacteremia, with the microorganisms causing this derived from the normal flora of the small bowel—streptococci, Enterobacteriaceae, anaerobes, and *Candida* species. Polymicrobial infection is the rule in these circumstances. This more occult result of vascular insufficiency is totally analogous to the ureteral leaks and stenoses developing after renal transplantation following vascular insufficiency of the ureter.[158–162,170–174,196–203]

As experience has been gained with liver transplantation, problems with the vascular anastomoses have become less common. The biliary anastomosis, however, remains the Achilles heel of liver transplantation. Whenever possible, the anastomosis of choice is a choledochocholedochostomy, which maintains the native sphincter of Oddi intact. When this is not possible for anatomic reasons, as in children with biliary atresia or ducts too small to safely carry out this anastomosis, or in adults with sclerosing cholangitis or other abnormalities of the extrahepatic biliary system, a choledochojejunostomy constructed with a Roux-en-Y technique that offers protection against microbial contamination from the GI tract can be employed. Although biliary leaks can develop with either anastomosis, obstruction is the major concern with the choledochocholedochostomy procedure, whereas reflux of organisms is the weakness of the choledochojejunostomy anastomosis. In either case, secondary infection is the consequence.[203,204]

The cardinal rule in the first few weeks following liver transplantation is that any episode of unexplained fever or bacteremia should be regarded as a manifestation of a technical problem involving the vascular tree, the biliary anastomosis, or deep wound infection until proven otherwise. Accordingly, such diagnostic procedures as abdominal CT scanning and/or ultrasound with Doppler to look for collections as well as vascular patency, cholangiography, and, when appropriate, hepatic angiography must be carried out expeditiously as well. Prompt surgical attack of technical problems under antimicrobial coverage can salvage these patients.[204]

Colonization of the bile of liver transplant patients with bacteria is the rule rather than the exception, and this can be easily documented by bile cultures in patients with T-tubes left in place to protect the biliary anastomosis. The colonizing flora usually consists of bowel streptococci, *Staphylococcus* species, and, intermittently, a variety of gram-negative bacilli. Such colonization by itself requires no therapy under normal circumstances. Indeed, patients developing symptomatic infection related to biliary colonization in the absence of biliary manipulation should be regarded as having an anatomic abnormality—obstruction or a leak—of the biliary tree, and should be investigated immediately. Occasionally, *Candida* species will colonize the biliary tree. Asymptomatic candidal colonization of the biliary tree can result in obstruction related to the formation of fungal balls, analogous to what can be observed in the urinary tract (*vide infra*). Accordingly, we treat such colonization preemptively, usually with fluconazole.

The asymptomatic colonization of the biliary tree becomes of importance under three circumstances: (1) biliary manipulations such as cholangiography can result in cholangitis and systemic sepsis; (2) liver biopsies can result in intraparenchymal collections of bile and blood, culminating in liver abscess formation; and (3) when the T-tube is removed, usually more than 2 months posttransplant, a bile leak with chemical and/or bacterial peritonitis can occur. We have found that administering single doses of broad-spectrum antibiotics prior to these procedures is quite helpful in preventing complications. Such prophylaxis is particularly important if the biliary anastomosis is a choledochojejunostomy.

In the first month following lung transplantation, the biggest concerns relate to the integrity of the bronchial anastomosis and postoperative pneumonia. Indeed, the great advances in lung transplantation have occurred as the surgical management of this anastomosis has improved (analogous to the ureteral and biliary anastomoses with renal and hepatic transplantation), including the provision of an omental wrap of the anastomosis. Avoidance of high-dose steroids to promote wound healing, care to preserve the vascular supply, and antimicrobial therapy to protect the anastomosis have all contributed to the success now being achieved.[205,206] Pneumonia, however, remains a significant problem for several reasons: Colonization of the native respiratory tract, particularly with gram-negative bacilli and *Aspergillus* species, provides an important source of infection for the new lung.[206,207] This is a particular problem in patients with cystic fibrosis.

4.5. Noninfectious Causes of Fever in the First Month Posttransplant

The most common noninfectious cause of fever in the first month posttransplant is allograft rejection. Approximately two-thirds of transplant recipients will have acute rejection in the first month posttransplant, with a little less than half of these having steroid-resistant rejection that requires antilymphocyte antibody therapy. Although less striking than in the precyclosporine era, fever may still be the first manifestation of rejection, particularly in this early posttransplant period. This is especially true in children, who tend to have more exaggerated temperature responses than adults. An important clinical point is that small children who receive an adult kidney may have significant rejection before changes in the serum creatinine level can be detected. Thus, an unexplained fever in the first few weeks posttransplant in a small child who has undergone renal transplantation should prompt either a renal biopsy or consideration of a steroid pulse. We have observed that in children who were febrile for 5–7 days prior to a rise in serum creatinine, if acute antirejection therapy was delayed until then, it was too late and the allograft was lost.[208]

The second most common noninfectious cause of fever in this time period is antilymphocyte antibody therapy. The first two or three doses of OKT3 (a monoclonal pan-T-cell antibody preparation) or antithymocyte globulin (a polyclonal pan-T-cell antibody preparation) are invariably associated with the release of a variety of cytokines, most notably tumor necrosis factor. In addition to fever and chills, some patients can develop hypotension, a febrile pulmonary edema syndrome, aseptic meningitis, encephalopathy, and other manifestations of a massive cytokine release. Some of this can be avoided with extra corticosteroids, diphenhydramine, and acetaminophen. As new anticytokine strategies become available, these inflammatory events should become more easily managed. In the great majority of patients, fever, chills, and malaise that disappear by the third dose are the only acute manifestations of antilymphocyte antibody therapy. The reappearance of fevers in the latter part of a 10- to 14-day course of therapy suggests an immunologic response to these animal proteins (murine in the case of OKT3 and equine or rabbit in the case of antithymocyte globulin) that may be limiting the antirejection effects of this treatment.[209,210]

Other noninfectious causes of fever to be considered include pulmonary emboli and drug reactions. Probably because of the anti-inflammatory effects of the immunosuppressive program, drug fevers and rashes are less common in transplant patients than in the normal population, although such adverse effects as bone marrow toxicity, hepatic toxicity, and renal injury are at least as common.

5. Infection 1 to 6 Months Posttransplant

Unless technical complications that can lead to life-threatening infection have occurred during the first month, the time period 1–6 months posttransplant is the critical period for the transplant patient in terms of the greatest risk of the infections that are unique to these immunocompromised individuals:

1. Immunosuppressive therapy is still at a relatively high level, particularly if significant amounts of antirejection therapy have been required. Even more important, the duration of immunosuppression is now sufficient that opportunistic forms of infection can occur.

2. Infections occurring during this period are usually quite challenging to treat with antimicrobial therapy, since most are caused by viruses, antibiotic-resistant bacterial species, and a variety of fungi.

3. Any technical errors lingering from the perioperative period (e.g., persistent anastomotic leaks, vascular compromise with tissue infarction) will, by definition, be serious. These require the presence of drains, catheters, and drainage tubes, for prolonged periods of time. The longer such foreign bodies are required, the greater the incidence of superinfection, and the more difficult to treat the infection becomes, usually evolving from an antibiotic-sensitive bacterial species to increasingly resistant gram-negative and fungal species.

4. The key factor, other than the technical issues that determine graft viability, in determining the patient's fate in this time period is the presence or absence of infection with the immunomodulating viruses (the herpes group viruses, hepatitis viruses, and, uncommonly, HIV).

5.1. Herpes Group Virus Infections in the Organ Transplant Recipient

The human herpes group viruses [CMV, EBV, varicella–zoster virus (VZV), herpes simplex virus types 1 and 2 (HSV-1 and HSV-2), and HHV-6, -7, and -8] share several characteristics that make them the most important group of microbial pathogens that affect organ transplant recipients[8,27]:

1. *Latency.* The term *latency* is used to designate the fact that once infected with these viruses, an individual is infected for life, even after evidence of active viral replication is no longer demonstrable. The viral genome is present in latently infected cells, but gene expression is limited, infectious virus is not produced, and the virus is hidden from attack by the host's immune system. Latent virus can be reactivated at a later date, either sponta-

neously or in response to exogenous influences. The laboratory marker for the presence of latent infection is the presence of circulating antibody ("seropositivity") in the absence of active viral replication. The different herpes group viruses differ in the stability of their latency, although all will be reactivated by immunosuppressive programs that include antilymphocyte antibodies, as well as other components (*vide infra*). CMV is latent in a more stable fashion, only being reactivated when the individual is exposed to certain forms of immunosuppression, as a result of an allogeneic reaction, during pregnancy, and on exposure to certain proinflammatory factors (*vide infra*). In contrast, EBV's latency is very unstable, with spontaneous reactivation occurring repetitively for variable periods of time in seropositive individuals. VZV and HHV-6, -7, and -8 seem to resemble CMV, requiring a specific event for reactivation to occur. In contrast, HSV-1 and -2 are more like EBV, being reactivated by a variety of "stresses" but also in conjunction with no apparent stimuli. Once viral reactivation occurs, then the extent of the viral replication, and the subsequent clinical effects, is amplified by the level of immunosuppression present.

2. *Cell association.* The term *cell association* means that these viruses are spread between individuals by intimate cell-to-cell contact, transfusion, or transplantation. Similarly, spread within tissues and systemically is accomplished by the direct contact of infected cells with other susceptible cells. This renders humoral immunity inefficient, and places particular emphasis on cell-mediated immunity for elimination of the infected cells. In particular, major histocompatibility complex (MHC)-restricted, virus-specific, cytotoxic T cells are the most important host defense against these viruses. The impact of different immunosuppressive agents and programs on the course of infection with these viruses is in large part related to the immunosuppressive program's effects on the functioning of the critical cytotoxic T-cell response. The MHC restriction of these T cells suggests that the transplant patient would have particular difficulty in eliminating virus-infected cells within allografts that are MHC disparate with the host. As will be discussed, this possibility may indeed be clinically important.

3. *Oncogenicity.* All herpes group viruses should be considered potentially oncogenic. Two types of oncogenic effects can be seen. First, and most important, is direct oncogenicity, as exemplified by the causation of posttransplant lymphoproliferative disorder (PTLD) by EBV and Kaposi's sarcoma by HHV-8. In addition, there can be more indirect effects. Thus, preceding CMV disease has been shown to increase the risk of subsequent EBV-associated PTLD 7- to 10-fold.[211,212]

4. *Clinical effects*. With regard to the clinical effects of the herpes group viruses, both *direct* and *indirect* effects must be considered. Direct effects include the various infectious disease clinical syndromes (e.g., pneumonia, hepatitis, tissue ulceration, gastroenteritis) that are a consequence of lytic viral infection, tissue invasion, and the host response to these events. Indirect effects refer to those events produced by cytokines, chemokines, and growth factors elaborated in response to viral replication. These indirect effects can be grouped into three categories: (a) immunomodulatory, such that the net state of immunosuppression is increased and opportunistic superinfection is promoted; (b) pathogenesis of allograft injury; and (c) oncogenesis. In many patients the indirect effects are at least as important as the direct.

5.1.1. CMV Infection in Organ Transplant Recipients

CMV has been, and continues to be, the single most important microbial pathogen affecting organ transplant recipients, contributing directly and indirectly to both morbidity and mortality in these patients. Evidence of CMV replication can be found in more than half of transplant recipients, although the rate of clinical illness is a fraction of this figure. There is a complex relationship between the effects of CMV replication and the two phenomena that are inherent in the transplant experience: allograft rejection and chronic immunosuppressive therapy. The end result is twofold: CMV, which latently infects more than half of the general population (the CMV seropositive population), will be reactivated from sites within the recipient, from the allograft itself, or from viable leukocyte-containing blood products; and the ability of the host to eradicate virus from any of these sources can be greatly impaired, leading to the possibility of prolonged or relapsing viral infection and a multitude of consequences. These consequences, the direct and indirect effects of CMV on the transplant recipient, can be grouped into four distinct categories[3,8,27,213]:

1. The direct causation of a variety of clinical infectious disease syndromes by the virus itself, ranging from prolonged fevers and a mononucleosis-like syndrome, pneumonia, hepatitis, and GI ulcerations and inflammation acutely, to a chronically progressive chorioretinitis.
2. The production of an immunosuppressed state that is over and above that caused by the immunosuppressive drugs being administered. This contributes significantly to the net immunosuppressed

state of the transplant patient and plays an important role in the pathogenesis of opportunistic superinfection caused by such organisms as *Pneumocystis carinii*, a variety of other fungi, *Nocardia asteroides*, and *Listeria monocytogenes*.
3. A probable role in the production of acute and chronic allograft injury, such that anti-CMV therapy may have a role in improving allograft function and longevity.
4. A possible role in the production of certain forms of malignancy.

5.1.1a. Epidemiology of CMV Infection in the Organ Transplant Recipient. There are three major sources, and one uncommon one, by which transplant patients may acquire active CMV infection[3,8,27,213–220]: from latently infected allografts from seropositive donors; from viable leukocyte-containing blood products that similarly harbor latent virus from seropositive donors; from reactivation of endogenous virus in seropositive transplant recipients; and, finally, by acquisition of the virus in the general community, as a result of intimate contact with an actively infected individual (this usually involves CMV-naive transplant recipients having sexual intercourse with such an individual, and acquiring primary infection via this route). Although this last is uncommon, it is important to emphasize that the manifestations can be every bit as devastating as primary infection acquired via the allograft, even though this community-acquired infection usually occurs many months posttransplant.

There are three major epidemiologic patterns of CMV infection in transplant recipients, each with its own rate of clinical illness[3,5,8,27,213–223]:

1. *Primary*. CMV disease occurs when the transplant patient has had no pretransplant experience with this virus (and is seronegative for CMV before the transplant), and is infected with virus carried latently in cells from a seropositive, latently infected donor. More than 90% of the time, the allograft is the source of such infections. In the remainder, viable leukocyte-containing blood products from seropositive donors are the source of primary infection. This is a particular problem in liver transplantation, when prodigious quantities of blood products may be required. Without anti-CMV therapy or the use of CMV-free blood products, ~15% of seronegative liver allograft recipients who receive allografts from CMV-seronegative donors develop symptomatic disease.[224] Similarly, in a large seroepidemiologic study of CMV infection involving 50 renal transplant centers around the United States, ~20% of seronegative recipients of kidneys

from seronegative donors who received blood transfusions seroconverted (and it would be expected that approximately two-thirds of these would develop symptomatic disease).[225] In contrast, among 79 seronegative individuals who received kidneys from seronegative donors and who were not transfused, not a single instance of seroconversion could be demonstrated.[225] Clinical disease following acquisition of primary CMV infection from blood products tends to be milder than when the allograft is the source, possibly because the burden of latently infected cells is significantly greater in the allograft.[226] Transmission of CMV via blood products can be prevented by the use of blood from CMV-seronegative donors or by the use of leukocyte filters.[8,27]

Approximately 90% of seronegative recipients who receive kidneys from seropositive cadaveric donors, as opposed to approximately 70% of seronegative recipients of kidneys from seropositive living related donors, seroconvert posttransplant. Presumably, this difference in attack rate is related to the increased level of rejection and the accompanying proinflammatory cytokines and other mediators elaborated in the process (*vide infra*), as well as the more intensive immunosuppressive therapy required with the transplantation of cadaveric as opposed to living related donors.[225] When the two kidneys from a seropositive donor are transplanted into two seronegative recipients, either both recipients develop symptomatic infection or neither does.[227] At present, there are no laboratory markers that delineate whether or not the organs of a particular seropositive donor are capable of transmitting the virus. Overall, approximately 50–60% of patients at risk for primary CMV infection, as defined by the donor being seropositive and the recipient seronegative, become clinically ill.[8,27,213,215–223]

2. Reactivation. CMV disease occurs when the transplant recipient is seropositive for CMV pretransplant, and reactivates endogenous latent virus posttransplant. It would appear that the great majority of patients who are seropositive pretransplant will show some evidence (serologic and/or virologic) of active CMV infection. Traditionally, it has been stated that 10–20% of individuals at risk for reactivation infection become clinically ill, although this can be greatly modified by the type of immunosuppression administered (*vide infra*) and whether or not the allograft donor is seropositive.[8,27,213,215–223,228,229]

3. Superinfection. CMV disease occurs when an allograft from a seropositive donor is transplanted into a seropositive recipient, and the virus that is reactivated is of donor rather than endogenous origin. The possible occurrence of this phenomenon had been a cause of much speculation for some time, as it had long been recognized that human CMV isolates in nature exhibit considerable genomic and antigenic heterogeneity.[27,213] Studies in other forms of CMV infection—in congenital CMV infection,[230] and in the AIDS patient[231]—have demonstrated that superinfection could occur. In two large studies in renal transplant populations, Fryd,[232] Smiley,[233] and their colleagues suggested that CMV superinfection was clinically important, since seropositive recipients of kidneys from seropositive donors had an outcome that was worse than if the kidneys came from seronegative donors. Utilizing DNA restriction enzyme analysis, several groups have now proven that CMV superinfection does occur commonly among allograft recipients. In at least 50% of cases, the virus that is reactivated following the transplantation of kidneys (and presumably other organs) from a seropositive donor into a seropositive recipient is of donor origin.[220–222] What is less clear is whether superinfected individuals are at increased risk of clinical disease from CMV than are those with reactivation of their own endogenous virus. Grundy *et al.*[222] reported that 40% of individuals with evidence of superinfection became symptomatic, whereas none of those with endogenous reactivation became symptomatic. Others have failed to find a difference between these two groups of patients in terms of the incidence of clinical disease.[234] Our own view is somewhere in between: superinfection has a somewhat greater impact than reactivation, but this difference is attenuated when intensive immunosuppression is administered and the incidence of clinical disease rises significantly in all seropositive individuals (*vide infra*).

The demonstration that superinfection occurs commonly has important clinical as well as scientific implications. Clearly, it increases the desirability of a CMV-seronegative donor for all recipients, not just seronegative ones. Similarly, the use of CMV-negative blood products or leukocyte filters when such products are administered would seem to be indicated for even seropositive individuals in order to prevent superinfection from blood administration. Finally, if natural infection cannot prevent symptomatic superinfection posttransplant, the challenge in developing an effective CMV vaccine appears to be quite daunting.

The level of CMV infection in the general community has important effects on the occurrence of CMV among transplant patients. In western Europe and North America, the level of seropositivity is 15% by age 2, 30% in young adults, and 50–60% in those over the age of 50, with higher rates among lower socioeconomic groups, male homosexuals, recipients of blood transfusions, and the sexually promiscuous. Thus, the incidence of CMV infection and the percentage of cases related to primary infection, superinfection, or endogenous reactivation will vary from center to center depending on the population being served.[213,235] Person-to-person spread of CMV among dialysis and transplant unit patients and personnel does not appear to occur under normal circumstances.[236]

If no antiviral therapy is administered to the transplant recipient, CMV infection will occur almost exclusively in the time period 1–4 months posttransplant, with a peak incidence ~5 weeks posttransplant.[237] Presently available antiviral preventive strategies offer incomplete protection, but, even when they fail, the incubation period is usually prolonged. For example, in a recent study comparing oral ganciclovir to oral acyclovir prophylaxis for 12 weeks after an initial 10-day course of intravenous ganciclovir in organ transplant recipients at risk for primary infection (donor seropositive, recipient seronegative), the mean time to disease when prophylaxis failed was 291 days for the ganciclovir patients and 212 days for the acyclovir patients.[238]

5.1.1b. Pathogenesis of CMV Infection. The critical first step in the pathogenesis of CMV infection is the reactivation of the virus from latency, whether the latent virus is present in the allograft or in the recipient's tissues. The CMV replication cycle has three distinct phases: (1) an *immediate early* phase, which lasts 3–4 hr, and results in the synthesis of proteins that regulate the subsequent expression of the remaining viral genes; (2) an *early* phase, which lasts 4–12 hr, and results in the synthesis of viral DNA polymerase (and other viral proteins); and (3) a *late* phase, which lasts 6–12 hr, during which structural viral proteins are produced, whole virions are assembled, and new infectious virus is released. The total replication cycle, then, takes an average of 18–24 hr, with the expression of the immediate early proteins being the key step in the initiation of this process.[239–242]

The key mediator in beginning this process is tumor necrosis factor-α (TNF), which combines with the TNF receptor of latently infected cells, resulting in a downstream signaling process that involves activation of protein kinase C and NFkB. The resulting activated p65/p50 NFkB heterodimer translocates into the nucleus and binds to the CMV immediate early enhancer region to initiate the process of viral replication.[242–246] The importance of this observation cannot be overestimated, as it explains why CMV infection and disease are linked to such widely disparate factors as sepsis, fulminant hepatic failure, allograft rejection, and the administration of such antilymphocyte antibodies as ATG and OKT3—all of these

are associated with the release of large amounts of TNF (and other proinflammatory factors) and hence will cause activation of latent virus.[8,242,247–251] It is likely that the observation that hypothermia in liver transplantation is associated with an increased risk of CMV infection is also mediated through TNF.[252]

There are two other signaling pathways that can be utilized to reactivate CMV from latency, although the TNF pathway appears to be the most important of the three: The first of these pathways is activated by the stress catecholamines, epinephrine and norepinephrine. Their effects on increasing concentrations of cyclic AMP result in immediate early enhancer/promoter stimulation and resultant viral reactivation. Similarly, proinflammatory prostaglandins, elaborated in the course of a variety of inflammatory processes, will also promote viral reactivation through the cyclic AMP pathway. In sum, inflammation, infection, and stress are the stimuli responsible for reactivating CMV from latency.[242]

Once actively replicating virus is present, the most important exogenous factor influencing the course of CMV infection posttransplant is the type and intensity of immunosuppression administered. Steroids, by themselves, appear to have minimal effects in terms of reactivating latent CMV. Thus, CMV infection in transplant patients was essentially unknown prior to the addition of such cytotoxic drugs as cyclophosphamide and azathioprine to the antirejection regimen.[8,27,213,253] Clinical observations over the past two decades have shown that, when antilymphocyte antibody therapy was added to conventional immunosuppression (whether azathioprine and prednisone in the 1970s, or cyclosporine- or tacrolimus-based immunosuppression in the 1980s and 1990s), the incidence of CMV disease increased strikingly. In this regard, it does not seem to matter whether the antilymphocyte antibody preparation employed is polyclonal (antithymocyte globulin, antilymphocyte serum, or antilymphocyte globulin) or monoclonal (OKT3). The incidence of viremia, and the incidence and severity of clinical disease, are increased in patients receiving antilymphocyte antibody therapy. In addition, the prophylactic benefits of such antiviral prophylactic programs as human leukocyte interferon, high-dose acyclovir, and hyperimmune anti-CMV immunoglobulin are attenuated by the use of antilymphocyte antibody therapies.[3,27,213,253–257]

Recent studies in the murine CMV model and in human transplant recipients provide a plausible explanation for these observations and reemphasize the importance of immunosuppression in modulating the effects of this virus. The two key steps in the pathogenesis of CMV infection are reactivation from latency and amplification and dissemination of actively replicating virus—the critical host defense against this last step being the previously mentioned MHC-restricted, CMV-specific, cytotoxic T cells. When equivalent antirejection regimens of cyclosporine, tacrolimus, antithymocyte globulin, anti-CD3 monoclonal antibody, rapamycin, or other immunosuppressive agents are administered to mice with either latent or active infection, very different effects are observed. Whereas cyclosporine, rapamycin, tacrolimus, and corticosteroids cannot reactivate latent virus, the antilymphocyte antibodies readily can; cytotoxic drugs such as cyclophosphamide and azathioprine are moderately potent in reactivating latent virus. In contrast, after active, replicating virus is present, cyclosporine, rapamycin, and tacrolimus are more potent in promoting viral replication and dissemination than are the antilymphocyte antibodies. Not surprisingly, then, the sequence in which immunosuppressive therapy is administered has an important effect on the course of CMV infection. The most dangerous scenario, unless an effective antiviral strategy is employed, is the reactivation of virus by antilymphocyte antibody therapy, followed by a cyclosporine- or tacrolimus-based immunosuppressive regimen that blocks the host's response to the now active virus.[5,8]

Observations in CMV-seropositive organ transplant recipients are consistent with these experimental results. When these patients receive only cyclosporine-based immunosuppression after transplantation, the incidence of clinical disease ranges between 10 and 20%. When OKT3 or polyclonal antilymphocyte antibody therapy is added to the program, the incidence of overt disease rises to as high as 60%. Thus, it appears that the host's ability to limit the replication of even small amounts of virus is blunted by current cyclosporine-based immunosuppressive regimens.[5,8,256]

The importance of cyclosporine- and tacrolimus-based immunosuppressive regimens in amplifying the extent of CMV (and other herpes group virus infections once reactivation from latency has occurred) is further emphasized by another clinical observation: In the pre-cyclosporine era, relapsing CMV infection was, with rare exceptions, essentially unknown; that is, patients became ill with CMV and either succumbed or recovered. In the present cyclosporine/tacrolimus era, patients almost universally require ganciclovir therapy to recover from symptomatic disease, and there is a 15–25% relapse rate of clinical disease following a course of treatment; that is, patients who become asymptomatic during therapy develop recrudescent symptoms 1–4 weeks after the completion of a 2- to 3-week course of ganciclovir. Presumably, small amounts of replicating CMV are amplified by

the calcineurin inhibitors until the virus reaches a level great enough to cause recrudescent clinical disease. We refer to the effects of cyclosporine and tacrolimus on replicating herpes group virus infection as "an *in vivo* PCR effect."[5,8] It is likely that viral load, both at the initiation of therapy (one measure of this being the presence of multiorgan disease) and at the completion of therapy, is a predictor of the risk of later relapse.[258]

The effects of the immunosuppressive program in amplifying the extent and effects of CMV replication are mediated largely through its effects on specific T-cell immunity directed against the virus. In particular, CMV-specific cytotoxic T lymphocytes (CTLs) act to limit the systemic viral load, the key determinant of the clinical effects of CMV replication (*vide infra*). In generating this response, helper T cells directed against CMV are essential both for recovery from viral invasion and to provide protection against such invasion (memory function). Recently, the complexity of this essential helper and cytotoxic T-cell response has been delineated. There is an expansion of gammadelta T cells in this process, and CTLs with varying specificities have been shown to be present. Particularly prominent in this array of T-cell responses are CTLs directed against the immediate early-1 (IE1) antigen and the late structural protein (pp65) that is the basis for the CMV antigenemia assay (*vide infra*).[259–264] Granzyme B (GrB) is produced by CTLs to induce apoptosis in target cells. Not surprisingly, then, the primary immune response to CMV infection is marked by the expansion of the numbers of CD8$^+$GrB$^+$CD62L$^+$ T cells and the appearance of soluble granzyme B in the peripheral blood.[265] Secondary immune responses to CMV appear to involve a recall response involving memory cells with the CD8$^+$CD28$^+$CD57$^+$ phenotype.[266]

As previously stated, the CTL response is MHC restricted. One can then ask the question as to how the transplant patient eliminates virus from an allograft that has a significant MHC discrepancy from that of the recipient. Two observations suggest that MHC restriction of the CTL response may be important: studies in renal allograft recipients have shown that persistent viruria can be present for years after transplantation (that is, persistent infection of the allograft is present)[267]; more recently, Fishman *et al.*[268] reported a high rate of relapsing CMV disease among patients with six antigen MHC donor–recipient mismatches. Traditionally, antibody responses to CMV have been utilized for diagnostic purposes, and have not been viewed as playing a major role in the patient's defense against the virus or recovery from clinical disease. A recent report documenting that the delayed acquisition of high-avidity anti-CMV antibody is correlated with prolonged bloodstream infection with the virus suggests that the contribution of humoral immunity to the host defense against CMV merits reexamination.[269] What is needed is a series of studies in which a careful assessment of both humoral and cellular immunity is carried out, with correlation of virologic parameters and clinical disease.

The cytokine response to CMV replication is an important determinant of the clinical effects of the virus and also its impact on such other processes as allograft rejection. For example, patients with symptomatic CMV disease have significantly greater evidence of a TNF response than those patients without symptoms from their CMV infection. Conversely, the major anti-inflammatory cytokine produced by monocytes, IL-10, is significantly increased in the asymptomatic patients compared to those with symptomatic disease. Thus, the balance between the proinflammatory TNF response and the anti-inflammatory IL-10 response is a significant determinant of the clinical effects of the virus.[270]

5.1.1c. Direct Clinical Effects of CMV in the Organ Transplant Recipient.   Whatever the category of CMV infection—primary, reactivation, or superinfection—its clinical and virologic manifestations are primarily seen in the time period 1–4 months posttransplant (unless the patient is receiving antiviral prophylaxis, in which case CMV disease can be observed more than 6 months posttransplant), with one delayed manifestation, chorioretinitis, occurring after that period. The effects of CMV in the different organ transplant populations are quite similar, with one major exception: CMV has far greater effects on the organ transplanted than on native organs. Thus, CMV hepatitis is a major problem in liver transplant patients, but is a relatively trivial issue in the other transplant populations; the attack rate for CMV pneumonia is far greater in lung and heart–lung recipients than in the other organ transplant populations; and CMV myocarditis is essentially only recognized in heart transplant patients.[258,271] The explanation for these observations is unclear, but several possibilities that are not mutually exclusive exist: There is a direct interaction or synergy between the effects of the virus and the effects of rejection on the allograft; the greatest viral burden, at least initially, is in the allograft where reactivation first occurs; and, finally, as previously discussed, the allograft is a privileged site for virus replication since the MHC-restricted, virus-specific, cytotoxic T cells will be unable to eliminate virally infected cells in the face of MHC mismatch.

As with most viral infections, CMV usually begins insidiously with constitutional symptoms of anorexia, malaise, and fever, often accompanied by myalgias and arthralgias. In many patients, unexplained fever and constitutional symptoms are all that the virus produces; in fact, prolonged fever is the most common recognizable clinical syndrome produced by CMV (approximately one-third of patients with clinically overt disease). These patients resemble normal hosts with CMV mononucleosis, even to the presence of 5–10% atypical lymphocytes on peripheral blood smear. The major difference lies in the usual absence of splenomegaly and lymphadenopathy in the organ transplant patient. In about one-third of patients who develop fever, a dry, nonproductive cough develops within a few days of the onset of the constitutional symptoms. Initially, dyspnea and tachypnea are not noted, but over several days progressive respiratory distress can ensue, although most patients with CMV pneumonia experience little respiratory distress at rest. On physical examination in patients with respiratory symptoms secondary to CMV infection, auscultation of the lungs is usually unrevealing. The best correlate on physical examination with the degree of respiratory embarrassment— hypoxemia on arterial blood gas determination and pneumonia on chest radiography—is the respiratory rate. Plugging of the lung by cytomegalic endothelial cells as well as the local inflammatory response to the virus appear to play an important role in the pathogenesis of CMV pneumonia.[213,217–219,255,272–277]

The radiographic manifestations of CMV pneumonia in the transplant patient may take a variety of forms. By far the most common form is a bilateral, symmetrical, peribronchovascular (interstitial), and alveolar process predominantly affecting the lower lobes.[219,278] Although a few renal, liver, and heart transplant patients with CMV pneumonia progress to total lung whiteout and respiratory failure,[279] in most individuals the lung involvement is relatively minor and would go unappreciated if a chest radiograph had not been obtained. The severe form of pneumonia is far more common in lung and heart–lung transplant patients. Less commonly, a focal consolidation more suggestive of bacterial or fungal disease,[219] or even a solitary pulmonary nodule may be caused by CMV.[280] Positive gallium[281] or indium-111-leukocyte[282] scans of the lungs have been reported in patients with CMV pneumonitis, although such information usually adds little to the diagnostic decision-making process in most patients.

An important point that cannot be overemphasized when considering the rate of progression of pneumonia in the nonpulmonary organ transplant patient is that CMV causes a subacute process that evolves over several days (in the lung transplant patient, this process can be greatly telescoped). The major differential consideration is to rule out *Pneumocystis carinii* infection, which presents in similar fashion and which is frequently present in addition to CMV. If acute respiratory deterioration over less than 12 hr occurs, superinfection with bacterial or invasive fungal agents should be considered rather than attributing such a deterioration to an exacerbation of the CMV infection. A relapsing form of CMV pneumonia has been reported, occurring when immunosuppression is reinstituted after recovery from serious CMV infection.[283]

In addition to actual invasive disease of the lung caused by CMV, careful studies by van Son and colleagues in The Netherlands have documented the presence of subtle abnormalities in pulmonary function in the majority of patients with CMV infection.[284,285] Concomitantly, evidence of complement activation, probably through the alternative pathway, can be found; it is appealing to speculate that these two events are related, just as subtle changes in oxygenation in the first few minutes of hemodialysis with a Cuprophane membrane are related to complement activation and resulting pulmonary leukosequestration.[286]

The second major organ system to be invaded with CMV in a fashion that can be life-threatening is the GI tract. Serious CMV hepatitis requiring intensive therapy is not uncommon in liver transplant patients (Fig. 2)[213,287–289]; similarly, pancreatitis severe enough to form "abscesses" may be produced by CMV in the pancreatic allograft.[290,291] In the other forms of organ transplantation, CMV infection of these organs is not a major clinical problem, although chemical abnormalities connoting infection at these sites are not uncommon.[8,213,219,292] Far more important clinically is the occurrence of infection of the gut itself. The consequences of GI CMV infection include diffuse inflammation with functional disturbances, hemorrhage, frank ulceration, perforation, and, possibly, the development of pneumatosis intestinalis. The stomach appears to be the most frequent site of symptomatic CMV infection, and appears to be associated with subjective complaints of nausea, a sense of abdominal fullness, and, occasionally, emesis and/or dysphagia. These symptoms have been correlated with the presence of CMV gastritis and inadequate gastric emptying, presumably the result of the infection. The Pittsburgh group has reported an incidence of upper GI CMV infection of 28% in liver transplant patients receiving cyclosporine-based immunosuppression, and a somewhat lower incidence (20%) in those treated with tacrolimus.[275,293–303] Unexplained abdominal pain, particularly midepigastric pain, in renal (and presumably other forms of solid organ) transplant recipients has been linked to the presence of active CMV infection in the upper GI tract.[303]

CMV infection at other sites in the upper GI tract is not uncommon, as well, with both esophagitis and duodenitis (including ulceration) being well documented.[300–304] A CMV-induced "pseudolymphoma" of the duodenum that responded to antiviral therapy has been reported in a renal transplant patient.[305] Colonic ulceration, often involving the right colon, presenting as GI hemorrhage or perforation, is a common manifestation of CMV infection.[292–297,299] Unusual forms of colonic CMV disease include pseudomembranous colitis,[306] and localized disease mimicking a neoplasm (including an apple core lesion on X-ray) or ischemic colitis.[307,308] Acute colonic pseudoobstruction (Ogilvie's syndrome) following liver transplantation has been linked to CMV, with resolution of the pseudoobstruction on treatment with ganciclovir.[309] Thus, it is clear that CMV can have significant effects on the function of both the upper and lower GI tract.

Uncommon effects of CMV on the GI tract in transplant patients include hemorrhoiditis[310] and cholecystitis.[311] Although CMV infection of the gut and of hepatic and pancreatic allografts can occur as part of a systemic febrile process, it is important to recognize that hematochezia, nausea and vomiting, hepatic enzyme abnormalities, and so on, in the absence of fever, leukopenia, or other manifestations of clinical CMV disease, can be the result of invasion of these viscera by this virus. Similarly, evidence of viremia (*vide infra*) may be absent in patients with CMV disease of the GI tract, requiring biopsy for diagnosis. Recognition of these clinical entities without some of the more common manifestations of CMV disease can lead to effective therapy with ganciclovir.[293–314]

Hematologic abnormalities are common during the course of CMV infection. For example, small numbers of atypical lymphocytes may be detected on examination of the peripheral blood smear. The most important effects, however, are on the white blood cell (WBC) and platelet counts. Leukopenia, not infrequently to counts less than 3000/mm^3, and/or thrombocytopenia, usually in the range of 30,000–60,000/mm^3 but sometimes even lower, occur in 20–30% of patients with CMV infection. Again, these hematologic findings may be the first manifestation of systemic CMV infection or be part of a multiorgan systemic process.[27,213–219] Idiopathic thrombocytopenic purpura has been reported as a consequence of CMV infection.[312]

Simmons *et al.*[275] have described what they have termed "the lethal CMV syndrome," which begins with fever and leukopenia (as does the more benign form of the illness), but progresses rapidly to include severe pulmonary and hepatic dysfunction, central nervous system (CNS) abnormalities, GI hemorrhage, and death. Death is

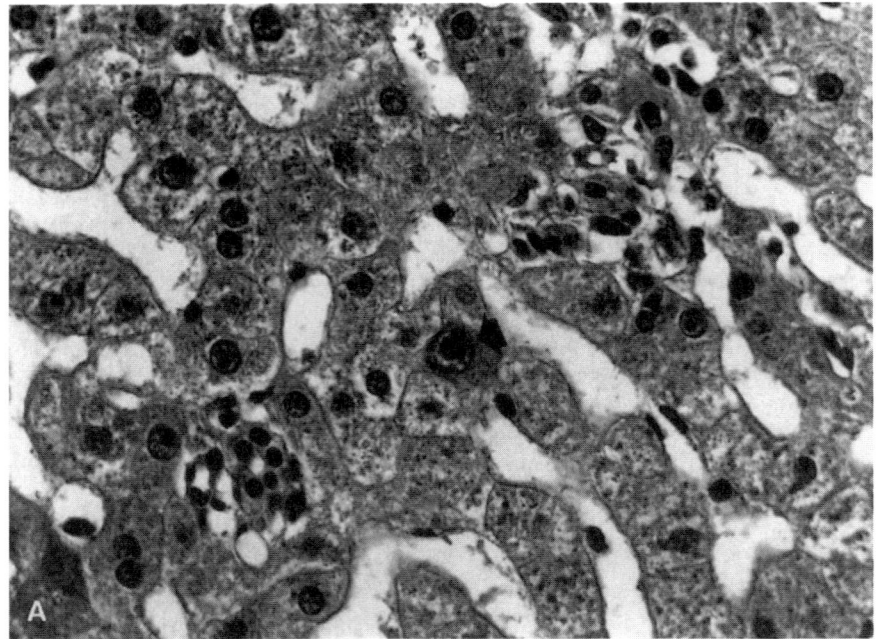

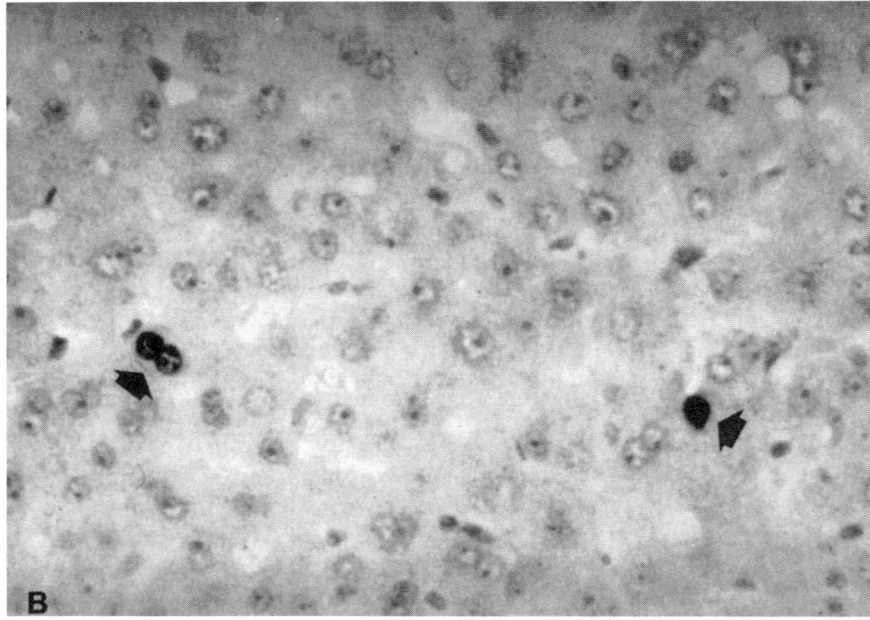

FIGURE 2. Liver biopsy of a liver transplant patient with cytomegalovirus hepatitis. (A) In addition to the focal sites of inflammation, a hepatocyte with the viral inclusion body typical of CMV infection is delineated by the arrow (hematoxylin and eosin stain, ×400). (B) Immunoperoxidase stain revealing three infected cells (arrows) in the absence of typical CMV inclusions.

usually caused by superinfection or bowel hemorrhage. Fortunately, the availability of ganciclovir therapy and the recognition of the possibility of this syndrome have resulted in early diagnosis and therapy, with few patients today developing the full-blown lethal syndrome. It is important to continue to emphasize, however, the potential for this constellation of events and the need for the clinician to be alert for early manifestations of severe disease. The addition of leukopenia to fever as a manifestation of CMV infection is often the first indication that serious clinical disease is developing, and prompt and aggressive therapy is mandatory.[8,27,213]

In renal transplant recipients, CMV infection has been associated with a number of processes that adversely affect the kidney: tubulointerstitial nephritis with cytomegalic inclusions within renal cells, particularly the tubular epithelium[223,313]; ureteral inflammation and/or necrosis[311,314]; hemolytic uremic syndrome[315]; thrombotic microangiopathy responsive to ganciclovir therapy[316]; and necrotizing and crescentic glomerulonephritis.[317] In addition, in renal transplant recipients, CMV infection has been associated with renal artery thrombosis[318]; whereas in liver transplant recipients, CMV infection has been associated with hepatic artery thrombosis.[319]

Uncommon infectious disease syndromes occurring in the organ transplant patient as a result of CMV infection include the following: endometritis[320]; epididymitis[321]; encephalitis[322]; transverse myelitis[323]; and skin ulcerations associated with an apparent cutaneous vasculitis.[294]

Chorioretinitis is the major late manifestation of CMV infection, usually being noted for the first time more than 6 months posttransplant. Although the retinitis may be asymptomatic at the time of discovery, most patients present with complaints of blurred vision, scotoma, and decreased visual acuity. Although symptoms are frequently restricted to one eye initially, progression to bilateral involvement is common. The initial retinal lesion on fundoscopic examination appears as scattered white dots or white granular patches without any characteristic distribution pattern. Irregular sheathing of the adjacent retinal vessels is common. This appearance of a gradually expanding, whitish, necrotic retinitis is thought to be distinctive for CMV. This fundoscopic picture of hypopigmented (white) areas surrounding atrophic retina corresponds to pathological findings in which the retinal pigment epithelium in the involved areas becomes so extensively necrotic that its capacity to proliferate is lost. The white, cordlike appearance of the involved retinal arterioles is thought to be secondary to the sloughing of infected endothelial cells from the retinal vessels, which leads to subendothelial hemorrhage and the collection of serofibrinous material in the same area (Figs. 3 and 4). Occasionally, retinal detachment or an anterior uveitis with secondary glaucoma may develop as the retinitis progresses, causing further loss of vision.[324–334] In one report from a heart transplant center, ~7% of patients had healed scars consistent with CMV retinitis or active CMV retinitis.[335] In our experience, the incidence of active retinitis is ≤ 1%. Rarely, CMV can cause acute retinal necrosis.[336]

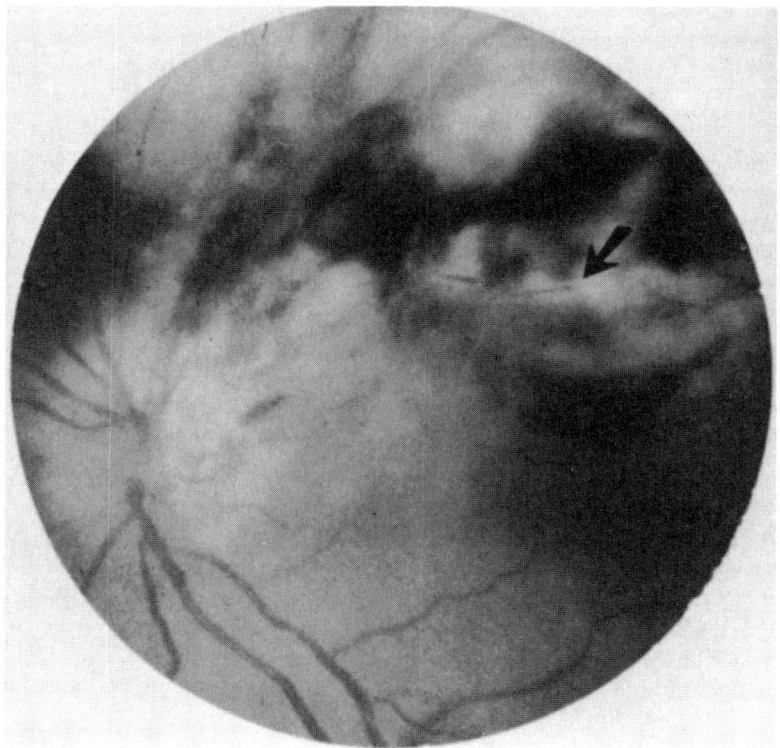

FIGURE 3. Fundoscopic appearance of cytomegalovirus chorioretinitis. Hemorrhagic infarction of the retina extends from the optic disk along the course of superotemporal vessels. A broad expanse of intense white retinal necrosis is largely obscured by extensive hemorrhage. Arteriole coursing through a zone of necrosis is attenuated and sheathed (arrow). (From Nicholson.[329])

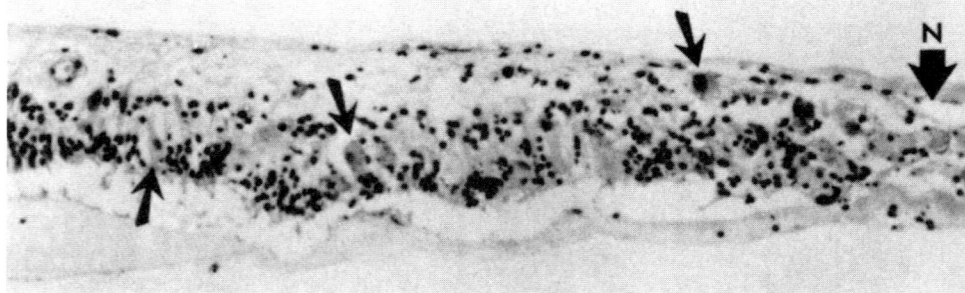

FIGURE 4. Histologic section at margin of retinal necrosis from the eye depicted in Fig. 3. Hypertrophied infected cells (smaller arrows) with nuclear and cytoplasmic inclusions are present in the region between completely necrotic retinal tissue (N) and that in which normal laminar retinal architecture is preserved. (Hematoxylin and eosin, ×175.) (From Nicholson.[329])

One other consequence of CMV infection in the transplant patient that bears mention is the possible impact on pregnancy. Congenital CMV infection has been documented in a child born to a woman who had suffered systemic CMV infection 3 years previously following her renal transplant. The mother had had an unspecified febrile illness during her first trimester of pregnancy, due either to a relapse in her infection or to superinfection. In any case, the child was severely affected, a tragic outcome that underlines the need for careful monitoring of successful transplant recipients for CMV replication before pregnancy is contemplated and during the course of the pregnancy, with prompt intervention if viremia is documented.[337]

5.1.1d. Indirect Clinical Effects of CMV Infection on the Organ Transplant Recipient.   CMV is an important contributor to the transplant recipient's net state of immunosuppression. Indeed, with the advent of effective antiviral chemotherapy against CMV (*vide infra*), it can be argued that the most important infectious disease effect of CMV on the transplant patient is its potentiation of superinfection. The clinical marker that appears to delineate those organ transplant patients most at risk for superinfection appears to be CMV-induced leukopenia. Severe leukopenia (WBC count < 1500/mm³) in conjunction with symptomatic CMV disease of greater than 5 days' duration, without ganciclovir and/or G-CSF therapy, has been associated with a greater than 50% mortality caused by superinfection, both of the lung and of the bloodstream. In the lung, *Pneumocystis carinii*, *Aspergillus* species, and a variety of gram-negative pathogens are the primary culprits. As far as the first two of these are concerned, alveolar macrophage dysfunction induced by CMV (in addition to the leukopenia in the case of *Aspergillus*) appears to be an important factor in the pathogene-

sis of superinfection with these organisms.[8,27,219,338–342] In addition, CMV appears to facilitate the colonization of the upper respiratory tract with gram-negative bacilli, with these serving as the reservoir from which gram-negative pulmonary infection is then derived.[343]

Septicemia with a variety of agents, but most particularly with *Listeria monocytogenes*, *Candida* species, and gram-negative organisms, is common in these patients with severe CMV-induced leukopenia. It is of interest that reports of infection from transplant centers not actively studying CMV disease have noticed a preponderance of both pneumonia and CNS infection in the period 1–4 months posttransplant, suggesting again a major role for this virus in the pathogenesis of serious infection. As with other clinical manifestations of CMV, such events appear to be more common in patients with primary as opposed to reactivation disease (in this context, all patients who are seropositive for CMV prior to transplant are regarded as having reactivation disease, with inadequate information currently available to distinguish between the clinical spectra caused by true reactivation as opposed to superinfection with a new strain of the virus). Again, viremia is a useful virologic marker to delineate those patients most at risk for this phenomenon.[213,344–349]

In addition to the abnormalities in leukocyte number and, possibly, function induced by the virus, a variety of other defects in host defense play a role as well. Humoral immunity, as measured by antibody response to the virus, appears to remain relatively intact in the majority of transplant patients with clinical CMV disease.[350] One possible detrimental effect of the humoral response to CMV has been suggested by Baldwin *et al.*[351–353] who demonstrated that IgM immune complexes appear in transplant patients in association with CMV infection, and that these have lymphocytotoxic properties. In addition, some CMV-induced rheumatoid factors may modify the immune response through anti-idiotypic activity. Cell-mediated immunity, whether measured by skin testing with Candida, SKSD, mumps, or intermediate-strength tuberculin recall antigens or by *in vitro* lymphocyte responsiveness to CMV antigens, other viral antigens, or a variety of mitogens, is markedly impaired.[350,354] It is noteworthy that patients with demonstrable cell-mediated immunity to CMV prior to transplantation, whether naturally induced or vaccine induced, lose this response in the first month posttransplant and may still be unresponsive more than 6 months posttransplant.[354,355]

Clinical recovery from CMV in the absence of specific antiviral therapy is largely dependent on the previously discussed activity of CMV-specific, MHC-restricted (class I MHC antigen), cytotoxic T cells (CD8$^+$) and NK cells, particularly the former. During CMV infection, expansion and activation of these lymphocyte populations occurs.[259–265,356–365] Van den Berg *et al.*[366] have demonstrated the importance of the dynamic relationship between viral burden (as quantitated by the antigenemia assay) and these lymphocyte responses. Whereas NK cells, as nonspecific effector cells, appear to be important in limiting viral replication and dissemination during the early phase of primary CMV infection, clinical recovery is determined by the ability to mount an activated, cytotoxic T-cell response commensurate with the level of virus present. In addition, development of this activated, cytotoxic T-cell response in the course of antiviral therapy appears to be a useful predictor that relapse will not occur following the cessation of antiviral therapy. This kind of analysis, which monitors both the replicating viral burden and the host's ability to defend against this burden, should permit a more precise approach to antiviral therapy. These findings also explain the occasional discrepancies that occur when clinical predictors are based solely on virologic measurements or lymphocyte measurements; what is really needed is an assessment of both.[366]

The mechanism by which CMV causes depressed cell-mediated immunity has received extensive investigation. At present, it would appear that CMV infection is associated with suppression of both monocyte and NK cell function, and that monocyte-induced suppression of lymphocyte function is the end result.[367–376] CMV infection of cells results in a downregulation of class I MHC antigens on these cells.[377–379] Since presentation of viral antigen by MHC class I antigens is of critical importance in mediating the key cytotoxic T-cell response to the virus, this may be an important mechanism by which the virus evades host defenses.[380] In addition, exhaustion of antiviral cytotoxic T cells may also play a role here.[381]

An accessible marker of disordered cell-mediated immunity induced by CMV involves the use of flow cytometry (and fluorescein-labeled monoclonal antibodies to lymphocyte cell surface markers) to characterize circulating T-lymphocyte subsets. In the normal individual, the ratio of CD4$^+$ (helper/inducer) cells to CD8$^+$ (cytotoxic/suppressor) cells is normally approximately 1.5–2; in patients with CMV and EBV infection, there is a marked reversal in this relationship, with the ratio now falling to 0.1–0.5, with clinical recovery being associated with normalization. The great majority of opportunistic infections occur in the subset of transplant patients with these changes in circulating T cells. Indeed, the occurrence of opportunistic infection in patients with normal circulating T cells and hence a reasonable net state of immunosuppression is an important clue to an excessive epidemiologic hazard in the environment (*vide supra*).[364,382]

Studies in the murine CMV model have supplemented these observations. If animals are given sublethal challenges with CMV, *Candida albicans*, *P. aeruginosa*, or other organisms, not much happens. If CMV is combined with each of the other organisms, then lethality approaches 100%.[383] In the murine model, CMV has been shown to reactivate latent *Toxoplasma gondii* infection in the lungs, producing an active pneumonia. Pathogenetically, it was suggested that a CMV-induced fall in the number of CD4$^+$ lymphocytes played a role in the reactivation of the protozoan, while the subsequent influx of CD8$^+$ cells was responsible for the active pneumonia that developed.[384] Of interest, evidence that HHV-6 activation is promoted by CMV infection (*vide infra*), particularly primary infection, in the transplant patient, suggests that this CMV-induced traffic in lymphocytes, and the elaboration of a variety of cytokines in conjunction with it, can play a role in the pathogenesis of a variety of secondary infections in the transplant recipient.[385]

In 1970, Simmons *et al.*[386] first suggested that CMV infection could in some fashion lead to allograft dysfunction. Proof of this relationship has been difficult to obtain, in part because of the ubiquity of CMV infection, and in part because of the lack of an adequate laboratory marker for distinguishing clinically important CMV infection from the asymptomatic or trivial. In addition, there has been a problem in most studies with an insufficient number of patients to permit adequate stratification that would control for such variables as type of donor, histocompatibility match, donor and recipient CMV status, and form of immunosuppression administered.[364,386–396]

Despite these difficulties there is now a compelling body of information suggesting that CMV can be involved in the pathogenesis of certain forms of allograft injury.

First, there are occasional cases in which renal dysfunction occurs with a biopsy picture that reveals cells laden with cytomegalic inclusions—typically in the tubular epithelium and interstitium, occasionally within the glomerulus. In these cases, treatment with ganciclovir can result in restoration of renal function.[223,313,397,398] However, such cases are few and far between, and clearly do not account for the increased incidence of renal dysfunction linked to CMV infection in several studies of renal transplant populations.[223,225,232,233,393,398–400] Particularly intriguing are studies in which "late rejection" was diagnosed in patients with asymptomatic CMV antigenemia—treatment with ganciclovir, but not immunosuppression, was effective in reversing the episode of renal dysfunction.[242,401,402]

If, then, under intensive immunosuppression, CMV can adversely affect renal allograft function, what is the mechanism? *In vitro* studies have shown that essentially all types of renal cells—glomerular, tubular, interstitial, and so on—can support the growth of this virus.[403] However, it is clear that if CMV is involved in the pathogenesis of allograft injury in more than a rare instance of disseminated infection, then direct infection, the deposition of immune complexes, and complement activation are not involved (although both circulating immune complexes and complement activation may occur not uncommonly in transplant patients with CMV infection).[27,223,404]

The renal lesion that has been the subject of the most discussion regarding possible linkage with CMV infection was studied by Richardson *et al.*[405] Patients were stratified on the basis of the presence or absence of CMV viremia, with renal biopsies obtained at the time of acute functional deterioration of the kidney being interpreted without knowledge of their CMV status. The results were striking. Those patients with nonviremic CMV disease and no CMV disease had renal biopsies that revealed the classic tubulointerstitial findings of acute cellular rejection; in contrast, those patients with viremic CMV disease were free of such changes and, instead, had a distinctive glomerular lesion on biopsy characterized by endothelial cell hypertrophy, necrosis, and loss with narrowing or obliteration of capillary lumina and the formation of finely fibrillary material between cells, and mild segmental hypercellularity. Immunofluorescent staining revealed deposits of immunoglobulin and the third component of complement within these glomeruli. Vascular injury was not uncommon in these biopsies, and no evidence of the virus could be found.

Since that original study, the following observations have been made by our group regarding this glomerulopathy[223,364,405,406]: In the original study, CMV viremia was the stratification marker that permitted us to separate out this lesion. Overall, CMV-infected recipients who are viremic have about twice the frequency of glomerulopathy as those who are nonviremic (58 versus 32%). Of 36 biopsies taken for renal allograft dysfunction, the glomerulopathy was observed in 12 of 25 (48%) with CMV and 0 of 11 with no virologic or serologic evidence of CMV ($p < 0.005$). Not unexpectedly, the glomerulopathy was highly associated with the previously described changes in circulating T-cell subsets that viral infection causes. When the infiltrating cells were characterized in patients with the glomerular lesion, there was a striking increase in the number of $CD8^+$ cells (compared with biopsies from patients with typical tubulointerstitial rejection); there was an increase in activated mononuclear phagocytes present; and the glomeruli from the glomerulopathy cases stained more intensely for MHC class I antigens than the tubules, in contrast to the typical rejection cases. Renal dysfunction associated with the glomerular lesion is far less likely to respond to antirejection therapy than classical rejection ($> 90\%$ in classical rejection versus $< 20\%$ for the glomerular cases). Finally, early intervention with ganciclovir therapy appears to have made this lesion far less common.

Since our description of this glomerular lesion, its existence has been confirmed by several groups. What has been controversial is the issue of whether this lesion is related to CMV infection. What all observers do agree on is the fact that this glomerular lesion only occurs in renal allografts. Studies in bone marrow transplant patients and recipients of other organ allografts, AIDS patients, and nonimmunosuppressed individuals with CMV infection who had renal tissue examined failed to yield evidence of this lesion. In sum, a histologic finding distinct from classical rejection has been described in renal allograft recipients that may be linked in some instances to CMV infection, but clearly has occurred in patients without evidence of CMV infection. As we look at the experience with the other allografts, it is clear that we must come up with a unifying hypothesis to link these two kinds of observation. In addition, the possibility that CMV infection could be involved in the pathogenesis of allograft injury that has a histologic picture more typical of classical rejection must also be kept in mind.[27,394–396,404]

In liver transplantation, this issue is even more complicated, since the direct and possibly indirect effects of CMV infection on the liver may be obscured by the impact of both classical allograft rejection and hepatitis C infection. However, Paya *et al.*[288] have observed among liver transplant recipients a group of individuals with pathological findings of focal necrosis and clustering neutrophils within the liver lobules, a picture distinct from that of typical hepatic allograft rejection. This pathological picture occurred in the setting of CMV viremia, but in individuals without histological or cultural evidence of CMV infection of the liver itself, a phenomenon similar to the glomerulopathy described in renal transplant patients. In addition, it has been postulated that CMV infection, when there is a one- to two-antigen match between donor and recipient, is associated with the disappearing bile duct syndrome in liver transplant recipients.[407] In patients with chronic rejection of the liver allograft, CMV infection of bile ducts and endothelial cells can be demonstrated.[408,409] Interestingly, CMV infection is associated with increased inflammation and severe bile duct damage in a rat liver allograft model as well.[410]

Even more important in terms of assessing the role of CMV in the pathogenesis of hepatic allograft injury are studies reporting an increased incidence of acute and chronic rejection, as well as mortality, in patients with CMV infection.[411–414] A particularly close relationship between rejection and CMV infection has been observed in those with a partial HLA class I match between donor and recipient.[415] An intriguing report from the University of Oregon has suggested that all CMV isolates do not carry the same risk for inducing rejection. In this analysis, CMV is divided into four genotypes on the basis of the envelope glycoprotein gB. In this study, CMV infection in general was associated with significantly reduced patient and graft survival rates at 1 and 5 years following orthotopic liver transplantation. Infection with CMV genotype gB1 was associated with a higher mean number of acute rejection episodes.[416]

Perhaps the most convincing clinical data linking CMV infection with allograft injury come from the cardiac transplant experience, where a number of studies have linked both acute rejection and accelerated coronary atherosclerosis in the allograft to both symptomatic and asymptomatic infection with the virus. In particular, CMV has been associated with what has been called *cardiac allograft vasculopathy*.[417–426] That vascular injury could be a consequence of systemic CMV infection should be no surprise: In the general population, CMV has long been hypothesized as playing a role in the pathogenesis of atherosclerosis, with evidence of herpesvirus infection at sites of atheromatous lesions in major arteries, as demonstrated by both DNA hybridization and immunoperoxidase staining for specific viral antigens. Epidemiologically, patients who required vascular surgery for atherosclerotic disease had a higher rate of CMV seropositivity than in matched control subjects.[426–429] Finally, a number of studies have compellingly demonstrated that CMV infection of both vascular smooth mus-

cle and endothelium is a regular occurrence during CMV infection, thus providing a mechanism for vascular injury—felt to be the foundation of chronic allograft injury.[426,430–436] Indeed, CMV-infected endothelial cells can be found in the circulation, presumably providing a means of viral dissemination as well as a possible measure of vascular injury at the capillary level.[435,436] As in liver transplantation (*vide supra*), different CMV strains appear to have differing abilities to affect vascular endothelial cells.[437]

In a similar fashion, CMV has been linked to the development of bronchiolitis obliterans in lung and heart–lung transplant recipients. This entity, the limiting factor in the long-term success of lung transplantation, as is the case with cardiac allograft vasculopathy, can occur without CMV on a pure "immunologic" basis. However, the incidence is increased with CMV, with at least some evidence suggesting synergy between virus and rejection in the pathogenesis of bronchiolitis obliterans. Relapsing CMV infection of the lung allograft, particularly in the presence of rejection, carries a particularly high risk for bronchiolitis obliterans.[426,438–445] The suggestion that CMV infection in general and relapsing or chronic infection in particular synergize with acute rejection in the pathogenesis of chronic allograft injury, has also been made in renal and liver transplant patients.[446,447] One potential link between rejection and CMV infection is the observation that both processes result in the release of endothelial cells, both infected and uninfected, potentially amplifying the extent of the infection and the immunologic stimulation from the allograft endothelium.[448]

Extensive effort has been devoted to determining the mechanisms by which CMV infection could result in acute and chronic allograft injury. In this effort, animal models that combine transplantation of tissues and organs across MHC barriers and CMV infection have provided important information that has greatly supplemented the information obtained from clinical studies. The essential challenge is to delineate how CMV could exert its effect when a pathologically identical process can occur in the absence of this virus.

Careful studies in a rat aortic allograft model demonstrated that early rat CMV infection (each species has its own specific virus, so human CMV cannot be utilized in rodent studies) was associated with inflammation of the endothelium, smooth muscle cell proliferation, and intimal arteriosclerotic alterations.[426,449,450] Similarly, rat CMV infection introduced into a rat heterotopic cardiac allograft model that included triple-drug immunosuppression (cyclosporine, methylprednisolone, and azathioprine) accelerated the development of cardiac allograft vasculopathy.[451] Utilizing a heterotopic, allogeneic, tracheal transplant model in rats, CMV was shown to cause enhanced MHC class II expression on the respiratory epithelium, airway wall infiltration with $CD4^+$ T cells and macrophages, and a 5-fold increase in luminal occlusion of the trachea—changes identical to "immunologically" caused rejection; that is, in this model bronchiolitis obliterans could be caused by either classical rejection or CMV.[452,453] In this model, platelet-derived growth factor appeared to play an important pathogenetic role in the evolution of these changes.[452] Of interest, increased immunosuppression decreased the changes induced by both processes (although increased immunosuppression could be expected to increase the level of viral replication), suggesting the possibility that CMV was acting through a signaling pathway similar to that involved with rejection.[426,449–452] Conversely, CMV-enhanced cardiac allograft vasculopathy is abolished by ganciclovir prophylaxis in the rat model.[451,454]

Recently, a series of studies exploring the mechanisms involved in the pathogenesis of CMV-induced allograft injury have extended our understanding of these processes considerably. First, it is clear that not only are key cells (e.g., vascular and bronchial smooth muscle cells, respiratory epithelium) infected, but there are consequences other than the generation of "cytomegaly" cells with inclusions or cytolysis. Thus, there is an increased display of MHC antigens, as well as activation of a variety of cell types involved in tissue injury—leukocytes, endothelial cells, and so on.[455–457] Particularly striking is the upregulation in the expression of proinflammatory adhesion molecules induced by CMV infection: CMV infection of human vascular endothelial cells results in the *de novo* expression of VCAM-1 and E-selectin, and a 200-fold increase of ICAM-1; in contrast, infection of vascular smooth muscle cells results in the *de novo* induction of ICAM-1, but has no effect on the expression of VCAM-1 or E-selectin. Of most importance, however, was the observation that uninfected cells in the same tissue culture system manifested the same findings as the infected cells, and these changes could be induced by a virus-free supernate from infected cells. Subsequent work demonstrated that the mediator of these effects was IL-1β released by CMV-infected cells, and the upregulation of the proinflammatory adhesion molecules on uninfected cells is due to the paracrine effects of this mediator produced in response to CMV infection.[458,459] The immediate early-1 antigen of CMV has been shown to upregulate IL-1β gene expression as well as that of another proinflammatory cytokine, IL-6.[460,461] Other adhesion molecules upregulated or induced *de novo* in re-

sponse to CMV infection are sialyl Lewis[x] and Lewis[x] antigens on human endothelial cells.[462]

Perhaps of greatest interest are studies of a rat kidney transplant model in which chronic allograft rejection occurred. The introduction of systemic CMV infection resulted in an increase in the level of inflammation within the kidney and accelerated and amplified the extent of the chronic rejection process. In this kidney model, as in the *in vitro* experiments, there was a significant and prolonged increase in VCAM-1 and ICAM-1 expression on the vascular endothelium of the renal allografts. In addition, the inflammatory cells infiltrating the allograft had a marked increase in their ligand adhesion molecules LFA-1 and VLA-4. These effects were secondary to the release of proinflammatory cytokines in response to CMV replication.[463,464] Interestingly, CMV infection and ischemic injury appeared to have similar and additive effects in contributing to the occurrence of "chronic rejection."[465]

Other mechanisms are possible as well. A variety of autoantibodies are produced as a consequence of CMV infection, including anti-endothelial cell antibodies.[466,467] Endothelial cells infected with CMV on the one hand manifest increased neutrophil and mononuclear cell adherence, and, on the other hand, provide a potent stimulus for a cytotoxic T-cell response that destroys uninfected bystander endothelial cells.[468–470] CMV also increases the production of the C-X-C chemokine and IL-8, thus causing neutrophil recruitment and enhanced neutrophil transendothelial migration.[471,472] A more direct form of immune injury precipitated by CMV infection also is possible. Sequence homology and immunologic cross-reactivity between an immediate early antigen of human CMV and the HLA-DR beta chain have been demonstrated.[473] In addition, CMV-infected cells produce a glycoprotein homologous to MHC class I antigens.[474] Thus, it is not unreasonable to speculate that immune injury triggered by the virus could be directed at cells that bear either the appropriate HLA-DR antigen or the particular class I antigen in a form of molecular mimicry.[404] The interaction between CMV infection, MHC antigens, and allograft injury may be even more complex. First, studies in the murine model have clearly shown that susceptibility to the virus is closely linked to the MHC locus (the H-2 complex in mice), with non-MHC genes also playing a role.[475,476] Several studies have suggested that DR-matched transplants may result in an increased incidence of CMV disease, and, possibly, chronic allograft injury. Thus, in evaluating the potential contribution of CMV infection to acute and chronic allograft injury, the MHC typing and the degree of match between donor and recipient may be important.[477,478]

CMV infection in transplant patients is associated with profound evidence of systemic immunologic/inflammatory activation: increased blood levels of IL-6, soluble forms of VCAM-1, and IL-2R, as well as a marked increase in the numbers of circulating $CD2^+CD8^+HLA^-DR^+$ cells.[479–482] As previously noted, activation of humoral inflammatory pathways can be demonstrated as well in transplant patients with CMV infection: the presence of circulating immune complexes, rheumatoid factor, and circulating products of complement activation C3d and C3a des-arg are easily demonstrated.[284,351–353] In addition, murine CMV infection has been shown in some mouse strains to cause an increase in alloreactivity.[483]

Perhaps the most compelling information to emerge linking CMV infection to allograft injury comes from reports that antiviral preventive programs aimed at CMV resulted in a decrease in the incidence of both acute and chronic rejection. Thus, oral valaciclovir and oral ganciclovir prophylaxis for the first 3–4 months posttransplant not only decreased the incidence of CMV infectious disease events, but also decreased the incidence of acute rejection episodes in both renal and liver transplant recipients.[238,484] In addition, there is preliminary evidence that in human cardiac transplant recipients, the administration of intravenous ganciclovir for a month decreased the incidence and severity of coronary artery atherosclerosis.[485,486]

In sum, an ever-expanding body of evidence has linked CMV infection to acute and chronic allograft injury. It would appear that a similar array of cytokines, chemokines, and growth factors are elaborated in the course of both CMV infection and classical rejection processes. Indeed, there is a bidirectional relationship between these two processes, with the one being influenced by the other, and both being influenced by the nature of the immunosuppressive therapy being administered. The fact that both of these can result in similar histologic pictures is not surprising—the mediators of allograft injury in both processes are similar; that is, although the initial stimuli may be different, the ultimate signaling pathway that effects injury is the same. A question that has long been debated has finally been answered: Which comes first, CMV or rejection? The answer is clearly either. What also becomes a reasonable hypothesis according to this view is that other infections (e.g., HHV-6) could have similar effects, and that ultimately the extent of allograft injury is the integration of a number of different processes that include ischemia, infection, and classical rejection, with room in this model for other as yet to be defined processes.

The final possible indirect effect of CMV to be considered is its possible role in the pathogenesis of malignancy in the transplant patient. Like other herpes group viruses, CMV must be thought of as a potentially oncogenic agent. Portions of the CMV genome have been shown to be homologous to the *myc* oncogene.[487,488] Not only is the intact virus a transforming agent in certain cell lines (the *in vitro* correlate of oncogenesis), but also specific cloned CMV DNA fragments are able to transform the NIH 3T3 cell line.[489] In other species, under special conditions, administration of the virus has been associated with the production of malignancy.[235]

What about humans? Rather weak associations between CMV and human colonic carcinoma and prostatic carcinoma have been made, chiefly on the basis of finding CMV DNA in the tumors of some patients with the former and a higher rate of serologic positivity for CMV in the sera of patients with the latter.[235] An association of testicular carcinoma with high antibody levels against VZV, herpes simplex virus, and CMV, with the strongest association being with the CMV titers, has been reported.[490]

The clearest relationship between CMV and malignancy is with EBV-associated posttransplant lymphoproliferative disease (PTLD) (*vide infra*). The incidence of PTLD is increased 7- to 10-fold in individuals with symptomatic CMV disease. Conversely, PTLD occurring in a transplant patient harboring latent CMV infection is almost uniformly associated with CMV reactivation. We would suggest that there is once again a bidirectional relationship between these two processes, with the link being cytokines, chemokines, and growth factors produced by the host in response to these processes. This hypothesis suggests that one could positively influence PTLD by antiviral strategies aimed at CMV (and possibly EBV).[211,491]

5.1.1e. The Diagnosis of CMV Infection in the Organ Transplant Patient. The diagnostic techniques utilized in evaluating the transplant patient for CMV infection can be divided into two general categories: serologic and virologic. Serologic techniques are most useful for assessing the past experience of donor and recipient with the virus, and thus for predicting risk of subsequent clinical disease (*vide supra*). Such techniques can also be utilized serially to delineate rises in titer or seroconversion as an indirect measure of the presence of active viral infection that is stimulating an immune response, although antibody response lags significantly behind the time when virologic diagnosis can be made. In this era of effective anti-CMV therapy, such a time lag is unacceptable, and thus the emphasis should be placed on virologic diagnosis of clinical disease, rather than waiting for serologic evidence of infection.[8,27,492] (As a general rule in immunocompromised patients, whose antibody response to microbial invasion may be attenuated, delayed, or totally abrogated, demonstration of replicating microbes in some fashion is far to be preferred to serial measurements of antibody levels.)

A variety of serologic techniques are currently available that measure the level of antibody to CMV in the serum. Essentially all of them utilize antigen extracted from fibroblasts infected with a laboratory strain of CMV (usually AD169). Although in theory the antigenic variation among different CMV strains in nature could affect the performance of a serologic assay, in practice this does not appear to be a major problem. Far more important in terms of the performance of a serologic assay is the method employed in extracting the antigen; for example, the development of the glycine-extracted complement fixing (CF) assay represented a major improvement in the sensitivity and reliability of the CF test.[492]

Although the CF assay has traditionally been considered the gold standard for testing for anti-CMV antibodies, today because of cost, speed, and sensitivity, most diagnostic laboratories have converted to immunofluorescence, ELISA, and latex agglutination systems for detecting anti-CMV antibodies. The advantages and pitfalls of these newer assays have been reviewed,[492] but one particular issue bears special emphasis here: Whereas there is a large clinical experience documenting the biologic meaning of a CF titer of $\geq$ 1:8 (a "true positive," meaning that the patient harbors latent or replicating virus), the meaning of similar titers with the newer, "more sensitive" assays is less clear. Thus, in a seroepidemiologic study of more than 1200 transplant patients in which multiple techniques were utilized in the analysis of each serum specimen, a small group of patients was identified that was negative by CF assay but positive in one or more of the newer assays at relatively low levels. Analysis of the clinical courses of these patients suggested that they were a heterogeneous group—a few with true-positive assays, and the remainder with false-positive assays. Therefore, the clinician needs to be aware of the performance characteristics of the assay being employed, as well as what the reliable cutoff level is that distinguishes true positives from more equivocal results.[225]

Most of the assays utilized to measure anti-CMV antibodies assess total antibody, both IgG and IgM. Commercially available CMV-specific IgM assays have added little to our diagnostic abilities: In primary infection, demonstration of replicating virus in blood and/or urine will usually precede the IgM response; some immunocompromised patients may not mount an antibody response even in the face of fatal infection; and IgM antibody can be present in asymptomatic reactivation infection. For all of these reasons, there is little to recommend the measurement of CMV-specific IgM for routine clinical use.[492]

Two other serologic techniques that bear mention here are the measurement of anti-CMV neutralizing antibody and the measurement of antibodies to specific viral proteins. Measurement of neutralizing antibody levels is a time- and resource-consuming research test that is impractical for routine clinical use, as the presence of measurable titers of neutralizing antibody activity does not guarantee either protection or recovery from clinical disease.[493] Studies with the murine CMV model have shown that monoclonal antibodies that were neutralizing *in vitro* were not necessarily protective *in vivo*, while other antibodies that were not neutralizing *in vitro* still could be protective *in vivo*.[494] Thus, the routine measurement of the titer of neutralizing antibody has no clear-cut clinical use at the present time.

Considerable work has been carried out to define the antibody response to specific CMV proteins. Thus, antibodies to various epitopes that mediate neutralizing activity, as well as antibodies to a variety of structural and regulatory proteins have been defined. Studies of a large number of patients have shown that there is considerable heterogeneity of response, and there is hope that one or more patterns of antibody response will correlate with the occurrence of clinical events. Such information would be of key importance both in defining protective immunity, and for the formulation of both an anti-CMV vaccine and an anti-CMV IgG product for the prevention and/or treatment of clinical disease.[492,495–501]

The cornerstone of CMV diagnosis, however, remains the direct demonstration of the presence of the virus in blood, respiratory secretions, urine, or tissues. Classical CMV virology involves the inoculation of specimens on fibroblast monolayers, with the endpoint for positivity being a cytopathic effect related to full replication of the virus, and the induction of visible changes in the monolayers. On average, this takes 1–2 weeks (and, thus, many replicative cycles are necessary before a cytopathic effect can be visualized), and may take upwards of 6 weeks if the viral burden is low.[492] A major advance in CMV diagnosis came with the development of the shell vial technique. In this technique, the clinical specimen is centrifuged onto the fibroblast monolayer, which assists adsorption of any virus present, resulting in a fourfold increase in the infectivity of the viral inoculum. Twenty-four to forty-eight hours later, at a time when there is no visible evidence of viral replication, the fibroblast monolayer is stained with a monoclonal antibody to the 72-kDa major immediate early protein of CMV, which can be demonstrated by immunofluorescence hours after inoculation. Urine and respiratory tract secretions are particularly well served by the shell vial technique. Unfortunately, cultures of buffy coat for evidence of viremia are less sensitive, because of leukocyte toxicity for the monolayer and problems with the centrifugation procedure; false-negative cultures of the buffy coat occur in as many as 50% of individuals. In addition, a single monoclonal antibody may yield falsely negative results because of antigenic variation among different isolates.[492,502–506]

Because of the importance of viremia in CMV diagnosis, particularly to facilitate early antiviral therapy, two other techniques have become the standard for the care of transplant patients: an antigenemia assay and PCR assays.[492,507] The and colleagues in The Netherlands have developed a direct immunoperoxidase assay on buffy coat preparations for the same-day diagnosis of viremia, utilizing monoclonal antibodies to a specific CMV antigen. Their assay utilizes monoclonal antibodies directed against a 65-kDa lower matrix phosphoprotein, a late structural antigen. At a time when granulocytes and monocytes positive for this antigen are easily demonstrated, only a fraction of these will show evidence of the immediate early antigen. It is now believed that the 65-kDa late antigen that is identified in the antigenemia assay represents antigen exogenously derived, probably from infected endothelial cells, which has been taken up by receptor-mediated endocytosis.[508]

Although these findings have importance in terms of understanding the pathogenesis of CMV infection in transplant patients, they also have practical diagnostic importance: Whereas studies with The's reagents have been uniformly successful, with a diagnostic sensitivity and specificity of > 95% in patients with clinical syndromes compatible with CMV,[508] other antigenemia assays, based on other monoclonal antibodies directed against other antigens, particularly the immediate early antigen, have been less useful.[509–511] Therefore, great care must be utilized in the choice of reagents used for the performance of this diagnostic test. The antigenemia assay has three advantages over previous diagnostic approaches: it is semiquantitative, giving a measure of the level of infection, viral load, as it occurs and as therapy is instituted; when performed prospectively, it, like the PCR assay for CMV, turns positive 4–7 days prior to the onset of clinical disease, opening up the possibility of preemptive therapy for those patients at major risk for clinical disease (not all patients with antigenemia become symptomatic, but sustained antigenemia, particularly a rising

titer, appears to be quite predictive); and, finally, it can serve as a means for monitoring the effects of antiviral chemotherapy.[508,512–519]

At the present time, the CMV antigenemia assay is the most accessible test for detecting viremia, determining viral load, and evaluating the need for and success of therapy. The availability of results in 1 day and the "low tech" nature of the test make it a useful assay for serial measurements. However, certain limits to the assay must be recognized. Although the assay has an excellent positive and negative predictive value when assessing patients with a mononucleosis-like syndrome and CMV pneumonia in non-lung transplant patients, there are situations in which the antigenemia assay has its limitations. Patients with significant CMV enterocolitis, proven on biopsy and responsive to ganciclovir therapy, will commonly have negative antigenemia assays. Lung transplant patients may have replicating CMV in their respiratory tracts, not infrequently with evidence of parenchymal invasion, in the face of a negative antigenemia assay. Finally, the blood specimen drawn for the assay needs to be processed within a few hours of attainment. A lapse of > 12 hr between blood draw and processing can greatly compromise the reliability of the results, a serious disadvantage when patients are being monitored at a distance from the transplant center—the assay does not "travel well."

Clearly, the future lies with PCR assays for CMV. The selective amplification of specific nucleic acid sequences, first introduced by Saiki et al. in 1985,[520] represents a major breakthrough in the application of molecular biology to diagnostics, including the diagnosis of CMV viremia. Although PCR assays require great technical skill, they have now been shown to be useful in the diagnosis of CMV viremia. Early concerns that active, replicating virus could not be differentiated from latent virus have not been borne out, provided the appropriate primers, number of amplification cycles, and specified stringency conditions are employed. In skilled laboratories, PCR and the antigenemia assay provide very similar information, although PCR may detect viremia a day or so earlier. In general, most laboratories that rely on PCR assays employ a qualitative PCR for screening, and then utilize a quantitative technique to determine the viral load, with serial measurements of viral load providing a dynamic picture of the interaction among virus, host defenses, and the antiviral strategy being employed. What is less clear at the moment is the precise format of the assay that is ideal and whether a whole blood, leukocyte-based, or serum assay provides optimal information. It is likely that these details will be worked out in the next few years, and that a "user-friendly" PCR assay will become the cornerstone of CMV diagnosis.[521–528]

Whether it is determined by antigenemia assay or by PCR, the key determination with CMV, as it is for HIV infection, is the viral load. Viral load determinations, preferably on blood but also on urine, provide critically important information for a number of purposes: viral load measurements have an important predictive value in determining subsequent risk of clinical disease; serial viral load measurements can form the basis for a preemptive antiviral strategy (*vide infra*); serial viral load measurements can be utilized to monitor the response to therapy; and viral load determinations at the conclusion of therapy can be utilized to predict the chance of relapsing clinical disease.[529–541]

5.1.1f. Clinical Management of CMV Infection in the Organ Transplant Patient.

Given the protean manifestations of CMV infection in the organ transplant recipient, it is not surprising that an extensive effort has been made to control this infection. This effort can be divided into two categories: the prevention of infection and disease, and the treatment of disease. Within the prevention category, there are three strategies that merit attention: those aimed at decreasing the risk of virus acquisition and reactivation; the induction, either actively or passively, of immunologic protection; and the utilization of antiviral drugs to prevent the clinical effects of viral replication. In fact, these strategies are not mutually exclusive, and are probably best used in combination. As one approaches the task of preventing the consequences of CMV replication, the key issue is what aspects of CMV infection is it desirable to prevent—the direct or the indirect effects of the virus? Although virtually all of the data that have been published in this regard have to do with the prevention of the infectious disease syndromes caused by the virus, the recent reports of antiviral therapy protecting against acute and chronic allograft injury suggest that the endpoints of anti-CMV intervention should be changed.[238,484–486]

There are two major sources of exogenous CMV infection for the transplant patient: viable leukocyte-containing blood products and the allograft itself. The first of these, transfusion-related infection, should be totally preventable, whereas the issues regarding allograft-transmitted infection are more complex. With the clear-cut evidence that CMV superinfection occurs in seropositive transplant patients (*vide supra*), and that this may have a greater clinical impact than endogenous reactivation, it is our belief that all transplant patients, not just seronegative ones, should receive only blood products that are free of CMV risk. This can be accomplished in two ways: (1)

screening all units of blood for the presence of CMV antibody (positive ones potentially harboring latently infected leukocytes), and utilizing only blood from seronegative donors for transplant patients, or (2) utilizing high-efficiency leukocyte filters to remove the viable leukocytes that harbor the virus. Both of these strategies work, in terms of preventing transfusion-related CMV infection; they differ in terms of cost and the possible prevention of other infections.[542–544]

Screening units of blood for CMV antibody can be accomplished at a cost of approximately $3 per unit; in contrast, a leukocyte filter costs approximately $21 per unit of red blood cells and a leukocyte filter for platelets costs approximately $40, and must be changed after the administration of six units of platelets through it. Therefore, costwise, the utilization of CMV-seronegative blood products is to be preferred, provided they are available. Theoretically, the removal of viable leukocytes might also be beneficial in decreasing the transmission of other infectious agents, such as EBV and HHV-6, through transfusions. Whatever strategy is employed, it is clear that the time has come when all blood products administered to transplant patients should be free of risk of transmitting CMV. This is particularly important in liver transplantation where a 16% incidence of primary infection resulting from transfusions has been documented.[192]

The issue of protective matching of donor and recipient so that an organ from a seropositive donor is not placed in a seronegative recipient is less clear-cut. Although eminently reasonable, such a policy would seriously curtail the potential donor pool. Data currently available suggest that the advantages of a living related transplant far outweigh any possible disadvantages from primary CMV,[224] so the only population to which protective matching could be applied would be patients awaiting cadaveric donor allografts. Even in these patients, both because of ethical issues related to the possible penalty for seropositive recipients who would then receive all of the organs from seropositive donors, and because of the desire not to further limit the already inadequate donor pool, our preference is not to carry out such protective matching. Instead, we would prefer to deploy an effective antiviral preventive program.

Two possible immunologic interventions against CMV have been evaluated in organ transplant patients: active immunization with a CMV vaccine and passive immunization with a variety of intravenous immunoglobulin preparations. Pioneering efforts by Plotkin and his colleagues have suggested that a CMV vaccine could have utility in organ transplant patients.[355,545–549] Their work with the live, attenuated Towne strain CMV vaccine demonstrated the following: administration of the vaccine to seronegative dialysis patients resulted in seroconversion, and if they received kidneys from seropositive donors, there appeared to be a decrease in the severity of the clinical disease that occurred, although there was no difference in the rate of viral infection. Of interest, graft survival at 36 months was improved in vaccinated recipients of cadaver kidneys compared with unvaccinated recipients. Pretransplant, a cell-mediated immune response to CMV could be demonstrated (although neither the humoral nor the cell-mediated immune response was as great in the uremic patient as in seronegative normal volunteers); however, this disappeared posttransplant in the face of immunosuppressive therapy. Reassuringly, reactivation of vaccine strain virus was not detected. Vaccination of seropositive patients prior to transplant had no discernible clinical benefit. Finally, efforts to develop a subunit vaccine have begun: In preliminary studies, normal human volunteers, administered purified glycoprotein B, the major envelope glycoprotein of CMV, developed both neutralizing antibodies and CMV-specific lymphocyte proliferation. The concern, here, is whether or not the antigenic variation among CMV strains that exists will prevent the development of an effective monovalent vaccine. Clearly, this effort needs to be continued, although as a practical strategy today, the vaccine option is not yet available.[355,545–549]

The administration of intravenous immunoglobulin preparations (IV-Ig) prophylactically to organ transplant patients is moderately effective in preventing CMV disease in organ transplant recipients (Table 4).[22] In saying this, one must recognize that the database is incomplete for a number of reasons: Since it is not clear what the critical viral antigen(s) is in terms of protective antibody, the definition of protective titers has not been possible. Studies in the murine model have underlined the difficulties involved: When various lots of an anti-murine CMV immunoglobulin preparation were studied, utilizing different techniques for titering the level of murine CMV antibody present, there was a discordance among the results obtained with the different assay systems, and none of them were reliably predictive of the level of antibody needed to protect against lethal challenge with the virus. Although titers of antibody that are protective *in vivo* cannot be measured in humans, it is very clear that there is a very poor correlation between antibody titers as measured by ELISA assay and *in vitro* neutralizing titers.[550] This is true both in individual patients, and when evaluating a variety of IgG preparations. Further complicating the issue is the antigenic heterogeneity of CMV strains in nature. Finally, it is very clear that different immunoglobulin preparations, both standard IV-Ig and hyperimmune anti-CMV, are very different from one another in terms of their anti-CMV content, and, in addition, there can be significant lot-to-lot variation.[551]

Given the challenges of defining and standardizing IgG preparations, and defining a dosage schedule for administering the preparation, what is remarkable is that they do have efficacy. Landmark studies by Snydman *et al.*[552–555] have demonstrated the following: Their hyperimmune anti-CMV IgG preparation significantly decreased the incidence of symptomatic disease in renal transplant recipients at risk for primary infection, although this benefit is modulated by the immunosuppressive therapy administered—prophylaxis with this agent decreased the incidence of primary disease by two-thirds

TABLE 4. Estimated Efficacies of Different Prophylactic Antiviral Strategies against CMV Infection in Different Forms of Organ Transplantation[a,b]

Type of transplant	Form of CMV infection	Antimicrobial strategy used	Estimated efficacy
Kidney	Primary	CMV hyperimmune globulin	2+
		High-dose acyclovir	2+
		CMV hyperimmune globulin + moderate-dose acyclovir	3+
	Secondary[c]	High-dose acyclovir	3+
		CMV hyperimmune globulin + moderate dose acyclovir	3+
Heart and/or lung	Primary	High-dose ganciclovir (1 month)	0
	Secondary[b]	High-dose ganciclovir (1 month)	4+
Liver	Primary	CMV hyperimmune globulin	0
	Secondary[b]	CMV hyperimmune globulin	3+

[a]Modified from Refs. 5 and 8.
[b]Unless otherwise noted, the regimens outlined were administered for a minimum of 3 months. Only semiquantitative assessments of efficacy are given, because of the recognition that the type of immunosuppression used will have a major effect on the efficacy of each of these regimens.
[c]Patients were not differentiated in the studies as to whether they had reactivation or superinfection; all patients seropositive for CMV prior to transplantation are grouped together.

when patients were being treated with azathioprine and prednisone, by one-half when immunosuppression was with cyclosporine-based programs, and is particularly attenuated when antilymphocyte antibody therapy is needed to treat rejection. In liver transplant patients, all treated with triple-drug immunosuppressive therapy (cyclosporine, azathioprine, and prednisone) ± antilymphocyte antibodies, this same group has noted significant protection against symptomatic disease in seropositive patients, in seronegative patients at risk for primary transfusion-related disease, but not in patients at risk for primary infection when the allograft is the source of the virus (donor seropositive, recipient seronegative).[192]

Utilizing a different anti-CMV hyperimmune globulin preparation, the Rotterdam group has reported prophylactic benefit in both renal and heart transplant recipients at risk for primary disease, but not for seropositive individuals.[556,557] In contrast, a Finnish group could not demonstrate protection in renal transplant patients at risk for primary disease, when they used a different globulin preparation and a different dosage schedule.[558] In a different study, CMV hyperimmune globulin appeared to attenuate the severity of clinical disease, but not its incidence, when administered in conjunction with antirejection therapy.[557] In a recent report, CMV immune globulin prophylaxis was reported to increase survival after orthotopic liver transplantation.[559] Finally, Steinmuller et al.[560] have reported that unselected lots of IV-Ig decreased the number of febrile days and the number of hospital days relating to CMV, as well as the number of CMV-related complications, when administered prophylactically to seropositive renal transplant recipients. Similarly, in a study restricted to pediatric renal transplant recipients, IV-Ig appeared to decrease both the incidence of CMV disease and the severity of that disease that broke through, although anti-CMV immune globulin was even more effective.[561,562]

In sum, then, the prophylactic administration of anti-CMV antibody is moderately effective, particularly in renal transplant patients. In addition to the issues previously raised, there are three other problems related to its use: cost (approximately $4800/patient in the Snydman study), the logistical difficulties involved in having to administer repeated intravenous doses of the globulin over a 4-month period, and, finally, the need to provide protection against other viruses, particularly EBV, and not just CMV.[22] Although Tsevat et al.[555] have delineated the cost-effectiveness of anti-CMV IgG ($29,800/life saved for those patients at risk for primary disease), our own feeling is that anti-CMV IgG is most useful when combined with an antiviral drug, when the dosage schedule

and cost can be made more "user-friendly," while the efficacy is increased.[22] However, with the advent of newer antiviral approaches to CMV prevention (vide infra), the routine use of intravenous IgG preparations in organ transplant patients without hypogammaglobulinemia or who are not undergoing plasmapheresis still remains to be defined.

High-dose oral acyclovir (approximately 3200 mg/day, with such high doses required because of the poor bioavailability of ≤ 10%) administered for 4–6 months posttransplant has also been shown to be moderately effective in preventing CMV disease (Table 4), again with attenuation of benefit in the face of antilymphocyte antibody antirejection therapy. As with immunoglobulin prophylaxis, acyclovir is more effective in seropositive individuals as opposed to those at risk for primary disease. What is perhaps most interesting here is that prophylactic efficacy occurs when peak levels of acyclovir in the blood are approximately 25 μmole/liter, and the average 50% inhibitory concentration in vitro is approximately twice that. This suggests that inhibition of the virus is most easily accomplished as it emerges from latency, when only small amounts of replicating virus are present. If so, then other antiviral strategies concentrating on this time point may also be effective.[22,563–568] Studies in which intravenous ganciclovir followed by oral acyclovir is compared to acyclovir alone have demonstrated that the combined regimen is more effective than acyclovir alone,[569,570] and the combination of immune globulin (particularly anti-CMV immune globulin) with acyclovir may be more effective than acyclovir alone.[571,572] With the immunosuppressive regimens now in use, it is fair to say that either high-dose acyclovir alone or immunoglobulin alone is less effective than regimens in which the acyclovir is combined with another modality, or a ganciclovir preparation or valaciclovir is administered alone.[573–575] Valaciclovir, the valine ester of acyclovir, has a bioavailability of 50–60% (in terms of the amount of acyclovir that enters the circulation). Not surprisingly, it far surpasses oral acyclovir in terms of its prophylactic efficacy, being the most effective anti-CMV prophylactic agent other than ganciclovir currently available.[434] In a pharmacoeconomic analysis, prophylactic valaciclovir was shown to be particularly cost effective in patients at risk for primary infection, decreasing by 5.5 days the number of inhospital patient days required; for seropositive individuals, there was little cost saving compared with placebo.[576]

Ganciclovir is the most potent anti-CMV drug currently available. When administered intravenously for 1 month to heart transplant recipients, it provided significant protection to seropositive recipients but none to those at risk for primary infection (D⁺R⁻).[577] In those patients whom the ganciclovir failed to protect (as is also true for oral ganciclovir, acyclovir, and valaciclovir), the incubation period is prolonged. Thus, in patients receiving prophylaxis with any of these agents, instead of the peak incidence being 1–4 months posttransplant (peak approximately 5 weeks), the CMV disease may occur 4–9 months posttransplant (6 weeks to 6 months after the completion of a prophylactic regiment). The exact timing will depend on the immunosuppressive program as well as whether or not systemic TNF release has occurred (as with infection, antilymphocyte antibody therapy, or rejection—*vide supra*). These events are more common the shorter the course of antiviral therapy administered.[577,578] In contrast, when intravenous ganciclovir is administered for 100 days or longer posttransplant, effective prophylaxis is achieved (although, at least in a liver transplant population, there was no overall effect on mortality).[579,580]

The key question has been whether or not oral ganciclovir, administered for 3–4 months, with its poor bioavailability of ≤ 10%, could provide effective prophylaxis without the need for prolonged vascular access required for intravenous prophylaxis. Currently available data show that at a dose of 1000 mg two or three times daily, significant protection is achieved.[581–584] Pharmacokinetic studies performed in patients with renal dysfunction and on dialysis have provided the following dosing guidelines: creatinine clearance > 50 ml/min, 1000 mg tid; creatinine clearance of 25–50 ml/min, 1000 mg every 24 hr; creatinine clearance of 10–24 ml/min, 500 mg each day; creatinine clearance < 10 ml/min or on dialysis, 500 mg every other day (after dialysis).[585] It should also be noted that although in animal studies ganciclovir has been noted to be teratogenic, thus far, at the doses cited, it appears to be free of these effects.[586]

A particular problem has been the prevention of CMV disease, especially CMV-pneumonitis, in lung transplant recipients. Sequential therapy with ganciclovir and nonselected IV-Ig for approximately 3 weeks, followed by a variety of doses of oral acyclovir was ineffective in preventing primary CMV disease.[587] More recently, prolonged courses of intravenous and/or oral ganciclovir or intravenous ganciclovir plus IV-IGg have been shown to have significant benefit in the protection of the lung transplant patient against CMV disease.[580,588–590]

An alternative approach to CMV prevention is a *preemptive* one. As described previously, there are two forms of preemptive therapy (*vide supra*): The first is based on the observation that CMV-seropositive individuals treated without antilymphocyte antibodies have a 10–15% incidence of CMV disease, while the 15% with steroid-fast rejection who receive antilymphocyte antibody therapy have a 65% incidence of CMV disease. Rather than administer the same prophylaxis to all seropositive transplant recipients, we have chosen to develop a program in which antiviral therapy is triggered by the need for antilymphocyte therapy. A daily dose of intravenous ganciclovir (5 mg/kg per day, with dosage correction for renal dysfunction) will decrease the incidence of CMV disease to ~20%; when this is followed by oral ganciclovir at a dose of 1 g two or three times per day, the incidence falls to zero.[581,591] Thus, an effective preventive strategy has been triggered by a clinical–epidemiologic observation, namely, the cytokine storm produced by antirejection antilymphocyte antibody therapy induces CMV disease at such a high rate that antiviral intervention, particularly one as effective as this one, is justified both clinically and pharmacoeconomically.

The second form of preemptive therapy is derived directly from the bone marrow transplant experience, in which monitoring patients for asymptomatic viral replication led directly to intravenous ganciclovir therapy, thus markedly decreasing the incidence of severe CMV disease.[592,593] The advent of quantitative virologic assays for viremia (antigenemia and quantitative PCR, *vide supra*) has made possible the monitoring of patients for early evidence of viremia and then developing algorithms for effecting preemptive therapy on the basis of this laboratory information. The appeal of this approach, once again, is to restrict the use of antiviral therapy to those who would most benefit from it. The disadvantages of this preemptive approach are logistic (getting blood specimens on a timely basis during the period of 2 weeks to 4 months posttransplant) and economic—although the cost of drug is decreased, the cost of the laboratory assays can be quite high. Despite these disadvantages this approach can work and is worthy of further attention.[446,530–541,594–597]

These prophylactic and preemptive regimens are likely to undergo further revision in the near future. The valine ester of ganciclovir has a bioavailability of ~50%, making possible the achievement via the oral route of blood levels of ganciclovir only possible up to now with intravenous therapy (much as was shown with the valine ester of acyclovir, valaciclovir).[598] This will permit the development of more efficacious regimens, although it is likely that the incidence of bone marrow toxicity will also be increased. Even more important, valganciclovir should be an ideal agent for testing whether or not eliminating CMV replication will protect the allograft from acute and chronic injury (*vide supra*).

In sum, major strides have been made in the prevention of CMV disease. It is likely that the regimen of the future will consist of some form of prophylaxis for patients at risk for primary disease, accompanied by preemptive therapy in association with antirejection therapy, and monitored by assays for presymptomatic viremia (which would also trigger preemptive therapy). Seropositive individuals would likely not require prophylaxis, but would receive the last two parts of the program (unless such prophylaxis can be shown to provide significant protection against the indirect effects of the virus, particularly allograft injury).[599]

The treatment of established clinical disease due to CMV is more straightforward. Ganciclovir at a dose of 5 mg/kg twice per day (with appropriate revision in the face of renal dysfunction) is quite effective in the treatment of CMV disease, even permitting the simultaneous treatment of rejection with increased immunosuppression.[600–604] However, the report of Duncan and Cook on the outcome of treating CMV pneumonia in heart transplant patients with ganciclovir issues an important caution.[605] Although therapy interrupted viral replication and resulted in initial clinical improvement, at 6 months follow-up, 70% of the patients had died because of late sequelae, predominantly cardiac allograft dysfunction. This report reemphasizes two lessons: Prevention is still far better than treatment of established disease; and, although ganciclovir therapy is quite effective for treating the direct infectious disease consequences of CMV infection, the indirect effects of the virus may not be managed quite as well.

Another important issue is the appropriate duration of the intravenous ganciclovir treatment. Many groups routinely utilize a fixed regimen of 2–3 weeks. However, in patients with a high viral load, such a duration of therapy may not be adequate to totally eradicate virus from the blood (and, presumably important tissue sites), with the continuing requirement for cyclosporine or tacrolimus then amplifying the residual virus, allowing for relapse to occur. Such relapse is particularly common in patients with a high viral load, an inadequate course of intravenous ganciclovir, high-dose immunosuppression, and a six-antigen mismatch between donor and recipient. Whereas primary ganciclovir resistance is essentially unheard of, relapsing infection is often associated with the development of ganciclovir resistance, requiring the use of the far more toxic foscarnet in management. One particular error in management that can lead to ganciclovir resistance is the initiation of oral ganciclovir while the patient still has evidence of a high viral load. Although oral ganciclovir is useful prophylactically, because of its poor oral bioavailability it is likely to induce resistance when used in a treatment mode. Quantitative assessment

of viral load (*vide supra*) before the end of intravenous ganciclovir therapy is of great use in this regard. Oral ganciclovir added after clearance of viremia is useful prophylactically then in preventing relapse. Relapsing infection with the risk of ganciclovir resistance is a particular issue in lung transplant patients, but can occur in any transplant patient with a high viral load and an inadequate course of intravenous ganciclovir. Thus, rather than a fixed regimen of intravenous ganciclovir for every patient with CMV disease, the course needs to be guided by viral load measurements and the recognition that unless immunosuppression can be decreased, more prolonged parenteral therapy is needed.[268,606–608]

Illustrative Case 4

A 47-year-old woman with end-stage liver disease caused by chronic biliary cirrhosis had undergone orthotopic hepatic transplantation 4 weeks previously. The donor had been CMV seronegative and the recipient CMV seronegative. Twenty-six units of red blood cells and 12 units of platelets were administered without leukocyte filters in the perioperative period; information was not available concerning the CMV serologic status of the donors of the blood products. The patient had a smooth peritransplant course, receiving standard triple-drug immunosuppression (cyclosporine, azathioprine, and prednisone) until the 11th posttransplant day when a low-grade fever and an elevation in the SGOT (from 25 to 65) and total bilirubin (from 1.7 to 2.5 mg/dl) were noted. Two pulse doses of methylprednisolone, 500 mg each, were administered intravenously over the next 48 hr. When the liver function tests failed to improve, a biopsy was performed, which revealed acute cellular rejection. OKT3, 5 mg/kg per day, was administered intravenously for 10 days, with prompt normalization of the liver function tests. No antiviral prophylaxis was prescribed. The patient remained well until 1 week later when, at a routine outpatient visit, her SGOT was again noted to be abnormal (three times the upper limits of normal), her alkaline phosphatase was twice the normal, and her bilirubin had doubled.

Evaluation of the patient consisted of the following: a negative physical examination, including no fever; a normal T-tube cholangiogram; a positive CMV antigen assay (with 650 positive cells noted on a buffy coat smear); and a liver biopsy which revealed CMV hepatitis (Fig. 2).

The patient was treated for 3 weeks with ganciclovir at a dose of 5 mg/kg twice daily, with improvement in her liver function tests first observed after 5 days of therapy. The patient then remained well for 26 days, when she returned with similar abnormalities in her liver function, fever, and malaise. Once again, a positive CMV antigenemia assay was noted (425 positive cells), and ganciclovir was reinstituted at the same dose for 6 weeks, with a drop in the antigenemia assay to 310 positive cells 7 days after the reinitiation of therapy. The antigenemia assay was documented to be negative prior to discontinuing the intravenous ganciclovir therapy. This was followed by 3 months of oral ganciclovir at a dose of 1000 mg twice daily for 3 months. In addition, there was a 50% decrease in the dose of her immunosuppressive drugs. The patient has remained well since then.

Comment. This is a case of primary, transfusion-acquired CMV infection that presented initially with hepatocellular dysfunction, in the absence of other symptoms, as the sole manifestation of disease. Biopsy was necessary to distinguish between rejection and infection, although the positive antigenemia assay argued strongly for infection. Indeed, the reason for the biopsy was to rule out the presence of dual processes. In this case, as in 15–20% of patients treated with ganciclovir, relapsing disease occurred. Fortunately, drug-resistant infection did not occur in this instance. By monitoring the viral load 1 week after reinstitution of the intravenous ganciclovir, a drug response could be documented, and foscarnet, with its attendant renal toxicities (renal failure, as well as excessive magnesium and calcium losses in the urine), could be avoided. Alternatively, if the patient had been more ill, it would have been appropriate to initiate therapy with foscarnet until testing of the virus isolate for ganciclovir resistance could be carried out. Because of the concerns regarding an increased risk of superinfection during the CMV infection, the patient was also maintained on anti-*Pneumocystis* prophylaxis, and protected while in the hospital from an ongoing nosocomial problem with *Aspergillus* infection by housing her in a HEPA-filtered room.

5.1.2. EBV Infection in Organ Transplant Recipients

The epithelial cells of the upper respiratory tract, particularly those of the oropharynx and the parotid duct, are the natural reservoir for EBV, with transmission of the virus occurring by means of intimate contact and the exchange of virus-laden saliva. Such infection is lytic in nature, resulting in the release of replicating virus, causing pharyngitis during infectious mononucleosis. Following acquisition of EBV, and the replication of the virus in the upper respiratory tract epithelium, long-lived B lymphocytes become infected as they travel through the lymphoid tissues of the oral cavity. Although the virus can be demonstrated in the cervical epithelium and semen, sexual transmission has yet to be proven. EBV can occasionally be transmitted through blood transfusion or bone marrow transplantation. The receptor for the virus on both epithelial cells and B lymphocytes is the CD21 molecule, which is also the receptor for the C3d component of complement.[609–611]

Whereas infection of epithelial cells results in viral replication, with release of infectious virions, infection of B lymphocytes usually results in latent infection without replication or release of virus. The consequences of such B-cell infection with EBV are transformation and immortalization. The nature of the EBV infection in these B lymphocytes is very different from that in the oropharynx: The virus exists in its circular episomal form (which is not susceptible to antiviral chemotherapy), and expresses only a few of the genes that EBV expresses during lytic infection—these genes are essential for the transformation of the B lymphocytes. Latent infection of the B lymphocytes by EBV is characterized by the differential expression of six EBV nuclear antigens (EBNAs), two

EBV-encoded RNAs (EBERs), and three latent membrane proteins (LMPs). Although a detailed examination of the role of these in the pathogenesis of PTLD is beyond the scope of this chapter, a few general comments can be made: EBNA-1 is important in maintaining the episomal form of EBV DNA within the dividing B cell and in the initiation of the expression of the other EBNAs. It also blocks HLA class I antigen expression, which may allow EBV-infected B cells to escape immune surveillance by CD8$^+$ cytotoxic T cells. EBNA-2 is essential for B cell immortalization, as are other EBNAs. EBERs appear to block the effects of interferon-χ, thus promoting the immortalization process. LMP1 may be considered an oncogene, with its ability to transform cells being due at least in part to its antiapoptotic effects.[612–618]

In the normal host, clinical disease is initiated by the infection of epithelial cells and B lymphocytes, which in turn incites an intense, CTL response to the B lymphocytes that have been latently infected, and become widely dispersed throughout the body, especially the liver, bone marrow, spleen, lymph nodes, and CNS. This immune response is responsible for most of the clinical manifestations of infectious mononucleosis, and is also the explanation for the lack of efficacy of antiviral chemotherapy in affecting this clinical syndrome.[615] Strains of EBV found in nature have been divided into two types, EBV-1 and EBV-2, on the basis of the gene that encodes EBV nuclear antigen (EBNA) 2A, a gene product involved in the immortalization of B lymphocytes. There is *in vitro* evidence that EBV-1 is a more efficient transformer of B lymphocytes than is EBV-2, and, at least in one series, infection with EBV-1 was particularly associated with the development of lymphoproliferative disease.[611,619,620]

Although most adults who are EBV seropositive harbor B lymphocytes latently infected with EBV that have the potential for unlimited growth, lymphoproliferation does not occur because of an active immunologic surveillance mechanism based primarily on MHC-restricted, EBV-specific, cytotoxic T cells. Suppression of this mechanism *in vitro*, or in such *in vivo* situations as EBV infection in mice with severe combined immunodeficiency, AIDS patients, and transplant patients results in outgrowth of these immortalized B cells and evident lymphoproliferation.[610,611,613,618,621] Primary EBV infection carries a significantly greater risk for PTLD than does reactivation infection (*vide infra*), presumably because of the lack of a preexisting immunologic surveillance mechanism in patients whose immunosuppressive therapy will attenuate their ability to develop such a response.[611,618]

Of probable importance, EBV-immortalized B cells produce growth factors, including lactic acid, IL-6, and

IL-10, that stimulate their own proliferation in an autocrine loop. IL-10, a pleiotropic cytokine produced mainly by activated Th2 cells, is of particular interest for several reasons: it has suppressive effects on macrophage, natural killer cell, and Th1 functions, including cytokine production, T-cell proliferation, and cytotoxicity; it increases the proliferation of activated B cells, their differentiation, and the secretion by these cells of immunoglobulin; in addition to T-cell production of IL-10, the protein product of the BCRF-1 region of the EBV genome exhibits extensive homology with the T-cell-derived IL-10, exhibiting similar ability to inhibit interferon-χ secretion, and suppress T-cell proliferation in response to antigenic and mitogenic stimuli. Thus, it has been hypothesized that these two forms of IL-10 play an important role in amplifying the extent of transformed B cells, a key step in the pathogenesis of PTLD.[610,618,622,623]

The question that may be asked is whether the elaboration of cytokines and growth factors in the course of other processes common in transplant patients (e.g., allograft rejection, the first few doses of antilymphocyte antibodies, or other infections) could also enhance the lymphoproliferative response. Consistent with this hypothesis is the observation that symptomatic CMV infection is a risk factor for the subsequent development of EBV-associated lymphoproliferative disease, increasing the risk of PTLD 7- to 10-fold.[212,624-628] In addition, there is a report that HCV infection is associated with a > 5-fold increased risk of PTLD among liver transplant recipients.[629] Since many of these effects are thought to be mediated by cytokines elaborated in response to these infections, it is not surprising that synergy among different processes can be observed in the promotion of PTLD. For example, the combination of pretransplant seronegativity for EBV, D^+R^- CMV status (thus, risk of primary infection with both of these viruses), and anti-CD3 monoclonal antibody treatment increases the risk of PTLD more than 500-fold.[212,624,628]

Since more than 90% of adults are seropositive for EBV, and far more transplants are performed in adults than children, the great majority of EBV infections in transplant patients represent reactivation infections, although EBV superinfection analogous to what is observed with CMV is quite possible. Such viral reactivation and excretion may occur without changes in antibody titer. In particular, heterophile antibodies do not appear with any regularity in transplant patients with either viral excretion or rises in specific antibody titer.[27] Primary EBV infection in transplant patients may occur in the community, or be acquired from the allograft or viable leukocyte-containing blood products. This is most common in pediatric transplant patients, the majority of whom are seronegative, with consequences ranging from asymptomatic infection to a mononucleosis syndrome to a rapidly fatal multiorgan lymphoproliferative disease similar to that seen in boys with the X-linked lymphoproliferative disorder that follows EBV infection. The infectious disease consequences of EBV reactivation closely resemble those observed with CMV, although, because of the ubiquity of CMV, the contributions of EBV to individual morbidity may be difficult to discern.[591,610,628,630-642] However, liver transplant patients with chronic hepatic allograft dysfunction resulting from primary EBV infection possibly acquired from the allograft have been described.[643] This entity appears to be more common than previously recognized, with a histologic picture of activated mononuclear infiltrates in the portal tracts associated with lobular disarray and sinusoidal lymphocytes arranged in linear beads and small aggregates. Demonstration of EBER-1 RNA in these mononuclear cells confirms the diagnosis.[642] The analogy to CMV hepatitis in the liver transplant recipient is obvious, and it is likely that other infectious disease syndromes currently attributed to CMV are actually related to EBV. It is also clear that EBV has immunosuppressive effects similar to those produced by CMV. Whether EBV infection has any influence on graft function is totally unknown. Another important unanswered question is whether dual infection with the two viruses has a greater impact than infection with either. The identification of symptomatic CMV infection as a risk factor for EBV lymphoproliferative disease suggests that such dual infection may be clinically important.

Far more important than EBV-induced hepatitis or mononucleosis in the transplant patient is the role this virus plays in the pathogenesis of the PTLD. The incidence of PTLD varies from 1–3% among kidney, heart, and liver transplant recipients to 7–33% among lung, intestine, and multivisceral transplants.[610,628,638,644,645] Approximately 90% of PTLD in organ transplant patients is of recipient origin; and a similar percentage shows evidence of EBV infection.[644] PTLD encompasses a range of pathologic entities, with the most useful classification system currently available being that put forth by the Society of Hematopathology, in which four distinct categories of PTLD are defined[645-647]:

1. *Lymphoid hyperplasia ("early lesions")*. Included in this category are plasma cell hyperplasia, lesions resembling infectious mononucleosis, and other forms of atypical lymphoid hyperplasia with preservation of the underlying architecture. These lesions are usually polyclonal

and often regress with reduction of immunosuppression.

2. *Polymorphic PTLD*. These are destructive lesions that infiltrate and destroy underlying tissue, with a wide range of B-cell maturation being present. Molecular studies suggest that virtually all of these tumors are monoclonal. Oncogene or tumor suppressor gene abnormalities are not a feature of these tumors. A lower percentage of these tumors respond to reduction of immunosuppression.

3. *Lymphomatous or monomorphic PTLD*. Most such tumors are of the diffuse large B-cell lymphoma subtype, although Burkitt-like lymphomas and mucosa-associated lymphomas (MALT lymphomas) may also occur as variants of B-cell disease. Abnormalities in the ras or p53 genes of the diffuse B-cell lymphomas are not uncommon. These tumors are monoclonal, with only a minority responding to decreased immunosuppressive therapy. In addition, such uncommon forms of PTLD as T cell, null cell, and NK cell lymphoma are placed in this category. A minority of these non-B-cell tumors are EBV positive, and response to decreased immunosuppression is uncommon.

4. *Other forms of PTLD*. Included in this category are such uncommon tumors as plasmacytoma, myeloma, and T-cell-rich/Hodgkin's disease-like large B-cell lymphoma. These tumors may or may not be EBV positive, and tend to be clinically aggressive.

In general, EBV-positive tumors occur in the first 6 months posttransplant; in contrast, although the EBV-negative PTLDs can occur as early as 6 months posttransplant, they occur at a mean time of >4 years posttransplant, accounting for the majority of the late PTLDs.[645–647]

A growing body of evidence suggests that there are two interacting factors that determine the incidence of PTLD in the transplant patient: the number of B lymphocytes infected and transformed by EBV, which is a manifestation of viral load; and the ability of the EBV-specific cytotoxic T-cell surveillance mechanism to eliminate these transformed cells, which is determined by the nature of the immunosuppressive therapy being administered and the cytokine, chemokine, and growth factor milieu that is present. These last are determined by such other factors as concomitant infections and the severity of rejection and the intensity of its therapy.[611,618,627,628]

In prospective, quantitative studies of oropharyngeal EBV shedding among renal and cardiac transplant patients, Preiskaitis *et al.*[611] reported that significantly higher levels of EBV shedding are observed in patients with primary, as opposed to reactivation infection, and in association with intensive immunosuppressive therapy with antilymphocyte antibodies or > 4 g of pulse methylprednisolone therapy. Similarly, the percentage of patients shedding is likewise modulated by the immunosuppressive regimen: in the general population, the incidence is approximately 20%; among transplant patients on maintenance immunosuppression, the figure rises to 30%; in association with a course of antilymphocyte antibody therapy, the incidence rises to approximately 80%.[611,630] Patients with the highest level of EBV shedding had the poorest serologic response to the virus, and the greatest risk for the development of PTLD. Finally, both acyclovir and ganciclovir were shown to totally suppress oropharyngeal shedding of EBV. It appears likely that the increased EBV replication noted in the oropharynx in these patients results in increased infection of circulating B lymphocytes. This may result directly in the polyclonal type of lymphomas or provide a larger pool of activated, replicating B lymphocytes for the cellular events that result in monoclonal tumors. Immunosuppressive therapy may thus promote the development of PTLD at two levels: increasing the incidence and level of oropharyngeal EBV replication (particularly by antilymphocyte antibody therapy), and blocking the surveillance mechanism (particularly by cyclosporine, but also by the rest of the immunosuppressive program).

Since this relationship between viral shedding and risk of PTLD was demonstrated, emphasis has shifted to correlations between levels of EBV DNA, as determined by PCR assay, in the blood and subsequent risk of PTLD. Although there is still lingering controversy about the optimum assay technique (lymphocytes, whole blood, or plasma), it is clear that serial measurements of EBV viral load in the blood can be utilized in a number of ways: a negative assay essentially rules out the possibility of PTLD in a patient with a clinically compatible syndrome; high levels of viral DNA have a high predictive value for either the presence of PTLD or the future development of it; and, perhaps most exciting, serial monitoring of EBV DNA levels may permit the adjustment of immunosuppressive therapy and/or the prescription of antiviral therapy so as to preempt the development of PTLD.[648–653]

Thus, the possibility exists that if one could interrupt EBV replication at a critical point in the posttransplant course (e.g., the replication induced by antirejection therapy), one could interrupt the oncogenic process by limiting the number of B lymphocytes that are transformed at a time when the surveillance mechanism has been particularly suppressed by high-dose immunosuppressive therapy.[654–661] Thus, a strategy akin to the preemptive program for CMV that was previously described, could have dual efficacy in also providing protection against the development of PTLD. A high priority must be given to determining if this strategy does indeed provide significant protection.

The clinical presentation of EBV-associated PTLD can be quite variable, and includes one or more of the following: unexplained fever (an FUO); a mononucleosis-type syndrome with fever, malaise, and lymphadenopathy, with or without pharyngitis/tonsillitis; a GI presentation, which can include GI bleeding, abdominal pain, gut perforation or obstruction; hepatocellular dysfunction; gallbladder disease presenting as acute cholecystitis; and CNS dysfunction such as seizures, change in state of consciousness, and focal neurologic disease. Not uncommonly, the allograft itself is frequently involved with the process, not particularly surprising given the MHC restriction of the EBV-specific cytotoxic T-cell surveillance mechanism—the MHC-disparate allograft representing a privileged immunologic site for the EBV-driven process to occur. Alternatively, PTLD can present as a hilar mass in renal transplant recipients resulting in hydronephrosis or renal vessel invasion (or a portal mass in liver transplant patients causing biliary obstruction). Finally, involvement of the bone marrow, lung, and virtually every other organ has been described. An important clinical point is that absence of adenopathy on CT scanning does not rule out PTLD, as this disease can be totally extra-nodal in nature.[638,662–664]

At present, optimal therapy for PTLD is unclear. An estimated 20–30% of patients improve with the cessation of immunosuppression, with a greater percentage of children (with PTLD as a consequence of primary EBV) responding to this maneuver than adults (virtually all of whom have PTLD as a consequence of reactivation of EBV). Although the evidence in favor of antiviral therapy with acyclovir, ganciclovir, or foscarnet is purely anecdotal,[627,665,666] most transplant centers do utilize one of these while decreasing immunosuppression. The concept justifying this approach is that although the EBV-transformed B cells present in the tumor are not exhibiting lytic infection, lytic infection is going on systemically "feeding the fire." Eliminating this element of the process, either directly or through modification of the resulting cytokine milieu, could add to the therapeutic program.[618,642,666] One subgroup of patients that does particularly well with PTLD is the group that presents with GI bleeding, obstruction, or perforation, and has a single tumor site resected. For the majority of patients,

however, in whom PTLD is discovered, further therapy is needed. Traditional therapies with antilymphoma chemotherapy, radiotherapy, and surgery have not yielded satisfactory results, and attention has turned to alternative approaches. In solid organ transplant patients with PTLD, these experimental therapies have included interferon-α plus nonspecific IgG, or the possible use of adoptive transfer therapy with the infusion of donor T lymphocytes or EBV-specific CTLs (although this last has had considerable success in bone marrow transplant patients with PTLD where donor cells are available, it is more problematic to utilize this approach in organ transplant patients where donor cells may not be available). What has been particularly exciting in recent years has been the experience with anti-B-cell monoclonal antibodies (directed against the CD20, CD21, or the CD24 antigens). More than 60% of patients, many of whom had failed earlier interventions, responded to such therapy, which was well tolerated. Other anti-B-cell antibodies (anti-CD19, anti-CD37, and anti-CD38) are also in development for this purpose. An interesting alternative approach is the use of a monoclonal antibody directed against IL-6. This was chosen because of IL-6's important role as a growth factor for EBV-induced B-cell proliferation. Complete remission was achieved in 5 of 12 patients so treated, with partial remission in another 3 patients, and stability in 1 patient; 3 patients showed no response to the therapy. These observations are important not only for defining a possible therapeutic option, but also for confirming the importance of the cytokine milieu in the evolution of PTLD.[627,665–673]

Our own practice at present is to decrease immunosuppression significantly (halting it completely and returning the patient to dialysis in the case of renal transplant patients) and to initiate systemic antiviral chemotherapy, while monitoring EBV DNA levels in the blood. Approximately 4–6 weeks is devoted to this effort. Progressive disease or failure to improve during this period indicates the need for further therapy. Our current preference is anti-B-cell antibody therapy, reserving more standard lymphoma chemotherapy for patients who fail this form of therapy. Clearly, prevention is much to be preferred, hence the emphasis on determining whether prophylactic or preemptive antiviral strategies, particularly at times of intensive immunosuppression, can prevent the evolution of this process.

Illustrative Case 5

A 47-year-old man underwent cardiac transplantation in treatment of an ischemic cardiomyopathy. The donor and recipient were seroposi-tive for both CMV and EBV. His immediate posttransplant course was marked by severe acute renal failure, resulting in the cessation of cyclosporine on the third posttransplant day and the initiation of a 14-day course of antithymocyte globulin. At the end of this time, he was doing well, with recovery of renal function and a well-functioning graft, with immunosuppression being maintained with cyclosporine, azathioprine, and prednisone. He was receiving no antiviral prophylaxis. Four weeks posttransplant, significant cellular rejection was diagnosed on routine endomyocardial biopsy, and, following two pulse doses of 500 mg of intravenous methylprednisolone, a 10-day course of OKT3 was initiated. Follow-up endomyocardial biopsy was negative, but 6 weeks posttransplant fever, leukopenia, hepatocellular dysfunction, and an interstitial pneumonia were noted, and CMV was isolated from the blood and on bronchoalveolar lavage. Ganciclovir and anti-CMV hyperimmune IgG therapy were instituted, with gradual clinical improvement. The patient received 4 weeks of therapy, and had returned home for the last week of this. He remained well until 6 months posttransplant, when he presented with fever, encephalopathy, hepatocellular dysfunction, and GI bleeding. PTLD was diagnosed on liver biopsy and on colonoscopic biopsy of a colonic mass.

Therapy was instituted by a 50% decrease in immunosuppressive therapy, followed by CHOP chemotherapy. In addition, full-dose ganciclovir therapy was administered. Unfortunately, the patient rapidly deteriorated, developing progressive encephalopathy, pancytopenia, and hepatic failure. At postmortem examination, PTLD involving the brain, liver, colon, small bowel, and infiltrating the cardiac allograft were found. Episomal EBV could be demonstrated quite easily in the tumor-bearing tissues. Samples of the tumor were demonstrated to manifest oligoclonality.

Comment. This cardiac transplant patient was heavily immunosuppressed, receiving two courses of antilymphocyte antibody therapy, without any concomitant antiviral preventive strategy. His sequence of clinical CMV disease followed by EBV-related PTLD was both expected and probably interrelated. Although management of the immunosuppression was expertly handled, the lack of an effective antiviral program in conjunction with the augmented immunosuppression virtually guaranteed the tragic course of events that occurred. Ideally, he would have received intravenous ganciclovir during the courses of antilymphocyte antibody, followed by 4 months of oral ganciclovir. It is likely that the CMV disease would have been prevented by this regimen. Such an approach would have decreased the risk of PTLD. However, because of the intensive immunosuppression required in this patient, monitoring of EBV blood DNA levels would have been of great interest prior to overt presentation of PTLD. Finally, given the gravity of the illness—rate of progression and disseminated tumor burden—early deployment of anti-B-cell antibody ± anti-IL-6 therapy would have been desirable.

5.1.3. Herpes Simplex Virus Infection in Organ Transplant Recipients

In the absence of antiviral prophylaxis, HSV is probably second only to CMV among viral agents causing clinical disease in the organ transplant patient. Virtually all of the infections caused by HSV are the result of reactivation of latent virus. Information currently available would suggest that approximately three fourths of patients with antibody to HSV pretransplant will excrete the virus in their throat washings, and approximately two thirds will demonstrate a fourfold or greater increase in their antibody titers. About one-half of the seropositive

patients and two-thirds of those who excrete the virus will develop visible mucocutaneous lesions. There appears to be no relationship among the presence, severity, or extent of such lesions, and the subsequent development of allograft dysfunction.[631,674]

By far the most common clinical manifestation of HSV infection in the organ transplant patient is herpes labialis, usually beginning by the second week posttransplant, peaking in severity by the end of the first month, healing over the next 2–6 weeks, and exacerbated and prolonged by acute antirejection therapy. Not only are such infections more prolonged than in the normal host, but they are considerably more severe: large, painful, crusted ulcerations that bleed or interfere with normal nutrition and require at least local analgesia. In some individuals, these lesions can interfere sufficiently with the handling of oral secretions to predispose to aspiration. Intraoral and esophageal infection may occur in association with the herpes labialis, particularly if the mucosa has been traumatized by endotracheal or nasogastric tubes. These should be avoided if possible in transplant patients who have active labial or intraoral infection. Both clinically and radiologically, herpetic esophagitis mimics the effects seen with candidal esophagitis. As in normal hosts, virtually all isolates from patients with oral HSV infection are type 1 (HSV-1). Although herpetic orolabial infection in the transplant patient can cause heaped-up, verrucous lesions of the lips, care must be taken in those individuals who fail to respond to conventional acyclovir therapy that a squamous cell carcinoma is not either simultaneously or solely present.[214,675,676]

In renal, cardiac, and liver transplant patients, HSV pneumonia is an uncommon event, usually occurring as a secondary bronchopneumonia in patients requiring prolonged intubation because of some other form of primary lung injury. The pathogenesis in these cases is related to the presence of replicating HSV that has been reactivated in the oropharynx, trauma to the mucosa by the endotracheal tube, and spread via the endotracheal tube to the lower respiratory tract. Recipients of lung and heart–lung transplants, however, if not receiving some form of acyclovir or ganciclovir prophylaxis, have a relatively high incidence of HSV pneumonia. Presumably the same pathogenetic mechanisms are involved here as previously stated, with the more prolonged intubation of these patients adding to the risk. In addition, by analogy to CMV infection, where the transplanted organ is more intensively affected than the native organ, it is not unreasonable to speculate that the lung allograft is more susceptible to inoculation with HSV than the native lungs of the other transplant groups. Because of this risk, our practice is to place all of our lung transplant patients on either ganciclovir or acyclovir prophylaxis in the first 4 months posttransplant (*vide infra*).[677]

Less commonly, anogenital infection may occur, caused predominantly by HSV-2. Unusually severe anogenital infection characterized by the presence of large coalescing, ulcerated lesions, without clear-cut vesicles, may be caused by HSV-2 in transplant patients who have been treated with particularly aggressive immunosuppressive therapy. In the evaluation of these lesions, routine bacteriologic cultures, Tzanck preparations, and morphologic evaluation will not yield the appropriate diagnosis. Viral culture or direct immunofluorescence studies with specific monoclonal antibodies directed against HSV—both performed on swabs of the lesion—will result in rapid diagnosis. Such lesions may become secondarily infected and act as a portal of entry for a variety of stool flora, resulting in local cellulitis and/or bacteremias, particularly if the patient is leukopenic.[678] We have also cared for a number of transplant patients with recurrent zosteriform lesions on the buttocks caused by HSV-2, and chronic chancrelike genital lesions caused by this virus have been reported.[679]

More severe HSV infections in the organ transplant patient are uncommon. In particular, although HSV may act as a secondary pathogen in an intubated patient with severe pneumonia caused by other agents, it is rarely a primary cause of pneumonia. The clinician evaluating an organ transplant patient with pneumonia who isolates HSV from respiratory secretions should be very circumspect in terms of accepting this as an adequate explanation for the pneumonitic process that is present.[680] Although several transplant patients have been reported with an illness characterized by severe oral lesions followed by high fever, fulminant hepatitis, GI bleeding, and disseminated intravascular coagulation because of HSV, this syndrome is exceedingly uncommon in the organ transplant setting.[681–685] Uncommonly, disseminated cutaneous infection with HSV, occurring at sites of previous skin injury such as burns or eczema (termed eczema herpeticum or Kaposi varicelliform eruption), may develop. In these cases, recovery without visceral dissemination is the rule, even without specific antiviral therapy. Anora *et al.*[686] reported an interesting case of a 34-year-old man in whom fever and then multiple vesicles around the sutures of his transplant incision developed about 2 weeks after receiving a kidney from his brother. HSV-2 was cultured from these vesicles, and an antibody response was documented. It is now apparent that on rare occasions the allograft may convey active HSV infection, which has resulted in disseminated disease in seronegative graft recipients.[687]

Similarly, CNS infection with HSV in the transplant patient is uncommon. Rarely, HSV-2 meningoencephalitis following anogenital infection may be noted, but more typical HSV-1 encephalitis as seen in the normal host is essentially unknown.[675] Dunn *et al.*[688] have suggested that the combination of HSV and CMV infections in the same individual was associated with a worse clinical outcome than either infection alone. In their experience at the University of Minnesota, concurrent HSV and CMV infection was associated with both increased patient mortality and increased allograft loss.

The advent of antiviral prophylaxis and therapy has had a remarkable effect on the occurrence and impact of HSV disease in organ transplant patients. Both acyclovir and ganciclovir are quite effective anti-HSV drugs, and essentially any preventive strategy utilizing these drugs that is effective against CMV and EBV will have a similar salutary effect on HSV infection. Symptomatic infection is easily treated with oral acyclovir, 200 mg five times per day for 7–10 days (prolonging the course of therapy, as needed, if intensive immunosuppressive therapy is being prescribed at the same time). Of importance, unlike in the AIDS patient, acyclovir-resistant HSV has not been a clinical issue in organ transplant recipients, even if the patient has received antiviral prophylaxis. An alternative to treatment of overt infection is the use of acyclovir prophylaxis in HSV-seropositive patients (in doses of 200 mg four times per day for the first month posttransplant), which is quite successful in preventing HSV disease. Since symptomatic HSV is relatively easy to treat, we prefer not to prophylax for this infection by itself, and tend to choose strategies primarily aimed at the other herpes group viruses (which, in most cases, will also provide significant benefit for HSV). Finally, the occasional patient with recurrent, symptomatic HSV infection may benefit from prophylaxis with acyclovir, 200 mg three or four times per day for periods of 6 months or longer.[22,679,689,690]

5.1.4. VZV in Organ Transplant Recipients

Three clinical syndromes are commonly recognized in organ transplant patients as being caused by VZV. First is typical localized dermatomal zoster resulting from viral reactivation which may involve two or three adjoining dermatomes, and even manifest a few sites of cutaneous dissemination at distant sites, but which is without evidence of visceral involvement. VZV can be isolated quite easily from these skin lesions, and rises in antibody to VZV can be demonstrated in the majority of such individuals. Traditionally, approximately 10% of organ transplant patients who remained on immunosuppression developed clinical zoster posttransplantation as a result of reactivation of VZV long dormant in dorsal root ganglia. Such infections were uncommon earlier than 2 months and later than 3 years posttransplant. With the widespread deployment of antiviral agents like acyclovir and ganciclovir, both of which have activity against VZV, the incidence of zoster has decreased, and those cases that do occur usually are later in the posttransplant course. Antiviral therapy with ganciclovir or valaciclovir will hasten the healing of the skin lesions. Such therapy is particularly indicated for zoster involving the face, the sacral dermatomes (as bladder and/or bowel function could be compromised), if significant underlying skin disease is present, or if high-dose immunosuppressive therapy is required to control rejection. Unlike the situation in individuals immunosuppressed by Hodgkin's disease and its therapy, visceral dissemination is uncommon in the organ transplant recipient.[22,675,691,692]

Second, disseminated VZV infection characterized by hemorrhagic pneumonia and skin lesions, encephalitis, pancreatitis, hepatitis, and disseminated intravascular coagulation may be observed as a consequence of primary VZV infection in organ transplant patients. In addition, there is some preliminary evidence suggesting that primary VZV infection in the organ transplant recipient may be associated with allograft injury in a fashion similar to that postulated for CMV (*vide supra*). If primary VZV infection is recognized sufficiently early, it can be effectively treated with high-dose intravenous acyclovir (10 mg/kg three times per day, with revisions of the dose for renal dysfunction). Rarely, VZV has been conveyed with the allograft to a seronegative recipient. More than 90% of adults are seropositive for VZV, and thus not at risk for primary VZV. Therefore, children are the prime concern. Our practice is to screen all transplant candidates serologically for VZV. Seronegative individuals should receive the varicella vaccine, with repeat serologic testing carried out to document seroconversion and protection. Seronegative transplant recipients, whether or not there has been a history of varicella vaccine administration prior to transplant, are instructed to promptly report all VZV exposures so that zoster immune globulin (ZIG) can be administered in a timely fashion. Oral valaciclovir prophylaxis is another option, although careful study of this approach in the transplant population has not been carried out. Although ZIG prophylaxis offers significant protection, in some individuals it will only attenuate the skin lesions while visceral infection spreads, thus delaying recognition and effective therapy.[693–697]

Recently, VZV-seropositive pediatric renal transplant recipients receiving an immunosuppressive program of cyclosporine, prednisone, and mycophenolate mofetil developed disseminated cutaneous varicella infection, with a generalized vesicular eruption without a dermatomal distribution. All of these episodes occurred in the first year posttransplant. None of the patients developed fever, respiratory, CNS, or other visceral disease, and had responded to acyclovir therapy. This syndrome was presumably due to reinfection in children whose immunity, despite seropositivity, was partially abrogated by the mycophenolate-containing regimen.[698]

Finally, a syndrome of unilateral pain without skin eruption associated with rises in specific antibody to VZV has been described in transplant patients and is presumably also caused by this virus. In addition, serial studies in transplant patients have demonstrated that asymptomatic rises in antibody titer to VZV may occur. This finding has been ascribed to an unstable relationship between virus and host, and is cited in support of the argument by Hope-Simpson that subclinical release of virus with resulting antigenic stimulation may maintain immunity to VZV.[692,697]

5.1.5. HHV-6 Infection in Organ Transplant Recipients

HHV-6 is a β-herpesvirus, closely related to both CMV and HHV-7. Indeed, there is 66% DNA sequence homology between CMV and HHV-6.[699] HHV-6 infects and replicates within a variety of leukocytes: most prominently CD4$^+$ T lymphocytes, but also including CD8$^+$ T lymphocytes, NK cells, macrophages, megakaryocytes, glial cells, and epithelial cells.[700,701] In addition, HHV-6 replication is a powerful stimulus for the elaboration of a broad array of proinflammatory cytokines, including TNF, IL-1, and interferon-χ. From this brief description of HHV-6's properties, one would predict that HHV-6 could cause febrile illnesses akin to those produced by CMV in transplant patients, that bone marrow dysfunction and encephalitis would also be within the range of illnesses produced by this virus, and that a variety of indirect effects would be produced by HHV-6 (e.g., promotion of CMV disease, contribution to the net state of immunosuppression and thus increasing the risk of opportunistic superinfection, and the potential for causing acute and chronic allograft injury).[700,702–704]

Primary infection with HHV-6 usually occurs in the first year of life; 90% of adults are seropositive for this virus and harbor latent virus. In the general population, HHV-6 has been shown to be the cause of roseola (exanthem subitum), a febrile exanthem of young children. In addition, HHV-6 has been linked to a mononucleosis syndrome, autoimmune disorders, lymphomas, necrotizing lymphadenitis, encephalitis, as well as multiple sclerosis and other demyelinating conditions.[700,701,705,706] HHV-6 infection has been documented, primarily in the period 1–6 months posttransplant, in 30–50% of organ transplant recipients. Although the transplantation of an allograft from a seropositive donor into a seronegative recipient carries a > 50% risk of virus transmission (primary infection), the great majority of infections are thought to be due to reactivation of latent virus.[700,707] Despite this information, several factors have rendered difficult the delineation of the clinical role of HHV-6 in transplantation: the fact that the clinical effects of this virus are probably very similar to those of such other viruses as CMV and EBV, whose peak level of replication is occurring during the same time period as HHV-6; the ubiquity of the other herpes group viruses in this patient population; and the fact that there can be serologic cross-reactivity between anti-CMV antibody and anti-HHV-6 antibody.[700,708–714]

At present, it is possible to say that the direct clinical effects of HHV-6 in transplant patients include fever with or without mononucleosis, interstitial pneumonitis, and hepatitis. Far more important are bone marrow suppression and CNS dysfunction (encephalitis). The CNS effects include mental status changes, seizures, and headache, with focal neurologic findings being rare. Evaluation of patients with HHV-6 CNS disease reveals less than dramatic results: minimal if any pleocytosis in the CSF; HHV-6 DNA demonstrable in the CSF of a distinct minority of patients; and usually negative neuroimaging studies. Rarely, low-attenuation white matter changes virtually identical to those seen with immunosuppression-associated leukoencephalopathy may be seen.[596,700,708–716] As previously mentioned, simultaneous infection with multiple herpesviruses, particularly HHV-6, CMV, and/or EBV, is a common event. It is likely, although not yet proven, that the patient with multiple infections will suffer greater direct and indirect consequences than if a single virus is replicating. Of interest is a study from the Royal Free Hospital in London in which HHV-6 infection was the leading factor correlating with both length of hospital stay and need for readmission.[717]

Although HHV-6 can be isolated by culture, and serologic assays are available, it is fair to say that reliable diagnosis of HHV-6 requires either antigenemia or PCR assay. The analogy to CMV is clear—serologic studies are primarily useful pretransplant to stratify risk; cultures require 5–21 days for results (and a shell vial assay is relatively insensitive); with antigenemia or quantitative PCR on blood being both timely and useful.[700]

The optimal management of HHV-6 infection remains to be defined. *In vitro* assays demonstrate susceptibility to ganciclovir and foscarnet analogous to that of CMV. Indeed, clinical benefit has been demonstrated with both of these drugs in patients with encephalitis.[715] It is likely that antiviral strategies aimed at CMV and employing ganciclovir are having a beneficial effect on the course of HHV-6 infection. However, optimal management, prophylaxis, and/or preemptive therapy remain to be defined. It might be predicted that in the future antiviral programs will not be aimed at a single virus like CMV; rather, programs will be directed at multiple agents simultaneously. For the present, a ganciclovir preparation appears to be the best means for accomplishing this.

5.1.6. HHV-7 Infection in Organ Transplant Recipients

HHV-7 resembles HHV-6 in its epidemiology, with primary infection being acquired commonly by the age of 5. It has a similar tropism for $CD4^+$ T lymphocytes, but unlike HHV-6, utilizes the CD4 molecule itself as its receptor.[700] The clinical effects of this "orphan virus" are currently unclear. Preliminary data from some,[718,719] but not all,[720] studies suggest that HHV-7 replication is a significant risk factor for clinical CMV disease. If one were to speculate on other possible clinical effects of this virus, it would be in the realm of indirect effects—that is, cytokines, chemokines, and growth factors produced in response to HHV-7 replication could influence the net state of immunosuppression (and the risk of other infections), the processes of allograft injury, and even the pathogenesis of certain forms of malignancy. Currently available evidence suggests that HHV-7 is ganciclovir resistant, and it is not clear what, if any, antiviral strategy should be employed for this virus.[721]

5.1.7. HHV-8 Infection in Organ Transplant Recipients

HHV-8 is a χ-herpesvirus which is the causative agent of Kaposi's sarcoma (KS). Most cases of KS in transplant patients represent instances in which viral reactivation from latency has occurred posttransplant. However, instances of primary HHV-8 infection in which an organ from a seropositive individual was transplanted into a seronegative individual are well documented, with subsequent development of KS in these individuals. Both corticosteroids and OKT3 administration appear to amplify the extent of HHV-8 replication. The incidence of KS in transplant recipients mirrors the seroprevalence of HHV-8 in a given population (with seropositivity equaling the presence of latent infection capable of being reactivated posttransplant). Thus, the incidence of KS in the United States is 0.5%, in Italy 1.6%, in Israel 2.4%, in South Africa 4%, and in Saudi Arabia 5.3%. The incidence of KS among HHV-8-seropositive transplant recipients approaches 25–30% in the first 3 years after transplantation. Genetic polymorphism in HHV-8 isolates has been demonstrated, but those causing KS are thought to belong to a single, or limited number of genotypes.[700,722–726]

Clinical manifestations of KS in transplant patients resemble those seen in other populations. Skin involvement with violaceous/blackish nodules is the most common presenting finding. However, visceral lesions involving the lungs, GI tract, bladder, and other sites can be found in ≥ 40% of transplant recipients with KS. GI manifestations include bleeding, abdominal discomfort, perforation, obstruction, and protein-losing enteropathy. Diagnosis of KS is made on biopsy; the diagnosis of HHV-8 infection is made by PCR assay of peripheral blood leukocytes, with viral load being an excellent predictor of the risk of developing KS. Optimal management is still being determined. The first step is to significantly reduce immunosuppression. There is suggestive evidence that such drugs as ganciclovir, foscarnet, adefovir, cidofovir, and HPMA [(S)-1-(3-hydroxy-2-phosphonylmethoxypropyadenine] may be useful in the management of high viral loads, and, possibly, overt KS. Clearly, an analogy can be made with EBV-associated PTLD and the concept of monitoring viral load, and then manipulating immunosuppression and/or preemptive antiviral therapy on the basis of these measurements. In patients not responding to the above interventions, more classical cancer chemotherapy is an option.[700,727–729]

5.2. Hepatitis in the Organ Transplant Recipient

The incidence of chronic liver disease in organ transplant patients has not changed significantly in 20 years, with 10–15% of successful transplant recipients being subject to the morbidity and mortality associated with what is usually progressive disease. In considering liver disease in this patient population, two types of etiologies must be considered: drug-induced hepatotoxicity; and virus-induced disease, particularly with a group of viruses that are subject to modulation by the immunosuppressive therapy being administered.[5] Two of the drugs that are a standard part of the immunosuppressive regimen, azathioprine and cyclosporine, can be injurious to the liver under certain circumstances. Of these drugs, azathioprine has received the most attention. At a dose of 2–4 mg/kg

per day, this drug is clearly hepatotoxic, with a histologic pattern consistent with chronic active hepatitis. However, at the doses of azathioprine utilized today (1–1.5 mg/kg per day), azathioprine is unlikely to be responsible for the development of chronic, progressive liver disease and cirrhosis. The substitution of cyclophosphamide for azathioprine in transplant patients with unexplained hepatocellular dysfunction is rarely associated with improvement. Similarly, although high-dose cyclosporine, particularly when administered intravenously, can cause hepatic injury, there is no convincing evidence that it is the cause of chronic, progressive liver disease in transplant recipients. Of importance, there is also little evidence that discontinuing azathioprine in transplant patients with proven viral hepatitis caused by hepatitis B will have a beneficial effect on the course of such infections.[5,730–735]

There are, however, unusual forms of liver injury that are caused by azathioprine administration that will respond to cessation of the drug: hepatic veno-occlusive disease (usually in men), peliosis hepatitis, perisinusoidal (Disse's space) fibrosis, and nodular regenerative hyperplasia. It has been postulated that azathioprine, even in the doses currently employed, can damage the endothelial cells lining the hepatic sinusoids and the terminal hepatic venules, producing this range of clinical disease.[736–738]

In addition to immunosuppressive agents, the transplant patient is subjected to a wide array of other pharmaceutical agents, most commonly antimicrobial drugs, antihypertensive agents, and diuretics. Although they are unusual causes of chronic liver disease, these should always be considered in the transplant patient. Perhaps the most common causes of hepatotoxicity have been the antituberculous medications isoniazid and rifampin, trimethoprim–sulfamethoxazole (particularly the sulfonamide component), alpha-methyl-dopa, and the azole group of antifungal drugs. As far as these last agents are concerned, it is reassuring that the initial experience with fluconazole has been so positive in organ transplant patients in terms of adverse interactions with the liver. It has been safely used, even in liver transplant patients, in the face of significant hepatic dysfunction, and its use has not been associated with significant hepatic injury in our experience of over 200 organ transplant patients treated with the drug.[5,179]

As previously noted, the herpes group viruses not uncommonly affect the liver. In particular, the rare instances of disseminated primary infection with HSV and VZV can have a major effect on the liver. In addition, the important effects of CMV and EBV, particularly on the transplanted liver, have already been reviewed. It is important to point out, however, that there is no evidence linking these viruses, or such other agents as adenoviruses (an unusual cause of fulminant hepatitis in the transplant patient),[739,740] to the pathogenesis of chronic liver disease that leads to end-stage liver failure—the entity under discussion here. Thus, the predominant causes of chronic hepatitis in transplant recipients are the classical hepatitis viruses. At present, five different viruses have been defined, each with a differing potential for injuring transplant patients (see Chapter 12)[86]:

1. *Hepatitis A virus* (HAV). HAV is a small RNA virus that is spread via the fecal–oral route. Although HAV can cause fulminant hepatitis that can require emergency liver transplantation and relapsing hepatitis can occur, chronic disease has never been shown to result from this virus.[86,741] Because viremia during the course of HAV infection is both transient and at a very low level, and because of the lack of chronic infection, HAV is rarely transmitted by blood transfusion, has not been documented to be transmitted by organ transplantation, and is not a significant problem among dialysis patients (unlike hepatitis B and C).[86,742–744] A number of patients with fulminant HAV infection have now successfully undergone emergency liver transplantation, although recurrent HAV infection of the allograft has been reported.[745]

2. *Hepatitis B virus* (HBV). HBV is caused by a DNA virus of the hepadna virus group. Unlike the situation with hepatitis A, HBV is present in a fully infectious form in the blood for prolonged periods (as long as 20 weeks in normal individuals with acute, self-limited infection; and as long as 20 years in patients with chronic infection). HBV infection is efficiently transmitted by transfusion or through the transplantation of organs from HBV carriers, as well as by such other routes as intimate mucosal contact between virus carriers and susceptible individuals. Just as the demonstration of HBsAg in the blood is a marker for infectious virus, antibody to HBsAg (anti-HBs) is a reliable marker for immunity to the virus. There have, however, been documented instances of anti-HBs-positive individuals reverting to HBsAg positivity with the initiation of immunosuppressive therapy.[746] HBV has an important impact on transplant patients in terms of the production of chronic active hepatitis and cirrhosis both before and after transplant, as well as in the pathogenesis of hepatocellular carcinoma in these patients. Although the liver is the major site of HBV replication, it is now clear that such extrahepatic reservoirs of HBV infection as mononuclear cells and bone marrow cells exist, and can account for the reinfection of liver allografts following transplantation (*vide infra*).[86,747–749]

3. *Hepatitis C virus* (HCV). HCV is an RNA virus that is the cause of the majority of parenterally transmitted and bloodborne cases of non-A, non-B hepatitis (NANBH). It is also likely to play the same role in transplant recipients. HCV is the hepatitis virus most likely to produce chronic, progressive liver disease. For example, in the normal population, 50% of individuals with posttransfusion HCV infection will have biochemical evidence of chronic liver disease at 12 months, with at least 40% of these having evidence of chronic active hepatitis and 10–20% having cirrhosis 5 years posttransfusion.[750,751] Thus, HCV is a not uncommon cause of end-stage liver disease necessitating liver transplantation, may cause progressive liver disease following transplantation, and is an important contributor to the net state of immunosuppression posttransplant (*vide infra*).[86]

4. *Hepatitis D virus* (HDV). HDV, sometimes called the delta virus, is a defective RNA virus whose replication requires the presence of HBV. When such coinfection occurs, there seems to be a higher rate of acute (including fulminant) and chronic hepatitis, as opposed to asymptomatic disease, than with HBV alone. Patients with chronic HBV infection who acquire HDV will frequently develop acute exacerbations, including fulminant hepatic failure. These events have been documented in transplant patients, particularly in countries such as Italy where HDV infection is common.[86,746,752]

5. *Hepatitis E virus* (HEV). HEV is the cause of enterically transmitted NANBH, a frequent cause of epidemic hepatitis in developing countries, and a major cause of sporadic hepatitis on the Indian subcontinent. HEV resembles HAV in terms of its fecal–oral mode of spread, its lack of ability to produce chronic liver disease, and its failure to have an impact on transplant patients.[86,753]

Thus, the two agents responsible for virtually all of the chronic liver disease in transplant patients are HBV and HCV, both viruses that are transmitted by blood products and by allografts from infected donors and whose effects are modulated by the immunosuppressive therapy being administered. It is estimated that ~20% of hepatitis occurring after transplantation remains unexplained, and that there are likely additional viruses causing hepatitis in these patients remaining to be described (so-called non-A, non-B, non-C virus or viruses).[111]

5.2.1. The Clinical Impact of HBV Infection on the Organ Transplant Recipient

The techniques currently employed in testing for HBsAg (and hence the presence of infectious virus) are so sensitive and specific that the current incidence of post-transfusion HBV infection is estimated to be approximately 0.002% per transfusion, with a comparable figure for organ transplantation. Because of this, the acute acquisition of HBV at the time of transplantation from either infected blood or a contaminated allograft is now extremely rare. This is fortunate for two reasons: The acquisition of HBV infection in the peritransplant period is associated with a markedly increased incidence of fulminant hepatitis; and HBV vaccine has a low rate of efficacy in dialysis patients, in those with end-stage liver disease and other debilitating illnesses, as well as posttransplant.[747,751,754–758]

The major problem with HBV infection in transplantation is the optimal management of the patient who is already infected with the virus prior to transplant, as virtually all such individuals will remain persistently infected posttransplant. Corticosteroid therapy seems to have a direct stimulatory effect on the level of virus replication, with a rapid increase in HBV DNA polymerase activity, HBeAg levels, HBV DNA, and HBsAg levels being observed. This stands in contrast to the gradual decrease in HBV production that is often observed in carriers of the virus not receiving immunosuppressive therapy.[746,759,760] The effects of the other immunosuppressive agents on HBV replication are currently unclear: In a short-term human hepatocyte tissue culture system, the amplifying effects of corticosteroids on HBV replication could be clearly demonstrated, with neither cyclosporine nor azathioprine having any effect[761]; in the woodchuck hepatitis virus model (which is felt to mimic many of the host–virus interactions of HBV in humans), the administration of cyclosporine during the incubation period increased significantly the incidence of chronic infection.[762] Thus, the *in vivo* effects may be greater than the tissue culture effects of these other agents. Paradoxically, liver function test abnormalities and even histologic evidence of disease activity may be blunted by the immunosuppressive therapy, despite the upregulation in viral replication. Unfortunately, despite this, liver injury is accelerated, and progressive liver disease can occur, even in the face of relatively normal biopsies until late in the course. Despite immunosuppression, a significant HBV-specific, MHC class II-restricted T-cell response is present. This is associated with both the local production and systemic release of interferon-χ and TNF, which is proportional to the degree of neuroinflammatory activity in the allograft.[86,763]

The effects of HBV on kidney and, presumably, other forms of nonhepatic transplantation should be considered in two time frames: the first 2 years posttransplant, and the late period after that. In the early period, HBV infection appears to have little clinical impact on the posttransplant course, with the possible exception of contributing to the net state of immunosuppression. However, beginning approximately 2 years posttransplant, patients with chronic HBV infection receiving chronic immunosuppression begin to develop serious illness resulting from progressive liver disease and/or the development of hepatocellular carcinoma.[748,764–772] As far as this last fact is concerned, the *HBx* gene of HBV, which codes for a viral transactivator, can induce the production of liver cancer in transgenic mice.[773] *HBx* appears to function through a complex signal transduction mechanism that activates a tumor promoter signaling pathway.[774] Presumably, the increased level of viral proliferation that is made possible by transplant patients' immunosuppressive therapy promotes this process, explaining in part the excessive incidence of hepatocellular carcinoma in this patient population.

The long-term effects of chronic HBV infection have been well chronicled in renal transplant patients. There is little reason to believe that these effects should be any different in the other forms of nonhepatic organ transplantation. By 8–10 years posttransplant, patients with HBV-associated chronic liver disease had an increased death rate from hepatic failure, hepatocellular cancer, and sepsis, with HBV-infected patients doing significantly worse than transplant patients with other forms of liver disease.[746,764–772] For example, Rao and Andersen reported that 38% of long-term renal transplant patients with HBV infection had chronic progressive hepatitis, 38% had chronic active hepatitis, 42% had evidence of cirrhosis, and 54% died of liver failure.[770] In contrast, only 17% of the HBsAg-negative patients had chronic progressive hepatitis, 14% had chronic active hepatitis, 19% had cirrhosis, and 12% died of liver failure. Thus, although NANBH in transplant patients is not a benign illness, HBV infection, in the absence of therapy, has carried a significantly worse prognosis. More recent reports, perhaps reflecting the use of cyclosporine-based immunosuppressive regimens that employ much lower steroid doses, have suggested that the 10-year prognosis for renal transplant patients with HBV is more optimistic.[87,775] In the case of renal transplantation, because of this profound impact of chronic HBV infection on long-term well-being and survival, arguments have been presented that such individuals would be better served by dialysis therapy rather than transplantation.[765–769] This has been controversial, to say the least, particularly in view of the more recent experience cited above and the dawn of a new era in which meaningful therapy of HBV infection is starting to become available (*vide infra*). One important point that bears further study is that preliminary evidence is available suggesting that the presence of HBeAg in the blood prior to renal transplant and/or histological findings of chronic progressive or active hepatitis delineate a group of patients at particularly high risk of progressive liver disease and excessive mortality after transplantation.[746,776] Our approach to this issue is to regard HBsAg positivity as a relative, rather than absolute, contraindication to transplantation in patients with endstage renal, heart, and lung disease. Such other factors as evidence of liver disease and HBeAg positivity (or, ideally, more specific markers of viral proliferation, such as HBV DNA levels), level of nutrition, other comorbid illnesses, quality of life, and, perhaps most important, the patient's own wishes should then enter into the decisionmaking.

The often dismal results of transplanting individuals with HBV infection have taken a significant turn for the better in recent years because of the emergence of new antiviral approaches. The first of these, α-interferon, has been shown to induce remission in approximately 40% of nonimmunosuppressed individuals with chronic HBV infection. However, in transplant patients, the rate of response is less than half of that. In addition, although α-interferon appears to be well tolerated in liver transplant recipients, experience in renal allograft recipients has indicated a significant risk of severe rejection in these patients. Hence, interferon therapy, at least with this agent alone, does not appear to be a promising avenue to follow.[777–780] Famciclovir, but not ganciclovir, has some activity against HBV, but clearly is not the answer to the problem.[781–785]

The advent of lamivudine represents the first major advance in this field. Lamivudine is a nucleoside analogue that is the most powerful inhibitor of HBV replication currently available, due to its effects on the viral polymerase/reverse transcriptase. The drug has been well tolerated and extremely effective not only in lowering HBV levels but also in improving hepatic function and histology, including patients with the devastating, fibrosing cholestatic hepatitis form of HBV disease. The major disadvantage of lamivudine is the emergence of drugresistant mutants (usually in a highly conserved region of the reverse transcriptase known as the YMDD motif). These mutants begin to emerge approximately 1 year after the drug is initiated, with an increasing incidence over time. The higher the viral load, the more likely and more quickly resistance mutants will emerge. After 1 year of therapy, the incidence of resistance is 14–31%; after 2 years, it is 38%. It is clear that multidrug programs will

be necessary to prevent this occurrence, analogous to what is required for the control of HIV infection. Such drugs in development as adefovir and entecavir offer great promise in this regard (see Chapter 12). Because lamivudine efficacy has a limited time span before the mutants appear and liver disease reemerges, there is currently great controversy about how best to use lamivudine— pretransplant, early posttransplant, or only treatment of symptomatic disease. Hopefully, the ability to use drug combinations in the near future will make this debate irrelevant.[785–802]

Liver transplantation in patients with HBV infection remains a particular challenge. Without antiviral intervention, there is almost universal recurrence of infection in the allograft despite perioperative hyperimmune globulin or α-interferon, and there have been a number of reports of excessive mortality in the first 6 months posttransplant, and even the occurrence of *de novo* hepatocellular carcinoma in the transplanted liver within 4 years of successful transplantation. Evidence of active viral replication at the time of transplantation has been particularly associated with poor results in some series.[749,803–809] Although fulminant HBV may occur posttransplant in these patients, the major cause of early death is sepsis, suggesting that the immunomodulating effects of this virus play an important role in the excessive early mortality.[805] The Boston Center for Liver Transplantation reported a 57% 1-year and 54% 3-year survival following liver transplantation for HBV infection without anti-HBV therapy, with most of the mortality resulting from infection in the first 6 months. A particularly poor result was obtained in patients with hepatocellular carcinoma and HBV infection (their mortality in the first year was > 80%).[810]

The first major advance in this area came from the use of prolonged immunoprophylaxis with anti-HBs hyperimmune globulin. Whereas high doses of this therapy restricted to the anhepatic phase of the operation and immediate perioperative period were ineffective, longterm immunoprophylaxis has had a major impact.[811–815] For example, Samuel et al.[811] have reported on an immunoprophylaxis program in which hyperimmune globulin was administered during the anhepatic phase of the liver transplant operation, daily for 6 days thereafter, and then as needed to maintain a circulating level > 100 IU/liter of anti-HBs. During an average follow-up of 20 months, all 110 HBsAg-positive patients became seronegative posttransplant, with circulating HBsAg reappearing in 22.7%— usually, approximately 9 months posttransplant. Overall 1-year survival was 83.6%. This prophylactic program was least effective in preventing HBV reinfection in patients with postnecrotic cirrhosis caused by HBV (59%

recurrence rate), more effective in those with combined HBV and HDV infection (13% recurrence), and was totally effective in those with fulminant hepatitis B, a population that has had fewer problems with HBV infection posttransplant in general. As in renal transplant patients, evidence of active viral replication (HBV DNA being a better marker than HBeAg) at the time of transplantation was highly predictive of HBV recurrence, a 96% rate of recurrence at 2 years.

In general, HBsAg-positive patients surviving liver transplantation for more than 6 months have usually done quite well for the next 12–24 months. Although the hyperimmune globulin approach is a significant advance, there are two remaining problems with it in addition to the lack of complete efficacy: cost—$25,000/year, although newer lower dose intramuscular regimens appear to offer comparable protection against reinfection at a significantly lower cost.[816] Of even greater concern is the selection of mutants with the globulin that are still pathogenetic but have changes in the surface antigen, thus obviating the effectiveness of the hyperimmune globulin approach.[817–821] Underlining the difficulties of single-modality therapy, there are now reports of patients for whom sequential therapy with immune globulin and then famciclovir and lamivudine, was associated with the development of mutants resistant to each of these in turn.[822,823] At present the combination of lamivudine and immune globulin appears to be the best available program for preventing reinfection of the transplanted liver with HBV.[824–828] A variation on this theme is the substitution of adefovir plus immune globulin when the patient has lamivudine-resistant infection prior to liver transplantation.[800]

5.2.2. The Clinical Impact of HCV Infection on the Organ Transplant Recipient

HCV is the major cause of NANBH in nontransplant patients, and appears to be the source of more than 80% of the progressive liver disease that occurs posttransplant.[86] End-stage liver disease caused by HCV is the leading indication for liver transplantation in Europe and the United States, accounting for more than one-third of all transplants.[829] In addition, chronic HCV infection, like chronic HBV infection, can result in the development of hepatocellular carcinoma.[830,831] HCV is spread via parenteral contact with blood with a high degree of efficiency. Thus, HCV infection posttransfusion (accounting for approximately 90% of cases of posttransfusion hepatitis) in intravenous drug users, in healthcare workers with occupational exposure to blood, and in hemodialysis patients is well recognized. In addition, as already discussed,

transplantation of an organ from an HCV-viremic individual is a highly efficient means of transmitting this virus. These epidemiologic risk factors account for only 50–60% of cases of acute hepatitis C, with sporadic cases occurring in the community. It is likely that sexual transmission, albeit with a low efficiency, accounts for many of these sporadic cases.[527,530,832–835] HCV appears to be the etiology only rarely of fulminant hepatitis, either sporadic or epidemic.[836,837]

The critical scientific step in unraveling the HCV story came when Choo *et al.*[838] successfully cloned a portion of the genome of this virus. The expression product of this clone has been used to devise a series of increasingly refined assays for detecting antibodies to this virus, as well as establishing the basis for the development of a PCR test for HCV RNA, which is now central to patient management. Currently available information on the antibody test may be summarized as follows: The antibody response to HCV takes a relatively long time to develop, even in individuals with normal immune responsiveness. Thus, although in one study, 80% of patients with posttransfusion NANBH developed anti-HCV by ELISA assay, the mean interval between the date of transfusion and anti-HCV seroconversion was 18 weeks; 61% of patients seroconverted within 15 weeks of transfusion, about 90% by 26 weeks, and one patient not until 12 months posttransfusion. In transplant patients, the development of the anti-HCV response is even more attenuated, with the degree of attenuation being proportional to the intensity of the immunosuppressive therapy being administered. Thus, anti-HCV testing would underestimate the presence of HCV infection among transplant patients. Consistent with the relatively weak immunizing effect of HCV are studies in chimpanzees suggesting that rechallenge with infectious virus, with either homologous or heterologous strains, had a high rate of reinfection—an observation of great potential importance for transplantation.[86,839–846]

In the general population, more than 50% of individuals with acute HCV infection develop chronic infection, with the course of the illness being extremely protracted in the majority (20–40 years). Approximately 20% develop serious end-stage liver disease and/or hepatocellular carcinoma as a consequence of the chronic HCV replication, with hepatocellular carcinoma being uncommon.[847] As with many infections in transplant patients, clinical consequences of HCV posttransplant are more common and occur over a shorter time span. LaQuaglia *et al.*[848] reported on the basis of studies in a renal transplant population from the precyclosporine era that more than 80% of individuals who develop liver function test abnormalities suggestive of NANBH maintain such abnormalities indefinitely. In the first 1–2 years post-transplant, acute clinical hepatitis caused by NANBH is quite unusual, and the major impact of such a process has been as an important contributor to the net state of immunosuppression as manifested by both an increased incidence of infection and an increased allograft survival. Beginning 3–5 years after renal transplantation, progressive liver disease begins to present in these patients, often culminating in all of the manifestations of end-stage liver disease. In renal transplant recipients, patients with chronic HCV infection have a similar mortality to those who are not infected for the first decade, but with a significantly increased mortality over the second decade (20-year survival for the HCV group, 63.9% versus 87.9% for the uninfected). As previously noted, both the incidence and the rate of progression of HBV-induced liver disease are greater than those for NANBH. However, because the incidence of NANBH is so very much higher, the net result is that more cases of serious liver disease are caused by NANBH in transplant patients than by HBV.[86,747,849–853] On the other hand, it would appear that patients with chronic HCV infection and end-stage renal disease fare better with renal transplantation than with dialysis. However, biopsy is recommended pre-transplant to rule out significant cirrhosis and to determine the need for antiviral therapy (*vide infra*). It is of interest that the combination of HBV and HCV appears to have a greater clinical impact than either of the viruses by themselves. On the other hand, the so-called hepatitis G virus has no impact on the course of patients with HCV infection (see Chapter 12).[850,854–858]

In liver transplant recipients who come to liver transplantation with HCV infection, reinfection of the allograft is the rule, with the source of the infection being HCV-infected peripheral blood mononuclear cells.[86,859,860] The consequences of HCV reinfection are controversial, with the incidence of adverse events being in part determined by the presence or absence of a variety of risk factors (*vide infra*) and, even more importantly, the duration of time over which these consequences are being assessed. In general, even up to 10 years, there is no effect of HCV reinfection on mortality following liver transplantation.[861] However, the following data from different series do suggest the potential for considerable morbidity: 42% of patients with recurrent HCV infection developed chronic hepatitis, with 25% of these going on to cirrhosis[862]; the King's College group reported that at 5 years, 20% of those transplanted for end-stage liver disease due to HCV had cirrhosis.[863]

A number of risk factors have been defined that are correlated with an increased morbidity and mortality from HCV reinfection, as well as with a shorter time course to end-stage liver disease. These include the following: *viral load*—the higher the viral load the greater the risk and the more accelerated the course; *intensity of immunosuppression*—in particular, higher doses (as pulse doses for acute rejection) and more sustained courses of corticosteroids, OKT3 therapy, and, perhaps, mycophenolate mofetil use have been correlated with marked increases in HCV viral load; *allograft rejection*—probably by two mechanisms: cytokines, chemokines, and growth factors produced in the rejection process will increase the level of circulating virus; and the increased immunosuppression required to treat the rejection will likely amplify the virus (a common error is to pare back immunosuppression so intensively that rejection is produced, thus leading to higher viral loads and an acceleration of the HCV infection far greater than if adequate immunosuppression had been provided); *HCV genotype*—genotype 1b infection has been associated with higher viral loads, a more accelerated course, and a poorer response to interferon-based treatment regimens; the presence of *quasispecies*—the greater the heterogeneity in the HCV population present, the greater difficulty the host has in containing the infection; *CMV viremia*—a risk factor for allograft cirrhosis after liver transplantation for hepatitis C (conversely, active HCV can trigger CMV disease; a bidirectional relationship between these two viruses mediated by cytokines and chemokines elaborated by the host in response to these infections); *iron overload*—iron overload decreases the host's cytotoxic T cell response against the virus, thus amplifying the level of virus; *donor and recipient class II HLA matching*—appears to promote reinfection and allograft injury; and *donor tumor necrosis factor gene*—the donor TNF-α promoter genotype may influence the inflammatory response to HCV infection of the graft and the extent of subsequent allograft injury.[854–879]

In addition to chronic active hepatitis and allograft cirrhosis, there are two relatively uncommon complications of HCV infection that can have devastating consequences for the transplant patient. The first of these is *fibrosing cholestatic hepatitis*. First described in patients with HBV infection and immunodeficiency (*vide supra*), this is a syndrome in which there are high levels of circulating HCV, rapidly progressive hepatic failure, mildly elevated serum aminotransferase level, an extensive periportal fibrosis, intense cholestasis, minimal inflammatory infiltrate, and no cirrhosis. The hepatocytes themselves are literally choked with exceedingly high levels of HCV which results in direct hepatocyte injury. This diagnosis requires immediate decrease in immunosuppression and the initiation of the best available antiviral therapy (*vide infra*).[880–882]

The second of these complications is glomerulonephritis, which can occur in both the general population with HCV infection, as well as the transplant recipient. Indeed, patients have been reported with severe nephrotic syndrome due to HCV, who improved with liver transplantation, and then relapsed when significant reinfection developed in the allograft. The most common form of this is membranoproliferative glomerulonephritis with or without mixed cryoglobulinemia. This is an immune complex-mediated condition, and, not surprisingly, other laboratory abnormalities commonly demonstrated include rheumatoid factor and hypocomplementemia. Of interest, a patient has been reported who developed fibrosing cholestatic hepatitis with cryoglobulinemia and a severe systemic vasculitis. A second form of glomerular disease reported in association with HCV is membranous glomerulonephritis. Other less common conditions linked to HCV are acute and chronic transplant glomerulopathy and thrombotic microangiopathy with anticardiolipin antibody. Management of these complications of HCV replication is difficult, and is focused on antiviral therapy directed against the HCV infection and supportive care of the renal disease.[883–889]

HCV infection has a considerable effect on the well-being and quality of life of patients following transplantation, particularly liver transplantation. At present, because of the availability of lamivudine and anti-HBs immune globulin, patients with HBV infection have less depression and an improved quality of life compared with those with HCV infection.[890–893] The first principle of managing HCV infection is to attempt to decrease immunosuppressive therapy and to treat such other conditions as CMV infection. Interferon-α and ribavirin individually have some activity against the virus. However, when combined, there is synergistic benefit, with ~50% of patients developing a meaningful response to such therapy. It must be emphasized, however, that these drugs have considerable side effects (interferon—fever, malaise, bone marrow dysfunction, and risk of rejection; ribavirin—hemolytic anemia), and new drugs are badly needed. The first of the new drugs will be pegolated versions of interferon-α, which permit weekly (as opposed to thrice weekly or more frequent injections) administration, have increased efficacy over the standard formulations, and have a similar adverse effect profile. In general, patients are more likely to respond, the lower the viral load.[800,894–900]

Illustrative Case 6

A 52-year-old woman with end-stage renal disease resulting from polycystic kidney disease underwent bilateral nephrectomy and cadaveric renal transplantation. Posttransplant, she had no clinical episodes of rejection, and by 1 year posttransplant, she had a serum creatinine of 0.7 mg/dl while being immunosuppressed with azathioprine 100 mg/day and prednisone 30 mg every other day. During the first year posttransplant, despite minimal immunosuppression and the absence of CMV infection, she had had bouts of *Listeria* sepsis, invasive pulmonary aspergillosis, and gram-negative sepsis caused by a perforated sigmoid diverticulum. The latter condition was treated surgically. Her liver function tests, which had been normal prior to transplant, began to show an SGOT level more than twice normal (without other abnormalities) by the third month posttransplant.

Over the next 5 years, her renal function remained excellent on the same immunosuppressive program. Her SGOT remained elevated at this same level (two to four times normal) as well. Percutaneous liver biopsy performed 2 years posttransplant revealed minimal changes of chronic persistent hepatitis. Substitution of cyclophosphamide for azathioprine had no effect on her SGOT level, and the azathioprine was restarted. Six years posttransplant, after a 2-month history of progressively increasing abdominal girth and edema, she presented with fever, rigors, and increasing abdominal pain. A diagnosis of *E. coli* peritonitis, with positive ascites and blood cultures, was made. Despite therapy with high-dose intravenous ampicillin and gentamicin (to which the organism was sensitive), as well as stress doses of steroids, the patient died on the second hospital day. At autopsy, far-advanced cirrhosis with esophageal varices and evidence of spontaneous bacterial peritonitis were noted. All tests for HBV were negative. The transplanted kidney appeared totally normal grossly and microscopically. Retrospective studies of stored sera revealed the following: the patient was anti-HCV negative pretransplant and throughout her posttransplant course; the patient was HCV RNA negative prior to transplant, and became positive, remaining so for the rest of her life, approximately 3 months posttransplant; the donor was anti-HCV and HCV RNA positive.

Comment. A tragic instance of chronic HCV infection, acquired from the donor at the time of transplantation, marred the course of an otherwise successful renal transplant. In the first year posttransplant, there were three notable events: several episodes of life-threatening extrahepatic infection, no rejection on minimal immunosuppression, and the development of mild elevations of the serum transaminase level. The first two of these presumably represent the contribution of HCV infection to the net state of immunosuppression. The "transaminitis" remained constant over the next 6 years and appeared to be of little consequence, particularly in view of the "benign" liver biopsy 2 years posttransplant. She then presented 6 years posttransplant with far-advanced cirrhosis and portal hypertension, and acute, spontaneous bacterial peritonitis that caused her death. This form of disease must be prevented.

5.3. HIV Infection in the Organ Transplant Recipient

Transplantation of an organ from an HIV-infected donor into an uninfected recipient is an extremely effective mechanism for transmitting the virus, with a transmission efficiency that approaches 100%. Indeed, the transplantation literature of the 1980s was replete with the whole gamut of clinical effects caused by HIV in this patient population: asymptomatic infection, mononucleosis, AIDS-related complex, recurrent opportunistic infection, Kaposi's sarcoma, and transmission of the virus to a spouse. With currently available techniques for screening prospective donors, virtually all transmission of the virus at the time of transplantation should be blocked.[901–904]

Having emphasized that appropriate screening of blood and organ donors for HIV positivity will significantly block the development of primary HIV infection posttransplant, it is only fair to state that HIV has been transmitted by an allograft from a donor who was seronegative for HIV by conventional testing (for antibody to HIV) but still harbored the virus. Presumably, such a donor is in the so-called "window period" when replicating virus is present but an antibody response has not yet appeared.[112] Studies in which PCR techniques have been used have shown that, in certain high-risk individuals, antibody-negative HIV infection that is transmissible can be present for several months.[112,113,903,904] Therefore, in addition to antibody screening, we would not, under normal circumstances, accept a donor with epidemiologic risk factors for HIV infection such as history of homosexual or bisexual behavior, history of intravenous drug abuse, hemophilia, promiscuous sexual history, incarceration in prisons with a high incidence of HIV, and so on. In such exceptional circumstances as a living related donor with one of these risk factors, PCR or viral culture techniques should be carried out in addition to routine antibody testing.[113]

When primary HIV infection is acquired posttransplant, it not uncommonly produces a mononucleosis syndrome, akin to that produced by CMV or EBV, approximately 6 weeks posttransplant. Most transplant patients seroconvert within weeks of this event, although we have reported a patient who remained HIV antibody negative for approximately 3 years posttransplant, until he developed rapidly progressive AIDS. The mean time to the development of overt AIDS in patients who acquire primary HIV infection at the time of transplant has been approximately 3 years.[113,901–904]

A far bigger issue than primary HIV infection posttransplant is the optimal management of patients with asymptomatic HIV infection who present with end-stage renal, hepatic, cardiac, or lung disease, and who otherwise would be deemed satisfactory transplant candidates. Currently available information suggests that HIV-positive individuals undergoing organ transplantation without highly active antiretroviral therapy (HAART) can be divided into three approximately equal groups in terms of clinical outcome: one third do very poorly, dying within 6 months of transplantation, primarily of infection; one third do very well, and are alive with a functioning allograft 5–7 years later; and one third have an intermediate result, developing clinical AIDS with a rapid downhill course 2–4 years posttransplant.[113,904–906]

Over the past two decades, it has been the general practice of virtually all transplant centers to exclude HIV-positive individuals from receiving an organ transplant. With the great success of HAART therapy, the prognosis for HIV-infected individuals has been changed markedly (see Chapter 12), and cautious efforts are being initiated to reopen this question. There are a number of hurdles to be overcome before this can become a routine step: drug interactions between anti-HIV drugs and the calcineurin inhibitors require careful attention; optimal dosimetry of both the immunosuppressive drugs and the AIDS drugs will need to be defined; and the ethical issues of the appropriate allocation of allografts will need to be carefully examined. Having said that, it would appear that cautious efforts for proceeding in this area are reasonable.[905–910] We have established the following criteria for proceeding in this area at the present time:

1. In general, we would restrict our transplant efforts in HIV-infected individuals to instances in which the organ is required to maintain life— hearts, livers, lungs, and kidneys only when dialysis is no longer an option. All such efforts should be regarded as clinical research requiring Human Studies Committee approval and Informed Consent.
2. Transplantation of HIV-infected individuals should only be carried out at centers with great expertise in both transplantation and HIV infection, and where these skills can be brought to bear in the day-to-day care of the patient.
3. HIV-infected patients accepted for transplantation should be on a stable HAART regimen, known to be compliant, and to have undetectable HIV viral loads.
4. HIV-infected patients accepted for transplantation should have no other contraindications for transplant, be well nourished, have a CD4 count $> 400/mm^3$, and have been free of infection for ≥ 3 years.

The report of the rapid development of a myelopathy after the transmission of HTLV-1 infection via a transfusion during heart transplantation emphasizes the care that must be taken to prevent the transmission of infection with the allograft or with blood.[911] In the present context, this case reminds us that in addition to screening for HIV-1, such other retroviruses as HIV-2, HTLV-1, and, probably, HTLV-2 should be part of the screening of potential donors.[911–913]

5.4. The Clinical Impact of Papovaviruses in the Organ Transplant Recipient

The papovaviruses are DNA viruses comprising two genera, the polyomaviruses and the papillomaviruses. The human polyomaviruses BK virus (BKV) and JC virus (JCV) infect most normal individuals during childhood, apparently without demonstrable clinical illness.[914–916] These agents are of interest for several reasons: immunosuppression results in the excretion of these agents and/or a specific antibody rise in the majority of transplant patients; BKV, JCV, and simian virus 40 (SV40) are closely related both antigenically and structurally; JCV is the cause of progressive multifocal leukoencephalopathy, a subacute, progressive demyelinating disease of the CNS; BKV is an increasingly important cause of interstitial nephritis in renal transplant patients; and all three agents are at least potentially oncogenic.[917–932]

BKV was first isolated by Gardner and colleagues in 1971 from the urine of a renal transplant patient 3 months posttransplant who presented with a ureteral stricture during an apparent rejection episode.[917] Since then, additional cases linking ureteral strictures and urinary papovavirus have been reported in renal transplant recipients.[915 917–919,921–924] Far more important has been the recognition of BKV as a not uncommon cause of progressive interstitial nephritis (an incidence as high as 5%, although evidence of asymptomatic BKV replication can be found in some 40% of renal transplant recipients). Renal biopsies from patients with kidney infection reveal the following: viral cytopathic changes (nuclear atypia, smudging, and inclusions), with necrosis and particular involvement of the tubular epithelium, and a dense, inflammatory infiltrate that can contain numerous plasma cells. Differential diagnosis considerations include calcineurin inhibitor toxicity and severe rejection. Definitive diagnosis can be made with immunoperoxidase staining with a monoclonal antibody specific for BKV, or electron microscopy of ultrathin sections from the biopsy. Patients' urine can be screened for BKV by cytologic examination (cytological evidence of BKV replication is made by the identification of so-called decoy cells, which contain intranuclear inclusions typical for polyomavirus) or by electron microscopic examination of negatively stained urine specimens. Negative results of a urine screen essen-

tially rule out BKV interstitial nephritis. Positive results, in the face of renal dysfunction, necessitate a biopsy for definitive diagnosis, as asymptomatic urinary shedding of BKV in the face of rejection is not uncommon.[929–932]

BKV interstitial nephritis is being recognized with increasing frequency. It has been associated with intense immunosuppression (particularly regimens that include tacrolimus ± mycophenolate) and significant rejection episodes. It has been suggested that damage to the tubular epithelium by rejection plays a role in the severe tubulointerstitial nephritis that develops. BKV nephritis has a poor prognosis, with progressive renal dysfunction being the rule. There is currently no antiviral drug effective for this entity. Increased immunosuppression will accelerate the process; decreased immunosuppression has been reported, in some patients, to be associated with an amelioration of the process. Practically, there is often difficulty in ruling out concomitant rejection as the immunosuppression is being tapered, with serial biopsies being necessary in the management of these patients.[929–932]

JCV was first isolated from the brain of a 38-year-old man with progressive multifocal leukoencephalopathy (PML),[933] and since then has been shown to be the cause of this illness (a demyelinating disease of the white matter of the brain, characterized by the development of progressive motor and sensory deficits, dementia, and death within 3-6 months). As with BKV, there is no known treatment, although decreased immunosuppression is strongly recommended. Polyomavirus infection has also been linked to pancreatic disease in the transplant recipient.[917–933]

Papillomavirus infection in transplant patients is of significance for several reasons: The most common manifestation in this population, as it is in the general population, is the production of warts. In the transplant patient warts may be so numerous as to be disfiguring, with the incidence and severity of warts in these patients being directly proportional to the intensity of the immunosuppressive therapy being prescribed. Such therapy appears to reactivate latent virus posttransplant. The significance of the warts goes beyond the cosmetic, however, as malignant transformation, particularly in sun-exposed areas, has been well documented. Human papillomavirus DNA has been demonstrated in the skin cancers of these patients, and it is likely that such cancers arise as a result of the combined effects of the virus, immunosuppression, and ultraviolet irradiation.[934–940] A particular subgroup of papillomaviruses, the epidermodysplasia verruciformis associated types, has been linked to the pathogenesis of cutaneous and anogenital squamous cell carcinomas in these patients, as well as to the presence of certain keratotic skin lesions that presumably represent premalignant lesions.[941–944]

Cervical papillomavirus, likewise, is more extensive in transplant recipients, is similarly modulated by the intensity of immunosuppressive therapy, and has been linked to the pathogenesis of cervical cancer in both the general population and among transplant patients. Such cancers are significantly increased in this patient population, commensurate with the increase in papillomavirus infection that is present at this site.[945,946]

5.5. Urinary Tract Infection in the Renal Transplant Patient

The most common form of bacterial infection affecting renal transplant recipients is urinary tract infection (UTI). The incidence of UTI in patients not receiving antimicrobial prophylaxis has been reported to vary from 35 to 79% in different series.[947–952] In addition, approximately 60% of the bacteremias observed in transplant patients have traditionally originated from this site.[953,954] The pathogenesis of such infections is only partially understood. In the earlier days of transplantation, UTI was particularly associated with two factors: technical complications associated with the ureteral anastomosis and UTI present prior to transplantation. Despite improvements in surgical techniques and the removal of native kidneys that were thought to be possibly harboring infection, the attack rate remained in the 35–45% range.[952,955]

The three major factors leading to this high incidence of posttransplant UTI appear to be the postoperative urinary catheter, the physical and immunologic trauma that the kidney suffers, and the immunosuppressive therapy that is administered. Now that bladder catheters are routinely removed 1–4 days posttransplant, it is unusual for overt infection to be documented while the catheter is still in place. However, the catheter tips not infrequently become contaminated, and provide a reservoir from which infection is derived. In animal models, the combination of bacteria inoculated into the bladder and trauma to the kidney will result in pyelonephritis, whereas bladder infection without renal trauma results only in a transient cystitis. It is logical that the kidney harvesting, transport, and then transplantation cause significant trauma to this organ, making it more susceptible to bacterial invasion—a susceptibility that is presumably exacerbated by the exogenous immunosuppressive therapy.[27,956–960]

UTI occurring in the first 3 months posttransplant is frequently associated with overt pyelonephritis, bacteremia (with even transient urosepsis having the potential for metastatic seeding in this patient population), and a high rate of relapse when treated with a conventional course of antibiotics. In contrast, UTI occurring at a later time is usually benign, can be managed with a conventional 10- to 14-day course of antibiotics, uncommonly is associated with a bacteremia or requires hospitalization, and has an excellent prognosis. Exceptions to this general pattern should be evaluated for such functional or anatomical abnormalities of the urinary tract as a stone, obstructive uropathy, or a poorly functioning bladder.[952,960]

The potential consequences of UTI and urosepsis in the renal transplant patient go beyond the direct infectious disease effects of the infection. It is now apparent that proinflammatory cytokines such as TNF, IL-1, IL-6, and others are released locally and systemically in response to bacterial invasion of the urinary tract.[762,763] As earlier discussed, this same array of cytokines is involved in two other processes of great importance in the transplant patient: allograft rejection and the reactivation from latency of CMV and other herpesviruses. Thus, clinically, it is not uncommon to see two events in the setting of transplant pyelonephritis, namely, measurable renal dysfunction and clinical CMV disease. The latter we refer to as "a second wave phenomenon," that is, 10–21 days after a febrile illness caused by UTI (and after a clinical response to effective antimicrobial therapy) fever reappears with demonstrable CMV viremia.

Not surprisingly, then, a great deal of effort has been expended in the prevention of UTI, with such significant progress that UTI and urosepsis have been largely eliminated in the transplant patient with the use of antibacterial prophylaxis. Trimethoprim–sulfamethoxazole (we use a dose of one single-strength tablet at bedtime, others use higher doses), trimethoprim, and ciprofloxacin (and presumably other fluoroquinolones) will all decrease the incidence of UTI to less than 10%, with the eradication of urosepsis unless obstruction to urine flow is present.[961–964] At this low dose of trimethoprim–sulfamethoxazole, nephrotoxicity caused by interactions with cyclosporine does not occur, although such other toxicities of trimethoprim–sulfamethoxazole as rash, Stevens–Johnson syndrome, bone marrow toxicity, and, rarely, interstitial nephritis can occur in these patients as well as the general population.[963,964] An additional benefit of trimethoprim–sulfamethoxazole prophylaxis is that it also provides significant protection against *P. carinii* pneumonia, *Nocardia asteroides* infection, sepsis caused by *L. monocytogenes*, and, probably, toxoplasmosis.[27,963,964] It is for this "extra protection" that we routinely utilize low-dose trimethoprim–sulfamethoxazole prophylaxis in the extrarenal transplant recipients as well. In a study comparing ciprofloxacin with trimethoprim–sulfamethoxazole for UTI prophylaxis in renal transplant recipients, ciprofloxacin was equal or superior to the standard regimen in terms of preventing UTI and in its side effect profile. However, 10% of the patients receiving ciprofloxacin developed *P. carinii* pneumonia, versus none of the patients receiving trimethoprim–sulfamethoxazole.[964] Therefore, our current practice is to utilize trimethoprim–sulfamethoxazole as our standard prophylaxis if the patient tolerates it; if some other regimen is necessary, we will combine ciprofloxacin with an alternative anti-*Pneumocystis* regimen, such as atavoquone.

One additional form of UTI that bears mention here is that relating to *Candida* species. Although asymptomatic candiduria is frequently also observed in the general population, in the renal transplant patient it can be particularly dangerous because of the potential for developing obstructing candidal fungal balls. This occurs most commonly in diabetic patients with poorly functioning bladders. Once such candidal fungal balls develop, ascending candidal pyelonephritis and candidal sepsis will ensue. For this reason, we preemptively treat even asymptomatic candiduria, preferably with fluconazole, or with low-dose amphotericin plus flucytosine if this fails.[27]

6. Infection in Organ Transplant Patients More Than 6 Months Posttransplant

One of the measures of the increasing success of organ transplantation in recent years is the fact that investigators are now looking at long-term survival and complications, rather than the 1-year statistics that have been standard for so long. As previously described, the patients who have had a good result from their transplant are primarily at risk for the usual community-acquired infections, particularly the respiratory viruses; in contrast, the "chronic n'er-do-wells" (patients who have received too much acute and chronic immunosuppression in an effort to salvage the allograft) are at special risk for the opportunistic bacteria (*Listeria monocytogenes* and *Nocardia asteroides*) and fungi (*Cryptococcus neoformans* and *Aspergillus fumigatus* in particular).[3,8,27]

6.1. Community-Acquired Respiratory Virus Infection in Organ Transplant Recipients

Respiratory virus infection circulating in the community can have a special impact on transplant recipients. Thus, during influenza epidemics, we have noted patients admitted with complications of influenza ranging from malaise, fever, and dehydration, to respiratory failure relating to influenza pneumonia, or complicating bacterial pneumonia caused by *Streptococcus pneumoniae*, *Haemophilus influenzae*, *Staphylococcus aureus*, and gram-negative bacilli. Although precise information on this subject is not available, we have the impression that complications of influenza are more common in this patient population. Trials of influenza vaccine in renal transplant patients have shown a lack of toxicity or adverse effects, but a disappointing level of efficacy. At present, there is no available information on the efficacy, side effects, or potential for cyclosporine interaction with amantadine prophylaxis in transplant patients. A high priority should be given to garnering this information.[8,965]

Respiratory syncytial virus (RSV) infection of a particularly virulent nature, causing an increased rate of pneumonia and mortality, has been reported in children following organ transplantation, particularly liver transplantation. In addition, we have documented RSV infection on lung biopsy of adult transplant patients with acute respiratory failure. Others have diagnosed RSV on bronchoalveolar lavage specimens from similar patients. Since aerosolized ribavirin therapy may have therapeutic efficacy for this infection, the possibility of RSV infection should be considered even in adult transplant patients with otherwise unexplained pneumonia.[5,23–25,966–968]

Adenoviruses not uncommonly infect the general population, producing asymptomatic disease or clinical illnesses such as upper and lower respiratory tract infection, conjunctivitis, and hemorrhagic cystitis. These viruses appear to have a particular propensity for causing disease in transplant patients: Diffuse interstitial pneumonia has been documented to occur by means of adenoviruses (types 34 and 35) previously not recognized; epidemic disease in a particular transplant population has been well documented; infection of the liver, lung, and GI tract by adenoviruses has been reported among pediatric liver transplant patients, with a mortality rate of 45%; and, finally, both tubulointerstitial nephritis and hemorrhagic cystitis have been documented to occur. The possibility that one or more of these adenoviruses could be oncogenic in this patient population is also worthy of investigation.[5,26–28,969–978]

Other community-acquired respiratory viruses reported to cause serious respiratory disease in the transplant patient include rhinoviruses and parainfluenza virus.[23,24,29] There are two general messages here: Transplant patients have a greater rate of complications from respiratory virus infection than do immunologically normal individuals, and it is worth a special effort to isolate them from such exposures, even in the late posttransplant period.

7. Infectious Disease Problems of Particular Importance in the Organ Transplant Patient

7.1. CNS Infection in the Organ Transplant Patient

Infection of the CNS is a significant cause of morbidity and mortality in the organ transplant patient, with an incidence of approximately 5%. Although the subject of CNS infection in the immunosuppressed patient is reviewed in more detail in Chapter 5, several points of particular emphasis in the transplant patient bear emphasis here. Four distinct patterns of infection may be observed in this population: (1) *acute to subacute meningitis*, almost invariably caused by *Listeria monocytogenes*, by far the most frequent cause of bacterial CNS infection in the organ transplant recipient; (2) *subacute to chronic meningitis*, usually caused by *Cryptococcus neoformans*, although such organisms as *Mycobacterium tuberculosis* and *Coccidioides immitis* can cause an identical clinical syndrome (subacute onset of fever and headache, sometimes associated with a decreased state of consciousness, with a lymphocytic pleocytosis and hypoglycorrhachia on CSF examination), and should be seriously considered if the appropriate epidemiologic history is obtained; (3) *focal brain infection* causing focal neurologic disturbances, which is occasionally caused by *Listeria*, *Toxoplasma*, *Cryptococcus*, *Nocardia*, or HHV-6, but is most commonly related to *Aspergillus* infection metastatic from a site of active pulmonary infection; and (4) *progressive dementia* resulting from progressive multifocal leukoencephalopathy, caused by the polyomavirus JCV. Together, *Listeria*, *Cryptococcus*, and *Aspergillus* account for more than three-fourths of the CNS infections occurring in the organ transplant patient.[348,349,979,980]

Each of these infections tends to occur at a particular time period posttransplant. The first month posttransplant is relatively free of CNS infection, unless an unusual epidemiologic exposure has occurred. Indeed, the major causes of CNS symptomatology in this time period are seizures secondary to cyclosporine, particularly the intravenous formulation, a severe OKT3 reaction (which can, not uncommonly, produce an aseptic meningitis picture), metabolic encephalopathy (especially in liver transplant patients with a poorly functioning allograft), cyclosporine- or tacrolimus-induced leukoencephalopathy, and a stroke (particularly in diabetics or in heart transplant patients with a poorly functioning allograft).[27,348,349,966,979–983]

If one then divides the time intervals into an early period 1–4 months posttransplant (corresponding to the period of maximal net immunosuppression relating to the combination of viral infection and sustained immunosup-

pression) and a later period, the following observations can be made: *Listeria*, *Nocardia*, *Toxoplasma*, and *Aspergillus* infection occur both in the early period and in the late period among the patients with failing allografts who have been intensively immunosuppressed ("the chronic ne'er-do-wells"). In contrast, cryptococcal infection is virtually always a late infection, again particularly in the chronic ne'er-do-wells.[3,27,348,349,979–981]

In the face of trimethoprim–sulfamethoxazole prophylaxis, *Listeria* and nocardial infection, and possibly *Toxoplasma* infection, are effectively prevented. The issue of *Toxoplasma* infection is not completely settled, as there is conflicting information available on the efficacy of trimethoprim–sulfamethoxazole against this infection.[980] In the setting of noncardiac organ transplantation, toxoplasmosis has not been a significant problem, and it is likely that trimethoprim–sulfamethoxazole provides adequate protection. In the special case of cardiac transplantation when the donor is seropositive and the recipient is seronegative, disseminated toxoplasmosis, with particular impact on the heart and the CNS, is a major problem (occurring in 10–20% of cardiac transplant patients, and an estimated 50% at risk of primary infection); current practice is to actively prophylax with pyrimethamine and sulfonamide, and not rely on trimethoprim–sulfamethoxazole. This regimen also supplies anti-*Pneumocystis* and antinocardial prophylaxis.[984–993]

It is important to emphasize that the presentation of CNS infection in the organ transplant patient may be very different from that in the normal host. In particular, the anti-inflammatory effects of the immunosuppressive therapy being administered may obscure the signs of meningeal irritation usually associated with meningitis in the normal patient. For example, in the Massachusetts General Hospital experience, only 60% of patients with *Listeria* meningitis had any evidence of meningeal irritation on physical examination, and, in many of these, the findings were subtle. Mild alterations in the state of consciousness were observed in 70% of patients; however, the most reliable combination of clinical findings for suggesting the possibility of significant CNS infection is the presence of fever and headache.[348,349,980,994]

Any transplant patient with an unexplained headache, especially if febrile, should undergo careful neurologic examination. If there is no evidence of papilledema or focal neurologic deficit, an immediate lumbar puncture should be carried out. If papilledema, focal neurologic deficit, or decreased state of consciousness is present, an immediate CT scan should be performed prior to the lumbar puncture. Magnetic resonance imaging (MRI) should be considered in the patient with spinal cord findings, the patient with focal neurologic findings or unexplained decreased mental status, and the patient with an abnormal CSF formula without an etiologic explanation—even in the face of a negative CT scan. The MRI scan is more sensitive than the CT, although more nonspecific findings may be observed. The CSF obtained at lumbar puncture should undergo the following evaluation: cell count, protein and sugar determinations, Gram's stain, acid-fast and fungal stains, fungal, mycobacterial, and bacterial cultures, and cryptococcal antigen testing. In addition, it is wise to save 2 ml of CSF for any later special studies (e.g., *Histoplasma capsulatum* or *Coccidioides immitis* antibody titers).[348,349,980,994]

7.1.1. *Listeria monocytogenes* Infection in the Organ Transplant Recipient

L. monocytogenes is a gram-positive bacillus that can produce a variety of clinical syndromes, the most important of which in the immunocompromised patient are bacteremia alone, meningitis, meningoencephalitis, and cerebritis without concomitant meningitis.[347,980,981,994–998] When parenchymal seeding of the brain occurs, *Listeria* has a particular propensity for involving the brain stem, causing a clinical syndrome akin to bulbar polio.[980,994,999] In addition, it has been reported to cause myocarditis in the cardiac transplant patient.[1000] In nonimmunosuppressed patients, *Listeria* is a well-recognized cause of bacterial endocarditis, and transplant patients should be assumed to be at comparable risk (i.e., transient *Listeria* bacteremia can seed abnormal heart valves). The portal of entry for *Listeria* is the GI tract, and, indeed, it is not uncommon for transplant patients with *Listeria* sepsis to report cramps and diarrhea as the initial manifestations of their infection.[348,980,994,1001] The *Listeria* gene that mediates penetration of the gut epithelial cells has been identified (it encodes internalin, a leucine-rich protein). Whether it or another protein is responsible for the remarkable CNS tropism of this organism remains to be determined.[1002]

Traditionally, listeriosis has been classified as a zoonosis, and, indeed, contact with animals and animal manure can result in human infection. However, for most patients, listeriosis is transmitted via contaminated food, particularly milk and cheeses, undercooked chicken and other meats, and uncooked vegetables (particularly ones grown in a manure fertilizer, as in a large epidemic in Canada in 1981, traceable to cole slaw from a farm that used sheep manure as fertilizer, with listeriosis having been documented in these sheep).[1003] In recent years, outbreaks of listeriosis have been particularly associated with mass-produced and distributed hot dogs and sau-

sages. In these instances, it is apparent that contamination of the machinery used to produce these foods greatly amplifies the epidemic risk.

It is important for the clinician caring for transplant patients to be aware that the bacteriology laboratory may mistakenly identify this organism initially as a diphtheroid, as a *Bacillus* species, or even as *Streptococcus pneumoniae*, when it is first isolated from a blood or CSF culture. In the context of a transplant patient, all such initial readings should be followed by further evaluation to rule out *Listeria*, with appropriate therapy instituted while this further evaluation is being carried out.[348,349,980]

All patients with documented *Listeria* bacteremia should undergo lumbar puncture to assess the possibility of CNS seeding. Even if the lumbar puncture is negative, the assumption should be made that subclinical CNS seeding has occurred. Watson *et al.*[1004] reported relapse of listerial infection with cerebritis in patients treated previously with a 10- to 14-day course of intravenous penicillin for documented bacteremia without, initially, evidence of CNS seeding. Both Watson's group and our own have observed a similar phenomenon in patients treated for 2 weeks for meningitis. Because of this pattern of relapse, we prefer to treat for at least 3 weeks with meningeal doses of antibiotics for documented listerial infection whether or not CNS involvement is initially documented. In addition, reinfection with separate episodes of listerial sepsis has been reported in transplant patients, presumably because of the effects of immunosuppressive therapy.[1005,1006]

Optimal antimicrobial therapy remains somewhat unclear. In nonallergic patients, penicillin or ampicillin in meningeal doses is the mainstay of therapy. In the transplant patient with varying degrees of renal dysfunction, we prefer to use ampicillin, because the alternate hepatobiliary route of excretion will prevent toxic accumulation of the drug while allowing full dosages to be used, ensuring that adequate therapeutic levels are reached.[348,349,980] In the laboratory, the combination of penicillin or ampicillin and an aminoglycoside leads to synergistic killing of the organism analogous to what is observed with enterococci.[1007] Whether this is clinically important is as yet unknown. Our practice is to combine ampicillin at a dose of 1.5–2.0 g intravenously every 4 hr with gentamicin at full therapeutic doses for the level of renal function present for the first 7 days of treatment and then finish the course of therapy with ampicillin alone. In penicillin-allergic patients, alternative therapy is usually employed. Although tetracycline, erythromycin, and chloramphenicol have all been suggested as possible alternatives in the penicillin-allergic patient, use of these bacteriostatic antibiotics in the treatment of systemic listeriosis has failed in a number of instances. In the past, we have preferred to desensitize penicillin-allergic patients with *Listeria* sepsis and/or meningitis and have used this approach successfully in a total of five patients with meningitis. With the clear-cut demonstration that trimethoprim–sulfamethoxazole can be successfully used in the treatment of life-threatening *Listeria* infection, we now would regard this as the preferred regimen in penicillin-allergic patients, reserving penicillin desensitization for the patient who cannot tolerate this drug as well.[1008,1009]

Illustrative Case 7

A 47-year-old man had undergone cardiac transplantation 8 months previously for congestive cardiomyopathy. His posttransplant course had been marked by recurrent rejection requiring repetitive pulse doses of steroids, as well as courses of antithymocyte globulin and OKT3. He received preemptive ganciclovir therapy to prevent viral infection during these antirejection treatments, as well as prophylactic trimethoprim–sulfamethoxazole for 6 months posttransplant. Because of continuing rejection activity, he had received high-dose steroids 1 month before, and he was placed on a tacrolimus, mycophenolate, and prednisone regimen instead of cyclosporine, azathioprine, and prednisone. The patient lived in rural Vermont on a working farm, although he himself had not been strong enough to work for some time.

The patient entered now with a 3-day illness characterized by abdominal cramps and diarrhea, fever and chills, and increasing headache. On physical examination, his temperature was 103.2°F, pulse 110, respiratory rate 22. He was a moderately ill-appearing man without nuchal rigidity or photophobia. Both the general physical examination and neurologic examination were within normal limits. CT scan was negative, but lumbar puncture revealed an opening pressure of 240 mm H_2O, 140 leukocytes (98% polys), protein 78 mg/dl, and sugar of 34 (peripheral blood sugar 69). Blood and CSF cryptococcal antigen were negative. Both blood and CSF grew *Listeria monocytogenes*. The patient was treated for 3 weeks with intravenous antibiotics: 2 g of ampicillin IV every 4 hr plus gentamicin, 80 mg IV every 8 hr for 1 week, and then with the ampicillin alone for an additional 2 weeks. He was markedly improved within 48 hr of initiating therapy, and was afebrile by the fifth hospital day. The patient made a complete recovery.

Comment. This is a classical case of systemic listeriosis, with CNS seeding, occurring 8 months posttransplant in a cardiac transplant patient who could be characterized as a "ne'er-do-well." His clinical presentation of a GI prodrome, and then the subacute onset of a febrile headache syndrome is characteristic of this infection. The epidemiologic history is probably relevant. The patient's response to intensive antimicrobial therapy was indeed gratifying. However, in the best of all possible worlds, this infection should have been prevented. As a "chronic ne'er-do-well," the patient should have been maintained on low-dose trimethoprim–sulfamethoxazole indefinitely. It is highly likely that this approach would have prevented this serious infection (*vide supra*).

7.1.2. *Cryptococcus neoformans* Infection in the Organ Transplant Recipient

C. neoformans is the single most common cause of CNS infection in the organ transplant patient, occurring almost exclusively in the late posttransplant period. It is the classic cause of subacute to chronic meningitis, often presenting after days to weeks of waxing and waning headaches and fever. Approximately one-third of patients will also note coughs related to the primary pulmonary portal of entry of such infection. Indeed, some of these patients will have simultaneous pulmonary and meningeal infection. The pathogenesis of cryptococcal disease consists of primary pulmonary infection following inhalation of aerosolized organisms, postprimary dissemination with seeding of skin, CNS (as well as the eye), the urinary tract, and the skeletal system, and the possibility of dormancy and later reactivation, again with the possibility of systemic dissemination. At least 20% of patients will have skin lesions early in the course of systemic cryptococcal infection, with this always connoting systemic infection with a high probability of simultaneous or subsequent CNS disease. As delineated in Chapter 3, skin lesions can be the first indication of systemic opportunistic infection, and careful search and biopsy of unexplained papules, nodules, and areas of atypical cellulitis can lead to the early diagnosis of systemic cryptococcal infection.[177,348,349,934,980,994,1010–1015]

Alternatively, cryptococcal infection can present as an asymptomatic pulmonary nodule discovered on chest X-ray or as sterile pyuria, with the organism identified on urine culture. This is particularly important in males, since the prostate is a not uncommon site of metastatic infection. Surgical manipulation of these sites carries a significant risk of bloodstream invasion and CNS seeding, even if it had not yet occurred. To prevent such events, a short course of antifungal therapy (usually with fluconazole; *vide infra*) should be considered.[177,934,1011–1016]

There have been two major advances in the management of cryptococcal disease in transplant patients: the development of the cryptococcal antigen test on blood and CSF, facilitating diagnosis; and the availability of fluconazole. The cryptococcal antigen test on blood is almost invariably positive in patients with systemic spread, and is particularly helpful in the evaluation of patients with undiagnosed skin lesions. The quantitative antigen test on CSF not only is useful in diagnosing meningitis, but also is of great help when measured serially in assessing the response to therapy.[177,934,1011–1016]

Standard therapy for cryptococcal infection has long been amphotericin with or without flucytosine. However, because of nephrotoxicity related to the interaction of amphotericin and cyclosporine, we (and others) have turned to fluconazole therapy, which can be administered both orally and intravenously. Based on the experience in both transplant patients and AIDS patients with cryptococcal infection, the following statements can be made: In the acutely, desperately ill patient, amphotericin remains the treatment of choice, despite toxicity issues, as amphotericin will gain microbiologic and clinical control more quickly than fluconazole; once such control has been achieved, the patient can be switched to fluconazole to eradicate the infection. In the more subacutely ill patient, we have used fluconazole as primary therapy successfully in a number of transplant patients, even in the face of preexisting renal or liver dysfunction. Toxicity from fluconazole has included a penicillinlike drug rash and the usual antimicrobial-associated upper GI distress, with only the drug rash leading to a change in therapy. To some extent (but not nearly as much as ketoconazole), fluconazole does block the metabolism of cyclosporine, requiring close monitoring of both cyclosporine and creatinine levels. One of the most difficult questions to answer is duration of therapy. Our approach to this, and essentially all fungal disease, is to treat until there is no evidence of the fungus (e.g., cultures are negative and the cryptococcal antigen has cleared), and then treat for an extra 2–4 weeks for added safety. Since, at that point, the drug is being administered orally, the extra therapy is no great hardship for most individuals.[177,1013,1016]

Note has already been made that cryptococcosis in the transplant patient usually occurs more than 6 months posttransplant, particularly in the "chronic ne'er-do-well" population. However, in recent years, we have observed an increasing number of patients coming to transplantation severely malnourished, a history of pretransplant immunosuppression, and an altered state of consciousness. The most common examples are patients with end-stage liver disease, in whom the altered mental status is attributed to hepatic encephalopathy. Posttransplant, when their neurologic status is still impaired, it is discovered that they have cryptococcosis. Retrospective testing of serum specimens drawn prior to transplant reveals that the cryptococcosis was present pretransplant. Unfortunately, the combination of pretransplant untreated cryptococcosis combined with a liver transplant operation, anesthesia, and exogenous immunosuppression carries a terrible prognosis, even with aggressive therapy. Therefore, evaluation of individuals with altered mental status for cryptococcosis (e.g., blood and CSF cryptococcal antigen testing) prior to transplant is strongly advocated.

Illustrative Case 8

A 55-year-old man had undergone cadaveric renal transplantation 14 months previously for chronic renal failure secondary to polycystic disease. Posttransplant immunosuppression was with azathioprine and prednisone. Despite mild chronic rejection that had been treated with multiple pulse doses of methylprednisolone, he had maintained a stable serum creatinine of 2.5 mg/dl. Four weeks prior to admission, several small papular lesions on an indurated base were noted on both arms but were not further evaluated. Over the last 10 days prior to admission, he had noted a bifrontal headache, first intermittently and then constantly, with increasing severity. Over the last 2 days, nausea and vomiting as well as fever developed, and he sought medical attention.

On physical examination, the temperature was 100.6°F and the blood pressure was 150/100. Skin examination revealed four to five nontender papular lesions on the dorsal surfaces of both wrists, with a small area of subcutaneous induration at the base of each papule. Fundoscopic examination revealed papilledema. There were no meningeal signs. Neurologic examination was otherwise unremarkable. A CT scan of the brain was nondiagnostic, and a lumbar puncture was performed that revealed an opening pressure of 300 mm H_2O, CSF protein 70 mg/dl, sugar 47 mg/dl, 24 leukocytes/mm³ (80% lymphocytes), a positive India ink preparation, and a cryptococcal antigen titer of 1:128 in the CSF. A skin biopsy of one of the papular lesions revealed cryptococci (Fig. 5). Both the CSF and skin specimens grew *C. neoformans*.

The patient was first treated with amphotericin B and flucytosine, and then with amphotericin B alone because of flucytosine-induced bone marrow toxicity in association with a serum creatinine climbing to 5.2 mg/dl. After a total dose of 2.5 g amphotericin, and with a negative cryptococcal antigen titer in the CSF, therapy was discontinued. Three months later, he presented again with relapsing cryptococcal infection. This time therapy was complicated by progressive renal failure and, despite combined amphotericin and flucytosine therapy, he succumbed to his infection.

Comment. This tragic result might have been avoided if the skin lesions had been biopsied 1 month before the onset of CNS symptoms, the diagnosis made, and appropriate therapy instituted earlier. We have made the diagnosis of systemic cryptococcosis on the basis of skin biopsy in more than a dozen patients now, and have cured virtually all of them before the development of CNS infection. This case example occurred before the availability of fluconazole. If this case presented today, a prolonged course of oral fluconazole would have been utilized, either by itself or following an initial course of amphotericin (< 500 mg), to gain control. Such an approach allows one to avoid premature termination of therapy because of toxicity, and, indeed, we have salvaged both patient and allograft in a number of instances with this approach.

7.1.3. *Aspergillus* Infection in the Organ Transplant Patient

Other than cryptococcal infection, the most common fungal agent affecting the CNS in the organ transplant patient is *Aspergillus*, particularly *A. fumigatus* (*A. flavus* is the second most common *Aspergillus* species causing invasive infection, with *A. niger*, *A. terreus*, and *A. nidulans* being quite uncommon). In this setting, *Aspergillus* appears to occur exclusively as a result of metastatic infection from a pulmonary (or, uncommonly, a nasal sinus) portal of entry. The usual sequence of events begins with a primary viral or bacterial pulmonary infection in a debilitated patient who has responded poorly to therapy, and this is followed by *Aspergillus* superinfection, which metastasizes to the brain via the hematogenous route within days of its invasion of the lung. This

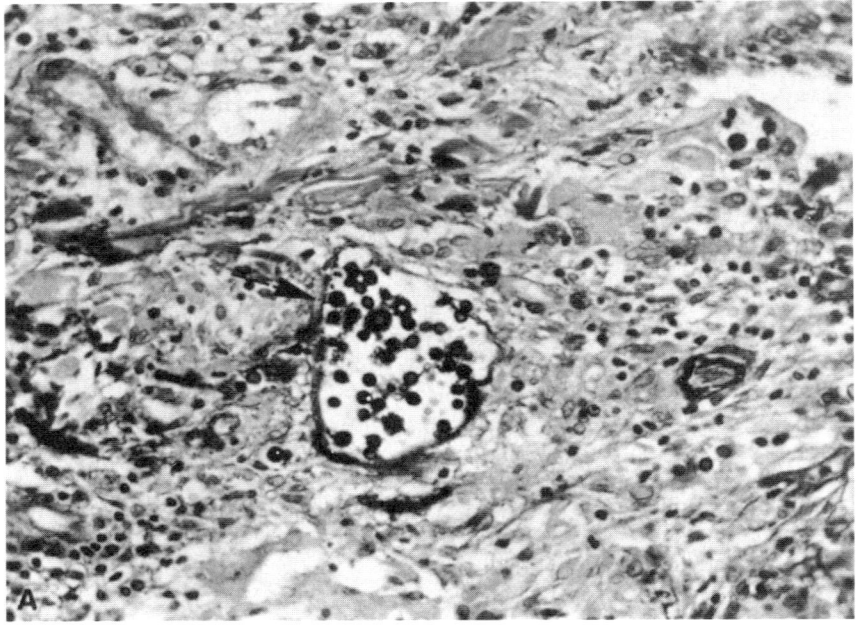

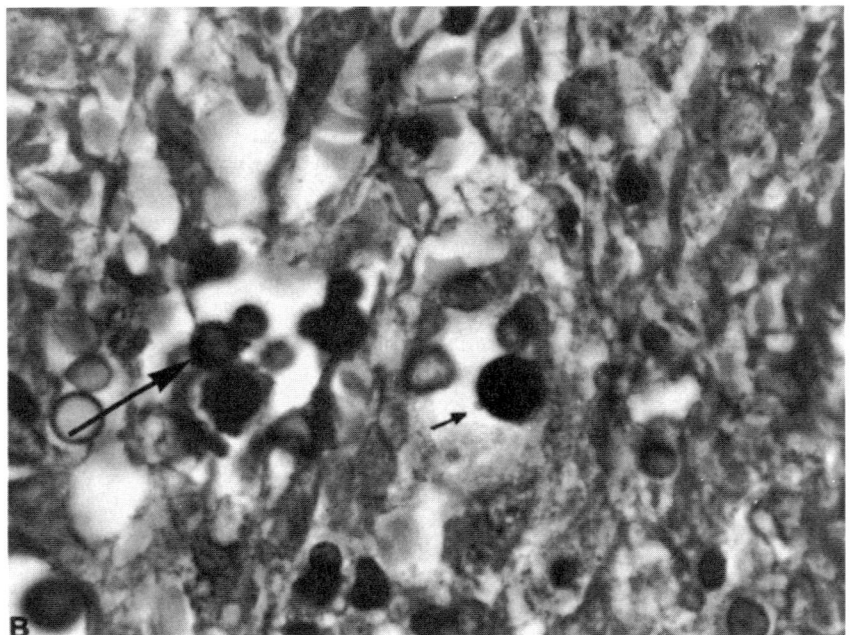

FIGURE 5. Skin biopsy of a patient with papular skin lesions caused by disseminated cryptococcal infection. (A) In the midst of the dermis were numerous cryptococci present focally as large aggregates (arrow) and singly surrounded by an infiltrate of histiocytes and lymphocytes. (Periodic acid–Schiff with diastase, ×400.) (B) Budding forms (small arrow) of *Cryptococcus neoformans*. Organisms with well-outlined capsules can be seen as well (large arrow). (Periodic acid–Schiff with diastase, ×1600.)

sequence of events is particularly common if the patient's primary pulmonary process has required prolonged intubation. Alternatively, the normal lung is invaded because of an unusually intense exposure to the organism by means of contaminated air, but again with dissemination via the bloodstream to the brain. This sequence of events is a result of this organism's propensity for invading blood vessels, which accounts for the three cardinal manifestations of *Aspergillus* infection—infarction, hemorrhage, and metastasis. The best way to deal with *Aspergillus* CNS infection is to prevent it either by preventing the pulmonary infection entirely or by recognizing it early and beginning high-dose antifungal therapy before hematogenous dissemination has occurred.[348,349,980,990,1016–1019]

The optimal therapy of invasive aspergillosis is currently in a state of flux, as data regarding a number of new agents are currently not available. Conventional amphotericin remains the gold standard of therapy, despite its toxicity. When dealing with CNS infection due to *Aspergillus* species, we, and others, commonly add a second drug such as rifampin, which has been shown to be synergistic with amphotericin *in vitro*, although translation of this into increased clinical efficacy has not been accomplished. Lipid-associated amphotericin preparations have come into increased use, and clearly are associated with less toxicity (both the acute fever, rigors, and hypotension that is mediated by cytokine release in response to amphotericin infusion; and nephrotoxicity). However, their expense is considerable, and it is not yet clear which amphotericin preparation is superior in terms of efficacy, particularly for CNS infection. Itraconazole is now available in an intravenous formulation as well as in a new oral formulation, which should make the pharmacokinetics and bioavailability more tractable, but once again efficacy data are incomplete, particularly in comparison with amphotericin preparations. The imminent availability of voriconazole and the new echinocandins will complicate decision-making further, until the database is more complete. At present, our practice is to initiate therapy with conventional amphotericin ± rifampin, switching to a lipid amphotericin formulation or voriconazole when toxicity issues become important. The echinocandins may play an important role in combination with regimens such as this.[980]

Nocardia asteroides as well as other fungal agents may follow a similar pattern (pulmonary portal of entry, with subsequent metastasis to the brain and elsewhere), although usually in less lethal fashion and with a greater chance that a single cerebral lesion may be identified and treated. As with all serious infections in the transplant patient, early diagnosis and aggressive therapy are the only hope for recovery.[348,349,980,994]

7.2. Bacteremia in the Organ Transplant Recipient

The organ transplanted has traditionally been the major source of posttransplant bacteremia. Thus, in the renal transplant patient, the urinary tract has been the portal of entry for 60–70% of bacteremias, and in the liver transplant patient, the biliary tree and infarcted liver following vascular anastomotic problems has been a major source of invasive infection. In the lung and heart patients, intravenous lines and pneumonia have been the major causes

of difficulties. In general, technical issues are the foundation of bacteremias occurring in the first 2 or 3 weeks posttransplant, as emphasized in Table 2. With better control of technical issues, and as the time posttransplant increases, listerial sepsis has been the most common cause of bacteremia in many transplant populations. However, with the widespread use of trimethoprim–sulfamethoxazole prophylaxis, this problem has become quite uncommon.[27]

At present, bacteremias in the organ transplant patient can be divided into three general categories: (1) those related to technical problems involving the allograft, intravenous lines, or, particularly in the liver transplant patient, collections of blood, lymph, bile, or urine that become secondarily infected causing soft tissue infection, peritonitis, and so on; (2) those related to breaks in the integrity of the GI tract, leading to inflammation of the gut wall and/or abscess or frank peritonitis; the most common example of this is sigmoid diverticulitis, which commonly leads to abscess formation and perforation; and (3) acute bacterial gastroenteritis, particularly that caused by *Salmonella* species.

Nontyphoidal salmonella infection, acquired, usually, by the ingestion of a contaminated foodstuff (particularly inadequately cooked chicken or other fowl), poses a particular hazard to the transplant patient. Whereas in the normal host, salmonella gastroenteritis is associated with a risk of bacteremia of less than 5%, in the transplant patient, more than 50% of individuals will have bloodstream invasion with this organism. In addition, metastatic infection as a consequence of such bloodstream invasion is the rule rather than the exception in the transplant patient. Seeding of the urinary tract, with positive urine cultures, is common. The cardiovascular tree, at sites of preexisting atherosclerotic lesions, aneurysms, fistulas, and so forth, is commonly involved. With involvement of the heart and/or vasculature, eradication of infection with antibiotics alone is often not possible. For this reason, prolonged antibiotic therapy is indicated in any transplant patient with documented salmonella bacteremia or evidence of metastatic seeding.[1020–1024]

Enterococcal infection, including vancomycin-resistant enterococci (VRE), is a particular problem in liver transplant recipients. Approximately 10–15% of liver transplant recipients will experience enterococcal bacteremia as a consequence of intravenous line-related sepsis, deep wound infection, or technical/anatomic problems with the liver transplant itself. In all types of organ transplant patients, enterococcal bacteremia occurs predominantly in the first month posttransplant, emphasizing the role of technical problems in the pathogenesis.[1025–1027]

7.3. Fungal Infections in the Organ Transplant Patient

Fungal infection in the organ transplant patient can be divided into two categories: pulmonary and/or disseminated infection with one of the geographically restricted systemic mycoses (histoplasmosis, coccidioidomycosis, or blastomycosis); and opportunistic infection with fungal species that rarely cause invasive infection in the normal host (*Candida* species, *Pneumocystis carinii*, *Aspergillus* species, *Cryptococcus neoformans*, the *Mucoraceae*, and others).

In the case of the systemic mycoses, two different patterns of disease are commonly observed: primary infection in the immunosuppressed patient with progressive disease and postprimary dissemination; and reactivation infection with secondary dissemination. In addition, in an occasional individual, reinfection, again with the potential for dissemination, may occur if immunosuppression has caused previous immunity to wane. The similarity to the pathogenesis of tuberculosis is obvious.[1028–1040] With the emphasis on pulmonary and disseminated infection with each of these entities, a variety of clinical presentations should lead to the consideration of these infections in the differential diagnosis: a subacute respiratory illness, with either focal or disseminated interstitial or miliary infiltrates on chest radiograph; a nonspecific febrile illness; or an illness in which metastatic aspects of the infection predominate (e.g., mucocutaneous manifestations in histoplasmosis and blastomycosis, or CNS manifestations in coccidioidomycosis). Amphotericin therapy remains the standard of care for these entities, although increasing experience with the azole antifungal agents, fluconazole and itraconazole, suggests that they are valuable for completing a course of treatment after clinical control has been achieved (see Chapter 6).[1028–1040]

A more common problem is the acquisition of opportunistic fungal infection posttransplant. Three patterns are observed: primary infection, usually of the lungs, occasionally of the nasal sinuses, most commonly by *P. carinii*, *C. neoformans*, or *A. fumigatus*; sequential and concurrent secondary infection, either of the lungs or via infected intravenous lines, by *Candida* species or *Aspergillus* species; and primary cutaneous infection following a break in the skin by *Aspergillus* species, *Paecelomyces*, *Penicillium*, *Pseudallescheria boydii*, and a variety of newly emerging fungal species. Two points bear reemphasis here. First, metastatic infection is the rule rather than the exception with these agents and the clinically similar *N. asteroides*. Therefore, a search for metastases should be carried out whenever a primary focus of infection is documented. Second, careful surveillance should be maintained for possible clustering of cases of opportunistic fungal infection, particularly *Aspergillus*. Such clustering suggests a major environmental hazard.[3,348,1039,1041–1044]

7.3.1. Candidiasis in the Organ Transplant Patient

Candidal species are commonly found on diseased skin, throughout the GI tract, and in the vagina. Overgrowth of these mucocutaneous surfaces occurs when the nutrient supply is increased because of elimination of the normal bacterial flora with antibiotics or because of metabolic factors (e.g., diabetes, corticosteroids, pregnancy). The most common manifestation of candidal infection in the transplant patient, then, is mucocutaneous overgrowth, which can cause one or more of the following clinical entities: oropharyngeal thrush, candidal esophagitis, vaginitis, intertrigo, and/or paronychia or onychomycosis. Topical therapy with nonabsorbable fungal agents such as clotrimazole or nystatin, when given together with antibacterial drugs in the transplant patient, will prevent the development of mucocutaneous infection and is usually effective in treating it when it occurs. If these agents do not produce a prompt therapeutic response, then fluconazole and, probably, the other azoles are quite effective.[177]

Once candidal overgrowth has occurred, the next step in the pathogenesis of invasive infection is penetration beyond the mucocutaneous barrier. In transplant patients, this often results from technical factors, e.g., contaminated intravenous lines or complicated liver transplantation with spillage of candidal organisms from the upper GI tract into devitalized tissue, hematomas, or ascites. If bloodstream invasion occurs, dissemination with the potential for visceral seeding will occur in 50% or more of transplant patients because of their immunosuppressed state (in contrast, transient candidemia in a normal host will disseminate in only 5% of instances). Therefore, every documented episode of invasive candidal infection in the transplant patient requires therapy. The clinical manifestations of disseminated candidal infection in the transplant patient are diverse, ranging from an acute septic picture to situations in which the metastatic site of infection is the predominant cause of symptoms—skin lesions, endophthalmitis, osteomyelitis, splenic abscess, meningitis, and so forth.

The therapeutic approach to invasive candidal infection in the transplant patient is similar to that for cryptococcal infection: initiating therapy with amphotericin if the patient is a therapeutic emergency, and then switching to fluconazole to complete the course of therapy while minimizing toxicity; primary therapy with fluconazole when the patient is more subacutely ill (a "diagnostic dilemma" as opposed to a "therapeutic emergency"). It should be emphasized that not all candidal species are equally susceptible to fluconazole, however. Fortunately, the most common species, *C. albicans* and *C. tropicalis*, are susceptible. In contrast, *C. krusei* and *C. glabrata* should always be considered fluconazole resistant, and amphotericin therapy is obligatory. There are currently insufficient data to evaluate the role of itraconazole, voriconazole, and the echinocandins in this situation.[27]

7.4. Mycobacterial Infections in the Organ Transplant Patient

One of the more controversial issues in the management of transplant patients is the approach to tuberculosis. The incidence of active tuberculosis in this patient population has been approximately 1%, compared with an overall rate in the United States at the same time of approximately 15 per 100,000.[1045–1053] The advent of the tuberculosis epidemic, including drug-resistant tuberculosis, in conjunction with AIDS, is in the process of increasing these figures further. One can predict that it is only a matter of time until the incidence of tuberculosis in transplant patients begins to rise as well, particularly in those urban areas where tuberculosis is having a major effect on the general community. Although the advent of drug-resistant infection will clearly complicate the process, it is important to emphasize that tuberculosis in a transplant patient can be effectively treated.

A variety of forms of tuberculosis have been observed in transplant patients, from cavitary disease of the lungs to miliary disease, from bowel disease to skeletal disease, and from skin to CNS disease. Particularly noteworthy have been instances in which the graft carried the infection from donor to recipient or when the graft was infected in the recipient as part of hematogenous dissemination of the organisms. An unusually high rate of bone and joint involvement has been noted in transplant patients with tuberculosis.[1045–1057] It is also worth emphasizing that epidemic disease can occur within a transplant program if prompt isolation of infected individuals is not carried out.[1058]

What has been surprising is not that tuberculosis has occurred in these immunocompromised patients, but rather that it has not occurred more frequently. For example, at the Massachusetts General Hospital, we have carried out organ transplantation in more than 1000 patients whose pretransplant tuberculin status was known. Of these, 127 were tuberculin positive and did not receive isoniazid prophylaxis either pre- or posttransplant. In only one of these was the tuberculosis reactivated, and this was promptly identified and treated successfully.

Why not use isoniazid prophylaxis routinely? The issue of isoniazid prophylaxis in transplant patients is a controversial one. On the one hand, the guidelines of the American Thoracic Society have traditionally recommended 1 year of isoniazid prophylaxis for individuals with positive tuberculin tests who are subjected to a prolonged course of immunosuppressive therapy. Recently, the Tuberculosis Committee of the Infectious Disease Society of America has advocated 9 months of isoniazid or 2 months of rifampin plus pyrazinamide for all individuals with positive tuberculin tests, although noting that because of toxicity issues judgment needs to be utilized in each individual.[1059] We and others have concluded that the risk of isoniazid hepatotoxicity is greater than the benefits of isoniazid prophylaxis in transplant patients with postive tuberculin tests and no other risk factors for tuberculosis. This is particularly true for liver transplant recipients.[1060–1062] The additional risk factors that would lead us to institute isoniazid prophylaxis in transplant patients include the following: recent tuberculin conversion; non-Caucasian racial background; the presence of other immunosuppressing conditions such as protein malnutrition; a history of active tuberculosis, particularly if it had been inadequately treated; and the presence of significant abnormalities on chest radiograph.

Current recommendations for the treatment of active tuberculosis include 6 months of therapy with isoniazid, rifampin, and pyrazinamide, with ethambutol added if antimicrobial resistance is deemed likely.[1059] Our own preference in transplant patients is to continue such therapy for a minimum of 6–9 months after all evidence of disease activity is no longer present. Particularly in liver transplant patients, such a regimen is often poorly tolerated. In those cases, innovative regimens such as the combination of ethambutol and ofloxacin are effective, but the duration of therapy with such less effective (but less toxic) regimens should be expanded to 2 years.[1062]

One other aspect of antituberculosis therapy in the transplant patient that bears comment is that antituberculous drugs can have an adverse effect on allograft survival through their effects on steroid metabolism. Isoniazid and rifampin cause the induction of hepatic microsomal enzymes that increase the catabolism of steroids. Similarly, rifampin, and possibly isoniazid, increase the metabolism of cyclosporine and tacrolimus by the hepatic cytochrome P450 enzyme system. Thus, patients will be underimmunosuppressed if only standard dosages are utilized. Fortunately, in the case of cyclosporine and tacrolimus, blood levels can be followed and dosages adjusted appropriately.

In addition to typical mycobacterial infection, atypical mycobacterial infection has been observed in transplant patients.[1063] This can be divided into two general categories: pulmonary, skin, skeletal, and disseminated infection related primarily to *M. kansasii*,[1048,1051,1053,1064,1065] but occasionally also to such other species as *M. chelonae*[1066] and *M. xenopi*[1067]; and skin infection alone with a variety of relatively less virulent mycobacterial species, including *M. marinum*,[1068] *M. haemophilum*,[1069] *M. thermoresistibile*,[1070] and *M. chelonae*.[1071,1072] In both categories, the atypical mycobacterial infection is often superimposed on previous tissue injury. In addition, rare instances of intestinal infection due to *M. avium/intracellulare* invasion analogous to that seen in AIDS patients have been reported in transplant recipients.[1073–1075]

Treatment of atypical mycobacterial infection is guided by antimicrobial susceptibility testing, with therapy usually initiated with a multidrug regimen that includes a macrolide (azithromycin or clarithromycin), ethambutol, and a third drug (clofazimine, rifabutin, or ciprofloxacin or another quinolone). Treatment should be for 2 years, preferably with a decrease in immunosuppression.[1075,1076]

8. Vaccine Use in Organ Transplant Recipients

Since prevention of infection is a primary goal in the management of transplant patients, an important consideration is the appropriate use of vaccines in this population of patients. There are two issues: safety and efficacy. When considering safety, two considerations come into play, namely, if a live virus vaccine is being considered, then the possibility of vaccine strain-induced disease is of primary concern; a general concern, whether the vaccine is a live virus vaccine or an inactivated or subunit vaccine, is the possibility of inducing rejection by the immunization process. In terms of efficacy, it can be generalized that the response posttransplant is attenuated, both in terms of the rate of seroconversion and in terms of the level of protective antibody produced if seroconversion indeed occurred (with the peak level of antibody obtained being a useful marker for predicting the duration of protection). Thus, from both a safety and an efficacy point of view, an important truism of transplant infectious disease is that all immunizations should be carried out before transplant (ideally, > 1 month pretransplant), and assessment of a transplant candidate's vaccine status, with correction of any deficits, is an important aspect of the pretransplant evaluation.

An important question is whether or not immunization with an indicated vaccine could induce rejection. In general, randomized controlled trials of such vaccines as influenza, hepatitis B, and tetanus toxoid have not revealed any statistically significant increased incidence of allograft rejection.[1077–1080] However, occasional reports have suggested that vaccine administration could have such an effect.[1079,1081] We have observed two renal transplant recipients who were > 12 months posttransplant, without previous history of allograft rejection, who proceeded to develop severe rejection (culminating in allograft loss) following tetanus toxoid administration. That these severe rejection episodes occurred in individuals with normal renal function (serum creatinine levels ~0.8 mg/dl) at a time when they should have been free from such a risk raises the question as to whether or not in an occasional patient vaccine administration could precipitate a rejection crisis. If so, then, what would be the mechanism and a rational way for the clinician to proceed?

The following hypothesis would be consistent with what has previously been presented in this chapter; that is, that cytokines, chemokines, and growth factors elaborated in response to a variety of stimuli could impact on the processes leading to allograft injury (*vide supra*). In this situation, it is well known that certain individuals, in response to the administration of particular vaccines, develop a flulike syndrome characterized by fever, inflammation at the administration site, and malaise that is mediated by the local and systemic release of proinflammatory cytokines in response to the vaccine. It is possible that individuals with particularly severe vaccine reactions would be the ones likely to develop allograft dysfunction due to the cytokine milieu that has resulted in response to the vaccine. This chain of events is analogous to what has been postulated to result from such infections as CMV, HHV-6, and, perhaps, others (*vide supra*). With these considerations in mind, appropriate vaccine use would include the following: use of vaccines when needed (*vide infra*), but careful screening for a history of previous vaccine reactions; all transplant patients receiving vaccines should be monitored postvaccination for allograft function, particularly those who have evidence of an inflammatory response to the vaccine.

As far as specific recommendations regarding the different vaccines, the reader is directed to the excellent review of Burroughs and Moscone.[1077] The following comments, however, bear particular emphasis:

1. Live virus vaccines have, in general, been shunned posttransplant. This is particularly important with the oral polio vaccine, where the administration of this vaccine to a transplant recipient or a family member (with subsequent person-to-person spread within the family) can result in vaccine-induced progressive meningoencephalitis. Accordingly, when polio vaccine is indicated for either a transplant patient or a family member, the inactivated polio vaccine is strongly advocated.[1077,1082–1084]

2. As far as other live virus vaccines are concerned, there is less of a consensus. In the case of the varicella vaccine, it has been safely used in other immunosuppressed patient populations, particularly seronegative leukemics and AIDS patients, although seroconversion rates have been lower than in the normal population. In one study of 17 pediatric renal transplant recipients, 65% developed protective antibody titers, with one child developing mild varicella postimmunization. As far as efficacy is concerned, 3 of these patients developed attenuated varicella 2–4 years later, suggesting some degree of protection.[1085] Our own experience has suggested a lower rate of seroconversion, often requiring multiple doses of the vaccine to achieve this, an ~10% incidence of mild varicella postimmunization, easily managed with acyclovir (*vide supra*).

3. Measles and mumps are clearly of increased severity in immunocompromised individuals, so immunization against these is clearly to be desired (there is less evidence of this for rubella, but most groups would use the combined measles–mumps–rubella vaccine when vaccinating against the other two viruses). Admittedly scanty data, particularly in other immunocompromised patients, suggest safety and possible efficacy.[1077] If possible, immunization should be followed up by serologic testing to demonstrate protection. This is particularly important for measles, where pediatric transplant recipients who are unvaccinated or whose immunity is not documented, should receive intramuscular immunoglobulin within 6 days of exposure to measles.[1077]

4. The use of inactivated or subunit vaccines needs to be individualized for the needs of the particular patient—use when clearly indicated, with close follow-up following vaccination. Pharmacoeconomic issues enter into consideration here as well. Thus, we routinely employ yearly influenza vaccination for heart and lung transplant recipients (after administering influenza and pneumococcal vaccines pretransplant), but regard it as more optional for kidney and liver transplant recipients, because of concerns about efficacy. More information in this area is badly needed.

9. Summary and Prospects for the Future

The advances in transplantation that have occurred over the past decade can be appreciated by comparing the discussion on the infectious complications in the four editions of this book. In the first edition, published in 1981, only bone marrow and kidney transplantation were discussed. In the second edition, published in 1988, liver and heart transplantation were extensively explored as well. In the third edition, published in 1994, issues peculiar to lung, heart–lung, and pancreatic transplantation were added to our consideration. Small bowel transplantation is now on the radar screen as well—the technical problems have been largely overcome, the immunosuppressive issues are being settled, and the major problems remaining to be overcome are infectious, particularly the translocation of aerobic gut flora (especially gram-negative) and candidal species across a mucosa whose barrier function is at least temporarily impaired; and the high incidence of EBV-related lymphoproliferative disease in these patients. The lessons learned with the other allografts should aid us here as well, and as we look to expand the horizons of transplantation by protecting patients from life-threatening infection, the following principles should be followed:

1. The infections that occur in transplant patients may be divided into three general categories: those related to technical complications, those related to epidemiologic exposures, and those related to viruses lurking in the graft recipient or his donor and that are rendered clinically manifest posttransplant. The modulation of these infections is accomplished by the dose, duration, and type of immunosuppressive therapy being administered.

2. The risk of infection in the transplant patient is determined by the interaction among three factors: the presence of technical/anatomical abnormalities, the epidemiologic exposures the patient encounters, and the net state of immunosuppression.

3. There is an expected timetable according to which particular infections occur at particular times in the posttransplant course. Exceptions to this timetable are usually the result of exposure to excessive environmental hazards. As one approaches new forms of transplantation or radically different immunosuppressive regimens, the obligation of the transplant infectious disease clinician is to define the timetable that fits this new situation.

4. The biggest challenge in approaching the infectious disease problems of the transplant patient is the

prevention and treatment of those viral diseases that contribute so broadly to the morbidity and mortality still associated with clinical human transplantation. The viruses of greatest importance are the herpes group viruses and the hepatitis viruses, with each of them modulated by immunosuppressive therapy. The advent of preemptive and prophylactic regimens for at least some of these infections holds promise that we can gain control and bring the promise of transplantation to reality for more patients.

5. Because of the impaired inflammatory response of this patient population, signs and symptoms of infection may be greatly muted. Physicians caring for such patients must be alert and aggressive in their approach to "minor" skin lesions or radiographic findings.

6. The therapeutic prescription for the transplant patient consists of an immunosuppressive regimen to prevent and treat rejection, and an antimicrobial strategy to make it safe. Just as we have learned to individualize the immunosuppressive program to fit the needs of the particular patient, we must be prepared to individualize the antimicrobial strategy in a similar fashion.

7. When considering the effects of a given infectious process on the transplant patient, both direct and indirect manifestations must be considered. The indirect effects are mediated by cytokines, chemokines, and growth factors elaborated by the patient in response to microbial invasion and replication. Among the indirect effects that have been defined are allograft injury, modulation of other infectious processes, and oncogenesis. In the next edition of this book, it is likely that we will be defining antimicrobial regimens in terms of their effects on the indirect as well as direct processes.

As we follow these precepts, it is important not to lose sight of the most important guideline of all: Although the challenge of caring for these patients is great, the rewards are even greater. We have come a long way in the care of these patients. Continued progress is on the horizon.

References

1. Hariharan S, Johnson CP, Bresnahan BA, *et al*: Improved graft survival after renal transplantation in the United States, 1988 to 1996 [see comments]. *N Engl J Med* **342**:605–612, 2000.

2. Sharing UNfO: Annual Report, the U.S. Scientific Registry of Transplant Recipients and the Organ Procurement and Transplantation Network 1999, pp. 7–8.

3. Rubin RH, Wolfson JS, Cosimi AB, Tolkoff-Rubin NE: Infection in the renal transplant recipient. *Am J Med* **70**:405–411, 1981.

4. Takemoto SK, Terasaki PI, Gjertson DW, Cecka JM: Twelve years' experience with national sharing of HLA-matched cadaveric kidneys for transplantation. *N Engl J Med* **343**:1078–1084, 2000.

5. Rubin R, Ikonen T, Gummert J, Morris R: The therapeutic prescription for the organ transplant recipient: The linkage of immunosuppression and antimicrobial strategies. *Transplant Infect Dis* **1**:29–39, 1999.

6. Jamil B, Nicholls K, Becker GJ, Walker RG: Impact of acute rejection therapy on infections and malignancies in renal transplant recipients. *Transplantation* **68**:1597–1603, 1999.

7. Miner JN, Brown M: Glucocorticoid action. In Austen K, Burakoff S, Rosen F, Strom T (eds): *Therapeutic Immunology*, ed. 2. Cambridge, Blackwell Science, 2001, pp. 103–116.

8. Fishman JA, Rubin RH: Infection in organ-transplant recipients [see comments]. *N Engl J Med* **338**:1741–1751, 1998.

9. Tolkoff-Rubin N, Rubin R: The purine antagonists: Azathioprine and mycophenolate mofetil. In Austen K, Burakoff S, Rosen F, Strom T (eds): *Therapeutic Immunology*, ed. 2. Cambridge, Blackwell Science, 2001, pp. 51–64.

10. Opelz G, Dohler B: Critical threshold of azathioprine dosage for maintenance immunosuppression in kidney graft recipients. Collaborative Transplant Study. *Transplantation* **69**:818–821, 2000.

11. Stolk JN, Boerbooms AM, De Abreu RA, van de Putte LB: Azathioprine treatment and thiopurine metabolism in rheumatic diseases. Introduction and first results of investigation. *Adv Exp Med Biol* **431**:487–493, 1998.

12. Sebbag L, Boucher P, Davelu P, *et al*: Thiopurine S-methyltransferase gene polymorphism is predictive of azathioprine-induced myelosuppression in heart transplant recipients. *Transplantation* **69**:1524–1527, 2000.

13. Sollinger HW: Mycophenolate mofetil for the prevention of acute rejection in primary cadaveric renal allograft recipients. U.S. Renal Transplant Mycophenolate Mofetil Study Group. *Transplantation* **60**:225–232, 1995.

14. Placebo-controlled study of mycophenolate mofetil combined with cyclosporin and corticosteroids for prevention of acute rejection. European Mycophenolate Mofetil Cooperative Study Group [see comments]. *Lancet* **345**:1321–1325, 1995.

15. Kahan BD: Cyclosporine [see comments]. *N Engl J Med* **321**:1725–1738, 1989.

16. Schreiber SL: Chemistry and biology of the immunophilins and their immunosuppressive ligands. *Science* **251**:283–287, 1991.

17. Shapiro R, Jordan M, Scantlebury V, *et al*: FK 506 in clinical kidney transplantation. *Transplant Proc* **23**:3065–3067, 1991.

18. Cao W, Mohacsi P, Shorthouse R, Pratt R, Morris RE: Effects of rapamycin on growth factor-stimulated vascular smooth muscle cell DNA synthesis. Inhibition of basic fibroblast growth factor and platelet-derived growth factor action and antagonism of rapamycin by FK506. *Transplantation* **59**:390–395, 1995.

19. Gregory CR, Huie P, Billingham ME, Morris RE: Rapamycin inhibits arterial intimal thickening caused by both alloimmune and mechanical injury. Its effect on cellular, growth factor, and cytokine response in injured vessels. *Transplantation* **55**:1409–1418, 1993.

20. Poon M, Marx SO, Gallo R, *et al*: Rapamycin inhibits vascular smooth muscle cell migration. *J Clin Invest* **98**:2277–2283, 1996.

21. Dominguez J, Mahalati K, Kiberd B, McAlister VC, MacDonald AS: Conversion to rapamycin immunosuppression in renal transplant recipients: Report of an initial experience [In Process Citation]. *Transplantation* **70**:1244–1247, 2000.

22. Cosimi A: Antilymphocyte globulin—A final look. In Morris P,

Tilney N (eds): *Progress in Transplantation*. Edinburgh, Churchill Livingstone, 1985, vol. 2, pp. 167–188.

23. Ortho Multicenter Transplant Group. A randomized clinical trial of OKT3 monoclonal antibody for acute rejection of cadaveric renal transplants. *N Engl J Med* **313:**337–342, 1985.

24. Hibberd PL, Rubin RH: Clinical aspects of fungal infection in organ transplant recipients. *Clin Infect Dis* **19**(Suppl 1)**:**S33–S40, 1994.

25. Soulillou JP, Cantarovich D, Le Mauff B, *et al*: Randomized controlled trial of a monoclonal antibody against the interleukin-2 receptor (33B3.1) as compared with rabbit antithymocyte globulin for prophylaxis against rejection of renal allografts [see comments]. *N Engl J Med* **322:**1175–1182, 1990.

26. Rubin RH, Tolkoff-Rubin NE: Antimicrobial strategies in the care of organ transplant recipients. *Antimicrob Agents Chemother* **37:**619–624, 1993.

27. Rubin RH: Infectious disease complications of renal transplantation [clinical conference]. *Kidney Int* **44:**221–236, 1993.

28. Sable CA, Hayden FG: Orthomyxoviral and paramyxoviral infections in transplant patients. *Infect Dis Clin North Am* **9:**987–1003, 1995.

29. Rabella N, Rodriguez P, Labeaga R, *et al*: Conventional respiratory viruses recovered from immunocompromised patients: Clinical considerations. *Clin Infect Dis* **28:**1043–1048, 1999.

30. Krinzman S, Basgoz N, Kradin R, *et al*: Respiratory syncytial virus-associated infections in adult recipients of solid organ transplants. *J Heart Lung Transplant* **17:**202–210, 1998.

31. McGrath D, Falagas ME, Freeman R, *et al*: Adenovirus infection in adult orthotopic liver transplant recipients: Incidence and clinical significance. *J Infect Dis* **177:**459–462, 1998.

32. Bridges ND, Spray TL, Collins MH, Bowles NE, Towbin JA: Adenovirus infection in the lung results in graft failure after lung transplantation. *J Thorac Cardiovasc Surg* **116:**617–623, 1998.

33. Simsir A, Greenebaum E, Nuovo G, Schulman LL: Late fatal adenovirus pneumonitis in a lung transplant recipient. *Transplantation* **65:**592–594, 1998.

34. Ghosh S, Champlin R, Couch R, *et al*: Rhinovirus infections in myelosuppressed adult blood and marrow transplant recipients [see comments]. *Clin Infect Dis* **29:**528–532, 1999.

35. Vilchez RA, Fung JJ, Kusne S: Influenza A myocarditis developing in an adult liver transplant recipient despite vaccination: A case report and review of the literature. *Transplantation* **70:**543–545, 2000.

36. Wittwer T, Pethig K, Heublein B, *et al*: Impact of chronic infection with chlamydia pneumoniae on incidence of cardiac allograft vasculopathy. *Transplantation* **69:**1962–1964, 2000.

37. Hopkins CC, Weber DJ, Rubin RH: Invasive aspergillus infection: Possible non-ward common source within the hospital environment. *J Hosp Infect* **13:**19–25, 1989.

38. Rubin RH: The compromised host as sentinel chicken [editorial]. *N Engl J Med* **317:**1151–1153, 1987.

39. Allo MD, Miller J, Townsend T, Tan C: Primary cutaneous aspergillosis associated with Hickman intravenous catheters. *N Engl J Med* **317:**1105–1108, 1987.

40. Kacmarek R, Kratohuil J, Dashevsky Y, *et al*: Performance of prototype portable HEPA-filtered positive pressure enclosures. *Respir Care* **37:**1368, 1992.

41. Turgeon N, Hovingh G, Fishman J, *et al*: Safety and efficacy of granulocyte-colony stimulating factor in kidney and liver transplant recipients. *Transplant Infect Dis* **2:**15–21, 2000.

42. Winston DJ, Foster PF, Somberg KA, *et al*: Randomized, placebo-controlled, double-blind, multicenter trial of efficacy and safety of granulocyte colony-stimulating factor in liver transplant recipients. *Transplantation* **68:**1298–1304, 1999.

43. Corales R, Chua J, Mawhortzer J, *et al*: Significant post-transplant hypogammaglobulinemia in six heart transplant recipients: An emerging clinical phenomenon. *Transplant Infect Dis* **2:**133–139, 2000.

44. Darouiche RO: Anti-infective efficacy of silver-coated medical prostheses. *Clin Infect Dis* **29:**1371–1377; quiz 1378, 1999.

45. Maki DG, Stolz SM, Wheeler S, Mermel LA: Prevention of central venous catheter-related bloodstream infection by use of an antiseptic-impregnated catheter. A randomized, controlled trial [see comments]. *Ann Intern Med* **127:**257–266, 1997.

46. Iseki K, Kawazoe N, Fukiyama K: Serum albumin is a strong predictor of death in chronic dialysis patients. *Kidney Int* **44:**115–119, 1993.

47. Meier-Kriesche HU, Friedman G, Jacobs M, *et al*: Infectious complications in geriatric renal transplant patients: Comparison of two immunosuppressive protocols. *Transplantation* **68:**1496–1502, 1999.

48. Meier-Kriesche HU, Ojo A, Magee JC, *et al*: African-American renal transplant recipients experience decreased risk of death due to infection: Possible implications for immunosuppressive strategies. *Transplantation* **70:**375–379, 2000.

49. Freeman RB Jr, Tran CL, Mattoli J, *et al*: Tumor necrosis factor genetic polymorphisms correlate with infections after liver transplantation. NEMC TNF Study Group. New England Medical Center Tumor Necrosis Factor [published erratum appears in *Transplantation* **68**(11)**:**1823, 1999]. *Transplantation* **67:**1005–1010, 1999.

50. Sahoo S, Kang S, Supran S, *et al*: Tumor necrosis factor genetic polymorphisms correlate with infections after renal transplantation. *Transplantation* **69:**880–884, 2000.

51. Chang FY, Singh N, Gayowski T, *et al*: Thrombocytopenia in liver transplant recipients: Predictors, impact on fungal infections, and role of endogenous thrombopoietin. *Transplantation* **69:**70–75, 2000.

52. Rubin R: Infectious disease problems. In Maddrey W (ed): *Current Topics in Gastroenterology: Transplantation of the Liver*. New York, Elsevier Science, 1988.

53. Johnson R, Rubin R: Respiratory disease in kidney and liver transplant recipients. In Shelhamer J, Pizzo P, Parrillo J, Masure H (eds): *Respiratory Disease in the Immunosuppressed Host*. Philadelphia, Lippincott, 1991, pp. 567–594.

54. Ettinger NA, Trulock EP: Pulmonary considerations of organ transplantation. Part I. *Am Rev Respir Dis* **143:**1386–1405, 1991.

55. Argenziano M, Catanese KA, Moazami N, *et al*: The influence of infection on survival and successful transplantation in patients with left ventricular assist devices. *J Heart Lung Transplant* **16:**822–831, 1997.

56. LiPuma J: Burkholderia cepacia complex: A contraindication to lung transplantation in cystic fibrosis. *Transplant Infect Dis* **3:**149–160, 2001.

57. Ramirez JC, Patterson GA, Winton TL, *et al*: Bilateral lung transplantation for cystic fibrosis. The Toronto Lung Transplant Group. *J Thorac Cardiovasc Surg* **103:**287–293; discussion 294, 1992.

58. Trulock EP: Lung transplantation. *Am J Respir Crit Care Med* **155:**789–818, 1997.

59. Walter S, Gudowius P, Bosshammer J, *et al*: Epidemiology of chronic Pseudomonas aeruginosa infections in the airways of

lung transplant recipients with cystic fibrosis. *Thorax* **52:**318–321, 1997.

60. Kanj SS, Tapson V, Davis RD, Madden J, Browning I: Infections in patients with cystic fibrosis following lung transplantation [see comments]. *Chest* **112:**924–930, 1997.

61. Nunley DR, Grgurich W, Iacono AT, *et al*: Allograft colonization and infections with pseudomonas in cystic fibrosis lung transplant recipients. *Chest* **113:**1235–1243, 1998.

62. Venuta F, Boehler A, Rendina EA, *et al*: Complications in the native lung after single lung transplantation. *Eur J Cardiothorac Surg* **16:**54–58, 1999.

63. Barker R, Mayes J, Schulak J: Wound abscesses following retroperitoneal pancreas transplantation. *Clin Transplant* **5:**403–407, 1991.

64. Gupta D, Sakorafas GH, McGregor CG, Harmsen WS, Farnell MB: Management of biliary tract disease in heart and lung transplant patients [In Process Citation]. *Surgery* **128:**641–649, 2000.

65. Papanicolaou GA, Meyers BR, Meyers J, *et al*: Nosocomial infections with vancomycin-resistant Enterococcus faecium in liver transplant recipients: Risk factors for acquisition and mortality. *Clin Infect Dis* **23:**760–766, 1996.

66. Newell KA, Millis JM, Arnow PM, *et al*: Incidence and outcome of infection by vancomycin-resistant Enterococcus following orthotopic liver transplantation. *Transplantation* **65:**439–442, 1998.

67. Singh N, Paterson DL, Chang FY, *et al*: Methicillin-resistant Staphylococcus aureus: The other emerging resistant grampositive coccus among liver transplant recipients. *Clin Infect Dis* **30:**322–327, 2000.

68. Winston DJ, Emmanouilides C, Kroeber A, *et al*: Quinupristin/dalfopristin therapy for infections due to vancomycin-resistant *Enterococcus faecium*. *Clin Infect Dis* **30:**790–797, 2000.

69. McNeil SA, Clark NM, Chandrasekar PH, Kauffman CA: Successful treatment of vancomycin-resistant *Enterococcus faecium* bacteremia with linezolid after failure of treatment with synercid (quinupristin/dalfopristin). *Clin Infect Dis* **30:**403–404, 2000.

70. Scowden EB, Schaffner W, Stone WJ: Overwhelming strongyloidiasis: An unappreciated opportunistic infection. *Medicine* **57:**527–544, 1978.

71. Morgan JS, Schaffner W, Stone WJ: Opportunistic strongyloidiasis in renal transplant recipients. *Transplantation* **42:**518–524, 1986.

72. Purtilo DT, Meyers WM, Connor DH: Fatal strongyloidiasis in immunosuppressed patients. *Am J Med* **56:**488–493, 1974.

73. DeVault GA Jr, King JW, Rohr MS, *et al*: Opportunistic infections with *Strongyloides stercoralis* in renal transplantation. *Rev Infect Dis* **12:**653–671, 1990.

74. Scoggin CH, Call NB: Acute respiratory failure due to disseminated strongyloidiasis in a renal transplant recipient. *Ann Intern Med* **87:**456–458, 1977.

75. Avagnina MA, Elsner B, Iotti RM, Re R: *Strongyloides stercoralis* in Papanicolaou-stained smears of ascitic fluid. *Acta Cytol* **24:**36–39, 1980.

76. Venizelos PC, Lopata M, Bardawil WA, Sharp JT: Respiratory failure due to *Strongyloides stercoralis* in a patient with a renal transplant. *Chest* **78:**104–106, 1980.

77. White JV, Garvey G, Hardy MA: Fatal strongyloidiasis after renal transplantation: A complication of immunosuppression. *Am Surg* **48:**39–41, 1982.

78. Weller IV, Copland P, Gabriel R: *Strongyloides stercoralis* infection in renal transplant recipients. *Br Med J (Clin Res Ed)* **282:**524, 1981.

79. Fowler CG, Lindsay I, Levin J, *et al*: Recurrent hyperinfestation with *Strongyloides stercoralis* in a renal allograft recipient. *Br Med J (Clin Res Ed)* **285:**1394, 1982.

80. DeVault GA Jr, Brown STD, Montoya SF Jr, *et al*: Disseminated strongyloidiasis complicating acute renal allograft rejection. Prolonged thiabendazole administration and successful retransplantation. *Transplantation* **34:**220–221, 1982.

81. Hirschmann JV, Plorde JJ, Ochi RF: Fever and pulmonary infiltrates in a patient with a renal transplant. *West J Med* **140:**914–920, 1984.

82. Schad GA: Cyclosporine may eliminate the threat of overwhelming strongyloidiasis in immunosuppressed patients [letter]. *J Infect Dis* **153:**178, 1986.

83. Schumaker JD, Band JD, Lensmeyer GL, Craig WA: Thiabendazole treatment of severe strongyloidiasis in a hemodialyzed patient. *Ann Intern Med* **89:**644–645, 1978.

84. Leapman SB, Rosenberg JB, Filo RS, Smith EJ: *Strongyloides stercoralis* in chronic renal failure: Safe therapy with thiabendazole. *South Med J* **73:**1400–1402, 1980.

85. Jones C: Clinical studies in human strongyloidiasis. I. Semeiology. *Gastroenterology* **16:**743–746, 1950.

86. Katkov W, Rubin R: Liver disease in the organ transplant recipient: Etiology, clinical impact, and clinical management. *Transplant Rev* **5:**200–208, 1991.

87. Chan PC, Lok AS, Cheng IK, Chan MK: The impact of donor and recipient hepatitis B surface antigen status on liver disease and survival in renal transplant recipients. *Transplantation* **53:**128–131, 1992.

88. Liang T, Baruch Y, Ben-Porath E, *et al*: HBV infection in patients with idiopathic liver disease. *Hepatology* **13:**1044–1048, 1991.

89. Delmonico F: Cadaver donor screening for infectious agents in solid organ transplantation. *Clin Infect Dis* **31:**781–786, 2000.

90. Madayag RM, Johnson LB, Bartlett ST, *et al*: Use of renal allografts from donors positive for hepatitis B core antibody confers minimal risk for subsequent development of clinical hepatitis B virus disease. *Transplantation* **64:**1781–1786, 1997.

91. Dickson RC, Everhart JE, Lake JR, *et al*: Transmission of hepatitis B by transplantation of livers from donors positive for antibody to hepatitis B core antigen. The National Institute of Diabetes and Digestive and Kidney Diseases Liver Transplantation Database. *Gastroenterology* **113:**1668–1674, 1997.

92. Uemoto S, Sugiyama K, Marusawa H, *et al*: Transmission of hepatitis B virus from hepatitis B core antibody-positive donors in living related liver transplants. *Transplantation* **65:**494–499, 1998.

93. Van Thiel DH, De Maria N, Colantoni A, Friedlander L: Can hepatitis B core antibody positive livers be used safely for transplantation: Hepatitis B virus detection in the liver of individuals who are hepatitis B core antibody positive. *Transplantation* **68:**519–522, 1999.

94. Dodson SF, Issa S, Araya V, *et al*: Infectivity of hepatic allografts with antibodies to hepatitis B virus. *Transplantation* **64:**1582–1584, 1997.

95. Douglas DD, Rakela J, Wright TL, Krom RA, Wiesner RH: The clinical course of transplantation-associated *de novo* hepatitis B infection in the liver transplant recipient [see comments]. *Liver Transplant Surg* **3:**105–111, 1997.

96. Dodson SF, Bonham CA, Geller DA, *et al*: Prevention of *de novo* hepatitis B infection in recipients of hepatic allografts from anti-HBc positive donors. *Transplantation* **68:**1058–1061, 1999.

97. Delmonico FL, Snydman DR: Organ donor screening for infec-

tious diseases: Review of practice and implications for transplantation. *Transplantation* **65:**603–610, 1998.

98. Pereira BJ, Milford EL, Kirkman RL, Levey AS: Transmission of hepatitis C virus by organ transplantation [see comments]. *N Engl J Med* **325:**454–460, 1991.

99. Pereira BJ, Milford EL, Kirkman RL, *et al*: Prevalence of hepatitis C virus RNA in organ donors positive for hepatitis C antibody and in the recipients of their organs [see comments]. *N Engl J Med* **327:**910–915, 1992.

100. Pereira BJ, Levey AS: Hepatitis C virus infection in dialysis and renal transplantation. *Kidney Int* **51:**981–999, 1997.

101. Fishman JA, Rubin RH, Koziel MJ, Periera BJ: Hepatitis C virus and organ transplantation. *Transplantation* **62:**147–154, 1996.

102. Miller J, Roth D, Schiff E: Hepatitis C and organ transplantation. *N Engl J Med* **328:**312–317, 1998.

103. Widell A, Mansson S, Persson NH, *et al*: Hepatitis C superinfection in hepatitis C virus (HCV)-infected patients transplanted with an HCV-infected kidney. *Transplantation* **60:**642–647, 1995.

104. Araya V, Rakela J, Wright T: Hepatitis C after orthotopic liver transplantation. *Gastroenterology* **112:**575–582, 1997.

105. Kliem V, van den Hoff U, Brunkhorst R, *et al*: The long-term course of hepatitis C after kidney transplantation. *Transplantation* **62:**1417–1421, 1996.

106. Bouthot BA, Murthy BV, Schmid CH, Levey AS, Pereira BJ: Long-term follow-up of hepatitis C virus infection among organ transplant recipients: Implications for policies on organ procurement. *Transplantation* **63:**849–853, 1997.

107. Ong JP, Barnes DS, Younossi ZM, *et al*: Outcome of *de novo* hepatitis C virus infection in heart transplant recipients. *Hepatology* **30:**1293–1298, 1999.

108. Everhart JE, Wei Y, Eng H, *et al*: Recurrent and new hepatitis C virus infection after liver transplantation [published erratum appears in *Hepatology* **30**(4)**:**1110, 1999]. *Hepatology* **29:**1220–1226, 1999.

109. Vargas HE, Laskus T, Wang LF, *et al*: Outcome of liver transplantation in hepatitis C virus-infected patients who received hepatitis C virus-infected grafts [see comments]. *Gastroenterology* **117:**149–153, 1999.

110. Feray C, Gigou M, Samuel D, *et al*: Incidence of hepatitis C in patients receiving different preparations of hepatitis B immunoglobulins after liver transplantation. *Ann Intern Med* **128:**810–816, 1998.

111. Pessoa MG, Terrault NA, Ferrell LD, *et al*: Hepatitis after liver transplantation: The role of the known and unknown viruses. *Liver Transplant Surg* **4:**461–468, 1998.

112. Simonds RJ, Holmberg SD, Hurwitz RL, *et al*: Transmission of human immunodeficiency virus type 1 from a seronegative organ and tissue donor [see comments]. *N Engl J Med* **326:**726–732, 1992.

113. Rubin RH, Tolkoff-Rubin NE: The problem of human immunodeficiency virus (HIV) infection and transplantation. *Transplant Int* **1:**36–42, 1988.

114. Limaye AP, Connolly PA, Sagar M, *et al*: Transmission of Histoplasma capsulatum by organ transplantation. *N Engl J Med* **343:**1163–1166, 2000.

115. Ooi BS, Chen BT, Lim CH, Khoo OT, Chan DT: Survival of a patient transplanted with a kidney infected with *Cryptococcus neoformans*. *Transplantation* **11:**428–429, 1971.

116. Gottesdiener KM: Transplanted infections: Donor-to-host transmission with the allograft. *Ann Intern Med* **110:**1001–1016, 1989.

117. Rubin R, Fishman J: A consideration of potential donors with active infection—Is this a way to expand the donor pool? *Transplant Int* **11:**333–335, 1998.

118. Johnston L, Chui L, Chang N, *et al*: Cross-Canada spread of methicillin-resistant *Staphylococcus aureus* via transplant organs. *Clin Infect Dis* **29:**819–823, 1999.

119. Nelson PW, Delmonico FL, Tolkoff-Rubin NE, *et al*: Unsuspected donor pseudomonas infection causing arterial disruption after renal transplantation. *Transplantation* **37:**313–314, 1984.

120. McCoy GC, Loening S, Braun WE, *et al*: The fate of cadaver renal allografts contaminated before transplantation. *Transplantation* **20:**467–472, 1975.

121. Doig RL, Boyd PJ, Eykyn S: *Staphylococcus aureus* transmitted in transplanted kidneys. *Lancet* **2:**243–245, 1975.

122. McLeish KR, McMurray SD, Smith EJ, Filo RS: The transmission of Candida albicans by cadaveric allografts. *J Urol* **118:**513–516, 1977.

123. Caballero F, Lopez-Navidad A, Domingo P, *et al*: Successful transplantation of organs retrieved from a donor with enterococcal endocarditis. *Transplant Int* **11:**387–389, 1998.

124. Anderson CB, Haid SD, Hruska KA, Etheredge EA: Significance of microbial contamination of stored cadaver kidneys. *Arch Surg* **113:**269–271, 1978.

125. Hayry P, Renkonen OV: Frequency and fate of human renal allografts contaminated prior to transplantation. *Surgery* **85:**404–407, 1979.

126. Majeski JA, Alexander JW, First MR, *et al*: Transplantation of microbially contaminated cadaver kidneys. *Arch Surg* **117:**221–224, 1982.

127. Spees EK, Light JA, Oakes DD, Reinmuth B: Experiences with cadaver renal allograft contamination before transplantation. *Br J Surg* **69:**482–485, 1982.

128. Owens ML, Wilson SE, Maxwell JG, *et al*: Major arterial hemorrhage after renal transplantation. *Transplantation* **27:**285–287, 1979.

129. Fernando ON, Higgins AF, Moorhead JF: Letter: Secondary haemorrhage after renal transplantation. *Lancet* **2:**368, 1976.

130. Weber TR, Freier DT, Turcotte JG: Transplantation of infected kidneys: Clinical and experimental results. *Transplantation* **27:**63–65, 1979.

131. Freeman RB, Giatras I, Falagas ME, *et al*: Outcome of transplantation of organs procured from bacteremic donors. *Transplantation* **68:**1107–1111, 1999.

132. van der Vliet JA, Tidow G, Kootstra G, *et al*: Transplantation of contaminated organs. *Br J Surg* **67:**596–598, 1980.

133. Slapak M: The immediate care of potential donors for cadaveric organ transplantation. *Anaesthesia* **33:**700–709, 1978.

134. Burgos-Calderon R, Pankey GA, Figueroa JE: Infection in kidney transplantation. *Surgery* **70:**334–340, 1971.

135. Moore TC, Hume DM: The period and nature of hazard in clinical renal transplantation. 3. The hazard to transplant kidney survival. *Ann Surg* **170:**25–29, 1969.

136. Schweizer RT, Kountz SL, Belzer FO: Wound complications in recipients of renal transplants. *Ann Surg* **177:**58–62, 1973.

137. Diethelm AG: Surgical management of complications of steroid therapy. *Ann Surg* **185:**251–263, 1977.

138. Muakkassa WF, Goldman MH, Mendez-Picon G, Lee HM: Wound infections in renal transplant patients. *J Urol* **130:**17–19, 1983.

139. Lee HM, Madge GE, Mendez-Picon G, Chatterjee SN: Surgical

complications in renal transplant recipients. *Surg Clin North Am* **58:**285–304, 1978.

140. Kyriakides GK, Simmons RL, Najarian JS: Wound infections in renal transplant wounds: Pathogenetic and prognostic factors. *Ann Surg* **182:**770–775, 1975.

141. Bruun JN: Post-operative wound infection. Predisposing factors and the effect of a reduction in the dissemination of staphylococci. *Acta Med Scand Suppl* **514:**3–89, 1970.

142. Belzer FO, Salvatierra O Jr, Schweizer RT, Kountz SL: Prevention of wound infections by topical antibiotics in high risk patients. *Am J Surg* **126:**180–185, 1973.

143. Tilney NL, Strom TB, Vineyard GC, Merrill JP: Factors contributing to the declining mortality rate in renal transplantation. *N Engl J Med* **299:**1321–1325, 1978.

144. Novick AC: The value of intraoperative antibiotics in preventing renal transplant wound infections. *J Urol* **125:**151–152, 1981.

145. Judson RT: Wound infection following renal transplantation. *Aust NZ J Surg* **54:**223–224, 1984.

146. Fisher MC, Baluarte HJ, Long SS: Bacteremia due to *Bacteroides fragilis* after elective appendectomy in renal transplant recipients. *J Infect Dis* **143:**635–638, 1981.

147. Morris PJ, Chan L, French ME, Ting A: Low dose oral prednisolone in renal transplantation. *Lancet* **1:**525–527, 1982.

148. Howard RJ, Condie RM, Sutherland DE, Simmons RL, Najarian JS: The use of antilymphoblast globulin in the treatment of renal allograft rejection. *Transplant Proc* **13:**473–474, 1981.

149. Palmer JM, Chatterjee SN: Urologic complications in renal transplantation. *Surg Clin North Am* **58:**305–319, 1978.

150. Leary FJ, Woods JE, DeWeerd JH: Urologic problems in renal transplantation. *Arch Surg* **110:**1124–1126, 1975.

151. Salvatierra O Jr, Olcott CT, Amend WJ Jr, Cochrum KC, Freduska NJ: Urological complications of renal transplantation can be prevented or controlled. *J Urol* **117:**421–424, 1977.

152. Kohlberg WI, Tellis VA, Bhat DJ, Driscoll B, Veith FJ: Wound infections after transplant nephrectomy. *Arch Surg* **115:**645–646, 1980.

153. Koehler PR, Kanemoto HH, Maxwell JG: Ultrasonic "B" scanning in the diagnosis of complications in renal transplant patients. *Radiology* **119:**661–664, 1976.

154. Schweizer RT, Cho S, Koutz KS, Belzer FO: Lymphoceles following renal transplantation. *Arch Surg* **104:**42–45, 1972.

155. Lorimer WS, Glassford DM, Sarles HE, Remmers Ar Jr, Fish JC: Lymphocele: A significant complication following renal transplantation. *Lymphology* **8:**20–23, 1975.

156. Belzer F: Technical complications after renal transplantation. In Morris P (ed): *Kidney Transplantation: Principles and Practice.* New York, Academic Press, 1979, pp. 267–284.

157. Townsend TR, Rudolf LE, Westervelt FB Jr, Mandell GL, Wenzel RP: Prophylactic antibiotic therapy with cefamandole and tobramycin for patients undergoing renal transplantation. *Infect Control* **1:**93–96, 1980.

158. Paya CV, Hermans PE: Bacterial infections after liver transplantation. *Eur J Clin Microbiol Infect Dis* **8:**499–504, 1989.

159. Lebeau G, Yanaga K, Marsh JW, *et al*: Analysis of surgical complications after 397 hepatic transplantations. *Surg Gynecol Obstet* **170:**317–322, 1990.

160. George DL, Arnow PM, Fox AS, *et al*: Bacterial infection as a complication of liver transplantation: Epidemiology and risk factors. *Rev Infect Dis* **13:**387–396, 1991.

161. Barkholt L, Ericzon BG, Tollemar J, *et al*: Infections in human liver recipients: Different patterns early and late after transplantation. *Transplant Int* **6:**77–84, 1993.

162. Korvick JA, Marsh JW, Starzl TE, Yu VL: *Pseudomonas aeruginosa* bacteremia in patients undergoing liver transplantation: An emerging problem. *Surgery* **109:**62–68, 1991.

163. Jacobs F, Van de Stadt J, Gelin M, *et al*: *Mycoplasma hominis* infection of perihepatic hematomas in a liver transplant recipient [see comments]. *Surgery* **111:**98–100, 1992.

164. Paya CV, Hermans PE, Washington JAD, *et al*: Incidence, distribution, and outcome of episodes of infection in 100 orthotopic liver transplantations. *Mayo Clin Proc* **64:**555–564, 1989.

165. Shaffer D, Jenkins RL, Karchmer AW, Monaco AP: Toxic shock syndrome complicating orthotopic liver transplantation—A case report. *Transplantation* **42:**434–437, 1986.

166. Rosman C, Klompmaker IJ, Bonsel GJ, *et al*: The efficacy of selective bowel decontamination as infection prevention after liver transplantation. *Transplant Proc* **22:**1554–1555, 1990.

167. van Zeijl JH, Kroes AC, Metselaar HJ, *et al*: Infections after auxiliary partial liver transplantation. Experiences in the first ten patients. *Infection* **18:**146–151, 1990.

168. Wiesner RH, Hermans PE, Rakela J, *et al*: Selective bowel decontamination to decrease gram-negative aerobic bacterial and candida colonization and prevent infection after orthotopic liver transplantation. *Transplantation* **45:**570–574, 1988.

169. Wiesner RH: Selective bowel decontamination for infection prophylaxis in liver transplantation patients. *Transplant Proc* **23:**1927–1928, 1991.

170. Castaldo P, Stratta RJ, Wood RP, *et al*: Clinical spectrum of fungal infections after orthotopic liver transplantation. *Arch Surg* **126:**149–156, 1991.

171. Colonna JOd, Winston DJ, Brill JE, *et al*: Infectious complications in liver transplantation. *Arch Surg* **123:**360–364, 1988.

172. Kusne S, Dummer JS, Singh N, *et al*: Infections after liver transplantation. An analysis of 101 consecutive cases. *Medicine* **67:**132–143, 1988.

173. Cuervas-Mons V, Barrios C, Garrido A, *et al*: Bacterial infections in liver transplant patients under selective decontamination with norfloxacin. *Transplant Proc* **21:**3558, 1989.

174. Wajszczuk CP, Dummer JS, Ho M, *et al*: Fungal infections in liver transplant recipients. *Transplantation* **40:**347–353, 1985.

175. Arnow PM, Furmaga K, Flaherty JP, George D: Microbiological efficacy and pharmacokinetics of prophylactic antibiotics in liver transplant patients. *Antimicrob Agents Chemother* **36:**2125–2130, 1992.

176. Arnow PM: Prevention of bacterial infection in the transplant recipient. The role of selective bowel decontamination. *Infect Dis Clin North Am* **9:**849–862, 1995.

177. Paya CV: Fungal infections in solid-organ transplantation. *Clin Infect Dis* **16:**677–688, 1993.

178. Hadley S, Karchmer AW: Fungal infections in solid organ transplant recipients. *Infect Dis Clin North Am* **9:**1045–1074, 1995.

179. Tolkoff-Rubin NE, Conti DJ, Doran M, DelVecchio A, Rubin RH: Fluconazole in the treatment of invasive candidal and cryptococcal infections in organ transplant recipients. *Pharmacotherapy* **10:**159S–163S, 1990.

180. Goldmann DA, Hopkins CC, Karchmer AW, *et al*: Cephalothin prophylaxis in cardiac valve surgery. A prospective, double-blind comparison of two-day and six-day regimens. *J Thorac Cardiovasc Surg* **73:**470–479, 1977.

181. Polk HC Jr, Lopez-Mayor JF: Postoperative wound infection: A prospective study of determinant factors and prevention. *Surgery* **66:**97–103, 1969.

182. Platt R, Munoz A, Stella J, VanDevanter S, Koster JR Jr: Anti-

biotic prophylaxis for cardiovascular surgery. Efficacy with coronary artery bypass. *Ann Intern Med* **101:**770–774, 1984.

183. Remington JS, Gaines JD, Griepp RB, Shumway NE: Further experience with infection after cardiac transplantation. *Transplant Proc* **4:**699–705, 1972.

184. Montgomery JR, Barrett FF, Williams TW Jr: Infectious complications in cardiac transplant patients. *Transplant Proc* **5:**1239–1243, 1973.

185. Pearl SN, Weiner MA, Dibbell DG: Sternal infection after cardiac transplantation. Successful salvage utilizing a variety of techniques. *J Thorac Cardiovasc Surg* **83:**632–634, 1982.

186. Gorensek MJ, Stewart RW, Keys TF, *et al*: Decreased infections in cardiac transplant recipients on cyclosporine with reduced corticosteroid use. *Cleve Clin J Med* **56:**690–695, 1989.

187. Petzold T, Feindt PR, Carl UM, Gams E: Hyperbaric oxygen therapy in deep sternal wound infection after heart transplantation. *Chest* **115:**1455–1458, 1999.

188. Canver CC, Patel AK, Kosolcharoen P, Voytovich MC: Fungal purulent constrictive pericarditis in a heart transplant patient. *Ann Thorac Surg* **65:**1792–1794, 1998.

189. Thomson D, Menkis A, Pflugfelder P, *et al*: Mycotic aortic aneurysm after heart–lung transplantation. *Transplantation* **47:**195–197, 1989.

190. Dowling RD, Baladi N, Zenati M, *et al*: Disruption of the aortic anastomosis after heart–lung transplantation. *Ann Thorac Surg* **49:**118–122, 1990.

191. Anthuber M, Kemkes B, Kreuzer E, *et al*: Mediastinitis and mycotic aneurysm of the aorta after orthotopic heart transplantation. *Tex Heart Inst J* **18:**186–193, 1991.

192. Palac RT, Strausbaugh LJ, Antonovic R, Floten HS: An unusual complication of cardiac transplantation—Infected aortic pseudoaneurysm. *Ann Thorac Surg* **51:**479–481, 1991.

193. Slater AD, Ganzel BL, Keller M, Robin GRD, Gray LA Jr: Repair of infected pseudoaneurysm with aortic arch replacement after orthotopic heart transplantation. *J Heart Transplant* **9:**230–235, 1990.

194. Phillips WS, Burton NA, Macmanus Q, Lefrak EA: Surgical complications in bridging to transplantation: The Thermo Cardiosystems LVAD [see comments]. *Ann Thorac Surg* **53:**482–485; discussion 485–486, 1992.

195. Edwards JE Jr, Bodey GP, Bowden RA, *et al*: International Conference for the Development of a Consensus on the Management and Prevention of Severe Candidal Infections [see comments]. *Clin Infect Dis* **25:**43–59, 1997.

196. Busuttil RW, Goldstein LI, Danovitch GM, Ament ME, Memsic LD: Liver transplantation today [clinical conference]. *Ann Intern Med* **104:**377–389, 1986.

197. Starzl TE, Ishikawa M, Putnam CW, *et al*: Progress in and deterrents to orthotopic liver transplantation, with special reference to survival, resistance to hyperacute rejection, and biliary duct reconstruction. *Transplant Proc* **6:**129–139, 1974.

198. Schroter GP, Hoelscher M, Putnam CW, *et al*: Infections complicating orthotopic liver transplantation: A study emphasizing graft-related septicemia. *Arch Surg* **111:**1337–1347, 1976.

199. Ho M, Wajszczuk C, Hardy A, *et al*: Infections in kidney, heart, and liver transplant recipients on cyclosporine. *Transplant Proc* **15:**2768–2772, 1983.

200. Calne RY, Williams R: Liver transplantation. *Curr Probl Surg* **16:**1–44, 1979.

201. Starzl TE, Iwatsuki S, Van Thiel DH, *et al*: Evolution of liver transplantation. *Hepatology* **2:**614–636, 1982.

202. Starzl TE, Putnam CW, Hansbrough JF, Porter KA, Reid HA:

Biliary complications after liver transplantation: With special reference to the biliary cast syndrome and techniques of secondary duct repair. *Surgery* **81:**212–221, 1977.

203. Starzl T, Demetris A: *Liver Transplantation: A 31 Year Perspective*. Chicago, Year Book Medical, 1990.

204. Bubak ME, Porayko MK, Krom RA, Wiesner RH: Complications of liver biopsy in liver transplant patients: Increased sepsis associated with choledochojejunostomy. *Hepatology* **14:**1063–1065, 1991.

205. Unilateral lung transplantation for pulmonary fibrosis. Toronto Lung Transplant Group. *N Engl J Med* **314:**1140–1145, 1986.

206. Cooper JD: Current status of lung transplantation. *Transplant Proc* **23:**2107–2114, 1991.

207. Dauber JH, Paradis IL, Dummer JS: Infectious complications in pulmonary allograft recipients. *Clin Chest Med* **11:**291–308, 1990.

208. Tolkoff-Rubin NE, Cosimi AB, Delmonico FL, *et al*: Diagnosis of tubular injury in renal transplant patients by a urinary assay for a proximal tubular antigen, the adenosine-deaminase-binding protein. *Transplantation* **41:**593–597, 1986.

209. Kreis H, Legendre C, Chatenoud L: OKT3 in organ transplantation. *Transplant Rev* **5:**181–199, 1991.

210. Chatenoud L, Ferran C, Legendre C, *et al*: In vivo cell activation following OKT3 administration. Systemic cytokine release and modulation by corticosteroids. *Transplantation* **49:**697–702, 1990.

211. Basgoz N, Preiksaitis JK: Post-transplant lymphoproliferative disorder. *Infect Dis Clin North Am* **9:**901–923, 1995.

212. Walker RC, Marshall WF, Strickler JG, *et al*: Pretransplantation assessment of the risk of lymphoproliferative disorder. *Clin Infect Dis* **20:**1346–1353, 1995.

213. Rubin RH: Impact of cytomegalovirus infection on organ transplant recipients. *Rev Infect Dis* **12**(Suppl 7):**S754–S766**, 1990.

214. Cheeseman SH, Rubin RH, Stewart JA, *et al*: Controlled clinical trial of prophylactic human-leukocyte interferon in renal transplantation. Effects on cytomegalovirus and herpes simplex virus infections. *N Engl J Med* **300:**1345–1349, 1979.

215. Betts RF, Freeman RB, Douglas RG Jr, Talley TE, Rundell B: Transmission of cytomegalovirus infection with renal allograft. *Kidney Int* **8:**385–392, 1975.

216. Ho M, Suwansirikul S, Dowling JN, Youngblood LA, Armstrong JA: The transplanted kidney as a source of cytomegalovirus infection. *N Engl J Med* **293:**1109–1112, 1975.

217. Suwansirikul S, Rao N, Dowling JN, Ho M: Primary and secondary cytomegalovirus infection. *Arch Intern Med* **137:**1026–1029, 1977.

218. Naraqi S, Jackson GG, Jonasson O, Yamashiroya HM: Prospective study of prevalence, incidence, and source of herpesvirus infections in patients with renal allografts. *J Infect Dis* **136:**531–540, 1977.

219. Rubin RH, Cosimi AB, Tolkoff-Rubin NE, Russell PS, Hirsch MS: Infectious disease syndromes attributable to cytomegalovirus and their significance among renal transplant recipients. *Transplantation* **24:**458–464, 1977.

220. Chou SW: Acquisition of donor strains of cytomegalovirus by renal-transplant recipients. *N Engl J Med* **314:**1418–1423, 1986.

221. Grundy JE, Super M, Lui S, Sweny P, Griffiths PD: The source of cytomegalovirus infection in seropositive renal allograft recipients is frequently the donor kidney. *Transplant Proc* **19:**2126–2128, 1987.

222. Grundy JE, Lui SF, Super M, *et al*: Symptomatic cytomegalovirus infection in seropositive kidney recipients: Reinfection

with donor virus rather than reactivation of recipient virus. *Lancet* **2:**132–135, 1988.

223. Rubin R, Colvin R: The impact of CMV infections on renal transplantation. In Racusen L, Solez K, Burdick J (eds): *Kidney Transplant Rejection*. New York, Marcel Dekker, 1998, pp. 605–626.

224. Snydman DR, Werner BG, Dougherty NN, *et al*: Cytomegalovirus immune globulin prophylaxis in liver transplantation. A randomized, double-blind, placebo-controlled trial. The Boston Center for Liver Transplantation CMVIG Study Group. *Ann Intern Med* **119:**984–991, 1993.

225. Rubin RH, Tolkoff-Rubin NE, Oliver D, *et al*: Multicenter seroepidemiologic study of the impact of cytomegalovirus infection on renal transplantation. *Transplantation* **40:**243–249, 1985.

226. Tolkoff-Rubin NE, Rubin RH: Recent advances in the diagnosis and management of infection in the organ transplant recipient. *Semin Nephrol* **20:**148–163, 2000.

227. Chou SW, Norman DJ: The influence of donor factors other than serologic status on transmission of cytomegalovirus to transplant recipients. *Transplantation* **46:**89–93, 1988.

228. Betts RF, Schmidt SG: Cytolytic IgM antibody to cytomegalovirus in primary cytomegalovirus infection in humans. *J Infect Dis* **143:**821–826, 1981.

229. Betts RF: Cytomegalovirus infection in transplant patients. *Prog Med Virol* **28:**44–64, 1982.

230. Huang ES, Alford CA, Reynolds DW, Stagno S, Pass RF: Molecular epidemiology of cytomegalovirus infections in women and their infants. *N Engl J Med* **303:**958–962, 1980.

231. Drew WL, Sweet ES, Miner RC, Mocarski ES: Multiple infections by cytomegalovirus in patients with acquired immunodeficiency syndrome: Documentation by Southern blot hybridization. *J Infect Dis* **150:**952–953, 1984.

232. Fryd DS, Peterson PK, Ferguson RM, Simmons RL, *et al*: Cytomegalovirus as a risk factor in renal transplantation. *Transplantation* **30:**436–439, 1980.

233. Smiley ML, Wlodaver CG, Grossman RA, *et al*: The role of pretransplant immunity in protection from cytomegalovirus disease following renal transplantation. *Transplantation* **40:**157–161, 1985.

234. Chou SW: Reactivation and recombination of multiple cytomegalovirus strains from individual organ donors. *J Infect Dis* **160:**11–15, 1989.

235. Ho M: *Cytomegalovirus: Biology and Infection*. New York, Plenum Medical, 1991.

236. Tolkoff-Rubin NA, Rubin RH, Keller EE, *et al*: Cytomegalovirus infection in dialysis patients and personnel. *Ann Intern Med* **89:**625–628, 1978.

237. Sagedal S, Nordal KP, Hartmann A, *et al*: A prospective study of the natural course of cytomegalovirus infection and disease in renal allograft recipients [In Process Citation]. *Transplantation* **70:**1166–1174, 2000.

238. Rubin R, Kemmerly S, Conti D, *et al*: Prevention of primary cytomegalovirus disease in organ transplant recipients. *Transplant Infect Dis* **2:**112–117, 2000.

239. Mach M, Stamminger T, Jahn G: Human cytomegalovirus: Recent aspects from molecular biology. *J Gen Virol* **70:**3117–3146, 1989.

240. Mocarski ES, Kemble GW, Lyle JM, Greaves RF: A deletion mutant in the human cytomegalovirus gene encoding IE1(491aa) is replication defective due to a failure in autoregulation. *Proc Natl Acad Sci USA* **93:**11321–11326, 1996.

241. Greaves RF, Mocarski ES: Defective growth correlates with reduced accumulation of a viral DNA replication protein after low-multiplicity infection by a human cytomegalovirus ie1 mutant. *J Virol* **72:**366–379, 1998.

242. Reinke P, Prosch S, Kern F, Volk H: Mechanisms of human cytomegalovirus (HCMV) (re)activation and its impact on organ transplant patients. *Transplant Infect Dis* **1:**157–164, 1999.

243. Fietze E, Prosch S, Reinke P, *et al*: Cytomegalovirus infection in transplant recipients. The role of tumor necrosis factor. *Transplantation* **58:**675–680, 1994.

244. Stein J, Volk HD, Liebenthal C, Kruger DH, Prosch S: Tumour necrosis factor alpha stimulates the activity of the human cytomegalovirus major immediate early enhancer/promoter in immature monocytic cells. *J Gen Virol* **74:**2333–2338, 1993.

245. Prosch S, Staak K, Stein J, *et al*: Stimulation of the human cytomegalovirus IE enhancer/promoter in HL-60 cells by TNF-alpha is mediated via induction of NFkB. *Virology* **208:**107–116, 1995.

246. Docke WD, Prosch S, Fietze E, *et al*: Cytomegalovirus reactivation and tumour necrosis factor. *Lancet* **343:**268–269, 1994.

247. Mutimer D, Mirza D, Shaw J, O'Donnell K, Elias E: Enhanced (cytomegalovirus) viral replication associated with septic bacterial complications in liver transplant recipients. *Transplantation* **63:**1411–1415, 1997.

248. Mutimer DJ, Shaw J, O'Donnell K, Elias E: Enhanced (cytomegalovirus) viral replication after transplantation for fulminant hepatic failure [see comments]. *Liver Transplant Surg* **3:**506–512, 1997.

249. Kutza AS, Muhl E, Hackstein H, Kirchner H, Bein G: High incidence of active cytomegalovirus infection among septic patients [see comments]. *Clin Infect Dis* **26:**1076–1082, 1998.

250. Ho M: Cytomegalovirus infection in patients with bacterial sepsis [editorial comment]. *Clin Infect Dis* **26:**1083–1084, 1998.

251. Brennan DC, Flavin K, Lowell JA, *et al*: A randomized, double-blinded comparison of thyroglobulin versus Atgam for induction immunosuppressive therapy in adult renal transplant recipients [published erratum appears in *Transplantation* 67(10):1386, 1999]. *Transplantation* **67:**1011–1018, 1999.

252. Paterson DL, Staplefeldt WH, Wagener MM, *et al*: Intraoperative hypothermia is an independent risk factor for early cytomegalovirus infection in liver transplant recipients. *Transplantation* **67:**1151–1155, 1999.

253. Dowling JN, Saslow AR, Armstrong JA, Ho M: Cytomegalovirus infection in patients receiving immunosuppressive therapy for rheumatologic disorders. *J Infect Dis* **133:**399–408, 1976.

254. Pass RF, Reynolds DW, Whelchel JD, Diethelm AG, Alford CA: Impaired lymphocyte transformation response to cytomegalovirus and phytohemagglutinin in recipients of renal transplants: Association with antithymocyte globulin. *J Infect Dis* **143:**259–265, 1981.

255. Rubin RH, Cosimi AB, Hirsch MS, *et al*: Effects of antithymocyte globulin on cytomegalovirus infection in renal transplant recipients. *Transplantation* **31:**143–145, 1981.

256. Hibberd PL, Tolkoff-Rubin NE, Cosimi AB, *et al*: Symptomatic cytomegalovirus disease in the cytomegalovirus antibody seropositive renal transplant recipient treated with OKT3. *Transplantation* **53:**68–72, 1992.

257. Gonwa TA, Capehart JE, Pilcher JW, Alivizatos PA: Cytomegalovirus myocarditis as a cause of cardiac dysfunction in a heart transplant recipient. *Transplantation* **47:**197–199, 1989.

258. Falagas ME, Snydman DR, Griffith J, *et al*: Clinical and epide-

miological predictors of recurrent cytomegalovirus disease in orthotopic liver transplant recipients. Boston Center for Liver Transplantation CMVIG Study Group. *Clin Infect Dis* **25:**314–317, 1997.

259. Reusser P, Cathomas G, Attenhofer R, Tamm M, Thiel G: Cytomegalovirus (CMV)-specific T cell immunity after renal transplantation mediates protection from CMV disease by limiting the systemic virus load. *J Infect Dis* **180:**247–253, 1999.

260. Sester M, Sester U, Gartner B, *et al*: Levels of virus-specific CD4 T-cells correlate with cytomegalovirus control and predict virus-induced disease following renal transplantation. In press.

261. Dechanet J, Merville P, Lim A, *et al*: Implication of gammadelta T cells in the human immune response to cytomegalovirus. *J Clin Invest* **103:**1437–1449, 1999.

262. Zeevi A, Morel P, Spichty K, *et al*: Clinical significance of CMV-specific T helper responses in lung transplant recipients. *Hum Immunol* **59:**768–775, 1998.

263. Asanuma H, Sharp M, Maecker HT, Maino VC, Arvin AM: Frequencies of memory T cells specific for varicella–zoster virus, herpes simplex virus, and cytomegalovirus by intracellular detection of cytokine expression. *J Infect Dis* **181:**859–866, 2000.

264. Gyulai Z, Endresz V, Burian K, *et al*: Cytotoxic T lymphocyte (CTL) responses to human cytomegalovirus pp65, IE1–Exon4, gB, pp150, and pp28 in healthy individuals: Reevaluation of prevalence of IE1–specific CTLs. *J Infect Dis* **181:**1537–1546, 2000.

265. Wever PC, Spaeny LH, van der Vliet HJ, *et al*: Expression of granzyme B during primary cytomegalovirus infection after renal transplantation. *J Infect Dis* **179:**693–696, 1999.

266. Hazzan M, Labalette M, Noel C, Lelievre G, Dessaint JP: Recall response to cytomegalovirus in allograft recipients: Mobilization of CD57+, CD28+ cells before expansion of CD57+, CD28− cells within the CD8+ T lymphocyte compartment. *Transplantation* **63:**693–698, 1997.

267. Cheeseman SH, Stewart JA, Winkle S, *et al*: Cytomegalovirus excretion 2–14 years after renal transplantation. *Transplant Proc* **11:**71–74, 1979.

268. Fishman JA, Doran MT, Volpicelli SA, *et al*: Dosing of intravenous ganciclovir for the prophylaxis and treatment of cytomegalovirus infection in solid organ transplant recipients. *Transplantation* **69:**389–394, 2000.

269. Lazzarotto T, Varani S, Spezzacatena P, *et al*: Delayed acquisition of high-avidity anti-cytomegalovirus antibody is correlated with prolonged antigenemia in solid organ transplant recipients. *J Infect Dis* **178:**1145–1149, 1998.

270. Nordoy I, Muller F, Nordal KP, *et al*: The role of the tumor necrosis factor system and interleukin-10 during cytomegalovirus infection in renal transplant recipients. *J Infect Dis* **181:**51–57, 2000.

271. Millett R, Tomita T, Marshall HE, Cohen L, Hannah HD: Cytomegalovirus endomyocarditis in a transplanted heart. A case report with *in situ* hybridization. *Arch Pathol Lab Med* **115:**511–515, 1991.

272. Calhoon JH, Nichols L, Davis R, *et al*: Single lung transplantation. Factors in postoperative cytomegalovirus infection. *J Thorac Cardiovasc Surg* **103:**21–25; discussion 25–26, 1992.

273. Peterson PK, Balfour HH Jr, Fryd DS, Ferguson RM, Simmons RL: Fever in renal transplant recipients: Causes, prognostic significance and changing patterns at the University of Minnesota Hospital. *Am J Med* **71:**345–351, 1981.

274. Simmons RL, Lopez C, Balfour H Jr, *et al*: Cytomegalovirus: Clinical virological correlations in renal transplant recipients. *Ann Surg* **180:**623–634, 1974.

275. Simmons RL, Matas AJ, Rattazzi LC, *et al*: Clinical characteristics of the lethal cytomegalovirus infection following renal transplantation. *Surgery* **82:**537–546, 1977.

276. Fine RN, Grushkin CM, Malekzadeh M, Wright HT Jr: Cytomegalovirus syndrome following renal transplantation. *Arch Surg* **105:**564–570, 1972.

277. de Maar EF, Kas-Deelen AM, van der Mark TW, *et al*: Cytomegalovirus pneumonitis after kidney transplantation is not caused by plugging of cytomegalic endothelial cells only. *Transplant Int* **12:**56–62, 1999.

278. Aafedt BC, Halvorsen RA Jr, Tylen U, Hertz M: Cytomegalovirus pneumonia: Computed tomography findings. *Can Assoc Radiol J* **41:**276–280, 1990.

279. Jeffery JR, Guttmann RD, Becklake MR, Beaudoin JF, Morehouse DD: Recovery from severe cytomegalovirus pneumonia in a renal transplant patient. *Am Rev Respir Dis* **109:**129–133, 1974.

280. Ravin CE, Smith GW, Ahern MJ, *et al*: Cytomegaloviral infection presenting as a solitary pulmonary nodule. *Chest* **71:**220–222, 1977.

281. Hamed IA, Wenzl JE, Leonard JC, Altshuler GP, Pederson JA: Pulmonary cytomegalovirus infection: Detection by gallium 67 imaging in the transplant patient. *Arch Intern Med* **139:**286–288, 1979.

282. Chinsky K, Goodenberger DM: Use of indium 111-labeled white blood cell scan in the diagnosis of cytomegalovirus pneumonia in a renal transplant recipient with a normal chest roentgenogram. *Chest* **99:**761–763, 1991.

283. Friedman HM, Grossman RA, Plotkin SA, Perloff LJ, Barker CF: Relapse of pneumonia caused by cytomegalovirus in two recipients of renal transplants. *J Infect Dis* **139:**465–473, 1979.

284. van Son WJ, van der Bij W, Tegzess AM, *et al*: Complement activation during an active cytomegalovirus infection after renal transplantation: Due to circulating immune complexes or alternative pathway activation? *Clin Immunol Immunopathol* **50:**109–121, 1989.

285. van Son WJ, van der Bij W, The TH, *et al*: Evidence for pulmonary dysfunction in all patients with symptomatic and asymptomatic cytomegalovirus (CMV) infection after renal transplantation. *Transplant Proc* **21:**2065–2068, 1989.

286. Cheung AK: Biocompatibility of hemodialysis membranes. *J Am Soc Nephrol* **1:**150–161, 1990.

287. Stratta RJ, Shaefer MS, Markin RS, *et al*: Clinical patterns of cytomegalovirus disease after liver transplantation. *Arch Surg* **124:**1443–1449; discussion 1449–1450, 1989.

288. Paya CV, Hermans PE, Wiesner RH, *et al*: Cytomegalovirus hepatitis in liver transplantation: Prospective analysis of 93 consecutive orthotopic liver transplantations. *J Infect Dis* **160:**752–758, 1989.

289. Alessiani M, Kusne S, Fung JJ, *et al*: CMV infection in liver transplantation under cyclosporine or FK 506 immunosuppression. *Transplant Proc* **23:**3035–3037, 1991.

290. Backman L, Brattstrom C, Reinholt FP, Andersson J, Tyden G: Development of intrapancreatic abscess—A consequence of CMV pancreatitis? *Transplant Int* **4:**116–121, 1991.

291. Klassen DK, Drachenberg CB, Papadimitriou JC, *et al*: CMV allograft pancreatitis: Diagnosis, treatment, and histological features. *Transplantation* **69:**1968–1971, 2000.

292. Luby JP, Burnett W, Hull AR, *et al*: Relationship between cyto-

megalovirus and hepatic function abnormalities in the period after renal transplant. *J Infect Dis* **129:**511–518, 1974.

293. Sutherland DE, Chan FY, Fourcar E, *et al:* The bleeding cecal ulcer in transplant patients. *Surgery* **86:**386–398, 1979.

294. Minars N, Silverman JF, Escobar MR, Martinez AJ: Fatal cytomegalic inclusion disease. Associated skin manifestations in a renal transplant patient. *Arch Dermatol* **113:**1569–1571, 1977.

295. Patel NP, Corry RJ: Cytomegalovirus as a cause of cecal ulcer with massive hemorrhage in a renal transplant recipient. *Am Surg* **46:**260–262, 1980.

296. Franzin G, Muolo A, Griminelli T: Cytomegalovirus inclusions in the gastroduodenal mucosa of patients after renal transplantation. *Gut* **22:**698–701, 1981.

297. van Son WJ, van der Jagt EJ, van der Woude FJ, *et al:* Pneumatosis intestinalis in patients after cadaveric kidney transplantation. Possible relationship with an active cytomegalovirus infection. *Transplantation* **38:**506–510, 1984.

298. Kaplan CS, Petersen EA, Icenogle TB, *et al:* Gastrointestinal cytomegalovirus infection in heart and heart–lung transplant recipients. *Arch Intern Med* **149:**2095–2100, 1989.

299. Escudero-Fabre A, Cummings O, Kirklin JK, Bourge RC, Aldrete JS: Cytomegalovirus colitis presenting as hematochezia and requiring resection. *Arch Surg* **127:**102–104, 1992.

300. Mayoral JL, Loeffler CM, Fasola CG, *et al:* Diagnosis and treatment of cytomegalovirus disease in transplant patients based on gastrointestinal tract manifestations. *Arch Surg* **126:**202–206, 1991.

301. Van Thiel DH, Gavaler JS, Schade RR, Chien MC, Starzl TE: Cytomegalovirus infection and gastric emptying. *Transplantation* **54:**70–73, 1992.

302. Sakr M, Hassanein T, Gavaler J, *et al:* Cytomegalovirus infection of the upper gastrointestinal tract following liver transplantation—Incidence, location, and severity in cyclosporine- and FK506-treated patients. *Transplantation* **53:**786–791, 1992.

303. Kaplan B, Meier-Kriesche HU, Jacobs MG, *et al:* Prevalence of cytomegalovirus in the gastrointestinal tract of renal transplant recipients with persistent abdominal pain. *Am J Kidney Dis* **34:**65–68, 1999.

304. Halme L, Hockerstedt K, Salmela K, Lautenschlager I: CMV infection detected in the upper gastrointestinal tract after liver transplantation. *Transplant Int* **11:**S242–S244, 1998.

305. Tilsed JV, Morgan JD, Veitch PS, Donnelly PK: Reactivation of duodenal cytomegalovirus infection mimicking a transplant lymphoma. *Transplantation* **54:**945–946, 1992.

306. Battaglino MP, Rockey DC: Cytomegalovirus colitis presenting with the endoscopic appearance of pseudomembranous colitis. *Gastrointest Endosc* **50:**697–700, 1999.

307. Crespo MG, Arnal FM, Gomez M, *et al:* Cytomegalovirus colitis mimicking a colonic neoplasm or ischemic colitis 4 years after heart transplantation. *Transplantation* **66:**1562–1565, 1998.

308. Diaz-Gonzalez VM, Altemose GT, Ogorek C, Palazzo I, Pina IL: Cytomegalovirus infection presenting as an apple-core lesion of the colon. *J Heart Lung Transplant* **16:**1171–1175, 1997.

309. Shapiro AM, Bain VG, Preiksaitis JK, *et al:* Ogilvie's syndrome associated with acute cytomegaloviral infection after liver transplantation. *Transplant Int* **13:**41–45, 2000.

310. Shutze WP, Kirklin JK, Cummings OW, Aldrete JS: Cytomegalovirus hemorrhoiditis in cardiac allograft recipients. *Transplantation* **51:**918–920, 1991.

311. Moudgil A, Germain BM, Nast CC, *et al:* Ureteritis and cholecystitis: Two unusual manifestations of cytomegalovirus disease in renal transplant recipients. *Transplantation* **64:**1071–1073, 1997.

312. McLaughlin K, Cruickshank M, Hollomby D, Jevnikar A, Muirhead N: Idiopathic thrombocytopenia after cytomegalovirus infection in a renal transplant recipient. *Am J Kidney Dis* **33:**e6, 1999.

313. Kashyap R, Shapiro R, Jordan M, Randhawa PS: The clinical significance of cytomegaloviral inclusions in the allograft kidney. *Transplantation* **67:**98–103, 1999.

314. Peretti N, Said MH, Bouvier R, *et al:* Cytomegalovirus infection may cause ureteral necrosis. *Transplantation* **69:**670–671, 2000.

315. Waiser J, Budde K, Rudolph B, Ortner MA, Neumayer HH: *De novo* hemolytic uremic syndrome postrenal transplant after cytomegalovirus infection. *Am J Kidney Dis* **34:**556–559, 1999.

316. Jeejeebhoy FM, Zaltzman JS: Thrombotic microangiopathy in association with cytomegalovirus infection in a renal transplant patient: A new treatment strategy [see comments]. *Transplantation* **65:**1645–1648, 1998.

317. Detwiler RK, Singh HK, Bolin P Jr, Jennette JC: Cytomegalovirus-induced necrotizing and crescentic glomerulonephritis in a renal transplant patient. *Am J Kidney Dis* **32:**820–824, 1998.

318. Pouria S, State OI, Wong W, Hendry BM: CMV infection is associated with transplant renal artery stenosis. *Q J Med* **91:**185–189, 1998.

319. Madalosso C, de Souza NF Jr, Ilstrup DM, Wiesner RH, Krom RA: Cytomegalovirus and its association with hepatic artery thrombosis after liver transplantation. *Transplantation* **66:**294–297, 1998.

320. Sayage L, Gunby R, Gonwa T, *et al:* Cytomegalovirus endometritis after liver transplantation. *Transplantation* **49:**815–817, 1990.

321. McCarthy JM, McLoughlin MG, Shackleton CR, *et al:* Cytomegalovirus epididymitis following renal transplantation. *J Urol* **146:**417–419, 1991.

322. Dorfman LJ: Cytomegalovirus encephalitis in adults. *Neurology* **23:**136–144, 1973.

323. Spitzer PG, Tarsy D, Eliopoulos GM: Acute transverse myelitis during disseminated cytomegalovirus infection in a renal transplant recipient. *Transplantation* **44:**151–153, 1987.

324. Aaberg TM, Cesarz TJ, Rytel MW: Correlation of virology and clinical course of cytomegalovirus retinitis. *Am J Ophthalmol* **74:**407–415, 1972.

325. De Venecia G, Zu Rhein GM, Pratt MV, Kisken W: Cytomegalic inclusion retinitis in an adult. *Arch Ophthalmol* **86:**44–57, 1971.

326. Porter R, Crombie AL, Gardner PS, Uldall RP: Incidence of ocular complications in patients undergoing renal transplantation. *Br Med J* **3:**133–136, 1972.

327. Wyhinny GJ, Apple DJ, Guastella FR, Vygantas CM: Adult cytomegalic inclusion retinitis. *Am J Ophthalmol* **76:**773–781, 1973.

328. Astle JN, Ellis PP: Ocular complications in renal transplant patients. *Ann Ophthalmol* **6:**1269–1274, 1974.

329. Nicholson D: Cytomegalovirus infection of the retina. In Pavan Langston D (ed): *Ocular Viral Disease.* Boston, Little, Brown, 1975, pp. 151–162.

330. Murray HW, Knox DL, Green WR, Susel RM: Cytomegalovirus retinitis in adults. A manifestation of disseminated viral infection. *Am J Med* **63:**574–584, 1977.

331. Merritt JC, Callender CO: Adult cytomegalic inclusion retinitis. *Ann Ophthalmol* **10:**1059–1063, 1978.

332. Carson S, Chatterjee SN: Cytomegalovirus retinitis: Two cases

occurring after renal transplantation. *Ann Ophthalmol* **10**:275–279, 1978.

333. Broughton WL, Cupples HP, Parver LM: Bilateral retinal detachment following cytomegalovirus retinitis. *Arch Ophthalmol* **96**:618–619, 1978.

334. Moeller MB, Gutman RA, Hamilton JD: Acquired cytomegalovirus retinitis. Four new cases and a review of the literature with implications for management. *Am J Nephrol* **2**:251–255, 1982.

335. Fishburne BC, Mitrani AA, Davis JL: Cytomegalovirus retinitis after cardiac transplantation. *Am J Ophthalmol* **125**:104–106, 1998.

336. Smit WM, Wagemans MA, Jansen CL, vd Horn GJ, Surachno JS: Acute retinal necrosis in a renal allograft recipient—An unusual manifestation of cytomegalovirus infection. *Transplantation* **55**:219–221, 1993.

337. Blau EB, Gross JR: Congenital cytomegalovirus infection after recurrent infection in a mother with a renal transplant. *Pediatr Nephrol* **11**:361–362, 1997.

338. Braun WE: Cytomegalovirus viremia and bacteremia in renal-allograft recipients [letter]. *N Engl J Med* **299**:1318–1319, 1978.

339. Rand KH, Pollard RB, Merigan TC: Increased pulmonary superinfections in cardiac-transplant patients undergoing primary cytomegalovirus infection. *N Engl J Med* **298**:951–953, 1978.

340. Chatterjee SN, Fiala M, Weiner J, *et al*: Primary cytomegalovirus and opportunistic infections. Incidence in renal transplant recipients. *JAMA* **240**:2446–2449, 1978.

341. George MJ, Snydman DR, Werner BG, *et al*: The independent role of cytomegalovirus as a risk factor for invasive fungal disease in orthotopic liver transplant recipients. Boston Center for Liver Transplantation CMVIG-Study Group. Cytogam, MedImmune, Inc. Gaithersburg, Maryland. *Am J Med* **103**:106–113, 1997.

342. Husni RN, Gordon SM, Longworth DL, *et al*: Cytomegalovirus infection is a risk factor for invasive aspergillosis in lung transplant recipients. *Clin Infect Dis* **26**:753–755, 1998.

343. Mackowiak PA, Goggans M, Torres W, *et al*: Relationship between cytomegalovirus and colonization of the oropharynx by gram-negative bacilli following renal transplantation. *Epidemiol Infect* **107**:411–420, 1991.

344. Rifkind D: *Pneumocystis carinii* pneumonia in renal transplant recipients. *Natl Cancer Inst Monogr* **43**:49–54, 1976.

345. Munda R, Alexander JW, First MR, Gartside PS, Fidler JP: Pulmonary infections in renal transplant recipients. *Ann Surg* **187**:126–133, 1978.

346. Gantz NM, Myerowitz RL, Medeiros AA, *et al*: Listeriosis in immunosuppressed patients. A cluster of eight cases. *Am J Med* **58**:637–643, 1975.

347. Schroter GP: Listeria monocytogenes and encephalitis [editorial]. *Arch Intern Med* **138**:198–199, 1978.

348. Hooper DC, Pruitt AA, Rubin RH: Central nervous system infection in the chronically immunosuppressed. *Medicine* **61**:166–188, 1982.

349. Rubin RH, Hooper DC: Central nervous system infection in the compromised host. *Med Clin North Am* **69**:281–296, 1985.

350. Rytel MW, Balay J: Cytomegalovirus infection and immunity in renal allograft recipients: Assessment of the competence of humoral immunity. *Infect Immun* **13**:1633–1637, 1976.

351. Baldwin WM, van Es A, Valentijn RM, *et al*: Increased IgM and IgM immune complex-like material in the circulation of renal transplant recipients with primary cytomegalovirus infections. *Clin Exp Immunol* **50**:515–524, 1982.

352. Baldwin WM, Claas FH, van Gemert GW, *et al*: IgM immune complexes, lymphocytotoxins, and rheumatoid factors in renal transplant recipients with CMV disease. *Scand J Urol Nephrol Suppl* **92**:9–13, 1985.

353. Baldwin WM, Claas F, van Gemert G, *et al*: Studies on lymphocytotoxins and rheumatoid factors in renal transplant recipients with cytomegalovirus disease. *Transplant Proc* **17**:616–617, 1985.

354. Linnemann CC Jr, Kauffman CA, First MR, Schiff GM, Phair JP: Cellular immune response to cytomegalovirus infection after renal transplantation. *Infect Immun* **22**:176–180, 1978.

355. Glazer JP, Friedman HM, Grossman RA, *et al*: Live cytomegalovirus vaccination of renal transplant candidates. A preliminary trial. *Ann Intern Med* **91**:676–683, 1979.

356. Reddehase MJ, Weiland F, Munch K, *et al*: Interstitial murine cytomegalovirus pneumonia after irradiation: Characterization of cells that limit viral replication during established infection of the lungs. *J Virol* **55**:264–273, 1985.

357. Bukowski JF, Warner JF, Dennert G, Welsh RM: Adoptive transfer studies demonstrating the antiviral effect of natural killer cells *in vivo*. *J Exp Med* **161**:40–52, 1985.

358. Quinnan GV Jr, Kirmani N, Rook AH, *et al*: Cytotoxic T cells in cytomegalovirus infection: HLA-restricted T-lymphocyte and non-T-lymphocyte cytotoxic responses correlate with recovery from cytomegalovirus infection in bone-marrow-transplant recipients. *N Engl J Med* **307**:7–13, 1982.

359. Rook AH, Quinnan GV Jr, Frederick WJ, *et al*: Importance of cytotoxic lymphocytes during cytomegalovirus infection in renal transplant recipients. *Am J Med* **76**:385–392, 1984.

360. Reusser P, Riddell SR, Meyers JD, Greenberg PD: Cytotoxic T-lymphocyte response to cytomegalovirus after human allogeneic bone marrow transplantation: Pattern of recovery and correlation with cytomegalovirus infection and disease. *Blood* **78**:1373–1380, 1991.

361. Biron CA, Byron KS, Sullivan JL: Severe herpesvirus infections in an adolescent without natural killer cells [see comments]. *N Engl J Med* **320**:1731–1735, 1989.

362. Carney WP, Rubin RH, Hoffman RA, *et al*: Analysis of T lymphocyte subsets in cytomegalovirus mononucleosis. *J Immunol* **126**:2114–2116, 1981.

363. Doody DP, Wilson EJ, Medearis DN, Rubin RH: Changes in the phenotype of T-cell subset determinants following murine cytomegalovirus infection. *Clin Immunol Immunopathol* **40**:466–475, 1986.

364. Schooley RT, Hirsch MS, Colvin RB, *et al*: Association of herpesvirus infections with T-lymphocyte-subset alterations, glomerulopathy, and opportunistic infections after renal transplantation. *N Engl J Med* **308**:307–313, 1983.

365. van Es A, Baldwin WM, Oljans PJ, *et al*: Expression of HLA-DR on T lymphocytes following renal transplantation, and association with graft-rejection episodes and cytomegalovirus infection. *Transplantation* **37**:65–69, 1984.

366. van den Berg AP, van Son WJ, Janssen RA, *et al*: Recovery from cytomegalovirus infection is associated with activation of peripheral blood lymphocytes. *J Infect Dis* **166**:1228–1235, 1992.

367. Craighead JE: Cytomegalovirus pulmonary disease. *Pathobiol Annu* **5**:197–220, 1975.

368. Green GM: The J. Burns Amberson Lecture—In defense of the lung. *Am Rev Respir Dis* **102**:691–703, 1970.

369. Rinaldo CR Jr, Black PH, Hirsch MS: Interaction of cytomegalovirus with leukocytes from patients with mononucleosis due to cytomegalovirus. *J Infect Dis* **136**:667–678, 1977.

370. Levin MJ, Rinaldo CR Jr, Leary PL, Zaia JA, Hirsch MS: Immune response to herpesvirus antigens in adults with acute cytomegaloviral mononucleosis. *J Infect Dis* **140:**851–857, 1979.

371. Carney WP, Hirsch MS: Mechanisms of immunosuppression in cytomegalovirus mononucleosis. II. Virus–monocyte interactions. *J Infect Dis* **144:**47–54, 1981.

372. Rinaldo CR Jr, Carney WP, Richter BS, Black PH, Hirsch MS: Mechanisms of immunosuppression in cytomegaloviral mononucleosis. *J Infect Dis* **141:**488–495, 1980.

373. Carney WP, Iacoviello V, Hirsch MS: Functional properties of T lymphocytes and their subsets in cytomegalovirus mononucleosis. *J Immunol* **130:**390–393, 1983.

374. Schrier RD, Rice GP, Oldstone MB: Suppression of natural killer cell activity and T cell proliferation by fresh isolates of human cytomegalovirus. *J Infect Dis* **153:**1084–1091, 1986.

375. Rook AH, Masur H, Lane HC, *et al*: Interleukin-2 enhances the depressed natural killer and cytomegalovirus-specific cytotoxic activities of lymphocytes from patients with the acquired immune deficiency syndrome. *J Clin Invest* **72:**398–403, 1983.

376. Charpentier B, Espinosa O, Martin B, *et al*: T cell immunity against cytomegalovirus modifies self-major histocompatibility complex antigens in kidney transplant recipients. *Transplant Proc* **17:**161–162, 1985.

377. Grundy JE: Virologic and pathogenetic aspects of cytomegalovirus infection. *Rev Infect Dis* **12**(suppl. 7)**:**S711–S719, 1990.

378. Grundy JE: Alterations of cellular proteins in human cytomegalovirus infection: Potential for disease pathogenesis. *Transplant Proc* **23:**38–41, discussion 41–42, 1991.

379. Barnes PD, Grundy JE: Down-regulation of the class I HLA heterodimer and beta 2-microglobulin on the surface of cells infected with cytomegalovirus. *J Gen Virol* **73:**2395–2403, 1992.

380. Joly E, Mucke L, Oldstone MB: Viral persistence in neurons explained by lack of major histocompatibility class I expression. *Science* **253:**1283–1285, 1991.

381. Moskophidis D, Lechner F, Pircher H, Zinkernagel RM: Virus persistence in acutely infected immunocompetent mice by exhaustion of antiviral cytotoxic effector T cells [published erratum appears in *Nature* **364**(6434)**:**262, 1993] [see comments]. *Nature* **362:**758–761, 1993.

382. Dummer JS, Ho M, Rabin B, *et al*: The effect of cytomegalovirus and Epstein–Barr virus infection on T lymphocyte subsets in cardiac transplant patients on cyclosporine. *Transplantation* **38:**433–435, 1984.

383. Hamilton JR, Overall JC, Glasgow LA: Synergistic effect on mortality in mice with murine cytomegalovirus and *Pseudomonas aeruginosa*, *Staphylococcus aureus*, or *Candida albicans* infections. *Infect Immun* **14:**982–989, 1976.

384. Pomeroy C, Filice GA, Hitt JA, Jordan MC: Cytomegalovirus-induced reactivation of *Toxoplasma gondii* pneumonia in mice: Lung lymphocyte phenotypes and suppressor function. *J Infect Dis* **166:**677–681, 1992.

385. Ward KN, Sheldon MJ, Gray JJ: Primary and recurrent cytomegalovirus infections have different effects on human herpesvirus-6 antibodies in immunosuppressed organ graft recipients: Absence of virus cross-reactivity and evidence for virus interaction. *J Med Virol* **34:**258–267, 1991.

386. Simmons R, Weil R, Tallent M, *et al*: Do mild infections trigger the rejection of renal allografts? *Transplant Proc* **2:**419–423, 1970.

387. Balfour HH Jr, Slade MS, Kalis JM, *et al*: Viral infections in renal transplant donors and their recipients: A prospective study. *Surgery* **81:**487–492, 1977.

388. David DS, Millian SJ, Whitsell JC, *et al*: Viral syndromes and renal homograft rejection. *Ann Surg* **175:**257–259, 1972.

389. Briggs JD, Timbury MC, Paton AM, Bell PR: Viral infection and renal transplant rejection. *Br Med J* **4:**520–522, 1972.

390. Lopez C, Simmons RL, Mauer SM, *et al*: Association of renal allograft rejection with virus infections. *Am J Med* **56:**280–289, 1974.

391. Betts RF, Freeman RB, Douglas RG Jr, Talley TE: Clinical manifestations of renal allograft derived primary cytomegalovirus infection. *Am J Dis Child* **131:**759–763, 1977.

392. May AG, Betts RF, Freeman RB, Andrus CH: An analysis of cytomegalovirus infection and HLA antigen matching on the outcome of renal transplantation. *Ann Surg* **187:**110–117, 1978.

393. von Willebrand E, Pettersson E. Ahonen J, Hayry P: CMV infection, class II antigen expression, and human kidney allograft rejection. *Transplantation* **42:**364–367, 1986.

394. Harmon J, Sibley R, Peterson P, *et al*: Cytomegalovirus viremia and renal allograft morphology: Are there distinct pathologic features? *Lab Invest* **46:**35A, 1982.

395. Herrera GA, Alexander RW, Cooley CF, *et al*: Cytomegalovirus glomerulopathy: A controversial lesion. *Kidney Int* **29:**725–733, 1986.

396. Boyce NW, Hayes K, Gee D, *et al*: Cytomegalovirus infection complicating renal transplantation and its relationship to acute transplant glomerulopathy. *Transplantation* **45:**706–709, 1988.

397. Shaver MJ, Bonsib SM, Abul-Ezz S, Barri YM: Renal allograft dysfunction associated with cytomegalovirus infection. *Am J Kidney Dis* **34:**942–946, 1999.

398. Pouteil-Noble C, Ecochard R, Landrivon G, *et al*: Cytomegalovirus infection—An etiological factor for rejection? A prospective study in 242 renal transplant patients. *Transplantation* **55:**851–857, 1993.

399. Cappel R, Hestermans O, Toussaint C, *et al*: Cytomegalovirus infection and graft survival in renal graft recipients. *Arch Virol* **56:**149–156, 1978.

400. Hornef MW, Bein G, Fricke L, *et al*: Coincidence of Epstein–Barr virus reactivation, cytomegalovirus infection, and rejection episodes in renal transplant recipients. *Transplantation* **60:**474–480, 1995.

401. Reinke P, Fietze E, Ode-Hakim S, *et al*: Late-acute renal allograft rejection and symptomless cytomegalovirus infection [see comments]. *Lancet* **344:**1737–1738, 1994.

402. Reinke P, Fietze E, Docke WD, Kern F, *et al*: Late acute rejection in long-term renal allograft recipients. Diagnostic and predictive value of circulating activated T cells. *Transplantation* **58:**35–41, 1994.

403. Ustinov JA, Loginov RJ, Mattila PM, *et al*: Cytomegalovirus infection of human kidney cells *in vitro*. *Kidney Int* **40:**954–960, 1991.

404. Rubin RH: The indirect effects of cytomegalovirus infection on the outcome of organ transplantation. *JAMA* **261:**3607–3609, 1989.

405. Richardson WP, Colvin RB, Cheeseman SH, *et al*: Glomerulopathy associated with cytomegalovirus viremia in renal allografts. *N Engl J Med* **305:**57–63, 1981.

406. Tuazon TV, Schneeberger EE, Bhan AK, *et al*: Mononuclear cells in acute allograft glomerulopathy. *Am J Pathol* **129:**119–132, 1987.

407. O'Grady JG, Alexander GJ, Sutherland S, *et al*: Cytomegalovirus infection and donor/recipient HLA antigens: Interdependent cofactors in pathogenesis of vanishing bile-duct syndrome after liver transplantation. *Lancet* **2:**302–305, 1988.

408. Lautenschlager I, Hockerstedt K, Jalanko H, *et al*: Persistent

cytomegalovirus in liver allografts with chronic rejection. *Hepatology* 25:190–194, 1997.

409. Evans PC, Coleman N, Wreghitt TG, Wight DG, Alexander GJ: Cytomegalovirus infection of bile duct epithelial cells, hepatic artery and portal venous endothelium in relation to chronic rejection of liver grafts. *J Hepatol* 31:913–920, 1999.

410. Martelius T, Krogerus L, Hockerstedt K, Bruggeman C, Lautenschlager I: Cytomegalovirus infection is associated with increased inflammation and severe bile duct damage in rat liver allografts. *Hepatology* 27:996–1002, 1998.

411. de Otero J, Gavalda J, Murio E, *et al*: Cytomegalovirus disease as a risk factor for graft loss and death after orthotopic liver transplantation [see comments]. *Clin Infect Dis* 26:865–870, 1998.

412. Rubin RH: Cytomegalovirus disease and allograft loss after organ transplantation [editorial; comment]. *Clin Infect Dis* 26:871–873, 1998.

413. Falagas ME, Snydman DR, Griffith J, Ruthazer R, Werner BG: Effect of cytomegalovirus infection status on first-year mortality rates among orthotopic liver transplant recipients. The Boston Center for Liver Transplantation CMVIG Study Group [see comments]. *Ann Intern Med* 126:275–279, 1997.

414. Falagas ME, Paya C, Ruthazer R, *et al*: Significance of cytomegalovirus for long-term survival after orthotopic liver transplantation: A prospective derivation and validation cohort analysis. *Transplantation* 66:1020–1028, 1998.

415. Ontanon J, Muro M, Garcia-Alonso AM, *et al*: Effect of partial HLA class I match on acute rejection in viral pre-infected human liver allograft recipients. *Transplantation* 65:1047–1053, 1998.

416. Rosen HR, Corless CL, Rabkin J, Chou S: Association of cytomegalovirus genotype with graft rejection after liver transplantation. *Transplantation* 66:1627–1631, 1998.

417. Grattan MT, Moreno-Cabral CE, Starnes VA, *et al*: Cytomegalovirus infection is associated with cardiac allograft rejection and atherosclerosis. *JAMA* 261:3561–3566, 1989.

418. Loebe M, Schuler S, Spiegelsberger S, *et al*: [Cytomegalovirus infection and coronary sclerosis after heart transplantation]. *Dtsch Med Wochenschr* 115:1266–1269, 1990.

419. Loebe M, Schuler S, Zais O, *et al*: Role of cytomegalovirus infection in the development of coronary artery disease in the transplanted heart. *J Heart Transplant* 9:707–711, 1990.

420. Weimar W, Balk AH, Metselaar HJ, Mochtar B, Rothbarth PH: On the relation between cytomegalovirus infection and rejection after heart transplantation. *Transplantation* 52:162–164, 1991.

421. Normann SJ, Salomon DR, Leelachaikul P, *et al*: Acute vascular rejection of the coronary arteries in human heart transplantation: Pathology and correlations with immunosuppression and cytomegalovirus infection. *J Heart Lung Transplant* 10:674–687, 1991.

422. McDonald K, Rector TS, Braulin EA, Kubo SH, Olivari MT: Association of coronary artery disease in cardiac transplant recipients with cytomegalovirus infection [see comments]. *Am J Cardiol* 64:359–362, 1989.

423. Everett JP, Hershberger RE, Norman DJ, *et al*: Prolonged cytomegalovirus infection with viremia is associated with development of cardiac allograft vasculopathy. *J Heart Lung Transplant* 11:S133–S137, 1992.

424. Koskinen PK, Krogerus LA, Nieminen MS, *et al*: Quantitation of cytomegalovirus infection-associated histologic findings in endomyocardial biopsies of heart allografts. *J Heart Lung Transplant* 12:343–354, 1993.

425. Koskinen PK, Nieminen MS, Krogerus LA, *et al*: Cytomegalo-

virus infection accelerates cardiac allograft vasculopathy: Correlation between angiographic and endomyocardial biopsy findings in heart transplant patients. *Transplant Int* 6:341–347, 1993.

426. Koskinen P, Kallio E, Tikkanen J, *et al*: Cytomegalovirus infection and cardiac allograft vasculopathy. *Transplant Infect Dis* 1:115–126, 1999.

427. Hendrix MG, Dormans PH, Kitslaar P, Bosman F, Bruggeman CA: The presence of cytomegalovirus nucleic acids in arterial walls of atherosclerotic and nonatherosclerotic patients. *Am J Pathol* 134:1151–1157, 1989.

428. Melnick JL, Adam E, DeBakey ME: Possible role of cytomegalovirus in atherogenesis. *JAMA* 263:2204–2207, 1990.

429. Orbaek Andersen H: Heart allograft vascular disease: An obliterative vascular disease in transplanted hearts. *Atherosclerosis* 142:243–263, 1999.

430. Melnick JL, Petrie BL, Dreesman GR, *et al*: Cytomegalovirus antigen within human arterial smooth muscle cells. *Lancet* 2:644–647, 1983.

431. Tumilowicz JJ, Gawlik ME, Powell BB, Trentin JJ: Replication of cytomegalovirus in human arterial smooth muscle cells. *J Virol* 56:839–845, 1985.

432. Smiley ML, Mar EC, Huang ES: Cytomegalovirus infection and viral-induced transformation of human endothelial cells. *J Med Virol* 25:213–226, 1988.

433. Waldman WJ, Adams PW, Orosz CG, Sedmak DD: T lymphocyte activation by cytomegalovirus-infected, allogeneic cultured human endothelial cells. *Transplantation* 54:887–896, 1992.

434. Hosenpud JD, Chou SW, Wagner CR: Cytomegalovirus-induced regulation of major histocompatibility complex class I antigen expression in human aortic smooth muscle cells. *Transplantation* 52:896–903, 1991.

435. Grefte A, van der Giessen M, van Son W, The TH: Circulating cytomegalovirus (CMV)-infected endothelial cells in patients with an active CMV infection. *J Infect Dis* 167:270–277, 1993.

436. Grefte A, Blom N, Van der Giessen M, *et al*: Ultrastructural analysis of the circulating cytomegalic cells in patients with an active cytomegalovirus infection: Evidence for virus production and endothelial origin. *J Infect Dis* 168:1110–1118, 1993.

437. Srivastava R, Curtis M, Hendrickson S, Burns WH, Hosenpud JD: Strain specific effects of cytomegalovirus on endothelial cells: Implications for investigating the relationship between CMV and cardiac allograft vasculopathy. *Transplantation* 68:1568–1573, 1999.

438. Milne DS, Gascoigne A, Wilkes J, *et al*: The immunohistopathology of obliterative bronchiolitis following lung transplantation. *Transplantation* 54:748–750, 1992.

439. Reinsmoen NL, Bolman RM, Savik K, Butters K, Hertz MI: Are multiple immunopathogenetic events occurring during the development of obliterative bronchiolitis and acute rejection? *Transplantation* 55:1040–1044, 1993.

440. Nakhleh RE, Bolman RM, Henke CA, Hertz MI: Lung transplant pathology. A comparative study of pulmonary acute rejection and cytomegaloviral infection. *Am J Surg Pathol* 15:1197–1201, 1991.

441. Scott JP, Higenbottam TW, Sharples L, *et al*: Risk factors for obliterative bronchiolitis in heart–lung transplant recipients [published erratum appears in *Transplantation* 52(2):388, 1991]. *Transplantation* 51:813–817, 1991.

442. Keenan RJ, Lega ME, Dummer JS, *et al*: Cytomegalovirus serologic status and postoperative infection correlated with risk of developing chronic rejection after pulmonary transplantation. *Transplantation* 51:433–438, 1991.

443. Smith MA, Sundaresan S, Mohanakumar T, *et al*: Effect of development of antibodies to HLA and cytomegalovirus mismatch on lung transplantation survival and development of bronchiolitis obliterans syndrome. *J Thorac Cardiovasc Surg* **116:** 812–820, 1998.

444. Ross DJ, Jordan SC, Nathan SD, Kass RM, Koerner SK: Delayed development of obliterative bronchiolitis syndrome with OKT3 after unilateral lung transplantation. A plea for multicenter immunosuppressive trials [see comments]. *Chest* **109:**870–873, 1996.

445. Girgis RE, Tu I, Berry GJ, *et al*: Risk factors for the development of obliterative bronchiolitis after lung transplantation. *J Heart Lung Transplant* **15:**1200–1208, 1996.

446. Humar A, Gillingham KJ, Payne WD, *et al*: Association between cytomegalovirus disease and chronic rejection in kidney transplant recipients. *Transplantation* **68:**1879–1883, 1999.

447. Evans PC, Soin A, Wreghitt TG, *et al*: An association between cytomegalovirus infection and chronic rejection after liver transplantation. *Transplantation* **69:**30–35, 2000.

448. Kas-Deelen AM, de Maar EF, Harmsen MC, *et al*: Uninfected and cytomegalic endothelial cells in blood during cytomegalovirus infection: Effect of acute rejection. *J Infect Dis* **181:**721–724, 2000.

449. Lemstrom KB, Bruning JH, Bruggeman CA, Lautenschlager IT, Hayry PJ: Cytomegalovirus infection enhances smooth muscle cell proliferation and intimal thickening of rat aortic allografts. *J Clin Invest* **92:**549–558, 1993.

450. Koskinen P, Lemstrom K, Bruggeman C, Lautenschlager I, Hayry P: Acute cytomegalovirus infection induces a subendothelial inflammation (endothelialitis) in the allograft vascular wall. A possible linkage with enhanced allograft arteriosclerosis. *Am J Pathol* **144:**41–50, 1994.

451. Lemstrom KB, Bruning JH, Bruggeman CA, *et al*: Cytomegalovirus infection-enhanced allograft arteriosclerosis is prevented by DHPG prophylaxis in the rat. *Circulation* **90:**1969–1978, 1994.

452. Koskinen PK, Kallio EA, Bruggeman CA, Lemstrom KB: Cytomegalovirus infection enhances experimental obliterative bronchiolitis in rat tracheal allografts. *Am J Respir Crit Care Med* **155:** 2078–2088, 1997.

453. Reichenspurner H, Soni V, Nitschke M, *et al*: Enhancement of obliterative airway disease in rat tracheal allografts infected with recombinant rat cytomegalovirus. *J Heart Lung Transplant* **17:** 439–451, 1998.

454. Lemstrom K, Sihvola R, Bruggeman C, Hayry P, Koskinen P: Cytomegalovirus infection-enhanced cardiac allograft vasculopathy is abolished by DHPG prophylaxis in the rat. *Circulation* **95:**2614–2616, 1997.

455. Arbustini E, Morbini P, Grasso M, *et al*: Human cytomegalovirus early infection, acute rejection, and major histocompatibility class II expression in transplanted lung. Molecular, immunocytochemical, and histopathologic investigations. *Transplantation* **61:**418–427, 1996.

456. Steinmuller C, Steinhoff G, Bauer D, *et al*: Analysis of leukocyte activation during acute rejection of pulmonary allografts in noninfected and cytomegalovirus-infected rats. *J Leukoc Biol* **61:**40–49, 1997.

457. Arkonac B, Mauck KA, Chou S, Hosenpud JD: Low multiplicity cytomegalovirus infection of human aortic smooth muscle cells increases levels of major histocompatibility complex class I antigens and induces a pro-inflammatory cytokine milieu in the absence of cytopathology. *J Heart Lung Transplant* **16:**1035–1045, 1997.

458. Dengler TJ, Raftery MJ, Werle M, Zimmerman R, Schonrich G: Cytomegalovirus infection of vascular cells induces expression of proinflammatory adhesion molecules by paracrine action of secreted interleukin-1beta. *Transplantation* **69:**1160–1168, 2000.

459. Knight DA, Waldman WJ, Sedmak DD: Cytomegalovirus-mediated modulation of adhesion molecule expression by human arterial and microvascular endothelial cells. *Transplantation* **68:**1814–1818, 1999.

460. Iwamoto GK, Monick MM, Clark BD, *et al*: Modulation of interleukin 1 beta gene expression by the immediate early genes of human cytomegalovirus. *J Clin Invest* **85:**1853–1857, 1990.

461. Iwamoto GK, Konicek SA: Cytomegalovirus immediate early genes upregulate interleukin-6 gene expression. *J Invest Med* **45:**175–182, 1997.

462. Cebulla CM, Miller DM, Knight DA, *et al*: Cytomegalovirus induces sialyl Lewis(x) and Lewis(x) on human endothelial cells. *Transplantation* **69:**1202–1209, 2000.

463. Lautenschlager I, Soots A, Krogerus L, *et al*: Effect of cytomegalovirus on an experimental model of chronic renal allograft rejection under triple-drug treatment in the rat. *Transplantation* **64:**391–398, 1997.

464. Kloover JS, Soots AP, Krogerus LA, *et al*: Rat cytomegalovirus infection in kidney allograft recipients is associated with increased expression of intracellular adhesion molecule-1, vascular adhesion molecule-1, and their ligands leukocyte function antigen-1 and very late antigen-4 in the graft. *Transplantation* **69:** 2641–2647, 2000.

465. van Dam JG, Li F, Yin M, *et al*: Effects of cytomegalovirus infection and prolonged cold ischemia on chronic rejection of rat renal allografts. *Transplant Int* **13:**54–63, 2000.

466. Toyoda M, Galfayan K, Galera OA, *et al*: Cytomegalovirus infection induces anti-endothelial cell antibodies in cardiac and renal allograft recipients. *Transplant Immunol* **5:**104–111, 1997.

467. Toyoda M, Petrosian A, Jordan SC: Immunological characterization of anti-endothelial cell antibodies induced by cytomegalovirus infection. *Transplantation* **68:**1311–1318, 1999.

468. Span AH, Van Boven CP, Bruggeman CA: The effect of cytomegalovirus infection on the adherence of polymorphonuclear leucocytes to endothelial cells. *Eur J Clin Invest* **19:**542–548, 1989.

469. Span AH, Endert J, van Boven CP, Bruggeman CA: Virus induced adherence of monocytes to endothelial cells. *FEMS Microbiol Immunol* **1:**237–244, 1989.

470. Waldman WJ, Knight DA, Adams PW: Cytolytic activity against allogeneic human endothelia: Resistance of cytomegalovirus-infected cells and virally activated lysis of uninfected cells. *Transplantation* **66:**67–77, 1998.

471. Craigen JL, Yong KL, Jordan NJ, *et al*: Human cytomegalovirus infection up-regulates interleukin-8 gene expression and stimulates neutrophil transendothelial migration. *Immunology* **92:**138–145, 1997.

472. Grundy JE, Lawson KM, MacCormac LP, Fletcher JM, Yong KL: Cytomegalovirus-infected endothelial cells recruit neutrophils by the secretion of C-X-C chemokines and transmit virus by direct neutrophil–endothelial cell contact and during neutrophil transendothelial migration. *J Infect Dis* **177:**1465–1474, 1998.

473. Fujinami RS, Nelson JA, Walker L, Oldstone MB: Sequence homology and immunologic cross-reactivity of human cyto-

megalovirus with HLA-DR beta chain: A means for graft rejection and immunosuppression. *J Virol* **62:**100–105, 1988.

474. Beck S, Barrell BG: Human cytomegalovirus encodes a glycoprotein homologous to MHC class-I antigens. *Nature* **331:**269–272, 1988.

475. Grundy JE, Mackenzie JS, Stanley NF: Influence of H-2 and non-H-2 genes on resistance to murine cytomegalovirus infection. *Infect Immun* **32:**277–286, 1981.

476. Allan JE, Shellam GR: Genetic control of murine cytomegalovirus infection: virus titres in resistant and susceptible strains of mice. *Arch Virol* **81:**139–150, 1984.

477. Blancho G, Josien R, Douillard D, *et al*: The influence of HLA A-B-DR matching on cytomegalovirus disease after renal transplantation. Evidence that HLA-DR7-matched recipients are more susceptible to cytomegalovirus disease. *Transplantation* **54:**871–874, 1992.

478. Manez R, White LT, Linden P, *et al*: The influence of HLA matching on cytomegalovirus hepatitis and chronic rejection after liver transplantation. *Transplantation* **55:**1067–1071, 1993.

479. Humbert M, Delattre RM, Fattal S, *et al*: *In situ* production of interleukin-6 within human lung allografts displaying rejection or cytomegalovirus pneumonia. *Transplantation* **56:**623–627, 1993.

480. Lebranchu Y, al Najjar A, Kapahi P, *et al*: The association of increased soluble VCAM-1 levels with CMV disease in human kidney allograft recipients. *Transplant Proc* **27:**960, 1995.

481. Grundy JE, Downes KL: Up-regulation of LFA-3 and ICAM-1 on the surface of fibroblasts infected with cytomegalovirus. *Immunology* **78:**405–412, 1993.

482. Siegel DL, Fox I, Dafoe DC, *et al*: Discriminating rejection from CMV infection in renal allograft recipients using flow cytometry. *Clin Immunol Immunopathol* **51:**157–171, 1989.

483. Grundy JE, Shearer GM: The effect of cytomegalovirus infection on the host response to foreign and hapten-modified self histocompatibility antigens. *Transplantation* **37:**484–490, 1984.

484. Lowance D, Neumayer HH, Legendre CM, *et al*: Valacyclovir for the prevention of cytomegalovirus disease after renal transplantation. International Valacyclovir Cytomegalovirus Prophylaxis Transplantation Study Group [see comments]. *N Engl J Med* **340:**1462–1470, 1999.

485. Valantine HA, Gao SZ, Menon SG, *et al*: Impact of prophylactic immediate posttransplant ganciclovir on development of transplant atherosclerosis: A post hoc analysis of a randomized, placebo-controlled study. *Circulation* **100:**61–66, 1999.

486. Valantine H: Role of CMV in transplant coronary artery disease and survival after heart transplantation. *Transplant Infect Dis* **1**(Suppl):25–30, 1999.

487. Spector DH, Vacquier JP: Human cytomegalovirus (strain AD169) contains sequences related to the avian retrovirus oncogene v-myc. *Proc Natl Acad Sci USA* **80:**3889–3893, 1983.

488. Gelmann EP, Clanton DJ, Jariwalla RJ, Rosenthal LJ: Characterization and location of myc homologous sequences in human cytomegalovirus DNA. *Proc Natl Acad Sci USA* **80:**5107–5111, 1983.

489. Nelson JA, Fleckenstein B, Galloway DA, McDougall JK: Transformation of NIH 3T3 cells with cloned fragments of human cytomegalovirus strain AD169. *J Virol* **43:**83–91, 1982.

490. Mueller N, Hinkula J, Wahren B: Elevated antibody titers against cytomegalovirus among patients with testicular cancer. *Int J Cancer* **41:**399–403, 1988.

491. Ho M, Miller G, Atchison RW, *et al*: Epstein–Barr virus infec-

tions and DNA hybridization studies in posttransplantation lymphoma and lymphoproliferative lesions: The role of primary infection. *J Infect Dis* **152:**876–886, 1985.

492. Chou S: Newer methods for diagnosis of cytomegalovirus infection. *Rev Infect Dis* **12**(Suppl 7):S727–S736, 1990.

493. Chou SW: Neutralizing antibody responses to reinfecting strains of cytomegalovirus in transplant recipients. *J Infect Dis* **160:**16–21, 1989.

494. Farrell HE, Shellam GR: Protection against murine cytomegalovirus infection by passive transfer of neutralizing and non-neutralizing monoclonal antibodies. *J Gen Virol* **72:**149–156, 1991.

495. Gretch DR, Kari B, Rasmussen L, Gehrz RC, Stinksi MF: Identification and characterization of three distinct families of glycoprotein complexes in the envelopes of human cytomegalovirus. *J Virol* **62:**875–881, 1988.

496. Britt WJ, Vugler L, Stephens EB: Induction of complement-dependent and -independent neutralizing antibodies by recombinant-derived human cytomegalovirus gp55-16 (gB). *J Virol* **62:**3309–3318, 1988.

497. Rasmussen L, Nelson M, Neff M, Merigan TC Jr: Characterization of two different human cytomegalovirus glycoproteins which are targets for virus neutralizing antibody. *Virology* **163:**308–318, 1988.

498. Rasmussen L, Matkin C, Spaete R, Pachl C, Merigan TC: Antibody response to human cytomegalovirus glycoproteins gB and gH after natural infection in humans. *J Infect Dis* **164:**835–842, 1991.

499. van Zanten J, van der Giessen M, van der Voort LH, *et al*: Cytomegalovirus-specific antibodies to an immediate early antigen and a late membrane antigen and their possible role in controlling secondary cytomegalovirus infection. *Clin Exp Immunol* **83:**102–107, 1991.

500. Marshall GS, Rabalais GP, Stout GG, Waldeyer SL: Antibodies to recombinant-derived glycoprotein B after natural human cytomegalovirus infection correlate with neutralizing activity. *J Infect Dis* **165:**381–384, 1992.

501. Chou S: Molecular epidemiology of envelope glycoprotein H of human cytomegalovirus. *J Infect Dis* **166:**604–607, 1992.

502. Chou SW, Scott KM: Rapid quantitation of cytomegalovirus and assay of neutralizing antibody by using monoclonal antibody to the major immediate-early viral protein. *J Clin Microbiol* **26:**504–507, 1988.

503. Gleaves CA, Smith TF, Shuster EA, Pearson GR: Comparison of standard tube and shell vial cell culture techniques for the detection of cytomegalovirus in clinical specimens. *J Clin Microbiol* **21:**217–221, 1985.

504. Swenson PD, Kaplan MH: Rapid detection of cytomegalovirus in cell culture by indirect immunoperoxidase staining with monoclonal antibody to an early nuclear antigen. *J Clin Microbiol* **21:**669–673, 1985.

505. Paya CV, Wold AD, Smith TF: Detection of cytomegalovirus infections in specimens other than urine by the shell vial assay and conventional tube cell cultures. *J Clin Microbiol* **25:**755–757, 1987.

506. Marsano L, Perrillo RP, Flye MW, *et al*: Comparison of culture and serology for the diagnosis of cytomegalovirus infection in kidney and liver transplant recipients. *J Infect Dis* **161:**454–461, 1990.

507. Dummer JS, White LT, Ho M, *et al*: Morbidity of cytomegalovirus infection in recipients of heart or heart–lung transplants who received cyclosporine. *J Infect Dis* **152:**1182–1191, 1985.

508. The TH, van der Ploeg M, van den Berg AP, *et al*: Direct detection of cytomegalovirus in peripheral blood leukocytes—A review of the antigenemia assay and polymerase chain reaction. *Transplantation* **54:**193–198, 1992.

509. Gerna G, Revello MG, Percivalle E, *et al*: Quantification of human cytomegalovirus viremia by using monoclonal antibodies to different viral proteins. *J Clin Microbiol* **28:**2681–2688, 1990.

510. Miller H, Rossier E, Milk R, Thomas C: Prospective study of cytomegalovirus antigenemia in allograft recipients [see comments]. *J Clin Microbiol* **29:**1054–1055, 1991.

511. Gerna G, Zipeto D, Parea M, *et al*: Monitoring of human cytomegalovirus infections and ganciclovir treatment in heart transplant recipients by determination of viremia, antigenemia, and DNAemia. *J Infect Dis* **164:**488–498, 1991.

512. van der Bij W, van Dijk RB, van Son WJ, *et al*: Antigen test for early diagnosis of active cytomegalovirus infection in heart transplant recipients. *J Heart Transplant* **7:**106–109, 1988.

513. van den Berg AP, van der Bij W, van Son WJ, *et al*: Cytomegalovirus antigenemia as a useful marker of symptomatic cytomegalovirus infection after renal transplantation—A report of 130 consecutive patients. *Transplantation* **48:**991–995, 1989.

514. van den Berg AP, Klompmaker IJ, Haagsma EB, *et al*: Antigenemia in the diagnosis and monitoring of active cytomegalovirus infection after liver transplantation. *J Infect Dis* **164:**265–270, 1991.

515. van den Berg AP, Tegzess AM, Scholten-Sampson A, *et al*: Monitoring antigenemia is useful in guiding treatment of severe cytomegalovirus disease after organ transplantation. *Transplant Int* **5:**101–106, 1992.

516. Erice A, Holm MA, Gill PC, *et al*: Cytomegalovirus (CMV) antigenemia assay is more sensitive than shell vial cultures for rapid detection of CMV in polymorphonuclear blood leukocytes [see comments]. *J Clin Microbiol* **30:**2822–2825, 1992.

517. Boland GJ, Ververs C, Hene RJ, *et al*: Early detection of primary cytomegalovirus infection after heart and kidney transplantation and the influence of hyperimmune globulin prophylaxis. *Transplant Int* **6:**34–38, 1993.

518. Koskinen PK, Nieminen MS, Mattila SP, Hayry PJ, Lautenschlager IT: The correlation between symptomatic CMV infection and CMV antigenemia in heart allograft recipients. *Transplantation* **55:**547–551, 1993.

519. Gaeta A, Nazzari C, Angeletti S, *et al*: Monitoring for cytomegalovirus infection in organ transplant recipients: Analysis of pp65 antigen, DNA and late mRNA in peripheral blood leukocytes. *J Med Virol* **53:**189–195, 1997.

520. Saiki RK, Scharf S, Faloona F, *et al*: Enzymatic amplification of beta-globin genomic sequences and restriction site analysis for diagnosis of sickle cell anemia. *Science* **230:**1350–1354, 1985.

521. van Dorp WT, Vlieger A, Jiwa NM, *et al*: The polymerase chain reaction, a sensitive and rapid technique for detecting cytomegalovirus infection after renal transplantation. *Transplantation* **54:**661–664, 1992.

522. Fox JC, Griffiths PD, Emery VC: Quantification of human cytomegalovirus DNA using the polymerase chain reaction. *J Gen Virol* **73:**2405–2408, 1992.

523. Bitsch A, Kirchner H, Dupke R, Bein G: Cytomegalovirus transcripts in peripheral blood leukocytes of actively infected transplant patients detected by reverse transcription-polymerase chain reaction. *J Infect Dis* **167:**740–743, 1993.

524. Kotsimbos AT, Sinickas V, Glare EM, *et al*: Quantitative detection of human cytomegalovirus DNA in lung transplant recipients. *Am J Respir Crit Care Med* **156:**1241–1246, 1997.

525. Abecassis MM, Koffron AJ, Kaplan B, *et al*: The role of PCR in the diagnosis and management of CMV in solid organ recipients: What is the predictive value for the development of disease and should PCR be used to guide antiviral therapy? *Transplantation* **63:**275–279, 1997.

526. Toyoda M, Carlos JB, Galera OA, *et al*: Correlation of cytomegalovirus DNA levels with response to antiviral therapy in cardiac and renal allograft recipients. *Transplantation* **63:**957–963, 1997.

527. Evans PC, Soin A, Wreghitt TG, Alexander GJ: Qualitative and semiquantitative polymerase chain reaction testing for cytomegalovirus DNA in serum allows prediction of CMV related disease in liver transplant recipients. *J Clin Pathol* **51:**914–921, 1998.

528. Pellegrin I, Garrigue I, Ekouevi D, *et al*: New molecular assays to predict occurrence of cytomegalovirus disease in renal transplant recipients. *J Infect Dis* **182:**36–42, 2000.

529. Humar A, Gregson D, Caliendo AM, *et al*: Clinical utility of quantitative cytomegalovirus viral load determination for predicting cytomegalovirus disease in liver transplant recipients. *Transplantation* **68:**1305–1311, 1999.

530. Cope AV, Sweny P, Sabin C, *et al*: Quantity of cytomegalovirus viruria is a major risk factor for cytomegalovirus disease after renal transplantation. *J Med Virol* **52:**200–205, 1997.

531. Cope AV, Sabin C, Burroughs A, *et al*: Interrelationships among quantity of human cytomegalovirus (HCMV) DNA in blood, donor–recipient serostatus, and administration of methylprednisolone as risk factors for HCMV disease following liver transplantation. *J Infect Dis* **176:**1484–1490, 1997.

532. Mendez J, Espy M, Smith TF, *et al*: Clinical significance of viral load in the diagnosis of cytomegalovirus disease after liver transplantation. *Transplantation* **65:**1477–1481, 1998.

533. Hassan-Walker AF, Kidd IM, Sabin C, *et al*: Quantity of human cytomegalovirus (CMV) DNAemia as a risk factor for CMV disease in renal allograft recipients: Relationship with donor/recipient CMV serostatus, receipt of augmented methylprednisolone and antithymocyte globulin (ATG). *J Med Virol* **58:**182–187, 1999.

534. Roberts TC, Brennan DC, Buller RS, *et al*: Quantitative polymerase chain reaction to predict occurrence of symptomatic cytomegalovirus infection and assess response to ganciclovir therapy in renal transplant recipients. *J Infect Dis* **178:**626–635, 1998.

535. Mas V, Alvarellos T, Albano S, *et al*: Utility of cytomegalovirus viral load in renal transplant patients in Argentina. *Transplantation* **67:**1050–1055, 1999.

536. Tong CY, Cuevas LE, Williams H, Bakran A: Prediction and diagnosis of cytomegalovirus disease in renal transplant recipients using qualitative and quantitative polymerase chain reaction. *Transplantation* **69:**985–991, 2000.

537. Barrett-Muir W, Breuer J, Millar C, *et al*: CMV viral load measurements in whole blood and plasma—Which is best following renal transplantation? *Transplantation* **70:**116–119, 2000.

538. Emery VC, Sabin CA, Cope AV, *et al*: Application of viral-load kinetics to identify patients who develop cytomegalovirus disease after transplantation. *Lancet* **355:**2032–2036, 2000.

539. Sia IG, Wilson JA, Groettum CM, *et al*: Cytomegalovirus (CMV) DNA load predicts relapsing CMV infection after solid organ transplantation. *J Infect Dis* **181:**717–720, 2000.

540. Emery V, Sabin C, Cope A, *et al*: Can the application of viral dynamics identify transplant patients destined to develop cytomegalovirus disease? *Lancet* **355:**2032–2036, 2000.

541. Griffiths P, Cope A, Hassan-Walker A, Emery V: Diagnostic

approach to cytomegalovirus infection in bone marrow and organ transplantation. *Transplant Infect Dis* **1:**179–186, 1999.

542. De Witte T, Schattenberg A, Van Dijk BA, *et al:* Prevention of primary cytomegalovirus infection after allogeneic bone marrow transplantation by using leukocyte-poor random blood products from cytomegalovirus-unscreened blood-bank donors. *Transplantation* **50:**964–968, 1990.

543. Preiksaitis JK: Indications for the use of cytomegalovirus-seronegative blood products. *Transfus Med Rev* **5:**1–17, 1991.

544. Sayers MH, Anderson KC, Goodnough LT, *et al:* Reducing the risk for transfusion-transmitted cytomegalovirus infection. *Ann Intern Med* **116:**55–62, 1992.

545. Starr SE, Glazer JP, Friedman HM, Farquhar JD, Plotkin SA: Specific cellular and humoral immunity after immunization with live Towne strain cytomegalovirus vaccine. *J Infect Dis* **143:**585–589, 1981.

546. Plotkin SA, Smiley ML, Friedman HM, *et al:* Towne-vaccine-induced prevention of cytomegalovirus disease after renal transplants. *Lancet* **1:**528–530, 1984.

547. Plotkin SA, Starr SE, Friedman HM, Gonczol E, Brayman K: Vaccines for the prevention of human cytomegalovirus infection. *Rev Infect Dis* **12**(Suppl 7)**:**S827–838, 1990.

548. Plotkin SA, Starr SE, Friedman HM, *et al:* Effect of Towne live virus vaccine on cytomegalovirus disease after renal transplant. A controlled trial [see comments]. *Ann Intern Med* **114:**525–531, 1991.

549. Spaete RR: A recombinant subunit vaccine approach to HCMV vaccine development. *Transplant Proc* **23:**90–96, 1991.

550. Rubin RH, Wilson EJ, Barrett LV, Medearis DN: The protective effects of hyperimmune anti-murine cytomegalovirus antiserum against lethal viral challenge: The case for passive-active immunization. *Clin Immunol Immunopathol* **39:**151–158, 1986.

551. Roy DM, Grundy JE: Evaluation of neutralizing antibody titers against human cytomegalovirus in intravenous gamma globulin preparations [see comments]. *Transplantation* **54:**1109–1110, 1992.

552. Snydman DR, Werner BG, Heinze-Lacey B, *et al:* Use of cytomegalovirus immune globulin to prevent cytomegalovirus disease in renal-transplant recipients. *N Engl J Med* **317:**1049–1054, 1987.

553. Snydman DR, Werner BG, Tilney NL, *et al:* A further analysis of primary cytomegalovirus disease prevention in renal transplant recipients with a cytomegalovirus immune globulin: interim comparison of a randomized and an open-label trial. *Transplant Proc* **20:**24–30, 1988.

554. Snydman DR, Werner BG, Tilney NL, *et al:* Final analysis of primary cytomegalovirus disease prevention in renal transplant recipients with a cytomegalovirus-immune globulin: Comparison of the randomized and open-label trials. *Transplant Proc* **23:**1357–1360, 1991.

555. Tsevat J, Snydman DR, Pauker SG, *et al:* Which renal transplant patients should receive cytomegalovirus immune globulin? A cost-effectiveness analysis. *Transplantation* **52:**259–265, 1991.

556. Metselaar HJ, Balk AH, Mochtar B, Rothbarth PH, Weimar W: Cytomegalovirus seronegative heart transplant recipients. Prophylactic use of anti-CMV immunoglobulin. *Chest* **97:**396–399, 1990.

557. Metselaar HJ, Rothbarth PH, Brouwer RM, *et al:* Prevention of cytomegalovirus-related death by passive immunization. A double-blind placebo-controlled study in kidney transplant recipients treated for rejection. *Transplantation* **48:**264–266, 1989.

558. Pakkala S, Salmela K, Lautenschlager I, Ahonen J, Hayry P: Anti-CMV hyperimmune globulin prophylaxis does not prevent CMV disease in CMV-negative renal transplant patients. *Transplant Proc* **24:** 283–284, 1992.

559. Falagas ME, Snydman DR, Ruthazer R, *et al:* Cytomegalovirus immune globulin (CMVIG) prophylaxis is associated with increased survival after orthotopic liver transplantation. The Boston Center for Liver Transplantation CMVIG Study Group. *Clin Transplant* **11:**432–437, 1997.

560. Steinmuller DR, Novick AC, Streem SB, Graneto D, Swift C: Intravenous immunoglobulin infusions for the prophylaxis of secondary cytomegalovirus infection. *Transplantation* **49:**68–70, 1990.

561. Flynn JT, Kaiser BA, Long SS, *et al:* Intravenous immunoglobulin prophylaxis of cytomegalovirus infection in pediatric renal transplant recipients. *Am J Nephrol* **17:**146–152, 1997.

562. Bock GH, Sullivan EK, Miller D, *et al:* Cytomegalovirus infections following renal transplantation—Effects on antiviral prophylaxis: A report of the North American Pediatric Renal Transplant Cooperative Study. *Pediatr Nephrol* **11:**665–671, 1997.

563. Balfour HH Jr, Chace BA, Stapleton JT, Simmons RL, Fryd DS: A randomized, placebo-controlled trial of oral acyclovir for the prevention of cytomegalovirus disease in recipients of renal allografts [see comments]. *N Engl J Med* **320:**1381–1387, 1989.

564. Balfour HH Jr: Prevention of cytomegalovirus disease in renal allograft recipients. *Scand J Infect Dis Suppl* **80:**88–93, 1991.

565. Fletcher CV, Englund JA, Edelman CK, *et al:* Pharmacologic basis for high-dose oral acyclovir prophylaxis of cytomegalovirus disease in renal allograft recipients. *Antimicrob Agents Chemother* **35:**938–943, 1991.

566. Vasquez EM, Sanchez J, Pollak R, *et al:* High-dose oral acyclovir prophylaxis for primary cytomegalovirus infection in seronegative renal allograft recipients. *Transplantation* **55:**448–450, 1993.

567. Gavalda J, de Otero J, Murio E, *et al:* Two grams daily of oral acyclovir reduces the incidence of cytomegalovirus disease in CMV-seropositive liver transplant recipients. *Transplant Int* **10:**462–465, 1997.

568. Rostaing L, Martinet O, Cisterne JM, *et al:* CMV prophylaxis in high-risk renal transplant patients (D+/R−) by acyclovir with or without hyperimmune (CMV) immunoglobulins: A prospective study. *Am J Nephrol* **17:**489–494, 1997.

569. Badley AD, Seaberg EC, Porayko MK, *et al:* Prophylaxis of cytomegalovirus infection in liver transplantation: A randomized trial comparing a combination of ganciclovir and acyclovir to acyclovir. NIDDK Liver Transplantation Database. *Transplantation* **64:**66–73, 1997.

570. Martin M, Manez R, Linden P, *et al:* A prospective randomized trial comparing sequential ganciclovir-high dose acyclovir to high dose acyclovir for prevention of cytomegalovirus disease in adult liver transplant recipients. *Transplantation* **58:**779–785, 1994.

571. Nicol DL, MacDonald AS, Belitsky P, *et al:* Reduction by combination prophylactic therapy with CMV hyperimmune globulin and acyclovir of the risk of primary CMV disease in renal transplant recipients. *Transplantation* **55:**841–846, 1993.

572. Stratta RJ, Shaefer MS, Cushing KA, *et al:* Successful prophylaxis of cytomegalovirus disease after primary CMV exposure in liver transplant recipients. *Transplantation* **51:**90–97, 1991.

573. Goral S, Ynares C, Dummer S, Helderman JH: Acyclovir prophylaxis for cytomegalovirus disease in high-risk renal transplant recipients: Is it effective? *Kidney Int Suppl* **57:**S62–S65, 1996.

574. Flechner SM, Avery RK, Fisher R, *et al:* A randomized prospec-

tive controlled trial of oral acyclovir versus oral ganciclovir for cytomegalovirus prophylaxis in high-risk kidney transplant recipients. *Transplantation* **66:**1682–1688, 1998.

575. Avery RK: Prevention and treatment of cytomegalovirus infection and disease in heart transplant recipients. *Curr Opin Cardiol* **13:**122–129, 1998.

576. Legendre CM, Norman DJ, Keating MR, Maclaine GD, Grant DM: Valaciclovir prophylaxis of cytomegalovirus infection and disease in renal transplantation: An economic evaluation. *Transplantation* **70:**1463–1468, 2000.

577. Merigan TC, Renlund DG, Keay S, *et al*: A controlled trial of ganciclovir to prevent cytomegalovirus disease after heart transplantation. *N Engl J Med* **326:**1182–1186, 1992.

578. Wreghitt TG, Abel SJ, McNeil K, *et al*: Intravenous ganciclovir prophylaxis for cytomegalovirus in heart, heart–lung, and lung transplant recipients. *Transplant Int* **12:**254–260, 1999.

579. Seu P, Winston DJ, Holt CD, *et al*: Long-term ganciclovir prophylaxis for successful prevention of primary cytomegalovirus (CMV) disease in CMV-seronegative liver transplant recipients with CMV-seropositive donors. *Transplantation* **64:**1614–1617, 1997.

580. Hertz MI, Jordan C, Savik SK, *et al*: Randomized trial of daily versus three-times-weekly prophylactic ganciclovir after lung and heart–lung transplantation. *J Heart Lung Transplant* **17:**913–920, 1998.

581. Turgeon N, Fishman JA, Basgoz N, *et al*: Effect of oral acyclovir or ganciclovir therapy after preemptive intravenous ganciclovir therapy to prevent cytomegalovirus disease in cytomegalovirus seropositive renal and liver transplant recipients receiving antilymphocyte antibody therapy. *Transplantation* **66:**1780–1786, 1998.

582. Gane E, Saliba F, Valdecasas GJ, *et al*: Randomised trial of efficacy and safety of oral ganciclovir in the prevention of cytomegalovirus disease in liver-transplant recipients. The Oral Ganciclovir International Transplantation Study Group [corrected] [see comments] [published erratum appears in *Lancet* **351**(9100)**:** 454, 1997]. *Lancet* **350:**1729–1733, 1997.

583. Ahsan N, Holman MJ, Yang HC: Efficacy of oral ganciclovir in prevention of cytomegalovirus infection in post-kidney transplant patients. *Clin Transplant* **11:**633–639, 1997.

584. Brennan DC, Garlock KA, Lippmann BA, *et al*: Control of cytomegalovirus-associated morbidity in renal transplant patients using intensive monitoring and either preemptive or deferred therapy. *J Am Soc Nephrol* **8:**118–125, 1997.

585. Pescovitz MD, Pruett TL, Gonwa T, *et al*: Oral ganciclovir dosing in transplant recipients and dialysis patients based on renal function. *Transplantation* **66:**1104–1107, 1998.

586. Pescovitz MD: Absence of teratogenicity of oral ganciclovir used during early pregnancy in a liver transplant recipient. *Transplantation* **67:**758–759, 1999.

587. Bailey TC, Trulock EP, Ettinger NA, *et al*: Failure of prophylactic ganciclovir to prevent cytomegalovirus disease in recipients of lung transplants. *J Infect Dis* **165:**548–552, 1992.

588. Gerbase MW, Dubois D, Rothmeier C, *et al*: Costs and outcomes of prolonged cytomegalovirus prophylaxis to cover the enhanced immunosuppression phase following lung transplantation [see comments]. *Chest* **116:**1265–1272, 1999.

589. Speich R, Thurnheer R, Gaspert A, Weder W, Boehler A: Efficacy and cost effectiveness of oral ganciclovir in the prevention of cytomegalovirus disease after lung transplantation. *Transplantation* **67:**315–320, 1999.

590. Gutierrez CA, Chaparro C, Krajden M, Winton T, Kesten S: Cytomegalovirus viremia in lung transplant recipients receiving ganciclovir and immune globulin. *Chest* **113:**924–932, 1998.

591. Cen H, Breinig MC, Atchison RW, Ho M, McKnight JL: Epstein–Barr virus transmission via the donor organs in solid organ transplantation: Polymerase chain reaction and restriction fragment length polymorphism analysis of IR2, IR3, and IR4. *J Virol* **65:**976–980, 1991.

592. Schmidt GM, Horak DA, Niland JC, *et al*: A randomized, controlled trial of prophylactic ganciclovir for cytomegalovirus pulmonary infection in recipients of allogeneic bone marrow transplants; the City of Hope-Stanford-Syntex CMV Study Group [see comments]. *N Engl J Med* **324:**1005–1011, 1991.

593. Goodrich JM, Mori M, Gleaves CA, *et al*: Early treatment with ganciclovir to prevent cytomegalovirus disease after allogeneic bone marrow transplantation. *N Engl J Med* **325:**1601–1607, 1991.

594. Egan JJ, Lomax J, Barber L, *et al*: Preemptive treatment for the prevention of cytomegalovirus disease: in lung and heart transplant recipients. *Transplantation* **65:**747–752, 1998.

595. Kusne S, Grossi P, Irish W, *et al*: Cytomegalovirus PP65 antigenemia monitoring as a guide for preemptive therapy: A cost effective strategy for prevention of cytomegalovirus disease in adult liver transplant recipients. *Transplantation* **68:**1125–1131, 1999.

596. Singh N, Paterson DL, Gayowski T, Wagener MM, Marino IR: Cytomegalovirus antigenemia directed pre-emptive prophylaxis with oral versus I.V. ganciclovir for the prevention of cytomegalovirus disease in liver transplant recipients: A randomized, controlled trial. *Transplantation* **70:**717–722, 2000.

597. Kunzle N, Petignat C, Francioli P, *et al*: Preemptive treatment approach to cytomegalovirus (CMV) infection in solid organ transplant patients: Relationship between compliance with the guidelines and prevention of CMV morbidity. *Transplant Infect Dis* **2:**118–126, 2000.

598. Pescovitz M: Oral ganciclovir and pharmacokinetics of valganciclovir in liver transplant recipients. *Transplant Infect Dis* **1**(Suppl 1)**:**31–34, 1999.

599. Jassal SV, Roscoe JM, Zaltzman JS, *et al*: Clinical practice guidelines: Prevention of cytomegalovirus disease after renal transplantation. *J Am Soc Nephrol* **9:**1697–1708, 1998.

600. Dunn DL, Mayoral JL, Gillingham KJ, *et al*: Treatment of invasive cytomegalovirus disease in solid organ transplant patients with ganciclovir. *Transplantation* **51:**98–106, 1991.

601. Cooper DK, Novitzky D, Schlegel V, *et al*: Successful management of symptomatic cytomegalovirus disease with ganciclovir after heart transplantation. *J Heart Lung Transplant* **10:**656–662; discussion 662–663, 1991.

602. van den Berg AP, Tegzess AM, Scholten-Sampson A, *et al*: Quo vadis? The clinical dilemma of simultaneous cytomegalovirus infection and steroid-resistant rejection. *Transplantation* **52:** 1081–1083, 1991.

603. Burns KD, Johnson-Whittaker L, Couture RA, Eidus L, Garber G: Successful treatment of renal allograft rejection in the presence of cytomegalovirus disease. A report of two cases. *Am J Nephrol* **10:**162–166, 1990.

604. de Koning J, van Dorp WT, van Es LA, van 't Wout JW, van der Woude FJ: Ganciclovir effectively treats cytomegalovirus disease after solid-organ transplantation, even during rejection treatment. *Nephrol Dial Transplant* **7:**350–356, 1992.

605. Duncan SR, Cook DJ: Survival of ganciclovir-treated heart trans-

plant recipients with cytomegalovirus pneumonitis [see comments]. *Transplantation* **52:**910–913, 1991.

606. Kruger RM, Shannon WD, Arens MQ, *et al*: The impact of ganciclovir-resistant cytomegalovirus infection after lung transplantation. *Transplantation* **68:**1272–1279, 1999.

607. Bienvenu B, Thervet E, Bedrossian J, *et al*: Development of cytomegalovirus resistance to ganciclovir after oral maintenance treatment in a renal transplant recipient. *Transplantation* **69:**182–184, 2000.

608. Alain S, Honderlick P, Grenet D, *et al*: Failure of ganciclovir treatment associated with selection of a ganciclovir-resistant cytomegalovirus strain in a lung transplant recipient. *Transplantation* **63:**1533–1536, 1997.

609. Sixbey JW, Nedrud JG, Raab-Traub N, Hanes RA, Pagano JS: Epstein–Barr virus replication in oropharyngeal epithelial cells. *N Engl J Med* **310:**1225–1230, 1984.

610. Straus SE, Cohen JI, Tosato G, Meier J: NIH conference. Epstein–Barr virus infections: Biology, pathogenesis, and management. *Ann Intern Med* **118:**45–58, 1993.

611. Preiksaitis JK, Diaz-Mitoma F, Mirzayans F, Roberts S, Tyrrell DL: Quantitative oropharyngeal Epstein–Barr virus shedding in renal and cardiac transplant recipients: Relationship to immunosuppressive therapy, serologic responses, and the risk of posttransplant lymphoproliferative disorder. *J Infect Dis* **166:**986–994, 1992.

612. Miller G: The switch between latency and replication of Epstein–Barr virus. *J Infect Dis* **161:**833–844, 1990.

613. Klein G: Viral latency and transformation: The strategy of Epstein–Barr virus. *Cell* **58:**5–8, 1989.

614. Gregory CD, Dive C, Henderson S, *et al*: Activation of Epstein–Barr virus latent genes protects human B cells from death by apoptosis. *Nature* **349:**612–614, 1991.

615. Pagano JS: Epstein–Barr virus: Culprit or consort? [editorial; comment] [see comments]. *N Engl J Med* **327:**1750–1752, 1992.

616. Cohen JI: The biology of Epstein–Barr virus: Lessons learned from the virus and the host. *Curr Opin Immunol* **11:**365–370, 1999.

617. Rowe DT: Epstein–Barr virus immortalization and latency. *Front Biosci* **4:**D346–D371, 1999.

618. Tanner J, Alfieri C: The Epstein–Barr virus and post-transplant lymphoproliferative disease: Interplay of immunosuppression, EBV, and the immune system in disease pathogenesis. *Transplant Infect Dis* **3:**60–69, 2001.

619. Sixbey JW, Shirley P, Chesney PJ, Buntin DM, Resnick L: Detection of a second widespread strain of Epstein–Barr virus. *Lancet* **2:**761–765, 1989.

620. Sixbey JW, Shirley P, Sloas M, Raab-Traub N, Israele V: A transformation-incompetent, nuclear antigen 2-deleted Epstein–Barr virus associated with replicative infection. *J Infect Dis* **163:**1008–1015, 1991.

621. Pisa P, Cannon MJ, Pisa EK, Cooper NR, Fox RI: Epstein–Barr virus induced lymphoproliferative tumors in severe combined immunodeficient mice are oligoclonal. *Blood* **79:**173–179, 1992.

622. Birkeland SA, Bendtzen K, Moller B, Hamilton-Dutoit S, Andersen HK: Interleukin-10 and posttransplant lymphoproliferative disorder after kidney transplantation. *Transplantation* **67:**876–881, 1999.

623. Vieira P, de Waal-Malefyt R, Dang MN, *et al*: Isolation and expression of human cytokine synthesis inhibitory factor cDNA clones: Homology to Epstein–Barr virus open reading frame BCRFI. *Proc Natl Acad Sci USA* **88:**1172–1176, 1991.

624. Basgoz N, Hibberd P, Tolkoff-Rubin N, *et al*: Possible role of cytomegalovirus disease in the pathogenesis of post-transplant lymphoproliferative disorder. American Society of Transplant Physicians, 12th Annual Meeting, Houston, 1993.

625. Manez R, Breinig MC, Linden P, *et al*: Posttransplant lymphoproliferative disease in primary Epstein–Barr virus infection after liver transplantation: The role of cytomegalovirus disease. *J Infect Dis* **176:**1462–1467, 1997.

626. Herbert D, Sullivan E: Malignancy and post-transplant lymphoproliferative disorder (PTLD) in pediatric renal transplant recipients: A report of the North American Pediatric Transplant Cooperative Study Group (NAPRTCS). *Pediatr Transplant* **2:**57, 1998.

627. Paya CV, Fung JJ, Nalesnik MA, *et al*: Epstein–Barr virus-induced posttransplant lymphoproliferative disorders. ASTS/ASTP EBV-PTLD Task Force and The Mayo Clinic Organized International Consensus Development Meeting. *Transplantation* **68:**1517–1525, 1999.

628. Cockfield S: Identifying the patient at risk for post-transplant lymphoproliferative disorder. *Transplant Infect Dis* 2001; **3** in press.

629. Hezode C, Duvoux C, Germanidis G, *et al*: Role of hepatitis C virus in lymphoproliferative disorders after liver transplantation [see comments]. *Hepatology* **30:**775–778, 1999.

630. Cheeseman SH, Henle W, Rubin RH, *et al*: Epstein–Barr virus infection in renal transplant recipients. Effects of antithymocyte globulin and interferon. *Ann Intern Med* **93:**39–42, 1980.

631. Spencer ES, Andersen HK: Clinically evident, non-terminal infections with herpesviruses and the wart virus in immunosuppressed renal allograft recipients. *Br Med J* **1:**251–254, 1970.

632. Strauch B, Andrews LL, Siegel N, Miller G: Oropharyngeal excretion of Epstein–Barr virus by renal transplant recipients and other patients treated with immunosuppressive drugs. *Lancet* **1:**234–237, 1974.

633. Chang RS, Lewis JP, Reynolds RD, Sullivan MJ, Neuman J: Oropharyngeal excretion of Epstein–Barr virus by patients with lymphoproliferative disorders and by recipients of renal homografts. *Ann Intern Med* **88:**34–40, 1978.

634. Grose C, Henle W, Horwitz MS: Primary Epstein–Barr virus infection in a renal transplant recipient. *South Med J* **70:**1276–1278, 1977.

635. Marker SC, Ascher NL, Kalis JM, *et al*: Epstein–Barr virus antibody responses and clinical illness in renal transplant recipients. *Surgery* **85:**433–440, 1979.

636. Salt A, Sutehall G, Sargaison M, *et al*: Viral and Toxoplasma gondii infections in children after liver transplantation. *J Clin Pathol* **43:**63–67, 1990.

637. Randhawa PS, Markin RS, Starzl TE, Demetris AJ: Epstein–Barr virus-associated syndromes in immunosuppressed liver transplant recipients. Clinical profile and recognition on routine allograft biopsy. *Am J Surg Pathol* **14:**538–547, 1990.

638. Stephaman E, Gruber S, Dunn D, *et al*: Posttransplant lymphoproliferative disorders. *Transplant Rev* **5:**120–129, 1991.

639. Randhawa PS, Yousem SA: Epstein–Barr virus-associated lymphoproliferative disease in a heart–lung allograft. Demonstration of host origin by restriction fragment-length polymorphism analysis. *Transplantation* **49:**126–130, 1990.

640. Jardine DL, Sizeland PC, Bailey RR, *et al*: Epstein–Barr virus infection acquired from a cadaveric renal transplant. *Nephron* **58:**359–361, 1991.

641. Canfield C, Hudnall S, Colonna J, *et al*: Fulminant Epstein–Barr virus associated post-transplant lymphoproliferative disorders following OKT3 therapy. *Clin Transplant* **6:**1–9, 1992.

642. Denning DW, Weiss LM, Martinez K, Flechner SM: Transmission of Epstein–Barr virus by a transplanted kidney, with activation by OKT3 antibody. *Transplantation* **48**:141–144, 1989.

643. Telenti A, Smith TF, Ludwig J, *et al*: Epstein–Barr virus and persistent graft dysfunction after liver transplantation. *Hepatology* **14**:282–286, 1991.

644. Weissmann DJ, Ferry JA, Harris NL, *et al*: Posttransplantation lymphoproliferative disorders in solid organ recipients are predominantly aggressive tumors of host origin. *Am J Clin Pathol* **103**:748–755, 1995.

645. Harris NL, Jaffe ES, Stein H, *et al*: A revised European-American classification of lymphoid neoplasms: A proposal from the International Lymphoma Study Group [see comments]. *Blood* **84**:1361–1392, 1994.

646. Harris NL, Ferry JA, Swerdlow SH: Posttransplant lymphoproliferative disorders: Summary of Society for Hematopathology Workshop. *Semin Diagn Pathol* **14**:8–14, 1997.

647. Nalesnik M: The diverse pathology of post-transplant lymphoproliferative disorder: Importance of a standardized approach. *Transplant Infect Dis* **3**:88–96, 2001.

648. Mutimer D, Kaur N, Tang H, *et al*: Quantitation of Epstein–Barr virus DNA in the blood of adult liver transplant recipients. *Transplantation* **69**:954–959, 2000.

649. Vajro P, Lucariello S, Migliaro F, *et al*: Predictive value of Epstein–Barr virus genome copy number and BZLF1 expression in blood lymphocytes of transplant recipients at risk for lymphoproliferative disease. *J Infect Dis* **181**:2050–2054, 2000.

650. Green M, Reyes J, Webber S, *et al*: The role of viral load in the diagnosis, management, and possible prevention of Epstein–Barr virus-associated post-transplant lymphoproliferative disease following solid organ transplantation. *Transplant Infect Dis* **4**:292–296, 1999.

651. Krieger NR, Martinez OM, Krams SM, *et al*: Significance of detecting Epstein–Barr-specific sequences in the peripheral blood of asymptomatic pediatric liver transplant recipients. *Liver Transplant* **6**:62–66, 2000.

652. Preiksaitis J: Epstein–Barr virus infection and malignancy in solid organ transplant recipients: Strategies for prevention and treatment. *Transplant Infect Dis* **3**:56–59, 2001.

653. Rowe D, Webber S, Shauer E, *et al*: Epstein–Barr virus load monitoring: Its role in the prevention and management of PTLD. *Transplant Infect Dis*. In press.

654. Malatack JF, Gartner JC Jr, Urbach AH, Zitelli BJ: Orthotopic liver transplantation, Epstein–Barr virus, cyclosporine, and lymphoproliferative disease: A growing concern [see comments]. *J Pediatr* **118**:667–675, 1991.

655. Swinnen LJ, Costanzo-Nordin MR, Fisher SG, *et al*: Increased incidence of lymphoproliferative disorder after immunosuppression with the monoclonal antibody OKT3 in cardiac-transplant recipients [see comments]. *N Engl J Med* **323**:1723–1728, 1990.

656. Patton DF, Wilkowski CW, Hanson CA, *et al*: Epstein–Barr virus-determined clonality in posttransplant lymphoproliferative disease. *Transplantation* **49**:1080–1084, 1990.

657. Nalesnik MA, Jaffe R, Starzl TE, *et al*: The pathology of posttransplant lymphoproliferative disorders occurring in the setting of cyclosporine A-prednisone immunosuppression. *Am J Pathol* **133**:173–192, 1988.

658. Crawford DH, Sweny P, Edwards JM, Janossy G, Hoffbrand AV: Long-term T-cell-mediated immunity to Epstein–Barr virus in renal-allograft recipients receiving cyclosporin A. *Lancet* **1**:10–12, 1981.

659. Bird AG, McLachlan SM, Britton S: Cyclosporin A promotes spontaneous outgrowth *in vitro* of Epstein–Barr virus-induced B-cell lines. *Nature* **289**:300–301, 1981.

660. Crawford DH, Edwards JM, Sweny P, Hoffbrand AV, Janossy G: Studies on long-term T-cell-mediated immunity to Epstein–Barr virus in immunosuppressed renal allograft recipients. *Int J Cancer* **28**:705–709, 1981.

661. Yao QY, Rickinson AB, Gaston JS, Epstein MA: *In vitro* analysis of the Epstein–Barr virus: Host balance in long-term renal allograft recipients. *Int J Cancer* **35**:43–49, 1985.

662. Hanto DW, Frizzera G, Purtilo DT, *et al*: Clinical spectrum of lymphoproliferative disorders in renal transplant recipients and evidence for the role of Epstein–Barr virus. *Cancer Res* **41**:4253–4261, 1981.

663. Kew CE 2nd, Lopez-Ben R, Smith JK, *et al*: Posttransplant lymphoproliferative disorder localized near the allograft in renal transplantation. *Transplantation* **69**:809–814, 2000.

664. Heller T, Drachenberg CB, Orens JB, Fantry GT: Primary posttransplant lymphoproliferative disorder of the gallbladder in a lung transplant patient presenting with acute cholecystitis. *Transplantation* **69**:668–670, 2000.

665. Oertel SH, Ruhnke MS, Anagnostopoulos I, *et al*: Treatment of Epstein–Barr virus-induced posttransplantation lymphoproliferative disorder with foscarnet alone in an adult after simultaneous heart and renal transplantation. *Transplantation* **67**:765–767, 1999.

666. Green M, Reyes J, Webber S, Rowe D: The role of antiviral and immunoglobulin therapy in the prevention of Epstein–Barr virus infection and post-transplant lymphoproliferative disease following solid organ transplantation. *Transplant Infect Dis* **3**:97–103, 2001 [review].

667. Hanto DW, Frizzera G, Gajl-Peczalska KJ, *et al*: Epstein–Barr virus-induced B-cell lymphoma after renal transplantation: Acyclovir therapy and transition from polyclonal to monoclonal B-cell proliferation. *N Engl J Med* **306**:913–918, 1982.

668. Pirsch JD, Stratta RJ, Sollinger HW, *et al*: Treatment of severe Epstein–Barr virus-induced lymphoproliferative syndrome with ganciclovir: Two cases after solid organ transplantation. *Am J Med* **86**:241–244, 1989.

669. Shapiro RS, Chauvenet A, McGuire W, *et al*: Treatment of B-cell lymphoproliferative disorders with interferon alfa and intravenous gamma globulin [letter]. *N Engl J Med* **318**:1334, 1988.

670. Rooney CM, Smith CA, Ng CY, *et al*: Use of gene-modified virus-specific T lymphocytes to control Epstein–Barr-virus-related lymphoproliferation. *Lancet* **345**:9–13, 1995.

671. Oertel SH, Anagnostopoulos I, Bechstein WO, Liehr H, Riess HB: Treatment of posttransplant lymphoproliferative disorder with the anti-CD20 monoclonal antibody rituximab alone in an adult after liver transplantation: A new drug in therapy of patients with posttransplant lymphoproliferative disorder after solid organ transplantation? *Transplantation* **69**:430–432, 2000.

672. Durandy A: Anti-B cell and anti-cytokine therapy for the treatment of PTLD: Past, present, and future. *Transplant Infect Dis* **3**:104–107, 2001 [review].

673. Fischer A, Blanche S, Le Bidois J, *et al*: Anti-B-cell monoclonal antibodies in the treatment of severe B-cell lymphoproliferative syndrome following bone marrow and organ transplantation. *N Engl J Med* **324**:1451–1456, 1991.

674. Korsager B, Spencer ES, Mordhorst CH, Andersen HK: Herpesvirus hominis infections in renal transplant recipients. *Scand J Infect Dis* **7**:11–19, 1975.

675. Rubin RH, Tolkoff-Rubin NE: Viral infection in the renal transplant patient. *Proc Eur Dial Transplant Assoc* **19**:513–526, 1983.

676. Ho M: Virus infections after transplantation in man. Brief review. *Arch Virol* **55:**1–24, 1977.

677. Smyth RL, Higenbottam TW, Scott JP, *et al*: Herpes simplex virus infection in heart–lung transplant recipients. *Transplantation* **49:**735–739, 1990.

678. Stone WJ, Scowden EB, Spannuth CL, Lowry SP, Alford RH: Atypical herpesvirus hominis type 2 infection in uremic patients receiving immunosuppressive therapy. *Am J Med* **63:**511–516, 1977.

679. Burkhart CG: Persistent cutaneous herpes simplex infection. *Int J Dermatol* **20:**552–554, 1981.

680. Ramsey PG, Rubin RH, Tolkoff-Rubin NE, *et al*: The renal transplant patient with fever and pulmonary infiltrates: Etiology, clinical manifestations, and management. *Medicine* **59:**206–222, 1980.

681. Anuras S, Summers R: Fulminant herpes simplex hepatitis in an adult: Report of a case in renal transplant recipient. *Gastroenterology* **70:**425–428, 1976.

682. Holdsworth SR, Atkins RC, Scott DF, Hayes K: Systemic herpes simplex infection with fulminant hepatitis post-transplantation. *Aust N Z J Med* **6:**588–590, 1976.

683. Elliott WC, Houghton DC, Bryant RE, *et al*: Herpes simplex type 1 hepatitis in renal transplantation. *Arch Intern Med* **140:**1656–1660, 1980.

684. Taylor RJ, Saul SH, Dowling JN, *et al*: Primary disseminated herpes simplex infection with fulminant hepatitis following renal transplantation. *Arch Intern Med* **141:**1519–1521, 1981.

685. Naraqi S, Jackson GG, Jonasson OM: Viremia with herpes simplex type 1 in adults. Four nonfatal cases, one with features of chicken pox. *Ann Intern Med* **85:**165–169, 1976.

686. Anora KK, Karalakulasingam R, Raff MJ, Martin DG: Cutaneous Herpesvirus hominis (type 2) infection after renal transplantation. *JAMA* **230:**1174–1175, 1974.

687. Dummer JS, Armstrong J, Somers J, *et al*: Transmission of infection with herpes simplex virus by renal transplantation. *J Infect Dis* **155:**202–206, 1987.

688. Dunn DL, Matas AJ, Fryd DS, Simmons RL, Najarian JS: Association of concurrent herpes simplex virus and cytomegalovirus with detrimental effects after renal transplantation. *Arch Surg* **119:**812–817, 1984.

689. Griffin P, Colbert J, Williamson E, *et al*: Oral acyclovir prophylaxis of herpes infection in renal transplant recipients. *Transplant Proc* **17:**84–85, 1985.

690. Pettersson E, Hovi T, Ahonen J, *et al*: Prophylactic oral acyclovir after renal transplantation. *Transplantation* **39:**279–281, 1985.

691. Rifkind D: The activation of varicella–zoster virus infections by immunosuppressive therapy. *J Lab Clin Med* **68:**463–474, 1966.

692. Luby JP, Ramirez-Ronda C, Rinner S, Hull A, Vergne-Marini P: A longitudinal study of varicella–zoster virus infections in renal transplant recipients. *J Infect Dis* **135:**659–663, 1977.

693. Feldhoff CM, Balfour HH Jr, Simmons RL, Najarian JS, Mauer SM: Varicella in children with renal transplants. *J Pediatr* **98:**25–31, 1981.

694. McGregor RS, Zitelli BJ, Urbach AH, Malatack JJ, Gartner JC Jr: Varicella in pediatric orthotopic liver transplant recipients. *Pediatrics* **83:**256–261, 1989.

695. Lynfield R, Herrin JT, Rubin RH: Varicella in pediatric renal transplant recipients. *Pediatrics* **90:**216–220, 1992.

696. Cohen JI, Brunell PA, Straus SE, Krause PR: Recent advances in varicella–zoster virus infection. *Ann Intern Med* **130:**922–932, 1999.

697. Fall AJ, Aitchison JD, Krause A, *et al*: Donor organ transmission of varicella zoster due to cardiac transplantation. *Transplantation* **70:**211–213, 2000.

698. Rothwell WS, Gloor JM, Morgenstern BZ, Milliner DS: Disseminated varicella infection in pediatric renal transplant recipients treated with mycophenolate mofetil. *Transplantation* **68:**158–161, 1999.

699. Lawrence GL, Chee M, Craxton MA, *et al*: Human herpesvirus 6 is closely related to human cytomegalovirus. *J Virol* **64:**287–299, 1990.

700. Sing N: Human herpesviruses-6, -7,and -8 in organ transplant recipients. *Clin Microbial Infect* **6:**453–459, 2000.

701. Agut H: Puzzles concerning the pathogenicity of human herpesvirus 6 [editorial; comment]. *N Engl J Med* **329:**203–204, 1993.

702. Flamand L, Gosselin J, Stefanescu I, Ablashi D, Menezes J: Immunosuppressive effect of human herpesvirus 6 on T-cell functions: Suppression of interleukin-2 synthesis and cell proliferation [published erratum appears in *Blood* **86**(1)**:**418, 1995]. *Blood* **85:**1263–1271, 1995.

703. Knox KK, Carrigan DR: *In vitro* suppression of bone marrow progenitor cell differentiation by human herpesvirus 6 infection. *J Infect Dis* **165:**925–929, 1992.

704. Carrigan DR, Knox KK: Human herpesvirus 6 (HHV-6) isolation from bone marrow: HHV-6-associated bone marrow suppression in bone marrow transplant patients [see comments]. *Blood* **84:**3307–3310, 1994.

705. Oren I, Sobel JD: Human herpesvirus type 6: Review. *Clin Infect Dis* **14:**741–746, 1992.

706. McCullers JA, Lakeman FD, Whitley RJ: Human herpesvirus 6 is associated with focal encephalitis. *Clin Infect Dis* **21:**571–576, 1995.

707. Dockrell DH, Prada J, Jones MF, *et al*: Seroconversion to human herpesvirus 6 following liver transplantation is a marker of cytomegalovirus disease. *J Infect Dis* **176:**1135–1140, 1997.

708. Ward KN, Gray JJ, Efstathiou S: Brief report: Primary human herpesvirus 6 infection in a patient following liver transplantation from a seropositive donor. *J Med Virol* **28:**69–72, 1989.

709. Kikuta H, Itami N, Matsumoto S, Chikaraishi T, Togashi M: Frequent detection of human herpesvirus 6 DNA in peripheral blood mononuclear cells from kidney transplant patients [letter]. *J Infect Dis* **163:**925, 1991.

710. Okuno T, Higashi K, Shiraki K, *et al*: Human herpesvirus 6 infection in renal transplantation. *Transplantation* **49:**519–522, 1990.

711. Gudnason T, Dunn D, Brown N, *et al*: Human herpesvirus 6 infections in hospitalized renal transplant recipients. *Clin Transplant* **5:**359–364, 1991.

712. Yoshikawa T, Suga S, Asano Y, *et al*: A prospective study of human herpesvirus-6 infection in renal transplantation. *Transplantation* **54:**879–883, 1992.

713. Sutherland S, Christofinis G, O'Grady J, Williams R: A serological investigation of human herpesvirus 6 infections in liver transplant recipients and the detection of cross-reacting antibodies to cytomegalovirus. *J Med Virol* **33:**172–176, 1991.

714. Chou SW, Scott KM: Rises in antibody to human herpesvirus 6 detected by enzyme immunoassay in transplant recipients with primary cytomegalovirus infection. *J Clin Microbiol* **28:**851–854, 1990.

715. Singh N, Paterson DL: Encephalitis caused by human herpesvirus-6 in transplant recipients: Relevance of a novel neurotropic virus. *Transplantation* **69:**2474–2479, 2000.

716. Dockrell DH, Smith TF, Paya CV: Human herpesvirus 6. *Mayo Clin Proc* **74:**163–170, 1999.

717. Griffiths PD, Ait-Khaled M, Bearcroft CP, *et al*: Human herpesviruses 6 and 7 as potential pathogens after liver transplant: Prospective comparison with the effect of cytomegalovirus. *J Med Virol* **59:**496–501, 1999.

718. Chan PK, Peiris JS, Yuen KY, *et al*: Human herpesvirus-6 and human herpesvirus-7 infections in bone marrow transplant recipients. *J Med Virol* **53:**295–305, 1997.

719. Tong CY, Bakran A, Williams H, Cheung CY, Peiris JS: Association of human herpesvirus 7 with cytomegalovirus disease in renal transplant recipients. *Transplantation* **70:**213–216, 2000.

720. Kidd IM, Clark DA, Sabin CA, *et al*: Prospective study of human beta herpesviruses after renal transplantation: Association of human herpesvirus 7 and cytomegalovirus co-infection with cytomegalovirus disease and increased rejection. *Transplantation* **69:**2400–2404, 2000.

721. Brennan DC, Storch GA, Singer GG, *et al*: The prevalence of human herpesvirus-7 in renal transplant recipients is unaffected by oral or intravenous ganciclovir. *J Infect Dis* **181:**1557–1561, 2000.

722. Mendez JC, Procop GW, Espy MJ, *et al*: Relationship of HHV8 replication and Kaposi's sarcoma after solid organ transplantation. *Transplantation* **67:**1200–1201, 1999.

723. Hudnall SD, Rady PL, Tyring SK, Fish JC: Hydrocortisone activation of human herpesvirus 8 viral DNA replication and gene expression *in vitro*. *Transplantation* **67:**648–652, 1999.

724. Gao SJ, Zhang YJ, Deng JH, *et al*: Molecular polymorphism of Kaposi's sarcoma-associated herpesvirus (human herpesvirus 8) latent nuclear antigen: Evidence for a large repertoire of viral genotypes and dual infection with different viral genotypes [published erratum appears in *J Infect Dis* **180**(5):1756, 1999]. *J Infect Dis* **180:**1466–1476, 1999.

725. Frances C, Mouquet C, Marcelin AG, *et al*: Outcome of kidney transplant recipients with previous human herpesvirus-8 infection [see comments]. *Transplantation* **69:**1776–1779, 2000.

726. Regamey N, Cathomas G: HHV-8: A newly recognized pathogen in transplantation [comment]. *Transplantation* **69:**1768–1769, 2000.

727. Neyts J, De Clercq E: Antiviral drug susceptibility of human herpesvirus 8. *Antimicrob Agents Chemother* **41:**2754–2756, 1997.

728. Hammoud Z, Parenti DM, Simon GL: Abatement of cutaneous Kaposi's sarcoma associated with cidofovir treatment [see comments]. *Clin Infect Dis* **26:**1233, 1998.

729. Grossi P, Baldunti F, Corona A, *et al*: Kaposi sarcoma following thoracic organ transplantation: Prevalence, correlation with human herpesvirus 8 and new therapeutic options. *Transplantation* **67:**S39, 1999.

730. Mozes MF, Ascher NL, Balfour HH Jr, Simmons RL, Najarian JS: Jaundice after renal allotransplantation. *Ann Surg* **188:**783–790, 1978.

731. Haxhe JJ, Alexandre GP, Kestens PJ: The effect of imuran and azaserine on liver function tests in the dog. Its relation to the detection of graft rejection following liver transplantation. *Arch Int Pharmacodyn Ther* **168:**366–372, 1967.

732. Starzl T, Marchioro T, Porter K, *et al*: Factors determining short and long-term survival after orthotopic liver homotransplantation. *Surgery* **58:**131–155, 1965.

733. Sparberg M, Simon N, del Greco F: Intrahepatic cholestasis due to azathioprine. *Gastroenterology* **57:**439–441, 1969.

734. Pirson Y, van Ypersele de Strihou C, Noel H, *et al*: Liver disease in transplanted patients. *Proc Eur Dial Transplant Assoc* **10:**434–445, 1973.

735. Ware AJ, Luby JP, Hollinger B, *et al*: Etiology of liver disease in renal-transplant patients. *Ann Intern Med* **91:**364–371, 1979.

736. Haboubi NY, Ali HH, Whitwell HL, Ackrill P: Role of endothelial cell injury in the spectrum of azathioprine-induced liver disease after renal transplant: Light microscopy and ultrastructural observations. *Am J Gastroenterol* **83:**256–261, 1988.

737. Read AE, Wiesner RH, LaBrecque DR, *et al*: Hepatic venoocclusive disease associated with renal transplantation and azathioprine therapy. *Ann Intern Med* **104:**651–655, 1986.

738. Katzka DA, Saul SH, Jorkasky D, *et al*: Azathioprine and hepatic venocclusive disease in renal transplant patients. *Gastroenterology* **90:**446–454, 1986.

739. Norris SH, Butler TC, Glass N, Tran R: Fatal hepatic necrosis caused by disseminated type 5 adenovirus infection in a renal transplant recipient. *Am J Nephrol* **9:**101–105, 1989.

740. Cames B, Rahier J, Burtomboy G, *et al*: Acute adenovirus hepatitis in liver transplant recipients. *J Pediatr* **120:**33–37, 1992.

741. Gimson AE, White YS, Eddleston AL, Williams R: Clinical and prognostic differences in fulminant hepatitis type A, B and non-A non-B. *Gut* **24:**1194–1198, 1983.

742. Szmuness W, Dienstag JL, Purcell RH, *et al*: Hepatitis type A and hemodialysis: A seroepidemiologic study in 15 U.S. centers. *Ann Intern Med* **87:**8–12, 1977.

743. Barbara JA, Howell DR, Briggs M, Parry JV: Post-transfusion hepatitis A [letter]. *Lancet* **1:**738, 1982.

744. Hollinger FB, Khan NC, Oefinger PE, *et al*: Posttransfusion hepatitis type A. *JAMA* **250:**2313–2317, 1983.

745. Fagan E, Yousef G, Brahm J, *et al*: Persistence of hepatitis A virus in fulminant hepatitis and after liver transplantation. *J Med Virol* **30:**131–136, 1990.

746. Robinson W: Hepatitis B virus and hepatitis delta virus. In Mandell G, Douglas RJ, Bennett J (eds): *Principles and Practice of Infectious Diseases*. Edinburgh, Churchill Livingstone, 1990, pp. 1204–1231.

747. Debure A, Degos F, Pol S, *et al*: Liver diseases and hepatic complications in renal transplant patients. *Adv Nephrol Necker Hosp* **17:**375–400, 1988.

748. Rao KV, Kasiske BL, Anderson WR: Variability in the morphological spectrum and clinical outcome of chronic liver disease in hepatitis B-positive and B-negative renal transplant recipients. *Transplantation* **51:**391–396, 1991.

749. Luketic VA, Shiffman ML, McCall JB, *et al*: Primary hepatocellular carcinoma after orthotopic liver transplantation for chronic hepatitis B infection. *Ann Intern Med* **114:**212–213, 1991.

750. Alter H: The chronic consequences of non-A, non-B hepatitis. In Seeff L, Lewis J (eds): *Current Perspectives in Hepatology*. New York, Plenum Press, 1989, pp. 83–97.

751. Service PH: Inter-Agency Guidelines for Screening Donors of Blood, Plasma, Organs, Tissues, and Semen for Evidence of Hepatitis B and Hepatitis C, 1991.

752. Rizzetto M: Hepatitis delta virus: Biology and infection. In Shikata T, Purcell R, Uchida T (eds): *Viral Hepatitis C, D, and E*. Amsterdam, Excerpta Medica, 1991, pp. 327–333.

753. Feinstone S: Non-A, non-B hepatitis. In Mandell G, Douglas RJ, Bennett J (eds): *Principles and Practice of Infectious Diseases*. Edinburgh, Churchill Livingstone, 1990, pp. 1407–1415.

754. Dusheiko G, Song E, Bowyer S, *et al*: Natural history of hepatitis B virus infection in renal transplant recipients—A fifteen-year follow-up. *Hepatology* **3:**330–336, 1983.

755. Scott D, Mijch A, Lucas CR, *et al*: Hepatitis B and renal transplantation. *Transplant Proc* **19:**2159–2160, 1987.

756. Stevens CE, Alter HJ, Taylor PE, *et al*: Hepatitis B vaccine in patients receiving hemodialysis. Immunogenicity and efficacy. *N Engl J Med* **311:**496–501, 1984.

757. Jacobson IM, Jaffers G, Dienstag JL, *et al*: Immunogenicity of hepatitis B vaccine in renal transplant recipients. *Transplantation* **39:**393–395, 1985.

758. Carey W, Pimentel R, Westveer MK, Vogt D, Broughan T: Failure of hepatitis B immunization in liver transplant recipients: Results of a prospective trial. *Am J Gastroenterol* **85:**1590–1592, 1990.

759. Nagington J: Reactivation of hepatitis B after transplantation operations. *Lancet* **1:**558–560, 1977.

760. Scullard GH, Smith CI, Merigan TC, Robinson WS, Gregory PB: Effects of immunosuppressive therapy on viral markers in chronic active hepatitis B. *Gastroenterology* **81:**987–991, 1981.

761. Lau JY, Bain VG, Smith HM, Alexander GJ, Williams R: Modulation of hepatitis B viral antigen expression by immunosuppressive drugs in primary hepatocyte culture. *Transplantation* **53:**894–898, 1992.

762. Cote PJ, Korba BE, Baldwin B, *et al*: Immunosuppression with cyclosporine during the incubation period of experimental woodchuck hepatitis virus infection increases the frequency of chronic infection in adult woodchucks. *J Infect Dis* **166:**628–631, 1992.

763. Marinos G, Rossol S, Carucci P, *et al*: Immunopathogenesis of hepatitis B virus recurrence after liver transplantation. *Transplantation* **69:**559–568, 2000.

764. Huang CC, Lai MK, Fong MT: Hepatitis B liver disease in cyclosporine-treated renal allograft recipients. *Transplantation* **49:**540–544, 1990.

765. Parfrey PS, Forbes RD, Hutchinson TA, *et al*: The clinical and pathological course of hepatitis B liver disease in renal transplant recipients. *Transplantation* **37:**461–466, 1984.

766. Parfrey PS, Forbes RD, Hutchinson TA, *et al*: The impact of renal transplantation on the course of hepatitis B liver disease. *Transplantation* **39:**610–615, 1985.

767. Weir MR, Kirkman RL, Strom TB, Tilney NL: Liver disease in recipients of long-functioning renal allografts. *Kidney Int* **28:**839–844, 1985.

768. Parfrey PS, Farge D, Forbes RD, *et al*: Chronic hepatitis in end-stage renal disease: Comparison of HBsAg-negative and HBsAg-positive patients. *Kidney Int* **28:**959–967, 1985.

769. Harnett JD, Zeldis JB, Parfrey PS, *et al*: Hepatitis B disease in dialysis and transplant patients. Further epidemiologic and serologic studies. *Transplantation* **44:**369–376, 1987.

770. Rao KV, Andersen RC: Long-term results and complications in renal transplant recipients. Observations in the second decade. *Transplantation* **45:**45–52, 1988.

771. Degos F, Lugassy C, Degott C, *et al*: Hepatitis B virus and hepatitis B-related viral infection in renal transplant recipients. A prospective study of 90 patients. *Gastroenterology* **94:**151–156, 1988.

772. Wedemeyer H, Pethig K, Wagner D, *et al*: Long-term outcome of chronic hepatitis B in heart transplant recipients. *Transplantation* **66:**1347–1353, 1998.

773. Kim CM, Koike K, Saito I, Miyamura T, Jay G: HBx gene of hepatitis B virus induces liver cancer in transgenic mice. *Nature* **351:**317–320, 1991.

774. Kekule AS, Lauer U, Weiss L, Luber B, Hofschneider PH: Hepatitis B virus transactivator HBx uses a tumour promoter signalling pathway [see comments]. *Nature* **361:**742–745, 1993.

775. Sengar DP, Couture RA, Lazarovits AI, Jindal SL: Long-term patient and renal allograft survival in HBsAg infection: A recent update. *Transplant Proc* **21:**3358–3359, 1989.

776. Fairley CK, Mijch A, Gust ID, *et al*: The increased risk of fatal liver disease in renal transplant patients who are hepatitis B antigen and/or HBV DNA positive. *Transplantation* **52:**497–500, 1991.

777. Perrillo RP, Schiff ER, Davis GL, *et al*: A randomized, controlled trial of interferon alfa-2b alone and after prednisone withdrawal for the treatment of chronic hepatitis B. The Hepatitis Interventional Therapy Group [see comments]. *N Engl J Med* **323:**295–301, 1990.

778. Kramer P, ten Kate FW, Bijnen AB, Jeekel J, Weimar W: Recombinant leucocyte interferon A induces steroid-resistant acute vascular rejection episodes in renal transplant recipients. *Lancet* **1:**989–990, 1984.

779. Weimar W, Kramer P, Bijnen AB, *et al*: The incidence of cytomegalo- and herpes simplex virus infections in renal allograft recipients treated with high dose recombinant leucocyte interferon: A controlled study. *Scand J Urol Nephrol Suppl* **92:**37–39, 1985.

780. Jain A, Demetris AJ, Manez R, *et al*: Incidence and severity of acute allograft rejection in liver transplant recipients treated with alfa interferon. *Liver Transplant Surg* **4:**197–203, 1998.

781. Kruger M, Tillmann HL, Trautwein C, *et al*: Famciclovir treatment of hepatitis B virus recurrence after liver transplantation: A pilot study. *Liver Transplant Surg* **2:**253–262, 1996.

782. Singh N, Gayowski T, Wannstedt CF, Wagener MM, Marino IR: Pretransplant famciclovir as prophylaxis for hepatitis B virus recurrence after liver transplantation. *Transplantation* **63:**1415–1419, 1997.

783. Wedemeyer H, Boker KH, Pethig K, *et al*: Famciclovir treatment of chronic hepatitis B in heart transplant recipients: A prospective trial. *Transplantation* **68:**1503–1511, 1999.

784. Jurim O, Martin P, Winston DJ, *et al*: Failure of ganciclovir prophylaxis to prevent allograft reinfection following orthotopic liver transplantation for chronic hepatitis B infection. *Liver Transplant Surg* **2:**370–374, 1996.

785. Rizzetto M, Marzano A: Posttransplantation prevention and treatment of recurrent hepatitis B [In Process Citation]. *Liver Transplant* **6:**S47–S51, 2000.

786. Jarvis B, Faulds D: Lamivudine: A review of its therapeutic potential in chronic hepatitis B [published erratum appears in *Drugs* **58(4):**587, 1999]. *Drugs* **58:**101–141, 1999.

787. Ben-Ari Z, Shmueli D, Mor E, Shapira Z, Tur-Kaspa R: Beneficial effect of lamivudine in recurrent hepatitis B after liver transplantation. *Transplantation* **63:**393–396, 1997.

788. Rostaing L, Henry S, Cisterne JM, *et al*: Efficacy and safety of lamivudine on replication of recurrent hepatitis B after cadaveric renal transplantation. *Transplantation* **64:**1624–1637, 1997.

789. Al Faraidy K, Yoshida EM, Davis JE, Vartanian RK, Anderson FH: Alteration of the dismal natural history of fibrosing cholestatic hepatitis secondary to hepatitis B virus with the use of lamivudine. *Transplantation* **64:**926–928, 1997.

790. Jung YO, Lee YS, Yang WS, *et al*: Treatment of chronic hepatitis B with lamivudine in renal transplant recipients. *Transplantation* **66:**733–737, 1998.

791. Andreone P, Caraceni P, Grazi GL, *et al*: Lamivudine treatment for acute hepatitis B after liver transplantation. *J Hepatol* **29:**985–989, 1998.

792. Nery JR, Weppler D, Rodriguez M, *et al*: Efficacy of lamivudine in controlling hepatitis B virus recurrence after liver transplantation. *Transplantation* **65:**1615–1621, 1998.

793. Perrillo R, Rakela J, Dienstag J, *et al*: Multicenter study of lamivudine therapy for hepatitis B after liver transplantation. Lamivudine Transplant Group. *Hepatology* **29:**1581–1586, 1999.

794. Mutimer D, Pillay D, Dragon E, *et al*: High pre-treatment serum hepatitis B virus titre predicts failure of lamivudine prophylaxis and graft re-infection after liver transplantation. *J Hepatol* **30:**715–721, 1999.

795. Bartholomew MM, Jansen RW, Jeffers LJ, *et al*: Hepatitis-B-virus resistance to lamivudine given for recurrent infection after orthotopic liver transplantation [see comments]. *Lancet* **349:**20–22, 1997.

796. Gauthier J, Bourne EJ, Lutz MW, *et al*: Quantitation of hepatitis B viremia and emergence of YMDD variants in patients with chronic hepatitis B treated with lamivudine. *J Infect Dis* **180:** 1757–1762, 1999.

797. Seehofer D, Rayes N, Berg T, *et al*: Lamivudine as first- and second-line treatment of hepatitis B infection after liver transplantation [In Process Citation]. *Transplant Int* **13:**290–296, 2000.

798. Puchhammer-Stockl E, Mandl CW, Kletzmayr J, *et al*: Monitoring the virus load can predict the emergence of drug-resistant hepatitis B virus strains in renal transplantation patients during lamivudine therapy. *J Infect Dis* **181:**2063–2066, 2000.

799. Fontaine H, Thiers V, Chretien Y, *et al*: HBV genotypic resistance to lamivudine in kidney recipients and hemodialyzed patients. *Transplantation* **69:**2090–2094, 2000.

800. Peters MG, Singer G, Howard T, *et al*: Fulminant hepatic failure resulting from lamivudine-resistant hepatitis B virus in a renal transplant recipient: Durable response after orthotopic liver transplantation on adefovir dipivoxil and hepatitis B immune globulin. *Transplantation* **68:**1912–1914, 1999.

801. Malkan G, Cattral MS, Humar A, *et al*: Lamivudine for hepatitis B in liver transplantation: A single-center experience. *Transplantation* **69:**1403–1407, 2000.

802. Mutimer D, Dusheiko G, Barrett C, *et al*: Lamivudine without HBIg for prevention of graft reinfection by hepatitis B: Long-term follow-up. *Transplantation* **70:**809–815, 2000.

803. Demetris AJ, Jaffe R, Sheahan DG, *et al*: Recurrent hepatitis B in liver allograft recipients. Differentiation between viral hepatitis B and rejection. *Am J Pathol* **125:**161–172, 1986.

804. Portmann B, O'Grady J, Williams R: Disease recurrence following orthotopic liver transplantation. *Transplant Proc* **18**(Suppl), 1986.

805. Freeman RB, Sanchez H, Lewis WD, *et al*: Serologic and DNA follow-up data from HBsAg-positive patients treated with orthotopic liver transplantation. *Transplantation* **51:**793–797, 1991.

806. Levy GA, Sherker A, Fung LS, *et al*: Relevance of hepatitis B viral DNA in assessment of potential liver allograft recipients. *Transplant Proc* **21:**3333–3334, 1989.

807. Demetris AJ, Todo S, Van Thiel DH, *et al*: Evolution of hepatitis B virus liver disease after hepatic replacement. Practical and theoretical considerations. *Am J Pathol* **137:**667–676, 1990.

808. O'Grady JG, Smith HM, Davies SE, *et al*: Hepatitis B virus reinfection after orthotopic liver transplantation. Serological and clinical implications. *J Hepatol* **14:**104–111, 1992.

809. Lavine JE, Lake JR, Ascher NL, *et al*: Persistent hepatitis B virus following interferon alfa therapy and liver transplantation. *Gastroenterology* **100:**263–267, 1991.

810. Eason J, Freeman RJ, Rohrer R, *et al*: Should liver allograft transplantation be performed for patients with hepatitis B? *Transplantation* **70:**397–401, 1994.

811. Samuel D, Bismuth A, Mathieu D, *et al*: Passive immunoprophylaxis after liver transplantation in HBsAg-positive patients. *Lancet* **337:**813–815, 1991.

812. Lauchart W, Muller R, Pichlmayr R: Long-term immunoprophylaxis of hepatitis B virus reinfection in recipients of human liver allografts. *Transplant Proc* **19:**4051–4053, 1987.

813. Grazi GL, Mazziotti A, Sama C, *et al*: Liver transplantation in HBsAg-positive HBV-DNA—Negative cirrhotics: Immunoprophylaxis and long-term outcome. *Liver Transplant Surg* **2:**418–425, 1996.

814. Tchervenkov JI, Tector AJ, Barkun JS, *et al*: Recurrence-free long-term survival after liver transplantation for hepatitis B using interferon-alpha pretransplant and hepatitis B immune globulin posttransplant. *Ann Surg* **226:**356–365; discussion 365–368, 1997.

815. Sawyer RG, McGory RW, Gaffey MJ, *et al*: Improved clinical outcomes with liver transplantation for hepatitis B-induced chronic liver failure using passive immunization. *Ann Surg* **227:** 841–850, 1998.

816. Burbach GJ, Bienzle U, Neuhaus R, *et al*: Intravenous or intramuscular anti-HBs immunoglobulin for the prevention of hepatitis B reinfection after orthotopic liver transplantation. *Transplantation* **63:**478–480, 1997.

817. Terrault NA, Zhou S, McCory RW, *et al*: Incidence and clinical consequences of surface and polymerase gene mutations in liver transplant recipients on hepatitis B immunoglobulin. *Hepatology* **28:**555–561, 1998.

818. Ghany MG, Ayola B, Villamil FG, *et al*: Hepatitis B virus S mutants in liver transplant recipients who were reinfected despite hepatitis B immune globulin prophylaxis [see comments]. *Hepatology* **27:**213–222, 1998.

819. Carman WF, Owsianka A, Wallace LA, Dow BC, Mutimer DJ: Antigenic characterization of pre- and post-liver transplant hepatitis B surface antigen sequences from patients treated with hepatitis B immune globulin. *J Hepatol* **31:**195–201, 1999.

820. Shields PL, Owsianka A, Carman WF, *et al*: Selection of hepatitis B surface "escape" mutants during passive immune prophylaxis following liver transplantation: Potential impact of genetic changes on polymerase protein function. *Gut* **45:**306–309, 1999.

821. Santantonio T, Gunther S, Sterneck M, *et al*: Liver graft infection by HBV S-gene mutants in transplant patients receiving long-term HBIg prophylaxis. *Hepatogastroenterology* **46:**1848–1854, 1999.

822. Tillmann HL, Trautwein C, Bock T, *et al*: Mutational pattern of hepatitis B virus on sequential therapy with famciclovir and lamivudine in patients with hepatitis B virus reinfection occurring under HBIg immunoglobulin after liver transplantation. *Hepatology* **30:**244–256, 1999.

823. de Man RA, Bartholomeusz AI, Niesters HG, Zondervan PE, Locarnini SA: The sequential occurrence of viral mutations in a liver transplant recipient re-infected with hepatitis B: Hepatitis B immune globulin escape, famciclovir non-response, followed by lamivudine resistance resulting in graft loss. *J Hepatol* **29:**669–675, 1998.

824. Markowitz JS, Martin P, Conrad AJ, *et al*: Prophylaxis against hepatitis B recurrence following liver transplantation using combination lamivudine and hepatitis B immune globulin. *Hepatology* **28:**585–589, 1998.

825. Yoshida EM, Erb SR, Partovi N, *et al*: Liver transplantation for chronic hepatitis B infection with the use of combination lamivudine and low-dose hepatitis B immune globulin. *Liver Transplant Surg* **5:**520–525, 1998.

826. Bain V: Hepatitis B in transplantation. *Transplant Infect Dis* **2:** 153–165, 2000.

827. Yao FY, Osorio RW, Roberts JP, *et al*: Intramuscular hepatitis B immune globulin combined with lamivudine for prophylaxis against hepatitis B recurrence after liver transplantation. *Liver Transplant Surg* **5:**491–496, 1999.

828. Angus PW, McCaughan GW, Gane EJ, Crawford DH, Harley H: Combination low-dose hepatitis B immune globulin and lamivudine therapy provides effective prophylaxis against posttransplantation hepatitis B [In Process Citation]. *Liver Transplant* **6:**429–433, 2000.

829. McCaughan G, Zekry A: Effects of immunosuppression and organ transplantation on the natural history and immunopathogenesis of hepatitis C virus infection. *Transplant Infect Dis* **2:**166–185, 2000.

830. Bruix J, Barrera JM, Calvet X, *et al*: Prevalence of antibodies to hepatitis C virus in Spanish patients with hepatocellular carcinoma and hepatic cirrhosis [see comments]. *Lancet* **2:**1004–1006, 1989.

831. Yu MC, Tong MJ, Coursaget P, *et al*: Prevalence of hepatitis B and C viral markers in black and white patients with hepatocellular carcinoma in the United States [see comments]. *J Natl Cancer Inst* **82:**1038–1041, 1990.

832. Esteban JI, Lopez-Talavera JC, Genesca J, *et al*: High rate of infectivity and liver disease in blood donors with antibodies to hepatitis C virus [see comments]. *Ann Intern Med* **115:**443–449, 1991.

833. Kao JH, Chen PJ, Yang PM, *et al*: Intrafamilial transmission of hepatitis C virus: The important role of infections between spouses. *J Infect Dis* **166:**900–903, 1992.

834. Prince AM, Brotman B, Inchauspe G, *et al*: Patterns and prevalence of hepatitis C virus infection in posttransfusion non-A, non-B hepatitis. *J Infect Dis* **167:**1296–1301, 1993.

835. McCashland TM, Wright TL, Donovan JP, *et al*: Low incidence of intraspousal transmission of hepatitis C virus after liver transplantation. *Liver Transplant Surg* **1:**358–361, 1995.

836. Wright TL, Hsu H, Donegan E, *et al*: Hepatitis C virus not found in fulminant non-A, non-B hepatitis [see comments]. *Ann Intern Med* **115:**111–112, 1991.

837. Woodfield DG: Acute post-transfusion fulminant hepatitis C virus infection: A case report. *Gastroenterol Jpn* **26**(Suppl 3)**:**221–223, 1991.

838. Choo QL, Kuo G, Weiner AJ, Overby LR, *et al*: Isolation of a cDNA clone derived from a blood-borne non-A, non-B viral hepatitis genome. *Science* **244:**359–362, 1989.

839. Kuo G, Choo QL, Alter HJ, *et al*: An assay for circulating antibodies to a major etiologic virus of human non-A, non-B hepatitis. *Science* **244:**362–364, 1989.

840. Alter HJ, Purcell RH, Shih JW, *et al*: Detection of antibody to hepatitis C virus in prospectively followed transfusion recipients with acute and chronic non-A, non-B hepatitis. *N Engl J Med* **321:**1494–1500, 1989.

841. Alter MJ, Hadler SC, Judson FN, *et al*: Risk factors for acute non-A, non-B hepatitis in the United States and association with hepatitis C virus infection. *JAMA* **264:**2231–2235, 1990.

842. Roth D, Fernandez JA, Burke GW, Esquenazi V, Miller J: Detection of antibody to hepatitis C virus in renal transplant recipients. *Transplantation* **51:**396–400, 1991.

843. Read AE, Donegan E, Lake J, *et al*: Hepatitis C in patients undergoing liver transplantation. *Ann Intern Med* **114:**282–284, 1991.

844. Prince AM, Brotman B, Huima T, *et al*: Immunity in hepatitis C infection. *J Infect Dis* **165:**438–443, 1992.

845. Farci P, London WT, Wong DC, *et al*: The natural history of infection with hepatitis C virus (HCV) in chimpanzees: Comparison of serologic responses measured with first- and second-generation assays and relationship to HCV viremia. *J Infect Dis* **165:**1006–1011, 1992.

846. Preiksaitis JK, Cockfield SM, Fenton JM, Burton NI, Chui LW: Serologic responses to hepatitis C virus in solid organ transplant recipients. *Transplantation* **64:**1775–1780, 1997.

847. Liang TJ, Rehermann B, Seeff LB, Hoofnagle JH: Pathogenesis, natural history, treatment, and prevention of hepatitis C. *Ann Intern Med* **132:**296–305, 2000.

848. LaQuaglia MP, Tolkoff-Rubin NE, Dienstag JL, *et al*: Impact of hepatitis on renal transplantation. *Transplantation* **32:**504–507, 1981.

849. Huang CC, Liaw YF, Lai MK, *et al*: The clinical outcome of hepatitis C virus antibody-positive renal allograft recipients. *Transplantation* **53:**763–765, 1992.

850. Ponz E, Campistol JM, Bruguera M, *et al*: Hepatitis C virus infection among kidney transplant recipients. *Kidney Int* **40:**748–751, 1991.

851. Hanafusa T, Ichikawa Y, Kishikawa H, *et al*: Retrospective study on the impact of hepatitis C virus infection on kidney transplant patients over 20 years. *Transplantation* **66:**471–476, 1998.

852. Rostaing L, Izopet J, Cisterne JM, *et al*: Impact of hepatitis C virus duration and hepatitis C virus genotypes on renal transplant patients: Correlation with clinicopathological features. *Transplantation* **65:**930–936, 1998.

853. Mathurin P, Mouquet C, Poynard T, *et al*: Impact of hepatitis B and C virus on kidney transplantation outcome [see comments]. *Hepatology* **29:**257–263, 1999.

854. Chan TM, Lok AS, Cheng IK: Hepatitis C in renal transplant recipients. *Transplantation* **52:**810–813, 1991.

855. Stempel CA, Lake J, Kuo G, Vincenti F: Hepatitis C—Its prevalence in end-stage renal failure patients and clinical course after kidney transplantation. *Transplantation* **55:**273–276, 1993.

856. Knoll GA, Tankersley MR, Lee JY, Julian BA, Curtis JJ: The impact of renal transplantation on survival in hepatitis C-positive end-stage renal disease patients. *Am J Kidney Dis* **29:**608–614, 1997.

857. Sterling RK, Sanyal AJ, Luketic VA, *et al*: Chronic hepatitis C infection in patients with end stage renal disease: Characterization of liver histology and viral load in patients awaiting renal transplantation. *Am J Gastroenterol* **94:**3576–3582, 1999.

858. Rostaing L, Izopet J, Arnaud C, *et al*: Long-term impact of superinfection by hepatitis G virus in hepatitis C virus-positive renal transplant patients. *Transplantation* **67:**556–560, 1999.

859. Muller H, Otto G, Goeser T, *et al*: Recurrence of hepatitis C virus infection after orthotopic liver transplantation. *Transplantation* **54:**743–745, 1992.

860. Bouffard P, Hayashi PH, Acevedo R, *et al*: Hepatitis C virus is detected in a monocyte/macrophage subpopulation of peripheral blood mononuclear cells of infected patients. *J Infect Dis* **166:** 1276–1280, 1992.

861. Boker KH, Dalley G, Bahr MJ, *et al*: Long-term outcome of hepatitis C virus infection after liver transplantation. *Hepatology* **25:**203–210, 1997.

862. Feray C, Gigou M, Samuel D, *et al*: The course of hepatitis C virus infection after liver transplantation. *Hepatology* **20:**1137–1143, 1994.

863. Gane EJ, Portmann BC, Naoumov NV, *et al*: Long-term outcome of hepatitis C infection after liver transplantation [see comments]. *N Engl J Med* **334:**815–820, 1996.

864. Charlton M, Seaberg E, Wiesner R, *et al*: Predictors of patient and graft survival following liver transplantation for hepatitis C. *Hepatology* **28**:823–830, 1998.

865. Rosen HR, Martin P: Hepatitis C infection in patients undergoing liver retransplantation. *Transplantation* **66**:1612–1616, 1998.

866. Feray C, Caccamo L, Alexander GJ, *et al*: European collaborative study on factors influencing outcome after liver transplantation for hepatitis C. European Concerted Action on Viral Hepatitis (EUROHEP) Group. *Gastroenterology* **117**:619–625, 1999.

867. Magy N, Cribier B, Schmitt C, *et al*: Effects of corticosteroids on HCV infection. *Int J Immunopharmacol* **21**:253–261, 1999.

868. Papatheodoridis GV, Barton SG, Andrew D, *et al*: Longitudinal variation in hepatitis C virus (HCV) viraemia and early course of HCV infection after liver transplantation for HCV cirrhosis: The role of different immunosuppressive regimens [see comments]. *Gut* **45**:427–434, 1999.

869. Charlton M, Seaberg E: Impact of immunosuppression and acute rejection on recurrence of hepatitis C: Results of the National Institute of Diabetes and Digestive and Kidney Diseases Liver Transplantation Database. *Liver Transplant Surg* **5**:S107–S114, 1999.

870. Rostaing L, Izopet J, Sandres K, *et al*: Changes in hepatitis C virus RNA viremia concentrations in long-term renal transplant patients after introduction of mycophenolate mofetil. *Transplantation* **69**:991–994, 2000.

871. Rosen HR, Shackleton CR, Higa L, *et al*: Use of OKT3 is associated with early and severe recurrence of hepatitis C after liver transplantation [see comments]. *Am J Gastroenterol* **92**:1453–1457, 1997.

872. Prieto M, Berenguer M, Rayon JM, *et al*: High incidence of allograft cirrhosis in hepatitis C virus genotype 1b infection following transplantation: Relationship with rejection episodes. *Hepatology* **29**:250–256, 1999.

873. Cotler SJ, Gaur LK, Gretch DR, *et al*: Donor–recipient sharing of HLA class II alleles predicts earlier recurrence and accelerated progression of hepatitis C following liver transplantation. *Tissue Antigens* **52**:435–443, 1998.

874. Neumann AU, Lam NP, Dahari H, *et al*: Differences in viral dynamics between genotypes 1 and 2 of hepatitis C virus. *J Infect Dis* **182**:28–35, 2000.

875. Pessoa MG, Bzowej N, Berenguer M, *et al*: Evolution of hepatitis C virus quasispecies in patients with severe cholestatic hepatitis after liver transplantation. *Hepatology* **30**:1513–1520, 1999.

876. Rosen HR, Chou S, Corless CL, *et al*: Cytomegalovirus viremia: Risk factor for allograft cirrhosis after liver transplantation for hepatitis C. *Transplantation* **64**:721–726, 1997.

877. Singh N, Zeevi A, Gayowski T, Marino IR: Late onset cytomegalovirus disease in liver transplant recipients: *De novo* reactivation in recurrent hepatitis C virus hepatitis. *Transplant Int* **11**:308–311, 1998.

878. Weiss G, Umlauft F, Urbanek M, *et al*: Associations between cellular immune effector function, iron metabolism, and disease activity in patients with chronic hepatitis C virus infection. *J Infect Dis* **180**:1452–1458, 1999.

879. Rosen HR, Lentz JJ, Rose SL, *et al*: Donor polymorphism of tumor necrosis factor gene: Relationship with variable severity of hepatitis C recurrence after liver transplantation. *Transplantation* **68**:1898–1902, 1999.

880. Toth CM, Pascual M, Chung RT, *et al*: Hepatitis C virus-associated fibrosing cholestatic hepatitis after renal transplantation: Response to interferon-alpha therapy. *Transplantation* **66**:1254–1258, 1998.

881. Delladetsima JK, Boletis JN, Makris F, *et al*: Fibrosing cholestatic hepatitis in renal transplant recipients with hepatitis C virus infection. *Liver Transplant Surg* **5**:294–300, 1999.

882. Abraczinskas D, Chung R: Allograft dysfunction and hyperbilirubinemia in a liver transplant recipient. *Transplant Infect Dis* **2**:186–193, 2000.

883. Davis CL, Gretch DR, Perkins JD, *et al*: Hepatitis C-associated glomerular disease in liver transplant recipients. *Liver Transplant Surg* **1**:166–175, 1995.

884. Morales JM, Campistol JM, Andres A, Rodicio JL: Glomerular diseases in patients with hepatitis C virus infection after renal transplantation. *Curr Opin Nephrol Hypertens* **6**:511–515, 1997.

885. Hestin D, Guillemin F, Castin N, *et al*: Pretransplant hepatitis C virus infection: A predictor of proteinuria after renal transplantation. *Transplantation* **65**:741–744, 1998.

886. Pascual M, Thadhani R, Chung RT, *et al*: Nephrotic syndrome after liver transplantation in a patient with hepatitis C virus-associated glomerulonephritis. *Transplantation* **64**:1073–1076, 1997.

887. Morales JM, Pascual-Capdevila J, Campistol JM, *et al*: Membranous glomerulonephritis associated with hepatitis C virus infection in renal transplant patients. *Transplantation* **63**:1634–1639, 1997.

888. Cantarell MC, Charco R, Capdevila L, *et al*: Outcome of hepatitis C virus-associated membranoproliferative glomerulonephritis after liver transplantation [see comments]. *Transplantation* **68**:1131–1134, 1999.

889. Baid S, Cosimi AB, Tolkoff-Rubin N, *et al*: Renal disease associated with hepatitis C infection after kidney and liver transplantation. *Transplantation* **70**:255–261, 2000.

890. Singh N, Gayowski T, Wagener MM, Marino IR: Vulnerability to psychologic distress and depression in patients with end-stage liver disease due to hepatitis C virus. *Clin Transplant* **11**:406–411, 1997.

891. Dickson RC, Wright RM, Bacchetta MD, *et al*: Quality of life of hepatitis B and C patients after liver transplantation. *Clin Transplant* **11**:282–285, 1997.

892. Bona MD, Rupolo G, Ponton P, *et al*: The effect of recurrence of HCV infection on life after liver transplantation. *Transplant Int* **11**:S475–S479, 1998.

893. Singh N, Gayowski T, Wagener MM, Marino IR: Quality of life, functional status, and depression in male liver transplant recipients with recurrent viral hepatitis C. *Transplantation* **67**:69–72, 1999.

894. Management of hepatitis C. *NIH Consensus Statement* **15**:1–41, 1997.

895. Sheiner PA, Boros P, Klion FM, *et al*: The efficacy of prophylactic interferon alfa-2b in preventing recurrent hepatitis C after liver transplantation. *Hepatology* **28**:831–838, 1998.

896. Cattral MS, Hemming AW, Wanless IR, *et al*: Outcome of long-term ribavirin therapy for recurrent hepatitis C after liver transplantation. *Transplantation* **67**:1277–1280, 1999.

897. Bizollon T, Palazzo U, Ducerf C, *et al*: Pilot study of the combination of interferon alfa and ribavirin as therapy of recurrent hepatitis C after liver transplantation [see comments]. *Hepatology* **26**:500–504, 1997.

898. Muramatsu S, Ku Y, Fukumoto T, *et al*: Successful rescue of severe recurrent hepatitis C with interferon and ribavirin in a liver transplant patient. *Transplantation* **69**:1956–1958, 2000.

899. Zeuzem S, Feinman SV, Rasenack J, *et al*: Peginterferon alfa-2a in patients with chronic hepatitis C. *N Engl J Med* **343**:1666–1672, 2000.

900. Heathcote EJ, Shiffman ML, Cooksley WG, *et al*: Peginterferon alfa-2a in patients with chronic hepatitis C and cirrhosis. *N Engl J Med* **343:**1673–1680, 2000.

901. Tzakis AG, Cooper MH, Dummer JS, *et al*: Transplantation in HIV⁺ patients. *Transplantation* **49:**354–358, 1990.

902. Erice A, Rhame FS, Heussner RC, Dunn DL, Balfour HH Jr: Human immunodeficiency virus infection in patients with solid-organ transplants: Report of five cases and review [see comments]. *Rev Infect Dis* **13:**537–547, 1991.

903. Patijn GA, Strengers PF, Harvey M, Persijn G: Prevention of transmission of HIV by organ and tissue transplantation. HIV testing protocol and a proposal for recommendations concerning donor selection. *Transplant Int* **6:**165–172, 1993.

904. Rubin RH, Jenkins RL, Shaw BW Jr, *et al*: The acquired immunodeficiency syndrome and transplantation. *Transplantation* **44:**1–4, 1987.

905. Ahuja TS, Zingman B, Glicklich D: Long-term survival in an HIV-infected renal transplant recipient. *Am J Nephrol* **17:**480–482, 1997.

906. Purgus R, Tamalet C, Poignard P, *et al*: Long-term nonprogressive human immunodeficiency virus-1 infection in a kidney allograft recipient. *Transplantation* **66:**1384–1386, 1998.

907. Hogg RS, Heath KV, Yip B, *et al*: Improved survival among HIV-infected individuals following initiation of antiretroviral therapy [see comments]. *JAMA* **279:**450–454, 1998.

908. Palella FJ Jr, Delaney KM, Moorman AC, *et al*: Declining morbidity and mortality among patients with advanced human immunodeficiency virus infection. HIV Outpatient Study Investigators [see comments]. *N Engl J Med* **338:**853–860, 1998.

909. Spital A: Should all human immunodeficiency virus-infected patients with end-stage renal disease be excluded from transplantation? The views of U.S. transplant centers. *Transplantation* **65:**1187–1191, 1998.

910. Schvarcz R, Rudbeck G, Soderdahl G, Stahle L: Interaction between nelfinavir and tacrolimus after orthoptic liver transplantation in a patient coinfected with HIV and hepatitis C virus (HCV). *Transplantation* **69:**2194–2195, 2000.

911. Gout O, Baulac M, Gessain A, *et al*: Rapid development of myelopathy after HTLV-I infection acquired by transfusion during cardiac transplantation. *N Engl J Med* **322:**383–388, 1990.

912. Sullivan MT, Williams AE, Fang CT, *et al*: Transmission of human T-lymphotropic virus types I and II by blood transfusion. A retrospective study of recipients of blood components (1983 through 1988). The American Red Cross HTLV-I/II Collaborative Study Group. *Arch Intern Med* **151:**2043–2048, 1991.

913. Perez G, Ortiz-Interian C, Bourgoignie JJ, *et al*: HIV-1 and HTLV-I infection in renal transplant recipients. *J Acquir Immune Defic Syndr* **3:**35–40, 1990.

914. Garner S: Prevalence in England of antibody to polyomavirus (B.K.). *Br Med J* **1:**77–78, 1973.

915. Shah KV, Daniel RW, Warszawski RM: High prevalence of antibodies to BK virus, an SV40-related papovavirus, in residents of Maryland. *J Infect Dis* **128:**784–787, 1973.

916. Brown P, Tsai T, Gajdusek DC: Seroepidemiology of human papovaviruses. Discovery of virgin populations and some unusual patterns of antibody prevalence among remote peoples of the world. *Am J Epidemiol* **102:**331–340, 1975.

917. Gardner SD, Field AM, Coleman DV, Hulme B: New human papovavirus (B.K.) isolated from urine after renal transplantation. *Lancet* **1:**1253–1257, 1971.

918. Coleman DV, Gardner SD, Field AM: Human polyomavirus infection in renal allograft recipients. *Br Med J* **3:**371–375, 1973.

919. Lecatsas G, Prozesky OW, Wyk Jv, Els HJ: Papova virus in urine after renal transplantation. *Nature* **241:**343–344, 1973.

920. Shah KV, Daniel RW, Zeigel RF, Murphy GP: Search for BK and SV40 virus reactivation in renal transplant recipients. *Transplantation* **17:**131–134, 1974.

921. Flower AJ, Banatvala JE, Chrystie IL: BK antibody and virus-specific IgM responses in renal transplant recipients, patients with malignant disease, and healthy people. *Br Med J* **2:**220–223, 1977.

922. Coleman DV, Mackenzie EF, Gardner SD, *et al*: Human polyomavirus (BK) infection and ureteric stenosis in renal allograft recipients. *J Clin Pathol* **31:**338–347, 1978.

923. Cheeseman SH, Black PH, Rubin RH, Cantell K, Hirsch MS: Interferon and BK papovavirus—Clinical and laboratory studies. *J Infect Dis* **141:**157–161, 1980.

924. Hogan TF, Borden EC, McBain JA, Padgett BL, Walker DL: Human polyomavirus infections with JC virus and BK virus in renal transplant patients. *Ann Intern Med* **92:**373–378, 1980.

925. Narayan O, Penney JB Jr, Johnson RT, Herndon RM, Weiner LP: Etiology of progressive multifocal leukoencephalopathy. Identification of papovavirus. *N Engl J Med* **289:**1278–1282, 1973.

926. Padgett B, Walker D: New human papovaviruses. In Melnick J (ed): *Progress in Medical Virology*. Basel, Karger, 1976, pp. 1–35.

927. Andrews CA, Shah KV, Daniel RW, Hirsch MS, Rubin RH: A serological investigation of BK virus and JC virus infections in recipients of renal allografts. *J Infect Dis* **158:**176–181, 1988.

928. Flaegstad T, Nilsen I, Skar AG, Traavik T: Antibodies against BK virus in renal transplant recipient sera: Results with five different methods indicate frequent reactivations. *Scand J Infect Dis* **23:**287–291, 1991.

929. Boubenider S, Hiesse C, Marchand S, *et al*: Post-transplantation polyomavirus infections. *J Nephrol* **12:**24–29, 1999.

930. Howell DN, Smith SR, Butterly DW, *et al*: Diagnosis and management of BK polyomavirus interstitial nephritis in renal transplant recipients. *Transplantation* **68:**1279–1288, 1999.

931. Binet I, Nickeleit V, Hirsch HH, *et al*: Polyomavirus disease under new immunosuppressive drugs: A cause of renal graft dysfunction and graft loss. *Transplantation* **67:**918–922, 1999.

932. Leventhal B, Soave R, Mouradian J, Cheigh J: Renal dysfunction and hyperglycemia in a renal transplant recipient. *Transplant Infect Dis* **1:**288–294, 1999.

933. Padgett BL, Walker DL, ZuRhein GM, *et al*: Cultivation of papova-like virus from human brain with progressive multifocal leucoencephalopathy. *Lancet* **1:**1257–1260, 1971.

934. Wolfson JS, Sober AJ, Rubin RH: Dermatologic manifestations of infections in immunocompromised patients. *Medicine* **64:**115–133, 1985.

935. Koranda FC, Dehmel EM, Kahn G, Penn I: Cutaneous complications in immunosuppressed renal homograft recipients. *JAMA* **229:**419–424, 1974.

936. Mullen D, Silverberg S, Penn I, *et al*: Squamous cell carcinoma of the skin and lip in renal homograft recipients. *JAMA* **229:**729–734, 1976.

937. Savin JA, Noble WC: Immunosuppression and skin infection. *Br J Dermatol* **93:**115–120, 1975.

938. Lutzner MA, Orth G, Dutronquay V, *et al*: Detection of human papillomavirus type 5 DNA in skin cancers of an immunosuppressed renal allograft recipient. *Lancet* **2:**422–424, 1983.

939. Ostrow RS, Bender M, Niimura M, *et al*: Human papillomavirus DNA in cutaneous primary and metastasized squamous cell carcinomas from patients with epidermodysplasia verruciformis. *Proc Natl Acad Sci USA* **79:**1634–1638, 1982.

940. Barr BB, Benton EC, McLaren K, *et al*: Human papilloma virus infection and skin cancer in renal allograft recipients. *Lancet* **1**:124–129, 1989.

941. Hopfl R, Bens G, Wieland U, *et al*: Human papillomavirus DNA in non-melanoma skin cancers of a renal transplant recipient: Detection of a new sequence related to epidermodysplasia verruciformis associated types. *J Invest Dermatol* **108**:53–56, 1997.

942. Arends MJ, Benton EC, McLaren KM, *et al*: Renal allograft recipients with high susceptibility to cutaneous malignancy have an increased prevalence of human papillomavirus DNA in skin tumours and a greater risk of anogenital malignancy. *Br J Cancer* **75**:722–728, 1997.

943. Bens G, Wieland U, Hofmann A, Hopfl R, Pfister H: Detection of new human papillomavirus sequences in skin lesions of a renal transplant recipient and characterization of one complete genome related to epidermodysplasia verruciformis-associated types. *J Gen Virol* **79**:779–787, 1998.

944. de Jong-Tieben LM, Berkhout RJ, ter Schegget J, *et al*: The prevalence of human papillomavirus DNA in benign keratotic skin lesions of renal transplant recipients with and without a history of skin cancer is equally high: A clinical study to assess risk factors for keratotic skin lesions and skin cancer. *Transplantation* **69**:44–49, 2000.

945. Alloub MI, Barr BB, McLaren KM, *et al*: Human papillomavirus infection and cervical intraepithelial neoplasia in women with renal allografts. *Br Med J* **298**:153–156, 1989.

946. Leptak C, Ramon y Cajal S, Kulke R, *et al*: Tumorigenic transformation of murine keratinocytes by the E5 genes of bovine papillomavirus type 1 and human papillomavirus type 16 [published erratum appears in *J Virol* **66(3)**:1833, 1992]. *J Virol* **65**:7078–7083, 1991.

947. Hinman F Jr, Schmaelzle JF, Belzer FO: Urinary tract infection and renal homotransplantation. II. Post-transplantation bacterial invasion. *J Urol* **101**:673–679, 1969.

948. Leigh DA: The outcome of urinary tract infections in patients after human cadaveric renal transplantation. *Br J Urol* **41**:406–413, 1969.

949. Martin DC: Urinary tract infection in clinical renal transplantation. *Arch Surg* **99**:474–476, 1969.

950. Prout GR Jr, Hume DM, Williams GM, Lee HM: Some urological aspects of 93 consecutive renal homotransplants in modified recipients. *J Urol* **97**:409–425, 1967.

951. Ramsey DE, Finch WT, Birtch AG: Urinary tract infections in kidney transplant recipients. *Arch Surg* **114**:1022–1025, 1979.

952. Rubin RH, Fang LS, Cosimi AB, *et al*: Usefulness of the antibody-coated bacteria assay in the management of urinary tract infection in the renal transplant patient. *Transplantation* **27**:18–20, 1979.

953. Myerowitz RL, Medeiros AA, O'Brien TF: Bacterial infection in renal homotransplant recipients. A study of fifty-three bacteremic episodes. *Am J Med* **53**:308–314, 1972.

954. Nielsen HE, Korsager B: Bacteremia after renal transplantation. *Scand J Infect Dis* **9**:111–117, 1977.

955. Pearson JC, Amend WJ Jr, Vincenti FG, Feduska NJ, Salvatierra O Jr: Post-transplantation pyelonephritis: Factors producing low patient and transplant morbidity. *J Urol* **123**:153–156, 1980.

956. Burleson RL, Brennan AM, Scruggs BF: Foley catheter tip cultures: A valuable diagnostic aid in the immunosuppressed patient. *Am J Surg* **133**:723–725, 1977.

957. Schaeffer AJ: Catheter-associated bacteriuria in patients in reverse isolation. *J Urol* **128**:752–754, 1982.

958. Schaeffer AJ, Chmiel J: Urethral meatal colonization in the pathogenesis of catheter-associated bacteriuria. *J Urol* **130**:1096–1099, 1983.

959. Hepinstall R: Experimental pyelonephritis: A comparison of blood-borne and ascending patterns of infection. *J Pathol* **89**:71–80, 1965.

960. Tolkoff-Rubin NE, Rubin RH: Urinary tract infection in the immunocompromised host. Lessons from kidney transplantation and the AIDS epidemic. *Infect Dis Clin North Am* **11**:707–717, 1997.

961. Tolkoff-Rubin N, Cosimi A, Russell P, *et al*: A controlled study of trimethoprim–sulfamethoxazole prophylaxis of urinary tract infections in renal transplant recipients. *Rev Infect Dis* **4**:614–618, 1982.

962. Fox BC, Sollinger HW, Belzer FO, Maki DG: A prospective, randomized, double-blind study of trimethoprim–sulfamethoxazole for prophylaxis of infection in renal transplantation: Clinical efficacy, absorption of trimethoprim–sulfamethoxazole, effects on the microflora, and the cost-benefit of prophylaxis. *Am J Med* **89**:255–274, 1990.

963. Maki DG, Fox BC, Kuntz J, Sollinger HW, Belzer FO: A prospective, randomized, double-blind study of trimethoprim–sulfamethoxazole for prophylaxis of infection in renal transplantation. Side effects of trimethoprim–sulfamethoxazole, interaction with cyclosporine. *J Lab Clin Med* **119**:11–24, 1992.

964. Hibberd P, Tolkoff-Rubin N, Doran M, *et al*: Trimethoprim–sulfamethoxazole compared with ciprofloxacin for the prevention of urinary tract infection in renal transplant recipients. *Online: J Curr Clin Trials* Doc. #15, 1992.

965. Hibberd PL, Rubin RH: Approach to immunization in the immunosuppressed host. *Infect Dis Clin North Am* **4**:123–142, 1990.

966. Cosimi AB, Colvin RB, Burton RC, *et al*: Use of monoclonal antibodies to T-cell subsets for immunologic monitoring and treatment in recipients of renal allografts. *N Engl J Med* **305**:308–314, 1981.

967. Peigue-Lafeuille H, Gazuy N, Mignot P, *et al*: Severe respiratory syncytial virus pneumonia in an adult renal transplant recipient: Successful treatment with ribavirin. *Scand J Infect Dis* **22**:87–89, 1990.

968. Doud JR, Hinkamp T, Garrity ER Jr: Respiratory syncytial virus pneumonia in a lung transplant recipient: Case report. *J Heart Lung Transplant* **11**:77–79, 1992.

969. Hierholzer JC, Atuk NO, Gwaltney JM Jr: New human adenovirus isolated from a renal transplant recipient: Description and characterization of candidate adenovirus type 34. *J Clin Microbiol* **1**:366–376, 1975.

970. Myerowitz RL, Stalder H, Oxman MN, *et al*: Fatal disseminated adenovirus infection in a renal transplant recipient. *Am J Med* **59**:591–598, 1975.

971. Keller EW, Rubin RH, Black PH, Hirsch MS, Hierholzer JC: Isolation of adenovirus type 34 from a renal transplant recipient with interstitial pneumonia. *Transplantation* **23**:188–191, 1977.

972. Stalder H, Hierholzer JC, Oxman MN: New human adenovirus (candidate adenovirus type 35) causing fatal disseminated infection in a renal transplant recipient. *J Clin Microbiol* **6**:257–265, 1977.

973. Lecatsas G, Prozesky OW, van Wyk J: Letter: Adenovirus type II associated with haemorrhagic cystitis after renal transplantation. *S Afr Med J* **48**:1932, 1974.

974. Harnett GB, Bucens MR, Clay SJ, Saker BM: Acute haemorrhagic cystitis caused by adenovirus type 11 in a recipient of a transplanted kidney. *Med J Aust* **1**:565–567, 1982.

975. Huebner R, Rowe W, Lane WT: Oncogenic effects in hamsters of human adenoviruses types 12 and 18. *Proc Natl Acad Sci USA* **48:**2051–2058, 1962.

976. Koneru B, Jaffe R, Esquivel CO, *et al*: Adenoviral infections in pediatric liver transplant recipients. *JAMA* **258:**489–492, 1987.

977. Yagisawa T, Takahashi K, Yamaguchi Y, *et al*: Adenovirus induced nephropathy in kidney transplant recipients. *Transplant Proc* **21:**2097–2099, 1989.

978. Wreghitt TG, Gray JJ, Ward KN, *et al*: Disseminated adenovirus infection after liver transplantation and its possible treatment with ganciclovir [letter]. *J Infect* **19:**88–89, 1989.

979. Hall WA, Martinez AJ, Dummer JS, *et al*: Central nervous system infections in heart and heart–lung transplant recipients. *Arch Neurol* **46:**173–177, 1989.

980. Tolkoff-Rubin N, Hovingh G, Rubin R: Central nervous system infections. In Wijdicks E (ed): *Neurologic Complications in Organ Transplant Recipients*. Boston, Butterworth Heinemann, 1999, pp. 141–168.

981. Singh N, Husain S: Infections of the central nervous system in transplant recipients. *Transplant Infect Dis* **2:**101–111, 2000.

982. de Groen PC, Aksamit AJ, Rakela J, Forbes GS, Krom RA: Central nervous system toxicity after liver transplantation. The role of cyclosporine and cholesterol. *N Engl J Med* **317:**861–866, 1987.

983. Wijdicks E: Neurologic manifestations of immunosuppressive agents. In Wijdicks E (ed): *Neurologic Complications in Organ Transplant Recipients*. Boston, Butterworth Heinemann, 1999, pp. 127–140.

984. Ryning FW, McLeod R, Maddox JC, Hunt S, Remington JS: Probable transmission of *Toxoplasma gondii* by organ transplantation. *Ann Intern Med* **90:**47–49, 1979.

985. Luft BJ, Naot Y, Araujo FG, Stinson EB, Remington JS: Primary and reactivated toxoplasma infection in patients with cardiac transplants. Clinical spectrum and problems in diagnosis in a defined population. *Ann Intern Med* **99:**27–31, 1983.

986. Rose AG, Uys CJ, Novitsky D, Cooper DK, Barnard CN: Toxoplasmosis of donor and recipient hearts after heterotopic cardiac transplantation. *Arch Pathol Lab Med* **107:**368–373, 1983.

987. McGregor CG, Fleck DG, Nagington J, *et al*: Disseminated toxoplasmosis in cardiac transplantation. *J Clin Pathol* **37:**74–77, 1984.

988. Nagington J, Martin AL: Toxoplasmosis and heart transplantation [letter]. *Lancet* **2:**679, 1983.

989. Hakim M, Esmore D, Wallwork J, English TA, Wreghitt T: Toxoplasmosis in cardiac transplantation. *Br Med J (Clin Res Ed)* **292:**1108, 1986.

990. Britt RH, Enzmann DR, Remington JS: Intracranial infection in cardiac transplant recipients. *Ann Neurol* **9:**107–119, 1981.

991. Luft BJ, Billingham M, Remington JS: Endomyocardial biopsy in the diagnosis of toxoplasmic myocarditis. *Transplant Proc* **18:**1871–1873, 1986.

992. Michaels MG, Wald ER, Fricker FJ, del Nido PJ, Armitage J: Toxoplasmosis in pediatric recipients of heart transplants. *Clin Infect Dis* **14:**847–851, 1992.

993. Wreghitt TG, Gray JJ, Pavel P, *et al*: Efficacy of pyrimethamine for the prevention of donor-acquired Toxoplasma gondii infection in heart and heart–lung transplant patients. *Transplant Int* **5:**197–200, 1992.

994. Conti DJ, Rubin RH: Infection of the central nervous system in organ transplant recipients. *Neurol Clin* **6:**241–260, 1988.

995. Ascher NL, Simmons RL, Marker S, Najarian JS: Listeria infection in transplant patients. Five cases and a review of the literature. *Arch Surg* **113:**90–94, 1978.

996. Schroter GP, Weil RD: Listeria monocytogenes infection after renal transplantation. *Arch Intern Med* **137:**1395–1399, 1977.

997. Niklasson PM, Hambraeus A, Lundgren G, *et al*: Listeria encephalitis in five renal transplant recipients. *Acta Med Scand* **203:**181–185, 1978.

998. Tilney NL, Kohler TR, Strom TB: Cerebromeningitis in immunosuppressed recipients of renal allografts. *Ann Surg* **195:**104–109, 1982.

999. Armstrong RW, Fung PC: Brainstem encephalitis (rhombencephalitis) due to Listeria monocytogenes: Case report and review. *Clin Infect Dis* **16:**689–702, 1993.

1000. Stamm AM, Smith SH, Kirklin JK, McGiffin DC: Listerial myocarditis in cardiac transplantation. *Rev Infect Dis* **12:**820–823, 1990.

1001. MacGowan AP, Marshall RJ, MacKay IM, Reeves DS: Listeria faecal carriage by renal transplant recipients, haemodialysis patients and patients in general practice: Its relation to season, drug therapy, foreign travel, animal exposure and diet. *Epidemiol Infect* **106:**157–166, 1991.

1002. Gaillard JL, Berche P, Frehel C, Gouin E, Cossart P: Entry of L. monocytogenes into cells is mediated by internalin, a repeat protein reminiscent of surface antigens from gram-positive cocci. *Cell* **65:**1127–1141, 1991.

1003. Schlech WFd, Lavigne PM, Bortolussi RA, *et al*: Epidemic listeriosis—Evidence for transmission by food. *N Engl J Med* **308:**203–206, 1983.

1004. Watson GW, Fuller TJ, Elms J, Kluge RM: Listeria cerebritis: Relapse of infection in renal transplant patients. *Arch Intern Med* **138:**83–87, 1978.

1005. Larner AJ, Conway MA, Mitchell RG, Forfar JC: Recurrent Listeria monocytogenes meningitis in a heart transplant recipient. *J Infect* **19:**263–266, 1989.

1006. Peetermans WE, Endtz HP, Janssens AR, van den Broek PJ: Recurrent Listeria monocytogenes bacteraemia in a liver transplant patient. *Infection* **18:**107–108, 1990.

1007. Moellering RC Jr, Medoff G, Leech I, Wennersten C, Kunz LJ: Antibiotic synergism against Listeria monocytogenes. *Antimicrob Agents Chemother* **1:**30–34, 1972.

1008. Scheer MS, Hirschman SZ: Oral and ambulatory therapy of Listeria bacteremia and meningitis with trimethoprim–sulfamethoxazole. *Mt Sinai J Med* **49:**411–414, 1982.

1009. Spitzer PG, Hammer SM, Karchmer AW: Treatment of Listeria monocytogenes infection with trimethoprim–sulfamethoxazole: Case report and review of the literature. *Rev Infect Dis* **8:**427–430, 1986.

1010. Agarwal A, Gupta A, Sakhuja V, *et al*: Retinitis following disseminated cryptococcosis in a renal allograft recipient. Efficacy of oral fluconazole. *Acta Ophthalmol* **69:**402–405, 1991.

1011. Schroter GP, Temple DR, Husberg BS, Weid RD, Starzl TE: Cryptococcosis after renal transplantation: Report of ten cases. *Surgery* **79:**268–277, 1976.

1012. Ellner JJ, Bennett JE: Chronic meningitis. *Medicine* **55:**341–369, 1976.

1013. Bennett JE, Dismukes WE, Duma RJ, *et al*: A comparison of amphotericin B alone and combined with flucytosine in the treatment of cryptococcal meningitis. *N Engl J Med* **301:**126–131, 1979.

1014. Hellman RN, Hinrichs J, Sicard G, *et al*: Cryptococcal pyelonephritis and disseminated cryptococcosis in a renal transplant recipient. *Arch Intern Med* **141:**128–130, 1981.

1015. Plunkett JM, Turner BI, Tallent MB, Johnson HK: Cryptococcal septicemia associated with urologic instrumentation in a renal allograft recipient. *J Urol* **125:**241–242, 1981.

1016. Conti DJ, Tolkoff-Rubin NE, Baker GP Jr, *et al*: Successful treatment of invasive fungal infection with fluconazole in organ transplant recipients. *Transplantation* **48:**692–695, 1989.

1017. Weiland D, Ferguson RM, Peterson PK, *et al*: Aspergillosis in 25 renal transplant patients. Epidemiology, clinical presentation, diagnosis, and management. *Ann Surg* **198:**622–629, 1983.

1018. Montero CG, Martinez AJ: Neuropathology of heart transplantation: 23 cases. *Neurology* **36:**1149–1154, 1986.

1019. Green M, Wald ER, Tzakis A, Todo S, Starzl TE: Aspergillosis of the CNS in a pediatric liver transplant recipient: Case report and review. *Rev Infect Dis* **13:**653–657, 1991.

1020. Rubin R, Weinstein L: *Salmonellosis: Microbiologic, Pathologic and Clinical Features*. New York, Stratton Intercontinental, 1977.

1021. Smith EJ, Milligan SL, Filo RS: Salmonella mycotic aneurysm after renal transplantation. *South Med J* **74:**1399–1401, 1981.

1022. Berk MR, Meyers AM, Cassal W, Botha JR, Myburgh JA: Nontyphoid salmonella infections after renal transplantation. A serious clinical problem. *Nephron* **37:**186–189, 1984.

1023. Samra Y, Shaked Y, Maier MK: Nontyphoid salmonellosis in renal transplant recipients: Report of five cases and review of the literature. *Rev Infect Dis* **8:**431–440, 1986.

1024. Dhar JM, al-Khader AA, al-Sulaiman M, al-Hasani MK: Nontyphoid salmonella in renal transplant recipients: A report of twenty cases and review of the literature. *Q J Med* **78:**235–250, 1991.

1025. Wade JJ, Rolando N, Hayllar K, *et al*: Bacterial and fungal infections after liver transplantation: An analysis of 284 patients. *Hepatology* **21:**1328–1336, 1995.

1026. Patel R, Badley AD, Larson-Keller J, *et al*: Relevance and risk factors of enterococcal bacteremia following liver transplantation. *Transplantation* **61:**1192–1197, 1996.

1027. George DL, Arnow PM, Fox A, *et al*: Patterns of infection after pediatric liver transplantation. *Am J Dis Child* **146:**924–929, 1992.

1028. King RW Jr, Kraikitpanitch S, Lindeman RD: Subcutaneous nodules caused by Histoplasma capsulatum [letter]. *Ann Intern Med* **86:**586–587, 1977.

1029. Kauffman CA, Israel KS, Smith JW, *et al*: Histoplasmosis in immunosuppressed patients. *Am J Med* **64:**923–932, 1978.

1030. Davies SF, Khan M, Sarosi GA: Disseminated histoplasmosis in immunologically suppressed patients. Occurrence in a nonendemic area. *Am J Med* **64:**94–100, 1978.

1031. Davies SF, Sarosi GA, Peterson PK, *et al*: Disseminated histoplasmosis in renal transplant recipients. *Am J Surg* **137:**686–691, 1979.

1032. Peterson PK, Dahl MV, Howard RJ, Simmons RL, Najarian JS: Mucormycosis and cutaneous histoplasmosis in a renal transplant recipient. *Arch Dermatol* **118:**275–277, 1982.

1033. Deresinski SC, Stevens DA: Coccidioidomycosis in compromised hosts. Experience at Stanford University Hospital. *Medicine* **54:**377–395, 1975.

1034. Bayer AS, Yoshikawa TT, Galpin JE, Guze LB: Unusual syndromes of coccidioidomycosis: Diagnostic and therapeutic considerations: A report of 10 cases and review of the English literature. *Medicine* **55:**131–152, 1976.

1035. Schroter GP, Bakshandeh K, Husberg BS, Well RD: Coccidioidomycosis and renal transplantation. *Transplantation* **23:**485–489, 1977.

1036. Cohen IM, Galgiani JN, Potter D, Ogden DA: Coccidioidomycosis in renal replacement therapy. *Arch Intern Med* **142:**489–494, 1982.

1037. Gallis HA, Berman RA, Cate TR, *et al*: Fungal infection following renal transplantation. *Arch Intern Med* **135:**1163–1172, 1975.

1038. Serody JS, Mill MR, Detterbeck FC, Harris DT, Cohen MS: Blastomycosis in transplant recipients: Report of a case and review. *Clin Infect Dis* **16:**54–58, 1993.

1039. Mayer JM, Nimer L, Carroll K: Isolated pulmonary aspergillar infection in cardiac transplant recipients: Case report and review. *Clin Infect Dis* **15:**698–700, 1992.

1040. Sridhar NR, Tchervenkov JI, Weiss MA, Hijazi YM, First MR: Disseminated histoplasmosis in a renal transplant patient: A cause of renal failure several years following transplantation. *Am J Kidney Dis* **17:**719–721, 1991.

1041. Benedict LM, Kusne S, Torre-Cisneros J, Hunt SJ: Primary cutaneous fungal infection after solid-organ transplantation: Report of five cases and review [see comments]. *Clin Infect Dis* **15:**17–21, 1992.

1042. Durand F, Bernuau J, Dupont B, *et al*: Aspergillus intraabdominal abscess after liver transplantation successfully treated with itraconazole. *Transplantation* **54:**734–735, 1992.

1043. Kusne S, Torre-Cisneros J, Manez R, *et al*: Factors associated with invasive lung aspergillosis and the significance of positive Aspergillus culture after liver transplantation. *J Infect Dis* **166:**1379–1383, 1992.

1044. Le Conte P, Blanloeil Y, Michel P, Francois T, Paineau J: Cutaneous aspergillosis in a patient with orthotopic hepatic transplantation. *Transplantation* **53:**1153–1154, 1992.

1045. Rattazzi LC, Simmons RL, Spanos PK, Bradford DS, Najarian JS: Successful management of miliary tuberculosis after renal transplantation. *Am J Surg* **130:**359–361, 1975.

1046. Oliver WA: Tuberculosis in renal transplant patients. *Med J Aust* **1:**828–829, 1976.

1047. Bell TJ, Williams GB: Successful treatment of tuberculosis in renal transplant recipients. *J R Soc Med* **71:**265–268, 1978.

1048. Ascher NL, Simmons RL, Marker S, Klugman J, Najarian JS: Tuberculous joint disease in transplant patients. *Am J Surg* **135:**853–856, 1978.

1049. Riska H, Kuhlback B: Tuberculosis and kidney transplantation. *Acta Med Scand* **205:**637–640, 1979.

1050. Vaz AJ: Miliary tuberculosis and the adult respiratory distress syndrome in a renal transplant recipient [letter]. *Chest* **75:**412, 1979.

1051. Lloveras J, Peterson PK, Simmons RL, Najarian JS: Mycobacterial infections in renal transplant recipients. Seven cases and a review of the literature. *Arch Intern Med* **142:**888–892, 1982.

1052. Coutts, II, Jegarajah S, Stark JE: Tuberculosis in renal transplant recipients. *Br J Dis Chest* **73:**141–148, 1979.

1053. Spence RK, Dafoe DC, Rabin G, *et al*: Mycobacterial infections in renal allograft recipients. *Arch Surg* **118:**356–359, 1983.

1054. Qunibi WY, al-Sibai MB, Taher S, *et al*: Mycobacterial infection after renal transplantation—Report of 14 cases and review of the literature. *Q J Med* **77:**1039–1060, 1990.

1055. al-Sulaiman MH, Dhar JM, al-Hasani MK, Haleem A, Al-Khader A: Tuberculous interstitial nephritis after kidney transplantation. *Transplantation* **50:**162–164, 1990.

1056. Higgins RM, Cahn AP, Porter D, *et al*: Mycobacterial infections after renal transplantation. *Q J Med* **78:**145–153, 1991.

1057. Sterneck M, Ferrell L, Ascher N, *et al*: Mycobacterial infection after liver transplantation: A report of three cases and review of the literature. *Clin Transplant* **6:**55–61, 1992.

1058. Sundberg R, Shapiro R, Darras F, *et al*: A tuberculosis outbreak in a renal transplant program. *Transplant Proc* **23**:3091–3092, 1991.

1059. Horsburgh CR Jr, Feldman S, Ridzon R: Practice guidelines for the treatment of tuberculosis [In Process Citation]. *Clin Infect Dis* **31**:633–639, 2000.

1060. Thomas PA Jr, Mozes MF, Jonasson O: Hepatic dysfunction during isoniazid chemoprophylaxis in renal allograft recipients. *Arch Surg* **114**:597–599, 1979.

1061. Higgins R, Kusne S, Reyes J, *et al*: Mycobacterium tuberculosis after liver transplantation: Management and guidelines for prevention. *Clin Transplant* **6**:81–90, 1992.

1062. Meyers BR, Papanicolaou GA, Sheiner P, Emre S, Miller C: Tuberculosis in orthotopic liver transplant patients: Increased toxicity of recommended agents; cure of disseminated infection with nonconventional regimens. *Transplantation* **69**:64–69, 2000.

1063. Novick RJ, Moreno-Cabral CE, Stinson EB, *et al*: Nontuberculous mycobacterial infections in heart transplant recipients: A seventeen-year experience. *J Heart Transplant* **9**:357–363, 1990.

1064. Cruz N, Ramirez-Muxo O, Bermudez RH, Santiago-Delpin EA: Pulmonary infection with M. kansasii in a renal transplant patient. *Nephron* **26**:187–188, 1980.

1065. Bolivar R, Satterwhite TK, Floyd M: Cutaneous lesions due to Mycobacterium kansasii. *Arch Dermatol* **116**:207–208, 1980.

1066. Trulock EP, Bolman RM, Genton R: Pulmonary disease caused by Mycobacterium chelonae in a heart–lung transplant recipient with obliterative bronchiolitis. *Am Rev Respir Dis* **140**:802–805, 1989.

1067. Weber J, Mettang T, Staerz E, Machleidt C, Kuhlman U: Pulmonary disease due to Mycobacterium xenopi in a renal allograft recipient: Report of a case and review. *Rev Infect Dis* **11**:964–969, 1989.

1068. Gombert ME, Goldstein EJ, Corrado ML, Stein AJ, Butt KM: Disseminated Mycobacterium marinum infection after renal transplantation. *Ann Intern Med* **94**:486–487, 1981.

1069. Davis BR, Brumbach J, Sanders WJ, Wolinsky E: Skin lesions caused by Mycobacterium haemophilum. *Ann Intern Med* **97**:723–724, 1982.

1070. Neeley SP, Denning DW: Cutaneous Mycobacterium thermoresistibile infection in a heart transplant recipient. *Rev Infect Dis* **11**:608–611, 1989.

1071. Heironimus JD, Winn RE, Collins CB: Cutaneous nonpulmonary Mycobacterium chelonei infection. Successful treatment with sulfonamides in an immunosuppressed patient. *Arch Dermatol* **120**:1061–1063, 1984.

1072. Cooper JF, Lichtenstein MJ, Graham BS, Schaffner W: Mycobacterium chelonae: A cause of nodular skin lesions with a proclivity for renal transplant recipients. *Am J Med* **86**:173–177, 1989.

1073. Kochhar R, Indudhara R, Nagi B, Yadav RV, Mehta SK: Colonic tuberculosis due to atypical mycobacteria in a renal transplant recipient [letter]. *Am J Gastroenterol* **83**:1435–1436, 1988.

1074. Patel R, Roberts GD, Keating MR, Paya CV: Infections due to nontuberculous mycobacteria in kidney, heart, and liver transplant recipients. *Clin Infect Dis* **19**:263–273, 1994.

1075. Munoz RM, Alonso-Pulpon L, Yebra M, *et al*: Intestinal involvement by nontuberculous mycobacteria after heart transplantation. *Clin Infect Dis* **30**:603–605, 2000.

1076. Diagnosis and treatment of disease caused by nontuberculous mycobacteria [published erratum appears in *Am Rev Respir Dis* **143**(1):204, 1991] [see comments]. *Am Rev Respir Dis* **142**:940–953, 1990.

1077. Burroughs M, Moscone A: Immunization of pediatric solid organ transplant candidates and recipients. *Clin Infect Dis* **30**:857–869, 2000.

1078. Loinaz C, de Juanes JR, Gonzalez EM, *et al*: Hepatitis B vaccination results in 140 liver transplant recipients. *Hepatogastroenterology* **44**:235–238, 1997.

1079. Blumberg EA, Albano C, Pruett T, *et al*: The immunogenicity of influenza virus vaccine in solid organ transplant recipients. *Clin Infect Dis* **22**:295–302, 1996.

1080. Admon D, Engelhard D, Strauss N, Goldman N, Zakay-Rones Z: Antibody response to influenza immunization in patients after heart transplantation. *Vaccine* **15**:1518–1522, 1997.

1081. Blumberg EA, Fitzpatrick J, Stutman PC, Hayden F, Brozena SC: Safety of influenza vaccine in heart transplant recipients. *J Heart Lung Transplant* **17**:1075–1080, 1998.

1082. Gershon AA: Immunizations for pediatric transplant patients. *Kidney Int Suppl* **43**:S87–S90, 1993.

1083. Pediatrics AAo: Poliovirus infections. In Georges M (ed): *Red Book: Report of the Committee on Infectious Disease*. Elk Grove Village, IL, American Academy of Pediatrics, 1997, pp. 424–433.

1084. Nkowane BM, Wassilak SG, Orenstein WA, *et al*: Vaccine-associated paralytic poliomyelitis. United States: 1973 through 1984. *JAMA* **257**:1335–1340, 1987.

1085. Zamora I, Simon JM, Da Silva ME, Piqueras AI: Attenuated varicella virus vaccine in children with renal transplants. *Pediatr Nephrol* **8**:190–192, 1994.

18

Surgical Aspects of Infection in the Compromised Host

A. BENEDICT COSIMI

1. Introduction

Steadily increasing numbers of patients with compromised immune responsiveness are being encountered in current surgical practice. The spectrum of these patients (Table 1) ranges from those with severely impaired host resistance, such as victims of acquired immunodeficiency syndrome (AIDS) or immunosuppressed transplant recipients, to those with more subtle defects, as occur in diabetics and those at the extremes of age. These patients should be expected to respond to inflammation and surgical stress quite differently than do normal individuals. Thus, it is not unusual to encounter as great as a 12-fold increase in postoperative sepsis and mortality[1] in patients whose acute condition is complicated by factors that diminish host defenses. The surgeon called on to evaluate a problem requiring possible surgical intervention in such patients must be aware of the special considerations that have been found to adversely influence their prognosis. Successful operative intervention is usually not possible following the solo approach employed for acute focal or traumatic conditions in the otherwise healthy individual. For compromised hosts, close collaboration and planning with the internist, infectious disease consultant, and anesthetist are essential. This chapter presents a number of guidelines found helpful in the diagnostic approach, preoperative preparation, intraoperative techniques, and

postoperative management of infectious conditions in these patients, illustrated by a review of typically encountered cases.

2. Diagnostic Approach

Since impaired immune responsiveness not only provides the soil for atypical infectious conditions but also masks the commonly expected signs and symptoms of the inflammatory process (Table 2), an unusually aggressive diagnostic approach must often be pursued in these patients. Their outcome is almost entirely dependent on the rapidity with which the correct diagnosis is established and specific therapy is instituted. In our experience with pulmonary infections in renal transplant recipients, for example, the overall mortality of patients diagnosed in the first 5 days of illness was 21%, as opposed to 65% for those whose diagnosis was not clarified until after 5 days of illness.[2] Others have emphasized that therapeutic delay for pulmonary infection in immunocompromised patients is, in fact, almost universally fatal and that the specific diagnosis is too often made only at postmortem examination.[3,4]

2.1. Diffuse Pulmonary Infiltrates in the Immunocompromised Host

Illustrative Cases 1 and 2

A 54-year-old renal allograft recipient (Case 1) was admitted with a history of low-grade fevers and dyspnea of 2 months' duration. His past history was complicated, including kidney transplantation 10 years earlier and then retransplantation 2 years earlier for recurrent membranous

A. Benedict Cosimi • Department of Surgery, Massachusetts General Hospital, Boston, Massachusetts 02114.

Clinical Approach to Infection in the Compromised Host (Fourth Edition), edited by Robert H. Rubin and Lowell S. Young. Kluwer Academic/Plenum Publishers, New York, 2002.

TABLE 1. Causes of Impaired Host Resistance

Patient's underlying condition
  AIDS
  Neoplasia
  Malnutrition
  Acute stress (burns, trauma)
  Metabolic factors (diabetes, uremia, obesity)
  Prematurity, advanced age
Iatrogenic
  Antineoplastic chemotherapy
  Immunosuppression (allograft or marrow recipients, autoimmune
    disorders)
  Splenectomy, perioperative transfusions

glomerulonephritis. He had been maintained on cyclosporine (CyA), azathioprine, and prednisone and had good allograft function. Prophylactic trimethoprim–sulfamethoxazole (TMP-SMX) therapy had been discontinued, according to protocol, at 6 months after the second renal transplant. Two months prior to admission, cough, low-grade fever, and malaise were first noted. Chest radiography showed only linear scarring at the bases. In view of negative sputum cultures, a presumptive diagnosis of viral upper respiratory infection was made and his prednisone and azathioprine dosages were reduced. Nevertheless, the respiratory symptoms persisted. Subsequently, a 5-day empirical course of azithromycin was administered, but his dyspnea continued to worsen. He was then referred to our institution for further evaluation. At the time of admission, he appeared acutely ill with fever and severe hypoxemia while breathing room air. Chest radiography showed bilateral hilar infiltrates consistent with infection or fluid overload. Intensive diuresis only marginally improved the respiratory function. Induced sputum smears and cultures remained nondiagnostic. By 48 hr after admission, tracheal intubation for ventilator-assisted respiration became necessary. On the second hospital day, a thoracoscopically obtained lung biopsy confirmed the presumptive diagnosis of *Pneumocystis carinii* pneumonia, which ultimately responded to high dosage TMP-SMX therapy.

A 60-year-old woman (Case 2) had been treated for 5 months with cyclophosphamide, methotrexate, and 5-fluorouracil for stage II breast cancer. She was now admitted with fever and respiratory distress of several days' duration. Chest radiography showed a focal consolidation in the right lung. When the transtracheal sputum aspirate yielded *Streptococcus pneumoniae*, ampicillin and gentamicin therapy were begun.

TABLE 2. Impaired Host Resistance: Consequences Affecting Surgical Management

Ineffective inflammatory reaction
Unusual presentation of common conditions
Unusual infectious problems
Increased tissue fragility
Impaired wound healing
Side effects of drug therapy for underlying condition (e.g., leukopenia,
  hyperglycemia, adrenal insufficiency)

Sequential radiographs showed progressive involvement of both lung fields. Further sputum cultures were not helpful. Despite the addition of erythromycin and TMP-SMX, her symptoms worsened, requiring tracheal intubation for assisted ventilation. An open lung biopsy revealed drug-induced pneumonitis with no evidence of significant infection. Following the addition of steroid therapy, the patient's condition rapidly improved, and she was discharged on a different adjuvant chemotherapy regimen for her breast cancer.

Comment. These cases emphasize the nature of the challenge the physician faces when assessing diffuse pulmonary infiltrates in an immunosuppressed patient. The differential diagnosis must include not only the usual bacterial or viral pathogens but also noninfectious causes, such as drug-induced pneumonitis,[5] pulmonary emboli or edema, and, most importantly, invasion by opportunistic agents such as *Pneumocystis* or *Legionella*.[6] The role of the surgeon in this assessment is to provide, as efficiently and safely as possible, appropriate secretions or tissues for study. Although empirical therapy is often begun immediately in these critically ill patients, one must expeditiously pursue clarification of the definitive diagnosis before, as in Case 1, the potentially treatable process has become life-threatening.

Spontaneously expectorated sputum specimens from immunocompromised patients are often nondiagnostic. We, therefore, favor obtaining an induced sputum as the initial approach to isolation of the infective agent. In our experience, this simple noninvasive procedure provides the specific diagnosis in approximately 50% of individuals with diffuse infiltrates, being most useful in patients with *P. carinii* pneumonia. If the induced sputum is nondiagnostic, early use of more invasive diagnostic approaches is essential. These include bronchoscopic lavage or transbronchial biopsy, percutaneous lung aspiration biopsy, and video-assisted thoracoscopic or even open lung biopsy. We favor early thoracoscopic biopsy, since the ample tissue obtained almost always provides the diagnosis, whereas the "less invasive" procedures not infrequently provide inadequate specimens and result in disappointing delays. Although some investigators continue to question the impact of lung biopsy in the management of these patients, many have concluded it is essential for early specific treatment. An unexpected diagnosis or change in therapy following lung biopsy has been reported in 45–50% of immunocompromised patients.[7,8] We have found that even open lung biopsy can be accomplished with relatively few complications using a limited intercostal incision through which the most obviously abnormal area is isolated using noncrushing clamps. After the wedge biopsy is obtained, the incised lung can be oversewn while the clamps maintain complete control of bleeding and air leakage. Similar results can be achieved using the automatic stapler for lung repair. Despite postoperative mechanical ventilation, bronchopleural fistula formation is seldom a problem.

Nevertheless, even this limited procedure can lead to

further complications in these fragile patients, so that most surgeons now favor thoracoscopic transpleural lung biopsy even for children.[9] This video-assisted procedure allows for an aggressive approach in obtaining tissue for diagnostic purposes without fear of the possible complications associated with standard thoracotomy. There is not much more morbidity than with transbronchial biopsy, but the tissue obtained is far superior, with an 85% successful diagnostic rate being achieved in patients with diffuse infiltrates.

2.2. Focal Pulmonary Infiltrates or Nodule(s) in the Immunocompromised Host

Illustrative Case 3

The patient was a 4-year-old male who underwent liver transplantation for biliary atresia. Induction immunosuppression included tacrolimus (FK506), azathioprine, and steroids. Primary nonfunction of the allograft necessitated emergency retransplantation on day 6. Excellent hepatic function was achieved by the second allograft, but methylprednisolone pulses and then a 10-day course of OKT3 monoclonal antibody were required for persistent rejection beginning on day 14. On day 21, a single colony of *Aspergillus fumigatus* was noted on routine surveillance sputum culture but the patient's chest X-ray revealed only minimal atelectasis in the right upper lobe. At that point, the patient's prophylactic antifungal therapy was changed from fluconazole to itraconazole. Over the following 10 days, several further sputum cultures became positive for *Aspergillus* and the right upper lobe pulmonary infiltrate progressed. Thoracic computerized tomography (CT) on day 31 revealed several nodular lesions in the right upper lung field. Liposomal amphotericin B therapy was begun and over the next 14 days the patient received a total of 60 mg/kg of this agent. Nevertheless, there was progression of the pulmonary process and head CT then revealed several lesions consistent with aspergillosis infection. Amphotericin B was therefore substituted for the liposomal preparation. Over the next 38 days, the patient received a total of more than 40 mg/kg of this agent. At that point, he developed seizures followed rapidly by brain death, presumably from hemorrhage into a cerebral lesion.

Comment. Invasive pulmonary aspergillosis has been reported to occur in 1–4% of liver transplant recipients.[10,11] At greatest risk are retransplant recipients, patients on high-dosage steroids, patients who have suffered vascular or biliary complications resulting in prolonged courses of antibiotics, and recipients whose transplant was required on an urgent basis or was accomplished with large-volume blood loss. Initial presentation may be only with unimpressive pulmonary infiltrates on chest X ray accompanied by low-grade fever. Bronchoscopy or even needle biopsy at this point are seldom diagnostic. Chest CT, however, may be useful in demonstrating pulmonary nodules at an early stage of the disease. Unfortunately, as in Case 3, there is often a delay in instituting appropriate medical therapy for invasive pulmonary aspergillosis and standard treatment in neutropenic patients or liver transplant recipients then fails in over 90% of the patients.[10,12] In view of these dismal mortality rates, some groups have turned to aggressive early surgical resection of suspected localized disease.[10] Survival rates of as high as 60% have been reported in patients undergoing resection of pulmonary masses within 7–10 days of the time that the diagnosis of aspergillosis was suggested clinically and confirmed by chest CT.

2.3. Gastrointestinal Complications of the Immunosuppressed State

Illustrative Case 4

A 61-year-old man was treated with azathioprine and prednisone after receiving a renal allograft from his daughter. With the onset of acute rejection 9 days after transplantation, the steroid dosage was sharply increased, and local graft irradiation was instituted. Renal function gradually improved, and tapering of the steroid dosage was begun. Six weeks after the transplant, the patient returned to his home in a neighboring state with normal allograft function while receiving azathioprine 125 mg/day and prednisone 20 mg/day. One week later, he complained of lower abdominal discomfort. He was evaluated at his community hospital, where a temperature of 101°F (38°C) was noted and the abdomen was described as slightly distended with moderate direct and rebound tenderness. Laboratory evaluation revealed a white blood cell (WBC) count of 11,500/mm³, hematocrit (Hct) of 33%, and no abnormalities of hepatic or renal function. Surgical consultation recommended routine abdominal radiographs. Since these were normal and the patient had remained reasonably stable, admission and simple observation were advised. During the succeeding hours, the patient complained of increasing abdominal distention and discomfort. By 12 hr after admission, he had developed oliguria, hypotension, and restlessness consistent with gram-negative septicemia. When transfer to a transplant center and emergency laparotomy were finally accomplished, diffuse fecal peritonitis secondary to perforated sigmoid diverticulitis was found. Despite resection of the involved bowel with construction of an end-colostomy, extensive irrigation and drainage of the abdomen, and intensive postoperative supportive measures, the patient subsequently expired.

Illustrative Case 5

A 48-year-old man with a history of α-1 antitrypsin deficiency had undergone bilateral lung transplantation 2 months earlier. Maintenance immunosuppression with CyA, steroids, and azathioprine had resulted in good pulmonary rehabilitation. He presented with a 2-day history of diarrhea followed by lower abdominal discomfort. Evaluation revealed him to be afebrile, in moderate distress from the abdominal pain, but otherwise stable except for definite lower abdominal direct and rebound tenderness. Hematologic and serum chemistry studies revealed mild leukocytosis. Plain radiographs and ultrasound examination of the abdomen were not remarkable. Paracentesis produced several milliliters of thin, yellow fluid containing multiple neutrophils on microscopic examination.

Abdominal CT with rectal contrast revealed colonic diverticulae and a small pelvic air-fluid level interposed between the bladder and the rectum. Emergency laparotomy performed within 4 hr of admission confirmed the preoperative diagnosis of perforated sigmoid colon which was treated by resection and end-sigmoid colostomy. Postoperatively, the patient recovered without incident. The colostomy was successfully closed as a subsequent procedure.

Comment. Since the possible occult sites of acute inflammatory processes in the abdomen or retroperitoneum are numerous, accurate evaluation in this situation is perplexing. Furthermore, because of the host's impaired response, typical findings of an acute abdominal catastrophe may be deceptively mild. In these compromised patients, active bowel sounds with continuing bowel movements and even diarrhea are not unusual, despite extensive peritoneal soilage. The impairment of the inflammatory response, unfortunately, also results in an inability to wall off the pathologic process so that irreversible disseminated sepsis may develop during even brief periods of hopeful observation. Most reports of colonic perforation in steroid-treated or uremic patients emphasize the paucity of symptoms, signs, and laboratory evidence of visceral perforation, leading to fatal delays in treatment or to an incorrect preoperative diagnosis in the majority of cases.[13,14] This complication has been reported so frequently that some authorities suggest that there is a direct adverse effect of steroids on normal colon. Certainly, inhibition of the normal inflammatory response, antifibroblastic activity, and atrophy of lymphoid elements of the bowel wall could interfere with normal barriers to invasive infection by intraluminal bacteria. Thus, perforation may occur whether the colon was previously diseased or not and, indeed, has been observed in apparently normal areas of the bowel.[15] This, plus the host's inability to adequately localize the process, results in extensive contamination. Early clinical recognition of this problem, therefore, is essential, and perforation of the colon should always be a prime suspect in any steroid-treated patient with fever and abdominal symptoms. Paracentesis has been particularly helpful in this diagnosis, often yielding purulent fluid despite absence of free air on plain radiographs. If the diagnosis remains in doubt or to better localize the process, contrast radiographic studies of the lower gastrointestinal (GI) tract should not be delayed. In general, a water-soluble contrast agent is preferred, but establishment of the diagnosis of perforated colon with barium studies has been reported without the serious complication of barium peritonitis.

Because of the often devastating consequences of perforated diverticulitis occurring posttransplantation, some centers have recommended that pretransplant diagnostic screening be performed on all potential candidates over 50 years of age.[16] Prophylactic colonic resection is then advised for individuals found to have significant diverticular disease. A more recent review of this issue has concluded that such an aggressive approach does not accurately predict posttransplant complications and should probably be abandoned.[17]

As more potent immunosuppressive drugs have been introduced into clinical protocols, other infectious agents such as cytomegalovirus (CMV) and Epstein–Barr virus (EBV) have been added as either primary or comorbid factors in the pathophysiology of GI complications.[18] CMV can present in a variety of forms ranging from asymptomatic infection, diagnosed by serologic or viral shedding studies, to symptomatic syndromes including a constellation of manifestations. These may include fever, myalgias, malaise, dyspnea, and leukopenia, and evidence of isolated organ involvement such as pneumonitis, retinitis, pancreatitis, hepatitis, GI hemorrhage or perforation, and even multiple organ system failure. Recently, several groups have emphasized that the pattern of CMV disease appears to be changing as it more frequently affects the GI tract in severely immunocompromised hosts.[19,20]

CMV infection can affect the GI tract from the esophagus to the rectum. The most common manifestations are abdominal pain, GI bleeding, and diarrhea. GI perforation is not infrequent. Some of the more interesting presentations of CMV GI disease are related to the development of pneumatosis intestinalis[21] or ischemic vasculitis.[22] The pathogenesis of these conditions is not well understood, but their early diagnosis, usually by endoscopy, so that specific treatment can be instituted is essential.

Illustrative Case 6

This patient was a 60-year-old man with severe encephalopathy and recurrent variceal bleeding resulting from chronic non-A, non-B hepatitis. He underwent semiemergent hepatectomy and orthotopic liver replacement following the most recent episode of variceal bleeding. Both donor and recipient were serologically positive for CMV. The operative procedure was complicated by significant blood loss requiring transfusion of 30 liters of blood products. Intraoperative anuria persisted into the early recovery period prompting the administration of OKT3 monoclonal antibody rather than cyclosporine as initial induction immunosuppression.[23] Following a 10-day course of OKT3 therapy, the patient's renal function recovered and he was subsequently maintained on CyA, azathioprine, and prednisone.

The patient's recovery, however, was further delayed by an episode of aspiration pneumonitis requiring reintubation and intensive antibiotic therapy. By 3 weeks posttransplantation, the patient was extubated and normal hepatic and renal function had returned. He began to complain of moderate abdominal pain and intermittent diarrhea but seemed to be slowly recovering. Abdominal CT revealed no significant abnormalities. Despite symptomatic and antacid therapy, there was no improvement; 10 days later, intermittent upper GI bleeding developed. Endoscopic examination revealed diffuse mucosal ulcerations which on biopsy confirmed tissue-invasive CMV. Despite institution of ganciclovir therapy at that point, the patient developed multiple organ failure and eventually expired.

Comment. CMV infection is perhaps the single most important infection affecting transplant recipients.[24] Typically CMV syndromes become manifest 1–4 months posttransplantation, but, as in this patient, these infections may occur earlier in hepatic allograft recipients. The important lesson of this case is that the surgeon evaluating vague abdominal symptoms must consider CMV infection as early as 3 weeks posttransplantation, even in the previously seropositive recipient. This epidemiologic form of the disease apparently results from reactivation of the patient's latent virus and results in symptomatic infection in approximately 20% of previously seropositive recipients.[24] A more aggressive diagnostic approach including early endoscopy in this patient with vague abdominal symptoms probably would have revealed the tissue-invasive CMV and led to more timely institution of potentially lifesaving therapy.[25]

Immunosuppressed allograft recipients are also at significantly increased risk for the development of virus-related malignant lymphoma and other lymphoproliferative disorders. The risk is increased when the patient is treated for rejection with repeated courses of high-dosage immunosuppression or with multiple agents. The interval between organ transplantation and the development of lymphoma is frequently short, often only a few months, when CyA or FK506 is part of the immunosuppressive regimen. In contrast, the average time to the development of lymphoma is usually over a year in allograft recipients treated with immunosuppressive regimens that do not include these calcineurin inhibitors. There is strong evidence for the association of EBV infection with post-transplantation lymphoproliferative disorders of B-cell lineage. Nearly all patients have serologic evidence of primary or reactivation infection with EBV.[26]

Patients receiving immunosuppression without CyA or FK506 are at greatest risk for extranodal disease, with a striking predilection for the development of lymphoma of the central nervous system.[27] In contrast, in patients receiving calcineurin inhibitors, the disease more often involves lymph nodes and CNS lymphoma develops much less frequently. The single most common extranodal site for the development of lymphoproliferative disease in patients treated with CyA or FK506 is the GI tract.[26,28] Both the small intestine and the large intestine can be involved, and the patients can present with bleeding, perforation, or even intussusception.

Illustrative Case 7

A 56-year-old woman underwent cadaver donor kidney transplantation 5 months prior to the present admission. She received immunosuppression with CyA, azathioprine, and prednisone. Her early postoperative course was complicated by several acute rejection episodes which were successfully reversed with high-dosage steroids and a 12-day course of OKT3 monoclonal antibody. One month after the most recent episode, she was readmitted with fever and pulmonary infiltrates.

CMV was cultured from urine and buffy coat blood. Treatment with ganciclovir was instituted and the patient's pulmonary status gradually resolved. She then began to note mild, diffuse abdominal pain and diarrhea without localizing peritoneal signs. Five days after the onset of these abdominal symptoms, a routine X-ray film of the chest revealed pneumoperitoneum. Emergency laparotomy disclosed a perforation of the transverse colon associated with an ulcerated mucosal mass. Right colectomy with terminal ileostomy and suture closure of the distal transverse colon were performed. The resected specimen revealed multiple mucosal lesions (Fig. 1), which on microscopic examination were consistent with B-cell lymphoma (Fig. 2). Despite withdrawal of immunosuppressive therapy, recurrent bleeding necessitated several subsequent bowel resections, all of which revealed evidence of lymphoma. The patient ultimately expired.

Comment. The features of this case are consistent with either CMV-mediated enteritis or the posttransplant lymphoproliferative disorder (PTLD) typically associated with EBV. The incidence of PTLD varies[29] according to the organ transplant and the type and degree of immunosuppression (Table 3). As in the case presented above, patients who develop these complications in the early posttransplant period have typically received an immunosuppressive regimen that included CyA or FK506, often in combination with one of the antilymphocyte serum preparations such as ATG or OKT3, during the previous several months.

Methods proposed for treating these lymphoproliferative disorders vary widely. They include surgical excision (especially of tumors localized to the GI tract), decreased immunosuppression, acyclovir therapy, immunoglobulin infusions, radiation therapy, and chemotherapy. Although favorable survival results have been reported,[30] most centers continue to observe a significant mortality rate in these patients often because of the insidious progression of the disease before it is finally diagnosed by catastrophic complications as occurred in our patient.

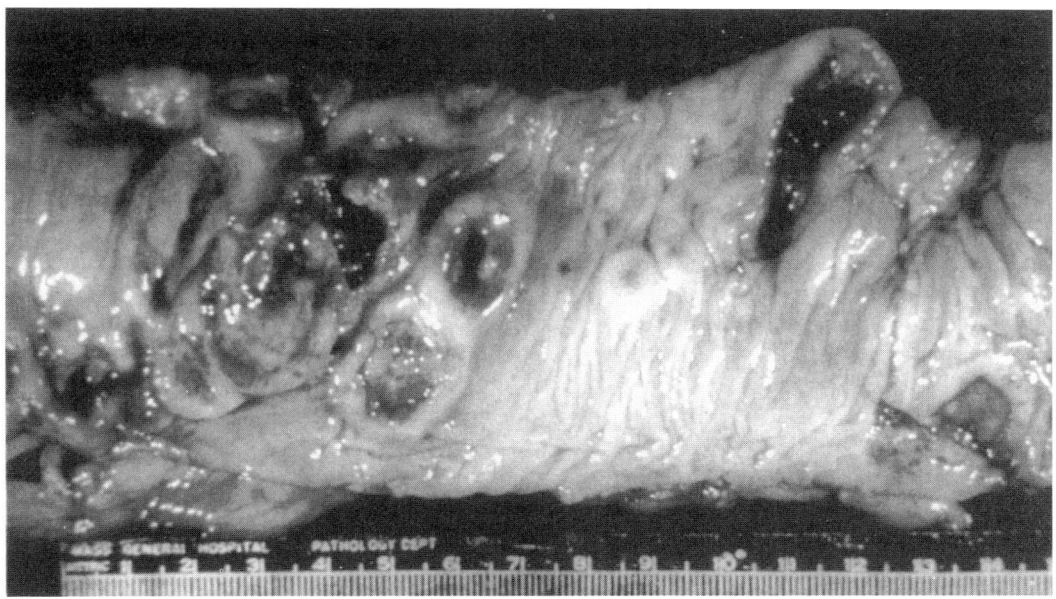

FIGURE 1. Colon resection specimen from Case 7. Multiple ulcerated tumors, penetrating the full thickness of the bowel wall in the larger lesions, are evident. (Reproduced from *N Engl J Med* **325:**191, 1991, with permission.)

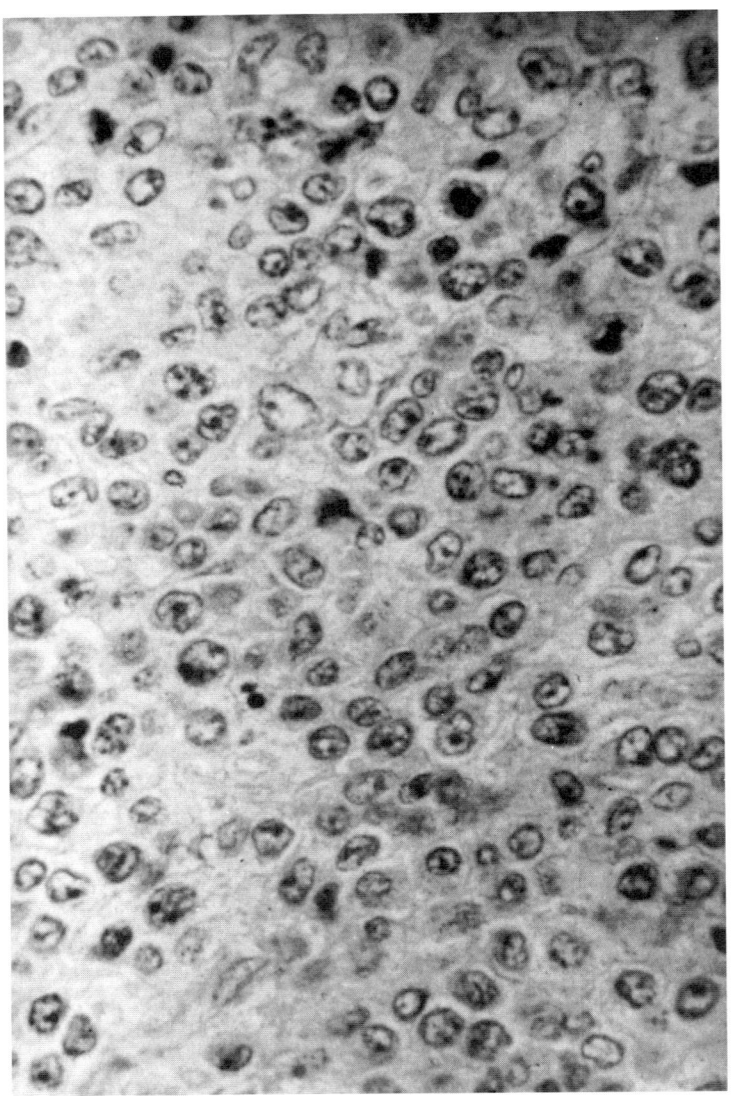

FIGURE 2. Microscopic examination of bowel wall tumor from Case 7. The lymphoma is composed of atypical lymphoid cells, including immunoblasts with prominent nucleoli, small and large noncleaved cells, and plasma cells. (Reproduced from *N Engl J Med* **325:**192, 1991, with permission.)

TABLE 3. Reported Incidence of PTLD in Recipients of Solid Organ Allografts

Allograft	Incidence of PTLD	
	Overall	Children (< age 5)
Kidney	1.3%	1.0%
Liver	3.0%	4% (CyA treated)
		6–20% (FK506 treated)
Heart	3.8%	Insufficient data
Heart/lung	4.6%	Insufficient data
Lung	7.9%	Insufficient data
Bowel	11%	31%

Another bowel complication being encountered with increasing frequency in the elderly or otherwise immunocompromised patient is intestinal infarction unrelated to major vascular occlusion. Although some investigators dispute the vascular etiology of the condition, it continues to be termed *nonocclusive mesenteric ischemia.*

Illustrative Case 8

A 46-year-old man received a cadaver donor renal allograft in the right iliac fossa, after which he was treated with cyclosporine and prednisone. He was continued on hemodialysis during a period of resolving acute tubular necrosis. During this interval, he received sodium polystyrene (Kayexalate) in sorbitol enemas on several occasions for treatment of hyperkalemia. Left-sided abdominal pain and tenderness and passage of grossly bloody stool occurred on the fifth posttransplant day. This was accompanied by worsening hyperkalemia despite hemodialysis. Plain radiographs of the abdomen were unremarkable. Sigmoidoscopy revealed pale rectal mucosa up to 18 cm, where the mucosa appeared dusky, edematous, and ulcerated. Biopsy revealed only severe inflammation. Barium enema revealed an area of narrowing in the distal sigmoid colon with irregularity and shallow ulcerations without perforation. At exploratory laparotomy, the left colon was severely ischemic with multiple areas of infarction despite pulsatile mesenteric vessels along the entire segment of involved bowel. A left colectomy with end-transverse colostomy and a distal mucous fistula were performed. Pathology of the resected specimen revealed prominent edema of all layers of the bowel with a diffuse, inflammatory infiltrate, extensive mucosal loss, and several areas of full-thickness infarction.

Postoperatively, the patient's immunosuppressive therapy was markedly reduced. He recovered rapidly and was discharged home with adequate allograft function 2 weeks later. He has subsequently had reestablishment of GI continuity and continues to do well with good renal allograft function.

Comment. Various etiologic mechanisms for this condition have been suggested, including steroid-induced vasculitis, connective tissue alterations which reduce mucosal resistance to bacteria, and the toxic effects of Kayexalate enemas.[31] Similar lesions have also been described in trauma victims who have suffered periods of hypotension.

Undoubtedly, numerous other factors, including host immuno-incompetence, postoperative blood volume changes, hypotension, marginally adequate blood supply secondary to atherosclerotic disease, uremia, and postoperative coagulopathies, are operative with differing degrees of importance in these individuals. The common denominator in all reports, however, is that delay in establishing the diagnosis in the compromised host almost invariably leads to a fatal outcome. In some patients, the acute symptomatology is preceded by reasonably prolonged periods of vague abdominal distress, malaise, and low-grade fevers. As in this patient, the development of difficult-to-control hyperkalemia or metabolic acidosis should serve as a diagnostic clue that intra-abdominal ischemic injury may be present. The surgeon should proceed with diagnostic evaluation on the premise that bowel pathology is likely unless proved otherwise. Confirmatory signs, such as guaiac-positive or bloody stool, demand immediate sigmoidoscopic examination which may reveal pale or cyanotic mucosa or areas of edema and ulceration. Barium enema in the early, less severe stages may show only "thumbprinting" secondary to hemorrhage into the bowel wall. This may progress to narrowing of the lumen and pseudopolyploid filling defects, indicating mucosal necrosis, and finally to perforation as total bowel wall infarction ensues. Obviously, hope of salvaging these patients is dependent on definitive surgical therapy during the early, more difficult to diagnose stages.

A GI complication that appears to be closely related to, if not a variant of, non-occlusive mesenteric ischemia is that of neutropenic colitis. Other common names for this condition are *typhilitis* or the *ileocolic syndrome.* Although most cases are associated with chemotherapy, the disease has also been described in transplant recipients and patients with aplastic anemia. The syndrome consists of fever, diarrhea, and abdominal pain—always in the setting of severe neutropenia. The pathogenesis is

believed to be that of chemotherapy-induced damage to the intestinal mucosa which, when coupled with neutropenia, allows bacterial invasion of the bowel wall. This then leads to variable degrees of necrosis with a predilection for the wall of the terminal ileum, cecum, and right colon. Several reviews have documented that this condition is the most common intra-abdominal disease discovered in neutropenic patients undergoing emergency laparotomy.[32,33] Of greatest importance to the surgical consultant has been the finding that many patients with this condition respond to bowel rest and antibiotics without the need for bowel resection.[34] Thus, careful sequential examinations are essential to distinguish the patients with resolving signs from those with a surgical condition. An excellent algorithm for such evaluation has been developed.[35]

2.4. Occult Intra-abdominal Sources of Fever and Infection

In patients with occult fever but no indication of an acute intra-abdominal catastrophe, the search for a surgically drainable collection often requires more sophisticated studies.

Illustrative Case 9

An 18-year-old patient with end-stage renal disease was being supported with hemodialysis while awaiting renal transplantation when he began having intermittent fevers in the range of $100.4–101.2°F$ ($38–39°C$). Five years earlier, he had undergone a right nephrectomy after several unsuccessful attempts to reconstruct a congenitally deformed drainage system. Renal function of the successfully reconstructed left kidney provided a marginal creatinine clearance, adequate to delay the need for dialysis during the succeeding $4\frac{1}{2}$ years. With the onset of the febrile course, there was occasional right-sided abdominal distress but no significant GI symptoms. CBT suggested a fluid collection in the right lower quadrant. The patient was taken to surgery with a preoperative diagnosis of periappendiceal abscess. An isolated, encapsulated 200-ml collection of white, purulent fluid that subsequently cultured *Staphylococcus aureus* was drained from the retroperitoneal site of the previous nephrectomy. A normal intraperitoneal appendix was removed as well. The postoperative and subsequent posttransplant course were benign.

Comment. Ultrasonography, CBT, or magnetic resonance imaging (MRI) have proved to be invaluable in the assessment of immunocompromised hosts with occult fevers. Most current reports favor ultrasound or CT-guided percutaneous aspiration of identified collections as the initial therapeutic approach.[36] Ultimately, however, exploratory laparotomy may be required as the definitive diagnostic and therapeutic maneuver.

Previous, apparently healed surgical sites should always be regarded with an index of suspicion in an immunocompromised patient. Hidden infections may be demonstrated months or even years after apparent uncomplicated healing. Needle aspiration or ultrasonography of areas of even minimal tenderness, erythema, or questionable fluctuance are often successful in revealing incredibly extensive underlying collections.

Obviously, an awareness of the likelihood of specific infectious complications aids the surgical evaluation. As noted, wound infections and colonic complications are seen commonly in the immunocompromised host. Another common but frequently overlooked condition is the occult perianal abscess which should always be carefully excluded, particularly in neutropenic patients.

Appendicitis is less common in immunosuppressed patients and can be difficult to diagnose even following perforation.[37] In neutropenic children receiving chemotherapy, abdominal pain and fever have been considered sufficient findings to justify immediate laparotomy with appendectomy being anticipated. For adults, we have found newer diagnostic tools, such as helical computed tomography (Fig. 3), to be very useful, providing rapid diagnosis with greater than 90% accuracy.[38,39]

Acute pancreatitis, although its etiology remains obscure, unfortunately, is not unusual in immunosuppressed patients.

Illustrative Case 10

This patient had received a cadaver donor renal allograft 4 months earlier. Excellent allograft function had been maintained with an immunosuppressive regimen including CyA and prednisone. The patient was admitted with complaints of abdominal pain, "bloating," and fever. He was found to have a serum amylase level of 515 units/liter ($N < 100$). With intravenous alimentation and conservative therapy, the symptoms resolved but the amylase level fell only to the 200 range. Ultrasound and CT studies identified no abscess cavities but suggested persistent edema of the tail of the pancreas. With each attempt to resume oral alimentation, recurrent pain and low-grade fever developed. Exploratory laparotomy revealed an edematous pancreas with multiple abscesses in the tail. Distal pancreatectomy provided complete relief of symptoms.

Four months later, the patient returned with similar findings while being maintained on cyclosporine and prednisone. Ultimately, subtotal pancreatectomy was required, again revealing multiple small abscesses in the gland. Following this, he has remained asymptomatic.

Comment. The incidence of acute focal pancreatitis in steroid-treated patients has been reported to be as great as 28.5% versus only 3.7% of matched but non-steroid-treated controls.[40] In allograft recipients, the incidence of clinically significant pancreatitis has been reported to range between 2 and 20%.[41,42] The mechanism whereby steroids could produce pancreatitis may be ductal ectasia and epithelial metaplasia, which favor obstruction. Perhaps of more significance is the fact that many of these patients have coincident severe infection, often of viral etiology. This may be the important underlying factor, with even direct viral infection of the pancreas being a possibility. Azathioprine has also been implicated as an etiologic agent for pancreatitis, emphasizing the probable multifactorial origin.[43] The mortality in these patients can be as high as 70%. The major factors responsible for this high mortality are the considerable delay in establishing the diagnosis and the unusually high incidence of later complications such as pseudocyst and abscess formation. In Case 10, the multiple pancreatic abscesses remained undetected for some time despite numerous radiographic studies. Fortunately, the process remained localized (presumably a benefit of the much lower dosages of steroids currently being used for allograft recipients) and could be managed by sequential surgical procedures.

As in colonic perforation, the surgeon is presented with the demanding role of providing early clinical recognition of the condition followed by direction of the necessarily aggressive management, which includes nasogastric suction; fluid, electrolyte, and caloric replacement; and further steroid reduction. Most importantly, prompt detection and appropriate, possibly repeated, drainage of developing collections is essential.

3. Preoperative Preparation

Once the need for surgery in an immunoincompetent patient has been established, the degree to which preoperative preparation can be extended is determined primarily by the nature and urgency of the surgical indica-

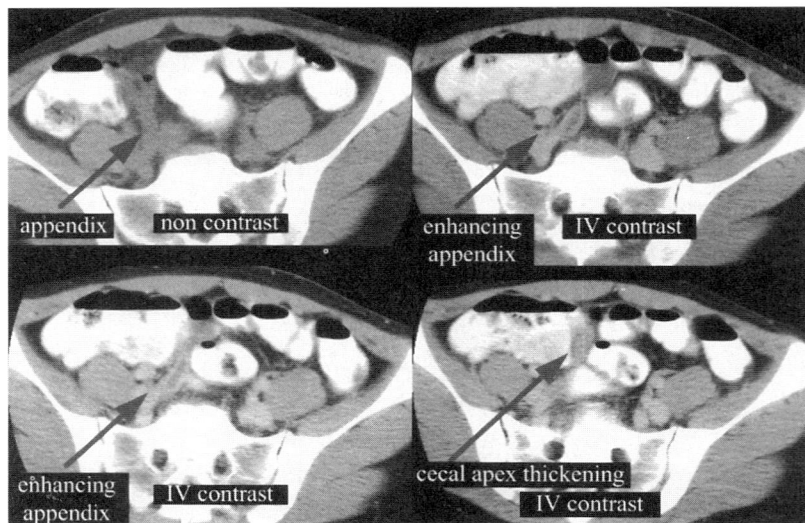

FIGURE 3. Selected images from helical abdominal CT in patient with lower abdominal pain. The upper images demonstrate a dilated appendix (>6 mm). Pericecal inflammatory stranding and apical thickening are noted in the lower images.

tion. Usual preoperative resuscitative measures including volume replacement, institution of antibiotic therapy, and correction of electrolyte imbalances follow the same guidelines employed for any acute surgical state. If the surgical indication is less urgent, a preoperative attempt to begin restoration of immunocompetence toward normal may be possible. Withdrawal of cytotoxic chemotherapy or high-dosage steroids will permit some recovery of depressed bone marrow and host-defense function and should reduce the risks of superinfection in the postoperative period. In addition, certain comorbid conditions unique to these patients must be considered.

3.1. Infection and Adrenal Insufficiency

Illustrative Case 11

A 27-year-old woman had been treated for several years with various immunosuppressive regimens, including steroids, cyclophosphamide, and azathioprine for an ill-defined vasculitis characterized by diabetes, blindness, and impaired renal function. Because of the need for frequent venipunctures and hemodialysis, a saphenous vein-to-femoral artery A-V fistula had been constructed for vascular access 6 months previously. The current admission followed the development of fever, chills, and pain in the area of the A-V fistula. On examination, she was found to be obtunded but arousable; temperature was 104.9°F (40.5°C); and blood pressure, which had been chronically elevated, was found to be 95/60. The soft tissues around the A-V fistula were erythematous and edematous, with a 3 × 5-cm tender fluctuant mass beneath the site of a recent venipuncture. The patient was begun on antibiotics and intravenous volume replacement. Under local anesthesia, the fluctuant mass was drained of 75 ml of purulent fluid containing numerous neutrophils and gram-negative bacteria. She nevertheless remained obtunded, febrile, and hypotensive, requiring vasopressor and massive fluid support. The diagnosis of relative adrenal insufficiency was eventually suggested. Increased doses of steroids were administered, resulting in satisfactory stabilization of the vital signs, defervescence, and improved mental function. The patient was continued on intravenous antibiotics and local wound care. Over the succeeding 2 weeks, there was complete resolution of the local infection, and the patient was discharged for continuing outpatient hemodialysis.

Comment. Of the many side effects suffered by patients receiving chronic steroid therapy (Table 4), one of the unique, admittedly unusual, conditions to keep in mind is occult adrenal insufficiency which may develop during periods of acute stress.[44] A hypotensive febrile crisis, abdominal findings, including severe ileus, and a variety of central nervous system (CNS) symptoms ranging from obtundation to mania may be rapidly reversed by administration of a single intravenous hydrocortisone bolus. However, the surgeon should avoid this crisis by advising perioperative steroid (hydrocortisone, 50–100 mg) supplementation, with maintenance doses continuing into the postoperative period. A typical schedule might be hydrocortisone, 50–100 mg q8h the day of surgery and 25 mg q8h the following day. By this time, resumption of the preoperative maintenance dosage should be adequate, provided continuing severe stress is not present.

Some investigators have reported that functional adrenal suppression is uncommon in allograft recipients receiving low-dosage baseline immunosuppression.[45] This group has concluded that continued administration of only the previously established maintenance steroid dosage through periods of physiologic stress is usually a safe approach, at least for this subset of patients. Nevertheless, they acknowledge that the steroid requirement may be qualitatively different for patients on higher-dosage steroids or with other diseases. We, therefore, continue to recommend the administration of "stress steroids" to all of these patients, recognizing that this brief period of modest-dosage glucocorticoid supplementation will have minimal detrimental effects on wound healing or infection control while completely preventing the possibility of clinical symptoms resulting from adrenal insufficiency.

3.2. Infection and Ketoacidosis

In addition to the acute infection, the immunocompromised patient's underlying condition must be considered and appropriately managed.

TABLE 4. Major Side Effects and Complications of Steroid Therapy

Side effect	Complications
Decreased phagocytosis	Increased infection risk
Glucose intolerance	Increased insulin requirement, infection
Decreased inflammatory response	Poor wound healing, increased infection
Lymphopenia	Increased viral infection risk
Capillary fragility	Breakdown of normal barriers
Dermatologic changes	Poor wound healing
Decreased mucosal resistance	Peptic ulceration, bacterial translocation
Catabolic effects	Myopathy, osteoporosis, fractures
Sodium retention	Edema, hypertension
Occult adrenal insufficiency	Hypotension, fever, ileus
? Local vasculitis	Pancreatitis, colonic ulceration

Illustrative Case 12

A 59-year-old diabetic patient had undergone successful renal transplantation for end-stage diabetic nephropathy, 3 years previously. He was admitted at the present time with fever and shaking chills of 2 days' duration and abdominal pain and vomiting for 1 day. Prior to these symptoms, he had been in good health with normal renal function while receiving maintenance dosages of CyA and prednisone.

On examination, he was found to be severely ill with a temperature of 103°F (39.5°C), pulse rate of 133/min, respiratory rate of 30/min, and blood pressure of 85/60. The renal allograft was not obviously enlarged but was slightly tender and the abdomen was moderately distended with direct and rebound tenderness, particularly in the epigastrium. Bowel sounds were hypoactive. The hematocrit was 43%, and the WBC 17,600/mm^3. Urinalysis revealed many white and red cells and gram-negative bacteria. Significant chemical abnormalities included a blood sugar of 595 mg/dl and CO_2 content of 14 mEq/liter. Abdominal and chest radiographs revealed only a distended stomach without evidence of free air.

The admitting diagnosis was gram-negative septicemia with consequent diabetic ketoacidosis. Although the ultrasound showed no stones, the source of infection was presumed to be acute, possibly gangrenous, cholecystitis. Urinary tract infection was also diagnosed. In preparation for proposed emergency laparotomy, intravenous antibiotics, rehydration, and insulin therapy were instituted. As the dehydration and ketoacidosis were corrected, the abdominal symptoms and signs rapidly resolved, and the patient's vital signs stabilized. Surgery was therefore deferred. Subsequent workup of the biliary tree revealed a normally functioning gallbladder without stones.

Comment. Diabetics may present with ketoacidosis in conjunction with a condition requiring surgery, in which case surgery must be delayed while vigorous treatment with insulin and intravenous fluids is initiated. It is important to recognize, however, that diabetic ketoacidosis itself may mimic an acute abdominal emergency. In this situation, treatment of the ketoacidosis may result in resolution of the symptoms and signs originally thought to require urgent surgical intervention. In the case reported above, the urinary tract infection apparently precipitated the ketoacidosis and the abdominal symptoms. On the other hand, the surgeon must not fall into the trap of ascribing all acute abdominal symptoms in diabetics to the pseudoperitonitis of ketoacidosis. During the few hours required to begin correcting the ketoacidosis with vigorous insulin and fluid therapy, the significance of the abdominal symptoms will usually become clarified.

The initial diagnosis of cholecystitis in Case 12 raises the issue of the most appropriate management of asymptomatic or mildly symptomatic cholelithiasis in immunocompromised patients. Some groups, for example, have recommended mandatory screening and prophylactic cholecystectomy, if cholelithiasis is demonstrated in patients awaiting transplantation.[46] More recent experience, however, has indicated that only a minority of patients with asymptomatic gallstones will become symptomatic after transplantation.[47] Thus, we do not routinely screen these patients or recommend cholecystectomy for incidentally discovered cholelithiasis. In those patients who do become symptomatic, laparoscopic cholecystectomy can typically be performed, without incident, even in the early posttransplant period.[48]

3.3. Infection and Malnutrition

Another factor, the importance of which has only recently received significant appreciation, is the role that malnutrition plays in the increased morbidity and mortality of surgical patients.

A number of studies have evaluated the ability of various clinical measurements to identify surgical patients at risk for nutritionally related postoperative complications. Such measures as serum albumin levels <3 g/dl and weight loss >20% have been suggested as specific determinants of significant malnutrition and increased likelihood of postoperative sepsis. Pre- and postoperative hyperalimentation (enteral or parenteral) has been recommended by some groups for surgical candidates meeting these criteria (e.g., cancer patients) if the proposed procedure can be safely delayed. Although improvement in various immune parameters after nutritional repletion can be demonstrated,[49] considerable controversy persists as to whether this regularly translates into decreased postoperative morbidity or which patients are truly appropriate candidates for hyperalimentation. A recent meta-analysis of randomized clinical trials in critically ill or surgical patients concluded that postoperative infectious complications may be reduced by hyperalimentation, but mortality rates are not influenced.[50] These observations emphasize that there is little evidence for the routine use of hyperalimentation for marginally nourished surgical candidates. Nevertheless, the well-documented evidence of poor anastomotic healing in severely protein-depleted experimental and clinical subjects does suggest the applicability of pre- and postoperative hyperalimentation for specific conditions, including enterocutaneous fistulas, burns, and upper GI malignancies.[49]

One must, of course, recognize the hazards associated with administration of parenteral alimentation, particularly in immunocompromised hosts. The most serious and prevalent complication is catheter-related sepsis. A suggested solution to this problem has been routine replacement of the catheter over a guide wire every 3–4 days. A controlled study, however, has demonstrated that this approach does not prevent infection.[51] In contrast, the use of central venous catheters impregnated with antimicrobial agents is associated with reduced rates of catheter-related bloodstream infection.[52] Thus, the use of single-lumen, antimicrobial-impregnated catheters that are restricted solely to parenteral nutrition administration and are placed and maintained by meticulous aseptic technique remains the most effective approach to limiting line-related infections.

Illustrative Case 13

A 27-year-old man was evaluated for complaints of crampy abdominal pain and bloody diarrhea. Barium enema was consistent with the diagnosis of regional enteritis of the distal ileum. Despite intensive medical therapy, intermittent fever, anorexia, and chronic diarrhea persisted over the succeeding 4-month period. The patient's weight fell from 165 to 130 lb (75 to 59 kg). Hospital admission was eventually required when increasing colicky pain, abdominal distention, and vomiting suggested partial bowel obstruction. Examination revealed a chronically ill young man who had a normal temperature and blood pressure. The abdomen was modestly distended with hyperactive, high-pitched bowel sounds in rushes, resulting in frequent watery stools. Significant laboratory studies revealed Hct 32%, WBC count 13,000/mm^3, and serum albumin 2.6 g/dl. Barium enema with reflux into the terminal ileum revealed marked thickening and irregularity of the bowel wall and a narrowed lumen with several enteroenteric fistulas, but no obstruction. Following resolution of the vomiting with nasogastric suction and intravenous fluids over the first 4 days, the patient was taken to surgery, for resection of all grossly diseased bowel, including the distal ileum and right colon. Over the subsequent 3 months, the patient was reexplored three times for subphrenic and pelvic abscesses and attempted closure of multiple enterocutaneous fistulas. Despite the eventual initiation of intravenous hyperalimentation, his weight continued to fall, to 105 lb (47.7 kg). Renal failure developed, prompting his transfer here for hemodialysis. At the time of admission, he was obtunded and wasted with a temperature of 102.5°F (39°C) and persistent anemia, hypoalbuminemia, and uremia. The partially disrupted abdominal incision was grossly purulent with evident tracking of the purulence into the perineum. Several persistent enterocutaneous fistulas were present.

During the subsequent 3-week hospitalization, the patient was treated with blood products, total parenteral nutrition, various antibiotics, multiple drainage procedures of abdominal and subcutaneous abscesses, suction drainage of the fistulas, and hemodialysis. Despite these measures, he had repeated positive blood cultures for gram-negative organisms, showed little evidence of wound healing, and finally expired 4 months after the initial surgical procedure.

Comment. Inflammatory bowel disease of the transmural type is frequently associated with partial bowel obstruction, fistulas, intra-abdominal abscess, and malabsorption. These patients are markedly malnourished, protein and electrolyte depleted, and severely catabolic. In this setting, preoperative hyperalimentation may be a valuable tool. Improvement in various immune parameters and remission of acute symptoms may be achieved. Even if only partial resolution occurs, the patient has become an improved operative candidate as nutritional deficiencies are partially corrected. One may assume that significant malnutrition exists in any patient who presents with weight loss of greater than 0.2% per day or 20% of normal, serum albumin of less than 3.0 g/dl, serum transferrin level of less than 200 mg/dl, or inability to respond to common delayed hypersensitivity skin tests. Such patients should be considered appropriate candidates for pre- and postoperative

hyperalimentation. With selective use of such therapy, many of the disastrous complications illustrated by this case should be avoided.

3.4. Preoperative Antibiotics

Preoperative administration of antibiotics to the compromised host may have therapeutic and prophylactic goals. Certainly, when a specific infectious process is suspected or has already been identified, or when a contaminated procedure such as bowel resection is planned, appropriate antibiotic coverage should be recommended. The importance of prophylactic antibiotics for "clean" surgical procedures in the compromised host remains more controversial. There is considerable evidence that antibiotic prophylaxis, even in the noncompromised host, reduces the risk of infections following extensive clean procedures, such as hip arthroplasty and cardiothoracic surgery.[53] The efficacy of this approach has been extended to even lesser procedures.[54] Not surprisingly, therefore, some transplant groups strongly recommend pre- and intraoperative broad-spectrum antibiotic coverage, particularly for diabetic patients. This therapy has been suggested to be a major reason for declining mortality risks for these patients. Our own experience without routine antibiotic coverage in renal allograft recipients, however, has indicated an extremely low rate of infection (less than 2%) of the primary transplant wound or following other uncontaminated procedures in these patients. Thus, prophylactic antibiotics have not been recommended for fear of selecting out resistant, more highly virulent, bacterial flora.

For potentially contaminated procedures, the status of the natural bacterial flora may suggest appropriate prophylactic antibiotics. For example, early studies of infection in cancer patients undergoing major surgical procedures suggested that pre- and postoperative antibiotic therapy for potential enteric pathogens greatly decreases the incidence of serious infection.[55] Based on such results, many surgeons advise the use of prophylactic antibiotics in all patients undergoing major surgical resections for malignancy. Specific recommendations obviously vary, but the antibiotic chosen is usually one with broad-spectrum coverage with particular consideration of the potential pathogens (e.g., oral, respiratory, enteric) expected to be encountered. The type of procedure to be undertaken is also an important determinant. For example, the indications for antibiotic coverage are greater if a major orthopedic or vascular procedure is to be performed versus a simple excision or node dissection.

It has been repeatedly demonstrated that there is a limited time period during which it is possible to augment the host's antibacterial mechanisms using antibiotics and that this period lasts only a few hours beginning at the time of bacterial contamination.[56] Thus, if they are to be used, it is essential that prophylactic antibiotics be administered before the incision is made. Following surgery, the course of antibiotic administration should be brief unless the host is colonized with resistant organisms.

Another approach, the precise value of which remains to be established, is that of selective decontamination of the GI tract. Studies have clearly demonstrated that intestinal flora can translocate to the peritoneal cavity, lymph nodes, liver, and spleen in patients with severe trauma or burns.[57] These observations have prompted evaluation of the efficacy of gut decontamination for preventing septic complications in variably compromised patients or experimental models.[58] Although a decreased incidence of nosocomial infections can be achieved, a decreased mortality rate has been more difficult to demonstrate. Continuing controlled trials will help to define the specific role of this approach to the antibacterial management of immunocompromised patients.[59]

4. Intraoperative Considerations

4.1. Choice of Anesthesia and Patient-Monitoring Techniques

Selection of anesthetic techniques in these patients is influenced primarily by any limitations resulting from the underlying disease. For example, in uremic individuals, agents eliminated via renal excretion should be avoided or dosages appropriately decreased. The use of average doses of galamine triethiodide, for example, in renal transplant recipients results in prolonged paralysis in 20% of patients. We have observed a similar problem following administration of pancuronium bromide, and therefore favor the use of agents such as atracurium to avoid protracted neuromuscular blockade.[60] This complication can be particularly serious because of the increased risk of pulmonary infection in immunosuppressed patients if prolonged endotracheal intubation is required. One important goal of anesthesia management, therefore, is to provide for return of adequate spontaneous respiratory activity and early postoperative extubation whenever possible.

Some patients, such as those with hepatic failure, are predisposed to developing hypoxia from atelectasis, due to ascites, and from intrapulmonary arteriovenous shunts.

In these individuals, careful preoxygenation and rapid sequence induction anesthesia are commonly used. An appropriately sized, low pressure cuff endotracheal tube is placed in anticipation of a possibly prolonged period of intubation. Adequate padding of the heels, sacrum, elbows, and head is provided to prevent pressure ulceration during the operation which not infrequently extends beyond 8 hr. Effective warming appliances are also essential and may even help to reduce the incidence of postoperative infection.[61]

The selection of anesthetic agents for liver transplantation is governed by the high-cardiac-output, low-peripheral-vascular-resistance state typically associated with end-stage liver disease. The preferred maintenance agent for hemodynamic stability is a narcotic anesthetic employing fentanyl and/or morphine. Because some patients may not tolerate administration of routine anesthesia, they are often given a benzodiazepine as well, to block memory. Lorazepam is frequently chosen, since this agent requires only glucuronidation for excretion.

The complexity of the procedure requires a number of extra lines. In adult recipients, access for infusion and monitoring is typically accomplished via a right radial arterial line, a large-bore intravenous site in the right arm, and introducer catheters in each internal jugular vein, one for placement of the pulmonary artery catheter and one for connection to the rapid infusion pump. Access sites in the legs are avoided since infusions would be unreliable during the period of vena cava occlusion.

Recognition that any indwelling line in these patients can be the source of serious bacteremia should emphasize that their use should never be allowed for only marginal indications, and that they should be removed as soon as clinical conditions permit. With the observation of such possibly obvious precautions, the administration of general or conduction anesthesia to immunocompromised patients can be provided without an unacceptably high incidence of complications.

4.2. Surgical Technique

The surgeon managing these patients must be prepared to constantly reevaluate and modify his usual techniques in order to deal with tissues (Fig. 4) that are weak, hold sutures poorly, and can be expected to require unusually prolonged healing periods.

Illustrative Case 14

A 71-year-old woman with leukemia received 6-thioguanine and cytosine arabinoside treatment for 5 months prior to the present hospital admission which was prompted by the sudden onset of crampy abdomi-

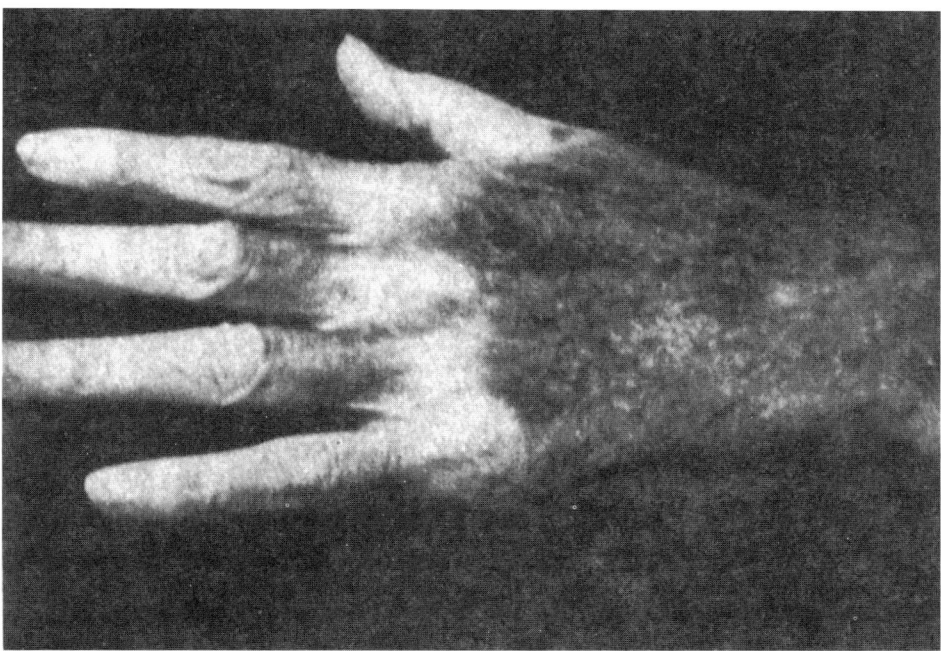

FIGURE 4. Typical thin, friable skin in a steroid-treated patient. Significant injuries after only minor trauma and poor wound healing following surgery can be anticipated in such tissues.

nal pain. On admission, her temperature was 104°F (40°C) and blood pressure 90/70. The abdomen was distended and tympanitic, and both direct and rebound tenderness were present. Following institution of antibiotics and rapid rehydration, the patient underwent emergency laparotomy. A volvulus of the distal ileum with dilation of the proximal bowel was found. A localized abscess surrounding a perforation of the antimesenteric border of the ileum was present as well. The volvulus was reduced, and no evidence of nonviable bowel was noted. The edges of the perforation were excised, and the defect was closed in two layers with chromic catgut and silk sutures. Postoperatively, the patient remained febrile with a quiet, distended abdomen until the fifth day, when a fecal fistula was noted at the inferior margin of the abdominal incision. Emergency reoperation revealed dehiscence of the bowel closure with local peritonitis. The involved small bowel segment was resected, and an end-to-end anastomosis constructed. The patient remained stable for several days but then developed fever, hypotension, and blood cultures positive for *Escherichia coli*. Her condition rapidly deteriorated, and she expired shortly thereafter. Postmortem examination revealed dehiscence of the bowel anastomosis and diffuse peritonitis.

Comment. In urgent GI tract operations in the compromised host, stomas should be considered in preference to anastomoses more frequently than in the usual surgical subject. The patient's chances of survival are enhanced if ileostomy or colostomy are constructed rather than a primary anastomosis which may result in fatal leakage when replaced in the abdomen.[62] Several months after construction of the stoma, when the patient has recovered from the acute event and host defenses have been at least partially restored, reestablishment of bowel continuity becomes a relatively minor surgical procedure. If sutured bowel must be left in the abdomen, nonabsorbable sutures must be used, and the surgeon should attempt to isolate and reinforce the closure, for example, with omentum, and provide proximal decompression via colostomy, gastrostomy, and so on. Similarly, tragic experience has emphasized that local drainage and proximal colonic diversion, which is adequate therapy for perforated diverticulitis in the usual patient, is totally inadequate following sigmoid perforation in the compromised host. In these patients, sepsis continues to spread, despite apparently complete local drainage, because the depressed inflammatory response is unable to provide satisfactory containment. Thus, exteriorization or primary resection of the involved bowel and construction of a descending colostomy and distal mucous fistula or turn in is the treatment of choice. With such an approach, one should expect at least an 80–85% salvage rate in these patients for a condition previously found to be almost universally fatal.

Copious intraperitoneal lavage should be used to reduce the residual contamination and limit the possibility of subsequent abscess formation. Some authors recommend antibiotic or antibacterial irrigations with continuing lavage into the postoperative period; however, saline alone is probably as efficacious and does not expose the patient to any side effects of the irrigation itself. Liberal drainage of all four quadrants and of the subphrenic spaces and pelvic pouch should be considered.

Following abdominal surgery in the immunocompromised host, the surgeon should anticipate the possibility of prolonged periods of gastric dysfunction. Drainage via gastrostomy tube is often employed to avoid the pulmonary and esophageal complications associated with nasogastric tubes. At the same time, consideration should be given to using the small bowel for early resumption of nutritional support. In fact, motility and absorption by the small intestine usually returns within hours of successful control of the septic process. Most patients with a reasonable length of normal small intestine can be given full nutrition via jejunostomy administration of elemental diet beginning in the early postoperative period. The elemental diet is absorbed in the proximal intestine without the need for digestion, thus providing an early return to anabolism without the risks of prolonged hyperalimentation via central venous lines. Because of the poor healing in these patients, special precautions must be taken during placement of cutaneous-enteric tubes. Unless the bowel around the enterostomy is carefully and completely sutured to the anterior abdominal wall, a satisfactory seal of the tract will not develop. Subsequent leakage of enteric contents around the tube and into the peritoneal cavity can be regularly anticipated. Similarly, poor tissue healing should be expected in all incised tissues.

Illustrative Case 15

A 10-year-old girl with end-stage liver failure and massive ascites was admitted because of fever and abdominal pain. Following the initial diagnostic impression of acute appendicitis, exploratory laparotomy was performed. Slightly cloudy ascites revealed white blood cells but no organisms on Gram's stain. Cultures subsequently yielded *E. coli*. The remainder of the examination, including the appendix, revealed no acute inflammatory process. A diagnosis of primarily infected ascites was presumed and continued antibiotic therapy was advised. The wound closure included all layer retention sutures because of the fear of poor healing. Postoperatively, continued ascites leakage around the sutures led to wound disruption, purulent peritonitis, and eventual death.

Comment. This patient's history emphasizes two points at which different surgical approaches could have changed her course. The correct diagnosis might have been made if a preoperative paracentesis had been performed. This would have avoided completely the surgical procedure. Once the laparotomy was performed, the possibility of wound complications was correctly recognized, but the inadequate nature of the approach taken for this problem is emphasized by the patient's postoperative course.

Because of the expected delayed healing, closure with interrupted nonabsorbable suture material is advisable. We have found fine stainless-steel sutures or synthetic monofilament products particularly appealing. Silk or cotton sutures are, of course, inadvisable in wounds at high risk of infection. Subcuticular skin closure (Fig. 5) also seems to decrease the incidence of complications since microabscesses, which frequently develop around transcutaneous sutures left in place for more than a few days, are avoided. Similarly, we rigorously avoid the use of all-layer sutures, especially in patients with ascites where direct communication between skin flora and peritoneum can occur along the suture tract because of the host's inability to effectively seal off the site. The value of topical application of antibacterial solutions into surgical wounds remains controversial. Some reports[63] seem to provide evidence that such irrigations lower the frequency of subcutaneous infections, especially in highly contaminated cases. Others suggest that normal saline is an equally effective irrigant.[64] In our renal transplantation practice, in which we do not use systemic prophylactic antibiotics but irrigate the incision with local antibiotics, a wound infection rate of less than 2% has

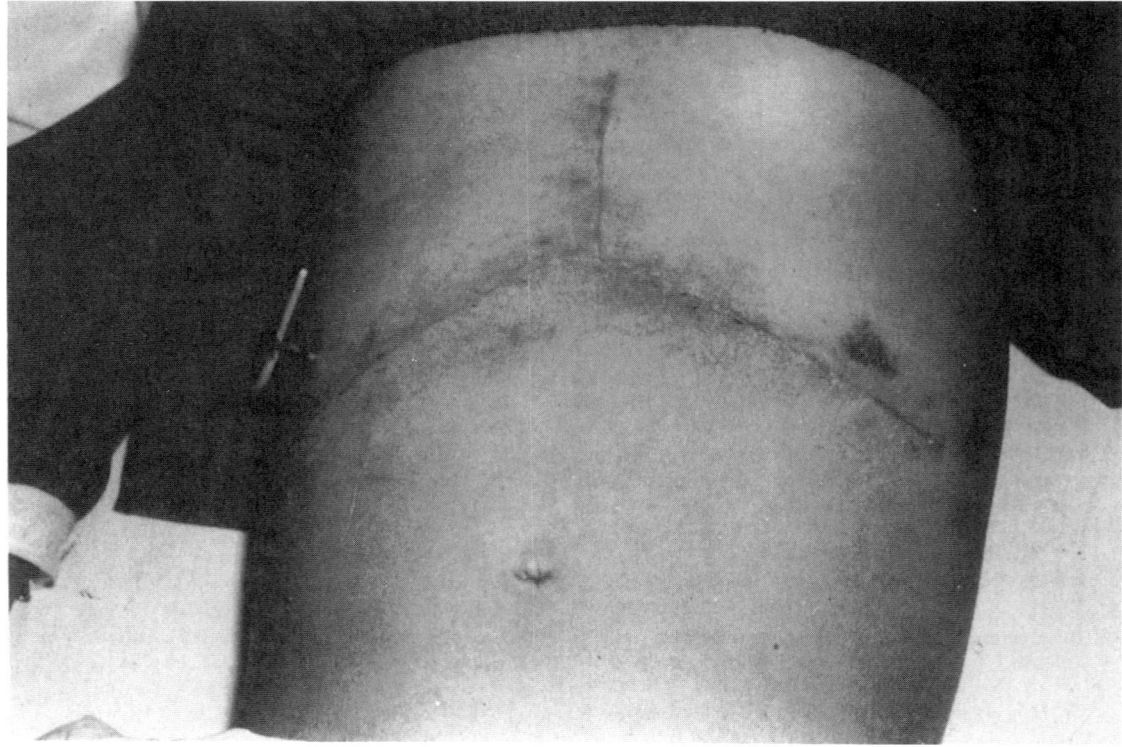

FIGURE 5. Abdominal wound closure following orthotopic liver transplantation. Avoidance of transcutaneous sutures helps to provide uncomplicated primary wound healing in over 95% of immunocompromised hosts in our experience. When utilized, the occluded T-tube, seen exiting in the right upper quadrant, is left in place as an intrabiliary stent for 4–6 months.

been achieved, an observation similar to that reported by others.[65] If the surgical procedure was one of marked contamination, delayed or secondary closure of the skin is in some instances safer than attempted primary closure.

Surgical management of certain conditions, usually unrelated to the patient's underlying immunologic defect, sometimes demands innovative intraoperative modifications. For example, more renal allograft recipients are surviving into their seventh and eighth decades, abdominal aortic aneurysms and aortoiliac occlusive disease are being encountered with increasing frequency. These vascular lesions typically develop in the aortoiliac segment proximal to the site of prior anastomosis of the donor renal artery. The standard technique of repair in this situation involves temporary interruption of blood flow to the allograft. Expeditious completion of this part of the procedure limits the period of warm ischemia and hopefully the degree of postoperative renal dysfunction.[66] However, the difficulty to be encountered during aneurysmectomy, the time required to complete the anastomoses, and the quality of collateral blood flow to the allograft cannot always be judged accurately prior to this procedure. Thus, some groups advise allograft protection, using temporary vascular bypass, to maintain partial perfusion to the pelvic vessels especially if the aortoiliac disease approaches or involves the area of the allograft vessel anastomosis.[67]

Illustrative Case 16

A 66-year-old man was admitted to the emergency ward with a 4-hr history of back and left lower quadrant pain. Two years earlier, he had received a cadaver donor renal allograft, placed into the left iliac fossa with end-to-side anastomosis of the donor renal artery to recipient external iliac artery. At the current admission, the Hct was 31% and serum creatinine was modestly elevated to 2.0 mg/dl from his typical baseline of 1.6 mg/dl. CBT scan demonstrated a 6.5-cm distal aortic aneurysm with probable retroperitoneal bleeding.

At emergency surgical exploration, proximal control of the ruptured but contained aneurysm was quickly achieved. The common iliac arteries were divided and cold heparinized Ringer's lactate solution was infused into the distal left iliac artery. Replacement of the aneurysm with a bifurcated prosthesis was completed in the usual fashion. The total cross-clamp time of the left iliac artery was 35 min.

The patient recovered well following this procedure, but allograft dysfunction persisted with the serum creatinine peaking at 4.5 mg/dl during the first postoperative week. Subsequently, his renal function gradually improved with a new stable baseline of 2.1–2.3 mg/dl.

Comment. The emergent procedure in this patient did not permit ready placement of a temporary arterial bypass system. Rapid completion of the operation and partial hypothermic perfusion were therefore relied on to protect the allograft. The subsequent renal dysfunction, although not life-threatening, undoubtedly worsens the prognosis for long-term function of this allograft. Such experience has led some groups to conclude that a temporary shunting technique should be used in these patients if it is feasible.[67] The proximal end of the shunt can be

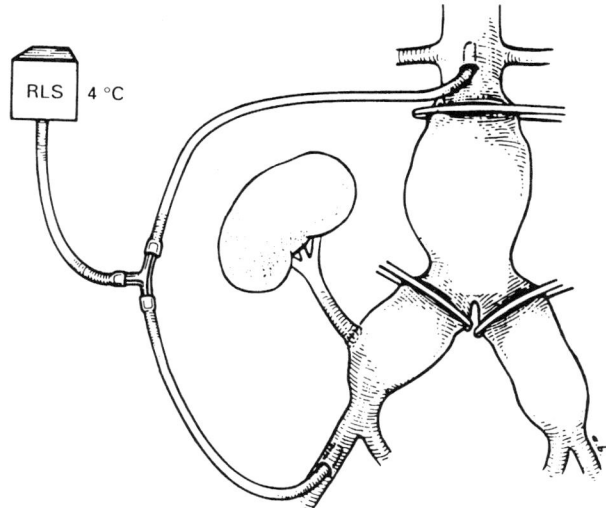

FIGURE 6. Temporary vascular bypass maintaining perfusion to renal allograft during aortoiliac surgery. The inflow in this case is from proximal aorta. Alternatively, the axillary or brachial artery could be utilized. The Y-connector allows for infusion of cold Ringer's lactate solution (RLS). (Reproduced from *Tex Heart Inst J* **17**:241, 1990, with permission.)

placed in the aorta above the aneurysm (Fig. 6). In some situations, the axillary or brachial artery can be used as the source of inflow. If the aneurysm is confined to the aorta, the distal shunt is placed into the common iliac artery on the side of the renal allograft. If there is aneurysmal involvement distally, however, the shunt is placed into the femoral or external iliac artery for retrograde perfusion of the allograft as depicted in Fig. 6.

4.3. The Effect of AIDS on Surgical Practice

Of increasing importance to the surgeon is the issue of transmission of the human immunodeficiency virus (HIV) within the medical setting. There are two aspects of this problem that are of special importance to the surgeon: the possibility that a surgeon will acquire HIV infection from percutaneous or mucosal inoculation with infected blood; and the possibility that an HIV-positive surgeon might transmit the virus to a patient during the course of an invasive procedure.

Currently available data suggest that the risk of HIV acquisition through percutaneous or mucous membrane exposure is less than 0.5%.[68] This has resulted in the report to the United States Centers for Disease Control and Prevention of over 150 documented or possible cases of occupationally acquired HIV infection in healthcare workers.[69] These observations and reports of a slowly rising incidence of undiagnosed HIV infection among patients presenting to urban hospitals for emergency care[70] underline the need for great care on the part of healthcare workers. In the operating room, increased attention to effective barrier precautions such as face shields and fluid-resistant gowns, the development of puncture-resistant glove materials, and improved surgical techniques to minimize percutaneous needle sticks will decrease the risk of HIV transmission.[71] Even with presently available gloves, data suggest that there is significant reduction in the volume of blood (and presumably the amount of virus) transmitted via needle stick when the needles pass through glove material before entering the skin.

As noted above, needle stick and other sharp injuries are of greatest concern. They account for greater than 80% of occupationally acquired cases of HIV infection among healthcare workers.[71,72] Whereas outside the operating room the majority of needle stick injuries involve nursing personnel, in the operating room more than 90% of such injuries involve surgeons or residents.[73] Percutaneous sharp injury occurs in approximately 7% of operative procedures. Most injuries to the surgeon involve solid suture needle sticks, usually occurring on the palmar surface of the fingers of the nondominant hand. These occur predominantly during the unsafe practice of manipulating the needle tip by blind palpation rather than by instrument control, as, for example, during vaginal hysterectomy or other gynecologic procedures deep within the pelvis.

If a recognized exposure to blood has occurred, a series of steps is recommended.[74] Although no data are currently available to document the efficacy of such procedures, it is not unreasonable to copiously rinse and decontaminate the local wound. The exposure should be reported to the responsible office within the hospital. The patient and the exposed healthcare worker should be evaluated clinically, epidemiologically, and by laboratory testing for HIV antibody and viral load (as well as for hepatitis B and C). In the case of HIV, this presently requires informed consent of the individuals involved. The Centers for Disease Control currently recommends follow-up HIV testing over the next 6 months, if the exposure indeed involved HIV-infected material. If unexplained fever, lymphadenopathy, rash, pharyngitis, aseptic meningitis, or lymphopenia (features of acute retroviral illness) develop, the healthcare worker should return immediately for further clinical and laboratory evaluation.

The question of whether or not early postexposure antiretroviral chemoprophylaxis should be initiated is generally answered affirmatively.[75] There are anecdotal reports of zidovudine prophylaxis failure when given as a single agent, but these typically represented relatively massive exposures and/or significant delay in initiating therapy.[76] Unfortunately, significant bone marrow and GI toxicity can be observed in individuals receiving antiviral prophylaxis, especially if combination therapy is utilized. Most centers, therefore, take the position that the final decision to initiate antiretroviral prophylaxis should still be made by the healthcare worker after complete disclosure of the available information. One group of individuals for whom chemoprophylaxis is currently not recommended are women who are pregnant or who might become pregnant in the near future.

The second issue regarding HIV infection and surgical practice is the possibility that the virus could be transmitted from an infected surgeon to a patient undergoing an invasive procedure. This theoretical possibility was brought to the public's attention when a well-publicized group of six individuals receiving dental care from the same dentist acquired HIV infection.[77] Subsequent extensive evaluation of over 23,000 individuals who had received surgical, medical, or dental care from HIV-infected individuals has identified only a single instance of probable transmission of HIV to a patient when other risk factors for HIV acquisition were not also present.[78,79] Therefore, the risk of transmission of HIV from a healthcare worker to a patient appears to be vanishingly small. As a result, routine screening of healthcare workers for HIV positivity is not currently recommended. Rather, the following recommendations have been made by the Centers for Disease Control[80]:

1. All healthcare workers should adhere to universal precautions, including the appropriate use of hand washing, protective barriers, and care in the use and disposal of needles and other sharp instruments. Healthcare workers who have exudative lesions or weeping dermatitis should refrain from all direct patient care and from handling patient-care equipment and devices used in performing invasive procedures.
2. Currently available data provide no basis for recommendations to restrict the practice of healthcare workers infected with HIV or hepatitis B virus who perform invasive procedures not identified as exposure-prone.
3. Exposure-prone procedures should be identified by medical/surgical/dental organizations and institutions at which the procedures are performed.
4. Healthcare workers who perform exposure-prone procedures should know their HIV antibody status.

5. Healthcare workers who are infected with HIV should not perform exposure-prone procedures unless they have sought counsel from an expert review panel and been advised under what circumstances, if any, they may continue to perform these procedures. Such circumstances would include notifying prospective patients of the healthcare worker's seropositivity.

5. Postoperative Management

5.1. Respiratory Management in the Immunocompromised Patient

As emphasized above, an important goal following postanesthesia stabilization of the compromised host is early extubation. We have observed superinfection pneumonitis in over 50% of immunosuppressed allograft recipients who require endotracheal intubation for longer than 96 hr. Such experience indicates that the importance of respiratory therapy in the early postoperative period cannot be overemphasized. Moderate areas of atelectasis or pooling of tracheobronchial secretions, which may be inconsequential in lower-risk surgical candidates, can provide the appropriate environment for sepsis in the compromised host. This is all too often followed by pulmonary failure, which is typically the initial event in a cumulative sequence of multiorgan failure leading to death. Thus, increased attention to positioning, early ambulation, oxygenation and humidification of inspired air, and aggressive respiratory therapy which includes judicious endotracheal suction or bronchoscopy to clear bronchial plugs is mandatory following extubation.

An important consideration related to the endotracheal intubation in immunocompromised patients has been the observation that the nasotracheal route may favor the development of sinus infection leading to serious complications. Thus, orotracheal intubation should usually be employed even when it is anticipated the tube will be in place for a number of days.

5.2. General Postoperative Care in the Immunocompromised Patient

Other postoperative considerations include careful monitoring of wounds for collections that may be deceptively asymptomatic. We recommend immediate needle aspiration of any suspicious area of even minimal induration or tenderness. Prompt ultrasound or CBT examination of the operative site is also indicated in any patient with signs of occult sepsis. Attempts to accelerate wound healing in these patients with any of the widely studied adjuvant agents (e.g., cartilage, zinc) are clearly not warranted. However, provision of adequate caloric and protein intake to achieve a positive nitrogen balance as early as possible is obviously appropriate. Similarly, addition of vitamins, particularly A and C, to the chronically malnourished patient cannot be severely criticized.

Guidelines for the administration of perioperative transfusions for all surgical patients have become increasingly conservative, as the risks of viral disease transmission, particularly CMV, hepatitis, and HIV, have become evident. In the immunocompromised patient, there is the additional concern that allogeneic blood transfusions themselves may produce a detrimental effect on the immune system.[81] This could be clinically advantageous to transplant recipients, and this effect had been incorporated into earlier immunosuppressive protocols to improve allograft survival.[82] The efficacy of currently used immunosuppressive agents has essentially eliminated any additional benefit of blood products so that deliberate transfusion is now seldom practiced. The importance of these observations, which were initially made in transplant recipients, raised the possibility that blood transfusions may be undesirable in other surgical patients because they could increase the risk of infectious complications or the more rapid growth or metastasis of residual tumor in patients undergoing resection for cancer.[83] In fact, retrospective clinical reviews have shown a significantly worse prognosis for patients who received perioperative transfusions during cancer surgery when compared with those who did not receive blood products.[84] Despite the obvious limitations of such retrospective analyses, these observations have significant implications, particularly for the surgeon caring for any immunocompromised patient. Increased efforts to use alternative methods of vascular volume support should be made. The arbitrary hematocrit level of 28–30% at which transfusion was previously thought necessary is no longer applicable. Where the threshold is reset obviously must be individualized on the basis of comorbidity risks. Many surgeons, however, accept a stable hematocrit in the 21–26% range unless there is evidence of inadequate tissue oxygen delivery or of the potential for localized ischemia, such as a history of coronary artery disease. There is little evidence that moderate anemia adversely affects the postoperative recovery, and iron supplementation, together with administration of recombinant erythropoietin, rapidly increases autologous erythrocytosis to restore the patient's hematocrit within a period of weeks. If transfusion is necessary, some groups recommend use of leukofiltered blood, since the immunomodulatory effects appear to be mediated by transfused allogeneic passenger leukocytes.[85]

The previous practice of administering granulocyte transfusions to prevent gram-negative sepsis in patients with severe neutropenia has been essentially replaced by widespread use of hematopoietic growth factors.[86] It is now evident that myeloid recovery can be significantly accelerated by both granulocyte-macrophage colony-stimulating factor (GM-CSF) and granulocyte colony-stimulating factor (G-CSF) and that such agents play an important role in the care of immunocompromised patients whose course is complicated by periods of neutropenia.[87] In fact, some groups have suggested that routine use of G-CSF, targeting a blood granulocyte count of 10,000–20,000/mm^3 in the perioperative period, can decrease the incidence of sepsis in immunocompromised patients.[88]

Other postoperative measures, such as vigorous patient isolation, are generally not required for immunosuppressed patients—except for those with severe neutropenia (granulocyte count less than 500/mm^3) or with major burns (see Section 5.3). Nevertheless, it is clear that significant environmental contamination in the hospital will be manifested first in these patients. Thus, any unusual incidence in the number of infections by organisms such as *Aspergillus* or *Legionella* must be immediately investigated and the contaminating source controlled.

5.3. Management of the Burn or Trauma Patient

The syndrome of sepsis and sequential postoperative organ failure may develop in a previously healthy immunocompetent host, particularly in the setting of major trauma such as burns or multiple injuries that produce chronic stress.[89]

Illustrative Case 17

A 9-year-old boy was admitted to the burn unit 9 days after suffering a flaming gasoline injury producing full-thickness burns over 80% of the body surface area. During the period between injury and transfer, the burns had been treated with AgNO$_3$ dressing; but, at the time of admission, early *Pseudomonas* sepsis of the thighs and buttocks was noted. These areas were debrided; the chest and abdomen were primarily excised and covered with autografts from the scalp. The patient was continued on AgNO$_3$ dressing and both enteric and parenteral hyperalimentation while being housed in the protective environment of a bacteria-controlled nursing unit. Initially, the take of the skin grafts was excellent, but by postgraft day 10, there was evidence of spread of *Pseudomonas* sepsis and partial graft loss. Multiple procedures of burn excision followed by coverage with allografts and autografts were successful in closing most of the trunk and proximal extremity wounds, but the patient's course was continually complicated by spiking fevers, positive blood cultures, and recurrent pulmonary infiltrates. By hospital day 35, GI dysfunction with ileus, distention, and vomiting severely

limited oral intake. On day 45, the patient developed respiratory insufficiency requiring mechanical ventilation after which he continued to deteriorate metabolically. He eventually died on day 54, at which time more than 90% of the burns had been successfully covered with autografts.

Comment. Despite major advances in the immediate resuscitation and surgical management of severe trauma victims, an unacceptably large fraction of these previously healthy, often young patients succumb in the later postoperative period, usually with multiple organ failure. Intensive evaluation of this syndrome has begun to clarify that the late onset of sepsis in these patients is enhanced by limitation of protein synthesis, possibly secondary to growth hormone resistance,[90] and excessive cytokine production. In Case 17, the *Pseudomonas* sepsis was already established at the time of admission. Despite institution of intensive efforts to eradicate the infection, it was never possible to regain control. The typical downhill course, inevitably leading to death, ensued even though the burns were nearly covered.

In patients with burns, as with all trauma, the goal must be to convert the open, contaminated wound to a closed, clean wound in the shortest time possible. This is especially important for major burns where continuing immune depression is produced by the wound itself.[91] In most cases, repeated skin grafting from uninjured donor areas is an adequate solution to the problem. However, if the area of normal skin available for donor sites is insufficient to provide reasonably prompt skin graft closure, we have again found that a much more aggressive approach is required. Following the serial evaluation of a number of techniques, we initially recommended immediate wound closure with skin allografts, whose survival was prolonged by immunosuppressive agents, as the most useful method of reducing fluid, protein, heat, and energy loss, controlling infection, and protecting underlying structures.[92] More recently, the addition of synthetic "skin" has revised this approach making it possible to remove allografts earlier and, thus, immunosuppressive agents are no longer required.[93]

In patients with massive burns, prompt excision and wound closure by a number of successive surgical procedures is begun usually on the first or second day after the burn injury when all available donor sites are harvested to cover a comparable area of excised eschar with autografts. Subsequent procedures consist of excision of eschar and closure with allograft skin or artificial skin (Fig. 7). Eventually, the allografts are excised stepwise and replaced with autografts as donor sites regenerate.

Patients with massive burns are usually nursed in some form of protective environment. We favor the bacteria-controlled nursing unit pictured in Fig. 8. The patient is protected against contact cross-infection by clear plastic access walls, and by a system of medical care delivery whereby the nursing and medical personnel do not directly enter the protected environment of the units. Autoinfection of the burn wound from the patient's respiratory or GI tract is reduced by meticulous aseptic precautions and the use of 0.5% aqueous silver nitrate dressings to the burn eschar until excised, to donor sites, and to newly grafted areas of the extremities.

The metabolic balance of these patients is carefully maintained with the goal of daily parenteral and enteric

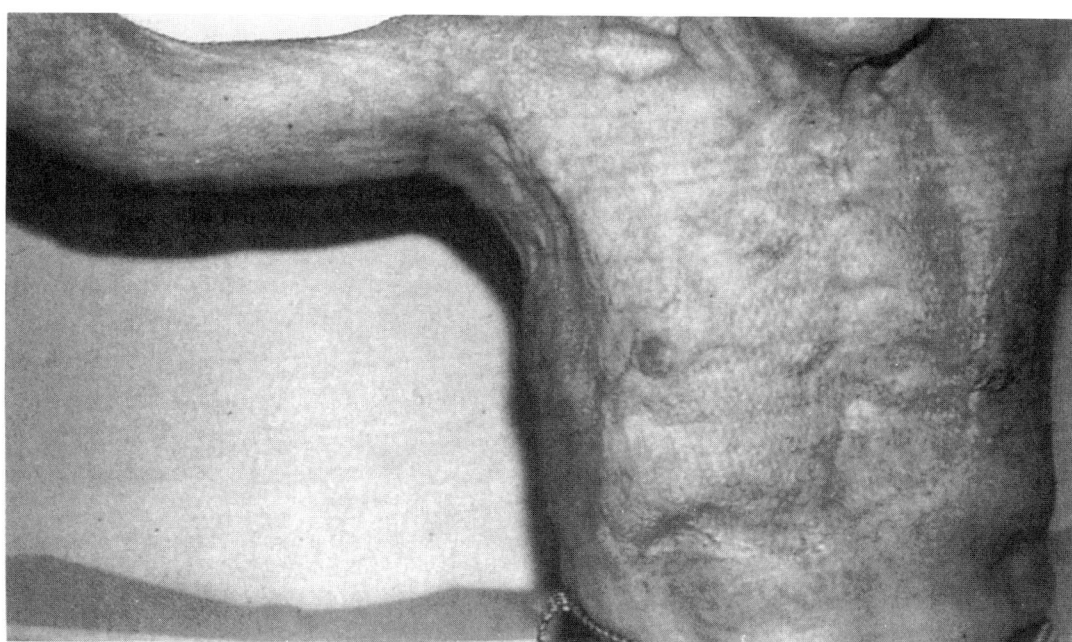

FIGURE 7. Appearance of full-thickness burn of chest and arm following treatment with artificial skin. The dermal portion of the artificial skin, made up of bovine collagen linked to a glycosaminoglycan, gradually biodegrades and is replaced by autologous "neodermis." The Silastic outer (epidermal) layer of the artificial skin is then removed and the neodermis is covered with thin meshed epidermal autografts or cultured epidermal cells. The result is cosmetically normal-appearing, supple, durable skin. (Reproduced from *Surg Rounds* 889, Oct 1992, with permission.)

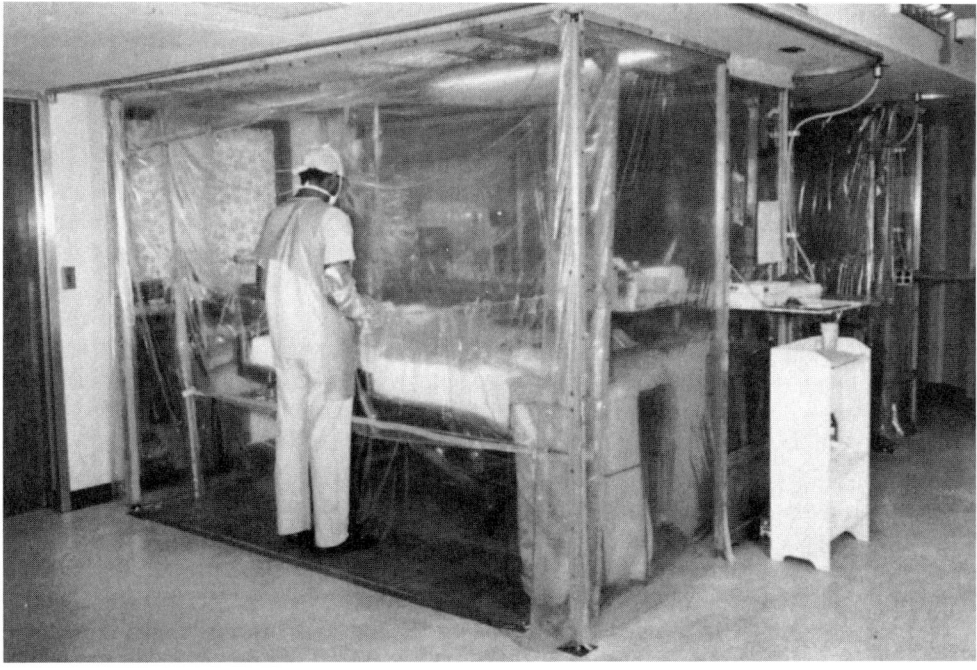

FIGURE 8. Bacteria-controlled nursing unit in which the patient area is provided with a continuous, pistonlike downflow of bacteria-free air at 90°F and 94% relative humidity to protect against energy loss as heat or evaporation as well as airborne cross-infection. Nursing and medical personnel access is via openings in the plastic walls.

intake of at least 3000 cal/m^2 and 3 g/kg of protein. Malfunction of the GI tract taking the form of gastric retention with ileus or diarrhea is an ominous sign. Its onset usually heralds severe malnutrition and sepsis.[94] Patients who expire usually develop diarrhea followed by persistent ileus and gastric retention that prevents oral alimentation. Surviving patients have occasional limited episodes of ileus, usually following an operative procedure, but GI tract dysfunction is not severe enough to interfere with nutrition and body weight is maintained throughout the course of the illness.

A large part of the effective control of bacterial infection demonstrated by these patients must be laid to the beneficial effect of prompt and almost complete wound closure and the maintenance of normal nutrition during periods of maximal stress. The contribution of the protective environment provided by the bacteria-controlled nursing unit is difficult to assess. Cross-infection is rarely seen in this group of highly susceptible patients, however, and late development of bacterial pneumonitis is seen only as a terminal episode in patients who fail to survive. Thus, although a dogmatic statement concerning the exact contribution of a protective environment to the immunosuppressed burn patient cannot be made, the evidence available indicates the contribution is probably significant.

Prior to the addition of immunosuppression to the routine treatment plan, which included prompt excision of dead tissue and immediate wound closure with autografts and allografts, the controlled environment of the bacteria-controlled nursing unit, intensive metabolic alimentation, and topical AgNO$_3$, patients with full-thickness burns of greater than 80% of the body surface suffered 100% mortality. With the addition of immunosuppression, which allowed continued wound closure by the allografts until they were electively replaced with autografts, we found it possible to salvage over half of this group of patients. Further advances, including the increasing use of synthetic "skin" without immunosuppression, have improved the care of these massively burned patients to the point that current mortality is around 33%.

Intensive postoperative support including nutritional management is also essential for preventing delayed sepsis and multiple organ failure in other postsurgical patients, in whom it is anticipated that oral alimentation will not be possible within 5–7 days. As noted in Section 3, many malnourished patients can be identified prior to surgery and will have already been started on a nutritional repletion program that is continued through the postoperative period. In previously healthy patients such as victims of major trauma, however, the extent of the postoperative catabolic response and increasingly nega-

tive nitrogen balance may not be appreciated until delayed wound healing, occult fevers, and early hepatic dysfunction become manifest. Unfortunately, at this point, institution of hyperalimentation is often ineffective. Thus, even in these acutely injured patients, it should be anticipated that although the body can initially employ its carbohydrate and fat reserves to meet the increased metabolism and negative caloric balance, these reserves are depleted within a few days of the injury or surgery. Thereafter, protein must be mobilized to provide these needs. Since there are no protein reserves as such, this immediate assault on the functional and structural capacity of the body must be prevented by early enteric or parenteral administration of adequate calories and amino acids.

5.4. Postoperative GI Bleeding in the Immunocompromised Patient

A particularly difficult postoperative problem is acute upper GI bleeding. Despite the use of antacids and H$_2$-receptor antagonists, this life-threatening complication continues to occur.

In transplant recipients, the incidence of bleeding gastroduodenal lesions has been reported to range from 1 to 10% with a mortality rate as high as 40%.[95,96] In the attempt to limit these complications, some groups have recommended prophylactic gastric surgery for patients with a history of ulcer disease or demonstrable hyperacidity. More recent reviews suggest that prophylactic surgery is not an effective approach to reducing the incidence of posttransplant gastroduodenal complications except for patients whose peptic ulcer disease has already proved to be recalcitrant to medical therapy.[97] Thus, immunocompromised patients who develop postoperative or posttraumatic gastroduodenal bleeding must often be managed surgically during their acute illness.

Illustrative Case 18

A 55-year-old woman with severe arthritis requiring chronic steroid therapy was admitted to the hospital with fever and lower abdominal pain. Bowel sounds were present, and the abdomen was nondistended, but lower abdominal direct and rebound tenderness was demonstrable. Urgent laparotomy revealed peritonitis secondary to perforated sigmoid diverticulitis which was managed with resection and end-sigmoid colostomy. A gastrostomy was performed, and extensive peritoneal lavage and drainage were instituted. The wound was managed with delayed primary closure.

The patient initially did well, but on postoperative day 7, black, guaiac-positive gastrostomy drainage was noted. This cleared with iced saline lavage. Endoscopy revealed diffuse superficial ulcerations of the distal stomach. Bright red bleeding developed 24 hr later. Selective angiography showed marked hypervascularity of the gastric mucosa. With somatostatin infusion, the bleeding was controlled for approximately 48 hr. However, active bleeding then recurred. Surgery was recommended but the patient and her family adamantly refused. Over the subsequent 7 days, repeated intra-arterial vasopressin infusions and embolic occlusion of the left gastric artery only intermittently controlled the slow but persistent bleeding. During this time, multiple blood cultures became positive for *Cryptococcus*. Culture of the removed angiography catheter similarly was positive for this organism. Despite amphotericin therapy and broad-spectrum antibiotic coverage, the patient remained febrile and lethargic, eventually developing pneumonia and respiratory failure leading to death.

Illustrative Case 19

An 18-year-old man suffered a severe crushing injury to the abdomen, pelvis, and legs in a motor-vehicle accident and required splenectomy, resection of devitalized bowel, and reduction and fixation of multiple fractures. Postoperatively, acute renal failure developed. He remained relatively stable, but intermittent fever occurred usually in association with hemodialysis treatments. On postinjury day 10, massive upper GI bleeding occurred. This was only partially controlled with iced saline lavage. Endoscopy revealed several prepyloric bleeding ulcerations. Cimetidine therapy was added, and the bleeding appeared to stop, only to recur 48 hr later. At exploration, a 300-ml foul-smelling hematoma was evacuated from the left subphrenic space. Vagotomy and distal gastrectomy were performed. The patient subsequently recovered without further bleeding or other significant complications.

Comment. Stress ulceration continues to provide a significant surgical challenge. Such ulcers characteristically bleed or perforate during the early postinjury or postoperative period. The incidence is markedly increased when significant complications, particularly sepsis, follow the initial procedure. In fact, the association of systemic or localized infection with acute gastroduodenal ulceration is so constant that a surgical axiom has been coined for these patients. It cautions that in the face of upper GI bleeding laparotomy is more urgently indicated for detection of the underlying sepsis than for control of the bleeding site. The occult subphrenic abscess in Case 19 emphasizes this point. Interestingly, this patient also exhibits another diagnostic clue occasionally observed, namely, the febrile episodes associated with hemodialysis. It has been suggested that these may be the result of expansion of the infected collection or release of cytokines during the rapid osmotic shifts associated with dialysis.

The pathologic condition of stress ulceration varies from diffuse erosive changes to a single punched-out ulcer with virtually no surrounding reaction. The progression of mucosal damage is closely related to the duration of sepsis. In patients studied serially, gastroduodenal disease becomes worse as sepsis is prolonged. On the other hand, mucosal lesions show a dramatic improvement when focal infection and septicemia are eradicated.[98]

The mechanism of sepsis-induced gastroduodenal lesion formation is unclear, but it may be favored by ischemia of the mucosa as it is with acute ulcers of other origins. It is possible that a vasoconstrictive process of the gastroduodenal mucosa is associated with the septic focus and is independent of systemic shock. In support of this, endoscopies have shown a pale, ischemic, marbled mucosa that strongly suggests an ischemic etiology.

Medical therapy of stress ulcers is often disappointing. The addition of H_2-receptor antagonists has improved these results, but again, ultimate patient survival can only be achieved if the accompanying sepsis is controlled as well.

The development of bipolar and laser coagulation devices has provided another approach to control for the patient who continues to bleed despite intensive medical therapy.[99] Emergency mesenteric angiography is usually undertaken only after endoscopy has been unsuccessful in identifying or controlling the bleeding site. If a bleeding site is defined by angiography, selective intra-arterial or peripheral intravenous infusion of aqueous vasopressin or somatostatin is sometimes successful. This is particularly true in patients with diffuse superficial gastric mucosal lesions in whom bleeding can often be controlled following infusion via the left gastric artery.[100]

In heavily immunosuppressed patients, especially in the presence of leukopenia, we have had little success with long-term angiographic control of GI bleeding. An obvious hazard is the development of sepsis from the indwelling catheter as occurred in Case 18. Thus, most groups advise surgical intervention if bleeding is not controlled within 24–48 hr following institution of treatment by selective angiography or somatostatin infusion.[101]

Surgical therapy in these patients also leaves much to be desired. The choice of operation is largely a matter of unsupported preference. Most authorities agree that vagotomy should be performed.[102] Gastric resection has been advised as well, since the incidence of rebleeding after vagotomy with only gastric drainage is usually higher. Of course, the greater morbidity and mortality following resection must also be taken into consideration.

5.5. Sepsis following Splenectomy

Illustrative Case 20

A 35-year-old man on chronic hemodialysis because of end-stage polycystic kidney disease was admitted to the hospital with fever and hypotension. Two years earlier, he had required left nephrectomy for infection. At that time, splenectomy was also performed because of intraoperative trauma to the spleen. He had been well until 1 week before the present admission when he noted the onset of occasional headaches and malaise. On the day of admission, he had been hiking on the beach, where he felt unusually hot. Swimming precipitated chills and he subsequently noted fever to 103°F (39°C). He was admitted to a local hospital with the possible diagnosis of heatstroke. He was observed there to be intermittently alert with a systolic blood pressure of 70 mm Hg and an oral temperature of 105°F (41°C). Ice packs were applied after which the temperature fell to 103°F (39°C), but the patient became increasingly obtunded and hypotensive. When transferred here, he was found to have generalized petechiae and ecchymoses. Despite intravenous pressors, antibiotics, and fluid resuscitation, he died 5 hr after admission. Blood cultures subsequently grew *Streptococcus pneumoniae*.

Comment. The occurrence of fulminant bacterial sepsis in splenectomized patients was first clearly demonstrated by King and Shumacker.[103] For many years, the risk was believed to be limited to infants or young children but subsequent studies have emphasized that splenectomy done at any age and for any reason increases the risks of death resulting from infection.[104]

Although well-opsonized bacteria are removed primarily by the liver, the spleen is responsible for removing nonopsonized blood-borne bacteria. The spleen also plays an important role in the humoral response, as the primary immunoglobulin response takes place in the spleen. Low levels of IgM have been observed after splenectomy in children. In addition, a reduced level of the complement factor, properdin, has been demonstrated after splenectomy. The spleen also serves as a major site of synthesis of tufsin, a basic tetrapeptide that coats polymorphonuclear leukocytes to promote phagocytosis. Children are clearly more susceptible to postsplenectomy sepsis than adults, especially during the first 6 to 12 months after splenectomy. In patients who underwent splenectomy for trauma, the risk of serious infection is approximately 1.5%, whereas asplenic patients with thalassemia have a risk of serious infection approaching 25%.[105] The mortality rate for overwhelming postsplenectomy infection is approximately 45%.[104]

The surgeon's role in the management of this condition should be on several fronts. First, it is clear that splenectomy is usually not required to achieve hemostasis following splenic disruption. Numerous reports have now documented the feasibility of splenic salvage procedures.[106] Second, if splenectomy is unavoidable because of exsanguinating hemorrhage or for definitive control of a specific disease process, the patient should be protected as much as possible by polyvalent pneumococcal vaccine and should be appropriately informed of the need for aggressive treatment of all infections.[104] Finally, the surgeon, who may be the first to evaluate these patients in an emergency ward, must keep this possibility in mind lest valuable time be lost, as in the case above, while nonspecific treatment for the fever is being administered.

Another concern of particular importance for surgeons results from the recent dramatically increased incidence of vancomycin-resistant enterococcus (VRE) colonization in hospitalized patients.[107]

Illustrative Case 21

A 5-year-old African-American child underwent orthotopic liver transplantation (OLT) for treatment of end-stage liver failure secondary to Alagille's syndrome. His previous history was significant for surgical repair of tetralogy of Fallot and multiple hospitalizations for antibiotic

treatment of recurrent otitis media, spontaneous bacterial peritonitis, and complications of chronic nutritional deficiency. At the time of admission for transplantation, there were no localizing signs of infection. Surveillance cultures proved to be unremarkable except for VRE in the patient's stool.

The transplant procedure using a reduced-size adult liver allograft was uncomplicated. The biliary tree was reconstructed with a Roux-en-Y choledochojejunostomy. Postoperatively, good allograft function was achieved, but fever and abdominal pain developed during the second postoperative week. Cholangiography demonstrated a bile leak and a small subhepatic fluid collection. At the time of exploration, a subhepatic hematoma, which subsequently proved to be colonized with VRE, was evacuated. The choledochojejunostomy was successfully revised. The patient subsequently developed a clinical picture consistent with intra-abdominal sepsis including renal failure requiring dialysis, hectic fevers, and persistent ileus. Despite treatment with chloramphenicol and synercid, cultures of the abdominal drainage fluid, bile, and stool were persistently positive for VRE. A therapeutic approach that included weekly laparotomy for drainage of loculated collections and debridement of any devitalized tissues was undertaken. Over the next 4 weeks, the patient's course gradually stabilized as the sepsis was controlled. He was gradually weaned from hemodialysis and he remains well with normal renal and hepatic function 2 years later.

Comment. This patient's postoperative course was initially complicated by biliary tract leakage, a continuing cause of significant morbidity and mortality following OLT.[108,109] Although many of these leaks can be managed nonoperatively, immediate surgery was recommended in this instance because of the known likelihood of VRE contamination. In fact, the finding at that point of a VRE-infected subhepatic hematoma presaged the patient's complicated subsequent hospital course.

In our experience, colonization of the biliary tree by VRE is almost universal following OLT in recipients whose baseline GI tract flora includes this organism. As a result, enterococcal infectious episodes are now being reported in as many as 20% of OLT recipients, even if no biliary leakage develops in the postoperative period.[110,111] We and others have found that aggressive open surgical debridement of devitalized or loculated collections is often required, since there is a much higher rate of unresolved infection following percutaneous drainage of intra-abdominal abscesses in patients with VRE.[112] In Case 21, the sites requiring this unusual (almost preantibiotic era) surgical approach included the subhepatic space around the biliary reconstruction as well as the resection margins of the reduced-sized liver allograft. As in this patient, pursuit of such aggressive surgical management will probably remain the only approach to salvaging patients with extensive intra-abdominal sepsis until an effective antimicrobial agent for VRE becomes available.

6. Conclusions

Surgical management of infection in the compromised host requires special considerations related to the diagnosis, preoperative preparation, intraoperative tech-

TABLE 5. Surgical Considerations for Management of Infection in the Immunocompromised Host

Aggressive diagnostic approach
  Recognition of impaired signs of inflammation and unusual clinical presentation
  Persistent search for occult infection
  Awareness of commonly occurring conditions
Expeditious preoperative preparation
  Systemic antibiotics
  Correction of malnutrition, hyperglycemia
  Prevention of adrenal insufficiency
  Reduction of immunosuppressive treatment
Special intraoperative attention
  Liberal use of lavage, drains, stomas versus anastomoses
  Provision for prolonged gastric drainage and enteric feeding
  Topical antibiotics
  Secure wound closure
  Limit surgical team's exposure to virus (e.g., AIDS)
  Limit homologous blood transfusions
Careful postoperative management
  Early extubation
  Early nutritional support
  Detection and aggressive management of occult septic complications
  Multispecialty control of GI bleeding
  Environmental precautions
  Hematopoietic growth factors

niques, and postoperative care (Table 5) of these fragile patients. With observation of certain guidelines as outlined here, it is possible to perform even highly complicated procedures with acceptable morbidity and mortality. The surgeon should consider appropriately aggressive evaluation and surgical intervention in these patients from the viewpoint that delay in performance of an indicated procedure, far from being the conservative approach, will more likely be the determining factor that prevents the patient's rehabilitation or even precludes survival.

ACKNOWLEDGMENTS. I am grateful for the excellent secretarial and editorial assistance provided by Cathy Padyk in the preparation of this manuscript.

References

1. MacLean LD: Delayed type hypersensitivity testing in surgical patients. *Surg Gynecol Obstet* **166**:285–293, 1988.
2. Ramsey PG, Rubin RH, Tolkoff-Rubin NE, *et al*: The renal transplant patient with fever and pulmonary infiltrates: Etiology, clinical manifestation, and management. *Medicine* **59**:206–222, 1980.
3. Lupinetti FM, Behrendt DM, Gilles RH, *et al*: Pulmonary resection for fungal infection in children undergoing bone marrow transplantation. *J Thorac Cardiovasc Surg* **104**:684-687, 1992.
4. Temeck BK, Venzon DJ, Moskaluk CA, Pass HI: Thoracotomy for pulmonary mycoses in non-HIV-immunosuppressed patients. *Ann Thorac Surg* **58**:333–338, 1994.
5. Rosenow EC, Limper AH: Drug-induced pulmonary disease. *Respir Infect* **10**:86–95, 1995.
6. O'Brien JD, Ettinger NA: Pulmonary complications of liver transplantation. *Clin Chest Med* **17**:99–114, 1996.
7. White DA: Pulmonary infection in the immunocompromised patient. *Semin Thorac Cardiovasc Surg* **7**:78–87, 1995.
8. Kramer MR, Berkman N, Mintz B, *et al*: The role of open lung biopsy in the management and outcome of patients with diffuse lung disease. *Ann Thorac Surg* **65**:198–202, 1998.
9. Rothenberg SS: Thoracoscopy in infants and children. *Semin Pediatr Surg* **7**:194-201, 1998.
10. Robinson LA, Reed EC, Galbraith TA, *et al*: Pulmonary resection for invasive *Aspergillus* infections in immunocompromised patients. *J Thorac Cardiovasc Surg* **109**:1182–1196, 1995.
11. Singh N, Arnow PM, Bonham A, *et al*: Invasive aspergillosis in liver transplant recipients in the 1990s. *Transplantation* **64**:716–720, 1997.
12. Denning DW, Stevens DA: Antifungal and surgical therapy of invasive aspergillosis: Review of 2121 published cases. *Rev Infect Dis* **12**:1147–1201, 1990.
13. Smith PC, Slaughter MS, Petty MG, *et al*: Abdominal complications after lung transplantation. *J Heart Lung Transplant* **14**:44–51, 1995.
14. Pirenne J, Lledo-Garcia E, Benedetti E, *et al*: Colon perforation after renal transplantation: A single-institution review. *Clin Transplant* **11**:88–93, 1997.
15. Scott-Connor CEH, Fabrega AJ: Gastrointestinal problems in the immunocompromised host. A review for surgeons. *Surg Endosc* **10**:959–964, 1996.
16. Flanigan RC, Reckard CR, Lucas BA: Colonic complications of renal transplantation. *J Urol* **139**:503–511, 1988.
17. McCune TR, Nylander WA, Van Buren DH, *et al*: Colonic screening prior to renal transplantation and its impact on post-transplant colonic complications. *Clin Transplant* **6**:91–96, 1992.
18. Van den Berg AP, Klompmaker IJ, Haagsma EB, *et al*: Evidence for an increased rate of bacterial infections in liver transplant patients with cytomegalovirus infection. *Clin Transplant* **10**:224–231, 1996.
19. Shrestha BM, Parton D, Gray A, *et al*: Cytomegalovirus involving the gastrointestinal tract in renal transplant recipients. *Clin Transplant* **10**:171–175, 1996.
20. Sakr M, Hassanein T, Gavaler J, Abu-Elmagd K, *et al*: Cytomegalovirus infection of the upper gastrointestinal tract following liver transplantation—Incidence, location, and severity in cyclosporine- and FK506–treated patients. *Transplantation* **53**:786–791, 1992.
21. van Son WJ, van der Jagt EJ, van der Woude FJ, *et al*: Pneumatosis intestinalis in patients after cadaveric kidney transplantation. *Transplantation* **38**:506–510, 1984.
22. Muldoon J, O'Riordan K, Rao S, Abecassis M: Ischemic colitis secondary to venous thrombosis. *Transplantation* **61**:1651–1653, 1996.
23. Delmonico FL, Auchincloss H, Rubin RH, *et al*: The selective use of antilymphocyte serum for cyclosporine treated patients with renal allograft dysfunction. *Ann Surg* **206**:649–654, 1987.
24. Rubin RH: Infectious disease problems. In Maddrey WC (ed): *Transplantation of the Liver*. Elsevier Science, Amsterdam, 1988, pp. 279–308.
25. Steck TB, Durkin MG, Costanzo-Nordin MR, Keshavarzian A: Gastrointestinal complications and endoscopic findings in heart transplant patients. *J Heart Lung Transplant* **12**:244–251, 1993.
26. Cao S, Cox K, Esquivel CO, *et al*: Posttransplant lympho-

proliferative disorders and gastrointestinal manifestations of Epstein–Barr virus infection in children following liver transplantation. *Transplantation* **66:**851–856, 1998.

27. Penn I, Porat G: Central nervous system lymphomas in organ allograft recipients. *Transplantation* **59:**240–244, 1995.

28. Starzl TE, Nalesnik MA, Porter KA, *et al*: Reversibility of lymphomas and lymphoproliferative lesions developing under cyclosporin-steroid therapy. *Lancet* **1:**583–587, 1984.

29. Armitage JM, Kormos RL, Stuart S, *et al*: Posttransplant lymphoproliferative disease in thoracic organ transplant patients: Ten years of cyclosporine-based immunosuppression. *J Heart Lung Transplant* **10:**877–887, 1991.

30. McDiarmid SV, Jordan S, Lee GS, *et al*: Prevention and preemptive therapy of posttransplant lymphoproliferative disease in pediatric liver recipients. *Transplantation* **66:**1604–1611, 1998.

31. Lillemoe KD, Romolo JL, Hamilton SR, *et al*: Intestinal necrosis due to sodium polystyrene (Kayexalate) in sorbital enemas: Clinical and experimental support for the hypothesis. *Surgery* **101:**267–279, 1987.

32. Koretz MJ, Neifeld JP: Emergency surgical treatment for patients with acute leukemia. *Surg Gynecol Obstet* **161:**149–151, 1985.

33. Vohra R, Prescott RJ, Banerjee SS, *et al*: Management of neutropenic colitis. *Surg Oncol* **1:**11–15, 1992.

34. Silliman CC, Haase GM, Strain JD, *et al*: Indications for surgical intervention for gastrointestinal emergencies in children receiving chemotherapy. *Cancer* **74:**203–216, 1994.

35. Wade DS, Douglass H, Nava HR, *et al*: Abdominal pain in neutropenic patients. *Ann Surg* **125:**1119–1127, 1990.

36. Civardi G, DiCandio G, Giorgio A, *et al*: Ultrasound guided percutaneous drainage of abdominal abscesses in the hands of the clinician: A multicenter Italian study. *Eur J Ultrasound* **8:**91–99, 1998.

37. Graffeo CS, Corinselman F: Appendicitis. *Emerg Med Clin North Am* **14:**653–671, 1996.

38. Rao PM, Rhea JT, Novelline RA, *et al*: Helical CT technique for the diagnosis of appendicitis: Prospective evaluation of a focused appendix CT examination. *Radiology* **202:**139–144, 1997.

39. Rao PM, Wittenberg J, McDowell RK, *et al*: Appendicitis: Use of arrowhead sign for diagnosis at CT. *Radiology* **202:**363–366, 1997.

40. Carone FA, Liebow AA: Acute pancreatic lesions in patients treated with ACTH and adrenal corticoids. *N Engl J Med* **257:**690–697, 1957.

41. Slakey DP, Johnson CP, Cziperle DJ, *et al*: Management of severe pancreatitis in renal transplant recipients. *Ann Surg* **225:**217–222, 1997.

42. Yanaga K, Shimada M, Gordon RDF, *et al*: Pancreatic complications following orthotopic liver transplantation. *Clin Transplant* **6:**126–130, 1992.

43. Foitzik T, Forgacs B, Ryschich E, *et al*: Effect of different immunosuppressive agents on acute pancreatitis. *Transplantation* **65:**1030–1036, 1998.

44. Hubay CA, Weckesser EC, Levy RP: Occult adrenal insufficiency in surgical patients. *Ann Surg* **181:**325–332, 1975.

45. Bromberg JS, Baliga P, Cofer JB, *et al*: Stress steroids are not required for patients receiving a renal allograft and undergoing operation. *J Am Coll Surg* **180:**532–536, 1995.

46. Girardet RE, Rosenbloom P, DeWeese BM, *et al*: Significance of asymptomatic biliary tract disease in heart transplant recipients. *J Heart Transplant* **8:**391–399, 1989.

47. Steck TB, Costanzo-Nordin MR, Keshavarzian A: Prevalence and management of cholelithiasis in heart transplant patients. *J Heart Lung Transplant* **10:**1029–1032, 1991.

48. Lopez Pa, Perrone SV, Kaplan J, *et al*: Laparoscopic cholecystectomy in heart transplant recipients. *J Heart Lung Transplant* **12:**147–149, 1993.

49. Daly JM, Lieberman MD, Goldfine J, *et al*: Enteral nutrition with supplemental arginine, RNA, and omega-3 fatty acids in patients after operation: Immunologic, metabolic, and clinical outcome. *Surgery* **112:**56–67, 1992.

50. Heyland DK, MacDonald S, Keefe L, *et al*: Total parenteral nutrition in the critically ill patient: A meta-analysis. *JAMA* **280:**2013–2019, 1998.

51. Cobb DK, High KP, Sawyer RG, *et al*: A controlled trial of scheduled replacement of central venous and pulmonary-artery catheters. *N Engl J Med* **327:**1062–1068, 1992.

52. Darouiche RO, Raad II, Heard SO, *et al*: A comparison of two antimicrobial-impregnated central venous catheters. *N Engl J Med* **340:**1–8, 1999.

53. Gyosens IC: Preventing postoperative infections: Current treatment recommendations. *Drugs* **57:**175–185, 1999.

54. Platt R, Zaleznik DF, Hopkins CC, *et al*: Perioperative antibiotic prophylaxis for herniorrhaphy and breast surgery. *N Engl J Med* **322:**153–160, 1990.

55. Bagley DH, Ketcham AS: Infections and their prevention in surgical cancer patients. In Burke JF, Hildick-Smith GY (eds): *The Infection-Prone Hospital Patient.* Little, Brown, Boston, 1978, pp. 175–181.

56. Classen DC, Evans RS, Pestotnik SL, *et al*: The timing of prophylactic administration of antibiotics and the risk of surgical-wound infection. *N Engl J Med* **326:**281–286, 1992.

57. Deitch EA, Baker T, Berg R, *et al*: Hemorrhagic shock promotes the systemic translocation of bacteria from the gut. *J Trauma* **27:**815, 1987.

58. Goris RJA, van Bebber IPT, Mollen RMH, *et al*: Does selective decontamination of the gastrointestinal tract prevent multiple organ failure? *Arch Surg* **126:**561–565, 1991.

59. D'Antonio D, Piccolomini R, Iacone A, *et al*: Comparison of ciprofloxacin, ofloxacin, and pefloxacin for the prevention of bacterial infection in neutropenic patients with hematological malignancies. *J Antimicrob Chemother* **33:**837–844, 1994.

60. deBros FM, Lai A, Scott R, *et al*: Pharmacokinetics and pharmacodynamics of atracurium during isoflurane anesthesia in normal and anephric patients. *Anesth Analg* **65:**743–746, 1986.

61. Paterson DL, Stapelfeldt WH, Wagener MM, *et al*: Intraoperative hypothermia is an independent risk factor for early cytomegalovirus infection in liver transplant recipients. *Transplantation* **67:**1151–1155, 1996.

62. Glenn J, Funkhouser WK, Schneider PS: Acute illness necessitating urgent abdominal surgery in neutropenic cancer patients: Description of 14 cases and review of the literature. *Surgery* **105:**778–789, 1989.

63. Sindelar WF, Mason GR: Irrigation of subcutaneous tissue with povidone-iodine solution for prevention of surgical wound infections. *Surg Gynecol Obstet* **148:**227–231, 1979.

64. Dire DJ, Welsh AP: A comparison of wound irrigation solutions used in the emergency department. *Ann Emerg Med* **19:**704–708, 1990.

65. Stephan RN, Munschauer CE, Kumar MS: Surgical wound infection in renal transplantation: Outcome data in 102 consecutive patients without peri-operative systemic antibiotic coverage. *Arch Surg* **132:**1315–1318, 1997.

66. Jivegard L, Blohme I, Holm J, *et al*: Abdominal aortic reconstruction without renal bypass in renal transplant patients. *Surgery* **106:** 110–113, 1989.

67. Jebara VA, Fabiani JN, Moulonguet-Deloris L, *et al*: Abdominal aortic aneurysmectomy in renal transplant patients. *Tex Heart Inst J* **17:**240–244, 1990.

68. Ippolito G, Puro V, Heptonstall J, *et al*: Occupational human immunodeficiency virus infection in health care workers: Worldwide cases through 1997. *Clin Infect Dis* **28:**365–383, 1999.

69. Bell DM: Occupational risk of human immunodeficiency virus infection in healthcare workers: An overview. *Am J Med* **102:**9–15, 1997.

70. Diettrich NA, Cacioppo JC, Kaplan G, Cohen SM: A growing spectrum of surgical disease in patients with human immunodeficiency virus/acquired immunodeficiency syndrome. *Arch Surg* **126:**860–866, 1991.

71. Nichols RL: Percutaneous injuries during operation: Who's at risk for what? *JAMA* **267:**2938–2939, 1992.

72. McCormick RD, Meisch MG, Ircink FG, *et al*: Epidemiology of hospital sharps injuries: A 14 year prospective study in the pre-AIDS and AIDS eras. *Am J Med* **91:**301S–307S, 1991.

73. Tokars JI, Bell DM, Culver DH, *et al*: Percutaneous injuries during surgical procedures. *JAMA* **267:**2899–2904, 1992.

74. McClinsey SC: Occupational exposure to HIV: Consideration for post exposure prophylaxis and prevention. *Nurs Clin North Am* **34:**213–225, 1999.

75. Puro V, Ippolito G: Issues on antiretroviral post exposure combination prophylaxis. *J Biol Regul Homeost Agents* **11:**11–19, 1997.

76. Lange JMA, Boucher CAB, Hollak CEM, *et al*: Failure of zidovudine prophylaxis after accidental exposure to HIV-1. *N Engl J Med* **322:**1375–1377, 1990.

77. Ciesielski C, Marianos D, Ou C-Y, *et al*: Transmission of human immunodeficiency virus in a dental practice. *Ann Intern Med* **116:** 798–805, 1992.

78. Robert LM, Chamberland ME, Cleveland JL, *et al*: Investigations of patients of health care workers infected with HIV. *Ann Intern Med* **122:**653–657, 1995.

79. Lot F, Seguier JXC, Fegueux S, *et al*: Probable transmission of HIV from an orthopedic surgeon to a patient in France. *Ann Intern Med* **130:**1–6, 1999.

80. Recommendations for preventing transmission of human immunodeficiency virus and hepatitis B virus to patients during exposure-prone invasive procedures. *MMWR* **40:**1–9, 1991.

81. Gascon P, Zoumbos NC, Young NS: Immunologic abnormalities in patients receiving multiple blood transfusions. *Ann Intern Med* **100:**173–177, 1984.

82. Opelz G, Terasaki PI: Dominant effect of transfusions on kidney graft survival. *Transplantation* **29:**153–158, 1980.

83. Quintiliani L, Pescini A, DiGirolamo M, *et al*: Relationship of blood transfusion, post-operative infections and immunoreactivity in patients undergoing surgery for gastrointestinal cancer. *Haematologica* **82:**318–323, 1997.

84. Wu HS, Little AG: Perioperative blood transfusions and cancer recurrence. *J Clin Oncol* **6:**1348–1354, 1988.

85. Bordin JO, Blajchman MA: Immunosuppressive effects of allogeneic blood transfusions: Implications for the patient with a malignancy. *Hematol Oncol Clin North Am* **9:**205–218, 1995.

86. Colquhoun SD, Shaked A, Jurim O, *et al*: Reversal of neutropenia with granulocyte colony-stimulating factor without precipitating liver allograft rejection. *Transplantation* **56:**755–758, 1993.

87. Crawford J, Ozner H, Stoller R, *et al*: Reduction by granulocyte colony-stimulating factor of fever and neutropenia induced by chemotherapy in patients with small-cell lung cancer. *N Engl J Med* **325:**164-169, 1991.

88. Foster PF, Mital D, Sankary HN, *et al*: The use of granulocyte colony-stimulating factor after liver transplantation. *Transplantation* **59:**1557–1563, 1995.

89. Tran DD, Cuesta MA, vanLeeuwen PAM, *et al*: Risk factors for multiple organ system failure and death in critically injured patients. *Surgery* **114:**21–30, 1993.

90. Singh KP, Prasad R, Chari PS, *et al*: Effect of growth hormone therapy in burn patients on conservative treatment. *Burns* **24:**733–738, 1998.

91. Nguyen TT, Gilpin DA, Meyer NA, *et al*: Current treatment of severely burned patients. *Ann Surg* **223:**14–25, 1996.

92. Cosimi AB, Burke JF, Russell PS: Transplantation of skin. *Surg Clin North Am* **24:**435–451, 1978.

93. Staley M, Richard R: Management of the acute burn wound: An overview. *Adv Wound Care* **10:**39–44, 1997.

94. Wolf SE, Rose JK, Desai MH, *et al*: Mortality determinants in massive pediatric burns. An analysis of 103 children with >80% TBSA burns. *Ann Surg* **225:**554–569, 1997.

95. Feduska NJ, Amend WJC, Vincenti F, *et al*: Peptic ulcer disease in kidney transplant recipients. *Ann J Surg* **148:**51–55, 1984.

96. Barsdaxoglou E, Maddern G, Ruso L, *et al*: Gastrointestinal surgical emergencies following kidney transplantation. *Transplant Int* **6:**148–152, 1993.

97. Bansky G, Do UH, Largiader F, *et al*: Gastroduodenal complications after renal transplantation: The role of prophylactic gastric surgery in hyperacid kidney allograft recipients. *Clin Transplant* **1:**209–213, 1987.

98. LeGall JR, Mignon FC, Rapin M, *et al*: Acute gastroduodenal lesions related to severe sepsis. *Surg Gynecol Obstet* **142:**377–380, 1976.

99. Donahue PE, Mobarhan S, Layden TJ, *et al*: Endoscopic control of upper gastrointestinal hemorrhage with a bipolar coagulation device. *Surg Gynecol Obstet* **159:**113–118, 1984.

100. Nusbaum M, Deren J, Chait A: Angiographic control of bleeding. *Contemp Surg* **23:**79–123, 1983.

101. Jenkins SA, Poulianos G, Coraggio F, *et al*: Somatostatin in the treatment of non-variceal upper GI bleeding. *Dig Dis* **16:**214–224, 1998.

102. Moody FG, Miller TA: Answers to questions on stress ulcer. *Hosp Med* **11:**33–56, 1983.

103. King H, Shumacker HB: Splenic studies: Susceptibility to infection after splenectomy performed in infancy. *Ann Surg* **136:**239–242, 1952.

104. Waghorn DJ, Mayon-White RT: A study of 42 episodes of overwhelming post-splenectomy infection: Is current guidance for asplenic individuals being followed? *J Infect* **35:**289–294, 1997.

105. Wilson SA, Johnson WD: Infections complicating surgical or functional splenectomy. In Grieco MH (ed): *Infections in the Abnormal Host*. Yorke, New York, 1980, pp. 848–865.

106. Clancy TV, Ramshaw DG, Maxwell JG, *et al*: Management outcomes in splenic injury: A statewide trauma center reviewed. *Ann Surg* **226:**17–24, 1997.

107. Fleenor-Ford A, Hayden MK, Weinstein RA: Vancomycin-resistant enterococci: Implications for surgeons. *Surgery* **125:**121–125, 1999.

108. Vicente E, Perkins JD, Sterioff S, *et al*: Biliary tract complications following orthotopic liver transplantation. *Clin Transplant* **1:**138–142, 1987.

109. Greif F, Bronsther OL, VanThiel DH, *et al*: The incidence, timing, and management of biliary tract complications after orthotopic liver transplantation. *Ann Surg* **219:**40–45, 1994.

110. Patel R, Badley AD, Larson-Keller J, *et al*: Relevance and risk factors of enterococcal bacteremia following liver transplantation. *Transplantation* **61:**1192–1197, 1996.

111. Papanicolaou GA, Meyers BR, Meyers J, *et al*: Nosocomial infections with vancomycin-resistant *Enterococcus faecium* in liver transplant recipients: Risk factors for acquisition and mortality. *Clin Infect Dis* **23:**760–766, 1996.

112. Linden PK, Pasculle AW, Manez R, *et al*: Differences in outcome for patients with bacteremia due to vancomycin-resistant *E. faecium* or vancomycin-susceptible *E. faecium*. *Clin Infect Dis* **22:**663–670, 1996.

Color Plates

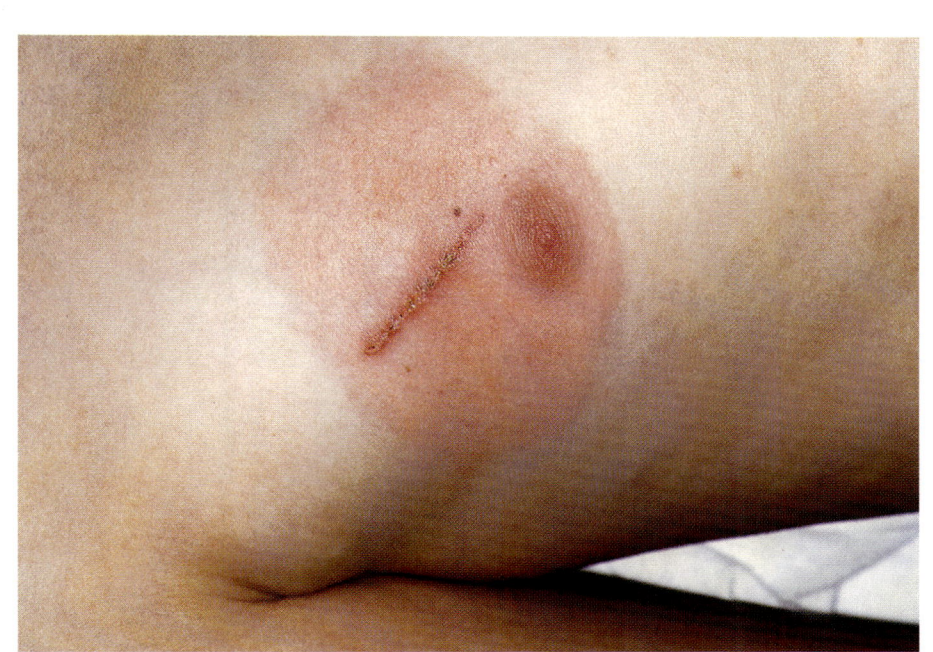

FIGURE 4. *S. aureus* cellulites in scratch. Early cellulites arising on the chest at the site of a cat scratch in an 8-year-old HIV infected girl.

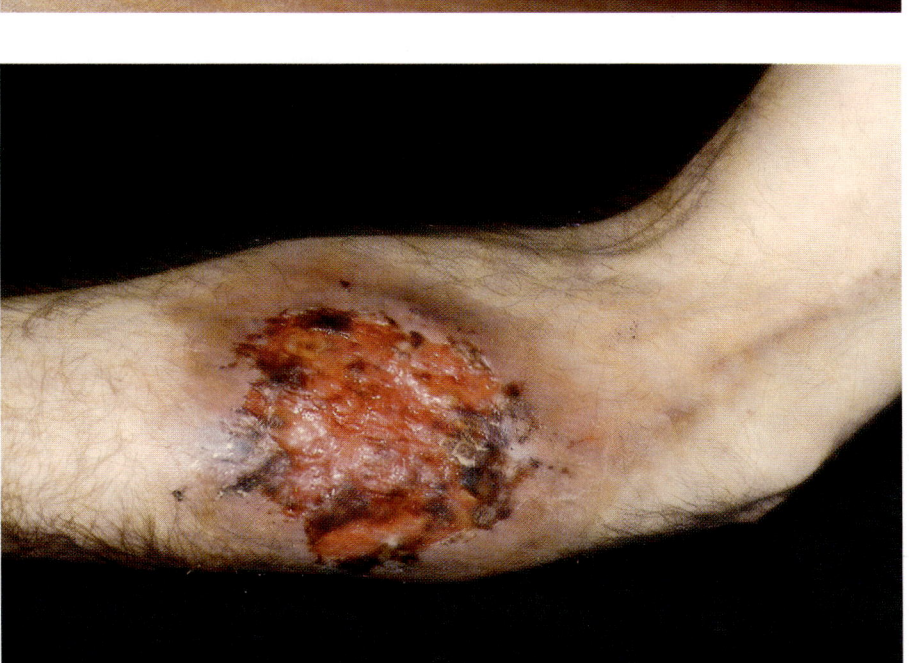

FIGURE 3. *S. aureus* cellulites in site of drug injection. Advance soft tissue infection on the forearm at the site of heroin injection. Lesion biopsy specimen also showed foreign body reaction.

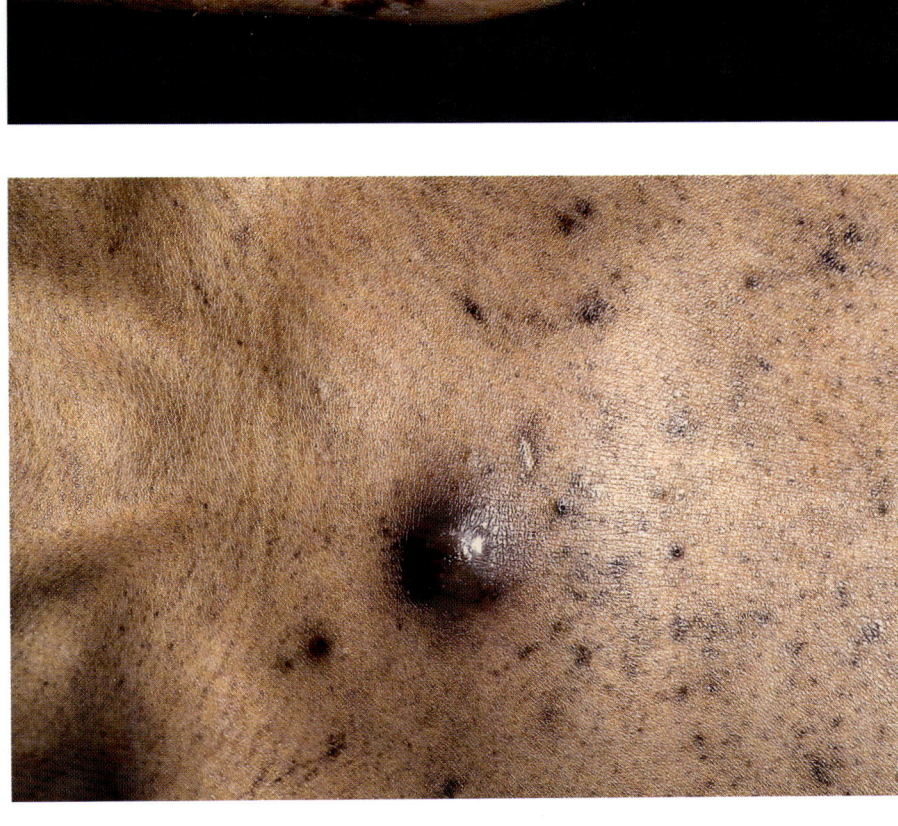

FIGURE 2. *Staphylococcus aureus*: Abscess arising in excoriation. This individual with HIV disease has pruritic eosinophilic folliculitis, excoriated chest lesions, resulting in secondary S. aureus infection with subsequent abscess formation.

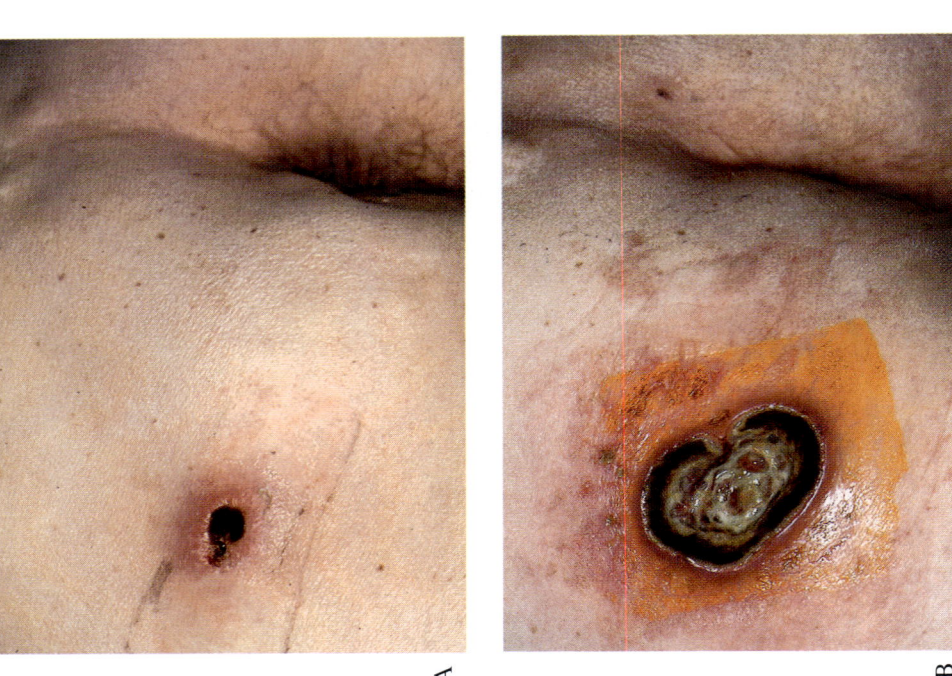

FIGURE 7. *Pseudomonas aeruginosa:* Ecthyma gangrenosum. (A) A very painful, ulcerated plaque on the buttock of a man with advanced HIV disease and neutropenia. Oral ciprofloxacin was given; an adverse drug eruption occurred and the drug was discontinued. (B) The ulcer enlarged and was associated with bacteremia. Neutropenia persisted and the patient subsequently died from *Pseudomonas* pneumonitis.

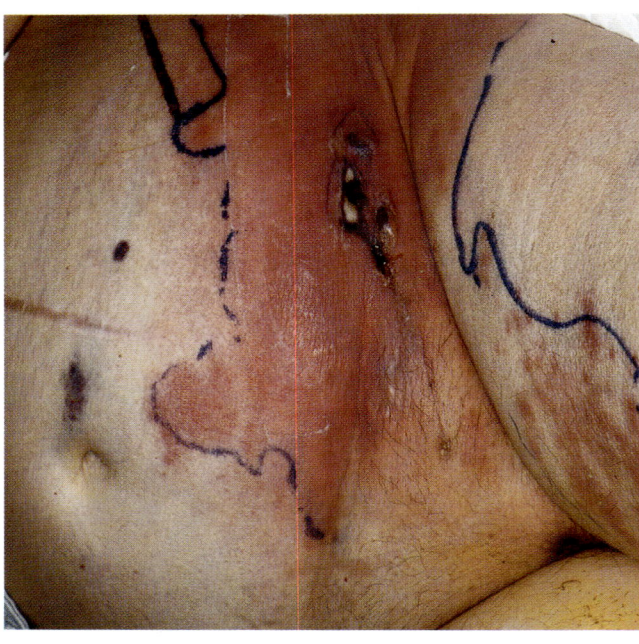

FIGURE 6. *Escherichia coli:* Cellulitis. Enlarging red, hot, tender plaque arising at a lymph node biopsy site in elderly patient with lymphoma.

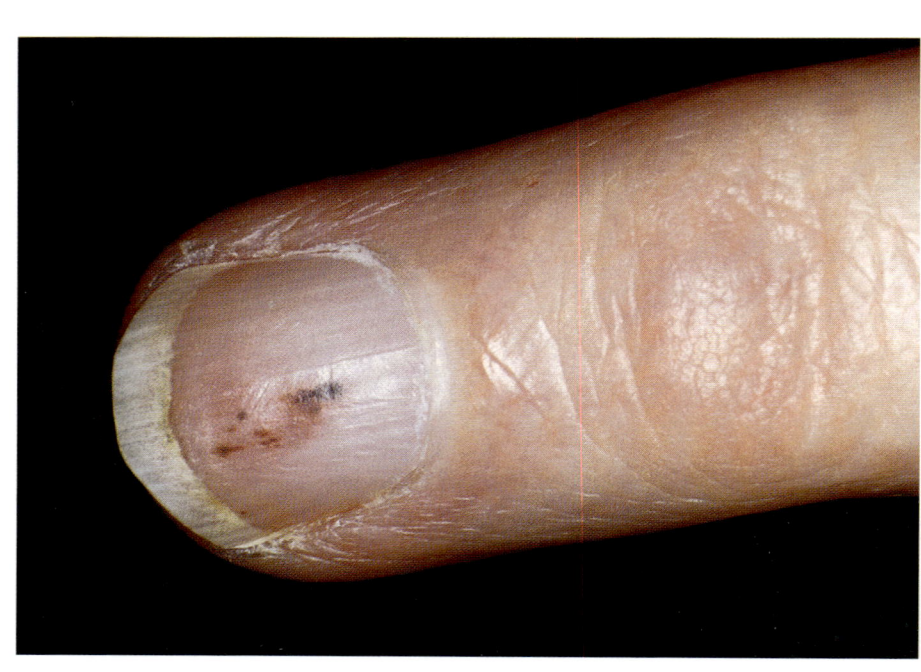

FIGURE 5. *Enterococcus:* Endocarditis with nail bed splinter hemorrhages. Splinter hemorrhages in the midportion of the nail bed in a diabetic with urinary tract infection, complicated by sepsis, endocarditis, and peripheral embolization.

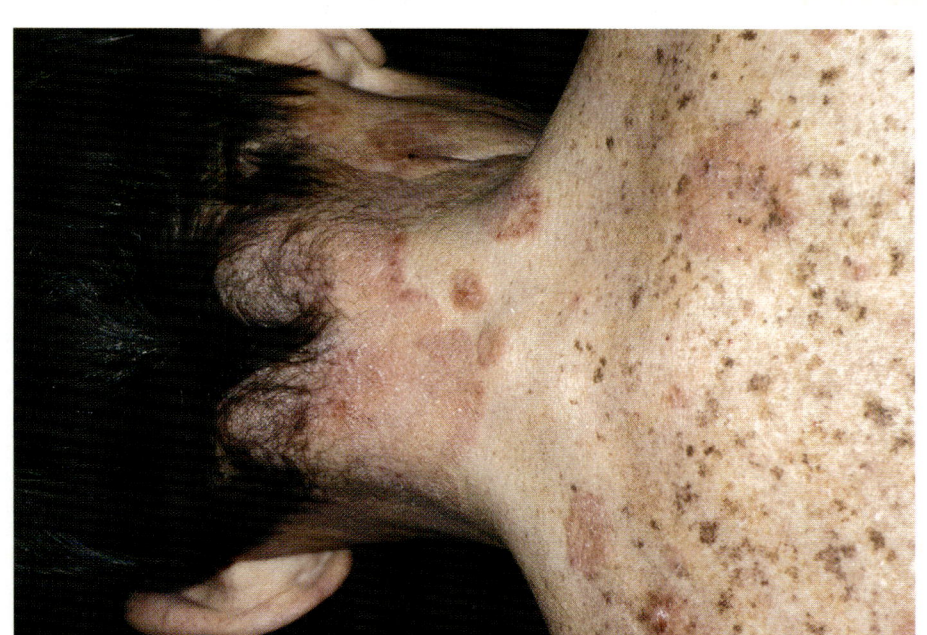

FIGURE 10. *Epidermal dermatophytosis:* Red scaling plaque on the neck are seen on the neck; lesions were widespread, involving 25% of the body surface area in a patient with advanced HIV disease. Tinea pedis and toe-nail onychomycosis was also present.

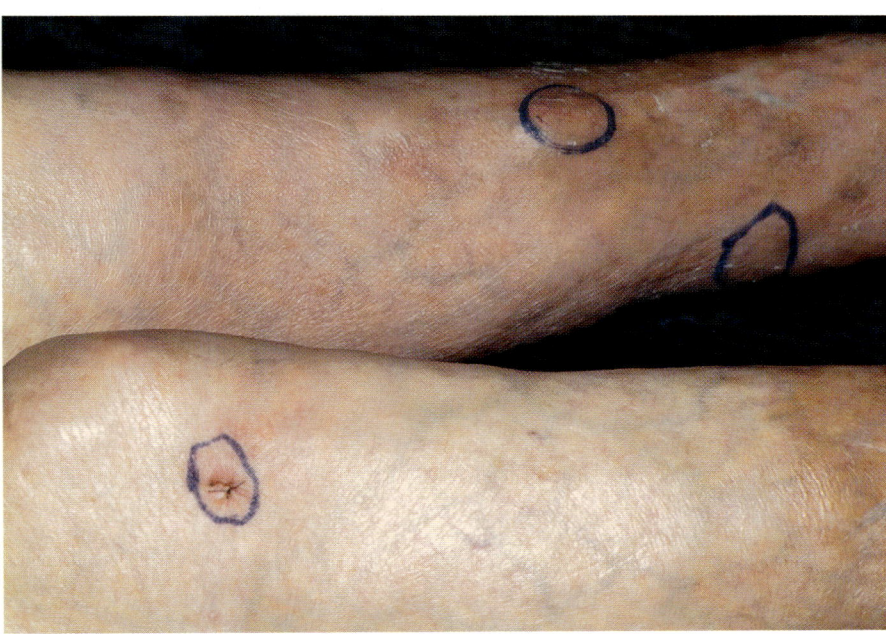

FIGURE 9. *Mycobacterium chelonae:* Cellulitis. Inflammatory, tender nodules on the legs of an 84-year-old woman treated with prednisone for asthma.

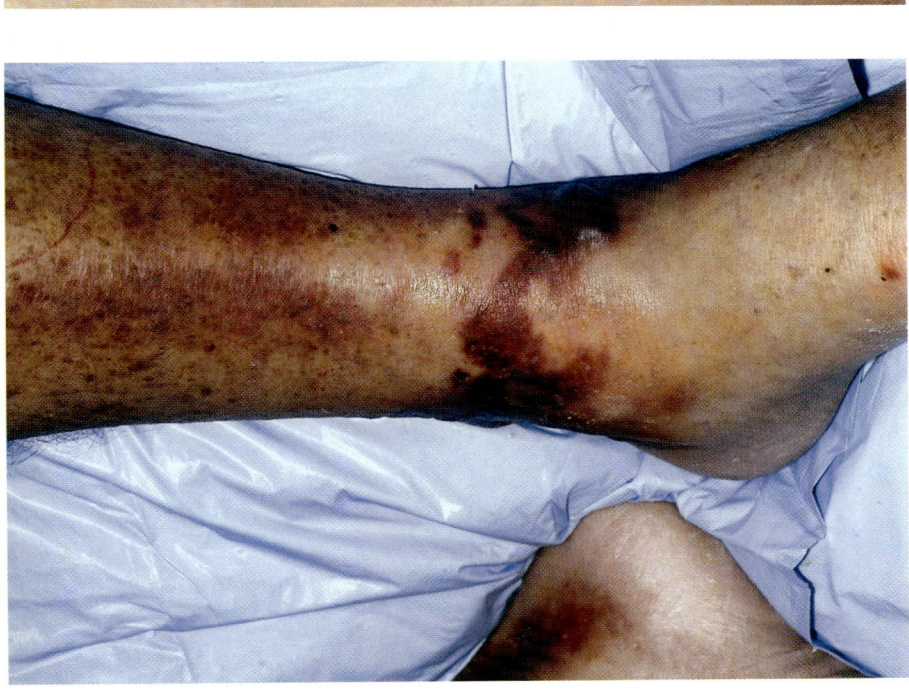

FIGURE 8. *Vibrio vulnificus:* Hemorrhage plaques and bullae on the lower legs and feet of a diabetic with cirrhosis. The patient had ingested raw clams with subsequent *V. vulnificus* gastroenteritis (asymptomatic), bacteremia, and bilateral lower leg soft tissue infection.

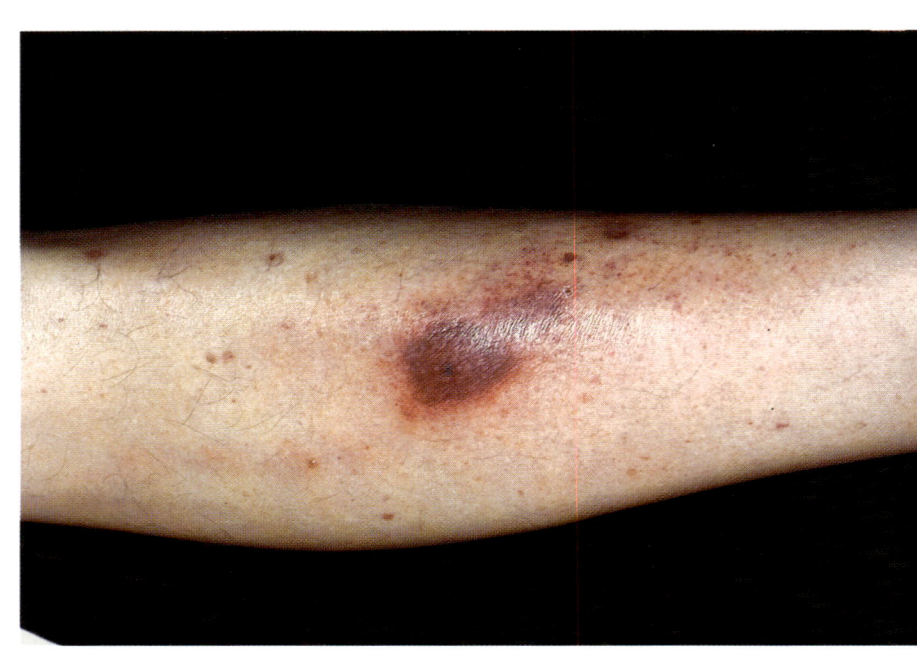

FIGURE 13. *Candidemia:* Disseminated *C. tropidalis* infection. A hemorrhagic nodule on the leg with multiple petechiae in an elderly patient with acute myellogenous leukemia and thrombocytopenia. *C. tropicalis* was isolated from the lesional skin biopsy specimen and blood.

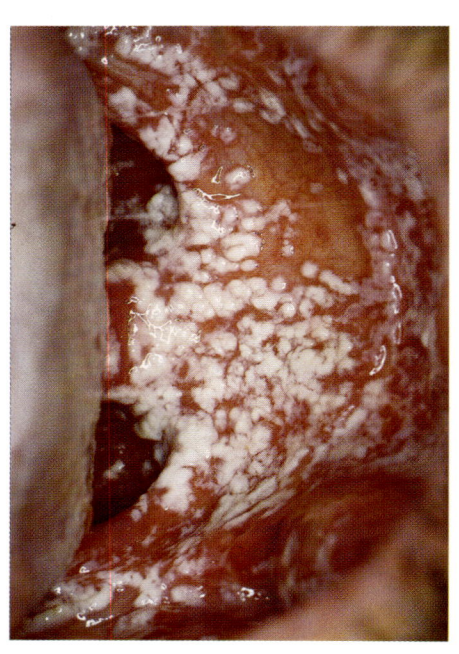

FIGURE 12. *Mucosal candidiasis:* Pseudomembranous or thrush. White colonizes of Candida are seen on the oral mucosa. The infant was being treated with ACTH.

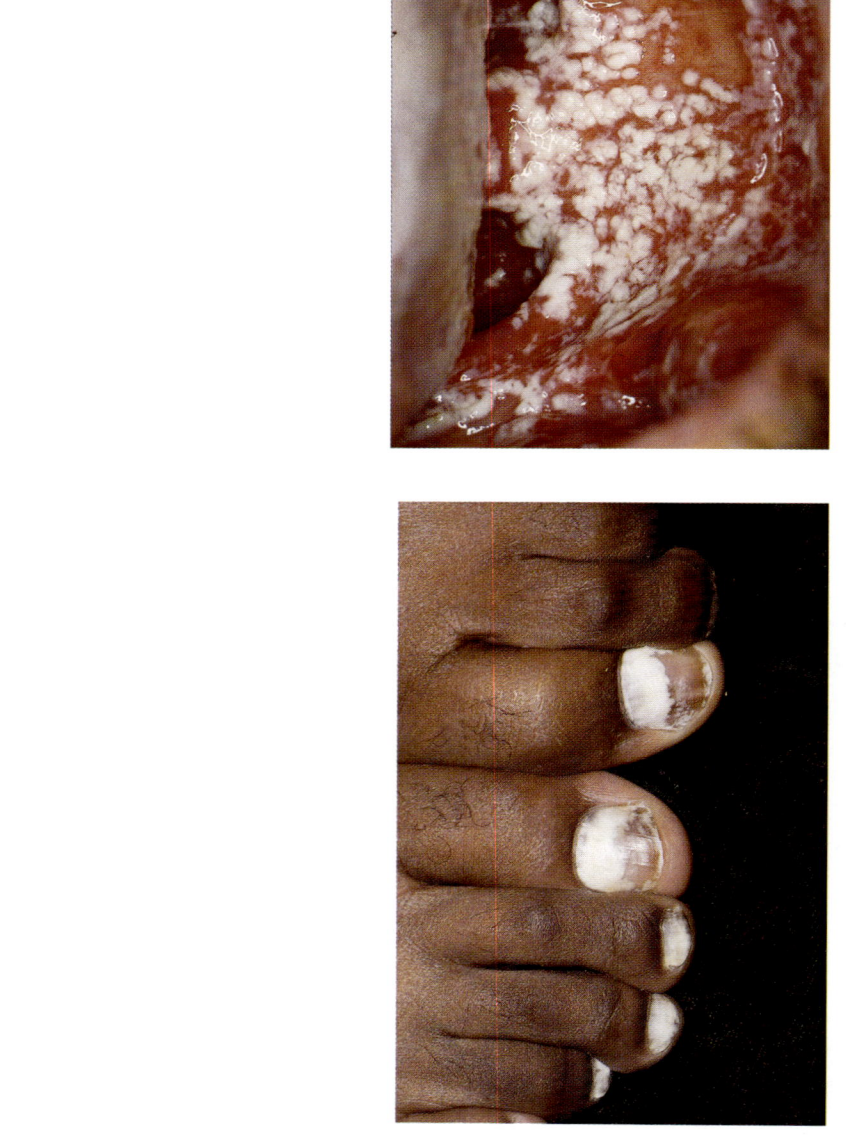

FIGURE 11. *Onychomycosis:* Proximal subungual. White nails are seen proximally on toenails of this HIV-infected black male. The dermatophyte *Trichophyton rubrum* track proximally over the dorsal nail plate to the underside surface of the nail. This variant of onychomycosis indicates significant immunocompromise, and usually occurs in moderately advanced HIV disease.

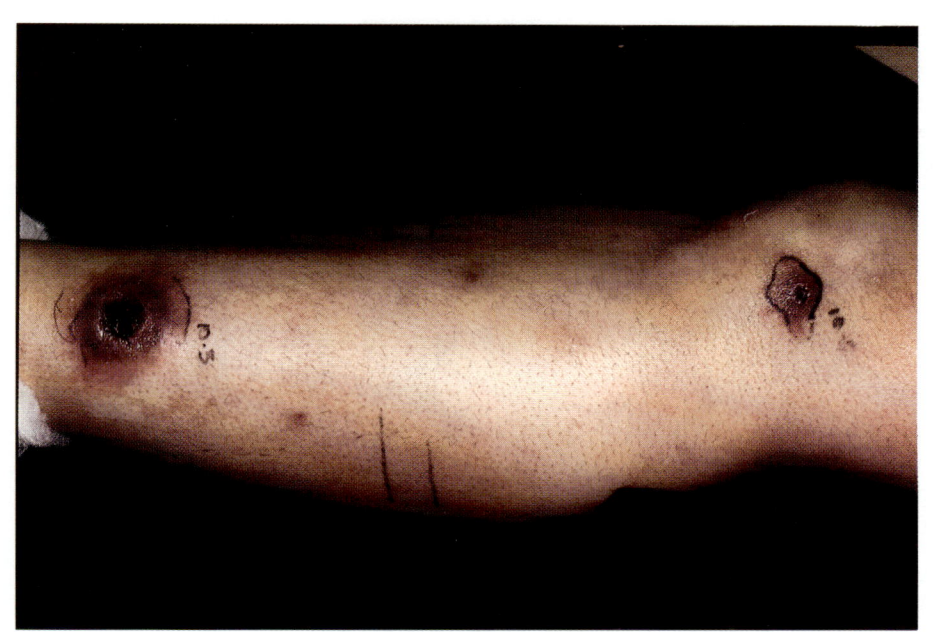

FIGURE 14. *Crypotococcosis*, disseminated. Multiple large molluscum contagiosumlike lesions on the face of a male with advanced HIV disease. Disseminated histoplasmosis and penicillinosis can also present with similar clinical findings.

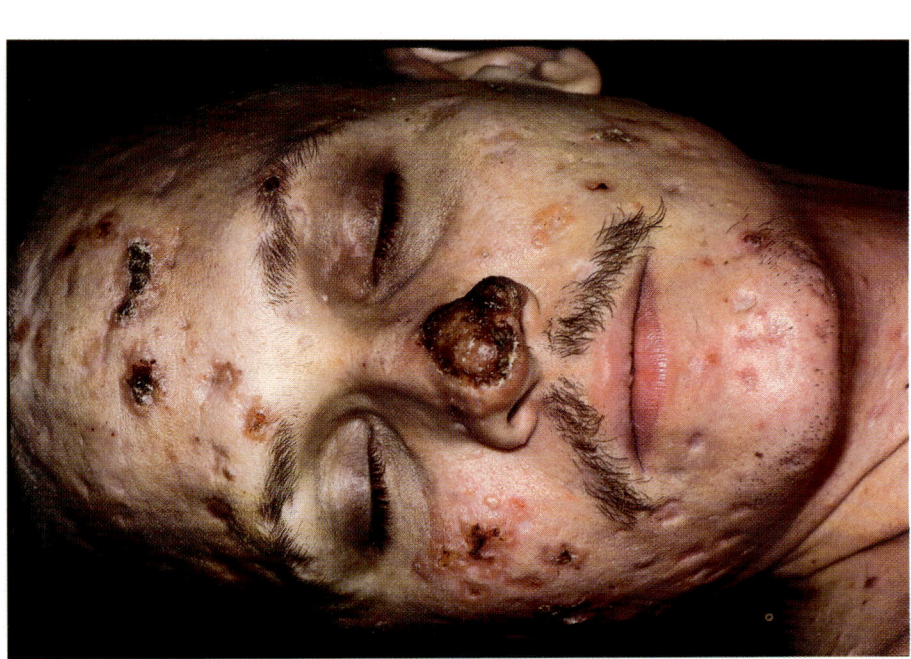

FIGURE 15. *Sporotrichosis*, disseminated. Multiple crusted ulcers on the face of an HIV-infected male with advanced HIV disease. Cutaneous lesions were disseminatee, and associated sporotrichoid infectious arthritis.

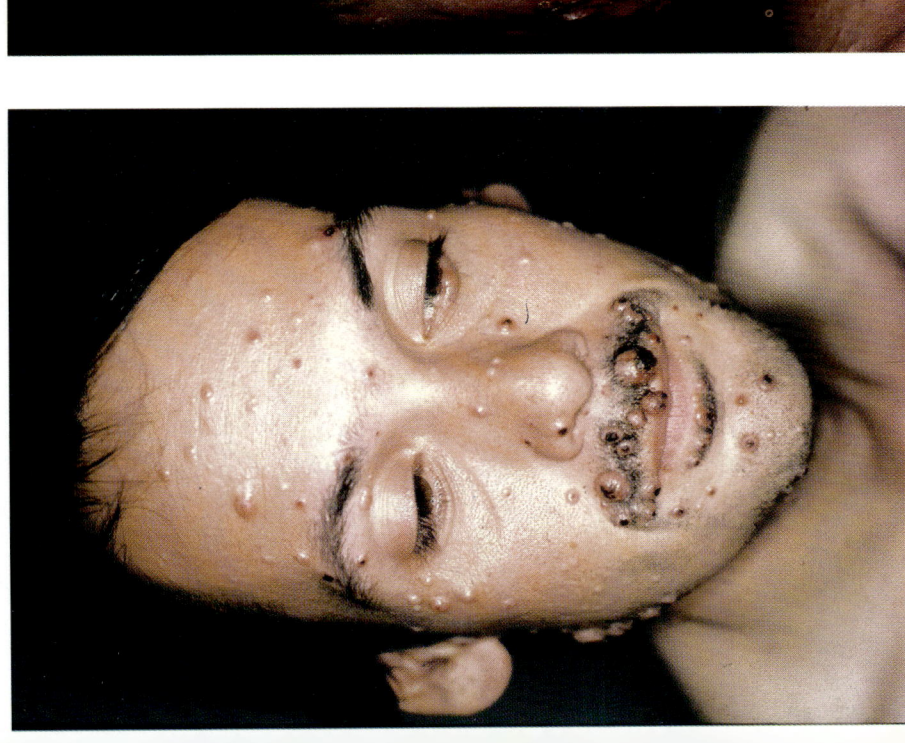

FIGURE 16. *Mucormycosis*, primary cutaneous. Two large crusted ulcers are seen on the leg. The patient was being treated for acute myelogenous leukemia.

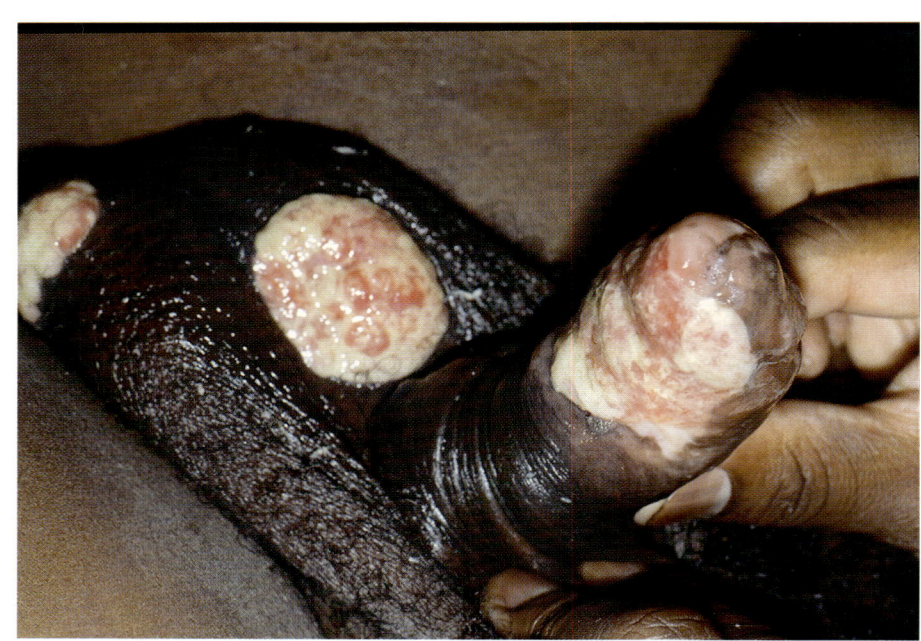

FIGURE 19. *HSV infection, acyclovir-resistant:* Chronic ulcer on penis. Large ulcers on the penis and scrotum of a male with advanced HIV disease failed to respond to intravenous acyclovir, foscarnet, or cidofovir.

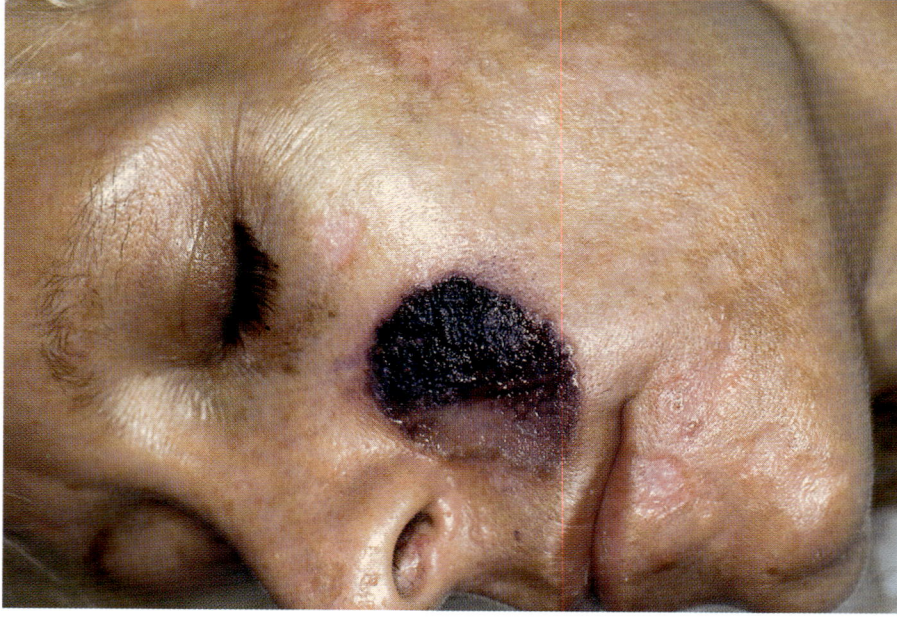

FIGURE 18. *HSV infection, acyclovir-resistant:* Chronic ulcer on face. A large ulcer painted with gential vioet and atrophic scars of the face of a woman with advanced HIV disease. Ulcers had been present for 1 year; the HSV isolated on culture was acyclovir-resistant. All lesions resolved with intravenous foscarnet.

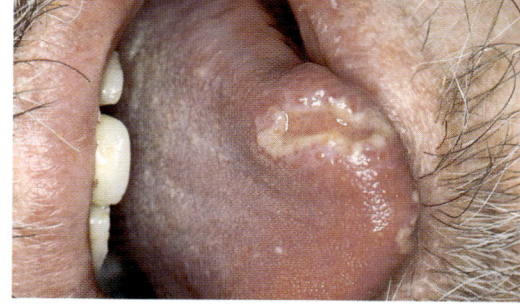
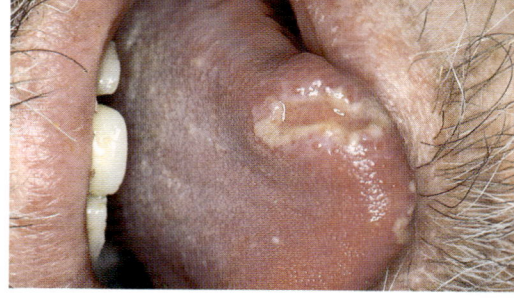
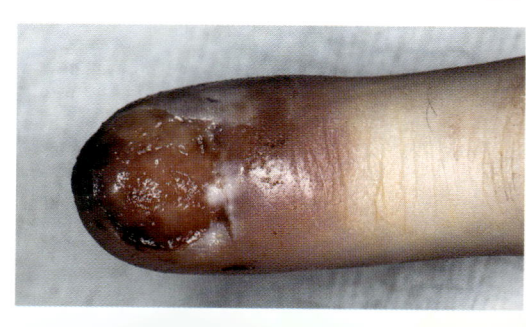
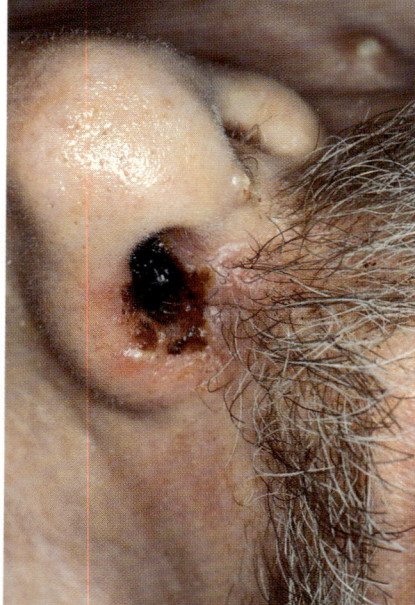
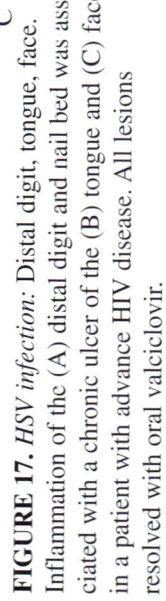

FIGURE 17. *HSV infection:* Distal digit, tongue, face. Inflammation of the (A) distal digit and nail bed was associated with a chronic ulcer of the (B) tongue and (C) face in a patient with advance HIV disease. All lesions resolved with oral valciclovir.

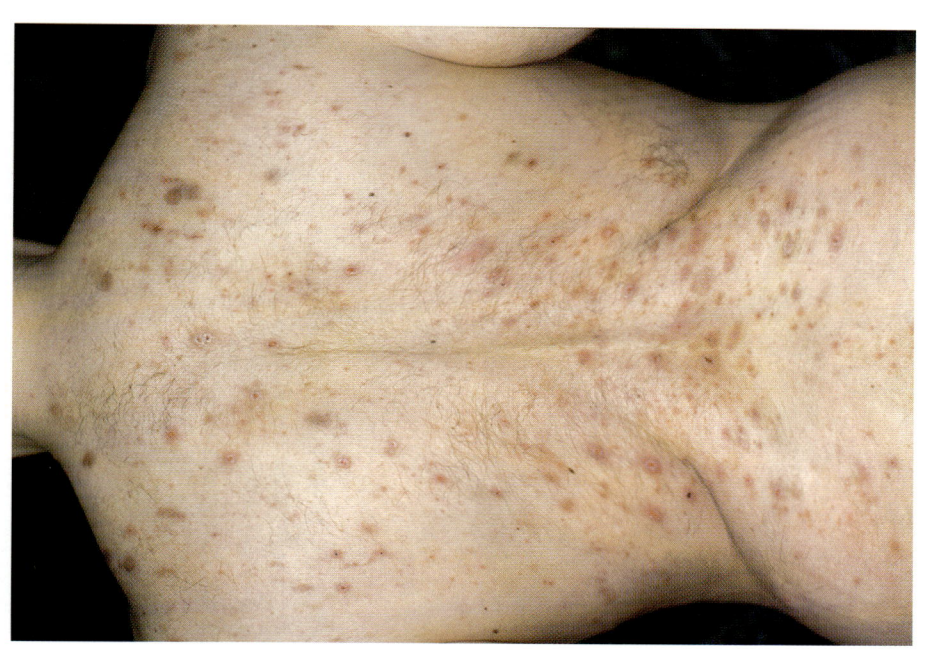

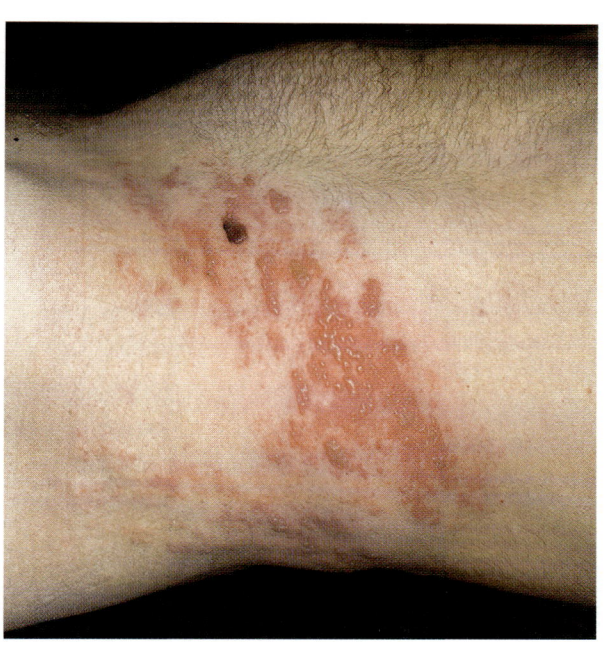

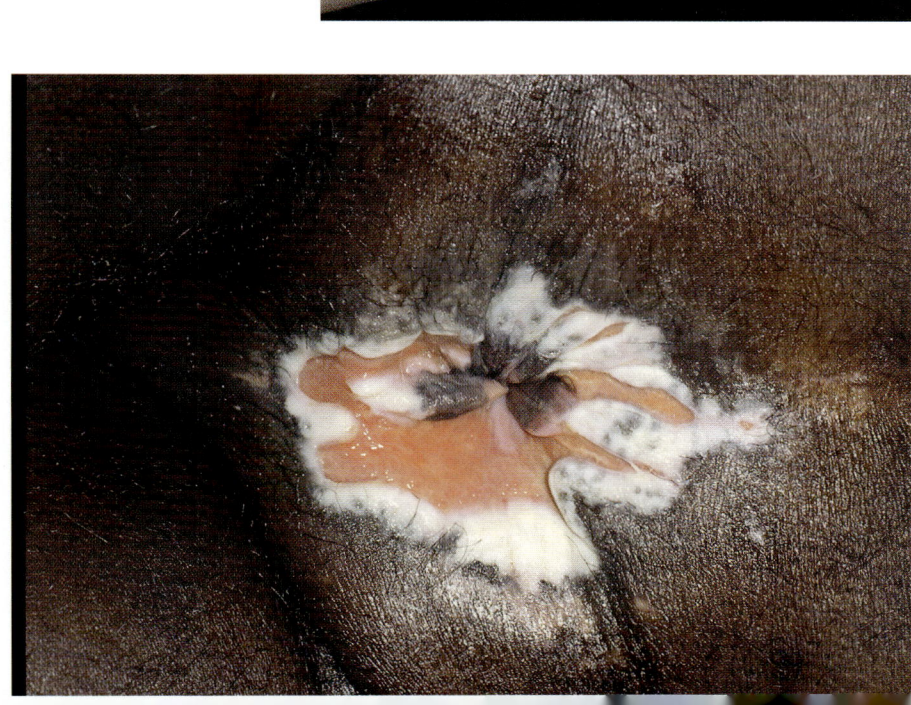

FIGURE 22. *VZV infection:* Disseminated infection. Disseminated VZV infection with >100 vesicles is seen on the back of a male with herpes zoster on the flank (not seen) and with advanced HIV disease.

FIGURE 21. *VZV infection:* Herpes zoster. This typical zosteriform vesiculobullus eruption involving three contiguous dermatomes on the flank was the indication for HIV testing in this HIV-infected male.

FIGURE 20. *HSV infection,* acyclovir-resistant: Chronic ulcer on perineum. A large chronic ulcer on the perineum of a male with advanced HIV disease failed to respond to oral and intravenous acyclovir.

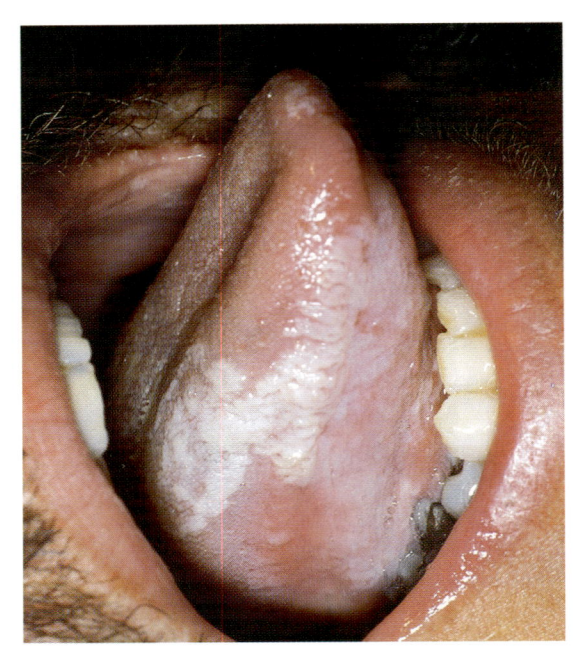

FIGURE 25. *EBV infection:* Oral hairy leukoplakia. A large asymptomatic plaque on the lateral, dorsal, and ventral tongue of a patient with moderately advanced HIV disease.

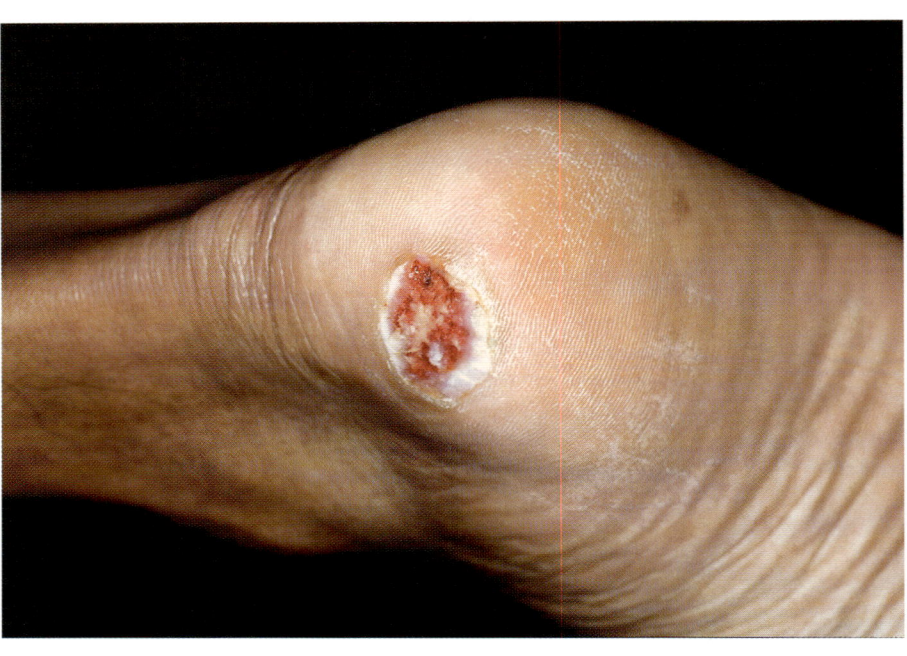

FIGURE 24. *VZV infection:* Local infection. A solitary very painful ulcer of the heel had been present for 3 weeks of a male with advanced HIV disease. The lesion resolved with oral acyclovir.

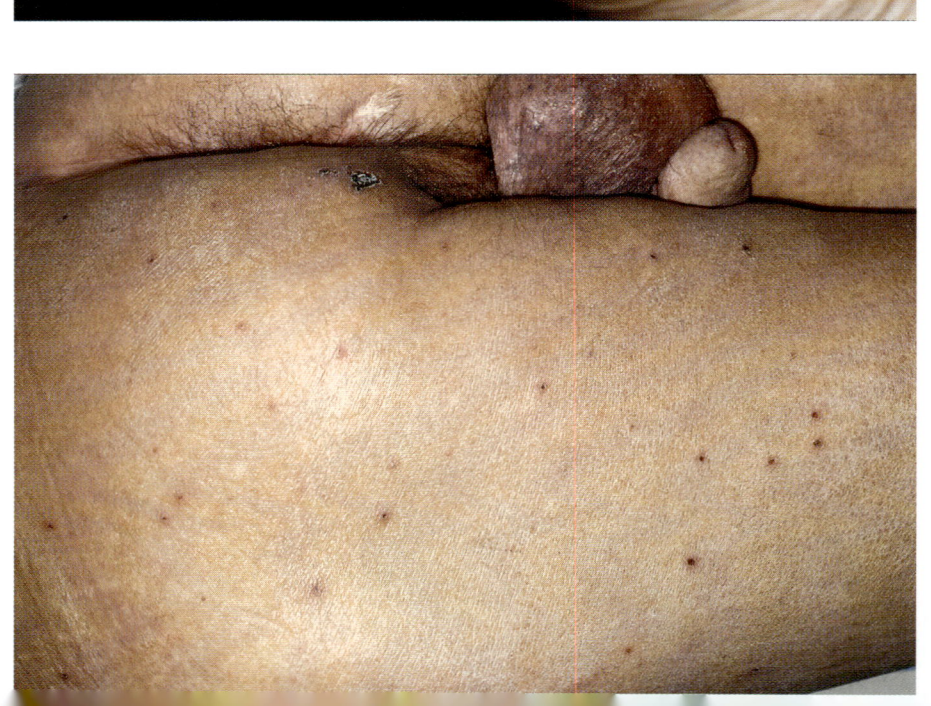

FIGURE 23. *VZV infection:* Disseminated infection. This patient with advanced HIV disease had approximately 40 vesicles and crustion erosion on the buttocks and thighs.

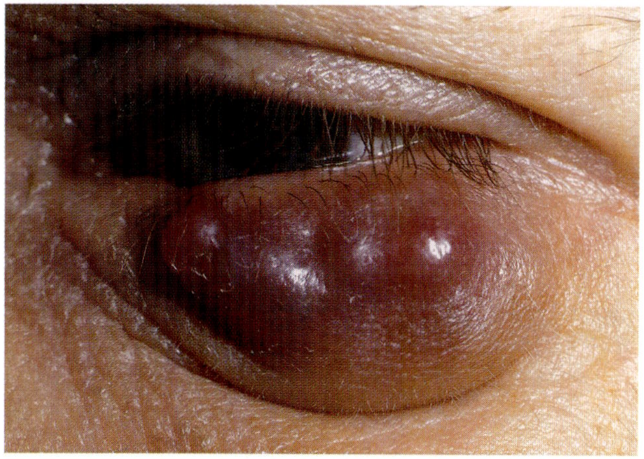

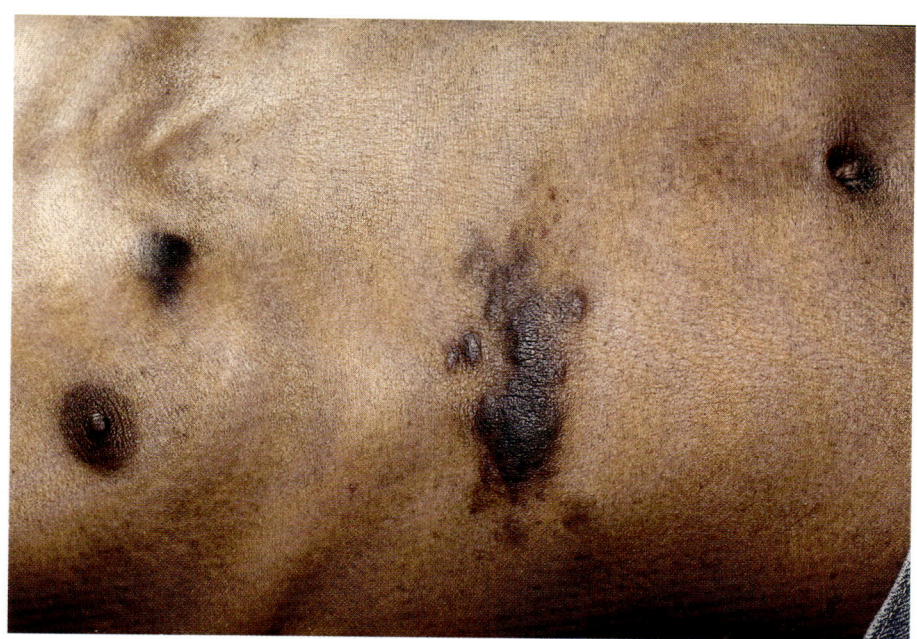

FIGURE 28. *Kaposi's sarcoma.* Large, confluent violaceous plaques on the trunk of an HIV-infected individual.

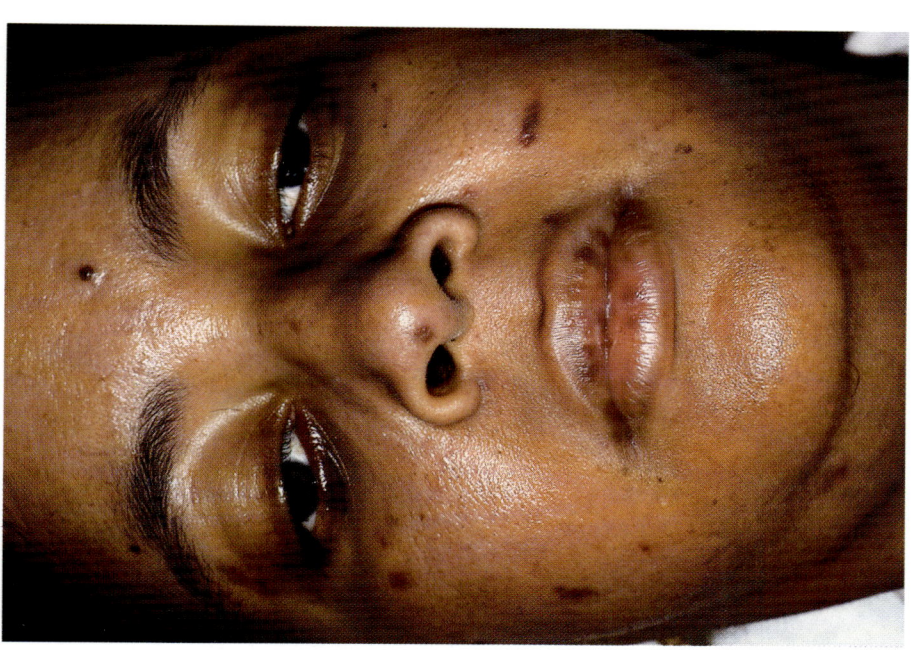

FIGURE 27. *Kaposi's sarcoma.* Multiple purple nodules on the face of a black HIV-infected male; asymptomatic but causing significant cosmetic disfigurement and stigmatization.

FIGURE 26. *Kaposi's sarcoma.* A large asymptomatic plaque on the lower eyelid of an HIV-infected male, cosmetically disfiguring.

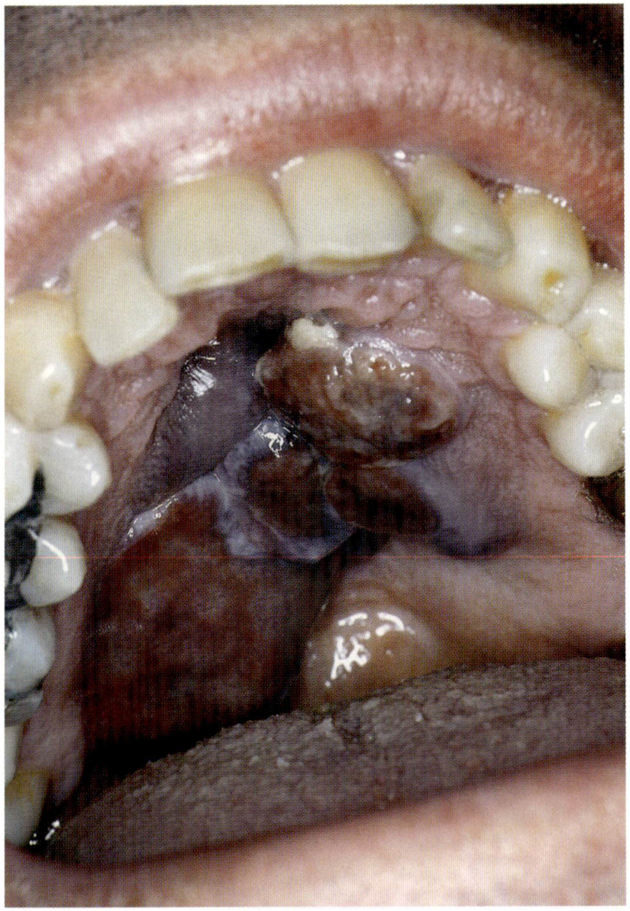

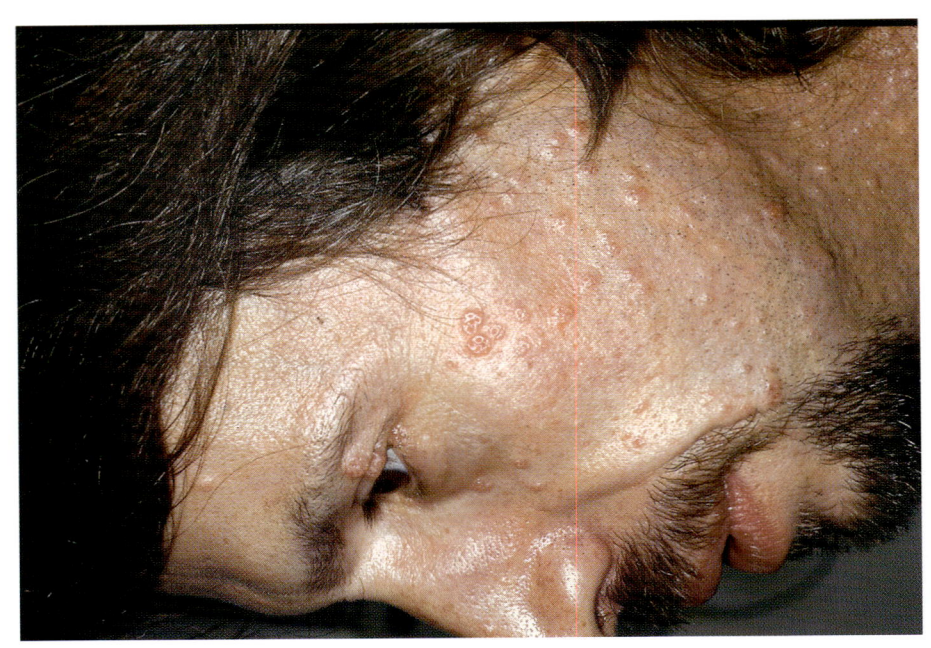

FIGURE 31. *Molluscum contagiosum.* Multiple, skin-colored papules on the face of an HIV-infected male.

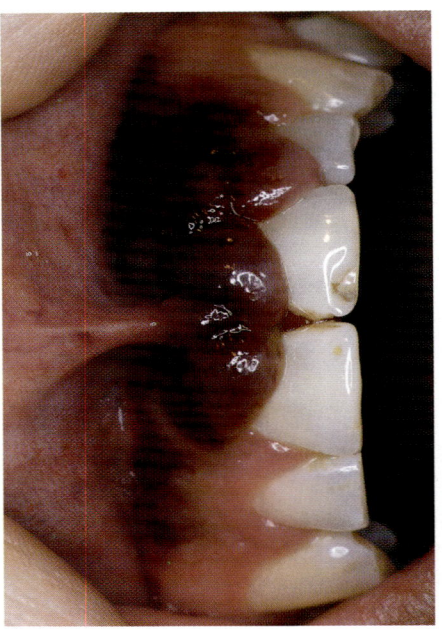

FIGURE 30. *Kaposi's sarcoma.* A large confluent plaque on the upper anterior gingiva of an HIV-infected individual.

FIGURE 29. *Kaposi's sarcoma.* Confluent, eroded, and painful tumors on the palate of an HIV-infected individual.

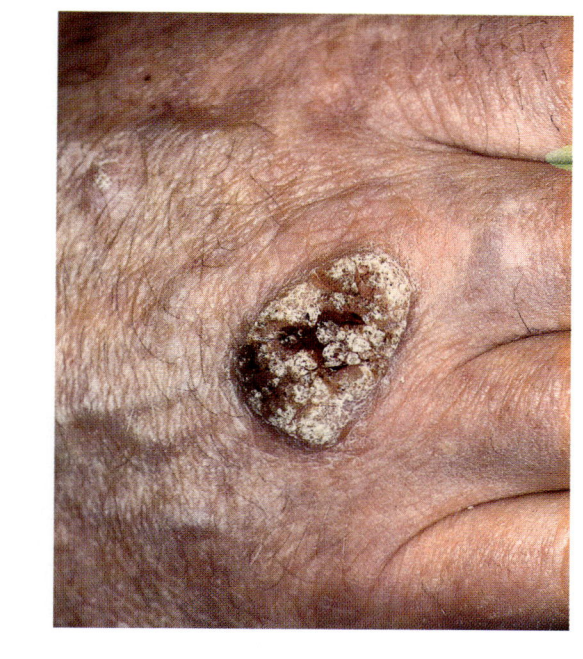

FIGURE 33. *HPV infection:* Verruca. A huge verruca vulgaris (common wart) on the dorsal hand of a renal transplant recipient.

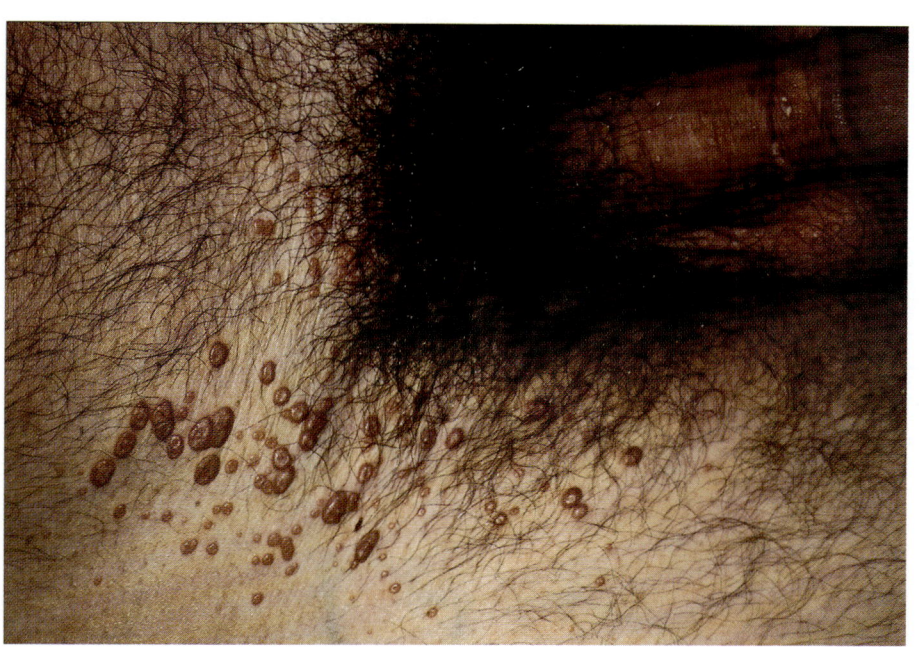

FIGURE 32. *Molluscum contagiosum.* Multiple, skin-colored papules and nodules, becoming confluent on the lower abdomen of a male with advanced HIV disease.

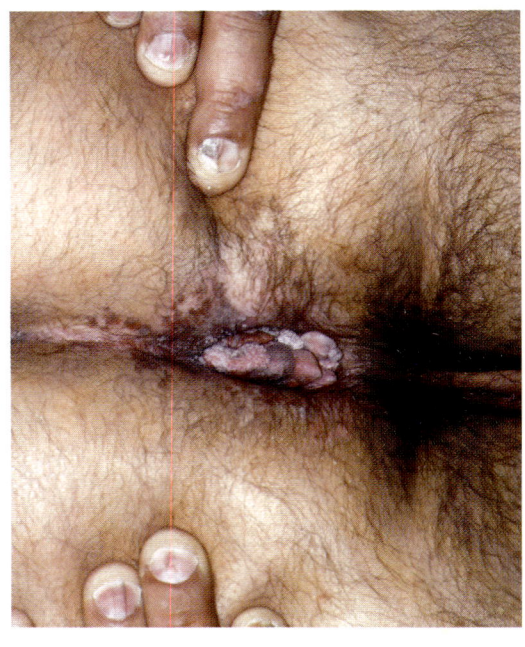

FIGURE 35. *HPV infection:* Squamous cell carcinoma. Squamous cell carcinoma *in situ* is seen on the perineum and periungual area on multiple fingers of male with advanced HIV disease. The squamous cell carcinoma on the left thumb and index finger had become invasive, necessitating amputation.

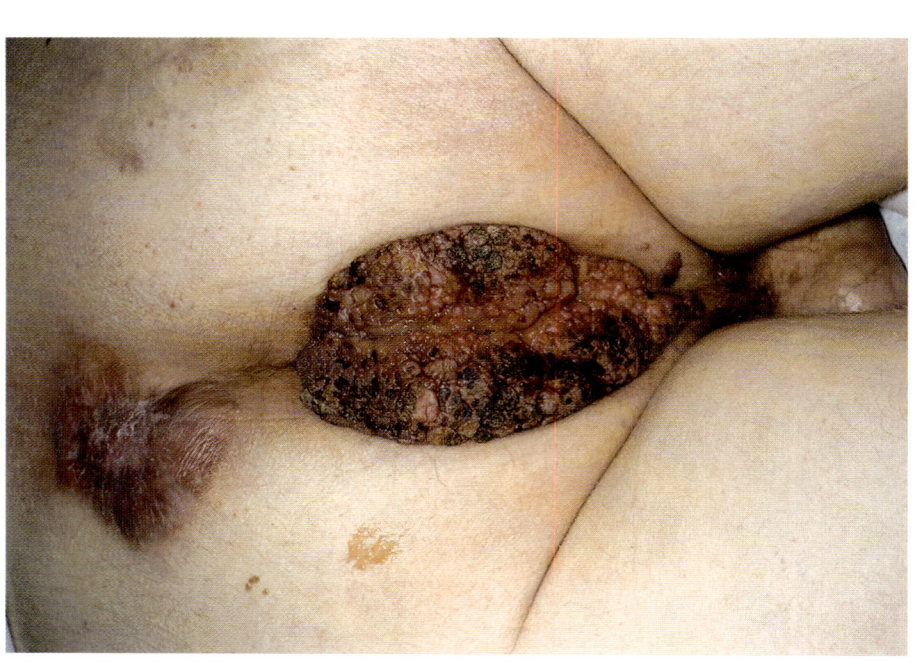
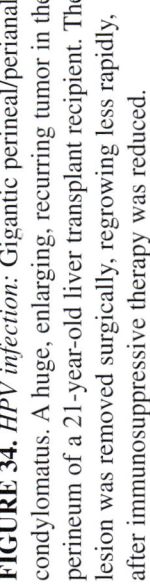

FIGURE 34. *HPV infection:* Gigantic perineal/perianal condylomatus. A huge, enlarging, recurring tumor in the perineum of a 21-year-old liver transplant recipient. The lesion was removed surgically, regrowing less rapidly, after immunosuppressive therapy was reduced.

Index

ISBN 0-306-46693-7

90000